W9-BHK-928

ESSENTIALS OF BIOLOGY

ESSENTIALS OF

BIOLOGY

Janet L. Hopson
University of California, Santa Cruz

Norman K. Wessells
University of Oregon

McGRAW-HILL CONSULTING EDITORS FOR THE LIFE SCIENCES

Howard A. Schneiderman — Monsanto Company and the University of California, Irvine

John H. Postlethwait — University of Oregon

McGRAW-HILL PUBLISHING COMPANY

New York St. Louis San Francisco Auckland Bogotá
Caracas Hamburg Lisbon London Madrid Mexico
Milan Montreal New Delhi Oklahoma City Paris
San Juan São Paulo Singapore Sydney Tokyo Toronto

Essentials of Biology

3 4 5 6 7 8 9 0 VNH VNH 9 5 4 3 2 1

ISBN 0-07-557108-0

Library of Congress Cataloging-in-Publication Data

Hopson, Janet L.
Essentials of biology / Janet L. Hopson, Norman K. Wessells.
p. cm.
Includes bibliographical references.
ISBN 0-07-557108-0
1. Biology. I. Wessells. Norman K. II. Title.
QH308.2.H67 1990 89-13511
574—dc20 CIP

Sponsoring Editor: Eirik Børve
Copyeditor: Janet Greenblatt
Project Managers: Hal Lockwood and Carol Dondrea, Bookman Productions
Production Supervisor: Pattie Myers
Senior Production Manager: Karen Judd
Text Designers: John Lennard and Renee Deprey
Cover Designer: Al Burkhardt
Illustrators: Dolores Bego, Martha Blake, Wayne Clark, Cyndie Clark-Huegel, Carol Donner, Marsha Dohrmann, Cecile Duray-Bito, Nelson Hee, Marilyn Hill, Illustrious, Inc., J & R Technical Services, David Lindroth, Paula McKenzie, Linda McVay, Elizabeth Morales-Denney, Victor Royer, Donna Salmon, Judy Skorpil, Vantage Art, Inc.
Photo Researcher: Stuart Kenter
Compositor and Color Separator: York Graphic Services, Inc.
Printer and Binder: Von Hoffman Press, Inc.
Cover Photo: Red-eyed leaf frog (*Agalychnis callidryas* on *Heliconia mathiasii*) by Michael Fogden
Cover Color Separator: Color Tech
Cover Printer: Phoenix Color Corporation

Manufactured in the United States of America

To my dear friends, with gratitude for their support and encouragement.
—J.L.H.

To Catherine, with thanks and love.
—N.K.W.

To Howard Schneiderman, with thanks from us both and from so many others for the contributions you have made to our lives, to science, and to the future.
—N.K.W. and J.L.H.

Prologue

At no time in history has the science of life been so visible and so important to human life and the future of our planet. Newspapers, magazines, and television feature biology prominently every day. Biological issues are discussed in Congress, in the courts, on Wall Street, at the World Bank, at the United Nations, and at summit meetings of heads of state, as well as in classrooms, laboratories, hospitals, and agricultural centers. We hear about viral diseases, especially AIDS. About repairing brain damage. About the ozone layer and skin cancer. The greenhouse effect. The disappearance of forests and the rapid loss of animal and plant diversity on our planet. Genetic engineering and frost-resistant strawberry plants. The costs—billions of dollars—of unraveling the genetic code of human beings (the so-called human genome project). How memory works. Organ transplants. Drugs that will prevent heart attacks. In vitro fertilization. The destruction of rain forests. Acid rain. Crops that require no insecticides. The extinction of dinosaurs after an asteroid crashed into the earth. Chemicals produced by plants that protect the plants from their enemies. The fate of whales. The language of wolves. The durability of cockroaches. The origin of humans. The future of our planet.

Essentials of Biology is a brief introduction to the worlds of biology. Its authors are superb guides for the journey. The book is a shorter version of the comprehensive introduction to biology that Norman Wessells and Janet Hopson successfully introduced recently. This new book has retained the clear explanations and exciting writing of the original, but is more selective in the subjects covered and has a vigorously revised and substantially improved art program. It is truly an essential book that will provide the reader with a powerful background for further studies in biology, medicine, agriculture, and the behavioral and social sciences, as well as the knowledge to function as an informed voter, consumer, and denizen of the planet Earth. Beyond that, *Essentials* will contribute to the reader's viewpoint: The world will look different; it will have more texture, more connectedness, a certain inner logic.

As consulting editors, we have contributed to the lively ferment that went into the book. It has been both challenging and exciting, and we are delighted with the finished product. We hope you will enjoy it.

Howard A. Schneiderman
St. Louis, Missouri

John H. Postlethwait
Eugene, Oregon

Preface

Biology, the study of life in all its manifestations, no doubt began with the first stirrings of curiosity in our early ancestors, taking root as these early humans applied their intelligence to the problems of tracking game and collecting plant materials. Biology became a more formalized intellectual endeavor soon after people could record their knowledge in pictures or words, and it continues today as a fundamental part of a good education.

Anyone can watch with interest and even inspiration as a bee lands on a fragrant flower. But the experience is far richer for the observer who understands that the shape of the insect and the shape of the blossom have evolved as complements to each other; that the flower is the plant's showy, tasty, fragrant advertisement, attracting animals that will inadvertently assist in the plant's reproduction; and that the bee has elaborate mechanisms for finding the flower and communicating its location to other members of the hive. The study of biology has vast practical applications, as well, in understanding our bodies and personal health, in grappling with the ethical questions that face us as citizens, and in sensing both our place in the web of interdependent living things and our need to help protect the delicate ecological balance that sustains us all.

For these reasons, the basic Principles of Biology course is a popular one on college campuses. Students who plan to pursue careers in life science, medicine, agriculture, and a broad range of other disciplines (listed in Chapter 1) usually take an introductory biology course, as do many nonmajors (students from unrelated fields) who are simply curious and want to learn the principles that underlie health, fitness, nutrition, genetic engineering, acid rain, the greenhouse effect, and dozens of other current topics.

At most schools, biology majors and nonmajors take introductory biology together, and these "mixed" courses can present the instructor with a real challenge. In particular, which textbook will give the majors the solid foundation of facts and concepts they need while providing the nonmajors with a source that is not overly detailed or presented at too high a level? We designed *Essentials of Biology* to address this need.

This book is a condensation of the well-received *Biology* that we first published in 1988. *Essentials of Biology* maintains the same authoritative selection of topics and the same reader-friendly presentation that has made *Biology* so popular for two-semester courses, but it also incorporates a number of new features that will make it successful for shorter courses or mixed enrollments.

Essentials follows the same levels-of-organization approach we took in *Biology*. The first part, From Atoms to Cells, discusses the building blocks of all matter; biological molecules, the stuff of cells and organisms; the flow of energy in living things; the parts of cells and how they function; and the central energy pathways of cellular respiration and photosynthesis that sustain all living organisms, directly or indirectly.

The second part, Like Begets Like, covers cellular reproduction, the mechanisms of heredity, and how genes and chromosomes control the daily functions of living cells. In this part, students see how genetics developed as a field, from the earliest studies of cell division to the latest applications of recombinant DNA; how researchers study human genetics, a subject of high student interest; and, in a block of chapters on development, how genes carry out their foremost task—controlling the formation of the embryo and young organism. As in *Biology*, development serves as the conceptual bridge between genetics and the remaining topics in the book, which are all at the level of whole organisms or their systems.

The third part, Order in Diversity, presents a clear picture of the wide spectrum of organisms and their basic characteristics. It starts with the origins of life on this planet and progresses through the kingdoms of organisms that emerged, describing the fascinating diversity of living things, their evolutionary relationships, and how each major group may have arisen.

The fourth and fifth parts, Plant Biology and Animal Biology, describe the physiology, or day-to-day functioning, of the most complex groups, the plants and animals, and how they interact with their environments and with each other.

The final part, Population Biology, introduces the sciences of evolution and ecology—from the way populations change over time, to the interactions of the earth's physical forces, to the way groups of organisms relate to those forces and to each other. An important part of this discussion is how environment, ecology, and behavior have shaped our own species' origin and history; how human evolution and activity have affected the earth's ecosystems; and how our future actions will continue to influence them—for better or worse—in the coming centuries.

While the general organization of topics in *Essentials* is the same as in *Biology*, we made some dramatic changes to create a shorter book useful for mixed audiences of student readers. We combined several chapters, reducing the total number from 51 to 46. We removed 25 to 40 percent of the material in each chapter by streamlining the prose; shortening the chapter introductions and endings by substituting point-by-point lists; removing some examples where fewer strong examples could make the same points; cutting some concepts and some detail, but leaving all those topics that our panel of academic reviewers felt were essential for students in a one-semester course; and designing hundreds of new and vastly improved figures that are closely coordinated to the text and can help students visualize and understand biological structures and processes more easily than through lengthier prose discussions. The result is a greater emphasis on essential concepts, a clearer presentation through words and illustrations, a greater proportion of space devoted to unifying themes and take-home messages, and a de-emphasis on detail.

In addition, we updated every chapter with relevant new research and applications of interest to students. We replaced many of the original boxed essays from *Biology* with new student-oriented subjects. And we added some new pedagogical features, including an advance organizer in the chapter introduction, underlined take-home messages for easy study and review, a very complete index to improve the book's utility as a reference, and the use of visual icons in many figures to help the student grasp the physical context for a structure or process (for example, see Guided Tour, page xxviii).

In all, *Essentials* maintains the same well-chosen and clearly explained subjects as *Biology*—thanks to our team of authors, consultants, contributors, and reviewers. But through vigorous revision and condensation, we have created an entirely new, up-to-date textbook with all the topics a biology major needs to know, presented in a way that will interest and give equal access to nonmajors.

We hope our strategy will be a winning one for the users of this book. And we hope our work will provide a foundation of knowledge from which the reader—whether future scientist or informed citizen in a nonscientific field—can understand the stream of discoveries sure to come in biology in the decades ahead, as well as to participate in the democratic process of regulating and utilizing the fruits of those discoveries.

Acknowledgments. We are indebted to hundreds of people for their help in undertaking and completing this project. None have done more than our consultants, Dr. John Postlethwait of the University of Oregon and Dr. Howard Schneiderman of the University of California, Irvine, and Monsanto Company, who were the primary formulators of the book's outline and organization. In addition, Dr. Postlethwait contributed heavily to the chapters on genetics, provided invaluable advice on matters large and small throughout the project's development, and designed virtually all of the new figures for the book.

Once again, we extend our warmest appreciation to those who contributed to *Biology*, as well as the reviewers from various colleges, universities, and institutions who provided critical feedback and recommendations for cutting and revising (see page xi). Their input was very important to our goals of authoritativeness and effective presentation, and it was much appreciated.

Many scientists, photographers, and artists contributed to the book's art program by allowing us to use or modify drawings and to print or reprint photographs. Their names and the figure references for their work appear in the Credits and Acknowledgments section at the end of the book.

A long and complex science book such as this, couched in readable prose and illustrated with hundreds of photos, drawings, charts, graphs, tables, boxes, and appendices, demands the tender loving care of a talented team of professionals at all stages of development. We are indebted to Hal Lockwood, Janet Greenblatt, Stuart Kenter, Carol Dondrea, Blake Edgar, Pattie Myers, Karen Judd, Bev Fraknoi, Lesley Walsh, Sandy Woods, Richard Lynch, Judith Levinson, and Renee Deprey.

Our deepest gratitude goes to Eirik Børve, our collaborating publisher, who has, at every step of the way, placed his intelligence, energy, and exceptional management skills behind our goal of teaching biology in the most effective manner possible. So much teaching and learning at the college level depends on carefully published textbooks, and we feel that Mr. Børve represents the very best in his field.

Finally, we express warmest appreciation to each other. A collaboration between a professional biologist and a widely published writer is truly a beneficial education for both parties.

Even with the careful contributions of our aforementioned friends and colleagues, errors of fact or interpretation may have found their way into the book. For these, we alone assume responsibility and stand ready to correct them.

We sincerely hope that the students and instructors who use this book will find it a stimulating introduction to the intricate, fascinating, and beautiful world of life on earth.

Janet L. Hopson and Norman K. Wessells

Academic Reviewers

Olukemi Adewusi, Ferris State University
John Alcock, Cornell University
Betty D. Allamong, Ball State University
Glenn Aumann, University of Houston, Central Campus
David Barrington, University of Vermont
Penelope H. Bauer, Colorado State University
Stanley Bayley, McMaster University
Barbara N. Benson, Cedar Crest College
Paul Biersuck, Nassau Community College
Sharon Bradish-Miller, College of DuPage
Osmond P. Breland, University of Texas, Austin
Louis Burnett, University of San Diego
T. E. Cartwright, University of Pittsburgh
Robert C. Cashner, University of New Orleans
Steve Chalgren, Radford University
Douglas T. Cheeseman, De Anza College
James S. Clegg, University of Miami
Mary U. Connell, Appalachian State University
Murray W. Coulter, Texas Tech University
Bradner Coursen, College of William and Mary
Larry Crawshaw, Oregon State University
Sidney Crow, Georgia State University
William Crumpton, Iowa State University
J. M. Cubina, New York Institute of Technology
R. Dean Decker, University of Richmond
Donald J. Defler, Portland Community College
Anthony Dickinson, Memorial University, St. Johns, Newfoundland
Gary Dolph, Indiana University at Kokomo
Warren D. Dolphin, Iowa State University
Marvin Druger, Syracuse University
Robert C. Eaton, University of Colorado
D. Craig Edwards, University of Massachusetts, Amherst
David W. Eldridge, Baylor University
James R. Estes, University of Oklahoma
Russell D. Fernald, University of Oregon
Michael Filosa, Scarborough College, University of Toronto
Conrad Firling, University of Minnesota at Duluth
Kathleen M. Fisher, University of California at Davis
Arlene Foley, Wright State University
Lawrence D. Friedman, University of Missouri at St. Louis
Larry Fulton, American River College
Arthur W. Galston, Yale University
Florence H. Gardner, University of Texas, Permian Basin
Lawrence G. Gilbert, University of Texas, Austin
Elizabeth Godrick, Boston University
Michael Gold, University of California at San Francisco
Jonathan Goldthwaite, Boston College
Judith Goodenough, University of Massachusetts, Amherst
Corey Goodman, Stanford University
D. Bruce Gray, University of Rhode Island
Margaret Hartman, California State University at Los Angeles
Robert Hehman, University of Cincinnati
Steven R. Heidemann, Michigan State University
Walter Hempfling, University of Rochester
Robert W. Hoshaw, University of Arizona
Dale Hoyt, University of Georgia
June D. Hudis, Suffolk County Community College
Robert J. Huskey, University of Virginia
Alice Jacklet, State University of New York at Albany
Robert J. Jonas, Washington State University
Pia Kallas-Harvey, University of Toronto
Jerry L. Kaster, University of Wisconsin at Milwaukee
Donald Kraft, Bemidji State University
T. C. Lacalli, University of Saskatchewan
Meredith A. Lane, University of Colorado
Joseph D. Laufersweiler, University of Dayton
William H. Leonard, Clemson University
Georgia Lesh-Laurie, Cleveland State University
Joseph S. Levine, Boston College
Joseph LoBue, New York University
Ellis R. Loew, Cornell University
V. Pat Lombardi, University of Oregon
Willliam F. Loomis, University of California at San Diego
Cran Lucas, Louisiana State University, Shreveport
Carl E. Ludwig, California State University, Sacramento
John H. Lyford, Jr., Oregon State University
Henry Merchant, George Washington University
Helen C. Miller, Oklahoma State University
William J. Moody, University of Washington
Frank L. Moore, Oregon State University
Randy Moore, Baylor University
Robert E. Moore, Montana State University
Dorothy B. Mooren, University of Wisconsin at Milwaukee
David Nanney, University of Illinois, Urbana
Maimon Nasatir, University of Toledo
Robert Neill, University of Texas, Arlington
Bette Nicotri, University of Washington
Herman Nixon, Jackson State University
Frank Nordlie, University of Florida, Gainesville
J. R. Nursall, University of Alberta
Ralph Ockerse, Purdue University
William D. O'Dell, University of Nebraska, Omaha

Nancy R. Parker, Southern Illinois University at Edwardsville
Rollin C. Richmond, Indiana University
Ezequiel Rivera, University of Lowell
Gerald G. Robinson, University of South Florida
Martin Roeder, Florida State University, Tallahassee
Thomas B. Roos, Dartmouth University
Ian Ross, University of California at Santa Barbara
Richard Russell, University of Pittsburgh
Roger H. Sawyer, University of South Carolina
Carl A. Scheel, Central Michigan University
John A. Schmitt, Ohio State University
Richard J. Shaw, Utah State University
Peter Shugarman, University of Southern California
Warren Smith, Central State University
David Stetler, Virginia Polytechnic Institute
Richard Swade, California State University, Northridge
Daryl Sweeney, University of Illinois, Champaign
Raymond Tamppari, Northern Arizona University
Harry D. Thiers, San Francisco State University
John Thomas, Stanford University
Sidney Townsley, University of Hawaii at Manoa
James Turpen, University of Nebraska Medical Center, Omaha
Joseph Vanable, Purdue University
Dan B. Walker, University of California at Los Angeles
Jack Ward, Illinois State University
Cherie L.R. Wetzel, City College of San Francisco
Larry G. Williams, Kansas State University
David Wilson, Miami University
Kathryn Wilson, Indiana-Purdue University
Leslie Wilson, University of California at Santa Barbara
Thomas Wilson, University of Vermont
G. A. Wistreich, East Los Angeles College
Daniel E. Wivagg, Baylor University
Keith H. Woodwick, California State University, Fresno
John Zimmerman, Kansas State University

Contents in Brief

1 The Study of Life 1

PART ONE FROM ATOMS TO CELLS 17

2 Atoms, Molecules, and Life 18
3 The Molecules of Living Things 35
4 Chemical Reactions, Enzymes, and Metabolism 58
5 Cells: Their Properties, Surfaces, and Interconnections 77
6 Inside the Living Cell: Structure and Function of Internal Cell Parts 98
7 Harvesting Energy from Nutrients: Fermentation and Cellular Respiration 117
8 Photosynthesis: Harnessing Solar Energy to Produce Carbohydrates 135

PART TWO LIKE BEGETS LIKE 153

9 Cellular Reproduction: Mitosis and Meiosis 154
10 Foundations of Genetics 170
11 Mendel Modified 185
12 Discovering the Chemical Nature of the Gene 199
13 Translating the Code of Life: Genes into Proteins 215
14 Bacterial Genetics, Gene Control, and Genetic Engineering 229
15 Human Genetics 251
16 Animal Development 269
17 Developmental Mechanisms and Differentiation 287
18 Animal and Human Reproduction 307

PART THREE ORDER IN DIVERSITY 325

19 The Origin and Diversity of Life 326
20 Monera and Viruses: The Invisible Kingdom 340
21 Protista: The Kingdom of Complex Cells 359
22 Fungi: The Great Decomposers 373
23 Plants: The Great Producers 388
24 Animals: The Great Consumers 409

PART FOUR PLANT BIOLOGY 447

25 The Architecture of Plants 448
26 How Plants Reproduce, Develop, and Grow 463
27 Exchange and Transport in Plants 483
28 Plant Hormones 498

PART FIVE ANIMAL BIOLOGY 513

29 The Circulatory and Transport Systems 514
30 The Immune System 533
31 Respiration: The Breath of Life 552
32 Digestion and Nutrition 570
33 Homeostasis: Maintaining Biological Constancy 592
34 The Nervous System 612
35 Hormonal Controls 629
36 Input and Output: The Senses and the Brain 649
37 Skeletons and Muscles 683

PART SIX POPULATION BIOLOGY 703

38 Evolution and the Genetics of Populations 704
39 Natural Selection 719
40 The Origin of Species 732
41 Ecosystems and the Biosphere 751
42 The Ecology of Communities 775
43 The Ecology of Populations 799
44 Behavioral Adaptations to the Environment 815
45 Social Behavior 832
46 Human Origins 846

Full Contents

Chapter 1 The Study of Life 1
What Is Life? 2
Life on Earth: A Brief History 5
Early Beliefs About the Origin of Life 5
A Modern View of the Origin of Life 7
Evolution: A Theory That Changed Biology and Human Thought 8
The Theory of Evolution 9
The Scientific Method 11
Biology, Society, and Your Future 12
Our Approach to Biology 14
Summary / Key Terms / Questions / Essay Question 15

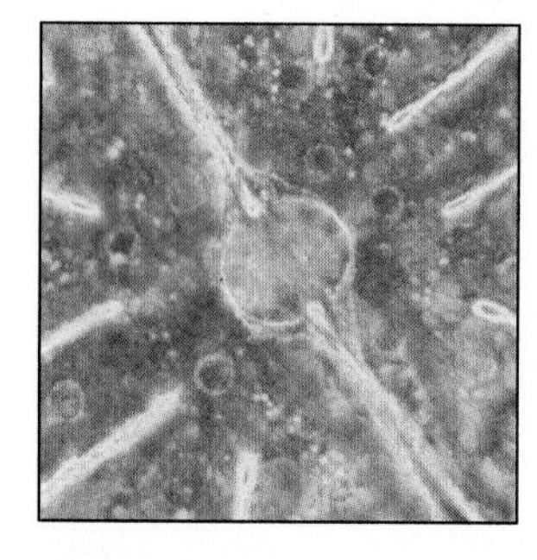

PART ONE

FROM ATOMS TO CELLS 17

Chapter 2 Atoms, Molecules, and Life 18
Elements and Atoms: Building Blocks of All Matter 19
The Elements of Life 19
Atomic Structure 20
Molecules and Compounds: Aggregates of Atoms 23
Chemical Bonds: The Glue That Holds Molecules Together 24
Bond Strength 26
Chemical Formulas and Equations 27
Water: Life's Precious Nectar 28
Physical Properties of Water 28
Molecular Structure of Water 29
Dissociation of Water: Acids and Bases 32
Atoms to Organisms: A Continuum of Organization 33
Summary / Key Terms / Questions / Essay Question 33
BOX: ATOMS IN MEDICINE 21
BOX: FREE RADICALS AND HUMAN HEALTH 25

Chapter 3 The Molecules of Living Things 35
The Fundamental Components of Biological Molecules 36
Carbon: The Indispensable Element 36
Functional Groups: The Key to Chemical Reactivity 38
Monomers and Polymers: Molecular Links in a Biological Chain 39
Carbohydrates: Sources of Stored Energy 39
Monosaccharides: Simple Sugars 40
Disaccharides: Sugars Built of Two Monosaccharides 40
Polysaccharides: Storage Depots and Structural Scaffolds 41
Lipids: Energy, Interfaces, and Signals 43
Fats and Oils: Storehouses of Cellular Energy 43
Phospholipids: The Ambivalent Lipids 45
Steroids: Regulatory Molecules 45
Proteins: The Basis of Life's Diversity 46
Amino Acids: The Building Blocks of Proteins 47
Polypeptides: Amino Acid Chains 48
The Structure of Proteins 49
Factors Causing a Protein's Specific Three-Dimensional Shape 52
Nucleic Acids: The Code of Life 54
Looking Ahead 56
Summary / Key Terms / Questions / Essay Question 56
BOX: PROTEIN FOLDING: A THREE-DIMENSIONAL PUZZLE 54

Chapter 4 Chemical Reactions, Enzymes, and Metabolism 58
The Energetics of Chemical Reactions 59
The First Law of Thermodynamics 59
The Second Law of Thermodynamics 60
Free Energy Changes 61
Equilibrium and Free Energy 62
Rates of Chemical Reactions 63
Temperature and Reaction Rates 64
Concentration and Reaction Rates 65
Catalysts and Reaction Rates 65
Enzymes and How They Work 66
Enzyme Structure 67
Enzyme Function 67
Enzymes and Reaction Rates 70
Metabolic Pathways 71
Control of Enzymes and Metabolic Pathways 72

Looking Ahead 74
Summary / Key Terms / Questions / Essay Questions 75
BOX: LUCIFERASE: A LUMINOUS ENZYME 70

Chapter 5 Cells: Their Properties, Surfaces, and Interconnections 77

How Cells Are Studied 78
Microscopy 78
Means of Studying Cell Function 80
Characteristics of Cells 81
The Nature and Diversity of Cells 81
Limits on Cell Size 83
The Cell Surface 84
The Plasma Membrane 84
Movement of Materials into and out of Cells: Role of the Plasma Membrane 87
Osmosis and Cell Integrity 89
Cell Walls and the Glycocalyx 92
Cell Walls 92
Glycocalyx 93
Linkage and Communication Between Cells 94
Multicellular Organization 95
Tissues, Organs, and Systems 95
Summary / Key Terms / Questions / Essay Questions 97
BOX: MICROSCOPES FOR THE LAST CELLULAR FRONTIER 79
BOX: THE SECRETS OF WINTER WHEAT 88

Chapter 6 Inside the Living Cell: Structure and Function of Internal Cell Parts 98

Cytoplasm: The Dynamic, Mobile Factory 99
The Nucleus: Information Central 99
Organelles: Specialized Work Units 100
Ribosomes: Protein Synthesis 100
Endoplasmic Reticulum: Production and Transport 101
Golgi Complex: Modifications for Membranes or Export 103
Vacuoles: Food and Fluid Storage and Processing 105
Coated Vesicles: Mediated Uptake and Transport 106
Lysosomes: Digestion and Degradation 106
Mitochondria and Plastids: Power Generators 107
The Cytoskeleton 109
Cellular Movements 111
Creeping and Gliding Cell Movements 111
Swimming Cell Movements 112
Internal Cell Movements 114
Summary / Key Terms / Questions / Essay Questions 115

Chapter 7 Harvesting Energy from Nutrients: Fermentation and Cellular Respiration 117

ATP: The Cell's Energy Currency 118
The Structure of ATP 118
ATP and the Harvesting of Energy 119
Oxidation–Reduction Reactions 119
Glycolysis: The First Phase of Energy Metabolism 121
Splitting Glucose: The Steps of Glycolysis 121
Fermentation 124
Cellular Respiration 124
Oxidation of Pyruvate: Prelude to the Krebs Cycle 127
The Krebs Cycle 128
The Electron Transport Chain 128
Mitochondrial Membranes and the Mitchell Hypothesis 128
The Energy Score for Respiration 130
Metabolism of Fats and Proteins 131
Control of Metabolism 132
Summary / Key Terms / Questions / Essay Questions 133

Chapter 8 Photosynthesis: Harnessing Solar Energy to Produce Carbohydrates 135

An Overview of Photosynthesis 136
The Two Basic Reactions to Photosynthesis 136
Chloroplasts: Sites of Photosynthesis 138
How Light Energy Reaches Photosynthetic Cells 138
The Nature of Light 139
The Absorption of Light by Photosynthetic Pigments 139
The Light-Dependent Reactions: Converting Solar Energy to Chemical Bond Energy 142
Electron Flow in Photosystems I and II 142
Photophosphorylation: Light Energy Captured in ATP 142
The Light-Independent Reactions: Building Carbohydrates 144
Oxygen: An Inhibitor of Photosynthesis 146
Reprieve from Photorespiration: The C_4 Pathway 147
The Carbon Cycle 149
Summary / Key Terms / Questions / Essay Questions 150
BOX: BACTERIAL GENES AND SOYBEANS 149

PART TWO

LIKE BEGETS LIKE 153

Chapter 9 Cellular Reproduction: Mitosis and Meiosis 154

The Nucleus and Chromosomes 155
- The Role of the Nucleus 155
- Chromosomes and DNA 156

The Cell Cycle 158
- Stages of the Cell Cycle 158
- Control of the Cell Cycle 159

Mitosis: Partitioning the Hereditary Material 159
- Prophase 160
- Metaphase 161
- Anaphase 161
- Telophase 162
- The Nature of the Spindle 162

Cytokinesis: Partitioning the Cytoplasm 163

Meiosis: The Basis of Sexual Reproduction 164
- Prophase I 165
- Metaphase I and Anaphase I 165
- Telophase I 165
- Meiotic Interphase and Prophase II 166
- Metaphase II and Anaphase II 166
- Telophase II 166
- Recombination in Meiosis 166

Asexual Versus Sexual Reproduction 166
- Asexual Reproduction 166
- Sexual Reproduction 167

Looking Ahead 168
- Summary / Key Terms / Questions / Essay Questions 168
- BOX: WHAT MAKES THE CHROMOSOMES DANCE? 162

Chapter 10 Foundations of Genetics 170

Early Theories of Inheritance 171

Gregor Mendel and the Birth of Genetics 172

Mendel's Classic Experiments 172
- Dominant and Recessive Traits 172
- Genetic Alleles 174
- Probability and Punnett Squares 174
- The Law of Segregation 176
- Test Crosses 177
- Dihybrid Crosses and the Law of Independent Assortment 177
- Incomplete Dominance 179
- Mendel's Ideas in Limbo: A Theory Before Its Time 179

Chromosomes and Mendelian Genetics 180

A Century of Progress 183
- Summary / Key Terms / Questions 183
- BOX: THOROUGHBREDS AND THOROUGH BREEDING 176

Chapter 11 Mendel Modified 185

How Genes Are Arranged on Chromosomes 186
- Crossing Over and Recombination 186
- Gene Maps 188

How Genes Act and Interact 190
- How Genes Act 190
- Genes with More Than Two Alleles 190
- Complementary Genes 193
- Epistatic Genes 193

Lethal Alleles and Pleiotropy 194

Mutation: One Source of Genetic Variation 196
- Summary / Key Terms / Questions 197

Chapter 12 Discovering the Chemical Nature of the Gene 199

Genes Code for Particular Proteins 200
- The One Gene–One Enzyme Hypothesis 200
- The One Gene–One Polypeptide Hypothesis 202

The Search for the Chemistry of the Gene 203
- Genes: Nucleic Acid, Not Protein 203
- Chemical Composition of DNA Revealed 206

The Research Race for the Molecular Structure of DNA 207

How DNA Replicates 209
- Is DNA Replication Semiconservative? 210
- DNA Replication in *E. coli* 211
- DNA Replication in Eukaryotes 212

Looking Ahead 213
- Summary / Key Terms / Questions / Essay Question 214
- BOX: THE PROPER MEASURE OF MAN? 209

Chapter 13 Translating the Code of Life: Genes into Proteins 215

Genetic Information Must Occur in Code 216
- The Genetic Code Is Colinear 216
- The Nature of the Genetic Code 217

DNA, RNA, and Protein Synthesis 218
- An Overview of Transcription 219
- Types of RNA 220

Protein Synthesis: Translating the Genetic Code 221
Step 1: Amino Acid Activation 221
Step 2: Initiation 222
Step 3: Elongation 222
Step 4: Termination 224
Cracking the Genetic Code 224
The Concept of the Gene Refined 227
Summary / Key Terms / Questions / Essay Question 227
BOX: A SHY GENETICIST AND JUMPING GENES 226

Chapter 14 Bacterial Genetics, Gene Control, and Genetic Engineering 229

Bacteria Exchange Genes Sexually and Asexually 230
Detection of Bacterial Phenotypes 230
Bacterial Gene Transmission 231
Use of DNA Transfer Processes to Map Prokaryotic Genes 234
Bacterial Genes: Subject to Precise Regulatory Control 234
The Operon Model of Gene Regulation 235
Catabolite Repression 236
Positive and Negative Control 238
Repressible Enzymes 238
Gene Regulation in Eukaryotes 240
The "Pieces" of Eukaryotic Genes 241
Eukaryotic mRNA: Special Processing 241
Recombinant DNA Technology 243
General Steps in a Gene Transfer Experiment 243
The Tools of the Genetic Engineer 243
Engineering a Bigger Mouse 244
Genetic Engineering: Promises and Prospects 247
Microbial Production of Valuable Gene Products 247
Genetic Alteration of Agricultural Species 247
Human Gene Therapy 248
Preventing the Proliferation of Dangerous Bacteria 249
Humanistic Concerns 249
Summary / Key Terms / Questions / Essay Question 249
BOX: ENGINEERING SUPERFISH 248

Chapter 15 Human Genetics 251

What Are the Methods of Human Genetics? 252
Constructing Pedigrees, or Family Trees 252
Karyotyping: Chromosomes Can Reveal Defects 252
Biochemical Analyses Reveal Mutations 254
Mapping Human Genes: Somatic Cell Genetics 255
Reverse Genetics 256
Nature, Nurture, and Twin Studies 256
Chromosomal Abnormalities: A Major Source of Genetic Disease 257
Abnormal Numbers of Autosomes 257
Abnormal Numbers of Sex Chromosomes 258
Translocations, Deletions, and Duplications 259
A Survey of Human Genetic Traits 260
Sex-Linked Traits 261
Recessive Traits on Autosomal Chromosomes 262
Dominant Traits and Variations in Gene Expression 264
Treating Genetic Diseases 266
Genetic Counseling 266
Summary / Key Terms / Questions / Essay Questions 267

Chapter 16 Animal Development 269

Production of Sperm and Eggs 270
Spermatogenesis 270
Oogenesis 271
Fertilization: Initiating Development 274
Sperm Penetration 274
Completing Meiosis 275
Barriers to Other Sperm 275
Cleavage: An Increase in Cell Number 276
Patterns of Cleavage 276
Cytoplasmic Distribution During Cleavage 278
Gastrulation: Rearrangement of Cells 279
Organogenesis: Formation of Functional Tissues and Organs 279
Embryonic Coverings and Membranes 282
Growth: Increase in Size 283
Growth in Embryos 283
Growth in Later Stages of Life 283
Regeneration: Development Reactivated in Adults 284
Aging and Death: Final Developmental Processes 285
Summary / Key Terms / Questions / Essay Questions 285

Chapter 17 Developmental Mechanisms and Differentiation 287

Determination: Commitment to a Type of Differentiation 288
Changes in Determination 288
The Stability of Determination Depends on the Cytoplasm 289
Differentiation: Building Cell Phenotype 290
The Causes of Differentiation 290
Characteristics of Differentiated Cells 291
Morphogenesis: The Organization of Cells into Functional Units 292
Single-Cell Movements 292
Moving Cells and Extracellular Substances 293

Cell Population Movements 294
Localized Relative Growth 295
Localized Cell Death 295
Deposition of Extracellular Matrix 295
Pattern Formation: The Vertebrate Limb 295
Regulatory Genes at Work During Pattern Formation 298
Gene Regulation in Development 299
DNA Processing 299
RNA Synthesis and Processing 299
Protein Synthesis 301
Cancer: Normal Cells Running Amok 302
What Causes Cancer? 302
Proto-Oncogenes in Cells 302
Viral Oncogenes Are Derived from Cells 303
Oncogene Products: Proteins Related to Normal Growth Factors 304
Summary / Key Terms / Questions / Essay Questions 305
BOX: AN ACNE DRUG AND DEVELOPMENTAL DEFECTS 297

Chapter 18 Animal and Human Reproduction 307
Reproduction in the Vertebrates: Anatomy and Strategy 308
The Male Reproductive System 309
Production and Transport of Sperm 309
The Male Sexual Response 310
The Role of Hormones in Male Reproductive Function 310
The Female Reproductive System 310
Production and Transport of the Egg 310
Female Genitals 312
The Female Sexual Response 312
The Role of Hormones in Female Reproductive Function 312
Origins of Sex Differences: Femaleness and Maleness 312
Hormonal Control of Sexual Development 313
Intersexes—Hormones Gone Wrong 314
Chromosome Imprinting and Development 315
Pregnancy and Prenatal Development 315
The First Trimester 315
The Second and Third Trimesters 317
The Placenta: Exchange Site and Hormone Producer 317
Birth: An End and a Beginning 318
Multiple Births 319
Milk Production and Lactation 321
Looking Ahead 322
Summary / Key Terms / Questions / Essay Question 323
BOX: IN VITRO FERTILIZATION: A GOING CONCERN 320
BOX: BIRTH CONTROL AND ABORTION 322

PART THREE ORDER IN DIVERSITY 325

Chapter 19 The Origin and Diversity of Life 326
A Home for Life: Formation of the Solar System and Planet Earth 327
The Emergence of Life: Organic and Biological Molecules on a Primitive Planet 328
The Formation of Monomers 328
The Next Step: Polymers 328
From Polymers to Aggregates 329
From Aggregates to Cell-Like Units 330
The Earliest Cells 332
Heterotrophs First, Autotrophs Later 332
The Emergence of Aerobes 333
Eukaryotes 333
The Changing Face of Planet Earth 334
Taxonomy: Categorizing the Variety of Living Things 335
The Binomial System of Nomenclature 335
Taxonomy and Darwin's Theory of Evolution 335
Taxonomy and Evolutionary Relationships 336
The Five Kingdoms 338
Summary / Key Terms / Questions / Essay Question 338
BOX: OUTER SPACE: THE SOURCE OF ORGANIC PRECURSORS? 329

Chapter 20 Monera and Viruses: The Invisible Kingdom 340
Monera: Tiny but Complex Cells 341
Bacterial Cell Structure 341
Bacterial Cell Movement 342
Bacterial Reproduction 343
Bacterial Metabolism 345
The Variety of Monerans 346
Subkingdom Schizomycete: The Bacteria 346
Cyanobacteria: The Blue-Green Photosynthesizers 348
Chloroxybacteria: Possible Ancestors to Chloroplasts 349
Archaebacteria: An Independent Kingdom? 349
Bacteria and Humans 350
Bacteria and Food Production 350
Bacteria and Disease 350

Viruses: Noncellular Molecular Parasites 351
Structure of Viruses 351
Reproduction of Viruses 352
Viruses and Disease 353
Viruses and Cancer 355
Evolution of Monerans, Viruses, and Eukaryotic Cells 355
Summary / Key Terms / Questions / Essay Question 356
BOX: BEATING VIRUSES AT THEIR OWN GAME 357

Chapter 21 Protista: The Kingdom of Complex Cells 359

Protist Classification by Life-Style 360
The Producers: Plantlike Protists 361
Euglenophyta 362
Pyrrophyta 362
Chrysophyta 363
The Decomposers: Funguslike Protists 364
Myxomycota: True Slime Molds 364
Acrasiomycota: Cellular Slime Molds 364
The Consumers: Animal-Like Protists 365
Mastigophora 365
Sarcodina 366
Sporozoa 369
Ciliophora 369
Caryoblastea 371
Microspora 371
Protistan Evolution 371
Summary / Key Terms / Questions / Essay Questions 371
BOX: MODERN METHODS FOR FIGHTING MALARIA 367

Chapter 22 Fungi: The Great Decomposers 373

Characteristics of Fungi 374
Fungal Growth Patterns 375
Fungal Life Cycles 376
Spores for Dispersal and Survival 376
Classification of Fungi 376
Chytridiomycota: The Interface Between Protists and Fungi 377
Oomycota 379
Zygomycota 379
Ascomycota 380
Basidiomycota 382
Deuteromycota (Fungi Imperfecti) 384
Lichens: The Ultimate Symbionts 384
Fungal Evolution 385
Summary / Key Terms / Questions / Essay Question 386
BOX: ELM EPIDEMIC IN DUTCH? 380

Chapter 23 Plants: The Great Producers 388

The Plant Kingdom and Its Major Divisions 389
The Basic Plant Life Cycle 390
The Algae: Diverse Aquatic Producers 390
Chlorophyta: Likely Ancestors of the Land Plants 391
Chlamydomonas and Other Volvocine Algae 391
Rhodophyta: Photosynthesizers in Shallow and Deep Waters 392
Phaeophyta: Giant Aquatic Organisms 393
Plants That Colonized Land 394
Nonvascular Land Plants: Bryophyta 395
Seedless Vascular Plants: Support and Transport, But No Seeds 396
Classification of the Seed Plants 398
Evolution of the Seed Plants 399
Gymnosperms: "Naked-Seed" Plants 399
Cycadophyta 399
Ginkgophyta: The Gingko Tree 400
Gnetophyta: The Gnetinas 400
Coniferophyta: The Familiar Conifers 400
Anthophyta: The Apex of Plant Diversity 403
Special Adaptations of Flowering Plants 403
Key to the Flowering Plant Life Cycle 404
Pollination and Pollinators: Key to Reproductive Success in Anthophyta 405
Fruits: An Important Evolutionary Strategy for Seed Dispersal 406
Looking Ahead 407
Summary / Key Terms / Questions / Essay Question 407
BOX: THE ORIGIN OF CORN: AN UNSOLVED MYSTERY 402

Chapter 24 Animals: The Great Consumers 409

Porifera: The Simplest Invertebrates 411
Radial Symmetry in the Sea 412
Cnidaria: Alternation of Generations in Animals 412
Ctenophora: The Comb Jellies 414
The Origins of Multicellularity 414
Organs and Bilateral Symmetry 414
Platyhelminthes 415
Nemertina: Two-Ended Gut and Blood Vessels 416
Nematoda: Emergence of the First Body Cavity 418
The Two Major Animal Lineages: Protostomes and Deuterostomes 420
Protostomes 420
Annelida: The Segmented Worms 420
Arthropoda: Exploiting Segmentation 422
Mollusca: Rearrangement of the Protostome Body Plan 427
Deuterostomes: A Loosely Allied Assemblage 428
Echinodermata: Radial Symmetry Encountered Once More 428
Hemichordata: The First Gill Slits 430

Characteristics of the Chordates 430
Gill Slits: Feeding and Respiration 430
Notochord: A Structure to Prevent Body Shortening 430
Muscle Blocks: The Basis for Swimming Locomotion 431
Nerve Cord: Coordination of Movements 431
The Nonvertebrate Chordates 431
Urochordata 431
Cephalochordata 432
Jawless Fishes: The First Vertebrates 432
Bony-Skinned Fishes and Modern-Day Descendants 432
Vertebrate Origins and the History of Bone 433
Jawed Fishes: An Evolutionary Milestone 433
Chondrichthyes: Fast-Swimming Predators 434
Osteichthyes: Adaptability and Diversity 434
Amphibians: Vertebrates Invade the Land 436
Amphibian Adaptations 436
Modern Amphibians 436
Reptiles: Adaptations for Dry Environments 436
Reptilian Adaptations 437
The Age of Reptiles 437
Modern Reptiles 439
Birds: Vertebrates Take to the Air 440
Mammals: Warm Blood, Hair, Mammary Glands, and a Large Brain 441
Mammalian Adaptations 442
Trends in Mammalian Evolution 442
Modern Mammals 443
Summary / Key Terms / Questions / Essay Questions 444
BOX: TICKS AND LYME DISEASE 425
BOX: ARE MASS EXTINCTIONS CYCLICAL? 440

PART FOUR

PLANT BIOLOGY 447

Chapter 25 The Architecture of Plants 448
Meeting the Challenges of Life on Land 448
Leaves: Living Collectors of Solar Energy 449
Leaf Structure 450
Leaf Adaptations 453
The Stem: Support and Transport 454
Stem Structure 454
Stem Vascular Tissue 455
Stem Adaptations 456
The Root: Anchorage and Uptake 457
Root Structure 457
Roots and Oxygen 460
Types of Roots 460
Root Adaptations 460
Looking Ahead 461
Summary / Key Terms / Questions / Essay Question 461
BOX: AGRICULTURE AND HUMAN CULTURE 450

Chapter 26 How Plants Reproduce, Develop, and Grow 463
Vegetative Reproduction: Multiplication Through Cloning 463
Vegetative Reproduction in Nature 464
Plant Propagation in Agriculture 465
Advantages and Disadvantages of Vegetative Reproduction 465
Sexual Reproduction: Multiplication and Diversity 466
Flowers: Ingenious Insurance for Fertilization 467
Pollen Production 468
Egg Production 469
Pollination and Fertilization 470
The Development of Plant Embryos 470
Seeds: Protection and Dispersal of the New Generation 472
Seed Maturation 473
Seed Dormancy 473
Germination and Seedling Development 474
Primary Growth: From Seedling to Mature Plant 475
Primary Growth of the Root 475
Primary Growth of the Shoot 475
Secondary Growth: The Development of Wood and Bark 478
Vascular Cambium 479
Cork Cambium 479
Plant Life Spans and Life-Styles 479
Looking Ahead 481
Summary / Key Terms / Questions / Essay Questions 481
BOX: WHAT PRICE ATTRACTIVENESS? 466

Chapter 27 Exchange and Transport in Plants 483
Plant Strategies for Meeting Basic Needs 484
Strategies for Gas Exchange 484
Strategies for Internal Transport of Liquids 484
Strategies for Coping with Wastes 484
Transport of Water in the Xylem 484
Principles of Water Movement 486
Theories of Upward Movement 487
Regulation of Water Loss 488
Transport of Solutions in the Phloem 489
What Is Translocated in the Phloem? 490

Kinetics of Transport in the Phloem 490
Mass Flow Theory: Transport in the Phloem 491
How Roots Obtain Nutrients from the Soil 492
Mineral Requirements and Uptake 492
Rate of Mineral Uptake 493
Nitrogen Uptake and Fixation 493
Mycorrhizae 494
Carnivorous and "Sensitive" Plants 495
Looking Ahead 496
Summary / Key Terms / Questions / Essay Question 496

Chapter 28 Plant Hormones 498

Auxins: Cell Elongation and Plant Movements 498
Effects of Auxins 500
How Auxins Work 501
Gibberellins: Growth Promoters 502
Cytokinins: Cell Division Hormones 503
Effects of Cytokinins 503
Cytokinins, Auxins, Tissue Cultures, and Plant Biotechnology 504
Abscisic Acid: The Growth-Slowing Hormone 505
Effects of Abscisic Acid 506
Interaction of Abscisic Acid and Other Hormones 506
Ethylene: The Gaseous Hormone 506
Interaction of Plant Hormones 507
Leaf Fall 507
Seed Germination 508
Control of Flowering 508
Photoperiodism 508
The Phytochrome Pigment System 509
Florigen: One Flowering Hormone or Many? 510
Looking Ahead 510
Summary / Key Terms / Questions 510
BOX: BRAVE NEW FOOD 505

PART FIVE ANIMAL BIOLOGY 513

Chapter 29 The Circulatory and Transport Systems 514

Some Basic Principles of Animal Physiology 514
Strategies for Transporting Materials in Animals 516
Relying on Diffusion: Limitations on Body Size and Shape 516
Circulatory Systems 516
The Vertebrate Circulatory System 518
The Heart 518
Blood Circuits 521
Structure of Arteries and Veins 521
Capillary Beds 523
Blood Pressure 524
Fluid Balance and the Diffusion of Materials 524
Regulation of Blood Flow 525
Blood Flow and Vessel Size 525
Control of the Heart's Output 525
Regulation of Blood Flow Through the Vessels 526
Brain Regulation of the Circulatory System 527
Blood: The Fluid of Life 528
Plasma: The Fluid Portion of Blood 528
The Blood's Solid Components 528
The Lymphatic System and Tissue Drainage 530
Looking Ahead 530
Summary / Key Terms / Questions / Essay Question 531
BOX: RISK FACTORS IN OUR NATION'S BIGGEST KILLER 522

Chapter 30 The Immune System 533

Components of the Immune System 534
White Blood Cells and Their Protein Products 534
Organs of the Immune System 535
Lines of Defense: Nonspecific and Specific 536
The Inflammatory Response 536
The Immune Response 537
Antibodies and Humoral Immunity 537
Antibody Structure: Variable and Constant Regions 538
Development of Antibody Specificity: The Clonal Selection Theory 539
Antibody Diversity: A Matter of Gene Shuffling 541
Monoclonal Antibodies: A Miracle for Medicine 542
T Cells and Cell-Mediated Immunity 543
Natural Killers: The Third Class of Lymphocytes 544
Autoimmune Disorders: A Breakdown in Tolerance 544
Immune Deficiency 545
Immune Phenomena and Human Medicine 546
Active Immunity: Protection Based on Past Exposures 546
Passive Immunity: Transfer of Antibodies 547
Allergy: An Immune Overreaction 548
Tissue Compatibility and Organ Transplants 549
Perspective 550
Summary / Key Terms / Questions / Essay Question 550
BOX: IMMUNITY AND THE MIND 549

Chapter 31 Respiration: The Breath of Life 552

Respiration and the Physics of Gases 553
Respiratory Organs: Structures for Efficient Gas Exchange 555
Wet Body Surface as a Site for Gas Exchange 555
Gills: Aquatic Respiratory Organs 556
Tracheae: Respiratory Tubules 558
Lungs: Intricate Air Sacs for Gas Exchange 558
The Human Respiratory System: The Hollow Tree of Life 559
The Air Pathway 559
How the Lungs Are Filled 560
Respiration in Birds: A Special System 561
Swim Bladders in Fish: Buoyancy Organs Derived from Lungs 562
Respiratory Pigments and the Transport of Blood Gases 562
Pigment Molecules 562
Control of Hemoglobin Function 564
Myoglobin: A Molecular Oxygen Reservoir 564
Carbon Dioxide Transport: Ridding the Body of a Metabolic Waste 565
Control of Breathing 565
How Animals Adapt to Oxygen-Poor Environments 567
Summary / Key Terms / Questions / Essay Question 568
BOX: THIN AIR AND THE HUMAN BRAIN 554

Chapter 32 Digestion and Nutrition 570

Animal Strategies for Ingestion and Digestion 571
Intracellular Digestion in Simple Organisms 571
Intracellular and Extracellular Digestion in Simple Animals 571
Extracellular Digestion in Complex Animals 571
The Human Digestive System 575
The Oral Cavity 575
The Pharynx and Esophagus 576
The Stomach 576
The Small Intestine: Main Site of Digestion 578
The Large Intestine 580
The Liver 581
The Chemistry of Digestion 581
Coordination of Ingestion and Digestion 584
Control of Enzyme Secretion and Gut Activity 584
Control of Hunger and Feeding 584
Control of Body Fat 585
Nutrition 585
Macronutrients: The Basic Foods 586
Micronutrients 587
Dietary Guidelines 589
Summary / Key Terms / Questions / Essay Question 590
BOX: SKIN COLOR, DRIFTING CONTINENTS, AND VITAMIN D 586

Chapter 33 Homeostasis: Maintaining Biological Constancy 592

Regulation of Body Fluids 593
Life in Fresh Water 593
Life in Salt Water 594
Life on Dry Land 595
Excretion of Nitrogenous Wastes 596
Excretory Systems in Invertebrates 597
Excretory Systems in Vertebrates: The Marvelous Kidney 598
Homeostasis and Temperature Regulation 604
The Impact of Temperature on Living Systems 604
Solutions to Temperature Problems 604
Temperature Regulation in Aquatic Organisms 605
Temperature Regulation in Amphibians, Reptiles, and Insects 606
Temperature Regulation in Birds and Mammals 607
Looking Ahead 610
Summary / Key Terms / Questions / Essay Question 610
BOX: ALL DRIED UP 594
BOX: SUPERCOOLED SQUIRRELS 610

Chapter 34 The Nervous System 612

Neurons: The Basic Units of the Nervous System 613
Neuron Structure: Key to Function 613
Three Primary Types of Neurons 614
How Neurons Signal 615
How the Action Potential (Nerve Impulse) Is Generated 615
How the Action Potential Is Propagated 618
Transmission of the Action Potential Between Cells 620
The Organization of Neurons into Systems 623
Simple Circuits 623
Complex Circuits 624
The Vertebrate Nervous System 624
Looking Ahead 627
Summary / Key Terms / Questions / Essay Question 627

Chapter 35 Hormonal Controls 629

Basic Characteristics of Hormones 630
Cellular Mechanisms of Hormonal Control 631
The Cellular Mechanisms of Steroid Action 632
Nonsteroids and the Cellular Mechanisms of Second Messengers 633
Endocrine Functions in Invertebrates 635
The Vertebrate Endocrine System 636
The Pancreas 636
The Kidneys 637
The Heart as an Endocrine Organ 638
The Adrenal Glands 638
The Thyroid and Parathyroid Glands 639
The Pituitary Gland 640

Hormonal Control of Physiology 644
Control of Blood Sugar Levels 644
Rhythms, Hormones, and Biological Clocks 644
Hormones and Development: Amphibian Metamorphosis 645
Hormones: New Types, New Sites, New Modes of Synthesis 646
Looking Ahead 647
Summary / Key Terms / Questions / Essay Questions 647
BOX: ANABOLIC STEROIDS AND THE INSTANT PHYSIQUE 633

Chapter 36 Input and Output: The Senses and the Brain 649

Sensory Receptors 650
Chemoreceptors 651
Gustation: The Sense of Taste 651
Olfaction: The Sense of Smell 651
Mechanoreceptors 652
The Statocyst and Inner Ear: Maintaining Equilibrium and Sensing Movement 654
The Ear and Hearing 656
Magnetism: Detecting the Earth's Magnetic Field 659
Thermoreceptors 659
Photoreceptors 660
The Vertebrate Eye 660
The Cephalopod and Arthropod Eyes 664
Structure and Function of Vertebrate Brains 665
The Hindbrain: Respiration, Circulation, and Balance 667
The Midbrain: Optic Tecta, Certain Instincts, and Some Reflexes 667
The Forebrain: Increasing the Complexity of Brain Functions 668
The Brain's Sentinel: The Reticular Activating System 669
The Neocortex of the Cerebrum: The "New Bark" 669
Exploring the Living Brain 671
Split-Brain Studies and the Cerebral Hemispheres 671
Electrical Stimulation of the Brain: Probing the Cortex 673
Electrical Recording to Explore the Brain: Sleep and Brain Waves 673
Brain Studies at the Level of the Neuron 674
Memory: Multiple Circuits at Work? 676
The Chemical Messengers of the Brain 677
Summary / Key Terms / Questions / Essay Questions 680
BOX: DOPAMINE AND ADDICTION: THE COCAINE CONNECTION 678

Chapter 37 Skeletons and Muscles 683

The Animal Skeleton: A Living Scaffold 684
Exoskeletons 684
Endoskeletons 685
Regulating Movement: Feedback Control of Muscle Action 690
Muscle Structures and Functions 692
Special Features of Skeletal Muscle 692
Muscle Structure 693
The Molecular Basis of Muscle Contraction 694
How Muscle Contraction Is Initiated 695
Turning Muscles Off 697
More on Muscle Action: Graded Responses and Muscle Tone 697
Smooth Muscle 698
Cardiac Muscle: Striated Tissue with Smooth Muscle Characteristics 699
Muscles in Evolution: The Contractile Systems of Invertebrates 700
Summary / Key Terms / Questions 701
BOX: RUNNING THE MILE: WHAT IS OUR SPEED LIMIT? 699

PART SIX POPULATION BIOLOGY 703

Chapter 38 Evolution and the Genetics of Populations 704

The Origins of Evolutionary Thought 705
Lamarck and the Inheritance of Acquired Characteristics 705
Darwin and Natural Selection 705
Variations in Genes: The Raw Material of Natural Selection 708
Looking for Genetic Variation: Protein Electrophoresis 708
Population Genetics: The Links Between Genetics and Evolution 710
The Gene Pool and Gene Frequencies 710
Genetic Equilibrium and the Hardy-Weinberg Law 711
Mechanisms of Evolution: Upsetting the Gene Pool Equilibrium 713
Nonrandom Mating 713
Genetic Drift 714
Mutation 716
Migration and Gene Flow 717
Looking Ahead 717
Summary / Key Terms / Questions / Essay Question 717
BOX: THE FABLE AND FACTS OF DARWIN'S FINCHES 707

Chapter 39 Natural Selection 719

The Darwinian and Genetic Meanings of Fitness 720
Natural Selection: How Phenotype Affects Genotype 722
Types of Selective Processes 723
Normalizing Selection: Preserving the Status Quo 723
Directional Selection: Changing Phenotypes 723
Diversifying Selection: Producing Variant Phenotypes 725
Maintaining Genetic Variation 726
New Views on Evolutionary Mechanisms 728
The Meaning of Adaptation 728
The Neutralist-Selectionist Debate and Molecular Evolution 728
Groups of Genes as Units of Selection 730
Looking Ahead 730
Summary / Key Terms / Questions / Essay Question 731

Chapter 40 The Origin of Species 732

How Biologists Define a Species 733
Preventing Gene Exchange 733
The Role of Isolating Mechanisms 733
The Role of Clines in Speciation 736
Becoming a Species: How Gene Pools Become Isolated 737
The Genetic Bases of Speciation 738
Regulatory Genes and Evolutionary Changes 739
Chromosomal Changes 739
Explaining Macroevolution: Higher-Order Changes 740
The Fossil Record 741
Parallel, Convergent, and Divergent Evolution 743
Extinction 745
The Punctuated Equilibrium Theory 747
The Role of Microevolution in Macroevolution 748
Looking Ahead 749
Summary / Key Terms / Questions / Essay Question 749
BOX: SORTING OUT SEXUAL SELECTION 735
BOX: SILVERSWORDS: HAWAII'S EXEMPLARS OF ADAPTIVE RADIATION 746
BOX: AN EPIDEMIC OF EXTINCTIONS: THE DISHEARTENING FACTS 747

Chapter 41 Ecosystems and the Biosphere 751

Ecosystems 752
The Biosphere 752
The Sun as a Source of Energy 753
The Sun, Seasons, and Climate 754
Temperature, Wind, and Rainfall 754
Biomes 757
The Habitats of Life: Air, Land, and Water 761
Air: The Atmosphere 761
Land 761
Water 762
Energy Flow in Ecosystems 763
Ecological Pyramids 764
Generating Biomass: Energy Relationships in the Ecosystem 765
The Efficiency of Energy Transfer 766
Cycles of Materials 768
The Water Cycle 768
The Carbon Cycle 769
The Nitrogen Cycle 770
The Phosphorus Cycle 771
Close-Up of a Lake Ecosystem 772
Looking Ahead 773
Summary / Key Terms / Questions / Essay Question 773
BOX: ECOLOGICAL TROUBLE IN THE TROPICS 758
BOX: ANTARCTIC FOOD WEBS: AN UNCONTROLLED EXPERIMENT 764
BOX: ACID RAIN AND DUST: AIRBORNE RECIPE FOR DISASTER? 767
BOX: HELP FIGHT GREENHOUSE WARMING 770

Chapter 42 The Ecology of Communities 775

Ecological Succession: A Basic Feature of Life 777
Primary Succession 777
Secondary Succession 778
Looking for the "Rules" of Succession 780
Animal Succession 780
The Structure of Communities 780
Physiognomy: What Does a Community Look Like? 780
Relative Abundance of Species 781
Species Richness: Taking Count of Species 781
The Niche 784
Adaptations for Defense 787
Mechanical Defenses of Animals and Plants 787
Chemical Defenses of Animals 787
Chemical Defenses of Plants 788
Warning Coloration and Mimicry 790
Adaptations for Escape 792
Escape in Space and Time 792
Camouflage 792
Symbiosis 794
Parasitism 794
Commensalism and Mutualism 795
Looking Ahead 796
Summary / Key Terms / Questions / Essay Question 797

Chapter 43 The Ecology of Populations 799

Population Growth 800
Exponential Growth 800
Carrying Capacity and Logistic Growth 801
Fluctuations in Population Size 801
The Effects of Age Structure 802
Reproductive Strategies: *r*- and *K*-Selection 803

Limits on Population Size 804
Factors Controlling Density 805
Population Fluctuation and Community Structure 809
How Populations Are Distributed 809
Human Populations: A Case Study in Exponential Growth 810
Looking Ahead 813
Summary / Key Terms / Questions / Essay Question 813
BOX: BIOLOGICAL PEST CONTROL: DOING WHAT COMES NATURALLY 808

Chapter 44 Behavioral Adaptations to the Environment 815

Reflex Behavior 816
Tropisms, Taxes, and Kineses 817
Instincts: Inherited Behavioral Programs 817
Open and Closed Programs 818
Triggering Behavior: Sign Stimuli and Innate Releasing Mechanisms 819
Physiological Readiness for Behavior 821
Habituation 821
Learning 822
Programmed Learning 822
Latent Learning 824
Complex Learning Patterns 825
Learning and Cultural Transmission 826
Complex Behavior: Navigation and Migration 827
Navigation 828
Migration 828
Behavior in Perspective 830
Summary / Key Terms / Questions / Essay Question 830
BOX: THE FLEXIBILITY OF ANIMAL BEHAVIOR 823

Chapter 45 Social Behavior 832

Behavior and Ecology 833
Communication 834
Insect Communication: The "Waggle Dance" of the Honeybee 835
Mating Behavior 835
Territoriality and Aggression 836
Altruistic Behavior 837
Kin Selection 838
Insect Societies 839
The Social Organization of the Honeybee 839
Social Systems of Vertebrates 840
Societies of Mammals 841
Life in the Herd 841
The Social Organization of Wolves 842
Primate Societies 842
Animal Behavior and *Homo sapiens* 844
Summary / Key Terms / Questions 844

Chapter 46 Human Origins 846

The Primates 847
Early Evolution of the Primates 849
Characteristics of the Primates 850
The Transition from Ape to Hominid 852
Molecular Relatedness 852
Bipedalism 853
Reproductive Behavior and Human Evolution 853
The Evolution of the Hominids 854
The Genus Homo 857
The Origin and Diversification of Recent Humans 860
Humankind and the Future of the Biosphere 862
Summary / Key Terms / Questions 864

Appendix A The Classification of Organisms A-1
Appendix B Some Useful Chemical Measurements A-5
Appendix C Suggested Readings A-6
Glossary G-1
Photo Credits C-1
Index I-1

List of Experiments

Experiment	Chapter	Page
Are cells surrounded by two layers of phospholipids?	5	84
Do proteins move about in the plasma membrane?	5	86
What are the roles of ER and the Golgi complex in moving cellular materials?	6	105
How do large molecules enter a cell?	6	106
Does genetic information reside in the cell nucleus?	9	156
Which cells transmit traits from parents to offspring?	10	171
What is the physical basis of heredity?	10	174
How can a test cross reveal an unknown genotype?	10	177
How are alleles of a gene distributed during gamete formation?	10	178
Where are the units of heredity located in a living cell?	10	180 ff.
Are genes located in a linear sequence along the chromosomes?	11	188
Are some alleles lethal, such as the one for short, kinked tails in mice?	11	194
Can environmental factors induce mutations?	11	196
Is there a link between specific genes and specific enzymes?	12	200 f.
Is there a link between genes and other proteins, as well?	12	202
What is the chemical nature of the gene?	12	203
Are genes made up primarily of DNA or protein?	12	204 f.
What is the molecular structure of DNA?	12	207
What is the basic mechanism for DNA replication?	12	210

Experiment	Chapter	Page
Does the order of subunits in a gene correspond to the order of subunits within a protein?	13	216
How is information coded in the genes?	13	218
What is the relationship between a given codon and a given amino acid?	13	225
How can we detect genetic variation in bacteria?	14	230
Can bacteria exchange genes?	14	231
What control mechanisms turn bacterial genes on and off?	14	235
What is the role of "extra" DNA in eukaryotes?	14	240
How do genetic engineers isolate the specific genes that code for desired proteins?	14	243
How does a researcher make thousands of duplicate copies of a gene and transfer them from one organism to another?	14	244
How can we locate particular genes on human chromosomes?	15	255
What is the genetic basis for Huntington's disease?	15	264
Can a cell's pattern of development be modified?	17	288
What factors govern a cell's developmental pathway?	17	289
What factors govern the way tissues and organs develop?	17	290
How could the simple building blocks of life have formed on the early earth?	19	328
How could aggregations of these building blocks have formed?	19	329
Could early RNAs have replicated themselves?	19	331
What is the composition of the nutrient fluid that flows from a leaf to other plant parts?	27	490

Experiment	Chapter	Page
Why does a plant bend toward the light?	28	499
What factor controls this bending?	28	499 f.
Does a plant flower in response to day length—the amount of light it receives in a 24-hour period?	28	508
Which parts of a plant actually measure day length?	28	510
How can a researcher fuse white blood cells and cancerous cells to produce antibodies of a given type?	30	542
How did an 18th-century doctor learn to prevent disease through vaccination?	30	547
What controls the breakdown of food in the small intestine?	32	584
What makes an animal stop eating when its stomach is full?	32	584
How does a small brain region act as the body's thermostat?	33	609
What happens to a person's speech and vision when a surgeon separates the two sides of his or her brain?	36	672
What happens when a researcher stimulates areas of the brain electrically?	36	673
How does the brain region involved in vision receive impulses and visual patterns?	36	674

Experiment	Chapter	Page
Can a researcher successfully transplant brain cells?	36	675
How can a researcher distinguish between different kinds of protein molecules?	38	709
Can natural selection produce fruit flies that only fly up or down?	39	722
Can a moth species undergo rapid evolutionary change when its environment changes rapidly?	39	725
Does competition between two species affect where and how each species may live?	42	808
How do plentiful resources affect a population's growth pattern?	43	825
Can the amount of available living space limit a population's size?	43	830
What factors govern the population sizes of predators and their prey?	43	831
Is a given behavior, such as egg rolling in the goose, instinctive or learned?	44	842
What triggers the instinctive feeding behavior in gull chicks?	44	843
Must an animal be physically primed before a stimulus can trigger an instinctive behavior?	44	845
How do birds navigate and migrate over long distances?	44	853

A Guided Tour to *Essentials of Biology*

Chapter Introduction

LITERARY QUOTE AND PHOTO

Appropriate literary quote and stunning photo to capture student interest and set the tone for the chapter.

26

How Plants Reproduce, Develop, and Grow

I took an earthenware pot, placed in it 200 lb of earth dried in an oven, soaked this with water, and planted in it a willow shoot weighing 5 lb. After 5 years had passed, the tree grown therefrom weighed 169 lb and about 3 oz. . . . Finally, I again dried the earth of the pot and it was found to be the same 200 lb minus about 2 oz. Therefore, 164 lb of wood, bark, and root had arisen from water alone.

Jean-Baptiste van Helmont,
Ortus Medicinae (1648)

A living castaway: After months at sea, a coconut palm sprouts and takes root just above the tide line on a Virgin Islands beach.

Flowering plants have flourished largely because of their innovations in sexual reproduction, the flowers, fruits, and seeds. Yet many plants within this group also exploit asexual *vegetative reproduction*, a kind of cloning that can involve stems, roots, or leaves and result in offspring that are genetically identical to the parent. The study of both modes of reproduction in flowering plants and of subsequent growth and development of new individuals helps reveal how plants differ from animals.

One aspect of plant development is particularly distinctive: Plants have perpetual embryonic centers that produce new organs throughout the life of the individual, whereas most animals form organs only as embryos. The plant's unique capacity for renewed growth and development throughout adult life is utilized during flowering and sexual reproduction as well as in modified form during vegetative reproduction.

Our goal in this chapter is to survey the entire range of plant parts, reproductive processes, and growth—from the drab to the glorious, from the asexual to the sexual, from pollen and eggs to the woody tissues of mature trees and bushes. The details of reproduction and growth help to characterize the flowering plants (and, in some respects, the conifers) and explain why they are such highly successful groups.

Our discussions will cover:

- Vegetative reproduction, or multiplication through cloning in nature and agriculture
- Sexual reproduction, and the roles of flower, pollen, sperm, egg, pollination, and fertilization
- The development of plant embryos
- Seeds and the dispersal of the new generation
- Germination of the seed and development of the new seedling
- Primary growth—increasing size and the addition of new tissues in the young plant
- Secondary growth—the development of wood and bark in older plants
- Plant life spans and life-styles

VEGETATIVE REPRODUCTION: MULTIPLICATION THROUGH CLONING

High in the Appalachian Mountains of West Virginia, there is a low, dense thicket of blueberry bushes nearly

UNIFYING THEME

Unifying themes to help tie together chapter facts and concepts.

ADVANCE ORGANIZER

A list of the chapter's major topics in their order of presentation.

Key to use of color appears on page xxxii.

Aids to Learning

BOXED ESSAYS

Boxed essays that present interesting experiments, new research findings, and newsworthy topics to help students understand biology's real-world applications and implications.

176 CHAPTER 10/FOUNDATIONS OF GENETICS

THOROUGHBREDS AND THOROUGH BREEDING

Horse racing is big business: Owners in 40 countries race more than half a million thoroughbreds, and fans wager billions annually. Ironically, while human Olympic sprinters continue to improve their speeds each year, horses are no faster now than they were 50 years ago. Have horses reached their inherent speed limit? Or is there another explanation for their lack of improvement, despite advances in nutrition and veterinary medicine?

Irish geneticists B. Gaffney and E. P. Cunningham recently developed a complicated method for analyzing change in 31,263 thoroughbreds over the past 25 years. Like other researchers, they found strong evidence of inbreeding, or unusually close genetic similarity based on matings between related horses. They expected this, because 80 percent of all thoroughbreds are descendants of just 31 horses brought to England from the Middle East and North Africa at the turn of the eighteenth century. Just as Mendel artificially selected and bred peas of a certain color, height, and shape, race horse breeders have selected and interbred these descendants to preserve competitive disposition, strong slender legs, and, above all, swift running speed. But did this practice of "breeding the best to the best" backfire and produce, instead, a "regression to the norm"—an averaging of traits and a loss from the population of alleles for both extreme slowness *and* extreme swiftness?

Figure A HAVE THOROUGHBREDS REACHED AN UPPER SPEED LIMIT?

Gaffney and Cunningham think not and report that thoroughbred populations still have considerable genetic variation. Based on their analyses, thoroughbreds should now be running about 2.5 percent faster than they did a quarter century ago and have definitely not approached their natural upper speed limit. Many observers think more practical factors are at fault: lax training methods (underexercising the valuable race horses); unsound breeding practices (pairing horses by their "papers" and family histories rather than their actual sizes, shapes, and speeds); and poor race track running surfaces. If the recent study is correct, we could be seeing faster horse races in the future. If it's wrong, the record-setting horses of the past, like Man O'War and Secretariat, could be legends forever.

green seeds in the F_2 generation. He knew that if you toss a coin in the air, it has an equal chance, or probability, of landing heads or tails. If you toss two coins, there are four possibilities: two heads; two tails; one head and one tail; or one tail and one head. Thus, the probability of two heads (or two tails) is 1 in 4, while the probability of one head and one tail is 2 in 4.

Figure 10-5b, step 3, shows a simple means (devised after Mendel's time) of displaying the probability of different allele combinations. This method, called a **Punnett square,** can be used for three, four, or more coin tosses or genetic traits. The various boxes in the Punnett square indicate the probabilities of seeing each of the possible head-tail combinations or, in the case of genetic factors, allele combinations. Thus, you can draw Punnett squares with the alleles from one parent along one side (representing the classes of possible gametes) and those from the other parent along the other side. By crossing each pair of alleles and filling in all the boxes of the Punnett square, you can see all the possible combinations and calculate the ratio of results.

The Law of Segregation

Since Mendel reasoned that each pea plant receives two alleles of each gene—one allele coming from each parent's gamete—he also reasoned that the number of alleles must be reduced as the parent produces gametes so that the offspring receives two alleles of each gene, not four. As Chapter 9 explained, just this kind of reduction occurs during meiosis and gamete formation and ensures that eggs, pollen, or sperm are haploid and carry just one allele of each gene.

By considering this separation and reduction of alleles during gamete production, one can understand the results of self-pollination within a heterozygote (*Yy*). In effect, this is equivalent to a cross between two *Yy* plants, each producing some gametes that carry the dominant *Y* allele for yellow seeds and some that carry the recessive *y* allele for green seeds. If the plants produce pollen (or sperm) and eggs with *Y* and *y* occurring in equal ratios, and if the gametes combine randomly at fertilization, one-quarter of the F_2 progeny will receive a *Y* from both

UNDERLINING

Underlined take-home messages that present or emphasize key concepts.

JAWED FISHES: AN EVOLUTIONARY MILESTONE 435

(a) (b) (c)

Figure 24-32 TELEOSTS, THE MOST DIVERSE FISHES.
The earth's oceans, rivers, and lakes teem with seemingly endless varieties of teleosts. (a) The sailfish (*Istiophorus platypterus*), with its long, bladelike upper jaw and huge dorsal fin that can be raised or lowered out of the way. (b) The masked butterfly fish (*Chaetodon semilarvatus*) of the Red Sea. (c) The rainbow wrasse (*Laproides phthirophagus*), a common reef fish in both hemispheres.

dramatically, and fins can be used as paddles, brakes, or even true gliding wings, as in the flying fish. We will discuss various aspects of fish physiology in later chapters.

The second subclass of bony fishes, the sarcopterygians, is older than the actinopterygians. It includes a few modern species with fascinating adaptations for breathing air and walking, plus extinct species that were the first vertebrates to crawl onto land more than 360 million years ago. In contrast to the thin, bony fins of teleosts, sarcopterygians have thick *lobed* fins with large bones and muscles. Certain sarcopterygians also have external and internal nostrils, or *nares*, and a good sense of smell.

Sarcopterygians include the living lungfish and coelacanths and the extinct rhipidistian fishes. *Lungfish* are rare freshwater fish that live in shallow rivers and lakes in Australia, South America, and Africa. They have both gills and lungs, relying on the former when their environment is wet and the latter when it is dry, and they are locked in a protective mud and mucus "cocoon."

Coelacanths are large (up to 1 m or so) primeval-looking fishes once thought to be extinct, but rediscovered off Madagascar in 1938 (Figure 24-33). They possess a fat-filled swim bladder for buoyancy, analogous to the shark's fat-storing liver, and pectoral and pelvic fins that move in a coordinated way during slow swimming, much as the forelegs and hind legs of land vertebrates move during walking.

The final group of fleshy-finned fish, the extinct ***rhipidistians***, were probably the ancestors of the land vertebrates. The muscular lobed fins of these fishes probably allowed them to "walk" along the bottom of shallow ocean bays and tidal flats. Rhipidistians lived during the Devonian period, about 350 to 400 million years ago, and may have pursued insects and other invertebrate food sources up onto land. The oldest fossilized animal tracks of vertebrates yet found were left by rhipidistians in 360- to 370-million-year-old sandstone formations on the Orkney Islands, off the northeastern coast of Scotland.

The evolution of the fishes was far from linear. During the Devonian period, the age of fishes, the waters literally teemed with a bewildering array of ostracoderms, cyclostomes, acanthodians, early actinopterygians, lungfish, and rhipidistians. The first amphibians also arose from ancestral fish and lived among this Devonian variety.

Figure 24-33 A COELACANTH (*Latimeria chalumnae*): A FLESHY-FINNED LIVING FOSSIL.
This rare fish closely resembles its ancient predecessors. The fish is about a meter in length and can move slowly forward using its thick-based fins, such as the posterior-dorsal one seen here.

Unique Art Program

ICONS

Extensive use of icons or small diagrams that help students fit the structure or process into its actual physical context in a living organism.

COLORS

The consistent use of assigned colors and shapes throughout the book to designate specific atoms, molecules, cellular structures, and processes.

DIAGRAMS

Extensive use of process diagrams that depict sequential biological events. Individual steps are numbered and keyed to step-by-step discussions in the text or figure legend.

Figure 8-4 THE STRUCTURE OF CHLOROPLASTS: SITES OF PHOTOSYNTHESIS.
Plant cells (a) contain chloroplasts, such as the one pictured in (b). This electron micrograph of a chloroplast (magnification 15,000×) reveals the internal stacks of membranes, the grana. Each granum (c) consists of flattened sacs called thylakoids; adjacent grana are interconnected by the thylakoid membrane (d). Most of the enzymes and pigments for the light reactions of photosynthesis are embedded in the thylakoid membranes. The stroma, a gel-like matrix, surrounds the grana. The enzymes for the light-independent reactions of photosynthesis as well as chloroplast DNA and ribosomes and other substances are located in the stroma.

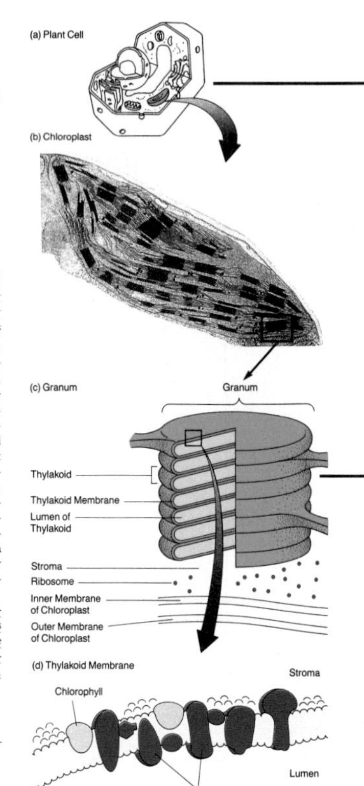

Chloroplasts: Sites of Photosynthesis

In eukaryotic cells, both phases of photosynthesis take place in chloroplasts, and the reactions depend on the unique structure of these organelles just as cellular respiration depends on the architecture of the mitochondrion (see Chapter 5). Chloroplasts have a variety of shapes, but most are elongated like minute bananas (Figure 8-4a and b). Chloroplasts are somewhat larger than mitochondria, and a typical plant cell contains from 20 to 80 of them (usually about 40). Two lipid bilayer membranes surround the chloroplast (Figure 8-4c); internally, a gel-like matrix called the *stroma* contains ribosomes, the machinery for protein synthesis, and DNA, which, in at least one species, forms genes that turn on and off in response to light. Some essential chloroplast proteins are encoded in the cell's nuclear DNA, synthesized in the cytoplasm, then moved into chloroplasts. The most prominent internal structures in chloroplasts are the stacks of flattened sacs called *grana* (meaning "grains"; see Figure 8-4c). Each flattened sac in a granum is called a **thylakoid** (from the Greek word for "sack"), and a *thylakoid membrane* surrounds the internal space, or *lumen*, of each sac (Figure 8-4d).

The chlorophyll, enzymes, and cofactors that participate in the light-dependent reactions of photosynthesis are embedded in the thylakoid membrane (see Figure 8-4d). Most of the enzymes that catalyze the light-independent reactions are found in the stroma, or matrix, surrounding the stacks of thylakoids.

HOW LIGHT ENERGY REACHES PHOTOSYNTHETIC CELLS

A browsing animal that eats a fresh green leaf from a bush is consuming light energy that may have left the sun just 8 minutes earlier. But what exactly is light energy, and how does it interact with the molecules of a living leaf?

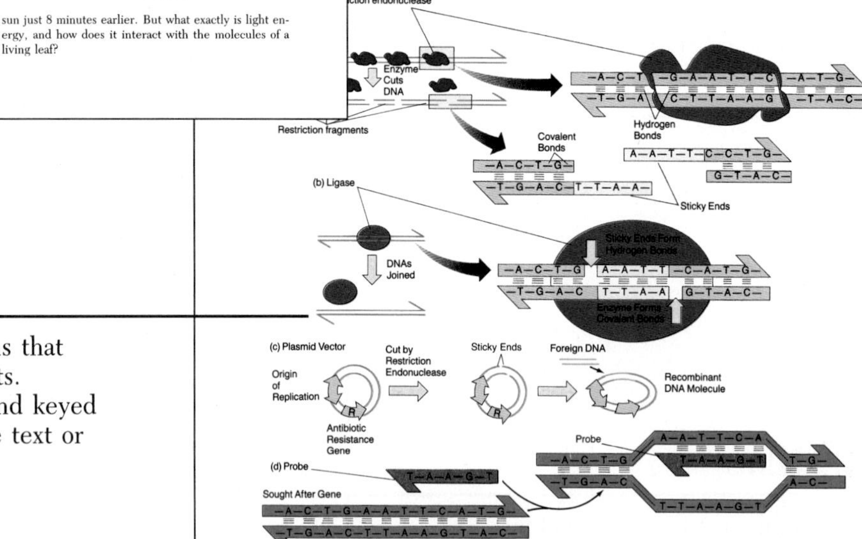

Figure 14-13 TOOLS OF THE GENETIC ENGINEER.
(a) Restriction endonucleases can cut DNA at specific sequences and leave overlapping "sticky" ends. (b) DNA ligase enzymes rejoin the sticky ends at complementary sequences. (c) New genes can be spliced into vectors, which are usually plasmids, to form recombinant DNA molecules. Vectors can carry such novel gene sequences into different host organisms. (d) Molecular probes are stretches of RNA with base sequences complementary to a desired gene, often part of a recombinant DNA molecule.

Molecular Probes

The final tool of the genetic engineer is a probe for locating recombinant DNA molecules, genes, or other desired pieces of genetic information. A **probe** can be a specially prepared stretch of RNA or DNA with a sequence complementary to the specific series of nucleotides in the sought-after gene (Figure 14-13d).

Let's see, now, how a researcher would use these tools to clone and transfer a gene.

Engineering a Bigger Mouse

Genetic engineers used the tools just described plus a few specialized techniques to isolate the gene for rat growth hormone, clone it, and transfer it into the giant mouse shown on page 229. Here's what they did, step by step.

A researcher took the DNA from rat cells and cut it with a restriction endonuclease into hundreds or thou-

Tools for Review

(a) Runner

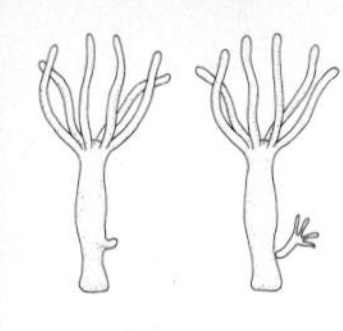

(b)

Figure 9-11 ASEXUAL REPRODUCTION IN PLANTS AND ANIMALS.
(a) Strawberry runners demonstrate a form of asexual reproduction used by plants. The plants grow new runners, or horizontal stems, which send their own roots down into the ground to form a new plant. (b) The hydra (here, magnified about 9×) is an aquatic animal that can reproduce asexually. Hydras can produce buds through simple mitotic division, and each bud pinches off and grows into a new full-sized hydra by further mitotic division.

based on novel combinations of traits from both parents. Spontaneous mutations can provide further variability, as well. The sexually reproducing organism in a sense gambles on giving its offspring a better hand; its genetic "cards" are "reshuffled" and "redealt" instead of being passed along in original form. Thus, new combinations of traits can arise much more rapidly in sexually than in asexually reproducing organisms and increase the chances of the species surviving sudden significant environmental changes.

Despite these great advantages, sexual reproduction does have drawbacks. An organism that cannot reproduce asexually—a mammal, for example—can never bequeath its own exact set of genetic material, no matter how successful, to its progeny, the way a prizewinning strawberry plant can pass along its hereditary complement to a clone. The very mixing process that created the successful gene combination in the adult works to dismantle it partially in the offspring. Researchers are currently trying to perfect techniques for cloning mammals so that the desirable traits of a prizewinning bull, let's say, could be reproduced in thousands of offspring, and not subject to the variability of sexual reproduction.

LOOKING AHEAD

The spectacular dance of the chromosomes during mitosis and meiosis, as well as the cycling of cells through periods of growth, synthesis, and division, allows for both fidelity and variability in the passage of genetic information from one cell generation to the next. Our next chapter explains how biologists unraveled those mysteries of inheritance.

SUMMARY

1. The nucleus is the repository of hereditary information, which is contained in DNA and organized in structures called *chromosomes*.

2. Chromosomes are made up of a substance called *chromatin*, which is a combination of DNA, *histones*, and nonhistone proteins. *Nucleosomes* are sites where DNA is wrapped about sets of histone molecules.

3. In most eukaryotic organisms, there are two copies of each chromosome; the two form a *homologous* pair. *Sex chromosomes* are the exception. Diploid cells and organisms contain homologous pairs of chromosomes. Haploid cells and organisms contain only one set of chromosomes.

4. The mitotic *cell cycle* consists of four phases: G_1, a period of normal metabolism; S, the phase of DNA replication plus metabolism; G_2, a brief period of further cell growth; and *M*, mitosis. The nonmitotic stages (G_1, S, G_2) are referred to collectively as *interphase*.

5. The events of mitosis can be divided into four phases: *prophase*, *metaphase*, *anaphase*, and *telophase*.

6. At the beginning of mitosis, two identical chromatids become associated with the *spindle* and align in the middle of the cell along the *metaphase plate*. In anaphase, the centromeres divide, and the two chromatids are drawn toward opposite poles.

7. The polar microtubules extend from the spindle poles and overlap at the equator. As the region of overlap decreases, the poles are moved farther apart. Spindle microtubules extend from the *kinetochores* in the centromere of the chromatids toward the spindle poles. These fibers shorten, pulling the chromatids toward the poles.

8. In animals, spindle formation is associated with centrioles; in most plants and fungi, it is associated with microtubule-organizing centers.

9. In animal cells, *cytokinesis*, or division of the cytoplasm, results from the

LOOKING AHEAD

Many chapters end with *Looking Ahead*, a short section that ties the main thread of the chapter to the discussions in the following chapter.

SUMMARY

A point-by-point recap of the chapter's main concepts and facts.

le ring of actin fila-
equator. In plants,
he building of a *cell*
s equator.

ial type of cell divi-
chromosome num-
tic prophase I, the
each homologous
gether in a process
called *synapsis*. In anaphase I, the homologous chromosomes separate to opposite poles. During meiosis II, each sister chromatid of the pair moves to one of the poles. The result is four haploid cells.

11. Meiosis allows for the random distribution of homologous parental chromosomes in offspring, as well as the genetic variability that results from *crossing over*.

12. Asexual reproduction, in which new organisms arise from mitotic processes, preserves an organism's "winning genetic formula" in a particular environment. Sexual reproduction, which involves meiosis and gamete production, ensures greater variability in offspring and ensures hereditary enrichment as new genes and gene families arise.

KEY TERMS

Key terms are listed to help students identify the chapter's most important vocabulary.

KEY TERMS

anaphase
autosome
cell cycle
cell plate
centromere
chalone
chromatid
chromatin
chromosome
clone
crossing over
cytokinesis
diploid
G_1 phase
G_2 phase
haploid
histone
homologous
interphase
karyotype
kinetochore
meiosis
metaphase
metaphase plate
mitosis
M phase
nucleosome
prophase
sex chromosome
S phase
spindle
synapsis
synaptinemal complex
telophase

QUESTIONS

Study questions, both short answer and essay questions, help students review and retain the most important information.

QUESTIONS

1. What part of the cell contains the hereditary information? Which structures contain this information? Which molecules make up these structures?

2. The cell cycle consists of four phases: G_1, S, G_2, and M. What occurs during each phase? Which has the most variable length?

3. What is the outcome of mitosis? Does each daughter cell receive identical chromosomes?

4. What is the outcome of meiosis? Does each haploid cell receive identical chromosomes?

5. Which of the following statements apply to mitosis, which to meiosis, and which to both?
   a. DNA replication occurs before this process starts.
   b. When the chromosomes first become visible, they are already doubled.
   c. Homologous chromosomes pair.
   d. Each daughter cell receives an identical complement of chromosomes.
   e. The centromere is the last part of the chromosome to divide.

6. Two kinds of microtubules separate the chromosomes during cell division. What are they, and how do they operate?

7. Describe the function of the centriole or microtubule-organizing center during cell division.

8. Describe the process of cytokinesis first in a dividing animal cell, then in a plant cell.

ESSAY QUESTIONS

1. Did you inherit equal numbers of chromosomes from your two parents? From your four grandparents? Explain.

2. Is a species more likely to survive in a changing environment if all its members are identical or if they are diverse? What is accomplished by sexual reproduction? What are the advantages of asexual reproduction?

For additional readings related to topics in this chapter, see Appendix C.

SUGGESTED READINGS

A list of classic and up-to-date references for additional reading beyond the text material occurs in a special appendix and is cited at the end of each chapter.

Key to Use of Color

Atoms

Oxygen

Hydrogen

Carbon or

Nitrogen

Phosphorus

Sulfur

Electrons

Atomic nucleus

Biological molecules

Carbohydrates

Lipids

Proteins or

DNA or

RNA

ATP

Cell parts

Nucleus

Cytoplasm

Mitochondria

Chloroplast

ER, Golgi complex

Energy pathways

Glycolysis

Fermentation

Krebs cycle

Electron transport chain

Light-dependent reactions

Light-independent reactions

1 The Study of Life

Sit down before fact as a little child, be prepared to give up every preconceived notion, follow humbly wherever and to whatever abyss nature leads, or you shall learn nothing.

Thomas H. Huxley, Letter (September 1860)

A kingfisher dives headlong into a rushing mountain stream and pulls out a young trout with its beak.

A tropical orchid growing on the trunk of a giant mahogany tree unfolds delicate and fragrant blooms to lure one type of pollen-eating beetle.

A rich growth of small, brownish white mushrooms springs up on elk droppings in a Wyoming meadow. The fungi, which are found only on the elk dung, are gathered by a homeward-bound fisherman, destined for tomorrow's breakfast with a rainbow trout.

An 18-year-old woman develops a condition of premature aging and a year later dies of this disease.

These examples, and hundreds of thousands more like them, are the subjects of **biology,** the study of life. The life histories of plants and animals represent the more traditional concerns of a science that dates back to

This lovely orchid grows as a parasite on a tree in Central America. Such fragrant, brilliantly colored flowers can lure insects of a certain size and shape to the plant. The insect visitors sip nectar or consume nutritious pollen grains, then wander off to another plant of the same species, inadvertently transferring pollen grains that may fertilize eggs in the next set of flowers.

Aristotle. In recent decades, biologists have acquired new tools and knowledge that enable them to engineer and explore life in unprecedented ways. Geneticists have learned to construct bacteria that manufacture human hormones such as insulin for the treatment of diabetes. Biochemists are learning to transfer hereditary information from bean plants to wheat and corn so that these critically needed food crops can generate their own fertilizer from nitrogen in the air. And physiologists are beginning to probe the circuits and functions of the most complex object in the known universe—the human brain, with its millions of interacting nerve cells. That fantastic organ has allowed us to amass a great wealth of information about the life process, and this book is an introduction to that body of knowledge.

Because the life sciences are in the midst of a scientific revolution, the modern biology student is presented with both a privilege and a challenge. The newcomer will learn facts and concepts not even imagined by the leaders of biology just a generation ago, and the amount to be learned by the beginner can seem overwhelming and grows more so each year. Our approach in this book is to present those concepts that promise to be enduring and so to provide a foundation for understanding biology as it continues to unfold. Despite its fast pace, modern biology presents a worthy challenge, and its student joins a venerable tradition of scholars who, for centuries, have been deeply curious about the process called life.

This chapter discusses:

- The basic characteristics of living things
- How organisms may have originated on our planet and diverged into millions of species
- The concept of evolution, a central tenet of biological science
- How biologists use the scientific method to investigate the living world
- The major recurrent themes in this book

WHAT IS LIFE?

What exactly is "life"? Everyone has an intuitive sense about what living things do. Many living things move: Fish swim, birds fly, and plants bend toward light. Most grow taller, wider, and heavier. Most produce eggs or seeds, or give birth to live young. Obviously, steel girders and pieces of wood fail to exhibit such tendencies, and we consider them to be nonliving. But what about dried-out seeds or virus particles? Are they living or nonliving? One way to answer this question is to construct a list of characteristics that put some boundaries around this elusive concept we call life.

Living things have a complex organization.
Living things take in and use energy.
Living things grow and develop.
Living things reproduce.
Living things show variations based on heredity.
Living things are adapted to their environments and ways of life.
Living things are responsive.

Most contemporary biologists define **life** as a particular set of processes that result from the organization of matter. The essential difference between the biologist's definition of life and the antiquated notion of a vital force is that to the biologist, life is not something *different* and *separate from* the organized processes and structures in a living thing. Instead, life is the sum of those processes and structures, whether observed in the diving kingfisher, in the radiance of an orchid, in the iridescent sheen of a hummingbird's feathers as it darts from flower to flower (Figure 1-1), or in the human brain at work. Because those processes and structures are so numerous and specific, it takes an entire book to define life adequately. So let's begin by considering the characteristics that best describe, if not precisely define, the living state.

Living things have a complex organization. Organisms have an intricacy of form—from the levels of atoms and molecules to those of cells, tissues, and entire organisms—that is not found in the nonliving world. Quartz crystals and snowflakes are organized assemblages of like atoms, but even the tiniest living cell contains thousands of complex substances arranged in special spatial relationships. That arrangement is the key to their coordinated function (Figure 1-2). Furthermore, the organism's life depends on the maintenance of this complex

Figure 1-1 LIFE'S BEAUTY AND DIVERSITY. The diversity of life is evident in the spectrum of shapes, colors, and textures in a violet-eared hummingbird, a jewel of the Brazilian jungle, and the delicate pink blossoms on which it feeds.

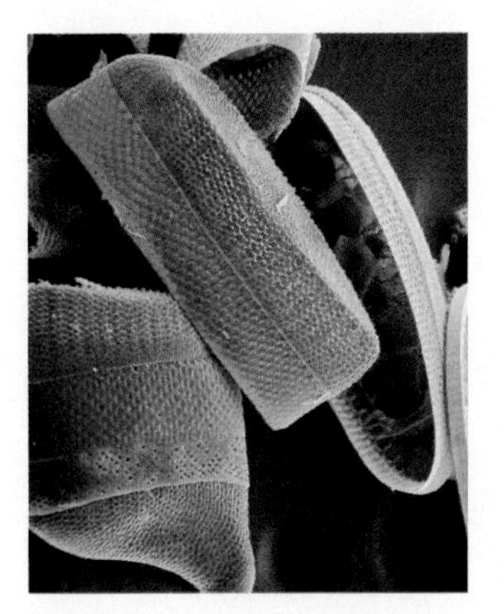

(a)

Figure 1-2 THE ARCHITECTURE OF LIFE. (a) Marine diatoms, magnified here about 450 times, are microscopic organisms with fragile shells that look like lacy blown glass and in fact are composed largely of silicon. (b) The broad antlers of this bull elk branch in a precise way, recalling the architecture of life on a macroscale.

(b)

Figure 1-3 SUNLIGHT: THE ENERGY THAT DRIVES LIFE. Pine needles, leaves on a maple, and tiny green organisms floating in the sea absorb sunlight's vital energy, then manufacture the nutrients that support virtually all life forms.

organization. It has been said that the body of an adult human contains about 85 L (20 gal) of water and about $6 worth of chemicals (see back endpapers for full names of weights and measures abbreviated in the book). Obviously, one cannot just stir the chemicals into solution and come up with a person, a pine tree, or a toadstool. Each organism's characteristics depend on the complex way the substances are arranged and organized in space.

Living things take in and use energy. Like elevators, sports cars, and other complex entities, living organisms tend to break down and gradually fall into disrepair unless their organized arrangement of substances is maintained. That continual maintenance depends absolutely on energy. So the organism must take in energy, most of which ultimately derives from the sun's light and heat (Figure 1-3). In plants, solar energy is converted into chemical energy and stored as nutrient molecules. In all organisms, including plants, energy from such nutrients is released and used for maintenance and for various life activities during a series of chemical events called *metabolism* (see Chapter 4).

Living things grow and develop. One of the most important activities supported by metabolism is *growth*, an increase in mass, size, or organization that is quite distinct from the simple addition of like molecules to a growing crystal (Figure 1-4). Organisms also *develop:* They become more complex and take on a series of new forms, such as when a fertilized egg cell develops into a chick inside a shell and then, after hatching, continues to develop into a hen (Figure 1-5).

Living things reproduce. All types of organisms generate offspring in a process called *reproduction*. Organisms can simply divide in two or can carry out a more elaborate process that includes courtship, mating, fertilization, internal incubation, and live birth. In every case, however, the organism begets a like organism: Redwood trees produce new redwoods, not pines; starfish, new starfish, not oysters; and so on. This reproductive fidelity depends on a set of chemical blueprints contained within each organism—the hereditary or *genetic* information, which specifies what form the new individual will take and, in part, how it will develop and grow.

Living things show variations. A basic feature of reproduction is that it results in *variation;* the offspring

Figure 1-4 GROWTH AND DEVELOPMENT.
A sprouting acorn (a) develops into a stalwart red oak tree (*Quercus coccinea*) (b). The growth and the change in physical form during maturation are easily observed characteristics of most living things.

(a)

(b)

Figure 1-5 LIKE BEGETS LIKE: REPRODUCTION.
Two stages in the development of a bird. The tiny early embryo just 13 days old (a) grows and develops all the organs found in the hatchling (b). Both stages are necessary for reproduction in this bird species.

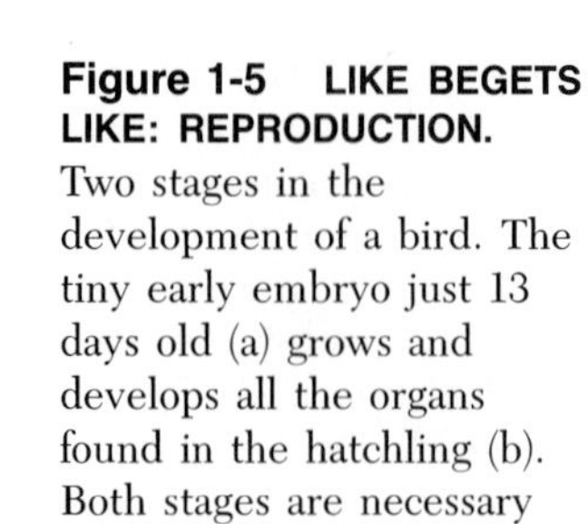

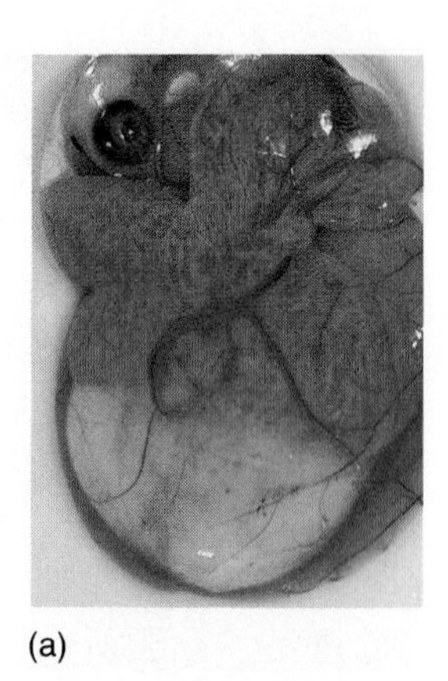

(a)

(b)

almost always differ in various ways from one or both parents (Figure 1-6). Change or reshuffling of the hereditary information and slight irregularities in the inheritance process occur normally. As we shall see, this variation is not always beneficial to the individual organism. Nevertheless, it is critical to the continuation of a species because it gives a lineage of organisms a better chance of surviving over vast spans of time as local and global environments are altered by natural forces. In other words, it permits evolutionary change.

Living things are adapted to their environments. Just as the emperor penguin can swim in icy antarctic seas (Figure 1-7), every living thing is organized and functions in a way that allows it to exploit and cope with its physical surroundings, including heat or cold, light or darkness, drought, salinity, predators, competitors, or other challenges it faces. To survive, an organism must be specifically suited to its environment and way of life. Such *adaptation* is the result of evolutionary change, the accumulation of inherited variations over time.

Living things are responsive. Flowers bend toward the sun. A baby pulls its hand away from a hot radiator. A trout darts away from a shadow cast on the edge of a

Figure 1-6 VARIATION BETWEEN GENERATIONS.
Offspring like these mixed-breed puppies are usually somewhat different from their parents—a fact that underlies the inevitable variations that are a basis for evolution.

Figure 1-7 ADAPTATION: COPING WITH THE ENVIRONMENT.
Organisms are physically and functionally suited to their surroundings. These flightless emperor penguins thrive in the frigid Antarctic, where robins, woodpeckers, or other inhabitants of temperate forests would quickly perish.

stream. A bacterium orients itself in a magnetic field, as does a migrating starling. A threatened squid releases an inky cloud and darts away to safety (Figure 1-8). These are examples of *responsiveness*, the ability to detect and adjust to certain features of the environment. It is an ability that allows short- and long-term adaptation and survival.

One other very special feature of living organisms is their *history*. Every living thing on earth today is a descendant of an organism that lived before it. Each is a member of an unbroken lineage stretching back in time to the era, billions of years ago, when life processes first became associated with organized sets of matter. Thus, to understand many characteristics of present-day organisms, we must have a knowledge of evolutionary history.

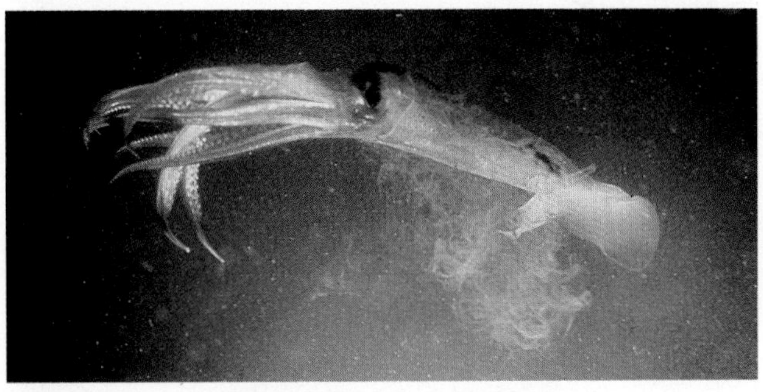

Figure 1-8 A CLOUD FOR SELF-DEFENSE.
When threatened, this boreal squid ejects a cloud of dark pigment and jet propels away.

LIFE ON EARTH: A BRIEF HISTORY

Biologists usually depict the evolutionary history of life forms as a branching tree, with the more complex modern groups at the ends of branches and the simpler groups that appeared earlier and gave rise to them along the trunk and limbs (Figure 1-9). This treelike picture of evolutionary history is based on fossil evidence, precise anatomies, comparisons of genetic blueprints of different organisms, and scientific studies of how life originated on earth, which we will discuss shortly. For many centuries, however, people's day-to-day observations led them to attribute the origin of life to **spontaneous generation.** The sudden, seemingly miraculous appearance of molds on bread, worms on aging meat, or toadstools on decaying wood led to the belief that life arises anew if conditions are conducive.

Early Beliefs About the Origin of Life

The belief in spontaneous generation held sway from well before the era of the Greek and Roman scholars until the seventeenth century, when an Italian named Francesco Redi published a careful refutation based on direct tests (Figure 1-10). He suspected that the maggots in meat actually arise from the eggs of flies that land on the spoiling food. To test his theory, Redi put fresh pieces of fish, veal, eels, and snakes into glass flasks, leaving some open and sealing others carefully. Flies visited the open flasks, and worms soon appeared on the meat; but in the sealed flasks, worms never developed. Redi concluded that maggots come only from eggs, and he generalized further that life forms originated only once and that all living organisms are direct descendants of preexisting individuals.

Despite Redi's well-founded arguments, the idea of spontaneous generation did not die gracefully. It was not until the mid-nineteenth century that French scientist Louis Pasteur finally laid the old idea to rest. Pasteur observed that the microscopic organisms living in a solution of food molecules (sugars, proteins, and water) can be killed by boiling the broth and will fail to reappear as long as the opening to the flask is heated (to sterilize it) and then sealed tightly. Pasteur showed that the broth in sealed flasks can remain uncontaminated for as long as 18 months. If the flasks are opened to the air, however, the solutions inside will teem with bacteria within a day or two. To Pasteur, this proved that microbes floating on dust particles in the air must enter the newly opened flasks and begin to multiply.

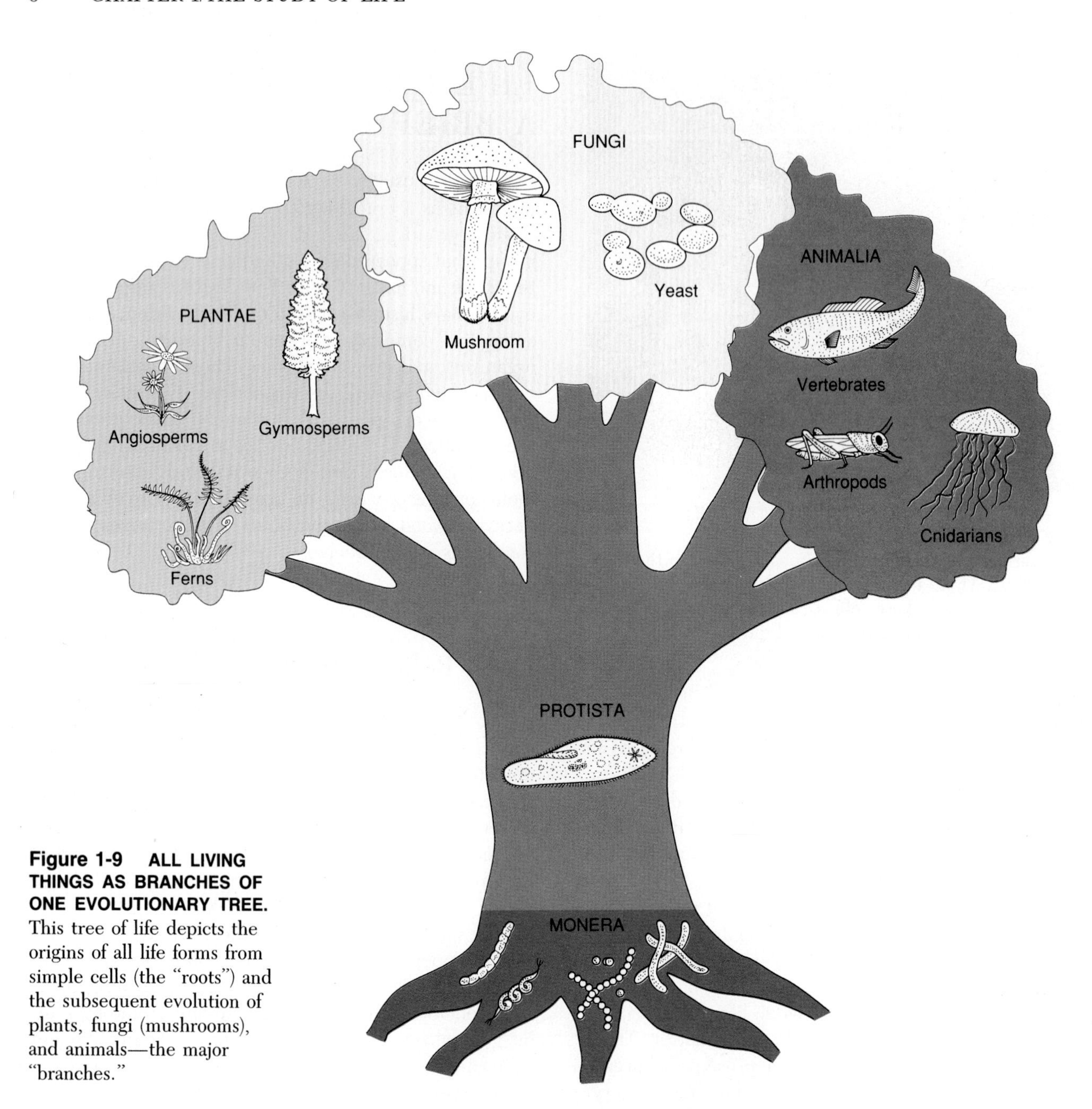

Figure 1-9 ALL LIVING THINGS AS BRANCHES OF ONE EVOLUTIONARY TREE. This tree of life depicts the origins of all life forms from simple cells (the "roots") and the subsequent evolution of plants, fungi (mushrooms), and animals—the major "branches."

Figure 1-10 FRANCESCO REDI. The myth of spontaneous generation was seriously damaged by Francesco Redi's careful experimentation, which showed that maggots do not arise from meat in sealed flasks.

In a brilliant move that finally defeated opposing theories about a "life-giving force" in the air, Pasteur designed a special type of flask with a long, downward-curving neck that allows air to reach the solution but traps dust particles and microbes in its lower part (Figure 1-11). He showed that a solution boiled in such a flask remains free of organisms despite the free exchange of air. Yet when the neck of the flask is removed, allowing dust to enter the solution, the growth of organisms becomes apparent within a matter of hours.

Pasteur's experiments proved that all life comes from preexisting life. However, this unbroken chain of organ-

Figure 1-11 LOUIS PASTEUR.
Pasteur showed that sterilized broth in a swan-necked flask remains free of microbes for days—even years. But when the curving glass neck is broken, the broth becomes clouded with bacterial growth within hours.

isms had some beginning, somewhere, at some time. One area of modern biological research (discussed at length in Chapter 19) is concerned with the origin of life and relies on data from paleontology, geology, astronomy, physics, and other fields.

A Modern View of the Origin of Life

In our solar system, the earth seems to be a uniquely hospitable place for life. The size of our planet and its distance from the sun dictate that gravitational force will hold an atmosphere near the surface. The atmosphere not only screens out much damaging ultraviolet light, but also helps maintain surface temperatures on earth within a range of about 0°–100° C. Within this range, the major constituent of most living organisms—water—is a liquid, not a gas or a solid. At temperatures much below 0° C and above 100° C, most life processes as we know them cannot take place.

Life's origins are intertwined with the history of the earth. About 4 billion years ago, our planet passed through a stage during which energy from the sun and from the heat of the earth's molten core helped complex chemicals form from simple atoms and molecules. Meteors bombarding the earth for millions of years in its early history may have delivered more such chemicals. Regardless of the source, those complex materials became the building blocks of living things (Figure 1-12). Biologists believe that aggregations of those building block chemicals led to the first organized systems with the characteristics of life we discussed earlier. These first *cells*, the fundamental units of all living things, could produce copies of themselves (Figure 1-13). However, imperfections in this hereditary process led to inevitable variations among the generations of cells that followed and set the stage for new kinds of organisms. Eventually, over vast spans of time, the major and minor branches on life's evolutionary tree arose.

Figure 1-12 FROM ORGANIC COMPOUNDS TO LIVING CELLS.
Organic chemicals, precursors of biological molecules, existed on our planet long before life arose. About 4 billion years ago, energy from the sun and from the heat of the earth's core aided the formation of complex chemicals, the building blocks of living things. Once living cells formed and proliferated, they slowly and irrevocably changed the face of the earth. This geyser basin with its hot sulfurous water—reminiscent, perhaps, of primal scenes on the early earth—teems with bacteria, algae, and other life forms.

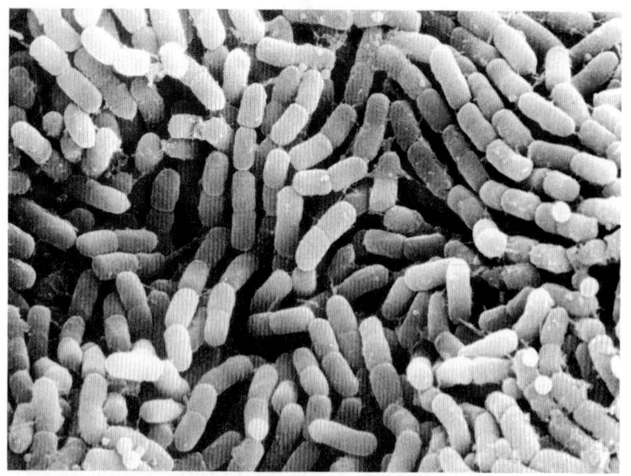

Figure 1-13 DESCENDANTS OF EARLY CELLS. Early cells may have resembled modern bacteria in fundamental ways. This bacterial colony with its sausagelike chains of cells is magnified about 2,000 times its normal size with the scanning electron microscope.

Through this process of hereditary variation, the first organisms to appear—various kinds of primitive bacteria—acquired new properties. Certain water-dwelling bacteria, for instance, began to release oxygen as a by-product of their metabolism, and over the course of hundreds of millions of years, huge quantities of oxygen accumulated in the atmosphere, blocking out deadly ultraviolet light from the sun. This protective covering allowed organisms to live near the surface of the sea or even in exposed moist places on land. Thus, life and earth began a process of interrelated change that continues today. Significantly, the physical attributes of the infant planet that led to life's origin from nonliving matter were *changed permanently* in the process and were no longer conducive to the spontaneous concatenation of living cells from nonliving matter.

The reproduction of the first cells and their descendants forged an unbroken chain of living organisms, each passing to its progeny the structures and processes of the living state. From the original types of single cells arose colonies of new cells that exhibited plantlike or animal-like characteristics. Those colonies, over many millions of years, gave rise to an enormous spectrum of organisms that inhabit water, land, and air. But despite the profound differences between, say, a bacterium growing on a cactus, a moss carpeting a rock, a tuna darting through deep ocean waters, and a stork sunning its broad wings, all living organisms continue to share fundamental chemical traits. Among these are the structure of the hereditary material, *deoxyribonucleic acid*, or *DNA* (Figure 1-14), and the fuel that drives life processes, *adenosine triphosphate*, or *ATP*.

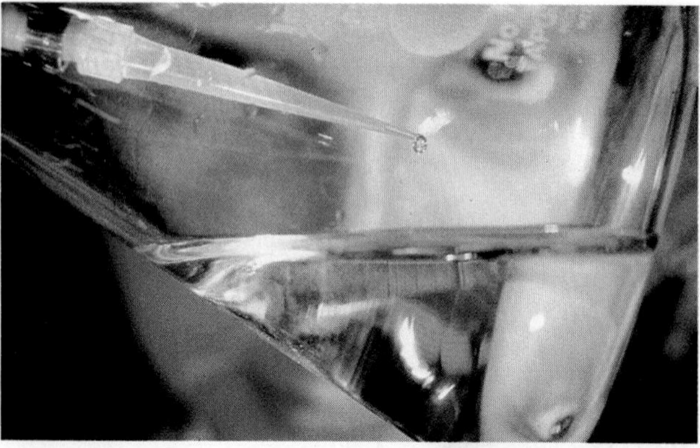

Figure 1-14 DNA: LIFE'S HEREDITARY MATERIAL. This drop of fluid contains DNA that has been extracted from living cells and purified. Through its use of such DNA, modern biotechnology is rapidly changing agriculture and medicine and expanding biological knowledge in unprecedented ways.

EVOLUTION: A THEORY THAT CHANGED BIOLOGY AND HUMAN THOUGHT

Most educated people accept the idea that today's creatures are descendants of yesterday's organisms. The idea that one group gives rise to another over time, however, was not always popular. One of the great intellectual adventures of human history led to our understanding that the immense variety of living things arose from simple, single-celled ancestors and that each new species was adapted to its particular habitat and way of life. The theory of evolution, proposed independently by the English naturalists Charles Darwin and Alfred Russell Wallace during the mid-nineteenth century, permanently changed the way people think about their origin and their place in the scheme of nature. It is, without a doubt, the most important unifying concept in biology. Neither Darwin nor Wallace, however, was the first to consider an alternative to the special creation of living things, the idea that all living things were created specifically for their place in nature and their usefulness to humankind. Their intellectual forebears included the French naturalists Georges-Louis Leclerc de Buffon and Jean Baptiste Lamarck, both of whom believed that variations among living creatures were the consequence of some mechanism of heredity.

Buffon is reputed to have believed in special creation, but he puzzled over such unexplainable, apparently functionless appendages as the two little side toes on a pig's foot, which never touch the ground. What purpose could these toes serve when compared with the functional and very similar toe bones in the limbs of most reptiles, mammals, and birds? Buffon concluded that the pig's extra toes, as well as other variations in the number and shape of limbs among vertebrates (animals with backbones), could be explained only if one assumed that all such functionless limbs had been inherited from ancestors whose limbs contained fully functional parts. Thus, one basic limb design must have been modified in different creatures for running, flying, digging, or swimming (Figure 1-15). All vertebrates, Buffon concluded, are descendants of a common ancestor. This was a remarkable idea for the 1760s, but it did not catch on because Buffon offered no evidence.

Lamarck believed that simple organisms had given rise to more complex ones in a natural progression of species, but he based his systematic evolutionary theory on two ideas that were later discredited: (1) Some natural force operates to purposefully move organisms up the ladder of complexity toward the place held by humankind, and (2) an organism can acquire a new characteristic during its lifetime and then pass it on to offspring. Lamarck's most famous example was the giraffe's neck, which the French naturalist claimed can grow longer as an individual animal stretches for leaves. The acquired trait of the longer neck, he said, can then be passed to the animal's young through the mechanisms of heredity. Although Lamarck's reasoning was faulty, his theory challenged the accepted belief that each form of life was immutable in form.

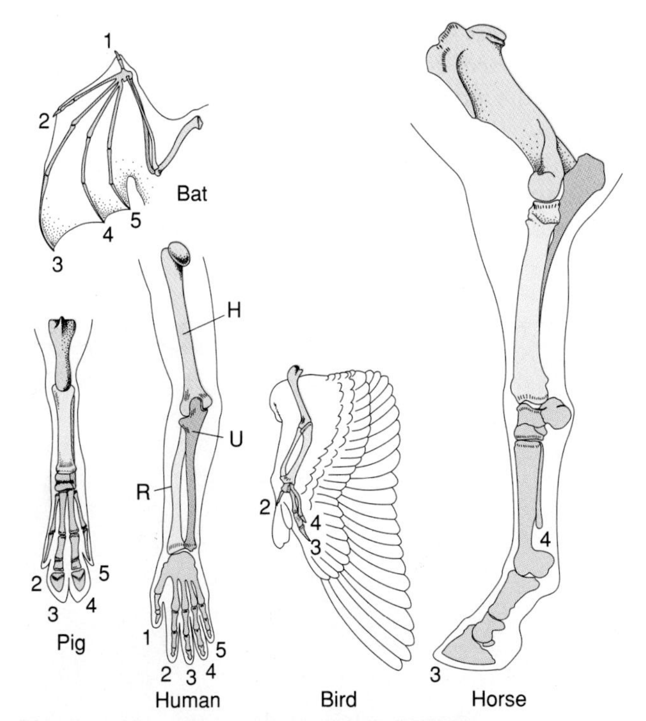

Figure 1-15 ONE DESIGN, MANY FORMS.
Buffon's reasoning led to the conclusion that the forelimbs of these five vertebrates (the bat, pig, human, bird, and horse) have been modified from a common ancestral condition. The letters and colors (H, bone; R, green, U, orange) show the position of the same bones in all five vertebrates.

As important as these early evolutionary theories were in shaping the ideas of Darwin and Wallace, certain geological evidence was even more provocative. Briefly, this evidence implied that the earth is much older than the 6,000 years that Christian scholars had calculated from the Bible. Earth's history has been marked by constant slow change, which is documented in the successive layers of rock that were deposited as the earth aged. Embedded in these layers are the remains of all sorts of extinct plants and animals that once inhabited the planet. These fossils sometimes reveal modifications in body organization and structure from one layer of rock to another. And finally, many of the animals and plants alive in the nineteenth century are not represented at all in the fossil record, implying that they did not live during ancient times (Figure 1-16).

Darwin carried along this assortment of biological, geological, and theological ideas when he signed on as the naturalist aboard the HMS *Beagle* in 1831 and began a 5-year exploratory voyage to South America and other continents. It was his careful observations of the natural world and much reading and reflecting after his return that led him some 20 years later to write his famous treatise, *On the Origin of Species by Means of Natural Selection.*

Unfamiliar with Darwin or his work, Wallace made his own trip to South America in 1848 and later sailed to Singapore and Borneo to collect biological specimens. He published three scientific papers on the mechanisms of evolution before Darwin made his work public. Then, in 1858, mutual friends presented Wallace's fourth paper simultaneously with an abstract of Darwin's book at a prestigious scientific meeting. Although Wallace independently devised most of the major tenets in Darwin's evolutionary theory, his work has long been overlooked. In recent years, however, it has started to receive well-deserved credit.

The Theory of Evolution

Darwin's book really offered two related theories: (1) the **theory of evolution,** which states that all living things have evolved from a common ancestor that diverged into millions of species by means of a gradual process of change and variation; and (2) the **theory of natural selection,** which states that natural events "select" organisms in such a way that the better-adapted individuals tend to survive and reproduce, whereas the less-adapted ones tend not to contribute to later generations. Darwin and

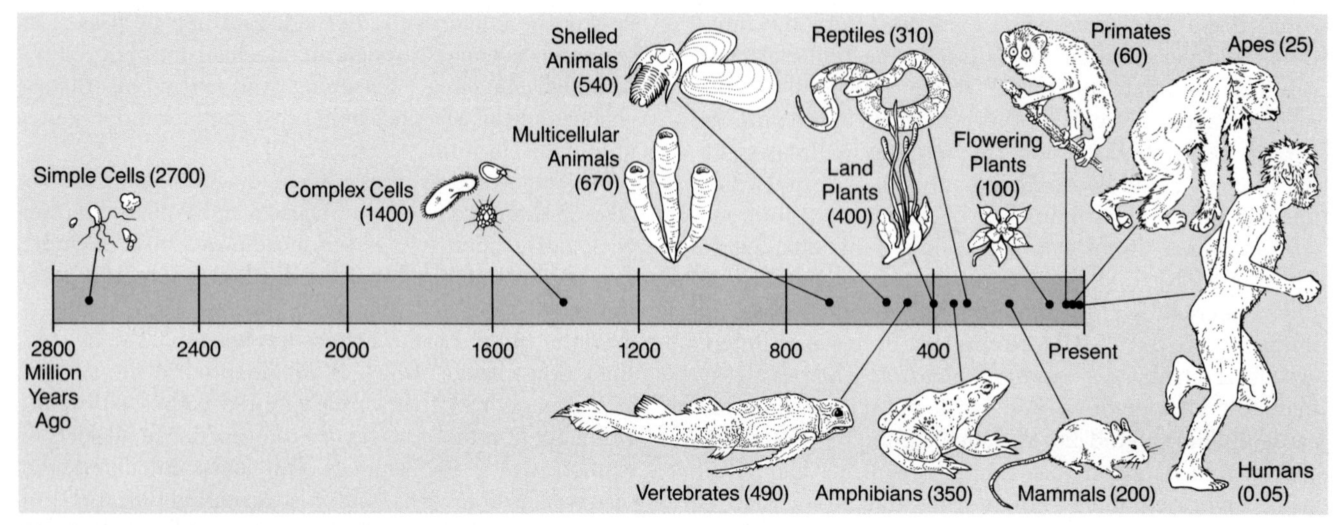

Figure 1-16 WHEN LIFE FORMS APPEARED IN THE FOSSIL RECORD.
This time line shows the enormous span between the appearance of the earliest cells, simple animals, plants, and complex mammals like apes and people.

Wallace both proposed that natural selection is the mechanism primarily underlying evolution.

Darwin cited important evidence from the selective breeding of farm animals. Since people began domesticating animals more than 10,000 years ago, they have selected the best milk producer, the best egg layer, the strongest burro or plow horse, the best watchdog. The farmer could slowly "improve" the breed for a desired characteristic by allowing a chosen pair of prize animals—but not others—to mate. This evidence alone convinced Darwin that a given kind of organism is not physically immutable (Figure 1-17). But if a farmer could act as an "artificial selector," causing a lineage of animals or plants to "evolve" in a certain way, then perhaps, Darwin reasoned, there is a "natural farmer" that selects in a nonpurposeful way certain plants and animals—but not others—for successful breeding. He called this hypothetical process *natural selection*. (Completely independently, Wallace also chose the term *natural selection* to describe the pressures of nature upon organisms' survival.)

Historians believe that Darwin was greatly influenced by the writings of Thomas Malthus, an English economist and clergyman. Malthus deduced that populations

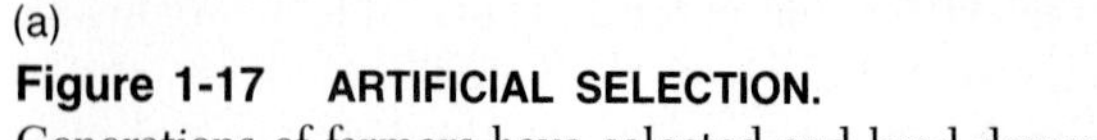

(a) (b)

Figure 1-17 ARTIFICIAL SELECTION.
Generations of farmers have selected and bred domestic cattle for specific traits, producing within the same species a remarkable range of variation. Scottish Highlanders (a) and Brahmans (b) are just two breeds of many dozens.

of organisms tend to increase in size because organisms produce an excess of offspring—hundreds of seeds from one flower, thousands of eggs from one pair of salmon, and so on. Such overproduction does not overwhelm the planet with organisms, however, because the hazards of living—harsh winters, disease, accidents, predation—wipe out most of the offspring (Figure 1-18). Darwin realized that such chance events could be the "natural farmer" working not just to reduce population size, but also to eliminate those individuals less fit than the survivors to evade predation and withstand environmental hardships.

Darwin recognized that the key issue was survival to reproductive age and then successful reproduction. Natural selection, he concluded, is the Grim Reaper that lets some individuals achieve reproductive success, while others die.

While Darwin's ideas were disconcerting to most of his nineteenth-century contemporaries, modern biologists use the theory of evolution to understand and organize an incredible array of facts and observations, including the obvious physical adaptations that all organisms show to their environments. Evolution explains both the unity and the diversity of life—why all organisms share many characteristics while at the same time showing great variation. And evolution gives us historical reasons for such vestigial structures as the pig's dangling toes.

There is so much accumulated evidence for the theory of evolution by natural selection, and it explains so much—from the properties of biological molecules to the characteristics of entire communities of diverse organisms—that almost all biologists view evolution as a scientific fact. But while the fact of evolution is believed to be beyond dispute, its mechanism (i.e., what causes it) is still under intensive investigation.

Figure 1-18 NATURAL POPULATION CONTROL. Elk populations are held in check partly by coyotes, which tend to attack the young, the old, and the sick.

THE SCIENTIFIC METHOD

Creative geniuses of many kinds have focused their attention on nature and given humankind great and lasting works: Ludwig van Beethoven's *Pastoral* Symphony; Claude Monet's *Water Lilies;* Henry David Thoreau's *Walden;* St. Augustine's *De Naturae Et Gratia (Of Nature and Grace).* There are, nevertheless, important differences between the ways artists, philosophers, and theologians approach nature and the way a scientist studies it. The nonscientist often works through inspiration, sentiment, or formal logic, while the scientist relies primarily on the **scientific method,** a kind of organized common sense that begins with observations of natural phenomena. Based on these observations, the researcher poses questions, frames hypotheses, designs experiments, gathers results, and formulates explanations.

Just as Darwin and Wallace began by observing the world around them, so all science begins with observations that stir the curiosity of scientists. The pea seedling grows rapidly; the oak tree, slowly. Why does the seedling grow fast? While making observations, scientists take measurements and establish basic facts. Then, in mulling over the results, they use **inductive reasoning**—go from the specific observations and facts to a general explanation—and formulate an educated guess, or **hypothesis,** that is a probable answer to the question. The framing of a scientific hypothesis is no less creative than painting a picture or developing an ethical theme. For many scientists, the wonderful excitement that comes at the moment of insight makes the entire process worthwhile. But it is the next step—*testing the hypothesis*—that sets science apart from other disciplines and enables scientists to produce accurate and enduring explanations of natural phenomena.

All scientists must design tests that could *disprove* a hypothesis, in case it is actually an incorrect guess. Figure 1-19 shows how a scientist tests the hypothesis that only the light of certain colors causes plants to develop flowers. To test the prediction, the scientist exposes sets of plants to different colors of light, then observes and records the results. If plants exposed to other colors of light (or no light) begin to flower, then the hypothesis is disproved and must be modified. Critical to the scientific method are **control experiments,** in which the scientist imposes conditions different from the ones set forth in the hypothesis. In our flowering plant experiment, the use of blue, green, and yellow light provides necessary controls. Only if many different approaches fail to disprove a hypothesis will it be considered accurate, and even then, the presumption of accuracy is tentative. Thus, the aim of a scientist is not to *prove* something, but to test it again and again in order to move toward a more and more accurate explanation.

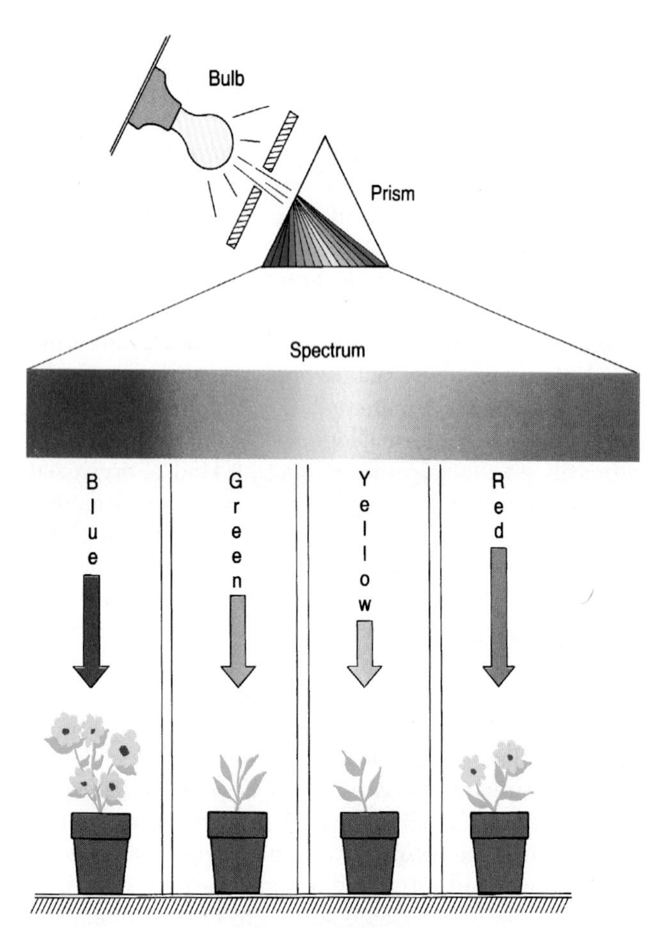

Figure 1-19 TESTING A HYPOTHESIS.
Beginning with the fact that white light is a combination of various colors (as a prism reveals), a plant scientist proposes a hypothesis such as this: *Red light causes plants to develop flowers.* Then he or she devises a test of the prediction. In this case, the scientist exposed plants to four colors of light, and found that only the plants receiving blue and red lights flowered. This helps support the original hypothesis that red light is required for plant flowering; but the hypothesis must be modified substantially because even better flowering occurs under blue light. More experiments will be necessary to explain why two colors of light induce flowering, not just one.

In addition, for any hypothesis to be considered tentatively accurate, the evidence in its favor must be *reproducible;* that is, an independent scientist must be able to repeat the same experiments and get essentially the same results. This is why a report in a scientific journal usually has a section called "Methods and Materials." In it, the scientist gives, in effect, a recipe that another scientist could follow to carry out an independent test of the hypothesis.

The final stage in the scientific method is the postulation of a **theory,** a general statement that is usually based on a number of tested hypotheses and is designed to explain a range of observations—even some not yet made. In this sense, a good theory is not just inductive, but can be used for **deductive reasoning,** a process of predicting new facts or relations for which new experiments can be designed and new information collected. For example, reexamine Figure 1-16. As paleontologists have searched the earth's rocks for fossils of vertebrates, they have found fossils of only ancient fishes in the oldest rocks, of amphibians in somewhat younger rocks, of the first reptiles in still younger rocks, and of the first mammals in even younger rock strata. This has helped to define the evolutionary sequence of backboned animals. But if paleontologists ever discovered mammalian fossils in the oldest rock layer, the entire theory would need to be revised. Furthermore, the fossil sequence leads to other predictions. One could predict, for instance, that the proteins or the molecules of the hereditary material (DNA) would be more dissimilar in a fish and a rabbit (a mammal) than in a fish and a frog (an amphibian). And indeed, molecular analyses help support these predictions.

When a theory such as the theory of evolution is tested again and again and corroborated in diverse ways, it is eventually accepted as a **natural law,** or a scientific fact. That is why we say that almost all biologists consider evolution to be far more than a theory. It is as much a fact as the physicist's theory that matter is built of particles called atoms and the astronomer's theory that the earth circles the sun.

The real impetus to the scientific method and the formulation of theories and principles is curiosity about nature. The truly creative scientist always keeps an eye open for the bizarre, the unexpected, the chance observation that may lead to sudden new insights. For instance, in 1928, when Sir Alexander Fleming noticed "halos" of killed bacteria around certain molds growing in culture dishes, he wondered why. His hypothesis that the molds produced an antibacterial agent led to the discovery of penicillin, an antibiotic that has saved millions of human lives.

BIOLOGY, SOCIETY, AND YOUR FUTURE

The scientific method has served as midwife to virtually all the great advances in biology. Despite its power, however, the scientific method has limitations; by itself, it cannot solve many important social problems. As science and technology advance, human society is faced with increasingly difficult ethical and moral choices. For example, it is now possible to determine during pregnancy whether a fetus is grossly deformed and will probably have a short and perhaps painful life, and it is also

Figure 1-20 SCIENCE AND MORALITY.
The monitoring of human fetuses, now routine, can reveal deformities, such as those caused by Thalidomide, long before the baby is due. But how should that knowledge be used?

possible to terminate most such pregnancies without risk to the mother (Figure 1-20). But should the parents choose to abort? And should society allow it? As another example, physicians can often prolong—sometimes for years—the lives of people whose conscious brain functions have been permanently destroyed by accident or disease. But should they do so? Or should the physician "pull the plug" after a certain amount of time? Finally, biologists soon will be able to transfer a number of specific, heritable traits from one person to another, such as replacing the defective gene (genetic material) that leads to sickle-cell anemia with a normal gene. But is it ethical to manipulate human genes?

Scientists and physicians can provide factual information and informed counsel about the problems and procedures just described, but the final ethical decision inevitably involves nonscientific considerations. Decisions such as these must be made by informed citizens or perhaps by governmental, professional, or legal groups. The scientific method is not directly applicable to ethical and moral questions. It can, however, provide a framework for rigorous, commonsense thinking that may help people make hard choices in these areas of human concern.

An understanding of science can have two other important consequences. Studying life science may lead to an interesting career in medicine, teaching, research, or applied biology (Figure 1-21). And regardless of one's professional path, biology is a source of the intellectual pleasure and perspective that make our lives richer and more enjoyable. To know how a tree functions or why migrating birds fly in a V formation can help us appreciate the beauty and complexity of the world around us (Figure 1-22).

(a)

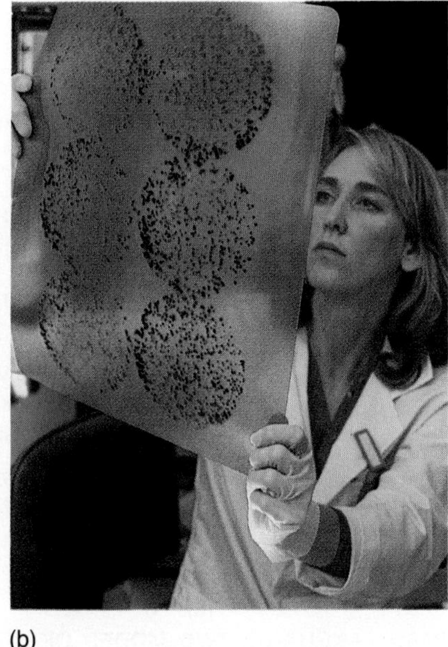

(b)

Careers in Biology

Biology teacher or researcher
Science writer, editor, photographer, illustrator
Wildlife manager, park ranger
Physician, osteopath, chiropractor, optometrist, podiatrist
Nurse
Medical technologist (occupational therapist, x-ray technician, etc.)
Dentist, dental hygienist, dental assistant
Veterinarian, veterinarian's assistant
Agricultural researcher, product salesperson
Landscape architect
Forestry or fisheries manager
Nutritionist, dietician
Laboratory technician
Staff member of aquarium, zoological park, botanical park
Water treatment engineer
Food science researcher, inspector, technician
Medical or science librarian
Biotechnology researcher, technician, product salesperson
Genetic counselor
Ecology researcher, field technician
Genetic researcher, lab technician
Epidemiologist
Marine biologist
Cell biologist
Microbiologist
Soil scientist
Environmental quality engineer

(c)

Figure 1-21 CAREERS IN APPLIED BIOLOGY.
Trained biologists work in dozens of areas from agriculture to zoology. (a) Here, an agricultural researcher works in the greenhouse, and (b) a biochemist uses recombinant DNA molecules in her cancer studies. (c) This partial list of careers indicates just some of the possibilities for those with training in biology. Labor statistics suggest that such jobs will increase 20 percent by the year 2000 and that biologists traditionally have a low rate of unemployment.

Figure 1-22 THE STUDY OF BIOLOGY—A WINDOW ON THE WORLD.
The patterns of nature are rich and varied—and more easily appreciated by the informed observer.

Figure 1-23 PHYSICAL LAWS AND THE PROPERTIES OF WATER.
Universal physical laws govern living and nonliving matter alike. The properties of water are the same whether it takes the form of snow in a Canadian meadow, water vapor issuing from a bison's nostrils, or hot fluid in an animal's coursing blood.

OUR APPROACH TO BIOLOGY

We divide our exploration of modern biology into six parts. Although each part covers a different area of biology, several themes run through the entire book.

1. Evolution, the foundation of modern biological thought, helping to explain both the unity and diversity of life on earth.
2. The central role of development in translating genetic instructions into a functioning organism. Through developmental biology, we can understand how the structures and life processes on which evolution operates actually arise in the individual and in the species.
3. The integration of the biological and physical worlds. Living matter obeys the same universal physical laws that govern nonliving matter, and these laws set limits on the characteristics of living things (Figure 1-23).
4. The coevolution of organisms with one another and the interplay of such biological evolution with changes in the earth's physical environment. Plants, animals, and other organisms evolve simultaneously, with a dependence on one another. But

GAGTTTTATCGCTTCCATGACGCAGAAGTTAACACTTTCGGATATTTCTGATGAGTCGAA
AAATTATCTTGATAAAGCAGGAATTACTACTGCTTGTTTACGAATTAAATCGAAGTGGAC
TGCTGGCGGAAAATGAGAAAATTCGACCTATCCTTGCGCAGCTCGAGAAGCTCTTACTTT
GCGACCTTTCGCCATCAACTAACGATTCTGTCAAAAACTGACGCGTTGGATGAGGAGAAG
TGGCTTAATATGCTTGGCACGTTCGTCAAGGACTGGTTTAGATATGAGTCACATTTTGTT
CATGGTAGAGATTCTCTTGTTGACATTTTAAAAGAGCGTGGATTACTATCTGAGTCCGAT
GCTGTTCAACCACTAATAGGTAAGAAATCATGAGTCAAGTTACTGAACAATCCGTACGTT
TCCAGACCGCTTTGGCCTCTATTAAGCTCATTCAGGCTTCTGCCGTTTTGGATTTAACCG
AAGATGATTTCGATTTTCTGACGAGTAACAAAGTTTGGATTGCTACTGACCGCTCTCGTG
CTCGTCGCTGCGTTGAGGCTTGCGTTTATGGTACGCTGGACTTTGTGGGATACCCTCGCT
TTCCTGCTCCTGTTGAGTTTATTGCTGCCGTCATTGCTTATTATGTTCATCCCGTCAACA
TTCAAACGGCCTGTCTCATCATGGAAGGCGCTGAATTTACGGAAAACATTATTAATGGCG
TCGAGCGTCCGGTTAAAGCCGCTGAATTGTTCGCGTTTACCTTGCGTGTACGCGCAGGAA
ACACTGACGTTCTTACTGACGCAGAAGAAAACGTGCGTCAAAAATTACGTGCGGAAGGAG
TGATGTAATGTCTAAAGGTAAAAAACGTTCTGGCGCTCGCCCTGGTCGTCCGCAGCCGTT
GCGAGGTACTAAAGGCAAGCGTAAAGGCGCTCGTCTTTGGTATGTAGGTGGTCAACAATT
TTAATTGCAGGGGCTTCGGCCCCTTACTTGAGGATAAATTATGTCTAATATTCAAACTGG
CGCCGAGCGTATGCCGCATGACCTTTCCCATCTTGGCTTCCTTGCTGGTCAGATTGGTCG
TCTTATTACCATTTCAACTACTCCGGTTATCGCTGGCGACTCCTTCGAGATGGACGCCGT

(a)

To be, or not to be: that is the question:
Whether 'tis nobler in the mind to suffer
The slings and arrows of outrageous fortune,
Or to take arms against a sea of troubles,
And by opposing end them. To die, to sleep—
No more—and by a sleep to say we end
The heartache, and the thousand natural shocks
That flesh is heir to! 'Tis a consummation
Devoutly to be wished. To die, to sleep—

(b)

Figure 1-24 GENETIC LANGUAGE AND HUMAN LANGUAGE: TWO MODES OF COMMUNICATION.
(a) This block of letters represents part of the genetic information in the DNA of a virus that infects bacteria. That information is contained in the order of four kinds of molecules called nucleotide bases, whose names are abbreviated here with the four letters A,T,G, and C. (b) This part of the most famous passage from Shakespeare's *Hamlet* exemplifies human language at its finest. Just as in genetic language, the information in human language is contained in the order of letters within words, and words within sentences.

in a broader sense, changes in the physical environment bring about changes in lineages of organisms, and organisms in turn alter the earth and its atmosphere.

The grand scheme of biological evolution is complemented in a few species, most notably our own, by **cultural evolution,** a process involving the transfer of information from generation to generation in a nongenetic way. Thus, we will discover how the two great languages—the language of heredity and the human language—interact in remarkable ways to accelerate the rate of change (Figure 1-24). The application of the scientific method, the invention and use of computers, and the manipulation of hereditary information are all parts of this cultural evolution. They give us the power to destroy the environment, other organisms, and ourselves or to improve the quality of life. Educated citizens, professional biologists, physicians, and others with knowledge of biological science can influence which choices are made and thus contribute to human cultural evolution and ultimately, the preservation of our planet.

SUMMARY

1. Living things have a complex organization, take in and use energy, grow and develop, reproduce, show variations based on heredity, are adapted to their environments and ways of life, and are responsive.

2. *Life* is a particular set of processes resulting from the organization of matter.

3. Redi and Pasteur proved that *spontaneous generation* of live organisms is not possible in the modern world. Nevertheless, the unique set of physical and chemical conditions that prevailed on the early earth was conducive to the spontaneous generation of the first living cells on our planet.

4. Living organisms have radically altered the lands, waters, and atmosphere, and the changing earth has affected how and where organisms may live.

5. The *theory of evolution* states that all living things have evolved from a common ancestor over the millions of years of earth's history.

6. The *theory of natural selection* states that natural events "select" organisms so that better adapted ones tend to survive to reproduce, while less-adapted ones tend not to contribute to subsequent generations.

7. The *scientific method* begins with observations. They stimulate curiosity, which leads to *inductive reasoning* and the development of a *hypothesis* to explain the facts and observations. Tests of a hypothesis include *control experiments,* which could support or disprove the hypothesis.

8. *Theories* are general statements designed to explain sets of related hypotheses that have withstood experimental verification. Theories lead to *deductive reasoning,* which involves making predictions about natural phenomena.

KEY TERMS

biology
control experiment
cultural evolution
deductive reasoning
hypothesis
inductive reasoning
life
natural law
scientific method
spontaneous generation
theory
theory of evolution
theory of natural selection

QUESTIONS

1. List the characteristics of life, and give an example of each from a different organism.

2. How did the experiments of Redi and Pasteur disprove spontaneous generation?

3. What do biologists think is a better explanation for how life originated?

4. How did the ideas of Buffon and Lamarck contribute to later evolutionary theory?

5. Define the theories of evolution and natural selection.

ESSAY QUESTION

1. Suppose that the lamp on your desk suddenly stopped giving off light. Design an experiment to determine why it stopped, and explain how each step of the experiment reflects the scientific method.

For additional readings related to topics in this chapter, see Appendix C.

Part
ONE

FROM ATOMS TO CELLS

From atoms come form, color, life itself—in this case, a desmid, a free-floating inhabitant of the sea surface.

2 Atoms, Molecules, and Life

Biology has been fortunate in discovering within the span of one hundred years two great and seminal ideas. One was Darwin's and Wallace's theory of evolution by natural selection. The other was the discovery by our own contemporaries of how to express the cycles of life in a chemical form that links them with nature as a whole.

Jacob Bronowski,
The Ascent of Man (1974)

Eighteenth-century chemists proved that all matter, living and nonliving, is composed of particles called atoms, and this discovery had a profound and permanent effect on the study of biology. In the decades that followed, biologists recognized that every organism contains the same two dozen types of atoms arranged in different ways. The glorious diversity of life on our planet—the millions of kinds of plants, animals, fungi, and microbes—could now be seen to stem from the myriad ways that specific atoms combine and interact.

Continued experimentation revealed that all life activities—from harvesting energy and breaking down food particles to moving, growing, and reproducing—depend, ultimately, on the chemical interactions between sets of atoms. Finally, biologists came to understand the central role of the water molecule, with its unique chemical properties, in the ongoing processes of life.

Ice crystals in the Olympic Forest: Assemblies of molecules of life's precious solvent, water.

By focusing on the basic building blocks of matter, this chapter lays a groundwork for all our remaining discussions of biological structure and function. The chapter covers:

- The structure of atoms and the ways in which these building blocks are assembled into more complex structures
- Molecules and compounds—aggregates of atoms with their own characteristics
- How atoms and molecules interact in *chemical reactions*
- The properties of water and its crucial role in living things
- The continuum of organization from atoms to living things

ELEMENTS AND ATOMS: BUILDING BLOCKS OF ALL MATTER

Two basic principles of chemistry emerged from the work of French chemist Antoine Lavoisier, English chemist John Dalton, and others in the late 1700s and early 1800s.

1. All matter, living and nonliving, is made up of **elements,** substances that cannot be decomposed by chemical processes into simpler substances. There are 92 chemical elements in nature, and 13 more have been created in the laboratory. Some examples of elements are hydrogen (symbolized H), oxygen (O), sulfur (S), gold (Au), iron (Fe), and carbon (C).
2. Each element is composed of identical particles called **atoms,** the smallest units of matter that still display the characteristic properties of the element. All the atoms in a brick of pure gold, for example, are identical to one another but different from all the atoms in a lump of carbon, an ingot of iron or samples of other elements. The properties of an element, such as the dense, shiny, metallic nature of gold or the dull black quality of carbon, are based on the structure of its individual atoms, as we shall see.

The Elements of Life

A natural question arose from the pioneering work of Lavoisier and Dalton: Are living things made up of the same elements as rocks, planets, and stars, or is our chemical makeup different? Living things, it turns out, display a special subset of the 92 naturally occurring elements in the earth's crust, but the elements occur in very different proportions (Figure 2-1). Fully 98 percent of the atoms in the earth's crust are the elements oxygen, silicon (Si), aluminum (Al), iron, calcium (Ca), sodium (Na), potassium (K), and magnesium (Mg), with the first

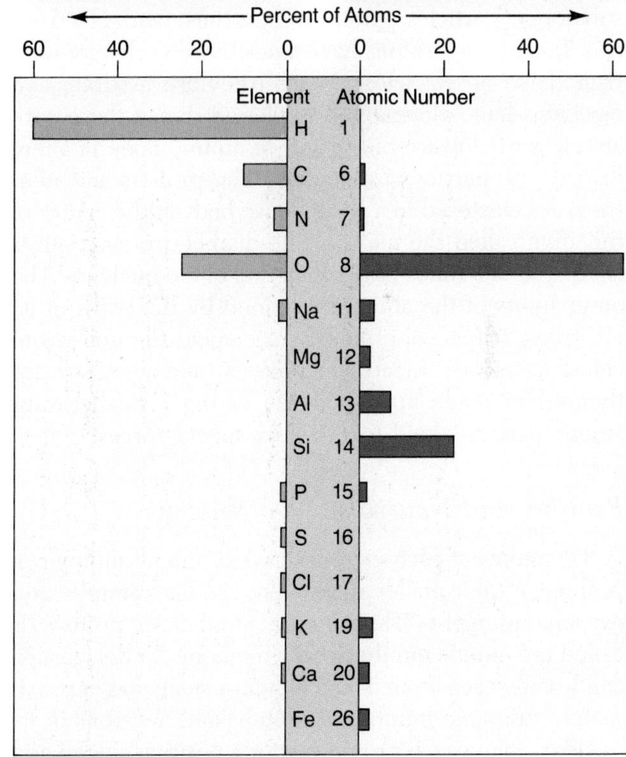

Figure 2-1 PRIMARY ELEMENTS IN THE HUMAN BODY AND THE EARTH'S CRUST
Because the human body consists mostly of water, hydrogen and oxygen are the most common elements in the human body. Hydrogen accounts for about 60 percent of the body atoms, and oxygen for more than 20 percent. Carbon contributes the majority of the remaining atoms in the body. The percent of iron and other elements in the human body is too small to be visible here. As the graph shows, Earth's crust is primarily oxygen, silicon, and aluminum—very different from the composition of living things.

three predominating. In a typical organism, however, 99 percent of the atoms are the markedly different subset carbon, hydrogen, nitrogen (N), and oxygen, with sodium, calcium, phosphorus (P), and sulfur making up most of the remaining 1 percent, plus a few other elements present in trace amounts.

Biologists are not certain why the chemical subsets of living and nonliving things are so different, but they do know that atomic architecture determines the physical properties of elements and, in turn, the properties of living organisms.

Atomic Structure

Atoms are extremely small: About 3 million atoms sitting side by side would probably cover the period at the end of this sentence. Physicist Gerald Feinberg once calculated that there are more atoms in the human body than there are stars in the known universe. Although minuscule in size, each atom is made up of three types of subatomic particles: protons, neutrons, and electrons (see Table 2-1). **Protons** have a positive (+) charge; **neutrons** have no electrical charge (they are neutral); and **electrons** have a negative (−) charge. Since these subatomic particles are only parts of atoms, none of them displays properties of elements. The protons and neutrons are clustered in a small dense body at the center of the atom called the *nucleus* (the diameter of an atom is about 100,000 times larger than that of the nucleus). The outer limits of the atom are defined by the paths of its electrons, which continuously race about the nucleus in cloudlike orbits. Electrons, protons, and neutrons are themselves made up of a dozen or more smaller subatomic particles held together by special forces.

Protons and Neutrons: The Nucleus

The atoms of each element have a unique number of protons in their nuclei: carbon has six, for example, and oxygen has eight. This unique number of protons is called the **atomic number** of the element. Under normal conditions, each atom of an element usually has approximately the same number of protons and neutrons in its nucleus. Since each proton carries a positive charge and neutrons have no electrical charge, the atom's nucleus has an overall positive charge equal in magnitude to the number of protons it contains; thus, the nucleus of a carbon atom has a charge of +6, and the nucleus of an oxygen atom has a charge of +8. Each proton and neutron has a mass of about 1 unit, called an atomic mass unit (see Table 2-1). The sum of an atom's neutrons and protons is called its *atomic mass*. Therefore, hydrogen, with one proton and no neutrons, has the atomic number 1 and an atomic mass of 1, while carbon, with six protons and six neutrons, has an atomic number of 6 and an atomic mass of 12. We will use the term **atomic weight** synonymously with *atomic mass*.

As uncharged particles, neutrons do not affect the chemical behavior of atoms. However, neutrons do impart a property to atoms that scientists have found particularly useful. Whereas the number of protons in atoms of a particular element always remains the same, the number of neutrons can vary. Most natural samples of elements are, in fact, mixtures of atoms that contain identical numbers of protons but different numbers of neutrons. These atoms have the same atomic number but different atomic weights and are called **isotopes** of the element. One familiar isotope is carbon-14 (written ^{14}C), which contains eight neutrons, not six, giving it an atomic weight of 14 instead of 12. Some isotopes, including ^{14}C, are *radioactive;* they emit energy that can be detected.

Modern medicine and biology would be in a vastly more primitive state were it not for the use of isotopes to chemically mark and trace biological molecules (see the box on page 21) and to determine the age of rocks containing fossil plants or animals.

Electrons, Atomic Orbitals, and Energy Levels

Within an atom, the electrons, the third type of subatomic particle, occur in numbers equal to the number of protons in the nucleus. An electrically neutral carbon atom, for example, has six electrons, and an oxygen atom has eight electrons. The negatively charged electrons are very much smaller than the protons and neutrons and are constantly in motion. Electrons add little to the mass of an atom but balance the protons' positive charge, leaving the atom electrically neutral (Table 2-1).

Table 2-1 PROPERTIES OF SUBATOMIC PARTICLES

Particle	*Mass (dalton)**	*Weight (g)*	*Electrical Charge*
Proton	1.00728	1.673×10^{-24}	+1
Neutron	1.00867	1.675×10^{-24}	0
Electron	0.000486	9.1095×10^{-28}	−1

*One dalton equals about 1.65×10^{-24} g, or one-twelfth the mass of an atom of carbon-12.

Electrons are so small that if we could collect and weigh them, just 1 g of electrons (about the weight of three aspirin tablets) would contain 10^{27} electrons (i.e., 10 multiplied by 10 a total of 27 times). This is a very large number; the radius of the known universe is 10^{25} m. The diameter of an electron is also vanishingly small—about 10^{-12}, or 0.000 000 000 001, cm across.

ATOMS IN MEDICINE

In 1895, German physicist Wilhelm Röntgen started a medical revolution with his discovery of x-rays. He found that this energetic form of radiation could pass through an opaque object, such as his own hand, expose a photographic plate, and thereby reveal dense structures inside the object, such as Röntgen's bones. Doctors suddenly had a window on the human body never before possible, and within a few weeks of Röntgen's great discovery, x-rays were being used to diagnose broken bones.

Today, another medical revolution is under way, based on a host of new and sophisticated imaging techniques. Like x-rays, these techniques rely upon the physical properties of atoms, and they promise to make the human body practically transparent to the diagnostician.

Computerized tomography, or CT, is an imaging technique that combines x-rays and computers. With CT, a physician can make high-resolution images based on a series of x-rays through cross sections, or "slices," of any part of the living body. CT can reveal tumors, blood clots, or other abnormalities in soft tissues as well as in bones (Figure A).

Nuclear magnetic resonance imaging, or NMR imaging, uses magnetic fields to alter the normal spin of the protons in certain atoms in the body and to change their natural magnetic fields. This temporary alteration causes the protons to give off a faint signal that can be detected and can reveal the presence and concentration of the chemical (Figure B). The signals created by the flipping of protons in hydrogen atoms, for example, enable doctors to watch blood (which is mainly H_2O) flow through vessels in tissues and organs. In this way, they can diagnose blood clots, tumors, clogged vessels, cancers in deeply embedded tissues, and damage to nerves, the brain, or joints.

The technique called positron-emission tomography, or PET, promises a still clearer window into the body and its biological processes. Radioactive isotopes that give off positrons (positively charged particles) can be injected into a patient, and the collisions between these particles and electrons can be detected with an array of scanners that surround the patient's body. In this way, the types and rates of important chemical reactions can be studied *as they occur* in the body. Physicians hope that PET will enable them to diagnose diseases of the brain and heart in their earliest stages and to probe the underlying causes of cancer and aging.

Other new imaging techniques include angiography, which uses dyes and x-rays to reveal blockages or defects in blood vessels, and intracranial embolotherapy, which allows physicians to detect and treat weak blood vessels in the brain.

These new methods—based on our increasingly sophisticated understanding of the atom—show how far both chemistry and medical technology have come since Wilhelm Röntgen first saw an image of the bones in his hand.

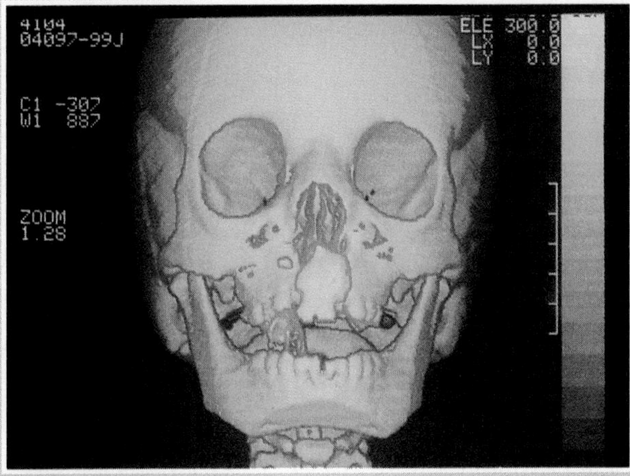

Figure A
CT scanning reveals details of the facial bones and muscles that cannot be seen on normal x-rays.

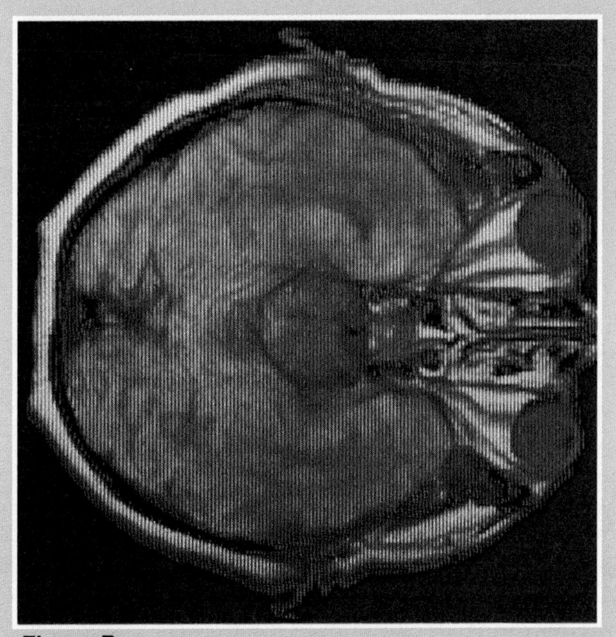

Figure B
An NMR image taken at the level of the eyes reveals both hemispheres of the brain, the lenses of the eyes, and even the cartilage in the ears.

Although the mass of an atom is concentrated in the protons and neutrons in the nucleus, the atom's chemical properties are based on the electrons. These tiny negatively charged particles occupy a *volume* as they zip about the nucleus. This volume is much larger than the nucleus alone: If the hydrogen nucleus could be magnified to the size of an orange, its single electron would be zipping around it in an orbit having a radius of one-third of a mile.

Danish physicist Niels Bohr, a pioneer in atomic structure, theorized in the early 1900s that electrons exist in energy levels around the nucleus. Bohr visualized electrons orbiting the nucleus like planets around the sun, (Figure 2-2a) and thought that each energy level could hold a certain number of electrons before the next level started to fill. For most biological molecules, the first three energy levels are the most important, and they hold two, eight, and eighteen electrons, respectively.

Within these energy levels, electrons actually do not move about in planar paths like planets around the sun, but instead zip about in three-dimensional **orbitals,** each with a specific shape and volume (Figure 2-2b). Electrons in energy level 1, for example, are usually found somewhere in a cloudlike spherical orbital (the 1*s* orbital) surrounding the nucleus. In the second energy level, a total of 8 electrons can be distributed among four orbitals. The first pair zip about in a spherical orbital called the 2*s*, which is slightly larger than the 1*s* (see Figure 2-2b), while the remaining three pairs of electrons move in three dumbbell-shaped orbitals called the $2p_x$, $2p_y$, and $2p_z$ orbitals. Within the third energy level, electrons move in spherical and dumbbell-shaped orbitals, but also in orbitals of more complex conformation. Figure 2-2c shows Bohr models for the elements hydrogen, helium (He), lithium (Li), carbon, and sodium and shows how their electrons are distributed into one, two, or three energy levels.

The system of energy levels and orbitals allows the negatively charged electrons to stay as close as possible

(a) Bohr Model of Energy Levels

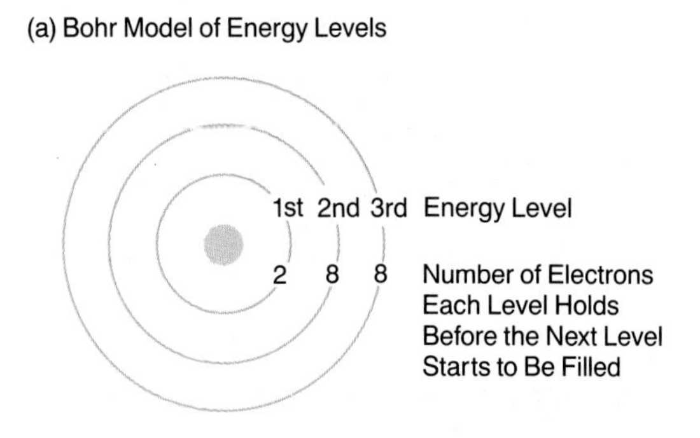

(b) Atomic Orbitals

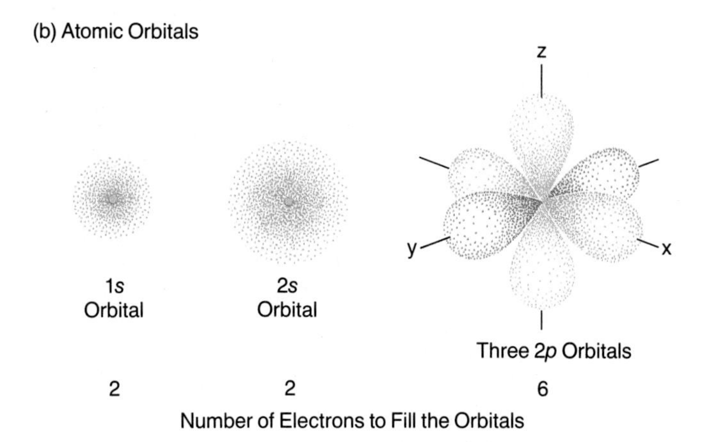

(c) Some Common Elements

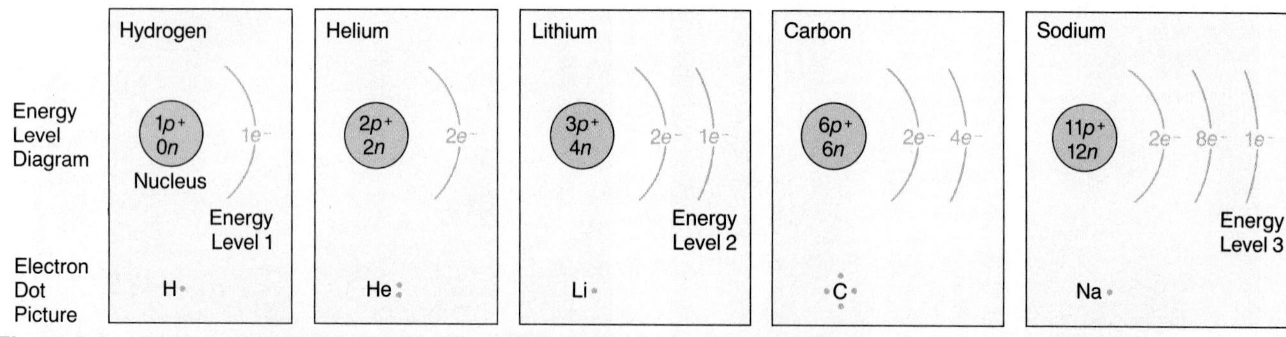

Figure 2-2 SIMPLE REPRESENTATIONS OF ATOMIC STRUCTURE: BOHR MODELS AND ATOMICAL ORBITALS. (a) Danish physicist Niels Bohr pictured electrons orbiting the atomic nucleus in energy levels, with level 1 holding a maximum of two electrons and both levels 2 and 3 a maximum of eight electrons each. (b) Within energy levels, electrons move about in three-dimensional cloudlike orbitals, the 1*s* and 2*s* orbitals and the dumbbell-shaped $2p_x$, $2p_y$, and $2p_z$ orbitals, arrayed at right angles to each other. The probability of finding an electron is greatest in the heavily shaded region nearest the nucleus and falls off as the distance from the nucleus increases. (c) Bohr models for the elements hydrogen, helium, lithium, carbon, and sodium show the number of electrons in each energy level at increasing distances from the nucleus. The electron dot diagram below each Bohr model shows the number of electrons in an atom's outermost energy level.

to the positively charged nucleus, yet still keep a maximum distance from one another. Chemists and physicists recognize a set of simple rules by which electrons distribute themselves among energy levels and orbitals:

1. An electron will occupy the lowest available energy level.
2. Orbitals in lower energy levels are filled completely before higher-level orbitals are occupied.
3. In any one energy level, a simpler orbital will be filled before an orbital of more complex shape is occupied.
4. Orbitals of similar shape at the same energy level must have one electron each before any of them can be filled.

To see how these rules apply to the atoms of specific elements, let's look at hydrogen, helium, lithium, and carbon once again, as well as a few other simple elements (Figure 2-3).

Hydrogen, atomic number 1, has a nucleus consisting of only one proton and no neutrons. Hydrogen's single electron occupies (but does not fill) the first energy level, closest to the nucleus. All orbitals in the second energy level remain unfilled as well. The next largest atom is helium, atomic number 2, with two protons and two electrons. Helium's two electrons complete the first energy level. Because this level is filled, helium is an extremely stable element; it does not react with other elements and is said to be *inert.*

Lithium, atomic number 3, has three electrons. The first two occupy the first energy level, and the third enters the second energy level (see Figure 2-3). Skipping ahead to carbon and oxygen, notice that carbon has four of its six electrons in the second energy level, while oxygen has six of its eight electrons in the second level, and that these electrons follow the orbital-filling rules listed earlier. In all elements heavier than sodium, electrons enter and continue to fill the third and higher energy levels in a specified order we won't discuss here. The important thing to remember is that the arrangement of the outermost electrons in the highest energy level accounts for an element's reactivity and brings about the formation of diverse substances. Those substances, in turn, give shape, size, color, and function to the myriad forms of living things.

MOLECULES AND COMPOUNDS: AGGREGATES OF ATOMS

In nature, atoms link up with other atoms in numerous ways. Two or more atoms bound together form a **molecule.** The bound atoms can be identical to each other, as in the O_2 and N_2 we breathe in the air or the carbon molecules found in the extremely hard crystal we call a diamond. Dissimilar atoms also can combine, however, as carbon and oxygen do to form the poisonous gas carbon monoxide (CO). Chemical **compounds,** such as CO or CO_2 (carbon dioxide), contain atoms of more than one element and can be decomposed into these elements. Molecules and compounds display properties not found in the constituent elements; thus, the gases hydrogen and oxygen form the compound water (H_2O), with its unique characteristics.

What holds atoms together, and what determines the chemical properties of molecules and compounds? The answers lie in the behavior of orbiting electrons.

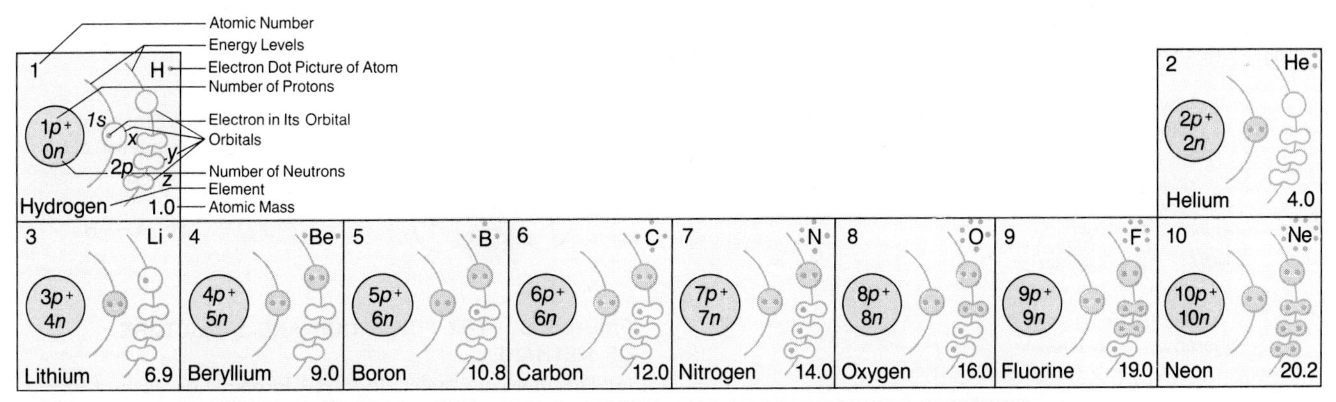

Figure 2-3 HOW ENERGY LEVELS AND ORBITALS ARE FILLED IN THE FIRST TEN ELEMENTS. Negatively charged electrons occupy atomic energy levels and orbitals in a specific order that causes them to stay as far away from each other as possible but as close to the positive nucleus as possible. This diagram shows the filling order in the first ten elements of the Periodic Table. For each element, a blue circle represents the nucleus, a tiny yellow dot an electron, an empty circle an unfilled spherical orbital, a filled-in circle a filled spherical orbital, an empty dumbbell an unfilled 2p orbital, and a filled dumbbell a filled 2p orbital.

Chemical Bonds: The Glue That Holds Molecules Together

Consider hydrogen gas, a trace component of the air around you. Hydrogen usually exists in the form of H_2 molecules, and is more stable than two separate hydrogen atoms. This is because two hydrogen atoms, each with its single orbiting electron, can *share* the electrons so that each effectively has two electrons in the 1*s* orbital, thereby completing the orbital and establishing the most stable arrangement (Figure 2-4a). As the atoms approach each other, each nucleus begins to attract the electron held by the other nucleus. Eventually, the electron clouds overlap and fuse into one **molecular orbital** (Figure 2-4b). Like an atomic orbital, a molecular orbital is most stable when filled by a pair of electrons. This shared orbital acts as a *chemical bond* between the two atoms and is like a strong spring in that it can be compressed or stretched to a certain extent without breaking. An atom can form as many bonds as there are unpaired electrons in its outermost orbital. The unpaired electrons involved in bond formation are called *valence electrons.*

The type of bond that forms between two atoms of hydrogen is called a *covalent bond.* There are two other types of chemical bonds, *ionic* and *polar,* and all three types differ in the way electrons are distributed around the nuclei of paired atoms.

Covalent Bonds

Because the bond between two hydrogen atoms involves shared valence electrons, it is called a **covalent bond.** Hydrogen can form such covalent bonds with other elements as well, including the biologically important element carbon. Carbon has four valence electrons and is "looking" for four additional electrons to fill its 2*p* orbitals, complete its second energy level, and give it maximum stability. The unfilled orbitals of four hydrogen atoms can fuse with the unfilled orbitals of one carbon atom to form four molecular orbitals with a shared electron pair in each, so that the one carbon and four hydrogens all achieve stable energy states with their outermost energy levels filled (Figure 2-5a). In the methane molecule that forms (CH_4), the four molecular orbitals point to the corners of a regular tetrahedron (a three-dimensional structure with four triangular sides) with the carbon atom at the center, because the four electron pairs tend to orbit the nucleus as far away from each other as possible (Figure 2-5b). Occasionally, the stable outermost molecular orbital of a covalently bonded atom loses an electron, and an unpaired electron continues to orbit the nucleus. If this highly energized molecular species (called a *free radical*) forms in a living thing, it can sometimes have destructive effects, as the box on page 25 explains.

The kinds of bonds formed in CH_4 are *single bonds,* meaning that only one pair of electrons is shared between two atoms (C and H; Figure 2-5c). But two atoms can share two or three pairs of valence electrons, forming *double* or *triple bonds.* Carbon atoms often form double bonds. For example, in ethylene (C_2H_4), a gas

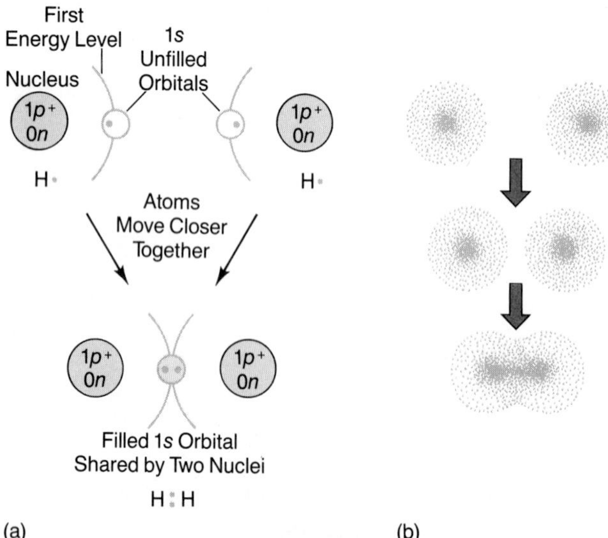

Figure 2-4 THE FORMATION OF A COVALENT BOND BETWEEN TWO HYDROGEN ATOMS.
When two hydrogen atoms come into extremely close proximity (a), their electrons fuse into one molecular orbital to form H_2, molecular hydrogen (b).

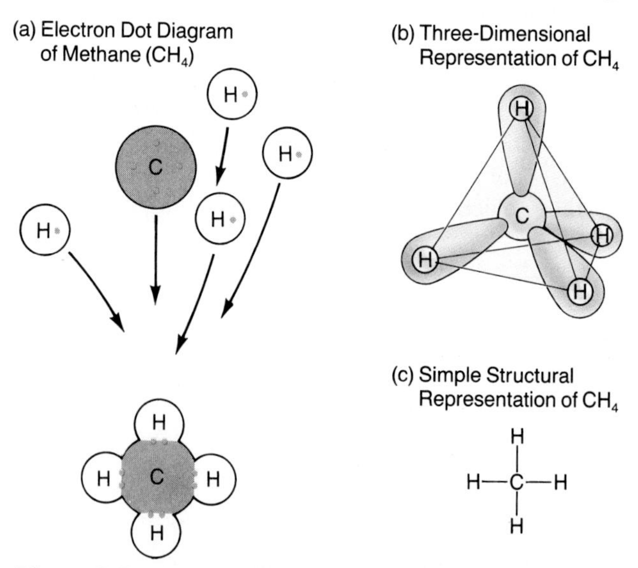

Figure 2-5 THE FORMATION AND STRUCTURE OF METHANE.
(a) The methane molecule (CH_4) forms when each of four hydrogen atoms form a covalent bond with one carbon atom. (b) The orbital structure shows the overlap of atomic orbitals and their tetrahedral configuration with the molecule. (c) A simpler structural representation shows the single bonds between the hydrogens and carbon as four single straight lines.

that aids fruit ripening, two carbon atoms share two pairs of electrons, and two hydrogen atoms also form single bonds with each carbon (Figure 2-6).

In single bonds, the two connected atoms can rotate around the bond much as wheels rotate on an axle. In double bonds, the connected atoms cannot rotate; in ethylene, for example, the two carbon atoms are fixed in space relative to each other. In compounds with triple bonds, such as N≡N (N_2, or nitrogen gas), three electron pairs are shared, the bond is stronger than a double bond, and the atoms are even more rigidly fixed in space—rigidity that can have biological consequences (see Chapter 3).

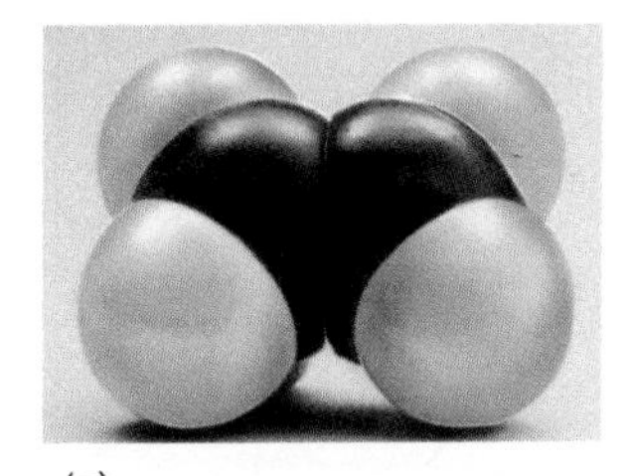

(a)

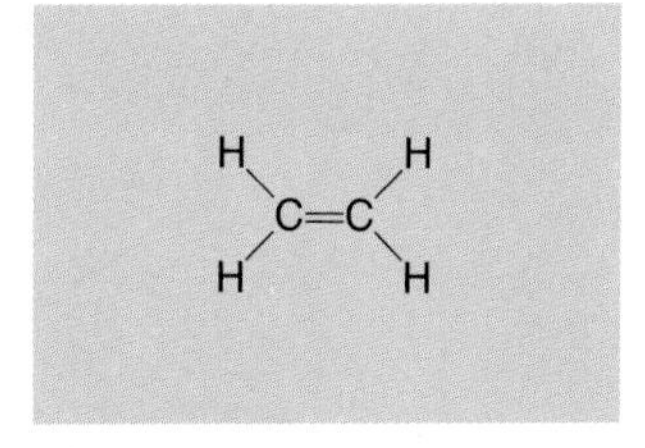

(b)

Figure 2-6 ETHYLENE HAS BOTH DOUBLE AND SINGLE BONDS.
(a) A space-filling model depicts the three-dimensional space occupied by the ethylene molecule (C_2H_4), including the electron orbitals, but shows little about the molecule's bonds. (b) This structural model reveals the double bond between carbon atoms.

Ionic Bonds

Whereas covalent bonds involve shared electrons, **ionic bonds** form when one atom gives up a valence electron and another atom adds the free electron to its outermost orbital, thereby holding the atoms together in an energetically stable unit. Lithium, for example, has three electrons: two 1*s* electrons and one 2*s* electron. Fluorine (F) has nine electrons, including two in two of its 2*p* orbitals and one in its third 2*p* orbital (Figure 2-7). When lithium and fluorine interact, lithium donates an

FREE RADICALS AND HUMAN HEALTH

Reactive chemical species called *free radicals* are being implicated in dozens of disorders, from wrinkled skin to cancer and premature senility. Free radicals, essentially molecules orbited by an unpaired electron, can form when molecular oxygen (O_2), hydroxyl ions (OH^-), nitrogen dioxide (NO_2), or other groups of covalently bonded atoms undergo certain types of chemical reactions and in the process lose an orbiting electron. In their new, unstable state, they can quickly steal an electron from a nearby molecule.

If the free radical steals an electron from a protein such as the collagen in connective tissue of joints or skin, the result can be increased wrinkling (Figure A) or intensified arthritis. If the electron donor is a molecule of fat in the delicate boundary that encloses a cell, a hole can form in that boundary, and the results can be damaged blood cells and damaged lung or eye tissue. If free radicals rob electrons from DNA, the "thefts" may alter genes and lead to cancer. And if a free radical attacks a nerve transmitter substance in the brain, it may contribute to Alzheimer's dementia.

The field of free radical chemistry is still in its infancy, but researchers have reached some tentative conclusions. Free radicals can form more easily during very heavy exercise; tobacco smoking; exposure to x-rays, strong sunlight, pesticides, cancer-causing substances, heavy air pollution, and certain kinds of drugs; and eating cured meats, fatty foods, and foods containing rancid fats. Avoiding these is clearly one way to reduce wear and tear by free radicals, and so is consuming more foods (not supplements) rich in vitamins A, C, and E, since these molecules scavenge unpaired electrons from free radicals before they can do damage.

Figure A
This Indian woman's skin is deeply furrowed from a lifetime's exposure to sun and the resultant formation of free radicals.

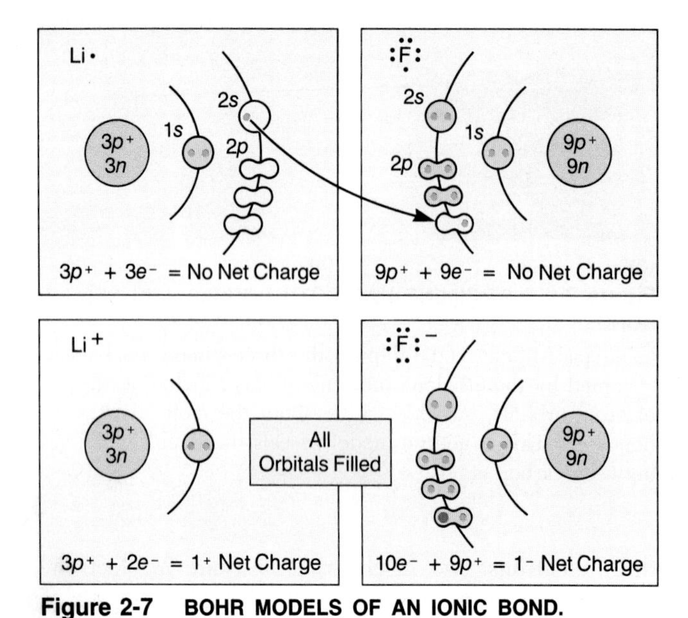

Figure 2-7 BOHR MODELS OF AN IONIC BOND.
Lithium donates an electron to fluorine. Fluorine then has a filled second energy level and takes on a net negative charge, while the lithium has a filled first energy level and takes on a positive charge—a stable configuration for each. Attraction between the positive and negative charges creates an ionic bond between the atoms.

electron to fluorine. As a result, the lithium atom becomes positively charged; it has three protons but is surrounded by only two electrons. At the same time, the fluorine atom, with nine protons and ten electrons, takes on a negative charge. The strong attraction between positive and negative ions gives rise to an ionic bond, and in the case of lithium and fluorine, to the ionic compound lithium fluoride (LiF). Charged atoms that have lost or gained electrons are called **ions** and are designated by a $^+$ or a $^-$: Li^+ and F^-, or O^{2-} and Mg^{2+}.

Ionic bonds hold together many common compounds, such as sodium chloride (NaCl), better known as table salt. When ionic compounds are dissolved in a solvent (e.g., when salt is dissolved in water for cooking spaghetti), they may *dissociate,* or break down, and free ions (Na^+ and Cl^-) would then be found in the solution. If the solution is dried by evaporating the water, the ions reassociate to form crystals in which the ions are once again bonded together. Ions play many important biological roles, as we shall see.

Polar Bonds

Covalent bonds and ionic bonds are like opposite sides of a coin when it comes to sharing electrons: Covalently bonded atoms share electrons equally, while ionically bonded atoms do not share electrons at all. Many bonds fall between these two extremes, however, and are characterized by a partial transfer of electrons. That is, atoms share the electrons, but the negative particles tend to spend more time orbiting one nucleus than orbiting the other, thus forming a **polar bond** (Figure 2-8). In a polar bond, the electrical charge from the cloud of moving electrons is asymmetrical; one atom is slightly negatively charged, and one is slightly positively charged.

The ability to attract electrons from other atoms in a molecule is called **electronegativity.** If a bond forms between two atoms with different degrees of electronegativity, the shared electron will spend most of its time near the nucleus of the more electronegative atom. In a polar molecule such as H_2O, for example, the electrons spend more time near the oxygen atom than near the two hydrogens; this leaves the hydrogen atoms somewhat positively charged and the oxygen atom somewhat negatively charged. This has powerful implications for the chemical properties of water, as we discuss later.

Bond Strength

Bonds may act as "chemical glue," but just how firmly do they hold atoms together? A bond can be strong or weak, but its strength, or **bond energy,** equals the amount of energy needed to break it. Chemists measure bond strength in **kilocalories** per mole (kcal/mole): 1 kcal is the amount of heat energy required to raise the temperature of 1 kg of water by 1° C, and the mass of a mole of a substance is the number of grams equal to the sum of the atomic weights of the constituent elements. For example, 1 mole of H_2O = 2 H (atomic weight 1) + 1 O (atomic weight 16) = 18. This quantity is the **gram molecular weight**. Thus, 18 g of H_2O = 1 mole of H_2O. One mole of any substance contains 6.023×10^{23} molecules—an enormous quantity known as *Avogadro's number*. Therefore, both 1 mole of water, which weighs 18 g, and 1 mole of O_2 molecules, which weighs 32 g (2 × 16), contain 6.023×10^{23} molecules.

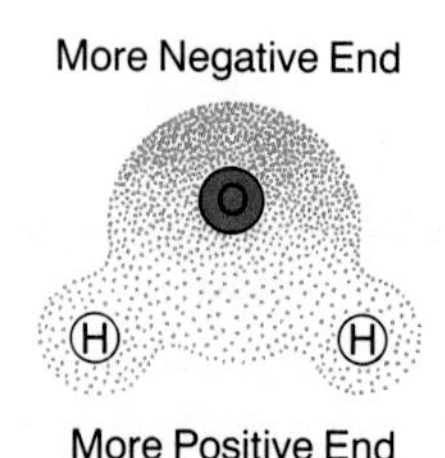

Figure 2-8 THE POLAR BONDS OF A WATER MOLECULE.
The electrons spend more time orbiting the oxygen nucleus than orbiting the two hydrogen nuclei, leaving the oxygen more negatively charged and the hydrogen more positively charged. This creates a polar molecule with negative and positive ends.

Table 2-2 SOME BOND ENERGIES

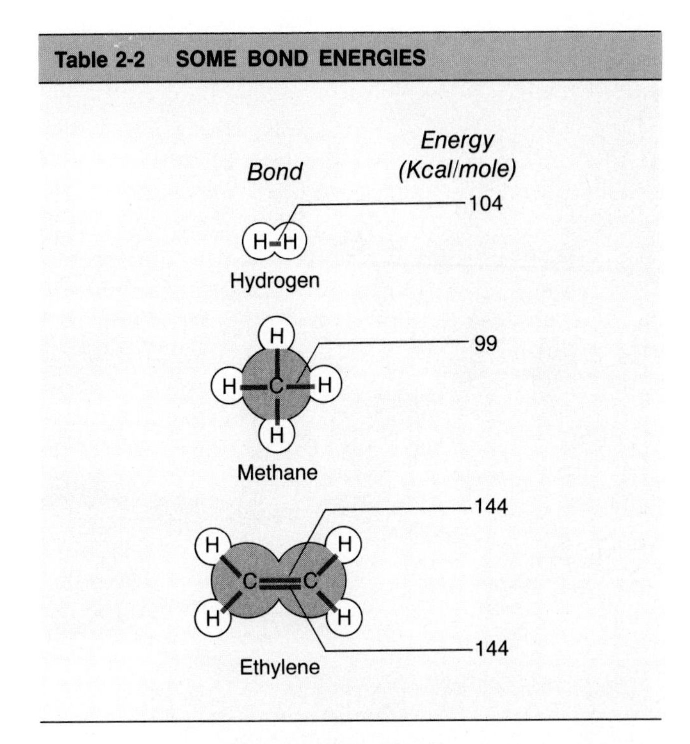

Table 2-2 lists the bond energies of various molecules we have discussed. Clearly, it takes a substantial amount of energy to break a stable bond. For example, the bond energy in a mole of hydrogen molecules (104 kcal in the 6.023×10^{23} H–H molecules constituting 1 mole) equals the energy needed to run a motor that could lift a 2-ton Cadillac 60 ft in the air or to keep a 100-watt light bulb glowing for an hour. The bond energies in molecules with more than two atoms are more complicated; for instance, it takes 120 kcal/mole to break the first O–H bond in H_2O, and 102 kcal/mole to break the second O–H bond. Both of the bonds in water, as well as all covalent and ionic bonds, are classified as **strong bonds.**

Weak bonds, which are easily broken by chance collisions or require very few kilocalories per mole to break, include two biologically important types: *hydrogen bonds,* which have a bond energy of about 4 kcal/mole, and *van der Waals forces,* which are very weak attractions that arise when any two atoms chance to come close together. We will encounter both kinds later.

Chemical Formulas and Equations

Chemists have created various shorthand systems for representing the composition and formation of molecules and compounds. Among them are molecular formulas, structural formulas, and chemical equations.

A **molecular formula** shows how many atoms of each type are present in a molecule and whether any of the atoms occur in certain common groups. Figure 2-9 shows the molecular formulas of water, carbon dioxide, methane, urea, and several other substances written below the molecular models. Notice that in the substance called urea, $(NH_2)_2CO$, two NH_2 groups are present along with one atom each of C and O.

A **structural formula,** such as that superimposed on each molecule in Figure 2-9, shows the approximate arrangement of the constituent atoms in space and the number of bonds between them. Chemists use molecular formulas to write chemical equations, which in turn show how molecules and ions interact with each other to form new substances in a process called a **chemical reac-**

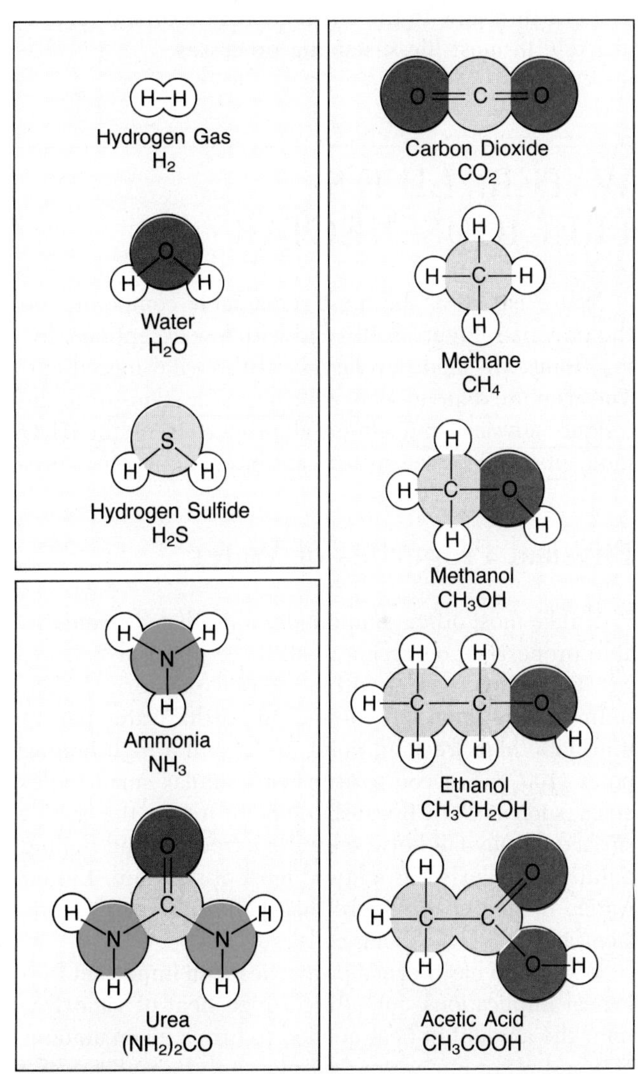

Figure 2-9 MOLECULAR MODELS AND STRUCTURAL FORMULAS HELP US VISUALIZE MOLECULES. The colored shapes shown here represent molecular models, while the letters and bonds within the shapes are structural formulas. The molecular formula is written below each model. Hydrogen compounds are grouped in the upper left-hand panel, nitrogen compounds in the lower left, and carbon compounds down the right side. Notice that hydrogens form one bond, nitrogens three, and carbons four.

tion. For example, they express the reaction between sodium hydroxide (NaOH) and hydrochloric acid (HCl) to yield sodium chloride (NaCl) and water this way:

$$NaOH + HCl \rightarrow NaCl + H_2O$$

The substances on the left side of the arrow are called **reactants,** and those on the right, which are formed as a result of the reaction, are called **products.** On both sides of a chemical equation, the total number of each kind of atom remains the same, but they can become distributed differently among the compounds as a result of the reaction process.

Let's turn now to the chemistry of water and its crucial role in most life-sustaining processes.

WATER: LIFE'S PRECIOUS NECTAR

Water is one of the most remarkable compounds in the universe (Figure 2-10), and with few exceptions, living things are made up largely of this vital ingredient. The absolute dependence of life on water stems from the unique physical and chemical properties of the H_2O molecule, and these, in turn, are based on its structure.

Figure 2-10 LIFE'S PRINCIPAL INGREDIENT: WATER. Living things are from 50 to 99 percent water, and the properties of the H_2O molecule are reflected in the structure and behavior of living matter. This vista from Yellowstone National Park shows water in its three physical states—as a liquid in the sparkling river, as a solid in the ground cover of snow, and as vapor rising from the geyser that soon condenses back into tiny liquid water droplets.

Physical Properties of Water

Unlike most other compounds, water has the remarkable property of occurring in all three physical states—gas, liquid, and solid—within the range of normal environmental temperatures found on earth. Water has an unusually high freezing point (0° C) and a high boiling point (100° C) in comparison with similar small molecules, such as CO_2. Because temperatures on the earth's surface usually fall between water's freezing and boiling points, water exists as a liquid most of the time. Liquid water, in fact, covers more than half of the earth's surface, and it fills all living cells.

Water also has thermal properties with important biological implications, including a high **heat of vaporization,** the amount of heat needed to turn a given amount of liquid water into water vapor (gas), and a high **specific heat,** the amount of heat needed to raise the temperature of 1 g of water by 1° C. (Water's heat of vaporization is 580 cal/g, and its specific heat is 1 cal/g/°C.) A high heat of vaporization means that for water to reach the gaseous state, it must absorb a great deal of heat from the surroundings. For many plants and animals, this property underlies a natural cooling system: Water evaporating from leaves, skin, or lungs uses up heat from the organism in the process of changing from liquid to gas.

Because of its high specific heat, water is slow to change temperature, and considerable amounts of heat must be added or removed to make the temperature change much. This property provides special insurance for living creatures, which often function best within a narrow temperature range. The seawater surrounding a kelp or a barnacle, as well as the water within the organisms, tends to heat or cool more slowly than the surrounding air or soil, and this provides a natural buffer against potentially damaging temperature fluctuations.

Water molecules also exhibit physical properties, such as strong cohesion, adhesion, tensile strength, and capillarity. **Cohesion** is the tendency of like molecules to cling to one another (such as water to water). **Adhesion** is the tendency of *unlike* molecules to cling together (such as water to the molecules of silicon dioxide on the walls of a drinking glass). **Tensile strength** is related to cohesion and is a measure of the resistance of molecules to being pulled apart. And **capillarity** is the tendency of molecules to move upward in a narrow space against the tug of gravity.

Let us see how these properties apply to some easily observed biological phenomena. You can observe the results of adhesion when you pour water into a flowerpot full of soil. The water sinks in and "wets" the soil, rather than remaining on top, because of water's inherent "stickiness," its tendency to adhere to the dissimilar molecules in the soil. Adhesion, cohesion, and capillarity also help explain why much of the water remains in the soil rather than running straight through. The molecules of water can resist the downward tug of gravity by first adhering to the surface of soil particles and then moving into tiny air spaces between them via capillarity. The water molecules enter the spaces and pull others along by means of cohesion, and thus remain in these tiny spaces rather than flowing out the bottom of the pot.

Adhesion, cohesion, and capillarity also explain why water moves upward in a thin glass tube. Water molecules on the periphery of the cylinder of liquid adhere to silicon dioxide molecules in the glass and move upward, pulling along the water molecules inside the column of water by means of cohesion. Fluid rises higher in a narrow tube than in a wide tube because in the wider vessel, a higher percentage of molecules lie inside the water column, away from the periphery and so are less liable to cohere to the glass. Significantly, in plants, water travels in exceedingly thin transport vessels from the tips of the roots through the stems to the highest leaves. Water can move so efficiently up these narrow vessels partly because of the features of capillarity, partly because water molecules adhere to sugar molecules in the walls of the vessels, and partly because the water molecules cohere to one another in unbroken chains. (We'll return to this water movement in Chapter 27).

Yet another of water's inherent physical properties is the formation of spherical droplets, as in faucet drips and raindrops. This occurs because of **surface tension**—the tendency of a liquid to minimize the surface area. The strong cohesion among water molecules causes them to be more attracted to one another than to air molecules, and thus the surface of a droplet acts like an elastic skin to confine the molecules into a shape with the smallest surface area—a sphere. Many small organisms, including the pond skater pictured in Figure 2-11, are able to exploit such surface tension, nimbly walking across pond or stream surfaces without breaking through.

Finally, water has a unique but familiar physical property with an important impact on biology: Its solid form, ice, floats. In almost every other substance, the solid state is denser than the liquid form; a cube of frozen CO_2 (also called "dry ice"), for example, will sink to the bottom of a beaker of liquid CO_2. Ice, however, is less dense than liquid water and floats in it. If solid water sank, lakes and seas would freeze from the bottom up, and many bodies of water would remain frozen nearly year round. Instead, the floating ice layer insulates the lower depths and allows many aquatic plants and animals to survive until the surface begins to melt.

Figure 2-11 WALKING ON WATER.
Surface tension allows the pond skater to walk on water. Cohesion between H_2O molecules creates an elastic "skin" that supports the weight of small insects such as pond skaters and water striders.

Molecular Structure of Water

Water is clearly a special substance, but what makes it so? The answer is a lesson in chemical structure and function: The properties of water derive from the structure of H_2O molecules, as well as from the bonds that form between them.

In an H_2O molecule, two covalent bonds form as the oxygen atom shares the two single electrons in its two outermost ($2p$) unfilled orbitals with the single electrons from two hydrogen atoms. Simultaneously, the two pairs of electrons not involved in the covalent bonds are repelled by the electron clouds of the covalent bonds and by each other. The four electron pairs, all repelling each

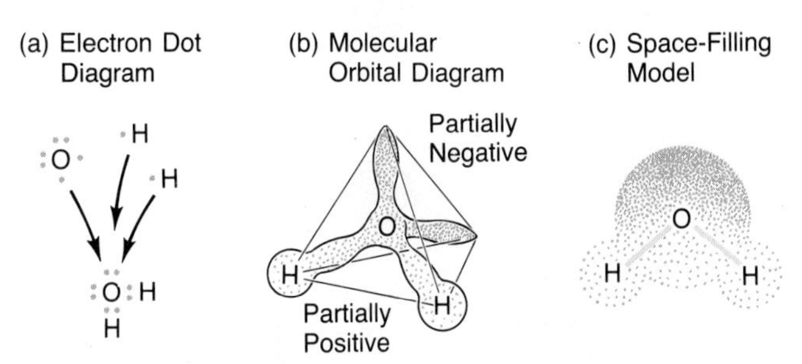

Figure 2-12 BASIC STRUCTURE OF A WATER MOLECULE.
(a) An electron dot diagram shows the covalent bonding of hydrogen and oxygen in a water molecule, while (b) a molecular orbital diagram reveals the molecule's tetrahedral shape, and (c) a space-filling model shows the shape the atoms occupy in space.

other, push out into a tetrahedron-shaped molecule, with two hydrogen atoms occupying two corners, the oxygen atom at the center, and the so-called lone pairs of electrons keeping to the other two corners (Figure 2-12). An H_2O molecule is electrically neutral, therefore, but it is also a polar molecule—that is, its charge is asymmetrically distributed. The side with the two hydrogen atoms has a positive charge, and the side with the lone pairs of electrons has a negative charge.

Hydrogen Bonding

Because of its polarity, one water molecule can interact with four other water molecules: two toward the negative side and two toward the positive side. The molecules lie as close to each other as possible without covalent or ionic bonds forming between the electron clouds of the five molecules (Figure 2-13). The attraction between the oxygen atom of one molecule and a hydrogen atom of another results in the formation of a weak bond called a **hydrogen bond**. The name *hydrogen bond* comes from the fact that, in a sense, one hydrogen atom is shared by two molecules. A hydrogen bond is almost twice as long as the O–H covalent bond in a single H_2O molecule and is 10 to 20 times weaker. The energy needed to break a hydrogen bond is 4–5 kcal/mole, which, in fact, is quite close to the average energy of motion of water molecules in liquid water. For this reason, only a fraction of the possible hydrogen bonds exist at any given time in liquid water, since hydrogen bonds that form in one instant are torn apart in the next. Hydrogen bonds play critical roles in the shape and function of many biological molecules, as Chapter 4 explains.

The nature of hydrogen bonds explains the special properties of water, including the ability of ice to float. In ice, water molecules are immobilized in a regular, latticelike arrangement with all possible hydrogen bonds formed and stable (Figure 2-14, left). The hydrogen bonds actually hold the water molecules farther apart from one another than in liquid water; thus, a piece of ice occupies a larger volume than an equivalent amount of liquid water, and is less dense. This explains why water pipes crack in winter and soft-drink bottles explode in the freezer: Water expands as it freezes. When the lattice of ice once again melts, the lattice structure collapses like a Tinkertoy tower from which the connecting rods are removed (Figure 2-14, right). The same number of water molecules can suddenly fit into a smaller volume, and the density of the liquid gradually increases until a peak is reached at about 4° C.

Hydrogen bonding also accounts for water's high specific heat and high heat of vaporization. When you heat a substance such as liquid nitrogen, which lacks hydrogen bonds, every bit of heat energy increases the motion of the molecules and thereby increases its temperature. When you heat water, however, much of the energy is used up in stretching or breaking hydrogen bonds rather than in increasing molecular motion. Thus, water is slow to increase in temperature, as its high specific heat indicates. The high heat of vaporization has a similar explanation: Before water can turn to vapor, all hydrogen bonds must be broken and the individual molecules set in rapid motion so they can break free of the fluid and "fly away" as gas molecules.

Finally, hydrogen bonding accounts for water's properties of cohesion, adhesion, tensile strength, and surface tension: Hydrogen bonds are the "glue" that holds

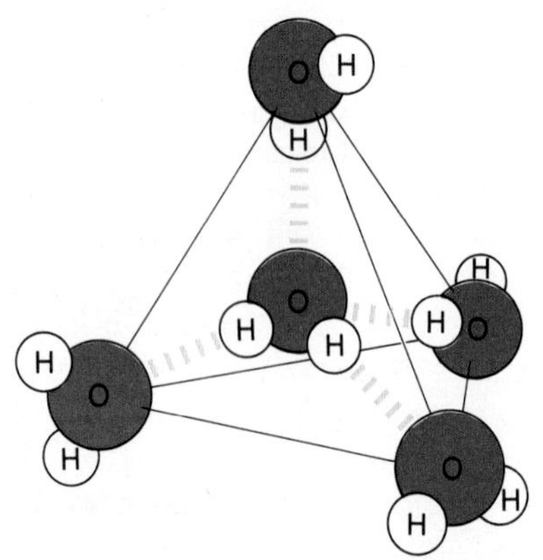

Figure 2-13 HYDROGEN BONDING AMONG WATER MOLECULES.
One water molecule can form hydrogen bonds (dotted lines) with four other water molecules. Negative and positive ends attract each other, but ends of like charge repel each other.

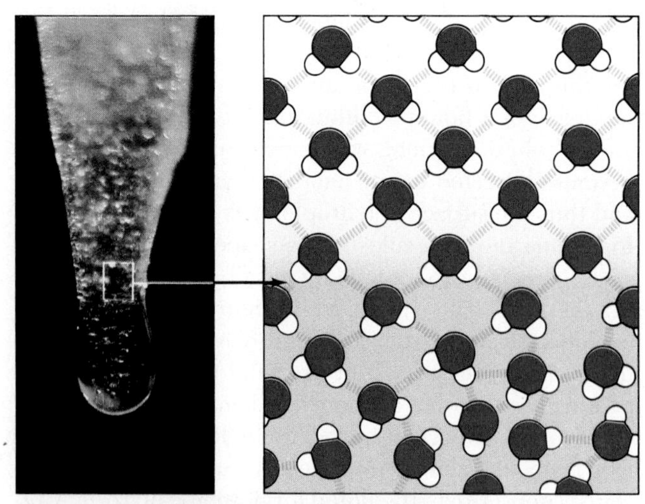

Figure 2-14 ICE HAS A LATTICELIKE STRUCTURE, WATER AN IRREGULAR CONFIGURATION.
Within an icicle, water molecules are held in a crystalline lattice (top, right). This regular array takes up more space than the randomly arranged molecules in liquid water (bottom, right), and this explains why an ice cube, icicle, or the frozen surface of a lake is less dense and thus floats.

together groups of water molecules, makes them resistant to penetration, and helps them adhere to other kinds of molecules. Not all substances form hydrogen bonds with water, however, and that is why water "beads" rather than wets a newly waxed car.

Water as a Solvent

More substances will dissolve in water than in any other liquid. This makes water the world's most efficient and widespread **solvent,** a substance capable of forming a homogeneous mixture with molecules of another substance. Substances that dissolve in solvents are called **solutes.** As a solute dissolves in water, the individual molecules of the solute disperse through the water and become surrounded by clusters of water molecules. The amount of solute in a solution is less than the amount of solvent. The amount of solute in a solution—expressed in grams per milliliter or moles per liter—is referred to as the *concentration* of the solution.

Ionic compounds, such as LiF and NaCl, provide the simplest examples of how solutes dissolve in solvents. When added to water, they tend to dissociate into ions, with the positive ions, called *cations* (Li^+, Na^+), being attracted to the negative side of water molecules, and the negative ions, called *anions* (F^-, Cl^-), attracted to the positive side of water molecules. In solution, tight clusters of water molecules surround the cations with negatively charged oxygen atoms pointing inward (Figure 2-15), while similar clusters surround the anions with positively charged hydrogen atoms pointing inward. In each case, the surrounded ion is said to be *hydrated*, and the groupings formed are called *hydration spheres.*

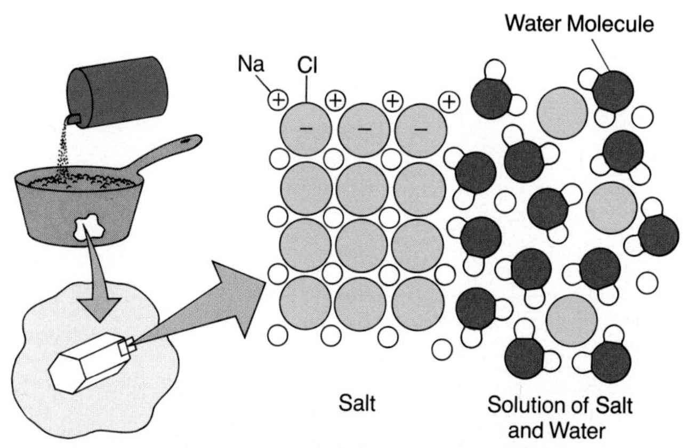

Figure 2-15 WHEN TABLE SALT DISSOLVES, HYDRATION SPHERES FORM.
In a solution of salt and water, Na^+ and Cl^- dissociate, and hydration spheres form, with charged ends of the water molecules oriented toward the positive or negative ions. The solid is shown in the left, the water and salt solution on the right. Hydrogen bonds among water molecules have been omitted for clarity.

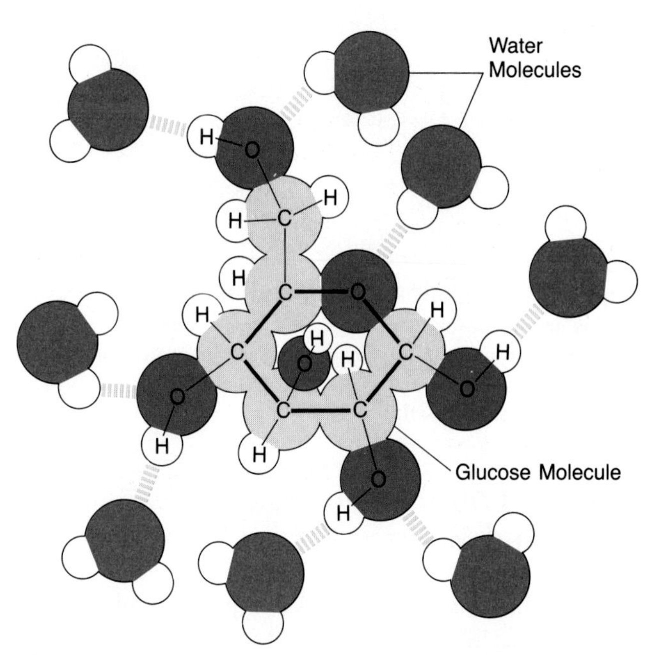

Figure 2-16 A GLUCOSE MOLECULE DISSOLVED IN WATER.
Sugars dissolve in water by forming weak hydrogen bonds with water molecules, again with positive and negative ends of the water molecules oriented toward charged regions of the glucose molecule.

Compounds can be categorized by their ability or inability to dissolve in water, and this property is very important to living things. Compounds that dissolve readily in water (such as NaCl and LiF) are called **hydrophilic** (water-loving) compounds. Polar molecules such as proteins, sugars, and many other large biological molecules are also hydrophilic, forming hydrogen bonds with water molecules (Figure 2-16). On the other hand, oils, waxes, and some plastics tend to be insoluble in water; therefore, they are called **hydrophobic** (water-fearing) compounds. These compounds have nonpolar covalent bonds (i.e., covalent bonds in which the bonding electrons are shared equally by both nuclei), and, as such, form neither hydrogen bonds nor purely electrical attractions with water molecules. If you drop oil on the surface of water, the two fluids form a boundary, or interface, because of their mutual insolubility. Cohesion tends to keep oil molecules with oil molecules and water with water, and little, if any, adhesion holds the drops together. In fact, depending on the amount of oil present, it will either stay together as a droplet or spread in a layer across the water surface, just as oil slicks form on the sea following an oil spill. The same principle is a basic feature of life; a hydrophobic layer of lipid molecules (types of fats) covers the surface of *every* living cell on earth. In fact, the bodies of such organisms as lizards, robins, and field mice can only retain water because hydrophobic compounds in their skin form a waterproof layer that inhibits water loss outward (Figure 2-17).

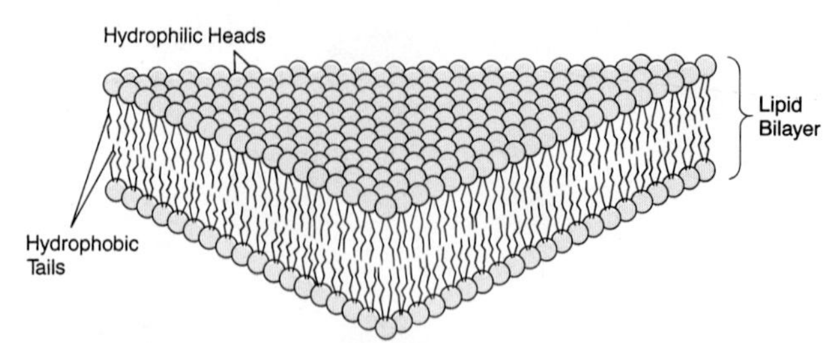

Figure 2-17 HYDROPHOBIC LIPIDS IN CELLS. Lipids are not soluble in water and form a water-impermeable membrane, or coating, at the surface of every living cell. Lipid layers also surround various specialized regions within cells.

Dissociation of Water: Acids and Bases

Water has one more remarkable property of interest to biologists: a slight tendency to "fall apart." Just as ionic compounds like NaCl tend to dissociate in water, water molecules themselves dissociate to a small extent: about 2 out of every 10^9 H_2O molecules spontaneously separate into hydrogen ions (H^+) and *hydroxyl* ions (OH^-). The hydrogen ions by and large associate with individual water molecules, thus forming *hydronium* ions (H_3O^+). (For all practical purposes, we can consider H^+ and H_3O^+ as equivalent in this discussion.) Like Na^+ and Cl^- in water, H^+ and OH^- become hydrated. Once dissociated, these ions can again combine, or associate, to re-form water molecules. Thus, two reactions proceed at the same time:

$$H_2O \rightarrow H^+ + OH^-$$

$$H^+ + OH^- \rightarrow H_2O$$

The number of hydrogen and hydroxyl ions in a sample of H_2O is equal and constant: in 1 L of pure water, there is 10^{-7} mole of H^+ and 10^{-7} mole of OH^-.

Adding a solute to water can change the concentration of H^+. Any substance that gives up, or donates, H^+ in solution, and thereby increases the H^+ content of the solution, is called an **acid**. Compounds that decrease the amount of H^+ in solution are called **bases.** Bases take up, or accept, H^+ from the solution (for instance the base OH^- would bind to H^+ to form water).

The concentration of H^+ in a solution is expressed in terms of **pH,** a scale running from 0 to 14. The pH of a solution is the negative logarithm of the concentration of H^+ in mole/L: $pH = -\log [H^+]$, where the brackets mean concentration. Thus, pure water, with an H^+ concentration of 10^{-7} mole/L, has a pH of 7. Pure water is said to be *neutral;* it is neither basic (*alkaline*) nor acidic. More H^+ means a *lower* pH. Hence acidic solutions have pH values less than 7. And because bases decrease H^+ concentration, they have pH values greater than 7.

Since the pH scale is a logarithmic scale, each unit represents a tenfold change in the concentration of H^+. A solution with a pH of 6, for example, has a concentration of hydrogen ions ten times greater than pure water, at pH 7.

Acids and bases are extremely important in biology because the chemical reactions that go on within living cells and tissues are sensitive to H^+ concentrations. Some biological materials, such as stomach fluid or the fruit of lemons and limes, are extremely acidic (Figure 2-18). Other fluids, such as the liquid surrounding human sperm or the seawater surrounding many aquatic organisms, are basic. Whether an organism's external environment is acidic or alkaline, however, the fluids in and around most living cells have an almost neutral pH, between about 6.5 and 7.5—a stability based on buffers.

Buffers are immensely important to cells and organisms because they help them to resist changes in pH when acids or bases are produced or added (as from food). Buffers are substances that bind H^+ when concentrations of H^+ are high and release H^+ when concentrations of H^+ are low. These activities tend to maintain a constant pH in the solution. Suppose such a solution has

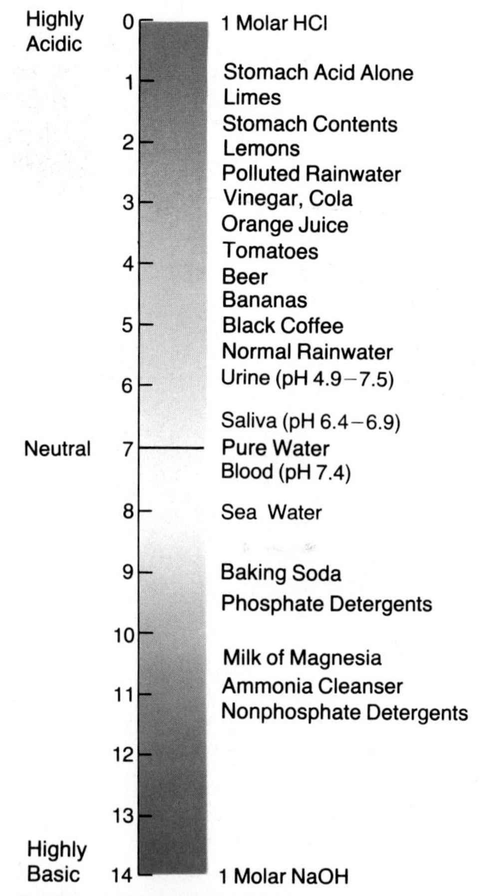

Figure 2-18 THE pH SCALE AND THE pH VALUES OF COMMON FLUIDS AND SUBSTANCES.

a pH of 6.3. If we add more H^+ to the solution, the buffer tends to absorb it, keeping the pH constant at about 6.3. If another test solution containing a buffer had a pH of 8 and a base, OH^-, was added, the buffer would release H^+, thereby neutralizing the added OH^- and maintaining the pH near 8. Chapter 3 will describe the buffering system in red blood cells.

ATOMS TO ORGANISMS: A CONTINUUM OF ORGANIZATION

We have seen in this chapter that the properties of atoms and molecules stem from the structure of those basic units of matter and their electron configurations. Throughout the remaining chapters, we will emphasize life's diversity and structural complexity, but the underlying importance of atomic structure remains, because all matter of significance to life is part of a continuum of organization. This continuum begins with subatomic particles—protons, neutrons, and electrons—and progresses to atoms, with their cloudlike orbitals; atoms are then built into molecules; small molecules are assembled into giant "macromolecules"; assemblages of macromolecules build parts of cells called organelles and membranes; those cell parts, when added together, form cells; cells put together yield tissues and organs; and, finally, assemblages of tissues and organs form individual organisms of all shapes and sizes. In Chapter 3, we move up the continuum of organization from atoms and small molecules to the large molecules found in living things.

SUMMARY

1. The universe is composed of 92 naturally occurring chemical *elements*, fundamental substances that cannot be broken down by chemical means. *Atoms* are the smallest units into which matter can be broken without changing its properties.

2. Each atom consists of a nucleus, which contains *protons* and *neutrons*, and negatively charged *electrons*, which are distributed in cloudlike orbitals around the nucleus. The number of positively charged protons determines an element's *atomic number;* the sum of protons and neutrons is an element's *atomic weight*, or atomic mass.

3. The number of electrons in an atom is equal to the number of protons, making the atom electrically neutral.

4. The distribution of electrons in orbitals is governed by a few basic rules: The simplest orbital of the lowest energy level is filled first; one electron must be present in each orbital of the same shape at one energy level before any of them can be filled; and so on.

5. *Molecules* are precisely ordered arrangements of atoms. *Compounds* are molecules made up of two or more types of atoms.

6. The atoms in molecules are held together by chemical bonds. In *covalent bonds*, electrons are shared by different atoms. In *ionic bonds*, electrons are transferred from one atom to another. Atoms that lose or gain electrons prior to the formation of an ionic bond are called *ions*, and the attraction between oppositely charged ions results in an ionic bond.

7. *Electronegativity* is a measure of the ability of atoms to attract other atoms in the same molecule. In molecules made up of atoms with different electronegativities, the bond is said to be *polar*.

8. The strength of chemical bonds is expressed as *bond energy* and is measured in *kilocalories per mole*. Covalent and ionic bonds are *strong bonds. Weak bonds* include hydrogen bonds and van der Waals forces.

9. *Molecular formulas* are used to represent the components of molecules and can be used to describe *chemical reactions* that involve *reactants* and *products. Structural formulas* show the approximate arrangement in space of the constituents of a molecule.

10. Water is crucial to life. It has a high melting point, high boiling point, high *specific heat*, and high *heat of vaporization*. Water molecules also show strong *cohesion, tensile strength, adhesion, capillarity*, and *surface tension;* in addition, ice floats. All these properties can be explained by the structure of the water molecule, which is characterized by a high degree of *hydrogen bonding*.

11. Water is an important *solvent:* it is capable of dissolving many kinds of *solutes*. Ionic compounds are examples of *hydrophilic* substances, which are readily hydrated. *Hydrophobic* substances, such as compounds with nonpolar bonds, are insoluble in water.

12. Water can dissociate into H^+ and OH^-. An *acid* is a substance that donates H^+ when dissolved in water; acidic solutions have pH values below 7. A *base* is a substance that accepts H^+; basic solutions have pH values above 7. A *buffer* binds or releases H^+, depending on the H^+ concentration in its vicinity. Buffers help stabilize the pH of solutions.

KEY TERMS

acid
adhesion
atom
atomic number
atomic weight
base
bond energy
buffer
capillarity
chemical reaction
cohesion
compound
covalent bond
electron

electronegativity
element
gram molecular weight
heat of vaporization
hydrogen bond
hydrophilic
hydrophobic
ion
ionic bond
isotope
kilocalorie
molecular formula
molecular orbital
molecule
neutron
orbital
pH
polar bond
product
proton
reactant
solute
solvent
specific heat
strong bond
structural formula
surface tension
tensile strength
weak bond

QUESTIONS

1. Draw atoms of two different elements and label the components.

2. Draw a molecule composed of two different elements and describe how it is held together.

3. Draw several molecules of liquid water, showing the bonds that form between the molecules.

4. Name one way to break some of the hydrogen bonds that link water molecules together.

5. What force is strong enough to break the covalent bonds in water and to separate each water molecule into hydrogen and oxygen?

6. Water forms a film on a clean drinking glass, but beads up on a greasy glass. It also forms beads on a plastic "glass." Which is more hydrophobic: glass or plastic?

7. What is an acid? A base? A buffer?

ESSAY QUESTION

1. Discuss the properties of water that make it a vital component of living organisms. How do hydrogen bonds contribute to these properties?

For additional readings related to topics in this chapter, see Appendix C.

Organic Molecules → Cont. H & C

P. 38

3 The Molecules of Living Things

This sameness of composition (encountered in all living beings from bacteria to man) is one of the most striking illustrations of the fact that the prodigious diversity of macroscopic structures of living beings rests in fact on a profound and no less remarkable unity of microscopic makeup.

Jacques Monod,
Chance and Necessity (1972)

Just as a novel is made up of words and words are made up of individual letters, the phenomenon we call life is written in a language of molecules and atoms. And just as atomic structure underlies the properties of molecules and compounds, the shapes and behaviors of biological molecules account for the physical characteristics and activities of living organisms.

Our present topic truly begins the study of biology, because it includes four major classes of large molecules that are unique to living things.

This chapter considers:

- Organic compounds—the carbon-containing molecules that make up the stuff of life
- Carbohydrates—sugars, starches, and related compounds central to cell energetics and structures
- Lipids—the fats, oils, and similar molecules with crucial roles in energy storage, waterproofing, and cell regulation
- Proteins—highly varied compounds that account for much of life's diversity and cellular structure and activity
- Nucleic acids—the information-bearing code of life that carries the instructions for inheritance as well as for a living thing's day-to-day operations

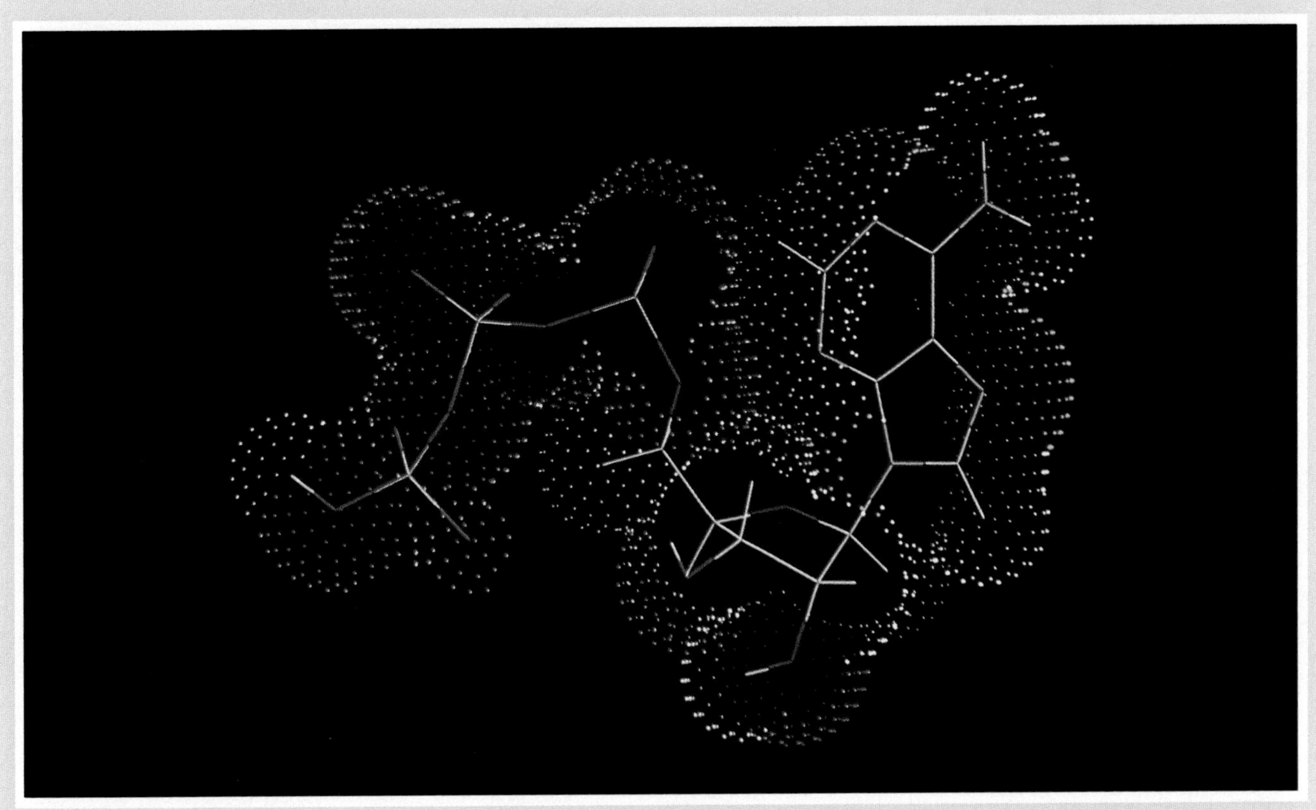

Fuel for the fire of life: A computer's view of ATP.

Later chapters will show how these diverse molecules are assembled into cell parts and how cells make up whole organisms. Here, however, we first encounter the transition zone between chemistry and biology, between the nonliving and the living.

THE FUNDAMENTAL COMPONENTS OF BIOLOGICAL MOLECULES

Carbon: The Indispensable Element

While some biological molecules are small and relatively simple, many of the carbohydrates, lipids, proteins, and nucleic acids are **macromolecules**—extremely large molecules with molecular weights of about 10,000 daltons or more. Most of the compounds that make up living things, however, share one thing in common: They contain carbon. In fact, life on earth cannot be separated from carbon and its unique chemistry. Any compound that contains carbon is called an **organic compound**; one without carbon is an *inorganic compound.* (A few simple molecules that contain carbon but no hydrogen, such as CO_2, are considered inorganic.) All biological molecules are organic, but not all organic molecules are manufactured by living things; some plastics, for example, are organic compounds not made within living cells.

The unique structure of the carbon atom ultimately accounts for the great diversity of molecules in living things. Since the carbon atom can form covalent bonds with up to four other atoms (see Chapter 2), it can bond to other carbon atoms and produce long, straight chains, branched chains, rings, and a variety of other shapes. Among these chains and rings are the backbones for the carbohydrates, lipids, proteins, and nucleic acids (Figure 3-1). Carbon atoms can also form covalent bonds with oxygen, nitrogen, hydrogen, sulfur, and occasion-

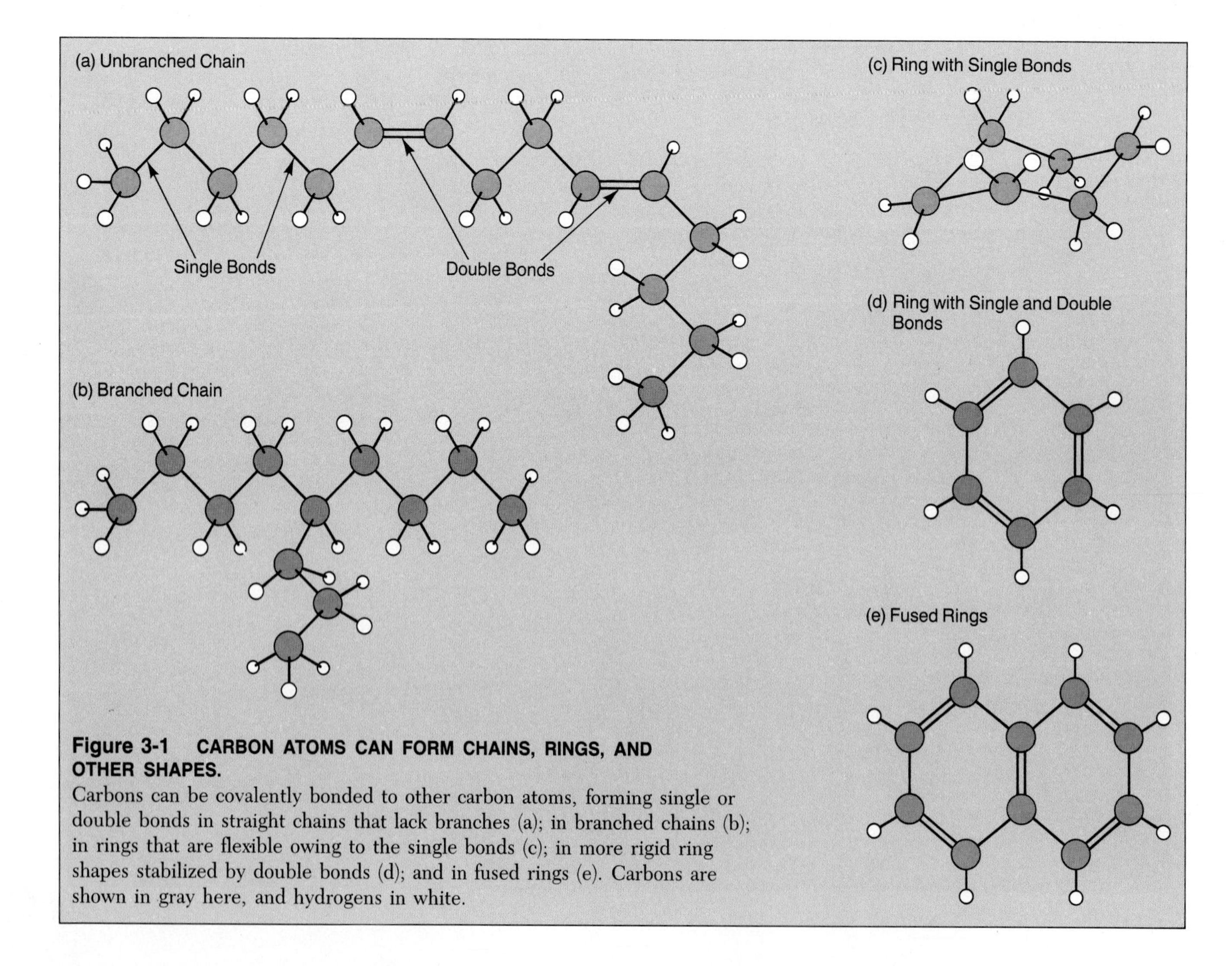

Figure 3-1 CARBON ATOMS CAN FORM CHAINS, RINGS, AND OTHER SHAPES.
Carbons can be covalently bonded to other carbon atoms, forming single or double bonds in straight chains that lack branches (a); in branched chains (b); in rings that are flexible owing to the single bonds (c); in more rigid ring shapes stabilized by double bonds (d); and in fused rings (e). Carbons are shown in gray here, and hydrogens in white.

ally phosphorus to produce a great variety of organic molecules. In all, carbon atoms can form the chemical skeletons for more than 1 million compounds—ten times more than the other 91 elements combined.

Carbon's ability to form bonds has a profound effect on the structure and function of biological molecules. The atoms in molecules are constantly jiggling, bending back and forth as the chemical bonds vibrate rapidly. Atoms rotate around the axis of a bond like spokes on a bicycle wheel whizzing around the hub. In small molecules, the effect of this rotation is not very great, but in a large organic molecule with a long chain of singly bonded carbon atoms, each of which is free to rotate relative to its neighbors, the molecule can sweep out a large volume as it assumes many shapes, or conformations. Figure 3-2 shows some of the possible conformations of a four-carbon chain. When carbon atoms are linked by double bonds, they are not free to rotate and change conformation. Consequently, all other atoms directly attached to double-bonded carbons lie in a single plane. Carbon rings, which occur in many important biological molecules, are typically flat if double bonds are present.

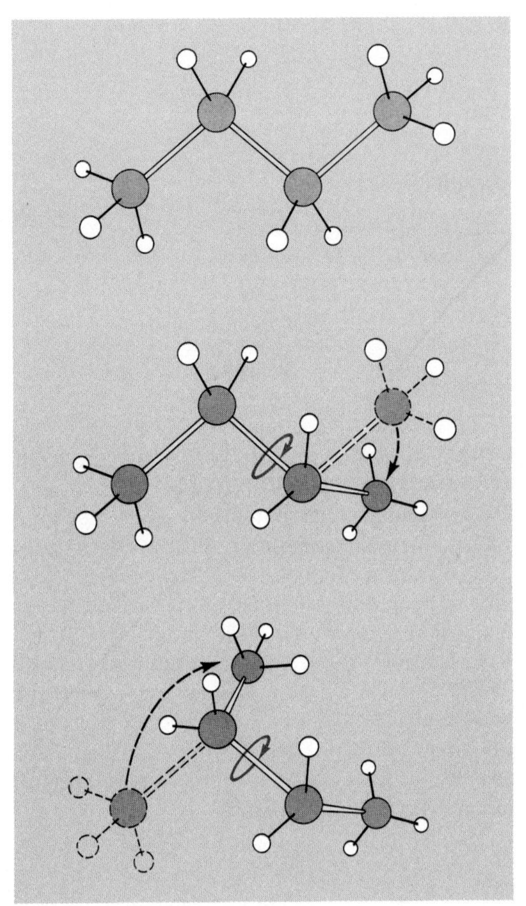

Figure 3-2 THE CHANGING CONFORMATIONS OF A FOUR-CARBON CHAIN.
A chain of four carbons can assume many possible shapes, or conformations, as it rotates around the carbon-carbon bonds.

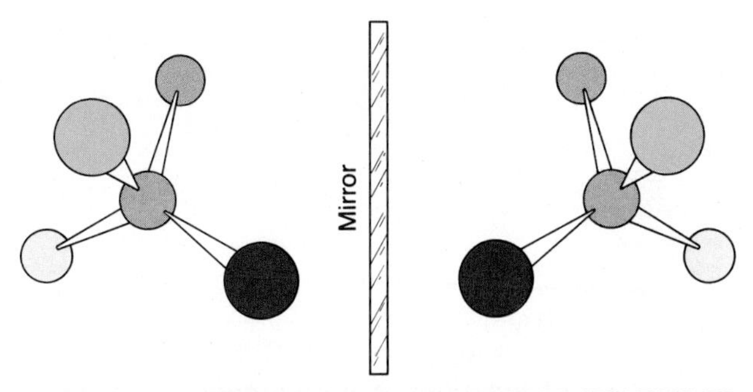

Figure 3-3 ENANTIOMERS: MIRROR-IMAGE MOLECULES.
These two molecules are enantiomers. Their chemical composition is the same; but since their atoms are arranged differently, their structure is different. They are mirror images of each other, like a person's right and left hands.

Another significant characteristic of carbon is the tetrahedral structure that results when it forms four single covalent bonds with four other atoms. In the methane (CH_4) molecule, for example, four hydrogen atoms are covalently bonded to carbon, and a tetrahedron forms (see Figure 2-5). Significantly, a molecule composed of one carbon atom with four different attached atoms can be arranged in two ways that are mirror images of each other. These mirror-image arrangements of atoms are called **enantiomers** (Figure 3-3). Enantiomers of a compound share the same chemical properties, but they will rotate light passing through them in opposite directions. Two compounds with exactly the same formula but different arrangements of atoms are called **isomers.** Isomers often have very different chemical properties.

If enantiomers of a molecule have the same chemical properties, why shouldn't we consider them equivalent? There are many important biological enantiomers, which can exist in either a right-handed, D form (from *dextro*, meaning "right") or a left-handed, L form (from *levo*, meaning "left"), and these forms can have different effects in biological systems. For example, a substance called L-adrenaline fits precisely in the pocket-shaped portions of certain molecules in the walls of blood vessels and causes those vessels to constrict. D-adrenaline, however, does not fit so well in the receptors and is some 12 times weaker than L-adrenaline in causing vessels to constrict. Although right-handed and left-handed molecules have the same chemical properties, only the L form is found in the molecules that make up proteins. This is true for every type of organism studied so far, from the simplest bacterium to the most complex plant or animal.

Compounds that have the same molecular formula but different properties are called **structural isomers.** Consider, for example, the isomers ethanol and dimethyl ether: both have the formula C_2H_6O. Ethanol

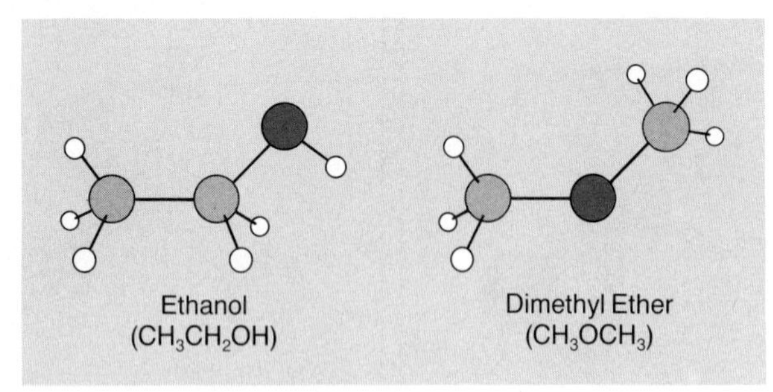

Figure 3-4 STRUCTURAL ISOMERS: SIMILAR ATOMS, VERY DIFFERENT PROPERTIES.
This diagram shows the molecular models and formulas for ethanol (ethyl alcohol) and dimethyl ether. Although both contain two carbon atoms, one oxygen atom, and six hydrogen atoms, the molecules have very different properties. Ethanol is a clear, intoxicating liquid, while dimethyl ether is a colorless gas.

(also called ethyl alcohol) is the colorless liquid that gives an intoxicating effect to wines, beers, and liquors. It is produced in yeast cells and some bacteria. Dimethyl ether, which is nonbiological in origin, is also colorless, but its boiling point is so low (−23° C) that it could boil and turn to vapor in a kitchen freezer. Dimethyl ether sometimes is used as an industrial refrigerant, but it never would be used in a cocktail. The arrangements of atoms in ethanol and dimethyl ether, shown in Figure 3-4, are quite different and account for the dissimilar properties of the two isomers.

Functional Groups: The Key to Chemical Reactivity

The real key to the chemical behavior of most biologically important molecules is the presence of specially arranged clusters of atoms called **functional groups.** The specific structure of each functional group imparts a similar chemical behavior to all molecules to which it is attached.

For example, the *hydroxyl group*, —OH (Figure 3-5a), has a role in the dissociation of water and the structure of ethanol. Ethanol and other carbon-containing compounds that have a hydroxyl group are classified as *alcohols*. Because of the hydroxyl group, alcohols tend to be polar (at least in the —OH region) and are often soluble in water, and they form weak hydrogen bonds. Thus, we can predict some aspects of a compound's chemical behavior simply by knowing that it has one or more —OH functional groups.

The *carboxyl group*, —COOH (see Figure 3-5a), is very common and usually exists as $—COO^-$ because it donates H^+ in solution. Like other substances that give up H^+ in solution, molecules containing the carboxyl

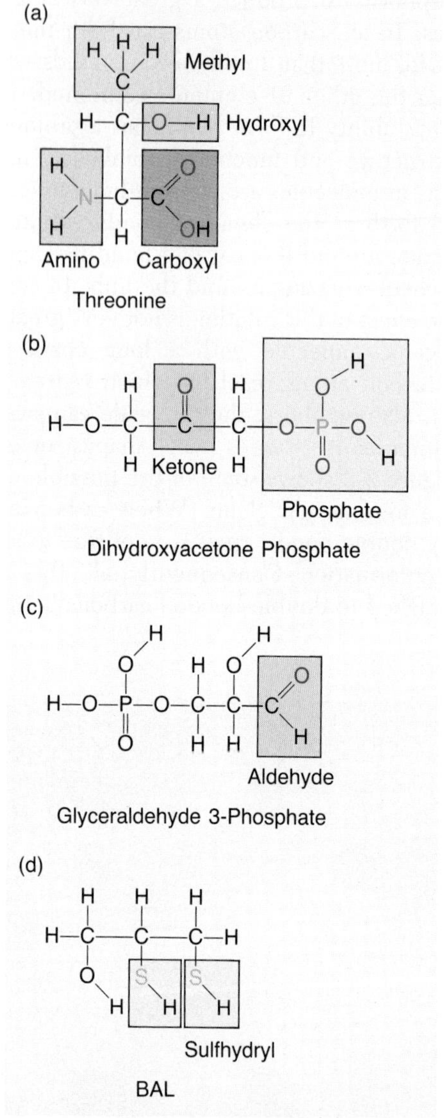

Figure 3-5 FUNCTIONAL GROUPS GIVE COMPOUNDS THEIR CHARACTERISTIC PROPERTIES.
The clusters of atoms highlighted here give complex molecules their predictable properties. (a) The amino acid threonine contains a methyl group (gray), a hydroxyl group (pink), an amino group (blue), and a carboxyl group (orange). (b) A molecule called dihydroxyacetone phosphate, which is a product of a cell's use of glucose, contains a ketone group (pink) and a phosphate group (green). (c) Glyceraldehyde-3-phosphate, also a product of glucose breakdown, contains an aldehyde group (pink). (d) And finally, a molecule that physicians use as an antidote to mercury poisoning, BAL, has two sulfhydryl groups (yellow).

group are acids, specifically *carboxylic acids*. Acetic acid, CH_3COOH, is a common carboxylic acid. Dissolved in water, it is the substance we call vinegar.

When a molecule contains both the —COOH and —OH functional groups, the characteristics of both are exhibited. For example, lactic acid, the substance that builds up in muscles and causes cramping during sustained periods of exercise, has both hydroxyl and carboxyl groups and displays properties of both an alcohol and a carboxylic acid.

The *amino group,* —NH_2 (see Figure 3-5a), is also a very common functional group and an important constituent of amino acids and proteins. The amino group may act as a base, accepting hydrogen ions to become ammonium ion, NH^{3+}. Urea, a waste product of metabolism, has two —NH_2 groups and is perhaps the most common amino compound.

The *ketone* (—C═O) and *aldehyde* (—CHO) functional groups (Figure 3-5b and c) characterize the class of biological compounds called sugars, the core constituents of carbohydrates. Both groups are polar and thus render the compounds soluble in water. In contrast, the *methyl group,* —CH_3 (see Figure 3-5a), which is an important component of lipids and oils, is hydrophobic and insoluble in water. The *phosphate group,* —PO_4 (see Figure 3-5b), is an equally important part of certain lipids as well as a component of nucleic acids and proteins. Another common functional group is the *sulfhydryl group,* —SH, which is a key to determining a protein's structure (Figure 3-5d). Two sulfhydryl groups on different parts of one large molecule, or even on separate "chains" of a molecule, may be joined together into a *disulfide group,* —S—S—. Disulfides stabilize the shape of many proteins.

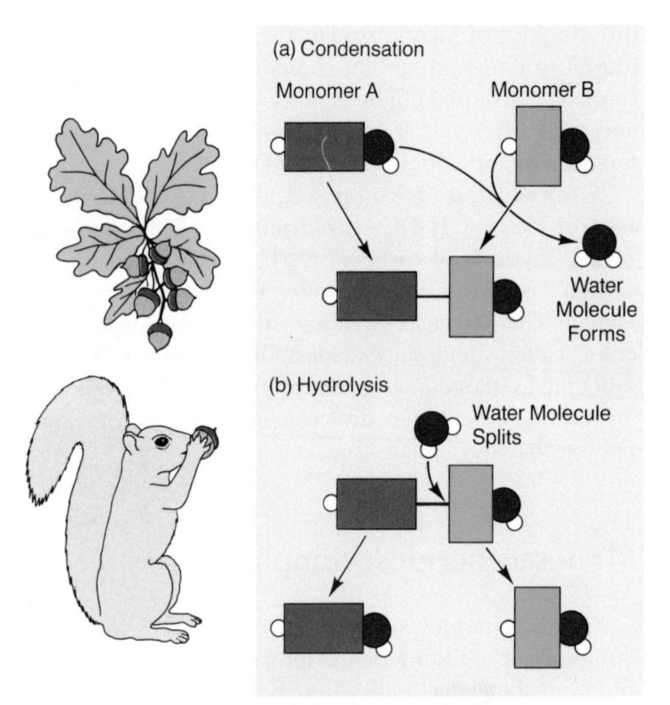

Figure 3-6 CONDENSATION AND HYDROLYSIS: MOLECULES JOINING AND SPLITTING APART.
(a) Inside an acorn, individual sugar molecules, such as monomers A and B shown here, join into carbohydrates by means of a condensation reaction. In the process, a water molecule forms. (b) Inside a squirrel's stomach, the carbohydrate is split by means of a hydrolysis reaction. During this event, a water molecule is split and the monomers are separated.

Monomers and Polymers: Molecular Links in a Biological Chain

The great diversity among biological molecules is due partly to the versatility of carbon bonding and partly to the various functional groups that can be attached to carbon atoms. But diversity also comes from the linkage of smaller units into **polymers** (meaning "many parts"). Polymers are long chains, much like strings of beads, made up of simpler units called **monomers** ("single parts") linked in a specific sequence by covalent bonds. The order in which monomers join is important, for it helps determine the function of biological polymers.

Polymers are formed by means of **condensation reactions:** As two monomers join and a covalent bond forms between them, one monomer loses an —OH group and the other loses a hydrogen atom (Figure 3-6a). The hydroxyl group and the hydrogen atom combine to produce a water molecule as a by-product of the condensation reaction. The opposite occurs when polymers are broken down into monomers: A hydrogen is added to one monomer and an —OH group is added to the adjoining monomer as they split apart. Because parts of water molecules are added to the monomers during this process, it is called **hydrolysis,** which means "splitting with water" (Figure 3-6b). Condensation takes place in cells inside an acorn, for example, as individual sugar molecules combine to form the more complex storage sugars called *carbohydrates.* Hydrolysis takes place in the stomach of a squirrel that later eats the acorn; the individual sugar molecules are removed one by one from the carbohydrate during the process of digestion.

Biological molecules, then, share some important features: a carbon "backbone," a chainlike polymer structure, and functional groups that impart characteristic properties as well as the ways in which the molecules are formed and broken apart. Let's survey the four classes now, including their roles in living organisms.

CARBOHYDRATES: SOURCES OF STORED ENERGY

Plants are largely constructed of **carbohydrates,** which include the sugars and starches, and this fact, plus the ubiquity of plants, explains why carbohydrates are the most abundant carbon compounds in living things. Immense offshore kelp beds, grassy plains and farmland, and the earth's vast tracts of forest can be thought of as huge living cities built of carbohydrates.

A carbohydrate is composed of C, H, and O in the ratio of 1:2:1 (CH_2O). This formula gives the group its name, "hydrate of carbon." Carbohydrates consist of a carbon backbone with various functional groups attached. The basic carbohydrate subunits are sugar molecules called **monosaccharides** ("single sugars"); they function as monomers that can be joined together to form the more complex **disaccharides** ("two sugars") and **polysaccharides** ("many sugars").

Monosaccharides: Simple Sugars

Monosaccharides serve as energy sources for living things and as building blocks for carbohydrate polymers and other biological molecules. Each simple sugar has a structure as based on a short carbon backbone, shown in Figure 3-7. A carbon-to-oxygen double bond may occur at the head of the chain in an aldehyde, or it may occur in the middle of the chain as a ketone group. Thc hydroxyl groups make the molecule polar and thus readily soluble in water (see Chapter 2).

The simplest monosaccharides are the three-carbon compound glyceraldehyde and its structural isomer, dihydroxyacetone, which share the same chemical formula, $C_3H_6O_3$, but have different configurations and properties. Several simple three-, four-, and five-carbon monosaccharides are involved in the cell's storage and use of energy. The five- and six-carbon monosaccharides are of the greatest importance biologically because they form the units in long carbohydrate molecules as well as participating in cellular energetics.

The universal cellular fuel burned by plants and animals for energy is a six-carbon monosaccharide called **glucose,** or grape sugar. Glucose can easily be degraded to yield its stored energy or readily built into large storage polymers. It can exist as a straight-chain compound (Figure 3-7a) or more commonly as a ring configuration (Figure 3-7c). Another six-carbon monosaccharide, *fructose* (Figures 3-7b and d), is a structural isomer of glucose and is common in sweet fruits and vegetables.

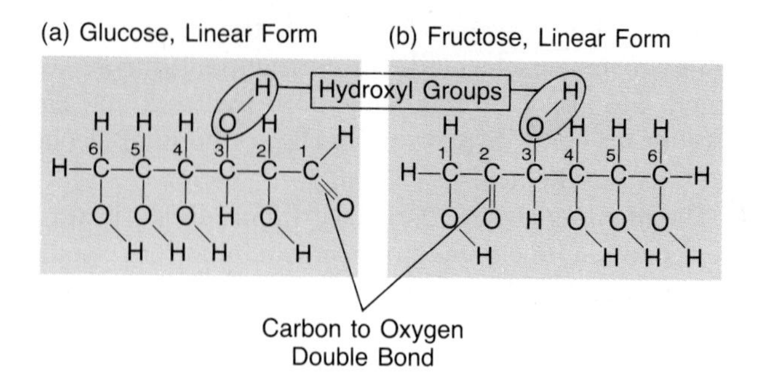

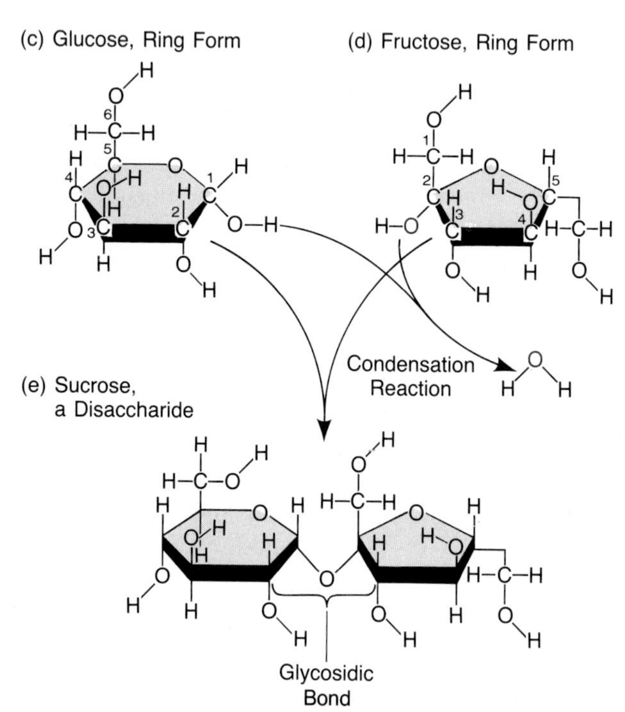

Figure 3-7 COMMON SUGARS IN LIVING THINGS: MONOSACCHARIDES JOIN TO FORM DISACCHARIDES. The monosaccharides glucose and fructose can occur as straight chains (a, b) or, more commonly, in ring form (c, d). Sucrose, or table sugar, is a disaccharide that forms when glucose and fructose join via a condensation reaction (e).

Disaccharides: Sugars Built of Two Monosaccharides

Real diversity in shape and properties can arise when monosaccharide monomers are linked into larger molecules. Two monosaccharides joined together form a disaccharide. For example, common table sugar, *sucrose* (Figure 3-7e), is made up of a glucose molecule linked to a fructose molecule. Sucrose occurs abundantly in honey and in the saps of sugarcane, maple trees, and sugar beets. Sucrose is the product of a condensation reaction in which a molecule of water is formed from two —OH groups as the two sugars join. The bond that links the two monosaccharides, C—O—C, is called a **glycosidic bond** (see Figure 3-7e).

Other important disaccharides include *lactose,* which is made up of galactose and glucose units and is the predominant sugar in milk, and *maltose,* the sugar that sweetens malted barley, containing two glucose molecules.

Polysaccharides: Storage Depots and Structural Scaffolds

Living organisms form the long-chain carbohydrates called polysaccharides by linking large numbers of single sugars, or monosaccharides, into polymers. In bacteria and in the cells of higher organisms, sugar polymers make up most of the walls and coatings that surround and protect the cells. And inside cells, large quantities of nutrients are stored as polysaccharides and later used to fuel cellular processes.

The most important polysaccharides are starch, glycogen, and cellulose. Other biologically significant polysaccharides include *chitin*, a major component of the shells of insects and crustaceans such as lobsters and crabs, and *glycosaminoglycans*, which are found on the surface of animal cells.

Starch

In most plants, the major nutrient reserve is **starch,** which is a mixture of two polysaccharides, *amylose* and *amylopectin*. Amylose occurs as long, straight polymer chains composed of glucose units joined by glycosidic bonds (Figure 3-8a). An amylose molecule is usually composed of at least 1,000 units of glucose, but this number can vary. Like amylose, amylopectin is made up of glucose units bonded together, but the chains have many branches joined to the larger molecule via cross linkages (Figure 3-8b). This results in separate chains of glucose units being bound together. An amylopectin polymer usually contains about 20,000 glucose monomers.

Inside a seed, a supply of starch will often support the early growth of the tiny plantlet that germinates. Starch also makes grains (the seeds of wheat, corn, rice, etc.) a rich source of nutrition for humans and other animals. Also on hand within the seed are *enzymes* (a class of protein that speeds up chemical reactions), which hydrolyze the glycosidic bonds in starch, releasing the disaccharide maltose and, in turn, glucose monomers to fuel the energy needs of the growing plant.

Glycogen

In animal cells, glucose is the chemical currency that provides energy for life processes and serves as the basic building block from which the subunits of lipids, proteins, and nucleic acids are built. Animals must store this raw material to avoid the necessity of nonstop eating. However, it must be stored in a more stable, less reactive form than free glucose molecules, since huge numbers of these would create many problems for living cells. That storage form is a polysaccharide called **glycogen,** made only in animal cells and containing several thousand glucose units. Compared to amylopectin, the branches in glycogen molecules are often shorter but more numerous (Figure 3-8c). This extensive branching allows the rapid breakdown and release of energy that animals often require (details in Chapter 37).

Cellulose

Cellulose is the fibrous structural material that gives strength and rigidity to plant cells and wood. Like starch, cellulose is composed of long chains of glucose units connected by glycosidic bonds, yet the two polysaccharides have very different properties: We eat starch in potatoes and bread, but we build houses from cellulosic wood. These physical differences are due entirely to the orientation in space of a single bond (Figure 3-8d). In starch and glycogen, the orientation of the bond allows the glucose chains to twist into compact spirals. In cellulose, the orientation of the bond prevents twisting, so that the chains are straight. This molecular rigidity explains the fibrous quality of cellulose and, in turn, the strength of leaves and stems and the hardness of wood. Reactive —OH groups projecting from both sides of the cellulose molecule allow it to link up with adjacent molecules, thereby forming stable, tough fibers.

Many organisms have the enzymes necessary to break the glycosidic bonds in starch and glycogen, but only a handful have those needed to break the bonds in cellulose. That is why a cracker will break down in your mouth, but wood or cotton never will, no matter how long it remains there. A few organisms, such as termites and cows, have microorganisms in their guts that produce the necessary enzymes for hydrolyzing cellulose; thus, these animals are able to feed on wood or on grass and leaves. Since cellulose passes through the human gut without being degraded, it provides no calories but does serve as beneficial fiber, or roughage.

We said at the beginning of this chapter that in the language of biological molecules, meaning lies in the order of the subunits and in overall molecular shape and function. Starch, glycogen, and cellulose all say the same thing: "glucose-glucose-glucose." Clearly, though, the three-dimensional shapes of these polymers (branched, spiraling, and straight chains) and the structural properties that are based on these shapes (softness, rigidity, etc.), provide additional information and varied functions to living things.

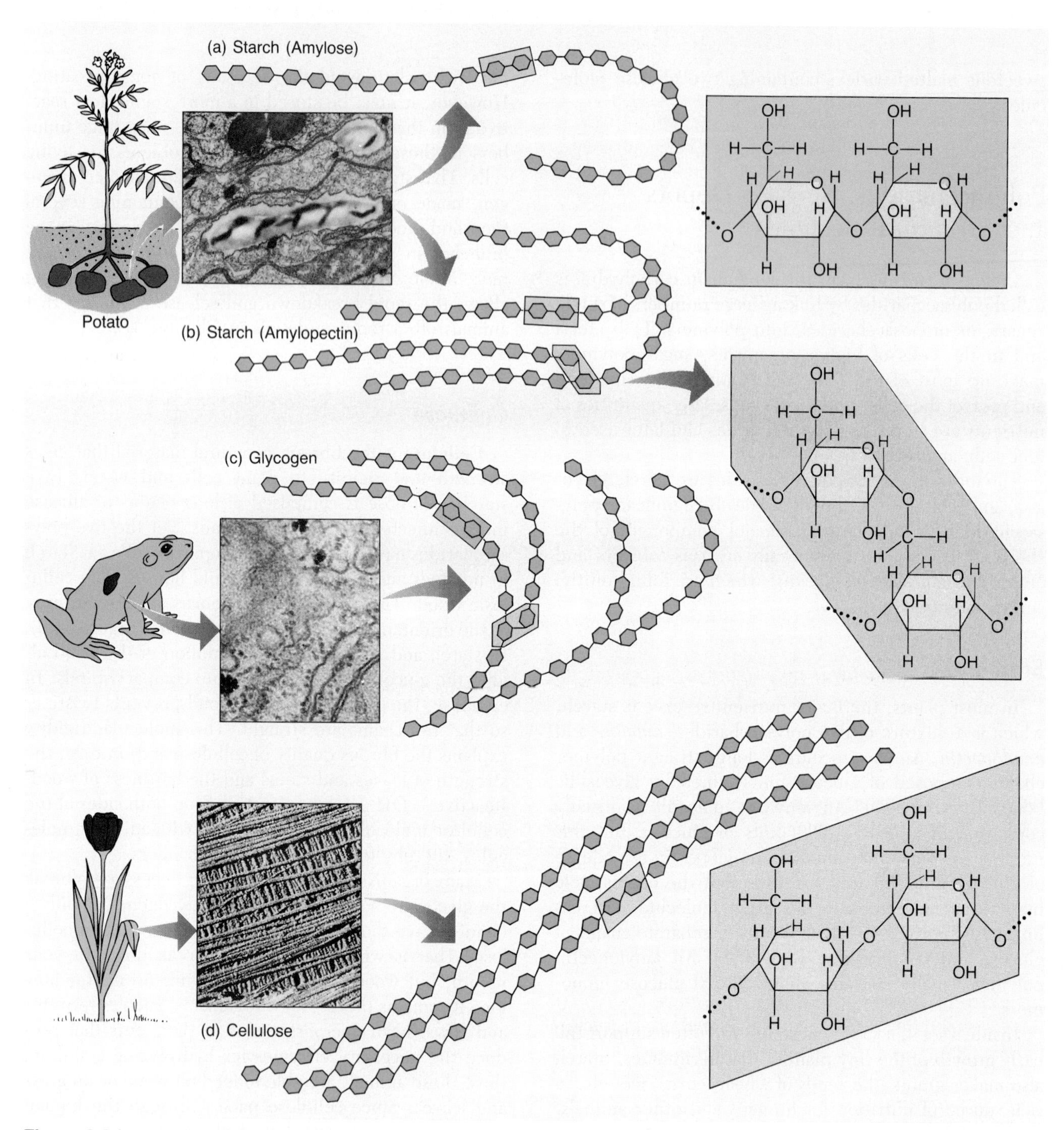

Figure 3-8 CARBOHYDRATES: CHAINS OF GLUCOSE SUBUNITS FORM ENERGY RESERVES AND BUILDING MATERIALS FOR LIVING CELLS.
Starch, such as the food reserves in a potato, is a mixture of amylose and amylopectin (a, b), both polysaccharide chains containing glucose subunits. Starch molecules cluster into granules or grains. An elongated granule can be seen clearly across the bottom of this portion of a plant cell under high magnification (about 5,000×). In amylose (a), glucose monomers form long straight chains, with each glucose unit linked to its neighbor by a glycosidic bond. Amylopectin (b) is also made up of linked glucose subunits, but it has occasional branches based on the glucose-to-glucose bands shown in the enlarged view. Glycogen (c) serves as a nutrient reserve for animals, much as starch does in plants. In a cell from a frog's liver, glycogen appears as dark rosettes (here magnified about 41,600×). Glycogen molecules are made up of long, branched chains of glucose units. The unbranched portions have glycosidic bonds (green box) as in amylose, while the branched chains are held together by cross-linkages (pink box) like those in amylopectin. Cellulose (d) is a structural polysaccharide that gives rigidity to plant cells such as those in leaves or stems. Cellulose forms fibers (shown here magnified 30,000×) and consists of straight chains of glucose units that gain rigidity owing to the orientation of the bonds that join the carbon atoms to the oxygen atoms in a bridge (blue box). The hydroxyl groups link with adjacent fibers, helping to form the stable, tough fibers.

LIPIDS: ENERGY, INTERFACES, AND SIGNALS

The second group of biological molecules, the **lipids,** make certain foods oily, keep us warm, and prevent the watery contents of cells from leaking out. Let us look at the three main types of lipids: fats and oils, phospholipids, and steroids.

Fats and Oils: Storehouses of Cellular Energy

Fats and oils are familiar substances that serve as nutrient reserves in animals and plants and are essential parts of the diet for most animals. Corn oil and olive oil (both yellowish liquids at room temperature) are typical lipids extracted from plant tissue. Lard and butter (whitish solids at room temperature) are made from typical fats.

Fats and oils contain two basic units: glycerol and fatty acids. **Glycerol** (Figure 3-9a) is a three-carbon alco-

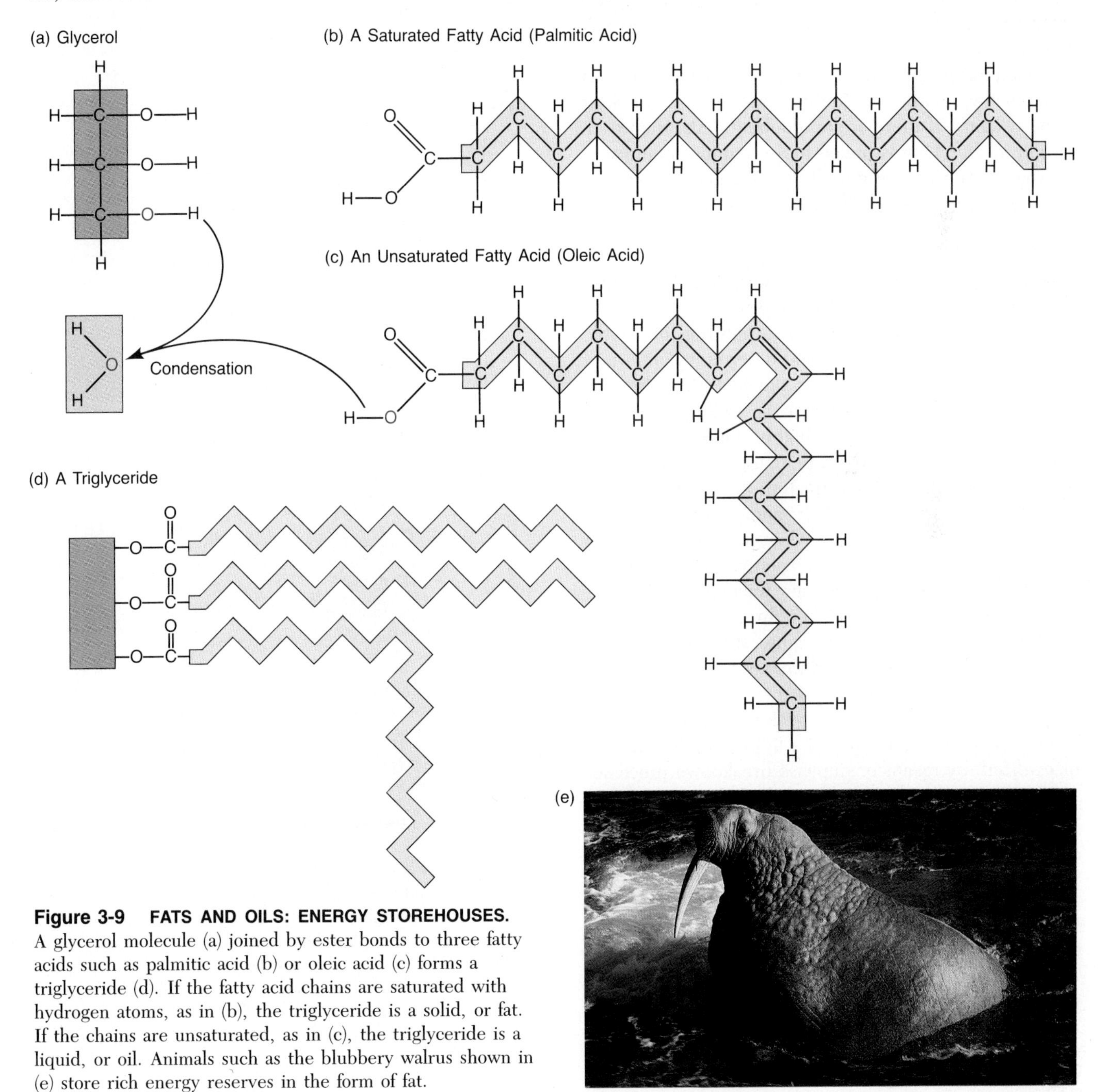

Figure 3-9 FATS AND OILS: ENERGY STOREHOUSES. A glycerol molecule (a) joined by ester bonds to three fatty acids such as palmitic acid (b) or oleic acid (c) forms a triglyceride (d). If the fatty acid chains are saturated with hydrogen atoms, as in (b), the triglyceride is a solid, or fat. If the chains are unsaturated, as in (c), the triglyceride is a liquid, or oil. Animals such as the blubbery walrus shown in (e) store rich energy reserves in the form of fat.

hol with three hydroxyl groups attached. Because —OH groups readily form hydrogen bonds with water molecules, glycerol is highly water-soluble. **Fatty acids** (Figure 3-9b and c) are long chains of carbon atoms attached to a carboxyl group (—COOH), which gives fatty acids their acidic properties. Because three fatty acids usually are joined to one glycerol molecule, fats and oils are called **triglycerides** (Figure 3-9d).

The entire triglyceride molecule is insoluble because the hydroxyl and carboxyl groups of the two kinds of units react with each other and thus are no longer available to form hydrogen bonds with water. The bonds that link the glycerol to the fatty acids in a triglyceride are **ester bonds,** which form between any alcohol and any carboxylic acid.

An important feature of the fatty acid molecules in lipids is the number of double bonds (C=C) in their carbon chains. If there are no double bonds, then each carbon in the chain is linked to a maximum number of hydrogen atoms (see Figure 3-9b). Such chains are said to be *saturated* with hydrogen atoms. In contrast, if some adjacent carbon atoms are linked by double covalent bonds, then the fatty acid is *unsaturated* because double-bonded carbons are not bound to the maximum number of hydrogens (see Figure 3-9c).

The degree of saturation (the proportion of double versus single bonds in the chain) determines important properties of the lipid. One is the melting point: the greater the saturation, the higher the melting point. For example, *neat's-foot oil,* a highly unsaturated lipid (found in the feet of reindeer, penguins, and cattle, from whose hooves it can be extracted) has a low melting point. It allows cattle, reindeer, and penguins to stand for days on ice or snow, and when applied to people's shoes, neat's-foot oil keeps them soft and pliable in cold weather. Oils from these animals' warm interiors, however, tend to be saturated. Therefore, these oils have a high melting point; they are quite stiff and inflexible when chilled and are poor lubricants.

Fats and oils are energy-rich compounds because they contain so many C—H covalent bonds. Many organic compounds, including the lipids, can be burned, or *oxidized,* by means of stepwise breakdown processes inside living cells. During the oxidation process, C—H and C—C bonds are broken and replaced by C—O bonds, the end product being CO_2. Fats and oils provide more energy than do sugars and proteins because they have fewer C—O bonds to begin with; thus, almost all the bonds in the molecule become oxidized, and energy is released as each C—H or C—C bond is broken. Because of this great capacity for storing energy, fats and oils serve as long-term nutrient reserves in plants and animals, while carbohydrates provide glucose and immediate fuels for most cellular processes. Thus, layers of fat or reservoirs of oil or other lipids often accumulate in organisms facing long periods with little or no access to renewed carbohydrate stores, as during hibernation, migration, seed and plant dormancy, and embryonic development (Figure 3-9e).

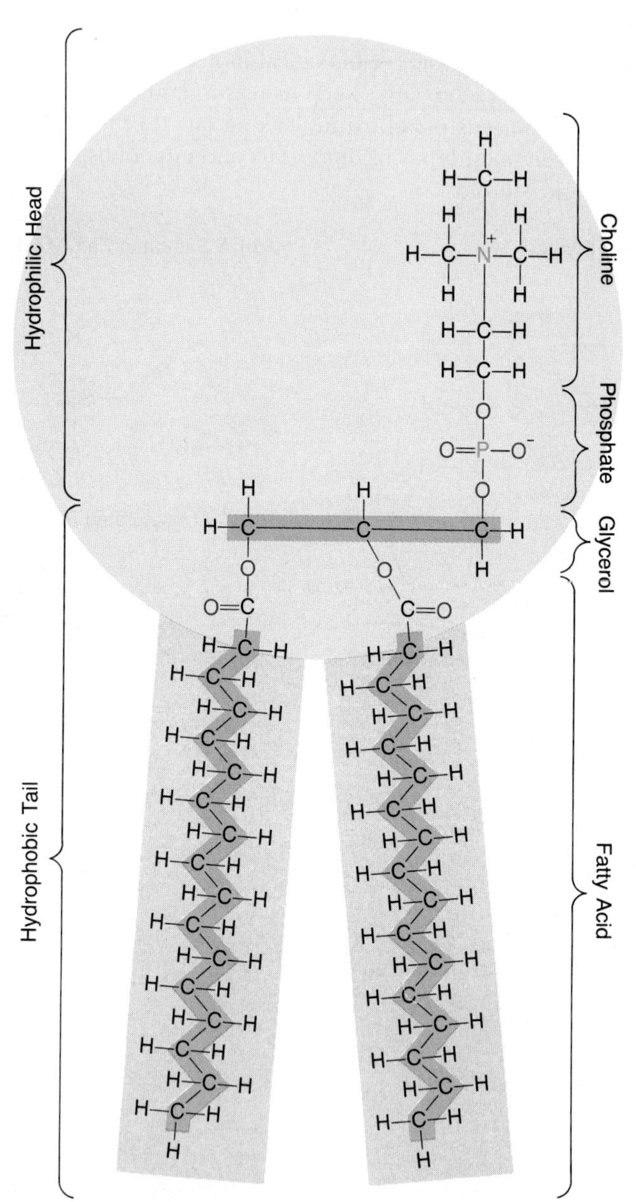

Figure 3-10 PHOSPHOLIPIDS: MOLECULAR HEADS AND TAILS.
Phospholipids are the major structural component of membranes and are responsible for waterproofing our bodies. The hydrophilic ("water-loving") head is made up of glycerol joined to a phosphate and a choline or other molecule. The hydrophobic ("water-fearing") tail is made up of fatty acid chains. Shown here is the phospholipid lecithin, a common ingredient in candy bars that acts as an emulsifying agent. It keeps droplets of cocoa butter and oil dispersed in moist surroundings (in this case, milk, which is mostly water, in the milk chocolate).

A variation on oils is the *waxes*, molecules made up of a long-chain alcohol linked to the carboxyl group of a fatty acid. Large numbers of wax molecules packed together form a waterproof outer layer on the leaves of plants, the bills and feathers of some birds, and other structures in living organisms.

Phospholipids: The Ambivalent Lipids

The **phospholipids** contain nitrogen and phosphorus as well as the carbon, hydrogen, and oxygen atoms in fats and oils. These additional elements give phospholipids their ability to maintain a cell's waterproof boundary. The main component of a phospholipid (Figure 3-10) is *glycerol phosphoric acid.* This is essentially glycerol with a phosphate group where one fatty acid would normally be found in an oil molecule (review Figure 3-9). Phospholipid molecules exemplify the importance of functional groups: The charged phosphate group ($—PO_4^{3-}$) constitutes a water-soluble (hydrophilic) region on a water-insoluble (hydrophobic) molecule, and this "ambivalence" has great biological significance.

As in fats and oils, the two fatty acids in phospholipids are joined to the glycerol with ester bonds. In many phospholipids, the phosphate group attached to the glycerol is also bound to a nitrogen-containing functional group (choline is a common one). The overall shape of a phospholipid molecule is something like a head with two long, thin tails streaming from the nape of the neck; the glycerol phosphate forms the hydrophilic head, while the fatty-acid chains form the hydrophobic tails (see Figure 3-10). The behavior of phospholipids is rooted in this shape. Since the tails are hydrophobic and the head is hydrophilic, the molecules have an ambivalent approach to water: If phospholipids are introduced into the zone between a layer of oil and a layer of water, the hydrophobic tails will orient themselves to extend into the oil, while the hydrophilic heads will face the water. Thus, the phospholipids form an interface, or separating layer, between the oil and the water (Figure 3-11). The fence-straddling behavior of phospholipids accounts for the structure of cell membranes, which are double-layered phospholipid barriers surrounding living cells, as Chapter 5 explains in detail.

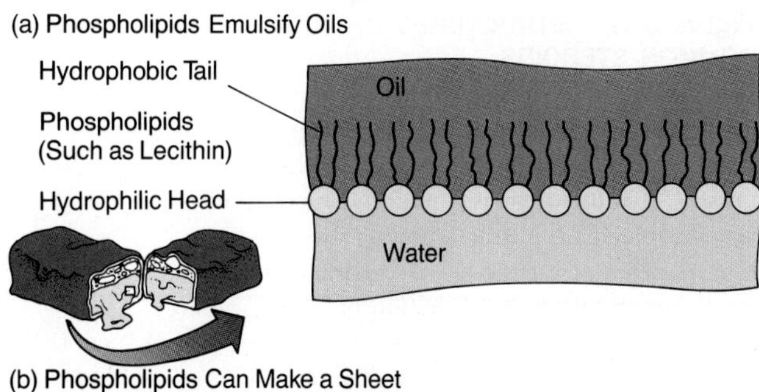

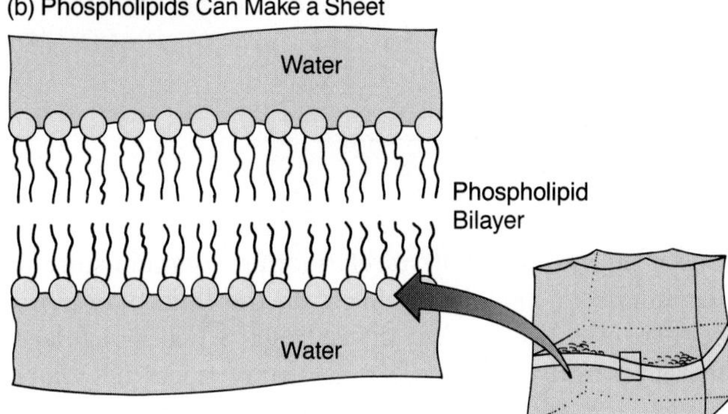

(c) A Phospholipid Bilayer Can Enclose a Water Droplet

Water

Water

Figure 3-11 PHOSPHOLIPIDS IN WATER: SINGLE LAYERS, DOUBLE LAYERS, AND DROPLETS.
(a) Phospholipids form a layer, with hydrophilic heads facing the water and hydrophobic fatty acid tails turned away from the water. (b) Phospholipids in water can form two layers, with their fatty acid tails facing toward each other and away from water. This forms a waterproof boundary. (c) Finally, phospholipid bilayers can form spheres around water droplets. This arrangement is reminiscent of the composition of a living cell, with its waterproof outer membrane.

Steroids: Regulatory Molecules

Steroids, a third type of lipid, are much less abundant in living cells than are fats, oils, and phospholipids, but they are no less important. Steroids contain interconnected rings of carbon atoms with functional groups attached, rather than glycerols and fatty acids (Figure 3-12). Steroids are nonpolar and hydrophobic; they are insoluble in water, can dissolve in oils or in lipid membranes, and can move into and out of many cells.

Some steroids act as vitamins (see Chapter 32), while others, such as estrogen and testosterone, are hormones

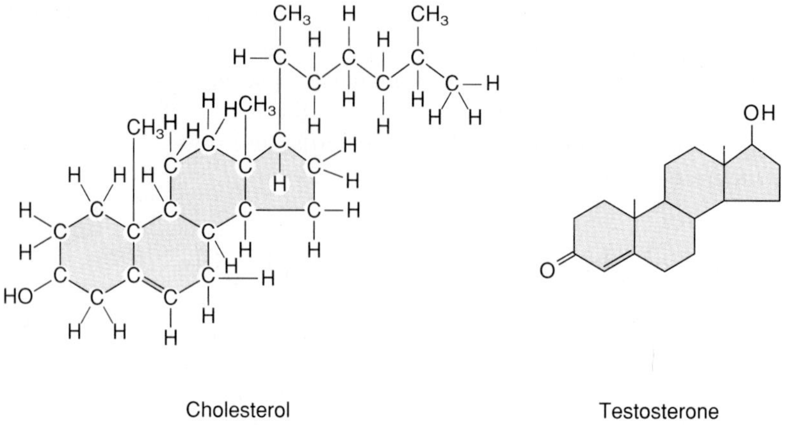

Figure 3-12 STRUCTURES OF SOME COMMON STEROIDS.
Cholesterol affects the fluidity of membranes, and biologists believe it is involved in the deposition of fatty materials within blood vessels. Testosterone is a male hormone that is responsible for the development and maintenance of secondary sex characteristics and for the maturation and function of accessory sex organs.

Cholesterol

Testosterone

(see Chapter 35). *Cholesterol,* another common steroid, has important beneficial effects on the fluidity of many cellular membranes. But this steroid, which is regularly manufactured in the body, may build up in the blood vessels and contribute to heart disease (see Chapter 29).

PROTEINS: THE BASIS OF LIFE'S DIVERSITY

When the water is removed from almost any living thing (except a woody plant), about 50 percent of the remaining matter is protein (Figure 3-13). **Proteins** are a class of diverse macromolecules that determine many characteristics of cells and, in turn, of whole organisms. Biologists often classify proteins according to the functions they perform. *Structural proteins,* for example, help to form bones, muscles, shells, leaves, roots, and even the microscopic cell "skeleton" that provides shape and allows cell movement. Other examples are *protein hormones,* which serve as chemical messengers; *antibodies,* which fight infections; and *transport proteins,* which act as carriers of other substances (the hemoglobin protein in blood, for example, transports oxygen). One of the most important kinds of proteins is the **enzymes,** which speed up, or catalyze, chemical reactions and thus underlie every biological activity.

Amino Acids: The Building Blocks of Proteins

All proteins are polymers constructed of subunits called **amino acids.** There are 20 types of amino acids in proteins. Thus, the biological language expressed in proteins is a huge vocabulary of complex "words" based on an alphabet (the 20 amino acids) that is almost as large as our own; like the 26 letters in the Latin alphabet, the 20 amino acids are capable of a virtually infinite number of combinations.

The meaning of a protein rests in the exact order of its amino acids. The order gives the molecule its special characteristics as a structural component in a cell, as an enzyme, as a carrier, or whatever. Within those categories, the order of amino acids gives the protein its unique properties as, for instance, a carrier of oxygen rather

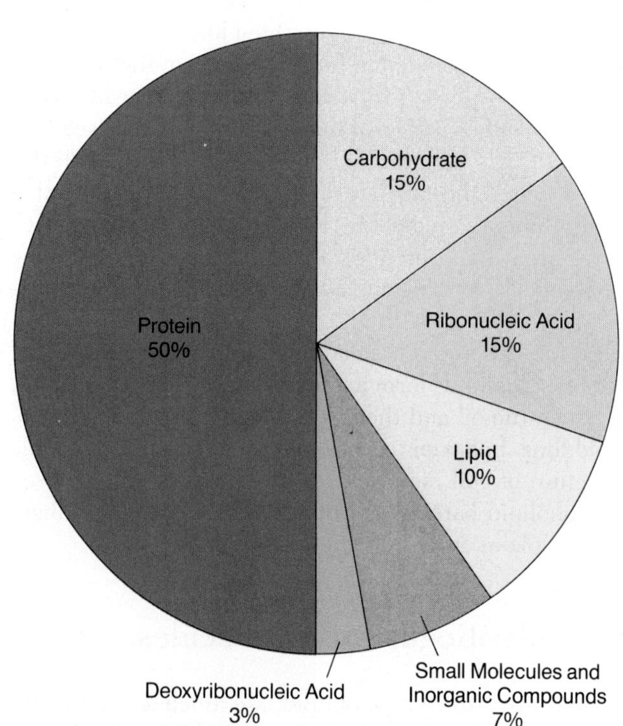

Figure 3-13 PROTEINS MAKE UP HALF OF A DEHYDRATED ORGANISM.
As this pie chart of molecular composition shows, dry organisms are about 50 percent protein and contain much smaller percentages of the other biological molecules.

than of potassium. We can see from the huge number of protein "words" that both individuality and diversity in nature stem largely from this class of biological molecules.

In each amino acid there are one central carbon atom, called the *alpha* (α) *carbon*, bound to a hydrogen atom, an amino group (—NH_2), a carboxyl group (—COOH), and a side chain, represented by R (Figure 3-14a).

In water, the carboxyl group may dissociate, giving up a hydrogen ion. Meanwhile, the amino group, a base, can accept a hydrogen ion. Thus, a molecule forms that has one positive group and one negative group, and those changes help determine protein behavior.

Each of the 20 amino acids commonly found in proteins has a distinctive R group. As Figure 3-14 shows, R can be as simple as a single hydrogen atom, as in glycine, or as complex as the double ring structure in tryptophan. Some amino acid side chains are hydrophobic; some are hydrophilic; and the others are ambivalent, or partially soluble in water. The properties of a protein are deter-

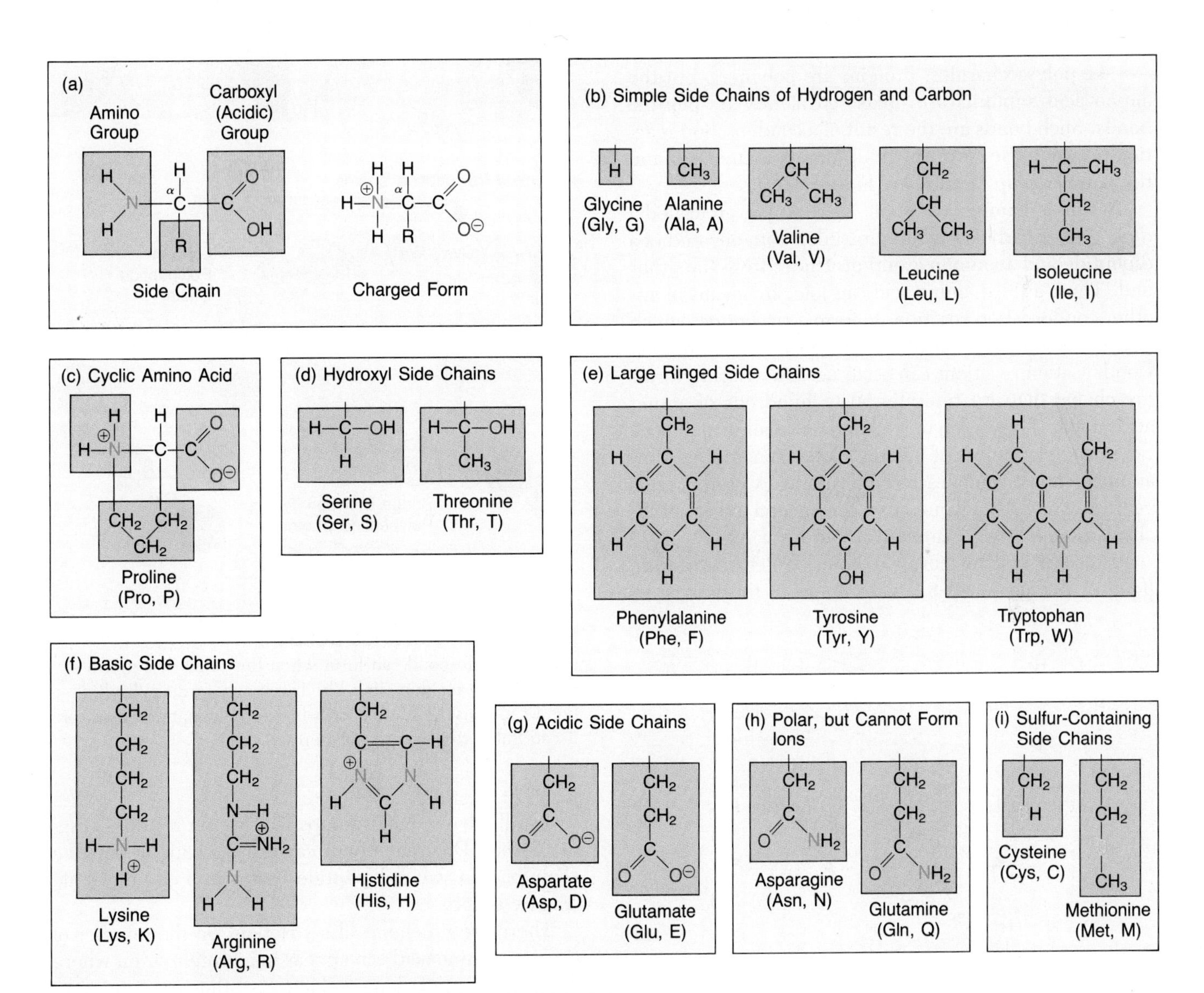

Figure 3-14 AMINO ACIDS: BUILDING BLOCKS OF PROTEINS.
Amino acids are the building blocks of proteins. This figure gives the structural formulas, conventional three-letter abbreviations, and more up-to-date single-letter designations for the 20 amino acids. (a) An amino acid consists of an amino group (blue), a carboxyl (acidic) group (pink), a hydrogen atom, and a unique side-chain R group (mauve) bonded to a central atom, which is known as the α-carbon. (b) Amino acids with simple side chains. (c) A cyclic amino acid, proline, in which the nitrogen atom and the α-carbon atom of the backbone form portions of a ring of atoms. (d) Amino acids with hydroxyl side chains. (e) Amino acids in which the side chains are large rings. (f) Amino acids with basic side chains. (g) Amino acids with acidic side chains. (h) Amino acids with side chains that are polar but do not form ions. (i) Amino acids with sulfur-containing side chains.

mined in good measure by the R groups of its constituent amino acids. Thus, for example, mostly hydrophobic side chains will make a protein quite insoluble.

Some proteins may carry various molecules or groups of atoms. Without these so-called *prosthetic groups*, the protein could not function properly. A prosthetic group can be a single atom, such as of copper or zinc, or a complex organic compound, such as the oxygen-binding, heme group found in the blood protein hemoglobin.

Polypeptides: Amino Acid Chains

Like polysaccharides, proteins are polymers, but the amino acid subunits are linked covalently by **peptide bonds.** Such bonds are the result of a condensation reaction between the carboxyl group of one amino acid and the amino group of another (Figure 3-15).

Two joined amino acid units, or residues, are called a *dipeptide*. A carboxyl group protrudes from one end of a dipeptide and an amino group protrudes from the other end. Thus, a third amino acid can join, by means of another condensation reaction, to form a *tripeptide*, which also has a carboxyl terminus and an amino terminus. Condensation reactions can occur again and again, forming chains that are typically 50 to hundreds of amino acids long. These polymer chains are called **polypeptides,** and regardless of length, each chain will have an amino terminus and a carboxyl terminus. A protein molecule can consist of one, two, or several polypeptide chains bound to one another in various ways.

A second kind of covalent bond, called a *disulfide bond* or *bridge* (—S—S—), results from the linking of two sulfhydryl (—SH) groups in cysteine residues (Figure 3-16a). Disulfide bonds can cause a kink or loop in a chain or join two polypeptide chains into one molecule (Figure 3-16b).

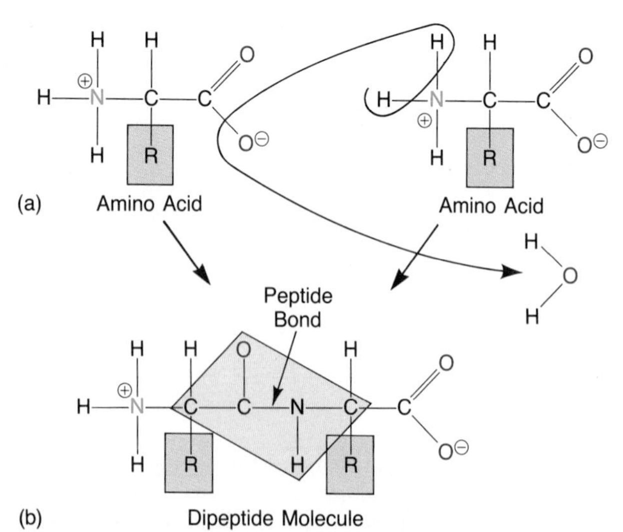

Figure 3-15 PEPTIDE BONDS: LINKS BETWEEN AMINO ACIDS.
(a) Two amino acids join via a condensation reaction during which a water molecule forms. (b) The resulting peptide bond links the two amino acids into a dipeptide.

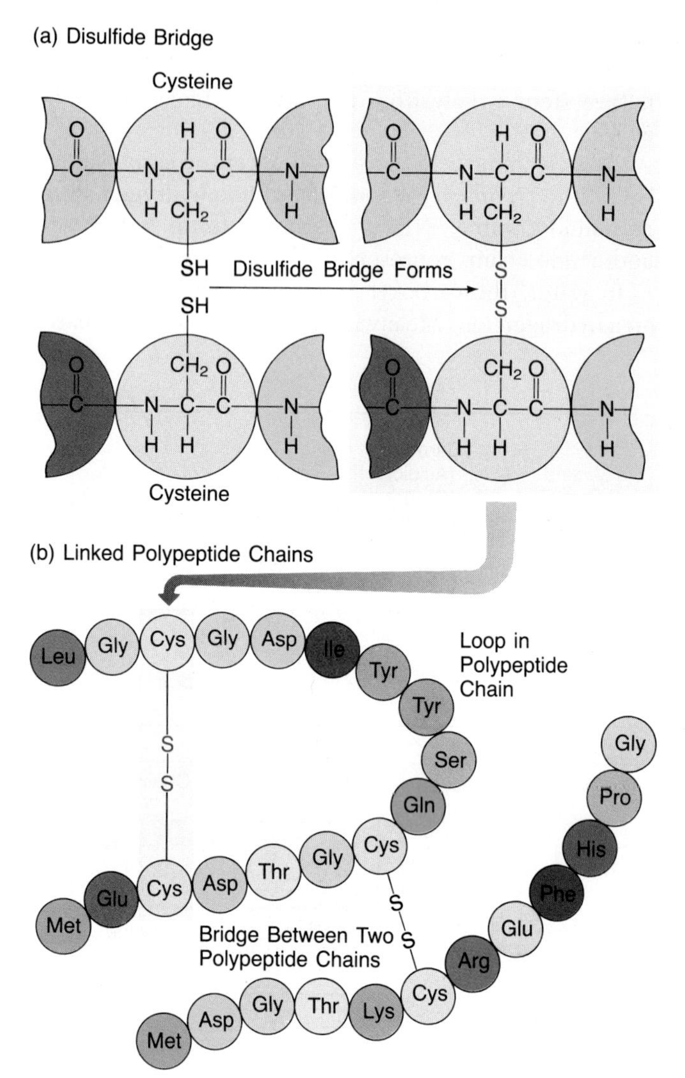

Figure 3-16 DISULFIDE BONDS.
(a) A disulfide bond can form when the sulfhydryl group of two cysteine residues link. (b) The newly formed disulfide bond can cause a loop to form in one polypeptide chain, or it can link two separate polypeptide chains.

There are no chemical restrictions on the number of times an amino acid can appear in a protein or on where it can be located along a chain, and there is nothing in the chemistry of the amino acids themselves that restricts the length of chains. Yet there are only 20 amino acids found in a seemingly endless variety of proteins. How can just 20 amino acids account for the millions of different kinds of proteins that biologists have observed in the living world? The answer lies in the *order* of the amino acids in polypeptide chains, and the possibilities are virtually unlimited. Consider the number of different polypeptides that could be formed by linking 50 amino acid molecules in a chain. In just the first two amino acid subunits along the chain, there are $20 \times 20 = 20^2$, or 400, potential combinations of the 20 different amino acids. Adding in the third amino acid, the possibilities grow to 20^3, or 8,000, possible tripeptides. Extrapolating further, the number of possible amino acid sequences in a polypeptide only 50 subunits long is 20^{50}, which can also be expressed as 10^{65}. In nature, a polypeptide with 50 amino acids would be a fairly short chain. Proteins usually contain between 100 and 10,000 amino acids and have molecular weights of 10,000 to 1 million daltons. Clearly, the number of potential amino acid sequences becomes astronomical.

These do not all occur in nature, however. There are millions of kinds of proteins (10^7 or 10^8) in the earth's living organisms, but nowhere near 10^{65}. Many of the potential amino acid sequences are not formed in real polypeptides, just as most potential combinations of the 26 letters do not form recognizable English words. The precise number and sequence of amino acids in a protein is dictated by an organism's hereditary material, as discussed in Chapters 12 and 13.

The Structure of Proteins

In the early 1950s, Frederick Sanger of Cambridge University made one of the great discoveries in the history of biology by exploring the precise amino acid sequence of the protein hormone called **insulin.** The innovative work of Sanger and his colleagues revealed that the order of amino acids is the key to the structure and function of proteins and that a specific sequence characterizes each type of protein. Before Sanger's discovery, biochemists had concentrated on measuring the relative quantities of the different amino acids in proteins. This would be equivalent to analyzing the meaning of the word *banana* by counting the number of *a*'s, *b*'s, and *n*'s rather than noting their order. Within a few years, biologists discovered that sequencing is fundamental not only to protein structure, but also to the hereditary information present in all cells, and, furthermore, that the se-

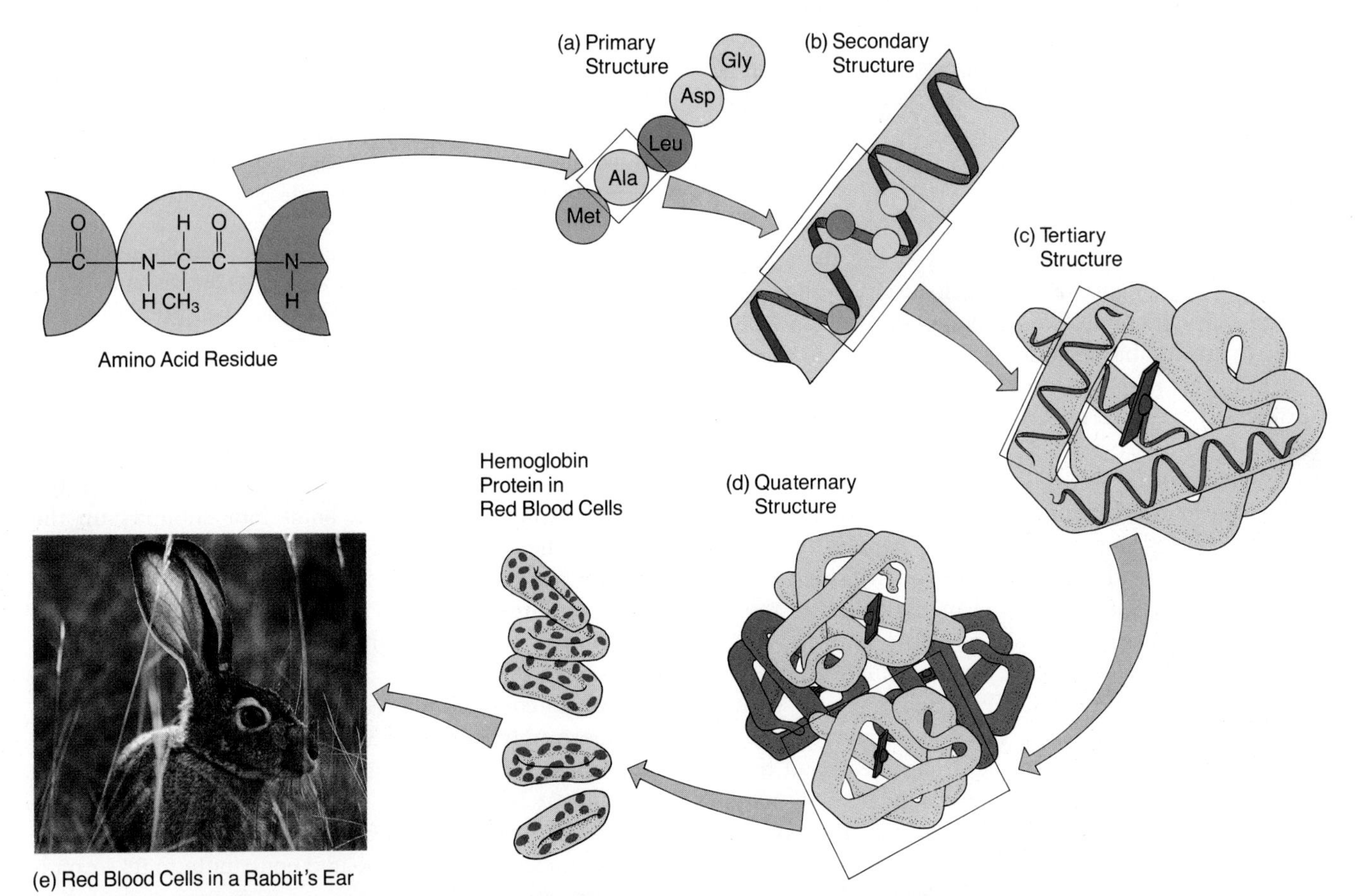

Figure 3-17 PROTEINS: FOUR LEVELS OF STRUCTURE.
Like all proteins, the blood protein hemoglobin has a linear sequence of amino acids, the primary structure (a); repeating helical coils, or secondary structure (b); and a folded three-dimensional shape, or tertiary structure (c). Hemoglobin also has a quaternary structure (d)—a three-dimensional arrangement of separate polypeptide chains. The millions of hemoglobin molecules in a red blood cell carry oxygen to some animals' tissues. (e) In this photo of a cottontail rabbit, red cells move through the visible blood vessels in the ears.

quence of the subunits in these informational molecules controls the amino acid sequence in proteins.

Primary Structure

The linear sequence of amino acids in a protein is referred to as its *primary structure* (Figure 3-17a). Whether a protein functions as an enzyme, a hormone, or a structural component of a cell depends on its primary structure. If just one amino acid is altered in a protein, an essential characteristic of the protein might change. For instance, the painful and debilitating human disease known as sickle-cell anemia results from the substitution of a single amino acid in hemoglobin, the oxygen-carrying protein within red blood cells.

Secondary Structure

The sequence of amino acids in a long polypeptide determines not just the biological meaning of the protein "word," but the way the chain twists, bends, and folds into a characteristic complex shape as well. This shape consists of a series of higher-order configurations called secondary, tertiary, and quaternary structure, which ultimately determine a protein's activity in a living thing. Figure 3-17 shows the four levels of structure in the hemoglobin protein and its role in a rabbit's blood supply.

The work of chemists Linus Pauling and Robert Corey, also in the 1950s, helped reveal this hierarchy of protein structure and function by focusing on the spatial relationships of amino acids in polypeptide chains. They knew that the subunits of some proteins tend to occur in regularly repeating patterns (like the bends in a coiled spring). These patterns form the *secondary structure* of the protein.

By studying these repeating configurations, Pauling and Corey found that atoms are *not* free to rotate around peptide bonds. This means that the peptide bond and the six atoms on either side of it remain in a single plane, even though atoms and side chains attached to the α-carbon may rotate (Figure 3-18a).

Pauling and Corey also examined how hydrogen bonds can form between —NH and —C=O groups along the main chain. These researchers had the flash of insight that marks great science: They reasoned that a polypeptide chain should assume a regular conformation that allows the maximum number of hydrogen bonds to form. They speculated that the inability of atoms to rotate about peptide bonds limits the ways a chain can fold to just a few basic shapes. These shapes include the repeating patterns called the α-helix and the β-pleated sheet, which form a protein's secondary structure.

In the **α-helix,** shown in Figure 3-18a, the long chain of amino acid residues is wrapped like a coiled telephone cord. The chain is held in position by hydrogen bonds joining the N—H group of one peptide bond with the C=O group of a peptide bond four subunits up the chain. The spiral α-helix is a more stable configuration than other conformations that do not allow such hydrogen bonding and it forms spontaneously in regions wherever the amino acid sequence (primary structure) allows a bond to form.

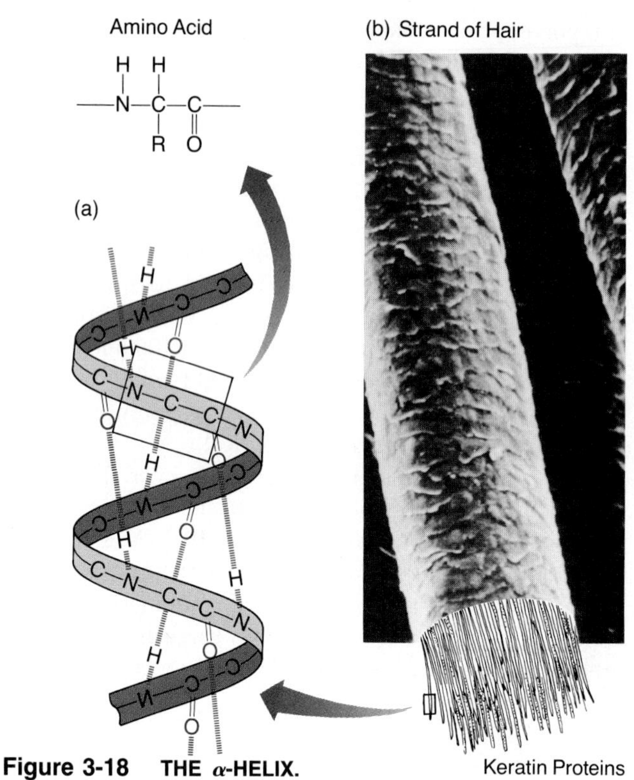

Figure 3-18 THE α-HELIX. The α-helix is formed by the coiling of a polypeptide chain, somewhat like the coiling of a telephone cord. The drawing in (a) shows the complete α-helix, with one portion of the C—N backbone enlarged. The hydrogen bonds are more or less parallel to the long axis of the helix. Each N—H group forms a hydrogen bond with the C=O group of the fourth amino acid up the chain. (b) A scanning electron micrograph of two hairs, magnified about 430 times, consisting of keratin protein in the α-helical configuration.

In the configuration called the **β-pleated sheet,** two or more polypeptide chains lying side by side become cross-linked by hydrogen bonds and form an accordion-like sheet of connected molecules (Figure 3-19). Both the α-helix and the β-pleated sheet repeat at intervals in a protein such as lysozyme and contribute to its secondary structure.

The protein keratin, found in mammalian hair, fingernails, horns, claws, and quills, and in birds' feathers, is composed mainly of α-helix regions, while in fibroin, a protein in the silk strands spun by silkworms and spiders, β-sheets are the common secondary structure.

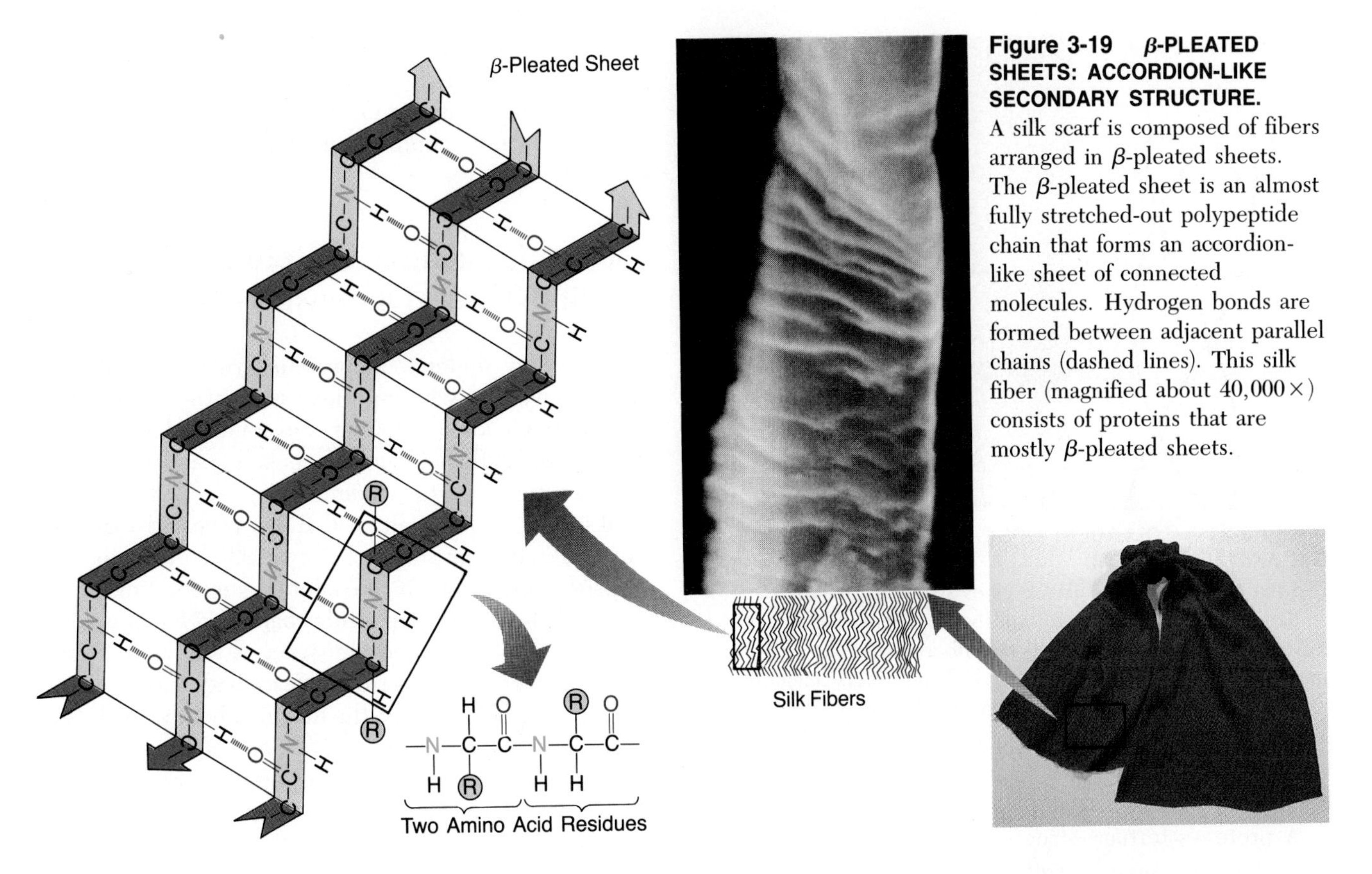

Figure 3-19 β-PLEATED SHEETS: ACCORDION-LIKE SECONDARY STRUCTURE.
A silk scarf is composed of fibers arranged in β-pleated sheets. The β-pleated sheet is an almost fully stretched-out polypeptide chain that forms an accordion-like sheet of connected molecules. Hydrogen bonds are formed between adjacent parallel chains (dashed lines). This silk fiber (magnified about 40,000×) consists of proteins that are mostly β-pleated sheets.

Tertiary Structure

Proteins have two more orders of organization above primary and secondary structure. An example of the next higher level, tertiary structure, is ribonuclease, a very stable protein that breaks apart RNA. Ribonuclease contains three α-helices and five β-sheets folded into the complex shape shown in Figure 3-20. The α-helices and β-sheets are connected by areas of *random coil* secondary structure—short sequences of amino acids that loop in a random way. Most of the amino acids in ribonuclease are stabilized in these helices and sheets by their own hydrogen bonds. The stretches of ribonuclease with no regular helix or sheet structure are extremely important: At each such region, there is a bend in the polypeptide chain. It is because of these bends that the helix and sheet sections fold into the *tertiary structure*—the characteristic three-dimensional shape shown in Figure 3-20.

Like ribonuclease, virtually all water-soluble proteins, including blood proteins, enzymes, and antibodies, form a compact, roughly spherical *globular* shape as a result of this tertiary folding. The outside of a globular protein has an irregular, crinkly surface studded with various chemical groups, while the molecule's internal core is relatively dry, since the portions of the polypeptide chain folded into the center tend to have hydrophobic side chains that repel water molecules.

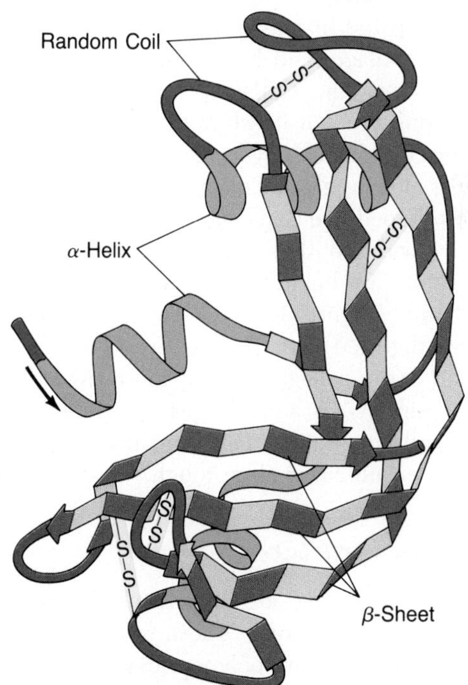

Figure 3-20 SECONDARY AND TERTIARY STRUCTURE IN THE PROTEIN RIBONUCLEASE.
This drawing shows ribonuclease with its three α-helices, its five β-sheets, and intervening regions of random coil, all folded into a characteristic three-dimensional shape. This shape allows the protein to bind to an RNA molecule and break it apart.

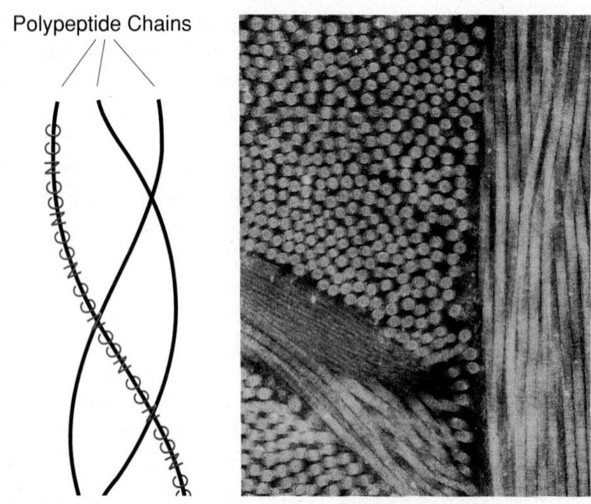

Figure 3-21 COLLAGEN: MOST ABUNDANT PROTEIN IN THE ANIMAL KINGDOM.
The structural protein collagen gives strength to skin, tendons, ligaments, cartilage, and bone. Thousands of collagen molecules are aligned into fibers, which are arrayed in parallel bundles. Here, collagen fibers (magnified about 30,000×) from a jawless fish called a lamprey can be seen in cross section (dots) and in longitudinal bundles.

A protein's tertiary structure is not always globular. Consider, for example, *collagen*, the structural protein that makes up the fibrous component of skin, tendons, ligaments, cartilage, and bone—fully 60 percent of all mammalian protein. Collagen makes leather tough, gives tendons the tensile strength of steel wires, and strengthens the corneal covering of the eye while leaving it nearly as clear as glass. Collagen protein is made up of molecules shaped like thin cigars some 200 times as long as they are wide. Within a given molecule, three polypeptide chains are held together in a triple helix by hydrogen bonds (Figure 3-21). Collagen is insoluble and not easily digested, but boiling breaks the hydrogen bonds and turns collagen to the more digestible substance gelatin. As an animal ages, covalent links form between collagen fibers. This explains why the meat from an older animal is often tough.

Quaternary Structure

Some proteins have a fourth level of organization, called quaternary structure. These molecules are made up of two, three, or more polypeptides, each folded into secondary and tertiary shapes and then intertwined in a complex multichain unit. The *quaternary structure* is the arrangement of these separate polypeptides in a three-dimensional shape held together by weak bonds. The hemoglobin molecule, which is composed of four heme-bearing polypeptide chains, has the quaternary structure shown in Figure 3-22.

Review Figure 3-17 to see how the precise linear sequence of amino acids, the side chains present, the particular foldings, and the intertwined polypeptide chains combine to yield a complex functioning protein with four levels of structure.

Factors Causing a Protein's Specific Three-Dimensional Shape

Once they understood protein structure, biologists began to wonder why proteins assume the shapes they do and remain kinked, folded, and conjoined rather than relaxing and intertwining like cooked spaghetti.

Experiments with ribonuclease, an enzyme that breaks down the nucleic acid RNA, revealed that proteins assume their shapes automatically, given the appropriate conditions in the solution surrounding them. Ribonuclease has a polypeptide chain of 124 amino acids, four disulfide bonds, and the structure shown in Figure 3-20. If a scientist treats ribonuclease with certain chemicals, it *denatures;* that is, the disulfide bonds break, and the enzyme loses its properties and its shape (Figure 3-23). The three-dimensional structure collapses during denaturation, and the chains of ribonuclease go limp. If the experimenter then removes the denaturing chemicals, the ribonuclease molecules regain their native (original) shape and activity. The conclusion: A protein's primary structure is sufficient to specify its three-dimensional structure and its *self-assembly.*

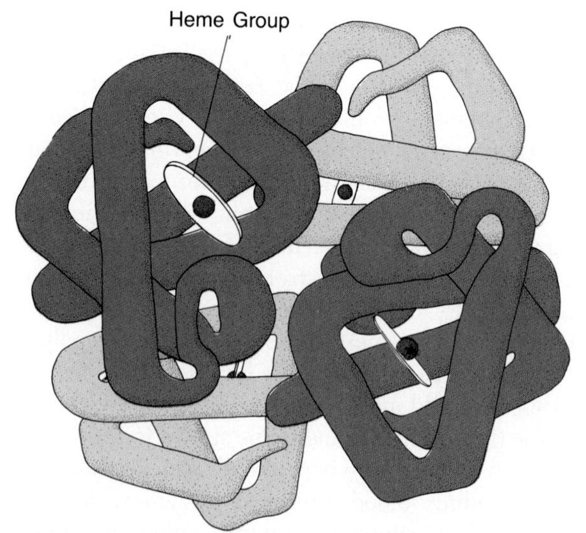

Figure 3-22 THE QUATERNARY STRUCTURE OF HEMOGLOBIN.
The quaternary structure of hemoglobin results when four polypeptides (α_1, α_2, β_1, β_2), each twisted into its characteristic tertiary shape, join by means of weak bonds. The heme prosthetic groups shown in red carry oxygen in this blood protein.

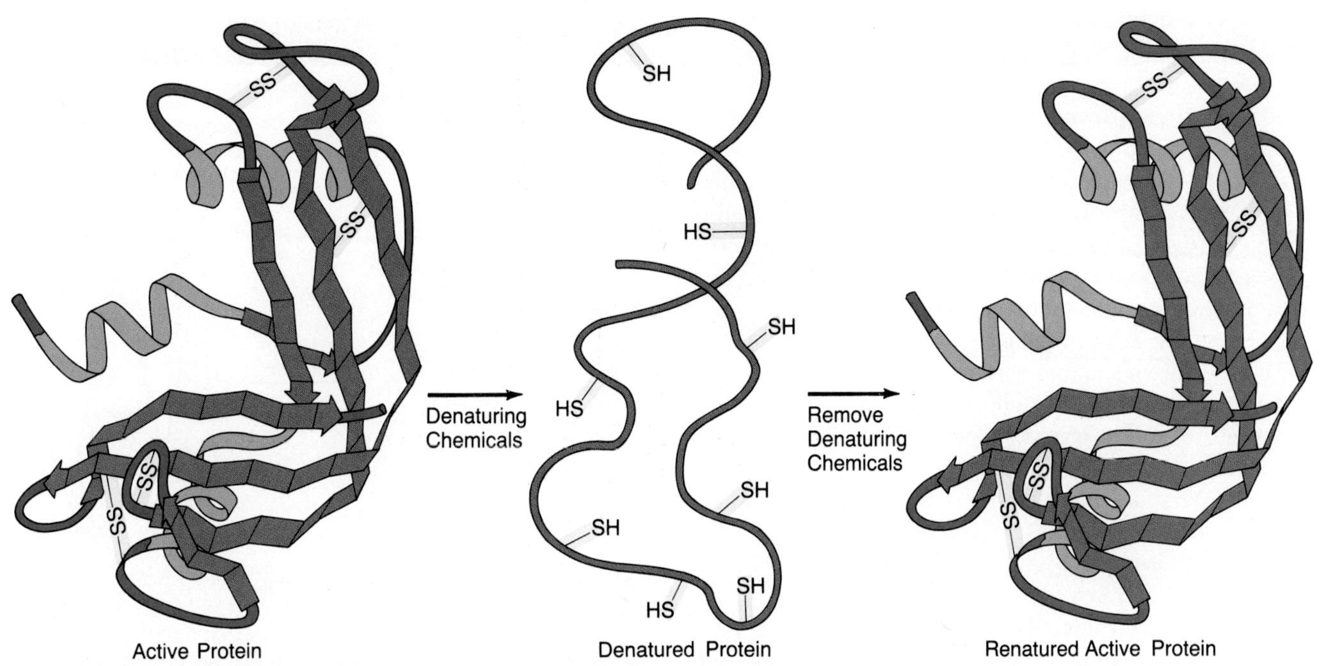

Figure 3-23 WHEN DENATURED, RIBONUCLEASE REASSEMBLES ITSELF.
When ribonuclease, which has four disulfide bridges, is denatured by chemical treatment, the disulfide bonds are broken and the molecule loses its shape. When the experimenter removes the denaturing chemicals, the enzyme spontaneously reassumes its native structure and regains enzymatic activity. This experiment shows that a protein's amino acid sequence is sufficient to specify much of its three-dimensional structure.

Other studies showed that chemical bonds and forces hold and stabilize the final shape of a protein. Most of the bonds that contribute to higher-order shapes in protein molecules are weak bonds of four types: hydrogen bonds, ionic bonds, van der Waals forces, and hydrophobic interactions.

We saw that hydrogen bonds help bring about secondary structure, including α-helices and β-pleated sheets. Ionic bonds can form between charged groups, such as the acidic or basic side chains of many hydrophilic amino acids. For example, the carboxyl group on the side chain of glutamic acid has a negative charge, while the amino group on the side chain of lysine has a positive charge. If glutamic acid and lysine are brought close together as a polypeptide chain folds, an ionic bond can form between them, stabilizing the loop or fold in the protein's shape.

If two parts of a molecule have reciprocal shapes and fit together closely, the electron clouds of the atoms in each part interact, and weak attractive forces, called *van der Waals forces,* are generated. Van der Waals forces can cause the parts of the molecule to pack more tightly, helping to hold the protein in its tertiary structure.

A fourth very important weak force that helps to hold a protein's shape once it has been attained is *hydrophobic interaction,* the tendency of hydrophobic side chains to tuck themselves inside the dry interiors of globular protein molecules. In the process, some hydrophilic side chains are pulled free of the hydrogen bonds they formed with water. As the hydrophilic side chains assume positions in the interior of the folded protein, they can reestablish hydrogen bonds, but this time with each other rather than with water. Thus, the main force responsible for the stability of globular proteins is hydrophobic interaction. Researchers are actively pursuing the rules that govern such protein folding so they can, in the future, design new proteins to order (see the box on page 54).

Because of the various types of weak bonds, proteins are usually on the brink of falling apart. If protein molecules were held in shape by strong covalent bonds, they would be enormously stable, but then the proteins would function poorly as enzymes or in carrying out other tasks essential to life. Reversible and rapid changes in shape are thus a hallmark of proteins and a necessity for life processes. Structural instability that stems from weak bonds is actually a major advantage and a precondition for protein functioning within living things.

Another advantage of weak bonds is that they allow self-assembly. A polypeptide can try many configurations before assuming the native form by chance; this form then persists because the maximum number of weak bonds form. Without weak bonds that are easily made and broken, this trial-and-error testing could not occur.

PROTEIN FOLDING: A THREE-DIMENSIONAL PUZZLE

With their growing skills in biological engineering, researchers are dreaming up artificial proteins that could help fight pollution, provide new medical tools, and even improve household detergents. Despite such promise, however, attempts to design new proteins have pointed out serious gaps in our basic understanding of protein structure and function.

Since the 1950s, biologists have known, in general, that a protein's amino acid sequence (primary structure) determines its helices and sheets (secondary structure) and its grooves, knobs, spikes, surface crinkles, and three-dimensional foldings (tertiary structure). What has eluded modern biologists is a detailed understanding of precisely how primary structure dictates higher-order protein folding, and with it, protein shape and activity.

The covert rules that underlie this folding have become a research goal in dozens of laboratories, but progress is frustratingly slow. Biologists are employing supercomputers and high-speed imaging devices to capture a protein's highly variable shape as it jiggles and moves in solution. So far, few guiding principles have emerged for why a given type of protein will bend in one spot and coil in another. But using the few available facts, William DeGrado and colleagues at DuPont in Wilmington, Delaware, have succeeded in designing a simple folding protein with four α-helices tightly packed and separated by three hairpin loops (Figure A). DeGrado hopes to afix a special binding site that will attach to metals so that the artificial protein could be used to clean up toxic wastes.

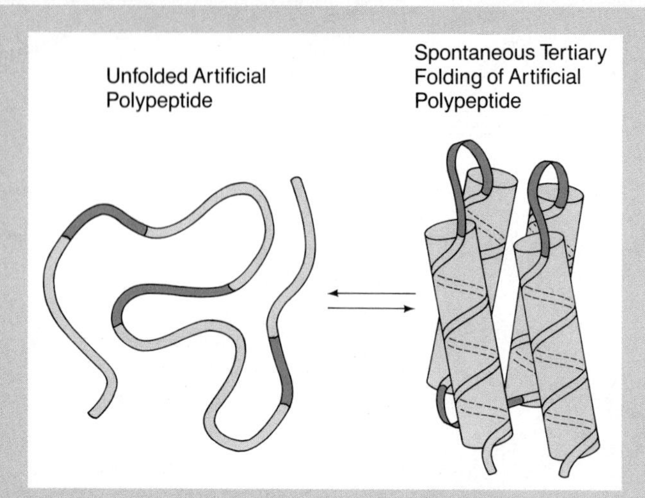

Figure A AN ARTIFICIAL PROTEIN.

Workers in other labs are trying to (1) modify insulin so that diabetics can absorb the tailored protein more quickly or slowly, depending on their medical needs; (2) modify a sticky protein from mussels for use as a medical adhesive; and (3) alter the protein subtilisin, an enzyme used in detergents, so that it will not degrade during clothes washing even if attacked by household bleach.

Biologists hope that by approaching the mystery of protein folding in two ways—through trial-and-error design and by seeking out underlying ground rules—they can, in a few years, create made-to-order proteins with environmental, agricultural, medical, and industrial uses.

NUCLEIC ACIDS: THE CODE OF LIFE

The final class of biological molecules, **nucleic acids,** are the information-bearing "code of life." Like proteins, nucleic acids have a specific linear sequence of subunits, a language of chemical "letters." These letters spell out instructions both for characteristics passed on to offspring and for translating that hereditary message into proteins that will be built into new cell parts, cells, and organisms. Nucleic acids are polymer chains made up of building blocks called **nucleotides.** Each nucleotide consists of a nitrogen-containing base, a five-carbon sugar, and a phosphate group (Figure 3-24). There are two types of nucleic acids: *deoxyribonucleic acid (DNA)* and *ribonucleic acid (RNA).* The sugar molecule in the nucleotides that make up DNA is a five-carbon monosaccharide called *deoxyribose;* a similar sugar, called *ribose,* is found in RNA (Figure 3-24a). These sugars become bonded to the nitrogen-containing bases through a condensation reaction, creating a *nucleoside.* The addition of a phosphate group to the sugar by means of another condensation reaction yields a nucleotide. Four different nitrogen-containing bases are found in DNA: *adenine, guanine, cytosine,* and *thymine* (Figure 3-24b). RNA also contains adenine, guanine, and cytosine; but instead of thymine, its fourth base is *uracil.* These are the "letters" that make up the nucleic acid alphabet, and although they are fewer than the 20 amino acids in proteins, they are just as capable of encoding diversity.

Condensation reactions join nucleotides together to form DNA and RNA in such a way that the sugar of one nucleotide is always bonded to the phosphate of the next

(a) Nucleotides

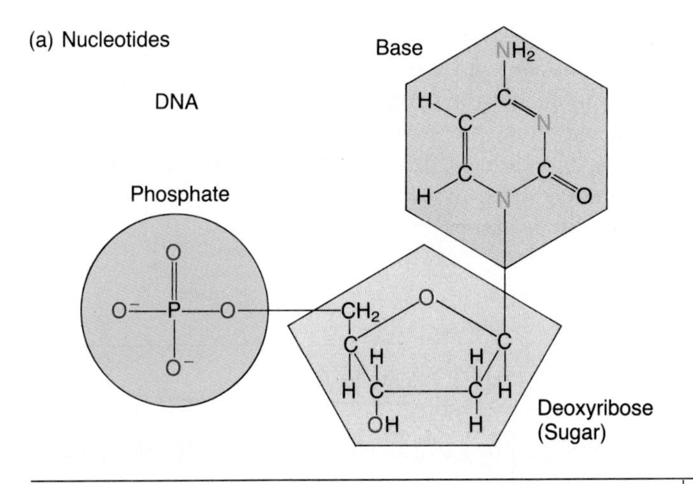

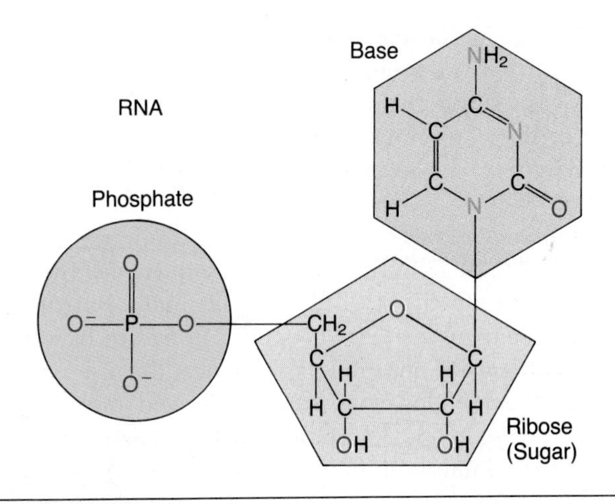

(b) Bases

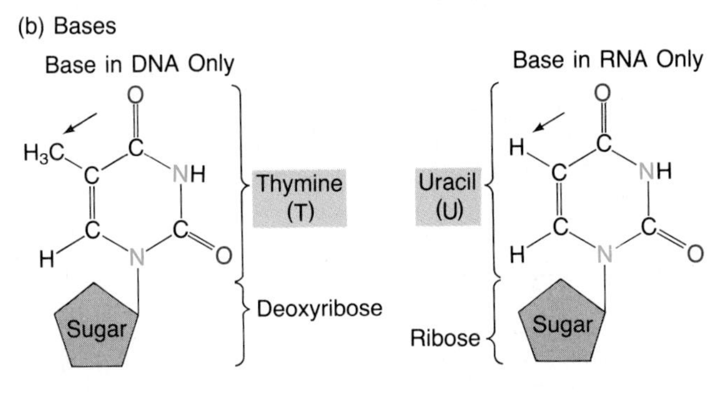

Bases in Both DNA and RNA

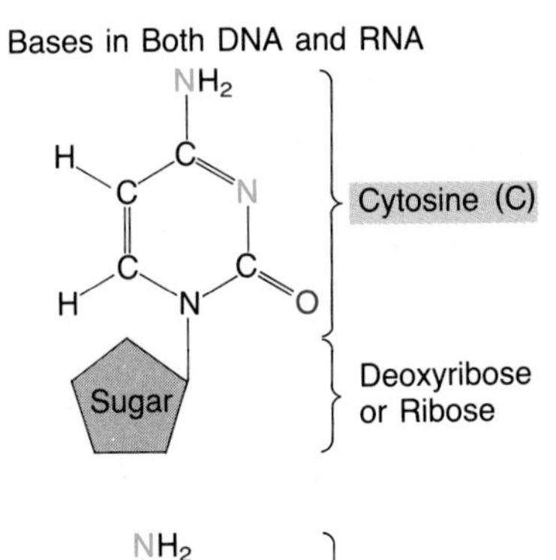

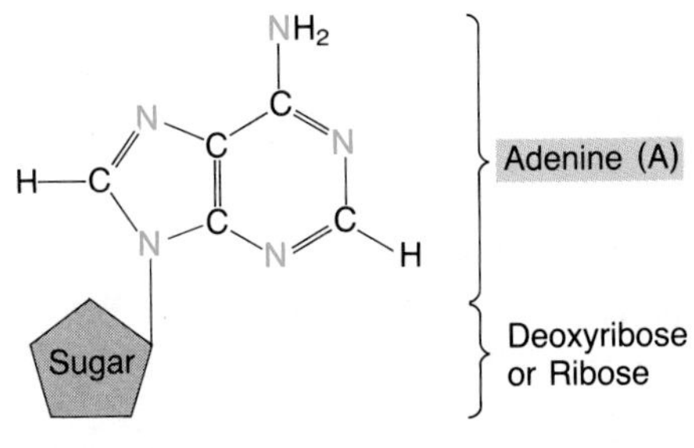

Guanine (G)

Sugar

Deoxyribose
or Ribose

(c) Full Molecules

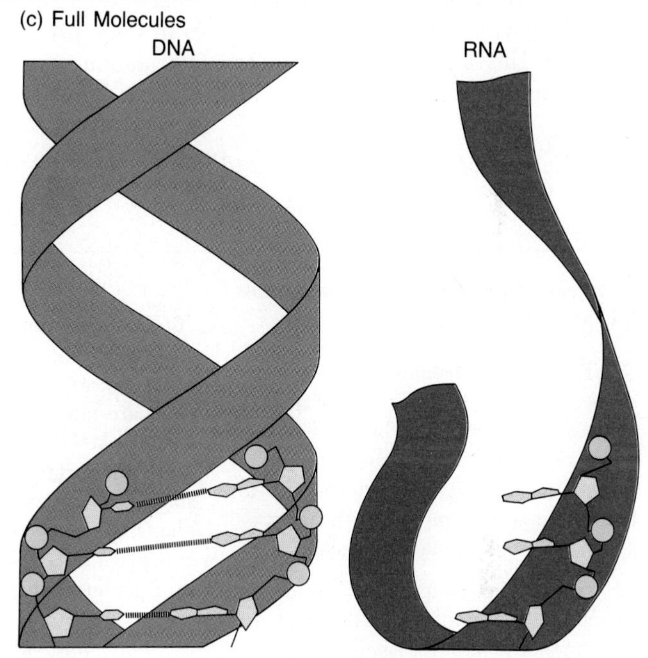

Figure 3-24 DNA AND RNA: NUCLEOTIDES, BASES, AND CHAINS.
DNA is the fundamental building block of genetic material. RNA also is involved in genetic processes. (a) Both DNA and RNA contain nucleotides made up of a phosphate, a base, and a sugar. (b) DNA alone has the base thymine. RNA alone has the base uracil. Both DNA and RNA share cytosine, guanine, and adenine. (c) Chains of nucleotides form the double helix of DNA and the single chains of RNA.

nucleotide in the chain (Figure 3-24c). DNA is composed of two such chains entwined in a double helix. RNA has a single chain. Just as polypeptides have amino and carboxyl terminal ends, nucleic acids have a hydroxyl group at one end and a phosphate group at the other.

The range of possible nucleic acids is infinite; that is, both the length of the chain and the sequence of the four nucleotides are theoretically without limit. But just as with proteins, nucleic acid "words" in nature have neither random size nor random sequence; the order of the letters (nucleotides) in each nucleic acid molecule is quite precise and is determined by a preexisting nucleic acid of the same type in a preexisting cell. As we shall see many times, life on earth is a process that only can be handed down directly from living things to their progeny, and this legacy is based on the information contained in the precise order of nucleotides in nucleic acids.

In its native state, DNA consists of two chains of nucleotides twisted around each other in a double-helix conformation and held together by hydrogen bonds. RNA molecules, in contrast, are made up of single chains that may fold into complex shapes or remain stretched out as long threads. The sequence of nucleotides in DNA ultimately determines the structures of every protein, in every living thing.

LOOKING AHEAD

Later chapters will trace the great intellectual adventure that led to the discovery of DNA, the meaning of the nucleic acid language, and how it controls both inheritance and the cell's day-to-day operations. Before that, however, we must see in more detail how life's macromolecular characters meet and interact—our subject in Chapter 4.

SUMMARY

1. The *macromolecules* of life (carbohydrates, lipids, proteins, and nucleic acids) are made up of carbon and are *organic compounds*. Carbon's unique properties allow it to bond with up to four other atoms and form the ring or chain skeletons of macromolecules.

2. *Isomers* are compounds with identical chemical formulas but different arrangements of atoms. *Enantiomers* are mirror-image molecules with the same properties, while *structural isomers* have quite different properties.

3. The specific structure of a *functional group* imparts a specific chemical behavior to another molecule. Examples are the hydroxyl group, the carboxyl group, the amino group, the methyl group, and the phosphate group.

4. Macromolecules are *polymers* formed by the linking of many *monomers* by means of *condensation reactions*. The splitting of polymers into their component monomers occurs through *hydrolysis*.

5. *Carbohydrates* are macromolecules that consist solely of carbon, hydrogen, and oxygen. *Monosaccharides*, such as glucose, can be combined into very large carbohydrate polymers.

6. *Disaccharides*, such as sucrose, are composed of two monosaccharides linked by a *glycosidic bond*. *Polysaccharides* are long polymers of monosaccharides linked by glycosidic bonds. The most important polysaccharides are *starch*, which is a storage material in plants; *glycogen*, a storage substance in animals; and *cellulose*, the fibrous structural material of plants.

7. *Fats*, *oils*, *phospholipids*, and *steroids* are important *lipids*. Fats and oils are *triglycerides* containing a *glycerol* molecule attached to three *fatty acid* chains by *ester bonds*. Phospholipids are hydrophilic at one end and hydrophobic at the other. Steroids have interconnected ring structures.

8. *Proteins* act as structural material, *enzymes*, chemical messengers, antibodies, and transport molecules. The monomers that make up proteins are the 20 amino acids.

9. *Amino acids* have a central carbon atom; a carboxyl group; an amino group; a hydrogen atom; and an R side chain, which determines the particular amino acid's properties. Amino acids join by means of *peptide bonds* to form *polypeptides*.

10. Every polypeptide has a unique amino acid sequence, dictated by an organism's hereditary material. This sequence is referred to as the protein's primary structure. A protein's three-dimensional shape is based on its secondary structure (such as *α-helices* and *β-pleated sheets*), its tertiary structure (precise folding patterns), and its quaternary structure (joining of several polypeptide chains).

11. Proteins have the property of self-assembly. The final shape of a protein is held and stabilized by weak hydrogen and ionic bonds, by van der Waals forces, by hydrophobic interaction, and sometimes by strong disulfide bonds.

12. *Nucleic acids*, such as DNA and RNA, are polymers of *nucleotides*, which consist of a nitrogen-containing base, a five-carbon sugar, and a phosphate group. DNA and RNA differ in three respects: Each contains a different sugar; one of their four bases is different; and DNA forms a double-helix shape, while RNA usually stays in single chains. The sequence of nucleotides in nucleic acids, which is inherited, determines the structures of proteins.

KEY TERMS

α-helix
amino acid
β-pleated sheet
carbohydrate
cellulose
condensation reaction
disaccharide
enantiomer
enzyme
ester bond
fatty acid
functional group
glucose
glycerol
glycogen
glycosidic bond
hydrolysis
insulin
isomer
lipid
macromolecule
monomer
monosaccharide
nucleic acid
nucleotide
organic compound
peptide bond
phospholipid
polymer
polypeptide
polysaccharide
protein
starch
steroid
structural isomer
triglyceride

QUESTIONS

1. Give an example of two isomers of a compound that have different properties.

2. Many polymers are formed by the linking of subunits in a condensation reaction. What small molecule is produced in this reaction?

3. What subunits make up cellulose? Starch? Glycogen? How do the structures of these three polymers differ?

4. Give an example of a saturated fat. Of an unsaturated fat. What is meant by the term *polyunsaturated*?

5. Give examples of lipids, and explain their functions in living organisms.

6. What is meant by the primary structure of a protein?

7. When a protein is denatured, what kinds of bonds are broken? When a protein is broken down into its component amino acids, what kinds of bonds are broken?

8. Give an example of a globular protein. Is it soluble or insoluble in water? Give an example of a fibrous protein.

9. How is the sugar in nucleic acids different from the sugar in starch?

10. Suppose you have a model-building kit that contains many nucleotides of four different kinds. How many different dinucleotides can you build? How many trinucleotides?

ESSAY QUESTION

1. Polysaccharides usually contain only a single kind of subunit, whereas polypeptides may contain 20 different kinds of subunits. Furthermore, the subunits of different polypeptides are arranged in different order. Explain why polypeptides have such complex structures and are so diverse. Why does each cell need so many different polypeptides?

For additional readings related to topics in this chapter, see Appendix C.

4

Chemical Reactions, Enzymes, and Metabolism

Laws of Thermodynamics:
1. *You cannot win.*
2. *You cannot break even.*
3. *You cannot get out of the game.*

Anonymous

When the Egyptian pharaoh Tutankhamen was interred in his gold sarcophagus in an underground tomb, slaves laid out a sumptuous ritual breakfast for the pharaoh's passage into the afterlife; 33 centuries later, when the tomb was opened, the breakfast was still there—completely dried out but still recognizable. Had Tutankhamen been alive to eat that breakfast when it was prepared, the digestive enzymes in his mouth, stomach, and intestines would have broken down the food molecules by lunchtime. Enzymes are the classic mediators of biological change, and change is central to life—change within atoms and molecules; fluctuations and adjustments within cells and tissues; and alterations in the organism itself. Except in a few rare cases, the organism that is no longer changing is no longer living.

Underlying every change in the living world are *chemical reactions,* transformations of sets of molecules into other kinds of molecules. Condensation reactions and hydrolysis reactions (see Chapter 3) are examples; molecules are transformed during these reactions, and as a result, complex chains are woven from simpler units or dismantled piece by piece. For these and most other

Sunlight, leaves, and browsing giraffe: Energy flows through all living things.

(a)

(b)

(c)

Figure 4-1 KINETIC, POTENTIAL, AND CHEMICAL ENERGY.
(a) A thundering waterfall has kinetic energy. This moving wall of water has tremendous energy to accomplish work, such as turning the wheels of a power generator or simply wearing away the rocks below. (b) This giant boulder, poised on a pinnacle, has the capacity to accomplish work, such as crushing cars, houses, and people. Potential energy can be stored for long periods; this rock has been in place for centuries. (c) A log, made up mostly of cellulose molecules, contains considerable amounts of chemical energy in its molecular bonds—energy that is released when the log burns.

reactions to take place within cells, however, energy must be expended, and our subjects in this chapter are life, energy, and the agents of change in biological systems.

Our specific topics include:

- *Energetics*, or the amounts and forms of energy that are part of chemical reactions
- How scientists measure free energy—the energy available to do biological work
- How concentration, temperature, and catalysts influence the rates of chemical reactions in living things
- The vitally important biological catalysts called enzymes, and how they work
- Metabolic pathways—orderly, interrelated patterns of chemical reactions controlled by enzymes

THE ENERGETICS OF CHEMICAL REACTIONS

As we saw in Chapters 2 and 3, molecules react in a specific way during a chemical reaction. In a hydrolysis reaction, for example, *reactants*, such as a dipeptide molecule and a water molecule, interact with each other and are converted into *products*, such as two amino acids, by means of the breaking and making of chemical bonds (review Figure 3-15). Regardless of type, all chemical reactions involve the transformation of energy from one of its many forms to another. *Light energy*, *heat energy*, and *electrical energy* are familiar forms from everyday life. **Kinetic energy** is the energy of motion, such as the energy generated by rushing water, a rolling rock, or moving molecules (Figure 4-1a). **Potential energy** is stored energy—the capacity to do work later. Water stored in a tank on top of a building, or a rock poised at the top of a hill, has potential energy (Figure 4-1b). **Chemical energy** is the energy stored in atoms and molecules and their bonds, so it is a kind of potential energy. The cellulose molecules in wood store significant amounts of chemical energy (Figure 4-1c).

The First Law of Thermodynamics

The different forms of energy share an important relationship described by the **first law of thermodynamics**, also called the law of conservation of energy (Figure 4-2a). This law states that energy can change from one form to another, but it can never be created or destroyed. Energy transformations take place every instant. Light energy strikes plant cells. Some is reflected,

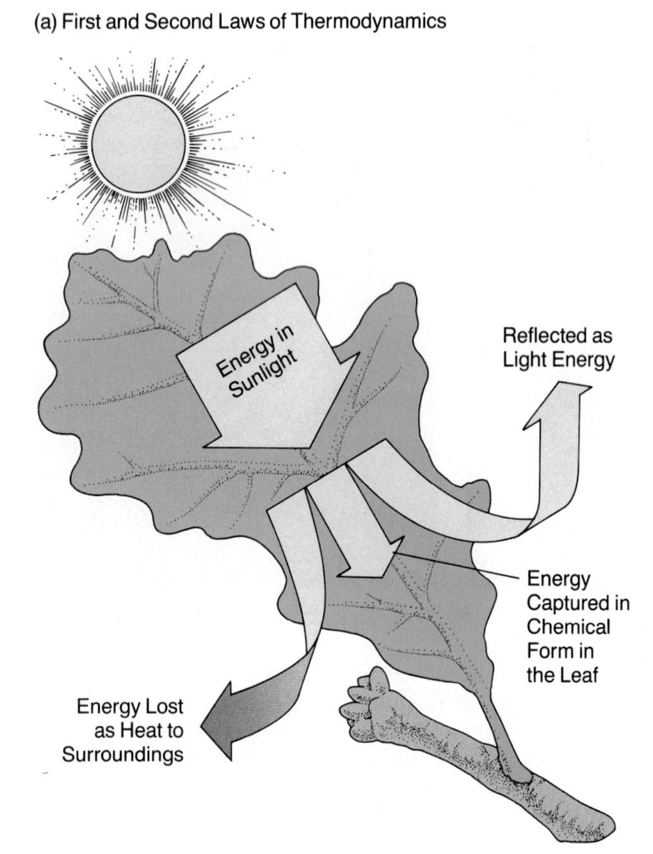

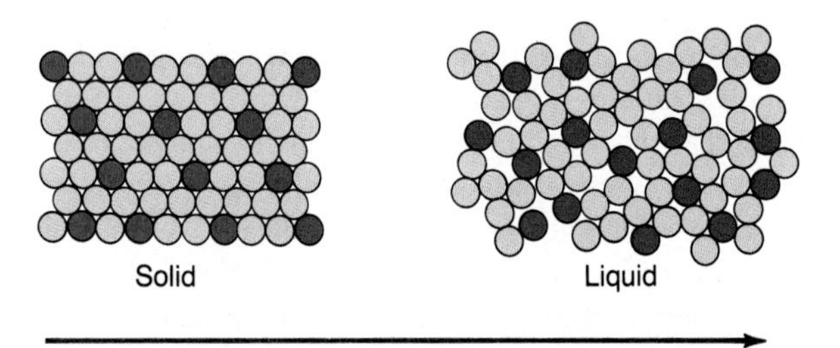

Figure 4-2 ENERGY CONVERSIONS AND ENTROPY: THE FIRST AND SECOND LAWS OF THERMODYNAMICS.
(a) An oak leaf in the sunshine illustrates both the first and second laws of thermodynamics. Light energy striking the leaf is converted to chemical energy by molecules inside the leaf. The process is not totally efficient, however; some of the energy is reflected as light, and some is lost as heat during the conversion. (b) When a solid is converted to a liquid, the geometric arrangement of the individual atoms breaks down. As the atoms become less ordered in the fluid, entropy increases.

but part is converted to chemical energy, which is stored in the bonds of carbohydrate molecules. And chemical energy is released as heat energy when wood is burned in a fireplace or when sugar is "burned" in a cell. In each case, energy changes form, but is never created anew or destroyed: It is *conserved.*

No energy transformation in the universe is 100 percent efficient. Whenever energy changes from one form to another, some portion is converted to heat energy and is lost; it is no longer available to do useful work. Heat energy can cause the rapid, random movement of molecules and thus contributes to the disorder of those molecules. For example, when proteins are digested in an animal's stomach and dipeptides are split into separate amino acids, some of the energy in the dipeptides and water molecules is released and lost as heat. This heat will simply cause molecules in a living organism or its surroundings to vibrate more rapidly for a while. Thus, the lost heat energy contributes no useful work.

The Second Law of Thermodynamics

Another universal principle, the **second law of thermodynamics,** reflects the fundamental fact that heat is a component of *every* energy conversion. This law states that the total energy in a system decreases inevitably as conversions take place and heat dissipates (see Figure 4-2a). An overall consequence of these inefficient conversions, then, is a tendency toward increasing randomness, or *disorder,* in the universe. For example, when a solid (such as ice) is converted to a liquid (such as water), the molecules become less ordered (Figure 4-2b). The amount of disorder in a system is known as **entropy.** As entropy increases and the usable energy in the universe continually decreases, the universe becomes more and more disordered. The heat that dissipates into the air around the blue flame of a gas burner is another example. This heat increases the random motion—the disorder—of the surrounding air molecules and thus the entropy of the system.

In light of these universal laws, how can the molecules, cells, and organs that make up living things exist and retain complex, well-ordered structures in a universe heading toward disorder? A protein, for instance, is much more than a disorganized pile of amino acids, and each of these subunits is more ordered than the same atoms free of one another. With its millions of proteins, a cell—let alone a tree or a fish—is an incredible oasis of order in a desert of entropy, the disorder that surrounds the cell in the universe.

The key to the maintenance of order in a living system, existing as it does within the disorder of the surrounding universe, is an input of energy. To stay alive and counteract the tendency of complex molecules and structures to degrade (and thus obey the second law), every cell must carry out chemical reactions involving a net influx of energy. The cell is an open system: It takes in energy but unavoidably frees waste products and heat back into its surroundings, heat that contributes to the disorder of the universe as the cell labors to keep its own integrity. The price of this process is dear: The cell must have enough energy in the form of nutrient molecules to drive the reactions that maintain the cell as well as to satisfy the second law of thermodynamics. Ultimately, the source of this energy for most organisms is the sun, as Chapters 7 and 8 will explain.

Eventually, some of the essential reactions that support life cease, and the cell or organism dies. The order of the living thing fades, disorder ensues, and entropy wins out. The animal or plant disintegrates back to "dust"—to the molecules and atoms of which it is formed.

FREE ENERGY CHANGES

A cell needs sufficient energy to power the thousands of simultaneous biochemical reactions that support life processes, as well as enough left over to satisfy the second law of thermodynamics. But just how much energy is enough?

Scientists use a set of equations to determine the maximum amount of energy available—or "free" to do work—in a given system. With such formulas, including the Gibbs equation, $\Delta G = \Delta H - T\,\Delta S$, they can track how

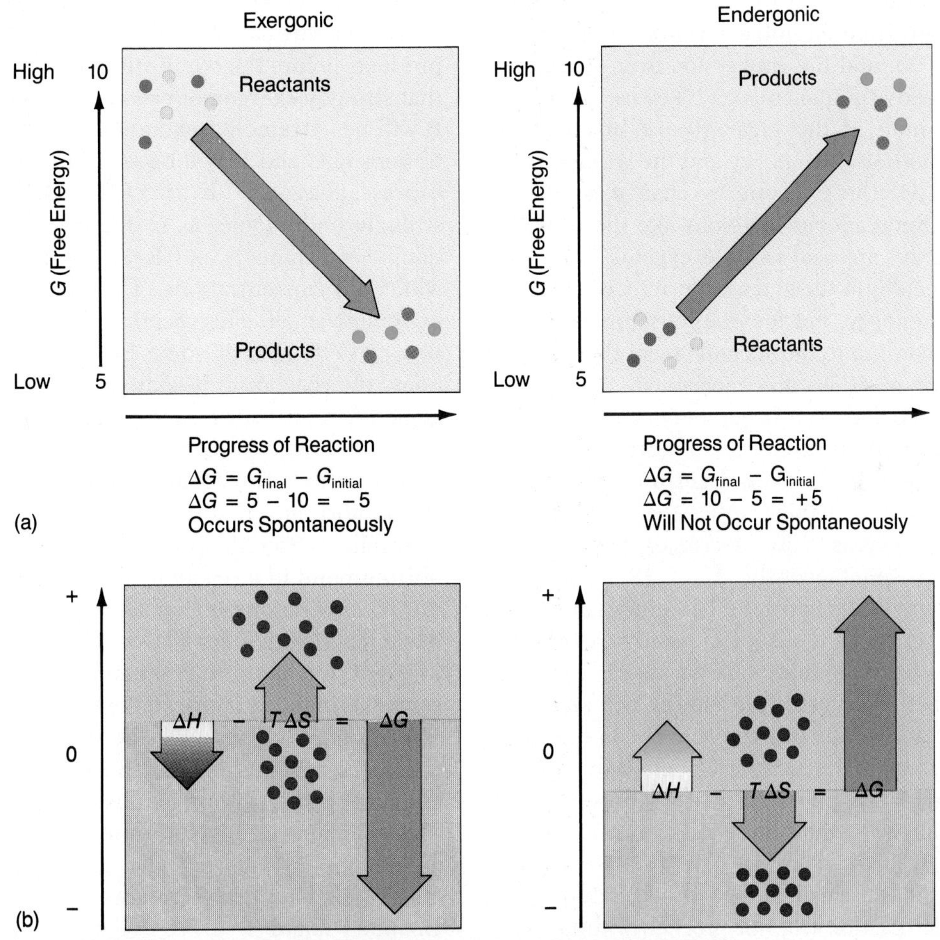

Figure 4-3 ENERGY CHANGES DURING EXERGONIC AND ENDERGONIC REACTIONS.
(a) In an exergonic reaction (left), the free energy (energy free to do work) in the products is less than in the reactants. Hence, ΔG is a negative number, and the conversion of reactants to products can proceed spontaneously. If, however, the free energy is greater in the products than in the reactants (right), then ΔG is a positive number, the reaction is endergonic, and it cannot proceed without additional energy input. (b) Enthalpy decreases during an exergonic reaction (left), meaning that the total energy trapped in the bonds of reactants and products drops. Since energy is released, the system now contains less, and it is less ordered; entropy thus increases. The reverse is true in an endergonic reaction (right); for the reaction to proceed, the total energy in the bonds must increase, entropy may decrease, and the system may become more ordered, at least temporarily. The diagrams in (b) represent just two of the possible outcomes, depending on whether a reaction is strongly or weakly endergonic or exergonic. For example, ΔH could be positive even in an exergonic reaction if $T\,\Delta S$ had an even larger value.

that free energy changes as a particular chemical reaction proceeds. The change in free energy, symbolized ΔG (delta G), is a function of (1) the change in total bond energies of reactants and products (designated ΔH, where H stands for *enthalpy*, or heat content); and (2) the change in the entropy, or disorder, in the reacting system (symbolized $T\,\Delta S$, where T equals temperature in degrees Celsius plus 273, and S equals entropy).

By measuring free energy changes during chemical reactions, scientists have determined that if ΔG is a negative number (i.e., if the free energy in the products is *less* than that in the reactants), then a given reaction can proceed spontaneously (the word *spontaneous* here really means "possible" and does not necessarily mean rapid, as we'll discover later). Figure 4-3a (left) shows that in a hypothetical reaction where the free energy of the system changes from an initial level of 10 to a final level of 5, $\Delta G = -5$, and the conversion from reactants to products proceeds spontaneously. Likewise, as a living cell burns 1 mole of the sugar glucose and in the process releases both usable energy and the waste products CO_2 and H_2O, the free energy change is $\Delta G = -686$ kcal/mole. Spontaneous reactions like these, with a net loss in energy, are said to be **exergonic** ("energy out"). Other molecules in the glucose-burning cell might trap some of that energy, but inevitably, some will also be lost as heat according to the second law of thermodynamics. Exergonic reactions are enormously important in living cells, since they can occur between reactants without any further input of energy. Figure 4-3b (left) shows how changes in enthalpy and entropy affect free energy in an exergonic reaction. Notice that as enthalpy drops and entropy increases, ΔG decreases, and the reaction can proceed spontaneously.

In a second type of reaction called an **endergonic** reaction (meaning "energy in"), ΔG is a positive number, and free energy is taken up, not released. As Figure 4-3a (right) shows, when the initial free energy is 5 and the final level is 10, $\Delta G = +5$, and the conversion of reactants to products cannot proceed spontaneously. Likewise, in the laboratory, experimenters can induce 1 mole of glucose to react with phosphoric acid to form water and glucose-6-phosphate—but only if they add energy to the system. The ΔG for this reaction is $+3.3$ kcal/mole, so the endergonic reaction will not proceed without energy input. As Figure 4-3b (right) shows, as the enthalpy in an endergonic reaction increases, the entropy decreases; an input of energy must make the system temporarily more ordered before the reaction will proceed. Like exergonic reactions, endergonic reactions are enormously important in biology. With beautiful symmetry and economy, exergonic and endergonic reactions are often linked in living things, as you will see shortly.

Equilibrium and Free Energy

You can watch an exergonic reaction in progress by simply mixing vinegar (an acid) with baking soda (a base) and observing the bubbling of the solution, which proceeds vigorously for a while, then less vigorously, and finally stops. What does this spontaneous cessation mean?

Chemists discovered long ago that as a reaction takes place, the free energy changes until $\Delta G = 0$. At that point, the reaction is said to be at **equilibrium;** no further *net* conversion of reactants to products takes place. This does not mean that the combined concentrations of two reactants, A and B, will be equal to the concentrations of the products, C and D. (Concentrations are expressed as moles per liter.) If the reaction is strongly exergonic, virtually all the reactants will be converted to products before the equilibrium point is reached, and for that strongly exergonic reaction, concentrations of A and B will be extremely small at equilibrium, while concentrations of C and D will be very large (4-4a). This can be expressed as $A + B \rightleftharpoons C + D$. If the reaction is strongly endergonic, as in Figure 4-4c, the concentrations of reactants A and B will be high at equilibrium, while the concentrations of products C and D will be extremely small. This reaction can be expressed as $A + B \rightleftharpoons C + D$. For some biologically significant reactions, the reaction is weakly exergonic or weakly endergonic. A more accurate representation for those reactions is $A + B \rightleftharpoons C + D$ (see Figure 4-4b), where the concentrations of reactants and products are more equal at equilibrium.

It follows from the preceding discussion that the equilibrium point of a reaction can be altered by controlling the concentrations of reactants and products. Imagine that a reaction $A + B \rightleftharpoons C + D$ goes to equilibrium. If we then decrease the residual quantities of A and B by removing them, or if we sufficiently increase the concentrations of products C and D, then the reverse reaction can occur and more A and B can be made. Controlling chemical equilibrium by adjusting the concentrations of reactants and products is a potentially valuable strategy for a living cell and one that will figure prominently in our discussions of respiration and photosynthesis (see Chapters 7 and 8).

Biological systems exhibit another strategy as well: They employ the energy freed during an exergonic reaction to drive an endergonic one. Such pairs of reactions are said to be **coupled.** Let's say the highly exergonic reaction $X \rightarrow M + N + \text{Energy}$ ($\Delta G = -10$ kcal/mole) is coupled to the highly endergonic reaction $D + \text{Energy} \rightarrow E + F$ ($\Delta G = +6$ kcal/mole). Even though the second reaction is endergonic when occurring alone,

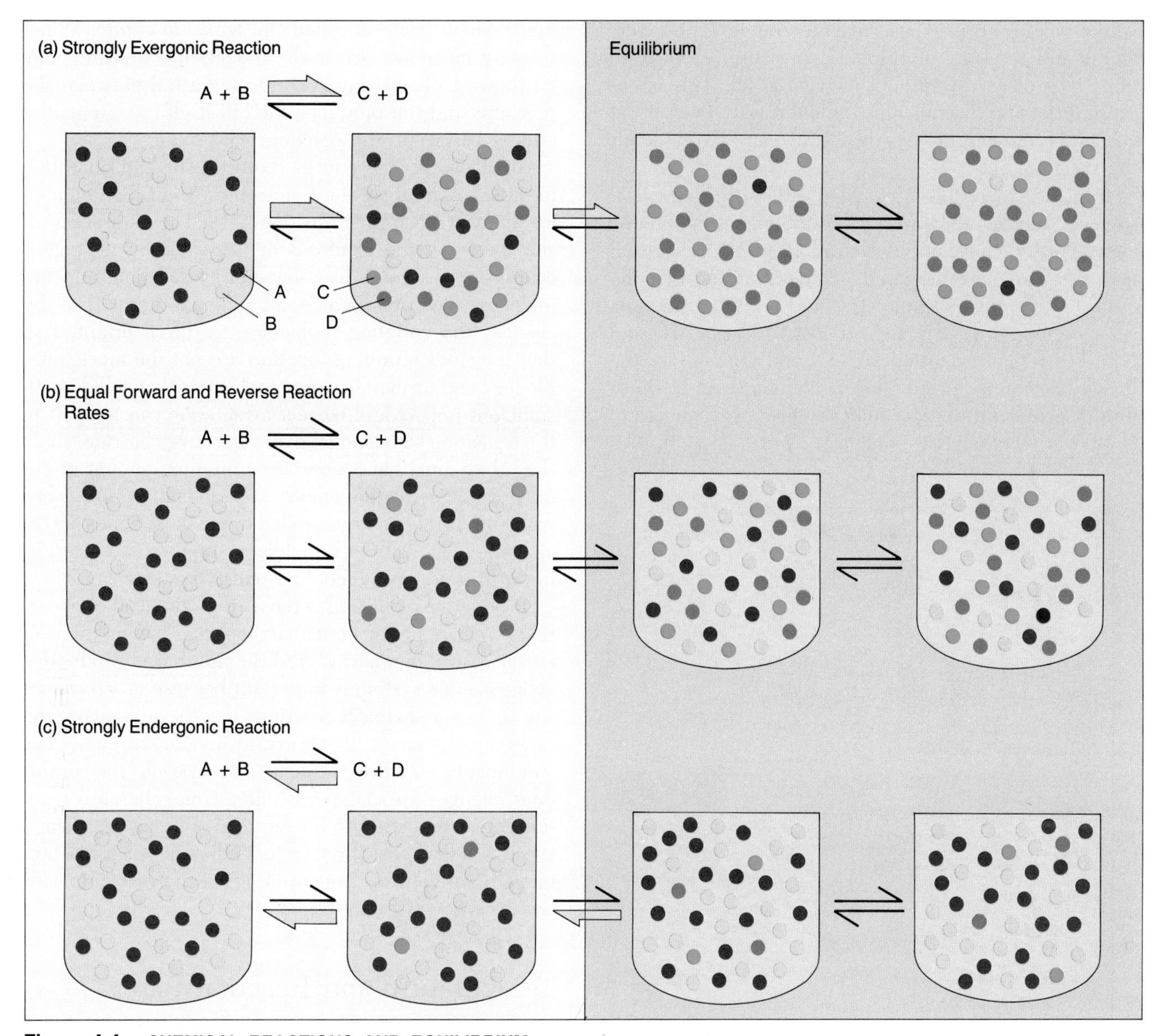

Figure 4-4 CHEMICAL REACTIONS AND EQUILIBRIUM.
(a) In a strongly exergonic reaction, there are more products (C and D) than reactants (A and B) at equilibrium. (b) When the reaction is only weakly exergonic, as in many biological reactions, reactants and products will be nearly equal. (c) In a strongly endergonic reaction, there will be more reactants (A and B) at equilibrium than products.

taken together, the coupled reactions are exergonic and proceed spontaneously.

Coupling is the basis for many of the interrelated reactions in living cells, such as those involved in the breakdown of nutrients and the building of proteins and fats. Because the products of a given reaction (including energy) generally participate in other reactions, true chemical equilibrium for a given reaction within a cell is rarely achieved. This is especially true for the sequential reactions of metabolism, as we shall see.

RATES OF CHEMICAL REACTIONS

Over many years, a bowl of sugar cubes or a large bone resting in a museum case will spontaneously break down into constituent molecules. The rate of breakdown will be so slow, however, that even after a few centuries, the cubes and the bone will look virtually unchanged.

Under what we consider normal environmental temperatures and sea-level atmospheric pressure, most reactions take place extremely slowly, if at all. This is because an energy barrier must be overcome, even if the change in free energy, ΔG, is negative and favorable for the reaction.

Consider two molecules, MN and OP (Figure 4-5, step 1). To react, they must collide. Some of the kinetic energy of the collision between molecules MN and OP goes into breaking the bonds of the atoms so that the entirely different molecules MO and NP can form. Just what occurs when MN and OP react to form MO and NP? For a fleeting instant as MN and OP collide, the bonds become distorted (step 2) and continue to do so until it is difficult to determine whether M is bonded to N or O or whether N is bonded to M or P (step 3). This intermediate state is called the *transition state.* When reacting molecules are in the transition state, they may be thought of as being at the top of a hill between the reactants and the products; the hill itself represents the energy barrier to the reaction.

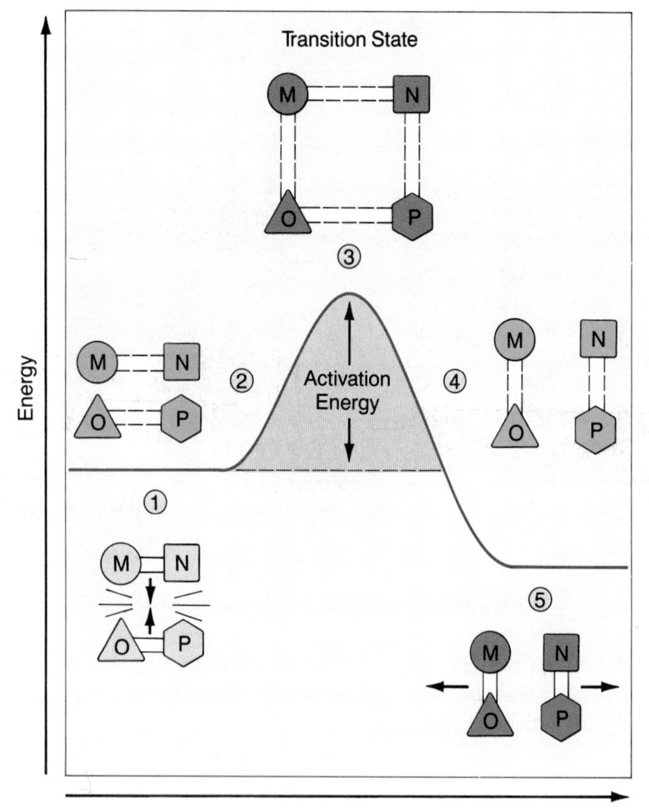

Figure 4-5 REACTING MOLECULES SURMOUNT THE ACTIVATION ENERGY "HILL."
When molecules react, the bonds between their atoms are rearranged in a series of steps. (1) Molecules MN and OP collide; (2) the kinetic energy of the collision begins to distort and stretch the springy bonds between atoms; (3) in the transition state, it is difficult to determine which atoms are bonded to each other; (4) after the reaction, the new bonds between the atoms of the products MO and NP momentarily remain distorted; (5) at the lowest energy level, the bonds between the atoms of the new product molecules, MO and NP, are stable.

Significantly, the energy (both kinetic and potential) in most colliding covalent molecules is not high enough for them to reach the transition state. The natural repulsion between the electrons of the colliding molecules cannot be overcome in a low-energy collision, and the molecules simply bounce off one another. Also, by chance, the colliding molecules are often oriented so that the sides bumping together are not the most suitable for bond formation. Only molecules that collide with sufficient impact and correct orientation can make it to the transition state and go over the energy barrier. To do this, they must have a certain minimum amount of energy, called **activation energy.** The activation energy of a molecule is the amount needed to break its bonds and to allow new bonds to form between atoms (step 4), resulting in the new products MO and NP (step 5).

Productive collisions—those that do lead to reactions—occur among *activated molecules* having the necessary activation energy. And the number of productive collisions in a system is important because *it determines the rate of a chemical reaction.* For the most part, the rates of reactions in living organisms are extremely fast and must be so if the organism is to carry out the continuous changes associated with life. Thus, chemical reaction rates are critical to the survival of living organisms, and as we shall see, three factors influence such reaction rates: temperature, concentrations of the reactants, and the presence of catalysts.

Temperature and Reaction Rates

Heat can supply the energy needed to activate molecules—to push them over the energy hill—so that a collision between them is productive. If an experimenter heats a population of molecules, they move about more quickly, and more high-energy molecules chance to collide (Figure 4-6). The reason is that heat energy dramatically increases both the frequency with which molecules collide and the force of the collisions. As a result, many more collisions attain activation energy levels, and the reaction accelerates. Even though the speed of some moving molecules goes up only about 3 percent with each 10° C rise in temperature, the rate of a chemical reaction goes up 100 to 300 percent; clearly, temperature affects the rate of a reaction far more strongly than it affects the speed of molecules.

Temperature changes affect every reaction differently because the activation energies differ between sets of molecules. One reaction might undergo a fourfold in-

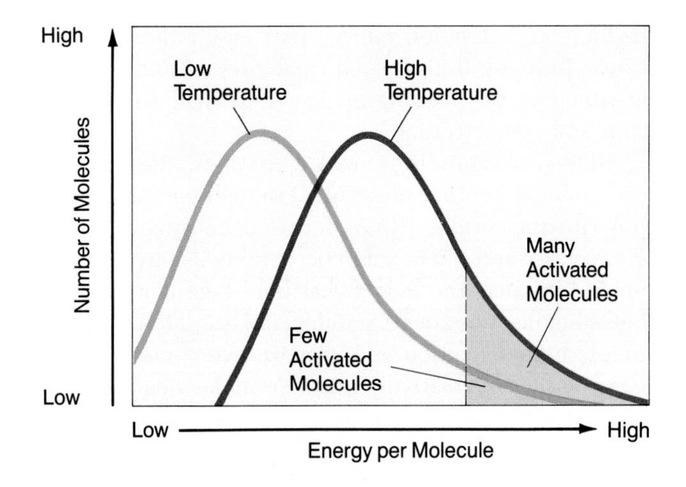

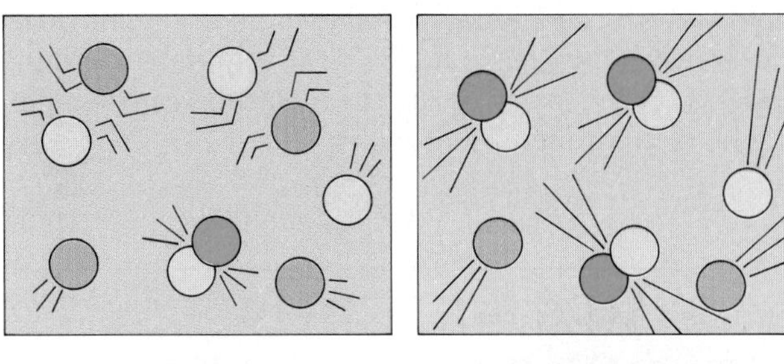

Figure 4-6 HIGHER TEMPERATURES ACCELERATE REACTION RATES.
A rise in temperature speeds reactions. At low temperature (blue curve), only a fraction of molecules in a given population has sufficient kinetic energy to react (blue area). At a higher temperature (red curve), the kinetic energy of all the molecules is higher, and thus a much larger fraction of molecules attains activation energy (red area). The explanation is straightforward: At lower temperatures, there are fewer productive collisions between reactants; at higher temperatures, more productive collisions.

crease in reaction rate in response to an increase in temperature, while a second reaction might undergo a sixteenfold increase. In most cells and organisms, however, normal internal temperatures are too low and too constant to promote very rapid chemical reactions.

Concentration and Reaction Rates

Quite logically, the more molecules there are in a given volume, the more collisions will occur among them—including ones of sufficiently high energy to bring about a reaction. And indeed, increasing the concentrations of reactants can speed their conversion to products (Figure 4-7). As we saw earlier, increasing concentrations also affect the rates of the reverse reactions. As reactants A and B are converted to products C and D, the concentrations of C and D increase; the "crowding" actually represents a decrease in entropy that can drive the reverse reaction ($C + D \rightarrow A + B$). Despite the effects of concentration on reaction rates, concentration alone, like temperature alone, is usually insufficient to accelerate reaction rates enough to sustain life in a cell or organism. The most important influence on rates of reactions in organisms rests with catalysts.

Catalysts and Reaction Rates

A **catalyst** is a molecule that increases the rate of a reaction without being used up during that reaction. Like a traditional matchmaker, it takes part in the reaction but emerges unscathed. Since it remains unaffected, a small amount of catalyst, used again and again, can speed up a reaction.

Catalysts have a far different effect on reaction rates than temperature and concentration. If you picture the energy hill that stands between reactants and products, then both temperature and concentration push reactants

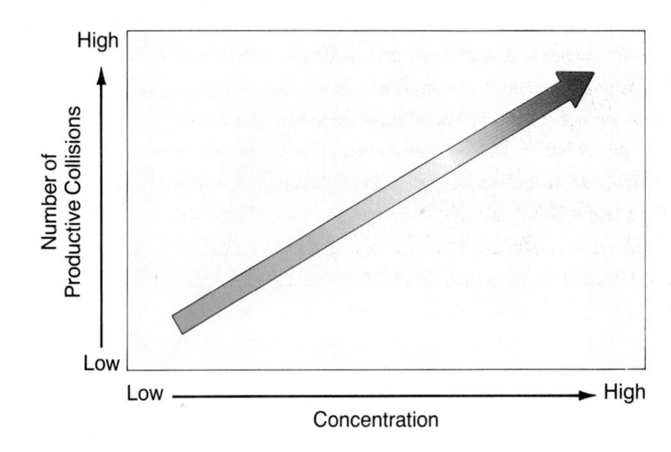

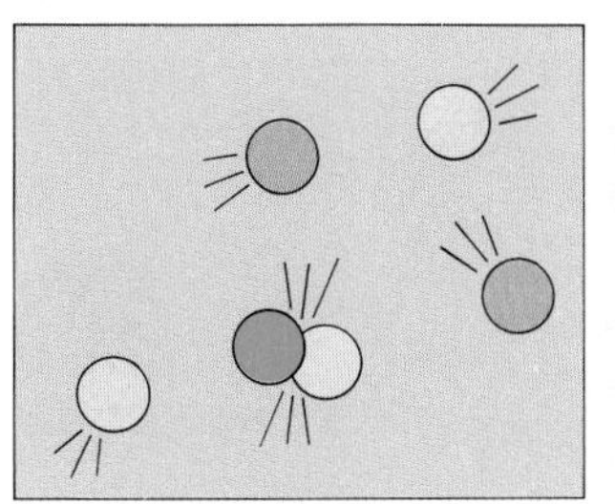

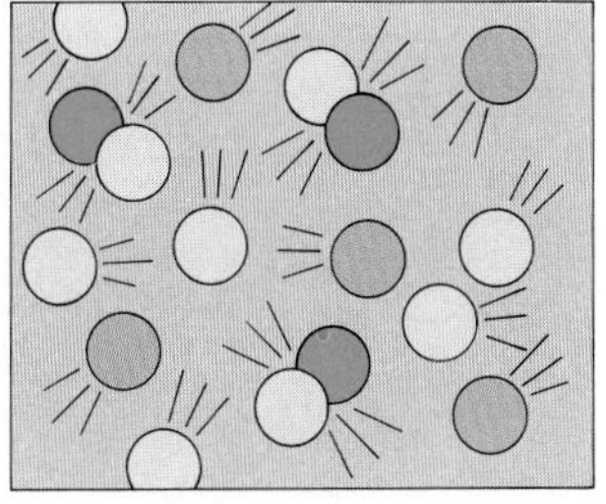

Figure 4-7 HIGHER CONCENTRATIONS ACCELERATE REACTION RATES.
Higher molecular concentrations lead to more reactions, as the linear curve reflects. Like dancers on a dance floor, the more molecules in a given area, the more likely they are to collide and subsequently react with one another.

up and over the energy hill. These factors work by increasing both the energy of colliding molecules and the frequency of their collisions, so that the molecules exceed the level of activation energy represented by the peak of the hill. Catalysts, however, actually lower the hill itself; *they reduce the activation energy necessary for the reaction to proceed* (Figure 4-8). This reduction has a dramatic effect on the rate of a reaction. For example, if the activation energy is reduced from 10 kcal/mole to 9 kcal/mole, many more collisions that would have been unproductive now have sufficient energy to overcome the energy barrier and thus to react. This 1 kcal/mole reduction yields a fivefold increase in the reaction rate. Remarkably, some biological catalysts can speed up cellular reactions a millionfold or more by lowering even further the activation energy barrier for those conversions. In living things, such catalytic proteins play a far greater role in speeding up reactions than do temperature and concentration.

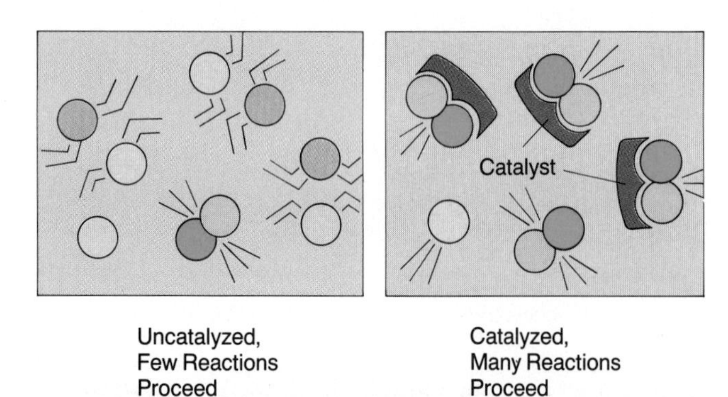

Figure 4-8 CATALYSTS ACCELERATE REACTION RATES. As biological catalysts, enzymes are the key to the chemistry of life. They lower the activation energy barrier, allowing many more reactions to proceed, and at a much faster rate, than would otherwise.

Biological catalysts are called *enzymes*; most enzymes are globular protein molecules, though certain kinds of polyribonucleotides (RNA) function enzymatically too. Enzymes—and their catalytic activities—are essential for life because the bonds that hold together most biological molecules are very stable and cannot be ruptured unless high activation energies are overcome. If organisms employed relatively unstable molecules that could react at low activation energies, their bonds would break spontaneously, and the result would be molecular chaos in the cell. Instead, most of the bonds in amino acids, lipids, sugars, and nucleic acid bases are stable, the macromolecules they make up are relatively stable, and the chemical reactions involving the molecules can be mediated and controlled by enzymes.

ENZYMES AND HOW THEY WORK

The study of enzymes began more than a century and a half ago. In 1822, William Beaumont, an army surgeon, treated a French Canadian soldier for a severe gunshot wound to the abdomen. The man survived and the wound healed, but it left a gaping hole in his upper abdomen. Through this "window," Beaumont was able to obtain samples of stomach fluids and to observe the secretion of digestive juices containing enzymes, noting their activity on foods. Seventy years later, researchers showed that enzymes from yeast cells could break down sugars even after the yeast cells had been disrupted. They coined the word *enzyme* during this experiment; to these researchers, an enzyme was simply a substance *en zyme*, from the Greek for "leavened," or "in yeast."

Since then, biologists have learned a great deal more about the roles of enzymes in living cells and about enzyme structure and function, including two unique characteristics. First, enzymes are specific. A given enzyme can act on only one type of compound or pair of reacting compounds, which is called its **substrate;** and it usually can catalyze only one type of reaction, such as condensation or hydrolysis. Enzyme names commonly end in "-ase" and reflect the type of substrate or reaction that they catalyze. For example, an enzyme that cleaves hydrogen from glucose-6-phosphate is called glucose-6-phosphate dehydrogenase. Second, enzymes can be controlled by the presence or absence of critical compounds. Let us now examine the structure, function, and unique characteristics of enzymes in more detail.

Enzyme Structure

Most enzymes are globular proteins, and the substrates on which they act are often much smaller molecules than the enzymes themselves (Figure 4-9). Each protein enzyme has a unique three-dimensional shape arising from its primary, secondary, tertiary, and (sometimes) quaternary structure (see Chapter 3). On the surface of each enzyme molecule is one small area (or sometimes a few areas) called the **active site.** The key to enzyme specificity is the shape of that site. The conformation of the active site complements the shape of the substrate(s) the enzyme acts on in much the same way that the keyhole of a lock fits around a key. It is this reciprocal matching of the three-dimensional shapes of active sites and substrates that accounts for enzyme specificity.

The active site is often a deep groove or pocket on the enzyme's surface. It is made up of amino acids from several parts of the enzyme molecule, brought into close proximity by the folding of the polypeptide chain. Thus, the active site is created by the enzyme's tertiary (and sometimes quaternary) structure. The active site may also contain a *prosthetic group* (see Chapter 3) that is essential to the enzyme's activity. Prosthetic groups may be atoms of zinc or magnesium, or they may be ring-shaped organic compounds that include metals similar to the heme group of hemoglobin.

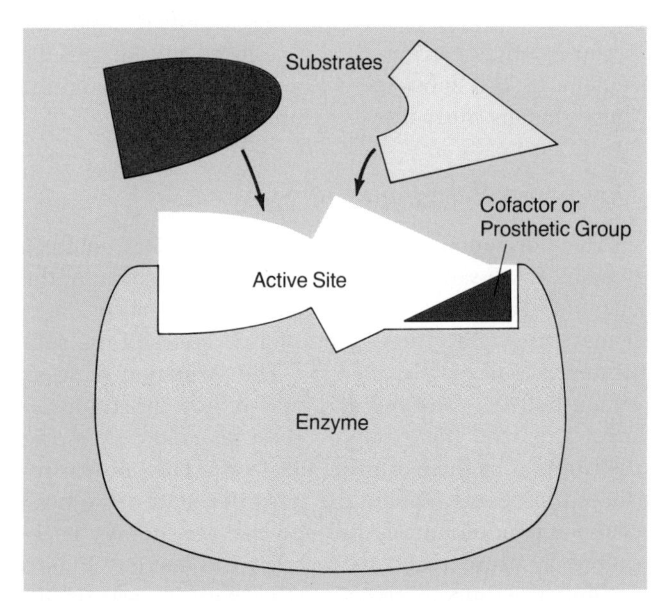

Figure 4-9 ENZYME STRUCTURE: GLOBULAR PROTEINS WITH ACTIVE SITES.
The precise folding of an enzyme's polypeptide chains creates a globular molecule (here, purple) with a cleft, or active site, that can bind substrates (here, red and yellow). The active site may contain a permanent prosthetic group or a temporary cofactor that is essential to the enzyme's activity.

Enzymes that lack permanently bound prosthetic groups may depend instead on **cofactors,** special compounds that bind temporarily to a site on the enzyme and take part in the reaction and the formation of products (see Figure 4-9). Most cofactors are *coenzymes,* small molecules associated with and essential to the activity of some enzymes. Coenzymes are always changed in some way as they interact with enzymes, and thus their original form must be restored by other reactions before they can function again. The compounds we call vitamins—essential substances that an organism cannot manufacture and so must consume in its diet—are often converted to coenzymes within cells, thereby enabling them to interact with enzymes (see Chapter 34).

Because an enzyme's activity depends on its precise three-dimensional shape, factors that affect its shape can also affect, or even destroy, enzyme activity. For example, excessive concentrations of Na^+ and Cl^- can break the ionic bonds that help maintain an enzyme's three-dimensional structure, thereby disrupting the integrity of the active site and its ability to bind substrate. Most enzymes have a *pH optimum,* or level of H^+ concentration at which they function best. Pepsin, for example, a digestive enzyme found in the human stomach, functions best at an acidic pH near 2, while chymotrypsin and trypsin, digestive enzymes found in the small intestine, function most efficiently in an alkaline environment at pH 8. Cells and organisms have many mechanisms that help keep salt and H^+ concentrations within narrow ranges and thus protect the structure and function of enzymes.

Enzyme Function

How do enzymes lower the activation energy barrier between reacting molecules? Enzymes function as catalysts by (1) forming complexes with the reacting molecules; (2) changing their own shapes slightly to improve the fit between enzyme and substrate; (3) increasing the local concentrations of the molecules; (4) orienting the molecules correctly so that the reaction can take place most efficiently; and (5) distorting the shape of the substrate molecules slightly, as well, thereby helping them reach the transition state (Figure 4-10a).

Enzyme Substrate Complexes

Just as other molecules must collide before they can react, reactants must collide with enzymes before a reaction can be catalyzed. The collision allows them to form the **enzyme-substrate (ES) complex,** a complex held together by weak bonds, and this complex is the essential first step in enzyme catalysis. In the ES complex, the substrates are positioned close together and in just the

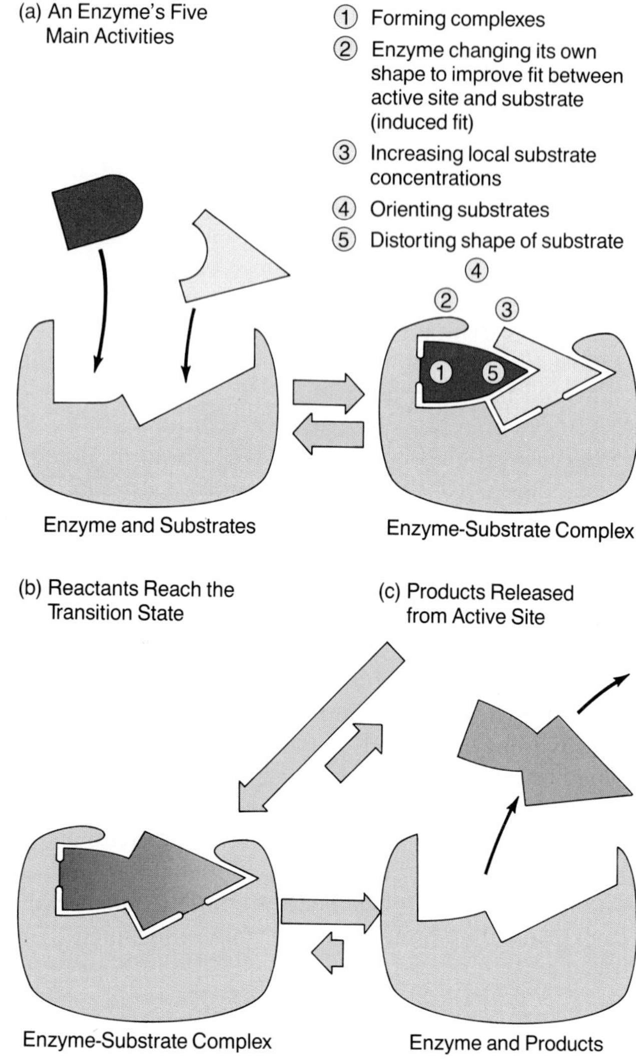

Figure 4-10 HOW AN ENZYME FUNCTIONS: FIVE ACTIVITIES THAT HELP CATALYZE REACTIONS.
(a) Once substrates collide with and bind to an enzyme's active site, the enzyme can perform the five functions listed and depicted here. (b) With bonds distorted, the reactants can reach the transition state, products can form, and these can be released from the active site, (c) leaving that site unchanged and ready to catalyze a new reaction.

right orientation so as to lower the activation energy barrier and facilitate the reaction between them, leading to the formation of products (Figure 4-10a, step 1, and b).

The number of weak bonds formed between substrate and enzyme determines the stability of the binding between a substrate molecule and an active site. Because weak bonds can form only when two atoms are very close together, the active site and the substrate must fit together well if such bonds are to form.

Change in Enzyme Shape

As the substrate binds to the active site, that site sometimes changes shape, allowing more weak bonds to form between enzyme and substrate. This change in shape improves the fit between active site and substrate, thereby increasing the binding *affinity*, or strength with which the substrate is bound to the enzyme. Such an adjustment in the shape of the enzyme is called **induced fit** (Figure 4-10a, step 2).

Increasing the Concentration of Substrate

Once the ES has formed, catalysis can take place, and the reaction can proceed. Several factors affect the reaction. First, the binding of two substrates to an enzyme brings them in close proximity to each other. In effect, then, enzymes greatly increase the effective *concentrations* of the substrates and thus the reaction rate itself (Figure 4-10a, step 3). For example, two substrates present in solution at concentrations of 10^{-4} M (molar concentration) may have effective concentrations of 4 M, or 40,000 times greater within the active sites of enzymes than in the solution. This is because the volume around and in the active site is so minute that the presence of substrate(s) in it raises the local concentration to a very high level.

Orienting the Position of Reactants

As mentioned earlier, the formation of an enzyme-substrate complex effectively orients the substrate molecules so that the reaction may proceed readily (Figure 4-10a, step 4). The position of the reactants is important because bonds can break and re-form only at specific regions of the substrates. For the reaction to occur, these regions must lie very close together.

Distorting the Shapes of Substrates

The formation of the enzyme-substrate complex (which, as we saw, sometimes changes the shape of the enzyme's active site) has a yet another important effect: it may distort the three-dimensional shapes of the substrates (Figure 4-10a, step 5). The formation of weak bonds between enzyme and substrates may strain the substrates, and this strain on their geometry alters the distribution of their orbiting electrons. This in turn reduces the energy level of the transition state—the peak of the activation energy hill—so that less energy is required to cause reactants to cross the barrier (Figure 4-10b). The rate of reaction, therefore, is greatly increased.

Completing the Catalysis

After the substrates have reacted, the product of the catalyzed reaction is released (Figure 4-10c), and the enzyme can reassume its initial conformation (if it has changed shape) and thus be ready to catalyze another reaction. This entire process can take place thousands of times a second. For example, the enzyme in red blood cells, carbonic anhydrase, can catalyze the reaction $CO_2 + H_2O \rightarrow H_2CO_3$ more than *600,000 times a second!* The same number of reactions would take many minutes without a catalyst. Carbonic anhydrase is exceptionally fast; most enzymes catalyze between 1 and 10,000 reactions a second.

Lysozyme: An Enzyme in Action

Figure 4-11 shows and explains how an individual enzyme, *lysozyme*, helps guard mammals against nose and eye infections and protects the embryo inside a bird or reptile egg. Lysozyme does this by hydrolyzing an essential polysaccharide component of bacterial cell walls, thereby weakening the invader's cell wall and causing it to lyse, or fall apart (hence the name *lysozyme*).

Lysozyme is a roughly spherical protein made up of a polypeptide chain of 125 amino acids (Figure 4-11a). The active site is a cleft long enough to accommodate six sugar units of the bacterial polysaccharide (Figure 4-11b). The sugars are held in the active site by hydrogen bonds. Each of six subsites within the active site corresponds to one of the six sugars, called A, B, C, D, E, and F. Only the covalent bond linking sugars D and E

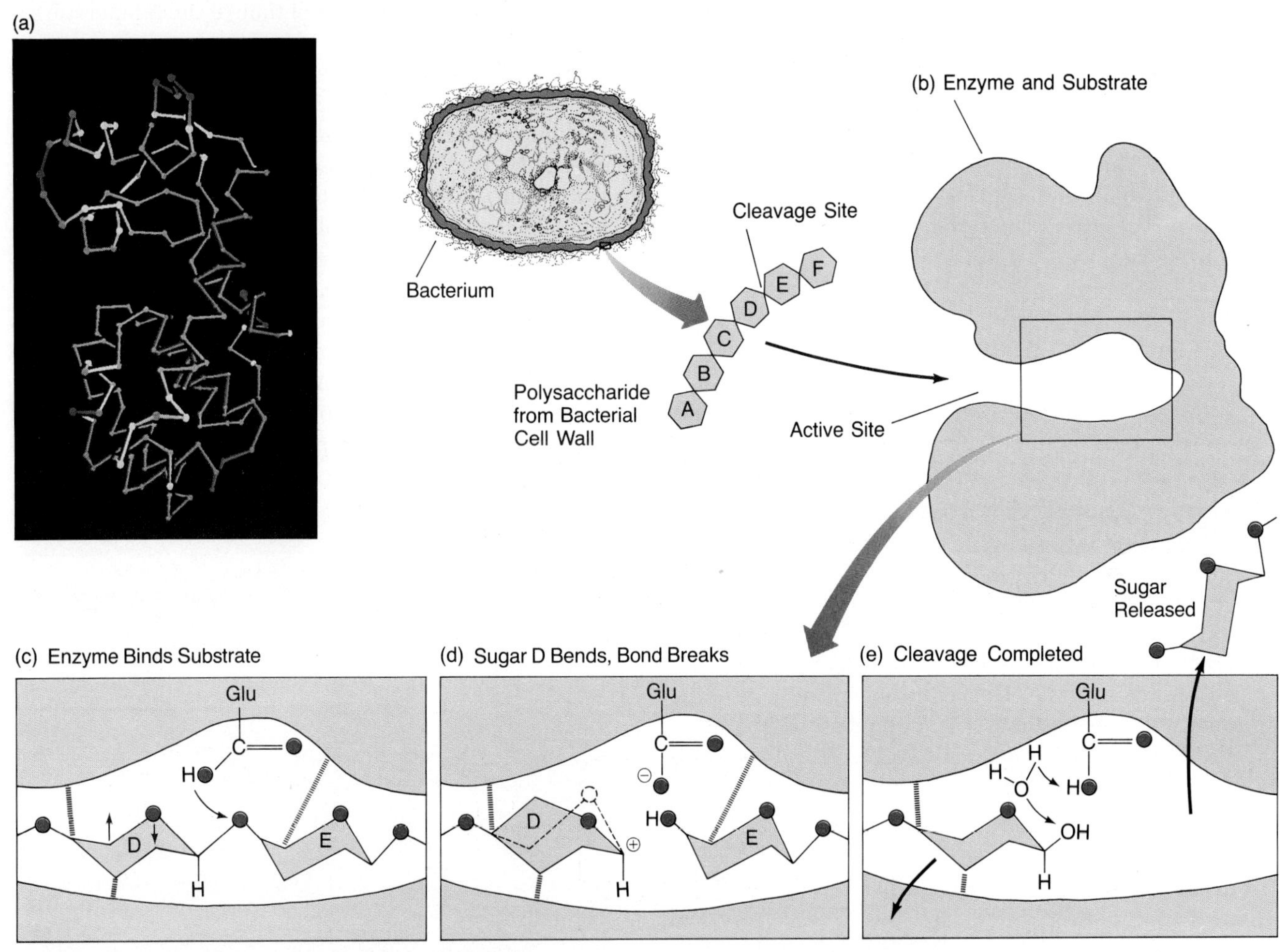

Figure 4-11 LYSOZYME: AN ENZYME IN ACTION.
Lysozyme protects animals from invading bacteria by dismantling a sugar chain in the bacterium's cell wall. (a) A computer-simulated model of lysozyme, showing the active site. (b) A drawing of lysozyme next to an invading bacterium reveals the enzyme's globular outer shape, the groovelike active site, and the chain of six hexagonal sugars from the bacterial cell wall that fits into the groove. Note the cleavage site between sugars D and E. (c) Sugar D has a chairlike shape as it begins to bind to lysozyme's active site. (d) However, the formation of weak hydrogen bonds causes that configuration to bend to a half chair or sofa shape. This distorted shape represents the transition state. By causing sugar D of the substrate to bend, the enzyme lowers the activation energy of the reaction, which can then proceed. Finally, a glutamic acid side chain (glu) in the enzyme's active site provides a hydrogen atom. This assists in the cleavage of sugars D and E. (e) The ions H^+ and OH^-, split from H_2O, neutralize the charged D sugar and the glutamic acid, and the cleavage is then completed. When enough sugar chains are broken in this way, the bacterial cell is lysed, and the invader dies.

is cleaved during the reaction (see Figure 4-11b). Therefore, the amino acid groups necessary to catalyze this particular reaction must be near subsites D and E on the enzyme molecule. Measurements show that sugar D is bound less tightly to the enzyme than are the other sugars. Sugar D, which has a ring shape formed by five carbon atoms and one oxygen atom, normally assumes a chairlike shape (Figure 4-11c). To fit into the active site of lysozyme, this ring must bend into a new configuration (Figure 4-11d); the distortion occurs because weak bonds form between the active site and the sugar molecule. The bent shape, called a "sofa" or "half chair" configuration, represents the transition state that surmounts the activation energy barrier. At this point, a glutamic acid side chain from the enzyme provides a hydrogen as the bond between sugars D and E is broken (see Figure 4-11d). This leaves charged intermediates. Sugars E and F then diffuse away from the enzyme, and water provides H^+ and OH^- to neutralize the charged forms (Figure 4-11e). This completes the cleavage of the substrate. When the lysozyme enzyme breaks enough of these bonds in the bacterial cell wall, the wall falls apart and the invader dies.

Enzymes and Reaction Rates

Although an enzyme like lysozyme can effectively lower the activation energy of a reaction, it cannot change an endergonic reaction into an exergonic one. It cannot, in other words, bring about reactions that are energetically unfavorable, nor does it change the concentrations of reactants and products at equilibrium. Moreover, an enzyme catalyzes the forward and reverse reactions to the same degree.

Among reactions that *are* energetically favorable, reaction speed can vary, depending on the concentrations of enzyme and substrate and depending on temperature.

The reaction rate is directly proportional to the amount of enzyme present. For example, if we have enough of the enzyme sucrase to catalyze the hydrolysis of 2.1×10^{-3} mole of sucrose in 10 minutes, then twice that amount of sucrase will break down twice as much sucrose in the same amount of time (if the temperature is held constant).

As with uncatalyzed reactions, temperature affects the speed of enzyme-mediated reactions. Because an enzyme reduces but does not eliminate the activation energy barrier, the additional energy provided by elevated temperature can further speed up a reaction. Figure 4-12 shows the effect of both increasing temperature and amount of functional enzyme on the rate of a catalyzed reaction (e.g., the reactions in yeast cells that produce CO_2 gas bubbles and cause dough to expand) as well as the effect of temperature and amount of enzyme on uncatalyzed reactions. As you can see from curve (a), temperature alone will speed the rate of uncatalyzed reactions. Increased temperature will also speed an

LUCIFERASE: A LUMINOUS ENZYME

The greenish yellow glow of a firefly on a summer night is produced by a fascinating enzyme called luciferase, which catalyzes the breakdown of a small molecule called luciferin. During this reaction, most of the energy is released as light rather than heat—hence the pleasant glow. Colonies of bacteria that contain a similar light-releasing enzyme inhabit four species of "flashlight fishes"—deep-sea fishes with glowing pockets directly beneath their eyes that act as beacons in the inky reaches of the deep ocean (Figure A).

Luciferase acts on luciferin only if energy is present in the form of adenosine triphosphate (ATP) molecules. This universal cellular fuel, which we will discuss in later chapters, provides the energy needed to drive the endergonic reaction

$$\text{Luciferin} \xrightarrow[\text{Luciferase}]{\text{ATP}} \text{Light} + \text{Heat}$$

By adding luciferase and luciferin to blood stored in blood banks, researchers can monitor useful shelf life: Red blood cells begin to degenerate and leak ATP molecules, these activate luciferase, and the blood begins to glow in the dark. Luciferase and similar enzymes are truly illuminating tools for biologists, as well as headlights for fish and taillights for fireflies.

Figure A
This flashlight fish has a pocket of luminous bacteria that lights its path through the darkness of the deep sea.

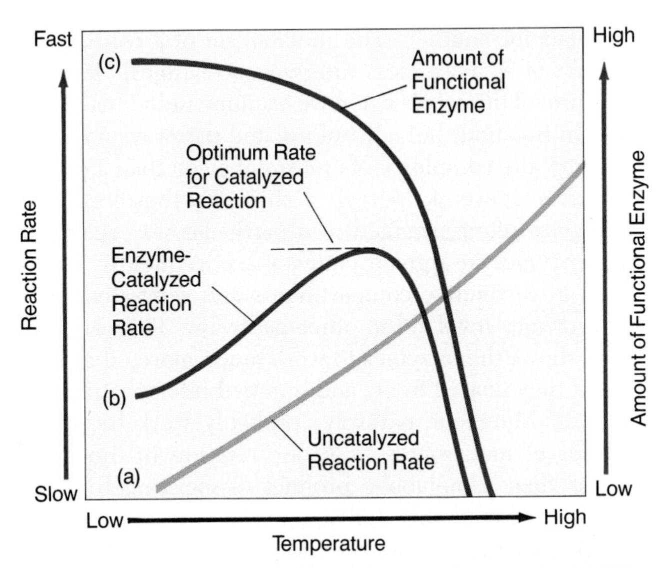

Figure 4-12 HOW TEMPERATURE AFFECTS THE RATES OF UNCATALYZED AND ENZYME-CATALYZED REACTIONS. (a) Increasing temperature will speed the rate of an uncatalyzed reaction in a linear way (blue curve). (b) Increasing temperature will also speed an enzyme-catalyzed reaction (red curve). Because an enzyme functions best within an optimal temperature range, however, the curve increases only to a point, then drops off as the rising heat denatures the enzyme's shape. (c) The purple curve reflects the influence of heat on enzymes. As temperature rises and the enzyme becomes denatured, the amount of functioning enzyme available to catalyze reactions declines and mimics the downward slope of the curve for reaction rate.

enzyme-catalyzed reaction curve (b), but only to a point. Many enzymes have an optimal temperature at which they can bind substrate, catalyze a reaction, and release the products most rapidly. At higher and lower temperatures, weak chemical bonds may be altered, and the enzyme's shape may be permanently altered to a nonfunctional shape.

Finally, increasing the concentration of a substrate will speed the rate of an enzyme-mediated reaction, but again, only to a point. If, for example, we put a given amount of enzyme in a solution and begin adding substrate, the reaction will go faster and faster in direct proportion to the amount of substrate added (Figure 4-12, curve c). But at some point, the rate will no longer increase at all, and will instead remain constant. Why? At the leveling-off point, so much substrate is present that the active site in every one of the enzyme molecules is occupied most of the time, and the enzyme is said to be *saturated* (Figure 4-13). No matter how much more substrate is added, the reaction cannot go faster. But if we add more of the enzyme, the rate of reaction again rises. The phenomenon of **enzyme saturation** sets enzyme-mediated reactions apart from noncatalyzed reactions, whose rates continue to increase with increasing concentrations of reactants.

The speed with which enzymes form ES complexes, catalyze specific reactions, then release products is critical to the life process. To appreciate how important it is, we have only to consider a calculation made by biologist David Kirk. He determined the enzyme urease breaks down in 1 second the same amount of urea that would hydrolyze spontaneously in 3 million years. No bacterium, fungus, plant, or animal could maintain itself and win the battle against entropy if its enzymes did not work so astonishingly fast to lower the activation energies of countless reactions.

METABOLIC PATHWAYS

Thousands of chemical reactions, all going on at once, support the activities that characterize and sustain life: growth, development, energy use, and responsiveness, among others (see Chapter 1). These chemical reactions take place in orderly, interrelated patterns that are con-

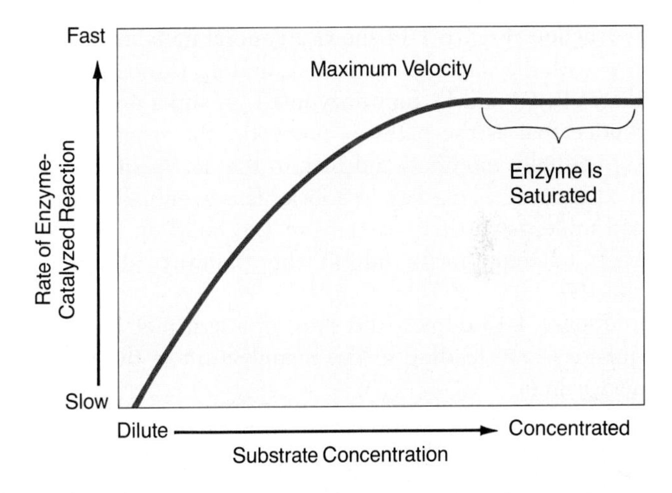

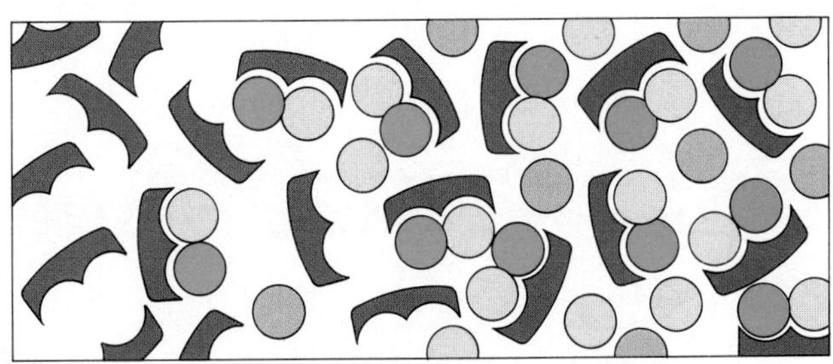

Figure 4-13 ENZYME SATURATION: ANOTHER MODULATOR OF REACTION RATE. Enzyme saturation limits reaction rates. An enzyme is "saturated" when the active sites of all the enzyme molecules are occupied most of the time. At the saturation point, the reaction will not speed up, no matter how much additional substrate is added to a solution.

trolled by enzymes. The combination of simultaneous, interrelated chemical reactions taking place at any given time in a cell is referred to as **metabolism.**

We have seen that endergonic reactions require energy, while exergonic ones release it. In cells, most energy-requiring reactions are involved in synthesizing needed biological molecules—amino acids, nucleic acids, fats, proteins, and carbohydrates. Together, these biosynthetic reactions are called **anabolism** (meaning "to build up"). Conversely, most energy-yielding reactions in cells break down molecules to obtain building blocks, release energy, or digest waste products. Together, these degradative reactions are called **catabolism** ("to tear down"). Energy-requiring anabolic reactions are often coupled to energy-releasing catabolic ones.

While our discussion of enzymes has focused on the way each catalyzes a particular reaction, many enzymes carry out the individual steps in long chains of reactions. As a result of this sequential enzyme activity, various compounds are progressively built up or broken down. These series of reactions are called **metabolic pathways.**

At each step of a metabolic pathway, an enzyme catalyzes a reaction that changes the starting material a little bit more—it adds a phosphate group, let's say, or removes an –OH group until eventually the final product is reached. Figure 4-14 shows a general metabolic pathway with the starting materials A_1 and A_2; three intermediates B, C, and D; four enzymes 1–4; and a product P. Notice that as the pathway proceeds, the reactants are sequentially modified and fit into the active site of the next enzyme like a key in a lock. The eventual product can undergo further reactions or can build up and feed back to temporarily inhibit the pathway (described shortly).

Figure 4-15 depicts the steps of a real metabolic sequence—one leading to the manufacture of the amino acid valine.

Metabolic pathways are interconnected; the products or intermediates of one pathway may be the starting materials for another. The metabolism of a cell can be thought of as a seamless interwoven chemical "fabric." We unravel this fabric when we examine individual pathways in isolation, but a list of isolated pathways no more indicates the complex metabolism of a cell than a pile of threads suggests a dress or a shirt. Furthermore, enzymes are often arranged in a particular way spatially: The enzymes in a given pathway are frequently segregated in distinctive compartments and separated from the enzymes involved in other pathways. This arrangement allows the enzymes to work much more efficiently than if they floated freely and bumped into substrates at random. Metabolic pathways probably work by direct transfer of metabolites from one enzyme to the next. Rather than a metabolic product dissociating from its enzyme and randomly diffusing to the next enzyme's active site, enzyme-to-enzyme complexes form and pass on the metabolite directly. This "bucket brigade" prevents large quantities of each intermediate in the pathway from building up. Chapters 7 and 8 will explore in much more depth the arrangement of enzymes in metabolic pathways.

Control of Enzymes and Metabolic Pathways

The fact that enzymes can be controlled is not only a distinguishing feature of these catalysts, but is crucial to their activity. Enzymes and metabolic pathways must be regulated if a cell is to function as an integrated whole and to have energy supplies and raw materials on hand when they are needed.

Cellular metabolism is subject to both external and internal controls. External controls include hormones—molecules produced in one part of the organism that can regulate enzyme activities within cells elsewhere in the body (see Chapters 30 and 35). Internal control mechanisms, such as the inhibitory feedback loop pictured in

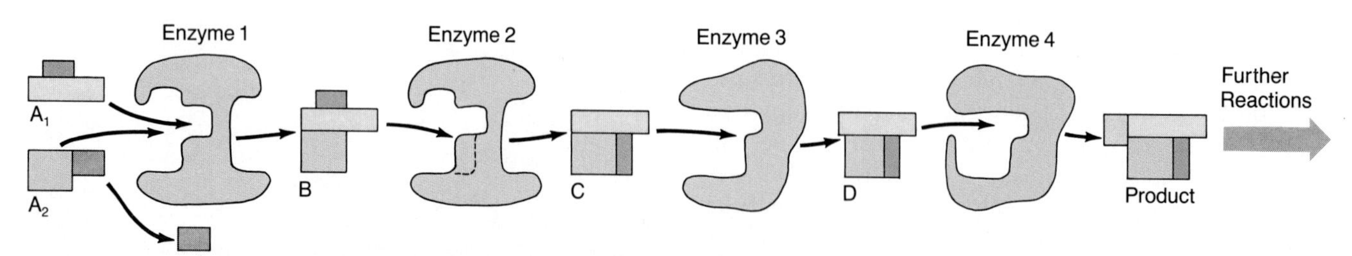

Figure 4-14 A GENERALIZED BIOSYNTHETIC PATHWAY AND ITS CONTROL BY FEEDBACK INHIBITION. Enzyme 1 catalyzes a reaction between starting materials A_1 and A_2. Enzyme 2 acts on B, the product of the first reaction, generating C. Enzyme 3 acts on C, further modifying its shape to D. And enzyme 4 adds a final tail to D, creating the product P. This can be incorporated into newly synthesized proteins, or it can build up in the cell. But if the product builds up too fast, it can feed back and inhibit the further activity of enzyme 1, temporarily blocking the pathway.

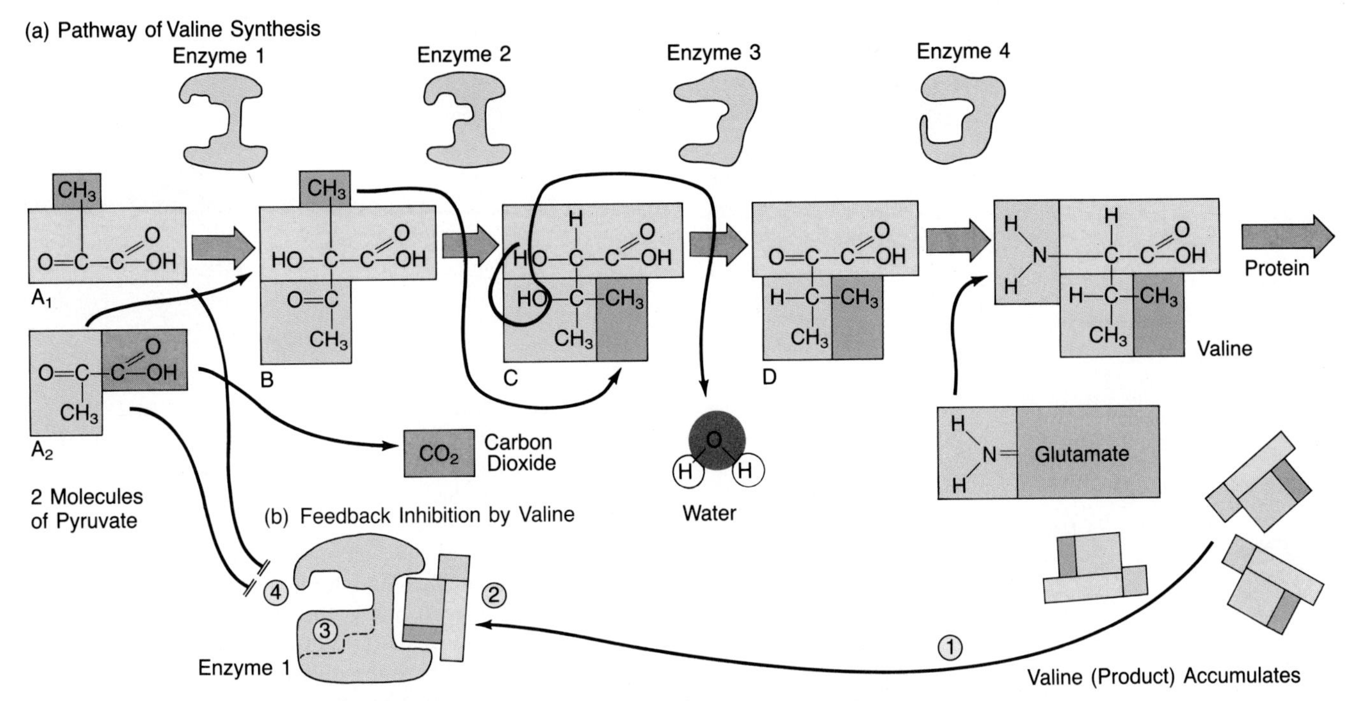

Figure 4-15 A SPECIFIC BIOSYNTHETIC PATHWAY: MANUFACTURING THE AMINO ACID VALINE.
When a cell makes valine, a four-step pathway leads from the raw material pyruvate to the finished amino acid. (a) Enzyme 1 catalyzes the joining of two pyruvate molecules. This step is called a decarboxylation, because CO_2 is removed. Enzyme 2 brings about a methyl migration, moving a methyl group from the top of B to the side. This requires the expenditure of an NADH molecule, which donates two hydrogens. Enzyme 3 catalyzes a dehydration—the removal of a water molecule from C. Finally, enzyme 4 causes an amino transfer; that is, it transfers an amino group from another animo acid, glutamate, to substrate D and forms the product valine. This can be incorporated into the polypeptide chain of a new protein, or it can build up and feed back to inhibit the activity of enzyme 1. (b) When the cell is producing valine faster than it incorporates the amino acid into proteins, valine diffuses through the cell cytoplasm (1) and binds to a regulatory site, the allosteric site of enzyme 1 (2). This distorts the shape of enzyme 1 (3), making it harder for the enzyme to function properly and thus inhibiting production of the next intermediate in the pathway (4). As a result, the rate of valine synthesis falls, and remains low until the cell uses its store of the amino acid.

Figure 4-14, operate within individual cells. Let us return to the synthesis of valine to see how one such internal feedback circuit controls the rate at which the cell produces valine (see Figure 4-15b).

It is advantageous for a cell to produce valine and other amino acids at the same rate as those building blocks are used up and incorporated into new proteins. Failure to control this rate would result in the stockpiling of unneeded amino acids or in a slowdown of protein synthesis because raw materials are in short supply. The system of regulation called **feedback inhibition** prevents either occurrence. Feedback inhibition allows the product of a biosynthetic pathway to inhibit the enzyme that catalyzes the first step in that pathway. A buildup of the amino acid valine, for example, reduces the rate at which this substance is synthesized, while a depletion of the amino acid removes the inhibition and increases the rate of its own production (see Figure 4-15b). Feedback inhibition is the basic means of internal control within cells, allowing the pathway product to feed back and control the rate of its own synthesis.

As in the valine pathway, feedback inhibition usually operates on the first enzyme in the pathway. Limiting amino acid synthesis at this point prevents the wasteful or dangerous buildup of intermediates that would occur if enzymes further down the pathway were the primary control targets. It is also most economical, because it prevents any waste of energy or of substrates along the pathway.

How can a small molecule like an amino acid control an enzyme's activity? Experiments show that the amino acid binds to the enzyme at a site other than the active site—a so-called *allosteric* binding site. The binding of a substance to an allosteric binding site causes the enzyme to change to a shape that is not compatible with active-site functioning, and so the enzyme shuts down (Figures

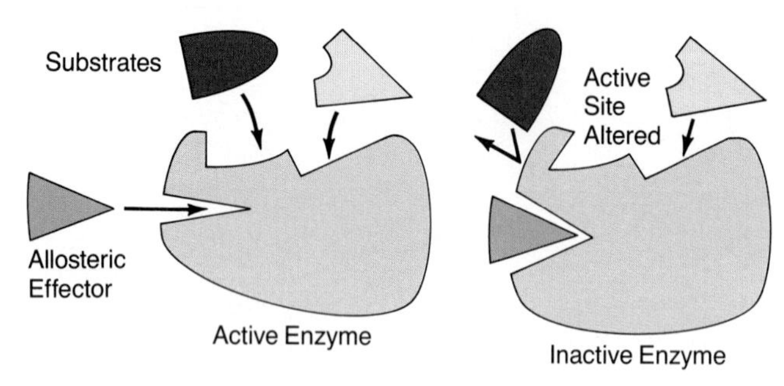

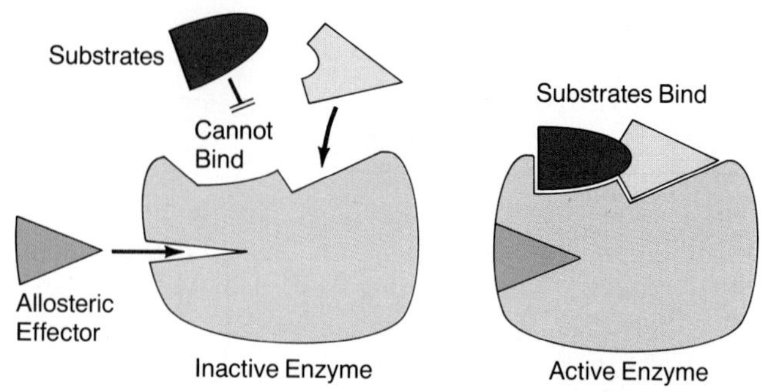

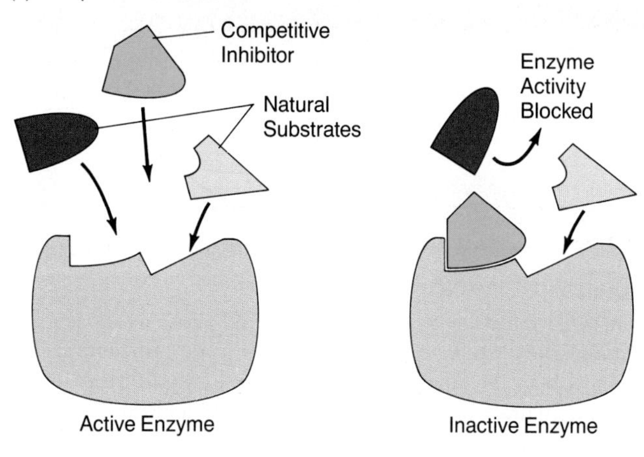

Figure 4-16 THE ACTIVITY OF ALLOSTERIC ENZYMES In addition to its normal active site, which binds substrate, an allosteric enzyme has another binding site for an allosteric effector, such as an amino acid. (a) When the effector is not bound to the enzyme, the active site and the substrate have complementary shapes, so binding and catalysis take place. But the binding of the effector to its site induces a shape change in the enzyme, distorting the active site so that it binds the substrate less tightly or not at all. This process is noncompetitive inhibition. (b) The binding of an allosteric effector does not always inhibit enzyme function. In some allosteric enzymes, an allosteric effector may induce a shape change in the active site that enhances the binding of the substrate and makes an inactive enzyme active. (c) A competitive inhibitor is shaped somewhat like the enzyme's normal substrate, with which it competes for binding in the active site. Cells do not regulate metabolism by means of competitive inhibition, but researchers apply the mechanism extensively in the design of drugs and medical treatments.

4-15b and 4-16a). This shape change is reversible, however, and the shutdown can be temporary; when the amino acid leaves the allosteric site, the enzyme reverts to its native shape, compatible with normal functioning. As the reaction pathway functions and more of the amino acid builds up, the excess product molecules will bind to allosteric sites on the first enzyme in the pathway, thereby inhibiting the enzyme and causing the pathway to slow and to come to a halt.

Enzymes whose activities are controlled in this way are called **allosteric enzymes** (meaning "other shape"), and the small molecule, such as a pathway product or a hormone, that binds to the allosteric binding site is called an *allosteric effector*. It should not be surprising to learn that if the binding of an allosteric effector can inhibit enzyme function, it also can activate an enzyme (Figure 4-16b). Some allosteric enzymes are inactive prior to the binding of the control substance, or activator.

Allosteric inhibition is part of a broader category of control called **noncompetitive inhibition.** In another type of enzyme control, called **competitive inhibition,** regulation involves the active site itself. A compound other than the normal substrate comes to occupy that active site and prevents binding of that substrate (Figure 4-16c). However, an increase in the concentration of the normal substrate can enable it to compete successfully for the active sites. Biologists have investigated many cases of competitive inhibition, and it seems likely that cells employ the "fine tuning" of noncompetitive inhibition much more often as a means of regulating metabolism.

LOOKING AHEAD

This chapter has shown how chemical reactions underlie the continuous changes associated with life and how these reactions depend on a favorable energy flow. While alive, organisms are the epitome of order, and their internal metabolic reactions are complex, controlled, and interrelated processes. Chapters 5 and 6 explore the orderly micro-universe within the living cell, the fundamental unit of life on earth and the site of life-sustaining energy transformations.

SUMMARY

1. Biological activities are based on *chemical reactions*—transformations of sets of molecules involving the making and breaking of chemical bonds.

2. *Energetics* refers to the study of energy changes during chemical reactions. Energy may be *kinetic* (energy of motion) or *potential* (stored energy). Energy in atoms, bonds, and molecules is called *chemical energy*.

3. The *first law of thermodynamics* says that energy can change form, but it can never be created or destroyed. The *second law of thermodynamics* says that during energy transformations, some usable energy is lost as heat, with an increase in *entropy*, or disorder.

4. *Free energy* (*G*) is a compound's usable energy, that is, the heat content of its bonds corrected for entropy and temperature. During an *exergonic* reaction, energy is released and ΔG is a negative number. During an *endergonic* reaction, energy input is required and ΔG is a positive number.

5. When a reaction is at equilibrium, $\Delta G = 0$. At equilibrium, there is no longer net conversion of reactants to products or products to reactants, though reactions in both directions continue to occur at the same rate.

6. For covalent bonds to break and for a reaction to take place, colliding molecules must achieve a minimum amount of energy, called the *activation energy*. At the top of the energy hill, activated molecules reach a transition state, and new bonds can begin to form.

7. The rate of a chemical reaction can be affected by temperature, concentrations of reactants, concentrations of products, and the presence of *catalysts*.

8. In living cells, enzymes serve as catalysts by reducing the activation energy necessary for biochemical reactions to take place. A given enzyme is specific: It can act on only one kind of compound or pair of compounds, called *substrates*, and it can catalyze only one type of reaction.

9. Each enzyme has an *active site*, a groove, or pocket, whose shape is reciprocal to the shape of a specific substrate(s). Factors affecting the weak bonds that maintain an enzyme's three-dimensional shape can alter its activity. In some enzymes, *cofactors* take part in the catalyzed reactions.

10. The first step in enzyme function is the formation of an *enzyme-substrate (ES) complex*. The active site sometimes changes shape in such a way that more weak bonds can form between enzyme and substrate; such a change is called *induced fit*. The binding of substrates to an enzyme can increase the effective concentrations of the substrates, orient the reacting substrate molecules, or change the shape of the substrate molecules. All such changes act to lower the activation energy.

11. Three factors—temperature, concentration of enzyme, and concentrations of substrates—can affect the rates of enzyme-catalyzed reactions. If enzymes become *saturated* with substrate the reactions show no further rate increase.

12. The combination of chemical reactions that take place in a cell is called *metabolism*. Biosynthetic reactions together make up *anabolism;* degradative and energy-yielding reactions make up *catabolism*. Series of reactions are called *metabolic pathways*.

13. Both external agents, such as hormones, and internal agents present within the cell can control metabolic pathways. Most internal control is by means of *feedback inhibition*, wherein the buildup of a product inhibits its further production. In *allosteric enzymes*, the regulatory agent binds to a site other than the enzyme's active site, changing the protein's shape and either activating or inhibiting enzyme function. This is a type of *noncompetitive inhibition*. Other types of enzymes exhibit *competitive inhibition*, wherein a substance binds to the active site and prohibits the binding of the enzyme's normal substrates.

KEY TERMS

activation energy
active site
allosteric enzyme
anabolism
catabolism
catalyst
chemical energy
cofactor
competitive inhibition
coupled
endergonic reaction
entropy
enzyme saturation
enzyme-substrate (ES) complex
equilibrium (chemical)
exergonic reaction
feedback inhibition
first law of thermodynamics
induced fit
kinetic energy
metabolic pathway
metabolism
noncompetitive inhibition
potential energy
second law of thermodynamics
substrate

QUESTIONS

1. Distinguish between potential, kinetic, and chemical energies in rocks, dammed-up water, and organic compounds such as sugars.

2. Use the second law of thermodynamics to explain why some energy is lost during every chemical reaction of every living cell. How does that loss relate to entropy?

3. How do exergonic and endergonic reactions differ, and why are the two types so frequently paired in living cells?

4. What is the activation energy of a

molecule? What sorts of things can affect activation energy and allow a reaction to take place? Which is most important in living cells?

5. Define enzyme, indicate the unique properties of enzyme-catalyzed reactions, and explain how an enzyme such as lysozyme works.

6. Metabolism = Anabolism + Catabolism. Explain.

7. Describe metabolic pathways and the ways that they can be controlled. What exactly is feedback inhibition?

ESSAY QUESTIONS

1. How are the terms *active site*, *induced fit*, and *ES complex* related?

2. Why is life an "oasis of order in a desert of disorder" in the universe? Does a living cell contradict the second law of thermodynamics? Why or why not?

For additional readings related to topics in this chapter, see Appendix C.

5
Cells: Their Properties, Surfaces, and Interconnections

The living cell is to biology what the electron and the proton are to physics. Apart from cells and from aggregates of cells there are no biological phenomena.

Alfred North Whitehead,
Science and the Modern World (1925)

In 1665, Robert Hooke, a 30-year-old physicist and amateur botanist, gazed in fascination through the lens of a primitive microscope. His subject was a thin sheet of tissue sliced from dried cork. The tiny, hollow honeycomb chambers he saw through the microscope reminded him of the small rooms, or cells, that monks inhabited in monasteries.

Other microscopists who later examined living plant and animal tissues found that every creature they studied consisted of what Hooke had called cells. Some organisms, such as bacteria, amoebae, and certain algae, are free-living single cells. However, most kinds of living things, from tiny freshwater colonies of plant cells to massive whales and redwoods, are composed of many cells that function together as a coordinated population.

In this and the next chapter, we move from the submicroscopic realm of atoms, molecules, and their chemical reactions to the basic units of life on earth—the mo-

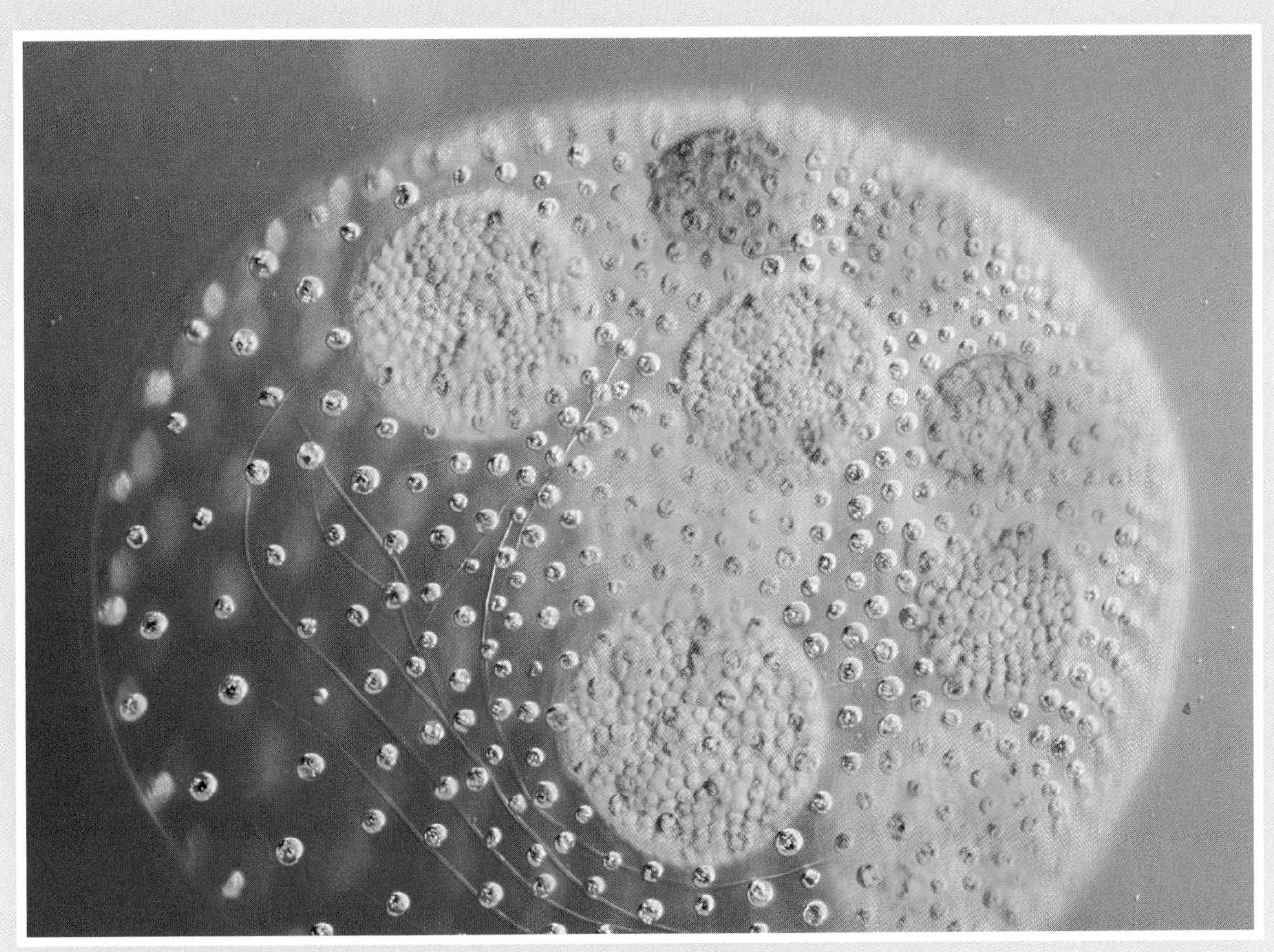

The borderline of multicellularity: Like miniature plants suspended in space, individual cells in the *Volvox* colony have minimal interaction with neighbors. The larger multifaceted green spheres inside are reproductive stages.

lecular "municipalities" known as **cells.** Our main topics in this chapter include:

- How biologists study cells
- The characteristics of cells, including their small size and general components
- The nature of the cell surface and passage of materials through it
- How cells are connected to and communicate with each other

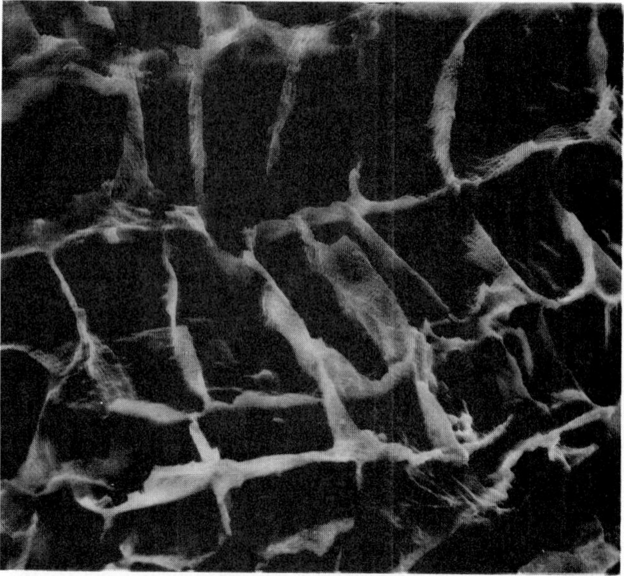

Figure 5-1 RIGID WALLS OF PLANT CELLS. This specimen, magnified about 600 times, reveals the compartments like those Anton van Leeuwenhoek saw in a thin slice of cork and which Robert Hooke called "cells." Modern scholars have discovered some of van Leeuwenhoek's well-preserved specimens, including algal filaments, cotton seeds, and the optic nerve of a cow.

HOW CELLS ARE STUDIED

Most cells are extremely small, and so it is not surprising that the discovery of these basic units of life followed closely after the invention of the microscope. Because sight is our primary sense, biologists still rely heavily on microscopes to detect cell structures. But as we shall see, biologists also employ many newer devices to take cells apart and learn how they function.

Microscopy

Much of what is known about cells was discovered by biologists peering into microscopes, just as Hooke and Anton van Leeuwenhoek, the seventeenth-century Dutch "father of the microscope," did so long ago (Figure 5-1). The history of cell biology parallels the invention of and improvements in microscopes, just as astronomy has advanced with the new and different types of telescopes.

The microscopes that Hooke, Leeuwenhoek, and other pioneers in cell biology employed were *light microscopes*, and modern versions are still used extensively in biological research, although not to magnify objects more than about 1,200 times normal size. This is because of the limitations of resolving power—the ability of the human eye to distinguish adjacent objects. If

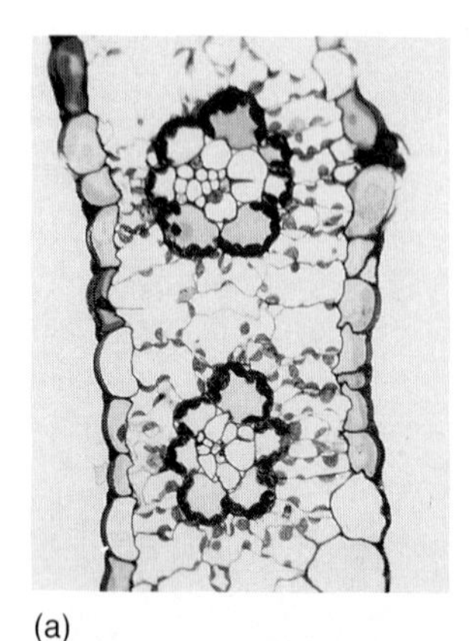
(a)

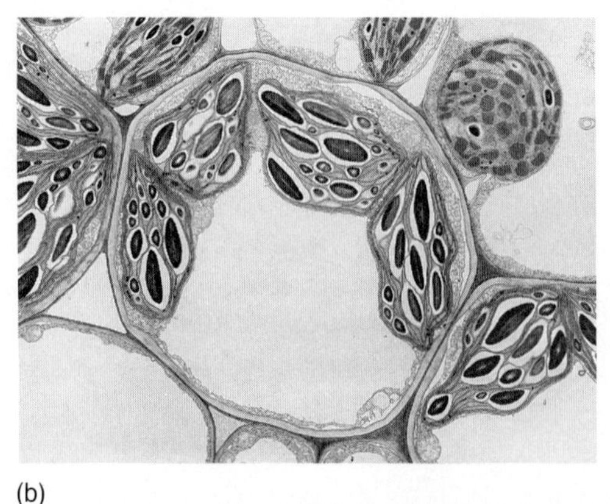
(b)

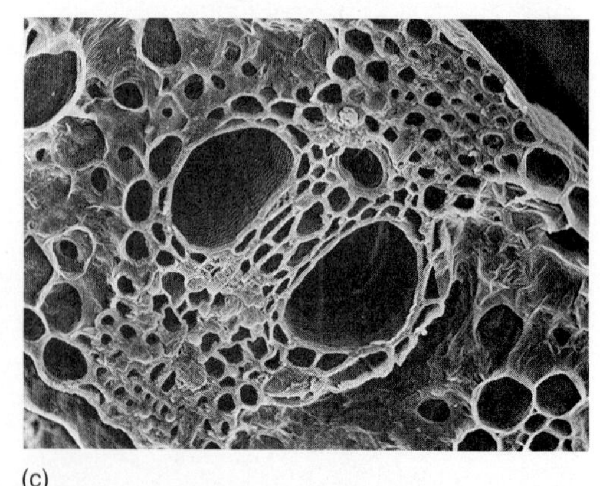
(c)

Figure 5-2 CLOSER AND CLOSER: CORN LEAF TISSUE VIEWED THROUGH DIFFERENT MICROSCOPES. The photograph in (a) shows a cross-section from a corn leaf as it looks through a light microscope (magnification 430×). Photograph (b) shows the starch granules and chloroplasts looking large and dark within a few corn leaf cells (16,700×). In (c) a scanning electron micrograph shows transport vessels as rigid, three-dimensional pipes and tunnels (740×).

MICROSCOPES FOR THE LAST CELLULAR FRONTIER

There is a revolution under way in the design and use of microscopes. Electron microscopes opened a new window on the cell beginning in the 1930s that led to spectacular insights. Biologists, however, would like to have even greater magnification, as well as instruments that could view cells in a more natural, three-dimensional state without harsh fixing procedures.

One relatively new machine called the *scanning tunneling microscope (STM)* can reveal individual atoms only 3 angstroms (12 billionths of an inch) across. It works on the principle of electron "tunneling," the tendency of electrons to jump between the tip of a fine metal needle and the surface to be studied. So far, scientists have used the STM to map the regular arrangement of atoms in benzene molecules (Figure A) and in crystals of silicon, graphite (carbon), and gold. But because the sample can be immersed in water, biologists hope to apply the STM to intact DNA molecules, cell membranes, and other biological targets.

A second instrument, called the *tandem scanning reflected-light microscope (TSRLM)*, uses a "confocal system": It illuminates and observes only a given slice or plane within a cell. This prevents shadows and glare from other structures in other planes from obscuring the particular organelle of interest and allows biologists to see the shapes and activities of very small organelles within living cells for the first time.

Still another microscope uses sound waves rather than light waves or beams of electrons and allows the scientist to detect density variations in different regions of the cytoplasm. Figure B shows two living fibroblast cells from a chicken's heart, moving about normally with nuclei, mitochondria, and small sacs, or vesicles, clearly visible.

Each new advance in microscopy, beginning with Hooke and van Leeuwenhoek, has brought new insights into the life of cells. These new instruments promise to do the same, but this time at the very limits of structure and function—the level of atoms, molecular assemblies, and organelles within healthy, active cells.

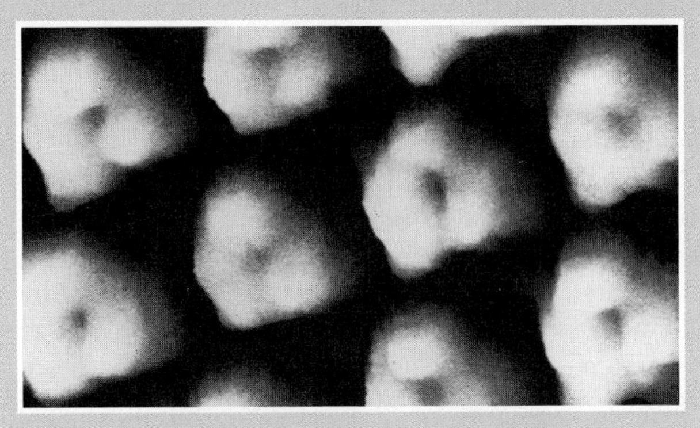

Figure A ATOM RINGS REVEALED. Individual carbon and hydrogen atoms show up clearly in benzene rings, photographed through a scanning tunneling microscope.

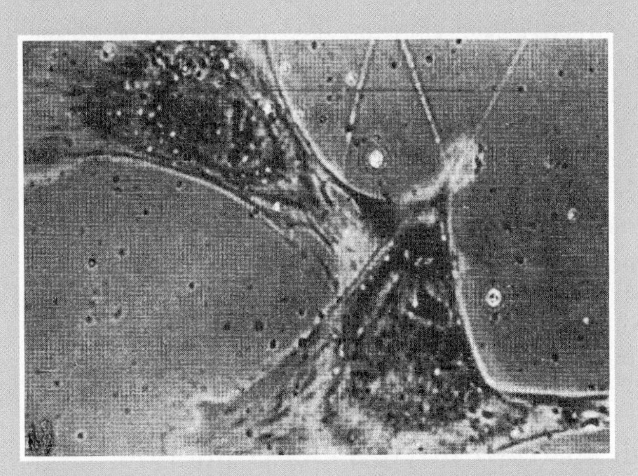

Figure B ORGANELLES IN LIVING CHICKEN HEART CELLS.

you move two black dots closer and closer together on a white background, they will eventually appear to have merged, because their proximity has gone below the resolving power of the unaided human eye (a distance of about 0.1 mm). Under a light microscope, two merged dots would be easily resolvable. The best light microscope is about 500 times better than the human eye at resolving dots, lines, or other small objects. Even so, resolving power is limited by the wavelength of visible light; the finest light microscope is limited to resolving objects spaced more than about 0.2 μm apart (Figure 5-2a).

Such limitations led scientists to design microscopes that use beams of electrons instead of light to irradiate specimens. These microscopes have a resolving power 100,000 times greater than that of the human eye. *Transmission electron microscopes (TEMs)* permit examination of a cell's interior (Figure 5-2b). They can send a beam of electrons through an ultrathin specimen slice and produce a negative image of the object on a small fluorescent screen.

Scanning electron microscopes (SEMs) bombard the surface of a specimen with electrons and allow scientists to see a three-dimensional image of the specimen's surface, with all its holes, ridges, spaces, and textures revealed (Figure 5-2c). The box on page 79 describes the latest generation of microscopes, which show cell structures and functions in even greater detail.

For a watery, translucent cell to be examined through a microscope and distinguished from the background, it must have contrast—distinct areas of light and dark. A biologist using a light microscope produces contrast in living cells by varying the light beam with prisms and other physical devices or by using dyes on dead cells. In electron microscopy, researchers stain specimens by soaking them in solutions of heavy metals, such as lead or uranium salts; the electrons will be absorbed more readily in denser areas of the object and thus look darker.

The electron gun and beam in a TEM or SEM must be operated in a high vacuum, and this necessitates special preparations. The operator normally fixes a specimen with preservative chemicals and then dries it. TEM specimens usually are embedded in a block of plastic and cut with an exceedingly sharp diamond into ultrathin sections (a pile of 1,000 such sections would be no thicker than this page). Alternatively, specimens may be quick-frozen and then literally cracked apart with a sharp blow. When coated with metal, these freeze-fracture preparations give good surface-relief views of cell parts. SEM specimens are fixed, dehydrated, and coated with a thin layer of gold (about 0.2 nm thick), which covers every nook and cranny of the surface and emits the electrons that yield a bright image of the specimen.

Means of Studying Cell Function

Viewing cells through a microscope is a good way to observe structures, but understanding a cell's living functions requires other instruments and techniques. One important method for determining function is *cell fractionation,* in which the experimenter grinds or breaks up cells and then analyzes their contents. Figure 5-3 shows how this is done with plant cells; in this case, a centrifuge separates cellular components from one another so that the researcher can, for example, study an enzyme that is active in one component (organelle) but not in others.

Another procedure is *radioactive isotope labeling* to trace ions, molecules, and individual chemical reactions in cells and organisms. For example, researchers make an amino acid with a radioactive isotope of carbon (^{14}C) or hydrogen (^{3}H) in place of its normal C and H atoms. Within minutes or even seconds of entering a cell, the labeled amino acid can be built into a new cellular protein, and the route the protein then follows can be traced by measuring the radioactivity of the various fractions.

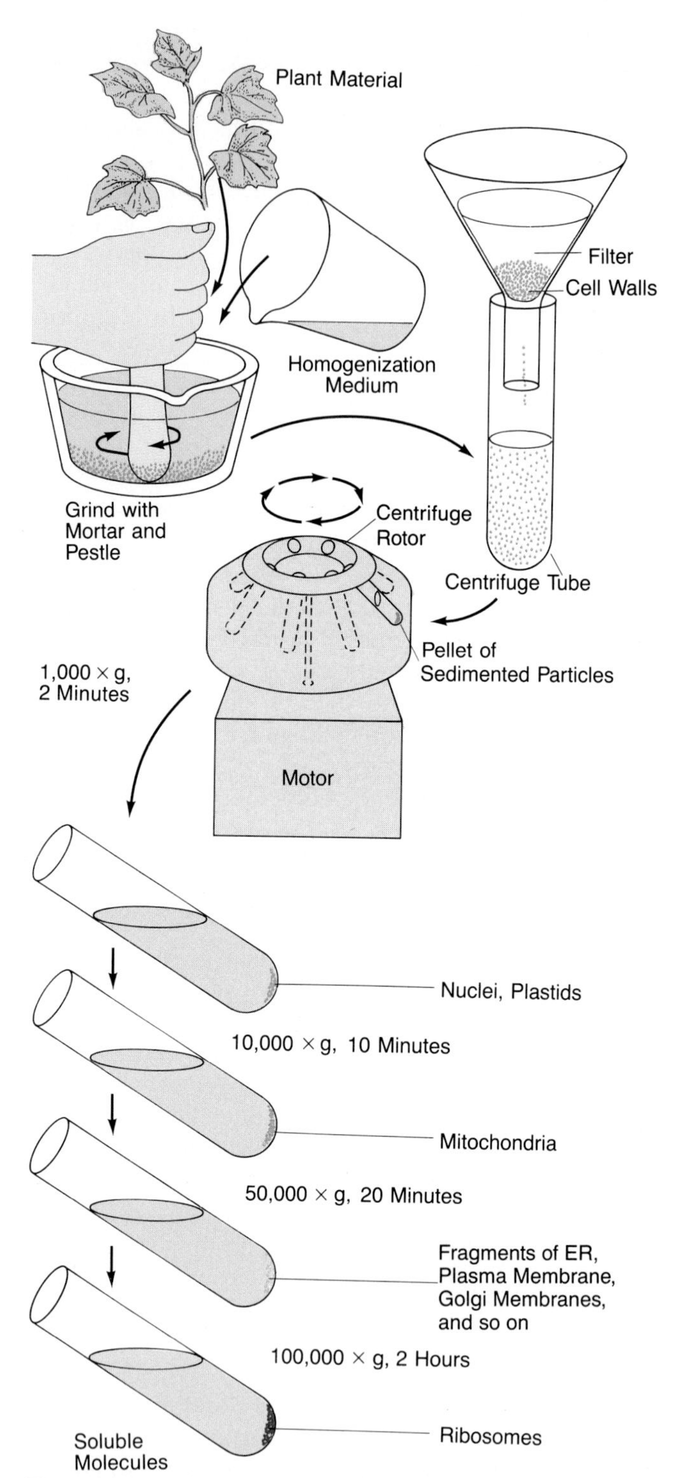

Figure 5-3 CELL FRACTIONATION TECHNIQUES: SEPARATING CELLULAR COMPONENTS.
In this case, plant leaves are homogenized and then filtered to remove the hard cell walls. Then the material is spun at progressively faster centrifugation speeds until a series of "pellets" forms. The label 1,000 × g signifies centrifugal force 1,000 times stronger than gravity. The first pellet contains the heaviest organelles, such as the nuclei; the last, high-speed run yields the smaller, lighter ribosomes.

Even more precise localization of the labeled, incorporated amino acid is possible using *autoradiography*, in which a researcher coats a thin section of the killed cell with a photographic emulsion exposes the emulsion to radioactive emission from the incorporated amino acid, just as light exposes the emulsion on photographic film; and locates the precise positions of the "hot" molecules.

CHARACTERISTICS OF CELLS

Perhaps the most important outcome of the three centuries of cell research that began with Robert Hooke is the modern **cell theory,** a set of statements that encapsulates the essential characteristics of cells. German biologists Lorenz Oken, Matthias Schleiden, Theodor Schwann, and Rudolf Virchow wove old ideas and observations about cells into a new synthesis, contributing the following ideas to the modern theory:

1. Cells are the basic units of life on earth. No organism has ever been found on earth that shows the attributes of life and yet is not composed of cells.
2. All organisms are constructed of cells. Every living thing on earth is either a single cell or a population of cells.
3. Except at the origin of life itself, all cells arise from preexisting cells. Cells arise only by division of living cells, never by aggregation of cell parts and cell chemicals ("from life comes life").

More recent research suggests two additions to the cell theory that apply to higher organisms and thus are not quite as general as the first three tenets:

4. Cells of multicellular organisms are sometimes interconnected, enabling the resultant populations to function as single units.
5. Cells of multicellular animals must stick to solid surfaces to divide, move, assume specialized shapes, and carry out necessary functions.

Like the theory of evolution by natural selection and the theory of the particulate nature of matter, the cell theory is a cornerstone of science. And just as the basic units of matter are atoms, the basic units of life are cells. But cells are not simply building blocks of life, as atoms are of molecules; the cells themselves *are what is alive in organisms*. The cell theory marked a major turning point for biology: By focusing on cells, biologists could begin to pose specific questions about how life operates and discover some profound answers.

The Nature and Diversity of Cells

Cells are the most highly ordered assemblages of dissimilar molecules on earth—perhaps in the universe. They can exist as discrete, free-living, single-celled species of bacteria, protozoa, and algae or as subunits making up the *tissues*, *organs*, and *organ systems* of multicellular organisms—the fungi, plants, and animals.

Types of Cells

There are thousands of kinds of cells, but all can be classified in two ways: (1) by fundamental elements of structure and (2) by the way they obtain energy.

Structurally speaking, cells are either **prokaryotes** or **eukaryotes.** The word *prokaryote* means "before nucleus," and it describes cells in which DNA is localized in a region but is not bounded by a membrane. In other words, a prokaryotic cell lacks a true, membrane-bound nucleus. All prokaryotes are independent, single-celled organisms, and they include the thousands of species of bacteria and cyanobacteria (blue-green algae).

A eukaryotic ("true nucleus") cell has a membrane-enclosed nucleus that houses the DNA in complex structures called chromosomes. In addition, eukaryotic cells contain several other types of membrane-bound internal structures, or **organelles.** Finally, eukaryotes have a special network of minute filaments and tubules called the *cytoskeleton*, which gives shape to the cell and allows movement. All single-celled organisms other than bacteria and cyanobacteria are eukaryotic cells, as are the basic units that make up all multicellular plants, animals, and fungi.

The second system of cell classification—methods of obtaining energy—categorizes cells as either **autotrophs** or **heterotrophs.** Autotrophs ("self-feeders") use light energy or chemical energy to manufacture their own sugars, fats, and proteins. A few types of bacteria are autotrophic, as are the more than 400,000 plant species on earth. Heterotrophs ("other-feeders") derive their energy by taking in foods in the form of whole autotrophs or other heterotrophs, their parts, or their waste products. Many kinds of bacteria and all the millions of animal and fungal species are heterotrophs.

Components of Eukaryotic Cells

While the presence of a true nucleus is the most obvious trait of a eukaryotic cell, a typical plant or animal cell has several kinds of organelles, some common to all eukaryotes and others characteristic of either plants or animals. Figure 5-4 depicts a generalized animal cell. That cell is enclosed by the *plasma membrane*, a flexible,

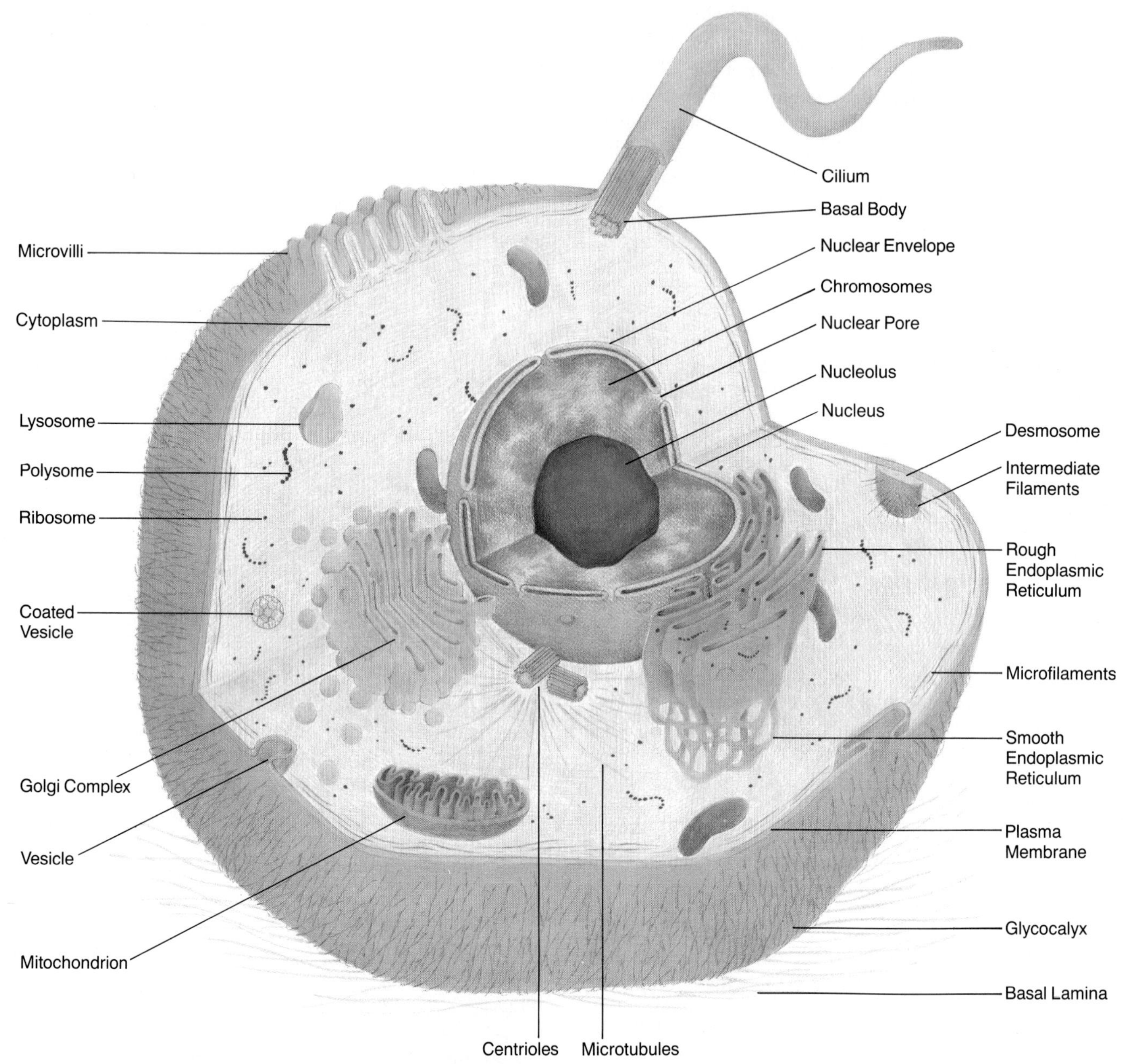

Figure 5-4 COMPONENTS OF A TYPICAL ANIMAL CELL.
The central nucleus contains hereditary material and a nucleolus, a site of ribosome manufacture. The watery cytoplasm has a cytoskeleton composed of microtubules, microfilaments, and other components. Cytoplasmic organelles include mitochondria, the energy powerhouses; rough endoplasmic reticulum, a site of protein synthesis; the Golgi complex, a place for modifying proteins; lysosomes, cellular digestive bodies; and centrioles, organizing bodies for microtubules. The cell surface includes the plasma membrane and attached glycocalyx. The surface may extend outward as cylindrical microvilli or as longer, motile cilia.

double layer on which are scattered a set of molecules, the *glycocalyx*, mediator of the cell's interactions with the environment. The semifluid interior of the cell, the *cytoplasm*, contains a *cytoskeleton* in which organelles are suspended. The cytoskeleton is a dynamic scaffolding composed of *microfilaments* and *microtubules* that support the cell's shape and often change it by their activity. The suspended organelles include the *mitochondria*, in which high-energy compounds are formed; the *ribosomes*, sites of protein synthesis; the *endoplasmic reticulum (ER)*, an assembly line for making certain types of proteins and fats; the *Golgi complex*, a depot where export materials are packaged; the *lysosomes*, which serve as digestion and waste-disposal systems for the cell; and the *nucleus,* the repository of the genetic material contained in chromosomes. The cell surface may possess one or more *cilia,* which act like tiny oars that move the whole cell or sweep materials past it, and it may also have fingerlike projections called *microvilli,* which usually facilitate absorption.

The components of a generalized plant cell are depicted in Figure 5-5. The plant cell's plasma membrane is surrounded by a protective, supportive *cell wall.* The cytoplasm contains not only the same sorts of organelles found in animal cells, but also sacs called *plastids,* including *chloroplasts,* the sites where nutrient molecules are built using the sun's energy, and *leucoplasts,* where nutrients are stored. The central portion of the plant cell, the *vacuole,* is a large drop of fluid enclosed by a membrane and filled with dissolved inorganic and organic substances. Neighboring plant cells may be interconnected by gossamer bridges of cytoplasm called *plasmodesmata.* Like animal cells, plant cells are extremely dynamic, and through a microscope, it is often possible to view the cytoplasm and nucleus streaming around and around the central vacuolar cavity.

Limits on Cell Size

One might assume that large organisms are built from large cells, and small organisms from small cells. However, measurements show that the difference between a whale and a mouse or between a redwood and a petunia lies in the total number of cells, not in the size of individual cells. A human, for instance, contains more than 100 trillion (10^{14}) cells, while a rat—possessing the same set of organs—has about 100 billion (10^{11}) cells. Individual cells do range in size, from the ostrich egg yolk, which is more than 3 cm in diameter, to small bacterial cells less than 0.2 μm, or 0.0000002 cm, in length (Table 5-1 and Figure 5-6). Most single cells, however, do not exceed a diameter of about 35 μm. Thus, 1,000 typical cells lined up neatly would just about span your thumbnail.

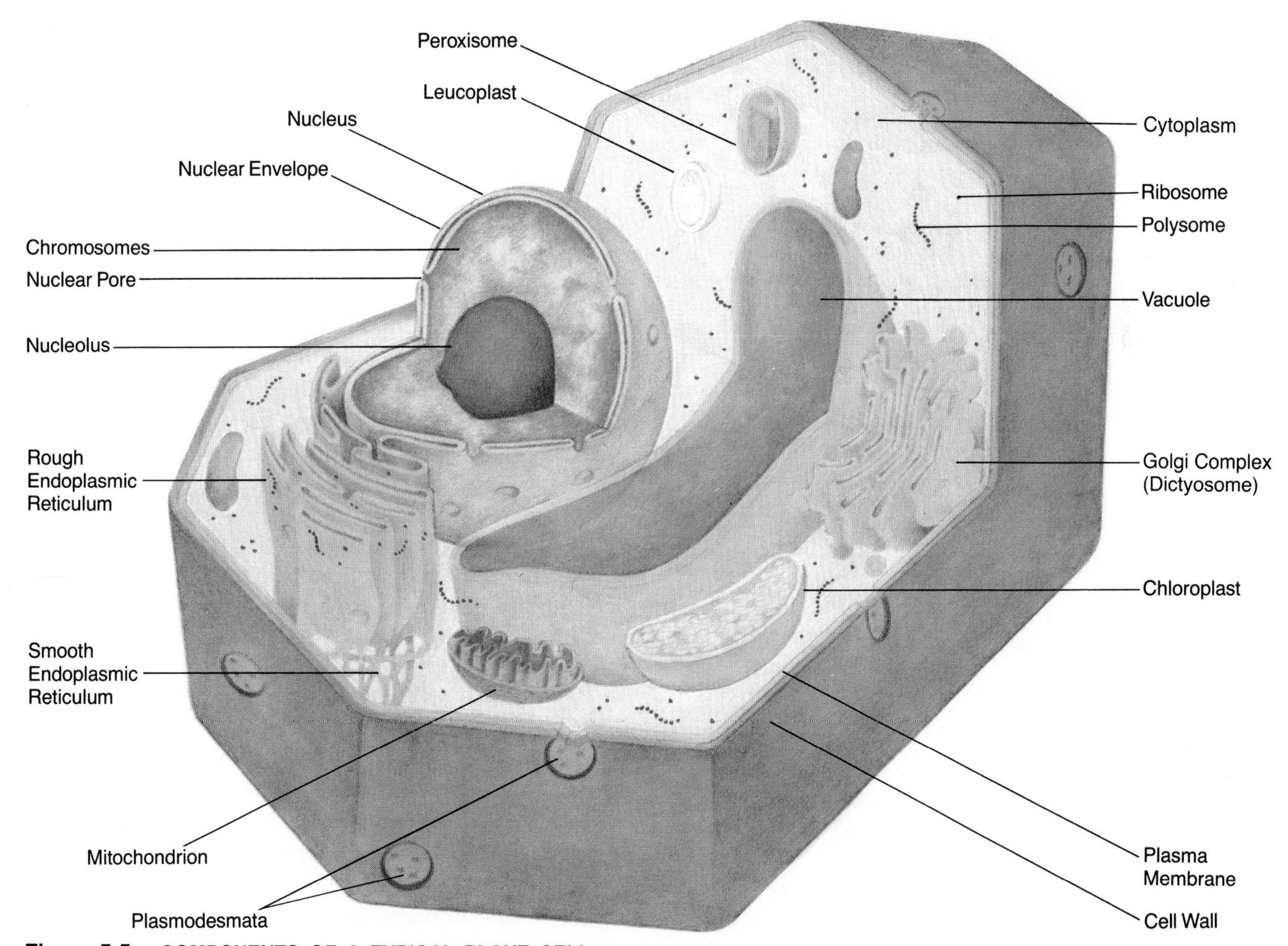

Figure 5-5 COMPONENTS OF A TYPICAL PLANT CELL.
Organelles present in animal cells are also found here. In addition, chloroplasts, sites of photosynthesis; leucoplasts, storage depots; and the large central vacuole are present. Plasmodesmata link plant cells together. Although it appears static here, the cytoplasm of a living cell streams around the edge of the central, fluid-filled vacuole.

Table 5-1 SIZES OF CELLS, ORGANELLES, AND MOLECULES

Cell, Organelle, or Molecule	*Diameter or Length (μm)*
Average plant cell	35.0
Average animal cell	20.0
Bacterium	2.0
Mitochondrion	1.0–8.0
Virus	0.02–0.2
Ribosome	0.025
Microtubule	0.022
Hemoglobin molecule	0.0064
Hydrogen ion	0.0001

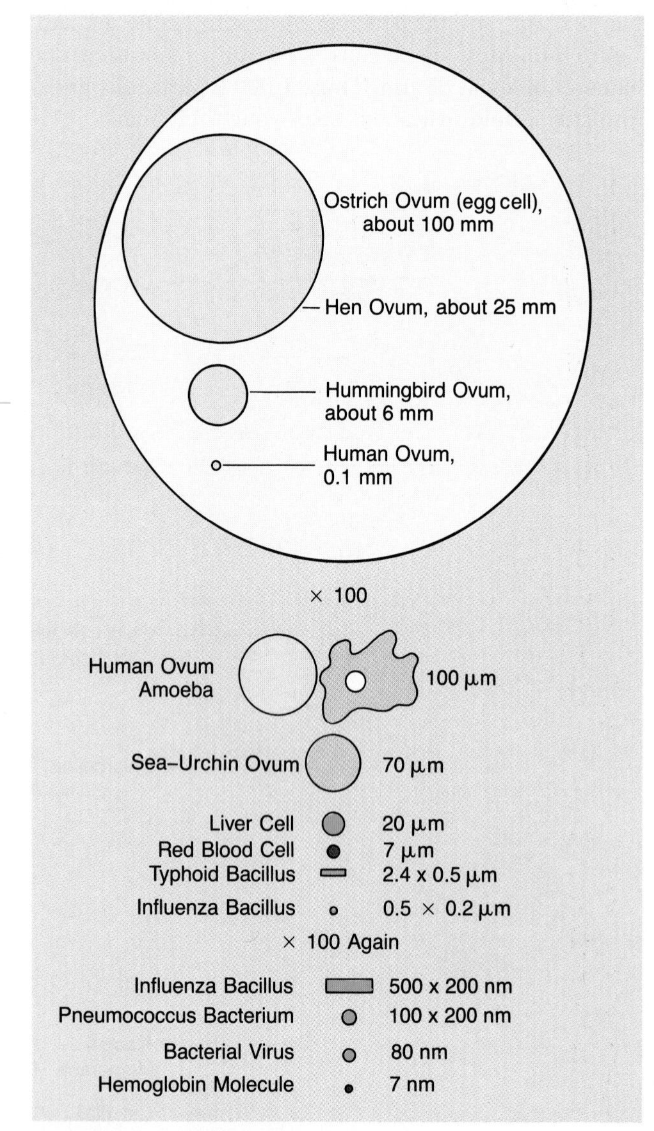

Figure 5-6 THE SIZES OF VARIOUS CELLS AND MOLECULES COMPARED.
A millimeter is 1/1000 of a meter; a micrometer is 1/1,000,000 of a meter; a nanometer is 1/1,000,000,000 of a meter.

The primary reason that spherical cells are rarely larger than 35 μm in diameter is that larger cells would not have enough surface area to exchange sufficient amounts of nutrients, gases, ions, and wastes with the environment. As Figure 5-7 shows, for each doubling in length, the cell's surface area is squared, but its volume is cubed. Conversely, as the cell gets smaller, its volume decreases faster than its surface area. Thus, the **surface-to-volume ratio** is greater in small cells. Since the cell surface is the site of interchange between the cell and the external environment, a larger relative surface area allows the cell to more readily absorb oxygen and nutrients for metabolism and to more easily excrete carbon dioxide and other waste products. Another advantage of greater surface area and small size in cells is that gases and nutrients have a shorter distance to travel from the edge to the center of the cell, where they are utilized for life processes.

Careful examination reveals that even extraordinarily large cells do not violate the principle of surface-to-volume ratio established for more common small cells. The yolk of an ostrich egg is an enormous single cell some 100 mm in diameter; however, most of the mass is relatively inert stored nutrient, and only a thin layer of cytoplasm surrounds the yolk.

The cell surface is clearly crucial, and as we will see, it is at once a raincoat, a guardian, and a sieve surrounding every living cell.

THE CELL SURFACE

Just as we see the brown paper on a package before discovering what is inside, the cell surface is the first part of a cell that we encounter. That surface is far more than a static wrapper, however; it is a dynamic barrier that governs all the traffic of materials into and out of the cell. Cells must also recognize and communicate with each other, especially within tissues and organs, and the cell surface governs these activities as well.

The Plasma Membrane

Despite the many kinds of walls and coatings that surround cells, every cell surface has one invariant component: the **plasma membrane,** or cell membrane, surrounding the watery cell contents, or cytoplasm.

In the 1930s, British biologists H. Davson and J. F. Danielli hypothesized that cells are covered by a thin, flexible envelope composed of two layers and that this **bilayer** contains phospholipid molecules and proteins. They believed—correctly, it turns out—that the hydrophilic, or water-loving, heads of the phospholipid mole-

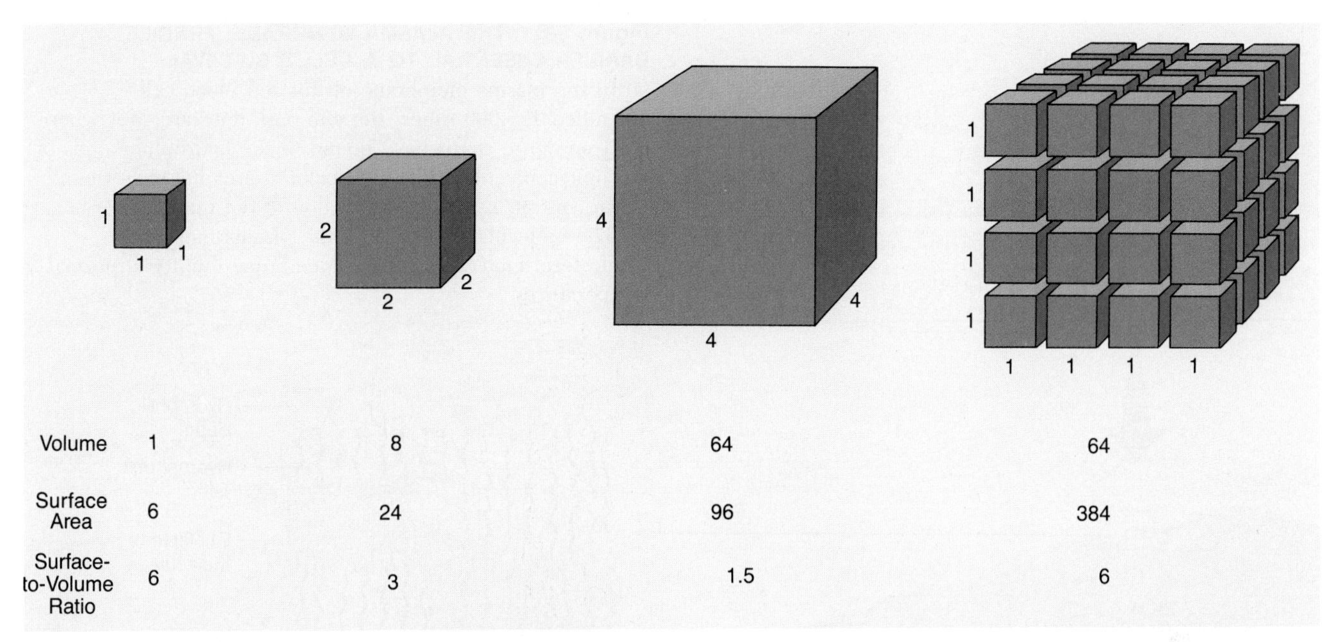

Figure 5-7 SURFACE-TO-VOLUME RATIO IN CELLS.
Subdividing a large cube demonstrates the concept of surface-to-volume ratio and shows why cells are small. As the length of each side of this cube doubles from 1 to 2 to 4, the volume increases more rapidly than does the surface area. Hence, the ratio of surface area to volume goes down. If the final large cube (4 × 4 × 4) is cut into 16 small cubes (1 × 1 × 1), the total surface area becomes four times as large. If these were cells, each of the small cells would be better able to exchange materials with its environment than would the one large cell.

cules are oriented outward in these layers and serve as a relatively waterproof boundary for the cell. This hypothetical lipid bilayer was corroborated by an early electron microscopist, who observed and photographed an 8 nm thick double line at the cell surface (Figure 5-8a). And later tests confirmed that plasma membranes and membranes surrounding many cell organelles consist of a bilayer of phospholipid molecules in which charged hydrophilic heads point outward and uncharged hydrophobic tails point inward (Figure 5-8b).

Additional tests have shown that membranes contain up to 100 different lipids and that the phospholipids in the inner and outer layers can differ so greatly that the bilayer is asymmetrical in composition and properties. The membrane's fluidity depends on the degree of saturation—that is, the proportion of double and single bonds—between adjacent carbon atoms in the fatty acid chains of certain phospholipids. The more double bonds in the chain, and thus the more unsaturated those chains are, the more flexible and fluid the membrane and the more easily substances can move laterally in the membrane. Research has also shown that some membrane lipids are linked to chains of sugars that project from the cell's surface. These so-called *glycolipids* may act as name tags, as in the human A, B, and O blood groups.

Contrary to Davson and Danielli's theory, membrane proteins do not form complete layers. Instead, they occur as **peripheral proteins**—individual molecules attached to the inner or outer membrane surface—or as **integral proteins,** embedded in the membrane. Many peripheral proteins are linked via sugars to lipids called phosphatidylinositols, which anchor the complex in the membrane. These complexes play a role in the adhesion of cells to other structures and in the signaling by hormones and other messenger molecules across membranes. Some integral proteins serve as sites where ions or molecules move across the membrane. Others have hydrophobic segments that span the membrane and are continuous with regions that protrude into the extracellular and intracellular spaces. Such proteins may moor the membrane to the cell's inner scaffolding (the cytoskeleton) and to molecules outside the cell such as collagen.

In 1972, biologists Jonathan Singer and Garth Nicolson proposed the **fluid-mosaic model** (Figure 5-9), which revolutionized our understanding of the dynamic barrier at the cell surface. It states that at normal biological temperatures, the plasma membrane behaves like a thin layer of fluid covering the cell surface. Individual lipid molecules move about laterally within the plane of the membrane, thereby contributing to the fluid state. Many of the proteins scattered through the membrane as in a patchwork or mosaic appear to be free to move laterally or diffuse in the lipid plane, like icebergs floating in

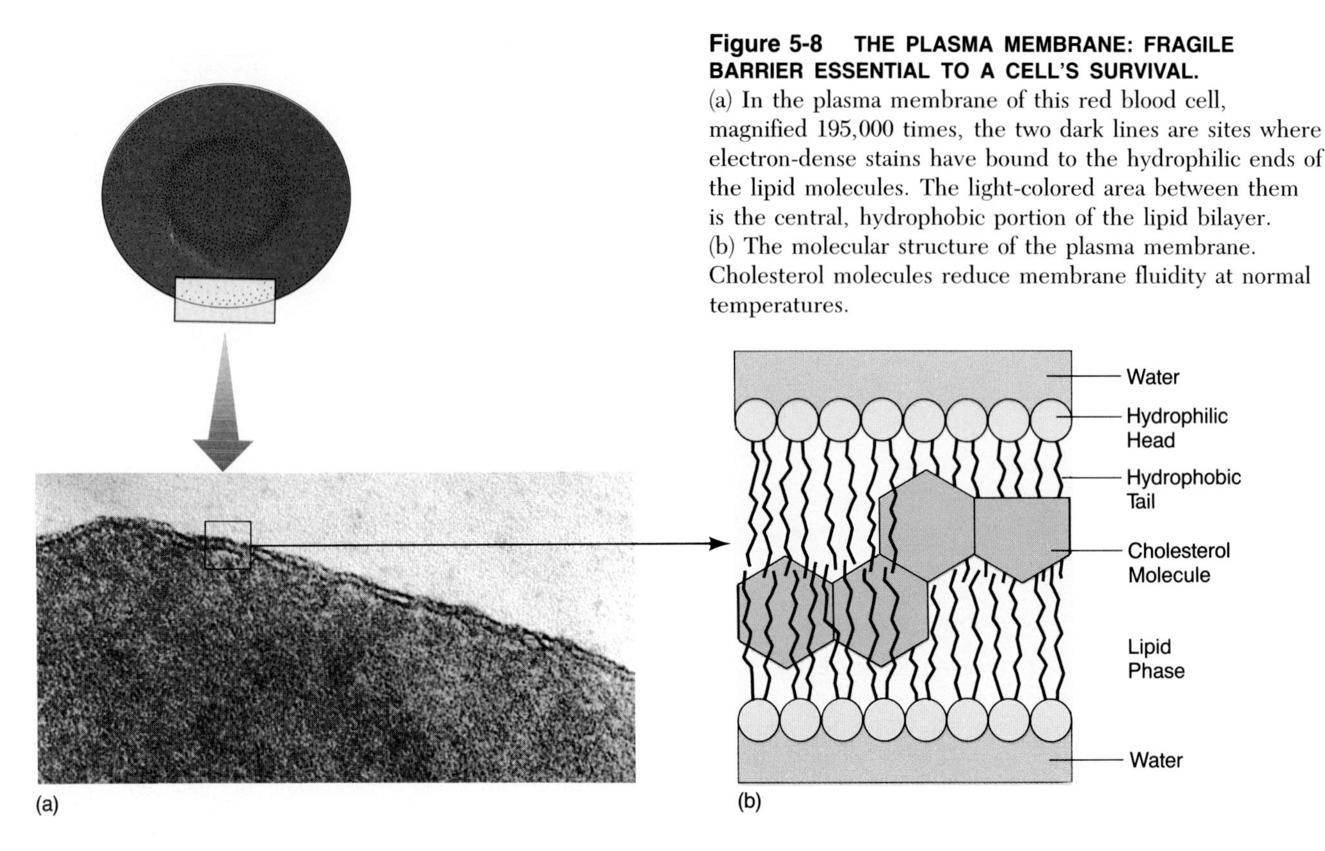

Figure 5-8 THE PLASMA MEMBRANE: FRAGILE BARRIER ESSENTIAL TO A CELL'S SURVIVAL.
(a) In the plasma membrane of this red blood cell, magnified 195,000 times, the two dark lines are sites where electron-dense stains have bound to the hydrophilic ends of the lipid molecules. The light-colored area between them is the central, hydrophobic portion of the lipid bilayer. (b) The molecular structure of the plasma membrane. Cholesterol molecules reduce membrane fluidity at normal temperatures.

a lipid sea. Not all proteins are free to diffuse, however, since some are linked to the cell's cytoskeleton or to adhesive materials outside the cell.

In 1970, researchers L. D. Frye and M. Edidin had revealed the movements of membrane molecules. Using human cells, they labeled certain sugar-protein surface markers with a dye that looks red under ultraviolet light. Next, they labeled surface markers of mouse cells with a green dye. Finally, they fused the cells, and observed as the red and green marked proteins intermingled and the

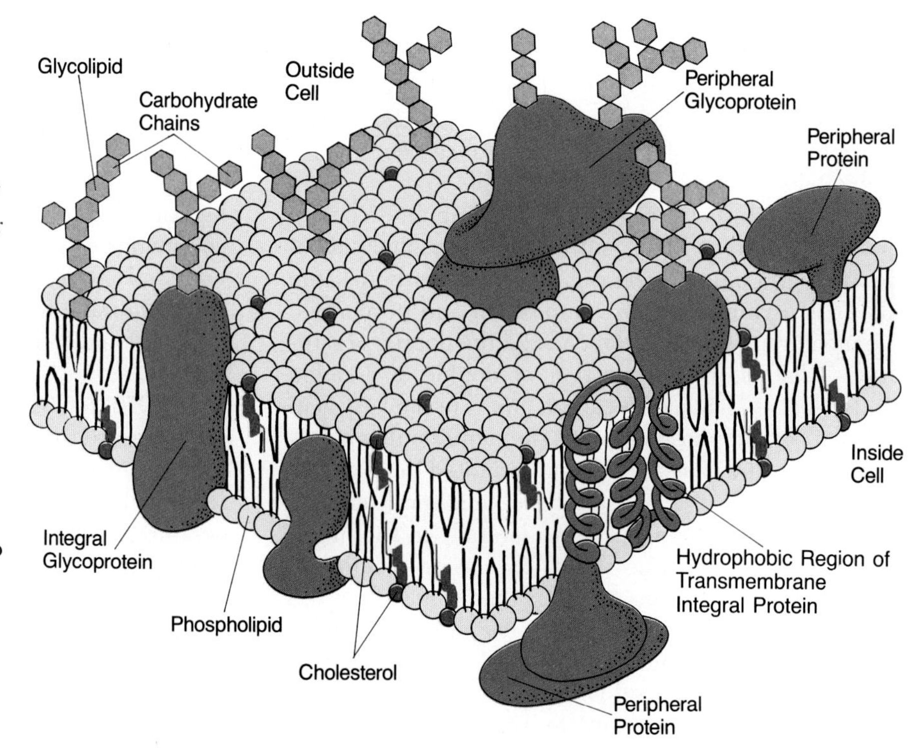

Figure 5-9 THE FLUID MOSAIC MODEL.
Globular integral proteins may protrude above or below the lipid bilayer and may move about in the plane of the membrane. Peripheral proteins are bound to integral ones. If membrane proteins are linked to other molecules, their diffusion in the membrane is slowed or prohibited. Notice that the charged hydrophobic regions of the proteins are positioned next to the charged hydrophilic head groups of the lipid molecules, while there are no surface charges on portions of protein in the core of the bilayer, where the uncharged hydrophobic lipid tails reside. Some peripheral proteins may be attached to fatty acid chains that anchor the protein in the bilayer (as shown at right).

entire cell surface became a mosaic of red and green dots. By later chilling the cells to reduce membrane fluidity, they observed that the colored markers stopped their movements and remained separated. This strong evidence helped lead Singer and Nicolson to their theory that membranes are fluid structures.

The thin membranes surrounding various organelles within cells are also composed of lipid bilayers, are fluid in nature, and may have scattered protein "icebergs." However, there are two major chemical differences between the plasma membrane and these other membranes: (1) Cholesterol, which reduces membrane fluidity at normal temperatures, is present in the plasma membranes of many eukaryotes, but not in organelle membranes, and (2) the kinds and percentages of various lipids differ. One consequence of these differences is that the membranes surrounding organelles are considerably more fluid than the plasma membrane. Regardless of differences, however, all membranes are lipid barriers that separate the cytoplasm from the watery world outside the cell or separate the fluid inside an organelle from the cytoplasm.

Movement of Materials into and out of Cells: Role of the Plasma Membrane

Life on earth evolved in water more than 3 billion years ago, and in a very real sense, the cells of multicellular organisms have never left that primal water. Inside all plants, animals, and fungi are salty, nutrient-filled liquids. Inside individual cells, *intracellular fluids* make up the fluid part of the cytoplasm. Outside of cells are *extracellular fluids* that bathe the surface of each cell like a shallow, miniature internal ocean.

Dissolved in a large animal's extracellular and intracellular fluids are several grams of salt ions, sugars, proteins, hormones, and other substances. However, some ions and molecules are maintained at higher concentrations outside the cell than inside, while others are more heavily concentrated inside the cell than outside. For example, there are 100 times more potassium ions inside human cells than in the blood outside those cells, but there are 100 times more sodium ions in the blood than in the intracellular fluid. A major activity at the cell membrane is regulation of these differences and the two-way flow of materials.

The plasma membrane, with its component lipids, acts like an oilskin raincoat, slowing or preventing the passage of polar ions or proteins (see Chapter 2) through the uncharged, hydrophobic interior of the lipid bilayer. Nevertheless, some ions and molecules are permitted to enter the cell via the passive processes of simple diffusion and carrier-facilitated diffusion and via the energy-requiring process of active transport.

Simple Diffusion

Diffusion is the tendency of a substance to move from a place of high concentration to one of lower concentration. Figure 5-10 demonstrates how diffusion takes place across a permeable membrane or porous partition. As the figure legend explains, salt (NaCl) poured into one side dissociates into Na^+ and Cl^-, and these ions move across the membrane from an area of higher concentration to an area of lower concentration.

In addition, the mobility of the water molecules is reduced in the immediate vicinity of dissolved ions because water molecules are attracted to dissolved polar substances, such as sugar and salt. In the system depicted in Figure 5-10, there is more drag on the right,

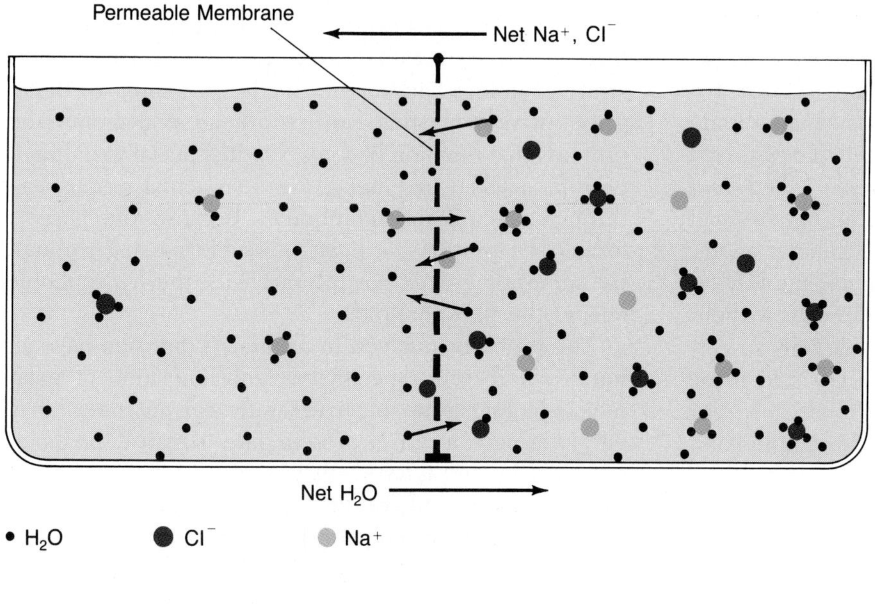

Figure 5-10 A SIMPLE DIFFUSION SYSTEM.
Solutions with high and low salt concentrations, separated by a partition perforated with many small holes. The more concentrated sodium and chloride ions (dissociated from NaCl) on the right strike pores in the permeable membrane more frequently than do the less concentrated ions on the left. As a result, more sodium and chloride ions move from right to left than the reverse. Water behaves the same way: It moves from a region of high concentration of water molecules (on the left) to a region of lower concentration (on the right). Net movements of water and of Na^+ and Cl^- thus occur in opposite directions until concentrations on both sides are equal.

THE SECRETS OF WINTER WHEAT

The United States and Canada are the "breadbaskets" of the world, exporting enormous amounts of wheat each year despite the fact that most of the wheat-growing areas are subjected to long, harsh winters and a foreshortened growing season. Wheat farmers in these regions grow a hardy grain plant called winter wheat, and until recently, no one knew what special characteristics allow it to survive the frigid winter months with roots frozen in the soil and leaves buried beneath deep snow. The answer, it turns out, lies in the plant's extraordinary cell membranes.

In a normal strain of wheat, as the cells freeze, water from the cytoplasm migrates outward and the cell membrane shrinks, causing the cell to shrivel. As the membrane shrinks, fat droplets appear to be squeezed from it into the cell's interior and to remain suspended within the cytoplasm. When the cell later thaws, water is reabsorbed and the cell expands again; but the fat droplets in the cytoplasm do not rejoin the membrane, and holes appear in the membrane where the droplets were located. The membrane eventually splits, and the cell's contents spill out. This explains why the frozen and thawed plant wilts and dies in the spring.

When the cells of a cold-tolerant plant, such as winter wheat, freeze, water migrates out and the membranes shrink, just as they do in the cells of a cold-sensitive plant. However, the lipids squeezed from the membrane do not move inward and float freely inside the cell's cytoplasm. Instead, those fat droplets are extruded outward and remain "tethered" to the membrane's outer surface in miniature spheres, filaments, and pockets. When the cell thaws, the fats are reabsorbed into the lipid bilayers. Because of this tethering and reabsorption, no holes form in the membrane, which does not split, and the cell remains intact. Thus, when the spring thaw comes, a winter wheat plant can resume active growth, having suffered no damage. Knowing this, plant breeders may be able to select and modify other staple food crops so they can grow better in colder areas and help provide badly needed food for the world's increasingly hungry population.

where the salt concentration is higher, than on the left, where fewer water molecules are encumbered. Thus, more water molecules on the left are free to encounter pores in the membrane and so to move from left to right.

The diffusion of ions and water molecules to and from areas of higher and lower concentration will continue until the concentrations on both sides are equal. Movement of individual ions and water molecules will continue, but at the same rate in both directions. Simple diffusion in living cells can be seen in the movement of lipid-soluble substances across the plasma membrane. Ether, a medical anesthetic, is a good example. Lacking a significant surface charge, ether molecules can dissolve freely in the hydrophobic lipids of the plasma membrane and then cross the membrane into the cell. The passage of molecules across the plasma membrane is called *lipid diffusion*. While lipid diffusion accounts for only a small percentage of molecules passing across cellular membranes, certain steroid hormones and lipid-soluble vitamins (including vitamin D) could not enter cells without it. Because cell membranes are permeable only to certain substances and not to others, they are said to be *semipermeable*.

What happens to substances that are not lipid-soluble? Diffusion of these molecules must occur through temporary and permanent openings in the plasma membrane—gates in the sea wall, if you will. Because the lipids of the plasma membrane are in a dynamic, fluid state, they are continuously moving laterally to some degree. Apparently, as this movement occurs, channels open temporarily to permit small polar molecules such as water or oxygen to enter or leave the cell's interior. Other polar substances may require more complex processes to move across a plasma membrane.

Carrier-Facilitated Diffusion

Some substances can cross the plasma membrane by linking up with so-called carrier proteins, much as a substrate combines with an enzyme (see Chapter 4). Biologists now believe that carriers are integral membrane proteins that function like pores, with special membrane proteins packed together to form an actual passageway through the plasma membrane. Because the pore is formed by proteins, a polar molecule passing through the membrane does not interact with the hydrophobic tails of the phospholipids.

The proteins involved in facilitated diffusion have an important feature: specificity. Only one kind of polar molecule (or group of structurally related ones) may enter via a given carrier protein pore. Because the molecules are moved into or out of the cell faster and more efficiently than could occur by simple diffusion, this process is called **carrier-facilitated diffusion.** No energy is required for this process. The direction of carrier-facilitated diffusion is wholly governed by the relative concentrations of the cargo substance inside and outside the cell: Movement is along a **concentration gradient,** from areas of higher concentration to areas of lower concentration.

The number and distribution of carrier molecules (or pores) in a membrane govern the rate of carrier-facilitated diffusion. If enough of a cargo substance is present to form complexes with all the available carrier molecules, the carrier becomes saturated, and the rate of carrier-facilitated diffusion reaches a maximum. This imposes limits on the speed with which cells can take in or extrude certain substances.

Discovery of a new class of molecules called *ionophores* has provided insight into yet another way that polar substances can cross a membrane. Produced by fungi, *ionophores* are doughnut-shaped molecules with an uncharged exterior (that can easily pass through the lipid bilayer) and a hole or slot to carry cargo—usually an ion. These fungal ionophores act as antibiotics that drastically upset the balance of ions inside bacterial cells and thus kill the cells before they can infect the fungus.

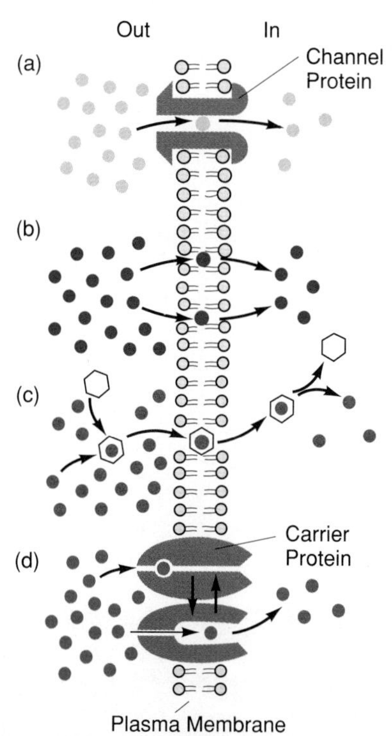

Figure 5-11 FOUR WAYS FOR SUBSTANCES TO PASS THROUGH A PLASMA MEMBRANE.
Molecules can move into or out of cells (a) through pores formed by pore proteins; (b) by diffusing directly through the lipid phase of the membrane (as urea or ethanol); (c) by being hidden inside an ionophore (as Ca^{2+}); and (d) by binding to a carrier, or transporter, protein that changes in conformation and releases the cargo (such as Cl^-) to the opposite side.

Diffusion has a profound effect on all life, even though it is a passive physical process that depends on the concentration of substances and the availability of membrane pores. Without it, the "cell-state" would quickly run out of many necessary commodities. But as we shall see, some molecules must be moved actively across membranes. The life of the cell and its commerce with the outside environment therefore hang in a delicate balance between energy-free passive diffusion and energy-costly active transport. Figure 5-11 summarizes the mechanisms of passive transport.

Active Transport

Transporting certain ions through the plasma membrane requires energy, particularly when this movement occurs *against* concentration gradients (from areas of low concentration to areas of high concentration). Figure 5-12 shows a simple model for **active transport:** A carrier protein binds its cargo, changes shape, and releases the cargo. ATP is consumed during each cycle. Enzymes that catalyze these processes hydrolyze ATP and so are referred to as ATPases. The cargo-pumping activity of such enzymes is truly impressive and is a primary reason why most cells require food and oxygen—they need to replenish ATP supplies for active transport.

Some active-transport carriers move a single cargo in a single direction. Such carriers are called *uniports.* Sometimes, however, two cargoes must be present for the port to function: *Symports* move two cargoes simultaneously in the same direction; *antiports* move two cargoes simultaneously in opposite directions. The Na·K·ATPase pump, for example, is a classic antiport; every time sodium ions are pumped out of a cell, potassium ions must pass in (Figure 5-12). If extracellular potassium is not present, the sodium pump halts.

A symport allows sugars or amino acids to enter cells. In this **co-transport,** or coupled transport, both sodium ions and an amino acid or sugar molecule bind to the carrier port, which discharges them inside the cell. Neither the sugar nor the amino acid can be transported alone. However, co-transport causes the level of sodium ions in the cell to rise. The ATPase pump therefore uses energy from ATP to pump out the excess salt that has entered with the sugar or amino acids (Figure 5-13). This relatively efficient process underlies the active uptake of sugars, amino acids, and other nutrients by the cells of the human gut.

Osmosis and Cell Integrity

We turn now to a major result of the transport of ions and molecules across the plasma membrane—a passive movement of water into and out of the cell. The diffusion

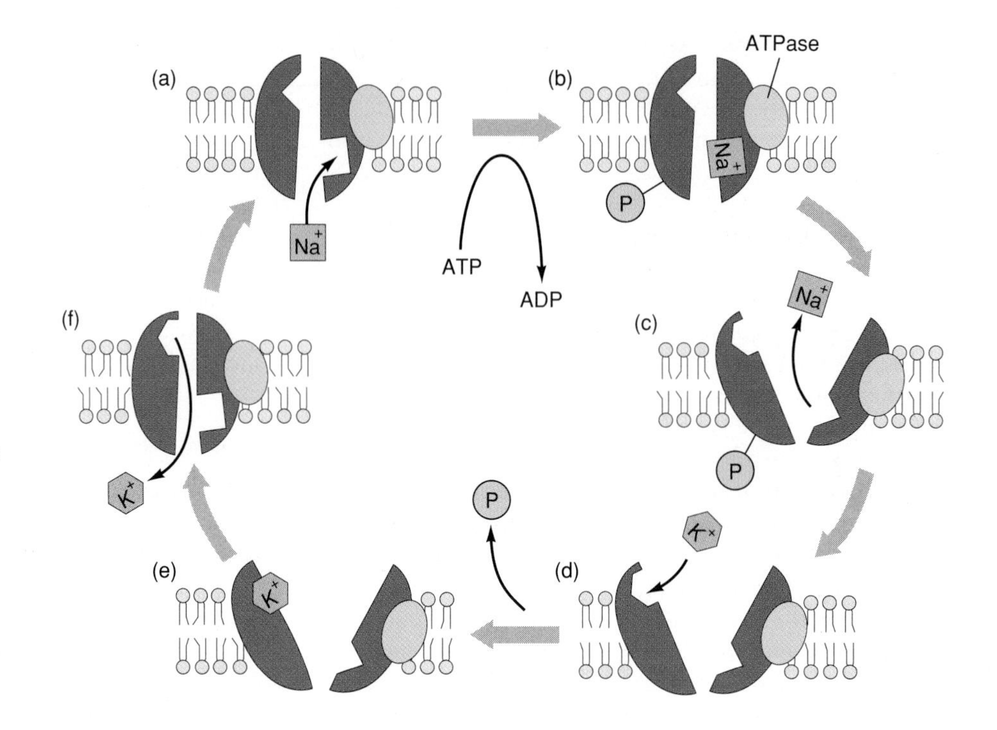

Figure 5-12 A MODEL FOR ACTIVE TRANSPORT.
One model for pumping Na^+ out of and K^+ into cells involves the Na · K · ATPase complex shown here. (a) After Na^+ binds, (b) phosphorylation occurs on the cytoplasmic side. (c) The protein changes in shape, releasing Na^+ to the exterior of the cell. (d) K^+ then binds; (e) as dephosphorylation occurs, (f) the protein reverts to its original shape, releasing K^+ into the cytoplasm. In other cases of active transport, two substances may be moved in the same direction, or only one substance may be transported into or out of the cell.

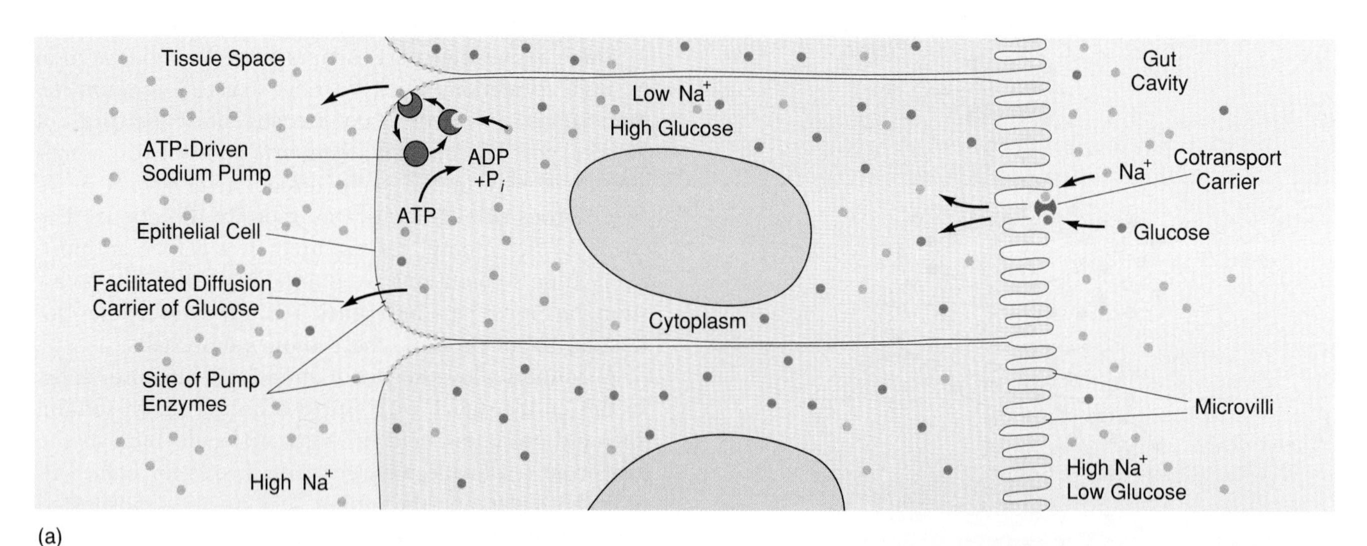

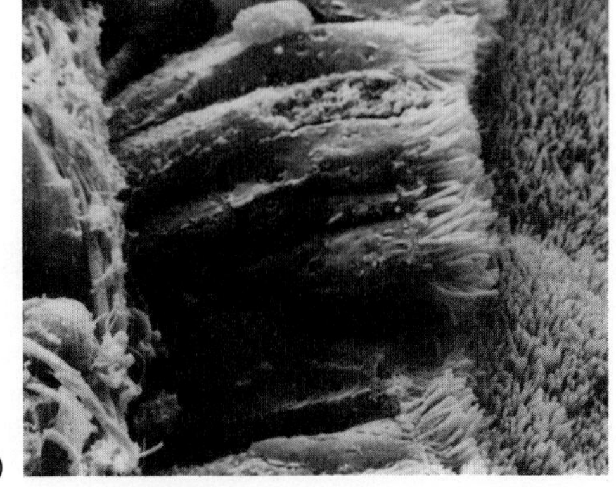

Figure 5-13 COTRANSPORT IN AN INTESTINAL EPITHELIAL CELL. (a) The ATP-dependent sodium pump at the left end of the cell lowers the concentration of Na^+ in the cytoplasm. As a result, Na^+ moves into the cell at the right end. However, Na^+ can pass in only when a sugar or an amino acid enters at the same time. The cotransport of salt and sugar takes place because the pump at the opposite end of the cell sets up an ion concentration gradient (low inside, compared with outside). For convenience, the pump and the carrier molecules are shown in the cytoplasm. In fact, those molecules probably are associated with the plasma membrane; the probable sites of the sodium pump enzymes are shown. (b) A scanning electron micrograph, magnified about 1,350 times, shows the many microvilli on intestinal epithelial cells; the cotransport carriers are associated with the microvilli.

of water through a semipermeable membrane like the cell membrane is called **osmosis.** It is a simple process that is nevertheless central to life on earth because if it fails, the cell swells and bursts or shrinks and dies.

Usually, water moves from a dilute solution to a more concentrated one. Thus, if the salt concentration inside a cell is high in relation to that outside a cell, water tends to cross the cell membrane and dilute the more concentrated solution inside. At the same time, the excess salt inside tends to move out of the cell. These processes continue until equilibrium is reached. The influx of water causes the cell to swell and exerts pressure on the surface. The force exerted outward by this increased internal water is called **osmotic pressure.** In plants and bacteria, the cell membrane is surrounded by a rigid wall, which permits the cell to swell only until the membrane pushes against it. This outward pressure against cell walls is called *turgor;* it is the reason why a well-watered plant is stiff and erect.

Figure 5-14 shows how the concentration of salt inside and outside a cell affects the movement of water and hence the volume of the cell itself. If the salt concentrations (*tonicity*) inside and outside animal and plant cells are equal, the extracellular fluid is said to be **isotonic** (the same) in relation to the intracellular fluid, and no net water movement occurs (Figure 5-14a). If the salt concentration outside the cell is less than that inside, the extracellular fluid is considered **hypotonic;** water tends to rush into the cell, and the cell swells (Figure 5-14b). This swelling is limited by the cell wall in bacterial and plant cells. But animal cells are bounded by only the cell membrane and a wispy glycocalyx and can actually swell so much that they burst. Finally, if the salt concentration outside the cell is greater than that inside, the extracellular fluid is said to be **hypertonic** relative to the cell contents. In this situation, water tends to leave the cell, and the cell shrinks (Figure 5-14c).

Many cells contain proteins and other molecules that cannot easily pass out through the semipermeable plasma membrane. These molecules attract water and so exert osmotic pressure. Their presence in the cell leads to a tendency for water to enter the cell continuously throughout its life. What, then, prevents swelling, overhydration, and the dilution of cell components? The answer lies in special membrane enzymes that pump ions out of the cell. For instance, the sodium pump (see Figure 5-12) is an enzyme system that continuously pumps Na^+ outward to lower the total internal ion content and raise the external ion concentration, thereby reducing the tendency for water to flow in. But the enzyme portion of the pump splits an ATP molecule each time sodium ions are extruded. Thus, the ATP supply must be continually replenished by metabolic processes. If a poison such as cyanide halts ATP production, the sodium pump halts, and the cell swells and dies.

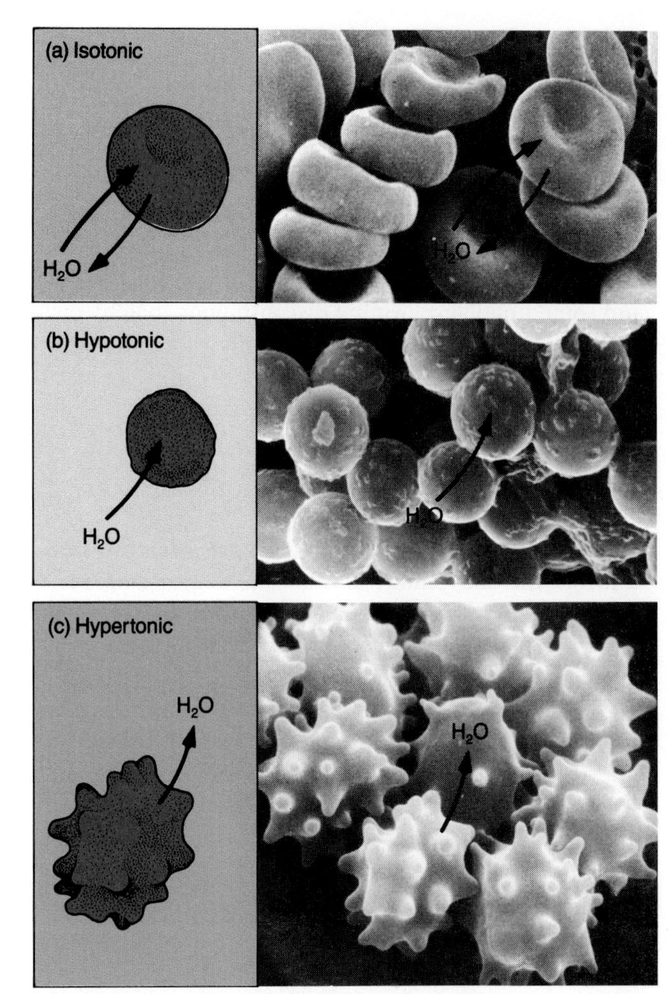

Figure 5-14 EFFECT OF SALT CONCENTRATION ON CELL VOLUME.
The concentration of salt in the fluid surrounding a cell may determine the cell's volume. (a) In an isotonic solution, with the same salt concentration as the cytoplasm, equal amounts of water enter and leave these red blood cells, so that their volume does not change. (b) A hypotonic (low-salt) solution results in an influx of water, which causes the cell to swell. (c) A hypertonic (high-salt) solution causes water to leave the cell and results in a shrunken appearance.

The tonicities of the tissue fluids that bathe cells must also remain fairly constant. The kidneys of complex animals are specialized to keep ion levels constant in the blood and, indirectly, in extracellular tissue fluid. As a result of the continuous balancing of ions and water, tissues are not subjected to overwhelming osmotic stresses, and the organism can survive.

In summary, the plasma membrane, by virtue of its lipid bilayer and associated proteins, regulates the passage of materials into and out of the cell and prevents the cytoplasmic fluid from leaking out. Although this dynamic barrier is vital to the life of the cell, it is not the only protective structure found at the cell surface, as our next section explains.

CELL WALLS AND THE GLYCOCALYX

Biologists rarely observe completely uncovered or unprotected plasma membranes in normal living cells. Molecular aggregates surround the plasma membranes of all cells, forming a rigid cell wall in bacteria, plants, and some fungi and forming discontinuous, patchy coating on the surface of animal cells. Both walls and patches help govern much of cell life.

Cell Walls

The plasma membrane, acting as a selective sieve, controls the movement of substances into and out of cells, but the rigid outer **cell wall** of plant, bacterial, and some fungal cells serves as protection and reinforcement.

The cell walls of plants and some fungi are made largely of *cellulose,* a high-molecular-weight polysaccharide that is arranged in multiple layers over the cell's plasma membrane. Plant cell walls have three portions: the middle lamella, the primary wall, and the secondary wall (Figure 5-15a). The *middle lamella* is the first layer to form when a plant cell divides into two new cells. It is composed largely of *pectin,* a gluelike polysaccharide that helps to hold adjacent cell walls together. *Primary walls* form on each side of the middle lamella as the new plant cells grow and mature. For a while, the primary wall of a plant cell remains flexible and stretchable; this enables the young cell to elongate and grow. The third layer, called the *secondary wall,* forms on the inner side of the primary wall after cell growth and the formation of new cells have stopped, particularly in the outer parts of plant tissue. The secondary wall tends to be very strong because the layers of cellulose within it lie at nearly right angles to one another, like the layers in plywood (Figure 5-15b). The cellulose molecules form tiny ropelike *microfibrils* that are glued together with a hardening substance called *hemicellulose. Lignin,* a complex carbon-containing substance, acts as a further stiffening agent for the secondary wall. Together, the cellulose and lignin in the walls of millions of individual plant cells make up the woody tissue of trees and parts such as pine cones and nuts.

The cell walls that surround bacteria are composed of polysaccharides, lipids, and short chains of amino acids and sugars. The amino acids and sugars, which are called *peptidoglycans,* link the components of the entire wall so

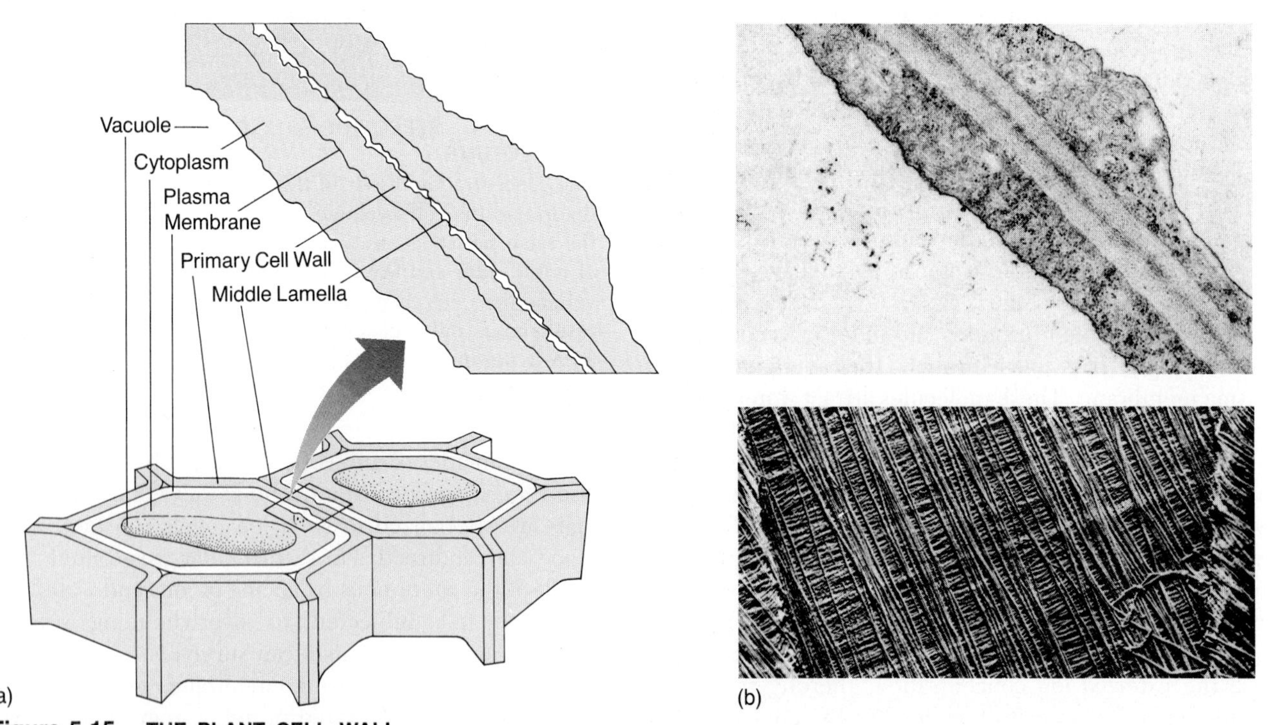

Figure 5-15 THE PLANT CELL WALL.
(a) A primary wall arising just after the division of a plant cell. The electron micrograph (magnified about 17,000×) and corresponding drawing show the middle lamella and the primary walls arising between two plant cells. The secondary wall forms later and is much thicker than the primary wall. (b) Ultrastructure of plant secondary cell walls (magnified about 9,300×). Cellulose microfibrils are welded together at almost right angles to one another within lignin and hemicellulose in woody tissue.

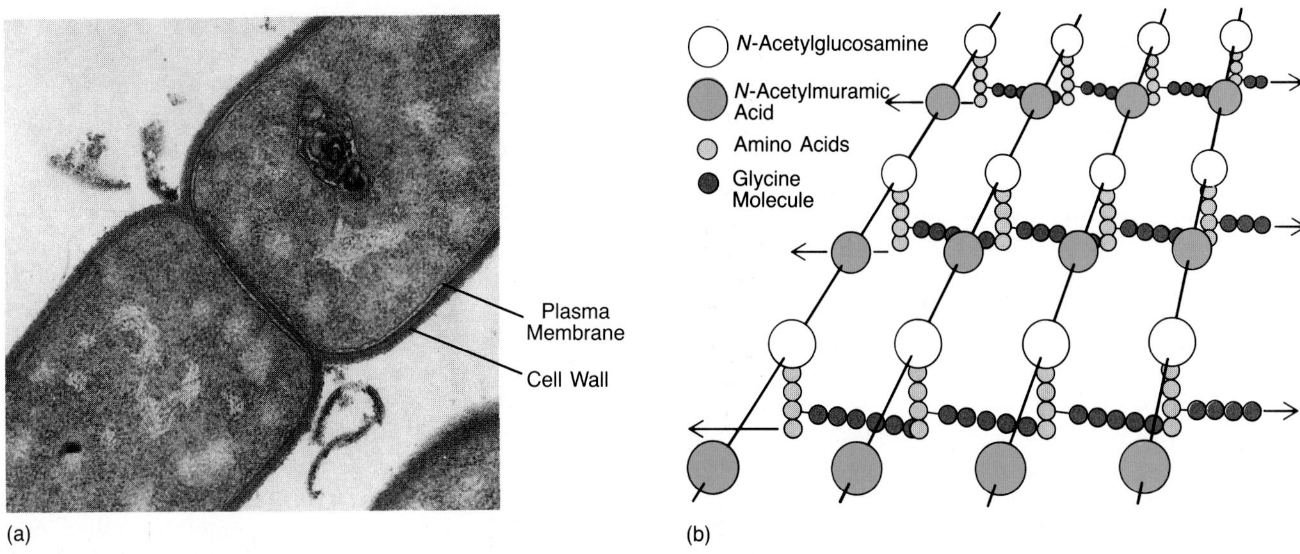

Figure 5-16 BACTERIAL CELL WALL.
(a) The bacterial gram-positive cell wall encircles, strengthens, and protects the cell (magnified about 60,000×). (b) The main wall components are an interconnected set of sugar polymers. The diagram shows a tiny portion of the wall, composed of peptidoglycans (short chains of amino acids and sugars). The small red spheres represent cross-linking glycine molecules. The entire wall can be considered a single macromolecule.

that it is quite rigid (Figure 5-16). This wall permits bacteria to invade other organisms and to withstand environmental conditions that would kill a typical animal cell. It also imparts to bacterial species their characteristic shapes (spheres, rods, commas, and spirals).

Fungal cell walls are complex combinations of cellulose, **chitin,** and other glucose polymer polysaccharides. Chitin—built from glucosamine, a nitrogen-containing sugar—is also a main ingredient of the external skeletons of lobsters, spiders, houseflies, and their kin.

Glycocalyx

Animal cells lack a continuous wall or rigid coat and thus are always potentially mobile and plastic. On their outer surface, however, they have patches of large molecules that act as glue and as molecular name tags. These molecules—glycoproteins and glycolipids—are referred to as the **glycocalyx,** or "sugar coat" (Figure 5-17). Gluelike glycocalyx molecules promote the adhesion of cells to one another and to external structures, such as the collagen fibers that twist like strong threads through the connective tissue of most organs. The complexly arranged and distinctively shaped groups of sugar molecules of the glycocalyx act as a molecular fingerprint for each cell type and are involved in cell recognition and in the coordinated responses of cells within both tissues and the organism, as a whole.

A variation of the glycocalyx is found on the surfaces of *epithelia,* cell populations arranged in sheets. Every sheet of epithelium is attached to a *basal lamina,* a feltlike layer that consists of highly ordered glycoproteins and a special type of collagen fiber.

Most cells of multicellular animals are attached to surfaces such as basal laminae, collagen fibers, or other cells. Contact and adhesion are essential if cells are to assemble an internal cytoskeletal scaffolding to form specialized shapes, divide, and move about. This need to adhere to a solid surface is called **anchorage dependence.** Thus, if one places animal cells in a fluid solution, they usually become spherical, cease to divide, and stop moving. Certain kinds of cancer cells have both an altered glycocalyx and reduced anchorage dependence, and these may figure in the cells' uncontrolled growth and ability to invade body tissues.

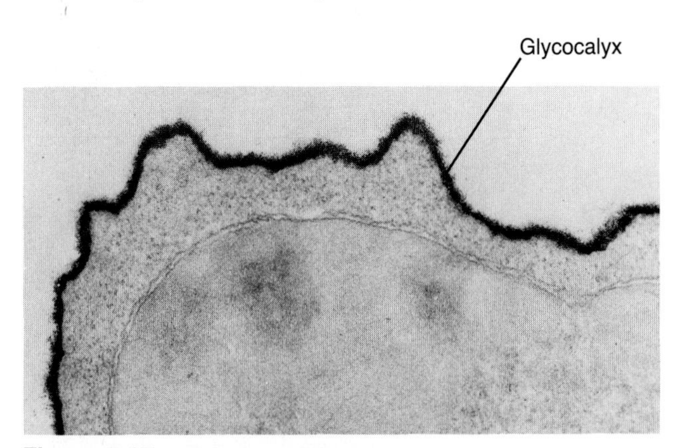

Figure 5-17 THE GLYCOCALYX.
A special stain, ruthenium red, reveals the glycocalyx on a mammalian cell (magnified about 30,400×). The process obscures the plasma membrane so that it can no longer be seen.

LINKAGE AND COMMUNICATION BETWEEN CELLS

In addition to outer walls or patches, cells have specialized structures on those surfaces that hold them firmly together in tissues, allow cells to communicate with one another and the environment, and prevent fluid leakage in certain tissues. Several types of junctions hold cells together and provide channels for intercellular communication: zonulae adherens, desmosomes, tight junctions, gap junctions, and plasmodesmata. Populations of animal epithelial cells provide an excellent place to study most of these junctions.

Zonulae adherens and desmosomes serve mainly to bind cells together (Figure 5-18). **Zonulae adherens** ("zones of adhesion") are sites of firm physical contact between cells. They are beltlike bands that run around most epithelial cells. In addition to linking adjacent cells, zonulae help control cell shape and serve as sites for insertion of important scaffolding filaments of the cytoskeleton.

Desmosomes are analogous to tiny spot welds, rivets, or buttons between cells. These small junctions are made up of unidentified molecules that apparently glue together adjacent plasma membranes. Desmosomes are particularly abundant in tissues subjected to mechanical stress, such as the outer layers of the human skin and the cervix of the uterus.

Tight junctions are seals that encompass the lateral surfaces of cells in epithelia and act a bit like rubber seals, forming barriers to fluid leakage. Tight junctions between adjacent epithelial cells of the bladder, for example, prevent urine from seeping between the cells back into the body tissue spaces.

The outer lipid layers of the plasma membranes of adjacent cells actually appear in the electron microscope to touch at tight junctions. The areas of contact are a network of ridges where proteins embedded in the two membranes' lipid bilayers are probably bound tightly together like the teeth of a zipper.

The primary communication junction between animal cells is the **gap junction,** a perforated channel that permits easy exchange of small molecules, ions, and electric currents across cell membranes and thus allows cells to communicate in the molecular, ionic, and electrical language they "speak" (Figure 5-19). These channels pass through the center of protein complexes called *connexons* that span both layers of the plasma membrane. The movement of ions and molecules between cells may help coordinate various cellular activities.

Biologists believe that many kinds of plant cells engage in intercellular exchange more easily than can animal cells. Bridges of cytoplasm called **plasmodesmata** (singular, *plasmodesma*) connect adjacent plant cells (Figure 5-20). These bridges normally arise as a plant cell divides; the two new cells fail to separate completely, leaving threads of cytoplasm, the plasmodesmata, and with them, direct means of intercellular communication and integration via the exchange of molecules, ions, and so on. If a gardener grafts the stem of a fragile but beautiful rose onto a hardy root stock, the graft "takes" only when plasmodesmata bridges become established between graft and host tissue cells.

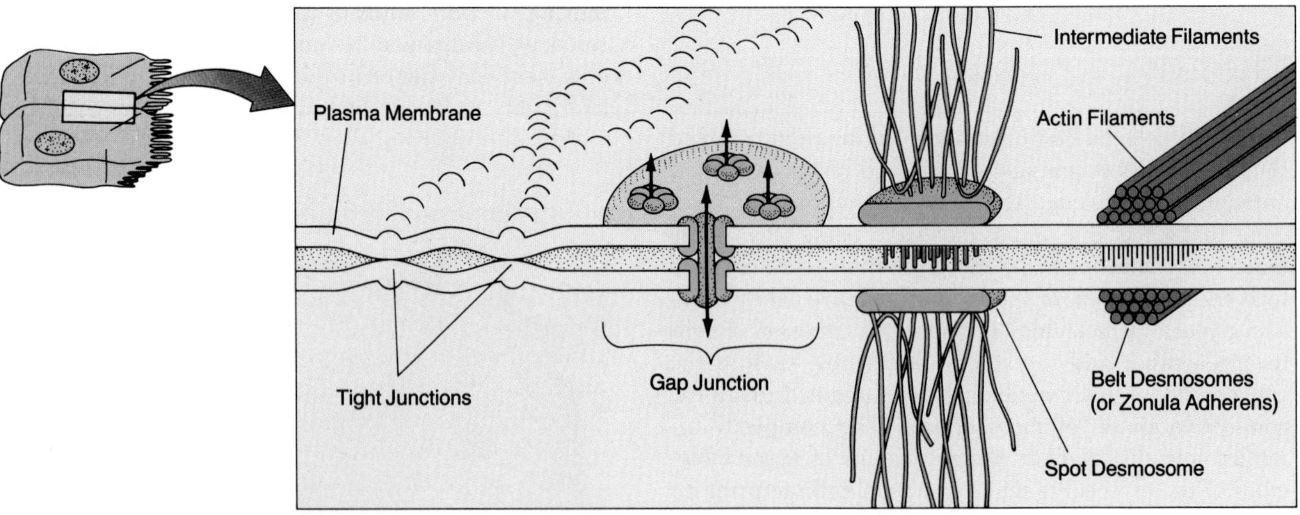

Figure 5-18 CELLULAR JUNCTIONS.
Cellular junctions occur in several forms and in a precise order. Zonulae adherens provide anchors for the cytoskeleton. Desmosomes, which run as bands around cells or as "spots," serve as strong "welds" between cells. Tight junctions can take several forms, but all serve as barriers to prevent leakage of substances into the space between cells. Gap junctions are patches where exchange of materials between cells can take place. The positions of the various types of junctions represented here are typical for many kinds of epithelial cells.

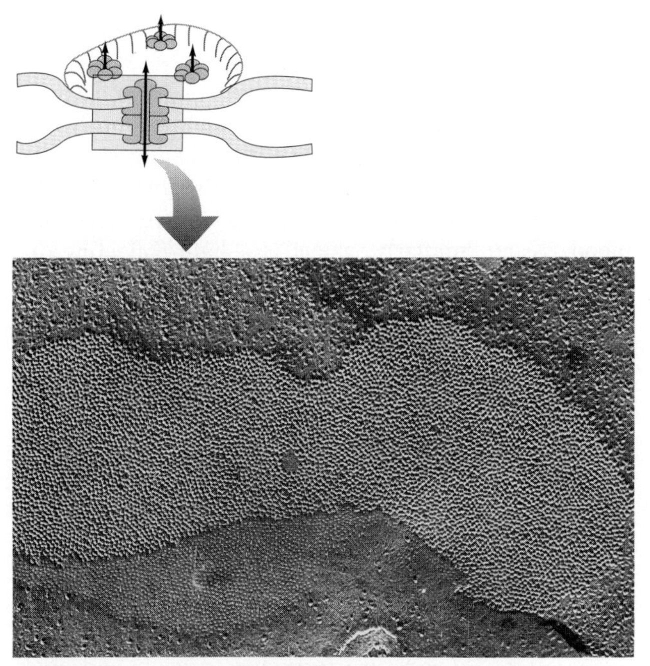

Figure 5-19 GAP JUNCTIONS ON THE LATERAL SURFACES OF EPITHELIAL CELLS.
This is a high-magnification electron micrograph (enlarged 61,500×) prepared by freeze-fracturing the surface of a mouse liver cell. It reveals closely packed particles making up the gap junction; each particle may be associated with a channel between the two cells.

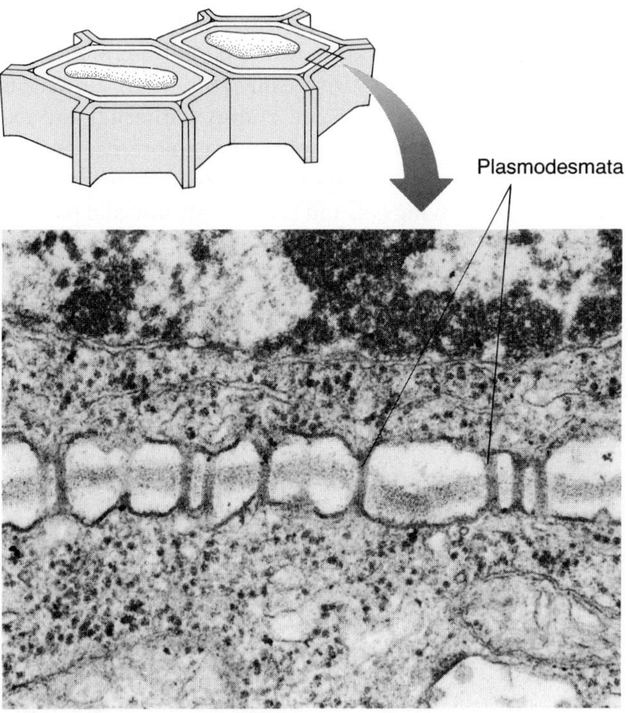

Figure 5-20 PLASMODESMATA—VITAL CHANNELS BETWEEN PLANT CELLS.
In this electron micrograph (magnified 50,350×), plasmodesmata can be seen interconnecting the cytoplasms of two young root tip cells from a timothy plant.

Recent studies of animal and plant tissue cells linked by gap junctions and plasmodesmata have led to a modification of the traditional cell theory. Biologists used to regard cells as units that are independent in structure and function. However, when linked by gap junctions, all the cells in a population, not the individual cells alone, become the unit of response and function. Thus, a regulatory molecule acting on one cell may trigger responses in adjacent cells because the "message" is passed through gap junctions. This remarkable property helps explain the coordination of cellular activity that is so essential in tissues and organisms made up of millions of cells.

MULTICELLULAR ORGANIZATION

Cell junctions are found in multicellular organisms and may have evolved as the earth's earliest single-celled organisms gave rise to multicellular colonies and, in turn, to organisms with millions of coexisting, coordinated cells. During this evolution, discrete subpopulations of cells within organisms began to carry out specialized tasks like support and movement. This division of labor led one of the founders of the cell theory, Rudolf Virchow, to compare the multicellular organism to a "cell-state," a country or state composed of individual cellular plumbers, delivery persons, managers, and other "citizens."

Tissues, Organs, and Systems

The individual cells in a multicellular organism are arranged into groups that function collectively as **tissues.** Tissues, in turn, are combined into *organs* and *organ systems*. The physiological division of labor in the "cell-state" depends on such arrangements.

Each tissue type is composed of cells performing the same or closely related functions. Epithelial tissue, for instance, consists of sheets of cells that cover organisms and organs and line cavities (Figure 5-21). Different types of epithelial tissue cover the human body and line the mouth, gut, lungs, heart, and blood vessels. Epithelial tissue also covers the leaves and stems of plants.

Epithelial cells can be squarish *(cuboidal)*, rectangular *(columnar)*, or flat *(squamous)*. Despite differences in shape, one feature common to all epithelial cells is that they usually adhere tightly to the similar cells on either side of them, joined by zonulae, desmosomes, and other junctions. Tight junctions between epithelial cells seal off a tissue from adjacent open spaces.

Another common tissue is *connective tissue*, the fi-

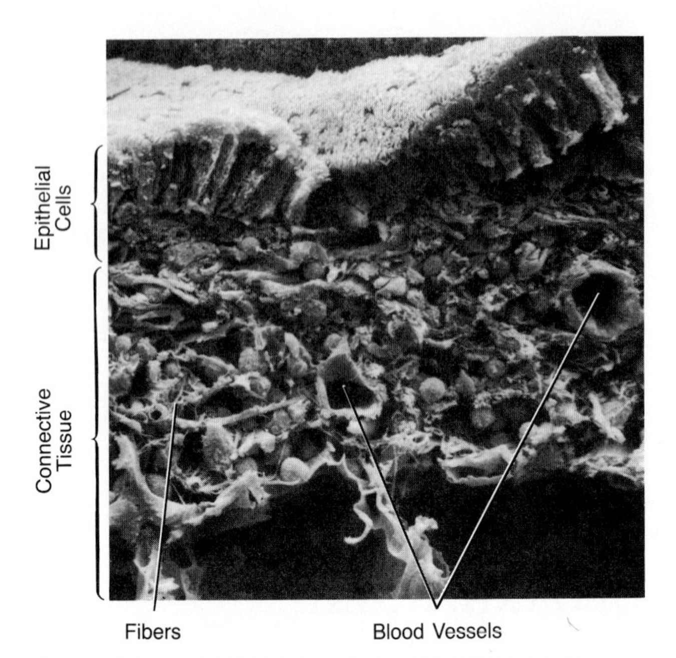

Figure 5-21 EPITHELIUM AND CONNECTIVE TISSUE. Typical animal tissues are epithelium and connective tissue (here magnified about 650×). The vertically oriented epithelial cells, visible at the top of this specimen of mammalian lung tissue, have many microvilli on their outer surface. The connective tissue below the epithelium has a variety of cell types plus tough fibers that give the tissue strength. Blood vessels are also visible.

brous component of most animal organs (see Figure 5-21). Related to connective tissue are bone and cartilage, tissues that are built on a fibrous base. Other types of tissue include muscle tissue, nervous tissue, and mesophyll, the light-absorbing green tissue inside leaves. Tissues in plants do not have formal boundaries; nevertheless, specialized groups of cells carry out tasks of support, transport, synthesis, and reproduction analogous to those in animals.

Organs are body parts composed of several tissues that work in concert to perform specific functions within the organism. For example, the skin is an organ that is made up of epithelial tissue, connective tissue, muscle tissue (such as the tiny muscles that move hairs), and blood tissue. Similarly, a leaf is a light-collecting and carbohydrate-producing organ made up of waterproof epithelia, conducting tissues, sugar-storing cells, and so on.

The final level of this hierarchy is the **organ system,** a collection of organs that carry out the various aspects of a complicated activity. In animals, the digestive system, for example, includes the diverse parts of the gut (esophagus, stomach, intestines, etc.) plus the salivary glands, liver, pancreas, and other organs. In plants, the water conduction system, for example, includes root hairs, vessels that extend up the root and stem into the leaf, and guard cells on the underside of the leaf that regulate evaporation. We'll return to tissues, organs, and organ systems in several later chapters.

While multicellularity and its division of labor allow greater functional efficiency, larger size, and a wider range of life-styles, there is a major price: Only certain cells in the reproductive organs contribute to the next generation, and the remainder of the organism dies. There is another price, as well: Single cells isolated from a plant or animal cannot live as independent organisms, but must rely on the unique internal environment found only within an intact organism. This environment is maintained through continual energy gathering and expenditure. Before we can profitably explore the energy flux between a cell and its environment, however, we must enter the living cell and examine its internal parts—the subject of Chapter 6.

SUMMARY

1. Most cells are minute, and so their discovery and subsequent examination followed the invention and development of light and electron microscopes. Microscopes offer the advantages of magnification and greater resolving power. Other methods of studying cells include cell fractionation, radioactive isotope labeling, and autoradiography.

2. The modern *cell theory* can be summarized as follows: Cells are the basic structural and functional units of life; all organisms are composed of cells; and all cells arise from preexisting cells.

3. Some cells are free-living, single-celled organisms, but others make up multicellular organisms. A *prokaryotic cell* lacks a membrane-bound nucleus, whereas a *eukaryotic cell* has one, as well as other *organelles* and a cytoskeleton.

4. *Autotrophs* can manufacture their own energy-rich organic compounds, while *heterotrophs* must obtain their energy by taking in autotrophs or other heterotrophs.

5. An animal cell is enclosed by the *plasma membrane*, which itself is surrounded by the *glycocalyx*, a patchy coating of molecules that mediates the cell's interaction with its environment. The cytoplasm, or semifluid interior of the cell, contains a cytoskeleton, various organelles, and the nucleus.

6. A plant cell has a plasma membrane, and surrounding it, a *cell wall,* which gives rigidity to the cell. The cytoplasm contains many of the same kinds of organelles found in an animal cell.

7. Cell size is limited by *surface-to-volume ratio.*

8. The plasma membrane is a lipid *bilayer* in which proteins are suspended. *Integral* proteins have hydrophobic exteriors and are found within the bilayer. *Peripheral* proteins are attached to either surface of the bilayer. The *fluid-mosaic model* describes how lipid molecules move about freely in the plane of the membrane.

9. *Diffusion*—the tendency of a substance to move from a region of high concentration to a region of low concentration—accounts for the movement of lipid-soluble substances across the semipermeable cell membrane. Substances that are not lipid-soluble enter the cell through carriers that function as pores. Some *carriers* simply *facilitate diffusion*, while others consume energy (ATP) to carry out *active transport* against the *concentration gradient*. Coupled transport of substances in the same or in opposite directions also occurs.

10. *Osmosis* is the passage of water through semipermeable membranes from a solution of low salt concentration to one of higher salt concentration. When the extracellular fluid is *isotonic* relative to the intracellular fluid, there is no net movement of water. When the extracellular fluid is *hypotonic*, water rushes into the cell. When the extracellular fluid is *hypertonic*, water leaves the cell, which shrinks. Water that enters a cell exerts a pressure outward; this is *osmotic pressure*.

11. Plant, bacterial, and some fungal cells are encased and protected by *cell walls*. In plants, these are largely *cellulose;* in bacteria, they are largely polysaccharides, lipids, amino acids, and sugars; in fungi, they include cellulose, *chitin*, and other polysaccharides. Animal cells possess a *glycocalyx* of sugars, lipids, and proteins.

12. *Zonulae adherens* and *desmosomes* provide firm structural linkage between some animal cells. *Tight junctions* seal plasma membranes together at ridges and so prevent leakage. At *gap junctions*, ions, small molecules, and electric currents can pass through pores. *Plasmodesmata* are large bridges between plant cells that may permit exchange of many substances.

13. Cells of multicellular creatures are organized into *tissues*, *organs*, and *organ systems* so that the activities of individual cells can be integrated into the whole functioning organism.

KEY TERMS

active transport
anchorage dependence
autotroph
bilayer (lipid)
carrier-facilitated diffusion
cell
cell theory
cell wall
chitin
concentration gradient
cotransport
desmosome
diffusion
eukaryote
protein
fluid-mosaic model
gap junction
glycocalyx
heterotroph
hypertonic solution
hypotonic solution
integral protein
isotonic solution
organ
organelle
organ system
osmosis
osmotic pressure
peripheral protein
plasma membrane
plasmodesma
prokaryote
surface-to-volume ratio
tight junction
tissue
zonula adherens

QUESTIONS

1. The cell theory is a cornerstone of biology. Explain the theory and why it is so central to science.

2. Draw typical prokaryotic and eukaryotic cells, and label the major parts of each.

3. Distinguish between the cell wall, the glycocalyx, and the plasma membrane. Which would you prefer to "wear" as a raincoat? Why? Which would best support a tree trunk? Why?

4. How does the surface-to-volume ratio help explain why cells are so small?

5. With labeled diagrams, show how various charged and uncharged substances can enter cells. Which processes require energy? Why don't the others?

6. Kidney disease might result in wild fluctuations of the salt concentration in an animal's extracellular fluids. How would this affect cells, and why?

7. How does tonicity relate to osmosis? What is osmotic pressure, and what is an example of its effects?

8. Prepare a table listing the types of animal and plant cell junctions, their structures, and their functions.

9. How are cells, tissues, and organs related? What is a heart? The skin surface? Red blood cells?

ESSAY QUESTIONS

1. What would happen if the mean temperature of all living things was suddenly lowered, making the plasma membranes of cells no longer fluid? Explain.

2. Are redwood trees and whales constructed from the same numbers of cells as rosebushes and minnows? Which has more and why?

For additional readings related to topics in this chapter, see Appendix C.

6

Inside the Living Cell: Structure and Function of Internal Cell Parts

The cell is the basic "module," the common denominator of all the immense variety of living forms. One must not forget, however, that all cells also play specialized roles over the entire range of diversity in biological form and function.

Ariel G. Loewy and Philip Siekevitz, *Cell Structure and Function* (1969)

Every living cell is like a miniature factory, whether that cell is a self-contained, independent organism, such as an amoeba, or just one tiny unit in a yellow dandelion petal or a gossamer dragonfly wing. In this living cellular factory, raw materials are turned into products to be stored, used in the cell, or exported. Like all factories, the cell consumes energy and creates wastes. But the outcome of its manufacturing process is not the production of gadgets to sell; it is the ephemeral quality we call life.

The cell is a model of organization within organization. The tasks of the cellular factory are carried out by a number of special units, the organelles. Organelles are built from orderly arrays of macromolecules, and these, in turn, are composed of small molecular subunits.

In a sense, cell organelles are the bridge between chemistry and biology, between molecules and life. Although no single organelle is alive, the precise arrangement of structures within the cell and the interrelated, simultaneous accomplishment of specialized functions by these structures account for the life process and, with it, for the cell's survival, reproduction, and function.

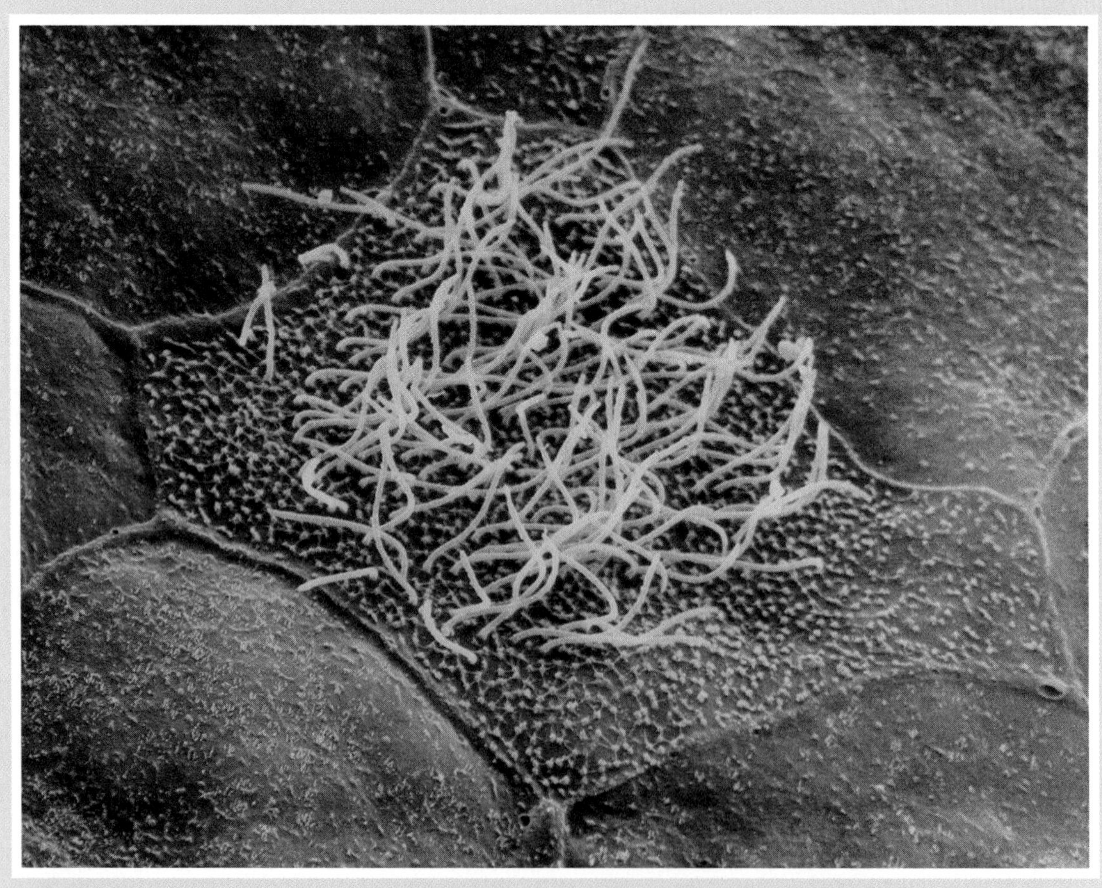

Cilia, microtubule-containing, hairlike structures, beat and propel fluid past these cells on the surface of a salamander embryo (magnified about 160×).

In this chapter, we pass beneath the cell surface and consider:

- The cytoplasm, or dynamic cellular ground substance
- Cellular organelles, structures with specific tasks in the cell's collective activities
- Cellular movements, based on the microscopic filaments and tubules of the cytoskeleton

CYTOPLASM: THE DYNAMIC, MOBILE FACTORY

Most of the mass of every prokaryotic and eukaryotic cell consists of **cytoplasm,** a semifluid substance bounded on the outside by the plasma membrane and viewed, by some, as being a bit like viscous bouillon filled with vermicelli. The organelles are suspended, each located in an ultrafine, dynamic, three-dimensional lattice (the "vermicelli") of filamentous proteins called the cytoskeleton or sometimes the cytomatrix. Soluble enzymes and proteins may be loosely attached to the cytoskeleton or to the surfaces of membranes. Dissolved in the cytoplasmic fluid (the "bouillon") are nutrients, ions, and other raw materials needed for the work of the cellular factory.

In the nineteenth century, biologists wondered whether the living state derives directly from properties of the translucent cytoplasm or whether the nucleus, with its genetic material, continuously instills eukaryotic cells with life. They wondered what would happen if the nucleus was removed from the cytoplasm: Would that semifluid substance die instantly or continue to exhibit lifelike activity for some period of time? To study this question, they and their twentieth-century successors developed various ways to remove the nucleus from large eukaryotic cells that could be manipulated fairly easily. Using modern experimental techniques, a researcher can suspend single cells in a nutrient fluid and treat them with a drug that weakens the cell surface and cytoskeleton. Then the living cells are centrifuged. Each nucleus literally pops out, surrounded by a thin halo of cytoplasm, and leaves the remainder of the cytoplasm intact within the plasma membrane. The nucleus soon degenerates. In contrast, the enucleated cytoplasm usually survives for many hours as a perfectly viable entity. It can move about normally, absorb and use nutrients and oxygen, synthesize proteins, excrete waste products, and so on. Not until the isolated cytoplasm needs further instructions from the genes that reside in its nucleus does the absence of that organelle become apparent. Only then does the enucleated cytoplasm gradually degenerate and die. These experiments show that most of the properties that we associate with life (see Chapter 1) are properties of the cytoplasm. Nevertheless, cellular reproduction depends on heredity, and thus the nucleus is critical to the maintenance of life in eukaryotic cells.

THE NUCLEUS: INFORMATION CENTRAL

The cell **nucleus** serves as the control room for most operations of the eukaryotic cell (Figure 6-1a). The largest cell organelle, the nucleus is often a rounded structure housing **chromosomes,** long strands of coiled DNA and proteins that contain **genes**—the basic blueprints for proteins. All cells depend on proteins for metabolism, shape, special functions, cell division, and other processes.

Genes are never directly used as patterns for the production of proteins. Instead, in all cells, genetic plans are copied on an intermediary molecule called *messenger RNA (mRNA).* In eukaryotic cells, mRNA bearing these genetic plans moves from the nucleus into the cytoplasm, where proteins are synthesized.

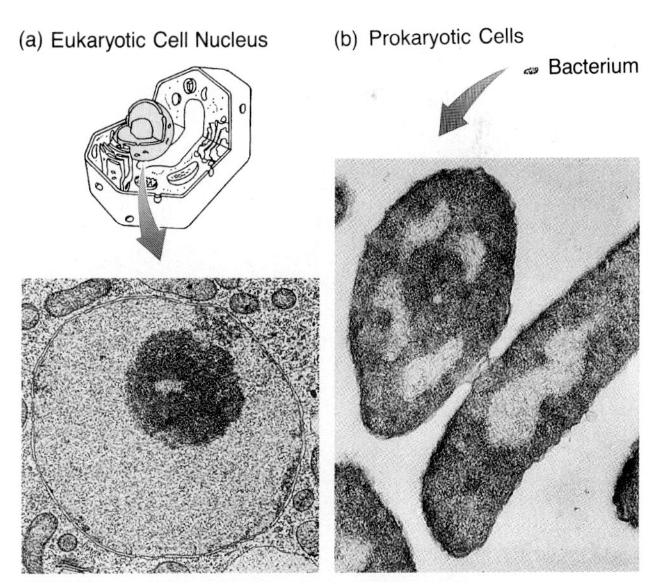

Figure 6-1 NUCLEUS VERSUS NUCLEOID: EUKARYOTIC AND PROKARYOTIC GENETIC CENTERS.
(a) The nucleus of a eukaryotic root tip cell can be seen here (magnification 14,500×). Note the double-layered nuclear envelope; the perforations in it are called nuclear pores. The large, darkened area in the nucleus is the nucleolus. (b) The light-colored, unbounded, irregular-shaped nucleoid regions are clearly visible in both of these mating prokaryotic bacterial cells, magnified about 40,000 times.

It is not clear why chromosomes are segregated in the nucleus of a eukaryotic cell (see Figure 6-1a), but perhaps the lengthy DNA molecules in chromosomes would become tangled or broken were they not segregated in a compartment. Alternatively, the separation of chromosomes from cytoplasm may allow more effective regulation of genes and a special processing of mRNA. In prokaryotic cells, one long, circular strand of DNA serves as a single chromosome. It is usually concentrated in one dense, unbounded area of the cell called the **nucleoid** (Figure 6-1b).

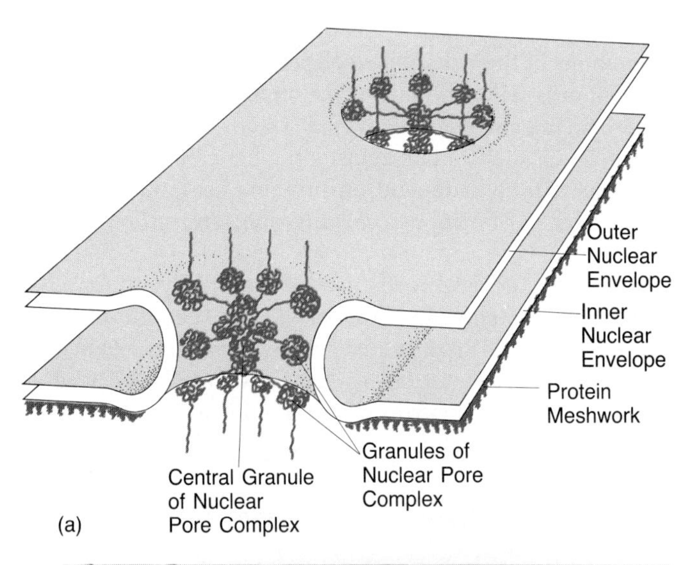

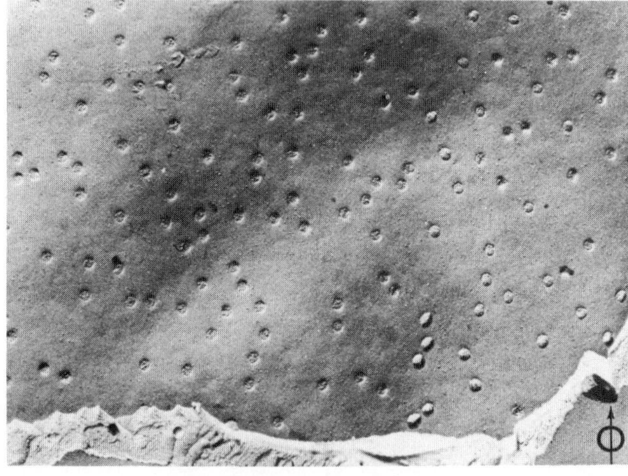

(b)

Figure 6-2 NUCLEAR PORES.
Nuclear pores allow materials to pass between the nucleus and cytoplasm. (a) An artist's interpretation of pore structure, revealing how the pore spans the two-layered nuclear envelope. Protein granules around the edges and in the center of the pore govern what passes through the pore. (b) In this freeze-fracture preparation of the nuclear envelope from a guinea pig cell, magnified about 14,250 times, the numerous craterlike nuclear pores look like dents on a golf ball.

In addition to chromosomes, the nucleus contains one or two organelles known as **nucleoli** (singular, *nucleolus*, meaning "a small kernel"). Just after a cell divides, one or two new nucleoli appear in association with specific chromosomes of each daughter cell. Within each nucleolus are minute fibers and granular components. These granules are precursors of *ribosomes*, organelles located in the cytoplasm that are the actual sites of protein synthesis.

The contents of the nucleus are isolated from the cytoplasm of eukaryotic cells by the **nuclear envelope,** a flattened, double-layered sac filled with fluid and perforated by pores (Figure 6-2). The membrane of the nuclear envelope is similar to the plasma membrane and the membranes of other organelles in that it is a lipid bilayer with associated proteins. Small molecules and ions can be transported through the double membrane, but large molecules (such as mRNA and proteins) and ribosomes can leave or enter the nucleus only through the nuclear pores. Precisely ordered globular and filamentous proteins make up the nuclear pore granules, some of which serve to seal off the pore and govern transport of large molecules into and out of the nucleus (Figure 6-2).

Biologists recently discovered that chromosomes are moored in a skeletal meshwork of protein attached to the inner layer of the nuclear envelope (see Figure 6-2) and probably do not drift about randomly inside the nuclear sap.

ORGANELLES: SPECIALIZED WORK UNITS

Suspended in the cell's semifluid cytoplasm are a variety of organelles—factory departments, if you will—each with a special function.

Ribosomes: Protein Synthesis

Ribosomes are perhaps the most numerous organelles within a cell. Anywhere from a few hundred to many hundreds of thousands of ribosomes may be present in the cytoplasm, either free inside the cell or attached to various membranes. Ribosomes are the sites at which amino acids are assembled into proteins, and their abundance is related to the importance of their function in the cell and the speed with which the cell can produce proteins.

Ribosomes of both prokaryotic and eukaryotic cells have a common structure. A complete ribosome is a submicroscopic particle only 25 nm—about one-millionth of

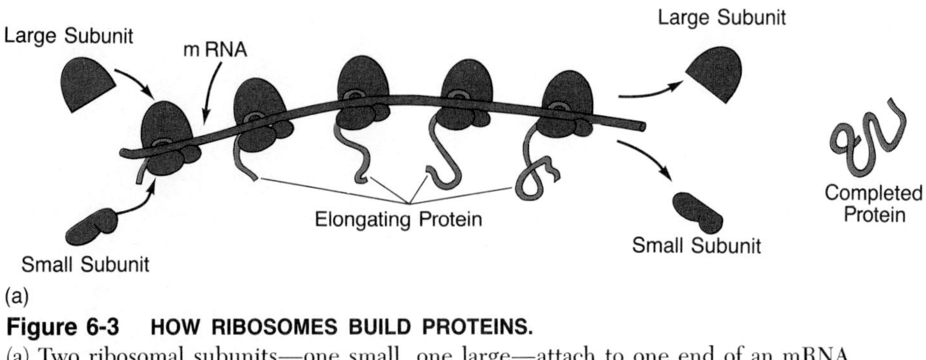

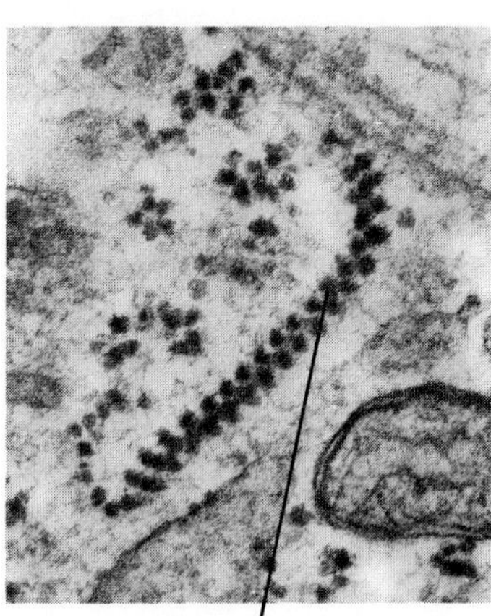

Figure 6-3 HOW RIBOSOMES BUILD PROTEINS.
(a) Two ribosomal subunits—one small, one large—attach to one end of an mRNA template. As each ribosome moves along the mRNA, "reading" the sequence coded in it, the new protein chain elongates. When the protein is released, the ribosomal subunits drop off the mRNA and reenter the pool of free subunits in the cytoplasm. (b) An electron micrograph of a free polysome (a group of ribosomes assembled on an mRNA molecule), magnified about 31,200 times, in the cytoplasm of a chick embryo's nerve cell.

an inch—in diameter. It weighs about the same as 100 to 150 average protein molecules and is composed of two globular subunits that differ in size and function. Each subunit contains large structural RNA molecules and many structural proteins. When an mRNA molecule leaves the cell nucleus, a large and a small ribosomal subunit join it at one of its ends (Figure 6-3a). Once the subunits combine, the single functioning ribosome moves along the length of the mRNA, as though reading the sequence of RNA instructions sent from the nucleus and translating the instructions into a protein. In accordance with each instruction on the mRNA molecule, a specific amino acid is linked to the end of an elongating protein chain by means of a peptide bond. This lengthening chain appears to protrude through a hole in the large ribosomal subunit. When the ribosome has finished reading the entire mRNA molecule and the protein copy is complete, the ribosomal subunits are released from the mRNA into the cytoplasm, ready to associate with another mRNA molecule and to participate in the next round of protein synthesis.

A number of ribosomes may move along a single mRNA molecule at the same time. One mRNA molecule plus several ribosomes is called a **polysome** (Figure 6-3b). Most *cellular proteins*—those released into the cytoplasm—are made on polysomes located free in the cytoplasm. Cellular proteins include common enzymes, some structural proteins, and certain membrane proteins. Some of the cytoplasmic proteins possess targeting signals that direct them to enter specific organelles, such as mitochondria or the nucleus. Cells also make two other types of proteins: *exportable proteins*, such as hormones, digestive enzymes, and structural proteins that become part of the cell wall, the glycocalyx, or the tissue spaces; and *membrane proteins*, which are incorporated into the plasma membrane and the membranes of organelles. Most exportable proteins and membrane proteins are made on membrane-bound polysomes associated with a special compartment, the endoplasmic reticulum. This ensures that exportable and membrane proteins do not mix with cellular proteins. It also allows the exportable and membrane proteins to be modified chemically and to be transported to specific sites in the cell, including sites of export.

Endoplasmic Reticulum: Production and Transport

In the mid-1950s, with the aid of the electron microscope, Rockefeller University researchers Keith Porter and George Palade discovered a lacy array of membranous sacs, tubules, and vesicles within the cytoplasm. They called this array the **endoplasmic reticulum** (meaning "intracellular web or network"), or **ER.** Further studies showed that the ER could be rough or smooth. The *rough endoplasmic reticulum (RER)* is studded with ribosomes, while the *smooth endoplasmic reticulum (SER)* is not. The business of these organelles is the synthesis, modification, and transport of materials.

Rough Endoplasmic Reticulum

The outer (cytoplasmic) surface of the RER is dotted with thousands of polysomes, each of which consists of an mRNA molecule loaded with a variable number of ribosomes. As protein is synthesized on these bound polysomes, the elongating peptide chains pass into channels, or *cisternae*, between the RER membranes (Figure 6-4). Once inside this membranous transport network, the proteins move to other organelles for packaging, storage, export, or modification.

In most proteins destined for export or for insertion in membranes, the first 20 or so amino acids of the protein

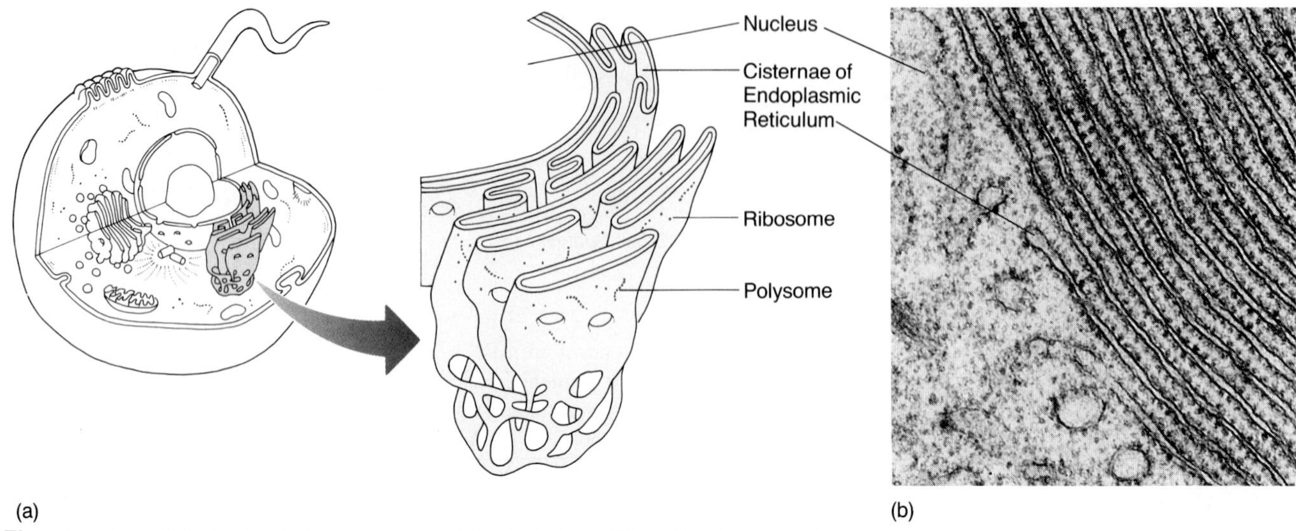

Figure 6-4 ROUGH ENDOPLASMIC RETICULUM: A PROTEIN-BUILDING FACTORY.
(a) This diagram shows an enlarged fraction of the RER. Stacked sacs serve as sites for ribosome attachment and channels for protein transport. A string of ribosomes is an individual polysome; elongating protein chains pass from the polysomes into the cavity of the RER. (b) This electron micrograph shows the RER in an animal cell, magnified about 4,000 times. Digestive enzymes and other exportable proteins are synthesized on these stacked membranes and are stored in the large, circular storage granules.

chain appear to act as a *signal peptide*, or address label. For some proteins, this label binds to a membrane receptor on the RER, thereby ensuring the passage of the elongating protein through the membrane and into the RER cisterna (Figure 6-5). Once released, the protein travels within the channel to other places in the cell. Some proteins are synthesized in the cytoplasm and then enter the RER, mitochondria, or other membranous organelles. These proteins literally unfold and expose their signal peptide region, allowing them to pass through the membrane.

Signal peptides allow a single type of ribosome to manufacture both cellular and exportable proteins. This means that cells do not have to possess two classes of ribosomes, one for reading only mRNAs with instructions for cellular proteins and one for reading only mRNAs with instructions for exportable and membrane proteins.

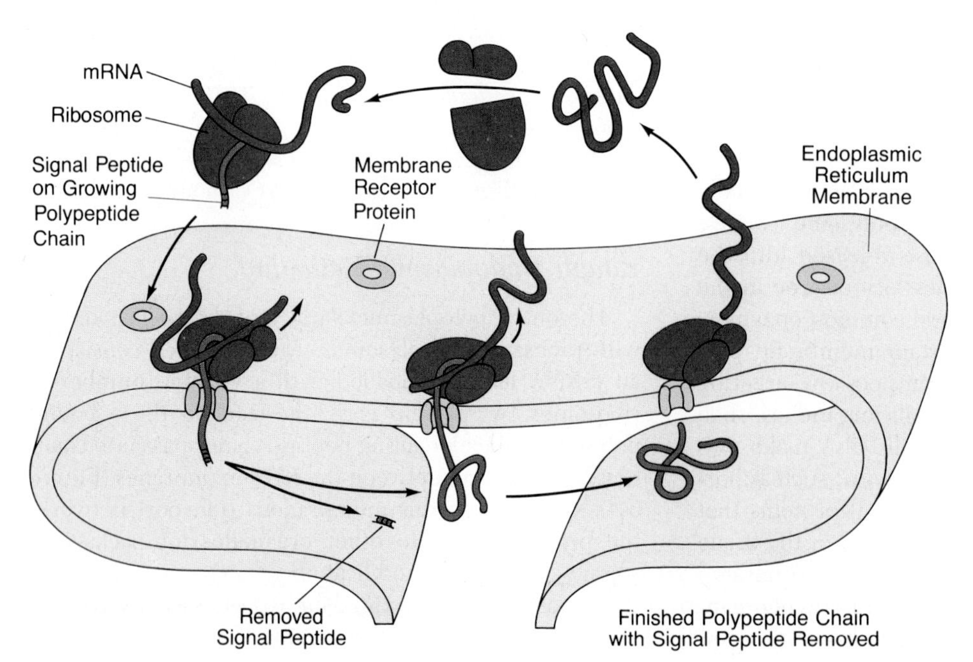

Figure 6-5 THE SIGNAL PEPTIDE HYPOTHESIS.
The signal peptide hypothesis explains how certain proteins may enter the RER. The two ribosomal subunits attach to an mRNA molecule. As the ribosome "reads" the instructions for protein synthesis, the signal peptide binds to a recognition site on the RER membrane; then the elongating polypeptide chain "grows" into the RER, ultimately to be released. Other proteins may be fully synthesized, unfold, and pass through the pore.

Proteins that have just been released into the RER cisternae can be modified by any of some 30 or 40 enzymes associated with the endoplasmic reticulum of different cells. These enzymes may add various chemical groups to newly synthesized proteins or may modify them in other ways.

In prokaryotic cells, signal peptides facilitate the correct export of proteins that can become part of the cell wall. To accomplish this task, signal peptides may cause polysomes to bind to the inner surface of the plasma membrane so that newly manufactured proteins pass through the lipid bilayer to become part of the cell wall or material on its outer surface. Other complete proteins unfold, and their signal peptide leads them through the membrane.

The RER of eukaryotic cells has another important function besides its role in the synthesis and transport of proteins: It apparently gives rise to the nuclear envelope. As we will see in Chapter 9, when a eukaryotic cell undergoes cell division, the nuclear envelope breaks into small vesicles that cannot be traced. After division, sacs of RER membrane surround the chromosomes of the two daughter cells and give rise to new nuclear envelopes.

To summarize, compartmentalization of the RER within the cytoplasm separates the proteins tagged by signal peptides for export or for insertion in membranes from the cellular proteins dissolved in the cytoplasm. Exportable proteins then move through the cell toward the outside world of the organism's tissue spaces, blood, or secretions.

Smooth Endoplasmic Reticulum

Most eukaryotic cells contain smooth endoplasmic reticulum, or SER, as well as RER. SER consists of a set of tubules or sacs that lack ribosomes and therefore are smooth (Figure 6-6). The SER and the enzymes associated with it carry out a variety of tasks, including the transportation, synthesis, and chemical modification of small molecules.

Radioactive tracers reveal that newly synthesized proteins may be transported from the RER to the SER en route to still another transport and packaging system, the Golgi complex. Besides being a cytoplasmic passageway, the SER is abundant in cells that synthesize fats and steroids. Finally, the SER is involved in the oxidation of toxic substances. If an animal swallows a toxic chemical such as carbon tetrachloride (dry-cleaning fluid) or a barbiturate, the SER in the liver cells becomes very prominent and active, functioning at maximum capacity until the toxic compound is broken down and excreted from the body.

Smooth endoplasmic reticulum also performs specialized functions in certain cell types. In liver cells, the SER contains a large quantity of an enzyme that helps modify glucose so that it can pass through the membrane and into the SER. Once inside the membranous channels of the SER, the sugar can be transported to the cell surface and out to needy cells throughout the body. Another specialized function occurs in skeletal muscle cells, where a special type of SER triggers muscle cell contraction in response to nerve impulses.

Golgi Complex: Modifications for Membranes or Export

What happens to the products of the ER when they leave this system of membranes with internal cavities? Transport vesicles, which are tiny membranous sacs, pinch off from the RER or SER and carry exportable molecules to the **Golgi complex,** a stack of saucer-shaped, baglike membranes surrounded by vesicles, as shown in Figure 6-7.

Discovered by Nobel Prize winner Camillo Golgi, the flat sacs of the Golgi complex seem to be stable structures that receive molecules brought in small transport vesicles from the RER or SER. There, the molecules are modified by Golgi enzymes. Sugars, lipids, phosphate groups, or sulfate groups may be added or removed, or the basic structure of the molecule may be altered. In animals, the molecule that undergoes such modification is usually a protein, a fat, or a steroid. In plants, it is a protein or a complex carbohydrate, such as cellulose, destined to be incorporated into the cell wall. New transport vesicles may pinch off from the side of one

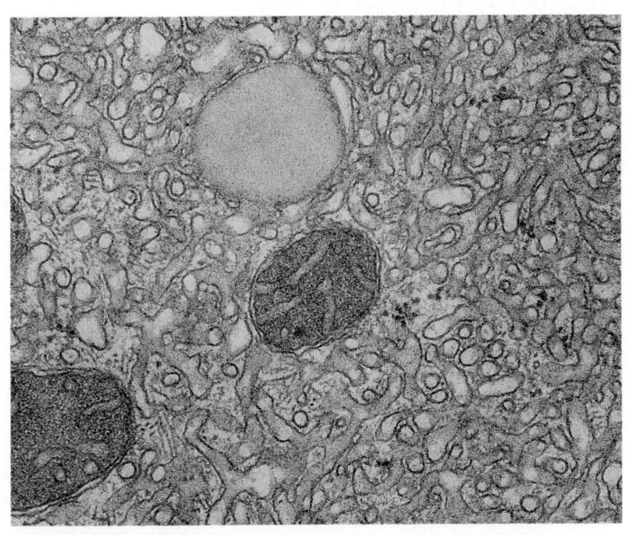

Figure 6-6 SMOOTH ENDOPLASMIC RETICULUM IN A NERVE CELL.
Membranous sacs are present without attached ribosomes in this animal cell's highly magnified smooth endoplasmic reticulum.

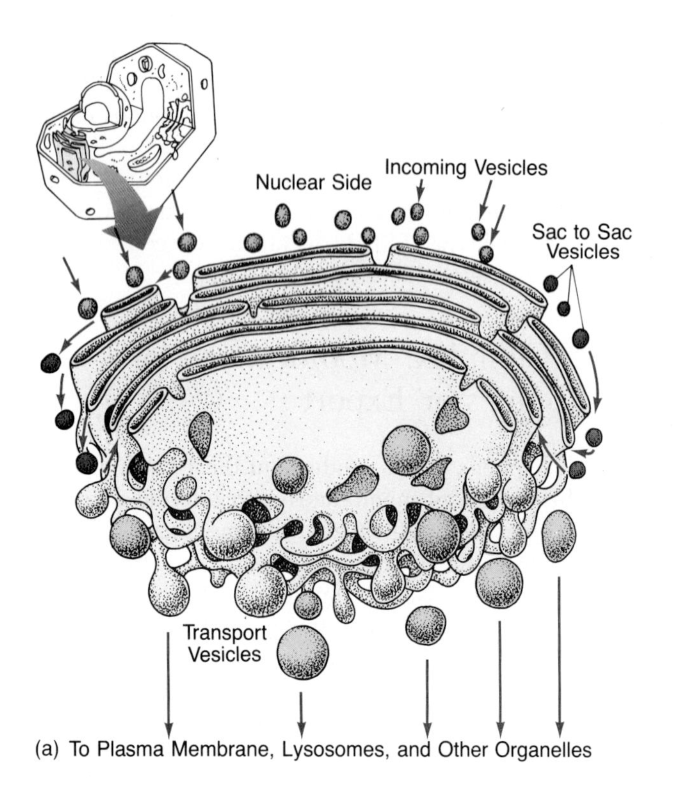

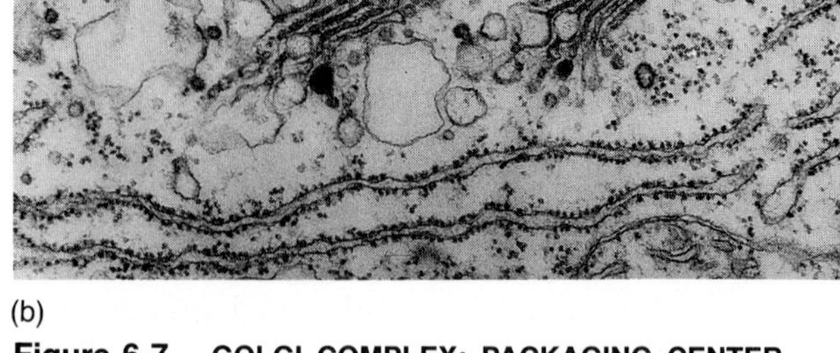

Figure 6-7 GOLGI COMPLEX: PACKAGING CENTER FOR THE CELL.
(a) Small vesicles containing newly synthesized proteins fuse with the edges of the flat Golgi sacs. Other small vesicles carry protein being processed from one sac to others. Finally, transport vesicles carry finished proteins to the plasma membrane, large storage vesicles, lysosomes, or other membranous organelles. (b) In this root cell from a radish plant, magnified about 15,000 times, the Golgi stacks are easily seen and are similar in appearance and function to those in animal cells. Although this organelle looks static, the small transport vesicles are constantly shuttling proteins to and fro.

Golgi sac and carry the contents to another Golgi sac for further modification. Then the cargo can be packaged in vesicles and transported from the Golgi sac, as Figure 6-8 shows.

Three groups of proteins pass through and are modified by the Golgi complex. The first are proteins of the nuclear envelope, the plasma membrane, and the membranes of other organelles. The second group pass to secretory granules such as those in a pancreas cell. The third group are enzymes of the lysosomes (discussed shortly). Probably through the presence of amino acid sequences called sorting or targeting domains, these

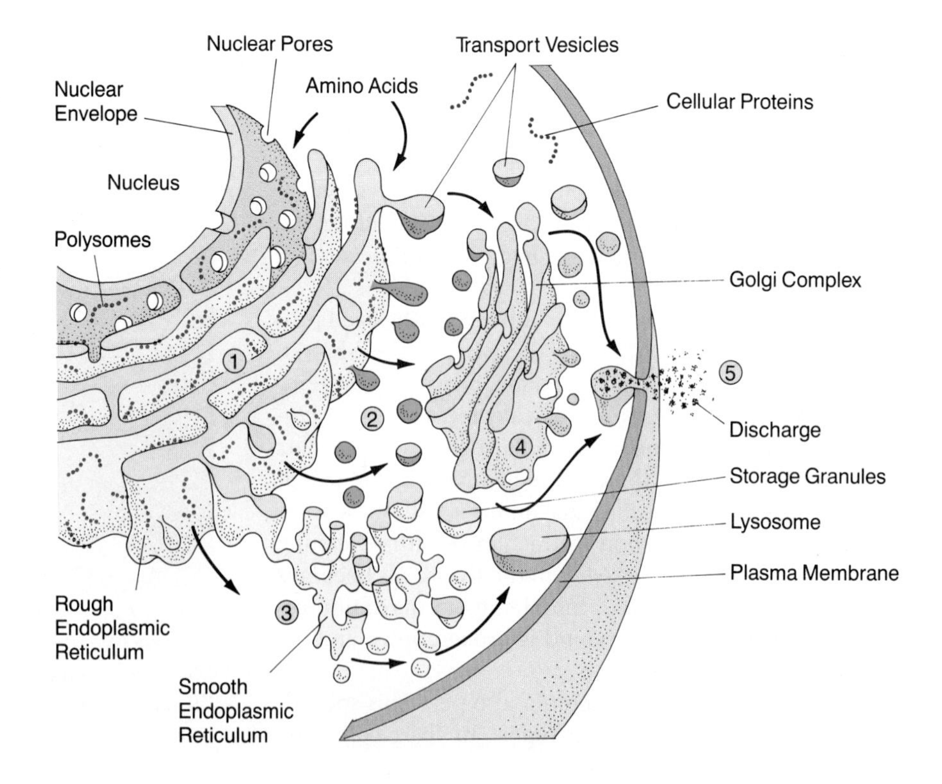

Figure 6-8 THE FLOW OF EXPORTABLE MATERIALS IN THE CYTOPLASMIC MEMBRANE SYSTEMS.
The nuclear envelope may be continuous with the RER (the site of considerable protein synthesis). The RER, in turn, contributes vesicles to the Golgi complex (the packaging site for exportable protein) or is continuous with the SER. Vesicles from the SER and from the Golgi complex fuse with the plasma membrane so that their contents may be expelled from the cell. Other vesicles carry digestive enzymes to the lysosomes.

proteins get sorted out in the specialized last sac of the Golgi complex, from which transport vesicles pinch off.

In 1964, George Palade and his colleagues at Rockefeller University confirmed the role of the endoplasmic reticulum and Golgi complex in processing, packaging, and transporting proteins from the cell. Employing autoradiography (see Chapter 5), the team found that labeled amino acids always appeared first over the RER (see Figure 6-8, step 1), then over vesicles near the RER (step 2), next over the SER and surrounding vesicles (step 3), then over the Golgi complex and large storage granules (step 4), and finally outside the cells (step 5). Clearly, the amino acids followed a cellular trade route that went: RER → SER → Golgi complex → secretory granules → cell exterior. This entire sequence, from RER to expulsion, is the main pathway for moving materials out of the cell.

Vacuoles: Food and Fluid Storage and Processing

We turn now from organelles that synthesize, transport, package, and export molecules to other organelles with functions analogous to eating, drinking, digesting, and excreting. Most important among them are **vacuoles** (from the Latin for "vacant"). Although vacuoles appear to be empty sacs, they are actually filled with fluids and soluble molecules, and they play a variety of roles.

Vacuoles are critically important to single-celled organisms such as amoebae. Such organisms face a perpetual influx of water because the tonicity of their cytoplasm is higher than that of the fresh water they inhabit. These one-celled creatures possess *contractile vacuoles*, membranous sacs that repeatedly accumulate cytoplasmic water, then contract, expelling it. The swelling of these vacuoles ultimately triggers the discharge of the contents, at which point the vacuolar and cell membranes fuse, and the contents of the vacuole are expelled.

The most prominent vacuoles appear in plant cells. Viewed at low magnification, many plant cells seem to contain little more than a huge central vacuole surrounded by a thin halo of cytoplasm (Figure 6-9). The vacuole serves as a water reservoir and as a storage site

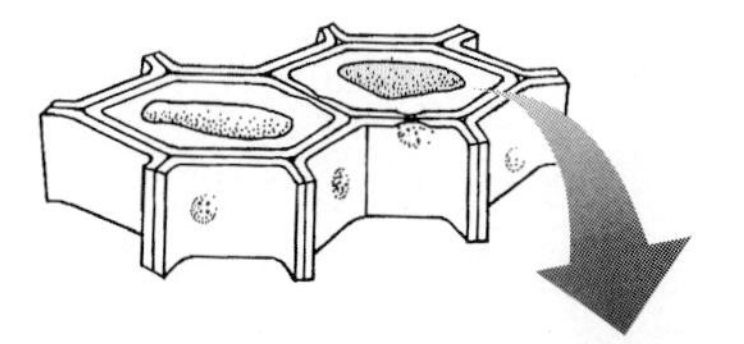

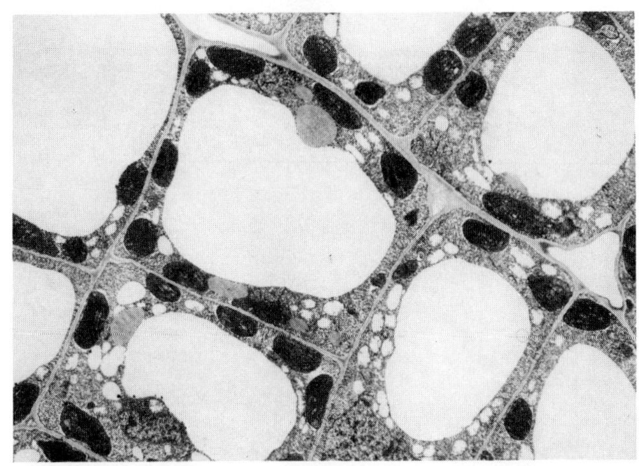

Figure 6-9 CENTRAL VACUOLES IN CELLS OF A TOBACCO LEAF.
The cytoplasm, with its many prominent chloroplasts and other organelles, streams, or circulates, around these large, central vacuoles (magnified about 2,000×).

for sugars, proteins, and the pigments responsible for the bright colors of many fruits and flowers. The fluid inside plant cell vacuoles also contributes to the turgor that keeps the cell stiff.

Many other cells also feed by means of vacuoles. Inside one-celled eukaryotes, membranous vacuolar sacs fill with food and pinch off from the large feeding organelles, commonly called gullets. This vacuolar feeding is a form of **phagocytosis,** the engulfment of particulate matter by animal cells (Figure 6-10). (Certain human white blood cells also ingest bacterial invaders by means of phagocytosis.) Food-containing vacuoles then receive digestive enzymes from the Golgi complex, and these degrade the food to component amino acids, lipids, and other nutrients.

Still another function of vacuoles is "cell drinking," or **pinocytosis.** A good example of pinocytosis occurs in narrow blood vessels called capillaries. Figure 6-11 shows

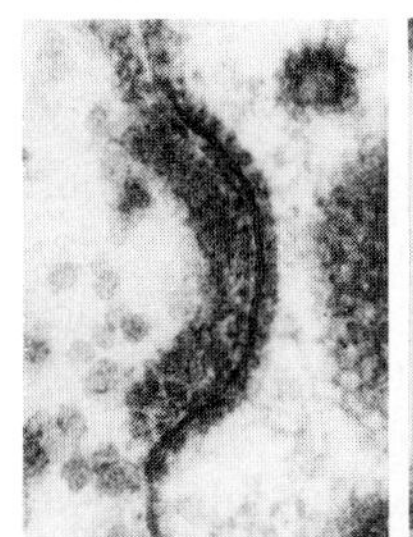

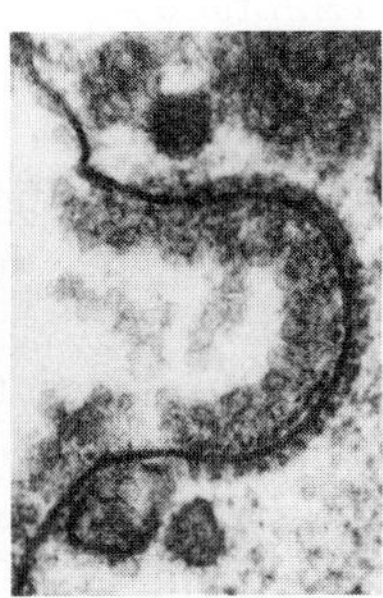

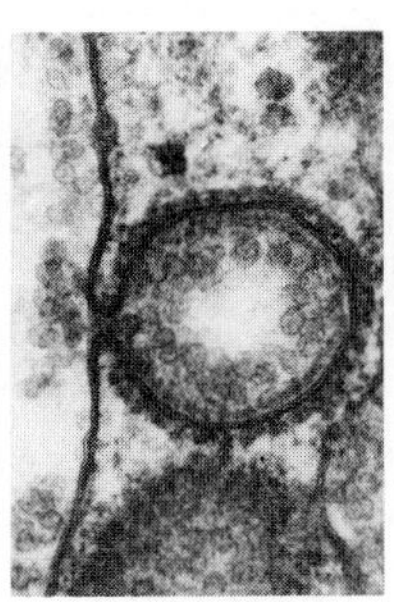

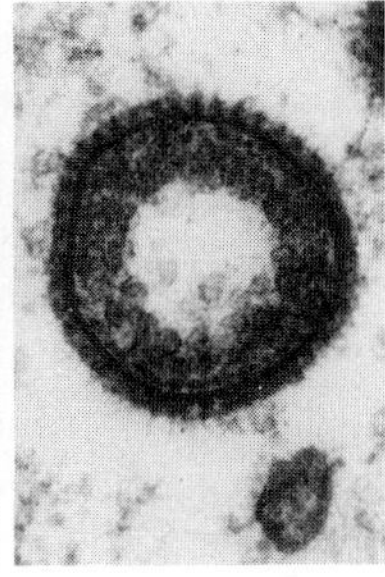

Figure 6-10 COATED PITS AND INCOMING CARGO.
Ingestion of extracellular substances by a coated pit (left). Protein particles are taken into a hen's egg cell to form yolk. The coated pit separates from the cell surface (middle) to form a coated vesicle (right). Yolk particles are visible inside, and clathrin dots the surface of the coated vesicle (all magnified about 60,000×).

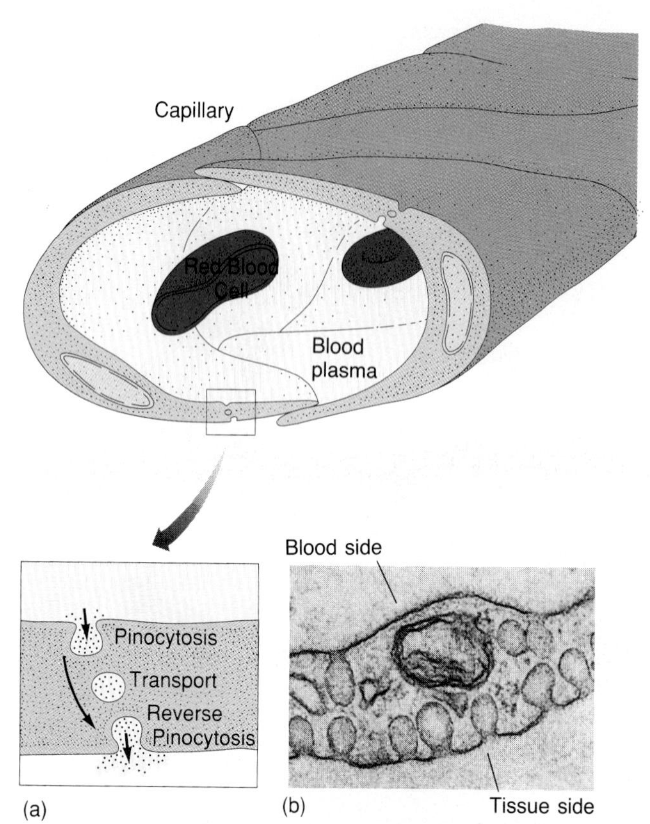

Figure 6-11 PINOCYTOSIS AND REVERSE PINOCYTOSIS IN A HUMAN CAPILLARY.
The drawings in (a) depict the events of pinocytosis, or fluid transport, across this capillary wall as recorded by the electron micrograph shown in (b) (magnification 15,000×).

how the membranes of cells that line capillaries pinch off vacuoles filled with yellowish, protein-rich blood serum, carry the fluid across the cytoplasm of the cell, then expel it from the capillary. The discharge, or "spitting out," of the fluid vacuolar contents is often called *reverse pinocytosis.* By coupling pinocytosis and reverse pinocytosis, capillary cells can transport serum from the blood vessel to the surrounding tissue space without exposing the serum components to the cytoplasm, where it might be chemically changed.

Phagocytosis and pinocytosis are examples of **endocytosis** (taking into the cell), while reverse pinocytosis is an example of **exocytosis** (putting out of the cell). Both mechanisms are an important adjunct to passive and active transport.

Coated Vesicles: Mediated Uptake and Transport

Just as George Palade traced cellular trade routes out of the cell, another group traced routes into the cell for molecules such as hormones and large proteins, which are too big to pass directly through the membrane. In the early 1980s, Ira Pastan, Mark Willingham, and other researchers at the National Cancer Institute studied an important type of endocytosis that occurs with relatively large molecules. By fluorescently labeling the molecules and following their passage into the cell with a television camera, videotape, and computer, Pastan and Willingham watched the molecules bind to receptor sites on the cell surface that are mobile in the semifluid membrane. These receptors can move laterally to specific regions called **coated pits**—indentations in the cell surface that are coated, or lined, with a protein called *clathrin* (see Figure 6-10). These pits pinch off tiny sacs called coated vesicles, which are moved through the cytoplasm toward the Golgi complex, the ER, or lysosomes. There the incoming cargo can be modified or degraded.

The team thus found the import route for large materials to be: molecule → mobile receptor site → coated pit → coated vesicle → Golgi complex, ER, or lysosomes. This is roughly the reverse of the export route and appears to involve simply a more complex passage through the membrane. Clathrin may also coat vesicles that transport materials from the ER to the Golgi complex, or from the Golgi to secretory vesicles or lysosomes. *Receptor-mediated endocytosis* has the great advantage of specificity: Only substances that bind to the membrane receptors will trigger importation.

Lysosomes: Digestion and Degradation

We saw that phagocytic vacuoles engulf food morsels, then receive an influx of digestive enzymes from the Golgi complex to help break down the ingested substance. But where do the enzymes come from? Some 50 kinds of digestive or hydrolytic enzymes can be formed in the RER and SER, then packaged by the Golgi complex into spherical membrane-bound bags called **lysosomes** which shuttle the enzymes to the vacuole. A lysosome fuses with a phagocytic vacuole and incorporates it into its structure, mingling the lysosomal digestive enzymes with the food morsels from the vacuole. The food contents in the lysosome can then be digested (Figure 6-12). The lysosomal membrane separates the rest of the cytoplasm from these powerful enzymes.

Besides helping to "feed" the cell, lysosomes play an important custodial role when cell components wear out. Lysosomal enzymes can degrade membranes, ribosomes, proteins, and a variety of other components and the subunits can then be returned to the cytoplasm for reuse. Sometimes, within injured or old cells, lysosomes also break open and free their enzymes, literally digesting the cell from the inside out.

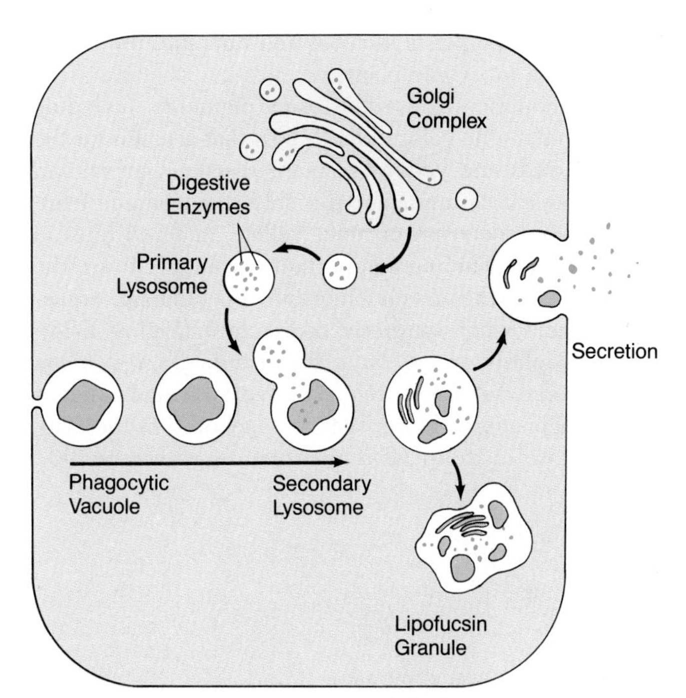

Figure 6-12 LYSOSOME LIFE HISTORY.
Lysosomes arise from the Golgi complex and fuse with vacuoles that have engulfed some foreign material. These fused structures persist for a varying length of time and then discharge their wastes or give rise to a very long-lived lipofucsin granule.

Most animal, plant, and fungal cells possess additional small membranous vesicles called **microbodies** (also known as *peroxisomes* and *glyoxysomes*). Like lysosomes, microbodies break down cellular waste products. About 1,000 microbodies are present in a typical mammalian liver cell. The microbodies of various animal and plant cells contain a number of different enzymes that, in general, carry out chemical processes called oxidation (see Chapter 7). One such enzyme is *catalase,* which splits hydrogen peroxide (H_2O_2) into water and oxygen and oxidizes an organic compound as it does so. Microbodies probably play a major role in cellular metabolism and detoxification.

Mitochondria and Plastids: Power Generators

Just like any manufacturing concern, the cell factory requires energy for its activities. Two types of organelles transform and store energy in forms usable by the cell: Mitochondria and plastids. **Mitochondria** (singular, *mitochondrion*) are sites of chemical reactions that harvest the energy from food molecules and generate high-energy compounds—such as ATP—that can be used directly to meet the cell's energy needs. **Plastids** are present in plant cells that use light energy to manufacture energy-rich carbon compounds, such as sugars, from simple inorganic raw materials.

Mitochondria

Mitochondria are the power plants of all eukaryotic cells. These organelles vary in shape from small spheres to long, sausage-shaped bodies about 1 μm wide and up to 8 μm long. Figure 6-13 shows the unique structure of a mitochondrion: It has a smooth, continuous outer membrane and an inner membrane thrown into folds called *cristae.* The outer mitochondrial membrane is

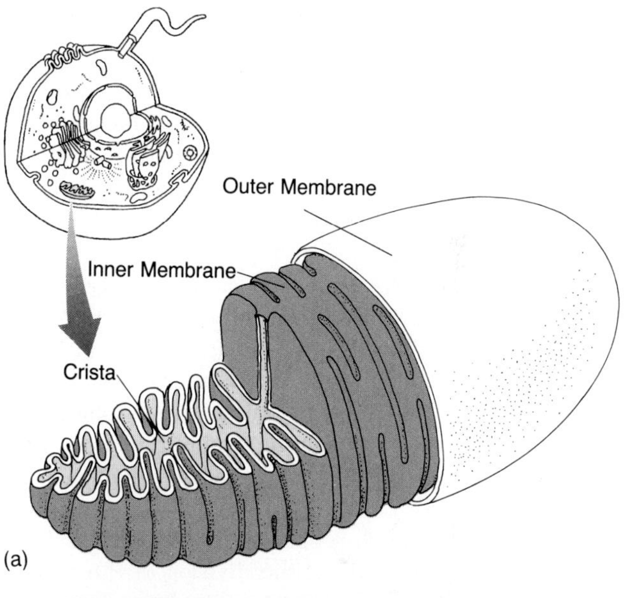

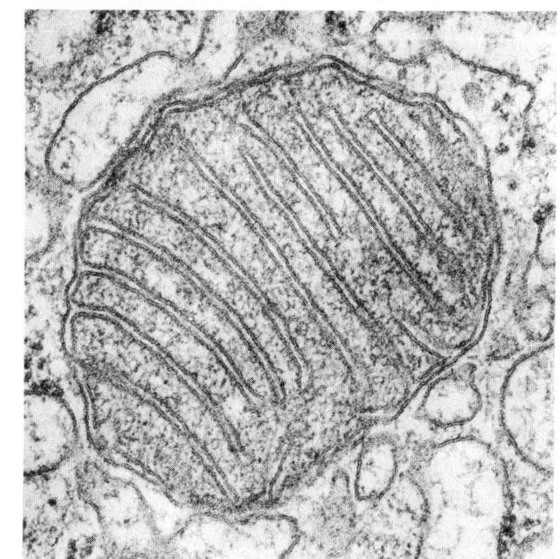

Figure 6-13 MITOCHONDRION: CELLULAR POWERHOUSE.
(a) An artist's depiction shows the cristae of the inner membrane in three dimensions. The inner matrix that bathes the cristae contains DNA, ribosomes, and many kinds of enzymes. (b) An electron micrograph reveals the membranous nature of the cristae, magnified about 54,000 times.

quite permeable, even to fairly large polypeptides. The cristae provide a large surface area on which many of the enzymes involved in generating ATP molecules are positioned. The proximity and spatial arrangements of these enzymes are critical to the cell's energy transformations (see Chapter 7). The number and size of mitochondria in a cell, as well as the number of cristae in each, depend on the cell's energy requirements. Cells with high energy requirements have numerous large mitochondria with many cristae, whereas cells with low energy needs have few or smaller mitochondria.

The cristae protrude into a semifluid **matrix** with a gel-like consistency that may stem from its high protein content. The matrix contains ribosomes constructed of RNAs and proteins that differ from those in the cytoplasm. Mitochondrial ribosomes are like those of prokaryotic bacteria. Every mitochondrion has its own genetic material in the form of DNA. In mitochondria of all eukaryotes, the genetic code is probably unique and differs from that of either eukaryotic or prokaryotic chromosomes.

Mitochondrial DNA codes for various mitochondrial proteins, including some of the subunits of the ATP-generating enzymes. Surprisingly, other subunits of those enzymes are encoded in the cell's nuclear DNA; the latter subunits are made on cytoplasmic polysomes and somehow enter the mitochondria, where they link up to the mitochondrial subunits to form the final enzyme molecules.

Mitochondria are self-replicating bodies; that is, they reproduce independently of the division of the rest of the cell. They can do so because they have their own genetic material. Indeed, it is very likely that mitochondria are descendants of unknown prokaryotes that invaded eukaryotic cells billions of years ago (see Chapter 19).

Plastids

A sure way to tell a plant cell from an animal cell is to look for plastids; every plant cell has at least one form of plastid in its cytoplasm in addition to the other organelles we have been discussing. Plastids turn plant cells into photosynthetic, carbohydrate-producing factories: They are responsible for capturing light energy to produce sugar and for storing sugar in the form of starch. The presence of plastids in the cells of plants allows them to produce their own food molecules from simple raw materials—CO_2, H_2O, and minerals.

There are two types of plastids. Those that lack pigments are the *leucoplasts*, and those that contain various pigments are the *chromoplasts*. The colorless leucoplasts serve as storage bins for starch, proteins, and oils that can be tapped by the plant as needed. Much of the food value of potatoes, carrots, and beets is stored in leucoplasts, and leucoplasts in seeds and nuts fuel the development of embryonic plants.

Chromoplasts can contain many pigments, including *carotenoids*—the colored molecules that account for the brilliant reds and yellows of maples, oaks, aspens, and other trees in autumn; for the colors of ripening fruit; and for the spectrum of pink, yellow, and red hues in flower petals. The most important chromoplasts are the **chloroplasts**, the green chlorophyll-containing organelles where photosynthesis takes place (Figure 6-14).

Chloroplasts are as large as an animal's red blood cells—some 3–8 μm in size. In a typical cell of a higher plant, 30 or more chloroplasts are anchored in the cytoplasm close to the plasma membrane (see Figure 5-5).

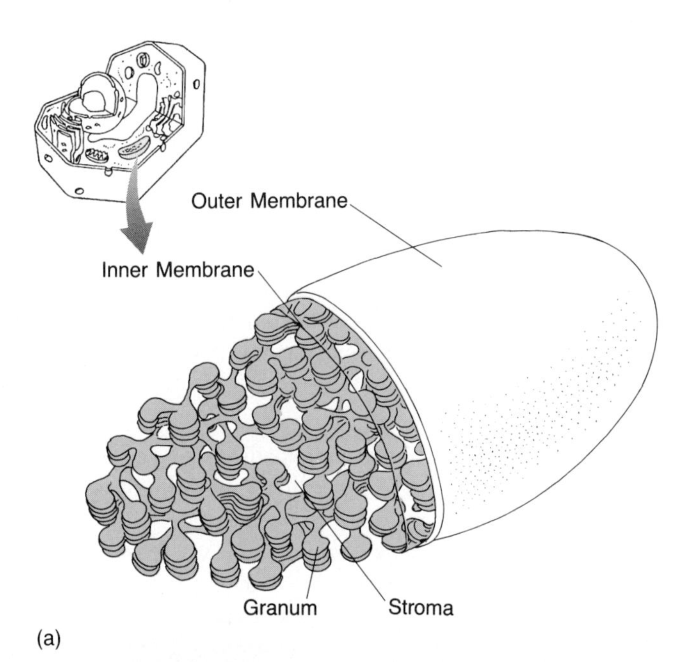

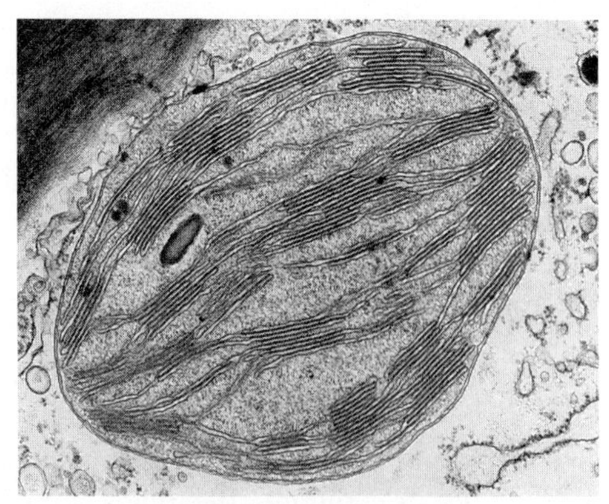

Figure 6-14 CHLOROPLAST: THE GREEN PLANT'S CARBOHYDRATE FACTORY.
A large number of stacks of flat sacs are visible in this chloroplast from the green alga *Vitella*. Each stack, or granum, has a huge surface area so that photons of light are likely to be captured, and thus energy can be used to build carbohydrates (see Chapter 8). The stroma fluid phase contains soluble enzymes, DNA, and ribosomes.

Two lipid bilayer membranes enclose each chloroplast, and much of the organelle's interior is filled with flattened membranous sacs, the *grana* (singular, *granum*), arranged much like stacks of coins (see Figure 6-14). The grana are surrounded by a matrix called the **stroma.** Enzymes required for photosynthesis are situated on the granum membranes, and their architectural arrangement is crucial to the photosynthetic process (see Chapter 8). Like mitochondria, chloroplasts are self-replicating organelles that contains circular strands of DNA but do require some proteins encoded by the cell's nuclear DNA and synthesized in the cytoplasm. All plastids arise from pigmentless *proplastids* and are to some extent interconvertible and able to lose or gain specific pigments. Some biologists trace the origins of plastids to prokaryotic cells that took up residence in some of the earth's earliest eukaryotic cells (see Chapter 19).

THE CYTOSKELETON

Another intricate unit bridges the mysterious gap between molecules and living cells: It is the **cytoskeleton,** the dynamic three-dimensional, weblike structure in the cytoplasm of all eukaryotic cells and in which the organelles are suspended. This intracellular scaffolding acts as both muscle and skeleton for the cell; allows the cell, its complex surface, and its organelles to move; and gives the cell its normal shape and keeps its parts in proper spatial relationship to each other. The cytoskeleton enables eukaryotic cells to carry out activities impossible for prokaryotic organisms.

The cytoskeleton is actually a convoluted latticework of microscopic filaments and tubules that seems to occupy most available space in the cell. Most abundant are

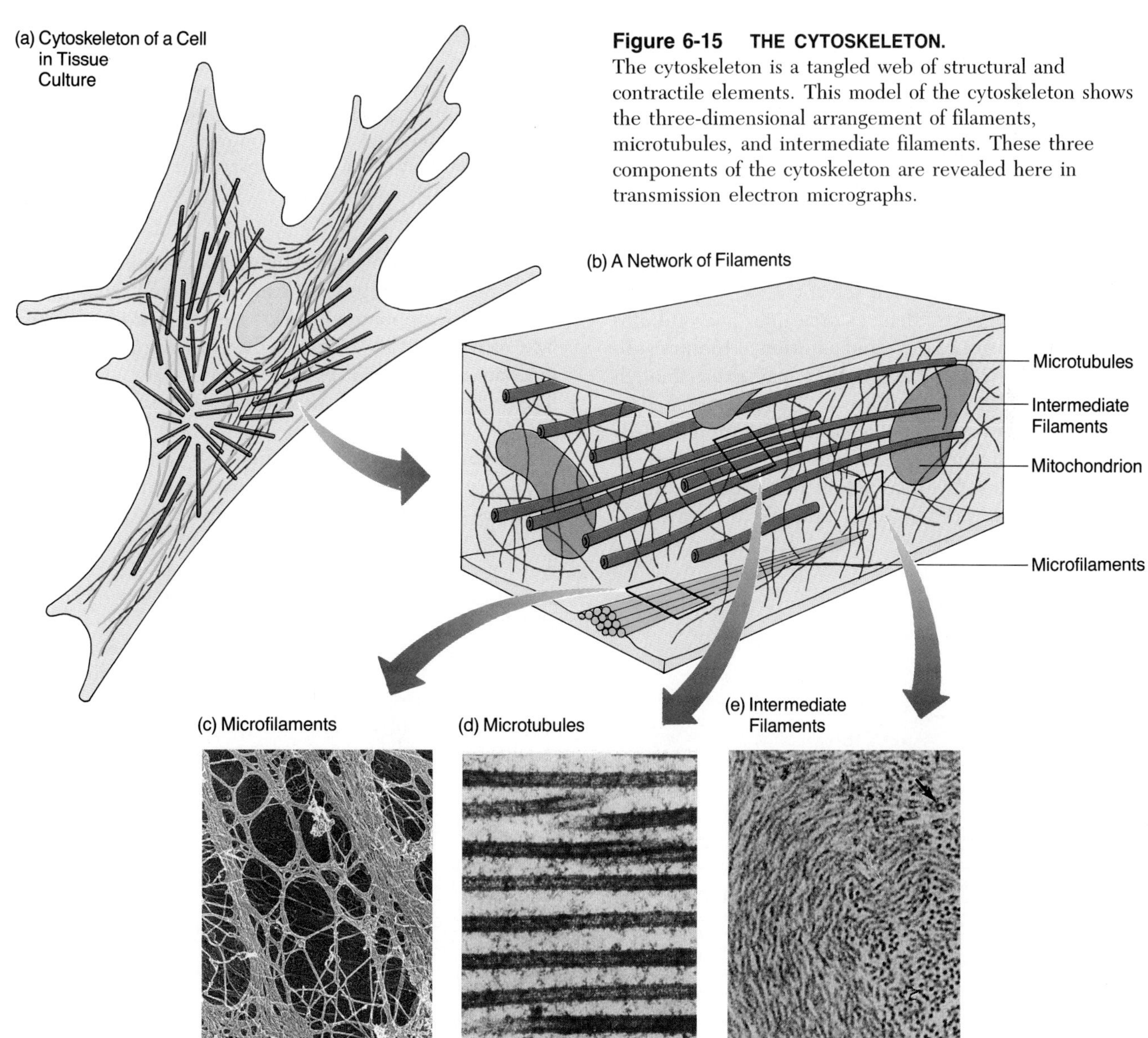

Figure 6-15 THE CYTOSKELETON.
The cytoskeleton is a tangled web of structural and contractile elements. This model of the cytoskeleton shows the three-dimensional arrangement of filaments, microtubules, and intermediate filaments. These three components of the cytoskeleton are revealed here in transmission electron micrographs.

the **microfilaments,** extremely fine, threadlike protein fibers only 3–6 nm in diameter (Figure 6-15). Composed predominantly of a structural protein called **actin** microfilaments are involved in many types of intracellular movements in plant and animal cells. Regular arrays of microfilaments in skeletal and heart muscle cells interact with other, thicker filaments made up of another contractile protein called **myosin.** This interaction brings about the contractions that move an animal's limbs or pump its blood. Muscle cells are another good example of cell specialization. Early eukaryotic cells used a basic contractile apparatus for activities such as cell division and circulating the cytoplasm. Only later when animal cells emerged did some cells possess the highly ordered arrays of actin and myosin which allow the exaggerated contractions that characterize true muscle cells.

Microtubules are long, cylindrical tubes 20–25 nm in diameter and thus are considerably wider than microfilaments. Microtubules are composed of subunits of the globular protein **tubulin** that are stacked in a spiral to form the long microtubules. Microtubules act as a scaffold that helps stabilize the shape of the cell (Figure 6-16). Microtubules are also the main component of the *spindle,* the apparatus that moves chromosomes when cells divide (see Chapter 9). And microtubules are arranged in geometric patterns inside the whiplike flagella and hairlike cilia that are used in certain kinds of cell locomotion, as we shall soon see.

Evidence suggests that cytoplasmic microtubules lengthen as a result of the rapid addition of tubulin subunits and that this polymerization may exert enough force to move the cell surface or internal organelles. Conversely, the rapid depolymerization of tubulin leads to shortening of microtubules and may cause them to pull on organelles. Cytoplasmic microtubules are highly dynamic structures (Figure 6-17) and may remain stable only if the ends are capped by special proteins or organelles. Single microtubules can be isolated from nerve cells, and if supplied with ATP, can glide along the surface of a glass slide. This is due to the presence of a **mechanoenzyme** (an enzyme that exerts mechanical forces, as does myosin) attached to the microtubule surface.

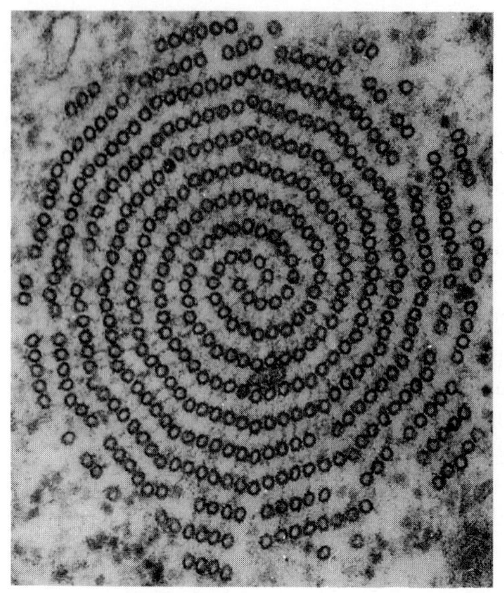

Figure 6-16 MICROTUBULES.
Microtubules are skeletal elements in cells that may be assembled in a variety of patterns. These regularly shaped and neatly arranged bundles in a protozoan's cellular extension (axopodium) are shown in cross section magnified about 17,000 times. For size comparison, note the portion of a mitochondrion to the left.

Intermediate filaments (about 10 nm in diameter) show up on electron micrographs as numerous wavy lines crisscrossing the interior of most kinds of animal

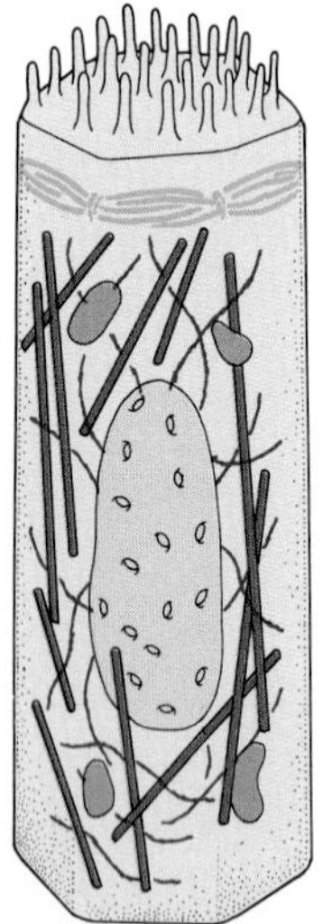
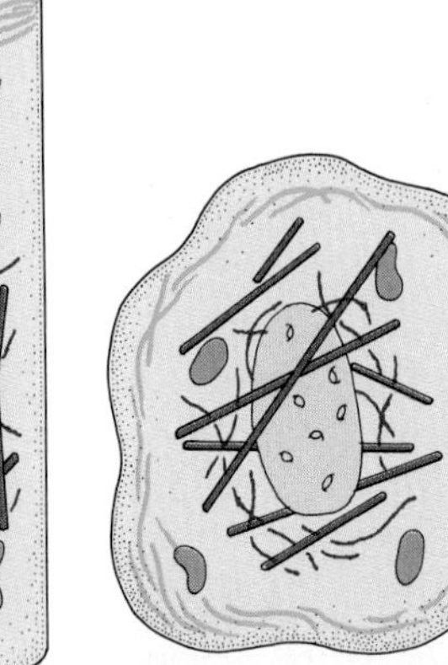
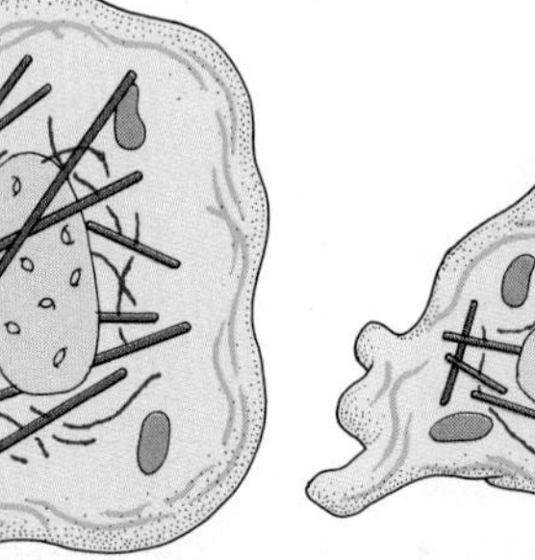
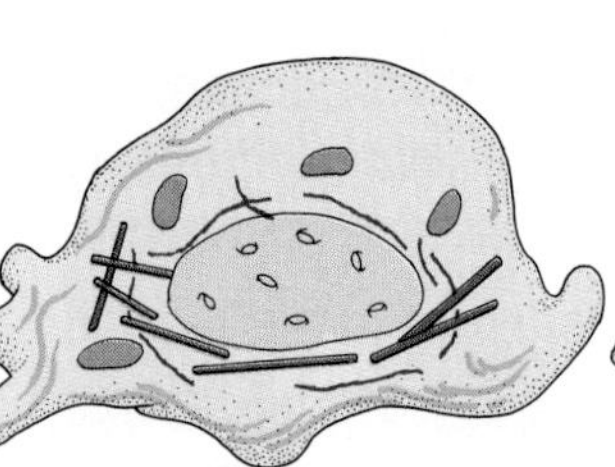
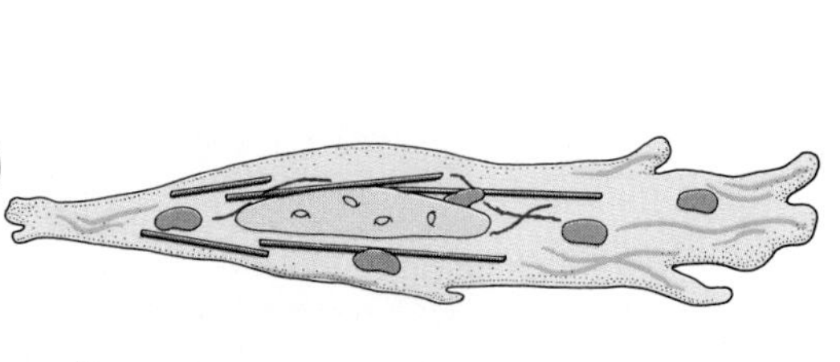

Figure 6-17 THE CYTOSKELETON OF AN EPITHELIAL CELL.
The organized cytoskeleton of an epithelial cell becomes disorganized when the cell is freed from the epithelial tissue. In fluid suspension, such a cell soon becomes spherical. On a solid surface, the cytoskeleton reorganizes in a flattened configuration, and the cell creeps away.

Table 6-1 COMPONENTS OF PROKARYOTIC, PLANT, AND ANIMAL CELLS

Component	*Prokaryote*	*Plant Cell*	*Animal Cell*
Cell wall	Present	Present	Absent
Glycocalyx	Absent	Absent	Present
Plasma membrane	Present	Present	Present
Cytoskeleton	Absent	Present	Present
Nucleus	Absent	Present	Present
Chromosomes	Single	Multiple	Multiple
Mitochondria	Absent	Present	Present
Plastids	Absent	Often present	Absent
Ribosomes	Present	Present	Present
Endoplasmic reticulum	Absent	Present	Present
Golgi complex	Absent	Present	Present
Vacuoles	Absent	Present	Present
Lysosomes	Absent	Often absent	Present
Cilia (9 + 2)	Absent	Absent in most	Present in some cells
Flagellum	Often present, unique type	Absent	Present in some cells
Centrioles	Absent	Absent in most	Present

cells. These filaments are made of a family of proteins called vimentins; those of nerves differ from those of muscle cells or of connective tissue cells. It seems likely that intermediate filaments impart tensile strength to the cytoplasm, since they are associated with the desmosomal spot welds between epithelial cells (see Chapter 5). Chromosomes within the nucleus may attach to three proteins of the intermediate filament family that form a structural meshwork.

The cytoskeleton is just as important a distinguishing characteristic of eukaryotic cells as is the envelope-bound nucleus or the discrete protein-containing chromosomes (Table 6-1). And without the ability of filaments and tubules to quickly disassemble and reassemble and to interact with mechanoenzymes, the cell and its parts would not be able to move, and many cellular functions would be impossible.

CELLULAR MOVEMENTS

Movement is a fundamental feature of life; many activities of living organisms, whether single-celled or multicellular, would cease were it not for movements generated by the cytoskeleton and contractile proteins. These structures enable the minute "rotors" of bacterial cells to propel them in search of food molecules; permit lashing tail movements to move sperm toward egg; allow an amoeba to creep toward a food source and engulf it; and enable vesicles to shuttle through the cytoplasm. Let us see how cellular movements are generated.

Creeping and Gliding Cell Movements

Free-living single-celled amoebae, as well as the great majority of animal cells, will glide and creep along solid surfaces. Even if immobilized for years in an organ, a cell could activate its locomotor machinery within minutes if freed from constraints. Thus, the primitive cellular capacity for movement persists in highly specialized cells as well as in free-living single-celled species.

Creeping and gliding movements are anchorage-dependent: They can take place only when there is a solid surface to which the animal cell can attach. Animal cells suspended in tissue fluids or culture liquids cannot swim. Figure 6-18 shows a mobile cell contacting and then moving across a solid surface. The cell's leading edge, protruding in sheetlike ruffles, thrusts forward, perhaps as actin and membrane components flow to the front. With each advance, the lower surface of the protrusion sticks to the substratum. Meanwhile, the cell surface itself moves backward as new actin microfilaments and membrane sections are assembled at the front. In some cells, the rear end of the cell is stretched rather like a rubber band. The rear end snaps free, the cytoplasm shifts forward, and the whole process starts again with new protrusions at the front end. In other cells, the cytoskeleton near the surface at the rear end of the cell may contract to force the cytoplasm forward.

If the cell's microtubules are experimentally dispersed by a drug, the entire surface of the cell protrudes with many active ruffles, making the full periphery a 360-degree "leading edge." In the absence of a single leading edge, the cell is effectively immobilized. Using genetic engineering techniques, researchers have produced slime mold cells that lack all myosin. Even without it, their lateral surfaces ruffle actively, and they can move across a substratum. These experiments show that actin is involved in the protrusion of a cell's leading edge and perhaps in snapping or pushing the contents of the rear end forward; myosin immobilizes the rear ends of some cells, just as microtubules immobilize the sides of other cells; thus, myosin and microtubules may channel movement in a single direction.

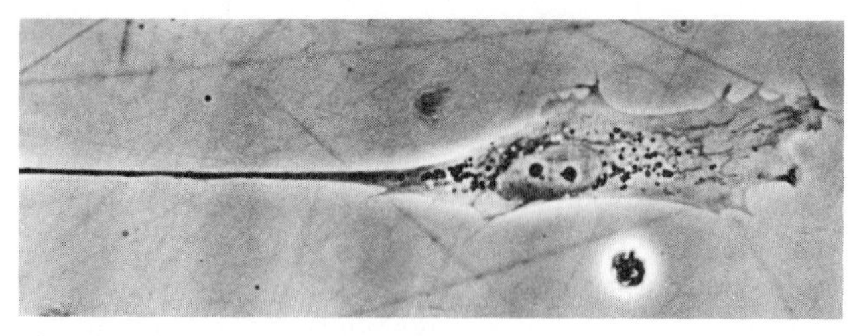

(a)

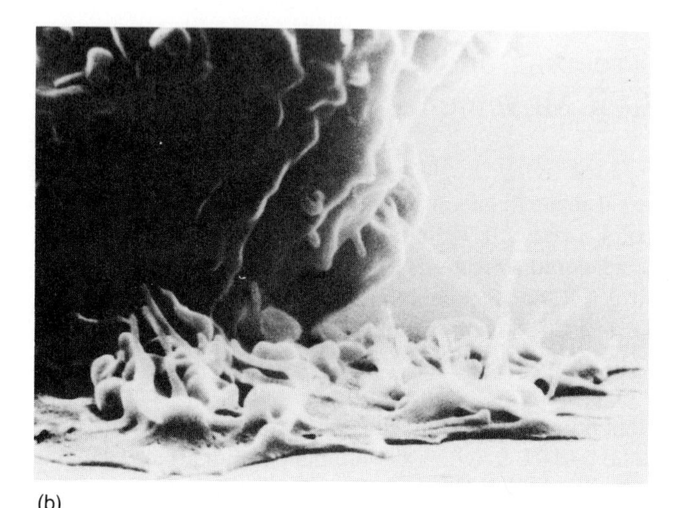

(b)

Figure 6-18 CREEPING CELL MOVEMENTS.
(a) A cell from a chick embryo (magnified about 450×) is stretched lengthwise as it creeps forward on a flat substrate. (b) The edge of a spreading cell (magnified about 10,500×), such as the epithelial cell drawn in Figure 6-17, ruffles and flutters dynamically and sends out fingerlike protrusions. A three-dimensional, space-filling network of actin microfilaments is attached to the inner, cytoplasmic side of the ruffling plasma membrane. At the lateral margins of the cell and within the lower cytoplasm, actin microfilaments often make up taut bundles or cables that are somehow stabilized by nearby microtubules.

Cells that creep or glide, such as amoebae and animal cells, appear to obey a few rules:

1. Cells must adhere to a solid surface to move.
2. Cells can be guided by the geometry of the surface on which they move; that is, they can glide or creep along bundles of fibers or along grooves, and in this way migrate from one part of the organism to another, especially in embryos.
3. The movement of cells can be turned on or off by internal or external signals.
4. Some cells exhibit **chemotaxis**, the ability to move toward or away from the source of a diffusing chemical. Slime mold cells (see Chapter 20), for example, creep toward the source of the molecule cyclic-AMP. Cyclic-AMP may cause myosin on the side of the cell nearest the source to disassemble. Actin can then create a leading edge at that site, while the cell's opposite side, with its assembled myosin, trails behind. Such cells move and congregate, beginning a new phase of the life cycle.

Swimming Cell Movements

While many eukaryotic cells can only creep along solid substrates, others can swim freely in liquid environments. These include the sperm cells of most animals and some plants, unicellular organisms such as *Paramecium* (Figure 6-19), spores of some fungi, certain algae, and many types of prokaryotic bacteria. Propulsion in a fluid is accomplished by means of flagella or cilia. **Flagella** (singular, *flagellum*) are fine, whiplike organelles that undulate to move a cell forward or backward. A cell usually has one or only a few flagella. **Cilia** (singular, *cilium*) are shorter than flagella, can number from a dozen to a few hundred, and usually beat in synchrony with each other.

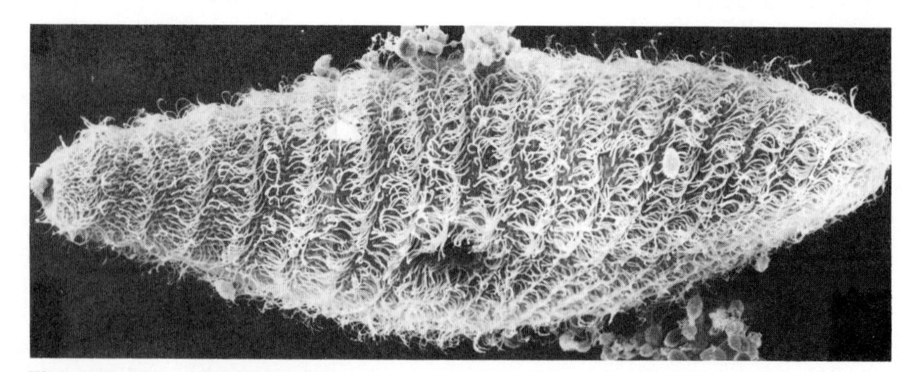

Figure 6-19 *PARAMECIUM.*
This single-celled eukaryotic organism, magnified about 400 times, is propelled by the synchronized beating of its many cilia.

Aside from length, motion, and number per cell, flagella and cilia of eukaryotes are physically identical. Both protrude from the cell surface and are covered by the plasma membrane. Both have the same intricate internal structure, shown in Figure 6-20: nine pairs of microtubules called *doublets* are arranged in a ring and extend the length of the cilium or flagellum. Two more microtubules run down the center of the doublet ring, and protein strands arranged much like the spokes of a wagon wheel interconnect the doublets and central singlets. Finally, each cilium or flagellum grows only from the cell surface at a site where a **basal body** is located. Basal bodies are one of a number of types of microtubule-organizing centers in animal and plant cells. As Figure 6-20 shows, a basal body has a distinctive internal arrangement.

How do cilia and flagella move and thus enable a cell to swim? Movement is based on tiny side arms that extend from one of the microtubules of each doublet. Each side arm is actually **dynein,** an ATP-cleaving mechanoenzyme. The outer, unattached end of each dynein side arm interacts with the surface of the adjacent microtubular doublet. When dynein binds and splits ATP, energy is released, and the dynein side arm apparently changes shape. The change in shape exerts a pushing force against the adjacent microtubular doublet and causes it to slide, just as a person poling a boat pushes against the river bottom with the pole. The movements of the microtubular doublet result in a slight bending of the cilium or flagellum. If thousands of dynein arms are splitting ATP and moving in sequence, the entire cilium or flagellum bends. The central microtubules act like a spring that resists the bending and controls the actual shape of the beating cilia or flagella.

The crucial role of dynein in the swimming movements of cilia and flagella was demonstrated in studies of certain Maori tribesmen from Samoa and New Zealand who were unable to impregnate their wives. Their sperm had straight tails and could not swim to reach and fertilize eggs, and careful analysis revealed that the dynein arms of the outer doublets were missing in each sperm's flagellum.

While flagella are virtually always involved in cellular swimming, cilia can play two distinct roles: swimming and sweeping fluids past stationary cells. In some free-living cells, coordinated beating of numerous cilia can propel the cells through a fluid environment. Such swimming is truly remarkable, because the viscosity of water to a microscopic organism would be the equivalent of a person swimming through tar or molasses. In certain other free-living cells, cilia sweep food particles toward sites of engulfment. And in some multicellular organisms, such as land vertebrates, thousands of cilia occur on the epithelial cells that line the trachea, the tube that carries air to and from the lungs. These cilia beat upward, sweeping mucus, dust particles, and contaminants out of the lungs and throat. Interestingly, in the sterile Maori males, the dynein deficiency prevents normal ciliary movements, and they frequently suffer respiratory diseases.

Many types of prokaryotic bacteria swim by means of flagella, but these flagella neither have the 9 + 2 internal structure shared by eukaryotic cilia and flagella nor are they covered by the bacterial plasma membrane. Bacte-

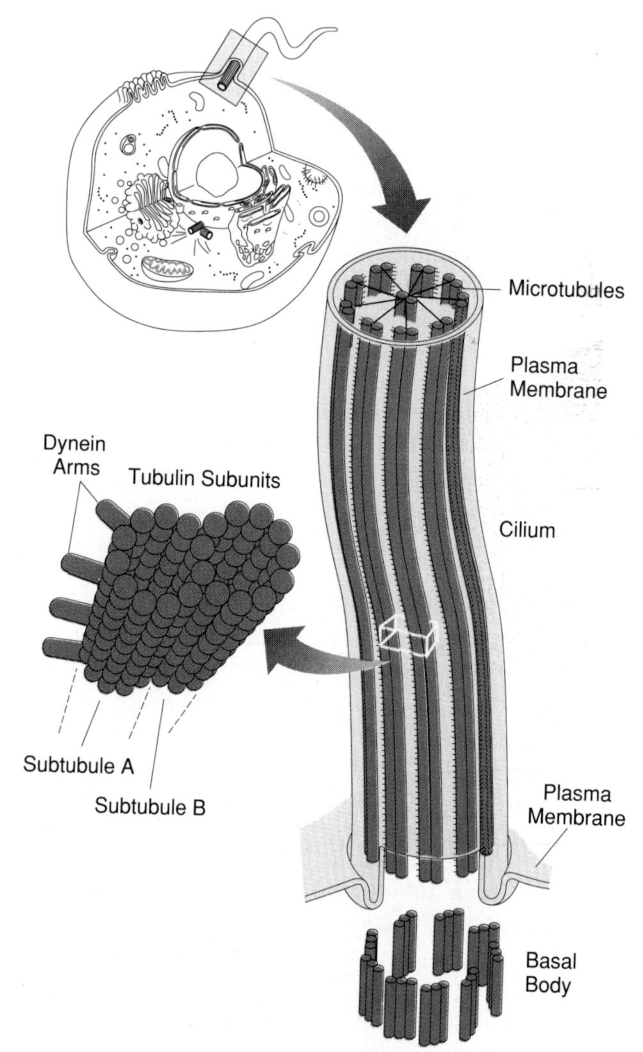

Figure 6-20 INTERNAL STRUCTURE OF CILIA AND FLAGELLA.
The dynein arms push against the adjacent microtubule doublet to cause bending. The two tubules in the center of the cilium or flagellum function as stiff springs. A basal body has nine microtubule triplets (instead of doublets) and lacks central singlets.

rial flagella are true molecular assemblages attached to the cell surface. These cylindrical protein strands do not bend back and forth, but rotate in miniature sockets, turning like propellers in a clockwise or counterclockwise direction to move the cell forward or backward.

Internal Cell Movements

All eukaryotic cells, whether mobile or not, need an internal transport system. We have already seen how "eating" and "drinking" vacuoles arise from the cell surface and shuttle inward to join lysosomes for "digestion" and how vesicles pinch off from the Golgi complex and move to different destinations in the cytoplasm.

Cytoskeletal microfilaments and microtubules are responsible for almost all major cytoplasmic movements. For example, most plant cells exhibit **cytoplasmic streaming:** Cytoplasm continuously flows around the central fluid-filled vacuole, circulating nutrients, proteins, and other cellular materials. Myosin proteins attached to vesicles and other organelles push against microfilament bundles in the plant cell and propel this streaming movement.

Most animal cells can move vesicles and other organelles outward toward the cell surface and inward toward the nucleus. Microtubular "railroad tracks" appear to be involved, whether they move the colored granules inside the pigment cells of a brilliantly colored tropical fish or certain organelles inside the long, spindly extensions of a nerve cell. Outward movements along microtubule tracks are due to the mechanoenzyme *kinesin*, which cleaves ATP for the energy needed to move particles. Another mechanoenzyme, MAP 1C, a kind of dynein, helps propel vesicles inward along microtubules. Clearly, a cargo's direction of movement is a property of the particular mechanoenzyme linked to it.

The dividing of animal cells into two new daughter cells also depends on actin microfilaments, myosin, and microtubules. Actin and myosin actually pinch the dividing cell in two (see Chapter 9). Microtubules of the spindle move the chromosomes. These microtubules are assembled from tubulin subunits near organelles called **centrioles,** which occur in pairs near the nuclear envelope (Figure 6-21). The role of centrioles can be observed if they are isolated from an animal cell and added to a solution of tubulin subunits: Spectacular elongation of microtubules occurs outward in all directions from the centriolar pair.

Most animal cells have centrioles, basal bodies, and spindles, and many can form cilia. In contrast, cells of higher plants usually lack centrioles, although they do have microtubule-organizing centers that facilitate the polymerization of spindle or cytoplasmic microtubules from tubulin. They also lack basal bodies and the cilia that grow from them. Biologists have not yet been able to explain why animal cells have centrioles and basal bodies, while higher plant cells do not.

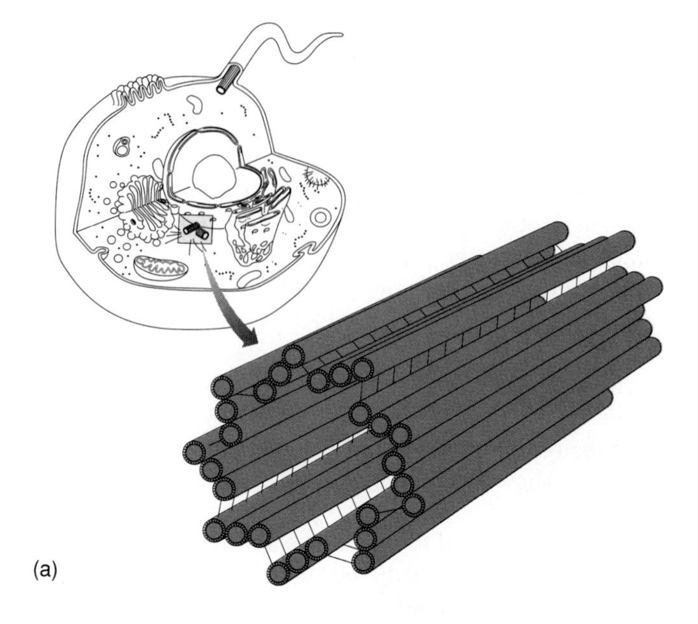

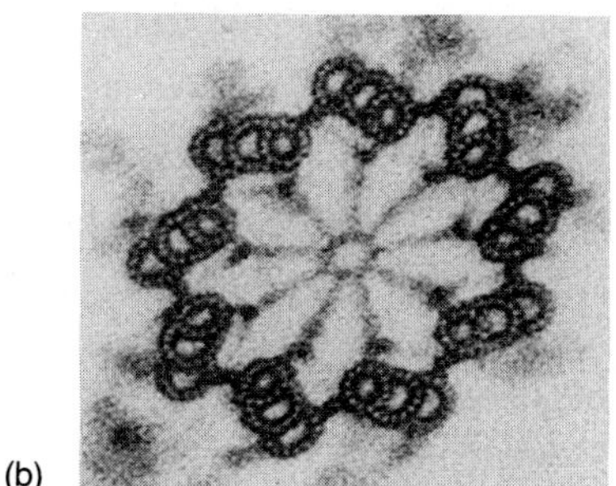

Figure 6-21 CENTRIOLES.
(a) Centrioles are made up of nine microtubule triplets, (nine groups of three microtubules each). (b) This electron micrograph, magnified about 160,000 times, shows a centriole in cross section. Spindle microtubules originate from centrioles. Each time a eukaryotic cell divides, its centrioles are duplicated so that each daughter cell receives a pair. The spindle moves the chromosomes during cell division.

In keeping with this chapter's theme, we can see that the organized aggregates of macromolecules forming the cytoskeleton and the cytoplasmic organelles are not themselves alive. Their coordinated activity, however, contributes to the ephemeral state we call life. So, too, other organelles that synthesize proteins, engulf food particles, or excrete waste molecules bridge the gap between molecules and life. When all are enclosed in the lipid bilayer plasma membrane—with its permeability, transport capabilities, and fluid properties—and when hereditary material is present, the result is the incredibly organized "factory" we call a cell, imbued with the state we call life.

SUMMARY

1. The *cytoplasm*, an organized semifluid substance bounded by the plasma membrane, contains organelles and molecules that are responsible for metabolism and many cell functions.

2. The *nucleus* of eukaryotic cells is delimited by the double-layered *nuclear envelope*, is the repository of genetic material (DNA) located on *chromosomes*, and is the primary center for the synthesis of RNA. *Nucleoli* are sites of ribosome manufacture in nuclei. Prokaryotes lack nuclei; their hereditary material is found in the *nucleoid* region.

3. *Ribosomes*, which are composed of two subunits of RNA and structural proteins, are the sites of protein synthesis. One mRNA molecule plus a number of attached ribosomes is called a *polysome*.

4. Polysomes free in the cytoplasm are sites of synthesis for many kinds of cellular proteins, which are released into the cytoplasm. Proteins made on membrane-bound polysomes are exportable or destined for membranes or lysosomes.

5. The *endoplasmic reticulum* is either rough or smooth. The rough endoplasmic reticulum (RER), whose outer surface is studded with polysomes, serves as a site of protein synthesis, modification, and transport. Smooth endoplasmic reticulum (SER) lacks attached ribosomes and is involved in the transport and synthesis of fats and steroids and detoxification.

6. The *Golgi complex* is the site at which molecules made in the ER are further modified and packaged for export or for delivery elsewhere in the cytoplasm.

7. *Vacuoles* can store engulfed fluids and nutrients or discharge fluids or wastes. Vacuolar feeding is a form of *phagocytosis*, while vacuolar drinking is termed *pinocytosis*. Taking things into cells is also called *endocytosis*, and expelling things, *exocytosis*.

8. *Lysosomes* contain a variety of digestive enzymes and degrade ingested materials and cellular debris. *Microbodies* are vesicles containing detoxifying enzymes.

9. *Mitochondria* are self-replicating organelles found in all eukaryotic cells. Each mitochondrion has an outer membrane and an inner membrane thrown into folds called cristae, where ATP is formed, and a central region, the *matrix*.

10. *Plastids* can be sites of photosynthesis in plant cells, or sites for storing a variety of nutrients and pigments. Both green chloroplasts and colorless leucoplasts arise from proplastids, contain DNA, and are self-replicating.

11. The eukaryotic cell's *cytoskeleton* is composed of *microfilaments*, *microtubules*, and *intermediate filaments*. These suspended organelles contribute to cell shape and allow cell movements.

12. Swimming cell movements are accomplished by means of *flagella* or *cilia*. Both have the same internal structure of microtubules, and arise from *basal bodies*. Gliding cell motion involves actin, myosin, and microtubules as well as adhesion and movements of the cell surface.

13. Internal cell movements may be caused by myosin pushing on actin cables or by mechanoenzymes such as kinesin or MAP 1C pushing on microtubules. Cell division depends on microtubules, microfilaments, and myosin. In animal cells, microtubules of the chromosomal spindle are assembled near organizing centers called *centrioles*.

KEY TERMS

actin
basal body
centriole
chemotaxis
chloroplast
chromosome
cilium
coated pit
cytoplasm
cytoplasmic streaming
cytoskeleton
dynein
endocytosis
endoplasmic reticulum (ER)
exocytosis
flagellum
gene
Golgi complex
intermediate filament
lysosome
matrix
mechanoenzyme
microbody
microfilament
microtubule
mitochondrion
myosin
nuclear envelope
nucleoid
nucleolus
nucleus
phagocytosis
pinocytosis
plastid
polysome
ribosome
stroma
tubulin
vacuole

QUESTIONS

1. Draw typical plant and animal cells, including their major organelles. Label the parts.

2. What are the parts and contents of a eukaryotic cell's nucleus?

3. What are the similarities and dissimilarities of the rough endoplasmic reticulum and free polysomes?

4. How are the RER and the Golgi complex related in structure and in function?

5. Various terms describe the taking in or passing out of materials from eukaryotic cells. List and define the terms.

Which occur in the wall of one of your blood capillaries?

6. Cells would be in grave difficulty if a mutation eliminated their lysosomes. Why?

7. Draw a mitochondrion and a chloroplast, label their parts, and then identify similarities and dissimilarities between them.

8. The cytoskeleton has several structures and functions. What are they? How are mechanoenzymes related to the structural components of the cytoskeleton?

9. How do cells creep, swim, or stream internally? What would adding an ATP poison do to such processes? Why?

ESSAY QUESTIONS

1. Why is contact with a solid surface important to basic cell functions?

2. How are vacuoles and various types of phagocytosis related to the transport of ions and other substances across the plasma membrane described in Chapter 5?

3. Recall the various types of junctions between cells in Chapter 5. What processes would need to go on if cell A synthesizes a small protein and (1) excretes it so it can be picked up by cell B; or (2) creeps to cell B and passes it directly?

For additional readings related to topics in this chapter, see Appendix C.

7

Harvesting Energy from Nutrients: Fermentation and Cellular Respiration

A living cell, like Lewis Carroll's Red Queen, has to run at top speed to stay in the same place. Without a constant input of energy, either from an outside source or from its own storage reservoirs, it will die.

Richard E. Dickerson, *Scientific American* (March 1980)

Energy flow and living things: A gerenuk nibbles leaves in the blazing East African sun.

Every second in the life of a cell, nutrients are being broken down to release energy. That energy is "spent" to build new cell parts and to fuel growth, maintenance, waste disposal, reproduction, movement, and other survival activities.

How do cells harness the energy to meet these high costs? The answer lies in the general flow of energy we began discussing in Chapter 4. Plants and certain kinds of microbes trap and convert solar energy in the process of photosynthesis, then store it in the chemical bonds of carbohydrate molecules. The cells of these light-gathering organisms as well as the cells of nonphotosynthesizers such as animals, fungi, and most microbes, release the energy stored in carbohydrates and "spend" it on their survival.

Our subject in this chapter is the precise way that cells trap and release chemical bond energy, step by step, in a series of chemical reactions. Those reactions are facilitated by enzymes, and the gradual release of energy prevents the kind of damage to the cell that might occur if all the bonds were broken at once to yield a burst of heat. The metabolic breakdown of nutrient molecules through a chain of mostly exergonic (energy-releasing) reactions is called *catabolism.* And much of the energy freed during catabolism is stored as molecular currency—specifically, ATP and other energy intermediates. The breakdown of carbohydrates continuously replenishes the cell's reserves of this molecular currency, and life activities continuously deplete it in a delicate balance. We study catabolism in this chapter and postpone photosynthesis until the next because all cells break down carbohydrates, whereas only certain kinds generate them through photosynthetic reactions. Moreover, catabolism evolved far earlier than photosynthesis; the latter is a biochemical specialization—albeit a crucial one that now supports virtually all living things, directly or indirectly.

The present chapter considers these subjects:

- The structure of ATP and the role of this molecular currency in the cell's energy harvest
- Oxidation-reduction reactions and the flow of electrons through metabolic pathways
- Glycolysis, the first phase of energy metabolism
- Cellular respiration—pathway to the biggest energy harvest
- How metabolic enzymes are arranged in mitochondrial membranes, and how this arrangement allows the high efficiency and high energy yields most cells require
- How organisms break down fats and proteins
- How cells ensure an available energy supply by controlling the rate of catabolism

ATP: THE CELL'S ENERGY CURRENCY

Cells work constantly to maintain a rich supply of the energy storage molecule **adenosine triphosphate,** or **ATP.** At any given instant, 10 billion molecules of ATP may be dissolved in the cytoplasm of a typical animal or plant cell. The energy stored in the relatively simple substance ATP is released when the molecule is cleaved into the related compound *adenosine diphosphate,* or *ADP,* plus inorganic phosphate (Figure 7-1a). While some of the energy is lost as heat during this reaction, some of it is released and conserved by being used to drive various endergonic reactions. In a luminescent fish, for example, the cells' burning of ATP may lead to an eerie underwater glow used in communication (Figure 7-1b and c).

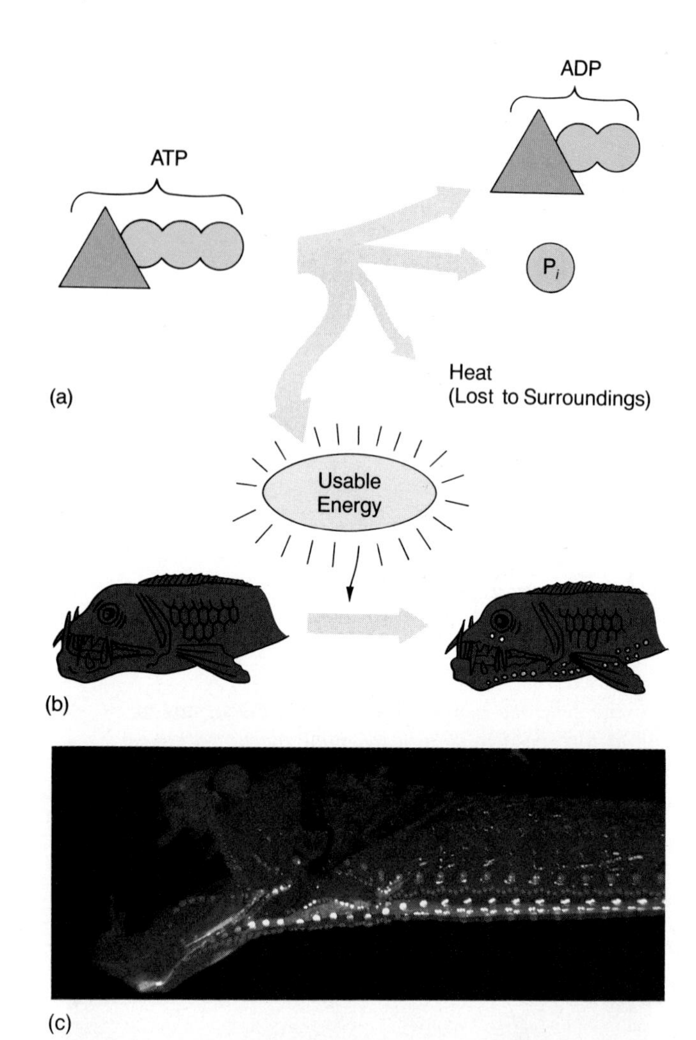

Figure 7-1 THE CLEAVAGE OF ATP RELEASES HEAT AND USABLE ENERGY.
(a) Cells contain a high-energy storage molecule, ATP, that can be cleaved to ADP and inorganic phosphate (P_i). In the process, some heat is released, but so is usable energy that can power many individual steps in biological processes. (b, c) Sloan's viperfish can communicate with potential mates through a deep-ocean light show—luminescent regions just under the skin. These dots, resembling tiny neon bulbs, are set aglow by the breakdown of millions of ATP molecules.

The Structure of ATP

The ability of ATP to store and release energy stems from its structure and that of the related molecules ADP and *AMP (adenosine monophosphate)* (Figure 7-2a). All three are nucleotides, the building blocks of nucleic acids. Each contains the purine base adenine forming a double ringlike portion; the sugar ribose in a five-sided ring formation; and a tail made up of phosphate groups—three groups in ATP, two in ADP, and one in AMP.

ATP's special ability to store and release energy is associated with the tail of phosphate groups. Adding phosphate groups to AMP or ADP requires energy; conversely, as phosphate groups are removed from ATP or ADP and transferred to other compounds, energy is released. Let's look first at the breakdown of ATP and the release of energy.

The terminal phosphate group can be transferred from ATP to a compound such as water (Figure 7-2b, step 1). The transfer of any phosphate group is called **phosphorylation.** This breakdown of ATP yields ADP plus inorganic phosphate (symbolized by P_i). The most common phosphorylation reaction in cells is just the reverse of this first reaction: An inorganic phosphate group is *added to* the ADP molecule, using up energy and forming ATP. This reaction *requires* about 8 kcal/mole.

Within simple conditions in a test tube, the cleavage (hydrolysis) of ATP to ADP releases its 8 kcal of energy per mole of ATP as heat. Cellular metabolism is usually not so wasteful; in many instances, the energy released when ATP is hydrolyzed is stored in other energy-intermediate molecules and used to power subsequent biological processes.

In a test tube (or rarely, in a cell), the ADP formed when ATP is cleaved can itself be cleaved further and the second phosphate group removed. This additional cleavage leaves the nucleotide AMP and once more releases an equivalent amount of energy—approximately 8 kcal/mole (Figure 7-2b, step 2). Yet another cleavage, however, removes the third phosphate group from AMP but liberates much less energy than the first two phosphates—only 2 kcal/mole (step 3). Since an ATP molecule's first two phosphate groups release so much free energy when cleaved, they are called *high-energy groups*, and the bonds linking them are known as *high-energy phosphate bonds.*

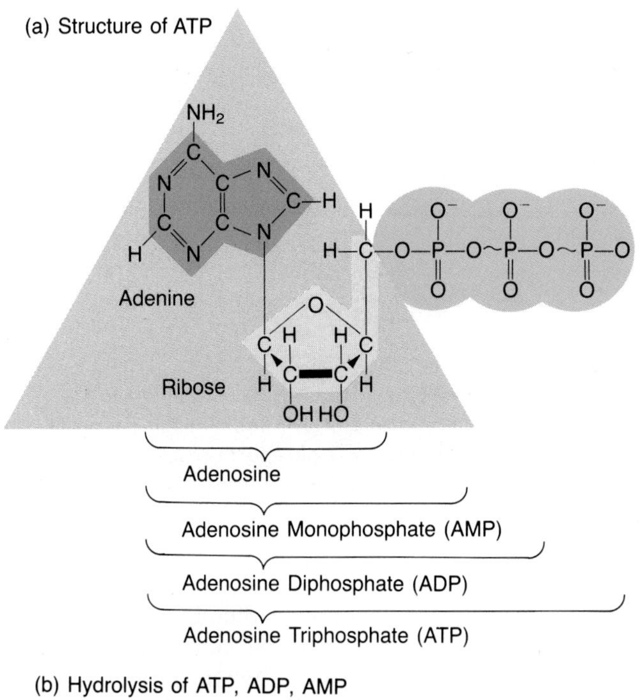

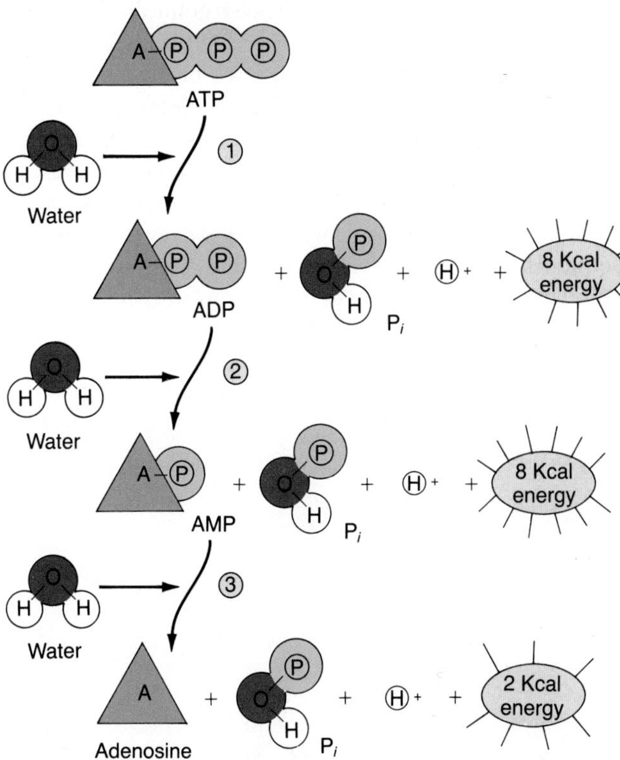

Figure 7-2 THE CELL'S HIGH-ENERGY FUEL, ATP. (a) ATP and the related molecules ADP and AMP each share an adenine base and the sugar ribose. ATP has three attached phosphate groups, however, while ADP has two and AMP just one. (b) When the terminal phosphate is split (hydrolyzed) from ATP and transferred to water (step 1), 8 kcal of energy is released. A similar release accompanies the cleavage of the terminal group from ADP (step 2), but the energy release is much smaller when the phosphate group is split from AMP (step 3).

ATP and the Harvesting of Energy

The continuous cleavage of ATP to ADP drives life processes, and thus the cell must form new ATP molecules every second. On the average, each of the 20 to 30 trillion cells in the human body cleaves from 1 to 2 billion ATP molecules into ADP every minute—the equivalent of about 90 lb (40 kg) of ATP a day! The only possible source of so much ATP is the continuous recycling of ADP and AMP molecules to ATP.

As the cell generates ATP molecules, how do these energy carriers drive typical endergonic reactions, such as the synthesis of the disaccharide sucrose from the monosaccharides glucose and fructose? For enzymes to catalyze the synthesis of this sugar, they require about 7 kcal of energy per mole of sucrose formed, and it is ATP that supplies this energy.

During sucrose synthesis, first glucose is phosphorylated; that is, ATP's terminal phosphate group is transferred to glucose, and the products are ADP and glucose-1-phosphate (Figure 7-3, left). In this reaction, some of ATP's chemical energy is stored in the new glucose-1-phosphate molecule. Second, if the proper enzyme is present, glucose-1-phosphate and fructose react to form sucrose and inorganic phosphate (Figure 7-3, right). In these two reactions, glucose-1-phosphate is a *common intermediate:* It is a product of the first reaction and a reactant in the second. And energy derived from ATP is conserved in this common intermediate in a form that can successfully drive the second, endergonic reaction. In most biological reactions, such common intermediates are the cell's means of transferring and spending the molecular energy currency of ATP to sustain life.

The energy transfer is never 100 percent efficient, however. Only about half the energy released in exergonic reactions is channeled into the recycling of ATP from ADP. The remaining energy will be wasted as heat. The same is true when ATP is later cleaved, as the phosphate is transferred to glucose or other molecules. Again, heat is lost. This inevitable freeing and loss of heat during ATP cleavage is another reflection of the second law of thermodynamics at work.

OXIDATION-REDUCTION REACTIONS

Exactly how is energy from nutrients channeled into ATP and not simply lost as heat? A cell's key metabolic reactions involve **oxidation-reduction reactions**—a flow of electrons from one molecule to the other that functions as a kind of energy current in the cell. Specifically,

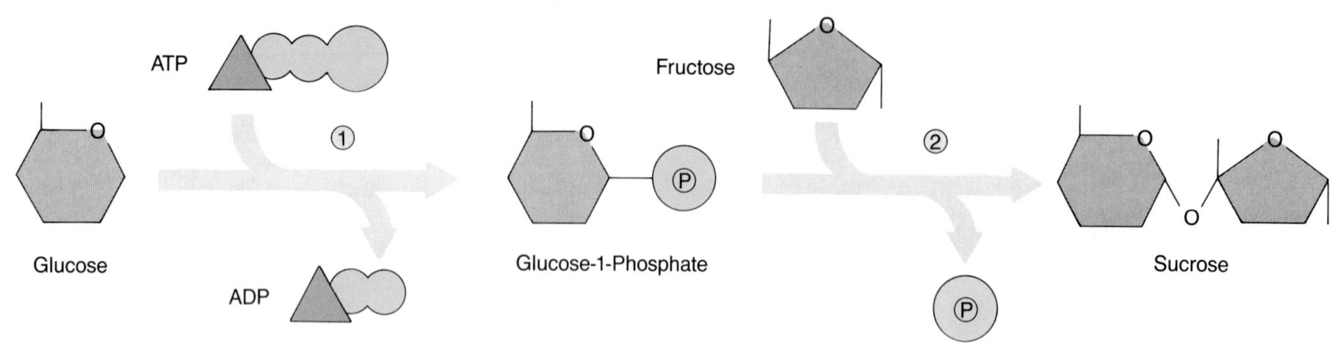

Figure 7-3 THE SYNTHESIS OF SUCROSE AND THE "SPENDING" OF AN ATP.
The building of a sucrose molecule from glucose and fructose requires the splitting of ATP and the transfer of energy to the reacting molecules. First, the terminal phosphate group of the ATP molecule is transferred to glucose. Some of the energy is now stored in glucose-1-phosphate, and the rest remains in the ADP molecule. Second, glucose-1-phosphate reacts with fructose to form the disaccharide sucrose, with inorganic phosphate left over. The high-energy form of glucose, glucose-1-phosphate, is a common intermediate for these two steps.

oxidation is the removal of electrons from an atom or a compound, while **reduction** is the addition of electrons and a concurrent decrease in net charge. These reactions always occur simultaneously, so that one partner molecule is oxidized and the other reduced (Figure 7-4a).

In many biological oxidation-reduction reactions, the electrons are transferred in the form of hydrogen atoms, which consist of one proton (H^+) and one electron (e^-) (Figure 7-4b). Like the transfer of electrons, the transfer of hydrogen atoms takes place in paired reactions.

Two particularly important coenzymes serve as electron carriers in many metabolic oxidations and reductions. (Coenzymes, recall from Chapter 4, are molecules that must be present for an enzyme to catalyze a reaction). The coenzymes are **NAD$^+$** (nicotinamide adenine dinucleotide; Figure 7-5a) and the structurally analogous **FAD** (flavin adenine dinucleotide). When NAD$^+$ is reduced to NADH (Figure 7-5b), two electrons and one proton are added to the molecule; the electrons can later be transferred during a chainlike series of reactions that provides much of a cell's energy supply of ATP.

These crucial coenzymes serve as energy intermediates, carrying electrons and hydrogen away from the sequential reactions of glycolysis and the first stages of cellular respiration and feeding them into marvelously

(a) Transferring an Electron

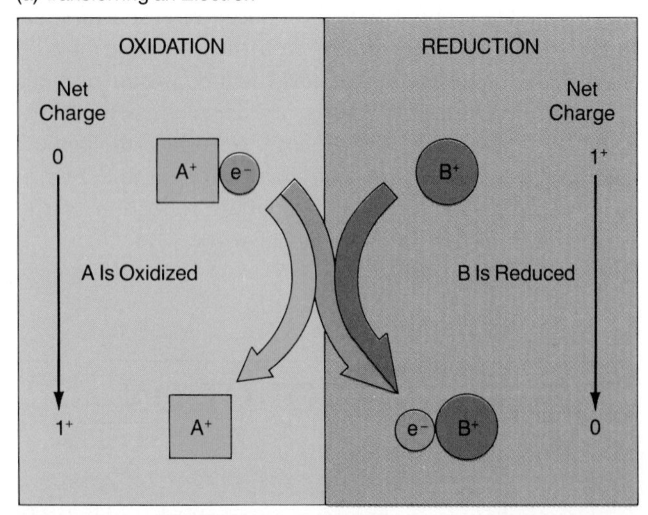

(b) A Hydrogen Atom Can Accompany Transferred Electrons

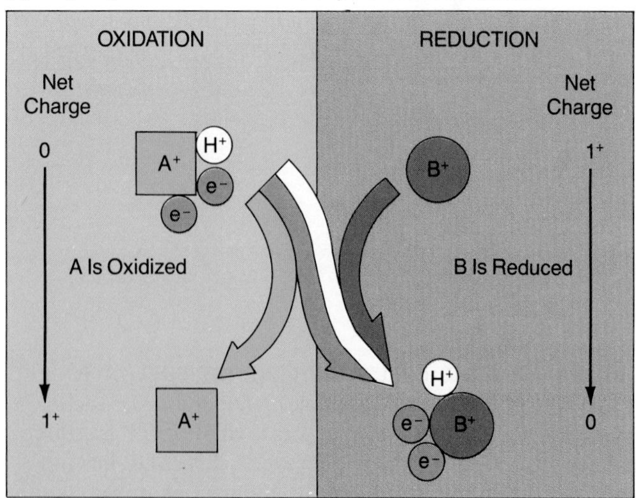

Figure 7-4 OXIDATION AND REDUCTION: PAIRED REACTIONS AND A FLOW OF ELECTRONS
Oxidation is the removal of electrons from an atom or a molecule, and reduction is the addition of electrons. (a) As compound A is oxidized, compound B is reduced, with the flow of electrons following the orange pathway. (b) The transfer of a hydrogen atom can be the equivalent of the transfer of electrons. Here, compound AH is oxidized, and a free hydrogen ion (H^+) forms and is transferred to compound B^+. B^+ is thus reduced to BH, and AH is oxidized to A.

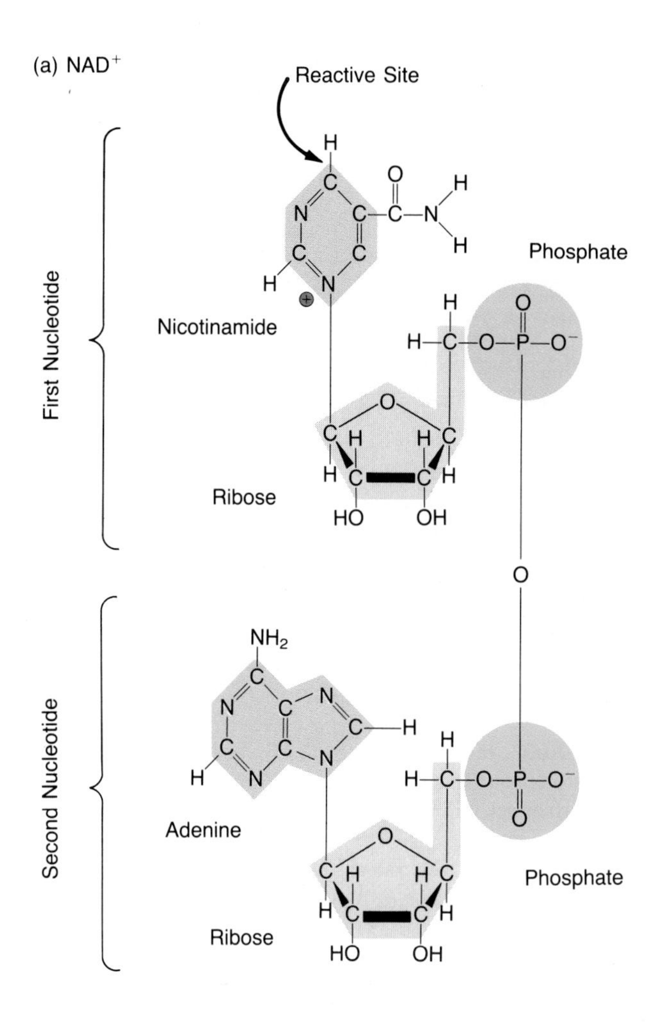

(b) Reduction of NAD+

NAD+
1+
NADH
0
NAD+ Is Reduced

Figure 7-5 THE ELECTRON CARRIER NAD$^+$.
(a) As the name implies, nicotinamide adenine dinucleotide (NAD^+) has one nucleotide containing the base nicotinamide, the sugar ribose, and an attached phosphate group; and a second nucleotide containing the base adenine, a ribose, and another phosphate group. (b) During the reduction of NAD^+, two electrons and one proton are transferred to the molecule, and the higher-energy molecule NADH results. NADH then feeds its electrons and protons into energy-requiring reactions that build ATP for the cell. FAD has a structure similar to NAD^+, but the sugars are ribitol, not ribose, and the reduction of FAD requires two hydrogen atoms, not one.

organized molecular machinery embedded in mitochondrial membranes. There the hydrogen drives the synthesis of ATP, which in turn runs the energy economy of the cells.

Let's begin, now, to explore the metabolic pathway called glycolysis, with its modest harvest of ATP.

GLYCOLYSIS: THE FIRST PHASE OF ENERGY METABOLISM

Glycolysis is the initial sequence of reactions used by virtually all cells to break six-carbon glucose molecules into two molecules of the three-carbon compound **pyruvate.** Glycolysis is a model for the stepwise buildup and breakdown of biological molecules and for the reaction pathways that underlie cellular activity.

Figure 7-6 provides an overview of cellular energy harvest and shows how glycolysis fits into it. Glycolysis and the subsequent steps of fermentation occur in the cytoplasm because the enzymes that bring about each reaction are dissolved in the watery cell fluid. Aerobic respiration takes place in the mitochondria in eukaryotic cells and along the inner surface of the plasma membrane in aerobic prokaryotes. Note the overall equation for the oxidation of glucose and its production of ATP.

Splitting Glucose: The Steps of Glycolysis

Glycolysis is the basis for energy metabolism in virtually all living creatures. During glycolysis, the six-carbon glucose molecule is broken down to two molecules of pyruvate in a series of nine reaction steps (Figure 7-7). Although biologists do not yet understand the exact mechanisms, recent studies show that pathway metabolites and coenzymes such as NADH are transferred directly from one enzyme to the next during glycolysis. This represents great cellular economy, because pools of intermediate compounds need not accumulate at each step in the metabolic pathway.

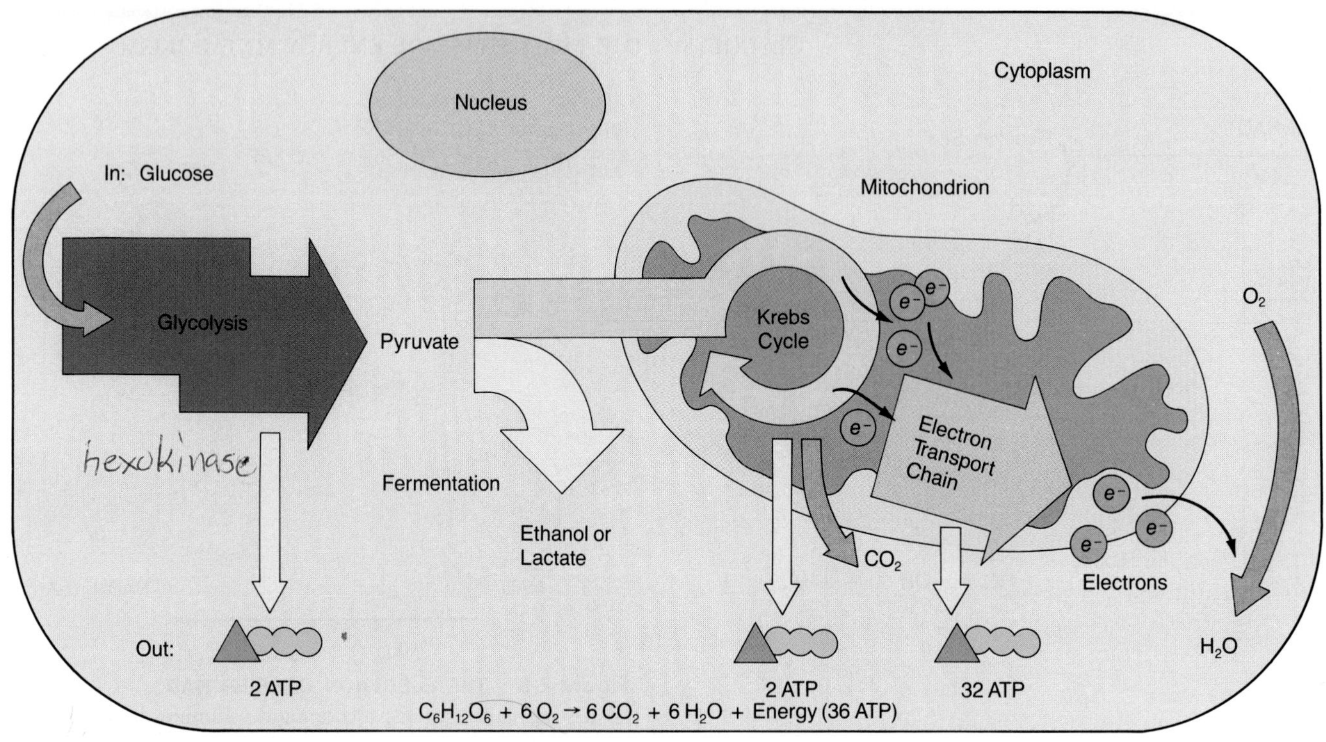

Figure 7-6 AN OVERVIEW OF CELLULAR ENERGY HARVEST: GLYCOLYSIS, FERMENTATION, AND AEROBIC RESPIRATION.
Glycolysis begins with the breakdown of glucose and ends with pyruvate. During the Krebs cycle, the first phase of aerobic respiration, pyruvate is converted to a high-energy compound. If the cell's environment lacks oxygen, however, pyruvate will undergo fermentation instead of entering the Krebs cycle, and the end product ethanol or lactate can form.

The high-energy compounds formed in the Krebs cycle are passed to the electron transport chain. Both of these processes take place in the mitochondria, where electrons are removed, and most of the ATP is generated from the initial molecule of glucose that entered glycolysis. Overall, for each glucose molecule ($C_6H_{12}O_6$), the cell uses 6 oxygen molecules and produces 6 carbon dioxides, 6 water molecules, and 36 molecules of ATP.

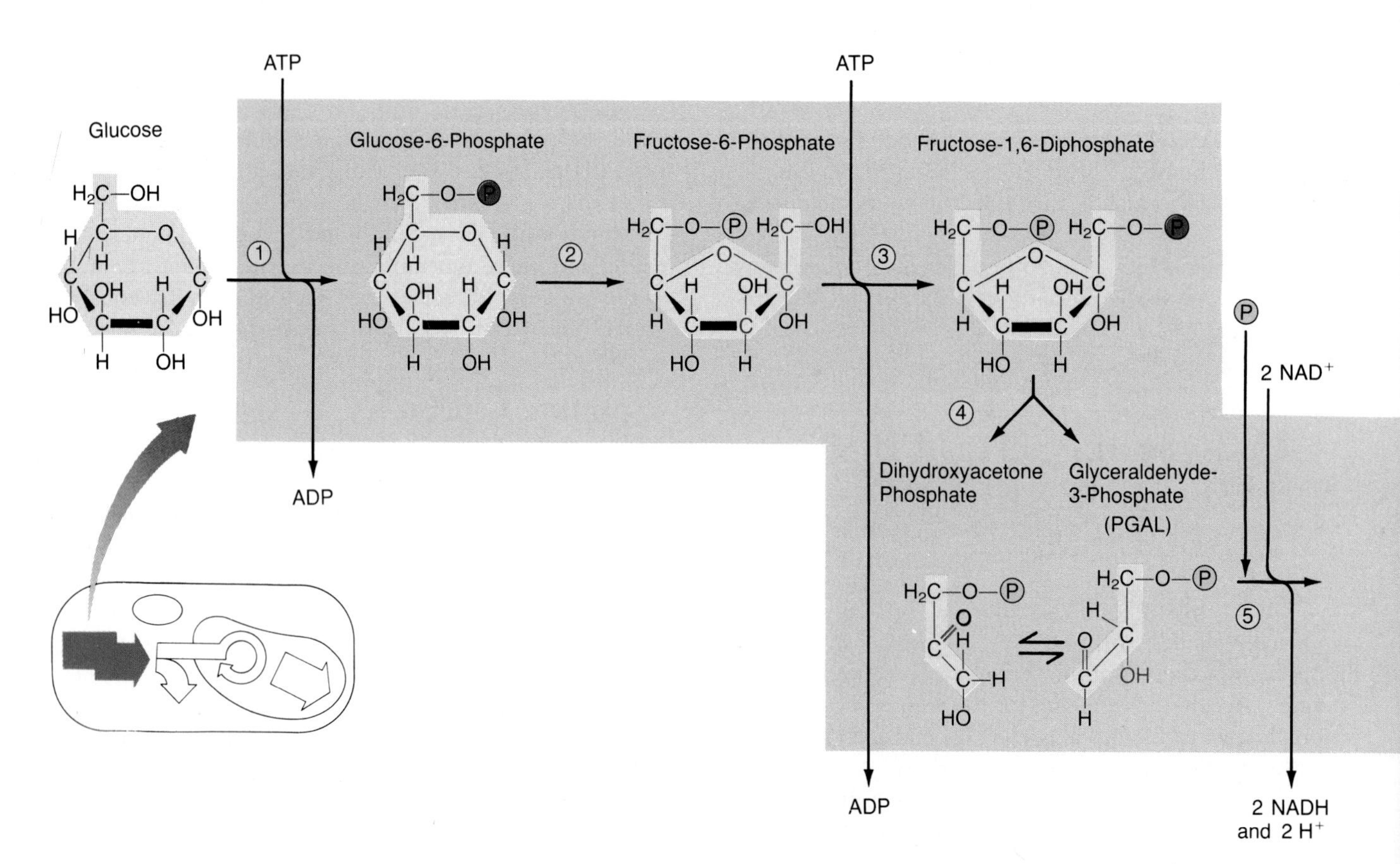

Step 1

The initiation of glycolysis requires a high-energy form of glucose. Thus, in step 1, a molecule of ATP is cleaved, forming both ADP and the *activated* molecule glucose-6-phosphate. A specific enzyme, hexokinase, catalyzes this reaction, just as specific enzymes catalyze *each* step in a metabolic pathway.

Steps 2 and 3

In step 2, the glucose-6-phosphate molecule undergoes a change in structure and is converted to its close relative, fructose-6-phosphate. During step 3, a molecule of ATP is cleaved (as in step 1), and its terminal phosphate group is transferred to the other end of the fructose carbon chain. We now have a molecule of fructose-1,6-diphosphate, which has been generated at the "cost" of two ATPs.

Step 4

During step 4, fructose-1,6-diphosphate is split into two similar three-carbon molecules, one called PGAL (phosphoglyceraldehyde, or glyceraldehyde-3-phosphate), and the other called dihydroxyacetone phosphate, which is converted immediately into another molecule of PGAL. From this point on, each step in the pathway must take place twice—once for each PGAL derived from the original glucose molecule.

Step 5

Steps 5 and 6 are key energy-capturing steps. During step 5, PGAL is both oxidized and phosphorylated. The aldehyde group $-\mathrm{C}\begin{array}{l}\diagup \mathrm{H}\\ \diagdown\!\!\diagdown \mathrm{O}\end{array}$ is oxidized when NAD^+ accepts two electrons, one of which is a hydrogen atom; NAD^+ becomes the reduced compound NADH, while the oxidized PGAL reacts with phosphate. A great deal of energy is released during the exergonic oxidation reaction involving PGAL and NAD^+; that energy is trapped immediately as a phosphate group joins PGAL to produce the high-energy molecule 1,3-diphosphoglycerate, or DPGA. Note that the phosphate needed to generate DPGA comes from inorganic phosphate (P_i) dissolved in the cell's cytoplasm, and not from ATP.

Step 6

In step 6, both 3-phosphoglycerate (3-PGA) and two molecules of ATP are formed as one phosphate group from each DPGA phosphorylates an ADP. This is possible because DPGA has an even higher energy content than does ATP. This last reaction "pays back" the earlier energy investment that set the stage for the exergonic reactions of glycolysis.

Figure 7-7 GLYCOLYSIS: METABOLIC PATHWAY FOR SPLITTING GLUCOSE.
The text describes the events of steps 1–9 in detail. Note that in steps 6 and 9, a total of four ATPs are produced, but that in steps 1 and 3, a total of two ATPs are used up for a net gain of two ATPs. Also notice that two NADHs are produced in step 5; these are used to make more ATP in the electron transport chain, a later phase of cellular respiration. Finally, notice that the pathway "jogs" at step 4, where fructose-1,6-diphosphate is broken into two molecules: dihydroxyacetone phosphate and glyceraldehyde-3-phosphate (PGAL). Since the dihydroxyacetone phosphate is immediately converted to glyceraldehyde-3-phosphate, the result is two PGALs. Both PGALs go through the rest of the glycolysis pathway. Thus, for each glucose molecule broken down, two pyruvate molecules form. New features at each step are shown in red.

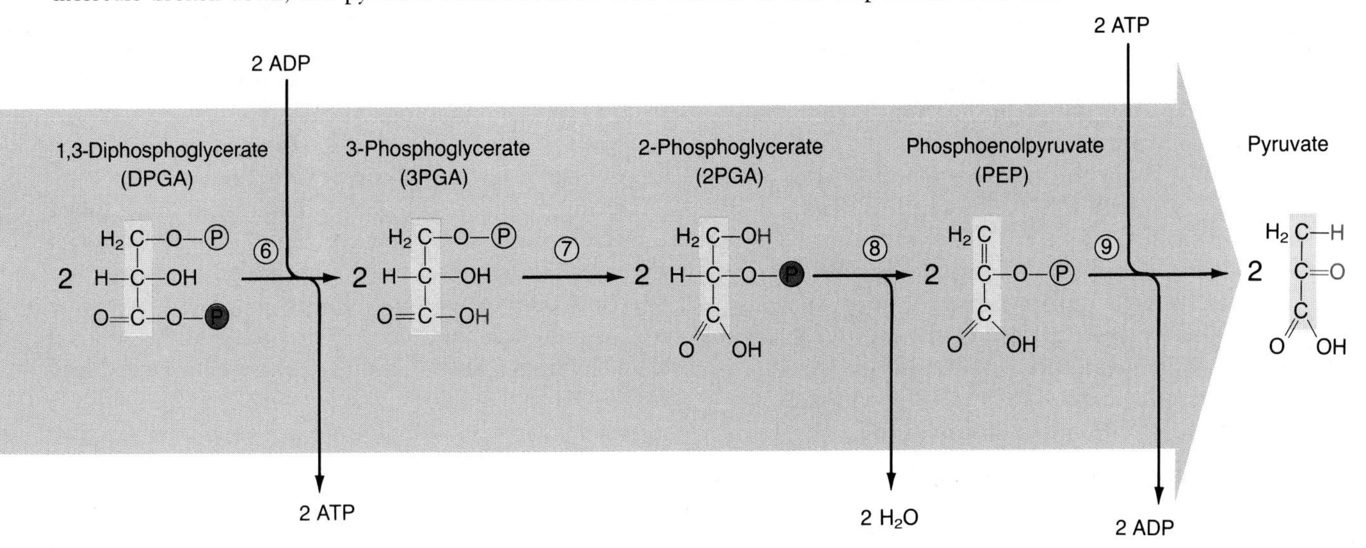

Steps 7–9

In steps 7 and 8, further reactions convert 3-phosphoglycerate to phosphoenolpyruvate (PEP). Like DPGA, PEP is a high-energy phosphate compound, and its energy is used to convert an ADP to ATP in step 9. Thus, for every molecule of glucose that enters the glycolysis pathway, four molecules of ATP are formed. However, since two ATP molecules are used up in steps 1 and 3, the net energy yield from glycolysis is two ATP molecules (see Figures 7-6 and 7-7, bottom). In addition, for each molecule of glucose, glycolysis generates two molecules of NADH and two molecules of pyruvate. The pyruvate molecules can be acted on further during the reaction sequences of fermentation or cellular respiration.

Fermentation

The biochemical point of **fermentation** is to allow the cell to carry out glycolysis—and generate ATP—even under anaerobic conditions, that is, when oxygen is absent from the environment. Two types of cells can carry out fermentation: *anaerobic cells* and *facultatively aerobic cells*. Anaerobic cells include certain types of bacteria that can survive only in the strict absence of molecular oxygen. Facultative aerobes include various other bacteria and yeasts, as well as animal muscle and certain other cells, that can ferment nutrients when oxygen is absent but carry out more efficient processes when oxygen is present.

Fermentation is a short pathway (just one reaction step) that follows the nine steps of glycolysis and acts on the pyruvate formed during that preliminary pathway. Depending on the fermenting organism, those additional steps can modify pyruvate to lactate, ethanol, or one of several other organic end products (Figure 7-8).

Curiously, the organism gains no further ATPs beyond the two already harvested during glycolysis. In fact, the point of fermentation is really to regenerate NAD^+, a limited commodity in cells but an essential cofactor for step 5 of glycolysis. During fermentation, pyruvate accepts electrons from NADH so that NAD^+ is regenerated. Without the regeneration of NAD^+, glycolysis would stop, and with it, the small ATP harvest of the anaerobic cell.

The conversion of pyruvate to lactate (Figure 7-8a) by certain bacteria helps turn milk into yogurt and many cheeses and helps create pickled foods, soy sauce, sourdough breads, and even chocolate. Anaerobic yeasts use a slightly different pathway (Figure 7-8b) to ferment the carbohydrates in fruits and grains to ethanol. This form of fermentation gives off carbon dioxide, and the released bubbles of CO_2 gas can cause dough to rise or give alcoholic beverages like champagne and beer their fizz.

Hard-working muscle cells can also carry out fermentation. Sometimes they use oxygen so fast that they become increasingly anaerobic and these facultatively aerobic cells can then begin to ferment pyruvate to lactate. Since lactate inhibits muscle function, a runner may end up with leg cramps. During so-called aerobic exercise, with its moderate exertion, oxygen flow to muscle cells is sufficient to prevent a build-up of lactate.

It may seem inefficient for a cell to carry out so many reactions when only certain ones lead to the formation of ATP, but there are two explanations. First, the intervening steps change the configurations of molecules, allowing energy-yielding reactions to occur. Second, energy release must be gradual and in small packets, since no biological molecule could absorb the large quantity of energy released by an explosive, single-step burning of glucose. In fact, the 686 kcal of potential energy contained in the C, H, and O bonds of 1 mole of glucose molecules (about 180 g) are released in many small steps by the "cold flame" of cellular metabolism.

CELLULAR RESPIRATION

The cells and tissues of multicellular plants and animals consume such large amounts of ATP that an animal capable only of fermenting food would have to find, eat, digest, and utilize almost 20 times more than a similar-sized animal whose cells carried out **cellular respiration**. This is because the end products of fermentation are organic molecules that still contain considerable energy, whereas cellular respiration breaks glucose down into small inorganic molecules and traps much of the released chemical energy in ATP.

During cellular respiration, the pyruvate molecules derived from glycolysis are shunted into a metabolic pathway called the *Krebs cycle,* which channels electrons to the *electron transport chain.* During this aerobic metabolism, oxygen accepts electrons and is reduced to H_2O, and 36 molecules of ATP are formed for each molecule of glucose consumed—18 times more than the energy harvest of glycolysis (review Figure 7-6).

During the first phase of cellular respiration, called the **Krebs cycle** (also known as the *citric acid cycle* and the *tricarboxylic acid cycle*), specific enzymes catalyze ten consecutive reactions, during which the pyruvate from glycolysis is oxidized to CO_2. Both NAD^+ and FAD act as hydrogen and electron acceptors and are reduced to NADH and $FADH_2$, respectively. As this pathway proceeds, three CO_2 molecules are produced for each pyruvate molecule, but only some of the energy is stored as ATP. The rest lies in the bonds of NADH and $FADH_2$ and is harvested during the final phase of respiration, the

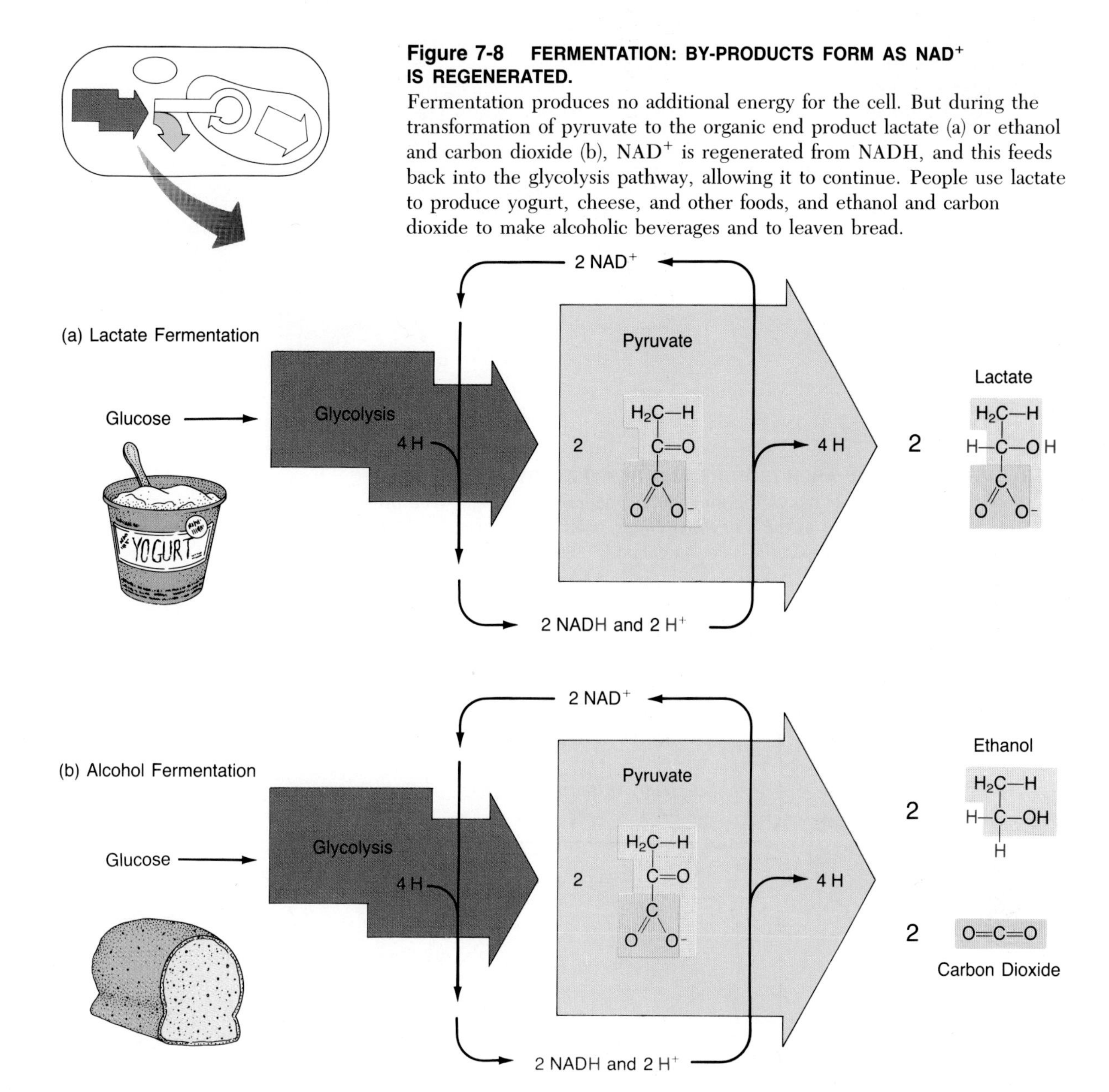

Figure 7-8 FERMENTATION: BY-PRODUCTS FORM AS NAD^+ IS REGENERATED.
Fermentation produces no additional energy for the cell. But during the transformation of pyruvate to the organic end product lactate (a) or ethanol and carbon dioxide (b), NAD^+ is regenerated from NADH, and this feeds back into the glycolysis pathway, allowing it to continue. People use lactate to produce yogurt, cheese, and other foods, and ethanol and carbon dioxide to make alcoholic beverages and to leaven bread.

electron transport chain. In this phase, the reduced compounds NADH and $FADH_2$ become oxidized, and their electrons are passed along a chain of oxidation-reduction steps to the final acceptor, O_2. This process releases a great deal of energy—in fact, most of the 36 ATPs.

In eukaryotes, both phases of cellular respiration take place in mitochondria. The Krebs cycle occurs in the mitochondrial matrix, the organelle's central region (Figure 7-9). Most of the enzymes that catalyze these reactions are in the matrix fluid, although they probably pass products directly from one to another, just as in glycolysis. In prokaryotes, the Krebs cycle enzymes are suspended in the cytoplasm. The proteins that bring about the second phase of cellular respiration, the reactions of the electron transport chain, are bound to the inner mitochondrial membrane in eukaryotes and to the plasma membrane in prokaryotes. Many of these proteins extend all the way through the lipid bilayer of the inner mitochondrial membrane, such that one part of each protein contacts the solution in the mitochondrial matrix, and the other part contacts the solution in the *outer compartment,* the fluid-filled space between the inner and outer mitochondrial membranes. This arrangement is essential to the formation of ATP, and we will see why shortly. The raw materials of respiration—pyruvate, oxygen, ADP, and inorganic phosphate—continuously diffuse into mitochondria. In turn, the waste products of energy metabolism—CO_2 and H_2O—and the energy storage product—ATP—diffuse outward into the cytoplasm. Mitochondria are truly the powerhouses of eukaryotic cells.

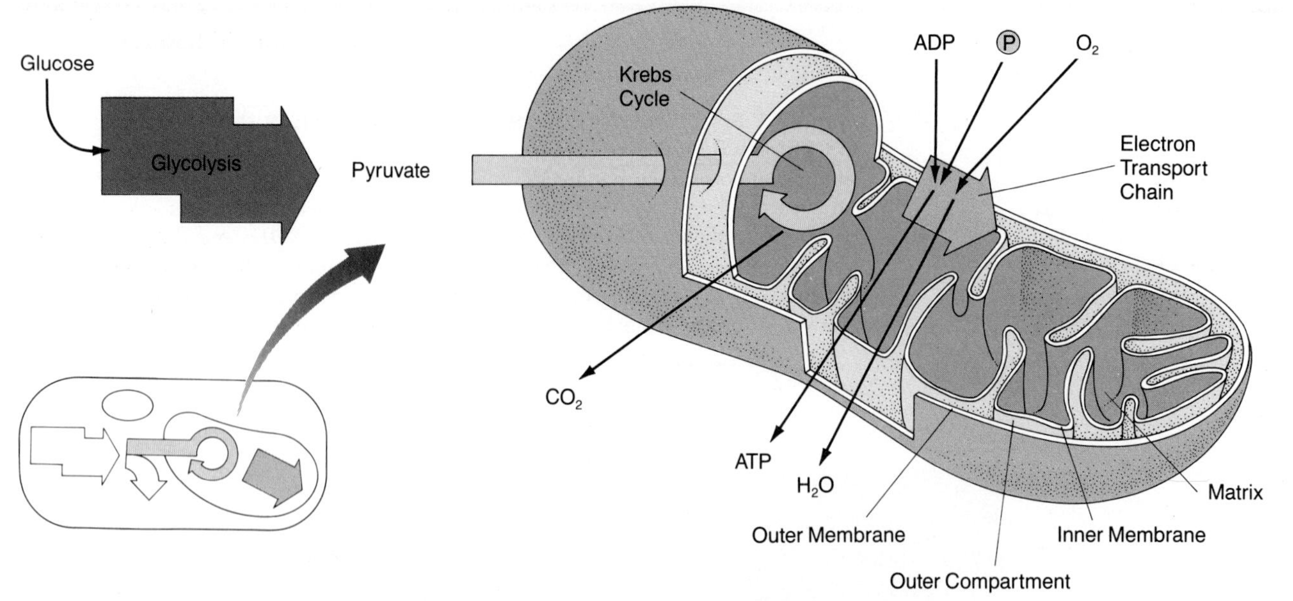

Figure 7-9 MITOCHONDRIAL ARCHITECTURE AND RESPIRATORY PROCESSES.
Respiration occurs within the foldings of the inner mitochondrial membrane. The processes in the electron transport chain are carried out by proteins embedded in the inner membrane itself. The raw materials of cellular respiration—pyruvate, ADP, inorganic phosphate, and oxygen—pass into the mitochondrion, and the end products exit.

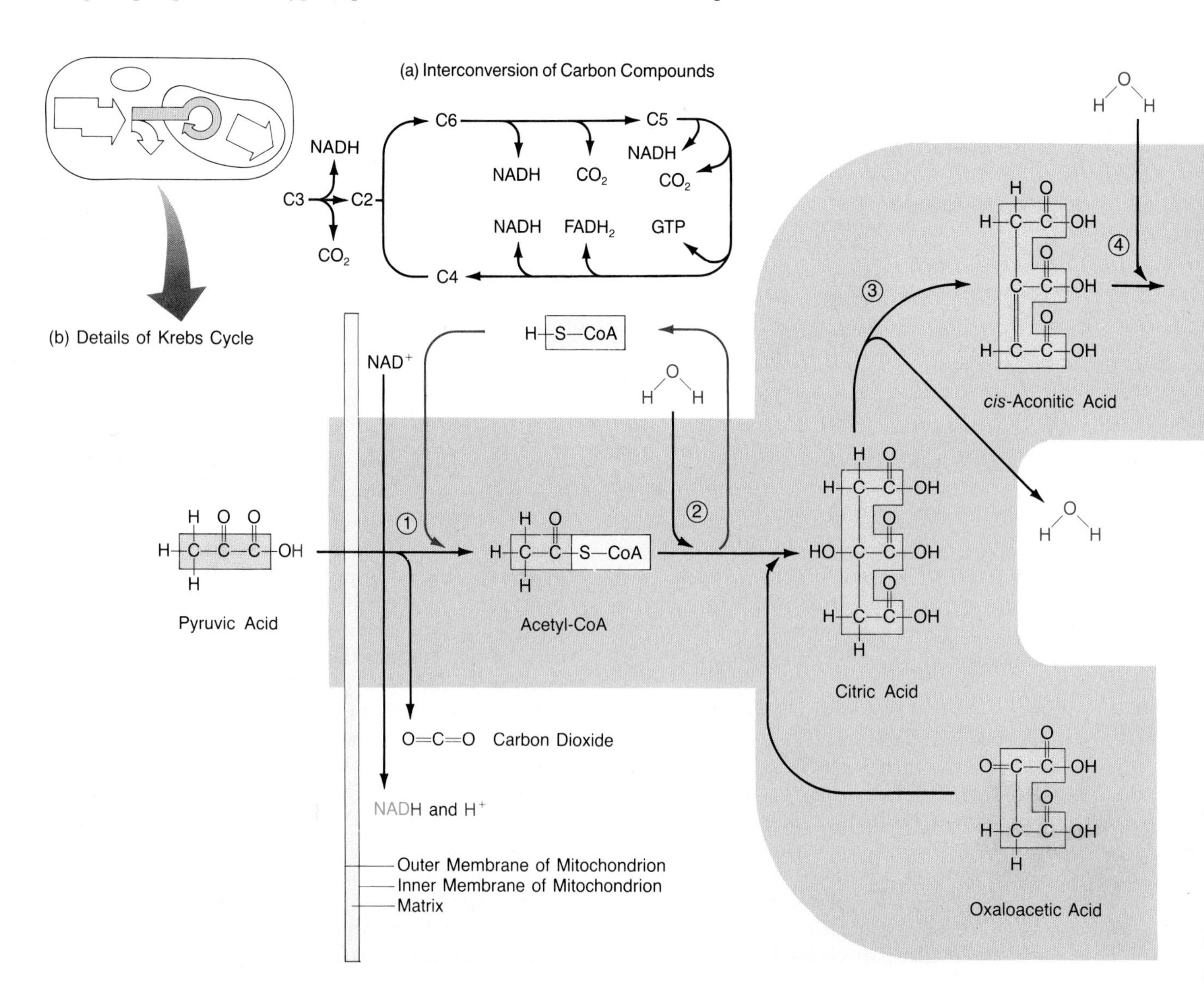

Figure 7-10 THE KREBS CYCLE: MOLECULAR CONVERSIONS AND A HARVEST OF ENERGY CARRIERS. (a) This very simple view of the Krebs cycle shows the number of carbons in each intermediate compound, as well as the NADH, $FADH_2$, and ATP produced. Note that a two-carbon compound enters the cycle and that two CO_2 molecules are generated each time the cycle "turns." The last four-carbon structure in the pathway joins an incoming two-carbon molecule to begin the cycle again. (b) This diagram shows the steps of the Krebs cycle in detail. Note the steps at which the energy carriers NADH, $FADH_2$, GTP, and ATP are produced. Also note the role of H_2O at various points of the cycle and the changes in the carbon skeleton of the intermediates. The cycle is not a closed system; it can be compared to a train in which some passengers (molecules) will ride the entire route (cycle), while other passengers will get on at the citrate station, and others might get off at the succinate station, for example. Figures 7-12 and 7-13 show in more detail how the Krebs cycle acts as a clearinghouse for metabolism.

Oxidation of Pyruvate: Prelude to the Krebs Cycle

Molecules of pyruvate generated by glycolysis pass easily through the highly permeable outer membrane of a mitochondrion and into the outer compartment. The molecules then move through the inner membrane by facilitated diffusion (see Chapter 5). Once in the mitochondrial matrix, a preliminary step takes place that precedes the Krebs cycle: Pyruvate is oxidized to CO_2 and an activated form of acetate (Figure 7-10b, step 1). In the process, NAD^+ is reduced to NADH, and the molecule of acetate becomes attached temporarily to a molecule of a coenzyme called *coenzyme A (CoA)*. Like ADP and ATP, this new compound, *acetyl-CoA*, is a high-energy compound; it conserves a large share of the energy available from the oxidation of pyruvate and with that energy drives the next reaction in the pathway, the first step of the Krebs cycle.

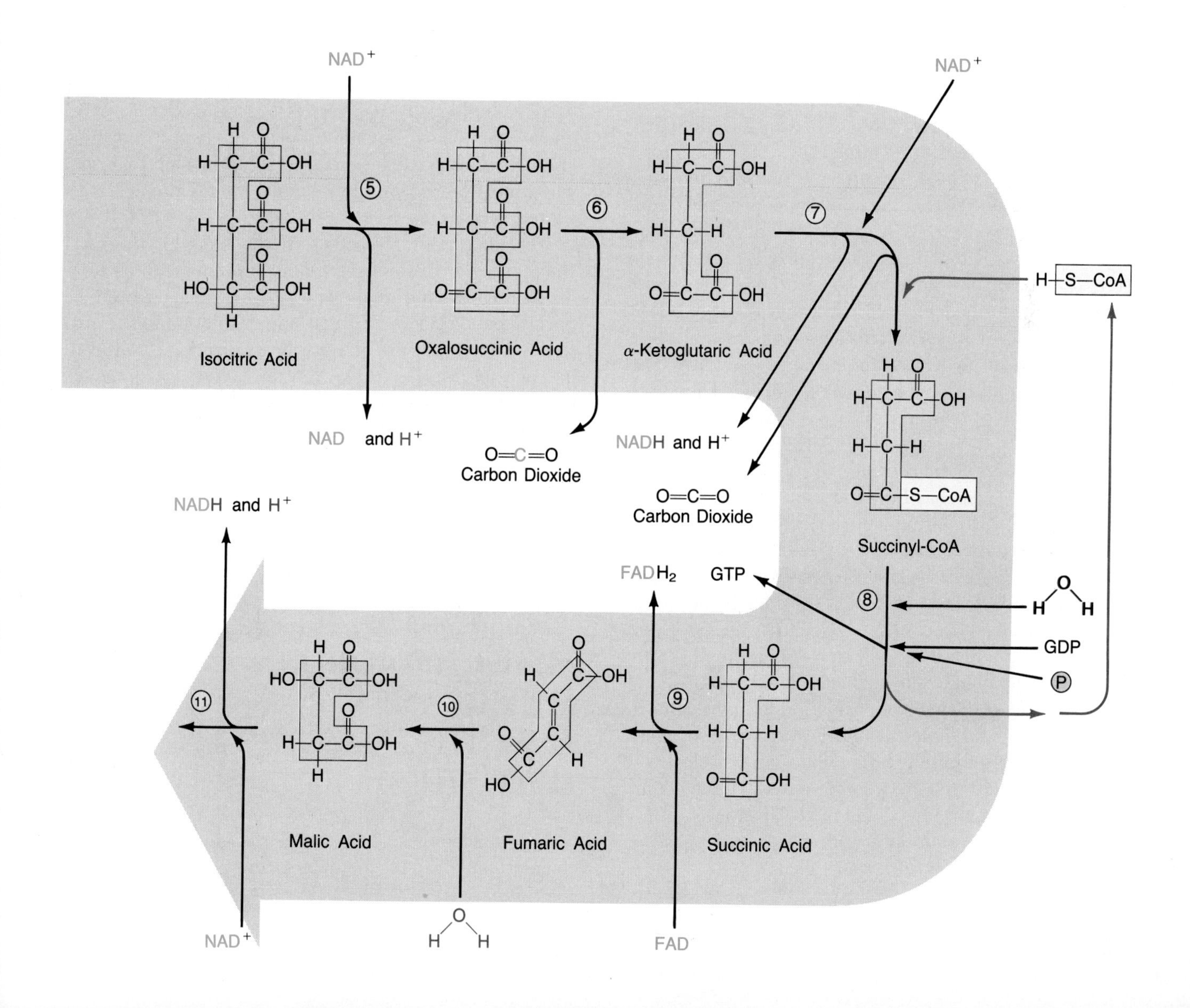

The Krebs Cycle

The Krebs cycle, whose steps were a masterpiece of biochemical detective work, was named after its discoverer, the British biochemist Sir Hans Krebs. Figure 7-10a shows the general conversions and products of that cycle, and Figure 7-10b shows the reaction steps in more detail.

The first true reaction of the Krebs cycle (Figure 7-10b, step 2) is actually a minicycle that links a two-carbon compound (the acetyl group of the acetyl-CoA molecule generated from the pyruvate formed during glycolysis) with a four-carbon compound (oxaloacetate, already present in mitochondria). A six-carbon compound, citrate, is formed; in the process, free acetyl-CoA is regenerated. The energy stored in acetyl-CoA drives this reaction.

The remaining reactions of the Krebs cycle remove two carbon atoms from the six-carbon sugar, generate two molecules of CO_2, and remove four pairs of electrons, which become stored in a number of energy-rich electron carriers. These carriers then provide the source of electrons that power the electron transport chain proteins to generate ATPs for the cell. At the three sites indicated in Figure 7-10b, (steps 5, 7, and 11), dehydrogenase enzymes transfer electrons to NAD^+ acceptors. At a fourth site (step 9), the enzyme succinic dehydrogenase, an integral part of the inner mitochondrial membrane, transfers electrons to FAD. And at the reaction steps where α-ketoglutarate is oxidized to succinate (steps 7 and 8), one molecule of guanine triphosphate (GTP) is also formed for each molecule of acetyl-CoA. Since two acetyl-CoAs result from each glucose molecule broken down in glycolysis, the Krebs cycle forms two GTPs for each glucose molecule metabolized. These two GTPs are later converted into two ATPs.

By step 11, the six carbons of the original glucose molecule have all been oxidized, and some of the energy stored in that sugar has been transferred to four molecules of ATP, two formed during glycolysis and two formed during the Krebs cycle. The total free energy change in the two phases is small: -62 kcal per molecule of glucose. In fact, most of the energy once bound up in glucose remains stored in NADH and $FADH_2$. Thus, during the first phase of cellular respiration, the tally from one original molecule of glucose is two molecules of NADH formed during the oxidation of pyruvate to acetyl-CoA and six molecules of NADH and two molecules of $FADH_2$ formed during the Krebs cycle. Add these energy carriers to two more NADH molecules generated during glycolysis, and the cell has a rich storehouse ready to be unlocked during the final phase of respiration, the electron transport chain.

The Electron Transport Chain

The electron transport chain is a metabolic cascade that begins by oxidizing the Krebs cycle products NADH and $FADH_2$, ends by reducing O_2 to H_2O, and in between harvests ATP (Figure 7-11a). A series of reaction steps drives this harvesting process and causes small amounts of free energy to be released gradually. For the first time during cellular respiration, oxygen plays a role, acting as an electron acceptor for both NADH and $FADH_2$.

As the two reduced energy carriers accept electrons, their oxidized forms, NAD^+ and FAD, are regenerated and thus ready to take part again in the reaction steps of glycolysis and the Krebs cycle. At the same time, protons set free when NADH is oxidized can join other protons in the watery matrix to provide the hydrogen atoms in H_2O, a final product of cellular respiration (see Figure 7-11a).

As NADH and $FADH_2$ are oxidized, the electrons they lose are passed along, like water in a bucket brigade, by a series of electron carriers (Figure 7-11b). Proteins along the electron bucket brigade gradually strip energy from the electrons, and this energy can then be used to build ATP in a process called **oxidative phosphorylation.** This "downhill" series of electron transfers gradually lowers the level of energy in the electrons, and when most of this energy is spent, the electrons are accepted by oxygen.

Certain carrier molecules in the electron transport chain are **cytochromes**—pigment proteins with an iron-containing *heme* at the center. Electrons donated by NADH pass from one cytochrome compound to the next as the compounds are alternately oxidized and reduced. As each pair of electrons from an NADH passes down the chain, hydrogen ions are pumped through the inner mitochondrial membrane. As the next section explains, those hydrogens reenter the mitochondrial matrix, generating three ATP molecules as they go. The oxidation of each molecule of NADH therefore generates three molecules of ATP.

MITOCHONDRIAL MEMBRANES AND THE MITCHELL HYPOTHESIS

The molecules that make up the electron transport chain are able to pass along electrons and generate ATP

Figure 7-11 THE ELECTRON TRANSPORT CHAIN: AN ENERGY CASCADE.
(a) As NADH and $FADH_2$ are oxidized, the electrons they lose are passed along a series of electron carriers (here represented by downward sloping yellow arrows). As the electrons move down the chain, their energy levels decrease until the electrons are finally accepted by oxygen, and H_2O forms. Oxidative phosphorylation is the process by which ATP is produced by means of this transport of electrons. (b) The electron carriers are large multienzyme complexes positioned within the mitochondrial membrane and represented here by blue or purple shapes. The large complexes include (from left to right) NADH-coenzyme Q reductase, coenzyme Q or ubiquinone (pink), cytochrome *c* reductase, cytochrome *c*, cytochrome *c* oxidase, and ATP synthetase. Although the complexes appear motionless in this drawing, they and the smaller carriers diffuse freely in the membrane; electron transfer only occurs when they chance to collide. In Mitchell's Chemiosmotic Coupling Hypothesis, the flow of protons toward the outer compartment of the mitochondrion is coupled to movement of electrons across each transport complex. The return of protons to the matrix via coupling factor (ATP synthetase) is linked to the formation of ATP.

(a) Aerobic Respiration

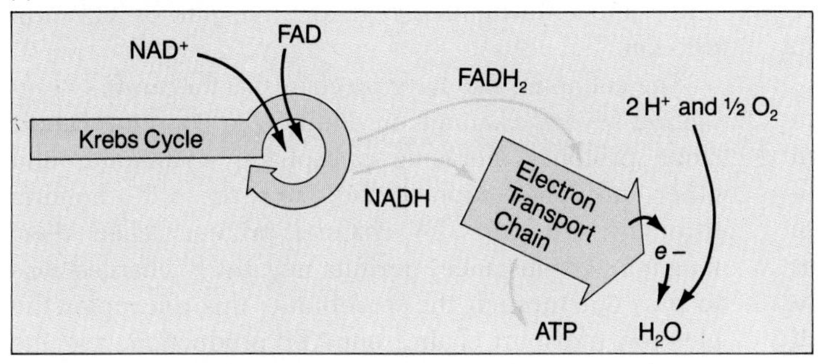

(b) Pathway of Electron Flow

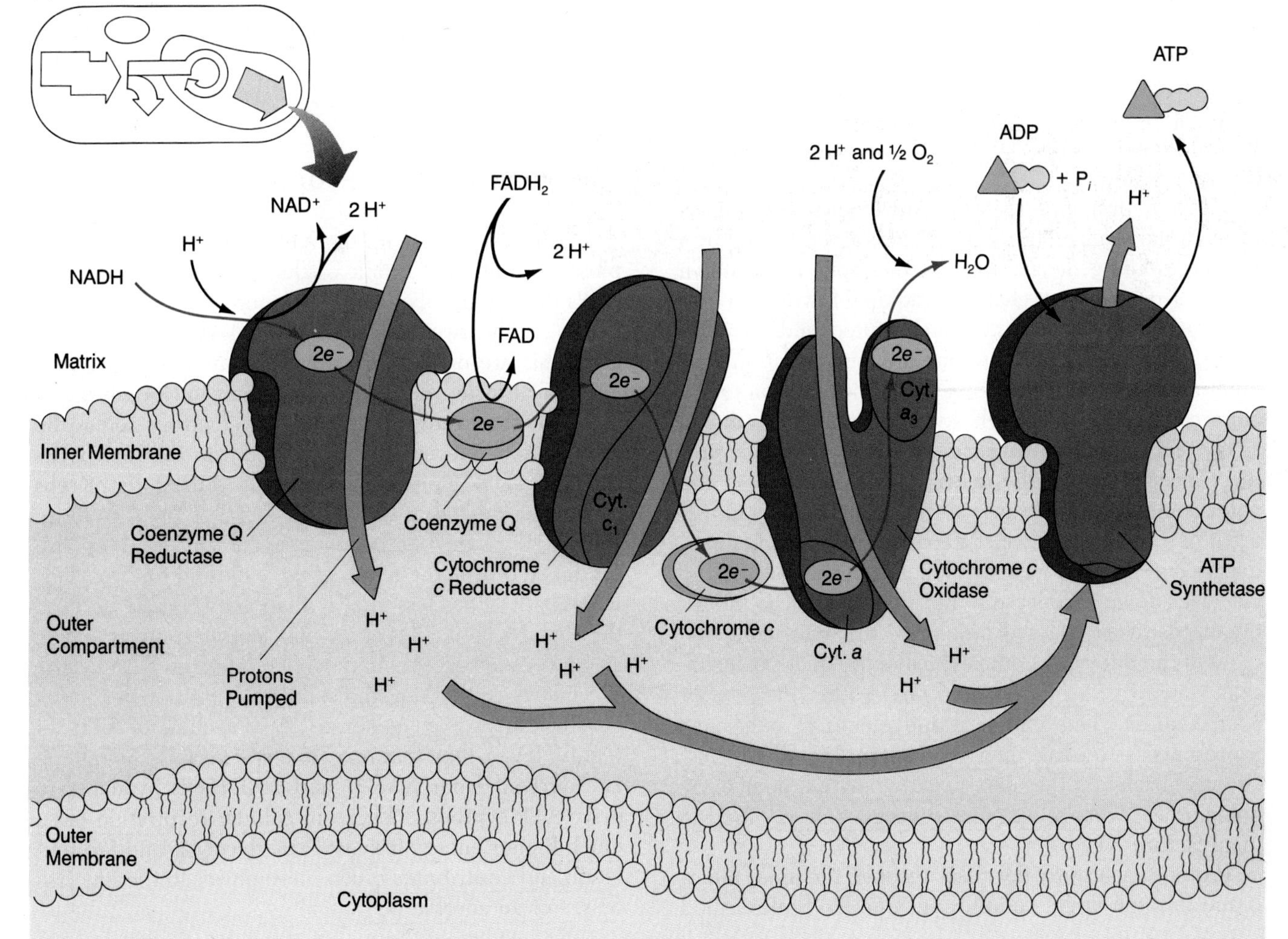

only because they are arrayed and embedded in a particular order in the inner membrane of the mitochondrion.

Although biologists are still investigating the complete mechanism of electron transport, they know that three of the energy carriers in the electron transport chain are parts of large multiprotein complexes, while the small molecule ubiquinone (also called coenzyme Q) and the single protein cytochrome *c* act as connecting carriers whose activities effectively link the larger complexes together (see Figure 7-11b). Each complex is an integral part of the inner mitochondrial membrane, and the passage of an electron pair through all three complexes and connecting carriers yields enough energy for the oxidative phosphorylation of about three ADPs to ATP.

The three large complexes constantly diffuse about in the membrane, moving distances six times their own diameters every 5 ms. The smaller ubiquinone and cytochrome *c* move ten times as fast. It is only when these electron carriers chance to collide that pairs of electrons can move from one complex to another. Despite these odds, electron pairs are received and donated by each multiprotein complex once every 5–20 ms.

For years, biochemists wondered whether the inner membrane of the mitochondrion is the site of ATP formation during respiration as well as the site of electron transport. The best current model, the **chemiosmotic coupling hypothesis,** proposed by British biochemist Peter Mitchell, states that ATP forms in association with this membrane.

Mitchell proposed that the movement of electrons through the electron transport chain is accompanied by a *proton-pumping* mechanism. As electrons move down the electron transport chain, a gradient of hydrogen ions (protons) is created across the inner mitochondrial membrane. About 12 to 13 protons are pumped from the inner compartment (matrix) of the mitochondrion to the outer compartment for each oxygen atom that is reduced as H_2O forms. This mechanism operates to keep the proton concentration in the matrix lower than that in the outer compartment. As we saw in Chapter 5, the movement of a substance against a concentration gradient requires energy. This energy comes from passage of the electrons down the electron transport chain. The pumping of positively charged protons to the outer compartment of the mitochondrion also leaves the matrix electrically negative in comparison with the outer compartment. This chemical and electrical imbalance represents potential energy: Protons tend to flow "down," back into the more negative matrix, in an exergonic process that releases free energy—energy used to manufacture ATP.

Protons apparently flow back through the inner membrane through large, complex enzyme molecules called **coupling factor** or **ATP synthetase** (see Figure 7-11b). Biologists believe that the protons cause the enzyme's catalytic site to release a newly formed ATP molecule, thereby opening the site to an ADP, which is then quickly converted to ATP.

This reaction acts as a major control point for ATP synthesis: The protons flow through the channels only as long as enough ADP enters the mitochondrion. This channeling mechanism may automatically adjust the rate of respiration to the speed at which ATP is used by the metabolizing cell, and hence the rate of ADP regeneration. Studies show, however, that in some cells, calcium entry into mitochondria regulates the rate of ATP production.

The cell must also have mechanisms for carefully regulating several channels in addition to coupling factor, since protons, inorganic phosphate, hydroxide, and other substances appear to enter the mitochondria through such ports. A channel protein called thermogenin, for instance, permits negatively charged ions to pass out through the membrane; this uncouples the electron transport chain from ATP production, and the energy, released as heat, can warm a human baby or a young rabbit.

While the electron transport chain and its relation to proton pumping is a complex topic, the activities of this chain constitute the real engine that drives most living cells.

The Energy Score for Respiration

Figure 7-12 tallies up the ATP harvest from the complete respiration of 1 mole of glucose. The cell gains two ATPs during glycolysis and two more during the Krebs cycle. As the two molecules of NADH formed during glycolysis are oxidized in the electron transport chain, four ATPs form; each "uses up" some of its energy in reactions that transport the NADH electrons across the inner mitochondrial membrane and into the matrix. As the 8 NADH molecules generated during the Krebs cycle are oxidized in the chain, a total of 24 ATPs form. Finally, the oxidation of two molecules of $FADH_2$ (also produced during the Krebs cycle) yields four more ATPs; $FADH_2$ provides less energy than does NADH because the $FADH_2$ molecules enter the electron transport chain at a lower energy level in the downhill series of transfers.

Theoretically, then, the cellular respiration of a molecule of glucose can generate 36 molecules of ATP. In reality, fewer than that may be made because not all the H^+ derived from NADH and $FADH_2$ is available to drive ATP synthesis. Even so, cellular respiration has a much higher energy harvest than glycolysis and fermentation and contributes much more power to the life processes of an aerobic cell.

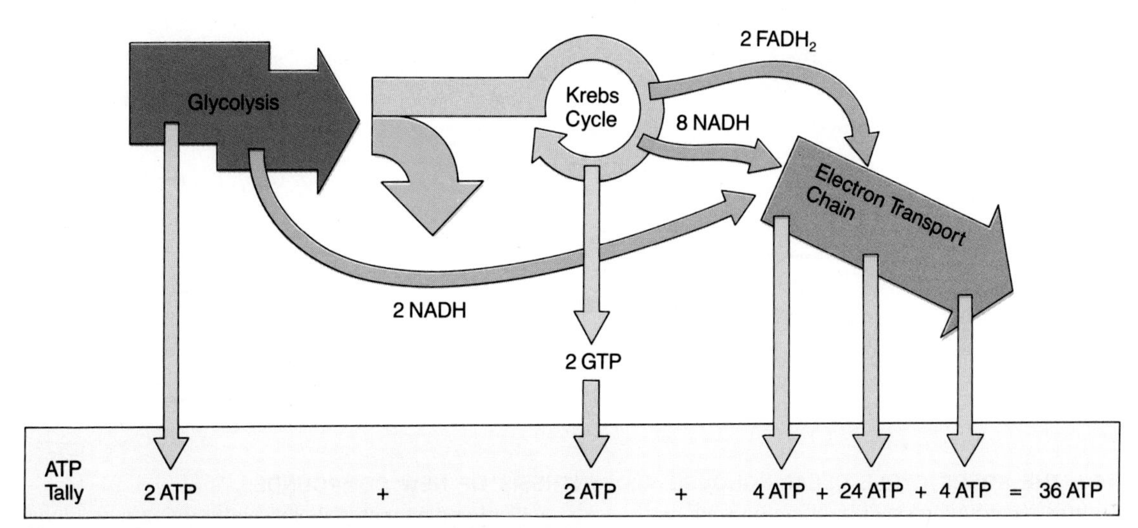

Figure 7-12 THE ATP TALLY FROM CELLULAR RESPIRATION.
The complete respiration of one molecule of glucose yields 2 ATPs and 2 NADHs from glycolysis and 2 ATPs, 2 NADHs, and 2 molecules of $FADH_2$ from the Krebs cycle. The electron transport chain generates 4 ATPs from the 2 NADH carriers of glycolysis. (The net yield is only 4 because moving each NADH from the cytoplasm into the mitochondrion costs the cell 1 ATP.) In addition, the electron transport chain produces 24 ATPs from the 8 NADH carriers generated in the Krebs cycle and 4 ATPs from the Krebs cycle's 2 $FADH_2$ molecules. The total harvest is thus 36 ATPs for each molecule of glucose metabolized in an aerobic cell.

METABOLISM OF FATS AND PROTEINS

Do the same energy pathways operate when an organism consumes fats or proteins instead of carbohydrates? Just as the cell breaks polysaccharides into glucose, and glucose into pyruvate, it also breaks down fats and proteins and converts them to the same compounds that serve as intermediates in the Krebs cycle. This is why the Krebs cycle can be considered a clearinghouse through which the building blocks of all organic energy sources—carbohydrates, proteins, and fats—are ultimately processed in aerobic cells (Figures 7-13 and 7-14).

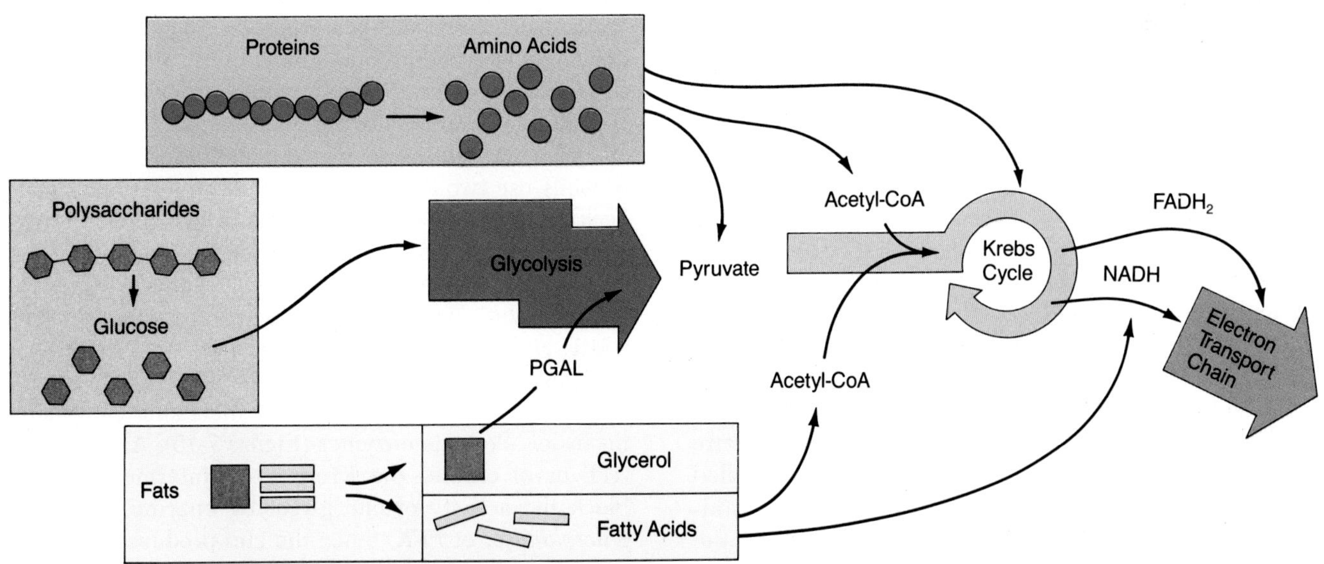

Figure 7-13 THE KREBS CYCLE CLEARINGHOUSE: HYDROLYSIS OF NUTRIENTS.
Just as pyruvate from the glycolysis of glucose molecules can enter the Krebs cycle, the subunits of proteins and fats can also be converted to acetyl-CoA or other Krebs cycle intermediates. Hence, the cell can convert these nutrients to ATP energy just as it can glucose.

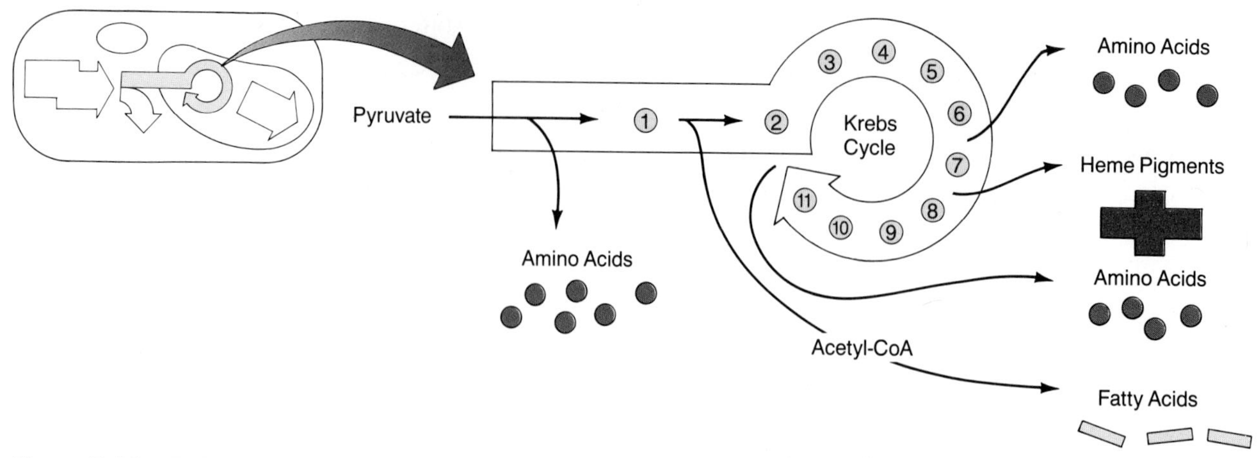

Figure 7-14 THE KREBS CYCLE CLEARINGHOUSE: BIOSYNTHESIS OF NEW COMPOUNDS. The cell requires a continuous supply of amino acids, fatty acids, and other raw materials for synthesizing new biological molecules and cell parts. Intermediates from the Krebs cycle can be drawn off and used directly or converted into these organic raw materials.

Through hydrolysis reactions, plant and animal cells may degrade proteins to their component amino acids (see Figure 7-13). These monomers can then be degraded further to pyruvate, to acetate or acetyl-CoA, or to intermediates in the Krebs cycle. Whatever the fragment's structure, its carbon skeleton is ultimately dismantled and oxidized to CO_2. On average, 1 g of protein yields about as much ATP as does 1 g of glucose.

Fats have a similar but more complicated fate. The breakdown of fats begins with hydrolysis of triglycerides to glycerol and free fatty acids. Glycerol is modified to glyceraldehyde-3-phosphate and enters step 4 of the glycolysis pathway. Fatty acids are broken down inside the mitochondrion to form acetyl-CoA (which feeds into the Krebs cycle) plus NADH and $FADH_2$, oxidized via the electron transport chain. One gram of fatty acid molecules provides 2.5 times more ATP than does 1 g of sugar or protein, which is why many animals and plants store some of their food reserves as fats or oils and is also why cutting down on fats helps a dieter shed pounds.

The Krebs cycle is important not just in the catabolism of nutrients, but also indirectly in the manufacture of proteins, fats, and carbohydrates for export from the cell or for structural roles inside the cell (see Figure 7-14). The building up process *(anabolism)* is also called **biosynthesis.** Pyruvate, acetyl-CoA, and Krebs cycle intermediates can serve as the starting compounds for the biosynthesis of glutamate and aspartate, for example, and these amino acids can be added to growing protein chains. The traffic of so many compounds through such a busy clearinghouse must be controlled in some way, and our next section explains how.

CONTROL OF METABOLISM

The rates at which molecules are broken down or built up in metabolic pathways are regulated in cells so that precious energy will be conserved and the products of the various pathways will be available when needed. Over the long term, genetic control mechanisms are employed to help regulate the amount of a particular enzyme. In the short run, the activity of the enzymes already produced is regulated too.

Most cellular controls are based on feedback inhibition, by which the product of a pathway affects the rate at which the pathway itself operates (review Figure 4–15). The concentration of the product thus has an inverse effect on the rate of its own production.

Cells use two distinct mechanisms to carry out feedback inhibition over metabolic pathways: allosteric enzymes and covalent modification of enzymes.

Allosteric enzymes (described in Chapter 4) have two binding sites. The first, the active site, binds the substrate, and the second binds an effector, a molecule that changes the activity of the enzyme. The glycolysis process involves a good example of metabolic control by means of allosteric enzymes (Figure 7-15). A buildup of ATP or of citrate, the Krebs cycle intermediate, can block the activity of one glycolytic enzyme, *phosphofructokinase*, or *PFK*. Since the end product of glycolysis, pyruvate, furnishes the carbon skeleton for the Krebs cycle, products of the Krebs cycle can build up and shut down production still further by interacting with the allosteric site on the PFK enzyme.

Covalent modification of enzymes involves a tempo-

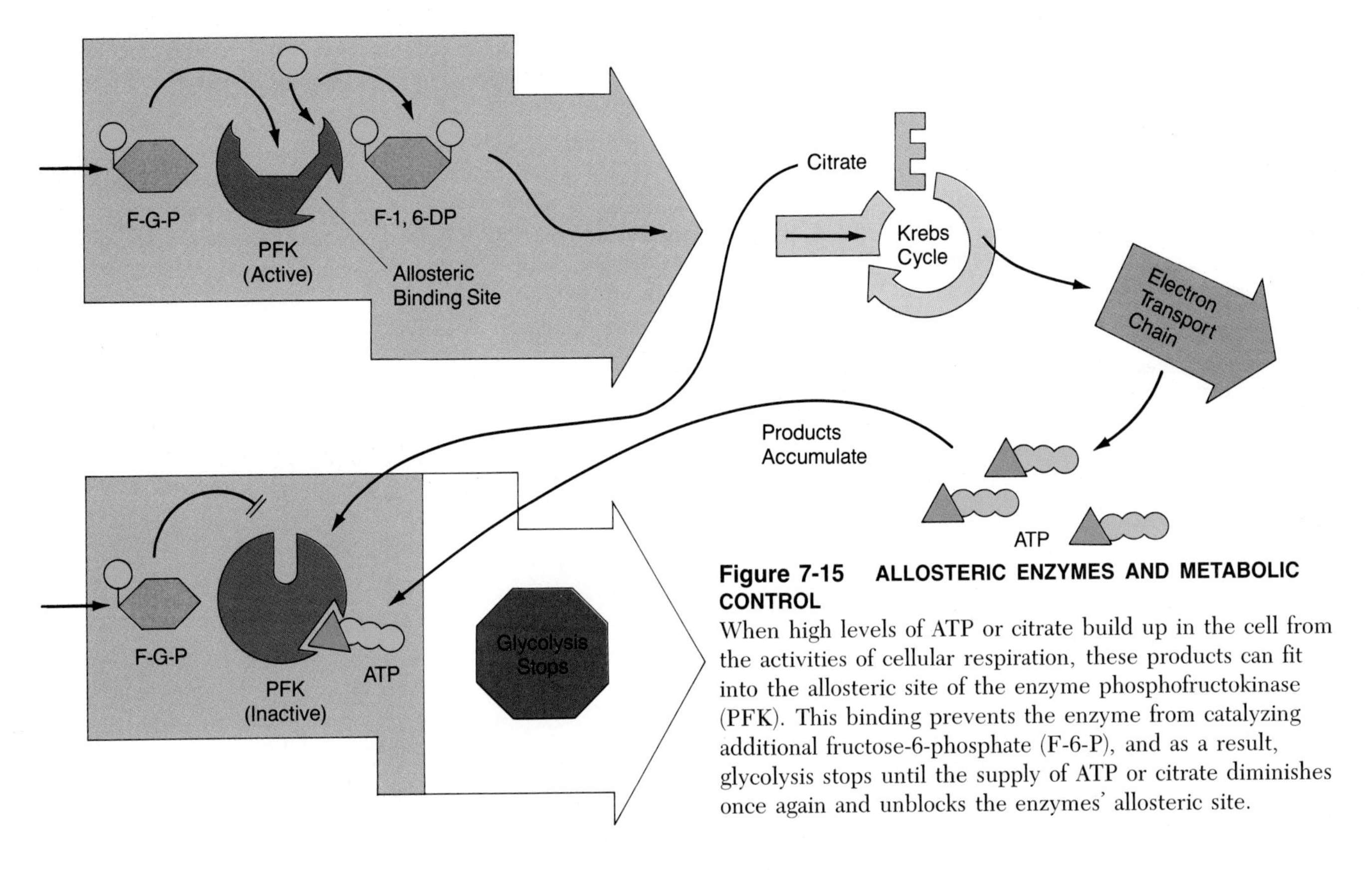

Figure 7-15 ALLOSTERIC ENZYMES AND METABOLIC CONTROL
When high levels of ATP or citrate build up in the cell from the activities of cellular respiration, these products can fit into the allosteric site of the enzyme phosphofructokinase (PFK). This binding prevents the enzyme from catalyzing additional fructose-6-phosphate (F-6-P), and as a result, glycolysis stops until the supply of ATP or citrate diminishes once again and unblocks the enzymes' allosteric site.

rary change in the chemical structure of these biological catalysts. Specifically, a chemical group attaches to the enzyme, inactivating it and reducing its catalytic function. Phosphate groups act this way on certain enzymes involved in respiration: When ATP accumulates, the enzymes become phosphorylated, and their activity is inhibited. When the level of ATP falls, another enzyme that functions especially well in the presence of abundant pyruvate removes the phosphate groups, and the inhibited enzymes are reactivated.

Dozens of anabolic and catabolic pathways function simultaneously in every cell and act at amazing speed. Every 5 seconds, for example, a yeast cell uses up and then regenerates *80 million ATP molecules*. The overall energy economy of a large multicellular organism such as a human, with 20 to 30 trillion cells, starts to seem impossibly complicated. Yet, metabolism goes on imperceptibly and automatically every second of an organism's life.

SUMMARY

1. Nutrient molecules are a temporary repository for solar energy trapped by plants and other autotrophs. Metabolic pathways in cells break down nutrients to release this energy in separate steps, each mediated by a specific enzyme.

2. *ATP (adenosine triphosphate)* is an energy storage molecule that transfers energy from exergonic reactions, such as nutrient catabolism, to endergonic reactions, such as synthesis of biological molecules.

3. All key metabolic reactions involve *oxidation*, the removal of electrons from an atom or a compound, coupled to *reduction,* the addition of electrons.

4. *Glycolysis* is the initial sequence of reactions used by virtually all cells to metabolize glucose. During the nine reaction steps of glycolysis, glucose is broken down to two molecules of *pyruvate*, a net of two molecules of ATP are formed, and two molecules of NAD^+ are reduced to NADH.

5. *Fermentation* is a reaction sequence beginning with glycolysis in which organic molecules rather than O_2 function as electron acceptors. Fermentation is important because it regenerates NAD^+.

6. During *cellular respiration,* glucose is oxidized completely to CO_2 and H_2O. The pyruvate generated from glycolysis is oxidized to an activated form of acetate, acetyl-CoA, which enters the *Krebs cycle*. That cycle produces two ATPs, as well as molecules of NADH and $FADH_2$.

7. NADH and $FADH_2$ are oxidized during the reactions of the *electron transport chain*, a series of electron carrier molecules embedded in the inner membrane of the mitochondrion. Electrons are passed from one of these pigment proteins to the next to the final acceptor, oxygen. During this sequence,

32 molecules of ATP are formed. In total, 36 ATPs form in an aerobic cell for each molecule of glucose consumed.

8. Mitchell's *chemiosmotic coupling hypothesis* is the current theory for how ATP is formed by *oxidative phosphorylation*. The flow of electrons down the electron transport chain in the inner mitochondrial membrane results in the pumping of protons outward through the membrane. Then, as protons pass back into the mitochondrial matrix through ATP synthetase, ADP is phosphorylated to ATP.

9. The subunits of amino acids and fats are converted to intermediates that can enter the Krebs cycle "clearinghouse." Various Krebs cycle intermediates can, in turn, serve as precursors for the biosynthesis of new biological molecules.

10. The control of metabolism allows the cell to store appropriate amounts of energy in ATP and raw materials. Cells employ allosteric enzymes and covalent modification of enzymes to control the rate of metabolism.

KEY TERMS

adenosine triphosphate (ATP)
biosynthesis
cellular respiration
chemiosmotic coupling hypothesis
coupling factor
cytochrome
electron transport chain
FAD
fermentation
glycolysis
Krebs cycle
NAD^+
oxidation
oxidation-reduction reaction
oxidative phosphorylation
phosphorylation
pyruvate
reduction

QUESTIONS

1. Explain how the cell's breakdown of glucose is similar to burning fuel in an engine. Explain a few important differences as well.

2. What is the function of each of the following in cell metabolism? ATP; NAD^+; FAD; Krebs cycle; cytochromes; oxygen.

3. Which of the following can proceed in the absence of oxygen? Glycolysis; fermentation; Krebs cycle; electron transport.

4. What extra metabolic step does fermentation accomplish?

5. Which molecules in a cell act as cyclic electron acceptors? What kinds of molecules act as terminal electron acceptors in fermentation? In respiration?

6. The electron transport system is embedded in membranes; glycolytic enzymes are located in the cytoplasm. Which set of molecules would have hydrophobic regions? Which would be easier to study in a test tube, using standard aqueous salt solutions?

7. Which is a more efficient fuel: a highly reduced molecule (like a fatty acid), a highly oxidized one (like carbon dioxide), or one in an intermediate state of oxidation (like a carbohydrate)?

8. If the respiratory system in a cell is poisoned so that electron transport is uncoupled from ATP production, what might happen to the energy produced?

ESSAY QUESTIONS

1. How does Mitchell's chemiosmotic pump work, and what relationship does it have to the concept of potential energy?

2. Yeast can ferment sugars and grow in the absence of oxygen. If a yeast cell were to lose its mitochondria, would it still be able to grow? Would it grow as fast as a normal cell? Explain.

For additional readings related to topics in this chapter, see Appendix C.

8

Photosynthesis: Harnessing Solar Energy to Produce Carbohydrates

The great invention of the plant kingdom . . . is chlorophyll. Whether it is a one-celled alga or a giant forest tree, a plant will contain this substance.

Anthony Huxley,
Plant and Planet (1974)

Every carbon atom in your body and every oxygen molecule you breathe once cycled through the tissues of a plant. The significance of plants, algae, and other photosynthetic organisms to the entire balance of life on earth cannot be overstated, because through photosynthesis, they convert solar energy to chemical energy in the form of carbohydrates. If plants suddenly stopped producing carbohydrates and liberating oxygen, much of the world's carbon would be oxidized to CO_2 gas within about 300 years, and the oceans and atmosphere would be devoid of free oxygen within about 2,000 years—mere instants in the vast stretch of geological time. Clearly, as aerobic, nonphotosynthetic organisms, we depend on plants, and they are dependably productive: They generate more than 150 billion tons of carbohydrates every year.

Both photosynthesis and cellular respiration are essential for the transfer of environmental energy to living things, but the two processes are chemical opposites.

Sunlight, air, the green of chlorophyll in Douglas firs: The components of photosynthesis, the sustainer of life on earth.

Photosynthesis traps solar energy, converts it to chemical bond energy, and generates carbohydrates, while, as we saw in Chapter 7, cellular respiration oxidizes carbohydrates, releasing chemical and heat energy.

Our coverage of photosynthesis in this chapter includes:

- The overall chemistry of the process, with its unique series of reactions mediated by enzymes
- How photosynthetic cells trap light energy in colored pigments and pigment complexes
- How the cell converts light energy to chemical energy in a series of light-dependent reactions
- How cells use that chemical energy to build carbohydrates in a series of light-independent reactions
- How photosynthesis in eukaryotic cells depends on the architecture of the chloroplast
- How too much oxygen can inhibit photosynthesis in a process called photorespiration
- How a special metabolic pathway can circumvent photorespiration in some drought-tolerant plants
- The global cycling of carbon through plants, animals, decomposers, and the atmosphere

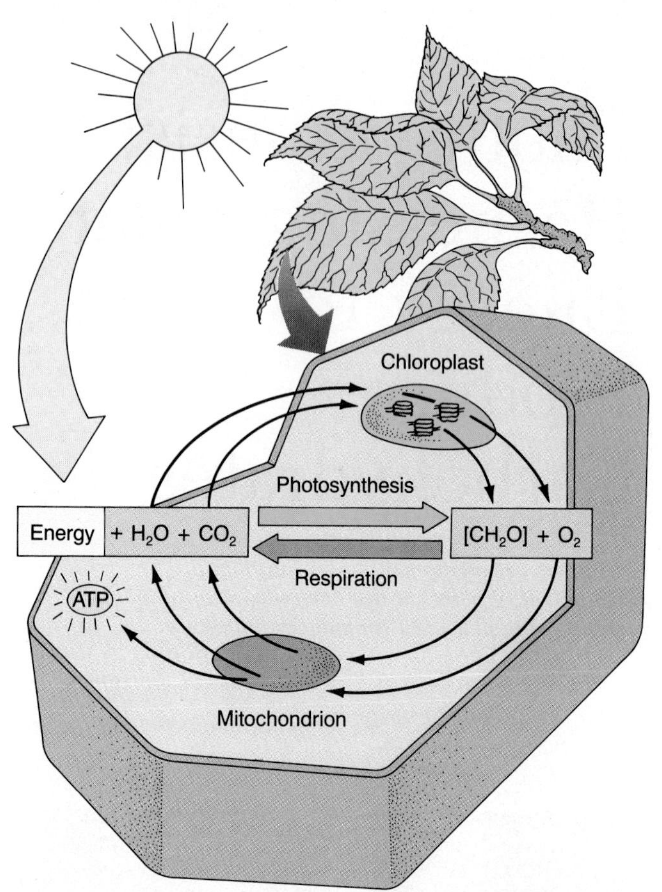

Figure 8-1 PHOTOSYNTHESIS AND RESPIRATION: BIOCHEMICAL OPPOSITES.
In a photosynthetic cell, light energy drives a series of reactions that convert CO_2 and H_2O to carbohydrates $[CH_2O]$ and O_2. This diagram shows the fate of the hydrogen, carbon, and oxygen atoms during photosynthesis and also shows that the reactions are the chemical reverse of respiration. In eukaryotic cells, photosynthesis takes place in chloroplasts, while aerobic respiration takes place within mitochondria.

AN OVERVIEW OF PHOTOSYNTHESIS

The metabolic process called **photosynthesis** occurs only in cells of green plants, algae, and certain protists and bacteria. During the process, energy from the sun is trapped and used to convert the inorganic raw materials CO_2 and H_2O to carbohydrates and O_2. A green pigment, *chlorophyll,* is the key to that process. We can summarize the process with the deceptively simple equation in Figure 8-1, where $[CH_2O]$ stands for carbohydrate. Notice that carbon dioxide is reduced to carbohydrate and that water is oxidized as molecular oxygen forms. Notice also that photosynthesis is chemically the reverse of cellular respiration. If we multiply each basic equation by 6 (because there are six carbon atoms in glucose), we get two new equations describing the synthesis and breakdown of glucose, $C_6H_{12}O_6$:

$$\text{Photosynthesis: } 6\,CO_2 + 6\,H_2O \xrightarrow{\text{light}} C_6H_{12}O_6 + 6\,O_2$$

$$\text{Respiration: } C_6H_{12}O_6 + 6\,O_2 \longrightarrow 6\,CO_2 + 6\,H_2O$$

As Chapter 7 explained, the oxidation of 1 mole of glucose to CO_2 during respiration is an exergonic process that releases 686 kcal of free energy. Not surprisingly, then, building glucose molecules from inorganic raw materials is highly endergonic and requires about 2,000 kcal per mole of glucose formed.

The Two Basic Reactions of Photosynthesis

Biologists first studying the general equation for photosynthesis assumed that since CO_2 and H_2O are raw materials for the process, CO_2 molecules must be cleaved somehow, O_2 released to the environment, and leftover carbon atoms joined to the H_2O molecules to form CH_2O. In 1929, C. B. van Niel of Stanford University studied photosynthetic bacteria that use the sulfur compound H_2S instead of H_2O during photosynthesis and release sulfur, not oxygen. By tracing their production of gaseous sulfur, van Niel found that the bacteria oxidize H_2S to sulfur plus hydrogen atoms and use the hydrogens to reduce CO_2 to carbohydrate (Figure 8-2a). During their photosynthesis, the cells also produce water, with oxygen coming from CO_2, since H_2S lacks oxygen. Generalizing from his work on bacteria, van Niel proposed that photosynthesis in plants could be described with the equation shown in Figure 8-2b.

Through scientific reasoning, van Niel was able to outline the two sets of reactions in photosynthesis. In the first set, water is oxidized; in the second, carbon dioxide is reduced. When water is split, O_2 is released to the environment; hydrogen atoms then reduce CO_2 to carbohydrate. But there was a great deal more to reveal, including the role of light in photosynthesis.

Light is clearly necessary for photosynthesis, since plants stop growing and eventually die if deprived of light for long periods. But the photosynthesis researchers who followed van Niel had to discover whether light is used directly during both sets of reactions or during just one of them. Their studies have shown that the first stage of photosynthesis, when water is split and O_2 released, is absolutely dependent on direct use of light energy; these reactions are therefore called the **light-dependent reactions** (Figure 8-3). The high-energy products of the first stage, ATP and NADPH (a coenzyme electron carrier), drive the second stage, during which CO_2 is reduced to carbohydrate. This second set of reactions, which can proceed whether light is present or not, is called the **light-independent reactions** (see Figure 8-3).

(a) Photosynthetic Sulfur Bacterial Cell

Light Energy + $2 H_2S + CO_2$ → $[CH_2O] + H_2O + 2 S$

(b) Chloroplast in Plant Cells

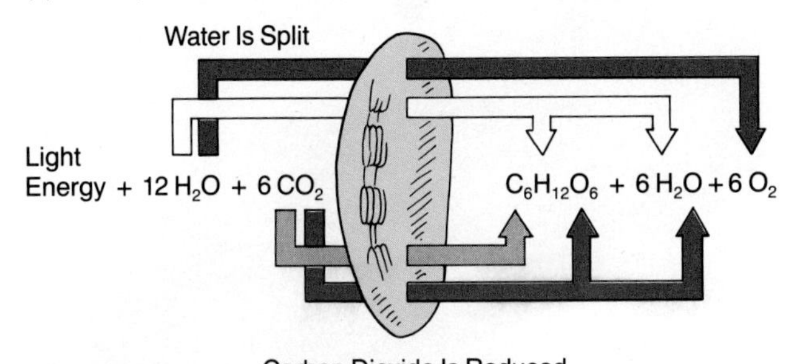

Figure 8-2 VAN NIEL'S EQUATIONS.
Van Niel's hypotheses and experiments led him to develop the equation in (a) to describe photosynthesis in sulfur bacteria and, by analogy, the equation in (b) to describe photosynthesis in plant cells. The yellow arrow traces the fate of sulfur, the gray arrows carbon, the white arrows hydrogen, and the red arrows oxygen.

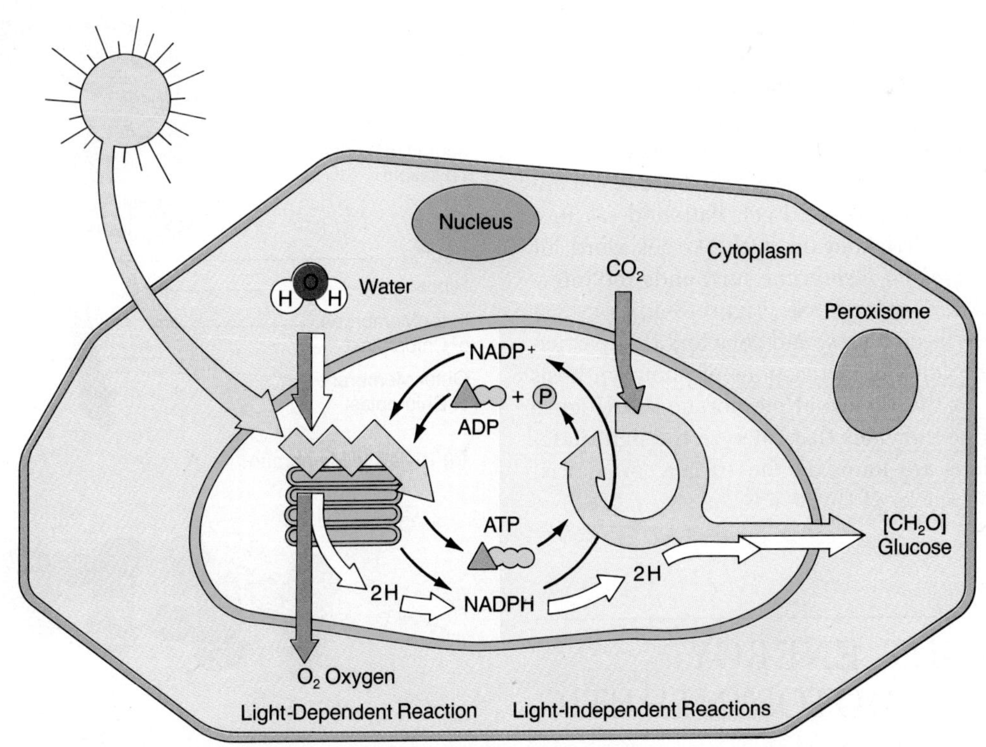

Figure 8-3 THE LIGHT-DEPENDENT AND LIGHT-INDEPENDENT REACTIONS OF PHOTOSYNTHESIS.
Energy from sunlight drives a series of light-dependent reactions that produce the energy carriers ATP and NADPH. Those compounds then power the light-independent reactions, which build carbohydrates such as glucose. During the first set of reactions, water is split and O_2 released; during the second set, CO_2 is reduced to carbohydrate.

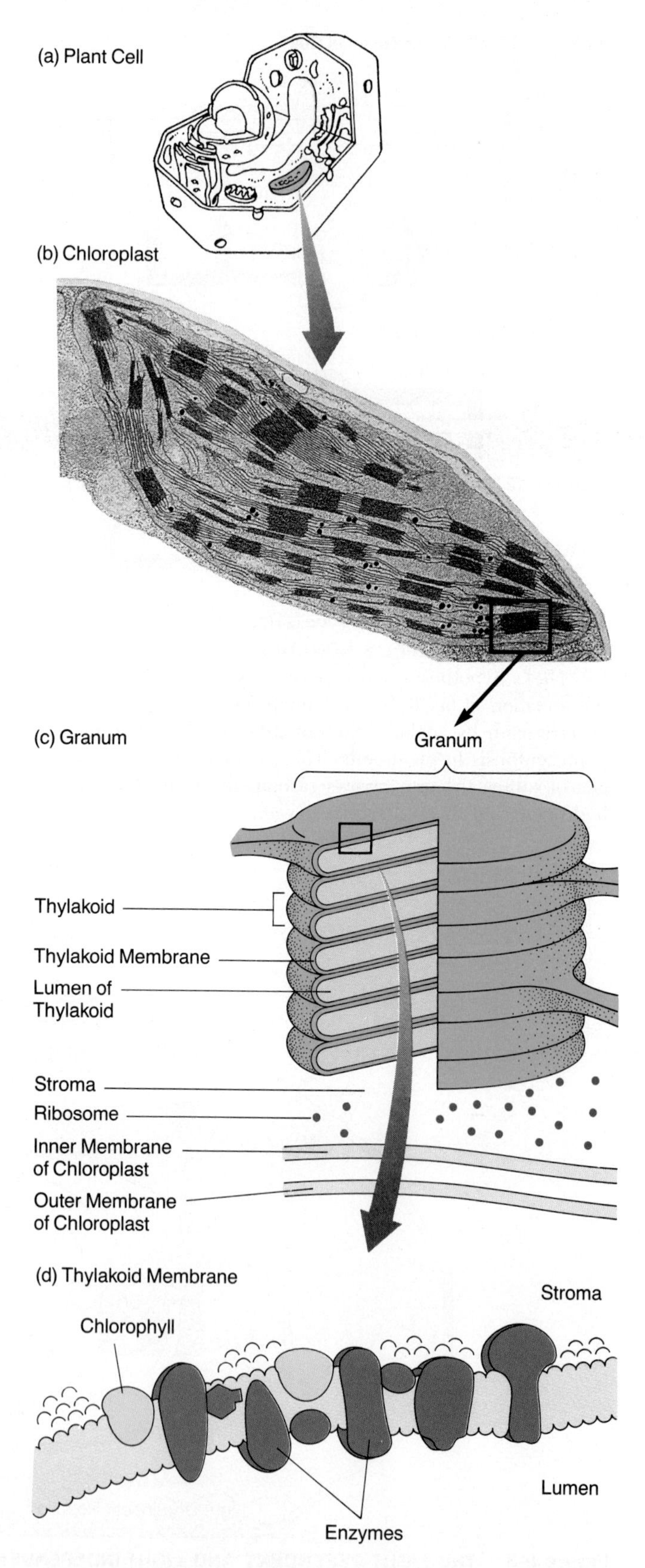

Figure 8-4 THE STRUCTURE OF CHLOROPLASTS: SITES OF PHOTOSYNTHESIS.
Plant cells (a) contain chloroplasts, such as the one pictured in (b). This electron micrograph of a chloroplast (magnification 15,000×) reveals the internal stacks of membranes, the grana. Each granum (c) consists of flattened sacs called thylakoids; adjacent grana are interconnected by the thylakoid membrane (d). Most of the enzymes and pigments for the light reactions of photosynthesis are embedded in the thylakoid membranes. The stroma, a gel-like matrix, surrounds the grana. The enzymes for the light-independent reactions of photosynthesis as well as chloroplast DNA and ribosomes and other substances are located in the stroma.

Chloroplasts: Sites of Photosynthesis

In eukaryotic cells, both phases of photosynthesis take place in chloroplasts, and the reactions depend on the unique structure of these organelles just as cellular respiration depends on the architecture of the mitochondrion (see Chapter 5). Chloroplasts have a variety of shapes, but most are elongated like minute bananas (Figure 8-4a and b). Chloroplasts are somewhat larger than mitochondria, and a typical plant cell contains from 20 to 80 of them (usually about 40). Two lipid bilayer membranes surround the chloroplast (Figure 8-4c); internally, a gel-like matrix called the *stroma* contains ribosomes, the machinery for protein synthesis, and DNA, which, in at least one species, forms genes that turn on and off in response to light. Some essential chloroplast proteins are encoded in the cell's nuclear DNA, synthesized in the cytoplasm, then moved into chloroplasts. The most prominent internal structures in chloroplasts are the stacks of flattened sacs called *grana* (meaning "grains"; see Figure 8-4c). Each flattened sac in a granum is called a **thylakoid** (from the Greek word for "sack"), and a *thylakoid membrane* surrounds the internal space, or *lumen*, of each sac (Figure 8-4d).

The chlorophyll, enzymes, and cofactors that participate in the light-dependent reactions of photosynthesis are embedded in the thylakoid membrane (see Figure 8-4d). Most of the enzymes that catalyze the light-independent reactions are found in the stroma, or matrix, surrounding the stacks of thylakoids.

HOW LIGHT ENERGY REACHES PHOTOSYNTHETIC CELLS

A browsing animal that eats a fresh green leaf from a bush is consuming light energy that may have left the sun just 8 minutes earlier. But what exactly is light energy, and how does it interact with the molecules of a living leaf?

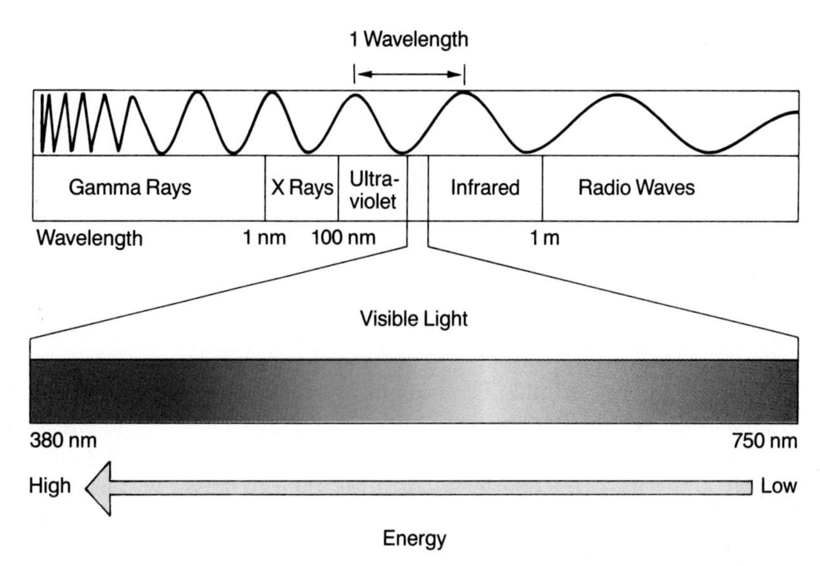

Figure 8-5 THE SPECTRUM OF ELECTROMAGNETIC RADIATION. The wavelength of electromagnetic radiation is measured in nanometers (1 nm = 10^{-7} cm). The shorter the wavelength, the higher the energy per photon. The spectrum of visible light is expanded here to show the relationship of color and wavelength. Only this small segment of the electromagnetic spectrum is visible to the human eye.

The Nature of Light

Visible light is a form of electromagnetic radiation. All such radiation has the properties of both particles and waves. Particles, or **photons,** are *packets* of light energy, and the distance they travel during one complete vibration is defined as the *wavelength* of that light. The intensity, or brightness, of a light beam is a measure of the total number of photons, or energy, per unit time.

The full range of light energies and wavelengths makes up just a small portion of the electromagnetic spectrum (Figure 8-5), which extends from high-energy gamma rays, with extremely short wavelengths, to low-energy radio waves, with long wavelengths, and includes ultraviolet (UV) light, which bees and certain other animals can see. Within the visible spectrum, wavelengths of about 400 nm appear violet to the human eye, while wavelengths of around 700 nm appear red.

Biological molecules, whether in the eye or in a living leaf, can capture the energy of photons for constructive work only when the wavelengths lie in the narrow range of the visible part of the spectrum.

The Absorption of Light by Photosynthetic Pigments

Photons of light can drive photosynthetic reactions only after pigment compounds within photosynthetic cells absorb those particles. Each **pigment** has distinctive properties based on its molecular structure. The principal pigment active in photosynthesis is **chlorophyll,** which appears green to the human eye because it absorbs light of most other wavelengths (violet, blue, orange, red) but *transmits* light with wavelengths in the green band of the visible spectrum (Figure 8-6). The

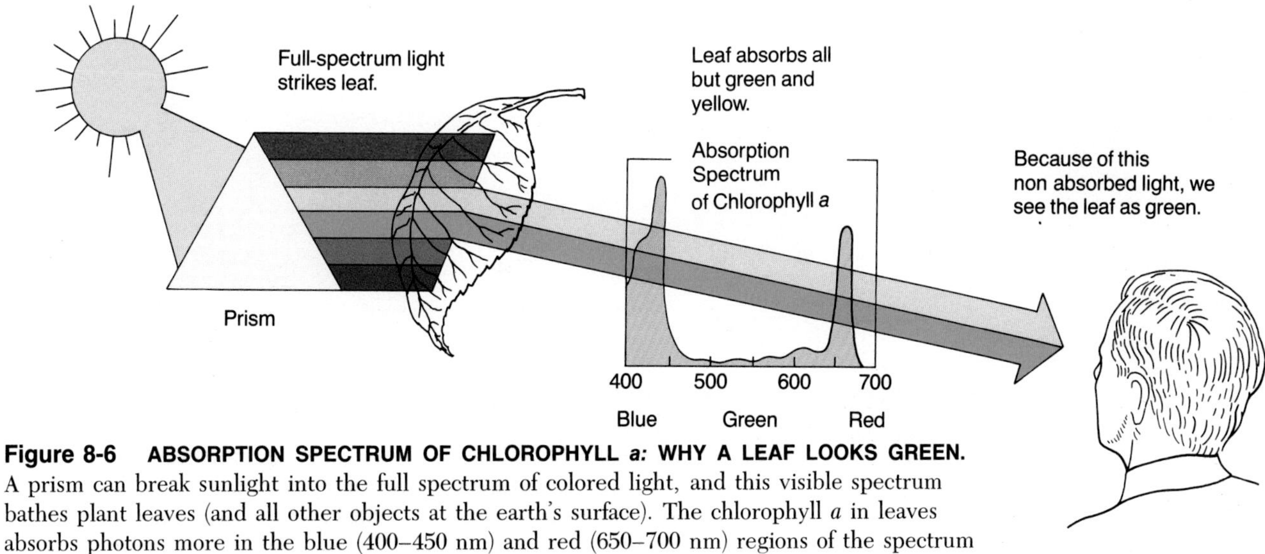

Figure 8-6 ABSORPTION SPECTRUM OF CHLOROPHYLL *a*: WHY A LEAF LOOKS GREEN. A prism can break sunlight into the full spectrum of colored light, and this visible spectrum bathes plant leaves (and all other objects at the earth's surface). The chlorophyll *a* in leaves absorbs photons more in the blue (400–450 nm) and red (650–700 nm) regions of the spectrum and less in the green (intermediate) wavelengths. As a result, a leaf appears green to the human eye because the green light is transmitted, while other wavelengths are absorbed.

absorption spectrum for chlorophyll *a* (found in all plants and cyanobacteria) is a graph of the molecule's ability to absorb various light wavelengths. Chlorophyll *a* absorbs a large fraction of the photons in both the red (650–700 nm) and blue (400–450) ends of the spectrum, but it does not absorb the green light of intermediate wavelengths.

When a pigment molecule absorbs a photon of light energy, the electron configuration of the absorbing molecule is changed: The light energy boosts an electron from its normal orbital around one atom in the molecule to a higher energy orbital. The pigment molecule with one electron in a higher energy orbital is now in an *excited* state, but this condition is fleeting. The molecule quickly loses its extra energy as heat; as fluorescence (photons with slightly longer wavelengths); or in one of two ways with significance for photosynthesis: passing excitation energy to neighboring molecules or transforming the excitation energy to chemical energy.

Types of Photosynthetic Pigments

Plants, algae, and photosynthetic bacteria have a rainbow of pigments in their cells, leaves, fronds, and fruits, many acting as *accessory pigments* to chlorophyll in the photosynthetic process. The primary photosynthetic pigment in green plants is chlorophyll *a* (Figure 8-7a). The molecule contains a ring portion with many alternating double and single bonds. With their shared electrons, these bonds are in constant flux, and this dynamism gives all biological pigments, including chlorophyll, the capacity to absorb low-energy photons in the visible spectrum.

Chlorophyll *a* is present in all eukaryotic plants and algae as well as in the prokaryotic cyanobacteria. A similar pigment, chlorophyll *b*, differs only in the presence of an aldehyde group rather than a methyl group at the top of the ring. Chlorophyll *b* is found in higher plants, in green algae, and in one prokaryote. Other classes of

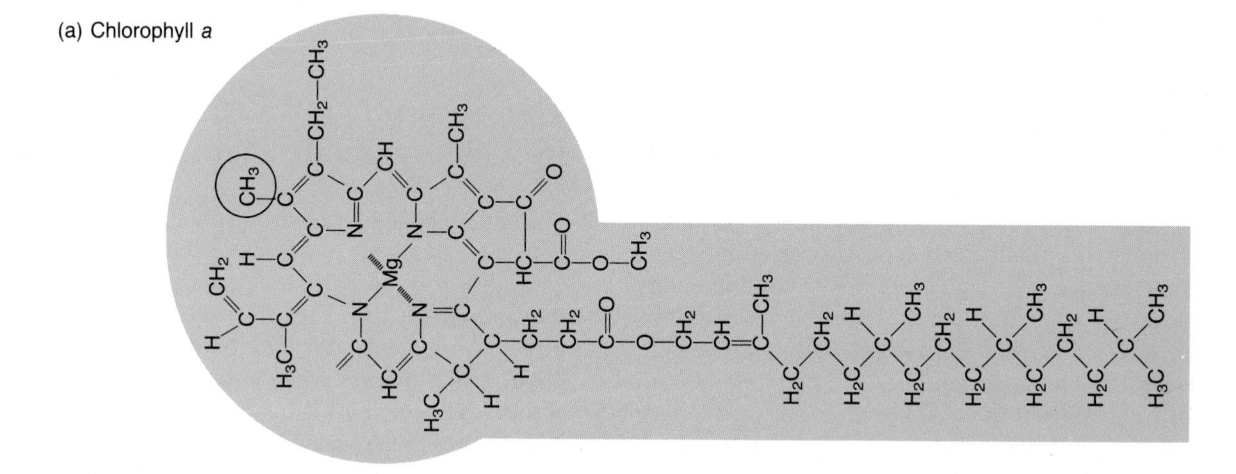

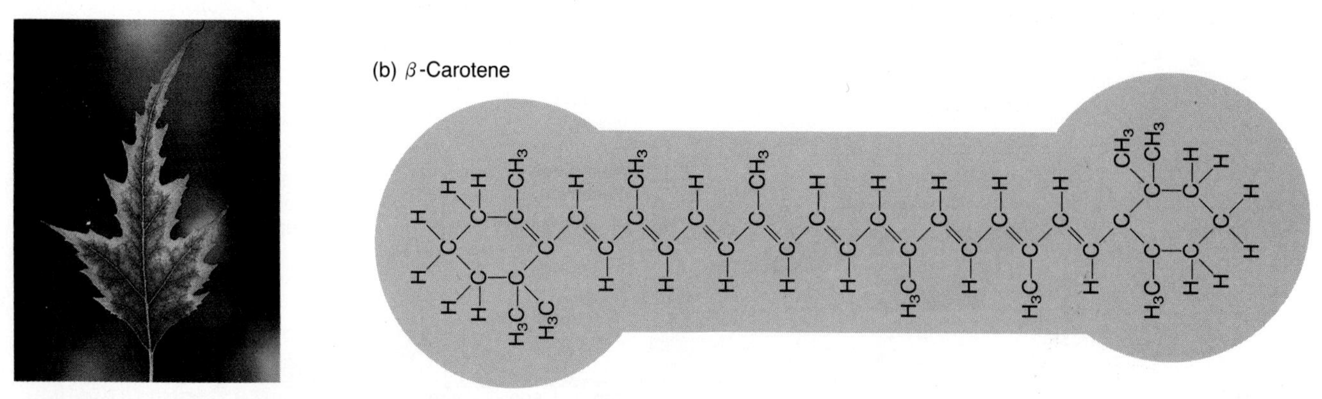

Figure 8-7 THE STRUCTURES OF CHLOROPHYLL *a* AND β-CAROTENE.
(a) Chlorophyll *a* pigment is found in virtually all photosynthetic organisms. The portion of the molecule responsible for absorbing light consists of four rings of carbon and nitrogen atoms surrounding a magnesium atom. Chlorophyll *a* is structurally very similar to chlorophyll *b* except that the methyl (CH_3) group in chlorophyll *a* (circled here) is replaced by an aldehyde (CHO) in chlorophyll *b*. (b) β-carotene is a bright yellow pigment found in leaves and stems; in red, orange, and yellow vegetables, fruits, and flowers; in the skins and skeletons of some animals; and in the feathers of some birds. β-carotene absorbs light in the 460–550 nm range and transmits the energy of this absorbed light to a photosynthetic reaction center.

chlorophyll with absorption spectra differing from those of chlorophylls *a* and *b*, occur in the brown, golden-brown, and red algae and in photosynthetic prokaryotes other than cyanobacteria (see Chapter 21).

In addition to their chlorophyll, all photosynthetic organisms contain one or more **carotenoid pigments** (Figure 8-7b). Carotenoids absorb photons of wavelengths from 460 to 550 nm; therefore, they appear to our eyes as red, orange, or yellow. These pigments are abundant in tomatoes and carrots, and they emblazon most autumn leaves after chlorophyll breaks down. Other pigments—purple and blue—that are neither carotenoids nor chlorophylls also participate in photosynthesis in some organisms.

Experiments have determined the precise wavelengths of light that are active in photosynthesis. This range of wavelengths is called the **action spectrum.** The action spectrum of photosynthesis is very similar to the absorption spectrum of chlorophyll; that correspondence helps establish that chlorophylls are the primary absorbers of light during photosynthesis.

Complexes of Photosynthetic Pigments

The green, yellow, and other photosynthetic pigments are associated physically and function in concert. On each thylakoid disk, hundreds of pigment molecules form compact aggregates called *antenna complexes* (Figure 8-8). Each such complex contains chlorophyll molecules, proteins, and sometimes red, orange, blue, or purple pigments, depending on the organism. Just as an antenna gathers radio signals, the molecules of an antenna complex function together to gather light energy. Because various colored pigments absorb light in different parts of the visible spectrum, their collective functioning means that the cell can gather light energy of a broader range of wavelengths than if chlorophyll *a* did the job by itself.

Because of the way pigment molecules in an antenna complex are arranged, when light strikes any one of them, its energy is funneled to a special chlorophyll *a* molecule, called a **reaction center chlorophyll** (see Figure 8-8). This funneling takes place because all the other associated pigments in a given antenna complex have higher excitation energies than the reaction center chlorophyll. Thus, they are able to pass the energy they absorb from high-energy photons "downhill" to the reaction center pigment. This pigment molecule can absorb only light with a slightly longer wavelength and lower energy content than a normal chlorophyll *a* molecule can absorb.

The energy transfer from other pigment molecules to the reaction center chlorophyll within the antenna complex is amazingly fast: Each molecule-to-molecule excitation event occurs in about 10^{-12} second, or 1,000 times

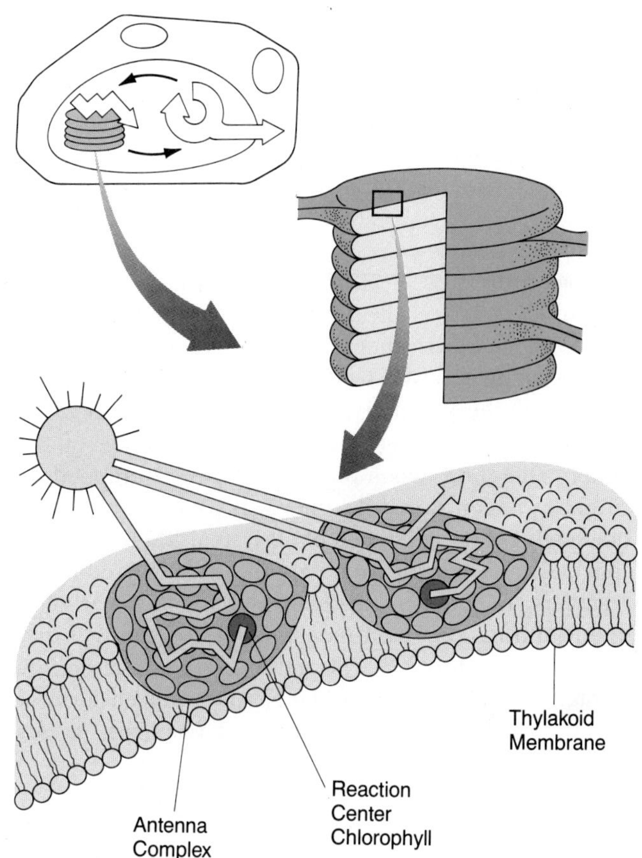

Figure 8-8 ANTENNA COMPLEXES: LIGHT-GATHERING PIGMENT CLUSTERS.
An antenna complex is a group of photosynthetic molecules that gather and transmit light energy to a special type of chlorophyll *a* molecule, the reaction center chlorophyll. Within each complex diagrammed here, the light green circles represent pigment molecules. After a photon is absorbed by one pigment molecule, the energy from the photon is passed along to adjacent molecules down the yellow path until it reaches a reaction center chlorophyll (dark green circle). The energy of a photon sometimes escapes from the complex without exciting a reaction center (upper right).

faster than it would take for the energy to be released as fluorescence or heat.

Within a plant's chloroplasts, the antenna complex with its chlorophyll molecules functions together with a few polypeptides and electron acceptor/donor molecules in the thylakoid membrane in what is called a **photosystem.** Green plants have two types of photosystems which transfer energized electrons from chlorophyll to more stable energy-trapping molecules. *Photosystem I* contains a reaction center chlorophyll that most strongly absorbs light with a wavelength of around 680 nm. Called P680 (for pigment 680), this chlorophyll passes electrons through several iron-sulfur centers to $NADP^+$, capturing the energy. *Photosystem II* has a reaction center chlorophyll, called P700, that absorbs light with a wavelength of 700 nm. The light-excited chlorophyll of photosystem II transfers its electrons through pheophytin (a molecule shaped like chlorophyll but lacking magnesium) to plastoquinone. The electrons are then shuttled into an electron transport system which pumps hydrogen ions across the thylakoid membrane, thus storing the energy.

The physical association of the reaction center, electron acceptor, and electron donor molecules in photosystems I and II is the key to the conversion of light energy to chemical energy during photosynthesis: The passage of an electron to the acceptor constitutes a reduction of the acceptor, while the contribution of an electron by the donor constitutes an oxidation of the donor. Therefore, when photons of light are trapped by antenna pigments, and the energy is passed to the reaction center chlorophyll in a photosystem, there can be a conversion to chemical energy via oxidation-reduction reactions. Eventually, this energy is stored in carbohydrates during the light-independent reactions. First, however, it is stored in ATP and in another high-energy storage molecule called NADPH (the reduced form of $NADP^+$) during the light-dependent reactions, as our next section explains.

THE LIGHT-DEPENDENT REACTIONS: CONVERTING SOLAR ENERGY TO CHEMICAL BOND ENERGY

The essence of the light-dependent reactions is the packaging of light energy in chemical compounds, during which H_2O is split and O_2 is released (review Figure 8-3). As H_2O is cleaved, electrons are donated and eventually flow in an energetically "uphill" direction to the storage compound NADPH. It is the energy from light that drives this overall uphill, endergonic set of reactions. Experiments reveal that photosystem II is responsible for the removal of electrons from H_2O and the release of O_2, while photosystem I brings about the reduction of the coenzyme $NADP^+$ to NADPH.

Electron Flow in Photosystems I and II

The light-dependent reactions of photosynthesis have been characterized as a zigzag, or Z, pathway (Figure 8-9). In the daytime, when light strikes the pigment molecules associated with photosystem II, energy is funneled through the antenna complexes, causing electrons to be ejected from P680 reaction centers (see Figure 8-9, step 1). Immediately, electrons donated by an H_2O molecule fill the holes. The water molecule is cleaved by a water-splitting enzyme, and an oxygen atom is released during the electron donation—the oxidation of H_2O (step 2). To yield an O_2 molecule, two H_2O molecules must be split.

The electrons ejected from the P680 reaction center chlorophyll molecule are passed to the electron acceptor *plastoquinone*, which is reduced (step 3). The high-energy electrons deposited in plastoquinone next descend an electron transport chain composed of a series of electron carrier molecules embedded in the thylakoid membrane between photosystems II and I (step 4).

The final electron acceptor in the electron transport chain is P700, the reaction center chlorophyll in photosystem I. Packets of light energy that are continually funneled to this molecule from its associated antenna complex cause electrons to be ejected sequentially. An electron passing down the electron transport chain from photosystem II can be accepted by P700 (step 5) and boosted to a very high energy state (step 6)—higher than when it started down the chain from plastoquinone, and much higher than when it left P680 a fraction of a second earlier. Four photons from P700 are needed to boost the four electrons from the original two molecules of H_2O to a high enough energy level to be accepted by the electron acceptor for photosystem I.

Biologists picture the boosting and reboosting of electrons from the P680 and P700 center molecules as an upward zigzag pathway, the Z scheme. The electron acceptor for photosystem I (the top of the "zag") is an iron-containing protein called *ferredoxin*, which becomes reduced as it accepts high-energy electrons boosted from P700 (step 7). Reduced ferredoxin is rapidly reoxidized, and the coenzyme $NADP^+$ is reduced to NADPH by protons originally generated during the splitting of H_2O (step 8). For each eight photons of light received by photosystems I and II and each four electrons boosted, two molecules of NADPH are generated:

$$4e^- + 2H^+ + 2\ NADP^+ + 8\ \text{photons} \longrightarrow 2\ NADPH$$

These two energy-storing molecules as well as ATP formed during electron transport are used during the light-independent reactions of photosynthesis. Thus, light energy has been trapped and converted to chemical bond energy, which can then be spent to reduce CO_2 to carbohydrates for the cell.

Photophosphorylation: Light Energy Captured in ATP

Recall from Chapter 7 that in the process of oxidative phosphorylation, free energy released by the electron transport chain is used to generate ATP. Photosynthetic

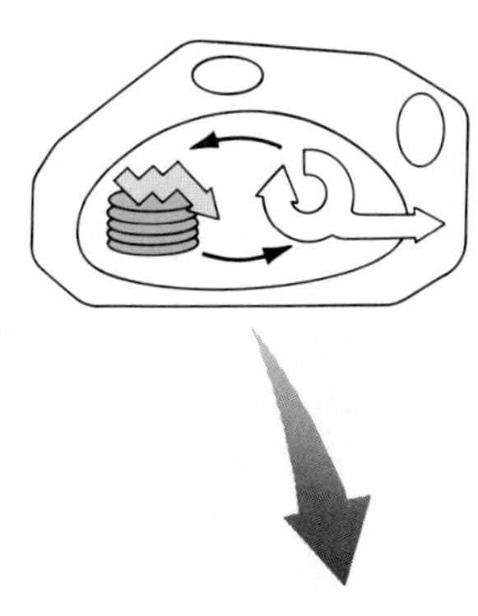

Figure 8-9 THE LIGHT REACTIONS OF PHOTOSYNTHESIS.
In this zigzag (Z) scheme, there is a net flow of electrons from water to $NADP^+$. The process occurs in two series of reactions. In the first, water donates electrons to a photon-activated P680 molecule in photosystem II. High-energy electrons are passed to plastoquinone (pQ), and from there, down an electron transport chain (cytochrome b–f) to plastocyanin (pC). As that occurs, H^+ is pumped through the thylakoid membrane, just as in the mitochondrial inner membrane, and the resulting electrochemical gradient generates ATP as protons pass through ATP synthetase. The electrons continue their energy descent to P700 of photosystem I, where another photon has provided the energy to activate the electrons to an even higher energy level. In the second series of reactions, the electrons pass to ferredoxin (Fd) and then are used to generate NADPH, which feeds into the light-independent, or carbon reduction, reactions of photosynthesis. Because the Z scheme results in ATP production, it is also called *noncyclic photophosphorylation*.

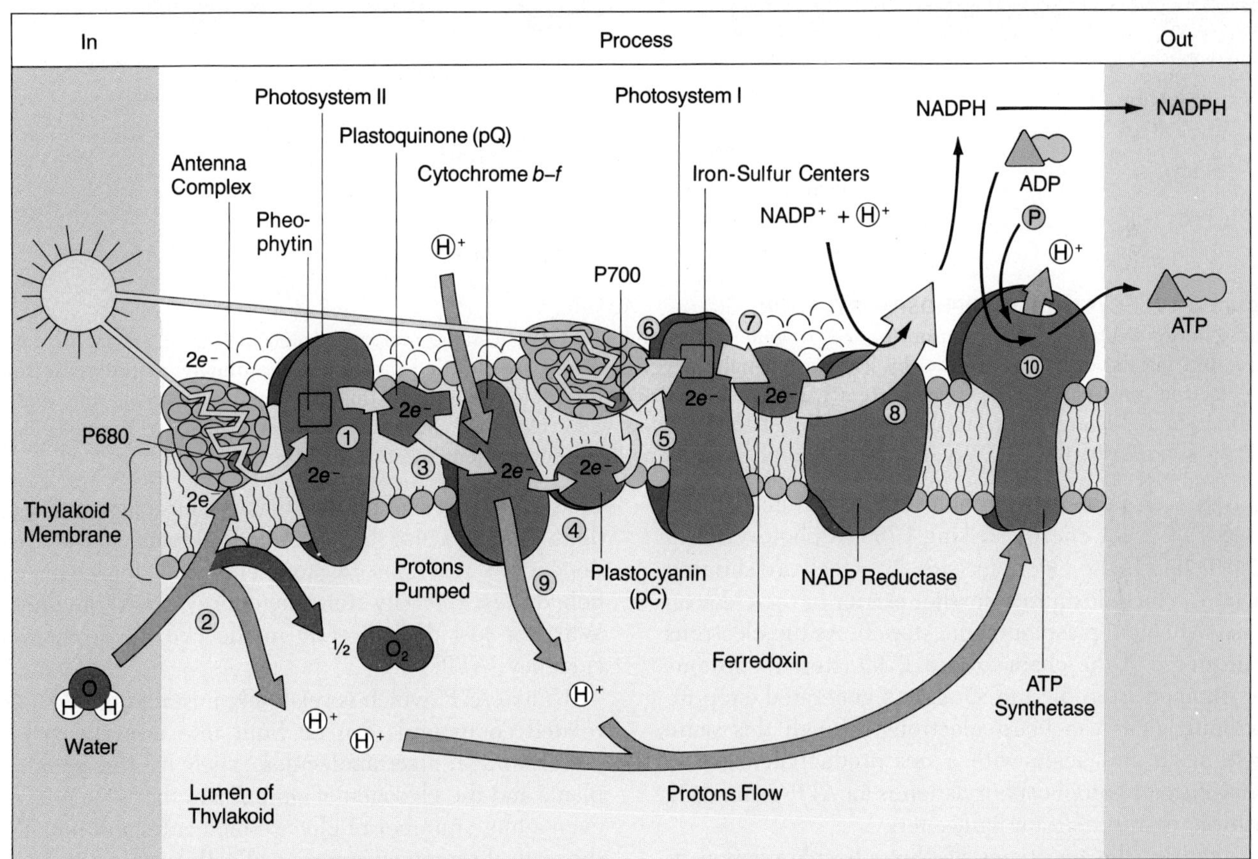

cells carry out a similar process called **photophosphorylation,** during which ATP forms as electrons excited by light energy move down an electron transport chain. Within the chloroplast, electron transport causes protons to be pumped through the membrane into the thylakoid inner space (Figure 8-9, step 9). This space is thus equivalent to the outer compartment of the mitochondrion, where protons also accumulate. Because the accumulation sets up a proton gradient across the membrane, protons flow back into the chloroplast stroma, passing through coupling factor proteins embedded in the thylakoid membrane (step 10). As protons flow out through the coupling factor channel, ATP is formed.

Biologists call the zigzag scheme of electron flow through photosystems II and I **noncyclic photophosphorylation.** But why "noncyclic"? In the zigzag scheme, electrons flow from H_2O to NADPH in a one-way, linear sequence through a series of pigments, proteins, and energy carriers. Apparently, noncyclic photophosphorylation is responsible for generating most of the ATP in the chloroplasts of photosynthetic organisms.

Plants can, however, produce a small additional supply of ATP by means of a marvelous trick called **cyclic photophosphorylation.** When the bulk of the $NADP^+$ molecules have been reduced to NADPH, electrons can be cycled from photosystem I back through the electron

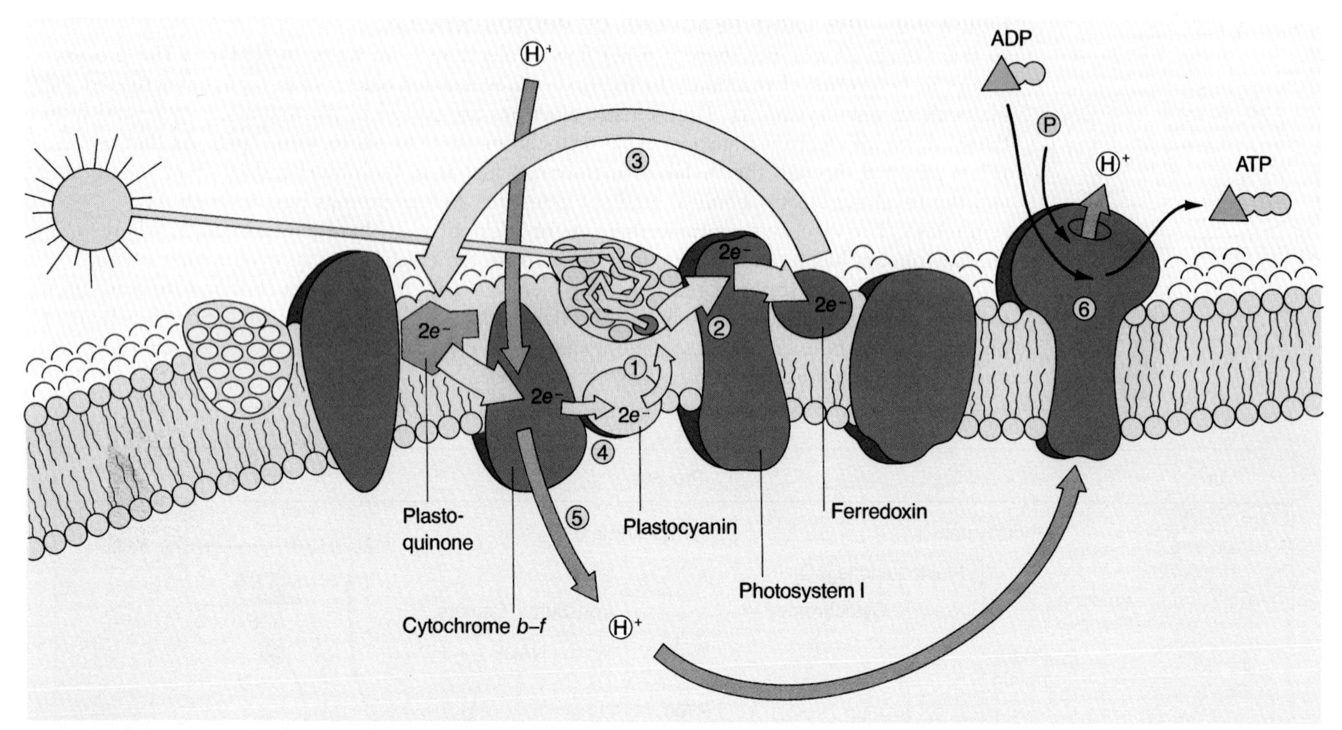

Figure 8-10 CYCLIC PHOTOPHOSPHORYLATION.
High-energy electrons can be reused (cycled) to generate additional ATP. The same molecules seen in Figure 8-9 are present here. But instead of being accepted by ferredoxin, high-energy electrons are shuttled back to plastoquinone: as they pass along the electron transport chain to photosystem I, their energy drives the proton pump and thus ATP synthesis. If photons continue to add energy to P700, this closed loop of electron flow will cycle again and again, producing many ATP molecules. If it does, however, the NADPH needed for the photosynthetic light-independent reactions will not be made from $NADP^+$.

transport chain between photosystems II and I (Figure 8-10). Light energy striking P700 in photosystem I (step 1) boosts the electrons (step 2), which are shunted back from ferredoxin to an earlier carrier in the electron transport chain (plastoquinone; step 3). As the electrons again descend the chain toward P700 (step 4), protons are pumped (step 5), and so ATP is generated (step 6). Incoming light can boost electrons through this same cycle again and again with a net production of ATP. Thus, *cyclic* photophosphorylation is an ATP-generating pathway driven *only* by light energy.

Together, the two types of photophosphorylation in the light-dependent reactions convert enough light energy to the chemical bonds in ATP and NADPH to fuel the second phase of photosynthesis, our next topic.

THE LIGHT-INDEPENDENT REACTIONS: BUILDING CARBOHYDRATES

The light-independent reactions of photosynthesis, which take place in the chloroplast stroma, constitute the second phase of photosythesis. These light-independent reactions use the organic products of the light-dependent reactions to store chemical energy in carbohydrates. But why store the energy in carbohydrates? Why not just deal directly in the cell's own chemical currency, ATP?

Unlike ATP, which is relatively unstable, glucose and related compounds can be built into densely packed, inert storage macromolecules, such as the starch of plants and the glycogen of animals. In these inert forms, even a huge number of glucose molecules will not affect the critical osmotic properties of cellular or extracellular fluids. Carbohydrates are also a good starting point for constructing the subunits of proteins, nucleic acids, and lipids. Equally important, carbohydrates are a highly efficient energy reservoir: When "burned," one molecule of glucose, which is smaller than one molecule of ATP, releases energy equivalent to about 36 ATPs. In effect, carbohydrates are much more compact and effective storage and transport units than is ATP; it is as though carbohydrates are "$100 bills," not commonly used in daily life, whereas ATPs are the "coins" used so much in the daily cellular economy.

Earlier we mentioned that the light-dependent reactions of photosynthesis generate energy-storing ATP and NADPH molecules and that these drive the light-independent reactions, which convert six CO_2 molecules to glucose, $C_6H_{12}O_6$. In the early 1950s, researchers Melvin Calvin and Andrew Benson at the University of California, Berkeley, worked out the reactions in this pathway and found, unexpectedly, that a three-carbon molecule, 3-phosphoglycerate (abbreviated PGA, for the acid form, phosphoglyceric acid), is the first *stable* intermediate in the pathway. They also discovered that a precursor of PGA, an electron acceptor for atmospheric CO_2, is a five-carbon compound called **ribulose bisphosphate (RuBP).** Subsequent research revealed that **ribulose bisphosphate carboxylase** catalyzes the first reaction in this cycle. This protein (which is composed of 16 polypeptide chains) is a key enzyme in a pathway so central to life on earth that it accounts for up to 50 percent of all the protein in chloroplasts. All in all, it probably is the single most abundant protein found in nature.

In recognition of Calvin and Benson's contributions, biologists call the light-independent reactions the **Calvin-Benson cycle** (Figure 8-11). During this series of

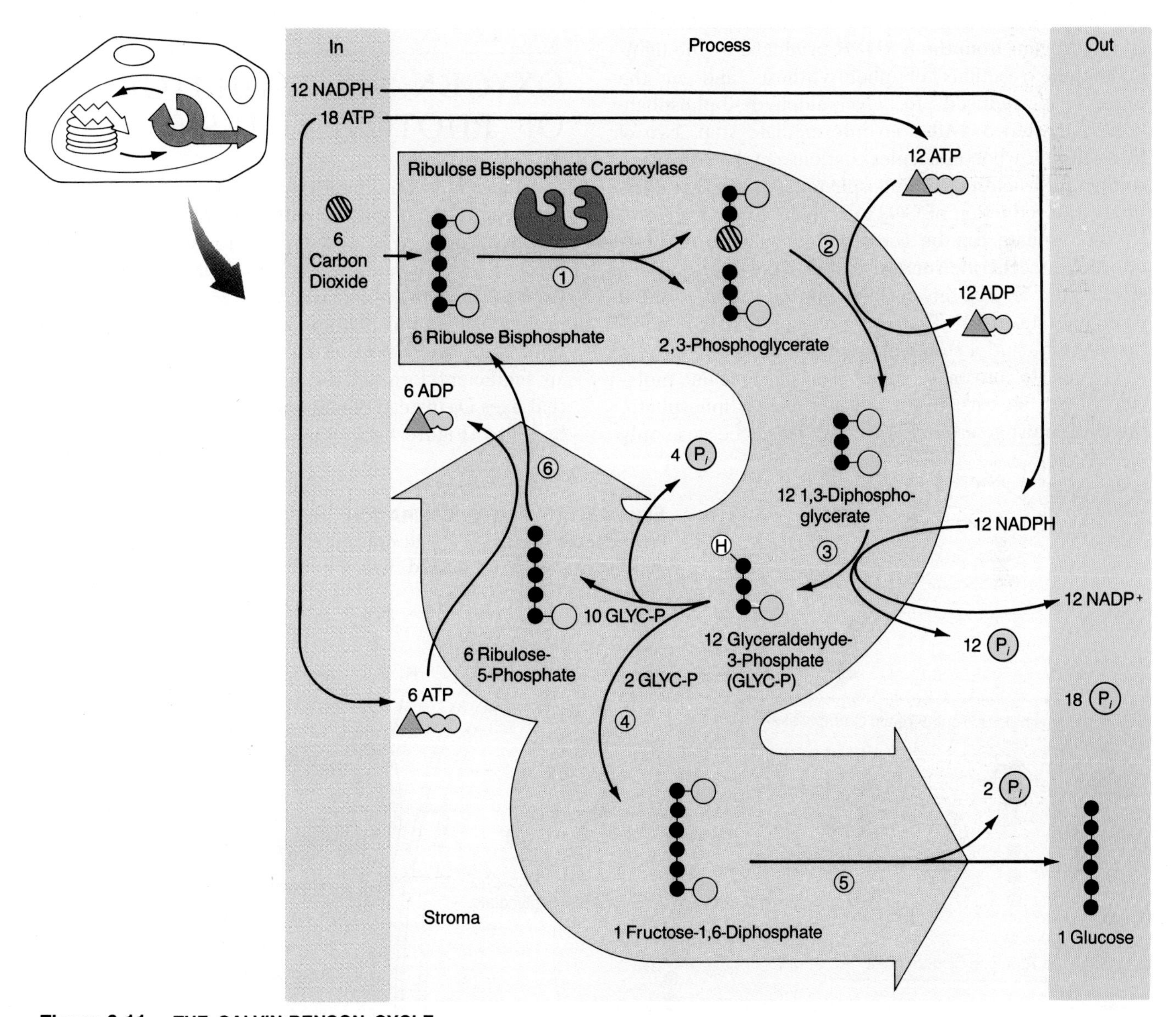

Figure 8-11 THE CALVIN-BENSON CYCLE.
The Calvin-Benson cycle is the set of light-independent reactions of photosynthesis that builds carbohydrates. The cycle must go around six times to convert six molecules of CO_2 to one molecule of the six-carbon sugar fructose diphosphate. For each turn of the cycle, ribulose bisphosphate must be regenerated from the other five carbons. The cycle is driven by energy from the light-dependent reactions in the form of NADPH and ATP. The dots in each intermediate here show the number of carbon atoms; the phosphates are shown in green. The left-hand column shows the input of materials, and the right-hand column the output of oxidized energy carriers, inorganic phosphate, and carbohydrates formed.

reactions, atmospheric carbon dioxide is first *fixed*, or incorporated into an organic compound, and then reduced to carbohydrate subunits.

In the first step of the cycle, the abundant enzyme ribulose bisphosphate carboxylase catalyzes the reaction of five-carbon RuBP with atmospheric CO_2 (step 1). An unstable six-carbon intermediate is formed and breaks down immediately to two molecules of 3-phosphoglycerate. In step 2, two ATPs are "spent" to phosphorylate this three-carbon sugar into an activated form, 1,3-diphosphoglycerate (DPGA). The next few steps of the Calvin-Benson cycle correspond to the steps of glycolysis, but in reverse. First, DPGA can accept electrons and hydrogens from the NADPH produced by the light-dependent reactions of photosynthesis and in the process is reduced to glyceraldehyde-3-phosphate (GLYC-P; step 3). After an intermediate step, two of those three-carbon molecules condense to the six-carbon compound fructose-1,6-diphosphate (step 4). That completes the reduction of CO_2 to carbohydrate. Fructose-1,6-diphosphate can be converted to glucose or built into the disaccharide sucrose or larger polymers such as starch (step 5). For this cycle to continue, it is essential to regenerate RuBP. That is done using GLYC-P, additional ATPs, and a series of reactions (step 6).

Overall, to convert six molecules of CO_2 to one molecule of the six-carbon sugar fructose-1,6-diphosphate, the cycle must go around six times. This is because only one carbon atom from RuBP plus CO_2 can be used for purposes *other than* regenerating the RuBP needed for the cycle to continue. The cost of six turns of the cycle is 18 molecules of ATP and 12 molecules of NADPH (Figure 8-11, left-hand column).

Although the energy expenditure is high, the ATP and NADPH produced by those first reactions fuel the light-independent reactions, with little waste of chemical energy. And thanks to the organization of chloroplasts, that energy is always in the right place at the right time.

OXYGEN: AN INHIBITOR OF PHOTOSYNTHESIS

Ironically, even though oxygen is essential for the aerobic respiration of glucose within the plant, too much O_2 can be harmful. When excited chlorophyll molecules are present, oxygen can assume high-energy states that are toxic to the cell. In normal plants, carotenoid pigments protect against this phenomenon, known as *photooxidation*. High levels of O_2 can also cause **photorespiration,** an inefficient form of the light-independent reactions that fixes O_2 instead of CO_2 and does not produce carbohydrates (Figure 8-12). On a day that is hot, dry, and

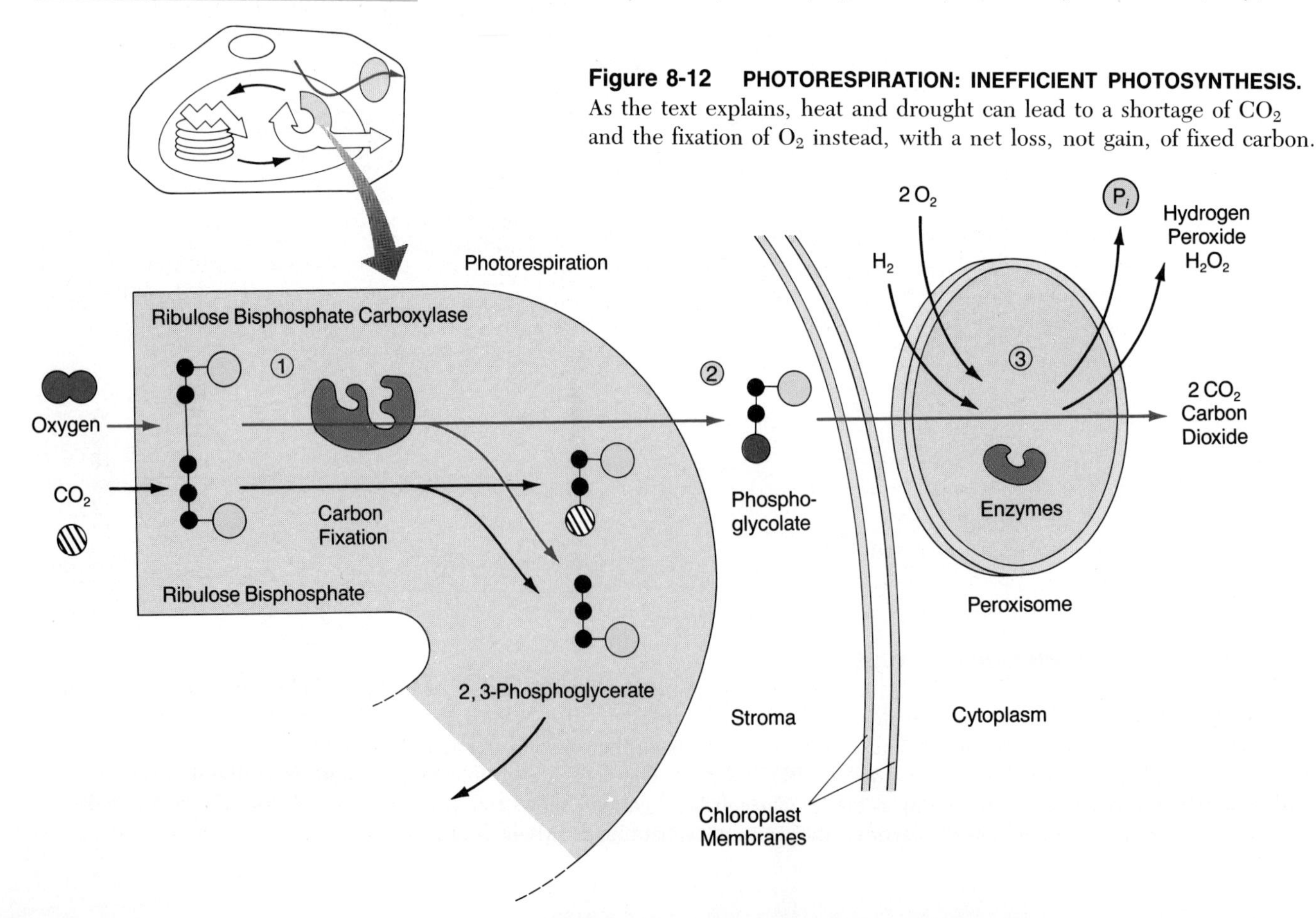

Figure 8-12 PHOTORESPIRATION: INEFFICIENT PHOTOSYNTHESIS. As the text explains, heat and drought can lead to a shortage of CO_2 and the fixation of O_2 instead, with a net loss, not gain, of fixed carbon.

sunny, tiny openings in a plant's leaves called *stomata* (singular, *stoma*) close to prevent water vapor from evaporating. Photosynthesis can continue at a greatly reduced rate, but the closure of these "portholes" prevents the waste product O_2 from escaping and most CO_2 molecules from entering.

Under these conditions, the important enzyme RuBP carboxylase begins to join O_2 to some molecules of RuBP in the first step of the Calvin-Benson cycle, instead of fixing only CO_2 (Figure 8-12, step 1). Thus, oxygen and carbon dioxide compete for the active site of the enzyme. Although this competition takes place to some degree even under normal conditions, the more oxygen that builds up in the plant (as on a hot, dry day), the more will be joined to RuBP.

When O_2 is added to RuBP, some *phosphoglycolate* forms in addition to 3-phosphoglycerate (step 2). Organelles called *peroxisomes* (see Chapter 6) can oxidize the phosphoglycolate molecules to CO_2, but without generating ATP or carbohydrate (step 3). Thus, CO_2, originally fixed at such great expense, is lost, and so the plant uses precious energy to metabolize phosphoglycolate and suffers a net loss of fixed carbon as a result.

(a)

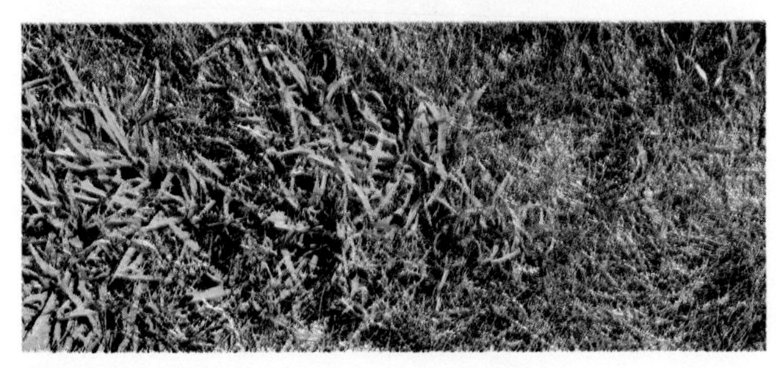
(b)

Figure 8-13 C_3 AND C_4 PLANTS.
(a) This lawn contains ordinary grass (C_3 plants), but also a few clumps of crabgrass (C_4 plants). (b) By midsummer, under conditions of minimal watering, most of the C_3 grass is dried and brown because C_3 plants cannot photosynthesize under hot, dry conditions. The C_4 crabgrass, however, is still green and growing, since it can carry on photosynthesis despite the blistering weather.

REPRIEVE FROM PHOTORESPIRATION: THE C_4 PATHWAY

Because of photorespiration, the majority of plants experience decreased carbohydrate production and thus slowed growth when the weather is hot and dry. These plants are called **C_3 plants** because the first stable intermediates in the Calvin-Benson cycle are three-carbon sugars. Another group, the **C_4 plants,** have a special leaf anatomy and a unique biochemical pathway that begins with a stable four-carbon intermediate. These adaptations, which had already evolved 5 to 7 million years ago, allow C_4 plants such as crabgrass, corn, and sugarcane to photosynthesize at a faster rate in hot, dry climates than can the C_3 plants (Figure 8-13).

Figure 8-14a and b compares the leaf anatomy of C_3 and C_4 plants. Both have a watertight, airtight "skin" called the *cuticle,* and both have *stomata,* through which gases and water vapor enter and exit the leaf. Both also have large *mesophyll cells,* which contain chloroplasts, and a network of veins, which transport water, nutrients, and minerals throughout the plant. In C_3 plants, the mesophyll cells are concentrated beneath the upper leaf surface but also occur in a spongy layer throughout the leaf. The mesophyll cells carry out all the photosynthesis in C_3 plants (Figure 8-14c). In C_4 plants, the veins are surrounded by two layers—an outer layer of mesophyll cells and an inner layer of *bundle-sheath cells*—both of which contain chloroplasts (Figure 8-14d). In these C_4 leaves, however, the Calvin-Benson cycle takes place only in the bundle-sheath cells because they contain all the RuBP carboxylase enzyme in the entire leaf. Carbohydrate production therefore occurs only in a layer of cells deep inside the leaf that is insulated from the high concentrations of oxygen in the air. In addition, the mesophyll cells actually carry out a kind of CO_2 "pumping" and thereby concentrate CO_2 and supply its carbon to the bundle-sheath cells in the form of a four-carbon compound.

The pumping of CO_2 by mesophyll cells begins with a special enzyme called phosphoenolpyruvate (PEP) carboxylase, which is present in the cytoplasm of the mesophyll cells in C_4 plants (Figure 8-14d). (It is absent in C_3 mesophyll cells; Figure 8-14c.) This enzyme fixes CO_2 molecules to molecules of phosphoenolpyruvate, a three-carbon compound, and forms the four-carbon compound oxaloacetate (Figure 8-14d, step 1). In many C_4 plants, this compound is transported into chloroplasts of the same mesophyll cells, where it is reduced by NADPH to another four-carbon compound, malate (step 2). Malate acts as a carbon "delivery person"; it diffuses through plasmodesmata (see Chapter 5) into the bundle-sheath cells, where it is cleaved to the three-

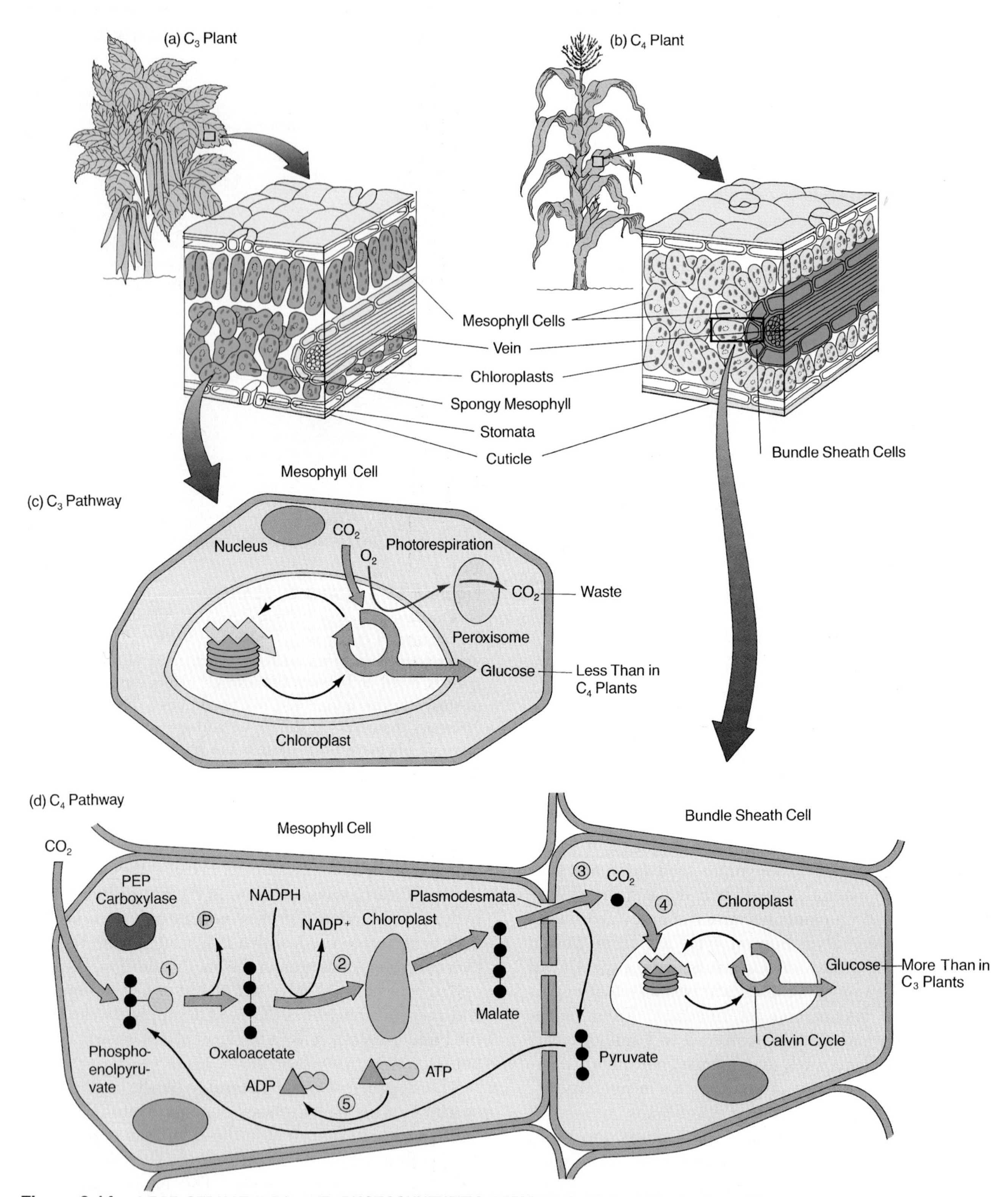

Figure 8-14 LEAF STRUCTURES AND PHOTOSYNTHETIC PATHWAYS IN C_3 AND C_4 PLANTS.
(a, b) Both C_3 plants and C_4 plants have mesophyll cells containing numerous bright green chloroplasts. However, the C_4 plants also have an inner layer of bundle-sheath cells. In C_4 plants, the light-independent reactions of photosynthesis take place only in these bundle-sheath cells, and the surrounding mesophyll cells act to "pump" CO_2 inward toward them. (c) The C_3 pathway is the same as shown in Figure 8-12, with O_2 inhibiting normal photosynthesis and little glucose formed. (d) The C_4 pathway involves some molecules that are not involved in C_3 photosynthesis but that are present during glycolysis and the Krebs cycle. The enzyme PEP carboxylase fixes CO_2 molecules to PEP, a three-carbon compound. The result is a four-carbon compound called oxaloacetate, which in turn is reduced to a different four-carbon compound, malate. Malate diffuses into the bundle-sheath cells, where it is converted to the three-carbon molecule pyruvate. In the process, CO_2 is released to take part in the Calvin-Benson cycle and the plant can continue to make carbohydrates even under hot, dry conditions.

BACTERIAL GENES AND SOYBEANS

In biology, the most esoteric-sounding research can sometimes lead to very practical applications. Take, for example, the recent genetic studies of *Rhodopseudomonas capsulata*, a so-called purple photosynthetic bacterium. This simple organism is anaerobic and probably captures sunlight much as the earliest photosynthetic cells did over 3 billion years ago. Surprisingly, the unfolding secrets of its photosynthetic machinery may someday help soybean farmers improve their yields.

By studying *R. capsulata*, researchers Barry Marrs and Douglas Youvan learned exact details about the genes that encode proteins in the cell's reaction centers and antenna complexes (Figure A). Colleague Edward Bylina then caused a mutation, or change, in the gene that encodes the protein that holds a quinone molecule—a molecule that acts in *R. capsulata* much as it does in plants (review plastoquinone in Figure 8-9). This mutation altered the protein in such a way that a common herbicide, atrazine, can no longer inhibit quinone's activity, block photosynthesis, and kill the bacterium.

Since the quinone-binding protein in *R. capsulata* is extremely similar to the equivalent protein in plants, the team is planning to mutate the protein's gene in important crop plants such as soybeans. Soybeans are sensitive to atrazine and will die if a farmer sprays a field with the herbicide to kill interspersed weeds. But if a genetic engineer inserted a new altered gene into bean chromosomes, perhaps the crop, like the mutated *R. capsulata* cells, could withstand atrazine spraying. Farmers could then eliminate herbicide-sensitive weeds without harming the herbicide-resistant soybeans and realize much higher soybean yields—and all because biologists studied the genetics of an ancient photosynthetic bacterium.

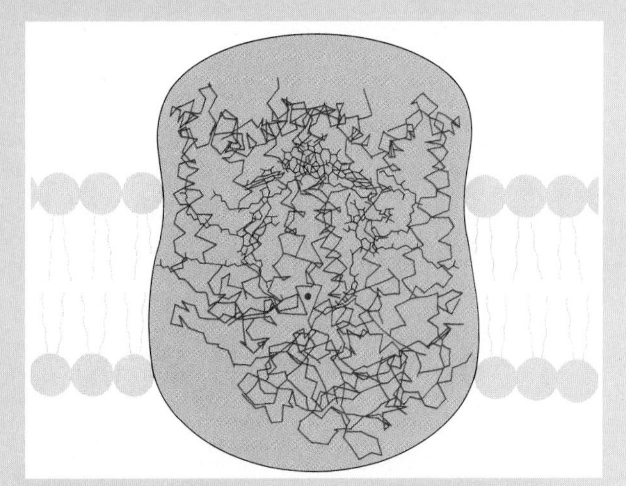

Figure A ANTENNA COMPLEX AND REACTION CENTER CHLOROPHYLL IN THE CHLOROPLAST MEMBRANE.

carbon compound pyruvate (step 3). In the process, CO_2 is released and can enter the Calvin-Benson cycle (step 4).

Once CO_2 is released from the malate "delivery person" and pyruvate is formed, this three-carbon "leftover" is transported back to the mesophyll cells, where it is transformed once again to PEP, ready for a new cycle of CO_2 incorporation and transport (step 5).

The ability of C_4 plants to thrive in arid conditions makes them superior candidates for agriculture, and they grow faster than C_3 plants under hot, dry conditions. Research is currently under way to transform such important C_3 crop plants as tomatoes, soybeans, and cereals into C_4 plants.

THE CARBON CYCLE

The emergence of photosynthetic organisms more than 3 billion years ago literally changed the face of the planet. By fixing carbon dioxide, these early bacteria helped reduce high levels of atmospheric CO_2, which can trap heat near the earth's surface in the so-called *greenhouse effect*. This reduction helped cool the surface, thereby increasing the size of the polar ice caps and lowering the sea levels. As the waters receded and land was exposed, new habitats opened up for organisms. By freeing oxygen in huge quantities, photosynthetic organisms revolutionized the atmosphere and everything bathed in it. That free oxygen created the potential for a great diversity of aerobic life forms with their large ATP needs. The simultaneous activities of photosynthesis and respiration established the **carbon cycle;** whereby photosynthetic organisms fix CO_2 into carbohydrate molecules. When these are metabolized by a plant or plant consumer, CO_2 is released back to the atmosphere (Figure 8-15). That replenishment makes enough CO_2 available for continuing rounds of photosynthesis in millions of acres of forests, grasslands, lakes, and sea surfaces: As CO_2 is fixed, O_2 is released; then as O_2 is fixed during aerobic respiration, CO_2 is released back to the atmosphere.

The continuous cycling of CO_2 maintains the delicate ecological balance on which most living organisms de-

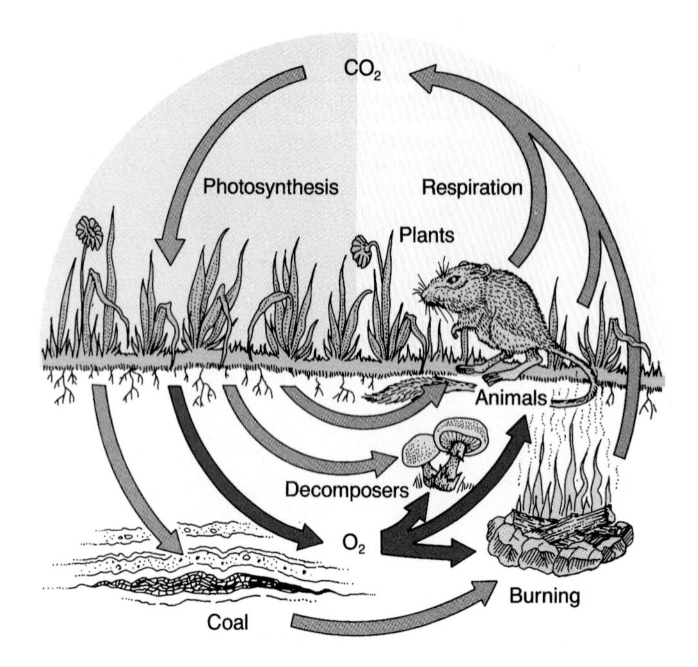

Figure 8-15 THE GLOBAL CARBON CYCLE. Photosynthetic organisms fix CO_2 into organic molecules, which are then used as energy compounds by both photosynthetic and nonphotosynthetic organisms. As organic molecules are "burned" in the fires of respiration, O_2 is used and CO_2 is released back into the atmosphere. The burning of fossil fuels derived from dead organisms also uses O_2 and releases CO_2. Decomposers include fungi and various microorganisms.

pend. Environmental insults like the unrestricted burning of fossil fuels and large-scale deforestation threaten that balance, as Chapter 41 will describe in more detail.

It is quite remarkable that reciprocal events such as photosynthesis and cellular respiration could produce such global effects. The carbon cycle demonstrates on the broadest scale the importance of metabolism and the importance of the flow of energy from the sun through the living world.

SUMMARY

1. *Photosynthesis* in green plants transforms molecules of water and carbon dioxide into carbohydrate and gaseous oxygen. Sunlight supplies the energy for this endergonic process.

2. The photosynthetic equation actually summarizes two partial reactions. In the first reaction, water is oxidized to produce ATP, NADPH, and oxygen. In the second, carbon dioxide is reduced, using ATP and NADPH to produce carbohydrate.

3. The reactions of the first stage of photosynthesis use light directly and thus are called *light-dependent reactions*, while the reactions of the second stage do not require light directly and thus are called *light-independent reactions*. Both series of reactions take place in the *chloroplast*. The *thylakoid membranes* are sites of the light-dependent reactions, and the central stroma the site of the light-independent reactions.

4. The light-dependent reactions convert solar energy to chemical energy. Several types of *chlorophyll* absorb photons of light. *Carotenoids* are also active in the process. Antenna complexes, composed of many light-harvesting pigment molecules and a *reaction center chlorophyll* molecule, ensure the high efficiency of light capture and conversion.

5. Reaction center chlorophylls are the sites where energy from photons energizes electrons, which then can be passed down electron transport chains that produce ATP and NADPH. In plants, reaction center chlorophyll P680 is associated with *photosystem II*, and reaction center P700 is associated with *photosystem I*.

6. The production of ATP from the transport of electrons excited by sunlight is called *photophosphorylation*. *Cyclic* and *noncyclic* photophosphorylation produce enough ATP to drive the light-independent reactions.

7. To form one new molecule of the six-carbon sugar fructose-1,6-diphosphate, six molecules of CO_2 must enter the light-independent reactions, or *Calvin-Benson cycle*. Regeneration of *ribulose bisphosphate* (RuBP) completes the cycle each time a CO_2 molecule passes through it.

8. The highly ordered arrangement of enzymes, pigments, and coupling factor within chloroplasts ensures the outcome of photosynthesis.

9. The oxygen produced in photosynthesis can harm chlorophyll-containing cells by forming compounds that are toxic to the cell or by contributing to an inefficient form of the light-independent reactions known as *photorespiration*.

10. *C_4 plants* reduce photorespiration by carrying out photosynthesis only in cells well insulated from high O_2 levels. They also use a novel method of CO_2 fixation that is not available to *C_3 plants*.

11. Photosynthesis works in tandem with respiration to foster the *carbon cycle*, an energy cycle that sustains most life on earth.

KEY TERMS

absorption spectrum
action spectrum
antenna complex
C_3 plant
C_4 plant
Calvin-Benson cycle
carbon cycle
carotenoid pigment
chlorophyll
chloroplast
cyclic photophosphorylation

electron acceptor
electron donor
granum
light-dependent reaction
light-independent reaction
NADP
noncyclic photophosphorylation
photon
photophosphorylation
photorespiration
photosynthesis
photosystem I
photosystem II
pigment
reaction center chlorophyll
ribulose bisphosphate (RuBP)
ribulose bisphosphate carboxylase
stroma
thylakoid

QUESTIONS

1. How did the study of photosynthesis in purple sulfur bacteria enable van Niel to understand photosynthesis in green plants?

2. What do the light-dependent reactions accomplish? The light-independent reactions? Which set of reactions is also called the Calvin-Benson cycle?

3. What is the carbon source for most plants? The energy source?

4. What is an antenna complex, and what is its function?

5. Would you expect to find chlorophyll *a* and β-carotene molecules embedded in membranes or dissolved in the cytoplasm? Explain.

6. The light-dependent reactions are often pictured as a zigzag pathway lying on its side. What molecule donates electrons to fill the holes in P680? As the electrons descend the diagonal path, some of their energy is captured. In what form? As the electrons start down a second diagonal, their reducing power is captured by what important coenzyme?

7. What is *cyclic* photophosphorylation?

8. Which abundant enzyme catalyzes the first step in the Calvin-Benson cycle? Which molecule contributes the electrons and hydrogen atoms that reduce sugars in the Calvin-Benson cycle? Which molecule was the original donor of the hydrogen atoms?

9. Do C_4 and C_3 plants share the same light-dependent reaction enzymes? Do both use the Calvin-Benson cycle? How are they different?

10. Describe our planet's carbon cycle.

ESSAY QUESTIONS

1. Why are pigment molecules so critical to photosynthesis? What would it mean if the molecules of the antenna complex appeared orange instead of green to the human eye?

2. Would you expect that cells containing chloroplasts would also contain the enzymes for glycolysis, the Krebs cycle, or oxidative phosphorylation? Explain.

For additional readings related to topics in this chapter, see Appendix C.

Part TWO

LIKE BEGETS LIKE

Equines of the same stripe: The principles of inheritance explain the striking family resemblance among these African zebras.

9
Cellular Reproduction: Mitosis and Meiosis

The cell cycle is the fundamental unit of temporal organization in the history of a population of well-nourished cells.

John J. Tyson and
Wilhelm Sachsenmaier,
Cell Cycle Clocks (1984)

Every day, a healthy person's bone marrow generates 2 million red blood cells, each with the same distinctive concave disk shape and the identical reddish color. The explanation for this unerring regularity is that in all forms of cellular and organismal reproduction, *like begets like.* Just as a single algal cell always gives rise to new algae, snails give rise only to new snails, not to sea urchins, and cardinals produce baby cardinals, not bluebirds, in the spring. This ability to breed true is based on a fundamental biological principle: When cells reproduce, they pass along instructions for constructing new cells that are almost identical to the parental cell. This legacy of information is apportioned equally and faithfully to each new cell and accounts for the striking resemblance from one generation to the next.

In an evolutionary sense, reproduction is the most important property of life; the continuation of a species depends on the replacement of at least some adult members with genetically similar new members that grow, mature, and in turn reproduce. Whether an organism is made up of a single cell or 20 trillion, its growth and reproduction depend on *cell division,* the separation of one cell into two. Biologists do not yet understand exactly why a cell divides when it does, but they do know a great deal about how the chromosomes, nucleus, and cytoskeleton of eukaryotic cells participate in cell division and how their actions help ensure the orderly distri-

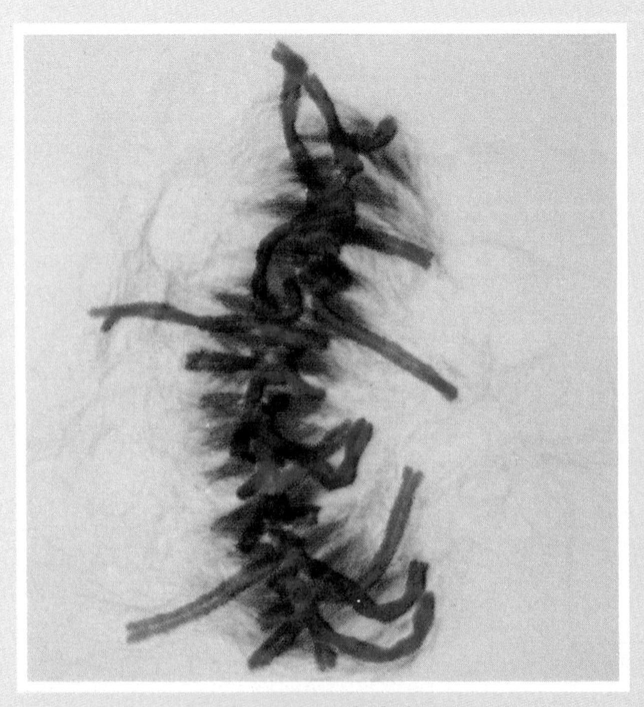

A plant cell in metaphase (magnified about 2,700×).

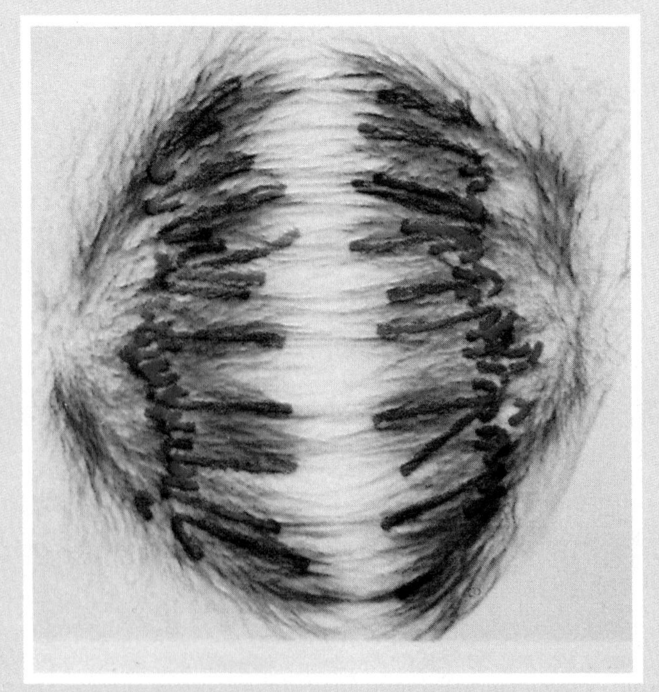

The same cell in anaphase (magnified about 2,700×).

bution of genetic material to the new generation. (We discuss the division of prokaryotic cells in Chapter 13.)

Our present discussions include:

- Proof that the nucleus and chromosomes contain blueprints for proteins and direct much of the cell's day-to-day functioning
- The stages of the cell cycle, including growth, DNA synthesis, and division into new daughter cells
- Mitosis, the partitioning of hereditary material into two cell nuclei
- Cytokinesis, the splitting of the cell cytoplasm so that two new individual cells form
- Meiosis, a special cell division within the organs that produce sperm, eggs, or spores
- The two forms of reproduction—asexual and sexual—and their evolutionary significance

THE NUCLEUS AND CHROMOSOMES

The most prominent organelle in the vast majority of eukaryotic cells is the *nucleus*, a spherical body with its own envelope dotted by special pores (review Figure 6-1). Within the nucleus lie the *chromosomes*, which contain the blueprints for the cell's proteins and thus, indirectly, for much of its function. These facts, while straightforward, came to light only after decades of research.

The Role of the Nucleus

As early as the 1870s, biologists could observe a sperm nucleus entering and fertilizing an egg. From this

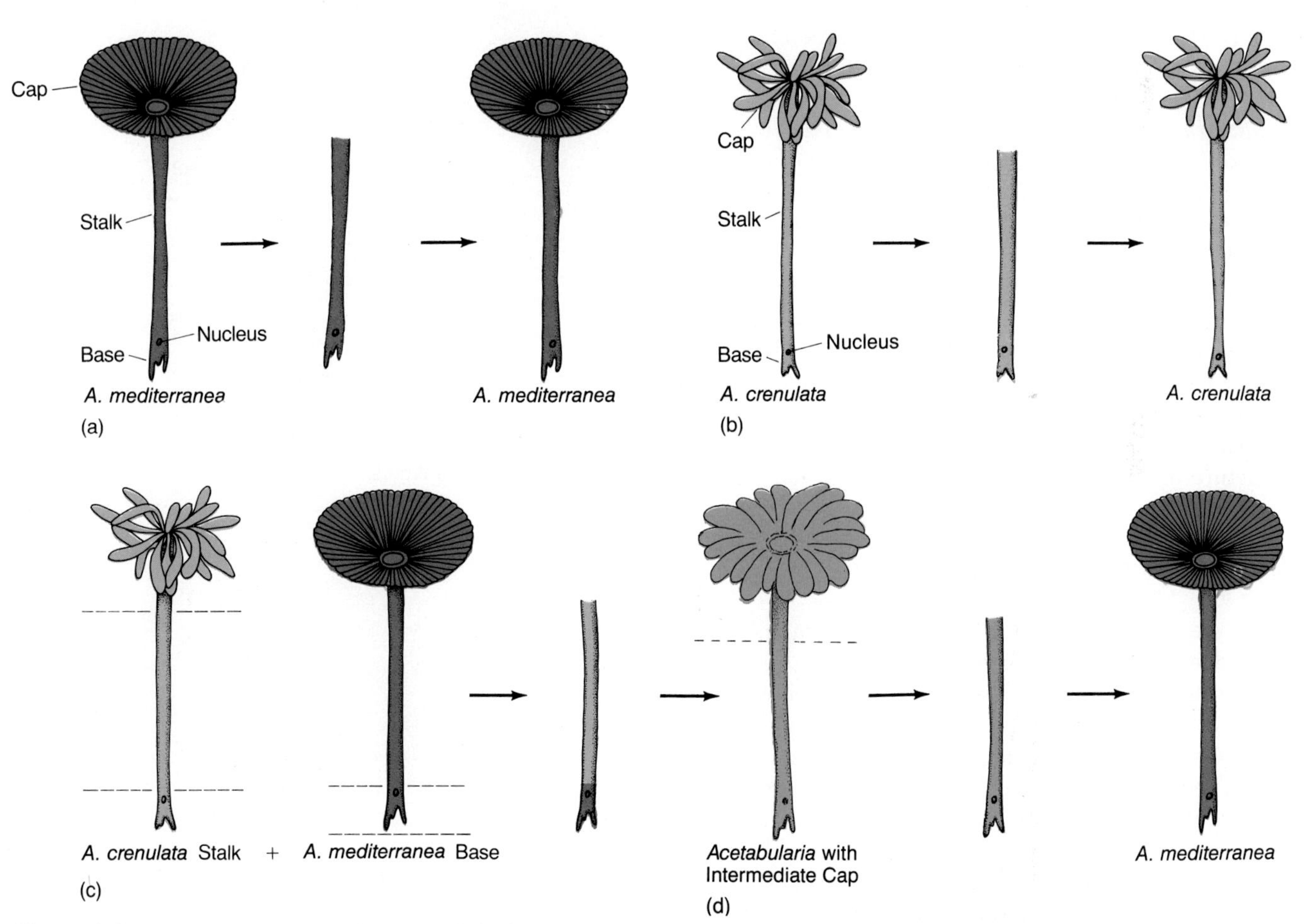

Figure 9-1 GENETIC INFORMATION CARRIED IN THE NUCLEUS.
Haemmerling's experiments with the single-celled alga *Acetabularia* revealed the essential role of the nucleus in heredity. (a) If a researcher cuts the upper stalk and cap from *Acetabularia mediterranea*, the cell regenerates the characteristic umbrella-like cap. (b) *Acetabularia crenulata* regenerates the characteristic petal-like cap if its stalk and cap are cut. (c) An *A. mediterranea* base with a transplanted *A. crenulata* stalk will regenerate a cap that is partly petal-like and partly umbrella-like. (d) If the experimenter then removes this partly intermediate cap, the algal cell regenerates an umbrella-like cap. Since the base, which contains the cell nucleus, came from an *A. mediterranea* cell, the experiment shows that the nucleus must contain information governing the cell's architecture.

they concluded that the nucleus must carry the male parent's genetic contribution to the embryo. Solid proof that the nucleus contains genetic information came decades later, however, when German biologist Joachim Haemmerling began in the 1930s to experiment with two species of *Acetabularia*, a unicellular ocean-dwelling green alga. Each *Acetabularia* organism is a large single cell (25–50 mm tall), with a rootlike base, a stalk, and a cap that varies in shape depending on species. *Acetabularia mediterranea* has an umbrella-like cap, whereas *A. crenulata* has a ragged, petal-like cap (Figure 9-1a and b).

Haemmerling tried cutting off the upper stalks and caps, and each species regenerated the appropriate umbrella- or petal-shaped cap (see Figure 9-1a and b). To determine whether the cytoplasm or the nucleus controlled the shape of the new cap, he transplanted a large *A. crenulata* stalk without its cap onto a small *A. mediterranea* base, where the nucleus is located. The new cap that formed was intermediate between the petal and the umbrella shapes (Figure 9-1c). Finally, by cutting off this new intermediate cap from the original *A. mediterranea* base, Haemmerling observed that the next and all subsequent regenerations produced *only* an umbrella-shaped *A. mediterranea* cap (Figure 9-1d).

Through these experiments, Haemmerling proved that the information from the nucleus present in the base ultimately governs the cell's architecture, including the shape of the cap (although residual information from the transplanted *A. crenulata* stalk could cause a transient effect on cap shape). Numerous experiments since Haemmerling's confirm that the nucleus is indeed the primary site of the information that determines the structure of the cell.

Chromosomes and DNA

If the nucleus contains genetic information, what form does it take? Early microscopists noticed that specific structures in the nucleus of a dividing cell will take up basic red or purple dyes. They named these structures **chromosomes,** meaning "colored bodies." Cells of each species possess a characteristic number of chromosomes: Human cells have 46; cotton plants, 52; turkeys, 82; and some ferns, more than 1,000. Some fascinating detective work eventually revealed that the information-bearing substance in chromosomes is DNA—deoxyribonucleic acid—an extremely long molecule that consists of two strands of nucleotides wound around each other in a helix. Chapters 12 and 13 will describe the search to understand DNA structure and how the molecule carries hereditary information. For now, to understand how cells divide, we need to concentrate on the structure and activities of chromosomes.

Under the light microscope, stained chromosomes may look like solid, flexible rods, as shown in Figure 9-2a. They are not solid structures at all, however. Instead, each is composed of a long, tightly coiled strand of DNA and associated proteins that knot up into "fuzzy rods" with a constriction called the **centromere** (Figure 9-2b).

If a chromosome were 1 cm (1/4 in.) long, the single slender DNA molecule coiled up in it would stretch the length of a football field. Clearly, the packaging of so much "string" into such a small "ball" must be very efficient. And biologists have discovered that stretches of the long, continuous DNA molecule wind around clus-

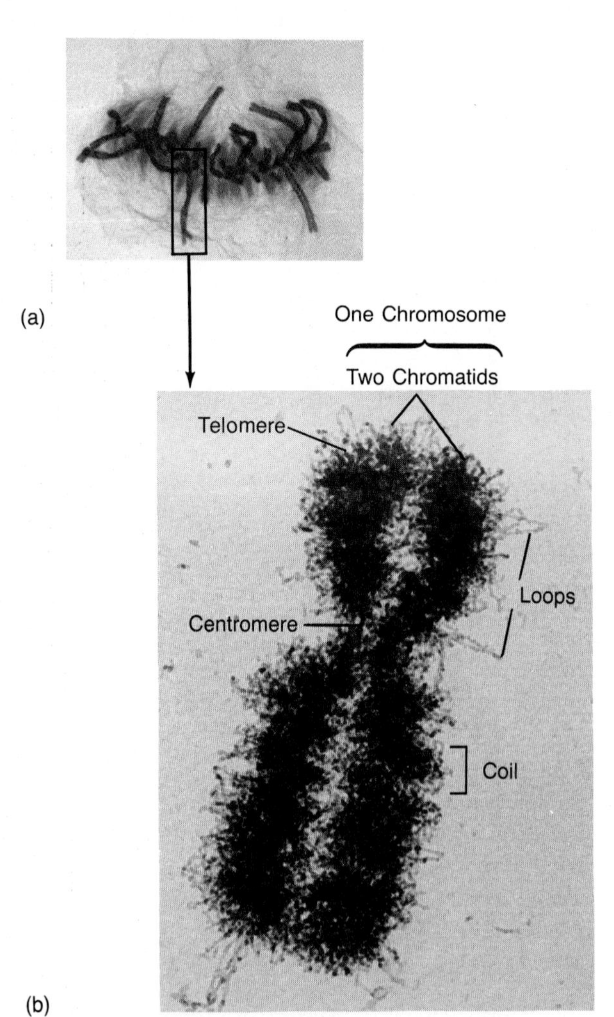

Figure 9-2 CHROMOSOMES IN THE NUCLEUS OF A EUKARYOTIC CELL.
(a) During cell division, chromosomes (magnified about 200×) appear to be solid bodies within the nucleus. (b) A higher magnification (about 20,000×) reveals that each body, or rod, is actually a coiled mass of threads, or strands of DNA and proteins. This chromosome has already been duplicated so that the two identical chromatids lie side by side, attached at the centromere, the narrow region of the chromatids.

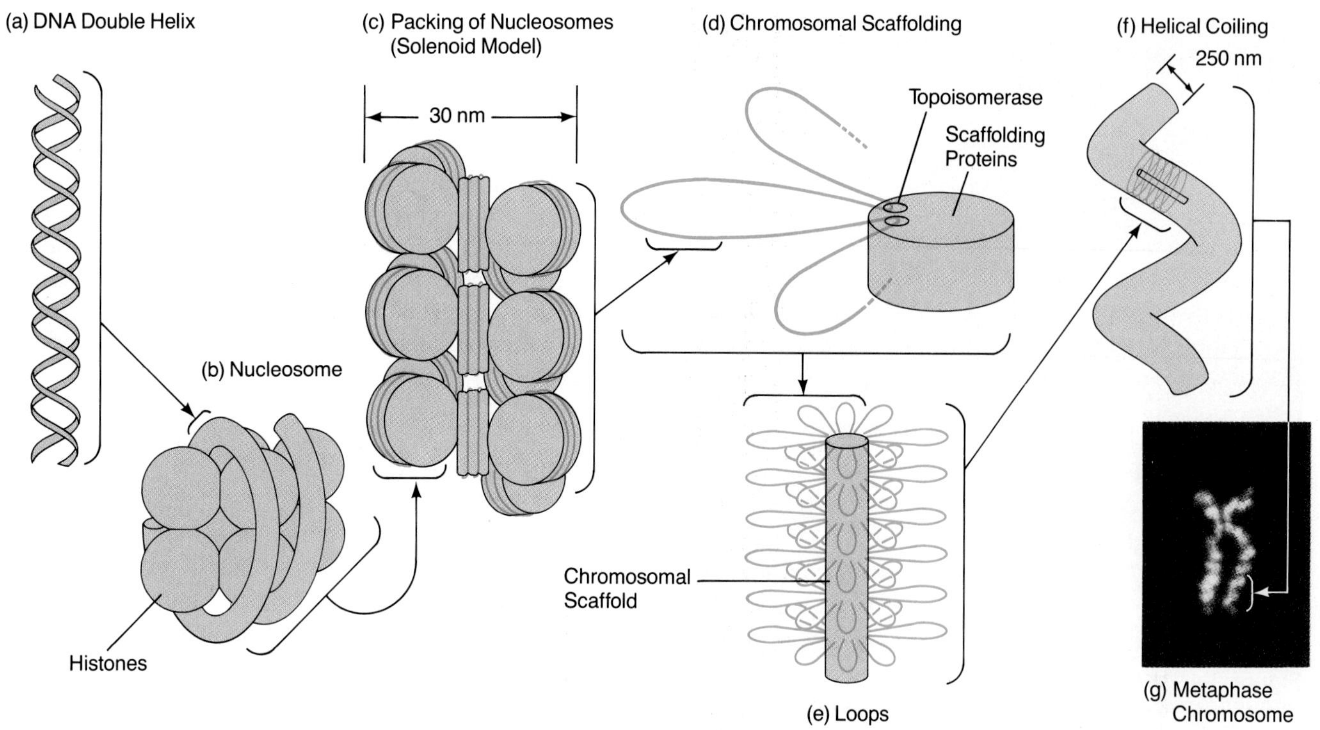

Figure 9-3 A CHROMOSOME UNWOUND: FROM DNA TO "COLORED BODY."
A chromosome contains highly coiled and supercoiled DNA strands. (a, b) The DNA molecule is wound around clusters of eight proteins called histones, forming nucleosomes. (c) Nucleosomes are packed into cylinders called solenoids. (d, e) Solenoids are thrown into loops attached by topoisomerase enzymes to a central scaffolding. (f) The loops and scaffold form a coiled rod, and (g) these coils are visible in chromosomes at metaphase, as this photo shows.

ters of positively charged proteins called **histones,** forming beadlike complexes, or **nucleosomes** (Figure 9-3a and b). They think that the nucleosomes interact with one another to form a solenoid, a flexible cylinder about 30 nm in diameter (Figure 9-3c) that is thrown into loops (Figure 9-3d) that in turn attach to a central scaffolding by means of a nonhistone protein called topoisomerase (Figure 9-3e). Together, the loops and scaffolding form a coiling rod about 250 nm in diameter (Figure 9-3f). The coils twist in opposite orientations in sister chromatids (Figure 9-3g). Thousands of genes lie along each individual DNA molecule, and the topoisomerase enzymes may help untangle knots in the DNA. Finally, the chromosomes of eukaryotes, from plants to protists to people, have special ends called *telomeres* composed of unusual repeating DNA sequences. These appear to protect the DNA during its replication.

Biologists call the combination of DNA, histones, and nonhistone proteins **chromatin.** When a cell is resting between cell divisions, the chromatin is attached to the *nuclear lamina,* a meshwork of intermediate-type filaments on the inner surface of the nuclear envelope. Chromosomes in a nondividing nucleus are unwound and tangled like a mass of yarn. As the time for cell division nears, individual chromosomes separate from the tangled mass and enter the supercoiled, condensed condition, in which they assume flexible rodlike shapes. This temporary coiling of chromatin makes chromosomes much more compact and permits the orderly allocation of genetic information during cell division.

Researchers can make a pictorial display of the number, sizes, and shapes of chromosomes in the coiled, condensed state. This display, called a **karyotype** (Figure 9-4), clearly shows that most cells have two copies of each chromosome, or **homologous** ("same-shaped") pairs. The only exception to this occurs with the sex-determining chromosomes of some animals. Humans, for example, have 22 pairs of **autosomes,** or non–sex chromosomes, and one pair of **sex chromosomes** in most cells of the body. Human females have homologous sex chromosomes (two *X* chromosomes), while males have a nonhomologous pair (one *X* chromosome and one stubby *Y* chromosome).

Organisms with cells containing two sets of chromosomes are called **diploid** (meaning "two"). Organisms with cells containing just one set of chromosomes are called **haploid.** In most plants and animals, the conspicuous adult is diploid, but the eggs, sperm cells, and sometimes a few associated cells are haploid. In typical eukaryotic, sexually reproducing organisms, diploid and haploid phases alternate during the *life cycle;* the life cycle of a species is defined as the passage from one adult

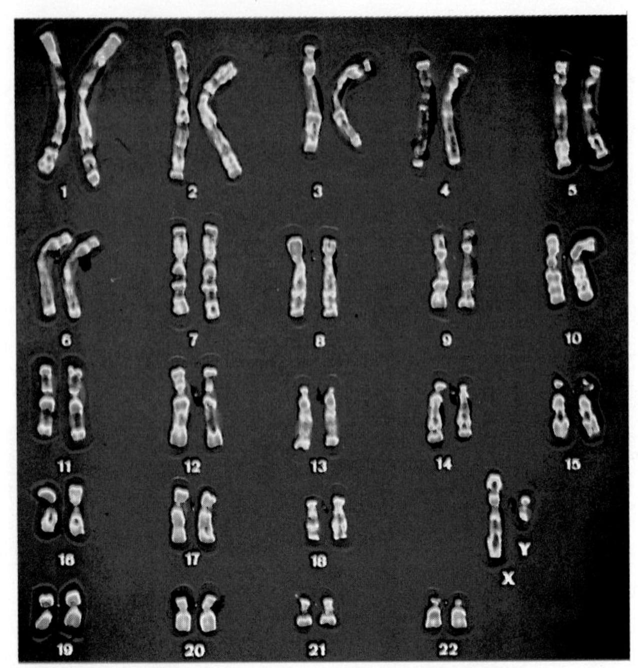

Figure 9-4 KARYOTYPE: AN ORGANIZED REPRESENTATION OF A SET OF CHROMOSOMES.
To produce a karyotype, a geneticist treats mitotic cells with a dye such as Giemsa that has an affinity for DNA and that fluoresces (glows) when viewed with ultraviolet light. The researcher then photographs the stained chromosomes, cuts the individual chromosomes from the picture, and arranges them roughly in descending order of size. In this human karyotype, there are 22 pairs of autosomes and one pair of sex chromosomes, an *X* and a *Y*. The presence of a *Y* reveals that the chromosome set comes from a male.

through the reproductive stages to the adult of the next generation. A sizable minority of plant species spend the adult phase of the life cycle as haploid organisms. For instance, among mosses, the adult you see growing in damp soil or on shaded logs is haploid, as Chapter 23 explains.

THE CELL CYCLE

Just as whole organisms pass through life cycles, individual cells pass through **cell cycles,** regular sequences during which cells grow, prepare for division, and divide to form two daughter cells, which repeat the sequence. Free-living single-celled eukaryotes, such as amoebae, are essentially immortal because of such cycles. As such cells divide, they distribute hereditary information to their daughter cells, which in turn distribute virtually the same information to their progeny, and so on for potentially millions of generations. Within multicellular plants and animals, some cells continue to grow and divide for the life of the organism. In plants, these include cells at the tips of roots, which probe ever deeper into the soil, and in animals, cells that line the small intestine of the digestive tract and continually produce daughter cells that are sloughed off as food is digested. Most cells in multicellular organisms, however, either slow the cell cycle greatly or "break out" of it altogether; they stop dividing and remain locked in one part of the sequence until either they die individually of old age or disease or the entire organism dies. These include muscle cells and nerve cells in animals, and cells that form sugar-conducting vessels in plants.

Stages of the Cell Cycle

Within the normal cell cycle, the period between cell divisions can be quite long, while the actual division phase takes place quickly. As Figure 9-5 shows, the cycle begins with a period of normal metabolism, called $\mathbf{G_1}$ (gap 1). During this phase, the cell synthesizes, assembles, and uses the components it needs for normal functioning. Next comes the **S phase** (synthesis phase), during which normal syntheses continue and, in addition,

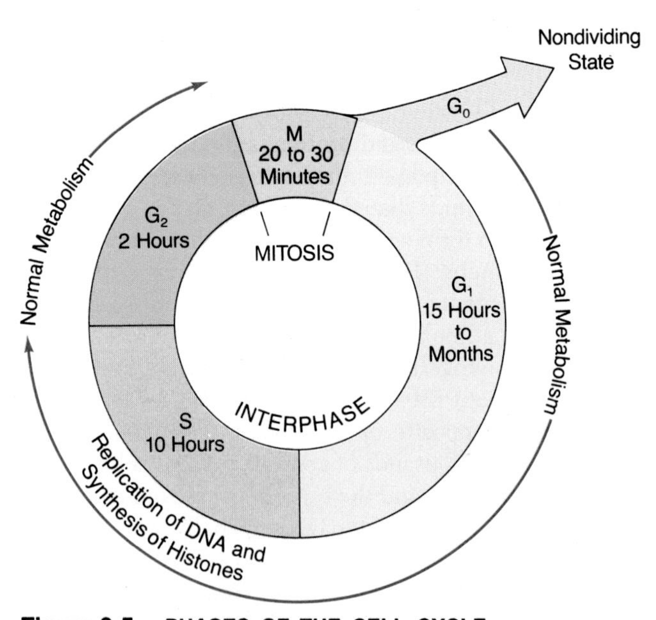

Figure 9-5 PHASES OF THE CELL CYCLE.
The cell cycle includes mitosis (designated M), a period of active division, and interphase, a period of nondivision, during which other processes take place. Interphase is usually divided into three phases. During G_1, normal components of the cell are synthesized and metabolized, often resulting in cell growth. Those processes continue during the S phase, but in addition, chromosomal DNA is replicated and histones are synthesized. During G_2, normal metabolism and cell growth continue until M is reached. Some cells remain for long periods in a nondividing state (G_0) after completing mitosis.

DNA is replicated and histones are synthesized. The amount of DNA in the nucleus doubles, and the histones and other chromosomal proteins synthesized in the cytoplasm pass into the nucleus, combine with the DNA, and form chromatin. Each chromosome in S phase now has two identical threads of DNA instead of the single thread present during the earlier G_1. The two daughter strands of a duplicated chromosome joined at the centromere are called a **chromatid.** At the end of the S phase, the cell enters $\mathbf{G_2}$ (gap 2), a brief period of normal metabolism and growth during which the cell manufactures more proteins and other substances. During G_1, S, and G_2, the nucleus is said to be in **interphase,** between nuclear divisions. Once a cell enters S, it is normally committed to proceed through S and G_2 and to divide during M.

When the cell enters the **M phase,** the period of **mitosis,** the chromosomes in the nucleus condense, move about in special ways, and become apportioned equally to the two new nuclei. This phase ends with *cytokinesis,* the actual separation of the cytoplasm and generation of two new daughter cells.

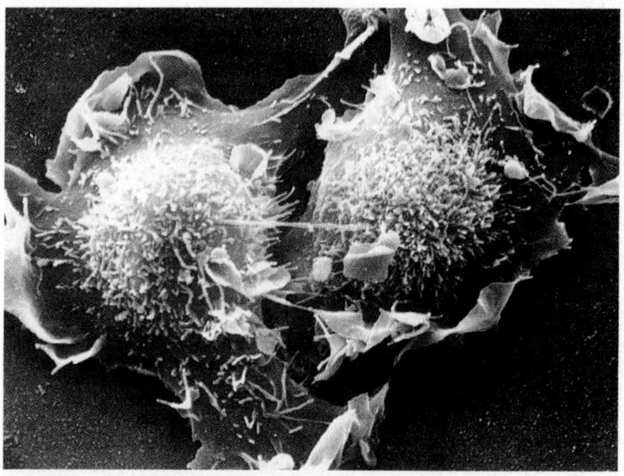

Figure 9-6 IDENTIFYING CANCEROUS CELLS. The division process in normal cells and cancer cells is identical, but division is not regulated normally in cancer cells. These dividing animal cells are magnified 1,300×, and researchers could use several tests to determine whether they are cancerous.

Control of the Cell Cycle

One of the most elusive problems in biology is understanding how the cell cycle is regulated. Most animal and some plant cells spend varying numbers of hours, days, or months in the G_1 phase before the S and M phases are initiated. And in nerve, muscle, and sugar transport cells, the functional adult cell remains in a permanent nondividing condition, sometimes called G_0. What prevents or triggers cell division at appropriate times?

Biologists have found that if they transplant nuclei from nondividing cells such as nerve cells into host cells in the S phase, within minutes the DNA replication characteristic of the S phase begins within the chromosomes of the transplanted nerve cell nuclei. Such experiments suggest that *the state of the cell's cytoplasm* controls the activity of the nucleus and helps determine its phase in the cell cycle.

What, then, determines the state of the cytoplasm? One factor is the cell's external environment. In multicellular animals, for instance, normal cells are *contact-dependent:* They can move and divide only if adhering to certain types of solid substrates (see Chapter 5). Another factor is intracellular calcium; certain types of cells must pump out calcium before they can pass through a mitotic cycle.

In both plants and animals, outside factors can stimulate or inhibit the cell cycle. In higher plants, for example, hormones called cytokinins stimulate mitotic activity in the root, stem, or leaf, while in complex animals, the hormones somatomedin, insulin, and epidermal growth factor can all stimulate cells to divide. Animal cells may also have inhibitors of cell division called **chalones.** Some biologists believe that chalones decrease in the area around an injury so that new cell divisions can repair the wound, then increase once the wound is healed to prevent an overgrowth of cells.

Understanding what controls the cell cycle is particularly important in the study of cancer. In some types of cancer, the normal controls on mitosis seem to be suppressed. While division is no faster, a higher proportion of tumor cells actively divide than do normal cells of the same type, and daughter cells continue to divide, cycle after cycle (Figure 9-6). Many cancer researchers and geneticists are currently focusing on the secrets of cell cycle control. One group, for example, recently found that yeast cells had a particular gene (the *RAD 9* gene) that can stop or delay the cell cycle when the cell has suffered potentially lethal radiation damage. This delay allows time for repair enzymes to fix the damage and suggests that the cell cycle may be under very precise genetic control.

MITOSIS: PARTITIONING THE HEREDITARY MATERIAL

Cells show an amazing fidelity from one generation to the next largely because mitosis, the shortest phase of the cell cycle, provides an intricate and highly accurate mechanism for apportioning genetic information equally to each daughter cell. During normal cell division, each of the two daughter cells must receive two copies of each chromosome type: two of chromosome number 1, two of number 2, and so on. It is the movements of chromosomes during mitosis that help guarantee these distributions.

The stages and events of mitosis in eukaryotic organisms have been described as a "dance of the chromosomes"—a series of spectacular movements during which the chromosomes jostle about in the middle of the nucleus and then, as if participating in an English country reel, split neatly into two groups and glide to opposite ends of the cell. The cell then separates into two, leaving each daughter cell with a full complement of chromosomes.

For convenience, biologists divide the continuous events of mitosis into four phases: *prophase, metaphase, anaphase,* and *telophase,* each illustrated in Figure 9-7.

Prophase

Recall that during interphase of the cell cycle—specifically during the S phase—the DNA of each chromosome is copied precisely (Figure 9-7a), and the original and the copy, joined at the centromere, are now called chromatids.

The first phase of mitosis (M), **prophase** ("before form"), starts after S and G2 are over. The tangled, yarnlike chromatin begins to condense during prophase, and the chromosomes become visible, each with its two identical chromatids. As prophase progresses, the coil-

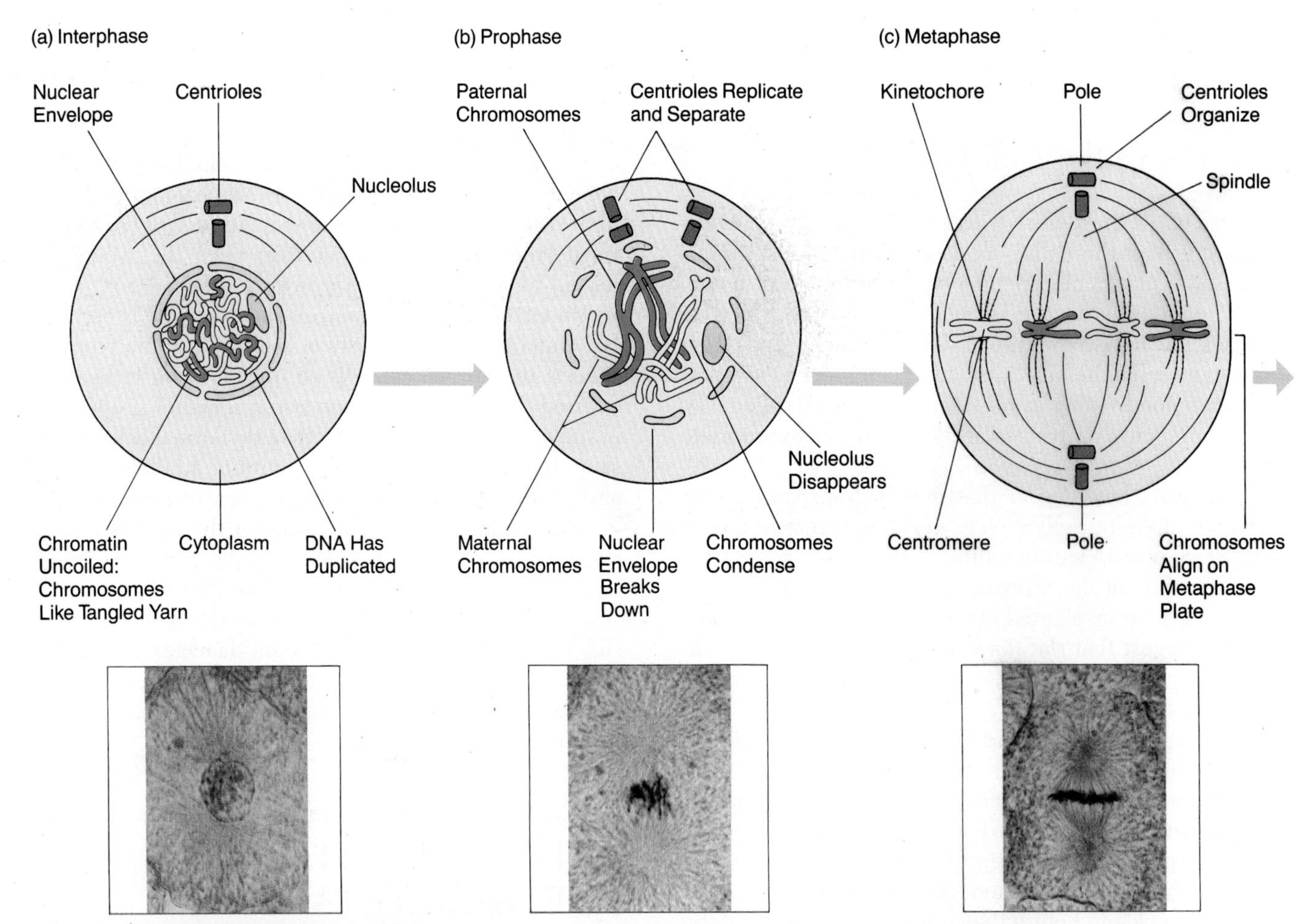

Figure 9-7 THE PHASES OF MITOSIS.
As the text describes, mitosis involves (a) interphase, (b) prophase, (c) metaphase, (d) anaphase, (e) telophase, and (f) cytokinesis. During these phases, chromosomes are replicated, pairs align at the cell's equator, they separate and move apart, and finally they become enclosed in the nuclei of newly formed daughter cells. For simplicity, the diagrams depict only two pairs of chromosomes. The accompanying photographs show mitosis in a plant cell.

ing of the chromosomes becomes tighter, the nucleolus disappears, the nuclear envelope usually breaks down, and the centrioles replicate (Figure 9-7b).

Metaphase

As prophase ends and **metaphase** ("middle form") begins, the fully condensed chromosomes become associated with the **spindle,** a football-shaped set of microtubules stretching across the center of the cell. In animal but not plant cells, additional arrays of microtubules called *asters* fan out in a star-shaped pattern from both *poles* (both ends of the dividing cell) (Figure 9-7c). Bundles of microtubules are attached at one end to specialized structures called **kinetochores,** which are adjacent to the centromeres of the chromosomes; these bundles extend toward the poles. The chromosomes appear to jostle about during early metaphase and eventually become arranged in a plane at a right angle to the spindle fibers in the center of the nuclear region. This plane is called the **metaphase plate.**

Anaphase

Anaphase ("again form") begins when the centromeres that link the sister chromatids split, thereby allowing each chromatid to behave as a separate chromosome. Some evidence suggests that a brief rise in cytoplasmic calcium signals this event. One chromatid from each pair moves toward each pole, with floppy arms lagging behind the centromeres (Figure 9-7d). Chromosome separation occurs in minutes, and anaphase ends as

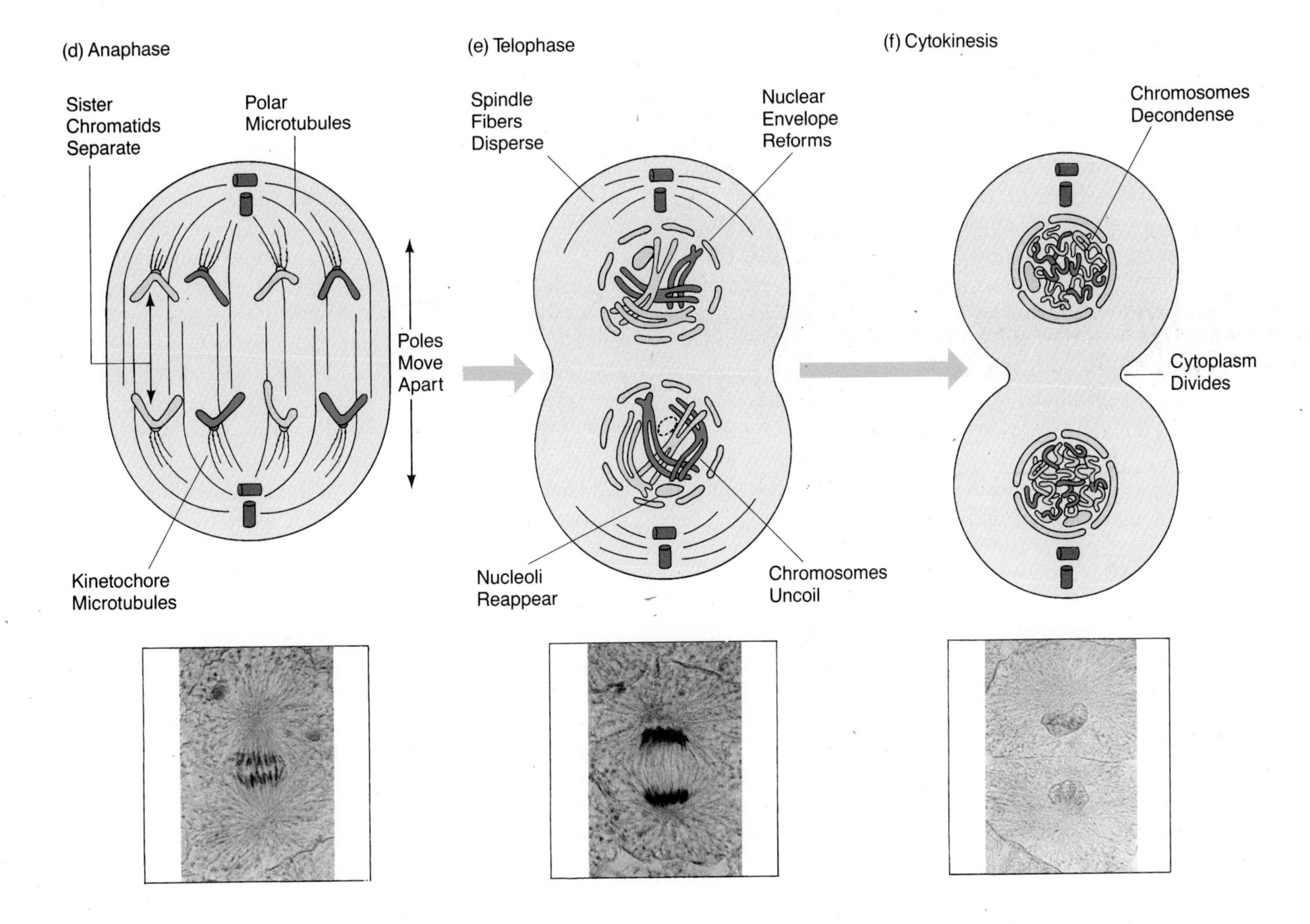

WHAT MAKES THE CHROMOSOMES DANCE?

The movement of the chromosomes to opposite spindle poles during anaphase guarantees that each new daughter cell will receive the hereditary instructions needed to direct its day-to-day survival, as well as its own eventual cell division. But exactly how do these bundles of coiled and packaged genetic material move from the cell's equator to the poles along the kinetochore microtubules?

Over the years, cell biologists have constructed two basic hypotheses: (1) that polar microtubules pull the chromosomes toward the pole like a fishing line reeling a fishing lure through a pond; or (2) that each chromosome moves along a stationary microtubule like a gymnast shimmying up a rope. To determine which, if either, is correct, a team at the University of Wisconsin devised a clever test.

Using live, dividing cells from pig kidneys, they stained the polar microtubules with fluorescent dye, and with a laser beam, bleached out a band of the dye halfway between the pole and the chromosomes (Figure A, top). With this system, they could allow anaphase to proceed and note the position of the marked bands. If the fishing line hypothesis were correct, then a bleached-out band would move toward the pole *along with* the chromosome (Figure A, middle). If the gymnast hypothesis were correct, the band would remain stationary while *the chromosome alone* moved toward the pole (Figure A, bottom). From several trials of the experiment, the team determined that the gymnast hypothesis is correct: Chromosomes move along stationary microtubules. They also determined that microtubules disassemble, one tubulin unit at a time, at the site where the chromosome attaches (at the kinetochore). Questions still remain about how this disassembly process allows chromosomes to move toward the poles. But the laser experiment helped explain in molecular terms a dramatic phenomenon that is central to cellular survival and heredity.

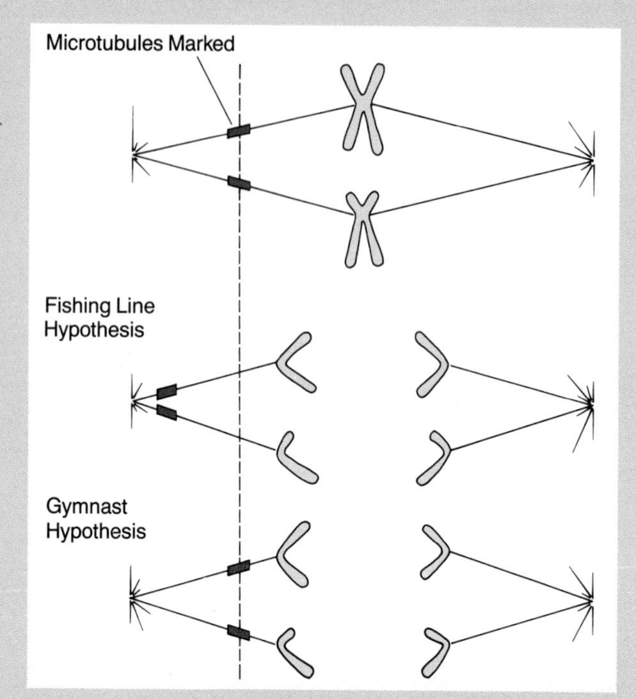

Figure A THE LASER BANDING EXPERIMENT.

the two sets of chromatids have parted and the two spindle poles move farther and farther apart.

Telophase

Telophase ("end form") starts as the spindle fibers disperse and nuclear envelopes begin to form around each set of chromosomes in a way that prevents cytoplasmic particles and soluble cytoplasmic macromolecules from becoming trapped inside the forming nuclei (Figure 9-7e). The chromosomes apparently become attached to the nuclear lamina. They uncoil once again into a tangle of chromatin, and nucleoli gradually reappear. Division of the cytoplasm then takes place (Figure 9-7f).

The Nature of the Spindle

Just as a dancing puppet is operated by moving strings, the dance of the chromosomes is choreographed and controlled by the action of the spindle fibers. Chromosomes are not alive and cannot move by themselves. The spindle's important task is ensuring that the paired chromatids separate properly and move in the right directions at the proper times.

In most cells, the spindle forms in the cytoplasm during prophase. It first appears as parallel sets of protein fibers, the microtubules, extending from the poles around the nucleus. Each half of the spindle forms as microtubules extend from each pole to the region of the metaphase plate, where the two halves of the spindle overlap (see Figure 9-7c). As the nuclear envelope disperses during prophase, the kinetochore microtubules extend from the kinetochores located next to the centromere of each chromatid.

During anaphase, two processes move the chromatids toward the poles. In a process driven by ATP, the microtubules slide past each other, and the overlapping of polar microtubules decreases (see Figure 9-7c and d).

With this, the poles move farther apart, and the two sets of chromatids are themselves pulled apart. Meanwhile, the kinetochore microtubules depolymerize near their attachment sites to the kinetochores; that disassembly process somehow allows the chromatids to move toward the pole, as the box on page 162 explains.

In animal cells, spindle formation is associated with the *centriole* (see Chapter 6), a cytoplasmic organelle similar to the basal bodies of cilia and flagella. During prophase, two pairs of centrioles move to opposite sides of the nucleus, marking the positions of the spindle poles. Forces generated by microtubule elongation apparently push the centrioles and asters to their final positions. Once in place, the centrioles probably help form the remainder of the spindle, especially at sites where tubulin polymerizes into microtubules.

In most plant and fungal cells, spindle formation proceeds as in animal cells, but the cells lack centrioles. Instead, they have regions called *microtubule-organizing centers* with a function analogous to that of the centrioles.

CYTOKINESIS: PARTITIONING THE CYTOPLASM

As the spectacular chromosome movements of mitosis end, the cell has two nuclei containing identical sets of chromosomes. **Cytokinesis,** the division of the cytoplasm, then takes place so that two individual cells result.

In animal cells, which are bounded only by a flexible plasma membrane, a contractile ring composed of actin filaments "pinches" the cell in two during cytokinesis. This beltlike array of filaments—part of the cytoskel-

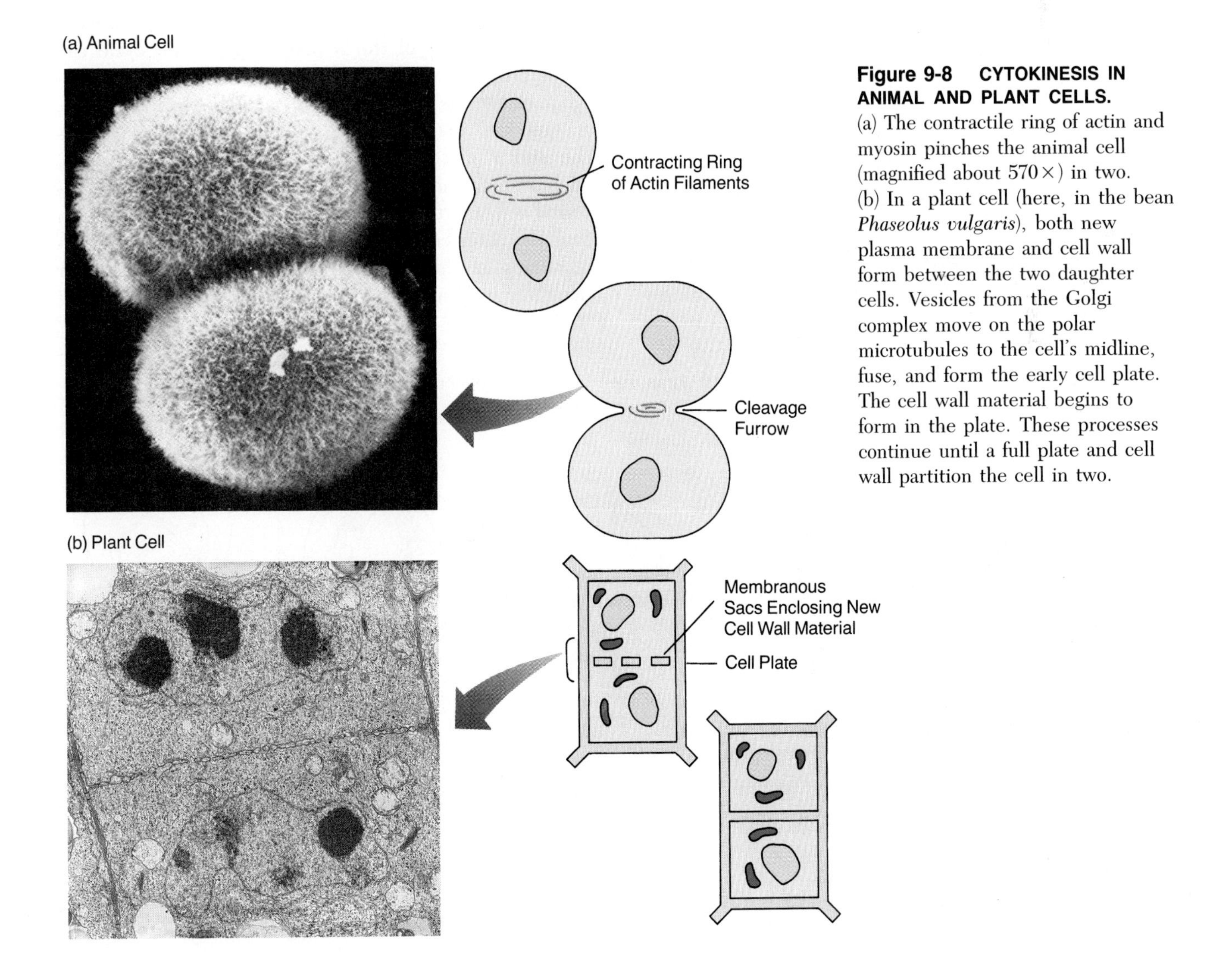

Figure 9-8 CYTOKINESIS IN ANIMAL AND PLANT CELLS.
(a) The contractile ring of actin and myosin pinches the animal cell (magnified about 570×) in two.
(b) In a plant cell (here, in the bean *Phaseolus vulgaris*), both new plasma membrane and cell wall form between the two daughter cells. Vesicles from the Golgi complex move on the polar microtubules to the cell's midline, fuse, and form the early cell plate. The cell wall material begins to form in the plate. These processes continue until a full plate and cell wall partition the cell in two.

eton—appears during telophase. The ring lies midway between the two spindle poles and asters, and actin filaments of the ring are linked to the inner surface of the plasma membrane so that their contractile activity pulls the cell surface inward tighter and tighter. This contraction forms a cleavage furrow, or division furrow (Figure 9-8a), that eventually separates the cells completely.

Plant cells, with their more rigid cell walls, cannot pinch in two. Instead, once the spindle has pulled the chromosomes apart, a different process constructs a **cell plate** across the cell at the equator and partitions the cytoplasm into two regions (Figure 9-8b). The cell plate is composed of membranous sacs that fuse with one another and gradually extend across the cell and outward toward the plasma membrane that lines the rigid cell wall on all sides. A cell wall begins to appear in the cell plate. The peripheral sacs fuse with the plasma membrane, thereby linking the new cell plate with the sides of the cell. As a result, each daughter cell is fully enclosed within its own intact plasma membrane. Following this, additional cell wall material is deposited in the region of the cell plate; a rigid cell wall forms; and each daughter cell is fully protected in its own tough shell.

MEIOSIS: THE BASIS OF SEXUAL REPRODUCTION

While single cells can divide in half to reproduce, few multicellular organisms can generate progeny by simply splitting in two. They rely, instead, on the production of spores or gametes (eggs or sperm) and the fusion of gametes at fertilization. Consider what would happen, though, if human gametes contained the full set of 46 chromosomes and offspring received a set from each parent: The progeny would have 92 chromosomes per cell, and future generations would have 184, 368, 736, and so on. But chromosomes do not accumulate this way because of a special kind of cell division called **meiosis,** which takes place only in the reproductive organs that generate sperm, eggs, or spores. In humans and most other animals, meiosis occurs only in certain cells of the ovaries and testes. In flowering plants, pollen grains (male gametophytes) are produced in anthers, and eggs are formed in an ovary at the base of each flower. Finally, as we will see, some plants spend most of their lives as haploid organisms; they may produce short-lived diploid stages prior to undergoing meiosis and producing a new generation of haploid organisms.

In yeast, at least, meiosis apparently begins when the product of one gene inactivates a protein kinase enzyme that inhibits meiosis. With that inhibition removed, DNA replication of the S phase of the cell cycle can occur. Two successive nuclear divisions take place, designated *meiosis I* and *meiosis II*. Each has prophase, metaphase, anaphase, and telophase stages. Whereas the similar phases of mitosis produce *two* daughter cells, each with the *same* number of chromosomes as the parental cell, the two nuclear divisions of meiosis result in *four* daughter cells, each with *half* the number of chromosomes as the parental cell. Moreover, these four daughter cells are not genetically identical, in part be-

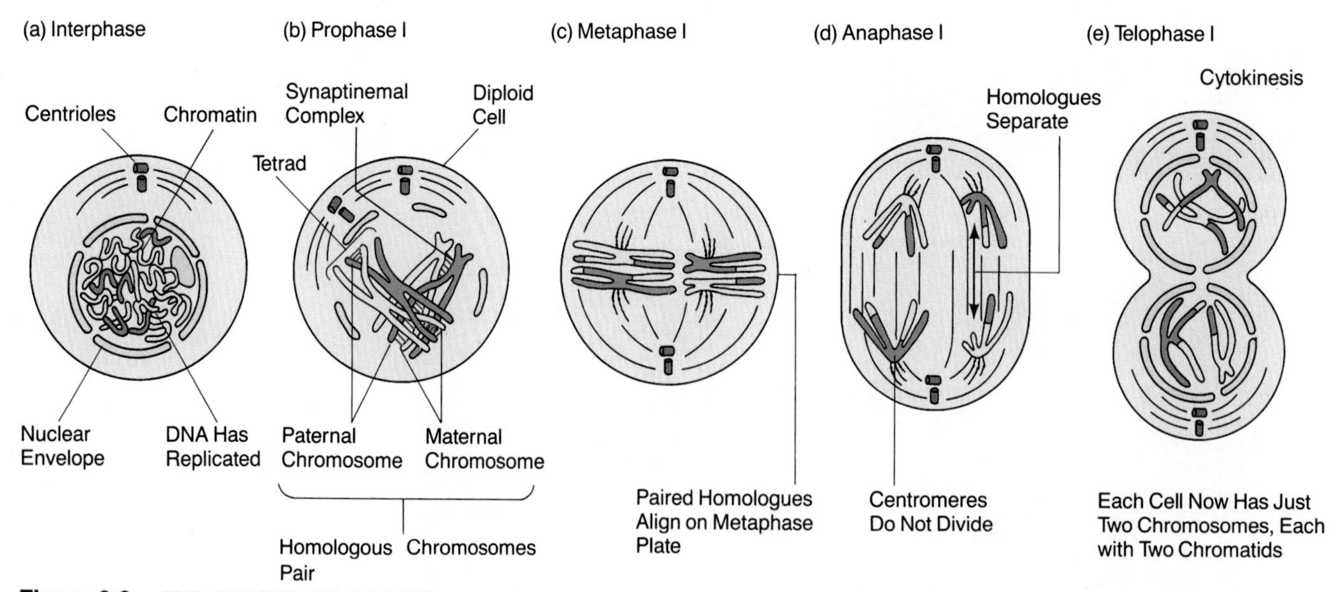

Figure 9-9 THE PHASES OF MEIOSIS.
As the text explains, two divisions ensue after the chromosomes are replicated: The first meiotic division reduces the number of chromosomes, whereas the second meiotic division results in the chromatids separating. Four haploid cells result.

cause a remarkable process called crossing over may exchange genetic material between chromosomes. Figure 9-9 presents the phases of meiosis.

Prophase I

Since DNA replication takes place during the S phase of the cell cycle, before meiosis begins, there are two chromatids for each chromosome at the start of prophase I of meiosis (Figure 9-9a). The sister chromatids of each chromosome are connected at the centromere, as in mitosis. Each chromosome in the nucleus of a diploid cell is part of a pair in which one chromosome is derived from the female parent and one from the male. These are called homologous pairs (each individual is a homologue), and each carries corresponding genes for corresponding traits. At the start of prophase I, the homologous chromosomes themselves come together and pair side by side to form a four-chromatid structure known as a *tetrad* (Figure 9-9b). The pairing of homologous chromosomes during prophase I is called **synapsis,** which literally means "union," and is unique to meiosis.

Synapsis of the homologous chromosomes is brought about by a bridging structure of proteins called the **synaptinemal complex,** which acts like a zipper to bring together in intimate association corresponding regions on the homologous chromosomes.

As in mitotic prophase, the nuclear envelope disappears in both plant and animal cells, and in animal cells, the centrioles separate during prophase I of meiosis. As prophase I comes to an end, the synaptinemal complex dissolves, and the coiling of the unzipped chromatids increases, making the individual chromosomes even more distinct. The complex holding together the tetrads is soon lost, but the pairing of the homologous chromosomes remains. It is at this stage that an "arrest" may occur, the cells remaining in this condition for hours, weeks, years, or decades, depending on the species.

Metaphase I and Anaphase I

In metaphase I, the homologous chromosomes remain associated in pairs as they align on the metaphase plate (Figure 9-9c). Recall that during anaphase of mitosis, the centromere of each chromosome splits, allowing the two sister chromatids to move to opposite poles. In contrast, during anaphase I of meiosis, the centromeres do *not* divide. Therefore, the *two* chromatids of each chromosome derived from the organism's mother move to one pole, and those from the chromosome derived from the individual's father move to the opposite pole (Figure 9-9d). Note that the maternal homologue may go at random to either pole. The important result of this unusual sequence of events is that the chromosome number in each newly forming nucleus is half the chromosome number of the parental cell.

Telophase I

During the next phase of meiosis, telophase I, nuclear envelopes enclose the chromosomes in nuclei, each

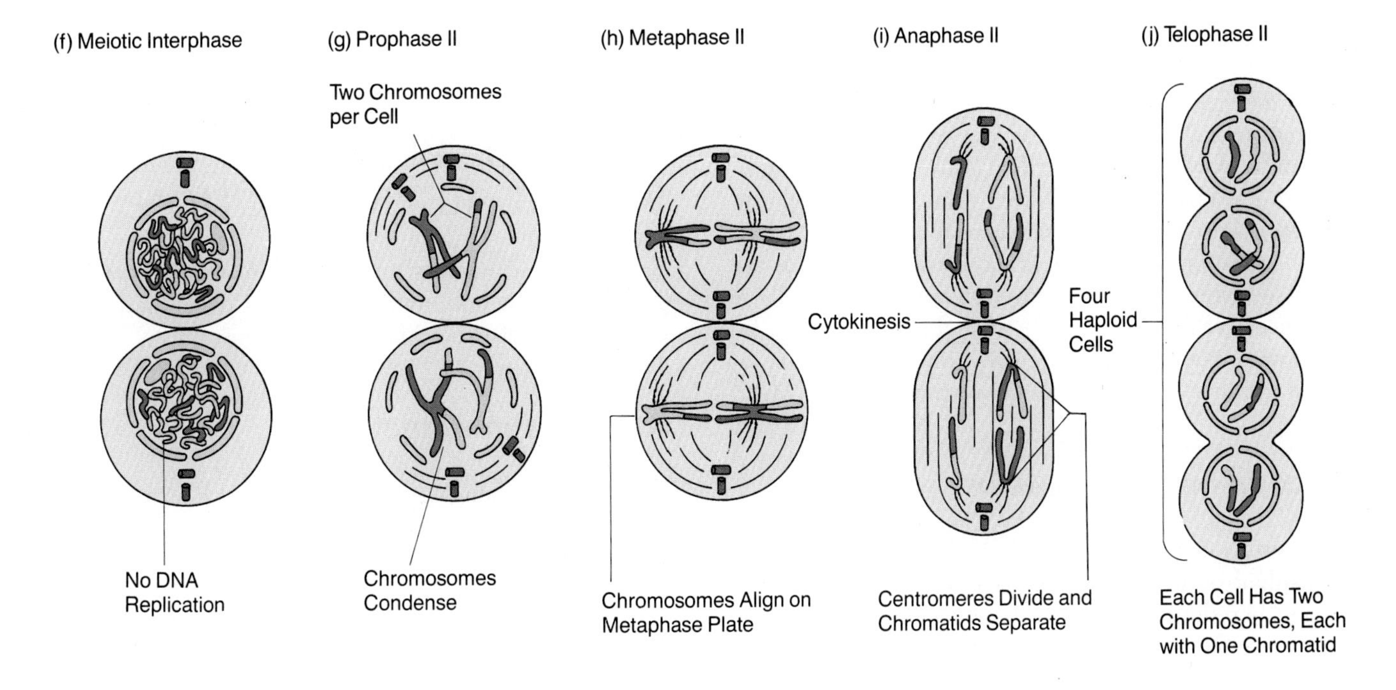

chromosome still consisting of paired chromatids because the centromeres did not divide in anaphase I (Figure 9-9e). For instance, during meiosis I in human sperm-producing cells, the original 46 chromosomes form 23 tetrads; on completion of telophase I, 23 chromosomes, each composed of 2 chromatids, are present in each daughter cell.

Meiotic Interphase and Prophase II

In most species, cytokinesis follows telophase I, and during the ensuing interphase, the chromosomes disperse in the nuclei of the two daughter cells (Figure 9-9f). The chromosomes recondense during prophase II of meiosis in most species (Figure 9-9g). In no case, however, does DNA replication occur in meiotic interphase.

Metaphase II and Anaphase II

In all species, as metaphase II begins, the chromosomes again align on a metaphase plate positioned by a new set of spindle microtubules (Figure 9-9h). At the beginning of anaphase II, the centromeres finally divide, and each sister chromatid moves to one of the poles of the spindle (Figure 9-9i).

Telophase II

Telophase II follows as nuclear envelopes re-form; the chromosomes expand to the interphase state; and cytokinesis is concluded. Four cells, each with a nucleus containing a haploid complement of chromosomes (23 in humans), have arisen from a single diploid nucleated cell (Figure 9-9j). Each of the four cells has only *one* of the four chromatids present in each tetrad as meiosis started. In male animals, these four cells can mature into sperm, each carrying the haploid ($1n$) chromosome number; in the male parts of flowers, they can develop into pollen grains. In female animals or female flower parts, the maturation of gametes is more complicated, as we shall see later, but a haploid egg nucleus eventually is produced. It is only when the nuclei from male and female parents unite that the correct diploid ($2n$) chromosome number is restored in the fertilized egg.

Recombination in Meiosis

There is much more to meiosis than just the diploid-to-haploid reduction. Meiosis is a primary means of ensuring hereditary variability among members of a species, and this variability is at the core of evolutionary change. Without such variability, organisms would be much less capable of adapting to changing environments, and the species would be less able to survive in the long term.

One of the most significant outcomes of meiosis is *the random distribution of parental chromosomes* (Figure 9-10a). In a newly fertilized human egg, or zygote, 23 chromosomes come from the mother and 23 from the father. When the individual who develops from that zygote grows into a man or a woman and produces sperm or eggs, the maternal and paternal chromosomes are distributed to the four daughter cells *at random* during meiosis I: Although every gamete receives 23 chromosomes, anywhere from none to all 23 can be of paternal or of maternal origin. This independent assortment of paternal and maternal chromosomes ensures that most individual gametes will receive new combinations of chromosomes and genes unlike those of either parent.

Another phenomenon that occurs during meiosis, called **crossing over,** contributes still further to genetic variability. During synapsis in prophase I, the homologous chromosomes—each composed of two identical chromatids—are "zipped together" by the synaptinemal complex. Then nonsister chromatids (i.e:, paternal and maternal ones) exchange corresponding pieces of genetic material (Figure 9-10b). This crossing over involves recombination nodules, large multienzyme aggregates that cut and stitch the maternal and paternal chromatids together. The visible points at which homologous chromosomes break and rejoin are called *chiasmata* (singular, *chiasma*). There is no apparent gain or loss of total genetic material during crossing over, just an equal exchange of corresponding chromosomal regions. Chapter 11 describes the genetic exchange of crossing over in more detail.

ASEXUAL VERSUS SEXUAL REPRODUCTION

Mitosis and meiosis make possible both asexual and sexual reproduction, and each type of reproduction has certain distinct evolutionary advantages.

Asexual Reproduction

Mitosis in a single-celled creature such as a yeast or an amoeba produces two daughter cells that function as independent and genetically identical organisms. Some multicellular eukaryotes can also reproduce by means of mitotic processes. Many plants undergo mitotic divisions that produce new plantlets on their horizontal stems;

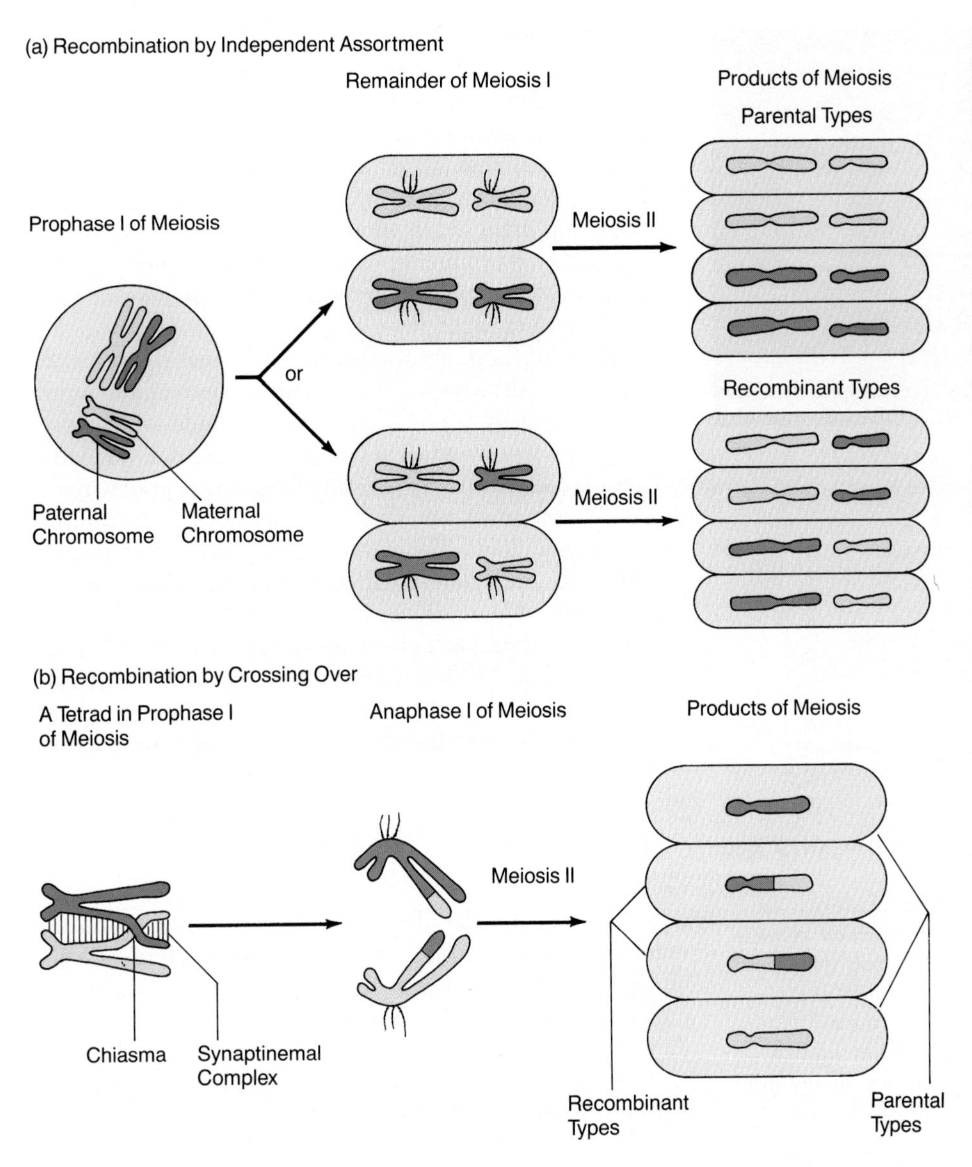

Figure 9-10 INDEPENDENT ASSORTMENT AND CROSSING OVER PRODUCE NEW GENETIC COMBINATIONS.
(a) During independent assortment, the chromosomes can align in different ways on the metaphase plate, and as meiosis proceeds, the resulting gametes can be either parental types (bearing the same chromosome combinations as mother or father) or recombinant types, with new chromosome combinations not seen in either parent. (b) During crossing over, a tetrad of homologous chromosomes begins to pair in early prophase I. They become tightly joined by proteins of the synaptinemal complexes, and portions of the chromatids exchange regions at a chiasma, or site of crossing over. As meiosis proceeds, the recombinant-type gametes can form, containing chromosomes with portions from each parent.

strawberries are an example (Figure 9-11a). Other plants produce new individuals from their roots or even their leaves. In simple animals such as the hydra, a relative of the jellyfish, buds grow by means of mitotic cell division, pinch off, and subsequently develop into a new hydra through further mitotic division (Figure 9-11b). These examples are all forms of *asexual reproduction*—reproduction without sex or meiosis.

Thanks to the regularity and precision of mitosis, asexual reproduction can produce many successive generations of organisms, all genetically identical to the parent. These offspring are often called **clones.** Asexual reproduction has several advantages. It is often more rapid than sexual reproduction, and it requires little or no specialization of reproductive organs. Most important, asexual reproduction preserves an organism's winning genetic formula. Every such organism gambles, or competes in the world, with the identical "cards" (genes) dealt to it by its parent. As long as environmental conditions remain stable, a genetically successful clone can enlarge rapidly and compete successfully with other organisms in the same environment.

Without the tremendous genetic variability bestowed by meiosis and sexual processes, however, a population of genetically identical organisms stands a greatly increased chance of being wiped out by one disease or by a single unusual environmental stress, such as a long drought. A line of asexually reproducing organisms can cope with changing conditions only through mutations, relatively rare events of spontaneous alterations in genetic material that may prove to be beneficial.

Sexual Reproduction

Sexual reproduction generates genetically unique progeny that have some traits from one parent, some traits from the other, and, significantly, some totally new traits

(a)

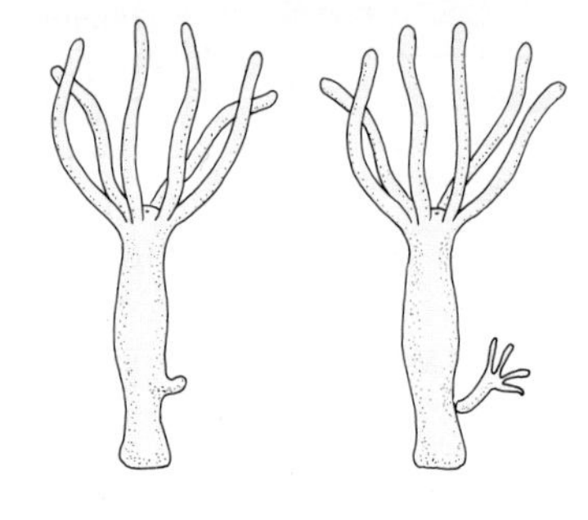

(b)

Figure 9-11 ASEXUAL REPRODUCTION IN PLANTS AND ANIMALS.
(a) Strawberry runners demonstrate a form of asexual reproduction used by plants. The plants grow new runners, or horizontal stems, which send their own roots down into the ground to form a new plant. (b) The hydra (here, magnified about 9×) is an aquatic animal that can reproduce asexually. Hydras can produce buds through simple mitotic division, and each bud pinches off and grows into a new full-sized hydra by further mitotic division.

based on novel combinations of traits from both parents. Spontaneous mutations can provide further variability, as well. The sexually reproducing organism in a sense gambles on giving its offspring a better hand; its genetic "cards" are "reshuffled" and "redealt" instead of being passed along in original form. Thus, new combinations of traits can arise much more rapidly in sexually than in asexually reproducing organisms and increase the chances of the species surviving sudden significant environmental changes.

Despite these great advantages, sexual reproduction does have drawbacks. An organism that cannot reproduce asexually—a mammal, for example—can never bequeath its own exact set of genetic material, no matter how successful, to its progeny, the way a prizewinning strawberry plant can pass along its hereditary complement to a clone. The very mixing process that created the successful gene combination in the adult works to dismantle it partially in the offspring. Researchers are currently trying to perfect techniques for cloning mammals so that the desirable traits of a prizewinning bull, let's say, could be reproduced in thousands of offspring, and not subject to the variability of sexual reproduction.

LOOKING AHEAD

The spectacular dance of the chromosomes during mitosis and meiosis, as well as the cycling of cells through periods of growth, synthesis, and division, allows for both fidelity and variability in the passage of genetic information from one cell generation to the next. Our next chapter explains how biologists unraveled those mysteries of inheritance.

SUMMARY

1. The nucleus is the repository of hereditary information, which is contained in DNA and organized in structures called *chromosomes*.

2. Chromosomes are made up of a substance called *chromatin*, which is a combination of DNA, *histones*, and nonhistone proteins. *Nucleosomes* are sites where DNA is wrapped about sets of histone molecules.

3. In most eukaryotic organisms, there are two copies of each chromosome; the two form a *homologous* pair. *Sex chromosomes* are the exception. Diploid cells and organisms contain homologous pairs of chromosomes. Haploid cells and organisms contain only one set of chromosomes.

4. The mitotic *cell cycle* consists of four phases: G_1, a period of normal metabolism; S, the phase of DNA replication plus metabolism; G_2, a brief period of further cell growth; and *M*, mitosis. The nonmitotic stages (G_1, S, G_2) are referred to collectively as *interphase*.

5. The events of mitosis can be divided into four phases: *prophase*, *metaphase*, *anaphase*, and *telophase*.

6. At the beginning of mitosis, two identical chromatids become associated with the *spindle* and align in the middle of the cell along the *metaphase plate*. In anaphase, the centromeres divide, and the two chromatids are drawn toward opposite poles.

7. The polar microtubules extend from the spindle poles and overlap at the equator. As the region of overlap decreases, the poles are moved farther apart. Spindle microtubules extend from the *kinetochores* in the centromere of the chromatids toward the spindle poles. These fibers shorten, pulling the chromatids toward the poles.

8. In animals, spindle formation is associated with centrioles; in most plants and fungi, it is associated with microtubule-organizing centers.

9. In animal cells, *cytokinesis*, or division of the cytoplasm, results from the

activity of a contractile ring of actin filaments at the cell's equator. In plants, cytokinesis involves the building of a *cell plate* across the cell's equator.

10. *Meiosis* is a special type of cell division that reduces the chromosome number by half. In meiotic prophase I, the two chromatids of each homologous chromosome come together in a process called *synapsis*. In anaphase I, the homologous chromosomes separate to opposite poles. During meiosis II, each sister chromatid of the pair moves to one of the poles. The result is four haploid cells.

11. Meiosis allows for the random distribution of homologous parental chromosomes in offspring, as well as the genetic variability that results from *crossing over*.

12. Asexual reproduction, in which new organisms arise from mitotic processes, preserves an organism's "winning genetic formula" in a particular environment. Sexual reproduction, which involves meiosis and gamete production, ensures greater variability in offspring and ensures hereditary enrichment as new genes and gene families arise.

KEY TERMS

anaphase
autosome
cell cycle
cell plate
centromere
chalone
chromatid
chromatin
chromosome
clone
crossing over
cytokinesis
diploid
G_1 phase
G_2 phase
haploid
histone
homologous
interphase
karyotype
kinetochore
meiosis
metaphase
metaphase plate
mitosis
M phase
nucleosome
prophase
sex chromosome
S phase
spindle
synapsis
synaptinemal complex
telophase

QUESTIONS

1. What part of the cell contains the hereditary information? Which structures contain this information? Which molecules make up these structures?

2. The cell cycle consists of four phases: G_1, S, G_2, and M. What occurs during each phase? Which has the most variable length?

3. What is the outcome of mitosis? Does each daughter cell receive identical chromosomes?

4. What is the outcome of meiosis? Does each haploid cell receive identical chromosomes?

5. Which of the following statements apply to mitosis, which to meiosis, and which to both?
   a. DNA replication occurs before this process starts.
   b. When the chromosomes first become visible, they are already doubled.
   c. Homologous chromosomes pair.
   d. Each daughter cell receives an identical complement of chromosomes.
   e. The centromere is the last part of the chromosome to divide.

6. Two kinds of microtubules separate the chromosomes during cell division. What are they, and how do they operate?

7. Describe the function of the centriole or microtubule-organizing center during cell division.

8. Describe the process of cytokinesis first in a dividing animal cell, then in a plant cell.

ESSAY QUESTIONS

1. Did you inherit equal numbers of chromosomes from your two parents? From your four grandparents? Explain.

2. Is a species more likely to survive in a changing environment if all its members are identical or if they are diverse? What is accomplished by sexual reproduction? What are the advantages of asexual reproduction?

For additional readings related to topics in this chapter, see Appendix C.

10 Foundations of Genetics

As a consequence of the application of Mendel's principles, that vast medley of seemingly capricious facts which have been recorded as to heredity and variation is rapidly being shaped into an orderly and consistent whole. A new world of intricate order previously undreamt of is disclosed. We are thus endowed with an instrument of peculiar range and precision, and we reach to certainty in problems of physiology which we might have supposed destined to continue for ages inscrutable.

William Bateson, *Mendel's Principles of Heredity* (1909)

Since the beginning of human civilization, people have observed the similarities between parents and offspring and have domesticated and selectively bred crop plants, pets, and livestock. Ten centuries of casual observations about heredity, however, failed to produce more than imaginative explanations for how it operates. Not until the mid-1800s did an extraordinary monk named Gregor Mendel discover the basic mechanism of heredity.

Heredity is a key feature of life as we know it, and its mechanisms ensure great stability to lineages of organisms, allowing them to remain adapted to their environments if the environments remain stable. At the same time, however, these mechanisms allow some genetic change and the potential to survive long-term environmental fluctuations. Mendel's experiments and the principles he discovered help explain not only this simultaneous genetic stability and flexibility, but also the dance of the chromosomes that we saw in our discussions of mitosis and meiosis in Chapter 9.

The fruit fly *Drosophila:* The solution to a thousand genetic puzzles.

In this chapter, we explore:

- The early theories of inheritance—incorrect, but based on common sense
- How Mendel's personal history led to the birth of genetics
- How Mendel's simple experiments revealed dominant and recessive traits, genetic alleles, and the laws of segregation and independent assortment
- How the work of Sutton, Boveri, Morgan, and others confirmed and extended Mendel's findings

EARLY THEORIES OF INHERITANCE

Around 400 B.C., Hippocrates made what may be the earliest attempt to explain the mechanisms of inheritance: the theory of **pangenesis**. According to this idea, each part of the body produces a characteristic "semen," or "seed," that somehow travels to the reproductive organs. On copulation, these seeds are released, and the combined fluids from the parents directly form the respective body parts of the offspring.

Despite the objections of no less a thinker than Aristotle, the theory of pangenesis persisted in one form or another for more than 2,000 years, until the late nineteenth century. At that time, a German biologist named August Weismann devised a clever experiment: He selected a few newborn mice three or four times a year for 7 years in a row, numbed their tails with ice, then lopped them off with a sharp knife.

During his long experiment, he removed the tails from the mice of 22 consecutive generations and found that invariably, tail-less mice sired normal offspring; the young never inherit the tail-less condition.

Weismann's results directly contradicted pangenesis and similar theories: Clearly, tails do not have to be present to pass out tail "semen." Weismann proposed an alternative explanation, the **germ plasm theory,** which holds that only hereditary information in the "germ plasm" of the gametes transmits traits to the progeny; other adult body cells (somatic cells) do not make a contribution. As Figure 10-1 illustrates, the fertilized egg gives rise to new lines of germ cells and somatic cells in the new organism, but since only the germ cells present in female ovaries or male testes divide and produce eggs or sperm, only the germ cells make a contribution to the offspring. In Weismann's theory, the germ plasm of the gametes is perpetuated generation after generation, whereas the somatic cells' sole function is to protect and nourish the germ cell lineage.

Although Weismann's theory emphasized the special nature of germ cells, eggs, and sperm, it shared an erroneous element with pangenesis and other antiquated theories: All agreed that heritable traits of both parents blend, or fuse, thereby losing the distinct characteristics of each. If the blending theory were correct, a purebred black dog and a purebred white dog, for instance, would always produce gray puppies, and over several generations, all the descendants would be a muddy gray and would be average in height, weight, and intelligence. This, of course, does not happen, but prior to the twentieth century, no one could explain heredity in a more satisfactory way. The belief in the "fluid" nature of heredity persisted up to and beyond the time Gregor Mendel made his momentous discoveries.

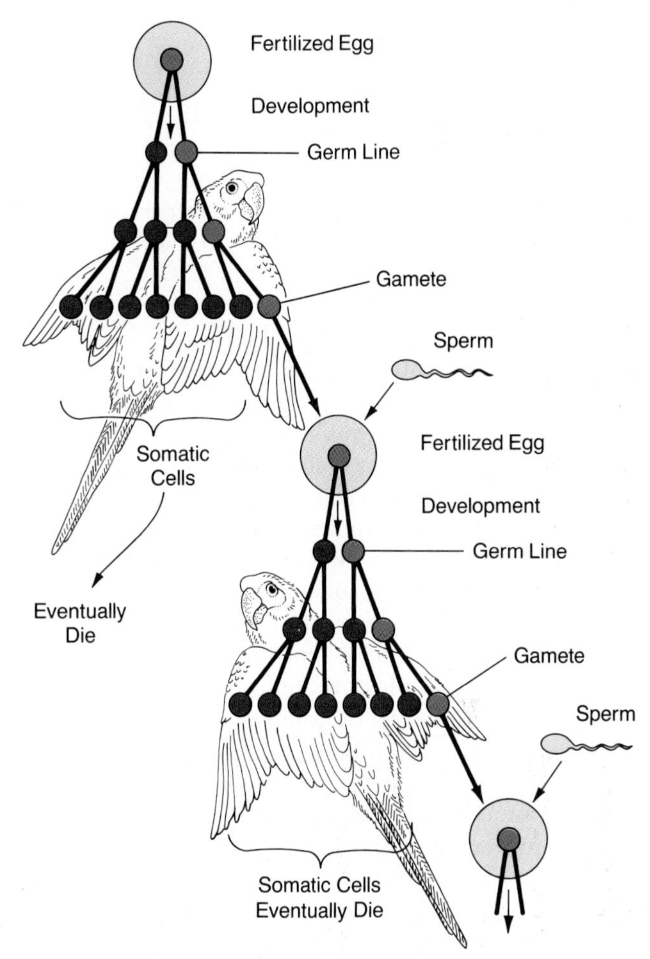

Figure 10-1 AUGUST WEISMANN'S GERM PLASM THEORY.
A fertilized egg gives rise to a new individual (here, a scarlet macaw) composed of two types of cells, germ cells and somatic cells. Germ cells (here, shown in blue) develop in the gonads and divide to produce the gametes, eggs or sperm. Somatic cells constitute all the other cells in the body; these make up its various parts and expire as a group when the organism dies. Only the germ cell lineage contributes directly to the next generation.

GREGOR MENDEL AND THE BIRTH OF GENETICS

Gregor Johann Mendel (Figure 10-2) was born in 1822 in a small Austrian village that is now part of Czechoslovakia. Although a bright student, Mendel came from a poor family and was unable to afford a university education. Instead, in 1843, he joined the Augustinian monastery at Brünn (now Brno, Czechoslovakia).

The Brünn monastery had long been a center of enlightenment and scientific thought, owing mainly to the efforts of Abbot Cyrill Napp. A proficient plant breeder as well as leader of Church affairs, Napp was intrigued by the question, What is inherited and how?

In 1851, Napp sent Mendel to study physics, mathematics (including the new field of statistics), chemistry, botany, and plant physiology at the University of Vienna. There Mendel encountered theories about the particulate nature of matter—the idea that all substances are made up of molecules and atoms. With Abbot Napp's encouragement, Mendel returned to the monastery and began a series of carefully planned experiments that eventually demonstrated the particulate nature of heredity. In 1866, Mendel presented the results of his experiments on the nature of inheritance to the Brünn Society for Natural History. However, his ideas were neither accepted nor understood in his lifetime. He died in 1884, convinced that it would "not be long before the whole world acknowledges" his discovery.

Figure 10-2 FATHER OF GENETICS: GREGOR MENDEL (1822–1884).

MENDEL'S CLASSIC EXPERIMENTS

Mendel carried out his experiments with the garden pea, a deliberate and, in retrospect, very wise choice. He obtained 34 distinct strains of peas from farmers and raised generations to select only *true-breeding strains*—those in which, for a given trait, each offspring is identical to the parent.

It was easy for Mendel to select and maintain true-breeding strains because the pea is self-fertilizing. That is, each single pea flower has both male and female reproductive organs enclosed by the delicate petals. Pollen grains, the sites of sperm nuclei, are formed in *anthers* (Figure 10-3). The anthers are near the *stigma*, the female structure that receives the pollen. Because the pea flower does not open fully, pollen usually reaches the stigma within the same flower and does not pass from one flower to another. Although self-fertilization is the normal mode, an experimenter can induce artificial cross-fertilization by carefully opening a flower bud and snipping off the anthers before they mature. The experimenter can then transfer pollen from the flower of a second pea plant to the stigma of the first to cause controlled fertilization with pollen from that second plant.

Mendel focused on the inheritance of several distinct traits—flower color, seed color, pod shape, and plant height, for example—rather than on the entire appearance of each plant in relation to that of the parents. Analyzing all of a given plant's traits at once would have been a statistical nightmare. So Mendel chose seven *characters*, or pairs of traits (Figure 10-4), and studied only one or two of those seven from any given plant and its progeny. Mendel was not the first to study genetics through plant-breeding experiments, but as we shall see, his education in mathematics and statistics largely made the difference between failure and success.

Dominant and Recessive Traits

Mendel's experiments had three unique and important features.

1. He studied characters of pea plants that offered just two clear-cut possibilities, such as tall or short stems, red or white flowers, or green or yellow seeds.
2. He traced and recorded the type and number of all the progeny produced from each pair of parent pea plants that he crossbred.
3. Finally, he followed results of each cross for two generations, not just one—a simple but profound decision that led to the downfall of the blending theory of inheritance.

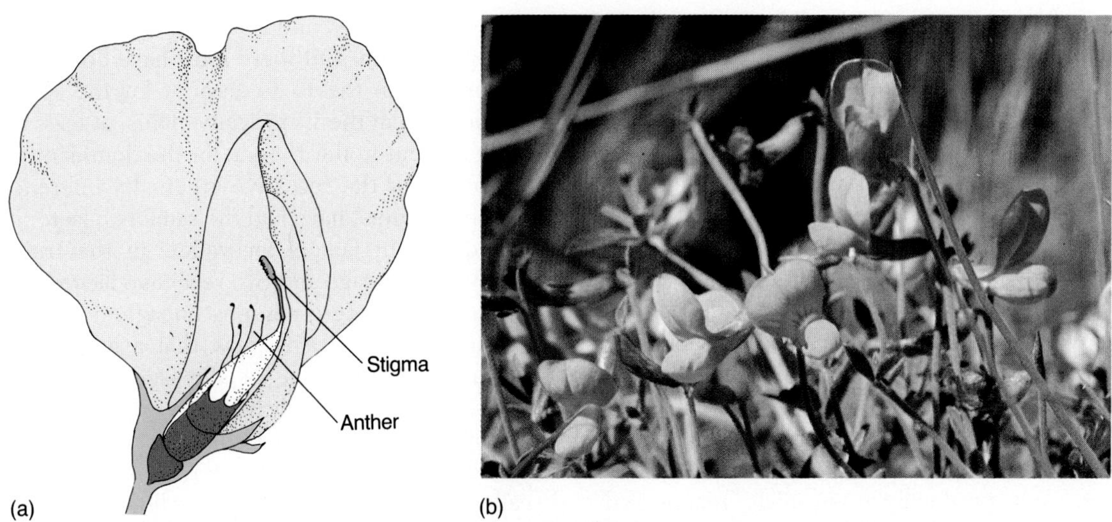

Figure 10-3 THE PEA FLOWER—MENDEL'S PRINCIPAL SUBJECT.
(a) Both the male and female parts of the pea flower are enclosed within a hoodlike petal, so the flowers are normally self-fertilizing. (Here, some outer petals have been removed to show the inner "hood.") Pollen from the male anthers can reach only the female stigma of the same flower. Mendel opened the bud of one flower before the pollen matured, removed the anthers, and fertilized the stigma with pollen collected from another flower; thus, he could control the cross. (b) Typical pea flowers.

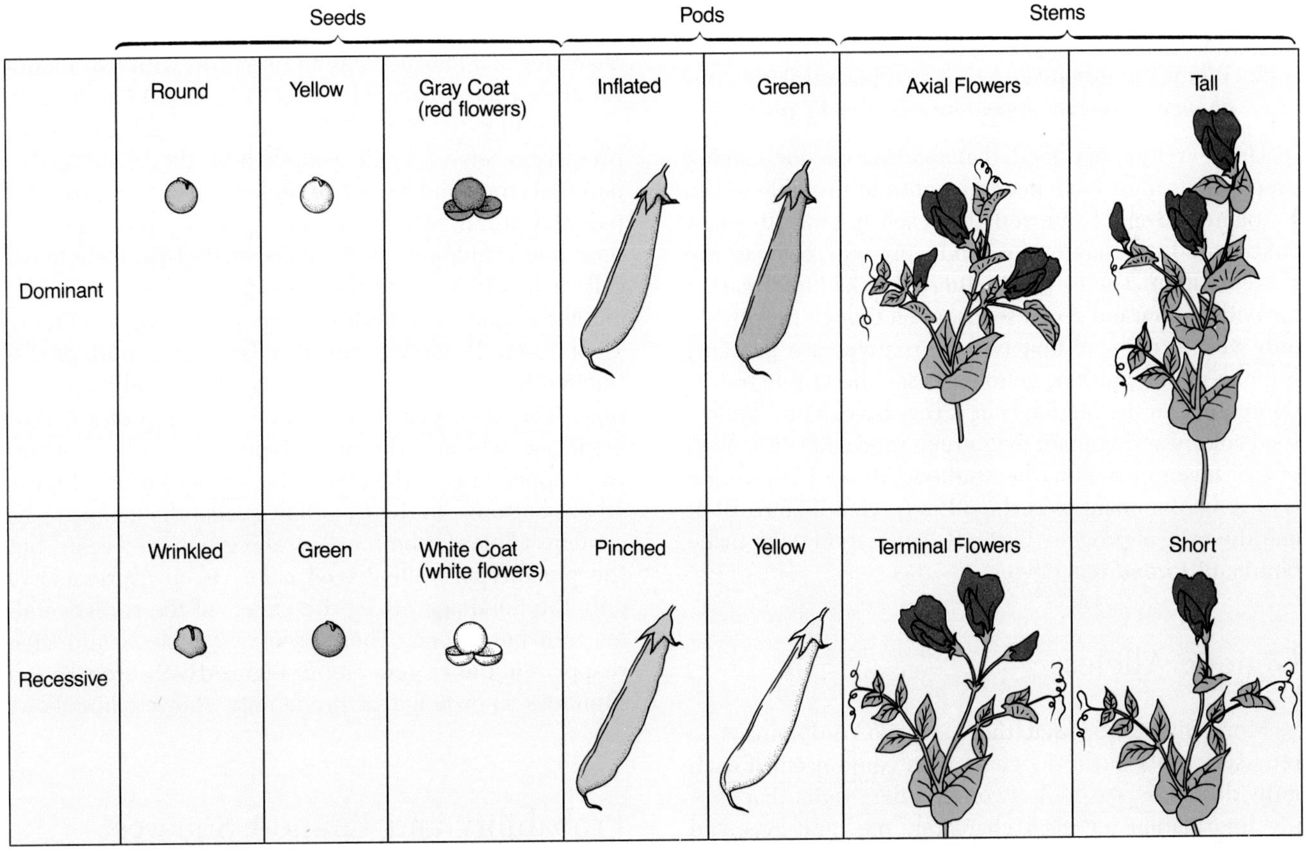

	Seeds			Pods		Stems	
Dominant	Round	Yellow	Gray Coat (red flowers)	Inflated	Green	Axial Flowers	Tall
Recessive	Wrinkled	Green	White Coat (white flowers)	Pinched	Yellow	Terminal Flowers	Short

Figure 10-4 THE PEA PLANT'S FAMOUS TRAITS.
Mendel studied seven clearly differentiated traits, or characters, of the pea plant in pure-breeding individuals of each strain. The seven traits and the dominant and recessive forms of each are shown here.

In a set of experiments on the inheritance of seed color, Mendel crossed pea plants from a yellow-seed strain with plants from a green-seed strain (Figure 10-5a, step 1). These plants are called the **P_1, or parental, generation,** and the progeny grown from the seeds of such a cross are designated the **F_1, or first filial, generation.** Mendel observed that in the cross between plants from green seeds and plants from yellow seeds, the progeny seeds were *only* yellow (step 2)—a fact that contradicted the blending theory of inheritance.

To see what would happen in future generations, Mendel allowed the F_1 generation yellow seeds to sprout, grow into mature plants, self-fertilize, and produce the next generation, which we now designate as the **F_2, or second filial, generation** (step 3). In Mendel's experiment, plants producing green seeds "reappeared" along with a great number of those producing yellow seeds in the second generation. Mendel's mathematical training now came into play. Among a total of 8,023 F_2 plant seeds, Mendel found that 6,022 were yellow and 2,001 were green. Since three-fourths of the plants produced yellow seeds and one-fourth produced green, this represented an almost exact 3:1 ratio.

Mendel noted two important points from this experiment:

1. Although the green-seed trait had disappeared in the F_1 generation, it *reappeared* in the F_2.
2. When the green-seed trait reappeared it was *unchanged* from its appearance in the P_1 plant.

He therefore reasoned that information for making green seeds must have been present but invisible in the F_1 plants. Mendel inferred that each original P_1 plant contributed information for producing seed color to the F_1 generation. The F_1 plants, therefore, had information for both yellow and green seeds, even though they made only yellow ones. When two alternatives are present, but one masks another, geneticists say the visible trait is **dominant** and the hidden trait is **recessive.** Thus, yellow seed color was dominant over green seed color. For each of the seven characters he studied, Mendel found one trait to be dominant over the other (review Figure 10-4) and the ratio of progeny in the F_2 generation to be three dominant to one recessive.

Genetic Alleles

Mendel deduced that the 3:1 ratio of dominant to recessive traits in the F_2 generation could occur if each individual possesses only two hereditary units that supply information for each character, one unit received from each parent. How did he reach this conclusion?

Mendel knew that a factor causing the dominant yellow trait was present in the F_1 generation, since the seeds were yellow, and yellow seeds also appeared in the next generation. But there must have been a second factor present in the F_1 to account for the appearance of green seeds in the F_2 progeny. Thus, at least two factors were present in the F_1: one for the dominant yellow and the other for the recessive green. By inference, the P_1 generation must have had the same two factors as well—two dominant factors for yellow in the true-breeding yellow-pea lineage and two recessive factors for green in the true-breeding green-pea lineage.

Modern geneticists know that a **gene,** a hereditary factor that regulates a specific aspect of an individual's looks, behavior, or biochemistry, determines each pair of traits such as green or yellow seed color or wrinkled or round seed shape. Each alternative form of a gene is called an **allele,** such as the yellow allele of the seed color gene or the wrinkled allele of the seed shape gene. One allele comes from each parent and is responsible for either a dominant or a recessive trait. We can assign the letter *Y* to the allele for yellow seed color and *y* to the allele for green seed color. (Geneticists often use a capital letter to represent a dominant allele and the lowercase form of the same letter to represent the recessive allele.) The true-breeding yellow parental strain has the alleles *YY*, and the true-breeding green parental strain, the alleles *yy*. Both true-breeding parents are said to be **homozygous** ("similar pair") for alleles of the seed color character; a *homozygote* is an organism with two identical alleles for a particular trait (Figure 10-5b, step 1).

According to Mendel's hypothesis, each of the F_1 progeny receive a single *Y* allele from its yellow-seeded parental strain and a single *y* allele from its green-seeded parental strain. All members of the F_1 progeny then have one *Y* and one *y* and are designated *Yy*. Geneticists call such pairs of genes **heterozygous** ("different pairs") and the organisms inheriting them *heterozygotes* (Figure 10-5b, step 2). Geneticists also draw an important distinction between an organism's *genotype* and its *phenotype*. The **genotype** is the genetic makeup of a cell or organism, whereas the **phenotype** is the cell or organism's appearance—the expression of those genes (Figure 10-5b, bottom). In the F_1 yellow seeds, the genotype *Yy* contains alleles for both yellow and green seed color, but the phenotype, yellow seed color, is an expression of only the dominant allele; the effects of the recessive allele remain hidden. These recessive alleles could then reappear in the F_2 generation (Figure 10-5b, step 3), and Mendel's knowledge of probability showed him how.

Probability and Punnett Squares

Mendel's superior knowledge of statistics and probability helped him predict the frequency of yellow and

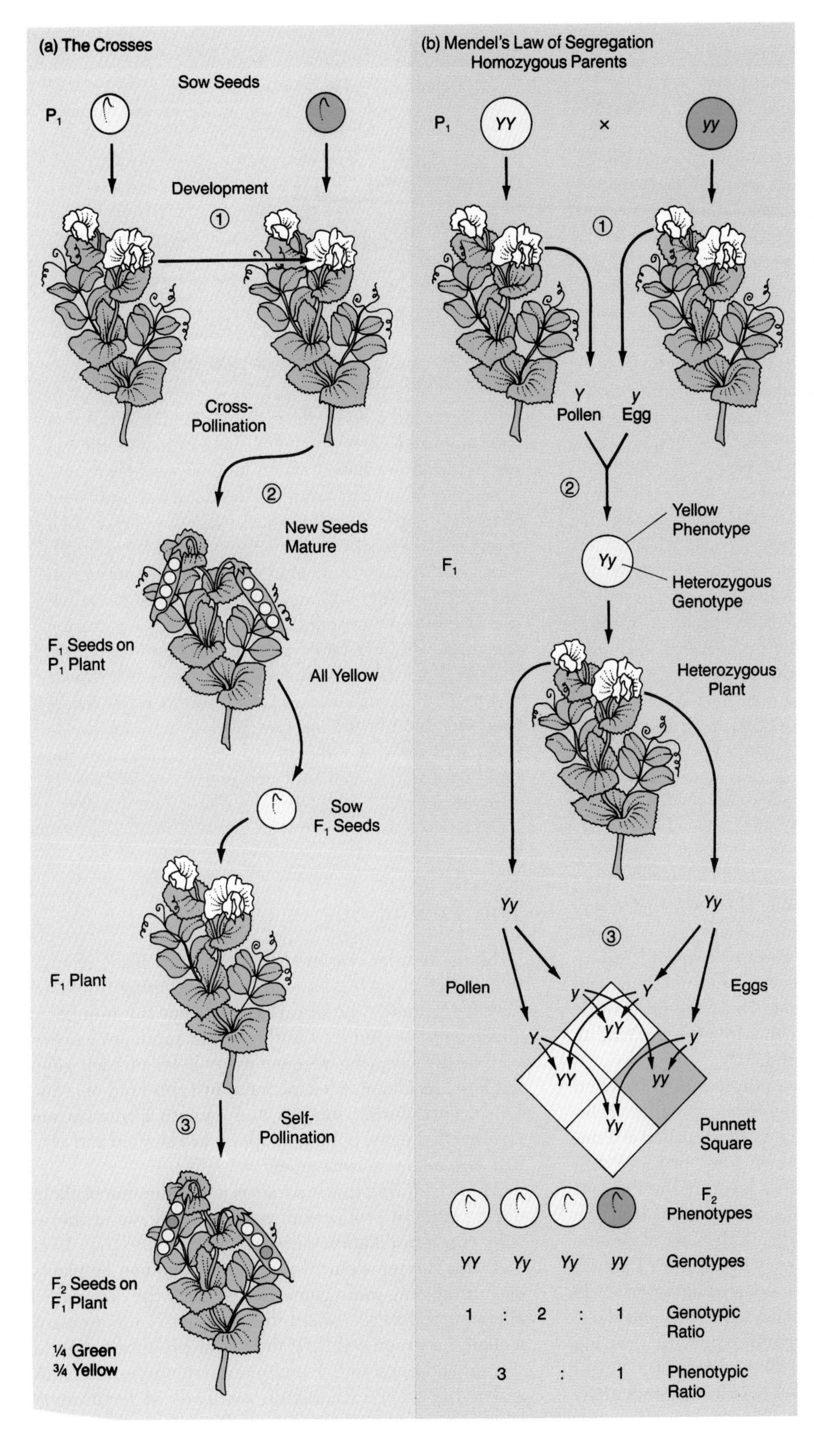

Figure 10-5 HOW SEED COLOR IS INHERITED IN PEA PLANTS.
(a) As the text explains, Mendel crossed plants in the P_1, or parental, generation grown from yellow seeds (peas) with P_1 plants grown from green seeds (step 1). All the progeny from this cross in the F_1 (first filial) generation bore yellow seeds (step 2). Mendel then planted the F_1 yellow seeds and allowed the resulting plants to self-pollinate and produce the F_2 (second filial) generation (step 3). He counted about three yellow seeds for every green seed in the F_2 generation pea pods. Since the green trait was absent in the F_1 generation but "returned" in the F_2, Mendel reasoned that it must have been present but hidden in the F_1 generation. (b) Mendel's law of segregation explains why we observe the results in (a). The P_1 plants were homozygous dominant (*YY*) and homozygous recessive (*yy*) for the seed color gene, and the F_1 progeny received one allele from each (*Y* from one, *y* from the other; step 1). Thus, the F_1 seeds were heterozygous (*Yy*), and all would appear yellow (step 2). Such heterozygotes can pass along two kinds of alleles in their gametes; half of the F_1 eggs will carry a *Y* and half will carry a *y*, as will the sperm (step 3). If one diagrams the possible combinations of these gametes in a Punnett square, both the phenotypes and genotypes of the F_2 seeds can be determined, along with their expected ratios.

THOROUGHBREDS AND THOROUGH BREEDING

Horse racing is big business: Owners in 40 countries race more than half a million thoroughbreds, and fans wager billions annually. Ironically, while human Olympic sprinters continue to improve their speeds each year, horses are no faster now than they were 50 years ago. Have horses reached their inherent speed limit? Or is there another explanation for their lack of improvement, despite advances in nutrition and veterinary medicine?

Irish geneticists B. Gaffney and E. P. Cunningham recently developed a complicated method for analyzing change in 31,263 thoroughbreds over the past 25 years. Like other researchers, they found strong evidence of inbreeding, or unusually close genetic similarity based on matings between related horses. They expected this, because 80 percent of all thoroughbreds are descendants of just 31 horses brought to England from the Middle East and North Africa at the turn of the eighteenth century. Just as Mendel artificially selected and bred peas of a certain color, height, and shape, race horse breeders have selected and interbred these descendants to preserve competitive disposition, strong slender legs, and, above all, swift running speed. But did this practice of "breeding the best to the best" backfire and produce, instead, a "regression to the norm"—an averaging of traits and a loss from the population of alleles for both extreme slowness *and* extreme swiftness?

Figure A HAVE THOROUGHBREDS REACHED AN UPPER SPEED LIMIT?

Gaffney and Cunningham think not and report that thoroughbred populations still have considerable genetic variation. Based on their analyses, thoroughbreds should now be running about 2.5 percent faster than they did a quarter century ago and have definitely not approached their natural upper speed limit. Many observers think more practical factors are at fault: lax training methods (underexercising the valuable race horses); unsound breeding practices (pairing horses by their "papers" and family histories rather than their actual sizes, shapes, and speeds); and poor race track running surfaces. If the recent study is correct, we could be seeing faster horse races in the future. If it's wrong, the record-setting horses of the past, like Man O'War and Secretariat, could be legends forever.

green seeds in the F_2 generation. He knew that if you toss a coin in the air, it has an equal chance, or probability, of landing heads or tails. If you toss two coins, there are four possibilities: two heads; two tails; one head and one tail; or one tail and one head. Thus, the probability of two heads (or two tails) is 1 in 4, while the probability of one head and one tail is 2 in 4.

Figure 10-5b, step 3, shows a simple means (devised after Mendel's time) of displaying the probability of different allele combinations. This method, called a **Punnett square,** can be used for three, four, or more coin tosses or genetic traits. The various boxes in the Punnett square indicate the probabilities of seeing each of the possible head-tail combinations or, in the case of genetic factors, allele combinations. Thus, you can draw Punnett squares with the alleles from one parent along one side (representing the classes of possible gametes) and those from the other parent along the other side. By crossing each pair of alleles and filling in all the boxes of the Punnett square, you can see all the possible combinations and calculate the ratio of results.

The Law of Segregation

Since Mendel reasoned that each pea plant receives two alleles of each gene—one allele coming from each parent's gamete—he also reasoned that the number of alleles must be reduced as the parent produces gametes so that the offspring receives two alleles of each gene, not four. As Chapter 9 explained, just this kind of reduction occurs during meiosis and gamete formation and ensures that eggs, pollen, or sperm are haploid and carry just one allele of each gene.

By considering this separation and reduction of alleles during gamete production, one can understand the results of self-pollination within a heterozygote (*Yy*). In effect, this is equivalent to a cross between two *Yy* plants, each producing some gametes that carry the dominant *Y* allele for yellow seeds and some that carry the recessive *y* allele for green seeds. If the plants produce pollen (or sperm) and eggs with *Y* and *y* occurring in equal ratios, and if the gametes combine randomly at fertilization, one-quarter of the F_2 progeny will receive a *Y* from both parents (*YY*), two-quarters will receive a *Y* from one parent and a *y* from the other (*Yy*), and one-quarter will receive a *y* from both parents (*yy*), as Figure 10-5b, step 3, shows. The genotypic ratio will be 1 *YY* to 2 *Yy* to 1 *yy*, but because the phenotype of *YY* and *Yy* is the same, the phenotypic ratio of yellow-seed progeny to green-seed progeny will be 3:1.

The results of Mendel's experiments on dominant and recessive inheritance led to what is now known as Mendel's first law: the **law of segregation.** According to this law, individuals carry two discrete hereditary units (alleles) affecting any given character. During meiosis, these two alleles segregate, or become separated, from each other. One allele for every character is then incorporated into each maturing gamete and is transmitted during random fertilization in an unaltered state to the next diploid generation. The new diploid individual has in every cell nucleus two alleles for each character, one from each parent.

Mendel's studies of all seven pea plant characters generated genotypes so close to the expected 3:1 ratio that some historians wonder whether he formulated the law of segregation before he carried out his experiments and consciously or unconsciously biased his own results. Regardless, Mendel's studies were extremely meticulous for his day, and he indeed proved experimentally the first principles of genetics.

Test Crosses

Mendel devised another set of tests to answer a knotty experimental problem, and in the process, found further support for his law of segregation. With every group of yellow seeds in the F_2 generation, some might be homozygous (*YY*) and some might be heterozygous (*Yy*), but which plants were which genotype? Mendel's solution was the **test cross,** mating the plant of known phenotype but unknown genotype to a plant that is homozygous recessive for the trait in question. He chose a homozygous recessive so that the contribution of the unknown parent, be it dominant or recessive, would be obvious in the phenotype of the offspring.

For example, if he grew a yellow seed (*Yy* or *YY*) from the F_2 generation into a plant and crossed it with a plant grown from a green seed (*yy*), the progeny would show one of two ratios: all yellow seeds or half yellow seeds and half green seeds (Figure 10-6). An F_2 yellow seed with the homozygous genotype *YY*, would donate only the *Y* allele, so when one of these combined with a *y* allele from the green-seeded plant, only yellow-seeded test cross progeny would form with the heterozygous genotype *Yy* (Figure 10-6a). An F_2 yellow seed with the heterozygous genotype *Yy*, on the other hand, would donate half *Y* alleles and half *y* alleles through its gametes. Thus, half the test cross progeny would be yellow (*Yy*), and the other half would be green (*yy*) (Figure 10-6b). Through the test cross, Mendel and later geneticists could certify the genotype of an F_2 individual.

Dihybrid Crosses and the Law of Independent Assortment

In a typical organism, each cell contains pairs of alleles for hundreds or thousands of traits, so let us go one

(a) Homozygous Yellow

If all the offspring have yellow seeds, the yellow parent must be *YY*

(b) Heterozygous Yellow

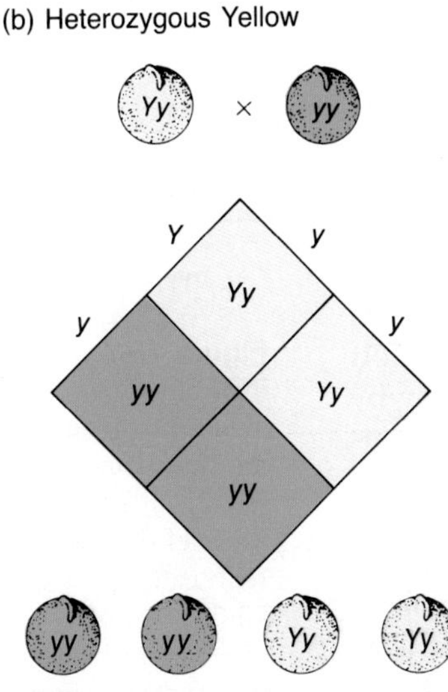

If half the offspring have yellow seeds and half have green seeds, the yellow parent must be *Yy*

Figure 10-6 TEST CROSSES: TRACING A PARENT'S UNKNOWN GENOTYPE.
Mendel invented the test cross to determine unknown genotypes. He crossbred a yellow-seeded plant of unknown genotype to a homozygous recessive for the trait in question, here *yy* (green). The phenotypic ratios of the test-cross progeny reveal whether the genotype of the yellow parent is *YY* or *Yy*.

step further and consider how two traits are inherited relative to each other.

Mendel designed experiments to study how two characters—shape and color—might interact during inheritance. After determining that round seeds are dominant to wrinkled seeds, he crossed a true-breeding strain with round yellow seeds (*RRYY* genotype) and a true-breeding strain with wrinkled green seeds (*rryy* genotype). In keeping with the dominance of the yellow-seed allele (*Y*) and the round-seed allele (*R*), the entire F_1 generation produced round yellow seeds.

Next Mendel performed a **dihybrid cross,** a cross between parents that are heterozygous (hybrid) for two characters, in this case, seed color and shape. Invariably, four phenotypes appeared in the F_2 generation, including two phenotypic combinations he had not observed in either parent (Figure 10-7). For this to happen, the alleles of the two genes for color and shape must have segregated *independently* of each other during gamete formation in the F_1 generation. That is, four types of eggs and four types of pollen were formed, not just two. Instead of gametes carrying only *RY* and *ry* as in the original parents, the alleles for seed color and seed shape must have reassorted independently so that different gametes carried *RY*, *ry*, *rY*, or *Ry*. Random recombination during fertilization would produce an F_2 generation with $4 \times 4 = 16$ combinations of alleles among the offspring. These 16 combinations would yield the 9 different genotypes and 4 different phenotypes shown in Figure 10-7 in a 9:3:3:1 ratio.

By knowing the rules, we can predict how many of the 16 possible allele combinations in a dihybrid cross will produce a given phenotype—say, round green seeds. Since the round allele (*R*) is dominant, three-fourths of the F_2 progeny will be round, and since green (*y*) is recessive, only one-fourth will be green. Because the two characters act independently, we can simply multiply their individual probabilities to determine their likelihood of occurring together: $3/4 \times 1/4 = 3/16$. If we want to know how many of the progeny will have wrinkled green seeds, we can multiply $1/4 \times 1/4$ (since both traits are recessive) and predict that only 1/16 will show the doubly recessive phenotype. This formula becomes very handy when we consider three, four, or more traits at the same time. Instead of drawing giant Punnett squares, we can predict phenotypes mathematically by simply multiplying the individual probabilities, no matter how many traits and alleles are involved.

Let's apply the method to three pea plant characters: seed color (*Y* or *y*), seed shape (*R* or *r*), and plant height (*T* for tall, *t* for short). The probability that a plant will be tall and have round yellow peas and thus have at least one dominant allele for each of three characters is $3/4 \times 3/4 \times 3/4 = 27/64$. The probability that a plant will be homozygous recessive for each of the three characteris-

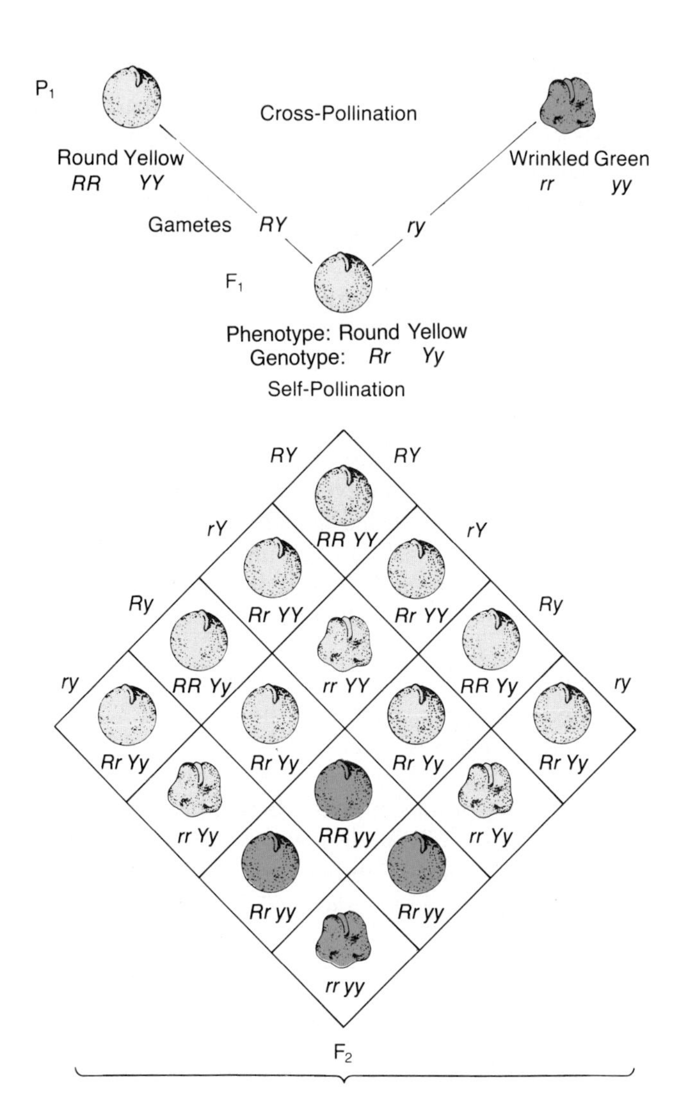

Genotypes	Phenotype	Phenotypic Ratio
R–Y–	Round Yellow	9/16
R–y–	Round Green	3/16
rrY–	Wrinkled Yellow	3/16
rryy	Wrinkled Green	1/16

Figure 10-7 DIHYBRID CROSSES: A CASE OF INDEPENDENT ASSORTMENT.
When pea plants with purebred round yellow seeds (*RRYY*) are mated to pea plants with purebred wrinkled green seeds (*rryy*), all F_1 progeny have round yellow seeds (*RrYy*). Mendel discovered that the genes for the two characters segregate, or "assort independently." They produce four kinds of gametes (*RY*, *Ry*, *rY*, and *ry*), and the F_2 generation displays four phenotypes in a 9:3:3:1 ratio. The two parental combinations of traits are represented, but two new recombinant types (round green seeds and wrinkled yellow seeds) also appear among the F_2 progeny.

tics (*yyrrtt*)—be short and have wrinkled green seeds—is 1/4 × 1/4 × 1/4 = 1/64. The probability of its being tall (dominant) with round (dominant) green (recessive) seeds is 3/4 × 3/4 × 1/4 = 9/64. And so on.

Mendel's dihybrid experiments led him to formulate what is now called Mendel's second law, the **law of independent assortment,** which states that alleles of different genes are distributed randomly to the gametes, and fertilization occurs at random. Chapter 11 discusses this law and its exceptions in more detail.

Incomplete Dominance

The characters and alleles Mendel used in establishing his two laws were lucky choices because they are so straightforward. For each of the seven characters that Mendel studied, he found that the dominant allele completely masked the recessive allele in the phenotype of the organism. But nature is not always so straightforward: The dominant-recessive relationship is not the only possible interaction between alleles of a gene. When a worker crosses a true-breeding red-flower strain of snapdragon (*RR*) with a true-breeding white-flower strain (*rr*), the flowers produced in the F_1 plants (*Rr*) are *pink*, not red (Figure 10-8a). This would seem to be a perfect case of blending inheritance and a contradiction of Mendel's laws. After further crosses, however, we can see that the law of segregation does hold true for snapdragon color. The expected segregation of flower color occurs when the pink F_1 snapdragons self-pollinate and produce F_2 progeny (see Figure 10-8a). The offspring have the usual 1:2:1 ratio, but each genotype has a distinctive and recognizable phenotype, since the heterozygote (*Rr*) is pink.

The "mixing" responsible for pink flowers occurs not at the level of genotype, since the alleles for red and white flower color remain unchanged, but at the level of the phenotype. In this plant, a single *R* allele may make half as much red pigment as two *R* alleles. This enables the white allele *r* (the absence of pigmentation) to "show through," and allows the human eye to perceive the flowers as pink. This phenomenon is called **incomplete dominance,** since both alleles exert an effect and jointly produce an intermediate phenotype. The palomino horse—the offspring of one white parent and one light chestnut or sorrel parent—is another example of this phenomenon.

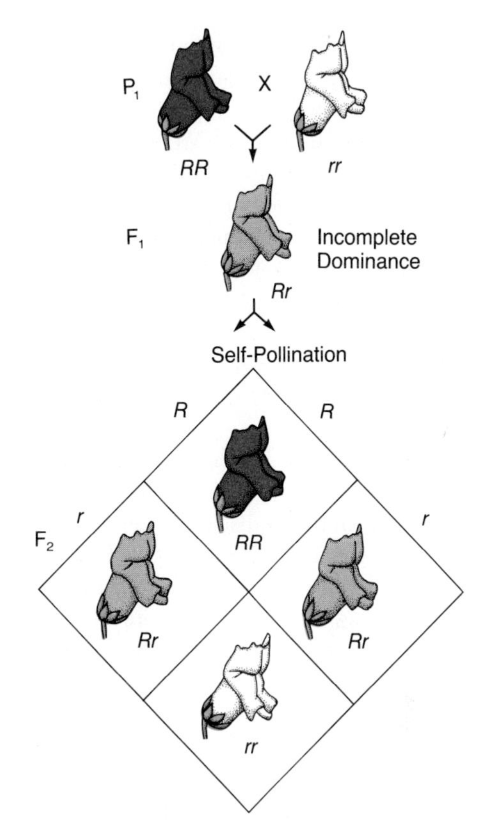

(b)

Figure 10-8 FLOWER COLOR IN SNAPDRAGONS: A CASE OF INCOMPLETE DOMINANCE.
(a) When red (*RR*) and white (*rr*) snapdragons are crossed, the F_1 generation has all pink flowers (*Rr*) because the red gene is incompletely dominant over the white, causing a plant with *Rr* alleles to be pink rather than red. In the F_2 generation, one-half of the plants have pink flowers, one-fourth have red flowers, and one-fourth have white flowers. (b) This gorgeous display depends on a number of genes and alleles for snapdragon color.

Mendel's Ideas in Limbo: A Theory Before Its Time

Despite the excellence of Mendel's research, as well as the clear presentation of his experimental results and his two fundamental laws of heredity before the Brünn Society for Natural History in 1866, Mendel's colleagues did not understand his paper. In particular, nineteenth-century biologists failed to grasp the mathematical presentation of results and the physical interpretations that Mendel provided.

Mendel's work was finally championed 35 years later when Hugo de Vries in Holland and Carl Correns in Germany achieved results identical to Mendel's in their own research. They discovered Mendel's paper in 1900 while preparing their data for publication, and gave him the well-deserved credit for priority. With the rediscovery of Mendel's laws, the science of genetics was under way.

CHROMOSOMES AND MENDELIAN GENETICS

Just 2 years after the rediscovery of Mendel's work, Walter Sutton in the United States and Theodor Boveri in Germany independently suggested that Mendel's hereditary units might be located on chromosomes. Sutton studied cells in the testes of grasshoppers and noted that as the chromosomes segregated during meiosis, the hereditary units were apparently reshuffled and independently assorted in the gametes, just as Mendel had predicted with his theoretical factors, the alleles of later geneticists. An abundance of careful observations led Sutton and Boveri to espouse the *chromosome theory of heredity*, the idea that the chromosomes carry Mendel's hereditary factors, that these segregate, and that parts can reshuffle during meiosis.

Figure 10-9 takes our understanding of chromosome movements during meiosis one step further and shows how hereditary factors are distributed during gamete formation. The male parent from the P_1 generation has donated *Y* and *R* alleles on the dark blue chromosomes, while the female parent has donated *y* and *r* alleles on the pale blue chromosomes (Figure 10-9a). When the F_1 individuals grow, mature sexually, and begin to produce gametes via meiosis (Figure 10-9b), the two pairs of homologous chromosomes may align in two alternative ways on the metaphase plate. One way (Figure 10-9c,

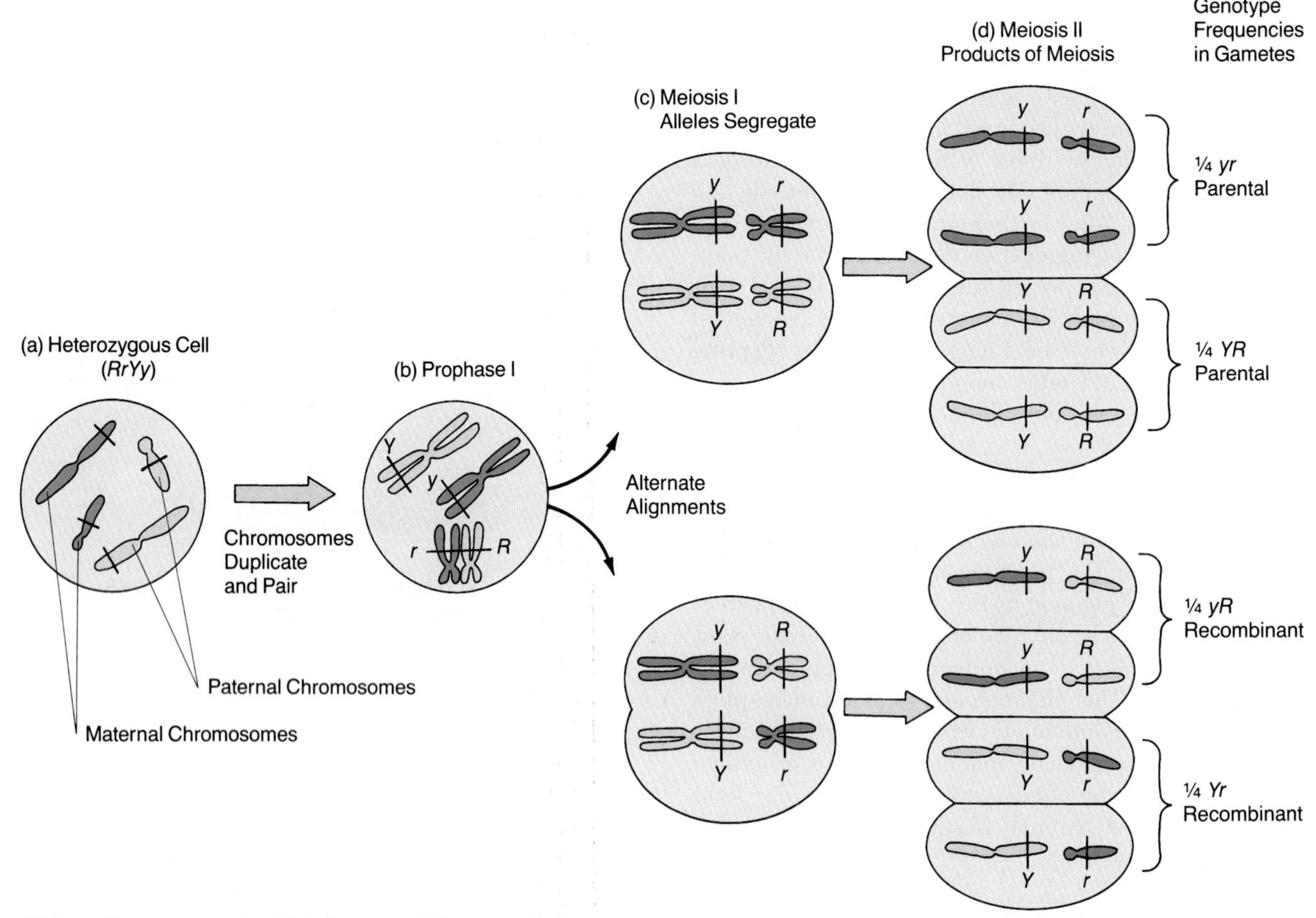

Figure 10-9 INDEPENDENT ASSORTMENT OF GENETIC TRAITS: A DIRECT CONSEQUENCE OF MEIOTIC ACTIVITY. The process of meiosis explains how alleles segregate during gamete formation, allowing independent assortment of genetic traits. Follow the movements of chromosomes bearing the alleles for seed shape and color in the pea plant. Paternal chromosomes are shown in pale blue, with alleles *Y* = yellow, and *R* = round noted, while maternal chromosomes are shown in dark blue, with alleles *y* = green and *r* = wrinkled noted. Follow the assortment of traits from the P_1 generation to the F_1 generation and through the F_1 generation's production of gametes by meiosis.

top), the segregation of chromosomes into gametes preserves the pairs of parental alleles (Y with R, y with r), producing *parental-type* gametes (Figure 10-9d, top). The other way (Figure 10-9c, bottom) segregates the alleles differently, leading to new allele associations (Y with r, y with R) or *recombinant-type* gametes (Figure 10-9d, bottom).

The work of Sutton and Boveri showed a correlation between chromosome movements and segregation patterns of Mendel's factors, but did not prove that the hereditary units lie on chromosomes. That proof came rapidly, however, with the experiments of Thomas Hunt Morgan and his students at Columbia University in the early 1900s. They worked with *Drosophila melanogaster* (Figure 10-10a), a type of fruit fly that is easily maintained in the laboratory, reproduces quickly, and has only four pairs of chromosomes—three pairs of autosomes, or non–sex chromosomes, and one pair of sex chromosomes (XX in females and XY in males). Earlier researchers had found that the male's X and Y chromosomes segregate at the first meiotic division, producing two types of sperm: (1) those with an X chromosome and three autosomes, and (2) those with a Y chromosome and three autosomes. Because the female has two X chromosomes, all eggs carry one X chromosome and three autosomes (Figure 10-10b). Thus, if an egg (with its X) is fertilized by a sperm with an X chromosome, it develops into a female fly (XX); if it is fertilized by a sperm with a Y chromosome, it becomes a male fly (XY). Morgan's elucidation of the fruit fly's sex chromosome system made it clear that the presence of a pair of differently shaped chromosomes residing in the cells of most animals correlates with the regulation of sexual phenotype, sexual physiology, and sexual behavior—a spectrum of specific gender-related hereditary traits.

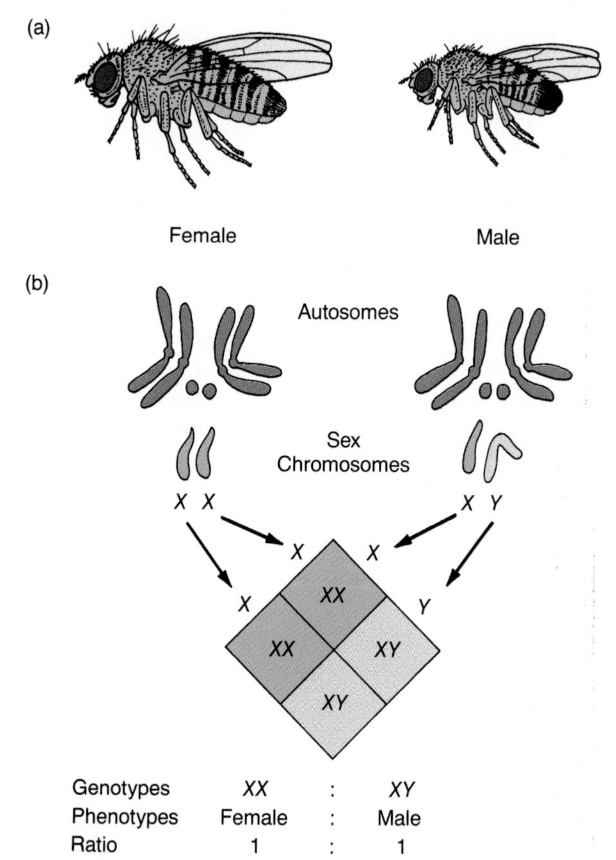

Figure 10-10 THE FRUIT FLY AND ITS FOUR CHROMOSOMES.
The fruit fly *Drosophila melanogaster* (a) is well suited for genetic studies because it has only four sets of chromosomes—three pairs of autosomes plus the sex chromosomes that determine the insect's gender (b). Females have two X chromosomes, while males have one X and one Y.

To see if chromosomes regulate other hereditary characters, Morgan's group studied eye color. Normal fruit flies have bright red eyes, but one day Morgan's group discovered a single white-eyed male in the colony. Morgan guessed that white eyes were a spontaneously occurring change (a *mutation*) of the gene for normal red eye color. (A *mutant* is an organism with a mutated gene, in this case, the gene for eye color. A mutated allele of a gene can be either dominant or recessive.) Morgan quickly discovered that this mutant trait did not act strictly according to Mendelian laws.

By crossing red-eyed females with the original white-eyed male, Morgan obtained an F_1 generation all showing red eyes; clearly, the white-eyed allele was recessive and the red-eyed allele dominant. However, when the red-eyed F_1 progeny were bred together, the F_2 generation did not turn out as expected. Rather, there were 3,470 red-eyed offspring and 782 white-eyed flies. Of these offspring, all the females had red eyes, while half the males had red eyes and half had white (Figure 10-11a).

After a series of crosses, recrosses, and careful analyses of the phenotypic patterns, Morgan figured out that this unusual inheritance pattern parallels the inheritance of the X chromosome. Morgan knew that males receive their single X chromosome from their mother, but females receive an X from each parent. Thus, an eye color allele carried on the X chromosome must be determining the male's eye color phenotype. If the X chromosome carries a dominant red allele (symbolized w^+), the male offspring will have red eyes; if it carries a white allele (w), the male will have white eyes (Figure 10-11b). This explanation would hold only if the Y chromosome carries *no* allele for eye color or *always* carries the white allele. Later studies showed the first assumption to be correct: The Y chromosome has no eye color allele to block the expression of the recessive w on the X it receives from its mother (X^wY). Since a female receives two X chromosomes, it shows the white-eye phenotype only if each X bears the recessive allele (X^wX^w).

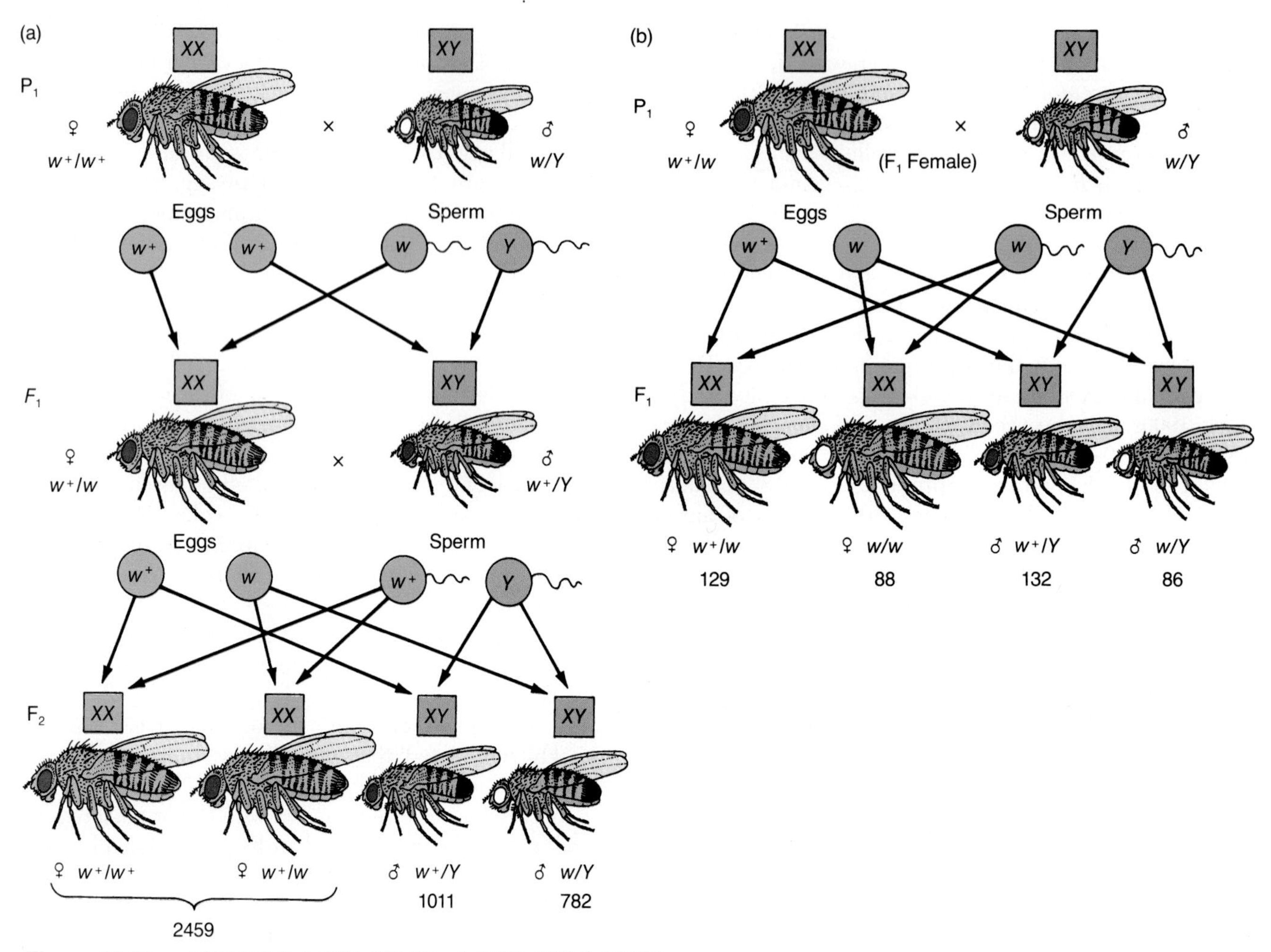

Figure 10-11 ***DROSOPHILA* EYE COLOR: A SEX-LINKED TRAIT.**
Eye color in fruit flies is not inherited according to simple Mendelian law. (a) All the F_1 progeny from a red-eyed female mated to a white-eyed male have red eyes. This proves that red is dominant to white. However, when F_1 males and females are crossed, the expected 3:1 phenotypic ratio of red-eyed to white-eyed flies is not produced. Instead, all females have red eyes, while half of the males have red eyes and half have white eyes. This result can be explained if the genetic factor for eye color lies on the *X* chromosome (inherited by a male offspring only from the mother) and if no corresponding allele occurs at all on the *Y* chromosome.

In the figure, w^+ represents the allele for red eyes, and w represents the allele for white eyes. Since the allele is carried on the *X* chromosome, a male will have only one allele. For females, with two alleles, w^+, or red eye, is dominant. (b) If a female inherits the white-eyed allele from both parents, she will have white eyes.

Morgan's experiments proved that the *X* chromosome carries a hereditary unit, the eye color gene. In so doing, they laid a solid foundation for the genetic research that followed over the next 75 years. Morgan's studies of eye color in fruit flies were also the first exploration of **sex-linked traits.** Geneticists have since identified thousands of such traits in various organisms, including several in our own species (as Chapter 14 explains).

Another peculiarity of the inheritance of white eyes in *Drosophila* led to yet another important discovery. In 1916, the geneticist Calvin Bridges observed that 1 in every 1,000 fruit flies displays an unpredicted phenotype: a white-eyed F_1 male resulting from the cross of a homozygous red-eyed female and a white-eyed male. As Figure 10-11a shows, that cross nearly always produces all red-eyed males. However, Bridges discovered that during meiosis in red-eyed females, the homologous *X* chromosome pair occasionally fails to segregate, or disjoin (Figure 10-12). He called this **nondisjunction** of the *X* chromosome. Some of the female's eggs therefore receive two *X* chromosomes, while others receive none at all. If sperm with an *X* chromosome fertilizes an egg

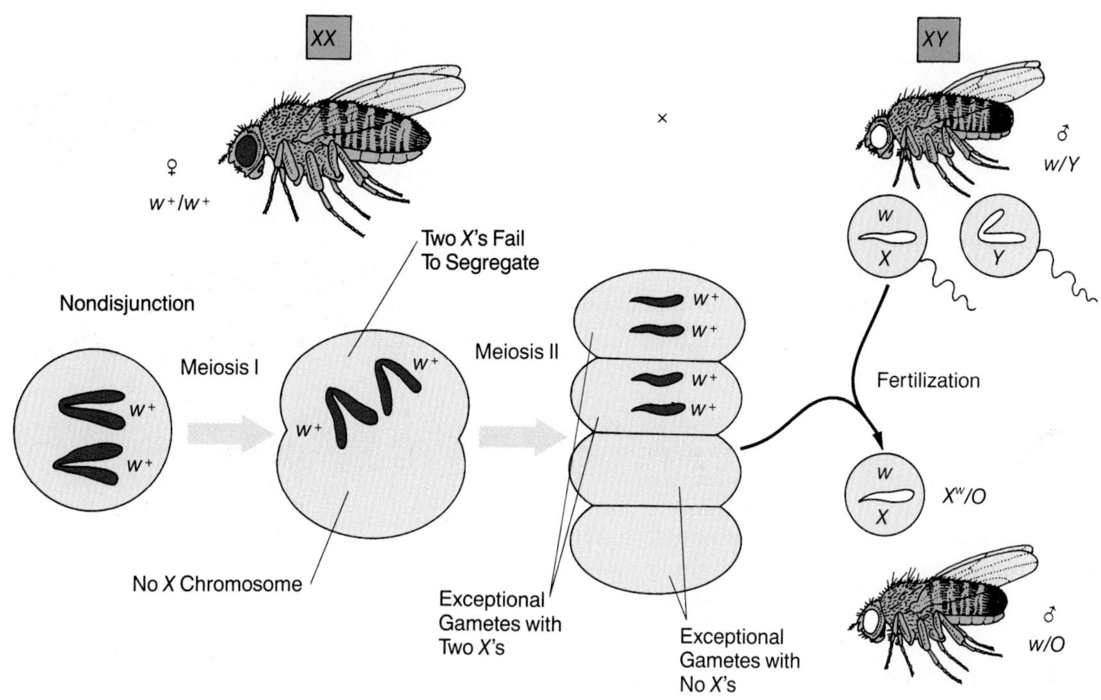

Figure 10-12 NONDISJUNCTION: A FAILURE TO SEGREGATE.
The homologous *X* chromosome pair in females sometimes fails to "disjoin" or segregate during meiosis, and thus one egg will receive two *X* chromosomes while another receives none at all. This situation is called nondisjunction. If an egg lacking an *X* chromosome is fertilized by a sperm that carries an *X* chromosome, the offspring will be a sterile *XO* male, and its eye color will correspond to the allele carried on the sperm's *X* chromosome, in this case, white eyes. This proves that the gene for eye color is associated with the *X* chromosome.

lacking an *X* chromosome, the offspring cannot be a female, since it lacks two *XX*'s and so is a sterile single-*X* (or *XO*) male; its eye color corresponds to the allele inherited on the father's *X*. Bridges's finding left no doubt that each chromosome carries a single specific allele for a given gene and was thus an additional foundation stone for the new field of genetics. Bridges's exploration of the nondisjunction phenomenon also led to an understanding of Down syndrome and other human genetic conditions and to the development of modern genetic screening techniques such as amniocentesis (see Chapter 15).

A CENTURY OF PROGRESS

The field of genetics is little more than a century old, yet it spans an enormous range of ideas—from pangenesis to gene manipulation in just a few decades. As we shall see in the next few chapters, however, just when most of the pieces of the puzzle seemed to be in place, new discoveries revealed wholly unsuspected levels of complexity in the hereditary material of eukaryotic organisms.

SUMMARY

1. One of the early theories of heredity, *pangenesis*, held sway until the 1880s, when Weismann formulated the *germ plasm theory*, which states that only gametes are involved in transmitting traits to progeny.

2. Mendel disproved blending theories of inheritance when he discovered that pea plants possess units of heredity, now called *genes*, for each of seven characters he studied. Mendel found alternative forms of the gene for each character; these are called *alleles*. One allele for a character can be *dominant* and the other *recessive*. The terms P_1 *(parental)*, F_1 *(first filial)*, and F_2 *(second filial)* label the generations of organisms crossed in genetic tests.

3. A *homozygous* individual has two identical alleles for a given character; a *heterozygous* individual has two dissimilar alleles (commonly, a dominant and a recessive allele) for a given trait.

4. *Genotype* is the genetic makeup, or gene content, of a cell or an organism; more precisely, it is the allele content. *Phenotype* is the physical appearance and properties of a cell or an organism and is an expression of the genotype.

5. According to Mendel's first law, the *law of segregation*, an organism inherits

one unit, or allele, for each trait from each parent; the separation, or segregation, of each allele pair occurs during meiosis.

6. Mendel performed *dihybrid crosses* to arrive at his second law, the *law of independent assortment,* which states that the alleles of genes governing different characters are inherited independently of each other.

7. A seeming exception to Mendel's laws is *incomplete dominance,* in which progeny show a phenotype intermediate between that of the parents. The parental types reappear in subsequent generations, however, bearing out Mendel's laws.

8. Sutton and Boveri advanced the *chromosome theory of inheritance,* which states that genes are carried by and inherited as parts of chromosomes. Morgan and his students proved this theory with their work on *Drosophila* sex chromosomes.

9. Bridges's work with *nondisjunction* of the *X* chromosome added the final proof that specific chromosomes carry specific alleles.

KEY TERMS

allele
dihybrid cross
dominant
F_1 (first filial) generation
F_2 (second filial) generation
genotype
germ plasm theory
heterozygous
homozygous
incomplete dominance
law of independent assortment
law of segregation
nondisjunction
pangenesis
phenotype
P_1 (parental) generation
Punnett square
recessive
sex-linked trait
test cross

QUESTIONS

1. What do geneticists call the units of heredity that segregate at meiosis? On what cellular structures are they found?

2. Let's call the dominant allele for red flower color (in peas) *W*, and the recessive allele for white flower color *w*. If a pea plant has the genotype *WW*, what is its phenotype? If the genotype is *ww*, what is the phenotype? What if the genotype is *Ww*?

3. What are the possible genotypes of a red-flowered pea plant? A white-flowered plant? Which of these genotypes are homozygous? Heterozygous? What gametes will each type produce?

4. When Mendel crossed pea plants that grew from round yellow seeds with pea plants that grew from wrinkled green seeds, what were the phenotypes of the F_1 hybrid seeds? When the F_1 hybrid seeds were grown, and the flowers self-pollinated, what kinds of seeds were produced? Were they like the original parents only, or were there also round green seeds and wrinkled yellow seeds? Explain this result.

5. A couple has two children. What is the probability that the first is a girl? A boy? What is the probability that the second is a girl? What is the probability that both are girls? Both boys? One of each sex?

6. Most living lobsters are brown, but a few are blue. Crosses have shown that the blue color is due to a recessive allele. Let's call the brown allele *B* and the blue allele *b*. What is the genotype of a blue lobster? What gametes would it produce? What is the genotype of a blue lobster's brown mother? What gametes would she produce? If a blue lobster has two brown parents, and the parents produce many offspring, what fraction of them would you expect to be blue? If a blue lobster is mated to its brown parent, what fraction of their offspring would have the genotype *bb*? What fraction would have the genotype *BB*? If two blue lobsters mate, what fraction of their offspring will be blue? Brown?

7. Wild red foxes occasionally produce a prized silver-black pup in a litter. When mated, such pups breed true. How would you explain this?

8. Consider two independent characters, normal skin color *C* versus albino *c* and normal blood *T* versus thalassemia (*t*) (anemia). Show by a Punnett square the expected ratio of a large number of children produced from parents heterozygous for both traits.

9. In the fruit fly, vestigial wings and scarlet eyes are caused by recessive genes on different chromosomes. If a vestigial scarlet male is crossed to a wild-type female, what will be the phenotype of the F_1? If the F_1 progeny mate among themselves, what would be the ratios of the F_2 offspring? Show the genotypes and phenotypes expected for each class. Use a Punnett square to analyze the F_2.

10. Brown eyes (*B*) are dominant to blue eyes (*b*). Show genotypic and phenotypic ratios of children from the following parental types:

a. $BB \times bb$
b. $Bb \times bb$
c. $Bb \times Bb$

In a family where the mother is blue-eyed (*bb*) and the father is brown-eyed (*Bb*), which genotypes would be possible for each offspring with the following phenotypes?

d. two children both blue-eyed: one boy, one girl
e. three brown-eyed children: two girls, one boy

What is the probability that a sixth child will be blue-eyed?

For additional readings related to topics in this chapter, see Appendix C.

11 Mendel Modified

Had Mendel studied coat color in dogs, eye color in fruit flies, or the genetics of Manx cats, he might never have devised his laws of inheritance; the rules might not have emerged from the exceptions.

John Postlethwait (1983)

Manx cats are a striking but peculiar breed of domestic cat, most notable for a stubby or missing tail and the difficulty in breeding the animals. Although a dominant mutation causes the Manx's short tail, the cat breeder will see *two* Manx kittens instead of three for each normal, long-tailed one. What accounts for this departure from Mendel's laws? Do Mendel's rules hold for peas and fruit flies but not for cats?

This chapter investigates inheritance patterns that appear to violate Mendel's laws. Mendel's work certainly revealed some basic relationships of heredity. But it was left for others, just a few decades later, to piece together more of the puzzle from logic and deduction, but also from chance observations of a new mutant trait and shrewd guesses about how to proceed experimentally.

The work of geneticists in the first half of the twentieth century helped explain many apparent contradictions between Mendel's laws and the hereditary patterns of organisms in nature. The nonlinear path of genetic discoveries illustrates in a clear and dramatic way how science actually advances much of the time.

A manx cat: Heterozygotes have special characteristics; homozygotes are dead.

In tracing the early history of genetics, this chapter will explain:

- How genes are arranged on chromosomes, and how the crossing over and recombination in meiosis causes gene rearrangements
- How genes act and interact within organisms
- How the alleles of some genes cause death or affect several physiological traits at once
- How different forms of the same gene arise through mutation and provide a source of genetic variation

HOW GENES ARE ARRANGED ON CHROMOSOMES

By establishing chromosomes as the bearers of genes, early geneticists gave a physical basis to the inheritance "factors" described by Mendel's laws. And later, of course, these genes turned out to be relatively short stretches of DNA that determine physical traits. Because all organisms have many traits and only a few chromosomes, geneticists became convinced that each chromosome must carry hundreds or thousands of genes. Indeed, later experimenters revealed that fruit flies have only 4 pairs of chromosomes but more than 5,000 genes, and humans have 23 pairs of chromosomes but about 50,000 genes.

The notion that chromosomes bear many genes helped explain some of Morgan's observations and those of other early geneticists. They found that whereas most genes obey the law of segregation, many pairs of genes do not follow the patterns predicted by Mendel's law of independent assortment. Some traits appear almost never to assort independently of one another. In a family, for example, traits *A* and *B* or traits *a* and *b* might tend to be inherited together, but traits *A* and *b* or traits *a* and *B* might only rarely appear together. The reason for this unexpected inheritance pattern is that many genes are arranged on few chromosomes, and some genes that lie on the same chromosome move together during meiosis and do not separate. Inheritance patterns among genes are described in terms of their **linkage,** or the degree to which genes on chromosomes are inherited together.

Consider a cross between two organisms, each of which has a pair of homologous chromosomes bearing alleles (or variants) of two genes, *A* and *B*. Either the dominant or the recessive alleles occur at a given location, or **locus,** on each chromosome. Of course, only one allele occurs at each locus on each homologous chromosome. Each parent in this cross has a homozygous, or matched, pair of alleles for each gene. That is, the mother is *ABAB*—homozygous dominant for both the *A* gene and the *B* gene. The father is *abab*—homozygous recessive for both genes.

Mendel's second law predicts that during meiosis in the F_1 hybrids (Figure 11-1a), the gene pairs will assort independently and gametes will form with four genotypes in equal proportions (25 percent of each type of gamete): two original, *parental* genotypes, *AB* and *ab;* and two novel, **recombinant genotypes,** *Ab* and *aB*. If such independent assortment occurred, it would indicate **nonlinkage;** that is, genes *A* and *B*, would be unlinked (Figure 11-1a). If genes on the same chromosome were always inherited together, gametes would form with only two genotypes (50 percent of each type): *AB* and *ab*. This would indicate **complete linkage;** both types of gametes would have the parental genotypes *AB* and *ab*, and no recombinant gametes would form (Figure 11-1b). Finally, if gametes formed with four genotypes (two parental, two recombinant) in unequal proportions, this would indicate **partial linkage**—an intermediate state between complete linkage and no linkage. For the gametes displaying partial linkage in Figure 11-1c, each parental genotype appears 49.5 percent of the time (49 + 0.5), which is more often than the 25 percent predicted by nonlinkage (independent assortment), but less often than the 50 percent predicted by complete linkage. Likewise, each recombinant genotype forms 0.5 percent of the time—less often than predicted by nonlinkage, but more often than predicted by complete linkage.

In nature, gametes reflect all three patterns. The seven characters that Mendel studied in peas, for instance, are not linked. They assort independently because they are on different chromosomes and move as independent units during meiosis. At the other extreme, a clear-cut instance of complete linkage occurs in *Drosophila* fruit flies, where gametes nearly always convey the parental arrangement of alleles for black body color and purple eye color. The genes for these traits lie very close to each other on the same chromosome and tend to be inherited together as a "package." Moreover, the genes for yellow body and white eyes are a case of partial linkage. Alleles for these genes lie a sufficient distance apart on the *X* chromosome that 99 percent of the time they are inherited in the parental combination, but 1.0 percent of the time, the alleles occur in different combinations from those found in the parents.

Crossing Over and Recombination

A fruit fly with yellow body and white eyes that arises in the F_2 from a cross between flies with yellow body and red eyes, and tan body and white eyes has a **recombinant phenotype.** Such novel phenotypes can be ex-

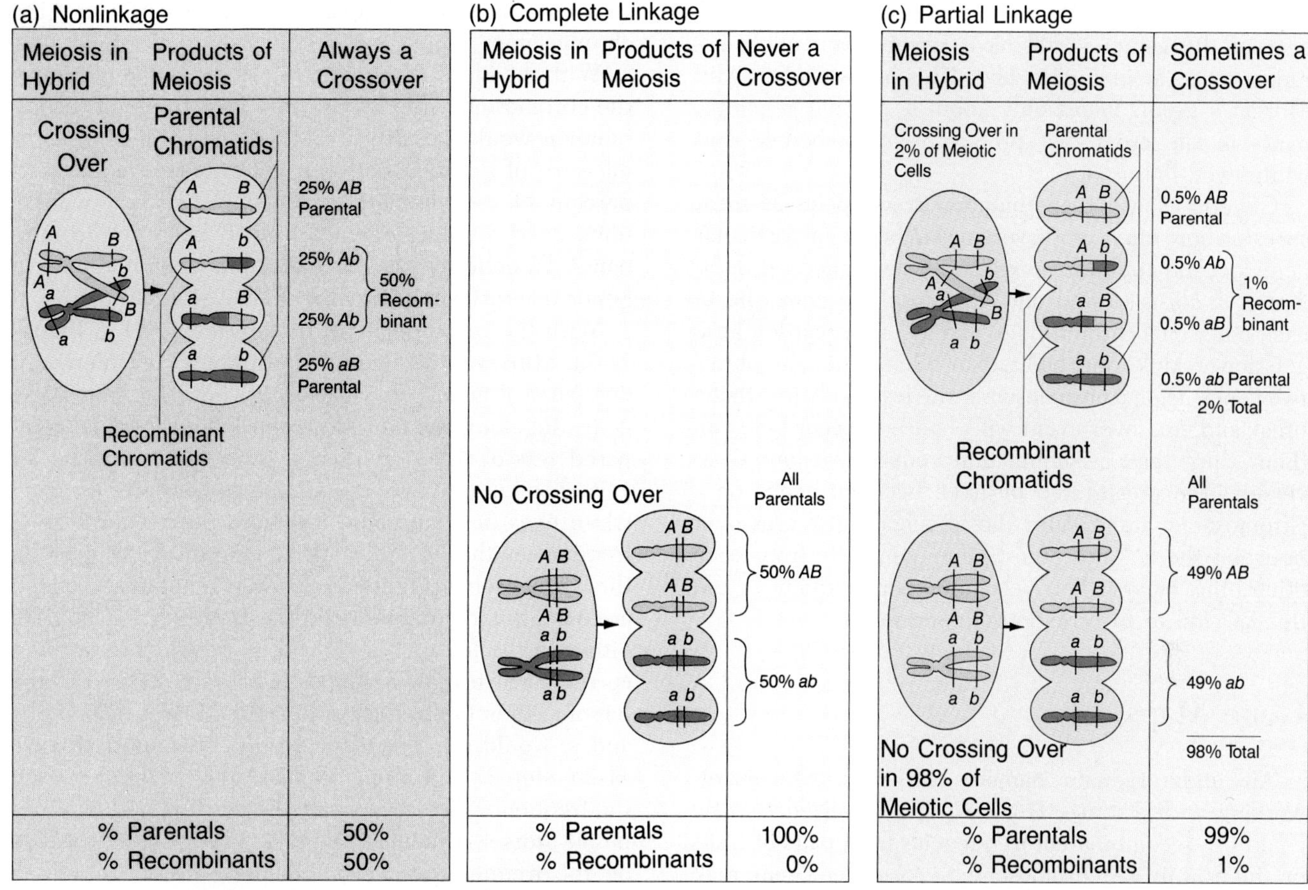

Figure 11-1 LINKAGE AND NEIGHBORING GENES.
(a) Genes *A* and *B* on the same chromosome can show different linkage patterns. Here the genes appear on maternal and paternal chromosomes in a hybrid. While chromosomes are closely paired during meiosis I, chromatids may exchange parts and result in a set of chromosomes in which alleles have recombined. Normal segregation of these chromosomes into gametes results in eggs or sperm that carry either the original set of parental alleles or the recombinant sets of alleles. With nonlinkage, the first possibility, genes that reside on the same chromosome may be inherited according to the law of independent assortment. In this example, parental and recombinant offspring occur in equal frequencies. Second, the two genes on the chromosome may always be inherited together and appear completely linked (b). Here, all the gametes are parental, and no recombinants occur. The third possibility is the intermediate pattern of partial linkage (c), in which some parental and some recombinant types occur, but in unequal percentages. Here, 1 percent of the gametes are recombinant types and 99 percent are parentals.

plained by *crossing over* during meiosis, the peculiar "trading" of pieces of chromosomes we discussed in Chapter 9. Although the genes for body color and eye color in *Drosophila* are physically linked on the *X* chromosome and cannot assort independently, crossing over during prophase of meiosis I can lead to a few recombinants.

During prophase I of meiosis in a female fruit fly, homologous chromosomes pair closely. Breakage of two nonsister chromatids then occurs by chance somewhere along their length, followed by a crossed rejoining of the broken chromatid arms derived from the original nonsister chromatids (see Figure 11-1b, top). If the break and repair happen to occur between the locus of the body color gene (locus *A* in Figure 11-1) and the locus of the eye color gene (locus *B* in Figure 11-1) some distance away on the same chromatid, the crossing over will produce two recombinant chromatids, one carrying *Ab* and the other *aB*. When these chromatids are then segregated into gametes, the resulting eggs will carry either the recombinant sets of alleles or the original sets of parental alleles from the two chromatids not involved in the exchange.

Crossing over between the sites of the *A* and *B* genes does not occur in every cell at prophase I of meiosis. In fact, it is a relatively rare event. If it happens in only 2 of every 100 cells going through meiosis I, then only 1 percent of the gametes will have recombinant allele arrangements, as in Figure 11-1. This helps explain why Morgan's group found only about 1 percent recombinants in the cross that produced yellow-bodied and white-eyed flies.

Geneticists use **recombination frequencies** as measures of how often crossovers occur between particular gene loci and apply them in a technique called *recombination analysis* to find out how close two genes lie to each other on a chromosome. Just as a short garden hose gets fewer kinks and knots than a long one, the closer two genes lie on a chromosome, the less likely it is that a break and crossover event will chance to occur between them; thus, their recombination frequency is low. Conversely, the greater the distance between genes on a chromosome, the greater the likelihood of a crossover between them. Thus, the farther genes lie from each other, the higher their recombination frequency, and this fact led to the creation of gene maps.

Gene Maps

An undergraduate named Alfred H. Sturtevant, working with Thomas Hunt Morgan, realized that he could use recombination frequencies from pairs of traits on the fruit fly's *X* chromosome to construct a **gene map** of where and how far apart those genes lie along the chromosome.

With the idea in mind that the closer two genes lie on the chromosome, the lower their frequencies of recombination would be, Sturtevant assigned 1 map unit, or measure of distance between genes, to be equal to 1 percent of recombinant gametes. Genes that were 5 units apart would therefore have 5 percent recombinants. Geneticists call the map units *centimorgans*, in honor of Sturtevant's mentor, T. H. Morgan.

With the recombination frequencies listed in Figure 11-2a, Sturtevant assigned map distances between various pairs of genes. In what order, however, did those genes lie along the chromosome? To find out, he compared sets of three genes at a time, such as those for vermilion-colored eyes (*v*), miniature wing size (*m*), and white eyes (*w*). He could tell from the recombination frequencies that *w* and *v* are 30 centimorgans apart on the chromosome (Figure 11-2b, step 1) and that *m* and *v* are 3 centimorgans apart (step 2). However, *m* could lie either to the *left* or the *right* of *v*. So which order was correct? Sturtevant reasoned that if *m* is located to the left of *v*, then the frequency of recombinants between *m* and *w* would be 30 units − 3 units = 27 units (Figure 11-2b, step 3). But if *m* is located to the right of *v*, then the frequency of recombinants between *m* and *w* would be 30 units + 3 units = 33 units. From the actual data on the frequency of recombinants between *w* and *m*

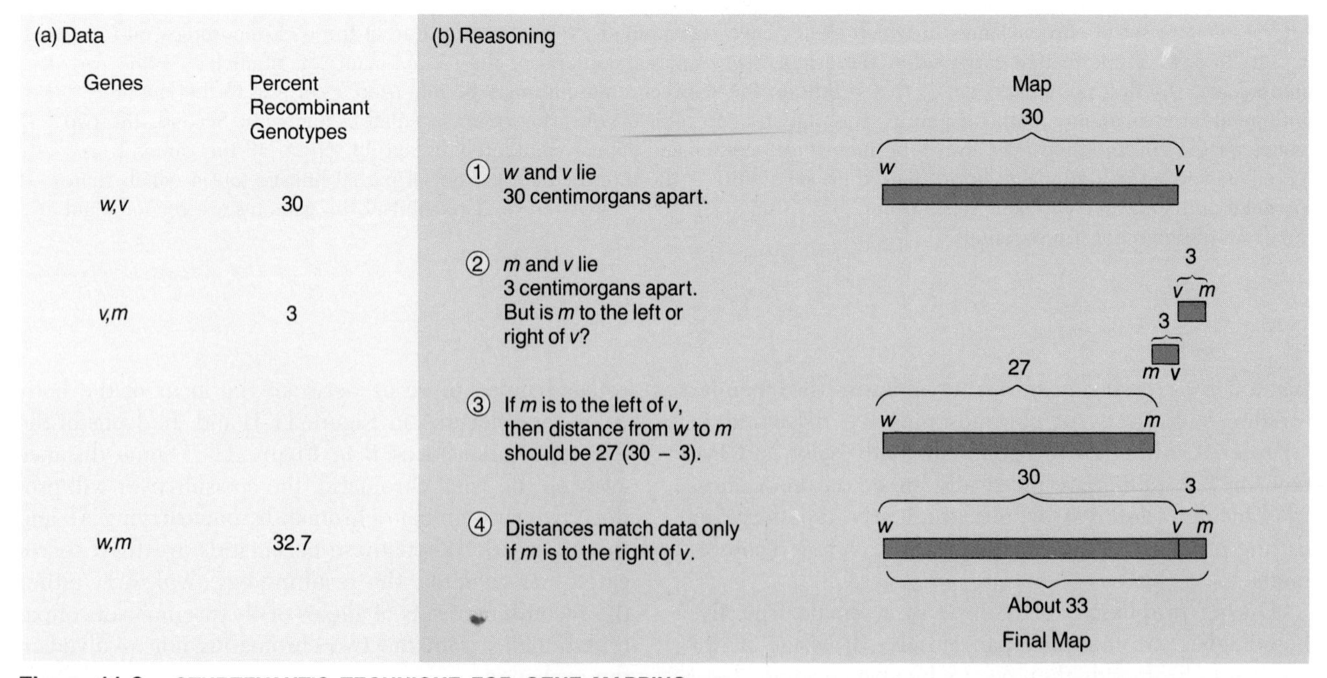

Figure 11-2 STURTEVANT'S TECHNIQUE FOR GENE MAPPING.
Sturtevant showed that genes on chromosomes have a linear relationship and lie in a specific order on a chromosome. The data in (a) shows the percent of recombinant genotypes between gene pairs *w* and *v*, *v* and *m*, and *w* and *m*. As the text explains, Sturtevant's comparisons (b) resulted in a mapping of the three genes to appropriate positions on the chromosome.

(32.7 units, or almost 33), Sturtevant concluded that the gene order is *w-v-m*, as in Figure 11-2b, step 4. Using this same technique, he was able to map five genes—*w*, *v*, and *m* as well as *y* (yellow body) and *r* (rudimentary wings)—occurring on the same *Drosophila* chromosome.

There was great excitement in Morgan's lab the morning that "young boy Sturtevant" presented his genetic map. His work served as the first proof that genes on chromosomes have a *linear sequence*—that is, they lie in "single-file" order along the chromosome. Sturtevant's linear model provided a framework for all the genetics research that followed. And the technique allowed Morgan and his colleagues to map genes on each of the four *Drosophila* chromosomes.

This early mapping work was further enhanced by a special type of chromosome. In certain insect cells (including some in *Drosophila*), DNA is replicated over and over, but the daughter strands fail to separate and the nucleus fails to divide. Consequently, as many as 1,000 DNA threads line up side by side, forming giant **polytene chromosomes** (Figure 11-3a). These are about 100 times as long as normal mitotic chromosomes and when stained reveal bands and interbands with sequences and sizes characteristic for each region of each giant chromosome. Researchers can letter and number the bands within regions to produce what is called a cytogenetic map of *Drosophila* chromosomes. They can also assign specific genes for striking mutations (Figure 11-3b) to particular bands on the stained polytene chromosomes and thus locate them on the map. If a gene is absent owing to a deletion, for instance, and a specific band also is absent, then the band is probably the site of the gene.

By placing a Sturtevant-style recombination gene map side by side with the *cytogenetic* (gene) *map*, as in Figure 11-3b, one can see a similar gene order but no exact correspondence of the distances between genes.

(a) Polytene Chromosome from Midge Fly Salivary Glands

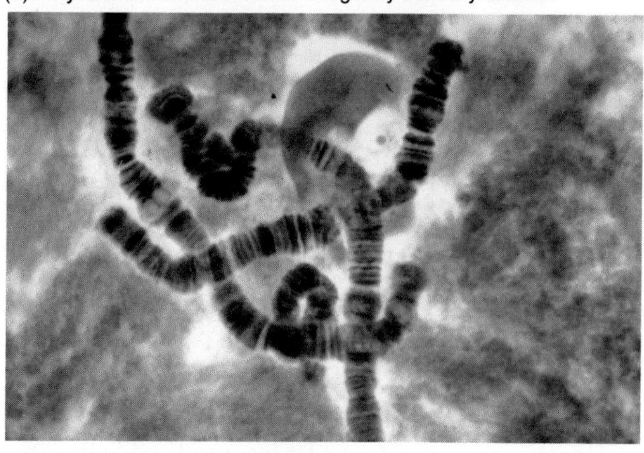

(b) Mapping Genes on Chromosomes

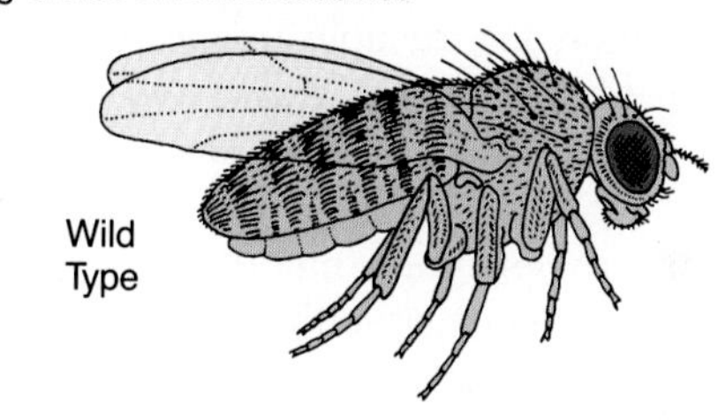

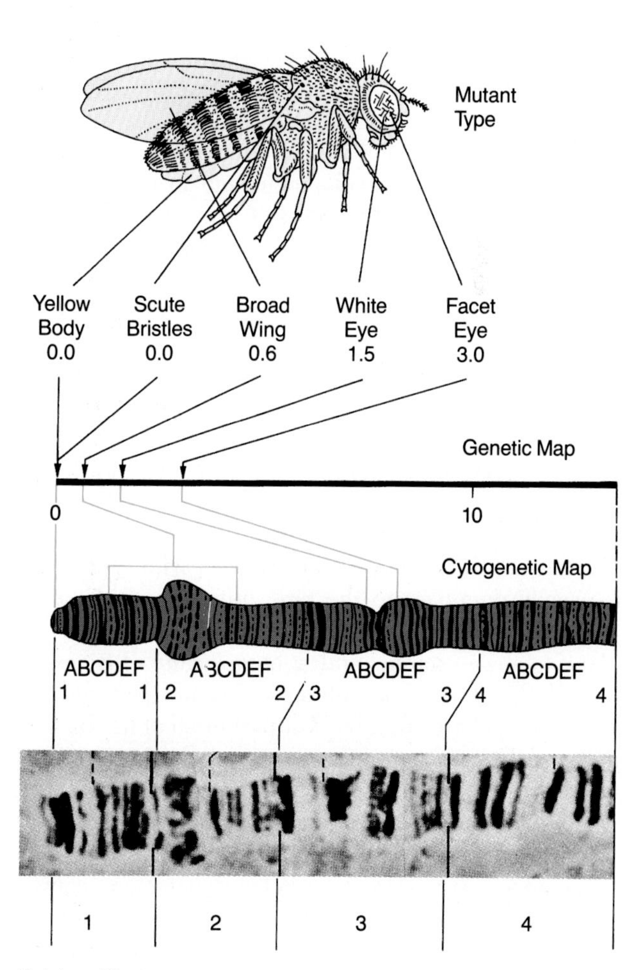

Figure 11-3 GIANT *DROSOPHILA* CHROMOSOMES: A BOON TO GENE MAPPING.
(a) These giant chromosomes from the salivary gland of a midge fly (magnified about 500×) have characteristic bands of chromatin that are large and distinct enough for researchers to study. Specific genes have been assigned to specific bands on such chromosomes. (b) Here, the mutant *Drosophila* genes for yellow body, scute bristle, broad wing, white eyes, and facet eye are mapped to the *X* chromosome. This figure shows genetic and cytogenetic maps of the end of the *X* chromosome containing those mutations (one-fifth of the entire *X* chromosome). A stained *X* polytene chromosome (below) shows the banding pattern. As the blue lines show, the distances on the genetic map do not represent exact distances on the chromosome.

Recombination is probably not an exact and simple function of distance along a chromosome, at least not in polytene chromosomes. Nevertheless, both mapping techniques support the notion of a linear arrangement of genes on chromosomes and allow geneticists to pinpoint gene positions and thus to probe more deeply into the mystery of how genes determine physical traits.

HOW GENES ACT AND INTERACT

Genes in nature often act far less predictably than the seven characteristics Mendel studied in pea plants. Human eye color, for example, ranges from pink (albinos) through blues, greens, grays, and browns, while human hair grows in a palette of shades from white to yellows, reds, browns, and jet black. Clearly, human eyes and hair are not simply one phenotype or another, the way that pea plants are either tall or short and peas are either yellow or green. So how do genes actually influence a pea plant's height or a person's eye color?

How Genes Act

The specific alleles of a gene often act during the organism's embryonic development and bring about the expression of contrasting traits. Genes act by causing specific proteins to be produced, and these proteins can play *structural* roles inside a cell, on a cell surface, or in the immediate environment of a cell. Proteins can act as *enzymes*, facilitating the myriad biochemical reactions that make up metabolism, or act as *regulators* that coordinate the timing and patterning of developmental processes. The interaction of many genes and many proteins during embryonic development culminates in a leaf, a hand, an eye, or a cat's tail.

Genes exist as either dominant or recessive alleles. The frequently encountered normal, or **wild-type, allele** is usually dominant to the rare mutant form. In a pea plant, for example, the dominant allele *R* causes the plant to make starch, and a homozygote (RR) makes enough that its seeds are plump and round (Figure 11-4). A recessive allele often has lost part or all of its ability to perform the function of the normal allele. In a homozygous recessive plant (*rr*), the protein controlled by the *r* allele is not enzymatically active; thus, the plant makes too little starch to make the seed plump, and so the seeds appear wrinkled. In a heterozygote, one copy of the dominant allele may provide enough of a given gene's normal function to support the development of a normal phenotype. Thus, in a pea plant heterozygous for round-seed trait (*Rr*), the single *R* allele encodes enough of a specific enzyme to be synthesized so that the plant manufactures enough starch to give the seeds a firm, round appearance. This explanation of gene action also helps us understand incomplete dominance, as in pink snapdragons (see Chapter 10). A red snapdragon with two *R* alleles makes two doses of the enzyme necessary for pigment production, and thus the flowers grow deeply pigmented. The heterozygotes with just one dose of the enzyme (*Rr*) make only half as much pigment and

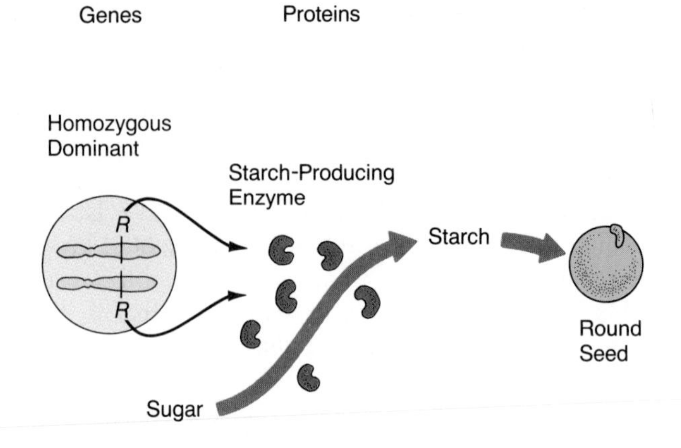

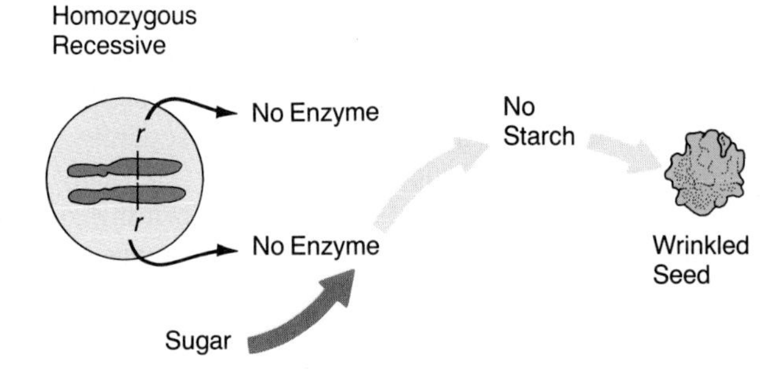

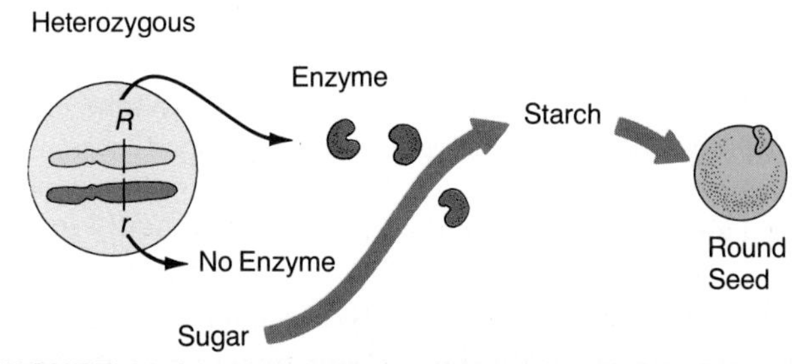

FIGURE 11-4 HOW GENES DETERMINE PHENOTYPE IN A PEA PLANT: ROUND AND WRINKLED SEEDS.
The gene for round seed (*R*) encodes an enzyme to produce starch, which fills the seed. In a homozygous dominant plant, the alleles cause abundant starch production (top). In a homozygous recessive, no starch is produced and the seeds are wrinkled (middle). In a heterozygote, the one active allele directs enough starch production to cause normal seeds (bottom).

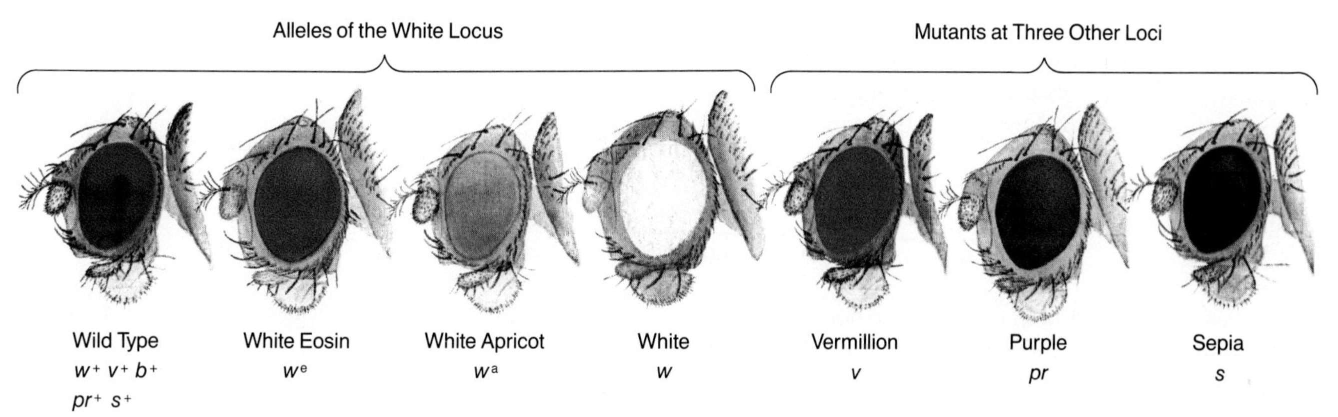

Figure 11-5 A PALETTE OF EYE COLORS IN *DROSOPHILA*.
The wild-type *Drosophila* eye is red, but a variety of sex-linked mutations of the white-eye locus or other loci can produce many eye colors, only some of which are shown here (with gene symbols below).

thus produce pink flowers. In such a case of incomplete dominance, the limiting factor in the phenotype is often the amount of active enzyme formed by the dominant allele.

Genes with More Than Two Alleles

Although Mendel studied traits with two alleles, a gene can have 3, 6, 12, or even 100 alleles, each of which may confer a slightly different phenotype for the same trait. In any individual of a diploid species, of course, only two of the possible alleles will be present. A group of alleles determining many forms of the same trait is called a **multiple allelic series.** In *Drosophila*, for example, dozens of sex-linked alleles of the white-eye gene bring about the palette of eye colors ranging from deep brown through brilliant red to pure white (Figure 11-5). And in humans, three alleles determine the A, B, AB, and O blood types.

Human Blood Type

Every day, the multiple allelic series for blood types affects human survival. If we become injured or ill, the type of blood each one of us can safely receive in a blood transfusion depends on our own blood type: A, B, O, or AB.

The ABO blood group gene has three alleles that code for the A, B, O, and AB blood types (Figure 11-6). One function of A, B, and O alleles is to govern the presence

Blood Type	Genotype	Antibodies Made	Recipient ↓	Donor				
				A	B	AB	O	
A	I^AI^A or I^AI^O	Anti B	A					
B	I^BI^B or I^AI^O	Anti A	B					
AB	I^AI^B	Neither Anti A or Anti B	AB					Universal Recipient
O	I^OI^O	Both Anti A and Anti B	O					
							Universal Donor	

Figure 11-6 BLOOD GROUPS AND TRANSFUSIONS.
This diagram shows four major human blood types (first column), the genotypes (second column), the antibodies a person's body will produce, depending on his or her own blood type (third column), and finally, the consequences for a blood recipient, depending on the donor's blood type (chart). If a person with type A blood receives type B, clumping occurs because the person's body produces anti-B antibodies. Note that AB is the universal recipient, and O is the universal donor.

or absence of specific molecular markers, or name tags, on the surfaces of red blood cells. Such name tags are *antigens* (substances that can evoke an immune response). Thus, every red blood cell of a type A person has A antigens but not B antigens; a type B person has red blood cells with B antigens only; a type AB person, both A and B antigens; and a type O person, neither A nor B antigens. The alleles that encode the A, B, and absence of antigens are called I^A, I^B, and I^O. Note that I^O is recessive and that both I^A and I^B are dominant to it. Furthermore, I^A and I^B alleles are fully expressed when they occur together. Hence, they are said to be **codominant,** and both phenotypic traits are present (an I^AI^B person has an AB blood type).

Medical difficulties can arise if a type A individual mistakenly receives a blood transfusion from a type B donor (or vice versa). This is because the body spontaneously makes proteins that will attack foreign blood group antigens—in other words, blood group substances that it does not possess. These proteins, called *antibodies,* act by binding to specific foreign molecules (antigens), causing them to clump together (see Figure 11-6). An immune reaction like this can be fatal.

Human *Rh blood group* represents a different gene affecting blood chemistry. About 85 percent of adult humans have a molecular marker called the Rh factor on their red blood cells and are classified as Rh positive (Rh^+). If an Rh^- person is exposed to Rh^+ blood, his or her immune system responds by making antibodies that will react with the Rh antigen factor on the red blood cells and cause a condition of blood cell clumping called Rh disease.

Rh disease can result from a blood transfusion, but more commonly, it follows the birth of an Rh^+ baby to an Rh^- mother (the father is Rh^+, hence the baby is heterozygous for the Rh factor; Figure 11-7). Problems do not usually occur with a first-born Rh^+ child to an Rh^- mother. However, some fetal cells containing Rh factor may cross the placenta, usually during the birth process (but sometimes before), and enter the mother's blood vessels (Figure 11-7, step 1). The mother's immune system retains a "memory" of the event in the form of cells primed to recognize the Rh factor and quickly produce antibodies against it (step 2). Later during the same pregnancy, or, more often, during a subsequent pregnancy, when the Rh^- mother is once again exposed to Rh^+ blood, these memory cells generate antibodies that cross the placenta and cause the fetus's red blood cells to clump (step 3). This creates a potentially fatal disease called erythroblastosis, which can damage or even kill the fetus or newborn.

Physicians can help prevent Rh-factor fatalities by replacing the newborn's Rh^+ blood with blood that does not carry the mother's antibodies against the Rh^+ factor. Physicians also use another means of preventing Rh disease: They inject the Rh^- mother with anti-Rh antibodies at the birth of an Rh^+ child (step 4); the antibodies clump any fetal red blood cells with Rh^+ markers that have entered her bloodstream, thus removing them and preventing her immune system from forming antibodies that could damage future babies.

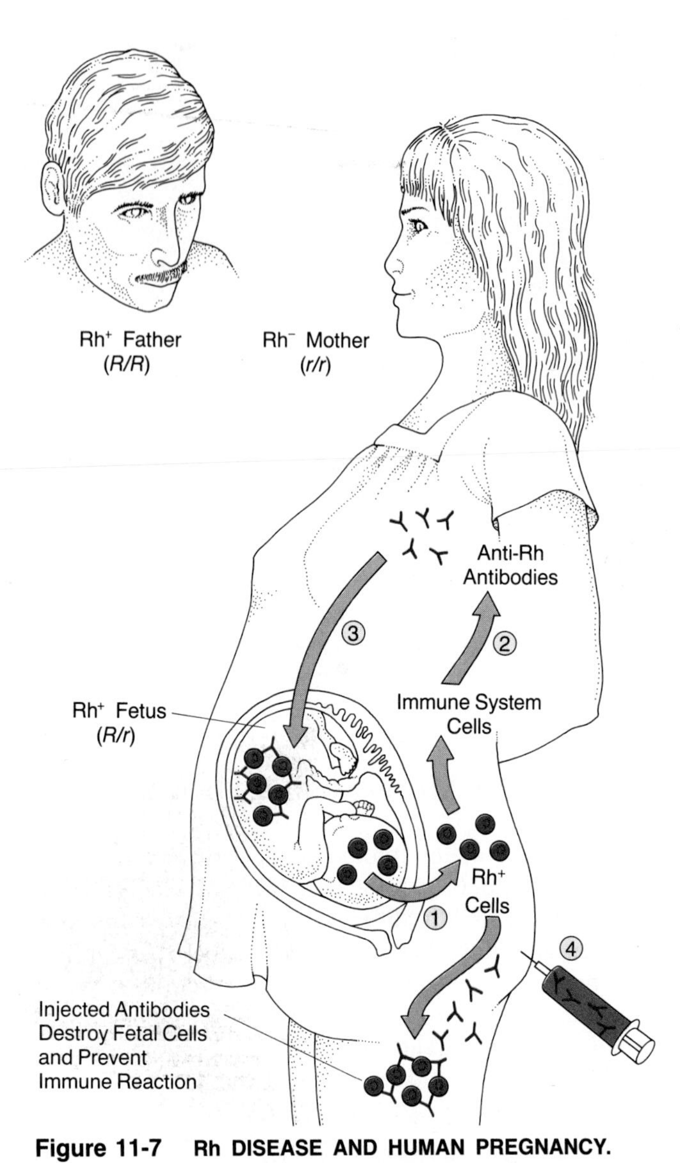

Figure 11-7 Rh DISEASE AND HUMAN PREGNANCY. As the text explains, a homozygous Rh^+ father and an Rh^- mother may produce an Rh^+ fetus, and if maternal and fetal bloods come into contact during pregnancy or birth (1), the mother's immune system will form anti-Rh antibodies (2). These can attack the present fetus or a subsequent fetus (3) unless a physician injects anti-Rh antibodies into the mother (4). These will attack any fetal blood cells circulating in the mother's system and prevent her body from making its own anti-Rh antibodies, thus protecting future pregnancies.

The Rh trait is an example of multiple alleles, because geneticists have found between 35 and 40 distinguishable phenotypes in Rh^+ individuals for the Rh factor. Physicians believe that the human genome contains many other multiple allelic series with more subtle effects that have not yet been recognized.

Complementary Genes

We have seen how several alleles of a single gene can affect a fruit fly's eye color or a human's blood type. Sometimes, however, two or more separate **complementary genes** can influence a single phenotype. The results of several such genes are shown in Figure 11-5. The products of complementary genes must act together to produce a given phenotype. For example, a fly needs both w^+ and v^+ to have a normal eye color.

The action of complementary genes also brings about human albinism. The couple in Figure 11-8 are both homozygous for recessive alleles causing the albino phenotype, which is characterized by the absence of pigment in skin, eyes, and hair and sometimes by the presence of crossed eyes. Their son, however, does not have the albino phenotype; he has normal pigmentation in his skin, eyes, and hair, and he does not have crossed eyes. Here's why. Each parent must be homozygous recessive for only *one* of the two complementary genes that act together to bring about albinism. The mother might be *AAbb*, normal for gene *A* but homozygous recessive for *b*, whereas the father *(aaBB)* lacks *A* but is normal for *B*. Since each lacks one critical enzyme in the two-step pigment-forming process, their hair, skin, and eye cells cannot generate pigment, and each shows the albino phenotype. Their son *(AaBb)*, however, apparently inherited one normal dominant allele of both genes, *A* from his mother and *B* from his father, and thus makes pigment normally.

Complementary genes help show that adult phenotype is the sum of biochemical pathways with many steps and the end product of a complex series of developmental events, each step or event dependent on products of a separate gene.

Figure 11-8 ALBINISM AND COMPLEMENTARY GENES. These two parents show albino traits to varying extents, yet their son is normal. Since normal pigmentation is controlled by two complementary genes (say, genes *A* and *B*), each parent must be homozygous recessive for one of the genes. The father could be normal for gene *A*, but aberrant for gene *B* (he is slightly pigmented), and the mother mutant for gene *A*, but normal for gene *B* (she is unpigmented). The son thus inherited one normal allele of each gene from each of his parents and has normal pigmentation.

Epistatic Genes

In a third type of gene interaction called **epistasis** (meaning "to stand over"), the effects of one gene over-

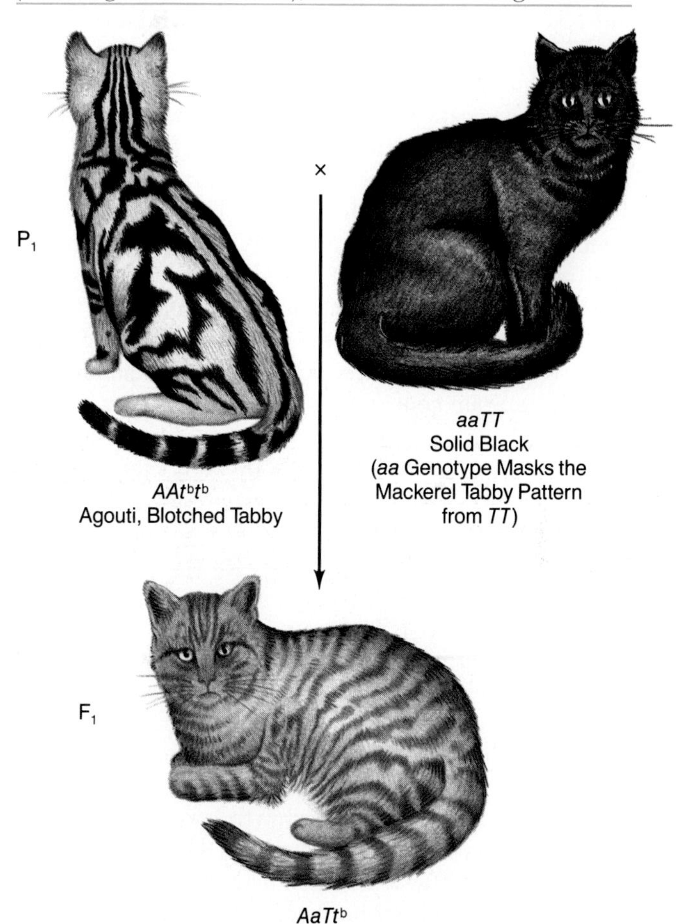

Figure 11-9 CATS' COATS: A DEMONSTRATION OF EPISTASIS.
When an agouti, blotched-tabby female (AAt^bt^b) is mated with an all-black male of genotype *aaTT*, all F_1 offspring are $AaTt^b$. They exhibit the dominant mackerel pattern even though neither parent appears to be mackerel. The reason for this unexpected coat pattern inheritance is epistasis—one gene masking the expression of a *different* gene for an entirely different trait. In this case, the male cat's coat color genotype (*aa*) is epistatic to the tabby pattern genotype (*TT*).

ride or mask the effects of other, entirely different genes.

Domestic cats, for example, have one gene that determines coat color (A) and another that determines coat pattern (T) (Figure 11-9). An AA or Aa cat is *agouti*, while an aa cat is pure black. Agouti is the typical grayish brown color of a wild rabbit, where each hair is gray with a yellowish band or tip. A cat with TT or Tt^b has a *mackerel-tabby* coat pattern, with vertical curving black stripes. A t^bt^b cat has a *blotched-tabby* coat pattern, with broad bands in swirls and whorls.

When a homozygous agouti, blotched-tabby female (AAt^bt^b) is mated to a pure-black male of genotype $aaTT$, all F_1 kittens are $AaTt^b$ (see Figure 11-9). These F_1 kittens represent a peculiar situation: All the offspring show the dominant mackerel-tabby pattern, yet neither parent appears mackerel. The only parent with a coat pattern has the recessive blotched-tabby phenotype. The kittens show this unexpected pattern of inheritance because the aa coat color genotype in the pure-black tomcat makes any coat pattern invisible. The coat color genotype aa is *epistatic to* the coat pattern gene T or t^b. Looking at a black cat, then, we really cannot tell if its genotype is $aaTT$, aat^bt^b, or $aaTt^b$. It could be a mackerel tabby or a blotched tabby disguised by epistasis.

It is easy to confuse epistasis with dominance, but in epistasis, one gene masks the expression of a *different* gene, while in dominance, one allele masks another allele of the *same* gene.

LETHAL ALLELES AND PLEIOTROPY

Genes perform such essential functions during an organism's development that the allelic forms of certain mutated genes—so-called **lethal alleles**—can lead to the organism's death. Depending on the essential function knocked out by the mutation, the organism can die as an embryo, later in embryonic growth and development, during infancy, or even at some time during adulthood.

In the late 1920s, a mutation named *brachyury* arose in a laboratory mouse, causing one of its offspring to have a particularly short, kinked tail. When this mutant mouse matured, biologists mated it to a normal mouse and found that half of the F_1 progeny had short, kinked tails (Figure 11-10). They concluded that the mating represented a test cross, that the original mutant was heterozygous for tail length (TT^+) and that the short-tail allele (T) is dominant to the normal, long-tail allele (T^+). Researchers next mated the short-tailed F_1 offspring with each other in a brother-sister mating. In keeping with Mendel's law of segregation, they expected to find a 3:1 ratio of the dominant short-tailed phenotype to the recessive long-tailed form. There was, instead, always a 2:1 ratio of short tails to normal tails (see Figure 11-10). After puzzling over these unexpected results, geneticists

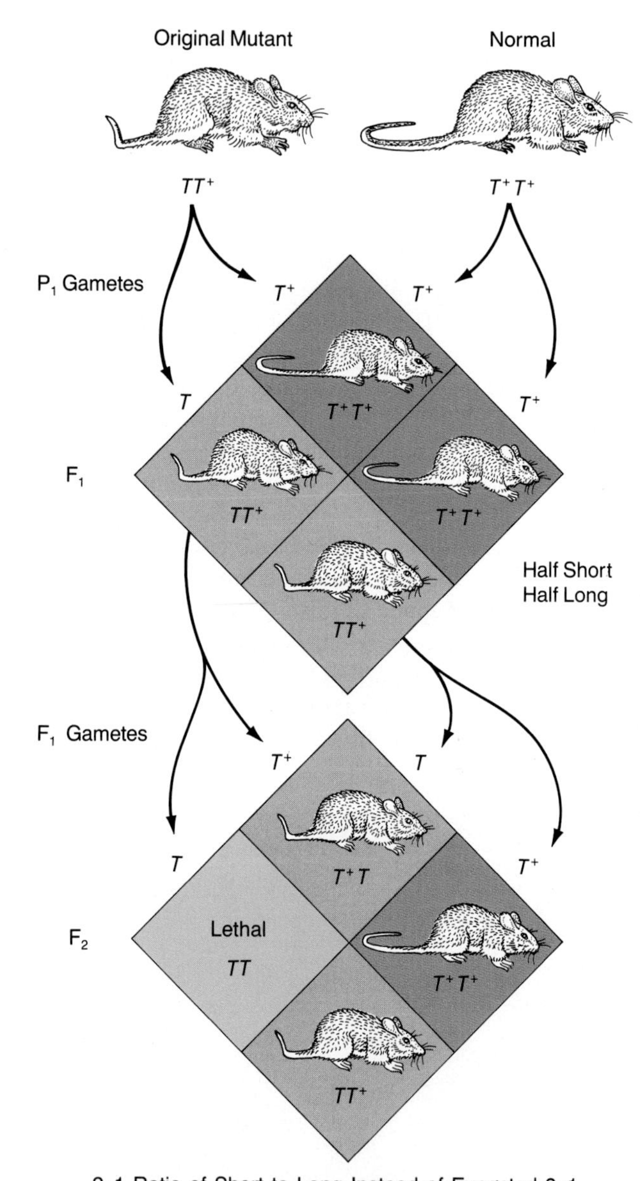

Figure 11-10 SHORT TAILS AND LETHAL ALLELES IN MICE.
A gene for tail length in mice displays an unusual inheritance pattern. Mice with the recessive genotype (T^+T^+) have normal-length tails, whereas heterozygotes (TT^+) have shorter, kinked tails. When crossed, mice with these genotypes produce half offspring with short tails and half offspring with long tails. When two short-tailed heterozygotes mate, the F_1 phenotypic ratio is 2:1 rather than the expected Mendelian ratio of 3:1 because mouse embryos with the homozygous dominant genotype (TT) die before birth.

realized that mice homozygous for the short-tailed trait must die early in development, and later researchers proved that such embryos do die and are resorbed in the mother's uterus. (Additional studies revealed a recessive lethal allele for tail length in Manx cats, and this explains the 2:1 phenotypic ratio mentioned at the beginning of this chapter.)

After studying lethal alleles in cats and mice, geneticists reasoned that in natural populations in the wild, most lethal genes are recessive. If they were dominant, then both homozygotes and heterozygotes would die. Dominant lethal genes do arise by mutation, of course, but disappear immediately owing to the death of both heterozygotes and homozygous dominants.

The Manx allele affects two phenotypes: When heterozygous, it causes a shortened tail; when homozygous, it causes the embryo's entire posterior to develop abnormally and the unborn animal to die. This correlation suggests that an individual gene can affect several traits. Mendel himself recognized this effect, now known as **pleiotropy**, when he observed that a single hereditary unit seems to determine simultaneously (1) whether a pea flower is red or white, (2) whether or not each leaf has red coloration where it joins the stem, and (3) whether the seed coat is gray or white.

Later researchers noticed other examples of single genes that affect several traits, including albinism in humans (review Figure 11-8). People homozygous for the albino gene not only have white hair, pink eyes, and pale skin, but also sometimes have crossed eyes. The allele causes defects in the way that nerves from the eye connect with targets in the brain, as well as failing to cause pigment production.

Another well-known example of pleiotropy is the gene that causes sickle-cell anemia, a human disease that usually results in death before age 20. Hemoglobin, the affected protein, is present in red blood cells and carries oxygen to the body tissues. Sickle-cell disease is caused by a mutated gene, Hb^S, for the production of hemoglobin. (The normal allele is Hb^A.) The Hb^S allele affects several traits and has at least five readily observed phenotypic conditions: (1) red blood cells shaped like sickles instead of normal disks (Figure 11-11); (2) severe and ultimately lethal anemia (reduced number of red blood cells); (3) pain in the abdomen and joints; (4) enlarged spleen; and (5) resistance to malaria in heterozygous individuals.

It is difficult to imagine that one mutation could cause such seemingly unrelated phenotypic characteristics. Each trait, however, can be traced back to the same fundamental defect: a *single* substitution of one amino acid for another in hemoglobin. The mutant abnormal hemoglobin molecules tend to join together to form long fibers of hemoglobin; these fibers then distort the shape of the red blood cell, causing it to sickle. Because the job of the spleen is to remove old and damaged red blood cells from the circulating blood, it becomes enlarged and overworked from removing the many sickled cells. This removal also leads to severe anemia, since there are fewer red blood cells left to carry oxygen. Finally, the shape of the sickled cells causes them to block blood flow in tiny capillaries, starving nearby body cells of oxygen and leading to pain in the abdomen and joints.

Figure 11-11 RED BLOOD CELLS, NORMAL AND SICKLED.
Normal red blood cells (bottom) are shaped like fat disks with a dent in each side. Sickled red blood cells (top) have elongated, irregular shapes because of abnormal hemoglobin molecules that have aggregated into fibers. These cells are magnified about 6,900×.

Individuals who are heterozygous for the sickle cell trait (Hb^AHb^S) produce both normal and abnormal hemoglobin molecules. Sickling occurs if malaria parasites are present in the red blood cells and use up much of the available oxygen. The spleen then removes and destroys the sickled infected cells while allowing the normal-shaped, uninfected cells to pass through unharmed. As a result, the spleen may become enlarged, the blood may become somewhat anemic, but the individual is resistant to malaria.

Despite its deleterious effects, the fact that the sickle-cell gene confers resistance to malaria explains why it was preserved during human evolution. In a malaria-infested region, individuals homozygous for the trait may die from the sickle-cell disease, and many of those homozygous for the normal gene may die from malaria. The sickle-cell heterozygotes, however, usually survive both conditions. The sickle-cell gene offers such marked protection from malaria that even with all its negative pleiotropic effects, it is sustained in human populations.

MUTATION: ONE SOURCE OF GENETIC VARIATION

How do different forms of the same gene arise and create 100 alleles for eye color in the fruit fly, as well as other less dramatic examples? New gene forms are the direct consequence of **mutation,** a change in the chemical structure of a gene or the physical structure of a chromosome.

Many mutations in the genetic information are deleterious to the organism, leading to a competitive disadvantage or to death. Occasionally, however, a genetic alteration lends the organism some survival advantage. The sickle-cell allele, for example, increases a heterozygous person's chances of surviving in a malaria region. When such advantages arise as a result of mutation, the organism may survive, pass the mutation to its offspring, and thereby spread the trait in the population. For this reason, mutation is one of the primary sources of biological diversity, providing the raw material for the evolutionary process.

Mutation is a word that encompasses several distinct processes. A **point mutation** alters the properties of a single gene and creates a new allele (Figure 11-12a). In the sickle-cell gene, a point mutation caused a substitution of one amino acid in hemoglobin, which in turn caused all the symptoms of the disease. Changes that alter the structure of chromosomes—**chromosomal mutations**—involve rearrangement of blocks of genes in the chromosome. These include *gene deletion; gene duplication; inversion*, or flipping, of genes; and *translocation*, or movement of a group of genes from one chromosome to another (Figure 11-12b–e).

A member of Thomas Hunt Morgan's team, Herman J. Muller, discovered a fundamental tool for probing genetic mutations. Muller found that certain physical and chemical factors in the environment can *induce* mutations by acting on the genetic material. He devised a system for measuring the **mutation rate** in *Drosophila*—that is, how frequently mutations arise naturally in a given population. Muller chose to measure the appearance of only sex-linked lethal mutations—mutations that cause the death of the affected fly before adulthood. He found that new lethal mutations arise spontaneously in about 2 of every 1,000 *X* chromosomes tested in each generation of flies. But because the *Drosophila X* chromosome contains perhaps 1,000 essential genes, the mutation rate of each individual gene is much lower—somewhere between 10^{-5} and 10^{-6} mutations per gene per generation.

Muller received the Nobel Prize for a subsequent discovery. When he exposed male fruit flies to x-rays and then tested their offspring for lethal mutations, he found that the frequency of mutations was greatly increased. Furthermore, the more powerful the x-ray dose, the more lethal mutations he could induce. Using Muller's techniques, later researchers discovered that certain chemicals called **mutagens** are also capable of causing mutations. The first of many chemicals shown to be mutagenic was mustard gas, a chemical-warfare agent widely used in World War I. Since then, scientists have identified hundreds of natural and synthetic mutagens.

Geneticists also discovered a link between mutagens and **carcinogens**—agents that can cause cancer. They

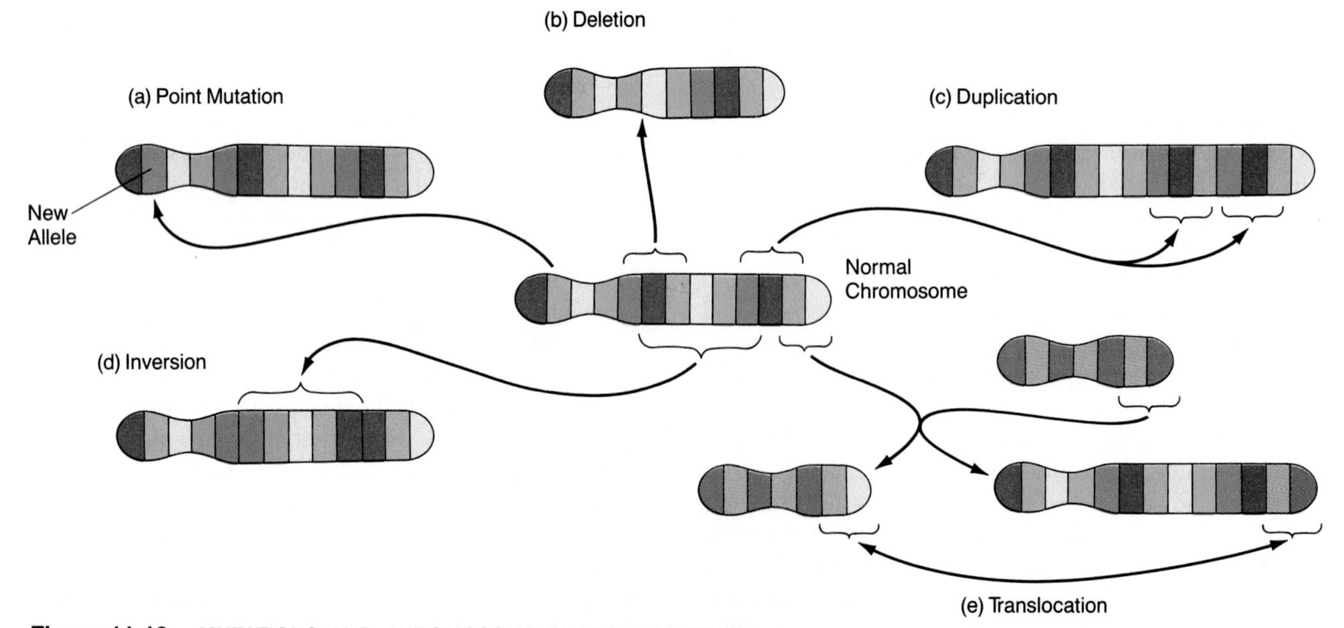

Figure 11-12 MUTATIONS AND CHROMOSOME REARRANGEMENTS.
Several processes can alter the structure or arrangement of genes on a chromosome. As the text explains, these include (a) point mutations, (b) deletions, (c) duplications, (d) inversions, and (e) translocations.

now frequently conduct tests that employ *Salmonella* bacteria to screen chemicals for possible mutagenic effects. More than 75 percent of the agents shown to be carcinogenic through long, complicated studies on animals also show up as genetic mutagens of bacteria in this fast, simple screening test. Therefore, experimenters use the *Salmonella* test, or **Ames test** (after its originator, Bruce Ames of the University of California at Berkeley), along with chromosome tests such as those Muller devised, to monitor the thousands of physical and chemical agents we are all exposed to regularly in our homes, in our workplaces, and out of doors.

It is easy to see that modern geneticists are standing on the shoulders of giants. Based on the profound theories and remarkable discoveries of such pioneers as Mendel, Weismann, Sutton, Morgan, Sturtevant, Bridges, and Muller, modern scientists have begun to unlock some of life's most complex secrets, including the chemical nature of the gene, our subject in Chapter 12.

SUMMARY

1. Genes on the same chromosome are said to be *linked* if they segregate together during meiosis; they are unlinked if they lie on separate chromosomes and segregate independently. Crossing over during prophase of meiosis I can cause pieces of chromatids to be exchanged, producing recombinant chromatids.

2. Genes that are close together on a chromosome recombine as a result of crossing over less frequently than do genes that lie farther apart.

3. Sturtevant showed that genes are arranged on chromosomes in a linear fashion. Subsequent analysis of the banding patterns of giant *polytene chromosomes* in *Drosophila* has yielded *cytogenetic maps*.

4. Genes control specific biochemical functions. Many dominant alleles are forms of genes that function normally, whereas most recessive alleles are forms of genes that have lost part or all of their ability to function.

5. Some genes exist in many alternative forms, including a *wild-type allele* and variations on it; such groups are called *multiple allelic series*.

6. *Complementary genes* are sets of genes whose products must act together to produce a given phenotype; thus, all genes must be present and active.

7. Masking of the phenotypic effects of one gene by an entirely different gene is called *epistasis*.

8. The presence of certain mutated alleles can lead to an organism's death. Such *lethal alleles* usually are recessive; heterozygotes for a lethal allele may show an unusual phenotype, as does the Manx cat.

9. One gene sometimes affects several traits; this phenomenon is called *pleiotropy*.

10. *Point mutations* alter the properties of single genes; changes that alter the structure of chromosomes are called *chromosomal mutations*. In nature, mutations occur randomly and spontaneously, but in low frequency; in the laboratory, they may be induced by physical or chemical factors (*mutagens*).

KEY TERMS

Ames test
carcinogen
chromosomal mutation
codominant allele
complementary gene
complete linkage
epistasis
gene map
lethal allele
linkage
locus
multiple allelic series
mutagen
mutation
mutation rate
nonlinkage
partial linkage
pleiotropy
point mutation
polytene chromosome
recombinant genotype
recombinant phenotype
recombination frequency
wild-type allele

QUESTIONS

1. Mendel crossed true-breeding peas producing round yellow seeds (*RRYY*) with peas producing wrinkled green seeds (*rryy*). All the F_1 seeds were round and yellow. What were the genotypes of the gametes in the F_1 flowers? Were there recombinant genotypes? Did the recombinant types represent 50 percent of the gametes and show independent assortment? Or did they represent less than 50 percent and show linkage of the two genes?

2. Many of the first mutations isolated by geneticists were sex-linked. Why would this have been true?

3. From which parent did Gregor Mendel inherit his *X* chromosome? Mendel's mother was a normal female with two *X* chromosomes, one from each parent. Her parents inherited them from their parents. Which of her four grandparents could *not* have contributed either of her *X* chromosomes? Which of Gregor's eight great-grandparents could *not* have been the source of his *X* chromosome?

4. What are homologous chromosomes? In what sense are the *X* and *Y* chromosomes homologous? In what sense are they not homologous?

5. Human ABO blood types are determined by three alleles at one locus: I^A, I^B, and I^O. I^A and I^B are codominant. Explain what this means. I^O is recessive. Is it possible for a normal individual to carry all three alleles? Explain. What gametes will someone with type AB blood produce? What gametes will someone with type O blood produce?

6. A fairly common sex-linked trait in humans is red-green color blindness. The allele for normal vision is dominant. In the United States, 6 percent of males are green color-blind, but fewer than 1 percent of females are. Explain why there are more color-blind males than females. Does every color-blind male have a color-blind parent? Does every color-blind female?

7. Severe combined immune deficiency (SCID), leaves infants with no immune system and subject to death by massive infection, usually within their first year of life. No female infant has ever been seen with this condition; explain why all victims of this disease are male.

8. Humans, mice, cats, and other animals have several dominant alleles that produce skeletal abnormalities (short, absent, or kinked tails, or short fingers and toes) in the heterozygotes. When two affected individuals mate, only two-thirds of their offspring are affected instead of the expected three-fourths. In one respect, these alleles are recessive. Explain.

9. Short-tailed sheep never breed true; they produce both short-tailed and long-tailed lambs in a 2:1 ratio. Matings of short-tailed sheep with long-tailed sheep produce equal numbers of short-tailed and long-tailed lambs. The short-tail trait is due to a mutant allele at one locus. Which allele is dominant? What genotypes are possible for a long-tailed sheep? For a short-tailed sheep?

10. If the map distance between *a* and *b* is 7 and between *b* and *c* is 46, what frequency of recombinant individuals would be expected between *a* and *b*, *b* and *c*, and *a* and *c*? If the only available information is the distance between *a* and *c*, what could be concluded about linkage between *a* and *c*?

The progeny from a mating of parents *AaBb* and *aabb* produce 96 offspring with *AB*; 84 with *ab*; 8 with *Ab*; and 12 with *aB*. What is the map distance between *a* and *b*?

11. A leghorn chicken with a large single red comb is homozygous for two unlinked recessive genes (*aabb*). It is mated with a hen, and the F_1 progeny are mated. The F_2 progeny display four comb types: walnut, rose, pea, and single, in a ratio of 9:3:3:1. Use a Punnett square to show the probable genotypes of each comb type and include the dominant alleles.

For additional readings related to topics in this chapter, see Appendix C.

12 Discovering the Chemical Nature of the Gene

The fundamental biological invariant is DNA.
Jacques Monod, *Chance and Necessity* (1970)

The discovery of the structure and activity of genes was one of the greatest intellectual adventures of modern science. The result of this century-long scientific exploration was no less than an understanding of the universal code of life, written in molecules of DNA. This search involved two avenues of inquiry pursued by geneticists and biochemists, working with little awareness or interest in each other's findings.

Geneticists were largely pursuing the mystery of gene activity. They knew that genes are copied precisely when chromosomes duplicate during cell division; that genes somehow bring about an organism's physical traits; and that genes occasionally change, or mutate, and give rise to new traits in subsequent cells or organisms. They did not know, however, *how genes act* to bring about these effects.

During the same decades, biochemists were studying a class of biological molecules that had been discovered

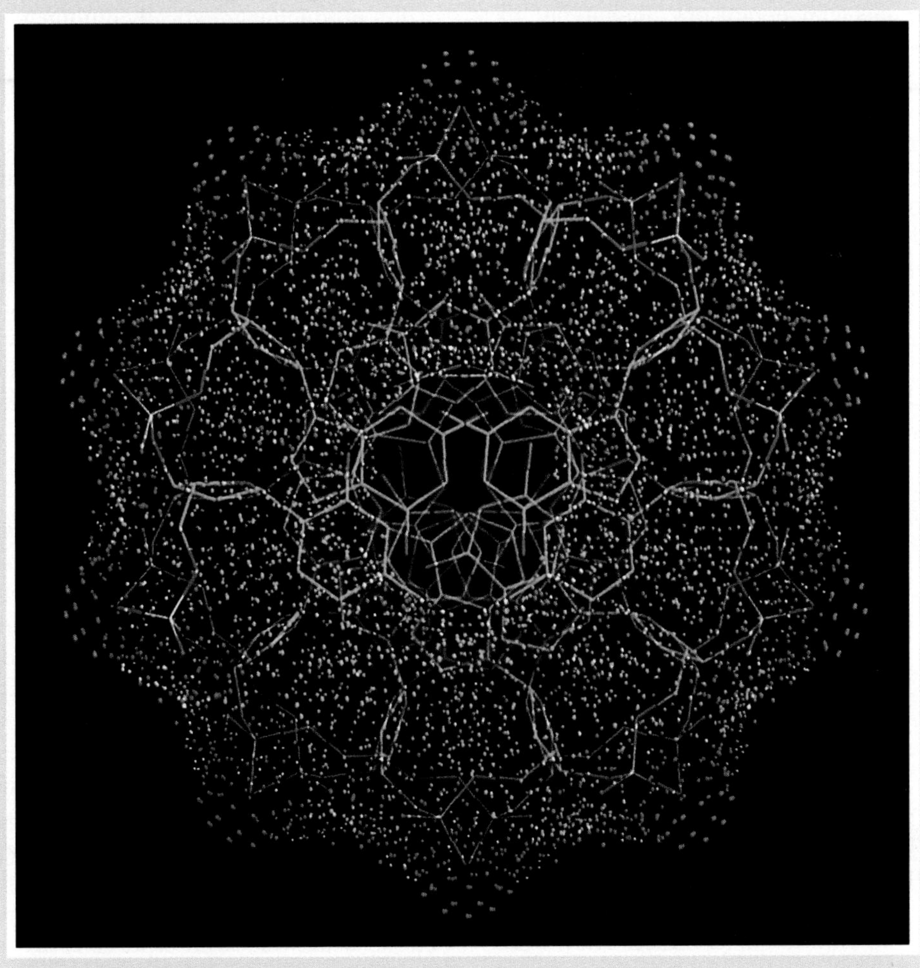

A computer-generated graphic of DNA, life's hereditary material, looking down through the center of the long molecule.

in the nucleus of cells—nucleic acids—and were trying to determine their properties and functions in living things. In 1953, with the work of James Watson and Francis Crick, the two lines of inquiry finally converged: By unraveling the structure of DNA, Watson and Crick explained the chemical nature of genes and suggested how genes are replicated and how they function.

This chapter retraces the historic experiments that unlocked the chemical code of life. It discusses:

- The proof that genes code for particular proteins—specifically, that one gene codes for one polypeptide
- The search for the chemistry of nucleic acids
- The race to reveal the molecular structure of DNA
- The way DNA is copied with amazing accuracy

GENES CODE FOR PARTICULAR PROTEINS

How do genes affect an organism's phenotype, its physical traits? Just 2 years after Mendel's work was rediscovered, the English physician Sir Archibald Garrod addressed this question by studying the human disease called alkaptonuria. Victims of alkaptonuria excrete a compound called *homogentisic acid* that turns their urine dark. This compound can build up and create vision problems and severe arthritis and can even turn the ears and nose black. Garrod hypothesized that the harmful compound builds up because the enzymatic reaction needed to break it down is blocked in victims of the disease.

While not stating it explicitly, Garrod implied that a defective gene is responsible for the absence of a particular enzyme, and thus he was the first to suggest that genes and enzymes are related and that genes are linked with specific chemical reactions in the body. Like Mendel's ideas, however, Garrod's work was ahead of its time and was not, in fact, addressed for several decades, until geneticists began to reexplore the same intellectual terrain.

The One Gene–One Enzyme Hypothesis

In the late 1930s, two geneticists, American George W. Beadle and Russian-born Boris Ephrussi, investigated particular genes in *Drosophila,* including the specific chemical reactions those genes direct. They were able to show that wild-type eye color in fruit flies depends on a series of biosynthetic reactions, each of which is controlled by a different gene. Beadle then went on with colleague Edward L. Tatum to discover that individual genes specify the activity of individual enzymes. In 1941, when Beadle began working with Tatum at Stanford University, the two decided to reverse the approach that had worked so well in early genetic studies: Rather than look for particular chemical reactions directed by particular genes in fruit flies, they decided to switch to an organism whose enzymes and chemical reactions were already understood and try to induce mutations in genes that would block these familiar reactions. They suggested that each gene might code for a particular enzyme—the so-called **one gene–one enzyme hypothesis.** But were they correct?

As their research model, Beadle and Tatum chose the red bread mold *Neurospora crassa.* This fungus grows quickly and easily from spores and is easy to maintain in the laboratory. *Neurospora* is haploid throughout most of its life cycle, and thus recessive mutated alleles are not masked by dominant alleles on the homologous chromosome. Moreover, *Neurospora* can reproduce asexually, producing genetically identical generations, or sexually, producing genetically unique generations. During sexual reproduction, the male and female gametes fuse, and the diploid zygote undergoes meiosis to produce the spores contained in tiny spore cases called *asci* (singular, *ascus*) within a cup-shaped fruiting body (Figure 12-1a). Inside those narrow asci, *ascospores* are produced with the order of the spores in each ascus reflecting the distribution of chromatids during meiosis (Figure 12-1b). Researchers can thus observe and analyze the products of meiosis directly (Figure 12-1c). Finally, and perhaps most important, *Neurospora* can grow on a simple synthetic medium containing no complex biochemicals but just a few inorganic salts, sugar, and the vitamin biotin (a vitamin of the B complex). From this so-called *minimal medium,* the fungus can synthesize all the more complex compounds its cells require.

Beadle and Tatum's plan was to cause mutations in *Neurospora* that would eliminate a mutant individual's ability to synthesize a specific biochemical it needed in order to grow. If they could link each mutated gene with an abnormality in one enzyme involved in synthesizing a nutrient, then perhaps they could establish the direct link between genes and enzymes.

The researchers first irradiated *Neurospora* cells with x-rays to induce genetic mutation (Figure 12-2a), then mated offspring from the irradiated culture with unirradiated (wild-type) cells to produce haploid ascospores. Beadle and Tatum cultured individual ascospores on a rich medium containing many vitamins, sugars, and amino acids that wild-type cells are able to make for themselves (Figure 12-2b). They took those ascospores

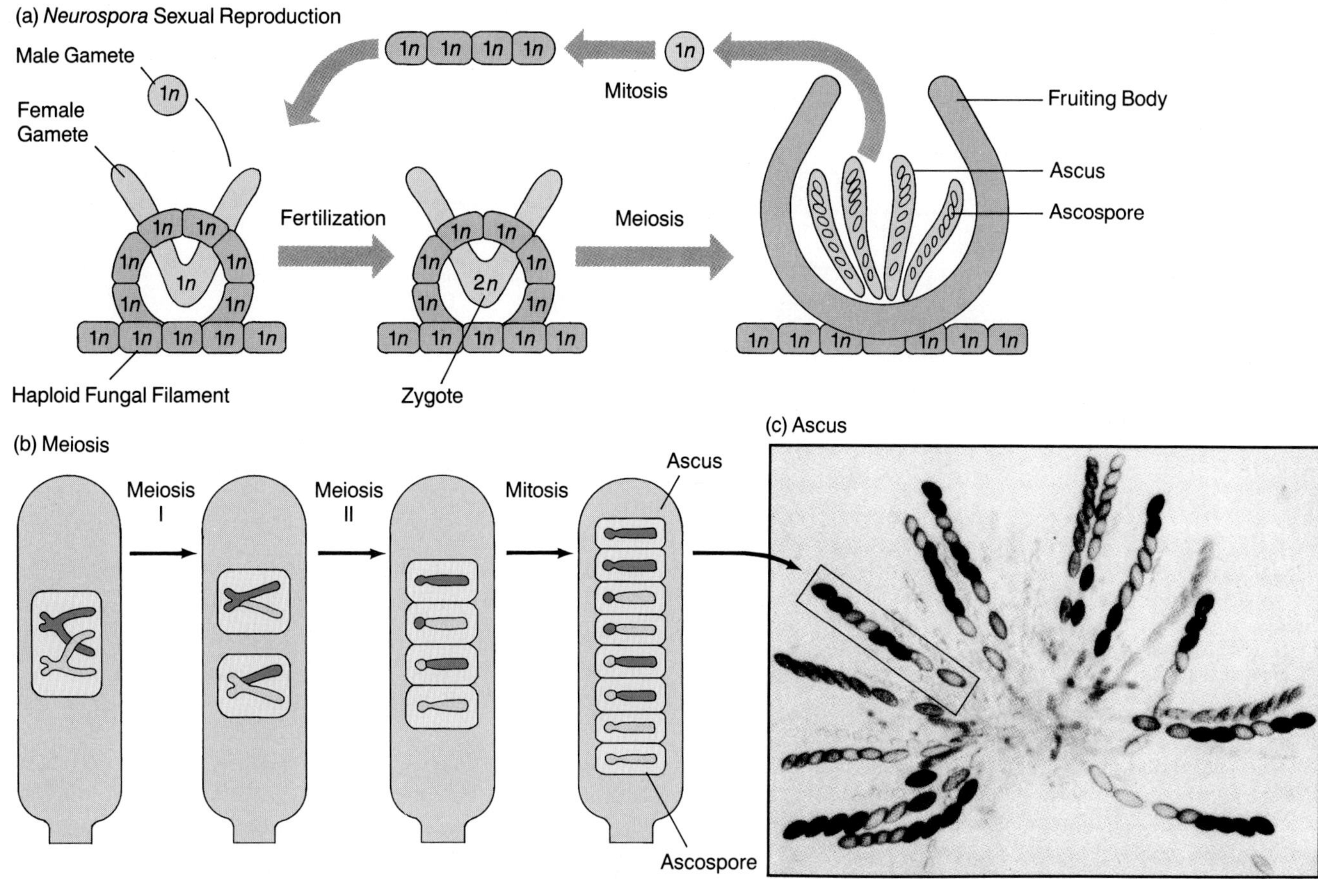

Figure 12-1 UNIQUE GENETIC SUBJECTS: ASCOSPORES OF THE RED BREAD MOLD, *NEUROSPORA CRASSA.*
(a) After haploid gametes of the red bread mold *Neurospora* fuse (left), meiosis takes place in the zygote, and a sexual fruiting body develops. This contains spore cases (asci), and inside of them grow ascospores. (b) *Neurospora* is ideally suited for genetic studies because the spores are arranged within the spore case in a linear order reflecting their origin during meiosis, an order that is maintained during subsequent mitosis. Because of this unique characteristic, an experimenter can deduce the genetic makeup of each zygote directly without statistical analysis. (c) This squashed fruiting body shows the eight spores in each case. Note the light and dark spores, reflecting the distribution of alleles during meiosis that led to those colors.

that grew on the rich medium and retested them in the kind of minimal medium that supports normal *Neurospora* cells (Figure 12-2c). Those that could not grow on the minimal medium clearly had mutated genes that interfered with some metabolic pathway for synthesizing some compound needed for growth.

Now they tested each mutant in a minimal medium supplemented with *one* specific complex molecule added to it—an amino acid, or a sugar, or a vitamin (Figure 12-2d). Most of the time, the mutant could not grow on the minimal medium plus one nutrient. If the spore could grow, however, the investigators knew they had supplied the nutrient that was missing as a result of the mutated allele. In this way, Beadle and Tatum isolated a number of mutant strains, each of which needed a single additional nutrient such as vitamin B_6 or the amino acid histidine in order to grow on a minimal medium (Figure 12-2d and e).

Finally, they crossed the mutant strains with normal *Neurospora*. If they found a 1:1 ratio of mutated to normal ascospores in a given ascus of the F_1 generation, they knew that only a single allele had changed and that each such mutant allele must interfere with the pathway for synthesizing the vitamin or amino acid by affecting a single enzyme. The one gene–one enzyme hypothesis appeared to be correct. In the years that followed the famous *Neurospora* studies, researchers found that occasionally, one nutrient can correct a defect in more than one mutated gene. This did not in itself contradict the one gene–one enzyme hypothesis, but it underscored the fact that a number of enzymes sequentially carry out different steps in each biochemical pathway (see Chapter 4). Thus, a mutation in any enzyme in the series can prevent the final product from forming. Not long after the *Neurospora* studies, new experiments expanded the one gene–one enzyme concept still further.

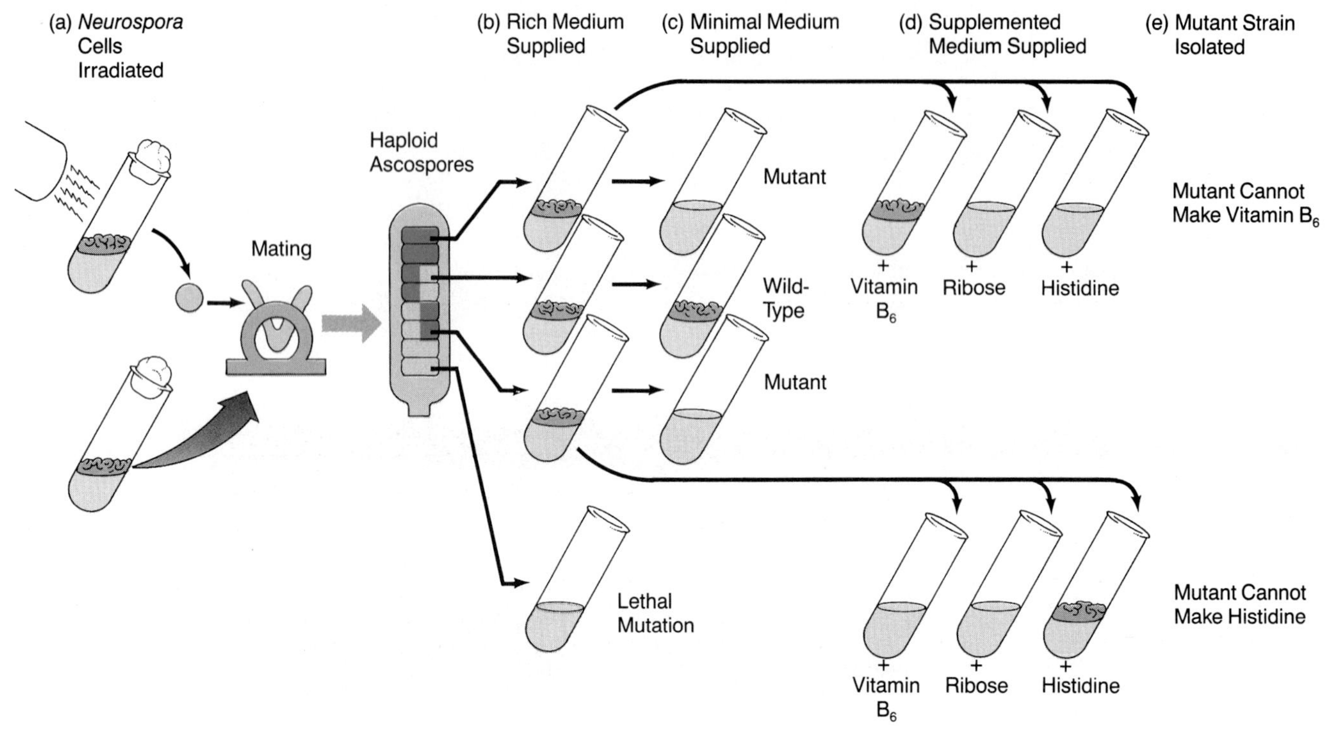

Figure 12-2 DETECTING *NEUROSPORA* MUTANTS.
Beadle and Tatum invented an ingenious method for detecting nutritional mutants in the red bread mold. Wild-type *Neurospora* ordinarily grows on a minimal medium, but nutritional mutants lack the ability to synthesize one or more nutrients. (a) Beadle and Tatum irradiated wild-type *Neurospora* to induce mutations and then mated the irradiated culture with an untreated wild-type culture to produce haploid ascospores. (b) They then introduced ascospores into a rich (complete) medium, where all grew. (c) When they transferred samples of each colony to a minimal medium, only the wild-type *Neurospora* grew. (d) In the final stage of the experiment, Beadle and Tatum transferred the mutant strain to supplemented media containing different vitamins. (e) Their results showed that one mutant strain could not synthesize vitamin B_6, and another could not synthesize histidine.

The One Gene–One Polypeptide Hypothesis

Beadle and Tatum's work raised a new question: Does the link with genes extend past enzymes to other kinds of proteins as well? In 1949, chemist Linus Pauling helped refine the one gene–one enzyme concept into its currently accepted form, the *one gene–one polypeptide hypothesis*.

Pauling and colleagues studied the role of the blood protein hemoglobin in the disease sickle-cell anemia (see Figures 11-11 and 12-3a). Pauling wondered whether the abnormal (recessive) allele could somehow affect the hemoglobin molecules in a way that caused red blood cells to sickle. Pauling's research group used *electrophoresis*—a technique that separates molecules on the basis of their electric charge—to compare hemoglobin from normal individuals, from homozygous recessives with a severe form of the disease, and from heterozygotes who had one abnormal allele but did not show disease symptoms.

Pauling's results, graphed in Figure 12-3b, confirmed that normal hemoglobin (HbA) has a different electric charge than mutant hemoglobin (HbS), migrates differently in an electric field, and hence is structurally different. This discovery suggested that the sickle-cell gene chemically alters the hemoglobin protein and automatically extended "one gene–one enzyme" to "one gene–one protein."

In 1956, British researcher Vernon Ingram, working at the Massachusetts Institute of Technology, modified this hypothesis yet again. Recall that normal hemoglobin consists of four polypeptide chains: two identical alpha (α) chains and two identical beta (β) chains (Figure 12-3c). Ingram discovered that in sickle-cell hemoglobin, both α chains are normal, but both β chains are mutant, with each abnormal β chain containing the amino acid valine in place of glutamic acid.

Clearly, if one gene codes for one protein, a single gene should have coded for both the α and β chains of the hemoglobin molecule. Research on another hereditary blood disease, thalassemia, also showed that a defect in just the α chain could destroy hemoglobin function. Such findings together suggested that two genes, one for each type of polypeptide chain, must be operating to produce a complete hemoglobin molecule. By the late 1950s, the theory of gene activity had been refined to the **one gene–one polypeptide hypothesis,** the idea that each gene brings about the formation of one polypeptide chain. We will see precisely *how* genes code for polypeptides in Chapter 13. In the meantime, let's see how researchers revealed the chemical nature of genes.

THE SEARCH FOR THE CHEMISTRY OF THE GENE

The study of nucleic acids had its start in 1869, when a young Swiss physician named Johann Friedrich Miescher began to collect human white blood cells from the pus on discarded bandages and to dissolve the cells' proteins with pepsin, a protein-digesting enzyme. Curiously, most of the contents of each cell nucleus remained intact during this enzymatic attack, suggesting that it was not a protein; Miescher called the material "nuclein." Others isolated nuclein from additional cell types, and by 1900, Miescher's compound was being called *nucleic acid.*

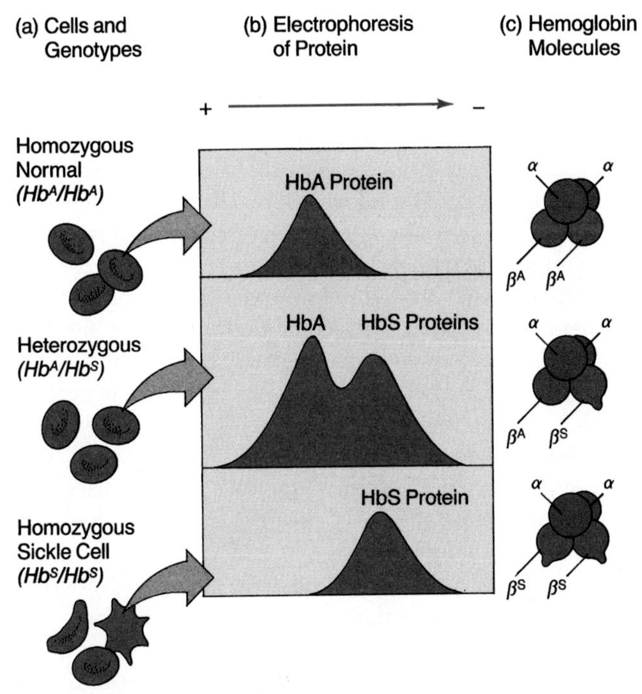

Figure 12-3 HEMOGLOBIN S FROM VICTIMS OF SICKLE-CELL ANEMIA COMPARED WITH NORMAL HEMOGLOBIN. (a) Hemoglobin from a normal person (Hb^AHb^A), from a heterozygous person with sickle-cell trait (Hb^AHb^S), and from a homozygous person with sickle-cell anemia (Hb^SHb^S) moves differently in an electric field. (b) During electrophoresis, normal hemoglobin moves closer to the positive pole, whereas the sickle-cell hemoglobin moves closer to the negative pole. (c) Normal hemoglobin has two α and two β chains. Hemoglobin from a heterozygote has one normal and one mutant, misshapen β chain. Hemoglobin S from a sickle-cell patient has two mutant β chains, and these cause the red blood cells to sickle.

A few years later, Robert Feulgen, a German chemist, developed a staining technique that dyes nucleic acids a deep red. Feulgen found that the red dye became concentrated in the chromosomes within the nucleus and concluded that chromosomes are largely made up of the newly identified nucleic acid. Some researchers speculated that nucleic acid might carry hereditary information, but others questioned whether the substance could convey such complex and variable information.

To confuse the issue, work in several labs had demonstrated that the nucleus also contains some protein. Because biologists knew that large protein molecules are built from about 20 types of amino acids and have great structural diversity, many in the early 1900s chose to believe that the genes themselves must be made of protein, not nucleic acid.

Genes: Nucleic Acid, Not Protein

The first real insight into the structure of genes came in 1928, when a British bacteriologist named Frederick Griffith was attempting to develop a vaccine against pneumococcus, the bacterium that causes pneumonia. In his work, Griffith inadvertently discovered a substance that could change one form of bacterium into another form with a different set of hereditary traits—a strange but obvious sign of genetic activity.

In that era, before the advent of penicillin and other antibiotics, pneumonia claimed thousands of lives each year. A vaccine seemed like the major hope for preventing these deaths, and Griffith studied two strains of pneumococci to try to develop such a vaccine. The *virulent,* or disease-causing, strain of pneumococci secretes a polysaccharide coat called a *capsule* around its cell wall. The capsule protects the bacterium from destruction by the host animal's immune system. Virulent pneumococci form large, smooth, glistening colonies and are designated S, for "smooth" (Figure 12-4a). The other strain Griffith studied lacks a capsule and is nonvirulent. These harmless cells form small, rough colonies and are designated R, for "rough."

When Griffith injected mice with live S-type pneumococci, they died from pneumonia (Figure 12-4b), while those he injected with live R-type bacteria survived (Figure 12-4c), as did those he injected with

S-strain pneumococci that had been heat-killed (Figure 12-4d). Curiously, though, Griffith found an unexpected fourth pattern in his well-designed controls: When injected with a mixture of dead S-strain and live R-strain bacteria, the mice died of pneumonia (Figure 12-4e). He found the tissues of the dead mice to be teeming with *live,* virulent S-strain pneumococci. How could this be? The live R-strain bacteria were harmless, and the dead S-strain bacteria certainly could not have come back to life! Griffith hypothesized that some material—perhaps hereditary material—from the dead S-strain cells must have transformed the harmless R-strain cells into the virulent S strain. Researchers soon found that an *extract* of the fluid surrounding the dead cells is sufficient to transform nonvirulent into virulent bacteria.

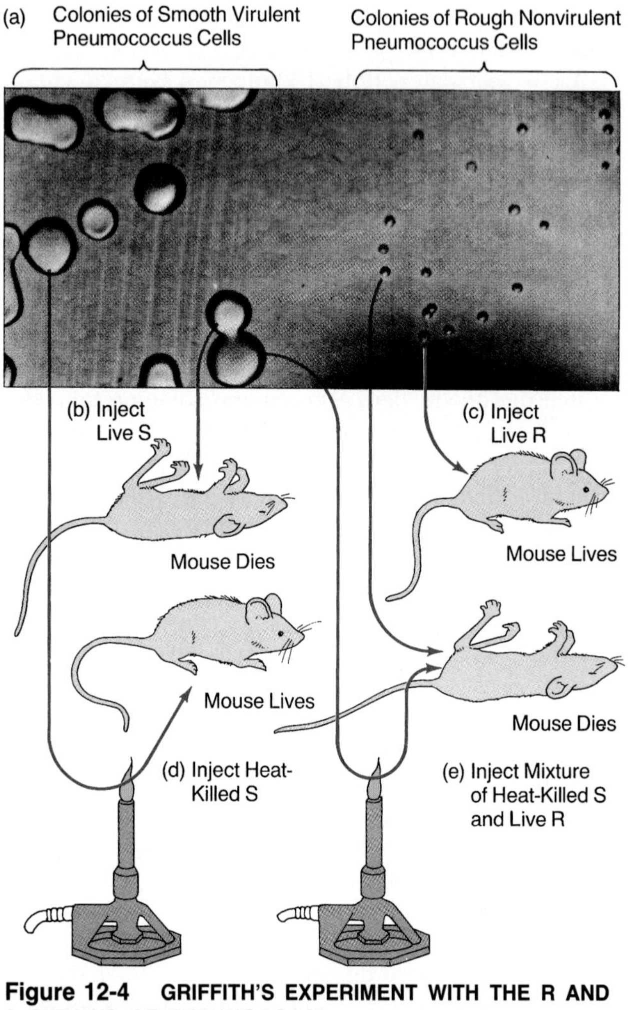

(a) Colonies of Smooth Virulent Pneumococcus Cells — Colonies of Rough Nonvirulent Pneumococcus Cells

(b) Inject Live S — Mouse Dies

(c) Inject Live R — Mouse Lives

(d) Inject Heat-Killed S — Mouse Lives

(e) Inject Mixture of Heat-Killed S and Live R — Mouse Dies

Figure 12-4 GRIFFITH'S EXPERIMENT WITH THE R AND S STRAINS OF PNEUMOCOCCI.
(a) The nonvirulent R (rough) strain of pneumococcus, on the right, grows on the same Petri dish with the virulent S (smooth) strain, on the left (both magnified about 1,640×). The S strain secretes a capsule around its cell wall that protects it from a host animal's immune system. (b) When a researcher injects a mouse with live S-strain bacteria, the rodent dies, whereas (c) a mouse injected with live R-strain bacteria lives. This shows that the S strain is virulent.
(d) When the S strain is killed by heat and then is injected, the mouse lives. (e) However, when the heat-killed S strain is mixed with the live R strain and injected, the mouse dies. A blood sample from the dead mouse shows that mixing the heat-killed S strain and the R strain transforms the live R strain into an S strain. Experiments years after Griffith's proved that the DNA of the heat-killed virulent strain was incorporated into the nonvirulent strain, rendering it virulent—and deadly.

In 1944, Oswald T. Avery, Colin MacLeod, and Maclyn McCarty at Rockefeller University set out to identify the substance in the extract. They gradually removed one chemical compound after another—first the proteins, then the carbohydrates, next the lipids—each time testing the ability of the material to transform nonvirulent pneumococci into virulent pneumococci. Finally, there was virtually nothing left in the extract but a fine, clear viscous thread that Avery could pick up on a glass stirring rod (review Figure 1-14). When dissolved, this threadlike substance could transform R-strain into S-strain bacteria. Chemical analysis showed the viscous substance to be millions of molecules of a nucleic acid called *deoxyribonucleic acid,* or *DNA.*

Despite the careful work of Avery's team in isolating DNA as the transforming factor in pneumococci, many biologists still could not accept that nucleic acids could carry complex genetic information. Thus, until the early 1950s, many continued to believe that proteins could serve as a richer information code.

At that time, Alfred Hershey and Martha Chase at the Carnegie Institution devised a definitive test for the chemical nature of genes and established beyond question that the carrier of hereditary information is indeed DNA.

The Carnegie researchers studied the intestinal bacterium *Escherichia coli* and a certain class of bacterial viruses, or *bacteriophages* ("phages" for short), that specifically infects and destroys *E. coli* cells. The experimenters knew that bacteriophages have a distinctive shape (Figure 12-5a); that they possess genes; and, significantly, that they are composed solely of protein and DNA. They also knew that bacteriophages attach to the outside of the *E. coli* cell (Figure 12-5b) and that within about 25 minutes, the bacterium bursts, releasing hundreds of complete, new bacteriophages. Hershey and Chase assumed that since the phage body remains outside the cell, the phages somehow send their genes in to direct the production of new virus particles. Because phages consist only of protein and DNA, whichever substance they injected into the bacteria had to be the genetic material.

The team had one chemical clue; they knew that DNA contains phosphorus, whereas most proteins do not, and that most proteins contain sulfur, whereas DNA

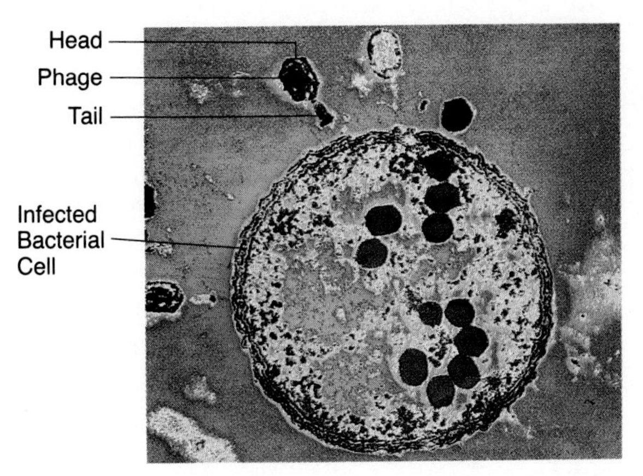

Figure 12-5
BACTERIOPHAGE: CRYSTALLINE PARASITES.
This electron micrograph (magnified about 100,000×) reveals the structure of a bacteriophage. The head contains DNA and sits above the tail. DNA passes down the core of the tail to enter a bacterium.

does not. By using radioactive isotopes of the two elements, they could label and trace the injected material. Hershey and Chase grew one set of *E. coli* cells and bacteriophages on a culture medium containing a radioactive isotope of phosphorus (^{32}P) and another set of cells on radioactive sulfur (^{35}S) (Figure 12-6a). The new phage particles that formed in one culture thus had ^{32}P-radioactively labeled DNA ("hot DNA") but had nonradioactive ("cold") proteins, whereas those that formed in the other culture had ^{35}S-radioactively labeled ("hot") proteins but "cold" DNA.

The researchers used both types of hot phages to infect unlabeled *E. coli* cells in isotope-free media (Figure 12-6b). Within a few minutes after mixing the phages

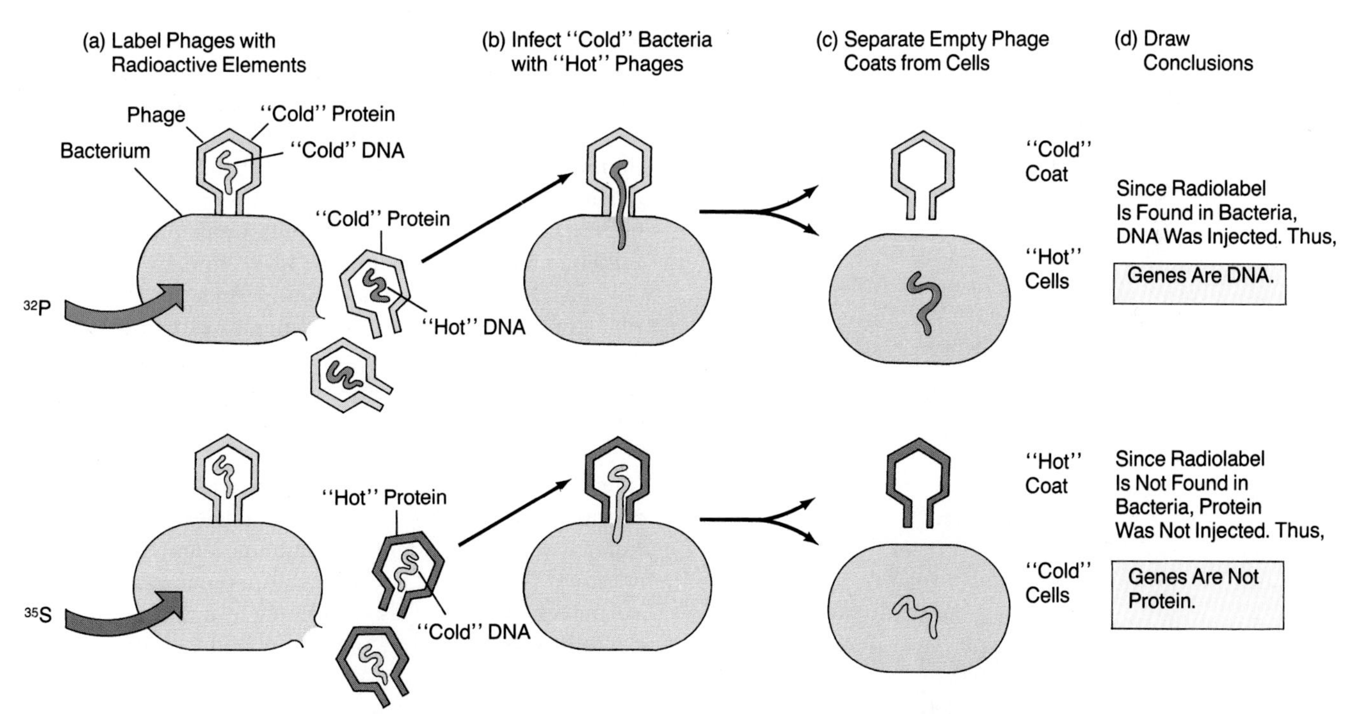

Figure 12-6 THE HERSHEY-CHASE BLENDER EXPERIMENTS: PROOF THAT THE GENETIC MATERIAL IS DNA, NOT PROTEIN.
Hershey and Chase employed bacteriophages, which consist only of DNA with a protein coat. Sulfur is a constituent of many proteins, but not of DNA, and phosphorus is a constituent of DNA, but not of most proteins. (a) The researchers used radioactively labeled sulfur and phosphorus to discover whether protein or DNA was incorporated into the bacteria, and then into the next generation of virus. (b) They infected "cold" bacteria with "hot" phages, waited for the phages to inject their DNA into the cells, then, with an electric blender, (c) they separated the empty phage coats and studied the cells to see which component the phages had injected. (d) From this they concluded that genes are made of DNA, not protein.

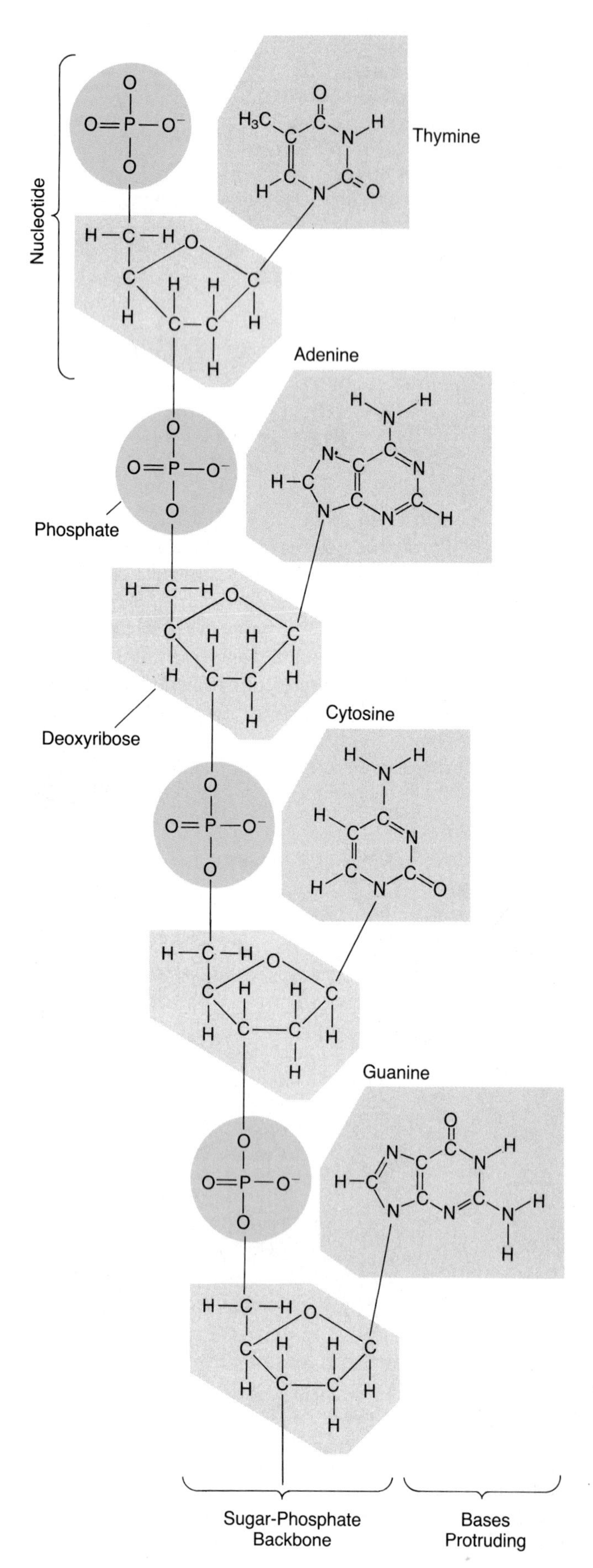

and the bacteria, the phages had inserted their genes, leaving behind empty viral coats that clung to the *E. coli* walls. Hershey and Chase agitated the two cultures in blenders to knock the empty virus particles from the cell surfaces. Then they separated the empty phage coats from the cells (Figure 12-6c). When they analyzed the cells and phage coats from the experimental run that used ^{32}P, they found the radioactive element associated with the *E. coli* cells, not with the empty "cold" phage protein coats, showing that DNA had entered the cells (Figure 12-6c and d). When they analyzed the coats and cells from the ^{35}S experimental run, they found radioactivity associated with the coats, but not the cells. Thus, phage protein had not entered the bacteria. Since phages inject only their genes, the researchers had proved that genes must be composed of DNA, not protein (Figure 12-6d).

Chemical Composition of DNA Revealed

A new question emerged from the knowledge that genes are made of DNA: How does the DNA molecule carry genetic information?

Years earlier, biochemist P. A. Levene had correctly identified the four bases in DNA—**adenine** (A), **guanine** (G), **cytosine** (C), and **thymine** (T)—each a ring structure composed of carbon and nitrogen (Figure 12-7). Adenine and guanine, double-ring structures, are called **purines;** thymine and cytosine, single-ring structures, are called **pyrimidines.** The molecule made up of a base plus the five-carbon sugar deoxyribose plus a phosphate group is called a **nucleotide** (see Figure 12-7). In each long molecule of DNA, the phosphate group links the five-carbon sugar of one nucleotide to the five-carbon sugar of the next nucleotide in the chain. This phosphate bonding creates a sugar-phosphate backbone, with the nitrogenous bases protruding (see Figure 12-7).

Another biochemist, Erwin Chargaff, made the additional finding that the four bases occur in differing proportions in the DNA of different species. The DNA in human cells, for example, contains about 30 percent adenine, 30 percent thymine, 20 percent guanine, and 20 percent cytosine, as compared to a sea urchin's 33 percent A and T and 17 percent G and C. The constant and variable features of DNA became known as **Char-**

◀ **Figure 12-7 THE STRUCTURE OF DNA AND ITS SUBUNITS.**
Each purine base (adenine or guanine) and each pyrimidine base (thymine or cytosine) is joined to deoxyribose sugar. A DNA molecule is built from chains of nucleotides, with the nitrogenous bases protruding as side groups from a phosphate-sugar backbone.

gaff's rules: First, A and T occur in equal amounts, as do G and C; second, the ratio of A and T to G and C is constant within a species but varies among species. To truly understand the mechanisms of inheritance at the molecular level, biologists had to account for these rules in the structure of DNA.

THE RESEARCH RACE FOR THE MOLECULAR STRUCTURE OF DNA

In the wake of these classic studies, a race began to uncover the precise structure of DNA, and the prize would soon go to two young researchers, the American biochemist James D. Watson and an English physicist turned molecular biologist named Francis Crick. Watson and Crick met at the Cavendish Laboratory at Cambridge University in England. Taking a theoretical route to the structure of nucleic acids, rather than designing and performing experiments, they carefully analyzed all the existing evidence on DNA and built models of cardboard and metal that could account for the observable facts.

Watson and Crick concentrated on four important pieces of evidence. First, DNA molecules are long, thin polymers containing four types of nucleotides linked by phosphate bonds. Second, [A] = [T] and [G] = [C]. Third, purified DNA forms a viscous solution like egg white that, when moderately heated, becomes nonviscous (more watery). Since moderate heating cannot break the covalent bonds in the sugar-phosphate backbone, a set of weaker chemical bonds must also help maintain the normal structure of DNA. Finally, Linus Pauling had found that hydrogen bonds (which *can* be broken by moderate heating) often hold polypeptide chains in the shape of an α-helix.

Evidence that DNA itself might have this same helical configuration came from photographs of crystallized DNA taken by Maurice Wilkins and Rosalind Franklin at King's College, London, using a technique called **x-ray diffraction** (Figure 12-8a). With this technique, researchers can determine the spatial arrangement of atoms in molecules—the lengths and angles of chemical bonds and the distances atoms lie from one another throughout the molecule. Franklin speculated that certain shadows and markings might mean that DNA is a helix, with the phosphate backbone on the outside and with a uniform diameter of about 2 nm (Figure 12-9a, b).

Armed with these pieces of evidence, Watson and Crick began trying to build three-dimensional models of DNA—models, they later wrote, ". . . superficially re-

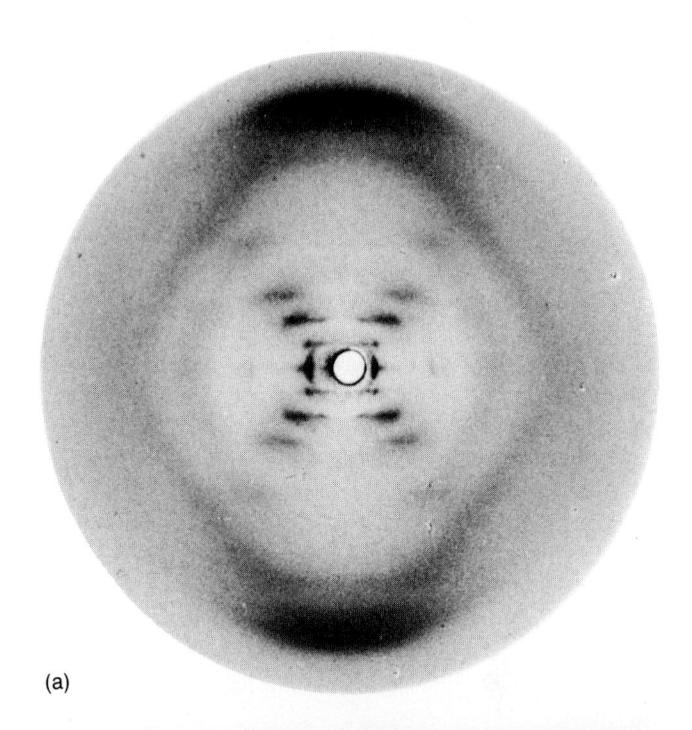

(a)

(b)

Figure 12-8 HISTORIC VIEWS OF THE DNA MOLECULE. (a) Photos of x-ray diffraction patterns such as this one produced by DNA fibers were interpreted to mean that DNA is a helix, with the phosphate backbone on the outside and the nitrogenous bases on the inside. (b) The discoverers of DNA structure, James Watson (left) and Francis Crick, in 1953 with one of the first models of DNA.

sembling the toys of preschool children," but accounting for the size of atoms, the lengths and angles of bonds, and so on (Figure 12-8b). Wilkins and Franklin's measurements of the x-ray diffraction photographs implied two repeating features of DNA, one occurring every 3.4 nm and the other every 0.34 nm. Watson and Crick surmised that the shorter measure, 0.34 nm, might represent the distances between the stacked bases of the nucleotides. Imagine the excitement and elation the young James Watson must have felt when one morning he suddenly saw that the cardboard pair of adenine and thymine has exactly the same shape as the pair of guanine and cytosine. That was the key! The two sets of base pairs could extend across the inside of a double helix, much like rungs on a ladder (see Figure 12-9a). It was obvious, then, that DNA is not a single helix, but *two* chains entwined about each other in a **double helix** (Figure 12-9b), with the phosphates and sugars arranged like the outer rails on a ladder, the nitrogenous bases fitting perfectly in place between the "rails," and with repeating features as Wilkins and Franklin had predicted (Figure 12-9c).

They also realized that if weak hydrogen bonds link A to T and G to C, they could explain the viscous-nonviscous behavior of DNA when heated. At low temperatures, the hydrogen bonds would cause the double helix to be stiff, and so the solution would be viscous. But at higher temperatures, the hydrogen bonds would be broken, and the two helices would come apart; as a result, the DNA would be less stiff, and the solution, too, would be less viscous and more watery.

The only combinations of paired bases that could fit together correctly and allow such hydrogen bonding were adenine-thymine and guanine-cytosine (Figure 12-9a)—a reflection of Chargaff's first rule: [A] = [T] and [G] = [C]. With marvelous insight, Watson and Crick had solved the mystery of the primary genetic molecule in all living organisms on earth.

Now Watson and Crick were in a position to deduce how the DNA molecule carries a wide variety of genetic information, and they did so, again correctly. The A-T and G-C base pairs can occur in *any* sequence along the length of the double helix; thus, the huge number of possible sequences can encode the genetic information

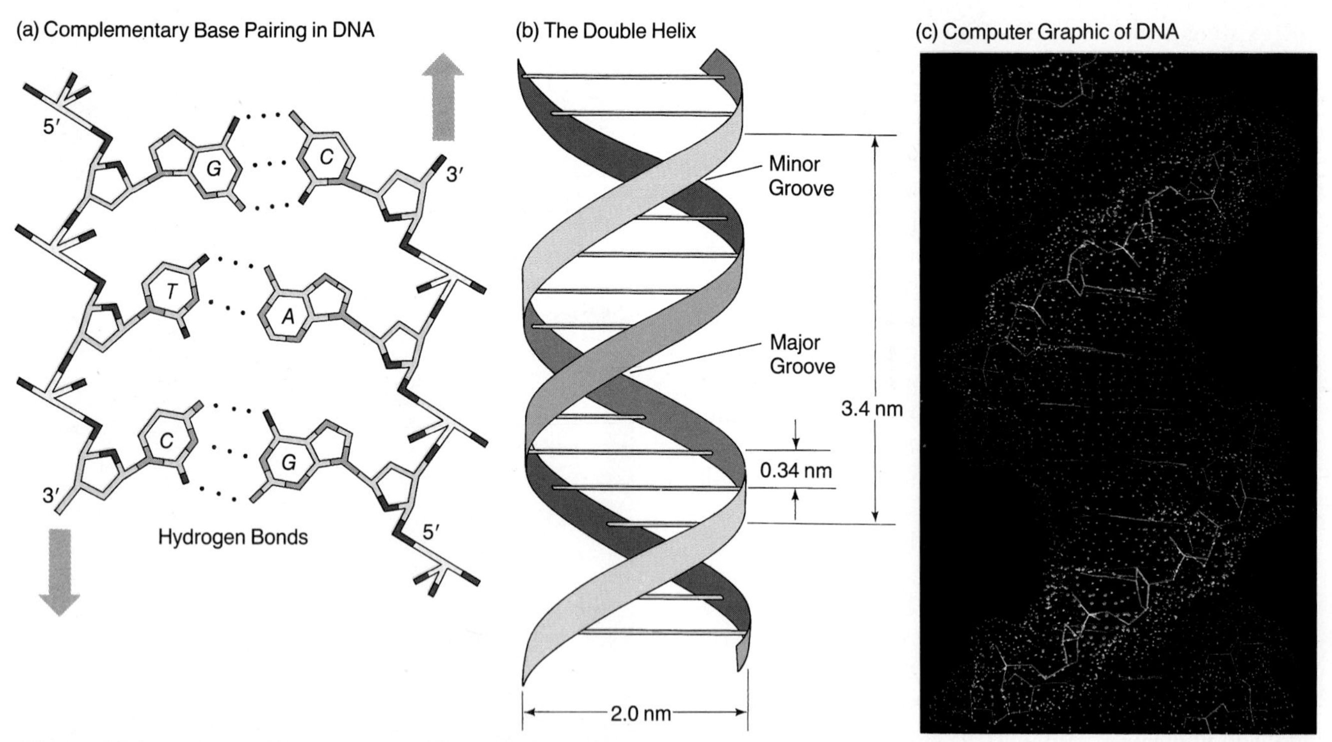

Figure 12-9 DNA DOUBLE HELIX: THE CODE OF LIFE.
(a) DNA consists of two sugar-phosphate backbones running in opposite orientations and arranged as a double helix, with the nitrogenous bases on the inside. These four bases are paired in just two possible combinations: adenine-thymine and cytosine-guanine. Notice that the two chains are arranged in opposite directions because the 5′ end is "up" on one chain and is "down" on the other chain. (b) Here the DNA double helix is depicted as two ribbons winding around each other and enlarged to show the base pairs as the rungs of a winding staircase between the helices. (c) Precise measurements of bond distances in DNA in varying states, plus computer modeling techniques, allow creation of computer-generated views of the genetic material.

THE PROPER MEASURE OF MAN?

Thirty-five years after his Nobel Prize–winning work on the DNA double helix, James Watson is heading an effort that promises to be far larger and, some say, more important than the original race for the structure of DNA: determining the exact sequence for all 3 billion bases in the human DNA. If the U.S. Congress decides to fund the so-called Human Genome Project fully, the effort will cost billions of dollars and take 15 years or more. But the prospect of such a huge and expensive undertaking has touched off vigorous controversy, and the eventual scope of the project is at present uncertain.

Figure A JAMES WATSON, HEAD OF THE HUMAN GENOME PROJECT, IN 1988.

Watson will direct those parts of the project that the National Institutes of Health oversees. Initially, geneticists plan to create a coarse, low-resolution physical map—a set of overlapping fragments of the total DNA arranged in order of location along the human chromosomes and placing the 1,000 or so currently known human genes in proper positions. In a planned second phase, they would locate and identify more of the 50,000 to 100,000 human genes. Eventually, researchers would determine the sequence of all 3 billion bases, even repetitive stretches and those with unknown functions.

Proponents point out the genome project's enormous potential for pinpointing the genes that cause some 3,500 human genetic diseases, including some forms of heart disease, cancer, obesity, diabetes, and Alzheimer's disease. This could help physicians diagnose genetic diseases accurately and might allow researchers to develop specific drugs and treatments. Critics, however, fear that the project would take funds away from more important research areas, and they wonder whether it is worth examining the entire DNA sequence to uncover the 3 percent that actually codes for proteins. Further, they fear that genome researchers would try to "copyright" rather than share their information so they could patent and profit from treatments for specific genetic diseases. And above all, some worry that the gene map could lead to "genetic discrimination," since mapping might reveal a person's inherent tendency toward cancer or heart disease, let's say, even though the individual might never develop the condition.

There are no answers as yet for such concerns, but the Human Genome Project is stimulating appropriately vigorous public discussion.

for the huge variety of proteins and enzymes responsible for each organism's phenotype. In our 46 chromosomes, for example, we humans have about 10 *billion* base pairs in the DNA. Just one sequence may characterize a species' DNA, so since there is a seemingly infinite number of base-pair sequences possible, it is not hard to understand the huge variety of organisms on earth.

Watson and Crick's model accounted both for the observed chemical and physical properties of DNA and for the information-carrying capacity of genes. In Crick's words, there is an "intrinsic beauty of the DNA double helix"—it is a "molecule which has style."

HOW DNA REPLICATES

The double-helix model went beyond even the observed chemical and physical properties of the molecule and its ability to carry complex genetic information. It also helped explain why the molecules normally replicate quite faithfully, but why they sometimes become mutated and then pass along this altered message to subsequent generations.

In the theoretical paper they published in 1953, Watson and Crick predicted a "copying mechanism" by

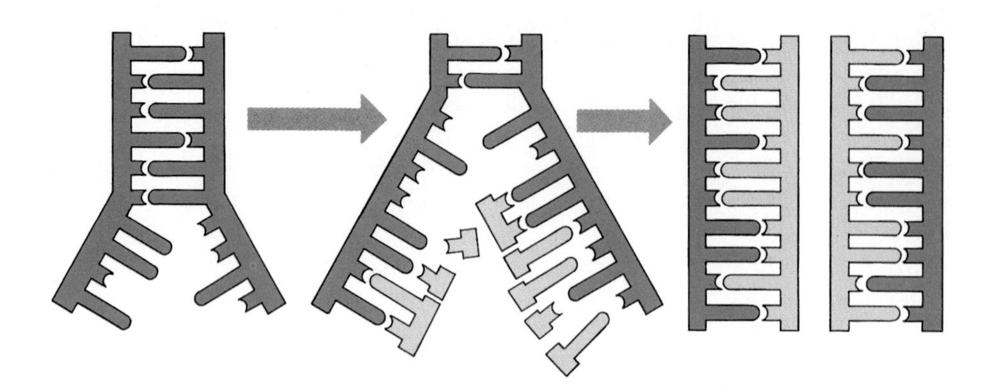

Figure 12-10 THE SEMICONSERVATIVE REPLICATION OF DNA. (a) The "unzipping" of a double-stranded parent molecule, followed by (b) pairing of bases complementary to the unwound strands, results in (c) the formation of two daughter molecules, each consisting of a parent template strand and a newly synthesized strand.

which the double helix might faithfully replicate itself. If the hydrogen bonds that connect adenine to thymine and guanine to cytosine were broken, the DNA molecule would "unzip" down the middle (Figure 12-10a). Because base pairs are complementary, these single strands of DNA could serve as *templates*, or patterns, on which new complementary strands could form. If free nucleotides existed in the surrounding medium, they would be linked by means of hydrogen bonds to the complementary bases on the two chains— free adenines to thymines, free cytosines to guanines (Figure 12-10b). Because of this complementary base pairing, the forming chain, when complete, would have the exact sequence as the original strands that had "unzipped." Thus, the "unzipping" of a double-helical DNA molecule would result in **DNA replication,** the formation of two daughter molecules, each with a "parent" strand and each with a new, "daughter" strand (Figure 12-10c). The two new double-stranded molecules would thus retain the original sequence of nucleotide base pairs found in the parent molecule and in this way preserve its encoded information from one generation to the next.

This copying mechanism could also explain gene mutation. If an error in base pairing occurred during DNA replication, so that an incorrect (noncomplementary) nucleotide became inserted into the forming strand, the new nucleotide sequence would be different from the original. This, in turn, would change the encoded information and could be transmitted to new generations when the cell later divided and its mutated DNA was replicated.

Is DNA Replication Semiconservative?

The Watson-Crick model for DNA replication became known as **semiconservative replication,** since each newly formed DNA molecule has one old strand conserved from the parent molecule and one entirely new daughter strand. Two other possibilities, however, had to be considered: that DNA replication is *conservative*, with one of the daughter DNA molecules conserved intact from the parent and the other daughter molecule completely new; or that it is *dispersive*, with both strands of both daughter DNAs consisting of interspersed fragments of new and old DNA.

Two young scientists at the California Institute of Technology, Matthew Meselson and Franklin W. Stahl, designed a test to determine which was true. They began by growing 12 consecutive generations of *E. coli* on a growth medium containing a heavy isotope of nitrogen, ^{15}N. By the twelfth generation, virtually all the normal nitrogen molecules (^{14}N) in the cells' proteins and nucleic acids had been replaced by heavy nitrogen. DNA molecules containing ^{15}N can be distinguished from those containing ^{14}N on the basis of their densities in a system called a *cesium chloride* (CsCl) *density gradient*. When a researcher centrifuges test tubes containing normal and heavy DNA and CsCl, heavy DNA forms a visible band at a lower level in the tube than does the normal, light DNA with ^{14}N (Figure 12-11a).

In the second step of the experiment, Meselson and Stahl took a batch of *E. coli* with 100 percent ^{15}N DNA and suddenly switched it to a growth medium containing only normal light nitrogen (^{14}N). After allowing the cells to go through only one cell division cycle (one period of DNA replication), they removed some cells in order to extract the DNA and measure its density, and they found that the DNA had a density exactly intermediate between heavy DNA and normal DNA (Figure 12-11b). This supported the hypothesis that each double-stranded molecule contained one ^{15}N strand and one ^{14}N strand, making their combined density midway between the two original densities (Figure 12-11b, bottom). Since there was no full-heavy DNA, this ruled out the "conservative" hypothesis, but left the "dispersive" hypothesis still viable. Measurements of DNA density after second and third cell division cycles, however, supported only

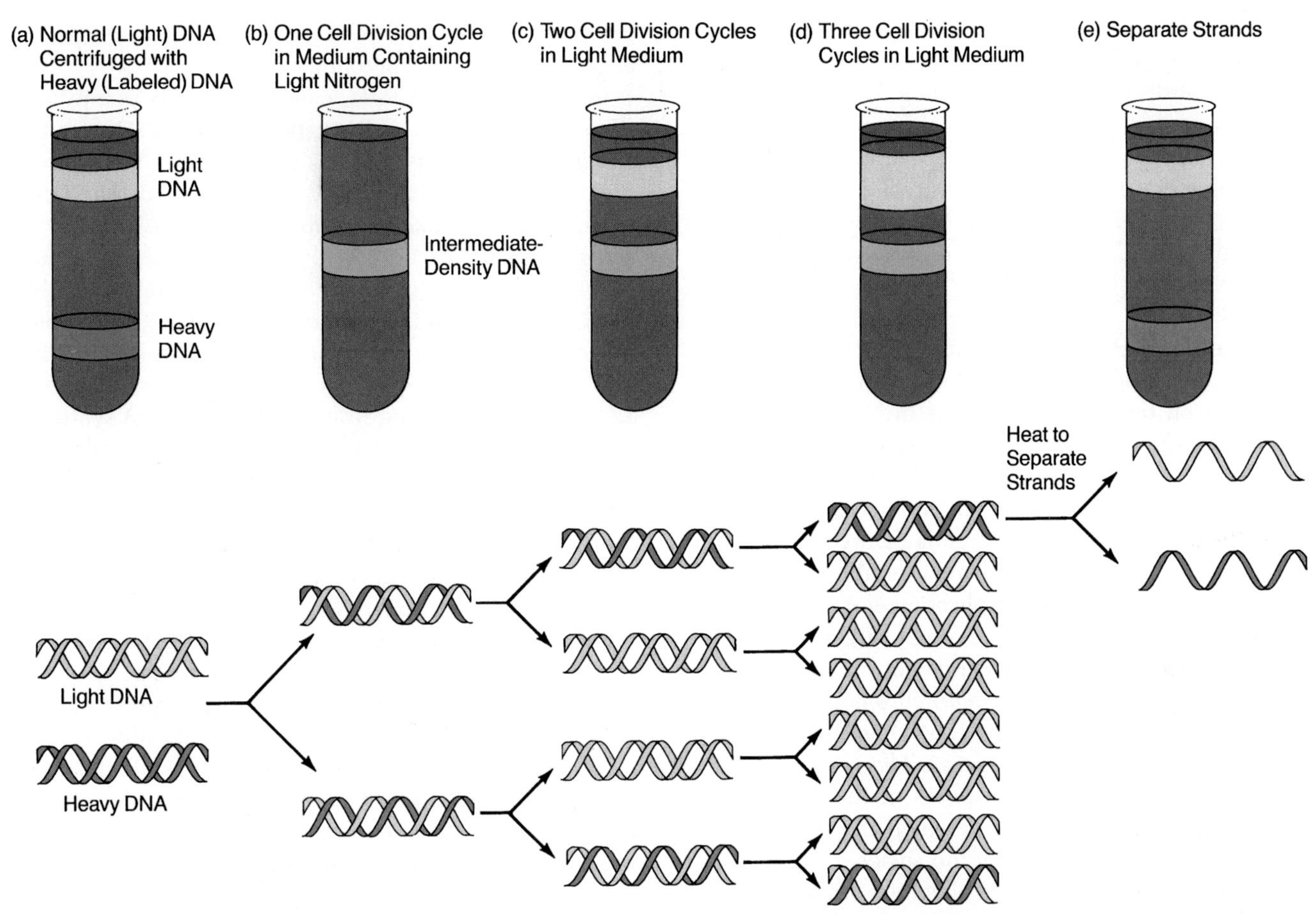

Figure 12-11 THE MESELSON-STAHL DNA REPLICATION EXPERIMENT: PROOF OF SEMICONSERVATIVE REPLICATION. Meselson and Stahl used DNA molecules, some "light" (containing normal ^{14}N nitrogen) and some "heavy" (containing heavy ^{15}N nitrogen) to elucidate DNA's mode of replication. (a) Centrifuging a mixture of DNA containing all-heavy strands and DNA containing all-light strands creates two bands in a CsCl density gradient. (b) The team allowed cells containing heavy DNA to grow on a nutrient medium containing light nitrogen for one division cycle and found DNA density intermediate between heavy and light. (c) After two cell division cycles, they saw one intermediate band and one light band. (d) After three cycles, they saw one intermediate band and a wide light band. (e) By heating and separating the strands, they once again saw the pattern for heavy and light DNA, proving that semiconservative replication does take place.

the semiconservative hypothesis (Figure 12-11c and d). Further, when Meselson and Stahl heated the hybrid DNA samples to break the hydrogen bonds and separate the DNA into single strands, each strand contained either *all* heavy or *all* normal nitrogen (Figure 12-11e). This showed that dispersive replication (with each strand having fragments of new and old DNA) was not occurring. The researchers concluded that the copying mechanism for DNA was indeed semiconservative, as Watson and Crick had theorized.

DNA Replication in *E. coli*

Studies by Meselson and Stahl and others probed the actual replication of DNA in living things. In *E. coli* bacteria, for example, DNA replication requires a series of specific enzymes and proteins. The enzyme **DNA polymerase** links free nucleotides as they line up on the template formed by the original strand of the parent molecule. And several other enzymes are involved in other aspects of the replication process. For instance, a special unwinding protein assists in the initial separation of the entwined helical strands; other enzymes repair damaged DNA or eliminate incorrectly paired bases; and still others join ends of broken chains. The biochemical complexity of the replication process probably evolved to ensure fidelity between the parent and the daughter DNA molecules. Because of the special enzymes, only 1 error occurs in every 1 billion to 10 billion base pairs, even though replication may occur at the rate of 1,000 nucleotides a second.

You may wonder why DNA is so stable in cells and why the same enzymes that catalyze its formation do not degrade it in a reverse reaction. The key to this stability is that, just as in ATP, phosphate groups are connected to the free nucleotide building blocks of DNA by high-energy bonds. When DNA polymerase helps attach the free nucleotide to a new, lengthening DNA strand, the two high-energy phosphates are released in a highly exergonic reaction. It should be no surprise, then, that DNA polymerase would require a great deal of energy to catalyze the reverse reaction and degrade DNA. Living cells thus expend much energy to make DNA, but the dividend is a greater stability of their precious genetic material.

Like other bacteria, *E. coli* has a single circular chromosome, and studies show that DNA replication starts at one point on the circle, called the origin of replication (Figure 12-12a), and proceeds around the circle in both directions at the same time like two base runners leaving home plate in opposite directions and meeting at second base. The points where the strands diverge for replication look like forks in the road and are called **replication forks.** This two-directional replication creates a bubble-shaped region that grows increasingly larger in both directions until the leading edges meet on the opposite side of the bacterial cell's circular chromosome.

Replication proceeds this way because the two strands of a double helix are *antiparallel,* or oriented in opposite directions. The phosphate in the sugar-phosphate backbone forms a link between the 5′ carbon of one sugar and the 3′ carbon of the adjacent sugar; thus, while one strand of the DNA molecule is oriented in a **5′ to 3′ direction,** the complementary strand has a 3′ to 5′ direction (Figure 12-13). DNA polymerase can add a new nucleotide only to one end of each growing chain; specifically, it can add the phosphate group on the 5′ carbon of the incoming nucleotide only to the free hydroxyl group on the 3′ carbon of the last sugar on the growing DNA chain. Therefore, a DNA chain lengthens only in the 5′ to 3′ direction.

At the replication fork, a short RNA acts as a primer (Figure 12-13a). To this RNA, newly formed DNA on the *leading strand* leads into the fork in the 5′ to 3′ direction, but the *lagging strand* is oriented so that DNA replication leads *away* from the fork (Figure 12-13b). To accommodate this asymmetry, DNA synthesis occurs differently on the two strands of the separating double helix. On the leading strand, synthesis is *continuous;* nucleotides are added constantly in the 5′ to 3′ direction as the replication fork moves along the strand (Figure 12-13c). On the lagging strand, however, DNA synthesis is *discontinuous* because the 5′ to 3′ direction leads away from the replication fork. Small DNA fragments called **Okazaki fragments** are synthesized in the 5′ to 3′ direction, but are not joined together right away. When an Okazaki fragment has grown to about 150 nucleotides, an enzyme removes the RNA primer and replaces it with DNA; another enzyme, *polynucleotide ligase,* joins the two Okazaki fragments (Figure 12-13d). By this time, the fork has moved further to the right, exposing more bases. A new Okazaki fragment forms and ultimately is joined to the previously formed fragment. This "back-stitching" process repeats over and over on the lagging strand as the replication fork progresses until the DNA molecule is completely replicated.

DNA Replication in Eukaryotes

DNA replication in eukaryotic cells with their separate nucleus and the more densely packed, protein-rich

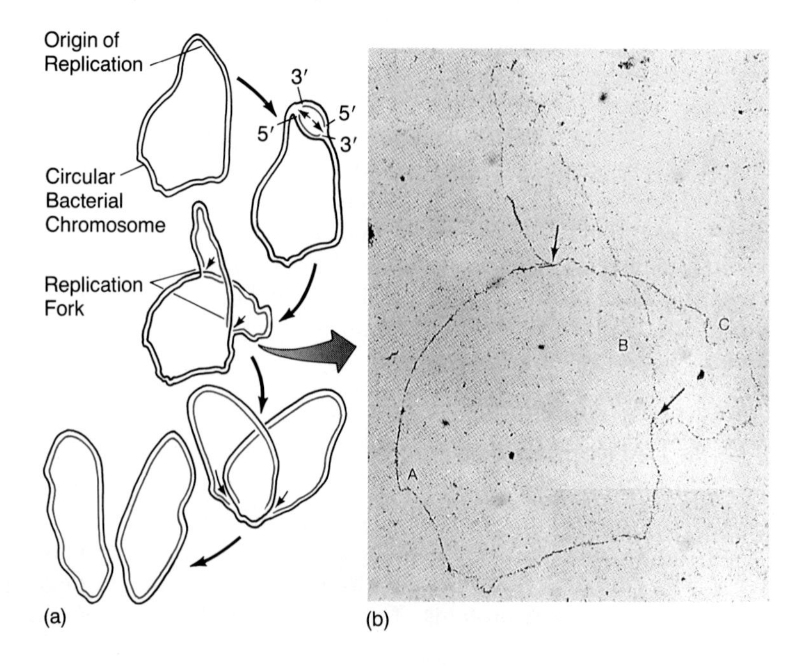

Figure 12-12 DNA REPLICATION IN *E. COLI.*
(a) When the DNA in an *E. coli* chromosome replicates, a bubble-shaped region forms and expands until the leading edges meet on the opposite side of the circle. The two strands separate at the origin of replication. The bubble formed by the replicating strands will expand around the circular chromosome until they meet on the other side. (b) An autoradiograph (produced by the exposure of a radioactive substance to photographic film in a dark place) of a replicating chromosome (magnified about 200,000×).

chromosomes, is similar to that in prokaryotes. The long DNA molecules of eukaryotes, which are supercoiled into chromosomes, have hundreds of sites from which replication proceeds in two directions simultaneously. This creates many replication bubbles (Figure 12-14) and ensures that the huge length of eukaryotic DNA can be replicated in just a few hours instead of the 45 or so days it would take if one replication proceeded from a single site. Studies show that DNA replication in eukaryotes is

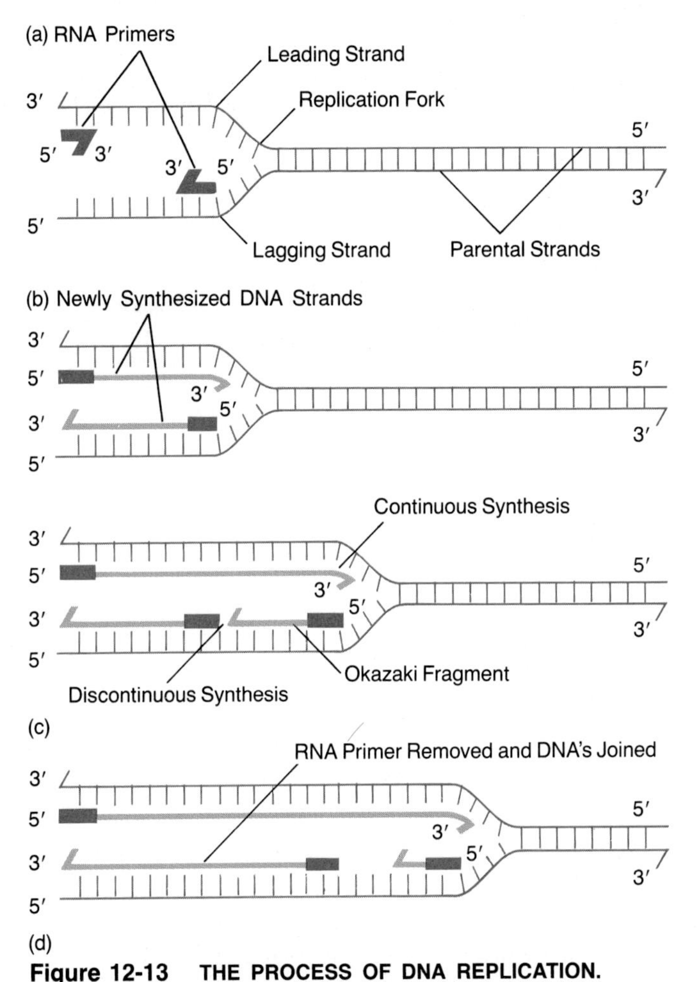

Figure 12-13 THE PROCESS OF DNA REPLICATION.
(a) DNA synthesis commences with the formation of a short RNA primer. (b) As the fork moves to the right, the *leading* strand is synthesized continuously in the 5′ to 3′ direction, or toward the moving fork. The *lagging* strand also is synthesized in the 5′ to 3′ direction, but away from the fork. (c) As more bases are exposed by the rightward movement of the fork, a second RNA primer forms on the lagging strand extending in the 5′ to 3′ direction away from the fork, generating an Okazaki fragment of DNA. (The leading strand continues to be synthesized as before.)
(d) The Okazaki fragment lengthens until it meets the previously formed fragment. Meanwhile, a new primer appears, and a new Okazaki fragment begins to form. This process is repeated over and over until the entire DNA molecule has been replicated.

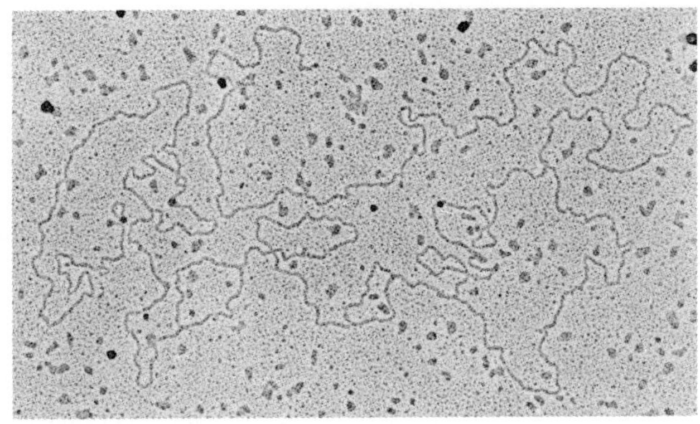

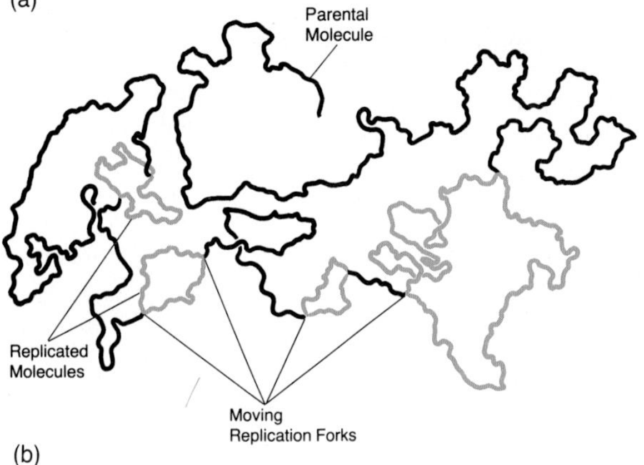

Figure 12-14 DNA REPLICATION IN EUKARYOTIC CHROMOSOMES: INITIATION AT MANY SITES SIMULTANEOUSLY.
In a eukaryotic cell, the replication of DNA takes place concurrently at hundreds of sites on a chromosome. This greatly decreases the time required for a chromosome to replicate its entire length of DNA. In this electron micrograph (a; magnified about 37,000×), and in the diagram of that same DNA (b), four regions of DNA replication can be seen (blue loops).

semiconservative, just as in prokaryotes. Replication appears to follow a universal mechanism in all DNA-containing organisms.

LOOKING AHEAD

From two separate lines of inquiry spanning just a few decades, we gained an understanding of the activity of the gene and the chemistry of DNA. This not only helped explain the link between genes and physical traits, but it also elucidated the mechanisms of heredity—the tendency of like to beget like. But the new field of molecular genetics was just beginning, and far greater understandings of cellular function were still to come, as Chapter 13 explains.

SUMMARY

1. Archibald Garrod, studying alkaptonuria, was the first to suggest that genes are responsible for specific chemical reactions in the body.

2. Working with *Neurospora*, Beadle and Tatum formulated the *one gene–one enzyme hypothesis*.

3. Working with hemoglobin protein, Pauling and Ingram refined this to the *one gene–one polypeptide hypothesis*.

4. Griffith's work on pneumococcal bacteria suggested that some substance in dead virulent strains can transform living nonvirulent strains, and Avery's team showed this substance to be DNA.

5. Using bacteriophages, which consist solely of proteins and DNA, Hershey and Chase proved that DNA alone is the carrier of genetic information.

6. DNA contains four bases: *adenine*, *guanine*, *cytosine*, and *thymine*. A and G are double-rings called *purines;* C and T are single-rings called *pyrimidines*.

7. *Chargaff's rules* state that [A] = [T] and [G] = [C] and that each species has a different ratio of [A] + [T] to [G] + [C]. Wilkins and Franklin's *x-ray diffraction* studies suggested a helical DNA shape with set distances between coils.

8. Watson and Crick constructed a three-dimensional model of the DNA molecule as a *double helix*, with two outer sugar-phosphate chains and with rungs formed by nucleotide pairs. Paired nucleotides, which always occur as A-T or G-C, are linked by hydrogen bonds.

9. Watson and Crick hypothesized that the sequence of the bases carries the genetic information. They also speculated that both accurate gene replication and occasional gene mutation can be explained if each DNA chain acts as a template for production of a new chain.

10. The Meselson-Stahl experiments confirmed Watson and Crick's hypothesis of *semiconservative replication*.

11. Because the enzyme *DNA polymerase* can join nucleotides only to the 3′ carbon of a growing DNA chain, DNA replication proceeds in a *5′ to 3′ direction* simultaneously along the two separating strands. The leading strand of a *replication fork* is synthesized as a continuous chain, and the lagging strand is synthesized discontinuously in short segments called *Okazaki fragments*.

KEY TERMS

adenine
Chargaff's rules
cytosine
DNA polymerase
DNA replication
double helix
5′ to 3′ direction
guanine
nucleotide
Okazaki fragment
one gene–one enzyme hypothesis
one gene–one polypeptide hypothesis
purine
pyrimidine
replication fork
semiconservative replication
thymine
x-ray diffraction

QUESTIONS

1. What is the genetic material in pneumococcus? In a bacteriophage?

2. What are Chargaff's rules? How did the Watson-Crick model for DNA explain Chargaff's rules?

3. Explain what is meant by semiconservative replication. If Meselson and Stahl had been able to separate the two DNA strands without breaking them, what would they have found after one generation? Two? What fraction of the DNA would have been in each band?

4. What feature of Watson and Crick's model suggested a method by which DNA could copy itself?

5. What is meant by polarity of DNA chains? Which subunits form the backbone of DNA? Which subunits form hydrogen bonds between the two chains?

6. In most prokaryotes, replication of the circular chromosome begins at just one point, while in most eukaryotes, DNA replication begins at many points. Why do eukaryotes need more?

7. Draw a diagram of replicating DNA with a single replication fork. Label leading strand, lagging strand, Okazaki fragments, RNA primer, and 5′ and 3′.

8. Four *Neurospora* mutants fail to grow on minimal medium, but they will grow on medium containing the amino acid arginine. The + mutants will grow when substances related to arginine are added to the medium, while the − mutants will not. What is the suggested metabolic pathway of arginine?

	Mutant 1	2	3	4
Minimal	−	−	−	−
Citruline	+	+	−	−
Ornithine	+	+	+	−
Arginine	+	+	+	+

ESSAY QUESTION

Why were Hershey and Chase's results accepted immediately, whereas those of Avery, MacLeod, and McCarty were not?

For additional readings related to topics in this chapter, see Appendix C.

13

Translating the Code of Life: Genes into Proteins

Proteins are, in the final analysis, the executors of each organism's inheritance.

John Cairns,
"The Bacterial Chromosome,"
Scientific American (1966)

By 1953, biologists knew that genes are made of DNA and that each gene carries the information for building a polypeptide. So momentous were these discoveries that they rapidly inspired important new inquiries, and within a decade, researchers had defined the *central theme of molecular biology:* that genetic information is stored in a linear message on nucleic acids and is expressed in a corresponding linear sequence of amino acids in proteins. They symbolized the correspondence this way: DNA → RNA → protein.

Researchers also discovered that the information in DNA is stored in a code—a code that unlocks the mechanisms by which genes direct the building of proteins and that underscores the evolutionary relationships between organisms at a biochemical level.

We will trace all these discoveries in this chapter as we discuss the following topics:

- Evidence that the language of DNA must be written in code, and the nature of that genetic code
- The functions of DNA and RNA within the living cell
- How the genetic code is translated into proteins
- How researchers learned to translate the genetic code
- How translating the code enabled researchers to refine their concept of the gene and solve new genetic questions

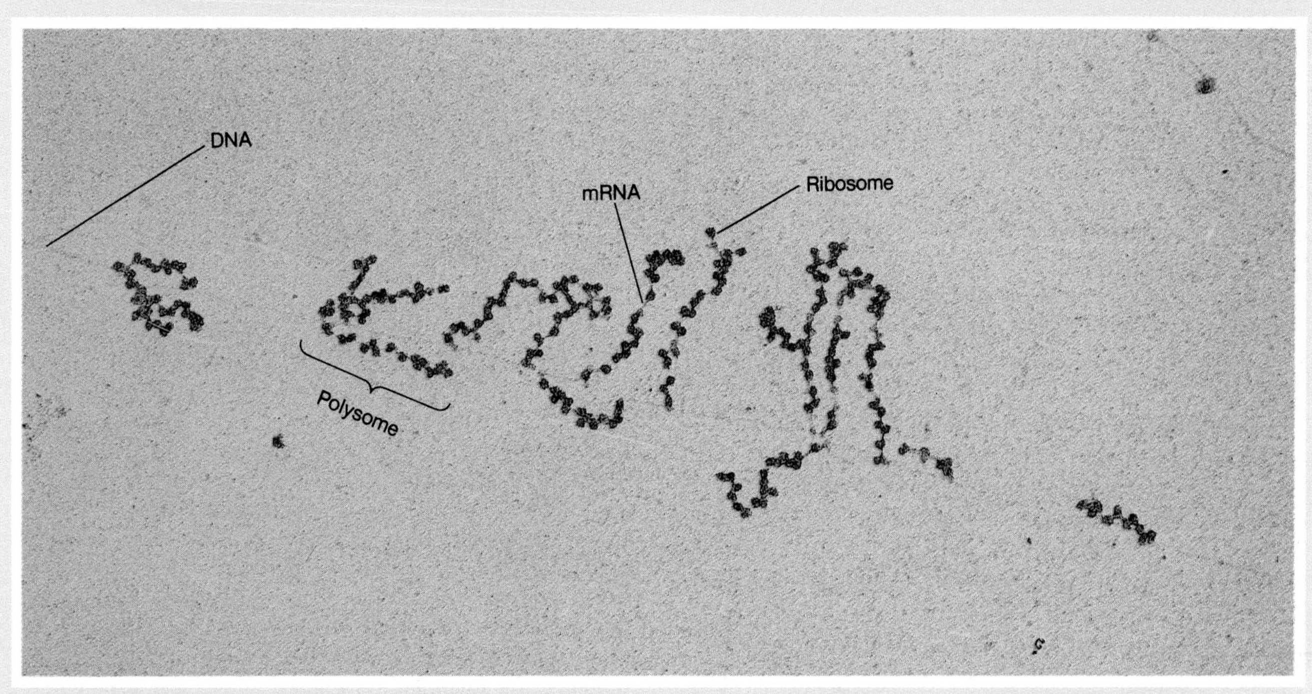

Heredity in action. In this photo (magnified about 64,000×), the genetic material DNA is being transcribed into the messenger, mRNA. Ribosomes on the mRNA mark sites where proteins encoded by the DNA are manufactured.

GENETIC INFORMATION MUST OCCUR IN CODE

Francis Crick once referred to nucleic acids and proteins as the "two great polymer languages." As we have seen, these "languages" have different alphabets: nucleic acids have a four-letter alphabet—the nucleotide bases A, T, G, and C—whereas the protein alphabet consists of the 20 amino acids (review Figure 3-14). In both languages, the sequence of "letters" conveys the meaning of the message, just as the words in this book convey meaning. But how do the alphabets correspond? The answer involves a *code* based on *sets* of DNA bases.

Logic shows why such a code must exist. If each of the 4 genetic "letters"—a C, let's say, or a G—coded for a single amino acid, the code could specify only 4 amino acids, not 20. Obviously, *sets* of nucleotide bases must correspond to a single amino acid. There could be 4^2, or 16, possible sets of two corresponding bases to serve as DNA code words (e.g., AT, TC, AG, GC) for specific amino acids. But this still falls a few short of the actual 20 amino acids found in proteins. Or instead, there could be 4^3, or 64, sets of *three* consecutive nucleotide bases (AAA, AAT, ATA, ATG, etc.)—more than enough to accommodate a language of 20 amino acids such as the one that occurs in the protein molecules produced by cells.

Soon after Watson and Crick identified the structure of DNA, biologists hypothesized that three consecutive nucleotides might code for a single amino acid, and they called the hypothetical sets **codons.** As we will see, these proved to be actual, not hypothetical.

The Genetic Code Is Colinear

Before any secret message—genetic or otherwise—can be translated from a cryptic language to a known language, a cryptographer must determine whether the letters in the coded message correspond in a *linear fashion* to the letters in the solution. Such an arrangement is said to be *colinear;* for example, in the code FDMDSHBR = GENETICS, each letter of the solution is represented in the code by the letter that precedes it in the Latin alphabet. A code that is not colinear could read backward or skip letters. So how does the genetic code operate?

To establish whether or not the genetic code is colinear, Charles Yanofsky and his colleagues at Stanford University performed experiments during the early 1960s comparing a gene and its corresponding polypeptide. *E. coli* produces an enzyme called *tryptophan synthetase,* which the cell needs to produce the amino acid tryptophan (abbreviated Trp). Yanofsky and his co-workers identified a large number of mutant *E. coli* strains with abnormal tryptophan synthetase activity. They also determined the sequence of the 268 amino acids in the enzyme's polypeptide chain in both normal and mutant cells. Each mutant polypeptide had an altered amino acid somewhere along the chain. Comparing the position of each altered amino acid and the genetic map location of each mutation, Yanofsky found a colinear correspondence.

Yanofsky also made a second important observation: Mutations at two distinct but nearby sites in the *trpA* gene (Figure 13-1) could change the same amino acid

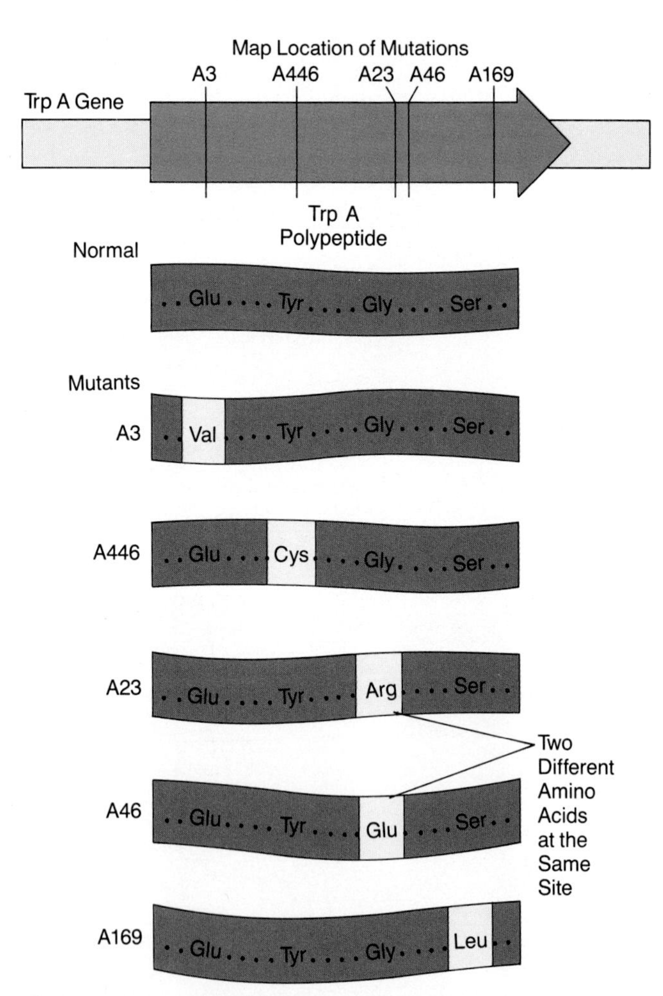

Figure 13-1 COLINEARITY REVEALED: MAP OF THE *trpA* GENES AND CORRESPONDING AMINO ACID SEQUENCE IN *trpA* POLYPEPTIDE.
The map at the top shows the locations of five mutations within the *trpA* gene of an *E. coli* bacterium. Immediately below lies the normal *trpA* polypeptide, with its regular amino acid sequence. Below that are five mutant polypeptides, each with an altered amino acid that corresponds to the location of the gene mutation. The A3 mutation, for example, has a single amino acid change of a Val for the wild-type Glu. Both the A23 and A46 mutations alter the amino acid Gly at the same position.

site, but in two different ways. He reasoned that the two mutations must reside within the *same* DNA codon but affect different nucleotides. Thus, the *trp* experiments proved two things: that nucleotides in a gene and amino acids in a polypeptide are colinear and, in keeping with the codon hypothesis, that more than one nucleotide codes for each amino acid.

The Nature of the Genetic Code

Geneticists quickly surmised that the **genetic code**—the molecular "grammar" relating nucleic acid bases to amino acids—might be colinear, sequential, and based on codons, in which case the order of amino acids in a polypeptide would depend on *the sequence of codons in DNA*. Experiments confirmed these suspicions and revealed four fundamental principles about the genetic code in virtually all organisms.

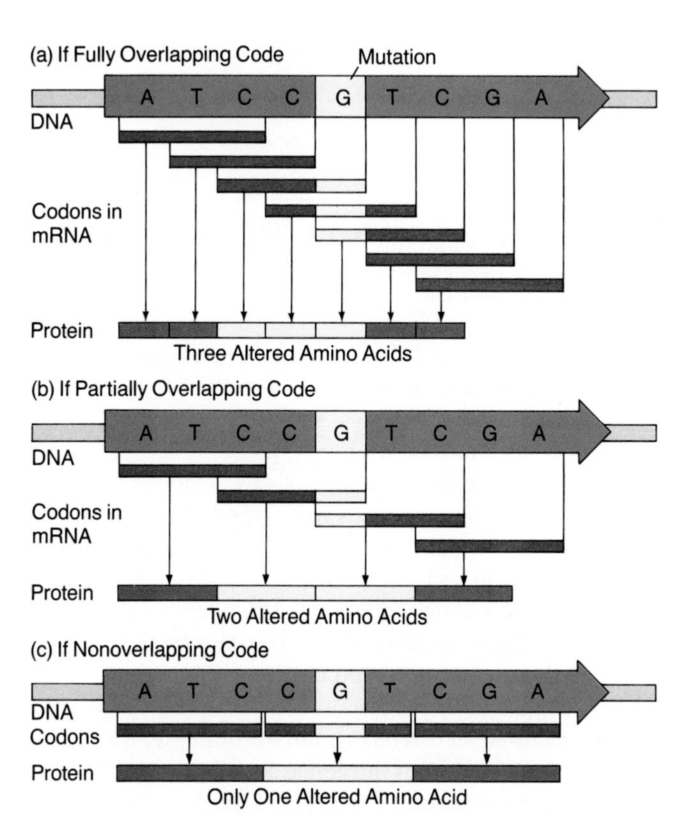

Figure 13-2 THREE WAYS OF READING A NUCLEOTIDE SEQUENCE.
(a) If a nucleotide codon consisted of three nucleotides that overlapped fully, then when a point mutation altered a single nucleotide, three amino acids would be affected by the change. (b) If the codons overlapped partially, a single point mutation would result in two incorrect amino acids. (c) If the nucleotide codons did not overlap at all, a single nucleotide substitution would affect only one codon and alter a single amino acid. This, in fact, is what biologists have observed experimentally.

1. *The Genetic Code Is Nonoverlapping*

It seemed likely that groups of three bases in a row on either strand of the double-stranded molecule might function as codons. However, it was conceivable that the sets of three might overlap, thereby functioning as an *overlapping code* (Figure 13-2a). In such a code, the alteration of a single nucleotide from, say, C to G, would affect three codons, and lead to three incorrect amino acids. In contrast, the same alteration in a partially overlapping code (Figure 13-2b) would lead to two incorrect amino acids, and in a nonoverlapping code, to just one incorrect amino acid (Figure 13-2c).

Biologists observed that mutations affecting just one nucleotide alter only a single amino acid. On the basis of this and other evidence, they determined the first principle of the genetic code: that it is *nonoverlapping*.

2. *The Genetic Code Is Deciphered by Reading Frames*

The helical backbone of the DNA molecule is a continuous chain of sugar and phosphate groups, spanned by an uninterrupted series of paired nucleotide bases,

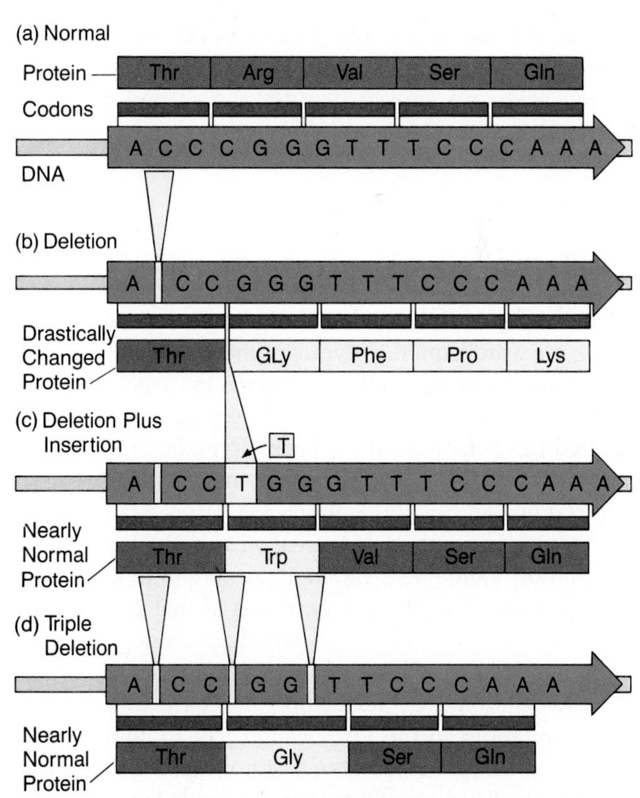

Figure 13-3 DELETIONS, INSERTIONS, AND READING FRAMES.
As the text explains, deletions and insertions can alter the reading frame and change the amino acids in a polypeptide, but certain numbers and combinations of these mutations can minimize the changes.

and not marked off into genes, codons, or other units of information. Yet somehow, inside a living cell, the information is read in appropriate units that code for individual amino acids.

Francis Crick conceptualized the reading of the genetic message in terms of a **reading frame.** If the nucleotide sequence shown in Figure 13-3a is read in three-nucleotide codons beginning at the first A, it encodes the amino acids threonine, arginine, valine, serine, and glutamine. If a chemical mutagen causes a deletion of the first C in the first codon of this nucleotide sequence, it will now code for threonine, glycine, phenylalanine, proline, and lysine (Figure 13-3b). The deletion would cause a shift in the reading frame from CGG in the second codon to GGG—a process that Crick called a **frameshift mutation.** The first *triplet* (reading group of three letters) is still normal, but the four subsequent ones are altered codons that specify different amino acids from those specified by the original triplets. Obviously, the resulting polypeptide would be radically changed.

If the *deletion* of a nucleotide can cause a frameshift, then a similar shift should occur following an *insertion* of one nucleotide. When Crick induced this insertion-type mutation in bacteriophages that had previously suffered a frameshift deletion mutation, he found that the original reading frame was restored (Figure 13-3c). Note that the only different amino acid is tryptophan; all the other amino acids are correct because the second mutation restored the normal reading frame and such a protein might function quite normally in some cases.

3. *The Genetic Code Is Degenerate*

A third principle is that the genetic code is *degenerate*—a word from the cryptographer's lexicon meaning that more than one codon can code for a given amino acid. Francis Crick's frameshift experiments demonstrated this principle. If the codon hypothesis is correct, and the code is read in triplets, then there are 64 possible codons but only 20 amino acids. Crick's team found that mutations to the DNA usually cause the insertion of a *different* amino acid rather than no amino acid at all (which would halt further synthesis of that protein). In contrast, if the code were nondegenerate, a single unique codon would specify each amino acid, 44 of the 64 codons would be noncoding *nonsense codons,* and most mutations to the DNA would halt protein synthesis, since it would stop at the first nonsense codon. Most or all of the 64 codons must therefore code for one amino acid or another rather than "nonsense." This has survival significance for the cell, since mutations will tend to cause substitutions of inappropriate amino acids but will not usually block protein synthesis entirely and lead to the cell's death.

4. *Codons Are Really Nucleotide Triplets*

Finally, Crick's frameshift experiments proved beyond doubt that codons are indeed triplets of nucleotides. Recall that one insertion mutation canceled out one deletion mutation and restored the original reading frame. Crick also showed that a second deletion or insertion following a first (in other words, two fewer or two additional nucleotides in a DNA chain) *fails* to restore a reading frame, while the deletion or insertion of *three* bases does restore it (Figure 13-3d). Where three nucleotides are deleted, the protein will be one amino acid too short and have one substitute amino acid (encoded by the second triplet in our example); where three nucleotides are inserted, the protein will usually be one amino acid too long. Experiments showed, then, that the genetic code is nonoverlapping, is read in appropriate frames, is degenerate, and consists of codons with three consecutive nucleotide bases. Nearly all organisms share a universal genetic language coded according to these principles, and our next topic is the way this language is translated into the equivalent but separate language of amino acids.

DNA, RNA, AND PROTEIN SYNTHESIS

The revelation of DNA's structure and coding principles had a profound ramification: an answer to how genes are translated into biological molecules—that is, how a genotype can produce a phenotype.

Knowing that the sequence of DNA codons specifies the sequence of amino acids in polypeptides, one might assume that the synthesis of proteins takes place right along one strand of the DNA double helix the way a machine part is stamped directly from a metal template. However, biochemists established in the 1960s that protein synthesis does not even occur in the nucleus of eukaryotic cells. Proteins are made only in the cytoplasm. In the mammalian red blood cell, for example, the entire nucleus is ejected from the cell as it matures, yet protein synthesis can continue for hours. Clearly, protein synthesis does *not* take place directly on the genes of nuclear DNA.

Researchers also found that when they break up prokaryotic and eukaryotic cells into homogenized mixtures containing no intact cells, the so-called *cell-free system* continues to incorporate amino acids into polypeptides. Moreover, this can take place in the absence of DNA: An enzyme that breaks up (hydrolyzes) DNA molecules will not stop the protein synthesis. If the experimenter adds an enzyme that digests *ribonucleic acid (RNA)* to the mixture, however, incorporation of amino acids into polypeptides ceases immediately. Thus, RNA must be a biochemical go-between in the transfer of information from DNA to proteins.

As we saw in Chapter 3, the class of nucleic acids called RNA is similar to DNA, except that each RNA molecule contains the sugar ribose instead of deoxyribose and is usually single-stranded rather than double-stranded. There is one additional difference: In RNA, the pyrimidine **uracil** (U) replaces thymine, which is found only in DNA. Thus, the RNA bases are A, U, G, and C, rather than A, T, G, and C.

If RNA somehow carried a copy of the genetic sequence inscribed on DNA out of the nucleus and into the cytoplasm, it might then be used as a template during protein synthesis to determine the sequence of amino acids in a polypeptide. This, in fact, turned out to be correct. Researchers, however, found that RNA is not a single homogeneous substance, but several different types of molecules. Let's take a closer look at how RNA is made, the process of **transcription,** and then look at how RNA is involved in constructing proteins, the process of **translation.**

An Overview of Transcription

The genetic information encoded in DNA is read out during the process of transcription into RNA. RNAs are single-stranded chains of nucleotides that are synthesized on single strands of the DNA molecule (Figure 13-4a). For transcription to occur as RNA (specifically

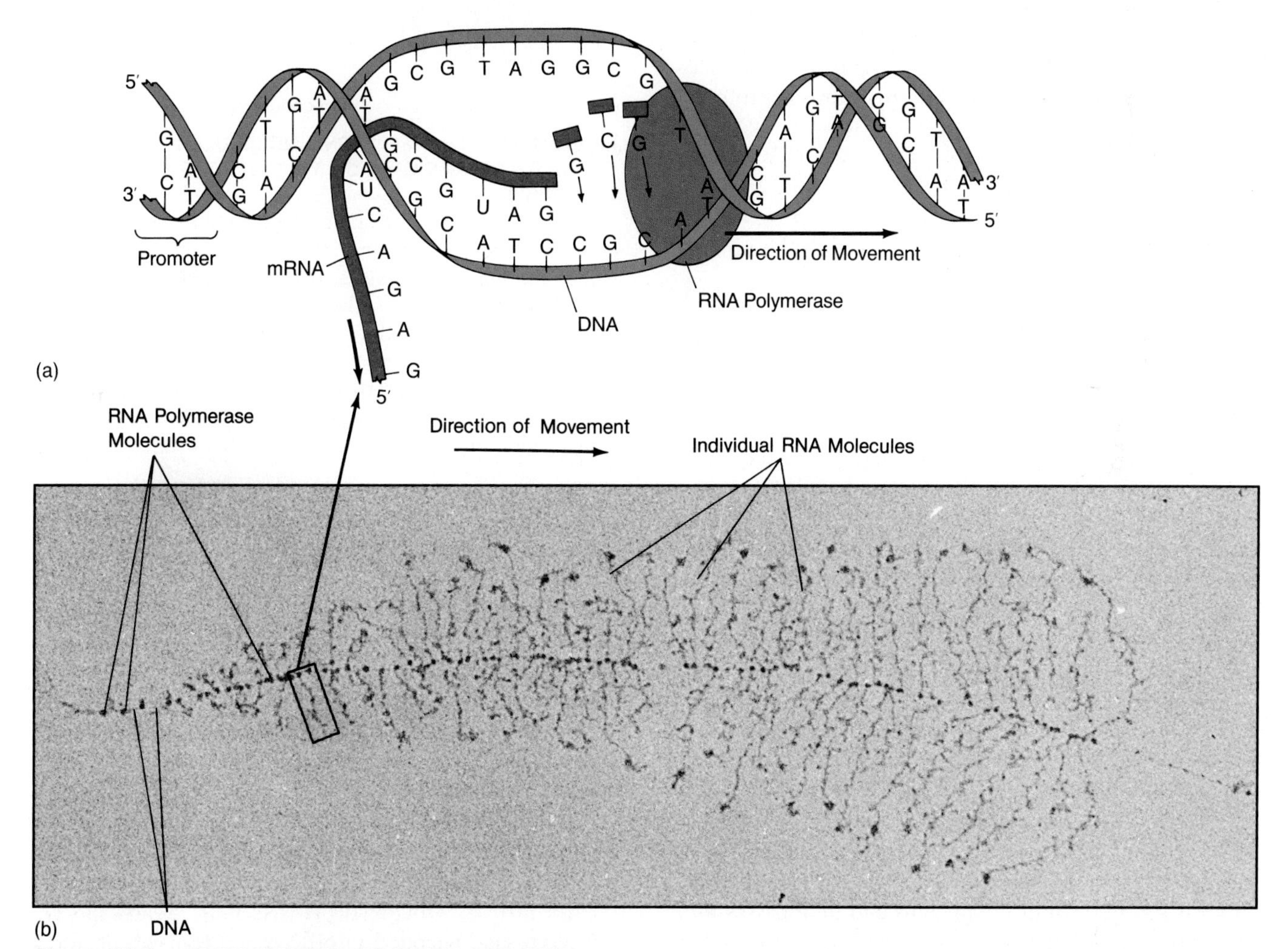

Figure 13-4 TRANSCRIPTION OF GENES INTO RNA.
(a) As the DNA double helix unwinds at the site of RNA polymerase attachment, mRNA is transcribed from one of the chains. The "matching" of complementary bases—U to A, G to C—is the true information transfer step. In another gene, the other strand of DNA might be transcribed. (b) Transcription is visible in this electron micrograph (magnified about 56,500×). DNA forms a continuous thread along which RNA polymerase molecules move; the farther they move along the DNA of these genes, the longer the chain of newly synthesized RNA that protrudes to the side. Many RNA chains are being synthesized simultaneously from this stretch of DNA.

messenger RNA) forms on a DNA template, an enzyme (RNA polymerase) links nucleotides floating free in the nucleus to the growing RNA strand in a sequence precisely complementary to that of DNA. The only exception is that uracil takes the place of thymine and occupies all locations opposite adenine (review Figure 3-24). An enzyme called *RNA polymerase* joins the high-energy, triphosphate form of the free-floating ribonucleotides, adding them to the growing chain in the 5′ to 3′ direction (see Chapter 12). This process generates an RNA that is a complementary copy of the DNA strand.

Near each gene, specific DNA sequences called *promoters* serve as sites where the RNA polymerase enzyme initially binds. Transcription of DNA into mRNA then proceeds from that site. In eukaryotic cells, the initial product of the transcription process is a primary transcript, and this RNA is subsequently altered by processes that we will describe in Chapter 17. Those alterations generate a mature RNA, which then passes out of the nucleus through a nuclear pore and into the cytoplasm to perform its function in the synthesis of proteins.

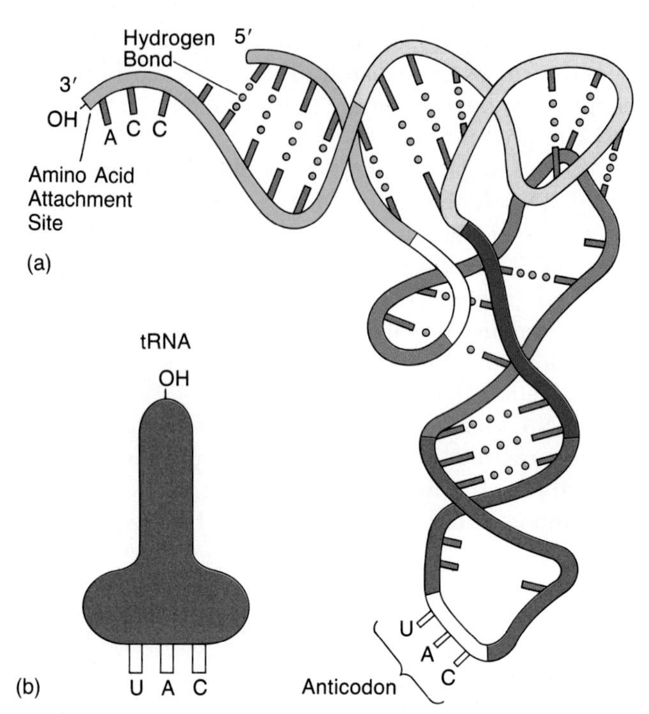

Figure 13-5 TRANSFER RNA MOLECULES: THE CARRIERS OF AMINO ACIDS TO THE SITES OF PROTEIN SYNTHESIS.
(a) Each tRNA codes for one of the 20 amino acids, and each has a looping structure stabilized by hydrogen bonds. Notice the unpaired triplet at the amino acid attachment site. This triplet, CCA, is the same for all tRNAs. The loop at the bottom contains the anticodon triplet sequence, which serves as the name tag for the amino acid that the tRNA specifies. For met-tRNA, the anticodon is UAC. (b) A simplified symbol, which appears in subsequent figures and which represents the various forms of tRNA.

Types of RNA

Three main types of RNA are involved in the process of protein synthesis.

Messenger RNA

Messenger RNA (mRNA) is a molecular emissary that carries information from the DNA in the nucleus to the sites of protein synthesis in the cytoplasm. The sequence of bases in mRNA reflects the triplets in DNA that encode amino acids. An mRNA will have at least three times as many bases as amino acids in the protein it encodes since three bases specify each amino acid.

Transfer RNA

The second type of RNA is **transfer RNA (tRNA),** small polynucleotide chains that transport individual amino acids to the sites of protein synthesis. At least one unique kind of tRNA joins specifically to each of the 20 kinds of amino acids. Transfer RNA molecules are small RNAs containing about 80 nucleotides, compared to 1,000 or more nucleotides in most mRNAs. Like mRNA, tRNA is synthesized in the nucleus and then passes into the cytoplasm; synthesis occurs on genes that code for the various types of tRNA. A key feature of all tRNA molecules is their looped shape, which results from the presence of a region where the RNA strand has folded back on itself to form a double strand.

Two especially important sites lie far apart on the tRNA molecule. At the 3′ end of the chain is the amino acid attachment site, where a specific amino acid attaches to the tRNA molecule. This site—a kind of molecular "hook"—always consists of the sequence 5′-CCA-3′ for all 20 kinds of tRNA. The tRNA then has a specific amino acid cargo "in tow," tugboat style, with a covalent bond. Some distance away on the tRNA molecule lies the **anticodon,** a kind of three-letter name tag with a specific set of bases for each amino acid (in Figure 13-5, 3′-UAC-5′ for methionine). Each type of anticodon can bind only to a specific, complementary three-base codon on an mRNA molecule (3′-UAC-5′ binds only to 5′-AUG-3′, for instance). Consequently, a tRNA can transport a particular amino acid to an appropriate site on the mRNA. Obviously, tRNAs help "read" the genetic message encoded in mRNA and translate it into a protein.

Ribosomal RNA

Ribosomes are made up of **ribosomal RNA (rRNA)** molecules, along with a variety of proteins. These tiny globular organelles can be likened to a movable vise that holds mRNA and tRNA molecules in just the right position so that amino acids on the tRNAs can be joined together. Ribosomes are made up of three types of RNA; genes for two of these types are located in the nucleolus (see Chapter 6), whereas multiple copies of the gene for the third type are found at many sites along various chromosomes. Proteins and rRNA molecules assemble in the nucleolus to form one large and one small subunit for each ribosome (Figure 13-6). These large molecular aggregates, or subunits, then migrate to the cytoplasm, where the two types of subunit attach to mRNA molecules.

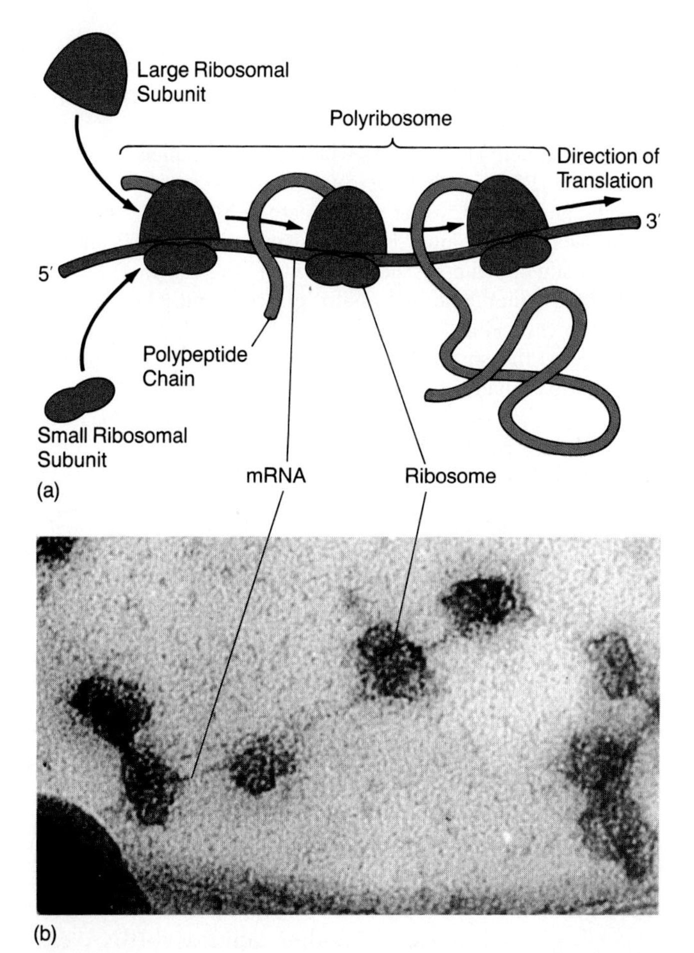

Figure 13-6 RIBOSOMES: MOVABLE SITES OF PROTEIN SYNTHESIS.
(a) Ribosomes are made up of ribosomal RNA (rRNA) and proteins, and they consist of two subunits that are unequal in size. When ribosomes are not engaged in protein synthesis, the subunits reside separately in the cytoplasm, joining together again for the next round of protein synthesis. The ribosomes function as movable vises that hold molecules in the right position so amino acids can be linked together into polypeptides during protein synthesis. (b) An electron micrograph (magnified about 560,000×) of a polyribosome from a red blood cell. A polyribosome (polysome) is a group of ribosomes on an mRNA.

In bacteria, ribosomes consist of one large and one small subunit designated 50S* and 30S; the complete organelle is called a 70S ribosome. In eukaryotic ribosomes, the subunits are somewhat larger (60S and 40S), and the complete organelle is an 80S ribosome. During protein synthesis, an mRNA molecule forms a complex with a small ribosomal subunit; then the large subunit joins, and if other factors are present, polypeptide synthesis can begin. Several ribosomes often move along a single mRNA molecule; the whole complex is called a *polysome*, or polyribosome (see Figure 13-6).

During the process of translating the genetic information from DNA into the amino acid sequence of a protein, the three types of RNA function together. Their activities in protein synthesis are our next subject.

PROTEIN SYNTHESIS: TRANSLATING THE GENETIC CODE

Hundreds of experiments in the 1960s unraveled the four general steps of protein synthesis: (1) **amino acid activation,** (2) **initiation** of protein synthesis, (3) polypeptide **elongation,** and (4) **termination** of protein synthesis. Experiments also found important differences between protein synthesis in prokaryotic and eukaryotic cells. In prokaryotes, transcription and translation take place simultaneously because the absence of a membrane-bound nucleus allows ribosomes and tRNAs to attach to an mRNA even before the synthesis of that molecule is complete (Figure 13-7). In eukaryotes, the processes are consecutive, since newly synthesized RNA must be altered, and the mRNA must leave the nucleus before protein synthesis can begin. Now let's consider the four steps.

Step 1: Amino Acid Activation

Amino acid activation precedes the translation of the nucleotide sequence of an mRNA into the amino acid sequence of a protein. This first step takes place in the cytoplasm and creates a pool of tRNA molecules with amino acids in tow, ready to take part in polypeptide

*S, the Svedberg unit, is a measure of size based on rates of sedimentation in a high centrifugal field. Sedimentation rates are not additive; hence, the complete bacterial ribosome is 70S rather than 80S (30 + 50).

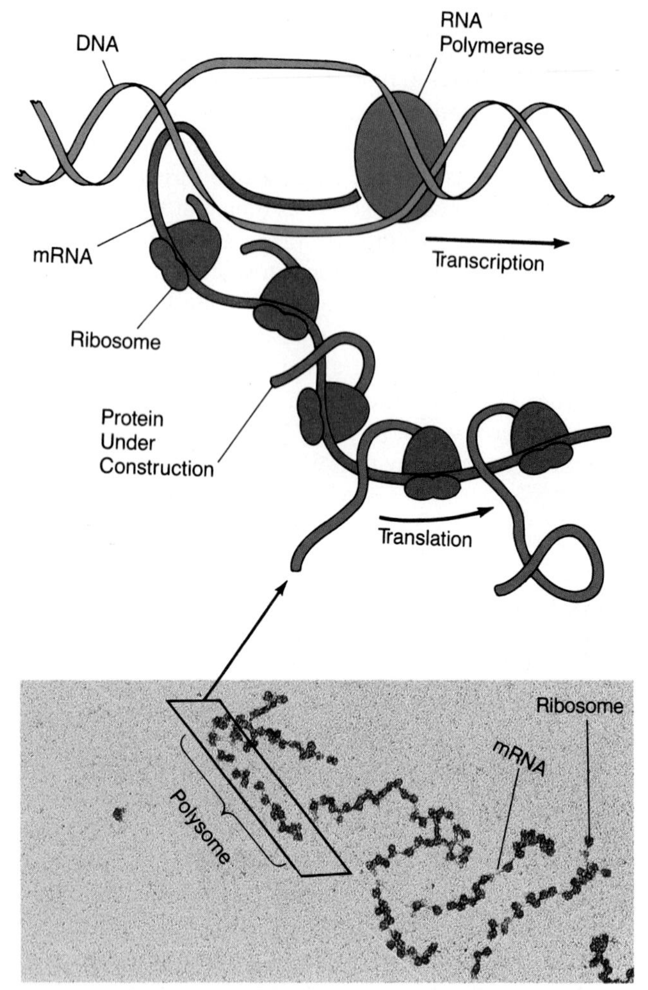

Figure 13-7 SIMULTANEOUS TRANSCRIPTION AND TRANSLATION.
In a prokaryote, as the genetic message is transcribed from DNA to mRNA (top, and review Figure 13-3), ribosomes attach to the forming mRNA, and translation (protein synthesis) begins simultaneously.

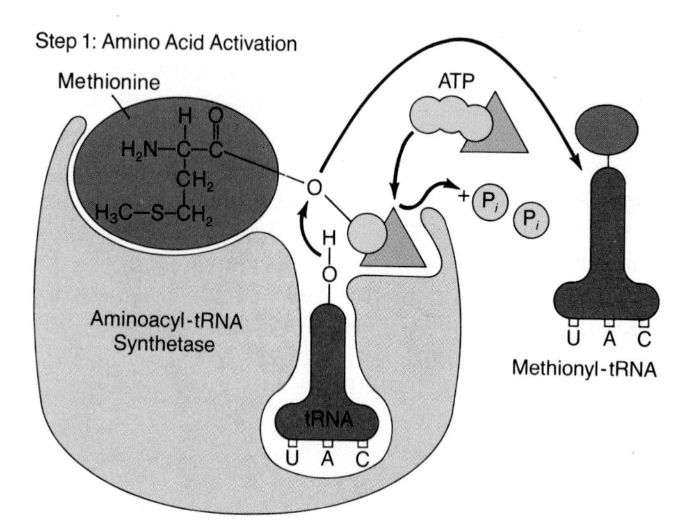

Figure 13-8 AMINO ACID ACTIVATION: FIRST STEP OF PROTEIN SYNTHESIS.
Amino acid activation occurs when enzymes called aminoacyl-tRNA synthetases catalyze the attachment of a specific amino acid to the appropriate tRNA. First, the amino acid (methionine, in this illustration) and ATP join specific sites on the synthetase enzyme and react. ATP is hydrolyzed, inorganic phosphate released, and AMP is joined to the amino acid. This adenylated amino acid then is joined to the specific tRNA that fits in another site on the synthetase. Finally, the newly created aminoacyl-tRNA molecule is released from the enzyme (along with AMP, not shown here). Energy from the ATP remains in the aminoacyl-tRNA molecule and helps to drive peptide bond formation.

synthesis (Figure 13-8). First, enzymes called **aminoacyl-tRNA synthetases** catalyze a reaction that attaches a specific amino acid (in this case, methionine) to an appropriate tRNA. The enzyme catalyzes the formation of a covalent bond between the carboxyl end (–COOH) of the amino acid and the 3′ end of the tRNA, forming aminoacyl-tRNAs, here a methionyl-tRNA. This is a high-energy bond resulting from the cleavage of ATP. The resulting activated amino acid contains some of the energy used later to help attach that amino acid subunit to a growing polypeptide chain.

Step 2: Initiation

Polypeptide synthesis actually begins with the attachment of a small ribosomal unit to a methionyl initiator tRNA (Figure 13-9a). This complex next joins a ribosome binding site on the mRNA molecule (Figure 13-9b). At this binding site is a special codon (AUG), which functions as a "start signal" for protein synthesis. Next, a large ribosomal subunit binds to the smaller one, forming a complete ribosome attached to the mRNA (Figure 13-9c). The ribosome has a groovelike site in which the initiator aminoacyl-tRNA nestles. This is called the **P site** (for "peptidyl"). The adjacent **A site** (for "aminoacyl") on the ribosome is vacant. Another aminoacyl-tRNA can bind to the mRNA codon located at the A site. Various tRNAs diffuse in and out of the A site randomly, until by chance an aminoacyl-tRNA approaches that has an anticodon complementary to the mRNA codon at the A site. This aminoacyl-tRNA then binds (Figure 13-9d).

Step 3: Elongation

Two amino acids are now sitting next to each other along the mRNA, each attached to a tRNA that is bound to the A or P binding site on the ribosome. At the start of

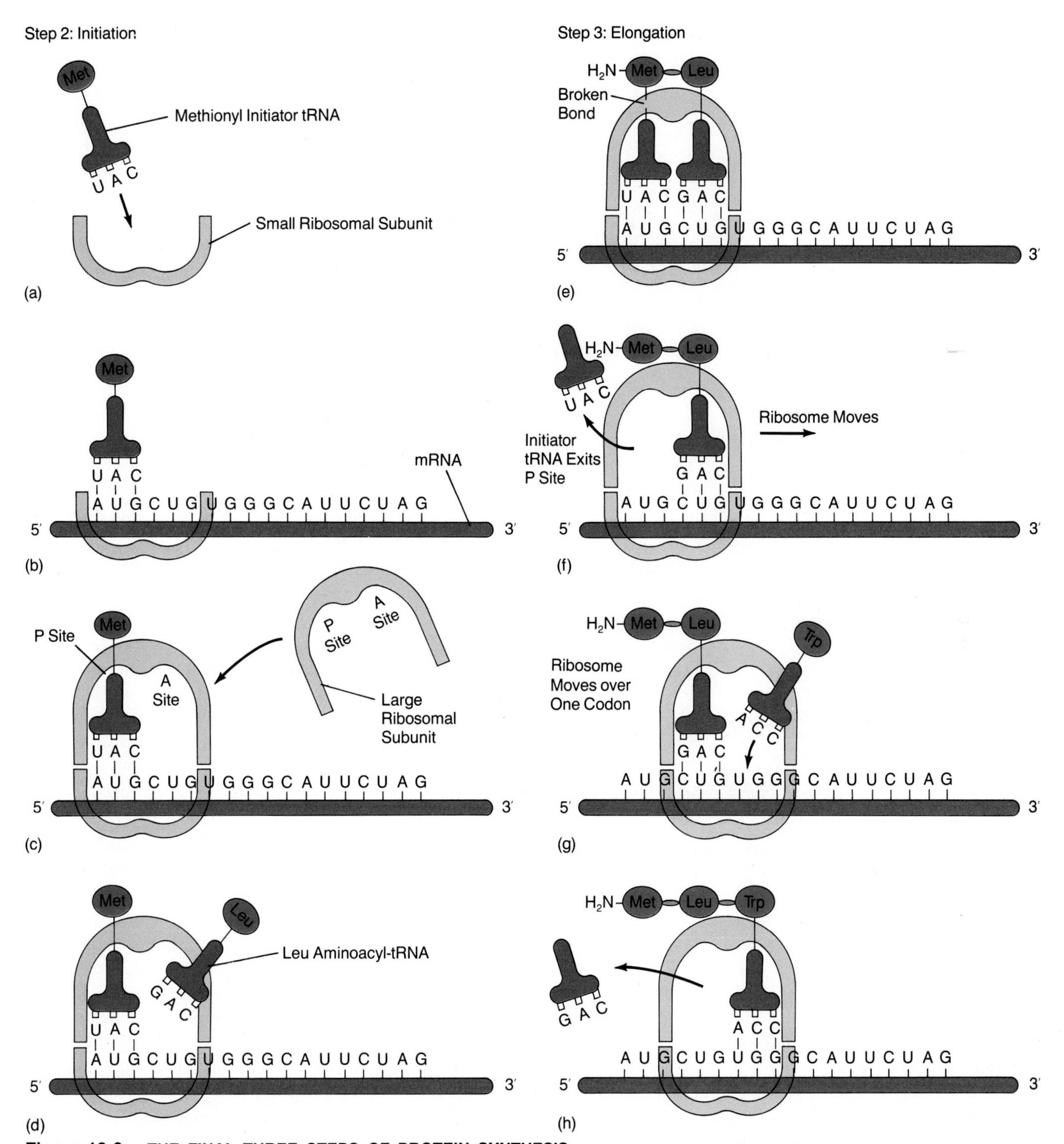

Figure 13-9 THE FINAL THREE STEPS OF PROTEIN SYNTHESIS.
After amino acid activation, steps 2–4 of protein synthesis take place. At the start, all required components are freely soluble in the cytoplasm of the cell. *Step 2, Initiation.* (a) Small ribosomal subunits each bind an aminoacyl-tRNA, including some that bind the initiator tRNA. (b) To start initiation, a small ribosomal subunit carrying an initiator tRNA binds to an mRNA so that its tRNA anticodon binds to an AUG sequence on the mRNA; the AUG is the start codon that codes for Met. (c) Next, a large ribosomal subunit with an A and a P site takes position so that the initiator tRNA is in its P site. The adjacent A site is open and can receive another aminoacyl-tRNA, (d) in this case one carrying leucine, since the leucine tRNA anticodon is complementary to the leucine codon (CUG) on the mRNA. *Step 3, Elongation.* (e) The enzyme peptidyl transferase breaks the bond between the first amino acid, Met, and its tRNA; a peptide bond is immediately formed between the first (Met) and second (Leu) amino acids, and the initiator tRNA exits the P site. (At this point, the N-terminal end of the peptide is indicated by H_2N–.) (f) The ribosome moves along the mRNA a distance of three nucleotides so the second tRNA molecule (carrying Met-Leu) is in the P site. (g) Then a new aminoacyl-tRNA with a Trp attached moves into the A site. Peptidyl transferase breaks the bond between the second amino acid (Leu) and its tRNA; (h) another peptide bond is formed between that amino acid (Leu) and the third one (Trp), adjacent to it. The ribosome then moves another three nucleotides, and the process is repeated. (Figure continues on next page.)

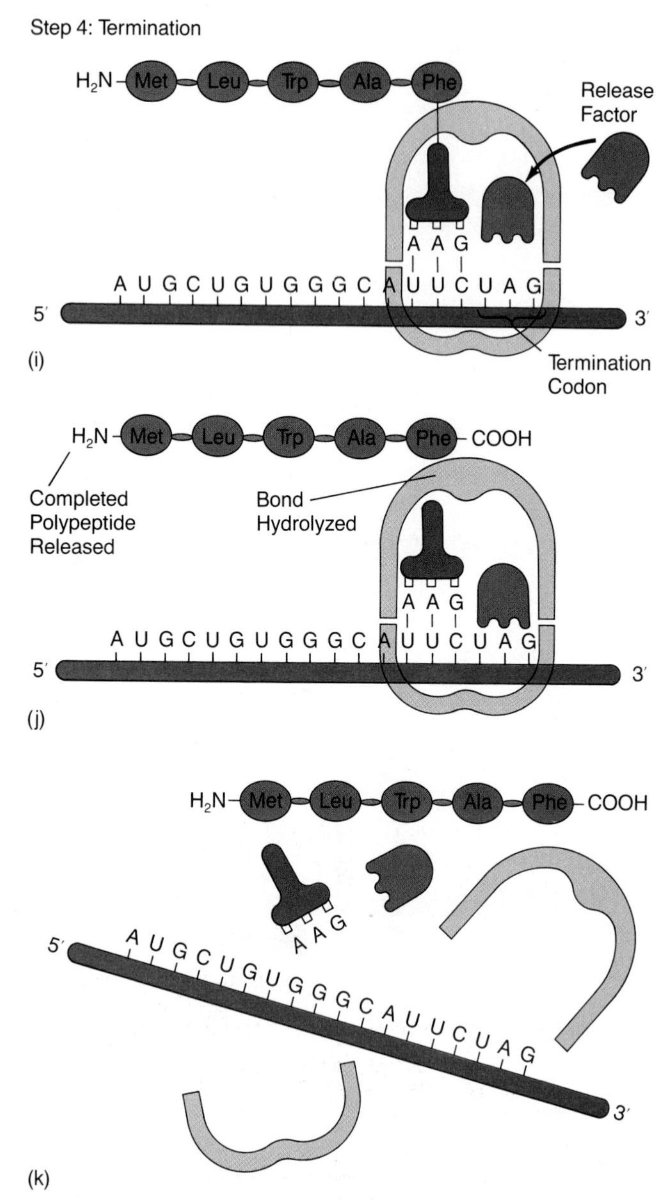

Figure 13-9 THE FINAL THREE STEPS OF PROTEIN SYNTHESIS. (Continued)
Step 4, Termination. (i) The ribosome ultimately reaches a termination codon on the mRNA. (j) The release factor binds to the stop codon and causes peptidyl transferase to release the completed polypeptide. (k) The ribosomal subunits dissociate from the mRNA.

this step, the enzyme **peptidyl transferase,** which is part of the large ribosomal subunit, breaks the bond holding the first amino acid (Met) to its tRNA and attaches that amino acid instantly by a peptide bond to the second amino acid, still bound to its tRNA at the A site (Figure 13-9e). The tRNA at the P site exits immediately, and the formation of the peptide bond creates a peptidyl-tRNA molecule, which then moves from the A site to the newly vacated P site, carrying the mRNA with it (Figure 13-9f). The distance moved is equivalent to three bases—that is, one codon. This process frees the A site on the ribosome so that still another aminoacyl-tRNA can bind with the third codon along the mRNA (Figure 13-9g). A new peptide bond then forms between the second and third amino acids, and the chain becomes one amino acid longer (Figure 13-9h).

Each step in the elongation process depends on the splitting of several high-energy phosphate bonds and makes protein synthesis an energetically expensive process for the cell.

Step 4: Termination

Elongation continues until a *termination codon* is reached on the mRNA molecule (Figure 13-9i). When this stop signal enters the vacant A site on the ribosome, a release factor binds and then hydrolyzes the peptidyl-tRNA to release the completed polypeptide (Figure 13-9j). The ribosomal subunits dissociate from the mRNA and are free to participate in another round of protein synthesis (Figure 13-9k).

As genetic information is translated into the amino acid sequence of a polypeptide, the small ribosomal subunit—where the mRNA codon and tRNA anticodon meet and pair—carries out the task of *ordering* the amino acids according to the base sequence in the mRNA message. The large subunit has the task of *linking* amino acids to the elongating polypeptide chain.

Figure 13-10 gives an overview of transcription, translation, and related processes in a bacterial cell. Protein synthesis occurs with amazing speed. A bacterial cell can generate an average-size protein, containing about 400 amino acids, in close to 20 seconds. And, of course, it can synthesize thousands of such proteins simultaneously on different mRNAs and ribosomes. In eukaryotes, the nuclear membrane physically separates the processes of transcription in the nucleus and translation in the cytoplasm. The speed and efficiency of protein synthesis in both cell types help explain how bacteria can produce a new generation in just 20 minutes and how many young plants and animals can grow so quickly.

CRACKING THE GENETIC CODE

By the early 1960s, biologists were in a position somewhat analogous to that of Egyptologists before 1799. The latter had by then discovered many examples of Egyp-

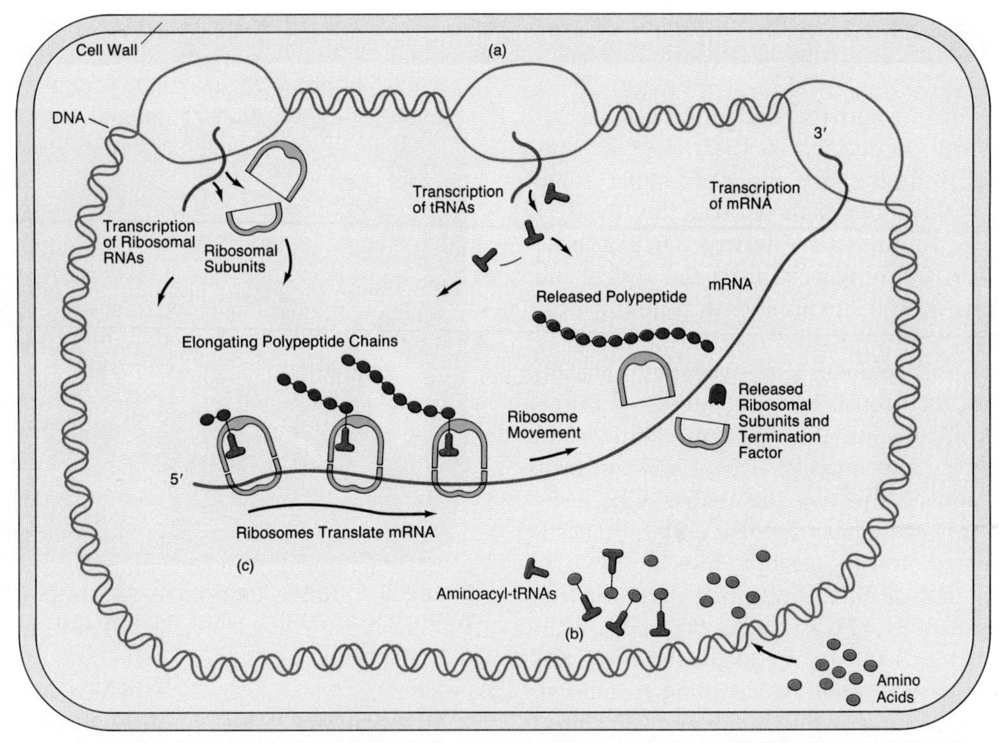

Figure 13-10 AN OVERVIEW OF THE TRANSCRIPTION AND TRANSLATION PROCESSES IN A BACTERIAL CELL. (a) The genetic message is transcribed from DNA to mRNA. Thus, the ribosomal RNAs and at least 32 different tRNA molecules specific for the 20 amino acids are transcribed from the DNA of the cell. Then mRNAs are transcribed. Even as the mRNA is synthesized, (b) activated amino acids on aminoacyl-tRNAs and small and large ribosomal subunits combine with the mRNA (c) to translate the message and synthesize proteins. Transcription can take place simultaneously at many locations on the DNA molecule, and translation can take place simultaneously at multiple sites on the mRNA molecule as ribosomes move along it. (For clarity, the various components are not drawn to scale.)

tian hieroglyphics and had determined that they represented a pictorial language. They could not, however, read an entire message in hieroglyphics, just as biologists knew of the genetic and protein "alphabets" but could not match a given codon (say, ATT or GTC) with a given amino acid. In the summer of 1799, a French soldier made an incredible chance find at a site in Egypt: the Rosetta stone, a large tablet inscribed with the same information in hieroglyphics and Greek that ultimately yielded the secrets of the Egyptian pictograms. Biologists, too, found the key to the genetic code through deep insight and hard work.

The first "crack" in the genetic code came as the unexpected result of a carefully controlled experiment. Two researchers at the National Institutes of Health, Marshall Nirenberg and J. Heinrich Matthaei, were studying protein synthesis in a test-tube solution containing materials extracted from cells. These ribosomes, amino acids, RNAs, and other substances in the cell-free system are capable of incorporating free amino acids into polypeptides despite the absence of living cells. As a control in one experiment, Nirenberg and Matthaei included a synthetic RNA called polyuridylic acid (poly-U), which is simply a long chain of RNA containing the base uracil—the equivalent of dozens of UUU codons: UUUUUUUUUUUUUUUUUU. . . . The researchers never guessed that the synthetic poly-U, their control, would direct the synthesis of a polypeptide. To their astonishment, they found that the special cell-free system containing poly-U generated a polypeptide, polyphenylalanine. Even more remarkable was the fact that although all 20 amino acids were available in the test tube, the peptide made contained *only* phenylalanine. This result indicated that the triplet UUU must be the codon that specifies the amino acid phenylalanine (Phe). The genetic message UUU UUU UUU was translated into the polypeptide Phe Phe Phe. The researchers soon created other synthetic messengers, including polycytosine (CCCCCCCCCCCC), which directed the synthesis of polyproline, and polyadenine (AAAAAAAAAAAA), which coded for polylysine. These findings suggested that CCC = proline and AAA = lysine.

Not long after this, others developed techniques for chemically synthesizing RNA triplets with all 64 possible

A SHY GENETICIST AND JUMPING GENES

In 1951, geneticist Barbara McClintock made a discovery for which, more than 30 years later, she received a Nobel Prize. Her discovery—that some genes are mobile rather than permanently fixed in place along the chromosome—was so startling in its day that it was largely ignored until the rest of the field caught up to McClintock's level of insight decades later.

McClintock was a pioneer in the genetics of maize, or Indian corn (Figure A). She was interested in the genetic basis for color variations in maize kernels and leaves, which can have numerous hues and patterns. McClintock noticed that the pigmentation patterns were usually inherited intact but that mutations occasionally occurred, causing abrupt shifts in the pigmentation phenotype of the offspring ears of corn. After 6 years of study, she arrived at the inescapable conclusion that certain control elements along the chromosome can *move* from one position to another and, in so doing, bring about the phenotypic shifts.

Specifically, McClintock found that a pair of genes along the ninth chromosome, the *Ac* (activator) and *Ds* (dissociation) genes, act in concert to turn on and off the genes that control color in the kernels. A signal from the *Ac* gene causes the *Ds* gene to jump to new positions along the chromosome, thus inactivating neighboring genes. This, in turn, causes abrupt changes in kernel pigmentation once the ears of corn develop.

Figure A NOBEL LAUREATE BARBARA McCLINTOCK WITH HER FAVORITE GENETIC SUBJECT.

Although McClintock published this work in 1951, it was neither comprehended nor accepted. Thus, the shy geneticist continued to study the movable, or "transposable," genetic elements of maize without further reporting her findings. Two decades passed before researchers using the modern techniques of molecular genetics discovered that "jumping genes" were ubiquitous in nature. Only then did the scientific community finally come to appreciate McClintock's data and conclusions.

codons and identifying their meanings. Figure 13-11 shows the codon dictionary they deciphered. This chart reveals the extreme degeneracy of the genetic code: six codons specify leucine; six code for serine; four for proline; and so on. Notice that only two amino acids, methionine and tryptophan, are represented by single codons. Also notice that three codons—UAA, UAG, and UGA—function as termination codons.

With the genetic code translated, researchers have been able to synthesize specific genetic sequences and proteins and alter the genes in living organisms, as Chapter 14 explains.

Researchers recently solved a long-standing puzzle about protein synthesis: If all tRNAs have the same cloverleaf shape, why does each one join to just a single kind of amino acid and bring it to the ribosome during protein synthesis? Geneticists at the Massachusetts Institute of Technology have found that an aminoacyl-tRNA synthetase enzyme recognizes a particular base sequence at a specific spot on a tRNA, then attaches the amino acid alanine to that molecule. The team thinks that in time, their group and others will decipher the recognition tags for all tRNAs. Some people have called the initial discovery the first step toward a "second genetic code," or the key to tRNA identity.

THE CONCEPT OF THE GENE REFINED

With their understanding of the genetic code and the way it is translated into protein, biologists could revise further their concept of the gene.

Recall that to Mendel, factors (his word for *genes*)

First position (5' end)	Second position: U	C	A	G	Third position (3' end)
U	Phe	Ser	Tyr	Cys	U
	Phe	Ser	Tyr	Cys	C
	Leu	Ser	Stop	Stop	A
	Leu	Ser	Stop	Trp	G
C	Leu	Pro	His	Arg	U
	Leu	Pro	His	Arg	C
	Leu	Pro	Gln	Arg	A
	Leu	Pro	Gln	Arg	G
A	Ile	Thr	Asn	Ser	U
	Ile	Thr	Asn	Ser	C
	Ile	Thr	Lys	Arg	A
	Met	Thr	Lys	Arg	G
G	Val	Ala	Asp	Gly	U
	Val	Ala	Asp	Gly	C
	Val	Ala	Glu	Gly	A
	Val	Ala	Glu	Gly	G

Figure 13-11 THE GENETIC CODE: ROSETTA STONE OF MOLECULAR GENETICS.
Each of the 20 amino acids is represented by at least one three-base codon. The genetic code is degenerate because most of the amino acids are specified by several codons. The degeneracy of the genetic code serves a protective function; the many codons for a single amino acid often are quite similar; for example, four of the six codons for leucine begin with CU, no matter what the third base. Because of this similarity, a point mutation in the third place will not lead to an incorrect amino acid being placed in a protein. Note that AUG, which codes for methionine, is a start signal for protein synthesis. There are three stop signals: UAA, UAG, and UGA.

were units of hereditary function that specify one of two alternatives for a given phenotypic characteristic. The study of mutations and alleles led later geneticists to picture the gene as a unit of mutation, as well.

When biologists further realized that genes occur at different locations on chromosomes and could recombine, they added the notion of the gene as a unit of recombination.

After formulating the one gene–one polypeptide hypothesis, molecular geneticists could further expand the definition of the gene to a unit of DNA that codes for a polypeptide or a structural RNA molecule. And the expansion continues to the present, as later chapters will explain. Despite the many facets of gene function, however, living organisms generally share the same genetic code. And the same basic form of translation into proteins underscores the unity of life and evolutionary relationships among all species.

SUMMARY

1. The sequence of gene *codons* is colinear with the amino acids in the corresponding polypeptide.

2. The *genetic code* is nonoverlapping, is translated in a fixed *reading frame*, and is degenerate; codons are nucleotide triplets. There are 64 possible codons to code for the 20 amino acids in organisms.

3. There are several types of RNA, all of which are made as complementary copies of the DNA nucleotide sequence. The copying process is called *transcription*. *Messenger RNA (mRNA)* serves as a template for protein synthesis. Each of the types of *transfer RNA (tRNA)* picks up a specific amino acid at its *amino acid attachment site;* another tRNA site, the *anticodon*, then can bond to the complementary codon of the mRNA. *Ribosomal RNAs (rRNA)*, in combination with a number of proteins, make up the ribosomes. They are sites where proteins are assembled.

4. Protein synthesis, or *translation*, involves four steps: (1) In *amino acid activation*, a specific amino acid becomes attached to a specific tRNA molecule; (2) at *initiation* of polypeptide synthesis, the small ribosomal subunit and the initiator tRNA bind the mRNA at the initiator codon, where they are joined by the large ribosomal subunit and the second tRNA; (3) *elongation* of the polypeptide occurs as the ribosome moves along the mRNA molecule from codon to codon and as peptide bonds are formed to link the amino acids into a polypeptide; (4) *termination* of protein synthesis takes place when a termination codon is reached on the mRNA molecule.

5. The small ribosomal subunit orders the amino acids according to the mRNA codon sequence, whereas the large ribosomal subunit links amino acids to the elongating polypeptide chain.

6. Nirenberg and Matthaei cracked the genetic code when they discovered that a long artificially prepared mRNA molecule composed solely of uracil directed the synthesis of a polypeptide that contained only phenylalanine. Later researchers uncovered the complete genetic code.

7. The gene has been defined as a unit of hereditary function, a unit of mutation, a unit of recombination, and a unit that codes for a polypeptide or a structural RNA.

KEY TERMS

amino acid activation
aminoacyl-tRNA synthetase
anticodon
A site
codon
elongation
frameshift mutation
genetic code
initiation
messenger RNA (mRNA)
peptidyl transferase
P site
reading frame
ribosomal RNA (rRNA)
termination
transcription
transfer RNA (tRNA)
translation
uracil

QUESTIONS AND PROBLEMS

1. A synthetic polymer of random sequence containing the bases U and G in a 5:1 ratio was made. In this polymer, the triplet UUU should occur five times more often than triplets containing two U's and a G, and 25 times more often than triplets containing one U and two G's; UUU should occur about 125 times more often than GGG. Nirenberg and Matthaei used this polymer to direct polypeptide synthesis in the cell-free system. Then they measured the proportions of the various amino acids in the polypeptide chains. Now look at the codon table (Figure 13-11) and determine which amino acid(s) was (were) most abundant. Which were present at about one-fifth that amount? Which were the least abundant?

2. Examine the genetic code (Figure 13-11). Do the codons have polarity? Is UUG different from GUU? Which way do we read them: 5′ to 3′, or 3′ to 5′? In which direction is mRNA translated into protein?

3. What two codons are present in the artificially constructed RNA molecule UGUGUGUGUG . . . ? When this RNA is used to direct protein synthesis in a cell-free system, what two amino acids would you expect to find in the polypeptides? What would be their sequence? In the repeating RNA molecule AUCAUCAUCAUC . . . , what three codons are present? Which amino acids do they code for? On genuine mRNA, how do the ribosomes "know" where to start translation?

4. Bacteriophage T4 lysozyme contains 164 amino acids in a single polypeptide chain. What is the minimum number of base pairs in the lysozyme gene? This is the base sequence of part of the mRNA for T4 lysozyme:
5′-AGGAGGUAUUAUGAAUAUAU UUGAAAUGUUACGUAUA . . . -3′
Write the base sequence of the DNA strand from which the mRNA was transcribed. Below it write the base sequence of its complementary strand. Does one of these match the mRNA? Now pretend you are a ribosome. Bind to the A- and G-rich ribosome binding site at the 5′ end of the mRNA, find the nearest initiation codon, and start translating the message into protein.

5. Suppose a mutation has occurred in the DNA in Problem 4 so that the underlined U is changed to C. How will that mutation affect the amino acid sequence?

6. This is the amino acid sequence of a section in the middle of the lysozyme molecule; the corresponding codons on the mRNA are written below:
. . . Thr-Lys-Ser-Pro-Ser-Leu-Asn-Ala-Ala-Lys-Ser-Glu-Leu-Asp
. . . ACA AAA AGU CCA UCA CUU AAU GCU GCU AAA UCU GAA UUA GAU . . .
A mutant strain #1 contains a frameshift mutation in the DNA corresponding to the underlined region: One of the A's is deleted. Write the mutant base sequence and the corresponding amino acid sequence for this region.

7. During protein synthesis, do the amino acids pair up with their corresponding codons in the mRNA? Or does tRNA pair with the codon?

8. Which of the following statements applies to mRNA? To rRNA? To tRNA? To all three? (1) Leaves the nucleus in eukaryotes after it is transcribed; (2) is transcribed from DNA; (3) contains an amino acid attachment site and an anticodon; (4) carries the blueprint for synthesis of a specific polypeptide; (5) acts as an adapter between an amino acid and its codon; (6) combines with a number of proteins to form a structure that acts as a movable vise that holds the blueprint and the subunits for protein synthesis.

ESSAY QUESTION

1. Define the word *gene*. Incorporate all the features of a gene that you have learned about so far.

For additional readings related to topics in this chapter, see Appendix C.

14 Bacterial Genetics, Gene Control, and Genetic Engineering

As the relentless revolution in genetics continues year after year, it becomes clear that we are learning enough about life's molecular dance to begin to become its choreographers.

Boyce Rensberger, "Tinkering with Life," *Science* 81 (1981)

A few years ago, a team of scientists from several universities set out to build not a better mousetrap, but a better *mouse*. The mice they produced grew twice as fast as normal, and while the utility of giant mice may seem obscure, it is a conceptually small step from bigger mice to faster growing cattle, sheep, rice, or corn.

The scientific team was carrying out **genetic engineering**, a set of procedures that depend on **recombinant DNA technology.** This technology involves the manipulation and transfer of genes from one organism to another in the laboratory and is the most exciting advance in genetics since Watson and Crick described the structure of DNA.

Recombinant DNA techniques grew directly from work on the genetics of bacteria and the mechanisms of gene control. Thus we begin our chapter with those subjects. Once geneticists understood more about genes in

Genetic engineering. The mouse on the left is normal; the giant mouse on the right contains the human growth hormone gene in all its cells.

bacteria, they were able to study gene activity and control in higher organisms, and eventually, they learned to transfer genes between species in the laboratory.

Geneticists traversed the path from Watson and Crick to recombinant DNA in just 20 years, and the next 20 hold significant promise for bettering human health, agriculture, and the environment. This chapter will describe:

- How genes are transferred between bacteria in nature
- How bacterial genes are controlled
- How genes are regulated in eukaryotic cells
- How researchers manipulate genes through recombinant DNA technology
- The promises and prospects for genetic engineering

BACTERIA EXCHANGE GENES SEXUALLY AND ASEXUALLY

Bacteria were such key organisms in the elucidation of DNA structure and protein synthesis that geneticists wanted to develop a gene map of the circular *E. coli* chromosome and use it to study the development of bacterial phenotypes. However, there were two significant stumbling blocks to mapping genes and their expression in bacteria: (1) Bacteria seemed too small to exhibit easily visible phenotypic traits; yet, to study how genes are passed from one generation to the next, one must have two pure-breeding strains that differ in some easily recognized trait. (2) In bacteria, the passage of genes from one generation to the next—so-called transmission genetics—differs from that in higher organisms. A bacterium does not undergo mitosis before it divides in two. Instead, the single, circular chromosome is duplicated, and one copy passes to each daughter cell during *fission*, the splitting of the cytoplasm in two. Bacteria also lack meiosis, but they do have a means of exchanging genes sexually. Therefore, Mendel's rules apply in principle to bacteria, but not in exact detail.

Detection of Bacterial Phenotypes

Because of these special obstacles, geneticists learned to use biochemical characteristics to study genetic variation in bacteria. Bacteria exhibit three kinds of biochemical phenotypes that can be detected and compared: (1) resistance to or sensitivity to certain antibiotics, (2) abnormal nutritional requirements, and (3) the inability to use certain compounds for growth.

Although normal *E. coli* cells cannot grow in the presence of antibiotics such as streptomycin, geneticists can use the procedures shown in Figure 14-1 to isolate mutants that are *resistant* to streptomycin. The experimenter spreads treated cells on *agar*, a rubbery culture medium derived from seaweed and, in this case, also containing a few nutrients and the antibiotic streptomycin. All the normal cells will die, but the one or two cells with the mutated gene for resistance to the antibiotic will survive, divide repeatedly, and form visible colonies on the surface of the agar.

Researchers use a different technique to identify mutant bacteria with nutritional deficiencies, including

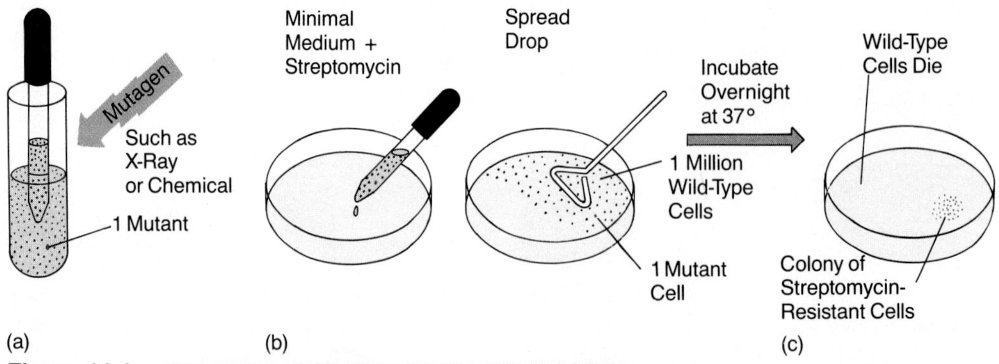

Figure 14-1 ISOLATING ANTIBIOTIC-RESISTANT MUTANTS.
(a) A researcher treats a suspension of normal bacteria with a mutagen to cause mutations. (The mutant cell is shown here in red). (b) The researcher then spreads the bacteria on a Petri dish containing agar with minimal medium plus the antibiotic streptomycin.
(c) Although the normal cells will die, any bacterial cells with the mutated gene for resistance to streptomycin will survive and multiply into colonies. Here the red cell divides into a red colony.

auxotrophs, or bacteria unable to make all their own necessary growth factors from the few chemicals present in a minimal medium.

Suppose, for example, that a bacterium has a deleterious mutation in a gene that codes for an enzyme the cell needs to make the amino acid tryptophan. Although such a mutated cell could not grow on a minimal medium, it could be "rescued" by the addition of tryptophan to the medium. Figure 14-2 shows and describes a procedure called *replica plating* that allows an experimenter to find such a mutant cell. By using a piece of velvet to pick up a dusting of colonies growing on a complete medium and "imprinting" them on a plate of minimal medium lacking tryptophan, the researcher can look for any colony that fails to appear on the deficient medium but does thrive on the complete medium. Such a colony can be assumed to be made up of mutant cells lacking a gene for tryptophan synthesis. Replica plating has revealed auxotrophs for hundreds of biochemicals.

A third type of mutant bacteria used in genetics research cannot use certain nutrients. Normal *E. coli* cells, for example, can grow just as readily on lactose as they do on glucose because they have enzymes that can convert lactose into the simpler glucose. Some mutant *E. coli*, however, have a defective gene and in turn a defective enzyme that prevents this conversion, leaving them unable to grow on a lactose medium. Such mutants proved invaluable, as we shall see.

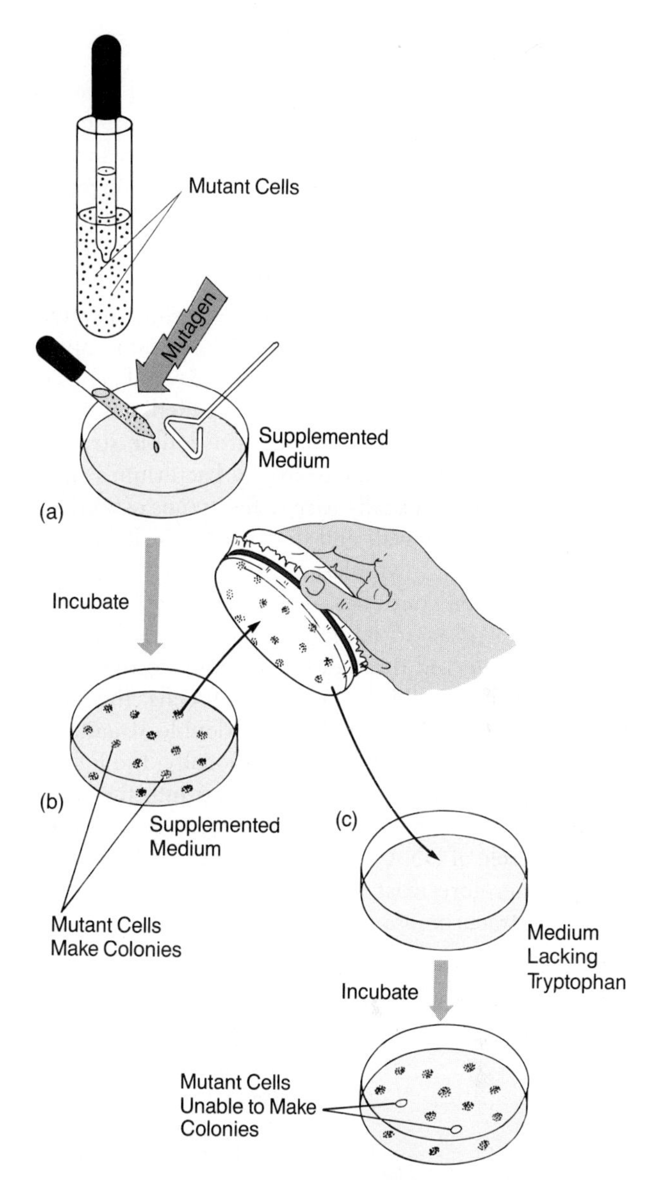

Figure 14-2 ISOLATING NUTRITIONAL MUTANTS BY REPLICA PLATING.
(a) To isolate an auxotroph (a mutant bacterial cell that cannot make all its own necessary growth factors), a researcher treats cells with a mutagen and then allows them to grow into colonies on supplemented medium along with normal bacteria. (b) The researcher next pushes a piece of velvet onto the plate, lifting some of the colonies, and (c) presses them onto the surface of a plate containing minimal medium lacking a specific nutrient, say, tryptophan. One can assume that any colony that is missing (that does not grow in a given spot on the minimal-medium plate) is an auxotroph (two colonies are missing in the figure). The worker can then retrieve those mutant colonies from the first plate and grow large numbers of the mutated auxotrophic cells.

Bacterial Gene Transmission

In 1946, Joshua Lederberg, a student, and his professor, Edward Tatum, performed an elegant experiment at Yale University demonstrating that "sex" occurs in bacteria.

Lederberg and Tatum worked with two strains of bacteria that were auxotrophic for different substances. To grow strain A, they had to add both the amino acid methionine and the vitamin biotin to the minimal medium. To grow strain B, they had to add the amino acids threonine and leucine. Since strain A could make threonine and leucine but not methionine and biotin, its genotype could be represented as met^- bio^- thr^+ leu^+, and strain B would then have the genotype met^+ bio^+ thr^- leu^-. Neither strain could grow on pure minimal medium, which lacks amino acids and vitamins. When Lederberg and Tatum mixed the two strains, they found that 1 in every 10 million cells could grow on minimal medium. The surviving cells must have had the genotype met^+ bio^+ thr^+ leu^+, since they could make all the amino acids and vitamins necessary for growth, and so they must have received some genes from strain A and some from strain B. The meaning of these results was clear: The bacteria were exchanging genes. Subsequent research has revealed three ways in which bacteria can exchange genes: *conjugation*, *transformation*, and *transduction*.

Conjugation: Direct Gene Transfer

Lederberg and Tatum discovered bacterial **conjugation,** during which DNA is transferred from one cell to another by direct contact. The donor cell, analogous to a male sexual organism, produces a *sex pilus*, a cytoplasmic bridge between the cells through which DNA is transferred to the recipient cell, analogous to a female organism (Figure 14-3a). Male cells contain the F ("fertility") factor, a piece of DNA that codes for the sex pilus. Male cells are therefore designated F^+, while recipient cells are designated F^-. The **F factor** is a set of genes on a small circle of double-stranded DNA called an **F factor plasmid** (Figure 14-3b). **Plasmids** are double-stranded DNA circles that are separate from a bacterium's single chromosome. They usually carry a few genes essential to the bacterium's survival, and they can replicate autonomously in the cell's cytoplasm.

As conjugation continues, one DNA strand of the circular F factor plasmid is nicked, or broken; this single strand is transferred through the sex pilus to the recipient F^- cell (Figure 14-3c), and in both cells, the single strands are replicated and become double-stranded F factor plasmids (Figure 14-3d). Occasionally, the F factor plasmid becomes inserted or integrated into the bacterial chromosome, so that the two form one double-stranded circle of DNA (Figure 14-3e). The F factor plasmid can therefore exist in two forms: free in the cytoplasm (extrachromosomal) or integrated into the bacterial chromosome. Genetic elements with the two modes of replication are called **episomes.**

A bacterial cell containing an F factor plasmid integrated into its circular bacterial chromosome is called an **Hfr cell** (high frequency of recombination cell), and when it conjugates with an F^- cell, *the integrated F factor pushes the entire bacterial chromosome into the recipient cell,* so that it now has two chromosomes, with two copies of many bacterial genes. If the donor and the recipient are genetically different, a researcher can detect recombination between the two **genomes** (the full set of genes on each organism's chromosomes).

It takes about 100 minutes for an entire strand of a bacterial chromosome to move through the pilus from an Hfr to an F^- cell. By interrupting mating after different periods of time, researchers can see which genes are transferred first and hence create a gene map (with units given in minutes) for the circular bacterial chromosome (Figure 14-4).

Transformation and Transduction: Indirect Gene Transfer

Genetic material can be transferred between bacteria even if the cells do not come into direct contact. During the process of **transformation,** DNA that has been released from one cell into the surrounding medium is taken up by another cell. The release of DNA is usually

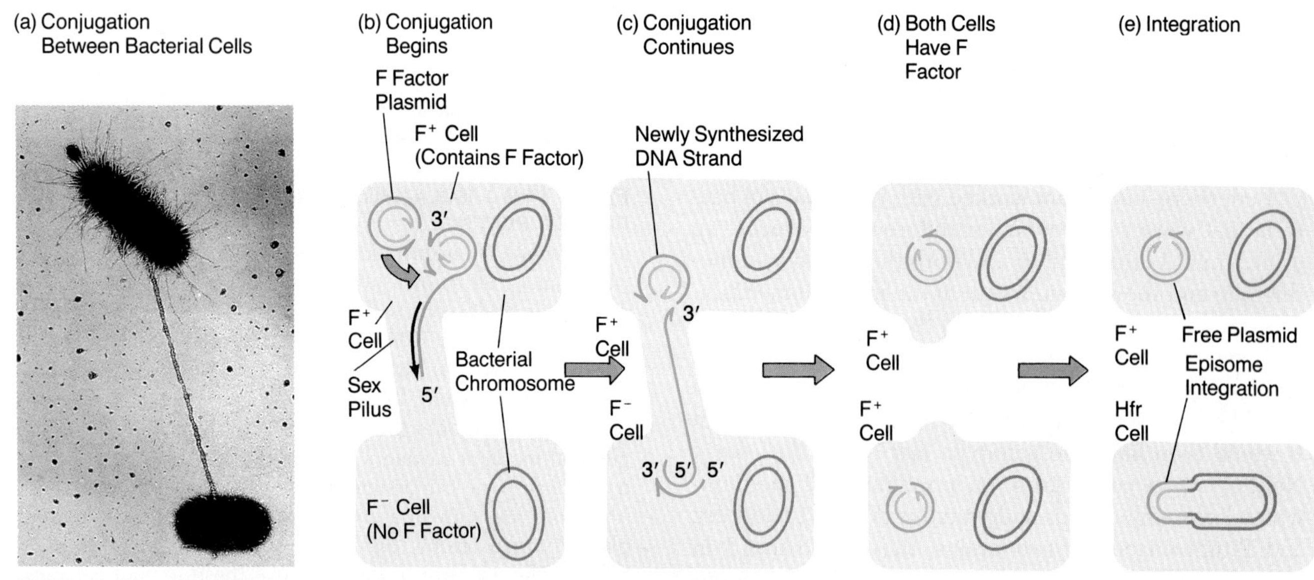

Figure 14-3 BACTERIAL CONJUGATION AND F FACTOR PLASMIDS.
(a) The *E. coli* cells shown in this electron micrograph (magnified about 18,000×) are joined by a sex pilus during conjugation. DNA is transferred from the male, or F^+, cell (top) to the female, or F^-, cell (bottom) through the pilus. (b) During bacterial conjugation, the fertility factor, or F factor, plasmid is transferred through the sex pilus to a recipient cell. (c) A single strand of the F factor plasmid is nicked, and that single DNA strand is transferred through the sex pilus to a recipient cell, which is then converted to an F^+ cell. (d) Both cells synthesize DNA strands complementary to the plasmid strand they contain.
(e) Sometimes, the transferred F factor becomes integrated into the recipient's DNA, changing it into an Hfr cell.

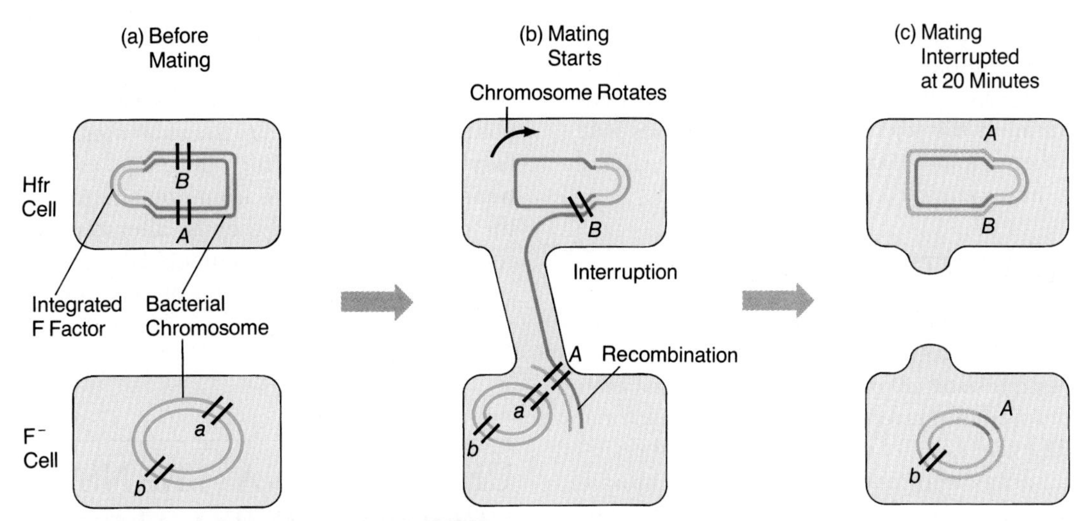

Figure 14-4 THE INTERRUPTED-MATING EXPERIMENT.
Interrupting the mating between bacterial cells at different time intervals allows geneticists to map the genes on the bacterial chromosome. (a) Hfr and F^- cells before mating, with alleles of the *A* and *B* genes on their circular chromosomes. (b) When F factor plasmid is transferred from an Hfr cell to an F^- cell, the migrating portion of the F factor plasmid seems to push the bacterial chromosome before it. (c) By interrupting the mating at regular intervals (here at 20 minutes) and thereby allowing only part of the genetic material to enter the F^- cell, geneticists can map the location of bacterial genes on the donor cell's chromosome. If the experiment goes on for 40 minutes instead of 20, the *B* allele will be transferred, too. Thus, the order in which, and time when, different genes enter the recipient cell correspond to their location on the chromosome.

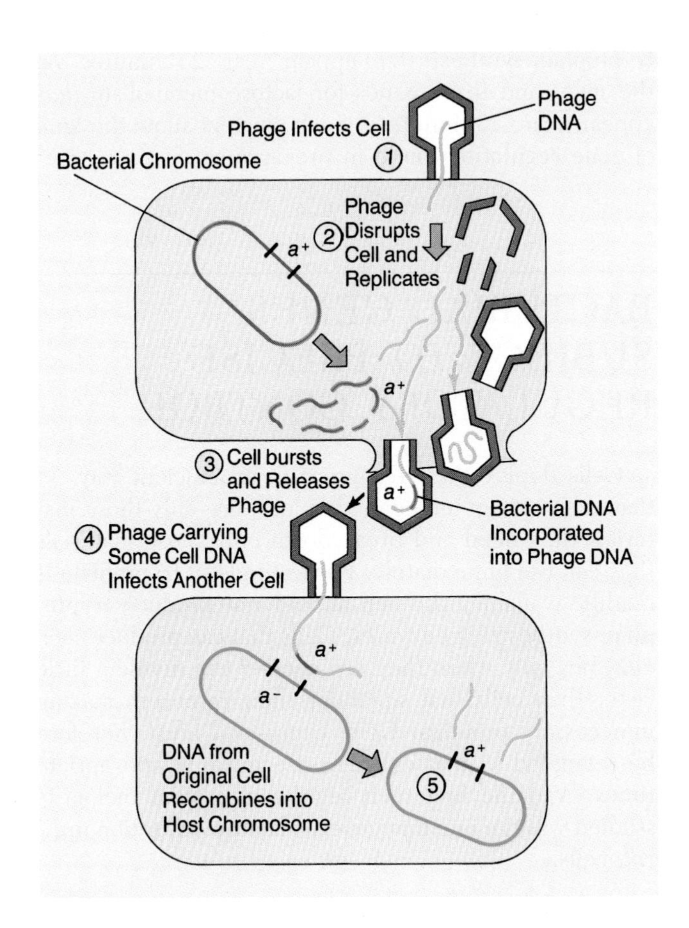

the result of cell disruption or cell death. This is the way the rough and smooth pneumococci exchange genes (see Chapter 12). In **transduction,** a virus carries DNA from one bacterium to another. Most *bacteriophages* (viruses that infect bacteria) consist of DNA surrounded by a protein coat. The steps in Figure 14-5 show that as new phages are being assembled in the cytoplasm of a host bacterial cell, phage coats encapsulate copies of viral DNA formed in the infected bacterium. Mistakes some-

◄ **Figure 14-5 PHAGE-MEDIATED TRANSDUCTION.**
When a phage infects a bacterium, phage DNA enters the host bacterial cell (1). The phage DNA replicates and is transcribed, and phage proteins begin to be synthesized. One result is that the host DNA breaks into fragments (2). DNA becomes packaged inside phage heads; occasionally, pieces of bacterial DNA, in addition to phage DNA, are enclosed in phages. Eventually, the cell bursts, and phage particles are released (3), some carrying bacterial DNA incorporated into their own. When such a phage infects a new cell (4), recombination may occur between the DNA of the donor bacterium carried by the phage and the host cell DNA (5). If the recombined donor DNA contains a new allele (here a^+), the new gene product produces an altered phenotype in the recipient cell and its progeny, and a researcher can confirm that transduction has taken place.

times occur, however, so that a coat encloses a piece of the bacterial chromosomal DNA along with phage DNA. The new phage can inject the bacterial DNA into another bacterium. If the injected DNA recombines into the DNA of the recipient bacterial cell, *phage-mediated transduction* has occurred.

Transposons Move Genes About

Another type of gene exchange in bacteria deserves mention. In 1955, a strain of the bacterium *Shigella dysenteriae* became resistant simultaneously to several antibiotics and caused a dangerous epidemic of bacterial dysentery. The genes, or **R factors,** that provided resistance to these drugs were not on the bacterial chromosome but were carried on "resistance" **R factor plasmids,** pictured in Figure 14-6. To make matters worse, some of these plasmids also carried genetic elements that allowed a rapid transfer of the R factor plasmids—and with them, multiple drug resistance—from one bacterial cell to another.

The genetic elements responsible for the transfer of these drug-resistant genes between plasmids are called *transposons.* They are DNA segments that owe their mobility to *insertion sequences,* short stretches of DNA that can insert a gene into a variety of chromosomal sites. Researchers, beginning with Barbara McClintock (see the box in Chapter 13, page 226) have discovered transposable genetic elements in eukaryotes as well as in prokaryotes. In corn, fruit flies, and other organisms, transposons may move from one site to another among the chromosomes and thereby alter gene activity. We'll return to transposable elements later and see how they may contribute to the origin of new genes and even of new species.

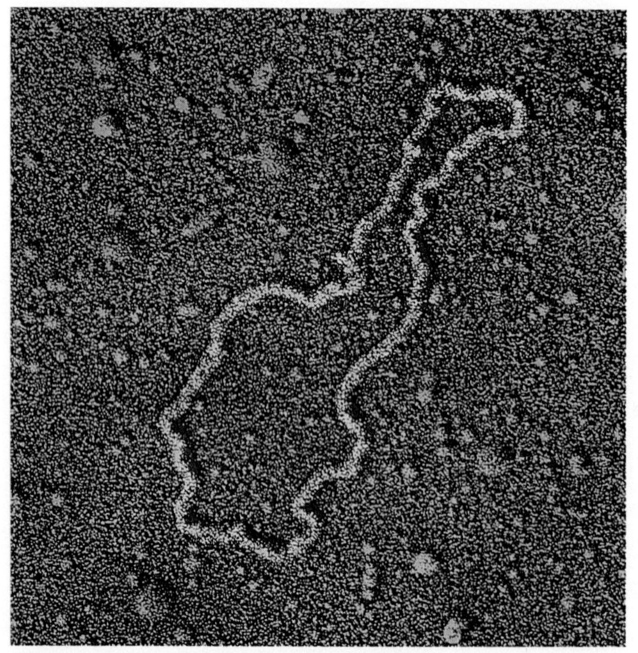

Figure 14-6 AN R FACTOR PLASMID. This circular DNA molecule (magnified about 82,000×) contains genes that confer resistance to antibiotics. Antibiotic resistance can be transmitted from bacterium to bacterium by transduction. During an epidemic, this type of infectious resistance can have dire medical consequences.

USE OF DNA TRANSFER PROCESSES TO MAP PROKARYOTIC GENES

Geneticists used conjugation, transformation, and transduction to help them map about 2,000 of the estimated 3,000 genes in *E. coli.* With the aid of newer technology, they recently completed the map. Notice in Figure 14-7 that the gene map for *E. coli* represents the organism's single circular chromosome, made up of a double-stranded DNA molecule. Notice, too, that genes for enzymes involved in related tasks are frequently grouped close together. For example, five genes for tryptophan synthesis (*trp*) appear near "25 minutes" on the map, and three genes for lactose metabolism (*lac*) appear near "10 minutes." Such clusters allow the kind of **gene regulation** found in prokaryotes.

BACTERIAL GENES: SUBJECT TO PRECISE REGULATORY CONTROL

Cells deploy their resources in an efficient way, so that at any given moment, they produce only those materials they need and break down only those materials they can use immediately. For example, if tryptophan is readily available, the cell should not produce tryptophan-synthesizing enzymes. Cells that can produce such enzymes only when they are needed can divide a little faster than cells that squander their resources making unnecessary proteins. Gene expression must therefore be regulated so that enzymes are made at appropriate times. And the first such regulatory system biologists studied was the one that governs lactose utilization in *E. coli* cells.

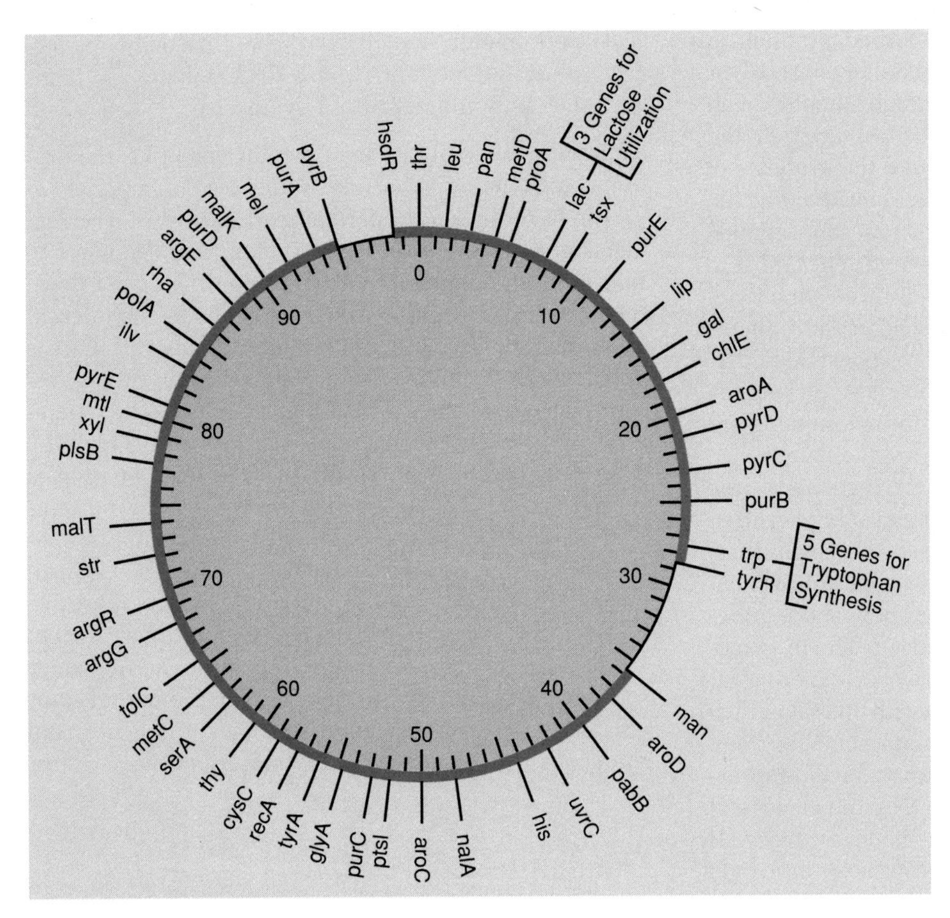

Figure 14-7 A GENE MAP OF *E. coli*.
The genes on this map represent only a small fraction of the 2,000 known genes of *E. coli*. Indeed, there is enough DNA in the circular chromosome to code for 3,000 polypeptides. The minutes around the inner circle signify times of entry into a cell during interrupted-mating experiments. Note that genes for enzymes involved in related tasks often occur close together on the genome—for example, the lac site, which is really three genes that code for polypeptides involved in lactose utilization.

The Operon Model of Gene Regulation

Several regulatory mechanisms can turn the supply of an enzyme on or off so the cell's precious energy is not wasted. Cofactors, for example, can block or allow certain enzymes' activity (review Figure 4-9), and feedback inhibition can block enzymes in metabolic pathways and prevent needless overproduction of the pathway's end product (review Figure 4-14). In prokaryotes, the amount of enzyme present in a cell is usually controlled at the level of mRNA synthesis. The details of regulation at this level were elucidated in the late 1950s and early 1960s, when French geneticists François Jacob and Jacques Monod studied the cluster of adjacent genes—or *operon*—that regulates how *E. coli* cells use the sugar lactose.

Regulation of Lactose Metabolism

Lactose is a disaccharide that is broken down by the enzyme β-galactosidase into its two constituent sugars, galactose and glucose. If no lactose is present in an *E. coli* cell's environment, then the cell will usually contain fewer than ten molecules of this enzyme. If the cell's environment suddenly contains lactose, however (as when you eat ice cream and the lactose in it reaches the *E. coli* cells in your intestine), each cell will soon contain several thousand molecules of the β-galactosidase enzyme needed to break down the lactose, as well as two other lactose pathway proteins. Clearly, the presence or absence of lactose regulates the amount of enzyme produced. Thus, lactose is called the *inducer*, and the three proteins are called *inducible enzymes* or inducible proteins.

Jacob and Monod noticed that the regulation of all three proteins seemed to be coordinated, so that when one of the proteins doubled in amount, so did the other two. Studies showed, however, that mutations can affect each protein individually, and thus each of the three proteins must be encoded by a separate *structural gene*—a gene coding for a polypeptide. The three genes occur in a linear cluster along the *E. coli* chromosome and are labeled *z* (for β-galactosidase), *y* (for β-galactoside permease), and *a* (for thiogalactoside transacetylase) (Figure 14-8a). A mutant *E. coli* cell that does not show β-galactosidase activity but shows activity in the other two enzymes specified by genes *y* and *a* can be designated $z^-y^+a^+$.

Jacob and Monod were surprised to find a mutant in which all three genes were affected simultaneously. This so-called *constitutive mutant* makes large amounts of all three enzymes even when lactose is absent from the environment; it clearly does not require the presence of the normal inducer (lactose) in order to make inducible enzymes. Jacob and Monod concluded that some genetic element distinct from the three structural genes must in some way govern the expression of all three simultaneously and, obviously, must be missing in the mutant. They labeled this regulatory element responsible for the inducibility of the enzymes as gene *i*. Wild-type bacteria are thus $i^+z^+y^+a^+$, and many constitutive mutants are $i^-z^+y^+a^+$.

Jacob and Monod wondered how the regulator gene *i* interacts with lactose, the inducer, to control the synthesis of the three enzymes, and devised this hypothesis: The *i* gene codes for a lactose **repressor** protein which, in the absence of lactose, goes to the *z*, *y*, and *a* genes and blocks, or represses, their expression. The researchers confirmed this hypothesis in a clever experiment. They used F^+ cells to create bacteria with one set of the *i* and *z* genes on the bacterial cell's chromosome, and one set on the inserted plasmid, each coding for the repressor and for β-galactosidase. One DNA molecule had i^-z^+, and the other had i^+z^-, and the heterozygous cells behaved as normal bacteria, making β-galactosidase only when lactose was present. From this the team concluded that the i^+ gene on one DNA molecule codes for some type of repressor molecule that can travel by diffusion to the z^+ gene on the other DNA molecule and somehow physically block the transcription of that gene and production of the enzyme β-galactosidase.

Jacob and Monod also hypothesized a genetic element near the *z* gene that could receive the signal (the repressor molecule) from the *i* gene: They called this regulator element the **operator,** designated by the symbol *o*. Indeed, mutations in the operator soon allowed them to map it to sites very near the *z* gene, and other experimenters showed that the operator does *not* code for a protein, as does the regulator gene *i*.

These experiments led Jacob and Monod to propose the operon model for gene regulation, illustrated in Figure 14-8a. There are four essential features of the operon model:

1. A *regulator* gene *i* that codes for a *repressor* molecule that can travel to another site and act as an "off" signal
2. A regulatory site, the *operator*, that receives the "off" signal from the repressor
3. A nearby set of protein-coding genes, *z*, *y*, and *a*, all of which are transcribed onto the same mRNA molecule
4. A **promoter site** adjacent to and partially overlapping the operator, where RNA polymerase binds DNA to initiate mRNA synthesis.

An **operon** is defined as the operator gene plus the protein-coding genes it controls, in this case, *o*, *z*, *y*, and *a*.

The lactose, or *lac*, operator works like this. The regulator gene makes a repressor protein that can move to the site of the operator gene, bind to it, and shut down the transcription of the structural genes in the operon. The repressor can exist in either of two shapes. If there is no lactose (inducer) available (Figure 14-8b), the repressor molecule assumes a form that can bind to the operator. This partially covers the promoter as well, and so the RNA polymerase is physically blocked from binding to the promoter. For this reason, no mRNA can be transcribed, and the three protein molecules cannot be synthesized by the cell. When lactose (the inducer) is present in the environment (Figure 14-8c), the repressor binds to this inducer and assumes an alternative shape. In this other conformation, the repressor can no longer recognize and bind to the operator. The promoter is thus left uncovered, RNA polymerase binds to the site, and the genes for the enzymes are transcribed.

Catabolite Repression

With the elegant repressor/operator mechanism, an *E. coli* cell makes only as much β-galactosidase as it needs to break down the available lactose. What would happen, however, if an *E. coli* cell were presented with both lactose and glucose? *E. coli* cells have a second mechanism that allows the cells to use glucose (a single-unit sugar, immediately useful for metabolism) whenever it is present in the medium, even if other energy sources are present, such as lactose (a disaccharide that must be cleaved). In the mechanism called **catabolite repression,** glucose represses the appearance of the enzymes that catabolize other sugars (Figure 14-9).

Catabolite repression depends on the fact that RNA polymerase binds to a promoter site much better if another protein, called *catabolite gene activator protein (CAP)*, is bound to a special DNA site nearby (this occurs when glucose is absent; Figure 14-9b). CAP acts as a guide, getting RNA polymerase started down the pathway of transcribing the *z* gene faster than if the polymerase had to find the way itself. When glucose is present, it indirectly causes CAP to change shape (Figure 14-9c). The altered CAP does *not* bind well to its site; in turn, RNA polymerase does not bind as well to the lactose promoter site. The synthesis of β-galactosidase is therefore reduced, and the cell cannot catabolize lactose as well. Actually, CAP does not bind directly to glucose.

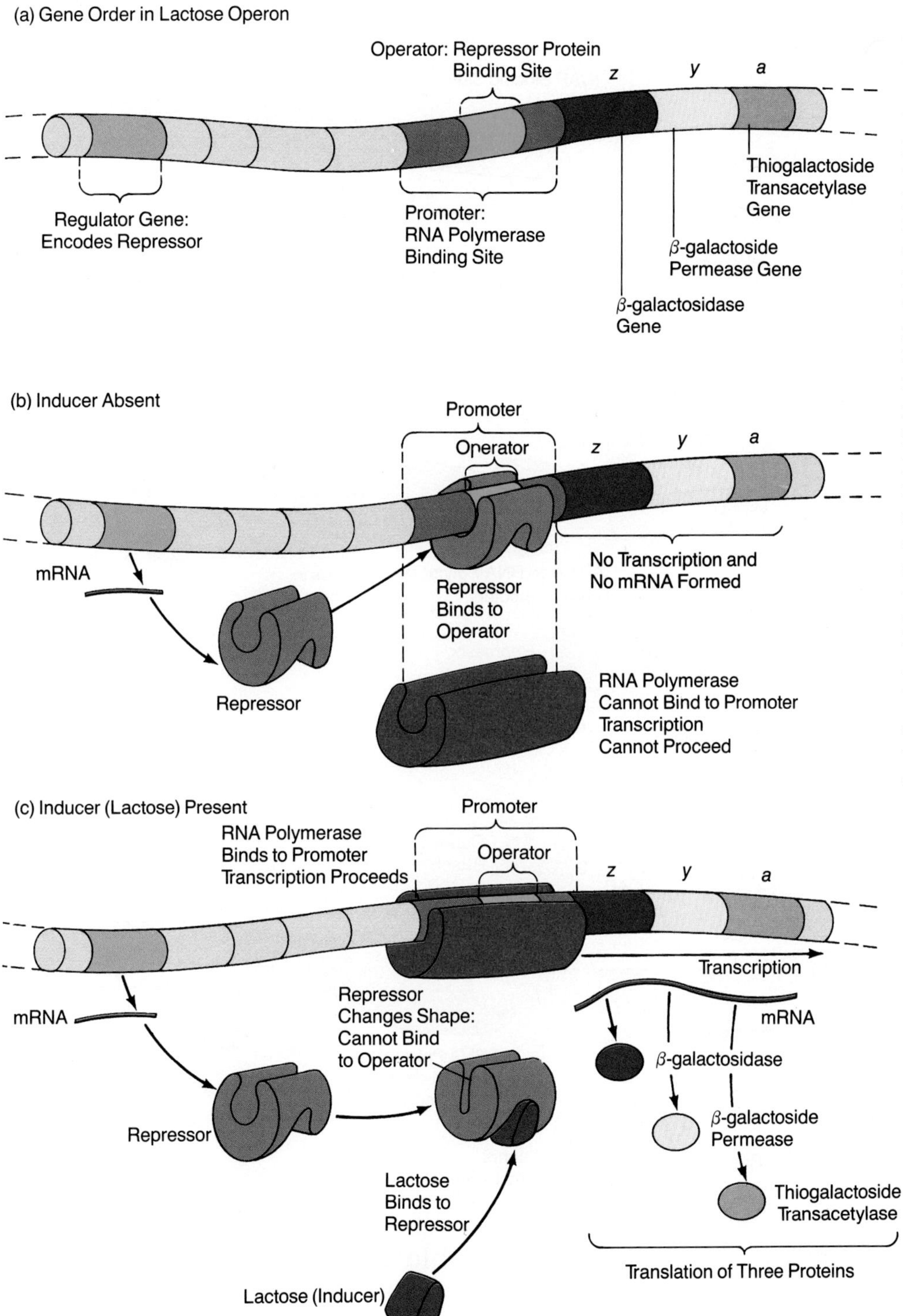

Figure 14-8 AN INDUCIBLE OPERON: THE LACTOSE OPERON. (a) The gene order. Note that the operator and the promoter overlap. (b) Inducer (lactose) absent. Repressor protein, coded by the *i* gene, is synthesized, recognizes the operator, and binds to it. This physically blocks the access of RNA polymerase to the promoter; hence, mRNA is not made, and lactose-metabolizing enzymes are not formed. (c) Inducer (lactose) present. When lactose is present, it functions as an inducer that binds to the repressor and causes it to change shape. This new shape does not allow binding to the operator. Thus, RNA polymerase has free access to the promoter, and transcription occurs, followed by translation to produce the enzymes. The cell can now use lactose as a source of energy and carbon.

Rather, CAP is sensitive to molecules of the chemical messenger *cyclic adenosine monophosphate (cAMP).* Levels of cAMP rise in the cell when glucose is absent; the cAMP binds to CAP and alters that molecule's shape.

We can see now that the β-galactosidase genes can operate at three levels of activity: low, medium, and high. These levels depend on the presence or absence of lactose and glucose and are governed by three separate proteins: the *lac* repressor, RNA polymerase, and CAP. Together, they render the cell exquisitely sensitive to external governing conditions, allowing it to use available nutrients in the most efficient way.

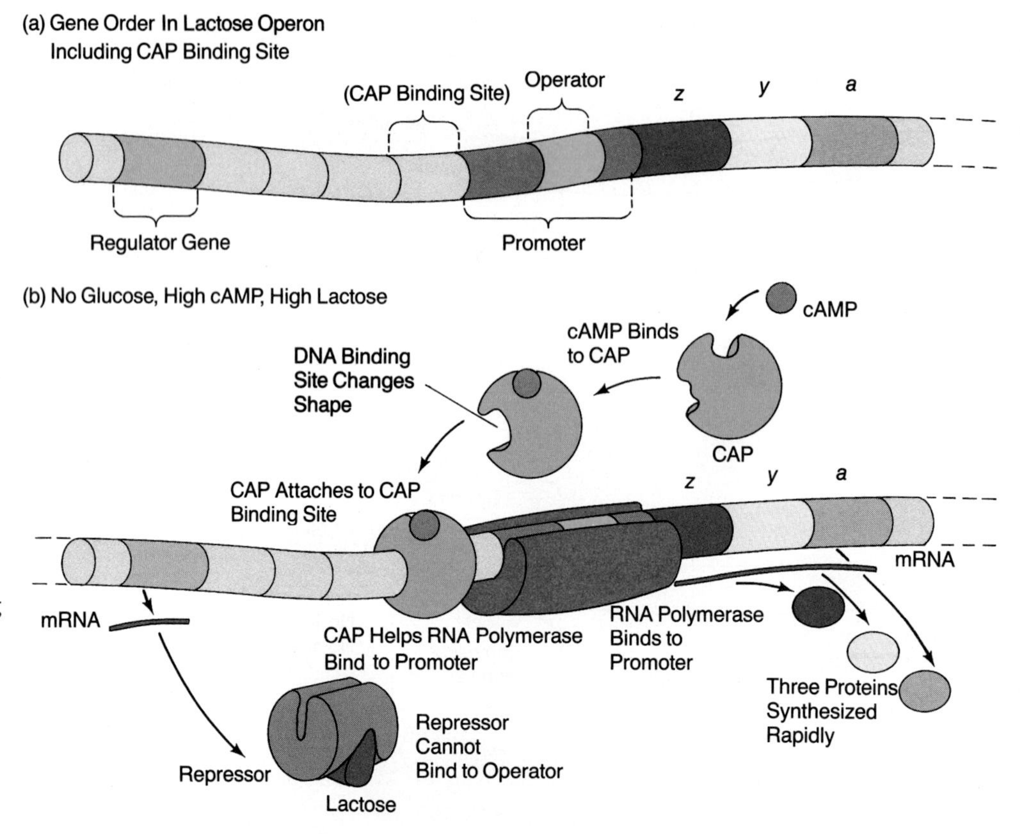

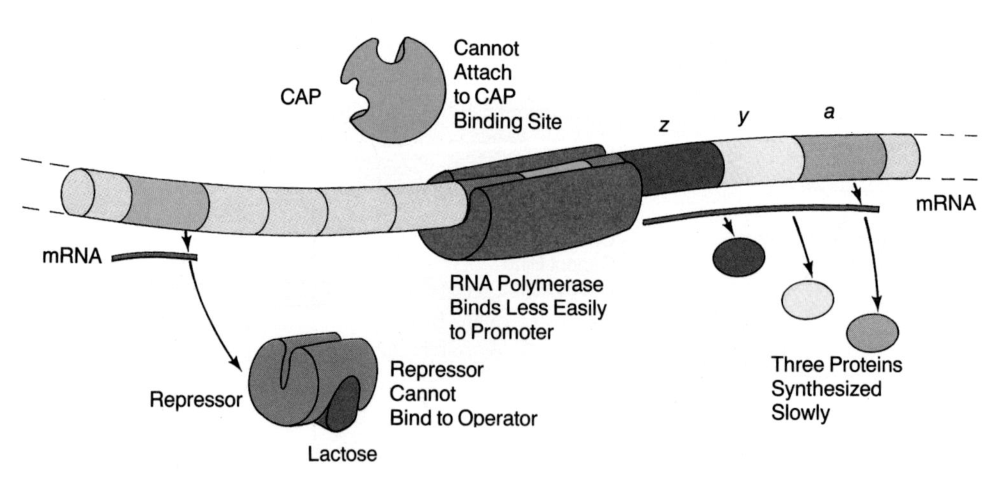

Figure 14-9 CATABOLITE REPRESSION. The formation of lactose-metabolizing enzymes is inhibited in the presence of glucose. (a) A site along the DNA called the CAP (catabolite gene activator protein) binding site is immediately adjacent to the promoter. (b) When glucose is absent, large amounts of cAMP are present in the cell. The cAMP binds to CAP, causing it to shift to a shape that can bind to the CAP binding site. This facilitates binding of RNA polymerase to the promoter, and transcription of the structural genes for lactose-metabolizing enzymes is speeded up. (c) When a large amount of glucose is present, the amount of cAMP is reciprocally small. CAP cannot bind to the CAP binding site, and RNA polymerase binds poorly to the promoter. Hence, little RNA is made, and few enzyme molecules are formed.

Positive and Negative Control

Despite the complex detail of these gene control systems in *E. coli*, there are really just two general modes of control at work, one positive and one negative. When CAP binds to the DNA, it acts like an accelerator pedal to speed the transcription rate. This is *positive control:* An action is followed by a positive response. On the contrary, when the lactose repressor acts by binding the operator, it slows the transcription rate like a brake pedal. This is an example of *negative control:* An action is followed by a negative response.

Repressible Enzymes

Just as some enzymes are considered *inducible* because an inducer substance like lactose helps turn on their production, cells also have **repressible enzymes.** For repressible enzymes, synthesis is blocked when their pathway end product is present, so that the cell makes the enzyme to synthesize a particular amino acid, vitamin, or nucleotide only if that substance is *not* already available in the medium. The tryptophan (*trp*) operon is a good example of repressible enzymes.

The *trp* operon is a repressible system consisting of

(a) Gene Order in Tryptophan Operon

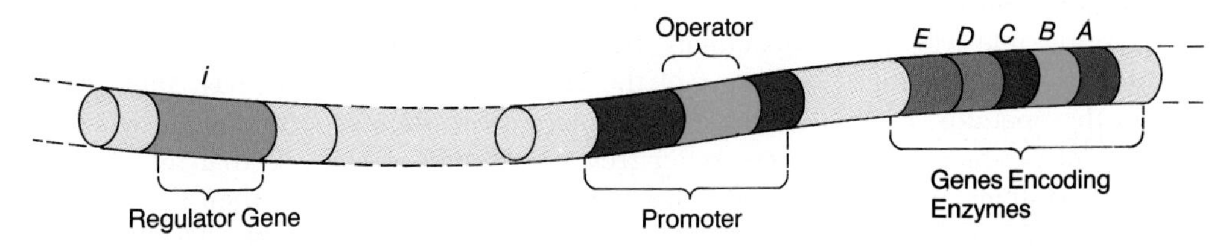

Figure 14-10 A REPRESSIBLE OPERON: THE TRYPTOPHAN *(trp)* OPERON. (a) The regulator gene lies far from the operon containing the five structural genes (*E*, *D*, *C*, *B*, *A*) that it controls. (b) When tryptophan is present, it binds to the repressor, causing the complex to assume a shape that can bind to the operator. This blocks access of RNA polymerase to the promoter; thus, transcription cannot occur and tryptophan-synthesizing enzymes are not made. (c) When tryptophan is absent, the repressor cannot bind to the operator, and the promoter is free to bind RNA polymerase. Transcription ensues, tryptophan-synthesizing enzymes are made, and the bacterium begins to supply itself with this amino acid.

genes for five enzymes that convert a particular starting substance to the amino acid tryptophan (Figure 14-10a). Transcription of the *trp* operon utilizes a repressor encoded by a gene some distance away from the *trp* operon. The *trp* repressor can exist in either of two shapes: One shape allows a repressor to bind the operator and block transcription; the other shape does not allow binding to the operator, so that gene expression can occur. If tryptophan is present (Figure 14-10b), it binds to the *trp* repressor, causing it to change shape. The resulting complex can bind to the operator and repress transcription (an example of negative control). In the absence of tryptophan, the repressor cannot bind, and enzymes for producing this necessary amino acid are synthesized (Figure 14-10c).

To summarize the control of inducible and repressible enzymes, both systems rely on the activity of allosteric regulatory proteins—that is, proteins that assume different shapes, depending on the binding of other substances (see Chapter 4). In an inducible system, when an inducer such as lactose binds to the repressor, the resulting complex cannot bind the operator, and lactose-degrading enzymes are formed. In a repressible system, when an end product such as tryptophan binds to the repressor, the complex does bind to the operator, and tryptophan-synthesizing enzymes are not produced.

The elucidation of these gene control systems in prokaryotes served as a guide for studying gene regulation in eukaryotes.

GENE REGULATION IN EUKARYOTES

Researchers attempting to probe gene control in eukaryotes were faced with two enigmas: Eukaryotes seemed to have too much DNA when compared with prokaryotes, and the majority of newly manufactured RNA seemed to be destroyed before it could leave the nucleus. These facts pointed to profound differences in the way prokaryotic and eukaryotic genes are organized and regulated.

Different organisms have very different amounts of DNA in their haploid genomes. As Table 14-1 shows, mammals have about 500 times as much DNA as *E. coli*, based on the numbers of nucleotide base pairs in each cell type, but they probably make only 50 times as many proteins as bacterial cells. The conger eel has almost 100 times as much DNA per haploid genome in each of its cells as do you and I, and a lily 200 times, but surely neither makes 100 times as many types of protein. So what is all this extra DNA doing in eukaryotes such as mammals, fish, and flowering plants?

Table 14-1 AMOUNT OF DNA IN HAPLOID GENOMES

Organism	*Base Pairs*
Plasmid	
pBR322	4.3×10^3
Virus, including bacteriophage	
SV-40 (a virus isolated from monkeys)	5.2×10^3
Lambda	4.6×10^4
Bacteria	
Diplococcus pneumoniae	1.8×10^6
Escherichia coli	4.1×10^6
Insect	
Drosophila melanogaster	1.8×10^8
Mammals	
Mouse	2.2×10^9
Human	2.8×10^9
Fish	
Conger eel	1.5×10^{11}
Plants	
Corn	6.6×10^9
Lily	3.0×10^{11}

Part of the answer to this question has come from DNA *dissociation/reassociation* experiments, in which a researcher treats a solution of DNA so that the double helices are separated into two strands; then the strands are allowed to reanneal (come back together) in a process called *hybridization*. The more copies of a gene present, the more easily the strands reanneal. A plot of the results of these experiments shows a single smooth curve in viruses and bacteria, indicating that nearly all DNA sequences are present just once per genome (Figure 14-11a). In mammals, however, the curve has three regions (Figure 14-11b), one reflecting rapid reassembly among multiple copies of a small number of DNA sequences. This *highly repeated DNA* consists of about 10 percent of the genome in mice—about 1 million copies of a repeating sequence some 300 bases long. Highly repeated DNA, sometimes called satellite DNA, is located in chromosomes mainly near the centromere and may play some role in aligning chromosomes during mitosis.

The middle portion of the curve reflects intermediate-speed reannealing and represents an additional 20 percent of mouse DNA made up of *moderately repeated DNA;* this includes DNA sequences that are present in 30 to 10,000 copies per haploid genome. At least some of these sequences specify the genes for rRNAs, tRNAs, histone proteins, and other familiar products. Each body cell of the clawed frog, *Xenopus*, for example, has about 800 copies of each gene coding for each of the large types of ribosomal RNAs.

The last portion of the mammalian curve shows slow reannealing in the remaining 70 percent of DNA, which reassociates as though there were only one copy per haploid genome. Geneticists call it the *unique DNA* and it includes most of the genes for proteins such as insulin, ovalbumin, and growth hormone.

Besides the presence of so much repetitive DNA, the eukaryotic genome is distinguished from the prokaryotic genome by the presence of *multiple-gene families*. Many eukaryotes contain families of genes—genes related in both structure and function. Clusters of genes are found for the five histone proteins, for example, and for the several different hemoglobin molecules. Scientists believe that these multiple-gene families arose by gene duplication during evolution. Once two diploid copies of an original gene (four alleles) had accumulated, a mutation in one pair of alleles would not harm the organism possessing it, since the other pair could maintain the original function. Thus, the base sequences of the two diploid copies were "free" to become different from each other, and new but related gene functions were able to develop.

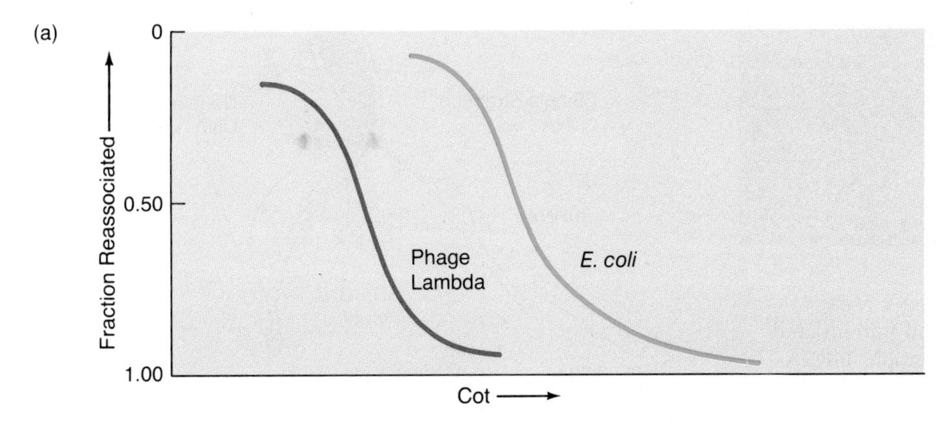

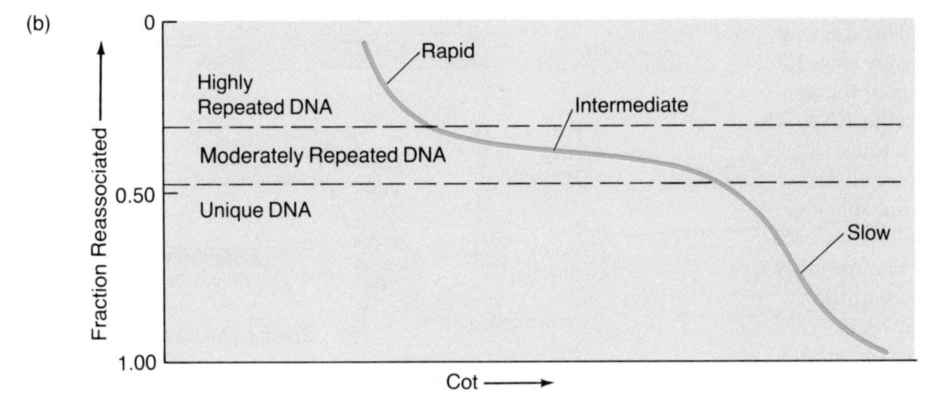

Cot = Initial DNA Concentration × Time for Reassociation

Figure 14-11 REASSOCIATION CURVES: DETERMINING THE AMOUNT OF REPETITIVE DNA AN ORGANISM CONTAINS.
(a) For simple situations such as phage lambda and *E. coli,* the curves are smooth, indicating that nearly all DNA sequences are present just once per genome. (b) Mammals, however, have huge numbers of copies of some DNA sequences (the rapid part of this graph), moderate numbers of other DNA sequences (the intermediate region), and single copies of some DNA sequences (the slow region).

The "Pieces" of Eukaryotic Genes

The second enigma—why most RNA is destroyed before leaving the eukaryotic cell nucleus—was explained by the discovery of two facts: (1) Most eukaryotic genes occur in "pieces," and therefore (2) the "pieces" of mRNA encoded by them must be joined together or processed in a special way.

A remarkable fact emerged when researchers allowed hemoglobin mRNA to bind to its complementary DNA sequence. A single mRNA molecule hybridized at distinct places on the gene, but intervening stretches of the gene did not hybridize with the mRNA. Instead, they formed loops of DNA (Figure 14-12a); evidently, nucleotides of these intervening stretches of DNA do not pair with mRNA nucleotides. Subsequent work has proved that these nonpairing regions are not represented in a functional mRNA molecule; thus, they do not code for protein. The nonpairing regions of the DNA are called **introns,** or *int*ervening sequences. The protein-coding parts of the gene are called **exons,** since they are *expressed* (Figure 14-12b).

Some genes have many introns; the collagen gene, for example, has 50, and the gene for fibronectin (a glue between cells) has 48, most of which are much longer than the exons of that gene. While still a major mystery, introns are characteristic of nearly all eukaryotic genes and are not usually found in prokaryotic genes.

Eukaryotic mRNA: Special Processing

In most eukaryotic genes, the exon regions are represented in the mRNAs that leave the nucleus and enter the cytoplasm, but the intron regions are not. How can this be, since the introns are transcribed along with exons into nuclear RNA molecules?

Experiments revealed that the introns are transcribed along with the exons of a gene into a single long RNA molecule called the *primary RNA transcript* or pre-mRNA (Figure 14-12c). Before this molecule leaves the nucleus, the portions of RNA corresponding to the introns are cleaved enzymatically from the primary transcript. Then the exon-encoded RNA sequences are spliced together to form an mRNA molecule. Five or more small ribonucleoproteins and several proteins form a complex called a spliceosome at the sites that must be spliced together, and the pieces are rejoined. After further processing, the spliced mRNA passes into the cytoplasm. Only when fully processed is it considered a mature mRNA molecule. Meanwhile, the excised RNA,

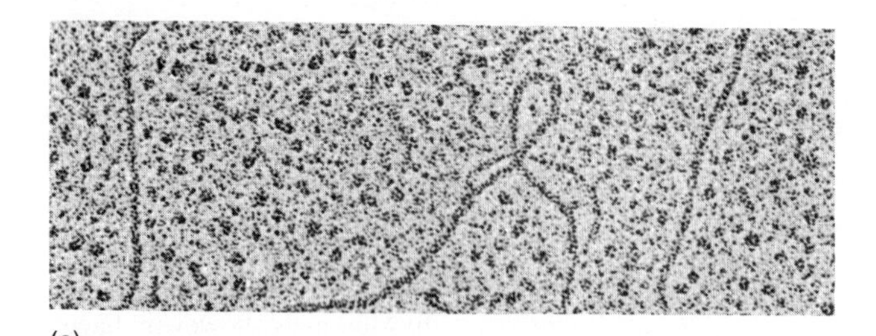

(a)

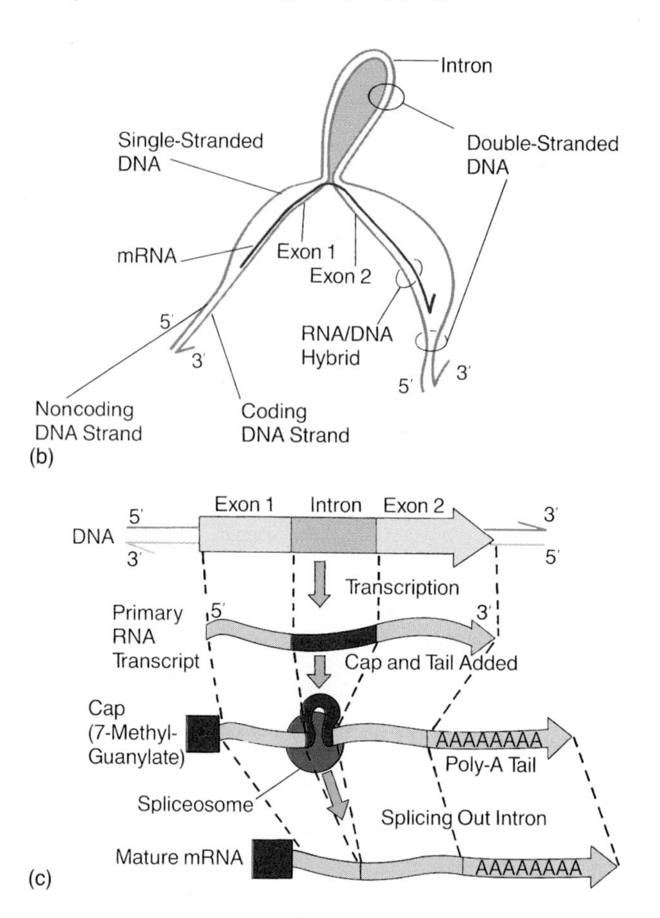

Figure 14-12 INTRONS AND EXONS. Eukaryotic genes have translated (exon) and untranslated (intron) segments. (a) To create this photograph, mRNA encoding globin (the protein part of hemoglobin) has been hybridized to a long sequence of DNA containing the globin gene (magnified about 73,000×). The vertical loop in the center is an intron in the globin gene. (b) This drawing interprets the photograph and shows an intron loop; the mRNA (red) binds only to the exon portions of the gene. (c) This drawing shows processing of the globin RNA. The full gene is transcribed into a primary RNA transcript. Then, processing includes capping, adding a tail, and splicing (at the spliceosome complex). During splicing, intron-coded regions (the loop here) are deleted as exon-coded ones are spliced together. The final mature mRNA is translated into hemoglobin after it passes into the cytoplasm. Both the cap and the poly-A tail remain untranslated when protein is synthesized on the mRNA. The cap and tail sequences are probably protective and may play a role in facilitating translation.

which corresponds to the introns, is degraded. The degradation of intron RNA molecules helps explain why so much newly synthesized RNA is destroyed rapidly and never leaves the eukaryotic nucleus.

The origins and basic functions of introns remain enigmatic. Researchers have identified some introns in viruses that actually code for viral enzymes. Others code for proteins that help remove introns from primary RNA transcripts. Still others called ribozymes catalyze their own removal from the primary transcript in the absence of protein enzymes. Some introns form loops, or lariats, reminiscent of bacterial plasmids. The elucidation of the origins of introns is likely to tell us something profound about the evolution of DNA function on earth.

Before the RNA becomes a mature message, a "cap" (often a methylated guanosine) is added to the 5′ end of the RNA, and a *poly-A tail* (100 to 200 adenosine residues) is added to the 3′ end. These new end regions may help protect the mature mRNA from degradation and may facilitate its translation.

How are eukaryotic genes, with their intron/exon structure, turned on and off at appropriate times? Molecular geneticists do not fully understand the mechanisms, but recent technological advances will help lead to answers. Scientists can quickly determine the sequence of bases in the DNA molecule. Geneticists now have two sophisticated instruments to automate parts of their research that used to require months of tedious effort. The first instrument, the DNA/RNA synthesizer, automatically produces gene segments with any sequence of nucleotides the researcher chooses. He or she simply types the sequence into a keyboard, and computer chips send signals to a set of valves. These open or close a set of reservoirs and allow tiny droplets of solution containing A, T, C, or G to reach a container. There, the bases are joined chemically, one by one, in the appropriate sequence, forming a gene segment of up to 30 nucleotides. The worker can link this segment to others made in the same way and eventually synthesize an entire functioning gene.

The second instrument is the protein sequencer. A researcher places a protein sample into the machine, and through a series of chemical reactions, amino acids are automatically "snipped off" one by one and identified by liquid chromatography. In about a day, instead of a month or more, the machine can reveal and record the entire amino acid sequence.

Using such automated sequencers, researchers have discovered certain base sequences at specific sites in eukaryotic genes that may act as promoters, determining where along chromosomal DNA transcription of a gene begins. Other sequences determine where along primary RNA transcripts processing enzymes will cleave transcripts to excise intron RNAs. Chapter 17 will return to the subject of eukaryotic gene control.

RECOMBINANT DNA TECHNOLOGY

In the early 1970s, molecular biologists developed a special set of laboratory techniques that built on their knowledge of bacterial genetics and gave them unprecedented power to manipulate genes in new ways. These techniques, collectively called *recombinant DNA technology,* accord molecular biologists the rather awesome ability to move genes from one chromosome to another. With this, they can create totally new genomes and altered organisms such as giant mice and bacteria that express human genes. Let's see how.

General Steps in a Gene Transfer Experiment

A biologist generally uses recombinant DNA techniques to find an interesting gene from either a bacterial or a eukaryotic chromosome and then move it into the genome of a different organism. For example, researchers have inserted human insulin genes into bacteria and then harvested quantities of the protein insulin for use as a human drug. The giant mouse is the product of another gene transfer in which experimenters moved the gene for rat growth hormone into normal mice, causing them to grow rapidly.

Finding the gene for a useful protein is harder than it sounds, because the gene is often a needle in a haystack of millions of genes and noncoding base sequences. To find this "needle," the researcher must (1) cut DNA from the donor cell into fragments, some containing whole functioning genes, and (2) make many copies of all these pieces of DNA by introducing the DNA into special vehicles (usually plasmids) and then inserting these vehicles into prokaryotic "factory" cells (usually bacteria such as *E. coli* that act as hosts). Once inside the host cells, the foreign genes can sometimes be expressed as foreign polypeptides. The researcher then must (3) identify the host cell that can suddenly make the polypeptide encoded by the desired gene (the "needle") and discard all the bacterial cells that express undesired genes (the "haystack") transferred from the original donor cell. Finally, he or she must (4) grow huge numbers of the host cells (each containing the desired gene) as a "gene factory." The experimenter can later remove the gene from these host cells, still in its plasmid vehicles, and transplant it by another set of steps into another host—say, a mouse, a frog, or a tomato plant.

The Tools of the Genetic Engineer

Molecular Scissors and Genetic Glue

In 1973, gene researchers Stanley Cohen at Stanford University and Herbert Boyer at the University of California at San Francisco identified a class of enzymes called **restriction endonucleases** and applied it to molecular genetics. Enzymes of this class recognize specific nucleotide sequences along DNA molecules and cut both complementary DNA strands of the double helix at those specific sites.

The interesting thing about restriction endonucleases is the *way* they cut the double-stranded DNA molecule: Many of them make a staggered cut in the two strands and leave "sticky" ends (Figure 14-13a). These sticky, or *cohesive,* ends are stretches of unpaired nucleotides that can form hydrogen bonds with other cohesive ends having the appropriate complementary sequence. Another class of enzymes called **DNA ligases** can act as molecular paste to rejoin the complementary cohesive ends of DNA fragments (Figure 14-13b).

Molecular Vectors

Molecular *vectors* are pieces of DNA that can carry a foreign gene into a host cell, where such genes are replicated along with the cell's own DNA. Geneticists typically use plasmids as vectors (Figure 14-13c). These naturally occurring circular strands of bacterial DNA have several features that make them ideal tools. First, they can be cleaved by restriction endonucleases to "cut the ring" so it opens and has two sticky ends. The experimenter can then splice in a gene from another organism, re-form the ring using ligase, and create a totally new genetic entity, a **recombinant DNA molecule** (see Figure 14-13c).

A second feature that makes plasmids an ideal vector is that the experimenter can introduce them into host cells via transformation, and the plasmids can carry a foreign gene into a host cell like a molecular Trojan horse. A third critical feature is that plasmids can replicate tens or hundreds of times inside a host cell and generate extra copies of the complete recombinant DNA molecule. Finally, since plasmids often carry genes for antibiotic resistance, they provide a means of screening for recombinant DNA molecules. We'll explain how this works shortly.

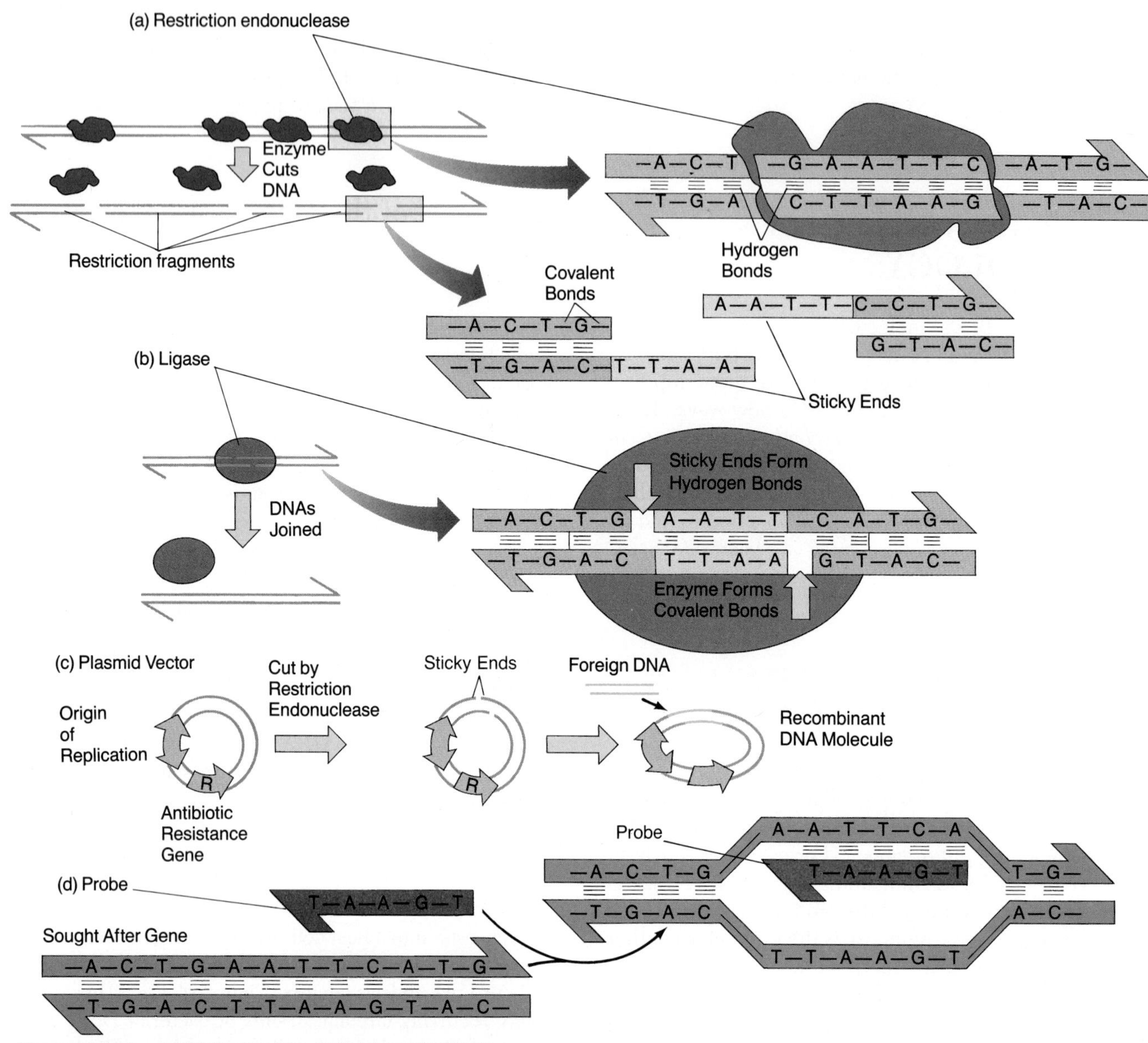

Figure 14-13 TOOLS OF THE GENETIC ENGINEER.
(a) Restriction endonucleases can cut DNA at specific sequences and leave overlapping "sticky" ends. (b) DNA ligase enzymes rejoin the sticky ends at complementary sequences. (c) New genes can be spliced into vectors, which are usually plasmids, to form recombinant DNA molecules. Vectors can carry such novel gene sequences into different host organisms. (d) Molecular probes are stretches of RNA with base sequences complementary to a desired gene, often part of a recombinant DNA molecule.

Molecular Probes

The final tool of the genetic engineer is a probe for locating recombinant DNA molecules, genes, or other desired pieces of genetic information. A **probe** can be a specially prepared stretch of RNA or DNA with a sequence complementary to the specific series of nucleotides in the sought-after gene (Figure 14-13d).

Let's see, now, how a researcher would use these tools to clone and transfer a gene.

Engineering a Bigger Mouse

Genetic engineers used the tools just described plus a few specialized techniques to isolate the gene for rat growth hormone, clone it, and transfer it into the giant mouse shown on page 229. Here's what they did, step by step.

A researcher took the DNA from rat cells and cut it with a restriction endonuclease into hundreds or thou-

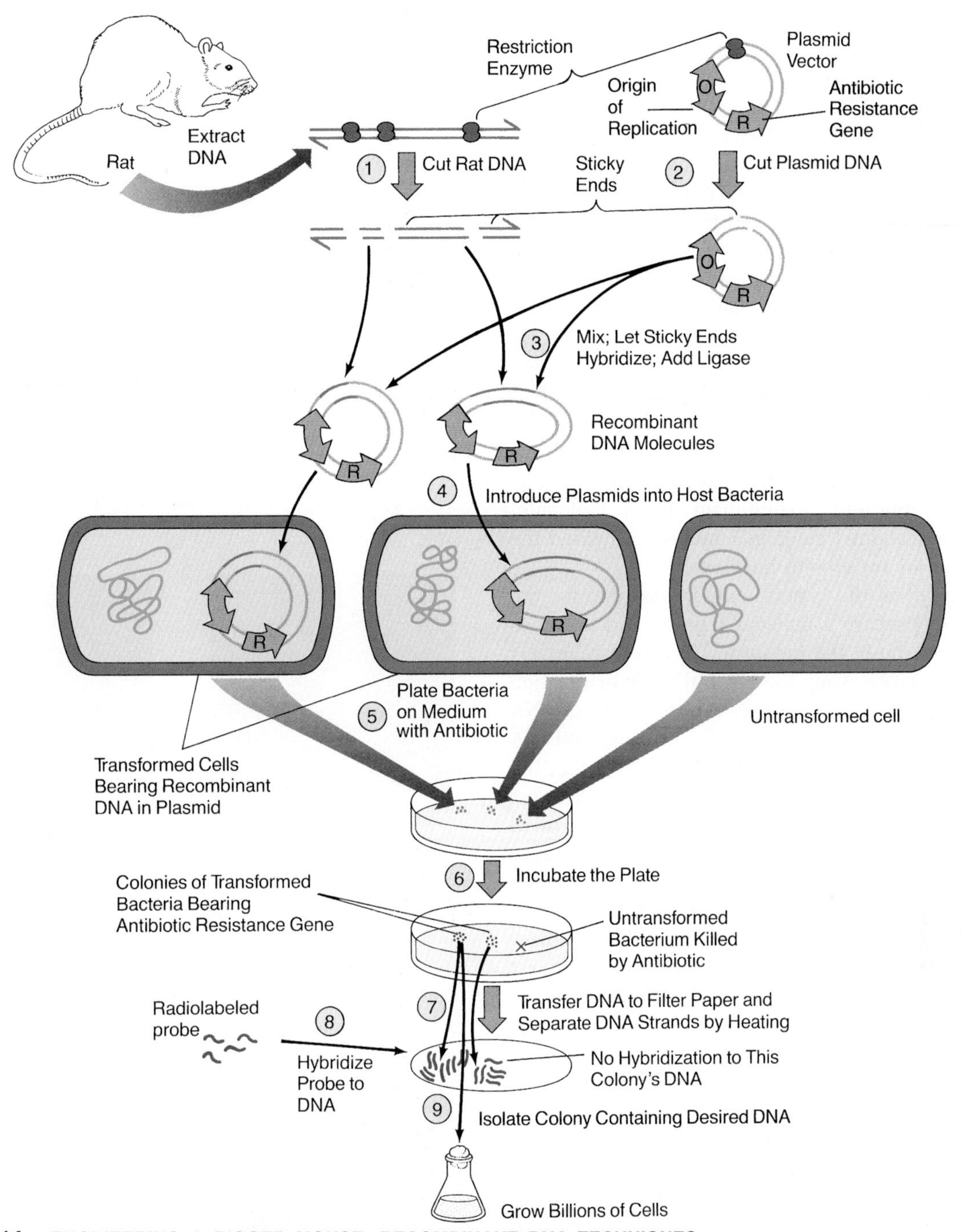

Figure 14-14 ENGINEERING A BIGGER MOUSE: RECOMBINANT DNA TECHNIQUES.
The creation of complementary sticky ends in the plasmid DNA and in the foreign DNA to be inserted is the key to this technology. Once the recombinant plasmid is created with the ligase enzyme, it is introduced into host cells. Millions of copies can then be made as both the cells divide repeatedly and plasmids are replicated in each of them. The text explains the specific steps from extracting and cleaving DNA (1) to preparing plasmid vectors containing recombinant DNA (2, 3); transferring, growing, and selecting bacterial host cells (4–6); and preparing a probe to find a desired gene (7, 8) and then isolating it (9).

sands of fragments, some containing complete, functioning genes (Figure 14-14, step 1). Since a restriction endonuclease cleaves DNA at specific sites, all the DNA fragments would have cohesive ends with the same specific complementary sequences, and thus the fragments could pair up with other DNA fragments cut by the same enzyme.

Next the researcher cut plasmids with the same restriction enzyme (step 2) and supplied DNA ligase to a mixture containing the rat DNA fragments and "cut" plasmids. Since the cohesive ends of the plasmids and the fragments cut by the same enzyme had matching complementary sequences, they could adhere to each other and be joined by the ligase (step 3). These steps spliced rat DNA into the re-formed plasmid rings.

Now the scientist mixed plasmids carrying foreign genes with bacteria lacking plasmids, waited for transformation to take place (step 4), and then spread the mixture on agar-filled Petri dishes containing a deadly antibiotic (step 5). Only transformed bacteria containing plasmids bearing antibiotic resistance genes and, coincidentally, the foreign DNA fragments—grew (step 6).

Transformed bacteria divided and redivided into **clones,** colonies in which every cell was genetically identical to the original parental cell. After several hours, the researcher had a set of Petri dishes dotted with hundreds of round bacterial colonies (the "haystack"), each clone containing a different recombinant DNA molecule. However, only about 1 clone in 1 million contained the desired gene for rat growth hormone.

At this point in the experiment, it was time for the researcher to prepare a molecular probe and use it to locate the "needle in the haystack." The researcher exploited such a probe using the **DNA/RNA hybridization** technique. The technique begins with tranferring DNA from the bacteria onto filter paper and heating the DNA to break the weak hydrogen bonds that hold together the molecule's complementary strands (step 7). The biologist then adds radioactively labeled mRNA for the growth hormone gene isolated from cells of a rat's pituitary gland (step 8). If any of the sequences on the mRNA probe are complementary to regions on the unzipped DNA strands, the two linear molecules will hybridize, revealing that the DNA sequence codes for the rat growth hormone gene. The researcher isolated clones of bacterial cells containing that stretch of DNA, grew the cells in mass culture, and eventually harvested billions of cells containing the desired gene (step 9).

Now for the transfer of the cloned gene to a new host. A different group of researchers used the techniques just outlined to locate and produce billions of copies of a mouse gene that codes for metallothionein, a protein that helps prevent poisoning from heavy metals such as zinc (Figure 14-15, step 1). Using a restriction endonuclease, they then cleaved the metallothionein gene and removed its regulatory region, which they joined to the gene for rat growth hormone they had cloned earlier (step 2). They injected this *fused gene* into a fertilized mouse egg (step 3), and it apparently became integrated into the chromosomal DNA. When cultured in a Petri dish, the egg developed normally into an early-stage mouse embryo, and a copy of the injected DNA was present in its cells (step 4). The researchers finally implanted the embryo into the reproductive tract of a foster mother and allowed it to develop normally (step 5).

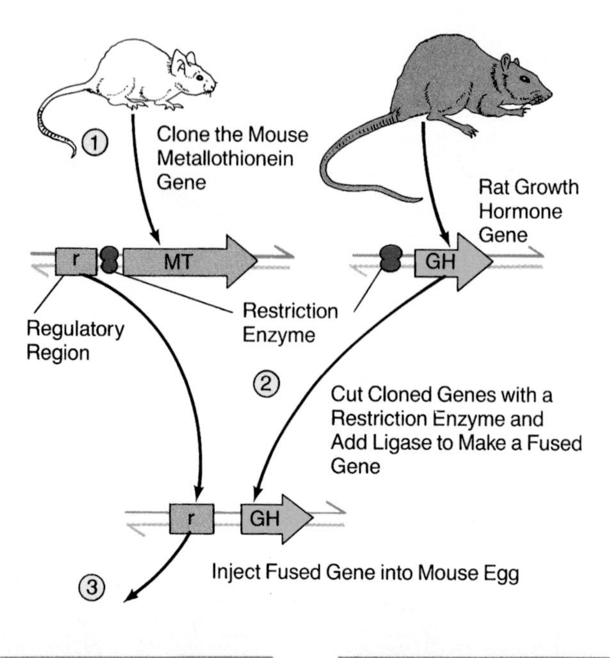

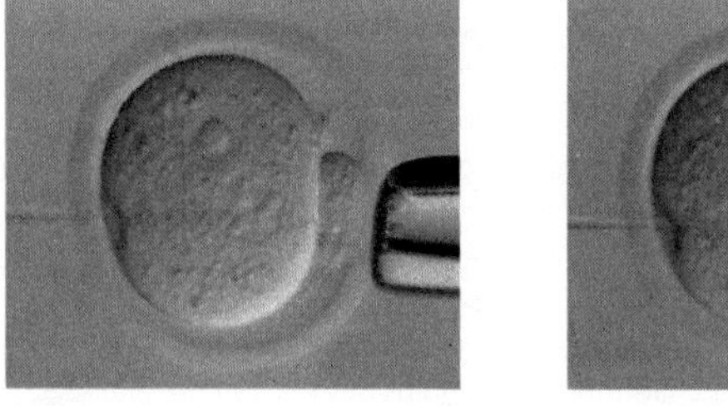

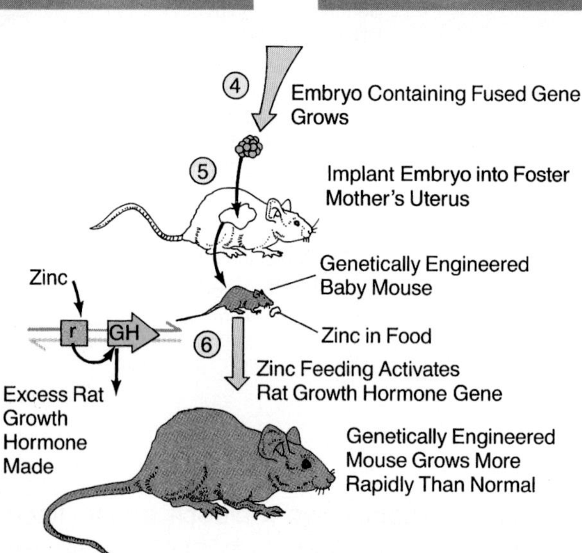

Figure 14-15 INCORPORATING NEW GENES INTO A MOUSE GENOME.
This illustration outlines the genetic engineering steps by which the rat growth hormone gene is placed under control of a mouse gene and then introduced into a fertilized egg. The growth hormone gene later acts in excess, and a giant mouse results. The text explains each step. The photo shows DNA being injected into a mouse egg. The zygote (magnified about 450×) is held in place with the pipette, seen at the right, while the injection microneedle on the left penetrates the male pronucleus.

After the genetically engineered mouse was born, they fed it zinc, and the regulatory region of the metallothionein gene turned on the adjacent sequence for rat growth hormone, and the mouse produced great quantities of growth hormone (step 6). With that substance abundantly present, the experimental mouse grew twice as fast as normal.

GENETIC ENGINEERING: PROMISES AND PROSPECTS

In a sense, people have practiced genetic engineering for thousands of years, allowing genes to recombine naturally in domesticated plants such as corn and wheat and in domesticated animals such as cattle and sheep, then choosing those individuals with the best new combination of desirable characteristics as parents for the next generation. The new technology of genetic engineering avoids chance mutation and random genetic recombination, altering a gene in a defined and desired way in the test tube, then introducing it into another organism's genome.

Geneticists have three basic strategies for employing recombinant DNA technology: (1) inserting a gene into a microorganism such as a bacterium or yeast, then harvesting a desired protein in large amounts; (2) inserting engineered genes into domestic plants or animals to improve their growth rate, protein quality, or other characteristics; and (3) inserting engineered genes directly into people to cure genetic disease. Let us look at the status of each application, as well as the ethical problems raised by some of these new approaches.

Microbial Production of Valuable Gene Products

One of the first applications scientists proposed for the new techniques was splicing into bacteria the genes for human insulin, the peptide hormone produced by the pancreas that helps regulate blood sugar levels. Diabetics secrete too little insulin, but many have been successfully treated with daily injections of insulin isolated from cattle, hogs, or other mammals. Insulin hormone from these sources, however, sometimes provokes adverse immunological reactions. In recent years, scientists at several corporations have used recombinant DNA techniques to genetically engineer bacteria that can produce human insulin, and the protein is now commercially available.

Many other proteins can be produced by means of bacterial synthesis. A partial list includes interferon, an antiviral and perhaps antitumor protein; growth hormone to treat people with genetic defects that retard growth and to treat burn victims; urokinase and tissue plasminogen activator enzymes that help to reduce blood clots and can be used to treat heart attack victims; tumor necrosis factor, used as an experimental drug to kill cancer cells; and atrial natriuretic factor, a hormone that promotes water and salt excretion and is used to treat people with heart failure and hypertension.

Geneticists have engineered yeasts to produce and secrete desired proteins into a liquid culture medium, thus making purification easier. As eukaryotes, yeasts can also modify proteins—for example, by adding sugars to them—which bacteria cannot do. Since vaccines often have carbohydrate parts, this might help researchers produce vaccines against hepatitis and other serious diseases. Photosynthetic cells are other possible single-celled hosts for eukaryotic genes. By inserting into cyanobacteria the genes for the enzymes that synthesize casein (milk protein) and lactose (milk sugar), we may someday replace milk cows. Such cyanobacteria could be grown in massive illuminated vats and would use the energy of sunlight to manufacture milk protein and sugar. The result would not be milk, but it could prove to be a marvelous dietary supplement for the world's hungry people.

Genetic Alteration of Agricultural Species

The giant mice we discussed earlier are prototypes for improving farm animals. Cattle engineered with growth hormone genes might grow faster and perhaps put on more muscle tissue (meat) for a given amount of feed consumed. Researchers do not yet know how these hormones might affect meat quality. Direct application of growth hormone also increases milk production.

Geneticists were surprised by the concentrations of growth hormone present in the blood of the engineered mice—sometimes 100 times higher than in cultures of genetically engineered bacteria. This suggests that perhaps someday we could carry out "genetic farming" of such valuable gene products as growth hormone or interferon in large animals instead of in bacteria.

The genetic engineering of plants will undoubtedly have the greatest impact of all on civilization. Although plants make their own food by photosynthesis, their proteins are usually low in certain amino acids, and this limits their food value for people. If plant scientists could introduce genes for animal-quality protein and these could be expressed successfully in plant crops, human nutrition in many parts of the world could be improved.

Researchers have already made considerable progress in the genetic engineering of plants. Scientists at Wash-

ENGINEERING SUPERFISH

Genetic engineers have been angling for another remarkable catch: "superfish" that grow faster, larger, and hardier than their fellow species. In 1987, Chinese researchers spliced human growth hormone into goldfish and loach (a type of carp), and the engineered fish grew four times faster than normal. The following year, American geneticists found an equivalent gene in rainbow trout, cloned it, and injected carp eggs with a fluid containing millions of copies of the gene. Of the 10,000 eggs they injected, only 20 carp appeared to express the trout gene, but they, too, grew far faster and more robust than normal carp (Figure A). In still other labs, researchers have transferred growth hormones from cattle into walleye and pink salmon and are searching for growth genes in the species themselves.

Figure A SUPER CARP HAVE GROWTH GENES FROM RAINBOW TROUT.

The list of potential piscine projects is long. Some researchers talk about combining genes for large size, frequent reproduction, and disease resistance into one superior species. Others would like to build pollution-fighting fish by splicing in the genes for metallothionein protein (which binds heavy metals). Some have discussed transferring the genes for glycoproteins (natural antifreeze in polar fish species) into common fish to allow them to live in colder waters. Finally, some researchers have daydreamed about finding the genes that underlie the salmon's homing instinct, so that tuna or other species could be engineered to return to their natal fish hatchery for harvesting. It remains to be seen whether some of these are viable research goals or just fish stories.

ington University in St. Louis have "borrowed" a plasmid from a soil bacterium called *Agrobacterium tumefaciens* that can infect certain broad-leafed plants and by means of the plasmid transfer pieces of its own DNA (so-called T-DNA) to the plant. They used a modified version of this vector to carry the gene for a yeast enzyme, alcohol dehydrogenase, into tobacco cells. Full tobacco plants grown from those cells all contained the yeast alcohol dehydrogenase gene, as did the cells of the F_1 generation grown from their seeds. In other experiments, researchers inserted a gene for a trypsin inhibitor into tobacco and produced plants with increased field resistance to insect pests. In 1988, plant scientists transferred genes into a cereal crop (corn) for the first time and transferred a corn gene into petunia, yielding flowers of novel color. Once foreign genes are introduced into new hosts with exciting new techniques such as these, experimenters can use standard agricultural practices to breed them into many other kinds of plants.

Plant genetic engineers predict they will soon be able to transfer genes for producing natural pesticides or for resistance to synthetic pesticides and herbicides. A farmer could then treat a growing field of engineered plants with high levels of such agents and keep the crop weed- and pest-free while not harming the crop plants themselves.

They also foresee the transfer of gene complexes that control drought or cold resistance, natural pest resistance, nitrogen fixation (a process by which atmospheric nitrogen is built into organic molecules), and other such traits. Indeed, farmers—and a hungry world population—can look forward in the not too distant future to plants with genes from many animal, plant, and microbial sources and with many improved qualities.

Human Gene Therapy

Serious genetic diseases, including sickle-cell anemia, thalassemia, cystic fibrosis, and hemophilia, affect about 1 percent of all children. These genetic conditions impose a large burden on the affected child, on his or her family, and on society. Chapter 15 will explain how physicians now diagnose many genetic diseases before birth so that prospective parents have the option to terminate pregnancy or to plan for a child with a disabling illness. The first trials are now under way to test the use of engineered genes in people, as that chapter explains.

Preventing the Proliferation of Dangerous Bacteria

As soon as recombinant DNA techniques became feasible, scientists acted to set strict safeguards to prevent accidents with recombined organisms. By the late 1970s, tight regulations were in place, and experimenters using high-containment procedures to prevent any chance of accidents deliberately cloned cancer-causing genes in animals and reintroduced them into sensitive animals. The results showed that the engineered bacteria used for recombinant DNA technology were safe to work with, given the precautions followed by competent bacteriologists. In fact, in some cases, it is safer to clone and study individual genes from pathogenic organisms or viruses in *E. coli* than to handle the intact organism or virus itself. Moreover, work on plasmids from *Agrobacterium tumefaciens* and other organisms suggested that plasmids have been transferring genes in nature for millions of years. Regulations governing recombinant experiments were subsequently relaxed.

Humanistic Concerns

Ultimately, society must address the ethical implications of genetic engineering, especially many questions surrounding human gene therapy. Adding normal genes to a human embryo or adult is not ethically different from doing kidney or bone marrow transplants, both long-accepted medical procedures. Introducing new genes into germ line cells, however, would alter the individual's progeny and thus directly influence human evolution. Geneticist Howard Schneiderman, a faculty member of the University of California and vice president of one of the largest chemical companies in the United States, is a strong advocate of both the new genetic technology and its responsible application:

> I believe—and many hard-headed scientists agree with me—that with the new biotechnology, almost anything that can be thought of can ultimately be achieved—new organisms, new limbs and organs, new treatments for disease, new ways of controlling pests, crops which produce their own pesticides, disease-free domestic animals, whole new industries that will sell products that even today cannot be imagined, let alone made. But if society is to reap social benefits from biotechnology and if industries are to realize financial rewards, we must understand and deal with not only the scientific and technical questions that confront us, but the social questions as well.*

Schneiderman emphasizes that the social, moral, and ethical issues posed by some applications of biotechnology can be resolved, as long as scientists work together with the press, legislators, and an informed public.

*Speech presented at a conference on Biotechnology: Research to Reality, October 26, 1981, in New York City.

SUMMARY

1. The mechanisms by which genes are passed from one generation to another differ between bacteria and eukaryotes. Prokaryotes do not undergo true mitosis and meiosis.

2. Bacteria can exchange genetic information in three ways. During *conjugation*, two cells make direct contact. The genes for forming a sex pilus are carried on a *plasmid* called the *F factor plasmid*, which can integrate into the chromosome. During *transformation*, DNA released from a dead or disrupted cell into the surrounding medium is taken up by another cell. During *transduction*, DNA is carried from one bacterium to another by a bacterial virus, or bacteriophage.

3. In prokaryotes, the process of *gene regulation* commonly includes the function of gene clusters called *operons*. These make it possible for limited numbers of regulatory molecules to simultaneously control the expression of sets of genes.

4. The lactose operon in *E. coli* is a cluster of adjacent genes *z*, *y*, and *a* involved in the cell's use of lactose. Regulatory gene *i* codes for a *repressor* substance, while *o* is the *operator* site on the operon DNA. The *promoter* site is adjacent to and partially overlaps the *o* gene.

5. If lactose is absent, the *i* gene makes a repressor that can bind to the operator. When so bound, the repressor partially blocks the promoter site and prevents transcription of the *z*, *y*, and *a* genes into an mRNA molecule.

6. *Catabolite repression* occurs if both lactose and glucose are available. The cell then uses glucose first. Glucose indirectly causes a catabolite gene activator protein (CAP) to change shape, and it no longer binds well to the promoter site. This slows the transcription of the *z*, *y*, and *a* genes and the synthesis of enzymes for utilizing lactose.

7. Lactose enzymes are *inducible*, since an inducer substance ultimately stimulates their production. *Repressible* enzymes are those in which the end product of their metabolic pathway builds up and blocks further synthesis of the enzyme.

8. Eukaryotic chromosomes have repetitive DNA sequences and a great deal of DNA that does not code for proteins. Within the genes themselves, there are regions that are expressed, *exons*, and intervening sequences that are not expressed, *introns*. Both exons and introns are transcribed into the primary RNA transcript; then the intron regions are cleaved out, and the exons are spliced together before a functional mRNA is made.

9. *Recombinant DNA technology* is a set of techniques for transferring genes

from one organism to another. *Restriction endonucleases* and *DNA ligases* are cut-and-paste enzymes that allow a researcher to cleave DNA in such a way that cohesive ends remain, and to rejoin DNA fragments into a repaired double helix. Vectors such as the F factor plasmid can carry foreign DNA into a host cell. A plasmid into which a foreign gene has been spliced is a new genetic entity, a *recombinant DNA molecule*.

10. Genetic engineering probably will be applied in three ways: (1) introducing foreign genes into microbes and harvesting the gene product, such as human insulin or interferon (this is already being done); (2) genetically altering livestock or food-crop species so they grow faster, are resistant to disease, cold, chemicals, or drought, and have improved protein quality; (3) replacing mutant human genes to cure genetic diseases such as sickle-cell anemia and hemophilia.

KEY TERMS

auxotroph
catabolite repression
clone
conjugation
DNA ligase
DNA/RNA hybridization
episome
exon
F factor
F factor plasmid
gene regulation
genetic engineering
genome
Hfr cell
intron
operator
operon
plasmid
probe
promoter site
recombinant DNA molecule
recombinant DNA technology
repressible enzyme
repressor
restriction endonuclease
R factor
R factor plasmid
transduction
transformation

QUESTIONS AND PROBLEMS

1. What is an operon? In which organisms would you expect to find operons? Draw a diagram of a specific operon, and describe how regulatory molecules control transcription in this operon.

2. A mutation sometimes occurs in the *lac* repressor gene so that the mutant repressor no longer can bind the inducer, but it can still bind to the operator. What phenotype would you expect this mutant strain to have? Suppose a different mutation changes the repressor so it can no longer bind to the operator, whether or not it is bound to inducer. What phenotype would this mutant strain have?

3. Would you expect promoters for many different genes to have the same base sequence or many different sequences (in other words, does RNA polymerase recognize a specific sequence)? If the *lac* promoter mutated so it could no longer bind RNA polymerase, what phenotype would the cell have?

4. What are F factors? What are F^+ and F^- cells? What are Hfr cells?

5. The genome of *E. coli* contains 4×10^6 base pairs. Assuming that the "average" gene contains 1,200 base pairs and most of the genome represents genes, how many genes might you expect *E. coli* to have? The genome of bacteriophage lambda contains 5×10^4 base pairs. How many genes would you expect to find? A human sperm or egg contains 3×10^9 base pairs. If most of the genome represented genes, how many genes would humans have? Eels have 1.5×10^{11} base pairs. Would eels have 50 times as many genes as humans? What kinds of evidence suggest that eukaryotes have lots of "extra" DNA?

6. A strain of *E. coli* contains plasmids that carry two genes for antibiotic resistance: one for resistance to streptomycin (Str^R) and the other for ampicillin (Amp^R). These genes make enzymes that are secreted by the cell and that destroy the antibiotics. The plasmid DNA has a single restriction site for restriction endonuclease *Eco* RI, in the Str^R gene, and a single restriction site for the restriction endonuclease *Hin* dIII, in the Amp^R gene. You wish to use this plasmid as a cloning vector for a piece of broccoli DNA. First you isolate lots of plasmid DNA and broccoli DNA; then you cut both kinds of DNA with the enzyme *Eco* RI. Next you mix the two kinds of DNA together. How will you join the broccoli DNA to the plasmid DNA? After joining the DNAs, you transform an antibiotic-sensitive strain of *E. coli* with this DNA and plate the cells on Petri plates containing ampicillin. What types of cells will grow? What types of cells will not grow? Some of the cells that grow will contain a plasmid that is carrying some broccoli DNA. A prokaryotic gene with the plasmid DNA in it will not make its product protein. Using that fact, how could you distinguish between colonies that contain a plasmid with inserted broccoli DNA from colonies that contain a plasmid with no broccoli DNA?

7. Gene transfer experiments were done using four different Hfr strains of *E. coli* (strains in which the F factor plasmid had integrated in different parts of the chromosome and possibly in different directions relative to the host cell chromosomes). Conjugation between the Hfr and F^- strains was interrupted at various times, and the following orders of gene transfer were found:

Hfr strain 1: A C T E R
Hfr strain 2: T E R I U
Hfr strain 3: A B M U I
Hfr strain 4: M U I R E

What is the order of all these genes on the bacterial chromosome?

8. What are some of the ways in which eukaryotic mRNA differs from that of prokaryotes?

For additional readings related to topics in this chapter, see Appendix C.

15 Human Genetics

The capacity to blunder slightly is the real marvel of DNA. Without this special attribute, we would still be anaerobic bacteria, and there would be no music.

Lewis Thomas, *The Medusa and the Snail* (1979)

Although we are all members of one species, human beings display a remarkable range of height, weight, coloration, facial features, and distinguishing traits such as baldness, cleft chin, and hairy ears. Pronounced variations like these among members of our own species as well as the underlying patterns of inheritance that explain them are the concerns of human genetics.

Most of the principles that we have discussed in our survey of genetics from Mendel to gene splicing apply equally to human genetics. And human genes are as subject to mutation and manipulation as those of other species. Until recently, however, the study of human genetics has been unusually difficult because human beings present at least three unique challenges as research subjects. First, geneticists can deliberately conduct matings between individuals of other species with interesting phenotypes and genotypes—but they cannot conduct such matings between humans. Second, humans don't produce a real F_2 generation, since human sisters and brothers rarely mate. Third, in most of the industrialized

Humankind: One species, many manifestations.

countries, a given human family is usually a very small statistical sample, with fewer than three children on average. For these reasons, human geneticists have always employed special study methods, and since the era of genetic engineering, now have even more powerful tools. Research in human genetics, using both traditional and leading-edge methods, has important implications for the treatment and prevention of many disorders through genetic analysis, screening, and counseling.

This chapter discusses:

- The special methodology of human genetics
- Chromosomal abnormalities, a major source of human genetic disease
- The most common human genetic traits
- How physicians treat genetic diseases
- How genetic screening and counseling can help couples at risk of passing on genetic defects

WHAT ARE THE METHODS OF HUMAN GENETICS?

Because of the unique nature of human subjects, geneticists must rely on indirect research methods such as pedigree analysis, karyotyping, biochemical analysis, somatic cell genetics, in situ hybridization, and twin studies. Let us consider each.

Constructing Pedigrees, or Family Trees

A **pedigree** is a formal representation that traces a genetic trait for all members of a family lineage over several generations. With the pedigree, a geneticist may be able to deduce whether the trait in question is inherited according to Mendel's laws. A dominant allele usually expresses itself (*AA*, *Aa*) and does not skip generations; that is, offspring will express the trait if one or both parents have the dominant allele. A recessive allele *can* skip generations and usually expresses itself only in homozygotes (*aa*).

Figure 15-1a shows a pedigree analysis for the occurrence of a dominant trait called *achondroplastic dwarfism* in a large family. Achondroplastic dwarfs have a developmental abnormality in which the long bones of the limbs do not reach normal size, although the bones of the trunk and head do.

In the first generation, a male dwarf married a normal-sized woman and produced seven children, including one pair of monozygotic (identical) twins and one pair of dizygotic (fraternal) twins. Four children were dwarfs (green symbols), and three were normal. Because the dwarf husband produced both dwarf sons and dwarf daughters, achondroplastic dwarfism must be carried on an autosome—a chromosome other than a sex chromosome (see Chapter 11). Geneticists noted that the trait is rare, that every affected child has an affected parent, and that about half the children of a dwarf are themselves dwarves. From this they reasoned that each mating is like a test cross where the dwarf is heterozygous for the dominant dwarf allele and the normal parent is homozygous recessive.

By using such pedigrees from a number of unrelated families whose members include achondroplastic dwarfs, geneticists have been able to confirm that the trait is determined by a dominant allele on an autosome. Furthermore, geneticists have discovered that the mutant allele for this trait is lethal when homozygous (see Chapter 11); the homozygous dwarf offspring of two achondroplastic dwarfs die before birth, so that only heterozygotes survive and become dwarfs.

Pedigree analyses are also useful tools for studying the genetic founder effect, wherein a large modern population derives from a small founding group. Geneticists studying black South Africans, for example, traced an autosomal dominant bone disease to "Arnold," a Chinese immigrant who had 70 descendants in four generations. In this disease, all of a person's teeth fall out by age 20.

Karyotyping: Chromosomes Can Reveal Defects

Another method for determining whether a given human trait might have a genetic cause is to examine the chromosomes themselves. This process, called **karyotyping,** involves staining chromosomes during mitotic metaphase, photographing them, and cutting out the images and arranging them on the basis of size and shape.

Karyotyping helped reveal several syndromes caused by chromosomal abnormalities. **Down syndrome,** for example, is a genetic disorder that occurs in 1 of every 600 newborns. This syndrome is characterized by mental retardation; heart defects; a short, stocky body; and distinctive eyelid folds, among other traits (Figure 15-2a). Karyotype analyses reveal that individuals with this syndrome have three copies of chromosome 21 (Down syndrome is also called *trisomy 21;* Figure 15-2b). The extra chromosome 21 is a result of **nondisjunction,** or the failure of homologous chromosomes to separate properly during meiosis. Geneticists have also traced inherited forms of Alzheimer's disease to a mutation on chromosome 21 and observed that abnormal quantities of a fibrous protein can accumulate in the brains of both aged Alzheimer's patients and young adult victims of Down syndrome.

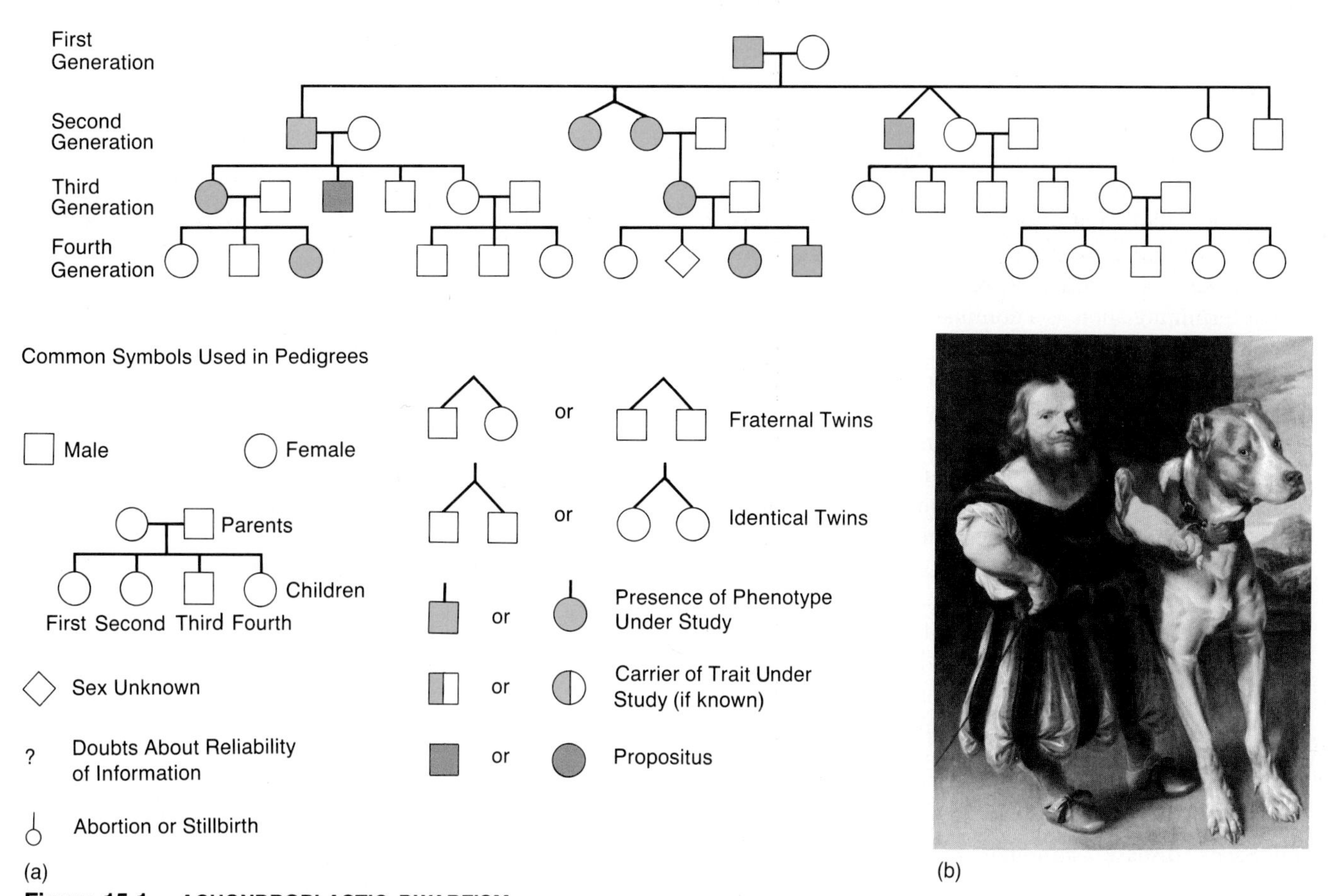

Figure 15-1 ACHONDROPLASTIC DWARFISM.
(a) A family pedigree for the occurrence of achondroplastic dwarfism. Individuals with the trait are indicated in green. The symbols in the legend are a key to reading the pedigree. (b) A portrait of an achondroplastic dwarf entitled *Giacomo Favorchi* and painted by the Dutch artist Karel van Mander around 1600.

(a)

1 2 3 4 5 6 XX
7 8 9 10 11 12
13 14 15 16 17 18
19 20 21 22

(b)

Figure 15-2 DOWN SYNDROME: AN EXTRA COPY OF CHROMOSOME 21.
(a) A sixteen-year-old girl with Down syndrome, or trisomy 21. (b) The karyotype of a human with Down syndrome. The extra copy of chromosome 21 (one of the smallest chromosomes) causes all of the deleterious traits characterizing this genetic disorder.

Biochemical Analyses Reveal Mutations

A geneticist's most direct tool for determining the genotypes that underlie human phenotypes is analyzing specific metabolic pathways. When the pathway links a single gene and a single gene product, such as an enzyme or antigen, a geneticist often can discover whether the determining allele is a dominant or a recessive mutation and how it is expressed.

One particularly deadly disorder that can be detected through biochemical analysis is **Tay-Sachs disease.** Children born with this disease are homozygous for a recessive allele that prevents them from producing *hexosaminidase A*, a digestive enzyme stored in the lysosome (see Chapter 6). Without this enzyme, lipids accumulate and damage the child's nervous system. Victims inherit one recessive allele from each parent—heterozygous *carriers* who do not show the trait themselves.

Figure 15-3 shows how the blood serum levels of hexosaminidase A differ among normal people, Tay-Sachs carriers (heterozygotes), and infants with the disease (homozygotes). Such analyses make carriers of the recessive allele fairly easy to identify, so that potential parents can learn their chances of producing a diseased child. Tay-Sachs disease, although rare, is most common

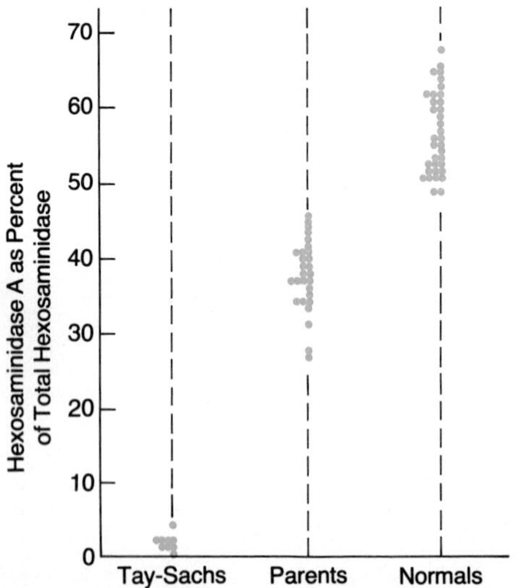

Figure 15-3 BIOCHEMICAL ASSAY OF TAY-SACHS DISEASE.
Tay-Sachs disease is caused by the body's inability to produce the enzyme hexosaminidase A, the result of a homozygous recessive allele. Geneticists can detect heterozygous carriers with a biochemical assay of the enzyme level in blood. As this graph shows, the percentage of hexosaminidase A is very low in homozygous victims of the disease, and the heterozygous carriers of the disease ("parents") show a value between those of normal individuals and victims.

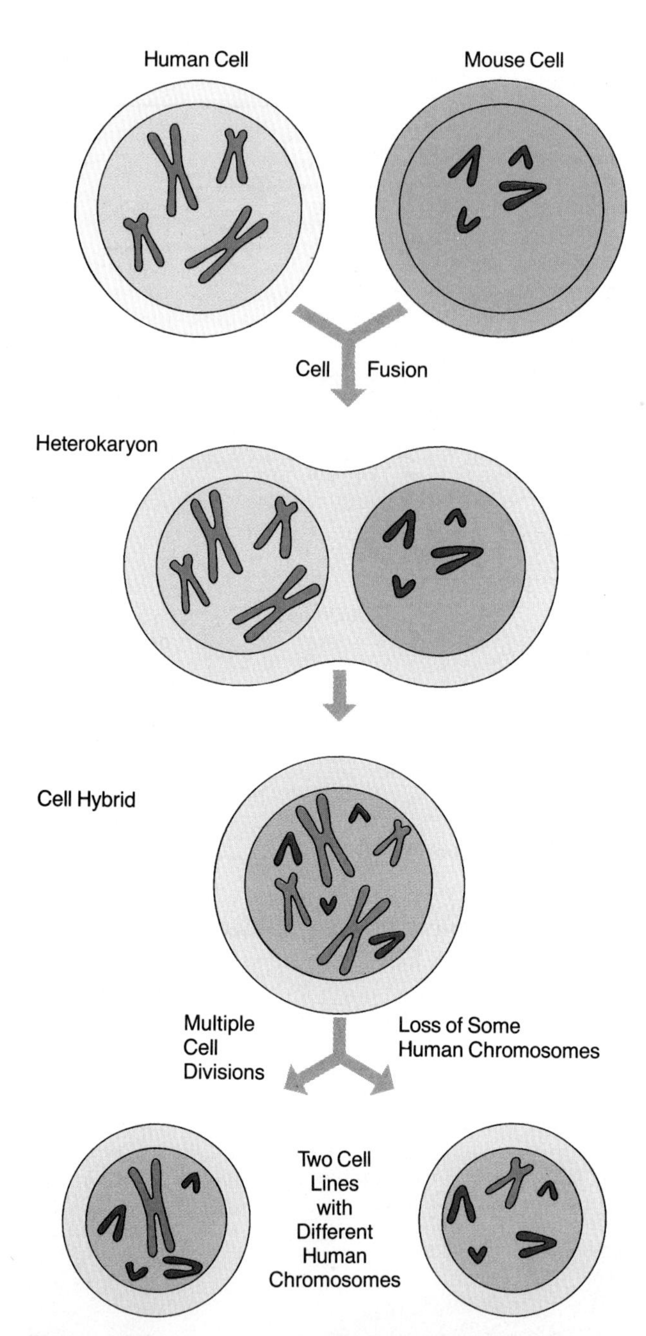

Figure 15-4 MAPPING GENES WITH SOMATIC CELL TECHNIQUES.
Researchers grow a human somatic cell in culture in the presence of a mouse cell. The chromosomes are easily distinguishable, since many human chromosomes have the centromere in the middle, whereas all mouse chromosomes have the centromere at the end. An inactivated virus binds to the surface of the cells and allows them to fuse. The result is a heterokaryon in which the two nuclei may later fuse. As the heterokaryons undergo multiple cell division cycles, they tend to lose human chromosomes. Ultimately, different subclones may contain different human chromosomes; by identifying which gene products are still made by a heterokaryon containing a specific human chromosome, a geneticist can assign a location for a gene on that chromosome.

among Jews from central Europe and their descendants, and in certain populations, up to 11 percent are carriers. The high incidence of Tay-Sachs and two related diseases involving lipid storage and lysosomes seems to have an evolutionary explanation analogous to the one for the sickle-cell gene and malaria. Historically, in Austria-Hungary, homozygous normals were more likely to die from tuberculosis, and homozygous recessives (with Tay-Sachs disease) died of the disease itself. Heterozygotes (Tay-Sachs carriers) apparently survived tuberculosis in greater numbers than noncarriers, and so the mutation was favored by natural selection.

Mapping Human Genes: Somatic Cell Genetics

In 1911, geneticists tracked the first sex-linked human trait, color blindness, to a gene located on the *X* chromosome. In 1971, a new generation of researchers developed a tool for mapping human genes, and within 10 years, they had assigned another 400 human genes to particular chromosomal locations. Today, gene mapping continues at a rate of about four genes per month.

The new method of chromosome mapping effectively bypasses human reproduction as a means of producing new genotypes because it relies on somatic cells, such as skin and liver cells, rather than on eggs and sperm. This field of research is called **somatic cell genetics.** In this approach, researchers grow two genetically distinct, diploid somatic cell types in tissue culture and fuse them by means of inactivated viruses that bind to the cell surfaces (Figure 15-4). The fused cell, with its two genetically distinct, diploid nuclei, is called a *heterokaryon* (meaning "different nucleus"). These two nuclei may fuse to make a tetraploid *cell hybrid,* which grows and divides normally to make a clone of cells.

Fortunate "accidents" sometimes occur as the somatic cell hybrids divide. Individual chromosomes may be lost from the cells. When mouse and human cells are fused to make hybrids, human chromosomes tend to become lost, so that the hybrids eventually have mostly mouse chromosomes plus one or a few human chromosomes of random type. One clone of cells might have human chromosomes 2, 12, and *X*, while another might have 7 and 9. By identifying which human genes or gene products are still found in the human-mouse cell hybrid with, say, its single remaining human chromosome, geneticists can be sure that particular genes are on individual human chromosomes.

Physicians can remove somatic cells from a developing fetus or from prospective parents, grow the cells in culture, and analyze them for the presence of a gene or DNA sequence called a *genetic marker* that is known to be closely linked to a mutant gene of interest. This allows them to predict or diagnose certain genetic defects.

Researchers have located markers with *restriction endonucleases,* enzymes that cleave DNA at specific sites (see Chapter 14), creating DNA fragments of different lengths. If the base sequence contains a mutation at one of these sites, the expected cleavage will not occur, and the DNA fragments will not be the predicted lengths (Figure 15-5). If such a site is very close to a deleterious allele, its presence serves to identify the allele, too.

Alec Jeffreys of Leicester University in England has exploited the power of restriction endonucleases to develop "DNA fingerprints" that are just as unique as the whorls on each person's fingertips. Within human DNA, there are large numbers of short, repeated sequences of bases called minisatellites, varying greatly in number from one individual to the next. Consequently, human DNA cleaved by certain restriction endonucleases yields immense polymorphism (variability) in the lengths of

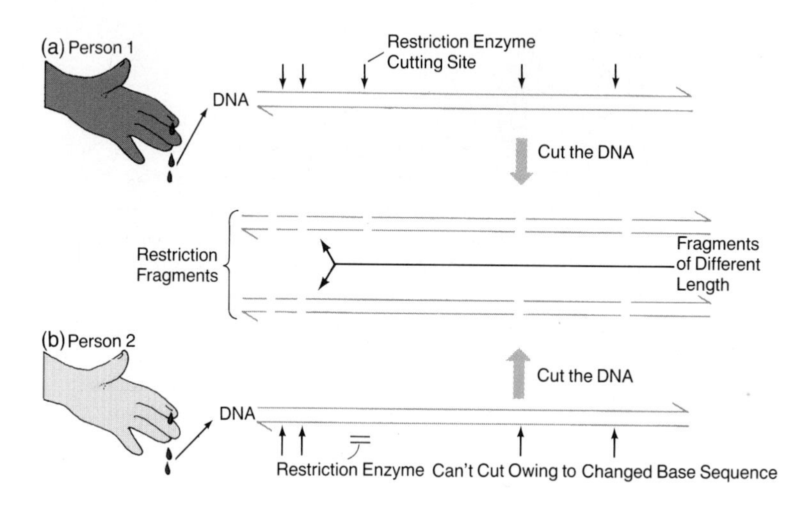

Figure 15-5 RESTRICTION ENDONUCLEASES: MAPPING HUMAN GENES.
(a) Person 1 has five characteristic cleavage sites on the DNA of a specific chromosome that will produce fragments of different sizes after cleavage. (b) Person 2 is a genetic variant with just four cleavage sites, since one has been eliminated by mutation. With fewer cutting sites, person 2 will have DNA fragments of different sizes. A geneticist can detect and compare the sizes of the fragments through gel electrophoresis. The researcher can then use the large amount of variability in sizes of restriction fragments to create unique "DNA fingerprints" and to map human genes.

restriction fragments. These **restriction fragment polymorphisms** have been dubbed **RFLPs** (pronounced "riflips"). Jeffreys found ways to probe this diverse array of RFLPs, and he concludes that the probability of two people having identical sets of restriction fragments is about 5×10^{-19}—essentially zero! Police and forensic pathologists can use "genetic fingerprints" to prove beyond doubt the identity of the human source of blood (even old, dried blood), of sperm in sexual assaults, or of biological parents in cases of disputed paternity.

Still another powerful new gene mapping technique called *in situ hybridization* creates probes from mRNA or DNA coding for a protein of interest such as sickle-cell hemoglobin. The probe will bind specifically to the chromosomal gene for that protein, and if the experimenter has prepared the probe with radioactive labels or tagged it with a fluorescent dye, the probe will reveal the site of the gene. Researchers are using this fast technique to map human chromosomes, including the genes for three color receptor proteins (opsins) in the human eye (see Chapter 38). In situ hybridization promises to allow the precise mapping of many genes coding for many proteins.

Reverse Genetics

With the new technologies of DNA sequencing, gene cloning, and controlled protein manufacture, geneticists are now carrying out "bottom up" or **reverse genetics.** Classical genetics starts with phenotypic traits and works back to the gene. Molecular biologists can now begin with long stretches of chromosomal DNA having an inheritance pattern that matches a given human disease; identify a specific gene in that DNA; determine what protein the gene encodes; and finally, learn the role of the protein in generating the disease. The first success with reverse genetics involved a gene on the human *X* chromosome that causes a disease abbreviated CGD in which victims suffer chronic severe bacterial infections. A long DNA segment from the *X* chromosome yields a gene that in turn produces a protein responsible for CGD. Reverse genetics is now being used to search for the genes and gene products that cause cystic fibrosis, Huntington's disease, and other deadly maladies.

Nature, Nurture, and Twin Studies

Yet another method of human genetics is the *twin study*, which can help researchers determine the influence of environment on gene expression.

The extent to which either genes or the environment affect a certain trait has come to be called the "nature-nurture" question. Obviously, the expression of an individual's genotype can be strongly affected by the environment in which he or she develops as an embryo, child, and adolescent. Factors such as nutrition, sanitation, disease, and availability of medical care can influence the eventual phenotypic traits of the adult. For example, even genetically identical monozygotic twins (derived from a single egg) can grow up looking fairly different if one ate poorly as a child or contracted a serious disease, while the other ate normally and stayed healthy. There are undoubtedly many subtle influences as well, such as presence or absence of siblings or parental neglect or affection.

One way geneticists approach the nature-nurture problem is to examine a given phenotypic trait in identical twins (Figure 15-6). Geneticists assume that any differences between two genetically identical individuals probably arise from environmental influences on gene expression. Consider the variation among individuals for height, weight, and abstract reasoning ability shown in Table 15-1. Identical twins who are reared by different sets of adoptive parents are much closer in height than are fraternal twins, regardless of whether they are reared together. Height, therefore, appears to have a strong genetic component little influenced by environment. However, weight is a different story: Identical twins reared separately show greater weight variation than do identical twins reared together; and both the former group and fraternal twins approach the variation seen in normal siblings. Therefore, environment, including

Figure 15-6 BEHAVIORAL SIMILARITIES IN IDENTICAL TWINS.
The twins in this picture were not told how to pose for the photographer. Notice that each set of twins has a characteristic posture and placement of hands. Genetics clearly plays a role in these examples of very complex behavior associated with tastes, body posture, and movement.

Table 15-1 DIFFERENCES BETWEEN TWINS* (Correlations: 1.0 perfect–0.0 no correlation)

	Identical Twins Reared Apart	*Identical Twins Reared Together*	*Fraternal Twins Reared Together*
Height	0.94	0.93	0.50
Weight	0.51	0.83	0.43
Abstract Reasoning Ability	0.58	0.66	0.19

*Based on a study reported in *Development, Genetics, and Psychology,* 1986, Lawrence Erlbaum Publishers, London, pp. 256, 257, 262.

learned eating habits, seems to affect weight to a much greater extent than it does height.

Some twin studies have shown that the kinds of traits measured by culture-free tests of abstract reasoning ability are more strongly influenced by environmental factors, such as parental stimulation and type of schooling, than by genotype. But other studies of identical and fraternal twins raised together or separately suggest a previously unappreciated role for the power of nature (rather than nurture) in shaping human personality traits such as activity level, anxiety, dominance, vocational interests, sociability, altruism, aggression, and even "traditionalism"—respect for discipline and authority and conservative social and political attitudes. These findings often disturb those who fear biological determinism. But the remarkable similarities between identical twins separated in infancy and raised apart argue too powerfully for a genetic basis to personality.

A case in point is Oskar Stohr and Jack Yufe. The twins had been separated at birth, one raised as a Catholic German and the other as a Jew in Trinidad. Some 40 years later, however, when they both arrived at a clinic to participate in a twin study, they were both wearing mustaches, wire-rimmed glasses, and blue, double-breasted shirts with epaulettes. They had nearly identical personalities, and both displayed such peculiarities as sneezing in elevators to surprise other people!

CHROMOSOMAL ABNORMALITIES: A MAJOR SOURCE OF GENETIC DISEASE

Using karyotyping and other techniques, geneticists have carefully examined the chromosomes of normal, malformed, and diseased adults, newborns, and embryos and determined that chromosomal defects underlie many specific disorders.

Abnormal Numbers of Autosomes

Deviation from the normal diploid number of 46 chromosomes is the most frequent human genetic defect. The most common deviation, called **aneuploidy,** is the absence of one or more chromosomes or the presence of one or more extra chromosomes. Aneuploidy often results in the death of the embryo at an early stage. In other forms of aneuploidy, portions of a chromosome are duplicated or deleted, changing chromosome content, not chromosome number.

Normal human development appears to require the entire complement of gene pairs represented on a diploid chromosome set. *About 45 percent or more of all human embryos* are spontaneously aborted during early pregnancy, and nearly one-quarter of these exhibit some deviation from the normal diploid chromosome number. As Figure 15-7 illustrates, *monosomics* have only one

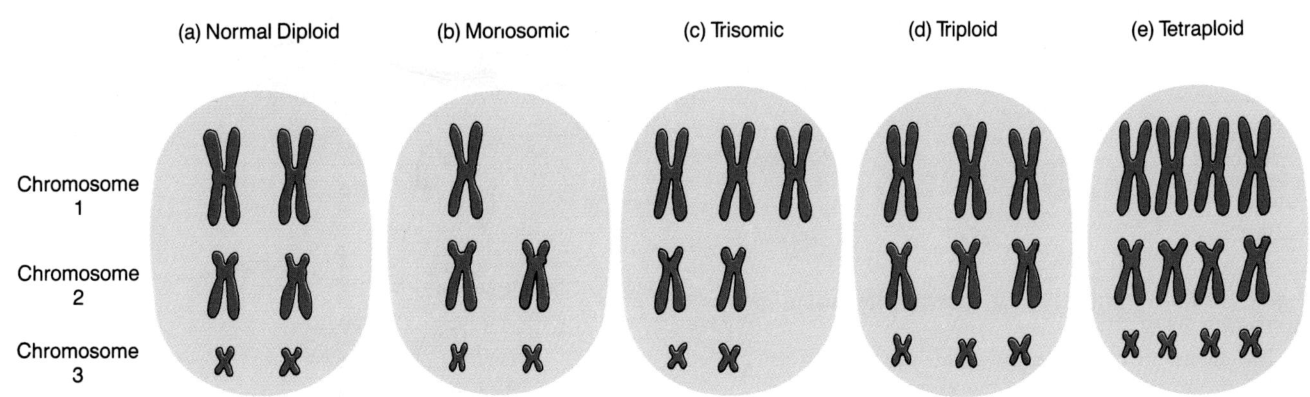

Figure 15-7 CHROMOSOME ABNORMALITIES.
(a) Just two pairs of homologous chromosomes constitute the normal diploid state of this cell line. (b) The monosomic cell lacks one chromosome 1. (c) The trisomic cell has an extra chromosome 1. (d) The triploid cell has three of each chromosome. (e) And the tetraploid cell has four of each chromosome.

partner of a given chromosome pair. *Trisomics* (such as individuals with trisomy 21) have three rather than two chromosomes of a given type. *Triploids* are individuals who have three entire sets of chromosomes (69 chromosomes) instead of two sets, and *tetraploids* have four sets (92 chromosomes).

Many forms of aneuploidy result from nondisjunction during meiosis in the mother (see Figure 10-12). As the mother's age increases, spontaneously aborted embryos show an increased incidence of aneuploidy (Figure 15-8). All human eggs become arrested in prophase of meiosis I while the female is herself still an embryo, and apparently, 40 or 50 years in the arrested state is simply too long; as the years pass, nondisjunction of chromosome 21 and others becomes more and more likely. Babies can survive with extra copies of certain chromosomes (21, 13, and 18), but display physical defects. As we have seen, trisomy 21 produces Down syndrome. *Trisomy 13*, which yields individuals with harelip, cleft palate, and various eye, brain, and cardiovascular defects, occurs in about 1 in 10,000 live births. *Trisomy 18*, which is observed in about 1 in 5,000 births, causes malformation in virtually every organ system. About 80 percent of babies born with an extra chromosome 18 are females, and virtually all affected infants of both sexes die within a few months.

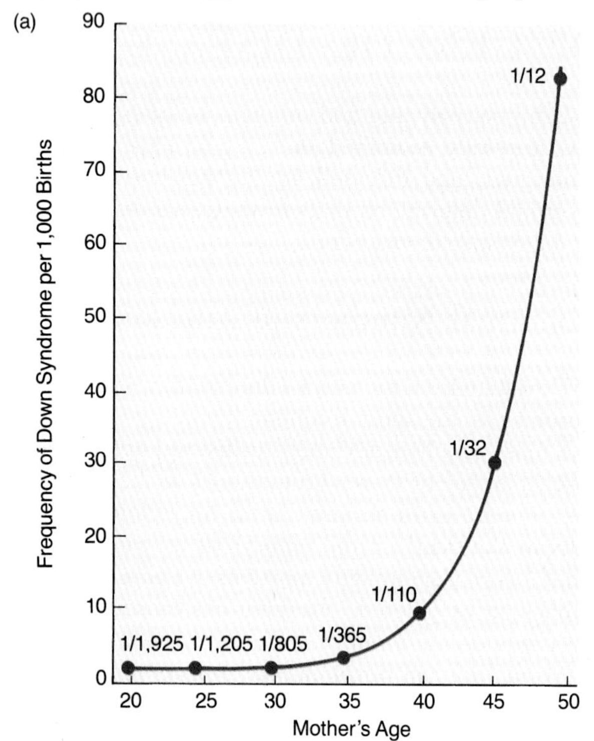

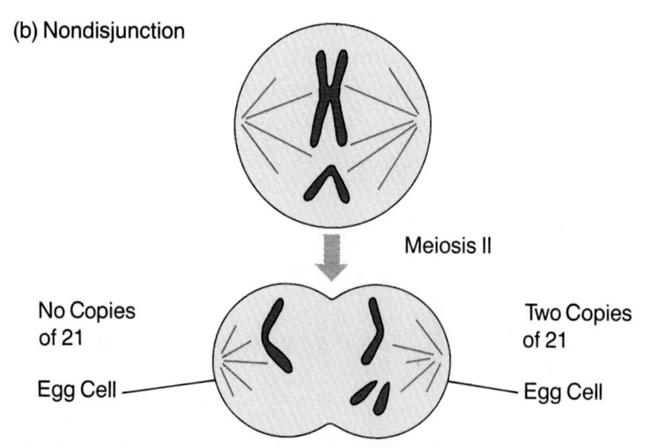

Figure 15-8 FREQUENCY OF DOWN SYNDROME AND MOTHER'S AGE.
(a) The danger of bearing a child after age 40 or so is evident from the steep rise in the numbers of Down syndrome babies as the mother's age advances.
(b) Aneuploidy in this case results from nondisjunction of chromosome 21, so that both copies are delivered to one gamete, while the other gets none.

Abnormal Numbers of Sex Chromosomes

Embryos with aneuploidy of the sex chromosomes frequently survive to birth and adulthood. Table 15-2 lists the various patterns that can occur. The presence or absence of the *Y* chromosome determines the route of sexual development in mammals, including humans. *XX* and *XO* embryos develop into females, whereas *XY* or *XXY* embryos develop into males.

Individuals with **Turner's syndrome** (*XO*), or monosomy of the *X* chromosome, based on nondisjunction in either parent have external female organs but degenerate ovaries lacking germ cells. Since ovarian hormones are absent, puberty does not occur, and secondary sexual characteristics do not develop.

Nondisjunction during meiosis can also produce *XX* eggs and either *XY* or *YY* sperm. Individuals with **Klinefelter's syndrome** (Table 15-2) have the external sex characteristics of males but are usually sterile. Some

Table 15-2 ABNORMAL NUMBERS OF SEX CHROMOSOMES

Sex Chromosomes	*Syndrome*	*Frequency at Birth*
Females		
XO, monosomic	Turner's	1 per 5,000
XXX, trisomic		
XXXX, tetrasomic	Metafemale	1 per 700
XXXXX, pentasomic		
Males		
XYY, trisomic	Normal	1 per 1,000
XXY, trisomic		
XXYY, tetrasomic		
XXXY, tetrasomic	Klinefelter's	1 per 500
XXXXY, pentasomic		
XXXXXY, hexasomic		

near-normal-appearing men who have small testes and are sterile have been shown to have two *X's* and no *Y* chromosome. Molecular cloning now reveals that despite their *XX* genotype, a short segment of *Y*-specific DNA is located at the end of one of the *X* chromosomes. This DNA includes the gene that causes testes to form instead of ovaries.

X Chromosome Inactivation

Regardless of the number of *X* chromosomes in a normal mammalian cell, only one is genetically active; the others become inactivated during embryonic development. The single *X* chromosome in every somatic cell of a male mammal but only one of the two *X* chromosomes in every somatic cell of a female mammal is active. A female's second *X* chromosome becomes a dark-staining, dense mass of chromatin that no longer functions and is called a **Barr body** (Figure 15-9). In normal (*XX*) females, every somatic cell has one Barr body; normal male (*XY*) somatic cells have none. Cells of *XO* females also lack Barr bodies, while Klinefelter males may have one Barr body.

Some regulatory mechanism shuts down the activity of a whole, specific chromosome in each female somatic cell, randomly inactivating the father's *X* chromosome in some cells and the mother's in other cells. The result is **mosaicism,** the condition in which some cells of an individual's body express one phenotype, while others express an alternative one. Every organ in a human female's body is thus a mosaic of cell clones with respect to her *X* chromosomes. If the inactivated *X* chromosome in a cell carries the only normal copy of a gene, and the active *X* carries a mutated allele of the gene, then that cell will have a mutant phenotype.

For example, the mutation anhidrotic ectodermal dysplasia deletes hair, teeth, and sweat glands. Women who are heterozygous for this *X*-linked trait are missing some teeth; have thin, sparse hair; and lack sweat glands in patches of their skin because the normal *X* has been inactivated to become a Barr body in these cells (Figure 15-10). Normal teeth, hair, and sweat glands occur only where cells have the wild-type allele on their active *X* chromosomes. The vacant sites occur where cells chance to have the mutant allele on their active *X* chromosomes. All women are mosaics, but the mosaicism is detectable only in heterozygotes for some genes.

Translocations, Deletions, and Duplications

Sometimes, only *part* of a chromosome rather than the whole structure is altered, lost, or moved during cell

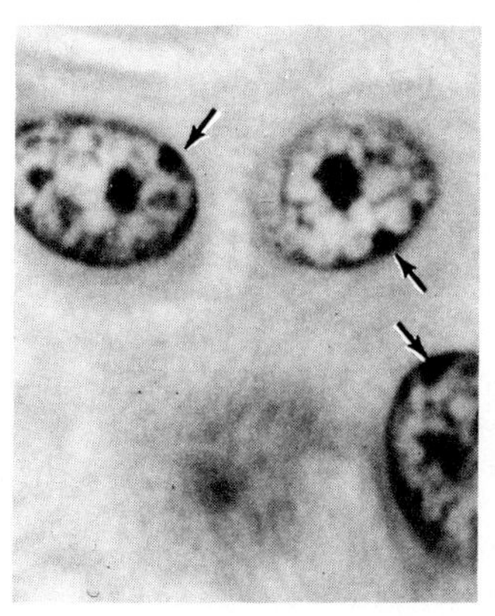

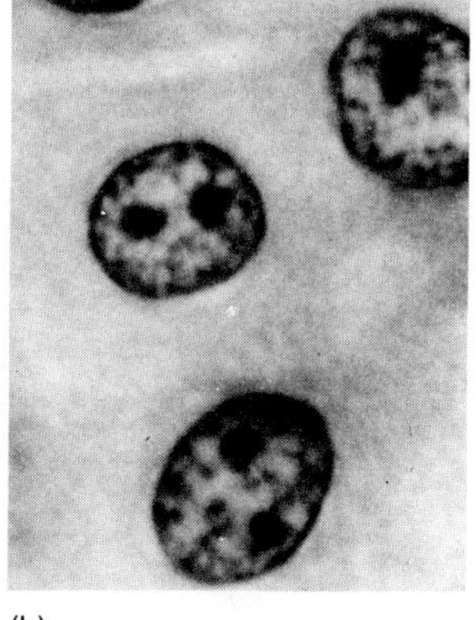

(a) (b)

Figure 15-9 BARR BODIES: CONDENSED X CHROMOSOMES.
(a) Stained nuclei from cells of a female mammal (magnified about 3,000×) reveal Barr bodies (arrows), compacted masses of chromatin comprising nonfunctional *X* chromosomes. Any individual with more than one *X* chromosome per cell has a Barr body for each inactive chromosome. (b) In cell nuclei from a male mammal (magnified about 3,000×), Barr bodies are not evident, since the single *X* chromosome is uncompacted. The large, darkly stained masses near the centers of both female and male nuclei are nucleoli.

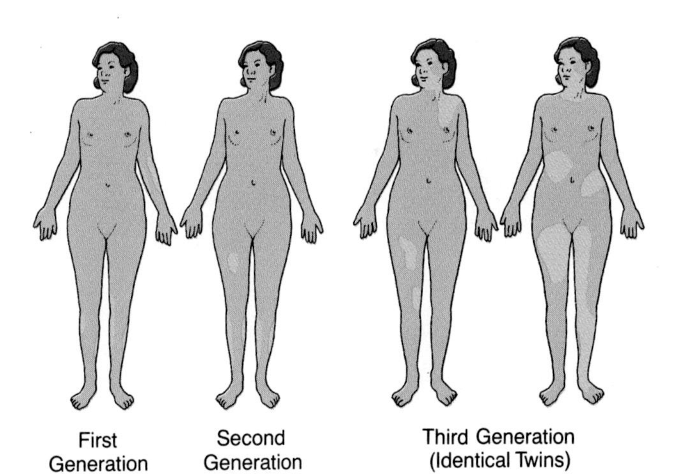

Figure 15-10 MOSAICISM RESULTING FROM X CHROMOSOME INACTIVATION.
If a woman's two *X* chromosomes carry different alleles, (i.e., if she is heterozygous for a given trait), different sets of alleles will be expressed in different cells as a result of *X* chromosome inactivation. If the activated *X* carries a mutated gene, the cell will have a mutant phenotype. This figure shows three generations of women from a family carrying the allele for anhidrotic ectodermal dysplasia. The colored areas indicate regions lacking sweat glands owing to the presence in those cells of an active *X* chromosome carrying the mutant allele. The two women in the third generation are identical twins. Their disorder is not expressed in identical patterns, since *X* inactivation is random.

division. These partial changes are the result of chromosomal *translocation, deletion,* and *duplication.*

Translocation

One form of Down syndrome arises when most of a chromosome 21 is translocated, or moved, to the tip of another large chromosome, usually chromosome 14 (Figure 15-11). The karyotype of an affected person will show the normal complement of 46 chromosomes, but one of the chromosomes 14 will bear the translocated portion of 21. By analyzing whether a couple's Down syndrome child has trisomy 21 or fused chromosomes 14 and 21, geneticists can predict their chances of producing another such child.

Some translocations can arise in certain somatic cells while the rest of the body cells retain normal chromosomes. Some patients, for example, suffer from a form of bone marrow cancer called chronic myelogenous leukemia. In 80 to 90 percent of these cases, the cancerous bone marrow cells show a translocation of part of chromosome 22 to the end of the long arm of chromosome 9, while the rest of the body cells show normal karyotypes. This specific translocation may arise spontaneously in a normal bone marrow cell and create a clone of cells that proliferates to form the cancer. Abnormalities like this are *somatic mutations*, since they occur in somatic, not germ line, cells and thus cannot be passed on to the next generation.

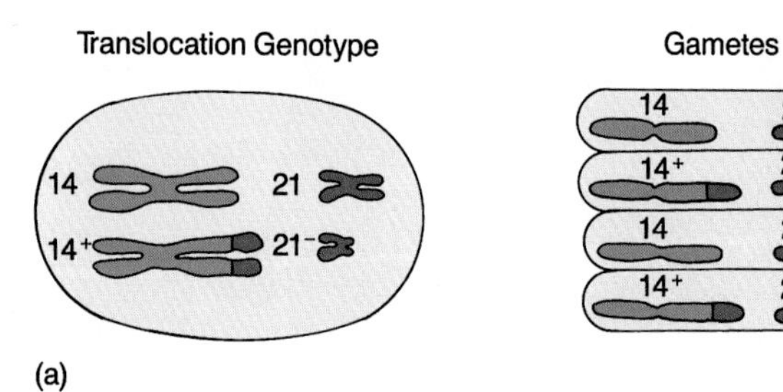

(a)

Copies of Long Arm of Chromosome 21		Phenotype of Zygote
in Gamete	in Zygote	
1	2	Normal
2	3	Down Syndrome
0	1	Lethal
1	2	Normal, but Carrier

(b)

Figure 15-11 TRANSLOCATION OF A PORTION OF CHROMOSOME 21 TO CHROMOSOME 14.
(a) This figure shows the nuclei of gametes from a female parent. (b) The first column shows the number of chromosomes 14 and 21 in each female gamete. The second column shows the number in each zygote following fertilization by a normal sperm (carrying one copy of each chromosome). The third column shows the zygote's phenotype.

Deletions and Duplications

Occasionally, parts of chromosomes are deleted entirely, usually in the germ cells of the parents. Physicians rarely see such chromosome abnormalities in live infants, since the loss of genes usually leads to severe defects and death in utero. A loss of material from the short arm of chromosome 5 leads to one exception: *cri du chat* ("cat's cry") *syndrome,* so called because the infant survives and has a distinctly catlike cry. The individuals affected by this rare disorder are mentally retarded but may survive through adolescence.

The most common inherited cause of mental retardation and learning disorders is so-called fragile *X* syndrome, a chromosomal defect in which the tip of the *X* chromosome can break off (Figure 15-12). Geneticists do not yet understand the inheritance pattern, but about 20 percent of the males and 66 percent of the females with a fragile *X* chromosome show no symptoms. Some speculate that *X* inactivation plays a role.

A SURVEY OF HUMAN GENETIC TRAITS

Like all other complex organisms, human beings have a host of normal traits determined by wild-type alleles and dozens of abnormal traits based on dominant and recessive mutations. Let's examine a number of the best-studied traits in each category.

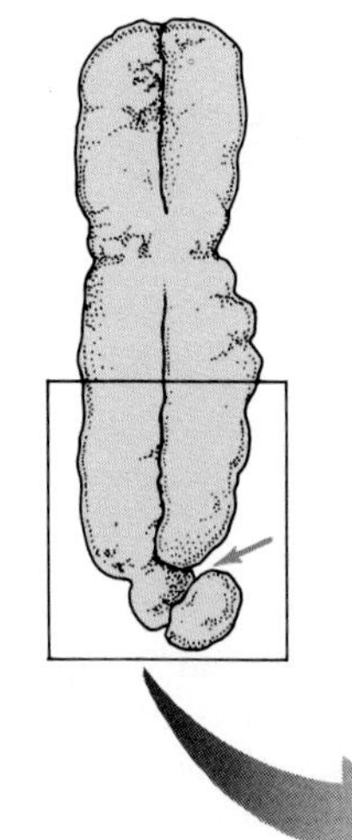

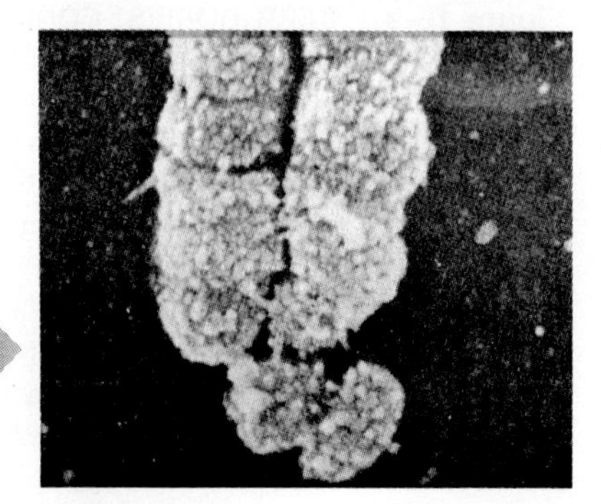

Figure 15-12 FRAGILE *X* CHROMOSOME.
The blue arrow shows the "fragile site" on a human *X* chromosome, where breakage is particularly likely to occur. Loss of this chromosome tip can lead to mental defects.

Sex-Linked Traits

Among the first human traits geneticists studied were those transmitted by human sex chromosomes. The fact that males are heterozygous, having one *X* and one *Y* chromosome, makes it relatively easy to identify sex-linked traits, since many genes carried by the *X* are not carried on the much smaller *Y*. The *Y* chromosome is transmitted directly from father to son in every generation and never to daughters. The only character that geneticists have unequivocally assigned to the *Y* chromosome is the trait of maleness itself. A father (*XY*) transmits an *X* chromosome to his daughter; similarly, a mother (*XX*) passes an *X* to any son. These *X* chromosomes can carry many traits and some significant mutations.

In the late eighteenth and early nineteenth centuries, physicians realized that red-green color blindness shows a regular pattern of inheritance. This type of color blindness affects about 8 percent of men in most Caucasian populations, but only 1 percent of women. If a color-blind man marries a woman with normal color vision, virtually all their children are normal. But if a color-blind woman marries a normal man, all the sons are color-blind and all the daughters are normal. The normal daughters of a color-blind mother can transmit their mutant, recessive allele to their sons (Figure 15-13a). Color blindness is a recessive mutation of a gene on the *X* chromosome; hence, it is an *X*-linked trait (Figure 15-13a–c). By employing recombinant DNA technology, geneticists have now discovered two genes on the *X* chromosome and one on an autosome that code for the proteins that absorb red, green, and blue light (see Chapter 38).

Females who are heterozygous for color blindness are actually displaying mosaicism due to *X* chromosome inactivation. Geneticists can tell this because in each eye, these women have both normal and abnormal cone cells (the cells in the retina responsible for color vision). Cells where the *X* chromosome is active and bears a wild-type allele develop as normal cone cells, while cells where the *X* is inactive or bears a mutant allele lack light-absorbing proteins and do not respond to colored light normally. Even though reduced in numbers, the normal cone cells are usually plentiful enough to produce normal color vision.

Another, much more serious X-linked trait is **hemophilia,** a disease in which the body fails to make a protein necessary for normal blood clotting. Hemophiliacs can bleed to death from even the slightest wound. About 1 in every 10,000 males shows hemophilia, and until the twentieth century, few victims survived adolescence.

For several generations, hemophilia has plagued the royal families of Europe descended from Queen Victoria of England. Geneticists have thoroughly investigated her pedigree (Figure 15-14a), and since none of her mother's, father's, or husband's relatives had the trait, a mutation probably arose during the meiotic divisions that produced the egg or sperm by which she was conceived or in an early mitotic division of her own germ line cells. The phenotypes of her descendants clearly show that she must have been heterozygous, having a normal dominant allele for the blood-clotting protein on one of her *X* chromosomes and a mutant recessive allele for hemophilia on the other *X* chromosome.

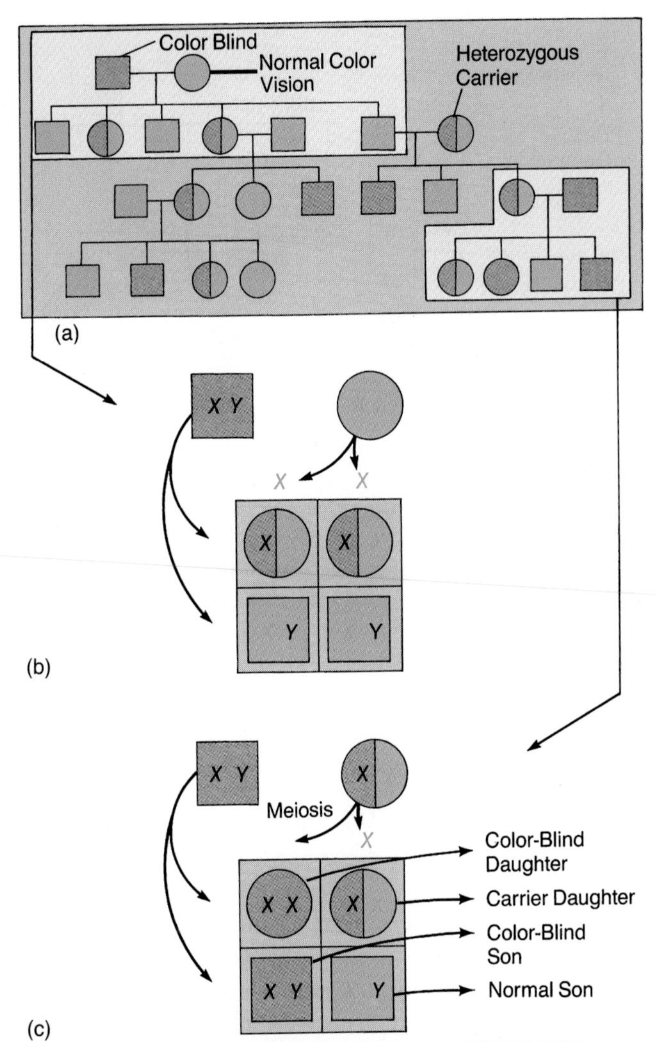

Figure 15-13 INHERITANCE OF COLOR BLINDNESS. (a) Typical pedigree of a family in which the father was color-blind and the mother was normal. The key in Figure 15-1 indicates symbols for normal individuals, affected individuals (here, color-blind), and carriers. (b) This chart shows how offspring inherit color blindness from such parents. (c) This chart shows how color blindness is inherited when the father is color-blind and the mother is a carrier.

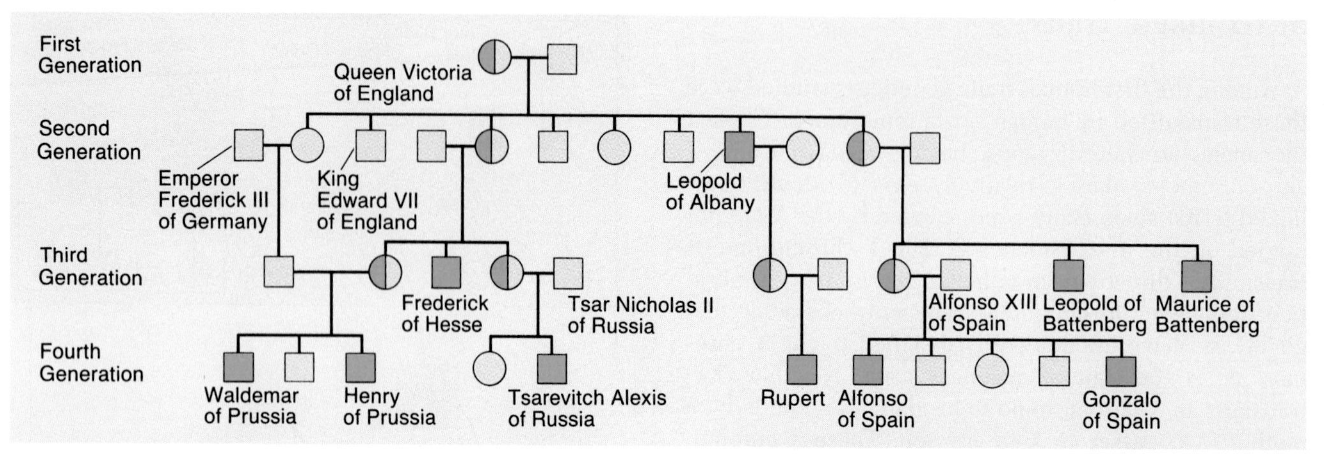

(a)

(b)

Figure 15-14 HEMOPHILIA IN THE ROYAL FAMILIES OF EUROPE.
(a) Pedigree for the hemophilia trait in descendants of Queen Victoria. The original mutation probably arose in the queen herself. (b) Queen Victoria with members of her immediate family in 1894.

Recessive Traits on Autosomal Chromosomes

Recessive alleles on autosomal chromosomes underlie most abnormal human traits. Some of the traits are harmless, such as tiny pits just above the ears or hairy elbows. Unfortunately, recessive alleles of autosomal genes also cause some of the most debilitating human abnormalities. Worse still, heterozygotes do not show the trait and are often unaware that they are carriers of the recessive mutation until they have produced a child with the defect.

Phenylketonuria (PKU)

Abnormal metabolism of the amino acid phenylalanine can cause a number of well-known genetic conditions (Figure 15-15). Newborn babies with the serious disease **phenylketonuria (PKU)** have abnormally high levels of phenylalanine in their blood serum and excrete high levels of phenylpyruvate (a breakdown product of phenylalanine) in their urine. The levels are high because the children make too little phenylalanine hydroxylase, the enzyme that converts phenylalanine to the harmless amino acid tyrosine (see Figure 15-15a). High levels of phenylalanine in the blood cause PKU victims to become severely mentally retarded. These children also have abnormal muscle tone and body movement, and their skin, hair, and eyes are pale owing to low levels of melanin, which comes from tyrosine. Until recently, about 1 percent of the severely retarded patients in mental institutions suffered from PKU.

Physicians now use blood screening tests to detect high levels of phenylalanine in newborns. If levels are high, they instruct parents to eliminate most sources of phenylalanine from the baby's diet, thereby preventing most or all of the PKU symptoms. The dietary treatment of PKU illustrates the effects of both nature and nurture on phenotype. A special diet excludes phenylalanine from the cellular environment, so the mutant genotype is not expressed, and instead, a normal phenotype develops.

Albinism

Another mutant recessive gene involving the same biochemical pathway as PKU causes **albinism,** a deficiency of pigment in the skin, eyes, and hair (Figure 15-15b). Albinism shows up in 1 person out of every

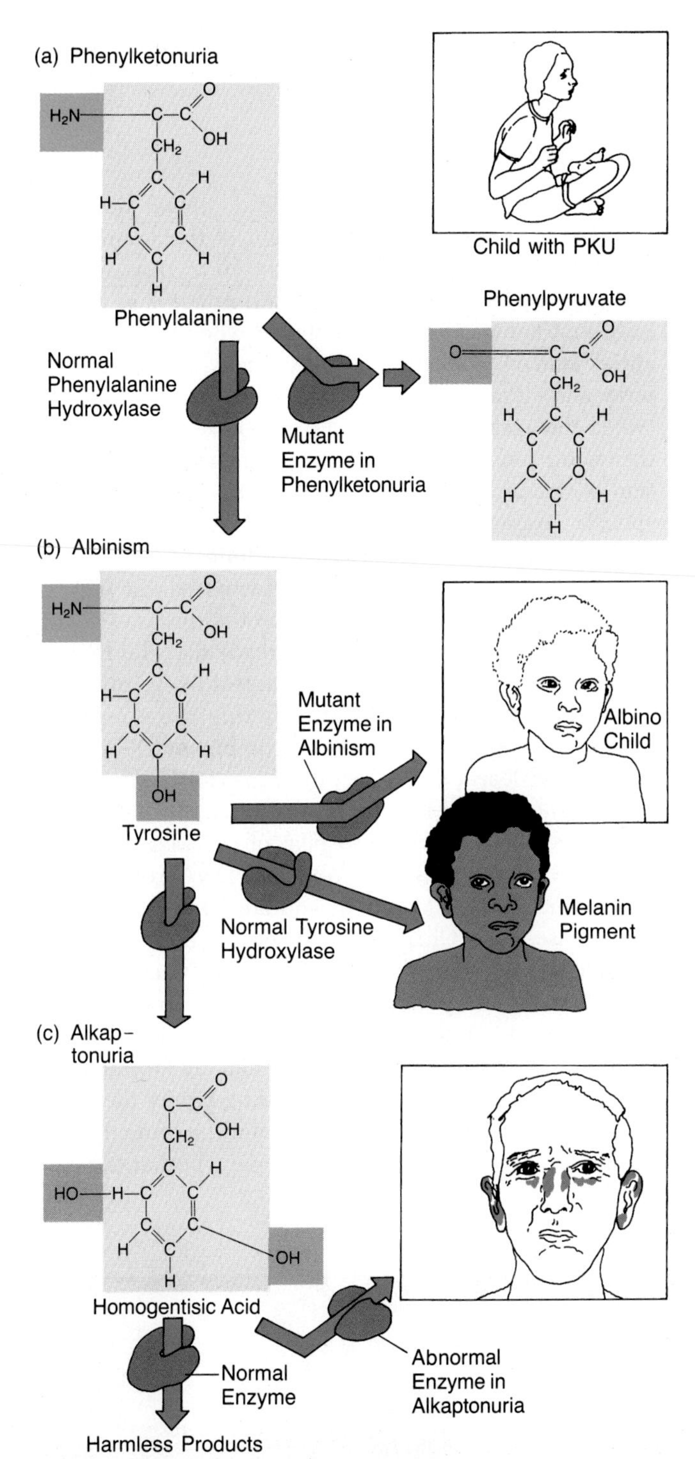

Figure 15-15 DEFECTS OF PHENYLALANINE METABOLISM AND RESULTING GENETIC DISORDERS. (a) Phenylalanine is normally broken down to tyrosine, but a mutant enzyme leads to a buildup of phenylpyruvate and the symptoms of PKU. (b) Tyrosine is normally converted to melanin, but a mutant enzyme leads to albinism, as in this male child. (c) In most people, tyrosine is converted to homogentisic acid, and this, in turn, is altered to a harmless compound. In those lacking the normal enzyme, however, homogentisic acid builds up and causes alkaptonuria, with its strange symptoms.

30,000, but it is 125 times more common among certain Native American tribes. Because of their recessive mutation, an albino makes too little of the enzyme tyrosinase, which begins the conversion of tyrosine to the dark pigment melanin. Melanin, in turn, creates most of the light and dark shades of the hair, skin, and eyes.

Lacking melanin pigment, an albino's skin is extremely sensitive to sunlight; exposure to the sun leads to roughness and wrinkling and to a high risk of skin cancer. Their eyes are also sensitive to light and appear pink because of light reflecting from underlying blood vessels. Most albinos exhibit *nystagmus*, an involuntary jerking of the eyes, and many suffer from eye defects, even near-blindness, that involve abnormal wiring of the nerves that extend from the eyes to the brain. No known treatment can overcome these enzymatic defects in albinos.

Yet another genetic condition stems from defects of this same metabolic pathway. Recall that Archibald Garrod found a buildup of homogentisic acid in victims of alkaptonuria (Chapter 12). A missing enzyme fails to break the compound down to a harmless product, and so the sufferers excrete black urine and can have dark coloration in nose and ears (Figure 15-15c).

Anemias

Two types of anemia (low red blood cell count), both caused by recessive genes, are quite prevalent in certain human populations. In malaria-endemic regions, the recessive allele for sickle-cell anemia occurs in high frequency because heterozygous carriers of the mutant allele are more resistant to malarial infection than are individuals with two normal alleles (see Chapter 11).

Thalassemias, another group of anemias prominent in malaria-ridden areas, are among the most serious health problems worldwide, claiming the lives of over 100,000 children each year. In the thalassemias, the body makes too little of either the α or the β chain of hemoglobin. Because the synthesis of only one chain is impaired, the other chain builds up within the red blood cells, causing them to burst.

New techniques for analyzing DNA sequences have revealed a point mutation in the β-globin structural gene of thalassemia patients. This mutation prevents the cell from correctly processing primary RNA transcripts, and so no final mRNA for β-globin passes to the cytoplasm, and the cell fails to make the crucial oxygen-carrying protein. Recently, molecular geneticists have cloned normal human globin genes and transferred them into the germ line cells of mice suffering from thalassemia. In the new generations of mice arising from these altered germ cells, the thalassemia was reduced or eliminated altogether. This is a model for how human gene therapy could conceivably correct the condition in people. We return to this topic later.

Dominant Traits and Variations in Gene Expression

Some mutations of human genes are dominant rather than recessive. The dominant alleles for cleft chin, baldness, or extra fingers and toes are harmless or at most bothersome. Other dominant alleles, however, can be lethal, such as the gene for Huntington's disease.

Pattern Baldness

A common dominant trait is pattern baldness, a condition in which heterozygous males (*Bb*) experience premature hair loss on the top and front of the head (Figure 15-16). Pattern baldness shows **sex-influenced inheritance:** Heterozygous males exhibit the trait, whereas heterozygous females may have thinning hair but do not become bald. High levels of testosterone, the male hormone, apparently are required for expression of the *B* gene; accidentally castrated males with the *Bb* genotype retain their hair unless they are treated with testosterone, in which case they become bald. Similarly, homozygous females become bald if they have an adrenal tumor that secretes testosterone.

Polydactyly

People with the mutant dominant allele for extra digits, or *polydactyly*, can have from five to seven digits on any hand or foot (Figure 15-17). The polydactyly gene shows **variable expressivity,** or different phenotypes in different affected people. Polydactyly also shows **incomplete penetrance,** in which the genotype is mutant but the phenotype is normal in certain individuals. Even a person affected with polydactyly usually has one or two normal limbs.

Huntington's Disease

Huntington's disease is a tragic dominant trait that involves a progressive deterioration of the brain cells, leading to muscle spasms and, after 10 to 15 years, the victim's death. Until recently, Huntington's victims have first known they were afflicted when their personalities started to change and their movements became jerky and clumsy—symptoms that appear sometime between the ages of 30 and 50. For the next 25 years or so, the victims suffer a progressive, irrevocable deterioration of the brain, leading to constant writhing, blurred speech, irrationality, and eventual death. Huntington's disease is caused by an autosomal dominant allele; thus, about half of the children of an affected parent will inherit the mutant allele and develop the disease (Figure 15-18). Since most victims have already had children by the time the disease begins to manifest itself, their offspring live in constant fear of suffering the same fate.

Recently, researchers using recombinant DNA techniques have learned to detect the genes for Huntington's disease and thus to identify probable victims far earlier than ever before. James Gusella, of the Massachusetts General Hospital in Boston, obtained DNA from mutant and normal white blood cells and added restriction endonucleases (see Chapter 14) to cleave the DNA into millions of fragments of different lengths. Gusella found that Huntington's patients always had a restriction fragment of short length, while the normal family members lacked the short fragment. Now, by looking for this short DNA fragment, or marker, in a young family member, researchers can predict whether or not the mutant allele for Huntington's disease is present and whether the per-

(a)

(b)

(c)

(d)

Figure 15-16 PATTERN BALDNESS IN THE ADAMS FAMILY.
(a) John Adams, the second president of the United States, fathered (b) John Quincy Adams, the sixth president, who, in turn, fathered (c) Charles Francis Adams, a diplomat and the father of (d) Henry Adams, a historian.

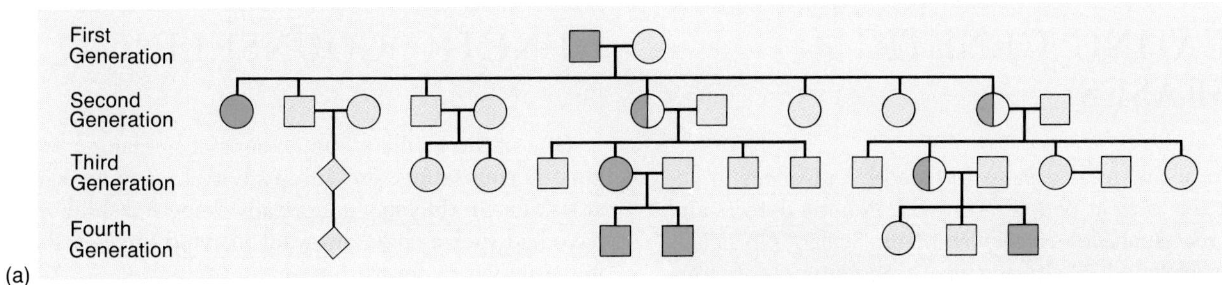

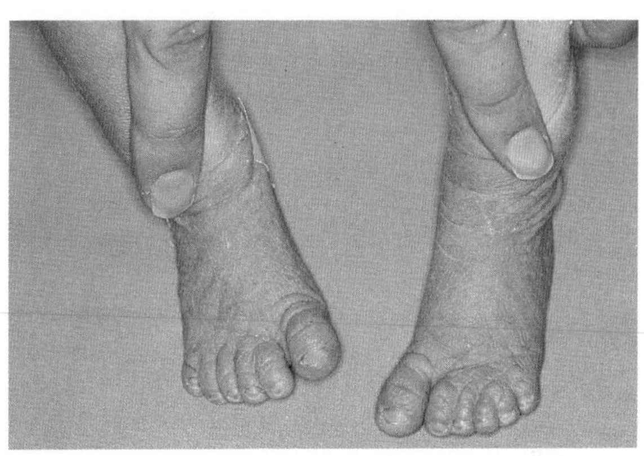

Figure 15-17 POLYDACTYLY—TOO MANY FINGERS OR TOES: AN EXAMPLE OF INCOMPLETE PENETRANCE.
(a) Polydactyly is caused by a dominant allele that is not always expressed. In this pedigree, three individuals (indicated by half-colored symbols) must have been carriers of the trait because they passed it on to their offspring. They did not exhibit the trait themselves, even though it is dominant. (b) A newborn with six toes per foot.

son is likely, decades later, to develop the disease. The knowledge itself is troubling, since there is currently no effective treatment, but the carrier may decide to remain childless and forego passing on the allele.

Using RFLPs, genetic researchers have pinpointed markers for several diseases besides Huntington's, including PKU; cystic fibrosis, a disorder that involves mucus buildup in the airways and is the most common fatal inherited disease among Caucasians; and Duchene muscular dystrophy, a disorder of progressive muscular weakness.

One goal of genetic research is to correct or replace defective genes like the Huntington's allele in the germ cells—an extension of the work that produced giant mice (see Chapter 14). Treating genetic diseases is one of the next frontiers of recombinant research.

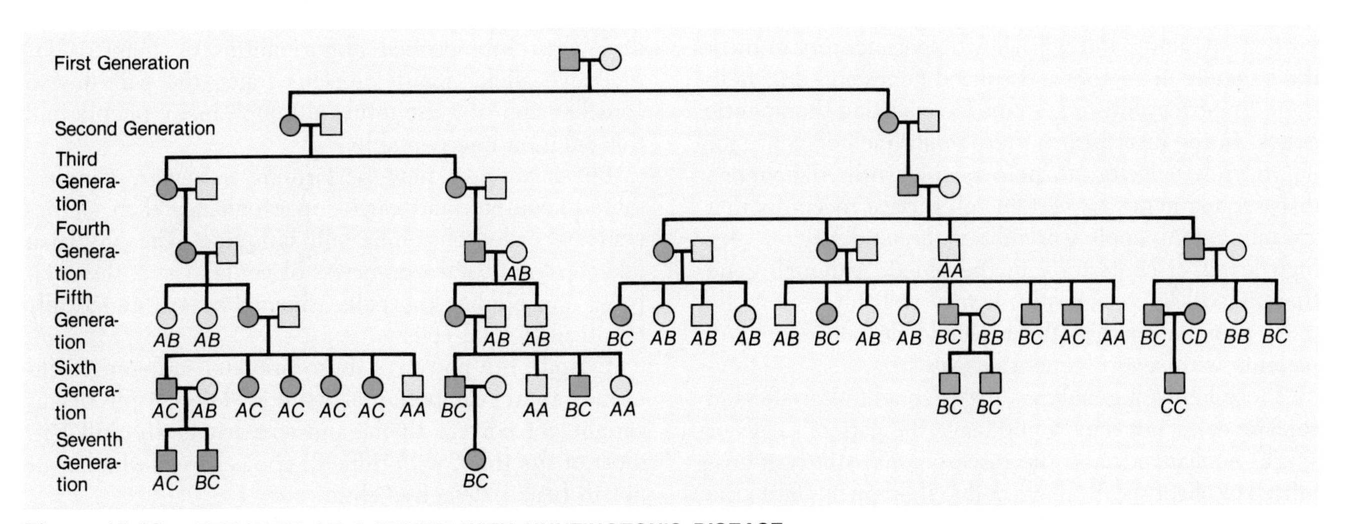

Figure 15-18 PEDIGREE OF A FAMILY WITH HUNTINGTON'S DISEASE.
The complete pedigree begins in the early 1800s and involves 3,000 individuals. Circles indicate females, and squares, males; open symbols represent normal individuals, and filled symbols, victims. Genotypes appear below surviving family members. Only people with the dominant *C* allele have the disease. Note that all victims have a parent who was also affected.

TREATING GENETIC DISEASES

Two of the great dreams fostered by advances in genetics are to treat people born with genetic defects and to correct such defects *before* birth. So far, physicians have succeeded in altering many phenotypes—for example, correcting visual defects with eyeglasses, correcting PKU or other metabolic defects through diet, and correcting physical defects such as extra fingers or toes through surgery.

Altering the genotype is a bigger challenge, but researchers are making rapid progress toward *gene therapy*—the insertion of normal genes into mutant cells to block genetic diseases. For example, physicians have plans for helping children born with a defect of the immune system called severe combined immunodeficiency. These patients lack the gene that codes for the enzyme adenine deaminase and cannot mount immune responses against any invading microorganisms. Physicians would like to help them by removing their bone marrow cells and transforming them—treating them so they take up the normal gene for adenine deaminase into their chromosomes. They will then reinject the transformed cells into the patient's bone marrow, where they will grow, express the gene, produce the normal enzyme, and give the child some immunity against diseases.

Human trials are planned, but their approval awaits the results of the very first test of gene therapy, set to begin before 1990. In that test, researchers will insert a "marker" gene for antibiotic resistance into the white blood cells of terminal cancer patients, using a disabled retrovirus as a vector. The genes will neither help nor hurt the patients, but will allow experimenters to track the behavior of vectors and spliced genes and thus help them to perfect future systems for inserting therapeutic genes. In the near future, these might include genes for nerve growth factor, to help people with Alzheimer's disease, and genes for certain cell surface receptors that are missing in people who inherit the tendency for very high levels of cholesterol in the blood. Although gene therapy will be very costly, it may well turn out to be more economical than the repeated hospitalizations of patients with severe genetic diseases.

These kinds of genotype "cures" might alter affected somatic cells, but they would leave the patient's eggs or sperm unchanged, and the disease could still be transmitted to offspring. Such germ line therapy lies far in the future, and even after the techniques are available decades hence, society will still have to consider the ethical aspects of such "human engineering."

GENETIC COUNSELING

One of the primary applications of human genetics is **genetic counseling,** providing advice to couples who are at risk for producing a genetically defective child or who have had such a child and want to avoid this heartbreaking situation in future births.

Consider the case of a couple whose child was born with a genetic defect. Before counseling begins, geneticists must identify the specific defect or medical syndrome through karyotyping, blood typing, and various immunological and biochemical tests. Next, they must construct a family pedigree. This pedigree may confirm the medical diagnosis and suggest a mode of inheritance as well as identify other family members who may be carriers of the mutant allele.

Next, counselors can assess a couple's statistical risk of producing another child with the same defect. If they suspect a single recessive or dominant gene, they can calculate the risk using Mendel's laws. If the defect is polygenic or the result of a chromosomal abnormality, they can estimate the degree of risk by using accumulated medical data. For most genetic defects, the only options open to couples are to avoid future pregnancies or to accept the risk and take a chance.

Physicians can detect many genetic defects prenatally with **amniocentesis,** during which they remove about 20 ml of the amniotic fluid surrounding the fetus through a needle and syringe. By centrifuging the fluid, they can recover some fetal epithelial cells and even a few fetal blood cells, then carry out tests for genetic defects on those cells. Table 15-3 lists a few of the 40 or so genetic defects that affect metabolic pathways and that can be detected using amniocentesis. If the geneticist discovers chromosomal abnormalities or homozygous recessive alleles for deleterious traits, the parents can consider aborting the fetus, although many people find this solution unacceptable.

Physicians are now performing a newer, simpler, safer technique (and earlier in pregnancy than amniocentesis) called **chorionic villi sampling.** The physician takes a sample of the embryo's placental tissue, then cultures and studies the cells, since they are genetically identical to fetal cells.

Despite the power of these prenatal screening techniques to reveal potential genetic defects, physicians cannot yet correct them, and so parents are still left, most of the time, with difficult choices over whether or not to bear a defective child.

The next section of this book explores the translation of genetic information, via the mechanisms of development, into an organism such as a human baby.

Table 15-3 SOME INHERITED BIOCHEMICAL DISORDERS THAT CAN BE DIAGNOSED PRENATALLY IN CULTURED FETAL CELLS

Error in the Metabolism of	*Disorder*	*Metabolic Defect*	*Brief Description of Phenotype*
Amino acids	Maple sugar urine disease (AR)*	Deficiency of enzymes needed in the breakdown of some amino acids	Poor development, convulsions, and early death; urine has maple sugar odor; diet therapy seems promising
Sugars	Galactosemia (AR)	Deficiency of an enzyme needed in the metabolism of galactose (derived primarily from milk)	Liver and eye defects, mental retardation, and early death if untreated; restrictive diet can control adverse symptoms
Lipids	Tay-Sachs disease (AR)	Deficiency of an enzyme needed in the breakdown of a complex lipid, allowing it to accumulate in nervous tissue	Progressive physical and mental degeneration, paralysis, blindness and death in infancy
Purines	Lesch-Nyhan syndrome (XR)	Deficiency of an enzyme involved in the metabolism of purines of purines	Mental retardation, muscular spasms, and compulsive self-mutilation; possible survival into adulthood
Complex polysaccharides	Hurler syndrome (AR) Hunter syndrome (XR)	Defects of connective tissue	Dwarfism, grotesque facial features, and mental retardation; Hunter syndrome is less severe

*AR = autosomal recessive, XR = X-linked recessive

Source: A. P. Mange and E. J. Mange, *Genetics: Human Aspects* (Philadelphia: Saunders, 1980, p. 581).

SUMMARY

1. Among the methods used to study human genetics are *pedigree* analysis, *karyotyping* of chromosomes, *biochemical analysis, somatic cell genetics, in situ hybridization,* and *twin studies.*

2. Most variable human characteristics reflect action of both genotype and environment during development, childhood, adolescence, and adult life.

3. In *reverse genetics,* experimenters go from the gene to the gene product that causes a human genetic disease. Such techniques also allow DNA fingerprinting by analyzing *restriction fragment polymorphisms (RFLPs).*

4. Some genetic disorders are attributable to chromosomal defects. *Aneuploidy* of autosomes involves losses or gains in chromosome number. Many forms of aneuploidy, including *Down syndrome* (trisomy 21), result from *nondisjunction* of one or more chromosomes during meiosis of the egg cells.

5. Aneuploidy of sex chromosomes can create individuals with *Turner's syndrome (XO), Klinefelter's syndrome (XXY),* or the *XYY* condition.

6. In mammalian cells, only one *X* chromosome is active; in normal (*XX*) females, the inactivated *X* can be seen in the nuclei of interphase cells as a mass of chromatin called a *Barr body.* Every female mammal is a *mosaic.*

7. Some genetic disorders arise from translocations, deletions, or duplications within chromosomes.

8. Among sex-linked traits in humans are color blindness, *hemophilia,* and anhidrotic ectodermal dysplasia.

9. Disorders of recessive alleles of autosomal genes that are associated with specific enzyme deficiencies include *phenylketonuria (PKU)* and *albinism.*

10. Dominant alleles determine such traits as pattern baldness, polydactyly, and Huntington's disease.

11. Efforts to treat genetic diseases focus on altering phenotype, but in the near future, will involve altering genotype.

12. In *genetic counseling,* couples are advised on the probability of producing a genetically defective child.

KEY TERMS

albinism
amniocentesis
aneuploidy
Barr body
chorionic villi sampling
Down syndrome
genetic counseling
hemophilia
incomplete penetrance
karotyping
Klinefelter's syndrome
mosaicism
nondisjunction
pedigree
phenylketonuria (PKU)
restriction fragment polymorphism (RFLP)
reverse genetics
sex-influenced inheritance
somatic cell genetics
Tay-Sachs disease
thalassemia
Turner's syndrome
variable expressivity

QUESTIONS AND PROBLEMS

1. A couple's son was born with hemophilia. There was no history of hemophilia in the husband's or wife's family. Is it likely that the son inherited the defective allele from his father? Explain.

2. Cystic fibrosis occurs once in every 2,000 live births. Victims are homozygous for this recessive allele. What is the probability that the child of a normal couple will suffer from cystic fibrosis? If a couple has an affected child, what is the probability that their next child will be affected?

3. A normal couple has a son with Duchenne muscular dystrophy, a sex-linked fatal disease affecting about one out of every 3,000 male infants. What is the probability that their next son will be affected? Their next daughter? What is the probability that their daughter will be a carrier? Which individuals have a one in four or greater chance of being carriers: The mother? The father? The father's mother? The mother's father?

4. How many chromosomes per cell does a Down syndrome (trisomy 21) victim have? A victim of Turner's syndrome (monosomy *XO*)?

5. Consider a woman who is heterozygous for the *X*-linked condition called anhidrotic ectodermal dysplasia; some patches of her skin lack sweat glands, and other patches are normal. Suppose she is also heterozygous for variants of G6PD (glucose-6-phosphate dehydrogenase), an enzyme encoded by a gene on the *X* chromosome. Many alleles of the gene produce variants of the enzyme. You examine the enzyme produced in different skin patches. One normal patch contains variant *A* and one abnormal patch contains variant *B*. Would you expect all the normal patches to contain variant *A*? Or would some contain variant *B*? Explain.

6. Assume that height is determined by three unlinked genes (*A*, *B*, and *C*) and that a person of average height (male: 5′9″; female: 5′4″) has three "tall" and three "short" alleles. Each tall allele adds 3 in. to the person's height. A person of average height could have the genotype *AaBbCc* or *AAbbCc* or *aaBBCc*. What would be the height of a woman with genotype *aabbcc*? The height of a man with genotype *AABBCC*? The height of a woman whose genotype is *aaBBcc*? What alleles would be present in the gametes she produces? What is the height of a woman whose genotype is *AaBbcc*? What alleles would be present in the gametes she produces?

ESSAY QUESTIONS

1. If people married at random with respect to height, skin color, hair color, blood type, nose length, and other traits, would differences disappear after a few generations? Explain.

2. Discuss the change in our ideas about the gene. Include the gene as a unit of heredity, of expression, of recombination, and of mutation.

3. Defend or attack the contention that the benefits of recombinant DNA research outweigh its potential disadvantages.

For additional readings related to topics in this chapter, see Appendix C.

16
Animal Development

There is no part of the future foetus actually in the egg, but yet all the parts of it are in it potentially. . . . The perfect animals are made by Epigenesis, or superaddition of parts.

William Harvey, *De Generatione Animalium* (1651)

At conception, you were a tiny sphere, one-tenth of a millimeter in diameter. Newly created by the union of sperm and egg, you were swept down a long, dark tunnel into the warmth and safety of the uterus. There, in the following 9 months, you enlarged thousands of times, to a length of 45–50 cm. You developed eyes, arms, legs, internal organs, and vocal chords capable of piercing cries. How did something so complex and wonderful happen so easily and naturally from a single cell? All animals begin as a single cell and go through an arduous period of development that includes cell divisions, cell specializations, and a shaping and unfolding of the new organism and its internal parts.

This chapter examines the basic processes of development seen in many animal species, including:

- How one cell becomes many
- How the cells are sculpted into organs

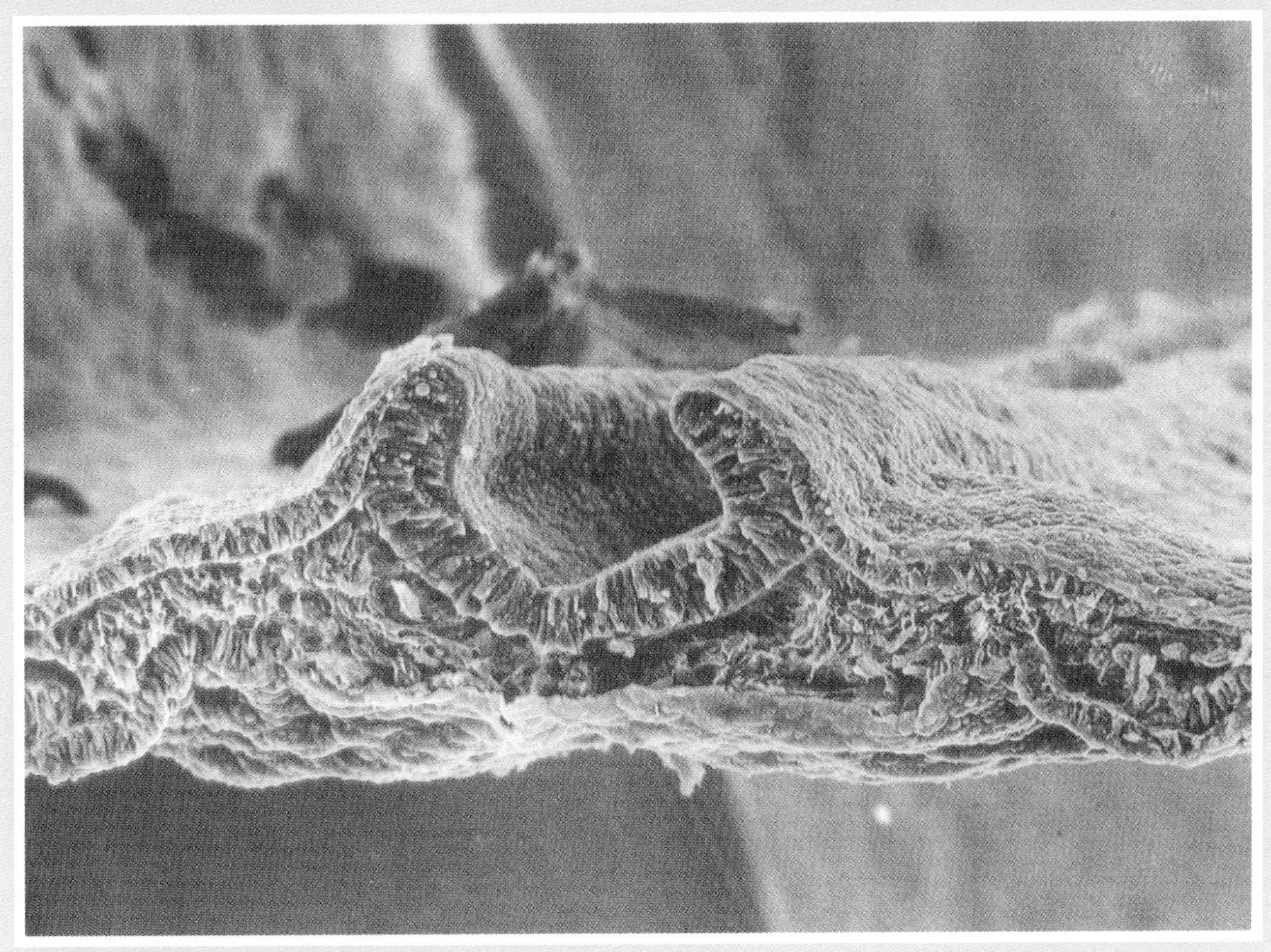

Development in process: A flat sheet of cells folds upward to form the hollow, tubular brain and spinal cord of a chicken embryo (magnified about 425×).

- How the new cells take on specialized functions
- How the new organism grows, matures, and ages

Chapter 17 considers the basic mechanisms underlying those processes, and Chapter 18 the way developmental principles apply to the human embryo, fetus, and child. Later chapters (particularly Chapter 24, on invertebrates and vertebrates), cover additional aspects of animal development, and Chapter 26 explores the development of plants in detail.

PRODUCTION OF SPERM AND EGGS

The coming together of two unique cells, the sperm and the egg, triggers the development of a new animal. How are such cells produced, and how can their union give rise to a new individual?

Sexually reproducing organisms manufacture sex cells called *gametes:* **sperm** in males and **ova** (eggs) in females. The process of sperm production is called **spermatogenesis;** the process of egg maturation is called **oogenesis.**

Spermatogenesis

Sperm are generated in male organs called **testes** (singular, *testis*). As Figure 16-1 shows, testes are composed primarily of hollow *seminiferous tubules.* These are twisted about like spaghetti in a dish, and sperm develop in the walls. Sperm originate from **gonial cells** *(spermatogonia),* which undergo repetitive cell divisions to yield huge numbers of cells that develop into the male gametes. More specifically, spermatogonia undergo mitotic divisions to produce *spermatocytes,* which then divide meiotically, reducing the adult male diploid ($2n$) genome to the haploid ($1n$) chromosome number (Figure 16-2). From the duplicated diploid state, two successive meiotic divisions produce four haploid *spermatids.* The heads of the spermatids are embedded in *Sertoli cells,* helper cells that may contribute vital materials to the maturing spermatids.

Each individual spermatid then matures, taking on a shape and function that will allow it to fertilize an egg. A tail begins to grow and protrudes toward the central cavity of the seminiferous tubule. The nucleus becomes more compact, and the Golgi complex gives rise to a storage granule called the *acrosome* (see Figure 16-2).

To successfully fertilize an egg, a sperm cell must be able to swim to it, penetrate its surface, and insert the haploid male genome into the egg cell. The motive force for swimming comes from the tail, or *flagellum,* which forms on one end of the spermatid. The flagellum undulates with a whiplike motion to drive the sperm through seawater, fresh water, or the fluids of the female's reproductive tract. The energy that powers the flagellum comes from the many mitochondria that are wrapped tightly around its base, as shown in Figure 16-2.

Sperm carry a compact payload in a small, streamlined head, containing only the acrosome and the nucleus. The acrosome carries enzymes needed to bore through the egg's protective layers and contains proteins used to produce a needlelike structure, also for penetrating to the egg. The nucleus becomes compact after meiosis is complete; all gene activity is repressed, and the

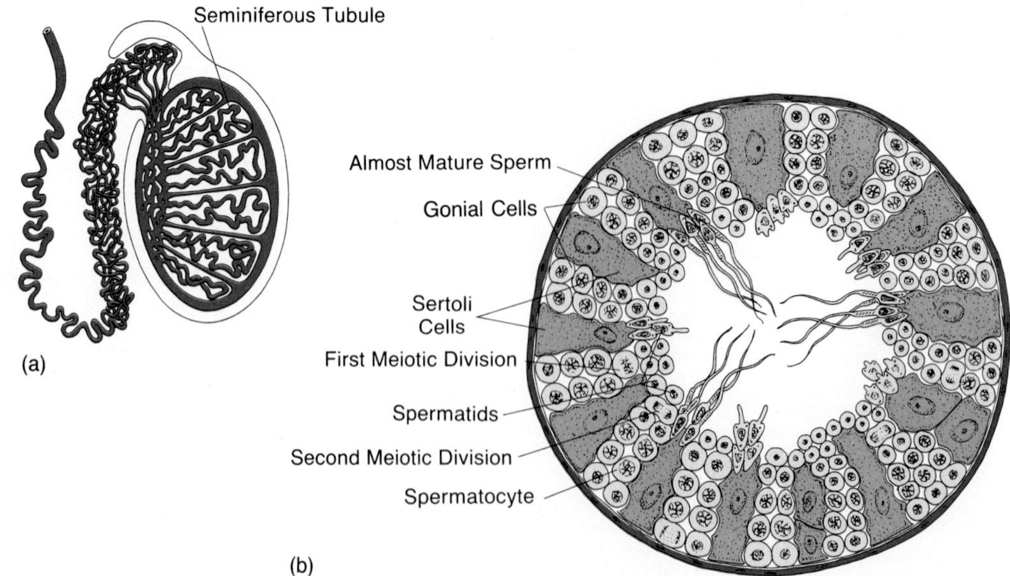

Figure 16-1 THE TESTIS: A LIVING FACTORY.
Year after year, the testis churns out hundreds of millions of sperm cells. (a) The coiled seminiferous tubules are visible in this cutaway diagram. Spermatogonia lining each tubule give rise to spermatocytes, which eventually mature into spermatids and then sperm with tails. (b) A simplified diagram of the seminiferous tubule wall, showing the various cell types and stages of sperm production.

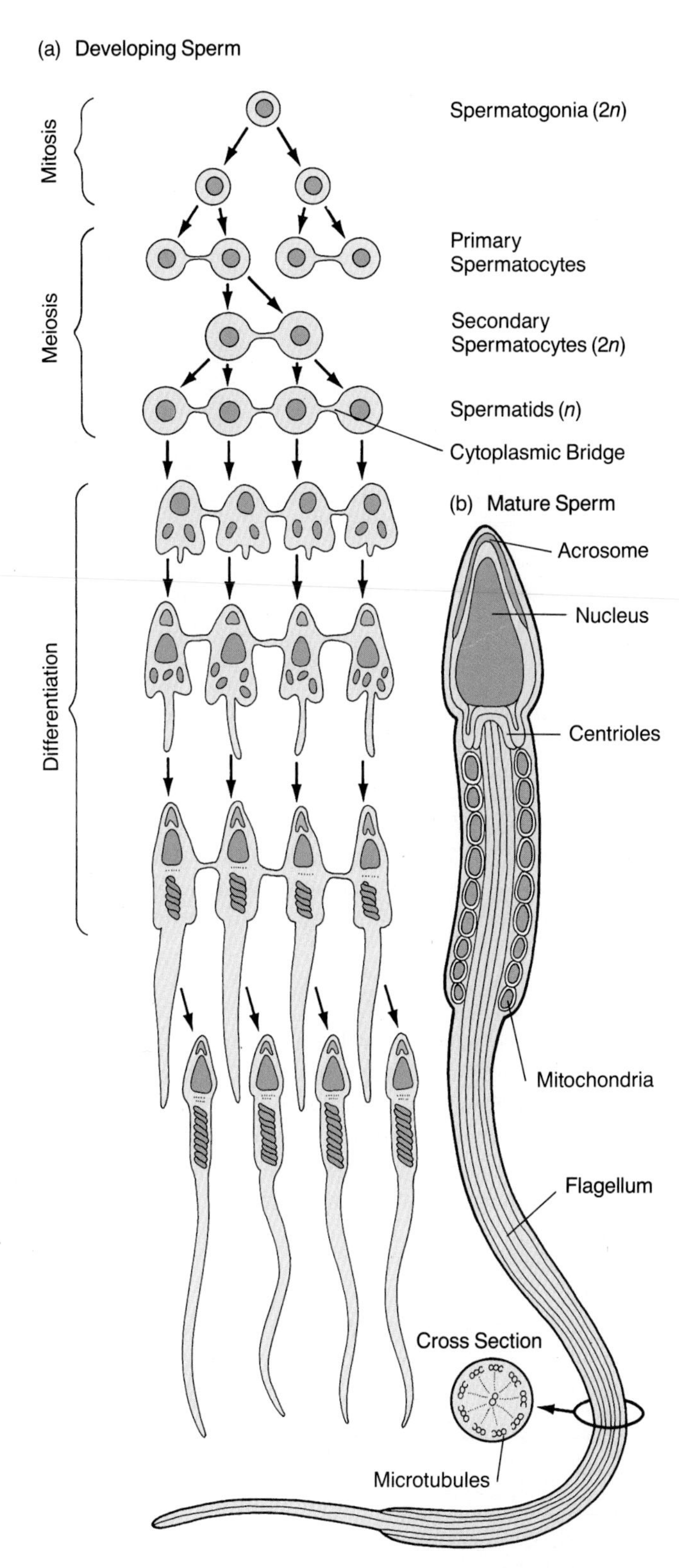

◀ **Figure 16-2 SPERMATOGENESIS.**
(a) During meiosis, the diploid ($2n$) spermatogonia produce haploid ($1n$) spermatids, which develop into mature sperm. This drawing shows a cohort of four spermatids, interconnected by bridges of cytoplasm. Spermatogonia are located peripherally, at the outer wall of the seminiferous tubule, and mature sperm are released into the tubule's central cavity (Figure 16-1). Huge Sertoli cells (not pictured here) can supply nutrients or other substances to aid the maturation process. (b) A mature sperm (enlarged view) is a streamlined, self-propelled torpedo, built to transport its genetic cargo a long distance, penetrate a tough barrier, and deliver its haploid genome safely inside the egg cytoplasm. These longitudinal and cross sections show several structures: the acrosome, storage site for the enzymes that will digest the egg's protective surface and allow sperm penetration; the nucleus, containing highly condensed chromosomes, which are compact and therefore require less energy for transport; the centrioles, involved in building the tail; and a spiral of mitochondria packed tightly around the microtubules, the backbone of the tail. Mitochondrial enzymes generate ATP, which fuels the sperm's whiplike tail movements.

haploid chromosomes are packed for transport in a crystal-like form.

Mature sperm cells are shed into the cavity of the seminiferous tubule and move down it passively; they begin to swim only after being expelled from the male's body.

Oogenesis

The female's **ovaries** generate ova, or eggs. In many female animals, the gonial cells (*oogonia*) complete their normal mitotic divisions early in life (in mammals, well before birth). The resulting cells, called *oocytes,* then enter a state of arrest in early meiosis, a pause that lasts for years in some species (Figure 16-3). In humans, meiotic arrest continues until puberty. The period of oocyte maturation and growth (see Figure 16-3) is followed by a final ripening, ovulation, and the first meiotic division. In women, about one egg per month matures and is released, and this pattern of ovulation continues throughout reproductive life. If fertilization occurs, the second meiotic division takes place, and development proceeds (Figure 16-4).

The egg provides not only genes, but also a place for them to reside and operate. In addition to containing the haploid set of female chromosomes, most animal eggs (1) contain nutrients to support early development, (2) contain regulatory molecules that can turn genes on and off or affect other aspects of embryonic development, (3) have protective layers, and (4) have mechanisms that enable them to respond appropriately to contact with sperm.

Egg Structure

Most animals have sperm cells about 50 μm long, but eggs vary tremendously in diameter and mass, depend-

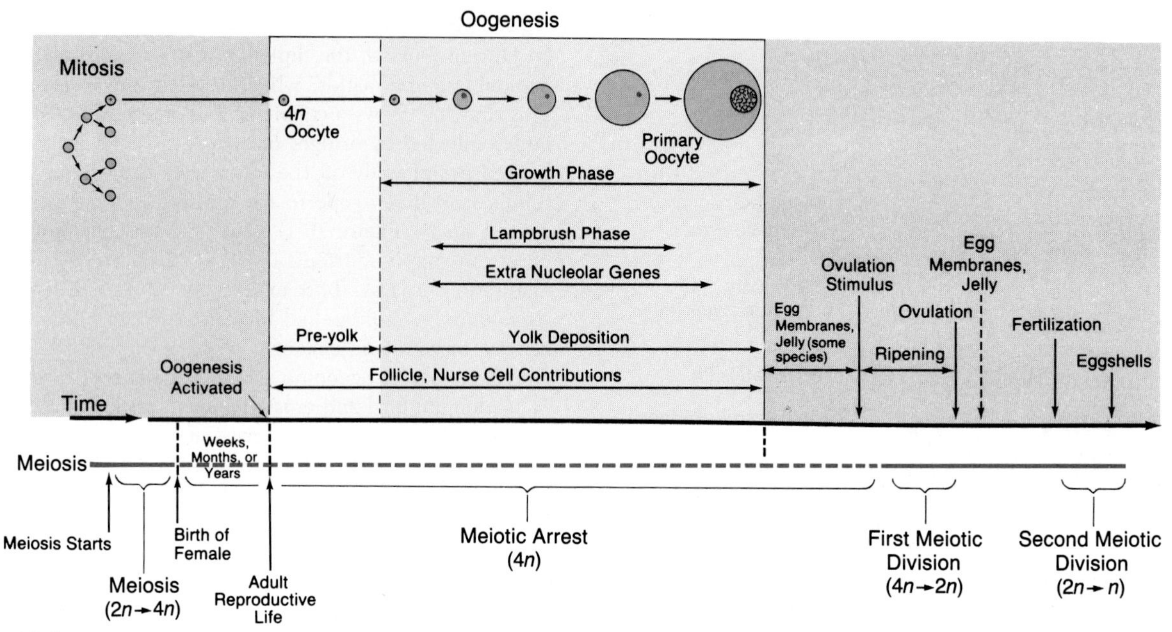

Figure 16-3 OVERVIEW OF OOGENESIS.
An overview of oogenesis in the frog *Xenopus*, including events in birds, mammals, and humans. Meiosis starts in the ovary of the female when she is an embryo. Then, in adult life, single eggs or groups of eggs in the ovary commence oogenesis. The total egg maturation period in *Xenopus* is 6 to 8 months; in other species, that time period might be several weeks, a month, or over a year long. Oogenesis, represented in the box, occurs during meiotic arrest. In some species, egg coatings and membranes are applied after oogenesis is complete and at various times prior to ovulation; in others, they are added in the oviduct after fertilization. Note that oogenesis starts only after the meiotic process has started and been arrested—hence chromosomes in the maturing diploid egg have doubled so the cell has four times the haploid amount of DNA, whereas the maturing spermatid (Figure 16-2) has completed meiosis and so has just one times the haploid (n) amount. The actual meiotic divisions in the egg occur only at ovulation and fertilization.

ing on species. A human egg is only about 0.1 mm across; a shark egg is 3 cm; and an ostrich egg is about 6 cm, making it the largest living single cell on earth. This divergence in size reflects wide differences in the quantities of stored nutrients.

Virtually all developing animal ova are surrounded by helper cells, either *nurse cells* or *follicle cells*. Nurse cells synthesize proteins and nucleic acids (mRNA and ribosomes composed of ribosomal RNAs and proteins) and transfer them to the cytoplasm of the developing oocyte, where they are stored for use after fertilization. Follicle cells also may transfer materials to oocytes and plaster various protective substances on their surfaces (Figure 16-5).

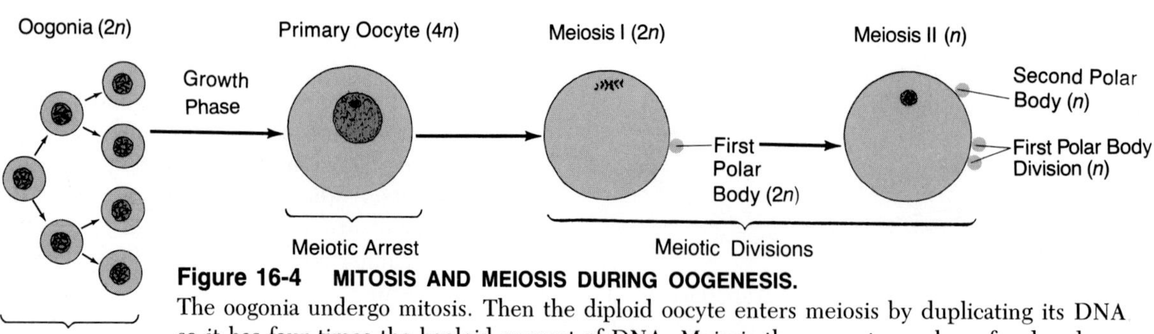

Figure 16-4 MITOSIS AND MEIOSIS DURING OOGENESIS.
The oogonia undergo mitosis. Then the diploid oocyte enters meiosis by duplicating its DNA so it has four times the haploid amount of DNA. Meiosis then arrests, perhaps for decades. Individual eggs then grow, as in Figure 16-3. Only in response to ovulation does the genome of the oocyte begin to be reduced to the haploid state. Fertilization is necessary in many species to trigger the second meiotic division. The first polar body usually divides in synchrony with the second meiotic division of the oocyte (or zygote in the case of fertilization). The three polar bodies degenerate.

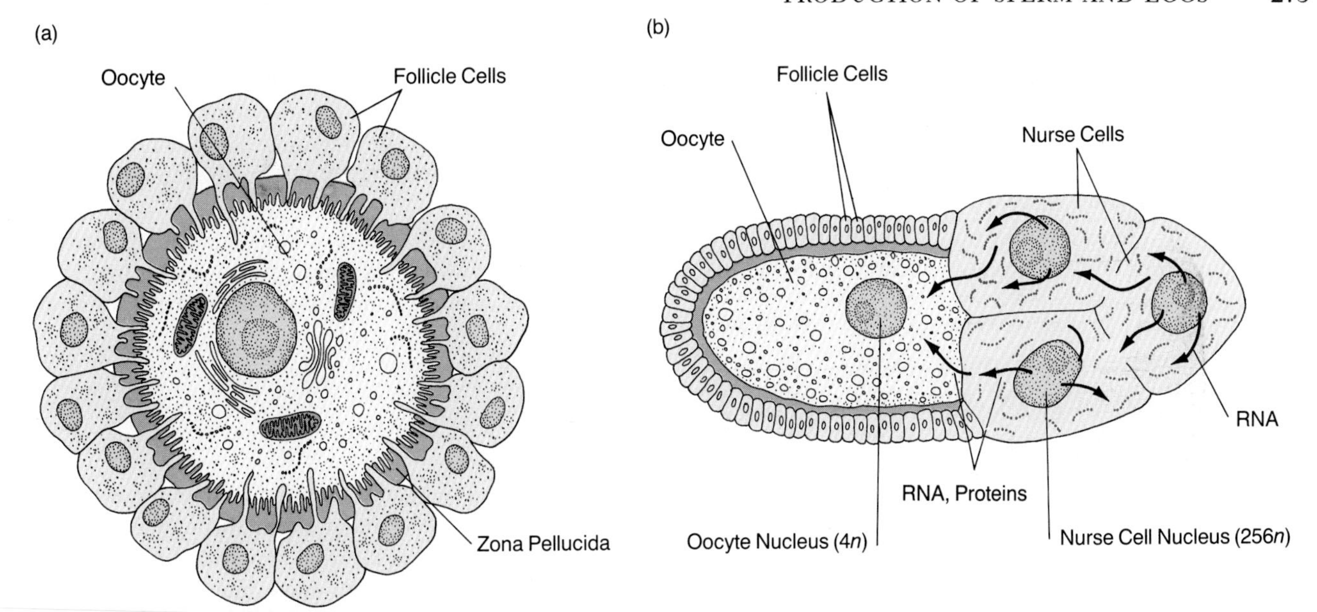

Figure 16-5 HELPER CELLS AND THE TRANSFER OF SUBSTANCES INTO THE OVUM. (a) An immature mammalian oocyte surrounded by follicle cells. The long, fingerlike projections from the follicle cells contact the surface of the egg and are sites for transferring substances to the egg cytoplasm. (b) An insect oocyte surrounded by small follicle cells. The egg's cytoplasm is directly open to the cytoplasm of the giant nurse cells at the top. RNA manufactured in the nurse cell nuclei may pass directly to the egg cytoplasm for storage. Or some of the RNA may be used in the nurse cell cytoplasm to synthesize proteins, which are then passed to the oocyte. The follicle cells also pass substances to the oocyte or add protective layers to its surface.

Other cells also help the egg produce the nutrients that it stores as the **yolk.** Yolk material, including proteins, carbohydrates, and lipids, usually is manufactured in digestive gland cells within the mother's body. In hens, for example, yolk protein is produced in liver cells and then is released into the bloodstream. In the ovaries, the yolk materials are preferentially absorbed from circulating blood into the developing oocyte, where they accumulate as the egg matures. Most of the yellow yolk in a hen's egg is made and deposited in this manner and nourishes the embryonic chick as it develops in isolation within the eggshell.

Finally, other cells produce a set of protective coatings around the egg. These coatings are built up by follicle cells or accumulate as the fertilized egg is transported through the female's reproductive tract. In hens, for example, the "white" and the shell of the egg are secreted by cells lining the **oviduct,** the tube through which the egg travels after leaving the ovary. The egg white, or *albumen*, is a special solution of salts and proteins that, among other things, protects the embryo from bacterial infection.

Manufacture of Molecules

As oocytes mature, some manufacture many of their own supplies, including a large quantity of RNA. During maturation, the meiotically arrested, duplicated chromosomes usually go through a period of great activity called the *lampbrush phase* (see Figure 16-3). The chromosomal backbone unravels at hundreds of sites, so that regions composed of certain genes loop outward from the main chromosomal axis. Large amounts of mRNA are made on each loop, processed (see Chapter 14), and passed to the egg cytoplasm. Here, most of the mRNA is stored in inactive form; only after fertilization is it employed in building proteins for the embryo.

Within the nucleus of a maturing egg, huge numbers of ribosomes are manufactured (10^{12} ribosomes in a single egg of the African clawed frog, *Xenopus*, for example) and are then passed to the cytoplasm for storage. Donald Brown of the Carnegie Institution of Washington calculated that even if all the chromosomal genes for the major classes of ribosomal RNA (rRNA) were functioning continuously at top speed, it would take 465 *years* for a frog oocyte to manufacture 10^{12} ribosomes. How, then, can this manufacture occur in just 5 or 6 months, the length of oogenesis in *Xenopus?* The answer lies in the special process of **gene amplification.** During this process, about 1 million extra copies of the ribosomal genes are made. These extra copies act as templates during oogenesis, so that the huge number of rRNA molecules and the resultant ribosomes may be generated in a reasonable length of time.

These ribosomes and mRNA from the oocyte nucleus—along with the RNA, yolk, and other substances manufactured by nurse cells or digestive gland cells—are *ma-*

ternal in origin. This means that genes and chromosomes in the mother's somatic (body) cells, as well as in the oocyte, code for the components in and around the egg. Thus, maternal genes and the spatial distribution of their products govern much of the course of embryonic development. After fertilization, the embryo's own genes, including paternal ones, begin to play a role.

FERTILIZATION: INITIATING DEVELOPMENT

The process that initiates development and unites the nuclei of male and female gametes is **fertilization.** Fertilization may be either external—in sea or fresh water—or internal—in the female's reproductive tract. But how do the gametes meet? How does a sperm "find" an egg?

Mating behavior is the main strategy that helps ensure the union of egg and sperm. For external fertilization, seasonal synchronization of *spawning* (release of gametes) is essential so that the relatively short-lived egg and sperm are released into water at the same time. Salmon, for example, scoop out a depression in a pebbly streambed; as the female lays thousands of eggs, the male pours forth sperm over them, and the chances of fertilization are increased.

Some aquatic organisms and most terrestrial animals carry out internal fertilization. This often involves the transfer of a *spermatophore,* a packet of sperm that can be stored in the female's body. Male squid, for example, insert the spermatophore with a tentacle, while a male salamander will leave the spermatophore on the ground and the female inserts it herself. Male sharks, insects, and mammals have a copulatory organ, the *penis,* which discharges sperm solution directly into the female's body without exposure to the environment. Mating rituals, coordinated by both the nervous and the endocrine systems, accompany the internal fertilization process in many organisms.

Once a sperm and egg do come together, the fertilization process involves several steps, including sperm penetration, completion of meiosis, and the raising of barriers to additional sperm.

Sperm Penetration

Eggs are universally immobile. Consequently, to meet the egg, sperm must swim through either body fluids if fertilization is internal or water if fertilization is external. Although some eggs may release specific chemicals to attract sperm, most do not. Thus, chance plays a large role in sperm meeting egg and helps explain why millions of sperm are released; the probability is low that any one sperm will encounter an egg.

Even after a sperm does contact an egg's surface, it must overcome several barriers before it can deliver its genetic cargo. First, it must penetrate the protective *jelly coat* that surrounds the egg; then it must pass through the *vitelline layer,* another protective coat; and finally, it must breach the plasma membrane itself.

The first contact of the sperm head with the egg's jelly coat triggers the remarkable and rapid **acrosome reaction.** This reaction consists of two parts: the release of the enzymes stored in the acrosome and the thrusting forth of a long, rigid filament that stabs through the jelly toward the plasma membrane. The enzymes digest a hole through the transparent jelly coat and the vitelline layer (Figure 16-6). Then the structural protein actin is rapidly rearranged into a rigid bundle of filaments, and

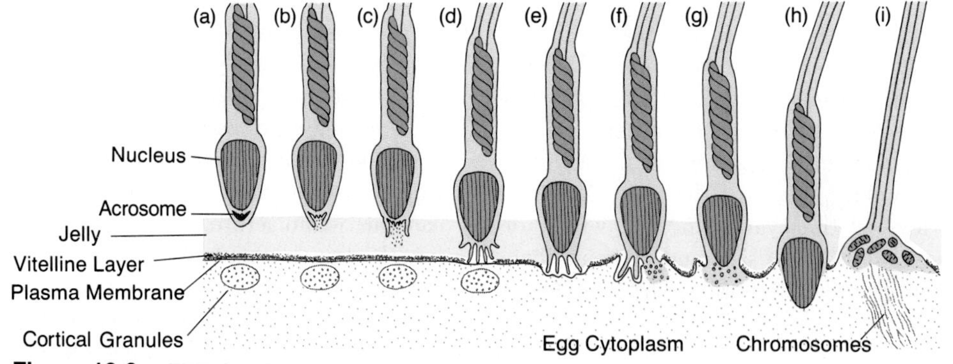

Figure 16-6 THE STAGES OF EGG FERTILIZATION BY A SINGLE SPERM. (a, b) The sperm approaches the egg, its front end touches the jelly coat, and the acrosome reaction is triggered. (c, d) The acrosome opens, and digestive enzymes pour forth to begin eating through the jelly coat and vitelline layer. (d, e) Fingerlike projections of the sperm surface protrude toward the egg. (f–h) The two cell surfaces fuse, and egg cytoplasm flows outward, engulfing the sperm head. (i) The sperm nucleus shows a rapid decondensation, or uncoiling, of its chromosomes.

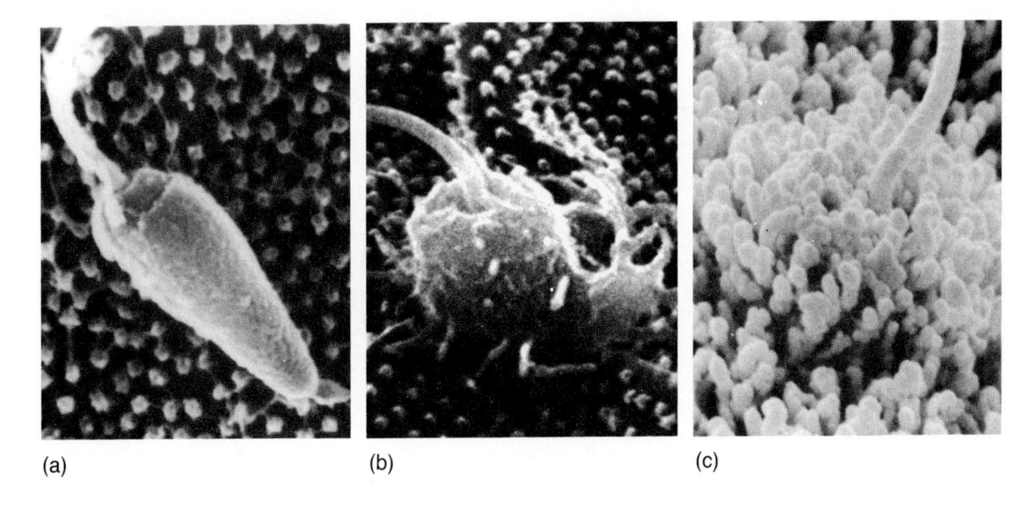

Figure 16-7 SPERM PENETRATING THE EGG SURFACE.
(a) Only a second or so after contact, this sea urchin sperm adheres to tiny projections on the vitelline layer. (b) The sperm head has penetrated halfway through the vitelline layer. (c) The vitelline layer has been removed, and the egg's plasma membrane is seen around the site where a sperm just penetrated. (Magnifications 5,000× to 15,000×)

this causes the front end of the sperm to protrude through the covering layers. The plasma membrane at the tip of the sperm can now bind to the ovum's surface, facilitated by a sperm membrane protein called *bindin* and a receptor molecule on the egg's plasma membrane.

The actual fusion of the sperm and the egg plasma membranes immediately triggers a series of processes at the egg's surface. In one such process, the sperm head is engulfed, as shown in Figure 16-7. Details vary among species, but in each, the haploid male nucleus moves into the egg cytoplasm, and almost immediately, its chromosomes begin to lose their inactive, crystal-like configuration. Microtubules then move the sperm and egg nuclei toward each other through the egg cytoplasm.

Completing Meiosis

Before most fully grown animal oocytes can be fertilized, the completion of meiosis must be initiated and the oocyte surface must undergo final maturation. Depending on species, the trigger for these events can be a chemical signal, such as the hormone *progesterone* acting on the unovulated mature egg (see Figure 16-3, "Ovulation Stimulus"), or it can be ovulation itself, in response to hormonal cycles or copulation. Frequently, ovulation stimulates only the first meiotic division, whereas the fusion of sperm and egg stimulates the second. The final reduction division then occurs, and only a haploid set of egg chromosomes remains to join those in the sperm nucleus (the other haploid set of female chromosomes is discarded in the second polar body). When the sperm and egg nuclei come together, the two sets of chromosomes mingle, yielding a diploid set and the genetic coding for a new organism. At this point, the first round of DNA synthesis can begin as the fertilized egg, now called a **zygote,** prepares for mitotic cleavage divisions.

Barriers to Other Sperm

Many sperm may reach the egg simultaneously and adhere to its surface, but only one may fertilize it. How is further sperm penetration prevented? The answer lies in the egg's multistep **cortical reaction** to contact with the sperm.

Fusion with a sperm's plasma membrane causes rapid changes in the electrical properties of the egg's plasma membrane. As a result, the egg loses its normal internal negative charge relative to the surrounding fluids and becomes positive. This blocks the attachment and entry of additional sperm. If the egg's internal negative charge is experimentally maintained, sperm after sperm can swim up and fuse with the egg. Conversely, if an unfertilized egg is artificially given a positive charge, not even one sperm can fuse with and enter the egg. Loss of normal negative charge lasts only long enough for other changes in the egg surface to create a permanent barrier to sperm entry.

Activation is the second step in the cortical reaction of the egg cell. Calcium is freed in the cytoplasm, possibly because the sperm causes a "second messenger" (inositol triphosphate; see Chapter 35) to be released. Sodium and hydrogen ions begin to move across the plasma membrane, and the intracellular pH rises. As the pH rises, the activity of enzymes and proteins is altered, the dormant egg cell begins to consume more oxygen, protein synthesis is initiated, and a variety of other metabolic events start up.

The final event of the cortical reaction is spectacular to see: the rapid elevation of the *fertilization membrane* from the surface of the egg cell. As Figure 16-8 shows, starting at the point of sperm-egg fusion, a "wave" involving the sequential release of calcium ions passes over the surface of the spherical egg. At the advancing front of the wave, cortical granules carry out exocytosis and fuse with the egg's plasma membrane, thereby discharging

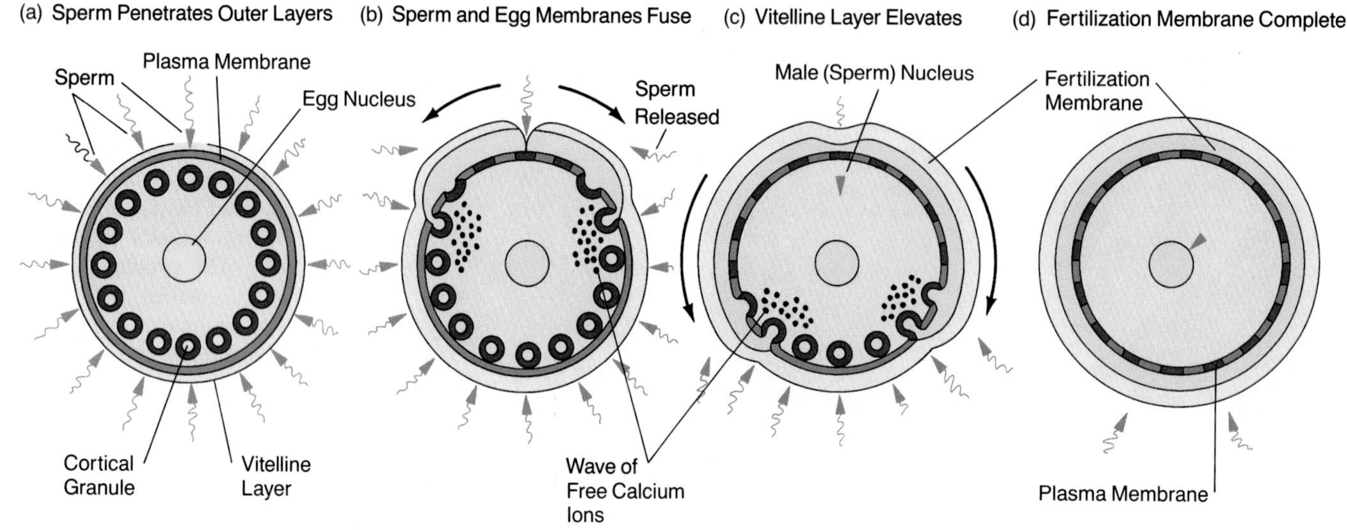

Figure 16-8 ELEVATION OF THE FERTILIZATION MEMBRANE.
(a) Numerous sperm surround the jelly-encased egg, but only one has succeeded in reaching the egg's plasma membrane. (b) Starting at the point where the sperm and egg plasma membranes fuse, cortical granules fuse with the egg's plasma membrane and discharge their contents. The activities of a discharged enzyme cause the vitelline layer to separate from the plasma membrane and destroy or modify sperm receptor-binding sites. (c) As the vitelline layer elevates from the egg surface and loses its sperm-binding sites, other sperm fall away. (d) An egg cell surrounded by a fertilization membrane to which no additional sperm can adhere. The final egg plasma membrane includes patches of membrane that originally encompassed each cortical granule.

their contents. Among these contents is an enzyme that separates the vitelline layer from the plasma membrane, permitting the vitelline layer to rise and form the tough fertilization membrane. This often takes place in just 10 to 20 seconds and leaves the egg floating free within the protective envelope of the fertilization membrane. It is no coincidence that this membrane is elevated by the time the internal electrical charge of the zygote has returned to its normal negative value; from then on, the fertilization membrane prevents the penetration of additional sperm.

Strangely enough, sperm and normal fertilization are not always required for reproduction. Insects, lizards, certain relatives of lobsters, and even turkeys can carry out **parthenogenesis:** a spontaneous activation of the mature egg followed by normal egg divisions and embryonic development—literally translated, a "virgin birth."

Parthenogenesis is relatively rare, and fertilization is a momentous event in the life of most individuals. Fertilization is not, however, the beginning or creation of life. Both sperm and egg cells are alive prior to fertilization and have undergone long and complex maturation processes; fertilization is the creation only of a new and unique diploid genome and cell. What's more, that cell might yield twins or quadruplets or might fuse with another embryo by chance and yield only some of the cells in the new individual. Clearly, fertilization neither creates life nor even "an individual" in every case.

CLEAVAGE: AN INCREASE IN CELL NUMBER

The major developmental event immediately following fertilization is **cleavage,** a special kind of mitosis, or cell division. Cleavage generates many small cells from the large single-celled zygote, and it distributes, in precise ways, yolk, mRNA, ribosomes, and other materials that were built into the egg during oogenesis. In most species, cleavage also produces a **blastula,** a sphere of individual cells, or **blastomeres,** with a central cavity.

Periodic increases in free cytoplasmic Ca^{2+} may trigger cleavage cell division cycles and those increases may occur when the second messenger inositol triphosphate releases the ions from intracellular storage sites.

Patterns of Cleavage

Cleavage patterns depend on several factors. One is the volume of yolk material incorporated into the egg. Figure 16-9 compares the cleavage patterns for a mammal's egg with little yolk, an amphibian's egg with somewhat more yolk, and a bird's egg with its massive store of yolk. The mammalian zygote cleaves completely through, forming an initial ball of cells called a *morula.* The morula becomes a *blastula* when a cavity, the *blas-*

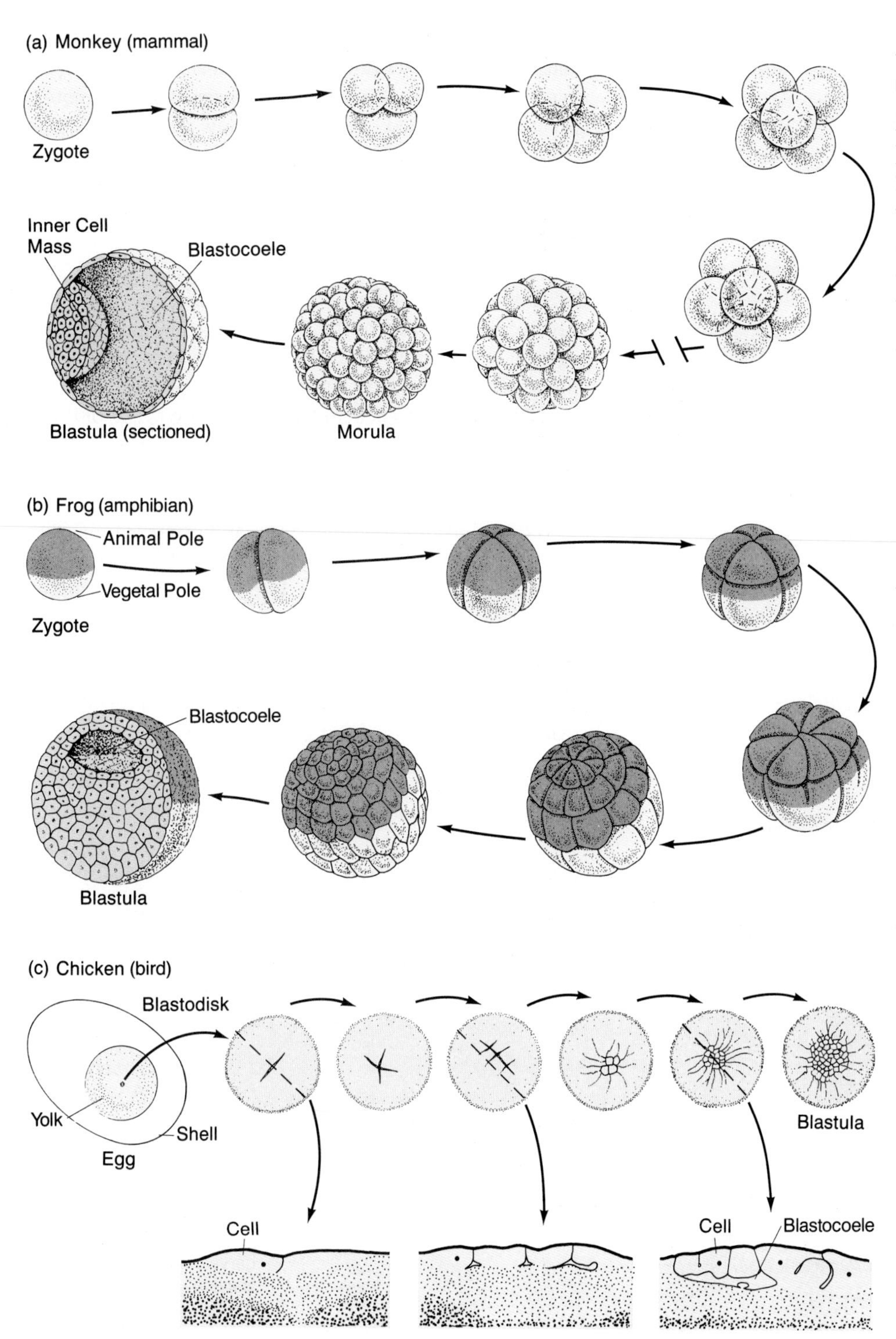

Figure 16-9 INFLUENCE OF YOLK ON CLEAVAGE PATTERNS IN EGGS.
(a) In a monkey embryo, which has very little yolk, the cleavage process occurs relatively independently in different blastomeres, producing arithmetic divisions (1, 2, 3, 4, 5, etc.) or geometric ones (1, 2, 4, 8, 16, etc.). The morula is a solid ball of cells that cavitates to form a blastula. The blastula consists of an inner cell mass and the surface layer. (b) Cleavage in an amphibian embryo, which has an intermediate quantity of yolk. Blastomeres at the darkly pigmented end, the animal pole, divide more rapidly than do those at the vegetal pole, where more yolk is located. (c) Cleavage in the flat disk of cytoplasm on a bird zygote as seen from the top and in cross section. The yolk is so large that it is not cleaved through, and only the tiny upper island of cytoplasm is shown. Each cleavage plane forms as though the surface of the dividing cell were sliced downward with a dull knife. The broken lines in the blastodisk drawings show the plane of the cross section below.

tocoele, forms within the mass of cells, each roughly the same size (Figure 16-9a).

In the frog or salamander egg (Figure 16-9b), cleavage proceeds more rapidly at the less yolky end (the *animal pole*) than at the more yolky end (the *vegetal pole*). The cells thus become smaller at the animal pole and the small blastocoele is offset toward it.

In a bird's egg (see Figure 16-12c), the massive yolk cannot be cleaved through completely, and cleavage divisions are restricted to one tiny island of cytoplasm at the animal pole. The result is a small disk of blastomeres, the *blastodisk*. Much later in development, cells at the edge of the disk spread and engulf the huge sphere of the yolk in a specialized embryonic membrane, the *yolk sac*.

Cleavage patterns depend not just on the amount of yolk present but also on an organism's evolutionary history. As Chapter 25 explains, there are two main lineages of animals. In one lineage (vertebrates, sea stars, and related organisms), cleavage is *radial,* with new blastomeres accumulating in even rows above each other. In other lineages (worms, clams, snails, and related organisms), cleavage is *spiral:* successive layers of newly arising blastomeres are offset, not evenly stacked.

Cytoplasmic Distribution During Cleavage

Cleavage has several critical outcomes, including, in some species, the precise distribution of cytoplasmic components to different blastomeres. This distribution probably takes place because substances are arranged in the zygote in such a way that cleavage restricts them to certain blastomeres. In a frog egg, for example, black pigment is concentrated at the animal pole, whereas the vegetal pole is white. This visible segregation represents a subtler distribution of organelles and chemical agents (ribosomes, mRNA, enzymes)—a distribution that may affect how different cell types develop in the embryo.

Experiments with the sea squirt, a urochordate and a relative of the vertebrates (see Chapter 24), have helped reveal this precise distribution. In sea squirt zygotes, two crescent-shaped regions—a yellow region and a gray region—appear in the minutes following fertilization (Figure 16-10). During subsequent cleavage divisions, the cytoplasm with yellow granules is segregated into certain blastomeres, the gray cytoplasm into others. Hours later, cells that arise from the yellow group give rise to the skeletal muscle of the little tadpole, while the gray group gives rise to the notochord, the main skeletal organ of the tadpole's tail. Biologists rearranged the zygote's cytoplasm experimentally, displacing yellow-crescent material to cells normally lacking it. All of these cells developed as muscle cells because the yellow-crescent determinants activated genes that specify muscle fibers. In other experiments, the researchers caused the redistribution of yellow, gray, and other cytoplasmic substances by centrifuging zygotes. Muscles, notochord, and other organs developed, but in abnormal locations and chaotic orientations. These experiments showed that muscle will develop wherever yellow cytoplasm ends up, and the notochord will form wherever gray cytoplasm is found. Redistributing cytoplasmic determinants clearly alters the resultant cell types. These regions of

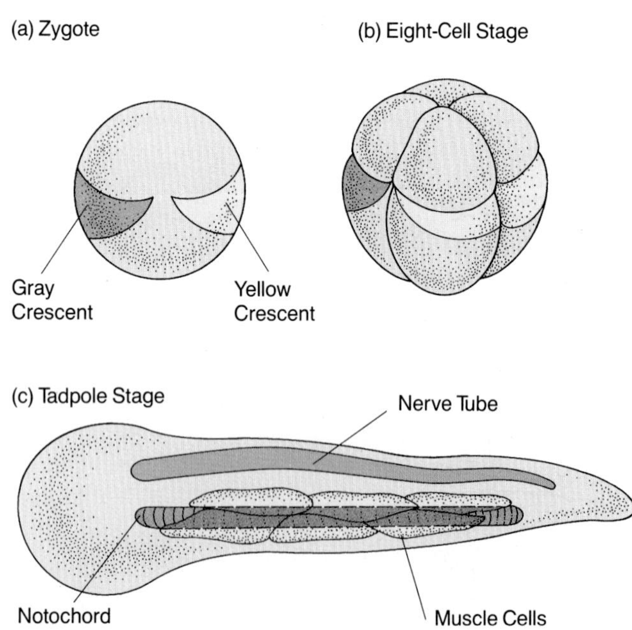

Figure 16-10 CYTOPLASMIC LOCALIZATIONS AND CELL FATES.
(a) The gray and yellow crescents in this sea squirt zygote are precisely segregated to certain blastomeres during cleavage. (b) At the eight-cell stage, two of the eight cells possess gray-crescent cytoplasm; two, yellow-crescent materials. (c) The tadpole contains a notochord derived from the gray-cell lineage and muscle cells from the yellow-cell lineage.

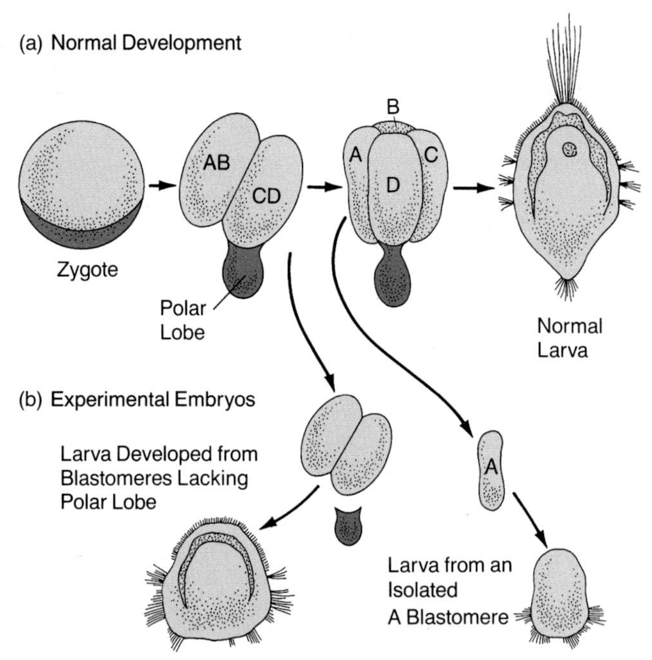

Figure 16-11 DEVELOPMENTAL INFORMATION AND THE POLAR LOBE SYSTEM OF A SNAIL EMBRYO.
(a) The lobe and its contents are restricted to one line of cells during cleavage; first CD, then D, and finally D's descendants. (b) A normal larva, a larva that developed after the polar lobe had been cut away early in cleavage, and a larva that developed from an isolated A blastomere. The maternally derived developmental "information" was not available to the A blastomere or to the embryo lacking the polar lobe; as a result, their development is deficient.

maternally originated cytoplasm must contain control agents for specific types of cell maturation. It is also clear that the normal distribution of the control substances determines whether the embryo's body develops correctly (i.e., with muscles and other organs formed at the correct sites). Biologists have concluded that the spatial arrangement of control substances is a form of developmental information built into a maturing oocyte.

More proof of this comes from the study of *polar lobes* (not to be confused with polar bodies) in some worm and snail embryos (and their relatives). As Figure 16-11 illustrates, the first cleavage divides the contents of the egg asymmetrically, so that the daughter cells have different developmental potentialities. The daughter cell (CD) receiving the polar lobe can go on to form a complete, normal embryo, whereas the daughter cell (AB) lacking the polar lobe forms an aberrant embryo.

In mammals and birds, developmental information is not partitioned into such obvious lobes or crescents; it actually arises as the cleavage process goes on. The position of a cell late in cleavage therefore determines what that cell will become: Those on the inside yield the embryo's body, while those on the outside form protective membranes and ultimately die.

GASTRULATION: REARRANGEMENT OF CELLS

Cleavage transforms one large cell into a sheet of many small cells, usually one cell layer thick, that surrounds either a cavity or a mass of yolk. A set of developmental processes called **gastrulation** then sculpts the blastula into a complex three-dimensional organism with inner, outer, and middle layers.

During gastrulation, some of the blastomeres located on the surface of the embryo move to the inside, producing the three-layered **gastrula.** As a result, cell populations originally separated from one another come into contact and interact, leading to normal development of most organs. The three *germ layers* that are produced (so called after *germinal,* for "origin") are the **ectoderm** (outer layer), **endoderm** (inner layer), and **mesoderm** (the layer between the ectoderm and the endoderm), each giving rise to specific tissues.

The quantity of yolk affects the movement of cell populations during gastrulation, just as it affects cleavage. In organisms with relatively little yolk, gastrulation movements are simple and involve a process of inpocketing called *invagination*, which resembles the dent you make by poking a finger into a soft rubber ball. This opening into the blastula is called the *blastopore*. Figure 16-12 compares gastrulation in *Amphioxus*, a simple aquatic relative of vertebrates whose eggs have little yolk; a frog egg, with intermediate quantities of yolk; and a bird's egg, with large quantities of yolk. In *Amphioxus*, gastrulation includes two processes: *invagination* of a sheet of cells to form the endodermal gut lining and the *migration* of individual cells to form mesoderm (Figure 16-12a).

In a frog's egg, a slit, the blastopore, marks the site near the embryo's equator where cells invaginate; the dorsal lip of the blastopore is composed of such moving cells (Figure 16-12b). Gastrulation in a bird's egg (Figure 16-12c) is a variation on the processes in fishes and amphibians. Cells located on the blastodisk surface move to the inside through a thickened portion of the embryo called the **primitive streak.** The first cells moving downward through the primitive streak become the lower endodermal layer; later ones become mesoderm. The primitive streak method of gastrulation originated in reptiles and is retained not just by birds but by mammals, even though they have virtually no yolk. This evidence helps support the hypothesis that mammals, like birds, descended from ancient reptiles (see Chapter 24).

ORGANOGENESIS: FORMATION OF FUNCTIONAL TISSUES AND ORGANS

Once gastrulation is complete, the next phase of development begins; it is **organogenesis,** the formation of the body's organs and tissues. During organogenesis, the cells brought inside the embryo or left on its surface during gastrulation become specialized to form the nervous system, the limbs, the kidneys, and other tissues. The ectoderm develops into the epidermis and its appendages (hair, skin, glands, and so on), the nervous system, and the sense organs. The mesoderm yields the connective tissues of most organs, the circulatory and immune systems, the kidneys and gonads, and the skeleton and muscles. The endoderm forms the lining of the gut and the epithelial portion of such internal organs as the liver, thyroid, lungs, and pancreas. Table 16-1 summarizes this process.

Even as the three germ layers arise, their cells are already committed to forming the cell types appropriate to each layer. Two major events, however, must still take place to elicit normal development of the various tissues and organs: **morphogenesis,** the change in shape of cells and cell populations, and **differentiation,** the maturation of cells so that they can perform separate functions, either in the embryo or in later stages of development.

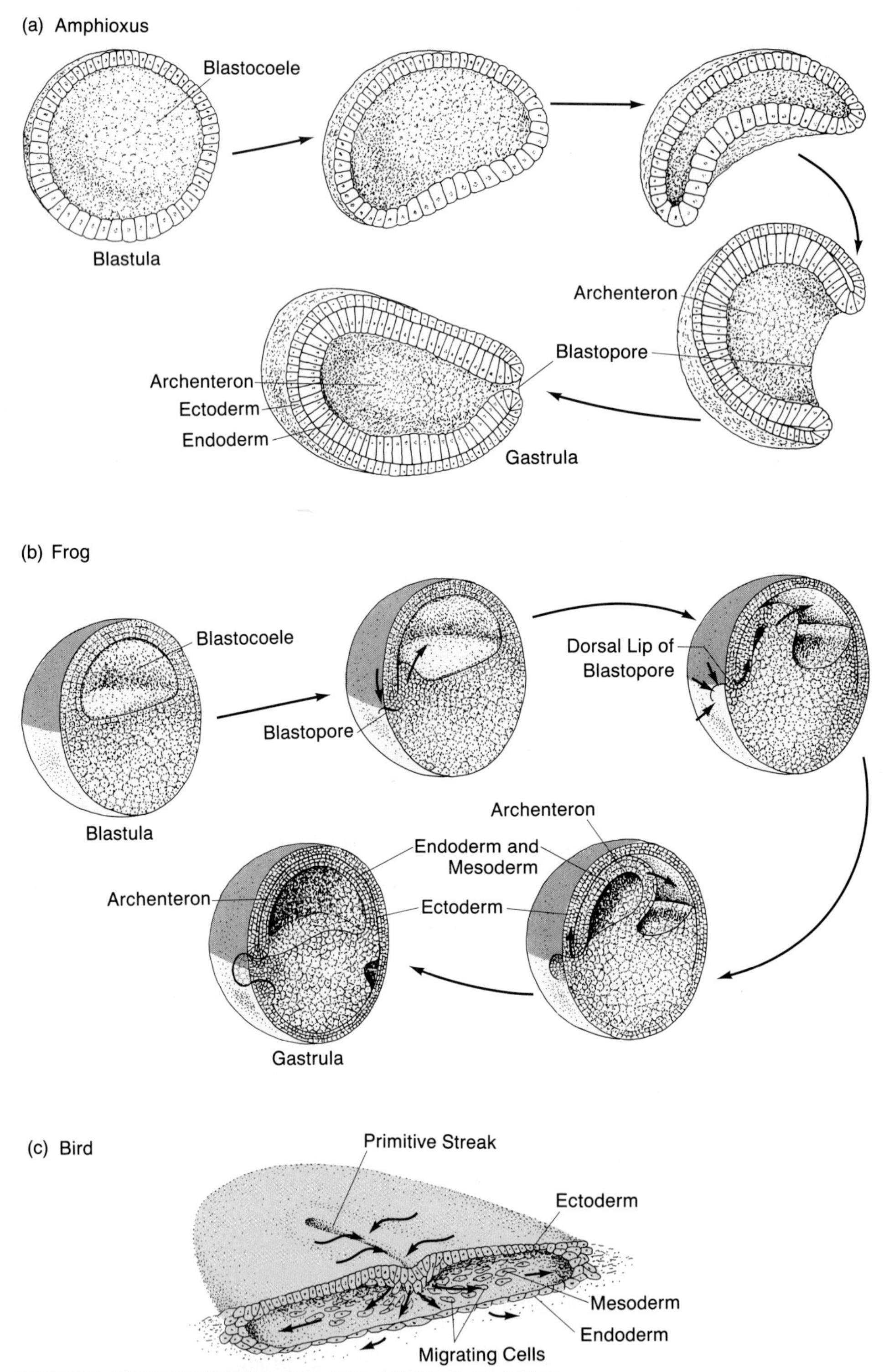

Figure 16-12 GASTRULATION PATTERNS AND THE AMOUNT OF YOLK.
(a) Gastrulation in an *Amphioxus* egg with very little yolk. The flattened plate of cells invaginates, forming two layers—the endoderm and ectoderm—and a cavity called the archenteron ("early gut"). This cavity will become the gut cavity.
(b) Gastrulation in a frog, whose eggs have an intermediate quantity of yolk. This frog embryo is cut in half so that the invaginating layer of cells (which begins near the junction of the smaller animal pole cells and the larger, vegetal pole cells containing most of the yolk) is visible as it moves gradually across the blastocoele space. Invaginated endoderm and mesoderm cells separate at a later stage. (c) Gastrulation in the flattened embryo of a bird. The tiny embryo rests on the huge sphere of yolk. Cells on the surface of the embryo move centrally toward the groovelike primitive streak and then migrate inward to give rise to the two inner cell layers: endoderm and mesoderm.

Table 16-1 FATE OF THE GERM LAYERS

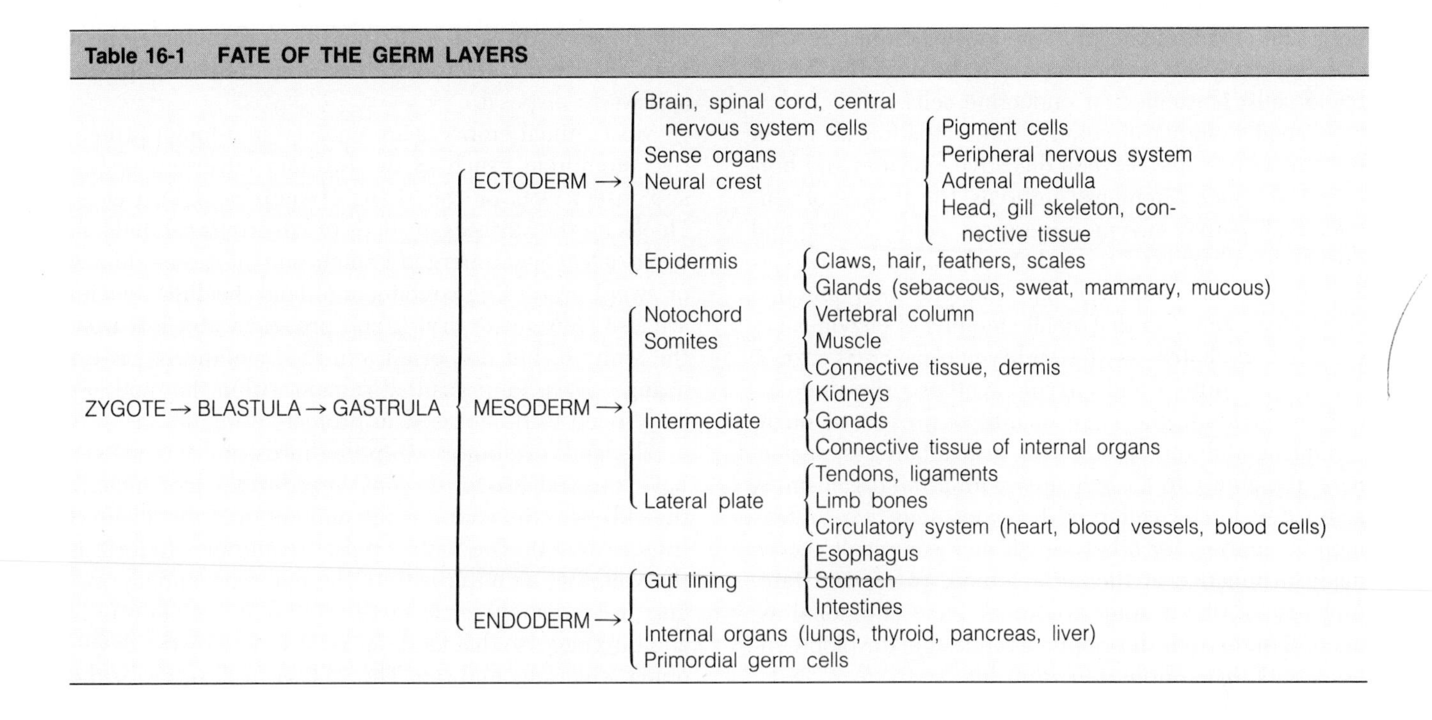

	Germ layer	Structure	Derivatives
ZYGOTE → BLASTULA → GASTRULA	ECTODERM →	Brain, spinal cord, central nervous system cells	
		Sense organs	
		Neural crest	Pigment cells; Peripheral nervous system; Adrenal medulla; Head, gill skeleton, connective tissue
		Epidermis	Claws, hair, feathers, scales; Glands (sebaceous, sweat, mammary, mucous)
	MESODERM →	Notochord, Somites	Vertebral column; Muscle; Connective tissue, dermis
		Intermediate	Kidneys; Gonads; Connective tissue of internal organs
		Lateral plate	Tendons, ligaments; Limb bones; Circulatory system (heart, blood vessels, blood cells)
	ENDODERM →	Gut lining	Esophagus; Stomach; Intestines
		Internal organs (lungs, thyroid, pancreas, liver)	
		Primordial germ cells	

Morphogenesis depends on the ability of cells and cell populations to assume specialized shapes and positions in the developing embryo. Take, for example, the process of **neurulation**—the formation of the neural tube, from which the brain and spinal cord later develop. Figure 16-13 shows the basic steps in this morphogenetic process. The cells start out in a flat and somewhat thickened sheet, called the *neural plate*. Through changes in cell shape, the right and left edges begin to fold upward toward each other. When they meet, the tissues fuse, and the resulting tube is pinched off from the adjacent ectoderm. The result is the hollow brain and spinal cord of the vertebrates. Through similar processes, the other cells of the embryo become arranged into tissues and organs that will be functional parts of the whole organism.

Through cell differentiation, cell groups become different from one another and from what they were earlier in embryonic development. They manufacture unique sets of molecules that account for their particular func-

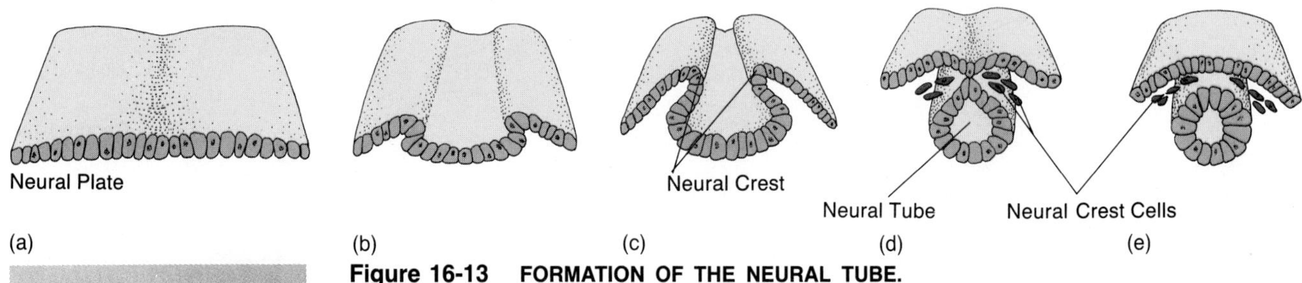

(f)

Figure 16-13 FORMATION OF THE NEURAL TUBE.
This sequence shows the morphogenesis of a small portion of spinal cord in cross section. (a) Morphogenesis starts with the neural plate, a flat sheet of cells on the surface of the embryo. (b, c) Changes in cell shape cause the plate to curve inward, so that the two sides move toward each other. (d) The cells at the edges of the neural folds come together, and the tissues fuse. (e) Finally, the neural tube separates from the overlying ectoderm. (f) A scanning electron micrograph of a stage between (c) and (d). From *Tissues and Organs: A Text-Atlas of Scanning Electron Microscopy* by Richard G. Kessel and Randy H. Kardon, W. H. Freeman and Company © 1979.

tion: sets of contractile proteins for heart muscle cells; chemical transmitters for nerve cells; hemoglobin for red blood cells; hormones for endocrine cells.

A good example of cellular differentiation occurs in the pancreas, where certain digestive enzymes and hormones are made. In mouse embryos, the amount of pancreatic digestive enzymes increases some 10,000-fold during development. Genes that code for these enzymes are turned off completely at first and then function slowly for a time, so that small amounts of enzymes are synthesized. Later, as the embryonic pancreas grows, the rates of mRNA and enzyme synthesis rise dramatically to generate the large quantities present at birth.

During differentiation, cells also attain *responsiveness*, the ability to be regulated within the organism. A given developing cell type becomes sensitive to hormones, neurons, or other signals and as a result can be made to function at the correct time, place, and rate. The rest of this chapter considers some additional aspects of embryonic development and some developmental events that continue in adult life.

EMBRYONIC COVERINGS AND MEMBRANES

An embryo is fragile; it must be protected from heat, drought, predators, the mother's immune system, and other dangers of its own particular environment. It cannot, however, be sealed off from the world completely; it must also have a means of gas exchange, nutrient uptake, and waste disposal.

Most animal embryos are encased in a tough protective membrane, such as the leathery coverings of fish and insect eggs or the shells of snake and bird eggs. Hooks on the surfaces of shells or coverings can hold an egg to a leaf or submerged branch, so that it is not blown or swept away. Egg containers of land-dwelling species are also resistant to drying and prevent water loss from the embryo, but do permit a limited amount of gas exchange. Even embryos that remain within the mother's body need to be encased in protective membranes.

The land vertebrates—reptiles, birds, and mammals—have evolved *extraembryonic membranes* that include the yolk sac, the chorion, the amnion, and the allantois (Figure 16-14). The first of these membranes to form is the **yolk sac,** an outgrowth of the embryo's endodermal gut. It completely encloses the sphere of yolk in reptile and bird eggs, so that food can be absorbed by the developing, enlarging embryo. The yolk sac wall serves as the first site of differentiation of red blood cells and thus provides the red cells and hemoglobin essential to oxygen binding and transport. The **chorion** and the **amnion** are originally continuous with each other but soon separate. The amnion fills with *amniotic fluid* and forms a cushionlike sac around the embryo. The chorion's functions are best understood in relation to the **allantois,** the fourth membrane, which arises as a pouchlike outgrowth of the embryo's posterior gut. Abundant allantoic blood vessels form and carry nitrogenous wastes away from the

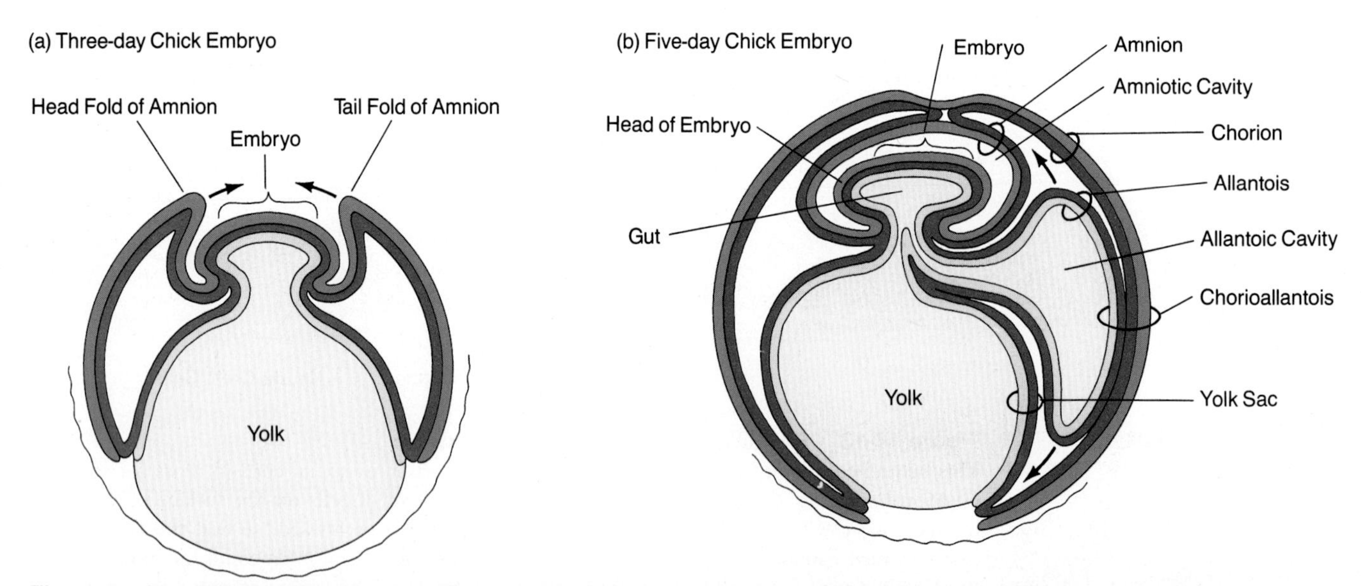

Figure 16-14 FORMATION OF EXTRAEMBRYONIC MEMBRANES DURING CHICK DEVELOPMENT. (a) A highly schematic drawing of the yolk sac membrane and the chorion encircling the spherical yolk mass in the developing chick. The amnion, or inner membrane, arises as it and the chorion travel over the top of the embryo and fuse. The amnion encloses the embryo in the fluid-filled amniotic cavity. As shown here, the allantois begins to expand. This pouchlike membrane is formed from the embryo's primitive hindgut. (b) The allantois enlarges greatly, and its outer wall fuses with the inner wall of the chorion; the result is the chorioallantois. The red lines indicate the position of mesoderm cells.

embryo to the allantois, which serves as a kind of organic garbage bag, storing such wastes until the reptile or bird hatches.

The allantois and the chorion become fused to form the *chorioallantois,* where oxygen diffusing inward through the egg white fluid is picked up and where carbon dioxide is given off from embryonic blood vessels.

In Chapter 18, we discuss the special function of these four extraembryonic membranes in humans and other mammals.

GROWTH: INCREASE IN SIZE

Development continues long past gastrulation, cell differentiation, and organogenesis. Until hatching or birth, an embryo *grows,* then continues increasing in size until adulthood.

Growth in Embryos

The rate of growth in an embryo can be astonishing: In just 22 days—from fertilization to birth—a rat embryo increases from one cell to *3 billion* cells. This developmental explosion is based largely on the simple, steady mitosis of cells rather than on the enlargement of individual cells. The fertilized egg divides into 2 cells; these 2 into 4, 8, 16, 32, 64, and so on. Thirty-one such cell generations precede birth in the rat. The tiny newborn, of course, continues its rapid growth into adolescence and adulthood: within 3 months, the rat pup's 3 billion cells become *67 billion.*

An extreme case of embryonic growth occurs in the blue whale. This marine mammal's egg is less than a millimeter in diameter and weighs a fraction of a gram, but the newborn calf weighs 2,000 kg and is 7 m long. That represents about a 200-millionfold increase in weight!

In many species, a finite amount of food available to the developing embryo limits its ultimate size. For instance, a chicken embryo inside its shell grows at the expense of the yolk and the "white," which gradually shrink as the number of cells increases. Hatching must occur soon after the yolk has been used up, or the embryo will starve to death. Similarly, in many other types of animal embryos, the only food to sustain development is stored in the egg from the start; hence, the late embryo at hatching can be no larger than the egg. Animals that carry out internal fertilization and incubation, however, do not have to depend on a limited food supply. Mammalian development provides the embryo in the womb with seemingly unlimited nutrients from the mother's body rather than from a store of yolk. Studies of mammalian evolution suggest that the general timing of mammalian birth is based not on limitations in the food supply, but on such things as the size of the fetal head relative to the birth canal and potential attack by the mother's immune system.

How do embryos that develop free of the maternal body, such as those of frogs and insects, get along with only intermediate or small quantities of yolk? Such animals have special growth phases in their life cycle. The embryo does not grow, but gives rise to a *larva* that can grow by feeding itself. Larvae typically lead very different lives from adults of the same species. The silkworm caterpillar, for example, feeds on plant leaves, but the moth flies about in search of flower nectar. A **metamorphosis,** a transformation controlled by hormones, intervenes between larval and adult stages. After growing manyfold, the caterpillar becomes a *pupa* that metamorphoses into the adult moth.

Growth in Later Stages of Life

Embryonic growth can be impressive, but the most spectacular growth stages of many species take place during the juvenile and adolescent phases of the organism's life cycle. It is then, for instance, that intensive cell division near the ends of the limb bones produces the typical adolescent growth spurt in humans. Specific hormones and growth factors regulate such spurts, as well as the eventual cessation of growth in the adults of most animal species.

A few animal species do continue to grow in adulthood. Lobsters, for example, enlarge in spurts year after year by casting off their old external skeletons and growing new, larger ones. Fish such as salmon also grow continuously, and like embryonic growth, such sustained adult growth depends on mitosis and increase in cell number.

Once an animal reaches its typical adult size, cell division slows, so that the number of new cells produced more or less compensates for older cells that die. The replacement of red blood cells in a normal adult human is a good example. Daily production of hundreds of millions of cells in the bone marrow (the core of many bones) occurs at a rate equal to the natural rate of cell death, so that the number of circulating red blood cells remains constant. This cell production is regulated by a protein called *erythropoietin,* secreted by the kidney whenever the blood flowing through the kidney is low in oxygen. After any loss of red blood cells (such as through bleeding), erythropoietin released by the kidney stimulates stem cells in the bone marrow to increase their already substantial mitotic activity. (**Stem cells** are cells that serve as a continuing source of a differentiated cell type, even for the lifetime of an organism.) This increased mitosis generates extra cells that differentiate into erythrocytes (red blood cells). Soon the concentration of red blood cells per milliliter of blood approaches normal values, oxygen levels rise and are detected by the kidney, and erythropoietin secretion falls to a resting level.

Biologists do not fully understand the exact mechanisms by which mitosis in adult tissues is regulated to keep the various populations of differentiated, functioning cells constant. The process of controlling growth during embryonic development is even more mysterious. But the fact that all the embryonic organs—eyes, limbs, heart, and so on—grow in proper proportion and stop enlarging at adulthood suggests that coordinated controls do exist.

Regeneration: Development Reactivated in Adults

Adult sea squirts and hydras can be produced from tiny buds; male elk grow new sets of antlers each year; whole new worms may regenerate from just pieces of the body; and some aquarium fish can sprout new fins to replace ones that have been bitten off. These are all cases of **regeneration,** a special type of growth in adults.

Salamanders are the best-studied organisms that regenerate lost parts. If, for example, the crystalline lens is removed from the eye of an adult salamander, a new one develops from the nearby iris, the pigmented ring of tissue that surrounds the pupil. Similarly, if the forelimb of a salamander is severed midway between the elbow and the wrist, the stump re-forms the missing forelimb, wrist, and digits within a few months (Figure 16-15). In each case, cells in the iris or in the stump tissue undergo a process of **dedifferentiation.** That is, they lose their functional phenotype (as pigment, bone, connective tissue, etc.). The dedifferentiated cells divide rapidly and generate a population of cells that will re-form the lost parts. The dividing daughter cells then carry out differentiation and morphogenesis, thereby forming a new lens or a new forelimb, wrist, and digits.

Before a salamander's limb can regenerate, a minimal amount of nervous tissue must be present in the stump near the wound site. Removal of these nerves prevents regeneration. Conversely, if extra nervous tissue is transplanted into a frog's forelimb (which normally cannot regenerate), a reasonable amount of regeneration occurs following amputation. Similar attempts are under way in mammals.

A related phenomenon often confused with regeneration is **compensatory hypertrophy,** a kind of temporary growth response that occurs in such organs as the liver and kidney when they are damaged. If a surgeon removes up to 70 percent of a diseased liver, the remaining liver tissue, arrayed in lobes, compensates by undergoing a very rapid rate of mitotic activity (greater than the fastest growing cancers) until almost all the original liver mass is restored. Similarly, if one kidney is removed, the other enlarges greatly to compensate. Note that the missing organ is not regenerated; the uninjured residual tissue merely increases in mass and cell number. By one hypothesis, levels of mitotic regulators fall in damaged tissues, and this drop allows mitosis to resume until normal cell numbers are restored.

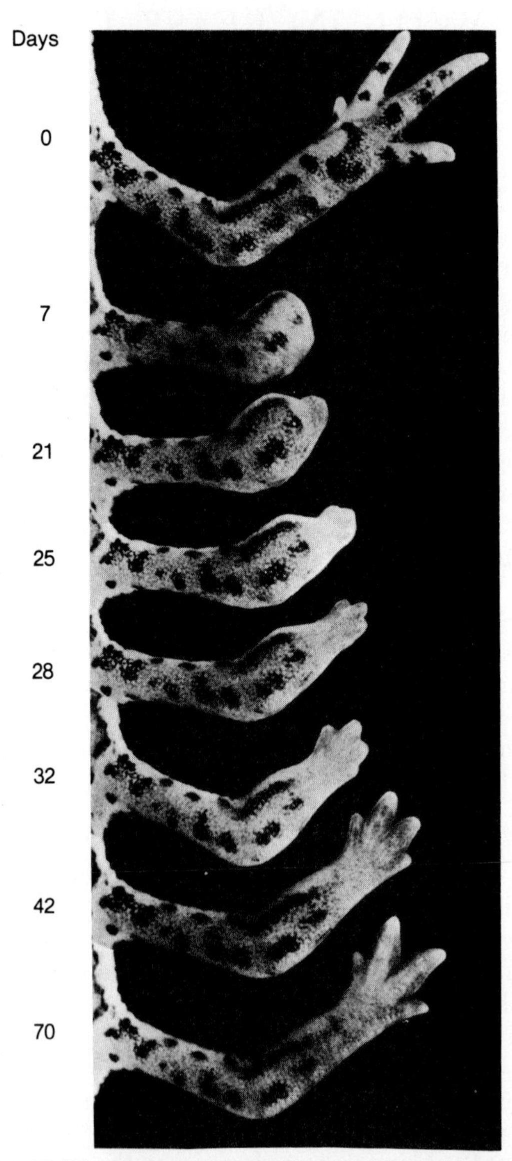

Figure 16-15 LIMB REGENERATION IN A SALAMANDER. Cells at the end of the stump dedifferentiate and proliferate intensively. They yield populations that once again differentiate into bone, muscle, tendon, skin, and so on. The numbers indicate days since the amputation.

AGING AND DEATH: FINAL DEVELOPMENTAL PROCESSES

"Death is but an end to dying," wrote the sixteenth-century French essayist Michel de Montaigne. By dying, he really meant *aging*, the time-dependent deterioration of many of the body's parts that actually begins early in life. Aging is a characteristic of multicellular organisms, both plants and animals, and in animals the germ line alone is "immortal."

Researchers are actively studying the causes of aging, including the graying of hair, the wrinkling of skin, and the deterioration of organs in humans. According to one theory, the fibrous proteins of the connective tissues degenerate. *Collagen*, which gives tensile strength to the connective tissues everywhere in the body, undergoes changes over time. Collagen fibers become less extensible, more brittle, and more subject to tearing. One result is the sagging skin of the older human. More important is impairment of functions in lungs, kidneys, heart, major blood vessels, and other vital organs as collagen and other connective tissue proteins deteriorate.

A second theory of aging suggests that cells are capable of only a limited number of divisions. Some highly controversial experiments have led to the idea that an intrinsic limitation in mitotic capacity ultimately leads to a depletion of new, healthy cells in vital tissues; thus, an insufficient number of cells may replace dying cells.

Other theories of aging are based on the decline of the immune system, and as Chapter 32 shows, decreased function of the crucial T-cell system following puberty may contribute to an increased frequency of cancers and debilitating infections and to increased *autoimmunity*—attack by the immune system on the body's own cells.

Yet another theory focuses on the way *lipofuscins*, or aging pigments, accumulate in cells. For unknown reasons, such pigments are not degraded or discharged from cells, so the volume builds up. For instance, lipofuscin makes up 5 percent of the volume of the heart muscle in an 80-year-old man. Perhaps these increasingly bulky lipofuscin granules begin to interfere with the proper functions of cells and so contribute to aging.

Whatever the mechanisms of decline, an aging animal eventually dies when some essential organ fails. Indeed, death may play a role in natural selection by removing older organisms that might compete for resources with younger breeding ones.

SUMMARY

1. *Sperm*, or male gametes, are specialized cells that develop in the testes through the process of spermatogenesis. Sperm can swim, penetrate the surface of an egg, and deliver the haploid male genome to the egg nucleus. The mature sperm cell has a head containing an acrosome and a compact nucleus, and it has a flagellum for swimming.

2. *Ova* (eggs), or female gametes, are specialized cells that develop in the *ovaries* through the process of *oogenesis*. Nurse cells, follicle cells, *oviduct* cells, and digestive glands help the egg to meet its specialized tasks by manufacturing proteins and nucleic acids, producing *yolk*, and adding protective coatings. Several processes take place in maturing oocytes, including gene amplification, the functioning of lampbrush chromosomes, and the importation of RNA from helper cells.

3. The process that unites egg and sperm is called *fertilization*. One step of fertilization is sperm penetration: The sperm head must pass through the several protective coatings of the egg. Fusion of the sperm and the egg plasma membranes triggers the three events of the *cortical reaction:* temporary loss of negative electrical potential of the egg, activation of the egg cell, and elevation of the fertilization membrane. Some species reproduce without normal fertilization by means of *parthenogenesis*.

4. *Cleavage* divides the fertilized egg, or *zygote*, into increasing numbers of cells, frequently in a precise manner that distributes developmentally significant control factors to different lines of *blastomeres* (cells) in the *blastula*.

5. *Gastrulation* is the rearrangement of cell populations to generate the three-layered *gastrula*. The three germ layers present after gastrulation are the *ectoderm*, *endoderm*, and *mesoderm*, which give rise to all body tissues and organs.

6. *Organogenesis*, the development of the body's organs and tissues, is achieved through two processes: *morphogenesis*, the attainment of special shapes of cells or cell populations, and cellular *differentiation*, the functional maturation of cells. Morphogenesis depends on cell locomotion and change in shape.

7. As they develop, some animal embryos are protected by a leathery or brittle outer covering. Reptile, bird, and mammal embryos have membranes that help provide nutrients, means of gas exchange, waste storage, and protection.

8. Growth results from mitotic activity of cells and accounts for the increase in mass of tissues, organs, and the whole organism. Once full growth is achieved, mitosis usually remains coupled to the rate of cell loss or death.

9. *Regeneration* can restore lost parts in some animal species. *Compensatory hypertrophy* is growth of certain organs in reaction to loss of tissue of like type.

10. Aging is the time-dependent deterioration of many of the body's organs and tissues.

KEY TERMS

acrosome reaction
allantois
amnion
blastomere
blastula
chorion
cleavage
compensatory hypertrophy
cortical reaction
dedifferentiation
differentiation
ectoderm
endoderm
fertilization
gastrula
gastrulation
gene amplification
gonial cell
mesoderm
metamorphosis
morphogenesis
neurulation
oogenesis
organogenesis
ovum
ovary
oviduct
parthenogenesis
primitive streak
regeneration
sperm
spermatogenesis
stem cell
testis
yolk
yolk sac
zygote

QUESTIONS

1. What is the function of a sperm cell? Draw one, indicate its size, label its structures, and describe their functions.

2. Do the same for an egg cell.

3. What fraction of the chromosomes in a fertilized egg is contributed by the sperm? By the egg? What fraction of the mitochondria is contributed by the sperm? By the egg? What fraction of the cytoplasm is contributed by the sperm? By the egg?

4. What is gene amplification? Give an example.

5. Describe the important events that take place during fertilization.

6. What is the significance of the yellow and gray crescents that appear in sea squirt zygotes soon after fertilization?

7. From which germ layer (ectoderm, mesoderm, or endoderm) is each of the following derived: hair, bones, nervous system, skin, muscles, outer layer of abdominal organs, germ cells, gut lining, connective tissue?

8. Describe the function of each membrane surrounding the embryo of a reptile, bird, and mammal.

ESSAY QUESTIONS

1. If you wanted to design a perfect birth control agent, what processes or properties of sperm, eggs, or fertilization would be good targets for your agent?

2. Are germ cells immortal? Are mitochondria? Explain.

For additional readings related to topics in this chapter, see Appendix C.

17

Developmental Mechanisms and Differentiation

Among the vertebrates as a whole—fishes, amphibians, reptiles, and mammals—there is some variation in cell type, but the key to the different organization of all these forms does not lie in the cells as such; it lies in how these basic building units are arranged in space during development.

Lewis Wolpert, "Pattern Formation in Biological Development," *Scientific American* (October 1978)

The human hand is exceptionally strong, graceful, and dexterous. No less remarkable, however, is the way that appendage develops from part of a single fertilized egg to a tiny bud to a paddle to the ultimate organic tool (Figure 17-1).

Chapter 16 described the processes underlying such development, including the cleavage of the fertilized egg to a mass of cells and the sculpting of specialized organs and tissues through cell rearrangement, morphogenesis, and differentiation. The question we address now is: What controls and coordinates this complex sequence of events? The real business of development is the translation of genotype into phenotype, and overall, this chapter will show how the sequential processes of development bridge the gap between genetic information and the three-dimensional living organism. Specifically, this chapter looks at:

- How individual cells differentiate and begin to function

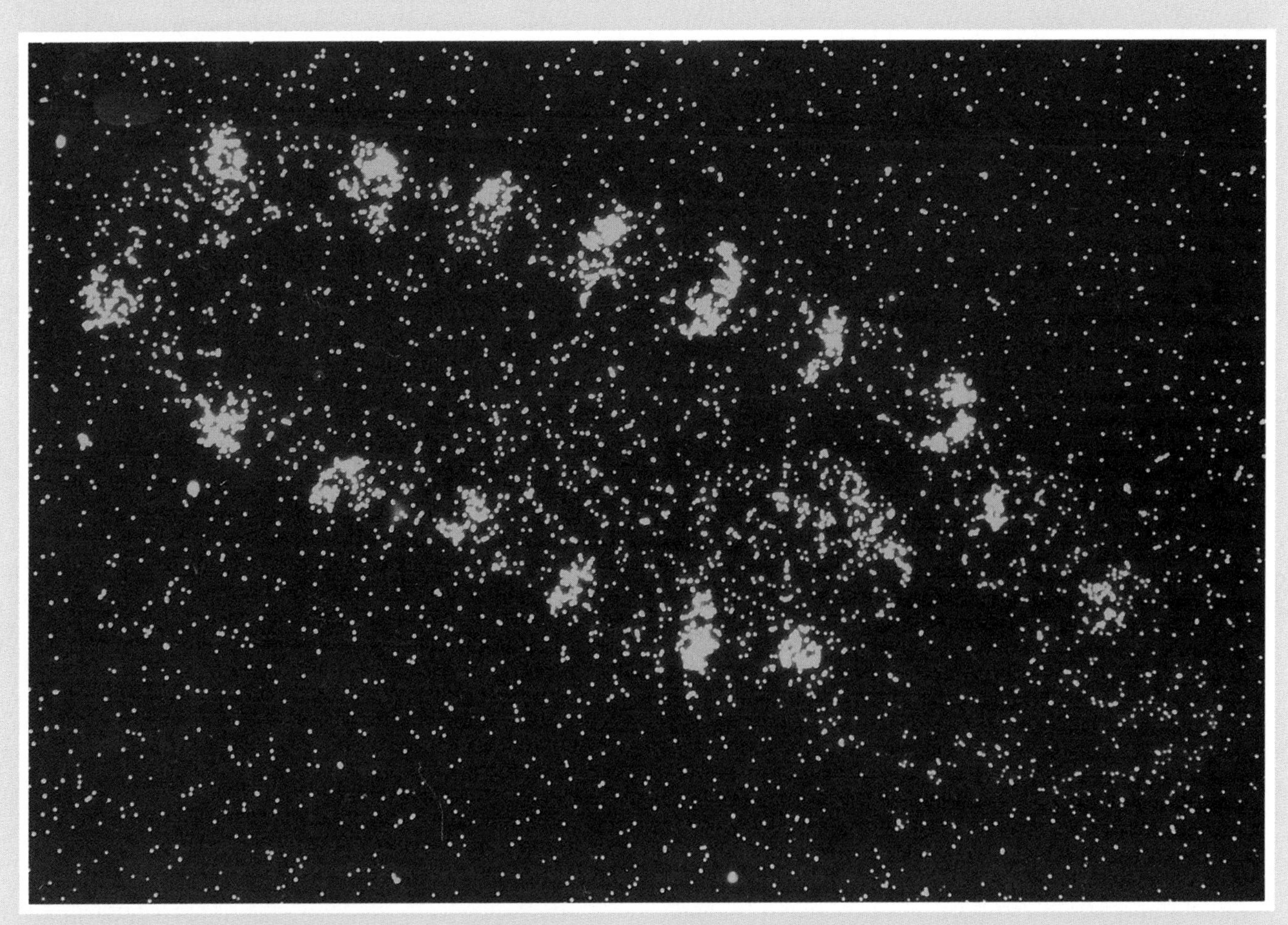

Sites of regulatory gene function in an early fruit fly embryo (magnified about 400×).

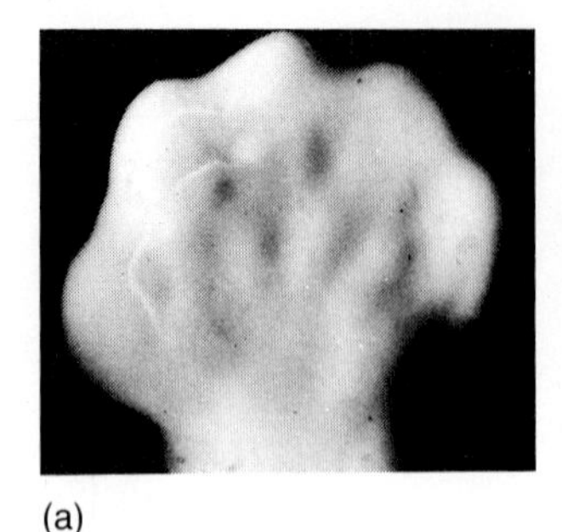
(a)

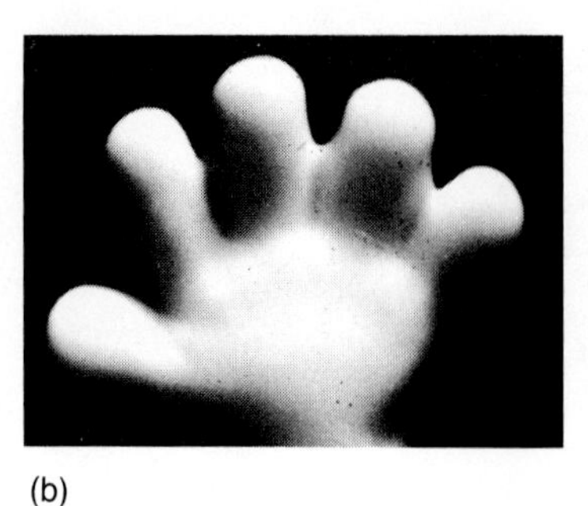
(b)

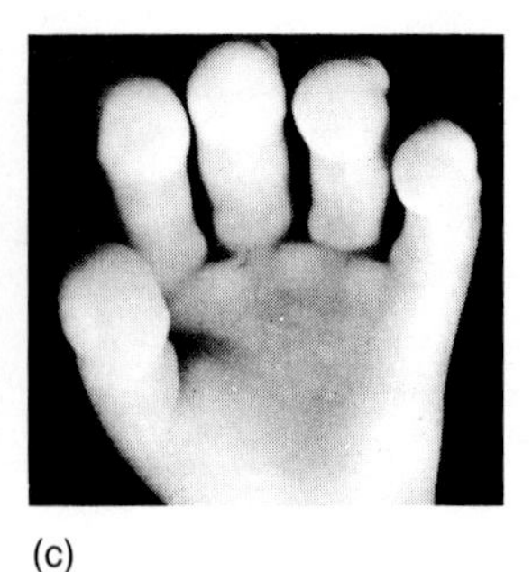
(c)

Figure 17-1 THE HUMAN HAND: FROM PADDLE TO SOPHISTICATED TOOL IN A FEW WEEKS. Gestational ages: (a) 5 weeks; (b) 6 weeks; (c) 7 to 8 weeks.

- How cells and cell populations are integrated so that each final organ will have the correct shape, size, and orientation
- The possible mechanisms of gene regulation during development

DETERMINATION: COMMITMENT TO A TYPE OF DIFFERENTIATION

Most cell lines in the embryos of higher animals undergo a remarkable process: Cells in such lines become *committed,* or *determined,* for one particular type of differentiation, such as to form bone, muscle, lung, or another specific differentiated cell type (Figure 17-2). During **determination,** cells commit to one final developmental pathway from among several alternatives.

The timetable for determination varies from cell type to cell type. In some cases, it takes place soon after fertilization. For example, as soon as the pole plasm (a mass of nucleic acid and protein) in a frog egg is incorporated into some blastomeres, these cells are determined as the germ line. Later, when the germ line cells and their progeny take up residence in the ovary or testis, they become the gonial cells, which will produce eggs or sperm. The pole plasm acts as a determinant in this case, just as the yellow- and gray-crescent materials and the polar lobe do in the sea squirt and the snail (see Chapter 16).

In most other cells, developmental options are reduced in a more gradual way. The covering layer of the embryo, the ectoderm, is committed at an early stage to forming the ectodermal lineage, or family, of cell types. Later, different subpopulations of ectoderm become the brain, the lens of the eye, the sweat glands, and many other types of cells (see Figure 17-2; review Table 16-1). In these cases, determination comes about through a series of steps: first as ectoderm, then as brain cells, and finally as one or another type of specific nerve cell used in seeing, moving, feeding, or some other brain function.

Once a subpopulation of cells is determined, the state is usually permanent. The pigment cells from the eye of a mouse or chick embryo are an example. These cells are already determined a few days before birth or hatching and have differentiated, so that they are black with pigment. If the pigment cells are removed from the embryo, separated from one another, and placed in a dish of nutrient medium, they may be induced to divide again and again, month after month, year after year. During this time, they remain determined, but they dedifferentiate; they are colorless, since their rapid division in the nutrient medium leaves no time for producing the black pigment. If the cells are eventually allowed to cease their rapid division and to differentiate once more, they again produce pigment granules. Why? The determined state of pigment cells has remained stable through these hundreds or thousands of divisions. Because the determined state is passed from one cell to its progeny during mitosis, it is clearly *inherited.* This stable control of which genes can be used prevents the cells of long-lived organisms from switching suddenly from liver to lung or eye cells. But mistakes sometimes do occur.

Changes in Determination

In Chapter 16, we saw that when researchers remove the lens from a salamander's eye, the nearby pigmented iris cells dedifferentiate, divide, and give rise to a population that differentiates anew as lens cells. In this case, the pigmented iris state *changes* to the lens state, demonstrating that determination can be modified.

Fruit flies provide another good example. Inside fruit fly larvae are little sacs called **imaginal disks,** each composed of cells determined to become a leg, an eye, a wing, a genital apparatus, or some other part of the adult fly. If, as shown in Figure 17-3, a researcher induces the genital imaginal disk cells to divide again and again, groups of these cells suddenly give rise to *leg* tissue. The

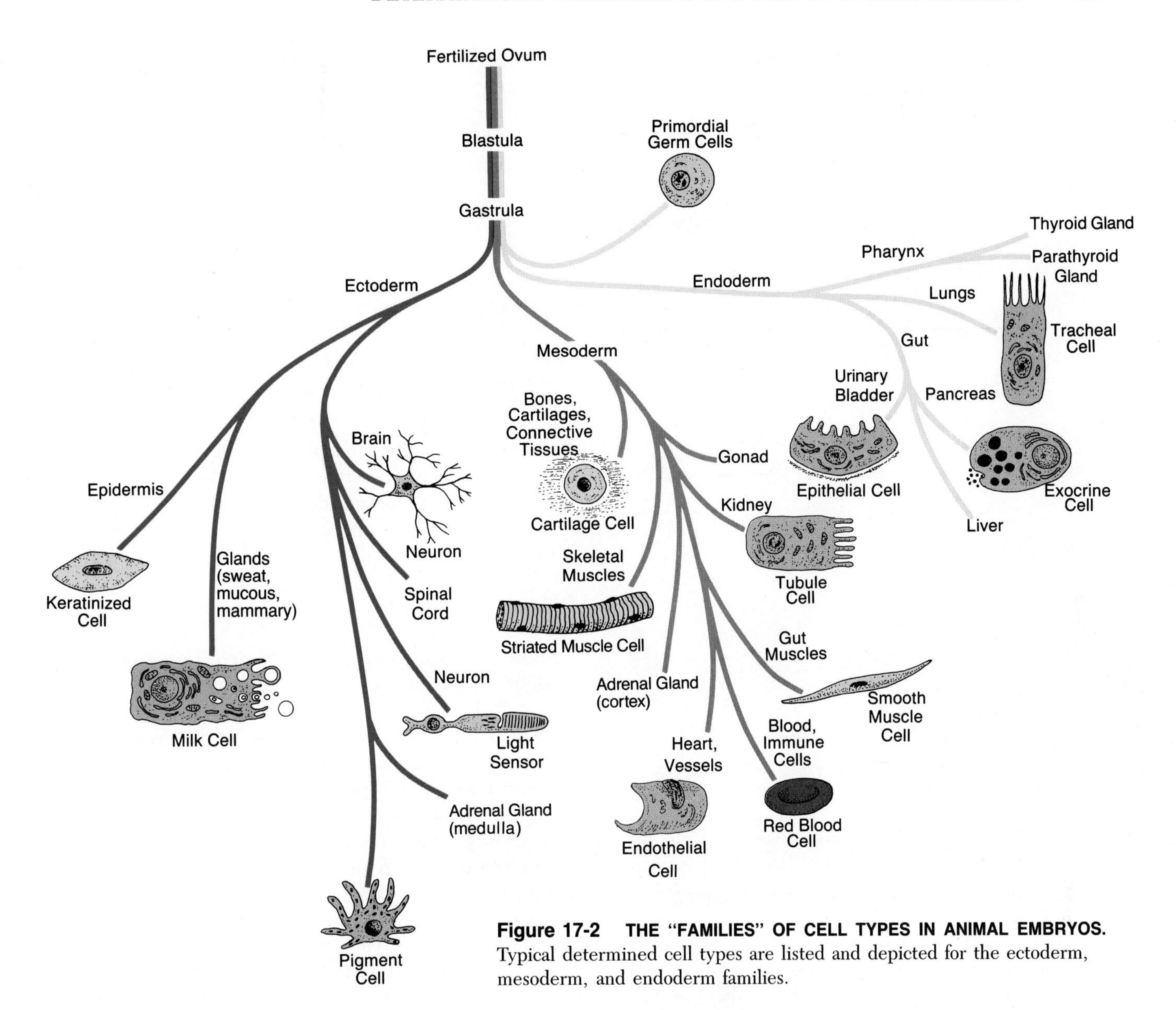

Figure 17-2 THE "FAMILIES" OF CELL TYPES IN ANIMAL EMBRYOS. Typical determined cell types are listed and depicted for the ectoderm, mesoderm, and endoderm families.

disk cells in such a case are said to be **transdetermined;** that is, they shift from one state of determination (genital) to another (leg). If divisions continue long enough, another transdetermination occurs, and in place of leg tissue, *wing* tissue forms. The reasons for transdetermination are still not understood. But related experiments on vertebrate embryos tell us more about determination.

The Stability of Determination Depends on the Cytoplasm

Robert Briggs and Thomas King at Indiana University studied determination through **nuclear transplantations.** They removed the diploid nucleus from a frog zygote and replaced it with a nucleus from one of various types of embryonic cells. If the transplanted nucleus came from a frog blastula cell, the injected zygote usually developed normally, even to the tadpole or adult frog stage. This suggests that the frog blastula cells, while partially determined, still carry a full set of genes capable of directing complete development.

John Gurdon used similar procedures on zygotes of *Xenopus*, the African clawed frog. He found that nuclei from fully differentiated gut cells and even from adult cell types could direct development of a zygote to a tadpole or adult with normal cell types. These results suggest that in a determined and differentiated cell, all the genes required for other determined states and for modes of differentiation are still present.

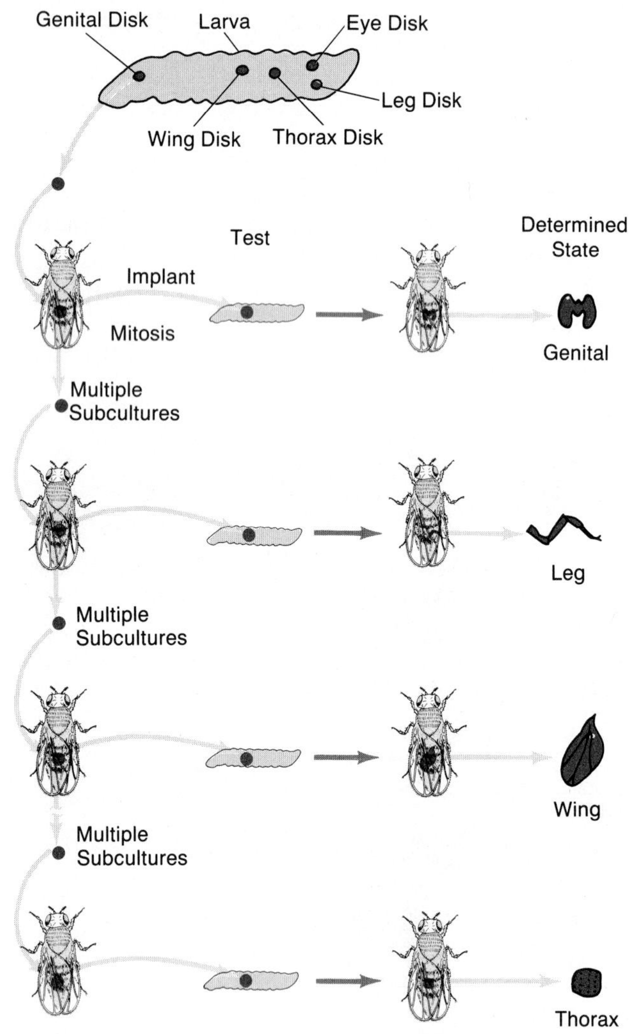

Figure 17-3 TRANSDETERMINATION SEQUENCES. A tiny fragment of a genital imaginal disk is removed from a fruit fly larva and is implanted in the body cavity of an adult female, where the fragment enlarges greatly. A tiny piece is "tested" by implanting it in another larva, which ultimately gives rise to an adult fly, in whose abdomen is the differentiated tissue derived from the implant. Another piece of the large imaginal disk fragment from the fruit fly abdomen is subcultured in another adult female. After the second fragment has grown large, a piece can be subcultured in still a third adult female—and so on, for serial culture. If the original genital disk cells divide enough times during a series of the subcultures, the disk cell progeny transdetermine into alternative structures. The sequence is not random: the first "mistake" is to form leg or antenna; from that state, the next change is to form wing; and that, in turn, can transdetermine to form thorax.

Nuclear transplantation experiments also revealed that the original state of determination is lost when the nucleus of a determined cell is removed and exposed to the cytoplasm of another cell type. Determination is an exceedingly stable condition, but it clearly depends on the cytoplasm of the determined cell type; exposure to a different cytoplasm "deprograms" the nucleus and permits it to participate in the entire developmental process. We will describe the probable molecular basis of determination later in this chapter.

DIFFERENTIATION: BUILDING CELL PHENOTYPE

Once determined to a particular developmental pathway, cells must undergo **differentiation,** or functional maturation. Let us consider the triggers for differentiation, then describe the unique characteristics of differentiated cells.

The Causes of Differentiation

In the 1920s, one of biology's greatest experiments identified a major developmental strategy for the control of cell differentiation and organogenesis. This mechanism is **tissue interaction,** which its discoverers called "induction." German biologists Hans Spemann and Hilde Mangold transplanted a mass of cells from the dorsal lip area of an amphibian blastopore to the surface of another gastrula (Figure 17-4). At the transplantation site, a second embryonic axis formed: A duplicate brain and spinal cord developed on the belly of the host embryo. Spemann and Mangold concluded that the implanted dorsal lip cells *induced,* or directed, the surrounding host ectoderm to become determined and to form new brain and spinal cord cells they normally would not have generated. Spemann called the dorsal lip tissue an *organizer,* since it appeared to organize a set of axial organs (brain, spinal cord, vertebral column) in a host embryo.

Subsequent experiments yielded similar results in birds and mammals. Mesodermal cells that migrate through the primitive streak of a bird or mammal embryo can be transplanted beneath ectoderm that normally gives rise to skin. Within a few hours, a brain or spinal cord begins to form from the nearby ectodermal cells in the embryo. Thus, Spemann's organizer activity operates in all land vertebrates.

A prime example of inductive tissue interaction occurs in the lens of the eye, which develops only where a hollow outpouching of the embryonic brain—the *optic vesicle*—contacts the head ectoderm, as shown in Figure 17-5. If a researcher surgically removes the vesicle from the developing embryo, the lens usually fails to

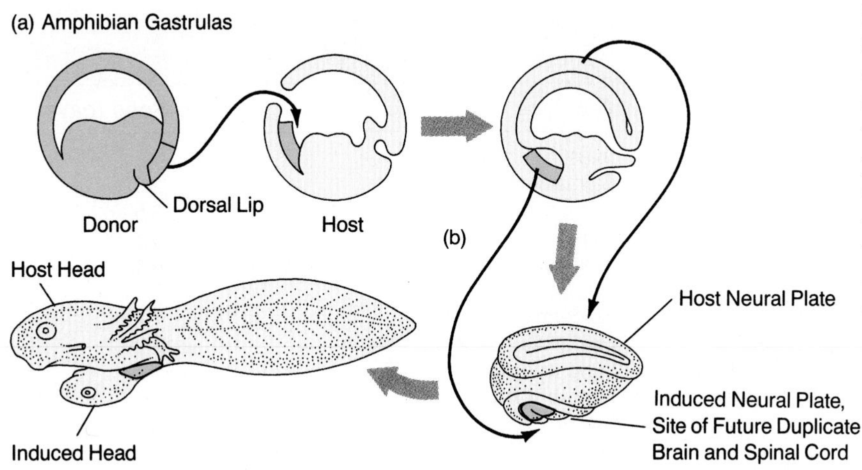

Figure 17-4 THE ORGANIZER-INDUCTION EXPERIMENT.
(a) Tissue from the dorsal lip of the forming donor blastopore is grafted into the blastocoele of a host. Gastrulation proceeds, and the implant is pressed closely against the host's belly ectoderm. (b) The ectoderm responds to the inducing tissue by forming a brain, and indeed, the entire head and anterior trunk may develop.

form on the side of the head from which the vesicle was removed. Furthermore, if that optic vesicle is transplanted so that it contacts ectoderm on the top of the embryo's head, ectodermal cells at that spot become determined as a lens and employ genes that they normally would not use.

Similarly, mesoderm commonly acts on an overlying ectoderm or underlying endoderm to support development of hair, mammary glands, lungs, thyroid, pancreas, teeth, and virtually every other organ of a vertebrate's body. Mesoderm is an inducer in these cases because the organ in question does not form if mesoderm is not present.

Inductive tissue interactions also ensure that developing organs will be arranged and spaced properly. The fact that a lens forms only where the optic vesicle happens to contact head ectoderm guarantees that the two organs will be aligned correctly. In fact, a broad area of the head ectoderm constitutes the *lens field*, an area in which all cells have the ability to respond to the inductive stimulus of the optic vesicle; it is as though the optic vesicle has a broad target, and a "hit" anywhere in it generates the lens. Induction as a control process therefore allows for a certain amount of developmental "noise"—a degree of imprecision of the sort one might expect in any living system.

Characteristics of Differentiated Cells

As it begins to differentiate, a cell takes on at least four specific characteristics that enable it to perform its particular roles in the organism.

1. A differentiated cell makes and uses a specific set of proteins that enables it to carry out its functions. It is this spectrum of proteins that really defines the differentiated condition of a cell type. Some proteins, such as the hemoglobin in red blood cells, carry out function,

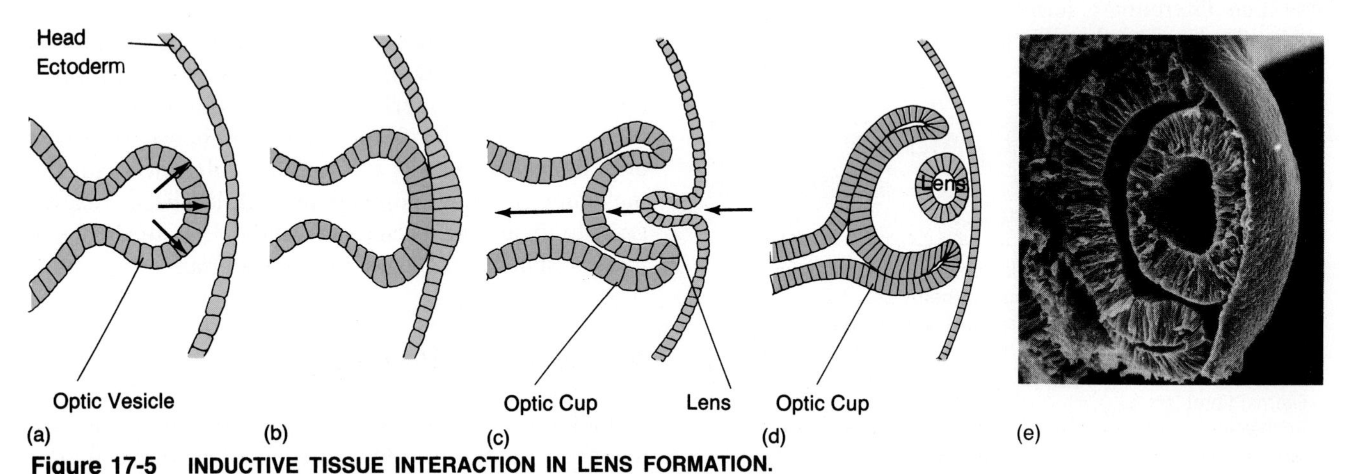

Figure 17-5 INDUCTIVE TISSUE INTERACTION IN LENS FORMATION.
(a) A lens forms from the embryo's head ectoderm at the site where the optic vesicle and the ectoderm come into close contact, as in (b). (c) Then, both the optic cup and the lens sink inward. (d, e) The hollow lens separates from the overlying ectoderm. This is visible in the diagram as well as the scanning electron micrograph (magnified 140×).

whereas others, such as tubulin in an epithelial cell, may govern shape (Table 17-1).

2. A differentiated cell is metabolically active. To maintain metabolic activity, cells usually have "housekeeping" proteins in addition to specialized proteins. These essential proteins include enzymes that help maintain general cell activity, as well as structural proteins like tubulin.

3. As it matures, a differentiating cell assumes a characteristic shape that enables it to function effectively in the tissue to which it belongs and in the organism as a whole. Mature cells can be flat, round, squarish, elongated, or disk-shaped, and specific proteins help determine these conformations. Obvious examples are tubulin, the building block of microtubules, and actin and myosin, which form some of the filament systems of the cytoplasm (see Chapter 6). New evidence suggests that such cytoskeletal proteins may differ slightly from cell type to cell type, with one actin gene directing a sperm cell to produce the actin of the acrosomal filament, another the actin in a muscle cell, and still another the actin in a nerve cell.

Ultimately, the shape of a differentiated cell helps govern its role in the organism. An appropriate shape and arrangement of organelles ensure that a secretory cell in a gland will secrete its product only from one end and into a cavity; that a muscle cell will apply contractile force at its ends and not at its center; and that a neuron will remain connected only to certain other cells.

4. A differentiated cell often ceases to undergo further cell division. Thus, differentiation is frequently a terminal condition. Red blood cells, skeletal muscle cells, and nerve cells are examples of differentiated cells that have completely lost the capacity for mitosis; once matured, they cannot divide into daughter cells.

The general cell division strategy in multicellular organisms is that (1) a determined cell divides, (2) its progeny then differentiate, forming a large number of protein molecules, and (3) these cells function for days, weeks, months, or years, depending on the kind of cell and the species of organism.

Table 17-2 lists average survival times of some human cell types. The range is enormous—from 1.3 days in the intestine to more than a year in the liver and a lifetime in the brain. It is remarkable that mitosis in each of these organs and cell types is closely coupled to the rate of cell loss. Without such coordination, organs and tissues could vary haphazardly in shape, size, and function.

Table 17-1 MAJOR SPECIFIC PROTEINS OF DIFFERENTIATED CELLS

Cell	*Protein*
Red blood cell	Hemoglobin (oxygen carrier)
Pancreatic B cell	Insulin (hormone)
Pancreatic exocrine cell	Trypsin (digestive enzyme)
Thyroid cell	Thyroglobulin (hormone storage)
Epidermal cell	Keratin (structural protein)
Dermal fibroblast	Collagen (structural protein)
Epithelial cell	Tubulin (structural protein)
Muscle cell	Myosin (contractile protein)
Pigment cell	Tyrosinase (pigment enzyme)
Immune B cell	Immunoglobulin (antibody)
Oviduct cell	Ovalbumin (egg white protein)

Table 17-2 AVERAGE LIFE SPANS OF HUMAN DIFFERENTIATED CELL TYPES

Cell Type	*Life Span (days)*
Intestinal lining	1.3
Stomach lining	2.9
Tongue surface	3.5
Cervix	5.7
Stomach mucus	6.4
Cornea	7
Epidermis—abdomen	7
Epidermis—cheek	10
Lung alveolus	21
Lung bronchus	167
Kidney	170
Bladder lining	333
Liver	450
Adrenal cortex	750
Brain nerve	27,375^{+} (75^{+} years)

MORPHOGENESIS: THE ORGANIZATION OF CELLS INTO FUNCTIONAL UNITS

Many diverse types of morphogenesis help sculpt an embryo's tissues and organs. Just a few different mechanisms, however, underlie these processes: (1) the movement of single cells, (2) the interaction between moving or stationary cells and extracellular substances, (3) the movement of cell populations, (4) localized relative growth, (5) localized cell death, and (6) the deposition of massive amounts of extracellular materials.

Single-Cell Movements

Individual cells may migrate from one site to another in embryos and thus establish populations that will form organs. *Neural crest cells,* for example, arise near the site where the neural tissue closes into a tube in vertebrate embryos (see Chapter 16) and follow regular path-

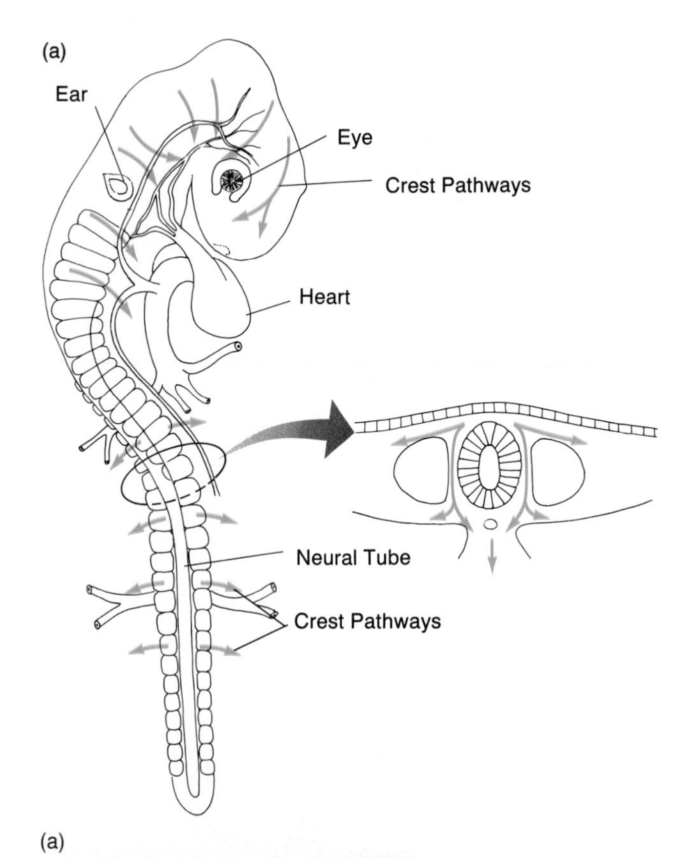

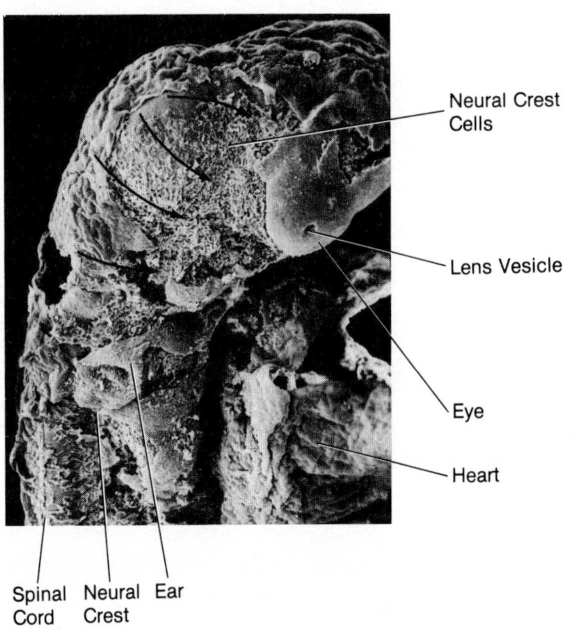

Figure 17-6 SINGLE-CELL MIGRATION.
(a) Green arrows show the routes of migration taken by neural crest cells in a chick embryo. Precise pathways are followed beneath the skin and around the neural tube and brain. (b) This scanning electron micrograph reveals neural crest cells in the head and neck of a chick embryo. The researcher has fractured away the ectoderm to show the underlying crest cells. Black arrows point out their routes of migration in the developing animal.

ways into the head and face, downward to the gut, and outward to the skin (Figure 17-6). These cells take up residence in a variety of locations, come under the influence of local controlling environments, and proceed to develop into a variety of tissues, such as the adrenal gland, pigment cells of the skin, most of the peripheral nervous system (sensory receptors and nerves that control the internal organs), and much of the face and head.

How do neural crest cells "know" where to travel and by what route? Studies show that the *adhesion* of cells to one another and to the extracellular environment help initiate, guide, and stop cell movements at correct locations. When pigment and cartilage cells are randomly intermingled, for example, they sort out into separate groups of like cells: The pigment cells stick to other pigment cells; the cartilage, to cartilage. This implies that cells have recognition markers on their surfaces—molecular name tags, so to speak—that help keep cells of similar type glued together in the body's tissues.

Moving Cells and Extracellular Substances

The locomotion of single cells is also affected by huge extracellular polymers composed of proteins and sugars. Some of these substances—fibronectin, laminin, collagen, and hyaluronic acid—provide a substratum, or ground substance, to which cells can stick and on which they can move, like a slug on a trail of mucus. A good example of this type of movement is seen in the developing cornea, the outermost portion of the eyeball, as Figure 17-7 depicts and explains.

The role of extracellular substances in morphogenesis is not just to enable cell movement, however. As moving cells interact with these substances, mechanical forces are also applied and can help shape the developing embryo. Far from being solid, extracellular materials in the living embryo comprise a deformable meshwork of fibers (collagens, fibronectins) and gel (hyaluronic acid) plus feltlike sheets of epithelial basement membranes (the sheets to which most epithelial layers attach). When a cell moves, it adheres to this substratum of substances and pushes backward so that it can move forward—just as the wheels of a dirt bike push backward and deform sand as the bike moves ahead. In this way, embryonic cells exert tension, or *traction*, on their substratum and can actually deform the network of fibers into parallel arrays. Biologists Albert Harris and David Stopak have shown that such *traction-induced* morphogenesis may account for the development of aligned collagen bundles, such as those in tendons and ligaments. In turn, individual muscles usually develop along tracts of such aligned fibers because the muscle-forming cells position themselves parallel to the fiber bundles.

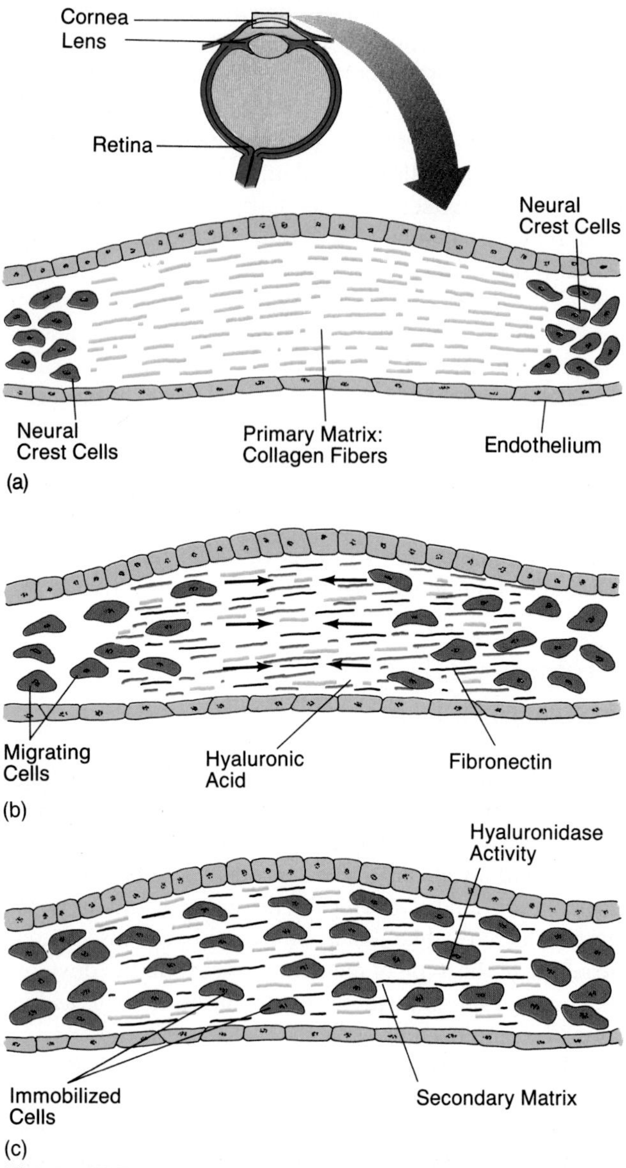

Cornea
Lens
Retina
Neural Crest Cells
Neural Crest Cells
Primary Matrix: Collagen Fibers
Endothelium
(a)

Migrating Cells
Hyaluronic Acid
Fibronectin
(b)

Hyaluronidase Activity
Immobilized Cells
Secondary Matrix
(c)

Figure 17-7 THE ENVIRONMENTAL CONTROL OF CELL MIGRATION.
(a) Neural crest cells migrate to the edge of the cornea (which is composed mainly of collagen fibers) but do not enter it. (b) A layer of corneal cells then manufactures and secretes hyaluronic acid. Fibronectin, a glycoprotein, also appears. These substances fill in the matrix of the cornea, and neural crest cells migrate in. (c) Eventually, the enzyme hyaluronidase is secreted, the hyaluronic acid is degraded, and the neural crest cells are left stranded in the corneal matrix, where they now differentiate into connective tissue cells.

Cell Population Movements

Entire populations of cells, such as epithelial sheets, may fold inward or outward during the formation of the lungs, kidneys, eyes, brain, and other organs. These foldings are caused by the activity of cellular cytoskeletal organelles, microtubules, and microfilaments, as well as by adhesion.

Microtubules may cause a cell to elongate; microfilaments may narrow one end of a cell. Figure 17-8 shows how such action may change the shape of a sheet of cells. For example, if cells are cemented together on their lateral surfaces by junctions (see Chapter 6), the narrowing of individual cells causes the entire population to change shape so that the sheet buckles or bulges. This sort of process leads to the formation of the neural tube (review Figure 16-13 and page 281).

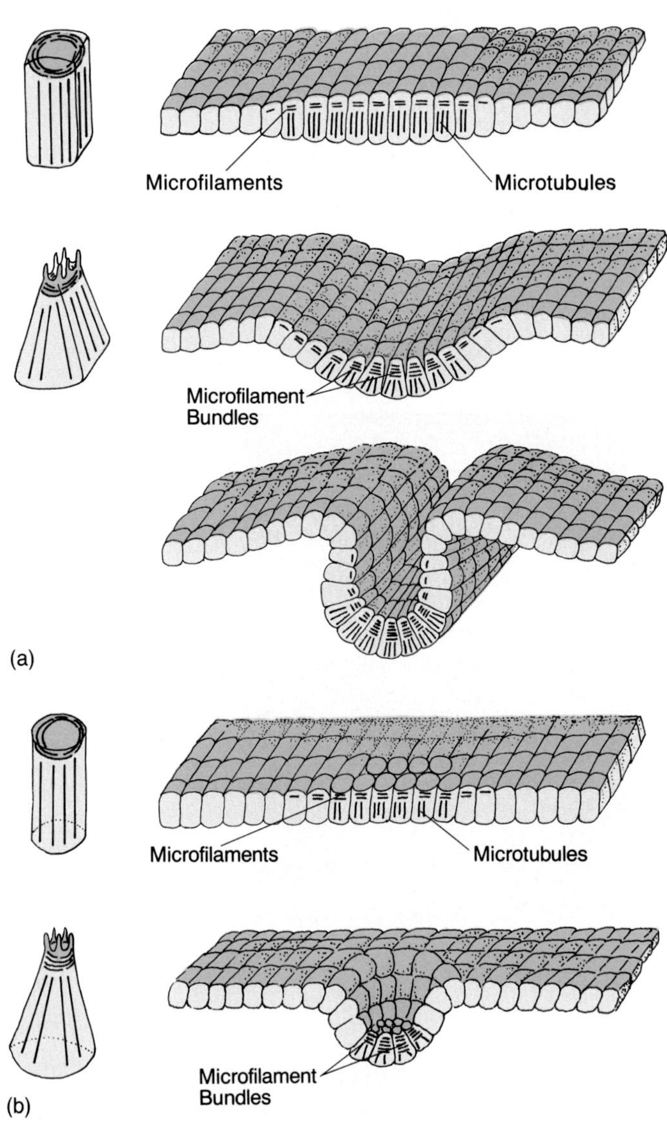

Figure 17-8 A BASIC PROCESS OF MORPHOGENESIS: NARROWING OF INDIVIDUAL CELLS.
(a) Cells may become wedge-shaped and so contribute to a sheet forming a groove; this occurs as the future central nervous system forms in vertebrates. (b) Cells may become conical and thus cause a pit to form; this occurs when the lens of the eye pinches off from the head ectoderm.

Localized Relative Growth

Another mechanism that underlies morphogenesis is the relative growth of specific groups of cells—growth that can be fast or slow, depending on the rate of mitosis at the site. In the earliest stages of vertebrate limb formation, mitosis slows in the surrounding body wall, while higher rates are maintained at the site of arm or leg formation. In addition, mitotic rates produce the branching of respiratory tubes in the mammalian lung. Right now, biologists know very little about how the increase or decrease in local mitotic rate is controlled.

Localized Cell Death

Ducks have webbing between their toes, but chickens and humans do not, because massive **cell death** occurs in the mesoderm between the toe bones of chicken and human embryos. As a result, the overlying skin ectoderm sinks inward between the toes (or fingers), and the digits are separated (Figure 17-9). Because comparable cell death does not occur in ducks, the mesoderm and overlying ectoderm form webs. Cell death occasionally fails to take place, and human and other vertebrate embryos sometimes have skin stretching between the fingers or toes.

Besides sculpting digits, cell death helps shape the shoulder, separate bones from one another, and eliminate large numbers of cells from the nervous system. This may seem a peculiar method of building organs, since tens of thousands of cells are first generated and then killed. Nevertheless, the strategy of cell death is a common means of achieving morphogenesis in organisms as different as insects and mammals.

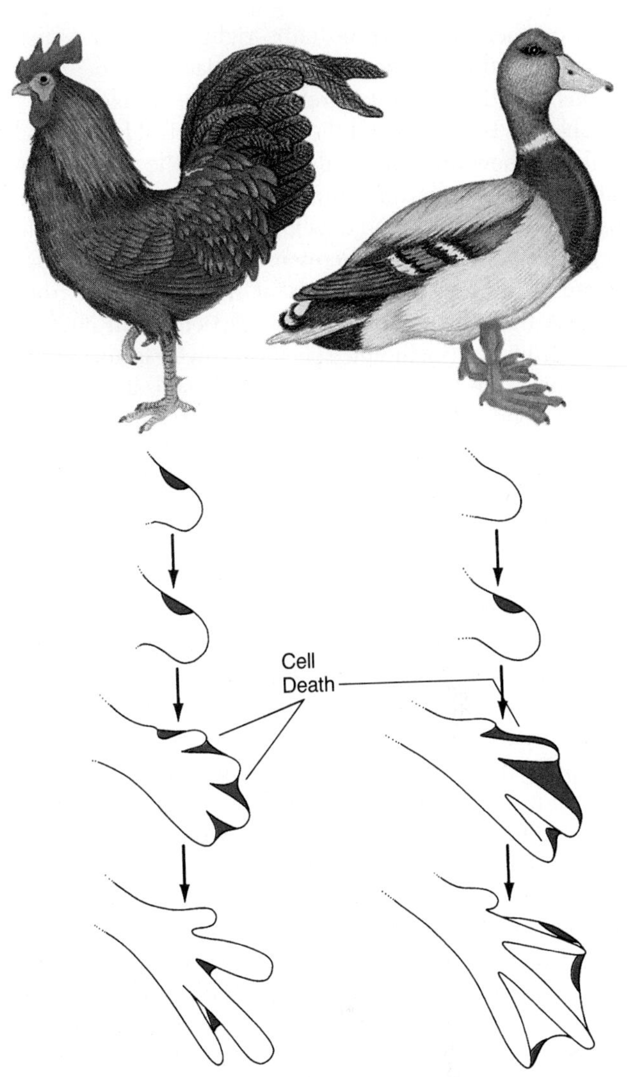

Figure 17-9 LOCALIZED CELL DEATH.
Compare the regions of cell death (red) in the feet of chick and duck embryos. The more limited areas of cell death in the duck's foot cause webbing to remain between the toes—all the better for paddling.

Deposition of Extracellular Matrix

Massive quantities of **extracellular matrix** contribute to the shape of the bones, cartilage, tendons, corneas, cavities of the eyes, and embryonic heart. Tough filamentous molecules such as collagen or huge, space-filling sugar polymers (called *glycosamino-glycans*) that absorb water are the main components of this matrix.

The six basic processes we have just described are employed, to varying extents, in the morphogenesis, or shaping, of each organ and tissue. Cell movement and traction may be particularly crucial for one developing organ; cell death, for another. Mutations operating on these basic morphogenetic mechanisms led to the new shapes of some organs and organisms that arose during evolution. Perhaps cell death failed to separate the toes of some ancient embryonic bird; it hatched with new webbed feet, it could paddle more efficiently, and it left behind web-footed progeny.

PATTERN FORMATION: THE VERTEBRATE LIMB

The mechanisms of morphogenesis we have just surveyed are integrated in the embryo to yield a complex organ or the basic body plan of an entire organism. **Pattern formation** is the gradual emergence of the body plan: one head; two, four, six, eight, or more legs, depending on the species; dorsal (upper), ventral (lower), anterior (head), posterior (tail), right, and left sides, again depending on the species; and placement of all the organs and their positions relative to the whole.

Chapter 16 introduced the topic of control factors and explained how their physical positions in the zygote help establish the embryo's future shape. The examples we discussed there included developmental determinants in a zygote's yellow and gray crescents or polar lobes. More recent studies of the embryos of African clawed frogs suggest that the distribution of particular mRNAs or their products (including specific growth factors) may determine the future dorsal-ventral organization of the animal's body. Biologists have amassed a fairly detailed picture of embryonic pattern formation in the vertebrate limb, so let's consider how the major axes form during limb development.

The first axis to arise in the forelimb of a developing vertebrate is the proximal-distal axis (proximal means near the main body, distal away from it): upper arm, forearm, wrist, and hand, with palm, thumb, and digits. (For simplicity, we will refer to parts of the human arm here, even though most experimental work has been done on embryonic chicken wings and mouse forelegs.) The different segments of the forelimb arise in an embryo in a sequence starting from the shoulder and progressing toward the digits. Lewis Wolpert, a leading British biologist, has proposed that the parts of the limb arise from the **progress zone,** a special set of mesodermal cells located just beneath the tip of the elongating limb bud. Biologists think that the first group of cells that remain behind the progress zone as it is moved passively outward during limb growth is assigned a "positional value" that may be called "upper arm." The next set to arise is assigned the positional value of forearm; the next is the wrist; and so on. Once a population of cells has assumed a positional value—say, upper arm—other processes cause the appropriate bones, muscles, tendons, and so on, to develop (Figure 17-10).

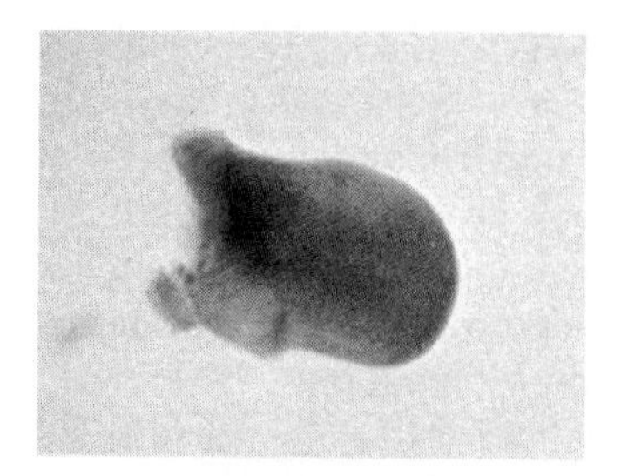

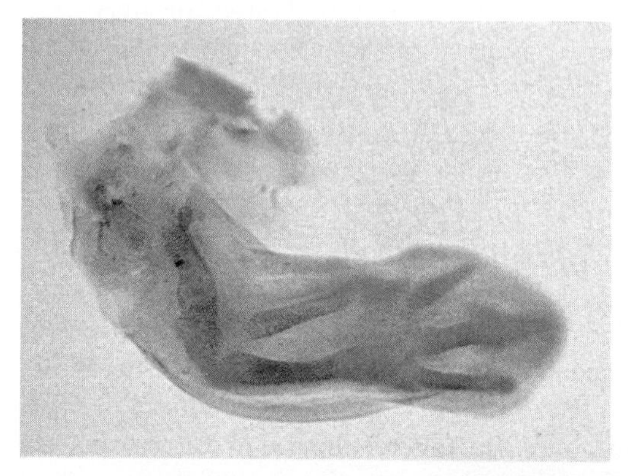

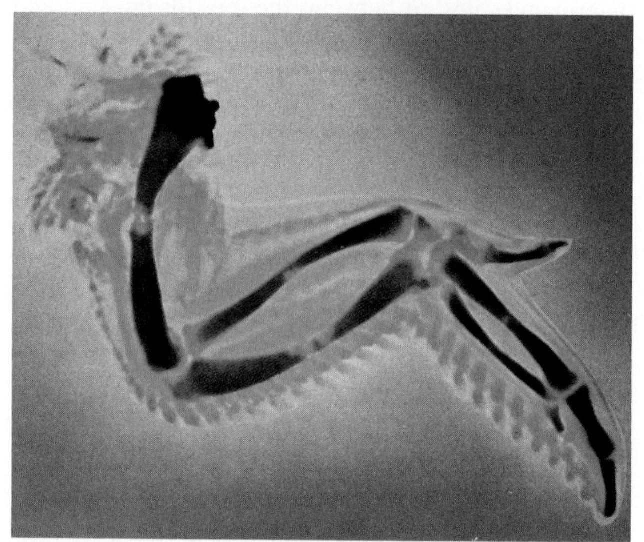

Figure 17-10 LIMB DEVELOPMENT IN A BIRD. The progress zone is located just beneath the tip of the bud, shown at the top. Already at that stage, the positional values for the upper wing and the forewing have been assigned. Immense growth of the wing occurs as the various bones, muscles, and other components of the wing differentiate.

What about the anterior-posterior (thumb–little finger) axis? At the posterior junction between the limb bud and the body wall is a region called the **zone of polarizing activity (ZPA).** The most posterior digit (the little finger, or fifth digit) forms nearest the ZPA.

If a researcher grafted an extra ZPA near the front of a young limb bud, extra digits would form, in reverse order, as shown in Figure 17-11. The digits of the resulting whole limb might have the sequence 543345—5 being the little finger and 3, the middle finger. The reason is that the implanted ZPA causes the digits that form near it to be posterior in type (5 and 4), while the host ZPA continues its normal action. So this limb would therefore have two "posteriors," and the duplicated and reversed set of digits would develop. Clearly, "posterior" can be defined as *near the ZPA.*

For years, biologists postulated that the ZPA must produce a *morphogen,* a substance that diffuses through the limb bud tissues and causes morphogenesis. Specifically, they thought that the morphogen might diffuse from the posterior region of the bud toward the anterior, and where it occurs in high concentrations cause posterior digits (digits 4 and 5) to form, while in lower concen-

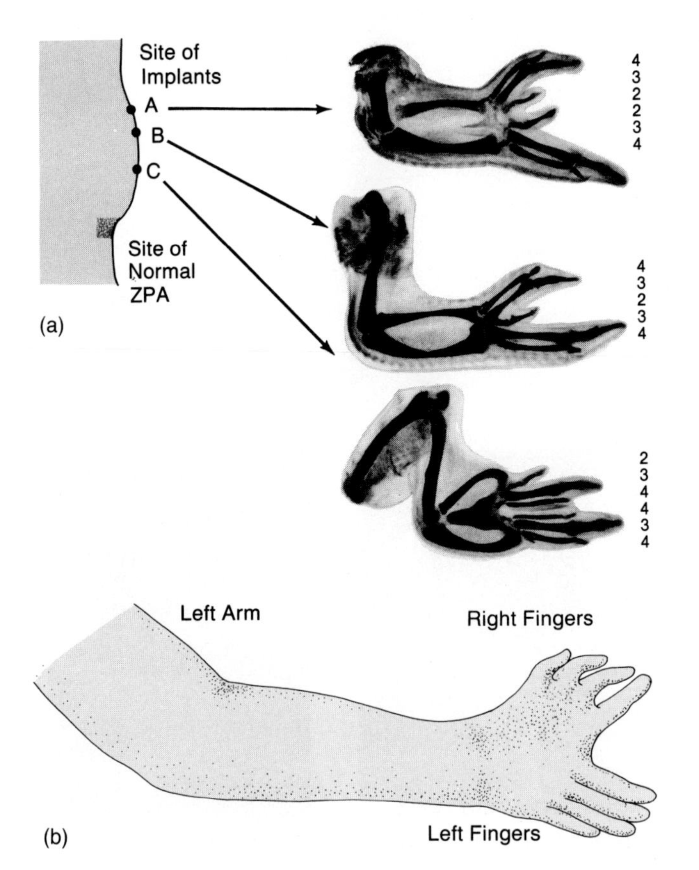

◀ **Figure 17-11 THE ZPA AND DIGIT REVERSAL.**
(a) ZPA tissue normally grows at the posterior part of the limb bud, where the bud meets the body wall. If extra ZPA tissue is grafted to sites A, B, and C of a chick embryo, it induces the formation of extra digits (whose numbers are shown to the right). These digits are always in inverted order, since the posterior-most digit is nearest to the ZPA. (b) An adult woman's hand on which the anterior, extra sets of digits formed instead of a thumb. Their order is reversed, just as though a ZPA activity acted as at site A in the chick experiment.

trations, farther from the source, allow anterior digits (3, 2, and 1) to form.

In fact, in 1988, researchers confirmed that normal chicken wing buds contain a derivative of vitamin A called *retinoic acid* and that this material indeed acts as a ZPA morphogen. They suggest that the cells in a developing limb can detect the presence of a retinoic acid gradient and form the different types of digits, bones, and parts of the limb in response. Not surprisingly, a new and powerful acne drug with a related structure can cause severe birth defects in people (see box below).

AN ACNE DRUG AND DEVELOPMENTAL DEFECTS

Most teenagers have acne, but ordinarily it is mild, and the blemishes disappear in early adulthood. A small percentage, unfortunately, develop a severe variety called cystic acne that can persist for decades and can leave the sufferer with a face or back full of deep pits and marks, not to mention deep psychological scars. In 1982, a Swiss pharmaceutical company released a drug called Accutane that can significantly help—even permanently cure—victims of cystic acne. But there is a catch: The drug can cause severe birth defects.

Accutane is a derivative of vitamin A called isotretinoin, and it has a structure similar to the morphogen retinoic acid (see above). Accutane, too, exerts profound influences on a developing fetus. If a woman becomes pregnant while taking Accutane, her baby may be born with a variety of defects, including stunted growth of the heart muscle, mental retardation, abnormally small jaws, and ears growing below chin level. Despite strident warning labels and the advice of physicians and pharmacists, hundreds of women have taken Accutane during pregnancy, and the U.S. Food and Drug Administration estimates that nearly 600 babies were born with Accutane-induced birth defects between 1982 and 1988.

One FDA official has compared Accutane to thalidomide, a sedative drug released in the early 1960s that caused limb malformations in about 6,000 babies before the drug was recalled from the market. Some observers would like to see Accutane recalled, as well, but many dermatologists oppose this, since the drug is by far their most effective tool for treating cystic acne. Researchers are continuing to study the mechanisms by which Accutane alters fetal development and to search for safer alternatives. In the meantime, the FDA recently directed the drug's manufacturers to print photographs of deformed newborns on the medicine's label. Only continued public education about the *teratogenic* (birth-defect-inducing) nature of isotretinoin will prevent tragic deformities and the equally tragic loss of a wonder drug that is changing people's lives for the better.

How, then, does the third axis—the dorsal-ventral (palm–back of the hand) axis—develop? An experiment that involves removing and rotating the embryonic ectoderm by 180 degrees reverses the dorsal-ventral orientation of the limb. Thus, biologists assume that in a normal limb bud, the ectoderm imposes the dorsal-ventral axis on all the limb parts.

The progress zone (proximal-distal), the zone of polarizing activity (anterior-posterior), and the ectoderm (dorsal-ventral) act together to coordinate the development of the limb. An error in one of their functions may lead to extra digits, missing parts, or incorrect orientations. But consider the long-term consequences of such errors if they have a genetic basis: They could provide the variations in limb structure that are the very source of evolutionary change. No doubt the evolution of the basic five-digit limb of land vertebrates into bat wings, horse legs, and whale flippers (see Chapter 1) was possible because mutations affected the control systems of limb axis formation. Thus, evolutionary alterations in body structure result from heritable changes in the developmental process of individual organisms.

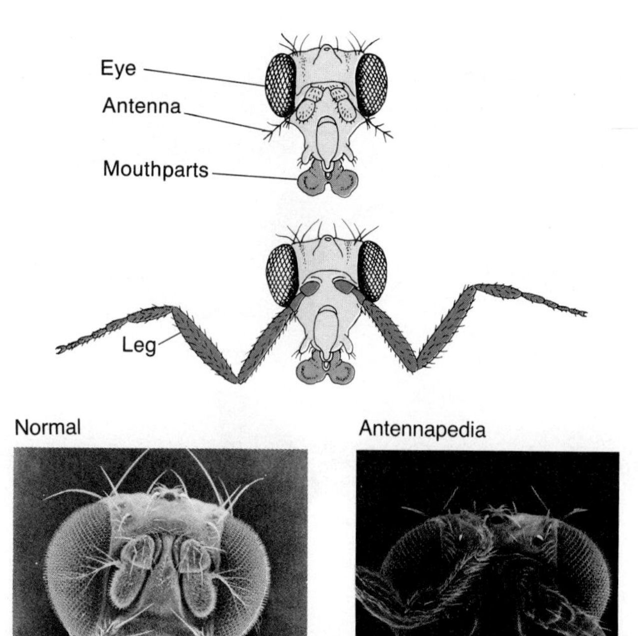

Figure 17-12 REGULATORY GENE MUTATION AT WORK. This is an extreme form of the antennapedia effect. Two full legs have grown from the sites on the head where antennae normally form.

REGULATORY GENES AT WORK DURING PATTERN FORMATION

Pattern formation of a single limb or of entire body axes must ultimately be based on gene activity, and the study of the regulatory genes that control pattern formation is one of the most exciting areas of biological research.

Researchers in the late 1970s and early 1980s found that in a bird, the difference between the upper arm bone and the upper leg bone (the humerus and femur, respectively), the thumb and the big toe, and the wing and the leg is shape alone—the way populations of differentiated cells are put together—not differences in specific structural proteins. These, in fact, are virtually identical, whether in bones, tendons, blood vessels, or other limb components. New work, however, suggests that a novel class of genes—the regulatory genes—determines shape by controlling morphogenetic processes. Regulatory genes apparently control batteries of structural genes, switching them on and off.

The best-studied regulatory genes come from the geneticist's workhorse, *Drosophila*. Biologists have identified a complex of genes in this insect called *antennapedia*, each of which can control the pattern formation of an appendage. A mutation in one such gene can cause a pair of perfect legs to grow out of the head in place of the two antennae (Figure 17-12). In a different *Drosophila* gene complex called *bithorax*, one mutation causes the fly to develop an extra thorax, or "chest" region, and an extra pair of wings. This and other evidence indicate that genes in the antennapedia and bithorax complexes act as simple switches that turn on and off batteries of genes that in turn control morphogenetic processes.

Researchers have found that at least 30 genes are expressed during egg formation in *Drosophila*, all affecting the anterior-posterior and dorsal-ventral axes of the zygote and future embryo. Experimenters can induce any region of a fruit fly egg to develop as the embryo's ventral surface, for example, just by injecting the product of the *Toll* gene. Other sets of genes (so-called *segmentation* genes) then determine the major body regions, and finally, *selector* genes (including the antennapedia and bithorax gene complexes) determine whether legs, wings, antennae, or other appendages will grow from a given region. The products of these regulatory genes can bind to other stretches of DNA and appear to act as switches that turn on and off the batteries of genes that actually carry out the shaping and formation of body parts during development.

A fascinating but still puzzling feature of regulatory genes such as those in the bithorax complex is the presence of a *homeobox*, a 180-base-pair DNA sequence. This same sequence has been found in sea urchins, frogs, mice, and humans. There is some evidence that the homeobox encodes a protein that in turn regulates batteries of genes involved in morphogenesis and differentiation. There is remarkably little difference in the amino acid sequences of homeobox proteins from one type of organism to another, and this suggests the conservation of developmental regulators during animal evolution.

GENE REGULATION IN DEVELOPMENT

Through their early experiments transplanting nuclei, biologists discovered that each determined body cell in a growing embryo usually retains all the original genetic material present in the zygote. Yet, once a cell is differentiated, it uses only some of those genes to manufacture the specific proteins that fulfill particular functions. One of the central research puzzles in developmental biology has been what turns on the correct genes for making that cell's characteristic protein.

Biologists have amassed a catalog of plausible mechanisms for the way genes are regulated during development. Here we present the most likely such mechanisms, organized according to the cellular level at which they operate: (1) the level of DNA processing, (2) the level of RNA synthesis and processing, and (3) the level of protein production and modification.

DNA Processing

Mechanisms of gene control that can occur at the level of DNA processing include gene amplification and gene rearrangement.

Gene Amplification

Classical genetics assumes that there are two copies of each gene in the usual diploid cell. As Chapter 16 explained, however, in the maturing oocytes of the African clawed frog, the genes that code for the structural RNA of ribosomes are *amplified* several thousandfold. As a result, huge numbers of ribosomes can be manufactured in reasonable amounts of time. But does this also mean that there are multiple copies of genes that code for proteins?

To answer this, biologists analyzed DNA in cells that make huge quantities of a single protein, such as the cells in silkworm glands, which produce silk. Most failed to find extra copies of the genes. The ovarian follicles of female fruit flies, however, make extra copies of three structural genes that code for eggshell proteins. Clearly, structural genes that code for proteins may themselves be amplified. While this case establishes the precedent, most differentiated cells do not in fact make extra copies of specific genes when they differentiate.

Gene Rearrangement

A unique mechanism in the cells of an embryo's immune system can shift pieces of genes around in differentiating cells to produce a variety of combined genes. We describe the details of this mechanism in Chapter 30 in our discussion of *antibodies*, proteins that help destroy foreign substances in the body. As we will see there, the process called **gene rearrangement** allows a relatively small number of gene pieces to be rearranged in a great variety of ways to generate millions of types of antibody proteins.

RNA Synthesis and Processing

Another way to state the central question in developmental biology is this: If all body cells contain all the genes inherited from both parents, how does each type of differentiated cell select and use only a tiny fraction of the full complement? The answer lies in **differential gene activity**—the functioning of some genes and not others.

Production of mRNA

Biologists have actually observed differential gene activity in the huge *polytene chromosomes* of certain insect cells (see Chapter 10). The characteristic banding patterns of such chromosomes result from a thousandfold duplication of DNA, with DNA helices aligned next to each other (Figure 17-13). When a specific genetic locus on a *Drosophila* polytene chromosome becomes active in synthesizing mRNA, the multiple copies of the DNA helices are somewhat unraveled from each other, and form a "puff." The unraveling allows the RNA polymerase enzyme to synthesize RNA.

Different puffs are active at different times during the development of a single *Drosophila* cell type. Certain sets of chromosome bands are puffed in salivary gland cells at one time, whereas a different set of bands is puffed at another time. What allows gene activity to be turned on and off at appropriate times?

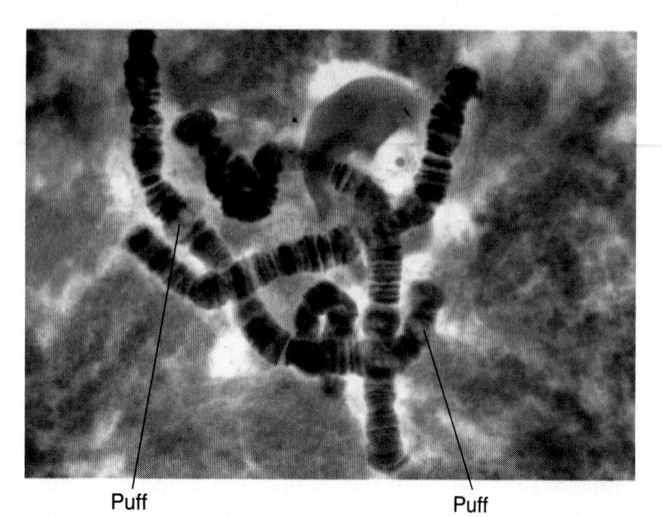

Figure 17-13 POLYTENE CHROMOSOMES OF INSECTS. This photograph made with a light microscope (magnified about 500×) shows genes at work. The many stained bands are unpuffed inactive genes. The light blue expanded "puffs" are sites of intense RNA synthesis.

To answer this, some researchers have been studying specific sites along the DNA that appear to regulate other genes. Sequences called *promoters* are located near the site where RNA polymerase binds to initiate the transcription of information from DNA to RNA. *Enhancers* are DNA sites that may be far removed from the promoter, but which increase the rate of initiation. Regulatory molecules, such as hormone receptors or so-called transcription factors, bind and thereby facilitate transcription of the gene. For example, when a steroid hormone binds to its receptor, the complex then binds to enhancers of specific genes. A transcriptional factor then binds nearby, allowing RNA polymerase to attach and synthesize RNA. Hence, researchers envision a multi-protein assembly that allows RNA polymerase to begin its transcriptional march along the gene.

From yeast to fruit flies to humans, many transcription factors and regulatory proteins share a curious feature: "zinc fingers" that bind to DNA. Fingers are amino acid sequences that contain zinc and bind to specific base sequences in DNA.

Some regulatory proteins lack zinc fingers but can still bind to DNA because they possess short amino acid sequences encoded by a homeobox. Referred to as "helix-turn-helix" structures, one of the α-helix units binds in the major groove of DNA. Repressor proteins in prokaryotes also have helix-turn-helix regions that bind to the operator DNA.

Biologists best understand the role of transcription factors in the genes that code for the class of ribosomal RNA called 5S. In *Xenopus,* the African clawed frog, there are two sets of 5S RNA genes in every cell, one called oocyte genes and the other somatic genes (Figure 17.14). Both sets are active in a maturing oocyte during oogenesis (review Figure 16-3). Later, however, in the cells of the embryo or adult, the oocyte 5S RNA genes become repressed, whereas the somatic genes remain active.

As Figure 17-14a shows, a transcription factor (A) with nine zinc fingers forms a stable complex with two other

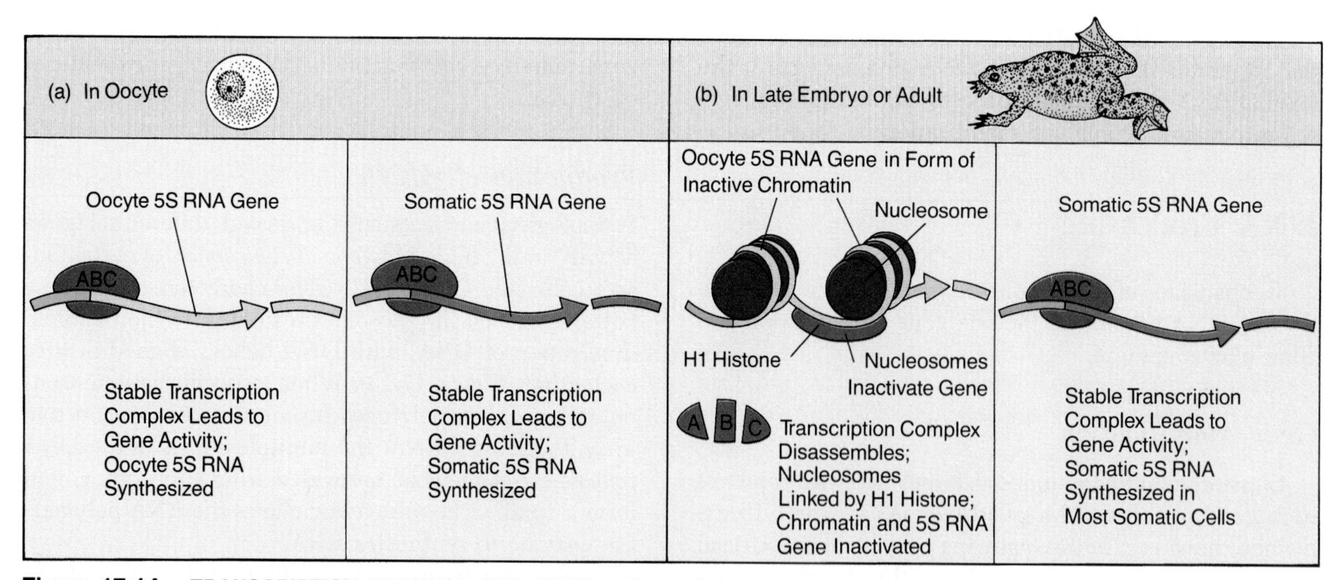

Figure 17-14 TRANSCRIPTION FACTORS AND GENE ACTIVATION. (a) In the oocyte of an African clawed frog, stable transcription complexes composed of three factors (A, B, C in a circle) are bound to the gene promoter regions of the oocyte and somatic 5S RNA genes, and both genes are transcribed. (b) In late embryonic and adult somatic cells, the stable transcription complex remains on the somatic 5S RNA promoter. But on the oocyte 5S RNA genes located in the same embryonic or somatic cells, the complex is unstable. The factors therefore leave those genes, and H1 histone can then link groups of nucleosomes into the inactive chromatin configuration.

factors (B and C), and all three bind to the promoter region of the somatic 5S RNA gene. Consequently, RNA polymerase can initiate transcription, and the oocyte nucleus manufactures copies of 5S RNA. Without this stable complex of A, B, and C transcription factors, transcription halts.

These fundamental findings give biologists the clearest picture so far of how determination, gene repression, and differential gene activity take place at the molecular level. Donald Brown of the Carnegie Institution in Washington, D.C., notes that any DNA not associated with stable transcription complexes will gradually be compacted into the repressed chromatin state by the protein histone H1. Apparently, in the *Xenopus* somatic and oocyte genes, differences in about three bases lead to the stable and unstable binding of transcription factors, respectively.

We can extrapolate from this example to other genes in developing cells: Regulatory genes may cause the cell to produce transcription complexes, which then maintain the genes in the active state and prevent their assembly into inactive, compacted chromatin. The determined state, with its characteristic heritable stability, may come about simply because transcription complexes keep certain genes in an active state, available for transcription, whereas other genes, by default, become compacted by H1 histone and are turned off.

Processing of RNA

Eukaryotic genes are separated into discrete sections (exons) interrupted by sequences of DNA called introns (see Chapter 14). Before a piece of RNA can leave the nucleus and act as an mRNA in translation, intron portions must be spliced out and special end sequences added. Cells that secrete the molecular glue called fibronectin (see Chapter 16) produce several different fibronectin proteins, depending on alternate splicing patterns of fibronectin primary RNA transcripts. This represents a case of *one gene–several polypeptides*, and it implies that the splicing enzymes present in a cell's nucleus are a key part of the machinery for regulating genes.

Storage and Stability of mRNAs

In most differentiating cell types, mRNAs are used soon after they enter the cytoplasm, but this is not necessarily the case. Some eggs can store a large quantity of inactive mRNA for months at a time, then use it rapidly during the cleavage period. A variety of plant seeds and pollen (see Chapter 27) also store inactive mRNAs ("masked messengers"). Certain mechanisms permit activation and use of these masked mRNAs at specific times in certain cells but not in others.

Not only can mRNAs be stored, but the molecules themselves can have long lifetimes. The mRNAs for specific proteins in differentiated cells—like ovalbumin in oviduct cells—have lifetimes of about 50 to several hundred hours. Since the mRNAs can be used as templates again and again, hour after hour, huge quantities of hemoglobin, ovalbumin, silk, and similar specific proteins may be manufactured on the long-lived mRNAs.

Protein Synthesis

There is yet another means by which gene expression can be regulated: *control of translation*, or protein synthesis. The presence or absence of other specific agents in the cytoplasm may affect protein synthesis. For example, a red blood cell will not synthesize *globin*, the protein part of hemoglobin, unless the cell has a supply of *hemin*, the precursor to the nonprotein part of the molecule, *heme*.

Yet another mechanism affecting the activity of specific proteins during differentiation is *modification* of newly manufactured proteins. This process sometimes involves chopping off and discarding pieces of a protein. Or it may entail the addition of sugars or phosphate groups to certain of the protein's amino acid side chains.

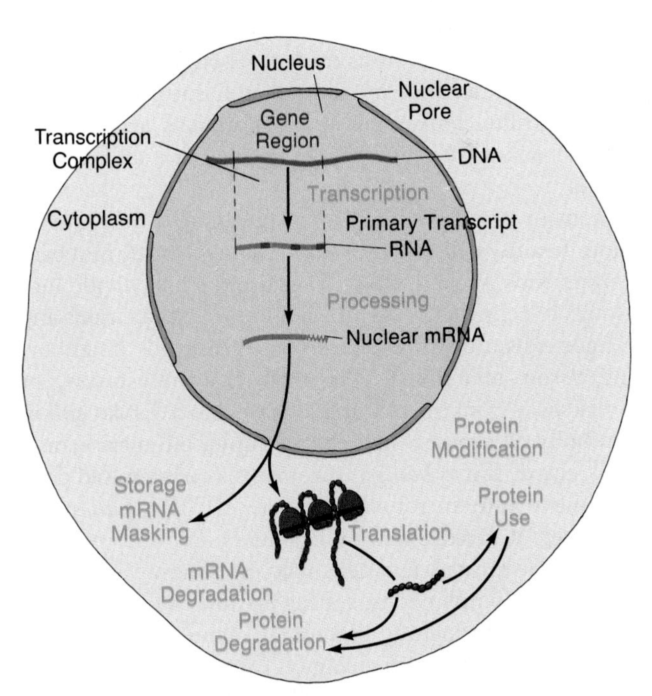

Figure 17-15 CONTROL POINTS FOR GENE EXPRESSION.
Each step or process labeled in green may serve as a potential control point in the overall expression of the genetic information on chromosomes as functional proteins.

One final process, *degradation*, regulates the levels of a specific protein in a cell. Short-lived cellular proteins have certain amino acids at their N-terminal (amino) end, while long-lived proteins, ones that persist for 20 hours or more, have other amino acids in the N-terminal site. These remarkable findings imply that cells have evolved a mechanism to ensure rapid turnover of some proteins and persistence of others.

To sum up what we know about gene regulation during development, differential gene activity is the basis for cellular differentiation and morphogenesis. Figure 17-15 summarizes the mechanisms that regulate gene expression during development.

CANCER: NORMAL CELLS RUNNING AMOK

Cancer is a scourge of plants and animals, and fossil evidence proves that it has been so for millions of years. The term *cancer* comes from the Greek word *karkinos*, meaning "crab," and derives from the large, red, claw-like arteries that feed the relentless growth of cancerous cells. In an animal with cancer, cellular growth controls function abnormally, and huge numbers of cancerous cells can accumulate, either as solid masses called **tumors** or as circulating cancer cells. Tumors known as **carcinomas** can arise in the epithelial sheets covering the outer and inner surfaces of the body, while tumors called **sarcomas** arise in connective tissues. Circulating cancers called **leukemias** and **lymphomas** arise in the blood-forming cells of the bone marrow and lymph nodes.

Cancer usually follows a set program: First, one or more healthy cells are altered, or *transformed*, into cancerous ones; second, these transformed cells divide into clones of cells that make up the tumor or the circulating cancer cells; third, the cancerous cells invade neighboring tissue; and fourth, the cells may **metastasize,** or break away from the tumor and spread to a distant site in the body. A breast tumor, for example, often arises in a milk gland. If it is *benign*, it remains as a lump and does not move into surrounding tissues. If it is *malignant*, however, it invades nearby ligaments, fat, and underlying muscle tissues, and cells may metastasize and lodge in adjacent lymph nodes, setting up new cancerous sites.

The transformation process in cancer cells brings about several profound changes. The most striking characteristic is unregulated growth; a cancerous tissue has no limit to size, unlike normal types of tissue. The surface properties and migrating behavior of cancer cells also change in some cases, as tumors produce motility factors that cause the cells to move about rapidly. In addition, cancer cells are abnormally long-lived and therefore tend to accumulate. The net result of uncontrolled mitosis and long-lived cells is an enlarging tumor, leukemia, or other malignant condition that disrupts normal tissues, ultimately to the point of killing the organism.

What Causes Cancer?

Clearly, the transformation of a healthy cell to a cancerous one is the key event in the development of cancer. But what causes it to take place? Scientists have spent many decades looking for *carcinogens*—cancer-causing agents—and have amassed a long list, including irradiation, chemicals, viruses, and genes.

Excessive exposure to ultraviolet light in sunlight, to x-rays, or to radioactive emissions (such as at Hiroshima or Chernobyl) can cause cancers. Thousands of chemicals are also carcinogens in some species. These include the hydrocarbon tars in cigarette smoke; substances in the blackened parts of charred meat; such dietary additives as red dye #2; and industrial solvents such as benzene. Dozens of viruses are also known to cause animal and plant cancers. Some of these viruses carry cancer-causing genes, or **oncogenes** (which we will discuss shortly), while others do not. Finally, abundant statistical evidence shows that families can pass on genetic tendencies toward cancer. Some families have histories of breast cancer among female members, while others have high rates of melanomas (deadly pigment cell cancers) or of prostate cancer among males.

Proto-Oncogenes in Cells

Researchers have shown that certain normal cellular genes may help initiate cancer. Every normal embryonic and adult cell has genes that code for the proteins and enzymes involved in mitosis, locomotion, adhesion to other cells, receptors for growth factors or hormones, and so on. Some of these normal cellular genes, however, also behave as **proto-oncogenes;** that is, they can cause cancers when altered in some way by mutations or chromosome translocations. In cancer patients with Burkitt's lymphoma, for example, a piece of chromosome 8 is abnormally attached to chromosome 14. As a result, mutations occur at a specific region of a proto-oncogene called *myc* (pronounced "mick"). In normal cells, the *myc* gene is on chromosome 8 and is inactive. In Burkitt's lymphoma patients, however, the *myc* gene is found at a new chromosomal site on 14 and becomes activated. These patients develop cancer and incur a high probability of death. Then, too, human lung cancer can be associated with a small deletion from chromo-

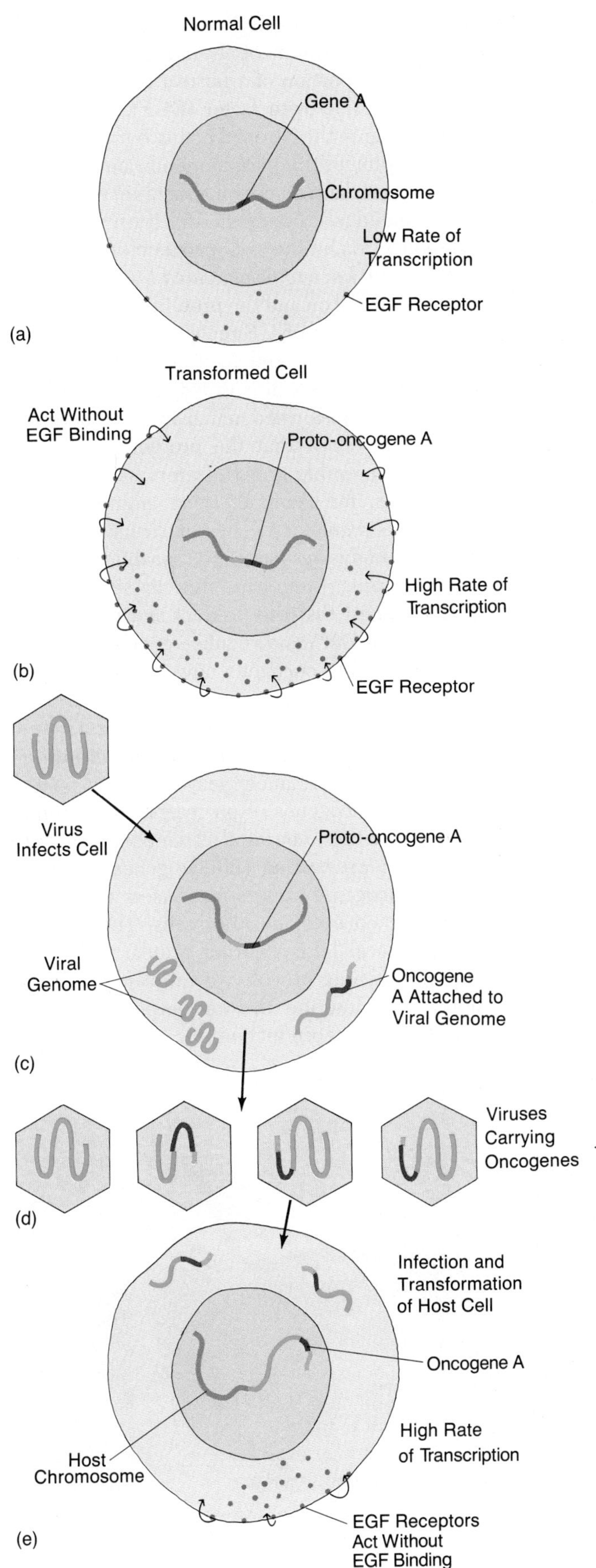

some 3; this alteration presumably eliminates a normal allele and leaves a mutant, cancer-causing allele to work. In still other cancers, another control region from viral genetic material becomes inserted near a cell's proto-oncogene; the result is transformation of the cell into the cancerous state. This is one way viruses lacking oncogenes themselves can still cause cancer. Cellular proto-oncogenes can be amplified, that is, can be present in multiple copies. This occurs in several *myc* genes as well as in the HER-2/*neu* gene. The apparent result of this amplification is a much greater likelihood that a woman will have a recurrence of breast cancer.

It is not farfetched to imagine a connection between the long list of carcinogens and the activation of the proto-oncogenes that may already be present inside normal cells. Chemicals, radiation, or viral infections may induce such mutations, chromosome translocations, or DNA insertions in proto-oncogenes, and these events, in turn, may trigger the cancerous transformation of a cell.

Viral Oncogenes Are Derived from Cells

Researchers made a startling discovery in the early 1980s: Cancer-causing genes in many viruses actually originated from proto-oncogenes in animal cells, and not in the viruses themselves. Biologists have studied several dozen viruses that carry oncogenes, including 20 that are *retroviruses* (viruses whose genetic material is a single-stranded RNA molecule). When a retrovirus, such as the Rous sarcoma virus, infects a host cell, a copy of the viral DNA is incorporated into the host cell's chromosomal DNA (Figure 17-16). There, the viral DNA oncogene is reproduced each time the cell goes through a division cycle. The viral genome itself may also be duplicated, yielding hundreds of new virus particles, each carrying a copy of the oncogene.

◀ **Figure 17-16 ONCOGENES, VIRUSES, AND CANCERS.** (a) A normal cell with gene A produces modest quantities of epidermal growth factor (EGF) receptors and grows slowly when EGF binds. (b) A change has occurred in the regulatory region for A, rendering it a proto-oncogene; many EGF receptors are made, and they function in the absence of EGF. The cell divides rapidly in an uncontrolled fashion. (c) A normal virus infects a cell. (d) By chance, copies of proto-oncogene A become attached to viral genetic material and are incorporated into viruses. Those cellular genes may mutate and become distinctive viral oncogenes. (e) When such viruses infect a normal cell, the oncogene may be incorporated into the host genome, transforming the cell. In this hypothetical case, its product is shown as the EGF receptors that function in the absence of EGF and cause uncontrolled cell division.

A natural assumption might be that oncogenes originate in the viruses that contain them. However, studies of 16 cancer-causing viruses show that the nucleotide sequence of the oncogene corresponds quite closely to those of 16 cellular proto-oncogenes. For example, the oncogene of Rous sarcoma virus (which causes tumors in chickens, abbreviated *src*, pronounced "sarc") is present in *normal* chicken cells. Surprisingly, genes almost identical to *src* are present in the DNA of humans, other mammals, and even fish. Thus, the *src* DNA sequence has been conserved during evolution and makes essentially the same protein in all vertebrates. Yet, the same gene, altered only slightly and injected by the virus, can cause cancer.

Biologists have concluded from these findings that the *src* oncogene originated in a cell, not in a virus, and that copies of it may have become inserted into the viral genome.

Oncogene Products: Proteins Related to Normal Growth Factors

Yet another recent finding helps explain the uncontrolled growth of a transformed cell—why, in other words, the cancerous cell runs amok and can lead to a tumor (Figure 17-17). The products of oncogenes usually become associated with the host cell's surface or cytoskeleton, are present in large quantity, and usually act as enzymes that catalyze the addition of phosphate groups to tyrosine residues in cellular proteins (hence, they are called **tyrosine phosphokinases**). Significantly, the addition of phosphate can alter all sorts of activities carried out by proteins; hence, an oncogene product—in plentiful supply—may literally trigger a cascade of events by phosphorylating a variety of cellular proteins.

Perhaps the most intriguing finding about oncogene products is that the protein of the *src* gene is almost identical to a major portion of a normal cell surface receptor for epidermal growth factor (EGF), a hormone that controls cell growth. Normally, this hormone must bind to the receptor in the plasma membrane and activate it (by phosphorylating a tyrosine residue) before cell growth is stimulated (see Figure 17-16). It turns out that the protein coded for by the *src* oncogene *acts as a receptor that does not have to be activated by EGF*. When the oncogene is turned on and the protein is produced, it functions like an activated EGF receptor *that cannot be turned off*. This puts the cell in a state of perpetual growth and division. The cell is thus tricked and shows all the features of a transformed malignant cell, including metastasis. Studies show that the protein products of other oncogenes resemble growth factors. A number of human oncogenes, for example, share amino acid sequences with the regulatory genes or structural genes that help specify body axes, segments, or differentiation in organisms ranging from fruit flies to sea urchins. These results suggest that the product of a proto-oncogene would bind to DNA and regulate gene expression, and products of a proto-oncogene now under investigation called *c-jun* do just that.

The pieces of the great cancer mystery are falling into place fairly quickly, and we can see that a predisposition toward a certain type of cancer may be based on the presence of proto-oncogenes or on rearranged chromosome pieces. Radiation, chemical carcinogens, and viruses may act on preexisting cellular genes or proto-oncogenes; or oncogenic viruses may insert the deadly baggage they picked up from other cells. The result is the same: the abnormal use of normal cellular molecules, and the life-threatening process we call cancer. In our next chapter, we describe the life-perpetuating processes of mating and reproduction.

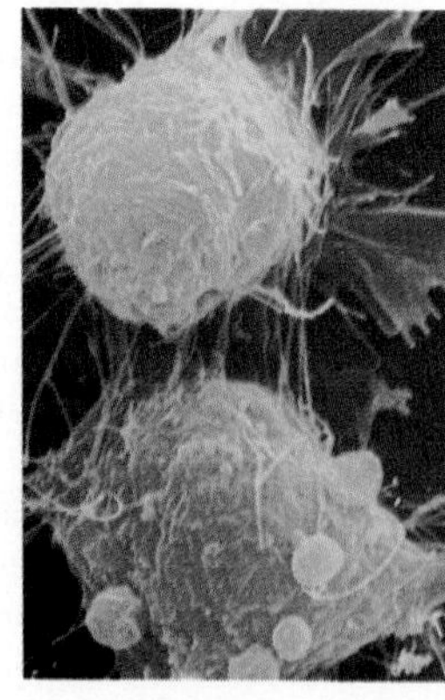

(a)

(b)

Figure 17-17 MALIGNANCIES: CELLS RUN AMOK.
(a) This neuroblastoma cell is the type of transformed cancerous cell often found in the adrenal gland tumors of children. (b) This mouse has a tumor of the mammary gland. If unchecked, such growths usually kill the organism.

SUMMARY

1. *Determination* is the commitment of a cell to a specific type of differentiation. Once a cell population is determined, the state is usually permanent and is passed on to the cells' progeny. The determined state can change as the result of exposure of the nucleus of a determined cell to the cytoplasm of a different cell type.

2. Determined cells undergo *differentiation*, or functional maturation. Differentiation may be triggered by inductive *tissue interactions*, or by regulatory factors (determinants) built into eggs.

3. A differentiated cell (a) makes and uses a specific set of proteins; (b) has "housekeeping" proteins to help keep it metabolically active; (c) assumes a characteristic shape; and (d) may cease to undergo further cell division.

4. Morphogenesis (at the cellular or organ level) can be produced by the movement of single cells; the interaction between moving or stationary cells and substrates; the movement of cell populations; localized cell growth; localized *cell death;* and the deposition of *extracellular matrix*.

5. Pattern formation includes development of an organ's or body's primary axes (proximal-distal, anterior-posterior, dorsal-ventral). In the arm, for example, the *progress zone* controls proximal-distal development, the *zone of polarizing activity* (*ZPA*) controls anterior-posterior development, and the ectoderm controls dorsal-ventral development.

6. Regulatory genes control batteries of other genes that are responsible for pattern formation. The body plan, its number of segments or limbs, and the shape and position of parts are controlled by regulatory genes. Some regulatory genes code for small proteins that bind to enhancers or promoters on DNA and thereby control activity of other genes.

7. Mechanisms of gene regulation that occur at the level of DNA include gene amplification (the making of extra copies of a gene) and *gene rearrangement*.

8. *Differential gene activity* involves transcriptional control and occurs in the synthesis and processing of mRNA. Proteins such as hormone receptors and transcriptional factors bind to enhancers and promoters to initiate transcription. The molecular basis of determination and of differential gene activity may involve the presence of stable transcription complexes on certain genes, whereas other genes are rendered inactive because H1 histone compacts their nucleosomes into inactive chromatin. Gene regulation may also occur in the processing of RNA transcripts (so that RNA sequences encoded by introns are deleted), including the rearrangement of exon-coded regions to yield different proteins from one gene. Messenger RNAs may be stored in inactive form as another means of regulating production of proteins.

9. Gene regulation at the level of protein synthesis and storage includes translational controls through the availability of necessary precursors, enzymes, energy, or cofactors; the modification of newly manufactured proteins; and the speeding up or slowing down of *protein degradation*.

10. Cancers are diseases in which normal cells are transformed into a state of sustained mitosis, long life, and altered surface properties. Cancerous cells can accumulate as *tumors* or circulate in the blood, and many can *metastasize*, and set up new sites in the organism.

11. Normal cells may contain *proto-oncogenes*, prospective cancer-causing genes, which when acted on by environmental factors (carcinogenic chemicals, radiation, or viruses) can lead to cancer.

12. Cancer-causing viruses may carry *oncogenes*, cancer-causing genes originally derived from animal or plant cells.

13. Cellular oncogenes may produce variations on molecules used normally in cell growth control. For example, the *src* gene codes for a molecule much like an epidermal growth factor receptor and tricks cells into uncontrolled cancerous growth.

KEY TERMS

cell death
determination
differential gene activity
differentiation
extracellular matrix
gene rearrangement
imaginal disk
leukemia
lymphoma
metastasis
nuclear transplantation
oncogene
pattern formation
progress zone
proto-oncogene
tissue interaction
transdetermination
tumor
tyrosine phosphokinase
zone of polarizing activity (ZPA)

QUESTIONS

1. Compare determination and differentiation. In what sense is determination heritable?

2. Some cells divide rapidly, some slowly, and some not at all. What are some examples of rapidly dividing cells? Slowly dividing cells? Nondividing cells? Is the rate of mitosis related to the survival time of the cells?

3. What developmental mechanisms give rise to morphogenesis? Give a specific example of each such mechanism.

4. At what stage in the development of an embryo are the axes of symmetry established? Describe how each axis is set up in the development of a forelimb.

5. What is a structural gene? What is a regulatory gene?

6. Each polytene chromosome contains about 1,000 DNA molecules lined up side by side. How do "puffs" relate to these multiple copies?

7. Give specific examples illustrating at least five methods by which expression of genes can be regulated.

8. Give an example of an inductive tissue interaction.

9. Proto-oncogenes have been found in yeasts, insects, fishes, birds, and mammals. They code for factors that are essential to normal growth and development. Explain how agents that damage DNA can cause cancer. Explain how viruses can cause cancer.

ESSAY QUESTIONS

1. Explain how mistakes in development (in nature or induced in the laboratory) can help us understand normal development. Give some specific examples.

2. How might mutations in regulatory genes generate a longer earthworm, an insect with six pairs of wings, or a human with nine fingers on each hand?

For additional readings related to topics in this chapter, see Appendix C.

18 Animal and Human Reproduction

The child lives inside its mother for nine months, floating weightlessly in a dark wet world of amniotic fluid. At delivery, it will literally be pressed and pushed out into a very different world. Not even the pearl diver returning to the surface experiences such a dramatic change.

David H. Ingvar, Stig Nordfeldt, and Rune Petterson, *Behold Man* (1974)

The individuals of every species, whether single cells or multicellular creatures, must reproduce. Although development is "center stage" in reproduction, the whole process includes the behavior, anatomy, and physiology of the adult males and females, whether sea urchins, frogs, or humans. We focus on human reproduction in this chapter because the subject has such high intrinsic interest and because biologists are so fully versed in its details.

The chapter describes:

- Reproduction in humans and other vertebrates
- The male and female reproductive systems and sexuality
- The origins of male and female gender
- Human pregnancy, prenatal development, birth, nursing, and birth control

The cord of life, the umbilical cord, carries this 4½-month human fetus's blood to and from the placenta, where oxygen and nutrients are added and carbon dioxide and wastes removed.

REPRODUCTION IN THE VERTEBRATES: ANATOMY AND STRATEGY

Throughout animal evolution, from invertebrate ancestors to the fishes, amphibians, reptiles, birds, and mammals, including humans, the sexual use of haploid male and female gametes has been preserved. That basic strategy of sexual reproduction provides a powerful selective advantage: *increased genetic diversity* due to crossing over and independent assortment (see Chapter 9) and to the random meeting of a sperm and an egg.

Many of the structures and processes that ensure successful human reproduction arose in our ancestors, the reptiles and early mammals. It was in the reptiles that a system was perfected for producing sealed, desiccation-resistant eggs, with the four basic embryonic membranes that still characterize every human embryo. And even though the yolk has been absent for perhaps 100 million years in our lineage, the reptilian gastrulation mechanism also is still present in human embryos. Early mammals apparently evolved with the ability to retain the developing embryo inside the female's body for long periods. During this gestation time, the embryo was nourished and supplied with oxygen, yet it was guarded from attack by the mother's immune system. And after birth, the ancient mammals nourished their newborns with milk from mammary glands, just as humans and all other mammals do today.

Nowhere are biology and behavior more closely linked than in reproduction and sexual behavior (Figure 18-1). The drive to reproduce truly dominates the lives of many vertebrates, as illustrated by the salmon's fateful spawning run and the seasonal matings of deer, bear, and whales. Females of most mammalian species come into heat, or **estrus** (the period of sexual receptivity), at the same time each year. The timing of each species' estrus has evolved so that either (1) the young are born when the environment and weather make their survival most likely (e.g., elk estrus and mating taking place in October with the young being born in April) or (2) the young are born early in the year (a time when conditions are not optimal) so that several months later, when the young have grown and are demanding the most milk, maximal forage will be available to the lactating mother.

Among the primates, the order to which humans belong, estrus follows several patterns. Individual female apes and monkeys tend to enter estrus asynchronously; this means that mating and births can take place over much of the year. Such females will mate only when in estrus, and the chances of achieving pregnancy are thereby increased. Human females show a less distinctive estrus phase and can reproduce year round. It is

(a)

(b)

(c)

Figure 18-1 MATING BEHAVIOR AND THE DRIVE TO REPRODUCE.
Mating strategies in (a) a terrestrial invertebrate, the damsel fly; (b) a terrestrial vertebrate, the rare Indian rhinoceros; and (c) a marine invertebrate, the spiny sea urchin, which spawns or releases eggs and sperm directly into seawater.

significant that human females can engage in sexual activity independently of reproduction; no longer is sexual behavior closely tied to ovulation. "Sex for pleasure" is an immensely important innovation that has physiological roots (which we explore in the next two sections), as well as dependence on the human brain's higher centers and the complex behaviors they control.

THE MALE REPRODUCTIVE SYSTEM

To participate successfully in reproduction, human males, like their reptilian and mammalian ancestors, must produce sperm and the necessary fluids to carry and protect the sperm, must discharge the sperm within the female reproductive tract so that the male gametes can reach the egg, and must have appropriate hormonal and behavioral mechanisms to support and control these processes.

Production and Transport of Sperm

The *testes* are the site of sperm production in the male reproductive system (see Chapter 16 and Figure 18-2). Most of the tissue in the testes consists of the *seminiferous tubules*, narrow conduits which, if unraveled from a single human testis, would extend more than twice the length of a football field. The ideal temperature for production of viable human sperm is about 34° C, which is lower than normal body temperature (37.5° C), and the testes hang suspended in the saclike *scrotum* outside the warmer abdominal cavity.

Mature sperm are moved from the seminiferous tubules into a storage tube, the *epididymis*, before they are released into another tube, the vas deferens. The *vas deferens* passes from the epididymis into the body cavity and curves around to ultimately join the urethra. It is in the vas deferens that sperm become motile in preparation for their journey up the female's duct system toward the egg. It is also the vas deferens of each testis that is cut and tied off in a *vasectomy*, the permanent male sterilization procedure.

Smooth muscle cells wrapped around the walls of the vas deferens contract, helping to move the sperm along, while the *seminal vesicles* and the *prostate gland* contribute nutrients such as sugars, fatty acids called **prostaglandins,** and additional substances that increase the pH so that the sperm suspension becomes alkaline. The resultant fluid is called **semen.**

Sperm and semen next pass through the *urethra*, the tube that leads through the **penis,** the male copulatory organ. The bulk of the penis is composed of three spongelike masses of tissue: two *corpora cavernosa* and one *corpus spongiosum*, as shown in Figure 18-2. The

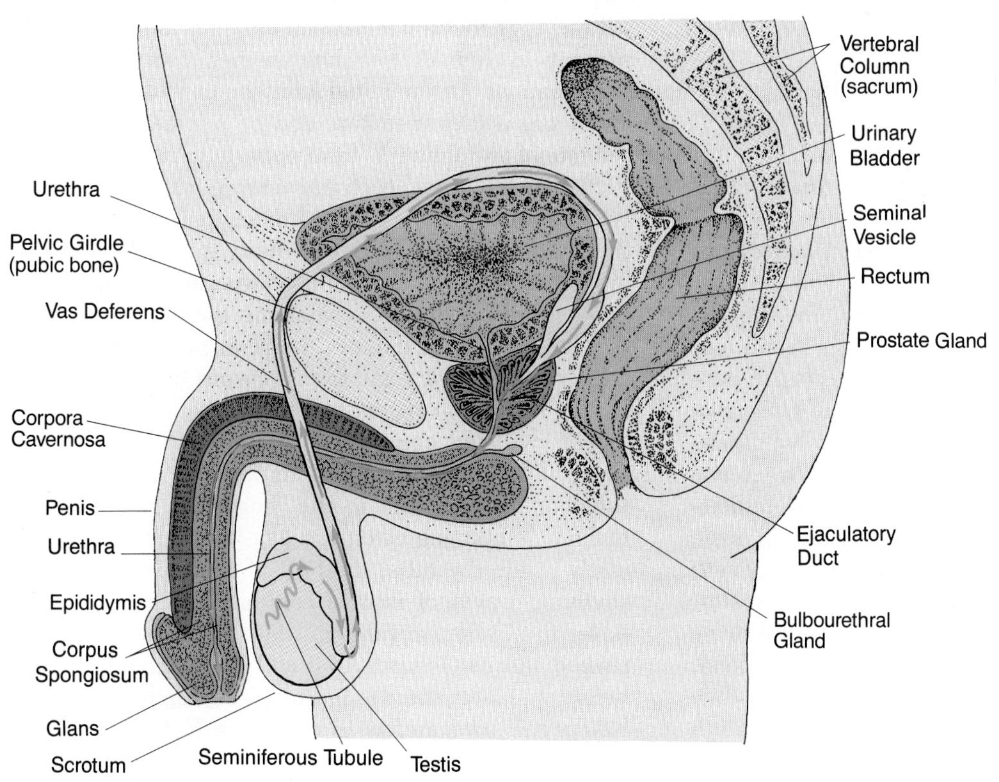

Figure 18-2 THE MALE REPRODUCTIVE SYSTEM. The route of sperm passage from the testis to the outside world is shown by the colored arrows.

end of the penis, called the *glans*, has a slightly larger diameter than the shaft and has an abundance of sensory nerve endings.

The Male Sexual Response

When the male is sexually stimulated, the spongy tissues of the penis fill with blood, the veins draining those tissues are squeezed shut, and the trapped blood causes *erection*, the swelling of the whole organ. Erection occurs in the first phase—the **excitement phase**—of the four-phase human sexual response identified by sex researchers William Masters and Virginia Johnson.

During erection, the *bulbourethral glands* secrete a slightly alkaline lubricating fluid into the urethra. This fluid neutralizes any residual acidic urine in the urethra and provides lubrication for movement of the penis in the vagina.

During the **plateau phase** of the male sexual response, blood pressure, heart rate, and breathing rate increase, and the testes enlarge from blood engorgement. Sexual excitement reaches its peak at the **orgasmic phase,** when a series of brief, rhythmic muscular contractions causes *ejaculation*—the jetlike expulsion of semen and sperm into the upper vagina. A typical male ejaculum consists of about 3–4 ml of semen, containing 300 to 500 million sperm. During the final phase, **resolution,** the penis returns to its unaroused size, and bodily processes such as heart rate and respiration gradually return to normal.

The Role of Hormones in Male Reproductive Function

Before a male can mature and function sexually, regulatory chemicals called *hormones* must come into play. Male sex hormones are collectively called **androgens.** The hormones that travel from the brain and pituitary gland to the testes (or ovaries in females) are the *gonadotropins*. The *hypothalamus*, a region in the brain (see Chapter 40), passes two releasing hormones (small peptides) to the *anterior pituitary*, an endocrine gland that lies at the base of the brain not far from the hypothalamus. These releasing factors trigger secretion of two gonadotropins, **luteinizing hormone (LH)** and **follicle-stimulating hormone (FSH),** into the blood. In male embryos and newborns, LH acts on the *interstitial cells* located between the seminiferous tubules of the testes. The interstitial cells respond to LH by manufacturing and secreting **testosterone,** an androgen, into the blood. The other pituitary hormone crucial to male reproductive function, FSH, acts on the seminiferous tubules to support sperm production. FSH is regulated by *inhibin* produced by the testes' sertoli cells.

As puberty approaches, testosterone causes various body changes, including the development of the **secondary sexual characteristics:** The penis, testes, and related glands enlarge and become sexually functional; body hair grows; the voice deepens; and general body shape changes as muscle and bone growth occurs. Without testosterone, most of the bodily features commonly considered masculine would fail to appear; like Peter Pan, the male would remain boyish forever.

THE FEMALE REPRODUCTIVE SYSTEM

Like the male, the adult human female must perform several tasks in order to participate successfully in reproduction. Her body must produce eggs, prepare the uterus for the embryo, participate in appropriate sexual behavior, meet the needs of the embryo as it develops, give birth to the child, nourish the newborn child, and develop hormonal and behavioral patterns that support these functions.

Production and Transport of the Egg

Human eggs, like those of other animals, are produced in the *ovaries* (see Chapter 16 and Figure 18-3). All the eggs that a woman will ever release are already present in her ovaries late in her own embryonic development. Of the initial 2 million or so oocytes, only about 400 will ever mature and be released, usually at the rate of one a month, from puberty to menopause. As Figure 18-3b shows, each egg matures surrounded by a layer of helper follicle cells. As the egg reaches maturity, the fluid-filled follicle bulges from the surface of the ovary. At a hormonal signal, **ovulation** occurs: The swollen follicle ruptures, releasing the egg. Soon after the egg has been expelled, the remnants of the follicle collapse in the ovary and form a **corpus luteum** ("yellow body," named for its appearance).

The ovulated egg must traverse a gap between the ovary and a tube with a fringed opening, the **Fallopian tube,** or *oviduct*. The motile, fingerlike projections at the tube's opening catch the egg, and once the egg is safely inside the tube, the beat of hairlike cilia and the rhythmic waves of muscle contractions move the egg down the Fallopian tube and into the muscular, pear-shaped **uterus** (Figures 18-3 and 18-4). About 10 days before ovulation, the uterine lining—the **endometrium**—has begun preparation to receive a fertilized egg by

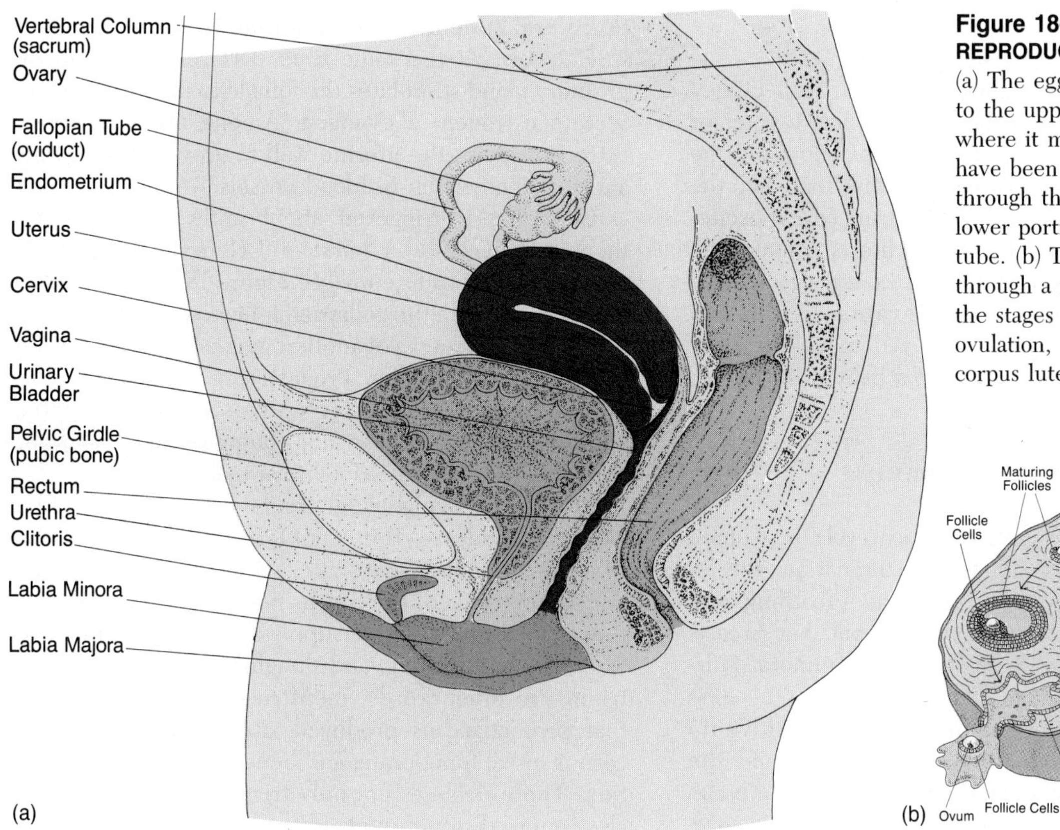

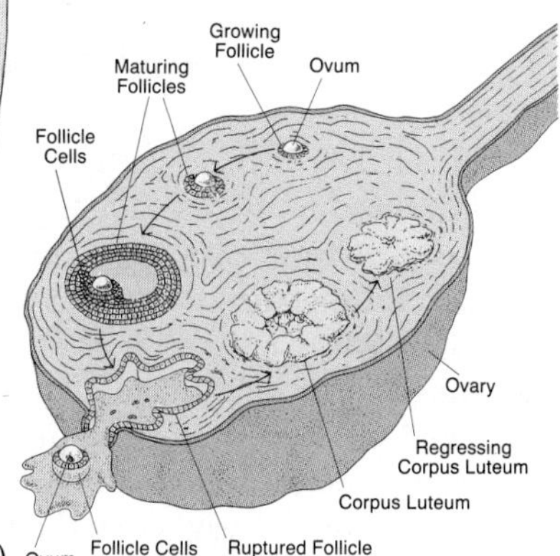

Figure 18-3 THE FEMALE REPRODUCTIVE SYSTEM.
(a) The egg passes from the ovary to the upper Fallopian tube, where it may meet sperm that have been carried from the vagina through the cervix, uterus, and lower portion of the Fallopian tube. (b) This idealized section through a mammalian ovary shows the stages of follicular maturation, ovulation, and formation of the corpus luteum.

thickening and filling with tiny blood vessels. If the egg is not fertilized, it passes through the uterus and is discharged, and the thick uterine lining subsequently is shed in the process called **menstruation.**

As the egg is being transported downward, various events may be easing the sperm's upward journey. Near the time of ovulation, an important change takes place in the **cervix,** the base of the uterus that also serves as the upper end of the **vagina,** the muscular passageway leading from the uterus to the outside of the body. The levels of various sex hormones present near the time of ovulation cause the mucus of the cervix to become much less viscous than it is during the rest of the monthly cycle. As a result, sperm can move through more easily en route to a chance meeting with the egg far up in a Fallopian tube. In addition, both the prostaglandins in the semen and **oxytocin,** a small peptide hormone released by the female's posterior pituitary gland during sexual intercourse, cause uterine contractions, which help move semen and sperm toward the Fallopian tubes and the egg. Tracts of cilia in the lower Fallopian tube may actually beat upward, thereby helping the swimming sperm to move up toward the egg. These various transport processes are rapid; it is estimated that human sperm may reach the upper third of the Fallopian tube in as little as 5 to 30 minutes following ejaculation. Even though 300 to 500 million sperm are released, the rigors of the journey, the dilution of alkaline semen by the acidic fluids of the female system, and the fact that about half the sperm may travel up the Fallopian tube lacking the egg mean that only a few hundred to a few thousand sperm may reach the level of the descending egg.

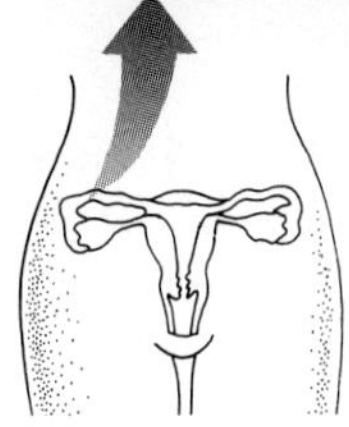

Figure 18-4 CILIA LINING THE FALLOPIAN TUBES.
Tiny, beating cilia on cell surfaces of the Fallopian tubes (magnified about 2,000×) propel the egg downward and perhaps propel the sperm upward.

Female Genitals

The human female external genitals, collectively called the **vulva,** consist of the *labia majora,* two major folds of skin; the *labia minora,* two inner folds; and the *clitoris,* a small, highly sensitive structure that, like the penis, becomes enlarged and erect during sexual excitement. The labia protect not only the clitoris, but also the openings of the urethra and vagina, the female organ of copulation. Because the walls of the vagina are thin and distensible, they can expand to encompass the erect penis and to permit the passage of a baby at birth.

The Female Sexual Response

The phases of the sexual response cycle in women parallel those in men. With sexual stimulation, the female's blood pressure, heart rate, and breathing rate increase during the excitement phase, and her breasts and nipples may also swell, as do the labia minora. During the plateau phase, the clitoris, which has become engorged, retracts upward so that its hypersensitive tip is better protected from further stimulation, and the outer third of the vagina swells and thickens into the *orgasmic platform,* which more effectively grips the penis.

A woman's orgasm is characterized by rhythmic contractions of vaginal muscles and intensely pleasurable sensations in the clitoral-pelvic area. Whereas men usually experience a sexually unresponsive refractory period, many women are capable of experiencing several orgasms without intervening rest periods. Following orgasm, the woman enters the resolution phase, during which the heightened physiological processes return to normal, and she may have a sensation of well-being and warmth.

The Role of Hormones in Female Reproductive Function

A variety of hormones secreted in various parts of the body coordinate reproductive cycles and sexual functioning in female mammals. These hormones regulate the **menstrual cycle,** the cyclic preparation of the uterus to receive an embryo, and the **ovarian cycle,** during which eggs mature and ovulation occurs (Figure 18-5).

The monthly preparation of the uterine lining for the fertilized egg normally begins at puberty with **menarche,** the first menstruation, and depends on several gonadotropins and **gynogens,** or female sex hormones. One gonadotropin, FSH, is secreted in minute quantities by the pituitary gland, and stimulates an ovarian follicle to commence intensive growth and the egg inside to mature. In females, inhibin from the ovary regulates FSH levels. At the same time, LH released from the pituitary gland stimulates the follicle to manufacture and secrete **estrogen,** a gynogen. Among other functions, estrogen causes the uterine wall to grow 2–6 mm thick and to become rich in blood vessels.

Ovulation is triggered about midway through the menstrual cycle by a burst of LH from the pituitary gland. The corpus luteum (see Figure 18-3b), the mass of cells formed from the collapsed follicle soon after the egg is released, secretes yet another gynogen, **progesterone,** in addition to estrogen. Progesterone promotes further development of the uterine wall, as well as preventing additional eggs from maturing, being released, and perhaps being fertilized following sexual activity.

The normal ovulated egg can be fertilized only within 24 hours of release. If it is not fertilized, it dies within a few days. About 11 days after ovulation, if the egg has not been fertilized, the corpus luteum begins to regress, thereby cutting off the supply of progesterone. This leads to the resorption and sloughing off of the endometrium—in other words, menstruation. Research shows that prostaglandins produced after ovulation decrease corpus luteal functioning and hence progesterone secretion. These declines not only trigger menstruation, but also renewed pituitary gland secretion of FSH and LH, and wish the latter, a new cycle starts.

Not until a woman reaches about 45 to 55 years of age do the ovaries lose their sensitivity to FSH and LH and stop making normal amounts of progesterone and estrogen, thereby ending the monthly menstrual cycles—all part of the **menopause.**

As in males, female secondary sexual characteristics develop at puberty in response to the presence of sex hormones. Estrogen secretion in females causes breast development, changes in body proportions and fat deposition, hair growth, initiation of ovulation and menstruation, and the female sexual response.

ORIGINS OF SEX DIFFERENCES: FEMALENESS AND MALENESS

Why do some embryos develop as females and others as males? The ultimate control of gender and sexual development resides in the chromosomes. Human females have two *X* chromosomes; males, an *X* and a *Y*. Females, therefore, lack the genes on the *Y* chromosome, and this difference has clear developmental consequences.

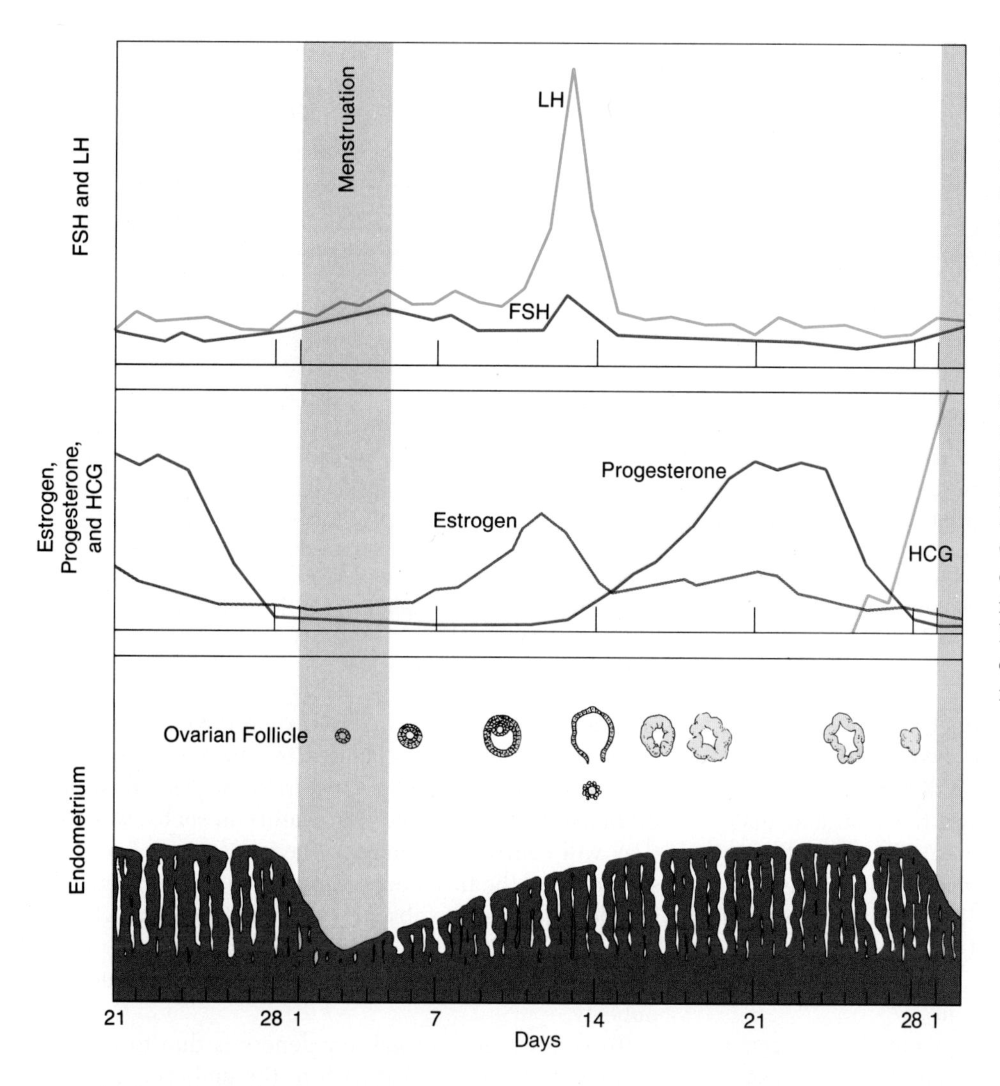

Figure 18-5 MONTHLY CYCLES OF HORMONE LEVELS IN THE HUMAN FEMALE. Pituitary FSH and LH control ovulation and production by the ovary of estrogen and progesterone. The endometrium grows thicker in response to estrogen and progesterone. If fertilization does not occur, progesterone levels fall, and the endometrium breaks down. If any embryo is present, it secretes human chorionic gonadotropin (HCG), which keeps the progesterone level high, and menstruation does not occur. Day 1 of each menstrual cycle begins when the menstrual flow starts and the endometrial lining is sloughed off. Beginning on day 5, the build up recommences.

In both male and female embryos, the early development of the gonads, internal duct system, and external genitals is identical. Figure 18-6a shows this *indifferent stage* of development.

An early step in gender development occurs when the gonial cells, the source of eggs or sperm, migrate to the indifferent gonads, each of which consists of a *cortex* on the surface and a *medulla* within. If the developing embryo has *XX* sex chromosomes, the cortex develops into the main tissues of the ovary. The cells surrounding each gonial cell become follicle cells, and the nearby connective tissue cells become interstitial cells that manufacture estrogen in response to LH. Conversely, if the embryo is *XY*, the cortex regresses, and the medulla develops into a testis. Gonial cells take up residence in the seminiferous tubules, where they begin sperm production later in life.

Depending on the direction of the indifferent gonads' development, the nearby tubes will develop a corresponding male or female pattern (Figure 18-6b and c).

If an indifferent gonad develops into an ovary, a nearby tube called the **Müllerian duct** develops into the Fallopian tubes, uterus, and upper portion of the vagina. If an indifferent gonad forms a testis instead, embryonic **Wolffian duct** forms the epididymis and the vas deferens for later sperm storage and transport. In females, the Wolffian duct regresses and disappears, whereas in males the Müllerian duct vanishes.

The external genitals also have an early indifferent stage when they are identical in the two sexes. A tiny swelling called the phallus forms the glans of the penis or the clitoris as development proceeds. Swellings surrounding the tiny phallus become either labia or scrotum, depending on the sex of the fetus. The differentiation of these external genitals depends on the direction taken by the developing gonads and the hormonal signals they produce and receive.

Hormonal Control of Sexual Development

Researchers have discovered that if they remove the indifferent gonads from a male and a female embryo,

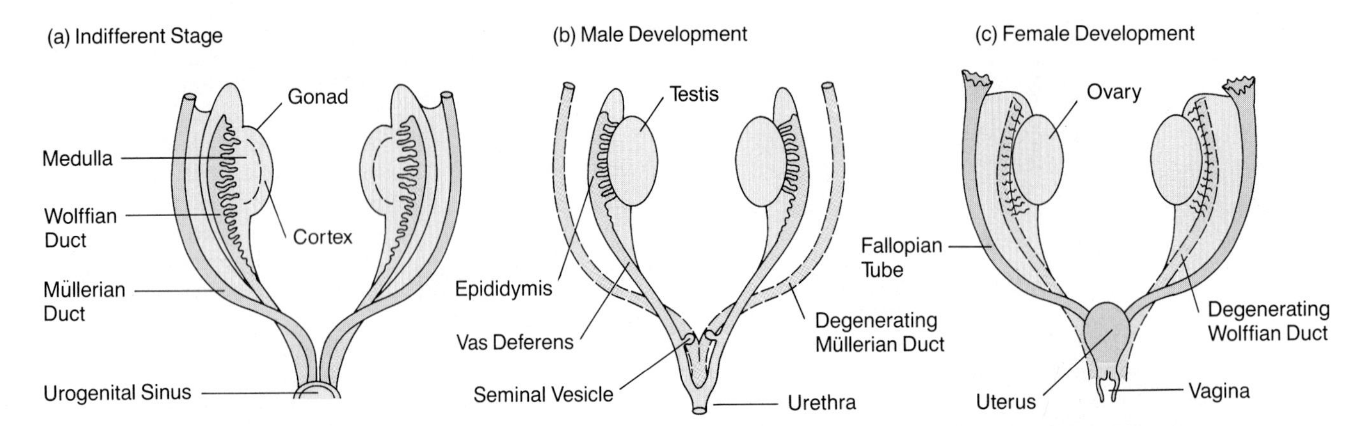

Figure 18-6 DIFFERENTIATION OF MALE AND FEMALE REPRODUCTIVE STRUCTURES IN HUMAN EMBRYOS.
(a) The indifferent stage, at about 5 weeks' gestation. (b) Male development begins at about the seventh week. In such embryos, the Wolffian ducts develop into the epididymis and vas deferens, while a nearby set of tubes called the Müllerian ducts degenerate. (c) Female development begins at about the eleventh week. Here, the Müllerian ducts develop into the Fallopian tubes, uterus, and vagina, while the Wolffian ducts degenerate. The indifferent-stage embryo has a tiny phallus surrounded by swellings. The phallus develops as the glans of the penis or the clitoris, depending on the embryo's sex, and the swellings become either scrotum or labia.

both will develop a **Müllerian duct** system (Fallopian tubes, uterus, and vagina). The two **Wolffian ducts,** which normally would yield the epididymis and the vas deferens, degenerate, even though the castrated male embryo has normal *XY* chromosomes in the cells of those structures. The embryos with gonads removed also develop female external genitals. Biologists therefore call the female body type the "neutral sex," since it develops in the *absence* of sex hormones, regardless of whether *XX* or *XY* chromosomes are present.

What, then, controls the development of male organs in a normal *XY* embryo? One gene and three substances play a role. A single gene, *testis determining factor* (TDF), or the so-called "sex switch," appears to regulate whether an individual will be male or female. If the gene is mistakenly transferred from a *Y* to an *X* chromosome, then presence of that factor in an *XX* individual will result in full maleness. Conversely, an *XY* person whose *Y* lacks the TDF gene will develop as a female.

When a normal *Y* chromosome with the TDF gene is present, the early gonad produces testosterone. This hormone causes each Wolffian duct to form an epididymis and a vas deferens, and some of the hormone is modified into another male hormone, *dihydrotestosterone,* which masculinizes the external genitals. Yet another hormone, **Müllerian inhibiting substance (MIS),** does what its name implies: It causes the Müllerian system to regress in males so that Fallopian tubes, uterus, and upper vagina do not develop. MIS, which is similar structurally to several mammalian and insect regulatory factors, apparently binds to DNA to accomplish these actions.

Sex hormones also affect the brain. Specifically, testosterone travels to the brain of a male embryo, where it is taken into and acted on by certain developing nerve cells. In the cells, it is changed by enzymes into estrogen. This estrogen causes clusters of nerve cells in the brain and lower spinal cord to mature in such a way that they will control such male behaviors as sexual responsiveness and the thrusting reflex of sexual intercourse. In female embryos, the absence of this estrogen effect permits the hypothalamus to develop in such a way that it can activate the menstrual and ovulatory cycles later, at puberty.

To summarize, normal femaleness is due to the absence of testosterone and MIS in the embryo, and normal maleness is due to the action of both testosterone and MIS in the embryo.

Intersexes—Hormones Gone Wrong

Errors in hormonal control of sexual development can produce individuals with both male and female sex characteristics, or **intersexes.** One type of intersex is the true **hermaphrodite,** an individual with both testes and ovaries. A more common intersex is the **pseudohermaphrodite,** a person having the gonads of one sex and the external genitals of the other.

While relatively uncommon, intersexes reflect the fact that all kinds of subtle gradations in sexual development and behavior can occur in humans; all of us passed through a neutral sex stage and went beyond it to varying degrees, depending on the intricate hormonal spectrum that shaped our bodies and brains. Prejudice against individuals who may not fit the stereotype of "male" or "female" has no biological or ethical justification.

Chromosome Imprinting and Development

The discovery of the "sex-switching" TDF gene helped biologists understand what switches on the production of male hormones in an individual with a *Y* chromosome and leads to male gender. A second line of research has revealed another fundamental difference between males and females at the chromosome level: a nonequivalency of the maternal and paternal genomes.

Through transplantation experiments, researchers have produced zygotes with two sperm nuclei or two egg nuclei. Curiously, even though each zygote has a diploid set of chromosomes, the embryos with only paternal chromosomes develop abundant trophoblast but almost no embryo body, while the embryos with only maternal chromosomes develop a more normal embryo body but very little trophoblast. All such embryos eventually die.

Clearly, the paternal alleles of particular genes must function in certain cells (those that make trophoblast); yet, a paternally derived *X* chromosome is always fully repressed in trophoblast cells, whereas paternal *X* chromosomes are active in some embryo cells. Maternal alleles must be expressed in certain cells (ones in the embryo's body); yet, a maternally derived *X* chromosome is always active in trophoblast cells. Researchers hypothesize that a parental imprinting process takes place on specific chromosomes and parts of chromosomes during oogenesis or spermatogenesis. Imprinting controls where and when alleles on such chromosomes can be used in development. Whatever the mechanism of the imprinting turns out to be, it should serve as a model for how genes are regulated during determination. At the very least, it already helps explain why mammals cannot reproduce parthenogenetically: Both maternal and paternal genomes are essential for a viable embryo.

PREGNANCY AND PRENATAL DEVELOPMENT

During pregnancy, the central process of human reproduction, the embryo grows to a full-term baby, nourished by the remarkable placenta. Pregnancy is arbitrarily divided into **trimesters,** periods of 3 months each. During the first trimester, beginning at fertilization, most of the embryo's organs take form, while during the remaining two trimesters, the developing young, now called a **fetus,** primarily grows.

The First Trimester

Once fertilized high in the Fallopian tube, the egg (now the zygote) follows the route shown in Figure 18-7. As the embryo is transported down the tube, it goes through several cleavages, becoming a solid ball of cells, the *morula* (Latin for "mulberry," which the cluster resembles). By the fourth day, this ball develops into the 50- to 100-cell blastula stage, called a *blastocyst,* with an outer layer of cells—the **trophoblast**—and an *inner cell mass* (see Chapter 16).

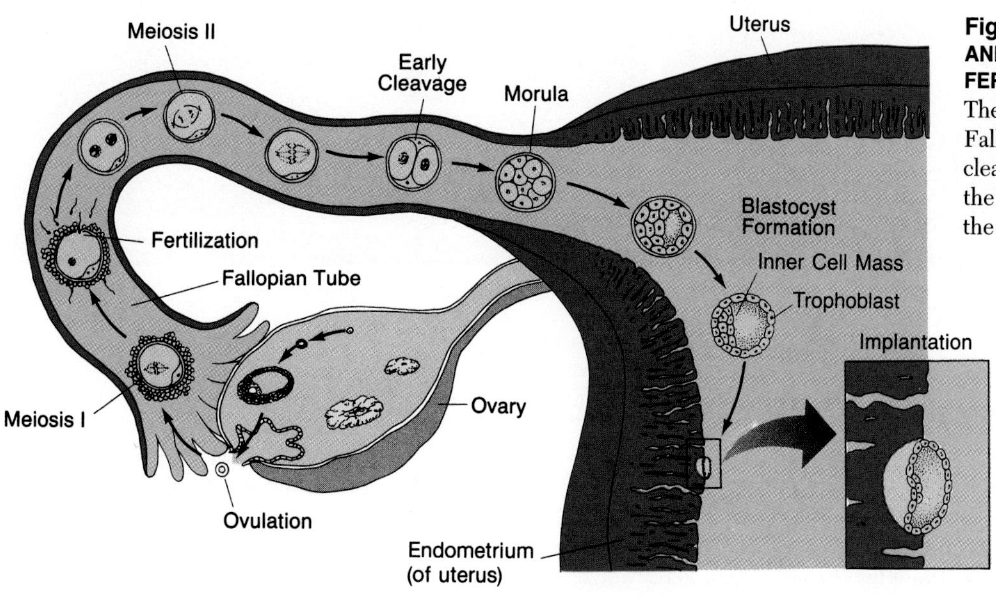

Figure 18-7 THE JOURNEY AND DEVELOPMENT OF THE FERTILIZED EGG.
The egg is fertilized in the upper Fallopian tube, undergoes cleavage while traveling down the tube, and finally implants in the endometrium of the uterus.

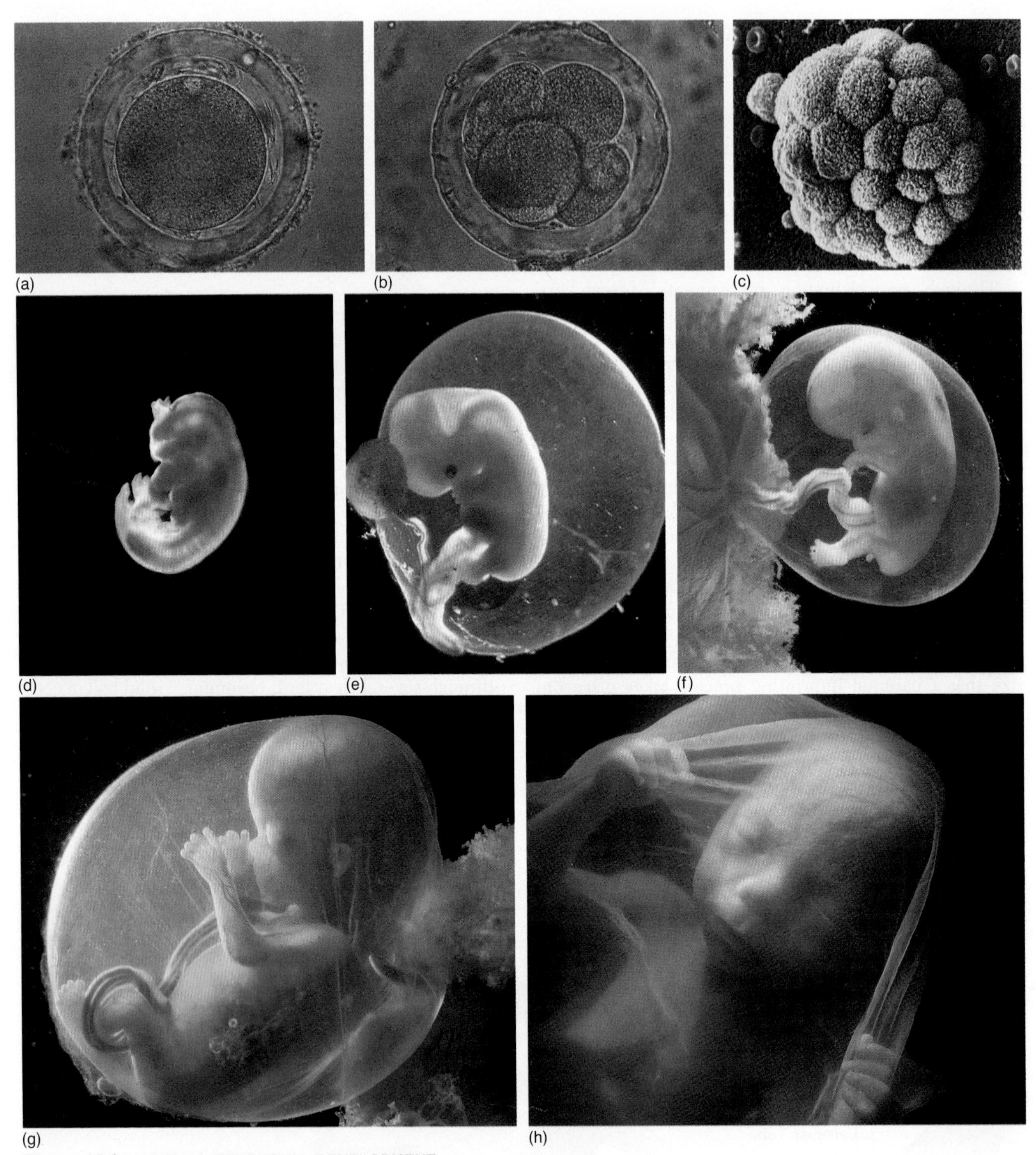

Figure 18-8 HUMAN EMBRYONIC DEVELOPMENT.
These are selected stages of the 9-month developmental odyssey that produces a baby. (a) A fertilized human zygote near the time of the first cleavage division. (b) The 4-cell stage. This embryo would still be traveling down the Fallopian tube. (c) The 64-cell stage. This blastocyst can implant in the endometrium of the uterus. (d) Between the fourth and fifth weeks of gestation. The major organs are forming; the bright red, blood-filled heart is just below the lower jaw and nearby gill slits (not visible). (e) The embryo at 6 to 7 weeks. The digits are beginning to form on the paddlelike limbs, the eye is pigmented and its clear lens can be seen, and the large brain is evident. The umbilical cord extends from the belly, and the fluid-filled amnion surrounds the embryo. (f) At 9 weeks. The fetus is recognizable as a primate, the limbs are elongated, and ears are clearly visible. (g) At about 3 months. The fetus is about 8 cm in length and weighs about 28 g. Its muscles begin moving the limbs and body. (h) At 4 months. The fetus is some 16 cm long, weighs 200 g, and grows rapidly until birth.

Next, the blastocyst enters the uterus, adheres to the uterine wall, and undergoes **implantation,** a process during which the trophoblast cells invade the endometrium and secure the embryo to the uterus. Implantation is usually completed 11 to 12 days after fertilization; from then on, the woman is considered pregnant.

The implanted embryo sends an important hormonal signal into the blood of the mother. The trophoblast cells secrete **human chorionic gonadotropin (HCG),** which stimulates the corpus luteum to continue secreting progesterone and estrogen. These hormones, in turn, do two things: They suppress release of FSH and LH by the pituitary gland, thereby preventing maturation and ovulation of another egg; and they stabilize the uterine wall, preventing menstruation and loss of the embryo. Home pregnancy tests detect HCG secreted in the urine.

Within the early embryo, only the inner cell mass gives rise to the embryonic body. These cells become arranged in a flat sheet, which subsequently undergoes gastrulation using a primitive streak, precisely as do embryos of reptiles and birds (see Chapter 16).

Once gastrulation has taken place, the remainder of the first trimester is devoted to organogenesis. The eyes, the brain and nervous system, the limbs, and most of the other organ systems develop, and the embryo's heart begins to beat. So many crucial regulatory events and inductive tissue interactions take place at this time that the embryo is particularly susceptible to foreign chemical agents (such as drugs, alcohol, or nicotine), and to diseases (such as rubella, or German measles). Rubella can cause severe abnormalities of the eyes, ears, brain, and heart of an embryo, while various drugs may lead to malformations in these or other organs (as do Accutane and thalidomide, as discussed in the box on page 297 in Chapter 17). As shown dramatically in Figure 18-8, the embryo grows substantially during the first trimester, and all its major internal and external organs develop.

The Second and Third Trimesters

Growth is spectacular during the second and third trimesters. The fetus's bones begin to harden, the brain adds millions of new cells, and the circulatory and respiratory systems prepare for the remarkable transition from being a fetus immersed in warm fluid to being an air-breathing newborn infant. By about the end of the fifth month, the fetus's circulatory and respiratory systems have developed enough to give it a chance of surviving if born prematurely—although only a moderate chance, even with intensive hospital care. During the eighth and ninth months, the fetus's weight doubles; organ development is completed, and all the essential organs will be fully functional at birth.

The Placenta: Exchange Site and Hormone Producer

The lengthy, intimate maternal-fetal relationship characteristic of mammals is possible because of embryonic membranes that originated in reptiles: the amnion, the yolk sac, the allantois, and the chorion (Figure 18-9). The latter two give rise to the embryonic parts of the **placenta,** the organ that sustains the embryo and fetus throughout pregnancy. In mammals, a fluid-filled sac, the *amnion*, surrounds, cushions, and protects the embryo. The rupture of the amnion, or "bag of waters," is often the signal that delivery is imminent.

The outermost membrane, the *chorion*, arises from the trophoblast, the early embryo's outer layer, which is involved in implantation (Figure 18-9a–c). The chorion grows thousands of fingerlike projections called **chorionic villi,** which provide a huge surface area for exchanging materials with the mother's endometrial blood vessels. Each villus eventually houses a tiny blood vessel, part of the embryo's vascular system (Figure 18-9d). This embryonic trophoblastic tissue and maternal endometrial tissue together make up the placenta, a roughly disk-shaped organ with an enormous surface area (167 m^2) for exchanging materials.

Many substances are exchanged at the placenta. Oxygen moves from maternal to fetal red blood cells. Carbon dioxide diffuses from fetal to maternal blood and is excreted by way of the mother's lungs. Nutrients pass from the mother's blood plasma into the fetus's. Waste products move from fetal to maternal blood plasma and are filtered away by the mother's kidneys. In addition, certain antibodies and hormones can pass across the placental barrier and act in the fetus or newborn.

The placenta remains connected to the abdomen of the fetus by the **umbilical cord,** in which the fetus's umbilical arteries and vein spiral about each other (see Figure 18-9d). The umbilical blood vessels are all that human embryos retain of the *allantois*, the fourth embryonic membrane of our reptilian and avian relatives, which utilize it as a storage site for wastes.

Besides being a site of exchange, the placenta serves as a barrier between fetal and material blood supplies and prevents certain kinds of molecules and cells from passing between mother and fetus. If the surface of the early placenta were not covered with a protective coating of molecules (one of which is thought to be HCG), the embryo, with its complement of paternal genes, might be recognized by the mother's immune system as foreign genetic tissue.

The placenta also secretes a variety of hormones for the embryo. Just when the embryo's production of HCG begins to drop, the placenta itself begins to manufacture and secrete large quantities of progesterone and estrogen. These hormones, as we saw, help sustain the endo-

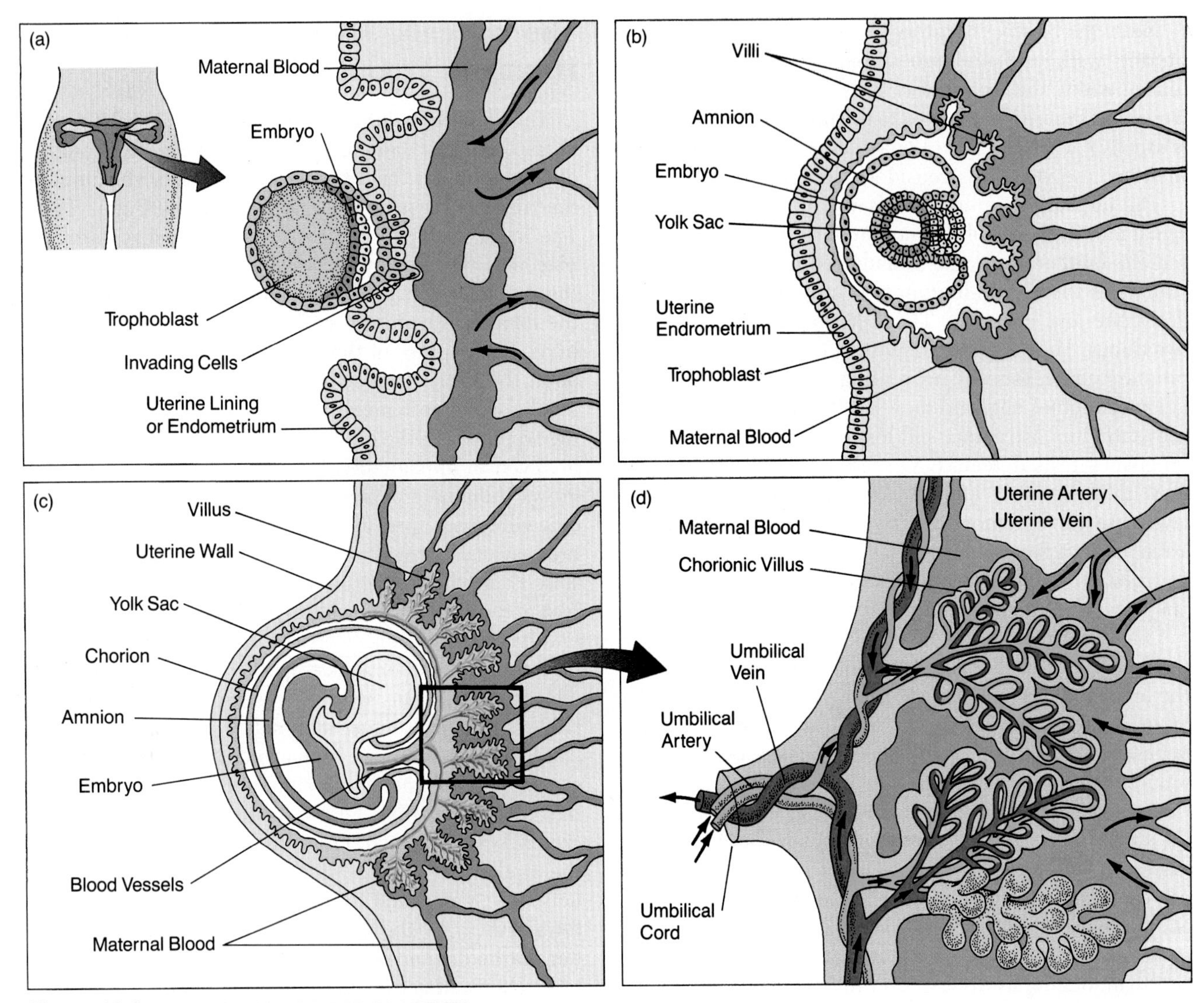

Figure 18-9 THE ORIGIN OF THE PLACENTA.
(a) The trophoblast implants in the endometrium. (b) The trophoblast is surrounded by uterine tissue, and trophoblastic villi of the placenta have started to form. (c) The number of villi and their surface area continue to increase to meet the needs of the enlarging embryo. (d) An enlarged view of mature chorionic villi housing blood vessels that are continuous with those of the umbilical cord.

metrial portion of the placenta and prevent a new cycle of ovulation and menstruation from taking place.

BIRTH: AN END AND A BEGINNING

About 266 days after fertilization, the human infant is born. Toward the end of pregnancy, **relaxin,** a hormone from the ovaries and placenta, acts to loosen the junction of the bones in the front of the mother's pelvis, so that they are better able to spread apart, allowing the baby to pass through without injury. The mother's uterine muscles are also keyed to act at this time, and the cervix is set to dilate, or open.

Changing hormonal levels are believed to initiate the process of birth, and the baby itself plays a key role in starting this chain of events. For birth to take place at the appropriate time, the fetal pituitary gland must secrete **adrenocorticotropic hormone (ACTH),** which stimulates the fetal adrenal glands to release steroid compounds. These compounds then apparently signal maternal cells in the placenta to manufacture and secrete two prostaglandins that are powerful stimulators of uterine muscle contractions; these contractions expel the fetus from the uterus. Biologists discovered the role of fetal hormones after noting that lamb fetuses will grow too large to be born if the mother sheep eats a certain

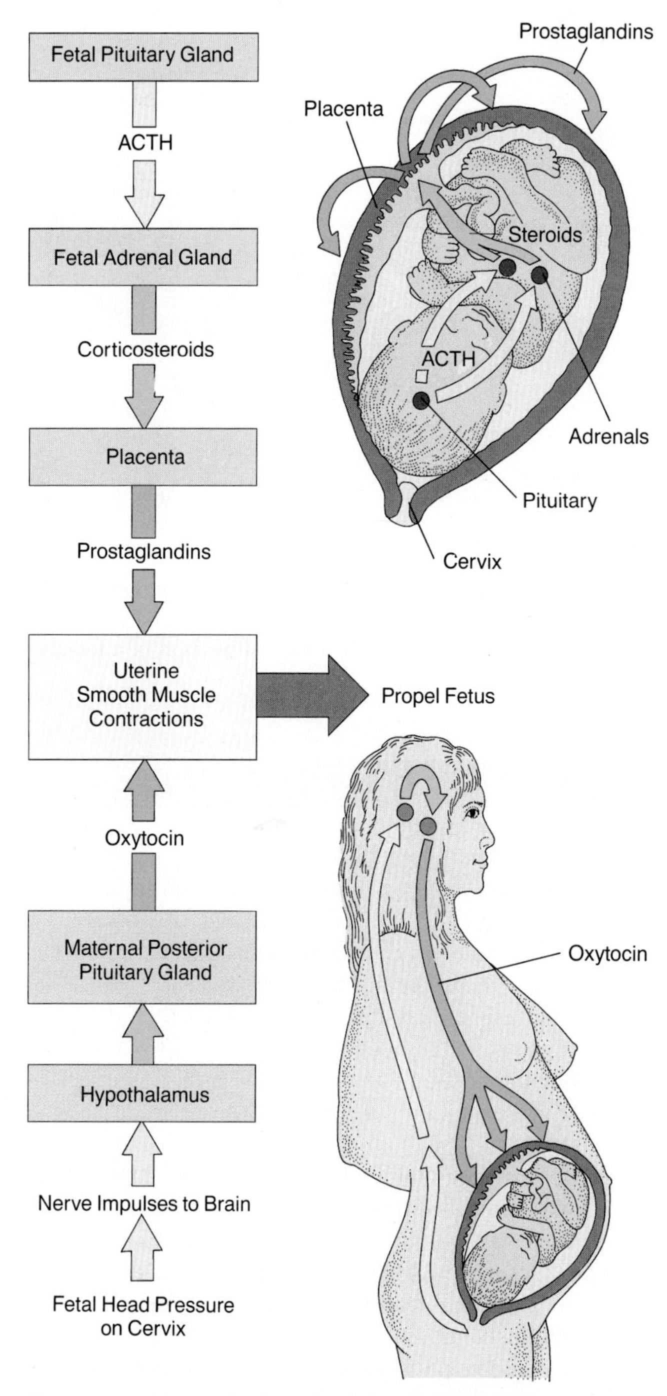

Figure 18-10 A HUMAN BIRTH IS TRIGGERED. Hormones from the fetus, the placenta, and the maternal brain must act together at the time of birth to propel the fetus along the birth canal and out into the world.

weed containing an alkaloid compound that interferes with the fetal pituitary and adrenal glands.

Other hormonal changes also precede birth, as Figure 18-10 shows. For example, the blood levels of progesterone, an inhibitor of smooth muscle contractions, drop as the time of birth approaches. Furthermore, the baby's head usually presses against the cervix, thereby triggering nerve impulses to the mother's brain. That causes the mother's hypothalamus to start releasing oxytocin from the posterior pituitary. Oxytocin works with the prostaglandins of the placenta to stimulate waves of muscle contractions in the walls of the uterus, forcing the baby downward on the dilating cervix. Ultimately, a series of very powerful contractions and a strong "pushing" by the woman advance the head and body of the infant through the cervical opening, down the vagina, and out into the world.

After the newborn emerges, uterine contractions continue and expel the placenta and attached membranes—now called the **afterbirth.** Following delivery, the medical team ties and cuts the umbilical cord. The baby is now an independent organism.

MULTIPLE BIRTHS

Twins are born about once in every 90 human births; triplets, once in every 7,500 human births; and quadruplets, only once in every 435,000 births. What biological processes are responsible for these relatively rare events? Human multiple births arise in two ways: (1) Monozygotic (identical) twins arise by fertilization of one egg followed by separation of the early cleavage-stage embryo into two (or more) developing systems (Figure 18-11a), whereas (2) dizygotic (fraternal) twins arise by fertilization of two eggs that happen to be ovulated during a single monthly cycle (Figure 18-11b).

While these events are rare in humans, they can be common in many other animals. The nine-banded armadillo of the southern United States and Mexico, for example, normally produces quadruplets. The fertilization of one egg by one sperm yields one cleaving embryo in which the inner cell mass develops four primitive streaks, or gastrulation sites, each of which produces a normal, genetically identical embryo.

Despite their genetic similarities, monozygotic siblings—whether armadillos or humans—are not precise physical copies of one another. The cytoplasm; the environment within the egg, embryo, or uterus; and the external environment after birth can all influence development and lead to variations among individuals.

Researchers have succeeded in producing a condition opposite to twinning by causing several embryos to fuse and yield a single individual. They can, for example, fuse a cleavage-stage embryo from a purebred strain of white mice with a similar embryo from a purebred brown strain, as shown in Figure 18-12. They then implant the aggregate of cells in the uterus of the host mother. When born 20 days later, the chimera (combined creature)

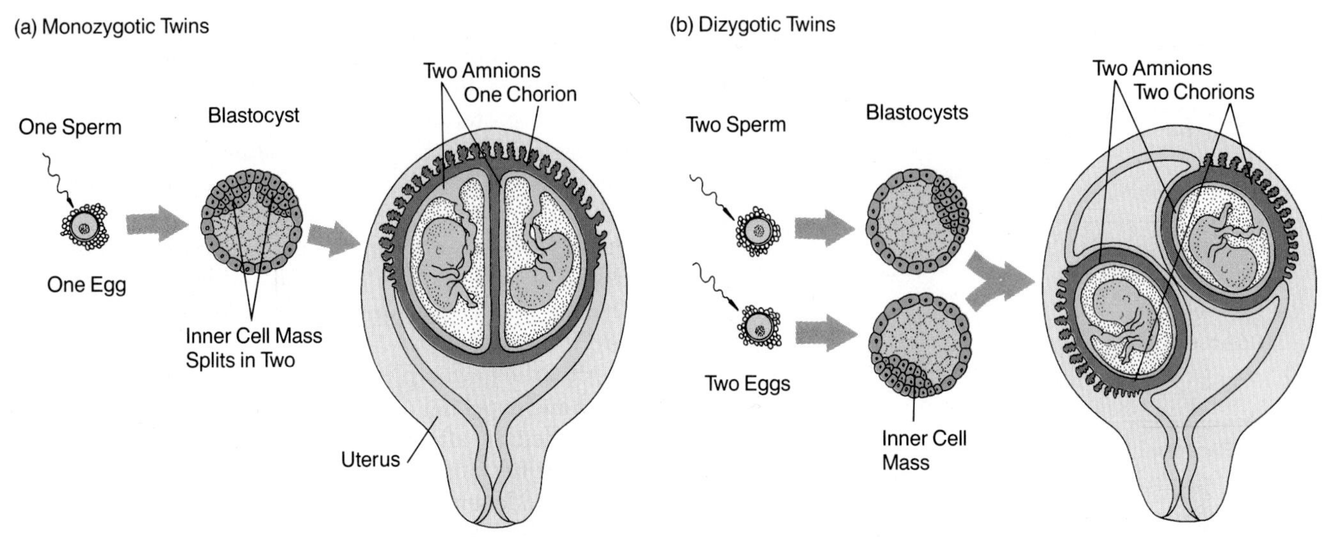

Figure 18-11 TWINNING IN HUMANS: MONOZYGOTIC AND DIZYGOTIC TWINS.
(a) Each inner cell mass is the source of an embryo. There is a single chorion (placenta) for identical twins. (b) The two chorions (placentas) of dizygotic twins normally remain separate. They may fuse, however, and as a result, the blood of the two embryos may mix. This situation can produce abnormalities in sex development, particularly a masculinization of a female embryo if its twin is a male.

IN VITRO FERTILIZATION: A GOING CONCERN

The first "test-tube baby," Louise Brown, was born to a British couple in 1978. A medical team accomplished this reproductive feat by removing an egg from the mother's ovary, combining it with the father's sperm in a glass dish, and reinserting the live, 2-day-old embryo into the mother's uterus. Since that historic birth, *in vitro fertilization* (literally, "fertilization in glass") has grown steadily from a bold experiment to an established medical procedure performed in clinics around the world. In vitro fertilization (IVF) is just one technique for overcoming infertility—a growing problem in our aging population. But the procedures associated with it have drawn considerable controversy.

Today, one couple in six cannot conceive without medical help, and the ranks of the infertile continue to grow as couples defer parenthood to their 30s and 40s and as venereal diseases and infections from IUDs cause the scarring of some people's delicate reproductive organs. An infertile woman can have missing or blocked Fallopian tubes, malfunctioning ovaries that fail to release eggs, or a cervix that prevents sperm from entering. An infertile man can produce too few or nonmotile sperm. And in some cases, the mother's immune system rejects and destroys the embryo, with its partially foreign (paternal) genes.

Physicians can now help nearly three-fourths of all infertile couples. Typically, they prescribe hormone-like drugs to induce ovulation, prevent immune rejection, or stimulate sperm production. If the drugs fail, surgery can often clear blocked passageways. IVF is an expensive procedure (more than $5,000 per attempt) and usually a last resort after other techniques have failed. It has a success rate of just 15 to 20 percent. For many couples, it offers the only possibility for reproduction, but for others, it poses ethical dilemmas.

At present, IVF technology is sophisticated enough to allow three procedures in addition to the external fertilization of a woman's egg by her husband's sperm: (1) The egg can be donated by a woman other than the wife, or the sperm can be donated by a man other than the husband; (2) an embryo can be produced from a couple's own gametes, but implanted in a second woman, a "surrogate mother," who carries the baby to term; or (3) a couple's embryos can be frozen, stored, and used later, whether by them or by an entirely different infertile couple.

So far, there are no federal laws prohibiting such procedures, but their moral and legal merits have been widely discussed by the public, the clergy, and medical ethicists. Each couple must decide how much intervention, and of what type, they are willing to accept in their attempts to conceive.

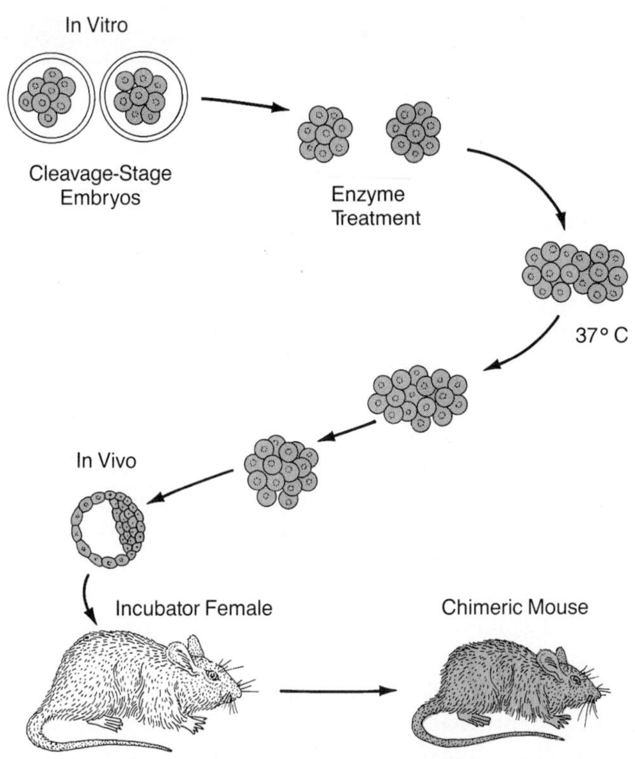

Figure 18-12 CHIMERIC MICE: ONE MOUSE FROM TWO. Two cleavage-stage embryos are fused and are implanted in the uterus of an incubator female hormonally induced into readiness for harboring an embryo. A normal embryo is born; seen here as an adolescent, all its organs have mixtures of cells from the brown and the gray parental strains.

proves to be a random mixture of cells from the white and brown strains. A similar mixed organism has been created with three embryos from six parents.

Experiments like these, as well as natural ones involving multiple births, demonstrate that it is not biologically accurate to regard fertilization as the step that "creates" a new individual. At fertilization, an already living system is given impetus to develop; but that system may form one, two, or more individuals or—experimentally—only portions of individuals. Morphological, physiological, and behavioral individuality appears only over the course of embryonic development and after birth. Fertilization creates only the potential for the individuality that develops later.

MILK PRODUCTION AND LACTATION

Once the baby is born and no longer is nourished by the placenta, the mother can take over feeding the infant by producing and secreting milk, a process known as **lactation. Mammary glands,** an evolutionary innovation of mammals, developed from glands in the skin and manufacture a highly nutritious mixture of fat, protein, and carbohydrate. Mammary glands mature in females at puberty, but not until a woman becomes pregnant does the final growth and maturation take place in the milk-secreting cells and ducts, in response to progesterone and estrogen.

In the fourth or fifth month of pregnancy, the mammary glands begin to synthesize and store small quantities of a remarkable yellowish fluid, **colostrum,** which will be the first food of the newborn. Colostrum contains an abundance of maternal antibodies that help protect the newborn from infection, as well as a high protein content, which combats diarrhea. A few days after birth, the infant's suckling and the pituitary hormone **prolactin**

Figure 18-13 NURSING IN HUMANS. The nursing relationship between a mother and her child is psychological as well as physiological.

BIRTH CONTROL AND ABORTION

All human societies have, at one time or another, developed mechanisms of birth control. The impact of too many children on the family, tribe, or population has led to methods as diverse as infanticide (permitting the newborn to die) and the sophisticated pharmacology of the "pill."

Table A lists the major methods of birth control, or *contraception*. Both men and women can use mechanical means of interfering with fertilization. A condom is a thin sheath designed to fit snugly over the erect penis and retain the sperm. A diaphragm is a round rubber dome filled with a sperm-killing jelly and inserted to cover the cervical opening. A cervical cap does the same job and is not removed except during menstruation. The intrauterine device, or IUD, is inserted by a physician into the uterus; there, the IUD interferes with the implantation process, but can be accidentally expelled or can cause infections.

Surgical methods of contraception are virtually permanent and, of course, are the most foolproof procedures. Whereas vasectomy of a male is a brief, minor operation, tubal ligation of a female is more complex; but new procedures make the operation easier and relatively risk-free.

Any process dependent on hormones, including ovulation, is subject to potential control by drugs. The "pill" is a minute dose of estrogen and progesterone that a woman takes daily between the fifth and twenty-sixth days of the menstrual cycle. The two hormones inhibit FSH and LH secretion and so prevent follicle growth and ovulation.

Nursing a baby causes a woman's pituitary gland to release the hormone prolactin, which can act as a natural suppressor of LH and perhaps FSH. Many societies employ extended nursing as a form of effective—but not foolproof—birth control.

To use the "rhythm" method of birth control, the couple must identify the time of ovulation and abstain from sexual intercourse for the period from 2 days before ovulation to ½ day afterward. A woman must take her temperature daily to detect the 0.5° F elevation that signals ovulation. Safety argues that the couple should probably avoid sexual congress for about a week around the time of ovulation.

Abortion is not an effective long-term form of birth control but has been practiced for centuries as a way to terminate occasional unwanted pregnancies. Induced abortion was described in Chinese writings from 4,000 years ago and by such philosophers as Aristotle and Plato. Almost three out of every five implanted embryos abort for "natural" reasons; hence, abortion is a fairly frequent phenomenon in nature.

Several types of abortions are performed surgically. Early in pregnancy, the cervix may be dilated, or gradually widened, so that an evacuation, or suction, device may be inserted to remove the embryo. Alternatively, a curette, or scraping instrument, may be used. Abortion of a fetus older than 16 weeks is a slightly more complicated process involving injection of a salt solution that triggers expulsion of the fetus.

Researchers are actively studying new contraceptive techniques, including "male pills" to prevent sperm production, "morning-after pills" to prevent implantation, and vaccines to temporarily mobilize a woman's immune system against HCG and a newly forming embryo. Someday, such methods will no doubt be available.

combine to stimulate the synthesis of true milk in the woman's breasts. Once the placenta has been separated from the newborn at birth and expelled from the mother, progesterone and estrogen from the placenta can no longer inhibit the release of prolactin. After milk production starts, an intimate physiological and behavioral relationship between mother and infant begins (Figure 18-13). The baby instinctively sucks on the nipple, sending sensory nerve impulses to the mother's brain that cause prolactin and oxytocin to be released from the pituitary gland. Prolactin stimulates more milk production, and oxytocin stimulates milk secretion.

LOOKING AHEAD

It is fitting that this portion of the book should end with the emergence of the new organism, because we now move from the processes of development that yield phenotype from genotype to a new section of the book. There we address a different question about life: not how the individual arises and develops, but how life itself began and how it then diversified into the myriad microbes, fungi, plants, and animals that populate our planet.

Table A BIRTH CONTROL METHODS

Device or Strategy Employed	Effect	Failure Rate*	Advantages	Disadvantages
		Methods Used by Female		
Abortion	Prevents completion of development	0	Effective	Expensive; morally objectionable to some
Tubal ligation (cutting and tying of oviducts)	Prevents egg transport	0–1	Effective	Irreversible
Oral contraceptives (the "pill")	Suppresses ovulation	0–3	Effective; reversible	Expensive; requires daily action; possible side effects
Intrauterine device (IUD)	Probably prevents implantation of embryo in uterus	0–3	Effective; reversible	May cause extensive bleeding; may be expelled and lost
Diaphragm	Inhibits sperm survival and/or transport	10–30	Safe; reversible	Must be inserted and removed regularly
Vaginal foams and jellies	Inhibits sperm survival and/or transport	8–40	No prescription required	Effective only if applied immediately before intercourse; messy
Diaphragm plus spermicidal	Inhibits sperm survival and/or transport	0–6	Safe; effective; reversible	Same as diaphragm
Vaginal douche	Inhibits sperm survival and/or transport	30–50	Inexpensive; can be used "after the fact"	Ineffective; may actually promote conception in some cases
		Methods Used by Male		
Vasectomy (cutting of vas deferens)	Prevents sperm release	0–1	Simple; effective	Irreversible
Condom	Prevents transfer of sperm to vagina	7–15	Simple; reversible; may prevent VD spread	Expensive; interrupts sexual activity; may leak or break
Premature withdrawal	Prevents transfer of sperm to vagina	15–25	No cost	Requires strong will; frustrating; ineffective
		Methods Used by Couple		
Rhythm method (abstinence from intercourse around time of ovulation)	Prevents contact of egg and sperm	15–35	No cost; acceptable to Roman Catholic Church	Requires extensive study and effort to be effective; ineffective if menstrual periods irregular

*Figures given are best and worst estimates of the number of undesired pregnancies per year per 100 couples using the method as their sole form of birth control.

SUMMARY

1. Sperm are produced in the seminiferous tubules of the testes, are stored in the epididymis, and then pass through the vas deferens, prior to being discharged through the urethra.

2. The male sexual response includes an *excitement phase, plateau phase, orgasmic phase,* and *resolution phase.*

3. *Androgens* such as *testosterone,* and pituitary gland hormones, namely, *luteinizing hormone (LH)* and *follicle-stimulating hormone (FSH),* govern sperm production, sexual maturation, and many aspects of male sexual behavior.

4. An egg released from the ovary during *ovulation* in response to a burst of LH passes down the *Fallopian tube* and, if fertilized, *implants* in the *endometrium* of the *uterus.*

5. The female external genitals include the labia majora, labia minora, clitoris, and openings to the urethra and *vagina.*

6. The female sexual response parallels the male's, with excitement, plateau, orgasmic, and resolution phases.

7. *Gynogens* (female sex hormones) control the monthly *menstrual* and *ovar-*

ian cycles. Two pituitary gland hormones are involved: FSH stimulates ovulation, and LH stimulates production of *estrogen* by the ovarian follicle. Estrogen causes the endometrium to thicken and mature. After ovulation, the follicle becomes a *corpus luteum* and begins to produce *progesterone,* which causes a final "ripening" of the endometrium and inhibits growth of additional follicles.

8. Both male and female embryos progress to the indifferent stage of sexual development, during which the embryo has both *Wolffian* and *Müllerian ducts.* Then, if testosterone and *Müllerian inhibiting substance (MIS)* act (as in normal males), the gonads, sex ducts, and genitals develop into the male type. If neither of these substances act, the structures develop into the female type.

9. Shortly after fertilization, the human zygote implants in the endometrium; the *trophoblast* cells secrete *human chorionic gonadotropin (HCG),* which helps maintain pregnancy.

10. The first *trimester* of prenatal development is characterized by formation of most of the body parts and organs. The second and third trimesters are marked by intense growth of the *fetus.*

11. Embryonic membranes, such as the amnion, protect the embryo and fetus as it develops. Embryonic trophoblastic tissue and maternal endometrial tissue form the *placenta,* the organ that sustains the embryo throughout pregnancy.

12. Birth results from hormonal and neural signals between the fetus and the mother and includes expulsion of the placenta from the uterus.

13. Multiple births arise from fertilization of two or more eggs (fraternal siblings) or of a single egg that cleaves into two or more developing systems (identical siblings).

14. The mother's *mammary glands* produce *colostrum,* the newborn's first food, but a few days after birth, commence *lactation,* stimulated by the infant's suckling and by the pituitary-gland hormone *prolactin.*

KEY TERMS

adrenocorticotropic hormone (ACTH)
androgen
cervix
chorionic villus
colostrum
corpus luteum
endometrium
estrogen
estrus
excitement phase
Fallopian tube
fetus
follicle-stimulating hormone (FSH)
gynogen
hermaphrodite
human chorionic gonadotropin (HCG)
implantation
lactation
luteinizing hormone (LH)
mammary gland
menarche
menopause
menstrual cycle
menstruation
Müllerian duct
Müllerian inhibiting substance (MIS)
orgasmic phase
ovarian cycle
ovulation
oxytocin
penis
placenta
plateau phase
progesterone
prolactin
prostaglandin
pseudohermaphrodite
relaxin
resolution
secondary sexual characteristic
semen
testosterone
trimester
trophoblast
umbilical cord
uterus
vagina
vulva
Wolffian duct

QUESTIONS

1. Why are the testes of many mammalian species located in a scrotum?

2. Describe the journey of an egg from the ovary to the uterus. Explain how three methods of birth control interfere with this journey.

3. Describe the journey of an individual sperm, and explain how three methods of birth control prevent or interrupt this journey, and explain how they work.

4. Explain how gender is determined in the embryo. What are indifferent gonads? Explain what is meant by the neutral sex. In what ways does the *Y* chromosome contribute to sex determination? The *X* chromosome?

5. If a zygote does not contain a *Y* chromosome, can it develop into an embryo? What if it does not contain an *X* chromosome? What is the consequence of parental imprinting of chromosomes?

6. Explain the terms *zygote, morula, blastula, embryo,* and *fetus.*

7. What tissues give rise to the placenta, and how does that organ function?

8. What is the role of human chorionic gonadotropin (HCG) in pregnancy? Where is this hormone produced? What makes detection of HCG a useful test for pregnancy?

ESSAY QUESTION

1. What is an individual? Are identical twins separate individuals? Is a chimera (formed by the fusion of two or more embryos) a single individual?

For additional readings related to topics in this chapter, see Appendix C.

Part
THREE

ORDER IN DIVERSITY

The expansive East African savanna is home to a wide variety of plants and animals, including thorn trees, native grasses, antelopes, giraffes, and wildebeests.

19

The Origin and Diversity of Life

We are indeed the stuff of which stars are made. Life may be so associated with carbon that at some point, we may be able to make a generalization that life is a property of the carbon atom.

Cyril Ponnamperuma,
The Origins of Life (1972)

How did life begin on earth? Most modern biologists believe that the splendid diversity of life forms on our planet—both alive and extinct—evolved from simple ancestral cells that lived billions of years ago and that arose, in turn, from nonliving substances by a process of **chemical evolution.** The exact details of that process are still not clear, but laboratory evidence suggests that early in our planet's history, small organic molecules coalesced in the ancient seas or soils, they began to interact in dynamic ways, cell-like structures formed, and eventually true cells, the basic units of life, emerged.

This chapter will describe current theories of life's origins, including:

- Earth's formation more than 4.6 billion years ago and its protracted geological history
- Theories and evidence for the evolution of living cells from nonliving matter

Mounds of photosynthetic cyanobacteria, so-called stromatolites, similar to 2.7-billion-year-old fossils.

- The changing face of planet earth, and how its alteration has influenced the evolution of diverse life forms
- How biologists using the science of taxonomy categorize the millions of types of organisms to reveal structural similarities and evolutionary relationships

Our study of life's origins, diversity, and taxonomy will help guide us through the next seven chapters and show that there is but a single continuum of matter that stretches from cosmic dust to rocks and planets to life in all its exquisite, interrelated forms.

Figure 19-2 OUTGASSING AT MOUNT SAINT HELENS. Explosive volcanic activity like this eruption from Mount St. Helens (Washington state) in 1980 helped create a second atmosphere early in earth's history (see text).

A HOME FOR LIFE: FORMATION OF THE SOLAR SYSTEM AND PLANET EARTH

As ancient as our world may be, there was a time when the earth and the sun did not exist. Cosmologists believe that the universe began 18 billion years ago with the **Big Bang** and that our sun formed about 5 billion years ago as massive clouds of hydrogen and other elements compressed into a ball of glowing gases. The sun's extraordinary heat (20 million° C at its core) caused various heavier elements to be made from atoms of hydrogen and helium, and clouds of this new heavier matter were expelled from the sun. Local clusters of gases, dust, and larger particles of matter eventually condensed and solidified into the planets of our solar system, including the planet earth (Figure 19-1).

The best current estimates suggest that the earth and its moon had aggregated as solid bodies by 4.6 billion years ago. A zone of light elements—primarily hydrogen and helium—shrouded the earth with its first atmosphere, but the new planet's weak gravity failed to hold it in place. The compaction of heavy elements, plus radioactive decay, generated immense heat inside the new planet, producing a molten metallic **core** that today is solid at its center. Intense upwellings of gases and heavier elements from the core resulted in widespread volcanic and hot-spring activity and the "outgassing" that helped form the second atmosphere (Figure 19-2). Besides all this internal activity, the earth was bombarded by millions of high-velocity planetesimals (bodies of all sizes from space). Their collective impact caused immense amounts of additional outgassing—accountable, some scientists believe, for the total H_2O content of the atmosphere and (later) the oceans. The outgassing may also have produced hydrogen cyanide (HCN) and oxidized carbon. The atmosphere that formed dates from about 4.4 billion years ago and at that time probably consisted of CO, CO_2, H_2, N_2, H_2O vapor, and little or no free oxygen. As eons passed, atmospheric temperatures fell, and water could then exist in liquid form, not just as vapor. Hot, torrential rains began falling, and the first oceans appeared.

Figure 19-1 FORMATION OF THE PLANETS. The planets condensed from gases and rings of matter at varying distances from the sun.

By about 3.9 billion years ago, some of the lighter nongaseous elements and compounds near the earth's surface had formed an outer rock "skin," the **crust,** which averages some 26–39 km in thickness and bears land masses surrounded by oceans. Global weather patterns became established, and rain falling over the continents resulted in erosion, an accumulation of salt and minerals in the oceans, and slow geological changes worldwide.

Given a history of formation somewhat similar to that of our neighboring planets, why did life emerge *here,* and not apparently on the others? One answer may lie in the earth's temperature, size, composition, and distance from the sun. Unlike Mars, the earth is not too cold for complex molecular processes to proceed spontaneously and for much of the water to remain liquid; nor is it so hot (as on Mercury and Venus) that complex organic polymers are degraded (denatured) and water can exist only as vapor. Moreover, smaller planets have gravities that are too weak, larger ones atmospheres that are too dense. Only earth, it seems, has acceptable ranges of temperature, gravity, and other factors for the type of life that originated here.

THE EMERGENCE OF LIFE: ORGANIC AND BIOLOGICAL MOLECULES ON A PRIMITIVE PLANET

With our reconstruction of earth's history, the stage is set for the drama of life to begin. Perhaps 600 million years have passed since the planet formed. Sterile continents rose above salty seas warmed by sunlight, by outgassing, and by the molten rock layers below the earth's crust. Volcanoes spewed out lava and released gases (especially CO_2 and H_2O vapor) into the atmosphere, which in turn was held like an invisible cloak by the forces of gravity. Visible light as well as intensive ultraviolet wavelengths penetrated the thin atmosphere and the oceans, perhaps to depths of nearly 20 m. And the atmosphere, rich in CO_2, trapped additional heat.

How could life have arisen in such a stark environment? The major current hypothesis holds that life arose spontaneously on the early earth by means of chemical evolution from nonliving substances. Scientists have re-created many of earth's original physical conditions in the laboratory and have discovered a great deal of evidence—geological, chemical, and biological—to support the hypothesis.

Each living organism is constructed of the same building blocks—a few types of amino acids, sugars, nucleic acids, and lipids. Therefore, these organic molecules themselves are the logical starting point in the search for life's origins. As early as the 1920s, researchers were posing fundamental questions about *monomers*, the small organic subunits of more complex molecules (see Chapter 3): Were such organic building blocks present on the early earth? And if so, how did they form?

The Formation of Monomers

In the 1950s, biochemists at the University of Chicago devised a clever experiment to test a 30-year-old hypothesis that monomers could have formed when gases in the ancient atmosphere were energized by heat, radiation, ultraviolet light, or massive displays of lightning. Using special laboratory apparatus, Harold Urey and his graduate student, Stanley Miller, re-created the hypothesized atmospheric conditions on early earth, filling an upper flask with four kinds of gases and the lower flask with a reservoir of water—a miniature ocean. To simulate an energy source such as lightning, they discharged electric sparks into the upper flask for an entire week. At the end of the experiment, the primal "sea" in the bottom flask had collected significant quantities of amino acids and simple sugars, as well as tarry residues.

Current geological evidence suggests that the Urey-Miller team had a misconception about the atmosphere that existed 4 billion years ago and that the early atmosphere was probably closer to the modern atmosphere minus free oxygen. Newer experiments re-creating those alternative conditions have yielded more than 100 kinds of organic monomers, and in fact, researchers have produced almost all the building blocks of living cells in laboratory "spark chambers." Additional evidence suggests that comets and meteorites may have delivered tons of monomers to the ancient earth (see the box on page 329).

The implication of these experiments is clear: Under conditions of heat, humidity, energy, and raw materials similar to those probably present on the earth billions of years ago, amino acids, sugars, fatty acids, and nucleic acid bases can readily form. Along with carbon compounds from space, this may explain the source of life's raw materials.

The Next Step: Polymers

At this point in our reconstruction of conditions and processes in the planet's history, the landscape is still lifeless, but an impressive array of organic monomers has either formed in the atmosphere or arrived from space and has been washed by rain into the soil as well as into warm lakes, rivers, and salty seas. What would be the next step toward life's origin? Biologists believe it was the spontaneous linking of monomers into polymers such as proteins and nucleic acids.

Until recently, many pictured the ancient oceans and seas as a rich "primordial soup" with a high concentration of organic molecules. This now seems unlikely, however, because studies show that ultraviolet light tends to degrade such molecules. Therefore, the ancient oceans and seas were probably a very dilute organic "broth," and the polymerization of monomers probably occurred only at sites of high concentration.

Recall from Chapter 3 that polymerization is a kind of "zipper chemistry" and that its individual steps are not necessarily complex, even if the final product is a large,

OUTER SPACE: THE SOURCE OF ORGANIC PRECURSORS?

Recent studies by astronomers and chemists suggest that organic precursors of life may have had an extraterrestrial origin. With the help of sophisticated radio and infrared telescopes, astronomers in the early 1970s discovered the presence of enormous clouds of organic molecules in the arms of our galaxy, which spiral outward. They have since identified more than 75 types of organic molecules in those stellar clouds, from simple CS, CN, CH, and CO to methyl and ethyl alcohol, formaldehyde, amino acid precursors, and straight-chain carbon compounds. Their conclusion: Organic molecules are ubiquitous in the universe and present in huge quantities.

Even given such a massive source, how could organic matter from space have ended up on earth early in its history? The answer, some contend, is that for almost 1 billion years, organic compounds showered down on our planet in the form of *carbonaceous chondrites*—small meteorites that contain a spectrum of organic molecules much like those detected in the galactic clouds. One team of astronomers has calculated the total mass of objects hitting the earth to have been at least ten times the mass of the present-day oceans. Large objects would have vaporized on impact, many geologists believe, but smaller objects, such as carbonaceous chondrites, would have survived—and along with them their load of complex organic molecules. Thus, most of the earth's carbon compounds were *delivered*, some contend, not made in the atmosphere or the oceans.

There is interesting confirming evidence for this hypothesis: Recent analysis of a meteorite that fell on Australia in 1969 revealed all five of the nitrogenous bases (adenine, guanine, cytosine, thymine, uracil) that make up DNA and RNA, as well as numerous amino acid precursors. Since the levels of radioactive hydrogen or deuterium in these organic molecules are twice the levels of similar molecules found on earth, scientists conclude that the organic molecules in the meteorite were indeed formed in space and were not simply contaminants picked up here.

There is growing speculation that interstellar objects could have seeded the primordial earth with millions of tons of organic matter. Questions still remain, however, about the role of this stardust in the chemical evolution that led to life. Many are hopeful that answers will come from future space missions to Mars, Saturn's moon Titan, and Jupiter's moon Europa, as well as from hurtling asteroids and comets.

complicated molecule. Where could polymerization have occurred if not in watery solutions? Current research favors the hypothesis that clay or rock surfaces were a likely site for polymer formation, since experiments confirm that the positive and negative charges within stacked sheets of clay or mica can act as simple catalysts to promote the linking of amino acids into polypeptides (so-called **proteinoids**). Hence, a watery solution is not essential to polymer formation. Researchers also believe that polynucleotides (early RNAs) might have formed on clay surfaces from individual bases. Evidence of such formation is especially significant, since biologists think that RNA was the first kind of self-replicating informational macromolecule to form.

Where would energy come from to drive organic polymerization of polypeptides in clays? Energy from sunshine or the earth's core or energy from ultraviolet light could have driven polymerization reactions. Alternatively, molecules of ATP, which form in Urey-Miller chambers, could have supplied the chemical energy for polymerization. If an experimenter mixes ATP with amino acids and various condensing agents, amino acid adenylates (the activated amino acids that take part in protein synthesis) form. Such molecules may then undergo a slow, spontaneous polymerization to form polypeptides. Even without ATP, however, and with simple heat applied to a batch of drying amino acids, polypeptides with 200 or more amino acid units will form spontaneously.

From Polymers to Aggregates

If, as many biologists believe, polypeptides and perhaps even short chains of nucleotides collected in clays, groundwater, or small pools, what might the next step toward life have been? Researchers have generated three model aggregates in the laboratory to study whether such groups of molecules would show properties of living cells.

In the 1960s, biologist Sidney Fox found that heating a mixture of dry amino acids and exposing it to water causes tiny spheres of proteinlike polymers to form, which he called **proteinoid microspheres** (Figure 19-3a). Each individual sphere has an outer layer of water and protein molecules and an aqueous interior that moves about rather like cytoplasmic streaming. These spheres can take up and concentrate other molecules from the surrounding solution, can fuse to form larger spheres, can shrink or swell osmotically, and behave as though

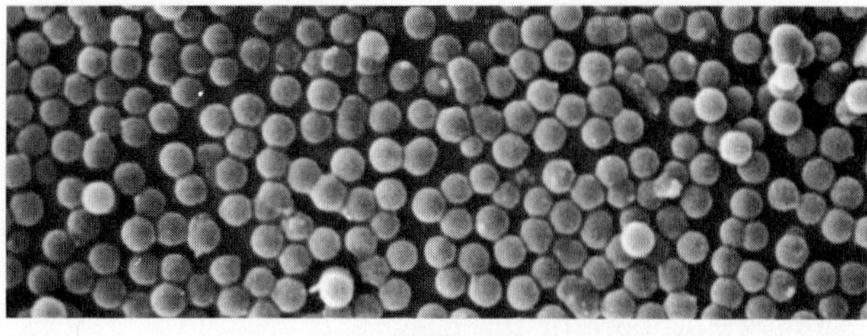

(a)

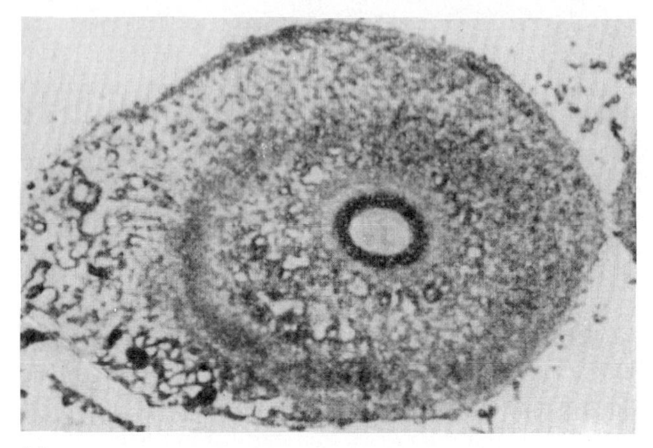

(b)

Figure 19-3 PROTEINOID MICROSPHERES AND COACERVATES.
(a) A group of proteinoid microspheres produced in the laboratory of Sidney Fox and seen magnified about 3,000 times with a scanning electron microscope. (b) A complex coacervate droplet magnified about 10,000 times. The inner portion of the proteinoid has been replaced with the lipid α-lecithin.

they have a selective barrier at their surface, even though no lipid is present.

Half a century earlier, Aleksandr I. Oparin found that solutions of various polymers derived from contemporary cells, such as proteins plus carbohydrates or proteins plus nucleic acids, form polymer-rich droplets—so-called **coacervates** (Figure 19-3b). Coacervates not only are reminiscent of tiny cells, but also, under certain circumstances, will divide into smaller spheres.

Other researchers have generated a third type of aggregate in the laboratory—the **liposome,** a spherical lipid bilayer that forms easily if phospholipids at the correct concentration range are shaken in an aqueous solution. Proteins may also be embedded in the lipid, just as they are in the fluid-mosaic plasma membrane of every living cell.

Liposomes, coacervates, and proteinoid microspheres are laboratory constructs, of course. But perhaps some aggregate with the combined properties of microspheres and liposomes was an ancient precursor of true cells. At the very least, these experiments show that polymers can form higher-order structures when energy is available; the chemistry of life is truly a self-ordering process.

From Aggregates to Cell-Like Units

While there is a profound difference between even the most complex aggregate of molecules and the living cell, the origins of reproduction, information transfer, and metabolism hint at the next set of steps in the appearance of life.

Origins of Reproduction and of Translation Machinery—RNA Leads the Way

Molecular reproduction of aggregates such as proteinoids or molecules such as RNA and DNA is the key to understanding how cellular reproduction started. One of the most intriguing aspects of proteinoids formed under laboratory conditions is their nonrandom composition: Certain amino acids within them tend to be linked, while others tend not to be associated. Different proteinoids can thus have acidic, basic, hydrophobic, hydrophilic, or other distinctive properties.

Significantly, many proteinoids formed in the laboratory have catalytic properties, and this suggests that biochemical pathways could have developed among the early polymers. As an amino acid chain folds into a three-dimensional shape, a catalytic site may form that is equivalent to an enzyme's active site.

RNA can also function as an enzymatic catalyst, and this contributes to the current view that RNA was the first informational macromolecule (Figure 19-4). By

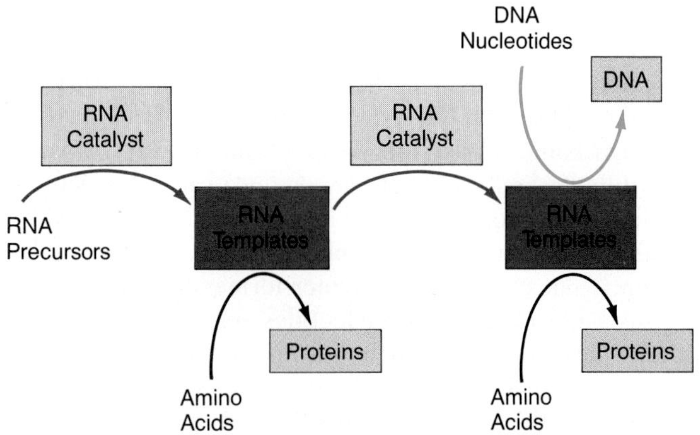

Figure 19-4 RNA: THE FIRST INFORMATIONAL MOLECULE?
The remarkable capacity of RNA to act as catalyst and template suggests that it may have served as the critical link in the origin of life. First, RNA could serve as a template that can be reproduced, including processes leading to RNA variation and evolution. Catalytic RNA, ribozymes, is the key to that process. Some RNAs might also serve as templates for protein manufacture. Finally, DNA copies of the RNA templates could have arisen, thereby yielding the extremely stable DNA genetic material common to most organisms on earth.

proving RNA's catalytic function, Thomas Cech and his colleagues at the University of Colorado revised the long-held notion that only protein enzymes can serve as biological catalysts. Additional studies confirmed that various sequences in tRNAs, mitochondrial RNAs, and nuclear RNAs can splice (cut) themselves out of longer RNA molecules. Furthermore, certain of the excised RNA pieces can function enzymatically as **ribozymes** to catalyze the construction of new RNA molecules on an RNA template. All this activity can occur without proteins, heat, ATP, or GTP added to drive the reactions.

Modern cells still employ the enzymatic activity of RNA to splice out introns from exons and to process precursor molecules into mature tRNAs, rRNAs, and mRNAs. Biologists speculate that long ago, catalytic RNA could have assembled new RNAs from nucleotides in warm pools or on clay surfaces. The mechanisms involved from the very start probably included self-removing and self-inserting catalytic RNAs that function just the way transposons move genes from one cell to another today (Figure 19-5). This activity is the equivalent to sex: transmission of genetic material from one cell to another. One can imagine, then, an "RNA world" of increasing complexity with self-replicating molecules having both functional (enzymatic) and informational regions, the probable presence of intron and exon regions, sexlike exchanges of base sequences, and finally a kind of natural selection favoring certain of those molecules more than others.

The key to reproduction and inheritance lies in the ability of informational molecules to replicate. Did these early RNAs form copies of themselves? When experimenters take spontaneously formed single-stranded RNA (or DNA) molecules and add them to a mixture of nucleotides, they can detect hydrogen bonding between the purine and pyrimidine bases of the free nucleotides and the complementary bases of the chain. When they add a condensing agent to the system, a slow polymerization of the bound nucleotides begins, and a complementary chain forms. Separation of these two chains yields two new templates for another round of "replication." Given the vastness of geological time, single- and double-stranded nucleic acids probably arose by just such processes.

Some of the RNA molecules that form in the test tube have hairpin shapes, like modern transfer RNA, and can be extremely stable. Furthermore, spontaneously produced proteins rich in basic amino acids will complex with the RNA molecules and form primitive ribosome-like structures. Perhaps, 4 billion years ago, such processes led to the development of the translation machinery—tRNAs, mRNAs, and ribosomes—to make proteins that are copies of RNA templates.

Although DNA is the modern cell's "code of life," it is usually regarded as the "last step" in the origin of informational macromolecules. RNA can serve as a template on which a complementary DNA strand can be assembled (using an enzyme called reverse transcriptase; see Chapter 20). Perhaps reverse transcription occurred long ago and the result was the extremely stable storage site for biological information, the DNA double helix, complete with introns and exons copied from equivalent regions in the more ancient RNA.

Origins of Metabolism

Biologists think that a process resembling natural selection may have led to the first metabolic pathways. Microspheres or naked genes in the primordial waters or on clay surfaces would have depended on an external supply of sugars, lipids, bases, and amino acids to reproduce and maintain structural integrity. As such naturally formed monomers were "consumed," first by nonliving aggregates and later by more and more cell-like struc-

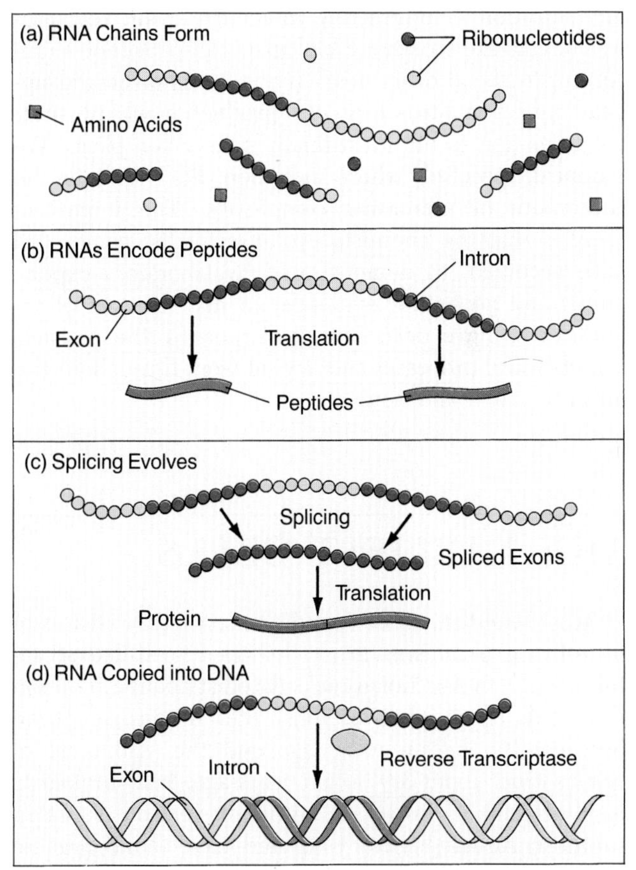

Figure 19-5 ACTIVITY IN THE EARLIEST GENES? Many biologists think that the earliest genes were RNA molecules (a) that could have contained intron and exon regions like those in modern genes (b, c) and that, through the activity of reverse transcriptase enzymes, were transcribed into functioning DNA molecules (d).

tures, those raw materials would have become increasingly scarce. Perhaps variations (mutations) in catalytic ribozymes or proteinoids allowed the use of a simpler and more abundant raw material, *Y*, instead of the rarer compound *X*. The mutant generations would obviously have benefited directly, just as they would have if a new enzyme, *y*, the chance product of a certain gene alteration, allowed them to form the rare and necessary *X* from the more common raw material *Y*:

$$Y \xrightarrow{y} X$$

While *Y* also might have become scarce in time, the formation of yet another enzyme, *z*, and the ability to carry out a second chemical conversion might still have allowed *X* to be made:

$$Z \xrightarrow{z} Y \xrightarrow{y} X$$

The first metabolic pathways, even forerunners of the Krebs cycle, might have evolved in this stepwise fashion. True metabolism was no doubt favored by compartmentalization of interacting molecular chain reactions within cell-like structures biologists call **progenotes** (ancestors to the prokaryotes). When metabolism did appear and when a true form of reproduction was present, the transition to life must finally have taken place. We cannot say precisely where and when "life" began on the continuum of increasing complexity. The important thing is that the chemical evolution of life could well have occurred, as demonstrated by laboratory experiments and geological evidence, in just the sort of sequence we have described, governed by the physical and chemical processes and laws at work throughout the universe, then and now.

THE EARLIEST CELLS

Most scientists agree that the oldest fossil remains of once-living organisms are rod-shaped, cell-like structures found in Australian rocks dated at about 3.5 billion years old. These fossils establish life's origins sometime between 3.5 billion years ago and the stabilizing of earth's crust about 3.9 billion years ago. Paleontologists have also found fossils in 3.4-billion-year-old rocks in southern Africa (Figure 19-6) and have found organic molecules in still older rocks (dated at 3.8 billion years) from western Greenland—molecules that may have had a biological origin. Significantly, geologists have detected in the 3.4-billion-year-old fossils spherical active cells that perhaps lived in colonies as cyanobacteria do today and carried out photosynthesis.

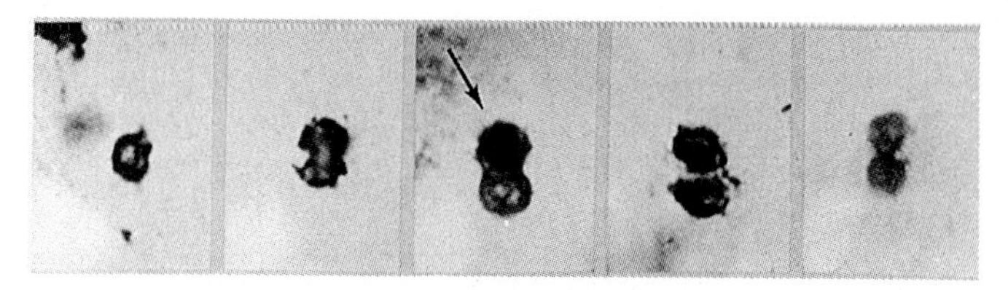

Figure 19-6 EARLY CELL-LIKE FOSSILS. These fossilized microorganisms were found in rocks more than 3.4 billion years old from the Swaziland System in southern Africa. The microfossils are arranged as if they are in various stages of cell division.

Heterotrophs First, Autotrophs Later

The first cells were in all probability *anaerobic*—able to survive in the absence of free oxygen—and *heterotrophic*—unable to make organic nutrients from simple inorganic precursors. These early *heterotrophs*, or "other feeders," consumed the amino acids and polypeptides, nucleotides, sugars, and other carbon-containing compounds that had formed spontaneously on earth or had been delivered in meteorites. As mentioned earlier, competition for increasingly scarce organic compounds would have favored evolution of metabolic pathways and probably also the use of ATP as an energy intermediary. True cells having such capacities soon would have depleted the store of monomers formed by nonbiological processes. Indeed, all life might have died out in a blitz of competition if it were not for one critical set of innovations: those leading to photosynthesis.

Perhaps as early as 3.8 billion years ago photosynthesis arose: structural proteins and enzymes inherited from earlier cells became capable of trapping energy from sunlight, probably at first to generate ATP by means of photophosphorylation (see Chapter 8) and later to store that energy in carbon compounds, the sugars. Cells with such attributes could have created their own nutrients from inorganic precursors and functioned as the first *autotrophs*, or "self-feeders."

Until recently, the emergence of autotrophs was dated to 3.4 billion years ago by the discovery of fossils with ^{12}C:^{13}C ratios very similar to those found in photosynthetic cells today; the same ancient rocks contain chlorophyll, the primary light-absorbing molecule of today's photosynthesizers. However, new carbon ratio measurements of sedimentary organic matter indicate that photosynthesis is much older than previously thought and that cells with extensive photosynthetic activity may have existed by 3.8 billion years ago. By 2.7 billion years ago, photosynthetic cyanobacteria were living in colonies and forming densely layered mats. Remains of these early cells formed **stromatolites,** minutely layered rock mounds like the ones pictured on page 326, which grow today in warm, salty inlets on the western coast of Australia.

The evolution of photosynthesis with its metabolic by-product, oxygen, had an immensely significant consequence for the earth and its inhabitants. Massive amounts of molecular oxygen began to accumulate in the atmosphere, and by 2 billion years ago, some of the atmospheric O_2 had been converted by sunlight to ozone (O_3) and had collected in a high-altitude layer, the **ozone layer.** This layer screened out much of the ultraviolet light that had been searing the earth's surface since the planet formed, and its protection allowed cells to live closer to the surfaces of oceans and lakes and even on the moist fringes of the continents.

The Emergence of Aerobes

The early autotrophs inadvertently created a potentially lethal condition for themselves, because the oxygen released during photosynthesis is poisonous to anaerobic cells, and it probably disrupted biochemical pathways that originated when the environment lacked free oxygen. Eventually, however, cells capable of living in the presence of oxygen and even of exploiting it metabolically emerged: **aerobic cells.**

In these aerobic cells, *cellular respiration* originated (see Chapter 7), allowing organisms to harvest 18 times more energy than by glycolysis and the additional steps of fermentation alone. The extra energy harvest of aerobic respiration permits much higher rates of growth and reproduction. In addition, some aerobic cells could now derive more energy by hunting down and digesting other cells than they used up in the hunting process.

Cellular respiration had another major consequence, as well: the beginning of the *carbon cycle.* During cellular respiration, carbon compounds are largely oxidized to CO_2. Thus, that gas can return to the atmosphere to help replenish the CO_2 reservoir used for photosynthesis.

Eukaryotes

Once aerobic cells appeared and the ozone layer screened much of the ultraviolet light from the sun's penetrating rays, the variety of organisms increased tremendously in the sea, in fresh water, and on land. But these cells were all *prokaryotic*—lacking a cytoskeleton, a nucleus, and other membrane-bound organelles (see Chapter 5). Not for another billion years or so would *eukaryotic* cells—having a true nucleus and complex organelles—appear. The oldest fossil eukaryote yet discovered lies embedded in rocks dated at 1.5 billion years old. Chapter 20 describes more about the possible origin of eukaryotes from prokaryotic ancestors.

Figure 19-7 summarizes the events that may have led to the origin of life and to the emergence of prokaryotic and eukaryotic cells.

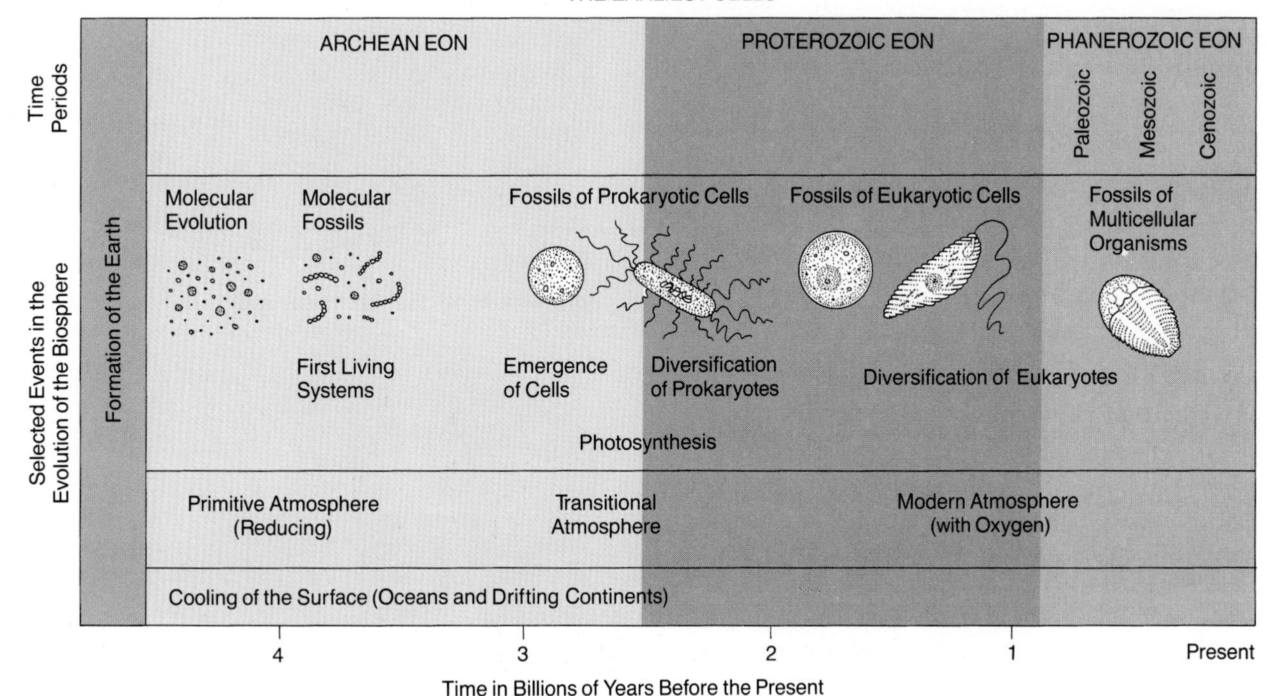

Figure 19-7 TIME PERIODS AND SELECTED EVENTS IN THE EARLY EVOLUTION OF LIFE AND THE BIOSPHERE. Following the process of chemical evolution early in earth's history, true cells arose. Ancient photosynthetic anaerobes released oxygen, an ozone layer formed, and the effective screen against ultraviolet radiation allowed other, more complex life forms to emerge, including the early aerobes and eukaryotes.

THE CHANGING FACE OF PLANET EARTH

The history of the earth is inextricably entwined with the history of life. Since the earth formed some 4.6 billion years ago, dynamic forces have continued to produce changes in landmasses, seas, and climate that in turn have altered the habitats of organisms and influenced biological evolution.

Our planet's outer molten metal core is surrounded by a hot, semisolid **mantle** and a lighter, solid crust. The crust, which is about 5 km thick over the seafloors and perhaps 33 km thick over the continents, is divided into massive plates. In a process called **continental drift,** these plates move slowly as convection currents in the mantle lead to upwelling and the addition of new crustal materials at certain sites.

The building of crustal plates, so-called **plate tectonics,** began more than 4 billion years ago. Since that time, about 15 plates (including 8 major plates) have formed and drifted about on the semisolid mantle. The upwelling of new materials near mid-ocean ridges produces new crustal areas and pushes the plates apart so that collisions occur. When the edges of two plates collide, one edge may be driven beneath the other, causing the Himalayas, Alps, Andes, and Rockies, and other mountain ranges to elevate.

By dating rock layers from the crustal plates and collecting other evidence, geologists have divided the past 550 million years into three *geological eras:* Paleozoic, Mesozoic, and Cenozoic. The transition time from one era to the next can be dated fairly precisely from the amounts of radioactive isotopes present in rocks. Subdivisions of geological eras, called *geological periods,* are dated in the same way. The immense span of time preceding the Paleozoic and during which life began is called the Precambrian period. (The geological time scale inside the book's front cover lists the eras and periods.)

Since the Paleozoic, the earth's landmasses have joined together into supercontinents, then separated once again. Figure 19-8 describes the formation of **Pangaea,** a single giant landmass some 250 million years ago. At the transition between the Paleozoic and Mesozoic eras, Pangaea broke apart. That time was also marked by the most massive extinctions of living creatures that ever

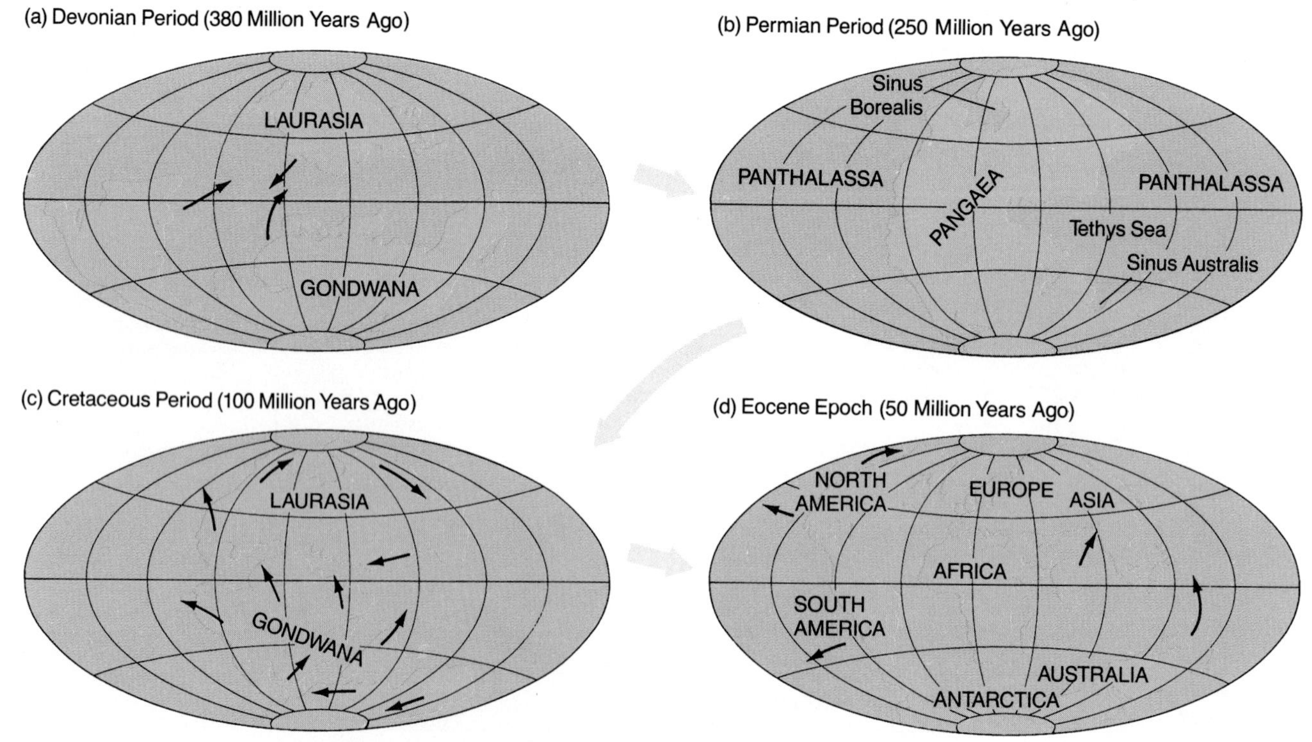

Figure 19-8 CONTINENTAL DRIFT OVER THE PAST 380 MILLION YEARS.
(a) Before and during the Devonian period of the Paleozoic era, 380 million years ago, a northern continent, Laurasia, and a southern continent, Gondwana, lay in separate hemispheres. (b) During the Permian period, Laurasia and Gondwana fused to form a single supercontinent, Pangaea. (c, d) During the Mesozoic era, Pangaea split apart, with Laurasia comprising what is now North America, Europe, and Asia. Gondwana later formed South America, Africa, Australia, and Antarctica. Drifting of continents has continued during the past 50 million years. Arrows indicate direction of drift.

took place on earth. Between 50 and 25 million years ago, the shifting landmasses took on the general continental configuration that we see today.

Accompanying these momentous plate movements were climatic changes that greatly affected living organisms. Sites on the drifting continents went from wet to dry, from cold to hot, and back again numerous times as their latitudes shifted and as their altitudes rose from seafloor to lofty peaks and were eroded again over time.

Such massive environmental changes influenced the evolution of specific groups of organisms, and as we shall see in subsequent chapters, the dominant species during each major geological period were in many ways shaped by the prevailing climates and terrains. To understand life and its evolution on earth, we must visualize both the vast, slow changes in land, sea, and weather that affect the evolution of species and the daily, seasonal, and cyclic changes that affect individual lives. Furthermore, we must remember that the earth itself changes continuously even today, and as its landscapes and climates alter, so must its living inhabitants.

TAXONOMY: CATEGORIZING THE VARIETY OF LIVING THINGS

Just as the earth's changes are chronicled in its rocks and recorded in the geological time scale, the evolutionary history of organisms is evident in the fossil record. The diversity of those fossils and also of living organisms has been systematically cataloged. Attempts to categorize the grand diversity of life predate Aristotle. But it was a Swedish botanist, Carl von Linné, who made the first major contribution to **taxonomy**—the classification of organisms.

The Binomial System of Nomenclature

In the mid-eighteenth century, Carl von Linné, who used the Latin form of his name, Carolus Linnaeus, wrote *Systema Naturae*, a landmark book organizing plants and animals in a new way. Linnaeus employed the **binomial system of nomenclature,** by which he assigned all organisms a name using two Latin words (Latin being the one language shared by all educated Europeans at the time). The first term denotes the **genus** (plural, *genera*), a grouping of very similar organisms to which the organism in question belongs. The second term denotes the organism's **species,** a specific group of closely related organisms within a genus. The full species binomial (two-part) name always includes genus and species; the modern domesticated horse, for example, is labeled *Equus caballus*, while its close relative, one of the zebras, is *Equus burchelli*.

Today, groupings do not stop at genus and species, but include broader *taxonomic groups*—arrangements of organisms into hierarchical classifications based on similarities (see Table 19-1 for examples). Similar genera are placed in the same **family**; similar families, in the same **order**; similar orders, in the same **class**; similar classes, in the same **phylum** (or **division** in botany); and, finally, similar phyla or divisions, in the largest, most inclusive category—the **kingdom.**

Taxonomy and Darwin's Theory of Evolution

Linnaeus had no alternative in the 1750s but to use a strictly morphological approach in defining his taxa: (singular, **taxon,** one category in a system of taxonomy). The more structural traits shared by different organisms, the closer their taxonomic relationship. But since that time, two major factors have altered the way biologists construct and use taxonomies. The first was Charles Darwin's theory of evolution. The second was the advent of new techniques to delineate the physiology, embryology, and biochemistry of living creatures—the hidden foundations on which overt structure depends.

Darwin's theory of evolution in response to natural selection (see Chapters 1 and 42) helped hone the defini-

Table 19-1 THE CLASSIFICATION OF ORGANISMS

Taxonomic Level	*Human*	*Honeybee*
Species	*Homo sapiens*	*Apis mellifera*
Genus	*Homo*	*Apis*
Family	Hominidae	Apidae
Order	Primates	Hymenoptera
Class	Mammalia	Insecta
Subphylum	Vertebrata	
Phylum	Chordata	Arthropoda
Kingdom	Animalia	Animalia
Taxonomic Level	***Corn***	***Mushroom***
Species	*Zea mays*	*Agaricus campestris*
Genus	*Zea*	*Agaricus*
Family	Poaceae	Agaricaceae
Order	Commelinales	Agaricales
Class	Monocotyledoneae	Basidiomycota
Subdivision*	Anthophyta	
Division*	Tracheophyta	Mycota
Kingdom	Plantae	Fungi

*Botanists use the terms *division* and *subdivision* instead of *phylum* and *subphylum*.

tion of species. According to evolutionary theory, one species has diverged into two when the group of organisms can no longer interbreed. One modern criterion for defining species, then, is *reproductive isolation*—the inability to breed successfully with other organisms, even closely related ones.

As we will see in later chapters, this definition of species applies more easily to certain sexually reproducing animals than it does to the microorganisms, fungi, and plants and animals that reproduce primarily by asexual means. Hence, the post-Darwinian definition of species helps categorize organisms in some, but not all, cases.

Progress in the fields of physiology, embryology, and biochemistry, as well as much more sophisticated morphological analyses have also influenced modern taxonomy. The sharing of metabolic pathways or properties and the similarities in nerve function, waste processing (such as from nitrogen metabolism), and properties of organs or tissues are among the kinds of evidence biologists use to define the degree of relationship among groups of organisms. Molecular taxonomists compare amino acid sequences of proteins or base sequences of DNA or RNA from different organisms, and with the latest rapid techniques for sequencing DNA bases, they will someday be able to analyze an organism's full haploid genome. Finally, by using electron microscopes and other sophisticated probes of tissue, cellular, and subcellular structures, biologists can observe physical differences between species at a much finer level.

Taxonomy and Evolutionary Relationships

Evolutionary thought and theory since the time of Darwin have created a new function for taxonomic classification; besides simply assigning organisms to specific groups, taxonomy provides a framework for studying evolutionary relationships.

Taxonomists have devised methods for tracing the family trees of living and fossil organisms. Some phylogenetic trees are based on the morphological evidence (structural similarities) used to define species. For example, molecular taxonomists have constructed phylogenetic trees based on amino acid differences in the respiratory pigment cytochrome *c*, as well as on other proteins or DNA itself.

Ideally, forms collected into one taxon—whether it be a species, family, phylum, division, or whatever—can be considered to have evolved from one ancestral species. Thus, all the robin species in the genus *Turdus* evolved from one ancestral *Turdus* species. They can be called **monophyletic,** meaning they share one ancestral source. A group of monophyletic organisms is called a **clade,** and the study of how closely related groups branched and separated from one another is called **cladistics.** An illustration of this view is provided in the *cladogram* in Figure 19-9. Like the more traditional taxonomy of Linnaeus, cladistics helps biologists study evolutionary relationships.

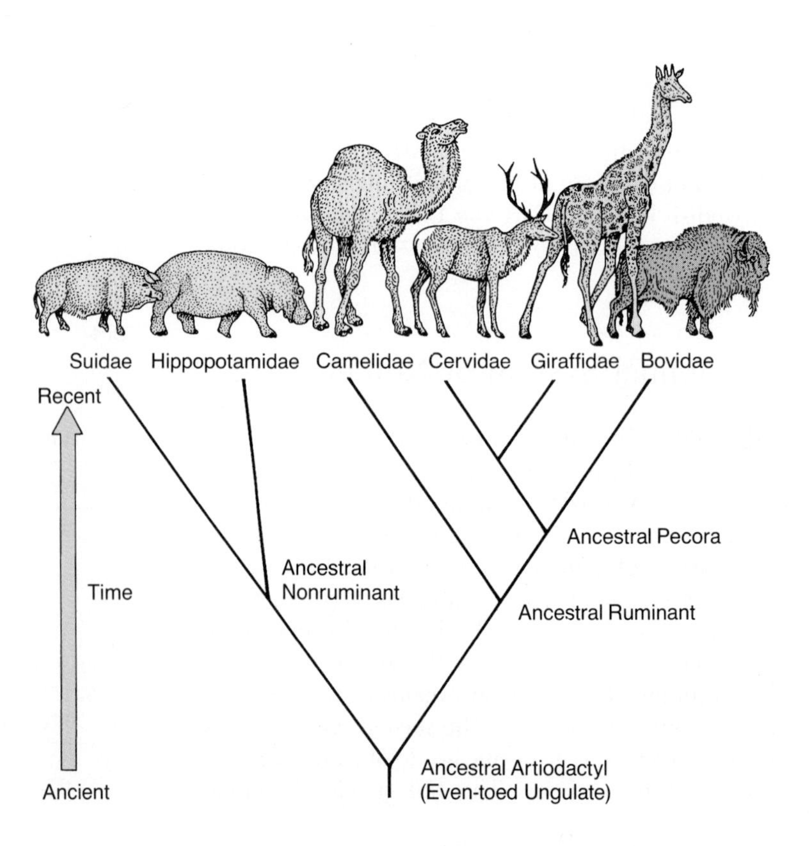

Figure 19-9 CLADOGRAMS: BRANCHING SEQUENCES OF RELATED GROUPS.
Here, one group of placental mammals is shown: those with an even number of toes, the artiodactyls. This clade includes two major divisions, the ruminants and nonruminants. Groups are arranged on the cladogram to reveal relative closeness of genetic relationship. Thus, the deer (Cervidae) are closer genetically to giraffes (Giraffidae) than either is to the cattle-buffalo-bison group (Bovidae). Hippos (Hippopotamus) diverged from the pigs (Suina) before that group diversified into warthogs, water hogs, pigs, and still other groups (not shown on the cladogram).

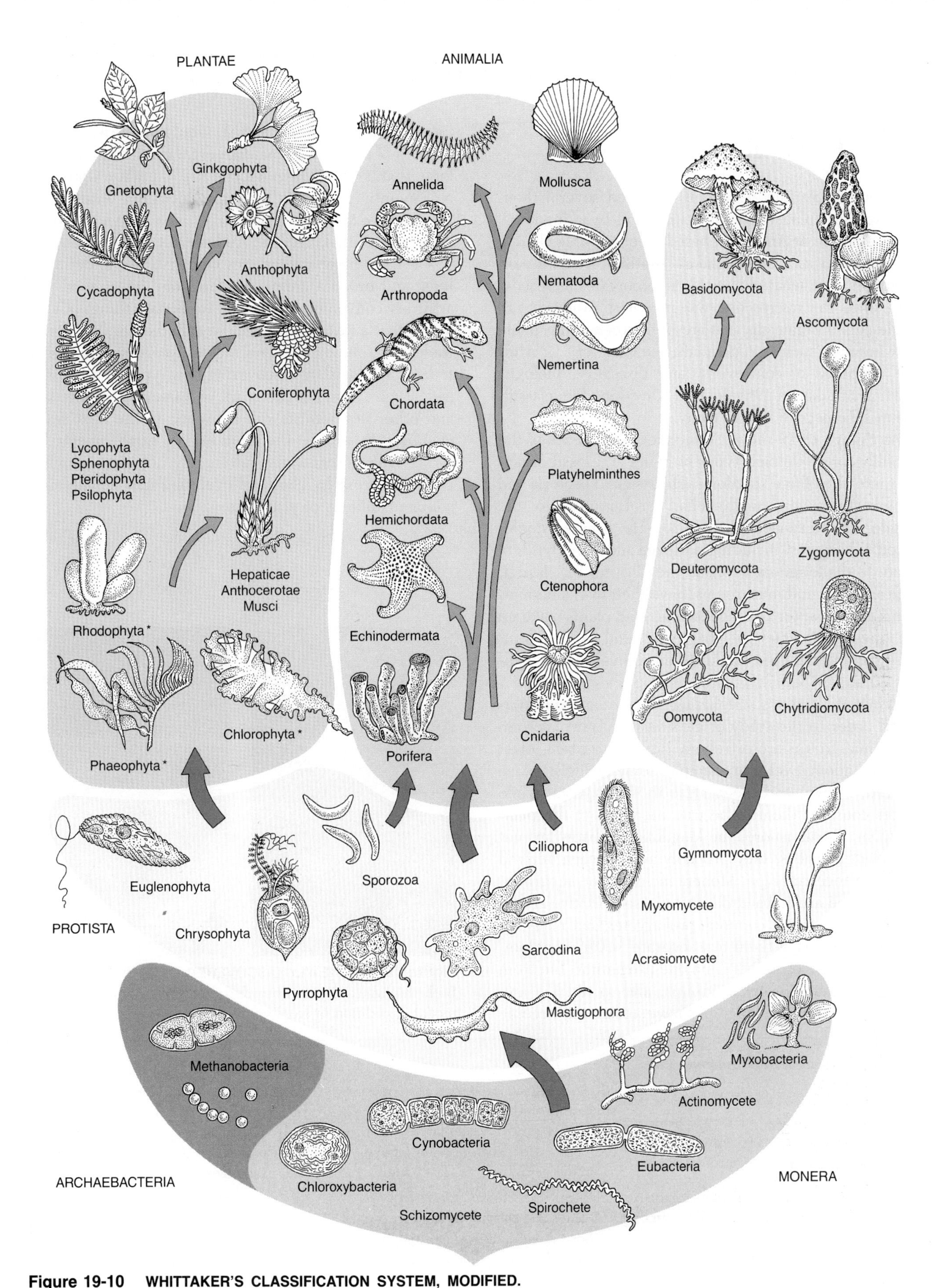

Figure 19-10 WHITTAKER'S CLASSIFICATION SYSTEM, MODIFIED.
Whittaker scheme, updated. The principal change is the addition of the Archaebacteria as a major group of prokaryotes, possibly a separate kingdom. Some biologists place these groups (the red, green, and brown algae) in the kingdom Protista.

THE FIVE KINGDOMS

Over his lifetime, Linnaeus classified several thousand species, placing them all in one of two kingdoms, either **Plantae** or **Animalia,** based on observable characteristics such as green pigment, motility, or food consumption. But what of the tiny microorganisms discovered since the microscope was invented? And what of single-celled organisms that are both green and capable of swimming (such as euglenoids; see Chapter 21)? After two centuries of controversy and discovery, biologist R. H. Whittaker in 1963 devised a five-kingdom classification scheme still in use today.

As Figure 19-10 shows, Whittaker placed each of the 2 million or so defined types of living and fossil organisms into one of five kingdoms. He assigned bacteria and cyanobacteria, the single-celled prokaryotes, to the kingdom **Monera** (see Chapter 20). He grouped single-celled eukaryotes, including protozoa and some types of algae, to the kingdom **Protista** (see Chapter 21). Fungi, such as molds and mushrooms, have their own kingdom, **Fungi** (see Chapter 22). Aquatic and land plants make up the kingdom Plantae (see Chapters 23 and 24), and multicellular animals form the kingdom Animalia (see Chapters 25 and 26).

While Whittaker's simplifying scheme has truly rendered order to the study of life's vast diversity, certain taxonomic groups are extremely difficult to place, even at the kingdom level. For example, some biologists place the red, brown, and green algae in the Protista, while others consider them to be true members of the plant kingdom. Furthermore, the kingdoms probably are not true clades; that is, they probably are not each derived from one ancestral type. Monera includes one subgroup, the so-called archaebacteria, that are so primitive and so different from more common prokaryotes that many biologists now assign them to a separate "sixth" kingdom. Fungi probably arose not from one progenitor, but from several separate protistan groups, and so did the animals. Finally, entire kingdoms have arisen and become extinct. Harvard geologist S. J. Gould has written about a group of possibly animal-like organisms, some a meter long, that lived worldwide prior to 570 million years ago. No members of this group, which he termed the Ediacaran experiment, appear to have survived. Were they a seventh kingdom? And were there still others?

Clearly, the characteristics biologists use to define so huge and broad a group as a kingdom are in good part arbitrary conveniences, helpful for certain things but not for all that a taxonomy might be called on to do. Nevertheless, taxonomy is by no means a dead science; even today, biologists are discovering new types of organisms (see Figure 19-11). The work of Linnaeus, Whittaker, and other taxonomists will allow us, in the next seven chapters, to retrace the evolution of earth's major life forms—all descendants, ultimately, of the primitive cells that arose by chemical evolution in a hostile environment billions of years ago.

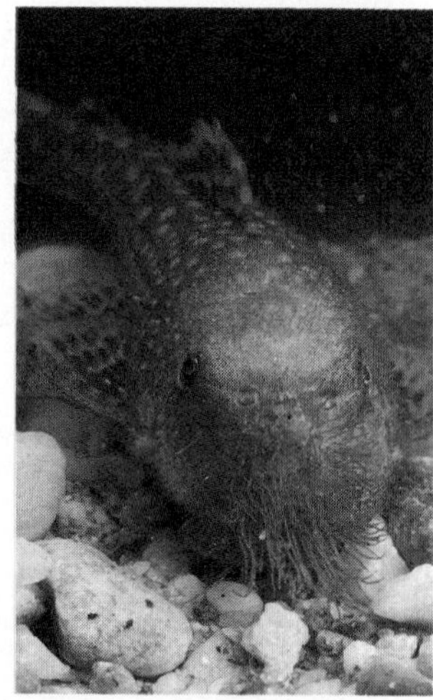

Figure 19-11 NEWLY DISCOVERED ORGANISMS. Both this black frog and bearded catfish were found on a recent expedition in tropic jungles.

SUMMARY

1. Biologists theorize that life arose spontaneously on the earth through normal chemical and physical processes.

2. Our solar system's planets arose from clouds of matter circling the sun. The earth's composition, size, temperature, and distance from the sun provided a special set of conditions compatible with the origin and maintenance of life.

3. Laboratory experiments demonstrate that under physical and chemical conditions probably similar to those on the early earth, such organic monomers as amino acids, sugars, fatty acids, and nucleic acid bases form.

4. The same conditions can cause some of those monomers to form polymers—*proteinoids* and nucleic acids. Polymers in solution can easily form aggregates with properties of precellular forms.

5. When supplied with appropriate nucleotides, certain RNAs act enzymatically as *ribozymes* and can build other RNA molecules as well as carry out sex-like exchanges of pieces of RNA.

6. Critical steps toward life included development of membranelike structures; biological informational macromolecules (probably RNA, then DNA); a genetic code; cell-like compartments;

and chains of metabolic reactions going on in those structures.

7. Organic molecules found in rocks dated at 3.8 billion years old may be the oldest fossil evidence of possible life. Fossilized cells occur in rocks that are 3.4 and 3.5 billion years old.

8. The first cells were almost certainly anaerobic heterotrophs that consumed organic materials. Autotrophs emerged later, producing their own nutrients and altering the atmosphere by releasing O_2. The *ozone layer* formed from this oxygen and reduced the ultraviolet light levels penetrating into shallow water and onto the land surface. Cellular respiration appeared after free O_2 accumulated in the atmosphere, and aerobic cells could harvest more energy from organic molecules.

9. Remains of the oldest eukaryotic cells have been found in rocks that are 1.5 billion years old. Single-celled prokaryotic and eukaryotic organisms were precursors to all later life forms.

10. Changes in geology, including *continental drift* due to *plate tectonics*, global climatic alterations, and volcanic and glacial activities, greatly affected the evolution of organisms and continues to do so today.

11. Biologists have adopted the *binomial system of nomenclature* as a means of categorizing the varieties of life on earth. Each type of organism has *genus* and *species* names and is further classified by a set of higher taxonomic categories: *family, order, class,* and *phylum (or division).*

12. *Taxonomy* reveals evolutionary relationships among organisms. *Clades* are taxonomic units whose members are derived from a common ancestor.

13. Phylogenetic trees have been used to depict evolutionary relationships graphically. The system used in this text arranges all organisms into five kingdoms: *Monera, Protista, Fungi, Plantae,* and *Animalia.*

KEY TERMS

aerobic cell
Animalia
Big Bang
binomial system of nomenclature
chemical evolution
clade
cladistics
class
coacervate
continental drift
core
crust
division
family
Fungi
genus (genera)
kingdom
liposome
mantle
Monera
monophyletic
order
ozone layer
Pangaea
phylum (phyla)
Plantae
plate tectonics
progenote
proteinoid
proteinoid microsphere
Protista
ribozyme
species
stromatolite
taxon
taxonomy

QUESTIONS

1. How did the earth's first atmosphere form, how has it changed, and what was responsible for that change?

2. What kinds of organic molecules have been produced in Urey-Miller spark chambers? What energy sources can be used in such experiments? What energy sources may have been available when organic molecules first formed on the Earth?

3. What are proteinoid microspheres, coacervates, and liposomes? In what ways are they analogous to living cells?

4. What roles did RNA play in the chemical evolution that preceded life on earth?

5. Is it more likely that the first living organisms on earth were heterotrophs or autotrophs? Aerobes or anaerobes?

6. What energy source is screened out by ozone? How did the presence of ozone affect the evolution of life on earth?

7. What is meant by a clade? Is each kingdom a clade? Is each phylum a clade? Each species? Are all living things on earth related?

8. Name the five kingdoms of organisms in Whittaker's classification system. To which kingdom do you belong? A mushroom? A pine tree?

ESSAY QUESTION

1. If pollutants suddenly triggered a chain reaction that destroyed all the ozone in the atmosphere, what would be the consequences?

For additional readings related to topics in this chapter, see Appendix C.

20 Monera and Viruses: The Invisible Kingdom

The most important discoveries of the laws, methods, and progress of Nature have nearly always sprung from the examination of the smallest objects which she contains.

Jean Baptiste de Lamarck,
Philosophie Zoologique (1809)

In one sense, the human body can be viewed as a set of small, interacting ecosystems inhabited by millions of microscopic residents. And the larger world around us is similarly inhabited: A few drops of water or a pinch of soil collected from anywhere on earth will teem with thousands to millions of individual bacteria.

Bacteria are the most prominent members of the kingdom Monera, which also includes *cyanobacteria*—blue-green photosynthetic cells—and *archaebacteria* (see Chapter 19). The 2,000 or more species of monerans play important ecological, practical, and evolutionary roles: They break down organic matter, recycle soil nutrients, ferment foods, manufacture carbohydrates by photosynthesis and chemosynthesis, and cause many animal and plant diseases. Humans harness monerans for food and drug production, leather and textile processing, and sewage treatment. And various lineages of ancient monerans probably gave rise to the other king-

Photosynthetic prokaryotic cells live in the hot water of these mineralized terraces in Yellowstone National Park.

doms of living things—the protists, fungi, plants, and animals. Quite literally, without monerans, we would not be here.

All monerans are **prokaryotes** (see Chapter 6). Monerans grow and reproduce quickly, and they display astonishing metabolic diversity, utilizing a wide range of energy sources and surviving in virtually all environments, from freezing to boiling, light to dark, high pressure to low.

The microscopic realm is also populated by *viruses*, complex packets of genetic material and protein that infect prokaryotic or eukaryotic cells, subvert their contents and metabolic machinery, and often kill the host cell in the process. Viruses are not true cells and do not fit neatly into any of Whittaker's five kingdoms. But their structural simplicity and possible origin from prokaryotic cells make this the appropriate place to consider viruses.

This chapter discusses:

- The basic moneran characteristics, including cellular structure, movement, reproduction, and metabolism
- The major groups of monerans—the bacteria, cyanobacteria, and archaebacteria—and their distinguishing traits
- The impact of bacteria on human food production and health
- The structure and reproduction of viruses and their ability to cause disease
- How monerans and viruses may have evolved

MONERA: TINY BUT COMPLEX CELLS

Each major type of moneran has unique characteristics, but all share certain features that are most easily exemplified by bacteria. Bacteria are usually 1–10 μm long (Figure 20-1), and many have volumes of about 1–5 μm^3. In contrast, the smallest eukaryotic yeast cells have volumes of about 20–50 μm^3, while the smallest algal cells are 5,000 μm^3.

Cellular architecture underlies the size differences between eukaryotes and prokaryotes. The lower limit on eukaryotic cell size is dictated by the cell's need to contain organelles such as the nucleus, one or more mitochondria, and Golgi bodies. For monerans, the lower limit on cell volume is set by the number and sizes of molecular assemblies: ribosomes, clusters of enzymes, and the circular DNA strand that serves as the cell's single chromosome. The smallest monerans (indeed, the smallest known living cells), called *mycoplasmas*, have no outer cell wall, contain half as much DNA as do larger bacteria, and usually are unable to move actively. Although monerans are single cells, many species can exist in complex communities and even form groups and collectively hunt prey.

Bacterial Cell Structure

One feature that is common to most monerans is a cell wall. Monerans live in hypotonic fluids (and therefore are hypertonic to them), and the cell wall resists swelling. This wall also gives shape, support, and protection to the plasma membrane and cytoplasm within. The typical bacterial cell wall is about 5–10 nm thick and is composed of huge carbohydrate and peptide chains called **peptidoglycan polymers** (or *murein*). Within the polymers, short peptides cross-link and stabilize the main chains of sugar molecules.

Some bacteria have walls made up of only a single thick layer of peptidoglycans outside the plasma membrane. Others, including *Escherichia coli*, have a second, outer layer of lipopolysaccharide that is a true lipid

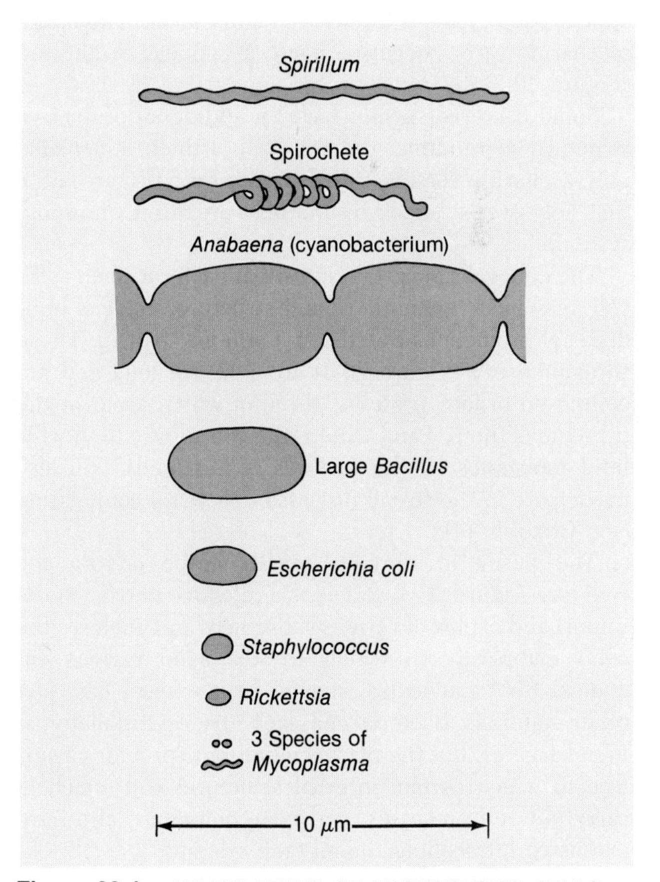

Figure 20-1 MAJOR TYPES OF PROKARYOTIC CELLS. Several of the prokaryotes we will discuss in this chapter are drawn to scale to show the range of sizes that exists.

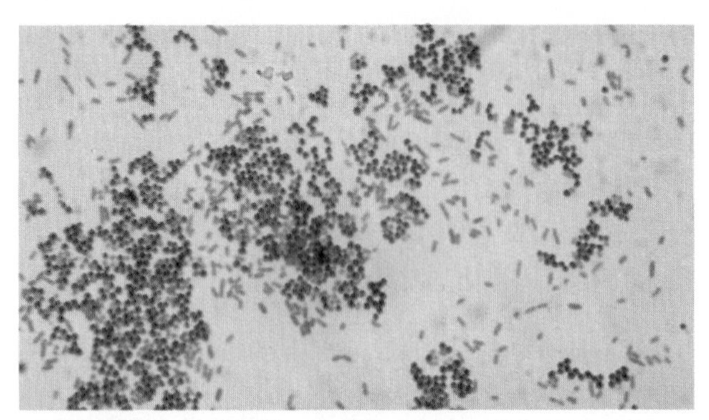

Figure 20-2 A MIXED SMEAR OF BACTERIAL CELLS STAINED BY THE GRAM PROCEDURE.
The gram-positive *Staphylococcus aureus* cells (magnified about 2,600×) have retained crystal violet dye and stain purple, whereas the *Escherichia coli* cells, which are gram-negative, do not retain crystal violet and instead show red dye.

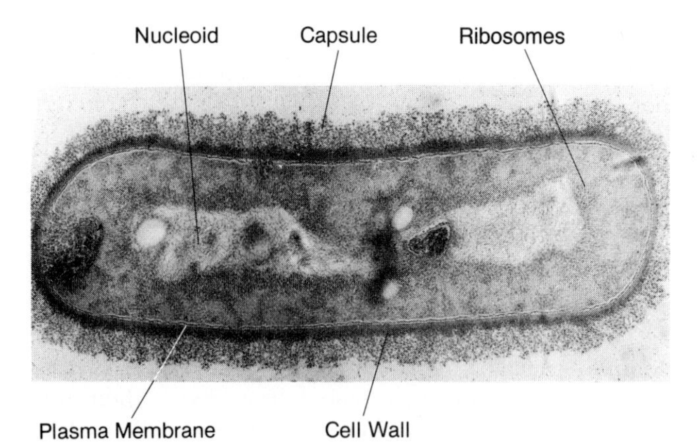

Figure 20-3 LIFE IN THE INVISIBLE REALM: A TYPICAL BACTERIAL CELL.
This electron micrograph of a bacillus, magnified about 29,000 times, shows the plasma membrane surrounded by the cell wall and capsule. The nucleoid region appears transparent, and the dense-staining cytoplasm contains many ribosomes.

bilayer surrounding a thick peptidoglycan layer. The traditional method of distinguishing among bacteria, *Gram staining,* exploits this difference in cell wall architecture, and after a series of procedures, those bacterial cells with a single peptidoglycan (so-called **gram-positive bacteria**) appear purple, and those with an outer layer of lipopolysaccharide (**gram-negative bacteria**) appear bright red (Figure 20-2).

Some bacterial species have an additional protective structure surrounding the cell wall: a thick, mucuslike polysaccharide coating called the *capsule* (Figure 20-3) that helps resist attack by the host organism's immune system.

The cell walls of certain intestinal bacteria, such as *E. coli,* may bear yet another outer structure, made of hundreds of projections called *pili* (Latin for "hairs"). These threadlike pili (Figure 20-4) are 1–2 μm long and are composed of four proteins, three of which occur at the tip. One of them, Pap G, helps the cell adhere to specific lipid molecules on the surfaces of host cells. Pili also participate in the sexual process of bacterial conjugation (see Chapter 14).

The plasma membrane beneath the prokaryotic cell wall (see Figure 20-3) serves as a selective barrier to the import and export of various substances and encloses the cell's cytoplasm, in which are dissolved various enzymes, RNA molecules, sugars, amino acids, fats, and other materials. Bacterial cells lack a true cytoskeleton as in eukaryotes, but the plasma membrane provides a surface to which certain internal structures can attach to carry out cell processes, and many monerans also have extensive intracellular membrane systems.

During DNA replication, for example, the circular chromosome present in the dense *nucleoid* region (see Figure 20-3) remains attached to the inner surface of the plasma membrane. Ribosomes associated with mRNAs that the cell exports also attach to the plasma membrane.

Projecting from the plasma membrane into the cell is a convoluted, whorled, membranous structure called the *mesosome*. The mesosome may be a source of new membrane during bacterial cell division; it may act as a crude mitochondrion, perhaps containing cytochromes arrayed in appropriate sequence for cellular respiration; and it may serve as a simple type of "chloroplast" in photosynthetic bacteria.

Bacterial Cell Movement

Some bacteria cannot actively propel themselves, but many others can move by means of a lovely piece of

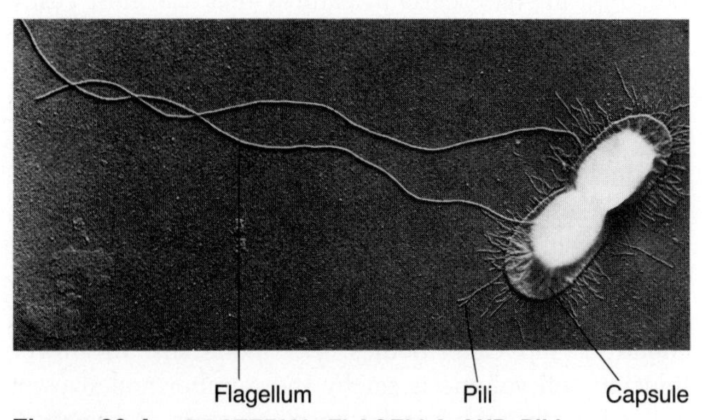

Figure 20-4 BACTERIAL FLAGELLA AND PILI: ORGANELLES OF LOCOMOTION AND EXCHANGE.
This dividing *Salmonella anatom* cell (magnified about 10,000×) has two flagella and hundreds of pili.

molecular engineering: the bacterial flagellum (Figures 20-4 and 20-5).

Bacterial flagella lack the microtubules found in eukaryotic flagella and cilia (see Chapter 6) and are, in fact, not much larger than one of those microtubules. Bacterial flagella also protrude beyond the plasma membrane and are truly extracellular.

Whereas eukaryotic flagella lash back and forth because of the bending of microtubule doublets, bacterial flagella *rotate* a bit like the propeller of a ship, clockwise propelling the cell forward and counterclockwise causing the cell to tumble chaotically. The bacterial flagellum is attached to the cell by means of a unique socket and motor composed of several dozen proteins that make up a shaft, or *filament;* a midpiece, or *hook;* a rotating rod extending through a protein disk; and finally, a ring of protein molecules (see Figure 20-5). The ring is a "motor" that responds to changing levels of a hydrogen ion gradient across the plasma membrane and so causes the attached flagellum (and with it, the cell) to rotate clockwise or counterclockwise.

Microbiologists have identified two other means of bacterial movement. In helically shaped bacteria called *spirochetes,* unusual protein fibrils called axial filaments lengthen and shorten, causing the cell to lash, snake, or bore forward. And minute rods called *myxobacteria* glide forward along a polysaccharide slime track that they secrete. While the gliding mechanism is still a mystery, experiments show that it cannot occur without a peculiar kind of sulfur-containing fatty acid called a *sulfonolipid.*

These three modes of movement allow bacteria to move appropriately toward or away from areas of high oxygen concentration, light, darkness, nutrients, toxic chemicals, and their own gaseous waste products. One photosynthetic bacterium, for example, automatically remains inside a lighted spot on a dark background, because ATP production from photophosphorylation (see Chapter 8) drops off sharply each time the cell enters the dark zone. This rapid decrease in ATP apparently causes changes in the H^+ gradient near the flagellum so that it reverses its direction of rotation, and the cell once more moves into the light.

Bacterial Reproduction

Bacteria can reproduce with astonishing speed—as often as once every 20 minutes! If unlimited space and nutrients were available to a single bacterium, and if all its progeny divided at this rate, after just 48 hours there would be some 2.2×10^{43} cells, weighing 4,000 times more than the planet earth. In nature, of course, bacterial growth is checked by limited nutrients and the buildup of waste products.

Bacteria can reproduce asexually, sexually, or through a survival process called spore formation. Let's look at each mode.

Asexual Reproduction: Budding and Binary Fission

Asexual reproduction can involve budding—producing outgrowths on an existing cell—but it usually entails **binary fission,** the division of prokaryotic cells into two virtually identical daughter cells.

Binary fission begins when the cell starts to elongate and the old cell wall starts to break down at a site close to

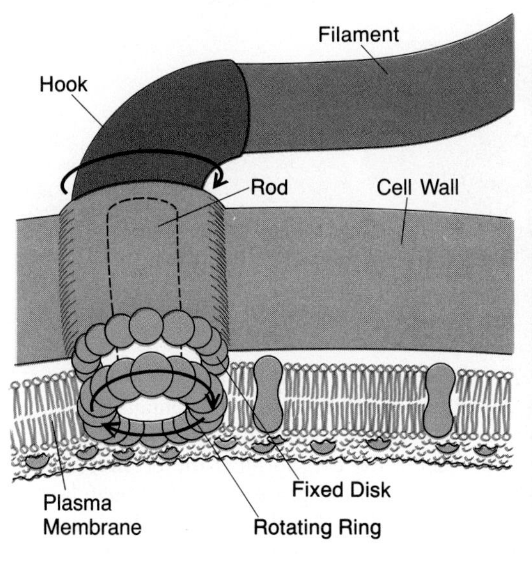

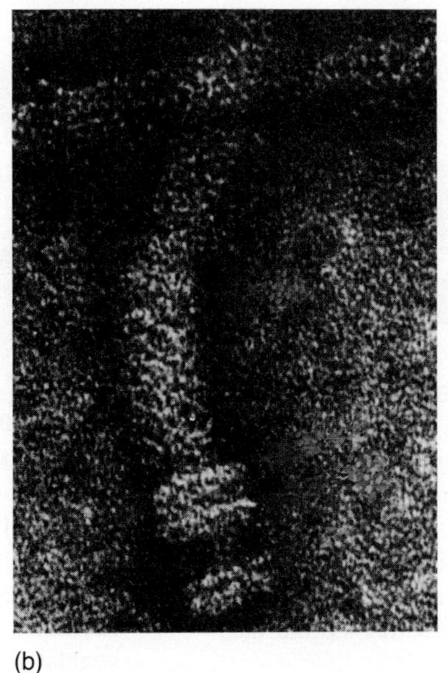

Figure 20-5 STRUCTURE OF THE BACTERIAL FLAGELLUM. (a) Unlike analogous eukaryotic flagella and cilia, the bacterial flagellum is a solid structure that extends through the cell wall and is anchored in the plasma membrane by protein rings (for simplicity, the four protein rings are shown as a disk and a rotating ring). The flagellum moves by means of the propeller-like rotation of the inner ring of proteins, which in turn rotates the attached rod, hook, and filament. Movement of hydrogen ions powers the rotation. (b) An electron micrograph of an *E. coli* flagellum (magnified about 422,000×). The filament and various protein rings can be seen.

the middle of the cell (Figure 20-6a). Just inside this region, the circular bacterial chromosome is attached to the plasma membrane. Enzymes soon nick open one of the two strands of DNA, which then unwind and separate. After DNA replication begins, the original attachment point to the plasma membrane becomes two points, new membrane is inserted between them, and they move apart. New cell wall materials begin to assemble at this location, forming a septum that extends inward until the cell has divided in two, with each daughter cell containing a circular chromosome attached to the plasma membrane (Figure 20-6b).

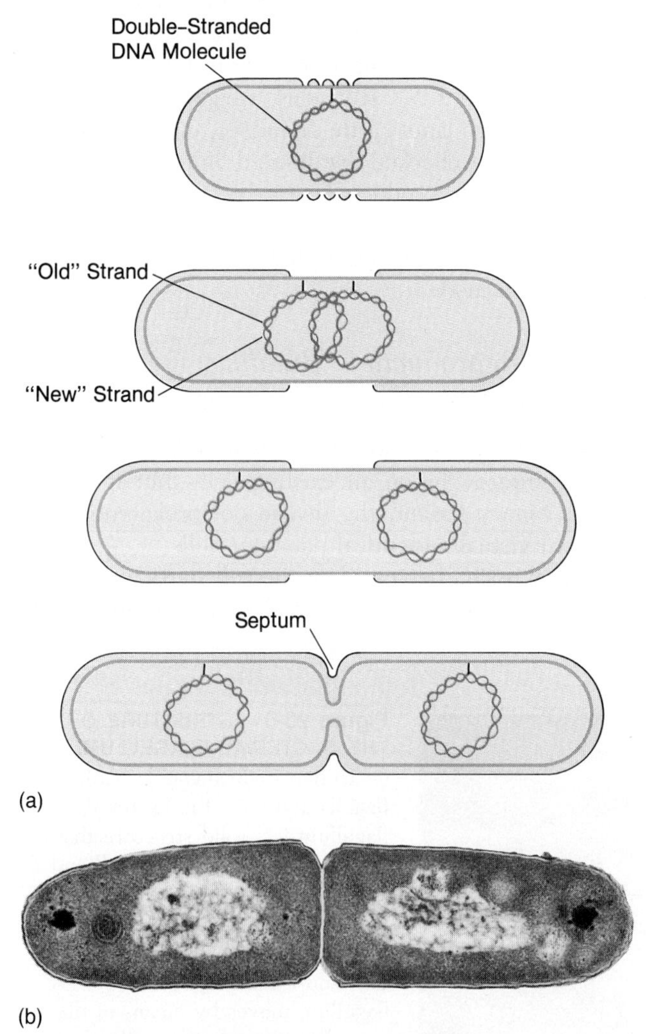

Figure 20-6 BACTERIAL CELL DIVISION.
(a) This series shows binary fission in a bacterial cell. As DNA replication begins, the cell wall ruptures, and new plasma membrane is added between the two sites where the DNA molecules are attached to the membrane. The plasma membrane then folds inward, and new cell wall material is added in the septum to complete the division of the cell. (b) A dividing bacterial cell after the septum has formed (magnified about 33,000×).

Sexual Processes

Although many bacteria apparently reproduce only through asexual binary fission, others carry out exchanges of genetic information reminiscent of sexual processes in eukaryotes. Chapter 14 discussed the three types of genetic exchange in bacteria: *transformation*, during which DNA fragments released from one cell are taken up by and incorporated into the DNA of another cell; *transduction* (see Figure 14-6), in which pieces of DNA from a host cell's chromosome are picked up by a virus and carried to the recipient cell; and *conjugation* (see Figure 14-3), which involves a donor male cell with a sex pilus, a recipient female cell, and the transfer of an F factor plasmid. The three sexual transfer processes are costly; in each, the donor cell dies. However, the evolutionary advantages of new gene combinations apparently outweigh the costs.

Spore Formation

The capability for both asexual and pseudosexual reproduction increases the chances of bacteria surviving under unfavorable conditions. In many species, environmental conditions can also trigger the formation of an **endospore,** a heavily encapsulated spore within another cell (Figure 20-7). The spore contains only those cellular components necessary to resume active metabolism once the cell is again in a favorable environment: genetic material, some ribosomes, enzymes to manufacture ATP and to carry out protein synthesis, amino acids, sugars, trace elements, and a minimum of water.

Endospores are truly an important evolutionary survival mechanism. They have been known to survive for

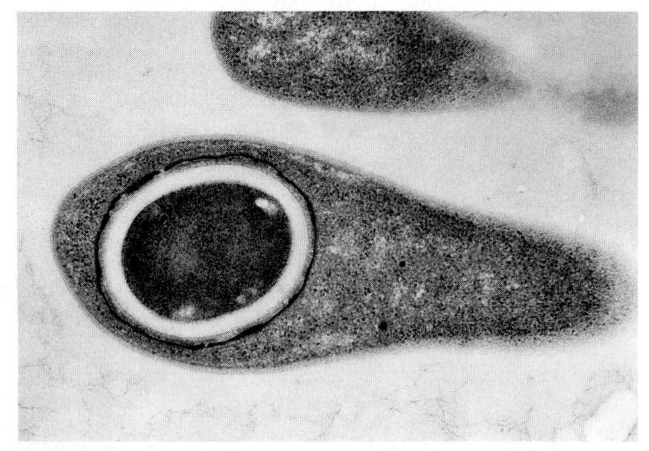

Figure 20-7 ENDOSPORE: A SURVIVAL CAPSULE.
This photo shows a *Clostridium tetani* cell containing a heavily walled endospore. A copy of the genetic material is enclosed in a portion of the cell, forming the prespore. A spore coat is added around the prespore, and ultimately, the original cell bursts, liberating the spore. Following a period of dormancy, the spore germinates to form a new vegetative cell.

over 1,000 years, for more than an hour of being boiled, for years of being locked in polar ice, and, in some species, even bombardment by ionizing gamma radiation.

Bacterial Metabolism

The versatile metabolism of bacteria is another key to their adaptation to wide-ranging environments and external stresses. *Heterotrophic bacteria* consume waste products and dead organic matter or exist as **parasites,** organisms that feed on living hosts, often harming them but usually not killing them. The parasitic species have tremendous medical impact, and those that break down organic substances are important to ecological cycles.

Autotrophic bacteria use energy and inorganic substances from the environment to manufacture the necessary organic building blocks and substrates for ATP production. They derive energy from either photosynthesis or oxidation of inorganic compounds. Thus, there are two types of autotrophs: *photoautotrophs* and *chemoautotrophs.*

Photoautotrophs include the green sulfur bacteria and two kinds of purple bacteria, sulfur and nonsulfur. Purple bacteria look purple because they have yellow and red carotenoids plus a bluish gray pigment called *bacteriochlorophyll,* which absorbs infrared wavelengths and thus can function in deep waters. Bacteriochlorophylls occur on membranous tubes, stacks of flattened sheets, or vesicles, not in the chloroplasts.

Bacterial photosynthesis is an anaerobic process. Both purple and green sulfur bacteria use molecules of H_2S (instead of H_2O) from mud sediments as electron donors for photosynthesis. Purple nonsulfur bacteria can grow in the presence of oxygen and use such organic compounds as alcohols and fatty acids.

Chemoautotrophs use some surprising substrates to obtain energy and can live in extremely harsh environments. These bacteria have the unique ability to oxidize inorganic compounds needed to build organic molecules. Thus, they trap energy released from such chemicals as hydrogen sulfide gas, elemental sulfur, ferrous iron, ammonium ions, and nitrites. Moreover, within the last few years, scientists have found in deep-ocean vents extraordinary chemoautotrophs that seem to thrive in environments several times hotter than those around thermal springs, such as the geysers in Yellowstone National Park, and that actually stop growing when "cooled" to the mere boiling point of water.

The researchers found the heat-loving bacteria in water samples obtained from tall sulfide chimneys almost 3,000 m below the ocean surface near Baja California. Remarkably, the bacteria divide at 250° C and cease reproducing when the temperature falls below 100° C. Thus, the growth of microbes is clearly not limited by temperatures that were long thought to denature proteins. Biologists still do not understand the kinds of structural proteins, enzymes, membranes, and other biological components that can withstand such intense pressures and temperatures.

One of the moneran's most important metabolic talents is *nitrogen fixation*—the process whereby bacteria take nitrogen from the air, oxidize it, and build it into amino acids and nucleotides. The process can involve free-living bacteria and cyanobacteria, as well as bacterial symbionts associated with the roots of various plants (Figure 20-8). (*Symbiosis* means "living together with mutual benefit.") The *Rhizobium* bacterium, for instance, fixes nitrogen in the roots of peas, clover, alfalfa, and other legumes. We will discuss the nitrogen cycle further in Chapter 44.

Regardless of their energy source, all bacteria have an extremely high rate of metabolism. Bacteria that ferment milk sugar (lactose), for example, break down 1,000 to 10,000 times their own weight of this substance each hour—a feat that would take a human 30 to 40 years.

Along with small cell size and the wide spectrum of available energy substrates, this metabolic speed helps explain the great success of the kingdom Monera. So does the broad range of oxygen levels they can tolerate. **Obligate anaerobes** grow only in the absence of free oxygen and employ only fermentation to generate ATP (see Chapter 7); **facultative anaerobes** can grow with or without oxygen; and **obligate aerobes** must have oxygen for metabolic processes. These variations in oxygen tolerance may reflect the evolution of microorganisms on a changing planet (see Chapter 19).

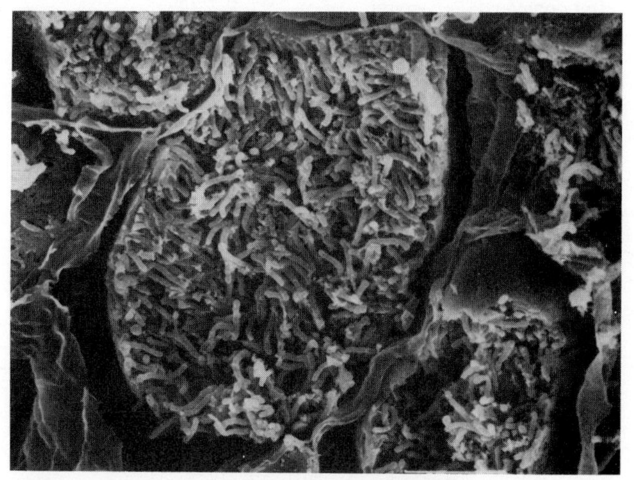

Figure 20-8 NITROGEN-FIXING BACTERIA.
Here a root nodule (magnified about 1,500×) is shown cracked open to reveal many symbiotic *Rhizobium* bacteria. These bacteria use sugars stored in plant roots as an energy source and in return provide the plant with a continuous supply of fixed nitrogen that can be used to build proteins.

THE VARIETY OF MONERANS

Monera is such a widely divergent kingdom that biologists disagree on which organisms should be considered members. They do agree, however, on two groups: the bacteria, or subkingdom *Schizophyta* and the *cyanobacteria* (formerly called blue-green algae). So let's begin our survey with them.

Subkingdom *Schizophyta:* The Bacteria

Historically, biologists assigned bacteria to taxonomic groups solely on the basis of shape. With the development of Gram staining and other techniques, however, bacteria, or **schizophytes** species, can be differentiated by their chemical, metabolic, and genetic differences, as well as by shape. Table 20-1 summarizes the major groups of bacteria we will consider and their key characteristics.

Eubacteria

Eubacteria, or "true" bacteria, are the most abundant and diverse group of prokaryotes, and many are named for their cellular shapes. A rod-shaped bacterium, for example, is called a *bacillus;* a sphere-shaped bacterium is called a *coccus;* and a helix-shaped bacterium is called a *spirillum.* Moreover, the rod-shaped agent that causes anthrax in animals and people is *Bacillus anthracis.* Members of the genus *Staphylococcus,* common inhabitants of the human body, form irregular clusters of

Table 20-1 MAJOR GROUPS OF SUBKINGDOM SCHIZOPHYTA

Phylum (Examples)	*Form*	*Mode of Movement*	*Mode of Reproduction*	*Mode of Nutrition*	*Ecological Role*	*Other Distinguishing Factors*
Eubacteria *(Escherichia coli, Streptococcus, Staphylococcus, Myobacterium tuberculosis, Clostridium botulinum)*	Rod, sphere, helix	Flagella, wriggling	Binary fission	Chemoautotroph, photoautotroph, heterotroph	Decomposer, symbiont, disease agent	Rigid cell wall, form endospores
Spirochaetes *(Spirochaeta, Treponema pallidum, Leptospira)*	Extremely long, flexible spiral	Twisting	Binary fission	Heterotroph	Symbiont (in mollusks), decomposer, disease agent (syphilis)	Obligate anaerobes (many forms)
Myxobacteria	Filament, large and short rods	Gliding (mechanism unknown)	Short filaments break off, binary fission	Heterotroph, chemoautotroph	Decomposer (especially of complex polysaccharides)	Flexible cell wall; some form fruiting bodies
Actinomycetes *(Streptomyces, Actinomyces)*	Branching multicellular filaments	Nonmotile	Spores, fragmentation, binary fission	Heterotroph	Decomposer (lipids, waxes), disease agent (tuberculosis, leprosy)	Many are moldlike in appearance
Rickettsiae, Chlamydiae	Small (0.5 by 1.1 μm)	Nonmotile	Binary fission	Heterotroph	Disease agent (typhus, Rocky Mountain spotted fever)	Intracellular parasites, rigid cell wall
Mycoplasmas	Smallest free-living cells	Nonmotile	Binary fission	Heterotroph	Disease agent (pleuropneumonia)	Plasma membranes contain cholesterol; no cell wall; many are intracellular parasites

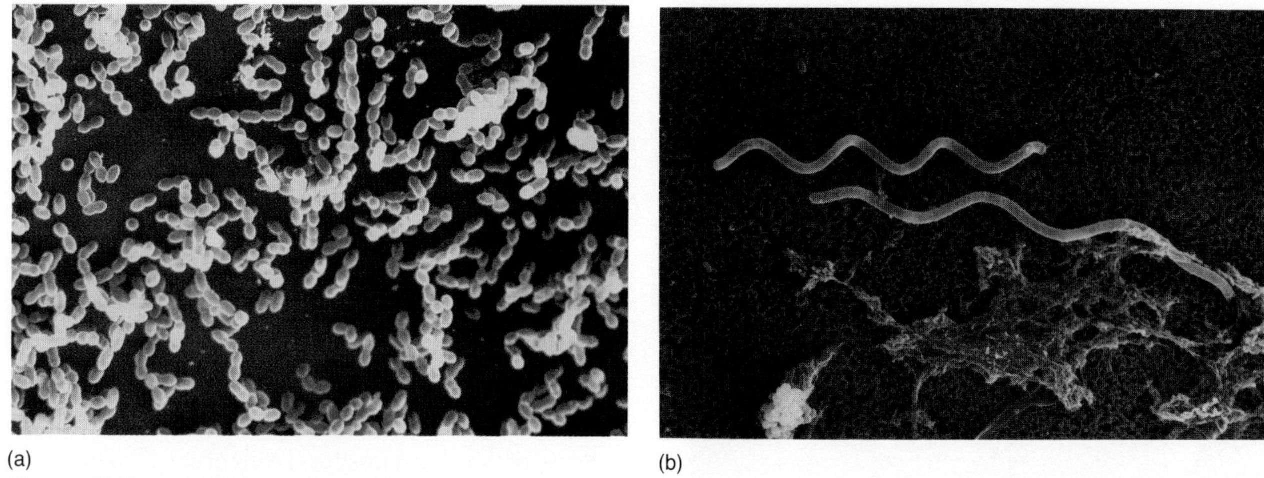

(a) (b)

Figure 20-9 COMMON EUBACTERIA.
These organisms include (a) spherical cocci *Streptococcus mutans* and (b) helical spirilla (both magnified about 1,500×).

spherical cells. The bacterium that causes strep throat, a species of *Streptococcus*, divides repeatedly along the same axis, forming a long chain of cocci (Figure 20-9a). Bacilli (and occasionally cocci) move by means of flagella. In contrast, many members of the genus *Spirillum* wriggle forward with a corkscrew motion (Figure 20-9b).

Spirochetes

The **spirochetes** are bacteria with a helical, or corkscrew, shape. Many live freely in aquatic environments and move by snakelike lashing movements, but parasitic species such as *Treponema pallidum*, the causative agent of syphilis, bore forward in a screwlike manner (Figure 20-10).

Myxobacteria

The **myxobacteria** are small, unflagellated, rod-shaped cells that, as we have noted, glide along slime tracks. They occasionally swarm together to form *fruiting bodies* (Figure 20-11), where some myxobacterial cells develop into *myxospores* capable of surviving harsh environmental conditions. Myxobacteria tend to live in soil, where they play an important ecological role in the breakdown and consumption of dead organic matter.

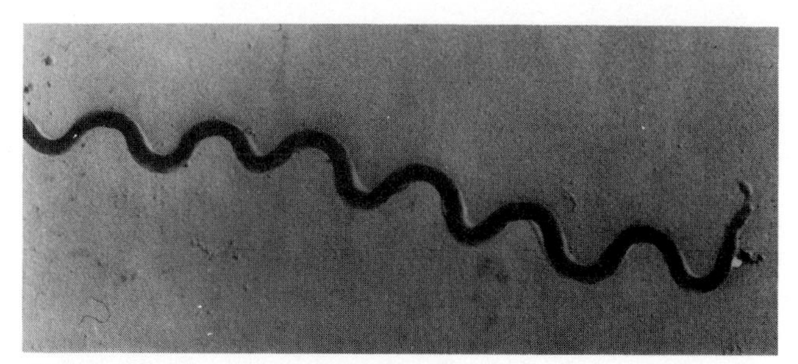

Figure 20-10 THE SPIROCHETE *Treponema pallidum*.
This spiral bacterium (magnified about 10,000×) is responsible for syphilis.

Figure 20-11 FRUITING BODIES FORMED BY MYXOBACTERIAL CELLS.
Myxobacteria *Chondromyces crocatus* secrete extracellular materials that compose branchlike stalks and then lie dormant in spherical clusters, or caps. The caps protrude above the soil surface and eventually burst. The released spores are then dispersed by wind. This electron micrograph is magnified several hundred times.

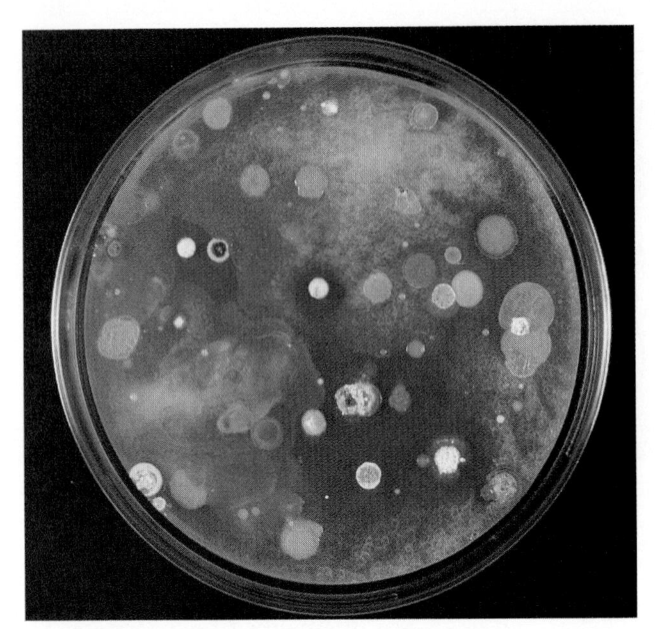

Figure 20-12 SOME ANTIBIOTIC-PRODUCING ACTINOMYCETES.
In the mixed bacterial culture shown here, clear zones free of bacterial growth can be seen around antibiotic-producing actinomycete colonies, such as *Streptomyces*.

Actinomycetes

Many types of diverse filamentous organisms called **actinomycetes** form the fourth schizomycete group. Actinomycetes form filaments because daughter cells remain attached to one another during binary fission, forming a multicellular network (Figure 20-12). The actinomycetes include species that can cause Hansen's disease (leprosy) and tuberculosis; species that make antibiotics such as streptomycin and erythromycin; and still other species that associate with the roots of nonleguminous plants and perform the important task of nitrogen fixation.

Rickettsias and Chlamydias

Rickettsias are among the tiniest prokaryotes and live as parasites in two alternating hosts: arthropods (such as lice, ticks, and fleas) and mammals or birds. Although rickettsias do not harm their arthropod hosts, they can be transmitted by means of the bites of fleas, ticks, or lice to mammals, where they can cause such serious diseases as typhus and Rocky Mountain spotted fever.

Chlamydias, even simpler prokaryotes, cause venereal infections, psittacosis (with its pneumonia-like symptoms), and trachoma, a leading cause of blindness.

Mycoplasmas

Mycoplasmas are the smallest living cells yet found. They lack rigid cell walls, grow to form filaments, globules, rings, and other shapes in colonies, and are usually aerobic. They cause pleuropneumonia and other diseases and are resistant to antibacterial agents such as penicillin.

Cyanobacteria: The Blue-Green Photosynthesizers

The second major group of monerans is the subkingdom Cyanophyta. These 200 species, formerly called blue-green algae and now referred to as **cyanobacteria,** live in a wide range of habitats, from salt and fresh water to soil.

Some of the oldest fossilized cells of any kind found on earth appear in sediments 2.7 billion years old and resemble modern cyanobacteria (see Chapter 19). Cyano-

(a)

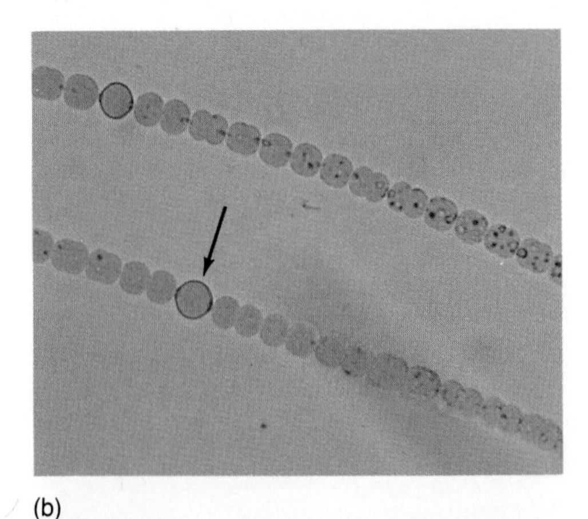

(b)

Figure 20-13 PHOTOSYNTHETIC POND SCUM.
(a) Colonies of cyanobacteria (blue-green algae) often form a scum on the surface of polluted ponds. (b) A common type of filamentous cyanobacteria is *Anabaena* (magnified about 75×). The arrow points to a heterocyst cell.

bacteria formed layered columns called stromatolites (see Chapter 19 opening photo on page 326) that may have released much of the free atmospheric oxygen, which in turn allowed animal life to evolve. Divers recently discovered living stromatolites up to 2 m high in the Bahama Islands.

Cyanobacteria are single rod-shaped or spherical cells that occur in clusters or in long filamentous chains (Figure 20-13). The cell walls, made of peptidoglycan polymers, are surrounded by a protective coating of gelatinous material that is often poisonous to grazers.

Cyanobacteria contain chlorophyll *a*, carotenoids, and red and blue pigments called *phycoerythrins* and *phycocyanins*, which together give the cells a range of hues from yellows and reds to violets and deep blues. The photosynthetic pigments are associated with layers of membranes called *thylakoids*, which are quite different from and simpler than the chloroplasts in plant cells.

Many cyanobacterial species can fix nitrogen because they have special thick-walled cells called heterocysts (Figure 20-14) that contain the oxygen-sensitive enzyme *nitrogenase*. Nitrogenase in heterocysts uses ATP generated by photosystem I (see Chapter 8) plus electrons to reduce N_2 to the fixed form NH_3 (ammonia).

The ability to carry out photosynthesis as well as nitrogen fixation means that many cyanobacterial species can survive on bare rocks, tree trunks, even on polar bear fur—harsh environments that provide only water, inorganic minerals, air, and light. A few species have particularly resistant cell walls that enable them to survive extreme environments—perhaps similar to those of early earth. Such extreme environments include thermal hot springs up to 75° C; alkaline water and soil up to pH 11; acidic sulfurous soil with a pH of 0.5; and highly concentrated salt solutions (25 to 30 percent NaCl). At least one company sells dried cyanobacteria as "living fertilizer" for arid lands.

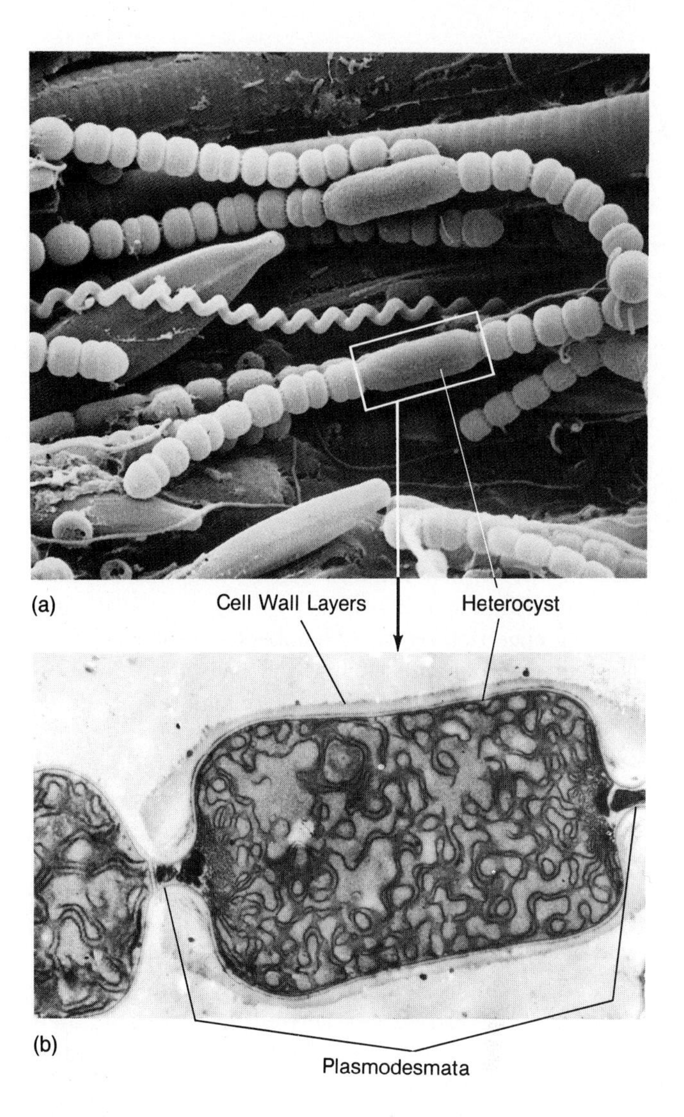

Figure 20-14 HETEROCYSTS IN *Anabaena*.
(a) In these *Anabaena* filaments (magnified about 1,500×), elongate, cylindrical heterocysts are interspersed between smaller, normal photosynthetic cells. (b) A heterocyst (magnified about 11,000×) has extra wall layers to impede oxygen entry and plasmodesmata to connect it to adjacent photosynthetic cells, thus facilitating the passage of fixed nitrogen or amino acids between cells.

Chloroxybacteria: Possible Ancestors to Chloroplasts

The recently discovered **chloroxybacteria** live in the bodies of small marine animals called tunicates or as free-living species in shallow lakes. Taxonomists created a new phylum, Chloroxybacteria (also called Prochlorophyta), for these light green, single-celled organisms, and some think chloroxybacteria constitute a separate subkingdom of monerans.

The cells are definitely prokaryotic, yet they contain chlorophylls *a* and *b* and yellow and orange carotenoids, and in these and other respects, they resemble chloroplasts of plants and green algae. Some biologists hypothesize that chloroxybacteria or similar green bacteria were engulfed by early cells, giving rise to chloroplasts.

Archaebacteria: An Independent Kingdom?

The fourth group of monerans are the **archaebacteria.** These organisms are so different from schizomycetes and cyanobacteria that some biologists prefer to classify them not as a subkingdom or phylum of monerans but as an independent kingdom. Archaebacteria include methane producers, sulfur-dependent species, and cells that tolerate very salty or hot environments.

The methane bacteria (**methanobacteria**) are anaerobes that derive energy from coupled reactions in which H_2 is oxidized to H_2O (yielding energy) and CO_2 is reduced to CH_4 (methane gas). Methanobacterial species can live in decaying vegetation in streams, lakes, and bogs, where they generate "swamp gas"; in the intestinal tracts of cattle, which belch the methane gas; thousands of fathoms deep on the seafloor; and even in hot sulfur springs or smoldering piles of coal mine tailings.

Sulfur-dependent archaebacteria (**sulfobales**) live by oxidizing or reducing sulfur, and many also thrive at high temperatures (80°–95° C). One remarkable species normally uses carbon dioxide as a carbon source and oxidizes sulfur with oxygen to provide an energy source. Microbiologists recently identified a genus intermediate between methanobacteria and sulfobales, called *Archaeoglobus*, which reduces sulfate and makes small amounts of methane.

The archaebacteria called **halobacteria** can survive in high salt environments such as salt pans and salt lakes. And the **thermoproteales** live at very high temperatures in volcanoes and equivalent locales.

Recent studies suggest that archaebacteria had an origin different from that of eubacteria and cyanobacteria. Archaebacteria have distinctive plasma membranes, several unique enzymes, and unique sequences in their transfer and ribosomal RNAs, and they lack cytochrome molecules (see Chapter 7). Significantly, their RNA polymerase and one of their ribosomal structural proteins resemble those of eukaryotes, and they have nucleosome-like structures as well as introns in their tRNA and rRNA genes, again like eukaryotes but unlike prokaryotes. Some biologists believe that some extremely primitive ancestral cell populations (the "urkaryotes") gave rise to three separate groups: the archaebacteria, the eukaryotes, and the monerans.

BACTERIA AND HUMANS

In addition to their ecological roles in soil nutrient cycles and in the breakdown of dead organic matter, monerans (bacteria in particular) are enormously important in the production of food and the spread of disease.

Bacteria and Food Production

Many of our favorite dairy products, including cheese, butter, sour cream, yogurt, and buttermilk, are produced when *Lactobacillus* species and *Streptococcus lacti* degrade the lactose in milk and release sharp-tasting lactic acid. We also employ lactic acid bacteria to ferment vegetables and fruits into pickles and sauerkraut, and acetic acid bacteria to produce vinegar from wine or fruit juices.

Of course, other bacteria can spoil food products, and so we must process and preserve foods with heat, pressure, and chemical additives and keep many foods frozen or refrigerated to retard bacterial growth.

Bacteria and Disease

A century ago, bacteria had a profound impact on daily life, as routine infections could be life-threatening, and improperly stored foods could easily poison the unsuspecting consumer. Not until the late nineteenth century, when scientists realized that microorganisms cause disease and spoil food, and German physician Robert Koch learned to distinguish **pathogens,** or disease-causing microbes, from harmless ones, could we begin to protect ourselves from the unseen agents (Figure 20-15).

Koch's studies and those of Louis Pasteur, who proposed the germ theory of disease (see Chapter 1), led people to adopt sanitation measures that have helped lengthen the human life span in this century. Table 20-2 lists a number of human diseases caused by bacteria and the primary ways these diseases are spread.

Bacteria cause actual disease symptoms by irritating or directly destroying the host organism's cells and tissues; by triggering exaggerated immune responses to the

Figure 20-15 A MILESTONE IN MEDICINE. The medical treatment of microbial diseases began with Robert Koch's experiments, which showed that pathogenic bacteria cause diseases such as cholera. Koch is seen here, to the left, in his tent near Lake Victoria in Africa during an expedition in 1906.

Table 20-2 SOME HUMAN BACTERIAL DISEASES

Disease	*Causative Agent*
Fecal contamination	
Typhoid fever	*Salmonella typhi*
Cholera	*Vibrio cholerae*
Dysentery (shigellosis)	*Shigella dysenteriae*
Food poisoning	*Clostridium botulinum*
	Salmonella species
	Staphylococcus species
Traveler's diarrhea	*Escherichia coli* strain
Animal bites	
Typhus	*Rickettsia typhi*
Q fever	*Coxiella burnetii*
Plague	*Pasturella pestis*
Tularemia	*Francisella tularensis*
Rocky Mountain spotted fever	*Rickettsia rickettsii*
Exhalation droplets	
Diphtheria	*Corynebacterium diphtheriae*
Tuberculosis	*Mycobacterium tuberculosis*
Meningitis	*Neisseria meningitidis*
Scarlet fever, rheumatic fever	*Streptococcus* species
Pneumonia	*Streptococcus pneumoniae*
Tonsillitis	*Streptococcus* species
Whooping cough	*Bordetella pertussis*
Direct contact	
Gonorrhea	*Neisseria gonorrhoeae*
Syphilis	*Treponema pallidum*
Puerperal fever	*Streptococcus* species
Wounds	
Tetanus	*Clostridium tetani*
Gas gangrene	*Clostridium perfringens*

presence of foreign cells; or by producing destructive toxins. These toxins can be released by living bacteria, in which case they are called **exotoxins,** or they may be liberated when bacterial cells die and burst, in which case they are called **endotoxins.** *Clostridium botulinum,* for example, the bacterium that causes botulism poisoning, releases a powerful exotoxin that can induce agonizing asphyxiation and death.

Physicians can effectively treat most bacterial diseases with antibiotics and other drugs, many of which are produced by other monerans or by fungi. Antibiotics may interfere with DNA, RNA, or protein synthesis; ribosome function; membrane integrity; or a variety of other essential structures or processes. Penicillin, for example, prevents synthesis and assembly of bacterial cell walls, but leaves animal or plant cells unaffected because their outer coverings are made of different substances.

Physicians can also prevent some bacterial diseases by *vaccination,* stimulating a person's immune system to set up a permanent defense response against a g[illegible] type of disease-causing bacterium or virus.

One critical problem now facing human medicine is the tendency of many types of bacteria to develop resistance to antibiotics as a result of chance mutations or the transfer of R factor plasmids that confer drug resistance. These drug-resistant cells can then quickly reproduce into large, infectious populations. Researchers continually grapple with this difficult and potentially dangerous problem.

VIRUSES: NONCELLULAR MOLECULAR PARASITES

Biologists sometimes consider viruses to lie on the threshold of life, straddling the shadowy line between living and nonliving. **Viruses** are not cells, but particles of genetic material and protein that can invade living cells, commandeer their metabolic machinery, and in this way reproduce. Viruses also evolve and may have originated from prokaryotic cells. Thus, while they are not true monerans, we consider viruses here.

In 1935, biologist Wendell Stanley purified the first virus by extracting a needlelike, crystalline material from the juice of 900 kg of tobacco leaves. When he dissolved these crystals and applied the solution to tobacco leaves, it could cause a serious disease. He named the particles—which were clearly not living cells but some sort of inert entity—*tobacco mosaic virus (TMV).* Later tests showed TMV to be rod-shaped particles with a protein coat around a core of RNA (Figure 20-16a).

Structure of Viruses

Not all viruses are rods. Others, such as the adenoviruses, which can cause human colds and animal tumors, are shaped like geodesic domes (Figure 20-16b and c). Still others, such as those that cause influenza, are somewhat globular, with spiky proteins over their surfaces. Viruses that infect bacteria, called *bacteriophages* (see Chapter 12), have a hexagonal head, a tail, and leglike tail fibers (Figure 20-17). The smallest viruses, or **viroids,** infect only plant cells and have just a single loop of RNA as genetic material, although it is not known to code for any protein. Host proteins apparently form the viroid's coat.

Regardless of overall configuration, each virus particle has an outer protein coat, or *capsid,* and a core of from five to several hundred genes made up of single- or double-stranded DNA or RNA. Together, the capsid and

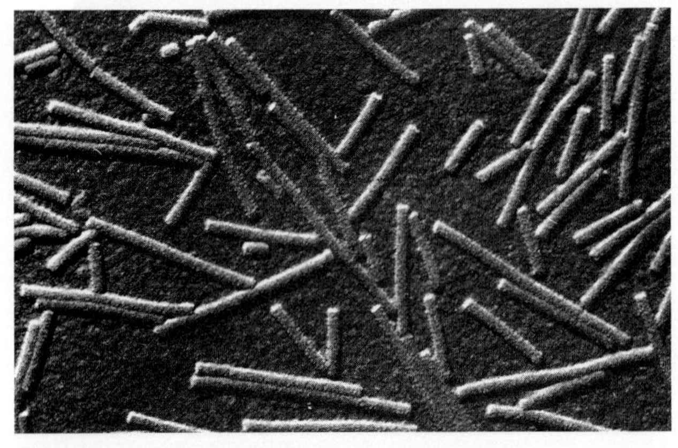

(a)

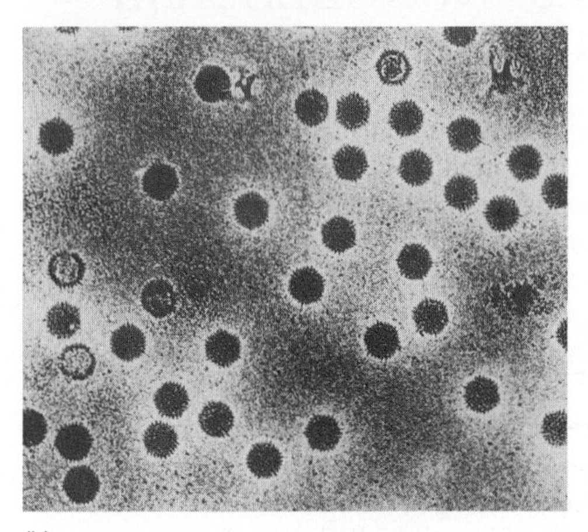

(b)

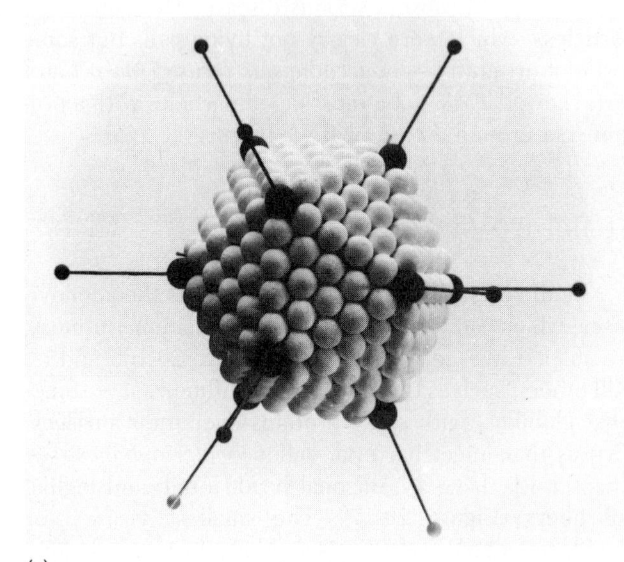

(c)

Figure 20-16 VIRUSES: A VARIETY OF PARTICLE SHAPES.
(a) The rod-shaped tobacco mosaic virus, TMV (magnified about 92,000×). (b) Negatively stained electron micrograph of adenovirus (magnified about 59,000×). (c) A model of the adenovirus shows some of its 20 sides, each an equilateral triangle.

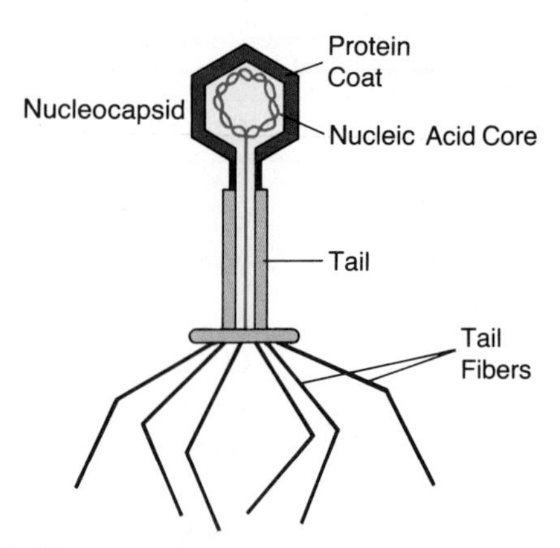

Figure 20-17 A TYPICAL VIRUS PARTICLE.
This bacteriophage has a head region, composed of a protein coat and nucleic acid core, and a tail. Tail fibers are involved in attachment to the bacterium that is to be invaded.

nucleic acid core are called the *nucleocapsid*, and in many viruses, an *envelope*, or compact membranelike structure composed of lipids and proteins, surrounds the nucleocapsid.

Reproduction of Viruses

To reproduce, viral nucleic acid must enter a host cell. Some viruses inject their genetic core into the host, leaving the protein capsid outside the cell as the viral genetic material begins functioning within. Certain animal viruses, such as the Semliki Forest virus, use an alternative strategy, binding to surface receptors on a host cell and triggering the cell to engulf the intact virion.

Many bacteriophages (viruses that infect bacteria) are *nonvirulent* on entering a host cell; that is, they do not kill their host, and instead, the DNA can become inserted into the host's chromosome as a *provirus* that is replicated along with the host's chromosome during each cell division cycle. During this process, called **lysogeny** (Figure 20-18a), the virus is *temperate* (inactive) because one of its genes blocks transcription of the other viral genes responsible for virulent behaviors.

Under certain environmental conditions, however, the lysogenic state breaks down, the viral genes are transcribed independently of the host chromosome, and viral proteins—including viral coat protein—are made. Mature viruses assemble, and the host cell bursts, releasing the newly replicated virus particles. This process of *virulent* (cell-killing) behavior is called the **lytic pathway** (see Figure 20-18b). Some viruses require proteins called *molecular chaperones* made by the host cell.

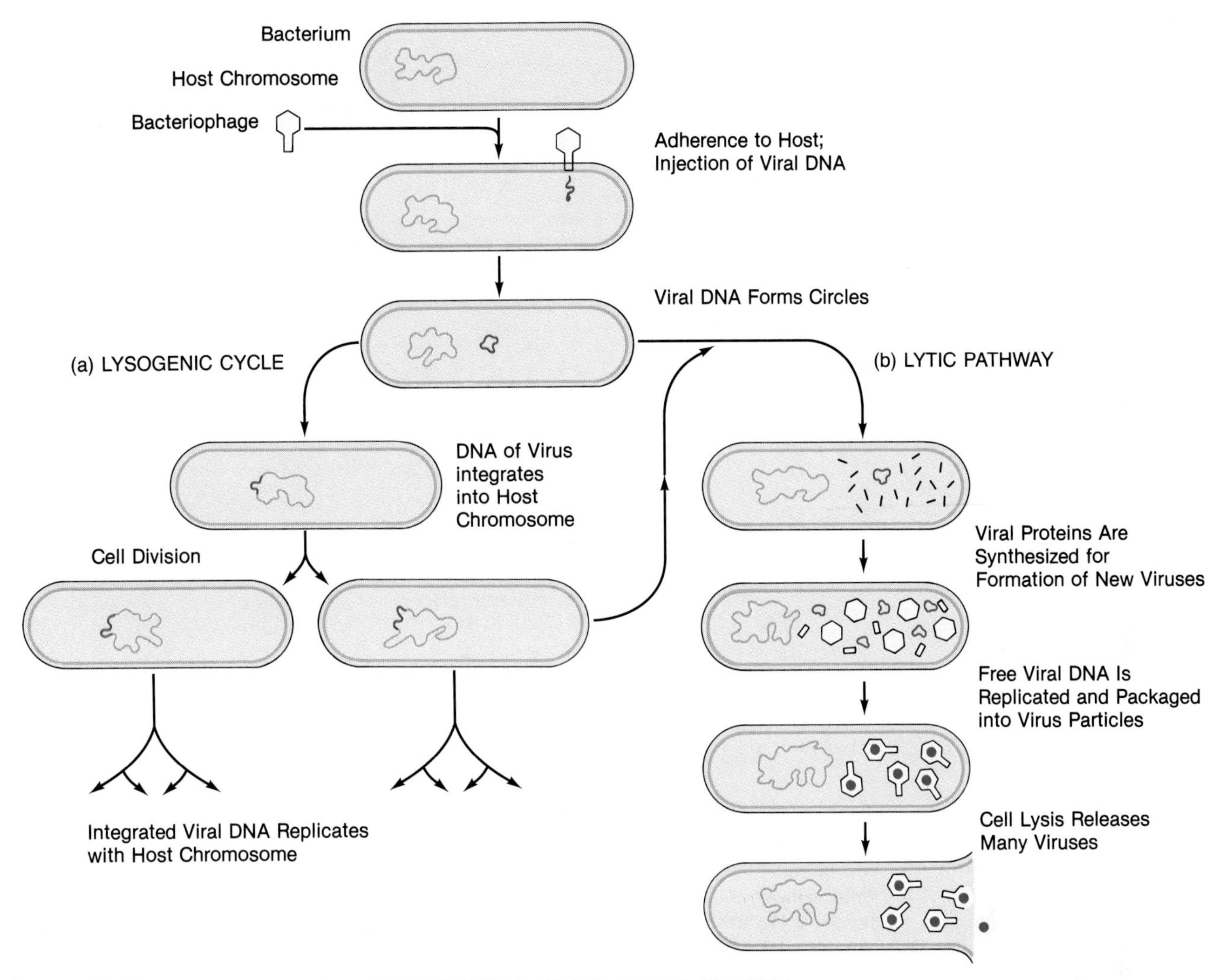

Figure 20-18 VIRAL LIFE CYCLE: REPRODUCTION AT THE HOST'S EXPENSE.
(a) During the lysogenic cycle, the viral DNA is integrated into the host cell DNA and is replicated each time the host cell divides. In this way, the viral genome (now called a provirus) is passed on to successive generations of bacterial cells. Proviruses may become virulent at any time and enter the lytic pathway. (b) In the lytic pathway, viral proteins are synthesized and assembled, along with viral nucleic acid, into numerous virus particles. The host cell is ruptured as the viruses are released.

Without them, proteins could not fold and assemble correctly as phage particles form; similarly, proteins entering mitochondria or chloroplasts do not fold and assemble properly if the chaperone protein is absent.

Details of viral replication depend on whether the viral genome consists of DNA or RNA. A DNA genome acts directly as a template for the synthesis of new viral DNA and mRNA—messages that are translated via the host's ribosomes to form viral proteins or enzymes needed for viral replication. In a *retrovirus*, the genetic material is RNA, and the infected host cell synthesizes a special viral enzyme called *reverse transcriptase* that makes DNA molecules complementary to the viral RNA genome (Figure 20-19). These DNA molecules can then be used for transcribing viral mRNAs and, in turn, viral proteins. Retroviruses cause many diseases, as our next section explains.

Viruses and Disease

Viruses cause various diseases in plants and animals, principally through cell lysis and toxin production. In the lytic cycle, once the virus replicates, the host cell is destroyed, and newly released particles can infect and destroy nearby cells or spread to other organisms. In contrast, some viruses cause the host cell to release lysosomal enzymes into its own cytoplasm, and these literally digest it from the inside out. Still other viruses can

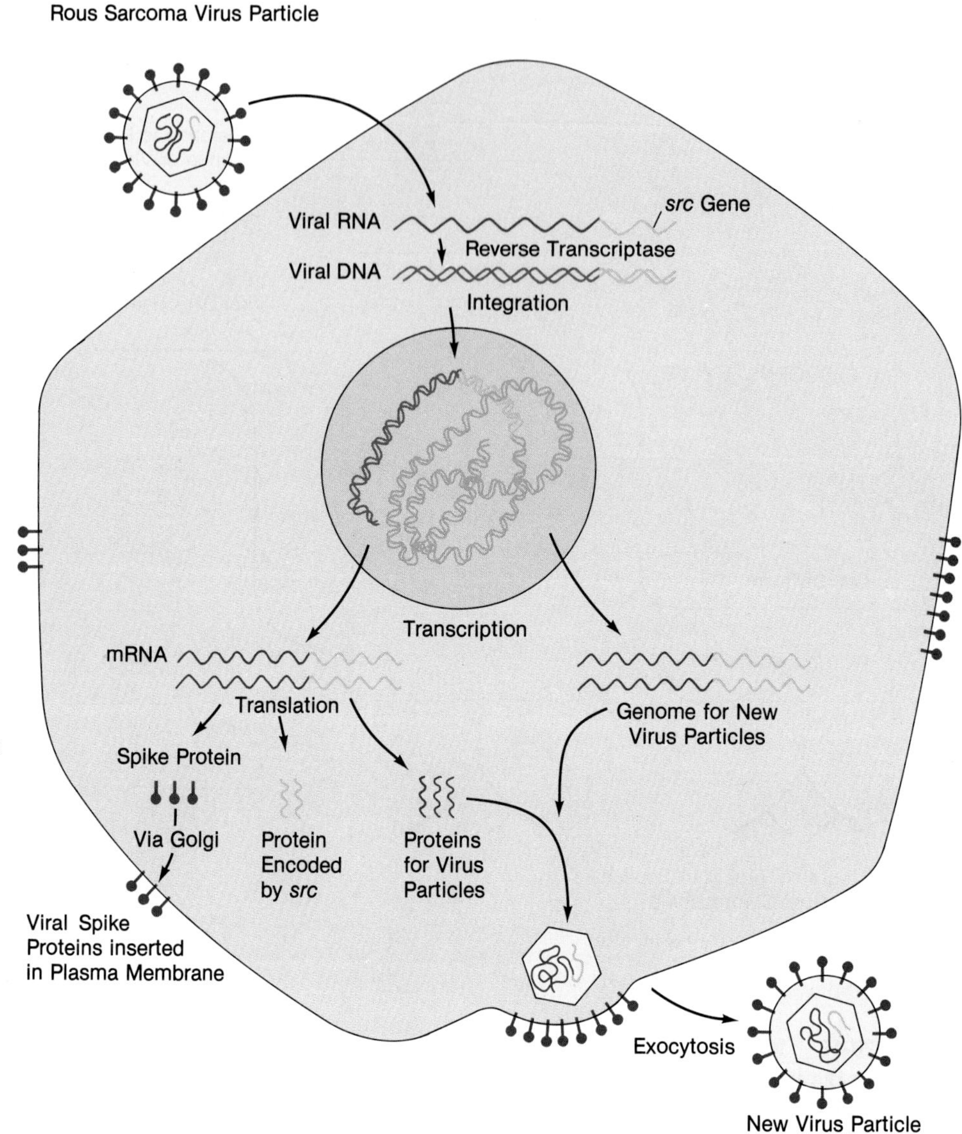

Figure 20-19 ROUS SARCOMA VIRUS. This RNA virus that induces cancer in chickens has in its genome the genetic information encoding viral coat proteins as well as the oncogene *src*, which codes for a protein believed to be involved in inducing the cancerous state in the infected cell. In this diagram, you can see that the viral genome is copied into DNA by the enzyme reverse transcriptase and then is integrated into the host cell's DNA. Transcription of the viral DNA yields RNA molecules that can be translated into viral coat proteins (as spike proteins) and the *src* product. The transcription of the viral DNA also serves to produce many copies of the viral RNA genome. The result of infection by the Rous sarcoma virus is the reproduction of the virus and the induction of the cancerous state in the host cell. After the viral genome and proteins have been assembled, the new virus particles exit from the host cell by outward budding (exocytosis).

cause disease by making toxins that inhibit cell metabolism. Table 20-3 lists several human viral diseases, including AIDS, which is caused by a retrovirus.

Frequently, a given virus will infect only certain tissues of a plant or animal. Cold viruses, for example, localize in the respiratory tract, not the heart or brain. Biologists have discovered that vulnerable cells carry molecules on their surfaces that can act as receptors; only when virus and receptor are a complementary fit will infection take place. Flowering plants (angiosperms) display such complementarity and are susceptible to viral infection, whereas conifers (pines and their relatives) do not and thus are resistant.

Why would cells possess receptors that can contribute to their own death? Biologists think that the surface molecules probably perform other functions for the cell and that only later did viruses evolve coat proteins that can bind to the cell surface molecules.

Viruses often evade a host's immune defenses because they evolve quickly and can lie dormant for long periods. Researchers found that over a 53-year period, one particular gene in the human influenza A virus averaged 1.73 nucleotide substitutions per year—a rate of evolutionary change 1 million times faster than in a mammal's germ line genes. And herpes simplex viruses type 1 and 2, which cause cold sores and genital herpes, respectively, can lie dormant for years, then become lytic when the person suffers physical or psychological stress.

Physicians have long sought effective antiviral treatments, since many of the chemicals that will poison a virus-infected cell will also harm normal, uninfected host cells. Researchers, however, have had a few successes, as the box on page 357 explains.

Table 20-3 SOME HUMAN VIRAL DISEASES

Disease	*Virus*
Measles	Paramyxovirus
Rubella (German measles)	Rubella virus
Atypical pneumonia	Paramyxoviruses, types 1–3; orthomyxoviruses, types A, B, and C
Common cold	Coryza viruses; rhinoviruses
Influenza	Orthomyxovirus, types A, B, and C
Hepatitis	Hepatitis A virus
Herpes	Herpes simplex, types 1 and 2
Mononucleosis	Epstein-Barr virus
Poliomyelitis	Poliovirus
Encephalitis	Semliki Forest virus
Mumps	Paramyxovirus
Smallpox	Smallpox virus
Rabies	Rhabdovirus
Dengue fever	Togavirus (flavivirus)
Yellow fever	Togavirus (flavivirus)
Acquired immunodeficiency syndrome	HTLV III (LAV)

Viruses and Cancer

As Chapter 17 explained, recombinant DNA technology has provided strong evidence linking viruses and cancer. Malignant cells often contain viral genetic material, either free or integrated into the cell's chromosomes. So-called *proto-oncogenes* are genes found in "normal" animal cells, including human ones, that are virtually identical to genes in the viruses that can cause cancers. Whereas the normal cellular gene product is beneficial to the cell, the presence of the viral oncogene alters normal cell behavior, the result being uncontrolled division and other abnormalities. As we saw in Chapter 17, treatments and, ultimately, prevention of cancers may emerge as we learn more about the perversion of normal cell processes by oncogenes.

EVOLUTION OF MONERANS, VIRUSES, AND EUKARYOTIC CELLS

Biologists have tried to piece together the evolutionary relationships among prokaryotes, viruses, and eukaryotes by using an internal biochemical record laid down in the structures and metabolic pathways of living species.

As we saw in Chapter 19, the first cells on earth were almost certainly simple anaerobic heterotrophs that lacked cell walls, flagella, and photosynthetic capability and that consumed preformed organic nutrients from the environment. Photosynthetic aerobes probably evolved from the early anaerobic cells and liberated free O_2 into the earth's atmosphere. Later, aerobic heterotrophs appeared.

The **endosymbiont theory** may explain how eukaryotic cells arose from these earlier prokaryotes (Figure 20-20). According to this theory, one or more simple progenitor cell types led to three early forms: the monerans, the Archaebacteria, and a prospective eukaryotic cell type without membrane-bound organelles but with the multiple chromosomes and other common biochemical attributes of eukaryotes. The eukaryotes then evolved by means of a mutually beneficial association, or *symbiosis*, with one or more types of bacteria. The prospective eukaryotes may have engulfed but not digested an efficient aerobic bacterium such as a purple eubacterium related to the rickettsias. This symbiont came to function in the host cell's metabolism and was the ancestor of the mitochondrion. The cell containing it could have been the progenitor of protistan, animal, and plant cells.

Although a controversial idea, some biologists think that undulating bacteria called spiroplasmas may have attached to the prospective eukaryotes and given rise to the flagella and cilia of various eukaryotic cell types. Finally, biologists generally agree that a second engulfment, this time of a photosynthetic aerobic prokaryote—similar to a chloroxybacterium—established a symbiont that evolved into the chloroplast.

Evidence in support of the endosymbiont theory includes the fact that all mitochondria and chloroplasts of eukaryotic cells possess the kind of ribosomes found in bacteria; tRNAs that differ from those in the cell's cytoplasm; and circular strands of DNA similar to bacterial chromosomes. In addition, these organelles reproduce independently during the cell division cycle. The evidence for the endosymbiont theory of the origin of eukaryotes is so persuasive that biologist F. J. R. Taylor has said, "The eukaryotic 'cell' is a multiple of the prokaryotic 'cell.'"*

Before the endosymbiont theory, some biologists thought that viruses may have evolved into the first prokaryotic cells. But did they lose their capacity to repro-

*F. J. R. Taylor, "Implications and Extensions of the Serial Endosymbiosis Theory of the Origin of Eukaryotes, *Taxon* 23 (1974): 5–34.

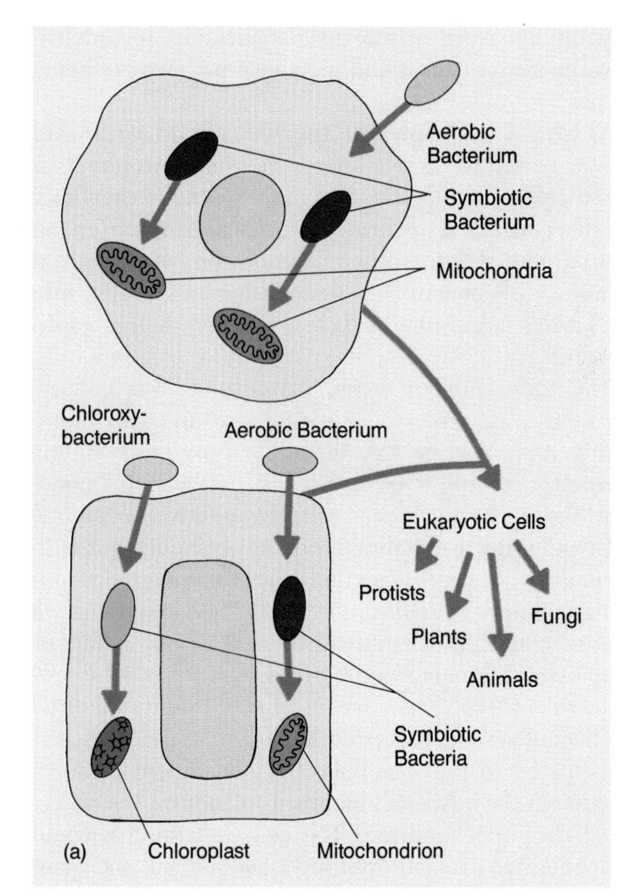

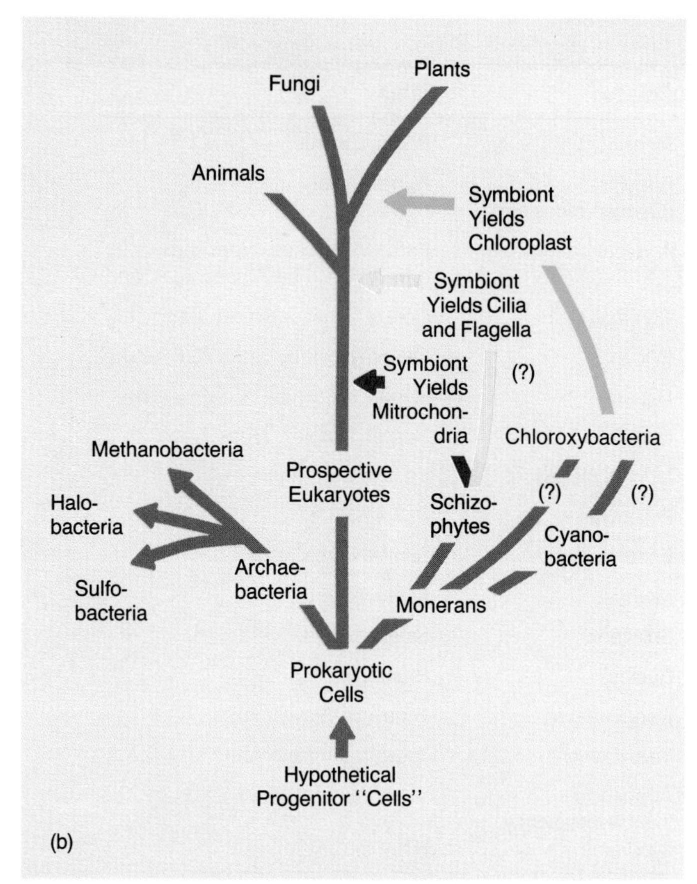

Figure 20-20 RELATIONSHIPS BETWEEN PRESENT-DAY PROKARYOTES, AND THE ORIGIN OF THE EUKARYOTIC CELL. (a) The endosymbiotic theory suggests that aerobic bacteria living in symbiosis with ancestral cells evolved into mitochondria, while symbiotic chloroxybacteria in later cells evolved into chloroplasts. The origin of cilia and flagella remains very uncertain. (b) This diagram shows separate lineages leading to archaebacteria and monerans.

duce? A newer, more easily defended theory suggests that viruses may be "escaped genes"—fragments of host cell chromosomes that survived at first without a protein coat. Perhaps the earliest viruses resembled modern viroids. Viroid RNA closely resembles one class of eukaryotic cell introns, implying that viroids are escaped introns or perhaps that both evolved from some common progenitor. Regardless of their origins, viruses have a special relationship with the invisible kingdom Monera, the ancestral kingdom of all other life on earth.

SUMMARY

1. The 2,000 or so species in the kingdom Monera are the smallest, oldest, and most numerous organisms on earth. All monerans are *prokaryotes.*

2. Prokaryotic cells are approximately one-tenth to one-hundredth the volume of average eukaryotic cells.

3. Bacterial cells have a rigid outer wall made of *peptidoglycan polymers.* Some also have additional outer layers of lipids and polysaccharides.

4. Bacteria lack a nucleus but have a nucleoid where the single DNA molecule is located.

5. Many bacteria move by means of a flagellum anchored in a freely rotating socket. Other motile bacteria have axial filaments or glide along a track of self-secreted slime.

6. Given unlimited space and nutrients, many types of bacteria can produce a new generation every 20 minutes. Bacteria reproduce asexually via budding or by *binary fission,* sexually through transformation, transduction, and conjugation; and some can form *endospores.*

7. Heterotrophic bacteria use organic compounds as substrates or exist as *parasites. Photoautotrophs* carry out photo-

BEATING VIRUSES AT THEIR OWN GAME

While it is difficult to kill an enemy that is noncellular, there are several "Achilles heels" in the viral reproductive process, and researchers are working to exploit them.

Tampering with Viral Enzymes

One of the few viral drugs currently available, *acyclovir*, resembles the nucleic acid base guanine. A viral enzyme converts the drug to an active form, and this substance then binds to DNA polymerase, preventing it from catalyzing the addition of nucleotides to DNA. In this way, acyclovir prevents the replication of the virus; a cell not infected by virus will not be affected by the drug. The drug is being used successfully to treat genital herpes, shingles, and cold sores. Another antiviral drug, AZT, is one of the few treatments now available for AIDS patients. It works by tying up the viral enzyme reverse transcriptase and thus preventing the transcription of RNA into DNA and then proteins. Unfortunately, there is increasing evidence that viruses are developing resistance to both acyclovir and AZT.

Tying up Viral Genes

Another approach is to make short stretches of RNA or DNA having a nucleotide sequence that is complementary to a section of the virus's own genetic material. These stretches can be introduced into infected host cells to bind the viral genes and prevent their being replicated or transcribed into proteins. Researchers have created sequences that strongly inhibit the activity of Rous sarcoma virus, which causes tumors in animals.

Fighting Viruses with Interferons

Part of the body's natural defensive reaction against viruses is the production of *interferons*, a group of small glycoproteins. Made and released by virus-infected cells, interferons act on neighboring cells to render them resistant to a wide range of viruses. Interferons produced through genetic engineering have proved effective experimentally in fighting the common cold.

Synthesizing Antiviral Vaccines

Although vaccines against pathogens such as poliovirus and rabies virus have been available for many years, they do pose a slight but real risk: Occasionally, the vaccine includes active virus particles, and patients contract the very diseases they sought to avoid. A new technique involves synthesizing short peptides that match the outermost regions of globular viral proteins and will safely trigger an immune reaction against the virus itself.

Targeting Antibodies

Researchers have also used the techniques of genetic engineering to prepare antibodies (immune system proteins) that recognize a surface feature on RNA rhinoviruses, which cause human colds. Specifically, the antibodies seek out one peptide at the base of a canyonlike groove on the viral capsid—a groove that normally binds to a receptor protein on human cells and allows infection. The engineered antibodies neutralize most types of rhinovirus and may someday help defeat the common cold.

synthesis. *Chemoautotrophs* oxidize inorganic compounds to obtain energy.

8. The subkingdom *Schizophyta* includes bacteria classified into the main phyla *eubacteria*, *spirochetes*, *myxobacteria*, *actinomycetes*, *rickettsias* and *chlamydiae*, and *mycoplasmas*.

9. *Cyanobacteria* are rod-shaped or spherical photosynthetic cells that often occur in filamentous groups. They contain chlorophyll *a*, carotenoids, and red and blue pigments. Some species fix nitrogen, and many can live in harsh environments.

10. The *chloroxybacteria* are prokaryotic cells that photosynthesize by means of chlorophylls *a* and *b*, as do green algae and higher plants.

11. *Archaebacteria* include methane producers (*methanobacteria*), extreme salt dwellers (*halobacteria*), sulfur-dependent species (*sulfobales*), and species that live at high temperatures (*thermoproteales*). Some biologists now classify these monerans separately as the kingdom Archaebacteria.

12. Many bacterial species aid in the production of human foods, but many others are major *pathogens*, or disease-producing agents.

13. *Viruses* are noncellular, metabolically inert particles composed of a core of RNA or DNA, a protein capsid, and sometimes a membranous outer envelope. Viral genetic material may become incorporated into a host cell's DNA, where it is replicated with the host chromosome. This is the process of *lysogeny*. Viruses may reproduce independently of the host cell and ultimately kill it; this is the *lytic pathway*.

14. According to the *endosymbiont theory*, the eukaryotes may have evolved from progenitors that engulfed aerobic purple eubacteria (the future mitochondria) and photosynthetic aerobic prokaryotes (future chloroplasts).

KEY TERMS

actinomycete
archaebacterium
binary fission
chemoautotroph
chloroxybacterium
cyanobacterium
endospore
endosymbiont theory
endotoxin
eubacterium
exotoxin
facultative anaerobe
Gram-negative bacteria
Gram-positive bacteria
halobacteria
interferon
lysogeny
lytic pathway
methanobacterium
mycoplasma
myxobacterium
obligate aerobe
obligate anaerobe
parasite
pathogen
peptidoglycan polymer
photoautotroph
prokaryote
rickettsia
schizophyte
spirochete
sulfobale
thermoproteale
viroid
virus

QUESTIONS

1. How could you use light to "guide" a flagellated, photosensitive bacterium through a tiny "maze" on a culture dish, and why would it work?

2. Draw a "family tree" for monerans, methanobacteria, and eukaryotic plant and animal cells.

3. Using two host cells, diagram the events during infection by an RNA and a DNA virus; include the lysogenic and lytic cycles.

4. What kind of prokaryote converts carbon dioxide into its own building materials? What energy source do such organisms use?

5. What features distinguish prokaryotes from eukaryotes?

6. Why do biologists believe that the first living organisms were anaerobic?

7. Review Chapter 1 and list the attributes of living organisms. Which of these attributes (if any) apply to viruses? Would you say that viruses are living organisms, nonliving particles, or incomplete organisms?

8. Draw and compare diagrams of a bacterial cell dividing and a eukaryotic cell in mitosis.

ESSAY QUESTION

1. Summarize the evidence supporting the endosymbiotic theory.

For additional readings related to topics in this chapter, see Appendix C.

21
Protista: The Kingdom of Complex Cells

The minute living Animals exhibited in . . . this Work, will excite a considerable Mind to admire in how small a Compass Life can be contained, what various Organs it can actuate, and by what different Means it can subsist. . . .

Henry Baker, *Employment for the Microscope* (1753)

The kingdom Protista comprises a wide range of mostly single-celled eukaryotes: plantlike cells that can swim; funguslike decomposers that creep toward food sources; animal-like cells that cannot move at all; cells that look like bells, eels, pincushions, fans, stars, shells, or snowflakes. While some protistan species form colonies at certain stages in their life cycle, the single-celled state predominates and thus their classification as true protists.

The transition from the kingdom Monera to the kingdom Protista, from prokaryotic to eukaryotic cells, proved to be a momentous one in the history of life on earth, for two reasons: The protists are far more complex in structure, function, behavior, and ecology than the monerans; and they evolved specialized organelles for feeding, digestion, excretion, respiration, coordination, and locomotion. As such, the early protists were, in a sense, the "testing grounds" for the life-styles of higher organisms.

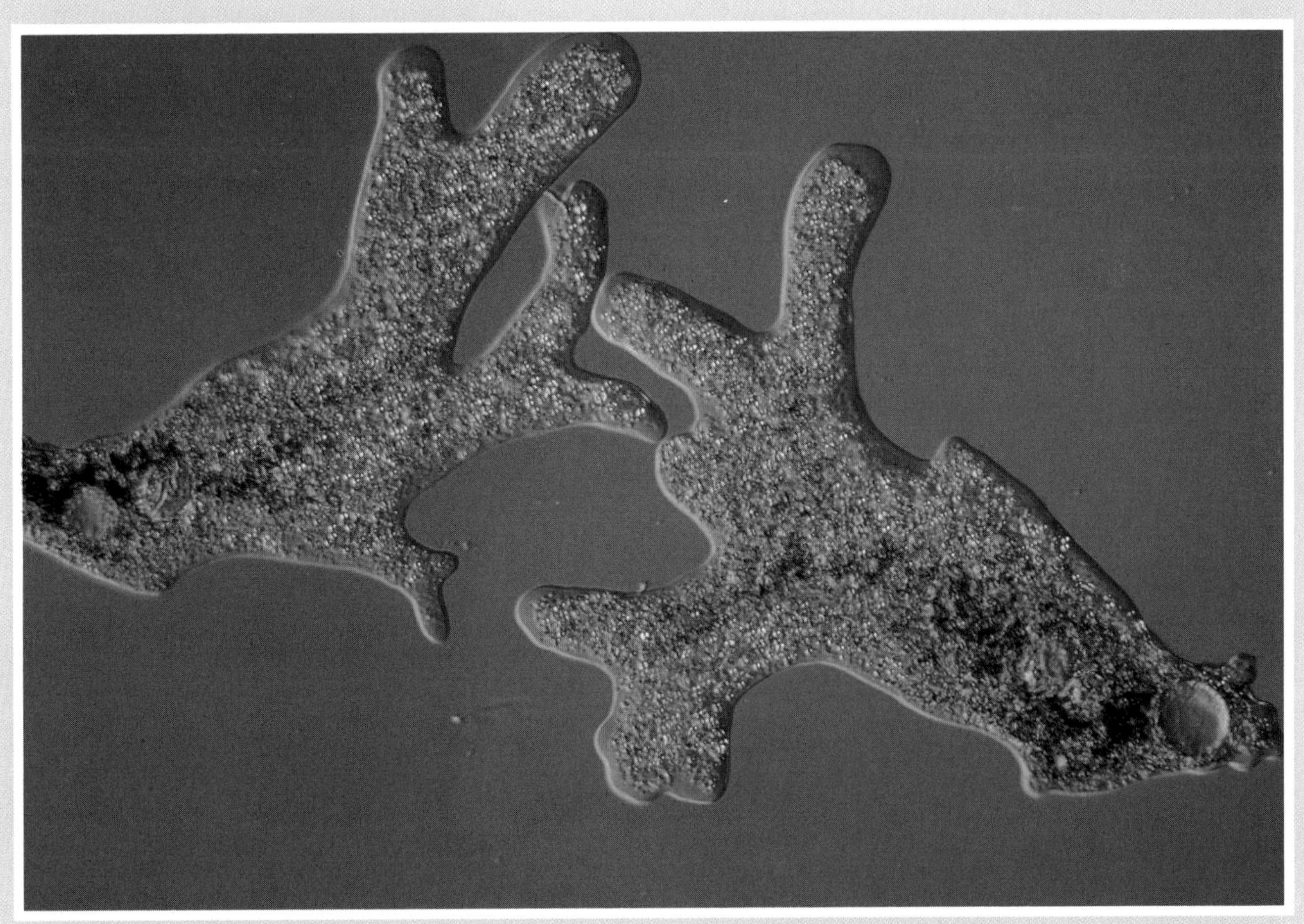

Like minute, many-armed fencers, two single-celled amoebae extend and retract their pseudopods in an encounter only seen by human eyes through a microscope.

This chapter describes:

- The classification of the protists by structure and general life-style
- The plantlike protists, important carbon fixers on a global scale
- The funguslike decomposers that live mainly in litter on the forest floor
- The animal-like consumers, complex motile cells that hunt down prey in microscopic realms
- The evolution of the protists from prokaryotic ancestors, and their possible descendants among the plants, fungi, and animals

PROTIST CLASSIFICATION BY LIFE-STYLE

Some protists are "mixed metaphors," displaying the photosynthesis of plants, for example, but the motility of animals. As a convenient method for categorizing the sometimes confusing species, biologists group protists according to three basic nutritional modes and resultant life-styles (Figure 21-1): (1) The *plantlike autotrophs* are "producers" that generate food molecules by means of chloroplasts and photosynthetic pigments; (2) the

Table 21-1 PROTISTAN PHYLA

Group or Phylum	*Representative Members*	*Means of Locomotion*	*Cell Wall*	*Habitat*	*Other Characteristics*
			Plantlike Protists		
Euglenophyta (euglenoids)	*Euglena*	Flagella	None	Mostly freshwater	Chlorophylls *a* and *b* plus carotenoids
Pyrrophyta (dinoflagellates)	*Gymnodinium* *Protogonyaulax*	Flagella	Rigid cellulose wall	Mostly marine; some parasitic or symbiotic	Chlorophylls *a* and *c* plus carotenoids
Chrysophyta (golden-brown algae, diatoms)	*Ochromonas*	Some with flagella	May have none; may have rigid pectin or pectin and silica wall	Freshwater and marine; some terrestrial	Chlorophylls *a* and *c* plus carotenoids
Algae*					
			Funguslike Protists		
Gymnomycota (slime molds)	*Dictyostelium*	Pseudopods	None	Terrestrial	Multinucleate plasmodium (true slime molds); multicellular slug (cellular slime molds)
			Animal-like Protists		
Mastigophora (zooflagellates)	*Trichonympha* *Trypanosoma*	Flagella; some with pseudopods	None	Mostly symbiotic or parasitic	Parasitic forms have complex life cycle with multiple hosts
Sarcodina (amoebae, foraminiferans, radiolarians)	*Amoeba* *Actinosphaerium*	Pseudopods; some with flagella	None	Marine and freshwater; terrestrial; some parasitic	Many secrete elaborate shells of silica or calcium compounds
Sporozoa	*Plasmodium*	Adult forms nonmotile	None	Parasitic	Complex life cycle with sporelike stage and multiple hosts
Ciliophora	*Paramecium*	Cilia	None	Marine and freshwater	Macronuclei and micronuclei; most complex single-celled organism
Caryoblastea	*Pelomyxa*	Amoeboid	None	Freshwater	Lack mitochondria
Microspora	*Vairimorpha*	Polar filament	None	Parasitic	Lack mitochondria; form unicellular spores; unique rRNA; most ancient offshoot of lineage to eukaryotes

*Some taxonomic schemes place Chlorophyta (green algae), Rhodophyta (red algae), and Phaeophyta (brown algae) in the kingdom Protista.

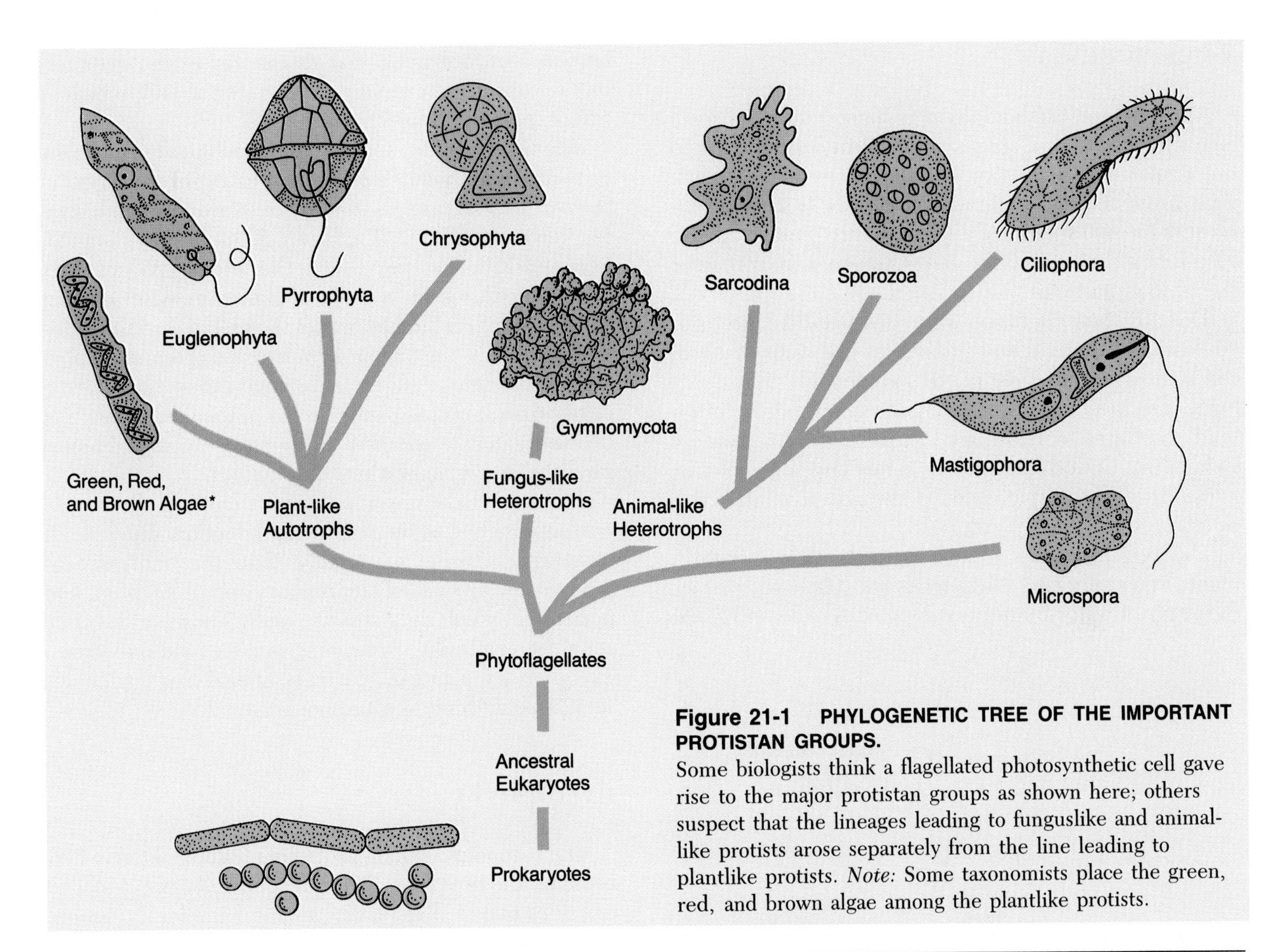

Figure 21-1 PHYLOGENETIC TREE OF THE IMPORTANT PROTISTAN GROUPS.
Some biologists think a flagellated photosynthetic cell gave rise to the major protistan groups as shown here; others suspect that the lineages leading to funguslike and animal-like protists arose separately from the line leading to plantlike protists. *Note:* Some taxonomists place the green, red, and brown algae among the plantlike protists.

funguslike heterotrophs are "consumers" and "decomposers" that feed on preexisting organic matter; (3) and the *animal-like heterotrophs* are "consumers" that feed on whole bacteria, other protists, or the cells of multicellular organisms.

Biologists presently disagree on how best to divide the 35,000 or more protistan species into specific phyla. All base their classifications on fine details of cellular structure and molecular constituents, especially among the animal-like protists. The life-style groupings we use here may or may not coincide with phyla recognized by different experts in the field, but they do help students identify important features.

The phyla listed in Table 21-1 are the main subgroups in the kingdom Protista. Despite nutritional differences, protists share the basic eukaryotic cell structure, including a membrane-bound nucleus; chromosomes made up of proteins and DNA; ribosomes and typical eukaryotic nuclear RNA processing; various membrane-bound organelles in the cytoplasm; and a cytoskeleton, providing cellular form and allowing cellular movement. Note that some taxonomists place the three major groups of algae—green, red, and brown—in the kingdom Protista.

THE PRODUCERS: PLANTLIKE PROTISTS

Immense numbers of microscopic autotrophic protists float near the surface of fresh and salt waters around the world. These masses of cells are one type of **phytoplankton,** a varied group of photosynthetic, usually single-celled species that generate huge quantities of atmospheric oxygen and biomass (the total amount of living matter in a given habitat). Most species of phytoplankton carry out photosynthesis, move by means of flagella, and are the first link in aquatic food chains: These primary producers are eaten by minute animal larvae and other small organisms, which in turn are eaten by larger invertebrate and vertebrate predators. The three most common phytoplankton phyla are *Euglenophyta*, or euglenoids; *Pyrrophyta*, or dinoflagellates; and *Chrysophyta*, the golden-brown algae and diatoms. If green, red, and brown algae are placed with Protista, then the photosynthetic protists also include very large multicellular organisms.

Euglenophyta

The phylum **Euglenophyta** is named after its most characteristic genus, *Euglena*—beautiful bright green unicellular organisms that swim about watery habitats, each propelled by a whiplike flagellum (Figure 21-2). Most of the 800 species of euglenoids are photosynthetic autotrophs, but some are heterotrophs that digest or engulf food particles.

Despite their inclusion with the plantlike protists, euglenoids lack rigid and prominent cell walls made of cellulose. Their absence permits euglenoids great flexibility of shape and allows the organisms, which are often mud dwellers, to travel by wriggling rather than by swimming. Underlying protein bands give a *pellicle*, or *periplast*, at the surface needed strength to contain the cell's contents.

The typical euglenoid cell is oval, with a slightly pointed posterior end. Euglenoids are able to counteract the entry of water by osmosis by means of a *contractile vacuole*—an organelle that discharges excess cytoplasmic water into a reservoir at the anterior end, which in turn expels the liquid from the cell.

The reservoir also serves as an attachment site for one or two microtubular flagella; the beating of the larger of the two flagella propels the tiny creature through its watery environment. A *Euglena* cell has a tiny, orange, light-sensitive eyespot called the *stigma*. When light stimulates this spot, a chain of processes is set off that triggers the flagellum to propel the cell toward the light.

Euglenoids build their carbohydrate products of photosynthesis into a unique substance called *paramylum*, rather than starch as plants do. As in the land plants that evolved later, euglenoid chloroplasts contain chlorophylls *a* and *b* and various carotenoids.

Biologists have observed mitosis and cell division in euglenoids, but never sexual reproduction. Euglenoids sometimes divide so rapidly that the independent growth and division of chloroplasts cannot keep up, and permanently colorless strains result. These strains then act as heterotrophs, consuming organic food molecules. The partially plantlike, partially animal-like euglenoids remain something of a taxonomic puzzle.

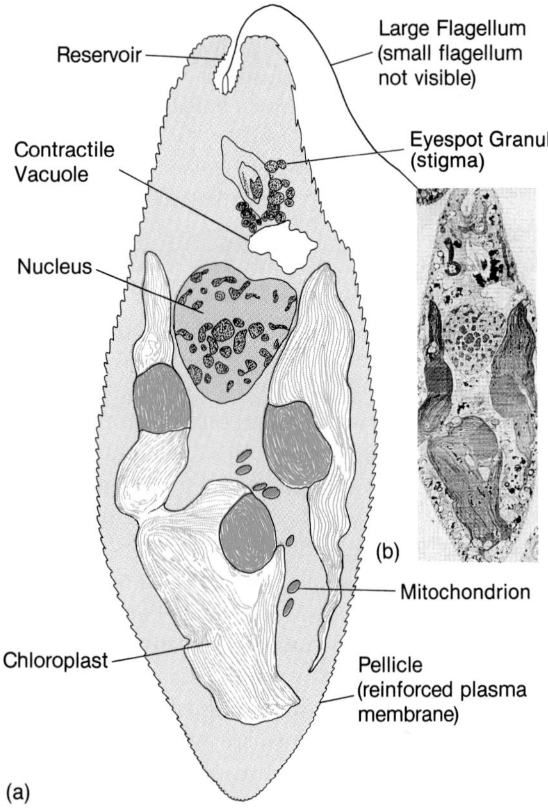

Reservoir
Large Flagellum (small flagellum not visible)
Contractile Vacuole
Eyespot Granules (stigma)
Nucleus
(b)
Mitochondrion
Chloroplast
Pellicle (reinforced plasma membrane)
(a)

Figure 21-2 EMERALD *Euglena:* A TYPICAL PROTIST. Various species of this protistan genus function in plantlike or animal-like ways. Euglenoids have typical eukaryotic features, including chloroplasts, mitochondria, and a nucleus. The drawing in (a) depicts the structures visible in the transmission electron micrograph (b) (magnified about 2,000×).

Pyrrophyta

The second phylum of protistan phytoplankton is **Pyrrophyta,** the fire plants. Their responsibility for so-called red tides makes "fire plants" an apt name for the group, but their common name, **dinoflagellates** (literally, "spinning cells with flagella"), is also quite descriptive. One flagellum runs around a circumferential groove like a belt and causes the cell to spin; another lies in a longitudinal groove, projects like a tail, and propels the cell forward or backward (Figure 21-3).

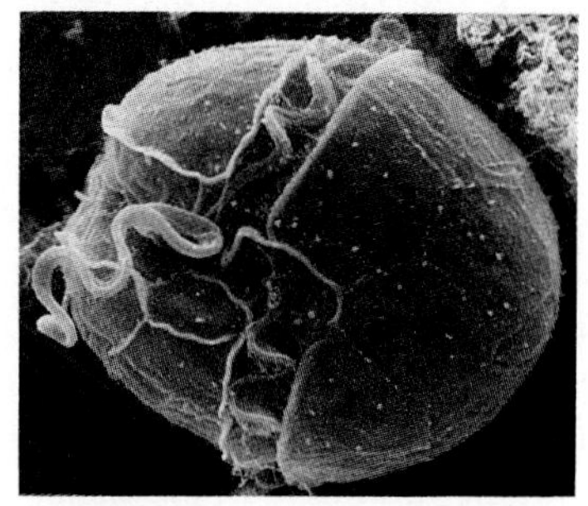

Figure 21-3 ARMORED PROTISTS: THE DINOFLAGELLATES. Dinoflagellates usually possess two flagella: The first, lying in a circumferential groove, causes the organism to spin, and the second, lying in a longitudinal groove, propels the organism forward or backward. These features are visible in the scanning electron micrograph (magnified about 1,600×) of a species called *Protogonyaulax catenella*.

Most of the 1,000 or so dinoflagellate species have contractile vacuoles, chloroplasts, and chlorophylls *a* and *c*, suggesting an independent evolution from the euglenoids, with their chlorophylls *a* and *b*. Members of Pyrrophyta also possess certain carotenoid compounds that cause the cells to appear gold, brown, or red, rather than green. Autotrophic dinoflagellates are very common types of phytoplankton and create tremendous amounts of both biomass and oxygen. Some photosynthetic dinoflagellates live symbiotically in the bodies of certain corals, sea anemones, flatworms, and giant clams. Some dinoflagellates lack chloroplasts and survive as free-living heterotrophs that can engulf organic matter and other cells. And a few species even live as true parasites within the bodies of various marine organisms, harming their hosts and deriving benefit, but giving nothing in return.

Like the euglenoids, dinoflagellates usually reproduce by asexual cell division, but some species generate cysts (resting stages) sexually. These germinate when conditions are appropriate.

Glowing Waves and Red Tides

Dinoflagellates cause two interesting phenomena in coastal areas. At night, the ocean may appear to glow with a greenish hue, and where disturbed, shimmer as though it were releasing a barrage of tiny sparks. At other times, offshore waters can turn rusty or bloody red, a phenomenon we call *red tides*. These tides contain toxins that poison fish, shellfish, and occasionally people.

Dinoflagellates sometimes cause waves to glow and sparkle because some species have an enzyme system that can emit light. Toxic red tides occur after dense populations (*blooms*) of certain species appear, such as *Gymnodinium* and *Protogonyaulax* (formerly called *Gonyaulax*). The toxins these cells produce are usually low-molecular-weight nonprotein compounds that act as nerve poisons or cause red blood cells to lyse (break open). These toxins can kill certain fish, barnacle, and clam species outright; in other species, the toxins build up in the tissues, perhaps poisoning a human consumer at some later time. For this reason, collection and consumption of shellfish are often banned in many coastal areas during warm summer months, the time of dinoflagellate blooms.

Chrysophyta

The *golden-brown algae* and *diatoms*, the most abundant phytoplankton organisms, are surely among the most beautiful members of the kingdom Protista. Diatoms have glasslike cell walls sculpted in thousands of geometric and faceted shapes. The other chrysophytes are a golden color. Both types are so common that a gallon of seawater may contain millions of them. Of the 6,000 to 10,000 species assigned to the phylum **Chrysophyta** by different taxonomic schemes, some live in fresh water and some in salt water; some are unicellular, and some live in colonies; most reproduce asexually, but some produce flagellated gametes that carry out sexual reproduction.

Golden-Brown Algae

The **golden-brown algae** have a variety of outer surfaces and locomotive structures. Some lack a cell wall and can creep along, much as amoebae and animal cells do (see Chapter 6). Others possess a cell wall made largely of *pectin* (rather than cellulose), a complex carbohydrate compound that acts as a glue between plant cell wall components. The pectin-walled algae usually have two flagella. Like the dinoflagellates, golden-brown algae contain chlorophylls *a* and *c* as well as carotenoids, especially *fucoxanthin*, the source of the golden color.

Diatoms

The photosynthetic **diatoms** in the oceans may be the greatest source of atmospheric oxygen, and they are delicately beautiful organisms. Their outer glassy shells are rigid and appear intricately sculpted with tiny holes, channels, and patterns (Figure 21-4) that allow nutrients, minerals, gases, and wastes to pass back and forth between the cell and the environment. After a diatom dies, its hard silica wall may sink and over time accumulate with other such cell walls into layers of a crumbly white sediment called **diatomaceous earth.** People make a gritty, abrasive, absorbent powder from these deposits and use it in swimming pool filters, toothpastes, detergents, fertilizers, and a range of other products.

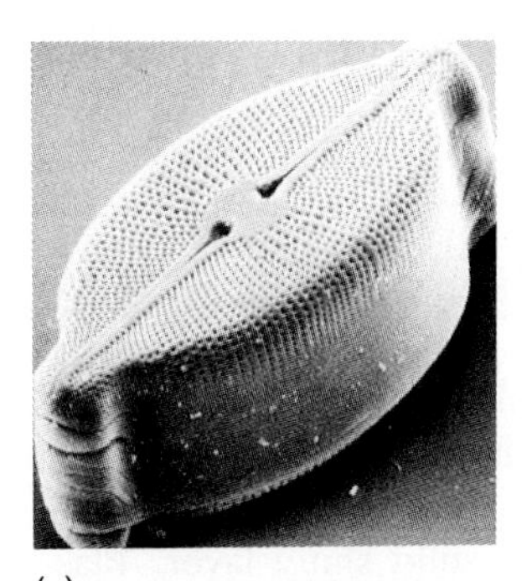

(a)

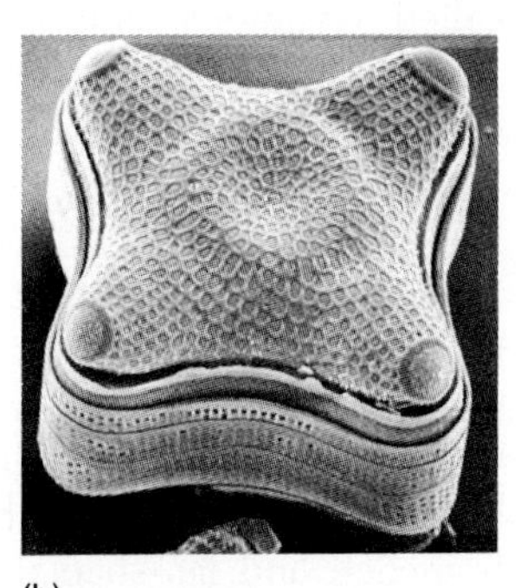

(b)

Figure 21-4 BIOLOGICAL ARCHITECTURE AT THE SINGLE-CELL LEVEL.
These diatoms, viewed with the scanning electron microscope, include (a) *Navicula perpusilla* (magnified about 1,150×) and (b) *Amphitetras antediluviana* (magnified about 600×).

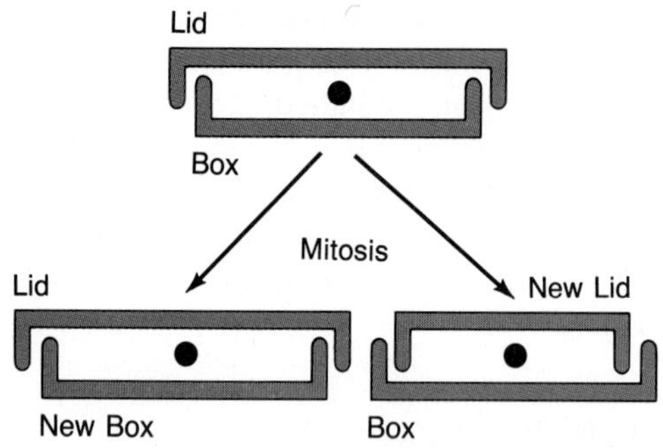

Figure 21-5 DIATOMS: DIMINISHING REPRODUCTION.
A diatom's cell wall fits together like a box with a tight-fitting lid. When a cell divides, the parts separate from each other, and new cell wall material is manufactured and assembled inside each original half. This unusual division pattern results in smaller and smaller cells until, after several divisions, the minute diatom cells can no longer undergo mitosis. Sexual reproduction is then triggered and the entire sequence can begin again.

Diatoms have cell walls constructed like tiny boxes with close-fitting lids. The cells reproduce through a bizarre process of building ever smaller "boxes." Figure 21-5 illustrates and describes this process.

THE DECOMPOSERS: FUNGUSLIKE PROTISTS

Several types of protists live in damp soil as funguslike heterotrophs that derive food and energy by breaking down fallen logs, leaves, small food particles, and other dead organic matter. One group of predominantly single-celled fungi, the Protomycota, straddles the line between protists and true fungi, and we discuss them in Chapter 22. Here we discuss the funguslike protists called **Gymnomycota,** or *slime molds,* including the **Myxomycota,** or true slime molds, and the **Acrasiomycota,** or cellular slime molds.

Myxomycota: True Slime Molds

The life cycle of the **true slime molds** has a mature stage characterized by the **plasmodium,** a whitish or brightly colored mass that may be either highly branched and fan-shaped or a solid slimy layer. Plasmodia may be tens of centimeters long and can weigh 40–50 g. Each plasmodium is actually a continuous cytoplasm containing many diploid nuclei that arise by mitotic division of existing nuclei. This creature may be thought of as one huge cell with many nuclei or as an "acellular" organism. The slimy myxomycote glides forward through rotting leaves, wood, or grass, ingesting woody debris, whole bacteria, or organic molecules decomposed by enzymes secreted onto the materials.

When the plasmodium encounters a nutrient-poor area or adverse environmental conditions, the slimy mass halts and sends up a series of vertical stalks bearing spore capsules (Figure 21-6). Inside these "fruiting bodies," or *sporangiophores,* the many diploid nuclei undergo meiosis, and soon the structures are bulging with rapidly produced haploid spores. Eventually, the spores are released, are dispersed by wind or rain, and may remain quiescent for months or years. Then, when conditions are favorable, germination occurs.

Germination produces a new generation of individual haploid cells with single flagella that swim about and may fuse to form a diploid zygote. Alternatively, the cells may lose their flagella and become **myxamoebae** that glide about feeding on plant debris or bacteria. Myxamoebae are classic protistan cells that divide, and they may survive for long periods if food is plentiful. Myxamoebae may also fuse to form diploid zygotes. A zygote arising by either mechanism develops into a new plasmodium.

The key characteristics to remember about the acellular slime molds are these: They carry out the sexual process and produce a huge, multinucleate cell mass.

Acrasiomycota: Cellular Slime Molds

The other type of slime mold, the **cellular slime mold,** also has an individual haploid stage, and these cells glide along the ground, engulfing food particles, debris from forest litter, or bacteria. They undergo nor-

Figure 21-6 FRUITING BODIES OF A TRUE SLIME MOLD.
Fruiting bodies or sporangia, resembling golf balls perched on tees and containing haploid spores, grow from the plasmodium or large, multinucleate cell of the true slime mold *Diachea.* In this photo, the sporangia rose from a rotting log.

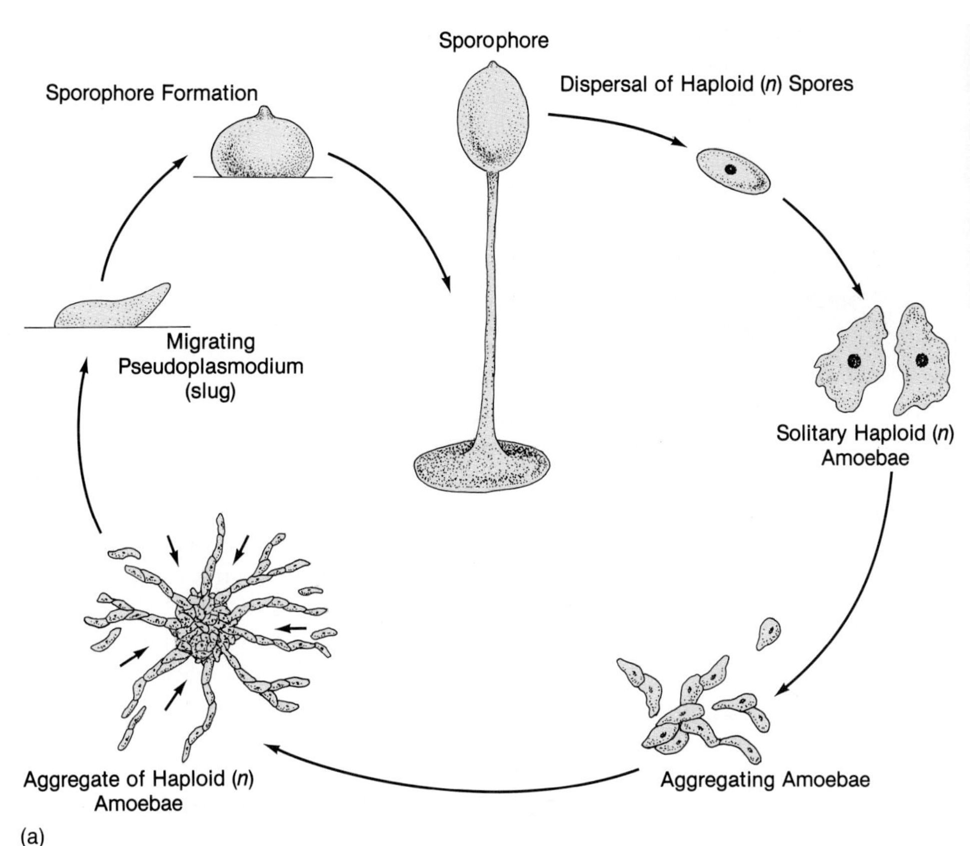

Figure 21-7 LIFE CYCLE OF THE CELLULAR SLIME MOLD *Dictyostelium*.
(a) Haploid spores are released from the fruiting body, or sporophore, and individual amoebae emerge. Solitary feeding amoebae are motile, but when starved, they begin to aggregate and form a migrating slug or pseudoplasmodium and finally a fruiting body. This species lacks a diploid stage.
(b) Aggregating amoebae (white dots) of *Dictyostelium* (magnified about 10×).

mal eukaryotic mitosis and cytokinesis, and large populations of single amoeboid cells can often result. When the food in an area is depleted, however, the amoebae begin to stream together into clumps (Figure 21-7). Each aggregate mass forms a sluglike **pseudoplasmodium** in response to a biochemical signal—pulses of the chemical cyclic AMP. This compound is released from the cell's posterior end and creates a "follow the leader" concentration gradient toward centers of aggregation.

The pseudoplasmodium moves forward in a coordinated fashion even though its thousands of cells are distinct individuals. Eventually, the pseudoplasmodium settles down and develops a stalked fruiting body. Tough-walled haploid spores with diverse genotypes are formed, are dispersed, germinate, and give rise to new individual haploid cells.

Cellular slime molds, then, are asexual and form a large mass by cell aggregation.

THE CONSUMERS: ANIMAL-LIKE PROTISTS

The **protozoans** ("first animals") are primarily animal-like in life-style, occur as single cells, have a variety of complex shapes, and carry out locomotion in diverse ways. (*Protozoan* is a general descriptive term, like *alga*, rather than a distinct taxonomic category.) Like animals, protozoans are consumers, cells that feed on other living cells or on food particles. Protozoans can be divided into four commonsense groupings, each distinguished by their means (or lack) of locomotion. The *Mastigophora* possess flagella; the *Sarcodina* display amoeboid movement; the *Sporozoa* are nonmotile in the adult form; and each of these groups includes a number of phyla. Members of the fourth group, the *Ciliophora*, have numerous cilia and form a true phylum.

Mastigophora

The Greek word *mastigos* means "whip," and indeed, many members of the phylum **Mastigophora,** or **zooflagellates,** move about by means of whiplike flagella. Some species, however, use their cytoskeleton for movement by extruding extensions of the cell surface called **pseudopods** ("false feet") instead of lashing flagella.

Zooflagellates are the simplest and perhaps most primitive of all protozoans. They lack cell walls, and most forms live as parasites or as harmless symbionts inside other organisms. A typical and economically sig-

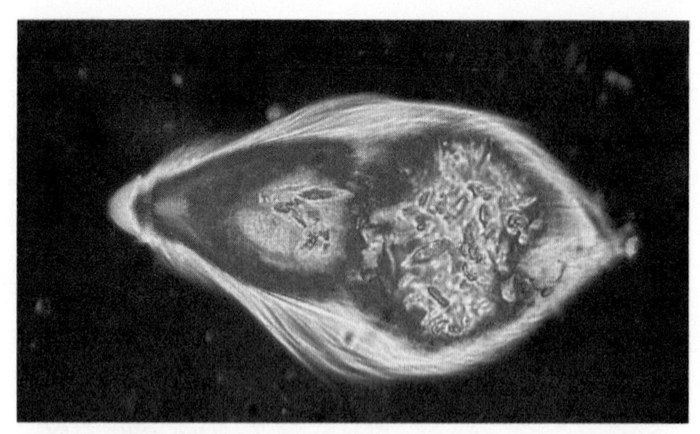

Figure 21-8 ***Trichonympha:*** **THE TERMITE'S HELPER.**
This zooflagellate (magnified about 380×) inhabits the termite's hindgut. The ability of *Trichonympha* to break down cellulose enables termites to use wood as a food source.

nificant zooflagellate is *Trichonympha,* a genus of cellulose-digesting protozoa that resides in the gut of termites (Figure 21-8). Without these digestive-tract organisms, termites could not use wood as food and would be unable to degrade fallen forest trees or tunnel through wooden floors and building foundations.

The most infamous zooflagellate is *Trypanosoma gambiense,* the cause of one form of African sleeping sickness. Researchers have studied sleeping sickness for more than a century and have found that 125 trypanosome species, including *Trypanosoma gambiense,* infect almost 400 species of mammals. Many of these zooflagellates are long, wriggling, eel-like cells, each with a flagellum and an undulating membrane on its dorsal surface (Figure 21-9). These cells have some

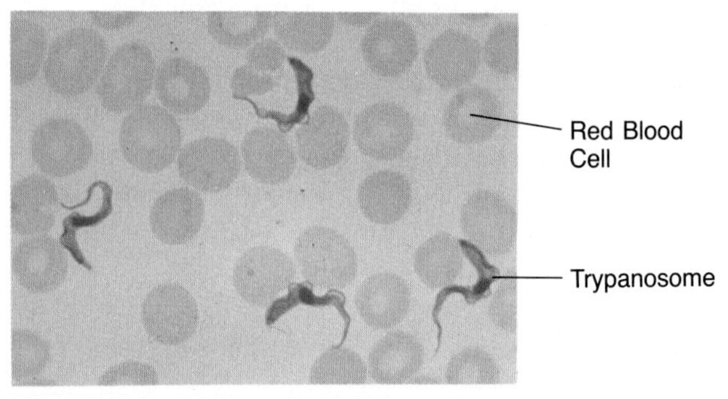

Figure 21-9 TRYPANOSOMES: DEADLY ZOOFLAGELLATES.
Trypanosomes (magnified about 700×) among red blood cells of an elderly missionary. Tsetse flies transmit *Trypanosoma gambiense* to mammals, including humans. Large numbers of the protozoans arise by mitosis; ultimately, some infect the brain and spinal cord, killing the host. "Stumpy" forms of the trypanosomes can be picked up by another tsetse fly; the organisms reproduce and mature in various parts of the fly's gut (hindgut, proventriculus, salivary glands) and then may be passed to a new mammalian host.

remarkable traits besides their shape and locomotion. For instance, the cells lack mitochondria for much of their life cycle, but have an unusual structure near the base of the flagellum that contains 100 "maxicircles" and up to several thousand "minicircles" of DNA. The maxicircles allow a mitochondrion to appear when the trypanosome is resident in one of its hosts, the tsetse fly. What's more, meiosis and fertilization seem to occur only when the trypanosome resides in a tsetse fly. The resultant F_1 generation has much more DNA than either parent—DNA that is slowly lost once the F_1 infects a mammal and its progeny divide and redivide within the host's bloodstream.

Trypanosome cells living inside the bloodstream of wild animals may release toxins, but these do not kill the host. When the same protists infect humans and domesticated animals, however, they can invade the brain and spinal cord and ultimately cause death. The protist has two hosts. Its *secondary host,* also called the transfer organism or the *vector,* is the tsetse fly, within which the trypanosomes develop, multiply, and infest the salivary glands. When the fly bites a mammal, the protozoans are passed through the insect's mouthparts into the *primary host's* body.

Modern molecular analyses reveal why *Trypanosoma gambiense* is so successful and so deadly. Each trypanosome cell is coated with millions of VSG proteins, or variable surface glycoproteins, anchored in the cell membrane. The host's immune system will gradually mount an attack against the particular VSGs on the surfaces of the clone of invading cells, but through a clever genetic switch, a few of the infecting trypanosomes will elude the host's immune system with a new temporarily unrecognized coat of VSG. The elusive protist can thus survive and reinfect new hosts in a changed form. A similar battle will be waged again and again, with the changing generations of parasites remaining one step ahead of the host's defensive system. Finally, trypanosomes invade the brain and kill the host.

These discoveries are a step toward designing better medical protection for the 1 million people who each year become victims of African sleeping sickness. But for the foreseeable future, they will only be adjuncts to the aerial spraying that has been used for decades to subdue the blood-sucking insect vectors.

Sarcodina

The second group of protozoans, the **Sarcodina,** range from the amoebae, with their constantly changing shapes, to beautiful hard-shelled organisms that resemble diatoms. The prototypal sarcodine is *Amoeba proteus,* with its irregularly shaped pseudopods that allow it to creep forward and engulf minute prey. Microtubules

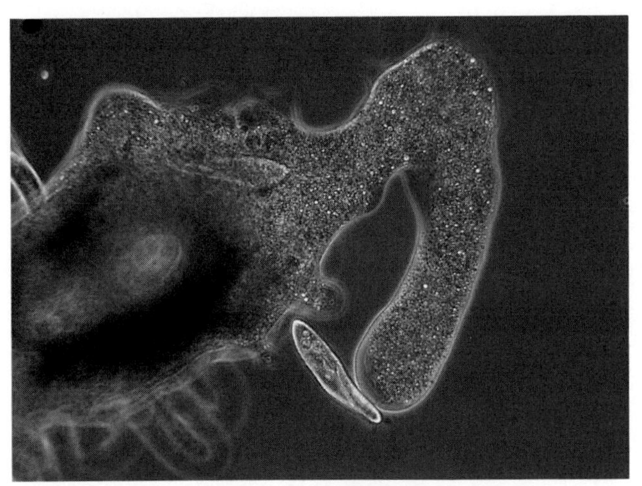

Figure 21-10 AMOEBA: THE PROTEAN PROTIST. Sarcodines such as *Amoeba* use pseudopods for locomotion and feeding. An amoeba (*Pelomyxa carolinensis*) capturing prey, which in this case is a ciliate, *Paramecium* (magnified about 100×).

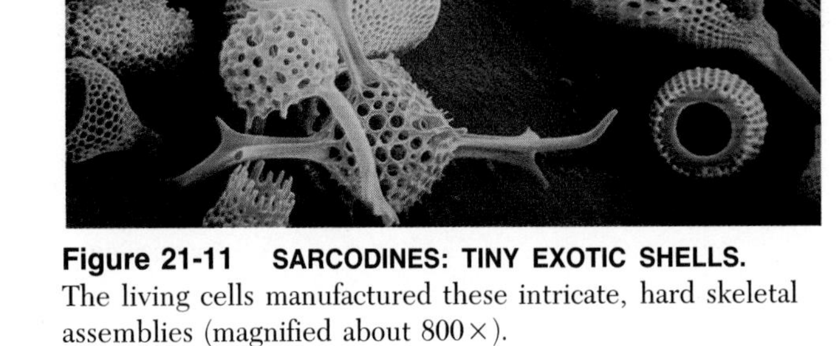

Figure 21-11 SARCODINES: TINY EXOTIC SHELLS. The living cells manufactured these intricate, hard skeletal assemblies (magnified about 800×).

and a dynein-like mechanoenzyme with ATPase activity move the amoeba's organelles and possibly its engulfed food materials rapidly about in the cytoplasm. Figure 21-10 shows a similar amoeba.

Several common sarcodines form hard, intricate shells of calcium or silicon salts about their soft cell bodies (Figure 21-11). In members of the genus *Actinosphaerium*, needlelike cytoplasmic extensions protrude through the shell, each "needle" supported by a complex double spiral of microtubules and containing actin and myosin. Food particles adhere to sticky secretions on these special pseudopods and are ingested.

MODERN METHODS FOR FIGHTING MALARIA

Malaria is considered by many to be the world's number one disease: Each year, more than 1 million African children under the age of five die from it; an additional 150 million people suffer the recurrent chills and fever of malaria; and almost 2 billion people in more than 100 tropical countries are in danger of contracting the disease. Despite decades of efforts to control its spread, with spraying programs and drugs such as chloroquine, malaria is making a dramatic resurgence. Some strains of malarial plasmodia have become drug-resistant by excreting chloroquine so rapidly that the antimalarial drug is ineffective. In many places, the *Anopheles* mosquitoes have also became resistant to pesticides. In addition, genetic recombination is unusually high between different strains of the parasites, and this source of genetic variability may also lead to new forms of resistance. To fight these growing threats, researchers are trying to understand how malarial parasites recognize the red blood cells they infect and destroy and to create vaccines that can immunize people against the disease in malaria-prone areas.

Merozoites seem to recognize two types of glycoproteins (sialoglycoproteins and glycophorins A, B, and C) on the surface of red blood cells, to bind to a sugar at the ends of some of the molecules, then to navigate through the membrane and multiply inside the red blood cell. Biologists are preparing monoclonal antibodies that would compete for and bind to the merozoite's normal receptor sites on the red blood cell and in that way prevent entry by the malarial parasites. A similar strategy has worked for a kind of trypanosome that attaches to fibronectin (see Chapter 6) and then invades human cells.

In parallel research, experimenters have developed a vaccine against the sporozoite stage of the life cycle and begun to test it in humans. Using recombinant DNA techniques, the researchers identified the main antigen on the parasite's surface that triggers a person's immune response. They then generated a benign synthetic version of the antigen and created a vaccine against it.

Two other major types of shelled sarcodines are the **foraminiferans** and the **radiolarians.** Both are marine protozoan and members of the **zooplankton**—nonphotosynthetic protists living at the ocean's surface. Radiolarians secrete delicately patterned silica-containing shells, through which they extend pseudopods that capture food and draw it inward for intracellular digestion. Foraminiferans secrete a calcium-containing shell that can look like a minute chambered nautilus, or spiral seashell. Either type can live as single cells, as colonies, or as fairly large multinucleate individuals.

Both radiolarian and foraminiferan shells fall to the ocean bottom after the cells die, where they can collect into sedimentary deposits and eventually form rocks. Radiolarians help form chert and other siliceous rocks, while foraminiferans contribute to limestone; the white

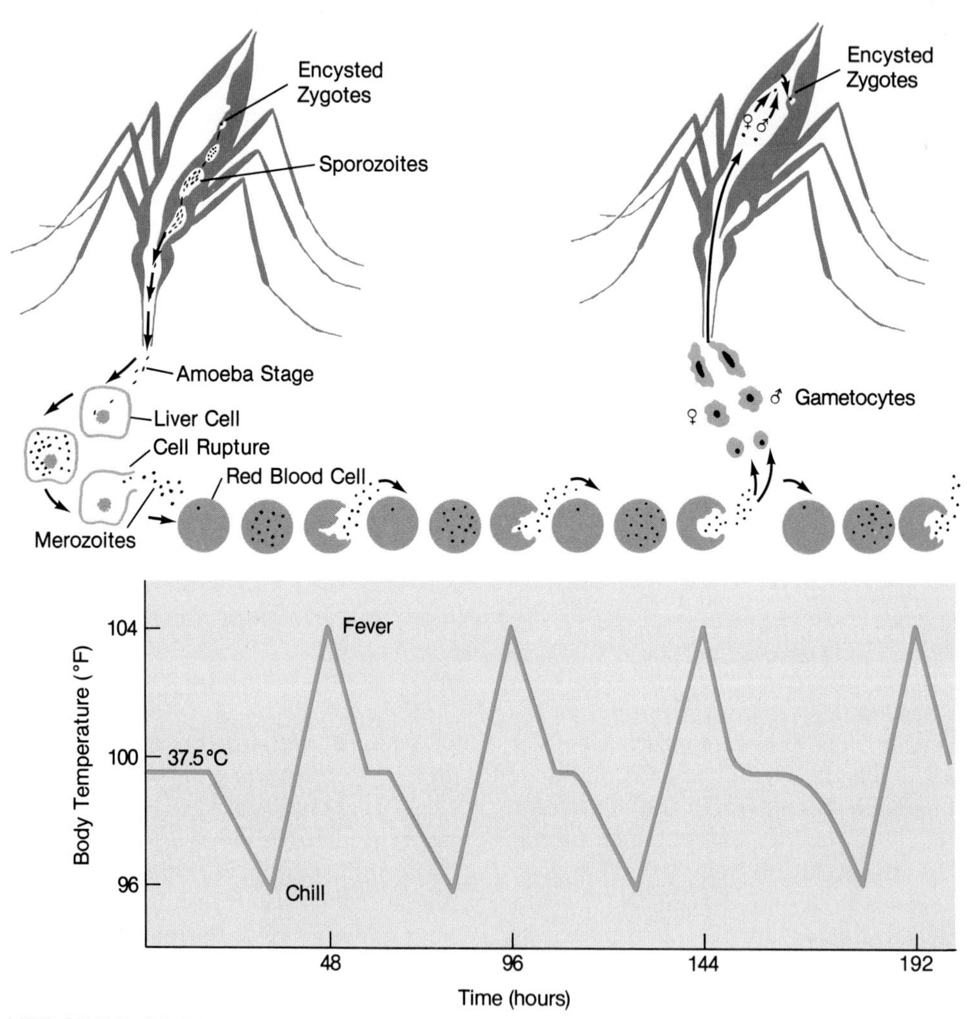

Figure 21-12 LIFE CYCLE OF *Plasmodium vivax.*
The life cycle of this and other malaria-causing sporozoan parasites of the genus *Plasmodium* begins when sporozoites enter the bloodstream of a human or other mammal along with the saliva of a female *Anopheles* mosquito. Sporozoites, which exhibit amoeboid movement, are harbored in the mosquito's salivary glands, and they enter the bloodstream along with a coagulant that prevents blood clotting and allows the mosquito to drink freely. The sporozoites develop in liver cells, reproduce asexually, and liberate tiny round merozoites into the blood. The merozoites enter red blood cells, reproduce synchronously at intervals of 48 or 72 hours, and then cause the cells to burst, or lyse. Toxins are liberated along with more merozoites, which enter other red blood cells, and the cycle begins again. This produces the cyclical patterns of chills and fevers seen in the graph. Note that the temperature peak corresponds to the bursting of thousands of red blood cells. The *Plasmodium* cells eventually become sexually reproducing male and female gametocytes, which can mature into sperm and egg cells only in the gut of a female *Anopheles.* If a malaria victim is bitten by such a female mosquito, some gametocytes enter the mosquito and fuse to form zygotes. Following fertilization, the zygotes bore into the mosquito's gut lining, where they develop into cysts. These cysts give rise to sporozoites, which ultimately are stored in the mosquito's salivary glands. The next blood meal for the mosquito injects the sporozoites into a new mammalian host, and the cycle starts again.

cliffs of Dover, England, for example, were formed by millions of generations of foraminiferans.

Sporozoa

A third grouping of protozoans is the **Sporozoa,** named for a sporelike stage of its life cycle. All sporozoans are parasites, inhabiting mammals, birds, fishes, insects, other invertebrates, and some plants. The adult forms are nonmotile, although immature cells can move about by flagella, pseudopods, or other means. The sporozoans most likely arose from both sarcodines and mastigophores.

The many sporozoan-induced diseases are typified by malaria, caused by *Plasmodium falciparum* and related *Plasmodium* species. Study of the *Plasmodium* life cycle, illustrated and described in Figure 21-12, helps reveal the complexities typical of sporozoan existence. These include different physical forms (sporozoites, merozoites, and gametocytes) carrying out different parts of the life cycle in the bodies of two kinds of hosts.

Ciliophora

The most complex single-celled organisms on earth are members of the phylum **Ciliophora,** commonly referred to as **ciliates.** These largely free-living aquatic protozoans derive their name from the many cilia distributed in rows, in bands, or uniformly across the cell surface (Figure 21-13).

Ciliates have several specialized organelles. The cell surface lacks a cell wall, but may have a pellicle of protective plates. Contractile fibers, called *myonemes,* and structural microtubules function like "muscles" and "bones" and somehow allow clumps or rows of cilia to beat with beautiful coordination so that the cell can swim forward or backward. The coordination and forward or backward beating of the cilia are controlled by currents of calcium ions and cyclic nucleotides (cyclic GMP). When Ca^{2+} enter a forward-swimming ciliate, for example, an action potential (the equivalent of a nerve impulse) is triggered. This reverses the direction of ciliary beating, and the cell begins to swim backward.

A typical ciliate such as *Paramecium* (Figure 21-14) can also produce trichocysts, toxin-bearing harpoonlike structures that help the cell adhere to substrates or that can be fired at prey or predators. The beating of the cilia around an oral groove, or cytostome, sweeps or draws prey species or other food particles into the cell. From the base of this tiny mouth, food vacuoles pinch off and join vesicles containing lysosomal digestive enzymes. An anal pore expels undigested wastes. Some marine ciliates consume algae, retain the cells' chloroplasts, and begin to carry on photosynthesis, supplying at least part of their energy needs.

Many protozoologists think that ciliates may be unique products of evolution. For one thing, they employ some exceptional codons of the genetic code (e.g., UAA and UAG code for glutamic acid rather than for a stop codon, as in all higher eukaryotes). They may thus

Figure 21-13 CILIATES: PROTISTS WITH MANY CILIA. Ciliates are the most complexly structured free-living single cells on earth. The cell shown here belongs to the genera *Didinium* (magnified about 270×).

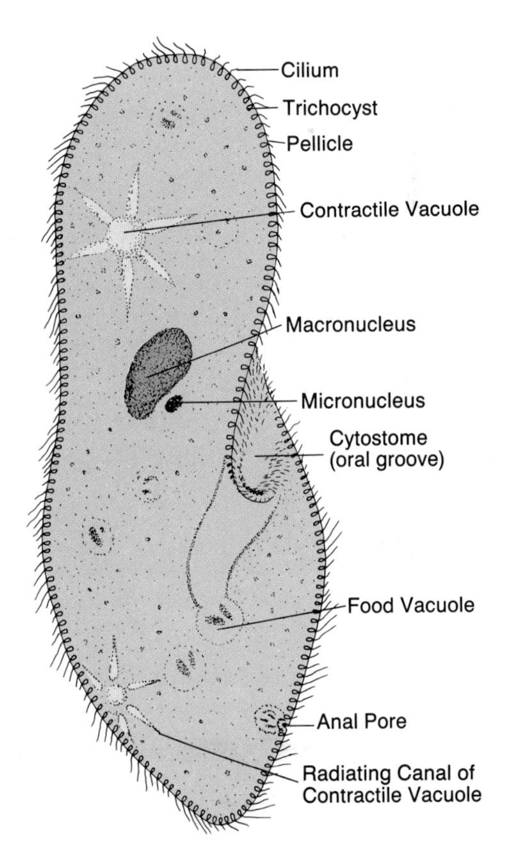

Figure 21-14 ANATOMY OF *Paramecium*. The major internal organelles are shown. The contractile vacuoles are involved in water discharge from the cell.

have arisen early in evolution when the genetic code itself was still highly ambiguous.

Beyond that, ciliates have certain complex structures found in no other species, including two types of nuclei. Each cell has at least one *micronucleus* and at least one *macronucleus.* Each micronucleus contains a diploid set of chromosomes. These chromosomes and the micronucleus function only during reproduction. Micronuclei therefore are not sites of RNA synthesis and are not involved in ongoing cellular metabolism. Macronuclei are highly polyploid, and their chromosomes are the sites of RNA synthesis. The macronuclei thus contribute to cellular metabolism and maintenance by providing the information needed to manufacture enzymes and structural proteins.

When a *Paramecium* cell divides in a kind of asexual reproduction, each micronucleus undergoes normal eukaryotic mitosis with a spindle apparatus. The macronucleus does not have a spindle; instead, chromosomes pass to daughter cells, apparently at random.

Ciliates also evolved an elaborate conjugation process involving the exchange of genetic information (Figure 21-15). During conjugation, ciliates align and join near their oral grooves. The diploid micronuclei of both cells undergo two meiotic divisions, and all the haploid nuclei except two per cell disintegrate. Then, in a remarkable exchange process, each cell passes one micronucleus to the other. The transferred micronucleus fuses with the resident one, restoring diploidy. Meanwhile, the macronuclei of both cells disintegrate. The diploid nuclei divide mitotically at least once. The conjugants now separate, and in each cell one product of the nuclear division remains diploid and becomes a micronucleus. The other product becomes polyploid and then an RNA-

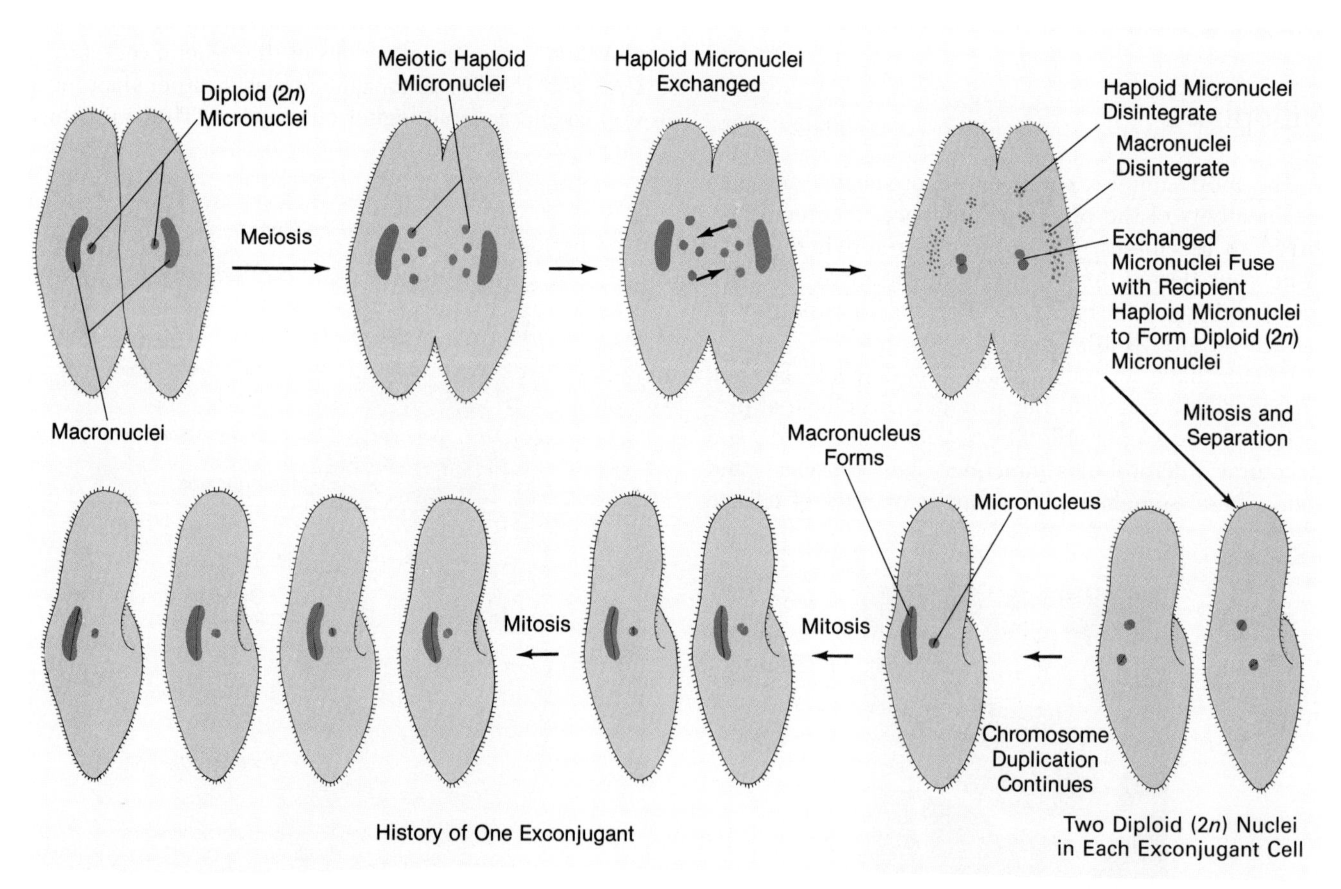

Figure 21-15 CONJUGATION IN CILIATES.
Conjugating individuals pair and undergo an elaborate exchange of micronuclear material. The micronuclei first undergo meiosis; then haploid micronuclei are exchanged. The macronuclei disintegrate, as do micronuclei that fail to fuse. As shown, the exchanged micronucleus fuses with a micronucleus of a recipient cell, yielding a diploid nucleus. New micronuclei then arise, and some enlarge to produce a highly polyploid macronucleus. Each exconjugant cell divides twice, producing four progeny cells. Thus, the original two cells yield eight progeny, each with chromosomes and genes from the parental cells. Conjugation may involve side-to-side contact between two cells or only contact near their ends.

producing macronucleus. This restores to the cell its typical nuclear organization.

Ciliates have provided developmental biologists with a clear example of cytoplasmic rather than nuclear inheritance. A piece of cell cortex and nearby cytoplasm cut from a *Stentor* cell can regenerate an entire new cell, as long as a micronucleus is included. No cilia will form, however, unless at least a portion of a row of cilia is also present. Clearly, the micronucleus containing the cell's chromosomal genetic information lacks some of the information needed to support development of the complete ciliary system.

Caryoblastea

The phylum **Caryoblastea** has only a single fascinating species. *Pelomyxa palustris*, an amoeba-like cell that lives in muddy pond sediments, is remarkable in two respects. First, it lacks mitochondria. It is inhabited by two types of symbiotic bacteria that conceivably play the role of mitochondria. Second, *Pelomyxa* never undergoes mitosis. Instead, after duplicating its chromosomes, *Pelomyxa*'s nucleus pinches in two in a process somewhat like bacterial binary fission (see Chapter 20). *Pelomyxa* is obviously an extremely primitive organism at the border between prokaryotes and eukaryotes, and for that reason, it is an object of curiosity and further study.

Microspora

An even more primitive phylum of protists, **Microspora,** contains parasitic species that infect a vast array of animals as well as other types of protistans. Microsporidia lack mitochondria, and their ribosomes and ribosomal RNA are so unusual that the cells may have arisen as a separate lineage long before any of the other eukaryotic lines originated. Perhaps microsporidia were already inhabiting animal and protistan cells with their own energy-generating mitochondria by the time oxygen built up in the earth's atmosphere.

PROTISTAN EVOLUTION

Some protozoologists believe that zooflagellates may have been ancestors of the other modern protozoan species with their broad spectrum of animal-like life-styles. The zooflagellates could have given rise to the sarcodines and to amoeboid forms by losing flagella. Ciliates are an ancient offshoot at the very least. The symbiotic and parasitic protozoans appear to have been derived from various free-living species. But where did the stem group of zooflagellates come from? Some biologists suggest that phytoflagellates (resembling euglenoids) evolved first; perhaps one line that lost its chloroplasts resulted in the animal-like protists. Another line, after losing chloroplasts, could have yielded the funguslike protists (review Figure 21-1). Still other lines may have yielded the green, red, and brown algae (Chapter 23).

If the entire kingdom Protista arose from flagellated phytoplankton, then what was the source of *those* cells? The best answer so far relies on the endosymbiont theory—the idea that moneran cells were colonized by mitochondrion-like and chloroplast-like prokaryotic symbionts. As we discussed in Chapter 20, there are strong arguments for this theory. But is not yet clear whether all protists arose from one type of stem cell or from several.

Regardless of their precise origin, protists reflect several evolutionary trends. One involves cellular diversity and the exploitation of different ways of life as producer, decomposer, or consumer. A second evolutionary feature of protists is diversity in locomotion. Pseudopods, flagella, cilia, or undulating membranous sheets propel these remarkable cells, sometimes in directed ways toward light, chemicals, or food. Some of the protists we have examined move by one mechanism; some, by another; and some, not at all. Nevertheless, there seems to be a trend toward complex, coordinated cell surface activity among some flagellated and ciliated forms.

Significantly, the eukaryotic protists with their specialized organelles and varied nutritional modes almost certainly gave rise to the multicellular fungi, plants, and animals, and our next few chapters will trace the interesting story of that descent.

SUMMARY

1. The kingdom Protista is composed primarily of single-celled eukaryotic organisms that function as producers, decomposers, or consumers of food materials.

2. The plantlike protists include the photosynthetic cells that make up *phytoplankton,* and in some taxonomic schemes, the green, red, and brown algae.

3. The euglenoids are mostly photosynthetic autotrophs that lack a rigid cell wall and are propelled by whiplike flagella.

4. The pyrrophytes, or *dinoflagellates,* are mostly photosynthetic autotrophs with unique flagella that cause them to spin as they swim. They are responsible for red tides.

5. The chrysophytes include the *golden-brown algae* and the *diatoms.* Both types contain the pigment fucoxanthin and have pectin in their cell walls. The cell walls of diatoms also contain silica.

6. Funguslike protists include the *true slime molds* and the *cellular slime molds*. The true slime mold forms a multinucleate *plasmodium*, whereas the cellular slime mold forms an aggregate of individual cells, or *pseudoplasmodium*.

7. The animal-like, consumer protists are the *protozoans*.

8. The mastigophores are *zooflagellates* that move about by means of flagella. Common types include *Trichonympha*, which inhabits the gut of termites, and *Trypanosoma*, which causes African sleeping sickness.

9. The sarcodines include amoebae, *radiolarians*, and *foraminiferans*. All move by means of *pseudopods*, and some have hard, intricate shells.

10. All sporozoans have a sporelike stage in their life cycle. The adults are nonmotile and are parasitic, such as the *Plasmodium* species that cause malaria.

11. The *ciliates* move via the coordinated beating of large numbers of cilia. They possess both micronuclei and macronuclei and can engage in a sexual conjugation process.

12. The single type of *Caryoblastea*, *Pelomyxa*, lacks mitochondria and does not undergo mitosis.

13. *Microspora* probably is the most primitive protistan phylum. Its members are parasites that lack mitochondria.

14. The many types of protists probably evolved from flagellated cells and were likely the stem groups for the multicellular fungi, plants, and animals.

KEY TERMS

Acrasiomycota
Caryoblastea
cellular slime mold
Chrysophyta
ciliate
Ciliophora
diatom
diatomaceous earth
dinoflagellate
Euglenophyta
foraminiferan
golden-brown alga
Gymnomycota
Mastigophora
Microspora
myxamoeba
Myxomycota
phytoplankton
plasmodium
protozoan
pseudoplasmodium
pseudopod
Pyrrophyta
radiolarian
Sarcodina
Sporozoa
true slime mold
zooflagellate
zooplankton

QUESTIONS

1. In what ways are some protists like plants? Like fungi? Like animals?

2. Name some examples of phytoplankton.

3. If a planetary disaster wiped out all the earth's phytoplankton, what would be the effect on fishes? On sea mammals? On seabirds? On the atmosphere?

4. Diatom cell division would seem to be a dead end because each division gives rise to a smaller daughter cell. How do diatoms overcome this problem?

5. How does the pseudoplasmodium of a cellular slime mold differ from the plasmodium of a true slime mold?

6. What group of protists is responsible for red tides? For chalk deposits? For diatomaceous earth? For cellulose digestion in termites? For African sleeping sickness? For malaria?

7. Give examples of ciliate structures that are analogous to legs, anchors, darts, muscles, and the mouth.

8. Describe some protistan methods of locomotion.

9. Diagram the stages of conjugation in ciliates. What are the functions of micronuclei? Of macronuclei?

ESSAY QUESTIONS

1. If a photosynthetic protist loses its chloroplasts, would you say that it is defective? Or would you say that it is cured of a parasite? Can the freed chloroplasts live on their own? Can the protist survive the loss of the chloroplasts? Can the protist generate new chloroplasts? Answer the same questions for a protist that loses its mitochondria.

2. African sleeping sickness and malaria both are evolving and staying "ahead" of human defenses. Explain how.

For additional readings related to topics in this chapter, see Appendix C.

22
Fungi: The Great Decomposers

Yeasts, molds, mushrooms, mildews, and the other fungi pervade our world. They work great good and terrible evil. Upon them, indeed, hangs the balance of life; for without their presence in the cycle of decay and regeneration, neither man nor any other living thing could survive.

Lucy Kavaler, *Mushrooms, Molds, and Miracles* (1965)

We leave the world of single-celled prokaryotes and eukaryotes now, and enter the realm of multicellularity, beginning with the fungi. Although the 175,000 or more species in the kingdom Fungi are an unheralded and often overlooked group of organisms, they are a critical link in the web of life on earth. Fungi are neither photosynthesizers, like plants, nor motile consumers, like animals. Instead, they derive nutrients by breaking down the waste products and remains of other organisms. Thus, they are the great decomposers, preventing our planet from being buried meters deep in the remains of dead plants and animals and returning essential inorganic and organic materials to the soil.

People have long had a close—if not entirely benevolent—relationship with the fungi. On the one hand, we eat many types and use others in the production of

Sturdy, helmeted decomposers of the forest floor: *Boletus luritus* mushrooms rise silently into the damp air.

bread, beer, wine, certain cheeses, and antibiotics. On the other hand, fungi are the single largest cause of plant diseases. They infect and consume stored foods, they can live as parasites in humans and other animals, causing ringworm, athlete's foot, yeast infections, and other diseases; and they can attack and decompose almost any organic material, from cloth and paper to leather and wood.

This chapter discusses:

- The general characteristics of the fungi, including their extracellular digestion, their peculiar structures and growth patterns, their use of spores for reproduction, and their life cycles, often dominated by the haploid phase
- The classification of the fungi into six main divisions, including the chytrids; water molds; black bread molds; yeasts, truffles, and morels; mushrooms, puffballs, and rusts; and the "imperfect" fungi. We also discuss lichens, the ultimate symbionts
- Fungal evolution—the predecessors and descendants of the great decomposers

CHARACTERISTICS OF FUNGI

Members of the kingdom Fungi have an innovative form of digestion, a unique structure, and several other uncommon characteristics that set them apart from microbes, plants, and animals.

Fungi are *heterotrophs* that obtain nutrients by secreting enzymes into their environment. The enzymes digest leaves, fruit, or other organic substances. The fungi then absorb the nutrient molecules that result from the breakdown of the organic matter. This process is called **extracellular digestion.** The particular battery of enzymes a given fungus secretes largely governs what the organism can digest as a food source. The fungus metabolizes the absorbed food, processes the nutrients, and stores them mainly as glycogen; this is similar to animals but not to plants, which store sugars as starch. One reason that the slime molds (see Chapter 21) are not classified as fungi is that they do not carry out extracellular digestion; cells of both types engulf food particles or absorb small molecules from the environment.

Fungi obtain nutrients from dead or living organisms. Those species that decompose dead matter are called **saprobes** and include bread molds, mushrooms, morels, and other fungi that grow on organic matter—from fallen trees to dung. Parasitic fungi obtain nutrients from living plants and animals and from most other unicellular and multicellular organisms. Finally, many fungi live as *symbionts* with the roots of higher plants, aiding the plants' growth and survival and deriving nutrients in return.

A fungus's basic structure is distinctive. Except for yeasts and a few other unicellular forms, fungi have a main *thallus* (body) composed of cellular filaments called **hyphae** (Figure 22-1). You can see individual hyphal filaments in molds that grow on bread or oranges. Mushrooms are actually a mass of hyphae packed tightly together. Each hypha, whether an individual strand or part of a mushroom, consists of a tubular cell wall surrounding a cytoplasm that contains many of the usual eukaryotic organelles plus one or more nuclei. Hyphae have no differentiated cell types, but some hyphae in certain species become specialized to form **rhizoids,** which anchor the fungus much like roots anchor a plant; some form **haustoria,** feeding structures that penetrate the living cells of other organisms and absorb nutrients from them; and some hyphae produce spores. This shows that fungi, as multicellular organisms, do display localized functional specialization.

The cell walls of most hyphae are made of *chitin* (a carbohydrate polymer composed of glucosamine) and other organic molecules. Some fungal hyphae are made up of true cells, while others are not. The difference lies in the presence or absence of cross walls, or **septa** (singular, *septum*), that segregate independent cells, each with at least one nucleus (Figure 22-2a). The hyphae of so-called "lower fungi" lack septa, and the cytoplasm streams freely through the branched hyphal channels (Figure 22-2b). Such fungi have multiple nuclei in one

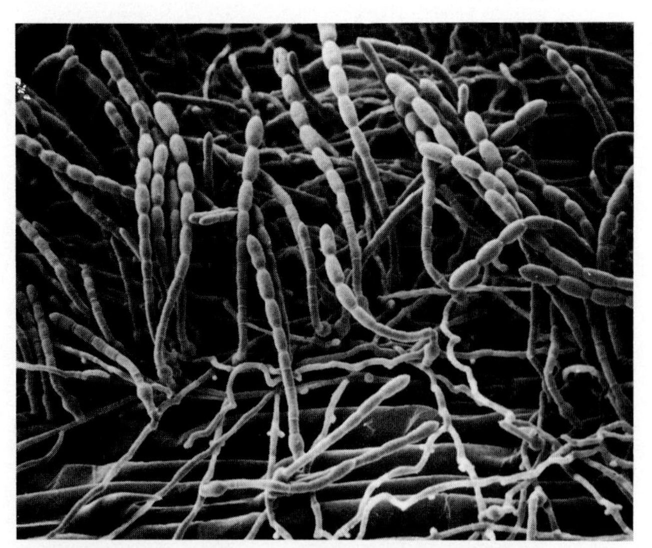

Figure 22-1 HYPHAE: BASIC FUNGAL UNITS. Hyphae of the fungus *Erysiphe graminis* growing on barley (magnified 600×). The small fibers at the bottom are the cytoplasmic tubes called hyphae. The structures with constrictions growing upward from the hyphae are conidiophores, reproductive structures.

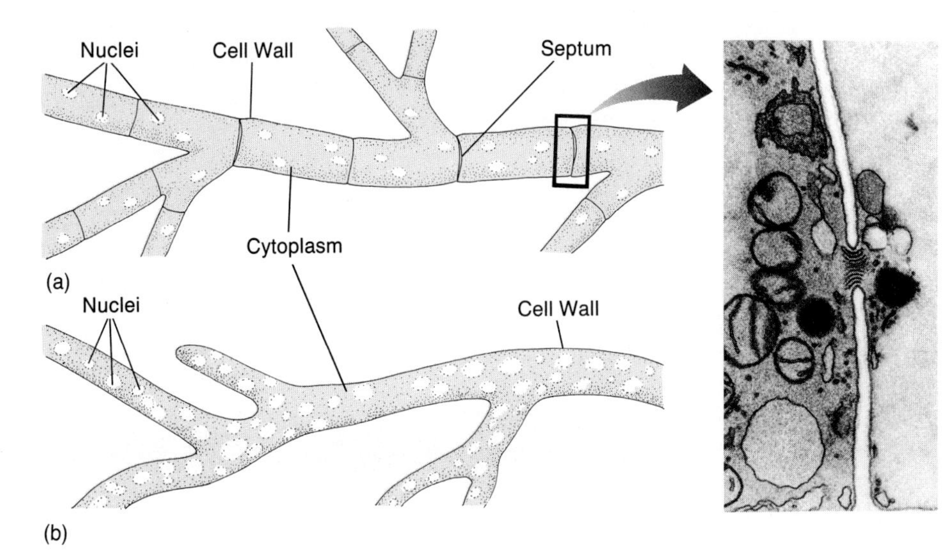

Figure 22-2 HYPHAE: WITH SEPTA AND WITHOUT.
(a) Septa delimit cell-like domains but are perforated with holes, as shown by the electron micrograph on the right (magnified 23,000×), allowing a variety of substances and organelles to pass between cells.
(b) In the absence of septa, a hypha is a single cytoplasmic domain.

mass of cytoplasm and are said to be **coenocytic.** The hyphae of "higher fungi" are *septate* (have septa). The electron microscope reveals cross walls, but even these are perforated by pores, which allow cytoplasmic continuity and streaming between separate cellular compartments.

Fungal Growth Patterns

Hyphae grow only by elongating at the tip and by forming branches some distance behind the tip. Although proteins and other essential macromolecules are synthesized throughout the cell, cytoplasmic streaming carries the products to the tips for assembly into new cell wall, plasma membrane, and so on. Such growth patterns generate a filamentous network, or **mycelium** (Figure 22-3). Continual growth at the tip is highly adaptive for fungal life-styles. As the fungus digests its way through a food source and depletes the available nutrients, the growing hyphal tip literally moves on, penetrating into undigested organic material with its fresh supply of nutrients. Fungal metabolic wastes also accumulate in regions around older hyphae, but such wastes do not harm them, because the tips grow onward.

The farther the tips of hyphal branches probe forward, the more nutrients are absorbed and the more hyphae are generated—sometimes with astonishing speed. At peak growth rates, a fungal colony may produce 200 m of hyphae—one-eighth of a mile—in just 5 hours.

Such prodigious growth depends on mitosis and rapid manufacture of cytoplasm. Fungal mitosis is unique in that it culminates in two nuclei that separate and are moved apart in the hyphal cytoplasm without cytokinesis. Some data also suggest that the fungi have unique chromosomes with only very small quantities of the histone proteins involved in DNA coiling (see Chapter 9).

As fungal hyphae grow through a rotting log, soil, or other substratum, hyphae from genetically distinct individuals of a species may contact each other, fuse, and give rise to a single cytoplasm with dissimilar nuclei, a so-called **heterokaryon. Heterokaryosis,** the process by which heterokaryons form, is a special feature of fungi. One of its great evolutionary advantages is that it brings dissimilar sets of genes into one mass of cytoplasm, and a new composite phenotype, combining features of more than one parental strain, may result.

Figure 22-3 A LIVING POWDER PUFF.
This fungal mycelium growing on a leaf derives nutrients by haustoria and uses them to rapidly build the whitish hyphae that make up the mycelium.

Fungal Life Cycles

Like all multicellular organisms, fungi exhibit life cycles. Some life cycle events involve asexual, vegetative reproduction, and others, the sexual processes of meiosis, and fertilization. In most fungal life cycles, the haploid stage is predominant. During sexual reproduction, fusion of the haploid nuclei yields the diploid stage; but meiosis soon follows, so that the major free-living organisms are haploid. In mushrooms and other complex fungi, however, a dikaryotic stage, equivalent to the diploid stages of plants and animals, alternates with the haploid portion of the life cycle. We'll consider several types of life cycles as we discuss the major divisions of fungi.

Spores for Dispersal and Survival

How can a nonmotile heterotroph like a fungus move to a new source of nutrients once it has consumed all the available nutrients in a given rotting log or moldy piece of bread? There must be a more mobile life stage to ensure dispersal to new sites, and there is: *spores*, the reproductive bodies fungi produce. Some spores are borne on specialized **aerial hyphae,** which grow vertically and discharge spores into the air in various ways. Other spores form internally, as in puffballs, and then are discharged explosively (Figure 22-4). And individual spores are quite variable: They can be composed of one or more cells, can have one or more nuclei, and can be formed sexually or asexually.

(a) (b)

Figure 22-4 SHOWERS OF SPORES: A FUNGAL STRATEGY.
(a) The discharge of spores from this earth star, *Geastrum triplex*, can be triggered by the slightest touch, even from falling raindrops. (b) This scanning electron micrograph magnifies the fungal spores from such a cloud by about 2,400 times. From *Living Images* by Gene Shih and Richard G. Kessel, Science Books International, 1982. Reprinted by permission of the present publisher, Jones and Bartlett Publishers.

Figure 22-5 THE GIANT PUFFBALL.
This *Lycoperdon giganteum* contains an astonishing 5×10^{12} spores by the best estimates. The chances of a spore landing at a favorable site, germinating, and giving rise to a new puffball are minute; hence the trillions of spores.

An individual fungus can produce a truly amazing number of spores. One average-size commercial mushroom can give rise to about 16 billion spores. The giant puffball is even more impressive: At maturity, its brown powdery interior is filled almost entirely with a loose network of as many as 7 trillion spores (Figure 22-5). Being very light, spores can be carried for hundreds and even thousands of miles, and they occur everywhere in our environment.

There are two main types of spores. *Dispersal spores* are usually short-lived, are produced in large numbers by an actively growing fungus, and can germinate quickly if environmental conditions are favorable. *Survival (resting) spores* are usually produced in lesser numbers when the fungus's growth is slowed by excessive heat, cold, or drought or lack of nutrients. They are derived sexually, they tend to be surrounded by a thick, darkly pigmented cell wall, and they must remain dormant for a while before they will germinate. Survival spores can remain viable for long periods, often years. A typical fungal species will form dispersal spores during its period of active growth and survival spores during unfavorable times.

CLASSIFICATION OF FUNGI

Fungi are difficult to classify because biologists do not know the evolutionary relationships between many members. Therefore, they rely on details of morphology, methods of reproduction, and modes of spore production to make their assignments. Mycologists (biologists who study fungi) place all members of the kingdom Fungi into six main divisions. Table 22-1 lists the primary characteristics of the six divisions. The 50,000 species of lower fungi make up three divisions: Chytridiomycota, Oomycota, and Zygomycota. (*Mycota* is based on the Greek word *mykes*, meaning "mushroom.") Their hyphae lack septa, and they form asexual spores in a special spore case. At certain stages, chytrids and oomycotes produce motile cells that rely on water for dispersal; hence the common name "water molds." The zygomycota share several characteristics with higher fungi and are probably closer to them in evolutionary terms than to the water molds.

The higher fungi include almost 100,000 species in the classes Ascomycota and Basidiomycota. Both groups are septate, and both reproduce sexually in a unique way—by the fusion of hyphae of different mating types to form spores.

The sixth class, Deuteromycota, or Fungi Imperfecti, is a grab bag of more than 25,000 species in which sexual reproduction either does not occur or has not yet been discovered.

Chytridiomycota: The Interface Between Protists and Fungi

Many mycologists believe that the 750 species in the class **Chytridiomycota** are the simplest and most ancient fungi. Some chytrids are aquatic, and some parasitize algae or other fungi. Most are unicellular, but a few species form chains of cells. Chytrids have a characteristic feature: They form two types of motile cells, one type formed asexually within a spore case, or **sporangium,** and gametes formed in separate male and female **gametangia** (gamete containers).

Most chytrid cells are haploid throughout the life cycle. The motile, diploid zygotes become dormant during unfavorable periods; then later, when conditions improve, they divide meiotically to form motile haploid *zoospores*, which settle onto appropriate substrata and grow into cell chains, somewhat resembling hyphae.

Table 22-1 CHARACTERISTICS OF THE SIX FUNGAL DIVISIONS

Division	*Usual Vegetative State*	*Asexual Reproduction (only mitosis involved)*	*Sexual Reproduction*		*Representative Member*
			Fusion of:	*Resulting in:*	
Lower Fungi (coenocytic)					
Chytridiomycota	Haploid	Flagellated spores in sporangia	Flagellated gametes	Resting spores	Chytrids
Oomycota	Diploid	Flagellated spores in sporangia	Gametes in gametangia	Oospores	Water molds, *Phytophthora infestans* (potato blight fungus), *Plasmopara viticola* (downy mildew)
Zygomycota	Haploid	Unflagellated spores in sporangiophores	Gametes in gametangia	Zygosporangium	*Rhizopus* (black bread mold)
Higher Fungi (septate)					
Ascomycota	Haploid	Conidia on conidiophores	Hyphae	Ascospores in an ascus	*Neurospora* (red bread mold), *Penicillium,* yeasts, truffles, morels, cup fungi, powdery mildews
Basidiomycota	Haploid, dikaryotic	None or conidia on conidiophores	Hyphal tip cells	Basidiospores on a basidium	Mushrooms, puffballs, bracket fungi, rusts, smuts
Deuteromycota (Fungi Imperfecti)	Haploid	Conidia on conidiophores	None known		*Aspergillus,* ringworm

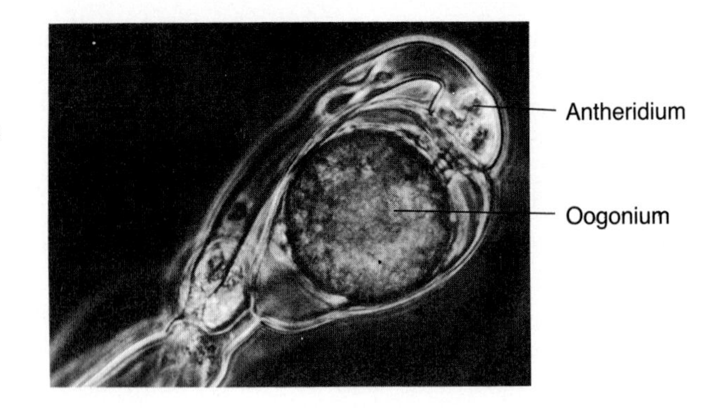

Figure 22-6 LOWER FUNGI: OOMYCOTA.
This photograph of *Sapromyces androgynus* (magnified 650×) shows structures for sexual reproduction—the male and female gametangia (the antheridium and oogonium)—in contact with each other.

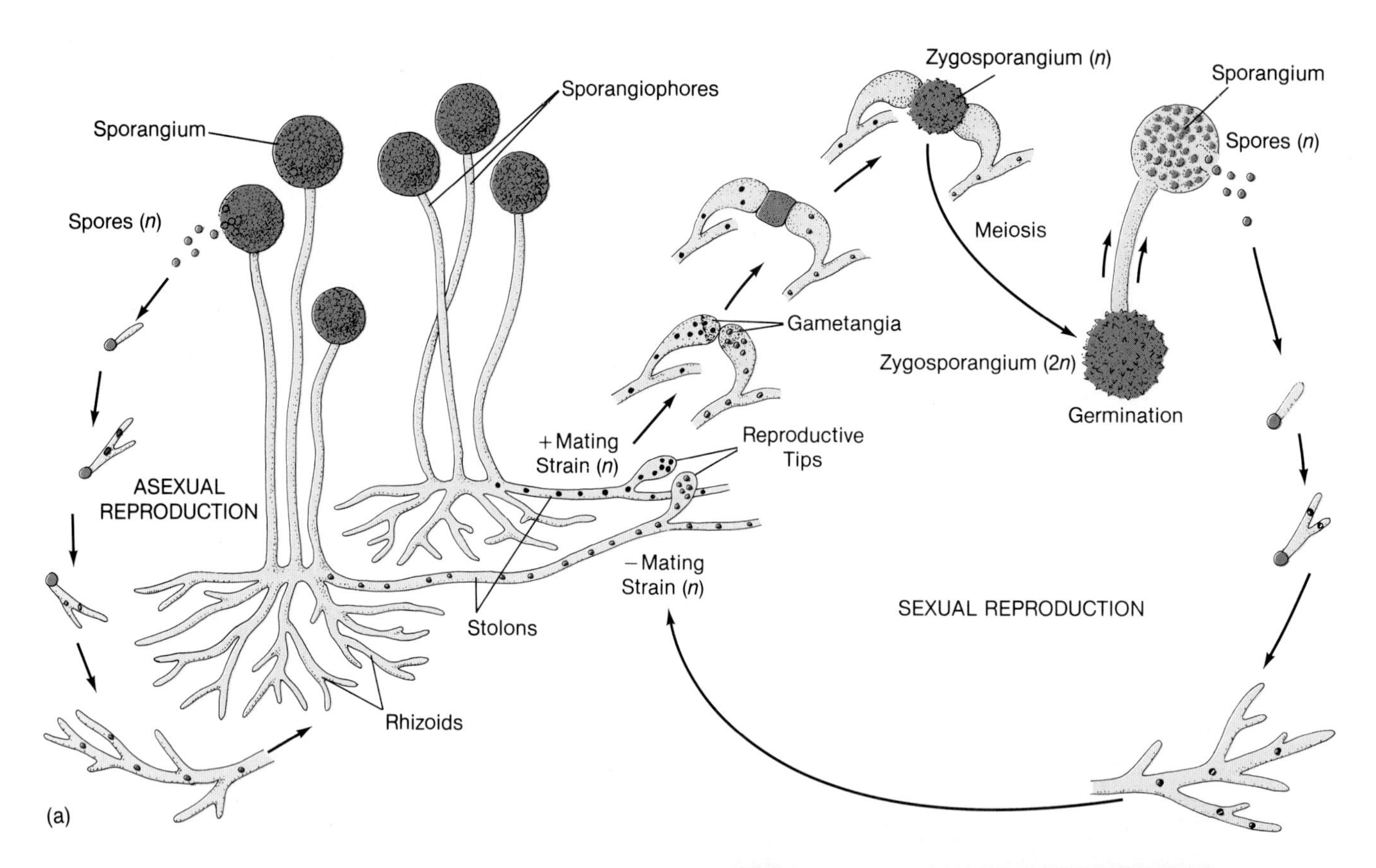

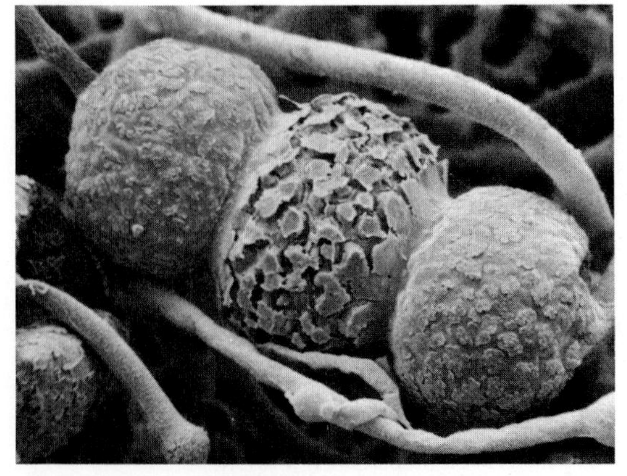

Figure 22-7 STAGES OF THE ZYGOMYCOTE LIFE CYCLE.
(a) Asexual reproduction (left side of figure) involves the formation of sporangia on the hyphae and discharge of haploid spores from those haploid sporangia. In sexual reproduction (right side of figure) opposite mating strains (+ and −) fuse at the tips of the hyphae. A zygosporangium forms and remains dormant for a period. Eventually, the two haploid nuclei fuse, meiosis takes place in the diploid zygote nucleus, and the zygosporangium germinates. A new sporangium grows upward from it and releases haploid spores that develop into mycelia. (b) Scanning electron micrograph of a mature zygosporangium between the adjacent reproductive tips (magnified 600×).

Oomycota

The oomycota are a diverse and unusual group. Some species in the division **Oomycota** are single-celled organisms, and others form branching hyphae. Some are aquatic, and others live in soil. Some are saprobes, and others are parasites. However, all oomycetes lack most or all cross walls in their hyphae and resemble colorless forms of certain algae (see Chapter 21). Like algae, but unlike other fungi, typical oomycota have cell walls containing cellulose instead of chitin. During most of the life cycle, oomycote nuclei are diploid, not haploid, and the chromosomes in those nuclei have basic histone proteins.

Oomycota are named after the large, immobile egg cells they produce (the prefix *oo* refers to eggs), and this is another alga-like feature. During sexual reproduction, both the egg and the male gamete (*protoplast*) develop in gametangia; these gametes eventually fuse, and a diploid oospore is formed. Figure 22-6 shows the gametangia of a typical oomycote.

In addition to the aquatic water molds, the oomycota include several parasites of terrestrial plants, and some of these have significantly altered human history. The oomycote called *Phytophthora infestans*, for example, causes late blight in potatoes. This fungus hit the potato crop in Ireland particularly hard between 1845 and 1847, when wet, cool weather provided optimal growth conditions. Since potatoes were the staple food of most of the Irish population at that time, the blight caused widespread famine, and more than 1 million people died. Another oomycote, *Plasmopara viticola*, causes downy mildew on grapes and almost destroyed the French wine industry around 1880 until chemists created the first fungicide for use on plants and saved some of the vines.

Zygomycota

The **zygomycota,** the third division of lower fungi, are entirely terrestrial, living usually as saprobes, digesting and absorbing dead plant and animal matter in the soil, or as parasites, feeding on a few species. The pin molds commonly found growing on fruit and bread are zygomycota, and a typical example is the black bread mold *Rhizopus*—a fungus with an interesting life cycle (Figure 22-7).

In the asexual reproduction cycle, spores released from sporangia germinate and send out *stolons*, or lateral stemlike hyphae that form a network; rhizoids, which absorb nutrients and anchor the mycelium; and *sporangiophores*, or stalks that bear the sporangia and produce and release new spores. Sexual reproduction occurs when the hyphae of two different mating strains (designated + and − because they are not usually morphologically distinguishable as male and female) contact each other at their hyphal tips. Cross walls form behind the zones of contact, forming gametangia that produce + and − gametes. These gametangia fuse, then their haploid gametes fuse, producing a cell that develops into a thick-walled survival stage, the *zygosporangium*. After a period of dormancy and when conditions are favorable, the haploid nuclei in the zygosporangium fuse to yield a true diploid zygote. Soon after, meiosis occurs, the zygote germinates, and a new sporangium is produced, borne on a single hypha. Once again, haploid spores inside this chamber are released, and those that chance to settle on a new food source germinate and begin a new life cycle.

Certain zygomycota, living deep in the soil play a critical ecological role as symbionts with the roots of most kinds of plants. Such associations of roots and the thin filaments of fungi are called **mycorrhizae,** which means "fungus roots." For many plant species, those individuals infected with mycorrhizae grow more successfully in poor soils, particularly soils deficient in phosphates, than do plants lacking the mycorrhizae. This is because the branching microscopic filaments of the mycelium fan out beyond the root, adding a huge amount of surface area for water and mineral-nutrient absorption (Figure 22-8). Mycorrhizae apparently supply the host plant with sub-

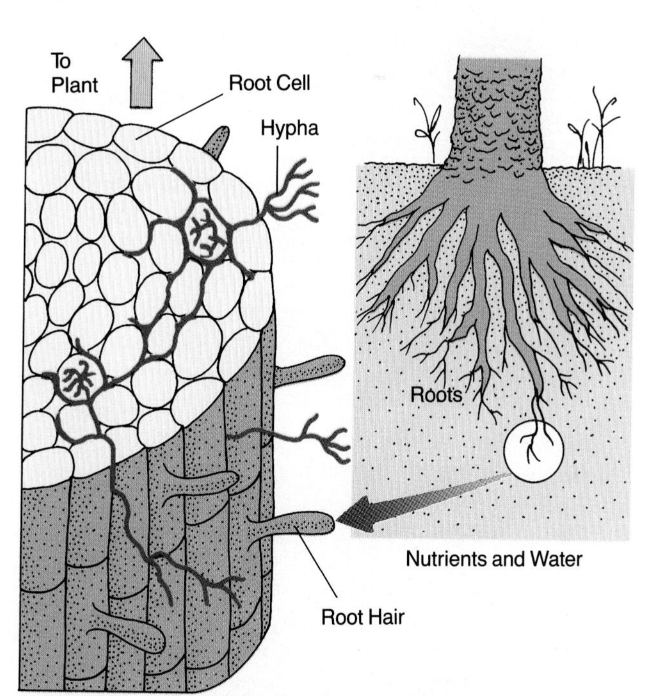

Figure 22-8 MYCORRHIZAE: FUNGUS-ROOT ASSOCIATIONS.
The large majority of land plants form associations with zygomycotes. Hyphae fan out beyond the plant's root cells and expand the surface area for absorbing water, minerals, and nutrients.

stantially more nutrients, especially phosphorus, than its roots alone can absorb. The plant more than repays its debt to the fungus by supplying it with photosynthetic products that are the raw materials for fungal cell metabolism. Almost all plant species can form mycorrhizae, and, indeed, it is estimated that 80 percent of all land plants develop mycorrhizae with one fungus or another. Fossils of some of the earliest land plants reveal mycorrhizal associations, and this suggests that the fungus-plant symbiosis may have helped ancient plants leave the water and colonize land.

Ascomycota

Ascomycota is the largest division of fungi, containing some 30,000 free-living species. An additional 18,000 species grow symbiotically with algae to form lichens

ELM EPIDEMIC IN DUTCH?

Plant researchers are applying some new and creative approaches to eradicating a fungal pest that has decimated the elm populations of North America: Dutch elm disease. This normally fatal tree disease causes leaves to yellow and wither and interferes with the critical passage of water upward through the trunk (Figure A). The fungus that causes Dutch elm disease is *Ophiostoma ulmi,* the infectious spores of which are carried from tree to tree by bark beetles tunneling through the tender wood just below the bark (Figure B). The boring of uncontaminated bark beetles can damage elm trees but not kill them—the fungus is the real pathogen, leaving millions of elms first leafless and then lifeless.

Figure A DUTCH ELM DISEASE. One elm survives, but the others have succumbed to the fungal disease. Normally, all the elms in an area of infection will die or must be destroyed.

Figure B BARK BEETLES TUNNEL IN AN ELM. Borings through the trunk destroy parts of the vascular system, but it is the fungus that kills the tree.

Dutch elm disease was first noticed in the Netherlands in 1919, and in 1930, it was reported in forests around Cleveland, Ohio. American biologists warned that the further importation of elm logs from Europe should be banned, but their advice was ignored. By the mid-1950s, biologists and forestry managers were using three weapons: an importation quarantine, the quick removal and destruction of diseased trees, and a host of chemical insecticides and fungicides to reduce populations of bark beetles and fungus. Despite these efforts, however, the disease had spread all the way across the North American continent by 1975.

In recent years, a new, highly pathogenic subgroup of the *O. ulmi* fungus has caused devastating disease outbreaks in North America and Southwest Asia and has stimulated the search for badly needed weapons to battle the epidemic. One promising tactic has been to create "trap trees" by injecting selected elms with a chemical called cacodylic acid and also spraying them with the attractants (**pheromones**) of female bark beetles. Cacodylic acid causes the elm to radiate the odor of a dying tree and attracts hungry beetles. The pheromones also draw large numbers of male bark beetles, and most of the attackers become trapped and perish on sheets of "fly paper."

A second technique employs a strain of the bacterium *Pseudomonas syringae,* which produces a natural *antimycotic,* or fungus-killing chemical analogous to antibiotics made by other prokaryotes. When released on a grove of trees, the *Pseudomonas* can kill fungal spores with remarkable effectiveness.

In still other research, geneticists have isolated and transferred a piece of double-stranded DNA called d^2 factor that inhibits growth of the deadly fungus and may allow experimenters to produce a biological control agent. Perhaps, someday, we can once again enjoy the tall shade trees that lined the streets and graced the yards of so many American towns.

(discussed in a later section). The name *ascomycota* means "sac fungus" and is derived from *ascus*, a type of spore sac. Most ascomycota are saprobes, but the class also includes many important parasites of living plants, such as those that cause Dutch elm disease, chestnut blight, and peach leaf curl. A few ascomycotes are unicellular (the yeasts), but most are composed of filaments in which the nuclei are haploid. The ascomycota, like other higher fungi, have septate hyphae.

Asexual reproduction involves the generation of spores called **conidia,** which come in many shapes and sizes and may be multicellular. Conidia develop not within enclosed sporangia, but usually on the tips of specialized aerial hyphae called *conidiophores* (Figure 22-9). The tiny, multitudinous conidia are blown about like dust and can land on new food sources. Such airborne conidia are often a source of allergy in humans.

Sexual reproduction involves the generation of *ascospores* in a little sac called an **ascus** (Figure 22-10). The fusing of hyphae from different mating strains brings together haploid nuclei, and a diploid stage forms—the only such stage in the ascomycete life cycle. The diploid cell immediately undergoes meiosis to produce eight tough-walled haploid ascospores. The ascus eventually bursts, freeing the ascospores.

The groups of asci that produce ascospores are usually surrounded by protective hyphae, forming a structure called the **fruiting body.** Fruiting bodies can be flask-shaped containers with a pore or channel, spherical containers with an opening, or disk- or cup-shaped. In the cup-shaped fruiting bodies, such as the cup fungi (see Figure 22-9), the asci are exposed on the upper, or inner, surface. A few related species, such as the morels and false morels, have stalked fruiting bodies crowned

(a)

Figure 22-9 THE ASCOMYCOTE LIFE CYCLE.
(a) A specimen of *Peziza aurantia*, a cup fungus or ascomycote. (b) During asexual reproduction (left), spores called conidia are produced at the ends of aerial hyphae, or conidiophores (review Figure 22-1). When dispersed, conidia form new mycelia. Mycelia may also be generated from ascospores through a type of sexual reproduction (right) that involves an unusual delayed fertilization. When the hyphae of two mating strains contact each other, the nuclei are brought together within the ovary-like ascogonium. Binucleate hyphae emerge from the ascogonium, each cell remaining dikaryotic, with separate nuclei originating from both the parental mating strains. At the tips of the dikaryotic hyphae, a special cell forms in which the nuclei fuse and yield a diploid nucleus. This cell and its progeny undergo meiosis and mitosis to produce eight ascospores, which then germinate and form new hyphae.

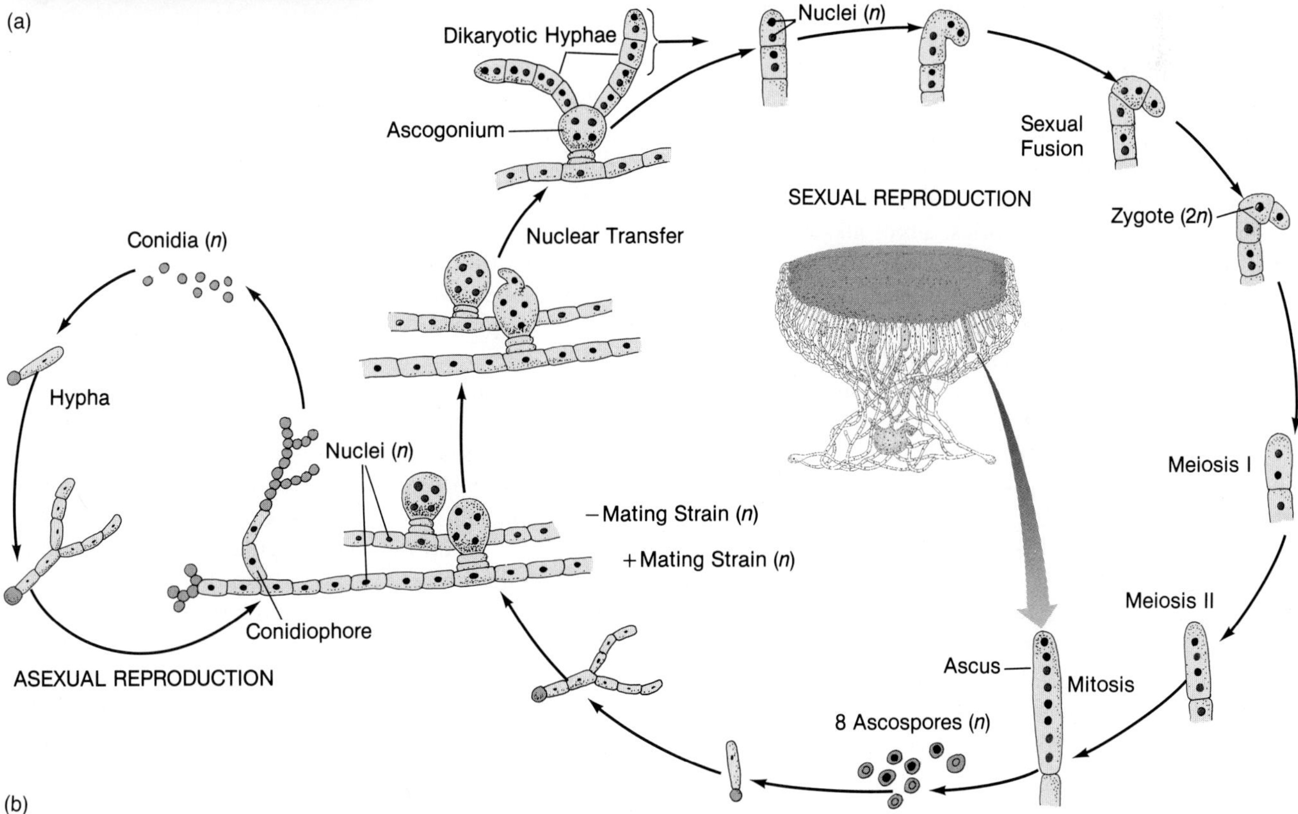

(b)

by bell-shaped, saddle-shaped, convoluted, or pitted tissue that contains the asci). The morels are highly prized as gourmet delicacies, but many of the false morels contain toxic compounds that can be fatal. The truffle is another prized edible ascomycote. Its dense underground fruiting body gives off pungent odors and flavors that attract squirrels, deer, pigs, and other animals—perhaps because truffles produce an odorous steroid called androstenol that can act as a sex attractant in some mammals. This production is a fascinating case of coevolution: The odors attract animals that get nutrients but also inadvertently help disperse the spores. Gourmet cooks find only a handful of the many truffle species palatable, but they so prize the cheesy-garlic bouquet of those few that they are willing to pay about $50 per ounce for fresh or canned truffles.

People also rely heavily on another group of ascomycota, the single-celled yeasts. Yeasts usually reproduce by asexual *budding*—the pinching off of small haploid cells from large older haploid ones. Sometimes, however, two cells unite, form a diploid nucleus, and eventually produce an ascus and ascospores. Yeasts such as *Saccharomyces cerevisiae* are enormously important in human food production because of their ability to ferment sugars and release ethyl alcohol and carbon dioxide. Vintners and brewers use yeasts to make wines and beer, and bakers employ them for leavening bread: The carbon dioxide that metabolizing yeasts give off causes bread dough to rise.

The parasitic *Claviceps purpurea,* is an ascomycote that infects rye grain, and causes a serious disease in humans and domesticated animals if bread is baked from contaminated grain. The fungal hyphae develop into a hard, tumorlike growth on rye called **ergot.** When ground with the rye, ergot releases a toxic chemical. Ergot poisoning is characterized by hallucinations, convulsions, uterine contractions, and tissue damage in the extremities. In small quantities, ergot alkaloids can be used to treat migraine headaches and to induce childbirth. Ergot is also the source of the hallucinogenic drug lysergic acid diethylamide (LSD).

Ascomycota include various *Penicillium* species, fungi that yield the antibiotic that has saved millions of human lives (Figure 22-10). In 1929, Alexander Fleming noted that a species of penicillin mold could inhibit the growth of colonies of the bacterium *Staphylococcus* on a Petri dish and that a diffusible substance coming from the fungus created the antibiotic effect. Researchers isolated this substance, and began to produce the drug **penicillin** commercially during World War II. Since then, biologists have discovered more than 4,000 antibiotics. These antibiotics are produced mainly by fungi and by the prokaryotic actinomycota (see Chapter 20). Annual world production is now more than 100,000 tons, and annual sales revenue in the United States is greater than $1 billion.

Figure 22-10 PENICILLIUM: A LIFESAVING ASCOMYCOTE.
Pharmaceutical chemists derive the lifesaving drug penicillin from the *Penicillium* mold, often seen growing on the surface of rotting citrus fruits such as this tangerine.

In addition to producing penicillin, species of *Penicillium* are used to ripen and flavor Camembert, Roquefort, and various other blue-veined cheeses.

Basidiomycota

Common mushrooms; coral fungi; puffballs; boletes, or pore mushrooms; and bracket, or shelf, fungi are members of **Basidiomycota,** the second division of higher fungi (Figure 22-11). The familiar mushroom is the conspicuous fruiting body of a huge hyphal mass that penetrates the soil or other substratum. A few basidiomycotes, however, such as the rusts and smuts that parasitize plants, do not form conspicuous fruiting bodies.

The basidiomycota differ from the ascomycota because of their **basidiocarp** (the "mushroom" we see on a forest floor) and because of their more lengthy dikaryotic stage in the life cycle. The basidiocarp of gilled mushrooms is actually a dense mass of hyphae with two nuclei per cell (so-called **dikaryotic** hyphae). Club-shaped **basidia** line the surfaces of the gills on the underside of the mushroom cap (Figure 22-12). In most species, each basidium bears four *basidiospores* on the tips of spikelike processes. Each basidiospore is haploid and results from meiotic divisions of a diploid cell at the tip of the basidium.

Figure 22-11 BASIDIOMYCOTES: FANTASTIC DIVERSITY IN SHAPE AND COLOR.
All basidiomycota are constructed of hyphae. (a) The coral fungus (*Clavaria* species), and (b) the amanita (*Amanita muscaria*), or fly agaric, demonstrate two ways hyphae can be welded into different shapes.

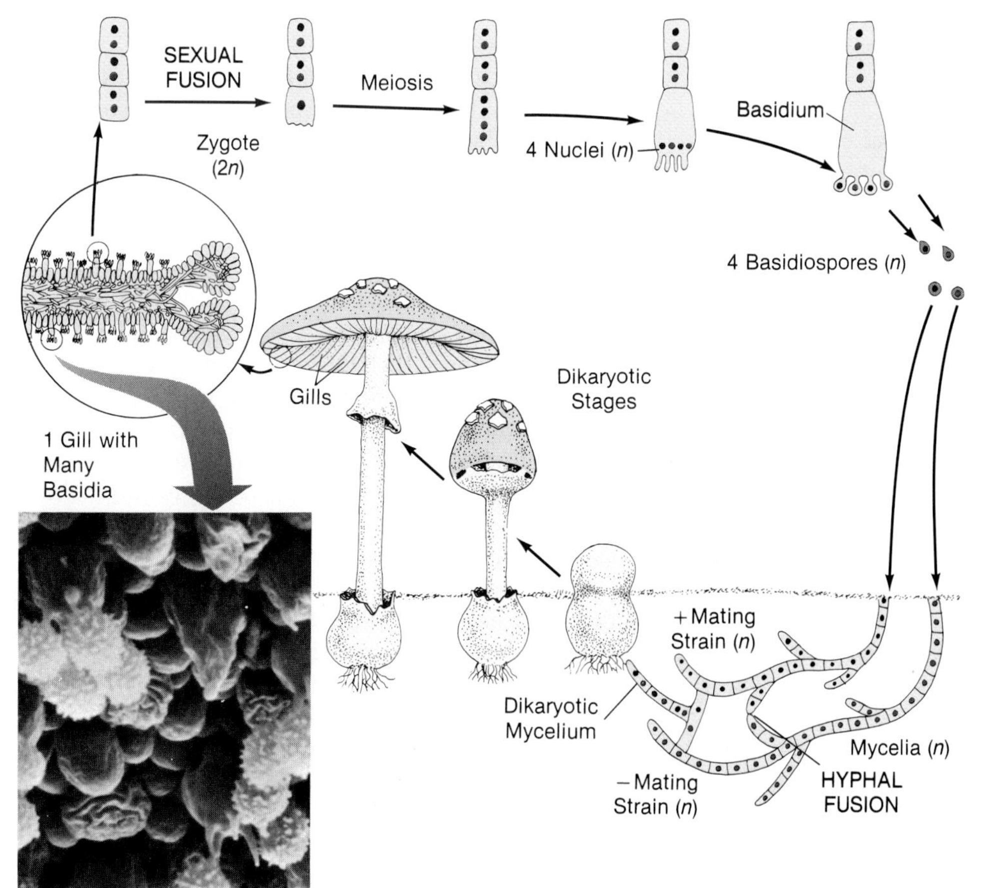

Figure 22-12 LIFE CYCLE OF A MUSHROOM, A TYPICAL BASIDIOMYCOTE.
The aboveground portion of a basidiomycote is the dikaryotic reproductive structure; below ground lies a mat of mycelia. Haploid mycelia of different mating strains make contact and fuse, giving rise to a "fusion hypha." This grows rapidly, producing the above-ground structure or basidiocarp, housing basidia. In each cell of the growing fusion hypha, the two parental nuclei remain separated, but the nuclei in the basidia eventually fuse and undergo meiosis, and basidiospores form. These are dispersed by wind or rain; they then germinate and once again produce underground mycelia. The insert shows basidiospores hanging from the basidium (magnified 2,200×).

Figure 22-12 shows and describes the life cycle of a typical basidiomycote. In general, a basidiospore germinates to produce a haploid mycelium, and when the mycelia of two different mating types make contact, they fuse. The nuclei remain separate, however, and thus the hyphae are dikaryotic. Haploid nuclei in the basidia (but not in other hyphal cells) eventually fuse to form diploid nuclei. Once fusion has occurred, meiosis follows, and the haploid basidiospores are produced.

Many mushroom species grow unseen for several years as huge dikaryotic mycelia in soil or wood until environmental conditions foster the development of fruiting bodies and basidiospores. Specific levels of moisture, hospitable temperatures, adequate carbon dioxide, and minimal light levels all must be available before fruiting bodies are induced. Once induced, the mushroom can grow very quickly. In the spring, for example, a lawnful of mushrooms can appear overnight. Some basidiomycotes, however, grow more slowly and under different kinds of conditions.

A few basidiomycotes reproduce asexually through dispersal of conidia, but most generate huge numbers of basidiospores through sexual processes, and these germinate quickly when they land on suitable food material.

Deuteromycota (Fungi Imperfecti)

Mycologists have described almost 25,000 fungal species that appear to lack modes of sexual reproduction. These they classify together as the **Deuteromycota,** or **Fungi Imperfecti.**

Most of the deuteromycota are known to reproduce asexually by means of conidia. Thus, they are related to the ascomycota and may have lost the ability to reproduce sexually during their evolution. A few species may have a sexual phase under certain environmental conditions never duplicated in the laboratory. When a sexual stage is discovered in the life cycle of one of these species, it is usually reclassified into an ascomycote genus or, more rarely, into a basidiomycote one.

Many of the most familiar and commercially important molds are deuteromycota, including species of *Aspergillus.* Food manufacturers use *Aspergillus niger* to produce citric acid for soft drinks, jams, jellies, salad dressings, and many other foods. They also use *A. tamarii* and *A. oryzae* to make soy sauce, sake, and a number of other fermented products.

One dangerous fungus in this genus is *Aspergillus flavus,* which produces an extremely potent compound, *aflatoxin,* which in high doses can kill farm animals or people and at lower levels can cause cancer. Stored peanuts are particularly susceptible to *Aspergillus* infection if they are not dried properly, but this fungus also attacks corn, millet, rice, and many other grains and seeds.

Other deuteromycotes are predators in a sense: They grow lassolike hyphae with which they can ensnare and digest tiny worms. One touch by the soil worm leads to an extremely rapid (0.1-second) osmotic change within the fungal cell, causing it to expand and trap the worm. Other hyphae soon grow into the prey and secrete digestive enzymes.

LICHENS: THE ULTIMATE SYMBIONTS

The gray, orange, and whitish encrustations we commonly see on rocks and trees are actually alive. They are **lichens,** composite organisms made up of one species of fungus and one or two species of algae (Figure 22-13). *Crustose* lichens are flat and cling tightly to the substratum; *fruticose* lichens grow upright or hang from a branch or rock; and *foliose* types look like leaves, having lobes that can be lifted from the rock, tree bark, or other surface on which they grow. Regardless of outer form, just below the surface of the lichen thallus, the algal species grows as a thin layer of single cells, entwined by the fungal hyphae.

There are some 18,000 lichen species, most containing their own unique fungal species. These are usually ascomycotes but are sometimes deuteromycotes or basidiomycotes. Only about 30 species of algae are found as part of these thousands of lichens, however. The most common algal components are members of the eukaryotic green alga genera *Trebouxia* and *Trentepohlia* or are prokaryotic cyanobacteria of the genus *Nostoc.*

Studies using radioactive tracer molecules such as $^{14}CO_2$ have shown that the algal partner fixes atmospheric carbon and provides the lichen with organic compounds (sugars), the products of photosynthesis. Some lichens can also fix atmospheric nitrogen and grow on almost any substratum. In fact, the majority of lichens derive most of their necessary water and minerals from rain water and air, allowing them to live where few other fungi could survive, including on bare rocks, barren soils, and tree trunks. This ability gives lichens an important ecological role: breaking down rocks and starting the process leading to soil production.

Biologists do not fully understand sexual reproduction in the lichens, but asexual reproduction involves fragmentation of the thallus into powdery bits that are carried by wind or splashing raindrops to a new site of attachment. Once in place, most lichens grow very slowly—from 0.1 to 10 mm per year in diameter. The low, relatively constant rates of lichen growth have al-

(a)

(b)

Figure 22-13 LICHENS.
Lichens are composite organisms formed from the close association between fungal hyphae and algae. (a) Crustose lichens are common sights on rocks or trees in areas free of heavy air pollution. (b) Fruticose lichens can resemble greenish beards hanging from limbs or ledges.

lowed them to be used for dating rocks and human artifacts. The giant carved statues on Easter Island in the Pacific, for instance, have been estimated to be 400 years old on the basis of lichen size and growth, and in other locations, some large lichens may be thousands of years old.

Mycologists have long debated whether lichens represent an association beneficial to both fungus and alga or whether the fungus is simply a parasite on its photosynthetic partner. The algal component of lichens do seem to benefit, however, by being able to inhabit an often dry, exposed ecological niche it otherwise could not exploit.

Lichens are so efficient at absorbing inorganic nutrients—including the sulfur dioxide component of air pollution—that they are sensitive indicators of air quality. The centers of industrialized cities are "lichen deserts," and the presence of even a few lichens, as in cities like St. Paul, Minnesota, signifies relatively clean air. Some experimenters are even designing preliminary monitoring systems that employ lichens for a fraction of the cost of sophisticated air-quality instruments.

FUNGAL EVOLUTION

How did the various fungal groups arise? Biologists believe that several lineages may have arisen from unicellular eukaryotes lacking chloroplasts. The oldest fossil filaments are Precambrian. If they are fungi, and not algae, then fungi must be among the earliest of all eukaryotic organisms. The early progenitor cells probably became coenocytic, thereby allowing larger cytoplasmic systems—hyphae—to evolve. Finally, in the higher fungi, the nuclei became at least partially walled off as more complex structures and ways of life evolved. The fact that zygomycotes, ascomycotes, and basidiomycotes lack basal bodies, centrioles, and microtubular flagella

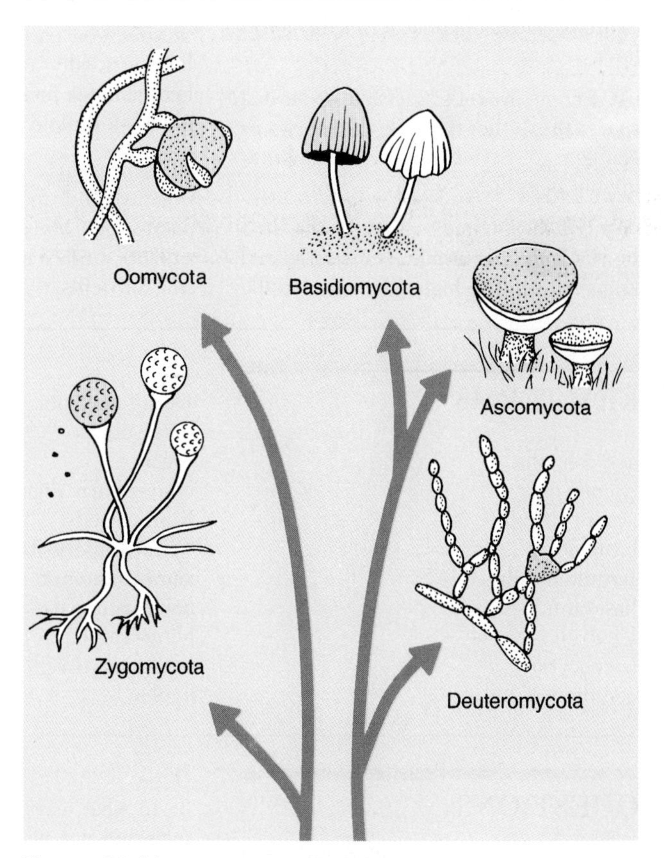

Figure 22-14 FUNGAL EVOLUTION.
One model showing the evolutionary relationships of the fungal classes, which arose from single-celled progenitors.

sets them off from the flagellated lower forms, the chytrids and water molds. In fact, some taxonomists classify chytrids and water molds in the kingdom Protista. This difference suggests that the former had a common ancestor. It also suggests that these higher fungi did not arise from the lower ones, but are independent descendants of prokaryotes (Figure 22-14). Finally, the presence of the dikaryotic state during the life cycles of ascomycotes and basidiomycotes links these groups as close relatives. Despite these differences, however, the fungi share a common successful strategy: decomposing organic matter produced by plants, animals, or algae. In Chapter 23, we turn to the primary producers of food on earth—the green plants.

SUMMARY

1. Fungi carry out *extracellular digestion* by secreting enzymes that digest organic matter, then absorbing the resulting nutrients derived from the food source. Fungi can be *saprobes*, parasites, or symbionts with plants.

2. Most fungi are composed of filamentous *hyphae*, whose cell walls contain chitin or, in some species, cellulose. Hyphae may have specialized anchoring, feeding, or spore-forming structures.

3. Hyphae can be *coenocytic* (multinucleate) or *septate* (having cross walls between nuclei). A filamentous network or dense packing of hyphae forms the *mycelium*.

4. Fungi produce asexual or sexual spores that either disperse the species or enable it to survive unfavorable environmental periods.

5. The lower fungi consist of the divisions *Chytridiomycota*, *Oomycota*, and *Zygomycota*. The higher fungi are in the classes *Ascomycota* and *Basidiomycota*. The sixth class—the *Deuteromycota*, or *Fungi Imperfecti*—apparently lack sexual reproduction.

6. The chytridiomycotes are aquatic or parasitic organisms with asexual spores and gametes propelled by flagella.

7. The oomycotes range from single-celled aquatic species to multicellular terrestrial plant parasites. They produce motile asexual zoospores; their nuclei are diploid through most of the life cycle; and sexual reproduction involves large, immobile egg cells.

8. The zygomycotes are terrestrial saprobes or parasites. They reproduce both asexually and sexually. Their haploid gametes fuse to form a zygosporangium—a diploid survival stage.

9. *Mycorrhizae* are associations of zygomycotes or basidiomycotes with plant roots. Mycorrhizae greatly expand a plant's surface area for absorbing water and nutrients.

10. The ascomycotes reproduce asexually, by means of spores called *conidia*, borne on *aerial hyphae*, or sexually, by producing ascospores in sacs called *asci*.

11. The basidiomycotes produce sexual spores, or basidiospores. The fruiting body, or *basidiocarp*, is constructed of *dikaryotic* hyphae in which each cell contains two nuclei, each of which arose from a different parent.

12. The deuteromycotes, which have no known sexual cycle, include many commercially important genera and some predatory fungi.

13. A *lichen* is a composite of a fungal species (usually an ascomycote) and a species of algae or cyanobacteria. The alga provides products of photosynthesis and probably is protected by the fungal hyphae from adverse environmental conditions.

14. The various fungal groups may have arisen independently from prokaryotes.

KEY TERMS

aerial hypha
Ascomycota
ascus
basidiocarp
Basidiomycota
basidium
Chytridiomycota
coenocytic
conidium
Deuteromycota
dikaryotic
ergot
extracellular digestion
fruiting body
Fungi Imperfecti
gametangium
haustorium
heterokaryon
heterokaryosis
hypha
lichen
mycelium
mycorrhiza
Oomycota
penicillin
pheromone
rhizoid
saprobe
septum
sporangium
Zygomycota

QUESTIONS

1. Why are the fungi called the great decomposers? What features contribute to their success?

2. In what ways do fungi resemble plants and not animals?

3. In what ways do fungi resemble animals and not plants?

4. The main "body" of a fungus is usually hidden underground or within living or decaying organisms. What is the structure and function of this body?

5. How do heterokaryons arise?

6. Fungi are nonmotile, yet their spores can disperse over thousands of miles. How is this accomplished?

7. Fungi can produce spores of two

types that have two very different functions. What are these functions?

8. What features distinguish the lower from the higher fungi?

9. What are the six divisions of fungi? Which of the following features are found in each division?
a. Motile gametes
b. Motile zoospores
c. Cellulose cell wall
d. Septate hyphae
e. Sexual reproduction
f. Mycorrhizal association with plant roots

ESSAY QUESTION

1. Do lichens represent a symbiotic relationship or a parasitic one? Give evidence to support your point of view.

For additional readings related to topics in this chapter, see Appendix C.

23

Plants: The Great Producers

The migration of plants from the seas to the dry land surfaces of the earth more than 400 million years ago ranks among the most important events in the history of the biological world.

Sara A. Fultz, in *Botany* (1983)

The majority of the earth's landscapes—whether alpine meadow, tropical rain forest, rolling farmland, tall stand of timber, scorched desert, urban park, or ocean surface—have a common denominator: Their primary biological component is plants. These photosynthetic organisms can be microscopic and single-celled, like many species of green algae. They can be stately giants like redwood and sequoia trees. Or they can bear flowers and fruit, in an endless variety of colors, fragrances, and forms.

Regardless of individual locale and appearance, plants are collectively the great producers: autotrophs that convert solar energy into fixed carbon that supplies their own energy needs and also supports most other living organisms, directly or indirectly. They also represent an evolutionary spectrum ranging from some of the earliest eukaryotic cells to the first large organisms that survived

". . . out of that fertile slime springs this spotless purity!" Water lilies. Henry D. Thoreau, *Journal*, vol. 283 (June 19, 1853).

on dry land to the diverse modern groups that we see so abundantly around us—and that we rely on so completely for our fruits, vegetables, beverages, spices, wood, grains, fibers, and many of our chemicals and drugs.

This chapter surveys that broad spectrum, including:

- The major characteristics of all plants and the main branches of the plant kingdom
- The basic plant life cycle, with its alternating generations
- The algae, diverse aquatic producers
- The challenges of land and its early colonization
- The nonvascular land plants—the mosses, liverworts, and hornworts
- The seedless vascular plants—the whisk ferns, club mosses, horsetails, and ferns
- How seed plants may have evolved
- The gymnosperms—plants with cones and seeds
- The anthophytes—plants with flowers, fruit, seeds, and a close interrelationship with animals

THE PLANT KINGDOM AND ITS MAJOR DIVISIONS

Because of its great diversity, the plant kingdom is far from easy to define. Plants usually are multicellular organisms; most contain chlorophyll and thus produce their own organic nutrients; they are usually stationary; and most have alternating haploid and diploid phases in the life cycle. Despite these general characteristics, some organisms classified as plants are single-celled; some lack chlorophyll; some are motile; and some show no alternation of phases.

Botanists have identified about 300,000 living plant species. Of those, more than 80 percent are the familiar flowering plants, which include all the major crops and hardwood trees and most ornamental plants; less than 5 percent are lower plants, while about 15 percent are pines and their relatives. Table 23-1 lists the major divisions in the plant kingdom and the approximate number of species in each. The divisions Chlorophyta, Rhodophyta, and Phaeophyta are referred to as algae, and many taxonomists now classify them as protists. The remaining divisions are considered land plants.

Some alternative classification schemes place the algae in the kingdom Protista because some algae are single-celled. Simple land plants that lack an internal system for conducting water and nutrients—a *vascular system*—are often placed in a single division, Bryophyta. These so-called **nonvascular plants** are small and low-growing, but are so physically distinct that our classification system places them in three classes: Hepaticae (liverworts), Anthocerotae (hornworts), and Musci (mosses).

All the remaining groups are *vascular plants*—plants with a specialized vascular system for transporting water

Table 23-1 MAJOR GROUPS OF THE PLANT KINGDOM*

Number of Living Species	*Division/Class*				
7,000	Chlorophyta (green algae)				Algae
4,000	Rhodophyta (red algae)				
1,500	Phaeophyta (brown algae)				
16,000	Bryophyta			Nonvascular Plants	Land Plants
6,000	Hepaticae (liverworts)				
100	Anthocerotae (hornworts)				
9,500	Musci (mosses)				
13	Psilophyta (whisk ferns)		Seedless Plants	Vascular Plants	
1,000	Lycophyta (club mosses)				
15	Sphenophyta (horsetails)				
12,000	Pteridophyta (ferns)				
100	Cycadophyta	Gymnosperms	Seed Plants		
1	Ginkgophyta				
3 (genera)	Gnetophyta				
550	Coniferophyta				
235,000	Anthophyta	Flowering Plants			
170,000	Dicotyledoneae				
65,000	Monocotyledoneae				

*Certain classification schemes other than the one we follow in this book categorize the divisions of algae differently. Some place the Chlorophyta, Rhodophyta, and Phaeophyta in the kingdom Protista, not in the kingdom Plantae.

and nutrients. Older classification systems generally placed all the vascular plants in one division (Tracheophyta), but since recent research has revealed significant differences among them, our system assigns them to nine divisions, four of which produce no seeds and five that do (see Table 23-1).

THE BASIC PLANT LIFE CYCLE

Despite their tremendous diversity, plants have just one basic type of reproductive life cycle that distinguishes it from the other four kingdoms of life. The plant life cycle is characterized by the *alternation* of a haploid phase and a diploid phase (Figure 23-1). Since gametes (eggs, sperm, or analogous structures) are formed during a plant's haploid phase, the haploid organism is called the **gametophyte,** or gamete-producing plant. Since spores are formed during the diploid phase, the diploid plant is called the **sporophyte,** or spore-producing plant. In some plants, both haploid and diploid forms are free-living green organisms of substantial size. In other plants, one place dominates and becomes the recognizable plant, while the other is reduced and inconspicuous and depends on the dominant one for moisture, nutrition, and physical support. Since haploid and diploid phases are often referred to as generations, the alternating life cycle of plants is called the **alternation of generations.**

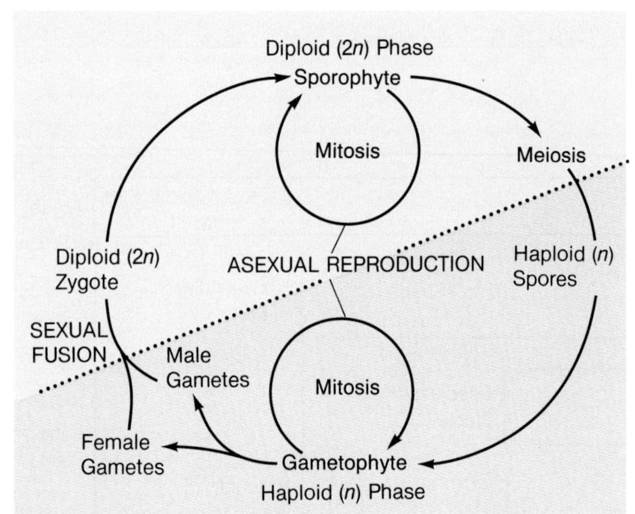

Figure 23-1 GENERALIZED PLANT LIFE CYCLE: ALTERNATION OF GENERATIONS.
Diploid and haploid generations alternate, separated by the events of meiosis and sexual fusion (the union of gametes). Plant life cycles differ mainly in the length of the two phases; for instance, the sporophyte stage can be very short. Both the sporophyte and gametophyte can reproduce asexually through mitotic divisions.

The plant life cycle differs from that of many animals in important ways. Plants can undergo direct, asexual reproduction by propagation from stems, roots, bulbs, leaves, or other specialized structures, whereas most animals cannot. Related to this is the absence of a plant germ line, the lineage of cells that yields gametes in animals; in plants, a given cell type can give rise to stems, leaves, flowers, and gametes. Plants have continuous organogenesis—production of new leaves, stems, and roots throughout adult life, whereas in most animals the number of eyes, limbs, and antennae is set, and no new organs arise after embryonic development. Plants can also enter an extended resting stage to survive periods of drought, cold, or other conditions unfavorable for growth and reproduction, whereas most animals cannot.

Let's begin our survey of the plant kingdom now with the ancestors to the plants, the three divisions of algae.

THE ALGAE: DIVERSE AQUATIC PRODUCERS

Algae display a stunning variety of shapes, sizes, and colors and are important in both evolutionary and ecological terms. **Algae** are an ancient group, and whether classified as plants or protists, they are believed to have been the ancestors of the other aquatic and land plants. The earliest organisms on earth, 3.5 billion years ago, were most likely anaerobic heterotrophs (see Chapter 19), but eventually, these organisms gave rise to single-celled autotrophs (much like modern chloroxybacteria) that produced their own food molecules through photosynthesis. Perhaps another billion years crept by before eukaryotic autotrophs emerged with their membrane-bound organelles, and these photosynthetic aerobes almost certainly gave rise to the first primitive algae, about 800 million years ago.

Over many millions of years, thousands of algal species evolved with the ability to gather sunlight at different water depths, depending on their specific chlorophylls and accessory pigments.

Huge numbers of algae inhabit the oceans and form an integral part of most aquatic ecosystems. Unicellular algae are primary producers in nearly all aquatic food chains. They produce nutrient molecules through photosynthesis and serve as food for a great variety of marine and freshwater nonphotosynthetic protists and animal consumers. Many algal species, especially large-bodied ones, also yield raw materials that people use as food, fuel, glue, and other products, as we shall see.

Despite their tremendous variation in size and shape, few algae exhibit any differentiation of specialized tissues. This is because their aquatic habitat provides support, prevents evaporation, and allows each cell to directly absorb nutrients and exchange gases with the watery environment. Nevertheless, algal reproduction and the relative development of haploid and diploid generations reveal significant evolutionary trends and help us understand specializations in the more complex plants.

Chlorophyta: Likely Ancestors of the Land Plants

Some ancient green algae growing in an ocean inlet or a freshwater pond were probably the ancestors of the lineage from which all land plants evolved more than 400 million years ago. Today, there are 7,000 species of green algae in the division **Chlorophyta.** A few inhabit the oceans, but most live submerged in fresh water or exposed on moist rocks, tree trunks, or soil. Most green algae are small, simple organisms shaped like filaments, hollow balls, or flat blades or sheets. Some marine forms, however, may range up to 8 m in length and 25 cm in width. Green algae store food as starch and possess both chlorophylls *a* and *b*—characteristics shared only with the land plants and major evidence for regarding green algae as land plant ancestors.

Siphonous Algae

Individual cells of green algae usually have a single nucleus, but some species have cells containing many nuclei in one large mass of cytoplasm. This group, the siphonous algae, includes plants like *Bryopsis,* which grows attached to rocks in clear, shallow water. A siphonous alga starts life as a single cell, but as it grows, the nucleus divides and redivides, while the cytoplasm does not. The result is a large coenocytic organism.

Chlamydomonas *and Other Volvocine Algae*

The best-known green alga is *Chlamydomonas,* an oval, motile, haploid alga propelled by two flagella. This common organism lives in moist soils and freshwater pools, and each haploid cell tends to contain a single large cup-shaped chloroplast; sometimes an eyespot, or *stigma;* and starch-producing organelles called *pyrenoids.*

Chlamydomonas can reproduce asexually by dividing into daughter cells that become motile **zoospores,** which mature into haploid adults. Under stressful environmental conditions, *Chlamydomonas* can reproduce sexually, by producing a thick-walled diploid zygote, or **zygospore,** that sinks to the bottom of a pond until favorable environmental conditions return. The zygospore then divides meiotically to produce haploid spores that develop into adults, completing the life cycle.

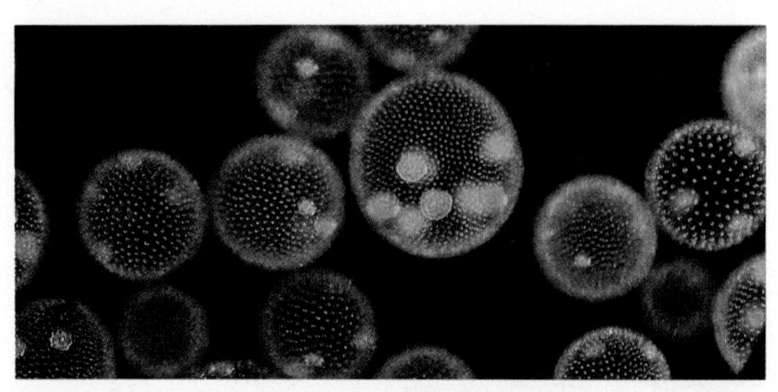

Figure 23-2 COLONIAL ALGAE.
Members of the volvocine line of colonial algae are made up of individual *Chlamydomonas*-like cells connected to each other by cytoplasmic strands. Each *Volvox* colony is a sphere with a single layer of cells embedded in a gelatinous matrix. As seen here, new daughter colonies form inside the hollow sphere (magnified about 15×).

Chlamydomonas is part of a lineage of algae in which individual cells sometimes function together in colonies of from 4 to 50,000 cells. The largest such colonies are *Volvox,* a bright green cluster that spins in the water and sparkles like the facets of an emerald (Figure 23-2). In fact, motile colonial green algae are called the *volvocine algae* after this jewel-like colony. In modern volvocine algae, an increase in colony size is accompanied by an ever-greater degree of cell coordination, interdependence, and specialization and by progressively more complex types of reproduction. *Gonium,* for example, consists of just 4 to 32 cells, and each can reproduce daughter colonies asexually or can form isogametes. In the larger colony *Pleodorina* (32, 64, or 128 cells), only the large posterior cells take a regular part in reproduction. Some species of this genus exhibit oogamy—the development of true oocytes, or large, immobile egg cells—in addition to producing swimming male gametes. *Volvox* is by far the most complex (see Figure 23-3), with 500 to 50,000 cells connected by tiny strands of cytoplasm. Only a few of the cells are exceptionally big and are specialized as oocytes. Like the other colonial algae, *Volvox* can also reproduce daughter colonies asexually inside the parent colony's central cavity.

Filamentous Algae

Another line of green algae, the **filamentous algae,** shows specialization in body form and in the alternation of generations. One common type is *Ulothrix,* a genus of mostly threadlike freshwater algae. While the threadlike chains of cells often reproduce asexually, adverse conditions can trigger sexual reproduction, during which one or more cells of the *Ulothrix* filament can act as a gametangium that produces flagellated isogametes. These fuse (as in *Chlamydomonas*) to form a diploid zy-

(a)

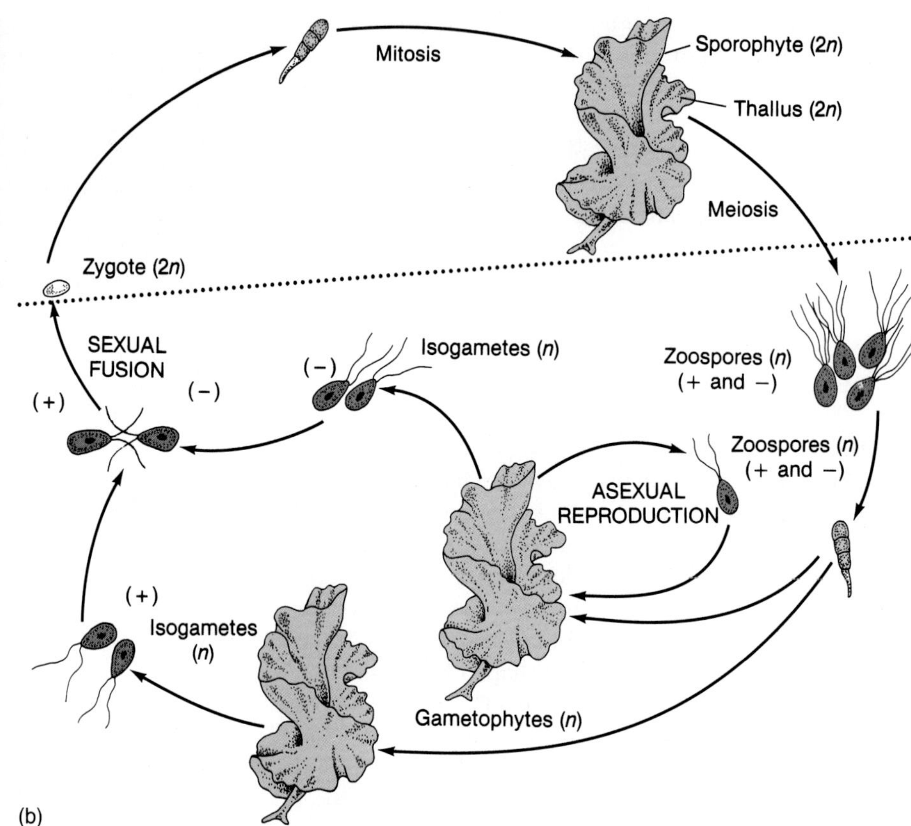

(b)

Figure 23-3 LIFE CYCLE OF THE SEA LETTUCE, *Ulva*.
(a) The sporophytes and gametophytes of *Ulva lectuca* are morphologically identical, consisting of ultrathin sheets of cells.
(b) Sexual fusion of isogametes. The cells fuse, yielding the zygote and in turn the new sporophyte.

gote, develop into a resting cell (the sporophyte), and eventually release four haploid zoospores that grow into new threadlike multicellular gametophytes, completing the cycle.

A multicellular green alga with an even more complex body type is sea lettuce, or *Ulva*, a delicate ocean-dwelling alga common in tidal pools (Figure 23-3). *Ulva* can reproduce either asexually or sexually and is the first alga we discuss in which both gametophyte and sporophyte are multicellular, as they are in the land plants. Each generation is a conspicuous, free-living organism—a nearly identical flat sheet of cells, a *thallus*, that undulates gracefully in the ocean's swells. This trait of alternating multicellular generations, as well as the production of chlorophylls *a* and *b* and other characteristics, leads botanists to believe that green algae of unknown types almost certainly gave rise to all the land plants.

Rhodophyta: Photosynthesizers in Shallow and Deep Waters

Most of the 4,000 or so species of red algae in the division **Rhodophyta** are small, delicate marine organisms that live in shallow or deep tropical waters. But several genera inhabit freshwater streams, lakes, and springs, and one genus even thrives in antarctic waters. Most are made up of many cells connected end to end in thin filaments (Figure 23-4) or side by side in thin, flat, leaflike sheets.

Red algae have a variety of pigments that absorb well all wavelengths of visible light. These pigments, called **phycobilins,** include the red *phycoerythrin* and the blue *phycocyanin* and enable some red algal species to live as the deepest photosynthesizers. Biologists long considered 150 m to be the maximal depth for any photosynthetic organism, but researchers recently found one species of red algae thriving at 268 m below the ocean surface. Red algae contain chlorophyll *a* and *carotenoids*, and some species also contain chlorophyll *d* rather than chlorophyll *b*. Red algae store carbohydrates in the form of *floridean starch*—layered starch granules built from glucose molecules and located in the cytoplasm, rather than in chloroplasts.

People extract materials from red algae for making laboratory *agar* and the stabilizing agent *carrageenan* used in paints, cosmetics, ice cream, and other applications.

Figure 23-4 THE EXQUISITE CELLULAR ARCHITECTURE OF A RED ALGA.
Most red algae have a filamentous, branched morphology as seen here (magnified about 60×).

Phaeophyta: Giant Aquatic Organisms

Many brown algae are fairly small, but the division **Phaeophyta** also includes the giants of the algal world—enormous kelps that can reach 100 m in length. Commonly called seaweed, the brown algae number about 1,500 species, many of which inhabit cool offshore waters. *Sargassum,* however, is found in warm tropical seas and sometimes forms huge floating masses.

The simplest brown algae have branched, filamentous bodies, but many have leaflike **fronds** that collect sunlight and produce sugars; a stemlike **stipe** that provides vertical support; and a **holdfast** that anchors the plant to the bottom, much like roots (Figure 23-5). Kelps also have localized regions of rapid cell division, similar to growth zones in vascular plants, as well as conducting tissue (a bit like the phloem in plants) that pipes the products of photosynthesis from parts of the alga receiving plentiful sunlight to those in deeper water. These structures represent important examples of parallel evolution, analogous to organs in complex plants.

Kelps growing in waters up to about 30 m deep overcome the problem of poor light penetration with long stipes and fronds that extend along the brightly illuminated ocean surface. Kelps also have unique brownish carotenoid pigments called *fucoxanthins* that absorb the deep-penetrating blue and violet wavelengths.

Brown algae store food reserves in the form of *laminarin,* a mixture of polysaccharides, lipids, and the sugar mannitol, and many species concentrate the important nutrients nitrogen, potassium, and iodine from the surrounding water. People sometimes farm kelps commercially for human food, livestock fodder, and fertilizer and extract useful materials such as alginic acid as a stabilizer, emulsifier, and waterproofing substance.

The life cycles of most brown algae, such as the well-known *Laminaria,* are similar to that of the green alga *Ulva* in that both alternating generations are multicellular. The brown alga *Fucus* and its close relatives, however, have a life cycle that resembles an animal's. In it, the main plant body, a greenish brown thallus, is diploid; it develops special structures at the tips of the

(a)

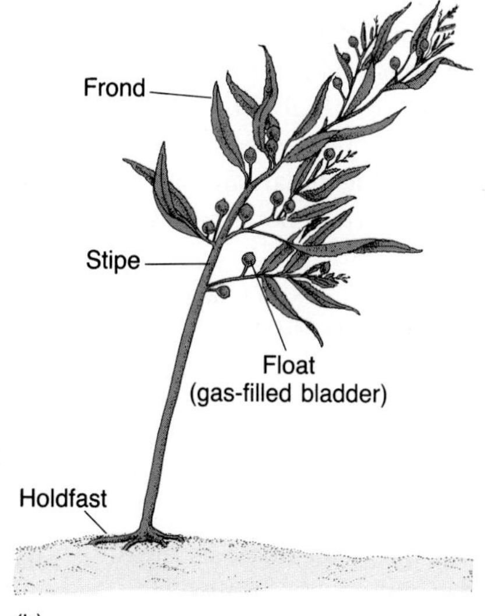

(b)

Figure 23-5 *Sargassum,* BROWN ALGAE.
(a) A sea otter finds protection and food in a giant kelp forest off the central California coast.
(b) The major structures of a specimen of a brown alga are indicated in this typical *Sargassum* and include the light-gathering fronds, the stipe, and the holdfast.

branches—sperm- or egg-producing organs that give rise to gametes. As in animals, the *Fucus*'s haploid phase is so reduced that these gametes are the only representatives. *Fucus* lacks the plant hallmark of alternating generations, perhaps underscoring their possible membership in the kingdom Protista.

PLANTS THAT COLONIZED LAND

The challenges of life on land and the diversity of distinctive habitats there allowed for greatly increased complexity and numerous adaptations in the descendants of the algae that left the oceans.

Plants apparently had colonized the land by the early Devonian period, about 400 million years ago (Figure 23-6), when the climate was warm and moist, episodes of mountain-building activity alternated with erosion, and the vast inland seas drained and refilled again and again. This may have placed a survival premium on the ability of any plant type to cope with changing environments. Plant biologists hypothesize that multicellular gametophytes arose that could produce gametes in response to adverse conditions and that when such gametes united, the resulting zygote could ride out the environmental stress. Some botanists think that such drought-tolerant resting zygotes, perhaps retained on the parental gametophyte body, gave rise to the sporophyte phase of the life cycle of the land plants.

Since land plants and green algae share the same photosynthetic pigments and storage products (chlorophylls *a* and *b* and starch granules), botanists believe that the drought-tolerant ancestors of the land plants were closely related to multicellular green algae (see Figure 23-6). All land plants share certain additional features: They have a relatively impervious jacket of cells surrounding the reproductive structures that acts to preserve a sterile environment inside; both gametophyte and sporophyte generations are multicellular; and openings for exchange of gas and moisture penetrate the *epidermis* (outer layer of the plant body).

Life on land presented stiff challenges not faced by aquatic plants, and desiccation was the first and most serious. A waterproof outer layer arose in most groups, and *stomata*—tiny pores through that impervious layer—allowed the controlled exchange of water vapor, carbon dioxide, and other gases. Such adaptations were already present in fossils of the genus *Cooksonia*, from the early Devonian. Second, support for the plant body also was essential, since without the buoyant effect of water, land plants are especially subject to gravity. Thick-walled support tissues arose first, and true vascular tissue later, capable of conducting water and nutrients to every cell of the multicellular plant body and providing strength for vertical support. Third, plants required a means of absorbing moisture and minerals from the soil and transporting them throughout the plant, and root systems evolved with large surface areas and the ability to absorb water, ions, and molecules. Finally, a new means of reproducing sexually evolved, since unprotected gamete cells could no longer swim or be transported in the water.

Recent fossil findings show that land fungi apparently coevolved with land plants. Fossils from the Triassic period show complex mycorrhizal structures (see Chapter 22), very much like those found in root cells today. The symbiotic pairing of fungus and root may have allowed both fungi and plants to survive and invade the relatively sterile soils of the early continents.

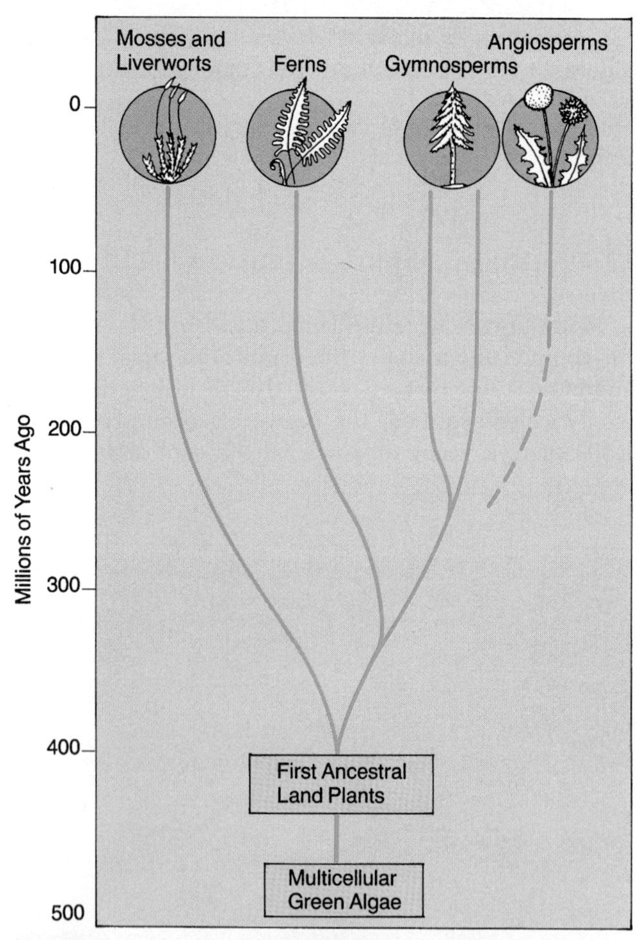

Figure 23-6 EVOLUTION OF THE MAJOR PLANT GROUPS.
The approximate time various plant groups arose is indicated on this geological time scale. The four types of gymnosperms—pines and relatives—are not closely related. The source of the flowering plants is not yet known.

Nonvascular Land Plants: Bryophyta

Liverworts, hornworts, and mosses are land plants that lack vascular tissue for conducting water and nutrients. Botanists have traditionally classified these nonvascular land plants in a single division, **Bryophyta,** but because the plants are so different from one another, some biologists divide them further into three separate classes: **Hepaticae** (liverworts), **Anthocerotae** (hornworts), and **Musci** (mosses).

All three types of nonvascular land plants are small and usually grow in moist places—in marshes and swamps and on damp rocks or logs, shady forest floors, stream banks, and sometimes the north side of trees. The early nonvascular and vascular plants probably diverged shortly after plants colonized the land and evolved with separate and very different solutions to the challenges of life out of water.

Hepaticae: The Liverworts

Liverworts are small plants that often grow horizontally over the soil surface as flat thalli or, in a few species, grow in water. Liverworts survive only in moist environments because (1) their stoma-like pores are not flanked by regulatory cells, and thus moisture escapes easily; and (2) while they are attached to the soil by hairlike rhizoids, they lack a true root system and usually absorb water and nutrients directly through the entire lower surface of the gametophyte. Most liverworts also lack water-conducting tissue.

The liverwort's conspicuous gametophyte generation bears a sperm-producing chamber, or **antheridium,** an egg-producing chamber, or *archegonium,* or both.

During rain or heavy dew, sperm are splashed or washed from a mature antheridium into the vicinity of an archegonium, swim down the neck of the chamber, and fertilize an egg. The resulting diploid sporophyte remains attached to the gametophyte during its entire growth, rather than being nutritionally independent and physically separate from the gametophyte, as in the green algae *Ulva* or *Ulothrix.* Spores produced by meiosis in the sporangium of the mature sporophyte are released and grow into new haploid gametophytes. Asexual reproduction is also fairly frequent in liverworts and involves the appearance of small *gemma cups*—analogous to flower buds—on the thallus.

Anthocerotae: The Hornworts

Hornworts are also inconspicuous residents of moist, shady places and are named after their spore cases. The gametophyte produces gametes in antheridia and archegonia, and when these gametes unite, the zygote develops into a sporophyte that remains attached to the gametophyte (Figure 23-7). The diploid sporophyte produces a long, cylindrical sporangium that splits as it dries and twists into "horn" shapes. This drying and twisting expels the spores, which develop into new haploid gametophytes. Hornworts have internal cavities filled with mucilage and inhabited by symbiotic cyanobacteria that fix nitrogen in organic compounds.

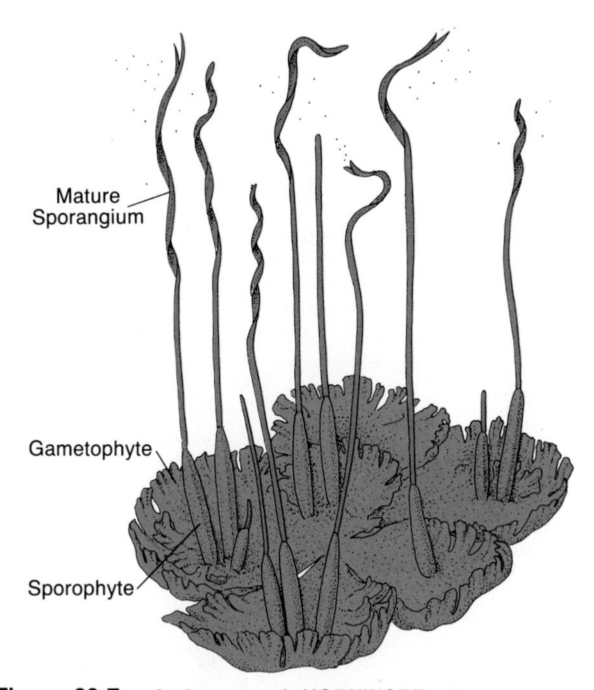

Figure 23-7 ***Anthoceros,* A HORNWORT.**
The hornwort sporophyte remains embedded in the gametophyte. Drying and twisting of the many sporangia create "horns," which split and release the spores.

Musci: The Mosses

By far the most numerous and familiar nonvascular land plants, mosses form velvety carpets and spongy beds almost anywhere the air is moist and clean. Moss spores germinate into a tiny threadlike protonema (Figure 23-8a) that bears a striking resemblance to a filamentous green alga, the probable ancestor of all land plants. The growing gametophyte forms the familiar moss plant that carpets moist soil and rocks (Figure 23-8b). As the plant reaches maturity within a few weeks, an antheridium or archegonium forms at the tip of each gametophyte. As in the other bryophytes, sperm must swim to the egg inside the archegonium, and thus reproduction requires heavy dew, rain, or spray from a nearby watercourse. A zygote inside the archegonium develops into a nonphotosynthetic sporophyte that resembles a minute golf club and depends on the gametophyte for nutrition.

Like liverworts, moss gametophytes develop rootlike rhizoids, but the entire lower portion of the plant, including the gametophyte's small leaflike structures,

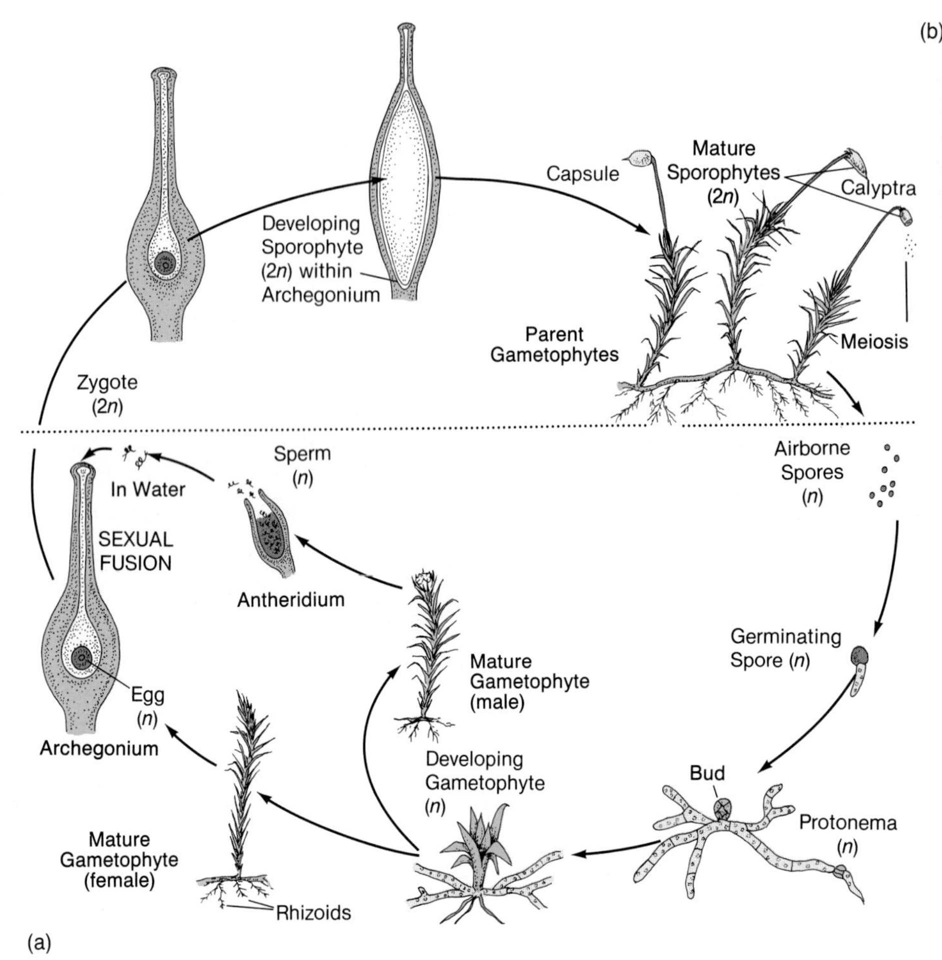

Figure 23-8 GENERALIZED LIFE CYCLE OF A MOSS.
(a) Leafy gametophytes develop from the haploid protonema and at the leaf tips produce antheridia and archegonia. On contact with dew or raindrops, the antheridia burst and release swimming sperm. After fertilization, the diploid zygote remains within the archegonium and divides mitotically. The sporophyte then develops, and following meiosis, produces haploid spores. The calyptra, the protective coating surrounding the capsule, falls off, exposing the capsule, which then ruptures, dispersing the spores. (b) The small, green, leafy structures are moss gametophytes (here, *Bryum capillare*). They produced the sperm and eggs that gave rise to the erect yellowish green sporophytes.

takes up water and minerals. Unlike liverworts, however, moss gametophytes have two specialized structures reminiscent of the major conducting vessels of vascular tissue in more complex plants, namely, elongated cells called *hydroids*, which aid in water transport, and *leptoids*, which transport nutrients.

The nonvascular land plants we have just described—the liverworts, hornworts, and mosses—are a fairly large and diverse group, but they are limited to small size and moist habitats by their swimming sperm, their lack of a true vascular system, and the absence of stomata that can close to prevent moisture loss. The next group of land plants evolved vascular tissues, which allowed larger size and survival in drier habitats. Nevertheless, they are still limited to seasonally moist areas because they reproduce by means of swimming sperm and disperse by means of spores, not seeds.

Seedless Vascular Plants: Support and Transport, But No Seeds

Colonization of the land was most successful for plants that developed an efficient internal transport system and vertical support tissue—the **vascular plants.** A special type of thick hollow cell called a tracheid makes up the vascular pipelines, lending vertical support and transporting water. Vascular plants share four unique characteristics: (1) two types of specialized conducting tissues, *xylem*, which conducts water, and *phloem*, which conducts sugars and other nutrients; (2) a layer of waterproofing material, *cutin*, on the portion of the plant aboveground to reduce water loss; (3) multicellular embryos that are retained within the archegonium; and (4) a diploid sporophyte that is the dominant stage in the life cycle. Four divisions of vascular plants produce spores, not seeds: **Psilophyta** (whisk ferns), **Lycophyta** (club mosses and quillworts), **Sphenophyta** (horsetails), and **Pteridophyta** (ferns).

Psilophyta: The Whisk Ferns

The whisk ferns are an ancient plant group whose members resemble *Rhynia* and some of the other early colonizers of land. The members of one surviving genus, *Psilotum*, are small, branching plants that resemble whisk brooms (Figure 23-9a). The sporophyte of *Psilotum*, which is the conspicuous generation, is com-

(a) (b)

Figure 23-9 ***Psilotum nudum,*** **A WHISK FERN, AND** ***Lycopodium complanatum,*** **A CLUB MOSS.**
(a) The stemlike and branchlike parts of this whisk fern are clearly visible. Primitive vascular tissue extends throughout such structures. (b) Spore-producing strobili are clearly visible at the top of these lycopods growing in the leaf litter on the shady forest floor.

posed of only a stem, with no true roots or leaves, although it does have rhizoids and scalelike appendages. Members of the other surviving genus, *Tmesipteris*, have leaflike appendages that may simply be flattened branches rather than true leaves. In both genera, a lateral stem called a *rhizome*, with hairlike rhizoids, serves as an absorptive and anchoring organ. Mycorrhizal fungi are usually present in the rhizoids and serve to increase nutrient availability to the whisk fern.

The whisk fern's sporophyte produces haploid spores in each sporangium, and these grow into small gametophytes that lack chlorophyll and live on decaying soil materials. At first the sporophyte is dependent on the gametophyte body for nourishment, but it eventually turns green and becomes autotrophic.

Lycophyta: The Club Mosses and Quillworts

The ancient club mosses and quillworts arose during the Devonian period, almost 400 million years ago. Club mosses and other lycophytes were probably the most numerous plants in the later Carboniferous period, and their decomposed and compressed tissue, collected in vast underground coal deposits, provides much of our fossil fuel.

The lycophytes have a conspicuous sporophyte and show the next advance in vascular plant architecture: *roots*, which not only anchor the plant, but also absorb water and minerals from the substratum, which can then move throughout the plant by means of conducting tissue. Club mosses and quillworts also have true leaves, since the appendages contain vascular tissue, and specialized **sporophylls** (spore leaves) which bear sporangia. In many species of the club moss genera *Lycopodium* and *Selaginella*, sporophylls are grouped on one axis to form a **strobilus,** or club-shaped, spore-producing organ (Figure 23-9b). Some lycophytes generate only identical **homospores,** which give rise to gametophytes that produce both antheridia and archegonia, while other species produce unidentical **heterospores**—either **microspores,** which differentiate into male gametophytes, or **megaspores,** which differentiate into female gametophytes. These plants still require water for reproduction, since sperm must swim to the archegonium in order to fertilize the egg.

Sphenophyta: The Horsetails

During the long Carboniferous period, when large portions of the earth's landmasses were covered by vast, steamy swamps, the horsetails were extremely common plants, growing globally in a wide range of sizes and forms and, like the lycophytes, forming many of our modern coal deposits. A single genus, *Equisetum*, remains today, and the common horsetail, or "scouring rush" (Figure 23-10), has stiff, abrasive stems containing silica crystals that people once used to scour pots.

The conspicuous plant is the diploid sporophyte. The shoots bear scalelike leaves in bright green whorls

Figure 23-10 ***Equisetum,*** **A HORSETAIL.**
This diminutive plant is a holdover from the Carboniferous swamps.

around joints in the rough stem. Many species produce a special reproductive stem called the *fertile shoot.* This tan or reddish stem is nonphotosynthetic and bears a terminal strobilus, which produces hundreds of identical spores in clusters of sporangia. The horsetails retain swimming sperm and depend on water for reproduction.

Pteridophyta

Ferns (division Pteridophyta) were also prominent during the Devonian and Carboniferous periods, when many types grew as large, treelike plants. In some tropical areas today, tall tree ferns still prosper, but the majority of the 12,000 modern fern species are low, graceful plants less than a meter tall. Ferns usually have lacy fronds, highly divided true leaves that grow from horizontal subterranean rhizomes and have vascular tissue that supplies water and nutrients to all parts of the leaf.

The leafy fern plant is the diploid sporophyte (Figure 23-11). In most species, sporangia are formed on the undersurface of the fronds and are grouped in structures called sori (singular, **sorus**). Virtually all fern species disperse similarly sized homospores into the air, sometimes forcefully, by a springlike mechanism in the sporangium wall. The haploid spores eventually settle and grow into tiny, thin, heart-shaped gametophytes, each of which is called a *prothallus.* The prothallus generally produces both antheridia and archegonia. It is free-living, photosynthetic, and has tiny rhizoids that imbibe water.

While ferns and other seedless vascular plants possess most of the basic characteristics required for success on land, their dependence on a film of water for swimming sperm has kept them from invading the drier continental areas. Only the next step in plant evolution, the seed, could break that final barrier to colonization of land. Keep in mind, though, that the hallmark of the plant kingdom—the alternation of generations—was present first in the algae and lower land plants, and these organisms are an important living chapter in biological history.

CLASSIFICATION OF THE SEED PLANTS

The five divisions of seed plants are **Cycadophyta** (cycads), **Ginkgophyta** (ginkgoes), **Gnetophyta** (gnetinas),

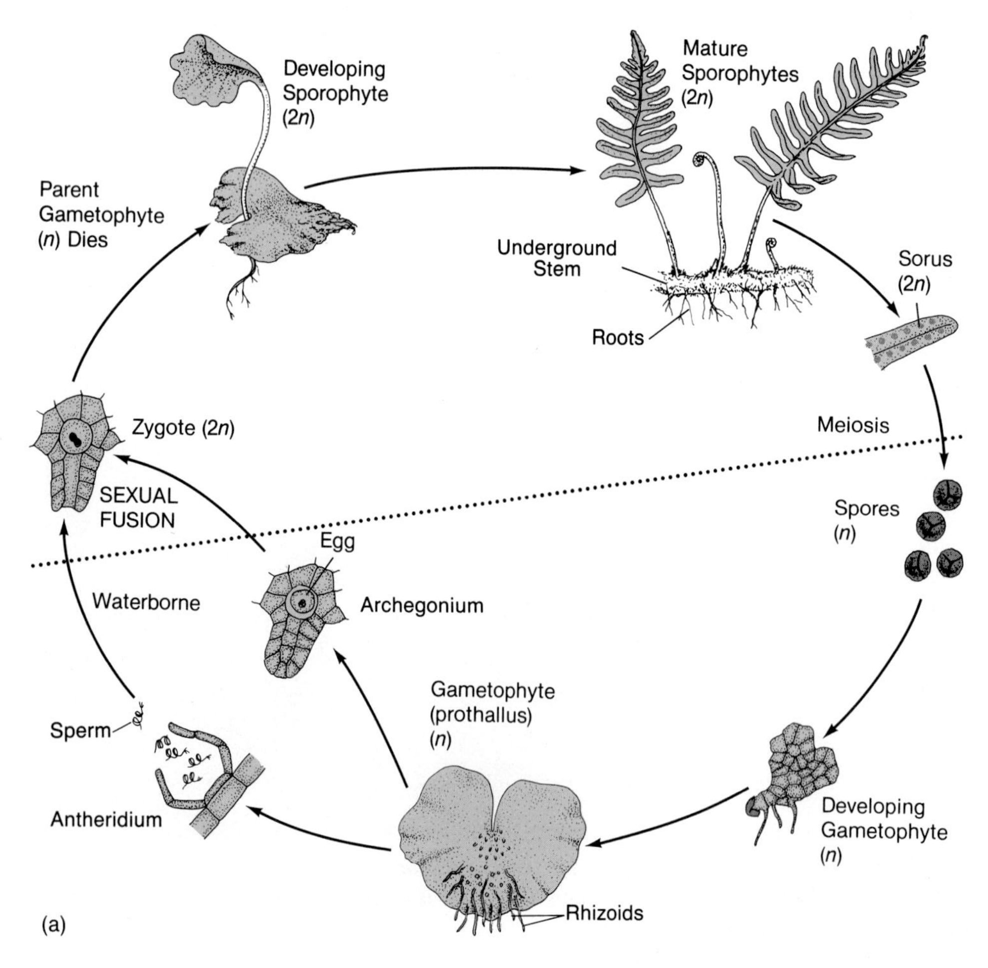

Figure 23-11 GENERALIZED LIFE CYCLE OF A TYPICAL FERN.
(a) Clusters of sporangia (called sori) underneath the mature sporophyte leaf rupture and liberate haploid spores. These develop into the gametophyte prothallus. Sperm produced in the antheridia fertilize the eggs within the archegonia. New sporophyte plants then develop, and the parental gametophyte prothallus withers and dies.
(b) The numerous sori are shown on the underside of this leaf of the fern *Polypodium vulgare.*

Coniferophyta (conifers), and **Anthophyta** (flowering plants) (see Table 23-1). Together, the members of the first four divisions are called *gymnosperms.* The fifth division—composed of the flowering plants—is the largest, with six times more members than all the other plant divisions combined. All seed plants share certain traits:

1. The *megasporangium,* the chamber in which meiosis takes place and the female gametophyte subsequently develops, is protected by one or two layers of skinlike tissue known as the *integument.*
2. Both male and female gametophytes are highly reduced, and the female gametophytes depend on the sporophytes for water, nutrients, and protection.
3. Sperm do not have to swim through a film of water to reach the egg, but reach it by other means.
4. Like ferns, most seed plants produce either needles or relatively large leaves with veins.

In addition, seed plants have the same type of life cycle and produce *pollen,* or movable male gametophytes; plant embryos; and, of course, **seeds**—complex units containing a developing embryo and stored nutrients.

EVOLUTION OF THE SEED PLANTS

It was during the long Carboniferous—the age of amphibians and coal-forming swamps—that the first seed-producing plants arose. Paleobotanists have found large numbers of fossilized ferns in layers of Carboniferous rock that appear to have produced seeds. These *seed ferns,* a now-extinct class of gymnosperms, were the first seed plants to evolve and dominated the continents for 70 million years.

Three more groups of gymnosperms appeared during the Permian: conifers, cycads, and ginkgoes—all likely descendants of a simple fernlike ancestor in a warm, swampy landscape. The earliest land plants, tied as they were to water for reproduction, were analogous to the amphibious animals; but these later groups of gymnosperms were more like the evolving reptiles—capable of withstanding drought and cold and of reproducing without abundant standing water. As the global climate cooled, the conifers, cycads, and ginkgoes gradually replaced the seed ferns. The "golden age of gymnosperms" stretched from 350 million to 100 million years ago and coincided with the rise and fall of the dinosaurs.

Beginning about 120 million years ago, flowering seed plants began to appear in higher, drier areas, apparent descendants of a low, shrublike gymnosperm. Within 20 million years, thousands of species of flowering plants had evolved and spread into the lowland areas formerly crowded by giant ferns, club mosses, cycads, ginkgoes, and conifers, populating much of the North American continent, while most of those more ancient species became extinct. Today, flowering plants dominate most ecological zones, and gymnosperms are more abundant only where the land is dry; the soil, poor; the average temperature, relatively low; or the elevation, high.

Let us begin our survey now of each group of seed plants.

GYMNOSPERMS: "NAKED-SEED" PLANTS

The term **gymnosperm** means "naked seed," and indeed, the cycads, ginkgoes, pines, spruces, redwoods, Douglas firs, and other gymnosperms have ovules and seeds that are exposed on the surface of the sporophyte. Despite this common trait, the major groups of gymnosperms are so different that they almost certainly evolved from separate kinds of ferns and are not closely related phylogenetically.

Cycadophyta

Cycads, or members of the division *Cycadophyta,* are a fascinating group of primitive seed plants native only to tropical and subtropical areas (Figure 23-12). A few spe-

Figure 23-12 THE CYCADS.
Cones of a cycad, the sago palm *Cycas media.* The tip of one cone is ruptured, revealing the individual red-colored seeds.

cies, like the sago palm, are popular house and garden plants. Cycads such as *Zamia pumila,* native to the sandy-soiled woodlands of Florida, resemble palms with their short, thick stems and crowns of finely divided leaves. True palms, however, are anthophytes.

Cycads grow as separate male and female plants, each producing a characteristic type of conelike strobilus near its apex. Because each adult cycad produces only one type of gametophyte, the species is said to be **dioecious** (literally, "two houses"). Cycads have a typical gymnosperm life cycle, which we will consider later in this chapter. Two particularly interesting details of the cycad cycle, however, are that cycad strobili and seeds can require up to 10 years to mature. And while the male **microgametophytes**—the pollen grains—are usually blown about by the wind, insects such as weevils are involved in pollen transfer in some species, inadvertently transporting pollen grains when they move from one strobilus to the next to feed. While animal-facilitated pollination is extremely common in flowering plants, it is rare in other seed plants, and this instance suggests that pollination may have originated in cycads millions of years ago.

Ginkgophyta: The Ginkgo Tree

Although **ginkgoes** grew abundantly throughout North America, Europe, and Asia some 25 million years ago, just a single species survives today: *Ginkgo biloba,* the "maidenhair tree" (Figure 23-13). The tree grows 25 m tall and produces graceful, fan-shaped leaves. It is **deciduous;** that is, like many anthophyte trees, its leaves fall in the autumn, after turning a luminous golden color.

The naked round seeds produced by the *ovulate,* or ovule-bearing, "female" tree have a fleshy outer layer that in late summer begins to overripen and produce butyric acid, a substance with an odor similar to that of rancid butter. For 5,000 years, Chinese healers have used ginkgo extracts as herbal medicines. Recently, American scientists isolated a complex compound from the leaves, ginkgolide B, which may help block asthma, counteract toxins, and help regulate blood pressure. Because ginkgoes are particularly resistant to air pollution, landscapers often plant the "male" trees in urban parks and along busy streets.

Gnetophyta: The Gnetinas

Gnetinas are a small but diverse group including tropical vines and trees, with large, leathery leaves, and densely branched shrubs that bear tiny leaves and grow in arid regions around the world.

Surely the oddest gymnosperms and one of the earth's strangest-looking plants is the gnetina called *Welwitschia mirabilis,* which grows in the Namib Desert of southwestern Africa. This plant has two large strap-shaped leaves that trail across the ground and blow in the wind, plus a huge, carrotlike taproot that penetrates several feet deep in the sandy soil of the Namib and can store gallons of water.

(a)

(b)

Figure 23-13 ***Ginkgo biloba:*** **THE ONLY SURVIVING SPECIES OF THE DIVISION GINKGOPHYTA.**
(a) A ginkgo growing in a New Jersey field. (b) A close-up view of the leaves and a ripe fruit of *Ginkgo.*

Coniferophyta: The Familiar Conifers

Cone-bearing plants, or **conifers,** grow in vast forests around the world and are often green year-round, although some, including certain larches, the bald cypress, and the dawn redwood, are deciduous (drop their leaves seasonally). In total, there are between 500 and 600 contemporary conifers species.

Conifers are our most important sources of paper, plywood, and other building materials. The larger species dominate millions of acres in cold, mountainous regions, and we often landscape with the smaller conifers. In addition, the tallest, oldest, and most massive plants on earth are conifers. One particular coast redwood *Sequoia sempervirens* growing in Northern California has reached almost 113 m (372 ft). One gnarled bristlecone pine *Pinus longaeva* in the high mountains of western Nevada has been estimated to be 4,900 years old. And the 2,500- to 3,000-year-old General Sherman Tree, a giant sequoia (*Sequoiadendron giganteum*) growing in California's Sierra Nevada range, has a circumference of 31.3 m (102.6 ft) at its base.

(a)

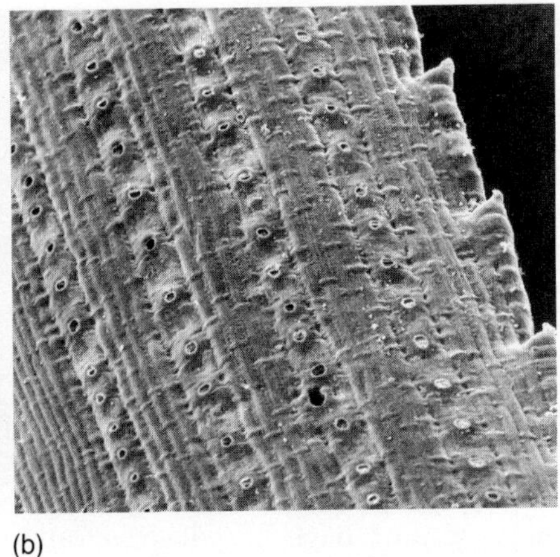

(b)

Figure 23-14 CONIFER LEAVES AND CONES.
(a) Conifer needles, two brown and green ovulate cones, and the small pollen-bearing cones (top) can be seen on this branch of a pine.
(b) The surface of a needle (magnified about 60×) from a Monterey pine is waxy and waterproof. The regular rows of stomata, looking like a mountain range of miniature volcanoes, are clearly visible.

Conifer leaves are often long, narrow, somewhat flattened needles with a thick, waterproof *cuticle*, or surface layer, that contains many stomata (Figure 23-14); all these modifications minimize water loss in dry environments. Needles also have internal vascular transport tissue and many contain *resins*—fragrant, gummy substances that discourage predators.

The life cycle of the pine, diagrammed in Figure 23-15, is typical of most conifers. The tall, conical adult tree is the diploid sporophyte and produces both male and

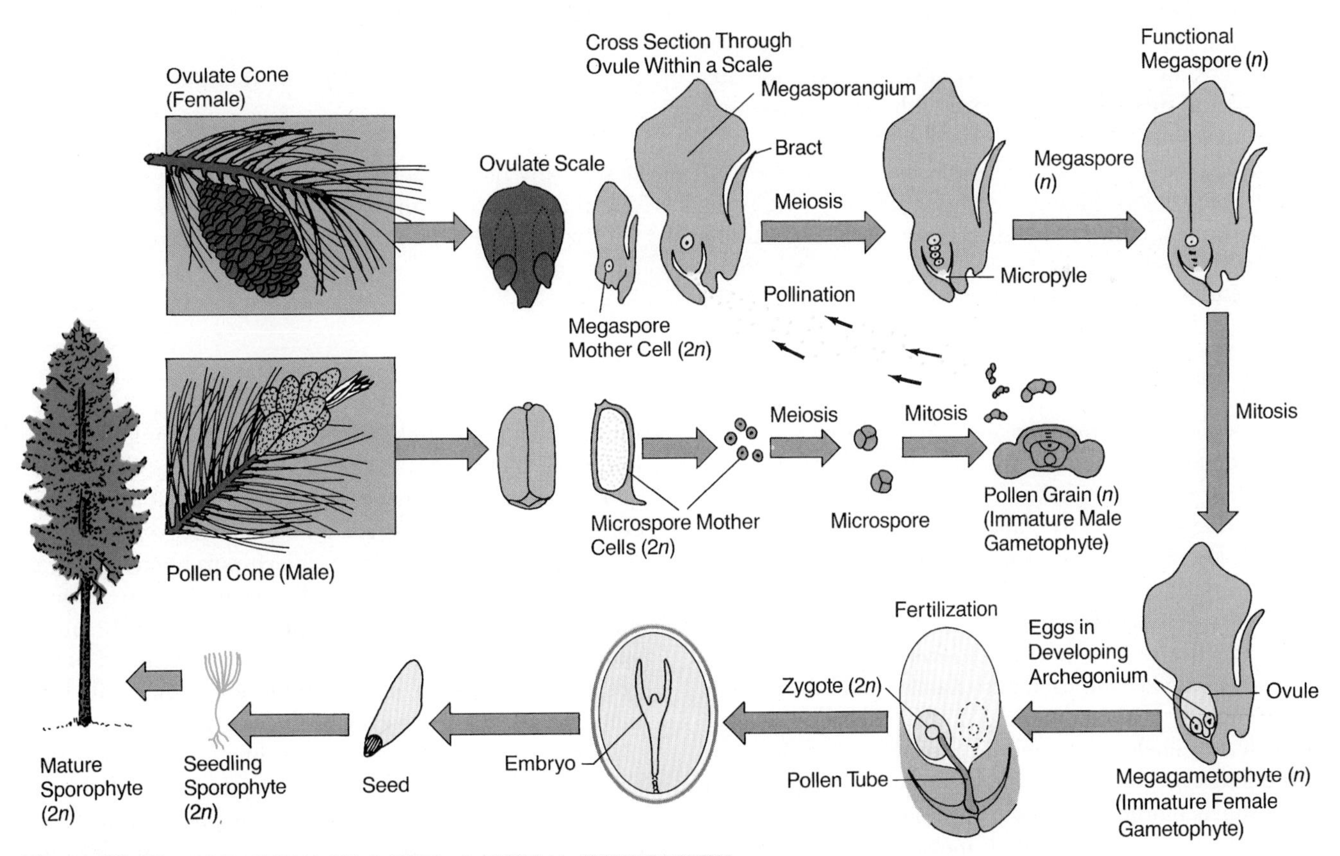

Figure 23-15 LIFE CYCLE OF A PINE, A TYPICAL GYMNOSPERM.
The mature sporophyte (tree) produces both ovulate (female) and pollen-bearing (male) cones. The pollen grains, or immature male gametophytes, develop from the microspores within the microsporangia of the male cones. When released, some pollen may chance to adhere to a female megasporangium (upper right). Each scale on a female cone is a megasporangium containing a megaspore mother cell. That cell yields haploid megaspores, which develop into archegonia, each housing one egg cell. A very slow process pulls the pollen grain's pollen tube through the micropyle. The pollen tube enables the sperm nuclei to fertilize the eggs within the archegonia. After fertilization, the ovule and its embryo within develop into the seed. The seed eventually germinates, and a seedling (sporophyte) is produced.

THE ORIGIN OF CORN: AN UNSOLVED MYSTERY

Corn is America's largest crop. It is also the most important food crop in the Western Hemisphere and a close third to wheat and rice as the largest annual harvest worldwide. Yet, the ancestry of this valuable staple is shrouded in mystery.

By the time Columbus landed in the New World, native Central American agriculturists had selected and bred more than 200 varieties of corn, including yellow, red, and blue field corn; yellow and white sweet corn; and popcorn. Many biologists agree that the indigenous people of the Americas must have begun selecting and breeding corn's progenitor between 8,000 and 15,000 years ago. But what *was* this progenitor?

Many biologists believe it was the grass called *teosinte*, which still grows wild in Mexico, Guatemala, and Honduras and bears a small spike of hard, triangular kernels (Figure A). Teosinte can form fertile hybrids with corn, and this indicates genetic closeness. Other biologists believe that a true wild corn species, now extinct, gave rise to both modern corn *and* teosinte. Still others hypothesize that the Indians crossbred the extinct wild corn with teosinte and that the resultant hybrid gave rise to corn.

The ancestry of corn may seem like an esoteric subject, but the question has relevance for the long-term survival of modern hybridized corn. Amazingly, more than 70 percent of the entire United States corn crop derives from only six parent lines. Thus, literally billions of corn plants are nearly identical—a lack of diversity that means that an insect, fungal, or viral pest could arise through mutation and devastate a large percentage of the crop in short order. If the ancestor to corn turns out to be teosinte instead of an extinct progenitor, then biologists would have a living reservoir of genes in the form of a sturdy Latin American weed that still grows wild and undergoes natural selection. Plant breeders hope that with advanced techniques of biotechnology, such genes could be transferred to modern hybrid corn to yield more vigorous plants.

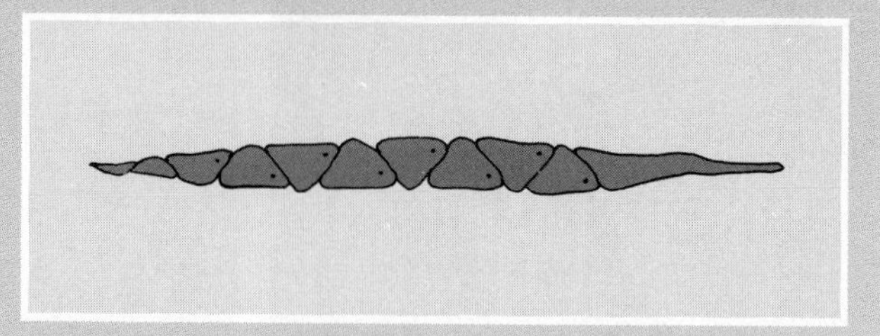

Figure A THE TEOSINTE SPIKE, ONE ROW OF HARD TRIANGULAR KERNELS.

Although corn's historical roots remain obscure for now, many biologists have called for the preservation of wild teosinte to protect what may be a priceless reservoir of wild genes until the mystery of corn's ancestry can be solved.

female cones (it is **monoecious**). A pine tree's ovulate cone is composed of tough, woody scales radiating from a central axis, and the "naked" ovules are thus protected. Below these scales are small, stiff structures called *bracts*. Ovulate cones range in length from 0.8 cm (white cedars) to as much as 61 cm (sugar pines) and also vary from woody to fleshy, as in the fleshy red berries of yews and fleshy blue berries of junipers.

Each **ovule** is composed of a chamber called the **megasporangium,** with an opening at one end called the **micropyle,** all enclosed in one or two outer coverings, or **integuments.** Within the megasporangium, a diploid **megaspore mother cell** divides meiotically to produce four relatively large haploid megaspores. The three megaspores nearest the micropyle degenerate, leaving one functional haploid megaspore, which divides mitotically and develops over a year or so into the multicellular, haploid female gametophyte, which is nonphotosynthetic, is nourished by the sporophyte, and develops into two or more archegonia, each containing one haploid egg.

Pollen-bearing pine cones are composed of soft tissue consisting of a central axis surrounded by sporophylls. Each sporophyll bears two **microsporangia,** inside of which diploid **microspore mother cells** give rise through meiotic divisions to haploid microspores. Each microspore divides mitotically to produce a four-celled **pollen grain,** an immature male gametophyte that will later mature and deliver sperm to the eggs.

Pollen-bearing cones produce huge numbers of pollen grains in the spring, often in golden clouds blown about by spring breezes. Individual pollen grains that chance to fall on an ovulate cone and sift down into the open scales become fixed to a sticky fluid secreted by the maturing ovule and exuded through the micropyle. This reception of pollen by the ovulate cone is called **pollination.** The sticky fluid later dries and contracts, thereby pulling the pollen through the micropyle and inside the ovule's integument. At the time of pollination and **germination,** or reactivated growth of the pollen grains, the megasporangium has not yet produced a megaspore. Megasporogenesis does not occur until about a month later.

Following pollination, the developing male gametophyte matures over the next 15 months, digesting a path through the integument and extending a pollen tube toward the developing female gametophyte and one of the archegonia that it will contain at maturity. When the pollen tube reaches the egg cell in the archegonium, much of its cytoplasm and two nonmotile haploid sperm nuclei are transferred into the egg. Fertilization occurs when one of the sperm nuclei fuses with the egg nucleus. The other sperm nucleus then degenerates.

In a typical conifer, well over a year passes between pollination and fertilization. During the second summer, the newly fertilized egg (zygote) develops into an embryo inside the ovulate cone. Although each **megagametophyte** contains several archegonia and the egg that each produces is often fertilized, usually just one embryo develops fully. The one successful embryo enlarges by mitotic divisions within the ovule to form a tiny plant, the new sporophyte, with an embryonic root and several embryonic leaves. The ovule's integument hardens and develops into the protective *seed coat.* The female gametophyte tissue remains as a food source for the developing embryo. By late in the second summer, the scales of the new woody ovulate cone open, and the mature seeds fall to the ground. The seeds that survive unscathed lie dormant through the fall and winter; the following spring, a small percentage germinate successfully and grow, completing the life cycle.

ANTHOPHYTA: THE APEX OF PLANT DIVERSITY

While gymnosperms are most common in cold, mountainous regions, the flowering plants (formerly called *angiosperms;* division **Anthophyta**) are the most common and conspicuous species in tropical and temperate regions. More than 95 percent of all living seed plants and over 80 percent of all the plant species on earth are flowering plants. They range in size from the giant *Eucalyptus marginata,* which may be more than 100 m tall, to some duckweeds like *Wolffia microscopica,* which is barely 1 mm across. And the flowering plants provide virtually all our food crops, our herbs and spices, beverages, much of our cloth, animal fodder, dyes, and medicines.

Angiosperms, the former name for the flowering plants, means "seed in a container," which refers to the specialized structure that encloses the ovules and the seeds of all members of this division. There are two major classes of flowering plants: The **dicotyledons,** or dicots (class Dicotyledoneae), number about 170,000 species, and the **monocotyledons,** or monocots (class Monocotyledoneae), total 65,000 species. The embryos of dicots have two nutrient-storing seed leaves (**cotyledons**), while the embryos of monocots have just one. Table 23-2 summarizes these and other differences.

Special Adaptations of Flowering Plants

Why have the flowering plants been so successful? First of all, they have the basic adaptations for life on land common to vascular plants: a waterproof covering of cutin plus stomata with guard cells to regulate gas exchange; a vascular system for transporting materials throughout the plant; specialized tissues for vertical support; and the production of seeds.

The flowering plants have some unique characteristics as well. The ovules and developing embryos of flowering plants are protected within a specialized structure that is part of the flower. **Flowers** are the key to the

Table 23-2 MONOCOTS AND DICOTS: DIFFERENCES OBVIOUS AND SUBTLE

Characteristic	*Dicotyledons*	*Monocotyledons*
Usual arrangement of flower parts (sepals, petals, stamens)	Multiples of fours or fives	Multiples of threes
Number of cotyledons (seed leaves)	Two	One
Usual pattern of leaf venation	Network	Parallel
Usual arrangement of vascular bundles in young stem	Circle	Scattered
Usual presence of secondary, or woody, growth	Present	Absent

success of these plants, since their colors, odors, and shapes attract insects, birds, and mammals that inadvertently transfer pollen from male to female flower parts. That free transportation makes the efficiency of fertilization far greater than fertilization that depends on chance distribution of pollen or male sex cells by wind or water, as in most gymnosperms and lower plants. Part of the flower's seed-protecting "container" develops into the **fruit,** a mature ovary or group of ovaries that surrounds the seeds and aids in their dispersal and protection.

Another crucial adaptation of the flowering plants is rapid growth. In a barley plant, for example, pollination, growth of the **pollen tubes,** and fertilization all take place in less than an hour. (Compare this to the year or more required in pines.) And some bamboo plants can grow as much as 3 ft in one day.

Rapid growth requires a high rate of photosynthesis, which in turn requires a relatively large leaf surface and explains why most flowering plants have broad leaves rather than needles. Large leaves lose more water, however, and flowering plants have specialized vascular cells called *vessel cells* as well as extensive systems of veins that aid in efficient water conduction (details in Chapters 25 and 27).

Key to the Flowering Plant Life Cycle

Of all the flowering plant adaptations, flowers are the most distinctive. A flower can be thought of as a central axis with four specialized whorls of distinctive parts: sepals, petals, stamens, and carpels (Figure 23-16). The *sepals* are small green leaves immediately below the flower. The *petals* are often large and colorful and attract animal **pollinators.** Male *anthers* (a bulbous structure at the tip of each *stamen*) are each a group of four chambers in which pollen grains develop. The female part, the *carpel*, has a *stigma*, the sticky top surface to which pollen grains adhere, and an enlarged base, or *ovary*, containing developing ovules. As in gymnosperms, the ovule of a flowering plant is a megasporangium surrounded by one or two integuments with a micropyle for fertilization.

An important feature of flowering plant reproductive strategy is the *endosperm*, a nutrient store that sustains the developing plant embryo. The endosperm develops within the ovary only after fertilization actually occurs, whereas in many gymnosperms, most of the food reserves are built up from female gametophyte tissue *before* fertilization. This gymnosperm strategy is less effi-

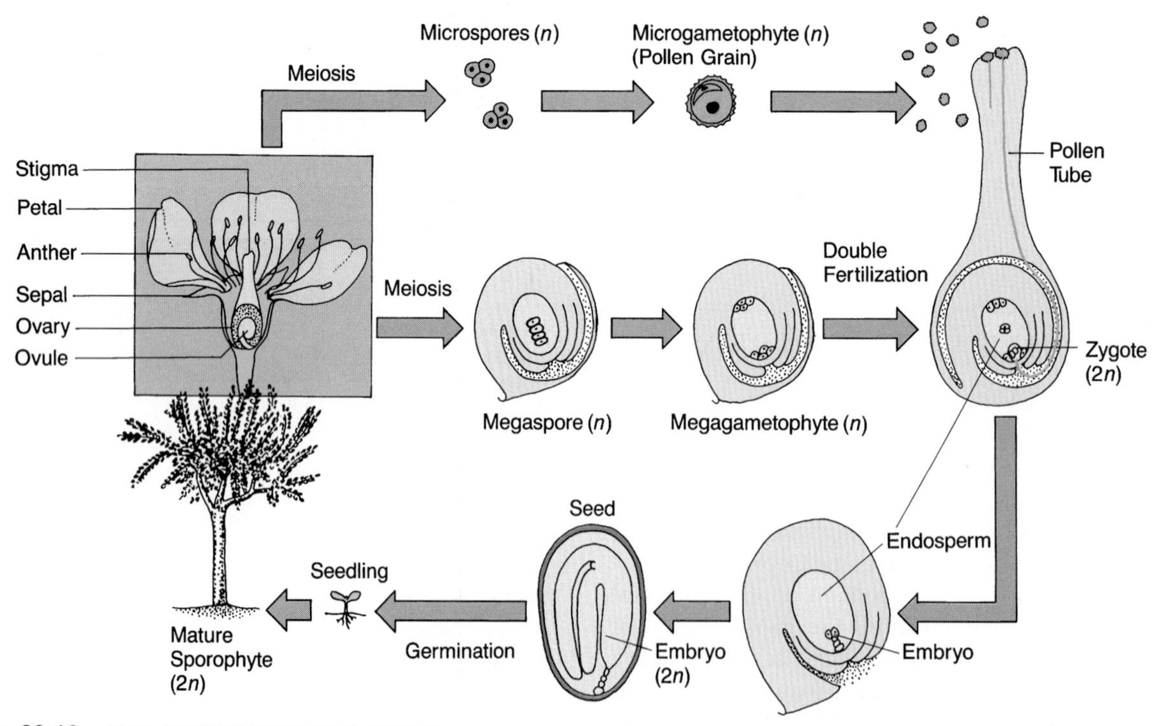

Figure 23-16 THE ANTHOPHYTE LIFE CYCLE.
The life cycle of a typical flowering plant is very similar to that of a typical gymnosperm (see Figure 23-15). The mature sporophyte plant produces flowers. Meiosis yields male (microspore) and female (megaspore) haploid cells, which develop into the microgametophyte (pollen grain) or megagametophyte. A process called double fertilization yields a diploid zygote and endosperm, which become enclosed in the seed (see Chapter 26). Germination and growth produce the mature sporophyte again.

cient, because the stored food is wasted if the egg is not fertilized.

Pollination and Pollinators: Key to Reproductive Success in Anthophyta

Flowers are more than protective enclosures for developing ovules and seeds: They help ensure that pollination will occur, and with it, greater reproductive success.

While some flowers are self-pollinated and others are wind-pollinated, most flowers attract animal pollinators that act as inadvertent couriers for the tiny pollen grains between separate flowers. Indeed, paleobotanists believe that flowers evolved as attractants for pollinators. Most pollinators benefit directly from the flowers they pollinate by consuming sugary nectar or protein-rich pollen. Unwittingly, their bodies become dusted with pollen grains, and they carry these from flower to flower and from plant to plant (even great distances apart) as they move around in search of food (Figure 23-17a). The grasses and oaks apparently reverted to wind pollination even though their ancestors were probably insect-pollinated. This is a curious reversion, however, since a wind-pollinated plant produces 1 million pollen grains for each successfully fertilized ovule, while an insect-pollinated plant produces only about 6,000.

A fascinating coevolution of flowers and their pollinators has produced specialized physical characteristics in both that help ensure the exchange of nutrients for the sake of pollination. Various species of bees, butterflies, moths, bats, and birds have mouthparts of the right shape and length to fit the flower parts of particular plant species. Plants, in turn, have specialized characteristics for attracting appropriate pollinators and ensuring that pollen is transmitted from flower to flower within the same species.

One striking example of coevolution can be seen in the interaction of scarlet gilia, which grows in the mountains of Arizona, with two pollinators, hummingbirds and hawk moths. In early summer, scarlet gilias produce long, red, tube-shaped flowers, which attract hummingbirds with their visual sensitivity to red and their long, narrow beaks and tongues. In late summer, hummingbirds migrate away, but the plants now produce pink and even white flowers—perfect for attracting hawk moths in the evening hours. Thus, the plants have evolved color shifting as an adaptation to the two types of pollinators. The result is a longer flowering season and more reproductive success for the plant species.

Among the flowering plants, such examples are legion. Flowers pollinated by bees, for example, are often blue or yellow and tend to be fragrant. Not surprisingly, bees can perceive blue and yellow but not red and can detect sweet odors from great distances.

Other flowers have evolved ingenious mechanical adaptations such as plates or certain petal shapes that help ensure pollination only by appropriate animals. Certain sage flowers, for example, have a plate blocking the path to the nectar. When a large bumblebee presses against this plate, two levers drop down and apply pollen to the bee's back, and when the bee visits the next sage flower, the transported pollen sticks to its curved carpel tip. In another example, the orchid *Ophrys speculum* has a petal that resembles a type of female wasp (Figure 23-17b). This lures a male wasp in search of a mate, and the insect, having landed on several flowers previously, inadvertently pollinates the rest he visits.

(a)

(b)

Figure 23-17 FLOWERS AND THEIR ANIMAL POLLINATORS.
The apparent coevolution of flowers and their animal pollinators—often birds, bees, beetles, and mammals—has involved the development of flower structures, colors, and odors linked to the body shape, color vision, and odor preference of animal pollinators. (a) Pollen from a banana plant dots the head of this fruit-eating bat. (b) A wasp attempts to mate with an orchid flower (species *Ophrys speculum*) that closely resembles a fuzzy female wasp.

(a)

(b)

Figure 23-18 EVOLUTIONARY TRENDS IN FLOWER SHAPES.
(a) Early flowers of the Chloranthacae may have resembled the modern liverleaf, or *Hepatica*. The flower is large and symmetrically arranged around the reproductive structures, has separate petals, and bears both male and female flower parts. (b) Evolutionary trends are visible in the orchid, which shows bilateral symmetry, has few petals, and bears fused reproductive organs (not visible). Other flower species may be imperfect (either male or female), may have fused petals, and may grow in a cluster called an *inflorescence*.

The fossil record shows that flowers have a much greater range of shapes today than they did 100 million years ago. Some botanists compare *Hepatica*, a modern woodland plant, with an orchid to show the sorts of evolutionary changes that took place in flowers during the Cenozoic (Figure 23-18).

Fruits: An Important Evolutionary Strategy for Seed Dispersal

Just as flowers increase the likelihood of pollination, so the shapes and sizes of fruits help ensure the dispersal of seeds from the parent flowering plant to new locations. A fruit is the product of a ripened ovary and often develops along with associated parts, such as the receptacle, the part of the stem to which the flower is attached. After fertilization, the wall of the ovary often undergoes modification to become a fruit wall, or *pericarp*. The pericarp can be thin and papery, as in the tiny fruits of the dandelion; hard and dry, as is the fibrous husk of the coconut; or fleshy and moist, as in cherries or blackberries (Figure 23-19).

It is sometimes advantageous for a parent plant simply to drop its seeds to the ground, directly below its outstretched branches. This often occurs in *annuals*, plants that live for just one year; since the sporophyte dies at the end of the yearly cycle, its growth site becomes vacant, and the dropped seeds have a good chance of attaining the foothold necessary for survival

(a) (b) (c)

Figure 23-19 FRUIT SHAPES AND SIZES.
The variation in shape, size, and type of fruit is important for the dispersal of the seeds contained within. The fruit of the dandelion *Taraxacum officinale* (a) is modified for dispersal by wind, while that of the coconut palm (*Cocos nucifera*) (b) is adapted to dispersal by ocean currents. (Note the pile of coconuts beneath the tree.) Blackberries (*Rubus rubrisetus*) (c) have seeds resistant to the acids in the digestive tracts of animals; thus, the seeds can be dispersed by the birds and mammals that eat the fruits. Soft, sweet cherry, peach, apple, or other luscious fruit tissues are digested, of course, and are the reason a variety of animals, including people, have evolved a "taste" for such fruits and berries.

when they germinate. For some types of annuals, however, and for *perennials*—plants that survive and reproduce over several years—it is more advantageous for the seeds to be dispersed from the parent plant and to germinate elsewhere, in a location where the new plants will not compete with the parent for nutrients, light, and moisture.

Structural modifications have evolved to facilitate seed dispersal by wind (e.g., the plumelike tufts of dry dandelion fruits); by water (the buoyant shell around a coconut embryo); or by animals (the brightly colored, fragrant, palatable fruits of the grape vine or peach tree). Hard, acid-resistant seeds can pass undamaged through an animal's digestive tract and thus be deposited at some distant location. And many fruits, such as that of the cocklebur, have hooked spines that "hitch a ride" to a new spot by clinging to mobile passersby.

LOOKING AHEAD

The fruits and flowers of anthophytes represent a high point in plant evolution and are the result of an immensely long progression of increasing specialization that began with the first algae. In our next chapter, we explore the origins and evolution of the animals, some of which came to act as plant pollinators, but all of which depend, directly or indirectly, on the ubiquitous primary producers—the plants.

SUMMARY

1. Virtually all members of the kingdom Plantae share four characteristics: multicellularity, chlorophyll, nonmotility, and alternating haploid and diploid generations.

2. Plants have a unique life cycle based on the *alternation of generations*. A diploid *sporophyte* divides meiotically to produce haploid spores that give rise to haploid *gametophytes*. Such gametophytes produce haploid gametes that fuse to produce diploid zygotes, which develop into sporophytes.

3. *Algae* are ancient, aquatic, make up much of the earth's biomass, and are classified either as protists or as primitive plants. Green algae (*Chlorophyta*) are mostly small freshwater organisms that contain chlorophylls *a* and *b*, store food as starch, and probably gave rise to the land plants. Red algae (*Rhodophyta*) are small, delicate, aquatic organisms that can live at a variety of depths, have *phycobilins* and other pigments, and store carbohydrates as floridean starch. Brown algae *Phaeophyta*, including giant kelps, contain pigments that can absorb light filtered through deep water and often have body parts analogous to those of land plants: leaflike *fronds*, a stemlike *stipe*, a rootlike *holdfast*, and sugar-conducting vessel cells.

4. Plant adaptations for terrestrial environments include "waterproofed" outer surfaces with pores that exchange moisture and gases; support tissue within the plant body; structures for absorbing moisture and minerals from the soil; vessels for transporting substances through the plant body; and mechanisms for fertilization and dispersion of spores and zygotes.

5. *Nonvascular land plants*—liverworts (*Hepaticae*), hornworts (*Anthocerotae*), and mosses (*Musci*) lack specialized vascular tissue, and the conspicuous plant body is a gametophyte with structures analogous to roots, stems, and leaves.

6. *Vascular land plants* have xylem and phloem tissue made up of *tracheid* cells, an outer layer of cutin to prevent water loss, and multicellular embryos retained on the diploid sporophyte. These plants include the whisk ferns (*Psilophyta*), club mosses and quillworts (*Lycophyta*), horsetails (*Sphenophyta*), and ferns (*Pteridophyta*). In these four divisions of vascular land plants, water must be present for reproduction to take place.

7. Seed plants are classified as Cycadophyta (the cycads), Ginkgophyta (the ginkgoes), Gnetophyta (the gnetinas), and Coniferophyta (the conifers), collectively known as *gymnosperms* ("naked-seed" plants). The remaining seed plants are classified as Anthophyta or flowering plants.

8. All seed plants produce *seeds*, the *ovule* containing the plant embryo and nutrients; produce multicellular female gametophytes (*megagametophytes*) in a *megasporangium* protected by sterile *integument* tissue; have reduced male (*microgametophyte*) and female gametophytes that are entirely dependent on the sporophyte; have sperm that do not require water to reach the egg; and have relatively large leaves and efficient vascular systems for internal transport.

9. The sporophyte (adult tree) of a typical conifer is *monoecious*, producing both male and female cones. Cycads, in contrast, are *dioecious*, growing as separate male or female plants. The ovulate (female) cone bears the multicellular female gametophyte, and later the eggs form. The male cone produces *pollen grains*, immature male gametophytes. Pollen grains are dispersed by wind, and fertilization takes place after sperm nuclei reach the egg through the pollen tube. The life cycle can take nearly 2 years to complete.

10. The flowering plants (division *Anthophyta*) are divided into *dicotyledons* and *monocotyledons* and produce *flowers* and *fruit*, the shapes and structures of which help ensure pollination and seed dispersal.

KEY TERMS

alga
alternation of generations
antheridium
Anthocerotae
Anthophyta
Bryophyta
Chlorophyta
conifer
Coniferophyta
cotyledon
cycad
Cycadophyta
deciduous
Dicotyledoneae
dioecious
filamentous alga
flower
frond
fruit
gametophyte
germination
ginkgo
Ginkgophyta
gnetina
Gnetophyta
gymnosperm
Hepaticae
heterospore
holdfast
homospore
integument
Lycophyta
megagametophyte
megasporangium
megaspore
megaspore mother cell
microgametophyte
micropyle
microsporangium
microspore
microspore mother cell
Monocotyledoneae
monoecious
Musci
nonvascular plant
ovule
Phaeophyta
phycobilin
pollen grain
pollen tube
pollination
pollinator
Psilophyta
Pteriodophyta
Rhodophyta
seed
sorus
Sphenophyta
sporophyll
stipe
strobilus
tracheid
vascular plant
zoospore
zygospore

QUESTIONS

1. What feature of the life cycle is unique to plants?

2. What are some important characteristics of the green algae? The red algae? The brown algae?

3. What is meant by parallel evolution? Give some examples of parallel evolution in the plant kingdom.

4. Name the major adaptations of the land plants.

5. Name and compare the nonvascular and the vascular land plants. What are some major differences in their basic structures and life cycles? What one feature ties the nonvascular land plants to moist environments?

6. What are the main characteristics of the seed plants?

7. Diagram the life cycle of a pine. How much time elapses between pollination and fertilization in a pine?

8. What characteristics would you look for in a newly discovered specimen to determine whether it was a seed plant or a nonseed plant? A gymnosperm or a flowering plant? A monocot or a dicot?

9. What are the biological roles of flowers and fruit?

10. Draw a flower and label the parts. Explain where the female and male gametophytes form.

ESSAY QUESTION

1. What are some facts that limit the size of nonvascular water plants? Nonvascular land plants? Vascular land plants? What are some advantages of large size? What are some disadvantages?

For additional readings related to topics in this chapter, see Appendix C.

24

Animals: The Great Consumers

From protozoans to sea stars, from squid to insects, invertebrates represent the essential diversity of animal life itself.

Robert H. Barth and Robert E. Broshears, *The Invertebrate World* (1982)

A distinct majority of multicellular species on earth today are animals. The marvelous diversity of the plant kingdom seems modest compared with the wealth of animal types: There are four animal species for every plant species. Of the 2 million kinds of animals, 97 percent are **invertebrates,** animals that lack backbones, while the remaining 3 percent have the vertebral column that is the hallmark of the **vertebrates.** As we will see, the distinction between invertebrates and vertebrates is an informal, not a formal, taxonomic one, and is left over from earlier classification schemes.

Invertebrates include a huge spectrum, from the bizarre to the beautiful, from the unfamiliar to the commonplace, and the animals are classified in dozens of phyla. While the vertebrates make up just one subgroup with one phylum, they too are a diverse lineage whose basic body design and physiology have given rise to the largest animals that ever lived, to long-distance fliers,

Splattered like a Jackson Pollock canvas, the nudibranch mollusk *Chromodoris banksi* crawls over a reef.

fast runners, and to creatures with highly complex behavior—our own species included.

Whether invertebrate or vertebrate, all animals are heterotrophs, consuming preformed organic matter rather than generating it through photosynthesis or chemosynthesis. They are all multicellular, and many lineages have the potential for large size; mobility; a stable, controlled internal environment; and relative independence from the harsh and changeable external environment.

One way to make sense of the fantastic diversity of the animals is to construct a rough family tree, giving a traditional overview of the relationships among the major phyla and their origins (Figure 24-1a). This type of overview suggests that the sponges branched off very early from the line leading to all other animals and that wormlike ancestors with internal body cavities (coeloms) gave rise to two major branches of animals, the *protostomes* (including annelid worms, mollusks, and arthropods, such as insects) and the *deuterostomes* (the echinoderms, such as sea stars, and the chordates, the phylum that includes the fishes, amphibians, reptiles, birds, mammals, and a few other kinds of animals). Recent analyses of nucleic acid and protein sequences yield a different sort of tree entirely (Figure 24-1b), in which cnidarians (simple, circular aquatic animals such as sea anemones) arose independently of all bilateral (two-sided) animals, including the flatworms, the annelid worms, arthropods, echinoderms, and chordates. We'll consider the evolutionary history of animals in greater detail as we go along—a history that spans over 600 million years and encompasses the "invention" of two-sided body symmetry, the head, the gut tube, the body cavity, body segments, jaws, bone, and dozens of other fundamental innovations.

Our survey of the animal kingdom includes the evolution and distinguishing characteristics of

- The sponges, the simplest invertebrates
- The radial animals, the cnidarians and ctenophorans
- The simple bilaterally symmetrical animals, the flatworms, flukes, nematodes, and relatives
- The protostomes, including the annelid worms, arthropods, and mollusks
- The deuterostomes, including the sea stars and re-

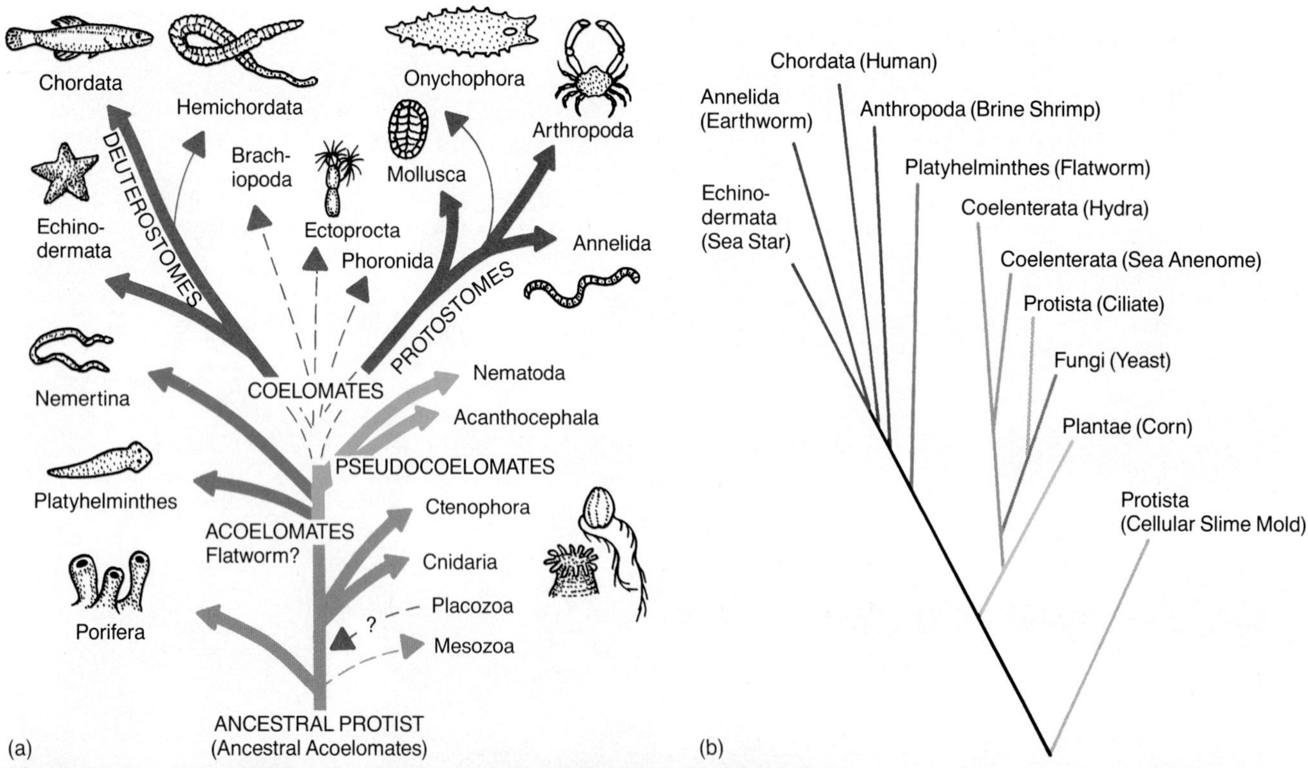

Figure 24-1 PHYLOGENETIC SCHEMES FOR THE EVOLUTION OF ANIMALS.
(a) All the major phyla discussed in this book appear in this traditional phylogenetic tree; there are many minor phyla, some of which are also included here. The phyla may be grouped on the basis of their body cavity (coelom) type, or they may be grouped as protostomes or deuterostomes, depending on whether their mouth or anus develops from the embryonic blastopore. These distinctions are indicated here by the different colored branches. (b) Taxonomists recently created a very different kind of evolutionary tree based on comparisons of nucleotide base sequences from 18S rRNAs.

lated echinoderms, and the chordates in general

- The simple nonvertebrate chordates, such as sea squirts and lancelets
- The earliest vertebrates, the jawless fishes
- The jawed fishes, including modern sharks and bony fishes
- The amphibians, first vertebrate invaders of land
- The reptiles, scaly egg-layers adapted to dry land
- The birds, the airborn vertebrates
- The mammals, warm-blooded animals with hair and milk

PORIFERA: THE SIMPLEST INVERTEBRATES

Our survey starts with the 5,000 species of sponges (phylum **Porifera**), aquatic creatures as small as a rice grain or as large as 2 m across. Sponges are **sessile:** They are permanently attached to rocks, pilings, and other hard surfaces, usually in relatively shallow salt water. They are also hollow **filter feeders:** They pump a stream of water, or *feeding current*, through hundreds of perforations called *incurrent pores*, then strain out food particles; then the water leaves the body cavity, the *atrium*, through one or more large *excurrent pores*, each of which is called an *osculum* (Figure 24-2). A sponge 1 centimeter in diameter and 10 centimeters high pumps 22.5 liters of water through its body in a day. Sponges are the most primitive animals with a true *tissue level of organization.*

As Figure 24-2 shows, the vaselike, often irregularly shaped body of a sponge has three tissues, or cell layers: (1) the outer *epithelium*, (2) an inner epithelial layer with embedded *choanocytes*, or collar cells, each containing a long flagellum surrounded by a delicate collar of microvilli; and (3) between those layers, the *mesenchyme* tissue made up mostly of *amoebocytes*, amoeba-like cells. *Pore cells*, perforated by the incurrent pores, open and close to regulate the flow of feeding current, while the beating of the choanocytes' flagella actually draws it in.

While sponges lack bones and muscles, amoebocytes can synthesize a type of fibrous internal "scaffolding" as well as *spicules*, chalky or glassy needles that provide rigidity, support, and protection from predators. In addition, specialized cells can contract like the smooth muscle cells of higher organisms, thereby regulating the diameter of the pores and channel openings. Such coordinated contractions near pore openings may somehow be governed by primitive nervelike cells.

Interestingly, a researcher can disperse a sponge into single cells by squeezing it through a fine mesh, and the individual cells can reaggregate and reorganize a new sponge just like the original. This ability may be related to the sponge's capacity for asexual reproduction, whereby the sponge forms buds that break away, settle onto rocks or pilings, and grow into new sponges. Sponges can also reproduce sexually, however, after amoeba-like cells undergo meiosis and form either sperm or eggs. Sperm liberated into the sea fertilize eggs in the mesenchyme of other sponges, and soon a ciliated, free-swimming or bottom-crawling larva develops, attaches to a substrate, and matures into a functioning sponge.

The unique features of sponge development and structure lead some biologists to classify sponges as a separate subkingdom, the *Parazoa*, and the rest of the animals in the subkingdom *Metazoa*. The more common placement of sponges in the phylum Porifera reflects their probable independent evolution from protistan ancestors and the belief that they are an evolutionary dead end, giving rise to no other group of multicellular organisms.

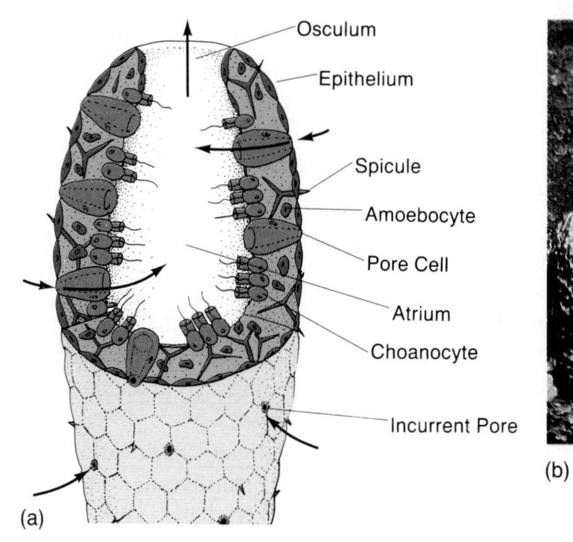

Figure 24-2 SPONGES: THE SIMPLEST MULTICELLULAR ANIMALS. (a) The structure of a simple sponge is shown in this diagram of a partially sectioned animal. The flow of the feeding current (shown by the arrows) is through the incurrent pores into the central cavity, the atrium. The beating of flagella on the choanocytes creates a current that drives the water out of the atrium through the osculum. (b) A cluster of tube sponges growing on a reef in the Caribbean Sea. Each large osculum is located at the top. The small black dots on the inner wall lining the atrium of each sponge are openings of pore cells.

RADIAL SYMMETRY IN THE SEA

Members of the phyla **Cnidaria** and **Ctenophora,** including jellyfishes, corals, comb jellies, and hundreds of other simple and often beautiful aquatic species, have a **radially symmetrical** body plan: Certain structures radiate in all directions from the center toward the circular periphery like spokes from the hub of a wheel. Radially symmetrical animals may be sessile or **pelagic** (drifting freely in the water), but all show a higher level of tissue organization than the sponges.

Cnidaria: Alternation of Generations in Animals

Cnidarians—the hydras, jellyfish, sea anemones, and corals—may be a distinct lineage that arose independently of the flatworms and all more complex groups, according to RNA sequence data. Cnidarians display an alternation of generations a bit like plants. Many species have a sessile stage with a hollow, elongated body, the **polyp,** which subsequently forms a pelagic **medusa,** a radial jellyfish drifting freely in the sea (Figure 24-3). Both forms have a three-layered body wall; a single opening, the mouth, surrounded by tentacles; and a central gut, or **gastrocoel,** also called the *coelenteron.* The phylum is sometimes called Coelenterata ("hollow gut") after this cavity.

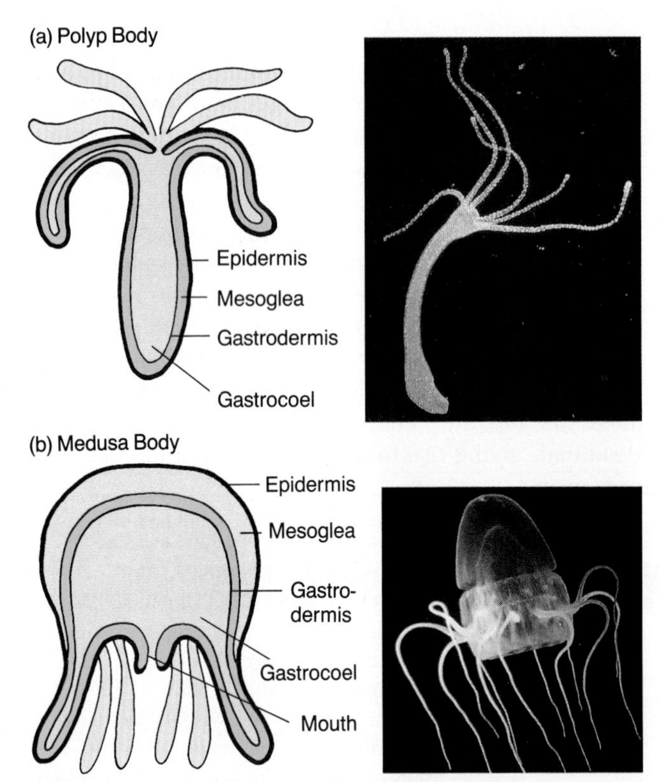

Figure 24-3 THE TWO BASIC BODY PLANS OF CNIDARIA.
(a) The polyp body form includes three tissue layers and tentacles protruding upward. Although an organism such as a hydra looks perfectly stable, new cells produced in growth zones move upward and outward in the tentacles and downward in the main body region. (b) The free-floating medusa is buoyant and swims by frequent pulsatile contractions; its tentacles hang downward in the water and, as in the polyp stage, contain numerous stinging cells.

Differentiated cells and tissues make up the external *epidermis* ("outer skin") and the inner *gastrodermis* ("stomach skin"). The middle layer of jellylike *mesoglea* ("middle glue") sometimes contains cells. The epidermis and whorl of tentacles bear stinging cells called *cnidocytes,* each housing *nematocysts,* remarkable organelles containing a coiled, harpoonlike filament in a hollow capsule (Figure 24-4). When triggered to discharge from the high-pressure capsule, the nematocyst filament may be barblike, ropelike, or sticky, and it is coated with toxins that immobilize prey.

Certain small fishes, such as the clown fish, swim among the tentacles of sea anemones and are stung but remain unharmed, having evolved immunity to the nematocyst toxin. An anemone attracts a particular fish species by giving off a synomone, a unique organic material that acts on the fish's nervous system to alter behavior, in this case, inducing symbiosis, or "joint living."

Cnidarians are the simplest creatures that employ **extracellular digestion.** After a cnidarian swallows a food morsel, digestive enzymes are secreted by gastrodermal gland cells, and partial digestion takes place in cells lining the gastrocoel. Nutritive cells can also ingest some tiny food pieces whole and digest them intracellularly through lysosomal enzymes. And some cnidarians derive nourishment from symbiotic algal cells living in their epidermis.

Cnidarians can capture and swallow prey because they have structures analogous to muscles and a skeleton. So-called *epitheliomuscular* cells in the inner gastrodermal layer have "tails" filled with microfilaments (see Figure 24-4) and can contract like tiny circular muscles, while such cells in the outer epidermal layer can contract like longitudinal muscles.

To do work effectively, muscles must pull against a fixed structure, and while cnidarians lack bone, water trapped in the gastrocoel acts as a **hydroskeleton,** a noncompressible mass that can change shape but not volume, like a squeezed balloon. Movement or locomotion based on muscle action opposing a hydroskeleton is a basic strategy among invertebrates.

Cnidarians show another evolutionary innovation: six types of *nerve cells* arranged in a primitive nervous system, the *nerve net* which allows the radially symmetrical

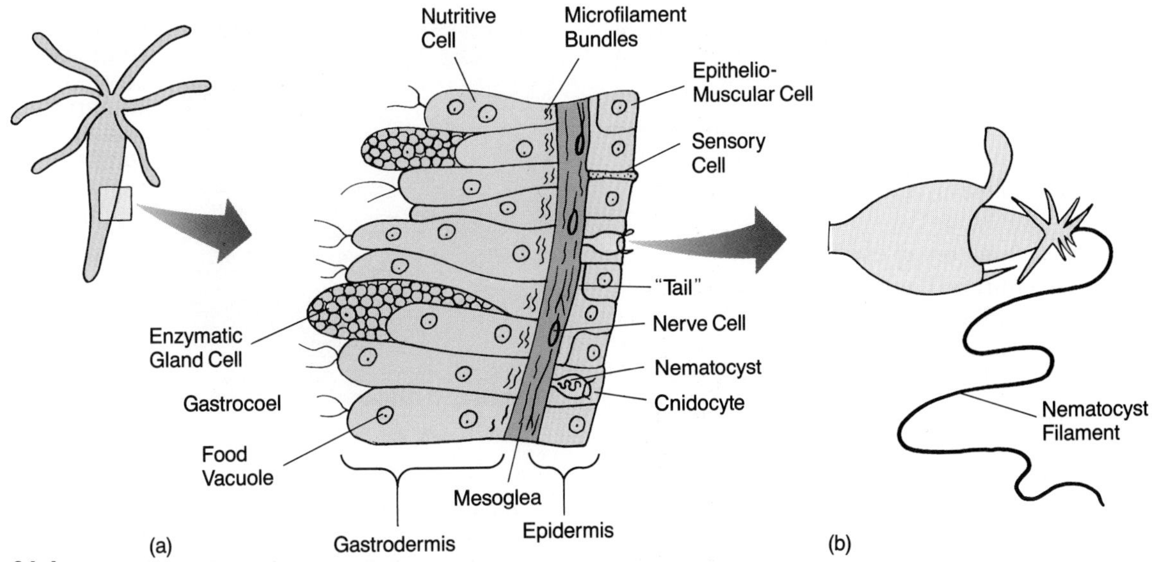

Figure 24-4 THE CNIDARIAN SPECIALIZED EPITHELIAL AND MESOGLEAL TISSUES.
(a) A detailed section of the body wall of *Hydra*. The outer epidermis consists of several cell types, including the epitheliomuscular cells, which are involved in the animal's movement, and cnidocytes, which contain the nematocysts for protection and trapping prey. The cells of the gastrodermis are concerned primarily with digestive processes. The middle mesoglea may lack cells, as shown here, or it may be cellular. (b) When a nematocyst is discharged, it can either entangle the prey or release a toxic substance that immobilizes the future food.

animal to swim, move its tentacles, or respond to stimuli from any side of its body.

Cnidarians also have a remarkable growth pattern that gives them a kind of perpetual rejuvenation. New cells arise continuously by mitosis in a growth zone just beneath the mouth and ring of tentacles, differentiate into the cells of body tissues, then slowly migrate toward sites where aged cells die and drop away. The hydra we see today will literally be a "new" creature 3 weeks later!

Reproduction in members of the phylum Cnidaria can be asexual or sexual. Polyp stages typically grow buds that mature into new polyps, and whole colonies can form in this asexual manner. But polyps can also develop buds that mature into medusae, the free-swimming sexual stage. These medusae develop true gonads, from which haploid sperm and eggs are released into the seawater. After fertilization, embryos develop into **planulae**—small, elongated, solid, cellular larvae propelled by cilia. Eventually, a hollow space appears in the planula's endoderm layer and forms the gut cavity, or coelenteron. The planula attaches to a hard substratum and gradually converts into a little polyp, thus completing the alternation of generations.

The degree to which the polyp or the medusa stage dominates the life cycle varies greatly in the cnidarian classes **Hydrozoa, Scyphozoa,** and **Anthozoa.** Hydrozoans, such as hydras, occur mainly in a sessile polyp stage and exist for only a short time as medusae and planulae. But other hydrozoans live only as medusae, and still others, such as the deadly Portuguese man-of-war, with its long, stinging tentacles, is a free-floating asexual polyp that produces groups of male and female medusae that look like bundles of deep-blue grapes.

Figure 24-5 BEAUTY IN THE DEEP.
This sea anemone (*Anthopleura* species), with its whorls of flexible tentacles, typifies the radial life style.

Scyphozoans, such as jellyfish, spend the majority of the life cycle as free-swimming medusae. The asexual polyp stage may be brief, inconspicuous, or absent. The tentacles of jellyfish—so ephemeral and gently billowing, yet so deadly to prey—can reach 40 m in length. Some Australian jellyfish have even caused human deaths.

Anthozoans, such as sea anemones and corals, skip the medusa stage entirely (Figure 24-5). They reproduce both by asexual budding and by sexual processes that yield planulae. Coral polyps live in colonies, and each animal secretes a hard limestone encasement around itself. Coral polyps can build up slowly into giant reefs tens or hundreds of kilometers long, but the living cnidarians inhabit only the narrow outer crust.

Ctenophora: The Comb Jellies

The second group of radially symmetrical animals probably arose independently of the cnidarians. In fact, the first fossilized ctenophorans, or comb jellies, some 400 million years old, are nearly identical to species alive today. All 90 living species are medusa-like and have eight unique rows of heavily ciliated cells—the combs—running from pole to pole on the transparent body (Figure 24-6). These cilia allow the organism to locomote, while sticky tentacles entrap food particles and move them slowly to the mouth. Comb jellies have nerve cells arrayed in a nerve net, and they have a sense organ that detects body position and coordinates the beating of the combs.

The unique combs, the absence of alternation of generations, and a dissimilar pattern of embryonic development all suggest that ctenophorans arose independently of the cnidarians.

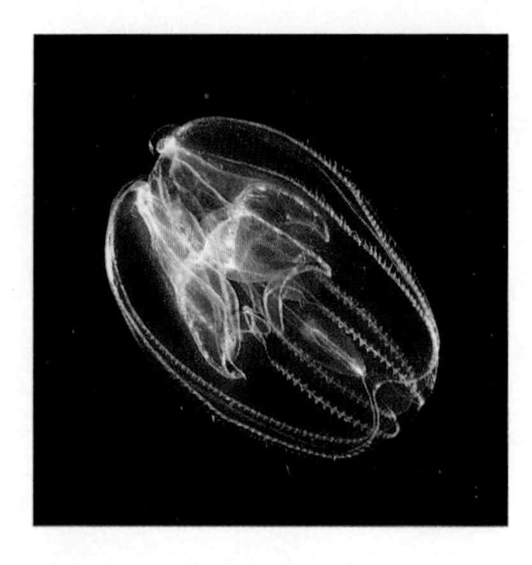

Figure 24-6 TRANSPARENT COMB JELLIES—INHABITANTS OF MARINE SURFACE WATERS. *Bolinopsis chuni*, a representative ctenophoran (comb jelly), swims through the warm waters of Sydney harbor, Australia.

THE ORIGINS OF MULTICELLULARITY

Biologists have two major hypotheses for how multicellularity arose in the ancestors of the radially symmetrical animals. The *syncytial theory* suggests that ciliated, single-celled protists containing many nuclei became "cellularized" when plasma membranes walled off those nuclei into separate cells. Most biologists, however, lean toward the *colonial theory*, the idea that flagellated protistan cells aggregated to yield either colonies resembling the hollow, ciliated blastula or gastrula stage of animal embryos (see Chapter 16) or solid organisms with a ciliated surface, resembling the planula larvae of sponges and cnidarians.

Zoologists hypothesize that the original multicellular colonies independently gave rise to creatures resembling the planulae of cnidarians and ctenophorans and in time the groups themselves. In contrast, they believe that the sponges probably arose from protozoa called *choanoflagellates*, which resemble the sponges' choanocytes.

Scientists in the 1880s described an intriguing three-layered marine worm, the **plakula,** which lives in the warm Mediterranean Sea and may be the most primitive free-living multicellular animal. Given the generic name *Trichoplax* and assigned to the phylum **Placozoa,** these minute animals glide over the seafloor, driven by surface flagella, and assume a variety of irregular shapes like an amoeba. *Trichoplax* has layers of distinct cell types and a middle zone much like the mesoglea of cnidarians, and it exhibits extracellular digestion in a temporary gutlike cavity.

Some zoologists suggest that the plakula is not far removed from the hypothetical planula that gave rise to radially symmetrical and other multicellular animals. Others argue that it is more like sponge planulae. The mystery awaits further research.

ORGANS AND BILATERAL SYMMETRY

The flatworms are the most primitive animals that carry out unidirectional locomotion. Their anterior end, containing the mouth, the major sense organs, and the primary integrating center of the nervous system, encounters new environments, food, and predators before other parts of the animal. This front-end specialization is called **cephalization,** or head formation.

A creature that moves forward in one predominant direction would generally experience the same thing on each side, and indeed, cephalized animals have also evolved the condition of **bilateral symmetry:** Their right and left sides are mirror images of each other. Most members of the animal kingdom have these characteristics.

Platyhelminthes

The flatworms, classified in the phylum **Platyhelminthes** (Figure 24-7), have both cephalization and bilateral symmetry. Their highest level of organization is the *organ level* rather than the tissue level. Like all the animals we will study from here on, including humans, the body and its organs develop from three embryonic layers: the ectoderm, mesoderm, and endoderm (Figure 24-8). The embryonic ectoderm becomes the surface epithelium, which is often ciliated and responsible for locomotion in many species. The mesoderm gives rise to true muscle fibers, excretory organs, and reproductive organs. The endoderm comes to line the digestive cavity. Flatworms lack respiratory and circulatory systems, but their thinness and flatness allow dissolved gases and nutrients to enter and move through, and wastes to leave the body by diffusion.

Turbellaria: Free-Living Flatworms

The members of the class **Turbellaria** include the most primitive bilaterally symmetrical invertebrates. The simplest of these, the *acoelomates*, are less than 1 mm long and lack a true gut and any hint of a body cavity. All turbellarians have an outer epidermis and a middle *parenchyma* (dense connective tissue) with true cells. In acoels, the mouth leads directly to the parenchyma. In other turbellarians, such as *Planaria* (Figure 24-9), a pharynx leads to endodermis tissue, specifically to lobed sacs called the gastrovascular cavity that hold food during digestion.

Planaria and other turbellarians have a more complex nervous system than do radially symmetrical animals,

Figure 24-7 MARINE FLATWORMS: BRILLIANT MARKINGS AND BILATERAL SYMMETRY.
This tiger flatworm, a marine turbellarian, inhabits the warm shallow waters around the Hawaiian Islands.

(a) Two- or Three-Layered: No Coelom (cnidarians, ctenophorans)

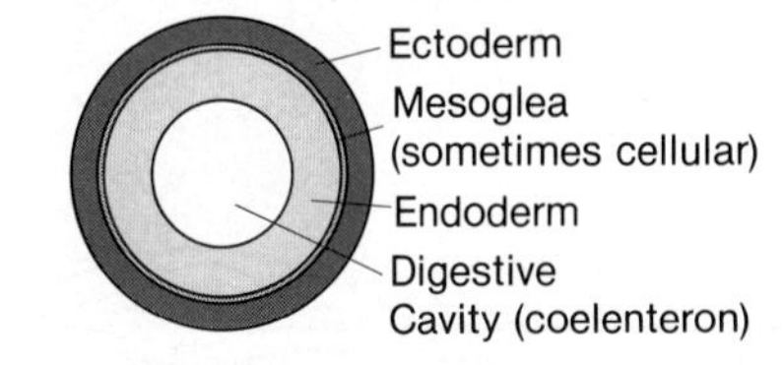

(b) Three-Layered: No Coelom (platyhelminthes)

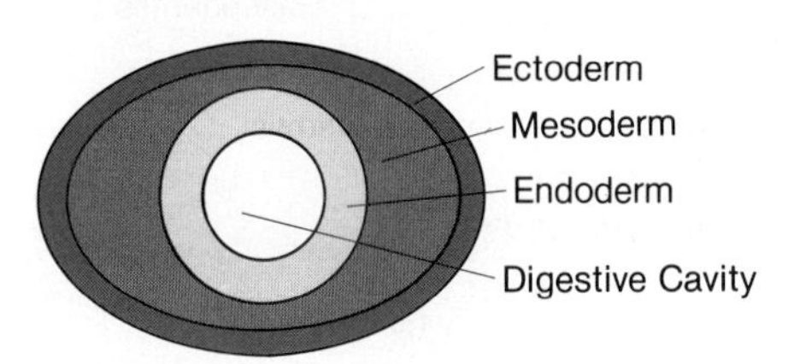

(c) Three-Layered: Pseudocoelom (nematodes)

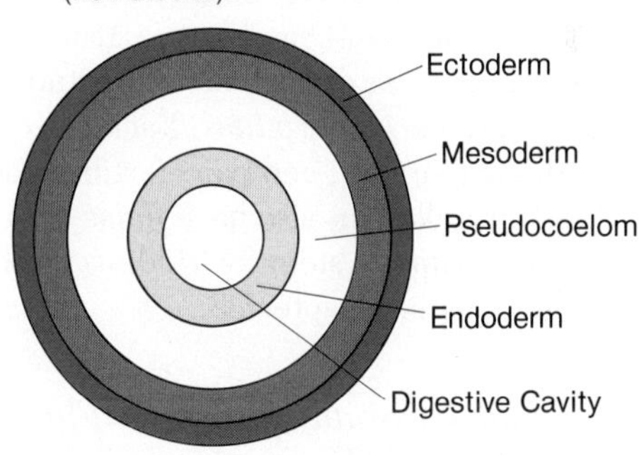

(d) Three-Layered: Coelom (annelids, mollusks, arthropods, chordates)

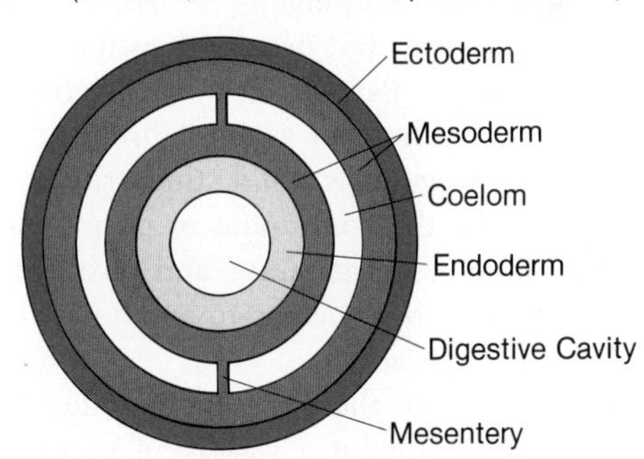

Figure 24-8 EVOLUTION OF THE BASIC ANIMAL BODY PLAN.
The evolution of body organization progressed from (a) a body with two epithelial layers plus mesoglea to (b, c) three distinct tissue layers surrounding a digestive cavity to (d) a coelom (body cavity) surrounded by mesoderm cells.

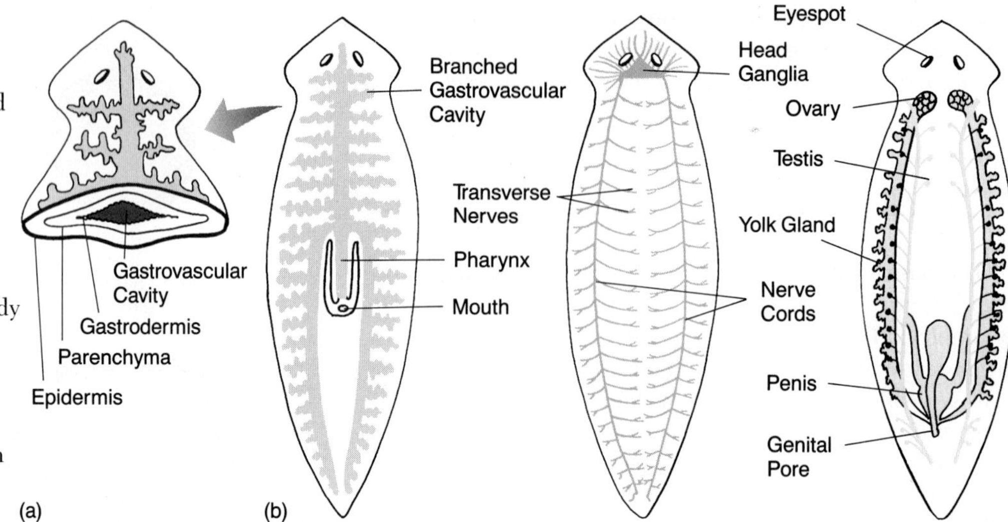

Figure 24-9
TURBELLARIANS: PRIMITIVE FLATWORMS.
Flatworms are unsegmented but have multiple organs. (a) Cross-sectional view through the region of the pharynx shows the three-layered, acoelomate body plan. The gastrovascular cavity is visible. (b) The body organization of a planarian flatworm has various organ systems, including the digestive, nervous, and reproductive systems shown here separately.

including two *eyespots* for sensing light, another organ that senses gravity, primitive brainlike organs called **ganglia** (singular, *ganglion*), and nerve trunk lines or cords down opposite sides of the body's longitudinal axis.

Turbellarians are *hermaphrodites*. Each has male and female reproductive organs, and pairs fertilize each other's eggs. Most turbellarians also have *flame cells* as part of an efficient excretory system for ridding the body of excess water (details in Chapter 33).

Trematoda and Cestoda: Parasitic Flatworms

Many flatworms are *parasites*, organisms that live on or within other organisms, derive nutrients from their hosts, but contribute nothing in return. The parasite may debilitate the host, but usually does not kill it (and thus lose its "home"). Parasites are usually transmitted from one host to another not as adults but as small, reproductive-stage organisms, and thus their survival strategy tends to be the continuous or periodic production of huge numbers of gametes and offspring, some tiny fraction of which will chance to find a new host and so perpetuate the species.

Flukes, members of the class **Trematoda,** are parasitic flatworms that inhabit a variety of vertebrate tissues. They range from less than 1 mm to 7 m in length and usually have suckers for attaching to the host organism and a covering, the *cuticle*, that resists attack by the host's enzymes. Flukes feed on and can extensively damage liver, muscle, lung, and other tissues. Many flukes have complex life cycles with alternating hosts. Blood flukes of the genus *Schistosoma* alternate between human and freshwater snail hosts and cause a severe human disease called *schistosomiasis,* which kills more people in the tropics than does any other human disease. Schistosomes have evolved cell surface markers immunologically identical to markers on the snail's cells. Hence, they are apparently not recognized as being "foreign" and are not killed by the cells of the human host's immune system (see Chapter 30).

Unlike flukes, tapeworms are members of the class **Cestoda.** These denizens of the vertebrate intestine have, in the course of evolution, lost their own gut and must absorb every nutrient from the host's digestive tract through their own epidermal wall, which is resistant to the host's digestive enzymes. The cestode's body is made up of a head unit, the *scolex*, followed by a continuously lengthening chain of units called *proglottids* (Figure 24-10), each of which is essentially a "packaged gonad" containing up to 80,000 eggs, as well as sperm. The proglottids mature, break off, and pass from the host organism in its feces. Proglottids can be picked up by an intermediate host, such as a pig. If a person eats inadequately cooked pork from an infected animal, the tapeworm can establish itself in a new human host.

Evolution of Flatworms

Genetic studies of RNA sequences show that flatworms, which lack an internal body cavity, are well separated from nearly all other bilaterally symmetrical animals, which do possess such a cavity, or coelom. Once it originated, the flatworm's grade of organization diversified. The turbellarians surely evolved into complex, free-living types and also gave rise to the parasitic trematodes and cestodes.

Nemertina: Two-Ended Gut and Blood Vessels

Members of the phylum Nemertina (or *Rhynchocoela*) display a major innovation: a one-way gut with

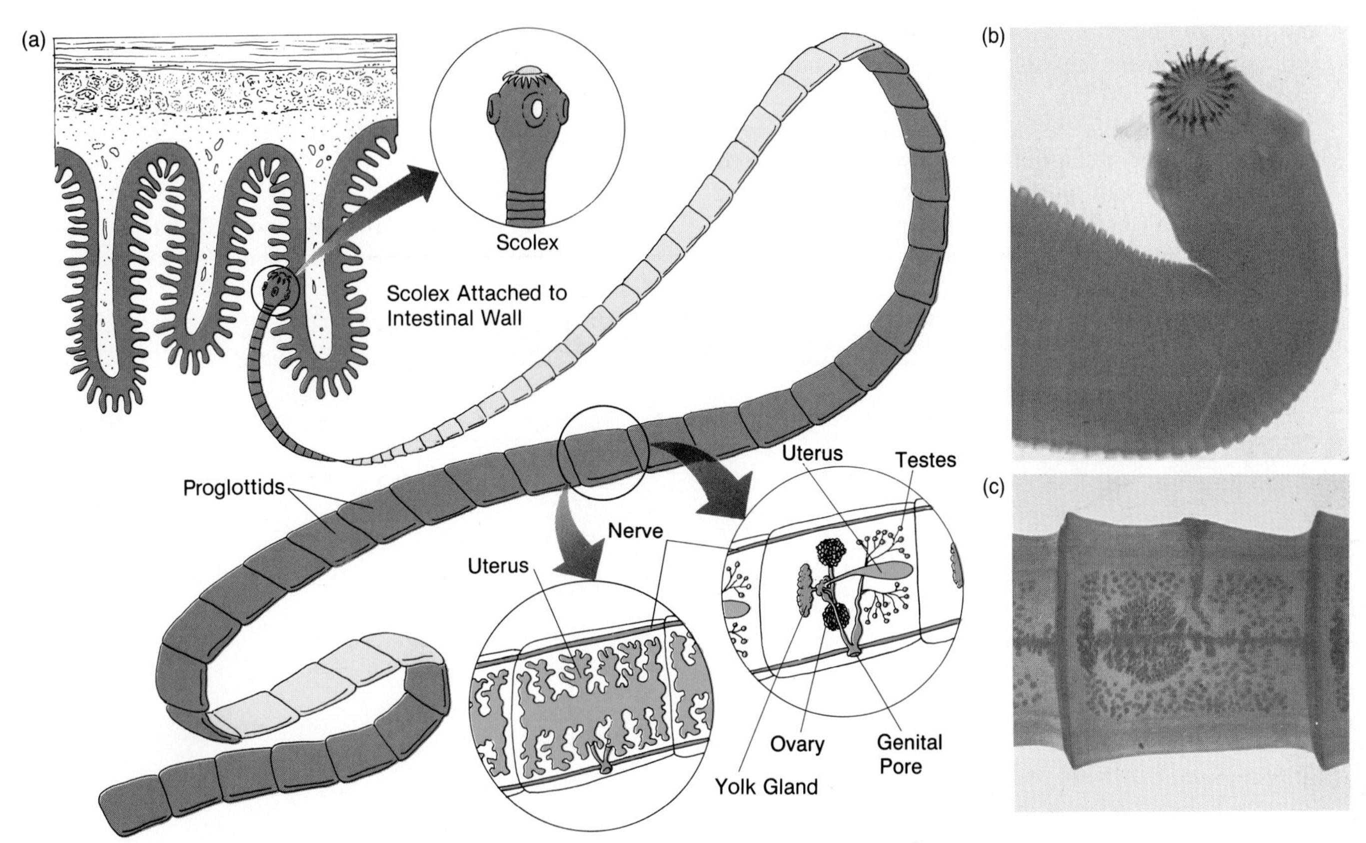

Figure 24-10 TAPEWORMS: EVOLUTIONARY DESIGN FOR ABSORBING NUTRIENTS. (a) The tapeworm's scolex, or head unit, is shown embedded in a host's intestinal wall. (b) This photo (magnified about 4×) of a stained tapeworm (*Taenia serrata*) shows the clublike scolex with its anterior ring of dark hooks and its many round suckers. The worm's body is made up of proglottid segments, each of which contains primarily reproductive organs. (c) In the drawing in (a), the circled enlargements of one proglottid show either male or female organs; both can be discerned in this photograph (magnified about 6×).

openings at both ends. So-called ribbonworms live on most marine coastlands, hidden under rocks, shells, and seaweed or burrowed into the sand and mud. Some are so thin and long (up to 30 cm) that they are called "bootlace" worms. The 650 species of these worms have locomotory, nervous, and excretory systems similar to those in flatworms and may have evolved from them.

Ribbonworms have a complete one-way *digestive system,* with mouth and anal openings at opposite ends, and with specialized regions—an esophagus and an intestine—which carry out different phases of digestion (Figure 24-11). In this one-way gut, partially digested materials are not mixed with undigested wastes. For the first time in animal evolution, the processes of digestion, absorption, and excretion are separated.

Finally, a major innovation for all later animals evolved first in ribbonworms: blood vessels that run along the body's longitudinal axis. These vessels convey a clear "blood" that lacks oxygen-carrying respiratory pigments such as hemoglobin, but the fluid does transport nutrients, dissolved gases, and metabolic wastes to

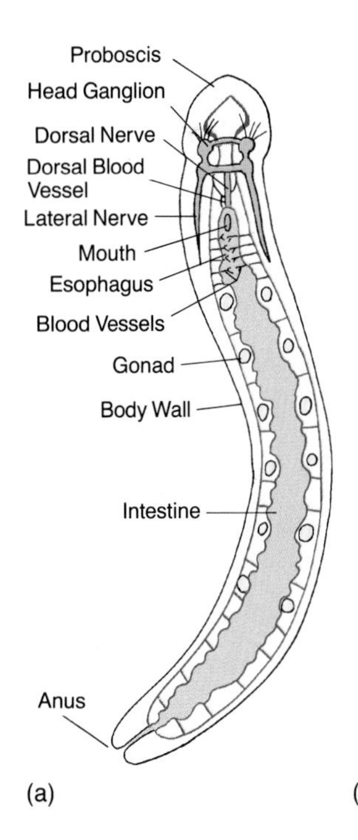

Figure 24-11 RIBBONWORMS: INNOVATIONS IN BODY FORM. The acoelomate ribbonworms (nemertines) have a circulatory system involving vessels and a two-ended gut. (a) The diagram shows the body organization of this three-layered acoelomate worm. (b) Specimens of the genus *Lineus* living on the Caribbean Sea floor off the coast of Panama.

and from all body regions. Movement of the body wall and contractions of the vessels cause the blood fluid to move slowly to and fro.

Nemertines display a spectacular capacity for regeneration. A fragment of a ribbonworm will regenerate the missing anterior end, posterior end, or both. Such fragmentation is in fact the common means of asexual reproduction in some ribbonworm species. Sexual reproduction also occurs, and the worms produce free-swimming larvae unlike those of any other invertebrate group.

Nematoda: Emergence of the First Body Cavity

Phylum **Nematoda,** the roundworms, has not only a one-way gut, but also a primitive body cavity separating

Table 24-1 CHARACTERISTICS OF THE MAJOR PRIMITIVE INVERTEBRATE PHYLA

Phylum	*Number of Species*	*Representative Animals*	*Body Cavity*	*Symmetry*	*Digestion*	*Circulation*	*Gas Exchange*	*Mode of Excretion*
Porifera	~5,000	Sponges	Ancestral acoelomate	Radial or asymmetric	Intracellular	Diffusion	Diffusion	Diffusion
Cnidaria	~10,000	Jellyfish	No coelom	Radial	Gut, with one opening (mouth); intra- and extra-cellular	Diffusion	Diffusion	Diffusion
Platyhelminthes	~13,000	Planarian	No coelom	Bilateral	Gut, with one opening	Diffusion	Diffusion	Flame cells and ducts
Nemertina (Rhynchocoela)	~650	Ribbonworms	No coelom	Bilateral	Complete gut, with two openings (mouth and anus)	Pulsating longitudinal blood vessels	Diffusion	Excretory canals with flame cells
Nematoda	~10,000+	Roundworms	Pseudocoelom	Bilateral	Complete gut, with two openings (mouth and anus)	Diffusion	Diffusion	Excretory canals
(Protostomes)								
Annelida	~8,000	Earthworms	True coelom	Bilateral	Complete gut, with two openings	Closed circulatory system	Diffusion	Nephridia
Arthropoda	~1,000,000	Insects	Coelom	Bilateral	Complete gut, with two openings	Open circulatory system	Tracheae, gills, or book lungs	Specialized excretory organs
Mollusca	~100,000	Clams	Coelom	Bilateral	Complete gut, with two openings	Open circulatory system	Gills	By pairs of nephridia, excreting wastes into the mantle cavity
(Deuterostomes)								
Echinodermata	~6,000	Sea Star	Coelom	Radial	Mouth, sac-like stomach, short intestine, and anus	Water vascular system	Usually by tube feet or specialized skin gills	No specialized excretory system; amoebocytes aid excretion
Chordata	~60,000	Mouse	Coelom	Bilateral	Complete gut, with two openings	Closed circulatory system	Gills, lungs	Kidneys

the body wall from the gut and allowing independent movement of the two. Roundworms are the most abundant animals on earth. There may be some 5 billion roundworms in the upper 3 in. of an acre of sandy beach or rich farm soil. Billions of nematodes are also found in every acre of ocean, lake, or river bottom. The 10,000 to 500,000 roundworm species feed on every conceivable source of organic matter.

Nervous Integration	*Reproduction*	*Distinguishing Characteristic*
Irritability of cells	Asexual, by budding; sexual (most hermaphroditic); larvae swim by cilia or crawl	Simplest animals
Nerve net	Asexual, by budding; sexual (sexes separate)	Nematocysts
Anterior ganglia; ladder-type system; simple sense organs	Asexual, by fission; sexual (hermaphroditic)	Simplest animals with true organs
Anterior ganglia; two nerve cords; simple sense organs	Asexual, by fragmentation; sexual (sexes separate)	Bootlace shaped marine worms
Simple brain; dorsal and ventral nerve cords; simple sense organs	Sexual (sexes separate)	Most abundant animals on earth
Ganglia and nerve cord	Asexual, by fragmentation, or sexual (sexes separate)	Many body segments
Brain with ventral nerve cord	Sexual (sexes separate)	Hard exoskeleton
Brain, ventral nerve cords, ganglia	Sexual (sexes usually separate)	Mantle; can have highly developed nervous system for invertebrates
Nerve ring and radiating nerves	Asexual, by regeneration, or sexual (sexes separate)	Stiff skeletal plates below skin
Brain and dorsal nerve cords	Sexual (sexes separate)	Gill slits, notochord, muscle blocks, hollow nerve cord

Nematodes are quite narrow in diameter. They lack both cilia and circular muscles around the body wall; thus, they can locomote only by bending and thrashing owing to contraction of longitudinal muscles. No discrete blood vessels are present. Table 24-1 lists characteristics of the nematodes and of the other primitive invertebrates.

Their great success and abundance are linked to an efficient digestive system with both mouth and anus and specialized gut regions for digestion and absorption. Unlike the situation in ribbonworms, the gut lies cushioned in a fluid-filled cavity, the **pseudocoelom** (meaning "false body cavity"). Rather than being packed inside a solid layer of cells and connective tissues, the pseudocoelom is bounded on one side by mesodermal tissue and on the other either by the gut tissue or by the epidermis (Figure 24-12). Like the coelom (true body cavity) of all higher animal phyla, the pseudocoelom allows the gut to be cushioned in fluid and the gut and body wall to move independently of each other.

Parasitic nematodes cause serious medical and agricultural problems. Disease-producing hookworms, pinworms, and filarial worms are members of this phylum. Of the 50 roundworm species that parasitize humans, one of the most common and dangerous is *Trichinella spiralis*, the agent of *trichinosis*, usually contracted by eating wild game or pork. If not killed by thorough cook-

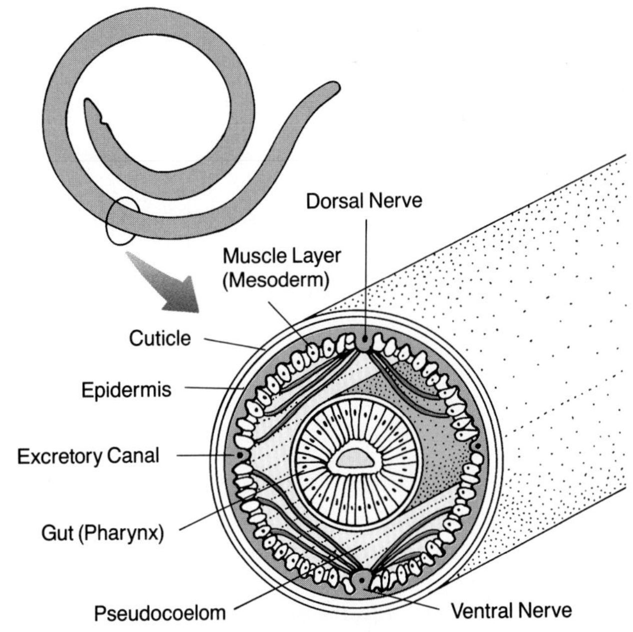

Figure 24-12 NEMATODES: A "FALSE BODY CAVITY," THE PSEUDOCOELOM.
A cross section reveals the organization of the nematode body. The body cavity surrounding the gut first appears as the pseudocoelom of the nematode worms, and it makes possible movement of the two-ended gut independent of the body wall.

ing of the meat, the worms infesting the animal's muscles produce larvae that bore into human muscles, especially those in the tongue and around the eyes, ribs, and diaphragm. Heavy infestations can nearly destroy these muscles, with potentially lethal consequences.

THE TWO MAJOR ANIMAL LINEAGES: PROTOSTOMES AND DEUTEROSTOMES

Biologists assign all animals more complex than roundworms to one of two assemblages, the protostomes or deuterostomes, and the embryology of these two lines tells us a great deal about their evolution. In one major animal lineage, the **protostomes,** the opening, or blastopore, of the future gut cavity (gastrocoel) becomes the embryo's mouth, and a separate orifice breaks through to become the anus. Protostomes (meaning "first mouth") include the segmented worms, arthropods, and mollusks. In the other animal lineage, the **deuterostomes** (meaning "second mouth"), the blastopore becomes the anus, and a second opening yields the mouth. Sea stars and their relatives, along with a few minor phyla and all vertebrates, display the deuterostome mode of development.

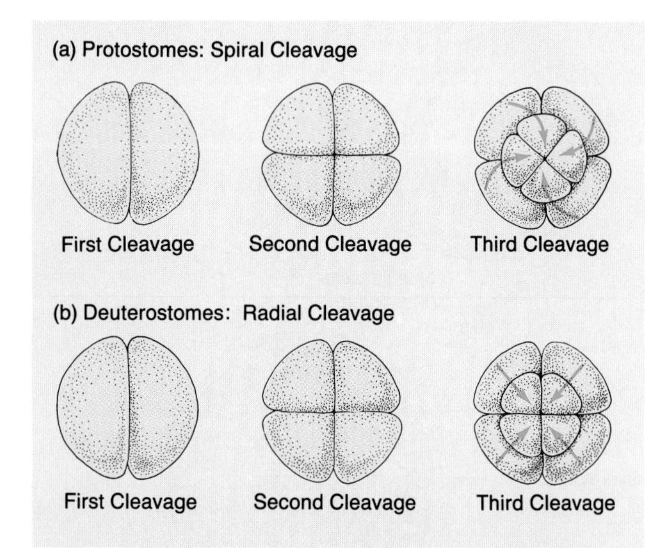

Figure 24-13 CLEAVAGE PATTERNS OF ANIMALS. The protostomes and the deuterostomes have different patterns of cleavage. (a) In the protostomes, cleavage is spiral; in the case shown, the third cleavage is in the clockwise direction, in terms of the position of the four small blastomeres, seen here from the animal pole of the embryo. (b) In the deuterostomes, cleavage is radial, with the four small, upper blastomeres arising directly above the larger, lower blastomeres. This difference in cleavage patterns is apparent from the third cleavage on.

Protostomes tend to show a spiral cleavage pattern, while deuterostomes tend to have radial cleavage (Figure 24-13; also see Chapter 16). Protostome cell lineages tend to receive specific developmental instructions very early, whereas in deuterostome embryos, developmental instructions arise more gradually. Despite certain important exceptions, in protostomes, mesoderm is usually derived from cells that migrate into the blastocoel near the blastopore, whereas in deuterostomes, pouches of mesoderm arise from the endoderm of the archenteron wall. Finally, the true body cavity (coelom) arises in many protostomes by a splitting of the mesodermal mass, whereas it develops directly from the cavity of the early mesodermal pouches in deuterostomes.

Studies of RNA base sequences provide evolutionary data on protostomes and deuterostomes, including evidence that chordates and echinoderms have been distinct lineages since the very early emergence of a true body cavity; that arthropods are very different from all other protostomes; and that, as biologists have long suspected, annelids and mollusks are related.

We have so far discussed the origins of radially and bilaterally symmetrical animals and the emergence of pseudocoeloms. Let us now see how true body cavities arose and trace the histories of the "eucoelomate protosomes" (those with coeloms)—the annelid worms, arthropods, and mollusks—as well as the deuterostomes—the echinoderms and chordates.

PROTOSTOMES

Annelida: The Segmented Worms

The familiar earthworm is among the 8,000 or so species in the phylum **Annelida,** or segmented worms. The word *annelid* means "tiny rings," and, indeed, the annelid's most striking external feature is the many body segments fused together in a linear arrangement (Figure 24-14a). Members of the three classes also have a true coelom and a closed circulatory system, that is, one consisting of a continuous set of vessels through which blood flows—the blood does not bathe tissue cells directly.

The 3,000 species of oligochaetes include earthworms and other soil and freshwater worms that range from tiny animals less than 1 mm long to giant tropical worms 2 m long and up to 40 mm in diameter. Almost all the 5,000 polychaete species are marine worms, many of which are brightly colored. Some build U-shaped tubular chambers in the sand and may bear a feathery crown (Figure 24-14b). The hirudineans are the 300 species of leeches that live in freshwater streams and lakes and in the foliage and leaf litter of moist tropical forests.

Figure 24-14 ANNELIDS: THE SEGMENTED WORMS. Segmented annelids have many forms. (a) Some, like the polychaete *Nereis*, consist of up to hundreds of nearly identical segments. (b) This Caribbean polychaete, *Spiro branchus*, has unfurled its spiraling crown of feathery tentacles, which act as respiratory gills and bear many cilia that help collect microscopic food particles. If irritated by nearby movement, the worm can quickly retract this showy crown and hide in its tube.

Segmentation: An Adaptation for Size and Specialization

Annelid worms show **segmentation:** Their bodies are organized into a series of segments, each containing the same key body structures, such as muscles, blood vessels, kidneylike excretory organs, and clusters of nerve cells, or ganglia. The evolutionary addition of similar units yielded longer and larger organisms, and certain individual segments could be highly specialized and bear intricate mouthparts, antennae, legs, or even wings. Thus, body segmentation also expanded the animal's ability to perform various specialized tasks.

Coelom and Circulatory System

Annelids are the first animal phylum with a true coelom: a fluid-filled cavity that is completely lined by mesoderm and in which many of the internal organs are suspended. The earthworm has discrete, separated chambers (Figure 24-15), and the fluid in each coelomic compartment, together with the fluid in the gut cavity, acts as a hydroskeleton: When body wall muscles contract in one segment of the body, the incompressible fluid transmits the force. Sequential contractions in the segments produce coordinated waves of elongation and swelling, during which bristlelike structures, or *setae*, on each segment are pushed against the soil and cause a net movement forward of the whole worm.

Annelids have another major adaptive feature: a *closed circulatory system,* in which blood circulates entirely within tubelike vessels, rather than partly circulating in vessels and partly bathing the body tissues themselves. Closed circulatory systems operate in some types of animals at high pressures and fast rates of flow, and thus can support high rates of metabolism and large body sizes.

In earthworms and other oligochaetes, oxygen and carbon dioxide are exchanged through the body surface itself. Polychaetes also have oarlike legs, or *parapodia*, which serve as sites for respiratory gas exchange.

Organs for Waste Removal

Diffusion alone cannot carry away the wastes of cellular metabolism in an earthworm, and instead, pairs of organs called *nephridia* ("small kidneys") collect wastes,

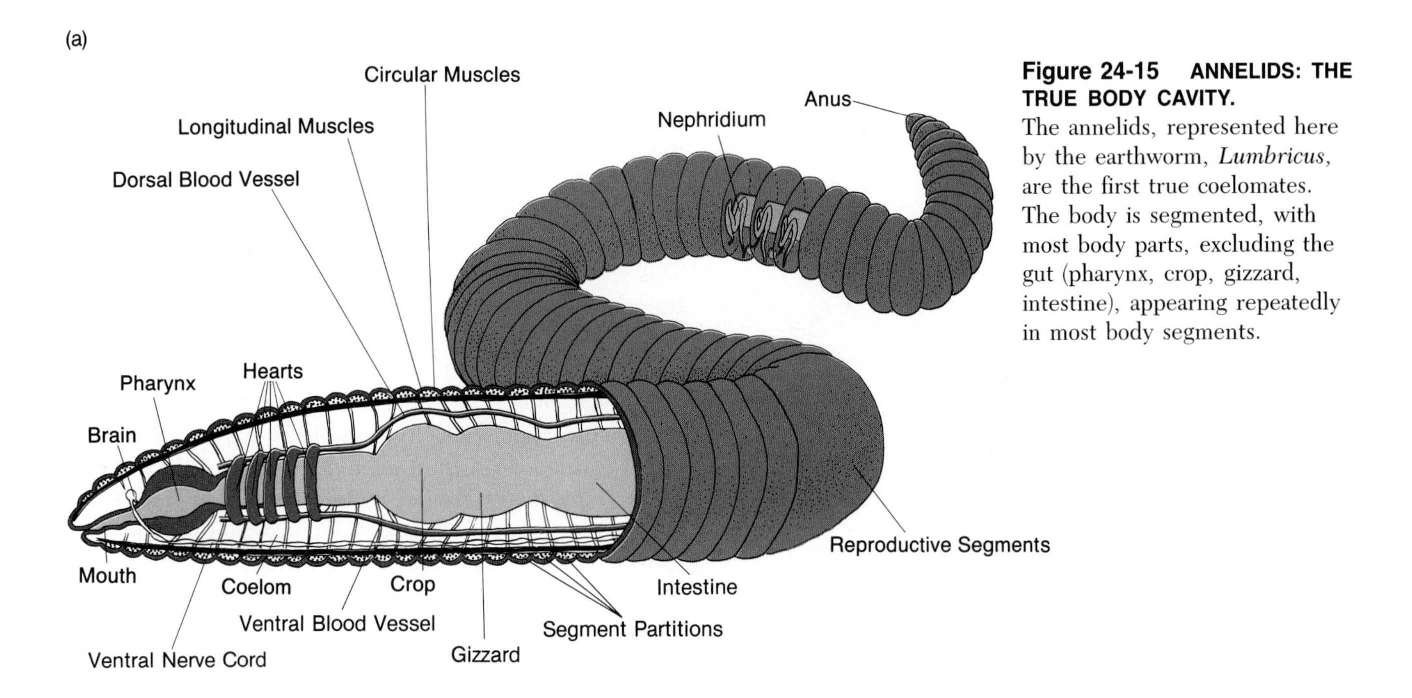

Figure 24-15 ANNELIDS: THE TRUE BODY CAVITY. The annelids, represented here by the earthworm, *Lumbricus*, are the first true coelomates. The body is segmented, with most body parts, excluding the gut (pharynx, crop, gizzard, intestine), appearing repeatedly in most body segments.

ions, and water in most body segments and excrete them. Chapter 33 considers these organs in more detail.

Annelid Reproduction: Asexual and Sexual

Polychaetes reproduce asexually when a single body segment breaks free and regenerates an entire worm. Alternatively, the entire body can break into separate segments, or local buds can form a set of segments to one side of the body. Both segments and buds develop into new adult worms. In some polychaete species, sexual reproduction takes place between male and female. In other species, and in oligochaetes, both mates are hermaphroditic and exchange sperm, as do the flatworms. The ciliated larva of aquatic polychaetes is called a **trochophore** and is just like the larva found in mollusks and other marine protostomes (Figure 24-16). Segments form from one end of the larva.

Arthropoda: Exploiting Segmentation

There are almost 1 million species in the phylum **Arthropoda** (meaning "jointed legs")—more than the combined number of species in all the other animal phyla. This huge diverse group of protostomes includes the crustaceans, insects, spiders, and mites, and fossil evidence shows that millipede-like arthropods may have been the first animals on land in late Ordovician times.

Arthropods share certain basic features with the annelids, including bilateral symmetry, segmentation, and a similar nervous system. However, they also display (1) highly modified segments and jointed appendages, (2) a blood-filled cavity, or **hemocoel,** instead of a coelom, and (3) a rigid **exoskeleton** that provides protection and serves as an attachment site for muscles.

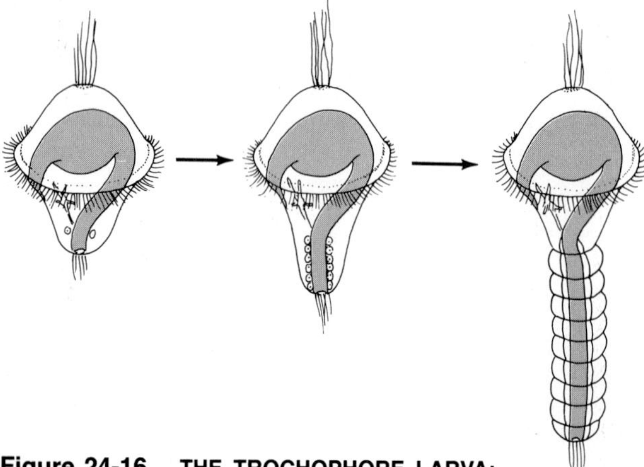

Figure 24-16 THE TROCHOPHORE LARVA: HALLMARK OF PROTOSTOMES.
Development of a worm from the trocophore larva. Repetitive developmental processes occurring at the posterior end of the larva generate the many body segments.

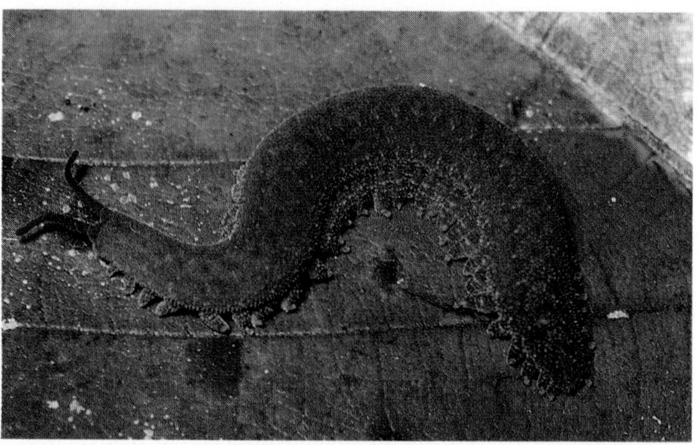

Figure 24-17 *Peripatus:* A LIVING FOSSIL.
Peripatus has the soft body of an annelid and the jointed limbs of an arthropod. This velvety invertebrate is found in the rain forests of Costa Rica.

One living group of velvety, caterpillar-like tropical animals appears to be intermediate between annelids and arthropods. Members of the most common genus, *Peripatus* (phylum **Onychophora**), have a soft cuticle and a segmented excretory system, as do the annelids, but they also have the jointed legs, hemocoel, and tubular respiratory system typical of arthropods (Figure 24-17).

The Advantages of a Hard Skeleton

The arthropod's exoskeleton is a tough cuticle that contains mostly chitin cross-linked to glycoproteins (sugar-protein complexes), but also waxes, as well as lipids that can seal body fluids in and water out. In one class of arthropods, the *crustaceans* (lobsters, crabs, shrimp, crayfish), the cuticle is impregnated with calcium carbonate crystals, making it very hard. In all arthropods, the cuticle remains thin and flexible at the leg and body joints, allowing the animal to move and bend.

Unlike the annelids and other worms, arthropods have discrete bundles of muscle tissue attached to the inner surface of the cuticle on each side of the joints. This arrangement provides leverage, so that muscle contractions can move appendages and body segments rapidly or forcefully.

The rigid exoskeleton is a nonexpandable "container" that could block the organism's growth. Thus, arthropods evolved the capacity to molt, or form a new, soft, larger exoskeleton inside the old one, then split and cast off the outgrown tight one at intervals throughout juvenile life and early adulthood (Figure 24-18). The new exoskeleton begins to harden by oxidation or by the deposition of calcium carbonate; thus, the arthropod is left with a soft exterior (which would make it vulnerable to predation) for as short a period as possible.

Figure 24-18 MOLTING HARD EXOSKELETONS. A cicada (*Tibecen* species) clings to a shed exoskeleton, its freed wings glinting electric blue.

High Metabolic Capacity: Vital for Rapid Movement

An arthropod's coelom is represented by only small pouches near the gonads. The embryo's blastocoel gives rise to the hemocoel, a cavity filled with the clear-looking *hemolymph* fluid that bathes most of the internal body tissues. Hemolymph transports gases, nutrients, and metabolic wastes, so that the needs of actively metabolizing and functioning tissue cells are met.

In general, arthropod tissues function at high rates of metabolism, especially when the body is warm, and high metabolism means high ATP requirement and oxygen demand at the cellular level. Insects have a branching network of blind-ended tubes called *tracheae,* which carry air deep into the body, near every cell. Mites, scorpions, and most spiders have *book lungs,* which are leaflike plates inside hollow chambers invaginated from the body surface. Aquatic arthropods, such as crabs and crayfish, have *gills,* which are highly branched, feathery structures over which water flows. (More on these respiratory organs in Chapter 31.)

Variations on the Arthropod Body Plan

The many attributes of the basic arthropod body plan have allowed a great evolutionary diversification of forms.

Members of the subphylum **Trilobita** were very abundant during the early Paleozoic era, from about 600 to 500 million years ago. Although all trilobite species are extinct, they left millions of marine fossils of flattened, oval-shaped animals with the *head, thorax,* and *abdomen* body regions. **Chelicerata,** the second subphylum of arthropods, includes horseshoe crabs, which often wash up on Atlantic coastal beaches; the ubiquitous *arachnids,* or spiders (Figure 24-19), mites, ticks, and scorpions (which were largely aquatic during the Paleozoic); and several rare and extinct groups. Chelicerates have only two body regions: the anterior *cephalothorax* (a fused head and thorax) and the posterior abdomen. This group's most distinguishing feature is the presence of *chelicerae:* pincerlike or clawlike mouthparts that replace the first pair of walking legs. The second pair of legs, *pedipalps,* also are used in feeding. The oversized pincers of scorpions are pedipalps. Chelicerate walking legs are attached to the cephalothorax; spiders, for example, have four pairs of walking legs on that body segment. Spiders also have silk glands for wrapping eggs and for building webs, as well as venom glands for stinging and paralyzing prey. A frightening outbreak of Lyme disease in the United States has been traced to deer ticks and black-legged ticks, both common chelicerates (see Box 24-1, page 425).

Crustacea, the third subphylum, includes some 25,000 species of shrimp, crabs, lobsters, barnacles, and sow bugs that are noteworthy for the great variety of their body forms (Figure 24-20). Crustaceans have two pairs of sensory antennae on the head, modified for touch and detection of chemicals (see Figure 24-20), and the *mandibles,* or jawlike mouthparts, for biting and chewing. The crustacean thorax is often covered by a hard, protective *carapace.*

Uniramia is a subphylum that includes centipedes, millipedes, and insects. Insects, the most abundant uniramians, have a head region; a thorax bearing three pairs of legs and sometimes one or two pairs of wings; and an abdomen. A tracheal system allows efficient gas exchange (see Chapter 31), and a pump—the heart—circulates the colorless hemolymph.

Figure 24-19 PREDATORY ARACHNIDS: LIFE IN A WORLD OF VIOLENCE. A pair of black widow spiders (*Latrodectus mactans*) hang suspended in their glistening web. The larger female has a red hourglass marking on the ventral surface of the abdomen, while the smaller male has a harlequin white and red pattern. Both produce a nerve poison that paralyzes their prey. The female will kill the male after mating.

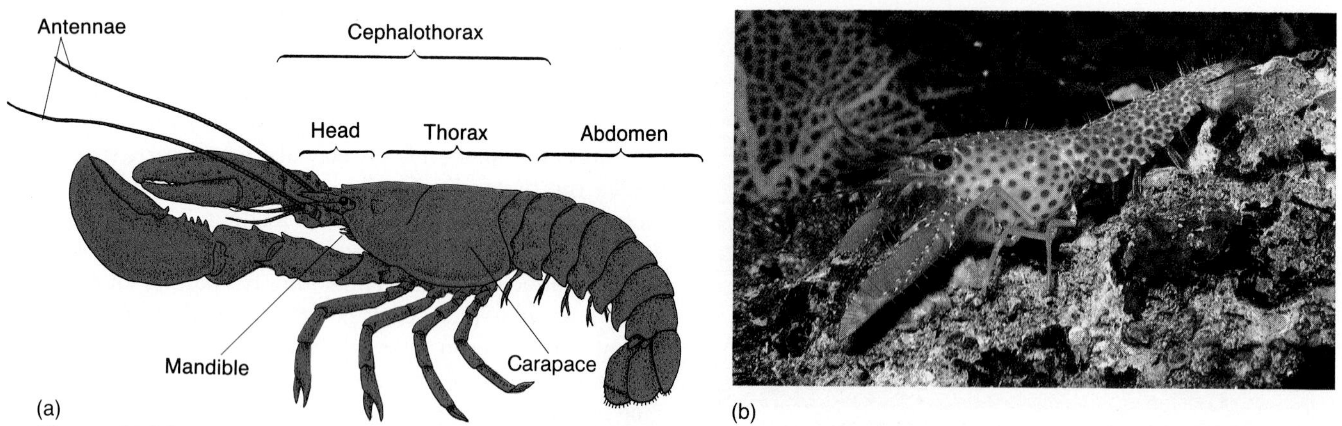

Figure 24-20 CRUSTACEANS: SPECIALIZED APPENDAGES ARE THEIR HALLMARK.
(a) Crustaceans, such as the American lobster, *Homarus*, have 19 pairs of appendages, including antennae, mouthparts, and legs specialized for feeding, walking, and swimming. (b) This tropical crustacean, the regal lobster *(Enoplometopus vanuato)*, is brilliantly colored and has jointed appendages for pinching, walking, swimming, and sensing the environment.

Figure 24-21 BIZARRE AND BEAUTIFUL INSECTS.
(a) A fully formed adult monarch butterfly *(Danaus plexippus)* emerges from its cocoon and rests. (b) This desert grasshopper from the Sudan is well camouflaged when at rest on these reeds. (c) Like construction cranes, these giraffe beetles *(Apoderus giraffa)* survey the landscape from on high. (d) Like a child's cutout figure, this leaf insect (*Phyllium* species) rests on its food plant.

(a)

(b)

(c)

(d)

TICKS AND LYME DISEASE

The summer of 1988 saw a major outbreak of Lyme disease, which can cause rashes, flulike symptoms, muscle weakness, arthritis, and—if untreated—serious damage to the heart and nervous system. Lyme disease is caused by a spirochete related to the syphilis agent. The carrier is a pair of common ticks. And the cure is, at this point, uncertain.

Ticks have always been a blood-sucking annoyance to people, pets, and livestock in outdoor settings. Since 1975, however, when deer ticks carrying the spirochete *Borrelia burgdorferi* were first detected around Lyme, Connecticut, the eight-legged relatives of spiders and mites have become even more of a threat. By 1989, more than 15,000 cases of Lyme disease in humans had been reported in 43 states, with most centering in the northeastern U.S., the upper Midwest, and the coastal zones of California and Oregon.

Deer ticks (species *Ixodes dammini*) in the East and Midwest have three hosts during their complex life cycle, and they suck blood from each host (Figure A). After hatching, a larva usually attaches to a white-footed field mouse, although the juvenile tick can also parasitize dozens of other kinds of mammals and birds. Since white-footed mice often harbor the spirochete in their blood, the tick larva tends to pick up the Lyme disease agent at this stage, then drop off to the ground. The following summer, the larva develops into a tiny nymph the size of a pinhead; it climbs up onto vegetation, then it grabs onto a second host brushing past. This new host can be a bird, a white-footed mouse, or a different mammal, including a human. As the nymph bites and drinks blood from its second host, it can transmit the spirochete. The nymph then drops off and develops further into an adult. Adults usually find their third and final meal on a deer. There they mate, lay eggs (if female), then die. In the Western states, the black-legged tick (*Ixodes pacificus*) carries the Lyme disease spirochete. While the tick tends to live on lizards and jackrabbits instead of deer and mice, its life cycle resembles the deer ticks' in other ways.

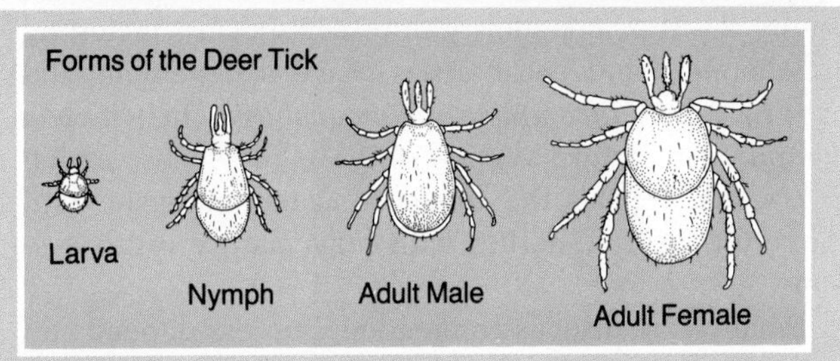

Figure A
Life cycle stages of the Eastern deer tick.

Because ticks are so abundant and their potential hosts are so widely varied, eliminating them is an unlikely solution to Lyme disease. In one study in a wooded area of Westchester County, New York, researchers found one infected tick in every 1.7 square meters of land area. And on a plot in central New Jersey, an insecticide spraying program in winter slashed tick populations that spring, but they returned to normal by fall.

There are some promising experimental approaches. One is the use of small traps that lure field mice to enter, then allow them to leave, carrying insecticide-laden cotton back to their nests where it kills ticks. Another is the deliberate release of tiny chalcid wasps, which parasitize and kill deer ticks. For now, however, many biologists and health officials agree that careful prevention when working or playing outdoors in Lyme disease areas is the safest approach. Wear light-colored, tightly woven slacks and a long-sleeved shirt; spray clothing with an insect repellant containing deet (N,N-diethyltoluamide); try to stay out of dense brush; check yourself often for ticks; and watch for early signs of Lyme disease, which can include a small red bump surrounded by a circular red rash, and/or fatigue, chills, headache, low-grade fever, and muscle and joint aches. Caught at an early stage, antibiotics can usually stop the infection.

Insecta: Diversity and Success

The 900,000 species in the class **Insecta** are amazingly diverse in appearance, and they thrive in virtually every terrestrial and freshwater habitat (Figure 24-21). Insects vary in size from tiny gall midges only 80 μm long to extinct giant dragonflies with a wingspread of 60 cm. Most contemporary insects are of a modest size that allows for exploitation of a vast array of microhabitats, such as the underside of leaves or the inner bark layer of trees.

An insect's hard, impervious cuticle is a successful adaptation that prevents water loss in terrestrial habitats and water gain in aquatic habitats. Another adaptation, the tracheal system, carries oxygen to all parts of the body. And the insect's striated (striped) muscles (see Chapter 37) are arranged to exert great force or to attain great speed and give insects great range and mobility. Such muscles permit various species to run extraordinarily fast, to carry many times the body weight, and to fly on wings that beat up to 1,000 strokes a second.

Versatile developmental strategies also contribute to the insect's success. Most insect species, including butterflies, beetles, and flies, undergo *complete metamor-*

phosis, transforming from a caterpillar, maggot, or other larva to an adult with a radically different body plan (see Figure 24-21a). A minority of insect species, including grasshoppers and cockroaches, have a developmental pattern called *incomplete metamorphosis*, in which a *nymph*, a miniature version of the adult, feeds, molts, grows, and repeats the sequence again and again until the adult finally emerges with fully mature organ systems.

Finally, the insect's tremendous success depends on complex sensory, endocrine, and nervous systems. Color vision and the capacity to see in the ultraviolet range are among the abilities of *compound eyes*, with their hundreds or thousands of facets. Many insects and other arthropods can also hear very well, owing to drum-like membranes on the abdomen or minute fringes of the antennae, which vibrate like a tuning fork. Antennae and other organs can also be involved in extremely sensitive chemical reception: Some male moths can detect chemical signals given off by females from a distance of over 1 km. Finally, a modest-sized brain receives sensory data and directs predominantly *stereotyped* behaviors, such as finding mates or collecting food, that is largely automatic and controlled by the activities of specific nerve circuits determined genetically and developmentally.

Social Insects and Their Complex Societies

Despite a small brain and stereotyped behavior patterns, complex social interdependence has evolved in the **social insects**, the orders *Isoptera* (termites) and *Hymenoptera* (ants, wasps, and bees). In social colonies, several generations coexist, and individuals in different *castes*, such as males, fertile females (queens), and infertile females (workers), perform different duties. Caste membership results from developmental and genetic processes: A queen bee develops from an immature female that is fed royal jelly, a special nutrient produced from pollen by worker bees. In the ants, wasps, and bees, males are haploid, and in one primitive Australian ant, *Myrmecia pilosula*, which has just one chromosome pair in the diploid state, each body cell in the adult male has just a single chromosome—the only case of this minimum number in all eukaryotes. We will learn more about the complex behavior of social insects in Chapter 45.

Arthropod Evolution

Fossils more than 570 million years old suggest that at least 24 and perhaps as many as 70 distinct types of arthropods had evolved by the start of the Cambrian period. Some biologists now conclude that the chelicerates, crustaceans, and uniramians arose independently of each other; however, developmental evidence supports the idea that annelids, *Peripatus*, and insects may be a real evolutionary lineage. On the one hand, it seems highly unlikely that the developmental pattern of crustaceans could have arisen from any annelid. This implies that they are probably an independent line; yet, data on 18S ribosomal RNA sequences, plus various body features, argue strongly that arthropods are a single lineage. The origins of the uniramia are of particular interest, since that group is the first type of invertebrate that arose on land. In fact, the earliest fossil centipedes, found in New York State in 1983, date from the Devonian period, 380 million years ago.

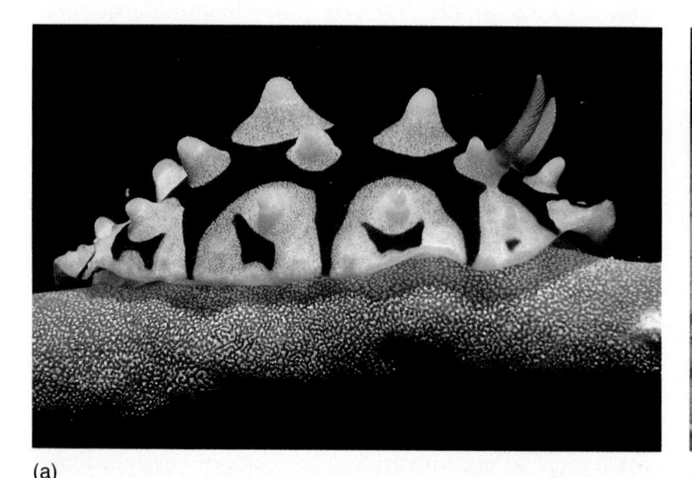

(a)

(b)

Figure 24-22 MOLLUSKS: A LARGE PHYLUM WITH STUNNING DIVERSITY.
Mollusks can be shell-less or shelled and can live in salt water, in fresh water, or on land. (a) Nudibranchs, or sea slugs, are among the ocean's most colorful animals. Projections from the dorsal surface can serve as gills or house extensions from the digestive system. (b) A giant octopus *(Octopus vulgaris)*, photographed from a safe distance in the Red Sea.

Mollusca: Rearrangement of the Protostome Body Plan

Of all the invertebrates, those with the most complex nervous systems and learned behaviors belong to the phylum **Mollusca**. The 100,000 living species of mollusks include slugs, snails, clams, scallops, octopuses, and squid (Figure 24-22). Approximately 35,000 additional species are extinct.

Because mollusks lack body segmentation, some biologists conclude that they arose from unsegmented flatworm tubellarians independently of the annelid and arthropod origin. However, mollusks do have a trochophore larva like that of annelids, and ribosomal RNA data suggest strong affinities between mollusks, annelids, and other protostomes.

As Figure 24-23 shows, the molluscan body has (1) a muscular organ called the *foot* used for locomotion and for tightly gripping rocks or other substrata; (2) a *visceral mass* containing the internal organs—heart, digestive system, excretory system; (3) a *head* housing the sensory organs, brain, and mouth; and (4) a *mantle*, a thickened fold of tissue covering the visceral mass. The space between mantle and visceral mass is the *mantle cavity;* respiratory gills hang down in this space in aquatic mollusks. Surrounding these organs in most mollusks is a calcium-rich shell that protects the animal, serves as an anchor for muscles, and supports the mantle folds, preventing the cavity around the gills from collapsing.

Mollusks have a true circulatory system with a heart; a few closed vessels that pass through the gills; and an open, blood-filled sinus, or cavity, that bathes internal tissues. They have an oxygen-carrying blood protein called hemocyanin. Muscles in the body wall push against the incompressible blood sinus, using it as a hemoskeleton. This causes the foot to protrude from the shell, as for locomotion.

Reproduction can involve fertilization of eggs in the seawater and development of trochophore larvae; internal fertilization and hatching of miniature "adults"; and other patterns.

Variations on the Molluscan Body Plan

There are four major classes of mollusks (plus the monoplacophorans), and while most follow the generalized body plan, some—like the octopus and squid—are very different.

Members of the class **Monoplacophora** are primitive mollusks once thought to have become extinct but "rediscovered" in 1957. Some zoologists interpret their oddly paired organs (six pairs of kidneys, for example) as a primitive type of segmentation, but others regard monoplacophorans as simple, unsegmented mollusks in which varying degrees of organ duplication took place.

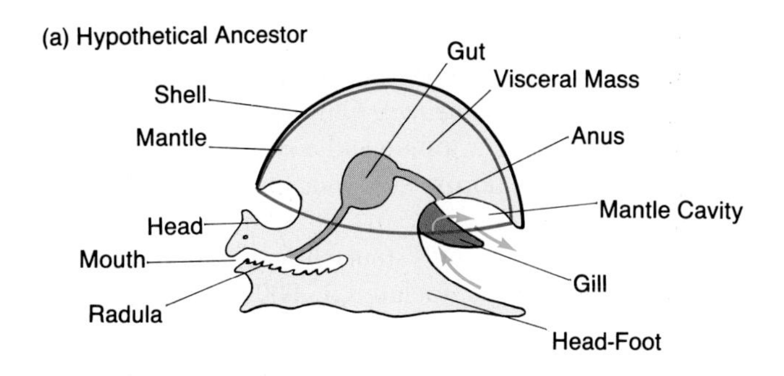

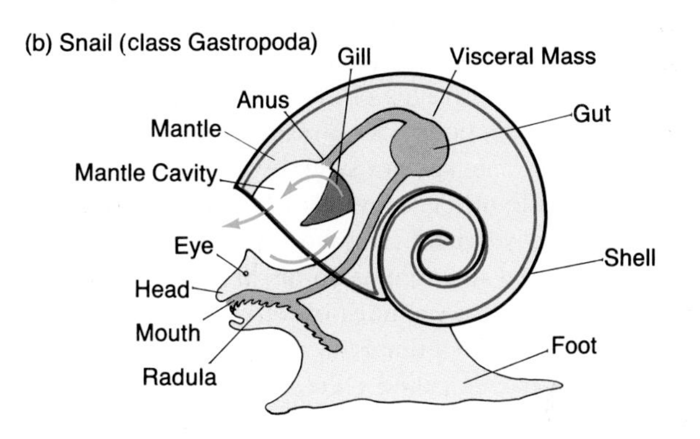

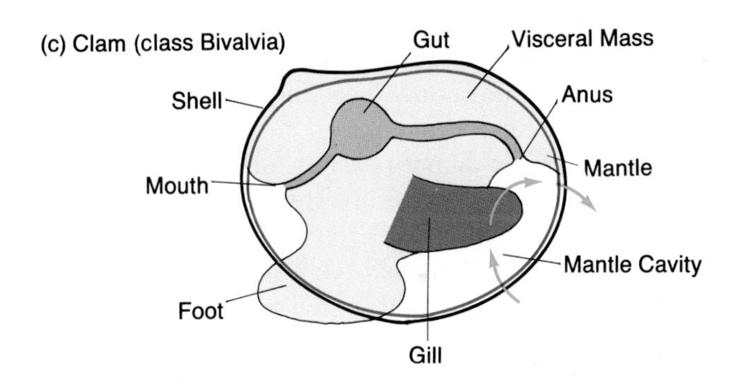

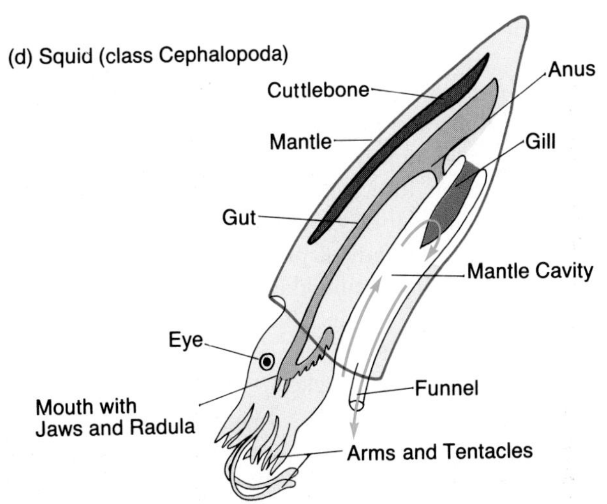

Figure 24-23 MOLLUSCAN BODY PLANS: VARIATIONS ON A THEME.
The relative position of the mantle cavity changes in each of these three classes. Water flow (blue) over the gills (purple) is shown.

Chitons—flattened oval marine mollusks with eight curving armored plates—make up the class **Polyplacophora** (or *Amphineura*). Chitons have a large foot capable of tenacious gripping and a *radula*, a tonguelike strap covered with horny, chitinous teeth that scrape off small bits of food.

Gastropoda (meaning "stomach-foot"), the most diverse class of mollusks, includes 55,000 species of snails, periwinkles, whelks, abalone, and related animals inhabiting saltwater, freshwater, and terrestrial habitats (see Figure 24-23b). All gastropods with shells display a characteristic coiling, or twisting, of the body that begins during the early cleavages of embryonic development. Like chitons, gastropods have a rasping radula to scrape off bits of food. Some species swim by means of an undulating foot, while others creep slowly along submerged rocks or crawl over land. In land dwellers, the mantle cavity is modified to serve as an air-breathing lung.

The class **Bivalvia** includes oysters, clams, scallops, mussels, and similar animals in which the shell is secreted in two halves called *valves* (see Figure 24-23c). The mantle extends downward on either side of the head and foot, enclosing them in a cavelike mantle cavity. Bivalve mollusks lack the radula and tend to be filter feeders with sievelike gills.

Cephalopoda (meaning "head-foot") includes squid, octopuses, chambered nautiluses, and a few other species that externally do not resemble other mollusks (see Figure 24-23d). The largest invertebrate on earth is the giant squid. In cephalopods, the foot is modified and terminates in the funnel, and the head is large and central and gives rise to eight to ten arms, or tentacles.

Members of the cephalopod lineage evolved as fast-moving hunters of the sea. Early species, most closely related to today's nautiloids, evolved gas-filled chambers inside their shells that provided buoyancy. In the squid, the shell is reduced to an internal *cuttlebone*—a light, flexible structure made of calcified chitin and permeated by tiny chambers and fluid that allows the squid to alter its buoyancy and depth in the water. Octopuses lack shells, external or internal. Most are not buoyant and live on the seafloor or on coral reefs; nevertheless, they can swim well in pursuit of prey for limited distances.

Squid have a narrow tubelike opening from the mantle cavity (the funnel), from which water can be forcibly expelled in a powerful jet, propelling the body in the opposite direction like a rocket.

The giant squid, the largest living invertebrate, is a denizen of the deep ocean that can reach 18 m (60 ft) in length and nearly 500 kg (1,100 lb) in weight. It has the largest eyes in the animal kingdom—complex organs that can equal the size of a car's headlight. Only a few dozen of these giants have washed up on beaches and have been studied. Octopuses and squid are capable of the most complex behavior of all invertebrates. One reason is their so-called camera eyes, which have high resolving power, like a vertebrate's eye. This is one of the best-known examples of *convergent evolution*, the appearance of a functionally equivalent feature in two very separate lineages.

Another reason for the complex behavior is brain size: A medium-sized octopus brain may contain nearly 170 million neurons (nerve cells) aggregated into one centralized, integrated mass. The uppermost lobes function, like the human cerebral cortex (see Chapter 40), as a learning center. As a result, octopuses can be trained to run mazes, to discriminate between shapes or colors, and to respond to rewards or to penalties such as a mild electric shock.

DEUTEROSTOMES: A LOOSELY ALLIED ASSEMBLAGE

We turn now to the deuterostomes, which include three very different-looking and ancient groups: the sea stars and their relatives (Echinodermata), the acorn worms (Hemichordata), and the animals with notochords (Chordata). Molecular data suggest that echinoderms and chordates have been distinct lineages for perhaps 700 million years, since the time that bilateral ancestors radiated into the major groups.

Echinodermata: Radial Symmetry Encountered Once More

Echinodermata (meaning "spiny skinned") is an apt phylum name for the sea stars (starfish), sea urchins, and their relatives. Most of the 6,000 species are relatively sedentary marine creatures that are radially symmetrical as adults but begin life as bilaterally symmetrical larvae. Some, like the sea cucumbers, are soft and cylindrical; others, like the sea lilies, are feathery; and many, like the sea urchins and sea stars, are well armored and multiarmed (Figure 24-24). All, however, possess a unique *water vascular system*, a modified portion of the coelom that acts as a hydraulic device and is involved in feeding, locomotion, respiration, and sensory perception.

In sea stars, special tubular extensions of the water vascular system (radial canals) end in rows of water-filled *tube feet*, each tipped with a sucker, on the ventral surface of the arms (Figure 24-25). At the end opposite from the sucker, each tube foot has a small muscular reservoir, the *ampulla*, which can contract, forcing more water into the foot and causing it to extend. Contractions of the ampullae, suckers, and muscles in the tube feet

Figure 24-24 ECHINODERMS: A WIDE ARRAY OF BODY FORMS SHARING FIVEFOLD SYMMETRY.
All echinoderms have five radiating arms or five markings on the body. (a) An ochre sea star *(Pisaster ochraceus)* feeds on mussels, barnacles or snails. (b) A sea apple *(Paracucumana tricolor)* is a rosy-hued sea cucumber with yellow tube feet and bright red branching gills. Many sea cucumbers are green and bumpy—as the name implies. (c) These Hawaiian slate-pencil sea urchins *(Heterocentrotus mammillatus)* have large, brilliantly colored spines for protection. Tube feet protruding below pull the animal along slowly.

are coordinated, allowing the sea star to move slowly across the seafloor.

Most echinoderms have a nervous system with a nerve ring near the mouth and main nerves that radiate outward to major portions of the body. Some echinoderms have eyespots at the tip of their arms that can sense light and shadow. Surprisingly, echinoderms lack a kidney system; they conform osmotically to seawater and thus are restricted to the marine habitat.

Echinoderms have small, porous but stiff skeletal plates, each one a separate calcite crystal ($CaCO_3$ and $MgCO_3$), lying in the mesoderm just beneath the outer epidermis. These plates, called *ossicles*, are unique in the animal kingdom and serve as a kind of armor.

Reproduction in echinoderms is by both asexual regeneration and sexual processes. A single sea star arm containing just a small portion of the central disk region will slowly regenerate the rest of the creature, although the process sometimes takes as long as a year. During sexual reproduction, following fertilization and radial cleavage, *tornaria* larvae form, resembling the larvae of the hemichordates, the next deuterostome phylum.

Biologists are not sure how the many different-looking echinoderms are related because much of the evolutionary divergence took place before the Cambrian period. The fossil record and other evidence suggest that sea stars and sea lilies make up one lineage, sea urchins and sea cucumbers a second, and brittle stars another. A

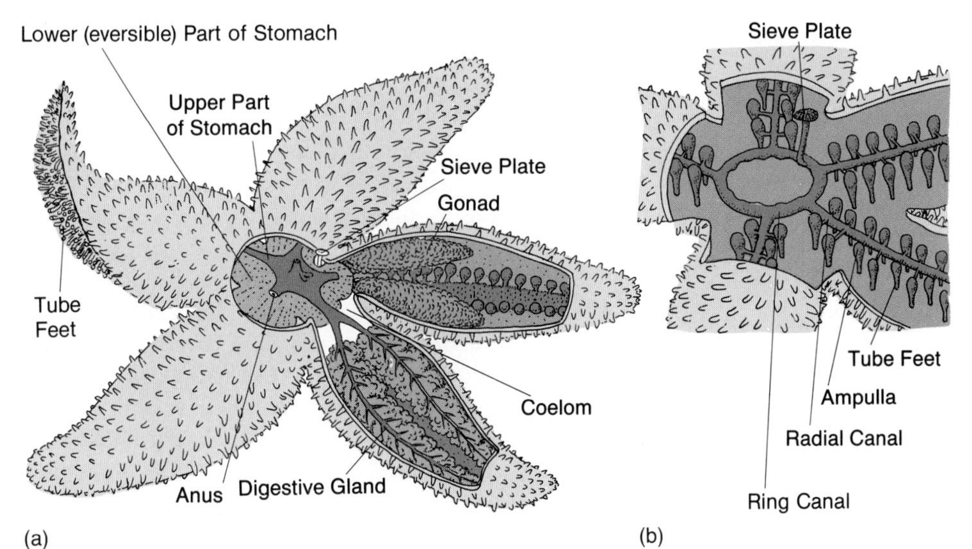

Figure 24-25 THE SEA STAR: AN EXAMPLE OF THE BASIC ECHINODERM BODY STRUCTURE.
(a) The digestive gland has been removed from one arm, revealing the gonad and one branch of the water vascular system. (b) The tube feet are external to the animal, while the other parts of the water vascular system, the ampullae and the ring canal, are internal. The sieve plate is the site where water enters or leaves the water vascular system.

newly discovered class, the Concentricycloidea, or sea daisies, has a medusa-like appearance, retains its embryos until they look like miniature adults, and is most closely related to sea stars.

Hemichordata: The First Gill Slits

It is in members of the invertebrate phylum **Hemichordata** that we see the most plausible candidates for creatures that might have yielded both echinoderms and chordates, our own phylum.

All hemichordates are marine animals; some are mobile and wormlike (the enteropneusts, or acorn worms), while others are sessile and bear tentacles (the pterobranchs). Both types are noteworthy for their paired sets of *gill slits,* openings through the lateral walls of the anterior gut. A feeding current driven solely by cilia enters the mouth and exits the body through the gill slits, gathering enough food to support the needs of acorn worms, which can reach lengths of 150 cm. Gill slits are the basic feeding mechanism of primitive chordates.

The wormlike enteropneusts have a dorsally situated nerve cord that arises by an infolding, as in vertebrates (see Chapter 16), but the groups share few other structures. The name *hemichordate* is actually a misnomer, since they lack a *notochord,* the primitive skeletal element of the chordates. Development in hemichordates yields a ciliated larva like the tornaria of echinoderms. An ancient pterobranch-like creature may have given rise to separate lineages leading to echinoderms and chordates.

We are now ready to consider the final major animal phylum, the chordates.

CHARACTERISTICS OF THE CHORDATES

The phylum **Chordata** encompasses animals as different as sea squirts and squirrel monkeys, but all chordates display the same four physical characteristics at some time during their life cycle: (1) *gill slits,* (2) a *notochord,* (3) a tail and blocks of muscles, and (4) a hollow *nerve cord* (Figure 24-26).

Gill Slits: Feeding and Respiration

Gill slits are paired openings through the lateral walls of the anterior gut in the region immediately posterior to

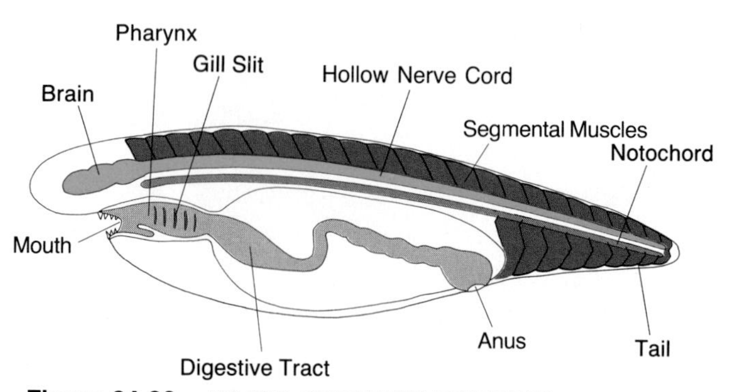

Figure 24-26 MAJOR CHORDATE FEATURES. Chordates are characterized by gill slits, a notochord (which may be incorporated into the vertebral column), a dorsal hollow nerve cord, and a postanal tail with segmental muscles.

the mouth cavity—the **pharynx.** The earliest chordate gill slits acted as exits for the feeding current that entered the mouth, and in living primitive chordates, tiny food particles become trapped in mucus produced by the pharynx and are passed to the rear portion of the gut for digestion.

Gill slits became associated with respiratory gas exchange in later chordates—the first vertebrate fishes. Today, some chordates (such as trout, sharks, and most other fishes) retain gill slits into adulthood; others (including reptiles, birds, and mammals) lose them in adult stages. But in every chordate embryo—including human embryos—gill slits and associated structures do develop (review Figure 18-8).

Notochord: A Structure to Prevent Body Shortening

All chordates have as their primary internal longitudinal skeletal element the **notochord,** a stiff but flexible rod that runs the length of the bilaterally symmetrical animal just ventral to the nerve cord, and which can bend but resists shortening. The notochord develops in every chordate embryo and is retained in many adults. The resistance to shortening allows chordates to swim in their unique and characteristic way, with blocks of muscles contracting to bend the body from side to side. Without the notochord, these contractions would shorten the body rather than propel it efficiently.

Virtually all modern vertebrates have a set of bones, the *vertebral column,* that develops around the notochord to protect the nearby nerve cord and major blood vessels and to provide additional stiffening and support for the body. This support is especially important in land vertebrates, and could not be met by the notochord alone.

Muscle Blocks: The Basis for Swimming Locomotion

All the ancient chordates (and many modern chordate species) lived in water and depended on swimming, made possible by the sweeping of a tail. Blocks of muscles, called **myotomes,** make up the bulk of the tail in a swimming chordate and also occur more anteriorly along both sides of the body. Each myotome can be regarded as a body segment, although not like that seen in annelids, where internal organs are often repeated in each segment. The key feature is that each myotome on each side of the body can contract as a unit; sequential waves of contraction from head to tail on opposite sides of the animal sweep the flank and tail one way and then the other, causing the body to push against a noncompressible material—the water—and propel the body forward.

Nerve Cord: Coordination of Movements

Sequential muscle movements require coordinated control. All chordates thus have a long, hollow **nerve cord,** the *spinal cord,* situated just dorsal to the notochord. Nerves run from the spinal cord to each myotome and to other parts of the body. The spinal cord acts as a switchboard for coordinating swimming, crawling, and walking, whether a brain is present or not.

Let us now survey the three subphyla that make up the phylum Chordata.

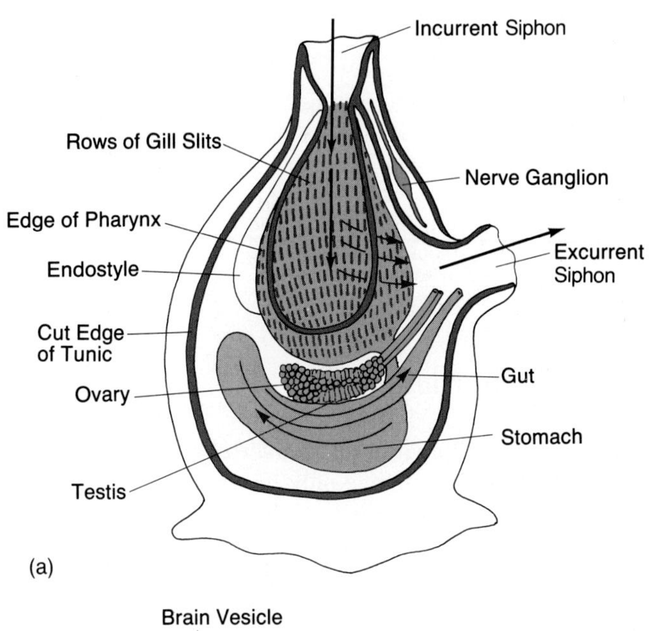

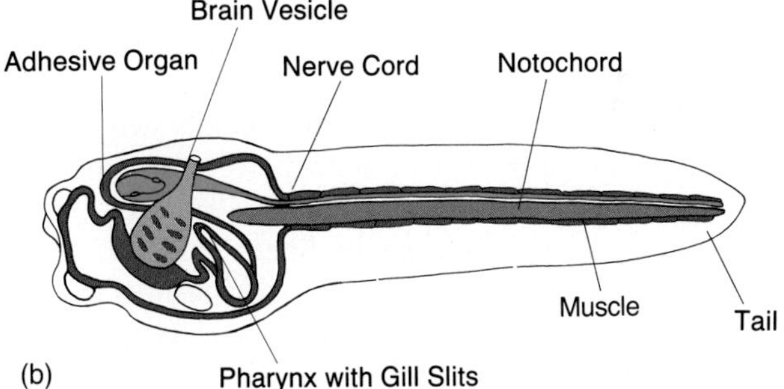

Figure 24-27 UROCHORDATES: EARLY CHORDATES. (a) This cut-away view of an adult urochordate shows how the feeding current passes through the gill slits and then out the excurrent siphon. (b) The chordate characteristics are apparent in the tadpole. Note in particular the dorsal nerve cord and the notochord. The tail muscles are not truly segmented in these organisms.

THE NONVERTEBRATE CHORDATES

Members of two of the three chordate subphyla—Urochordata and Cephalochordata—are nonvertebrates: Although they have a notochord and gill slits, they lack a bony skeleton and a vertebral column.

Urochordata

Members of the subphylum **Urochordata**—the commonest type being the sea squirts—are small marine organisms that usually spend their adult lives permanently attached to rocks or other hard surfaces. Urochordates have an outer tough and leathery *tunic* surrounding the body walls and pharynx, a large basket-shaped structure hanging within the body cavity and perforated by hundreds of ciliated gill slits (Figure 24-27a). Beating cilia draw in a feeding current through the mouth, or incurrent siphon. Food particles become mired in a mucus sheet in the pharynx, and these particles pass into the gut, where they are digested and the nutrients are absorbed.

While the hollow nerve cord, notochord, and tail are absent in the adult, they are present in the animal's free-swimming larval stage, the *tadpole* (Figure 24-27b). Tadpoles of most species never feed; within a day or so of hatching, they undergo metamorphosis, during which the tail is retracted and its parts are degraded. Concomitantly, the immature pharynx begins to develop, and the gill slits begin to function. The remarkable urochordates are like primitive fish as larvae, but are unlike other chordates as adults, and are considered nonvertebrates because they lack a bony skeleton or vertebral column.

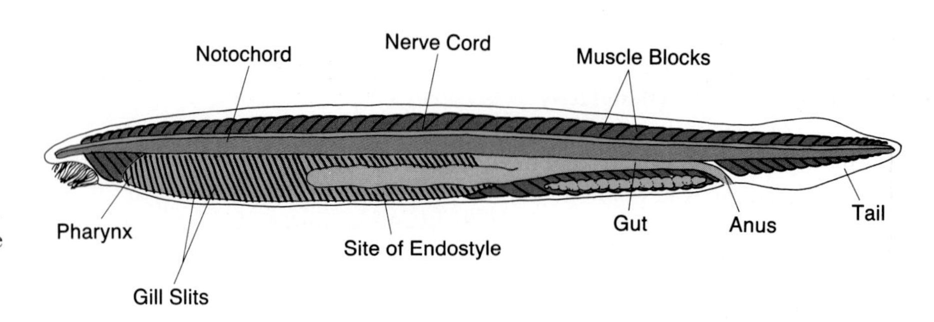

Figure 24-28
CEPHALOCHORDATES: A VERTEBRATE RELATIVE.
The typical chordate characteristics are easy to spot in this adult lancelet (*Branchiastoma lanceolatum*). Note the dorsal hollow nerve cord (lacking a brain), the notochord, the gill slits, the segmental muscle blocks, and the postanal tail.

Cephalochordata

A second subphylum of nonvertebrate chordates, **Cephalochordata,** contains only one type of organism, the lancelet (formerly called amphioxus). It is the adult body of this small marine animal, which lives half-buried in the sand of shallow ocean bays and inlets, that exhibits all four chordate characteristics (Figure 24-28). Lancelets are filter feeders; their long pharynx, perforated by more than 100 ciliated gill slits, sieves and traps food particles on a mucus sheet. Although lancelets swim only sporadically, the adults have a notochord, a nerve cord (but no brain), myotomes, and a tail. The lancelet notochord contains unique musclelike fibers, and this suggests that the lancelet lineage was an early offshoot of the line that led to the vertebrates.

JAWLESS FISHES: THE FIRST VERTEBRATES

The earliest members of the subphylum Vertebrata were fishes of the class **Agnatha,** which means "jawless" (Table 24-2). They arose more than 540 million years ago, lived as filter feeders into late Devonian times, and included the *ostracoderms,* or bony-skinned fish. Two major "innovations" distinguish them from earlier chordates: their feeding current was powered by muscles rather than by cilia; and their bodies were covered by plates made of bone lying within the skin, although they still lacked a vertebral column.

Bony-Skinned Fishes and Modern-Day Descendants

Fossils of many different kinds of ostracoderms show that each type had a mouth and a series of paired muscle-powered gill slits that could pump more water (containing suspended food particles) than could be propelled by the ciliated gills of the more primitive chordates. More food intake meant that these jawless fishes could attain larger body size.

The ostracoderms also had a bony brain case, the head shield, surrounding a complex brain with the three parts found in every subsequent vertebrate—the forebrain, midbrain, and hindbrain—as well as *cranial*

Table 24-2 MAJOR LIVING CLASSES OF THE SUBPHYLUM VERTEBRATA

Common Name of Group	*Class Name*	*Approximate Number of Species*	*Distinguishing Characteristics*
Fishes	Agnatha Chondrichthyes Osteichthyes	50 625 30,000	Skull; bones; jaws; fins (paired appendages)
Amphibians	Amphibia	2,600	Legs (extending sideways); fully functioning lungs; chambered heart
Reptiles	Reptilia	6,500	Dry, scaly skin; expandable rib cage; amniotic eggs (leathery); legs (extending below body)
Birds	Aves	8,600	Feathers; air sacs; amniotic eggs (hard-shelled); warm-blooded
Mammals	Mammalia	4,100	Milk and mammary glands; body hair; placenta (most)

Figure 24-29 A LIVING REPRESENTATIVE OF AGNATHA, THE LAMPREY.
A close-up view of a lamprey, *Lampetra fluviatilis*, shows the seven gill slit openings.

nerves leading from the brain to the gill slit muscles, eyes, and ears. Clearly, the sense organs and nervous system that are so vital to all vertebrate life were fully formed in these first known vertebrates.

The ancient armored ostracoderms are survived today by the jawless lamprey (Figure 24-29) and hagfish, both of which are classified in the subclass *Cyclostomata*, which means "round mouthed." Instead of bony plates, cyclostomes have skeletons of *cartilage*, a pliable tissue formed of cells embedded in a fibrous, extracellular matrix composed largely of collagen and sugar-protein polymers. Lampreys are parasites that feed on other fish by attaching to an animal's side with the suction-cup action of their mouths, cutting a wound in the host's skin with their rasping tongue, and drinking the victim's blood. The very distantly related hagfish use their circular, jawless mouths to feed on various invertebrates and on dead animal and plant matter that falls to the ocean floor. The extinct conodonts, once abundant in ancient seas, are structurally related to hagfish.

Vertebrate Origins and the History of Bone

Zoologists now believe that hemichordates may have developed larvae with gill slits and a chordate-type tail, and that these little creatures yielded three distinct lines: the urochordates, with their sessile adult stage; the cephalochordates, with their unique notochord; and the first jawless vertebrate ostracoderms.

It is only in the last group that bone appeared, providing scientists with the fossil record of these extinct fishes. Bone consists of *osteocytes*, or bone cells, embedded in a matrix of organic fibers composed of collagen molecules, on which grow crystals of an inorganic material called *apatite*, made of calcium and phosphate. Living bone is a highly dynamic tissue that serves as the body's main reservoir of calcium and phosphate ions, releasing one or both types of ions when blood levels are low, and storing the ions when blood levels are high. Its hardness also makes bone a protective encasement, a supportive skeleton, and a site of attachment for muscles.

Biologists long theorized that bone evolved because of its hardness, but a newer theory suggests that the role of bone in storing phosphate (so crucial for nucleic acid synthesis and ATP production) may have been the key evolutionary factor.

JAWED FISHES: AN EVOLUTIONARY MILESTONE

Even though the earliest vertebrates had a chordate tail, bones, and a "new" mode of feeding (muscular gill slits), they were still limited to filtering tiny bits of food. The evolution of jaws, however, led to a revolution in body size and feeding styles.

The most ancient jawed-fish fossils are members of the class Acanthodii, an extinct group that left fossils dating back to the Ordovician period, 435 million years ago. Strong evidence suggests that the jaws in acanthodians and their later relatives, the placoderms, arose from the two or three anterior sets of skeletal gill support pieces called *branchial arches*, as well as from nerves and muscles of the filter-feeding apparatus of the jawless ostracoderms (Figure 24-30).

Jaws literally transformed the world's ecology. Jawed fishes could eat other fish, graze on large marine algae, or crush worms, clams, and other invertebrates in their mouths, and in so doing attain a wide range of body sizes. Complex food chains also emerged in which herbivorous fish grazed on plants, carnivorous fish ate the herbivores, and those predators were themselves eaten by still larger carnivorous fish.

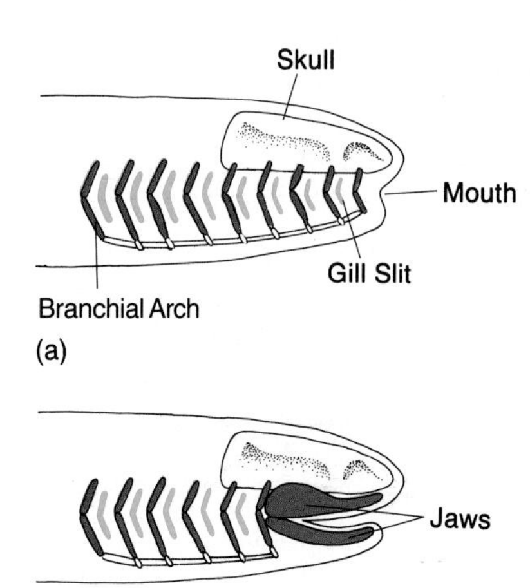

Figure 24-30 THE EVOLUTION OF JAWS.
Jaws probably arose from the anterior branchial arches of the gill system. The anterior branchial arch skeletal pieces are believed to have become altered to form the upper and lower jawbones. As that occurred (a), the number of gill slits decreased (b).

Scientists surmise that gills came to be used for both feeding and respiration in the jawless ostracoderms and that as the jaws took over the feeding function in the acanthodians, the gills remained as respiratory exchange sites.

While the skull and skeleton were bony in these early jawed fishes, the adults had flexible notochords rather than complete vertebral columns. Fossils do show small, separate bony arches over the spinal cord that probably led to the vertebral column in later jawed fishes.

The early jawless fishes were probably limited to movement along the sea bottom, sucking up mud and food particles like self-propelled vacuum cleaners. But the ancient acanthodians showed another structural advance: *fins*, or swimming appendages. At first little more than sharp spines, these evolved into highly sophisticated, movable structures that allowed acanthodians and their descendants to maneuver more easily and swim at a range of depths, speeds, and directions.

It is of great evolutionary significance that two pairs of fins, so-called *pectoral* (anterior) and *pelvic* (posterior) fins, are supported by *girdles*, sets of bones or cartilage that provide a firm base to which muscles can attach and against which the paired fins can move. As we will see, fin girdles were forerunners of the shoulder and hipbones of land vertebrates.

(a)

(b)

Figure 24-31 CHONDRICHTHYES: CARTILAGINOUS FISHES.
(a) This streamlined black-tip shark *(Carcharhinus maculipinnis)* is a voracious predator. Note the gill slit openings posterior to the eye. (b) A blue-spotted stingray *(Taeniura lymma)* rests quietly on this Red Sea reef off Egypt. The greatly expanded pectoral fins are almost like wings that flap through the water.

Chondrichthyes: Fast-Swimming Predators

One ancient lineage of jawed, heavily armored fish, the *placoderms*, gave rise to **Chondrichthyes** (see Table 24-2). This class, which contains modern sharks, skates, and rays (Figure 24-31), is distinguished by a lightweight but strong skeleton made only of cartilage.

Cartilaginous fishes have huge, oil-storing livers that lend buoyancy, which in turn leads to speed and maneuverability and contributes to the success of these fishes as voracious predators. The fins also contribute to success, allowing control of swimming direction, as with the shark's pectoral, pelvic, and tail (*caudal*) fins. Pectoral fins provide "lift" at the front end that counteracts lift produced by the caudal fin, which thrusts water backward and downward as it sweeps from side to side during swimming. The angle of the pectoral fins can direct the body upward, downward, or sideward as the animal pursues prey or swims in leisurely fashion.

In skates and rays, the pectoral fins are greatly enlarged into "wings" that literally flap through the water, gracefully propelling the animal.

While remarkable and successful predators, the cartilaginous fishes never diversified to the same extent as did their bony cousins, the Osteichthyes.

Osteichthyes: Adaptability and Diversity

Fishes with bony skeletons, members of the class **Osteichthyes,** arose almost 400 million years ago from bony, jawed acanthodians (see Table 24-2). Two subclasses now predominate: *Actinopterygii,* or spiny-finned fishes, and *Sarcopterygii,* or fleshy-finned fishes. The actinopterygians include the infraclass *Teleostei,* almost 30,000 species of familiar, modern bony fish ranging from freshwater trout to sea bass, from brightly colored reef fishes to giant tuna, from eels to sea horses (Figure 24-32).

Teleosts owe their great diversity in part to the presence of a *swim bladder,* an internal gas-filled organ that can change volume and allow the animal to be neutrally buoyant at nearly any depth. Since pectoral fins and tail are not essential for lift, body and fin shapes can vary

(a) (b) (c)

Figure 24-32 TELEOSTS, THE MOST DIVERSE FISHES.
The earth's oceans, rivers, and lakes teem with seemingly endless varieties of teleosts. (a) The sailfish *(Istiophorus platypterus)*, with its long, bladelike upper jaw and huge dorsal fin that can be raised or lowered out of the way. (b) The masked butterfly fish *(Chaetodon semilarvatus)* of the Red Sea. (c) The rainbow wrasse *(Laproides phthirophagus)*, a common reef fish in both hemispheres.

dramatically, and fins can be used as paddles, brakes, or even true gliding wings, as in the flying fish. We will discuss various aspects of fish physiology in later chapters.

The second subclass of bony fishes, the sarcopterygians, is older than the actinopterygians. It includes a few modern species with fascinating adaptations for breathing air and walking, plus extinct species that were the first vertebrates to crawl onto land more than 360 million years ago. In contrast to the thin, bony fins of teleosts, sarcopterygians have thick *lobed* fins with large bones and muscles. Certain sarcopterygians also have external and internal nostrils, or *nares*, and a good sense of smell.

Sarcopterygians include the living lungfish and coelacanths and the extinct rhipidistian fishes. *Lungfish* are rare freshwater fish that live in shallow rivers and lakes in Australia, South America, and Africa. They have both gills and lungs, relying on the former when their environment is wet and the latter when it is dry, and they are locked in a protective mud and mucus "cocoon."

Coelacanths are large (up to 1 m or so) primeval-looking fishes once thought to be extinct, but rediscovered off Madagascar in 1938 (Figure 24-33). They possess a fat-filled swim bladder for buoyancy, analogous to the shark's fat-storing liver, and pectoral and pelvic fins that move in a coordinated way during slow swimming, much as the forelegs and hind legs of land vertebrates move during walking.

The final group of fleshy-finned fish, the extinct *rhipidistians*, were probably the ancestors of the land vertebrates. The muscular lobed fins of these fishes probably allowed them to "walk" along the bottom of shallow ocean bays and tidal flats. Rhipidistians lived during the Devonian period, about 350 to 400 million years ago, and may have pursued insects and other invertebrate food sources up onto land. The oldest fossilized animal tracks of vertebrates yet found were left by rhipidistians in 360- to 370-million-year-old sandstone formations on the Orkney Islands, off the northeastern coast of Scotland.

The evolution of the fishes was far from linear. During the Devonian period, the age of fishes, the waters literally teemed with a bewildering array of ostracoderms, cyclostomes, acanthodians, early actinopterygians, lungfish, and rhipidistians. The first amphibians also arose from ancestral fish and lived among this Devonian variety.

Figure 24-33 A COELACANTH *(Latimaria chalumnae)*: A FLESHY-FINNED LIVING FOSSIL.
This rare fish closely resembles its ancient predecessors. The fish is about a meter in length and can move slowly forward using its thick-based fins, such as the posterior-dorsal one seen here.

AMPHIBIANS: VERTEBRATES INVADE THE LAND

Fossils show that early fishlike animals of the class **Amphibia** (see Table 24-2) lived as land vertebrates perhaps as long as 350 million years ago and gave rise to the familiar amphibians—frogs, toads, and salamanders. The oldest amphibian fossils yet found were *ichthyostegids*, animals up to 100 cm long. These shared many traits with the rhipidistian fishes, including similar teeth and skull bones; intact notochords; small bony scales in the skin similar to fish scales; and rows of sensory structures called the *lateral line*. However, they also possessed some unique adaptations for life on land: (1) a skeletal and muscular system better adapted to locomotion on land; (2) skin conducive to gas exchange; (3) the ability to return to watery habitats for reproduction; and (4) better sensory equipment for receiving airborne sound waves. Thus, the ichthyostegids are almost a perfect "missing link."

Amphibian Adaptations

Lacking the buoyant effect of water to support body weight, every land animal faces two primary problems: support and locomotion. Fossils show that the earliest amphibians already had large forelimb and hind limb bones like those found today in modern amphibians, reptiles, birds, and mammals. These limbs had powerful muscles to help lift the belly off the ground and to walk, albeit slowly and with the limbs oriented more outward than forward.

The notochord of the early amphibians was surrounded by U-shaped bones, and the spinal cord was protected by bony arches called *neural arches*. Only in later amphibians did the complete vertebral column arise, with each individual vertebra a fused bone made up of a *centrum* (derived from the U-shaped bone) and a neural arch. In adults, the notochord is pinched into pieces that yield *disks*, the shock-absorbing pads between vertebrae.

In a land vertebrate, such as one of the early amphibians, the full set of vertebrae form a strong "suspension girder," while the attached arches form a hollow canal that surrounds and protects the spinal cord. The body weight hangs on the "girder," is transferred through the pectoral and pelvic girdle to the limbs, and from the limbs to the ground.

Another challenge for life out of water involved gas exchange and drying out in the air. Nearly all amphibians possess true lungs but have only an inefficient mechanism of swallowing movements to fill and empty them. To compensate for this inefficient air pumping, amphibians have a scaleless skin (with embedded blood vessels) that releases most of the body's excess carbon dioxide and absorbs about half the oxygen the animal needs. Since gas exchange takes place by diffusion through a layer of water, much of an amphibian's skin surface must remain moist at all times. This allows water to continually evaporate from the skin surface and to carry away body heat with it. Consequently, no amphibian has a high body temperature or a rapid rate of metabolism, and nearly all are limited to moist habitats and behaviors that minimize evaporation.

Amphibians are tied to moist habitats for another equally compelling reason: Their eggs have the same kind of clear, jellylike covering as some fish eggs and can easily desiccate, or dry out. Most modern amphibians must return to water to reproduce, no matter how widely they range on land.

Modern Amphibians

Modern amphibians (Figure 24-34) include the tailless frogs and toads (order *Anura*, derived from one ancestral line; the tailed salamanders and newts (order *Urodela*), of separate derivation; and the legless salamanders (order *Apoda*). Like their ancestors, most modern amphibians live in or near water and as adults have a strictly carnivorous life-style; they consume insects, worms, and other animals but usually not plant matter.

Most frogs, toads, and salamanders breed and lay their eggs in water, although some use such damp places as wet earthen burrows or small watery pools in trees. The embryos usually develop in the water, becoming tadpoles that swim about and feed for varying lengths of time, depending on the species. Details of the metamorphosis from tadpole to frog or salamander appear in Chapter 35.

REPTILES: ADAPTATIONS FOR DRY ENVIRONMENTS

About 40 million years after the first amphibians crawled onto the muddy shores, creatures about the size of small dogs were coexisting with the amphibians. These new animals were the *Cotylosauria*, the first reptiles, and are the ancestral group from which the dinosaurs and the modern representatives of the class **Reptilia** stemmed (see Table 24-2), a class that includes lizards, snakes, iguanas, crocodilians, turtles, and tortoises. The plants and invertebrates of the Carboniferous landscapes presented a diverse and rich food source for any land vertebrates that could resist desiccation and venture inland into higher, drier areas.

(a)

(b)

Figure 24-34 MODERN AMPHIBIANS.
(a) Golden toads *(Bufo periglenes)* live only along the Continental Divide in the Costa Rican Cloud Forest Reserve. Here, a male (right) sits astride a larger, mottled female as a second male (left) looks on. (b) This red spotted salamander *(Notophthalmus viridescens)*, like all amphibians, must remain in moist habitats so that its skin won't dry out.

Reptilian Adaptations

Ancient and contemporary reptiles shared certain key adaptations that greatly expanded their range of habitats. These include shelled eggs, dry and scaly skin, and mechanisms for excretion that greatly restrict water loss.

The reptilian egg is essentially a pool of water and a supply of food for the developing embryo—all surrounded by membranes and a shell that is either leathery and collagenous or brittle and made of calcium carbonate. It is referred to as a *cleidoic*, or "sealed off," egg, but the shell is perforated by several thousand tiny pores, which allow oxygen to pass into and carbon dioxide to pass out from the embryo. Since these eggs minimize moisture loss, they need not be laid in water. Thus, like seed plants, reptiles are free from dependence on standing water for reproduction.

Reptilian eggs are larger than amphibian eggs and contain enough yolk to sustain the developing embryo until it can survive on land. Thus, when the baby reptile hatches, it is not a larva, but a free-living animal capable of breathing, moving in search of food, and resisting desiccation. Because fertilization must occur before the tough shell is laid down, reptilian mating behavior culminates in internal fertilization, as does that of the reptiles' descendants—the birds and mammals. Some reptiles evolved the bearing of live young (viviparity) rather than the laying of eggs.

The reptile's dry, scaly skin—composed of dead, dry cells containing keratin protein, plus lipids that block water loss—protect the animal once hatched. The skin can be dry and impervious to water loss because reptiles have an expandable rib cage, a set of rib bones around the chest region that fills and empties the lungs with great efficiency. As a result, the lungs are the sole site of oxygen and carbon dioxide exchange, and the dry-skinned animals can inhabit drier inland areas.

Finally, reptiles evolved with enzymes and kidney functions that allow them to excrete a pasty guano instead of watery urine, and this conserves a large quantity of water otherwise lost (details in Chapter 33).

The Age of Reptiles

The entire Mesozoic era—dominated by flying, swimming, and running dinosaurs—has been called the age of reptiles. The cotylosaurs first appeared about 310 million years ago, and by the end of the Paleozoic era, some 85 million years later, they had radiated into a host of reptilian groups that inhabited the shifting continental landmasses. One early line of the cotylosaurs gave rise in the late Permian period to the *therapsids*, an equally diverse group that contained the direct ancestors of the mammals (Figure 24-35).

Toward the end of the Permian period, about 230 million years ago, some 96 percent of all animal types—invertebrates and vertebrates—were wiped out. One important line of reptiles, the *thecodonts*, did survive, however, and eventually gave rise to the dinosaurs and to the modern crocodilians (see Figure 24-36). The early thecodonts were similar to small lizards with very short forelimbs; they probably ran on their two hind legs and thus were *bipedal* (two-legged). Some early dinosaurs were bipedal, but later ones were four-legged herbivores and carnivores.

Nearly all dinosaurs were large as adults, weighing thousands of kilograms in some cases, and they may have been highly active, warm-bodied creatures. Some biologists argue that the heat was self-generated, as in mod-

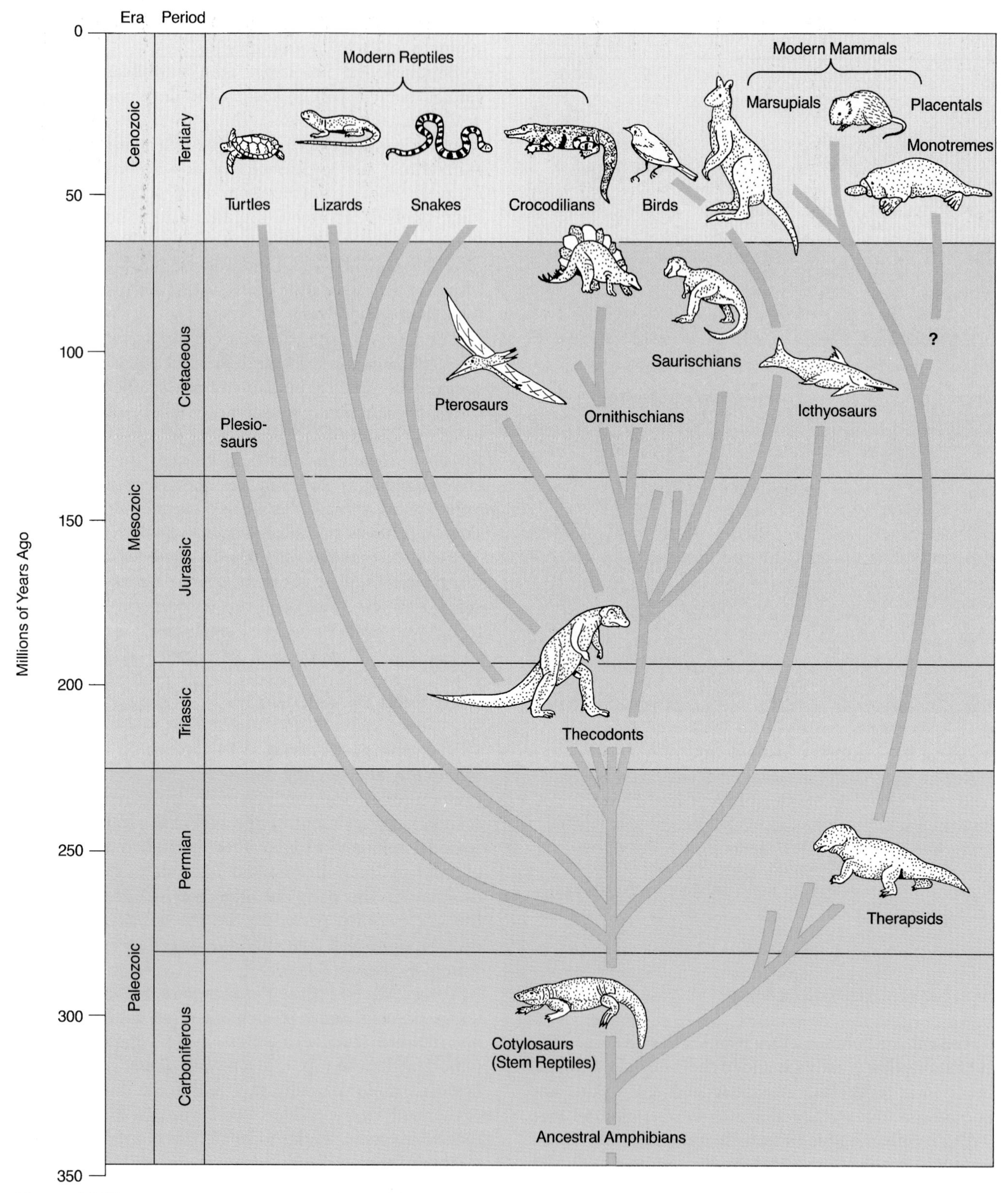

Figure 24-35 THE EVOLUTIONARY RELATIONSHIPS OF THE REPTILES AND THEIR DESCENDANTS, THE BIRDS AND MAMMALS.
This complex family tree of the reptiles is really like viewing a bare maple tree in mid-winter—large numbers of reptilian types are omitted, and only the "main branches" are shown here. The dinosaurs—ornithischians and saurischians—were especially diverse. The flying pterosaurs and birds are very distinctive lineages. The therapsid-mammal lineage is an ancient one. New evidence suggests that monotremes are closely related to placentals and marsupials.

(a) (b) (c)

Figure 24-36 SOME LIVING REPTILES: ANCIENT SURVIVORS IN MODERN HABITATS.
Most reptiles share certain traits, such as dry skin, an expandable rib cage, and cleiodic eggs, but their body forms and lifestyles are quite diverse. (a) These salt-water crocodiles *(Crocodylus porosus)* are heavily armed—their teeth are viciously sharp, their jaws are powerful, and their heavy tail can lash sideways to deliver a crushing blow. (b) A baby Galápagos tortoise *(Testudo elephantopus)* breaks from its egg. (c) The aqua-colored chameleon (*Chamaeleo* species) is a small tree-dwelling reptile that feeds on insects, snatching them with a long, sticky tongue.

ern warm-blooded birds and mammals, but considerable evidence suggests that many species lived in warm Mesozoic forests and swamps, absorbed heat from sunlight during the day, and retained a good deal of the heat overnight.

Many reptilian lineages reinvaded the water or took to the air. *Ichthyosaurs* were beautiful, graceful creatures with features that parallel those of modern sharks and porpoises (see Figure 24-35). And the *pterosaurs* were flying vertebrates that descended from the thecodonts (see Figure 24-35). A few were the size of pheasants or large bats, but most had large wing spans, some up to 15 m (48 ft), with huge sheets of skin acting as flight surfaces. The pterosaur's chest and leg bones were light, and these animals could probably glide well and flap fairly effectively, but walked clumsily on land and lost large amounts of moisture through their wings. Their aerial niche was later filled by the birds and bats, which evolved quite independently from the pterosaurs and each other.

At the close of the Cretaceous, 65 million years ago, the last dinosaurs disappeared. Their demise was not as sudden as some have proposed: Paleontologists working in Montana recently found bones from dinosaurs that were alive 41,000 years after the purported final extinction. And the dinosaurs did not die out alone; many types of plankton and tropical reef organisms also became extinct.

Intense volcanic activity in the late Cretaceous period probably depleted atmospheric ozone and led to a prolonged acid rain. This could have lowered sea pH, altered oceanic chemistry, and killed plankton. In addition to volcanic activity, continental drift and mountain building during this period might have caused temperature extremes, both daily and seasonal, that created severe problems for dinosaurs and contributed to their eventual extinction.

Another hypothesis is that an enormous object—perhaps a meteorite—slammed into the earth in the late Cretaceous period, vaporized on impact, and threw an immense cloud of ash and dust into the atmosphere. Global fires apparently followed, burning perhaps 10 percent of the surface biomass and further contributing to the dense soot clouds, which could have lowered temperatures by blocking sunlight.

Whatever the geological events and their consequences, many kinds of animals alive in the late Cretaceous period, at the end of the Mesozoic era, did survive. The survivors included the smaller "cold-blooded" reptiles—turtles, lizards, snakes, and crocodiles—as well as mammals, birds, amphibians, teleost fishes, lungfish, insects, worms, many other invertebrates, plants, fungi, protists, and monerans. It was the mammals, with their fur insulation, capacity to generate body heat, and placentas that so rapidly radiated at this time. Perhaps competition from mammals, the intense volcanic activity, and sensitivity to fluctuating and cool temperatures all contributed to the extinction of the dinosaurs.

Modern Reptiles

Three orders of reptiles survived the extinctions at the end of the Mesozoic era. The order *Crocodilia* is closest to the dinosaurs and radiated during the Cenozoic (beginning about 65 million years ago) into the alli-

ARE MASS EXTINCTIONS CYCLICAL?

Several recent theories hold that mass extinctions such as the demise of the dinosaurs occur in regular, predictable cycles rather than as single catastrophic events. If true, then the extinction of many species may be due more to bad luck than bad genes.

Two paleontologists from the University of Chicago, David Raup and John Sepkoski, made a startling discovery in 1983. While graphing the patterns of extinction among 9773 genera of marine species over the last 250 million years, they saw about ten peaks, each about 26 million years apart (Figure A). These peaks suggested that some agent, at work over extremely long but regular intervals, could periodically disrupt biological systems on earth, wiping out large numbers of organisms and opening up a race among the survivors to repopulate the blighted areas.

Geologists and astronomers began searching for solar or galactic phenomena that might occur on a 26-million-year cycle and suggested that showers of comets might periodically strike the earth, throw up huge clouds of dust and debris, block sunlight, drastically reduce the levels of photosynthesis, and lead to mass extinctions. Luis and Walter Alvarez at the University of California at Berkeley have proposed that this sort of event led to the demise of the dinosaurs.

Subsequent analyses by other astronomers and paleontologists have challenged the cyclical nature of the mass extinctions. More important, however, is the concept that unpredictable catastrophes play a major role in shaping the evolutionary history of life on earth. Organisms have not evolved defenses against such drastic events, partly because organisms have life cycles that are minute in comparison with events that occur at intervals of tens or hundreds of millions of years.

Thus, natural selection would favor organisms with attributes that allow survival in the cold and reduced light following such an event, just as it continually selects organisms in the intervals in between catastrophes. Whatever the regular or irregular intervals, mass extinctions may have opened the way for equally massive radiations of surviving species and thus may help explain the overall evolution of life on earth.

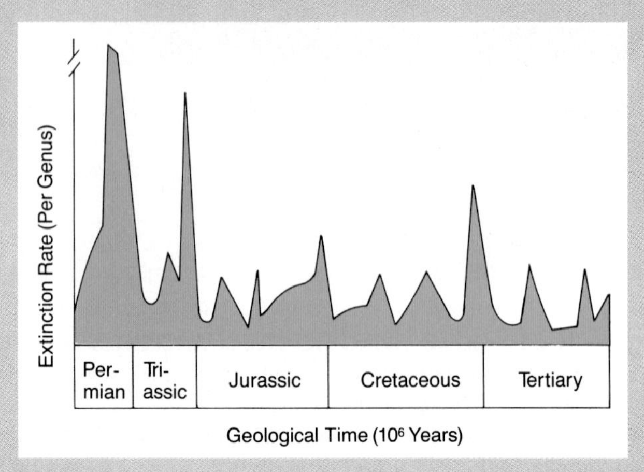

Figure A
Mass extinctions at 26-million-year intervals.

gators of China and the United States; the caiman of Central and South America; and the crocodiles of Asia, Africa, Australia, and Central America (Figure 24-36a).

Tortoises and turtles (order *Chelonia*) arose from an ancient stem group that survived both the Permian and Cretaceous extinctions. Their protective *carapace,* or shell, of bone or leathery skin and their production of cleidoic eggs contributed to their success (Figure 24-36b).

The order *Squamata* includes lizards, snakes, and iguanas (Figure 24-36). Living along the ocean shore and in jungles, temperate regions, and deserts, lizards are among the most versatile of the reptiles. Snakes first appeared late in the Mesozoic era and are essentially legless lizards with modifications for locomotion and for killing and swallowing prey. Many snakes manufacture venom in their salivary glands and can swallow prey whole with jaws that open extraordinarily wide because they are not attached to the rest of the skull.

BIRDS: VERTEBRATES TAKE TO THE AIR

Members of the class **Aves** (Latin for "birds"; see Table 24-2) probably arose in the Triassic period, about 225 million years ago. The first known bird, *Protoavis,* already had light, hollow bones for flight, but was like a small, meat-eating, fast-running, bipedal dinosaur in other respects.

The much-studied fossils of another ancestral bird, *Archaeopteryx,* reveal a key feature of avian physiology and flight—feathers. Because they are made of dead cells, feathers become dehydrated and are exceedingly light. Thus, they can be large and form the flight surface of the wings and tail without weighing much and without being susceptible to water or heat loss.

Feathers probably originated as overlapping, fringed extensions from the ends of reptilian scales that trapped a layer of stagnant air next to the skin to keep excess heat *out*. Once present, feathers could have come to serve the opposite purpose in cooler environments: to prevent heat loss from the body. Thus, they were probably a precondition for the development of *homeothermy*, or warm-bloodedness, in birds as well as structures for flight.

A bird's skeletal and muscular system is equally well adapted to flight. The bones are particularly light and strong; the large pectoral, or breast, muscles powerfully raise and lower the wings; the lower legs and feet are little more than scaly skin covering bone and tendon, and these can be moved rapidly during landing or while grasping prey or can be folded against the body like airplane landing gear so they will not impede flight. Finally, although birds lack forepaws to manipulate objects, their highly flexible and somewhat elongated necks allow them to use their beak to build nests, crack seeds, or spear tasty insects or minnows.

Biologists speculate that wings may have evolved when birds' fast-running *bipedal* ancestors began to hold out their forelimbs while running to escape from predators and glided to the safety of a tree branch. Forelimbs became true wings, of course, only after feathers evolved for insulation and had elongated to provide extra surface area. In all other airborne vertebrates—bats, flying foxes, flying squirrels, and the extinct pterosaurs—sheets of skin became the wing surfaces, and early flight involved a gliding rather than a flapping motion.

Today, Aves is a large, highly diverse class with 8,600 species of water birds, songbirds, and birds of prey, including huge flightless runners, deep divers, and migrators that travel thousands of miles each year (Figure 24-37). All bird species lay cleidoic eggs, and none evolved the capacity to bear live young.

MAMMALS: WARM BLOOD, HAIR, MAMMARY GLANDS, AND A LARGE BRAIN

A true mammal can be distinguished by the production of milk for its young in mammary glands, a high internal body temperature, and the presence of hair. Mammals include the primates, bats, whales, cats, and many other familiar animals. The history of the class **Mammalia** (see Table 24-2), began with the therapsids in the distant Mesozoic era, the age of reptiles. It is difficult to say exactly where those mammal-like reptiles end and the first mammals begin, but 180-million-year-old fossils show characteristic mammalian bone structures.

Small mammals lived alongside the dinosaurs for millions of years, yet did not become predominant until the Cenozoic dawned. The explanation may lie in the efficiency of food utilization. A grassy meadow that supported 11 cow-sized dinosaurs could only support one modern cow. A mammal turns 90 percent of its food energy into heat to keep the body warm, while reptiles, in contrast, turn 60 to 90 percent of food energy into new reptilian biomass. With this metabolism, dinosaurs had a distinct advantage in the warm Mesozoic. But once global climates cooled in the Cretaceous period, small insulated mammals that could generate high internal body temperatures could better survive and exploit the new environmental conditions.

Fossils of the primitive and fascinating **monotremes** date back to 100 million years ago. Their modern relatives, the duck-billed platypus of Australia and two species of spiny anteaters, or echidnas, of Australia and New Guinea retain certain reptilian traits (Figure 24-38). In

(a)

(b)

(c)

Figure 24-37 ADAPTIVE RADIATION OF BIRDS.
The diversity of modern birds is reflected in their bodies and behaviors. (a) A spruce grouse (*Dendragapus canadensis*) cocks a wary eye and ear for danger lurking in the thickets. (b) Two huge waved albatrosses (*Diomedea* species) "presenting" during their courtship ritual on a South Sea island. (c) Like a skilled weaver at work, a thick-billed weaver finch (*Amblyospiza albifrons*) perches on its roosting platform and weaves its nest in Zululand, South Africa.

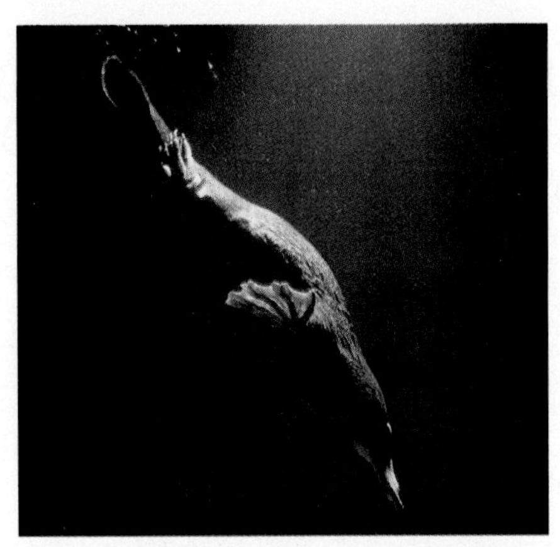

Figure 24-38 THE DUCK-BILLED PLATYPUS. The duck-billed platypus *(Ornithorhynchus anatinus)* is a monotreme, a primitive mammal that both lays eggs and nurses its young with milk produced in mammary glands.

particular, these strange monotreme species lay small leathery eggs, but, like all other mammals, nurse their young with milk after hatching.

The stem groups of other modern mammals appeared later in the Cretaceous period. These mouselike creatures, resembling modern tree shrews, radiated into the marsupials and the placental mammals. **Marsupials**—pouched animals such as the kangaroo and opossum—apparently originated and diversified in North America. Later, dozens of species evolved on the then-isolated continents of South America and Australia. The **placentals** diversified into hundreds of rodents, cats, hooved animals, and others on the continents of North America, Europe, Asia, and Africa.

Mammalian Adaptations

Mammals show several important physical innovations. Their high, constant body temperature is possible because hair serves as an excellent form of insulation; stored fat or blubber acts as both insulation and a source of heat energy; metabolic pathways produce heat (see Chapter 33); and efficient respiratory and circulatory systems deliver large quantities of oxygen needed for the generation of heat.

A second significant physical advance in mammals entailed a set of new reproductive structures and processes based on old reptilian designs. In monotremes, the fertilized egg, and then the developing embryo, is maintained in the uterus for about 2 weeks as the corpus luteum of the ovary (see Chapter 18) secretes female steroid hormones. Various membranes and a shell are then built around the embryo, and the female platypus or echidna lays a shelled, reptilian-like egg. Once the primitive newborn chips its way out, it begins to suckle milk from its mother's mammary glands.

These gestation processes are modified further in marsupials. The embryo develops in the uterus for about 3 to 4 weeks, before the tiny, bean-shaped marsupial is born. The embryo cannot grow any larger because a placenta does not develop in most marsupial species. At birth, the blind marsupial—still little more than a fetus—crawls from the birth canal through the mother's abdominal fur and into the *marsupium,* or pouch. There it finds the teat of a mammary gland, attaches to the tip of it, and proceeds to feed, grow, and develop for up to 5 months or more (Figure 24-39a).

A foal, a human baby, or a newborn whale is huge compared to a marsupial, because the placenta meets the growing embryo's needs for gas, nutrient, and waste exchange. The placental embryo is also protected from attack by the mother's immune system and so can grow far larger than the marsupial embryo. A blue whale, for example, can weigh 2,700 kg (3 tons) at birth—27 million times more than the largest newborn marsupial. Moreover, the placental mammal is much more mature at the time of birth—the foal can stand, and the whale can swim.

Trends in Mammalian Evolution

A fantastic diversification has taken place among mammals during the 65 million years since the Cenozoic era (the age of mammals) began. That diversity is based on several factors:

1. *Increasingly sophisticated temperature regulation.* This allows the marsupials and placentals to live in a broader range of habitats than the monotremes.
2. *Increased body size.* Within nearly every mammalian order, there is a trend toward larger size, apparently related to heat balance: The larger the body, the lower the ratio of surface area to body mass.
3. *Diversification of tooth shape.* Mammals have evolved a wide array of tooth shapes and functions that correlate with food sources: wide, flat molars and premolars (bicuspids) for grinding grains and grasses or for crushing bones; sharp, conical canines for slicing and tearing flesh; and flat, chisel-like incisors for gnawing on woody plants or grazing on grasses.
4. *Elongation and specialization of limbs.* Mammals have developed diversified limb structures that permit many different locomotory styles and behaviors. These limbs include the elongated leg, ankle, and toe bones of the antelope and jack rabbit; the powerful arms of primates; the bat's wings; and the porpoise's streamlined flippers.

(a)

(b)

(c)

Figure 24-39 MARSUPIALS: RADIATION INTO MANY TYPES.
Although all marsupials share reproduction using the pouch and mammary glands, they differ widely in life-styles. (a) The development of young marsupials such as this brush tail opossum *(Trichosurus vulpecula)* is completed in the maternal pouch, where the tiny young remains attached to and is fed by way of a teat. At birth, the marsupial is bean-sized and has few well-developed features. After several weeks in the pouch, the baby is seen with its head poking out of the marsupium. (b) Kangaroos are fast-hopping grazers. Note the joey (or young) of this red kangaroo *(Macropus rufus)* dangling its head and front limbs outside the pouch, while its hindquarters rest safely inside. (c) The koala spends most of its time in trees, usually eucalyptus, feeding on leaves. This young koala *(Phascolarctos cinereus)* clings to its mother, though it still may return to the pouch for milk.

5. *Increased brain size.* All mammals have an enlarged *neocortex,* the region of the forebrain that controls complex learned behavior (details in Chapter 36).

Modern Mammals

Taxonomists group mammals largely on the basis of their teeth and head bones, placing the monotremes in the infraclass *Prototheria* ("first mammals"), the marsupials in the infraclass *Metatheria* ("middle mammals"), and the placental mammals in the infraclass *Eutheria* ("true mammals").

The surviving prototherians (monotremes)—the duck-billed platypus and the spiny anteaters—are more primitive in some ways than their mammalian cousins: Besides laying eggs, they have pelvic girdles and middle-ear bones that are intermediate in form between those of reptiles and those of other mammals. Platypuses do have specialized electroreceptors to detect food underwater, and they maintain a body temperature close to 32°C despite swimming in near-freezing water.

The kangaroos, koalas, wombats, bandicoots, and other metatherians (marsupials) have a range of life-styles as broad as that of the placental mammals (Figure 24-39b and c). The group radiated first in North America and later in Australia and South America during the Cenozoic era. All North American marsupials became extinct 15 million years ago, but the opossum reentered about 75,000 years ago. Despite their success in isolation, many Australian marsupials are now threatened by imported placentals, especially rabbits, dogs, and cats.

The eutherians (placental mammals) are the most diverse infraclass of mammals (Figure 24-40). Here is a listing of only ten orders and a few characteristics of each:

1. *Insectivora:* shrews, moles, hedgehogs; small, nocturnal, insect-eating mammals.
2. *Chiroptera:* bats and flying foxes; mostly night fliers that use echolocation for navigation.
3. *Primates:* lemurs, tarsiers, monkeys, apes, and humans; limbs, sense organs, and other adaptations for life in the trees.
4. *Rodentia:* mice, rats, beavers, porcupines, squirrels; chisel-like teeth for gnawing wood, nuts, and grains.
5. *Lagomorpha:* rabbits and hares; rapid running and prolific reproduction.
6. *Cetacea:* whales, dolphins, and porpoises; large-brained marine mammals with streamlined bodies.
7. *Carnivora:* dogs, bears, cats, skunks, weasels; a diverse group of meat-eating predators. Suborder *Pinnipedia* includes seals, sea lions, and walruses.
8. *Proboscidea:* Asian and African elephants; distinguished by their versatile appendage, the trunk.
9. *Perissodactyla:* horses, zebras, tapirs, rhinoceroses; ungulates (hooved animals) with an odd

(a) (b) (c)

Figure 24-40 PLACENTAL MAMMALS: FAMILIAR FUR-BEARERS, WITH A MULTITUDE OF BODY FORMS AND LIFE-STYLES.
(a) The long, powerfully muscled arms of this orangutan of Borneo *(Pongo pygmaeus abelii)* allow the intelligent and agile creature to move safely through the trees far above the forest floor. (b) Alaskan fur seals *(Callorhinus ursinus)* are streamlined fish eaters, highly maneuverable in the water, but awkward on land. These marine mammals inhabit coastal areas in the far northern and southern latitudes. (c) Bats are the most diverse mammals, making up one-quarter of all mammalian species; the greater horseshoe bat *(Rhinolophus ferrum-equinum)* shown here uses echolocation (a type of sonar) to find flying insect prey at night.

number of toes (one or three) and special fermenting digestion.

10. *Artiodactyla:* cattle, sheep, deer, camels; ungulates with an even number of toes (two or four) and a special stomach chamber for digestion.

The human species, dating from only about 150,000 years ago, is but one newly sprouted twig on the tree of animal life. We return to our species' evolutionary history in Chapter 46. With our survey of life's variety now complete, we can explore in Part Four and Part Five how members of two major kingdoms, Plantae and Animalia, survive day to day.

SUMMARY

1. Multicellular animals probably arose either from aggregates of flagellated protistan cells or from organisms that resembled the *planula* larva of cnidarians. Multicellularity allows animals the advantages of large size, greater mobility, a more stable internal environment, and greater independence from the external environment.

2. Of the 2 million species of animals alive today, 97 percent are *invertebrates,* which lack a backbone, and 3 percent are *vertebrates.*

3. The sponges, the members of the phylum *Porifera,* are hollow, *sessile* filter feeders organized into three layers: the epithelium, the mesenchyme, and the flagellated choanocytes. Sponges probably arose independently of all the other animal phyla and did not give rise to any other group of animals.

4. Animals that are sessile or *pelagic* (free floating) tend to be *radially symmetrical,* while those that move in one predominant direction tend to be *bilaterally symmetrical* and *cephalized* (possess a head).

5. Many members of phylum *Cnidaria* (hydras, jellyfish, and corals) show a kind of alternation of generations between the polyp and the medusa. Cnidarians are distinguished by their stinging cells, the cnidocytes. They also possess a *hydroskeleton,* carry out *extracellular digestion,* and have a nerve net, a primitive kind of nervous system.

6. The comb jellies (phylum *Ctenophora*) are free-floating medusae with eight rows of ciliated cells—the combs.

7. The *plakula* is classified in the phylum *Placozoa.* It is uncertain whether it is related to sponges or is on the main evolutionary line toward all higher animals.

8. Flatworms (phylum *Platyhelminthes*) are bilaterally symmetrical and have a gut with one opening. Planarians contain *ganglia,* aggregations of nerve cells; nerve cords; and eyespots. The species of the classes *Trematoda* (flukes) and *Cestoda* (tapeworms) parasitize vertebrates.

9. Ribbonworms, members of the phylum *Nemertina,* are the most primitive animals with a complete, one-way digestive system, including a mouth and an anus, and with blood vessels.

10. *Nematoda,* the phylum of roundworms, lack blood vessels and circular muscles, but have a *pseudocoelom* surrounding a one-way gut. Although most roundworms are free-living, the parasitic species cause plant and animal diseases, including trichinosis.

11. The animals with a *coelom*—a fluid-filled cavity surrounded by mesoderm in which many of the internal organs are suspended—are divided into two lineages: the *protostomes* and the *deuterostomes*. The most important difference is that the embryonic blastopore becomes the mouth in protostomes, while the blastopore becomes the anus in deuterostomes.

12. The worms of the phylum *Annelida* are characterized by the *segmentation* of the body into units, each of which contains muscles, blood vessels, excretory nephridia, and ganglia.

13. The phylum *Arthropoda* includes spiders, crustaceans, insects, and mites. It is by far the most diverse animal phylum, with 1 million species. Arthropods usually have specialized body segments, a rigid *exoskeleton*, an open circulatory system, complex sense organs, and a nervous system that allows complicated, rather stereotyped behavior.

14. Members of the phylum *Mollusca* have variations on a four-part body plan: the foot, the visceral mass, the head, and the mantle. The class *Bivalvia* (clams, oysters, mussels) and some species in the classes *Gastropoda* (snails and abalone) and *Cephalopoda* (chambered nautiluses) have shells. The cephalopods—especially the squid and octopuses—have the largest and most complex brain among invertebrates.

15. Members of the phylum *Echinodermata*—sea stars, sea urchins, sea cucumbers, and brittle stars—tend to be radially symmetrical as adults. They have a unique water vascular system that serves as a hydroskeleton.

16. The phylum *Hemichordata* consists of the mobile, wormlike enteropneusts and the sessile, tentacled pterobranchs. Adults of these marine animals possess gill slits.

17. The phylum *Chordata* includes three subphyla: *Urochordata*, *Cephalochordata* (nonvertebrate chordates), and *Vertebrata*.

18. At some stage of their life cycle, all chordates possess *gill slits*, a *notochord*, segmental body and tail muscles (*myotomes*), and a dorsal hollow *nerve cord*.

19. The first-known vertebrates, jawless fishes of the class *Agnatha*, used muscle-driven gill slits for filter feeding. These agnathan ostracoderms also possessed bone.

20. The early jawed fishes, members of the extinct class *Acanthodii*, could ingest larger pieces of food and had paired fins that helped control movement in the water.

21. *Chondrichthyes* include modern sharks, skates, and rays. *Osteichthyes* include the teleosts and other spiny-finned fishes and the fleshy-finned lungfish, coelacanths, and rhipidistians.

22. Modern amphibians include the frogs and toads, salamanders, and legless salamanders and show such adaptations for life on land as a strong vertebral column; strong limb muscles, limb bones, and pectoral and pelvic girdles; and the ability to respire across the moist skin.

23. Members of the class *Reptilia* have dry, relatively impermeable skin and an expandable rib cage; a water-conserving excretory system; and the cleidoic egg, which resists desiccation. Modern reptiles include crocodiles, turtles, lizards, and snakes.

24. Birds, class *Aves*, have feathers for insulation and flight surfaces and bones, muscles, and limbs modified for flying. They maintain a high, constant temperature and generate heat internally.

25. Mammals, including the monotremes, marsupials, and placentals, are insulated with hair or fat and can maintain a high, constant body temperature, as can birds. All mammals feed their newborn and young with milk from mammary glands.

KEY TERMS

Acanthodii
adaptive radiation
Agnatha
Amphibia
Annelida
Anthozoa
Arthropoda
Aves
bilateral symmetry
Bivalvia
cephalization
Cephalochordata
Cephalopoda
Cestoda
Chelicerata
Chondrichthyes
Chordata
Cnidaria
coelom
Crustacea
Ctenophora
deuterostome
Echinodermata
exoskeleton
extracellular digestion
filter feeder
ganglion
gastrocoel
Gastropoda
gill slits
Hemichordata
hemocoel
hemoskeleton
Hirudinea
hydroskeleton
Hydrozoa
Insecta
invertebrate
Mammalia
marsupial
medusa
Mollusca
Monoplacophora
monotreme
myotome
Nematoda
Nemertina
nerve card
notochord
Oligochaeta
Onychophora
Osteichthyes
pelagic
pharynx
placental mammal
Placozoa
plakula
planula
Platyhelminthes
Polychaeta
polyp
Polyplacophora
Porifera
protostome
pseudocoelom
radial symmetry
Reptilia

Scyphozoa
segmentation
sessile
social insect
Trematoda
Trilobita
trochophore
Turbellaria
Uniramia
Urochordata
Vertebrata
vertebrate

QUESTIONS

1. Which of the phyla discussed in this chapter include individuals with radial symmetry? Bilateral symmetry?

2. What are the alternating generations found in many species of the phylum Cnidaria? Do these stages differ in genetic makeup, chromosome number, life-style, or form? Which form produces gametes? Describe briefly the three classes of cnidarians.

3. What are the major evolutionary innovations seen in each of the worm phyla?

4. In what ways do arthropods differ from annelids? What are the subphyla in the phylum Arthropoda?

5. Describe some of the adaptations to land among the subphylum Uniramia.

6. Describe the major classes in the phylum Mollusca. Which contain individuals with a radula? With external shells? With no shells? With twisted or coiled shells? With a large head and brain? With large eyes?

7. Describe the water vascular system of echinoderms. What are the functions of this unique system?

8. Most animals that are bilaterally symmetrical move in a head-first direction. What are some examples of bilaterally symmetrical invertebrates described in this chapter?

9. What features are common to all three subphyla of chordates?

10. What are the functions of gill slits?

11. How does bone differ from cartilage? What are some advantages of each?

12. How did jaws evolve? How did jaws change the life-style of the early fishes?

13. The cartilaginous fishes include what present-day fishes? What sort of "flotation device" do they possess? What sort of "flotation device" do the teleosts possess? What are some advantages of each device?

14. What problems did amphibians encounter in living on land? Why are they restricted to moist environments?

15. What are some of the innovations of reptiles that freed them from dependence on a very moist environment?

16. What innovations aid birds in flying?

17. What unique attributes are shared by all mammals?

18. What features distinguish the monotremes, marsupials, and placentals? What are some features that distinguish each group? Can you name a monotreme that is common in the United States? A marsupial?

ESSAY QUESTIONS

1. Discuss parasitism as an evolutionary strategy as seen in worms. What are its advantages and disadvantages?

2. Mammals have a much more rapid metabolism than do reptiles. What are some advantages of a rapid metabolism? What are some disadvantages?

For additional readings related to topics in this chapter, see Appendix C.

Part
FOUR

PLANT BIOLOGY

This lone flowering plant survives in a coastal sand dune despite extremes of temperature, sunlight, and moisture.

25 The Architecture of Plants

Of the theory of vegetables, or of the growth, propagation, and nutriment of vegetables, our knowledge is only slight and superficial. A close inspection into the structure of plants affords the best ground for reasoning on this subject, and, indeed, every thing beyond it is little better than mere fancy or conjecture.

George Le Clerc de Buffon,
Natural History (1821)

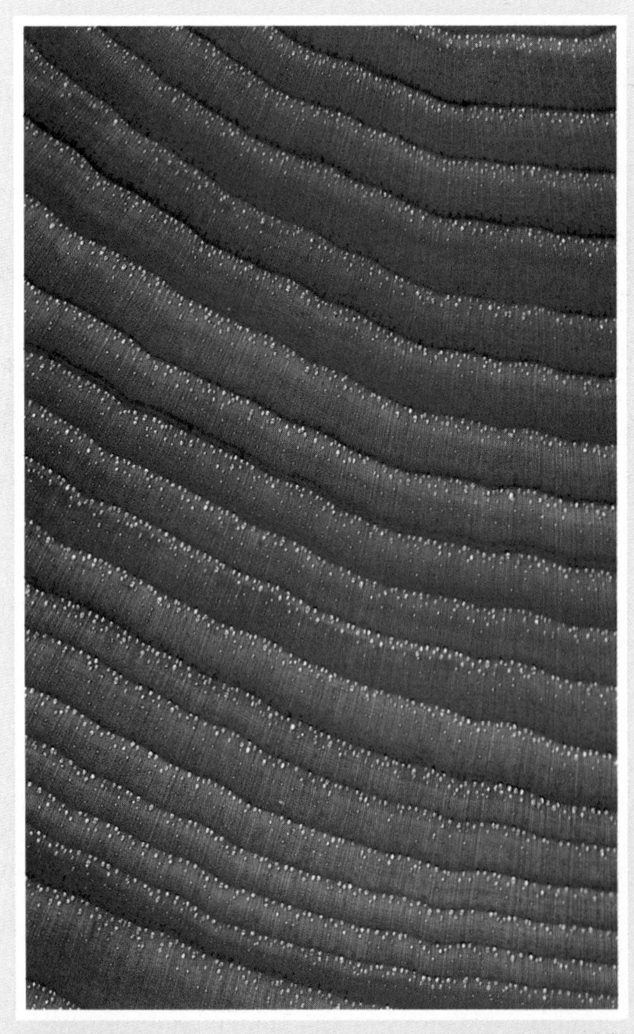

Bitternut hickory rings: living historical records of past years' growth.

Seed plants—gymnosperms and flowering plants—have an inspiring diversity of sizes, shapes, structures, colors, and life-styles, in a host of species from delicate violets to furrowed giants like ponderosa pine or Douglas fir. Despite their diversity, however, the ubiquitous and ecologically important seed plants share a number of characteristic physical structures, including leaves, stems, roots, and reproductive parts. As with all living things, such shared physical structures are the basis of function. And the study of plant anatomy is a foundation that allows us to more fully understand plant physiology—how plants grow, reproduce, collect solar energy, transport materials, absorb water, and resist attack from predators.

With its emphasis on plant form and function, this chapter begins a unit on plant physiology. In general, we will consider the main structures of seed plants and the host of variations that allow them to survive in a nearly endless variety of habitats. Specifically, the chapter discusses:

- Plant structures that overcome the challenges of life on land
- Leaves, living collectors of solar energy
- The stems, which provide vertical support and a passageway for material transport
- The root, which anchors the plant and allows the uptake of water and minerals

MEETING THE CHALLENGES OF LIFE ON LAND

The ancestors to the plants—the algae—lived in shallow seas and bodies of fresh water about 600 to 800 million years ago. The first plants to invade the moist fringes of the barren continents encountered the danger of desiccation; a higher intensity of sunlight than that penetrating watery habitats; rapid daily and seasonal shifts in temperature; lack of buoyant support for the plant body; buffeting by rain and wind; and the limited availability of water, needed by swimming sperm to reach and fertilize eggs.

The leaves, stems, and roots we will discuss in this chapter, as well as the seeds, flowers, and fruits discussed in the next, evolved over tens of millions of years, and together they enabled plants to adapt to this harsher terrestrial existence and to diverge into more than 300,000 species of land plants, most of which are flowering plants. Because they are so prevalent and ecologically important, we will focus mainly on the flowering plants in this and the next three chapters. The two major classes of anthophytes, the monocots (such as Bermuda grass, banana plants, and daffodils) and the dicots (such as potatoes, maples, and sunflowers), have significant anatomical differences in their roots, stems, leaves, and flowers, as listed in Table 23-2. In this chapter, we will encounter three plant tissue systems: the *dermal tissue*, *ground tissue*, and *vascular tissue* (Figure 25-1 and Table 25-1).

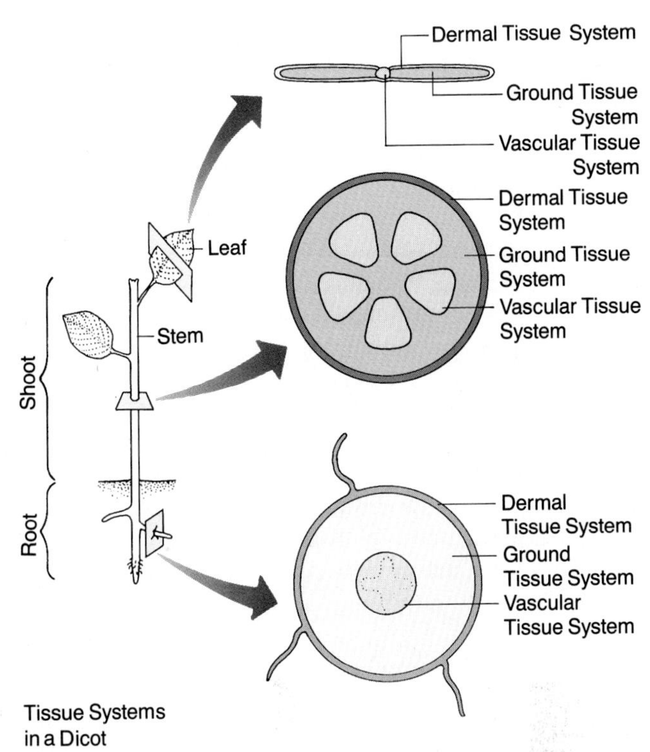

Figure 25-1 TISSUE SYSTEMS IN THE MAJOR PLANT ORGANS.
The many differences in the way cells are arranged can only be seen by looking within the leaves, stems, and roots at the cells that make up the three basic tissue systems of plants: the dermal tissue, ground tissue, and vascular tissue.

LEAVES: LIVING COLLECTORS OF SOLAR ENERGY

Plants trap solar energy in their specialized light-collecting organs, the **leaves.** Human solar engineers would be hard pressed to match the all-around performance of even the simplest leaf: Leaves need little more than light, water, carbon dioxide, and a few minerals to function; leaves often move to track the sun's motion; and light and gases move into the leaf while only some water is lost. What mechanical device could compare to this efficiency, or to the graceful leaves of a willow tree or quaking aspen?

Table 25-1 MAJOR TISSUE AND CELL TYPES OF FLOWERING PLANTS

Tissue System	*Tissue*	*Common Cell Types*	*Description*	*Function*
Dermal	Epidermis	Epidermal cells, guard cells, hair cells	Small living cells; secrete cuticle; flank stomata; discourage predators	Protection of internal cells; prevents water loss; preserves sterility
Ground	Parenchyma	Parenchyma cells	Small living cells; usually thin cell walls	Photosynthesis and storage
	Collenchyma	Collenchyma cells	Elongated living cells; thick cell walls	Support
	Sclerenchyma	Sclerenchyma fibers and sclereids	Usually extremely elongated cells; thick, reinforced cell walls; usually dead at maturity	Protection, support, rigidity, hardness
Vascular	Xylem	Tracheids and vessel elements	Hollow, thick walls of dead cells	Xylem tubes conduct water and minerals
	Phloem	Sieve tube elements	Elongated living cells lacking nuclei; thick cell walls; sieve plates with pores connect successive cells to form sieve tube	Phloem tubes conduct sugars and other nutrients
		Companion cells	Small, elongated living cells with nuclei; adjacent to sieve tube elements in phloem	Involved in transport by phloem's sieve tube elements

AGRICULTURE AND HUMAN CULTURE

Few of us realize that much of human culture is a direct outgrowth of agriculture—the deliberate cultivation of plants. Until about 10,000 years ago, all humans lived as roving hunter-gatherers. Then hunter-gatherer tribes in several parts of the world abandoned their nomadic life-styles in favor of domesticating useful plants. They cultivated rice in Southeast Asia, wheat in the Middle East, maize and beans in North and Central America, and potatoes in the Andean highlands of South America. After harvest, these crops were easy to store for long periods, and they provided carbohydrates, the major source of calories, as well as many of the vitamins and minerals humans need to stay healthy. These starchy foods became the staff of human life, and their domestication led to many changes in human society.

The advent of agriculture led to the establishment of the first permanent settlements and allowed people to remain in one area and produce enough food for their needs. With these original settlements came the need for architecture and engineering, to create buildings, streets, water systems, and bridges. People also needed systems of arithmetic to keep track of their food supplies and population. A large percentage of Egyptian hieroglyphics was devoted to tabulating food stores and agricultural productivity (Figure A), and many Aztec writings are also thought to be vast ledgers tabulating the annual harvests of corn and beans.

Agriculture is closely linked to early astronomy and religion. Calendars were probably created so that farmers could predict the proper times for planting and harvesting. Religions, too, were based on agriculture: Many cultures worshiped the sun and soil fertility—an indication that early farmers knew a great deal about the requirements of plants.

Figure A ANCIENT AGRICULTURE IN EGYPT. Hieroglyphics depict the harvesting of cereal grain crops (middle panels), as well as healthy orchards under cultivation (bottom panels).

Finally, agriculture had direct consequences for our species' social interactions and family size. Communities of people needed to cooperate to terrace fields, dig canals and irrigation ditches, and store food supplies, and this almost certainly led to expanded social skills and perhaps even to castes and social classes. And because families needed more hands to help farm the land, the human population increased. Ironically, this marked the beginning of the population explosion that now strains the earth's resources and ecology—including the soil and mineral resources needed to produce sufficient food for a global population now approaching 5 billion.

The leaf's efficiency at collecting solar energy—the amount of energy it harvests relative to the amount of light impinging on a given surface area—depends on the size, shape, and position of the leaf, as well as on the photosynthetic process itself (see Chapter 8). Most dicots have leaves with two distinct portions: an enlarged, usually flattened region called the **blade** and a stemlike portion called the **petiole,** which connects the leaf blade to the stem of the plant. Plants are often identified by their leaf shapes, and the range is stunning—from long and narrow to rounded, smooth, lobed, toothed, fluted, finely divided, or needlelike. Figure 25-2 shows just a few examples.

Leaves are a successful compromise between the need to collect sunlight and the need to prevent water loss from the plant. In most cases, the broader and flatter the blade, the more light it collects, and the more surface area there is for evaporation to occur. A completely sealed, watertight coating on the leaf could prevent such evaporation; however, it would also prevent the exchange of carbon dioxide and oxygen, so critical to photosynthesis.

Leaf Structure

The leaf structure must somehow accommodate the need for a large, light-absorbing surface area; the need for waterproofing to prevent excess evaporation; the need for gas exchange; and the need for a transport system to carry water and nutrients between the leaves and the rest of the plant. Figure 25-3 illustrates the structures of a typical leaf in cross section. Let us examine each, starting from the outermost and working inward.

Figure 25-2 LEAVES: A VARIETY OF SHAPES TO MEET A VARIETY OF NEEDS.
Leaves have two major parts, a blade and a petiole. Leaf blades exist in many complex shapes and are adapted for the conditions under which the plant lives. (a) A colorful leaf, such as that of the coleus plant, can attract insects to the tiny purple flowers. (b) The needles of this fir tree (*Abies grandis*) are long and narrow. This shape helps reduce evaporative water loss by diverting rather than catching wind. (c) Aquatic plant leaves, supported by the water and not subjected to tearing by the wind, can grow to huge proportions. These water lily leaves (*Victoria amazonica*) are 1–2 m in diameter.

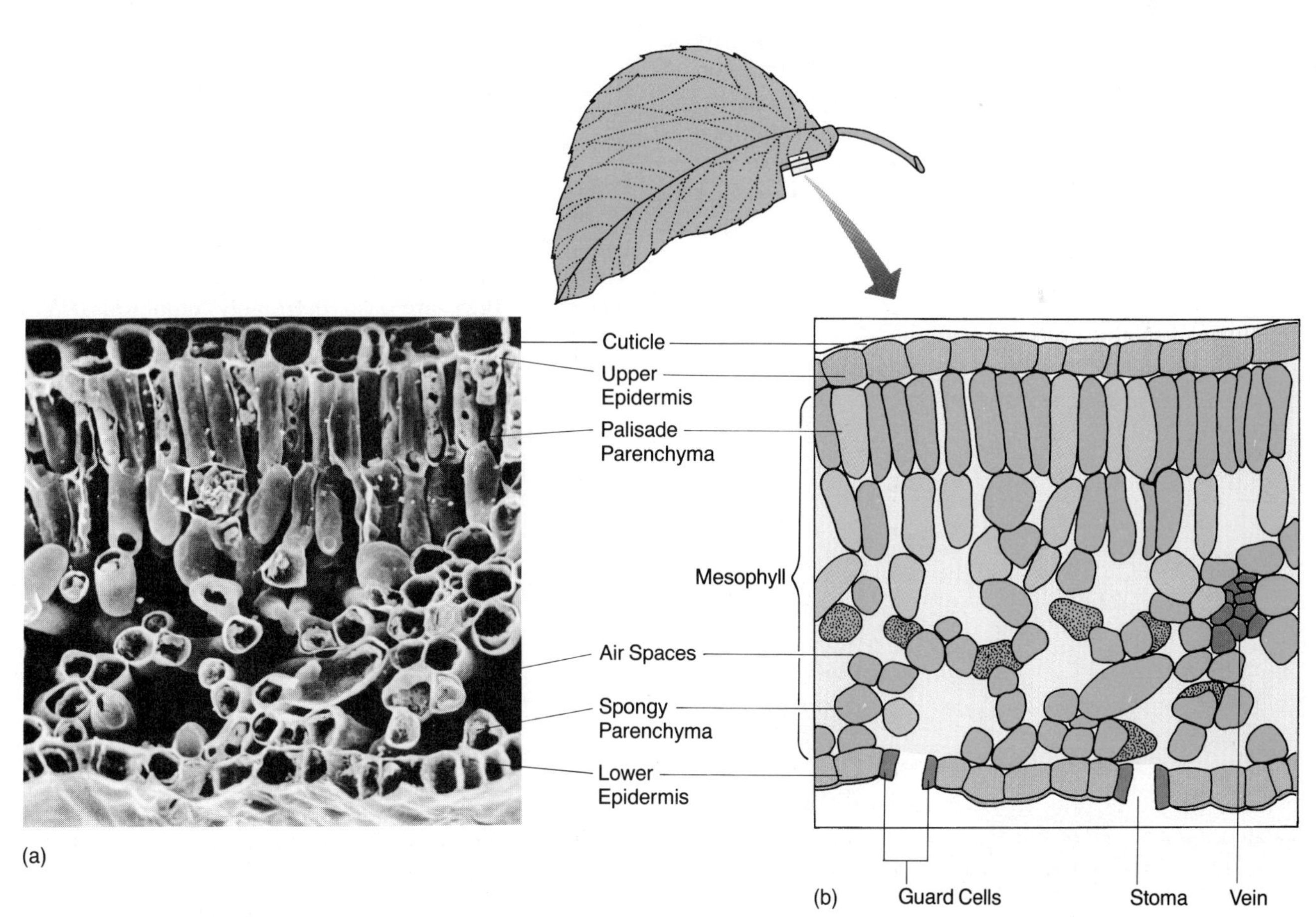

Figure 25-3 ANATOMY OF A LEAF.
(a) This scanning electron micrograph shows an apple leaf in cross section, magnified about 400 times. (b) The various tissue layers and some of the cell types that make up the leaf. The upper epidermis, with its overlying cuticle, seals and guards the upper leaf surface. Palisade parenchyma cells contain many chloroplasts (not shown) and are the major site of photosynthesis. Air spaces permeate the lower spongy parenchyma. A very small vein is visible in cross section (left side of leaf). The lower epidermis and its cuticle seal off and protect the underside of the leaf. The guard cells on either side of each stoma (not shown in the micrograph) control the movement of gases into and out of the plant.

Epidermis: Outer Protection

Leaves are protected on both their upper and lower surfaces by the **epidermis,** a layer of cells, usually one cell thick, that cuts down on moisture loss and helps prevent invasion by microorganisms. In some leaves, certain epidermal cells develop hairs that hold water near the leaf structure, and reduce water loss. The guard hairs that give an African violet its velvety feel or dot the lower surface of a live oak leaf may also make the leaf a difficult place for insects to land or walk (Figure 25-4). In most leaves, epidermal cells secrete a transparent waxy material called **cutin,** which forms an outer layer, or **cuticle,** over the leaf; this seals moisture in the leaf tissues. In arid regions, the cuticle can be quite thick and can give an entire plant, such as a blue spruce or eucalyptus, a whitish, grayish, or bluish hue. Whether a cuticle is thick or thin, it is essentially transparent to incoming light. The epidermal cells below the cuticle are semitransparent, allowing light to pass through to the inner cellular layers. Most epidermal cells lack chlorophyll; thus, they do not carry out photosynthesis.

The epidermis and the cuticle are extremely effective at reducing water vapor loss. Too tight a seal, however, would prevent the normal exchange of oxygen and carbon dioxide, and thus the epidermis is dotted with thousands of tiny pores called **stomata** (singular, *stoma*). Each stoma is bounded by a pair of **guard cells,** which regulate the opening and closing of the pore and thus affect the movement of gases into and out of the plant (Figure 25-5). Guard cells are the only photosynthetic epidermal cells and are usually kidney-shaped in dicots and dumbbell-shaped in monocots. The opening or near-total closure of the stomata regulates the passage of air and water vapor in the aboveground parts of the plant, as Chapter 27 explains in more detail.

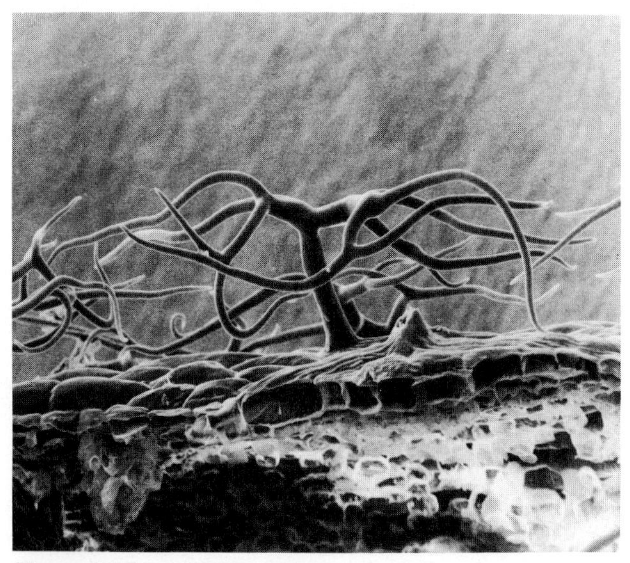

Figure 25-4 GUARD HAIRS.
These incredibly shaped guard hairs are found on the lower surface of a live oak leaf (*Quercus agrifolia*) (magnified about 125×). Their true diminutive size can be appreciated by noting the sizes of the epidermis cells of the leaf's lower surface.

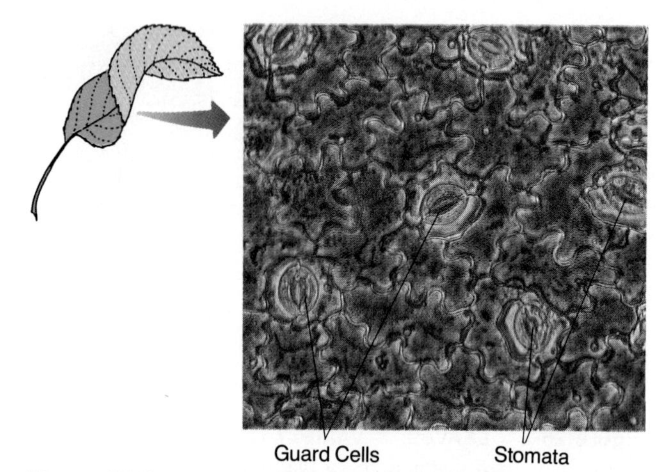

Figure 25-5 REGULATING GAS EXCHANGE.
Stomata regulate the movement of oxygen, carbon dioxide, water vapor, and other gases into and out of the leaf. In this photo micrograph of a morning glory leaf (magnified 200×), the guard cells appear to have closed the stomata, thus reducing gas movement in and out of this flowering plant's leaf.

Stomata are extremely small—only about 100 μm long—but they are also extremely numerous. With about 19,000 stomata per square centimeter, a rose leaf might have nearly 250,000 stomata on its lower surface and about 60,000 on its upper surface. Collectively, stomata represent the major site of water loss from the leaf: Fully 90 percent of a plant's water loss occurs through the stomata. For every carbon dioxide molecule that enters a leaf through an open stoma, several hundred water molecules exit from that same opening (again, details in Chapter 27).

Mesophyll: Photosynthetic Cell Layers

The leaf's main photosynthetic tissue, the **mesophyll** (see Figure 25-3) lies inside the upper and lower epidermis. Mesophyll is bright green because the cells have many chlorophyll-containing chloroplasts, suspended in a clear cytoplasm. The leaf mesophyll tissue is made up of **parenchyma** cells, which are thin-walled and have large vacuoles. Monocots have a single type of parenchyma cell, dicots two types: the **palisade parenchyma** and the **spongy parenchyma,** arranged in two layers (see Figure 25-3). The layers contain cells with specialized shapes, but photosynthesis takes place in both layers.

Palisade parenchyma cells are rod-shaped and are arranged vertically in a layer one or two cells deep just below the translucent upper epidermis—an ideal location for absorbing light energy. Spongy parenchyma cells are irregularly shaped and are loosely arranged in a layer directly below the palisade parenchyma. The ample space around them allows for more contact between their surfaces and gases.

Much of the air entering through the stomata can permeate the spongy parenchyma layer through an air space just interior to each stoma and through smaller channels that branch into the layer like numerous twisting passageways off a main cave (see Figure 25-3a and b). Thus, carbon dioxide, oxygen, and other gases can pass quickly into and out of the spongy parenchyma cells by diffusing across the moist cell walls and plasma membranes. As in a mammal's lung, the total cell surface area for the exchange of gases is enormous in the loosely packed spongy parenchyma layer.

Veins: Pipelines for Transport and Support

If you hold a thin leaf up to a strong light, you can see a pattern of veins inside the leaf blade somewhat reminiscent of the ribs in a kite or a fan. The veins in leaves act not only as a structural framework that helps maintain the shape and stiffness of the blade, but also as a transport system that is continuous with the vascular system in the rest of the plant. The rigid-walled *xylem* cells carry water and minerals and strengthen the veins, while the sievelike *phloem* cells transport sugars and other products of photosynthesis.

In dicots, smaller and smaller veins branch off the major veins in a netlike array that permeates nearly every region of the leaf. The largest vein in a dicot leaf, usually located at the center of the blade, is called the **midrib** (Figure 25-6a). The midrib passes through the petiole, which attaches the leaf to the stem, and joins the vascular systems of leaf and stem. In many monocots, such as iris or corn, veins extend in parallel lines from the petiole to the leaf tip, and midribs are not so apparent (Figure 25-6b).

Leaf Adaptations

From the brilliant coleus to the fragile baby tears, it is obvious that leaves come in an artful palette of colors and a range of shapes and sizes that generally represent adaptations to the demands of a particular environment or season (see Figure 25-2). In tropical rain forests, for example, richly colored leaves often help attract pollinators; and extremely large leaves compete well for the dim light filtering through the forest canopy, yet evaporation is minimal, since the environment is moist.

Where light levels are high and moisture levels are low, leaf adaptations are quite different. Pine needles, for example, are long and thin and have a thick epidermis and cuticle. The needle form not only reduces water loss, but also prevents wind damage, allowing the force of the wind to pass between the leaves.

Plants such as the woodland violet must contend with two distinctly different environments during the same growing season. In early spring, intense sunlight reaches the forest floor where violets emerge, since the majority of the woodland trees are still leafless. The first leaves of the violet plant, so-called sun leaves, are small and well adapted to intense sunshine and efficient light gathering. As leaves emerge on oaks, maples, and other canopy trees and shade the ground, violets produce shade leaves, which are larger than sun leaves, have more surface area, and are a darker green color owing to more chlorophyll. The shade leaves absorb and use more of the dim light that reaches the plant during the summer, and the sun leaves wither and die.

Some leaves are so modified to reduce water loss that they are no longer even recognizable as leaves. For example, the water-storing leaves of the ice plant are so thick that they appear to be stems, and many familiar vegetables—such as cabbage, Brussels sprouts, and onions—consist of thick starch-storing leaves tightly compressed into "heads." Spines are the ultimate leaf modification: No photosynthesis takes place at all, but neither does water loss.

(a) (b)

Figure 25-6 VEIN PATTERNS IN DICOT AND MONOCOT LEAVES.
(a) A large vein called a midrib usually runs down the center of a dicot leaf, as in the large tropical leaf shown here. Smaller veins branch out in a network from the midrib. These veins are subdivided into smaller and smaller passageways, carrying materials within easy diffusion range of every leaf cell. (b) In a monocot, such as these cattail leaves, the veins run in characteristic parallel lines from the petiole to the leaf tip.

Clearly, leaf architecture is well suited to meet the demands of the surrounding environment and to absorb sunlight, minimize water loss, maximize gas exchange, and transport water and nutrients.

THE STEM: SUPPORT AND TRANSPORT

In vascular land plants, **stems** carry out the two critical functions of support and transport. The stem supports the leaves, holding them aloft toward the light. Developmental processes ensure that the leaves grow on the stem in an arrangement that interferes as little as possible with each leaf's light-gathering function.

The stem's dual functions of support and transport are served primarily by vascular tissue, which connects the leaves to the absorptive roots anchored in the soil. The vascular tissue is specialized for moving materials back and forth and is also reinforced to lend strength to the entire plant. In Chapter 27, we consider woody plants with their mature vascular tissue. Here we consider the stems of **herbaceous** dicots and monocots, which normally have short, thin, soft, and nonwoody stems.

Stem Structure

If you cut a cross section of the stem of a dicot, such as a tomato plant, you would see four concentric zones of tissue: epidermis, cortex, vascular tissue, and pith (Figure 25-7).

Epidermis

Stem epidermis is similar to leaf epidermis; it is a layer, one cell thick, of usually nonphotosynthetic cells. The cells secrete a waxy cuticle that prevents water loss, and stomata dot the stem epidermis. Stem epidermis is always present in young plants, but older plants often lack this outer layer.

Cortex

Just inside the epidermis lies the **cortex.** Most of the cortex cells are parenchyma cells—large, thin-walled, regularly shaped structures similar to those in leaf mesophyll. Stem parenchyma cells are usually considered unspecialized, but they can differentiate into a variety of cell types, including **collenchyma** and **sclerenchyma.** Collenchyma cells have particularly strong primary walls (see Chapter 5), thickened at the corners, which lend

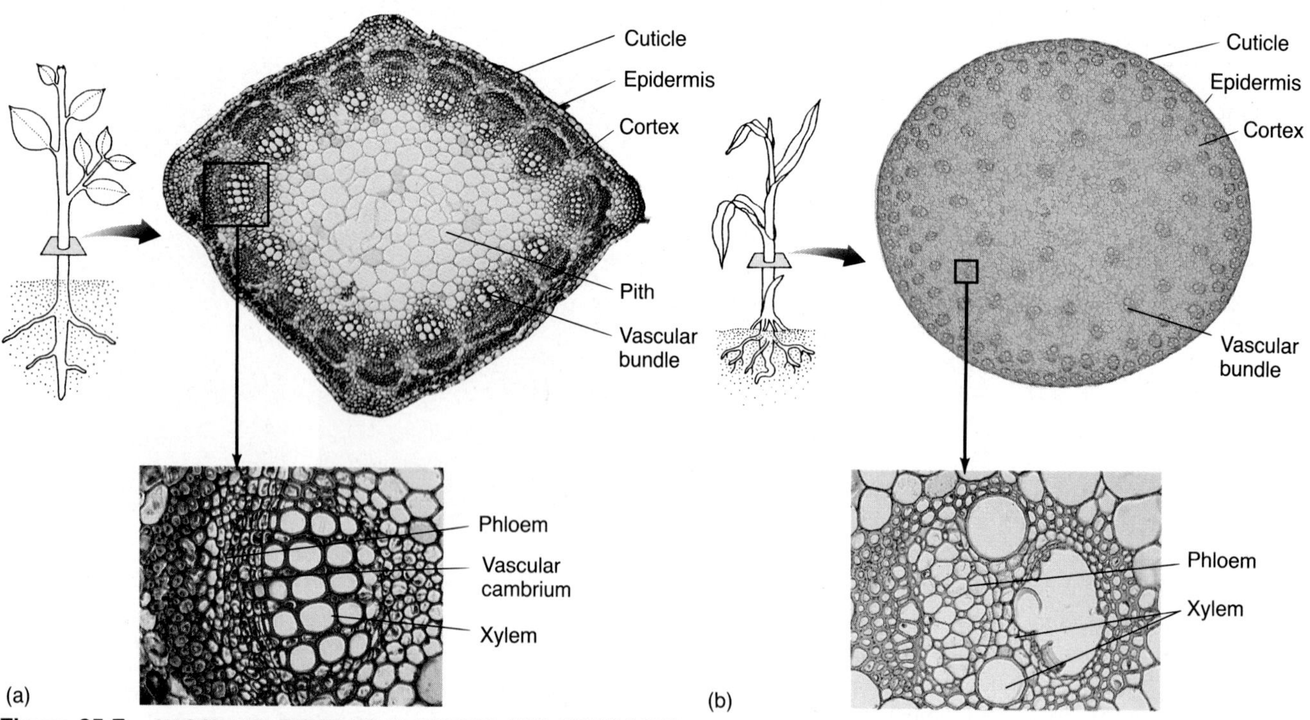

Figure 25-7 VASCULAR BUNDLES IN DICOTS AND MONOCOTS.
(a) The cross section of a dicot (here, alfalfa, magnified about 20×) shows the epidermis, the adjacent cortex, and the vascular bundles arrayed around the central pith. Each vascular bundle has a central growth zone (the vascular cambium), with xylem toward the center of the stem and phloem toward the periphery. (b) In the monocot stem (here, corn, magnified about 6×), the epidermis surrounds the ground parenchyma. The vascular bundles are scattered throughout this parenchyma, but within each bundle, phloem is external to xylem, as in dicots.

extra support to the stem. The cells are usually found in sheets or vertical bands just inside the epidermis. In celery, this cell type makes up the strings of the stalk and contributes to the pleasant crunch you hear when taking a bite. Botanists believe that sclerenchyma cells enable plant organs to withstand various strains, such as gusts of wind, and strengthen the plant so that as it grows, it can better support the ever-increasing weight of the leaves and lengthening stem.

There are two types of sclerenchyma cells: fibers and sclereids. Both have strong secondary walls composed of cellulose that is deposited as the cell matures. **Fibers** are extremely long cells and tend to occur in bundles or cylinders, often toward the periphery of the stem or root. Plant fibers make up the strong filaments in hemp, sisal, jute, and flax. A strengthening substance called **lignin** is present, and gives the cells their hardness and strength. Specialized lignin-containing "stone cells" or **sclereids** can be present elsewhere in the plant and, for example, give pears their gritty texture and nutshells and peach pits their stoniness.

Vascular Tissue

Moving farther inward, the stem zone lying just interior to the cortex is the **vascular tissue,** made up of two types of tissue, the *xylem* and *phloem,* which support the entire plant and conduct materials throughout it. In a herbaceous dicot such as alfalfa, xylem and phloem occur together in a ring of distinctive **vascular bundles,** with phloem exterior to xylem in each bundle (see Figure 25-7). A layer of cells called the **vascular cambium** separates the xylem from the phloem in each bundle.

Pith

The fourth and final zone of stem cells, the **pith,** occurs in the center of a dicot stem (see Figure 25-7). Pith is a storage tissue composed of large, thin-walled parenchyma cells. It provides much of the diameter of a young green stem, but as the stem ages, it plays a progressively less important role.

Monocots have the same four stem structures as in herbaceous dicots, but with some modifications (see Figure 25-7). The epidermal tissue is quite similar in monocots and dicots, but the monocot vascular bundles tend to be scattered throughout a tissue called the **ground parenchyma,** rather than arranged in a ring, and thus there are no discrete zones of cortex and pith.

Stem Vascular Tissue

Xylem and phloem tissues are critical to support and transport in every vascular plant, so let us look at them more closely.

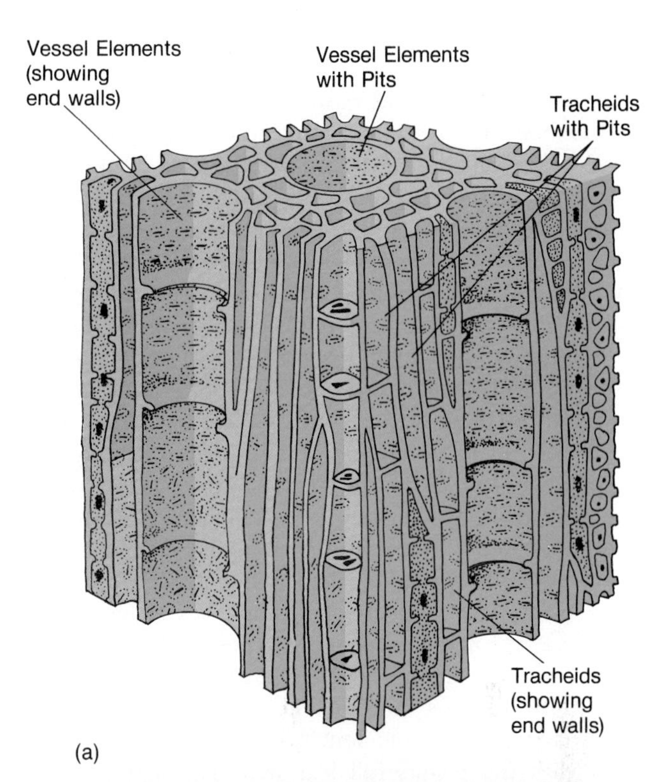

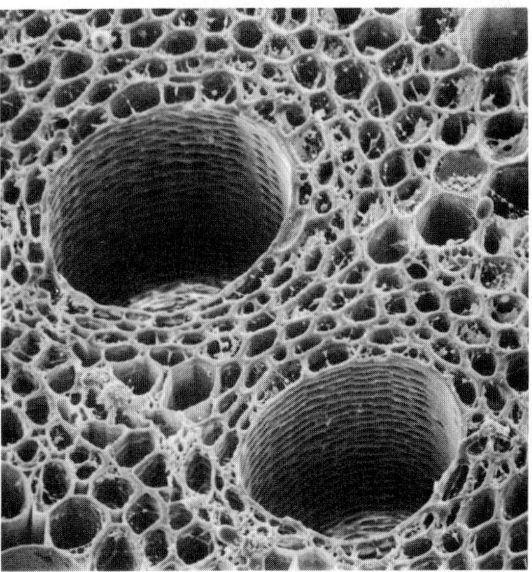

Figure 25-8 THE ORGANIZATION OF VASCULAR TISSUE: XYLEM.
(a) In xylem tissue, there are two major cell types, tracheids and vessel elements. In a woody plant, both cell types lose their cytoplasm and remain as dead empty cell walls. Both are arranged in columns end to end, but the vessel elements lose their end walls, while the tracheids retain angled end walls that are thin and perforated. Together, the columns of both cell types form the plant's water-conducting xylem tissue. (b) This scanning electron micrograph (taken of root tissue very near the stem; magnified about 320×) shows two vessel elements extending vertically (large holes). Perforations can be seen in their side walls and end walls. The smaller cylinders on each side of the vessel elements are tracheids.

Xylem

Xylem is a conducting tissue made up of cells stacked end to end like sections of pipe; these cells, although dead when functional, transport water and minerals, usually from the roots to the stem, leaves, and fruit.

In flowering plants, there are two types of xylem cells, tracheids and vessel elements (Figure 25-8). **Tracheids** are long, tapering cells that overlap, forming thin tubes within the vascular bundles. The walls of the tapered ends are incomplete, sievelike, and perforated by pits. Water is pulled through these pits as it moves through the xylem (details in Chapter 27). **Vessel elements** are much shorter and broader than tracheids and have either blunt end walls with large perforations or no end walls at all. A string of vessel elements, stacked open end to open end, functions like an open pipeline and so allows a freer flow of water than do tracheids. Most flowering plants have both tracheids and vessel elements, while the xylem tissue of other types of vascular land plants (whisk ferns, horsetails, ferns, and conifers) contains only tracheids. The total amount of stem tissue—and hence of vessel elements and tracheids—is correlated with the plant's full leaf area: The more water-carrying "pipelines," the more leaves the plant tends to have.

As a flowering plant grows and matures, xylem cells die, and thick walls of cellulose, lignin, and other polymers remain. The end walls of each vessel element are digested away completely, creating a pipeline channel for water movement only one cell diameter wide but thousands of cells long. As tracheids develop, the tapered end walls of the dying cells are incompletely digested, leaving porous sieves. Like vessel elements, mature tracheids are dead; they no longer contain cytoplasm.

Phloem

Phloem is a tissue specialized for carbohydrate transport, and its cells remain alive in order to transport nutrients. In flowering plants, each column of phloem is composed of **sieve tube elements** stacked vertically end to end to form the **sieve tube** (Figure 25-9). The end walls, or **sieve plates,** of each sieve tube element are perforated by many pores. At maturity, the sieve tube elements lose their nuclei but retain living cytoplasm. These strands of cytoplasm line the inner cell surface and pass through the sieve plate connecting the adjoining sieve tube elements. Each mature sieve tube element is associated with a companion cell, a nucleated cell derived from the same precursor cell as the sieve tube element. The companion cell is believed to serve as the supplier of macromolecules—proteins, enzymes, and so on—required for the long-term survival of the sieve tube elements.

The phloem tubes occur within the same vascular bundles as the xylem conduits, but they transport sugars and other organic materials away from the sites of manufacture in photosynthetic tissue and toward the parts of the plant (roots, woody stems) that do not carry out photosynthesis. In spring, a reverse process can occur in some plants: Sugars stored in stems or roots during the winter begin to be transported upward to help fuel the rapid growth of leaves and new stems. The phrase "the sap is rising" refers to this phenomenon, and when it occurs in sugar maples, people sometimes tap the tree and harvest the rising "sap" to make a rich, sweet syrup. Chapter 27 explains phloem function in detail.

Stem Adaptations

Besides transporting material and supporting the plant, stems can serve as storage organs or protect the plant from predators. For example, the stems of sugar cane, rhubarb, and broccoli store carbohydrates produced in the leaves. And the potato, which people often

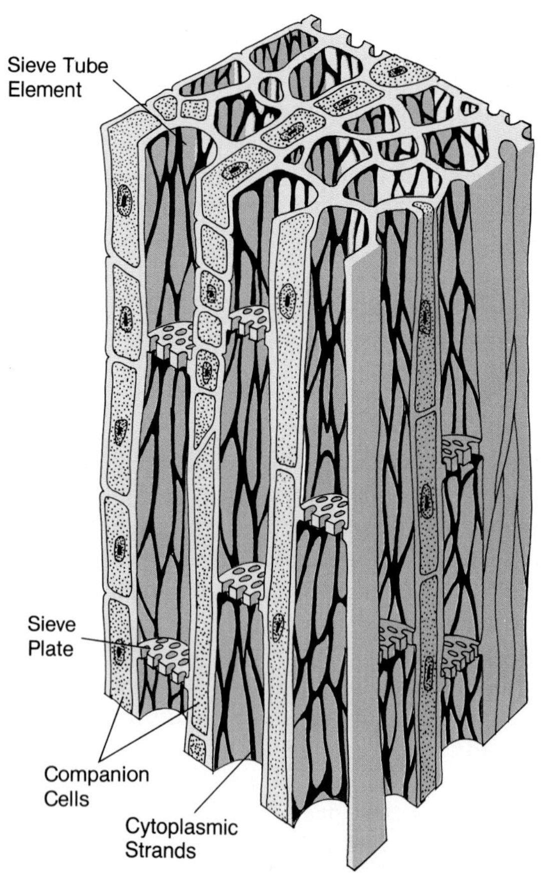

Figure 25-9 THE ORGANIZATION OF VASCULAR TISSUE: PHLOEM.
Phloem is composed primarily of sieve tube elements, with their perforated end walls (sieve plates). The closely associated living companion cells are visible, with their cytoplasm and nuclei.

Figure 25-10 CACTUS SPINES: MANY FORMS, TWO ROLES.
These long, sharp spines are sufficiently close together to protect the *Opuntia* species cactus from predation. The whitish color of the spines reflects sunlight and helps keep down the plant's internal temperature.

think of as a root, is actually an enlarged undergound stem modified for starch storage.

The stems of plants such as roses or blackberries bear prickles, and these discourage predation by browsing animals. Many cactus species grow spines (modified leaves) and hairs that also reduce predation (Figure 25-10). And the thick, photosynthetic, water-storing stems of cactuses and succulents are yet another set of stem adaptations.

THE ROOT: ANCHORAGE AND UPTAKE

An extensive root system spreads underground below most plants and may represent 60 to 80 percent of the organism's biomass in some tundra and prairie plants. Tree roots usually extend to a radius far larger than the crown, and an alfalfa plant less than 2 m tall can have roots at least 6 m deep. **Roots** have two key functions in the overall life of the plant: (1) They absorb water, minerals, and oxygen from the soil; and (2) they anchor the plant firmly in the soil or, if the plant is a vine, to the vertical surface it may be climbing. In many plants, roots also store energy reserves for the plant in the form of starch. The vascular system that serves the stem and leaves is continuous in the roots; all the "pipes" are therefore in place for transporting water and mineral nutrients from the soil to the rest of the plant. Vascular tissue also strengthens the root, augmenting the anchoring function.

Root Structure

Roots have concentric zones of tissue, including the epidermis, cortex, endodermis, pericycle, and stele (Figure 25-11a).

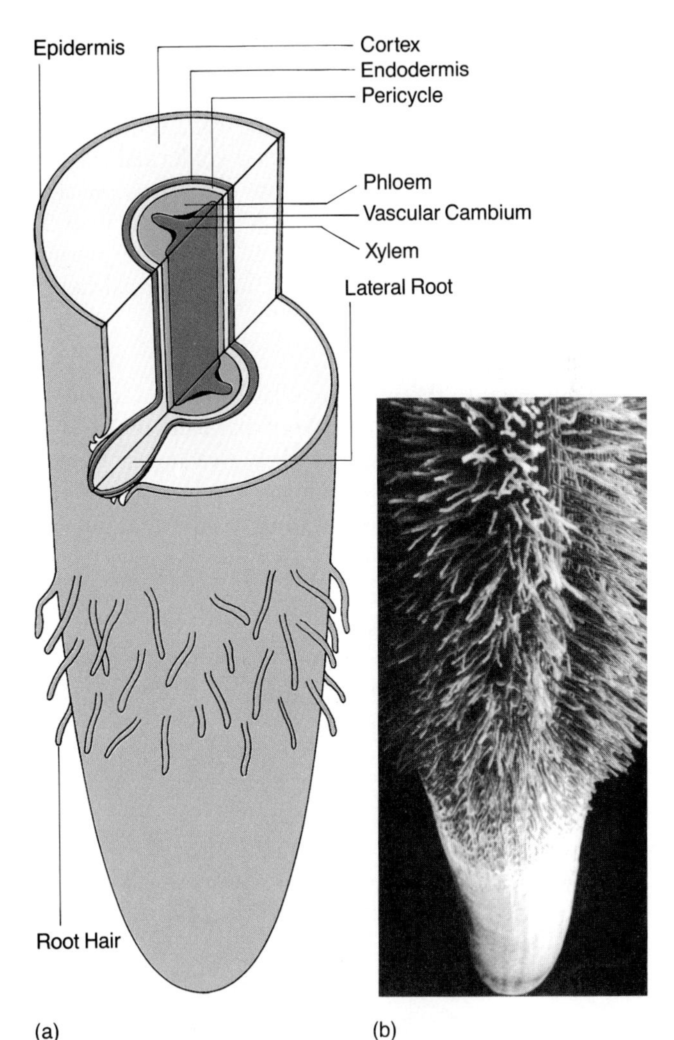

Figure 25-11 MAJOR TISSUES OF A ROOT.
(a) The various cell layers and root hairs are evident in this cross section of a root. Water and ions entering root hairs must ultimately pass through the endodermis cell cytoplasm before entering the pericycle and thus the stele. (The stele includes the pericycle, xylem, and phloem.) The pericycle gives rise to lateral roots, one of which is shown here just starting to form. (b) Root hairs are extensions of single root epidermal cells. Their enormous number on this radish root creates a huge surface area for absorbing water and minerals. From *Living Images* by Gene Shih and Richard G. Kessel, Science Books International, 1982. Reprinted by permission of the present publisher, Jones and Bartlett Publishers.

Epidermis

Roots have an outermost zone of cells, the epidermis, that is one cell thick and sometimes lacks a waterproof coating. The root is an organ of water **absorption;** most of the actual uptake occurs through fine, hairlike extensions called **root hairs,** extending from individual epidermal cells (Figure 25-11b). Collectively, root hairs increase the absorptive surface area of the root tip enormously. Botanists have estimated that the number of root hairs on the root system of a 4-month-old rye plant exceeds 14 billion and that the surface area equals 401 m^2—the floor area of a large house.

Roots continuously grow into the soil, sending out new root hairs from the newly formed epidermal cells in the probing root tip. Because solute concentrations are higher inside the epidermal cells and root hairs than in soil water, water moves by osmosis across the plasma membrane of root hairs and into the epidermal cells.

Cortex and Endodermis

Just internal to the epidermal layer lies the root cortex, made up of nonphotosynthetic parenchyma cells—large, thin-walled structures with big vacuoles. Cortex cells often store a large amount of starch as an energy reserve: The enlarged roots of the carrot, sweet potato, and turnip plants are familiar examples.

The cortex lies between the epidermis, where water and minerals enter the root, and the stele (the central vascular cylinder), where these materials are transported to other parts of the plant. Water and minerals must therefore traverse the cortex before entering the stele. There are two pathways for this passage of substances through the cortex: the apoplastic pathway and the symplastic pathway (Figure 25-12a). The **apoplastic pathway** is a compartment of sorts that consists of all the extracellular spaces between cells as well as the spaces inside cell walls through which water and minerals can freely traverse *without crossing any plasma membranes.* Conversely, the **symplastic pathway** is composed of the collective cytoplasms of all the individual cells, plus the cytoplasmic connection between them. Recall from Chapter 5 that the cytoplasmic regions of adjoining plant cells are connected by small channels called plasmodesmata and that the cytoplasm of each cell is surrounded by a plasma membrane. In most roots, the membrane-bound symplast extends from the cytoplasm within the root hair through the cortex cells and ends next to the xylem. Once a mineral ion or a water molecule moves across the plasma membrane, no matter where it crosses, it is within the symplastic pathway, and because of the plasmodesmata, the molecule has free access to the entire symplastic compartment. As materials traverse the cortex in this way and reach the region of the

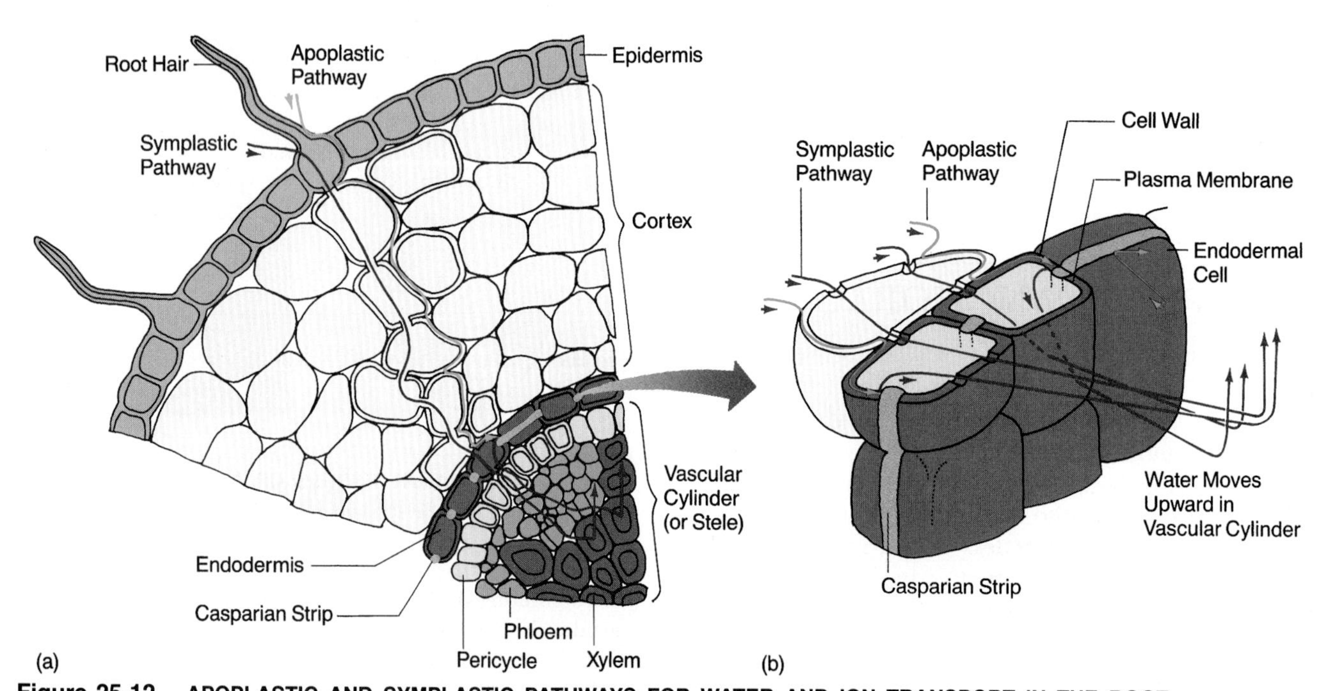

Figure 25-12 APOPLASTIC AND SYMPLASTIC PATHWAYS FOR WATER AND ION TRANSPORT IN THE ROOT.
(a) The root cross section shows the two pathways of water and mineral uptake. Materials moving through the apoplastic pathway do not cross plasma membranes, but instead move between cells and just inside cell walls. Most water and ions move this way, but a small percentage cross plasma membranes and diffuse through the interconnected cytoplasms of the root cells; this is the symplastic pathway. (b) The Casparian strip blocks water and mineral movement, making these nutrients flow through the cytoplasm of the endodermal cells. The endodermis and Casparian strip also prevent backflow of ions from the stele.

symplast adjacent to the xylem, they may cross the plasma membrane once again, leaving the symplast to enter the xylem, where they are transported upward toward the stem or leaves.

Despite the efficiency of water and mineral passage through the symplastic pathway, experiments show that more water passes through the apoplast, which ends at the innermost layer of the cortex, the **endodermis,** rather than extending directly to the vascular elements. Endodermal cells form a tight ring that separates the cortical cells from the vascular tissues and act as a barrier that regulates the lateral flow of water and nutrients from the apoplast to the vascular cylinder. Here is how the mechanism works: endodermal cells are closely packed together, and certain walls where they abut each other are impregnated with **suberin,** a substance impermeable to water. The special suberin-waterproofed walls of endodermal cells are referred to as **Casparian strips** (Figure 25-12b). The side walls of endodermal cells, which face the outer cortex in one direction and the inner vascular cylinder in the other, are not coated with suberin. Thus, for water and minerals in the apoplast to gain access to the vascular tissues, they must enter the symplastic pathway and thereby pass *through* the endodermal cell cytoplasm. Once in the endodermal cells, ions can be moved by active transport toward the xylem. The water follows osmotically. This restriction of water and ion movement to a route through the endodermal cells provides an opportunity for selective uptake of various mineral ions dissolved in water. By acting as a barrier to apoplastic movement, the endodermis can serve as the critical regulator of substances passing from the soil into the root and to the rest of the plant. Furthermore, the endodermis, with its Casparian strips, prevents backflow of ions from the xylem out of the root. Chapter 27 returns to this important subject.

Pericycle

Just interior to the endodermis is another circular zone of cells, the **pericycle,** which surrounds the xylem and phloem. Water and ions from the endodermal cells first enter the pericycle before moving toward the xylem. Pericycle cells are capable of generating new roots and thus are vitally important for the expansion of the root system. Branches of existing roots, or **lateral roots,** arise in the pericycle, grow through the cortex, rupture the epidermis, and emerge from the surface of the root. This difference in the internal origination of branches represents a major distinction between roots and stems.

Stele

The central cylinder of vascular tissue in a root is called the **stele.** This core includes the tissues interior to the endodermal layer: pericycle, xylem, phloem, and (in monocots) root pith. In dicots, the core also includes vascular cambium. In angiosperms, root xylem and phloem are made up of tracheids, vessel elements, and sieve tube elements, as in stems and leaves. The xylem cells in young dicot roots often form a central cross-shaped region with phloem bundles between the arms of the cross, as shown in Figure 25-13. In dicots, the number of "arms" is usually 2 to 4, while in monocots, that number can reach 20. In the roots of many mature monocots, the xylem and phloem bundles alternate to form a ring that separates pericycle cells from those of the pith, which serve a storage function similar to those in stem pith. The xylem vessels in roots are the beginning "pipe sections" in the long, continuous water conduits that stretch from roots to leaves.

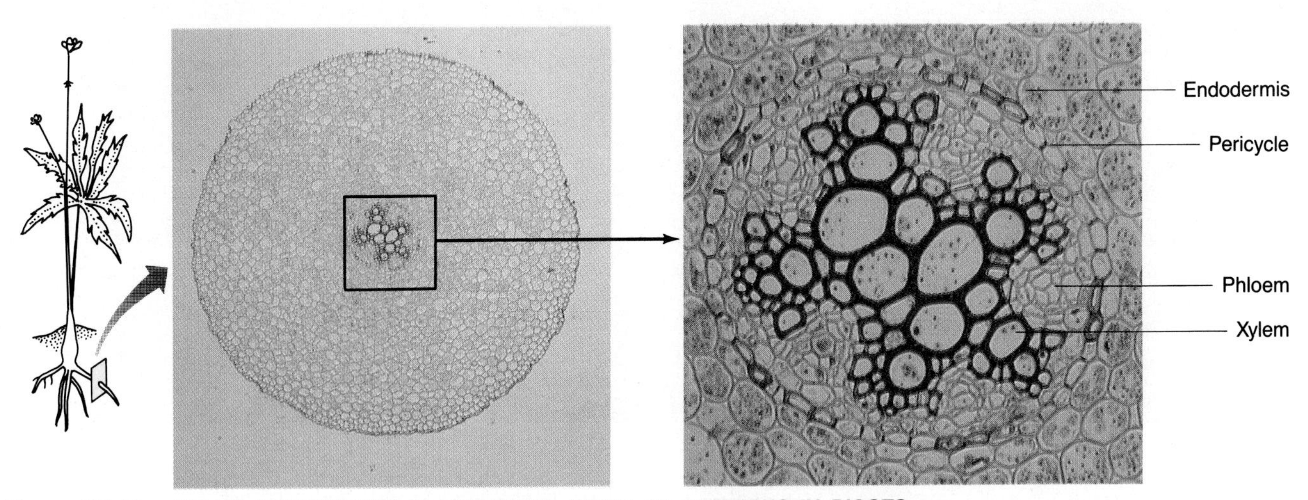

Figure 25-13 XYLEM AND PHLOEM IN THE ROOT: REGULAR PATTERNS IN DICOTS.
Here a buttercup root section (*Ranunculus* species) shows the large central xylem vessels with nearby bundles of phloem vessels. The pericycle and endodermis are also clearly visible; the large-celled cortex surrounds the endodermis.

Roots and Oxygen

While stems and leaves are usually surrounded by air, roots are usually surrounded by soil; thus, the amount of air in the soil often determines how successfully the root system, and hence the plant, will grow. Because root epidermal tissue lacks a cuticle, oxygen from the soil can diffuse easily into the internal tissues of the cortex. Root cortex cells, in turn, are loosely packed, allowing oxygen to move freely toward the stele. If soils are deficient in air and oxygen, owing to waterlogging or tight compaction, the root tissue can become anaerobic and eventually die.

Types of Roots

The anchorage roots provide is crucial to plant survival, and two patterns of root growth have evolved to facilitate anchorage as well as absorption: Most conifers and dicots have a taproot system, while most ferns and monocots have a fibrous root system.

Plants with a **taproot system** have one main root extending up to 1 m or more undergound (Figure 25-14a). Taproots allow the plant to withstand adverse weather as well as to gather water from far below the surface. In addition to the large taproot itself, such plants usually have numerous lateral roots, smaller branches off the main axis that allow the plant to absorb water and nutrients from a large soil area. Taproots such as carrots and turnips, which are biennials, store large quantities of starch to support the continuing growth of the shoot during the plant's second year of life.

Plants with a **fibrous root system** have many equal-sized roots arising at the junction of stem and root and fanning out through the soil (Figure 25-14b). In some plants, such as corn, the primary root that emerges from the seed dies, and new **adventitious** roots sprout from the base of the stem. As fibrous roots penetrate the ground, they produce lateral roots and form a dense mat that anchors and supports the plant. Because such roots spread in all directions, plants with a fibrous system are difficult to remove from the soil. Grasses have fibrous root systems that prevent soil erosion by literally tying the soil together.

Fibrous root systems can provide a tremendous surface area for absorbing water and minerals. The total length of the fibrous roots of the 4-month-old rye grass mentioned earlier was more than 600 km—the distance from Los Angeles to San Francisco—even though the stems averaged only 38 cm in height. The surface area of the root system of this plant was 639 m^2—130 times greater than the surface area of the shoot and leaves.

Root Adaptations

Plants that live in unusual places often have highly specialized roots. The mangrove trees in swampy saltwater habitats, for example, may develop dozens of **prop roots:** roots that sprout downward from the stem and

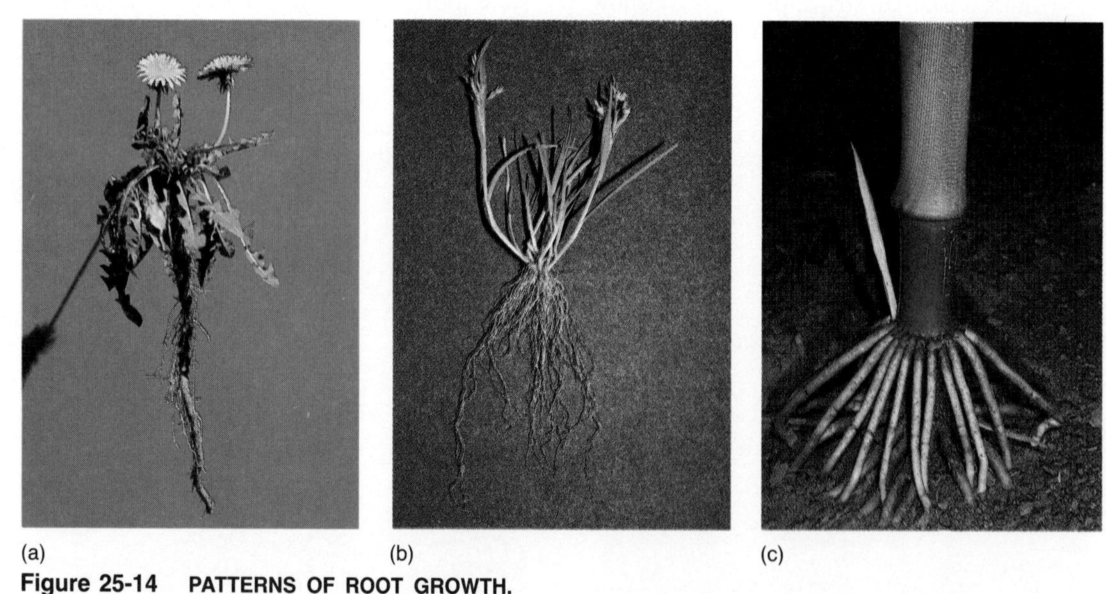

(a) (b) (c)

Figure 25-14 PATTERNS OF ROOT GROWTH.
(a) The dandelion's taproot system and (b) the fibrous root system of this grass plant demonstrate that though the former has a very strong central root, both provide great surface area for absorbing water and minerals. (c) The prop roots on this corn plant lend support and absorption.

exist partly in the air and partly in the waterlogged soil (Figure 25-14c). These prop roots can anchor the trees against hurricane-force winds despite the soft substratum.

Plants that live in the air also have special root adaptations. "Air plants," or **epiphytes,** usually grow on other plants rather than in the soil. An orchid growing on the trunk of a jungle tree can absorb water vapor from the air through its exposed roots. Vines have also developed specialized roots: Ivy stems sprout small roots with "suckers" at each root tip that help the plant cling tenaciously to a wall or a tree trunk.

Finally, we have seen that carbohydrate storage is another specialized root function in sweet potatoes, beets, carrots, radishes, turnips, and similar plants.

LOOKING AHEAD

By surveying the basic elements of plant architecture—the cells and tissues of the leaves, stems, and roots—we can appreciate the close interplay of form and function in the plant world and see how particular cell and tissue types allow plant organs to collect sunlight, exchange gases, prevent moisture loss, absorb water and minerals, store nutrients, conduct materials, and anchor the plant to its substratum. A plant's day-to-day existence depends on the smoothly integrated functioning of all its parts, and the next three chapters explore that functioning, beginning with how plants reproduce, develop, and grow.

SUMMARY

1. Vascular seed plants have evolved leaves, stems, and roots and a system of vascular conduits for carrying water, minerals, and nutrients throughout the plant.

2. The *leaf* is the plant's primary sunlight collector, and its efficiency depends largely on the size, shape, and position of the *blade*. The *epidermis* secretes a waterproofing material (*cutin*) in a layer, the *cuticle*, and contains specialized pores, or *stomata*, through which gases diffuse. *Guard cells* regulate the opening and closing of stomata.

3. Photosynthesis takes place in the *mesophyll*, a tissue made up of *parenchyma* cells and containing chloroplasts.

4. Leaf *veins* transport materials between the leaf and the rest of the plant and act as structural "girders" for the leaf.

5. The *stem* supports the plant and contains the main vascular tissues. An *herbaceous* stem consists of outer epidermis; a photosynthetic zone of *cortex* cells, including *collenchyma* and *sclerenchyma fibers* and *sclereids* for vertical and radial strength; *vascular tissue* containing xylem and phloem; and, in dicots, a central storage area of *pith*.

6. There are two types of *xylem* transport cells: *tracheids* (found in all vascular plants) and *vessel elements* (found only in flowering plants).

7. *Phloem* is composed of *sieve tube elements* stacked end to end to form the *sieve tube* and associated with living, nucleated *companion cells*.

8. Xylem and phloem occur together in *vascular bundles*. These can be in rings or scattered throughout the *ground parenchyma*.

9. *Roots* anchor the plant, absorb water and minerals, and transport the absorbed substances to the stem. Root structures include the epidermis, with its *root hairs;* the root cortex, including parenchyma cells and the *endodermis;* and the *stele*, or central cylinder of vascular tissue.

10. Water and minerals may move through the root cortex in either the *apoplastic* or *symplastic* pathways. The Casparian strip forces water and ions to pass through endodermal cell cytoplasm.

11. Roots can grow in a deep *taproot system* or a branched *fibrous root system* composed of *adventitious roots*.

KEY TERMS

absorption
adventitious root
apoplastic pathway
blade
Casparian strip
collenchyma
companion cell
cortex
cuticle
cutin
endodermis
epidermis
epiphyte
fibrous root system
ground parenchyma
guard cell
herbaceous
lateral root
leaf
lignin
mesophyll
midrib
palisade parenchyma
parenchyma
pericycle
petiole
phloem
pith
procambium
prop root
radicle
root
root hair
sclereid
sclerenchyma
sieve plate
sieve tube
sieve tube element
spongy parenchyma
stele
stem
stoma
suberin
symplastic pathway
taproot system
tracheid
vascular bundle
vascular cambium
vascular tissue
vein
vessel element
xylem

QUESTIONS

1. What is the major function of a leaf?

2. In what ways is a leaf a "compromise"?

3. What is the functional relationship between the leaf epidermis and the leaf cuticle?

4. Sketch a stoma and guard cells. Does your sketch indicate the presence of chloroplasts? Where? What is the function of the guard cells?

5. Sketch and label a cross section of a dicot stem and of a dicot root.

6. What tissue is usually at the center of a dicot root? At the center of a dicot stem? Where is the endodermis in such roots, and what is its primary function?

7. Why do roots need a large surface area? What accounts for the enormous surface area of many roots?

8. Describe parenchyma tissue. Where is parenchyma tissue located in leaves? In stems? In roots?

ESSAY QUESTION

1. Trace the path followed by a molecule of water from the soil through the root, stem, and leaf of a vascular plant, until the molecule enters the atmosphere. Name the various cells and tissues along the pathway. Can a water molecule pass from root to leaf without passing through cytoplasm?

For additional readings related to topics in this chapter, see Appendix C.

26

How Plants Reproduce, Develop, and Grow

I took an earthenware pot, placed in it 200 lb of earth dried in an oven, soaked this with water, and planted in it a willow shoot weighing 5 lb. After 5 years had passed, the tree grown therefrom weighed 169 lb and about 3 oz. . . . Finally, I again dried the earth of the pot and it was found to be the same 200 lb minus about 2 oz. Therefore, 164 lb of wood, bark, and root had arisen from water alone.

Jean-Baptiste van Helmont,
Ortus Medicinae (1648)

A living castaway: After months at sea, a coconut palm sprouts and takes root just above the tide line on a Virgin Islands beach.

Flowering plants have flourished largely because of their innovations in sexual reproduction, the flowers, fruits, and seeds. Yet many plants within this group also exploit asexual *vegetative reproduction,* a kind of cloning that can involve stems, roots, or leaves and result in offspring that are genetically identical to the parent. The study of both modes of reproduction in flowering plants and of subsequent growth and development of new individuals helps reveal how plants differ from animals.

One aspect of plant development is particularly distinctive: Plants have perpetual embryonic centers that produce new organs throughout the life of the individual, whereas most animals form organs only as embryos. The plant's unique capacity for renewed growth and development throughout adult life is utilized during flowering and sexual reproduction as well as in modified form during vegetative reproduction.

Our goal in this chapter is to survey the entire range of plant parts, reproductive processes, and growth—from the drab to the glorious, from the asexual to the sexual, from pollen and eggs to the woody tissues of mature trees and bushes. The details of reproduction and growth help to characterize the flowering plants (and, in some respects, the conifers) and explain why they are such highly successful groups.

Our discussions will cover:

- Vegetative reproduction, or multiplication through cloning in nature and agriculture
- Sexual reproduction, and the roles of flower, pollen, sperm, egg, pollination, and fertilization
- The development of plant embryos
- Seeds and the dispersal of the new generation
- Germination of the seed and development of the new seedling
- Primary growth—increasing size and the addition of new tissues in the young plant
- Secondary growth—the development of wood and bark in older plants
- Plant life spans and life-styles

VEGETATIVE REPRODUCTION: MULTIPLICATION THROUGH CLONING

High in the Appalachian Mountains of West Virginia, there is a low, dense thicket of blueberry bushes nearly

1 km in diameter. Even more startling than its size is its derivation: The bushes are all connected by underground stems and were derived from the same initial plant. Thus, they are a single clone of genetically identical plants, illustrating **vegetative reproduction,** a process in which identical new plants emerge from the parent's body.

Like these blueberries, many flowering plants can multiply asexually, or vegetatively, as well as through sexual reproduction. Recall that in flowering plants, the *sporophyte,* the conspicuous, diploid, spore-producing plant (with its leaves, stems, and roots), alternates with the *gametophyte,* the small gamete-producing generation (see Chapter 23). For most kinds of flowering plants in most situations, the sexual phase is retained in the life cycle, since it ensures the spread of new, genetically unique individuals.

For some flowering plants, however, there is a distinct advantage in multiplying asexually—going directly from one sporophyte to the next, relying on mitosis, rather than meiosis, as the basis for reproduction. If an individual plant grows successfully at a particular spot, it is, in a sense, adapted to that place on the earth's surface. Hence, if the plant can reproduce vegetatively, it can spread rapidly into similar adjoining areas. There are some inherent disadvantages to this form of reproduction, but first let us discuss details of the process.

Vegetative Reproduction in Nature

A wide array of plant parts, including modified stems, leaves, and roots, can give rise to new individuals through vegetative reproduction.

Rhizomes, stolons, runners, corms, and tubers are all stem structures that can lead to plant reproduction. **Rhizomes** are subterranean stems that grow laterally from the main shoot. Periodically, a new set of roots and aerial stems emerges from the rhizome; then leaves develop, and a genetically identical plant becomes established. Blueberries spread by means of rhizomes, as do cattails and bamboo. One cattail can send out rhizomes that produce more than 60 new shoots in a single summer. And some bamboo forests in Southeast Asia are reported to be more than 160 km long and yet contain only the plants of a single clone, all connected by a vast network of rhizomes.

Stolons are branches of aerial stems that grow laterally, touch the ground, and put out roots and stems at those sites. A troublesome weed, the water hyacinth reproduces vegetatively via stolons, and botanists estimate that in a year, 10 water hyacinths can reproduce into more than 600,000 identical plants! Strawberries commonly reproduce through stolons (also called *runners*), as does a fast-growing vine called kudzu that now covers fields and forests throughout the South with billows of leafy vines (Figure 26-1a).

(a)

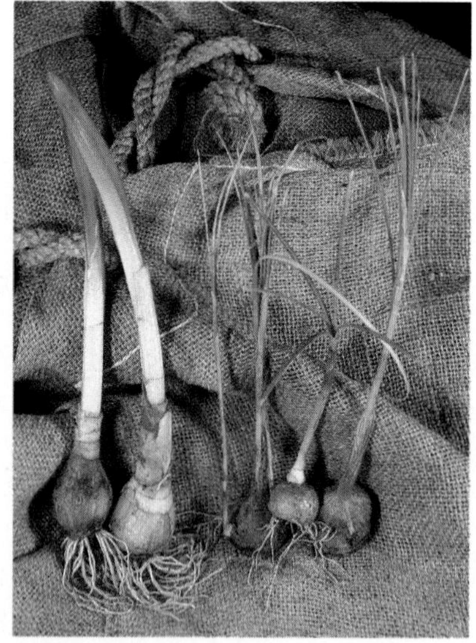

(b)

(c)

Figure 26-1 VEGETATIVE REPRODUCTION.
(a) Kudzu vines *Pueraria lobata* propagate vegetatively by runners and can grow nearly 30 cm (about 12 in.) a day in the summer, blanketing trees and fields with millions of leaves. (b) The daffodil (*Narcissus pseudonarcissus*) corms on the left are compressed, nutrient-filled parts of stems. The saffron crocus (*Crocus sativus*) bulbs on the right are modified leaves. (c) The *Kalanchoë daigremontiana* plant (also called mother of thousands) produces genetically identical plantlets in the margins of leaves. Each has tiny roots and can fall to the ground and quickly become established near the base of the parent plant.

Corms are dense underground stem structures involved in vegetative reproduction as well as carbohydrate storage. Corms are usually smaller than bulbs (which are actually modified leaves) and are solid (Figure 26-1b). Iris, tulip, and grape hyacinth grow from bulbs; crocus and gladiolus, from corms. Tiny *bulblets* can form at the base of a bulb, and *cormels* at the base of a corm; they remain attached to the original bulb or corm, but can grow separate new plants.

Tubers are another type of modified stem that can participate in vegetative reproduction. The common white potato is a tuber and is the swollen tip of a stolon. Each tuber bears many "eyes," or buds, which can sprout new stems and roots.

Unlikely as it may seem, leaves, too, can be reproductive structures. In the *Kalanchoë* plant, dozens of tiny plantlets—complete with leaves and roots—are produced along the margin of each leaf (Figure 26-1c). Begonias and African violets can also reproduce asexually via leaves. Finally, roots can give rise to new plants. Lilac, poplar, elderberry, and many types of grasses send out horizontal roots from which new stems and roots can emerge. Whole groves of aspens, turning to gold in the crisp autumn air, give evidence of being clones linked by root suckers.

Plant Propagation in Agriculture

Farmers propagate many kinds of plants with seeds. Sometimes, however, it can take several years for a species to grow from seedling to flower- or fruit-producing adult. More important, the genetic recombination that occurs during sexual reproduction can cause some desirable characteristics to be lost or masked in the offspring. And some hybrids, including pineapple and Marsh grapefruit, do not reproduce sexually at all.

Vegetative propagation is both faster and more certain than seed propagation, since the offspring are clones of the parent. Consequently, farmers, gardeners, and others who raise plants use various techniques to produce plants vegetatively. They break off bulblets and cormels, for example, and replant them individually to grow new tulips or crocuses. They cut tubers into pieces, with one "eye" per piece, and plant these to yield new individuals. And they take stem cuttings, or *slips*, from sugarcane and pineapple plants and treat them with plant hormones to induce them to sprout roots and hence form new plants.

Another important and widely used technique for propagating woody plants is *grafting*, attaching a stem cutting from a desirable plant—a Peace rose, for example, or a red Delicious apple—to the root stock of a hardier variety of rose or apple. Grafting is such a dependable means of propagation that most commercial fruit grown in the United States comes from trees reproduced in this way.

Researchers at several universities and biotechnology firms are devising ways of coating somatic embryos (tiny green embryos derived from cells other than fertilized eggs) with a transparent organic jelly like that on a fish egg and then encapsulating both embryo and gelatinous layer in a thin polymer jacket that will biodegrade once in the soil. With this technique, huge numbers of genetically altered tomato or other plants could be produced from a single, original hybrid cell and then be planted directly in the fields (Figure 26-2).

These encapsulated embryos promise to be a valuable tool since the gelatinous coating around the somatic embryo can be tailored to suit the needs of the infant plant with nutrients, hormones, pesticides, and fungicides to speed growth and prevent damage. Such made-to-order capsules may indeed be the bridge between lab and field, between the genetic revolution and the agricultural revolution of the 1990s.

Advantages and Disadvantages of Vegetative Reproduction

Vegetative reproduction has distinct advantages both in agriculture and for plants in nature. However, the presence of identical offspring with a single genotype can be a major disadvantage if it leaves the clone more susceptible to disease or to changes in the environment.

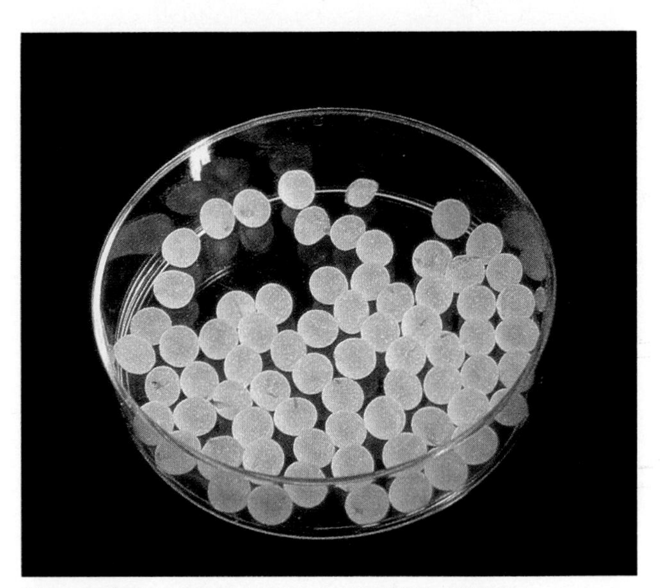

Figure 26-2 SPACE-AGE "SYNTHETIC SEEDS." An organic jelly is used to coat alfalfa embryos generated from callus on undifferentiated plant tissue.

WHAT PRICE ATTRACTIVENESS?

Flowers are an anthophyte's insurance against genetic oblivion. Many produce bright colors and strong scents to attract animal pollinators, as well as sweet nectar and protein-rich pollen that compensates the unwitting agents. The race to attract the right visitor at the right time is critically important, and some plants have evolved particularly costly mechanisms for attraction and reward. *Arum* lilies and ground-flowering protea are two cases in point.

Like the burning of a perfumed candle, an *Arum* lily sends a pungent plume of odorous compounds into the air by heating the central club-shaped flower part up to 15° C higher than the ambient air (Figure A). The flowering period can be short-lived, but during the 8- to 10-hour period, one species burns its starch reserves rapidly, and its metabolic rate equals that of a flying hummingbird! Mitochondria in the hot-headed flower have an alternate electron transport chain that allows energy to be released as heat instead of trapped as ATP.

Ground-flowering protea are odd-looking flowers from the southern parts of Africa and Australia that are unique in attracting mice, shrews, and other small mammals as pollinators. These animals are mainly seed eaters, and protea make up a very small proportion of their diet. So why do they even bother to visit the scrubby plants? The answer, says researcher Delbert Wiens, is "junk food": The flowers use considerable energy to produce an intensely sweet nectar that is nearly half sucrose—a nectar that activates the rodent's natural sweet tooth. The small visitor gets its irresistible reward—but only after probing so deeply among the curving flower parts that it picks up a load of sticky pollen on its fur (Figure B).

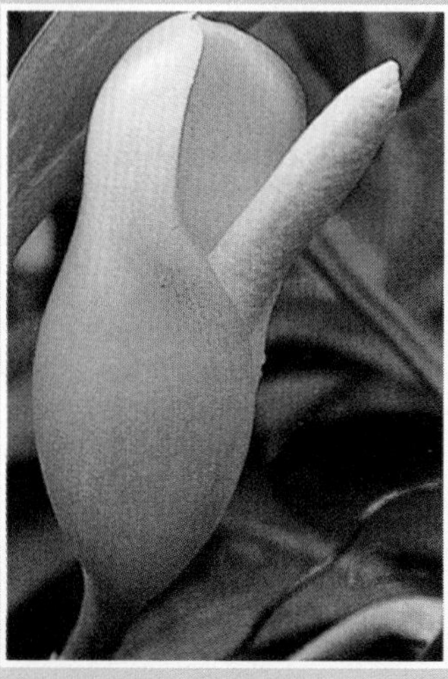

Figure A
Arum lily: A hot-headed flower.

Figure B
Protea: "Junk food" for rodents.

Farmers often grow vast fields of genetically identical strawberries, for example, and each plant tends to produce the same size and flavor of fruit. A single viral infection, however, could destroy the farmer's entire crop. Clones also tend to flower at the same time. Huge stands of bamboo, for instance, can reproduce vegetatively for up to a century, then suddenly reproduce sexually by flowering and then die in the same brief period—greatly altering the local ecology and animal life in the area.

The prevalence of vegetative reproduction among the flowering plants suggests that in many species, the risks are offset by the ability to spread quickly with genetic fidelity and to reproduce any mutated genes that arise in plant growth zones—so-called somatic mutations.

SEXUAL REPRODUCTION: MULTIPLICATION AND DIVERSITY

The ubiquity of flowers and fruit remind us that most plants reproduce sexually, producing haploid male and

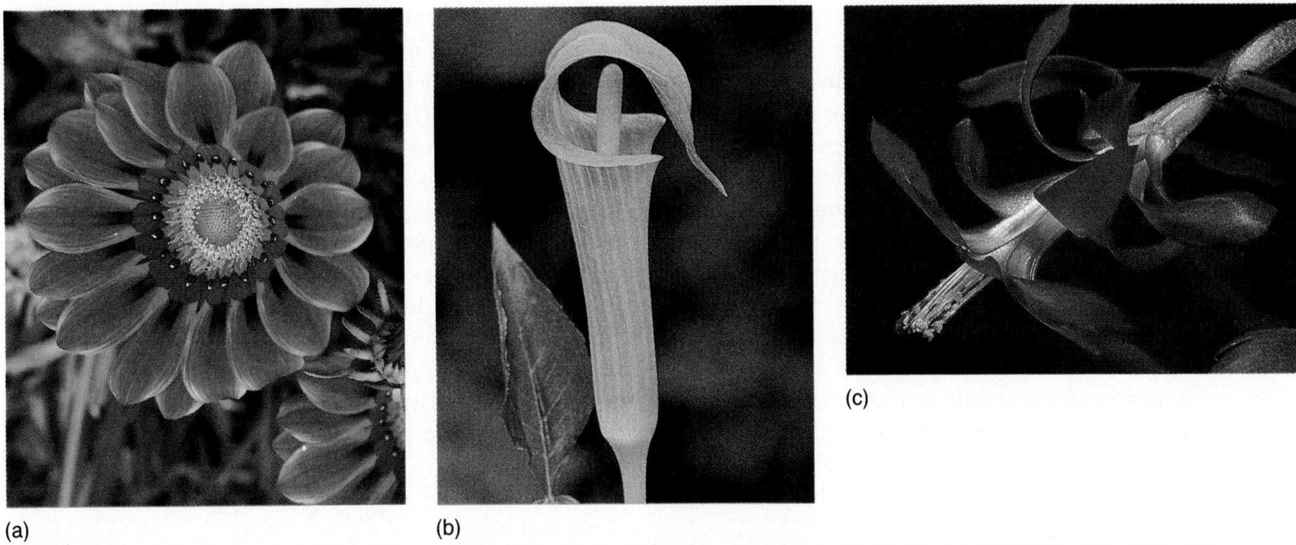

Figure 26-3 FLOWERS: SEXUAL STRUCTURES WITH A SEEMINGLY INFINITE VARIETY OF SHAPES, COLORS, AND ODORS.
(a) Each *Gazania* flower is actually composed of a large number of very small flowers, whose parts cannot be seen at this magnification. The *Gazania* is thus a "composite" head of flowers. Because individual flowers within this head mature over several days, different pollinators may fertilize the egg cells. (b) Only the tip of the female pistil (a group of fused carpels) is visible in this jack-in-the-pulpit (*Arisaema atrorubens*). (c) Multiple whorls of colored petals characterize this Christmas cactus (*Schlumbergira* species).

female gametes that fuse to create new gene combinations in the zygotes. Let us look at how these processes take place inside flowers.

Flowers: Ingenious Insurance for Fertilization

Each of the quarter of a million anthophyta species has a flower uniquely its own in at least some minor detail. Flower appearance is the most obvious, as *Gazania*, jack-in-the-pulpit, and Christmas cactus demonstrate so well (Figure 26-3). Fragrances differ widely, too, as do modes of pollination—including wind or particular animal species (review Chapter 23). Nevertheless, all flowers are based on the same organizational plan and an effective "design" for producing gametes and protecting them throughout fertilization and seed development.

The typical flower consists of four whorls of parts: sepals, petals, stamens, and carpels (pistils) (Figure 26-4). The *sepals*, or outer ring of floral parts, are often green and photosynthetic, although they can be petal-like, as in tulips and lilies. Sepals are attached to the *receptacle*, the enlarged end of the stalk that supports the flower. Collectively, the sepals are called the **calyx.** The *petals*, the next whorl of modified leaves (collectively called the **corolla**), are often brightly colored—from white to blackish purple and every hue in between—and serve to attract pollinators to the flower. The flower's showy exterior parts (calyx and corolla) are called the **perianth.** The interior parts—*stamens* and *carpels*—are the sexual structures involved in producing pollen and embryo sacs.

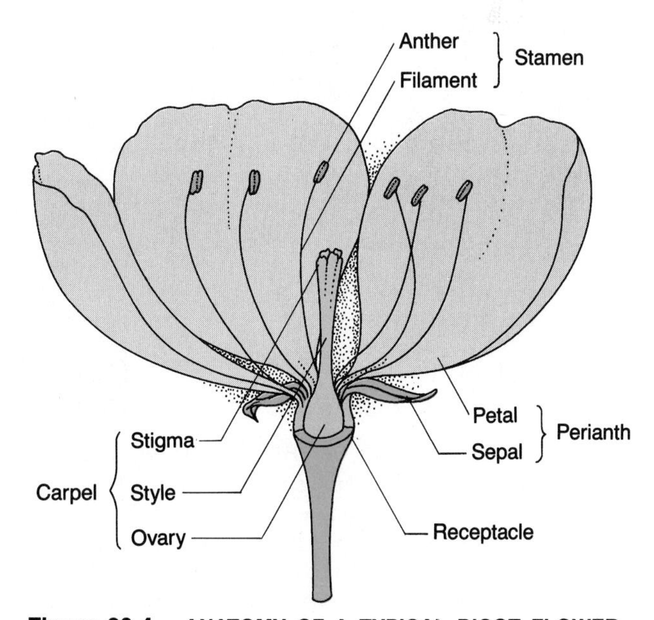

Figure 26-4 ANATOMY OF A TYPICAL DICOT FLOWER.
The stamen consists of the anther and filament. The carpel consists of the ovary, style, and stigma. Several fused carpels constitute a pistil.

Pollen Production

The pollen-bearing organs, the **stamens,** are the plant's male flower parts. They produce the microspores that develop into pollen. Most stamens consist of two parts: the stalklike *filament,* attached to the receptacle; and the *anther,* borne on top of each filament (Figure 26-5a). Typically, each anther contains four *microsporangia,* or *pollen sacs,* chambers in which the pollen grains form.

Pollen sacs are "factories" that generate enormous quantities of pollen grains. Each pollen sac is composed of two cell types: *peripheral cells* and *microspore mother cells.* The peripheral, or outer, cells form a sac of tissue around the developing pollen grains (Figure 26-5b). The innermost layer of peripheral sac cells, the *tapetum,* nourishes the developing pollen grains. In the center of each microsporangium is a large number of microspore mother cells, which are destined to give rise to pollen grains through a complex process.

The first step in the development of pollen grains is **microsporogenesis** (Figure 26-5c). Each diploid microspore mother cell divides meiotically, producing four haploid microspores, each of which gives rise to a single pollen grain.

The second step is **microgametogenesis,** during which the microspores differentiate into functional pollen grains (see Figure 26-5c). Maturation of pollen involves mitotic division of each haploid microspore. This step occurs in two parts. First, the haploid nucleus and adjacent cytoplasm divide once mitotically to form two cells: the *tube cell* and the *generative cell,* which is en-

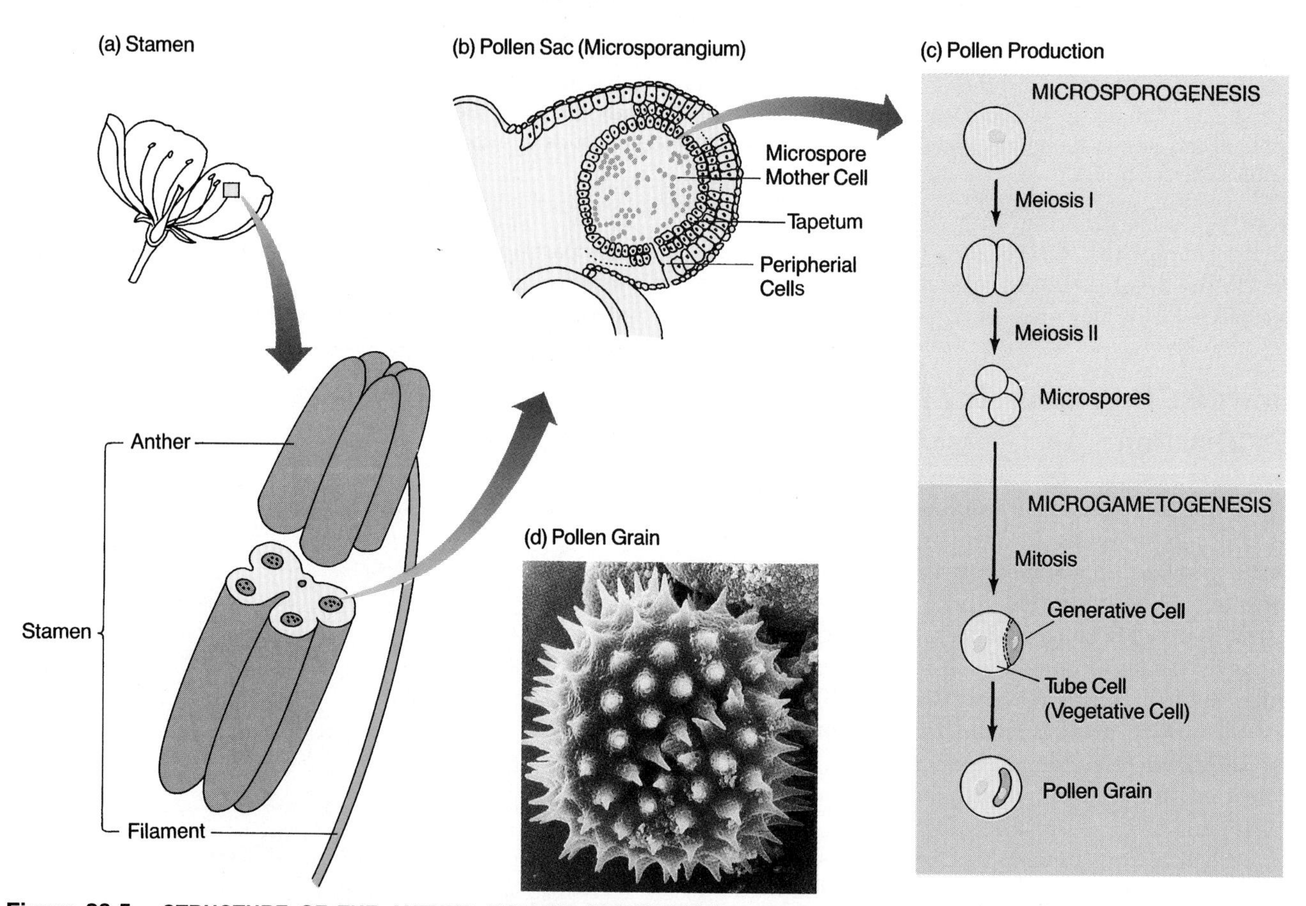

Figure 26-5 STRUCTURE OF THE ANTHER, POLLEN PRODUCTION.
(a) The anther is the pollen-producing structure of the stamen. A detailed cross section through the anther reveals four pollen sacs (microsporangia). (b) The inner layer of each pollen sac is the tapetum, a set of cells that secretes nutrients into the pollen sac. Microspore mother cells fill each pollen sac. (c) Pollen production has two parts. During microsporogenesis, each microspore mother cell in a microsporangium undergoes meiosis to form four haploid microspores. During microgametogenesis, each haploid microspore undergoes mitosis, forming two differentiated cells: a tube cell and a generative cell. The tube cell then envelops the generative cell and forms a pollen grain that contains the generative cell and that has a specialized wall. (d) The often elaborate outer protein and sugar coats of pollen grains induce allergic reactions in some people. Different species of plants produce distinctively different pollen. Flowers related to asters (*Aster chilensis*) have "spiky" pollen (here, magnified about 1,350×).

veloped by the tube cell. These two cells, surrounded by a single cell wall, constitute the pollen grain. The tube cell will eventually form a *pollen tube*, and the generative cell will divide to yield the two haploid sperm nuclei that fertilize an egg.

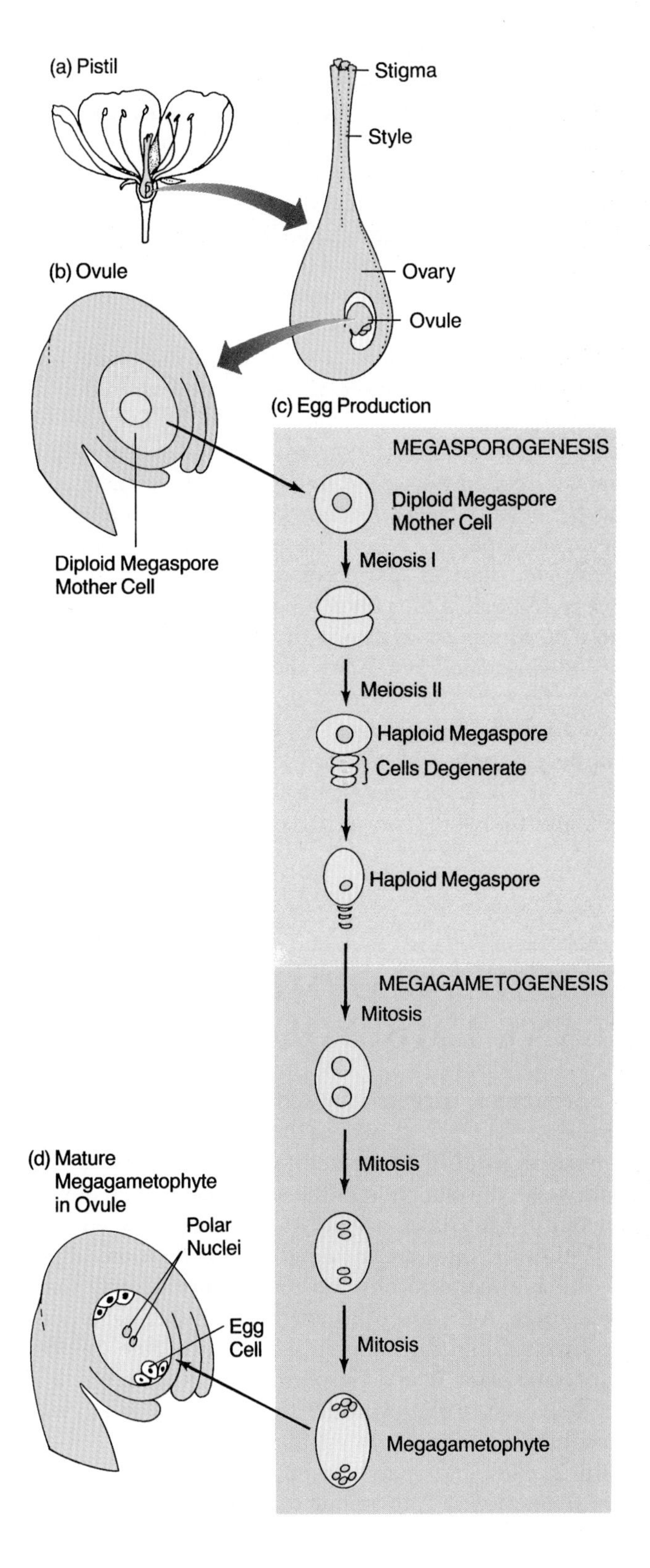

During pollen maturation, *sporopollenin*, an extremely hard carotenoid polymer, along with other polysaccharides and proteins, is deposited in the pollen cell wall, creating spikes, knobs, cavities, and craters (Figure 26-5d). Each pollen surface pattern is unique and species-specific and acts as a kind of structural signature. Unfortunately for many humans, the spikes, knobs, and other projections of pollen grains contain proteins that can induce the strong allergic reactions we call hayfever.

Egg Production

Like pollen, eggs are produced and protected within the flower, specifically in the *pistil*, a central vaselike structure consisting of one carpel or several fused carpels (Figure 26-6a). The pistil itself is the site of egg cell production and of fertilization. The pistil consists of three parts: the *stigma*, the pistil's broadened sticky top surface, to which pollen adheres; the *style*, the slender "neck" that connects the stigma to the ovary; and the *ovary*, at the base of the pistil, the actual site of egg development. The ovary and certain accessory structures mature into the fruit. Within the ovary, some plant species, such as the peach, have only one *ovule*, or embryo sac (Figure 26-6b), and thus develop just one seed, which rests inside the structure we know as the pit. The fruits of many other plants can have several ovules. String beans and pea pods are good examples, and each bean or pea within the ripened ovary is a separate ovule.

Megasporogenesis takes place within the ovule (Figure 26-6c); during this process, a diploid *megaspore mother cell* undergoes meiotic division to produce four haploid cells. Of these, three degenerate, and the fourth becomes the megaspore, with a haploid nucleus. **Megagametogenesis** follows, during which the female gamete is produced and readied for fertilization (see Figure 26-6c). The megaspore undergoes three mitotic divi-

◀ **Figure 26-6 OVULE STRUCTURE, EGG CELL PRODUCTION.**
(a) Each ovule in a flower's ovary contains (b) one megaspore mother cell that will ultimately give rise to the egg cell. (c) Egg production has two phases. During megasporogenesis, the megaspore mother cell in an ovule divides meiotically to form four haploid daughter cells. Three of these degenerate, leaving one megaspore with a haploid nucleus. Megagametogenesis follows and involves three mitotic divisions with no cytokinesis. (d) Within the mature megagametophyte (inside the ovule), the resultant eight haploid nuclei are then walled off in separate cells. These eight include the egg cell and the central cell (usually binucleate) that will form the endosperm. (Note that in various anthophyta, the central cell may contain anywhere from 1 to 14 nuclei.)

sions of the nucleus alone; in many cases cytokinesis does not occur. The result is a single cell, the *megagametophyte*, or *embryo sac*, which usually contains eight haploid nuclei. These eight nuclei separate into two groups of four at opposite ends of the megagametophyte (Figure 26-6d). Then, as if in a dance, one nucleus from each group of four migrates to the center of the cell, leaving three nuclei behind at each end. The two centrally located nuclei are called the **polar nuclei.** The other six nuclei become separated from this central region as plasma membranes form and dissect the original megagametophyte's cytoplasm into a central binucleate cell (with the polar nuclei), three small cells at one end, and three small cells at the opposite end. One of the six cells becomes the haploid egg cell, and it remains flanked by two other living cells. The three cells at the opposite end degenerate after fertilization.

Pollination and Fertilization

Once the egg cells inside flower ovaries have matured and are ready for fertilization, the corolla unfurls, often creating a showy display of bright petals. Because each flowering plant species has its own internal "timetable" for egg maturation that is influenced by environmental conditions, the flowering of various species can occur throughout the year, from the earliest crocus pushing up through the snow to chrysanthemums flowering among drifts of fall leaves.

As we saw in Chapter 23, flowers can function as advertisements to pollinators. The colors, shapes, and odors of the petals and other flower parts entice the proper pollinator to a reward of sugary nectar or protein-rich pollen grains, and the animal inadvertently ferries a load of pollen to the next flower. Small, drab flowers with reduced or absent petals and no fragrance are often wind pollinated; the trend in these flowers is toward long, exposed stamens that bear masses of pollen that can easily be blown about or enlarged, sticky stigmas.

Whatever its form, once the flower opens, the stigma is exposed, the pollen is released from the anthers, and the pollen grains are blown or carried to an exposed stigma on the same or another flower (Figure 26-7a). The molecules that make up the surface of the pollen grain interact with proteins and polysaccharides on the stigma, and if they "recognize" each other by a series of reactions, the pollen grain is stimulated to begin growing—to *germinate* (Figure 26-7b). This specificity of molecular matching helps prevent cross-fertilization between unrelated plants. Recall that the pollen grain contains a tube nucleus and a generative nucleus. When the pollen grain germinates, a pollen tube, which is a filamentous extension of the tube cell, grows through the hard wall of the grain. This tube then begins to grow through the stigma and style, toward the ovary (Figure 26-7c). As the pollen tube continues to grow downward, the generative cell divides mitotically to produce two sperm cells, with their haploid nuclei, which remain just behind the growing tip. Ultimately, the tip of the elongating pollen tube passes through the *micropyle*, a specialized opening at the surface of one end of the ovule (Figure 26-7d).

The growing pollen tube penetrates the highly protective ovary in what can be truly an Olympian performance. Corn pollen, for instance, sticks to one of the silks (the hairlike stigmas and styles at the end of ears of corn), and in just 24 hours, the pollen tube enters the silk, and the cell composing the tube elongates rapidly, grows down the full length of the strand, and reaches the kernel (ovary with ovule)—an amazing distance of up to 30 cm in large ears of corn.

Following pollen tube growth, flowering plants undergo a complex **double fertilization** process unique to this group of organisms (Figure 26-7e). The five participants in double fertilization are the two sperm nuclei from the pollen grain, the nucleus of the egg cell, and the two polar nuclei of the large central cell of the female megagametophyte. When the pollen tube enters the micropyle, the two sperm cells are released through a pore in the wall of the pollen tube. One sperm fuses with the egg, forming a zygote with a diploid nucleus. The second sperm cell penetrates the large central cell containing the two polar nuclei. The three nuclei fuse, forming a triploid ($3n$) nucleus. This triploid endosperm cell will develop into *endosperm*, the primary source of nutrition for the embryonic plant, familiar to us as coconut milk and the white chewy parts of popcorn.

THE DEVELOPMENT OF PLANT EMBRYOS

Fertilization triggers four simultaneous processes in flowering plants: (1) growth of the triploid endosperm as a nutrient tissue for the embryo; (2) formation of the embryo; (3) development of the seed coat; and (4) development of fruit tissue around the developing seed.

Within the endosperm, rapid mitotic divisions allow the storage tissue to develop much more quickly than the embryo. A process of "laying food by" begins, ensuring that an abundant supply will be available for the embryonic plant. The endosperm absorbs nutrients that are delivered to the ovule by the parent plant's xylem and phloem.

In grasses and many other monocots, endosperm forms the largest component of the mature seed. The

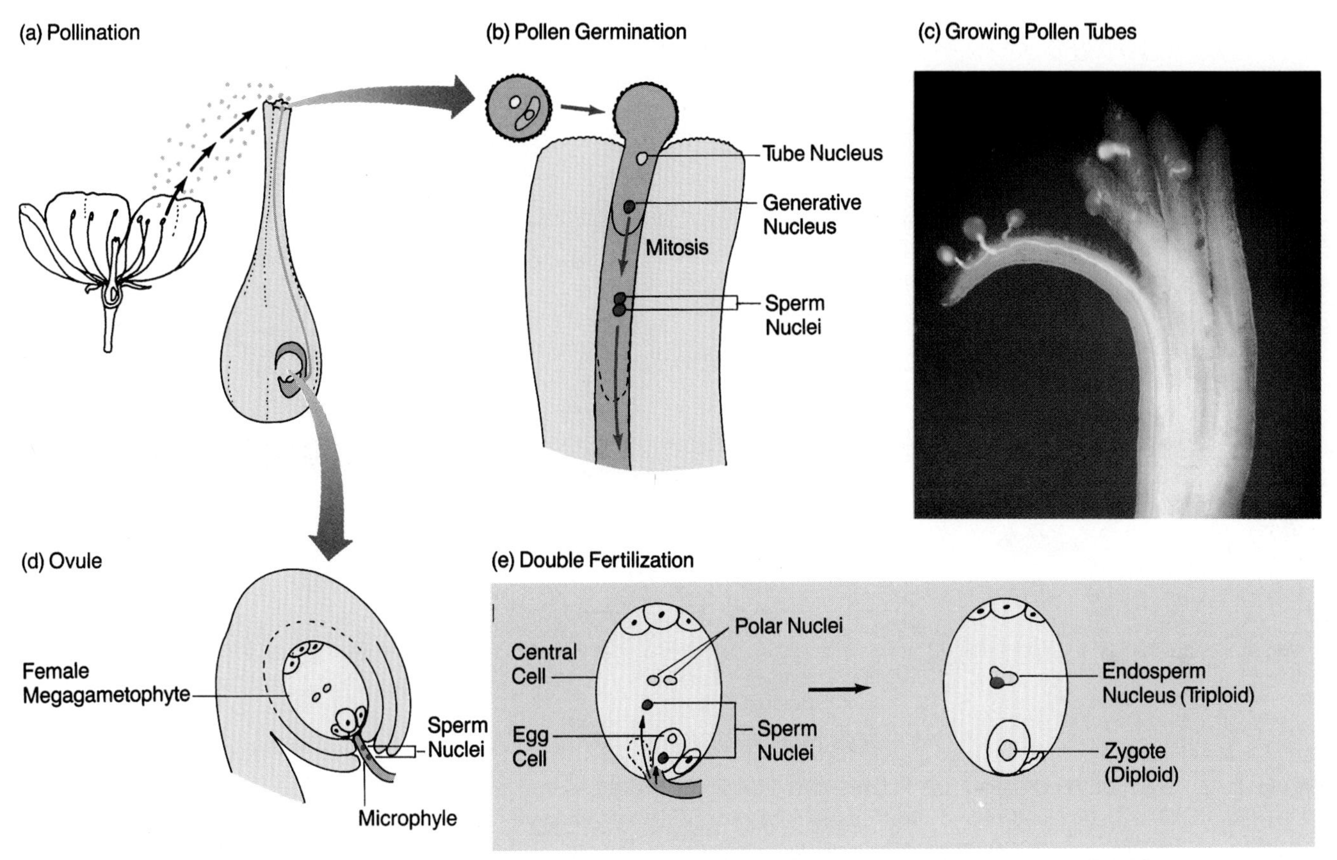

Figure 26-7 POLLINATION, POLLEN GERMINATION, POLLEN TUBE GROWTH, AND DOUBLE FERTILIZATION. (a) Pollination occurs when pollen lands on a stigma. (b) As the pollen germinates, the generative cell undergoes mitosis to form two haploid sperm cells that pass down the elongating pollen tube. (c) Here, a photo of a geranium stigma shows the growth of a pollen tube 1 cm long in just 20 minutes. (d) When the pollen tube enters the micropyle of a mature ovule, (e) a double fertilization takes place. One sperm cell fuses with the egg cell, forming the zygote. The other fuses with the central cell containing the polar nuclei to form the triploid endosperm nucleus. This endosperm nucleus divides to produce many triploid endosperm nuclei; as that occurs, the zygote divides and begins to form the embryo.

seed itself is composed of a tough coat surrounding the stored nutrient tissue and the embryonic plant. Endosperm forms the starchy part of a kernel of corn, for example, which is a member of the grass family. The endosperm of wheat comprises the stored starch from which we make flour. Grass embryos consume most of the food reserves of the endosperm after they burst forth from their seed coats in the spring. In contrast to such monocots, the embryos of most dicots, such as peas or beans, consume the endosperm while developing and store the nutrient reserves in two large, plump embryonic leaves, the *cotyledons*.

Mitotic divisions of the diploid zygote produce the monocot or dicot embryo. The embryo arises from cells at the upper end of the zygote (Figure 26-8b). The **suspensor,** a column of cells that connects the embryo to the ovule wall, arises from cells at the opposite end. Mitosis continues in both cell populations; the upper cells produce a ball of cells, the *globular-stage embryo*. The suspensor feeds the embryo by passing nutrients from suspensor cells and from the endosperm to the embryo's cells.

Cells in the globular embryonic stage differentiate into **inside cells** and **outside cells.** The small outside cells form the **protoderm,** the embryonic epidermis. Protoderm also is present in the growth centers of roots and shoots (where it is part of the apical meristems, the growth sites). The original protoderm is one cell thick, and it remains that way as it expands, increasing the embryo's surface area to allow for internal cell division and growth.

Some of the inside cells form the layer of cells that gives rise to the embryo's vascular tissue (see Chapter 25). In dicots, two bumps emerge at the top of the embryo above the procambium. These two bumps, which give the embryo a heart-shaped appearance, will form the two **cotyledons,** or seed leaves, characteristic of dicots (see Figure 26-8b). In monocots, there is one cylin-

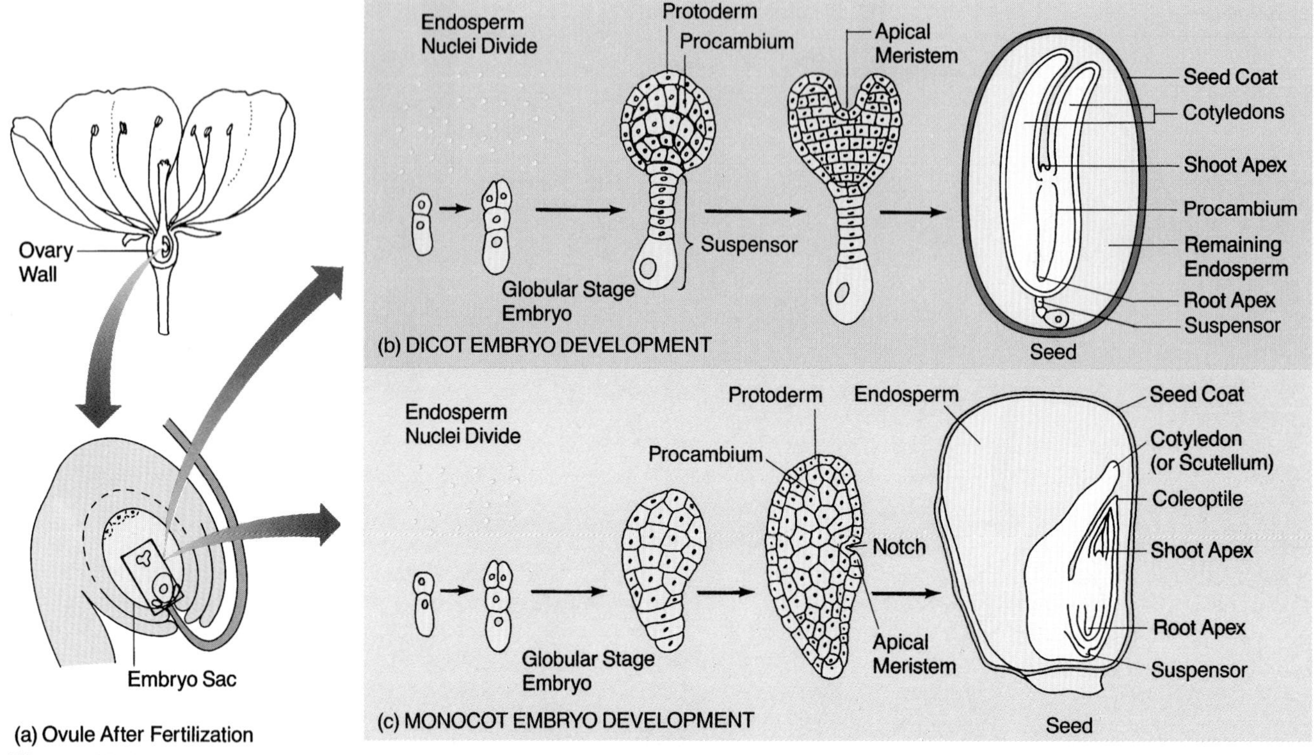

Figure 26-8 THE DEVELOPMENT OF FLOWERING PLANT EMBRYOS.
(a) Development of both monocot and dicot embryos begins inside the ovule after fertilization. Mitosis of the zygote forms cells that will develop into the embryo and the suspensor. Cells at the end of the suspensor form the globular embryo that soon has inside cells and outside cells. The outside cells form a protoderm (forerunner of epidermis), and the inside cells form procambium, which gives rise to vascular tissue. (b) At this point, dicot development diverges from monocot development. In dicots, the two cotyledons gradually take form, leaving the apical meristem located between these "seedling leaves." (c) In later stages of monocot development, a notch forms on one side of the cylindrical embryo, marking the site of the apical meristem. The single cotyledon, the scutellum, is located above and the root below the apical meristem.

drical cotyledon, the **scutellum,** which arises directly from the end of the embryo opposite the suspensor (Figure 26-8c).

Wedged between the two dicot cotyledons, or at one side of the monocot's cotyledon, is the **apical meristem.** A meristem is an organizing center of undifferentiated, actively dividing cells. Meristems are found only in plants and form zones where cells for new organs can be generated throughout the life of the plant.

The apical meristem lies at the growing tip of the shoot and is responsible for generating—directly or indirectly—all the cells for a plant's leaves, stems, branches, and flowers. Although the apical meristem is the true apex of the plant, it may appear to be lower than a dicot's two cotyledons because they elongate very rapidly. In monocots, the apical meristem is found in a notch on the side of the embryo; from that position, it is reoriented in the seedling so that the apex points upward (see Figure 26-8c).

At the opposite end of the embryo, another important precursor tissue forms: the **root meristem.** Like the apical meristem, this zone of undifferentiated, rapidly dividing cells will continue to generate new cells, but these will grow and mature into functioning root tissue. Formation of the shoot and root meristems defines the growth axis of the plant; that is, it establishes the polarity of root and shoot critical for normal plant development.

SEEDS: PROTECTION AND DISPERSAL OF THE NEW GENERATION

The embryo develops as part of the seed, and this, in turn, usually develops within the fruit. *Seeds* are the mature ovules and protect the embryo and the endosperm until the appropriate time for germination. In many species, the establishment of the embryo's meristems is a crucial stage of development; once this occurs, the embryo stops growing, and the *seed coat* forms from the ovule walls (see Figure 26-8). This coat can be thin and papery, as in the "skin" on a peanut, or stony, as on a Brazil nut. But it always serves as a tight, protective seal around the embryo (and often the endosperm as well), and once the seed coat has formed, the fruit begins to enlarge around the seed.

Most mature *fruit* tissue is derived from the wall of the ovary at the base of the flower, although some fruits, such as the strawberry, are derived from the enlarged receptacle, below the ovary. The ovary wall can become fleshy and succulent, as in cucumbers, citrus, pears, tomatoes, and cherries; it can be leathery, as in pea pods and the twisted pods of a locust tree; or it can be papery and dry, as in the "squirts" that flutter down from maple or elm trees.

The fruit tissue that surrounds the seeds helps protect them from drying out during early development and, through a number of ingenious adaptations, eventually aids in their dispersal. The fruit can split open, forcibly ejecting the seeds or allowing them to blow about like the papery seeds and gossamer plumes of a milkweed plant (Figure 26-9a). Succulent fruits such as apples and tomatoes often attract hungry animals that swallow the seeds and later deposit them unharmed (Figure 26-9b). Finally, a number of fruits and seeds have hooks, barbs, or sticky exudates that allow them to hitch a ride on an animal's body (Figure 26-9c).

Seed Maturation

Many kinds of seeds must mature by storing nutrients and dehydrating. The seeds of most staple food crops, such as peanuts and beans, have a special period for weight gain in the embryo and accumulation of food in the cotyledons and endosperm. Such food reserves are crucial, since the seedling must survive on stored food until it becomes photosynthetic and self-sufficient.

Before the cavity within the seed coat is totally filled, weight gain in the cotyledons and endosperm is *wet weight*—that is, mostly water. However, after the cotyledons or endosperm fills the seed cavity completely, the stage of *dry weight* accumulation occurs, during which the water in the cells is replaced progressively with oils, starch, and proteins. The adult legume plant may look yellow and pathetic at the time of seed maturation because the entire sugary photosynthetic output is diverted to the ovarian tissue and seeds.

During the late stages of seed maturation, the embryo begins to dehydrate—in many species, almost completely—yet it remains alive. The early embryo is about 90 percent water and 10 percent dry matter (cell walls, membranes, nutrients, etc.). As the seed matures, the synthesis and storage of dry matter increase the proportion of these materials to about 50 percent. After this state is attained, dehydration accelerates to 10 percent wet weight and 90 percent dry. Few other living tissues can withstand such a water loss and still survive.

Seed Dormancy

How does the plant embryo survive in a dehydrated condition? It becomes **dormant.** The cells of the dormant embryo respire at a very low rate and carry out only a small amount of metabolic work. In this resting state, the seed can survive unfavorable environmental conditions, such as a cold winter, a prolonged drought, or both. In some species, the embryo may remain in this resting state for years. Weed seeds excavated in Denmark, for example, germinated after 1,700 years.

(a)

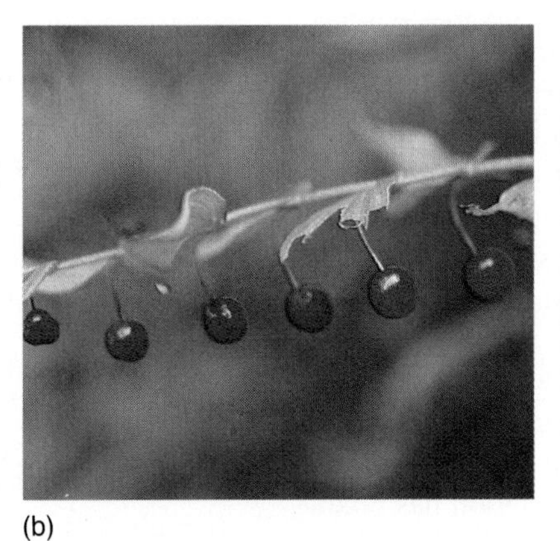
(b)

(c)

Figure 26-9 SEEDS AND FRUITS: ADAPTATIONS FOR SURVIVAL.
(a) The wind will soon carry this last milkweed seed (*Asclepias syriaca*) from its pod to some distant site. (b) The sweet, fleshy fruits of the twisted stalk rose (*Streptopus roseus*) attract hungry birds, bears, or other animals; the seeds inside the fruit pass through the animal's digestive tract intact, to be deposited elsewhere. (c) When spring comes and snow melts, the burrs of this burdock (*Arctium minus*) will adhere to the fur, feathers, or clothing of a passing animal and ensure transport of the seeds to new sites.

Most seeds have dormancy mechanisms besides dehydration, since the addition of water and the appropriate temperature for growth would automatically end dormancy even if it were a temporary, mid-winter warm spell. To prevent premature germination, the seeds of many plants must be exposed to several weeks of cold at or below a certain temperature such as 4°C.

But how can a dormant seed "keep track" of time? The answer lies in cellular chemistry. Some seed coats are filled with compounds that inhibit germination and that are gradually inactivated over time. These compounds act like a biochemical hourglass, requiring the embryo to remain dormant until a certain number of months have passed—a number that guarantees germination at a time of year when the chances for survival are statistically greatest. For the seeds of many plants in temperate climates, prolonged cold gradually inactivates the inhibitors, while for other plants from mild climates, a long, hot summer is needed to "burn off" the inhibitors.

Some seeds have extremely hard (sclerified) seed coats that must be partially abraded away by blowing over rough ground or passing through a bird's digestive system, where acidic digestive juices etch its surface or small pebbles fracture it and grind it. The abrasion process allows germination to begin.

GERMINATION AND SEEDLING DEVELOPMENT

When conditions are conducive, dormancy ends and germination begins like an erupting volcano. Growth of the tiny plant resting inside the seed resumes when ground temperatures are right for the particular species, sometimes when light strikes the seed, and when water is present. The quiescent embryo imbibes, or takes up, water at a rapid rate, restoring the water content of 80 to 90 percent and causing the embryo's cells and tissues to swell. Imbibing water also establishes conditions of tonicity and solubility that are conducive to enzyme activity; as a result, the embryo's sluggish metabolism soon speeds up, allowing cell division to resume in the meristematic regions. The pressure caused by the swelling and renewed growth of the embryo cracks and weakens the encapsulating seed coat, and within hours, the root cells elongate, and the future root pokes out through the ruptured seed coat. At this point, germination has taken place, and the embryo begins a race to establish itself as a photosynthetic seedling before it runs out of the nutrients stored in the endosperm or cotyledons.

The first structure to emerge from the rupturing seed coat is the **radicle,** the short length of root that becomes the primary root, and is part of the **hypocotyl.** The radicle is *positively gravitropic,* which means literally "movement toward gravity"; hence, it grows downward into the soil. As that occurs, the **epicotyl,** or future stem, emerges from the soil and begins to grow upward.

The many thousands of small cells in embryonic root, shoot, and cotyledon undergo a remarkable elongation process as part of germination. Individual cells may lengthen 10 to 100 times, so that a small segment of tissue only 1 mm long could elongate to 100 mm. This burst of cell elongation pushes the root several centime-

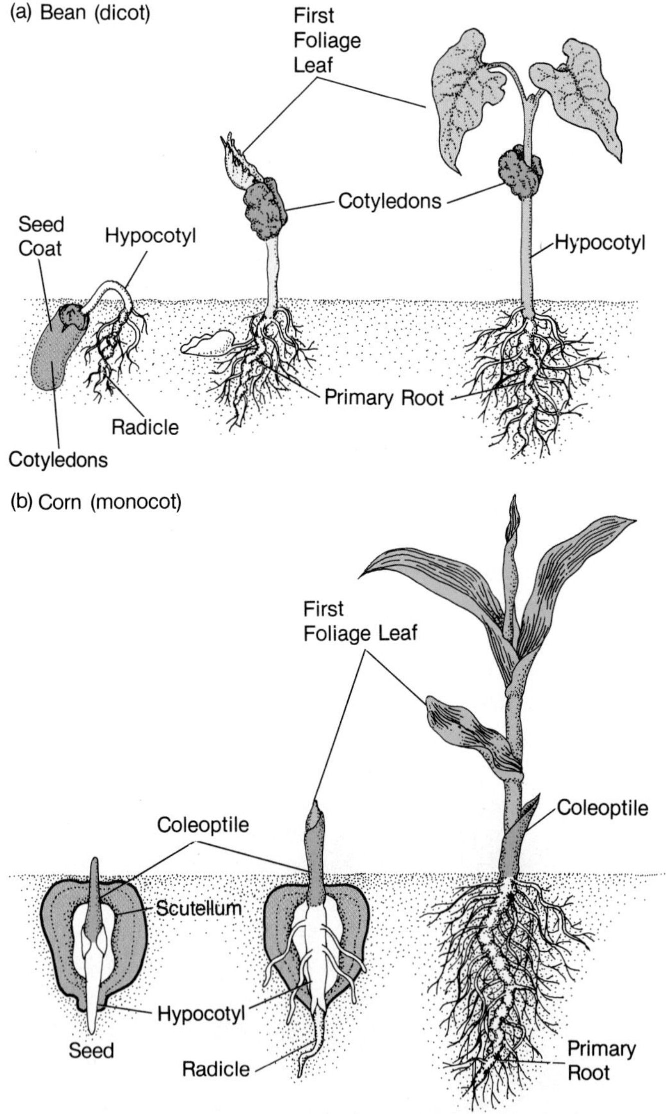

Figure 26-10 THE NEW PLANT: GERMINATION AND SEEDLING GROWTH IN A MONOCOT AND IN A DICOT. In both (a) dicots and (b) monocots, seedling growth begins with the emergence of the radicle (root sprout) and the hypocotyl. Whereas the dicot cotyledons of many species will emerge from the soil as the hypocotyl elongates, the monocot's scutellum remains within the seed coat with the endosperm to nourish the seedling. In a monocot, the coleoptile leads the shoot through the soil.

ters deep into the soil and propels the shoot upward toward air and sunlight.

Early seedling growth differs somewhat between monocots and dicots, as Figure 26-10 shows. In both, the radicle and hypocotyl emerge first. However, monocots have a specialized organ for shoot growth, the **coleoptile** (Figure 26-8c). This protective sheath around the embryonic leaves pushes upward through the soil toward the light. Once exposed to light, the rest of the shoot tissues can also grow toward light (see Chapter 28). Recent studies show that light can directly activate certain plant genes involved in development.

Soon after the monocot's coleoptile emerges from the soil, it is split apart. The rapid expansion of the embryonic leaves within pushes these young leaves into the sunlight; they then turn green, and begin photosynthesis. Although the monocot's shoot has erupted, the rest of the seed remains underground, and the endosperm and scutellum stay within the seed coat to nourish the growing seedling.

Mobilizing stored food becomes every seedling's top priority. Seeds of barley, a monocot, are a good example. Barley endosperm stores both starch and protein in dead, centrally situated cells. Surrounding that food cache is a layer of living cells called the **aleurone,** which comes from the ovule wall. When a barley seed germinates, the embryo makes or releases a hormone that stimulates protein synthesis in the aleurone layer. An aleurone-produced enzyme, *amylase,* is secreted into the endosperm at a very high rate, resulting in the rapid conversion of starch to sugars. These sugars are used to manufacture new cell walls during the rapid cell division and elongation following germination.

In some dicot seedlings, such as beans, a slightly different sequence of events takes place. Rather than relying on a central store of endosperm, the seedling's two food-storing cotyledons are ferried up through the soil by the elongating shoot and, once they have broken ground, are stimulated by sunlight to turn green and become the new plant's first photosynthetic organs.

Whether in a monocot or dicot, germination results in the building and elongating of new cells and, in time, in the lengthening of root and shoot. The seedling is on its way to becoming a mature plant.

PRIMARY GROWTH: FROM SEEDLING TO MATURE PLANT

As development proceeds in the established seedling, the root and shoot meristems generate cells, which grow and differentiate into specialized cell types. Most of the size expansion in stems and roots of adult plants is the result of cell elongation in zones just behind the meristematic regions. Thus, cells arising by division in the meristem elongate dramatically to yield real growth in size. This elongation or linear growth at the tips of the root and shoot is called **primary growth.** In contrast, expanding girth or lateral growth occurs differently and is called **secondary growth.**

Primary Growth of the Root

Primary growth in the tips of young roots occurs rapidly and continuously. The primary roots of maize plants, for example, can elongate up to 11 cm per day! Each tip has a root meristem that produces several tissues. Cell divisions at the front of the meristem yield **root cap cells** (Figure 26-11), which act as a shield to protect the delicate meristem and secrete a slime called *mucigel.* This slippery material eases the root's passage through the soil, as well as facilitating the uptake of phosphate and iron and promoting symbiotic relationships with certain beneficial soil bacteria. As the root probes the soil, the root cap cells are continually sloughed off or crushed, so that new ones must be generated continuously.

Root tips are often damaged by even gentle handling. As a result, they cease functioning, and root growth stops. Since the youngest root tissues, near the tips, are sites of root hair formation and thus of maximal water and mineral absorption, the simultaneous injury of many root tips can cause severe water stress in a plant.

There are three major zones in a rapidly growing root: the **meristematic region,** the **region of elongation,** and the **region of maturation and differentiation** (see Figure 26-11a). The first two zones allow the root's vertical growth; the third, its maturation. The meristematic region is a zone of mitotic cell division where growth results from an increase in the number of cells. The region of elongation, just behind the meristem, contains mitotically active precursors of root epidermal cells and a central core of cells that become vascular tissues and cortex. New cells that are left behind the advancing meristematic region show growth by an increase in size rather than an increase in number. Once the new cells have increased in length 10- to 100-fold, some of them begin to differentiate into individual cell types, such as xylem and epidermal cells that form root hairs.

Primary Growth of the Shoot

Just as in the root system belowground, primary growth of the visible parts of the plant depends on meristematic activity and cell elongation in growth regions. New leaves, stem, and branches are generated by the apical meristem, which sits at the very top of the plant,

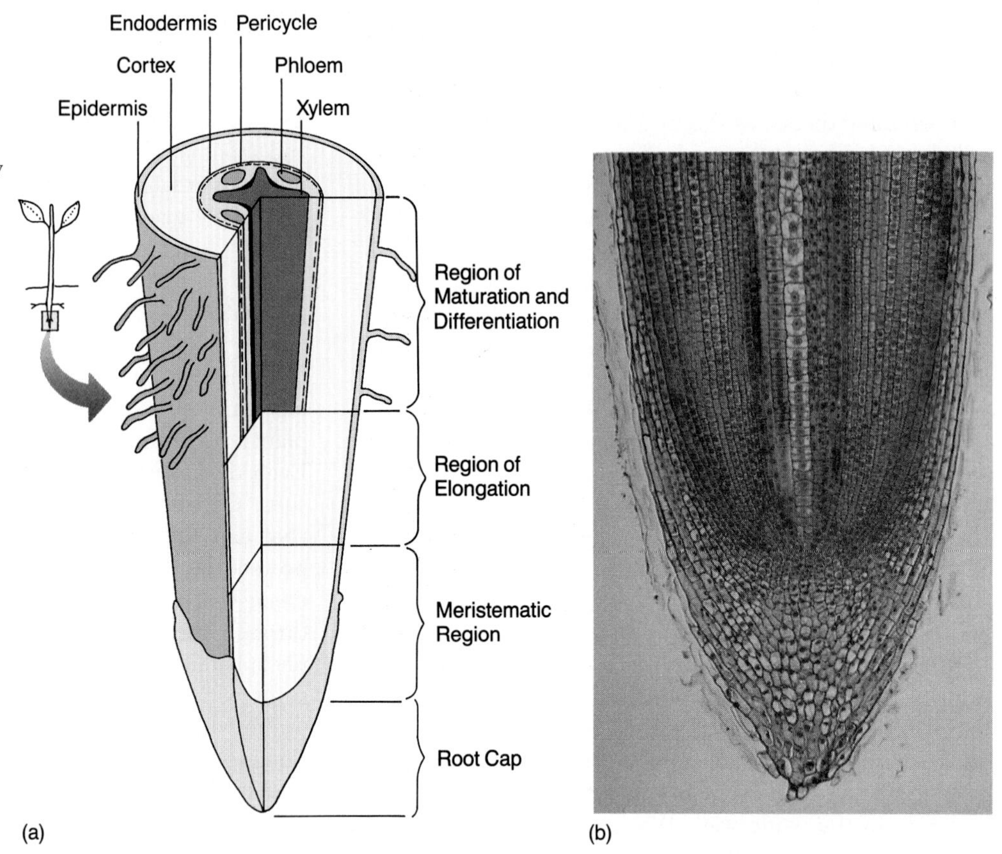

Figure 26-11 PRIMARY GROWTH IN THE ROOT. (a) Three major zones contribute to root growth. New cells arise by mitosis in the meristematic region. They increase in size, and especially in length, in the region of elongation. Finally, the cells differentiate into epidermis and vascular tissue cells in the region of maturation and differentiation, where root hairs extend from epidermal cells. (b) The delicate meristem of this onion root (*Allium* species) is protected by root cap cells (magnified about 85×). These large cells with tough walls are continually abraded and replaced as the root tip pushes through the soil.

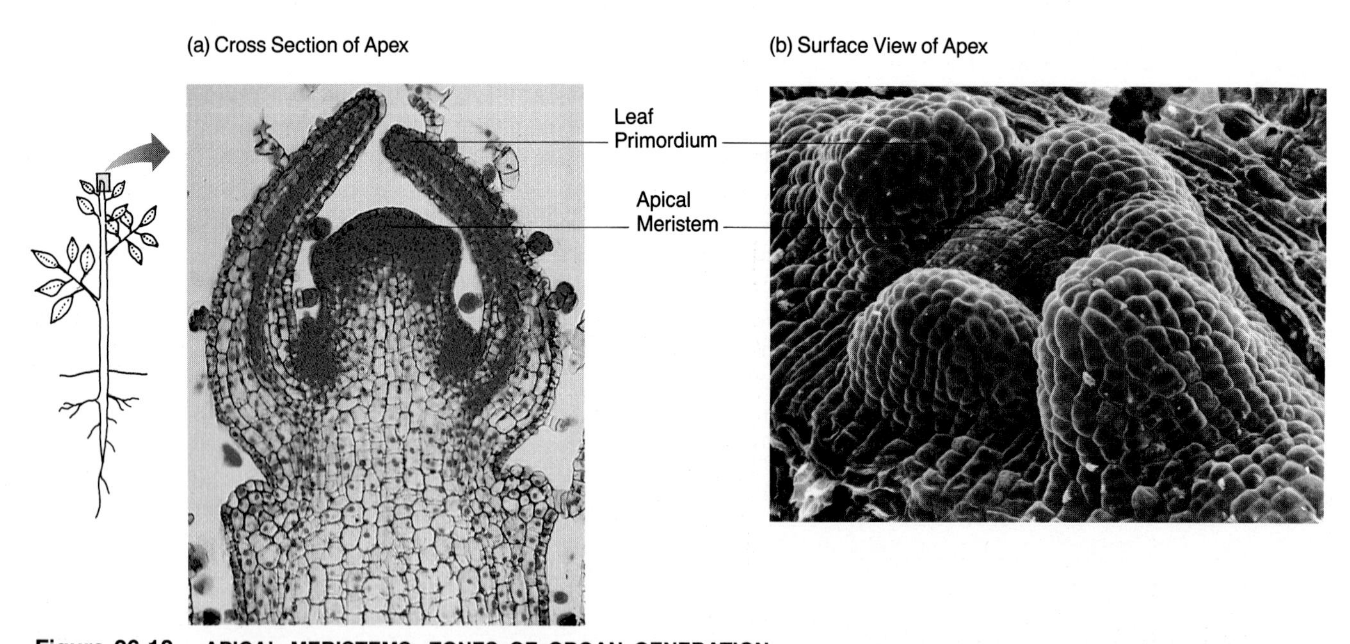

Figure 26-12 APICAL MERISTEMS: ZONES OF ORGAN GENERATION. (a) This dicot apical meristem (magnified about 16×) is a growth center giving rise to cells that will form new leaves and stem. The immature leaf primordia of this coleus (*Coleus* species) are folded over the apical meristem, protecting it. Bud primordia may later become active meristems and sites of stem outgrowth as branches. (b) This scanning electron micrograph from a different plant, a succulent, *Graptopetalum paraguageuse* (magnified about 28×), shows the apical meristem of a shoot as a dome in the center. The meristem cells give rise to the leaf primordia, four of which are seen here in various stages of development.

normally covered and protected by many immature leaves folded around it (Figure 26-12). As a new leaf forms, a group of meristematic cells on the apical meristem divides, creating a bulge on the side of the meristem called the **leaf primordium,** which resembles a swollen mound at first and then a flattened pad with the start of recognizable leaf features, including a midrib and lobed or toothed leaf margins.

After a leaf primordium forms on one side of the apical meristem, the next primordium forms a precise distance away, a specific fraction (one-half, one-third, one-quarter) of the way around the cylindrical stem. This process occurs again and again, all new leaves forming at these regular intervals around the stem. The result can be an *alternate* pattern or an *opposite* pattern of leaves around the plant stem (Figure 26-13).

The position at which a leaf arises is called the **node,** and the stem tissue between leaves is the **internode.** The length and growth rate of the internodal tissue can vary greatly. In a head of cabbage, for example, internodes are extremely short, and the leaves are virtually stacked on top of each other. In many other plants, such as corn, the internodes are quite long, and the leaves are far apart. Besides giving rise to new leaves, the apical meristem produces the plant's stem tissue—epidermis, cortex, phloem, xylem, and pith.

The apical meristem has yet another task besides initiating leaf primordia and producing stem tissue: It is the site of formation of **bud primordia** (see Figure 26-13). These primordia form in the angle between the base of developing leaf primordia and the apical meristem and later may grow into branches or form flowers. The angular junction of leaf primordium and shoot apical meristem is called the *axil.* Hence, bud primordia are said to be **axillary.** As the leaf primordium develops and is displaced from the shoot apex, the axillary bud may continue developing and, depending on species, may eventually come to look like the shoot apical meristem and itself initiate new leaf and bud primordia. It is only because of this phenomenon that large, lateral branches of a tree or bush can themselves give rise to numerous smaller branches and they, in turn, to still smaller ones.

The dormant buds at the tips of maple twigs in the spring are another type of bud, the **terminal buds,** enclosing the shoot apical meristem and several sets of well-developed leaves waiting to begin growth when conditions are appropriate. Perhaps the most familiar bud produced by the apical meristem is the **flower bud,** enclosing future petals and reproductive parts.

The presence of meristems truly distinguishes embryos of higher plants from those of higher animals. In general, embryos of the more complex animals develop the same number of organs as in the adult (two arms, one liver, etc.); these organs are constructed during embryonic development and must last for the animal's lifetime. In plants, on the other hand, the major life strategy involves periodic generation of new organs. As leaves become less efficient with age, these light-collecting organs are replaced by new, more efficient leaves. As shrubs or trees increase so spectacularly in size, the total number of leaves, branches, and limbs increases greatly. And roots grow continuously, too, so there is perpetually a fresh zone of absorptive root hairs. A healthy plant, therefore, is always growing and making new organs—leaves, roots, and stems—except when dormant during cold or very hot weather. The plant embryo simply contains the first examples of these organs and, by means of the perpetual growth centers, the meristems (embryonic tissue), the plant retains the potential to make many more.

(a)

(b)

Figure 26-13 LEAF ARRANGEMENTS RESULT FROM PATTERNS OF MERISTEM ACTIVITY.
(a) If a meristem shows several discrete active regions, which occur in a spiral pattern along the stem, the resulting leaf arrangements will be alternate, as in this European fly honeysuckle. (b) If both sides of the apical meristem are highly active at the same time, the resultant leaves lie opposite each other, as in this whorled loosestrife.

Because of this potential, the dividing cell populations are said to be *totipotent:* Each cell can give rise to a complete new plant or any of its parts. This tendency helps explain why plants do not need (and do not have) germ lines; totipotent cells derived from the meristem are located in the forming flower and give rise to the microspore or megaspore mother cells, which yield sperm or eggs.

SECONDARY GROWTH: THE DEVELOPMENT OF WOOD AND BARK

Wood is one of the most common and useful materials to humans, but what is wood, exactly, and how does it fit into the sequence of events we called plant development?

We have seen that all the tissues in a young seedling arise by primary growth from root and apical meristems. As a nonherbaceous plant matures and grows taller, its stem also begins to grow laterally, increasing its diameter—its girth. This thickening of the stem, or **secondary growth,** enables the plant to withstand the added load of branches and leaves, as well as wind, rain, gravity, and other environmental forces. The new, secondary tissues are generated by two growth regions, collectively called **lateral meristems.** The regions are specifically the *vascular cambium* and the *cork cambium.*

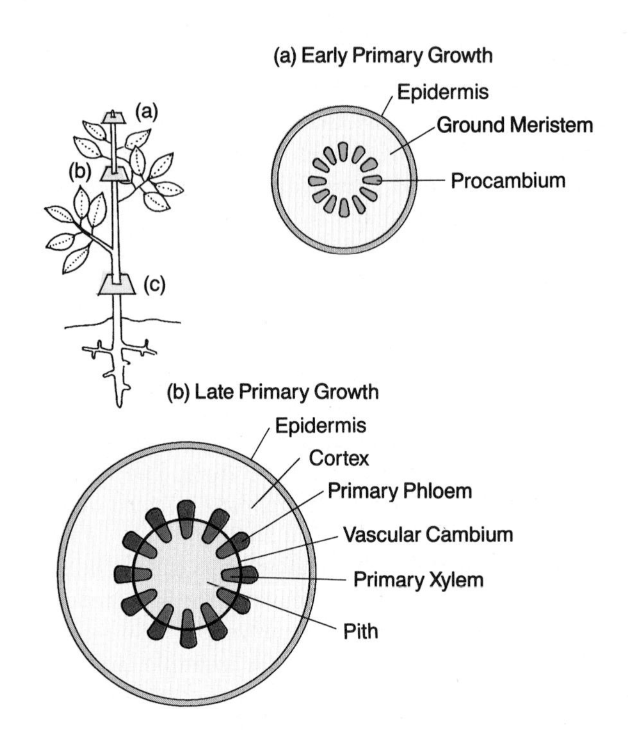

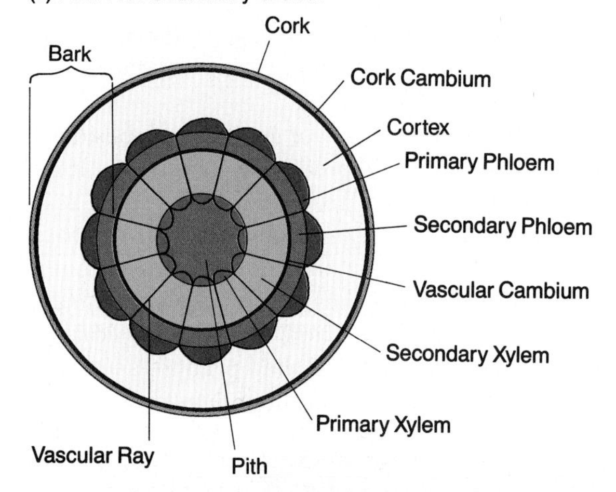

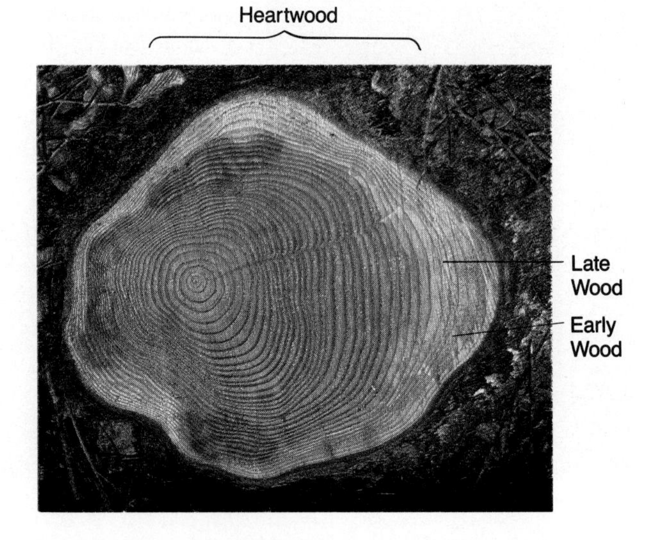

Figure 26-14 SECONDARY GROWTH EMERGES FROM PRIMARY, FORMING WOOD, BARK, AND GROWTH RINGS. Developing stem tissue in a woody dicot. (a) Early primary growth. (b) Late primary growth, with vascular tissue bundles arising from the vascular cambium and surrounded by cortex. (c) With the first year of secondary growth comes expanding girth, the addition of secondary xylem and phloem, and an outer layer of bark. The vascular rays are lines of parenchyma cells involved in lateral transport of minerals and water. The phloem rays conduct the products of photosynthesis in the vertical transport system. (d) Forty years of secondary growth in a larch. During the late summer, the water-carrying xylem cells are becoming dormant, and few are produced from the cambium. These are represented by narrow dark rings, the late wood. In the spring, the vascular cambium is reactivated and begins to produce many large xylem cells. These account for the lighter-colored broad xylem bands, the early wood. Note the darker heartwood in the center, a region composed of older dead xylem. The outer rings of younger growth are called sapwood because they are composed of xylem that still actively conducts water. Note the asymmetrical shape of the rings and the trunk itself; perhaps environmental factors led growth to be faster on the left than on the right side of this tree.

Vascular Cambium

We have discussed the critical transportation roles of the vascular tissues, xylem and phloem, as well as their importance in providing physical support (see Chapter 25). As a plant matures, individual cells of both xylem and phloem cease to transport materials; phloem elements usually function for only a year or two before dying, and xylem elements often fill up with materials that diminish the vital flow of water and nutrients. The cylinder of generative cells called the vascular cambium overcomes the problem of vascular tissue maturation by producing new xylem and phloem cells and, in the process, adding girth to the plant. The vascular cambium is a cylindrical zone of actively dividing cells that lies between the xylem and the phloem in both the stem and the root (Figure 26-14a and b). Mitotic divisions in the vascular cambium produce new, undifferentiated cells. Those on the inner side of the cambium become secondary xylem, or wood, while those on the outer side differentiate into secondary phloem, one component of *bark* (Figure 26-14c).

Each spring, the vascular cambium, dormant over the winter, is reactivated and begins producing new xylem and phloem. In many plants, spring xylem cells are large in diameter and constitute **early wood** or **spring wood,** while summer xylem cells are much smaller and form the **late wood,** or **summer wood.** The difference in cell size probably results from the availability of water: Larger cells form in the moist spring; smaller ones, in the drier summer. The difference in cell size between early and late wood is quite apparent and results in the distinct rings we see in cross sections of tree trunks (Figure 26-14d). A count of tree rings actually reveals the number of times the vascular cambium has been activated to produce spring wood, the lighter part of each ring. The rings of younger xylem, nearer the periphery of the trunk or stem, are called the **sapwood;** sapwood xylem still functions in water conduction. A darker region, at the center of the trunk, is composed of nonconducting xylem and is called **heartwood.** This is the older xylem that is often clogged with various substances and no longer transports water and nutrients. The growth rings of trees literally retain an organic history of the harsh winters, wet springs, and dry summers through which the organism has survived. Plant scientists can tell, for example, that trees with uneven rings (as in Figure 26-14d) grew at an angle rather than straight up; resisting gravity caused more growth on one side than on the other.

Since the diameters of the stems, branches, and roots increase each year as plants grow older, transport *across* the tissue—**radial transport**—becomes as much of a problem as **vertical transport** between roots and leaves. While the cells of vertical transport tissue, xylem and phloem, arise from **fusiform initial cells,** squarish cells called **ray initial cells** extending from the center of the stem toward the periphery produce **vascular rays,** which function as the system of lateral transport (Figure 26-14c). Rays help distribute minerals and water laterally to the sapwood and store excess starch.

Cork Cambium

The other type of lateral meristem that produces secondary growth in plants is the **cork cambium,** a layer of cells just beneath the epidermis (see Figure 26-14c). As the trunk and roots increase in diameter owing to the activity of the vascular cambium, the original epidermis and cortex become too small to encompass the expanding trunk or roots and split open. Cell divisions in the cork cambium then produce a secondary layer, the **cork,** which replaces the epidermis. Cork cells become impregnated with suberin, the same waxy substance that helps prevent water loss from young stems and roots. Such waterproofing, of course, also inhibits the passage of oxygen and carbon dioxide into and out of the plant. Therefore, in cork-covered stems and roots, only certain dark-looking spots in the cork, called **lenticels,** allow gas exchange.

The cork cambium develops from the cortex in stems. In certain trees, if one layer of cork cambium is destroyed, the secondary phloem parenchyma cells can give rise to a new one. This is why the bark of some old trees is layered, resulting in bark patterns characteristic of particular species. People harvest the outer cork layer from cork oak trees for bulletin boards and bottle stoppers.

The familiar **bark** of the woody stem includes all the tissues outside the vascular cambium: the cork, cork cambium, cortex (lost after one to three years), and the secondary phloem (see Figure 26-14c). The outer bark (cork tissue) is dead, while the inner bark is alive. Most trees cannot withstand the loss of the bark layer because the phloem is lost, too. Thus, the removal of a ring of bark all the way around the tree by deer or other animals (so-called *girdling*) is often fatal.

PLANT LIFE SPANS AND LIFE-STYLES

Plants have different growth habits, and anyone who has purchased flower or vegetable seeds probably has encountered the terms *annual, biennial,* and *perennial.*

In **annual plants,** the seed germinates in the spring, the plant grows, the flowers and seeds develop, and the plant dies—all within a single growing season (Figure 26-15a). The entire cycle is repeated annually, with only new sets of seeds surviving from one year to the next. Many weeds, grains, flowers, and garden vegetables are annual plants.

Annual plants may live for just a few weeks or for nearly an entire year. Much of their energy goes toward flowering and producing seeds. Thus, the plant body grows only large enough to support the maturation of the seeds; very little secondary growth or wood and bark production occur. As soon as the seeds appear on an annual, the plant typically begins to age and die.

Biennial plants are much less common than annuals and require 2 years from the time the original seed germinates to the production of a new generation of seeds. Parsley and foxglove are typical biennials (Figure 26-15b). The plant body grows and becomes established in the first season; in the second year, the plant flowers, produces seeds, and dies soon thereafter.

Annuals and biennials die in just a year or two because they cease growing after the apical meristem is converted into a flower. When the apical meristem is making leaves and internodes, the meristematic tissue is not used up, since it continuously regenerates itself through cell division. The stimulus to produce a flower, however, causes a transformation of the meristem from vegetative to sexually reproductive growth, and the whole meristem is "sacrificed" to make the flower. With no dividing cells left to make more leaves or stems, the annual or biennial plant is doomed.

Perennial plants, in contrast, live for many years, and some, such as trees, have large bodies with a considerable degree of secondary growth (Figure 26-15c). Other perennials, such as grasses and herbs, may not be particularly woody or large. Perennials often require several years of growth before they can flower and produce fruits and seeds. Flowering in perennials does indeed "consume" apical meristems; however, this does not affect the plant's survival because flowering typically involves only a small subset of the lateral branches, not the entire complement of meristems. The perennial plant can produce leaves and flowers for many years from different subsets of lateral branches arising each year. Most woody perennials have a typical life span of 10 to 50 years, although many trees can survive for hundreds or thousands of years.

Annuals, biennials, and perennials share the potential for having new, chance mutations in meristem cells incorporated into flowers formed from such meristems and so into pollen and eggs. Thus, one branch of a large tree or even one flower on a herbaceous plant could produce gametes with a different genotype than the rest of the plant. This is potentially a major source of genetic variability and points to a profound difference between plant and animal development: In general, plants tolerate a far greater level of genetic imbalance.

Recall that in animals, alterations of normal chromosome number are often fatal (see Chapter 15), and to be passed on to future generations, mutations must occur in germ line cells, since somatic mutations do not enter sperm or eggs and so make no direct contribution to evolution. In plants, however, polyploidy (extra chromo-

Figure 26-15 ANNUALS, BIENNIALS, AND PERENNIALS.
(a) Annual plants, such as these poppies (*Papaver* species), live just one growing season; the seeds they produce survive to repeat the cycle. (b) Biennial plants, such as foxglove (*Digitalis purpurea*), require 2 years from seed germination to production of the next generation of seeds. The plants then die at the end of the second season. (c) Perennials, such as these blossom-covered plum trees (*Prunus cerasifera*), can survive indefinitely because their flower production does not use up a significant portion of the plant's apical meristem tissue.

some sets) is quite common: About half of all flowering plants thus far studied are polyploid. What's more, plants have a haploid phase, the gametophyte, in which a single copy of the genome directs complex development and physiology. Some plant scientists have even begun to think of a single large plant—with the capacity to tolerate genetic imbalance and to pass to new generations those novel genotypes arising in meristems—as being more similar to a population of animals than to an individual animal.

LOOKING AHEAD

Thus far in our discussion of plant biology, we have surveyed plant structures and seen how they grow, develop, and reproduce. In the next chapter, we will see how plants live day to day—in particular, how they absorb and transport water, minerals, and sugars—and thus, in part, how they cope so successfully with the earth's many and varied environments.

SUMMARY

1. During *vegetative reproduction*, plants genetically identical to each other and to the parent arise from the parent's body.

2. Vegetative reproduction can involve *rhizomes, stolons or runners, bulbs, corms, tubers, leaves,* and *root suckers.*

3. Sexual reproduction in anthophytes involves *flowers* that attract the appropriate pollinator. Flowers are modified leaves consisting of sepals, petals, stamens, and carpels (pistils).

4. Pollen production involves *microsporogenesis*, the production of haploid microspores, and *microgametogenesis*, during which each microspore divides mitotically to form the tube cell and generative cell, encased together in the pollen grain.

5. Egg production involves *megasporogenesis*, the formation of a haploid megaspore, and *megagametogenesis*, mitotic divisions of the megaspore to produce two *polar nuclei*, the egg, and five other cells.

6. When a pollen grain adheres to a stigma, the pollen is stimulated to grow, or *germinate*. A pollen tube grows through the stigma and style and enters the ovule; two sperm are released, and a double fertilization generates the diploid zygote and the triploid endosperm cell.

7. The zygote divides, and one cell gives rise to the *suspensor*, which attaches the embryo to the ovarian wall. The other cell derived from the zygote gives rise to a ball of cells—the globular-stage embryo—which remains attached to one end of the suspensor.

8. The endosperm serves as the source of nutrients for the embryo inside the seed coat. (Dicots generally store nutrients in the cotyledons.)

9. *Seeds*, the mature ovules, develop within the *fruit*, which is usually derived from the wall of the ovary. Fruits protect the seeds and aid dispersal. The embryo within the seed enters *dormancy*.

10. Germination begins when the seed takes up water and swells and meristematic activity commences. The seed coat cracks, and via cell elongation, the seedling emerges.

11. Endosperm nourishes monocot seedlings, as do newly photosynthetic leaves. Nutrient reserves in the cotyledons nourish dicot seedlings; the cotyledons may become the first photosynthetic leaves.

12. *Primary growth* involves cell elongation at the tips of roots and shoots. A rapidly growing root has a *meristematic region* (zone of cell division), a *region of elongation* (zone of cell-size increase), and a *region of maturation and differentiation* (zone of tissue cell specialization).

13. The *apical meristem* produces new leaves and branches. A leaf arises from a *leaf primordium*, a bulge on the side of the meristem. Branches arise from bulges called *bud primordia*.

14. *Secondary growth*, the thickening of the stem, involves *lateral meristems;* the *vascular cambium*, which generates xylem and phloem; and *cork cambium*, which produces a layer of *cork*.

15. *Wood*, or secondary xylem, in a thick stem shows annual growth rings. Functional xylem near the stem or trunk periphery is called *sapwood*, the central *heartwood* is nonfunctional xylem.

16. Waterproof cork replaces the epidermis in a thick trunk. *Lenticels* allow gas exchange through the cork. *Bark* is made up of cork, cork cambium, and secondary phloem.

17. *Annual plants* mature rapidly, produce seeds, and die within a single growing season. *Biennial plants* do the same over a 2-year period. *Perennial plants* produce new meristems each year and are long-lived.

KEY TERMS

aleurone
annual plant
apical meristem
axillary bud
bark
biennial plant
bud primordium
calyx
cleavage division
coleoptile
cork
cork cambium
corm
corolla
cotyledon
dormancy
double fertilization
early wood
epicotyl
flower bud
fusiform initial cell
heartwood
hypocotyl
inside cell
internode

lateral meristem
late wood
leaf primordium
lenticel
megagametogenesis
megasporogenesis
meristematic region
microgametogenesis
microsporogenesis
node
outside cell
perennial plant
perianth
polar nucleus
primary growth
protoderm
radial transport
radicle
ray initial cell
region of differentiation
region of elongation
rhizome
root cap cell
root meristem
root sucker
runner
sapwood
scutellum
secondary growth
spring wood
stamen
stolon
summer wood
suspensor
terminal bud
tuber
vascular ray
vegetative reproduction
wood

QUESTIONS

1. Give examples to show how plants use roots, stems, and leaves for vegetative (asexual) reproduction.

2. What advantages does a plant species gain from vegetative reproduction? From sexual reproduction? Which has the greater evolutionary advantage for the species?

3. Describe the roles of the following in the life cycle of a flowering plant: pollen tube, polar nuclei, double fertilization, seed coat.

4. Briefly describe each of the following parts of a flowering plant's embryo: protoderm, cotyledon, scutellum, apical meristem.

5. Where are meristems found in the plant body? What role does each type of meristem play in the growth, development, or reproduction of the plant?

6. What is the physical and metabolic condition of a dormant seed? How may seed dormancy be broken? How is the breaking of dormancy regulated in the seed?

7. List, in chronological order, the major steps in seed germination and seedling development.

ESSAY QUESTIONS

1. Compare and contrast primary plant growth from apical meristems and secondary plant growth from lateral meristems.

2. Explain the steps in the formation of a pollen grain (male gametophyte) and an ovule (female gametophyte) in a typical flowering plant.

For additional readings related to topics in this chapter, see Appendix C.

27
Exchange and Transport in Plants

From its Office, which is To feed the Trunk . . . the *sap* must also, in some Part or other, have a more especial motion of Ascent.

Nehemiah Grew, *The Anatomy of Plants: With an Idea of the Philosophical History of Plants* (1682)

In 1628, the English physician and scientist William Harvey described how the pumping heart can circulate an animal's blood. Naturally enough, botanists in Harvey's time began to search for an analogous pumping mechanism and accompanying transport vessels in plants. While they found no pump or circulation in plants, they did discover, by the 1720s, that water travels from the roots to the leaves.

It remained for twentieth-century plant scientists to work out the details of how plants obtain gases, nutrients, minerals, and water and how these materials are transported throughout the organisms. As this chapter proceeds, it will become clear that basic physical properties and events—adhesion, cohesion, evaporation, osmosis, and others—underlie the movement of fluids so essential to the continuing life of every plant.

A grisly fate for a fly stuck to sweet-smelling nectar. Action potentials like those in an animal's nerve cells traveled down the tentacles of this sundew (*Drosera rotundifolia*), causing them to bend and entrap the hapless insect so that it can be digested and its nitrogenous compounds absorbed by the carnivorous flower.

We consider these topics:

- Basic plant strategies for exchanging gases, transporting liquids, and handling wastes
- How water is transported in the xylem
- How sugar solutions are transported in the phloem
- How roots take up nutrients from the soil
- Unusual plants that trap prey or move when touched

PLANT STRATEGIES FOR MEETING BASIC NEEDS

Plants have evolved "energetically inexpensive" strategies for exchanging gases, moving food and water, and eliminating wastes (Figure 27-1); these processes do not depend on a large-scale expenditure of ATP or other high-energy compounds that are metabolically costly to form. Thus, plants tend to use passive processes—processes that operate by the physical rather than the chemical laws of nature.

Strategies for Gas Exchange

Plants exchange gases passively with the environment. Gases enter and leave the green parts of a plant through thousands of stomata. In the nongreen parts of the shoot system, limited gas exchange occurs through the epidermis or lenticels. The roots, too, take up oxygen from the soil.

All the normal gases in air diffuse into and out of the stomata on leaves and green stems. But there is net movement of carbon dioxide inward, since that gas is used up in photosynthesis. And, of course, excess oxygen liberated by photosynthesis diffuses back out into the atmosphere. Once carbon dioxide or other gases have entered the leaf or stem, they need diffuse only short distances through intercellular air spaces in order to reach all tissue cells. Thus, there is no true respiratory or circulatory system of the sort common in many animals to carry gases long distances within plants.

Strategies for Internal Transport of Liquids

Within vascular plants, watery fluids move in two sets of transport vessels: the *xylem*, which transports water and mineral nutrients; and the *phloem*, which transports solutions containing organic molecules. Water continuously enters the xylem in the roots, passes upward, and is lost through the leaves; sugars and other nutrients enter the phloem from their site of manufacture in the leaves and other photosynthetic sites and are removed in the meristems, stems, and roots either for use or for storage in nonphotosynthetic sites. These movements depend on passive physical processes without the extensive, coordinated cell activity common in animals.

Strategies for Coping with Wastes

All organisms produce waste products, but plants tend to have fewer excretory problems than do animals. Nitrogenous wastes, which accumulate in animals, are not a major problem in plants. Plants acquire nitrate (NO_3^-) from the soil and manufacture their own amino acids. Nitrogen is usually a limited nutrient in the soil for reasons we will discuss later, so plants are rarely favored with an excess. Moreover, the continuous growth of leaves and roots demands a steady supply of amino acids and other nitrogen-containing organic molecules. Plants are often so short on nitrogen that much of their new growth is supported by recycling amino acids and other nitrogen compounds from older tissue. A yellowing leaf often is undergoing a process of dissolution, during which its existing proteins are broken down to amino acids, shipped to the growing portions of the plant, and incorporated into new proteins. Plants recycle most wastes, incorporating their breakdown products into new cytoplasmic components.

Plants do make some waste products toxic to their own cells, such as tannins and phenolics (both small molecular derivatives of phenol, the basic ring compound of lignin). Plants have no means of excreting such products through the leaves or stem, however, and therefore, some specialized cells act as storehouses for the noxious products. Likewise, the central cell vacuoles or the resin ducts of a conifer tree may sequester resins and tars, and this long-term storage protects the plant from the toxins. Toxins themselves help discourage predators (details in Chapter 42).

TRANSPORT OF WATER IN THE XYLEM

Drying and wetting can have dramatic effects on a plant, causing it to droop and wilt or become stiff and erect. The principles of osmosis and water movement across membranes help explain why.

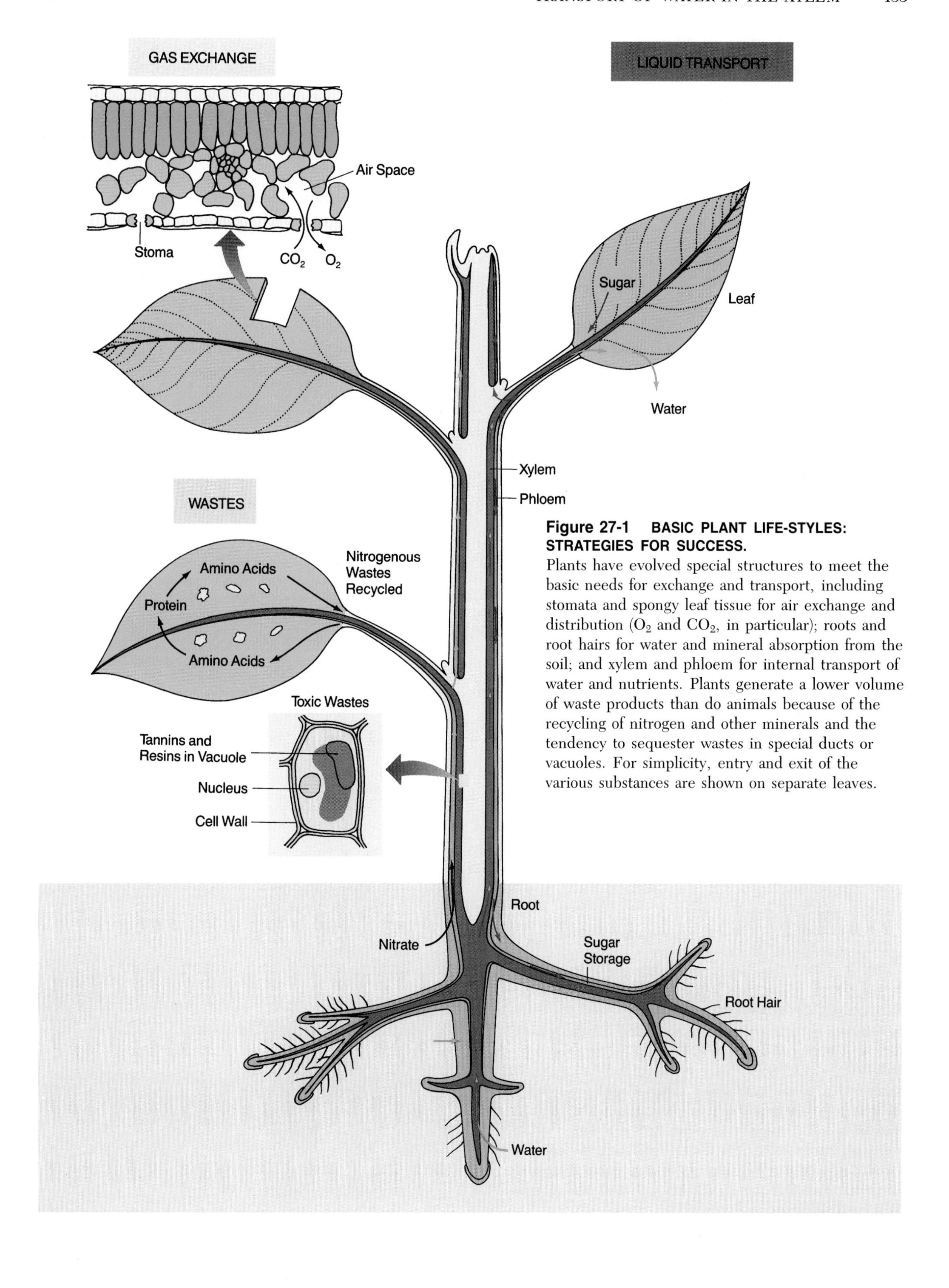

Figure 27-1 BASIC PLANT LIFE-STYLES: STRATEGIES FOR SUCCESS.
Plants have evolved special structures to meet the basic needs for exchange and transport, including stomata and spongy leaf tissue for air exchange and distribution (O_2 and CO_2, in particular); roots and root hairs for water and mineral absorption from the soil; and xylem and phloem for internal transport of water and nutrients. Plants generate a lower volume of waste products than do animals because of the recycling of nitrogen and other minerals and the tendency to sequester wastes in special ducts or vacuoles. For simplicity, entry and exit of the various substances are shown on separate leaves.

Principles of Water Movement

If a cell has a high solute concentration (and thus a low water concentration) relative to the solution surrounding it, water will move into the cell, from an area of high concentration from to one of low, by *osmosis*. The force causing water to move can be measured, and biologists call it the *osmotic pressure* or **osmotic potential** (Figure 27-2a). As water moves inward, the cell's cytoplasm and vacuole begin to swell. But as the interior expands, it pushes outward against the cell wall, and the cell becomes *rigid*, or stiffened and distended (Figure 27-2b). The force a turgid cell exerts on its cell wall is called *turgor pressure*. A high turgor pressure blocks additional water entry into the cell. Thus, osmotic pressure and turgor pressure are opposing forces—equivalent to blowing up a balloon inside a rigid box. At equilibrium, the turgor pressure blocking the inward movement of more water precisely counterbalances the osmotic pressure forcing water to pass inward (Figure 27-2c).

To quantify a cell's water balance, plant physiologists have developed the concept of **water potential,** the turgor pressure plus the osmotic pressure. By convention, the water potential of pure water is set at zero. Any solutes dissolved in the water make it less pure, and the value of the water potential becomes a negative number. Normal plant tissues have a negative water potential, say, -5 bars (-5 kg/cm^2); this results from a small positive turgor pressure forcing water out (say, $+5$ bars) added to a larger negative osmotic potential (-10 bars). Consequently, water tends to enter the cell.

If a plant cell with a water potential of zero (water tends to neither enter nor leave) is placed in a hypertonic (very salty) solution, water will leave the cell and pass into the medium, causing the cell to become *plasmolyzed* (Figure 27-2d). In plasmolyzed cells, the cytoplasm and vacuole shrink because of water loss and pull away from the rigid cell wall. If the cell is placed in a *hypotonic* (low-salt) medium, the cell takes up water from the medium, thereby becoming turgid. In the cells in a wilted house plant, the water potential is more negative than that of soil; thus, these cells quickly take up water and become turgid (Figure 27-2e).

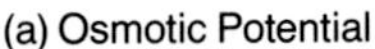

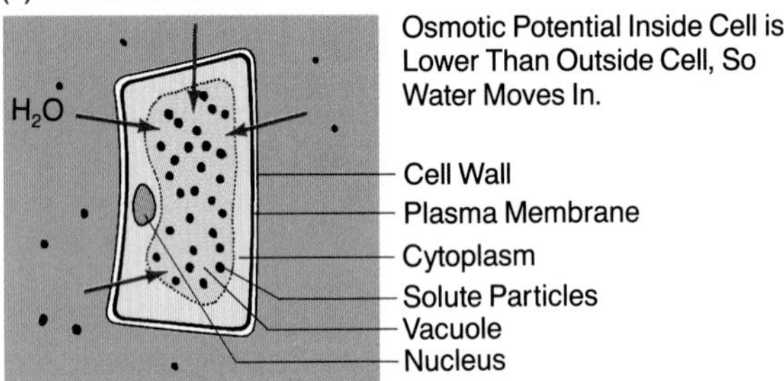

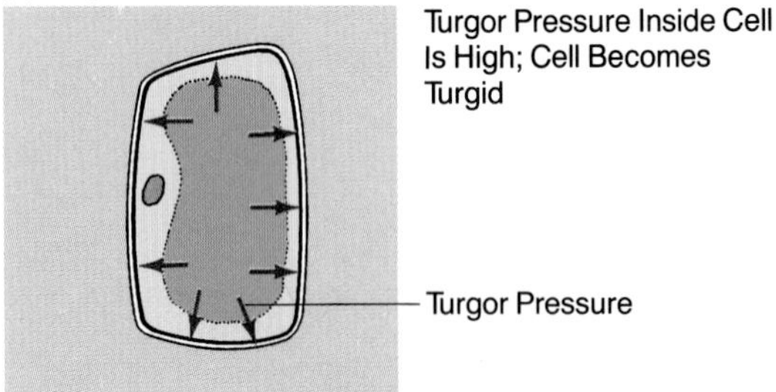

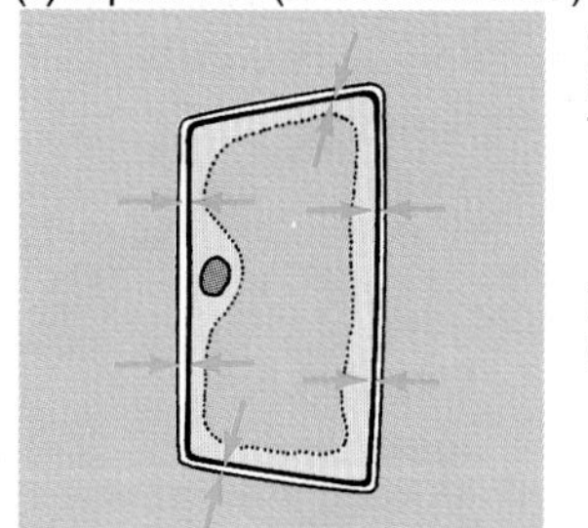

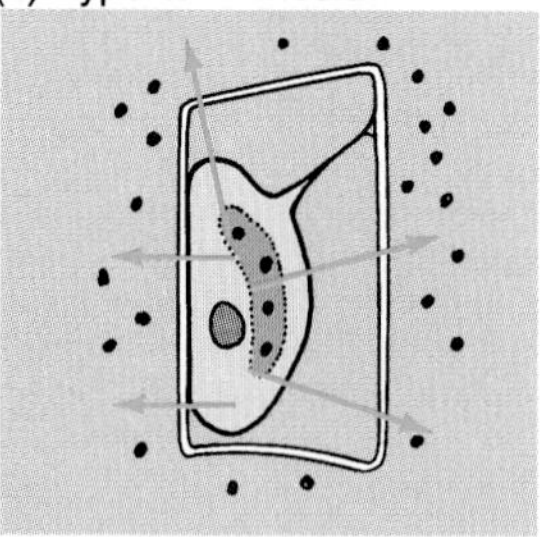

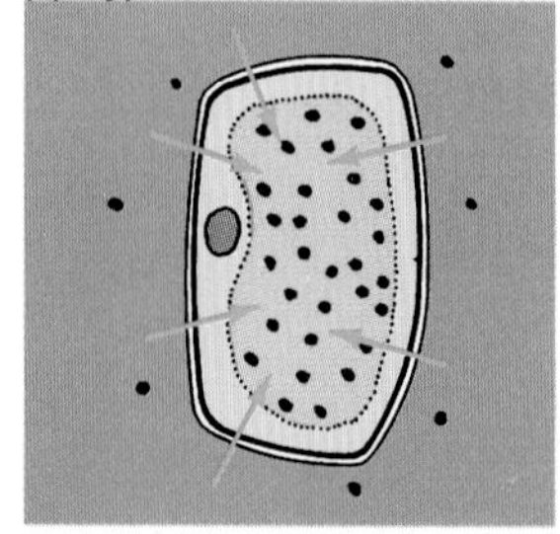

Figure 27-2 SWELLING AND SHRINKING: BASIC RESPONSES OF PLANT CELLS.
(a) Osmotic potential causes water to diffuse into a plant cell. (b) Turgor pressure pushes outward against the cell wall. (c) At equilibrium, these two opposing forces are balanced. (d) Plasmolysis occurs when a cell in a hypertonic medium loses water and the volume occupied by the cytoplasm and its vacuole is reduced. The plasma membrane pulls away from the cell wall, and the cell shrinks. (e) If the cell is returned to a solution of higher water potential (hypotonic), it takes up water and returns to its normal shape and volume.

Theories of Upward Movement

How does water get from the roots to the top of a tall tree? Plant scientists have proposed several theories based on root pressure, capillary action, and transpiration.

Root Pressure

According to the **root pressure theory,** water pressure builds up in the roots and pushes water upward toward the leaves. Indeed, researchers have demonstrated that plants such as the tomato have considerable root pressure; they removed the stem just above the roots and attached a pressure-measuring device. If the experimenters kill the roots by heat or other damage, the pressure subsides and the water ascent stops.

In nature, root pressure is demonstrated by **guttation,** the formation of water droplets on pores at the edges of leaves in some plants (Figure 27-3).

While low root pressures might suffice for small plants, tremendous pressure would be required to force water to rise to the top of a large tree. Yet, if a branch is sawed off, water does not rush out of the xylem. Root pressure, therefore, cannot account for water movement in the stems of most plants.

Figure 27-3 GUTTATION: SEEPAGE OF WATER FROM A LEAF.
When minerals become highly concentrated in the root xylem, so much water enters and passes upward in xylem vessels that water is forced out of the leaf margins, as in these strawberry leaves (*Fragaria ananassa*).

Capillary Action

A second suggestion for how water might move in plants is **capillary action.** This phenomenon causes water molecules to creep upward by adhering tightly to the thin strands of cellulose and other polysaccharides that compose the walls of xylem tubes. However, water can rise this way only about 0.5 m, and so capillary action is also insufficient to explain water movement in larger plants and trees.

Transpiration

The best hypothesis for how water makes its extensive upward climb in plants is the **transpiration-pull theory** (also called the **cohesion-adhesion-tension theory**): the idea that water is pulled up through the xylem from the top of the plant rather than pushed up from the roots. When researchers cut or puncture a xylem vessel in a stem, they can detect air moving into the stem to fill the space vacated by water that has moved upward in the stem above the cut. This implies that the water is pulled through the plant from above. Other experiments reveal that the rate of fluid movement in the xylem is greatest when the plant is in bright sunlight and lowest at night, when evaporative water loss slows.

Such observations have established that **transpiration**—evaporative water loss through the leaves—is crucial to water movement upward in the xylem (Figure 27-4). Water is constantly lost when stomata on the surface of leaves and stems are open, allowing gas exchange with the atmosphere. The loss can be explained in terms of the water potential equation. Water in the atmosphere is at a much lower concentration (a more negative water potential) than it is in the interior of the leaf. Thus, when the stomata are open, water moves toward the atmosphere. As water vapor departs the leaf by the process of transpiration, more water moves from the root to the leaf to replace the lost water.

In nature, transpiration does follow these principles but can be a bit more complicated. In a larch tree, for example, water transpired from the needles each morning comes largely from the sapwood of twigs and branches; then those reservoirs are refilled by water from the soil in afternoon and evening.

The transpiration-pull theory depends on the properties of water molecules (see Chapter 2). In the xylem, the water column exists as a long liquid "chain" held together by the high cohesion among individual water molecules—enough to support a thin column of water 100 m high. When a water molecule in the leaf's air

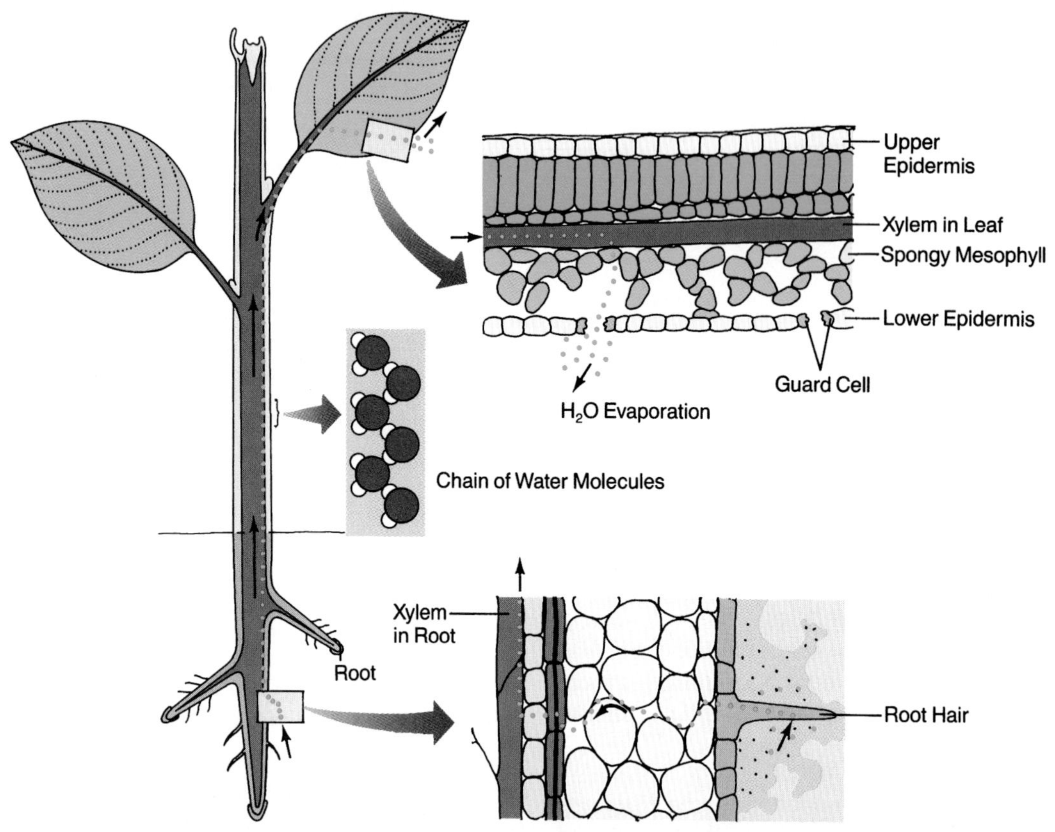

Figure 27-4 TRANSPIRATION: KEY TO UPWARD WATER MOVEMENT IN PLANTS.
When the air surrounding a plant is not saturated with water vapor, water will evaporate from the leaves. Water is taken up through the roots of the plant by osmosis and moves up the stem to the leaves, where it evaporates into the intercellular spaces of the leaf. Then water vapor exits the leaf through the stomata. Water moves up because of its cohesive properties; that is, diffusion from cells into the leaf's gas-filled spaces sets up a transpirational pull, which helps pull the column of water upward. Adhesion of the water column to the walls of xylem vessels also plays a role in the process.

channels changes from the liquid to the gaseous state and transpires through a stoma, the entire column in that xylem tube is drawn upward a tiny amount, and another water molecule then moves into the roots from the soil (see Figure 27-4). As a result of one water molecule pulling on another, the entire water column in the xylem is stretched, like a piece of elastic. Plant physiologists say that water in the xylem is under *tension*.

What happens if an ax blow, a hard freeze, or the snapping of a branch breaks the upward moving chain of water molecules? Air bubbles form near an injury, enlarge, and move up the water column, eventually blocking the cut vessel's entire diameter. The nearby plant tissue usually responds by sealing off the injured section with resins and tars to prevent water seepage or infection. The pits on the side walls of the xylem vessels allow water rising from the roots to move laterally and thus bypass the sealed area and continue its upward movement in nearby uninjured vessels.

Regulation of Water Loss

Plants regulate the amount of water vapor lost from the leaves—and, in turn, the upward water movement through the roots and stems—by the opening and closing of stomata. The two guard cells bordering a stomatal pore have thick walls on the side next to the stomatal opening and much thinner, more flexible walls on the opposite side (Figure 27-5a). Stomata open and close in response to the movement of ions across the guard cell membranes. When ions such as potassium (K^+) and chloride (Cl^-) enter the guard cells (see Figure 27-5a), water follows by osmosis (Figure 27-5b). (Experiments show that there is about a 20-fold increase in K^+ in guard cells when they swell.) Resultant turgor pressure causes the cells to swell and bulge sideways, pushing the thin, flexible walls outward and pulling the thick, inner walls apart (Figure 27-5c). Because this bulging occurs on both sides of the pore, the pore opens wider, permitting gas

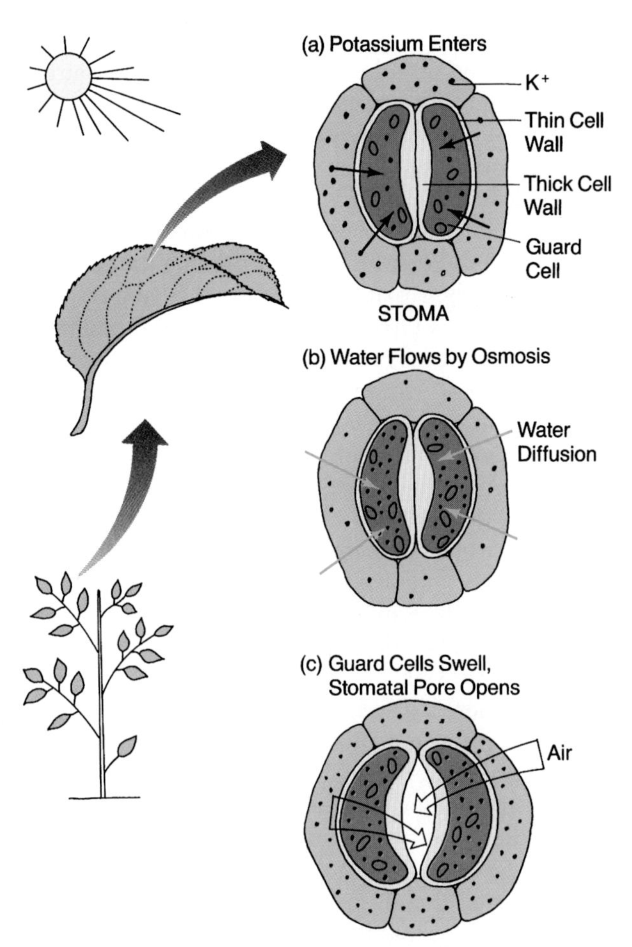

Figure 27-5 STOMATA: OPENING AND CLOSING THE GUARDIAN GATES.
(a) When turgor is low in both guard cells, the stoma is closed. (b) When ions such as K^+ and Cl^- enter the guard cells, water follows by osmosis, (c) increasing the turgor pressure within the cells and causing them to bulge sideways, pulling their thick inner walls apart. This movement opens the pore, and air containing carbon dioxide can enter, or water vapor and oxygen can exit.

exchange by diffusion: Water vapor leaves, net CO_2 flows inward, and net O_2 flows outward.

What causes the guard cells to gain or lose potassium ions and therefore to open or close stomata and, in turn, to regulate transpiration and photosynthesis? Experimental findings imply that low water potential within the leaf is a strong environmental signal causing guard cell shrinking. This fits with the so-called midday closure of stomata biologists observe in many plants; stomata close when leaf cells lose turgor because more water is exiting the stomata than entering through the roots. How is a leaf's water potential communicated to the guard cells? Certain evidence hints strongly that the plant hormone *abscisic acid* (see Chapter 28) accumulates in leaves subjected to water stress. Furthermore, abscisic acid applied to guard cells causes them to close their stomata. However, there must be more to the regulation of stomata.

Recent studies show that a drop in soil water content or in air humidity can somehow trigger stomatal closures even before severe water loss occurs in the leaves. In such a *feedforward* response, an environmental cue (lowered humidity) acts on a control element (stomata) independent of the process being controlled (transpiration).

Stomatal pores also open when the concentration of CO_2 drops inside the leaf, as when light intensity is high and photosynthesis occurs rapidly. Finally, guard cells have pigments that absorb blue or red light; when stimulated, these pigments activate processes that may affect K^+ movement and, in turn, the degree of guard cell swelling. The intact plant in nature is subjected to varying light, humidity, and other conditions that can affect guard cell behavior, and clearly, all these interrelated environmental factors may help regulate guard cells.

Gas exchange across the leaf surface also depends on (1) the concentration gradients of gases from inside to outside of the leaf, (2) the diffusion properties of the gases, and (3) the wind speed across the leaf surface (faster-moving air carries water vapor away more rapidly). The relative humidity of air is usually well below 100 percent, so that evaporation and transpiration are favored. There is usually little absolute difference between the low CO_2 levels outside and inside the leaf. However, as CO_2 is used up by photosynthesis within the leaf cells, a *sink*, or absorption region, is created within the leaf toward which more CO_2 diffuses.

Despite the elegant regulatory machinery of stomata, water loss due to transpiration can be staggering. A corn plant in full sun can lose five times the combined weight of the water in its own leaves each day. And transpiration also affects the movement of sugar in the phloem.

TRANSPORT OF SOLUTIONS IN THE PHLOEM

The transport of solutes in the phloem, or **translocation,** involves the movement of sugars and other organic molecules from the photosynthetic leaf cells to the leaf's phloem, and from there throughout the plant. The phloem is a living tissue composed primarily of sieve tube elements, which have lost their nuclei but retain some cytoplasm, and companion cells, which retain their nuclei.

What Is Translocated in the Phloem?

Nature has provided botanists with the perfect tool for sampling the contents of the phloem: aphids (Figure 27-6a). No manufactured needle is as fine and as accurate a probe as the *stylet*, or feeding tube, of an aphid. These small insects feed by boring into individual sieve tube elements of stems or leaves, then sucking in nutrient solutions as they flow in the phloem.

To study the fluid contents of phloem, plant physiologists allow aphids to attach to a plant and penetrate sieve tube elements with their stylets (see Figure 27-6a). The aphids are then anesthetized with ether, and their bodies are cut away, leaving only the stylet protruding from the phloem (Figure 27-6b). The phloem fluid oozes out of the stylets, and the researcher can collect and analyze it.

Studies of this sort have revealed that in most plants, the fluid transported in the phloem is a concentrated solution, with up to 30 percent dissolved materials. The major component is sucrose, but other sugars and organic compounds can be found, along with low concentrations of amino acids and other nitrogenous compounds that support protein synthesis in other parts of the plant.

Kinetics of Transport in the Phloem

Researchers study the *kinetics* of phloem transport—the direction and rate of material flow—by adding radioactive tracers to the phloem, either by injection or by enclosing a photosynthetically active leaf in an airtight bag and then introducing radioactive CO_2. Within a few minutes, the leaf chloroplasts incorporate the radioactive carbon dioxide into sugars, which begin to be exported from the leaf cells. The movement of radioactive sugars can be traced by removing pieces of leaf, stem, and root tissue at various intervals and determining the amount of radioactive carbon that has reached the sample tissue.

Experimenters typically observe that the radioactive compounds exit the leaf, enter the stem, and are translocated *downward*. However, a fraction of the newly synthesized compounds also moves up the stem. Developing fruits—growing bean pods, for example—receive most of the photosynthetic product from nearby leaves. This observation has led to the concept that leaves are *sources* of carbon compounds and that growing regions, storage tissues, and other parts of the plant are *sinks* for those compounds. Experiments have also shown that the plant expends ATP energy to "load" at the source and to "unload" at the sink.

Translocation is clearly more complex than transpiration, since phloem is a two-way transport system in which organic compounds can flow either up or down the plant, depending on the location of strong sinks, and in which different organic molecules can be translocated at different rates.

The translocation of sugars can be very fast. Radioactive tracers reveal that the first few sugar molecules that exit a leaf may move at a rate of more than 200 cm an hour, although the average is 50–100 cm an hour. This

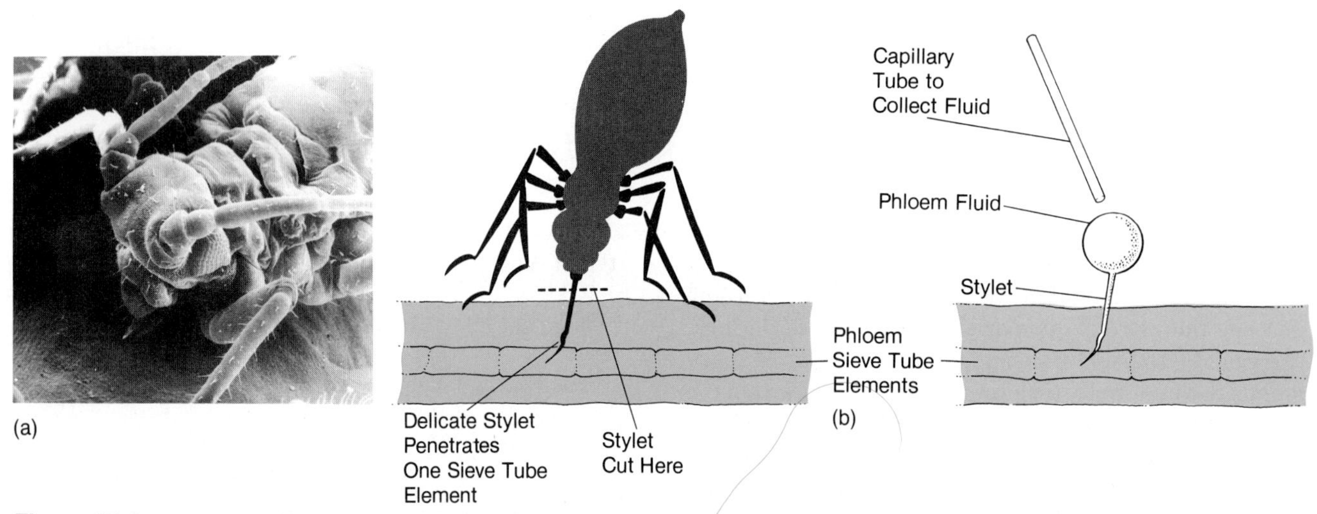

Figure 27-6 APHIDS: BIOLOGICAL TOOLS FOR STUDYING PHLOEM.
(a) Aphids have a delicate feeding tube called a stylet that they plunge into a sieve tube element like a hypodermic needle into an animal's vein. (b) Researchers employ the aphid's stylet by severing the body of a feeding aphid and leaving the stylet in the sieve tube. The phloem contents continue to seep out of the stylet and can be collected and analyzed. An aphid (magnified about 50×) is shown with its feeding tube inserted into a marigold stem.

bulk flow, or mass movement, of phloem fluid is quite rapid for a system that has no apparent pump and is carrying a viscous sugar solution through thin, partially blocked pipelines composed of sieve tube elements with perforated end walls—the *sieve plates.* What accounts for this rapid transport?

Mass Flow Theory: Transport in the Phloem

The best explanation is the **mass flow theory.** Sugar produced by photosynthesizing cells in leaves is loaded into the sieve tube elements by the companion cells, creating a high solute concentration (Figure 27-7, step 1). The phloem is thus hypertonic relative to neighboring xylem cells, and water moves by osmosis from the xylem to the phloem (step 2). This increases the phloem cell's water content and creates a high turgor pressure within the phloem. But the phloem cells cannot expand past a certain size because of their rigid cell walls. The only open route to relieving the increased pressure is through the perforated sieve plates (step 3). Therefore, the mass flow theory suggests that the elevated water pressure in the leaf is relieved by the flow of water and organic substances through the leaf phloem and then through the stem phloem.

Experiments show that translocation occurs down a gradient of turgor pressure. The turgor pressure in the phloem is highest in the small vascular bundles of the leaf, where sugars are pumped into the sieve tube elements by the companion cells. As the sugary phloem contents are carried out of the leaves, the turgor pressure in the phloem gradually falls (step 4). This is because nonphotosynthetic cells of the stem actively take up disaccharides from the passing phloem fluid to support their own cellular metabolism. This loss of sugar renders the phloem solution progressively more dilute. Turgor pressure falls steadily as the phloem's fluid contents move downward toward the roots or upward toward the meristems, and more and more sugar is removed. The final carbohydrates are removed from the phloem in the roots (step 5). The water in the phloem, now at a high water potential, moves into the xylem to be recirculated toward the top of the plant (step 6). This is a remarkably efficient strategy: Water transferred from xylem to phloem in leaves creates the turgor pressure in the phloem required to move the sugar out of the leaf. Transport both up the xylem and down the phloem is ultimately traceable to the transpiration from the leaves, which carries water upward in the first place.

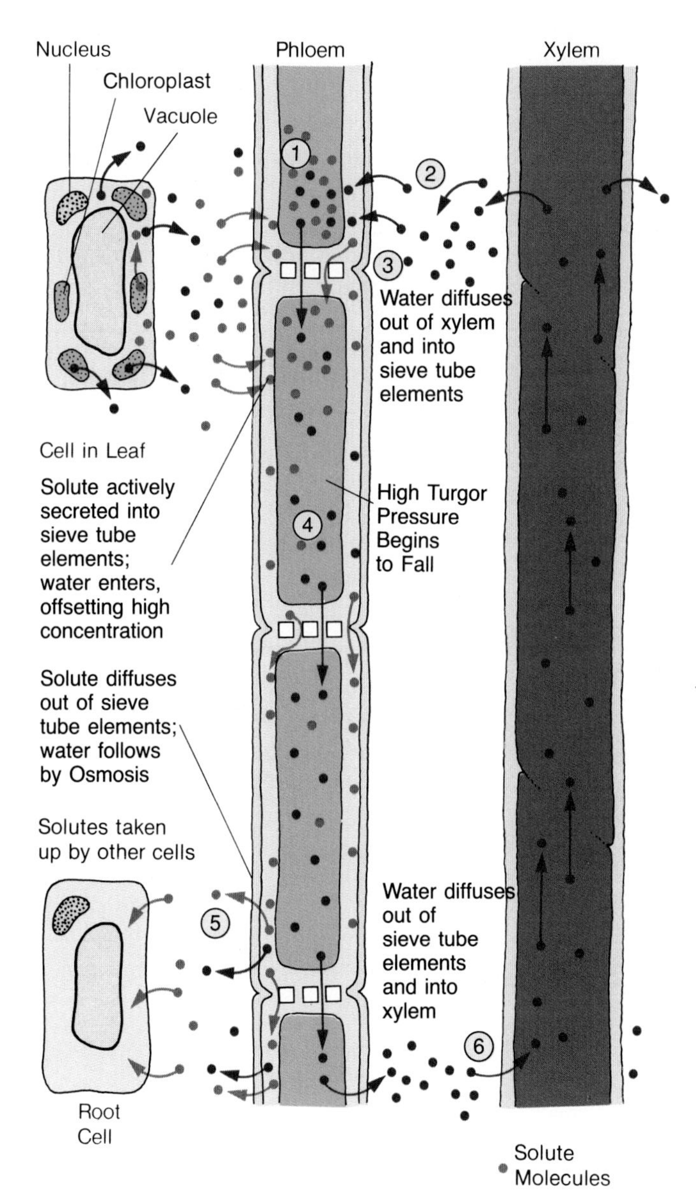

◀ **Figure 27-7 THE MASS FLOW THEORY AND NUTRIENT TRANSPORT.**
Sugar, which acts as a solute, is produced by photosynthesizing cells and passes into the sieve tube elements of the phloem. As more solutes enter those tubes, the resultant higher solute concentration causes water to flow into the sieve tube elements from neighboring xylem cells. Since the volume of the sieve tube elements is restricted by their cell walls, the continuously enlarging fluid volume in the sieve elements is reduced by passage of fluid through the sieve plates. As a result, a flow of water and nutrients is set up through the leaf phloem and into the stem phloem and thus to the root phloem. This translocation of water and nutrients occurs down a water pressure gradient from the leaves to the roots. Solutes are supplied to stem and root cells, as shown. Water that passes down the phloem passes out of the sieve tube elements and then joins the water that enters root hairs. It then moves upward in the xylem, completing the cycle.

Some plant physiologists think that turgor pressures in the leaf phloem are not sufficiently different from those in the stem and root phloem to create a flow of sugars. While the mass flow theory is the best explanation so far, it may well be modified by future research.

HOW ROOTS OBTAIN NUTRIENTS FROM THE SOIL

Houseplants in old leached soil often grow poorly, but a few drops of commercial plant food sets everything to right. So what *is* plant food?

Mineral Requirements and Uptake

Complete plant foods contain **minerals,** inorganic substances such as nitrogen, potassium, and phosphorus, plus traces of iron, zinc, manganese, and magnesium—essential nutrients that the plant cannot obtain through gas exchange or photosynthesis. Table 27-1 lists the minerals that are *macronutrients* (required in large amounts) and those that are *micronutrients* (required in smaller amounts). If a plant is deficient in any one of these mineral nutrients, it will exhibit specific symptoms. A plant deficient in nitrogen, for example, often has small, yellowed leaves, while one deficient in phosphorus can have stunted, purple leaves.

The minerals required for plant growth originate from the earth's rocks, eroded over time into small pieces or dissolved in soil water. Although the water enters the roots by simple diffusion in response to a water potential gradient, mineral uptake may occur in two ways: passive or active. (You can review these mechanisms in Chapter 5.) Active transport is a more reliable means of accumulating minerals and ions than passive, since it can import materials against a concentration gradient, moving even trace quantities of minerals from the soil into the cytoplasm.

Cells of the root cortex, epidermis, and endodermis can actively transport mineral ions from the dilute solu-

Table 27-1 ESSENTIAL MINERALS FOR HIGHER PLANTS

	Major Functions	*Typical Compounds*
Macronutrients		
Nitrogen (N)	Major ingredient in many compounds	Amino acids, purines, pyrimidines, porphyrins, hormones
Phosphorus (P)	Energy transfer; structural component of DNA and other compounds	Sugar phosphates, ATP, GTP, nucleic acids, coenzymes, many proteins
Potassium (K)	Osmotic relations; protein conformation and stability; stomata; enzyme cofactor	Potassium ion
Magnesium (Mg)	Enzyme activation; pigments and ribosome stability; nucleic acid synthesis	Chorophyll
Calcium (Ca)	Enzyme activation; component of cell walls; permeability of membranes	Calcium pectate
Sulfur (S)	Active groups in enzymes and coenzymes	Amino acids, coenzymes, vitamins
Iron (Fe)	Active groups in enzymes and electron carriers	Cytochromes, peroxidases
Micronutrients		
Copper (Cu)	Enzymes, photosynthesis	Enzymes for oxidation reactions; plastocyanin
Manganese (Mn)	Photosynthesis; cellular respiration	Cofactor for many enzymes
Molybdenum (Mo)	Nitrogen fixation; nitrate reduction	Nitrate reductase, nitrogenase
Zinc (Zn)	Enzyme activation	Carbonic anhydrase, auxin synthesis
Sodium (Na)	Enzyme activation; osmotic relations	Sodium ion
Cobalt (Co)	Nitrogen fixation	Vitamin B_{12} in nitrogen-fixing microorganisms only
Silicon (Si)	Structural	Hydrated silicon dioxide
Chlorine (Cl)	Photosynthesis; ion and electrical balance	Chlorinated alkaloids
Boron (Bo)	Sugar transport	Borate ion

tions that diffuse into the apoplastic space between cell walls into the cell's cytoplasm, the symplastic pathway (review Figure 25-12). Ions absorbed into cells of the cortex and epidermis follow the symplastic pathway; eventually, they reach the root's endodermis by traveling through the plasmodesmata—the cytoplasmic connections between plant cells—and are transferred from the endodermis into the xylem.

The xylem carries the dissolved minerals passively upward, presumably because of the transpiration-pull effect; in this way, the nutrients are distributed throughout the plant. Leaf and stem cells then remove ions from the xylem solution using active transport pump enzymes similar to those used by the root cells to remove ions from soil water.

Rate of Mineral Uptake

The rate at which plants can take up various minerals depends on the concentration of the minerals in the soil and on the soil's pH. Some soils are rich in inorganic nutrients, whereas others are deficient in one or more. Most plants grow best in a slightly acidic soil, in which mineral nutrients are most soluble. At more extreme high or low pH values, most of the mineral nutrients in soil are tied up as insoluble salts and are not available to plants.

Soil composition affects how plants take up minerals. Sandy soils have mainly large particles and large air spaces, but retain little water and therefore provide few dissolved minerals. Clay soils have very fine particles, little air space, and hold much more water and dissolved mineral ions. In soils with little organic material—decaying vegetation, or compost, for example—calcium is often tied up in an insoluble salt, calcium carbonate. In soils with more organic material, calcium ions form complexes with organic acids and remain available to plants. Farmers and gardeners use both organic compost and mineral-containing fertilizers to recondition and replenish soils.

Nitrogen Uptake and Fixation

Plants have a special problem obtaining nitrogen. It is a crucial component for the synthesis of protein, RNA, and DNA, yet the element mainly occurs in an inaccessible form: in the atmosphere as the gaseous molecule N_2. While available nitrogen is usually the most important single factor limiting plant growth, plants cannot take in molecular nitrogen directly and instead must take in nitrate (NO_3^-) from the soil and transport it to the leaves, where an enzyme called *nitrate reductase* converts the oxidized form of the element back to reduced nitrogen, or ammonium ion (NH_4^+), that can be further modified and built into biological molecules. A few kinds of herbaceous plants, however, can use NO_3^- directly, converting and incorporating the ion into organic compounds in root cells, then transporting them upward.

How does atmospheric N_2 become oxidized into NO_3^- in soil? Such conversion of N_2, or **nitrogen fixation,** can take place by nonliving processes, by free-living soil microorganisms, or by certain bacteria that live symbiotically with plant roots.

Nitrogen fixation by nonliving processes takes place in the atmosphere when the energy from lightning breaks the strong bonds between nitrogen atoms in N_2, allowing nitrogen oxides (NO_2^-, NO_3^-) and ammonia (NH_3) to form spontaneously. During rainstorms, these compounds are washed down into the soil, where the nitrates can be taken in directly by plants (Figure 27-8, step 1).

Free-living soil microorganisms carry out some biological nitrogen fixation, converting N_2 to NH_4^+ (step 2). Nitrogen-fixing species of the bacterial genera *Azotobacter* and *Clostridium*, among others, and certain ammonifying cyanobacteria can reduce N_2 into NH_3 by means of the enzyme *nitrogenase*. Species of *Nitrosomonas* and *Nitrobacter* carry out the process of **nitrification,** by which they oxidize NH_3 to NO_2^- and NO_3^-, the latter being directly available for absorption by plant roots (step 3).

The richest source of biologically fixed nitrogen is bacteria living symbiotically in the roots of certain plants. *Legumes*, a large group of plants that includes peas and beans, can form symbiotic relationships with particular species of *Rhizobium*, a genus of symbiotic, nitrogen-fixing bacteria that forms NH_4 (step 4). The host plant then uses the NH_4 to form amino acids and other compounds. Although rhizobia can live freely in the soil, most do not fix nitrogen unless they are living in the roots of a legume.

Rhizobia pass into a root hair through a channel or *infection thread*, then enter the cortex of the root, begin to divide rapidly, and build up a large population. Soon they invade many cells of the cortex. At the same time, the root cells near the invaded hairs proliferate and create **root nodules,** small, hard lumps on the root surface that encapsulate the nitrogen-fixing factory (see Figure 27-8).

Within the root nodule, the bacteria change their metabolism and appearance and become **bacteroids,** no longer able to survive outside the root nodules. The bacteroids make large amounts of the enzymes required for nitrogen fixation. Rhizobial nitrogen-fixing enzymes require a nearly anaerobic environment in which to reduce nitrogen. Although the root cells of the nodule are aerobic, they secrete a pigment-protein complex called *leghemoglobin* that can bind oxygen very tightly and that can keep the free oxygen concentration near the bacteria

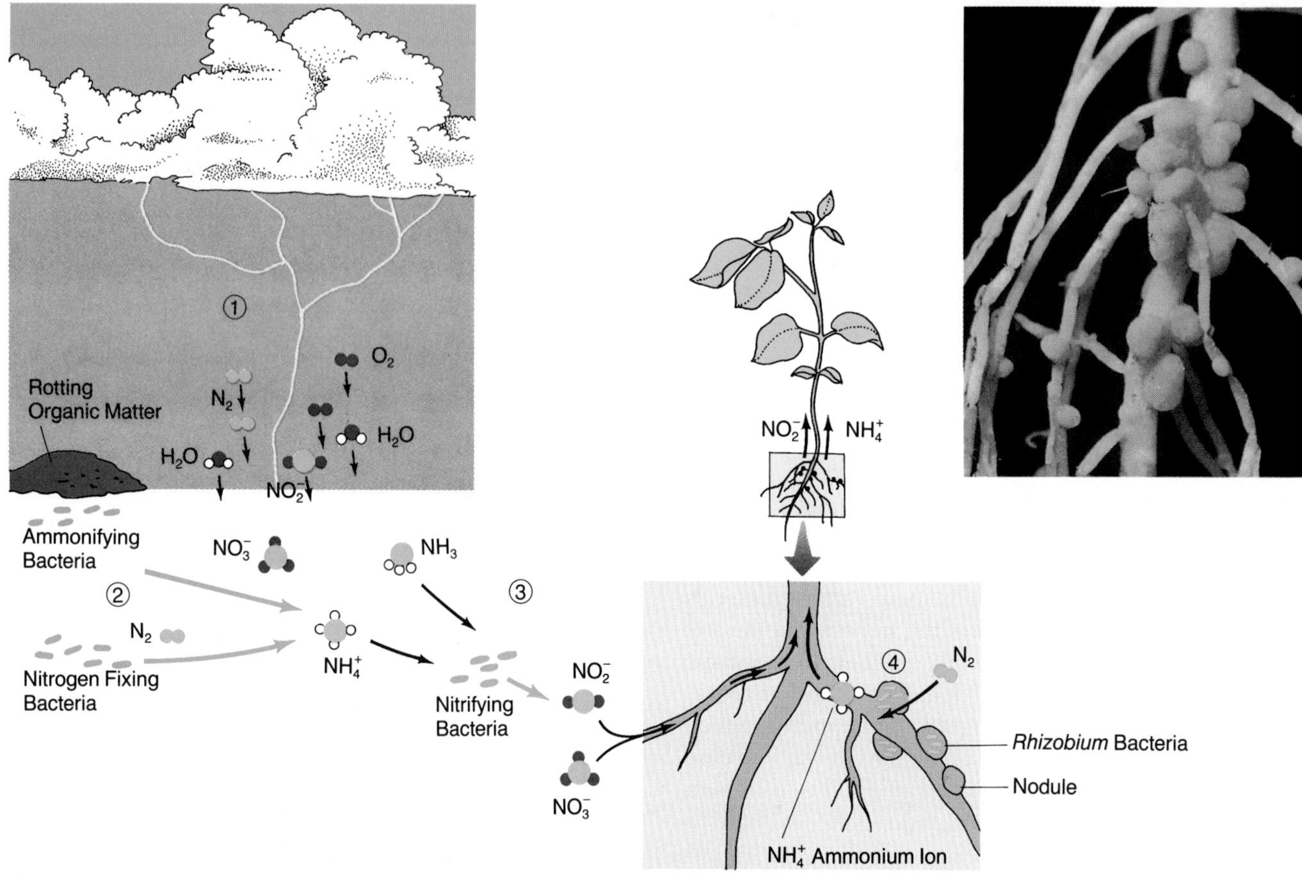

Figure 27-8 NITROGEN FIXATION: NONBIOLOGICAL AND BIOLOGICAL.
As the text explains, lightning flashes can fix atmospheric nitrogen nonbiologically (step 1). Ammonifying and nitrogen-fixing bacteria can also fix nitrogen biologically (step 2). Nitrifying bacteria can oxidize NH_3 and make it available to plant roots (step 3). Moreover, *Rhizobium* bacteria can invade a leguminous plant, and a mutually beneficial relationship can be established in which *Rhizobium* supplies the legume with nitrogen, while the legume supplies the bacterium with carbohydrates (step 4). The photo shows nodules on the roots of a soybean plant.

low enough to allow nitrogen fixation. Root nodules look pinkish-red because of their high concentration of leghemoglobin.

The leguminous plant supplies the bacteroids with carbohydrates. The bacteroids then metabolize these compounds to reap the large amount of energy needed to reduce N_2 to NH_3. In an inadvertent exchange for their "room and board," the bacteroids produce NH_3 as a waste product, which the plant then assimilates and uses as a source of reduced nitrogen. The basic interaction between legume root and *Rhizobium* is clearly successful for both parties. Studies done in 1988, however, show that nitrogen fixation is even more efficient in a plant such as alfalfa if a second type of soil bacteria infects the nodules and releases a toxin that inactivates a plant enzyme. While data are still coming in on this newly uncovered process, researchers have observed that the doubly infected alfalfa grow twice as fast as those with *Rhizobium* alone.

Certain nonleguminous plants have microbial symbionts that fix nitrogen. Western mountain lilac, alder shrubs, and the eastern sweet gale are typical examples; they can survive in rocky soils containing little nitrogen. Some types of water-dwelling ferns have cyanobacteria of the genus *Anabena* living in specialized compartments in their leaves, making nitrogen fixation possible in those organs (Figure 27-9). Chinese farmers fertilize rice plants by cultivating water ferns in their flooded fields.

Mycorrhizae

Roots in their natural environment have still other associations with soil microorganisms. As we said in Chapter 22, most land plants have mycorrhizal associations with specific types of fungi. The fungi invade or cover the roots with threadlike hyphae, often completely surrounding them like a sheath. This extends the roots' absorptive surface area for water and provides needed nutrients, since fungi are more efficient than plant root hairs at absorbing mineral nutrients from the soil. In

Figure 27-9 SYMBIOSIS BETWEEN FERNS AND CYANOBACTERIA.
A frond of the water-dwelling fern *Azolla caroliniana* (magnified about 10×). Inside many parts of the frond are found cyanobacteria capable of fixing nitrogen.

fact, mycorrhizae-covered roots often lack root hairs altogether and depend on the fungi to handle absorption.

CARNIVOROUS AND "SENSITIVE" PLANTS

Carnivorous plants, which tend to live where levels of available nitrogen are low, meet some of their nitrogen requirements in an unusual way: They trap and digest insects. The pitcher plant is a "passive trapper": Prey slide down a slippery chute into a pool of digestive fluid. The Venus flytrap and the sundew are "active trappers": Modified leaves spring shut to ensnare unsuspecting prey. The prey's arrival triggers the osmotically driven secretion of digestive enzymes; then the process is reversed and the enzymes—plus nutrients from the victim's body—are reabsorbed and distributed in the plant's vascular system. Despite its inherent fascination, this is getting nitrogen the hard way, since digestive enzymes are "costly" to manufacture and secrete, and special absorption mechanisms are required.

Perhaps the most curious thing about carnivorous plants is how they move rapidly enough to catch their highly mobile prey. Most plants, if touched, do not respond visibly. However, carnivorous plants and a few others *do* respond to touch by means of *action potentials*—changes in electrical potential that trigger changes in turgor pressure in certain cells and that travel from the stimulated area, such as a leaf, toward the base of the leaf and the stem. (We will see analogous processes in animal nerve cells in Chapter 34.)

The sundew (see chapter-opening photo) attracts insects via sweet nectar secreted on the tips of sticky, club-shaped tentacles that jut upward from the leaves. When a fly lands on one of the tentacles, it sets off an action potential that travels down the tentacle, stimulating closure of the tentacles and entrapment of the fly. The Venus's flytrap works on a similar principle.

Although it does not feed on insects, *Mimosa pudica* also responds to touch by means of action potentials. When a leaf is touched, an electrical stimulus travels quickly down the leaf to the leaf base, where cells in the axil of the petiole rapidly lose water so that their turgor pressure falls. Because the normally turgid axil cells hold the leaf upright, this water loss causes the leaf to droop rapidly (Figure 27-10a). If the touch stimulus is strong enough or is repeated often enough, the action potential

(a)

(b)

Figure 27-10 SENSITIVITY WITHOUT NERVES: A SPEEDY DEFENSIVE REACTION.
(a) The *Mimosa pudica* responds to touch with a fast drop in turgor pressure. An action potential is produced on touch; turgor pressure drops as a result, and the leaf droops. (b) If the touch is repeated, the action potential travels to other parts of the plant; eventually, all the leaves on the plant droop. This response may have been an evolutionary response to prevent predation. Mimosa are also bitter tasting—another protection from animals.

travels to other leaves and soon the entire plant looks wilted (Figure 27-10b).

The folding response of the mimosa may be an evolutionary adaptation to avoid being eaten; if a grazing deer or a hungry insect touches the mimosa bush, the plant folds up and the leaves are more difficult to find.

Animals move their legs, wings, or jaws by harnessing the contractility of the actin-myosin protein system—not through turgor pressure changes, as in plants. Plant cells do contain actins and myosins, but they are involved in cytoplasmic streaming, not in moving individual cells or organs relative to each other. Cell movement is no doubt absent because of the rigid, all-encompassing cell walls that forever anchor the individual plant cell in one spot, but these do provide strength, protection, and capillarity.

LOOKING AHEAD

Plants and their cells are characterized by the use of passive physical forces and minimal expenditure of energy. The transpiration-pull theory and the coupling of upward water flow in the xylem and the nutrient distribution system in the phloem are good examples of this unwitting parsimony. The result is efficient distribution of water, minerals, photosynthetic products, and other nutrients through a plant as tall as a redwood or as specialized as a desert cactus. But such varied activities must be coordinated, as our next chapter explains.

SUMMARY

1. Gases are exchanged through stomata, the epidermis itself, through lenticels, and across root surfaces.

2. Plants usually cannot excrete their own toxic wastes; thus, they store some indefinitely and break others down and recycle them for use in new cells and tissues.

3. Water transport in the xylem relies on several principles of water movement, including osmosis and *water potential*, which in a cell is equal to turgor pressure plus *solute (osmotic) potential*.

4. The *root pressure theory* states that water pressure building up in the roots due to osmotic pressure pushes water upward.

5. According to the *transpiration-pull theory*, *transpiration*—water loss through leaves—causes water to be pulled up through the xylem as a result of the cohesion of water molecules to each other and the adhesion of the molecules to the cell walls. More water molecules are taken in by the roots.

6. Regulation of water loss is governed primarily by the opening and closing of the stomata, which is a response to movement of ions across guard cell membranes.

7. *Translocation* of sugars and other organic substances through the phloem generally proceeds from sites of synthesis in leaves and stems (sources) to sites of use or storage in meristems, stems, roots, and flowers (sinks).

8. Transport of solutions in the phloem is well explained by the *mass flow theory*. Water has a tendency to flow from the xylem to the phloem in leaves because of the high solute concentration there due to photosynthesis; this drives the leaf phloem's contents into the stem's phloem, from where the contents move downward or occasionally upward (to sinks).

9. Plants require several *minerals*, inorganic substances dissolved in soil water, taken up by the roots, and transported in the xylem. Essential nutrients may be passively or actively transported by root cells.

10. The rate at which plants can take up various mineral nutrients depends on the concentration of the nutrients in the soil and on the soil pH.

11. *Nitrogen fixation* can be carried out through nonbiological means (e.g., lightning flashes) and by both free-living and symbiotic microorganisms. Plants must depend primarily on nitrogen-fixing microorganisms to convert atmospheric N_2 into NH_3 (and NH_4^+). Some microbes carry out *nitrification*, converting NH_3 to NO_2^- or NO_3^-. In legumes, symbiotic nitrogen-fixing bacteria invade root hairs and become encapsulated in *root nodules*.

12. *Carnivorous plants* can obtain nitrogen by trapping and digesting insects. Trapping sometimes involves quick movement of leaves brought about by electrical action potentials, which trigger rapid loss of turgor pressure in cells that control the position of the leaves.

KEY TERMS

bacteroid
capillary action
carnivorous plant
cohesion-adhesion-tension theory
guttation
mass flow theory
mineral
nitrification
nitrogen fixation
osmotic potential
root nodule
root-pressure theory
translocation
transpiration
transpiration-pull theory
water potential

QUESTIONS

1. Explain briefly how each of the following physical phenomena may play a role in the upward movement of water in the xylem: osmosis, capillary action, adhesion, cohesion, transpiration.

2. What experimental evidence suggests that transpiration is a primary process contributing to water movement in the xylem?

3. Explain how the movement of potassium (K^+) and chloride (Cl^-) ions into and out of guard cells regulates the opening and closing of stomata and thus the loss of water vapor from leaves.

4. Explain the difference between *transpiration* and *translocation*.

5. Describe how phloem cells may operate to cause flow of the phloem contents.

6. How do plant physiologists use (a) aphids and (b) the radioactive tracer $^{14}CO_2$ to study the activity of the phloem? What have these studies revealed about the phloem contents and its movement?

7. What accounts for the several types of symbiotic relationships between plant roots and bacteria or fungi?

8. How are waste products handled by plants?

9. Name the seven macronutrients and the nine micronutrients essential for optimum plant growth.

ESSAY QUESTION

1. Why is nitrogen fixation essential to the growth of higher plants? How are various bacteria involved in nitrogen fixation? Describe the very important symbiotic relationship between leguminous plants and bacteria of the genus *Rhizobium*.

For additional readings related to topics in this chapter, see Appendix C.

28
Plant Hormones

. . . the faint illumination of a narrow stripe on one side of the upper part of the cotyledons of Phalaris *determined the direction of the curvature of the lower part. . . . These results seem to imply the presence of some matter in the upper part which is acted on by light, and which transmits its effects to the lower part.*

C. Darwin and F. Darwin, *The Power of Movement in Plants* (1888)

Lofty survivor: A towering redwood tree in northern California's Jedediah Smith State Park.

There is a spectacular stand of virgin redwood trees in a park north of San Francisco, and in it sits a cross-sectional slab about 10 ft across cut from one towering tree. Starting at the center of the slab and then every foot or so across it, park rangers have painted and labeled individual growth rings. The center label shows that the tree was a small sapling in A.D. 909. It was a strong young tree in 1066, the year the Battle of Hastings was fought. It was sturdy and tall when the Magna Carta was signed in 1215, and it had grown into an impressive pillar standing in an untouched forest by 1492. The tree was a venerable giant by 1776, and it had added another 12 in. of girth, ring by ring, by the time loggers cut it down in 1930 to build California bungalows.

During its millennium of growth, the tree withstood innumerable storms, periods of drought, cold snaps, raging fires, swarms of insects, attacks by airborne fungi, and endless combinations of sunlight, temperature, water, and nutrients. Without a system for controlling growth and internal functions to meet these environmental challenges, the tree never could have had such a long life. Neither, for that matter, could a zinnia live out its short life span from seed to adult to seed in your summer garden. Indeed, all plants require coordinated control over growth and internal processes, and the substances that provide much of this control are **plant growth regulators,** or **hormones**—substances manufactured in minute quantities in one part of the plant that produce effects in other parts. Most plant tissues can make and respond to each of the five major classes of plant hormones, and each type has many effects.

This chapter discusses each hormone class:

- The auxins, which stimulate cell elongation, cell division, and the control of various processes
- The gibberellins, involved in shoot growth, bud and seed dormancy, and other activities
- Cytokinins, hormones that help induce cell division and delay aging in plant tissues
- Abscisic acid, a plant hormone that suppresses growth, accelerates aging, and in some plants helps promote the falling of leaves and fruit
- Ethylene, the fruit-ripening hormone
- How plant hormones interact to bring about leaf fall, seed germination, and flowering

AUXINS: CELL ELONGATION AND PLANT MOVEMENTS

Plants move in response to environmental signals: Leaves bend toward light, roots grow downward under the influence of gravity, some flowers trace the path of the sun across the sky each day, and others close tightly at night and open up in daylight. These usually imperceptibly slow movements are called **tropisms** and are accomplished primarily by changes in cell size and sometimes number.

In about 1880, Charles Darwin and his son Francis performed the first recorded experiments on tropisms. The Darwins wondered why plant stems display **phototropism** (bending toward the light), and to study the phenomenon, they grew some oat seeds on the windowsill. When the coleoptiles (the sheaths covering the embryonic shoot, see Figure 26-10) emerged, the Darwins put tiny transparent glass caps on the tips of some coleoptiles, black caps on others, cut some tips off, and covered the coleoptiles on some seedlings below the tip (Figure 28-1). The glass-capped coleoptiles grew toward a light source, and so did the seedlings with stems masked. In contrast, the black-capped ones grew straight up, as did the decapitated ones. The elder Darwin concluded that only the coleoptile tip is able to respond to light and that it produces a substance or transmits a signal that causes the stem to grow toward the light.

Some 40 years later, Dutch botanist F. W. Went carried the tests further. He set out to isolate this signal by removing the tips from coleoptiles that had grown in the dark and placing them on plates of gelatin-like agar. Went carried out all his operations in the dark so he could tell whether a chemical signal or light itself was causing an effect. Assuming that the chemical signal might move out of the tip into the agar, he cut a small block of the rubbery material and placed it symmetrically on top of a decapitated coleoptile. He observed that the coleoptile grew straight up (Figure 28-2a), and he

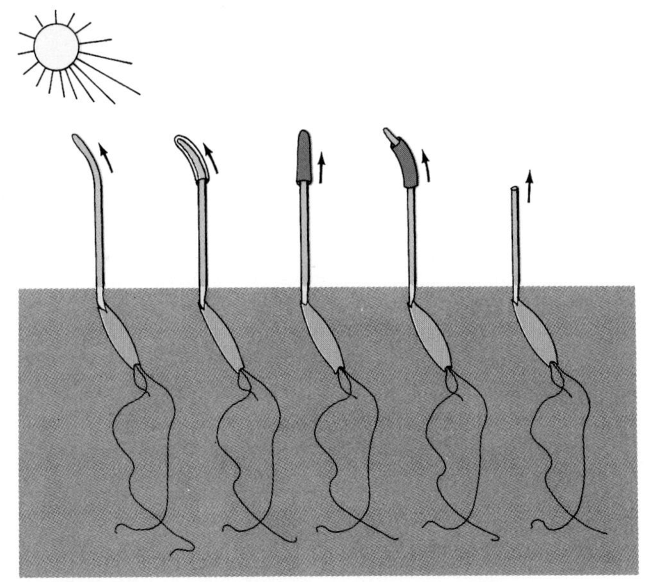

Figure 28-1 THE DARWINS' DISCOVERY OF PHOTOTROPISM.
Charles and Francis Darwin masked parts of grass and oat seedling coleoptiles and found that the tips had to be exposed to light if directed growth (phototropism) was to occur. A transparent glass cap did not interfere with the response, but an opaque cap over the tip did inhibit the phototropism. Covering the stem had no effect.

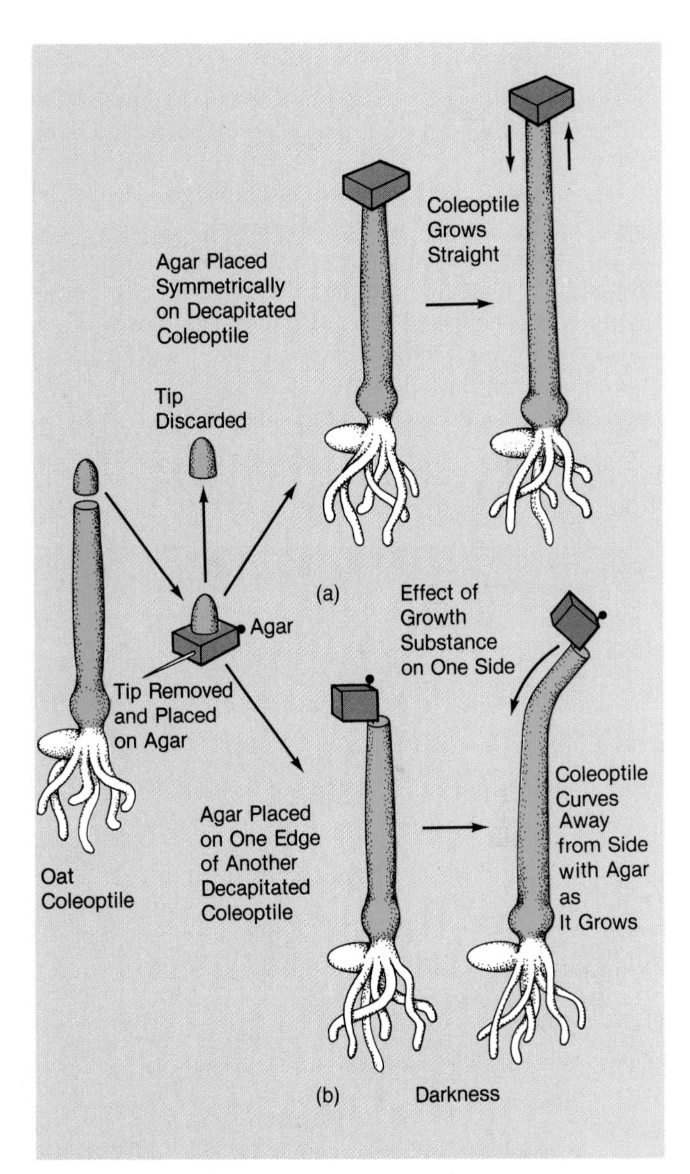

Figure 28-2 WENT'S AGAR BLOCK EXPERIMENT.
Went showed that a diffusible growth factor controls seedlings' phototropism. Went cut the tips from a group of dark-reared seedlings and placed the tips on agar blocks, hoping that the "signal" would move into the agar block. (a) It did, and when the agar block was placed over the end of a decapitated coleoptile, the coleoptile grew straight upward. (b) Went then placed the agar containing the "growth factor" on one side of the coleoptile or the other, and the coleoptile elongated more on the side with the block; in other words, the plant curved away from the side with the block. Went gave the name auxin to the growth factor.

concluded that a substance diffused from the agar and stimulated the coleoptile to grow.

Carrying his work still further, Went decapitated another coleoptile and placed an agar block off center at one edge (Figure 28-2b); he observed that the coleoptile grew faster on the side with the attached agar block, so that the stem naturally bent away from the block. Since this bending took place in the dark, Went concluded that light must affect the distribution of the signal agent in intact coleoptiles, so that one side curves toward the light. Went called the substance **auxin,** from the Greek "to increase."

Later researchers purified the substance from the seedling tips and identified it as indole-3-acetic acid (IAA; Table 28-1), a hydrophobic molecule that can easily pass through the plasma membrane of cells. Those researchers confirmed Went's original hypothesis: When a plant is exposed to light, auxins produced near the tip accumulate on the side of the stem away from the light and cause it to grow quickly, thus bending the plant toward the light source. Others discovered that auxins promote growth by stimulating cell elongation; as the cells on the nonilluminated side lengthen, the stem must curve toward the shorter (illuminated) side (see Figure 28-1). And still others learned that the group of related hormones called auxins can influence cell enlargement, axillary bud development, fruit development, leaf abscission, cambial activity, and the growth of adventitious roots.

Effects of Auxins

Surprisingly, while auxins stimulate cell enlargement in some tissues, they suppress it in others. The main shoot of many plants produces auxins that inhibit the growth of lower buds, resulting in **apical dominance,** the tendency of the main shoot of a plant to predominate over all the others (Figure 28-3a). An A-shaped plant—a fast-growing young pine tree, for example—exhibits

Table 28-1 PLANT GROWTH REGULATORS (HORMONES)

Name	*Structure*	*Origin*	*Some Functions*
Auxin	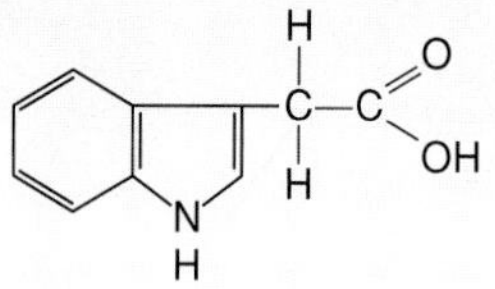 (indole ring with N–H, attached $-CH_2-$ with C(=O)OH)	Young growing regions	Growth promoter: stimulates cell elongation; inhibits lateral buds; prevents leaf and fruit drop; orients root and shoot growth
Gibberellic acid	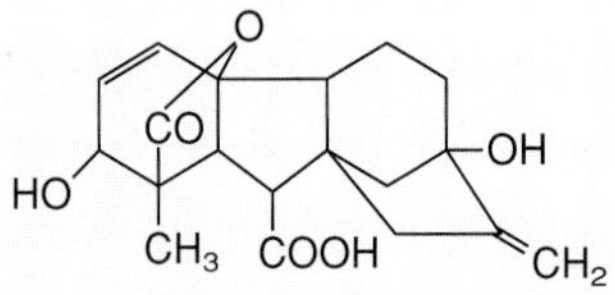 (ring structure with O, CO, HO, OH, CH_3, COOH, CH_2)	Young growing regions	Growth promoter: stimulates cell elongation; promotes grass seed germination
Cytokinin	(a) (furan ring, CH_2—NH, purine ring with N, N, N, N–H) (b) 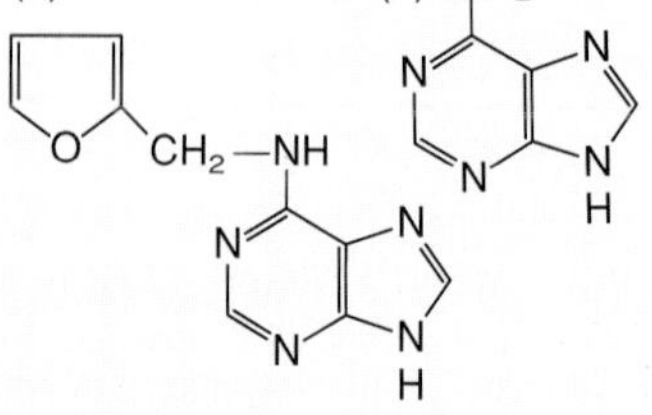 NH_2 (purine ring with N, N, N, N–H)	Roots	Growth promoter: stimulates cell division; blocks leaf senescence
Abscisic acid	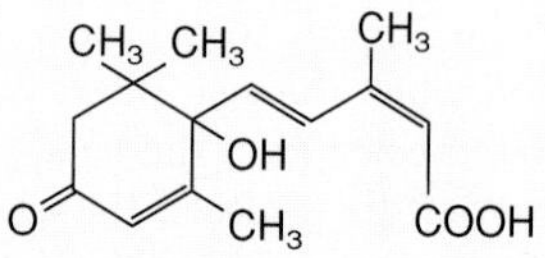 (ring with O, CH_3, CH_3, OH, CH_3; side chain with CH_3, COOH)	Cells under stress	Growth inhibitor; maintains dormancy
Ethylene	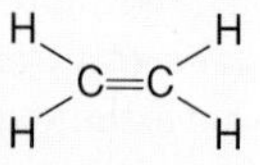 $H_2C{=}CH_2$ (H, H, C=C, H, H)	In regions of high auxin concentration	Maturation promoter: stimulates lateral expansion of elongating cells; promotes fruit ripening and leaf and fruit drop

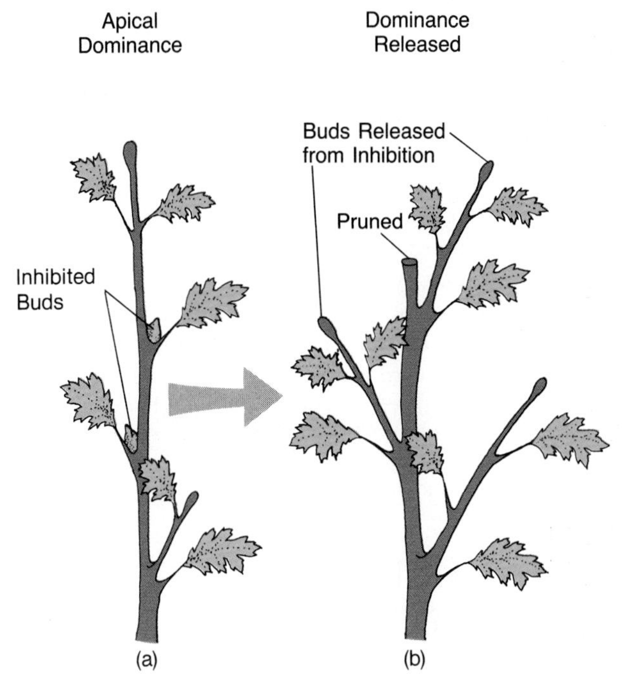

Figure 28-3 APICAL DOMINANCE: THE INFLUENCE OF AUXIN.
(a) Apical dominance is the tendency of the main shoot of a plant to predominate over lateral growth. However, if the apical meristem has grown far enough from the lower axillary buds (as at bottom), the buds may begin to grow. (b) Apical dominance can be overcome by pruning (cutting off) the main growing shoot. When this is done, axillary buds produce branches.

strong apical dominance, while a short, squatty, box-shaped plant, such as a boxwood shrub with many actively growing lateral branches, has weaker apical dominance. To promote the growth of side branches in a normally A-shaped plant, gardeners often prune, or pinch back, a plant's main growing tip. This releases the axillary buds from the inhibition of apical dominance exerted by auxin produced in the main stem (Figure 28-3b).

Auxins are also instrumental in fruit development. Young embryos often produce auxins that promote the development of fruit tissue around the seed. The amount of the hormone often determines how large the fruit will ultimately grow. The strawberry is a good example. Each dark spot on a strawberry is a seed, and the conical red strawberry "fruit" we eat results from the fusion of the many tiny fruits surrounding the seeds (Figure 28-4a). Removing all the developing ovules except one causes a misshapen strawberry to form with succulent flesh around only the remaining seed (Figure 28-4b).

Auxins and another class of plant hormones, the *gibberellins,* help determine a plant's response to gravity. If you place a normal seedling horizontally rather than vertically, the elongating shoot tip will soon bend upward, and the growing root tip, downward. These responses to gravity are called **gravitropisms.** Experiments reveal that the two hormones build up on the lower side of a shoot lying horizontally, causing greater cell elongation to occur on the lower surface and thus the tip to bend upward, away from the earth. When the growing shoot is oriented vertically once again, hormone concentrations equalize across the shoot, and straight vertical growth proceeds. Gravitropism is thought to be somewhat more complex in roots than in shoots; we discuss it a bit later.

How Auxins Work

In the many years since the Darwins' experiment with coleoptiles, botanists have found indirect evidence to suggest that when pigment molecules on the illuminated side of the stem receive light, auxins are transported down the opposite, dark side of the stem. There, as we saw, auxins promote cell elongation, causing the seedling to curve toward the lighted side. Despite years of searching, biologists have never succeeded in isolating the pigment and know just one additional fact about it: Light at the blue end of the spectrum (wavelengths shorter than 500 nm) most effectively stimulates the pigment molecules.

Botanists would like to know whether auxin is preferentially transported to the dark side of a stem when the pigment receives blue light or whether auxin is transported to *both* sides and is simply destroyed on the illuminated side. To study the path and rate of auxin movement, researchers applied a radioactively labeled auxin solution to the cut end of a stem. Evidence suggests that auxins move downward in association with plasma membranes or cell walls, not through xylem and

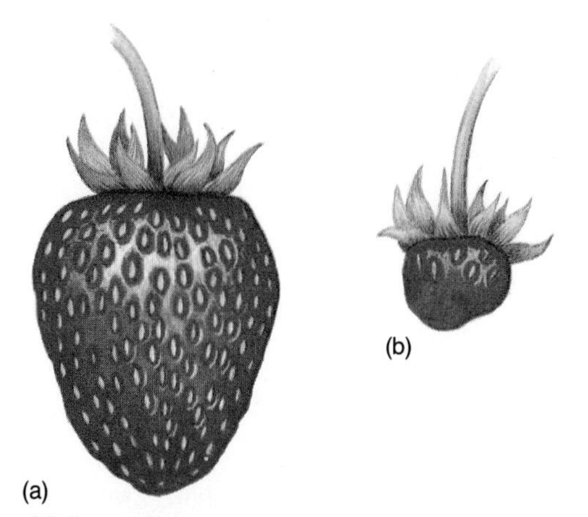

Figure 28-4 AUXIN AND STRAWBERRIES.
Auxins released by strawberry ovules promote fruit development around the seeds. (a) A normal strawberry. (b) A strawberry from which all but one of the seeds were removed.

phloem—a process called *polar transport*. In stems kept in the dark, the downward polar transport is equivalent on all sides of the stem. In stem sections illuminated on just one side, however, most of the auxin appears to be transported across the stem to the dark side by an unknown mechanism.

Despite the mysteries surrounding auxin transport, plant biologists do know how auxin promotes cell elongation and thus stem or root curvature. When auxin reaches a responsive cell, it indirectly causes loosening of the cell wall; the cell's internal turgor pressure then stretches the wall and allows the cell to lengthen—sometimes more than tenfold. Auxin indirectly loosens the cell wall by inducing the cell to secrete hydrogen ions. These protons lower the pH within the cell wall and perhaps activate enzymes that break some bonds within the walls, allowing cellulose fibers to slip past each other and thus allowing turgor pressure to push the wall into a new shape.

Recent research suggests that acid (H^+) release may be involved in gravitropisms, as well as in cell elongation. Specifically, plant physiologists have found that tiny amounts of acid are released from the zone where cells elongate in a vertical root or shoot. In a shoot placed horizontally, acid is secreted from the lower surface, which then curves upward. Acid is also secreted from the upper side of a root placed horizontally, and the root soon curves downward (Figure 28-5a).

Auxin appears to affect the release of hydrogen ions, which in turn bring about cell wall loosening, cell elongation, asymmetrical growth in shoots and roots, and plant tropisms in response to light and gravity.

There may well be other mechanisms involved in gravitropism, however. A recent study shows that dense starch granules, or amyloplasts (also called statoliths), sink to the lower surface of root cells in a root placed horizontally (Figure 28-5b, top). This apparently causes calcium ions to be released, and both Ca^{2+} and auxin are pumped out of those cells. As a result, more auxin flows along the lower surface of the root than along the upper, differential cell elongation takes place in the upper surface, and the root tip curves downward (Figure 28-5b, bottom). Even without its root cap, though, a horizontal root curves downward if researchers place an agar block that contains auxin or free Ca^{2+} along the lower surface.

GIBBERELLINS: GROWTH PROMOTERS

A disease that affects one of humankind's most important staple crops, rice, led to the discovery of a second class of growth-promoting hormones, the **gibberellins.** Rice seedlings normally grow into chest-high mature plants, but sometimes, the seedlings grow abnormally tall and spindly, with stems so frail that the wind eventually knocks the plant over.

Japanese scientists studying so-called "foolish seedling disease" in the 1930s discovered not only the cause of the disease, the fungus *Gibberella fujikuori*, but also an entirely new class of plant hormones, the gibberel-

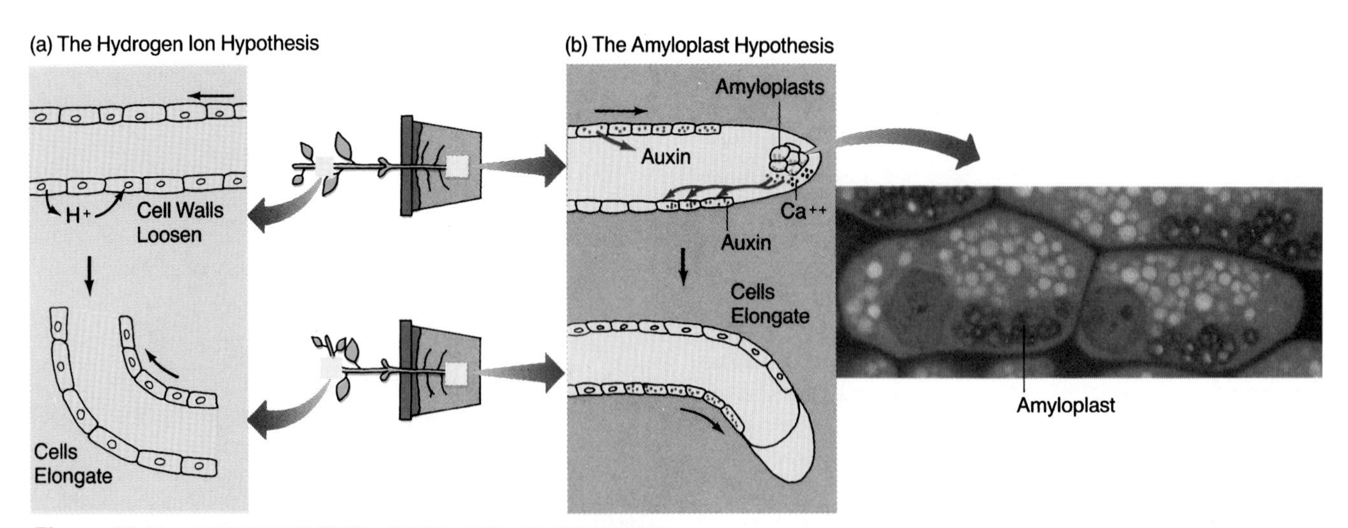

Figure 28-5 HYDROGEN IONS, AUXIN, AND GRAVITROPISM.
(a) Shoot growth patterns and patterns of H^+ release show that acid is released on the side of the shoot that elongates as the shoot curves away from gravity. The ions may play a role in the hormone-induced loosening of cell walls and thereby allow this negative gravitropism. (b) Amyloplasts may play a role in the positive gravitropism of roots, sinking to the lower surface of root cells and causing Ca^{2+} and auxin release and the subsequent elongation in the upper cell surfaces that leads to downward root growth.

Figure 28-6 TOWERING CABBAGES: A RESULT OF GIBBERELLIN TREATMENT.
Cabbage ordinarily has almost no internodal elongation, despite the presence of a low level of gibberellins. Instead, as seen at the left, the leaves are packed in a tight head. Treatment with higher levels of gibberellin induced the plants on the right to elongate, producing spectacular stems instead of tight rosettes.

lins, which cause elongation of plant internodes (the stem regions between leaves). By now, researchers have identified more than 70 related gibberellin compounds that occur naturally in higher plants. The most active appears to be gibberellic acid (GA_3; see Table 28-1). Today, biologists also know that unlike auxin, gibberellins increase *both* the size and number of internodal cells, and as a result, increase stem length.

Plant scientists probed the action of gibberellins by applying the hormones to genetic dwarf plants. In some of the plants, the internodes elongated—the more gibberellins, the longer the internodes grew (up to a maximum hormone concentration. This finding helped show that some dwarf plants make too little gibberellin and that the hormones do indeed lengthen internodes.

Gibberellins also regulate various aspects of plant development. Many biennials and some annuals grow as a *rosette*, a cluster of leaves in a tightly packed roundish head and with almost no internodal elongation. When a rosette plant such as a cabbage is ready to flower, it *bolts*, or sends a shoot high in the air. A small dose of gibberellic acid will induce bolting, and a large dose can induce both stem elongation and subsequent flowering or even create a 10-foot-tall cabbage (Figure 28-6)! In the rosette stage, gibberellin concentrations are too low to trigger bolting and flowering.

Like plant auxins, gibberellins can influence the growth rate of apples, peaches, and some other fruits and can substitute for the presence of a developing embryo, the source of the normal stimulant for fruit growth. Many fruits that are responsive to treatment with gibberellins are not sensitive to auxins, and vice versa.

Although gibberellic acid and auxins have similar effects, auxins and gibberellins are bound by different responsive cells in the plant. Nevertheless, both hormones are required for normal growth, and recent studies show that both may function by activating enzymes that release sugar polymers called oligosaccharins from plant cell walls. These, in turn, may act as second messengers to regulate plant genes and affect various phases of growth and development.

CYTOKININS: CELL DIVISION HORMONES

A third class of hormones, the **cytokinins,** influences plant growth by regulating cell division. Cytokinins are structurally similar to the nucleic acid base adenine (see Table 28-1).

Plant biologists discovered one of the first cytokinins in coconut milk, which is a type of liquid endosperm that promotes the growth and cell division of virtually all cultured plant tissues because of its cytokinin content.

Effects of Cytokinins

When cytokinins alone are added to a culture of undifferentiated plant cells, nothing happens. When auxin alone is added, the cells elongate. When both hormones are added, the cells divide. These observations led biologists to the conclusion that the ratio of cytokinin to auxin determines the rate of cell division in plants. If the concentration of cytokinin is higher than that of auxin, the rapidly dividing cells tend to differentiate into shoot or leaf tissue. If the concentration of auxin is higher,

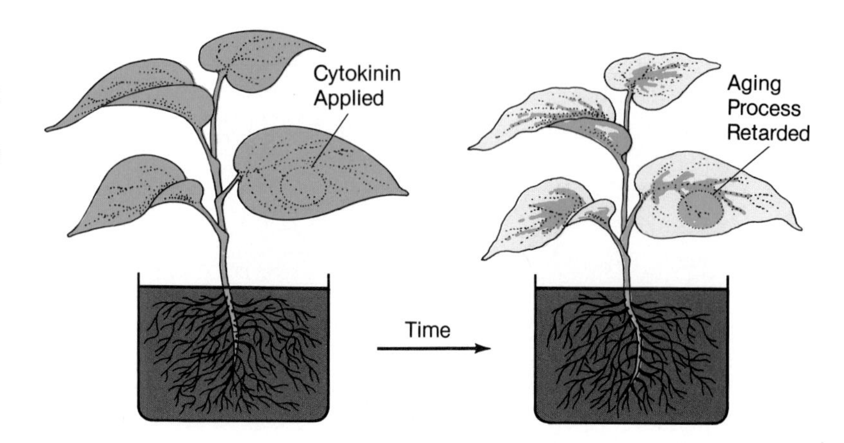

Figure 28-7 CYTOKININS: INHIBITORS OF SENESCENCE IN A LEAF.
Cytokinins not only promote cell division but also delay aging. When cytokinins are applied to an area of a leaf, that spot remains healthy while the remainder undergoes normal senescence. Cytokinins may act by promoting transport of nutrients to the treated area, thereby slowing the degradation of treated leaf tissue.

roots or disorganized lumps of cells called callus tissue tend to form. Clearly, some balance of the two is needed for normal cell growth and division.

Cytokinins play another role: They delay aging, or *senescence*. When leaves begin to die, they lose much of their chlorophyll, their proteins are degraded into amino acids, and many of their organic constituents are exported to the stem. If an experimenter applies cytokinin to one spot on a senescing leaf, the yellow leaf will sport a single bright green region where the hormone retards senescence (Figure 28-7). Cytokinin may spare leaf tissue from aging by promoting nutrient transport to the area. A decrease in cytokinin in older leaves may lead to the senescent state because once used up, vital minerals and organic nitrogen compounds are not replenished.

Cytokinins, Auxins, Tissue Cultures, and Plant Biotechnology

Forestry and agricultural researchers have worked hard to perfect plant tissue culture techniques so they can achieve vegetative reproduction in lines where sexual reproduction might lead to the loss of desired traits (see Chapter 25). Cytokinins and auxins are the two major hormones used in culturing plant cells and tissues. Most plant species can be grown and maintained this way for periods of up to 30 years.

With certain plant species, researchers are able to grow an entire plant in culture from one cell, or *protoplast* (Figure 28-8). The ability to regenerate a whole organism from a single cell is a reflection of the cell's *totipotency* (see Chapter 26).

Since plant meristems proliferate constantly (or at least seasonally) in nature and continually produce new organs, it is no surprise that plant cells in culture organize new meristems and produce either root or shoot, depending on the hormonal environment. If conditions favor extremely fast growth and cell division, **callus tissue** results. If such a disorganized lump is given appropriate hormones, it can form many organized structures, each growing from a meristem that develops on the surface of the lump. Recently, a team of English plant researchers succeeded in regenerating rice plants from rice protoplasts. This achievement had great significance, because in general, the staple cereal crops humans depend on have not been subject to genetic modification with the techniques of biotechnology.

(a)

(b)

Figure 28-8 ENTIRE PLANTS GROWN IN CULTURE FROM SINGLE CELLS, OR PROTOPLASTS.
Researchers place tiny pieces of tissue or single cells (protoplasts) from apical meristems under sterile conditions on agar plates containing a medium with all the nutrients needed by the plant. (a) These petri dishes show varying stages of growth of cultured corn and sunflower clones. (b) This plant was raised from a protoplast.

Despite the initial difficulties, researchers are working on several fronts to pioneer new techniques of plant engineering and thus to substantially improve cereals and other food crops (see the box below). The efforts to engineer better plants are crucial; observers predict that during the next 30 years we will have to grow as much food as was produced during *all of previous human history* if we are to stave off massive hunger.

Some geneticists have been combing the plant kingdom for species with desirable genes—genes for traits such as salt tolerance, cold tolerance, insect resistance, firmer fruit, nitrogen fixation, oil production, and increased protein yield. Others have been working on a scraggly weed in the mustard family called *Arabidopsis*, which may become the fruit fly of molecular botany and a useful model in which to develop methods for transferring genes. So far, those methods include the transport of gene fragments into cells via bacterial vectors (see Chapter 14); the fusion of two different kinds of protoplasts; the passage of genes into cells through artificial, electrically produced pores in their plasma membranes; and even the removal of DNA from one cell and its injection into another using extremely fine needles.

ABSCISIC ACID: THE GROWTH-SLOWING HORMONE

One could almost predict the existence of a hormone that suppresses a plant's natural tendency to grow, as might be needed during droughts or cold winters. And indeed, there is one: **abscisic acid.** This name is derived from the word *abscission*, which means "cutting off" or

BRAVE NEW FOOD

Agricultural researchers are hard at work in fields and laboratories to improve the flavor, hardiness, yields, and nutritional value of the foods we eat. Here's a smorgasbord of foods that researchers are currently engineering through the techniques of modern genetics:

- Soybeans, corn, and sugarbeets outfitted with genes that allow the plants to withstand high levels of herbicides. The idea is that farmers can spray whole fields of such plants to kill weeds but leave the crops unharmed. With fewer weeds to compete for nutrients, space, and sun, crop yields improve substantially.
- Super hardy tomato, potato, and cucumber plants with genes for resistance to plant viruses so they are better protected from viral attack.
- Tomato plants with genes that make their leaves impervious to attack by caterpillars.
- Tomatoes that can stay plump and firm for weeks after being picked.
- QPM, or Quality Protein Maize—corn with a protein content that approaches that of milk. Normal corn is missing two essential amino acids (tryptophan and lysine), and people who get most of their food calories from corn often suffer protein deficiencies. QPM, now being tested in Guatemala, China, Brazil, and Senegal, may help address this serious problem.
- Avocados with a creamier texture and less water content.

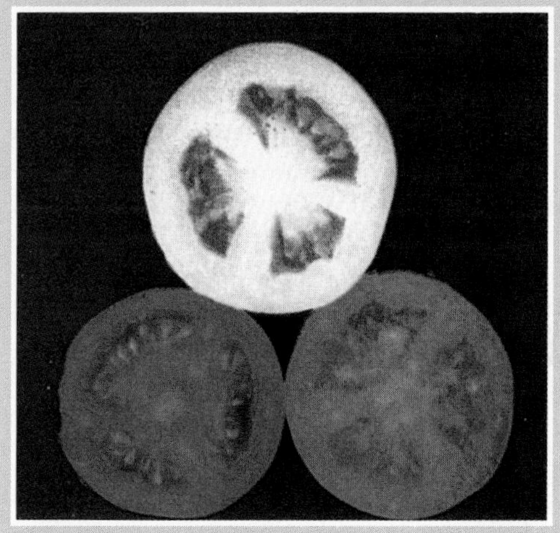

Figure A
New tomato varieties with a longer shelf life

- Strawberry plants with the genes to withstand frost damage, yet still produce tasty fruit.
- Lettuce that resists downy mildew, a common fungal parasite.
- Even beans that don't cause flatulence, so that more people will find these high-protein legumes acceptable.

Our current gene manipulation techniques are so powerful, and new techniques are being developed so rapidly, that one can hardly predict what other brave new foods of the near future there may be. It is safe to say, however, that the "shopping list" will continue to grow.

"removing," but is not nearly as apt a term as the alternative name *dormin*.

Abscisic acid is found in dormant bulbs and seeds and in some fruits, leaves, and other tissues. Its presence can trigger changes that prepare a plant for leaf drop, for winter dormancy, or simply for general slowed growth. In some plants, abscisic acid helps control flowering. In all plants, abscisic acid works in a delicate balance with growth-promoting hormones to govern appropriate responses to seasonal and short-term environmental changes.

Abscisic acid was discovered twice. It was studied first in the 1940s, when a group of British scientists prepared an extract from birch leaves that could arrest the growth of seedlings of birch trees and other plants. Then in the 1960s, scientists at the University of California at Davis isolated sufficient amounts of the hormone from cotton bolls (fruits of cotton plants) to determine the compound's chemical structure (see Table 28-1).

Effects of Abscisic Acid

Abscisic acid has a remarkable range of effects. It stimulates buds to form a set of outer leaves that become tough, protective bud scales in preparation for winter dormancy. It acts as a general inhibitor, especially in response to stress. For instance, drought causes the concentrations of abscisic acid to rise in leaf tissue and, in turn, causes leaf stomata to close (see Chapter 27). Abscisic acid is often an antagonist to the positive effects of the other plant hormones. Whereas gibberellic acid stimulates the production of hydrolyzing enzymes in endosperm, for example, abscisic acid inhibits this response, thereby acting as a check on the release of the seed's food energy and in turn the embryo's growth.

Abscisic acid is active in other ways in seed dormancy. It builds up in maturing seeds, for instance, and suppresses root and shoot elongation in the embryo. *Stratification* (treatment of seeds at low temperatures) appears to be necessary for the breakdown of abscisic acid in many seeds. Presumably, the same thing occurs in seeds "asleep" under snow or ice as they await the spring thaw to germinate.

Finally, abscisic acid accelerates senescence and in some plants promotes (although it is not primarily responsible for) **abscission,** the normal separation of leaf, fruit, or flower from the plant. In cotton bolls, for instance, abscisic acid levels are highest in the ovary bases at the time of normal fruit drop. Abscisic acid normally is produced in the leaves, probably in the chloroplasts, is shipped throughout the plant, and accumulates in older tissues. As a result, senescence and dormancy are accelerated in those tissues. A drop of abscisic acid solution on a healthy green leaf causes a yellowed, aged spot to form rapidly—just opposite to the effects of cytokinin.

Interaction of Abscisic Acid and Other Hormones

Abscisic acid interacts with the growth-promoting hormones to regulate plant responses. Almost all of abscisic acid's growth-suppressing effects can be offset by one of the other plant hormones. Perhaps the most confusing feature of plant hormone biology is that each of the growth-promoting hormones counteracts only *a few* of abscisic acid's inhibiting effects. Although the specific interactions are complex, the overall meaning of this give-and-take system is clear: The various plant hormones work together to control growth so that a plant even as tall as a towering redwood can meet the numerous physical challenges it faces during any given season by slowing down or speeding up growth at appropriate times.

ETHYLENE: THE GASEOUS HORMONE

The common expression "One rotten apple spoils the barrel" is an accurate observation: One ripening fruit does produce a volatile compound that accelerates the ripening of nearby fruit. This volatile gas is the hormone **ethylene** (see Table 28-1). The presence of ethylene in air causes increased respiration, which in turn leads to the changes in fruit composition that transform a hard, acidic, inedible fruit into a sugary ripe one.

A plant's ethylene production can be triggered by localized increases in auxin concentration and perhaps by other physiological factors as well. Ethylene released into the air can then act on the whole fruit or plant; its increase is accompanied by a sharp rise in metabolic activity and a hastening of the condition we call *fruit ripening*. Fruit dealers often have fruits picked while still green, then treat them later with ethylene to ready them for market. Conversely, dealers sometimes want to delay the ripening process and so employ a special low-pressure storage chamber that is continuously flushed with air to remove any ethylene released by fruits or flowers. In such an environment, an apple slice remains juicy, without browning, for days, and roses and carnations can be kept fresh for 6 months.

In addition to promoting fruit ripening, ethylene influences plant growth. Ethylene inhibits the transport of auxin across the stem or root—a movement important during the plant's tropic responses to light and gravity. Because ethylene is produced in areas of high auxin concentration yet inhibits its transport, ethylene may serve as part of a negative-feedback loop that prevents excess auxin accumulation. The ratio of ethylene to auxin seems to influence whether cells elongate or expand radially in response to auxin. Because auxin can stimulate ethylene formation, many of the effects originally attributed to auxin may actually be due to ethylene.

In 1988, researchers found some evidence that plant cells may have receptors for ethylene by studying a mutant form of the weed *Arabidopsis thaliam*. A mutation in the plant's *etr* gene apparently affects a receptor for the gaseous hormone. Significantly, the mutant plants are insensitive to ethylene in leaf senescence, seed germination, cell elongation, and other responses.

INTERACTION OF PLANT HORMONES

All five major types of plant hormones interact in complex ways to control the physiology and structure of the entire plant. Biologists know where each is produced: auxins and gibberellins in the leaves and young growing regions of the shoot; cytokinins in the roots; abscisic acid in cells experiencing stress, particularly water stress; and ethylene wherever auxin concentrations are high.

Although there is no general theory for how the five types of hormones interact, biologists do accept the principle that two or more positive hormonal signals are required for plant growth, but one negative signal can prevent further growth. For example, cell elongation requires both auxins and gibberellins, two growth-promoting signals, while ethylene or abscisic acid can act as a single negative signal to halt growth processes. For every example of growth inhibition by abscisic acid, however, combinations of auxin, gibberellin, and cytokinin (each stimulating different aspects of growth) can reverse the negative impact. The evolution of this complex control system provides plants with a flexible repertoire of responses for meeting daily challenges as well as seasonal ones such as leaf fall and seed germination.

Leaf Fall

The red, gold, and orange leaves of autumn create one of the most enchanting signs of the changing seasons, and the falling of those leaves has a complex, integrated underlying chemical and physical basis. At the base of the leaf petiole, flower stalk, and fruit stem, a special zone of cells called the **abscission layer** forms in response to hormonal signals (Figure 28-9). When the production of auxin declines in the leaf blade, for example, the abscission layer begins to form. An increase in ethylene production also appears to promote such formation. Thus, auxin is an inhibitor of abscission, while

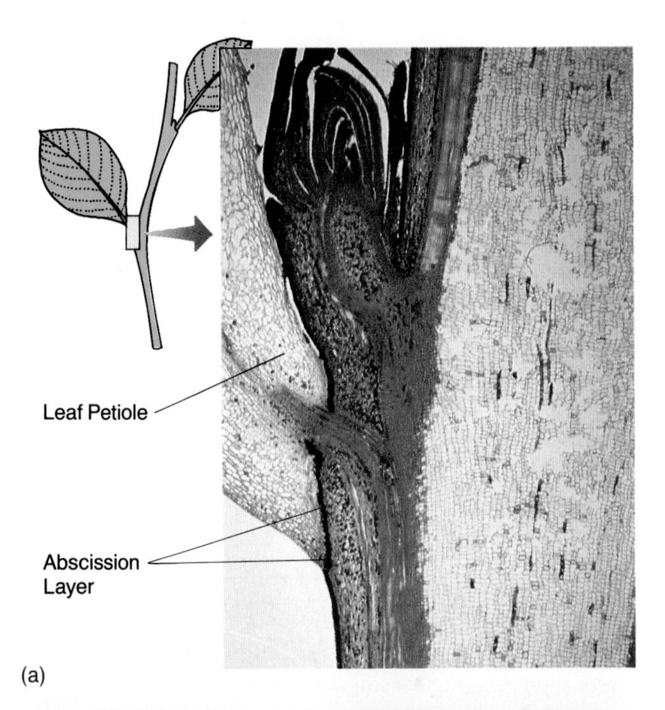

(a)

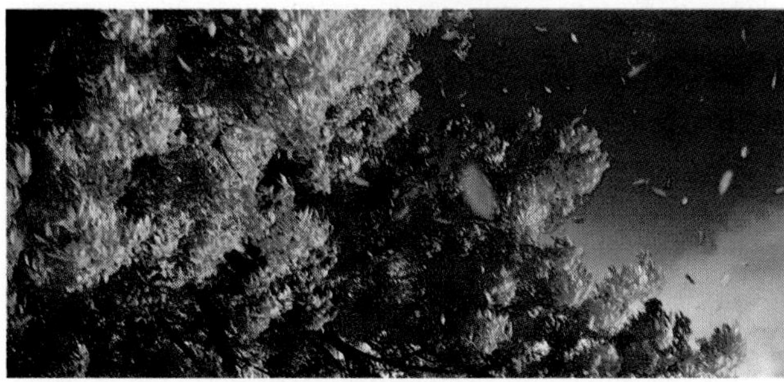

(b)

Figure 28-9 FORMATION OF ABSCISSION LAYERS: A RESPONSE TO HORMONAL SIGNALS.
The abscission zone is a layer of structurally weak cells in or near the petiole. Enzymes are synthesized that weaken the cell walls in this zone to such an extent that the leaf's own weight or a gust of wind can cause the leaf to break free. A *decrease* in the production of auxin and an *increase* in that of ethylene cause the abscission layer to form. (a) In this view of a maple branch (*Acer* species), the vertical black line on the left is the abscission layer; the leaf's petiole extends toward the left, and an axillary bud is seen at the top, left. (b) The result of the weakening abscission layer is a fluttering fall leaf.

ethylene is a promoter. The cell walls in the abscission layer are broken down, creating a weak area. The weight of the leaf or a gust of wind can cause the leaf to separate from the stem and fall. The same basic mechanism occurs in flowers and fruits.

Seed Germination

The two hormones needed for seed germination are gibberellic acid and abscisic acid. A first critical step in barley seed germination, for example, is mobilization of starch stores in the endosperm. This begins when the embryo secretes gibberellic acid, which in turn stimulates synthesis of the messenger RNA for the enzyme α-amylase. The aleurone layer then secretes α-amylase, and this enzyme digests the starchy endosperm (see Chapter 26). Abscisic acid stimulates ribosomal RNA synthesis, and so both hormones appear to act at the level of gene transcription in the nuclei. Scientists hypothesize that in nature, abscisic acid content rises if the seed experiences water stress and thus prevents the seed from germinating and using its nutrient stores when conditions are unfavorable for the seedling to become established.

CONTROL OF FLOWERING

As the changing seasons reveal, most temperate-zone plants flower in spring or summer, with each species seeming to have an internal calendar that tells it when to bloom: Crocuses often break through the snow, tulips emerge a few weeks later, and chrysanthemums don't unfurl until early autumn. The timing of a plant's reproductive cycle is critical to the pollination of its flowers and hence to the production of fruits and seeds at appropriate times, and recent research suggests that an interplay of ethylene and gibberellin may be involved in flower opening. The young flower bud seems to produce high levels of ethylene, which prevents premature expansion. As the corolla matures, however, it generates less ethylene, allowing gibberellin levels to climb and floral opening to follow quickly. The key to all this activity is timing. But how does a plant "measure" the seasons so as to flower at an appropriate time?

The answer involves environmental cues that influence hormonal activity, which in turn can trigger and support life cycle events. In many long-lived species such as trees, the coming of spring light cycles and temperatures induce hormone levels to rise, and with them, last year's growth of floral buds to break dormancy, develop further, and open. Quite a different strategy is employed by some annual plants, such as zinnias. For them, physical factors influence the speed of vegetative growth, and after a fixed amount of such growth, the plant produces floral buds. How, then, does the environment trigger the activity of plant hormones so that flowering occurs at appropriate times for the various species?

Photoperiodism

In 1920, W. W. Garner and H. A. Allard showed that the number of hours of light in a given day—the day length, or **photoperiod**—is crucial in determining whether many plants will flower. In the continental United States, the day length in a 24-hour period changes from about 9 hours to 15 hours between the winter and summer solstices, thus providing an accurate measure of the changing seasons.

Garner and Allard classified plants into three groups, depending on the effects of photoperiod on flowering: *short-day* plants; *long-day* plants; and *indeterminate*, or *day-neutral*, plants. Short-day plants require a short photoperiod—fewer than some critical number of hours for each species—for flowering to be triggered. Long-day plants require a long photoperiod—more than some critical number of hours for each species. The flowering of day-neutral plants appears to be unaffected by the number of hours of light in a day.

Researchers Karl Hamner and James Bonner later modified this concept by means of a very simple experiment. When they exposed cocklebur (a short-day plant) to a 10-hour (short) day and a 14-hour night interrupted by a brief flash of light, they found that the plants did not flower (Figure 28-10). Consequently, they determined that it is the length of the *night* period that is critical to a short-day plant. Applying hindsight, short-day plants should really have been called long-night plants. Hamner and Bonner also showed that long-day plants actually require a short, uninterrupted night; in other words, they are short-night plants. If, for instance, a long-day plant is kept in just a few hours of light and many hours of darkness, it will not flower. If that long night is interrupted by a brief period of light, however, flowering occurs because the plant is tricked into behaving as if the interrupted long night were actually a short night.

Once the day length concept had been proposed, botanists observed the effects of the photoperiod on flowering in many kinds of plants and found that the number of treatment periods required to induce flowering can vary widely. Some short-day (long-night) plants—rice and cocklebur, for example—require only a single long night to induce flowering. Some long-day (short-night) plants—such as dill weed, spinach, and white mustard—require

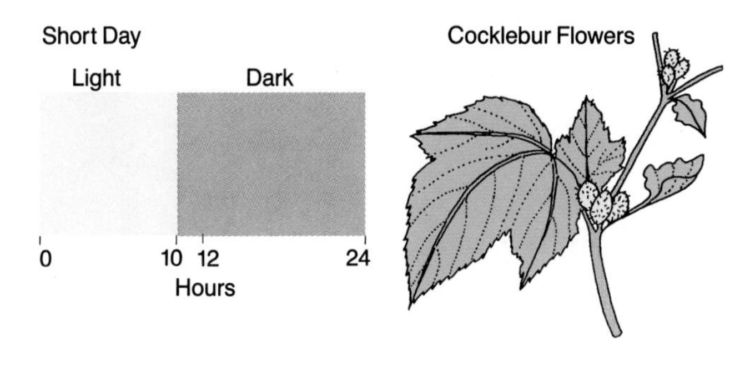

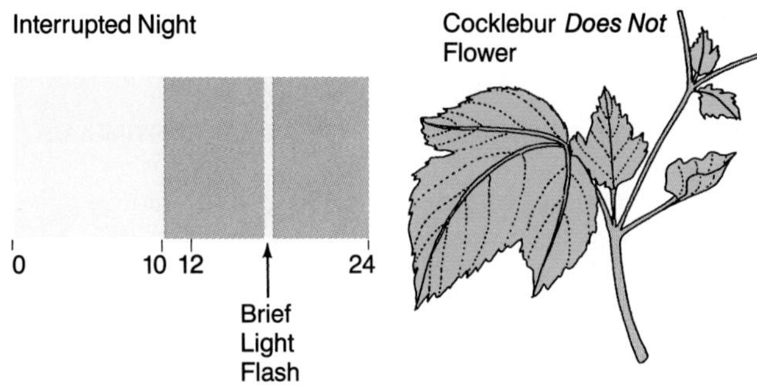

Figure 28-10 FLOWERING: A RESPONSE TO SPECIFIC PHOTOPERIODS.
Experiments with a cocklebur in controlled-light chambers show that plants must experience nights of certain lengths before flowering is induced. For the cocklebur, the minimal night length is 9 hours. No matter what time the night begins, or how long the day is, only an *uninterrupted* period of darkness, at least 9 hours long, will induce flowering. When that time period is interrupted (here, midway through the night), the cocklebur plant does not flower.

only a single short night. Other plants, however, require several days with the appropriate hours of light or darkness. Plants such as winter wheat, soybean, and orchids also require *cool* nights in addition to the appropriate photoperiod before they will flower. (Most plants do not have such temperature requirements, however.) Finally, there are day-neutral plants—such as cucumber, peas, corn, and onion—whose flowering appears unaffected by day length.

The Phytochrome Pigment System

Interrupting the required long night of so-called short-day plants with a flash of light gave plant researchers a simple experimental system for probing the mechanism by which plants measure photoperiods and led to the discovery of a pigment called **phytochrome,** which contains a simple protein as well as a light-absorbing pigment molecule.

Phytochrome exists in two chemical forms—phytochrome red (P_r) and phytochrome far-red (P_{fr})—that provide the plant with a basis for detecting the length of the dark period. Each form can be converted into the other, but each absorbs a different wavelength of light. P_r absorbs red light (about 660 nm), and the absorbed light energy converts the phytochrome molecules to the P_{fr} form. P_{fr} absorbs far-red light (about 730 nm) and converts the pigment-protein complex back to P_r (Figure 28-11). During daylight hours, the P_r absorbs light energy and is converted to P_{fr}. Sunlight is much more energetic in the red part of the spectrum than in the far-red part, so by the end of the day, about 95 percent of the phytochrome molecules have been converted to the P_{fr} form. When P_{fr} is present in plant tissues, more than 40 different enzymes increase in activity in various cell types of leaves or stems, and in some plants flowering can result. Recent studies suggest that P_{fr} may activate a protein that binds to a special "light switch"—a light-responsive element on the DNA—and that the binding may turn on genes for one or more of these enzymes, including ribulose-1,5-bisphosphate carboxylase, the key to carbohydrate production during photosynthesis (see Chapter 8). During the night, P_{fr} is then enzymatically converted back to P_r (see Figure 28-11).

It would be easy to conclude from the experiments that the time-limiting dark reversion of P_{fr} to P_r is the plant's biological clock; however, that reversion takes only about 3 hours in both short-day and long-day plants. Plant biologists suspect that some other as yet unknown factor acts as an internal clock that somehow regulates daily and seasonal patterns of overall activity within the plant. Phytochrome pigment itself appears to be in-

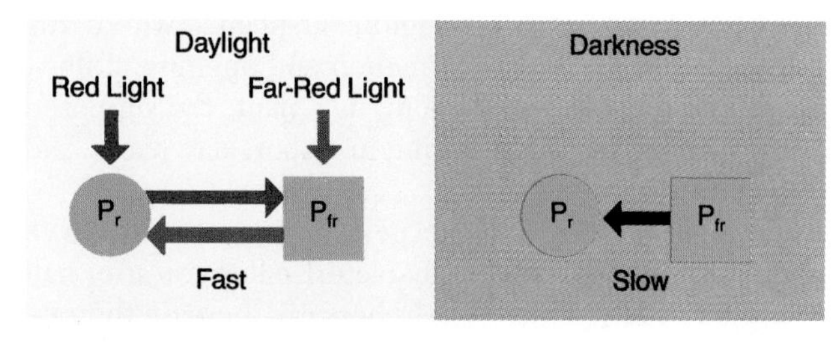

Figure 28-11 LIGHT ABSORPTION BY PHYTOCHROME PIGMENT: THE PLANT'S MEANS OF DETECTING LIGHT AND DARK.
Phytochrome exists in two forms: phytochrome red (P_r), which absorbs light at about 660 nm, and phytochrome far-red (P_{fr}), which absorbs light at the far-red end of the spectrum, about 730 nm. During daylight hours, the P_r absorbs light energy and is converted to P_{fr}. P_{fr} is unstable and is converted back to P_r during the night or, more quickly, by far-red light. Somehow, the amounts of each form of the pigment allow plants to measure the number of hours of day and night.

volved in a range of plant responses besides flowering, including seed germination. Lettuce seeds, for example, require light to germinate. If planted too deep in the soil, where light cannot penetrate, the seeds fail to germinate. Exposing seeds to red light allows their germination. This exposure—even a brief one—converts all the phytochrome pigment within the seed to the far-red form. If, however, the seeds treated with red light are then exposed to just a few minutes of far-red light, they will not germinate.

Florigen: One Flowering Hormone or Many?

A Russian plant physiologist named M. H. Chailakhian, working in the 1930s, wondered which parts of plants are actually involved in measuring the photoperiod. By exposing the apex of a chrysanthemum, a short-day plant, to one photoperiod and the remaining leaves to another, he found that only the young leaves below the apex, not the apical meristem, are involved in detecting photoperiod with phytochrome. Without its young leaves, a plant cannot measure day length; but with just one or a few leaves, it can.

In another experiment, Chailakhian induced one set of plants to flower with the appropriate light conditions, then grafted a piece of an induced plant to each of several noninduced plants—all of which then flowered. He observed that the flowering stimulus can travel across a graft union, from plant to plant, and can convert a flowerless apex to one with flowers (Figure 28-12).

Chailakhian suggested that a "flower-making" hormone, which he called **florigen,** was being transmitted from the leaves of the induced plants to the flower bud meristem tissue of the noninduced plants, where the hormone stimulated the formation and opening of flowers. Even if produced in a long-day plant, the supposed hormone can induce flowering in a short-day plant, and vice versa.

Despite these grafting experiments, scientists have yet to purify or chemically characterize florigen after half a century. The fact that researchers can increase the rate at which florigen moves out of the leaves suggests that a signal does appear to pass from one place to another. The current view is that not one but a number of compounds, including perhaps gibberellins, may interact to promote flowering.

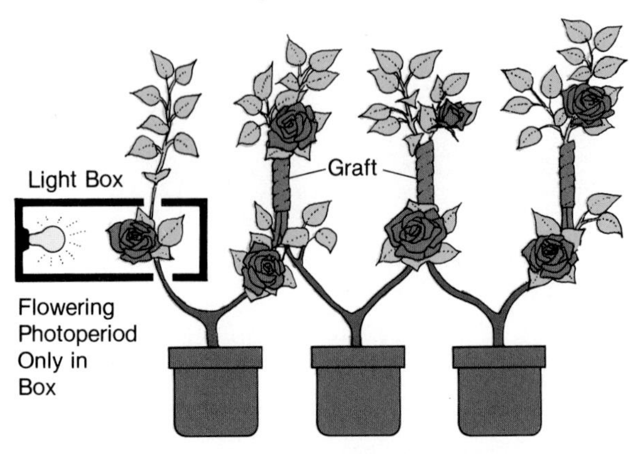

Figure 28-12 FLORIGEN: THE MYSTERIOUS STIMULUS FOR FLOWERING.
Florigen can move through a series of grafts to induce flowering in plants that have not been exposed to the appropriate photoperiod. Although the exact nature of florigen is not known, this experiment, performed by M. H. Chailakhian, showed that the flowering stimulus can travel from a leaf given the appropriate photoperiod (at left) through a series of grafted plants, to stimulate flowering in all of the plants.

LOOKING AHEAD

The reproductive behavior, growth, and internal processes of plants are complex and well coordinated. Much of that coordination is based on hormones, natural internal cycles, and physical factors—such as water, temperature, and light. A range of hormones—triggered and distributed in response to directional light, gravity, day or night length, and other environmental factors—helps allow plants to produce new leaves, shoots, roots, or flowers, sometimes over decades or centuries. Our next part explores analogous animal processes that coordinate and control daily and seasonal survival.

SUMMARY

1. Plant *hormones* are substances produced in extremely low concentrations in one part of a plant that bring about effects in another part. Most plant tissue can both make and respond to each of the hormones.

2. *Auxins* are generally produced in the plant's growing apex and regulate plant *tropisms*, or movements, by stimulating cell elongation. Auxins influence *phototropism* and *gravitropism*.

3. Auxins may also inhibit growth, producing such effects as *apical dominance*, the tendency of the main shoot of a plant to predominate over all the others.

4. Auxins stimulate cell elongation by promoting the secretion of hydrogen ions from the cell, lowering the pH in the cell wall, and facilitating enzymatic weakening of the cell walls.

5. *Gibberellins* are a class of more than 70 growth-promoting hormones that cause cell and stem elongation, bolting in the rosette stage of long-day plants, and increase in fruit size.

6. *Cytokinins* are a class of hormones that, in combination with auxins, promote cell division. Cytokinins also delay cell aging, or senescence, possibly by promoting nutrient transport to cells.

7. *Abscisic acid* is a hormone that suppresses plant growth, helps maintain dormancy of seeds and tissues, and speeds leaf senescence and, in some plants, *abscission*.

8. *Ethylene* accelerates fruit ripening by stimulating tissue respiration and plays a role in growth patterns by inhibiting auxin transport.

9. The five types of hormones interact in complex ways to control leaf fall, seed germination, and other life cycle events. In general, two or more positive hormonal signals are required for growth, while one negative signal can prevent growth.

10. Induction of flowering depends on *photoperiod*, or day length. Plants detect day length with a pigment-protein complex called *phytochrome*.

11. Flowering is most likely controlled by several chemical—perhaps hormonal—factors. In addition to the hypothetical hormone *florigen*, plant age, water availability, temperature, and season also seem to be involved.

KEY TERMS

abscisic acid
abscission
abscission layer
apical dominance
auxin
callus tissue
cytokinin
ethylene
florigen
gibberellin
gravitropism
hormone
photoperiod
phototropism
phytochrome
tropism

QUESTIONS

1. Construct a table that lists the five major types of plant hormones, and include the primary modes of action at the cell level and actions on the entire plant.

2. How do we know that certain of auxins' effects occur outside cells, whereas gibberellins' effects occur within the nuclei of aleurone cells in seeds?

3. Why would abscisic acid be better named *dormin*?

4. Explain the basic mechanisms of photoperiodism in flowering.

5. How is each form of phytochrome converted into the other by light, and what happens when that occurs?

6. Why do researchers suspect that florigen is not gaseous? That ethylene is?

7. Describe how a leaf separates from the stem. What hormones may be involved?

8. Why is the temperature compensation of the plant photoperiod system an important adaptation for most plants on our planet?

9. What is the generally accepted principle of hormone interaction in plants?

For additional readings related to topics in this chapter, see Appendix C.

Part FIVE

ANIMAL BIOLOGY

On the lava rock shores of an island in the Galápagos Archipelago, a flightless cormorant basks in the sun near a marine iguana and a Sally lightfoot crab.

29 The Circulatory and Transport Systems

I began to think within myself whether it [the blood] might have a sort of motion, as it were, in a circle.

William Harvey, *The Circulation of the Blood* (1628)

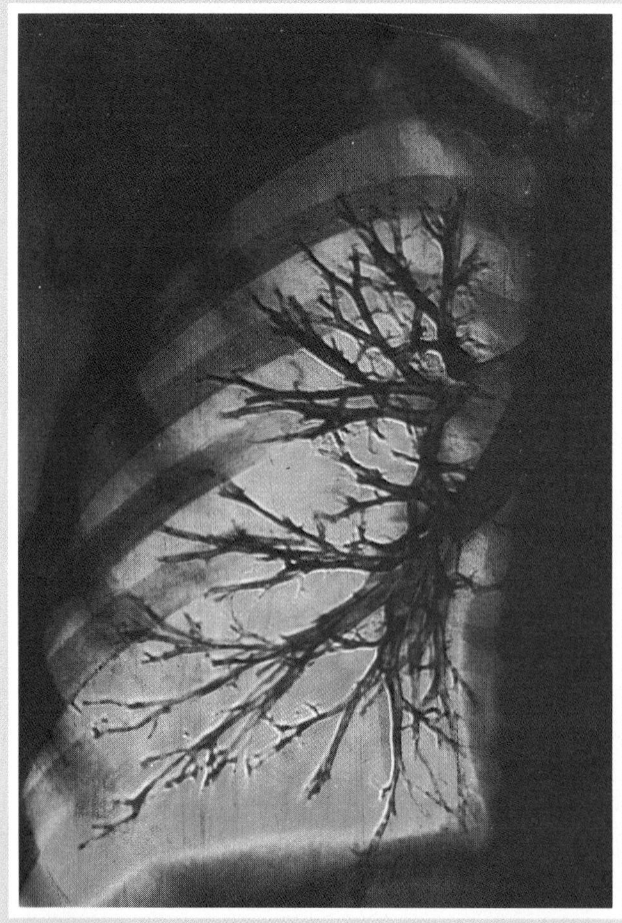

The branching "tree" of arteries in the right human lung as revealed by x-ray photography after injection of a radiopaque dye.

Like all animals, a human is a multicellular heterotroph—and a large one, at that. The adult human body contains about 75 trillion cells, and each of them has the same basic needs as an independent, free-living aerobic cell: extracting energy from the environment; receiving a supply of nutrients, oxygen, and other materials; ridding itself of certain organic molecules, CO_2, and other wastes; and maintaining *homeostasis*, a constant internal state, despite environmental fluctuations.

Because a human liver cell, let's say, is several centimeters deep in the body's interior, it is utterly dependent on a transport system capable of bringing in needed materials and carrying away wastes. Indeed, nearly every human cell lies very close to a branch of an elaborate transport network called the **circulatory system.** This system includes (1) an energetic, muscular pump, the *heart*; (2) a set of hollow vessels, the *arteries* and *veins*, which ramify throughout the body; and (3) a "microcircuitry" of extremely fine vessels, the *capillaries*, interwoven in virtually every body tissue; and (4) the *blood*, or transport fluid. Other kinds of transport systems are common in the animal kingdom, too, but all are analogous strategies.

Transport systems are an appropriate place to begin our series of chapters on animal anatomy and physiology, the study of animal structures and how they work. We will encounter transport systems again and again as we discuss how animals carry out certain types of defense against disease, distribute oxygen, digest food, excrete wastes, deliver hormones, move their bodies, and regulate their internal temperature and physiological environment.

In this chapter, we consider:

- Some basic principles of animal physiology
- Two solutions to the problems of material transport, simple diffusion and a circulatory system
- The structures of the vertebrate's circulatory, or cardiovascular, system and how they function
- How an animal's blood flow is controlled during activity or rest
- The fluid of life—blood—and another important fluid—lymph

SOME BASIC PRINCIPLES OF ANIMAL PHYSIOLOGY

As Chapter 24 explained, the animal kingdom is amazingly diverse; animals live in an endless variety of habitats and are subjected to wide variations in humidity, light, temperature, nutrients, pH, and other physical factors. In the face of such fluctuations and often harsh external conditions, an animal's central physiological problem is maintaining **homeostasis,** a steady yet highly dynamic internal state. This steady state requires the maintenance, within certain upper and lower boundaries (depending on species and individual differences), of internal temperature, and of pH, ion, nutrient, and gas (O_2, CO_2) levels.

The chapters in this part of the book focus on the many physiological systems responsible for homeostasis. The circulatory or other transport system supplies the material needs of every cell. The immune system defends the body against attack. The respiratory system brings oxygen into the body and carries away carbon dioxide. The digestive system supplies nutrients and removes leftover organic wastes. The excretory system removes nitrogenous wastes and helps balance water, pH, and ionic levels. The musculoskeletal system supports the body (in most animal species) and allows it to move. And the hormonal and nervous systems help coordinate animal behavior and all the activities of homeostasis.

Just as in plants, an animal's physiological functions depend on and are closely intertwined with its anatomical structures. Chapters 29 through 37 present those structures in a range of simple and complex animals, focusing most attention on humans and other vertebrates. Again, as in plants, those structures rest on a hierarchy of organization from molecules to cells, tissues, organs, organ systems, and, ultimately, whole organisms. While we will encounter many unique organ types in the different physiological systems, it is useful to note before starting that they tend to be composed of the four general kinds of tissues: epithelial, connective, nervous, and muscle.

Epithelial tissue is a single or multilayered sheet of cells bounded on one side by attachment to a ruglike basement membrane (Figure 29-1a). Epithelial tissue usually forms an interface between two dissimilar environments, such as the intestinal wall and the gut cavity, or the bladder wall and the bladder cavity. Epithelial cells contribute to an organism's homeostasis by regulating exchanges between such adjacent environments, and they can be differentiated for protection (as in the epidermis of the skin); for absorption (as in the lining of the intestine); for secretion (as in the mammary glands); or for sensory function (as in the ear or eye).

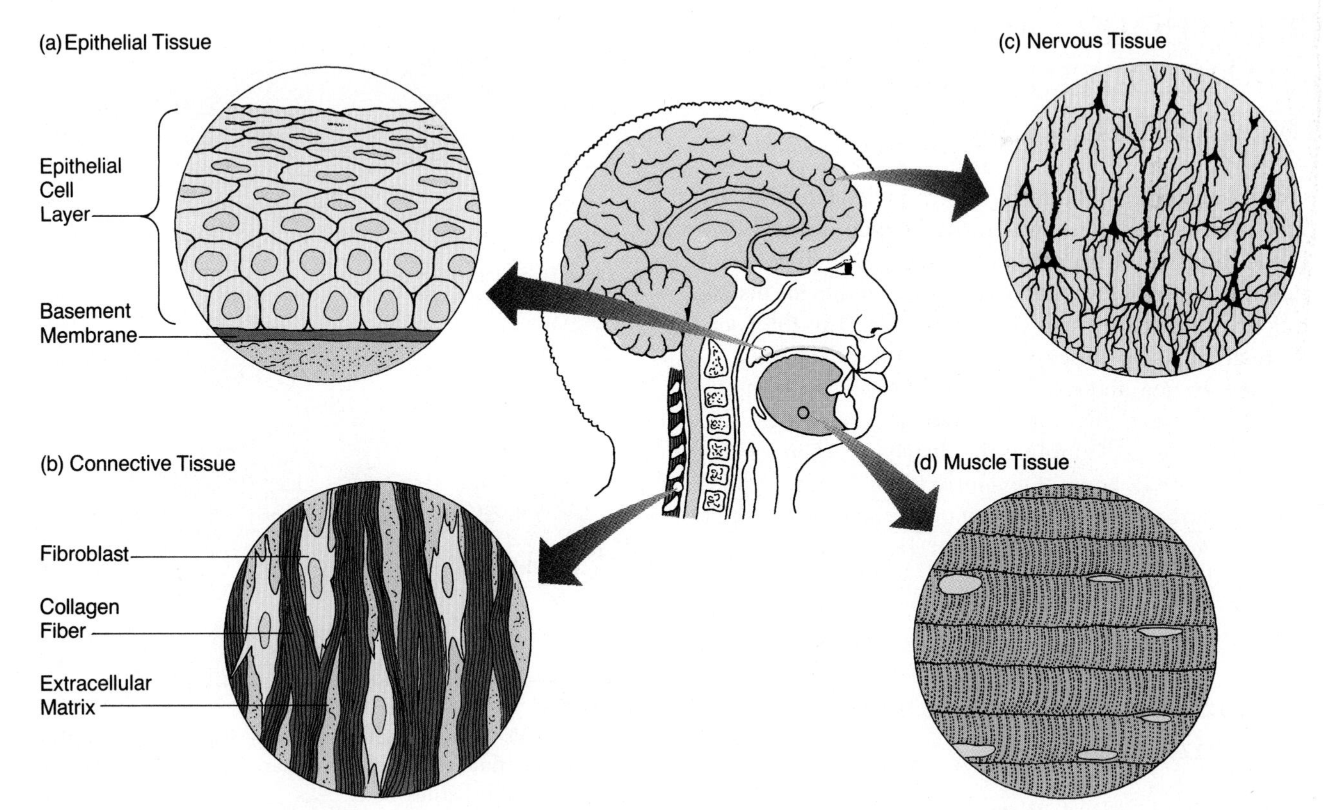

Figure 29-1 ANIMAL TISSUES.
(a) Epithelial tissue forms sheets that act as interfaces between dissimilar environments. (b) Connective tissue binds or supports other tissues. (c) Nervous tissue relays messages and helps integrate and coordinate body functioning. (d) Muscle tissue contracts, allowing cells, organs, or whole animal to move.

Connective tissue lies just below the basement membrane to which epithelia are attached; it binds other tissues together and supports flexible body parts (Figure 29-1b). The common connective tissue cells called *fibroblasts* found in skin and internal organs like the lungs and kidneys produce an extracellular matrix of nonliving materials such as the proteins collagen and elastin. Other connective tissue types include adipose (fat), which provides insulation and energy storage; cartilage, tendons, and bones, which provide structural support for the body and (along with muscles) allow movements; and tissue fluids and blood, which transport ions, molecules, and dissolved gases. (We cover two other tissue types, nervous tissue and muscle tissue (Figure 29-1c and d), in detail in Chapters 34 and 37.)

In this chapter, we consider the connective and epithelial tissues of the *vascular system* (the heart and blood vessels), which is more responsible for the homeostasis of vertebrate animals than any other system. This is because blood carries materials to and from virtually every region of the body and serves as a means of monitoring overall levels of activity and health. Let us begin, however, with the general problems and principles of animal transport systems.

STRATEGIES FOR TRANSPORTING MATERIALS IN ANIMALS

The transport of vital materials in single-celled organisms, such as monerans and protists, depends largely on simple diffusion of materials from areas of high concentration to those of low. Diffusion also accounts for distribution within the cytoplasm of most cells. But diffusion is a relatively slow process: In the absence of stirring, it can take oxygen about 3 hours to diffuse 1 cm through water, and since the rate for diffusion varies as the square of distance, doubling the distance requires fourfold longer, or 12 hours. Obviously, diffusion can suffice as a transport mechanism in living organisms only when the distances that materials must be carried are extremely short—a fact with consequences for both small animals and large.

Relying on Diffusion: Limitations on Body Size and Shape

Multicellular animals that rely solely on diffusion for transport must be porous like a sponge or extremely flat and thin like a flatworm. In Porifera (sponges), for example, channels carry seawater or fresh water throughout the body (see Figure 24-2). Internal cells not directly bathed by the water are rarely more than 1 mm from it, so materials diffuse through a layer at most two to three cells thick. In the flatworm *Planaria*, the body is so flat and thin that every cell is close enough to either the body surface or the fluid-filled gut cavity to receive and exchange materials by simple diffusion (see Figure 24-12). Similarly, some cnidarians have such a thin, gossamer body wall that despite an often large body size, diffusion alone can also supply their needs.

Circulatory Systems

All multicellular animals that are not porous, flat and thin, or very thin-walled have a separate system for circulation that moves masses of fluid through the organism in a circular flow called **bulk flow.** Circulatory systems are either open or closed. **Open circulatory systems** are found in insects, certain mollusks, and several other kinds of invertebrates. In the open system of an insect, for example, a fluid such as hemolymph flows anteriorly through the body in a major vessel or vessels, drains

(a) Open Circulatory System

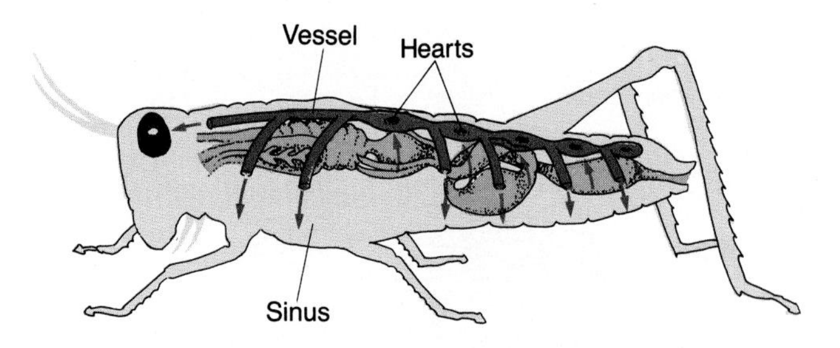

(b) Closed Circulatory System

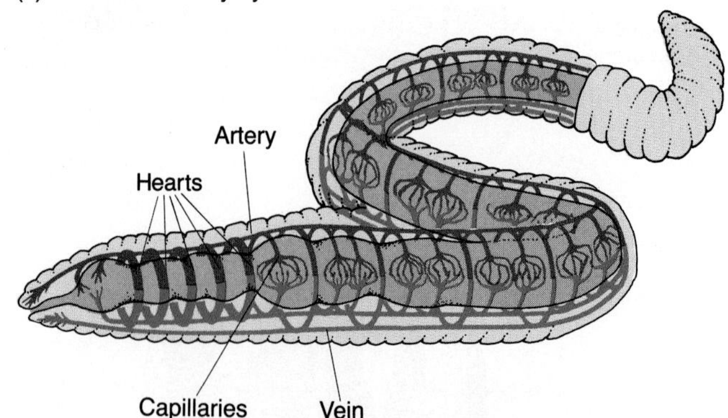

Figure 29-2 OPEN AND CLOSED CIRCULATORY SYSTEMS.
(a) In animals with an open circulatory system, such as the grasshopper, fluid pumped by a heart (or a series of hearts) exits the vessel, bathes body cells, and drains back into the vessel. (b) In animals with a closed circulatory system, such as the earthworm, the blood is contained within a system of vessels and is pumped through the system by one or more hearts.

through slits in the vessel wall into large open spaces called *sinuses,* and seeps in a posterior direction, bathing tissues or organs directly as it goes (Figure 29-2a). The fluid in the sinuses eventually reenters the vessel through openings in its wall, and the circuit is completed.

In a **closed circulatory system,** such as in earthworms, squid, octopuses, and all vertebrates, blood moves through a continuous set of interconnected vessels: the arteries, capillaries, and veins (Figure 29-2b). **Arteries** are relatively large vessels that carry blood away from the heart. The smallest arteries are continuous with **capillaries,** tiny, thin-walled vessels interwoven throughout the various body tissues. Nutrients, oxygen, and ions carried in the blood diffuse through the walls of the capillaries into the extracellular fluid that bathes the outside of tissue cells, and from there, those substances enter the cells. Metabolic wastes (carbon dioxide, lactic acid, ions, etc.) leave the cells, enter the extracellular fluid, and diffuse from it into the capillaries. Capillaries are continuous with the smallest **veins,** vessels that carry the blood toward the heart. Veins, in turn, feed the blood through one or more pumps back into the arteries, completing the circuit.

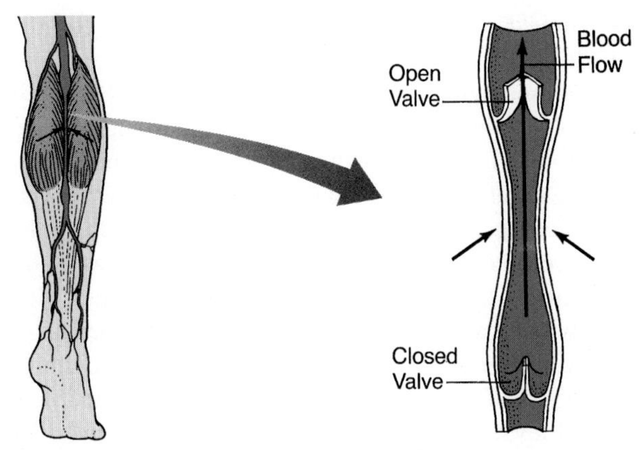

Figure 29-3 VALVES.
Valves in veins ensure that blood flows toward the heart by preventing backflow. The movement of venous blood is caused by the heart's pumping action as well as by compression of veins due to the contractions of nearby skeletal muscles (colored arrows). Here, as the calf muscles contract, the vein is compressed; the upper valve remains open while back pressure closes the lower valve, thereby preventing flow toward the foot.

Why Fluids Flow in One Direction in Circulatory Systems

All circulatory systems share certain features (including contracting muscles, pressurized fluid, and valves) that cause the fluid to flow in a circuit rather than in random directions.

First, the walls of some vessels contain muscle tissue that contracts sequentially in *peristaltic waves* to sweep fluid along.

Second, specialized regions of cardiac muscle tissue, or **hearts,** act as pumps. Vertebrates have one large heart, whereas insects and other invertebrates can have several small hearts that act like booster pumps for the circulating fluid. The rhythmic contraction of these pumps and the peristaltic waves of contraction in the walls of vessels push against the volume of circulating fluid and create pressure, the **blood pressure.** In a closed system, the pressure is often quite high, and blood flow is rapid, and this serves the needs of rapidly metabolizing cells. In an open system, in contrast, blood fluids in the large open sinus have low pressure and move slowly, and the animals tend to have sluggish metabolism. The fast-moving insects are an exception, and their open circulatory system is not responsible for gas transport. Instead, oxygen is supplied to an insect's tissue cells through a unique system of hollow air pipes, the *tracheae,* which branch throughout the body and deliver large amounts of oxygen to actively metabolizing tissues (details in Chapter 31).

Body movements help massage vessels and "stir" blood fluids. And the flow of fluid into and out of local regions of the circulatory system is controlled by tiny muscles—*sphincters*—that act as floodgates to shut off or open up vessels, depending on the need for materials in a local tissue. Sphincter muscles encircle vessels and decrease their diameter as they contract.

Finally, blood or other circulating fluid flows in a one-way circuit because *valves,* flaps of tissue that extend into the vessels from the inside walls, close when fluid begins to flow in the reverse direction (Figure 29-3). Valves prevent backflow in both vessels and the heart.

Circulating Fluids and the Diffusion of Materials

Animals with circulatory systems differ in how materials are carried to their tissue cells. In an animal with an open circulatory system, the **hemolymph** bathes the outside of tissue cells and forms the animal's *extracellular fluid.* This colorless extracellular fluid is roughly similar in composition to the fluid within individual cells (the *intracellular fluid*) and consists of water with dissolved nutrients, ions, gases, and some macromolecules. Many of these substances move from the extracellular fluid into tissue cells whenever their concentrations are higher outside the cells than inside—a result of diffusion down concentration gradients (see Chapter 5). In turn, substances move out of the intracellular fluid, across the plasma membrane, and into the extracellular fluid whenever their concentrations are higher inside the cell than outside.

An animal with a closed circulatory system has a third fluid, blood, in addition to its intracellular and extracellular fluids. Blood flows through the closed network of arteries, capillaries, and veins and consists of two parts: a watery component, the *plasma*, which contains organic molecules, ions, and dissolved gases; as well as a solid component, the *cells*, among which are the red blood cells, which carry most of the oxygen in the circulatory system. Materials in the blood diffuse first through the capillary walls into the extracellular fluid, and from there, they may diffuse into the interiors of cells. For instance, oxygen carried by red blood cells diffuses through the thin capillary walls into the extracellular fluid since the level of oxygen in the extracellular fluid is lower than in the blood. From there, the oxygen diffuses passively into tissue cells because the oxygen concentration within those cells is lower than in the extracellular fluid.

Figure 29-4 summarizes the delivery of materials to and from animal cells, using the transport of oxygen and carbon dioxide as examples. Whether a system is open or closed, net diffusion occurs when the concentrations of substances differ within adjacent fluids. Moreover, many types of molecules cannot enter or leave the tissue cells unless they are actively pumped across the plasma membrane by means of certain enzymes.

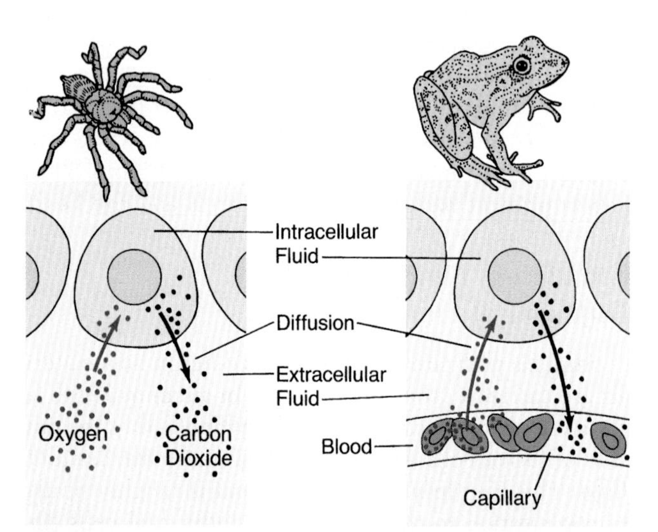

Figure 29-4 THE EXCHANGE OF MATERIALS TO AND FROM ANIMAL CELLS.
(a) In an open circulatory system, oxygen diffuses from the extracellular fluid into the cells' intracellular fluid, and carbon dioxide diffuses in the opposite direction. (b) In a closed circulatory system, oxygen carried along in the bloodstream diffuses out of the capillary, into the extracellular fluid, then into a tissue cell, whereas carbon dioxide moves in the reverse direction.

THE VERTEBRATE CIRCULATORY SYSTEM

In the 360 years since English physician William Harvey discovered that blood circulates rather than ebbs and flows from the heart, physiologists have learned a great deal about the structures and functions of the vertebrate circulatory system. We will first study the heart—the muscular pump behind the circulatory flow—and then consider the circuit itself—the 60,000 miles of arteries, capillaries, and veins that pervade the body.

The Heart

William Harvey's experiments proved that the heart of a land vertebrate is an energetic, fist-sized pump with thick muscular walls that contract forcefully, expelling blood into the arteries. Harvey also showed that mammalian and avian hearts are double pumps, divided in half longitudinally (Figure 29-5), and that the two circuits in the closed circulatory systems of birds and mammals both begin and end in the heart. Blood moves from the right side of the heart through the lungs and back to the left side of the heart (one circuit) and then from the left side of the heart to the rest of the body, returning once more to the heart's right side (second circuit).

The evolution of the heart helps us understand this double construction. In *Amphioxus* (see Chapter 24), the cephalochordate relative of the vertebrates, the heart is merely a specialized portion of a single major blood vessel, with waves of contraction "kneading" the blood fluid along. Many vertebrate fish species have the same sort of contractile vessel, but it is divided into four chambers that come one after the other in posterior-to-anterior sequence: the *sinus venosus*, the *atrium*, the *ventricle*, and the *conus arteriosus* (Figure 29-6a). Evidence indicates that during the course of evolution, this basic four-chambered heart arose as a means to elevate blood pressure in a stepwise fashion. As **Starling's law** explains, the more the cardiac muscle is stretched, the more vigorously it responds and the larger the volume of blood that can be pumped per contraction. Thus, stretched heart chambers can supply more blood to an active animal.

In land vertebrates, the four basic chambers seen in the fish heart have been reduced to two—the atrium and the ventricle—by the loss of the sinus venosus and the conus arteriosus. The atrium and ventricle are split in half to varying degrees. This separation into right and left heart chambers permits blood to flow through two separate circuits: the **pulmonary,** or **lung, circulation** (right heart → lungs → left heart), where the blood picks up O_2 and dumps CO_2, and the **systemic,** or **body, circu-**

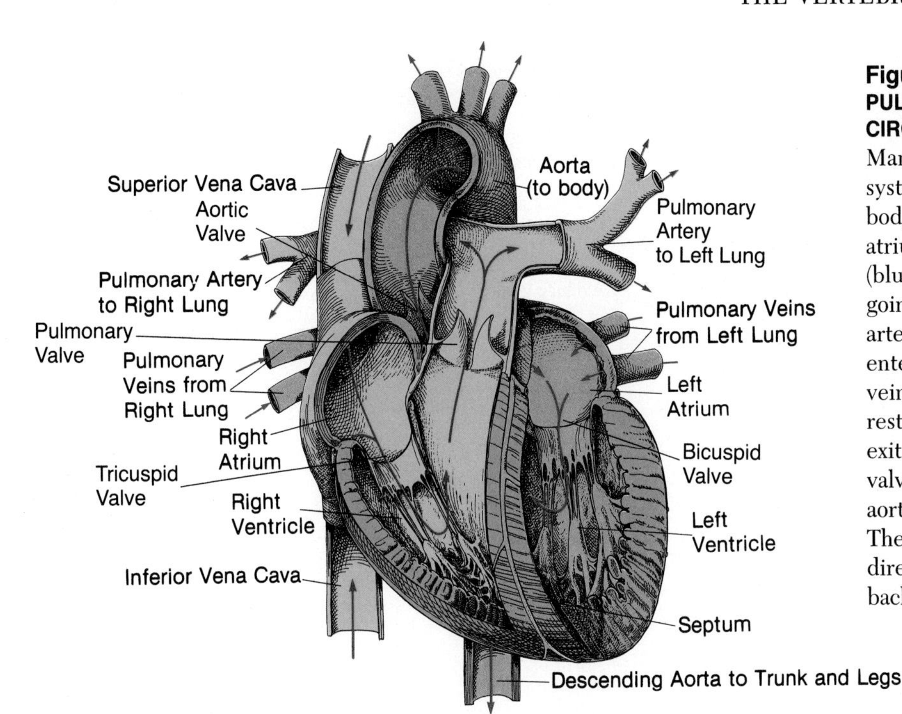

Figure 29-5 THE HUMAN HEART: PULMONARY CIRCULATION AND SYSTEMIC CIRCULATION.
Mammals have a "double" circulatory system. The veins that carry blood from the body tissues to the heart enter the right atrium. The venous, deoxygenated blood (blue) passes into the right ventricle before going to the lungs through the pulmonary arteries. The oxygenated blood (red) then enters the left atrium via the pulmonary veins. The blood begins its circulation to the rest of the body through the aorta, which exits from the left ventricle. Note the four valves between human heart chambers (the aortic, pulmonary, tricuspid, and bicuspid). These valves assure the proper forward direction of blood flow and prevent a backward flow.

lation (left heart → body → right heart), where the oxygen is delivered to body tissues and CO_2 is collected.

Adult amphibians, which carry out gas exchange through both the skin and the lungs, have a heart with an atrium and a ventricle (Figure 29-6b). The amphibian atrium is split in half, but the ventricle is not. This arrangement allows a small proportion of blood from the right atrium (which carries a low level of oxygen) to mix with blood from the left atrium (which carries a high level of oxygen) when the two blood supplies pass through the ventricle. Blood is routed so that the most oxygen-rich portion goes to the brain, but even that is somewhat lowered in oxygen content because of the mixing in the ventricle. In reptiles, which (except for crocodiles), depend solely on lungs for gas exchange, the splitting of the heart's two chambers is carried a step further: The ventricle is partially separated so that deoxygenated blood in the right side of the heart remains

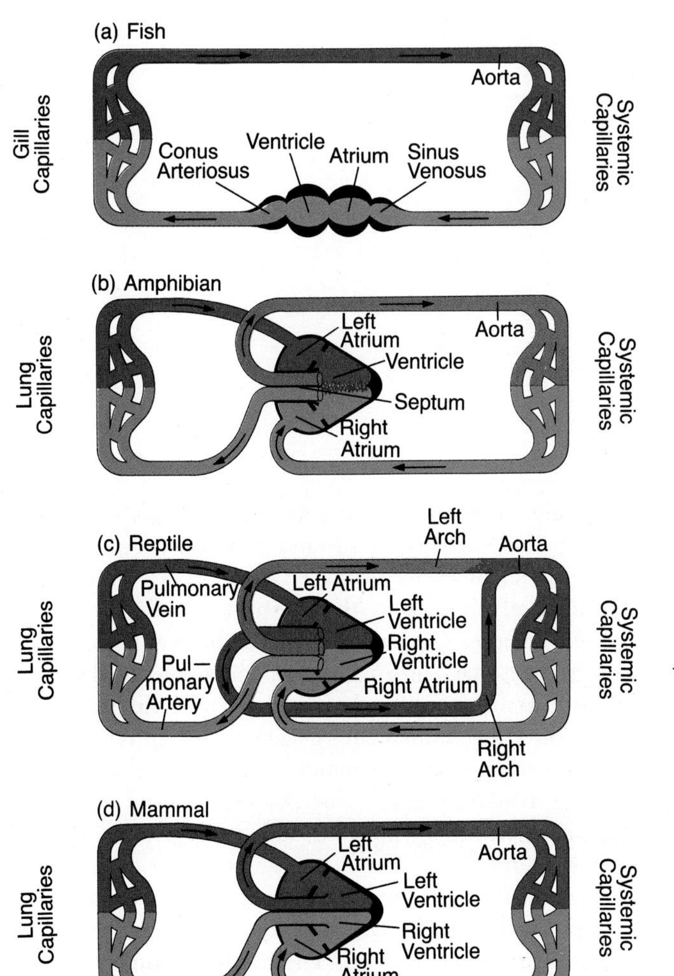

◀ **Figure 29-6 EVOLUTION OF THE HEART AND CIRCULATORY SYSTEM IN VERTEBRATES.**
(a) Many fish have four separate heart chambers. (b) Blood is mixed in the single amphibian ventricle because it receives both oxygenated blood from the lungs via the left atrium and deoxygenated blood from the body tissues via the right atrium. (c) In reptiles, the septum between the ventricles is usually incomplete, and the base of the aorta is split into two vessels that lead to the right and left arches of the aorta. (d) In mammals and birds, the ventricle is fully separated into two halves, and only one aorta exits the heart; the drawing is of a mammal.

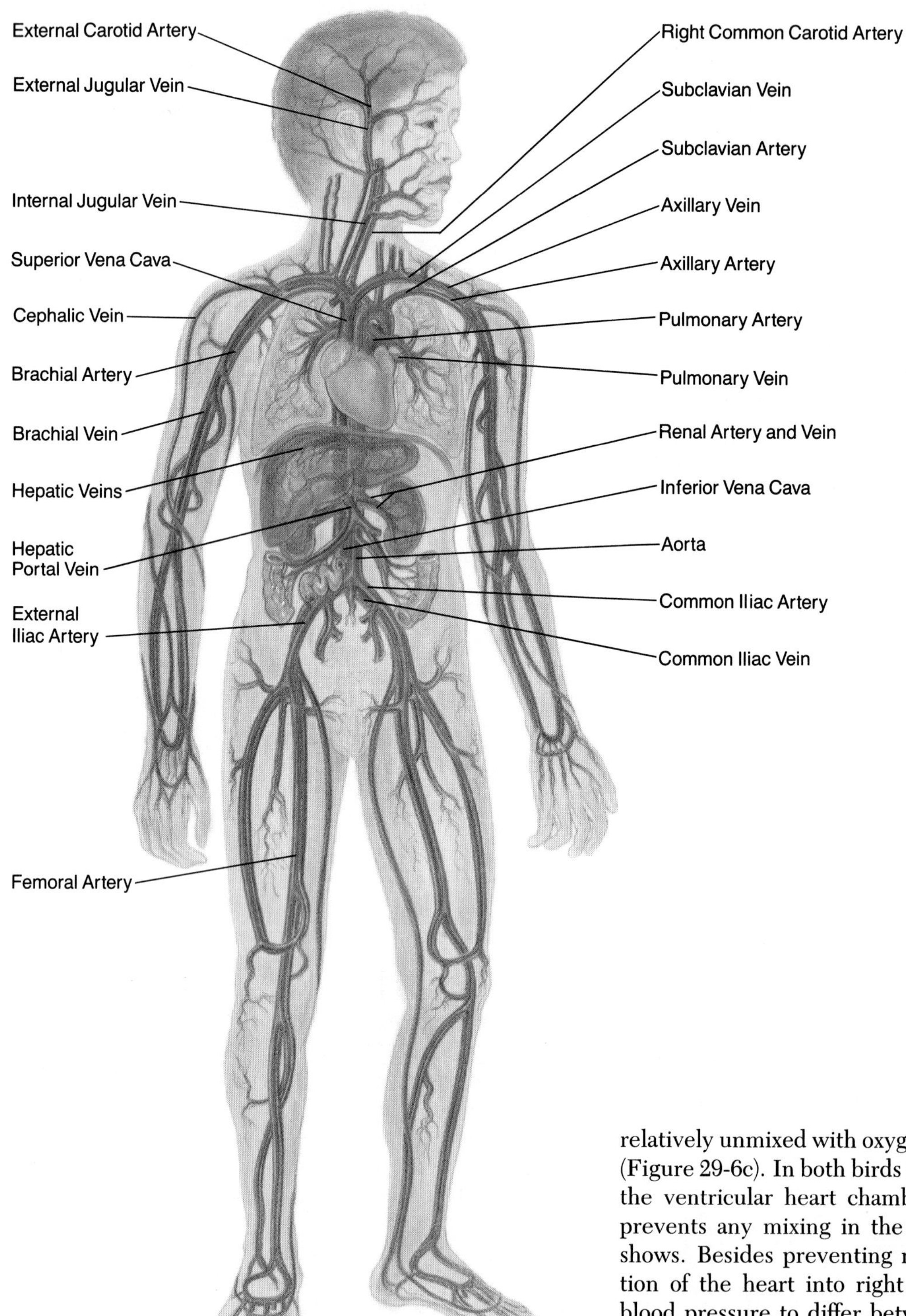

Figure 29-7 THE HUMAN (MAMMALIAN) CIRCULATORY SYSTEM, SHOWING THE MAJOR ARTERIES AND VEINS. The pulmonary circulation and the systemic circulation consist of major arteries and veins leading to and returning from all parts of the body. The myriad smaller arteries, arterioles, capillaries, venules, and smaller veins are linked within the tissues and organs to constitute a closed circulatory system.

relatively unmixed with oxygenated blood in the left side (Figure 29-6c). In both birds and mammals, separation of the ventricular heart chamber is complete: A septum prevents any mixing in the ventricle, as Figure 29-6d shows. Besides preventing mixing, the complete partition of the heart into right and left halves allows the blood pressure to differ between the two sides.

No bird or mammal could sustain its high rate of metabolism, high oxygen demand, and high level of activity without a high-pressure, fast-flowing internal transport system in which oxygenated and deoxygenated bloods are never mixed.

The result of the evolutionary process we have described is a marvelous organ that pumps blood throughout an organism's lifetime. The adult human heart, for example, pumps a teacupful (7 fl oz) of blood with each three beats, and every 15 minutes, it pumps enough blood to fill a 20-gallon gasoline tank. In a day, the contraction of this fist-sized organ could fill 70 barrels, and over a lifetime, *18 million barrels.* In a day, an adult's 5 L of blood circuits the body 1,440 times.

Blood Circuits

Figure 29-7 shows the full circulatory pathway that the blood takes through the human body. Blood carried to the body tissues by the arteries delivers oxygen and nutrients to the capillary beds. Then bluish blood with low oxygen tension moves into tiny veins, continues on into larger ones, and eventually reaches the two largest veins in the body: the *posterior vena cava,* which carries blood from the legs and most of the body, and the *anterior vena cava,* which carries blood from the head, neck, and arms. In humans, these are called the *inferior vena cava* and the *superior vena cava,* respectively. The two venae cavae empty into the right atrium of the heart. The muscular walls of this chamber contract, and the blood is forced from the right atrium into the right ventricle (review Figure 29-5). Contraction of this chamber, in turn, sends the oxygen-depleted venous blood through the pulmonary arteries to the lungs. In the lungs, arteries branch to finer and finer arterioles, and the blood eventually passes through capillary beds, where carbon dioxide is released and oxygen is taken up.

The newly oxygenated blood then passes into the pulmonary veins and is carried to the left atrium of the heart. The contraction of the left atrium sends blood past a valve into the left ventricle, from which it is pumped into the **aorta.** This largest artery in the body feeds into a number of major arteries, including the *coronary arteries,* which serve the heart muscle itself. Other major arteries branching from the aorta supply the head, shoulders, and arms, the gut, the liver and pancreas, the kidneys, the pelvic area, and the legs. Each major artery gives rise to a treelike array of progressively smaller arteries until finally the arterioles and capillaries are reached. In these smallest vessels, blood is carried very close to every cell in every tissue in the body. After passing through the capillary beds, blood passes into the "mirror image" of the arterial system—the treelike array of venules (tiny veins), larger veins, and so on, until it arrives once again where it started—the venae cavae.

One aspect of blood circuitry merits special mention: *Portal vessels* carry blood from one capillary bed to another. For example, a major vessel, the *hepatic portal vein,* connects the capillary beds in the wall of the intestine—the site of nutrient absorption—directly to capillary beds in the liver. Blood leaves the liver via the *hepatic veins,* which join the posterior (inferior) vena cava en route to the heart.

Structure of Arteries and Veins

The structure of arteries and veins allows them to function as important reservoirs for blood pressure and blood volume. Arteries have relatively thick walls, composed of an inner lining called the *endothelium,* a middle layer of smooth muscle cells, and an outer layer of connective tissue, which contains collagen and springy elastin fibers (Figure 29-8). In general, the closer an artery is to the heart, the more muscular and elastic are its walls.

The largest artery in the body, the aorta, illustrates how muscles and elastin fibers work together to transport blood. Each time the heart's muscular left ventricle contracts, a pulse of oxygenated blood is driven at high pressure along the aorta. This pulse of blood causes the aortic walls to expand slightly, and potential energy is therefore stored in the expanded walls. The elastic walls of the aorta then rebound, contracting inward, exerting force on the blood and helping to propel it. Overall, the expansion and contraction of the major arteries take much of the high-pressure "shock" out of the strong pulses of blood that leave the heart and help propel blood through the body.

Like arteries, veins have a three-layered wall of endothelial cells, smooth muscle, and connective tissue (see Figure 29-8). With less smooth muscle, vein walls are much more flexible than arterial walls. This flexibility

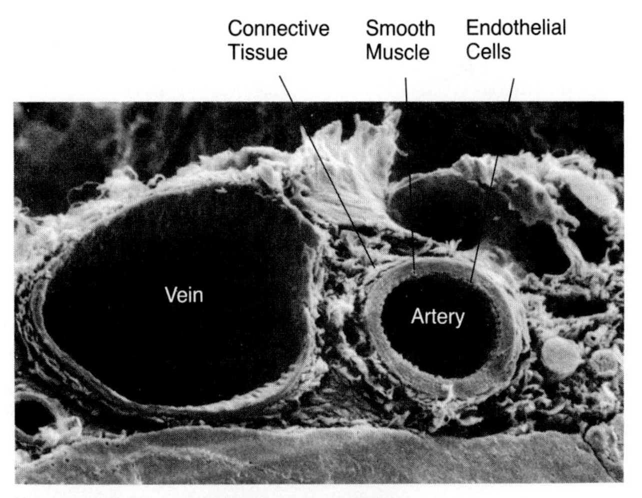

Figure 29-8 ARTERIES AND VEINS.
Scanning electron micrograph (magnified about 170×) of a vein (left) and its companion artery (right) in cross section. Arterial and venous walls consist of three layers: endothelium, smooth muscle, and connective tissue. In general, veins have thinner walls and larger diameters than arteries and are much less rigid. From *Tissues and Organs: A Text-Atlas of Scanning Electron Microscopy* by Richard G. Kessel and Randy H. Kardon, W. H. Freeman and Company © 1979.

RISK FACTORS IN OUR NATION'S BIGGEST KILLER

America's biggest killer is *atherosclerosis,* a kind of "hardening of the arteries," a buildup of yellowish fatty deposits called *plaque* inside the arteries, beginning in late childhood and continuing throughout life. Like corrosion that builds up in water pipes, plaque can accumulate until it seriously impedes the flow of blood (Figure A). Reduced blood flow, in turn, can lead to the formation of a blood clot that obstructs an artery completely—especially one in the heart or brain—and cause a heart attack or stroke. In the U.S., atherosclerosis accounts for half of all deaths each year and has inspired a great deal of research into risk factors and preventive measures.

Diet

In recent years, researchers have demonstrated that a diet high in cholesterol can increase one's risk for developing atherosclerosis. Within the bloodstream, cholesterol travels attached to carrier lipoproteins, two types of which are the HDLs (high-density lipoproteins) and the LDLs (low-density lipoproteins). Biologists sometimes picture HDLs as molecular vacuum cleaners that take up cholesterol from cells and plaques and ultimately deliver it to liver cells, where the fatty material is incorporated into bile (and where new cholesterol molecules are synthesized). People who exercise regularly; maintain ideal weight; eat unsaturated vegetable and fish oils; and limit foods such as coconut and palm oils, organ meats, shrimp and lobster, egg yolks, fatty meats, and rich dairy products tend to have higher levels of HDLs, less circulating cholesterol, lower LDLs, and, in turn, a lower risk of developing atherosclerosis. For those with elevated cholesterol levels in the blood, certain medicines can help. These include niacin and Lovastatin. The latter drug disrupts cholesterol synthesis by liver cells so that more cholesterol is pulled out of the blood by the liver and secreted as bile salts. The best policy, however, is still to minimize saturated fat intake and adopt other aspects of a healthy life-style.

Family History

A tendency to develop atherosclerosis can be inherited; if your mother, father, brother, or sister died of a heart attack or stroke before age 65, you are at increased risk.

Personal Habits

Both smoking and a sedentary life-style are risk factors for developing atherosclerosis. Among other effects, smoking is believed to cause damage to make the thin endothelial layer that lines the arteries susceptible to damage. It is at such points that plaque begins to form and then accumulates. Research on nonhuman primates also reveals that regular exercise helps maintain coronary arteries with larger diameters, regardless of diet.

Medical Conditions

Both high blood pressure and diabetes increase one's chances of developing atherosclerosis. As with smoking, these conditions can damage arteries and lead to sites of plaque formation.

Behavior

Psychologists and psychiatrists have found common behavior patterns in many victims of heart disease. The most often cited is the "type A personality," a constellation of behaviors and attitudes that includes competitiveness, aggressiveness, a sense of constant time pressure and heavy responsibilities, and sometimes hostility and vengefulness.

Considering all these risk factors, the most prudent course is obviously the balanced life-style so long advocated by health professionals: Reduce intake of saturated fat, exercise regularly, avoid smoking and heavy drinking, and reduce stress.

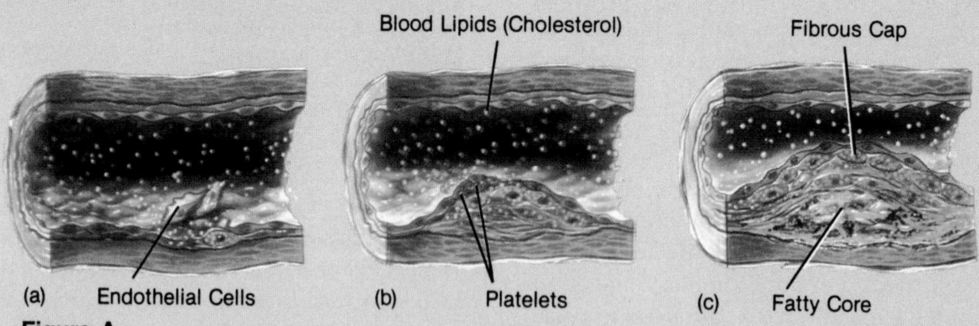

Figure A

(a) The start of a plaque deposit. (b) Blood platelets amass, cause a cap of cells to form above the plaque, and isolate the plaque within the arterial wall. This leads to the narrowing of the entire artery. (c) If the cap breaks and lipids from the fatty core combine with blood-clotting factors, a clot can form, block the artery, and lead to a stroke or heart attack.

allows veins to distend and increase greatly in volume, so that they act as a *volume reservoir*, holding up to 70 percent of the total blood in the body. The pliability of veins also permits the contraction of skeletal muscles that surround veins in the arms and legs to compress the veins and so to push blood along past the one-way valves toward the heart (see Figure 29-3).

Capillary Beds

Capillaries, the very narrowest extensions of the arteries, are about 1 mm long and about 3–10 μm or less in diameter—small enough so that red blood cells have to squeeze through one at a time (Figure 29-9). Large white blood cells often temporarily block their passage. Each capillary has extremely thin walls composed primarily of a single layer of endothelial cells. Beds of capillaries permeate virtually all tissues of the body, and it is here that the real action takes place—the transfer of materials carried by the blood to the extracellular fluid, and from there, to the individual body cells. Conversely, substances from cells pass to the extracellular fluid and then into the capillaries.

A human's pervasive capillaries, if joined end to end, would be a major part of the vascular system's length, some 60,000 miles or roughly 2.4 times around the earth! In most tissues, no cell is more than three or four cell diameters from a capillary. Capillary beds have a complex structure, including main channels and smaller channels in which blood flow is controlled by tiny sphincter smooth muscle cells that encircle the entrance to local networks of capillaries. These muscles act like miniature floodgates that are contracted about 80 percent of the time; they open and close intermittently to allow blood to flow through the capillary periodically.

A capillary's thin walls are the key to its function. Substances can diffuse in either direction between the blood and the extracellular fluid through a capillary's single layer of endothelial cells (Figure 29-10). This movement of substances can take one of four routes: Fluid can be ingested by tiny pinocytotic vesicles on one side of the endothelial cells and discharged on the other side (Figure 29-10, route 1); fluid and other substances can diffuse through narrow clefts between endothelial cells at sites lacking tight junctions (route 2); the inner and outer plasma membranes of individual endothelial cells

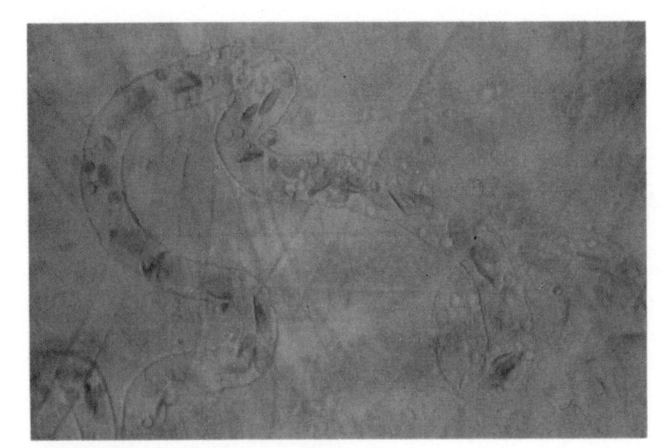

Figure 29-9 CAPILLARIES.
Red and white blood cells (magnified about 550×) are visible in a capillary in the webbed foot of a living frog.

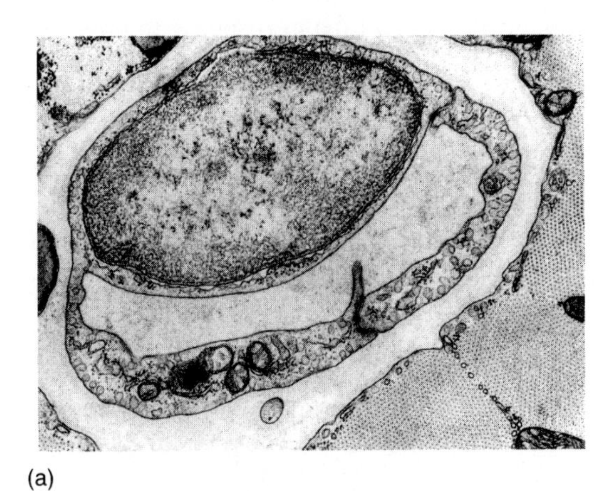

(a)

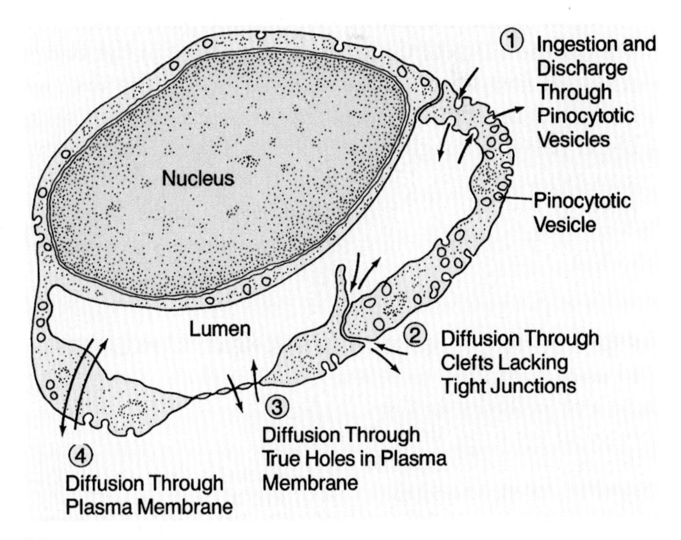

(b)

Figure 29-10 CAPILLARY WALL: KEY TO FUNCTION.
The capillary wall consists of a single layer of endothelial cells enclosing the narrow central cavity. (a) The electron micrograph (magnified about 14,000×) shows a capillary wall with an uninterrupted endothelium. (b) The contrasting type of capillary has an endothelium interrupted by minute pores closed only with a thin piece of basal lamina. The routes of transport—1, 2, 3, and 4—are described in the text.

may fuse at specific sites to form true holes through the cells, such as occurs in kidney capillaries, which filter blood or generate urine (route 3); and lipid-soluble materials can diffuse directly across the plasma membrane, through the cytoplasm, and out the other side (route 4). The route taken by a substance depends on its chemical properties and the capillary's architecture.

Blood Pressure

You have no doubt had your blood pressure taken with an inflatable cuff wrapped tightly around your upper arm. What was being measured was the force that the blood pushes outward against a large artery's wall. Since fluid flows through a tube in response to pressure differences between the two ends, blood moves through the circulatory system as a result of pressure differences between the two sides of the heart as well as within the capillary beds. The measurement of blood pressure is made in millimeters of mercury (mm Hg)—that is, the height to which a column of liquid mercury would rise if pushed on by the blood at that pressure. During rest, human blood is at its highest pressure, about 120 mm Hg, as it leaves the left ventricle during contraction of the heart, or **systole.** Blood in the heart is at its lowest pressure, about 80 mm Hg, during relaxation of the heart muscle, or **diastole.** Blood pressure of 120/80 is considered normal, or average, in human adults.

Blood pressure readings taken in one part of the body are not the same as those taken elsewhere in the body. As blood flows through the ever-narrowing arteries, the increased resistance lowers the pressure. By the time blood passes through the capillaries, the pressure averages perhaps 10 or 15 mm Hg. If it were much higher than 30 mm Hg, the fragile capillaries might burst.

In birds and mammals, blood pressure is substantially lower on the right side of the heart (which supplies the lungs) than on the left; it drops even lower—to perhaps 7 mm Hg—in the lungs, owing to the fantastic system of branching capillaries. This, plus a layer called surfactant (see Chapter 2), means that less fluid passes from the blood through the thin capillary walls into the lungs, and thus carbon dioxide and oxygen do not have to diffuse through thick layers of liquid during gas exchange.

Fluid Balance and the Diffusion of Materials

The movement of critical substances from the blood to the extracellular fluid and then to the cells, and vice versa, is based on two kinds of pressure: blood pressure and osmotic pressure. The net effect of blood and osmotic pressures is the movement of fluid *out* of the capillaries and into the surrounding extracellular fluid. Let us see why. Blood pressure at the arterial end of capillary beds, about 30 mm Hg, is higher than in the surrounding tissue; thus, water is driven out of the blood plasma through the narrow clefts between the endothelial cells of the capillary walls and into the extracellular fluid. At the same time, *osmosis* (see Chapter 5) leads to water movement in the opposite direction. This is because the blood contains proteins that give it an osmotic pressure of about 25 mm Hg. This is called **colloidal osmotic pressure** and is caused mainly by proteins, or colloids (Figure 29-11). Tissue fluids outside the capillaries lack most of these colloids and so have an osmotic pressure of only about 4 mm Hg. The difference ($25 - 4 = 21$ mm Hg) is the net osmotic pressure representing a net tendency for water to enter the blood at the high-pressure, or arterial, end of capillaries. This net pressure inward, however, is more than offset by the blood pressure pushing outward against the capillary walls—about 30 mm Hg. The result

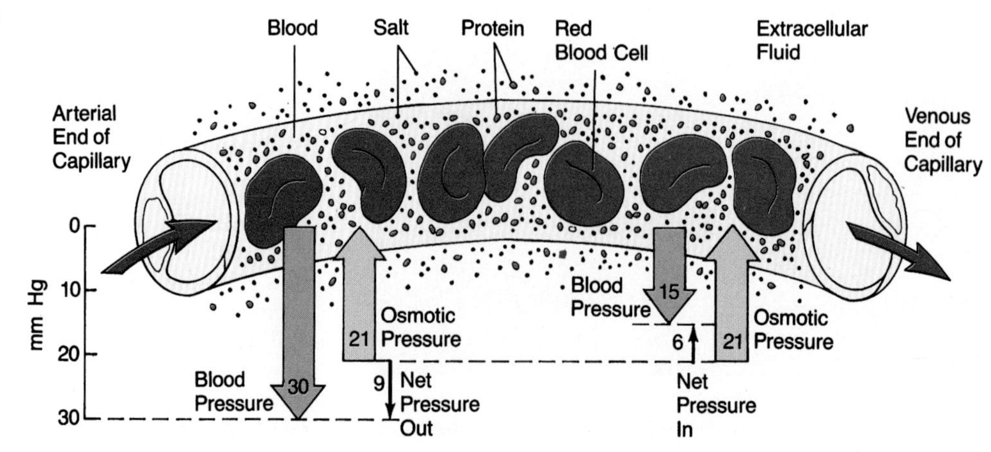

Figure 29-11 FLUID MOVEMENTS NEAR CAPILLARIES. In many capillaries, there is a net flow of fluid from the blood to the tissue fluid. This depends on the hydrostatic pressure of blood and the osmotic pressure differences between blood and extracellular fluid. Because the hydrostatic pressure falls as blood travels through a narrow capillary, water tends to exit at the high-pressure end and enter at the low-pressure end. Any net loss drains to the lymphatic system.

is a net outward pressure in the arterial-end capillaries equal to 30 − 21, or 9 mm Hg. It is this net outward pressure that causes the watery contents of the blood to move out through the capillary walls and into the extracellular fluid.

What happens, then, at the low-pressure, or venous, end of the same capillaries? The osmotic pressures here are the same as at the arterial end, giving rise to a net inward pressure of 21 mm Hg. However, resistance and the huge total surface area of the capillary walls lowers the blood pressure to about 15 mm Hg or lower by the time the blood reaches the venous end of a capillary. Subtracting 15 mm Hg outward pressure from 21 mm Hg inward pressure gives 6 mm Hg net inward pressure. As a result, fluid at the venous end of a capillary moves back into the capillary from the tissues.

Can you see the puzzle here? A net pressure of 9 mm Hg pushes fluid out at the high-pressure end of a capillary bed, whereas a net pressure of only 6 mm Hg pushes fluid back in at the low-pressure end. Since the two pressures are not in balance, there is a continuous net movement, every second of every day, of fluid out of the capillaries. So why doesn't all the fluid gradually leave the blood and accumulate as extracellular fluid? The reason is that fluid is absorbed in a special system of lymphatic collecting ducts (discussed later) and returned to the blood vessels.

REGULATION OF BLOOD FLOW

A strenuously exercising animal needs the ready transport of materials to and from the actively metabolizing muscle cells, while an animal at rest after a meal needs greater blood flow to the stomach and intestines. The rate at which materials are transported is closely tied to the speed of the flowing blood, which is based on vessel size, the frequency of heartbeats, and other factors.

Blood Flow and Vessel Size

The rate of blood flow at any spot in the circulatory system depends on the *total* cross-sectional area of the vessels at that spot. Just as water slows down on entering a large pipe and speeds up again on entering a smaller one, blood velocity depends on vessel size. The aorta and venae cavae have smaller cross-sectional areas in comparison with the combined cross sections of the arterioles, capillaries, and venules added together. This means that the aorta and venae cavae are "constricted" regions, whereas the total capillary cross-sectional area is like a wide place in a pipe or like a wide, shallow region of a river. The rate of blood flow is therefore fastest in the aorta and venae cavae and slowest in the capillaries.

In a given individual vessel, the velocity of blood flow varies directly as a function of vessel radius. Specifically, reduction of a vessel's radius by ½ reduces the rate of blood flow through it to 1/16 the original speed. Thus, in the body's biggest pipe—the aorta—blood flows about 330 mm per second, whereas it moves about 1 mm per second in an average capillary.

As blood leaves the capillaries and passes into the veins, the total cross-sectional area through which it flows is reduced, so the rate of blood flow accelerates even though the blood pressure remains low. As we mentioned earlier, return flow of blood to the heart is aided by the kneading action of body muscles on the veins and by one-way valves scattered along the length of many veins. You have no doubt heard of soldiers fainting after standing at attention too long. This is because inactivity allows large quantities of blood to collect in the leg veins; not enough blood is returned to the heart to be pumped to the brain, and without enough oxygen in the brain, the soldier grows dizzy and faints. The situation is usually quickly rectified, however, by the control system that regulates activity of the heart and blood vessels.

Control of the Heart's Output

Unlike other types of muscle contractions, heart muscle contractions—heartbeats—are generated by a specialized conducting system in the heart itself. The heart has a *myogenic beat*—one that will continue even if every nerve serving the heart is severed.

The basic source of the myogenic heartbeat lies in every heart muscle cell, since each cell can beat spontaneously. However, a special strip of modified heart muscle cells beats a bit faster and so sets the pace. This strip of cells, the heart's own pacemaker, is called the **sinoatrial node,** or **S-A node** (Figure 29-12). These cells can conduct electrical impulses of the same type carried by nerve cells. A special set of internodal fibers carries impulses to another node, which is located at the base of the right atrium near the ventricles; this is the **atrioventricular node,** or **A-V node** (see Figure 29-12). Finally, another special set of conductive fibers, the *bundle of His* (pronounced "hiss"), extends from the A-V node into the muscular walls of both ventricles, so that a wave of contraction can be triggered synchronously in all the ventricular muscle cells, and a heartbeat can result.

How are heartbeats triggered? At regular intervals,

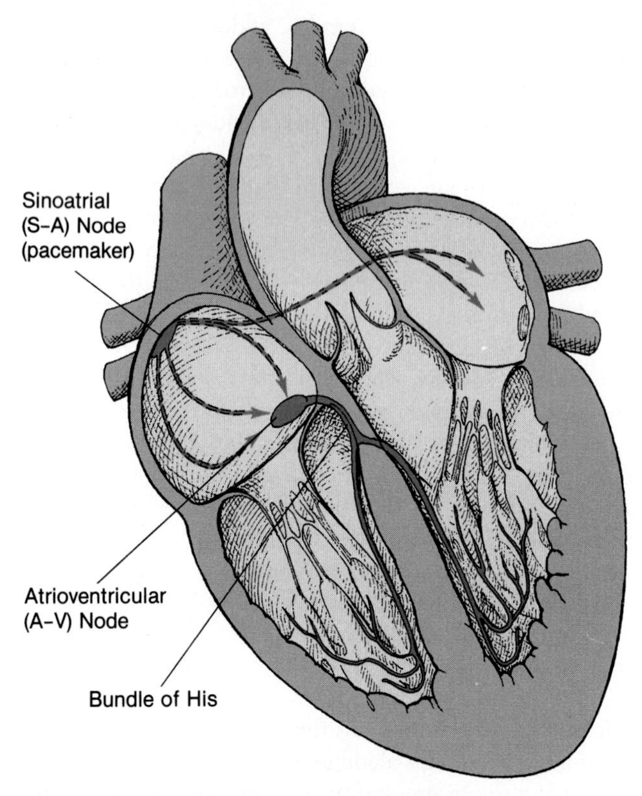

Figure 29-12 MYOGENIC HEARTBEAT.
The coordinated contractions of the heart are regulated by the pacemaker, the sinoatrial (S-A) node. This region of specialized excitatory and conductive fibers initiates electrical impulses that spread rapidly across the two atria, which then contract almost simultaneously. When the wave of excitation reaches the atrioventricular (A-V) node after a brief pause, it travels down the bundle of His to right and left bundle branches, then to all parts of the ventricular walls, stimulating the "power pumps" to contract.

usually about 1 second apart in humans, the S-A node initiates an electrical impulse by spontaneously *depolarizing;* that is, the electrical charges inside and outside the cells reverse. This impulse spreads over the internodal fibers and causes the atrial muscle cells to contract, thereby propelling blood into the ventricles. The impulse originating in the S-A node travels toward the A-V node. By then, the ventricles are already filled with blood. After a 0.1 second delay, the A-V node fires, sending impulses down the bundles of His to most of the muscle cells of the ventricular walls. The A-V depolarization impulse spreads virtually instantaneously from one heart muscle cell to the next because the cells are interconnected by numerous gap junctions that readily pass electrical current. Thus, the whole of the large ventricular muscle mass contracts at once, just slightly later than the contraction of the atria.

The electrical activity associated with each contraction is easily measured using an electrocardiograph. Beating of the heart also makes the characteristic "lub-dub" sound that can be heard with a stethoscope. This sound is due to valve closure at the onset of systole (contraction) and diastole (relaxation). The "lub" corresponds to the closing of the two valves between the atria and the ventricles, while the "dub" corresponds to the closing of the pulmonary and aortic valves. A new view of the heart is that the organ's expansion during diastole causes a suction pump action that pulls blood from the veins into the heart. Networks of connective tissue fibers between heart cells may assist in this organ expansion.

Although the heartbeat is intrinsic to the S-A and A-V nodes, the *rate* of contraction is often slowed or speeded by nerve signals in order to meet changing demands. If the body needs less, the beat rate and stroke volume can be lowered. Nerve impulses liberate a chemical called *acetylcholine*, which acts on the S-A and A-V nodes to slow the heartbeat. Conversely, when more cardiac output is needed, different types of nerves release the chemical *norepinephrine* (also called *adrenalin*), which acts on the nodes and the heart muscle cells to speed and strengthen ventricular contraction.

A slow heart rate is sufficient to meet the body's resting demands, but as an animal begins to exercise, the inhibition can be overcome by shutting off the release of acetylcholine. The heart then speeds up. If exercise becomes even more vigorous, the second type of nerve releases norepinephrine, and the heart rate speeds even more. Thus, mammals have a two-step system for increasing cardiac output to meet peak demands for oxygen and nutrients.

Regulation of Blood Flow Through the Vessels

The amount of blood flowing into a given capillary bed or into the veins is also under the direct control of the nervous system. In the brain and heart, most of the capillaries are opened, so these master organs get a steady supply of oxygen and nutrients. Elsewhere in the body, however, sphincter muscles can tighten and thereby cut down blood flow through a particular capillary bed. At any one instant, most of these sphincters are contracted, and only 30 percent or so of the body's capillaries are open to blood flow. It would be very dangerous if all opened simultaneously, since the capillaries can hold 1.4 times the total volume of blood in the body. Only by closing off a high proportion of capillaries at any given time can blood be kept at adequate pressures and volumes in the remainder of the circulatory system.

Contraction of the precapillary sphincters and the walls of vessels in response to nerve impulses is called **vasoconstriction.** (Figure 29-13a). Vasoconstriction creates a dynamic control over capillary beds in a local area. First one bed is closed, then another when the first reopens, and so on, so that local cellular needs are met, but the total volume of blood accumulated in the full set of capillaries is never too great. Vasoconstriction also helps stabilize blood flow during a change in body position, such as standing up after lying in bed. To cut down on blood pooling in the leg veins, vasoconstriction partially closes those veins and also shuts down some capillary beds, forcing blood back toward the heart. Likewise, as a giraffe lowers its head to drink or an antelope lifts its head to sniff for predators, coordinated vasoconstriction must take place.

Researchers recently discovered a peptide in the lining of the aorta called endothelin, which has nearly the same amino acid sequence as deadly sarafotoxins from snake venom. The snake toxins specifically attack hormones in heart and blood vessel cells, and researchers wonder whether endothelin may cause a natural toxicity that leads to vasoconstriction and hypertension.

Capillaries close and open in response to local tissue needs. The opening is called **vasodilation** (Figure 29-13b). Vasodilation allows blood to flow into muscle tissue capillary beds so that a muscle's oxygen demand is met, into skin capillaries so heat can be released from the body, and so on. One cause of vasodilation is EDRF, endothelium-derived relaxing factor. This substance is released locally and counteracts constriction so that flow rates are balanced within vessel networks.

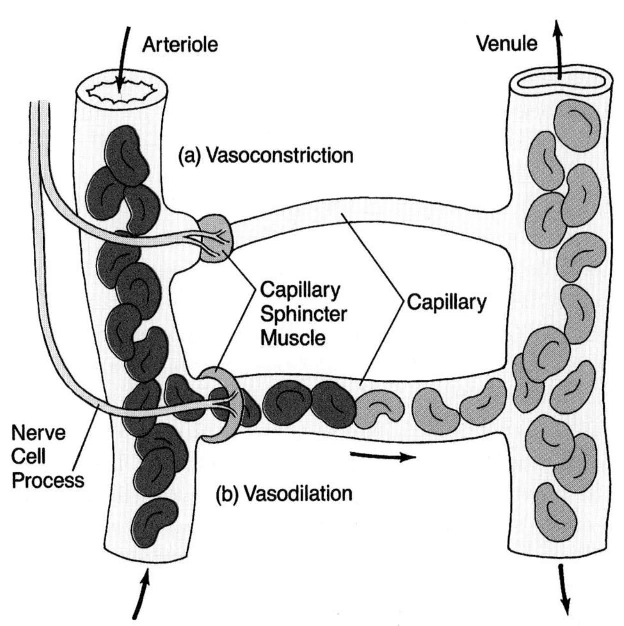

Figure 29-13 VASOCONSTRICTION AND VASODILATION REGULATE BLOOD FLOW.
(a) Vasoconstriction involves contraction of the capillary sphincters and restricted blood flow. (b) Vasodilation involves relaxation of the capillary sphincters and increased blood flow.

Brain Regulation of the Circulatory System

We have seen that nerves can regulate the heart and blood vessels, but what controls the nerves? An area of the brain stem called the **vasomotor center** receives input from a variety of sites in the body, including higher brain centers, but it primarily regulates blood pressure. The vasomotor center coordinates the heart and blood vessels and in combination with local capillaries keeps the levels of essential substances or properties within normal ranges. In general, this is accomplished by activating vasoconstriction or vasodilation to control blood flow. During vigorous exercise, for example, the vasomotor center, the respiratory control center, and the hypothalamus coordinate an increase in breathing rate, heartbeat, blood flow, and sometimes heat loss. A coordinated response also diverts blood from the skin to the body core when a mammal becomes chilled. The vasomotor center interacts with brain centers that regulate breathing and temperature to produce these coordinated responses. When an animal is in danger, the vasomotor center can also trigger release of the hormones *epinephrine* and norepinephrine. These hormones mobilize the "fight-or-flight" response and raises the cardiac output by causing vasodilation in the heart and skeletal muscles and vasoconstriction in the intestine and many other areas.

The functioning of the circulatory and respiratory systems is beautifully coordinated by the vasomotor and respiratory centers so that the blood pressure and the levels of oxygen, carbon dioxide, and hydrogen ions in the blood remain relatively constant whether exercise is heavy, light, or absent. One might suppose, for example, that the carbon dioxide levels would rise or the oxygen levels fall *before* the heart rate changes. Instead, the heart rate changes *first,* so that despite heavy exercise, levels of hydrogen ions, carbon dioxide, and oxygen in the blood remain normal. Muscle and joint movements occurring during exercise are reported to the brain by sensory nerves. The vasomotor and respiratory centers then become activated, and, in turn, speed heart rate, increase return of the blood through the veins, and coordinate the various responses that maintain steady levels of ions and blood gases *during* the exercise, not lagging behind it.

BLOOD: THE FLUID OF LIFE

The remarkable life-sustaining **blood** is a dynamic solution containing ions, nutrients, waste products, hormones, several kinds of cells, and other substances. The solid fraction of blood—blood cells and platelets—makes up about 40 to 50 percent of the blood's volume and is suspended in a yellowish fluid called plasma.

Plasma: The Fluid Portion of Blood

Plasma is water that contains various types of dissolved substances, including some oxygen, carbon dioxide, and nitrogen as well as ions and nutrient molecules. Positive and negative ions (cations and anions) make up about 1 percent of plasma by weight. The cations include sodium (Na^+), potassium (K^+), calcium (Ca^{2+}), and magnesium (Mg^{2+}). The anions include chloride (Cl^-), bicarbonate (HCO_3^-), sulfate (SO_4^{2-}), and phosphate (HPO_4^{2-}, $H_2PO_4^-$). Because the compound NaCl is the most common substance in plasma, blood is often compared to dilute seawater.

Also dissolved in plasma is a variety of small organic molecules, including glucose, amino acids, nucleic acid bases, certain fats, cholesterol, and vitamins. These molecules can be derived from food digested in the gut, or they can be secreted by cells, particularly in the liver. In addition, plasma carries away the wastes of cellular metabolism, such as lactic acid, urea, uric acid, and ammonia.

Proteins make up about 8 percent of the total volume of blood plasma. Together with salts, the proteins in the blood bring about the colloidal osmotic pressure mentioned earlier in this chapter. Another general effect of blood proteins is to act as a buffer that minimizes changes in blood pH.

The Blood's Solid Components

Blood Cells

The most numerous blood cells suspended in plasma are the red blood cells, or **erythrocytes.** A single milliliter of mammalian blood contains about 5 million of them, and at any given time, 25×10^{12} red blood cells are circulating through the human body. Each erythrocyte is disk-shaped, with a thickened, relatively thick periphery and a thinner, extensible central region that can bulge forward like a spinnaker sail as the cell moves through a narrow capillary (Figure 29-14). Throughout

Figure 29-14 BLOOD CELLS.
Erythrocytes (red blood cells) are the most numerous. In mammals, they do not have nuclei or mitochondria and are seen as biconcave disks, magnified about 2,250 times in this scanning electron micrograph. Leukocytes (white blood cells) consist of several types; the cells with bumpy surfaces visible in the midst of the erythrocytes are leukocytes. From *Tissues and Organs: A Text-Atlas of Scanning Electron Microscopy* by Richard G. Kessel and Randy H. Kardon, W. H. Freeman and Company © 1979.

life, cellular differentiation occurs in the *bone marrow,* the blood-forming tissue of a mammal. During this process, red blood cells lose their nuclei but retain a few other organelles. The cytoplasm in a mature red blood cell is filled with hemoglobin molecules—300 million per cell. *Hemoglobin* is the agent responsible for binding and carrying oxygen in the blood.

Less numerous, but equally important, are the white blood cells, or **leukocytes,** which are involved in the body's defense system. The average healthy person has about 7,500 leukocytes per milliliter of blood (see Figure 29-14); but that number may rise rapidly in response to infection or disease. The several types of white blood cells are **monocytes,** which mature into scavenger cells called *macrophages;* **granulocytes,** which are involved in inflammatory and allergic reactions; and **lymphocytes,** immune cells that attack foreign substances, such as bacteria and viruses, and that sometimes secrete *antibody* molecules—proteins that bind the foreign substances (details in Chapter 30).

Throughout adult life, blood cells continually die and are replaced. Red blood cells survive about 120 days. Every second of your adult life, about 2 million new blood cells are produced in the central marrow of your bones, and an equivalent number die. The fatty bone marrow tissue contains many *stem cells* that give rise to either red or white blood cells. Growth inducer proteins and differentiation inducer proteins regulate the sequence and number of red and white cells that develop in bone marrow. One specific type of marrow stem cell, the *megakaryocyte,* gives rise to masses of membrane-bound living fragments called **platelets,** which play an important role in blood clotting.

Platelets and Blood Clotting

A modest cut or scrape does not normally endanger life from loss of blood because blood *clots,* forming a semisolid mass that seals off the walls of ruptured blood vessels. When a rupture occurs, blood plasma, blood cells, and platelets leak into the surrounding tissue spaces. Contact between platelets and tissue components such as collagen fibers causes some platelets to break open and leak several agents that reduce blood flow to the wound site and start a cascade of events (Figure 29-15a). Chemical events in this cascade act on one of the globular proteins contained in blood plasma, *prothrombin,* and lead to the formation of millions of *thrombin* molecules if calcium ion levels are sufficient. Thrombin then acts on another plasma protein, *fibrinogen,* catalyzing its conversion to billions of *fibrin* molecules. Fibrin makes up the fibrous strands that compose a clot and gives the clot tensile strength. Platelets and some blood cells pouring from the ruptured blood vessel are trapped within the three-dimensional fibrin meshwork as the clot forms (Figure 29-15b). The contraction of cytoskeletal proteins within platelets pulls the clot into a firmer, tougher configuration.

The sequence of clotting just described is highly simplified, since still more factors are actually involved. Nevertheless, we can see that clotting is a good example of *amplification* in a biological system: A relatively small event triggers a stepwise series of larger and larger events. Fortunately, tissue factor, the start of it all, is not normally found on the surfaces of cells lining blood vessels or the heart, and thus clot formation within intact vessels is unlikely.

A number of diseases or genetic defects may affect any step in this complicated series of reactions and lead to poor clotting or the complete inability to clot. Hemophiliacs ("bleeders"), for example, lack one critical factor in the clotting process. In other persons, clots tend to form and drift through the vascular system. If the clots lodge in a capillary bed of the heart, they can trigger a heart attack; when they travel to the brain, they can cause a stroke. In 1987, genetic engineering produced medically useful quantities of an enzyme that occurs naturally in trace amounts; this *tissue plasminogen activator* digests clots and literally can reverse some heart attacks if it is injected within a few hours.

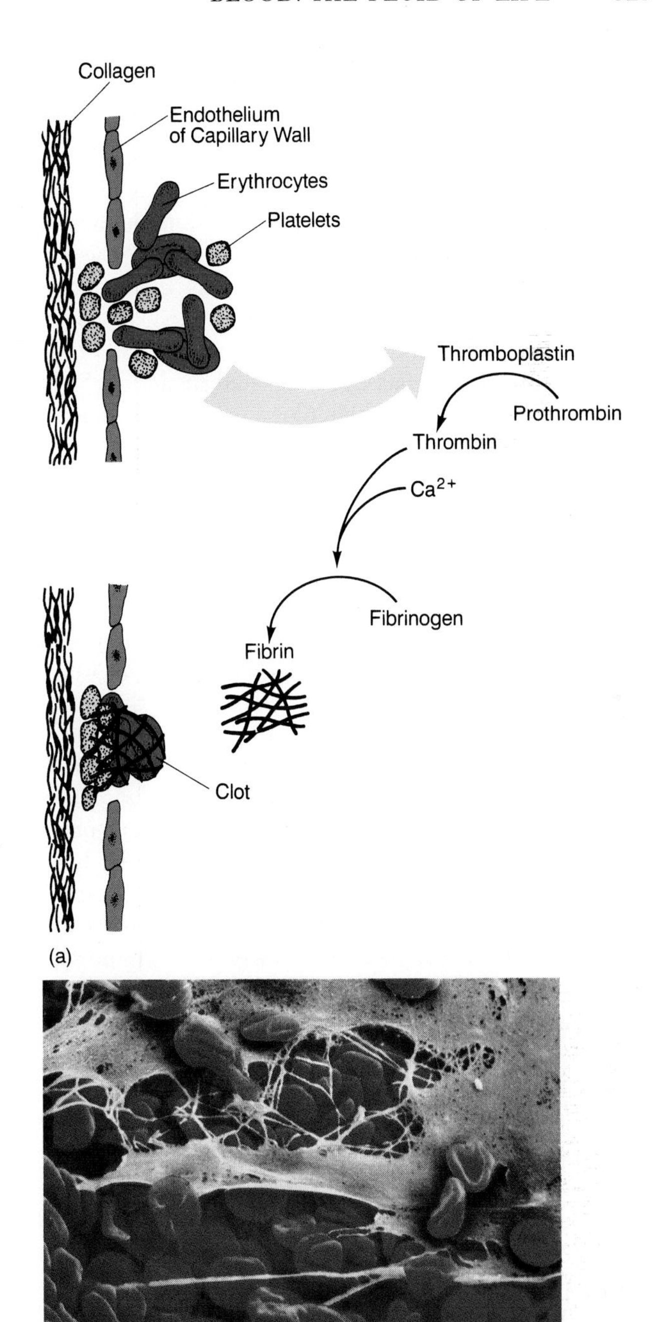

Figure 29-15 PLATELETS AND BLOOD CLOTTING. The steps in blood clotting. (a) The leakage of blood platelets and erythrocytes out of the capillary leads to their contacting collagen. The platelets adhere, spread on the collagen, and release the contents of their granules, thereby initiating a sequence of reactions leading to the formation of fibrin strands and a clot. See text for details. (b) A network of fibrin (magnified about 1,250×) that covers a wound site and traps red blood cells and platelets (not visible here).

THE LYMPHATIC SYSTEM AND TISSUE DRAINAGE

A second system of vessels, called the **lymphatic system,** runs roughly parallel to the venous half of a vertebrate's circulatory system, drains excess extracellular fluid that bathes body cells, and houses important parts of the immune system. (Lymph vessels in the wall of the intestine also absorb fats from digested foods.)

As we saw earlier, the high blood pressure present at the arterial end of capillary beds in mammals and birds causes a net flow of water from the blood into the spaces between cells. The lymphatic system drains this fluid and returns it to the venous system to prevent tissue swelling.

The human lymphatic system is shown in Figure 29-16. Tissue fluid enters the system through pores in tiny, thin-walled, blind-ended *lymphatic vessels* reminiscent of blood capillaries. Some proteins that have leaked from the blood plasma into the tissue spaces may enter the lymphatic vessels. These vessels merge into larger and larger lymphatic vessels, and the fluid, called *lymph,* flows through them away from the tissues and toward the heart. At various sites in the body, the lymphatic vessels pass through *lymph nodes,* compact meshworks of connective tissue that filter particles from the lymph and produce some of the cells of the immune system. The tonsils, for example, are lymph nodes. Lymphatic vessels continue to merge into larger and larger channels and ultimately join at the *thoracic duct* and right lymphatic duct, which pour lymph into major neck veins. In humans, about 11 ml of lymph drains into the blood each hour, although during certain illnesses, the flow may reach the fantastic rate of 900 ml per kilogram of body weight per hour.

Lymph is propelled by several mechanisms. In mammals, body muscles compress the lymphatic vessel walls, kneading the lymph along, while valves like those in veins prevent backflow. In addition, larger lymphatic vessels have smooth muscle along their walls that contracts in waves and thus propels the watery lymph solution.

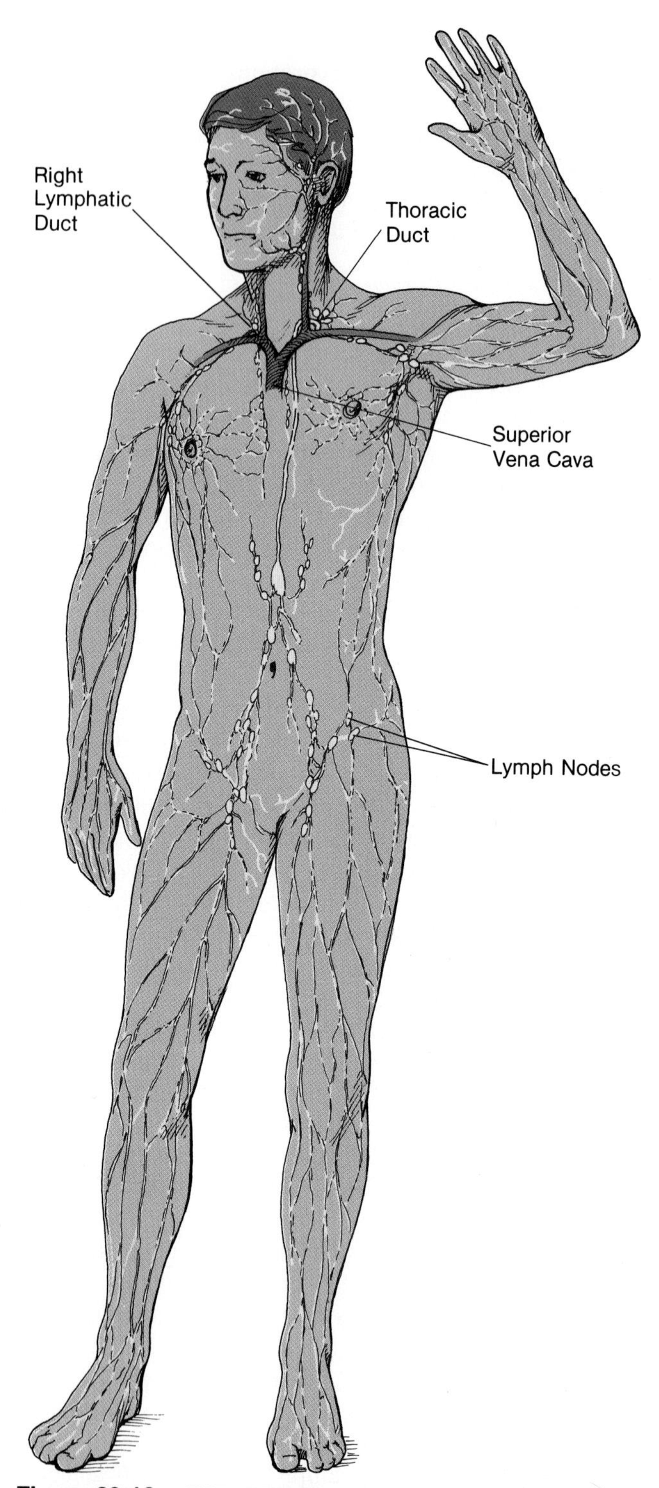

Figure 29-16 THE HUMAN LYMPHATIC SYSTEM. The lymphatic system is an auxiliary system of vessels that drains via the thoracic duct and the smaller right lymphatic duct into major veins that join the superior vena cava. Lymph is filtered as it percolates through the lymph nodes located at many sites in the body.

LOOKING AHEAD

The lymphatic system is remarkable and critical to the survival of warm-blooded creatures. Equally remarkable, however, are the circulatory system itself—the complex and highly coordinated transport network that provides animal cells with the materials they need for metabolism—and the immune system, our next subject.

SUMMARY

1. Diffusion alone can transport commodities into and out of cells over short distances, but long-distance transport requires energy-dependent bulk flow.

2. *Open circulatory systems* are present in many multicellular invertebrates; *closed circulatory systems* occur in some invertebrates and in all vertebrates. *Blood* moves through a closed system in *arteries, capillaries,* and *veins.*

3. Fluid circulates because muscles contract in vessel walls and the *heart;* body muscles knead vessels and augment blood flow; sphincter muscles control the flow to and from local areas; and valves prevent backflow.

4. This fish heart has four chambers. Land vertebrates retain two chambers, the atrium and the ventricle. In birds and mammals, those chambers are completely divided into right and left halves associated with separate *pulmonary (lung)* and *systemic (body) circulation.*

5. The major arteries of land vertebrates evolved from the ventral aorta and *aortic arches* of fishes. In mammals and birds, the carotid arch becomes the carotid arteries; the systemic arch, the aorta; and the pulmonary arch, the pulmonary arteries.

6. In mammals and birds, deoxygenated blood circulates from capillaries to *venules,* to veins, to the heart's right atrium, to the right ventricle, and then into the pulmonary arteries and lungs. Oxygenated blood passes through the pulmonary veins to the heart's left atrium, to the left ventricle, to the aorta, to major arteries, to smaller arteries, and finally to the capillaries in body tissues.

7. Substances can move across capillary walls owing to pinocytosis, diffusion, and passage through holes.

8. *Blood pressure,* the force of blood against a vessel's wall, is highest during contraction of the heart, or *systole,* and lowest during heart relaxation, or *diastole.*

9. Net fluid loss from capillaries is counterbalanced by absorption of tissue fluid by the lymphatic vessels. The *lymphatic system* transports this fluid back to the circulatory system.

10. The vertebrate heart has a self-generated, myogenic beat produced by intrinsic nodes and a specialized electrical conducting system. The heartbeat can be slowed down or accelerated by nerves, hormones, or local autoregulation.

11. The *vasomotor center* in the brain stem helps control the rate of heartbeat as well as blood flow through peripheral vessels. The firing of nerves causes *vasoconstriction* and *vasodilation,* as do local autoregulation processes.

12. *Blood* contains plasma, red and white blood cells, and platelets. *Plasma* carries ions, maintained in a balanced concentration, as well as small organic molecules, proteins, hormones, vitamins, and wastes. Red cells carry hemoglobin and oxygen. White cells are involved in the body's immune response. Platelets are fragments of cells involved in blood clotting.

KEY TERMS

aorta
aortic arch
artery
atrioventricular (A-V) node
blood
blood pressure
bulk flow
capillary
cardiac output
circulatory system
closed circulatory system
colloidal osmotic pressure
connective tissue
diastole
epithelial tissue
erythrocyte
granulocyte
heart
hemolymph
homeostasis
leukocyte
lymphatic system
lymphocyte
monocyte
open circulatory system
plasma
platelet
pulmonary (lung) circulation
sinoatrial (S-A) node
Starling's law
systemic (body) circulation
systole
vasoconstriction
vasodilation
vasomotor center
vein

QUESTIONS

1. What is the function of the circulatory system? The heart? The large arteries and veins? The capillary beds?

2. What role does diffusion play in single-celled, small, and large animals in the distribution of gases and nutrients to individual cells?

3. What is the main difference between open and closed circulatory systems?

4. Compare the hearts of fishes, amphibians, reptiles, birds, and mammals.

5. What is the function of aortic arches in fishes? What are portal vessels? How are veins and arteries similar in structure? How are they different?

6. Imagine you are a red blood cell in a capillary of your smallest left toe. Describe your journey through the bloodstream and back to that toe.

7. Where does the lymph originate? Where does it join the blood system?

8. Explain the source of excess tissue fluid.

9. What controls the heart's intrinsic myogenic beat? What other nodes and fibers are involved in regulation of heartbeat, and how do they function? What role does the nervous system play in regulating the heart activity?

10. Why is vasoconstriction vital for every mammal? What happens if it fails?

11. Where do mammalian blood cells form? What are stem cells? What are the functions of the red cells? White cells? Platelets?

12. Diagram the formation of a clot at a wound site.

ESSAY QUESTION

1. When severe blood loss occurs after an accident, what other body changes take place? Would water be a good replacement for the lost blood? Salt water (saline)? Plasma? Explain.

For additional readings related to topics in this chapter, see Appendix C.

30 The Immune System

The remarkable capacity of the immune system to respond to many thousands of different substances with exquisite specificity saves us all from certain death by infection.

Martin C. Raff,
"Cell Surface Immunology,"
Scientific American (May 1976)

Every environment on earth is literally teeming with bacteria, viruses, yeasts, molds, toxins, and other harmful substances, some of which can cause potentially fatal diseases in plants and animals. Yet most humans, as typical animals, live in fairly good health for six or seven decades, suffering diseases for only short periods and recovering fully. The reason is that we are each defended by an **immune system**—a standing army of 1 trillion white blood cells, 100 million trillion special protein molecules called *antibodies,* and a few small organs. Together, the components of the immune system protect the body from foreign substances, such as bacterial or viral invaders, and from abnormal cells that arise within, including some that could lead to cancers. The dangerous epidemic of Acquired Immune Deficiency Syndrome (AIDS)—a viral disease that destroys a person's immune system and usually proves fatal—reveals just

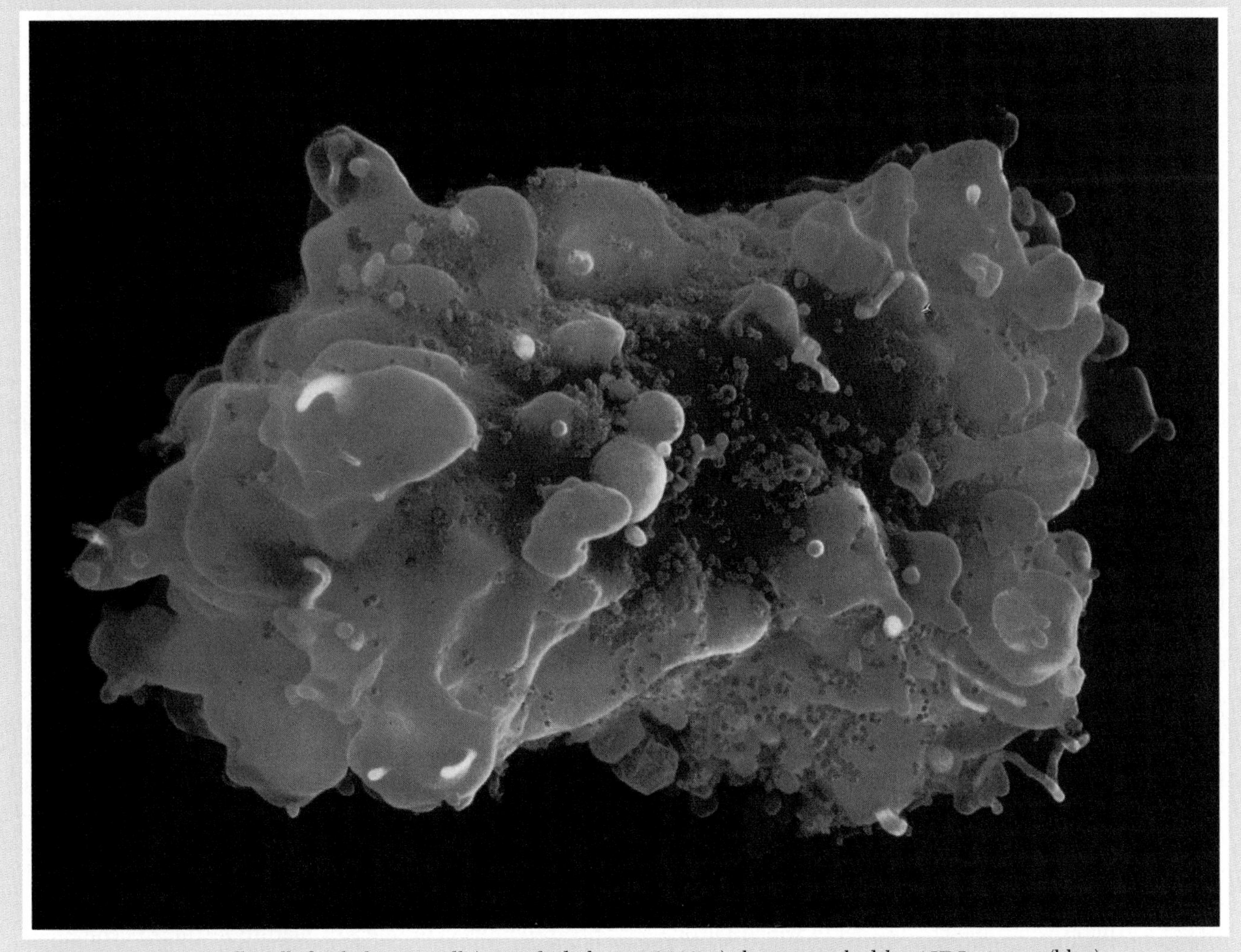

An immune system cell, called a helper T cell (magnified about 27,000×), being attacked by AIDS viruses (blue).

how vulnerable each of us can be to infections and cancers if the immune system is disrupted.

The body has various lines of defense, some general and some highly specific in their recognition of and response to foreign substances. We shall see that the mechanism for recognizing foreign and "domestic" body substances depends on molecular fit—the binding of molecules with reciprocal shapes—just as the critical functioning of enzymes depends on molecular fit. And we shall see that this immense diversity of molecular fit depends on a unique process of gene shuffling that occurs in two types of white blood cells. Immune responses are a truly amazing evolutionary adaptation to life among potentially harmful microbes and molecules.

This chapter describes:

- The cells and organs of the immune system
- Nonspecific and specific defenses, and how they operate
- The structure, production, and activity of antibodies—diverse defensive molecules that attack invaders
- White blood cells called T cells, and how they work
- White blood cells called natural killers, and their role in defending the body
- Autoimmunity: how the immune system sometimes attacks the body's own cells
- Immune deficiency—the terrible consequences of a missing, impaired, or destroyed immune system
- Immunity and human health, including vaccinations, allergies, and organ transplants

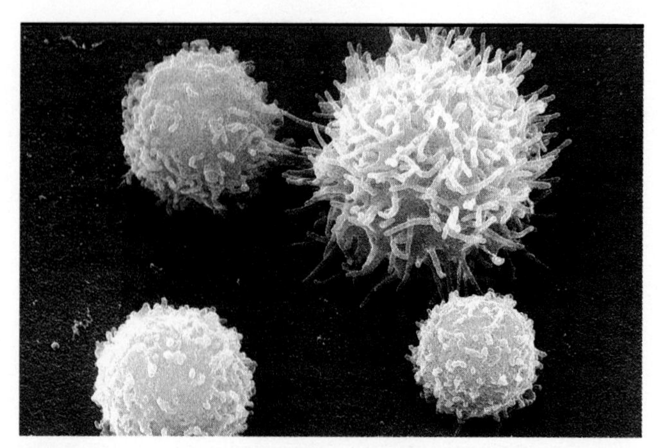

Figure 30-1 CELLS OF THE IMMUNE SYSTEM. Scanning electron micrograph (magnified about 2,000×) of several lymphocytes (round cells) and a macrophage (top right). These two cell types interact to initiate many immune responses.

COMPONENTS OF THE IMMUNE SYSTEM

The cells and organs of the immune system are the body's "standing army" for defense, although some of the structures also play other roles in the body. The components of the immune system include several classes of blood cells; organs such as the thymus, spleen, and lymph nodes; and soluble circulating proteins, or antibodies.

White Blood Cells and Their Protein Products

Although there are two main kinds of blood cells, red and white, fully 99 percent of the circulating cells are red blood cells. Nevertheless, the remaining 1 percent consists of 1 trillion white blood cells, all of which participate in the body's defense. One class of white blood cells, the *granulocytes*, characterized by granules in their cytoplasm, takes part in the body's inflammatory response to infections—a nonspecific response. The other two classes of white blood cells are **macrophages** (mature monocytes) and lymphocytes (Figure 30-1). Macrophages engulf and digest most foreign particles and initiate many immune responses by trapping foreign substances and "presenting" them to lymphocytes. There are two major kinds of *lymphocytes:* **T cells** and **B cells.** Whereas T cells congregate at a site of infection and mount a direct attack on foreign organisms or tissues, B cells do not localize at the infection site; instead, they synthesize proteins called antibodies, which perform the attack function. The effect of B cells, therefore, is indirect. A third type of lymphocyte is the large, granular white blood cell called a **natural killer (NK) cell,** which can attack certain viruses and virus-infected tumor cells.

The Varied Sources of White Blood Cells

In all adult vertebrates, stem cells in bone marrow give rise to red blood cells, as well as to various types of white cells (see Chapter 29). This is why physicians sometimes transplant bone marrow into patients who have received massive doses of radiation, which destroys bone marrow stem cells. In healthy bone marrow, after T-lymphocyte stem cells arise, these pre-T cells migrate to the thymus, where they mature further. (The "T" in "T cell" stands for "thymus-derived".) During this maturation process, they become committed or determined. B-lymphocyte stem cells arise in the bone marrow or liver of mammalian embryos. In avian embryos, these pre-B-lymphocyte lineages mature further (become determined) in the bursa of Fabricius, an appendix-like organ near the posterior gut. Subsequently, bursa-derived B lymphocytes (hence the "B" in "B cells") and their equivalents in mammals become localized in **peripheral lymphoid tissues,** including the lymph nodes, spleen, gut-associated appendix, tonsils, adenoids, and intestinal Peyer's patches (Figure 30-2). T-cell lineages can sometimes be found in the spleen, as well.

In 1988, researcher Irving Weissman and colleagues at Stanford University purified bone marrow stem cells for the first time, and found that they can give rise to single or multiple types of blood cells, depending on environment. Stem cells placed in the spleen, for example, yield red cells and several white cell types, while those placed in the thymus yield only T cells.

Organs of the Immune System

The discrete organs of the immune system are the lymph nodes, spleen, and thymus (see Figure 30-2).

Lymph Nodes

The lymphatic vessels help drain excess tissue fluid (see Chapter 29). These vessels also carry white blood cells to and from the tissues; large pores in the walls of lymphatic capillaries allow white cells to pass into tissue spaces at the site of an infection. The tiny, nearly translucent lymphatic vessels appear to merge into thickened masses—pinkish gray beads of tissue called **lymph nodes** (see Figure 30-2)—which occur at many sites in a mammal's body, with clusters in the neck, armpit, abdomen, and groin. Lymph nodes are dynamic tissue masses, containing millions of lymphocytes that constantly arrive and populate the nodes while older cells die or pass through the lymphatic vessels as they return to the circulatory system.

Lymph fluid is filtered as it penetrates a lymph node; particulate materials such as bacteria and debris from dead cells are removed through this filtering. Macrophages take up residence in the lymph nodes and ingest foreign particles as they float by. Such macrophages may also bind foreign substances onto long, fingerlike protrusions of their cell surface. T lymphocytes resident in the node may then "recognize" the foreign substance carried by the macrophage, proliferate, and migrate through the lymphatic vessels to the site of the infection, where the foreign substance originates. When an infection occurs in the body (say, a sore throat), the nearest lymph nodes (those in the neck) often swell greatly and become tender as filtering increases and lymphocytes proliferate. Lymph nodes near a tumor tend to trap cancer cells that have broken free from the tumor; thus, the nodes must often be irradiated or removed along with the tumor itself.

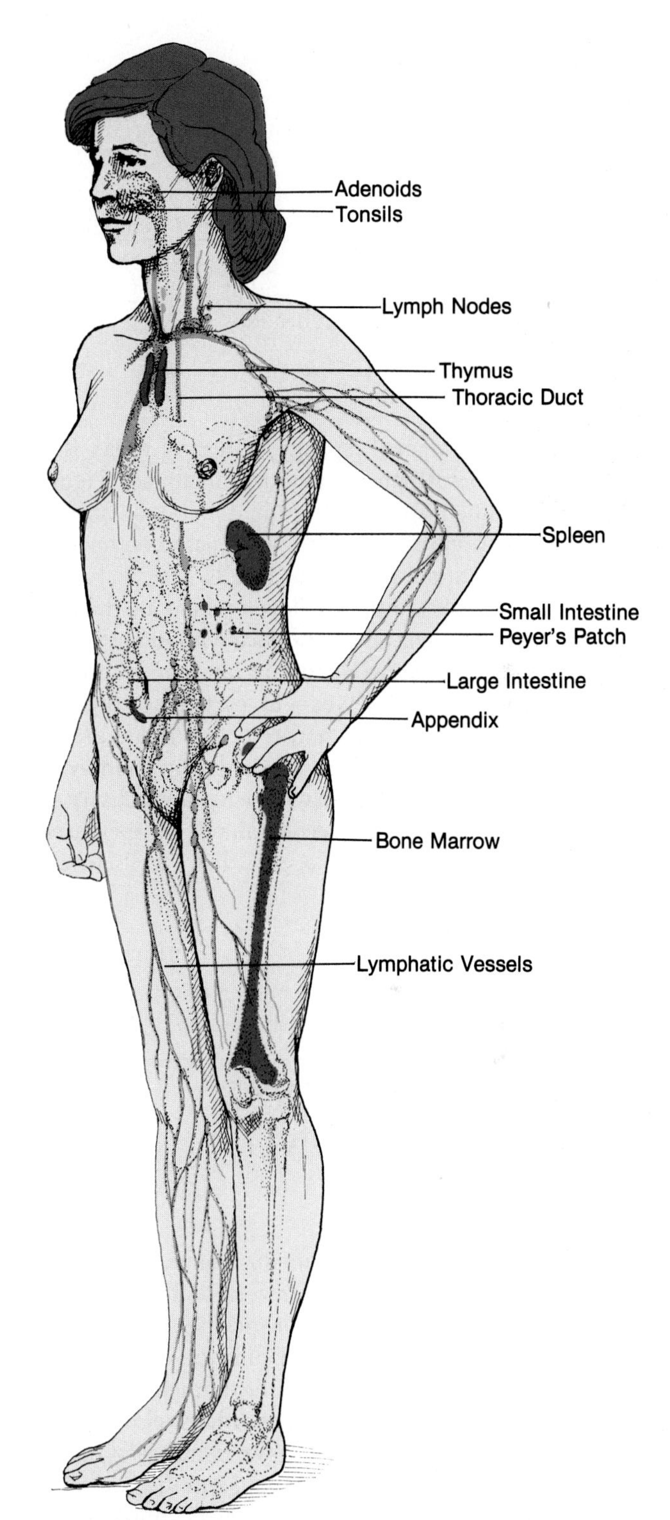

Figure 30-2 THE IMMUNE SYSTEM: THE BODY'S GUARDIAN.
The primary organs of the immune system are the lymph nodes (pinkish gray), spleen, thymus, and bone marrow. Marrow forms inside of most bones. Lymph nodes are associated with lymphatic vessels, allowing the lymph to percolate through the nodes.

Spleen

The **spleen,** an oval, flattened, deep-purple organ the size of a small beefsteak (in a person or other large mammal), is located near the liver (see Figure 30-2). The spleen's main functions are to store red blood cells, destroy aged red cells, and recycle iron from hemoglobin molecules. However, the spleen also contains many lymphocytes and macrophages, and it filters lymph in much the same way that lymph nodes do.

Thymus

The **thymus** is a small, spongy, spherical organ that sits behind the breastbone in adult mammals (see Figure 30-2). Determined stem cells from the bone marrow migrate to the thymus, where they are acted on by hormones to differentiate further as T cells.

The thymus is essential to development of the immune system. Humans and mice born without a thymus lack the capacity to carry out certain immune responses dependent on T lymphocytes—in particular, fighting viral and bacterial infections and rejecting tissue grafts. The thymus is largest and most active during childhood. It shrinks after puberty, and with age, the activity of T lymphocytes decreases, accompanied by increased susceptibility to infections by agents that T cells would normally combat.

LINES OF DEFENSE: NONSPECIFIC AND SPECIFIC

When a foreign organism or substance enters the body of a mammal, bird, or other vertebrate, two types of defenses can be mounted: nonspecific and specific. *Nonspecific* defense mechanisms function in a similar way, regardless of the nature of the foreign substance they serve to eliminate. The skin and the sticky mucous membranes, for example, impede the entrance of bacteria or foreign molecules into the body, and the sweeping action of cilia along the air passages in mammals helps prevent the bacteria from adhering to and penetrating body cells. Similarly, secretions such as tears, mucus, and sputum contain *lysozyme*, an enzyme that digests the cell walls of bacteria. If foreign substances do gain entrance to the body, they are ingested and destroyed by macrophages in the blood, liver, and lymph nodes. Moreover, another nonspecific response, inflammation, can be mounted to fight invaders.

The Inflammatory Response

The site of a wound becomes red, puffy, and tender because of an **inflammatory response.** If you get a large splinter in your finger, nearby granulocytes immediately release *histamine* (Figure 30-3, step 1). This molecule increases the permeability of local capillaries and relaxes the sphincter muscles surrounding the capillaries at the wound site. This increases blood flow to the area and causes fluid, serum proteins (serum is the fluid that exudes from clotted blood), antibodies, platelets (see Chapter 29), macrophages, and complement proteins

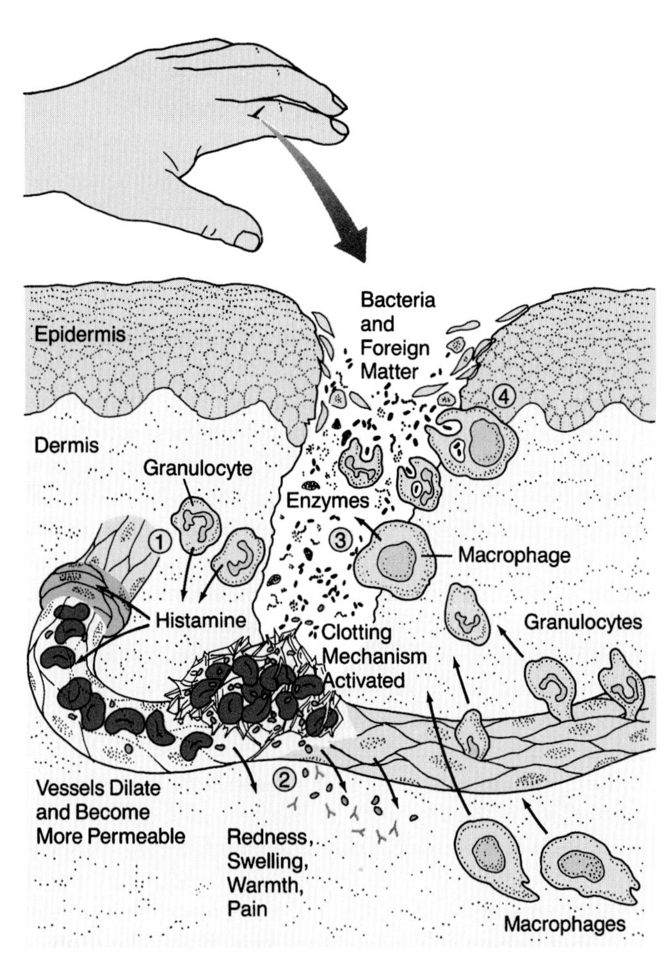

Figure 30-3 INFLAMMATION: THE NONSPECIFIC DEFENSE MECHANISM.
Penetration of the first line of the body's defense, the epidermis of skin, and introduction of a foreign agent (e.g., bacteria) elicits inflammation. The inflammatory response involves the migration of granulocytes and macrophages to the site of the wound. Granulocytes release histamine (step 1), while macrophages engulf the foreign particles and damaged cellular material (steps 2–4). This may be easily accomplished, and healing occurs rapidly; or the result may be a more general bacterial invasion if bacteria at the wound site succeed in invading lymphatic vessels or blood vessels.

(which help burst foreign cells) to leak from damaged blood vessels into the intracellular spaces (step 2). Puffiness results. The clotting mechanism (see Chapter 29) is activated to prevent undue loss of blood. Some of the substances released at wound sites act as attractants for macrophages, and those cells release enzymes that destroy and inactivate many kinds of bacteria introduced into the wound (step 3). As the inflammatory response proceeds, a large number of macrophages accumulate and ingest bacteria as well as debris from cells that have been killed around the puncture (step 4). Eventually *pus* may accumulate; this yellowish substance is an aggregation of macrophages, granulocytes, tissue cells, dead cells, dead or dying bacteria, and extracellular fluid.

The inflammatory response is the basic defense mechanism in most invertebrates and in chordates. Mollusks and arthropods, for instance, have inflammatory responses that can destroy most invading foreign particles. Insects also produce families of antimicrobial, lytic peptides. It is only in the chordates, however, that the highly specific defense system, the immune system, operates as well.

The Immune Response

Specific defense mechanisms are the exclusive province of the immune system: T and B lymphocytes act in response to specific invading organisms, and NK cells respond to viruses, to transplanted tissues, and perhaps to other foreign agents. The immune system reacts in a unique and specific way to each species of bacterium or virus. That is why, for example, a child who has the mumps becomes "immune" to subsequent infection by the mumps virus but not to the agents that cause measles or whooping cough. We describe *specificity* in more detail later. Here, we discuss three additional differences between the immune response and the nonspecific defense mechanisms: *diversity*, *memory*, and *tolerance*.

Diversity

Despite the enormous diversity of foreign agents that can invade the body—including newly mutated strains of bacteria and virus and even newly synthesized chemicals—the vertebrate immune system can usually respond to the challenge with a "tailor-made" defense.

Memory

The immune system retains a "memory" of the foreign agents it has reacted to in the past. The body of a child with chicken pox, for example, will usually mount a full *primary response* to the invading virus after about a week or so. If the child is exposed, even years later, to this highly contagious disease, a *secondary response* takes place—a faster, more vigorous reaction to the second exposure. The immune system "remembers" the earlier exposure to the same infectious agent, and this memory renders the individual *immune* to the disease. Nonspecific defenses show no hint of such memory.

Tolerance

Immunity also depends on tolerance—that is, the immune system's capacity to recognize the body's own molecules and cells, to distinguish "self" from "nonself." Nonself—foreign cells, tissues, or substances—is reacted against vigorously, whereas self—one's own cells, tissues, and molecules—is generally not reacted against; instead, it is *tolerated*. Cells of the immune system may also be able to recognize and attack cancerous cells—"self gone wrong." However, the ability to distinguish self from nonself sometimes goes awry, resulting in attack against the body's own healthy cells (autoimmunity, described later).

The immune response to a nonself substance can be *humoral*, involving B cells and circulating antibodies, or it may be *cellular*, involving T cells. Let us see how.

ANTIBODIES AND HUMORAL IMMUNITY

In 1888, American researchers E. Roux and D. Yersin observed that laboratory animals with immunity to diphtheria toxin had a substance in their blood that could neutralize the toxin. Because this substance was found in the blood fluid, immunity conferred by such substances was termed **humoral immunity,** based on the ancient term for body fluids, *humors*. The substances that act against foreign bodies such as bacteria are now known to be globular proteins called **antibodies,** produced by B lymphocytes in response to foreign entities in the body. Such foreign substances are called **antigens,** and they include bacterial toxins, viral coat proteins, and surface proteins and carbohydrates that label cells. An antigen is any substance that elicits an immune response.

More than 50 years ago, researcher Karl Landsteiner found that antibodies usually "recognize" small clusters of atoms arranged in a precise shape perceived as "foreign" to the body. These clusters are called *antigenic determinants,* and recognition results in antibodies binding to them. Such antigenic determinants occur in literally millions of configurations on the surfaces of every virus, bacterium, and protozoan and on the cell surfaces of every plant, fungus, and animal on earth.

Using synthetic antigens produced in the laboratory, Landsteiner found that animals could make antibodies against these new compounds to which they could never possibly have been exposed and that the *binding sites* on the antibody molecules must be highly specific, just as are the active sites on enzyme molecules. Because there are millions of possible antigens, a single animal must somehow be capable of making millions of different antibody molecules with great specificity in recognizing antigens. How is this possible? And how do the antibodies recognize the antigens? The answer lies in antibody structure.

Antibody Structure: Variable and Constant Regions

There are several classes of antibodies, but the basic structure of an antibody molecule has the shape of the letter Y (Figure 30-4). Each antibody molecule is composed of four polypeptide chains: two identical **heavy chains** and two identical **light chains,** all linked by *disulfide bonds* (see Chapter 3). Each "arm" of the Y consists of one light chain and part of one heavy chain; the hinge region allows the arms to move; and the "stem" is formed by the rest of the two heavy chains. Each arm region of the Y serves as an antigen-binding site.

Comparing two different antibodies from mouse blood, one directed against tetanus toxin, the other against diphtheria toxin, we would find that one end of the heavy chains is identical in both antibodies, and one end of the light chains is identical in both: These are the **constant regions** of the polypeptide chains (see Figure 30-4a). But the amino acid sequences at the opposite ends of the light and heavy chains in the arms of the diphtheria antibody would differ from those of the tetanus antibody. These are the **variable regions** of the polypeptides. Thus, the end of each arm of the Y-shaped antibody molecule is composed of the variable region of a light chain and the variable region of a heavy chain, which together form a binding site for an antigen.

Each arm has an identical pocket, or crevice, between the two variable regions with a configuration complementary to that of the antigen to which the antibody binds (Figure 30-4b). Slight differences in the amino acid sequences of variable regions account for slight differences in the shape and properties of the binding site, giving different antibodies their "lock-and-key" specificity for differently shaped antigens. These amino acid differences yield an animal millions of different antibodies, and arise via a remarkable process of "gene shuffling" (described later) during lymphocyte maturation. Careful analysis reveals that three **hypervariable regions** situated at intervals in the variable portions of both light and heavy chains actually account for the different binding site specificities of antibodies.

Whereas the hypervariable regions determine an antibody's binding specificity, the constant region determines the antibody's general *class.* There are five major classes of antibody proteins, or **immunoglobulins:** IgM, IgG, IgA, IgD, and IgE (Figure 30-5a), each with specific functions. Early in an immune response, B cells synthesize IgM, each molecule of which looks like a cluster of five Y-shaped antibody molecules joined together by their constant chains and by another protein chain (Figure 30-5b), and which is capable of binding large aggregations of antigen. Later in the same immune response, B cells synthesize and secrete a large quantity of

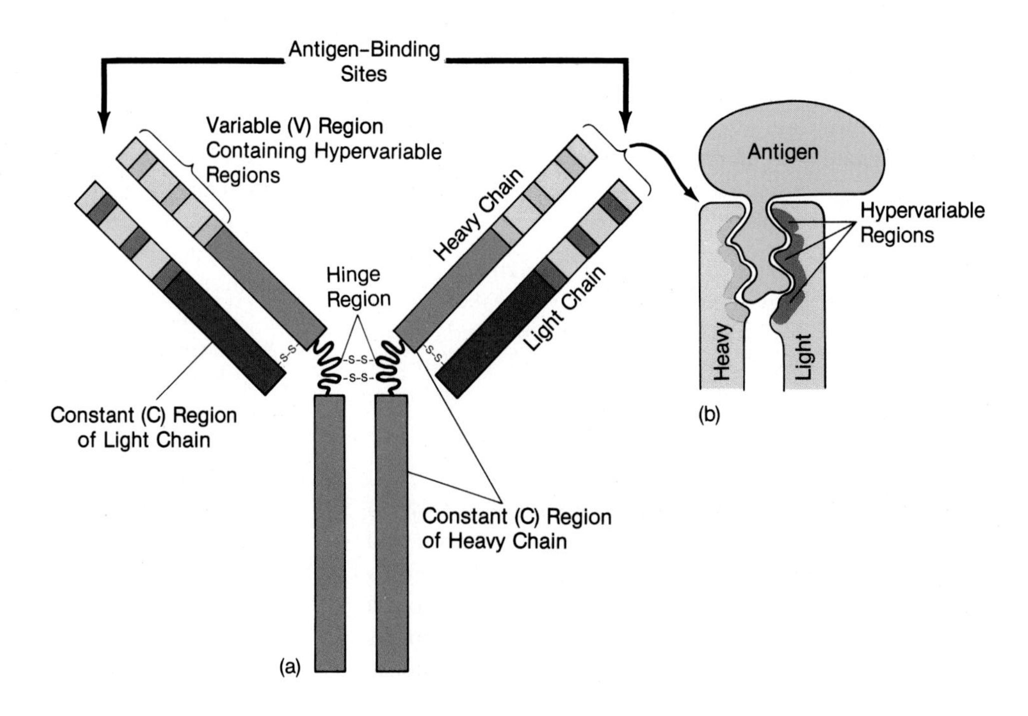

Figure 30-4 STRUCTURE OF THE ANTIBODY MOLECULE.
(a) The heavy and light chains are held together by disulfide bridges. The antibody's antigen-binding sites are associated with variable regions of the polypeptide. (b) In the hypervariable regions, amino acid sequences vary from one antibody type to another; as a result, the shape and properties of the binding sites vary, as do their specificity for antigens.

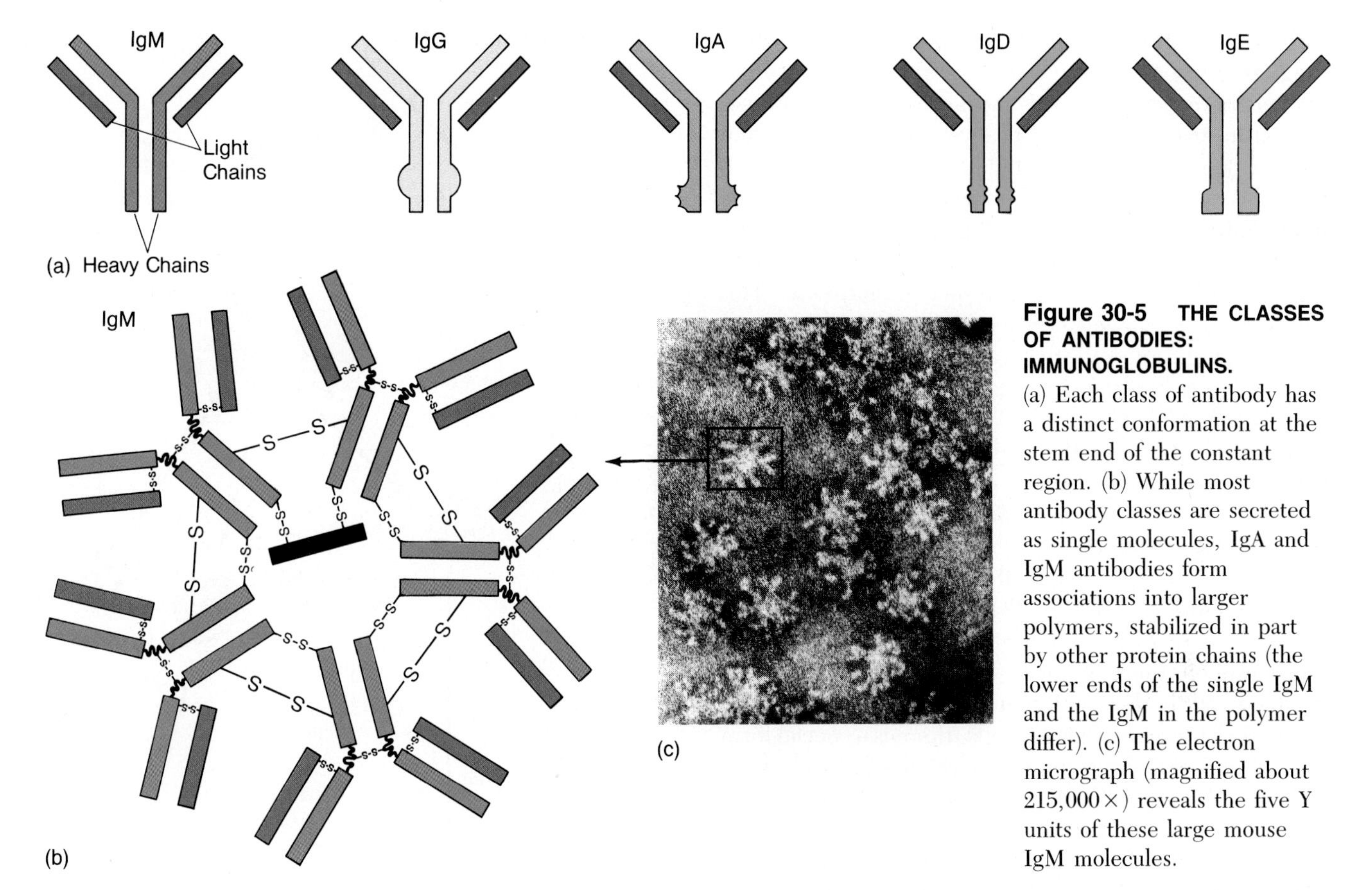

Figure 30-5 THE CLASSES OF ANTIBODIES: IMMUNOGLOBULINS. (a) Each class of antibody has a distinct conformation at the stem end of the constant region. (b) While most antibody classes are secreted as single molecules, IgA and IgM antibodies form associations into larger polymers, stabilized in part by other protein chains (the lower ends of the single IgM and the IgM in the polymer differ). (c) The electron micrograph (magnified about 215,000×) reveals the five Y units of these large mouse IgM molecules.

Y-shaped IgG antibodies, which are much more stable than IgM and can persist for long periods.

IgG and IgM antibodies help protect animals from invaders by three mechanisms: precipitation, agglutination, and the complement response. *Precipitation* of large aggregates, or clumps, of antigen-antibody complexes may prevent the antigen access to cells and thus inactivate it. Antibody molecules may bind to antigens on the surface of invading foreign cells and cause the cells to stick together in groups—a process called *agglutination*. Both precipitation and agglutination can stimulate phagocytosis by macrophages to clean up the antigen-antibody complex.

The binding of IgG or IgM antibodies to a cell can also activate the **complement response,** a series of enzymatic reactions involving complement proteins that causes the invading cell to burst. Some of the complement proteins bind to the antigen-antibody complex on the surface of the invading cell. A cascade of reactions occurs in which one complement protein enzymatically activates another; finally, a protein called *perforin* assembles to form open pores in the invader's plasma membrane. As a result, the cell usually swells and dies.

The third class of immunoglobulins, the IgA molecules, are present in tears, mucous secretions, and saliva and attack the foreign microbes that enter the body's orifices. IgA reacts with the surface antigens on bacteria and reduces the likelihood of bacterial invasions. IgD is found on the surfaces of lymphocytes and plays a role in their activation. IgE molecules are bound to mast cells and play an important role in allergic responses (described later).

Sharks and bony fishes have B cells and IgM, but not the other immunoglobulin classes, and the IgM molecules have only 500,000 differently shaped binding sites for antigens, compared with perhaps 18 billion in mammals. Many biologists think that IgM is the oldest immunoglobulin type, and that the other classes evolved in amphibians, reptiles, birds, and mammals.

Development of Antibody Specificity: The Clonal Selection Theory

What are the cellular and genetic bases for the specificity, memory, diversity, and switching of antibodies from IgM early in the immune response to IgG later?

In the 1950s, Sir Macfarlane Burnet, Niels Jerne, and other immunologists proposed the **clonal selection theory** to integrate and explain all these characteristics of the immune response. B cells originate from embryonic stem cells that become determined and that proliferate in the bursa of birds and in the bone marrow of mammals. During these developmental processes, gene shuffling takes place. A stage called the pre-B cell becomes irreversibly committed (by means of exon rearrangements) to making only one heavy-chain variable region (Figure 30-6, step 1). At a later stage, called the **naive B cell,** the cell becomes further committed to only one light-chain variable region through still more rearrangements (step 2). As a result, the particular heavy and light chains can combine to form a specific antibody. This committed naive B cell and all of its progeny constitute a *clone* of identical cells that can manufacture antibody with only one binding site specificity (Figure 30-6). The early commitment processes are said to be antigen-independent because the antigen itself is *not* present when the pre-B and naive B cells become committed. Within the embryo's bone marrow or bursa, millions of different committed pre-B cells arise, each with a single specificity, and each before the organism has ever encountered foreign antigens. When it does so later, it can respond to millions of different antigens because for any given antigen, there is at least one preexisting clone that produces antibodies with some capacity to bind to it.

On the surfaces of naive B cells, there is a special form of IgM and IgD, each immunoglobulin molecule with precisely the same binding sites (step 3). Naive B cells are commonly found in the lymph nodes of adults and sometimes in the blood as they disperse from bone marrow to the nodes. The 100,000 or so surface IgM and IgD molecules on such cells may act as *receptors* for an antigen. Suppose a foreign microbe enters the body and ultimately lodges in a lymph node; its antigenic determinants bind only to naive B cells whose surface immunoglobulins happen to have specificity for that antigen (step 4). This process is called *clonal selection* (Figure 30-6). The particular committed B cell is being "selected" for maturation, but others that do not bind are not. The "selected" naive B cell divides to generate a clone (step 5). Rapidly, a series of divisions occurs in the B cell and its progeny, and they differentiate and begin synthesizing antibody molecules (step 6). After about 4 days, these **plasma cells** are actively secreting an immense number of antibody molecules (up to 2,000 per second for the 3- to 4-day life of the cell). Since one resting immature B cell gives rise to many plasma cells of precisely the same specificity, this process is termed *clonal expansion.*

When the resting B cell is stimulated to divide by binding an antigen, it actually gives rise to two types of cells: a huge number of actively secreting plasma cells and a smaller number of **memory cells** (step 7). Memory cells circulate in the blood or reside in the lymph nodes, ready to provide a rapid and effective secondary immune response against future exposures to the antigen for which they are specific (details to come).

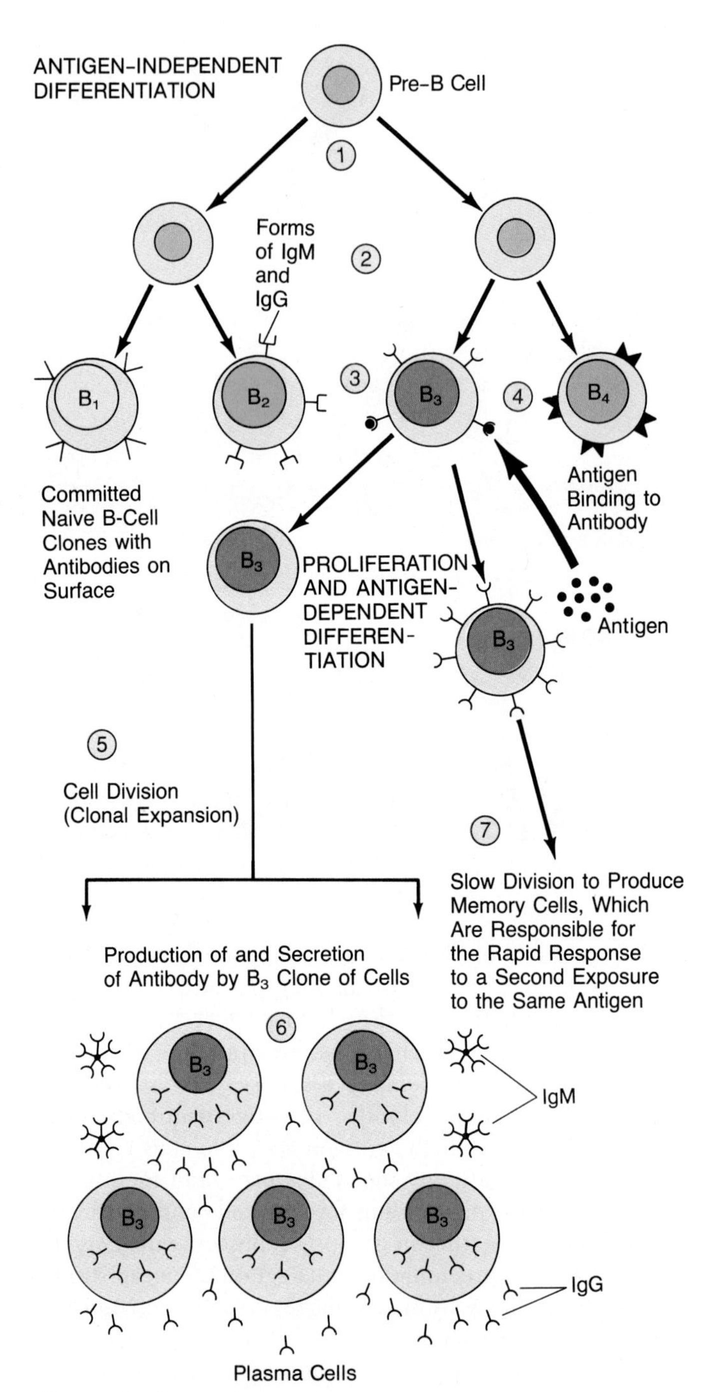

Figure 30-6 CLONAL SELECTION THEORY.
During the development of the immune system, millions of different committed naive B cells are produced. Antigens need not be present for this to occur. Each committed naive B cell has unique antibody receptors on its surface. In the course of life, the chance appearance of an antigen that binds to a particular naive B cell causes that cell to divide and differentiate, forming a clone of cells that produce identical antibodies that bind the stimulating antigen. If a different antigen happened to appear, a different naive B cell would be "selected," and a different clone would be generated. The "selected" B-cell clone also gives rise to a set of memory cells that can carry out secondary responses.

Antibody Diversity: A Matter of Gene Shuffling

Antibody diversity presents an intriguing paradox: An individual bird or mammal can have literally millions of different kinds of antibodies, each with a uniquely shaped binding site based on variable sequences of amino acids. If each variable region were coded for by its own gene, an organism would need millions of separate genes to make antibodies; yet most mammals have a total of about 1 million genes, and only a few code for immunoglobulins. In recent years, geneticists have discovered a unique process of shuffling among segments of certain genes in the nuclei of the early stages of B cells. This process accounts for the immense diversity that is the key to the body's defense against a host of invaders.

Data from studies of immunoglobulin genes in mice and people suggest that there are about 7,500 possible amino acid sequences for the light-chain variable regions and perhaps 2.4 million for the heavy chain. Because a given cell expresses one light-chain variable region and one heavy-chain variable region, the total number of different binding sites that can be formed is 7,500 × 2.4 million, or 18 billion possible combinations. No one is sure whether that many actually form or whether certain sequences are favored.

Recall from Chapter 14 that the gene sequences that code for most eukaryotic polypeptides consist of exons (expressed regions) separated by introns (intervening sequences) of DNA. The genes coding for immunoglobulin chains also occur in pieces; however, the many sequences that code for the variable regions are separated from the sequences that code for the constant regions by long stretches of DNA. When the pre-B and naive B cells become committed to making a single type of antibody, DNA rearrangements occur: One of the segments coding for the heavy-chain variable region moves next to a sequence coding for the heavy-chain constant region; likewise, the sequence coding for the light-chain variable region moves next to a sequence coding for the light-chain constant region (Figure 30-7). These two rearrangements of DNA give rise to one complete light-chain gene and one complete heavy-chain gene. Extremely specific nuclear enzymes, endonucleases, actually cleave DNA at precise spots to allow the recombination process to occur.

A key to antibody diversity is that the process of moving any one of the large number of variable sequences next to a single constant sequence is *random*. This occurs for both the light and the heavy chains. Thus, a great variety of constant-variable combinations are made, each in a different pre-B or naive B cell, and the cell becomes committed to making one particularly shaped binding site.

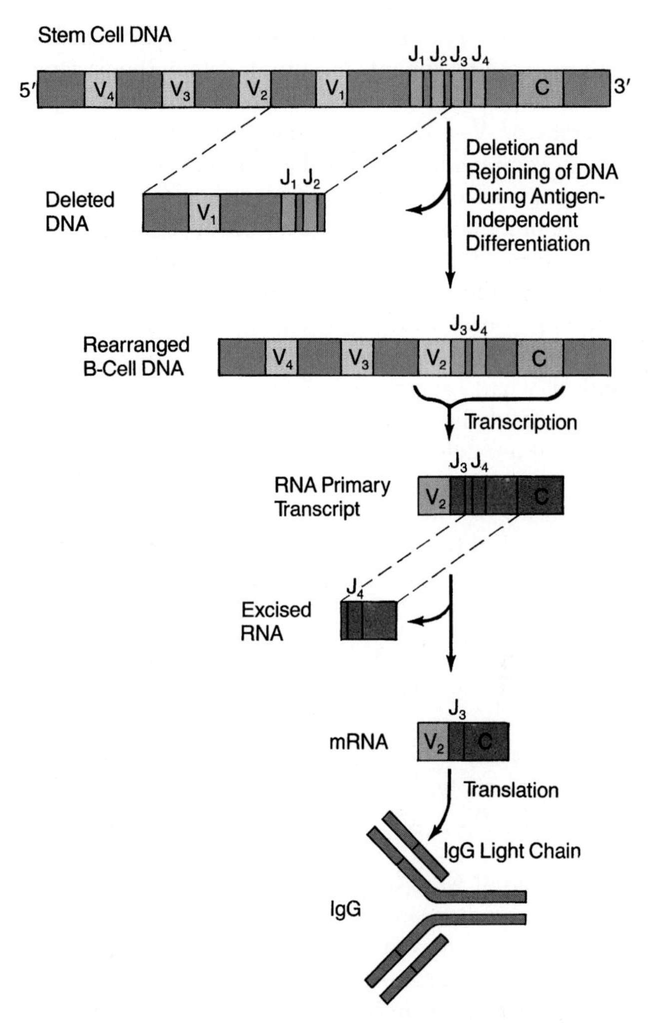

Figure 30-7 GENE SHUFFLING AND ANTIBODY DIVERSITY.
Antibody diversity is largely the result of the shuffling of gene sequences that code for both heavy and light chains. This figure shows the shuffling process during the formation of a gene for an immunoglobulin light chain. Great variability is achieved by combining different V, J, and constant regions. "Extra" DNA remains in the final gene, so the RNA primary transcript from the gene must be spliced to yield the final mRNA. As a result, a piece of RNA is excised.

It turns out that the gene sequences that code for the variable region also occur in several pieces. The sequences that give rise to the light-chain variable region occur in two pieces: the V (for "variable") segment and the J (for "joining") segment. For the heavy chain, the complete variable region is assembled from three different gene segments: the V, the J, and the D (for "diversity"). There are multiple copies of each of these segments, each coding for a slightly different amino acid sequence. For example, the human chromosome that carries the heavy-chain genes has about 80 V segments clustered together, at least 50 D segments clustered together, and 6 J segments in a group. When the variable-region sequences are moved next to a constant-region sequence, 1 V, 1 D, and 1 J segment are randomly selected, combined, and moved as a group. Based on this random association of different segments, a total of 80 × 50 × 6, or 24,000, different heavy-chain variable regions can be formed. In addition, when these three segments are joined, the joining can occur in about 100 slightly different ways, yielding even more total variations (24,000 × 100 = 2.4 million). Similarly, the different kinds of V and J segments in light-chain sequences yield at least 750 light-chain possibilities. The fact that there are about ten different ways to join the V and J segments greatly increases—to 7,500—the number of possible variable regions that can be formed.

To put all this together, there are at least five factors involved in generating antibody diversity in mammals:

1. Many V, D, and J segments are encoded in the genome and are passed from generation to generation by the germ cells.
2. When a B cell passes through the pre-B and naive B stages, it becomes committed, and the various segments for both heavy chains (V, D, J) and light chains (V, J) are randomly combined.
3. The joining of the segments can occur in slightly different ways to give rise to additional variation.
4. The binding sites are formed by random combinations of different heavy and light chains.
5. Finally, a process of actual somatic mutation appears to go on in the portions of genes that code for the hypervariable regions of heavy and light chains. Thus, new DNA base sequences not inherited from either parent suddenly appear in the variable-region DNA during the origin of pre-B cells, as well as later during clone development.

Such gene shuffling has been compared to filing slightly different notches in billions of keys so that each one opens a different lock. In this case, the matching of key and lock allows for the defense of the organism against a diversity of foreign intruders.

Monoclonal Antibodies: A Miracle for Medicine

Understanding B-cell clones and antibody production has led to the recent development of a procedure that promises to be as important to human medicine as the discovery of penicillin: the production of **monoclonal antibodies,** highly specific antibodies produced by a single clone of B cells.

Normally, even when a single antigen enters the body, different parts of it are recognized by different naive B cells, each determined to form antibodies with slightly different binding sites. As a result, not one but a whole set of B-cell clones is selected, and a mixture of antibody molecules is produced. In 1977, two researchers at Cambridge University, Georges Kohler and Cesar Milstein, discovered a way to obtain large amounts of just a single type of antibody.

First, they injected a mouse with an antigen (Figure 30-8, step 1). Next, they removed the mouse's spleen and obtained B cells from it, some of which were directed against the antigen that had been injected (step 2). Then they fused individual spleen B cells with a special kind of "immortal" cell called a *myeloma*—a cancer cell that divides very rapidly (step 3). Each resultant fused cell, called a *hybridoma,* expresses the characteristics of the two cell types: Each hybridoma clone can synthesize a huge amount of a single IgG antibody directed against a specific antigen and can divide again and again to yield a clone of millions of tiny, living factories, all producing this same antibody in huge quantity (step 4). Researchers use specific techniques to identify the particular hybridoma clone that makes the desired antibody.

The product of these hybridoma factories, then, the monoclonal antibody, is a pure reagent directed against a specific target (step 5). Medical researchers can use monoclonal antibodies to identify specific biological molecules and structures, such as hormones and hormone receptor sites on cell membranes. Researchers have used monoclonal antibodies in the basic study of diabetes, cancer, heart disease, rheumatoid arthritis, allergies, and the function of the brain.

Molecular identification also allows the diagnosis of diseases and conditions. Several commercial pregnancy tests for use in hospital laboratories and at home are now based on the ability of targeted monoclonals to detect the presence of very tiny amounts of human chorionic gonadotropin and thus to verify pregnancy as early as 10 days after conception.

Monoclonal antibodies have also been prepared against a large number of viruses (including the AIDS-causing virus), bacteria, fungi, and parasites to aid in the diagnosis of infectious diseases. Laboratory technicians, for example, can determine if a person has gonorrhea, herpes, or chlamydia infection with a high degree of accuracy (85 to 99 percent) in only 15 to 20 minutes (not the 3 to 6 days previously required). Thus, appropriate treatment can begin almost immediately.

Monoclonals can also be used to diagnose a host of other conditions, including anemia, pituitary insufficiency, infertility, and prostate and liver cancers. A variation on this targeting is the discovery of *abzymes,* antibodies whose binding site catalyzes a specific reaction involving the bound antigen.

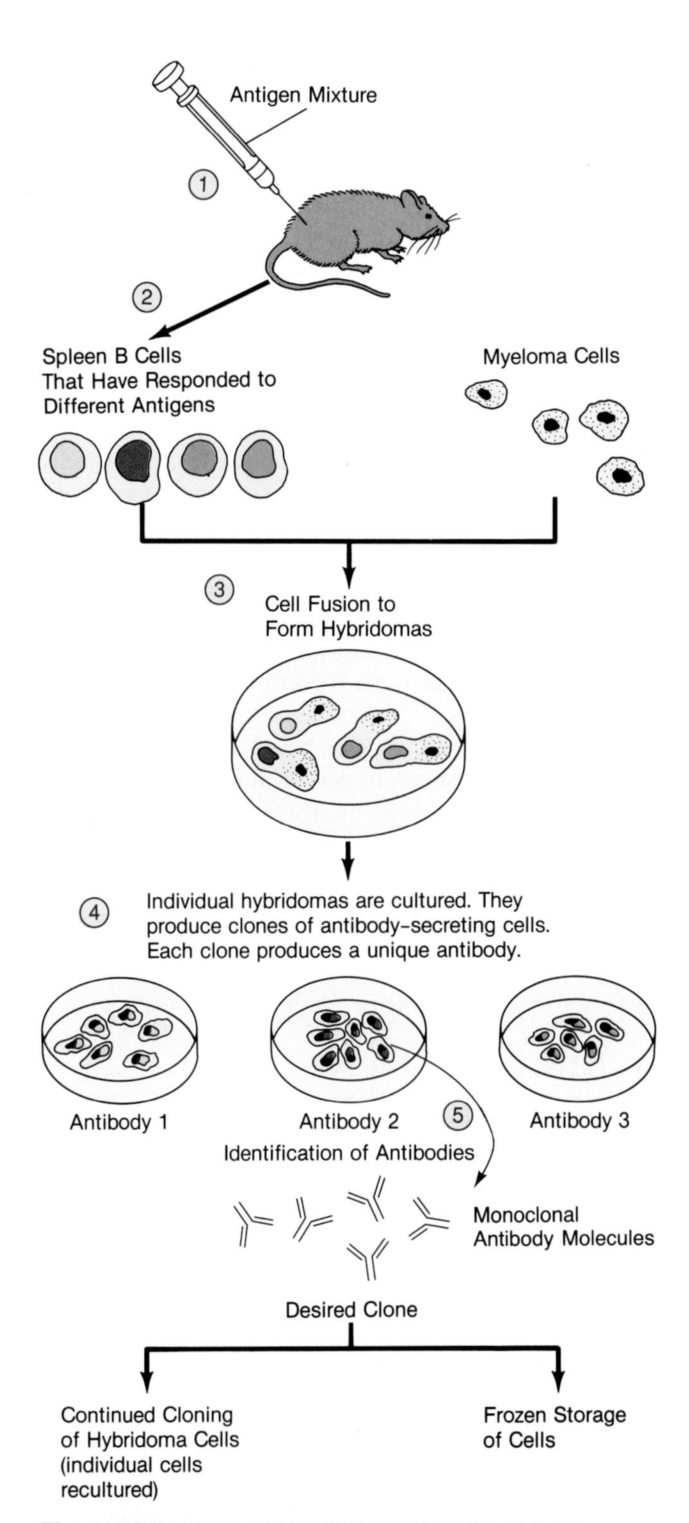

Figure 30-8 THE PRODUCTION OF MONOCLONAL ANTIBODIES.
Antigen-stimulated spleen cells are fused with myeloma cells, yielding "immortal" hybridomas. Each of them secretes a single, "monoclonal" antibody. Once the hybridoma secreting the desired antigen is identified, it is cloned to generate millions of antibody-secreting cells that yield the huge quantity of a single antibody needed in medicine or science. Some hybridoma cells may be stored frozen and later cloned for antibody production.

Finally, physicians have used monoclonals to help counter the rejection of transplanted organs, to rectify immune deficiency diseases, and to fight cancers by delivering lethal materials directly to specific cancer cells.

T CELLS AND CELL-MEDIATED IMMUNITY

The other major type of immune response besides humoral immunity is **cell-mediated immunity**—the direct attack of foreign cells or substances by T lymphocytes. Unlike B cells, T cells do not manufacture and secrete a large amount of antibody protein. Instead, they have antibody-like receptors on their surfaces that enable them to recognize and respond to specific antigens and then to interact with cells directly.

Immunologists divide T lymphocytes into three main types, each with distinctive surface labels: *killer T cells,* which attack foreign cells and substances directly; *helper T cells,* which assist certain B cells in producing antibodies and assist killer T cells in mounting an attack; and *suppressor T cells,* which seem to play a role in control of the immune system and in tolerance (Figure 30-9).

Committed immature T cells undergo a clonal selection process similar to that of B cells, recognizing the foreign antigen to which the T cell can respond and dividing into a large clone of T cells. Those cells with killer properties accumulate at the site of an infection, where they attack antigen-bearing bacteria, virus-infected cells, tissue transplants, parasites, and cancer cells. The attack may involve activating the complement response or synthesizing and inserting pore-forming proteins (perforins) in the foreign cell's plasma membrane so that it dies.

Macrophages play a role in bringing the antigen and immature T cell together. They do this by "presenting" the foreign antigen to the immature T cell. To become active, a cytotoxic T cell or a helper T cell must recognize both the foreign antigen *and* a surface macrophage marker molecule. Specifically, a foreign protein antigen binds to a macrophage and is phagocytized by the macrophage. The antigen is then chopped up, and a fragment of it is "presented" on the surface of the macrophage in a complex with an MHC molecule marker. This presentation of the antigen-MHC complex activates the T cell, as does the macrophage's secretion of a **lymphokine,** a lymphocyte-activating protein (see Figure 30-9b). As a result, the helper T cell secretes other lymphokines that activate killer T cells and possibly naive B cells with the identical antigenic specificity. These lymphokines are in a sense amplifying signals that mobilize

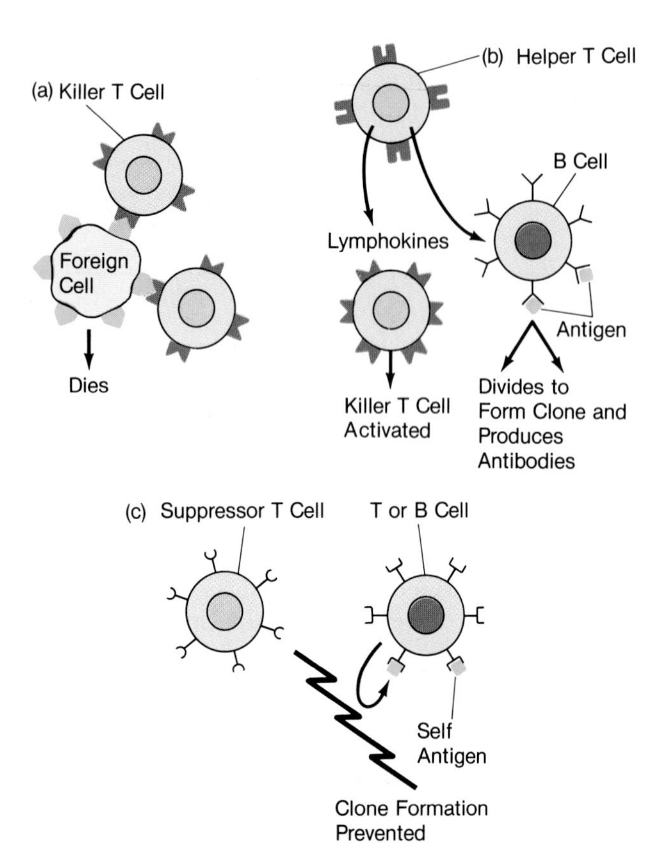

Figure 30-9 TYPES OF T CELLS.
(a) Killer T cells are activated by macrophages, and after forming a clone, directly attack a foreign antigen. (b) Helper T cells augment the function of B and T cells by secreting lymphokines that enable B cells to carry out a full set of responses to antigens and that enable killer T cells to attack foreign antigens. (c) Suppressor T cells apparently slow and halt the production of B or T cells once an immune response successfully eliminates foreign antigens. Suppressor T cells also may help maintain tolerance of self antigens. Thus, as shown here, they may block formation of active T or B cell clones. If this inhibition ceases, autoimmune disease may arise.

more and more killer T cells or B cells to respond to the foreign invader. One such substance is tumor necrosis factor, which is involved in inflammation, fever induction, and regression of certain tumors. It does the latter by killing cancer cells directly or by causing blood vessels supplying the tumor to rupture, thereby cutting off oxygen to the mass. In general, for a given kind of antibody to be made against an antigen, both a helper T cell and a B cell must recognize that antigen as foreign.

Finally, immunologists think that suppressor T cells may play a role in turning off immune responses or at least in limiting their magnitude and duration (see Figure 30-9c). Another, more widely investigated role of suppressor T cells is the active prevention of immune responses against self antigens. Thus, suppressor T cells may be involved in the tolerance process.

NATURAL KILLERS: THE THIRD CLASS OF LYMPHOCYTES

In the early 1980s, a third class of lymphocytes was discovered: natural killers (NK) (Figure 30-10). These large, granular lymphocytes, which arise in bone marrow, appear to lack binding sites for antigens on their surfaces. However, extensive lysosomal qualities are associated with their granules. NK cells, which make up about 10 percent of the body's lymphocytes, can rapidly attach to and lyse tumor cells while leaving normal cells unharmed. They apparently use perforins to build pores in tumor cell membranes, thereby causing cell death. NK cells are also found in the epithelial layers of the lungs, cervix, intestine, and male reproductive tract. In these locations, NK cells may participate in surveillance for foreign agents and can directly attack viruses and virus-infected cells.

Immunologists are not sure how NK cells can distinguish potentially cancerous cells or how they relate to T and B cells. Nevertheless, intensive research is showing that NK cells are certainly a line of defense against cancer and probably take part in many other immune responses.

AUTOIMMUNE DISORDERS: A BREAKDOWN IN TOLERANCE

Under certain circumstances, the normal phenomenon of tolerance breaks down, and the immune system can attack the body's own cells in a process called an

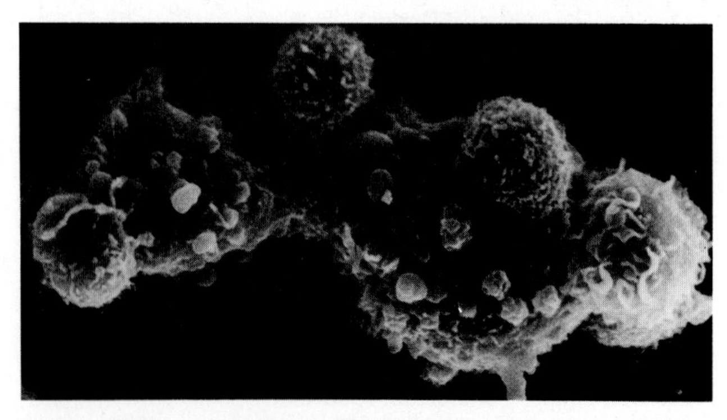

Figure 30-10 NATURAL KILLER (NK) CELLS.
Two large target cells with a number of smaller NK cells bound to their surface (magnified about 2,800 ×). The small bulbous extensions from the surface of the target cells are indicators of their imminent death.

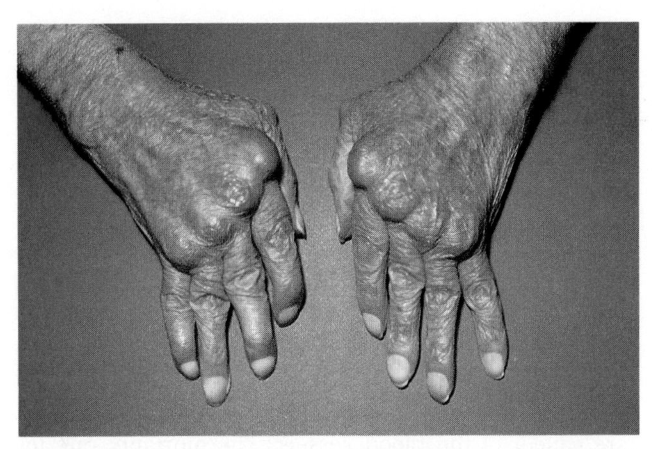

Figure 30-11 AUTOIMMUNE RESPONSE: A BREAKDOWN IN TOLERANCE.
Tolerance appears to be an acquired characteristic of the immune system based on the presence of antigen(s) during the immune system's development. Arthritis involves an attack by the body's immune cells on connective tissue in the joints or other organs.

autoimmune response. The result can be devastating diseases, such as arthritis and lupus erythematosus (Figure 30-11).

Tolerance, the recognition of one's own tissues as self rather than as nonself, begins in the embryo. Research suggests that during late embryonic development, the immune system takes an inventory of normal antigens already present in the body; all such antigens are thereafter defined as self. If transplanted foreign cells bearing their unique antigens on their surface are present at the time of the tolerance inventory, then they, too, will be considered self forever after.

Tolerance appears to be based on a failure of self antigens to stimulate development of clones of T or B cells. One explanation is *clonal deletion*, in which clones of maturing B and T cells with reactivity against the millions of self antigens may be selectively destroyed or inactivated during embryonic development. Another mechanism leading to tolerance may simply be the physical masking of certain antigens on the surface of some normal cells so that those antigens are not exposed to the immune system when it is developing. Tissues inside the eye and cells inside the seminiferous tubules of the testes, for example, are apparently not exposed to cells of the developing immune system. Penetrating wounds of the eye or testis in adulthood can be followed by a massive immune attack on the internal tissues. It is as though those masked cells were never inventoried during embryonic development and are mistakenly recognized as nonself by immune cells that see them for the first time during adult life.

Suppressor T cells may also bring about tolerance by transferring it. If one takes a mouse that is tolerant to foreign antigen A, removes some of the mouse's T cells, and injects them into another mouse, the second mouse immediately becomes tolerant to antigen A. It will not be able to mount an immune attack on antigen A, even though it should be able to.

When the elaborate system of tolerance breaks down, a number of grave medical problems can result. Either T or B cells may suddenly begin to attack normal cells or cell products and may produce a life-threatening situation. Victims of the disease myasthenia gravis suffer paralysis and sometimes die because their B cells generate antibodies that attack an enzyme essential to the functioning of the nerve-muscle junction. Another disease, lupus erythematosus, strikes women in their late teens and can attack internal organs and cause swelling and inflammation of the joints and connective tissue, similar to that in rheumatoid arthritis.

The most common autoimmune disease is arthritis, which affects more than 30 million people in the United States. Rheumatoid arthritis is a severe inflammation of connective tissue that can affect the joints, heart, lungs, blood vessels, and spleen. Numerous studies are currently under way to determine how the immune system causes this and other types of arthritis and to improve treatment of the diseases. Current treatments for autoimmune diseases include immunosuppressant drugs and, in severe cases, blood filtering to remove lymphocytes.

IMMUNE DEFICIENCY

The importance of the immune system is demonstrated most dramatically by individuals born without certain components of the system and by people who lose their immune response through disease.

A few individuals have even been born with complete **immune deficiency**—no immune response at all. Most victims die from infections shortly after birth. However, physicians quickly diagnosed the condition in one baby born in Texas in 1971 and confined baby David to a germ-free chamber. He survived there for more than a decade to become the longest-living person lacking both T- and B-cell systems. He left the chamber after receiving some promising treatments, but died shortly thereafter.

In 1981, a startling new immune deficiency disease started to show up: *acquired immune deficiency syndrome*, or *AIDS*, which is the result of a viral infection. Most of the early cases in the United States involved young homosexual men, users of intravenous drugs and their sexual partners, hemophiliacs, and persons from regions of Africa where the disease is believed to have originated. In Africa today, nearly 10 percent of sub-Saharan black populations may be infected, and the totals worldwide may exceed 5 to 10 million people. A high proportion of the victims have the human immunodeficiency virus (called HIV) in their bodies—a virus with an incubation time of up to 9 years or more. The AIDS virus invades macrophages and multiplies, but does not kill those cells. Instead, it interferes with the macrophages' important immune roles, and it causes the cells to secrete substances that may kill the victim's brain cells (Figure 30-12, right). The virus can also invade the central nervous system directly, and it invades and kills helper T cells so they cannot help killer T cells or B cells to mount attacks against foreign antigens (Figure 30-12, left). Although killer and suppressor T cells and B cells are present in AIDS victims, the deficit in helpers prevents their normal function. Because human AIDS victims show vastly reduced B- and T-cell responsiveness, their bodies are exceedingly susceptible to a host of pathogens—bacteria, protozoa, molds, yeasts, viruses, toxins—encountered in everyday life. Human victims sometimes suffer numerous localized and systemic infections and often succumb to pneumonia or to Kaposi's sarcoma, a rare cancer of the blood vessels. The mortality rate for AIDS has been more than 70 percent, and an intense research effort is now under way to learn more about the cause and treatment of this disease. Recent data suggest that the HTLV virus mutates at an alarmingly fast rate, with some of its gene sequences changing more than 30 percent in just 20 years. Even if researchers are able to produce an effective AIDS vaccine, this mutation rate suggests that the drug won't remain effective for long. AIDS is certainly the most dangerous disease to arise since the black plague in medieval times, and more than any other disease, it demonstrates the critical role of a normal immune system in day-to-day survival.

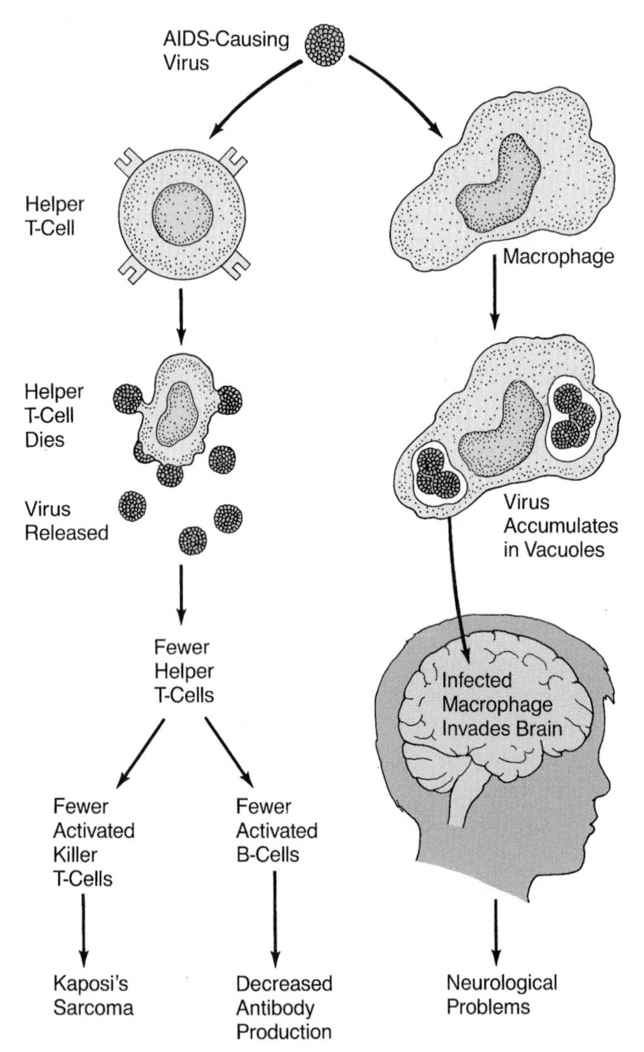

Figure 30-12 AIDS: THE NEW PLAGUE.
When the HIV virus infects helper T cells, it kills the infected immune cells. Without them, the body generates fewer activated killer T cells and B cells, and the result can be the cancer called Kaposi's sarcoma and a decrease in the production of antibodies to fight invaders—including the HIV virus. When the HIV virus infects macrophages, the virus accumulates in vacuoles, and such infected macrophages can invade and damage the brain and nervous system.

IMMUNE PHENOMENA AND HUMAN MEDICINE

Under normal circumstances, the immune system protects rather than attacks the body and, in fact, can be stimulated in various ways to provide still broader defense. Let us see how.

Active Immunity: Protection Based on Past Exposures

Most adults who had mumps, measles, chicken pox, and rubella (German measles) as children are immune to catching them again. Why is that? The answer is that a reservoir of memory cells specific to each particular disease-causing agent persists throughout life, able to mount a quick defense (secondary response) should the pathogens return. This type of immunity—based on prior exposure to pathogens—is called **active immunity** (Figure 30-13a).

Luckily, active immunity can be conferred by vaccinations and booster shots as well as by contracting (and surviving) diseases themselves. A vaccine contains a dead or attenuated (harmless) form of the virus or bacte-

(a) Active Immunity

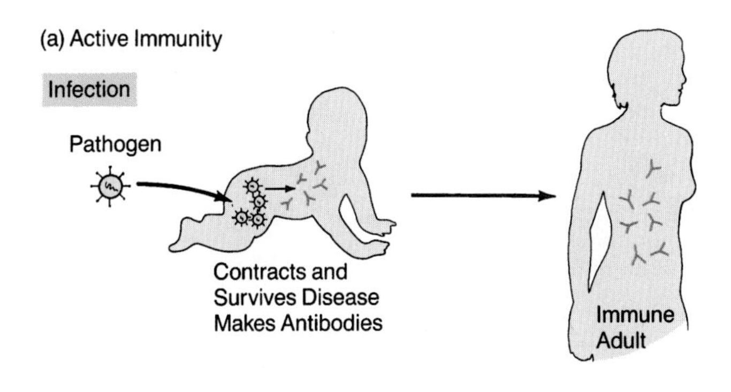

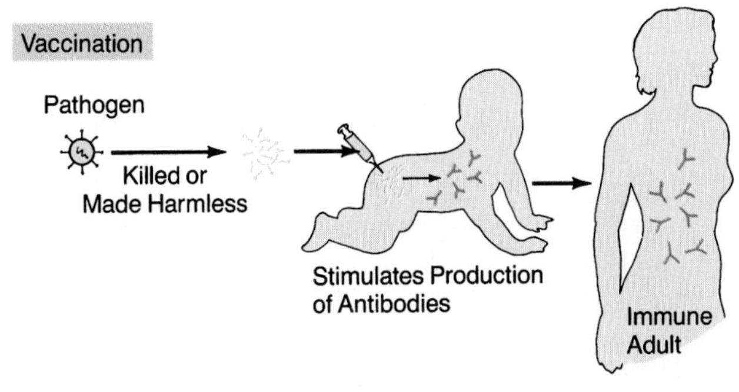

(b) Passive Immunity

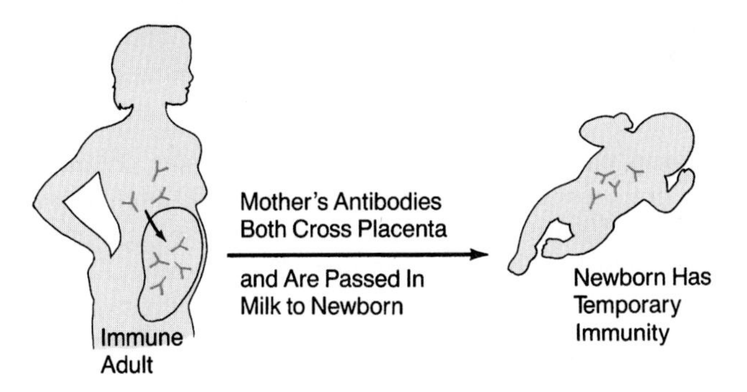

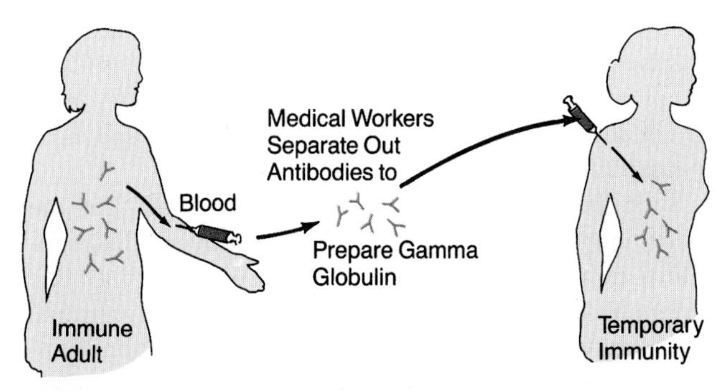

Figure 30-13 ACTIVE AND PASSIVE IMMUNITY.
(a) Physicians can stimulate active immunity by vaccinating patients with killed viruses, which stimulate a mild immune response. The resulting memory B cells can respond quickly to a new challenge by the same virus later in life.
(b) Physicians can also confer passive immunity by transferring preformed antibodies from one person (or laboratory animal) to another person.

rium responsible for polio, tetanus, diphtheria, whooping cough, or certain other communicable diseases. These agents, in turn, stimulate the generation of memory cells without actually causing the disease.

The concept of vaccination dates to the late eighteenth century and the English physician Edward Jenner. Jenner noticed that some milkmaids had developed scars on their hands and arms after milking cows infected with cowpox—scars that resembled those of the often deadly human disease smallpox. Following a hunch, Jenner gathered some pus from a cowpox sore on a milkmaid's hand, dipped a needle in it, and made a scratch on a young boy's skin. Several months later, Jenner gave the boy a dose of smallpox microbes, but no disease developed. (Such an experiment on humans would never be allowed today!) The boy had acquired immunity as a result of the cowpox vaccination. Jenner went on to vaccinate thousands of Londoners, protecting virtually all of them from smallpox.

Despite their benefits to medicine, vaccines are not without risk. If the vaccine maker does not completely inactivate a pathogen, the recipient can actually develop polio or another serious disease. A solution to this problem is currently under development: synthetic vaccines. Using new techniques, scientists construct short peptides that mimic a small region of a virus's protein coat; these peptides act as antigens that stimulate antibody production and confer active immunity. Guinea pigs injected with a short synthetic peptide that mimics part of the antigen from the foot-and-mouth disease virus developed immunity to even huge, lethal doses of the active virus. An experimental malaria vaccine based on synthetic peptides has fully protected monkeys from the disease and may be the basis for a vaccine against malaria that could save millions of human lives. And work is proceeding on similar vaccines against diphtheria, hepatitis virus, and other pathogens of humans and domesticated animals.

Passive Immunity: Transfer of Antibodies

In some mammals, certain antibodies can cross the placenta from the maternal blood into the fetal blood and so confer immunity to the fetus and newborn. This is an example of **passive immunity,** in which antibodies are not stimulated within an individual but are provided indirectly (Figure 30-13b) and can function for a few days or weeks. Other antibodies are passed to the newborn in colostrum (the first milk) and can protect the nursing infant until its own immune system begins to operate more fully (see Chapter 18).

A third form of passive immunity results from injecting a patient with *gamma globulin,* a mixture of IgG mol-

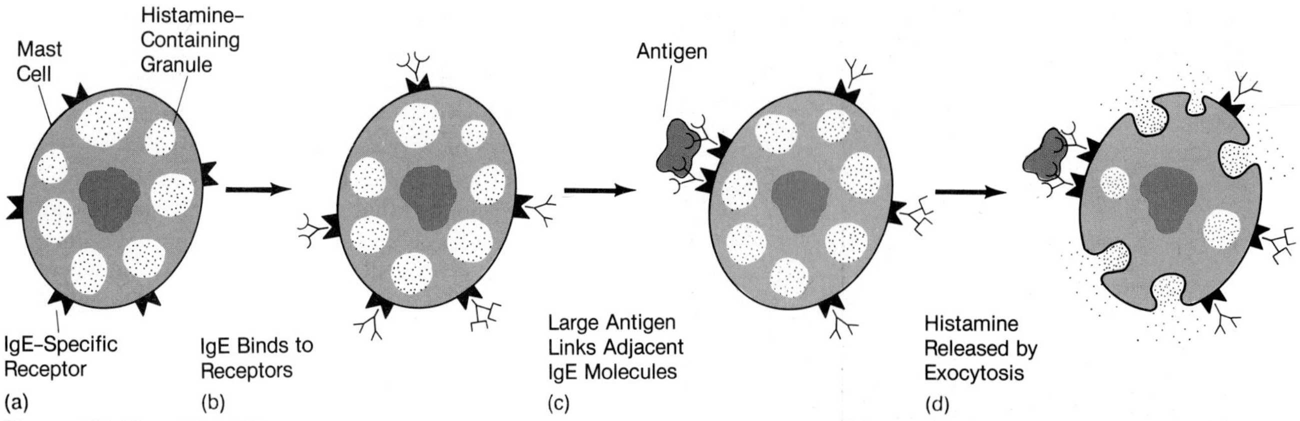

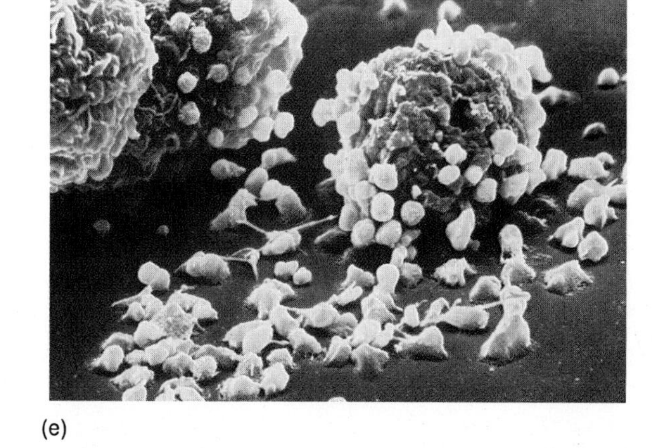
(e)

Figure 30-14 ALLERGY.
Mast cells release histamine by exocytosis in response to large antigenic agents, such as pollen grains, mites, and cat dander. (a) The mast cells have cell surface receptors for IgE and therefore (b) can bind a range of IgE molecules specific for different antigens. (c) A substance such as a pollen grain has bound to antibodies on this mast cell's surface, (d) triggering massive release of the cell's granules. (e) A mast cell (magnified about 2,500×) in the process of releasing its granules.

ecules directed against diverse antigens. Some of these IgG molecules have binding sites specifically directed against potential disease-causing agents and can protect the patient even though they are not produced by his or her own immune system. For example, physicians often recommend gamma globulin injections for people traveling to countries where hepatitis and certain other diseases are endemic.

Passive immunity can also be provided against the bite of certain venomous snakes by injecting *antivenins* (once called antivenoms). These contain massive amounts of IgG specifically directed against the particular snake venom antigens. The antibodies in the antivenin can go to work immediately to neutralize the snake toxin and prevent it from harming the victim further.

Allergy: An Immune Overreaction

Sensitivity to dust, cat dander, ragweed and other pollens, strawberries, rhubarb, mussels, and countless other "foreign" agents results from the action of the class E immunoglobulins. IgE molecules are bound to the surface of *mast cells* (Figure 30-14a and b), which are a type of granulocyte abundant in connective tissue, especially in the skin, lymph nodes, intestines, lungs, and membranes of the eyes, nose, and mouth. When the airborne particles contact the IgE molecules on mast cells in the lung or tracheal tissue (Figure 30-14c), these granulated cells are triggered to release their granules explosively by exocytosis (Figure 30-14d and e). Mast cell granules contain heparin, which helps prevent blood clotting, and histamine, which increases capillary wall permeability. The outpouring of these substances results in leakage of fluid from capillaries into tissue spaces; as a result, tissues swell, especially in the respiratory tract. Coughing, sneezing, and mucus secreting also occur—all aimed at expelling the foreign invader. There is also a loss of systemic blood pressure because of capillary leakage. A severe reaction of this type is called **anaphylaxis** or **anaphylactic shock.**

Such localized symptoms may be irritating, but the result is positive—the invader is removed. Unfortunately, in more than 35 million Americans, this mast cell system tends to overreact, producing **allergies**—chronic reactions induced by any number of benign environmental stimulants, from pollen grains to milk, eggs, wheat, peanuts, shellfish, and berries. Because mast cells are so widespread in the body, allergic reactions are correspondingly diverse, ranging from the sneezing and coughing of hay fever to the severe wheezing and breathing difficulties of asthma, to such digestive reactions as diarrhea, vomiting, or cramping. The most serious allergic reactions trigger anaphylactic shock, a life-

IMMUNITY AND THE MIND

A bell rings and a dog salivates, anticipating a helping of dog food. This is the classical example of a conditioned response: an automatic physiological reaction to a cue received in the brain, based on repeated training—in this case, the sound of a bell followed by a food offering. The Russian psychologist Pavlov discovered this response at the turn of the century, and 30 years later, a Russian emigré to Paris conducted an even more intriguing series of experiments based on conditioning.

S. I. Metalnikov stimulated the immune systems of rabbits and guinea pigs by injecting bacteria, and simultaneously blaring trumpets, scratching the animals' paws, or raising the temperature in their cages. Eventually, the sensory stimuli alone caused increased immune activity in the animals, even with no injection of bacteria.

Despite Metalnikov's early suggestion of a link between the immune and nervous systems, most biologists have considered the systems distinct and separate. Newer research, however, is pointing once again to a strong and direct link between immunity and the brain.

Glenda MacQueen and colleagues at McMaster University in Ontario injected rats with egg white protein (albumin) to stimulate an allergic response—the production of IgE antibodies and a massive discharge from the mast cells in the rat's mucous membranes. Simultaneously, the researchers exposed the rats to a humming fan and flashing strobe lights. Eventually, after several repetitions of the injections and the sensations, the rats experienced an allergic response when they simply heard the fan or saw the lights. Here, a modern research team using carefully controlled experiments found evidence of the same links that Metalnikov saw half a century ago: Somehow, the central nervous system can influence the immune system—perhaps even turn it on and off.

Some physicians have long suspected that emotions like anxiety and grief can depress the immune system and leave the body more vulnerable to attack. Does an immune system–nervous system link underlie such effects? In addition, physicians can't explain certain autoimmune conditions like Crohn's disease, ulcerative colitis, irritable bowel syndrome, and asthma, in which the body's defenses turn against parts of the body itself. Could episodes of these illnesses start with sensory and emotional signals? Continued study of a possible link between mind and immunity may someday provide answers.

threatening state during which the mast cells in many tissues discharge the contents of their granules simultaneously, leading to reduced blood flow to the brain and heart muscle and difficulty in breathing. This can be brought about by bee or spider toxins or drugs such as penicillin in an individual who is severely allergic to one of those substances.

Treatment of mild allergies usually includes administration of antihistamines, drugs that counteract the effects of histamine released from mast cells. Allergy researchers are currently investigating more fundamental ways to disarm the "firing pin," IgE, by chemically "tying it up" or suppressing its production. Some researchers predict that within a few years, allergies will be very effectively treated or stopped entirely.

Some fascinating new studies suggest a link between allergies, other immune overreactions, and the nervous system, as the box above explains.

Tissue Compatibility and Organ Transplants

Since about 1960, the transplantation of organs such as kidneys has become an increasingly important medical procedure. However, since all body tissues are labeled antigenically and can be recognized as self or nonself, the recipient's immune system is likely to reject a piece of skin, a kidney, or a transplanted heart unless the donor and the recipient are closely related. Researchers have, in fact, discovered antigenic protein labels called *histocompatibility* antigens (*histo-* means "tissue"), as well as a group of genes that code for these labels. These so-called **major histocompatibility (MHC) complexes** contain at least 4 closely linked genes in humans and as many as 50 alleles in total—which explains why two people chosen at random are so unlikely to have the same constellation of alleles and, in turn, compatible tissue types.

Obviously, transplantations do not go on in nature. What, then, is the biological role of histocompatibility genes? Antigen recognition and stimulation of lymphocytes to carry out immune responses may be their major role. Histocompatibility genes code for proteins that appear on most cell surfaces throughout the body. In immune system cells (T and B lymphocytes and macrophages), these proteins are involved in the ability to recognize foreign substances. For instance, when macrophages accumulate at a wound site and devour debris, they also stimulate the proliferation of T cells. For this stimulation to occur, the histocompatibility proteins on the macrophages and T cells must match. In addition, the histocompatibility proteins function as receptors that bind the foreign antigen in a remarkably slow and still not understood process. A T cell then recognizes the complex composed of the foreign antigen bound to the histocompatibility protein (as we saw in our discussion of cell-mediated immunity). In fact, the whole basis for distinguishing self from nonself has been linked to the fact that histocompatibility proteins may bind self molecules more tightly than foreign molecules.

Some histocompatibility complex genes specifically regulate the level of the immune response; hence, they are called *Ir* genes. Animals with deficient functioning of the *Ir* genes have defective helper and suppressor T cells. Perhaps the *Ir* genes regulate the entire immune response, and with it the body's day-to-day protection.

Histocompatibility genes also may affect human reproductive success. Couples that share many cell surface antigens (i.e., histocompatibility alleles) tend to have a high rate of spontaneous abortion. Early in pregnancy, the mother's immune system may fail to recognize the embryo as containing foreign genes and proteins. In the absence of this recognition, a blocking process that normally shields fetal antigens from the mother's immune system is not initiated. Consequently, her lymphocytes eventually attack the placenta, and the embryo is rejected.

PERSPECTIVE

With its numerous cells and organs, the immune system is a sort of microcosm of biology, acting independently at times and in close coordination at other times. Yet, the immune system also has unique properties—in particular, the gene shuffling that leads to specific targeted antibodies for fighting invaders. Because the immune system has both universal and special features and has important implications for medicine as well, biologists will continue to investigate it intensively.

SUMMARY

1. The *immune system* includes white blood cells, lymph nodes, spleen, thymus, and antibody molecules.

2. White blood cells include granulocytes, *macrophages*, *T* and *B lymphocytes*, and *natural killers (NK)*. T lymphocytes arise in the bone marrow and mature in the thymus; B lymphocytes arise in the bone marrow or liver of the embryo and in peripheral lymphoid tissues of adults.

3. *Lymph nodes* are small, thickened masses of tissue that filter lymph and contain millions of lymphocytes and macrophages.

4. The *spleen* is a small, oval organ that stores macrophages, lymphocytes, and red blood cells.

5. The *thymus* is a spongy, spherical organ that produces a hormone and is a site of T-lymphocyte differentiation.

6. Nonspecific defense mechanisms include physical barriers and the *inflammatory response*. Specific immune defense mechanisms involve B lymphocytes and antibodies in *humoral immunity;* T lymphocytes in *cell-mediated immunity;* and the natural killer (NK) cells.

7. *Antibodies* are globular blood proteins produced by B lymphocytes in response to the presence of an *antigen*. An antigen is any substance that elicits an immune response.

8. Type IgG antibody molecules each have two identical *heavy chains* and two identical *light chains*, each with *constant* and *variable regions*. The binding site is formed between the *hypervariable regions* of light and heavy chains. The variable regions determine an antibody's specificity; the constant region determines an antibody's class: IgG, IgM, IgD, IgA, or IgE. Antibodies, killer T cells, and NK cells may kill a target cell by assembling pores in the cell's plasma membranes.

9. The *clonal selection theory* states that during early development, cells pass through pre-B and naive stages during which they become committed to making antibodies with only one specifically shaped binding site. Foreign substances stimulate naive B cells with appropriate binding sites to divide rapidly and repeatedly. The resulting clone of *plasma cells* actively secretes an immense number of antibody molecules targeted to the antigen. *Memory cells* also remain available throughout the individual's life.

10. The immense diversity of variable regions (and hence antibody binding specificities) is based on the rearrangement of gene segments coding for heavy- and light-chain variable and constant regions.

11. A hybridoma created by fusing an antigen-stimulated B cell and a myeloma cell can generate *monoclonal antibodies*, all specific for a given antigen.

12. Killer T cells attack foreign substances; helper T cells produce lymphokines that activate antibody production and killer T-cell attack; and suppressor T cells control the immune system and turn off the immune response.

13. Natural killers are lymphocytes that attack tumor cells, viruses, and virus-infected cells.

14. *Autoimmune response* diseases, such as arthritis, result from a breakdown in tolerance. Tolerance depends on the recognition of one's own tissues or molecules as self rather than nonself; tolerance arises as the immune system develops.

15. *Active immunity* is defense based on exposure to disease pathogens or vaccination. *Passive immunity* results from the transfer of antibodies.

16. *Allergies* represent a reaction of the immune system's mast cells to a foreign substance such as pollen or a particular food.

17. Transplanted organs are usually rejected because of histocompatibility antigens on the surfaces of all tissue cells.

KEY TERMS

active immunity
allergy
anaphylaxis (anaphylactic shock)
antibody
antigen
autoimmune response
B cell (B lymphocyte)
cell-mediated immunity
clonal selection theory
complement response
constant region
heavy chain
humoral immunity
hypervariable region
immune deficiency
immune system
immunoglobulin
inflammatory response
light chain
lymph node
lymphokine
macrophage
major histocompatibility complex
memory cell
monoclonal antibody
naive B cell
natural killer (NK) cell
passive immunity
peripheral lymphoid tissue
plasma cell
spleen
T cell (T lymphocyte)
thymus
variable region

QUESTIONS

1. Describe each type of cell that is part of the immune system, and explain its functions.

2. Describe the organs of the immune system, and explain their functions.

3. What are antibodies? Which cells synthesize them, and how? Does each antibody-producing cell make only one kind of antibody? A few kinds? Many kinds? Draw a diagram of an IgG antibody, and label all of its parts.

4. Describe some features of specific defense.

5. What is clonal selection? Explain its relation to a primary response and a later secondary response.

6. Discuss the factors that contribute to antibody diversity, from genes through antibody assembly.

7. What are hybridomas? What are monoclonal antibodies? Do they differ from ordinary antibodies?

8. Severe combined immune deficiency (SCID) is an *X*-linked disease in which the individual lacks B and T cells. What sort of prognosis would you expect for an individual with SCID? What is AIDS? What, specifically, is wrong with the immune system of AIDS victims?

9. What is an antigen? What is meant by active immunity? Passive immunity? Autoimmunity? Tolerance?

10. What is anaphylactic shock? How can it result from a bee sting or from eating egg yolk in an allergic person?

ESSAY QUESTION

1. The study of abnormalities or diseases often leads to an understanding of normal biological processes. Discuss an example from the immune system.

For additional readings related to topics in this chapter, see Appendix C.

31

Respiration: The Breath of Life

Like a fountain, the living organism retains its improbable configuration by borrowing sources of energy from the world around it and by conferring and re-conferring organisation upon the matter which is ceaselessly flowing through it. And, in order to do this, to exploit the energy resources of the substances it borrows from the outside world, the cell must have oxygen.

Jonathan Miller, *The Body in Question* (1979)

"Breath of life" is more than just a well-turned phrase; we and the vast majority of other multicellular organisms are utterly dependent on a supply of oxygen. This ubiquitous element is needed to meet the collective demands of our individual cells as they consume oxygen during the transfer of stored energy from nutrients to ATP, the universal energy currency. Eighteenth-century French chemists held that "life is combustion." In all eukaryotes, the flame of life will not burn for long without oxygen.

The process by which organisms exchange gases with the environment is called called **respiration.** Respiration includes the intake of oxygen, the transport and delivery of oxygen to cells, and the removal and release of carbon dioxide. A whole organism's respiration is not to be confused with cellular respiration (see Chapter 7), during which mitochondria manufacture ATP as glucose and other organic molecules are metabolized.

Most animals have special sets of respiratory organs, such as gills or lungs, as well as cilia or muscles, to help

A column of ocean spray and water vapor rises from the blow hole of a gray whale *(Eschrichtius robustus).*

pump in a steady supply of air or of water containing dissolved oxygen. This chapter considers the respiratory organs in detail and how they function in concert with the circulation of blood to ensure a constant supply of oxygen to all cells, whether the organism is resting or exercising strenuously.

The chapter discusses:

- The physics of transporting gases and how those principles affect an organism's oxygen supply
- The kinds of respiratory organs and systems that have evolved in the animal kingdom, including the structural trends toward expanded surface area; maintenance of a wet surface for gas exchange; and efficient ventilating or pumping mechanisms
- The human respiratory system with its branching air pathways and bellowslike lungs
- The complex respiratory systems of birds
- The special buoyancy organs of fish
- How hemoglobin and other respiratory pigments help to transport blood gases
- Physiological controls over breathing and respiration
- How animals adapt to oxygen-poor environments

RESPIRATION AND THE PHYSICS OF GASES

Why are the cabins of commercial airliners pressurized? The answer is that air pressure affects the efficiency of every breath we take and, in turn, the intake of oxygen so critical to life. Like other physiological systems, the respiratory system demonstrates how life processes must conform to the chemistry and physics of matter.

Air pressure is created by the weight of the gases surrounding the earth in a layer several hundred miles thick. Every square inch of the ocean and land surface near sea level is pressed on by a column of air 1 in. by 1 in. by about 200 mi high. This column weighs about 14.7 lb; thus, the pressure of this weight of gases, called **atmospheric pressure,** is 14.7 lb/in.^{2} at sea level. The standard way to measure atmospheric pressure is to see how high that column of air will push a column of water in a U-shaped tube (about 10 m at sea level) or a column of mercury, Hg (about 760 mm at sea level) (Figure 31-1).

Since air contains a mixture of different gases, the full measurement of 760 mm Hg is the total pressure of that mixture, with each gas contributing only part of the column's weight and the pressure it exerts. Oxygen occupies approximately 21 percent of the air by volume and thus exerts a pressure of 0.21×760, or 160 mm Hg.

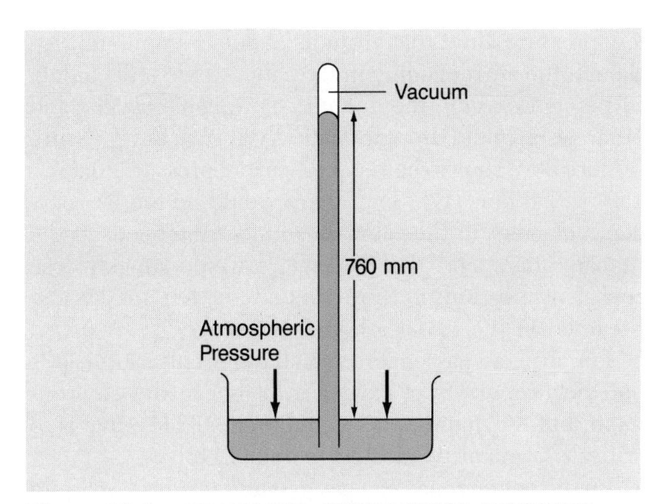

Figure 31-1 MEASURING ATMOSPHERIC PRESSURE. At sea level, the atmospheric pressure will support a column of mercury 760 mm high.

Likewise, nitrogen and carbon dioxide exert pressures of 593 mm Hg and 0.2 mm Hg, respectively. These figures represent the **partial pressures** exerted by each gas and are abbreviated P_{O_2}, P_{N_2}, P_{CO_2}, and so on.

Returning to the commercial airliner in flight at 35,000 ft, let us say, what would happen if a cabin wall were punctured? The air in the cabin, maintained artificially at an air pressure similar to that of Denver (about 5,000 ft above sea level), would rush out into the thin,

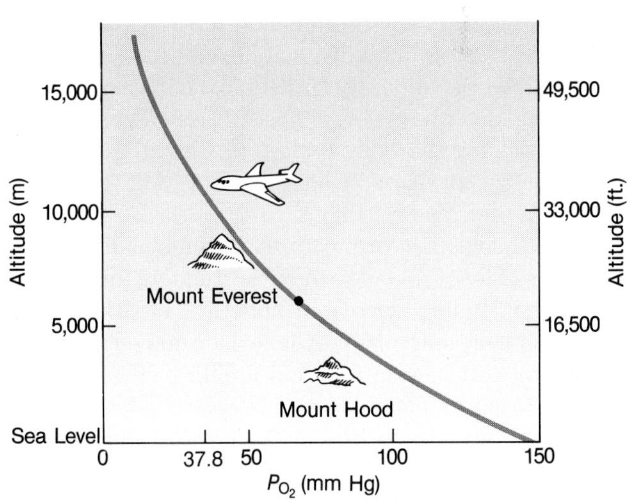

Figure 31-2 PARTIAL PRESSURE OF OXYGEN AND ALTITUDE. Most unacclimated humans will lose consciousness because of the low oxygen levels at altitudes around 6,000 m (60 mm Hg or less). Thus, airplane cabin pressures are maintained at the equivalent of 1,500 m (5,000 ft). At the summit of Mt. Everest, a person acclimated to high altitude can survive for a few hours breathing air with oxygen at a partial pressure of about 40 mm Hg.

low-pressure air at that altitude (about 180 mm Hg). Although the air remaining in the cabin would still contain 21 percent oxygen, the *quantity* of available oxygen and other gases would drop dramatically at such low pressure (Figure 31-2 shows the reduced partial pressure, $0.21 \times 180 = 37.8$ mm Hg). Soon there would be too few oxygen molecules in the cabin air to meet the needs of the human brain, and the passengers would quickly lose consciousness unless they donned oxygen masks and breathed in the gas at a higher pressure.

For airplane passengers and all other eukaryotes, it is not the *proportion* of oxygen available in the environment, but the *quantity* delivered to the tissues, that is so critical—a quantity affected by four factors.

First, the tendency of a gas to enter an adjacent liquid increases as the partial pressure of the gas rises. In a glass of carbonated soda, the partial pressure of carbon dioxide (P_{CO_2}) in the liquid is higher than in the air above it. That is why the CO_2 diffuses out of the liquid and forms gas bubbles that rise to the surface. Likewise, if P_{CO_2} in a cell is higher than in the surrounding extracellular fluid, the dissolved CO_2 molecules will diffuse outward from the cell to that fluid (Figure 31-3). Since CO_2 levels in the extracellular fluid build up so that P_{CO_2} is higher than in the blood, CO_2 diffuses into the blood plasma and is carried away. Conversely, since P_{O_2} in air is high, O_2 diffuses from air to blood, from blood to tissue fluid, and from tissue fluid to cell; this is because P_{O_2} is successively lower in each of these compartments in living animals.

The solubility of gases in liquids is a second important factor governing O_2 and CO_2 availability and movement in organisms and cells. **Solubility** is defined as the amount of a gas that will dissolve in a specified volume of liquid at specified temperatures and at a pressure of 1 atmosphere (760 mm Hg). The solubility of gases in liquids is usually quite low. For example, in 100 ml of blood plasma, there is only 0.5 ml of O_2, whereas in 100 ml of air, there is 42 times more (21 ml). The low solubility of gases in liquids has had several effects on the evolution of large, complex organisms, the most significant being the evolution of specialized pigments, such as

THIN AIR AND THE HUMAN BRAIN

One small but important branch of physiology deals with the effect of high altitude on animals—including human mountain climbers (Figure A), and many researchers have studied what happens to the human brain at extreme altitudes of 5,800 m (18,850 ft) or more. At such heights, an organism faces bitter cold, high winds, low humidity, and high levels of solar and ultraviolet radiation. By far the most important physical challenge, however, is *hypoxia*, low levels of oxygen reaching the body tissues; this is caused by low barometric pressure, which drops steadily with increasing height (see Figure 31-2).

Physiologists have measured a number of distinct effects of low P_{O_2} and extreme altitude on the human body, including decreased appetite, breathlessness even at rest, muscular fatigue, exhaustion on attempting any sort of exertion, and a falling off of mental capacity and memory.

As a climber presses on to greater and greater heights, brain tissue is further deprived of oxygen, and bizarre mental states often occur. Some people suffer frank hallucinations, seeing highway equipment, skiers, trees, and dead mules on a mountain's summit.

On a daring assault of Mt. Everest, Peter Habeler and Reinhold Messner carried no oxygen, and Habeler recalls being overtaken at the summit by a treacherous sense of euphoria that no harm could befall him. This feeling passed quickly, and Habeler suddenly feared that he was suffering severe brain damage. He commenced sliding thousands of feet down the mountainside and reached base camp in less than an hour. Unfortunately, Habeler did notice carryover effects after the ascent, including lapses of memory and nightmares. In fact, physiologists are concerned that a number of climbers have sustained massive nerve cell death in the brain.

Figure A
Climbers laboring upward, during a scientific expedition to Mt. Everest.

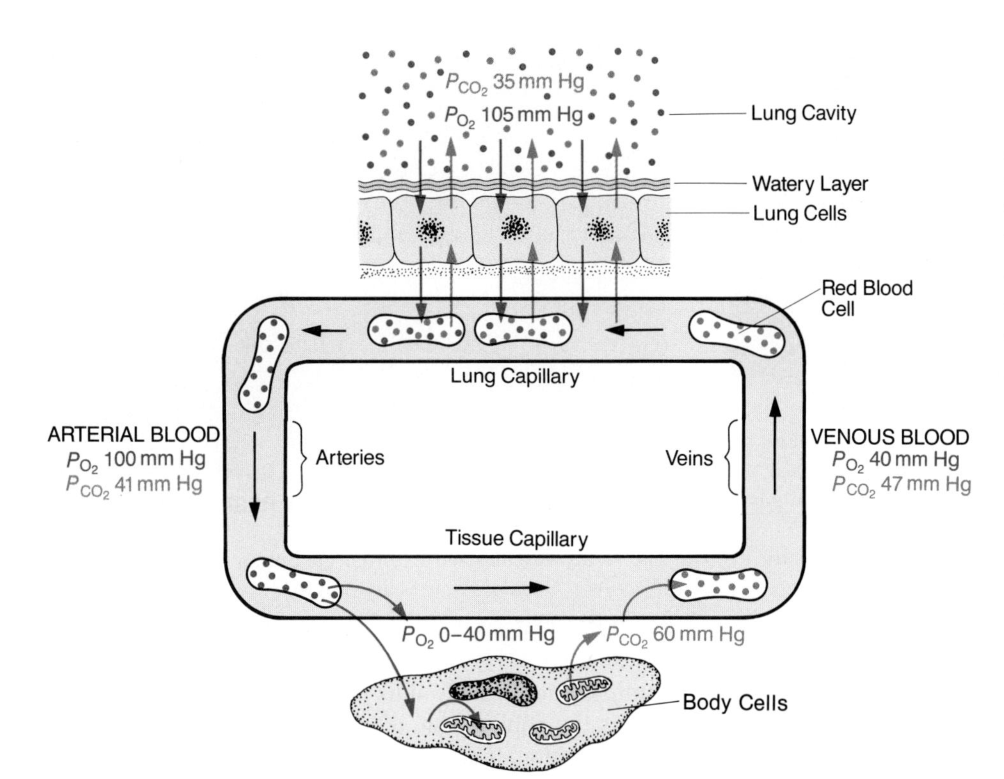

Figure 31-3 GAS EXCHANGE IN THE LUNGS AND TISSUES. The delivery of oxygen (red) to body cells and the elimination of carbon dioxide (blue) from lung cells involve diffusion into and out of the alveolar lung cells, tissue fluids, capillaries, and red blood cells. Oxygen from the lungs diffuses into the blood, and CO_2 diffuses in the opposite direction. When oxygenated blood passes through a tissue capillary, O_2 diffuses out of the blood into the extracellular fluid and then into nearby cells; again, CO_2 moves in the opposite direction.

hemoglobin, that can carry a large amount of O_2 in the blood.

The temperature of the surrounding environment, of the respiratory organs, of blood (if the organism has any), and of the cells is a third set of crucial determinants of gas contents and the ease with which dissolved gases diffuse into and out of cells. If, for instance, a cold-water fish, such as a trout, moves into warm water (which holds less dissolved oxygen), the fish will be unable to take in as much oxygen as before through its gills, even though its rate of metabolism rises in the warm water and its cells require more oxygen. Clearly, it must quickly escape this situation or perish.

The fourth factor that influences the amount of gas in a fluid is the rate of diffusion. For O_2 to enter an organism's tissues or cells, it must dissolve in the fluid surrounding the cell and then it must diffuse through that fluid to the cell surface. Gases diffuse much more slowly in liquid than in air, severely limiting the distance that cells can be from a source of O_2. Organisms dependent solely on diffusion of O_2 and CO_2 through the body surface must be very small, flat, or porous, like a sponge (see Chapter 29). Larger, thicker organisms use bulk flow to transport gases and other substances. Breathing air into lungs and pulling water across gills are methods of bulk flow that bring volumes of gaseous or dissolved O_2 to special respiratory surfaces, where O_2 can be picked up by the blood and carried to all tissue cells.

RESPIRATORY ORGANS: STRUCTURES FOR EFFICIENT GAS EXCHANGE

A primary characteristic of animals is the presence of a gas exchange organ with a large enough surface area that sufficient intake of oxygen and release of carbon dioxide can occur to meet the needs of all body cells. Let us survey four types of respiratory systems that have evolved in animals: the wet body surface of small organisms; the *gills* of aquatic creatures; the *tracheae* of insects and related animals; and the *lungs* of terrestrial vertebrates.

Wet Body Surface as a Site for Gas Exchange

The cell membranes of single-celled and colonial organisms, such as prokaryotes, fungi, sponges, and small flatworms, can provide adequate exchange of oxygen and carbon dioxide by means of diffusion alone. In earthworms, marine worms, lungless salamanders that inhabit ice-cold mountain streams, and certain kinds of frogs, respiration involves both diffusion and bulk transport. Oxygen diffuses through the moist body surface and into

Figure 31-4 SKIN AS A RESPIRATORY ORGAN. The Lake Titicaca frog (*Telmatobius culcus*), which lives submerged for weeks in the depths of Lake Titicaca, has small lungs, but relies when submerged exclusively on gas diffusion through the skin for obtaining oxygen and eliminating carbon dioxide. The folds of skin on the body and legs provide increased surface area for gas exchange.

Figure 31-5 GILLS: ORGANS FOR RESPIRATION IN WATER. The richly vascularized external gills of an albino mudpuppy (*Necturus maculosus*) are easy to see.

blood capillaries that lie just below the skin surface, and carbon dioxide follows the reverse pathway, moving from body cells to extracellular fluid, to blood vessels, to skin capillaries, and then diffusing out through the moist skin (Figure 31-4). Recent studies show that frogs can regulate blood flow to the skin and hence the exchange of O_2 and CO_2. The rates of metabolism are low enough in these animals that no specialized, highly efficient respiratory organs are needed. However, there is an ecological price: Such organisms can only live where their skins will stay moist constantly. In addition, they may be vulnerable to predators, since their skins cannot be covered with impermeable armor. With the other types of respiratory organs—gills, tracheae, and lungs—the gas exchange surface remains moist and permeable, but the animal itself develops an impermeable outer layer.

Gills: Aquatic Respiratory Organs

Many kinds of aquatic animals have specialized gas exchange organs called **gills** that provide both a large surface area and a very short distance for oxygen and carbon dioxide to diffuse between the surrounding water and the blood (Figure 31-5). Thus, gills have a rich supply of blood separated from the outside water by just a few thicknesses of cells of the capillary wall and gill epithelium. Blood in the capillaries picks up O_2 that diffuses the short distance from the water into the gill epithelial cells and then through the capillary wall. Carbon dioxide (often moving as HCO_3^-, bicarbonate) diffuses in the opposite direction, that is, down its concentration gradient.

Fish gills have a special design feature in which water and blood fluids flow in opposite directions, improving oxygen uptake. A trout, for example, performs repeated pumping movements of its jaws and opercula (an operculum is the flaplike side of a bony fish's head that covers the gill region) that pull water into the mouth and force it past the gills and out the opercular openings. At the rear of the mouth cavity, in the pharynx, water passes sideways over the gills' gas exchange surfaces. Simultaneously, blood in gill capillary beds flows in the opposite direction—from the lateral edges of the gills toward the medial (Figure 31-6). This opposite flow of two fluids separated by a permeable interface is called a **countercurrent exchange system** (Figure 31-7). Countercurrent flow leads to far more complete uptake of O_2 from water to blood than would occur if both flowed in the same direction. Without countercurrent exchange, a fish could theoretically extract only 50 percent of the dissolved O_2 at best; with it, some fish can extract up to 85 percent of the O_2 dissolved in seawater.

Aquatic creatures must expend substantial energy pumping a large amount of water over their respiratory surfaces so they can "harvest" enough O_2. A trout, for example, must pump its operculae 400 times for each single breath of air taken by the lungfish, and the trout must expend 10 to 20 percent of its total body energy to ventilate its gills, compared to the lungfish's 1 to 2 percent. In some aquatic animals, locomotor movements aid the pumping of water through the gills: Fast-moving fishes such as mackerel swim with their mouths open in "ram-jet" fashion, causing a large amount of water to move over their gill surfaces. Gills represent a distinct evolutionary advance over simple diffusion through wet body surfaces. As animals began to colonize dry land, gills lost their usefulness, since their exquisitely thin, pliable exchange surfaces collapse together when not supported by water, preventing efficient exchange of gases. In land animals, tracheae and lungs evolved.

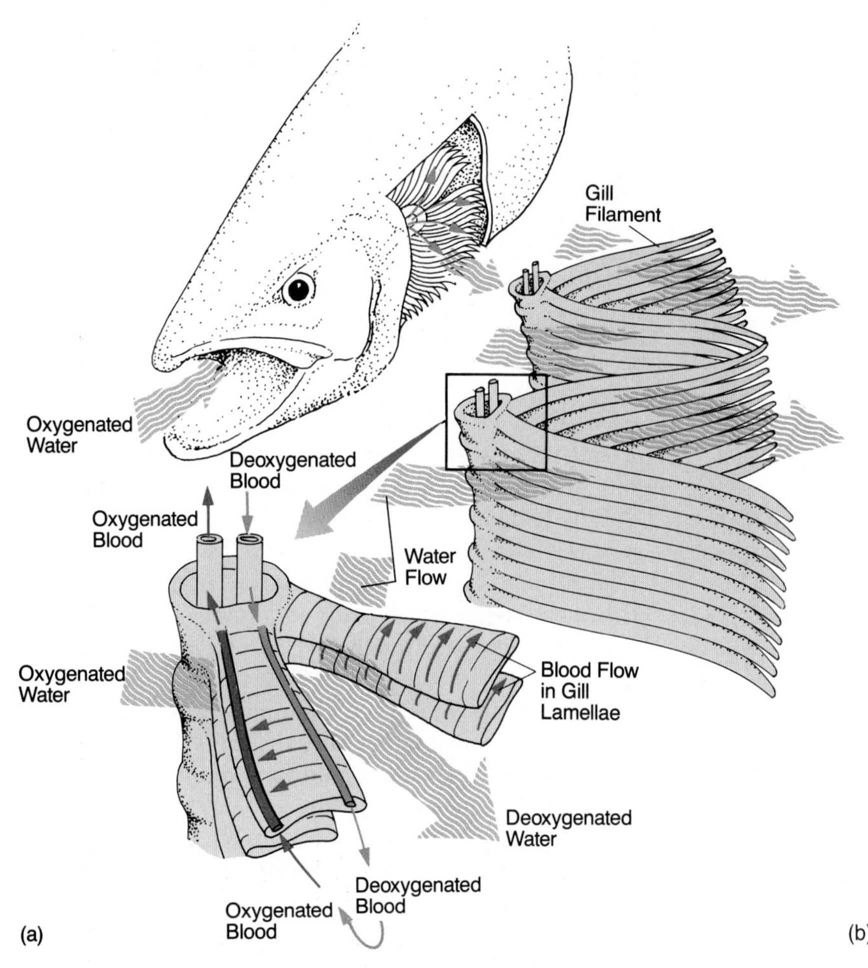

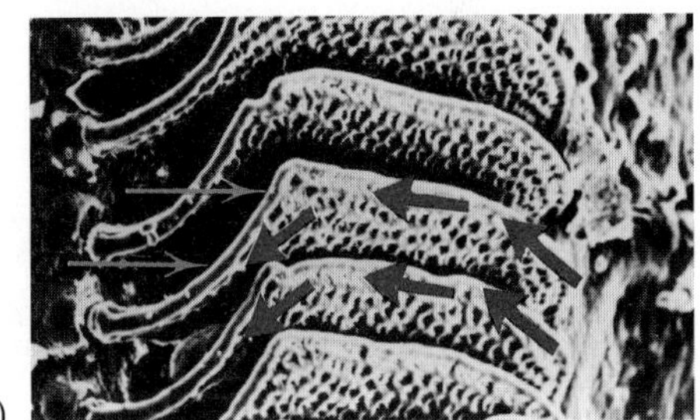

Figure 31-6 GAS EXCHANGE IN THE FISH GILL.
(a) Water containing dissolved oxygen enters the mouth cavity and then passes between the gill filaments. The transfer of oxygen from the water to the blood takes place at the gill filaments, where the capillaries in the gill lamellae are located. The flow of water between adjacent gill filaments (turquoise arrows) is in the opposite direction from the flow of blood (red arrows). This type of countercurrent exchange system facilitates maximum extraction of oxygen from water. (b) This scanning electron micrograph (magnified about 56×) shows a plastic cast of the capillary beds found in a set of gill lamellae, with their huge lacy surface area for O_2 and CO_2 exchange. Blood flows from right to left (red arrows), and water (turquoise arrows) in a countercurrent direction.

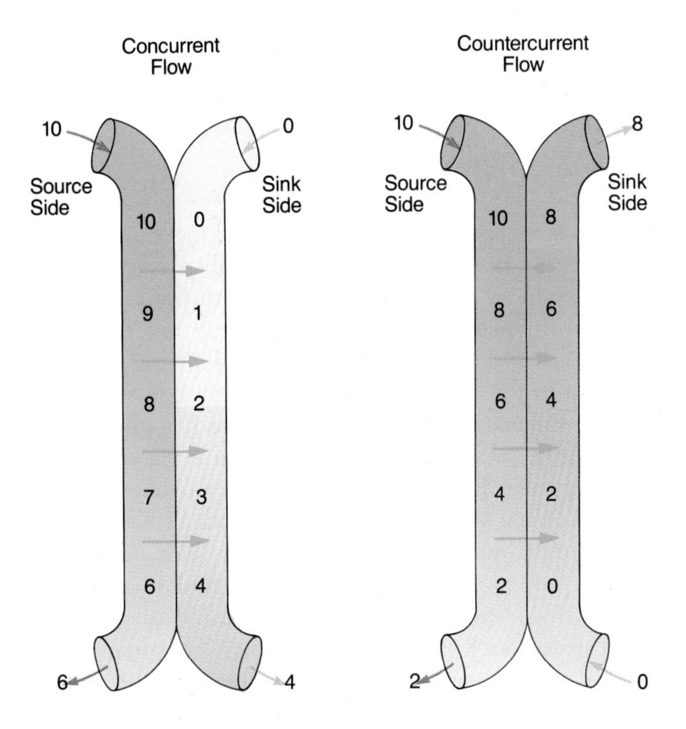

Figure 31-7 COUNTERCURRENT FLOW: AN EFFICIENT TRANSFER METHOD.
Compare the degree to which units of a commodity—such as oxygen, lactic acid, or heat—are transferred when fluids flow in the same (concurrent) direction and when they flow in opposite (countercurrent) directions. In this example, only four of ten units are transferred in the concurrent exchange system (40 percent efficiency), whereas eight of ten units are transferred in the countercurrent exchange system (80 percent efficiency). The source is the site from which a commodity moves, and the sink is the site to which it goes.

Tracheae: Respiratory Tubules

Insects and most terrestrial arthropods possess **tracheae** (singular, *trachea*): air-filled hollow tubes that branch and rebranch into a fine network of air passages that penetrate the animal's body and deliver air directly to tissues deep in the interior (Figure 31-8a). In some species, air sacs connected to the internal passages serve as temporary air reservoirs. Like gills, tracheae provide a large surface area for gas exchange and a short diffusion path to the capillaries and tissue fluids.

On examining almost any insect, one can see a set of tiny "portholes" that run in a line down the animal's side. These portholes are closable vents called *spiracles*, which lead into the tracheae. The tiniest tracheae deep in the tissues are called *tracheoles*, and they eventually reach close to every body cell (Figure 31-8b). The tip of each tracheole, sometimes called an "air capillary," is usually filled with a watery fluid. When an insect is resting, air travels into the open spiracles, moves throughout the body in the tracheae, and reaches the fluid-filled portion of the tracheoles. Oxygen diffuses into the fluid, then through the walls of the tracheoles, and finally into the hemolymph, which bathes the surfaces of most tissue cells as it moves through the insect's open circulatory system.

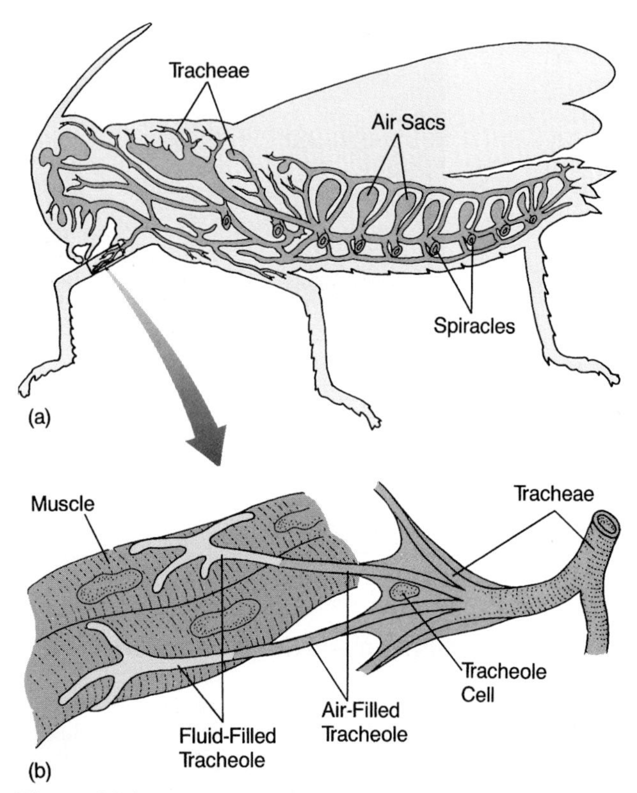

Figure 31-8 INSECT TRACHEAE: HOLLOW TUBULES FOR EFFICIENT RESPIRATION.
The respiratory system of insects consists of tracheae, which open to the exterior by way of spiracles. (a) The tracheae branch to form smaller, hollow tracheoles, which convey gases to and from all parts of the body. Ventilation is based mostly on compression of the air sacs through muscular action of the abdomen. (b) The terminal portions of tracheoles are filled with fluid through which gases must diffuse. In active tissues, such as muscles, some of this fluid is drawn osmotically out of the tracheole tip; this, in turn, brings more oxygen-containing air closer to the tissue cells.

Table 31-1 OXYGEN USE UNDER CONDITIONS OF REST AND ACTIVITY

Organisms	*Weight*	*Oxygen Consumption (ml/kg/hr)*
Butterfly	0.3 g	600 at rest; 100,000 while flying
Mouse	20.0 g	2,500 at rest; 20,000 while running
Human male	70 kg	200 at rest; 4,000 while working maximally

When an insect is flying, jumping, or running and its cells are metabolizing rapidly, the fluid drains from the tracheole tips, and air is able to penetrate farther into the empty tracheoles and deliver O_2 faster to the tissues because the diffusion pathway through the residual liquid is shorter. Tracheal tubes also carry CO_2 away from the hemolymph and tissue cells and release it through the spiracles.

In active insects, such as bees and grasshoppers, pumping action brought about by muscle contractions as the animal flies, runs, or jumps further increases O_2 supplies just as the insect becomes more active. Table 31-1 shows just how dramatically the need for oxygen can increase during activity.

Lungs: Intricate Air Sacs for Gas Exchange

Lungs are hollow, usually branched internal respiratory organs that are connected through passageways to the outside air and form interfaces with the circulatory system. In many vertebrates and some invertebrates, lungs arise during embryonic development as outgrowths of the gut wall. Mature lungs vary in structure. Frogs, salamanders, and lungfish have smooth-walled, balloonlike sacs with a relatively small surface area for gas exchange. Reptiles have more complex lungs, with a larger surface area. And birds and mammals have highly convoluted lungs with tortuous inner folds, branches, tubes, and sacs capable of meeting a high demand for O_2 and having a surface area up to 40 times greater than the total surface area of the skin. Irrespective of surface area, the actual gas exchange site of all lungs is the watery layer covering the innermost layer of lung cells. Since that moisture is subject to evaporation, the drying out of the lungs is an obstacle to life in the air.

Once evolved, air breathing had physiological advantages over gill respiration, since air holds 40 times more O_2 than does water, and since an animal with lungs consumes far less energy respiring than does an animal with gills. However, the situation is somewhat different for CO_2 transport. CO_2 is much more soluble in water than is O_2, and it diffuses in water more rapidly. Because of this high solubility and fast diffusion, many aquatic animals that take in O_2 through the lungs still give off CO_2 through the skin or gills. The CO_2 diffuses from abundant skin capillaries across the skin and directly into the water. Even while an amphibian is sitting on a riverbank, its skin is so moist that CO_2 can diffuse directly out into the air.

Animals with a body covering that resists desiccation, such as reptiles, birds, and mammals, exchange nearly all O_2 and CO_2 across the respiratory surfaces of the lungs and have ventilatory mechanisms to bring large quantities of air to those surfaces.

THE HUMAN RESPIRATORY SYSTEM: THE HOLLOW TREE OF LIFE

As long as life persists, the demand for oxygen never ceases, and so our very lives depend on our complex respiratory system. As we discuss the air pathways and ventilation of human lungs, keep in mind that the basic structures and strategies apply to all mammals, birds, and reptiles.

The Air Pathway

As air enters the human body, it is filtered, humidified, and warmed by different parts of the upper respiratory tract (Figure 31-9). The *nostrils*, the familiar openings to the nose, lead into nasal passageways that are lined with small hairs. These hairs act as filters, preventing large, airborne objects from penetrating into the

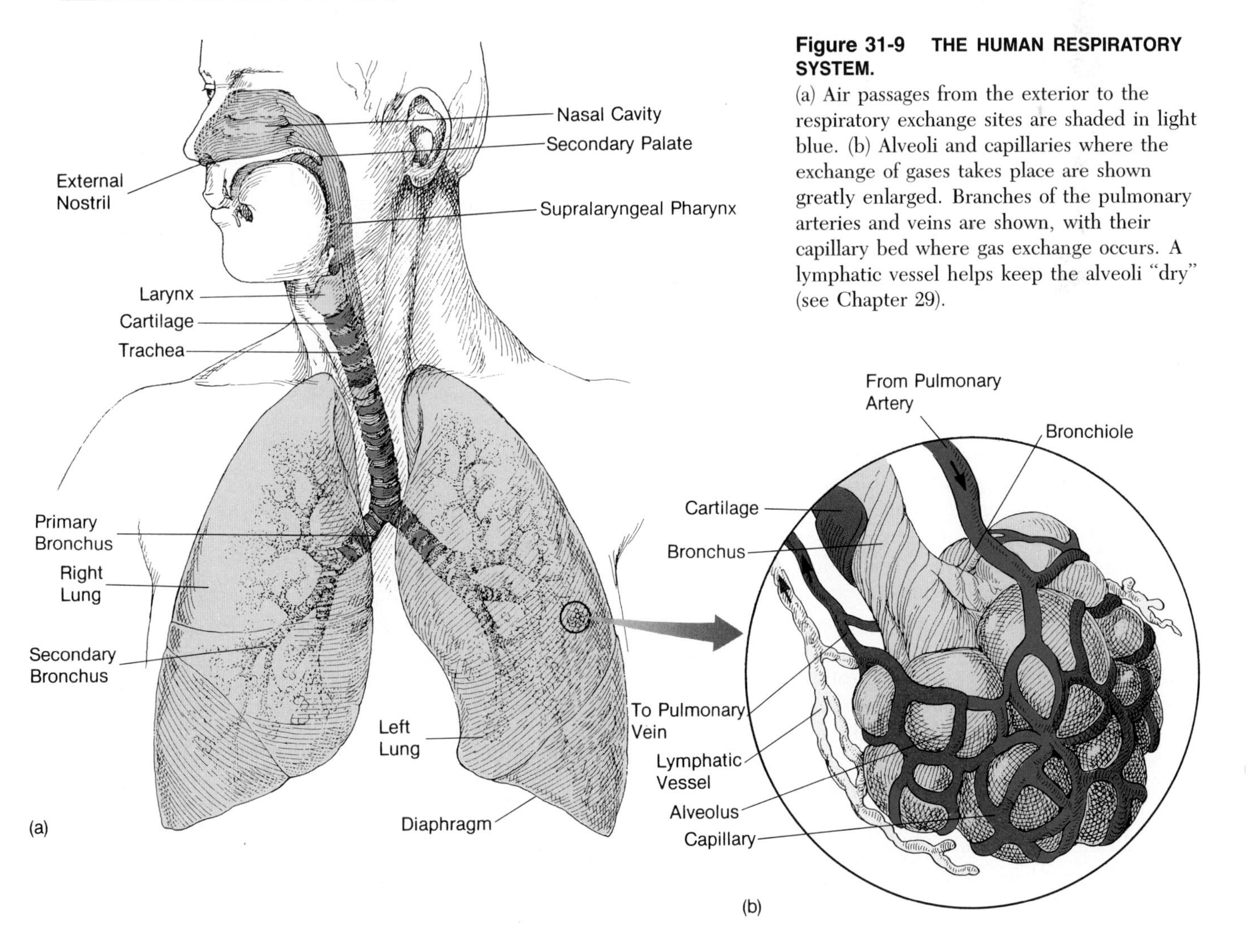

Figure 31-9 THE HUMAN RESPIRATORY SYSTEM.
(a) Air passages from the exterior to the respiratory exchange sites are shaded in light blue. (b) Alveoli and capillaries where the exchange of gases takes place are shown greatly enlarged. Branches of the pulmonary arteries and veins are shown, with their capillary bed where gas exchange occurs. A lymphatic vessel helps keep the alveoli "dry" (see Chapter 29).

piratory system. Nasal cavities warm and humidify the air and collect airborne particles on a mucous layer. Tiny cilia protruding from epithelial cells lining the nasal cavities sweep the mucus and trapped particles back toward the throat for swallowing.

The nasal and mouth cavities are separated from each other by a shelf called the *secondary palate;* food being chewed passes below it toward the esophagus, while air entering the nasal cavities passes above it toward the back of the mouth. Air, liquids, and food pass through a chamber behind the tongue, the **pharynx.** To prevent food or liquid from entering the windpipe, or **trachea,** a flap of tissue, the *epiglottis,* and two nearby vocal cords execute well-coordinated movements during swallowing. These movements temporarily seal off the **larynx,** the voice box or first part of the trachea and lung system, so that food and liquids cannot enter the air passages.

The larynx, or voice box, is the main site of sound production in most mammals. Two ligaments called *vocal cords* can be stretched to varying extents across the opening to the larynx like the sheet stretched on top of a kettledrum. Small muscles vary the position and tension of the two vocal cords, and as air is exhaled from the lungs and passes between them, the resulting vibrations create basic sounds.

The trachea is a long tube leading to the lungs. The inner surface of the trachea is lined with ciliated epithelial cells that produce mucus, while the outer wall contains rings of cartilage a bit like the wire helix that keeps a vacuum cleaner hose open. The trachea branches into two primary **bronchi** (singular, *bronchus*), which in turn branch into secondary bronchi, which branch still further into small *bronchioles.* These bronchial tubes form a system of hollow air ducts that resemble an upside-down tree, with branches that end in clusters of tiny blind-ended cavities called **alveoli** (singular, *alveolus*) (Figure 31-9b). The branching air passages and tiny cup-like alveoli together make up the lungs, a mass of moist, delicate tissue enclosing miniature air spaces. Adult human lungs contain about 300 million alveoli, with a total surface area for gas exchange of 75–160 m^2—over 40 times the surface area of the skin and as large as half a tennis court. Surrounding each alveolus is a jacket of capillaries. Lymphatics are also present and help keep the lungs dry (see Chapter 29).

It is across the thin-walled alveoli that O_2 is delivered to the blood and CO_2 is released from it. Once in the alveoli, oxygen from the air diffuses a distance of just 0.2–2.0 μm to reach the blood in the capillaries, about the same distance separating seawater and blood in a mackerel's gills. To reach the blood plasma, the O_2 must pass through: (1) a thin film of fluid containing a soapy material called *surfactant,* which lowers the fluid's surface tension (see Chapter 2) and helps prevent the alveoli from collapsing; (2) the thin layer of alveolar epithelial cells; and (3) the endothelial cells of the capillary walls. (Two basement membranes are also present in those walls.) Once in the blood plasma, most O_2 passes into red blood cells, is bound to hemoglobin, and is transported throughout the body. In many premature newborns, the immature lung cells do not yet secrete surfactant; without it, the alveoli collapse, and the infant suffocates. Hormonal treatments can speed lung cell maturation in such newborns.

How the Lungs Are Filled

Humans and other mammals fill and empty their lungs by means of a group of bones and muscles that work together like a bellows or suction pump. The evolution of **ventilation**—the process of filling and emptying internal respiratory organs—will help us understand the marvelous mechanism in humans. Lungfish and amphibians employ a *force pump system:* The closing of valves in the nostrils and glottis (the opening into the trachea) and the coordinated swallowing movements of the lower surface of the mouth cavity push air backward into the lungs (Figure 31-10). But the amphibian's force pump is so inefficient that, as we noted earlier, the skin must serve as a major auxiliary site of gas exchange.

In land vertebrates, a *suction pump action* operates because of an innovation first found in reptiles, the *expandable rib cage.* When a lizard expands its rib cage, its chest volume increases, pressure within the lungs is lowered, and air is drawn into the lungs by a kind of suction. With this efficient suction pump, all gas exchange can be

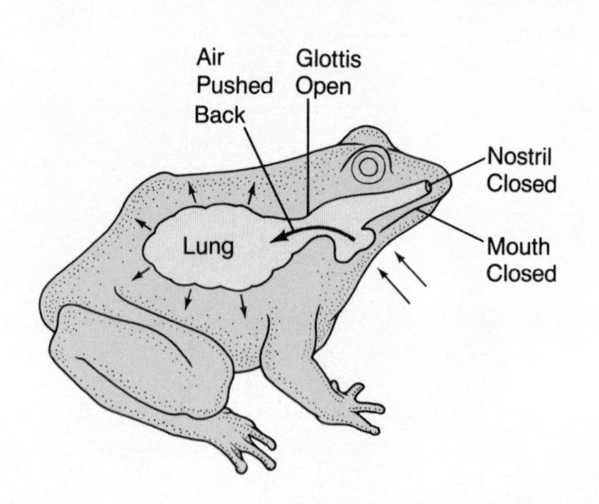

Figure 31-10 AMPHIBIAN VENTILATION: A FORCE PUMP SYSTEM.
As the mouth floor is lowered, a frog takes in air through the open nostrils. As the figure shows, the frog then closes its nostrils, raises the mouth floor, and the external air is forced from the cavity through the glottis and into the lungs.

restricted to the lungs, the skin can be dry, and life is possible away from watery habitats.

Birds and mammals also use the rib cage suction pump, and mammals have an auxiliary sheet of muscle, the **diaphragm.** The human lungs lie within the *thoracic cavity*, a space enclosed by the ribs and separated from the abdominal cavity (which lies below) by the diaphragm (Figure 31-11). In an erect human, the ribs curve forward from the backbone and meet at the breastbone, or *sternum*, and a diagonal set of muscles, the **external intercostals,** stretches between each pair of ribs. Both the external intercostals and the diaphragm contract during the inhalation of a breath of air. As the external intercostals shorten, the rib cage is raised up and out. Simultaneously, the diaphragm, which is dome-shaped at rest, flattens and pulls downward toward the feet. During inhalation, these muscle actions expand the volume of the thoracic cavity, lowering the air pressure inside the cavity slightly below atmospheric pressure. As a result of this natural suction, air flows in. At rest, exhalation is entirely a passive process brought about by relaxation of the external intercostals and the diaphragm. This relaxation results in a decrease in chest volume and an expulsion of air.

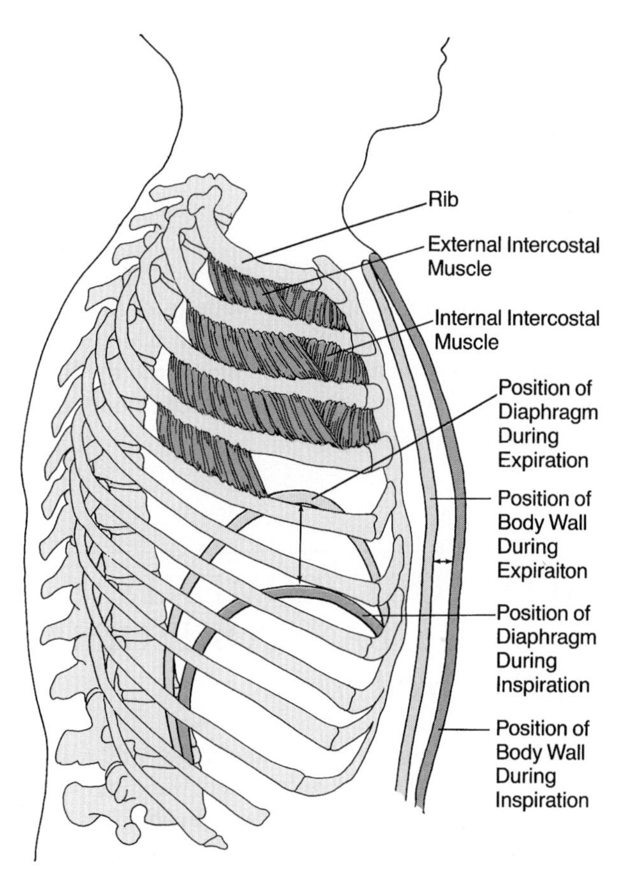

Figure 31-11 VENTILATION IN MAMMALS: A SUCTION PUMP SYSTEM.
The movements of the diaphragm and rib cage bring air into the lungs or force it out again. The downward movement of the diaphragm as it contracts, coupled with the upward and outward movement of the ribs as the external intercostal muscles contract, increases the volume of the thoracic cavity, and air flows into the lungs. (The black arrows show the changes in position of diaphragm and body wall during inspiration.) During exhalation (expiration), the relaxation of the diaphragm and external intercostal muscles (which is sometimes aided by contraction of the internal intercostal muscles) returns the diaphragm and ribs to their resting position. The volume of the thoracic cavity decreases, and air is forced out of the lungs. Only a portion of the intercostal musculature is shown.

During active exercise, however, a second set of muscles between the ribs, the **internal intercostals,** contracts and forcibly expels more air from the lungs. Larger volumes of air also are inhaled. A normal inhalation takes in a volume of air (the **tidal volume**) equivalent to only 10 percent of the total 5 L volume of the trachea and lungs, or about 0.5 L. Maximal exhalation and inhalation can boost that figure to 80 percent of total lung capacity, called the **vital capacity.** Some residual air always remains in the mammalian lung system to help keep the lungs from collapsing, to buffer the blood, and to minimize wide fluctuations in blood gases.

RESPIRATION IN BIRDS: A SPECIAL SYSTEM

Birds have a special system of *air sacs* that makes fresh air almost continuously available to the sites of gas exchange. This provides the animals with the large quantities of oxygen needed for long-distance, high-altitude flights—even a flight over Mt. Everest.

The paired lungs of birds are relatively small and have the consistency of dense sponges. However, nine or more hollow air sacs are attached to the lungs and fill much of the body cavity (Figure 31-12a). These **air sacs** are like balloons that lighten the body and serve as reservoirs for air that will later enter the lungs.

When a bird inhales, air in the posterior air sacs flows into the tiny *air capillaries* of the lungs, the only sites where gas exchange can occur (Figure 31-12b). At the same time, fresh air entering the mouth passes through the trachea and, without O_2 exchange, collects mainly in the posterior air sacs. During exhalation, some of this still fresh air passes forward from the posterior sacs into the air capillaries for gas exchange. Experiments show that air flows continuously, during both inhalation and exhalation, in one direction through the lungs. Therefore, O_2 and CO_2 exchange occurs in the lungs' air capillaries during both inhalation and exhalation.

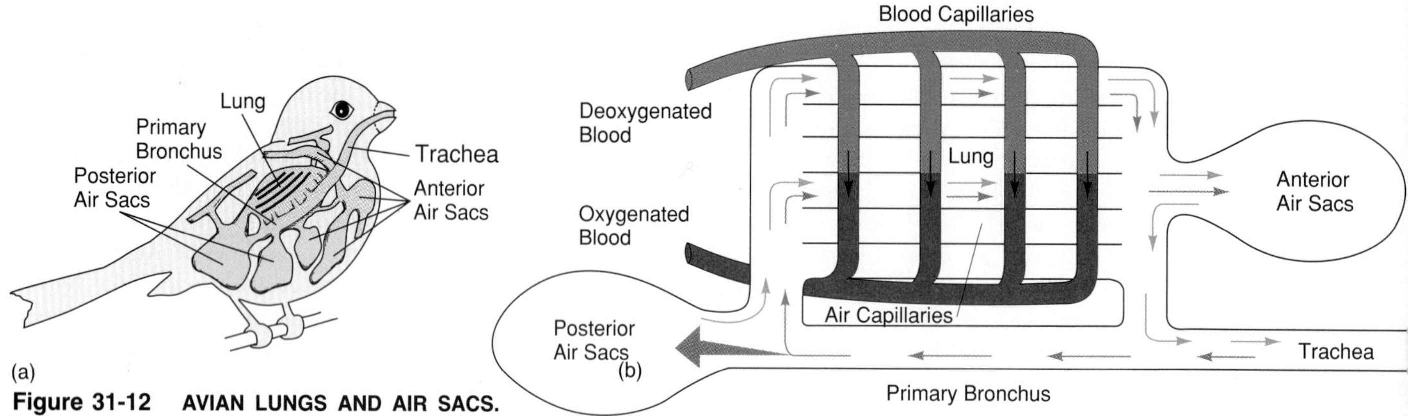

Figure 31-12 AVIAN LUNGS AND AIR SACS.
In birds, oxygenated air passes over the respiratory surfaces in the lungs during both inhalation and exhalation. (a) The position of the sacs and lungs. (b) Inhaled air passes through the primary bronchus, and most enters the posterior air sacs as well as the lung (red arrows). During this passage, air that was in the posterior sacs moves through the lung's air capillaries (red arrows) so that exchange takes place. During exhalation, the air in the posterior air sacs passes through the lung en route to the external environment (turquoise arrows). Again, exchange goes on during exhalation. Blood flow is shown diagrammatically to move in crosscurrent fashion to the direction of air flow.

The combination of continuous unidirectional air flow and a modified type of countercurrent exchange, along with other physiological systems, enables birds to do the intense work of flying, even at high altitudes, and still maintain a high body temperature. Rapid breathing also contributes to such performance; a Venezuelan hummingbird, the sparkling violet-ear, breathes 330 times per minute at sea level and 380 times per minute at high altitudes. No mammal can accomplish such a feat.

In addition to augmenting O_2 uptake in the lungs, a bird's many air sacs lower the body weight relative to its size. Branches of the air sacs even extend into the large wing and thigh bones, making them strong but light in weight. Both features help a bird stay aloft while doing much less work and using less energy.

SWIM BLADDERS IN FISH: BUOYANCY ORGANS DERIVED FROM LUNGS

Bony fishes such as trout, tuna, and sea bass have an organ called the swim bladder, which is somewhat analogous to a bird's air sacs. The swim bladder is derived from lungs and was probably possessed by early types of bony fishes (teleosts). A swim bladder is a gas-filled sac much like a long, enclosed balloon running below the spine. It is a hydrostatic organ—one that enables the fish to adjust its buoyancy so it will not sink at a given water depth, thus allowing it to expend less energy while hovering effortlessly in one spot. The volume of the swim bladder can be increased by the *red gland*, which allows O_2 and other gases to pass from the blood into the bladder, or decreased by the *oval gland*, which removes gas from the bladder (Figure 31-13).

RESPIRATORY PIGMENTS AND THE TRANSPORT OF BLOOD GASES

Gills, tracheae, and lungs are amazingly well suited to deliver masses of air or oxygenated water into the body and carrying away waste carbon dioxide, but special pigment molecules do most of the actual transporting.

Pigment Molecules

We know that oxygen has a low solubility in water (only about 0.5 ml of O_2 will dissolve in 100 ml of water), which means that blood plasma (see Chapter 29) cannot carry nearly enough O_2 to meet the total needs of body cells if their metabolism is high. The evolutionary solution to this problem is special **respiratory pigments,** molecules often contained in the blood cells. These complex pigment molecules bind O_2 reversibly, taking it up when the partial pressure of O_2 is high and releasing it when P_{O_2} is low. In some invertebrates, the primary

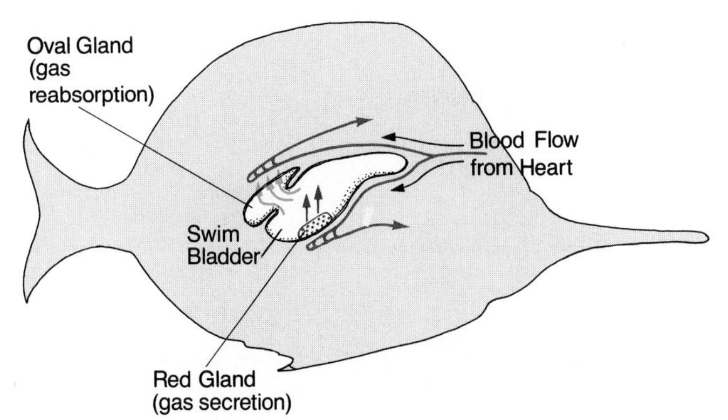

Figure 31-13 FISH BUOYANCY AND THE SWIM BLADDER.
Buoyancy in many bony fishes (teleosts) is controlled by adding oxygen to or removing oxygen from the swim bladder. This exchange is achieved by two glands: the red gland, which secretes oxygen into the swim bladder (dark red arrows), and the oval gland, which reabsorbs oxygen from the swim bladder (blue arrows). These glands are associated with nets of capillaries.

respiratory pigment is the copper-containing protein *hemocyanin.* In vertebrates, various marine worms, the mud-dwelling larvae of some flies, and even certain beans, the pigment is the iron-containing protein *hemoglobin* (Hb).

A hemoglobin molecule is composed of four polypeptide chains, each encasing a ringlike, iron-containing *heme* group (review Figure 3-25). Oxygen binds specifically between the iron atoms and a particular amino acid in the protein part of the four polypeptide chains. Recent research shows that specific amino acids lying near the iron atom help govern oxygen binding. Hemoglobin with four bound O_2 molecules is called *oxyhemoglobin,* HbO_2.

Hemoglobin is manufactured in immature red blood cells and gives these cells their red color. In vertebrates, hemoglobin molecules are always retained within red blood cells. The retention of hemoglobin in red blood cells has several advantages:

1. The hemoglobin molecules are always close to enzymes and other factors in the red blood cell's cytoplasm that maintain or vary the pigment's binding properties (as we shall soon see).
2. Bound within cells, hemoglobin does not add to the colloidal osmotic pressure of blood plasma; if hemoglobin did circulate freely, osmosis would tend to draw fluid more strongly from tissues.
3. Because a red blood cell is about the same diameter as a capillary, the squeezing of these "sacs" of hemoglobin through capillaries stirs up layers of plasma and makes exchange of O_2, CO_2, nutrients, and wastes more efficient.

Hemoglobin picks up O_2 in regions where the P_{O_2} is high and releases it in regions where the P_{O_2} is low, which neatly explains why the hemoglobin in blood picks up O_2 from the lungs and releases it in the tissue capillary beds. The release of O_2 follows the splitting, or *dissociation,* of HbO_2 into Hb and O_2. The tendency for hemoglobin to bind O_2 under different partial pressures is represented by an S-shaped **oxygen dissociation curve** (Figure 31-14). In the human lung, where P_{O_2} is about 105 mm Hg, about 97 percent of the heme groups in the hemoglobin molecules will bind O_2. When this happens, the hemoglobin is said to be 97 percent *saturated.* Moving left on the graph, we find the typical P_{O_2} in resting tissues of large mammals to be about 30 mm Hg. At this P_{O_2}, the hemoglobin can be only 50 percent saturated. The difference between the two saturation figures—97 − 50, or 47 percent—represents the amount of bound O_2 that is released by the blood when it travels from the lungs to typical capillary beds. If exercise or intensified metabolism in a tissue causes the local P_{O_2} to drop from 30 mm Hg to 15 or 10 mm Hg, more HbO_2

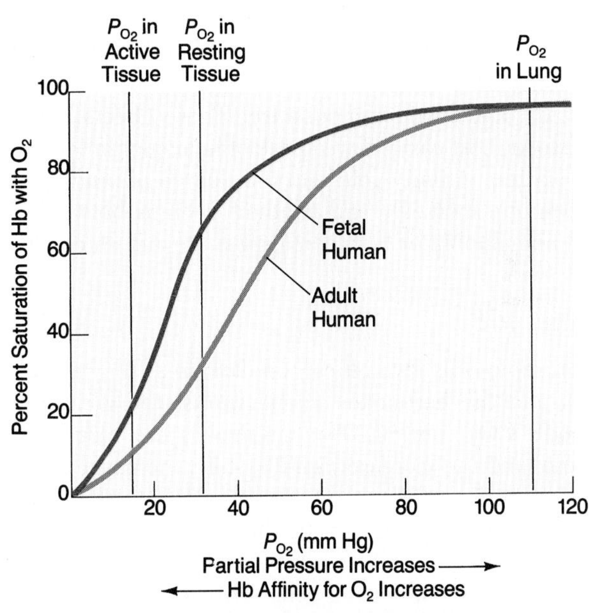

Figure 31-14 OXYGEN DISSOCIATION CURVE.
Oxygen dissociation curves reflect the amount of oxygen that is bound to hemoglobin at different partial pressures of oxygen. Curves to the left reflect hemoglobins with greater affinity for oxygen than those for which the curves are to the right. A curve to the right indicates that a lower percent saturation of hemoglobin with oxygen will prevail at any particular partial pressure of oxygen in the tissues; this means that more oxygen is given up or delivered to the tissues. Fetal hemoglobin (left) has a greater affinity for oxygen than does maternal hemoglobin (right). Oxygen that is released from the maternal hemoglobin in the placenta will be taken up by the fetal hemoglobin and carried to the fetal tissues.

will dissociate (97 − 15, or 82 percent), releasing a larger proportion of O_2 to the deprived tissue cells.

The steep slope in the S curve has important consequences: First, it means that even a modest increase in oxygen partial pressures at the gills or lungs results in rapid loading of O_2 by hemoglobin; second, and conversely, it means that a relatively small drop in P_{O_2} in metabolizing tissues, in the vicinity of the 50 percent loading point on the oxygen dissociation curve, will result in a massive unloading of oxygen from hemoglobin.

Note that the curve is shifted to the left in mammalian fetuses. A fetus faces something of an oxygen crisis: It has no direct access to O_2, but at the same time has a substantial O_2 demand owing to its rapid metabolism and growth. Because fetal hemoglobin has a higher affinity for O_2 than does that of the mother, O_2 tends to leave the mother's blood as it passes through the placenta and enters the fetal red cells, where it binds to the higher-affinity fetal hemoglobin.

Control of Hemoglobin Function

The binding of oxygen by hemoglobin is affected by several environmental factors, including blood pH, the presence in red blood cells of organophosphate compounds, and blood temperature. These effects are understood most easily if we speak of hemoglobin's *affinity* for oxygen. Affinity refers to the degree of binding between one molecule and another, such as between O_2 and hemoglobin or between a substrate and an enzyme.

Blood pH

Aerobic cells give off carbon dioxide as they metabolize, and the harder an animal works, the greater the blood's CO_2 content. This gas makes the blood more acidic, and the increased acidity shifts the oxygen dissociation curve to the right (Figure 31-15). This means that at a given tissue P_{O_2}, the hemoglobin can bind less O_2, and thus more O_2 is released where it is needed by the actively metabolizing cells. This shift to the right is an aspect of the **Bohr effect.** In the lungs, just the opposite occurs, and as a result, more CO_2 is given off into the air. Thus, the Bohr effect brings about enhanced uptake of O_2 by hemoglobin as the red blood cells pass through the lungs and an enhanced release of O_2 as they pass through tissues.

Blood Temperature

The temperature of blood is yet another influence on the affinity of hemoglobin for oxygen; as temperature increases, O_2 affinity decreases. Within blood flowing through a warm area, such as an actively working muscle, oxygenated hemoglobin releases O_2 more easily than in a cooler area.

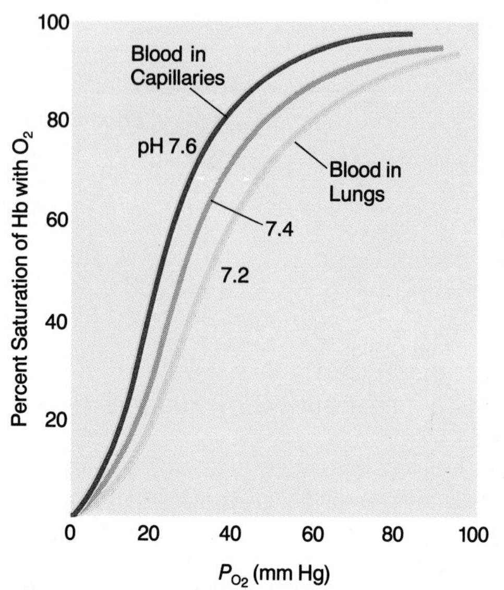

Figure 31-15 THE BOHR EFFECT: BLOOD PH INFLUENCES ON THE BINDING OF OXYGEN TO HEMOGLOBIN.
A lowering of blood pH causes a shift of the oxygen dissociation curve to the right and reflects a reduced affinity of hemoglobin for oxygen. Because of this shift, oxygen will be released from hemoglobin under conditions encountered in the tissue capillaries. This helps ensure the release of oxygen where it is required. Conversely, in the lungs, carbon dioxide is released, H^+ concentration is reduced, pH rises, and the curve shifts to the left. In this high-affinity state, hemoglobin binds oxygen more avidly. The pH values shown are hypothetical.

Carbon Monoxide

The heme groups of hemoglobin molecules can bind a substance of similar shape and properties to oxygen: carbon monoxide (CO). Carbon monoxide binds more avidly to hemoglobin than does O_2 and increases the affinity of hemoglobin for both CO and O_2 molecules that are bound. The hemoglobin then cannot release any O_2 as it passes through the tissue capillaries. This explains why people exposed to CO, such as from automobile exhaust, can be asphyxiated, even though O_2 is present in the air.

Myoglobin: A Molecular Oxygen Reservoir

An animal's body can store oxygen by means of a special respiratory pigment called **myoglobin,** which is abundant in some muscles. The structure of myoglobin (Figure 31-16) is roughly similar to that of one of the four globin subunits that make up a hemoglobin molecule.

Each myoglobin molecule contains a single heme group and iron atom. Myoglobin has a higher affinity for O_2 than does hemoglobin. Therefore, when blood flows through a muscle capillary bed, O_2 is transferred from hemoglobin to myoglobin. When muscle cells are resting or are moderately active, P_{O_2} remains higher than 20 mm Hg, and myoglobin retains its O_2. But when the animal exercises so hard that the blood alone cannot supply all the O_2 needed, P_{O_2} will fall below 20 mm Hg in muscle cells, and O_2 will dissociate from myoglobin, making O_2 available in the cells. Myoglobin also serves as a "ferryboat," carrying O_2 to the mitochondria, where it is quickly used in metabolism.

Abundant myoglobin gives muscles a red hue. In sharks and mackerel, for example, a red strip of muscle fibers running down the fish's sides is the primary muscle tissue that contracts when the fish swims at normal speeds. The abundant O_2 from the myoglobin in these muscles helps support and sustain this activity. The rest of the fish's body muscles (which appear white due to lack of myoglobin and to a paucity of capillaries), go along for the ride—they contract only when a burst of speed is needed. Similarly, the red breast muscles of ducks can support sustained flight, whereas the white breast muscles of domestic turkeys, low in myoglobin, allow for only short bursts of activity.

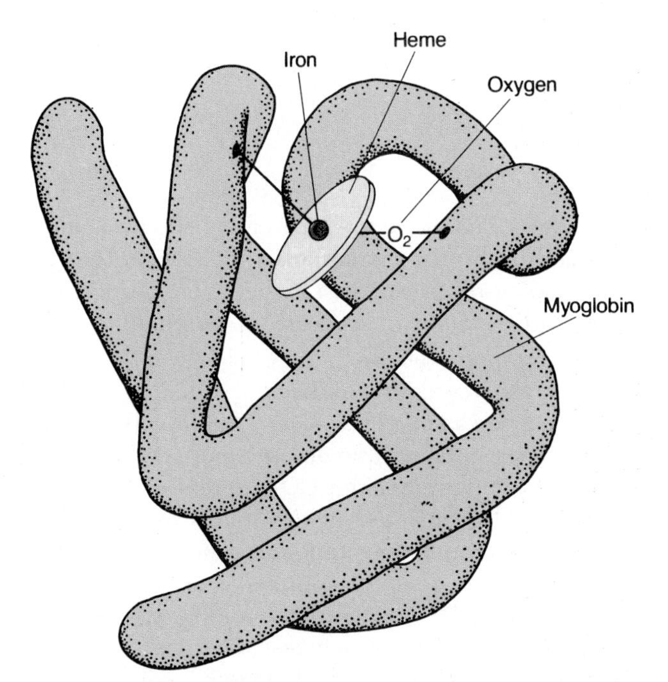

Figure 31-16 MYOGLOBIN: THE STORAGE PIGMENT. Myoglobin consists of a single polypeptide chain with one heme group for binding oxygen. This storage pigment is present in certain muscles, where it binds oxygen at low partial pressures and surrenders its oxygen to the muscle cells when oxygen levels are severely depleted owing to intense contractions.

Carbon Dioxide Transport: Ridding the Body of a Metabolic Waste

The metabolic waste product carbon dioxide diffuses out of cells and into the capillaries because P_{CO_2} is higher in the tissues than in the blood at the venous end of the capillary bed (Figure 31-17a, step 1). Some CO_2 dissolves in the blood plasma and is carried to the lungs unchanged. Some dissolved CO_2 reacts slowly with water in the plasma to form carbonic acid and then dissociates into HCO_3^- and H^+ (step 2). Plasma proteins and hemoglobin act as *buffers* to bind the H^+ and prevent wide swings in pH (step 3). In terrestrial vertebrates, most of the CO_2 diffuses into the red blood cells, where the enzyme *carbonic anhydrase* greatly speeds up the initial reaction (step 4):

$$CO_2 + H_2O \rightarrow H_2CO_3$$

Then, during the secondary reaction, H^+ is released:

$$H_2CO_3 \rightarrow H^+ + HCO_3^-$$

The presence of H^+ lowers the pH, triggering the Bohr effect in the red blood cells. The resulting acidity could affect enzymes and proteins within the red blood cells and plasma, but it does not because hemoglobin acts as a buffer (step 5). Because of such buffering in the plasma and red blood cells, the blood pH remains at about 7.35, even though a large quantity of CO_2 is being carried.

In the air within the lungs, P_{CO_2} is lower than in the blood. This lower P_{CO_2} shifts the chemical equilibrium of the circulating blood (Figure 31-17b): The hemoglobin side chains release H^+ (step 6), which can then reassociate with the HCO_3^- that diffuses into the red blood cells from the plasma. Carbon dioxide bound to hemoglobin ($HbCO_2$) also dissociates as it binds O_2 (step 7). This once again yields H_2CO_3, which in the presence of carbonic anhydrase dissociates into H_2O and CO_2. The CO_2 diffuses out of the red blood cells into the plasma (step 8), then through the walls of the capillaries in the lungs, and finally into the alveoli. Then, during exhalation, the CO_2 is expelled through the bronchial passages and out through the nose or mouth.

CONTROL OF BREATHING

A person's normal rhythmic breathing cycle is automatic. But what regulates it? A region in the brain of all vertebrates, the *medulla*, contains the *respiratory center*, a group of nerve cells that automatically triggers periodic contractions of the muscles that bring about ventilation (in mammals, the intercostal muscles and

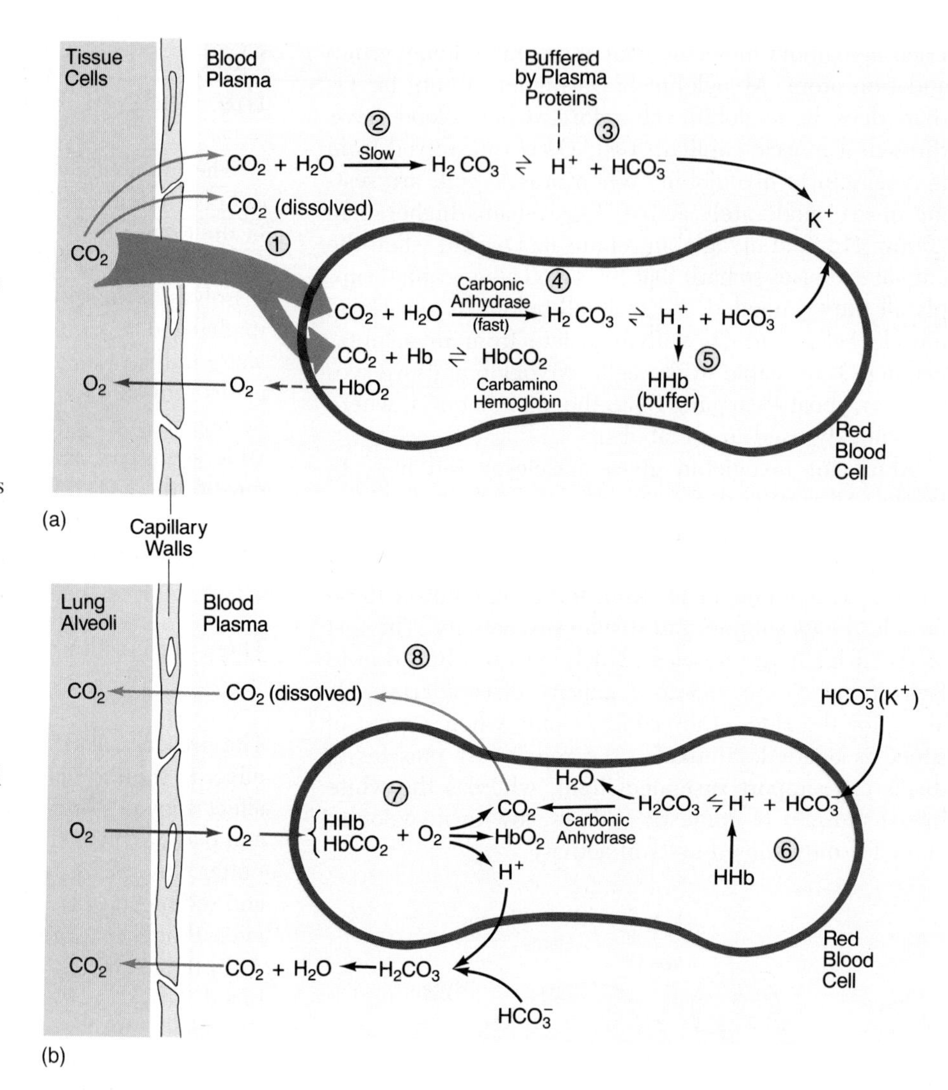

Figure 31-17 TRANSPORT OF CARBON DIOXIDE IN THE BLOOD.
Most carbon dioxide in the blood is transported as HCO_3^-, especially the large amount formed rapidly in red blood cells as a result of the activity of the carbonic anhydrase enzyme. Some CO_2 is bound to hemoglobin (carbamino hemoglobin), and a small proportion moves as dissolved gas. The reactions occurring in the blood capillaries (a), where CO_2 is taken up, are reversed in the lungs (b) as CO_2 is eliminated into the alveoli. A key feature of this system is that it operates to keep blood H^+ levels low and pH near 7.35. If H^+ builds up in blood plasma or red blood cells, breathing speeds; this eliminates CO_2 faster and shifts the various reactions so that more H^+ combines with HCO_3^-. Note that O_2 diffuses into the red blood cells in lung capillaries and binds to hemoglobin (HbO_2) after HHb has surrendered its H^+. In the tissues, the HbO_2 gives up O_2, which diffuses into tissue cells.

diaphragm; Figure 31-18). The repeated and synchronous signals from the respiratory center regulate an organism's ventilatory movements and hence its breathing rate so that P_{CO_2} and P_{O_2} in the blood remain constant.

The respiratory center of aquatic vertebrates responds mainly to the O_2 level in the body, because O_2 tends to be in short supply underwater and because CO_2 diffuses away so easily. In air, O_2 is abundant, and CO_2 removal is the problem; thus, in terrestrial vertebrates, the CO_2 level mainly drives the respiratory center. In fact, the center's rate also varies to help keep blood pH constant. If blood pH grows too acidic, the animal breathes faster and exhales more CO_2 from the body; conversely, if blood pH becomes too alkaline, the animal breathes more slowly.

The respiratory center receives information on P_{O_2} from chemoreceptor nerve endings in the *aortic bodies* found in the wall of the aorta and from nerve endings in the *carotid bodies* located in the carotid arteries (see Figure 31-18). (*Pressure receptors* also report blood pressure based on heart rate during activity.) Chemoreceptor nerve endings in the walls of the medulla monitor P_{CO_2} and pH in the extracellular fluid that bathes the nerve endings in that portion of the brain. Changes primarily in P_{CO_2} and pH stimulate the respiratory center to slow or speed the rhythmic discharge of nerve impulses to the ventilatory muscles.

When a person takes fast, deep breaths without exercising (hyperventilates), he or she expels a great deal of CO_2, thereby lowering the CO_2 content of the blood; this raises blood pH and in turn raises the pH of the brain extracellular fluid within the tissue near the respiratory center. As a result, respiratory center nerve firing is slowed and the breathing rate drops, so that P_{CO_2} builds up and pH falls again to normal ranges. If a person exercises vigorously, blood and tissue P_{O_2} will drop, and P_{CO_2} will increase (pH drops), triggering more frequent respiratory center nerve firing and hence faster, deeper breathing. These mechanisms govern breathing rates even if a person tries to overcome their control: Breath holding is invariably interrupted by an involuntary gasp for air, while hyperventilation ends in a fainting spell,

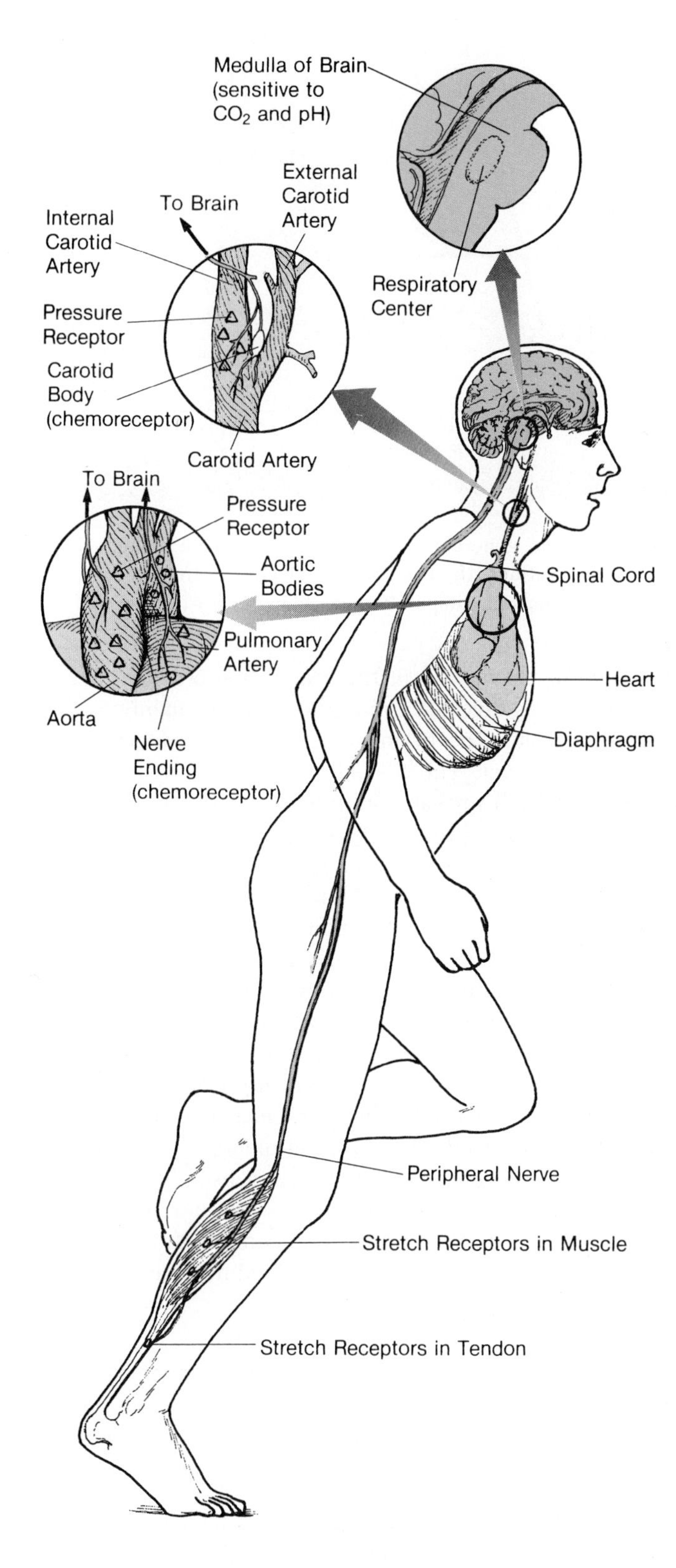

◀ **Figure 31-18 RESPIRATORY RATE AND CHEMORECEPTORS AND MECHANORECEPTORS.** Chemoreceptors are sensitive to O_2 levels (carotid body, aortic bodies, and pulmonary artery) as well as to CO_2 levels and pH (medulla). Mechanoreceptors measure stretch in certain body muscles and tendons, and pressure receptors in the blood vessels also participate in respiratory control, especially during exercise. The sites of these various sensory receptors are shown. The respiratory center itself is situated in the medulla.

with a normal, slower breathing rate quickly restored.

New experiments show that at the start of vigorous exercise, the breathing rate increases quickly—well before CO_2 builds up. Conversely, when exercise is suddenly stopped, the breathing rate drops before normal P_{O_2} is restored. The explanation for these phenomena is that special receptors monitor the degree of contraction of limb muscles. These stretch receptors in the tendons and joints of the arms and legs actually trigger a change in breathing rate to fit the level of exercise, and not simply to correspond to blood chemistry. This response is called the **mechanoreceptor reflex.** It ensures that blood gases are kept at normal levels during and after exercise, thereby avoiding the undesirable consequences for blood and body chemistry of letting P_{O_2}, P_{CO_2}, and pH levels fluctuate. Instead, the changes are anticipated and are compensated for ahead of time.

Other research shows that improper brain control of breathing may explain sudden infant death syndrome, which kills 8,000 babies per year and accounts for about half of all infant deaths in the United States. Some biologists believe that slow development of the medulla and its respiratory center may impair the arousal response needed to wake a child if breathing stops.

HOW ANIMALS ADAPT TO OXYGEN-POOR ENVIRONMENTS

An out-of-breath hiker in a high-mountain area may suffer altitude sickness: a spectrum of maladies that includes headache, extreme fatigue, nausea, and breathlessness. These symptoms result from low atmospheric P_{O_2}, which causes a drop in P_{O_2} in the blood and a shifting of the oxygen dissociation curve to the right, making it easier for hemoglobin to release O_2 in tissues, but harder for hemoglobin to bind O_2 in the lungs. In birds and certain mammals at high altitudes, however, the response is quite different. High-flying birds and the llamas of the South American Andes, for example, have oxygen dissociation curves that are shifted to the left, resulting in better binding of O_2 in the lungs—not worse (Figure 31-19).

A person who lives at sea level and goes to the mountains for several months gradually does become accli-

(a)

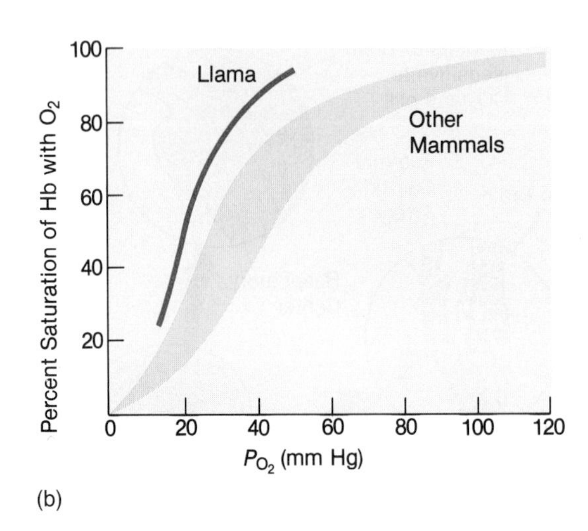

(b)

Figure 31-19 LIFE AT HIGH ALTITUDE.
(a) Animals that live at high altitudes, such as these llamas, alpacas, and sheep, generally have hemoglobin with a higher affinity for oxygen than the hemoglobin of similar animals living at lower elevations. (b) This fact is reflected in the oxygen dissociation curve. The llama's curve is located to the left of curves for mammals of comparable size living at lower altitudes.

mated to high elevations. The number of red blood cells increases, carrying more O_2, and the rate of breathing and heart rate also increase to facilitate gas exchange and delivery.

A person who lives an entire lifetime in the high mountains will develop enlarged ventricles of the heart for driving large volumes of blood containing extra red cells. The Quechua Indians of the high Andes, for example, tend to be small in stature, yet they have large chests enclosing significantly larger hearts and lungs than those of South American Indians living at lower elevations.

Permanent adaptations can also be seen in deep-diving seals, whales, and other marine mammals that may be deprived of the opportunity to breathe for up to an hour. Most have a huge number of very large red blood cells, large quantities of myoglobin in their muscles, and hemoglobin with an oxygen dissociation curve shifted to the right, making it easier to release oxygen to the tissues during a dive.

Marine mammals have two other adaptations that are part of the **diving reflex,** a complex set of responses to swimming downward in the sea. When a whale dives, blood flow is diverted from other regions mainly to the heart muscles, the brain, and, to a lesser degree, the skeletal muscles. The skeletal muscles can incur a large **oxygen debt:** They can continue to contract for locomotion even after all stored O_2 is depleted. The energy for such contractions comes from glycolysis (see Chapter 7), so lactic acid is formed and builds up in the muscle tissues. Contrary to what might be predicted, marine mammals probably do not surface (see chapter-opening photo) because they run out of oxygen; instead, they apparently come up when falling blood glucose levels get very low. The second part of the diving reflex, the slowing of the heart rate (also called *bradycardia*), occurs in all vertebrates. Thus, the heart of a marine mammal or of a terrestrial vertebrate slows automatically when the nostrils are immersed; conversely, a fish's heart slows when the animal is removed from water and held in air.

Our next chapter, on digestion and nutrition, considers the source and regulation of an animal's glucose levels in more detail.

SUMMARY

1. *Respiration* is the process by which organisms exchange O_2 and CO_2 with the environment.

2. Air exerts a weight on the earth's surface that is measured as *atmospheric pressure*. Individual gases in the air—N_2, O_2, and so on—exert *partial pressures*. Gases have different solubilities in liquids, and this has consequences for O_2 and CO_2 movement in living things.

3. Simple diffusion of respiratory gases can meet the needs of only very small or very flat organisms. In large organisms, bulk flow processes are required to distribute sufficient quantities of O_2 and CO_2 to the body tissues.

4. Respiratory organs can be external, as are *gills*, or internal, as are *tracheae* and *lungs*.

5. Gills take many forms but always function as a site for exchange of dissolved gases with water.

6. Insects have a system of tiny, hollow tubes called *tracheae* that transport air deep into the body. There, O_2 dissolves in fluid and passes into the hemolymph, which bathes cell surfaces.

7. Lungs are hollow, branched internal respiratory organs that receive air from the outer environment; they can be smooth and saclike or very convoluted.

8. In mammals, air is filtered, humidified, and warmed as it passes through

the nasal passages, mouth cavity, and *pharynx*. The epiglottis and vocal cords seal off the *larynx*, or entrance to the *trachea* and lungs, during swallowing.

9. In mammals, the respiratory tree includes the trachea, which branches into two primary *bronchi*, then secondary bronchi and bronchioles, before ending in cuplike *alveoli*, where gas exchange takes place.

10. The *diaphragm* and the muscle-powered expandable rib cage of mammals change the volume of the thoracic cavity and draw in air or push it out through a suction pump system with elastic recoil.

11. *Air sacs* lighten a bird's body and are part of a reservoir system that allows the lungs to be flushed completely with fresh air.

12. A *swim bladder* is a hydrostatic organ that enables a bony fish to adjust its buoyancy.

13. Hemoglobin is an iron-containing *respiratory pigment* that binds O_2 when P_{O_2} is high (as in lungs) and releases it when P_{O_2} is low (as in tissues).

14. Blood pH, organophosphates in the red blood cells, and blood temperature alter the binding affinity of hemoglobin for O_2. By shifting the *oxygen dissociation curve*, they increase the efficiency of O_2 binding in the lungs and O_2 release in the tissues.

15. *Myoglobin* is a respiratory storage pigment found mostly in muscle tissue; it maintains a reserve supply of O_2 that can be released under conditions of high O_2 demand and low local P_{O_2} levels.

16. Carbon dioxide is transported in blood plasma and red blood cells, mostly as the bicarbonate ion HCO_3^-. Hemoglobin is the blood's primary buffer and carries H^+.

17. The brain's respiratory center responds primarily to changes in P_{CO_2} and pH and, to a lesser degree, to changes in P_{O_2} in the peripheral chemoreceptors. Rates of breathing during exercise are controlled primarily by *mechanoreceptor reflexes*, and not by altered blood chemistry.

18. Different animals adapt to oxygen-poor environments through a variety of mechanisms, including production of more red blood cells, faster breathing and heart rates, enlarged heart ventricles and lungs, shunting of blood to vital organs, and the *dive reflex*.

KEY TERMS

air sac
alveolus
atmospheric pressure
Bohr effect
bronchus
countercurrent exchange system
diaphragm
diving reflex
external intercostal
gill
internal intercostal
larynx
lung
mechanoreceptor reflex
myoglobin
oxygen debt
oxygen dissociation curve
partial pressure
pharynx
respiration
respiratory pigment
solubility
swim bladder
tidal volume
trachea
ventilation
vital capacity

QUESTIONS

1. What is atmospheric pressure at sea level? How high (in mm) will this pressure push a column of mercury (Hg)? At sea level, what is the partial pressure of oxygen (P_{O_2}) in mm Hg?

2. Name four important physical factors that affect the amount of O_2 available to an organism's cells at a given time.

3. What four major types of respiratory systems have evolved in the animal kingdom?

4. What is meant by countercurrent flow?

5. Through what structures do insects take in air? How is air delivered to an insect's body cells?

6. Contrast the avian respiratory system with the mammalian one.

7. Name three important respiratory pigments found in the animal kingdom, and name the animal groups in which they appear.

8. What chemical and physical factors control hemoglobin function?

9. How is CO_2 carried in mammalian blood? What is the role of carbonic anhydrase in CO_2 transport?

ESSAY QUESTION

1. On the same graph, draw oxygen dissociation curves for normal human hemoglobin (Hb), human Hb plus DPG, fetal human Hb, and myoglobin. Explain the significance of each curve.

For additional readings related to topics in this chapter, see Appendix C.

32 Digestion and Nutrition

Now good digestion wait on appetite, And health on both!

William Shakespeare,
Macbeth (III.iv)

The bright green aphid sucking juice from a plant stem; the coyote tearing the soft belly of a jackrabbit; the college student munching an apple. All are carrying out one of an animal's most basic strategies: feeding—taking in prefabricated nutrients for metabolism and growth.

Feeding usually involves both **ingestion**—the taking of food pieces into the body—and **digestion**—a multistep process that includes the mechanical and chemical breakdown of food pieces into large organic molecules, the breakdown of these molecules into smaller compounds, and the absorption of these compounds and their transformation into molecules that can be used in cellular metabolism.

This chapter describes:

- Modes of ingestion and digestion in the simplest heterotrophs and in more complex invertebrates and vertebrates

Galápagos gourmet: A giant tortoise (*Geochelone elephantopus abingdoni*) grazes on a Pinta Island cactus.

- The human digestive system, with its 23 ft of intestines and its huge surface area for absorbing nutrients
- The chemistry of digestion and the absorption of nutrients
- The coordination of ingestion, digestion, and nutrient levels so that an organism has a constant supply of energy, even if it feeds only periodically
- Nutrients—the carbohydrates, fats, proteins, vitamins, minerals, and water that animals must consume, and the roles these essential substances play in cellular metabolism and health

ANIMAL STRATEGIES FOR INGESTION AND DIGESTION

One might call feeding the universal pastime, one that is reflected in the fascinating spectrum of mechanisms for ingesting and digesting food among simple and complex organisms.

Intracellular Digestion in Simple Organisms

Many kinds of protozoa, as well as internal parasites and certain marine invertebrates, take in nutrients or food particles directly through the body walls. Nutrients can be absorbed or engulfed directly by body wall cells and are then broken down by *intracellular digestion*—digestive enzymes acting within each cell.

Parasites such as tapeworms are bathed in partially digested nutrients in their host's intestines, and actively transport such nutrients across the plasma membrane of their body cells. Certain free-living marine worms and other aquatic invertebrates absorb dissolved glucose, amino acids, and other organic molecules directly from seawater and soft bottom sediments. Protozoa such as *Paramecium* have an ingestion organelle, the gullet, lined with beating cilia that sweep in tiny food particles (see Figures 21-13 and 21-14). When a particle arrives at the base of the gullet, a food vacuole pinches off and carries the particle into the cytoplasm. Smaller vesicles within the cytoplasm fuse with the vacuole and transfer digestive enzymes to it. The enzymes degrade the food particle, hydrolyzing it into sugars, fats, amino acids, nucleic acid bases, inorganic ions, and so on. These nutrient molecules then move from the food vacuole into the cytoplasm. Finally, the food vacuole travels to the inner side of the cell surface, fuses with a special area called the anal pore, and excretes undigested debris.

Intracellular and Extracellular Digestion in Simple Animals

Animals that rely on intracellular digestion alone must have bodies that are very small or very flat. Larger size and more complex body organization are possible for animals such as flatworms and cnidarians (like jellyfish) because they have an internal gut chamber called the *gastrovascular cavity* (see Figure 24-3). Within such cavities, ingested water and food particles and dissolved gases can travel passively throughout the body.

In the flatworm *Planaria* and in cnidarians such as hydra, the gastrovascular cavity is a blind-ended chamber with a single opening through which food enters and wastes exit (see Figure 24-12). Once flatworms or cnidarians ingest food particles, enzymes are secreted into the gut, and the particles are partially broken down into macromolecules. This sort of breakdown process, taking place within the *lumen*, or hollow part of the gut chamber, rather than within individual body cells, is called *extracellular digestion*. In the flatworm and cnidarians, extracellular digestion is incomplete; cells lining the gut phagocytize the macromolecules, and within these gut cells, the nutrients are degraded further by enzymes.

Extracellular Digestion in Complex Animals

In most animals, including fish-eating water bugs, plant-eating manatees, and Galápagos tortoises (see opening photo), complete extracellular digestion takes place in a two-ended gut called the **alimentary canal,** which extends from mouth to anus. Food makes a long, one-way passage through the lumen (central cavity) of this tube. The alimentary canal, also called the *digestive tract*, or *gut*, has discrete regions that carry out specialized tasks: breaking up food, storing it temporarily, digesting it chemically, absorbing digested nutrients, reabsorbing water, storing wastes, and finally, eliminating wastes.

The Earthworm Gut

The digestive tract of the earthworm serves as a simple example of the two-ended gut with specialized regions and functions (Figure 32-1). The worm's mouth operates like a suction tube to draw in dirt and tiny food particles. After passing through the *pharynx*, ingested materials travel into a tubular passageway, the **esophagus,** where an alkaline secretion is added. The soil and food are stored in a thick-walled chamber, the **crop,** then pass to the **gizzard,** a larger muscular chamber for processing the food into smaller pieces. In the gizzard, hard, toothlike projections from the inner walls function

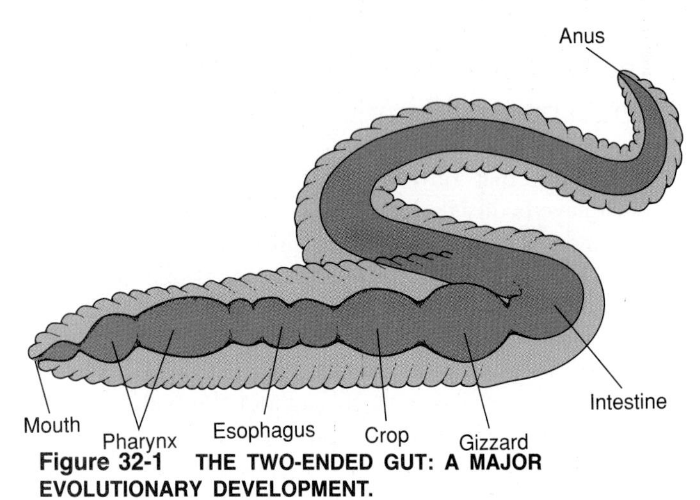

Figure 32-1 THE TWO-ENDED GUT: A MAJOR EVOLUTIONARY DEVELOPMENT.
Most bilaterally symmetrical animals have a digestive tract with separate openings at the mouth and anus. This earthworm gut also has specialized regions that carry out different functions.

like tiny millstones to pulverize and grind the food. Such pulverizing increases the food's total surface area so that extracellular enzymes can more effectively break it down into sugars, fats, amino acids, and other nutrients. The enzymatic breakdown of food and the absorption of nutrients take place in the earthworm's **intestine,** the long, tubelike remainder of the alimentary canal. Secretory cells in the intestine wall produce and excrete digestive enzymes into the lumen. The inner wall of the intestine has a single large infolding that increases the surface area for nutrient absorption. In addition, each epithelial cell in the gut lining has tiny hairlike projections, called **microvilli** (singular, *microvillus*), that expand the gut's absorptive surface even further. Blind-ended pouches called **ceca** (singular, *cecum*) extend from the intestine and hold food for extended periods of time, so that enzymes have time to act and absorption can be fairly complete. Most water is reabsorbed by the terminal portion of the intestine, and finally, fairly dry waste matter is excreted through the anus.

Insects: Variations on the Basic Plan

The earthworm's basic digestive structures and processes are also characteristic of snails, lobsters, sea cucumbers, spiders, and relatives of each of those animals. Insects, with their range of specialized mechanisms for ingesting and digesting, display interesting variations on that simple model. Many insects feed on plant or animal juices and so have a piercing, needlelike proboscis that can penetrate flowers, stems, leaves, or skin and suck in only fluid (Figure 32-2). Other insects, such as termites and horseflies, have strong mandibles for cutting tiny chunks of wood or for biting animal prey. Various insects have ceca or other storage chambers in the gut that store food and obviate the need for continuous feeding. Female mosquitoes, for example, store mammalian blood in thin-walled expendable ceca. The blood from a single feeding can be held for up to a week in these reservoirs and passed into the alimentary canal as needed. As in the earthworm, the terminal portion of the insect intestine reabsorbs water, so that large quantities are not lost during excretion of wastes.

Feeding Styles of Meat-Eating Vertebrates

Meat-eating vertebrates have alimentary canals analogous to those of earthworms and insects, but the mouth is usually outfitted with sharp teeth (Figure 32-3). Exclusive meat eaters, or **carnivores,** tear other organisms into edible-sized pieces—often big chunks that they swallow hastily. Some carnivores, such as alligators and toothed whales, have only pointed, conical teeth for piercing and tearing. Mammals, such as cats and dogs, typically have piercing *canines* in the front of the mouth, to rip apart flesh; broader *molars* with raised points in the rear, to grind flesh and bones into small pieces; and teeth of other shapes. Meat eaters can swallow less pulverized food than can plant eaters, since animal flesh is not surrounded by the tough cellulose walls found in plants.

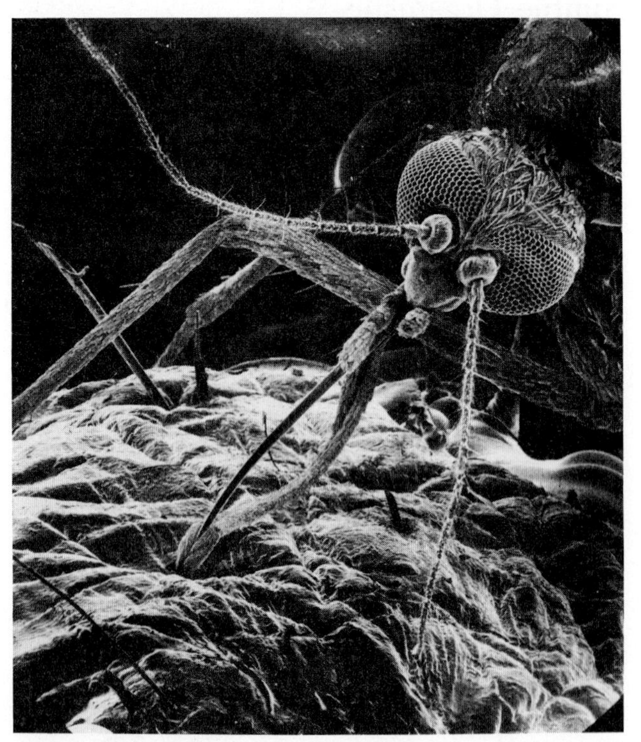

Figure 32-2 INSECT MOUTHPARTS.
Insects have an array of mouthparts that are adapted for biting, sucking, and chewing. This mosquito's syringelike proboscis (magnified about 22×) is an ideal shape for piercing this human skin and sucking blood.

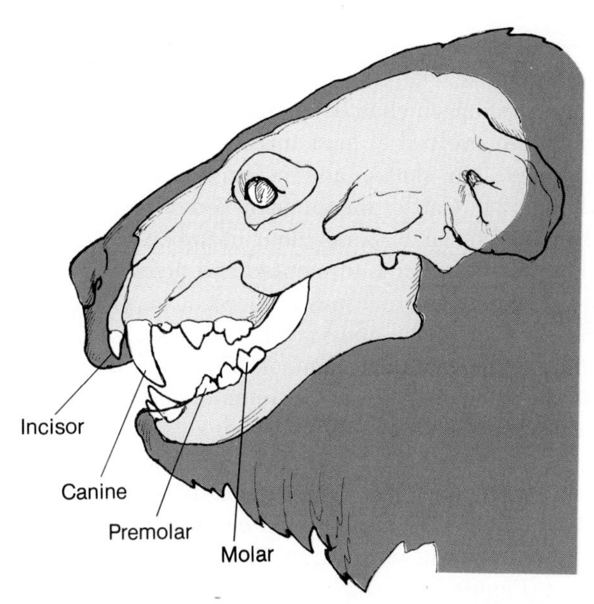

Figure 32-3 CARNIVORE TEETH: FOR TEARING AND CRUSHING.
Teeth, such as a lion's, puncture, tear, and crush the flesh and bones of the animal's prey. But such teeth are of little use in grinding tough plant materials; thus, carnivores cannot graze on tough grasses and shrubs.

The tongue and muscular walls of the pharynx and esophagus propel the meat chunks through the upper alimentary canal and into the **stomach,** an elastic-walled sac that can expand to hold the large, infrequent meals consumed by carnivores. Partial digestion occurs in the stomach, and then the food is passed into the *small intestine* where further digestion and nutrient absorption occur. Less time is required to digest animal tissues than plant tissues, so a carnivore's intestine tends to be shorter than an herbivore's.

Not all large animals have teeth. Birds lack teeth, and this probably reduces body weight for flight. When they eat meat (worms, insects) or plant material (seeds), the food passes into the muscular gizzard, where the food is ground into fine particles between pebbles. The so-called baleen whales also lack teeth. Baleens include the largest animals on earth, the blue whales. They and their relatives, such as finback, humpback, and gray whales (Figure 32-4a), survive by filtering krill, small crustaceans, and plankton from seawater. A 90 ft blue whale can collect up to 1.5 tons of food a day on fringed sheets of baleen that hang down from the upper gums. Baleen (Figure 32-4b) is nothing more than hardened, shredded sheets of the gum epithelial layers (a bit like the ridges on the roof of a person's mouth) that provide an evolutionary solution to a common problem for aquatic animals—removing small food particles from a dilute medium.

Feeding Styles of Plant-Eating Vertebrates

Most of the earth's large vertebrates—elephants, giraffes, buffalo—are exclusively plant eaters, or **herbivores,** with similar equipment for ingesting plant matter: chisel-like *incisors* in the front of the mouth, to gnaw or slice off plant materials, and broad, flat molars in the back, to crush and grind plant cell walls and fibers for long periods of time (Figure 32-5).

To appreciate the remarkable gut structures of large herbivores, recall from Chapter 24 that the "cold-

(a)

(b)

Figure 32-4 FILTER FEEDING: A COMMON FOOD-GATHERING TECHNIQUE THAT ALLOWS THE VERY BIG TO FEED ON THE VERY SMALL.
(a) The California gray whale (*Eschrichtius robustus*) sieves tons of krill daily on baleen plates. (b) The plates, which split at their ends into broomlike bristles, hang from the palate in the roof of the mouth and strain out the tiny prey.

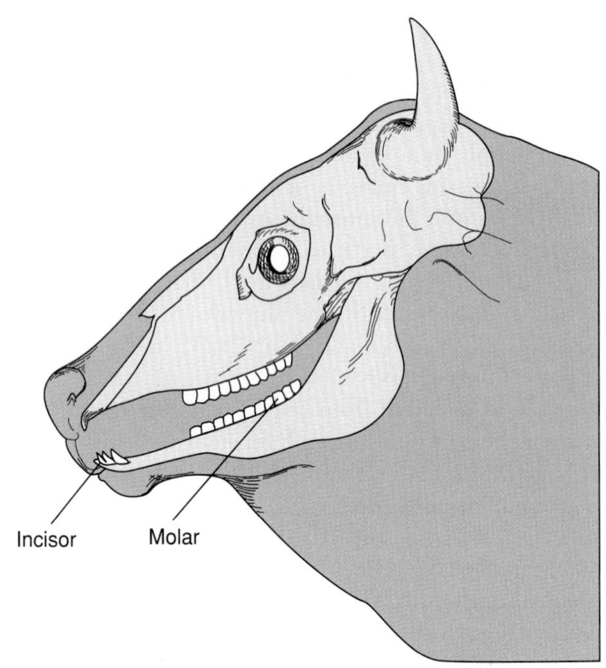

Figure 32-5 HERBIVORE TEETH: FOR GRINDING.
This ox skull provides an example of the chisel-like incisors that herbivores use to slice off plant materials. The large set of molars is used in grinding tough plant cell walls.

blooded" reptiles do not generate their own body heat and, as a result, eat very infrequently: A reptile weighing about as much as an adult human needs the equivalent of only two or three of our meals per week. Furthermore, a reptile's intestine is very short compared with a mammal's, since digestion and absorption in the reptilian gut can go on slowly. The mammalian way of life, with its high body heat and great food consumption, depends partly on the evolution of a very long intestine. And herbivores show an even greater specialization of the long intestinal tract than do other mammals.

This specialized gut helps explain how herbivores weighing hundreds of kilograms can thrive on a diet of nothing but grasses and leaves. Herbivores have symbiotic microbes that break down cellulose even though the animals themselves lack enzymes to digest plant matter directly. Thus, herbivores reap calories from plant cellulose—a trick we humans cannot do. When animals lack enzymes or symbiotic microbes that can digest cellulose, they simply pass cellulose through the alimentary canal as undigested "roughage." In contrast, the stomach chambers of cows, the large intestine of the horse, and the termite's gut all harbor special monerans or protists that do have such enzymes, called cellulases. Protists in the termite gut secrete cellulase into the insect's gut lumen; enzymatic breakdown of cellulose takes place there, and the insect can then absorb some of the cellulose subunits.

Extra stomach chambers containing symbiotic organisms have evolved at least three times in mammals: in ruminants, in sloths, and in certain marsupials (see Chapter 24). Cattle, sheep, deer, and other hooved herbivores have an elaborate four-chambered stomach in which plant matter is fermented and digested. Three of these chambers are considered ceca, or storage sites, and the fourth, the *abomasum*, is the equivalent of the stomach of most other mammals (Figure 32-6), since it is the only chamber that secretes digestive enzymes. Hooved herbivores are called *ruminants* after the largest of the three ceca, the **rumen.** When a cow eats grass, the partially chewed plant matter passes from the esophagus into the first cecum, the *reticulum*, and then slowly into the rumen, along with a large volume of alkaline (pH 8.5) saliva (Figure 32-6, step 1). A cow's daily saliva production exceeds 200 L, and the fluid's alkalinity greatly decreases the acidity of the stomach. In a large cow, the digestive chambers hold 40 gal of this semiliquid food and saliva. Immense numbers of anaerobic bacteria and ciliate protozoa that live in the rumen and reticulum fer-

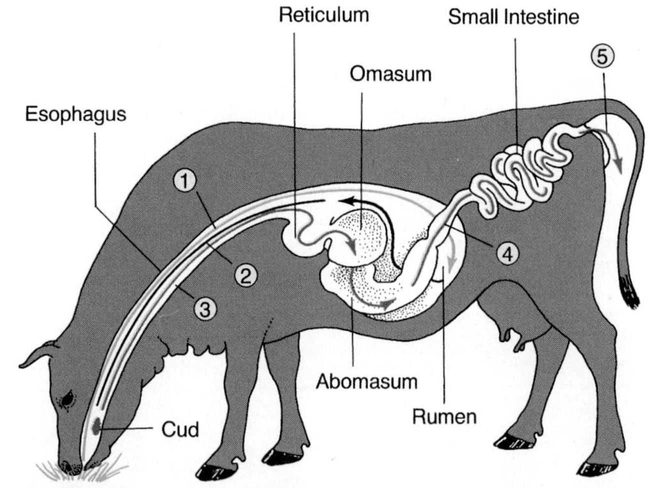

Figure 32-6 THE RUMINANT'S FOUR-CHAMBERED STOMACH.
In this cow's digestive system, the green arrows show the passage of ingested food and the gray arrows show the passage of rechewed food through the chambers of the stomach. During the first passage (step 1), food enters the rumen, where it is broken down by microorganisms. This cud is then regurgitated (step 2; gray arrows), rechewed, and swallowed; it then moves into the reticulum, omasum, and abomasum (true stomach) (steps 3 and 4), where chemical digestion by enzymes takes place. The digested food (brown arrow) then enters the small intestine and eventually exits the anus (step 5). (For clarity, the route of partially chewed food from the reticulum to the rumen is not shown.)

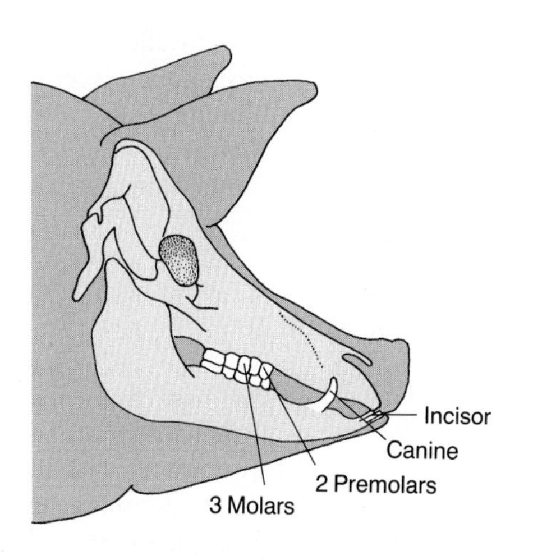

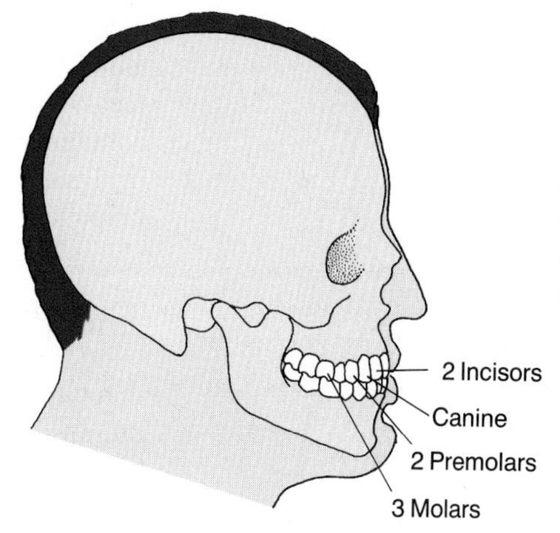

Figure 32-7 OMNIVORE VERSATILITY IN TEETH—AND THUS IN DIET.
Some mammals can be omnivorous because they have teeth for slicing, puncturing, tearing, and grinding. Hence, they are able to consume both animal flesh and plant materials.

ment the grass into its component sugars, fatty acids, and amino acids. These molecules and peptides may be absorbed into the ruminant's bloodstream. As these processes occur, large quantities of methane gas (CH_4) and CO_2 are formed and belched out. Each year, 20 percent of the CH_4 released to the earth's atmosphere comes from domestic ruminants; this amounts to more than 100 trillion kilograms!

Cows can regurgitate from the rumen undigested plant matter, called *cud,* and chew and grind any fibrous material left intact during the first round of digestion (step 2). The cud is swallowed and returned to the rumen, while the smaller and more soluble materials are passed on to the third cecum, the *omasum,* where vigorous churning further breaks up the food and allows more digestion (step 3). From the omasum, the food matter moves to the abomasum, where yet more digestion and absorption take place (step 4). Bacteria in all four chambers can combine nitrogenous compounds, such as ammonia and urea, with other molecules to form amino acids that the microbes can use for their own protein synthesis and that the cow can absorb in significant quantities along with B vitamins and other nutrients produced by the bacteria. In addition, millions of these microbial cells die each day, are themselves digested, and provide their ruminant hosts with proteins, carbohydrates, and nucleic acids. Ultimately, the largely digested food matter passes along the lengthy small and large intestines, where further digestion and water absorption take place (step 5), and the wastes are excreted.

In contrast to cows, horses and their relatives exist on cellulose-rich foods but lack the ruminant's special stomach chambers. Symbiotic bacteria are present, but mostly in the posterior part of the large intestine. Although that region of the intestine can absorb some of the nutrients from degraded food, the bacteria themselves are not killed and digested, as in cows, but are lost from the body in the feces. Consequently, a horse's digestive system is not nearly as efficient as a ruminant's at obtaining nutrients from food and from dead symbionts, and therefore, a horse must eat more than a cow of similar weight. In addition, ruminants can consume large amounts of plant matter, then retire to safe, sheltered places away from predators to regurgitate and chew their cud.

THE HUMAN DIGESTIVE SYSTEM

Humans, many primates, bears, and pigs are **omnivores:** animals that consume both plant and animal matter (Figure 32-7). While Eskimos consume mostly meat, blubber, and organs, and Hindus consume only vegetable foods, most people are fully omnivorous, eating a range of foods. Our digestive system (Figure 32-8) therefore has the mechanical and chemical ability to process many kinds of food.

The Oral Cavity

The human mouth is "guarded" by a pair of structures, the upper and lower *lips,* highly vascularized muscular flaps with an abundance of sensory nerves. They help retain food as we chew it and are important structures for facial expressions and human speech.

The oral cavity itself contains the tongue and teeth, including four groups in the adult: the chisel-shaped *incisors,* used for biting and cutting; the pointed *canines,* used for shredding and tearing; the relatively small *premolars* (or bicuspids), used for grinding; and the large, flattened *molars,* also used for grinding (see Figure 32-7). Tooth enamel, the hardest material in the body,

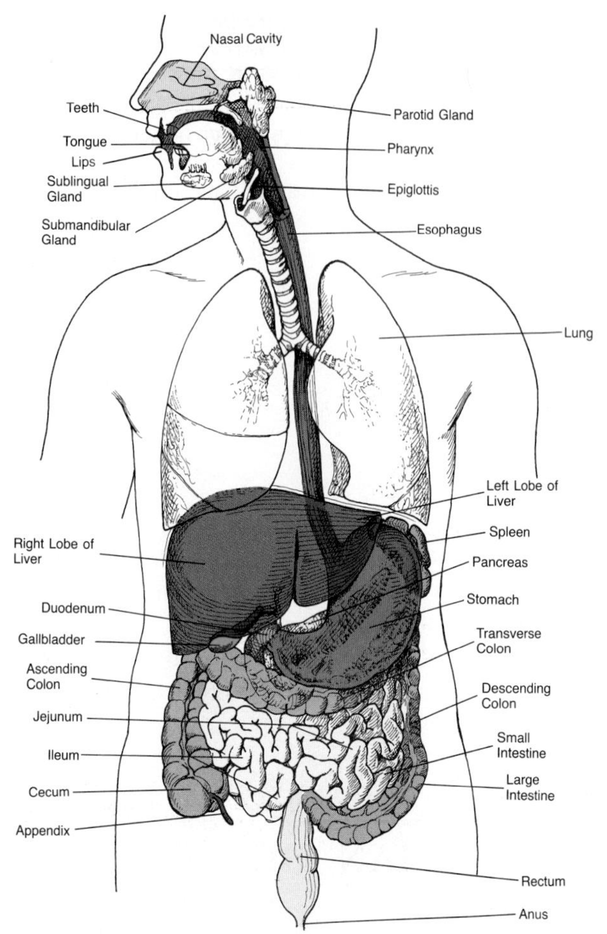

Figure 32-8 HUMAN DIGESTIVE SYSTEM: THE PERSONAL FOOD PROCESSOR.
The major organs of the human digestive system are shown here. The liver has been partially cut away to show the stomach and the pancreas. The gallbladder, small and large intestines, and other organs are easily seen. The sublingual, parotid, and submandibular glands are salivary glands.

plus our amazingly strong jaw muscles, allow us to mechanically process even hard nuts and small bones.

The *tongue*—an innovation seen only in chordates—is a muscular organ that moves and manipulates food during chewing (see Figure 32-8). It also monitors the texture and taste of foods and helps to form words in humans. The posterior part of the tongue shapes chewed food into a **bolus,** a moistened lump, which is then swallowed.

The oral cavity is continuously bathed by *saliva,* a watery fluid secreted by three pairs of **salivary glands**—the submandibular, sublingual, and parotid glands—which produce 1–1.5 L of fluid each day in an adult. Saliva begins the humidification of air passing through the mouth toward the lungs; moistens food, thereby contributing to bolus formation; and carries various molecules and ions necessary to digestion. Swallowing is aided by the presence in saliva of *mucus,* slippery secretion of many tiny glands in the mouth lining. Saliva also contains *amylase,* a parotid enzyme that begins the hydrolysis of starch into sugars, plus another enzyme that helps kill bacteria entering the mouth.

The Pharynx and Esophagus

The foods and liquids we swallow pass from the mouth into the pharynx, the thin-walled chamber at the back of the mouth that leads to both respiratory and digestive tracts. The epiglottis and vocal cords temporarily seal off the opening to the trachea so that swallowed food will enter the esophagus and not the tracheal air passages (Figure 32-9). Initiation of the swallowing reflex can be voluntary, but most of the time, it is involuntary. When swallowing begins, sequential, involuntary contractions of skeletal muscles in the walls of the pharynx and upper esophagus propel the bolus or liquid toward the stomach. Coordination of this muscle contraction and the movement of the epiglottis normally prevents aspiration of food into the trachea.

The gut contains four layers (Figure 32-10): the innermost layer, or *mucosa* surrounding the lumen; around it, the *submucosa,* a zone of blood and lymphatic vessels, nerves, and mucus-secreting glands embedded in fibrous connective tissue; the *muscularis externa,* a zone with circular and longitudinal smooth muscle layers and nerves; and an outermost layer, the *serosa,* made up of more fibrous connective tissue and a moist epithelial sheet called the *peritoneum.* This peritoneal covering lines the entire body cavity and all exposed internal organs and encompasses the coelom (see Chapter 24). Double sheets of peritoneum, the *mesenteries,* hang down from the body wall, and like a hammock, suspend the esophagus, stomach, and other gut parts.

How do the four layers of the gut wall function to move swallowed food? A wave of contraction in the smooth muscle cells of the muscularis externa and mucosa constricts first one region of the esophagus and then the next. Such waves of contraction, called **peristalsis,** propel the food bolus through the esophagus toward its junction with the next chamber, the stomach. There, a circular muscle, or sphincter, opens periodically to allow food to move from the esophagus into the stomach, and then closes to prevent backflow of the stomach contents.

The Stomach

The human stomach is a J-shaped expandable bag in which the initial breakdown of proteins takes place in the presence of a highly acidic solution containing water,

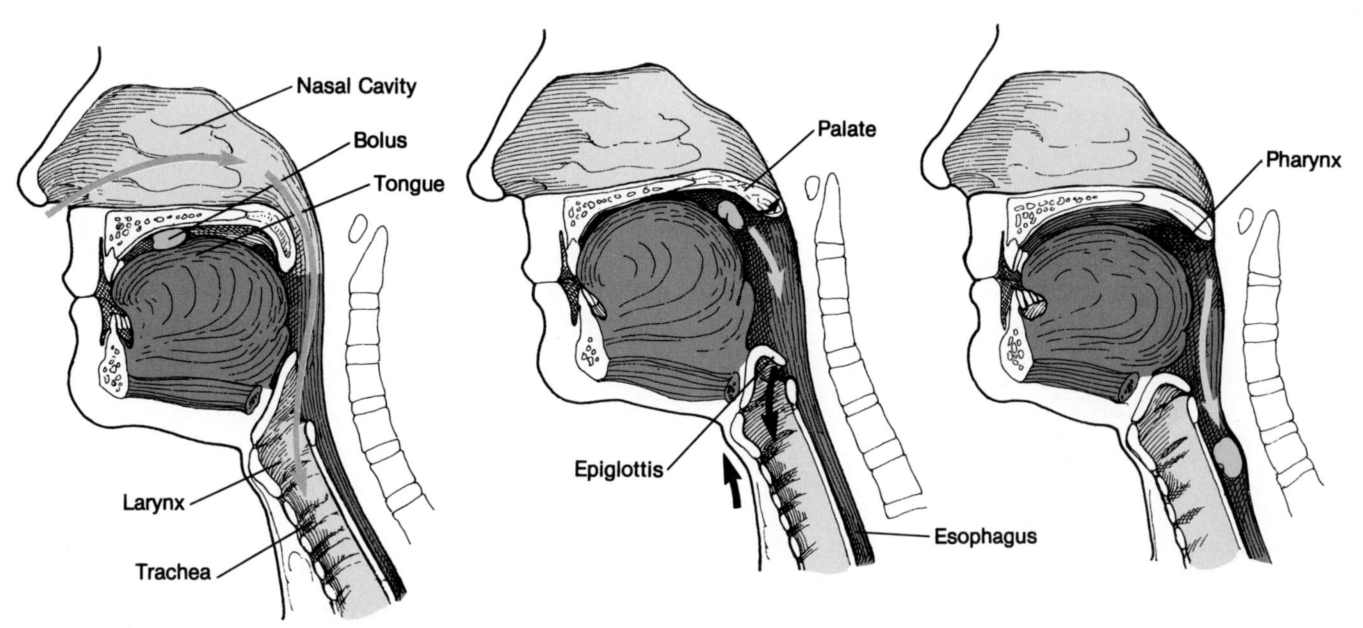

Figure 32-9 SWALLOWING: A PRELUDE TO DIGESTION.
(a) Prior to swallowing, the chewed and partially digested food begins to form a bolus. (b, c) The tongue and epiglottis move in precise coordination during swallowing to press food backward and downward. The epiglottis closes during swallowing and prevents food from entering the respiratory system. The blue arrow represents air; the red arrow the bolus's path.

mucus, enzymes, and hydrochloric acid (HCl). As Figure 32-11 shows, the stomach mucosa contains thousands of *gastric pits*, tiny crevices lined with *chief cells* (which secrete *pepsinogen*, a precursor of the enzyme **pepsin**), *parietal cells* (which secrete a solution containing HCl), and mucus-secreting cells.

Mucus coats the inner lining of the stomach and protects it from the corrosive effects of digestive enzymes

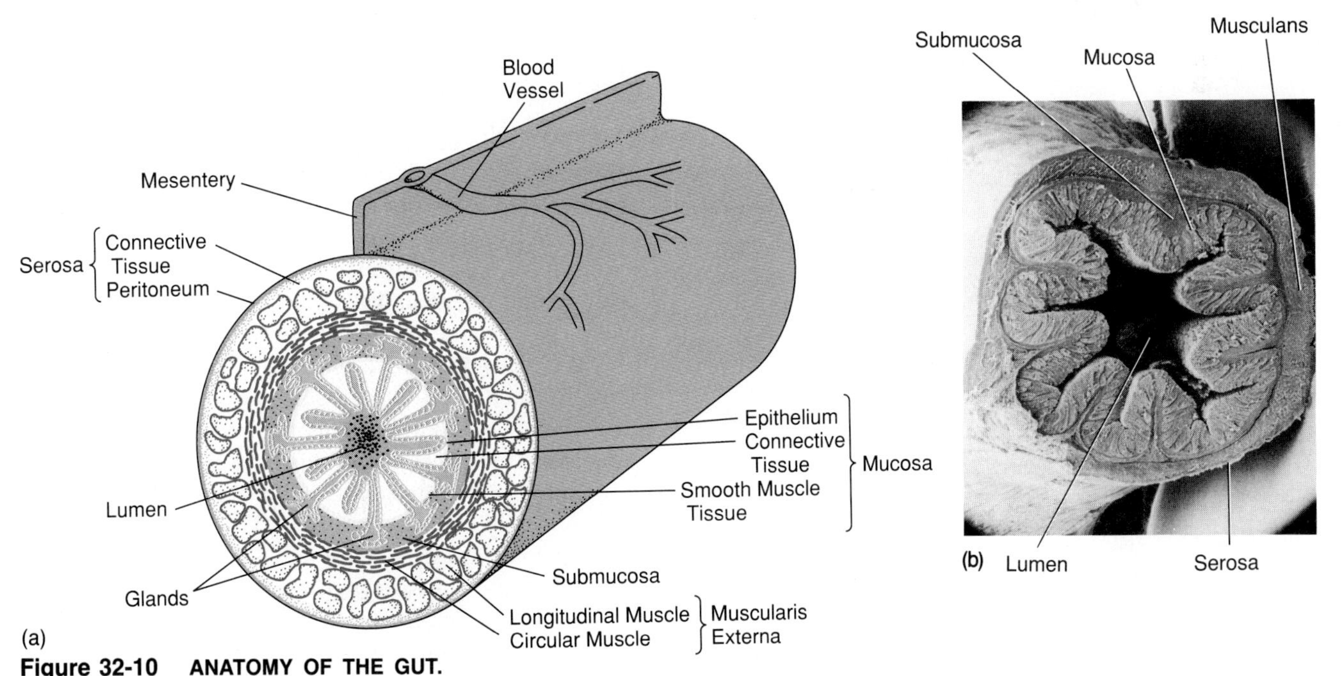

Figure 32-10 ANATOMY OF THE GUT.
(a) The mammalian digestive tract has several functional layers. The organization of these layers varies somewhat in different regions of the tract. (b) This scanning electron micrograph cross section of the colon (magnified about 20×) shows the mucosa (surrounding the lumen), the submucosa, and the longitudinal muscle.

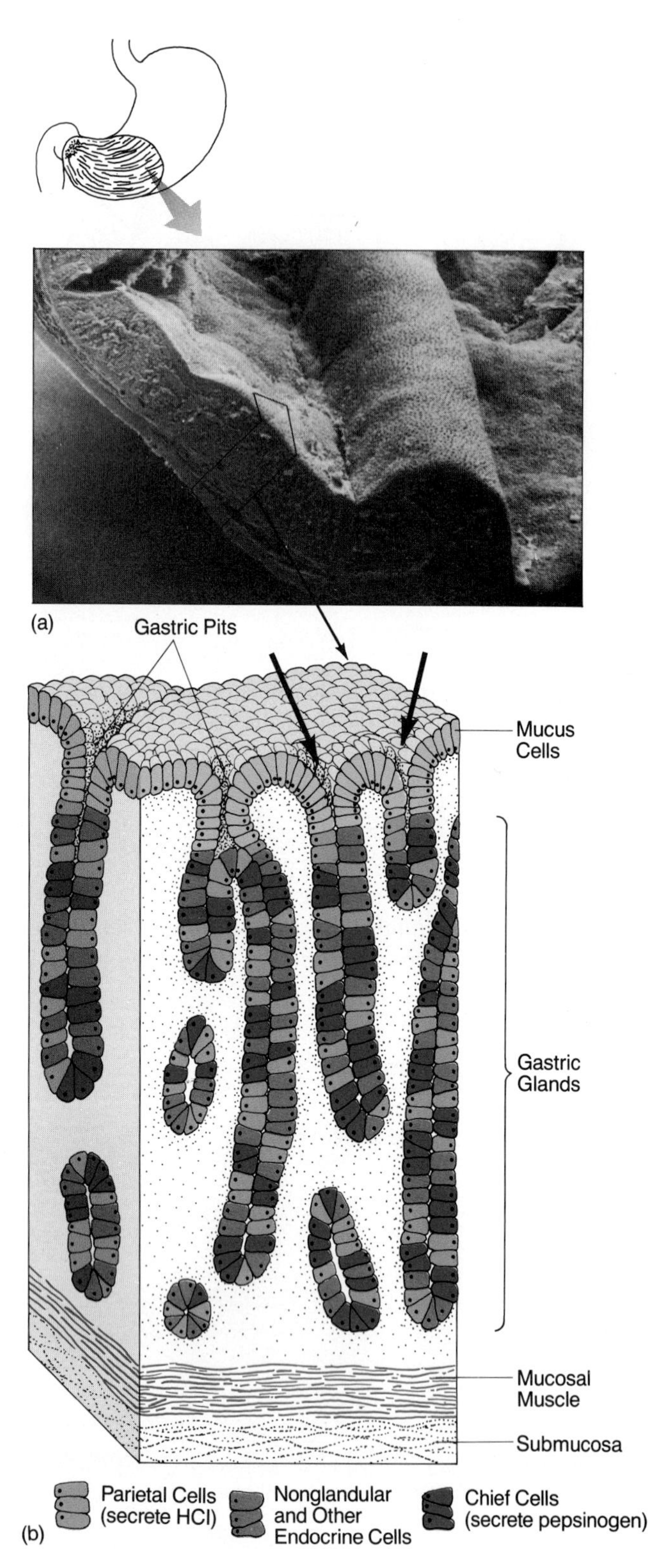

Figure 32-11 THE STOMACH AND ITS INNER SURFACE: LAYERS AND FOLDS.
(a) Openings of thousands of gastric pits are clearly visible as black dots on this scanning electron micrograph of the stomach wall (magnified about 22×). (b) The epithelium of the stomach contains numerous gastric glands and mucus-secreting cells. The mucus protects the stomach lining from digesting itself.

and acidic stomach juices. Rapid replacement of stomach-lining cells also protects against the acid, as does the secretion of acid only when food is present. At a pH of 1.5 to 2.5, gastric juice is the most acidic substance in the body. If the stomach's mucous lining is destroyed or reduced, the pepsin and HCl can begin to digest the stomach lining itself, producing a lesion called a *peptic ulcer*. Stress is thought to be one factor, resulting in the production of excess stomach acid when no food is in the stomach.

The act of eating and the presence of food induce the gastric pits to secrete HCl (as H^+ and Cl^-) and pepsinogen. The H^+ causes pepsinogen to be cleaved into pepsin, and pepsin, in turn, cleaves additional pepsinogen into more pepsin. As H^+, pepsin, mucus, and fluid mix with and begin to degrade the food, smooth muscles in the stomach wall contract and vigorously churn the mixture to prepare it for digestion in the small intestine. About 3 or 4 hours after a meal, the contents have become a semiliquid mass called **chyme** with the consistency of cream. The chyme spurts through the ringlike valve at the lower end of the stomach, the *pyloric sphincter*, and from there it enters the small intestine.

The Small Intestine: Main Site of Digestion

Most digestion of food and absorption of nutrients and water take place in the **small intestine,** a lengthy stretch of gut between the stomach and the large intestine (see Figure 32-8). A human's small intestine is only about 4 cm (1½ in.) in diameter, but it is 7–8 m (20–25 ft) long. This long length provides a large total surface area for absorbing nutrients, but a more significant enlargement comes from convolutions and minute projections of the inner gut surface (Figure 32-12). On the ridges and folds of the inner intestinal wall, thousands of tiny, fingerlike villi project from each square centimeter of the mucosa, giving it the appearance of velvet to the unaided eye. Both the folds and the villi are covered by epithelial cells, each bearing numerous microvilli. These minute projections are packed at a density of 200,000 per square millimeter and extend the surface area of the intestinal folds and villi by a factor of 20. The inner wall of the human small intestine thus has a total surface area of some 250 m^2—the size of a tennis court.

Recent experiments reveal a high turnover in the epithelial cells covering the villi; new ones are continuously produced near the bases, while old ones drop off the ends. The result is that daily about *17 billion* cells weighing about 250 g pass into the intestinal lumen.

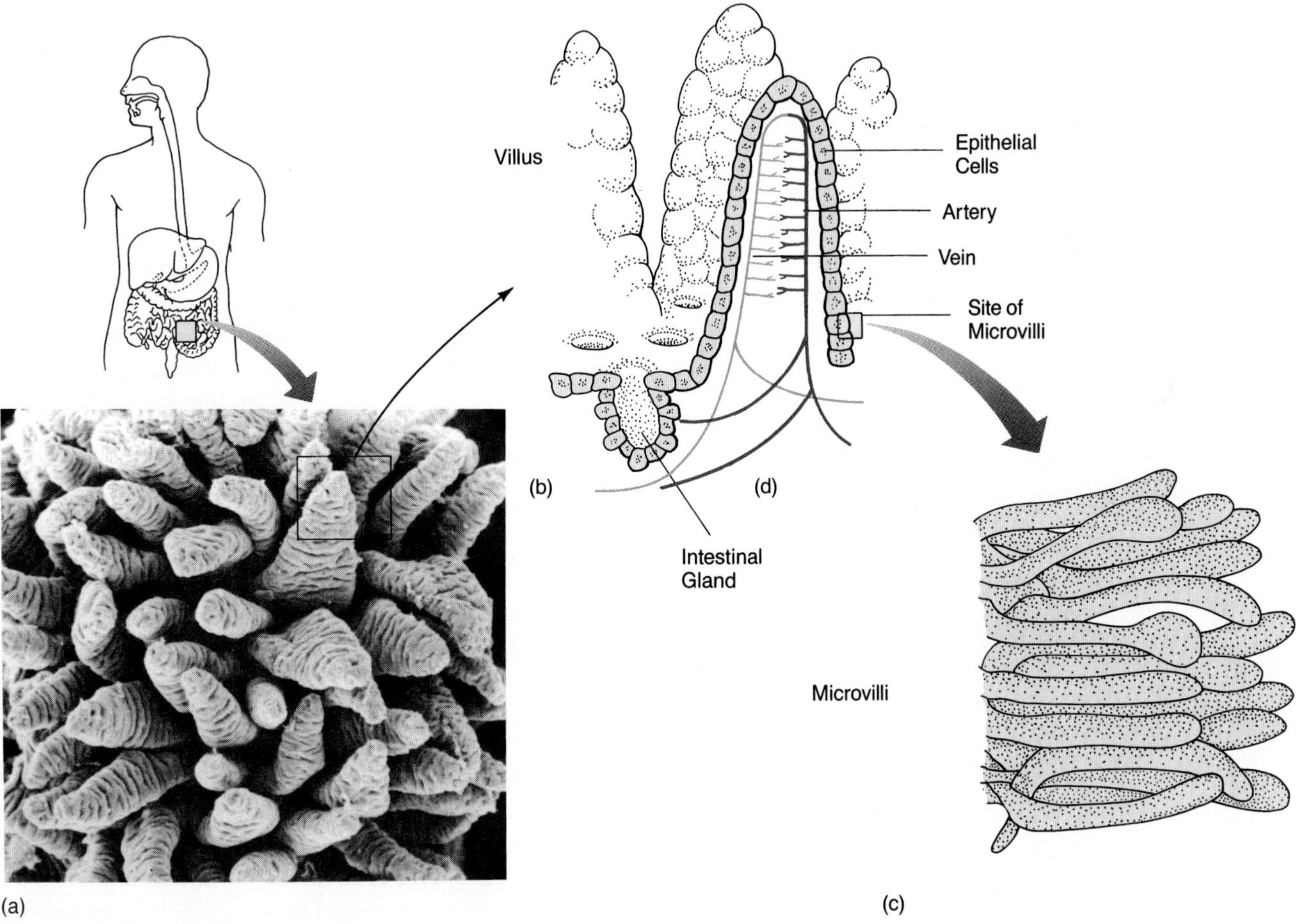

Figure 32-12 INTESTINAL VILLI.
The small intestine has a huge absorptive surface area owing to fingerlike, cylindrical villi. (a) This scanning electron micrograph shows the villi on the inside surface of a monkey's small intestine (magnified about 53×). (b) Absorbed nutrients pass through the epithelial cells into blood capillaries and lymphatic vessels, both of which are in the core of each villus. The villus surface area is further expanded by minute, fingerlike projections called microvilli (c) that are found covering the outermost lumenal end of each epithelial cell. The microvilli are too small to be seen here.

The first 30 cm of the human small intestine make up the *duodenum*, a region devoted solely to digestion. The next 3 m segment is the *jejunum*, and the final 4 m segment, the *ileum;* both carry out absorption of nutrients. The duodenum contains many digestive enzymes. Some of these enzymes are secreted by glands in the duodenal mucosa; others are secreted in the **pancreas** and flow into the duodenum through the pancreatic ducts. The pancreas is a glandular (secretory) organ that is leaf-shaped and that sits just ventral to the stomach (see Figure 32-8). The enzymes it secretes break down fats, proteins, carbohydrates, and nucleic acids. Another duct leads toward the duodenum from the **gallbladder,** a small organ beneath the liver (see Figure 32-8). The gallbladder is a storage and concentrating sac for the greenish fluid called **bile,** which is produced in the liver. Bile is very alkaline and contains pigments, cholesterol, and bile salts that act rather like detergents to emulsify fats (mold them into droplets suspended in water) and aid in fat digestion and absorption.

Both pancreatic juice and bile contain abundant bicarbonate ions (HCO_3^-), which neutralize the acidity of the chyme flowing from the stomach into the duodenum. The pH rises from about 2 to about 7.8—an optimum, slightly alkaline pH for the activity of pancreatic enzymes.

In the jejunum and ileum, amino acids, sugars, fatty acids, nucleic acid bases, minerals, and water are absorbed from the chyme across the surfaces of the epithelial villi (Figure 32-13). Much of this absorption involves active transport, and as Chapter 5 explained, large amounts of ATP must be expended to maintain the so-

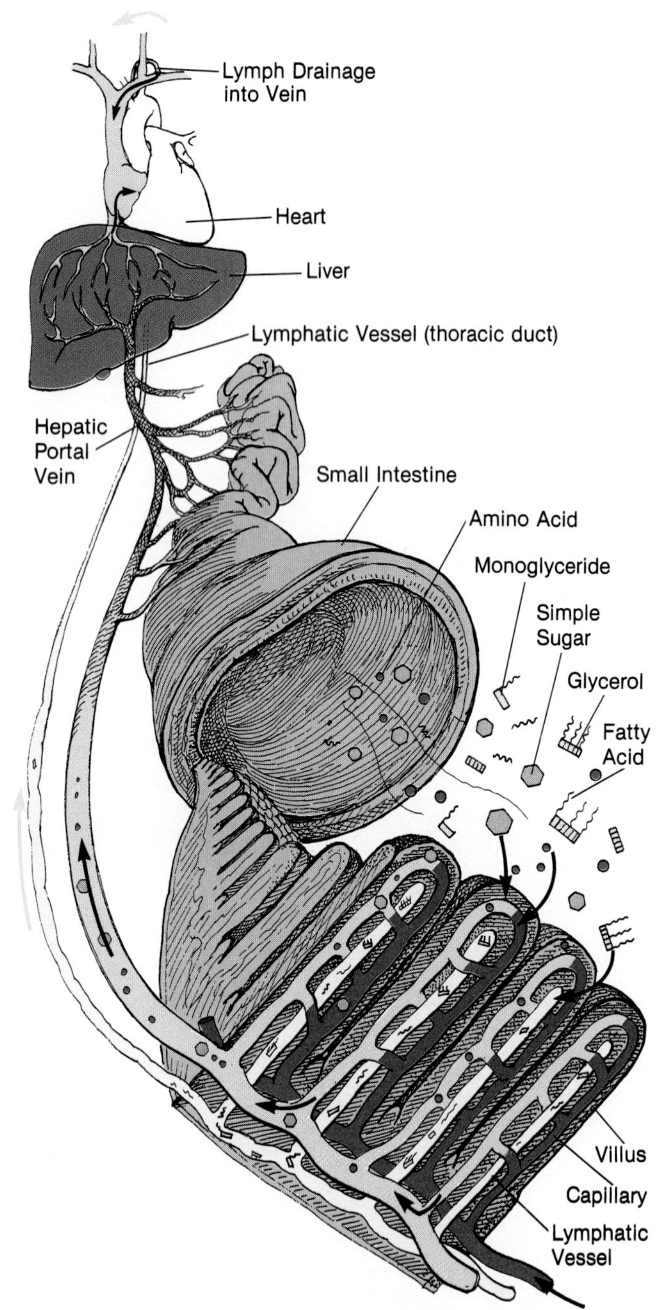

Figure 32-13 ABSORPTION OF DIGESTED NUTRIENTS IN THE SMALL INTESTINE.
Sugars and amino acids pass through the intestinal epithelium into the blood capillaries, and most fats pass into the lymphatic vessels. The absorbed sugars and amino acids are rapidly transported to the liver, whereas the absorbed fats enter the general circulation through the interconnection of the lymphatic and venous systems.

dium gradient that causes nutrients to move across the intestinal wall. Once absorbed, amino acids and sugars pass into blood capillaries inside the villi and drain into the hepatic portal vein, which flows to the liver (Figure 32-13). There, macrophages destroy invading bacteria. A large proportion of the nutrients are taken into the liver cells, where they act as raw materials for the synthesis of proteins, glycogen, nucleic acids, and other substances.

Some lipids are absorbed undigested through the plasma membranes of the epithelial cells; however, most fats are broken down into fatty acids and monoglycerides (and some glycerol) before being absorbed. Once inside the cells of the intestinal wall, triglycerides are resynthesized and coated with phosphoproteins to form minute droplets called *chylomicrons.* These droplets pass out of the cells and into lymphatic vessels called *lacteals,* which are located at the core of each villus, and from there, move into the bloodstream and are transported throughout the body.

The small intestine also absorbs a great deal of water along with the nutrients. Each day, about 10 L of fluid pass into the gut. A person drinks about 1.5 L in the form of water, soft drinks, coffee, and so on, and the other 8.5 L pass into the gut as saliva, gastric juice, pancreatic secretions, and other fluids. The small intestine absorbs about 9 L of liquid, and the large intestine absorbs most of the rest.

The Large Intestine

The human **large intestine** is a sizable segment of the gut, with a diameter of about 6.5 cm and a length of 2 m. Because the inner surface has few convolutions and has no villi or microvilli, the surface area is only 1/30 that of the small intestine. The large intestine is joined to the small intestine near a blind-ended sac, the cecum. The human cecum and its fingerlike extension, the *appendix,* are nonfunctional as storage sites and are probably evolutionary remnants. From its junction with the small intestine, the large intestine forms a squarish configuration around the folds of the small intestine (see Figure 32-8). Also called the *colon,* it ascends the right side of the abdominal cavity, crosses to the left, and then descends (with sections called, respectively, the ascending, transverse, and descending colon). The descending colon terminates in the rectum, which leads to the *anus,* the exit point of the entire alimentary canal.

The large intestine absorbs water and minerals and forms the feces. As chyme is moved along by peristaltic waves, minerals diffuse or are actively transported from the chyme across the epithelial surface of the large intestine. Water follows osmotically and is returned to the lymph and blood. When water absorption is disrupted, diarrhea results; the waste passes too quickly and is eliminated as watery feces. When fecal matter moves too slowly and becomes excessively dehydrated, constipation results.

Many bacterial species, including *E. coli,* exist symbiotically in the large intestine. Housed in a warm, moist environment, millions of these cells can consume food materials that resisted digestion previously. At the same time, they secrete amino acids and vitamin K, which the host absorbs along with minerals and water.

The last portion of the large intestine is an expandable storage chamber, the *rectum*, which holds feces until they are eliminated. Feces are composed primarily of water and of great numbers of dead bacteria, undigested food such as plant fibers, debris from sloughed-off cells, and other waste products. Strong peristaltic contractions of the large intestine's smooth muscle layer and a defecation reflex lead to the expulsion of the semisolid feces.

The Liver

The largest and one of the most versatile organs in the human body is the **liver,** an intermediary between digestion and the organism's metabolic needs. It is a reddish brown, lobed mass that weighs about 3–3.5 lb and sits just under the diaphragm (see Figure 32-8). The liver contains millions of cells called *hepatocytes,* which help regulate the nutrient content of the blood as it flows through the lobed organ. Hepatocytes also manufacture several blood proteins, including prothrombin—an enzyme involved in blood clotting (see Chapter 29)—and albumin—a plasma protein.

The mammalian liver is a major site for the uptake and conversion of a wide variety of substances, including toxic ones. Excess amino acids are absorbed in the liver and converted to urea, which is then excreted in the urine. Hemoglobin from dead red blood cells is collected in the liver and converted to bile pigments. These pigments, *bilirubins* (red) and *biliverdins* (green), color the bile and the feces. Certain enzymes in hepatocytes break down toxins such as alcohol and other drugs. If the exposure is high-level or chronic, the cells can be damaged in the process, and a scarring disease called *cirrhosis* can result.

Besides its other functions, the liver produces *somatomedins,* agents that mediate bone growth; it stores fat-soluble vitamins; it acts as a reservoir for glycogen; and it is involved in maintaining normal blood sugar levels. Clearly, the condition of the liver is critical to the health of the entire body.

THE CHEMISTRY OF DIGESTION

Between the time an animal eats food and excretes undigested wastes, its body must absorb nutrients. The chemical part of the digestion process is the breakdown of foods into small absorbable units by enzymes such as those listed in Table 32-1.

Chemical digestion breaks the bonds linking monomers into large protein, carbohydrate, and lipid polymers. These reactions most often involve hydrolysis, with enzymes working in tandem to break first one molecular bond and then another.

Let us take the breakdown of starch (amylose and amylopectin) as an example of the chemical digestion process. Amylase in the saliva begins the hydrolysis of amylose (Figure 32-14). Amylase secreted by the pancreas into the small intestine continues that process, but at an alkaline pH, yielding large amounts of the disaccharide maltose. Much disaccharide passes into epithelial cells on the villi of the small intestine and is digested by eight disaccharide-splitting enzymes. In addition, the enzyme *maltase,* located in the luminal membrane of cells in the wall of the small intestine, cleaves maltose to produce two molecules of the monosaccharide glucose. This sugar is then absorbed through the intestinal wall via active cotransport with Na^+ (see Chapter 5).

The digestive enzyme *lactase* specifically degrades lactose, the carbohydrate in milk. It is synthesized by baby mammals but not by most adults, since they usually cease drinking milk at the time of weaning. Interestingly, many adult humans can drink unlimited quantities of milk, while others develop cramps, stomach gas, diarrhea, and vomiting if they drink more than a little. These latter individuals are said to be *lactose-intolerant.* Anthropologists and geneticists have found that the ability of adults to drink and digest milk evolved within the last 5,000 years or so, and only in populations that traditionally maintained dairy herds, such as cultures in northern Africa, the Near East, and Europe. About 70 to 100 per-

Table 32-1 SOME MAJOR DIGESTIVE ENZYMES

Source	*Enzyme*	*Substrate*
Salivary glands	Amylase	Starch, glycogen
Stomach	Pepsin	Proteins
Pancreas	Amylase	Starch, glycogen
	Lipase	Lipids
	Trypsin	Proteins
	Chymotrypsin	Proteins
	Deoxyribonuclease	DNA
	Ribonuclease	RNA
Small intestine	Maltase	Maltose
	Lactase	Lactose
	Sucrase	Sucrose
	Aminopeptidase Carboxypeptidase Tripeptidase Dipeptidase	Peptides
	Nucleases	Nucleotides

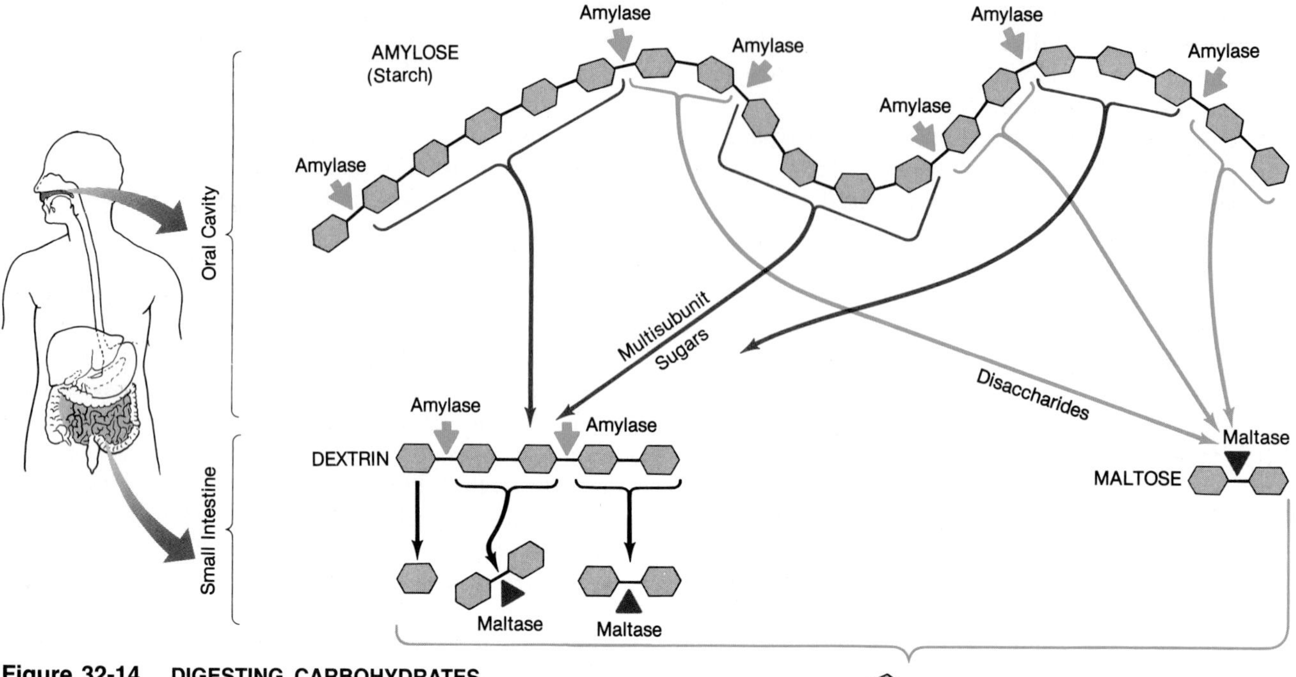

Figure 32-14 DIGESTING CARBOHYDRATES. The original starch polymer is broken into disaccharides and larger polysaccharides termed dextrins. The latter are broken into shorter molecules, depending on the point at which amylase hydrolyzes bonds. Ultimately, only monosaccharides and disaccharides remain.

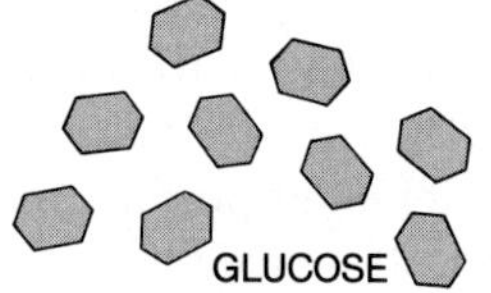

cent of individuals whose ancestors come from dairying regions produce lactase as adults, while 70 to 100 percent of adults from traditionally nondairying regions do not.

The enzymatic breakdown of fats is similar to that of carbohydrates, but since fats are insoluble in water, they must first be emulsified by bile salts so they remain suspended in the chyme (Figure 32-15). Then enzymes such as pancreatic *lipase* hydrolyze the triglycerides into fatty acids and glycerol. Those subunits are absorbed by the small intestine and are transported via lymph vessels to the neck, where they enter the blood and pass to the liver and other organs.

Proteins make up the most formidable class of molecules for chemical digestion because of their great structural variation and their 20 different subunit amino acids. Enzymes such as pepsin, trypsin, and chymotrypsin, collectively called *endopeptidases*, cleave proteins into smaller molecules, splitting bonds at very specific places in the middle of protein molecules (Figure 32-16). That step generates many more amino and carboxyl ends than existed in the uncleaved protein. Those ends become substrates for the *exopeptidases* (including aminopeptidase, dipeptidase, and carboxypeptidase), which remove amino acid subunits one or two at a time.

Protein digestion starts in the stomach when pepsin, an endopeptidase, cleaves food molecules wherever five particular amino acids occur along protein polymers. The pancreatic endopeptidases work next, and finally, exopeptidases, secreted by cells in the wall of the small intestine, complete the job of freeing amino acids. Some dipeptides (and even tripeptides and higher polymers), however, are absorbed by the microvilli and are digested intracellularly.

Additional kinds of food molecules—RNAs, DNAs, and so on—are broken down by other digestive enzymes working alone and in series. Together, the enzymes listed in Table 32-1 ensure that virtually all ingested nutrients are reduced to an absorbable state.

If enzymes can degrade almost any kind of food, why don't they attack and digest the very cells that generate them? First, the entire gut secretes mucus, which helps resist enzymatic attack. Second, digestive enzymes are packaged in storage granules in the RER and Golgi complex (see Chapter 6). Third, many digestive enzymes are synthesized as inactive precursors called **zymogens,** which must be cleaved into an active enzyme and a leftover polypeptide. For example, when pepsinogen is secreted into the stomach, H^+ (from HCl) cleaves a region from one end of the zymogen to yield pepsin. For the

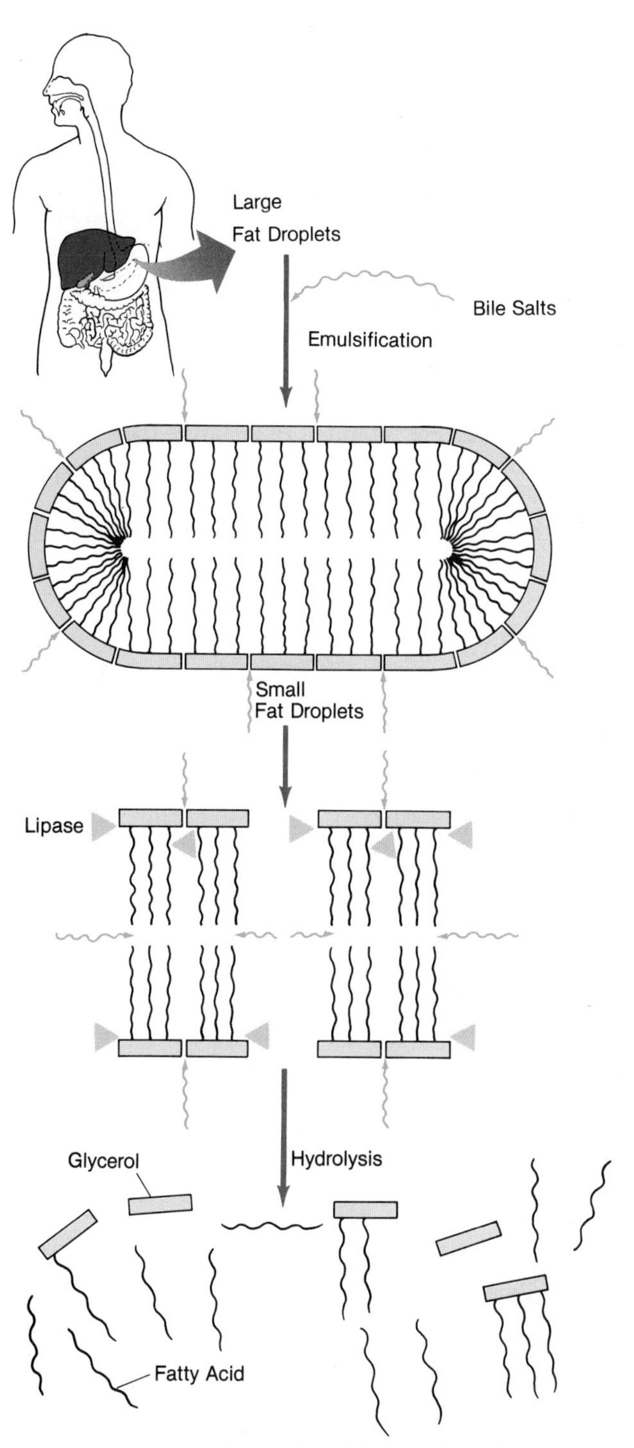

Figure 32-15 DIGESTING AND ABSORBING FATS. Large fat droplets in the chyme are dispersed by bile salts. Small fat droplets form and are attacked by lipase from the pancreas. The breakdown products (glycerol, fatty acids) cross the plasma membrane of the intestinal epithelial cells that cover the surface of intestinal villi. There the fatty acids and glycerol are reassembled into lipids and released in association with protein. The lipid-protein aggregates enter lacteals or occasionally capillaries in the villus.

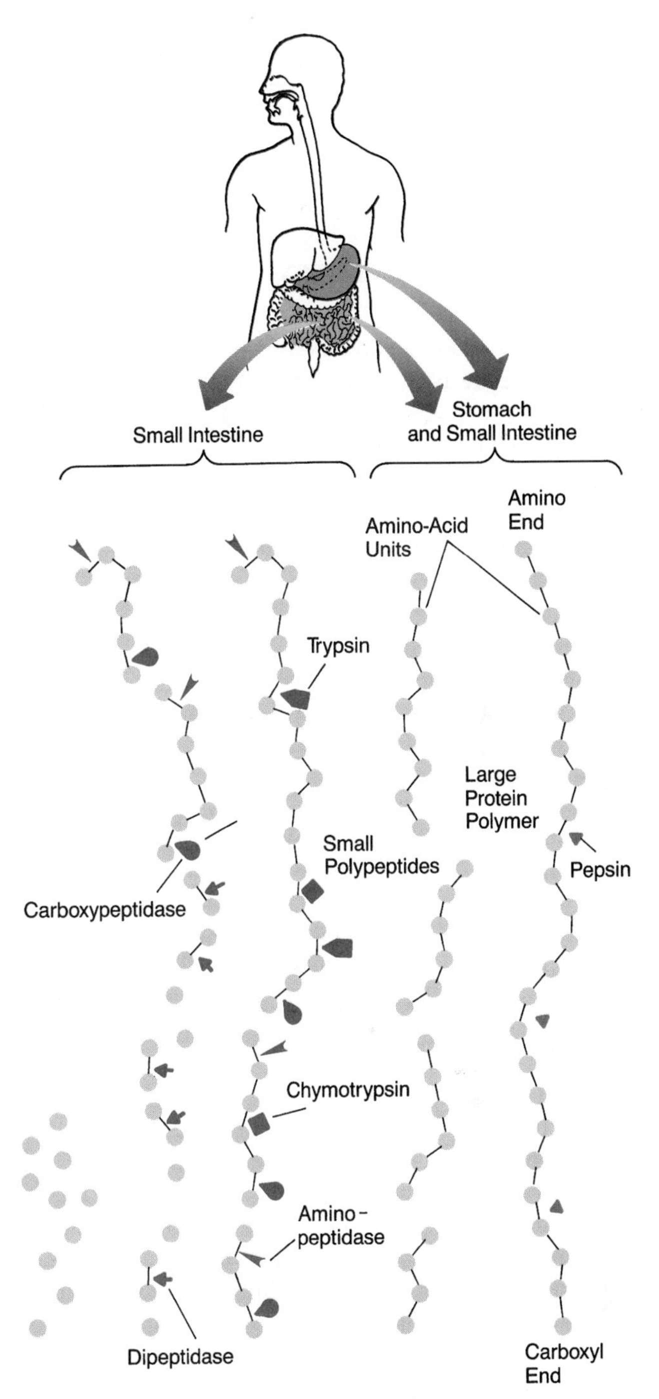

Figure 32-16 DIGESTING PROTEINS. Pepsin, trypsin, and chymotrypsin are endopeptidases (in red); aminopeptidase, carboxypeptidase, and dipeptidase are exopeptidases (in orange). All of these are enzymes that hydrolyze different peptide bonds. (Water is added during the hydrolysis.) Aminopeptidase, carboxypeptidase, and dipeptidase free individual amino acids. Note in this sequence the sites where the different kinds of enzymes can hydrolyze peptide bonds.

pancreatic zymogens, trypsin is the common activator (as, for example, chymotrypsinogen), cleaving off a small part of each precursor to trigger activity.

COORDINATION OF INGESTION AND DIGESTION

Ingestion and digestion are obviously highly efficient ways to supply body cells with the specific molecules they need for metabolism, cell growth, and repair. But controls are needed so that the animal senses when to *ingest* and the body when to *digest*.

Control of Enzyme Secretion and Gut Activity

If you are like most people, simply thinking about whatever most delights your palate will bring on a flow of saliva and gastric juices, ready to digest the imaginary delicacy. What controls the release of enzymes when feeding is imminent?

Investigations have revealed that impulses from the brain stimulate secretion of saliva and gastric juices at the sight or smell of food (Figure 32-17, step 1). A branch of the vagus nerve stimulates cells in the stomach to secrete a digestive hormone called **gastrin** into the bloodstream (step 2). It is gastrin that causes the gastric glands in the stomach wall to secrete HCl and pepsinogen (step 3). When food reaches the stomach, it stimulates further gastric secretion.

When chyme from the stomach reaches the duodenum (step 4), the acid in chyme causes the digestive hormone *secretin* to be secreted, and the proteins and fats in chyme cause intestinal cells to release the digestive hormone *cholecystokinin (CCK)* (Figure 32-17, step 5). Both hormones slow contractions of the gut's smooth muscle layer so that food will be passed along more slowly and fats can be fully digested and absorbed. In addition, secretin causes the pancreas to secrete bicarbonate, which neutralizes the acid in chyme (step 6), and CCK causes the pancreas to secrete fluid rich in protein-digesting enzymes (step 7).

Intestinal cells secrete at least six additional hormones to control the stomach and pancreas. One of the most interesting is **vasoactive intestinal peptide (VIP)**, a hormone secreted by the duodenum when fats are present in its lumen (step 8). VIP increases secretion of pancreatic juice and inhibits secretion of gastric juice (and gastrin), among other actions. VIP, secretin, CCK, and several other hormones are synthesized not only in the intestine, but also at specific sites in the brain, and are considered neuropeptides (see Chapter 36).

Control of Hunger and Feeding

What makes an animal seek food when it needs nutrients and stop eating when it has taken in enough? Some biologists believe that an organism senses hunger after cells in the hypothalamus or in the liver respond to lowered levels of glucose, amino acids, or fats in the blood. The animal then starts to hunt, graze, or perhaps open a refrigerator. Eating then distends the stomach and begins to raise the levels of those nutrients in the blood; as a result, the animal stops eating.

Using a feeding blowfly as an experimental model, researcher Vincent Dethier discovered that cells are stretched as an animal's stomach fills, and this stretching sends an inhibitory signal to the brain. When he cut nerves between the stomach and the brain, inhibitory

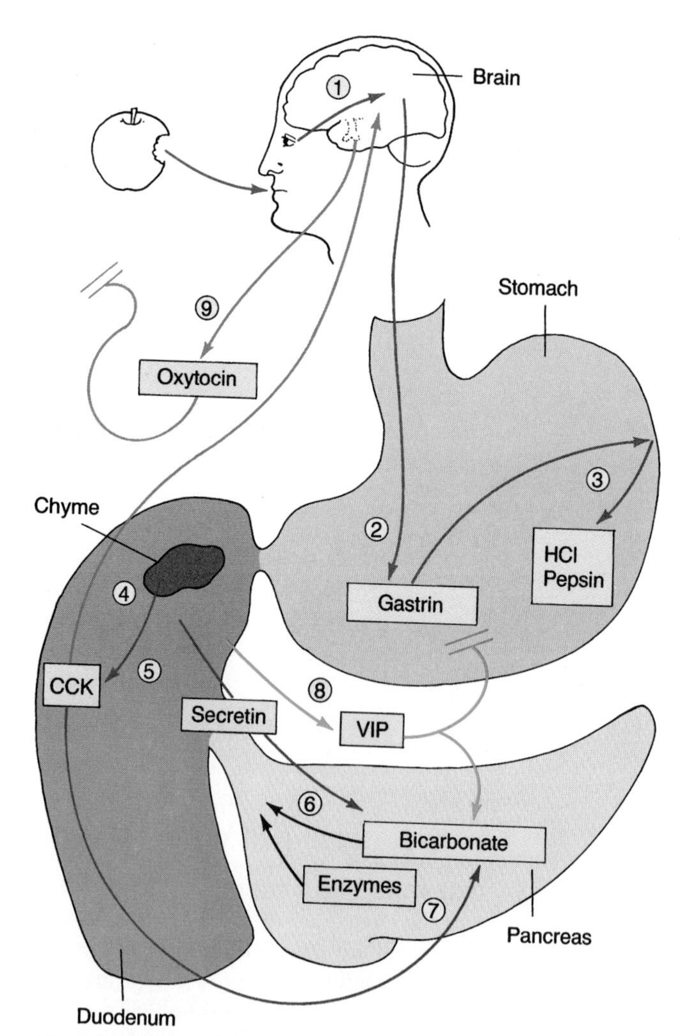

Figure 32-17 HORMONE CONTROL OF DIGESTIVE JUICE SECRETION BY THE GUT.
Impulses from the brain, as well as the presence of food in the gut, stimulate the regulatory processes that affect stomach, pancreatic, and intestinal secretions. The text explains the sequence of events, step by step.

signals could not be sent and the fly continued to eat until it nearly burst.

Experiments with rats, chickens, and various mammals have located the feeding control mechanism in the brain's hypothalamus. Certain animals that receive surgical lesions in a part of the hypothalamus will eat compulsively, while lesions to another part cause the animal to cease eating altogether. Information about nutrient levels travels from liver cells to the brain and is then integrated in the hypothalamus, eventually resulting in a sensation of hunger.

The brain hormone oxytocin also plays a key role in stopping ingestion. Cholecystokinin (CCK) seems to trigger the brain's pituitary gland to secrete oxytocin (Figure 32-17, step 9). Injecting CCK into a rat that has just begun to feed causes a massive release of oxytocin into the blood plasma and brings the feeding to an immediate halt. Some biologists hypothesize that when an animal has eaten enough to be full or when it eats nauseating food, the vagus nerve (which carries information from the stomach to the brain) causes the hypothalamus to release CCK. That release, in turn, causes the pituitary gland to secrete oxytocin. By still unknown mechanisms, these high levels of oxytocin influence the animal's behavior—it stops eating.

Control of Body Fat

If eating is the universal pastime, trying to lose weight is becoming a close second—at least for many North Americans. The amount an organism eats is of course strongly related to its weight, and so is the composition of its diet. Research shows that when we eat fats, the body can convert them into stored body fat more easily than it can the carbohydrates or proteins we consume. This, plus the risk of clogged blood vessels, explains why agricultural researchers are using genetic engineering techniques to develop leaner cows and pigs. In one recent study, experimenters inserted genes for bovine (cow's) growth hormone into pigs. The pigs grew faster on less feed and their meat had only one-third as much fat as that of a normal pig.

Recent evidence shows that other factors besides total food intake and the leanness or fatness of the diet are also involved in a person's weight. Weight researchers have developed the concept of a *set point*, a genetically determined level for each individual's body weight and, particularly, for the proportion of fat his or her body stores. This set point is analogous to an animal's body temperature, which is automatically maintained within a very narrow range. Tests show that in most people and in various other mammals tested, when the caloric intake is greatly increased, the body does not gain much weight, but burns the excess calories and dissipates them as body heat. When food intake is experimentally lowered, the reverse happens: Body weight and fat stores are maintained, and the available calories are used more efficiently. Research also suggests that increased physical activity is the only successful way to change the set point and alter the amount of fat the body stores over an extended period.

The number of fat cells in the human body may influence the set point. Fat cells probably arise early in life; childhood fat intake influences the number of fat cells in the adult. Once present, fat cells remain fairly constant in number, but they can vary greatly in size; in obese animals, for example, they are four to five times larger than normal. Fat cells may signal the brain to control eating by releasing a protein called adipsin into the blood. When the animal is fasting, more adipsin is released, and fat is mobilized. Animals with genetic obesity have lower levels of adipsin.

Under certain environmental stresses, the set point may shift temporarily—probably following changes in hormone production. Birds or mammals that migrate or hibernate seem to experience seasonal changes in set point. First, the appetite and the tendency to store fat increase; then, after hibernation or migration begins, the "thermostat for feeding" is lowered, and the body fat is slowly burned. In the northern parts of the United States and Europe, a surprisingly high percentage of people experience SADS (seasonal affective disorder syndrome): They crave excessive amounts of carbohydrates during the winter and suffer depression, poor sleep, and other symptoms. The hormone melatonin and the neurotransmitter serotonin are implicated in this carbohydrate craving, and exposure to strong artificial light can help to ease the symptoms for many until the days lengthen naturally.

Aberrations of appetite and feeding seen only in humans, *anorexia nervosa* and *bulimia nervosa*, are characterized by minimal food intake over long periods or by excess eating and induced vomiting. Research suggests deep psychological reasons for the two conditions, but new data also show that after a meal, bulimics (which include about 4 percent of young adult women in the United States) secrete only half as much CCK—the hormone that induces satiety—as normal eaters.

NUTRITION

We humans, along with most other heterotrophs, must take in proteins, carbohydrates, fats, water, and some minerals—the so-called **macronutrients**—in substantial amounts. Heterotrophs need smaller amounts of vitamins and trace inorganic minerals, the **micronutrients.**

SKIN COLOR, DRIFTING CONTINENTS, AND VITAMIN D

Anthropologists believe that our own species, *Homo sapiens*, originated in Africa and that the first populations were dark-skinned. Why, then, did some human populations evolve later with light skins? The answer, strangely enough, apparently involves drifting continents and vitamin D.

During the long Mesozoic era, the continents drifted to their current positions, and the tropical Gulf Stream began to flow northward, past Newfoundland, Greenland, and Iceland, and toward northern Europe, making these far northern latitudes green and habitable. As a result, human populations were able to migrate from the lower latitudes into these areas. And it was during that migration and early habitation perhaps 75,000 years ago that lightly pigmented skin, eyes, and hair apparently evolved.

Interestingly, the absence of dark pigmentation is directly related to the human body's production of vitamin D. Human skin cells manufacture the active form of vitamin D, 3-dehydrocholesterol, when they are exposed to certain ultraviolet wavelengths in sunlight. A person making or consuming too little vitamin D is very likely to develop *rickets,* a disease in which the bones become brittle, misshapen, and easily broken.

Vitamin D and rickets have a straightforward connection to human skin color. In a tropical region, a light-skinned person can absorb too much ultraviolet light and convert too much precursor to vitamin D. The results can be a condition called *hypervitaminosus,* in which too much calcium is present in the blood plasma, kidney stones build up, extra pieces of bone form in the tendons, and other problems occur. Among dark-skinned people living in tropical areas, the relatively large amounts of black and brown melanin pigments in their skin and hair absorb the ultraviolet wavelengths from the sun and prevent their bodies from producing a damaging oversupply of the vitamin D precursor (not to mention avoiding skin cancers). Since the earth's first *Homo sapiens* were dark-skinned peoples in tropical regions of Africa, this physiological mechanism almost certainly operated in them.

Figure A
Dark skin and light skin: A matter of melanin.

In dark-skinned migrants to northern Europe, however, such concentrated melanin pigments would have absorbed much of the ultraviolet light (since sunlight is weaker there much of the year), and this would have prevented them from producing enough vitamin D to avoid rickets. Many biologists believe that strong selection pressure would have operated for lighter and lighter skins, and a pale blond exposed to the Scandinavian sunlight for just one hour a day can manufacture enough vitamin D to prevent rickets. The only far-northern peoples with heavily pigmented skins are the Eskimos. And they have traditionally eaten large quantities of fish oils—a good source of vitamin D.

How poignant and tragic it is that the evolution of skin colors appropriate to latitude and sunlight levels should have led to prejudice, racial discrimination, and cruel persecution over the centuries.

Macronutrients: The Basic Foods

In the gut, macronutrients are broken down into subunits that are absorbed, transported in the bloodstream to tissue cells, and used as building blocks, as inputs to various metabolic pathways, or as substrates for storage of chemical energy in ATP. The energy value of food is equivalent to the **calories** produced when the food is oxidized completely. One calorie is the amount of heat required to elevate the temperature of 1 g of water by 1° C. The kilocalorie (1,000 calories, or 1 Calorie) is the unit of measure used most often by nutritionists. When 1 g of protein or 1 g of carbohydrate is metabolized, about the same amount of heat is released—around 4 kilocalories per gram. In contrast, fat yields more than 9 kilocalories per gram, and this explains why animals' bodies store energy reserves in this form.

The subunits of macronutrients may be stored within

liver, muscle, lung, or other types of cells for future needs. Because the individual subunit molecules would exert an osmotic effect in the cytoplasm, they are converted to large polymers for storage. Carbohydrates are built into glycogen granules (see Figure 3-11), and fats can be stored as oil droplets. Interestingly, protein is very rarely stored. Apparently, it takes less energy to store carbon as carbohydrate or fat and then convert those substances to amino acids than it does to manufacture and store large reservoirs of protein directly. When protein is digested, liver enzymes remove the amino groups from amino acids in a process called *deamination.* The carbon atoms are then passed via acetyl coenzyme A to carbohydrates or fats for storage.

Virtually every animal must consume a variety of foods to fulfill its dietary requirements, since many single foods lack specific amino acids or fatty acids and since animals cannot synthesize all of them. Most animals, for instance, can synthesize only about half of the 20 different amino acids and so must obtain the others in the diet. Adult humans cannot synthesize isoleucine, leucine, histidine, lysine, methionine, phenylalanine, threonine, tryptophan, and valine; nutritionists call these the nine **essential amino acids.**

The typical, omnivorous human diet in the United States contains poultry, meat, fish, eggs, and milk, each of which usually contains all the essential amino acids. In contrast, the proteins in grains, vegetables, and fruits are usually deficient in one or more of the essential amino acids. For example, beans contain lysine, tryptophan, and cysteine, but are low in methionine, whereas corn is a source of methionine but is low in lysine, tryptophan, and cysteine. The Indians of central Mexico have thrived on their traditional diet of corn tortillas and beans because they eat them both in the same meal and so get a full set of amino acids for protein synthesis.

Some animals cannot synthesize certain fats. Rats and humans can convert carbohydrates to various types of fatty acids and lipids for storage or use in cell membranes. However, neither organism can synthesize enough of the common fatty acid *linoleic acid*, an essential component of cell membranes. If a person or rat fails to eat enough of this and two other essential fatty acids in the form of corn, safflower, or other vegetable oils, the result will be nerve cell degeneration, severe malfunction of other cell types, and perhaps even death.

Micronutrients

The micronutrients, the vitamins and trace minerals used in the body's enzymatic reactions, are usually small ions and molecules that the body cannot synthesize rapidly (if at all). Thus, they too are essential parts of the diet.

Vitamins

Vitamins are organic molecules that function as coenzymes or cofactors of enzymes (see Chapter 4). The need for vitamin C was first discovered in 1536. The significance of this vitamin was noted in the eighteenth century. At that time, British sailors on long sea voyages ate little besides salt pork, biscuits, and rum. These men frequently suffered *scurvy,* with its poor healing of wounds, excruciating joint pain, anemia, loose teeth, and connective tissue diseases. Eventually, people noticed that lemons, limes, and sauerkraut (all of which we now know contain vitamin C, or ascorbic acid) prevented the symptoms. Without this vitamin, cells cannot secrete collagen, the most common protein in the body, and scurvy symptoms result.

In general, identifying essential vitamins has been difficult for two reasons. First, vitamin needs vary among organisms. Cats and dogs, for example, can synthesize their own vitamin C, whereas humans and monkeys cannot; thus, early research results based on studies of laboratory animals were often quite confusing. Second, the quantities of a vitamin needed tend to be so tiny that it is extremely difficult to prove that the compound is essential. For example, 1 oz (28.3 g) of vitamin B_{12} can supply the daily need of 4,724,921 people! What is more, bacteria residing in the animal gut synthesize vitamins such as B_{12} or K and provide a built-in vitamin source to confuse experimental results still further. Nevertheless, nutritionists have developed a list of essential vitamins, their roles in the body, and deficiency symptoms (Table 32-2).

Vitamins are classified as either water-soluble or fat-soluble. **Water-soluble vitamins,** such as C and the B vitamins, are transported as free compounds in the blood and serve as coenzymes in chains of metabolic reactions (see Chapter 4). **Fat-soluble vitamins,** such as A, D, E, and K, are transported in the blood as complexes linked to lipids or proteins. These vitamins play more specialized roles in particular tissues and physiological activities—vitamin A in the formation of visual pigments, vitamin K in a vertebrate's blood-clotting mechanism, and so on. The body easily excretes excess amounts of water-soluble vitamins in the urine, but fat-soluble vitamins cannot pass so easily from the system and can accumulate in toxic levels.

Minerals

Minerals are inorganic molecules that provide ions critical to the functioning of many cellular enzymes or proteins (as iron for hemoglobin). When a zebra, giraffe, or cow licks a salt block or when a person eats a banana (K^+ source), an essential mineral need is being filled. Table 32-3 lists the roles and sources of the major mineral nutrients essential to mammals.

Table 32-2 VITAMINS

Vitamin	*Distribution*	*Function*	*Deficiency Symptoms*	*Primary Sources*
Water-soluble				
B_1 (thiamine)*	Absorbed from gut; stored in liver, brain, kidney, heart	Formation of coenzyme involved in Krebs cycle	Beriberi, neuritis, heart failure	Organ meats, whole grains
B_2 (riboflavin)*	Absorbed from gut; stored in kidney, liver, heart	Cofactor in oxidative phosphorylation	Photophobia, skin fissures	Milk, eggs, liver, whole grains
B_6 (pyridoxine)*	Absorbed from gut; one-half appears in urine	Coenzyme in amino acid and fatty acid metabolism	Dermatitis, nervous disorders	Whole grains
B_{12} (cyanocobalamin)*	Absorbed from gut; stored in liver, kidney, brain	Nucleic acid synthesis; prevents pernicious anemia	Pernicious anemia, malformed red blood cells	Organ meats, synthesis by intestinal bacteria
Biotin	Absorbed from gut	Protein synthesis; CO_2 fixation; amine metabolism	Scaly dermatitis, muscle pains, weakness	Egg white, synthesis by digestive-tract flora
Folic acid (folacin, pteroylglutamic acid)	Absorbed from gut; utilized as taken in	Nucleic acid synthesis; red blood cell formation	Failure of red blood cells to mature, anemia	Meats
Niacin	Absorbed from gut; distributed to all tissues	Coenzyme in hydrogen transport (NAD, NADP)	Pellagra, skin lesions, digestive disturbances, dementia	Whole grains
Pantothenic acid	Absorbed from gut; stored in all tissues	Forms part of coenzyme A	Neuromotor and cardiovascular disorders	Most foods
C (ascorbic acid)	Absorbed from gut; little storage	Vital to collagen and ground substance	Scurvy, failure to form connective tissue	Citrus fruits
Para-aminobenzoic acid (PABA)	Absorbed from gut; little storage	Essential nutrient for bacteria; aids in folic acid synthesis	No symptoms established for humans	
Fat-soluble				
A (carotene)	Absorbed from gut; stored in liver	Visual pigment formation; maintains epithelial structure	Night blindness, skin lesions	Egg yolk, green and yellow vegetables, fruits
D_3 (calciferol)	Absorbed from gut; little storage	Increases calcium absorption from gut; bone and tooth formation	Rickets (defective bone formation)	Fish oils, liver
E (tocopherol)	Absorbed from gut; stored in fat and muscle tissue	Maintains red blood cells in humans	Increased fragility of red blood cells	Green leafy vegetables
K (naphthaquinone)	Absorbed from gut; little storage	Stimulates prothrombin synthesis by liver	Failure of blood-clotting mechanism	Synthesis by intestinal flora, liver

*B complex vitamins

As Chapter 29 explained, blood plasma and body fluids contain many types of mineral ions. Since significant amounts of minerals are carried away in the sweat, urine, and feces, an animal must replenish its mineral supply regularly. An adult human, for example, must consume about 3 g of NaCl daily. Deficiencies of Na^+, Cl^-, and K^+ seriously disturb the osmotic balance of the blood, body fluids, and cells and can disrupt nerve impulses (see Chapter 34). Calcium is also essential for nerve function, bone structure, cell movements, and contraction of cytoplasmic filaments. Phosphate, particularly in the form of ATP, is the cell's vital energy currency. Manganese is needed for normal structure of bones and tendons. Iron is the critical element that

Table 32-3 MINERALS

Mineral	*Function*	*Primary Sources*
Essential Minerals		
Calcium	Component of bone and teeth; essential for normal blood clotting; needed for normal muscle, nerve, and cell function	Milk and other dairy products, green leafy vegetables
Chlorine	Principal negative ion in interstitial fluid; important in fluid balance and in acid-base balance	Most foods, table salt
Magnesium	Component of many coenzymes; balance between magnesium and calcium ions needed for normal muscle and nerve function	Many foods
Phosphorus	As calcium phosphate, an important structural component of bone; essential for energy transfer and storage (component of ATP) and for many other metabolic processes; component of DNA, RNA, and many proteins	All foods
Potassium	Principal positive ion within cells; influences muscle contraction and nerve excitability	Many foods
Sodium	Principal positive ion in interstitial fluid; important in fluid balance; essential for conduction of nerve impulses	Most foods, table salt
Sulfur	Component of many proteins; essential for normal metabolic activity	Meat, fish, legumes, nuts
Trace Minerals		
Cobalt	Component of vitamin B_{12}; essential for red blood cell production	Meat, dairy products (strict vegetarians may suffer from cobalt deficiency)
Copper	Component of many enzymes; essential for melanin and for hemoglobin syntheses	Liver, eggs, fish, whole-wheat flour, beans
Fluorine	Component of bone and teeth	Some natural waters; may be added to water supplies
Iodine	Component of hormones that stimulate metabolic rate	Seafood, iodized salt, vegetables grown in iodine-rich soils
Iron	Component of hemoglobin, myoglobin, cytochromes, and other enzymes essential to oxygen transport and cellular respiration	Meat (especially liver), nuts, egg yolk, legumes (mineral most likely to be deficient in diet)
Manganese	Activates many enzymes; as arginase, an enzyme essential for urea formation	Whole-grain cereals, egg yolk, green vegetables
Zinc	Component of at least 70 enzymes, including carbonic anhydrase; component of some peptidases and thus important in protein digestion; may be important in wound healing and fertilization	Shellfish (oysters), meats, liver

binds O_2 in hemoglobin, myoglobin, and cytochromes. Iodine occurs in thyroxine, a hormone that regulates metabolic rate and heat production in mammals.

Dietary Guidelines

A great deal of information and misinformation is published today about the human diet, but in fact, vitamin and mineral deficiencies are rare in the American population, and obesity is a much more common problem than malnutrition. Table 32-4 presents the U.S. government guidelines for good nutrition. In addition, recent evidence strongly suggests that regular aerobic exercise is important to maintaining ideal weight, reducing stress, and preventing constipation, atherosclerosis (fatty deposits in the arteries), and muscle degeneration.

Table 32-4 MODIFIED UNITED STATES GOVERNMENT GUIDELINES FOR GOOD NUTRITION

1. *Eat a variety of foods daily,* including fruits and vegetables; whole grains, enriched breads, and cereals; meats, fish, and poultry; and dried peas and beans.
2. *Maintain ideal weight.* Increase physical activity; reduce fatty foods and sweets; avoid too much alcohol; lose weight gradually.
3. *Avoid fats, saturated fats, and cholesterol.* Eat fish, poultry, lean meats, dry peas and beans; use eggs and organ meats sparingly; minimize high-fat dairy products; reduce intake of fats on and in all foods; trim fats from meats; broil, bake, or boil—don't fry; read food labels for fat content.
4. *Consume adequate starch and fiber.* Substitute starches for fats and sugars; eat whole-grain breads and cereals, fruits and vegetables, dried beans and peas, and nuts to increase fiber and starch intake.
5. *Avoid excess sugar.* Limit consumption of sugar, syrup, honey, candy, soft drinks, and cookies. Select fresh fruits or fruits canned in their own juices; watch for sucrose, glucose, dextrose, maltose, lactose, fructose, syrups, and honey on food labels.
6. *Avoid excess sodium.* Reduce salt in cooking and at the table. Cut back on potato chips, pretzels, salted nuts, popcorn, condiments, cheese, pickled foods, and cured meats. Watch food labels for sodium or salt content.
7. *Drink alcohol in moderation if at all.* Limit alcoholic beverage intake (including wine, beer, and liquors) to one or two drinks per day.

*Adapted from *Dietary Guidelines*, U.S. Department of Agriculture; U.S. Department of Health and Human Services, 1979.

SUMMARY

1. Feeding involves both *ingestion* and *digestion*—chemical breakdown of food.

2. Animals' digestive organs vary in complexity, depending on the size of the organism, its nutrient needs, and whether it is an *herbivore* (exclusive plant eater), a *carnivore* (exclusive meat eater), or an *omnivore* (both plant and meat eater).

3. The *alimentary canal* is the two-ended gut that extends from mouth to anus.

4. The teeth and tongue prepare the food for swallowing. The *salivary glands* secrete saliva, which moistens food and contains amylase.

5. Food moves through the *pharynx* to the *esophagus*, and *peristalsis* in esophageal walls propels it toward the stomach.

6. The *stomach* is commonly the site of further mechanical breakdown of food pieces. Protein digestion begins as *pepsin* attacks the food *bolus*.

7. Alkaline enzymatic digestion and absorption of nutrients occur in the *small intestine*. Chyme flows from the duodenum (a digestive region) into the jejunum and the ileum (absorptive regions). *Villi* and *microvilli* increase the small intestine's absorptive area manyfold.

8. The *pancreas* manufactures digestive enzymes that are secreted into the duodenum. The *liver* is a primary site for the use, modification, or storage of absorbed nutrients. It also secretes *bile*, which is stored and concentrated in the *gallbladder*.

9. The *large intestine*, made up of the colon and the rectum, absorbs most of the water and minerals not absorbed by the small intestine. It stores, modifies, and transports feces.

10. In vertebrates, the relative length of the small intestine correlates with whether or not the animal generates heat internally. Symbiotic bacteria (and sometimes protozoa) live in most animals' intestines and in the specialized stomach *ceca* of ruminant mammals.

11. During chemical digestion, enzymes work in sequence to break polymers into smaller units and those molecules, in turn, into monomer subunits.

12. Proteins are digested by endopeptidases and then by exopeptidases. Complex carbohydrates such as starch or glycogen are broken down into smaller and smaller units and ultimately into monosaccharides. Fats are commonly hydrolyzed to fatty acids and glycerol before being absorbed.

13. Nerves and hormones help regulate ingestion and digestion. Eating triggers the initial digestive processes in the mouth (saliva secretion) and stomach (pepsin and HCl secretion).

14. The hormones secretin and CCK control the secretion of pancreatic enzymes and juice, as well as the release of bile from the liver and gallbladder. Food movement out of the stomach and through the intestine is coordinated by the *enterogastric reflex* so that there is ample time for digestion and absorption.

15. Proteins, carbohydrates, fats, and some *minerals* are *macronutrients*.

16. *Micronutrients* include *water-soluble* and *fat-soluble vitamins* and trace minerals.

KEY TERMS

alimentary canal
bile
bolus
cecum
calorie
carnivore
chyme
crop
digestion
enterogastric reflex
esophagus
essential amino acid
fat-soluble vitamin
gall bladder
gastrin
gizzard
herbivore
ingestion
intestine
large intestine
liver
macronutrient

micronutrient
microvillus
mineral
omnivore
pancreas
pepsin
peristalsis
rumen
salivary gland
small intestine
stomach
vasoactive intestinal peptide (VIP)
vitamin
water-soluble vitamin
zymogen

QUESTIONS

1. Describe ingestion and digestion.

2. Briefly outline the steps in ingestion and digestion as carried out by *Paramecium, Planaria,* and an earthworm.

3. Explain extracellular digestion, and give several examples.

4. What are some problems faced by a large herbivore, such as a cow, in getting and processing food?

5. Define *herbivore, carnivore,* and *omnivore,* and give an example of each.

6. Name the organs of the human alimentary canal, and describe the role that each plays in digestion.

7. Name two major organs of the human digestive system that are *not* part of the alimentary canal, and describe the role that each plays in digestion.

8. How are carbohydrates, fats, and proteins broken down during digestion?

9. Give several examples of how the chemical and physical activities of the digestive system are coordinated.

10. List the nine essential amino acids in the human diet, and explain why they are "essential."

ESSAY QUESTION

1. All heterotrophs must possess certain general behavioral and chemical adaptations to acquire organic molecules. List some of these adaptations.

For additional readings related to topics in this chapter, see Appendix C.

33 Homeostasis: Maintaining Biological Constancy

Stability of the internal environment is the condition of free life.

Claude Bernard, *The Way of a Medical Investigator* (1865)

The bodies of animals are more than 50 percent water, and yet kangaroo rats, bluegills, and sharks can live entire lifetimes drinking little or no water. How is this possible?

All complex animals have a salty "internal sea" of body fluids. In many vertebrates, these fluids stay within a narrow range of values for water and salt concentration appropriate to each species, despite the animal's diet or activities, while in many vertebrates, the values fluctuate along with those of the environment. How do the values remain so constant in the more complex animals?

Large mammals live nearly everywhere, from the polar ice caps to equatorial deserts to temperate forests, and they often encounter dramatic temperature shifts within a given day, season, or year. Yet they maintain a high internal temperature that rarely fluctuates. How do they do it?

The answer to all three questions is that animals have marvelous mechanisms for **homeostasis**—the maintenance of a relatively constant internal environment despite fluctuations in the external environment. This chapter deals with three separate but interrelated homeostatic systems that enable animals to survive the var-

A husky's fur insulation: one means of maintaining internal constancy.

iations in salinity, water availability, and temperature on our planet: the **osmoregulatory system,** which governs the levels of water and salt in body fluids; the **excretory system,** which eliminates several kinds of metabolic wastes, usually in the urine; and the **thermoregulatory system,** which maintains body temperature in a bird or mammal and/or governs its responses to shifts in environmental temperature.

This chapter explores:

- Mechanisms for regulating internal salt and water balance, including special adaptations of animals that live in particularly dry or salty environments
- How wastes are filtered out of the body fluids and excreted, and the role of one of the most complex organs ever to evolve—the kidney
- How certain animals cope with changes in environmental and body temperatures and are thus able to survive in the earth's wide range of habitats

REGULATION OF BODY FLUIDS

People driving on a freeway, shrimp scudding along in the sea, and mayflies darting above a field in summer are all, in a sense, complex, mobile pools of water. Water is the chief component of the three body fluids common to animals with circulatory systems (intracellular fluid, extracellular fluid, and blood or hemolymph). But many organisms live in environments where the balance between water and salt is unfavorable to life processes. For this reason, the ability to maintain a stable internal fluid environment by means of **osmoregulation** was one of the most important evolutionary innovations in life's history.

As Chapter 5 explained, *osmosis* is the tendency of water to move through semipermeable membranes, depending on the relative concentrations of osmotically active substances, so-called **osmolytes** (ions, small organic molecules, proteins), on the two sides of the membrane. Thus, water moves into or out of a cell, depending on the relative concentrations of such substances in the intracellular and extracellular fluids. The majority of invertebrates and vertebrates cannot tolerate very high salt concentrations in their body fluids because salts can harm protein structure and function. Instead, their bodies display what is called the **compatible osmolyte strategy;** they have, in addition to salt ions, various organic molecules (including amino acids, sucrose, or glycerol) that raise the total osmotic pressure of the body fluids close to that of seawater, yet leave enzymes and structural proteins unharmed (hence "compatible").

The ratio of different salts in body fluids is also a key to cellular health. This *salt balance* commonly involves higher concentrations of potassium ions in the intracellular than in the extracellular fluid and higher concentrations of sodium ions in the extracellular than in the intracellular fluid. Maintaining an internal environment that is constant in osmotic pressure and salt balance requires specialized osmoregulatory mechanisms, which vary widely according to an organism's habitat.

Life in Fresh Water

Freshwater organisms have body fluids that contain more dissolved salts and organic molecules than the water in which they live (Figure 33-1a). Thus, the surrounding water is *hypotonic* (see Chapter 4) relative to their body fluids and tends to move constantly into the

(a) Freshwater Teleosts

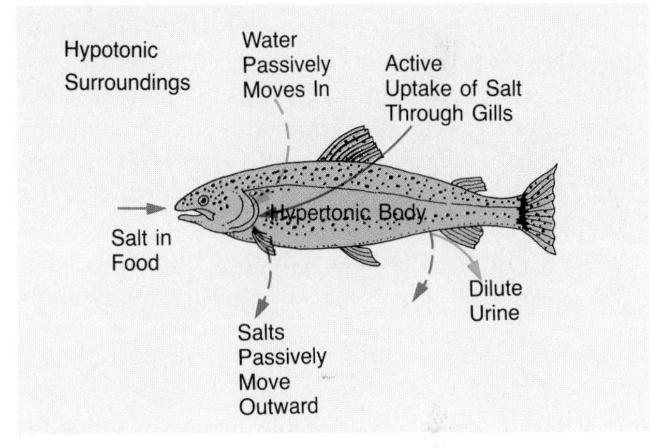

(b) Marine Teleosts

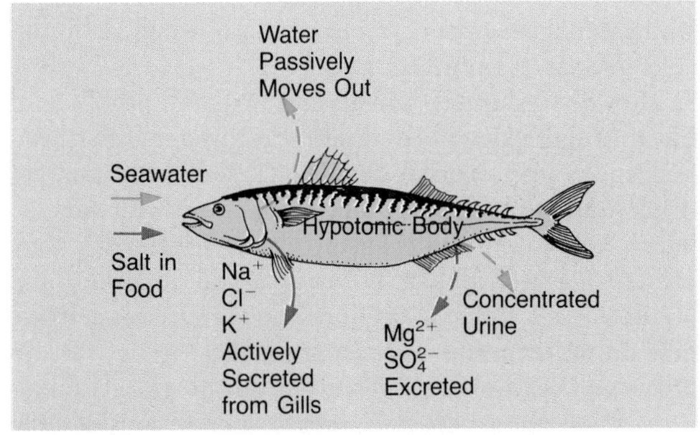

Figure 33-1 WATER AND SOLUTE EXCHANGE IN FRESHWATER AND MARINE TELEOSTS.
(a) To compensate for the tendency to take in water, freshwater teleosts excrete a large volume of watery urine and actively take up the salts from the water passing through the gills. (b) Marine fishes tend to lose water to the environment and to become too "salty." They obtain water by drinking and minimize water loss by excreting a small volume of highly concentrated urine. Solid arrows indicate active movement of water and solutes; broken arrows indicate passive movements.

ALL DRIED UP

Most organisms die if they lose too much water, but a few intriguing species can dry up without ill effect. The tardigrade, or "water bear," for example, is normally 85 percent water by weight, but that can drop to 2 percent. In this state, the microscopic invertebrate looks like a desiccated seed and can be frozen, heated, or blasted with radiation without altering its apparently lifeless state. Just add water, however, and the tardigrade unfurls like an inflated beach toy and ambles away (Figure A).

Researchers John Crowe, Lois Crowe, and colleagues at the University of California at Davis have studied how tardigrades, bacteria, fungal spores, yeasts, seeds, insect larvae, and other organisms survive drying out. They found that the sugar trehalose, and to a lesser degree, glucose and sucrose, can act as a "chemical place keeper" that prevents damage to membranes and proteins as the organism dehydrates. When water leaves the surface of a normal cell membrane, for example, the membrane's organization may change from a fluid state (see Chapter 5) to a more solid gel-like state. When the membrane is rehydrated, the phospholipid heads may no longer be aligned side by side, and thus they may allow leaks that eventually lead to the cell's death. The Crowes think that in extremely drought-tolerant organisms, trehalose molecules may become wedged between the phospholipids, replacing the water molecules as the membranes begin to dry; thus, the sugar maintains the proper spacing until the water is restored.

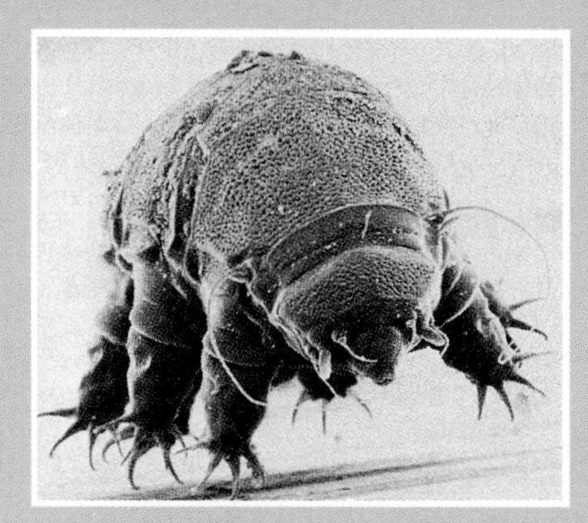

Figure A REHYDRATED TARDIGRADE.

Researchers would like to use trehalose (or create even better substitutes) to more effectively preserve biological materials such as baker's yeast, brine shrimp, hemoglobin, pollen, seeds, even human eggs and sperm, and thus extend their "shelf life" in the dehydrated state.

animal's body across all permeable surfaces, causing excessive *hydration*, or bloating. At the same time, the organism's own dissolved salts tend to move outward, while the small organic molecules are retained by the cells' plasma membranes.

How do freshwater animals cope with the problems of bloating and losing needed salt? Freshwater fishes have four major adaptations for coping: (1) They almost never drink water; (2) their bodies are coated with mucus, which helps stem the constant influx of water; (3) they excrete a large volume of water as dilute urine; and (4) special salt-absorbing cells on the gill surfaces actively take up salt from the fresh water passing through. Pump enzymes (ATPases; see Chapter 4) in the plasma membranes of these special cells expend considerable amounts of ATP energy to continuously pump salt into the body.

These gill cells represent a lovely case of integrated physiological function, because the transport of Na^+ and Cl^- into the cells is coupled to the excretion of the waste products ammonium (NH_4^+) and carbon dioxide in the form of HCO_3^- plus H^+. The pump enzyme requires a counter ion: To transport Na^+ in one direction, another positively charged ion must move in the opposite direction. Removal of NH_4^+ is the organism's main means of excreting nitrogen (more details to come). The CO_2 is the product of oxidative metabolism by body cells. The H^+ comes from water in the reaction

$$H_2O + CO_2 \rightarrow H_2CO_3 \rightarrow H^+ + HCO_3^-$$

The controlled elimination of this H^+ helps regulate body fluid pH. Furthermore, as the HCO_3^- leaves through the gills, Cl^- enters. The homeostasis of salts in these freshwater organisms is thus linked to the excretion of major waste products—nitrogen and CO_2—as well as to the regulation of pH in body fluids.

Life in Salt Water

Organisms living in salt water face radically different challenges based on the high salt levels of the environment. Many marine invertebrates—such as lobsters, sea stars, cnidaria, and annelid worms—have body fluids with osmotic pressure close to that of seawater. Such animals are essentially *isoosmotic* (osmotically the same) relative to the salty ocean; there is little net movement of water into or out of the animal's tissues. Many other marine creatures, however, such as clams and bony fishes, have internal osmotic pressures roughly similar to those of their freshwater relatives and hence much lower than that of seawater (Figure 33-1b). Seawater is *hypertonic* ("higher tonicity") relative to the fish's body fluids, so water tends to leave its tissues, resulting in the threat of *dehydration* and a simultaneous influx of salts, which could disturb the normal osmotic and salt balances.

As one might expect, marine bony fishes must combat dehydration and dump excess salt, and they do this via adaptations nearly opposite to those of freshwater fishes: (1) They drink seawater frequently to replace water lost across gills or permeable body surfaces; (2) they excrete certain salt ions (Mg^{2+}, SO_4^{2-}) in a small volume of concentrated urine; and (3) they excrete Na^+ and Cl^- by means of salt-secreting cells on the gills. These salt-secreting cells have a pump enzyme with ATPase activity; other features work with the pump enzyme to move Na^+ from inside the cells outward into the salty seawater against a concentration gradient.

Sharks, skates, and rays also live in the salty oceans, but instead of pumping out salt and drinking water, as do other marine fishes, they have two kinds of organic molecules in their body fluids that raise the osmotic pressure to a level slightly higher than that of seawater: urea, a by-product of nitrogen metabolism, and trimethylamine oxide (TMO). Because urea and TMO render its body fluids slightly hypertonic, a shark or skate actually *gains* water from the sea across permeable surfaces. Research shows that urea alone in high concentrations denatures proteins and inhibits enzymes; in contrast, TMO stabilizes proteins and activates enzymes. The two together, in a proper ratio, counteract each other and thus raise the osmotic pressure without interfering with proteins. This has been termed the **counteracting osmolyte strategy.** A number of other fishes and invertebrates have evolved similar pairs of counteracting osmolytes.

Despite this useful mechanism, the shark must still rid itself of excess salt that enters across the gills and in its food, and it does so by secreting salts in the urine and also through a posterior *rectal gland.*

Life on Dry Land

Like marine animals, terrestrial animals tend to dehydrate—not via osmosis, however, but by evaporation of body fluids into the surrounding "ocean of air." Land vertebrates display various adaptations for preventing moisture loss, for taking in water, and for eliminating or conserving salt, depending on the saltiness of the animal's food and water.

Amphibians take up water through the skin and across the wall of the urine-storing **bladder,** a balloonlike storage organ (Figure 33-2). A hormone from the brain can cause water to enter the body through the skin when the animal is on a moist surface or is immersed in water. The bladder of a frog, toad, or salamander is an important water and salt reservoir; the drier the environment, the larger the bladder and the more urine it can store. If the animal begins to dehydrate, a hormone causes water to be physiologically "reclaimed" from the bladder and returned to the extracellular fluid. In other land vertebrates, the bladder is largely a one-way receptacle.

Figure 33-2 WATER AND SALT UPTAKE IN AN AMPHIBIAN.
A poison arrow frog *(Dendrobates histrionicus)* on a moist leaf. Salts may be absorbed actively, especially through the belly skin in contact with the ground or other substratum, such as a leaf. Water is stored in the bladder, from which it can be recovered; water may also enter the body through the skin by passive processes.

For most terrestrial vertebrates, the lungs represent a potential major site of water loss. In many mammals, the nasal cavities function as a *countercurrent exchange system* to combat such loss. During inhalation, air passing through the nasal cavities picks up heat from adjacent tissues, causing the temperature of these tissues to fall; that's why a dog has a cold nose. The inhaled air is further warmed and humidified in the lungs. Then, as it passes back out during the next exhalation, the warm, moist air flows over the cooler nasal surfaces. As the air cools, much of the water vapor condenses out on the nasal surfaces and so is not breathed out of the body. This mechanism saves 50 percent of the water that would be lost via respiratory evaporation in kangaroo rats, camels, and giraffes—all residents of arid areas.

In most terrestrial vertebrates, the primary means of regulating the osmotic balance of body fluids is by controlling the amount of water and salt lost in the urine. Thus, the kidneys are the primary regulatory organs (details later).

In desert and marine birds and reptiles, special organs supplement the kidneys. These animals often build up high salt concentrations in their bodies because they consume salty foods or seawater and lose water through evaporation and in the urine and feces. To rid themselves of this excess salt, these animals have *salt glands,* special secretory organs near the eye or in the tongue that remove excess NaCl from the blood and secrete it in a solution of tearlike drops. Sensors in the wall of the bird's heart apparently monitor the osmotic pressure of the blood and send nerve signals to the brain that trigger activation of the salt glands. Thus, the animal can drink seawater and yet, by excreting some of the water and all of the salts, achieve a net water gain.

If terrestrial vertebrates without salt glands—including humans—drink seawater, they ultimately *lose* body water. Our kidneys generate urine that is less salty than seawater, so for every liter of seawater a person drinks, 1.75 L of urine must be produced to maintain the appropriate salt balance in the body fluids, and the precious water is depleted.

EXCRETION OF NITROGENOUS WASTES

When the body breaks down proteins, nucleic acids, and other nitrogen-containing compounds, *nitrogenous,* or nitrogen-bearing, wastes form. As we metabolize proteins, for example, enzymes in the liver remove the amino groups (NH_2) from amino acids in a process called *deamination.* The remainder of the molecules (keto acids) can be converted to sugars and lipids for storage. The released NH_2 groups combine with free hydrogen ions to form NH_3 (ammonia) and NH_4^+ (ammonium ions) (Figure 33-3). Ammonia is a very strong and toxic compound, as a small whiff of glass cleaner suggests, and so organisms must transport or store it in very dilute solution. A single gram of nitrogen reduced to NH_3 must be dissolved in about 400 ml of water to be flushed from the body without harming cells along the way.

The basic problem for organisms handling nitrogenous wastes is simple: the need to safely flush out toxic wastes versus the need to conserve water. When water is abundant—as in typical freshwater fishes—the answer to the problem is equally simple: "Dilution is the solution to pollution." But when water is precious—as it is to many marine and most terrestrial animals—the dilution strategy creates problems of severe dehydration and osmotic imbalance. Those animals solve the problem by detoxifying the nitrogenous wastes or by making them insoluble.

Both freshwater and saltwater fishes overcome the problem of NH_3 toxicity by deaminating amino acids in the gills, rather than in the liver, and releasing NH_3 directly into the respiratory water current as it leaves the

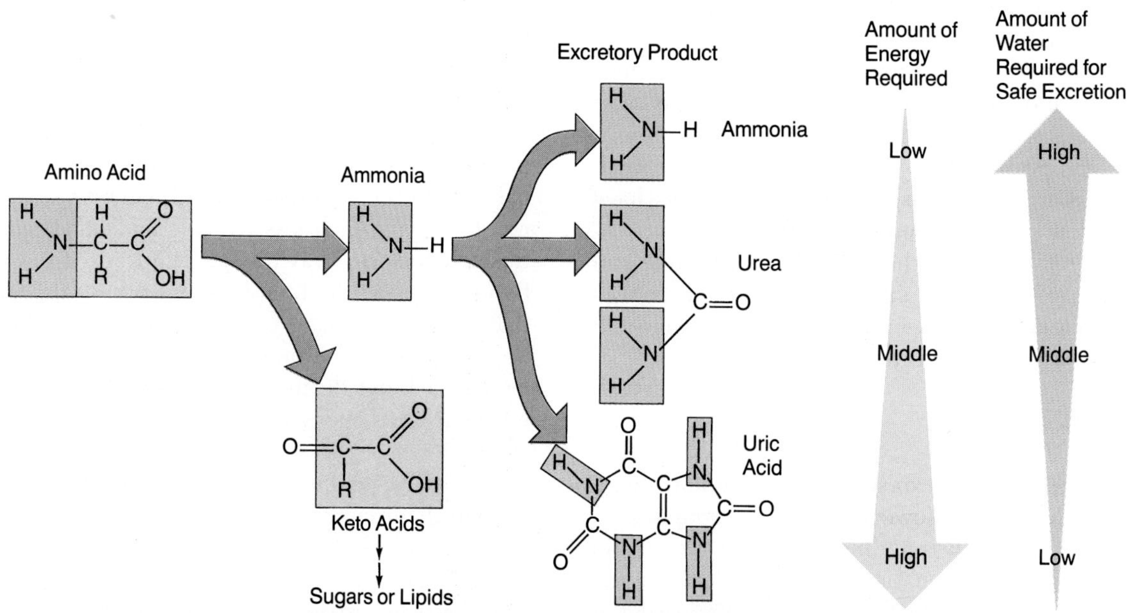

Figure 33-3 THE BREAKDOWN OF AMINO ACIDS INTO NITROGENOUS WASTES.
Nitrogenous wastes form mainly when amino acids are deaminated. The first breakdown product is ammonia, and keto acids, common intermediates in glycolytic pathways and the Krebs cycle, are also formed. Many aquatic animals excrete ammonia directly. In terrestrial animals, ammonia is converted into either urea or uric acid. These processes require energy. This diagram shows the relative toxicity of the three nitrogenous waste products. The more toxic a compound, the more water is required to dilute and excrete it. Note that only one nitrogen atom is excreted per molecule of ammonia, whereas two are present in urea and four in uric acid.

body. Consequently, body cells never really come in contact with the NH_3, so it need not be diluted and no body water is lost excreting it.

In land animals, enzymes reduce the toxicity of nitrogenous wastes; specifically, liver enzymes manufacture urea or uric acid (see Figure 33-3). **Urea** is less toxic than NH_3 and can be transported safely in far less water—1 g of urea in 50 ml of water instead of 400 ml. But its production requires more energy than leaving NH_3 in its more toxic form. In many land animals, urea is removed from the body by the kidneys and is eliminated in fluid urine.

Land animals faced with water shortages often excrete nitrogenous wastes in the form of **uric acid** (see Figure 33-3). In this form, 1 g of nitrogen can be transported and excreted in only 10 ml of water. Uric acid is usually formed in the liver, in a lengthy pathway that requires more than a dozen enzymes and a considerable amount of energy. As uric acid solutions move through the excretory tract, more and more water is reabsorbed until the compound crystallizes and precipitates out of solution, and the remaining water can be reclaimed. Guano, the white semisolid paste excreted by birds and reptiles, is largely uric acid crystals.

In humans and some other animals, uric acid may be formed from the breakdown of nucleic acids (rather than from deamination of amino acids). If too much uric acid is produced, *gout* can develop. In this condition, uric acid crystals form in fingers, toes, and other joints, causing severe pain. Table 33-1 lists the major nitrogenous wastes of various marine and terrestrial vertebrates, as well as other aspects of their osmoregulation.

Let us turn, now, to the organs and associated structures that rid the body of wastes, ions, and excess water. The same basic kind of "plumbing"—a tubule with the capability to filter and reabsorb water, ions, and molecules—evolved independently in at least three groups: the annelids, arthropods, and chordates.

Excretory Systems in Invertebrates

The simplest excretory structure is the protist's *contractile vacuole* (see Figure 21-2), which primarily expels excess water, but also ejects nitrogenous wastes. Small multicellular invertebrates, such as flatworms, have more complex structures for excretion, tubules with *flame cells* (Figure 33-4a). At the next level of complexity are the **nephridia,** or excretory organs, of the earthworm (Figure 33-4b). The nephridium is composed of a tube with a funnel-like opening into the body cavity called the *nephrostome;* a coiled *tubule;* an expanded storage portion of the tubule, the *bladder;* and an exit pore through the body wall, the *nephridiopore*. Water, salts, and wastes in the body fluids can enter the nephrostome directly and pass down the tubule. Alternatively, substances in the hemolymph (blood) can move from the blood capillaries that are entwined around the coiled tubule and pass across the tubule wall into the fluid. Water, salt ions, and organic substances may be reabsorbed from the fluid passing through the tubule, depending on the body's needs.

Grasshoppers and most other insects have several narrow, blind-ended sacs that arise at the junction of the

Table 33-1 MAJOR STRATEGIES FOR MAINTAINING WATER AND SALT BALANCE

Animal	*Environment Concentration Relative to Cells*	*Urine Concentration Relative to Blood*	*Major Nitrogenous Waste*	*Key Adaptation*
Freshwater teleost fish	Hypotonic	Very hypotonic	NH_3	Absorbs salts through gills
Marine teleost fish	Hypertonic	Isotonic	NH_3	Secretes salts through gills
Marine chondrichthian fish	Isotonic	Isotonic	NH_3	Urea, TMO, and rectal gland
Amphibian	Hypotonic	Very hypotonic	NH_3 and urea	Absorbs salts through skin
Marine reptile	Hypertonic	Isotonic	Urea and NH_3	Secretes salts through salt glands
Marine mammal	Hypertonic	Very hypertonic	Urea	Drinks some seawater
Desert mammal	—	Very hypertonic	Urea	Manufactures metabolic water
Marine bird	—	Weakly hypertonic	Uric acid	Drinks seawater; uses salt glands
Terrestrial bird	—	Weakly hypertonic	Uric acid	Drinks fresh water

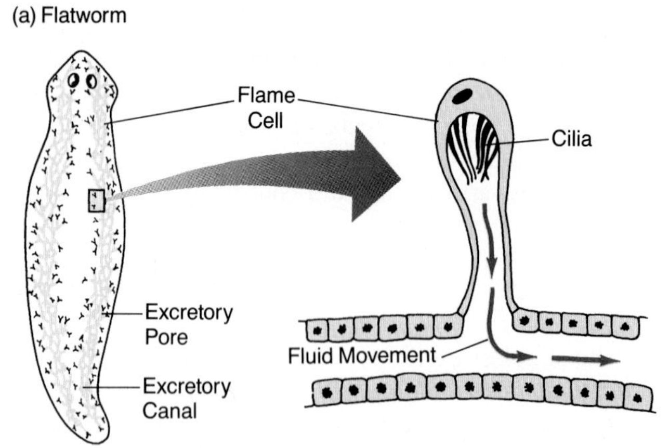

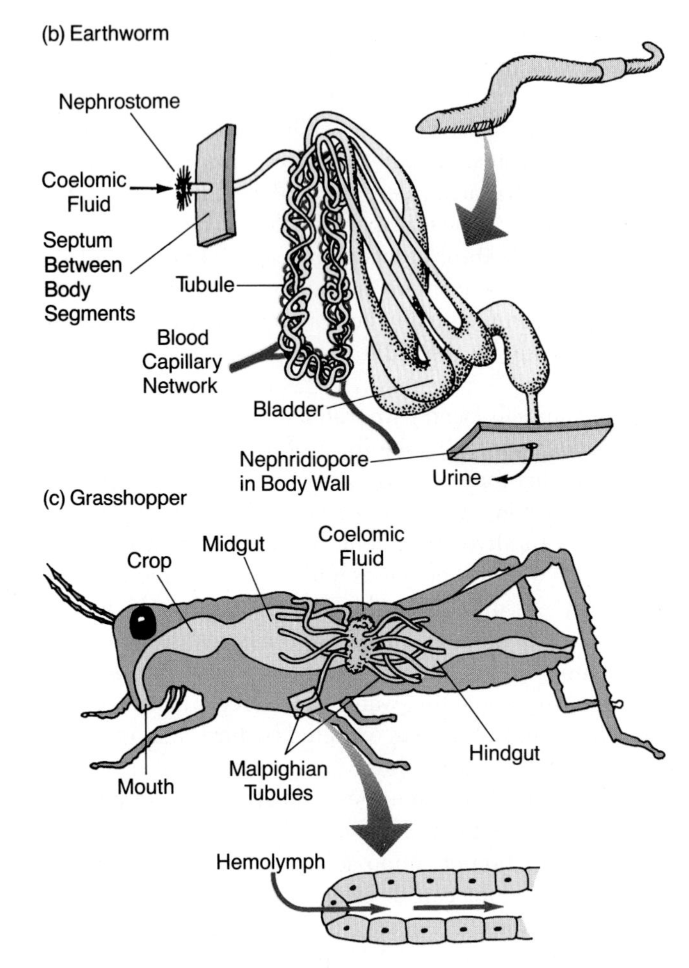

Figure 33-4 EXCRETORY ORGANS IN TYPICAL INVERTEBRATES.
(a) Planarians have excretory systems with a branching network of ducts. Excretory fluid is discharged into these ducts by the flame cells, and wastes then exit the body through excretory pores. (b) In the earthworm, wastes pass into the coelomic fluid, which then enters the excretory organ, the nephridium, through the nephrostome. The excretory fluid (urine) then collects in the bladder and is discharged from the body via the nephridiopore. (c) In insects, excretory products in the coelomic fluid pass through the walls of the many Malpighian tubules (arrows), and urine is discharged into the hindgut, where it is eliminated with the solid fecal waste.

midgut with the hindgut. These sacs, called **Malpighian tubules** (Figure 33-4c), are bathed on the outside by hemolymph; the cells of the tubule walls absorb uric acid, K^+, and other substances from the hemolymph and pass them into the tubule by means of active transport. A small amount of water passively accompanies the solutes because of osmosis. The solution drains into the gut and accumulates in the hindgut. There, much of the water and any salts needed by the body are reabsorbed and returned to the hemolymph. Thus, the insect's Malpighian tubule, like the worm's nephridium, is the site of nitrogenous waste removal and a stabilizer of water and salt balance in the body.

Excretory Systems in Vertebrates: The Marvelous Kidney

In all reptiles, birds, and mammals, the key to excretion, water and salt balance, and regulation of pH in body fluids is the **kidney,** a blood-filtering and waste-excreting organ. The functional unit of the kidney, the **nephron,** is a tubule in which secretion and reabsorption of ions, molecules, and water take place. In freshwater fishes, the nephrons produce copious urine, but in ocean fishes, they produce just a dribble. Amphibian kidneys, called *mesonephric kidneys,* are generally similar in structure to those of freshwater fishes and usually produce a large volume of urine. It is in reptiles that a new kind of kidney develops in the embryo—the *metanephros,* which matures into the adult kidney with many nephrons. Individual mammalian and avian nephrons have a unique section—the loop of Henle—which offers the unprecedented ability to concentrate a hypertonic urine and, in turn, to conserve water for the body.

In the human, the kidney is a dark purplish organ owing to its immense blood supply. It is about 10 cm long and 7 cm wide, and it has a distinctive shape—rounded on one side and indented on the other (Figure 33-5a). Each of the two human kidneys is made of more than 1 million nephrons, and together they perform an astonishing task: Every 24 hours, all the blood in the human body passes through these two compact organs and their many nephrons some 500 to 600 times! During each pass through the kidneys, the blood's salt and water contents are balanced, and wastes are removed. Of the

700–800 L of blood moving through the human kidneys daily, about 200 L of fluid leave the blood and enter the nephrons, but only 1.5 L are excreted as urine.

Each kidney has an outer **cortex** region in which the nephron's main portion is located (Figure 33-5b). Internal to the cortex is the **medulla,** where the loops of Henle reside. Ducts passing through the medulla open into the kidney's central cavity, the **pelvis,** which drains urine.

Each nephron is a long looped unit (unfolded, it would be 30–38 mm long) composed of several parts (Figure 33-5c): a complex, branched, ball-shaped mass of blood capillaries called the *glomerulus,* which is surrounded by a double-walled cup called the *Bowman's capsule;* a slender tube called the *proximal convoluted tubule* extending from the Bowman's capsule; a hairpin-shaped portion called the *loop of Henle;* and a final twisted section called the *distal convoluted tubule,* which joins the *collecting duct.* Many collecting ducts run together like tributaries into a river and then drain into the kidney's *pelvis.* This cavity narrows and leads into the *ureter,* a duct from each kidney that carries urine to the *bladder,* the muscular-walled, balloon-shaped storage organ. A single duct, the *urethra,* exits from the bladder and carries urine out of the body.

All vertebrate kidneys operate with the same two-step strategy: (1) They produce a primary filtrate that contains all the blood plasma's small molecules and ions; and (2) they reabsorb most of the water and all the substances needed by the body, while allowing urea, toxic substances, and other ions and molecules to pass out in the urine. Different parts of each nephron carry out these tasks by means of four physiological mechanisms: filtration, reabsorption, secretion, and concentration.

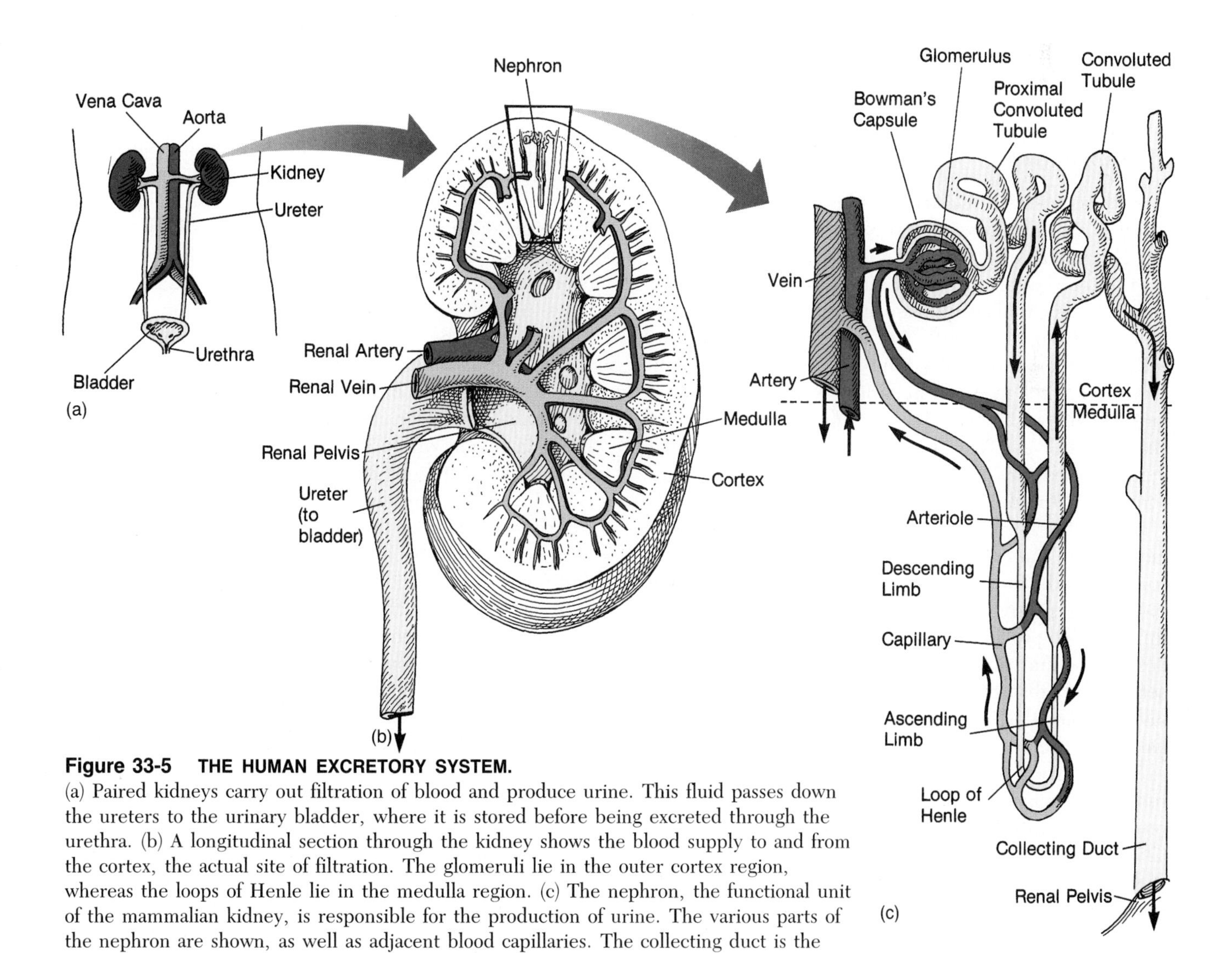

Figure 33-5 THE HUMAN EXCRETORY SYSTEM.
(a) Paired kidneys carry out filtration of blood and produce urine. This fluid passes down the ureters to the urinary bladder, where it is stored before being excreted through the urethra. (b) A longitudinal section through the kidney shows the blood supply to and from the cortex, the actual site of filtration. The glomeruli lie in the outer cortex region, whereas the loops of Henle lie in the medulla region. (c) The nephron, the functional unit of the mammalian kidney, is responsible for the production of urine. The various parts of the nephron are shown, as well as adjacent blood capillaries. The collecting duct is the site of final concentration of the urine.

Filtration

Blood reaches each kidney from a major vessel, the renal artery, and enters the glomerulus of each nephron under high pressure (about 70 mm Hg). This pressure actually *pushes* a solution of water, salts, urea, traces of protein, and other molecules through holes in the walls of the glomerular capillaries and the Bowman's capsule (Figure 33-6). The filtration mechanism is so efficient that about 25 percent of the water and solutes entering the glomerulus passes across into the Bowman's capsule. This filtration is entirely passive and depends completely on blood pressure. In moving from the blood to the cavity of the nephron, all substances must pass through a *basement membrane*, a thick layer composed of sugar polymers and proteins and which includes a basal lamina (see Chapter 29). The basement membrane acts as a filter and ensures that most of the large molecules in blood plasma do not pass into the cavity of the Bowman's capsule. The solution that does go through is called the **glomerular filtrate,** and it contains useful substances, such as glucose, water, salts, and amino acids, as well as urea.

Reabsorption

Both active (ATP-requiring) and passive processes are involved in the recovery of the useful substances from the glomerular filtrate (see Figure 33-6). Fully 75 percent of the sodium and potassium ions and amino acids and 100 percent of the glucose and vitamins are actively removed from the filtrate, are secreted by the tubule cells into the extracellular fluid surrounding the nephrons, and enter nearby capillaries. Negatively charged ions, such as Cl^- and HCO_3^-, follow the positively charged Na^+ and K^+ passively into the extracellular fluid. Most of this reabsorption occurs in the proximal convoluted tubule.

As solutes are removed from the filtrate, it becomes more dilute, and the extracellular fluid becomes more concentrated. To correct this imbalance, water leaves the filtrate passively by osmosis as the filtrate moves through the proximal convoluted tubule and into the descending limb of the loop of Henle. This mechanism—active salt reabsorption with water following passively—saves a dramatic 75 percent of the water from the original filtrate before it reaches the loop of Henle.

The filtration and reabsorption activities of the nephron clearly maintain water and salt balance, as well as removing wastes. If the blood has excessive levels of K^+, Cl^-, or certain other ions, the nephron's tubule cells reabsorb fewer ions from the filtrate, and more are subsequently lost in the urine. If the concentration is too low, the tubule cells reabsorb more ions and return them to the blood, so fewer are excreted. The homeostatic function of the nephrons and kidneys is obviously crucial and continuous, balancing salt and water despite fluctuations in the animal's diet and environment.

Secretion

A process the reverse of reabsorption also takes place: Materials—particularly hydrogen ions—diffuse from the blood in the capillaries into the tubule cells, which actively secrete them into the filtrate. Secretion of H^+ occurs in both the proximal and the distal convoluted tubules and serves to rid the body of the excess acid that accumulates during normal feeding and metabolism. (see Figure 33-6). Excess K^+ is also secreted this way from the distal tubules and collecting ducts, and so are certain drugs and toxins. A current controversy surrounds the ethics of mandatory drug testing for athletes, pilots, and others, and the issue really starts here: Without this secretion process, the drugs would not appear in the urine in detectable quantities.

Concentration

A mammal's ability to excrete wastes and salts with minimum water loss depends on the fourth mechanism, concentration. We saw that the process of reabsorption in the proximal convoluted tubule removes both salt and 75 percent of the water, but it does not significantly change the concentration of salts and urea. As the filtrate flows through the descending limb of the **loop of Henle,** it becomes further reduced in volume and more concentrated—not because of active transport of either water or salts in this region, but because the descending limb is permeable to water, and it passes through a "brine bath," a region where extracellular fluid is highly concentrated (see Figure 33-6). Thus, water leaves the filtrate in the descending limb by passive diffusion, and the concentration of the filtrate rises three- to tenfold.

Several factors account for this brine bath of high salt and urea concentration around the lower portion of the loop of Henle. The filtrate passes through the descending limb, moves into the hairpin turn of the loop, and then climbs the ascending limb. In the ascending limb, it passes through a thick-walled section, where it is currently thought that Cl^- is actively transported out of the tubule, with Na^+ following passively (see Figure 33-6, step 1). Here, in contrast to the proximal convoluted tubule, water cannot follow passively because the cells of the ascending limb are impermeable to water. Thus, the salt concentration in the extracellular fluid becomes quite high, approaching an osmotic pressure three times that of seawater! This salt is free to reenter the descending limb of the loop of Henle (step 2) and cycle back to the ascending limb (step 3). Therefore, the salt makes a cycle: ascending limb to extracellular fluid to descending

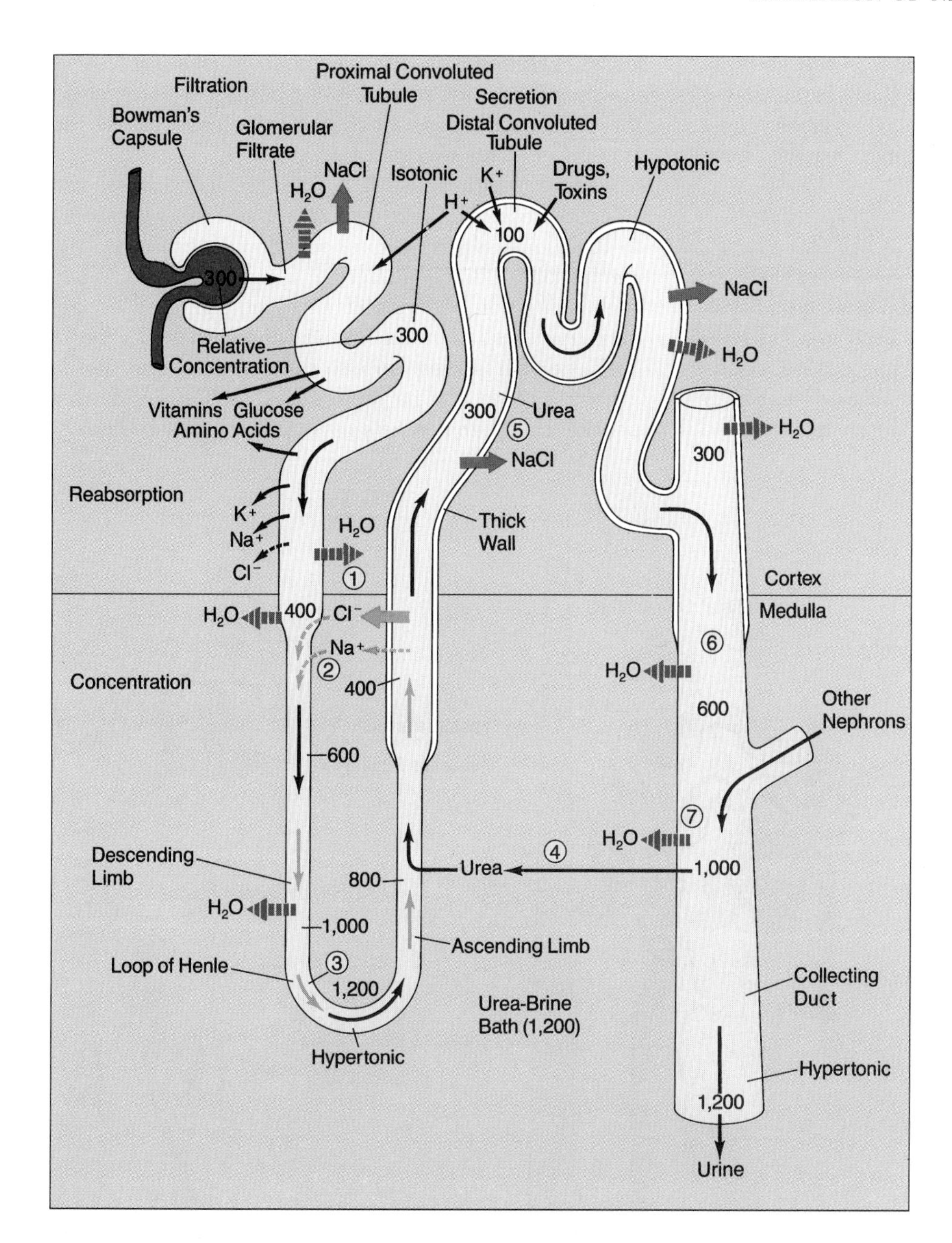

Figure 33-6 URINE PRODUCTION IN THE MAMMALIAN NEPHRON. This figure follows the movement of filtrate through the nephron, described in the text step by step. Active transport of ions is shown with solid arrows; passive transport with broken arrows. The hypothesized active transport of Cl^- in the thick-walled portion of the ascending limb of the loop of Henle is the key to NaCl accumulation in the fluid bathing the loop of Henle. Passive movement of urea from the collecting duct builds a very high urea concentration in the fluid surrounding the loop. This, along with the NaCl, gives rise to the extracellular "urea-brine bath." As filtrate passes through the loop bathed with this solution, much water can be recovered and carried away by the blood. Water also may diffuse out of the collecting duct, and a reduced volume of hypertonic urine is thereby produced. The numbers at intervals along the tubules provide a measure of the relative concentration of the filtrate.

limb to ascending limb again. The opposite flow in the descending and ascending limbs acts as a countercurrent exchange that sets up a gradient of extracellular salt concentration (see the concentration numbers in Figure 33-6). The osmotic pressure of that extracellular brine bath is made still higher because urea moves out of the collecting ducts and builds up (step 4); thus, a more accurate description is "urea-brine bath."

As the filtrate moves up the ascending limb, it becomes temporarily less concentrated as salt is secreted (step 5). But it then passes from the distal convoluted tubule through the **collecting duct,** which itself plunges back through the urea-brine bath in the kidney's medulla and passes close by the loop of Henle (step 6). Because the walls of the collecting duct can be made permeable to water by hormone action, water leaves the filtrate (step 7), enters the extracellular fluid urea-brine bath, and diffuses back into the blood. This further concentrates the filtrate remaining in the collecting tubule and reduces its volume to form a highly concentrated, low-volume excretory product—the *urine*. The urine moves down the ureter into the bladder, where it is held until the fullness of this organ stimulates *micturition*, or urination.

Because of the loop of Henle's countercurrent exchange mechanism, the kidneys of humans, other mammals, and birds, can reclaim up to 99 percent of the water in the original glomerular filtrate. Furthermore, the longer the loop of Henle, the more concentrated the urine that can be made and the more water that is saved: Desert mammals, like kangaroo rats, have very long loops of Henle; aquatic mammals, like beavers, have very short ones.

Regulation of Kidney Function

When a person has been drinking a large quantity of water, the kidneys excrete the extra fluid in a more dilute urine. Conversely, when a person eats very salty food, the kidneys conserve water and balance the higher salt concentrations. What regulates kidney function and allows such homeostatic adjustments to take place? Several hormones are responsible, some of which control the concentration of salt in the urine, and others the amount of water excreted.

Some conditions, such as reduced salt intake, can cause the osmotic pressure in the blood vessels leading to the glomeruli to be unusually low or can cause the amount of salt in the glomerular filtrate to be low. When such a situation occurs, cells near the glomeruli release an enzyme called *renin* into the blood plasma (Figure 33-7, step 1); nerves can also cause this release. Renin acts on a blood plasma protein (α-2-globulin) to form a peptide called *angiotensin I*, (step 2), while another en-

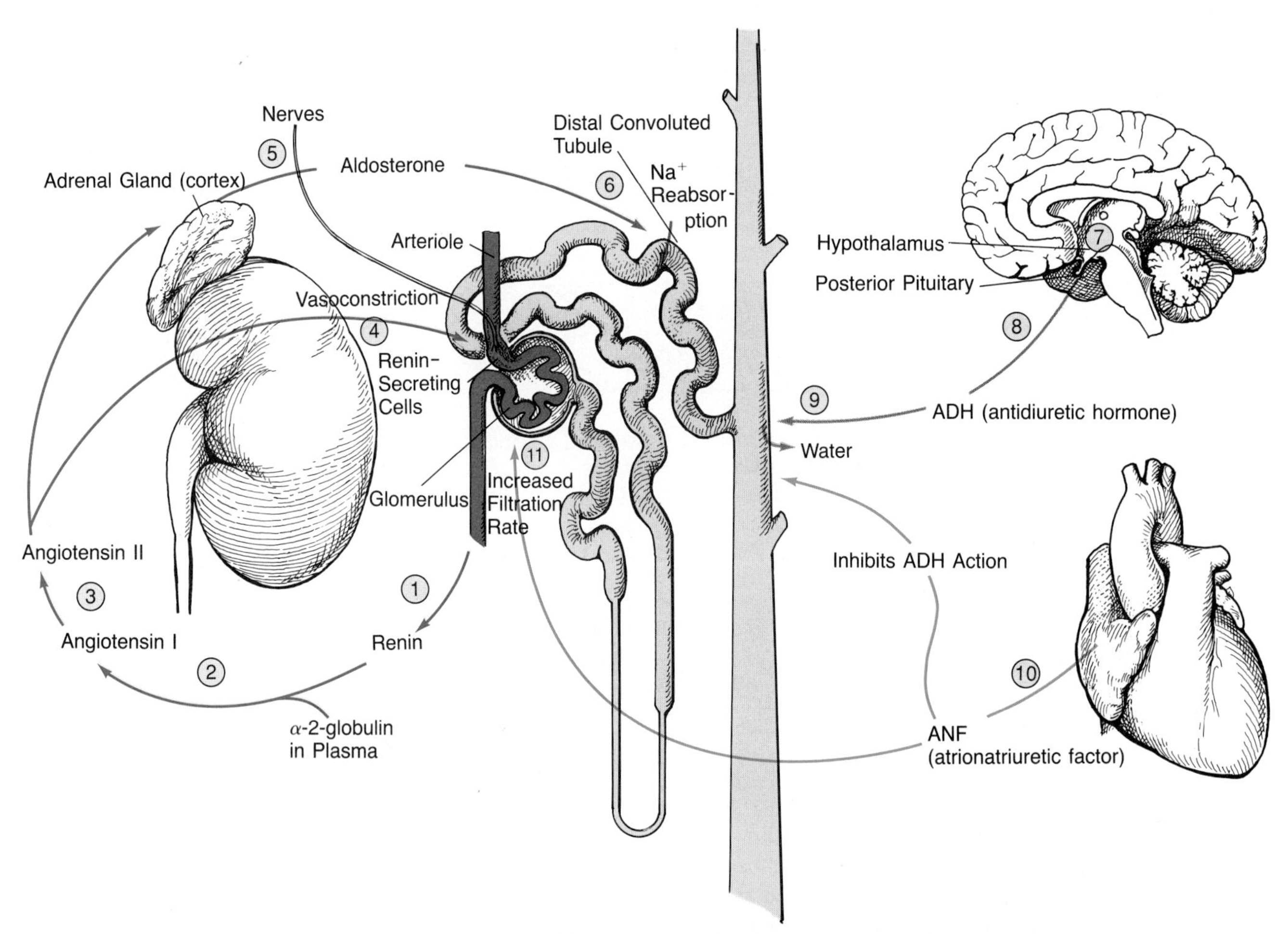

Figure 33-7 HORMONAL CONTROL OF REABSORPTION IN THE NEPHRON AND COLLECTING DUCT.
As the text explains, step by step, decreased blood pressure or low salt in the blood result in the secretion of renin by cells located next to the walls of the arterioles approaching glomeruli. (Nerves leading to such cells may also control this secretion.) Renin initiates a series of reactions in the blood that produces angiotensin II; that compound may cause vasoconstriction of the arteriole to the glomerulus, thereby decreasing glomerular filtration; angiotensin II also causes aldosterone release, which promotes Na^+ reabsorption (and, indirectly, water reabsorption). Hypertonic blood traveling to the brain's hypothalamus triggers ADH secretion; as a result, water is reabsorbed at the collecting duct, and the blood tonicity falls. Heart atrial cells secrete atrionatriuretic factor (ANF) in response to high blood pressure; ANF acts directly to increase glomerular filtration and inhibits ADH action. As a result, more water and Na^+ are lost. Thus, these various hormones counteract each other and permit precise control of blood volume, tonicity, and pH.

zyme in turn converts to *angiotensin II* (step 3). This circulating peptide hormone triggers constriction of the blood vessels leading into the glomeruli, so that less blood enters them and the volume of glomerular filtrate is reduced (step 4). Angiotensin II also acts on the adrenal gland, a gland that sits on top of the kidney, to trigger the release of the steroid hormone **aldosterone** (step 5), which travels in the blood to the kidneys and acts on the distal convoluted tubules, so that more salt is reabsorbed before the filtrate passes into the collecting ducts (step 6). Also, more water is reabsorbed osmotically, thereby correcting the low blood pressure that may have triggered renin release in the first place. Aldosterone also causes salt to be reabsorbed in the salivary glands, sweat glands, and colon. A series of events regulates aldosterone secretion into the blood.

A second hormone that regulates kidney function is **antidiuretic hormone (ADH),** also called *vasopressin.* Neurosecretory cells in the hypothalamus fire nerve impulses when the osmotic pressure of the blood plasma is too high (step 7). The impulses travel along axons of neurosecretory cells to the posterior pituitary gland, where they cause secretion of ADH into the blood (step 8). ADH binds to cells in the walls of the kidney's collecting ducts and increases their permeability to water (step 9). The higher the ADH level, the more water passes osmotically out of the filtrate and back into the extracellular fluid, and the less urine is produced. Most of the reclaimed water eventually reenters the blood plasma, diluting it and thereby lowering its osmotic pressure. This decrease is then detected by cells in the hypothalamus, and soon less ADH is released. Just the opposite occurs if the osmotic pressure of the blood is too low (such as after imbibing too much beer): ADH level falls, less water is reclaimed in the collecting ducts, and copious urine is produced. (The alcohol in beer has the added effect of inhibiting ADH itself.) Different levels of ADH counteract osmotic pressures at both extremes, thus helping to maintain a homeostatic normal condition.

Changes in blood pressure can also increase or decrease ADH secretion. Drinking water increases blood volume and ADH secretion drops, less water is reabsorbed, more urine passes, and the blood pressure decreases. If you lose water as the result of sweating or bleeding, blood pressure drops, ADH secretion increases, urine production decreases, and the blood pressure once again rises.

Another hormone, **atrionatriuretic factor (ANF),** also called atriopeptin, can act as an antagonist to the ADH mechanism. Cells in the walls of the heart's atrial chambers monitor blood pressure, and when the walls are stretched excessively owing to increased blood volume (pressure), the heart muscle cells process a precursor hormone and release ANF (see Figure 33-7, step 10). ANF causes both water and Na^+ to be excreted in urine, leading to a fall in blood pressure, and it also acts on the blood vessels supplying and draining the glomeruli, increasing glomerular filtration rates (step 11). Moreover, ANF acts directly on collecting duct cells to inhibit the water retention effects of ADH. Thus, ANF and ADH may be viewed as opposites: ANF causes water loss; ADH causes water retention. Researchers have noticed that people suffering congestive heart failure have a puzzling response to ANF. Although such individuals have high blood pressure, abnormally high Na^+ and water retention, and make and release high levels of ANF, the kidneys do not respond properly to the ANF: Na^+ and water are retained, and the heart is not protected from high blood volume and pressure.

ANF also regulates the amount of fluid in the brain and spinal cord cavities. This is an important adaptation, since the brain lacks lymphatic drainage and is especially sensitive to edema (water accumulation).

Foods such as coffee, tea, and cranberry juice, as well as certain over-the-counter and prescription drugs, suppress the secretion of ADH and thus act as *diuretics,* causing increased urination.

Not surprisingly, changes in water concentration due to salty foods, sweating, urination, or blood loss can stimulate the *drinking center* in the hypothalamus and lead to thirst. Physicians often warn people with high blood pressure to avoid salty foods and thus water retention, but studies show that most people have a preference for excess salt—a preference stemming mainly from habit. Additional studies reveal, however, that those on low-salt diets eventually stop craving salt and that salt restriction early in life can lead to a lifetime of healthier eating habits.

As a homeostatic organ, the kidney also helps to regulate the pH of body fluids in reptiles, birds, and mammals. Recall from Chapter 31 that the rate of breathing is a major factor in controlling the pH of blood and body tissues. The kidney, too, keeps the blood plasma at a fairly constant pH of 7.35 to 7.4 by automatically adjusting the type and number of ions leaving the body so that the basic chemical composition of the body fluids stays within a set range of values.

Proper kidney function is so important for ridding the body of nitrogenous wastes and maintaining pH and salt balance that diseases of these organs are often life-threatening. Inborn defects, infections, tumors, kidney stones, and exposure to toxic chemicals can all diminish kidney function and allow wastes to build up in the blood. Victims of kidney disease or failure often must be placed on an artificial kidney or dialysis machine, which removes wastes from the blood, and they sometimes require kidney transplants.

As one might suspect, the kidney's activities are energy-expensive, since many aspects of reabsorption and secretion rely on active transport processes involving ATPase pump enzymes. Supporting these homeostatic activities hour after hour, day after day, costs the body a substantial portion of its free oxygen, its food energy, and its supply of ATP.

HOMEOSTASIS AND TEMPERATURE REGULATION

The earth's environments vary dramatically in temperature, as well as in moisture and salinity, and many animals have homeostatic mechanisms involving the nervous, endocrine, circulatory, and other organ systems that govern **thermoregulation,** their internal and external responses to temperature.

The Impact of Temperature on Living Systems

Temperature has been a strong source of selective pressure on animals because the rates of chemical reactions vary with temperature, usually doubling for each 10° C rise. Superficially, this might mean that enzymes would catalyze reactions much faster in a mouse, with a normal body temperature of 37° C, than in a snail, at 5° C. In fact, all enzymes have *temperature optima,* temperatures at which they catalyze reactions most efficiently. Enzymes have evolved so that they function optimally in each animal's typical habitat. For example, a digestive enzyme in a trout may function just as rapidly at 10° C as does the analogous enzyme in a desert lizard at 34° C. In a sense, enzyme evolution helps overcome the simple relationship between reaction rates and temperature.

Extremes of temperature can also disrupt basic biological molecules and structures, such as proteins and membranes. If body temperature rises to between 45° C and 50° C, proteins begin to denature; further rises can literally destroy the enzymes that sustain metabolism. When temperatures fall, reactions run more slowly, and weak bonds can rupture, causing enzymes to dissociate and cease functioning. Furthermore, low temperatures cause the lipids of plasma membranes and organelles such as mitochondria to change from a fluid to a solid state, particularly in birds and mammals, and to interfere with ion pumping, mitochondrial activity, and other fundamental processes.

Solutions to Temperature Problems

Animals have evolved effective strategies for coping with fluctuations and extremes of temperature. Some animals occupy a habitat where the temperature remains constant; some have a body temperature that fluctuates along with their surroundings and have physiological processes adapted to fluctuating temperatures; and some generate heat internally and thus maintain a constant body temperature despite fluctuations in external temperatures. Animals in the first category live in the deep ocean or deep lakes, where the temperature remains near 4° C all year long. Animals in the second and third categories have a much wider distribution in aquatic and terrestrial habitats.

Birds and mammals tend to feel warm to a person's touch and so were traditionally called "warm-blooded." Most other animals feel cool and were called "cold-blooded." But these labels are inadequate, since the body temperatures of many so-called cold-blooded animals at times exceed those of mammals and birds. The terms *poikilotherm* and *homeotherm* are more precise. **Poikilotherms** are animals with a variable body temperature, one that tends to track the temperature of the surrounding environment (Figure 33-8). Most fishes, amphibians, reptiles, insects, and other invertebrates are poikilotherms. **Homeotherms,** in contrast, have relatively constant body temperatures; birds and mammals are examples (see Figure 33-8). Even these terms and

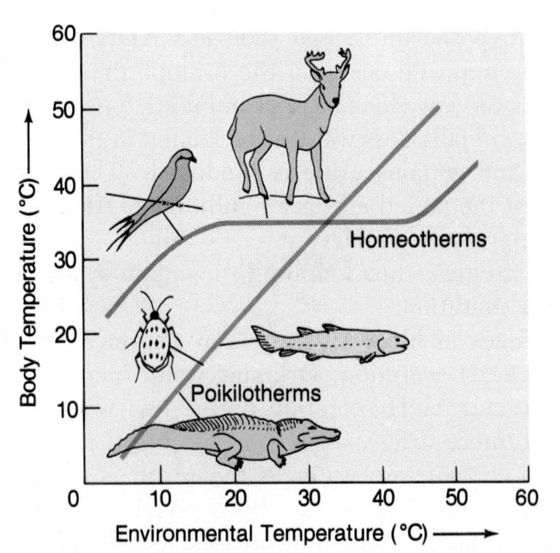

Figure 33-8 HOMEOTHERMS AND POIKILOTHERMS: THE RELATIONSHIP BETWEEN BODY TEMPERATURE AND ENVIRONMENTAL TEMPERATURE.

In homeotherms, such as birds and mammals, body temperature remains more or less constant over a range of environmental temperatures. In contrast, the body temperature of poikilotherms, such as insects, fishes, and amphibians, is usually within a degree or two of the environmental temperature.

definitions, however, are inadequate for all cases. A lizard tethered to a stake would function as a poikilotherm, body temperature fluctuating with the environment. In nature, however, the same lizard is a "behavioral homeotherm"; it can maintain a temperature of between 33° C and 35° C all day long by moving into and out of the sun and carrying out other behaviors. Alternatively, a brittle star or sea cucumber living on the seafloor, where the water temperature varies little, might have a body temperature of 2° C to 4° C its whole life; that superb homeothermy is based on the stability of its environment, not the homeothermy a wood thrush or mule deer displays by generating heat internally and maintaining body temperatures between 37° C and 39° C.

Animals can be classified in yet another way. **Ectotherms** derive most of their body heat from the environment rather than from their own metabolism. Reptiles are a good example. **Endotherms**—birds and mammals—produce their own body heat. Of course, chemical reactions in the cells of ectotherms do release a small amount of heat. However, their bodies are not well insulated, and the heat dissipates into the environment. Most endotherms have bodies insulated by fur, feathers, or fat and so can retain heat, enabling them to maintain high body temperatures (near 40° C) and to keep at least the deep body temperature, the so-called *core temperature*, quite constant. The body temperature of some homeothermic endotherms can drop radically, however, when they are inactive (a state called torpor) or in the winter (hibernation). Such drops in core temperature are frequently seen in very small birds and mammals.

Organisms have a veritable arsenal of responses to the heat and cold of their surrounding environments. These responses can be ecological (where and how the animal lives), metabolic (how its cells and tissues function), behavioral (how the entire organism responds to changes in temperature), structural (anatomical solutions, such as insulation), and physiological (specific responses such as shivering or sweating). Let us consider temperature regulation in various animal groups.

Temperature Regulation in Aquatic Organisms

The aquatic environment is remarkably stable. Water gains and loses heat slowly and conducts heat poorly (see Chapter 4); thus, in the absence of turbulence, water temperatures at a given depth in oceans, lakes, and rivers tend to fluctuate very little on a daily basis, to remain constant within each depth zone of a deep body of water, and to change slowly, if at all, from season to season. Given these stable surroundings, most aquatic invertebrates and vertebrates are ectothermic poikilotherms: They derive heat from the environment, and their body temperatures fluctuate slowly with that of the surrounding water.

Because the environment is stable, the body temperatures of sponges, marine worms, crayfish, squid, shrimp, and most fishes are quite unchanging as long as the organism remains at a given depth. In many aquatic organisms, the enzymes that carry out metabolic reactions function optimally and very rapidly at the water temperature where the organism spends most of its time—a tidy outcome of evolutionary processes. In the laboratory, young sockeye salmon will cluster where the water is held at the optimal temperature for their growth, heart rate, and swimming speed. And some antarctic animals have the sugars sorbitol, glycerol, and trehalose in their blood, acting as "antifreeze" that allows them to swim about all winter in supercooled (subfreezing) waters.

In an aquatic environment where organisms are fairly limited to specific depths and activity levels, a predator that can move more quickly and range widely through various zones can successfully harvest prey. Diving mammals and birds, such as seals and cormorants, take advantage of this principle. And curiously, certain fishes, such as mako sharks and tuna, are "warm-bodied"; they have evolved mechanisms that permit at least parts of the body to be maintained at a high temperature and activity level (Figure 33-9). The mako shark and the tuna are both endothermic homeotherms: They derive heat from their own metabolic activities, not from the external environment, and they maintain a fairly constant temperature around the spinal cord and brain. These tissues may be 5° C to 15° C higher than the surrounding water temperature and sometimes may approach the typical mammalian core temperature of 37° C. Although makos and tunas are not insulated, they can conserve the heat generated by their major red muscles during swimming (see Figure 33-9), whereas the more common poikilothermic sharks and fishes constantly lose heat through their skin and gills. Makos and tunas "trap" heat with a kind of countercurrent heat exchange system built from blood vessels running between the body core and the skin. Specifically, like many animals that live in cold or hot climates, makos and tunas have a "miraculous net" of blood vessels, or **rete mirabile,** in which blood flow in countercurrent vessels can be altered to regulate heat loss or gain (see Figure 33-9).

The important difference between tunas and other fishes is the tuna's endothermic source of body heat and the high core temperature. The selective advantage of a high body temperature has been profound for these fish, allowing their muscles to contract three times more strongly than cooler-bodied fishes and enabling them to swim faster and search for prey in more varied ocean depths than can other predatory fishes.

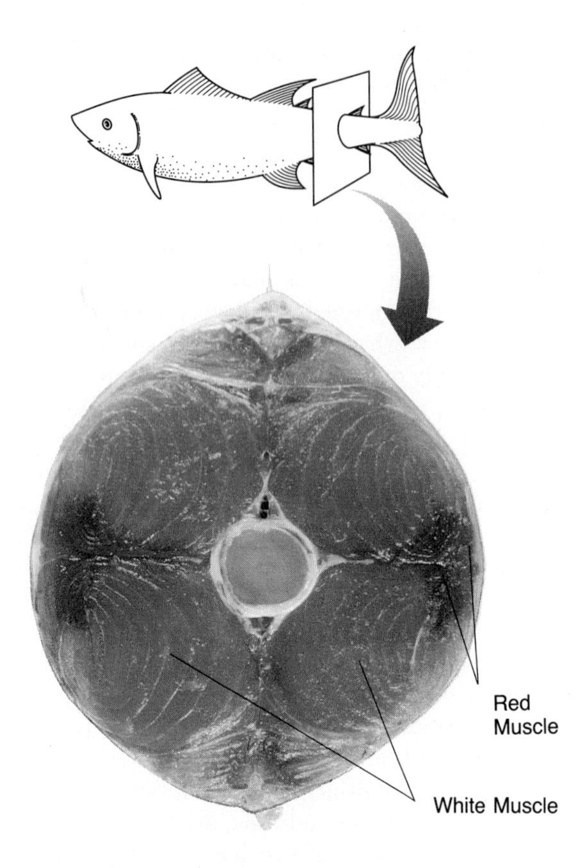

Figure 33-9 WARM-BODIED FISHES.
In a few fishes, including this tuna, bands of red muscle possess a capillary network known as the rete mirabile. The heat generated by these muscles is not lost because it is transferred in the rete from hot venous blood passing outward to cold arterial blood passing inward from the body surface. This enables the animal's core temperature to remain higher than the environmental temperature.

Temperature Regulation in Amphibians, Reptiles, and Insects

Surrounded by air, land animals must cope with marked temperature fluctuations. Most terrestrial animals are ectothermic poikilotherms: They derive heat from the environment, and their body temperatures vary with external temperatures. Let's consider the amphibians, reptiles, and insects.

Amphibians, such as frogs, take in heat from the sunshine and water and absorb warmth radiating from nearby soil and rocks. In cold climates, frogs and salamanders have an extremely low rate of metabolism in winter and are quite inactive. Even when more active in the warmer seasons, however, they have a persistent thermoregulatory problem: Amphibians must exchange carbon dioxide and oxygen across a moist skin surface, and this moisture layer acts as a natural cooling system. Fully 580 calories of body heat are consumed and lost with each gram of water that evaporates from the skin. This continuous evaporative cooling means that the body temperature of a frog or salamander in air is often lower than the environmental temperature, even on a warm day. No wonder they are abundant in the tropics but rare in most dry or cold areas.

Reptiles, with their dry, scaly skin, do not lose body heat through respiration-related surface evaporation. Terrestrial reptiles seek out sources of heat—direct sunlight, warm rocks, soil—to raise their body temperature. Early in the morning, for example, a lizard sits in the sun; its skin darkens in response to hormones, making it better able to absorb infrared radiation in sunlight. The lizard may also do "push-ups," pumping its legs up and down to generate heat, and it may crawl out of the sun and grow pale if it overheats. Marine iguanas will also dive into the sea to cool themselves and to feed (Figure 33-10). Although lizards and other reptiles are classified as poikilotherms and ectotherms, they are actually homeotherms that can maintain a fairly constant body temperature by means of their own behavior. As true ectotherms, however, they do cool off at night and become considerably less active after sunset.

Invertebrates, with their vast diversity of life-styles and habitats, are mostly ectotherms and poikilotherms. Their adaptations are mostly metabolic—enzymes, biochemical functions, and membrane components that remain functional within the range of temperatures they

Figure 33-10 TEMPERATURE REGULATION IN REPTILES.
By combining a rapid heartbeat with vasodilation, the Galápagos marine iguana *(Amblyrhynchus oristosis)* gains heat rapidly as it sunbathes. The skin darkens, allowing much radiation to be absorbed. When the iguana dives into the cold seawater to feed, it reduces its heart rate and constricts blood vessels in the skin; the combined result is less heat loss.

normally encounter in burrows, tropical trees, or wherever they live.

The metabolic means by which a few kinds of invertebrates raise their body temperatures represent a form of endothermy. Certain species of moths, flies, bees, dragonflies, and beetles, for example, weighing only 100 mg to 2 g, can maintain temperatures of 35° C to 40° C. All such insects have high temperatures only when active; they are called *heterotherms*. In addition, insects may have some kind of insulation; bumblebees and the sphinx moths, for instance, have hairlike bristles on the thorax that houses the rapidly contracting wing muscles. These muscles generate a large amount of heat during flight, and the insulating bristles help retain this heat within the segment. If the insect's internal temperature climbs too high, its heart beats faster, and extra hemolymph (blood) flows to the hairless abdomen so that heat radiates away.

Still another mechanism—a preflight "warm-up"—heats up a cool insect's body. The warmed moth then maintains a remarkably high core temperature of 41° C, even as air temperatures fluctuate between 3° C and 35° C. A bumblebee's flight muscles must reach 35° C before it can generate enough lift to fly. Its warm-up involves shivering, producing frequent minute muscle contractions that break down ATP and generate heat, which is retained in the thorax. Bumblebees are one of the few insects that have an additional capacity to generate heat metabolically, and it allows them to fly and forage on cool, damp days, when other types of bees are huddled together for warmth in their hives.

Temperature Regulation in Birds and Mammals

The most active and behaviorally complex animals—the birds and mammals—have large repertoires of activities and can live in habitats all over the world because they are endothermic homeotherms: They maintain constant body temperatures of 35° C to 42° C with internally generated heat.

A vertebrate homeotherm's first prerequisite for maintaining constant temperature is *insulation*. Feathers and fur trap a layer of stagnant air next to the body and reduce heat loss from the underlying skin (Figure 33-11). Animals with sparse hair, such as whales and humans, get insulation from layers of fat. Insulation can also keep heat out: Large mammals living in the hottest, driest deserts invariably have very thick, light-colored, almost silvery fur on their backs. The back fur of camels and Merino sheep, for instance, can reach surface temperatures of 70° C and 85° C, respectively, while the underlying skin remains at 40° C.

Figure 33-11 FEATHER INSULATION.
The thick layer of down feathers keeps these Emperor penguin chicks snug inside their own sleeping bags. Feathers, as well as hair on huskies and some other mammals, can provide such effective insulation that the animal's metabolism does not have to be turned up unless temperatures go far, far below freezing.

When a mammal or bird becomes chilled, its fur or feathers are held erect by tiny muscles so that the layer of air trapped next to the skin becomes thicker. In mammals, this process is called **piloerection.** In the warmer months, some mammals have thinner pelts or mat down their fur by licking it so that body heat can be used up by evaporation.

In many birds and mammals, overheating is combated by evaporation of body fluids. Mammals and birds will often pant or take short, shallow breaths to speed evaporation in the lungs. Some mammals also have *sweat glands*, which release a watery salt solution that carries away heat by evaporation. Sweating involves loss of both water and salt and can lead to severe problems if prolonged. A person doing hard physical labor on a hot day can lose up to 4 L of sweat per hour.

Yet another important mechanism for regulating body heat involves vasoconstriction and vasodilation—the opening and closing of capillary beds (Figure 33-12). Vasoconstriction of the skin capillaries prevents most hot blood from passing through them; the temperatures of the extremities and skin surface drop, while those of the central nervous system, heart, and other vital organs remain warm. Vasodilation increases blood flow to the skin, so that heat can be lost more easily, either by *radiation* (transfer of heat waves to the environment) or by *convection* (movement of warmed air away from the skin). The uninsulated feet and flippers of penguins, walruses, and many other arctic animals have rete mirabile countercurrent exchange nets. In these animals, arteries carrying hot blood from the body core toward the ex-

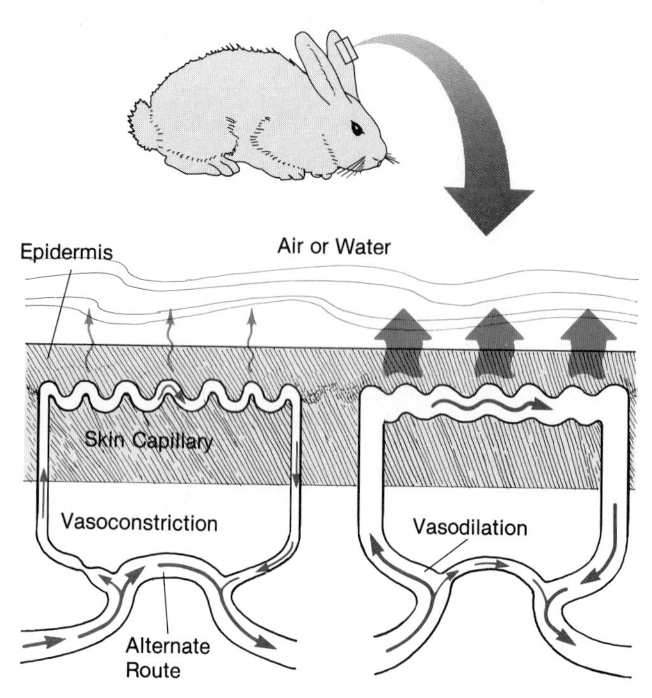

Figure 33-12 CONTROLLING HEAT LOSS BY THE EXTREMITIES.
Vasoconstriction and vasodilation effectively reduce and promote, respectively, the loss of heat to the environment by controlling blood flow to the extremities. Vasoconstriction reduces the flow of blood to an extremity, and hot blood passes back to the body core. Thus, the loss of heat is low. Vasodilation promotes heat loss by increasing blood flow to an extremity and allowing heat from the body core to be carried off by convection.

tremities lie next to veins carrying cold blood from the feet or flippers back toward the core (Figure 33-13). Heat is transferred from the arterial blood to the venous blood and is not carried to the surface, where it would be lost. Thus, the body core can remain warm.

Another function of the rete mirabile is the cooling of the brain when the body is subject to overheating. The eland, a cowlike animal of dry, hot African regions, has a rete mirabile in the walls of its nasal passages. Evaporation during inhalation and exhalation cools the venous blood in the walls of these chambers; that blood flows counter to the hot arterial blood approaching the brain and carries away some of the heat. As a result, an eland can keep a cool head—a cool brain, that is—under a blistering sun.

Mammals and birds also have behavioral mechanisms: They sun themselves or seek shade, as temperature fluctuations dictate. Small homeotherms sometimes share burrows for warmth, and larger range animals huddle together in the wind. Migration to warm climates and hibernation enable many kinds of birds and mammals to survive through the winter. And people alter their clothing, dwellings, and activities to compensate for changes in the weather and climate.

Heat Production in Birds and Mammals

How do endotherms generate their own internal heat in the first place? There are four main sources of such heat generation, or **thermiogenesis:** (1) muscle contraction, (2) ATPase pump enzymes, (3) brown fat, and (4) metabolic processes. Each time a muscle contracts, millions of ATP molecules are hydrolyzed, releasing heat as a by-product. Voluntary muscular work—such as running, jumping, flying, or stamping the feet—thus generates heat, as does the involuntary muscular process of *shivering*. Enzymatic activity, brown fat, and metabolic processes are sometimes collectively called *nonshivering thermiogenesis*, since they do not involve muscle contraction, voluntary or involuntary.

Birds and mammals have a unique capacity to generate heat by using enzymes of ancient evolutionary and physiological vintage—the basic ATPase pump enzymes found in the plasma membranes of all cells. Chilling leads to the release of the hormones thyroxine from the thyroid gland and norepinephrine from certain nerve endings. Their release leads to increased sodium pumping by the enzymes in the plasma membranes of liver, fat, and muscle cells, and with it, increased heat generation.

The third source of heat production is **brown fat,** a special type of fat deposited beneath the ribs and shoulder blades and around some major blood vessels in new-

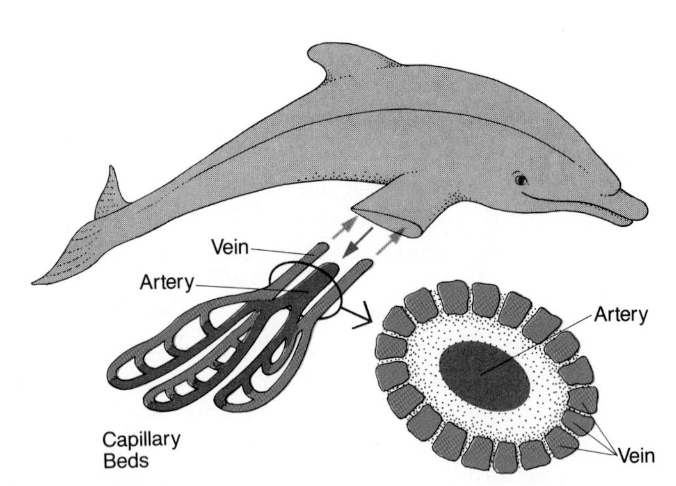

Figure 33-13 COUNTERCURRENT MECHANISMS: BIOLOGICAL HEAT EXCHANGERS.
In animals that live in environments less extreme than the Arctic, arteries and veins can be arranged in a way that conserves heat in the appendages. In the dolphin flipper, for instance, the main artery is surrounded by veins. Heat from the artery is transferred to the veins, which then carry warmed blood back to the body core.

born mammals and in adult mammals adapted to cold climates or that hibernate. Heat is generated when the brown fat cells oxidize fatty acids. Blood flowing near active brown fat is heated and then carries heat to the brain and the heart muscles, in particular. The brown color comes from immense numbers of mitochondria, with their iron-containing cytochromes.

The fourth source of thermiogenesis is metabolism. Mammals and birds have a *high basal metabolic rate* that is controlled largely by thyroxine, and the intense metabolic activities give off heat as a by-product.

The Hypothalamus: The Body's Thermostat

How are changes in the core and surface temperatures of an animal's body measured so that the organism can respond to cold with shivering and to heat with panting, and not vice versa? The answer for mammals and birds is that the portion of the brain called the hypothalamus functions as a thermostat with a specific setting, or **set point,** for each species; the familiar body temperature of 98.6° F, or 37.5° C, is the human set point.

If an experimenter surgically destroys the hypothalamus of a ground squirrel or guinea pig, the animal functions like a poikilotherm—its internal temperature rises and falls along with that of the external environment. Other researchers implanted in a rat's hypothalamus a *thermode,* an extremely fine probe that can be heated or cooled to precise temperatures. Throughout the experiment, the rat's body temperature remained the same. However, when the implanted thermode lowered the temperature of the hypothalamus from 38° C to 36.5° C, the animal began to shiver, produce more heat metabolically, and experience vasoconstriction and piloerection. When the researchers warmed the thermode, the temperature of the hypothalamus increased, and the rat's heat-dissipating arsenal was activated, causing the rat to pant, sweat, and exhibit vasodilation. Physiologists consider the hypothalamus to be a mammal's **thermoregulatory center.**

Just as breathing rate increases at the start of exercise in an anticipatory manner, temperature responses usually take place very fast; you begin almost immediately to shiver if you walk into a meat locker or to sweat if you sit in a sauna. This is because nerve impulses, especially from cold receptors in the skin and from cold and heat receptors in the walls of major blood vessels and in internal organs, send data to the brain and trigger local thermoregulatory processes long before the blood, the body core, or the hypothalamus change temperature at all. This anticipation helps keep the core temperature constant.

A homeotherm's set point can itself be altered at times. At the onset of a fever, for example, the set point rises. This is usually a result of circulating compounds called *pyrogens;* bacteria release some pyrogens, while white blood cells release others in response to infections or to other systemic conditions. Pyrogens act on the hypothalamus to raise the setting of the natural thermostat, so a person with fever shivers, even though his or her temperature is elevated.

During the winter, various homeotherms—including chipmunks, ground squirrels, marmots, and skunks—go into **hibernation,** during which their set points fall to about 5° C (Figure 33-14). They remain in their dens or burrows throughout the coldest months in a state of very low metabolism, with slowed heart and breathing rates. Mammals prepare for hibernation by building up fat reserves and growing heavy winter pelts.

During the months of hibernation, the kidneys continue to cleanse the blood, and the animal awakens periodically to excrete wastes. If a hibernator is disturbed or if the air temperature grows so cold that the animal is in danger of freezing, emergency neuronal circuits arouse

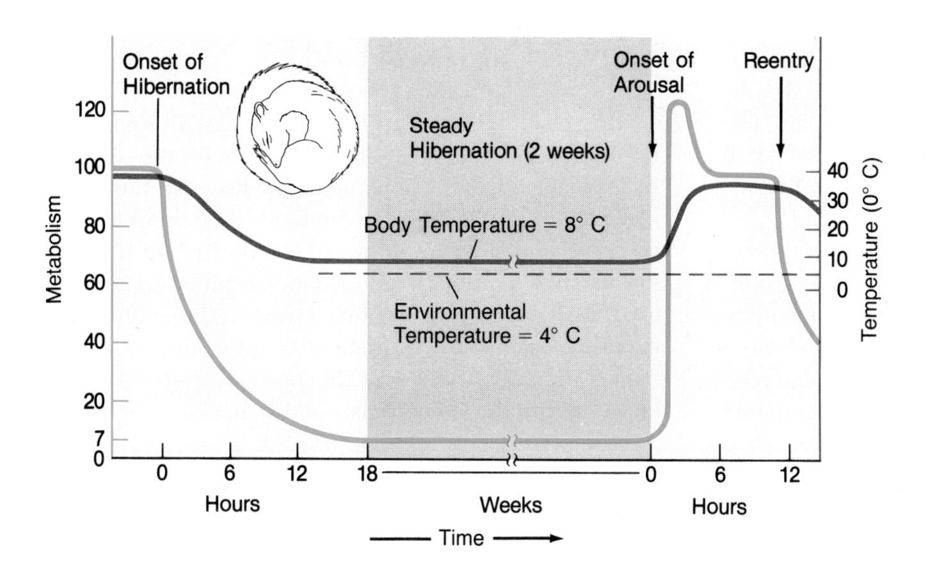

Figure 33-14 HIBERNATION AND AROUSAL IN GROUND SQUIRRELS. The body temperature of a ground squirrel (red) falls close to the environmental temperature after metabolism (blue) is turned down during hibernation. After 2 weeks, the ground squirrel arouses, and this increased activity allows its blood to be cleansed by the kidneys. After a few hours, metabolism and temperature drop once again.

SUPERCOOLED SQUIRRELS

Few animals face conditions as extreme as the arctic ground squirrel (Figure A). This small mammal, with its finite capacity to store energy in fat reserves, must somehow survive through the dark and frigid winters in Alaska and northern Canada that last eight months or more, by hibernating in small straw-lined burrows beneath a meter or more of frozen dirt, ice, and snow.

How does a small animal (weighing just 450 to 1135 g, or 1 to 2.5 lb) remain alive for so long while eating nothing and steadily burning off its limited fat reserves? Brian M. Barnes at the University of Alaska in Fairbanks recently found out. Barnes captured a dozen arctic ground squirrels, implanted miniature temperature sensors in their abdomens, then allowed them to dig burrows and hibernate for the winter. Surprisingly, he found that their blood became *supercooled;* that is, it dipped below the freezing point of water but remained liquid. Experimenters have artificially chilled a few small mammals to the supercooled state, but inevitably, if left in this condition for more than an hour, the blood has frozen and the animal has died. The arctic ground squirrel, however, can remain subzero for at least three weeks, and show no ill effects.

Supercooling is just one strategy for surviving long freezing winters. Some insects freeze partially but generate compounds that cause ice crystals to form outside of cells, where they do less damage. Some fish, insects, and other animals produce antifreezes that prevent ice from forming at all so blood remains fluid. And some mammals have a thermostat that rouses them from hibernation when the body temperature falls too low. Barnes speculates that supercooling provides significant energy-saving advantages for hibernators in bitter cold environments. He estimates, in fact, that a supercooled squirrel might expend only one-tenth as much energy to remain at −3° C as it would to stay above 0° C.

Figure A
An arctic ground squirrel emerging in May after its long hibernation.

the animal and it warms to nearly 37° C. Research shows that the cell membranes of cold-tolerant species are less permeable to ions of various types than the membranes of nonhibernators. At low temperatures, ATP synthesis drops, ion pumping slows, and nonhibernators can die of respiratory failure or cardiac fibrillation. Since hibernators' cells are geared to lower rates of ion passage (probably by having fewer ion channels), their slowed physiological processes can continue to sustain life. The box above explains the specific mechanisms that keep arctic ground squirrels alive while hibernating in the bitter cold winters of Alaska and northern Canada.

A variation on hibernation is **torpor,** a temporary lowering of the set point. Bats and hummingbirds, for example, must enter torpor every day when they are inactive. This is because they have small body masses and large surface areas and cannot stay warm without feeding constantly and maintaining high metabolic rates to keep one step ahead of heat loss from the body surface. While resting, they simply cannot stay warm, so they enter torpor, and their normal set points drop to nearly match the temperature of the surrounding air.

LOOKING AHEAD

This chapter showed how an animal's internal environment is carefully maintained by homeostatic systems. In particular, we saw how metabolic fires can burn for a lifetime but can be adjusted moment by moment to meet external demands, and how the composition of the salty internal sea is remarkably stable, despite variations in food and water consumption, waste elimination, and moisture or salt loss. Our next chapter focuses on the body's electrical network, the nervous system, which helps coordinate homeostasis and behavior.

SUMMARY

1. Mechanisms for *homeostasis*—the maintenance of a relatively constant internal environment—include the *osmoregulatory system*, the *excretory system*, and the *thermoregulatory system*.

2. Freshwater fish tend to lose salts and to take in water, and therefore excrete much dilute urine, and take up salts through the gills.

3. Marine fish tend to take in salts from ocean water and to lose water, and thus they excrete a small volume of concentrated urine and salt ions from the gills.

4. Amphibians can absorb water across the skin and bladder wall. Desert and marine reptiles and birds have salt glands to secrete excess NaCl.

5. Animals excrete nitrogen as ammonia, *urea,* or *uric acid,* depending on water available for safe transport.

6. In reptiles, birds, and mammals, *kidneys* perform *osmoregulation* and secretion of nitrogenous wastes, and regulate the pH of body fluids. The unit of kidney structure and function is the *nephron.*

7. To make urine, kidneys produce a primary filtrate and reabsorb most of the water and needed substances, while allowing wastes and unneeded substances to exit. Four physiological mechanisms are involved: filtration, reabsorption, secretion, and concentration.

8. *Antidiuretic hormone (ADH), aldosterone,* and *atrionatriuretic factor (ANF)* are all involved in controlling salt and water balance.

9. *Thermoregulation* helps maintain body temperature despite variations in environmental temperature.

10. *Poikilotherms* generally derive heat from the environment and have body temperatures that track the environment. *Homeotherms* have relatively constant core body temperatures.

11. *Ectotherms* derive most of their body heat from the environment. *Endotherms* generate their own body heat and have high body temperatures.

12. The high, constant body temperature in birds and mammals depends on insulation, on vasoconstriction and vasodilation, on panting, sweating, and other specific behaviors and structures.

13. The hypothalamus regulates drinking, feeding, and certain aspects of kidney function and acts as a *thermoregulatory center.* An animal's set point can change during fever or *hibernation.*

KEY TERMS

aldosterone
antidiuretic hormone (ADH)
atrionatriuretic factor (ANF)
bladder
collecting duct
compatible osmolyte strategy
cortex
counteracting osmolyte strategy
ectotherm
endotherm
excretory system
glomerular filtrate
hibernation
homeostasis
homeotherm
kidney
loop of Henle
Malpighian tubule
medulla
nephridium
nephron
osmolyte
osmoregulation
osmoregulatory system
piloerection
poikilotherm
rete mirabile
set point
thermiogenesis
thermoregulation
thermoregulatory system
torpor
urea
uric acid

QUESTIONS

1. Identify the substances or body conditions regulated by the osmoregulatory, excretory, and thermoregulatory systems. What three body fluids are involved in this regulation?

2. Which of the following describe freshwater fishes and which describe marine fishes?
   a. Live in a hypotonic environment.
   b. Live in a hypertonic environment.
   c. Drink large amounts of water.
   d. Drink little water.
   e. Excrete a small volume of concentrated urine.
   f. Excrete dilute urine.
   g. Take up salts by active transport across the gills.
   h. Excrete salts by active transport across the gills.

3. Describe briefly the counteracting osmolyte strategy that operates in sharks and related fishes to raise the osmotic pressure of body fluids.

4. Identify two groups of animals that excrete nitrogenous wastes as ammonia, two groups that excrete uric acid, and two that excrete urea.

5. Sketch the nephron and label the glomerulus, Bowman's capsule, proximal convoluted tubule, distal convoluted tubule, and collecting duct. What other part of the nephron is found in mammals and birds?

6. Identify the four basic processes kidneys carry out. What blood components do *not* enter the Bowman's capsule?

7. Describe the countercurrent exchange activity in the proximal and distal convoluted tubules and its effect on the nephron.

8. Identify three hormones that regulate kidney function; discuss the source, mode of action, and effects of each.

9. Reptiles are said to be "behavioral homeotherms." Explain.

10. How is heat generated in birds and mammals?

ESSAY QUESTION

1. Describe how some insects, reptiles, birds, and mammals maintain a high body temperature. Why do very small birds and mammals often pass the night in a state of torpor?

For additional readings related to topics in this chapter, see Appendix C.

34 The Nervous System

Nervous systems are undoubtedly the most intricately organized structures to have evolved on earth.

Roger Eckert and David Randall, *Animal Physiology* (1983)

If you have ever tried to catch a crayfish with your hand, then you've probably found this little crustacean to be much faster than you are; it will suddenly flip its powerful abdomen, dart effortlessly out of your grasp, then swim a few extra strokes to complete its escape. Even such a "simple" behavior as this requires the rapid coordination of several body parts—a coordinating role played by the nervous system.

The many physiological processes we have discussed so far must be coordinated and controlled if an organism is to function in a unified way. Such coordination, or combining of elements into a harmonious whole, is called *integration*. Integration of the activities of various cells, tissues, and organs in a multicellular organism comes about through the functioning of two control systems—the nervous and endocrine (hormonal) systems, which work via the same general principles: They use molecular, ionic, or electrical signals for cell-to-cell com-

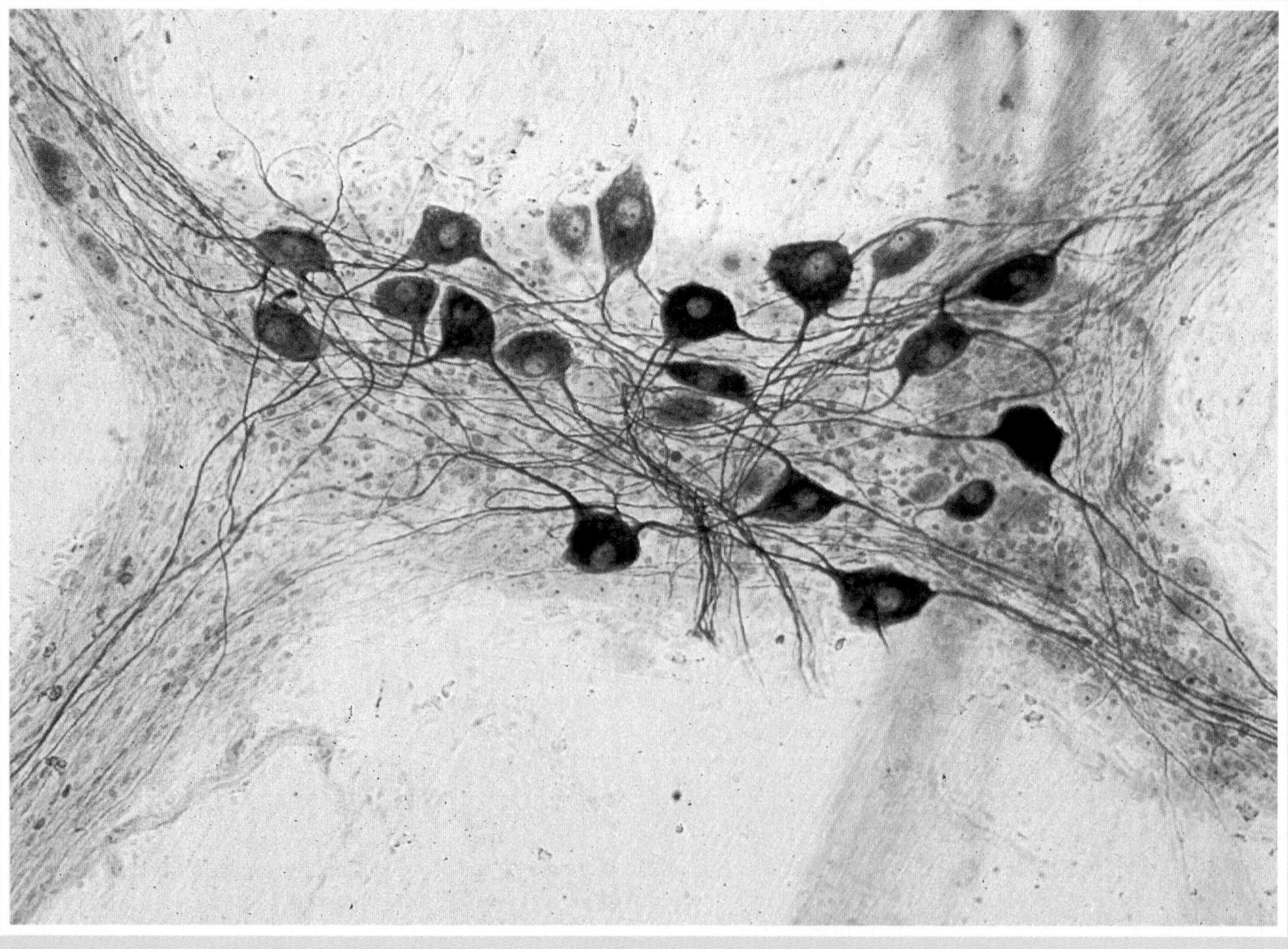

Neurons: a delicate network for electrical and molecular "cross-talk," here in an animal's gut wall.

munication; and they trigger changes in special *effector* cells and organs, which bring about physiological changes or behaviors in the organism. Traditionally, biologists drew sharp distinctions between the endocrine system, characterized by slow-acting hormones, and the nervous system, characterized by fast-acting electrical signals. Modern research, however, has blurred the former boundaries between the two by identifying literally dozens of "local hormones" that are secreted by nerve cells and that modulate the behavior of other neurons or neuronal circuits. The overlap and similarities between the two systems will become increasingly apparent in this chapter and the next.

We first encountered nervous systems in Chapter 24. In each nervous system, whether a simple nerve net or a highly developed spinal cord and brain, the basic unit is the individual nerve cell, or *neuron*, arranged by the millions or billions into chains, nets, or other combinations. No matter how complex, the function of every such network depends, in the end, on the activities of this unique cell type. For this reason, the chapter discusses:

- The general structure and individual types of neurons
- The activity of neurons—how they generate and transmit electrical signals
- How neurons are organized into nerve systems of varying complexity

NEURONS: THE BASIC UNITS OF THE NERVOUS SYSTEM

The nerve cell, or **neuron,** is essentially a cellular switching device—an energy *transducer.* When acted on by chemicals, heat, pressure, or a variety of other energy forms, the neuron transduces that initial input, the stimulus, into an electrical signal—the **nerve impulse,** or **action potential.** Neurons not only transduce the energy of a stimulus, but also conduct information in the form of these nerve impulses and communicate with other cells in the nerve network. Their unique structure explains how.

Neuron Structure: Key to Function

Despite their wide variety of shapes, nearly all neurons have four basic structures, each critical to function: an "antenna" to receive messages; a "cable" to transmit messages; special endings of the cables to allow communication with the next cell in a circuit; and a maintenance factory to keep all the rest in good repair. In a typical neuron (Figure 34-1), these four structures are the dendrites ("antennas"), the axon ("cable"), the synaptic terminal (cable ending), and the cell body (maintenance site).

Most **dendrites** are relatively short, multibranched, often spinelike extensions of the cell surface. Collectively, they provide a very large surface area for receiving information, which they transmit to the surface of the cell body, from where it may or may not be passed down the axonal cable. Thus, the more branched the dendrites, the more separate inputs a neuron can receive. Some kinds of neurons, however, have only a single dendrite, and a few types have none—they receive information messages directly on the surface of the cell body from other neurons.

The *cell body* contains the nucleus and most of the kinds of organelles found in all somatic cells (see Chapter 6). Protein synthesis and many metabolic activities are carried out mainly in the cell body, with products pass-

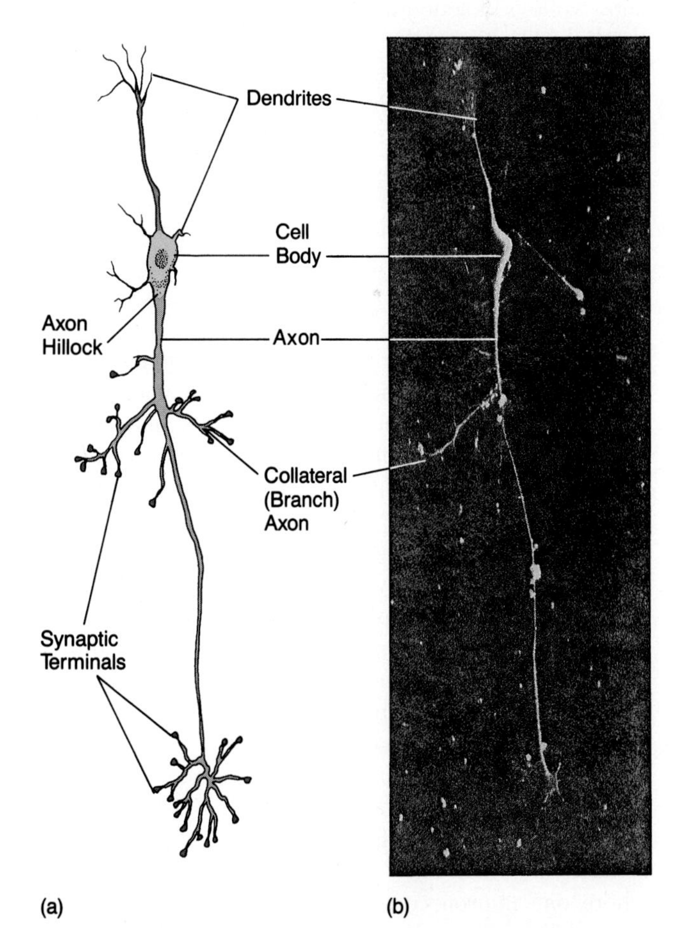

Figure 34-1 THE MAJOR PARTS OF A NERVE CELL. (a) A typical neuron consists of dendrites, the cell body, the axon, and synaptic terminals. (b) A scanning electron micrograph (magnified about 400×) of an embryonic neuron growing in a culture dish.

ing to the more specialized portions of the neuron for their maintenance. Polyribosomes, for example, sit just internal to sites where dendrites receive signals from other neurons, and ribosomes and mRNA are transported from the cell body outward to the dendrites, allowing protein synthesis to occur on those polyribosomes.

The **axon** is the neuronal cable that transmits signals in the form of action potentials (nerve impulses) from one point to another in the nervous system. The plasma membrane is the actual site of the transmission process. The axon contains a complex cytoskeleton and many mitochondria. It also functions as a "pipeline" that carries molecules manufactured in the cell body to the axonal endings and that transports other molecules from the axonal endings back to the cell body. The area where the axon joins the cell body is called the *axon hillock*, and it is thought to be the site where action potentials are generated. Individual axons often branch to form *collateral axons*, which can link up to many different nerve cells or target cells, such as muscle cells. These collateral axons allow volleys of action potentials to be transmitted to all the linked cells simultaneously.

Axons vary in diameter and length. In mammals, axons can be 1–20 μm in diameter. The largest axons in the animal kingdom, found in the giant squid, are up to 1 mm across. Since an action potential's traveling speed varies with the diameter of the axon, differences in size are important to nervous system function. An axon's length can range from only tens of micrometers to 4 m or more; some of a giraffe's axons, for example, extend from the spinal cord down to the feet.

Axons are not themselves what we commonly refer to as nerves. **Nerves** are actually bundles of many axons. A single human optic nerve, for example, may contain 1 million axons. The axons in a nerve can run parallel to each other or can be intertwined like the strands of a telephone cable.

The many axons in a nerve can carry independent messages simultaneously. Thus, sensory neurons in the skin can receive heat, pain, or other kinds of information, which are then transmitted to the spinal cord over certain axons within a nerve, while commands may be traveling from the brain to the muscles in a different subset of axons in the same nerve.

The fourth major neuron structure is the **synaptic terminal,** at the end of the axon (see Figure 34-1). A synaptic terminal and the surface of the adjacent target cell it contacts together form a synapse. *Synapses* are sites where one neuron communicates either chemically or electrically with a target cell, which can be another neuron, a muscle cell, or a secretory cell. The synaptic terminals contain packets of the chemical used in cell-to-cell communication, the *neurotransmitter*, as well as secretory machinery and specialized membranes. Laboratory techniques that allow researchers to visualize the same neuron over several weeks in a living mouse reveal that synapses are dynamic, changing slowly in number and distribution—a fact that may help explain the plasticity of nerve circuits and systems we will see in later chapters.

Animal nervous systems also contain **glial cells,** or simply **glia.** Depending on the animal species and the site within the nervous system, there may be as many as ten times more glia than neurons. Glia provide mechanical support and metabolic aid to neurons; in some species, they can sustain an axon for months, even if it is severed from its cell body by injury. Glial cells wrap layers of *myelin*, or electrical insulating material, around individual axons. This myelin insulation allows extremely rapid transmission of action potentials and thus much faster functioning by myelinated nerves and nervous systems. Researchers have recently discovered one type of glial cell that may help protect the nervous system from autoimmune attack by presenting antigens to white blood cells much as macrophages do. Malfunctioning of these so-called microglial cells could conceivably play a role in autoimmune diseases like multiple sclerosis.

Three Primary Types of Neurons

Differences in the various locations and proportions of dendrites and axons help distinguish neurons involved in sensation, movement, and brain activity. Three major classes of neurons deserve our attention: receptor neurons, effector neurons, and interneurons.

Receptor (sensory) neurons are specialized energy transducers. Each receptor neuron is sensitive to a particular type of stimulus, such as light, pressure, heat, or a specific chemical, and once received by the dendrites, this triggers a change in electrical activity that travels as an impulse down the axon (Figure 34-2). Some sensory organs have vast numbers of receptor cells: An eye can have 100 million; an ear, 20,000. Such receptor cells lack axons and pass their information to true sensory neurons, which carry it to interneurons or, occasionally, to motor neurons.

Effector (motor) neurons transmit messages to muscles, causing them to contract, and to glands, causing them to secrete (see Figure 34-2). Whether in response to specific stimuli or to higher-order commands from the brain, everything an animal does—each eye blink or growl—is the direct consequence of coordinated activity in some set of effector neurons. Humans have about 3 million effector neurons.

Interneurons receive information from receptor neurons, sensory neurons, or other interneurons, process it, and send commands on to the effector neurons (see Figure 34-2). Interneurons are arranged in circuits that vary tremendously in complexity and in the number of interneurons they contain, the most complex being tracts of thousands of neurons in a mammalian brain. The interneuron circuits are the sole sites for coordinating sets of motor neurons and thus the movement and activity of body parts. Interneuron circuits are also the seat of higher-order processes, such as learning and memory. Thus, interneurons are the integration sites of the nervous system. Approximately 98 percent of the 10 billion cells in the human nervous system are interneurons.

The three basic types of neurons are built into a vast variety of circuits. In some, a thousand neurons may converge on just one; or the branching dendrites of one neuron may connect to a thousand others; or neurons may form loops that feed back on themselves.

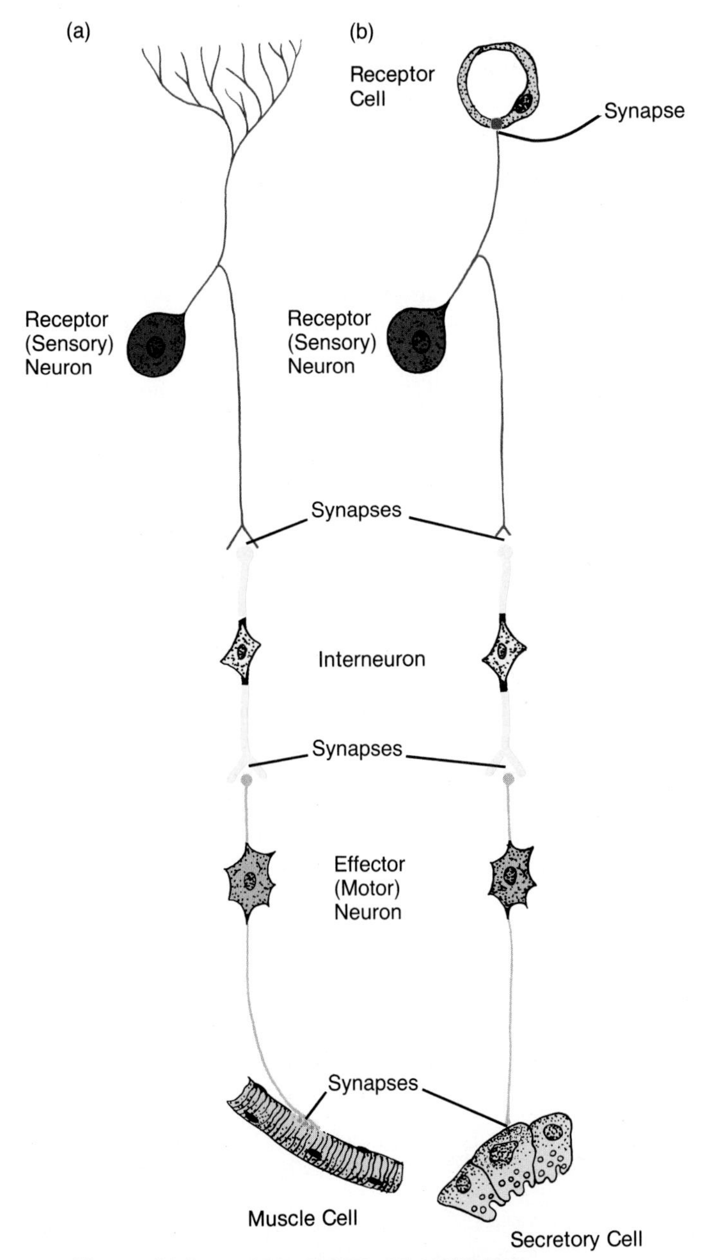

Figure 34-2 BASIC TYPES OF NEURONS.
(a) The simplest type of sensory neuron also functions as a receptor neuron. Sensory neurons form synapses with interneurons or, more rarely, with motor neurons. (b) Some sense organs have special receptor cells that form synapses with sensory neurons. Interneurons both conduct and process information. Effector (motor) neurons form synapses with muscle cells or secretory cells.

HOW NEURONS SIGNAL

Nerve impulses, or action potentials, are the "language" of the nervous system—a kind of Morse code by which neurons communicate with one another and with other cells. German physiologist Julius Bernstein, working near the turn of the century, demonstrated that the nerve impulse is an *electrochemical* event during which ions move across the nerve cell's plasma membrane, causing the normally negatively charged cell to lose its charge.

Not until 40 years later, however, when J. Z. Young discovered two giant axons in the squid, did the mysteries of nerve action start to unravel. Squid axons, which are 1 mm in diameter and nearly 50 times bigger than the largest human axons, are large enough to make experimentation relatively easy. Young's contemporaries, English physiologists Alan L. Hodgkin and Andrew F. Huxley, learned how to insert electrodes into the squid axon and measure the movement of ions (Figure 34-3). They soon realized that changes in the permeability of the axon's plasma membrane and the subsequent movement of ions produce a nerve impulse and transmit it along the length of the axon.

How the Action Potential (Nerve Impulse) Is Generated

To understand action potentials, we have to consider the normal electrical state of the neuron, the so-called resting potential. We also need to see how the resting potential changes to generate the action potential and how the cell returns to the resting state.

Every living cell, whether it is the photosynthetic cell of a plant or the nerve cell of an animal, has a **resting potential,** a state of electrical charge in which the inside of the cell is electrically negative relative to the outside of the cell. The resting potential is usually due to the relative numbers of Na^+, K^+, and Cl^- on the two sides of the membrane and to the relative permeability of the plasma membrane to those ions.

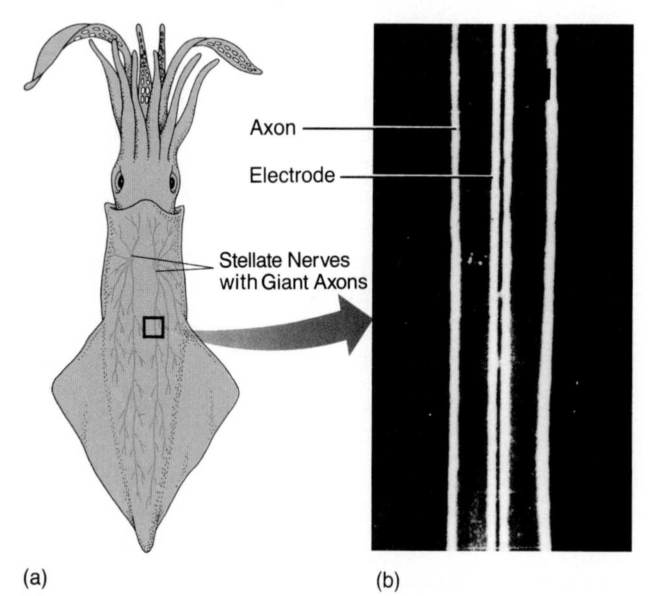

Figure 34-3 GIANT AXONS OF THE SQUID.
The squid's thick, rapidly conducting axons have revealed a great deal about the basic nature of nerve impulses.
(a) Organization of squid nerves. The stellate nerves contain the giant axons. (b) A microelectrode runs down the center of a giant axon (magnified only about 15×).

Biologists have found that a neuron's plasma membrane contains a very active Na·K·ATPase pump enzyme that shuttles Na^+ out of and K^+ into the cell when ATP is hydrolyzed (Figure 34-4). This establishes a concentration gradient for Na^+, so that as it accumulates outside the cell, it tends to leak back in. Na^+ can only leak back slowly, however, owing to properties of the ion channels through which the Na^+ must pass as it crosses the hydrophobic plasma membrane. The pumping also establishes a concentration gradient for K^+; that ion moves down its gradient and leaks out of the cell more easily than Na^+ moves into the cell because of properties of the ion channels through which K^+ moves. As a result of these differences in Na^+ and K^+ movement, a net positive charge builds up outside the cell and a net negative charge inside. At equilibrium, the net outward flow of K^+ equals the new inward flow of Na^+, and the cell is at its resting potential and in a **polarized** state.

A third ion, Cl^-, moves passively across the cell membrane. Because it is attracted to the net positive charge outside the cell and is repulsed by the net negative charge inside the cytoplasm, Cl^-, like Na^+, tends to accumulate outside the cell.

Resting potential is measured in millivolts (mV); a millivolt is one-thousandth of a volt. In most neurons, the resting potential is about -70 mV, reflecting the difference between the net negative charge inside the cell and the net positive charge outside. The high Na^+ concentration outside the neuron is like a wall poised to fall inward; and the high relative K^+ concentration inside is like another wall poised to fall outward. Thus, the polarized, resting potential is a state of stored energy.

Note that if a cell's resting potential rises from -70 mV toward 0 mV, the cell becomes *less* polarized and is in fact said to be **depolarized** (it has lost some of its polarization). Conversely, if the cell becomes more negative (say, to -90 mV), it is **hyperpolarized.**

The ion channels in the neuron's plasma membrane are critical to setting up the resting potential and to transmitting volleys of electrical signals, or action potentials—the main business of neurons. Ion channels specific to each type of ion may open or close, making the membrane temporarily permeable to the specific ion. Changes in membrane potential may trigger the transient opening of "gates" in the neuron's membrane: **voltage-gated channels** may open when a certain mem-

Figure 34-4 THE NEURON'S ION PUMP.
The Na·K·ATPase pump enzyme utilizes energy from ATP to transport Na^+ out of the neuron and K^+ into it. This mechanism helps establish the resting potential of eukaryotic cells. The pump also restores the normal relative concentrations of Na^+ and K^+ after many action potentials have passed along an axon. Ion channels (protein pores) through the membrane permit relatively faster K^+ leakage outward and slower Na^+ leakage inward; these processes also contribute to the resting potential.

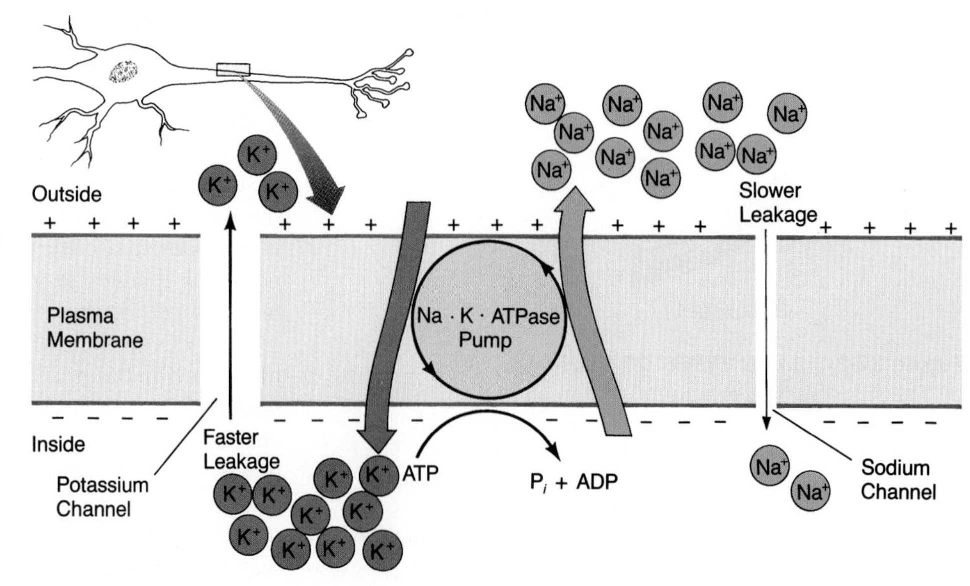

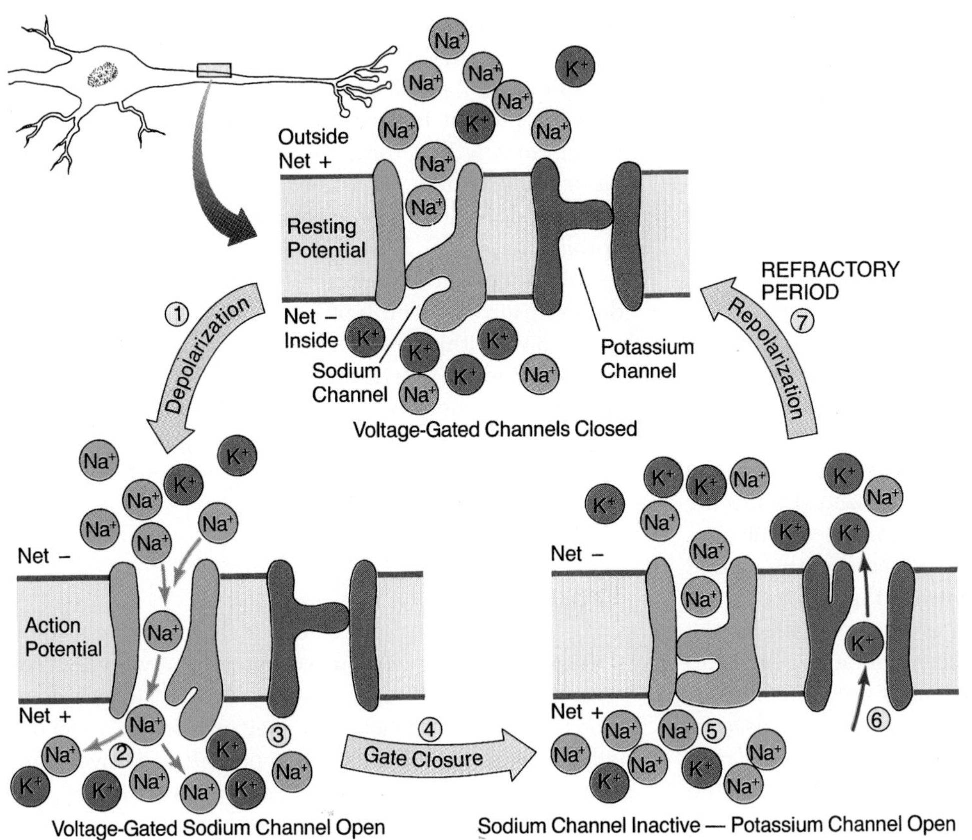

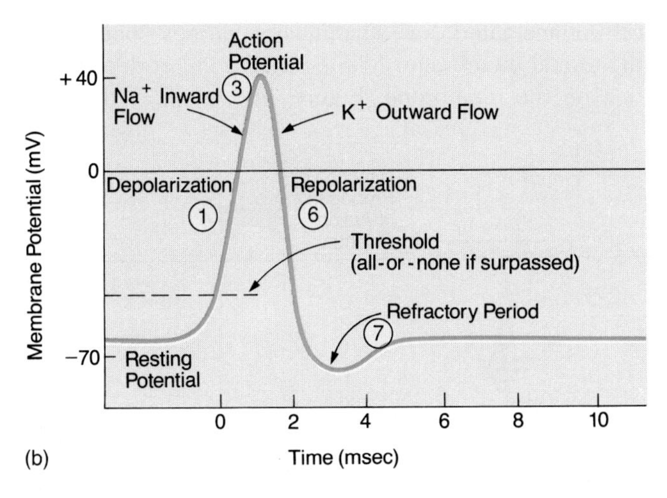

Figure 34-5 THE IONIC AND MOLECULAR BASES FOR AN ACTION POTENTIAL.
(a) Voltage-gated Na^+ channels (gates) are closed as the resting potential is maintained. On depolarization above a threshold value, the Na^+ channels open, permitting Na^+ to flow into the cell. (b) The effects are seen in the oscilloscope recording as the full action potential is attained. The Na^+ channels then close suddenly and enter an inactive state; at the same time, K^+ leaks outward through its channels, thereby lowering the potential below the original resting potential level. The refractory period terminates as the potential returns to the resting level and the Na^+ channels revert to the closed but active state.

brane potential is reached or as a membrane depolarizes. Specific chemicals can do the same thing: **Chemical-gated channels** open when a specific chemical is present.

Whereas all cells have resting potentials, neurons and a few other cell types (including muscle cells and certain sensory cells) are unique in that they are *excitable*, that is, the cell membrane responds in an explosive way to an electrical or chemical signal (the so-called stimulus) by generating an action potential. In such excitable cells, ion channels (gates) that are selective for Na^+ open, allowing Na^+ to pass through the membrane at the point of electrical or chemical stimulation and allowing an action potential to follow. Experimenters have probed the events surrounding ion channel opening by depolarizing a nerve cell, raising the potential of the neuron from −70 mV to about −30 mV (Figure 34-5, step 1). This change from the resting potential opens the voltage-gated Na^+ channels in the membrane, allowing some of the excess Na^+ in the extracellular fluid to rush into the cell (step 2). Enough Na^+ moves rapidly into the cell to depolarize the cytoplasm just inside the membrane to perhaps +40 mV at the site where the channels opened (step 3). The Na^+ channels remain open for only about 0.5 millisecond before slamming shut (step 4), as the pore proteins of which the channel is built enter an inactivated state (step 5). The result is the transient entry of relatively few Na^+, but even that small number is sufficient to generate the action potential at that particular site on the membrane.

During our experiment, the total change between the resting potential (−70 mV) and the peak positive voltage as Na^+ rushes in (+40 mV) is the action potential of the neuron. When the change from resting potential to action potential is recorded using electrodes, the resulting display on an oscilloscope shows the action potential as a spike (Figure 34-5b). When the positive charge inside the cell reaches a specific level, voltage-gated K^+ channels open, and K^+ rushes out of the cell down its concentration and electrical gradient (Figure 34-5, step 6). The exodus of K^+ means that less total positive charge remains inside the cell; thus, the potential falls rapidly toward the resting potential level. Indeed, the charge is briefly reduced below the resting potential of −70 mV. During this **refractory period** (0.5 to 2 milliseconds), the membrane is unable to react to additional stimulation. Although the refractory period is short, it has great biological significance because it limits the number of action potentials that can be generated per second. During this short time period, the inactivated Na^+ channels return to their original closed but ready configuration, and the potassium channels close, preparing the system to work again if an appropriate stimulus causes them to do so (step 7). After the action potential has fired and the membrane has repolarized at a specific site, there is a local imbalance to be corrected: The concentrations of K^+ and Na^+ have shifted slightly, so that there is now a bit more K^+ outside the cell and a bit more Na^+ inside the cell than is characteristic of the resting potential. Although the accumulations inside and outside from a single action potential are insignificant, they add up after thousands of action potentials pass down an axon. What restores the original ionic balance to the cell? That is the task of the battery charger, the Na·K·ATPase pump enzyme we discussed earlier. This highly active enzyme consumes a large amount of ATP as it pumps Na^+ out and K^+ in. It should not be surprising to learn that when a giant squid axon is poisoned with cyanide, which prevents ATP production, there is *no* immediate effect on action potentials. Since their firing does not consume ATP directly, thousands of impulses can be generated before the absence of Na·K·ATPase pump enzyme activity takes effect.

An important feature of all electrically excitable cell membranes is that they initiate action potentials in an *all-or-none* fashion. If a neuron or other cell type receives a depolarizing stimulus below a certain voltage, nothing happens. However, if the stimulus reaches a specific **threshold** voltage (−30 mV in our example, but different for each cell type), the Na^+ channels begin to open, and within a fraction of a millisecond, the full action potential is triggered. No matter how large the stimulus becomes, the same action potential is obtained. This all-or-none character of action potentials allows their transmission for long distances without fading out.

How the Action Potential Is Propagated

Messages in the nervous system typically involve volleys of action potentials traveling long distances down axons running within a turkey's leg or a person's arm or a jellyfish's tentacle. For an action potential to travel along an axon, it must be *propagated*, or disseminated, the length of that axon, a bit like fire burning along a fuse (but not consuming it); axons are reusable "fuses."

Impulse propagation depends on the presence of typical voltage-gated Na^+ channels at various spots along the axonal membrane. At the initial point of depolarization on the membrane (Figure 34-6, step 1), the Na^+

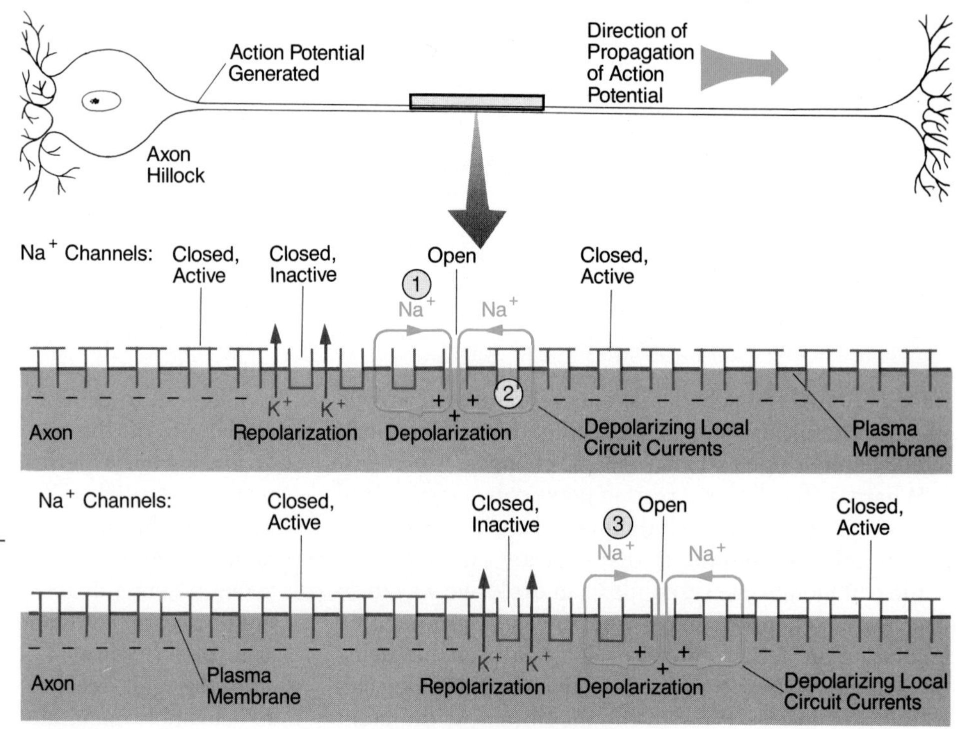

Figure 34-6 PROPAGATION OF AN ACTION POTENTIAL IN AN UNMYELINATED AXON. The action potential is normally generated at the axon hillock. Here we see the action potential at two sites next to each other. Propagation from one site to the next depends on local circuit currents that depolarize above threshold the adjacent membrane and thereby open voltage-gated Na^+ channels there. As the action potential sweeps along the axon, the voltage-gated channels successively open, close and become inactive, and become active again at each site. (Voltage-gated K^+ channels are not shown.)

channels are caused to open, and an all-or-none action potential is initiated. The local influx of Na^+ causes minute **local circuit currents** to flow (step 2), thereby depolarizing beyond threshold the adjacent portion of the membrane so that its Na^+ channels open (step 3). This sequence is the start of propagation. The process continues down the axon, as one after another Na^+ channel opens, in domino fashion, slams shut into the inactive state, and then returns to the closed condition. The action potential travels at speeds of 1–100 m per second and does not fade out or damp in intensity for an obvious reason: At each new site, the depolarizing local circuit current causes a surplus of Na^+ channels to open so that a full all-or-none action potential is produced.

Another feature of a propagating action potential is its unidirectionality; from the site of initiation, usually the axon hillock, the impulse travels down the axon. The brief refractory period during which the Na^+ channels are inactive and K^+ flows outward accounts for the inability of the action potential to double back on itself toward the cell body.

The speed with which an axon can propagate an action potential varies from only a few centimeters to 120 m per second. One variable that affects velocity of propagation is axon diameter. Conduction velocity increases only as the square root of the diameter; that is, increasing the diameter 16 times yields only a fourfold increase in speed, but this can still ensure an animal's survival. Fish, for example, have two large neurons, the Mauthner cells, whose firing and rapid conduction allows the animal to flip violently and forcefully away from the lunging beak, claws, or fingers of an attacker.

Temperature also affects impulse propagation: Rising temperature speeds propagation. Because of their high body temperatures, birds and mammals can have very narrow axons (1 m or so) that still conduct rapidly.

The typically small-diameter axons of vertebrates can conduct impulses rapidly for another reason: They are insulated by **myelin sheaths.** Myelin is the product of special glial cells (called **Schwann cells** in peripheral nerves and **oligodendrocytes** in the brain and spinal cord) that form layer upon layer of specialized plasma membrane that is wrapped around the axons of many vertebrate neurons like the plastic or rubber insulation around an electrical wire (Figure 34-7).

At regular intervals along a myelinated axon, the myelin sheath is interrupted, and tiny sections of axonal membrane, called the **nodes of Ranvier,** are exposed (see Figure 34-7). Na^+ channels are abundant at these nodes,. but not in the stretches sheathed with myelin. Because of this structure, the impulse moving along a myelinated axon jumps from one node to the next about as fast as it would travel from one site of axonal membrane to the next in an unmyelinated nerve cell. This process of jumping from node to node is called **saltatory propagation.**

The myelin sheath makes a remarkable difference: The squid's giant unmyelinated axon (the largest known) is 1 mm in diameter and conducts at 20 m per second. Yet, a myelinated frog axon, 500 times smaller in diameter, can conduct twice as fast.

To sum up, the action potential, or nerve impulse, in all types of animals is an electrochemical event involving the rapid depolarization and repolarization of the nerve cell membrane. The process begins at the point where the membrane is depolarized beyond a threshold value and continues along the axonal membrane. Each depolarization-repolarization cycle initiates a new cycle at the adjacent point on the axonal membrane (or at the next node of Ranvier in a myelinated axon). Now let's see what happens when an impulse reaches the synaptic terminal at the end of an axon.

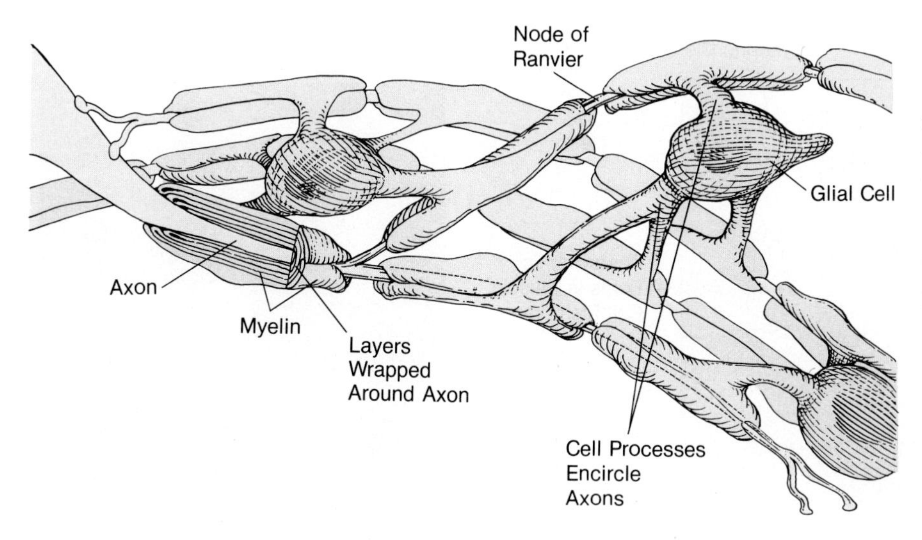

Figure 34-7 MYELIN: THE FATTY INSULATION OF AXONS. In myelinated nerves, cell processes from glial cells wrap around several axons and form myelin sheaths composed of multilayers of plasma membrane. Unmyelinated regions—the nodes of Ranvier—occur at regular intervals along each axon; the action potential jumps from one node to the next.

Transmission of the Action Potential Between Cells

As you read this page, millions of action potentials are being transmitted from one neuron to another in your brain across junctions called **synapses.** A synapse is usually located at the end of an axon or collateral axon of a so-called *presynaptic* neuron (a neuron whose action potential moves toward the synapse). The dendrite of the *postsynaptic* cell lies on the opposite side of the synaptic junction and receives electrical or chemical stimulation during **synaptic transmission,** the crossing of the synapse by the signal. Postsynaptic neurons usually form synapses with dozens, hundreds, or thousands of presynaptic neurons and therefore can receive independent signals from a variety of sources in any brief time interval.

There are two types of synapses: electrical and chemical. In *electrical synapses,* the plasma membranes of the communicating cells actually touch, facilitating ion transfer (Figure 34-8a). The action potential travels from one neuron to the next via gap junctions, through which ions, small molecules, or electric currents may pass rather freely from cell to cell (see Chapter 5). Because transmission at a gap junction is extremely fast, electrical

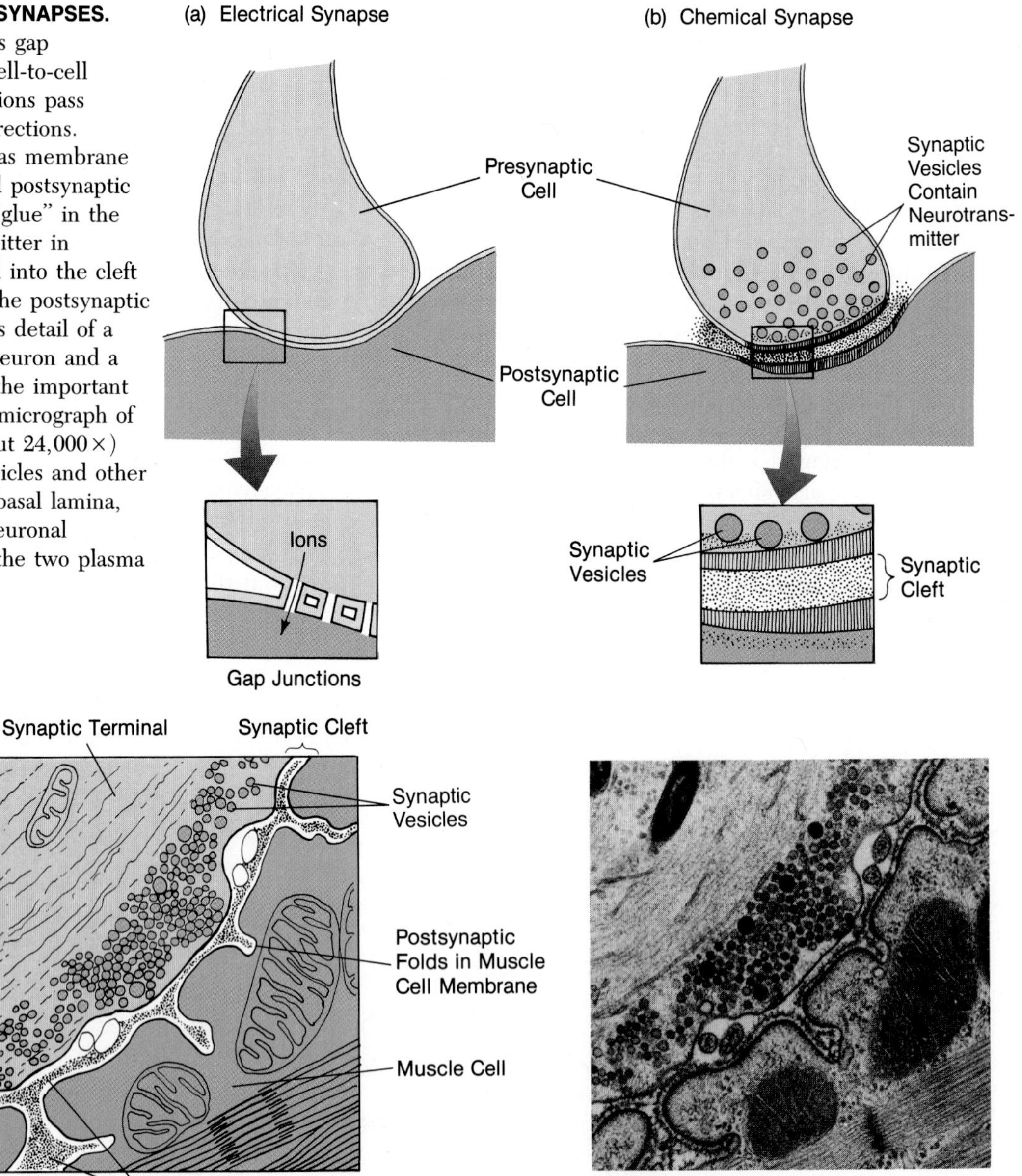

Figure 34-8 TYPES OF SYNAPSES. (a) An electrical synapse has gap junctions that are sites of cell-to-cell communication. Such junctions pass action potentials in both directions. (b) The chemical synapse has membrane thickenings on the pre- and postsynaptic surfaces, plus a molecular "glue" in the synaptic cleft. Neurotransmitter in synaptic vesicles is released into the cleft and binds to receptors on the postsynaptic surface (not shown). (c) This detail of a synapse between a motor neuron and a skeletal muscle cell shows the important structures. (d) An electron micrograph of the synapse (magnified about 24,000 ×) shows how the synaptic vesicles and other structures actually look. A basal lamina, equivalent to the glue in neuronal synapses, is seen between the two plasma membranes.

synapses are found where fast impulse transfer is important, as in the muscle cells of the vertebrate heart. Most electrical synapses are *nonrectifying* (bidirectional); impulses can move across them in either direction.

In *chemical synapses*, electrical impulses are transduced into chemical signals instead of being transmitted directly (Figure 34-8b). Chemical messengers, called **neurotransmitters,** are released from the presynaptic cell and diffuse across the space between the presynaptic and the postsynaptic cell membranes. The chemical message is then transduced back into an electrical voltage change in the action potential of the postsynaptic cell. Chemical synapses are generally slower at transmission than are electrical synapses, but they have a greater capacity to convey information because of the many kinds of possible neurotransmitters and their various speeds of release and breakdown. Because only presynaptic cells release neurotransmitters, chemical synapses transmit information in only one direction; they are *rectifying* (unidirectional) synapses.

In chemical synapses, the membranes of the two cells are not in direct contact; they are separated by a microscopic space—the *synaptic cleft.* Just inside the presynaptic cell membrane are a large number of tiny sacs, called *synaptic vesicles,* which contain neurotransmitter. Dozens—perhaps hundreds—of different chemicals serve as neurotransmitters in various neurons. Among the better understood neurotransmitters are acetylcholine, epinephrine, norepinephrine, dopamine, and serotonin. While different types of neurons release different neurotransmitters, neurons of a single type generally secrete the same single neurotransmitter or the same mixture of several. For example, all motor neurons that drive human skeletal muscles secrete the neurotransmitter acetylcholine.

The Role of Neurotransmitters

When an action potential reaches a synaptic terminal, it depolarizes the presynaptic cell membrane. This depolarization opens voltage-gated Ca^{2+} channels, allowing a small number of Ca^{2+} to flow into the cytoplasm of the terminal from the extracellular fluid surrounding it (Figure 34-9a). The Ca^{2+} promotes the fusion of synaptic vesicles with the presynaptic cell membrane, and the membrane opens to release neurotransmitter from the vesicles into the synaptic cleft (Figure 34-9b).

Neurotransmitter molecules diffuse across the narrow synaptic cleft in a fraction of a millisecond. Some reach and bind to *receptor molecules* embedded in the postsynaptic cell membrane (Figure 34-9c). These receptors are part of the chemical-gated channels that are opened only by specific neurotransmitters. The specificity of neurotransmitter-receptor binding ensures that only the "right" signal can trigger the postsynaptic cell and perhaps propagate the action potential.

When the ion channels associated with the receptors open, ions flow along their concentration gradients into the postsynaptic cell's cytoplasm, and the channels remain open for only about 400 microseconds. The influx of ions through the many channels at a synapse can have a positive (excitatory) or a negative (inhibitory) effect on the postsynaptic cell membrane.

At an **excitatory synapse,** synaptic transmission depolarizes the receiving (postsynaptic) cell. Cation channels (ones that pass positively charged ions) open in the post-

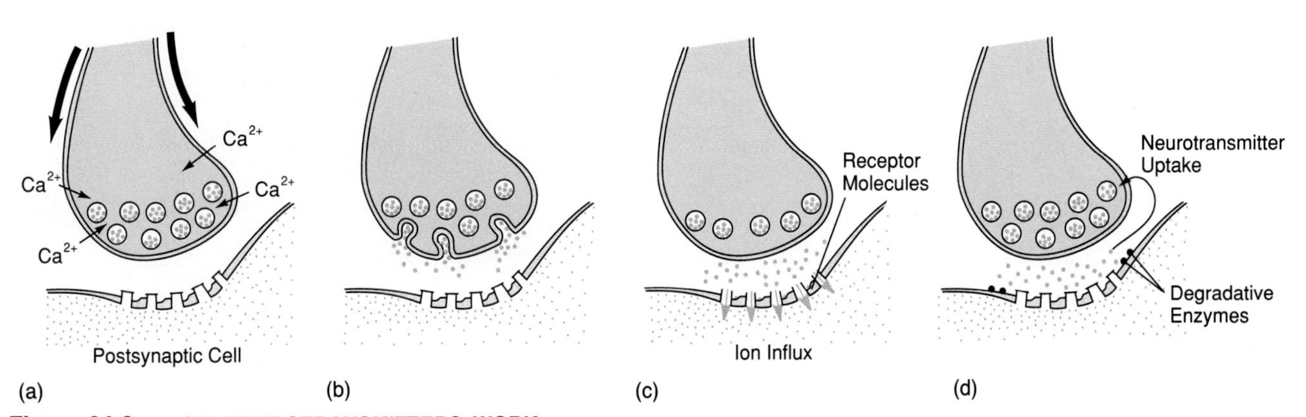

Figure 34-9 HOW NEUROTRANSMITTERS WORK.
(a) Arrival of the action potential at the synaptic terminal triggers Ca^{2+} entry into the presynaptic cell cytoplasm. (b) This, in turn, triggers fusion of presynaptic vesicles with the plasma membrane, thereby releasing neurotransmitter. (c) Binding of the neurotransmitter—say, acetylcholine—triggers the entry of ions into the postsynaptic cell cytoplasm. In response to this binding, Na^+, K^+, and Ca^{2+} enter at excitatory synapses, and Cl^- enters or K^+ leaves at inhibitory synapses. (d) After a brief time, the neurotransmitter is degraded by an enzyme or is taken back into the presynaptic terminal.

synaptic cell membrane; Na^+, K^+, and some Ca^{2+} flow in; and the electrical potential changes a bit from −70 mV toward 0 mV. This depolarization is called an excitatory **postsynaptic potential (PSP).** It opens any nearby voltage-gated Na^+ channels and may cause the receiving (postsynaptic) cell to fire.

An **inhibitory synapse** is one in which the action of synaptic transmission is to *prevent* depolarization of the receiving cell. The neurotransmitter "unlocks" channels selective for ions such as K^+ or Cl^-. If K^+ flows out or Cl^- flows in, the cytoplasm inside the membrane becomes more negative in its potential; it hyperpolarizes, making the potential even farther from threshold. This is an inhibitory postsynaptic potential. As a result, the receiving cell is much less likely to generate an action potential—it is "inhibited."

Neurotransmitters are removed from receptor sites and the synaptic cleft in a speedy and efficient way, and this allows the postsynaptic cell to receive newly arriving messages. Neurotransmitter molecules are removed from the cleft in two ways: They are either rapidly broken down by enzymes on the postsynaptic cell surface or taken back into the presynaptic terminals (Figure 34-9d). Even the transmitter molecules bound to receptor molecules are released and diffuse back into the cleft; they, too, are degraded or taken up. Many nerve gases and insecticides are designed to inactivate the postsynaptic enzymes that break down neurotransmitters. Since the neurotransmitters do not disappear, they continue to stimulate the postsynaptic cell, which in turn causes muscles to contract, leading to paralysis and death.

Summation and the Grand Postsynaptic Potential (GPSP)

We have seen that an action potential can travel down a neuron, cause a neurotransmitter to be released at an excitatory synapse, and trigger a postsynaptic potential (PSP). Is this enough, however, to initiate a true action potential in the postsynaptic cell? The answer is no, and the reason is that there are too few voltage-gated Na^+ channels on the dendrites and cell body of most postsynaptic cells to allow the initiation of action potentials. However, most nerve cells have many synapses on their dendrites and cell body (Figure 34-10), some inhibitory, some excitatory, and at any one time, a number of these may be active. The result is the **grand postsynaptic potential (GPSP)**, which is roughly the additive effects of

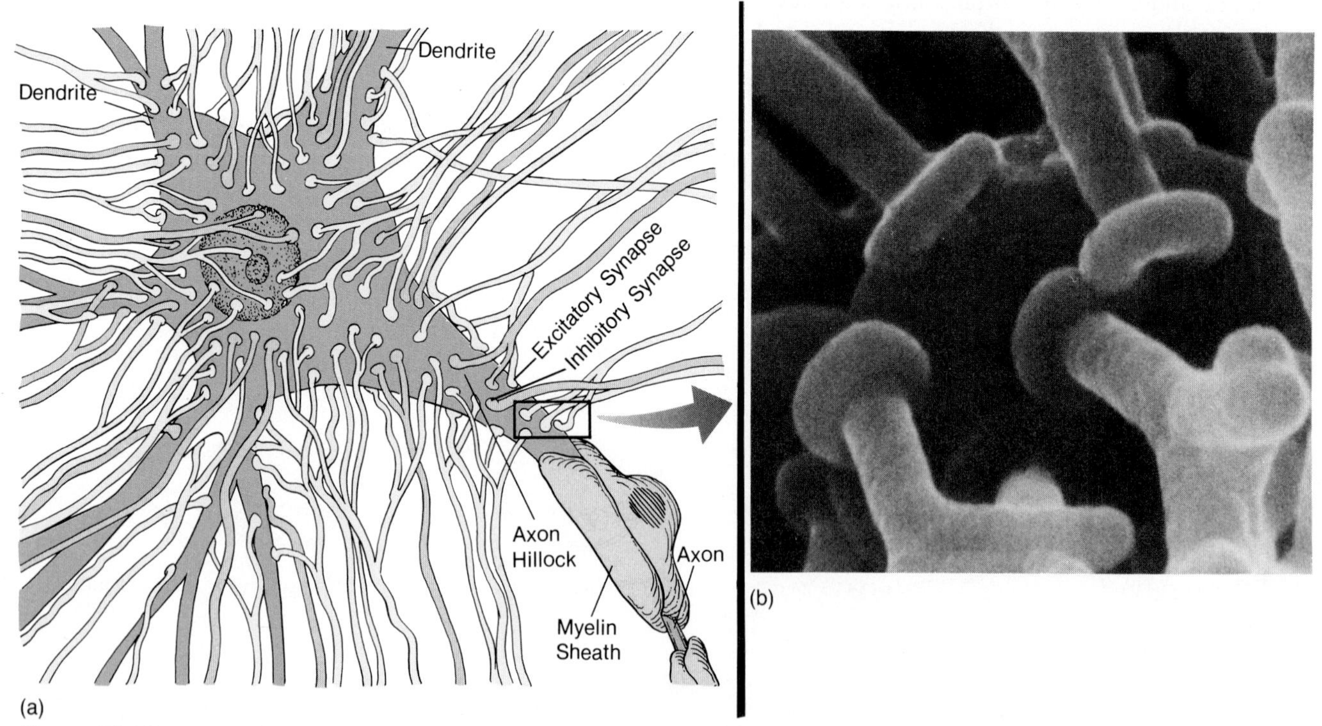

Figure 34-10 THE GRAND POSTSYNAPTIC POTENTIAL AND SUMMATION.
The body of this spinal cord neuron is virtually covered with excitatory (green) and inhibitory (red) synaptic terminals. (b) A scanning electron micrograph (magnified about 23,000×) shows such terminals. At any one time, varying numbers of the excitatory and inhibitory synapses fire to generate excitatory and inhibitory postsynaptic potentials (PSPs). They "add" together as the GPSP, and when their total exceeds threshold for this neuron, action potentials are generated at the axon hillock.

all the excitatory and inhibitory PSPs over any short time span. If many excitatory synapses but few inhibitory synapses are active at a given time, the combined PSPs will probably cause enough depolarization to generate an action potential. The adding together of individual PSPs is called **summation.** Summation occurs over both space and time. *Spatial summation* takes place when a number of excitatory and inhibitory PSPs at different sites are combined. *Temporal summation* is the cumulative effect of a number of PSPs arising at different times. If a volley of action potentials comes down a presynaptic neuron fast enough, the potential does not have time to fall back to resting potential levels. The potential might continue to increase until the grand finale is reached and a GPSP causes the postsynaptic cell to fire.

Within an animal's nervous system, action potentials pass along a chain of neurons—a GPSP being reached in one, then the next, then the next—until an effector cell, such as a muscle or secretory cell, is reached. Thousands, millions, and indeed tens of millions of such processes go on in nervous systems of varying complexity every second. From this combined activity come behavior, muscle contraction, secretion—in short, the functioning of the animal organism.

Let us turn now from the individual neuron and nerve impulses to systems of neurons within functioning organisms.

THE ORGANIZATION OF NEURONS INTO SYSTEMS

Nervous systems vary in complexity from simple networks in organisms with the least complicated behavior to incredibly intricate living computers—brains—in organisms with the most complicated behavior (Figure 34-11).

Simple Circuits

All living things are *irritable*; that is, they respond to chemical, mechanical, electrical, and other stimuli. The simplest unicellular organism may move away from a noxious chemical or the touch of a glass rod. Such a stimulus usually affects the cell surface directly or indirectly, often changing the proportions and distributions of ions. Nervous systems are really just complex variations on this basic irritability and ionic response.

In animals, specialized cells evolved to receive and respond to stimuli. Sponges have a few cells that resemble neurons, but they have no true nervous system. The radially symmetrical cnidarians, such as jellyfish and hydras, have **nerve nets,** lacking a centralized control center, but often having a concentration of the network

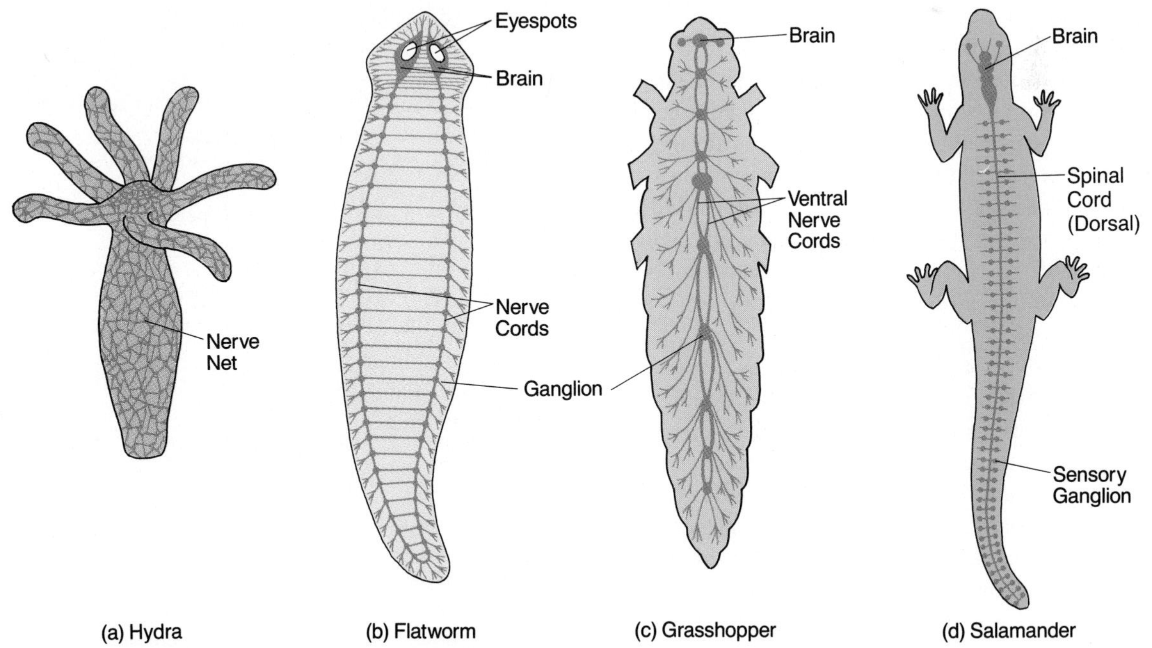

Figure 34-11 REPRESENTATIVE NERVOUS SYSTEMS.
(a) Cnidarians, such as this hydra, may have a simple nerve net that includes a concentration of network neurons around the mouth. (b) Flatworms may have a simple network or two nerve cords plus nerve centers called ganglia and a primitive brain formed of aggregations of neurons. (c) Arthropods, such as a grasshopper, have two ventral, solid nerve cords and brain ganglia above the anterior gut. (d) The salamander, a representative vertebrate, has a dorsal, hollow spinal cord and brain.

near the mouth (Figure 34-11a). These animals have two circuits, one with rapid-conducting neurons that control muscular movements, the other with slow-conducting neurons that control feeding. The synapses in such nerve nets can conduct in either direction and so are nonrectifying—an important survival mechanism in a radially symmetrical organism that receives stimuli from all sides.

Complex Circuits

The next major components of the nervous system to evolve after the nerve net were **reflex arcs** and nervous control centers. Reflex arcs begin with sensory input. This may occur in a sensory receptor cell, which receives light, pressure, heat, and so on, and alters that physical signal to an electrochemical form that activates a sensory neuron. Alternatively, the sensory neuron itself may function both as receptor (as for odor, in the human nose) and as neuron to transmit information on the first leg of the reflex arc. The simplest reflex arcs are two-neuron circuits linking a sensory neuron directly to a motor neuron. This type of circuit is quite rare, although some are still present in primates (e.g., the knee-jerk reflex in humans). Most reflex arcs have one or more interneurons that relay sensory input from sensory neurons to motor neurons and stimulate the motor response. Besides relaying information, interneurons may process it—and that is the basis for all complex animal behavior. Thus, interneurons may be arranged in circuits of varying complexity that converge, diverge, or feed back on themselves. The more interneurons present in a circuit, the greater the opportunity for increasingly complex information processing and output. In the most complex circuits, multiple excitatory and inhibitory outputs are generated, and behavior can no longer be referred to as a reflex, since processes such as learning may be involved (see Chapter 36). The evolution of the brain and of complex behavior among animals is the evolution of interneuron properties and circuitry.

Aggregations of the cell bodies of neurons responsible for coordinating functions or behavior in bilaterally symmetrical animals are called **ganglia** (singular, *ganglion*). Ganglia vary greatly in size and complexity; *cerebral ganglia*, or brains, are the most complex ganglia of all. In general, cerebral ganglia are composed primarily of interneurons, whereas ganglia localized in the body segments may contain higher proportions of motor neurons.

The centralization of the nervous system in bilaterally symmetrical animals is apparent from the ganglia as well as from formation of one or more nerve cords. Seen even in flatworms and ribbon worms (Figure 34-11b), central nerve cords become prominent features in relatively complex organisms, such as the protostomes (annelids, arthropods, mollusks; Figure 34-11c). In a grasshopper, for instance, ganglia are located along the animal's ventrally placed nerve cord, often one in each segment of the body. Usually, these ganglia control local activity in the body segment, such as leg or wing movements. As Chapter 24 explained, chordates also possess a single longitudinal nerve cord, but this one develops dorsally and does not possess segmental ganglia along its length (Figure 34-11d).

The Vertebrate Nervous System

Vertebrates' complex nervous systems underlie their courtship, group hunting, and other complex behaviors. The vertebrate nervous system is divided into two major parts: the central nervous system and the peripheral nervous system (Figure 34-12). The brain and spinal cord form the **central nervous system (CNS),** the integration and control system that ultimately regulates nearly all functions of the body. The **peripheral nervous system (PNS)** innervates all parts of the body and carries information to and from the central nervous system.

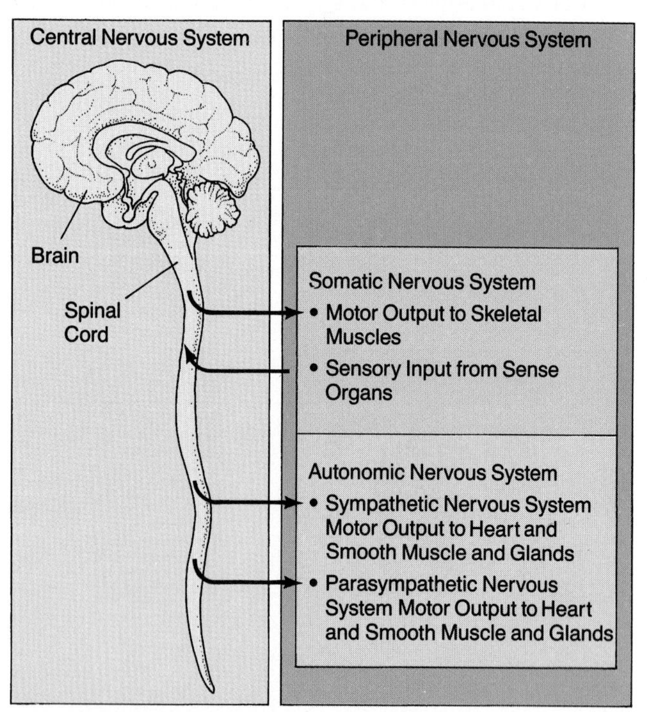

Figure 34-12 SIMPLIFIED SCHEME OF THE VERTEBRATE NERVOUS SYSTEM.
The nervous system consists of the central nervous system (CNS), which is composed of the brain and spinal cord, and the peripheral nervous system (PNS). The PNS includes the somatic nervous system, serving the body musculature, in particular, and the autonomic nervous system, which has parasympathetic and sympathetic divisions that serve most body organs in an antagonistic fashion.

The vertebrate's brain is the primary integration and control organ of the body. It receives information in the form of nerve impulses from major sense organs in the head as well as from other body regions; the latter information is relayed through the spinal cord. The brain, in turn, communicates with the rest of the body (1) via the cranial nerves, which exit the brain through openings in the skull, (2) again by relaying information through the spinal cord to the PNS, and (3) by controlling various hormone-secreting organs, such as the pituitary gland. The brain's millions or billions of neuronal cell bodies tend to be located near the organ's surface, and as a result, the exterior of the brain appears gray and is called *gray matter*. Tracts of myelinated axons tend to be found centrally, and because the lipids in myelin appear white, the interior brain region is called *white matter*.

The spinal cord, extending posteriorly from the brain, is hollow and filled with cerebrospinal fluid, which also fills the brain's cavities. In the simplest sense, the spinal cord can be thought of as a relay network between the brain and the motor and sensory nerves running to and from the body posterior to the head. But, in fact, the spinal cord is an integrating switchboard capable of coordinating such complicated activities as walking and running. A cross section of the spinal cord shows the centrally located gray matter, made up of the cell bodies of neurons surrounded by white matter composed of myelinated axons (Figure 34-13). Like the brain, the spinal cord is composed mostly of information-processing interneurons. The axons carry information to the brain or to neurons extending out from the spinal cord to body tissues. Motor neurons are another important neuron type in the spinal cord; they control much of the body musculature (see Chapter 37).

The second major division of the vertebrate nervous system, the peripheral nervous system, is made up of receptor cells; sensory neurons, which carry sensory information inward to the central nervous system; and motor neurons, which carry control signals outward to the muscles and organs in various portions of the body, such as the fingers or the stomach. The PNS includes the major sense organs as well as paired *sensory ganglia* (also called dorsal root ganglia) arrayed on each segment of the bony vertebral column (see Figure 34-13). Sensory neurons of the trunk have their cell bodies located in the sensory ganglia and have two extensions. The dendrite extends from the cell body to the periphery, where it may synapse with a sensory receptor cell or, depending on its particular set of functions, may itself monitor some sensations; thus, heat, cold, pressure, pain, and so on, are reported via volleys of impulses moving toward the spinal cord over the sensory axon. The axon of each sensory ganglion neuron extends into the spinal cord, where it synapses with interneurons that feed the information up, down, or across the spinal cord or to effector cells.

Functionally, the peripheral nervous system is divided into the somatic nervous system and the autonomic nervous system (see Figure 34-12). The **somatic nervous system** includes sensory and motor parts that control the muscles of the body that we think of as being

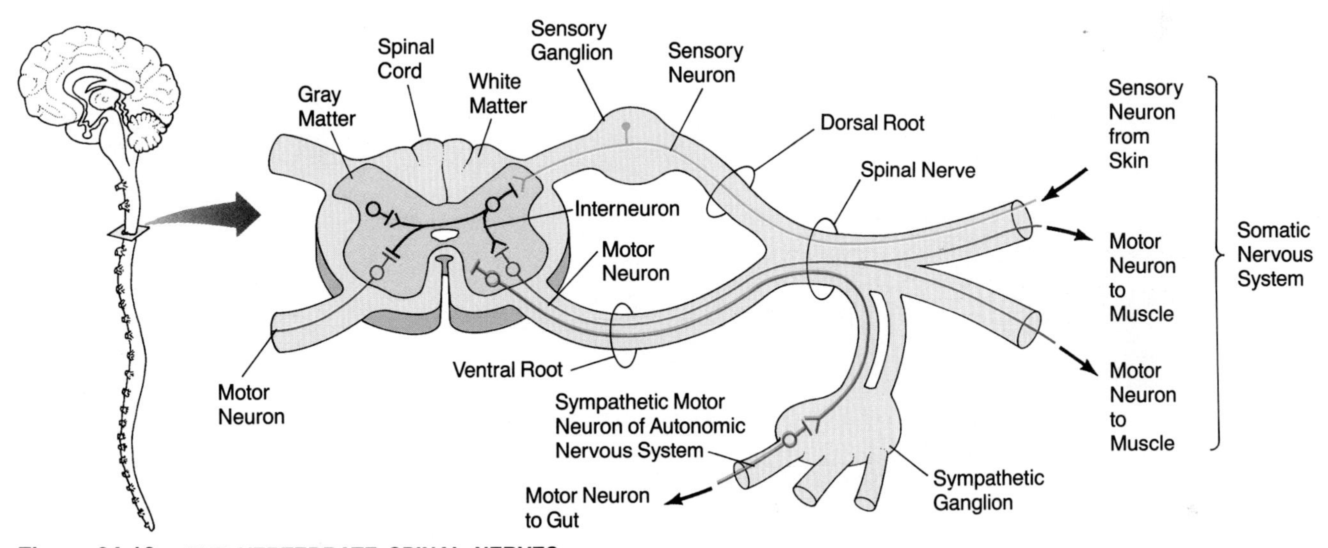

Figure 34-13 THE VERTEBRATE SPINAL NERVES.
This cross section of the vertebrate spinal cord reveals the relationship between white and gray matter. Sensory neuronal cell bodies are in the sensory ganglion located on the dorsal root of the spinal nerve. Motor neurons in the cord send axons over the ventral root and then out the spinal nerve to body muscles. Other ventral root axons go to the sympathetic ganglia, where they drive sympathetic neurons that innervate a variety of organs. Note the interneurons in the cord; one synapses with a motor neuron on the right and also sends a collateral axon across to the left side, where a synapse with a different neuron is formed.

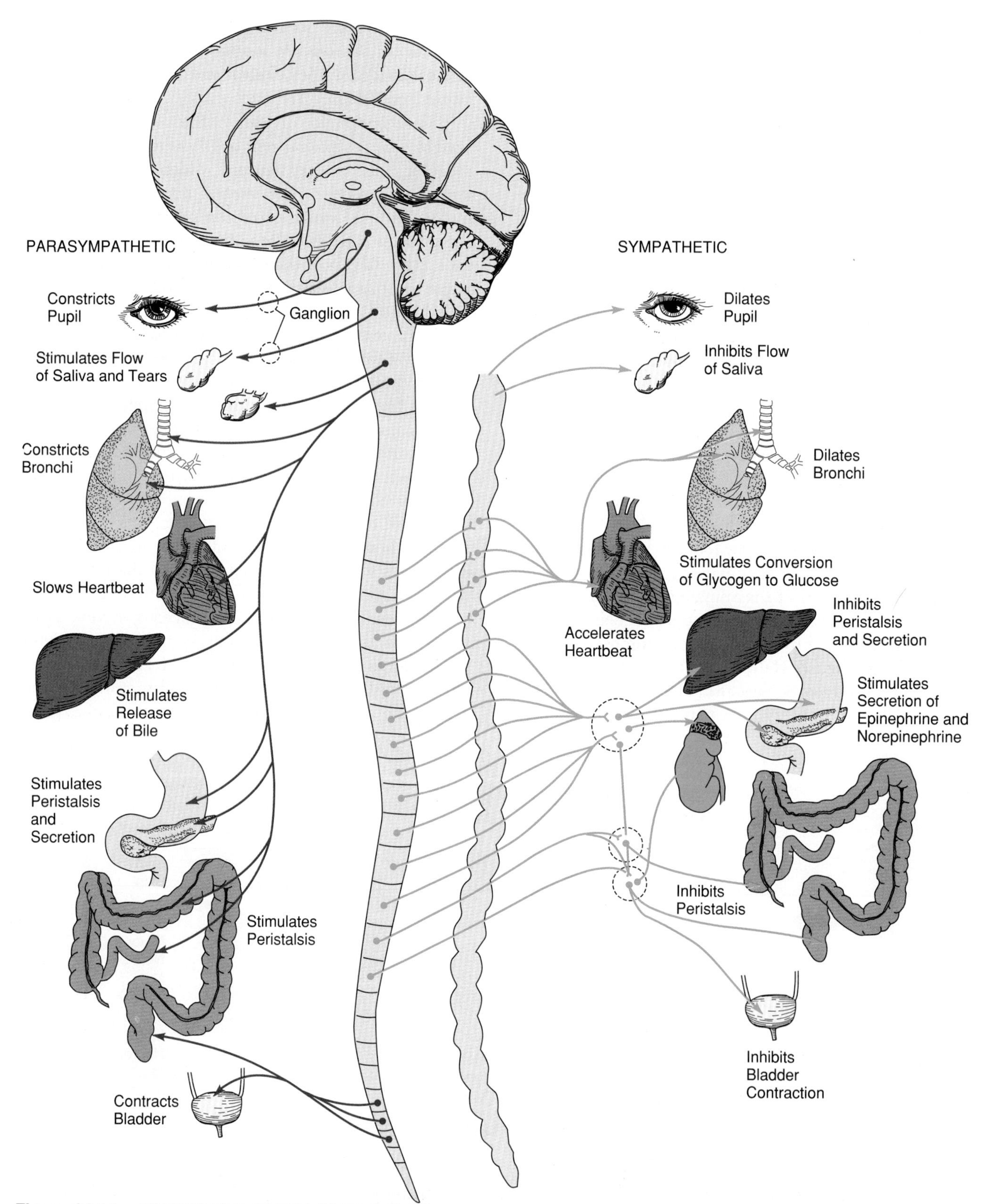

Figure 34-14 INNERVATION OF THE BODY ORGANS.
The autonomic nervous system consists of two sets of motor systems that generally work antagonistically. Each organ receives parasympathetic and sympathetic innervation. In general, sympathetic nerves are stimulatory and parasympathetic ones inhibitory on the organs. For convenience, only one sympathetic chain and one set of parasympathetic nerves are shown; in fact, both systems are paired, occurring on both sides of the body.

voluntary, including those of the arms, legs, neck, throat, lips, tongue, and eyelids. The **autonomic nervous system** is solely a motor system and acts generally without a human's awareness to control physiological functions such as heart rate, digestion, and excretion. The organs affected by the autonomic nervous system include glands (e.g., salivary glands and pancreas) and the smooth muscles of the blood vessels. Using the techniques of biofeedback, a person can learn to influence certain autonomic functions voluntarily.

The autonomic nervous system, in turn, has two divisions: the sympathetic and the parasympathetic (Figure 34-14). The two divisions often act antagonistically. In many cases, the **sympathetic nervous system** stimulates cell or organ function, whereas the **parasympathetic nervous system** inhibits it. However, there are many exceptions, especially among the parasympathetic neurons that stimulate such processes as secretion. The two systems, in general, employ different neurotransmitters. Sympathetic neurons usually secrete norepinephrine, which tends to speed up physiological processes; parasympathetic neurons normally secrete acetylcholine, which tends to slow down the processes. For example, sympathetic impulses to the heart are excitatory, speeding the heartbeat and increasing the volume of blood pumped, while inhibitory parasympathetic impulses act to slow down the heart rate and lower its output.

In vertebrates, pairs of spinal nerves extend laterally from the spinal cord at the level of each vertebra (see Figure 34-14). Each nerve contains sensory and motor fibers (another term for axons) of the somatic nervous system, as well as motor fibers of the sympathetic nervous system. An animal that suffers damage to the spinal cord nerves will often lose sensation or mobility in some part of its body. Researchers, however, recently learned to encourage the regrowth of neurons into the spinal cord. In a rat with a damaged spinal nerve to the foot, they implanted tiny paper bridges coated with astrocytes, a type of fetal nerve cell, and the adult's neurons grew to reconnect with the spinal cord. Perhaps a similar technique will one day help people with spinal injuries.

In the head, 12 pairs of cranial nerves carry sensory and/or motor axons between the brain and various organs. Thus, the olfactory organs, eyes, and ears send information to the brain via cranial nerves. Alternatively, parasympathetic axons that run in other cranial nerves serve the heart, lungs, and other organs. Whereas sympathetic ganglia are located far from the organs they serve, parasympathetic ganglia of the cranial nerves are usually found near the organs they control.

We have encountered the functions of the various parts of the vertebrate central and peripheral nervous systems many times in the preceding chapters on physiology. Later chapters on the endocrine system, sensory system, effector organs, brain, and behavior will complete the picture.

LOOKING AHEAD

Nerves alone do not control an animal's body: Our next chapter explains how the endocrine system works with the nervous system to coordinate the full range of physiological functions and behavioral responses. Moreover, Chapter 35 shows how the endocrine and nervous systems are really variations on a single theme. Both utilize chemical signals, receptors, and transduction to generate specific behavioral responses.

SUMMARY

1. The nervous system works rapidly by transmitting nerve impulses, but neurons also secrete slower-acting hormonelike compounds that modulate nervous system function.

2. The basic unit of all nervous systems is the *neuron*. Neurons are specialized cells that receive, process, and transmit information. Most neurons have four types of structures: *dendrites*, the cell body, *axons*, and *synaptic terminals*. Anatomical *nerves* are bundles of many axons.

3. There are three main types of neurons: *receptor (sensory) neurons*, which are special energy transducers; *effector (motor) neurons*, which transmit messages to muscles and glands; and *interneurons*, which are information processors between sensory and effector neurons.

4. Most neurons transmit signals called *action potentials (nerve impulses)*. To evoke an action potential, a stimulus must depolarize the cell beyond a certain *threshold* level above the cell's *resting potential*. Action potentials depend on *voltage-gated channels* in the neuron's membrane that are sites of ion movement.

5. During an action potential, channels open and Na^+ flows into the cell, causing depolarization. The channels then slam shut and are in an inactive state. The action potential spike then falls as K^+ moves out of the cell at that site. Soon, the channels return to the closed-but-active state, enabling them to respond to the next depolarization event.

6. The propagation of action potentials along an axon involves *local circuit currents* that open new voltage-gated Na^+ channels. The speed with which an axon can propagate a nerve impulse depends on the axon's diameter, on temperature, and on the presence or absence of insulating material. The *refractory period* due to the temporary inactivity of closed

Na^+ channels ensures that the nerve impulse does not move back up the axon. Axonal currents jump from one *node of Ranvier* to the next in nerve cells wrapped in *myelin sheaths.* Such *saltatory propagation* is much faster than the propagation in unmyelinated fibers of the same diameter.

7. The site where messages are transmitted from one neuron to another cell is called a *synapse.* Chemicals called *neurotransmitters* are stored in the synaptic vesicles in the axon's presynaptic terminals; arrival of an action potential at the terminal triggers release of a neurotransmitter into the synaptic cleft between cells. At *excitatory synapses,* this initiates a small depolarization event (the *PSP,* or *postsynaptic potential*). At *inhibitory synapses,* this increases polarization (a negative PSP).

8. Spatial and temporal *summation* of numerous excitatory and inhibitory postsynaptic potentials (PSPs) results in a *grand postsynaptic potential (GPSP),* which, if above the threshold for the neuron, causes it to generate impulses.

9. The simplest sort of circuit in nervous systems is the *reflex arc.* Aggregations of neuronal cell bodies are called *ganglia.*

10. The vertebrate nervous system consists of the *central nervous system (CNS)*—the brain and spinal cord—and the *peripheral nervous system (PNS),* which innervates all parts of the body and transmits information to and from the CNS. The PNS is divided into the *somatic* and *autonomic nervous systems.* The somatic system includes sensory and motor components. The autonomic system includes only effector neurons that control the body's involuntary physiological functioning. It has two subdivisions that work antagonistically: the *sympathetic* and the *parasympathetic.*

KEY TERMS

action potential (nerve impulse)
autonomic nervous system
axon
central nervous system (CNS)
chemical-gated channel
dendrite
depolarization
effector (motor) neuron
excitatory synapse
ganglion
glial cell
grand postsynaptic potential (GPSP)
hyperpolarization
inhibitory synapse
interneuron
local circuit current
myelin sheath
nerve
nerve net
neuron
neurotransmitter
node of Ranvier
oligodendrocyte
parasympathetic nervous system
peripheral nervous system (PNS)
polarization
postsynaptic potential (PSP)
receptor (sensory) neuron
reflex arc
refractory period
resting potential
saltatory propagation
Schwann cell
somatic nervous system
summation
sympathetic nervous system
synapse
synaptic terminal
synaptic transmission
threshold
voltage-gated channel

QUESTIONS

1. Which two control systems integrate the functions of multicellular animals? Why has the boundary between these two systems recently become blurred?

2. In what way does a nerve cell act as a transducer of energy?

3. Name and sketch a typical nerve cell, and label its parts.

4. What are the three basic functional types of nerve cells? Briefly distinguish among them.

5. Define the resting potential of a nerve cell in terms of ions and electrical charges; explain how this potential is maintained.

6. Describe an action potential (nerve impulse), and tell how it is propagated along a nerve cell. How do the states of Na^+ channels help explain the action potential?

7. What variables influence the speed of conduction of the action potential?

8. Identify the two major types of synapses, and describe briefly how they function. How are excitatory and inhibitory PSPs related to a GPSP, and that, in turn, to initiation of an action potential by a postsynaptic neuron?

9. Identify and describe the major parts of a vertebrate's central nervous system (CNS).

10. Identify and describe briefly the two functional divisions of the peripheral nervous system (PNS).

ESSAY QUESTION

1. Discuss the role of each of the following in the propagation of action potentials along axons and across synapses: Na^+, K^+, Cl^-, Ca^{2+}, myelin, acetylcholine, ATP, Na·K·ATPase pump.

For additional readings related to topics in this chapter, see Appendix C.

35 Hormonal Controls

The functioning and the survival of complex organisms would hardly be conceivable, were it not for the existence of these regulatory chemical interactions between cells, tissues, and organs.

Jacques Monod, *Chance and Necessity* (1966)

In 1771, an English surgeon named John Hunter inserted a rooster's testis into a hen's abdomen, then recorded changes in the hen's appearance as it gradually developed large roosterlike tail feathers and many other male secondary sex characteristics.

John Hunter's research was an early exploration of the **endocrine system,** a set of organs, tissues, and cells that secrete substances called *hormones.* Not for a century, however, did physiologists actually define hormones as chemicals that are secreted into the blood by specific endocrine glands and that travel to target cells in other parts of the body, where they change and regulate the

Sex hormones at work. Both the bright red throat skin and the display behavior of this male great frigate bird *(Fregata minor)* on a Galápagos island result from the action of androgens, male sex hormones.

activities or growth patterns of specific cells and thereby help coordinate an organism's physiology.

Each hormone plays a different part in an animal's biology: Some hormones trigger and regulate growth; others control sexual maturation or reproductive activity; still others control the secretion of digestive juices after a meal, maintain the chemical balance in blood, or alter an animal's behavior with the changing seasons. The endocrine system and the nervous system together integrate and control all of an animal's fundamental life processes.

This chapter discusses:

- The characteristics of vertebrate hormones
- How hormones exert their effects on target cells
- How hormones regulate the life cycles of invertebrates
- The major organs and hormones of the vertebrate endocrine system
- How hormones regulate blood sugar levels, daily (circadian) rhythms, and other life processes
- How hormones control metamorphosis in a developing frog
- The frontiers of hormone research, including some newly discovered hormones that serve as powerful chemical messengers in the brain

BASIC CHARACTERISTICS OF HORMONES

Every vertebrate possesses two types of glands. *Exocrine glands* secrete substances into ducts that in turn empty into body cavities or onto body surfaces (Figure 35-1a); sweat glands, mammary glands, and salivary

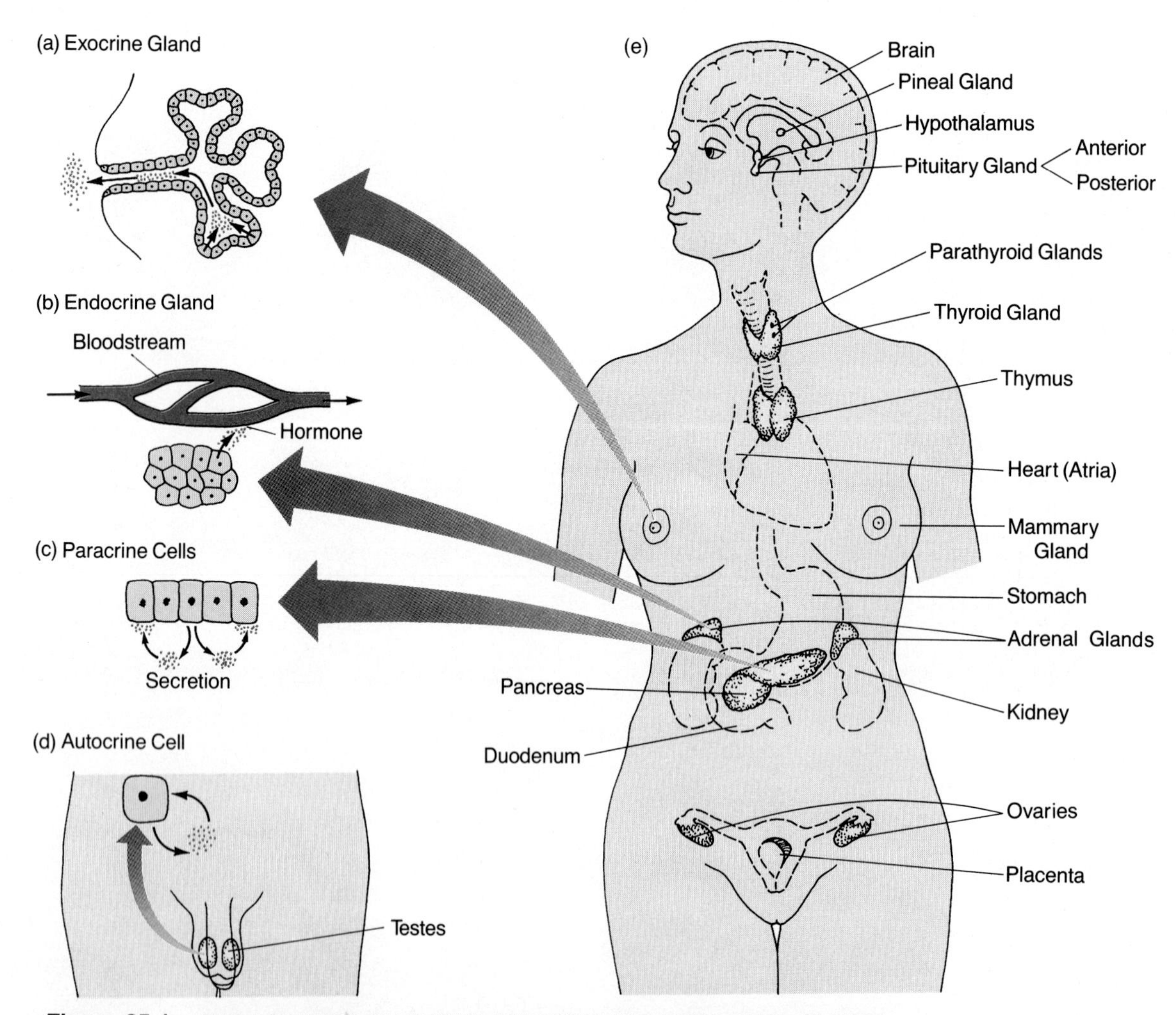

Figure 35-1 THE LOCATION OF HUMAN EXOCRINE AND ENDOCRINE GLANDS. (a) Exocrine glands secrete milk, sweat, or other substances into ducts. (b) Endocrine glands secrete hormones into the tissue space, and from there into the bloodstream. (c) Paracrine cells secrete hormones that affect adjacent cells. (d) Autocrine cells regulate their own activities via hormonal secretions. The major vertebrate endocrine glands in (e) the adult female and (d) male.

glands are examples. *Endocrine glands* have no ducts and instead secrete their chemical products—hormones—directly into the tissue space next to each endocrine cell (Figure 35-1b). Some of these hormones may then diffuse into the bloodstream and be carried throughout the body. Figure 35-1e shows the major vertebrate endocrine glands. Vertebrates also have *paracrine* cells; the hormones they secrete diffuse short distances, affect target cells in adjacent tissues, and do not enter the bloodsteam (Figure 35-1c); examples include certain cells of the intestinal tract that liberate a substance that causes adjacent cells to secrete enzymes. Finally, the *autocrine* cells secrete regulatory substances that act on the secreting cells themselves (see Figure 35-1d).

The dozens of hormones fall into three broad chemical categories. Many hormones are *proteins*—either short-chain polypeptides or more complex proteins. This group includes insulin and oxytocin (shown in Figure 35-2a and discussed later). Others are *amines*—metabolic derivatives of amino acids (Figure 35-2b). The amine hormones include *epinephrine* (adrenaline), which increases pulse rate and affects other physiological changes when a person experiences rage or fear. The third group of hormones are the *steroids*—complex molecules derived from cholesterol—and they include the sex hormones estradiol-17β and testosterone (Figure 35-2c).

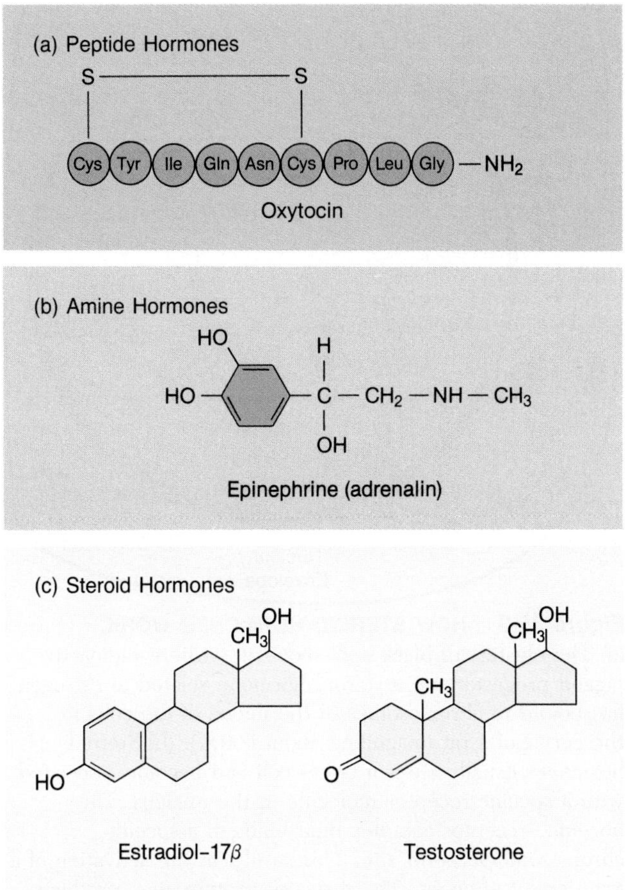

Regardless of hormone type, a few principles apply to all hormonal activity:

1. A given hormone affects only *target cells*. Target cells are responsive to hormones because they possess *receptor molecules* made of protein in their membranes, cytoplasm, or nucleus, as well as *transduction machinery* that the *hormone-receptor complex* activates. This machinery then carries out the cellular response. Cells that lack receptors for a particular hormone do not respond to that hormone.
2. Each target cell in the body is regulated by only certain hormonal signals because it manufactures one set of hormone receptor molecules, but not others.
3. Different cells may respond in different ways to the same hormone. Such diversity in target-tissue responses is possible because the transduction machinery of different target cell types "reads" the hormone signal in different ways. Characteristics of target cells, rather than of hormones, therefore define the specificity of hormonal action.
4. Some hormones, such as those that maintain homeostasis in body fluids, are present much of the time, whereas others, such as the "fight-or-flight" hormone epinephrine, appear only when needed.
5. The amount of a circulating hormone is usually governed by negative feedback control; a falling off of hormone levels in the blood stimulates additional secretion, and a rising of those levels inhibits its further secretion.
6. Once hormones bind to receptor molecules, they are usually broken down rapidly. Without such a finely tuned recovery mechanism, target cells could not be sensitive to changing levels of the hormones that regulate their activities.

With these principles in mind, let us see, in general, how cells respond to a regulatory signal.

(a) Peptide Hormones

Cys Tyr Ile Gln Asn Cys Pro Leu Gly —NH_2

Oxytocin

(b) Amine Hormones

Epinephrine (adrenalin)

(c) Steroid Hormones

Estradiol-17β Testosterone

Figure 35-2 REPRESENTATIVE HORMONES.
Hormones usually fall into three categories: (a) proteins and small polypeptides, (b) amines, and (c) steroids.

CELLULAR MECHANISMS OF HORMONAL CONTROL

Once a receptor molecule on the cell surface or within the nucleus of the target cell has bound to a hormone, the hormone's chemical signal is translated into new or different rates of cellular activity through one of three mechanisms:

1. A hormone can allow other substances to enter or leave the target cell more quickly.
2. A hormone can stimulate a target cell to synthesize enzymes, proteins, or other substances.
3. A hormone can prompt the target cell's machinery to activate or suppress existing cellular enzymes.

Each of these effects is accomplished by an intricate series of chemical events at the cell surface, within the cytoplasm, and sometimes in the nucleus. The precise sequence depends on the type of hormone.

The Cellular Mechanisms of Steroid Action

Steroid hormones must gain entry into a target cell in order to act, and the hydrophobic character of these molecules allows them to diffuse through the target cell's plasma membrane. Once inside, the steroid binds to receptor molecules located in the cytoplasm or (usually) the nucleus—receptors that are specific to the cell type and steroid. Cells of the uterus, for example, contain receptors for estradiol-17β, one of the estrogen family of hormones (Figure 35-3b, step 1). When estradiol enters the cell, it diffuses into the nucleus and forms a hormone-receptor complex (step 2), which is then chemically activated (step 3). Biologists think that the hormone-receptor complex then attaches to acceptor sites on the chromosomes (step 4). The binding of the complex to the chromosomes activates specific genes to transcribe their messenger RNAs (step 5); subsequently, these mRNAs are translated into protein (step 6). Thus, a target cell's specific response to steroid hormones is the production of specific proteins, which may act in the cell or be excreted (steps 7 and 8). The physiological effects of a steroid injection are usually seen after 6 to 24 hours. The binding to chromosomes occurs within about 5 minutes, and new mRNAs can be detected within 15 minutes, but it takes several hours for concentrations of new proteins to build up.

The exquisite specificity of steroid hormone action results from the two binding steps: first, the binding of hormone to receptor, and second, the binding of hormone-receptor complex to acceptor sites on the chromosome. Researchers are currently focusing on the proteins that bind to acceptor sites. Steroid receptors appear to have DNA binding domains like the polypeptide loops called zinc fingers in transcription factors (see Chapter 17), and the acceptor sites on the chromosomes can be short stretches of DNA with specific base sequences near hormone-activated genes.

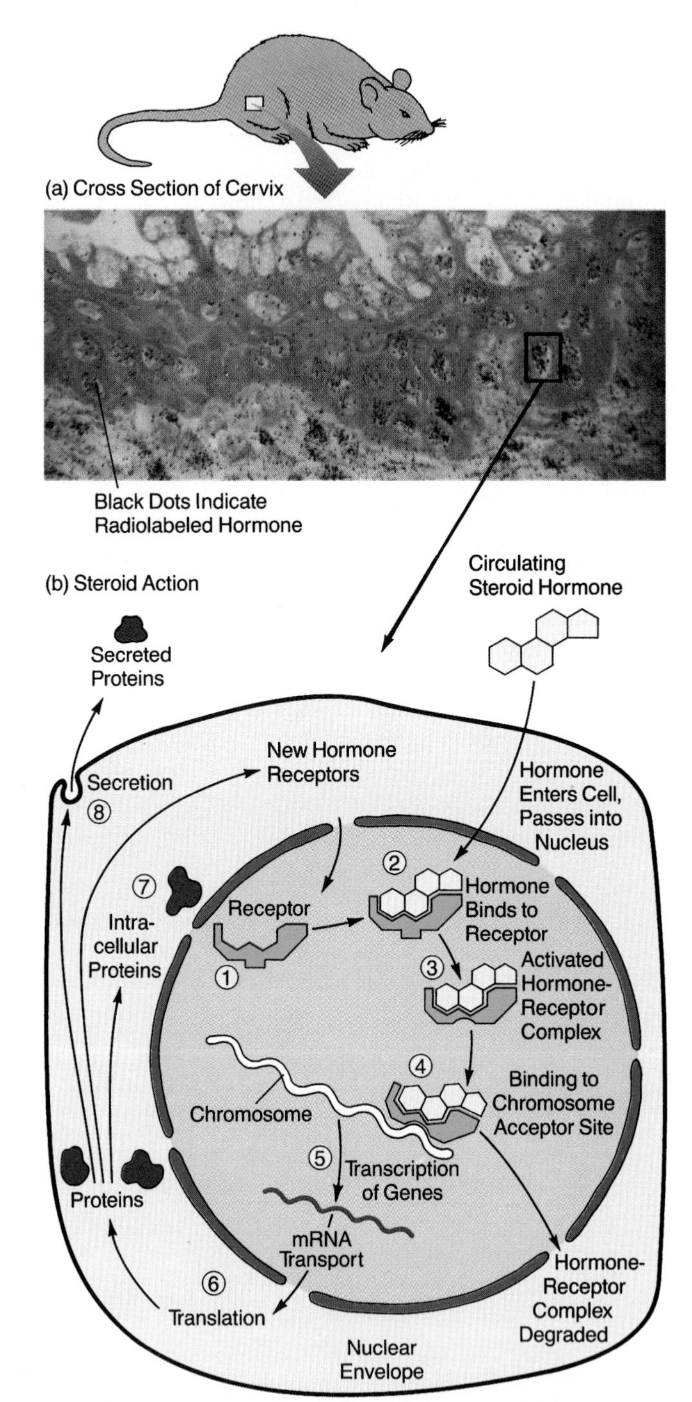

Figure 35-3 HOW STEROID HORMONES WORK.
(a) The clusters of black dots show sites where radioactively tagged progesterone, a steroid hormone related to estrogen, has bound to chromosomes in the nuclei of the cells in the cervix of a rat (magnified about 600×). (b) Steroid hormones usually enter a target cell and become associated with a specific receptor molecule in the nucleus. The hormone-receptor complex then binds to a specific chromosome acceptor site. This results in the activation of a gene or set of genes. The resultant protein gene products can be secreted into the circulation, can lead to cellular growth or new cellular activities, or can serve as receptor molecules for the steroid.

ANABOLIC STEROIDS AND THE INSTANT PHYSIQUE

For 72 hours, Canadian track star Ben Johnson was officially the fastest man in the world. During the 1988 summer Olympics, Johnson ran the 100-meter dash in 9.79 seconds (Figure A), and for three days, millions of sports fans celebrated his world's record and his gold medal. But then the results came in from a routine blood test taken right after the race: Johnson's blood contained an anabolic steroid called stanozolol. Olympic officials rescinded his gold medal, and awarded it to second place American runner Carl Lewis.

Stanozolol is a synthetic form of testosterone that, like the natural male hormone, promotes anabolism or the building of tissue. Weight lifters began using anabolic steroids in the 1950s, and now the practice is widespread. Recent studies suggest that 7 to 8 percent of professional football players use steroids, as do at least 5 percent of college athletes, and 6.6 percent or more of high school boys.

For years, physicians denied the very real anabolic effects of the drugs, even though males could see them firsthand in the locker room and on the playing field. This may help explain why so many men and boys have ignored warnings about the equally real risks: In addition to promoting muscle development, artificial androgens also suppress natural testosterone. This suppression can lead to shrunken testes and enlarged breasts. But paradoxically, the drugs can also confer the masculinizing effects of testosterone in exaggerated proportions. These can include heavy hair growth, acne, and premature baldness, as well as damage to the kidneys, liver, heart, and to psychological symptoms such as depression, anxiety, hallucinations, or paranoia. At the very least, a man or boy can become psychologically dependent on anabolic steroids and be unable to accept a smaller physique once he stops using them. Clearly, the benefits of the "instant body" are short-lived, while the risks can be long term, even deadly.

Figure A
Winner and loser: The heavily muscled Ben Johnson sprints to his temporary world record.

In mammals, steroids such as estrogen and progesterone also function in the brain, where they affect both development and function. Besides regulating genes, the steroids may alter the polarization-repolarization activity of neurons, changing the way the neuronal circuits function and thus altering sexual behavior. Athletes have discovered that a class of synthetic testosterones called anabolic steroids can build up the muscles, but can also have numerous harmful effects on brain, heart, and other organs, as the box above explains.

Nonsteroids and the Cellular Mechanisms of Second Messengers

At least one nonsteroid hormone *(thyroxine,* produced by the thyroid gland) works inside target cells, much as steroids do. For most nonsteroids, however, the signal is relayed into the target cell by chemical go-betweens, so-called *second messengers.*

Cyclic Nucleotides: The First Set of Second Messengers Discovered

In the early 1960s, Earl W. Sutherland, a scientist at Western Reserve University, won a Nobel Prize for discovering the second-messenger phenomenon. By studying the way the hormone epinephrine (Figure 35-2b) stimulates liver cells to convert glycogen to glucose, he found that epinephrine can transmit its regulatory information *without entering the target cell.*

When epinephrine binds to a receptor molecule embedded in the plasma membrane (Figure 35-4a, step 1), the receptor changes shape and binds the so-called G-protein (step 2). G-protein then binds to the nucleotide GTP (guanosine triphosphate; step 3), which allows the G-protein–GTP complex to activate *adenyl cyclase,* an enzyme associated with the membrane (step 4). This enzyme, in turn, catalyzes increased production of *cyclic adenosine monophosphate*—or **cyclic AMP (cAMP)**—from ATP that is normally found in cells in low concentrations (step 5). Only a small number of bound

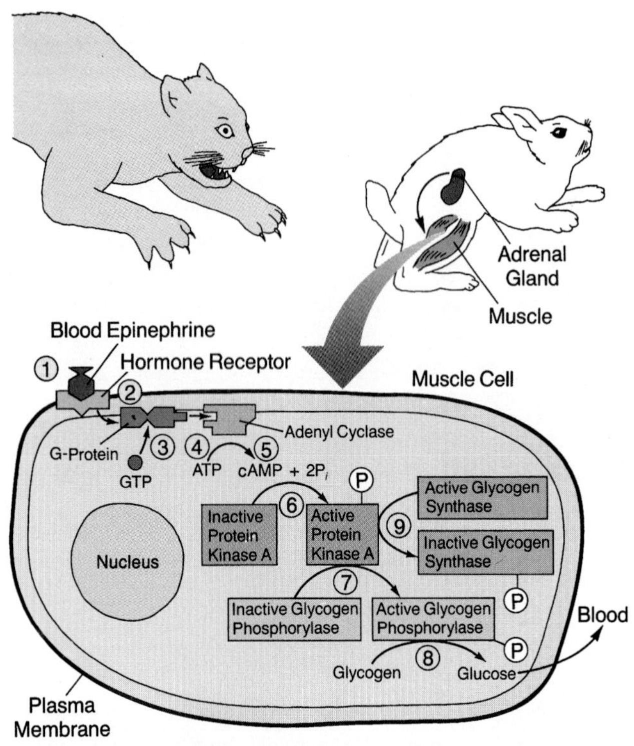

Figure 35-4 cAMP: HORMONAL ACTION BY MEANS OF A SECOND MESSENGER.
The liberation of glucose in a rabbit's muscle cell in response to epinephrine during the fight-or-flight response involves a series of reactions that are initiated by the cAMP system. Note that the activated protein kinase A activates one enzyme (phosphorylase) and inactivates another (glycogen synthetase).

hormone molecules are needed to cause catalytic enzymes to form a large amount of cAMP; thus, cyclic AMP is an *amplified* form of the hormone's message. Even a tiny amount of hormone can have major effects because of this cAMP amplification. Many hydrophilic hormones—such as proteins, polypeptides, and some amines—act by stimulating cAMP production.

Once produced, cAMP acts as an intermediary, or **second messenger,** setting in motion a cascade of chemical interactions. First, cAMP combines with an inactive form of *protein kinase A*, a cytoplasmic enzyme (step 6), thus activating protein kinase A so that it catalyzes the addition of phosphate groups to other cellular proteins or enzymes. This phosphorylation may be the final step in the sequence; as it takes place, the phosphorylated proteins or enzymes are either activated or inactivated, and so carry out the hormone-specific responses. In some cases, however, the phosphorylation activates still another kinase that then activates another enzyme in the chain. This adds one more amplification step. An example of such an amplification cascade occurs after epinephrine binds to skeletal muscle cells; cyclic AMP is produced and protein kinase A is activated (step 6). Protein kinase A phosphorylates glycogen phosphorylase (step 7), activating that enzyme to break down glycogen into glucose (step 8). Protein kinase A also phosphorylates glycogen synthetase, inactivating that enzyme and thereby preventing it from synthesizing glycogen from glucose (step 9). Both processes help to elevate the levels of free glucose in the muscle cells. Calculations of one such sequence indicate that an amplification of up to 10^8 may occur; in other words, the binding of even one molecule of epinephrine can produce 100,000,000 molecules of glucose!

After the second messenger has done its job, another enzyme, phosphodiesterase, degrades cAMP to AMP, which is used to resynthesize ATP. The cell thus returns to its original state, in which it can once again be sensitive to new hormone molecules that may bind to the cell surface.

In recent years, researchers have discovered another second-messenger compound, *cyclic guanosine monophosphate*—or **cyclic GMP (cGMP).** In heart cells, cGMP is a second messenger for the neurotransmitter acetylcholine, which slows muscle cell contraction—an action opposite to the speeding of contractions by epinephrine and cAMP. This pairing helps explain the antagonistic actions of parasympathetic and sympathetic nerves on the heart's contractile rate (see Chapter 34).

Inositol Triphosphate and DG: Other Second Messengers

Researchers in the early 1980s discovered a second-messenger system that may be even more widespread than cAMP, since it helps trigger cell responses to a variety of neurotransmitters, hormones, and growth factors. Numerous substances bind to receptor molecules, with the common effect of releasing Ca^{2+} in the cytoplasm, but these ions do not act as second messengers. Instead, that role is filled by **inositol triphosphate,** or $\mathbf{IP_3}$ (inositol is a six-carbon sugar related to glucose).

The plasma membranes of cells contain small amounts of a phospholipid called phosphatidylinositol 4,5-bisphosphate, or PIP_2 (Figure 35-5, step 1). When epinephrine, for example, binds to what is called an alpha receptor in the plasma membrane, PIP_2 is immediately cleaved into IP_3 plus **diacylglycerol,** or **DG** (step 2). Both substances act as second messengers.

IP_3 instantly triggers the release of Ca^{2+} from cellular storage sites (step 3), and those ions act as a third messenger, causing all sorts of responses, including contraction of the actin and myosin systems, polymerization of macromolecules, and activation of enzymes. Meanwhile, DG activates *protein kinase C*, (step 4), which phosphorylates a variety of proteins and activates a different set of responses than those brought about by cAMP and pro-

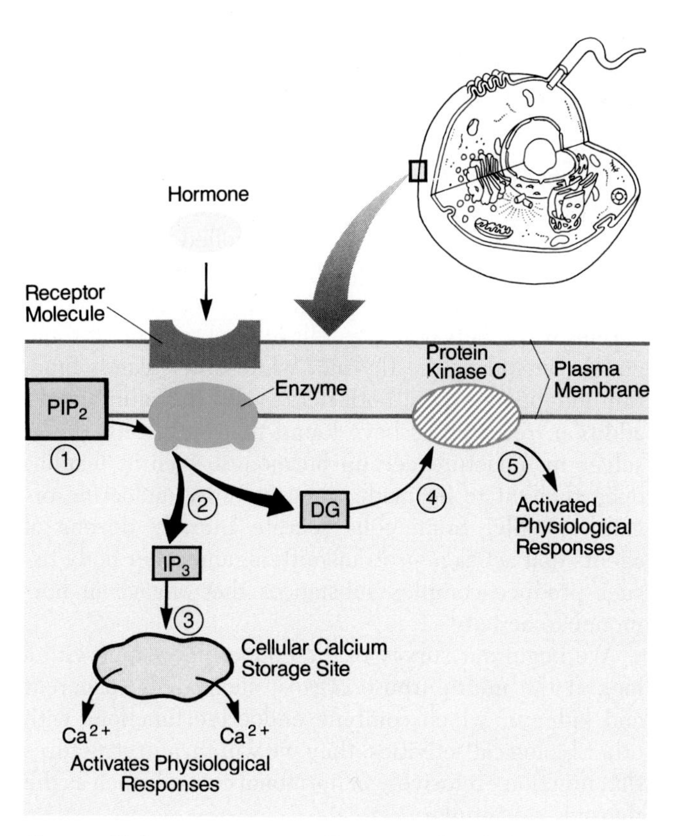

Figure 35-5 IP_3 AND DG: THE NEW SECOND MESSENGERS.
When the hormone or other signaling molecule binds to its receptor molecule, an enzyme cleaves PIP_2 into the second messengers IP_3 and DG. IP_3 liberates Ca^{2+} from storage sites, while DG activates protein kinase C. A cascade of cellular events can follow: division, motility, secretion, and so on.

tein kinase A (step 5). Protein kinase C is activated by insulin secreted by beta cells in the pancreas, aldosterone secreted by the kidney's adrenal cortex, and epinephrine secreted by the kidney's adrenal medulla.

The discovery of IP_3 and DG as second messengers helps explain important events in the activation of sperm and eggs, in the stimulation of cancer cells to divide, and in important aspects of brain function. Together with cAMP and cGMP, they enable biologists to understand the molecular basis of communication between body cells and regulatory systems. To see how hormones regulate an entire process, we now turn to the invertebrates.

ENDOCRINE FUNCTIONS IN INVERTEBRATES

Among the invertebrates, hormones control most aspects of molting, reproduction, and metamorphosis, as well as other life cycle events. As Chapter 16 explained, certain insects develop into caterpillars—larvae that feed, grow, and pass through a series of stages, or instars. Finally, they pupate, enter a cocoon, and undergo the radical developmental changes that produce a moth or butterfly—each transition under hormonal control. Metamorphosis from pupa to adult moth in the silkworm *Hyalophora cecropia* is controlled by *brain hormone* (Figure 35-6). This peptide hormone is produced by neurosecretory cells of the brain. Such cells in invertebrate and vertebrate brains are electrically excitable and can conduct action potentials; yet, the "transmitter" they secrete is a hormone, not a conventional neurotransmitter. When a silkworm secretes brain hormone, the substance sets in motion a series of chemical changes. Brain hormone directly stimulates the *prothoracic gland* to secrete a steroid hormone called **α-ecdysone.** The α-ecdysone is then converted to 20-hydroxyecdysone, which triggers the cellular changes of molting and metamorphosis.

It turns out that ecdysones trigger molting (the change from one instar to the next), but another hormone actually determines what the larva changes into. The *corpora allata* are paired endocrine glands located behind the brain that secrete **juvenile hormone (JH)** during the caterpillar instar stages. A sufficiently high level of circulating JH prevents metamorphosis from beginning as the larva molts from one stage, or instar, to the next in response to ecdysones. As long as the levels of both hormones remain high, larva-to-larva molts will occur (see Figure 35-6). When other metabolic changes cause the JH level to fall below a certain threshold, but ecdysone levels still remain high, the larva pupates. In the pupa, JH secretion stops altogether; but since ecdysone (and therefore 20-hydroxyecdysone) is still abundant owing to the presence of its regulator, brain hormone, the pupa finally develops into an adult. Note that 20-hydroxyecdysone causes the series of molts to occur, whereas the levels of juvenile hormone determine which pattern of development will take place during those molts—to larva (high JH), to pupa (low JH), or to adult (no JH).

All insects produce the ecdysone family of hormones, as do many other arthropods; for example, 20-hydroxyecdysone also stimulates molting in shrimp and other crustaceans. Periodic molting throughout adult life allows the animal to grow during the brief period between the shedding of the old, hard exoskeleton and the hardening of the new, larger one.

Biologists have discovered a variety of hormones among the hundreds of thousands of invertebrate species. Particularly widespread are a number of peptides now known to be secreted by human and other vertebrate brain neurons as neurotransmitter peptides (see Chapter 34). Such chemicals influence the life processes of protozoa, simple invertebrates such as sponges, some plants, and perhaps even some prokaryotes. Clearly, the

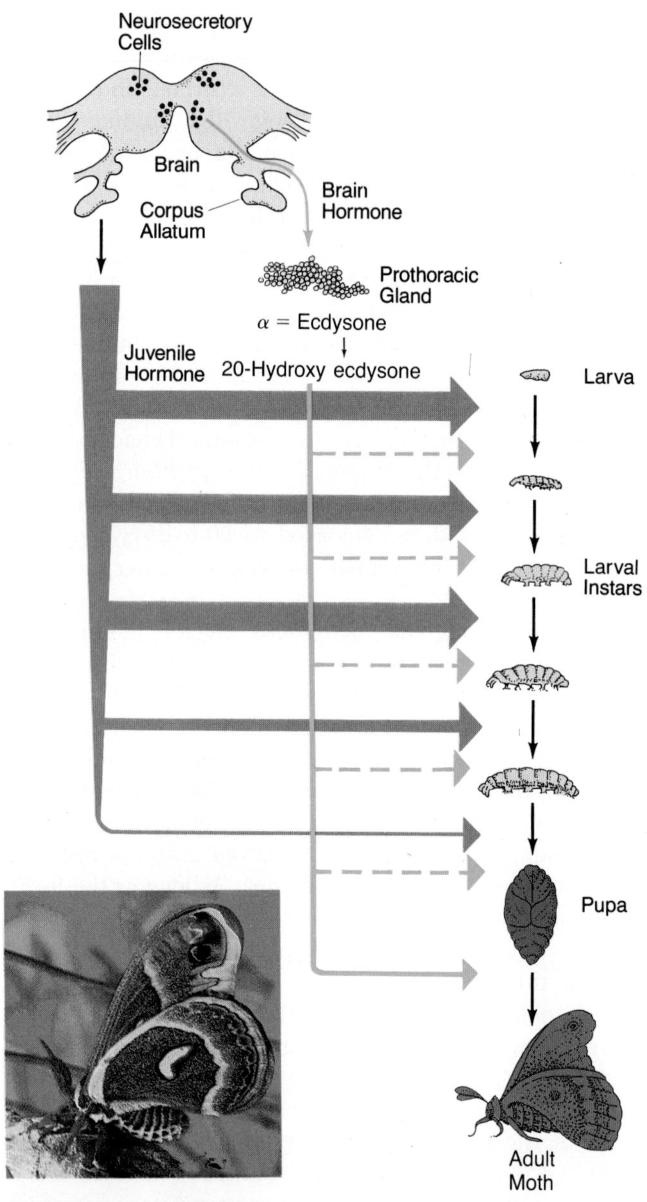

Figure 35-6 HORMONE CONTROL OF INSECT METAMORPHOSIS.
In the final metamorphic event, the pupa is transformed into an adult. This is due to the action of ecdysone, produced by the prothoracic gland in response to brain hormone. The ecdysone is present during much of a larva's life and is responsible for molts—from larva to larva, larva to pupa, and pupa to adult. But it is the level of juvenile hormone that determines what the molting-stage insect develops into. The width of the orange arrows reflects juvenile hormone levels. Note that the amount of juvenile hormone is less over time; thus, the pupa and then the adult develop.

hormones we know best in vertebrates originally evolved as regulatory molecules in much simpler organisms, where their original tasks were no doubt quite different.

THE VERTEBRATE ENDOCRINE SYSTEM

For many years, biologists thought that the endocrine systems of all vertebrates (including humans) consisted of a series of separate glands all controlled by a "master" gland (the pituitary). Today, however, they believe that hormone-producing cells and tissues are controlled in various ways. Sets of nerve cells in the brain direct some glands, including the thyroid, while other glands function independently of both nerves and the pituitary. In addition, researchers have found that brain cells themselves manufacture certain hormones, such as insulin, once thought to be made exclusively by endocrine organs. Finally, brain cells secrete literally dozens of agents that act as neurotransmitters, and other body tissues produce complex substances that engage in hormonelike activity.

We begin our survey of the endocrine system with a look at the multipurpose organs, such as the pancreas and kidneys, which combine endocrine functions with other biological activities; then we will turn to structures that function exclusively in hormonal control, such as the thyroid and pituitary.

The Pancreas

The *pancreas* is an organ near the stomach that secretes digestive enzymes and bicarbonate ions into the intestine and has endocrine functions as well (see Figure 35-1). Scattered throughout the pancreas are spherical clusters of cells and adjacent blood vessels called **islets of Langerhans.** Endocrine cells in these islands produce **insulin** and **glucagon,** hormones that regulate the glucose levels in the blood, and **somatostatin,** a hormone that inhibits the secretion of insulin and glucagon and glucose absorption by the intestine (Figure 35-7).

When an animal eats, it digests and absorbs nutrients, and the level of glucose in its blood rises (Figure 35-7, step 1). Normally, *β cells* in the pancreatic islets of Langerhans respond to an elevated level of blood glucose by secreting insulin (step 2), which has several effects (step 3): Insulin facilitates the passage of glucose into many cells, where it fuels cellular metabolism; it also causes transcription of a number of genes and acts on the liver, where it increases the rate at which glucose is converted into glycogen and the rate at which amino acids are converted into protein. Insulin also stimulates the synthesis of glycogen in skeletal muscles, and causes adipose (fat) tissue to increase fat storage.

Insulin's action causes the glucose level in the blood to fall (step 4), but when it drops below a certain thresh-

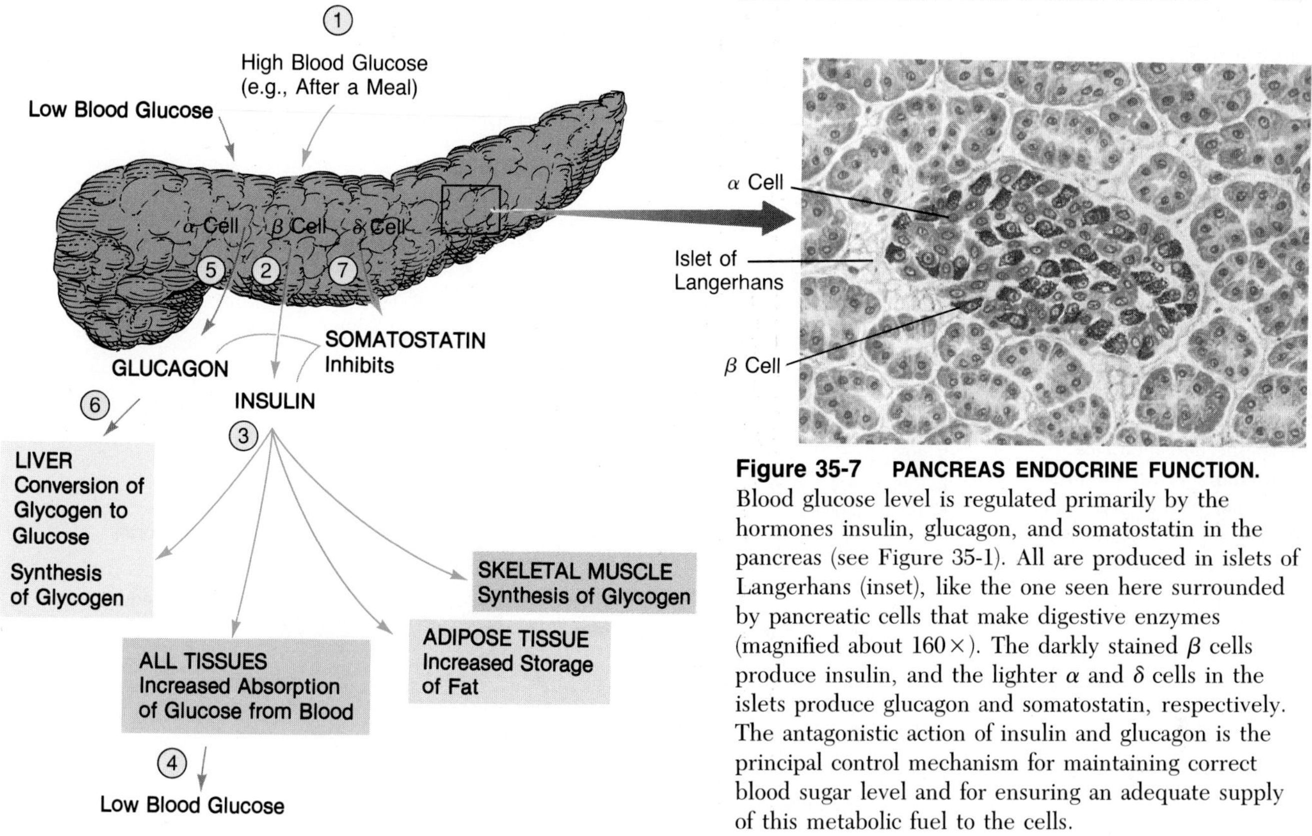

Figure 35-7 PANCREAS ENDOCRINE FUNCTION. Blood glucose level is regulated primarily by the hormones insulin, glucagon, and somatostatin in the pancreas (see Figure 35-1). All are produced in islets of Langerhans (inset), like the one seen here surrounded by pancreatic cells that make digestive enzymes (magnified about 160×). The darkly stained β cells produce insulin, and the lighter α and δ cells in the islets produce glucagon and somatostatin, respectively. The antagonistic action of insulin and glucagon is the principal control mechanism for maintaining correct blood sugar level and for ensuring an adequate supply of this metabolic fuel to the cells.

old, pancreatic *α cells* begin to secrete glucagon (step 5), which accelerates the conversion of liver glycogen to glucose (step 6) via the second messenger cAMP. Insulin and glucagon are thus antagonists, since one causes glycogen synthesis and the other glycogen breakdown. The third pancreatic hormone, somatostatin, is secreted by *δ cells* (step 7). Somatostatin inhibits the secretion of both insulin and glucagon and absorption of glucose across the intestinal wall. Somatostatin is released when glucose or amino acid levels in the blood rise, and by its effects, it lowers those levels.

Both positive and negative feedback loops regulate glucose levels in blood. Signals from the hypothalamus stimulate parasympathetic nerves (see Chapter 34) and increase insulin secretion, while other hypothalamic signals trigger sympathetic neurons, which cause α cells to secrete glucagon and β cells to cease insulin secretion. Finally, both types of nerves may affect the δ cells that secrete somatostatin.

When any of these complex mechanisms goes awry and the pancreas does not secrete sufficient insulin, the level of glucose in the blood and urine rises. The result is *diabetes mellitus*. Diabetes can also result if an enzyme destroys insulin too rapidly or if insulin receptors are abnormally few in number. Diabetics tend to lose body water because of excessive urination, and they may lose weight and tire easily. Human diabetics usually take insulin injections or oral doses of other recently developed drugs that act like insulin, causing the glucose to move into the body cells and thereby reducing the blood glucose levels.

The Kidneys

The kidneys, too, are multipurpose organs: They not only filter the blood and manufacture urine, but also secrete hormones. One of these hormones is **erythropoietin,** a protein that stimulates the production of red blood cells in bone marrow. If a mammal loses more than a small amount of blood, the kidneys produce more erythropoietin, and the increasing levels in the bloodstream trigger heightened production of red blood cells for the next few days.

The kidneys also contain enzymes that convert vitamin D_3 in the blood into a hormone with the cumbersome name **1,25-dihydroxycholecalciferol.** This substance regulates the uptake of calcium in the small intestine; without it, the gut cannot absorb calcium from food.

A third hormone is produced indirectly when certain cells in the kidney cortex respond to low levels of salt (hypotonicity) in the blood; the cells secrete the enzyme *renin* into the blood (Figure 35-8). Renin boosts the production of a polypeptide, *angiotensin II,* which in turn causes cells in the cortex of the nearby adrenal glands to

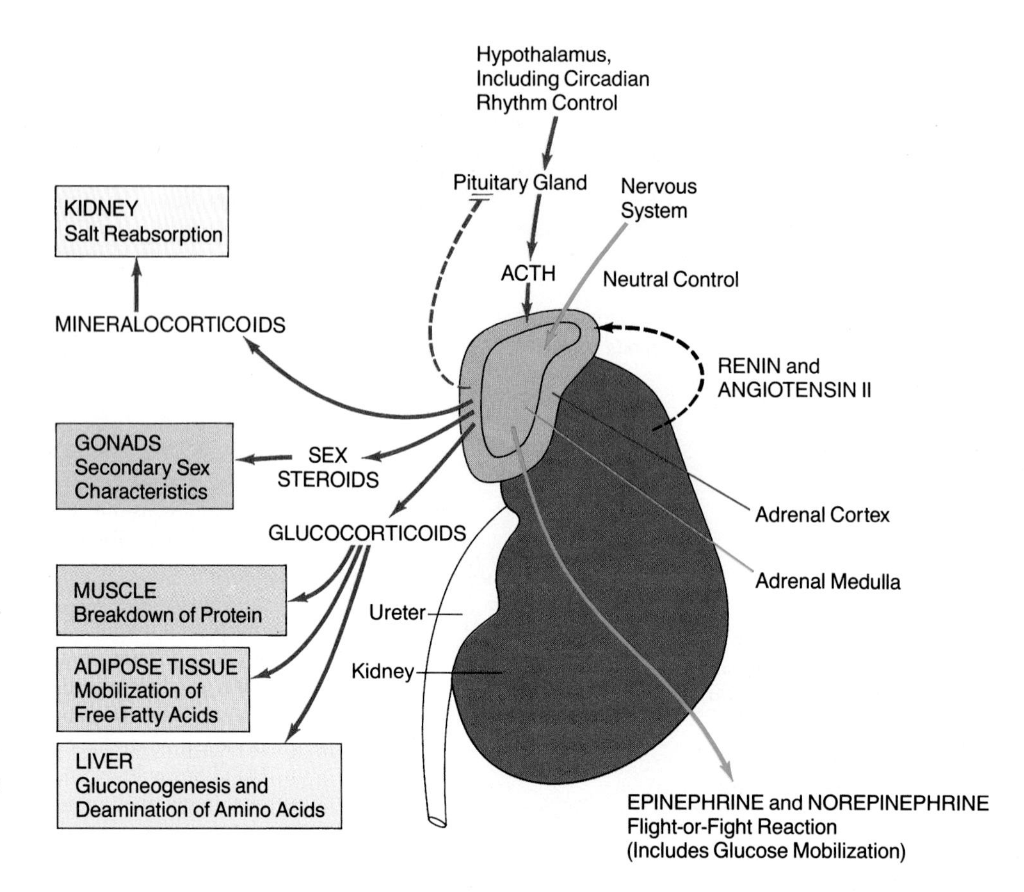

Figure 35-8 ENDOCRINE ACTIVITIES OF THE ADRENAL GLANDS. The adrenal cortex produces glucocorticoids, mineralocorticoids, and sex steroid hormones. Glucocorticoid production is regulated by adrenocorticotropic hormone (ACTH) secreted by the pituitary gland in response to hypothalamic releasing hormone and feedback regulation. (The rhythmic circadian system also affects ACTH release.) Aldosterone (a mineralocorticoid) is regulated by the renin-angiotensin system, as shown. The secretion of epinephrine by the adrenal medulla is controlled by the nervous system. Sex steroids are under control of the pituitary gland; they were discussed in Chapter 18.

secrete the steroid hormone **aldosterone.** As Chapter 33 explained, aldosterone increases the amount of sodium (and hence water) recovered from the urine by cells of the distal convoluted tubules of nephrons, and thus salt levels rise in the blood.

The Heart as an Endocrine Organ

Research has shown that heart cells produce substances that increase urine production when blood pressure is too high and the atrial walls are stretched. Muscle cells in the atrial walls secrete *atrionatriuretic factor (ANF),* which causes the kidneys to excrete copious amounts of sodium and water by countering the effects of antidiuretic hormone (ADH) on collecting ducts and also by dilating the blood vessels leading to the glomeruli (see Chapter 33). Together, ANF and ADH either raise or lower blood pressure and volume to help maintain fluid homeostasis—a key to the organism's internal physiology. ANF also acts in the brain to help govern the volume of fluid in the cavity of the central nervous system.

The Adrenal Glands

Located anterior to each kidney is another endocrine organ, the **adrenal gland.** An outer *adrenal cortex* secretes dozens of steroid hormones, and an inner core, the *adrenal medulla,* secretes two kinds of amine hormones.

Hormones of the Adrenal Cortex

Adrenal cortex cells produce steroid hormones, known as **corticosteroids** (see Figure 35-8). One subset of these, the **glucocorticoids,** stimulates the production of glucose from proteins and carbohydrates. In mammals, the most prevalent glucocorticoid is *cortisol.* Although involved in maintaining blood glucose levels, the glucocorticoids differ from insulin in their long-term action; even when an animal is not taking in food, they ensure that an adequate supply of glucose will be available as a cellular fuel. A person who lacks glucocorticoid hormones develops *Addison's disease,* an illness characterized by weakness and weight loss. Glucocorticoids also increase the amount of fat stored in the body and reduce inflammation. People often use *cortisone,* a well-known anti-inflammatory derivative of cortisol, to treat swelling, athletic injuries, arthritis, insect bites, and even poison ivy.

A second group of corticosteroids is the **mineralocorticoids.** In addition to stimulating kidney tissue to retain sodium ions, aldosterone triggers the excretion of potassium (see Figure 35-8). An excess of aldosterone causes edema (water retention in tissues), high blood pressure, and a severe loss of potassium. If potassium loss continues over a prolonged period, the supply of potassium in muscles is depleted and paralysis results (details in Chapter 37). Conversely, if aldosterone levels are low, an animal eliminates large quantities of urine, and its blood pressure drops precipitously. Untreated, an aldosterone deficit quickly leads to death.

Recent studies explain a long-standing puzzle: Many body tissues have receptors for both mineralocorticoids and glucocorticoids, yet some target tissues respond to just one of the classes of hormones, and other tissues respond to neither. It seems that a specific dehydrogenase enzyme may break down and deactivate one or both classes of hormone, depending on the tissues. Thus, the presence of the enzyme, not the receptor, may govern the target tissue specificity.

The adrenal cortex secretes a third subgroup of corticosteroids: the sex steroids—the *androgens* (male sex hormones) and *estrogens* (female sex hormones). These sex steroids affect initial development of sex-related organs and secondary sex characteristics (see Chapter 18 and Figure 35-8). The hormones also act during and after puberty to regulate maturation and sexual activity. Under normal conditions, the sex steroids are produced mainly by the gonads. On occasion, the adrenal gland overproduces androgens or estrogens and induces an intermediate biological gender with mixed male and female secondary sexual characteristics.

Hormones of the Adrenal Medulla

Located inside the adrenal cortex, somewhat like the pit of a peach, is the adrenal medulla. This structure is essentially an addendum to the sympathetic nervous system and receives direct nervous input (see Chapter 34). Its endocrine cells, called **chromaffin cells,** secrete primarily **epinephrine** (adrenaline; see Figure 35-2b) and some **norepinephrine** (noradrenaline, which is the common neurotransmitter of the sympathetic nervous system). When released by neurons at synapses, these amine hormones serve as neurotransmitters. When released as hormones by adrenal medulla cells, they produce the so-called fight-or-flight reaction (see Figure 35-8). In any vertebrate reacting to an emergency, nerve impulses from centers in the brain stimulate the adrenal medulla to release epinephrine into the blood. The epinephrine then mobilizes the tissues and organs crucial to self-defense or escape and inactivates others. For example, it causes respiration to increase, the pupils to dilate, and blood vessels leading to the brain and skeletal muscles to open; simultaneously, however, epinephrine causes blood vessels in the skin and kidneys to constrict and hence more blood to be shunted to the brain, muscles, and rapidly beating heart. These differences are due to different receptors and transduction machinery in the target cell types; as a result, one hormone can act in opposite ways.

The Thyroid and Parathyroid Glands

The **thyroid gland** is a butterfly-shaped organ located near the esophagus and trachea in air-breathing vertebrates (Figure 35-9). The thyroid produces the hormones **triiodothyronine** (with three iodine atoms), or simply **T_3; thyroxine** (with four iodine atoms), or **T_4** (see Figure 35-2b); and **calcitonin,** a protein hormone involved in calcium balance (see Figure 35-9). The T_3 and T_4 hormones are derivatives of the amino acid tyrosine and seem to act similarly on target cells.

Thyroxine's primary function is to set an animal's metabolic rate by regulating the rate at which cells consume O_2 to synthesize ATP. It also speeds the breakdown of complex carbohydrates and the synthesis of proteins and

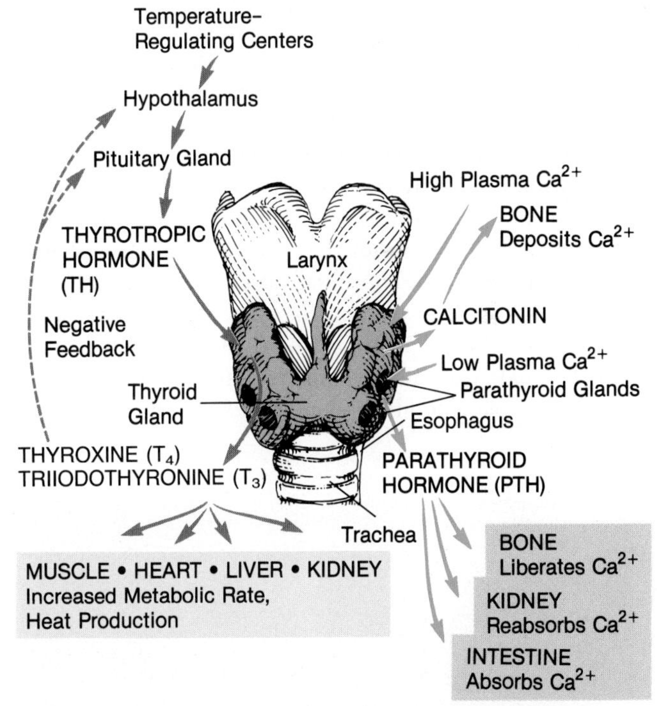

Figure 35-9 THYROID AND PARATHYROID FUNCTION. Metabolic rate (and heat production in mammals) is regulated by the thyroid hormones, which are released into the bloodstream in response to thyrotropic hormone (TH) from the pituitary gland. Plasma calcium levels are regulated through the action of parathyroid hormone (PTH), which is produced by the parathyroid glands, and calcitonin, which is produced by specialized cells that lie between the thyroid follicles. Note that the hypothalamus and neurons are not involved in calcium regulation by those hormones.

slows the rate at which fats are burned. In addition, thyroxine increases heat production in mammals by activating the ATPase pump enzymes on certain cell surfaces to cleave much ATP and thus step up heat release. Thyroxine also acts in the central nervous system, and human brain cells have a novel receptor for it unlike others found in the body.

Both T_3 and T_4 are secreted in response to a third hormone, known as **thyrotropic hormone,** or **TH** (see Figure 35-9). TH is produced in the anterior pituitary gland and travels through the blood before binding to receptor molecules on thyroid cells. When blood levels of thyroxine rise, TH secretion decreases in a negative feedback loop, and as a result, T_3 and T_4 secretion falls. When blood levels of thyroxine are low, more TH is secreted, and thyroid cells are stimulated to increase their output of T_3 and T_4. In addition, the brain can cause TH release and thereby thyroid hormone secretion.

If the negative feedback loop fails to function properly, the levels of thyroid hormones in the blood can be too high or too low, resulting in a *hyperthyroid* condition, with accelerated metabolism, weight loss, and hyperactivity, or a *hypothyroid* condition, with slowed metabolism, weight gain, and sluggishness. Hypothyroidism in a developing fetus can result in *cretinism,* and afflicted children suffer greatly retarded physical and mental growth.

Goiter is another response to a disturbance in the thyroid-pituitary negative feedback loop, in this case caused by too little iodine in the diet. Without iodine, the thyroid gland cannot synthesize T_3 and T_4, and as a result, the anterior pituitary continually secretes a large amount of TH. This "overdose," in turn, causes the thyroid gland to enlarge (Figure 35-10). The advent of iodized table salt has made this once-common ailment relatively rare.

The thyroid gland has yet another function: regulating calcium storage in the body. When calcium levels in the blood rise, small clusters of specialized cells in the thyroid secrete calcitonin, a protein hormone that decreases the calcium levels. Calcitonin inhibits the activity of *osteoclasts*—cells that normally break down bone and release calcium (see Chapter 37). In addition, calcitonin stimulates bone cells (osteoblasts) to bind calcium ions into bone salts (see Figure 35-9).

Just the opposite occurs when hormones of the **parathyroid glands** are secreted. These four tiny bean-shaped glands situated on the surface of the thyroid secrete **parathyroid hormone (PTH)** (see Figure 35-9). PTH, acting when the calcium level in the blood falls, increases both the number and the activity of osteoclasts. They degrade bone salts, freeing Ca^{2+} into the blood plasma.

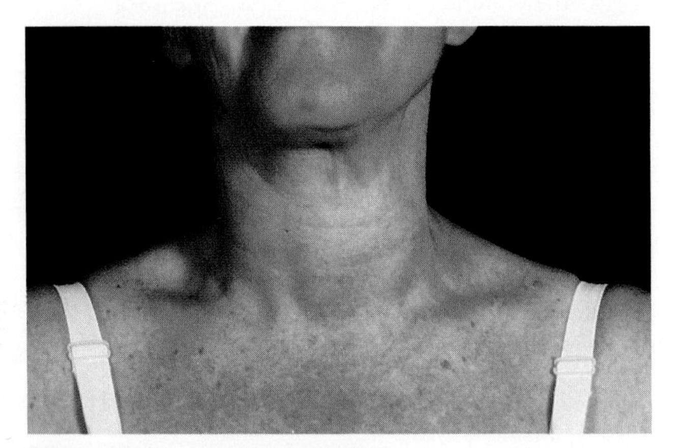

Figure 35-10 THYROID GOITER.
The thyroid is enlarged greatly because of continuing TH release from the anterior pituitary, despite the inability of the thyroid to make its hormones.

The Pituitary Gland

Nestled in a cup-shaped depression at the base of the skull lies the **pituitary gland.** It is no larger than a small marble (about 1.3 cm in diameter) in humans, but was once thought to be the vertebrate's "master" endocrine gland. Research has shown, however, that the pituitary is more like a go-between than a master, receiving regulatory orders directly from the brain and transmitting them to other endocrine organs. The pituitary consists of two lobes—the *anterior pituitary* and the *posterior pituitary*—that together produce at least ten hormones (Figure 35-11). The anterior pituitary must receive normal hormonal signals to do its work, while the posterior pituitary is derived from brain tissue and is a site of neurosecretory cell endings.

Hormones of the Anterior Pituitary

The anterior pituitary secretes several peptide hormones in response to signals received from the hypothalamus. These substances either regulate hormone secretion by other organs or act on target tissues directly. They fall into three families of structurally related proteins (Table 35-1). The first family includes the glycoprotein hormones: thyrotropic hormone (TH) and the gonadotropins—*follicle-stimulating hormone (FSH)* and *luteinizing hormone (LH).* TH regulates thyroid function. FSH and LH trigger gamete production in ovaries and testes, and they also regulate the gonads' secretion of estrogens and progesterone (high levels in females) and testosterone (high levels in males).

The second family of protein hormones includes **prolactin** and **growth hormone (GH).** In fish, prolactin acts as a regulator of water and salt balance. In some birds, prolactin stimulates esophageal cells to produce a cheesy "crop milk," which is fed to hatchlings. In mammals,

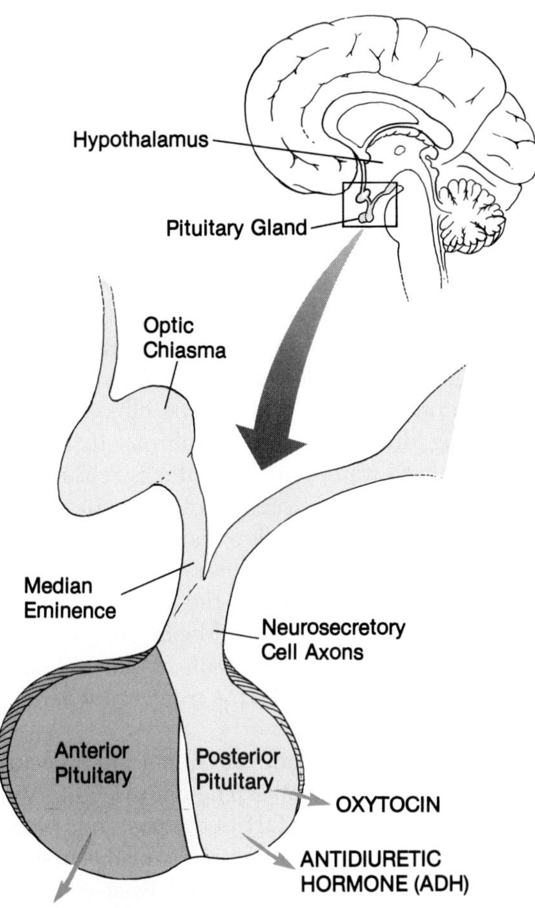

Figure 35-11 PITUITARY GLAND FUNCTION.
The pituitary rests in a bony cavity at the base of the skull. Its secretions are controlled by the hypothalamus of the brain. The pituitary has two distinct regions, the anterior pituitary and the posterior pituitary. The posterior pituitary is connected directly to the hypothalamus and is true brain tissue. The anterior pituitary is not brain tissue and is controlled by chemicals carried in the blood from the nearby brain.

different sets of cells in the anterior pituitary secrete four prolactin proteins that differ slightly in structure. In females, prolactin stimulates the synthesis and secretion of milk from mammary gland tissue that has first been "primed" by estrogen and other steroid hormones secreted during pregnancy (see Chapter 18). In some cases, both infertile men and women have been found to have high levels of prolactin in their blood, and both have often responded favorably to a drug (bromocryptine) that inhibits prolactin secretion.

The anterior pituitary also secretes *growth hormone (GH)*, which acts both in adults and in maturing animals. GH affects cells in adult, nongrowing tissues in several ways: It augments the incorportion of amino acids into protein, promotes the use of fats instead of carbohydrates for energy, and causes the liver to catabolize glycogen to glucose, thereby raising blood glucose levels for metabolism. But the hormone's best-known effect—stimulating growth during maturation—occurs indirectly, as it triggers cells in the liver to secrete small peptide hormones called **somatomedins.** One type of somatomedin stimulates the formation of cartilage, especially in the so-called growth plates near the ends of long bones in arms and legs. This is the basis for growth in such bones and occurs during the "growth spurts" that all land vertebrates experience in adolescence. Another

Table 35-1 HORMONES OF THE ANTERIOR PITUITARY

Hormone	*Target*	*Action*
Thyrotropic hormone (TH)	Thyroid gland	Increases release of thyroxine
Follicle-stimulating hormone (FSH)	Seminiferous tubules (male), ovarian follicles (female)	In male, causes production of sperm; in female, stimulates maturation of follicle
Luteinizing hormone (LH)	Interstitial cells of ovaries or testes	In female, induces final maturation of follicle, ovulation, and formation of corpus luteum; in male, induces secretion of androgens
Prolactin	Mammary glands	Increases synthesis of milk proteins and growth of mammary glands
Growth hormone (GH)	Bone, fat, other tissues	Increases synthesis of RNA and proteins, transport of glucose and amino acids, lipolysis, and formation of antibodies
Corticotropin (ACTH)	Adrenal cortex	Increases secretion of steroids in adrenal cortex
Melanophore-stimulating hormone (MSH)	Melanophores and melanocytes	Increases synthesis of melanin, causes darkening of skin
Lipotropin (LPH)	Fat and other cells	Hydrolysis of fats

type of somatomedin promotes cell division in connective tissues so that they may grow, too. Somatomedins and insulin have quite similar amino acid sequences, suggesting that the genes for these hormones may have arisen from a single ancestral gene during evolution.

Individuals with a GH deficiency will be abnormally short, a condition called *pituitary dwarfism.* An excess of GH, on the other hand, causes a pituitary disorder called *pituitary gigantism,* which is characterized by great height but normal body proportions. When surplus growth hormone is secreted after the growth spurt of puberty, the result is *acromegaly,* a condition in which the bones and tissues of the face and joints become enlarged.

In the mid-1980s, a California biotechnology firm succeeded in isolating the gene for human growth hormone, cloning it in bacteria, and producing enough synthetic human GH to market as a pharmaceutical. Clinical trials with the substance showed that hypopituitary dwarf children (those with a demonstrated deficiency of their own growth hormone) grew an average of 4 in. the first year and 3.5 in. the second—far more than untreated dwarf children grew.

Members of the third family of anterior pituitary hormones are polypeptides. It includes **corticotropin** (or adrenocorticotropic hormone, **ACTH), melanophore-stimulating hormone (MSH),** and **lipotropin (LPH).** ACTH regulates corticosteroid production by the adrenal cortex. MSH affects skin pigmentation. Its effects are best understood in lizards, in which it causes the skin to darken by stimulating the dispersion of small granules of brown pigment (melanin) within cells called *melanophores.* Darkened skin can absorb solar heat and warm the animal; if the lizard grows too hot, the MSH secretion drops and the skin pales. Lipotropin has quite a different function. It causes the hydrolysis of fat into free fatty acids and glycerol, making these important precursors available to cells.

Hormones of the Posterior Pituitary

The posterior lobe of the pituitary gland secretes only two hormones: **oxytocin** and **antidiuretic hormone,** or **ADH** (also called *vasopressin*) (see Figure 35-11). These chemicals are synthesized not by the pituitary itself, but by neurosecretory cells in the nearby hypothalamus. They are then secreted into the bloodstream by way of neurosecretory cell endings in the posterior pituitary as well as elsewhere in the brain.

Oxytocin (see Figure 35-2a) is best known as a stimulator of contraction in smooth muscle cells. This occurs, for instance, in a mammal's mammary glands, so that milk is expressed through the nipple. Oxytocin also stimulates the contraction of uterine smooth muscle during labor or orgasm. It turns out that oxytocin synthesis is not restricted to hypothalamic neurosecretory cells. The hormone is also produced in the thymus, where its role in the immune system is not defined, and in the corpus luteum of the ovary. In reptiles, amphibians, fishes, and birds, a relative of oxytocin, *arginine vasotocin (AVT),* plays an analogous biological role. When a female blue jay is ready to lay an egg, for example, AVT starts the contractions of her oviduct.

The second posterior pituitary hormone, ADH, has a very different function from oxytocin; it decreases the production of urine. It exerts its effect by causing the collecting ducts of kidney nephrons to become more permeable to water, so that water is reclaimed into the blood from the urine (see Chapter 33). In the absence of ADH, a person's urine flow increases tenfold, resulting in a condition known as *diabetes insipidus.* ADH also has a second function, which is the source of its alternative name, vasopressin. It raises blood pressure by stimulating smooth muscles in the walls of blood vessels to constrict. The release of ADH is triggered whenever blood pressure falls.

Biologists have made an important new discovery about pituitary hormones. The neurosecretory cells that secrete oxytocin and ADH in the posterior pituitary also send axons to sites in other parts of the brain. The researchers do not yet understand what either hormone does when it is released locally in the midst of brain neurons, but effects on local blood flow and oxygen availability seem likely.

The Hypothalamus-Pituitary Connection

The hypothalamus governs nearly all pituitary functions and, through them, hunger, thirst, body temperature, emotional reactions to stress, sexual stimulation, and many other essential processes controlled by hormones.

Embedded in the hypothalamus are grapelike clusters of neurosecretory cells that manufacture both ADH and oxytocin (Figure 35-12, step 1). Carrier proteins transport the hormones along the nerve cell axons to axonal endings in the posterior pituitary (or to other sites in the brain), where they are then stored (step 2). When the neurosecretory cell bodies are stimulated to initiate action potentials (see Chapter 34), those nerve impulses pass down the neurosecretory cell axons and cause the hormones' release into the blood (step 3), just as neurotransmitters are released at the axonal endings of neurons.

Other hypothalamic neurosecretory cells (step 4) have axons with a different destination; they extend to an area

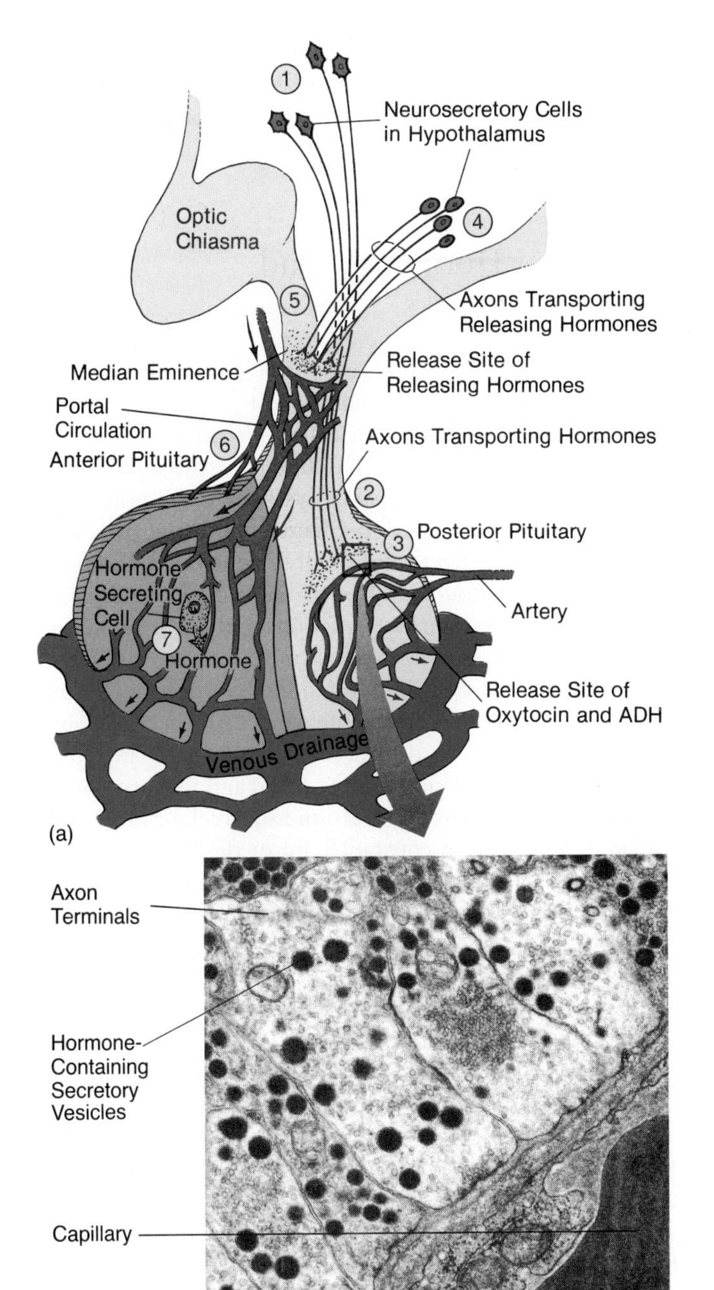

Figure 35-12 ROUTES OF HORMONES IN THE PITUITARY.
(a) The releasing hormones produced by the hypothalamus control the production and secretion of anterior pituitary hormones. The neurosecretory cells discharge their releasing hormones into a capillary system that extends from the median eminence to the anterior pituitary. (b) Terminals of neurosecretory cells in the posterior pituitary gland are seen at the upper left in this electron micrograph (magnified about 18,000×). The large black spots are secretory vesicles; each contains oxytocin or antidiuretic hormone. The wall of a blood capillary fills the lower right corner; secreted releasing hormones pass through the holes in the capillary wall into the plasma.

near the base of the hypothalamus called the *median eminence*, where, on appropriate stimulation, they secrete a variety of small peptide hormones (step 5). These substances are carried through a specialized system of capillaries to the neighboring anterior pituitary (step 6); there, they regulate the release of anterior pituitary hormones (step 7). Those hypothalamic neurohormones that stimulate hormone secretion are called **releasing hormones (RHs)**, while those that inhibit it are termed **release-inhibiting hormones (RIHs).**

Table 35-2 shows that every hormone secreted by the anterior pituitary is controlled by an RH or an RH-RIH combination. For example, prolactin-release-inhibiting hormone (PRIH)—consisting of either a peptide dubbed GAP or the neurotransmitter dopamine—must cease to be secreted when a female mammal suckles her young in order for prolactin to be released in response to prolactin-releasing hormone (PRH). It is as though there are two locks to the prolactin door: The inhibitory signal must cease, and the stimulatory signal must start.

Many of the releasing hormones are secreted not only in the median eminence but also at other sites in the

Table 35-2 RELEASING AND RELEASE-INHIBITING HORMONES OF THE HYPOTHALAMUS

Hormone	*Action*
TH-releasing hormone (TRH)	Stimulates TH release
Gonadotropin-releasing hormone (GnRH)	Stimulates FSH and LH release
Prolactin-release-inhibiting hormone (PRIH)	Inhibits prolactin release
GH-release-inhibiting hormone (GRIH)	Inhibits GH release; interferes with TH release
GH-releasing hormone (GRH)	Stimulates GH release
Corticotropic-releasing hormone (CRH)	Stimulates ACTH release
MSH-release-inhibiting hormone (MRIH)	Inhibits MSH release

brain to which hypothalamic neurosecretory cells send their axons. Some of these substances act on brain cells in spectacular ways. For instance, TRH triggers subdued behavior, heat production, neuronal electrical activity, and several other processes that have nothing to do with the pituitary. Such surprising discoveries suggest that many RHs, RIHs, and other classic hormones participate, in ways not yet known, in a range of normal brain functions. Clearly, the nervous and endocrine systems are not discrete: Neurons secrete hormones, hormones affect neurons, and this dual system controls an animal's behavior and its physiology.

HORMONAL CONTROL OF PHYSIOLOGY

As hormones integrate and control an animal's body functions, one of their primary roles is to maintain homeostasis. Let's first consider the homeostatic mechanism governing blood sugar levels and then take a look at an animal's biological rhythms and cycles.

Control of Blood Sugar Levels

What happens when the human body receives a sugar "jolt" like a candy bar? Within minutes, the sucrose in the candy is converted to glucose, which enters your blood plasma. If the glucose level is high enough, β cells in the islets of Langerhans react by secreting insulin. At the same time, secretion by the adrenal cortex of glucocorticoids decreases. Circulating insulin binds to receptors on skeletal muscle, fat, and liver cells and stimulates them to take up glucose. As a result, the level of glucose in the blood then falls back toward normal.

In contrast, what happens when a mammal cannot find food for several days? During such fasting, blood glucose levels fall below normal, producing a condition known as *hypoglycemia*. As a result, glucagon from pancreatic α cells stimulates the conversion of liver glycogen into glucose, and this is then released into the blood.

The liver's supply of glycogen is only sufficient to compensate for 12 to 24 hours or so of fasting. Thereafter, growth hormone from the pituitary and glucocorticoids from the adrenal cortex stimulate the synthesis of glucose from amino acids in the liver (a process called *gluconeogenesis*). In addition, GH, glucagon, and possibly lipotropin cause fatty acids and glycerol to be converted into glucose. As fatty acids in the liver are oxidized, ketone bodies are produced. These diffuse into the blood and can then be converted to acetyl coenzyme A in muscle cells for synthesis of ATP. In this way, the fasting animal's muscle cells continue to obtain the energy they need to contract.

Rhythms, Hormones, and Biological Clocks

Certain hormonal effects on physiology occur as automatic internal cycles. Animals, plants, and even some microbes function on cycles called *circadian* ("about a day") rhythms. Circadian rhythms are intrinsic daily fluctuations in the way various systems in the organism function. As we shall see in Chapter 44, some complex organisms also have internally controlled *annual rhythms;* the migrations of birds and mammals, for example, depend on hormonal actions and occur on automatic, annual cycles. External cues, such as the daily cycle of light and darkness due to the earth's rotation, or the changes in season due to the earth's tilt, may "set" or entrain circadian or annual clocks, but the internal rhythm itself is a property of the brain and its hormones. In other words, if a fruit fly, ground squirrel, or human is isolated for weeks in constant light or darkness, cycling will persist; sleep cycles, hunger cycles, mitotic cycles in various tissues, and all sorts of other events will continue normally despite the absence of the standard cues of daylight and darkness.

An organism's biological clock has a genetic basis. Strains of fruit flies, for instance, can be isolated that have short-day (say, 18-hour) clocks or long-day (say, 30-hour) rhythms. The genes responsible for such characteristics presumably do their work in the brain. A diagnostic feature of all such circadian cycles is that their timing is temperature-compensated. This makes sense, since a salamander or fruit fly may be exposed to radically different temperatures during a typical 24-hour day; if their internal clocks ran faster at high temperatures and slower at low ones, timekeeping would be chaotic. But as shown by Princeton University biologist C. S. Pittendrigh, the internal clocks are compensated, so that despite varying temperatures, they run at a constant rate.

How are circadian rhythms controlled? In complex animals, many hormones are released at preset times during each 24-hour period. Epinephrine levels in human blood and body fluids, for example, are high during the day and low at night. Since epinephrine inhibits cellular mitosis, most mitotic activity in body tissues—including the growth of hair and fingernails—takes place while we sleep.

One source of rhythmic control in many animal species is the **pineal gland,** a pealike nugget of tissue located in the roof of the brain. The pineal gland releases the hormone **melatonin,** a derivative of the amino acid tryptophan, usually when an animal's environment is dark (in humans, it is most abundant in the blood between 11 P.M. and 7 A.M.). Melatonin apparently acts on a variety of neurons and hormone-secreting cells, functioning as a pacemaker signal to synchronize other body activities.

When whole pineal glands or even a few cells from pineal tissues are cultured in a dish in the dark, they release melatonin on a cyclic schedule, just as they do when still embedded in the brain of an animal in an environment with normal changes of light and darkness. Thus, cells themselves can have intrinsic rhythmic properties. Other experiments have revealed how the extraordinary built-in rhythmic activity of the brain and pineal gland is related to the animal's "biological clock"—its intrinsic, roughly 24-hour timetable of biological activities. Suppose, for example, that a bird is kept in constant light or darkness for an extended period. Despite the absence of external signals, the pineal gland continues to function on a cycle approximating 24 hours in length, and most body functions continue to cycle as well. If the pineal gland is removed, however, the bird's internal "schedule" disappears, and its biological functioning becomes arhythmic. Conversely, if a pineal gland is transplanted into an arhythmic bird lacking one, the animal immediately adopts the precise rhythm of the implanted organ.

Recent studies suggest that paired regions of the hypothalamus called the **suprachiasmatic nuclei (SCN)** are important for the mammalian biological clock. Through exceptionally delicate microsurgery, researchers can separate neurons of this region from the rest of the brain. Like cultured pineal cells, these SCN neurons continue to function on a circadian cycle. However, every other cyclic activity of the mammal's brain, endocrine system, and body ceases (including that of melatonin secretion by the mammal's pineal gland). The pineal gland exerts its influence by secreting melatonin; the suprachiasmatic nucleus sends its regulatory messages via nerve axons to other parts of the brain. But new techniques reveal axons extending from pineal cells in hamsters (a mammal) into the brain near the hypothalamus. Other work demonstrates melatonin receptors in the human SCN, as well as the successful use of melatonin to correct jet lag, synchronize sleep-wake cycles in the blind, and alleviate disorders of sleep and biological rhythms. These and other findings suggest that the suprachiasmatic nuclei may, as pacemakers, drive mammalian cycles, but that the evolutionarily ancient pineal cells may still be involved in activating or entraining hypothalamic neurons and so influencing cyclic activities.

HORMONES AND DEVELOPMENT: AMPHIBIAN METAMORPHOSIS

Hormones influence the maturation and division of specific cell types in embryos (see Chapter 16), and this hormonal control of development persists as different kinds of endocrine cells function through the early stages of an animal's life. The metamorphosis of a tadpole into an adult frog is an excellent example of this developmental control in action.

When metamorphosis occurs, the tadpole's body undergoes enormous changes. Hind legs begin to develop, and forelimbs soon follow. The tail disappears, literally digested away by lysosomal enzymes. A new visual pigment appears, one suitable for light traveling through air, not water, and the tadpole's hemoglobin is altered to a type compatible with air breathing. At the same time, lungs begin to develop, and the gills are later degraded. The gut shortens as an adaptation for future carnivorous feeding, and enzymes for digesting protein are synthesized to allow digestion of animal food. Because of a change in liver enzymes, the kidneys start to secrete mostly urea, instead of the more toxic ammonia—a substance that must be passed into large volumes of water. The end result of this developmental magic is a frog whose physiology is radically different from that of the tadpole that preceded it.

The brain and the thyroid gland are two main sources of the hormones that control the tadpole's impressive transformation. The sequence starts when the hypothalamus begins to secrete TRH (thyrotropic hormone–releasing hormone; Figure 35-13, step 1). Pituitary cells release thyrotropic hormone in response (step 2), and that hormone, in turn, triggers the secretion of thyroxine by the thyroid gland (step 3). At the same time, the tadpole's pituitary, aided by the secretion of PRIH (prolactin-release-inhibiting hormone) from the hypothalamus (step 4), slows secretion of prolactin (an antagonist of thyroid hormones) from the pituitary (step 5). As thyroxine circulates through the tadpole's bloodstream, it is bound by abundant receptor molecules in tissues affected by metamorphosis—the tail, legs, liver, and others.

But what causes the hypothalamus to initiate the secretion of the hormones responsible for metamorphosis? Factors in the tadpole's environment that make survival in the water more difficult, such as crowding, falling oxygen content in the water, or limitation of the food supply, cause the tadpole's brain to initiate metamorphosis.

Some salamander species remain at the larval stage if their pond is not crowded and has adequate oxygen and

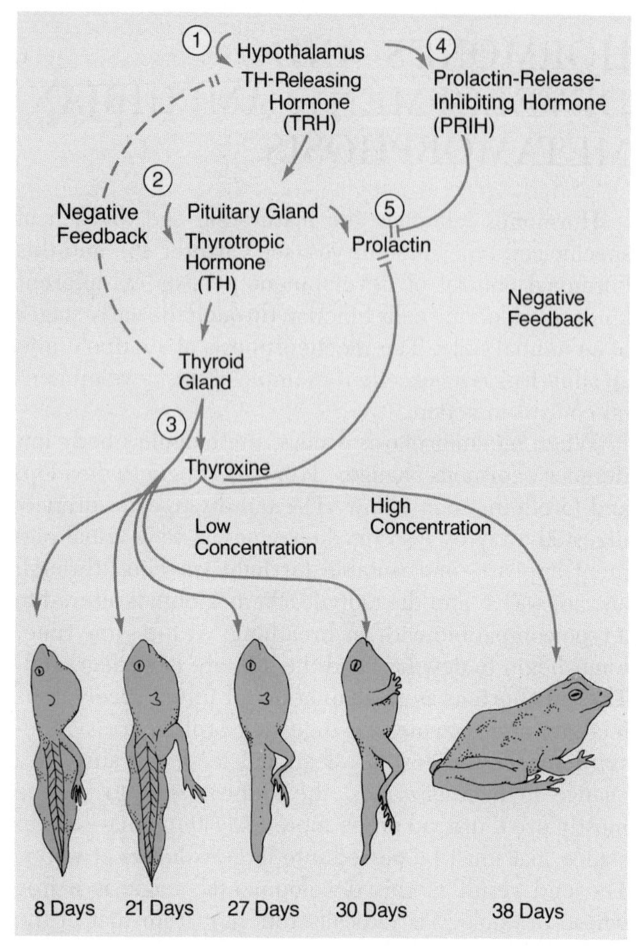

Figure 35-13 AMPHIBIAN METAMORPHOSIS: FROM TADPOLE TO FROG.
During metamorphosis, the tadpole loses some anatomical structures as new ones develop. Biochemical changes, largely the result of thyroid hormone action, are associated with the metamorphosis.

plentiful food. Their gonads develop, however, and the larvae reproduce. This phenomenon, in which a juvenile form reproduces, is called **neotony.** In a subsequent year, if food in the water becomes very scarce, the hypothalami of the larvae in that generation trigger metamorphosis, and adult salamanders emerge.

HORMONES: NEW TYPES, NEW SITES, NEW MODES OF SYNTHESIS

Endocrinology is one of the most exciting areas of biology today, and among the more recently identified control agents are the **prostaglandins,** a fascinating group of chemicals derived from fatty acids. Once thought to be present only in semen, prostaglandin-like substances are in fact produced by virtually all cells (usually from membrane phospholipids).

Unlike most hormones, prostaglandins do not circulate in the blood, but appear in increased quantity in local regions where tissues are disturbed. They are potent stimulators of smooth muscle contractions and thus may be involved in birth, miscarriage, or blood vessel constriction in the kidney and other tissues. An excess of prostaglandins acting on human cerebral blood vessels causes the vessels to dilate and press against nerves, producing a headache. Aspirin is an effective reliever for headache pain because it inhibits prostaglandin production.

The *brain peptides,* also fairly recently discovered, have revolutionized our view of the relationship between the endocrine and nervous systems. Researchers have identified these substances in the brains of mammals and other vertebrates. Oxytocin (described earlier) is released from axonal branches of hypothalamic neurons at various specific sites in the brain. Another surprising brain peptide is CCK, a hormone long thought to work only in the gut (see Chapter 32). Certain axons of single neurons in the cerebral cortex synapse on tiny blood vessels, where they release CCK. The hormone then affects local blood flow and the electrical activity of neighboring nerve cells. Conceivably, this relates to the role of CCK in establishing vivid memories. Still another peptide, VIP (an intestinal polypeptide), is released by other cerebral cortex neurons.

Some biologists postulate that hormones such as CCK and oxytocin help to maintain normal relationships between nerve cell activity, blood flow, and cellular metabolism in the brain. Brain peptides probably also exert relatively long-lasting effects (many seconds, minutes, or longer) on the polarized state of neurons. Thus, traditional neurotransmitters do their work in milliseconds, while a brain peptide, perhaps released from the same neuron, has a much more sustained action. This leads to the view that the nervous system is composed of hard-wired circuits that are "tuned" or modulated by brain peptides and certain neurotransmitters. As a result, nerve circuits can be flexible and adaptable, characteristics we associate with learning.

Most brain peptides have ancient evolutionary histories; they are found not only in complex mammals and invertebrates, but even in unicellular organisms and in plants lacking nervous systems. Evidence suggests that such substances originated as chemical vehicles for intercellular communication. Yeast cells, for instance, possess a mating pheromone, *α factor,* with an amino acid sequence very similar to that of the mammalian gonadotropin-releasing hormone. Some researchers believe that as nervous systems evolved, such communication molecules took on roles as neurotransmitters, and as endocrine tissues arose, they came to function as hormones.

While modern research techniques have given us many surprises about hormones, none is more novel than multiple-hormone production from single genes. For example, newly isolated enzymes can chop up one large naturally occurring polypeptide called POMC (pro-opiomelanocortin) in pituitary and hypothalamus cells to yield ACTH, β-lipoprotein (a hormone that causes fat cells to break down lipids), three forms of MSH, several brain peptides (called endorphins; see Chapter 36), and other peptides. In other cells, one polypeptide can be cleaved to yield GnRH and GAP, the prolactin-release-inhibiting peptide. Based on these and other examples, it is clear that the one gene–one polypeptide rule worked out so elegantly by geneticists in the 1960s must be modified in hormone-producing cells: One gene can code for several polypeptides, each with its own functions. The gene that codes for POMC is, in fact, turned on in a variety of body cell types, and different regulatory agents in those cells control the gene and the resultant hormonal peptides.

LOOKING AHEAD

Having considered two interrelated networks of physiological control, the nervous and endocrine systems, we turn, in the next chapter, to the sensory systems that take in data and to the brain that integrates and interprets those data.

SUMMARY

1. Hormones are chemical messengers that are secreted by cells and glands of the *endocrine system* into the blood and that influence the activities of cells in specific target tissues.

2. A hormone affects only target cells specific to it. Different cells develop different transduction mechanisms and so may respond differently to the same hormone.

3. Chemically, most hormones are proteins (or short-chain polypeptides), amines, or steroids.

4. Steroid hormones bind to specific receptor molecules in the nuclei of target cells. The hormone-receptor complex then binds to acceptor sites on the chromosomes and stimulates specific genes to synthesize messenger RNA, which is then translated into protein.

5. *Cyclic AMP* and *cyclic GMP* are second messengers for many protein, polypeptide, and amine hormones. These and other second-messenger systems function as amplification cascades, so that binding of a tiny amount of hormone can ultimately trigger a massive biochemical response.

6. Many protein hormones, neurotransmitters, and growth factors and some electrical stimulation act via *inositol triphosphate (IP_3)* and *diacylglycerol (DG)* as second messengers, which in turn activate major cellular responses.

7. Two important hormones in arthropods are *juvenile hormone (JH)*, which regulates the stages of development, and *ecdysone* hormones, which are responsible for metamorphosis, molting, and other processes.

8. β cells in the pancreatic *islets of Langerhans* secrete *insulin*, which lowers blood glucose level. Pancreatic α cells secrete *glucagon*, which raises blood glucose level. *Somatostatin*, a hormone secreted by pancreatic δ cells, inhibits the secretion of insulin and glucagon and the absorption of glucose across the intestinal wall.

9. The kidney secretes *erythropoietin*, which stimulates the production of red blood cells in bone marrow, and *1,25-dihydroxycholecalciferol*, which regulates calcium absorption in the small intestine. *Aldosterone*, made by the adrenal glands, regulates sodium and (indirectly) water retention in the kidney.

10. Heart cells secrete atrionatriuretic factor (ANF) in response to elevated blood pressure. This hormone causes vasodilation and urine production, thereby lowering blood pressure.

11. The cortex of the *adrenal glands* produces *corticosteroids* (the *glucocorticoids* and *mineralocorticoids*) and the sex steroids (androgens and estrogens).

12. *Chromaffin cells* in the adrenal medulla secrete *epinephrine* and *norepinephrine*, which mobilize crucial organs and tissues in the fight-or-flight reaction.

13. The *thyroid gland* secretes *triiodothyronine (T_3)* and *thyroxine (T_4)*, which regulate the rate of oxygen consumption and metabolism in cells and also may play a role in generation of heat. A third thyroid hormone, *calcitonin*, lowers blood calcium level, whereas *parathyroid hormone (PTH)*, secreted by the *parathyroid*, elevates blood calcium.

14. The anterior pituitary gland secretes *thyrotropic hormone (TH)*, *corticotropin (ACTH)*, *follicle-stimulating hormone (FSH)*, and *luteinizing hormone (LH)*; these hormones regulate endocrine functions in the thyroid, adrenals, and gonads. The anterior pituitary also secretes *prolactin*, *growth hormone (GH)*, *melanophore-stimulating hormone (MSH)*, and *lipotropin (LPH)*.

15. The posterior pituitary secretes *oxytocin* and *antidiuretic hormone (ADH)*, or vasopressin. ADH causes water reabsorption in kidneys, whereas oxytocin stimulates smooth muscle cell contraction.

16. Anterior pituitary cells are regulated by the hypothalamus through *releasing hormones* and *release-inhibiting hormones*. These substances pass from the median eminence, near the base of the hypothalamus, to the anterior pituitary to exert their effects.

17. The *pineal gland*, which secretes the hormone *melatonin*, generates circadian rhythms in birds, and the *suprachiasmatic nucleus (SCN)* of the hypothala-

mus plays the same role in mammals. Light/dark cycles or other cyclic cues can "reset" the brain's internal pacemaker cells and so the organism's biological clock.

18. Metamorphosis in amphibians is a dramatic example of hormonal control of a major maturational process.

19. Derivatives of fatty acids called *prostaglandins* are synthesized when local areas are disturbed, and they stimulate smooth muscle contraction. Brain peptides are released at specific sites in the brain and appear to have special, relatively long-lasting effects on neurons.

KEY TERMS

adrenal gland
aldosterone
antidiuretic hormone (ADH)
calcitonin
chromaffin cell
corticosteroid
corticotropin (ACTH)
cyclic AMP (cAMP)
cyclic GMP (cGMP)
diacylglycerol (DG)
1,25-dihydroxycholecalciferol
ecdysone
endocrine system
epinephrine
erythropoietin
glucagon
glucocorticoid
goiter
growth hormone (GH)
inositol triphosphate (IP_3)
insulin
islet of Langerhans
juvenile hormone (JH)
lipotropin (LPH)
melanophore-stimulating hormone (MSH)
melatonin
mineralocorticoid
neotony
norepinephrine
oxytocin
parathyroid gland
parathyroid hormone (PTH)
pineal gland
pituitary gland
prolactin
prostaglandin
release-inhibiting hormone (RIH)
releasing hormone (RH)
second messenger
somatomedin
somatostatin
suprachiasmatic nucleus (SCN)
thyroid gland
thyrotropic hormone (TH)
thyroxine (T_4)
triiodothyronine (T_3)

QUESTIONS

1. What are the functional and chemical similarities between the endocrine and nervous systems?

2. What do biologists call substances produced by the endocrine system that exert influence on target cells elsewhere? Name the three chemical categories into which these substances fall, and give an example of each from vertebrates.

3. List the six principles and the three mechanisms of hormonal action.

4. What are the basic differences between the way steroid hormones and nonsteroid hormones act on target cells?

5. Explain the roles of juvenile hormone and ecdysone in the life cycle of a cecropia moth.

6. Describe the role played by pancreatic hormones in glucose metabolism. What other glands and hormones affect the metabolism of glucose?

7. Explain in terms of glands, hormones, and target cells what biologists mean by the fight-or-flight reaction.

8. From what you know of the chemical structure of thyroid hormones, explain how the lack of dietary iodine interferes with basic metabolism.

9. The pituitary gland is now considered to be a "go-between" in endocrine functions. Explain.

10. Give examples of circadian rhythms. What part of the brain controls circadian rhythms in birds? In mammals?

ESSAY QUESTIONS

1. Define the term *homeostasis*, and discuss the role of the endocrine system in the maintenance of homeostasis.

2. Discuss the various ways in which the endocrine system and the nervous system are variations on one theme.

For additional readings related to topics in this chapter, see Appendix C.

36

Input and Output: The Senses and the Brain

" . . . weary after a dull day with the prospect of a depressing morrow, I raised to my lips a spoonful of the tea in which I had soaked a morsel of the cake. No sooner had the warm liquid, and the crumbs with it, touched my palate than a shudder ran through my whole body, and . . . immediately the old grey house upon the street, where [my aunt's] room was, rose up like the scenery of a theatre . . . and the whole of Combray and of its surroundings, taking their proper shapes and growing solid, sprang into being, town and gardens alike, from my cup of tea."

Marcel Proust, *Remembrance of Things Past* (1922)

Proust's unlocking of a childhood memory—complete with emotions and visual details—through the door of the senses—the taste and smell of tea and cakes—is not just a famous literary passage. It points up two of the most characteristic and interesting animal traits: the ability to perceive the environment and then to integrate and analyze this sensory data. Sensing and integrating can involve simple receptors and neural centers, or it can involve elaborate sense organs, such as the eye, ear, nose, and tongue, together with a fantastically intricate organ of coordination, such as the human brain. At either extreme—and all along the range of neural complexity that lies between flatworms, say, and a French novelist—the sensing and integrating allows the animal to take in information on food, mates, danger, or internal states and to respond in a way that promotes survival.

The sensory process begins when specialized sensory receptor cells in the eyes, skin, or other body parts respond to chemical, mechanical, or electromagnetic en-

Eyes, ears, nose, whiskers, fingers: sensory sharing and reacting between two animal friends.

ergy. When energy acts on a receptor neuron, the flow of ions through the cell membrane increases or decreases, and this sends a sensory message to the nervous system. If the animal is complex, huge numbers of neurons with even more numerous axons, dendrites, and synapses will be woven into circuits. In the human brain, for example, at least 100 billion neurons and a trillion support cells are divided subpopulations with specific functions, such as receiving and analyzing images, sounds, or other kinds of sensory data; coordinating movements; storing new memories; and producing spoken language. The combined activities of all these brain regions lead to the higher-order processes that are so distinctly human: talking, reading, writing, playing music, studying biology, and remembering childhood scenes.

To approach the rich and intertwined subjects of sensory activity and the brain—much of that material at the frontiers of modern biology—we divide our chapter into the following discussions:

- Sensory receptors, the specialized neurons that register environmental data
- Chemoreceptors, those sensory neurons responsible for taste and smell
- The mechanoreceptors, which detect pressure, position, movement, and sound
- The thermoreceptors, which detect heat
- The photoreceptors, which detect light and allow vision
- The structure and function of the brain—site of sensory integration in vertebrates
- The living brain and its amazing capabilities
- Memory, the brain activity that underlies learning
- Neurotransmitters and neuropeptides, the chemical messengers of the brain

SENSORY RECEPTORS

Most multicellular animals have receptor cells that sense features of the organism's internal or external environment. These receptor cells act as transducers, transforming the energy of the stimulus—the sound waves from a tinkling bell, let's say—into a form of electrical activity, which is then transmitted to the central nervous system. During this transduction process, the original energy stimulus is often *amplified;* nerve impulses traveling from the eye to the brain, for instance, may have 100,000 times the energy of the light that stimulates each of the eye's receptor cells.

As Chapter 34 explained, some sensory neurons function as receptor cells, detecting and transducing energy themselves, whereas the receptor cells in certain special sense organs are separate and synapse with sensory neurons. In both types of sensory receptor cells, the dendritic end has unique structures and molecules that respond to light, chemicals, mechanical deformation, or other stimuli. Most such responses alter ion flow along the cell and elicit a particular pattern of electrical activity at the opposite end of the cell—which may communicate directly with dendrites of a postsynaptic cell.

Neurons then convey information about the stimulus encoded in this electrical pattern to the central nervous system—usually the brain—where the information is translated into a particular sensory perception, such as the tinkling of a bell. The environmental information detected and sent by the receptor cell is just the beginning. The way the receiving brain center *processes* the message is what determines the nature of the sensory perception resulting from a stimulus.

Most animals' sensory systems are more sensitive to new information than to background sensations. Receptor cells have a spontaneous level of electrical activity—a *baseline level*—that signals that a particular environmental condition is constant. When the environmental stimulus grows or recedes, the receptor cells register this new information via the increase or decrease in electrical activity. This mechanism allows an animal to distinguish between constant conditions that may be of little interest and changing ones that could affect its survival.

The brain can also distinguish "old" and "new" information. For instance, a person's sensory system and brain adapt to the presence of a newly donned pair of socks in less than a second; at first you sense their "touch," but you quickly become unaware of their presence. Receptors may continue to send in sensory information about the socks' presence, but certain brain circuits filter out this excess sensory input. If a pebble gets into your sock, however, you would notice this new condition immediately.

Sensory receptor cells can occur singly or can be clustered in **sense organs,** such as ears, eyes, or taste buds. These organs often have intricate anatomical parts that efficiently funnel stimuli—sound waves, light rays, or aromatic molecules—to the actual sensory receptor cells. **Chemoreceptors** detect specific molecules and ions and are prominent in taste and smell perception. **Mechanoreceptors** monitor pressure, stretching, and bending (shearing) forces and can sense body motion, sound waves vibrating in air or water, body acceleration or deceleration, and perhaps the pull of gravity. **Thermoreceptors** sense infrared heat waves or temperature changes in the skin or brain. Most animals have also evolved **photoreceptors,** which are sensitive to visible or ultraviolet light.

CHEMORECEPTORS

Virtually every animal on earth shares the ability to sense chemicals. The two types of chemical perception that biologists understand best are **gustation** (taste) and **olfaction** (smell).

Gustation: The Sense of Taste

The human tongue is dotted with 10,000 or so **taste buds,** many of which are located as part of larger swellings called *papillae* (Figure 36-1a and b). Each taste bud includes clusters of *taste* (gustatory) *receptor cells* (Figure 36-1c), which detect relatively high concentrations of dissolved substances and provide information about the flavor of a pickle, say, or a peach.

The tongue has receptors for four basic flavors—*sweet, bitter, sour,* and *salty*—located in specific zones (see Figure 36-1a). We can distinguish considerably more than four tastes, however, including such subtle flavors as honey versus table sugar, even though both are sweet. Several cell types make up each taste bud, and this may explain our range of taste capabilities. Also, while some taste receptor cells bind only one class of stimulus molecule, others can respond to several.

The outer ends of taste receptor cells have numerous microvilli (and thus a large crinkled surface area) on which the taste receptor molecules are located. Sensory neuron endings extend into each taste bud, cupping the receptor cells like fingertips holding an orange. When a tasteable material such as denatonium, the most bitter substance yet discovered, binds to receptor molecules on the microvilli, a second messenger causes internal calcium to be released. Sodium and potassium ions flow into the taste receptor cell, and neurotransmitter is released at its opposite end. As this occurs, there is a net flow of ions across the epithelium. These events cause the adjacent postsynaptic sensory neuron to increase the frequency with which it generates and propagates nerve impulses, which the brain then receives and interprets as particular tastes. In vertebrates, taste may allow an animal to discriminate among foods so that it can accept and eat safe ones and reject potentially harmful ones.

Olfaction: The Sense of Smell

Whereas taste receptors primarily detect concentrated molecules in an animal's food or water, olfactory receptors are sensitive to amazingly low concentrations of odorants in the environment, and they convey survival information about food, predators, potential mates, even imminent changes in the weather.

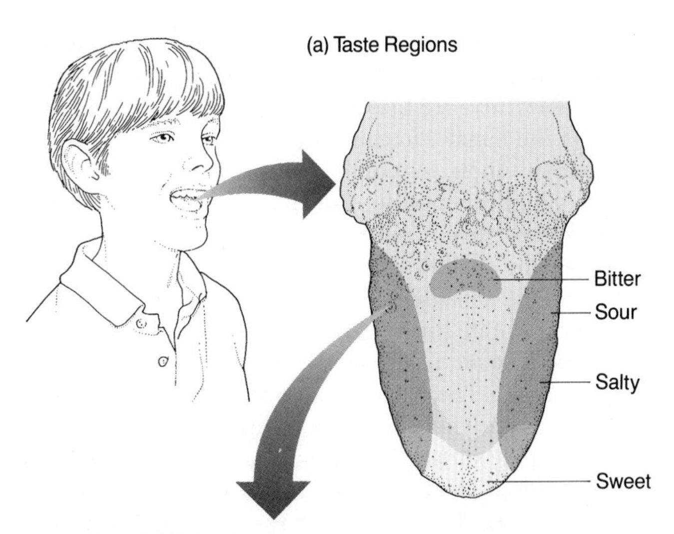

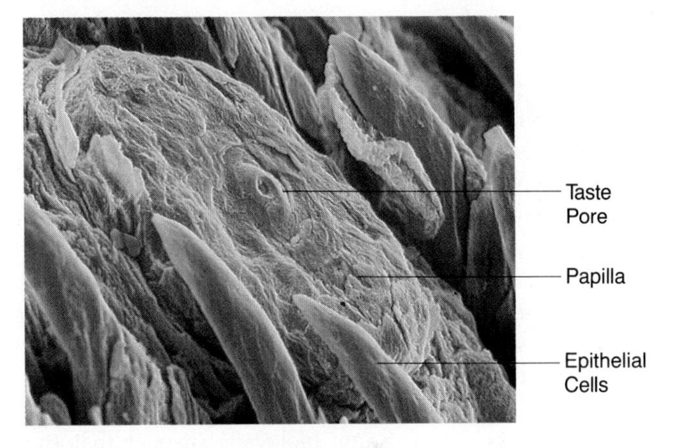

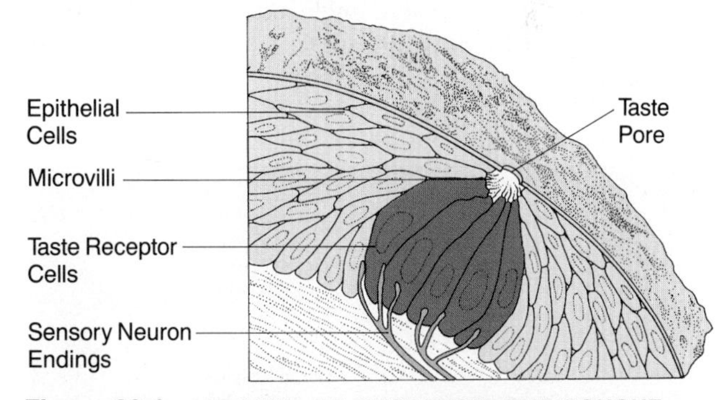

Figure 36-1 ORGANS OF GUSTATION: THE TONGUE AND TASTE BUDS.
(a) Taste buds are distributed on the tongue in zones with sensitivity to sweet, bitter, sour, and salty flavors. These regions overlap each other somewhat. (b) This scanning electron micrograph (magnified about 520×) shows the surface of a papilla on the tongue from above. The dot in the center is a taste pore that opens onto the taste bud cells below. (c) The diagram shows the nerve endings that synapse with receptor cells in a taste bud. From *Tissues and Organs: A Text-Atlas of Scanning Electron Microscopy* by Richard G. Kessel and Randy H. Kardon, W. H. Freeman and Company © 1979.

Consider, for instance, the 20,000 receptor cells in the fringed antennae of the male silk moth (Figure 36-2). These allow the male to detect *bombykol*, the sex pheromone of the female, in concentrations as low as 1 molecule per 10^{17} molecules of the gases in air. Similarly, the salmon's highly sensitive olfactory capacity allows it to migrate hundreds of miles up mountain river systems, following the shallow streams where the fish spent the early months of its life.

In a vertebrate's nasal cavities, the olfactory receptors are embedded in a thin sheath of olfactory epithelium (Figure 36-3). These receptors are true neurons; the outer end of each receptor is composed of microvilli, which increase surface area and contain the actual receptor molecules. The inner end is an axon, which carries nerve impulses. In land vertebrates, odorant molecules entering the nasal cavities bind to an odorant-binding protein (OBP) that spews out of tiny nearby glands. This protein then ferries the incoming molecules to the olfactory neurons. The binding process so concentrates the odorants that a human can, for example, smell one molecule of green pepper in one trillion air molecules.

Precisely how do olfactory receptors work? One hypothesis holds that the microvillar surface of each receptor cell is a patchwork of receptor molecules, which apparently have binding sites with precise configurations. The three-dimensional shape of a molecule, in turn, determines whether it will bind to given receptor molecules. Adding a methyl (CH_3) group to a certain ring compound, for example, changes its odor from spearmint to wintergreen. Some biologists suspect that the olfactory epithelium contains only about ten types of receptor cells and that each single odorant might attach with a different binding affinity to a subset of these cell types. The specific pattern of output from the unique combination of olfactory neurons might then equal the particular odor the animal perceives.

Figure 36-2 THE FEATHERY ANTENNAE OF THE MALE SILK MOTH.
Male silk moth antennae are extremely sensitive receptors for the female pheromone bombykol. This sensitivity enables male silk moths to home in on females over great distances.

Once an odorous substance like wintergreen binds to a glycoprotein receptor molecule, a series of events takes place that causes a second messenger (cyclic AMP or GMP) to bind to and open channel proteins in the cell membrane. This permits Na^+ to enter the receptor cell and change its state of polarization (see Chapter 34). As additional odor molecules bind, nerve impulses are generated either more rapidly or less rapidly than the baseline rate. Thus, the response in the olfactory receptor cells is *graded* (see Chapter 34), as is common in many sensory systems. The changed patterns of impulses travel to the *olfactory bulb* of the brain (see Figure 36-3), where information processing begins. Interestingly, nerve tracts from the olfactory bulbs run directly to two brain regions involved in memory and mood, the hippocampus and the limbic system. This may help explain why odors can sometimes trigger elaborate memories and moods.

MECHANORECEPTORS

Mechanoreceptors can detect touch, sound, electrical and magnetic fields, and gravity. Nevertheless, all these detections depend on the same physical principle: distortion of the shape of the mechanoreceptor cell and its plasma membrane. Consider a vertebrate's skin, for example, with its two types of mechanoreceptors, *Meissner's* and *Pacinian corpuscles*, both sensitive to tactile pressure (Figure 36-4). A Pacinian corpuscle resembles an onion in which layers of cells surround the dendritic ending of a mechanoreceptor neuron. Firm pressure applied to the skin stretches the Pacinian corpuscle out of its usual shape, the neuron's baseline firing pattern is altered, and the brain registers pressure. The club-shaped Meissner's corpuscles detect light pressure (see Figure 36-4).

Skin also appears to have bare nerve endings that function as *pain receptors* and *thermoreceptors* (see Figure 36-4).

Two additional kinds of receptors take in information on environmental changes and body positions. Mammals and birds have mechanoreceptor cells wrapped around the base of hairs and feathers, and arthropods have the receptors inside sensory hairs. Gusting wind, flowing water, or other disturbances bend the hairs or feathers, distort the receptor cells, and generate new patterns of sensory impulses. Similarly, joints, tendons, and muscles have stretch receptors, or **proprioceptors,** that help

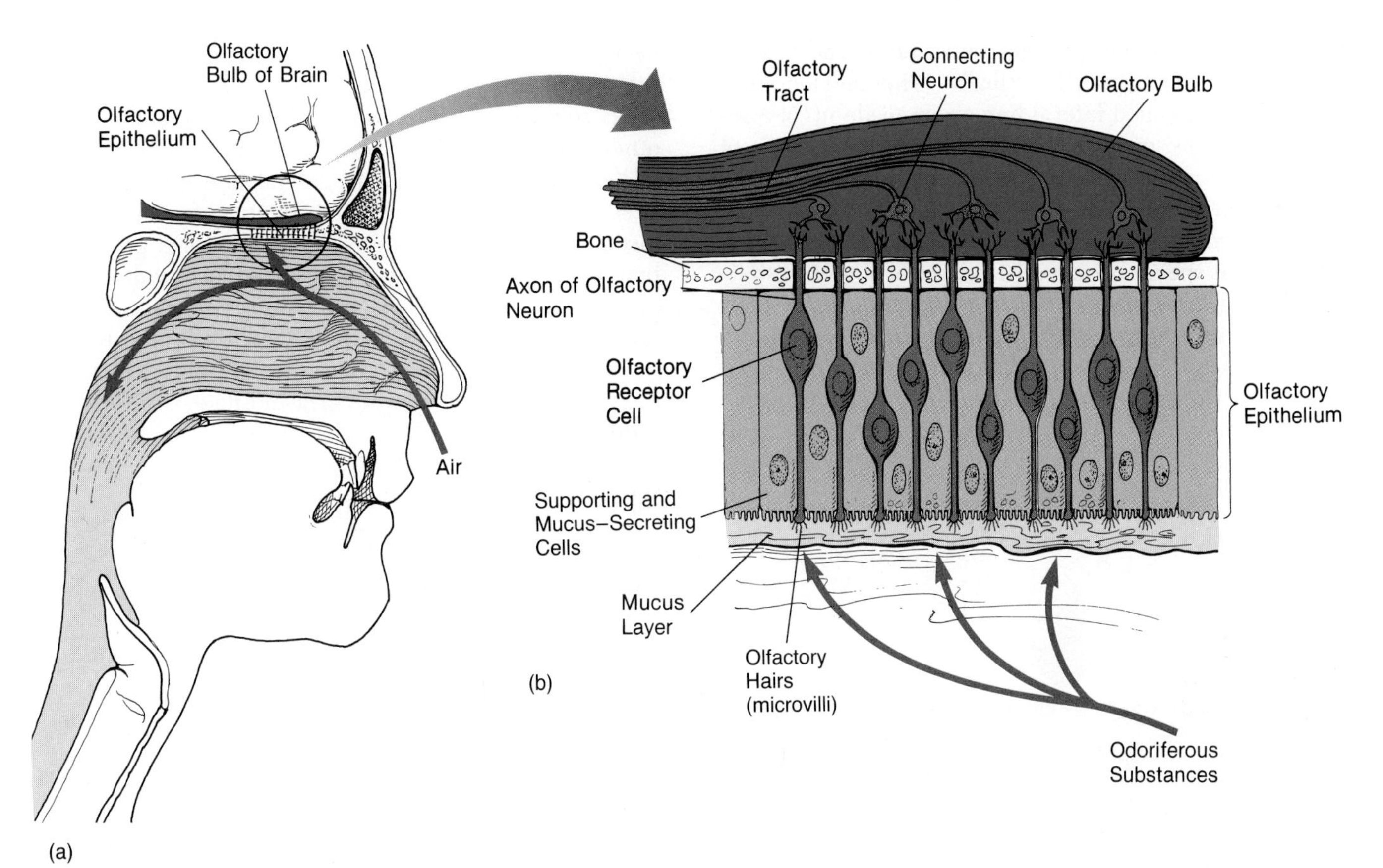

Figure 36-3 THE SOURCE OF SMELL: OLFACTORY RECEPTORS IN THE NOSE.
(a) Olfactory receptor cells are located in the epithelium of the nasal cavity; each receptor projects olfactory hairs into the thin layer of mucus that lines the nasal passage. (b) When an odor molecule binds to the receptor, a chain of events involving second messengers ensues; the chain culminates in the relay of patterns of nerve impulses to the olfactory bulb of the brain. From there, other neurons carry the information to higher brain centers.

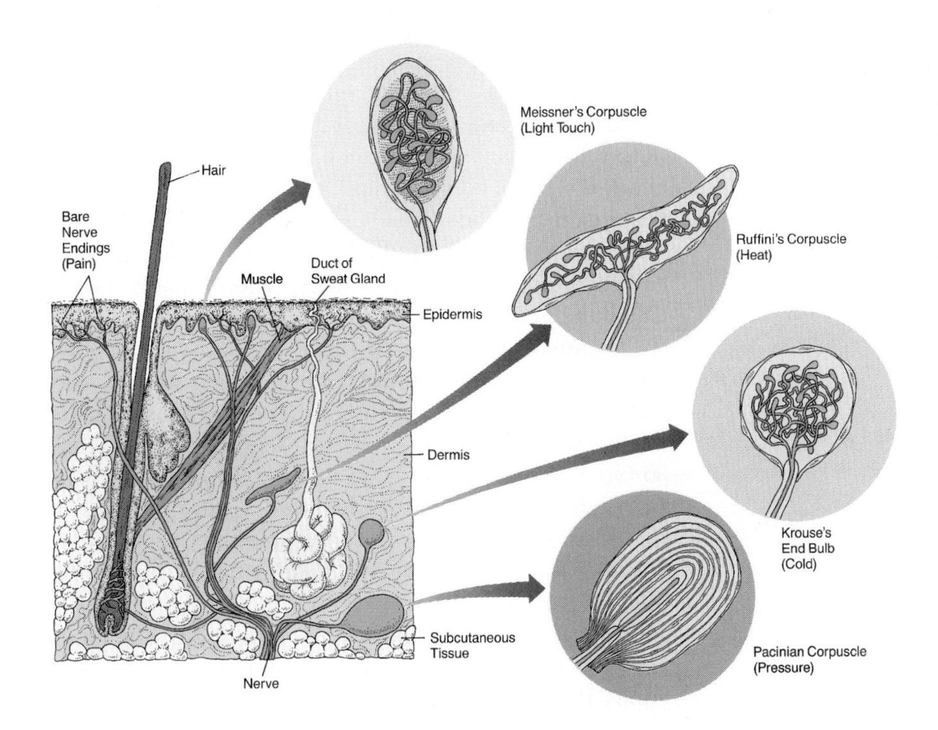

Figure 36-4 SENSORY RECEPTORS IN THE SKIN.
This longitudinal section through a piece of mammalian skin shows the major structures and their sensory capacities. Bare nerve endings in epidermis and dermis report pain. Meissner's corpuscles report light touch (pressure). Pacinian corpuscles report firmer pressure. Ruffini's corpuscles report heat, while Krause's end bulbs report cold. Pacinian corpuscles are classic mechanoreceptors in which onion-shaped layers of cells and fibers surround a mechanoreceptor dendrite; firm pressure downward on the skin, or body or limb movements that stretch or compress the skin, deform the "onion" and thus act on the sensory nerve endings.

detect limb position, muscle contractions, and related processes (see Chapter 37). Finally, fishes and amphibians have a so-called **lateral line organ** made up of a series of cavities, or pitlike depressions, along the side of the body (Figure 36-5). As the organism swims or water flows along its sides, sensory hairs embedded in flexible jellylike *cupulae* (tiny cups) are bent one way or the other. This speeds or slows the release of neurotransmitter by the hair cells and triggers adjacent sensory neurons to send impulses more frequently or less frequently to the brain. This, in turn, allows the fish or amphibian to detect the presence of nearby objects or organisms, sound vibrations, movements through the water, and even minute variations in water temperature or salinity. Sharks, catfish, mormyrids (weak electric fish; sometimes sold as "elephant fish" in pet shops), and certain other species have modified mechanoreceptors (electroreceptor cells) in the lateral line pits that lack cilia and can detect weak electric currents. Sharks, for instance, can detect minute electric currents produced by the muscles of prey species, while mormyrids generate a stream of weak electric pulses from an electric organ in the tail, which is then detected by the lateral line organ. When a rock, plant, or other object disrupts the field, the fish can locate or avoid it, even in murky water.

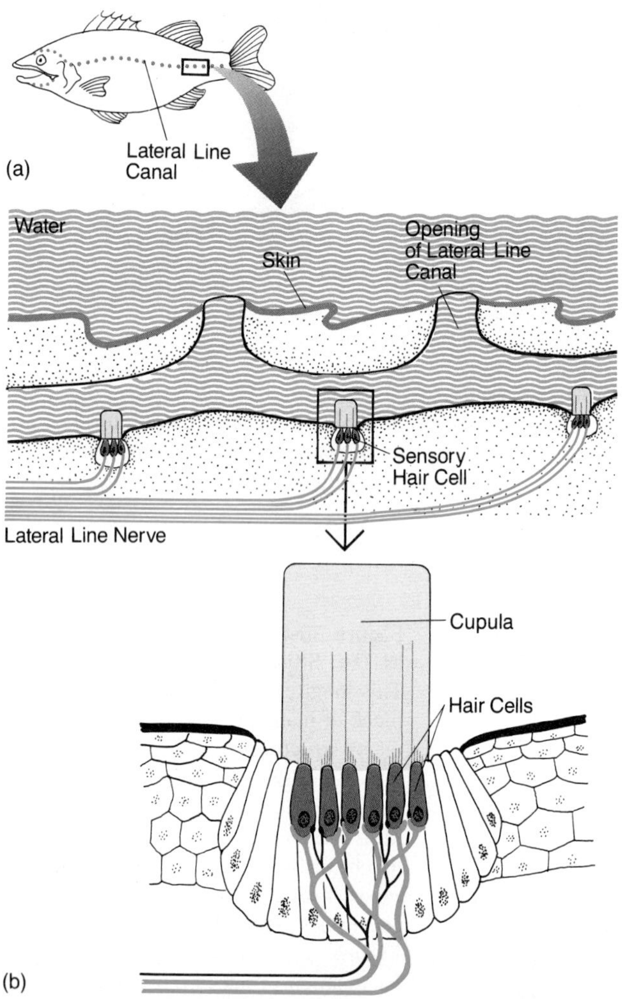

(a) Lateral Line Canal

Water
Skin
Opening of Lateral Line Canal
Sensory Hair Cell
Lateral Line Nerve

Cupula
Hair Cells

(b)

Figure 36-5 SENSATION IN THE SEA: THE LATERAL LINE ORGAN.
(a) Sensory cells associated with the lateral line systems of many fishes and amphibians detect changes in the velocity of the water currents around the animal, as well as other environmental properties. (b) The hair cells generate nerve impulses in adjacent sensory neurons faster if the cupula is bent in one direction and slower if it is bent in the opposite direction, thus signaling the direction and relative speed of water flow.

The Statocyst and Inner Ear: Maintaining Equilibrium and Sensing Movement

In both vertebrates and invertebrates, the ability to detect up from down and to sense motion depends on the deformation of sensory hair cells of the sort present in the lateral line.

In invertebrates, these sensory hair cells surround structures called **statoliths**—masses of calcium carbonate crystals, sand grains, or other dense materials. Like a yogi on a curved bed of nails, each statolith rests on the hairs within a fluid-filled sac called a **statocyst**. When the animal moves, the relatively heavy statoliths shift position too, bending some of the projections of underlying sensory hair cells. This bending, in turn, alters the electrical activity of those cells, the release of neurotransmitter from their axons, and the sensory messages that reach the central nervous system.

For example, the statoliths of a jellyfish adrift in the sea roll from side to side, relieving pressure on some sensory hairs and applying it to others. The changing output from many receptor neurons conveys information about the statocyst's position, and hence the position of the organism's body, relative to the center of the earth.

In vertebrates, fluid-filled chambers within the inner ear carry out identical functions. These chambers are called the *utriculus, sacculus,* and *lagena* (Figure 36-6a). (In mammals, the lagena is called the *cochlea.*) Crystalline structures called **otoliths** (literally, "ear stones") rest on cupula material that overlies patches of sensory hairs on the bottom or side walls of each chamber (Figure 36-6b). Like the statoliths of invertebrates, the otoliths shift and bend the sensory hairs to register changes in the position of an animal's head and thus of its body. Otoliths also detect motion, shifting backward as the body moves forward, and vice versa.

Finally, tubelike **semicircular canals** within the inner ear measure angular (curving) accelerations and decelerations of the head and body and allow animals to detect

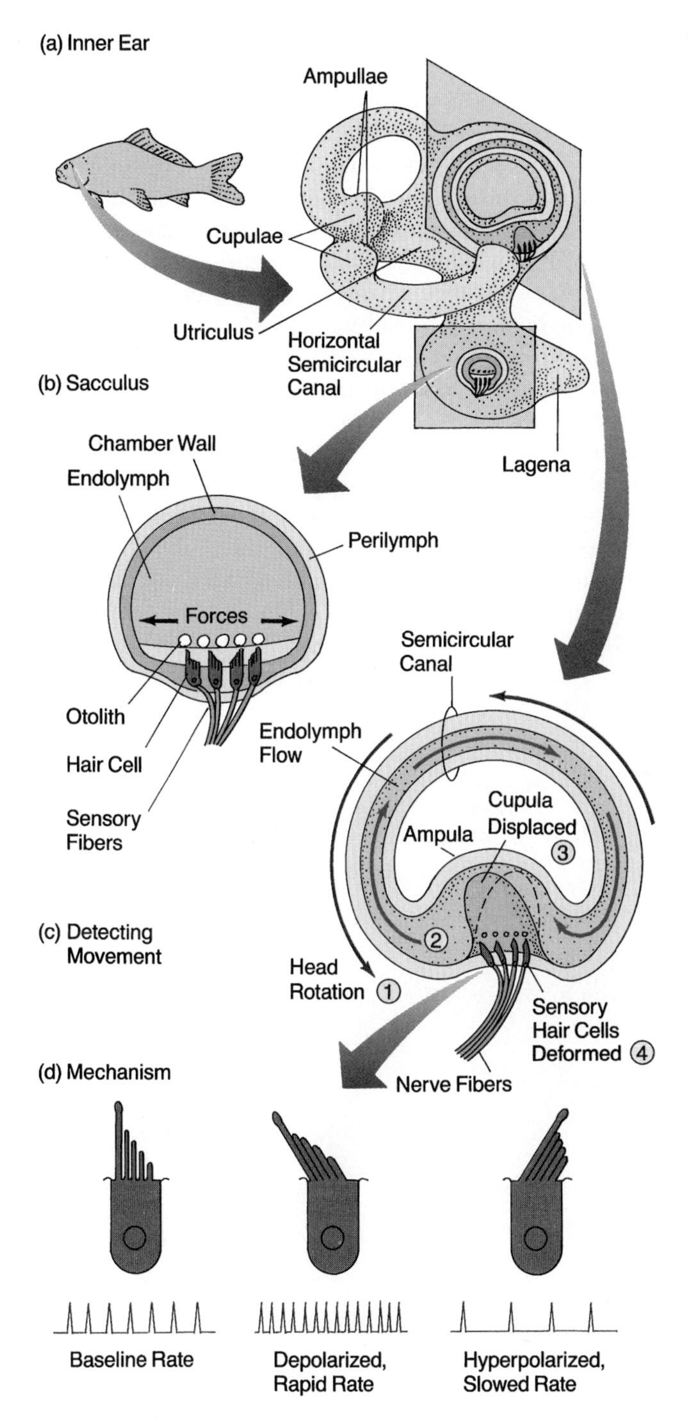

◀ **Figure 36-6 MECHANORECEPTORS OF THE INNER EAR: SENSING ANGULAR ACCELERATIONS AND DECELERATIONS.**
(a) The hollow, fluid-filled canal and chamber system of a fish's left inner ear. Each of the three semicircular canals possesses a cupula attached to a set of sensory hair mechanoreceptor cells. (b) Each of the three inner-ear chambers—the utriculus, the sacculus (shown here), and, in some species, the lagena—contains sensory hair cells. The hairs of these cells are embedded in a gelatinous cupula that is covered with otoliths. Body position and gravitational force are constantly monitored by the sensory hair cells as the otoliths shift due to body movement. (c) As the head moves in a counterclockwise direction (1), the walls of the canal move in the same direction (red arrows). Endolymph fluid in the canal moves in the opposite, clockwise direction (2, blue arrows), thereby displacing the cupula (in this case to the left, 3). That displacement in turn bends the "hairs" on the sensory hair cells (4) and changes the rate at which they cause generation of nerve impulses. If the animal's head, and hence the canal, stopped moving and rotated clockwise instead, the fluid would flow counterclockwise, bending the cupula in the opposite direction and changing accordingly the rate at which impulses are generated. (d) In the resting state, the cilia are not bent, and a baseline nerve impulse rate is transmitted over the nerve fibers (left). When the cilia are bent in the direction of the largest cilium, the cell depolarizes, and the impulse rate rises (center). When the cilia are bent in the opposite direction, the cell hyperpolarizes, and the impulse rate falls below baseline (right).

the extraordinary range of movements possible in three dimensions.

On each side of a vertebrate's head, there are three semicircular canals, one oriented horizontally and two oriented vertically. The three are more or less at right angles to each other (see Figure 36-6a). Each canal is filled with a fluid, called **endolymph,** which moves freely into and out of the ends of each canal, something like water sloshing through a pipe (Figure 36-6c). At one end of each canal is a hollow endolymph-filled chamber, or *ampulla*. Extending across each ampulla, much like a hinged door or gate, is a gelatinous cupula; one end of the cupula is attached to the hairlike extensions of mechanoreceptor cells. How does this system of pipes, fluid, and sensors operate?

In a series of experiments, Werner Loewenstein, a physiologist at the University of Miami, discovered a key mechanism. When a vertebrate turns its head (Figure 36-6c, step 1), endolymph flows into a semicircular canal on one side and out of the corresponding canal on the opposite side (step 2), and both flows stimulate sensory hairs (steps 3 and 4). The bending of the hairs in one direction causes the cells to release neurotransmitter more rapidly; in the opposite direction, more slowly (Figure 36-6d). When the sensory hairs are not bent at all, the cells release neurotransmitter at the baseline rate. The rate of neurotransmitter release governs the rate at which postsynaptic neurons generate nerve impulses that travel to the brain. The animal's brain interprets that information along with sensory data from the ear's other semicircular canals, otolith chamber systems, and eyes, and the result is continuous perception of the animal's position in space and its linear and angular movements.

The Ear and Hearing

In many vertebrates, sound is detected in the inner ear. An animal's ability to hear, however, depends on the anatomy of the outer and middle ears.

Whether an external ear is large or small, furry or fleshy, it is designed to gather high-frequency vibrations (sound waves) traveling through the air and to conduct them inward so that they strike the **tympanic membrane,** or *eardrum,* separating the outer and middle ears (Figure 36-8). This tightly stretched membrane vibrates when sound waves strike it, and it, in turn, sets in motion similar vibrations in a chain of three tiny bones in the middle ear—first the *incus* (sometimes called the hammer), next the *malleus* (anvil), and finally the *stapes* (stirrup).

As the incus, malleus, and stapes vibrate in sequence, they conduct sound waves through the air-filled middle ear to the **oval window,** a thin, taut inner-ear membrane in the upper wall of the **cochlea.** As this chain of events proceeds, the sound waves are amplified, setting up vibrations that are powerful enough to overcome the inertia of an inner-ear fluid, the *perilymph.*

The mammalian cochlea is the primary structure involved in hearing, and it is essentially a set of three tubes, coiled one atop another. Every time sound waves set the oval window vibrating (Figure 36-8a, step 1), pressure waves arise in the perilymph in the cochlear chamber behind the oval window (step 2). Each wave travels through the upper chamber of the cochlea (step 3) and, when it reaches the end, passes through an opening into the lower chamber (step 4). From there, the waves sweep along the lower chamber, ending up at the **round window** at the chamber's base, which bows outward with each vibration to relieve the pressure (step 5).

Pressure waves in the perilymph affect the position of the **basilar membrane,** a thin sheet of tissue that lies along the base of the middle cochlear chamber. Resting on the basilar membrane and directly beneath a second

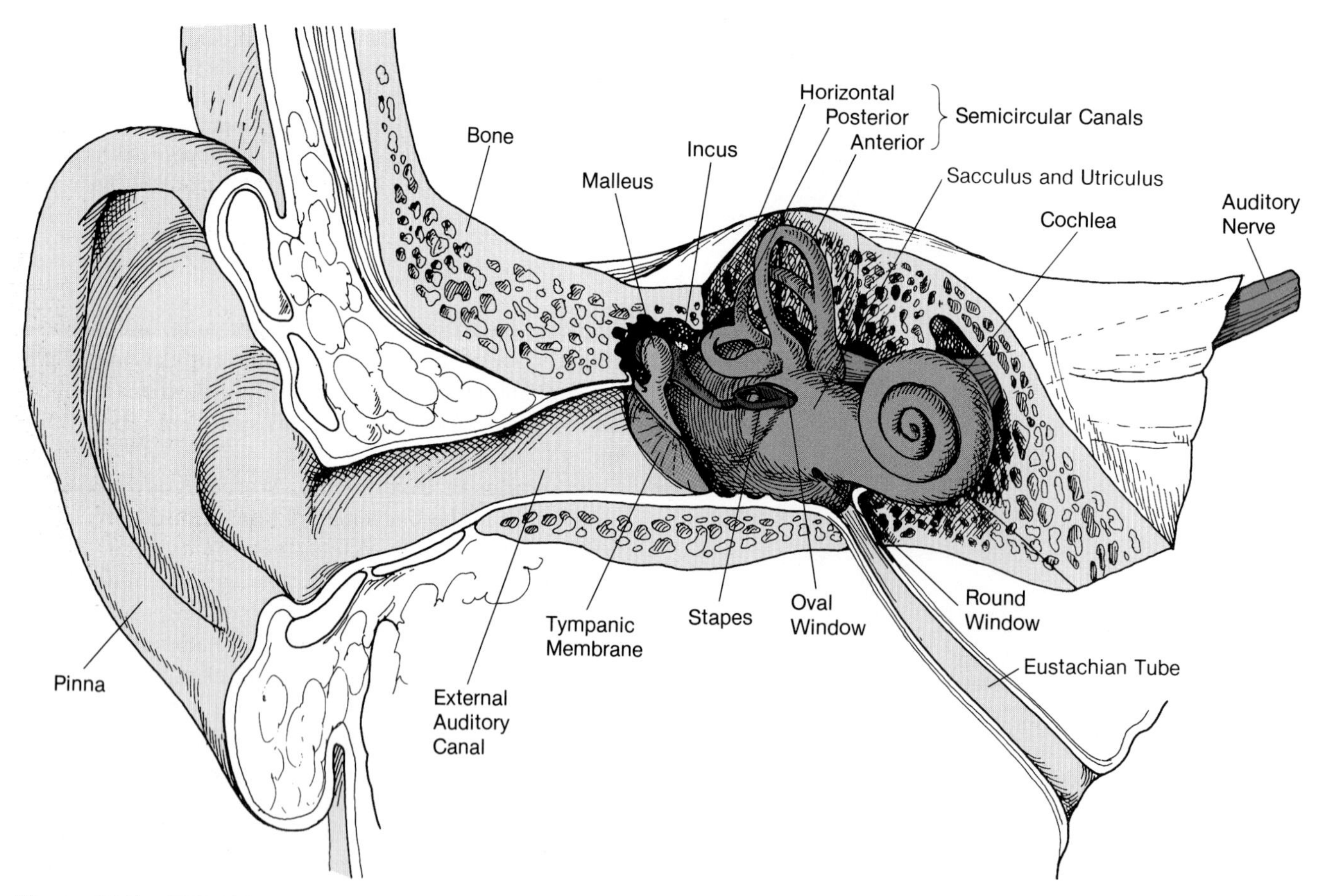

Figure 36-7 THE EAR.
The sense of hearing depends on the anatomy of the ear. The ear has three sections—the external ear, the middle ear, and the inner ear. Sound is channeled through the external ear inward to the tympanic membrane. Sound vibrations pass along the middle-ear bones, vibrate the oval window, and set up similar vibrations in the fluid of the inner ear. Those vibrations pass into the snail-shell-shaped cochlea, where they are detected by sensory hair cells. Pressure is relieved at the round window. The Eustachian tube connects the spaces around the inner-ear structures with the mouth cavity.

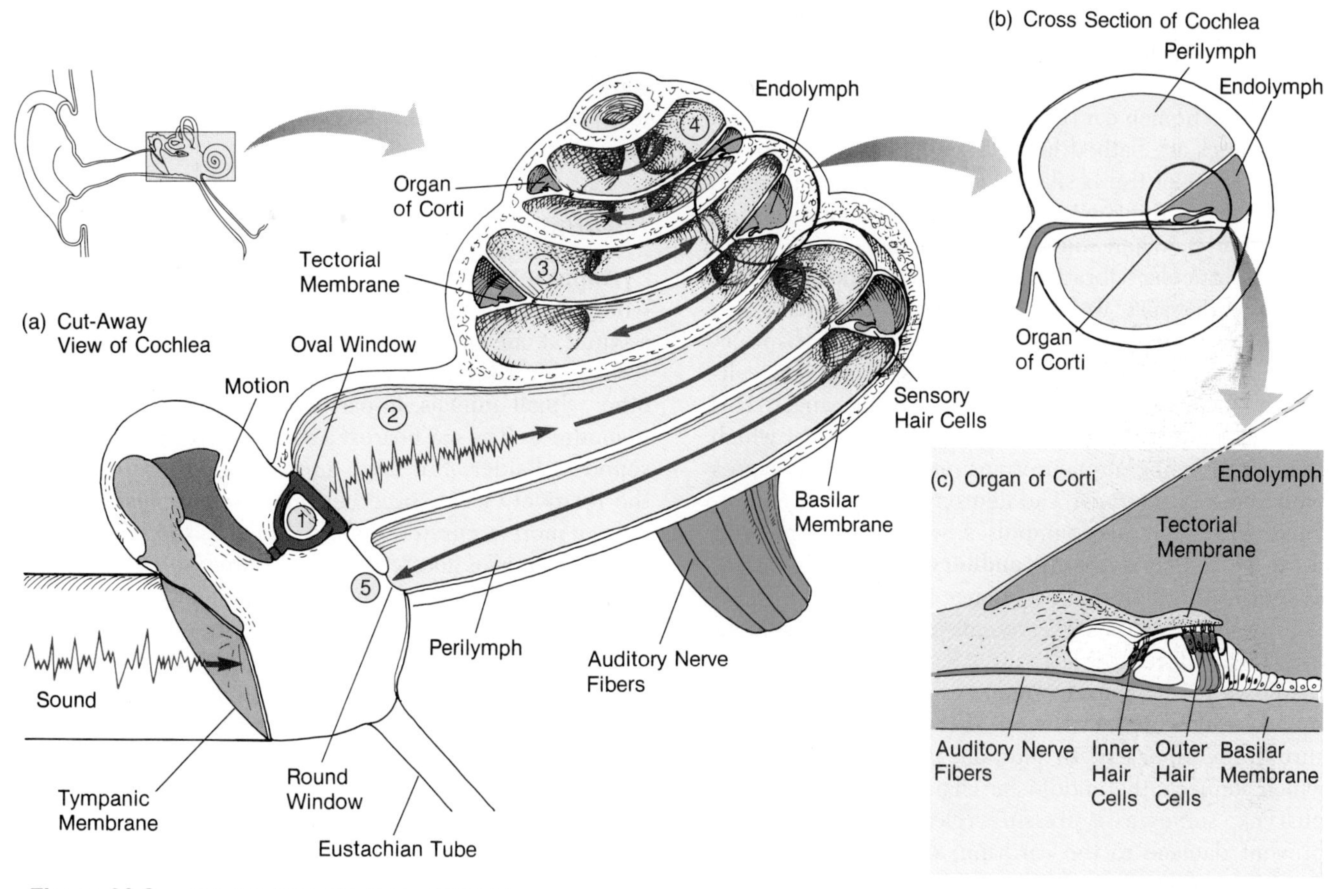

Figure 36-8 STRUCTURE OF THE COCHLEA.
The cochlea of mammals is the space-saving spiral structure in the inner ear that is involved in hearing. (a) This section of a model cochlea shows the tympanic membrane, middle-ear bones, oval window, and round window as sound striking the tympanic membrane sets up compression waves that travel in the perilymph fluid of the cochlear chambers (arrows). (b) A cross section of the cochlea reveals the middle chamber which contains (c) the organ of Corti with its sensory hair cells that rest on the basilar membrane. The tectorial membrane rests on the sensory hairs themselves. Compression waves traveling outward from the oval window in the upper chamber, and then back in the lower chamber toward the round window, displace the basilar membrane. That, in turn, moves the sensory hair cells relative to the tectorial membrane, resulting in altered release of neurotransmitter by the hair cells and new nerve impulse patterns over the auditory nerve to the brain, interpreted as sound. (d) This scanning electron micrograph shows the organ of Corti with the overlying tectorial membrane removed (magnified about 1,700×). Each V-shaped set of structures is the group of so-called stereocilia—the sensory hairs—that arise from a single underlying mechanoreceptor hair cell. When sound waves vibrate perilymph in the cochlea, the cells visible here would vibrate up and down (above and below the plane of this page), and thus the stereocilia would be bent as they moved relative to the overlying tectorial membrane. From *Tissues and Organs: A Text-Atlas of Scanning Electron Microscopy* by Richard G. Kessel and Randy H. Kardon, W. H. Freeman and Company © 1979.

(d) Surface View of Hair Cells

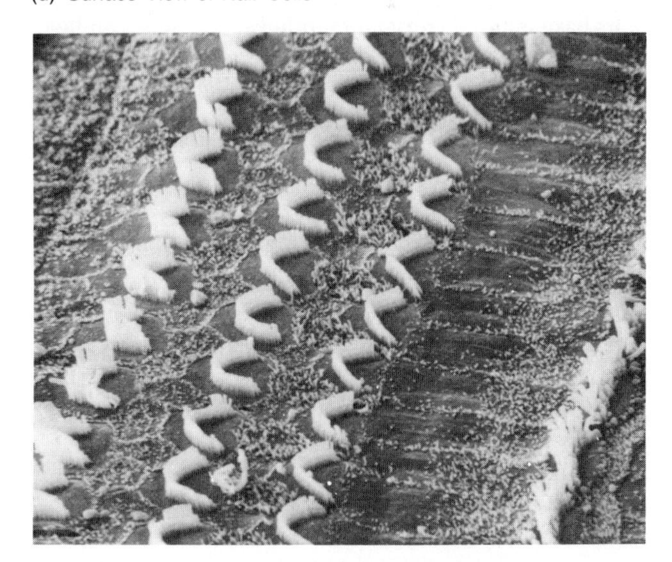

layer is the **tectorial membrane** of the **organ of Corti,** composed of thousands of sensory hair cells—about 17,000 in a human ear (Figure 36-8b–d). These cells and membranes are bathed in endolymph fluid. The spatial arrangement of the basilar and tectorial membranes in the cochlea is the key feature of the ear's sound-sensing architecture. When sounds enter the ear—each sound with a particular vibrational frequency—the vibrations generate pressure waves that displace particular sites along the basilar membrane. Wherever this displacement takes place, microvilli on the overlying hair cells are bent against the tectorial membrane at an angle. This angle of bending, in turn, determines the rate at which the hair cells liberate neurotransmitter at their synapses with auditory neurons. The neurons are thereby stimulated to generate nerve impulses, which are propagated down axons that run in the **auditory nerve** to the brain. The result of this simple sequence is an auditory "message" telling the brain that sounds of particular frequencies have entered the ear.

Whenever the oval window vibrates inward, the round window flexes outward, thereby equalizing pressure in the inner ear. In addition, the *Eustachian tube,* which connects the middle-ear chamber with the upper pharynx, serves as a pressure release site, helping to prevent damage to the eardrum, oval window, round window, and other ear parts when an animal is exposed to very loud noises.

The range of frequencies to which an ear is sensitive varies with the length of the cochlea and the properties of its basilar and tectorial membranes. A human can hear frequencies as low as 30 to 50 hertz and as high as 20,000 hertz; bats and porpoises may detect echoing sounds of 100,000 to 140,000 hertz.

Extremely loud noises—one loud note at a rock concert, for example—can permanently or temporarily deafen an animal, displacing or damaging some of the sensory hair cells so that instead of standing erect, they flop over like wet noodles. Other sources of damage include high doses of antibiotics, such as gentamicin, neomycin, tobramycin, and streptomycin, which can cause the stereocilia to clump and thus fail to function, and middle-ear infections caused by bacteria and viruses, mumps, measles, and meningitis.

As many as 200,000 people in the United States are deaf because of inner-ear damage. Often, their condition is mistakenly called "nerve deafness" when, in fact, the nerve cells can function but their hair cells can no longer transduce. But some of these people can now be helped because medical researchers have designed devices called *multichannel cochlear implants* to bypass the damaged hair cells and deliver electrical signals directly to the nerves. The patient wears a tiny microphone connected to a signal processor with electrodes running directly into the cochlea, where hair cells and sensory nerve cells reside. Sounds picked up by the microphone can then be converted to electrical signals that stimulate inner-ear nerves directly and enable the patient to hear once more.

Locating Sounds

Animals use two cues to detect the direction from which a sound comes: the relative loudness (intensity) of a sound at one ear or the other and the difference in time it takes for a sound to register in one ear and then the other. Small animals, in particular, measure differences in loudness: The ear nearest a sound source registers the sound as louder. For large animals, including humans, the relatively wide space between the ears makes time delay a more accurate gauge. In general, though, highly visual organisms locate a sound source not by hearing alone but by flicking their eyes immediately toward the general direction from which a new sound comes. Another indication that hearing and seeing are closely related in some mammals is the presence of two topographically aligned brain "maps" of auditory and visual space. When experimenters surgically redirect an animal's eye, the visual brain map shifts, and the auditory map shifts as well. Similar shifts no doubt occur during a young mammal's growth, as its head changes shape. Although the distance between eyes and ears is altered, the animal's ability to localize sights and sounds is maintained.

Echolocation: "Seeing" with Sound

Small nocturnal animals, such as shrews and bats, and aquatic mammals, such as porpoises and whales, are often active under conditions of reduced light and low visibility. They have evolved the ability to use sound to sense their way in the dark.

The sound-based system that bats use to navigate and hunt in the dark is known as **echolocation.** As the animal flies, it produces a series of sound pulses at the rate of about 20 per second, each with a frequency ranging from 40,000 to 100,000 hertz—well above the maximum frequency audible to humans—and each with the equivalent intensity of a subway train roaring past a platform. Such pulses of sound bounce off moths, tree branches, and other obstacles and produce faint echoes that the bat, with its often huge ears (Figure 36-9), is able to gather, hear, and analyze.

Once the bat senses an object of interest, the rate of sound pulse production increases to 50 or 80 per second, and the returning echoes provide an amazingly detailed acoustic picture of the darkened forest or cave through which the bat rapidly flies and navigates.

Porpoises, whales, and other marine mammals generate echolocating clicks in their larynx or in the system of

Figure 36-9 ECHOLOCATION.
Bats navigate by generating ultrasonic sound pulses that are inaudible to the human ear. Echoes of the pulses provide information about an object's position, direction of movement, and even composition. The huge ear pinnae help this pallid bat (*Antrozous pallidus*) to gather these faint echoes and may aid in directional location of the echo source.

air-filled tubes leading to the blowhole. In addition, such animals have an array of adaptations for gathering directional information, including isolation of the inner ear in a foamy cushion so that sound may reach the cochlea only from the eardrum. Perhaps most remarkable, however, are the eerily beautiful moans, whistles, squeaks, and other noises that sea mammals use to communicate and, apparently, coordinate highly complex group behavior.

Magnetism: Detecting the Earth's Magnetic Field

Few modern biological discoveries have been more surprising than the finding, in the 1970s, that birds, bacteria, and perhaps many other types of organisms orient their bodies to the earth's magnetic field. How can living things perceive a force as elusive as magnetism?

Many birds and at least one mammal—the porpoise—have a small number of tiny crystals of magnetite (a form of iron oxide) in tissues near the brain. Some humans have such crystals in the delicate bones of their sinuses. Likewise, *magnetobacteria* have magnetite crystals in their cytoplasm and swim toward or away from the poles of an applied magnetic field. How some higher organisms use magnetic particles to sense the direction of the lines of magnetic force generated by the earth's core remains a subject of research. One hypothesis suggests that the crystals in a pigeon's head, for example, function like tiny compass needles, and mechanoreceptors might detect their relative movement. Other work indicates that in the rat, changes in magnetic field can alter enzyme activity and levels of melatonin in the pineal gland. Still other work suggests that regularly arrayed visual pigment molecules in the eyes' light-sensing layer (also present in pineal glands) undergo certain changes as the head and eyes move relative to the earth's magnetic field. They could allow the animal to sense magnetic fields, but not by a mechanoreceptor mechanism. Whatever the mechanisms, homing, migration, and many other directional animal behaviors depend on detecting magnetic fields.

THERMORECEPTORS

Detecting magnetic fields is relatively rare, but nearly all animals can respond to variations in temperature. In various vertebrates, nerve endings in the skin and tongue alter their firing pattern when the local temperature changes. Warm receptors generate nerve impulses faster as temperature rises; conversely, cold receptors generate impulses faster when the temperature of the nerve endings drops. No one knows just how these cells function, but snakes that hunt their food by detecting temperature changes may provide clues.

Figure 36-10 THERMORECEPTORS IN A NIGHT HUNTER.
The facial pit of a bamboo viper, a member of the pit viper family, is visible as a dark line just in front of the eye. A pit viper's infrared detectors, located in its facial pits, help it find prey in the dark.

Even on a moonless night, with its prey remaining perfectly still, a rattler can detect a mouse's whereabouts and strike with deadly aim. This flawless marksmanship depends on *infrared detectors*, organs located in the snake's *facial pits* that detect the long infrared wavelengths of radiant-heat energy. All pit vipers have such pits (Figure 36-10).

While they probably function in a similar manner, a snake's facial pits are supersensitive compared to the warm and cold receptors of an insect's antennae or a person's skin; they can detect a temperature increase of only 0.002° C. Consequently, a pit viper can sense, within half a second, a mouse whose body temperature is 10° C warmer than its environment if the mouse is within range of about 40 cm.

PHOTORECEPTORS

Animals have an amazingly diverse array of light-sensitive structures capable of detecting visible wavelengths of light. Some protists, for example, have clusters of light-sensitive molecules in eyespots that allow the cells to move toward or away from a light source (see Chapter 21). Flagellates, whose light-sensing pigment is masked on one side, will perform a kind of miniature ballet, rotating toward a light source shined from one side on a culture dish. The same light-absorbing pigment (called opsin) in protist cells is found in organisms as diverse as archaebacteria, unicellular algae, fruit flies, goldfish, and cows, and all these pigment proteins share very similar amino acid sequences. The gene for opsin probably originated billions of years ago in prokaryotes, reflecting the antiquity of the capacity to absorb light energy.

In complex animals, accessory cells form a lens that gathers and focuses light on the underlying light receptors. As the different but wonderfully complex eyes of vertebrates, mollusks, and arthropods all reveal, such an arrangement allows an animal to detect *lower intensities* of light, to sense *direction*, and to sense *motion* in nearby objects.

The Vertebrate Eye

Vertebrate eyes, and those of octopuses and squid, function much like cameras: A single lens focuses light images on a sheet of light receptors (equivalent to the film), and the image is then "developed" by neuron processing in the visual cortex of the brain.

Typical "camera eyes" are roughly spherical (Figure 36-11) and are mostly surrounded by a tough, protective sheet of connective tissue called the *sclera* and by the *choroid*, a layer containing blood vessels that carry oxygen and nutrients to the eye and remove carbon dioxide and wastes. Pressurized fluid-filled spaces—the *aqueous humor* and *vitreous humor* chambers—help give the eye its shape. The **cornea,** the tough, transparent portion of the sphere through which light enters, is a specialized skin tissue composed mainly of aligned collagen fibers.

When light enters the eye, it passes through the cornea and aqueous humor. It may then be halted by the **iris,** a pigmented ring of tissue (blue, brown, green, gray, or intermediate shades in the human eye) that can be opened or closed by radially oriented muscles under reflex control. Or the light may pass through the iris's dark central opening—the **pupil**—to the **lens,** a structure built of highly ordered cells packed with structural proteins called crystallins. Each of four different crystallins is identical to a metabolic enzyme. During animal evolution, lens proteins may have been selected solely because they formed transparent crystals.

Light rays traveling through the lens and the pupil are focused on the rear surface of a light-sensing epithelial layer, the **neural retina.** An important component of the neural retina in primates, birds, and some reptiles is the **fovea,** a region about 1 mm in diameter where maximal *acuity* (high-resolution viewing) is achieved. In all vertebrate eyes, there is an optic disk, or "blind spot," where nerve fibers of the neural retina exit from the eye over the **optic nerve,** which goes to the brain. Finally, surrounding the neural retina is the **pigmented retina,** a layer that absorbs or reflects light that chances to pass unabsorbed through the neural retina.

Forming an Image: Focusing the Eye

When we focus a camera, we actually move the lens toward or away from the film, and the eyes of fishes, amphibians, and snakes focus in just this way. In birds and mammals, however, both cornea and lens participate in the focusing process.

First, the curvature of the cornea causes light rays to bend to varying degrees. This corneal focusing is usually sufficient to provide a sharp image of distant objects. As focusing takes place, the lens is held taut in a somewhat flattened position by the elastic ligament of the *ciliary body*, which supports it (see Figure 36-11). When the object being viewed is near, however, the cornea cannot bend the light enough to focus the image on the retina, so the *ciliary muscles* contract, allowing the lens to round up. This increases the angle of bending of the light rays, bringing the image into sharp focus. Birds also bend the cornea to help this process. In land vertebrates, this process—called *accommodation*—goes on constantly as an animal shifts its gaze from one object to another at differing distances. The nearsightedness and farsightedness so common in human eyes stems from abnormalities in the shape or functioning of the cornea

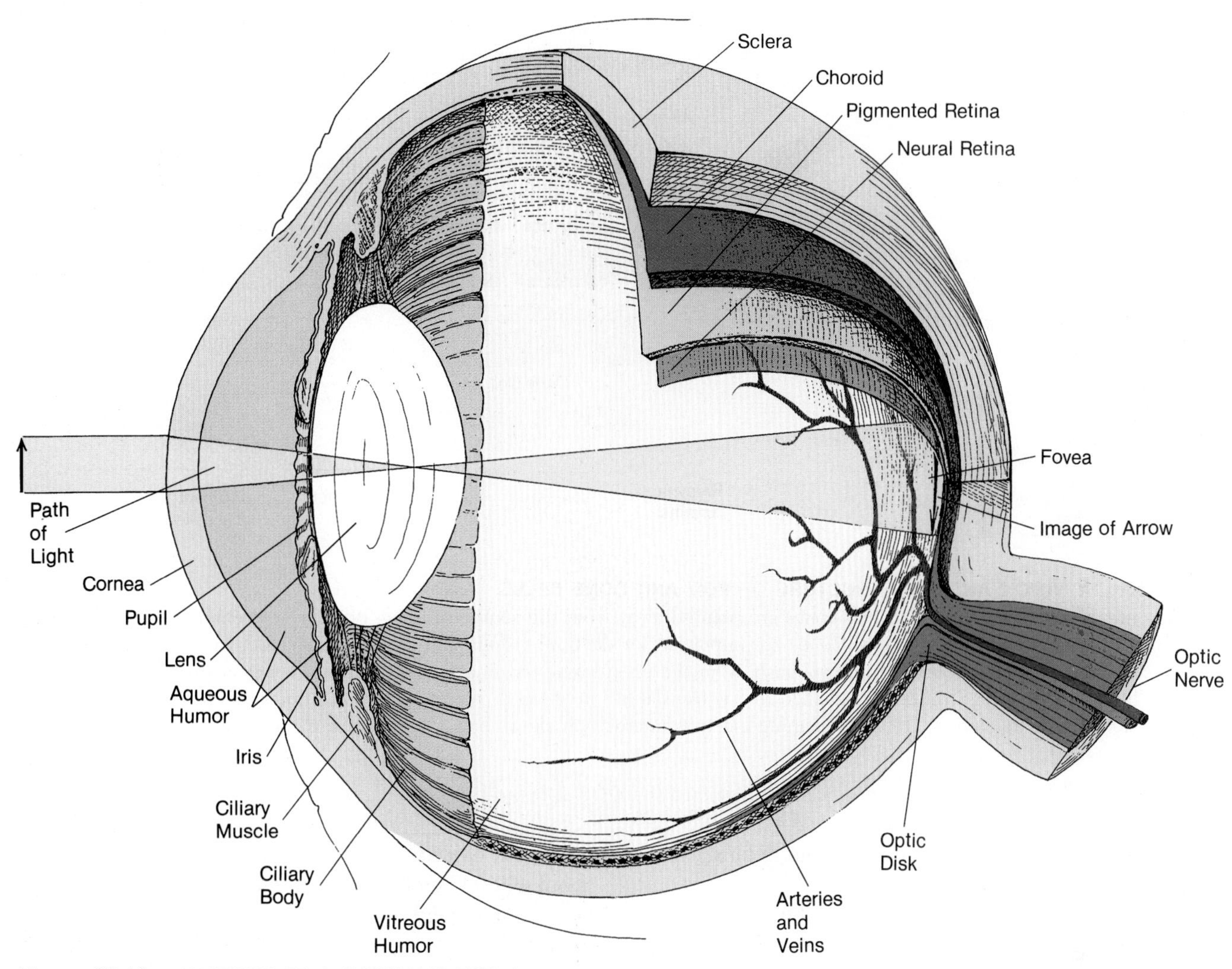

Figure 36-11 ANATOMY OF A MAMMAL'S EYE.
Light—here, the image of an arrow—passes through the cornea and the lens and is focused on the neural retina. The eye's various layers are described in the text. The fovea is the site of highest activity; the optic disk region is a blind spot.

or lens—abnormalities that can usually be corrected with artificial lenses.

The Retina

The neural retina, on which an image is focused, is really part of the wall of the brain that is specialized for vision. The highly complex neural retina contains the visual receptor cells, or photoreceptors, as well as a variety of nerve cells. In many vertebrates, including humans, the eye contains two types of photoreceptors, **rods** and **cones,** which are shaped as their names imply (Figure 36-12) and which contain different sets of light-absorbing pigments. These pigments are assembled in the membranes of a highly modified cilium, the *outer segment,* which points away from the center of the eye—that is, away from light.

Rods are highly sensitive: They respond even when the amount of light entering the eye is very low, such as at night or in deep water. Cones, on the other hand, react only to higher light intensities and so are most active in daytime light levels. Cones also provide high visual acuity and color vision.

The sharpness of an animal's vision depends on the density of cones in each eye's fovea. Each human fovea has about 160,000 cones per square millimeter, compared to a hawk's roughly 1 million cones in the same region. A hawk therefore has a visual acuity some eight times better than a person's, enabling it to detect a small mouse in the grass from high in the air.

Whereas humans and other animals that are active both day and night have both rods and cones in their retinas, nocturnal creatures and organisms that live in low-light environments, such as deep-sea fish, tend to have predominantly "rod eyes." Such eyes are marvelously adapted to life in the dark, but they usually lack

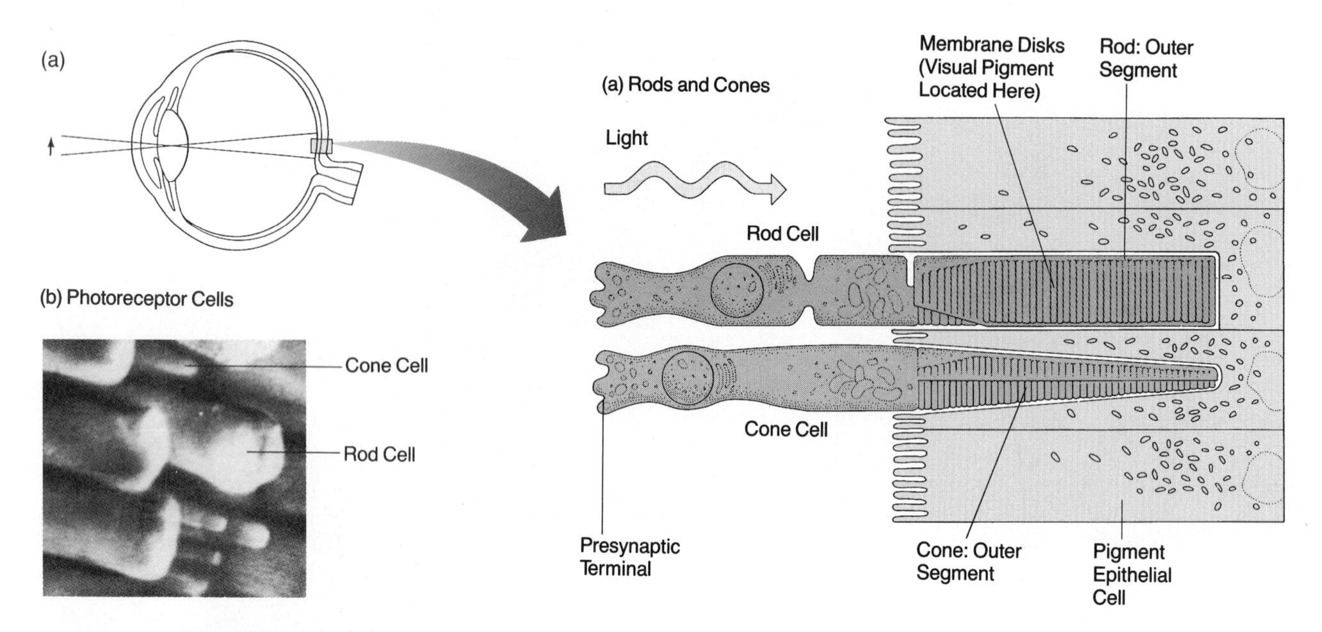

Figure 36-12 VISION AND THE STRUCTURE OF ROD AND CONE CELLS.
(a) The outer segments of both rods and cones are pointed away from the source of light. Rods are extremely sensitive to dim light and are important in night vision. Cones are important for daytime vision, sharp visual acuity and, in some animals, color vision. Neurotransmitter is released spontaneously from the presynaptic terminals when rods and cones are in the dark. The effect of light is to slow that release—a change that is translated into altered patterns of nerve impulses sent to the brain by other cells of the neural retina. (b) This scanning electron micrograph shows how the rods dwarf the much smaller cones.

color vision and the ability to resolve fine details, both processes carried out by cones.

Both rods and cones possess millions of molecules of a visual pigment called **rhodopsin** (Figure 36-13b). This pigment is composed of a lipoprotein, **opsin,** and the light receptor molecule itself, **retinal.** Derived from vitamin A, retinal is a small molecule with a unique property: When it is activated by a photon of light, it undergoes a radical change in shape, forming a structural isomer of the original molecule. This is the sole role that light plays in vision; retinal's shape change converts the energy of a photon into molecular motion.

Imagine an eye shielded from light for several hours. The retinal portion of all the rhodopsin molecules in the rods and cones will be bent in an extremely unstable configuration called 11-*cis* retinal (Figure 36-13a). Retinal in this shape sits snugly in a pocket on the surface of the opsin molecule (forming rhodopsin). If a light is flashed at the eye, photons will strike some rhodopsin molecules. If the 11-*cis* retinal molecule captures sufficient light energy of the correct wavelength, it changes to the configuration called all-*trans* retinal (Figure 36-13c). But the all-*trans* molecule fits poorly in the pocket on opsin, and so the opsin immediately changes in volume and position in the lipid bilayer membrane in the rod or cone and becomes activated.

A single photoactivated rhodopsin molecule instantly triggers a cascade of steps resulting in the splitting of about 400,000 molecules of the second messenger cyclic GMP per second (see Chapter 35). The loss of cGMP hyperpolarizes the visual receptor cell, the initial electrical event in vision. This sequence has strong correlations with events during olfaction and hormonal action (see Chapter 35).

As a result of this hyperpolarization, nerve impulses are generated that carry the message to the brain, and a flash of light is detected.

Seeing in Color

The "colors" we perceive in a shimmering rainbow or a peacock's bright plumage are sensations produced by different wavelengths of the electromagnetic spectrum.

The accepted theory of color vision, the **trichromatic theory,** states that there are three classes of cones, one type maximally sensitive in the blue wavelengths of visible light, another in the green, and the third in the yellow or red. Each type of cone responds to lesser extents in other parts of the spectrum. Color of any given hue—say, blue-green—evokes a characteristic level of activity from each of the three cone types; orange light would yield a different ratio of activities among the cones, as would red light. It is the complex activity patterns of the

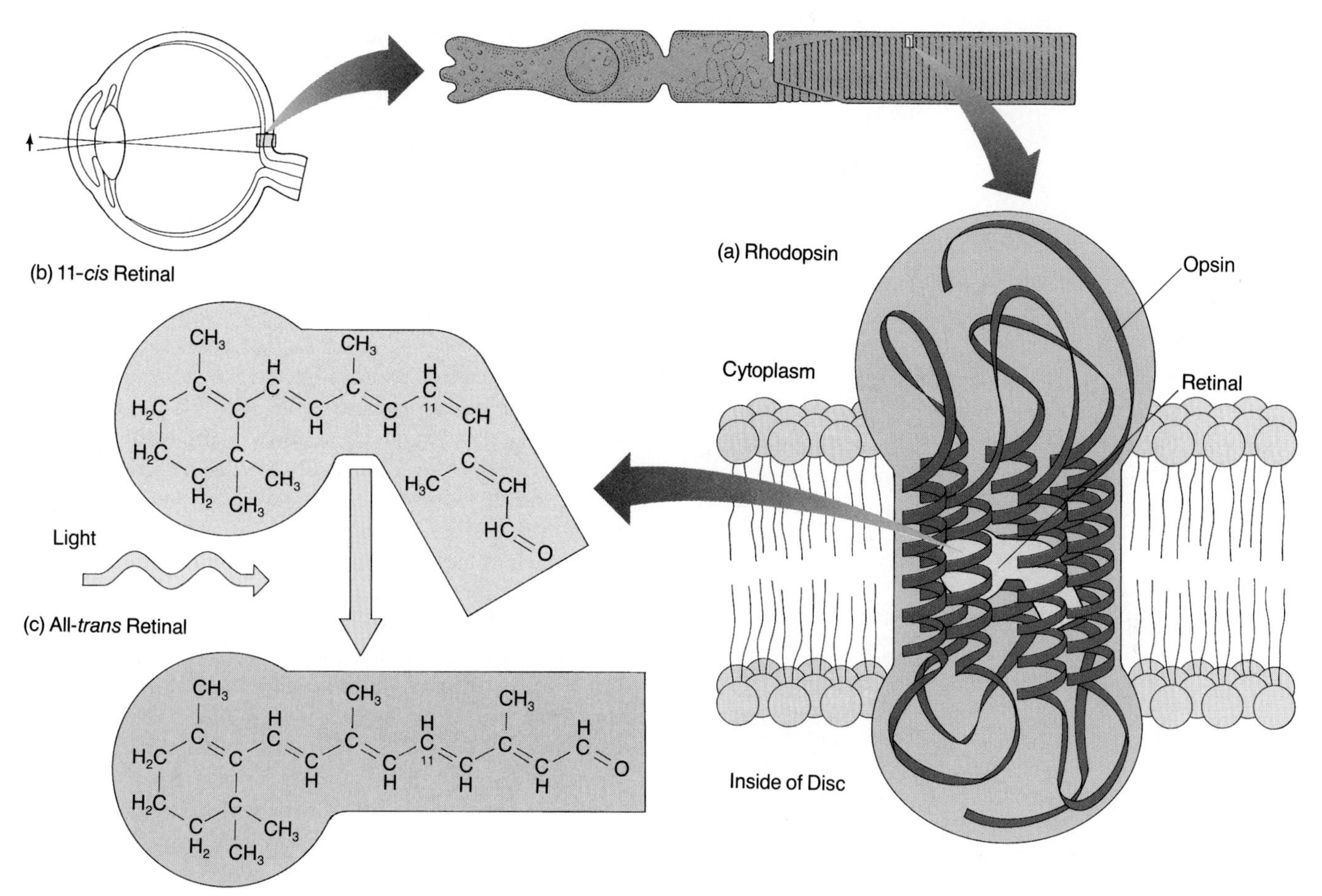

Figure 36-13 AS LIGHT STRIKES RHODOPSIN, THE STRUCTURE OF RETINAL CHANGES. When light strikes a rhodopsin molecule (a), its component 11-*cis* retinal (b) changes configuration to all-*trans* retinal (c). An enzyme later converts the molecule back to the resting *cis* formation. The effect of light on retinal triggers a chain of events that culminates in hyperpolarization of the rod or cone.

three types of cones that the brain interprets as the colors of the spectrum.

In 1986, researchers using recombinant DNA techniques proved the correctness of the trichromatic theory. They found that human males have a separate gene for blue-sensitive opsin; another one for red-sensitive opsin; and one, two, or three for green-sensitive opsin, depending on the individual. The red and green genes occur only on *X* chromosomes, and males lacking one or the other display one of the two types of human color blindness. Women, with their two *X* chromosomes, are almost always heterozygous for these recessive defects and are therefore carriers, while not color-blind themselves.

The opsin genes may still be evolving in humans: Researchers have found a number of men with hybrid red and green genes of the sort that might arise through mistakes in crossing over during meiosis. Other work shows that mutations upstream from the green and red genes result in "blue monochromacy," a condition in males in which all colored light looks gray.

Visual Processing in the Retina and the Brain

As images of the world are focused on the retina, rods and cones hyperpolarize to varying degrees, and the result is a stream of neural information that is turned into the sense of vision—an extraordinary transformation.

Rods and cones form synapses with several types of neurons, most frequently with **bipolar cells.** These bipolar cells, in turn, synapse with **ganglion cells,** neurons whose axons extend along the inner surface of the eye cavity, exit the eye at the optic disk, and travel via the optic nerve to the brain (Figure 36-14).

The axons of ganglion cells are the sole information channels from eye to brain. This means that the wiring of rods, cones, bipolar cells, and other nerve cells in the retina determines the quality of vision. For instance, in the primate fovea (see Figure 36-11), there may be a simple one-to-one hookup from a cone to a bipolar cell to a ganglion cell. This means that there is point-by-point reporting to the brain of precisely which cones are struck

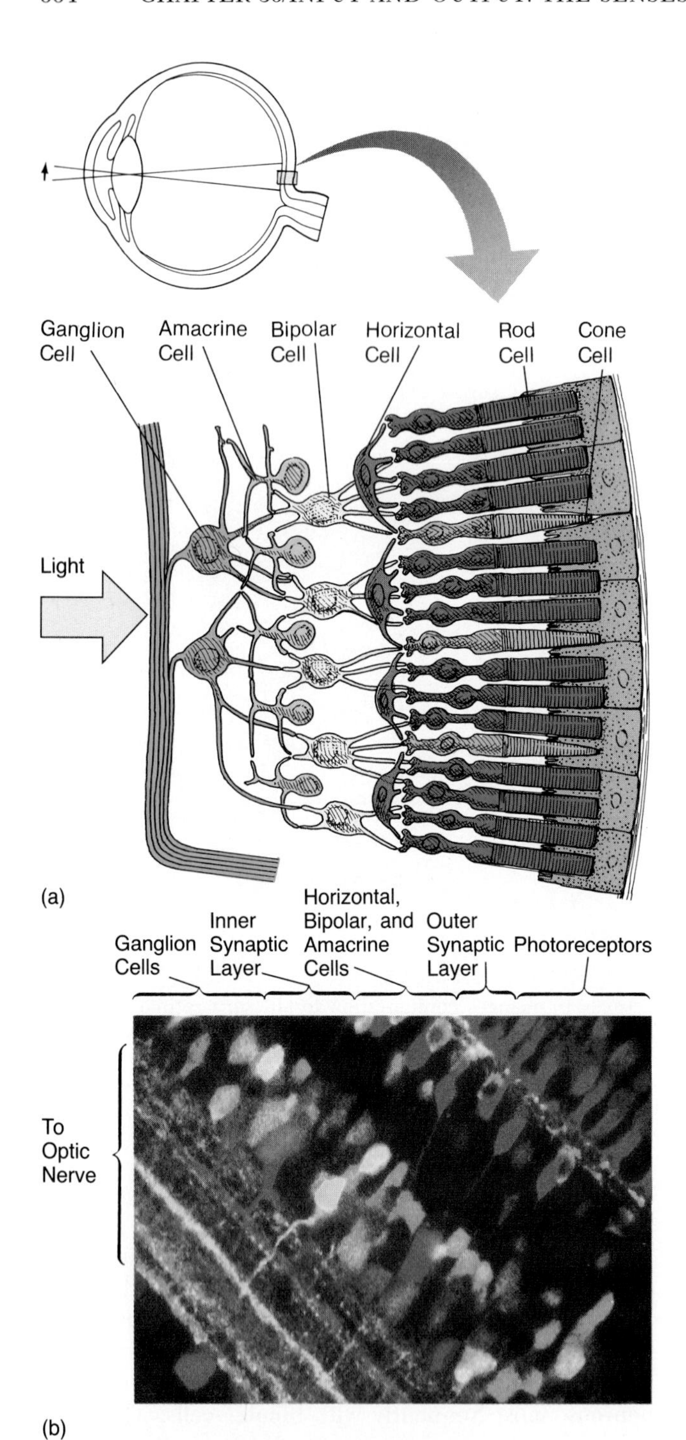

Figure 36-14 LAYERS OF THE RETINA.
(a) This drawing shows the hierarchy of visual processing by the retina of the eye. Rods and cones absorb light. They then form synapses with either horizontal cells or bipolar cells; bipolar cells continue the processing by synapsing onto amacrine cells and ganglion cells. Ganglion cells send nerve impulses to the brain via the optic nerve. Note that a single ganglion cell receives input from many rods and cones. (b) This confocal light micrograph shows an ultrathin section of a chick's retina, treated with fluorescently labeled antibodies to show the different cell types.

by photons. In most parts of the retina, however, many rods (or cones) synapse with a single bipolar cell, and a number of bipolar cells synapse with each ganglion cell. In the human eye, some 131 million rods and cones are connected via bipolar cells to only 1 million ganglion cells. The full set of rods or cones joined in this way to a single ganglion cell makes up the **visual field** of that ganglion cell; the photons hitting any of the rods or cones connected to the ganglion cell can thus trigger it to generate a nerve impulse.

Two other types of neurons that participate in information processing in the retina are *horizontal cells,* which interconnect rods or cones with bipolar cells, and *amacrine cells,* which interconnect sets of bipolar and ganglion cells (see Figure 36-14). This latticework of cross connections within the neural retina permits the visual field pathways of different ganglion cells to influence each other, producing heightened contrast and thus information about an object's contours.

Thousands of ganglion cells in the retina are wired to respond to all sorts of concave, convex, straight, and other edges; their combined activities translate within the brain of a frog, let's say, into the animal's visual world, with its insects, water lilies, and other frogs. Furthermore, such cells allow the frog to identify moving insects as opposed to still objects, and so to capture food.

Like all other senses, vision ultimately is the product of integrating steps in the brain. The large optic nerve from each eye carries axons of ganglion cells to the *optic chiasma,* where, in lower vertebrates, all the axons cross to the opposite side of the brain; in primates and some other higher vertebrates, up to half of the axons from each eye do not cross to the opposite side of the brain at the chiasma (Figure 36-15). The information from the eyes is ultimately routed to the brain's right and left **visual cortices** (singular, *cortex*), where the most complex visual processing takes place.

Biologists know more about the visual cortex and the manner in which it processes information than about the functions of any other part of the brain, and we return to this important subject later in the chapter.

The Cephalopod and Arthropod Eyes

The paired eyes of an octopus and the multiple eyes of a scallop have structures and functions strikingly like those of a vertebrate eye, including the relative locations of the cornea, pupil, lens, retina, and other parts. Despite some differences in focusing, neural processing, and anatomy, the evolution of camera eyes in both vertebrates and certain marine invertebrates has resulted in a large surface area to array the visual pigments and in a neuron network for initial processing of visual information.

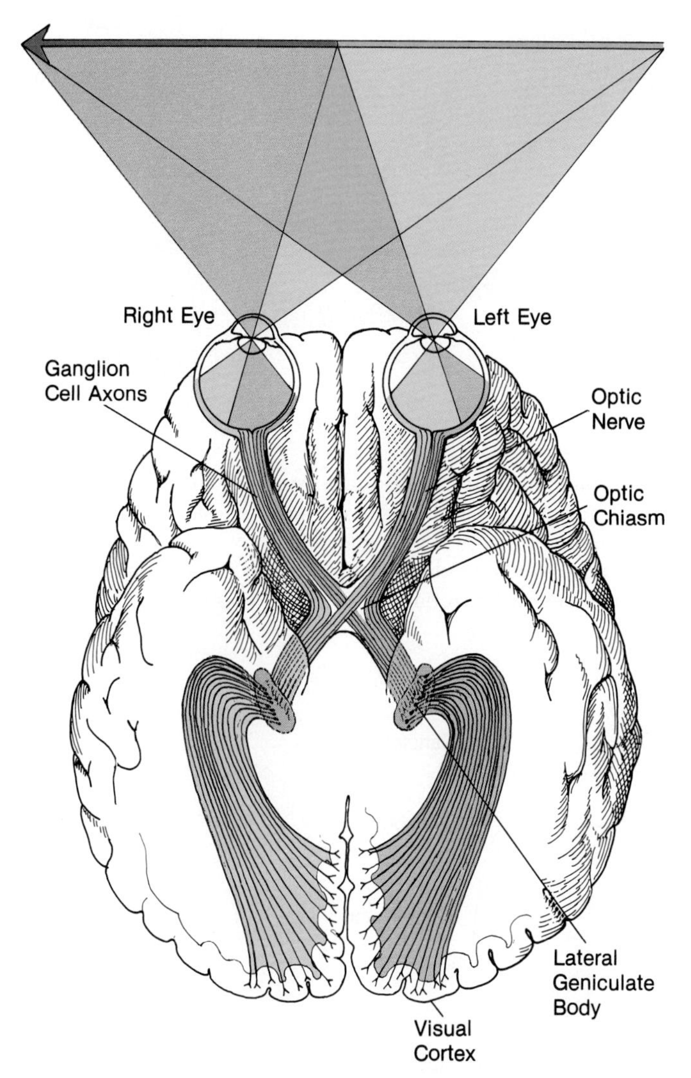

Figure 36-15 THE ROUTES OF VISUAL INFORMATION. In primates, the left halves of both neural retinas are connected by ganglion cell axons to the left side of the visual cortex, while the right halves of both eyes send signals to the right side of the brain. Trace the routing of ganglion cell axons on this drawing (viewed from below). Information, upon reaching the brain, is processed in each lateral geniculate body, and from there is passed to the visual cortex for final processing and interpretation.

By contrast, arthropods, such as crayfish, spiders, and insects, have **compound eyes,** composed of many separate optic units called **ommatidia** (Figure 36-16). Each ommatidium is oriented at a slightly different angle from the others and so is aimed at a different part of the visual world. In the ommatidia of a horseshoe crab, light passes through a lens into a cluster of sensory cells *(retinular cells)* that have visual pigment molecules spread on numerous microvilli (called collectively the **rhabdomere**). When the visual pigment molecules are stimulated, second messengers may act, the retinular cells depolarize, and the single *eccentric cell* of the ommatidium generates and propagates nerve impulses to the brain.

Each retinular cell in an ommatidium "sees" a relatively large patch of the world, and high visual acuity is not possible. Instead, arthropods see a grainier, fuzzier version of their surroundings than do species with camera eyes. At the same time, however, there is good evidence that the arthropod's compound eye acts as a collection of lenses, parabolic mirrors, and light guides that can refract, focus, and detect very low light levels and provide good depth perception. These qualities allow an animal like the nautilid shrimp to aim and snap its lethal pincers more quickly than any other animal can strike.

STRUCTURE AND FUNCTION OF VERTEBRATE BRAINS

In many invertebrates, a dominant mass of neurons—a "superganglion," or brain—sits near the major sense organs in the anterior part of the body, or head. The basic vertebrate brain evolved from such simpler invertebrate nervous systems, and even in the earliest known vertebrate fossils (jawless fishes), the central nervous systems have all the complex structures—brain divisions, cranial nerves, complex sense organs, and other major components—that are found in a modern shark or trout.

In contemporary vertebrate embryos, the developing brain consists of three large swellings: the forebrain, the midbrain, and the hindbrain, which in turn merges into the spinal cord (Figure 36-17a). The **forebrain** includes an anterior region, the telencephalon, and a posterior region, the diencephalon. The telencephalon includes the olfactory bulbs and the future site of the *cerebrum.* The diencephalon gives rise to the dorsally situated thalamus, the ventrally placed hypothalamus, and part of the pituitary gland. The **midbrain** forms the mesencephalon and gives rise to the optic tectum in fishes and amphibians and to the superior and inferior colliculi in mammals. The large **hindbrain** has two major parts: the metencephalon, which becomes the *cerebellum* and the *pons,* and the myelencephalon, which forms the *medulla oblongata*—the site of connection between brain and spinal cord.

As the brain continues to develop, its internal cavity (the interior of the neural tube) is partitioned into four interior chambers, the *ventricles.* Also arising during development are protective sheets of connective tissue called *meninges,* which wrap the brain and spinal cord and line the inside of the skull. A liquid called **cerebrospinal fluid** fills the space between the meninges, as well as the ventricular chambers and the intercellular spaces

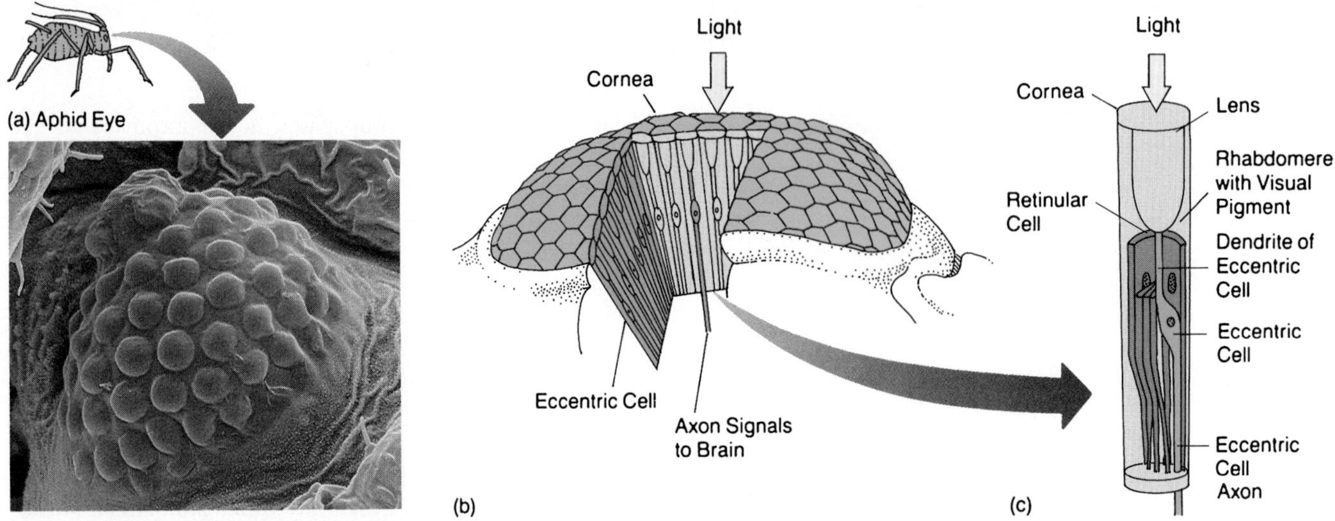

Figure 36-16 VISION IN A COMPOUND EYE.
(a) An aphid's compound eye, magnified with an electron microscope about 500 times. Each of the "spheres" in this clump is the outer end of an ommatidium, a separate optic unit. In comparison, a fruit fly eye would have over a thousand ommatidia. (b) Each ommatidium in an insect eye samples 2 to 3 degrees of the animal's visual world. Depending on which ommatidia are stimulated by a change in light, an insect can locate and respond to the source of change. (c) Close-up of an ommatidium of the horseshoe crab, showing the single eccentric cell neuron, which serves all the retinular cells of this optic unit.

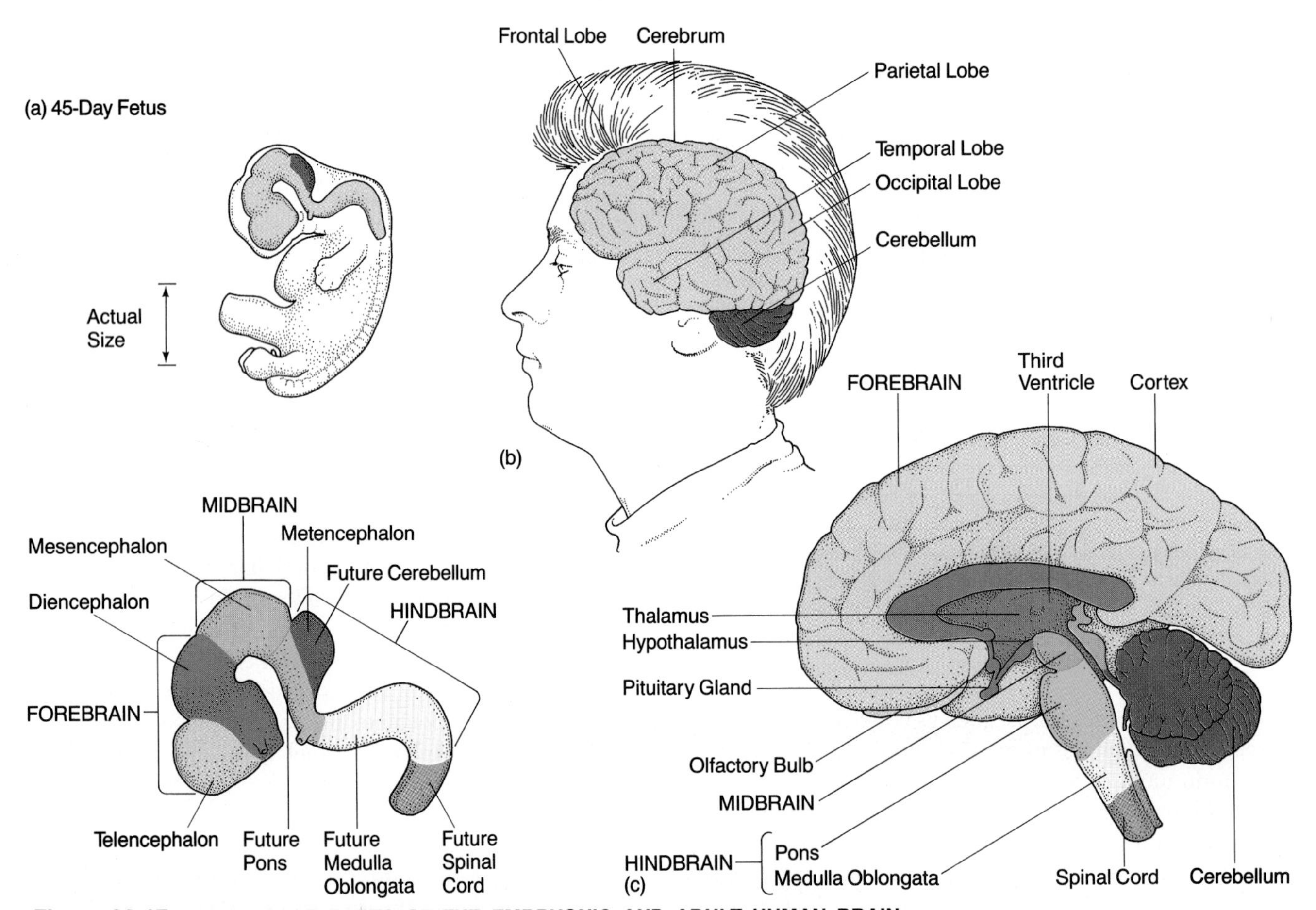

Figure 36-17 THE MAJOR PARTS OF THE EMBRYONIC AND ADULT HUMAN BRAIN.
(a) The major parts of the embryonic human brain include the forebrain, midbrain, and hindbrain. Each gives rise to specific subdivisions and regions. (b) This drawing shows a surface view of the adult brain. The major areas of the cerebrum of the forebrain are called lobes (frontal, temporal, parietal, and occipital). (c) This drawing shows a view of the adult brain split down its midline. The hindbrain's major regions are shown in green, purple, and pink; the midbrain is orange; and the forebrain is blue.

in the walls of the brain and spinal cord. Cerebrospinal fluid is vital to normal brain functions; among other activities, it transports nutrients, oxygen, carbon dioxide, and wastes, and the ions that allow nerve cells to generate and conduct impulses diffuse through it. Table 36-2 lists all the parts of the adult vertebrate brain, and the next few subsections describe them one by one.

The Hindbrain: Respiration, Circulation, and Balance

The spinal cord serves as the central nervous system's main "trunk line" for the transmission of sensory and motor information between the body and the brain. The cord and brain merge at the **medulla oblongata,** a thickened stalk at the base of the brain (Figure 36-17c). Just anterior to the medulla is the **pons,** as well as a dorsal swelling, the **cerebellum.** Together with the midbrain, the medulla and pons compose the *brain stem,* forming the base on which the cerebellum and forebrain rest. Distributed in these brain regions are bundles of nerve axons called *tracts* and clusters of neuronal cell bodies called *nuclei.*

The medulla oblongata controls respiration, swallowing, vomiting, and other basic physiological functions.

Table 36-2 PARTS OF THE ADULT VERTEBRATE BRAIN

Division	*Functions*
Medulla Oblongata	Regulates respiration, heartbeat, and blood pressure; reflex center for swallowing, sneezing, coughing, and vomiting; relays messages between brain and spinal cord
Pons	Connects and integrates various parts of the brain; helps regulate respiration
Cerebellum	Controls posture and muscle tone; helps maintain equilibrium and coordinates movements
Midbrain	Superior colliculi mediate visual reflexes; inferior colliculi mediate auditory reflexes; muscle tone and posture information integrated by red nucleus
Hypothalamus	Controls body temperature, appetite, and fluid balance; secretes releasing hormones; produces oxytocin, ADH; helps control emotional and sexual responses
Thalamus	Relay center between spinal cord and cerebrum; messages are interpreted within thalamic nuclei before being relayed to the cerebrum
Cerebrum	Site of intelligence, memory, language, and human consciousness; controls some motor activities

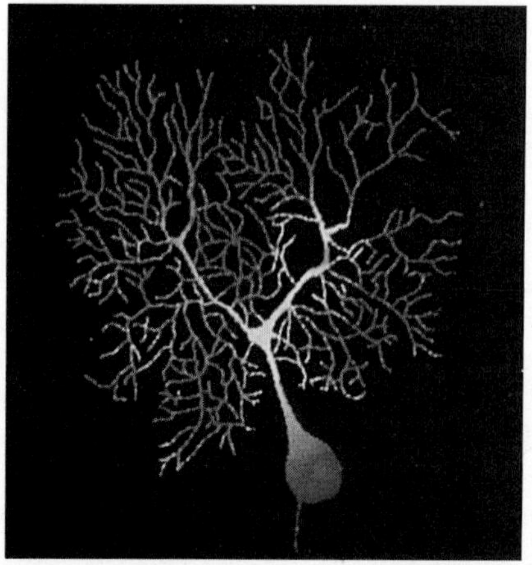

Figure 36-18 PURKINJE CELL: A COMPUTER MODEL. Each type of neuron has a unique morphology. This computer-generated version of a Purkinje cell (neuron) has thick branching dendrites and a long, fine axon (bottom) extending downward from the cell body.

This brain region receives sensory information from the spinal cord and from cranial nerves (see Chapter 34), as well as instructions from brain centers such as the hypothalamus of the forebrain (the seat of control of temperature regulation, osmoregulation, etc.). Some of the sensory information coming to the medulla is routed via the pons to the nearby cerebellum. The cerebellum modulates a vertebrate's balance and stance and some locomotor movements. The sensory nerves from the inner ear's gravity and acceleration/deceleration detectors are major sources of input to the cerebellum. Similarly, special receptors in muscle cells and joints provide a constant flow of additional information about the position of the head, trunk, and limbs (see Chapter 37).

The incredibly complicated tasks of the cerebellum are carried out in good part by giant, many-branched **Purkinje cells.** Arranged in complex networks, each Purkinje cell receives signals across an estimated 200,000 synapses on its highly branched dendrites (Figure 36-18). The output of Purkinje cell networks is sent via various tracts and nuclei down the spinal cord and causes motor neurons to increase or decrease the varying tensions required in neck, limb, trunk, and leg muscles as a terrestrial vertebrate stands, sits, or moves in complex ways or as an aquatic or airborne vertebrate swims or flies.

The Midbrain: Optic Tecta, Certain Instincts, and Some Reflexes

The midbrain lies just anterior to the pons and cerebellum (see Figure 36-17a). In fishes and amphibians, the midbrain's dorsal region, the *optic tectum* (*tectum* means "roof" or "covering"), is the primary site for processing sensory impulses from the eyes. In reptiles and their descendants (the birds and mammals), the optic tectum relays information from the auditory and visual systems to the forebrain. It also has relinquished most of its control of behavior to the increasingly complex forebrain.

The Forebrain: Increasing the Complexity of Brain Functions

The forebrain of an amphibian or fish is a simple, smooth-surfaced region, traditionally thought to be a processing center for sensory impulses from the *olfactory bulbs*. The olfactory bulbs form the anterior ventral portion of most vertebrate brains, reflecting the importance of odor as a sensory modality. As reptiles and birds evolved, the size and complexity of the anterior dorsal part of the forebrain, or **cerebrum,** expanded, and the activities it regulates came to include complex instinctive behaviors. Warblers can migrate using constellations of stars as a compass, and weaverbirds can construct elaborate nests of woven grass because of their forebrain neuronal circuitry. Among mammals, the forebrain's cerebrum has grown to massive proportions. We discuss the resulting behavioral capabilities shortly.

The posterior part of the forebrain, arising from the diencephalon (see Figure 36-17), coordinates much of the organism's physiology. The ventrally situated *hypothalamus* regulates the autonomic nervous system and numerous physiological functions, including panting and sweating in response to overheating and producing and eliminating urine in response to overhydration. In addition, cell bodies of neurosecretory cells reside in the hypothalamus, and some of these cells (located in an area called the *supraoptic nucleus*) extend their axons to the nearby posterior pituitary, where the hormones oxytocin and antidiuretic hormone (ADH) are stored and released (see Chapter 35). Other axons terminate at the median eminence, where their endings secrete releasing hormones for the many anterior pituitary hormones. An-

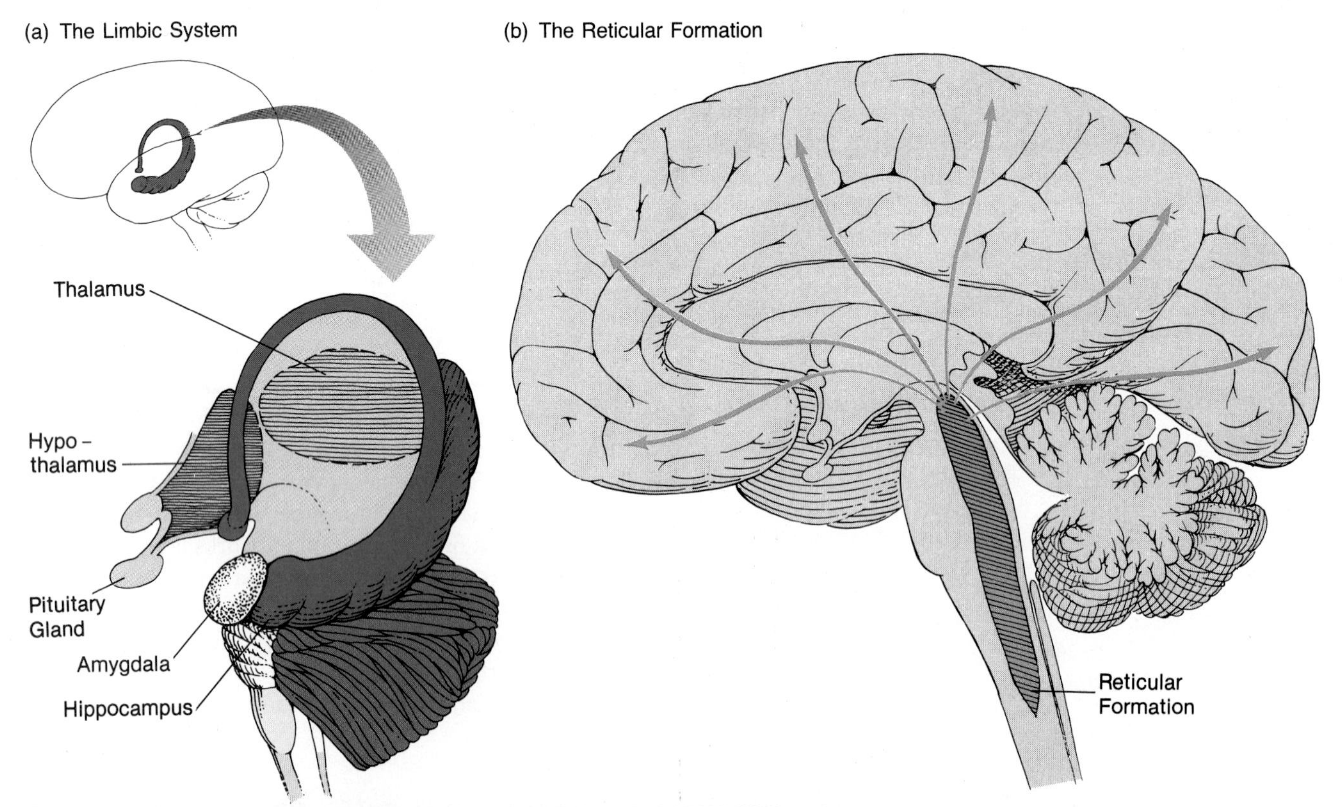

Figure 36-19 THE LIMBIC SYSTEM AND THE RETICULAR FORMATION.
(a) The limbic structures lie deep beneath the cortex. Besides contributing to the regulation of temperature and blood pressure, various sites in the limbic system are involved in feeding, fighting, fleeing, and sexual reproduction, as well as moods and emotions such as pleasure, playfulness, anger, and embarrassment. (b) The reticular formation is a diffuse network of neurons located in the brain stem. This system monitors information from the sensory systems of the body, and via the reticular activating system (RAS), tracts that lead to the thalamus may activate the higher cortical centers of the brain. The RAS filters sensory input and either dampens or amplifies the signals.

other important region of the hypothalamus contains neural cell body clusters responsible for gonadotropin release and sexual behavior in mammals. Still another portion of the hypothalamus, the suprachiasmatic nucleus (SCN), is the seat of a mammal's "biological clock" (see Chapter 35), and from it, signals originate that coordinate a wide variety of body activities on roughly a 24-hour cycle. Among these regular activities are release of hormones, mitotic activity of body cells, hunger, and wakefulness.

Dorsal to the hypothalamus is the *thalamus.* In land vertebrates, this area integrates most kinds of sensory information and channels it to appropriate parts of the nearby *cerebral cortex* (described later). The thalamus also receives and processes signals from the cerebrum and, in turn, sends signals to the cerebellum, where balance, posture, and movements are coordinated.

Emotions are largely controlled by the **limbic system** (Figure 36-19a). Part of the thalamus, much of the hypothalamus, and the *amygdala* and *hippocampus* deep within the forebrain make up the limbic system. When researchers use electrodes to stimulate portions of the limbic hypothalamus, a variety of physical states or emotions are evoked in the animal: hunger, thirst, pain, or even complex emotions such as sexual desire, anger, rage, and pleasure. When a person experiences strong emotion, the limbic hypothalamus sets in motion a coordinated set of related physiological reactions—blushing, sweating, a more rapid heartbeat, and so on—in response to such experiences. When the limbic system's regulation of mood and desire becomes abnormal, behaviors such as pathological overeating or undereating may be generated. The limbic system contains a reward, or pleasure, center and an avoidance, or punishment, center. These centers are probably important contributors to the "unconscious" drives that control aspects of behavior in all mammals, including humans. Two other limbic system structures—the forebrain's amygdala and hippocampus—are involved in the highly important process called *short-term memory*, the temporary storage of memories of recent events that are not yet stored permanently elsewhere in the forebrain.

The Brain's Sentinel: The Reticular Activating System

Embedded in tissues of the medulla and cerebrum are portions of an elaborate but diffuse network of neurons, the **reticular formation** (Figure 36-19b). This network receives information from each of the body's sensory systems: Its axons ascend from the medulla through the midbrain to the thalamus and descend to the spinal cord, where they inhibit or amplify incoming sensory information or modify motor instructions traveling down the cord.

The **reticular activating system (RAS)** is composed of the reticular formation plus its tracts that lead to the thalamus. In humans, it is sometimes called the "gateway to consciousness." Why? Studies have shown that an electrical stimulus applied to the RAS will cause a sleeping cat to become aroused and alert, while drugs that inactivate the RAS produce sleep or, in humans, loss of consciousness. The RAS is like a filter that somehow tunes out the higher brain centers. The RAS is also responsible for arousing the higher centers, so that the animal awakens and responds or becomes aware of sensory input. During waking hours, the RAS probably acts like a filter in helping the higher centers pay attention to new sensory inputs important for survival while ignoring the flood of other information that is merely "noise" for the system.

The Neocortex of the Cerebrum: The "New Bark"

During evolution, the mammalian brain enlarged tremendously, and millions or billions of additional neurons came to make up the area of the cerebrum called the **neocortex,** or *neopallium*—literally, "new bark." For billions of new neurons to be arranged in the typical configuration of the neocortex—a sheet of nerve tissue with an average thickness of 2.5 mm, containing six precisely ordered *layers* of neurons—creases and furrows developed in the neocortex, forming hills (*convolutions*, or *gyri*) and valleys (deep *fissures* or shallow *sulci*) of gray matter. If the living hills and valleys of the human neocortex were pulled taut into a smooth-surfaced configuration like a shrew's, the human skull would have to be the size of an elephant's to accommodate our complex brain.

The six layers of neurons are composed of **columns** about 20 μm in diameter, each containing up to 100 neurons (Figure 36-20). The human cortex is estimated to consist of perhaps 400 million columns, with a total of more than 10 billion neurons. To add to this complexity, at least 60 distinct classes of neurons have been described within the various layers and columns. Moreover, in the visual cortex, groups of columns lying next to each other form *slabs*, whose function we describe later.

The neocortex has a specialized geography (Figure 36-21). Three large regions receive sensory input: one from vision (the visual cortex), one from hearing (the auditory cortex), and the third from bodily sensations (the sensory cortex). A fourth region acts in motor control (the motor cortex). Within each cortical region, local groups of neurons are associated with specific parts of

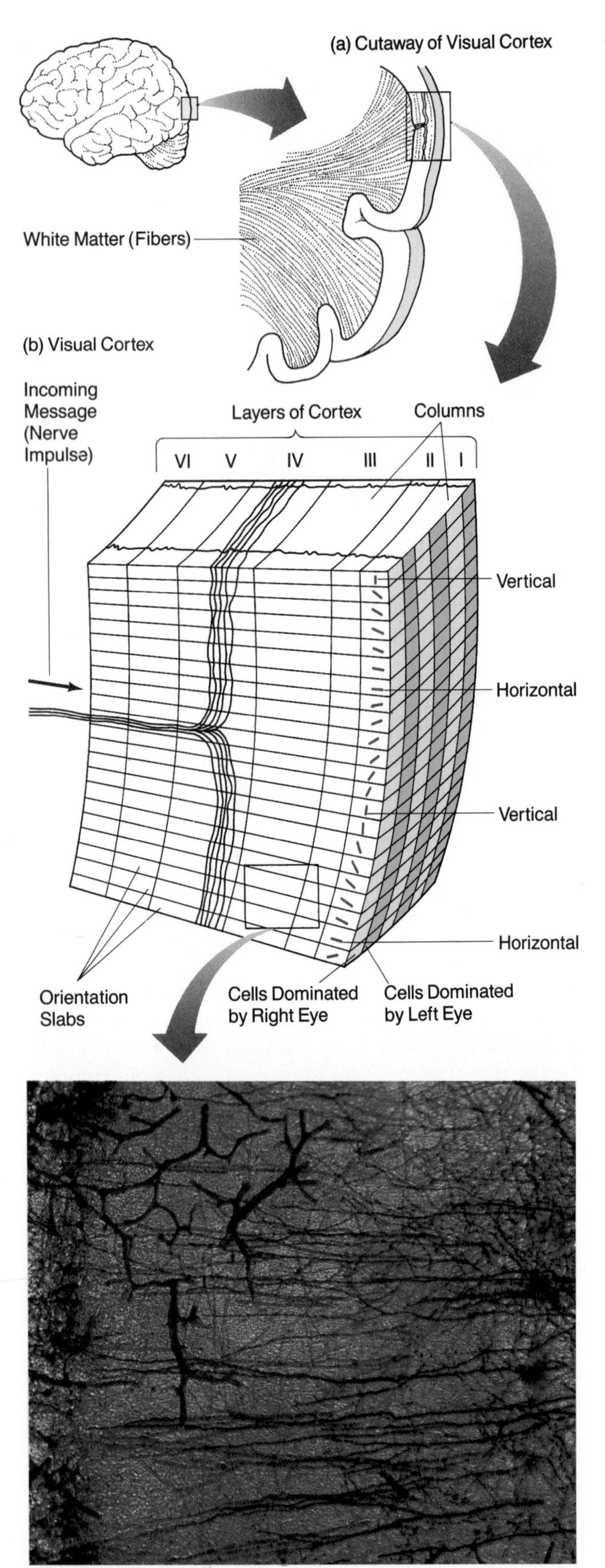

Figure 36-20 OCULAR DOMINANCE AND ORIENTATION SLABS.
(a) A cut portion of the visual cortex, indicating the position of the six layers. (b) The light and dark blue stripes on the surface of the cortex to the right of the numbered layers show which of the layer IV and other neurons are driven by either the right or left eye. The expanded view also shows the orientation slabs; the small lines in layer I reveal the angles that a straight line must have in order to be perceived by complex neurons in that slab. Thus, the slab marked "horizontal" could see the horizontal upper or lower edge of a picture frame (explained in text), while the slab marked "vertical" could see the frame's vertical sides.

the body. For example, on the *sensory cortex*, each portion of the body's sensory surface—lips, fingers, and so on—is represented by a set of neurons responsible for processing information from that body area. The tracts of axons carrying sensory information cross the central nervous system, so that the left side of the body is reported to the right sensory cortex, and the right side of the body to the left sensory cortex. In effect, there is a maplike representation of the whole body on the sensory cortex, referred to as a *homunculus*, or "little person" (Figure 36-22a). In similar fashion, researchers have traced out maps of the visual, auditory, and motor cortices.

Experimenters worked out the motor homunculus by stimulating various regions of the *motor cortex* and noting which muscles twitch or contract (Figure 36-22b). In general, motor sites on each side of the cortex have the same spatial arrangement as muscles on the opposite side of the body. Distortions in the relative size of body

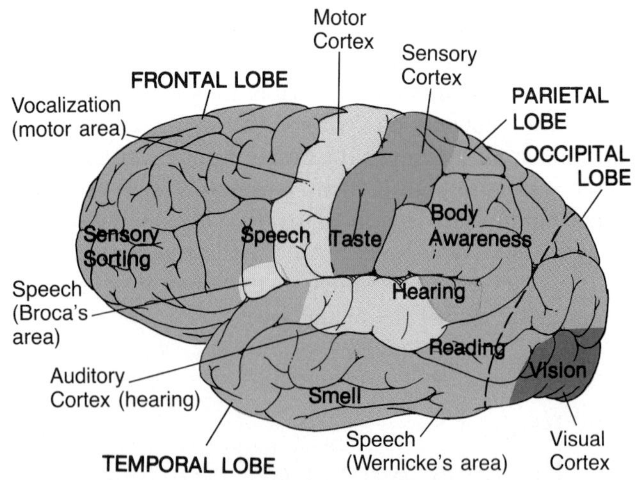

Figure 36-21 THE CORTICAL LOBES.
The portion of the cerebral cortex that is designated the associational cortex is divided into the frontal, temporal, parietal, and occipital lobes. Each lobe possesses specific areas of function, some of which are indicated. Together, these areas of function constitute the so-called higher centers of the brain. Mapping of the areas of higher function has been achieved through various means—for example, electrical stimulation.

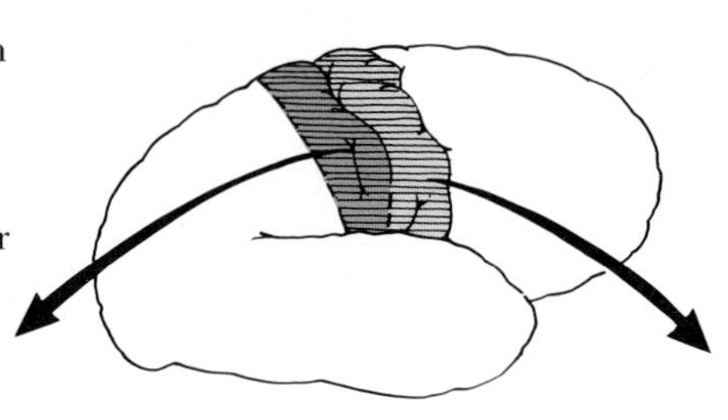

Figure 36-22 THE SENSORY CORTEX AND MOTOR CORTEX. Cross sections made through a brain in the planes shown here reveal the motor and sensory cortices. (a) By inserting electrodes in the sensory cortex at various sites, the experimenter can detect electrical activity while touching the area of the body whose sensory input projects to that spot. Using this procedure, biologists have defined the sensory homunculus.

(b) Conversely, stimulating the various parts of the motor cortex with an electrode causes constriction of muscles driven by that region of the cortex. For instance, using the stimulating electrode at the site shown here would cause the left hand to curl up. This process maps the centers of control of body movements and defines the motor homunculus.

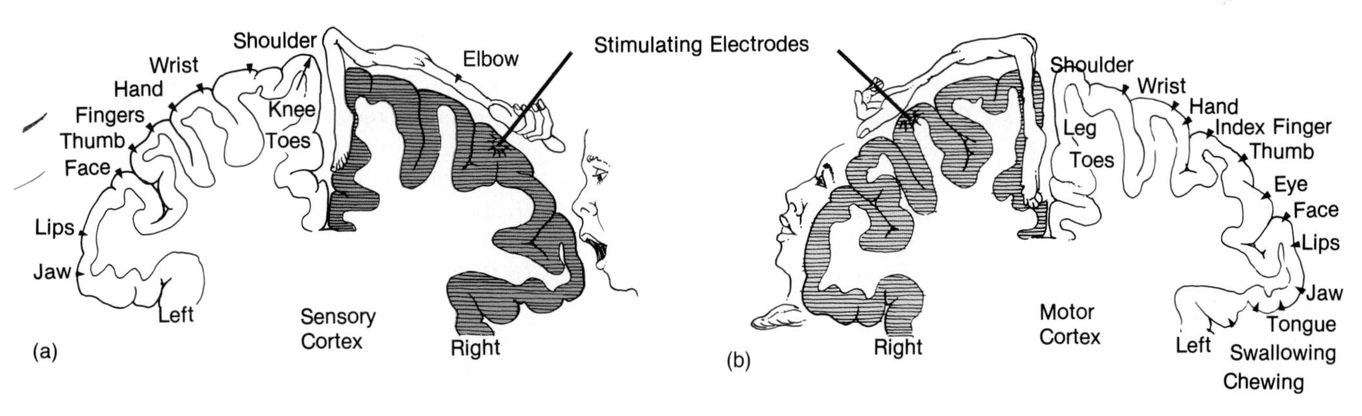

parts in both sensory and motor homunculi reflect the differential density of nerves in various body regions; for instance, the highly innervated, sensitive lips loom grossly large, while the tough skin of the elbow claims only a tiny bit of cortical space (see Figure 36-22a).

The four areas of cortex "assigned" to vision, hearing, body senses, and motor functions make up a relatively small proportion of a primate's neocortex. The vast remaining cortical area is said to be "unassigned." What is the role of the billions of neurons and columns in this large area? Surprisingly, some dozen representations of the visual cortex map and a half dozen each of the somatic and auditory maps are present. If, for example, an experimenter destroys the neurons corresponding to the left thumb on the sensory cortex, the corresponding "left thumb" region of the duplicate sensory map disappears even though neurons in *that* region are untouched. This and other results show that the multiple representation maps are dynamic and change with use. The duplicate maps and the remaining cortex is called the *associational cortex*. This unassigned tissue is the site where the brain's most elusive characteristics and abilities arise—memory, learning, language, and personality. Neurologists have assigned anatomical names to portions or *lobes* of the associational cortex: the temporal, frontal, parietal, and occipital lobes, as well as functional areas of the lobes where memory, speech, reading, and higher-order brain phenomena are generated. Our next section traces key experiments that helped reveal the brain's deepest secrets.

EXPLORING THE LIVING BRAIN

To learn about the brain, researchers destroy specific areas of tissue, stimulate neurons with electrodes, record the brain's electrical activity, and study individual neurons.

Split-Brain Studies and the Cerebral Hemispheres

Most severe lesions, or damage to specific regions of the brain, result in clear behavioral deficiencies. For instance, damage to the left side of the brain in the temporal lobe might lead to speech impairment, while a lesion in the dorsal cortex may cause loss of motor functions, such as the ability to move the legs. In the 1920s and 1930s, neurosurgeons began to make systematic discoveries of brain function while attempting to help patients suffering from severe epilepsy. One of the few techniques that became available several years later was surgically splitting the two hemispheres of the cerebral cortex—an induced lesion with dramatic consequences.

A gap, the *median fissure*, normally separates the brain's two hemispheres; hidden deep within the fissure is a giant bundle of 200 million axons, the **corpus callosum**, which extends from one hemisphere to the other, linking the two sides of the brain. The split-brain operations sliced through this giant axonal bundle, and as a result, the epileptic seizures themselves became much rarer and less severe, and the patients retained almost all their former mental faculties save some impairment of memory.

In the early 1960s, brain researcher Roger Sperry and his colleagues conducted studies that helped determine how the splitting affected brain activity. The experimenters knew from previous studies that images in the left visual field of both eyes are conveyed to the right side of the brain, and images in the right visual field are sent to the left hemisphere (Figure 36-23). Sperry and colleagues also knew that the brain's adult language functions are divided; for most humans, the left brain is specialized for speech. Sperry showed split-brain patients two different groups of objects simultaneously, including a ring in the right visual field, "reported" to the left brain, and a key in the left visual field, "reported" to the right brain. Next, he presented the written words "ring" and "key" independently to each visual field. No matter how hard the subjects tried, they could say "ring" (the object "seen" by the left hemisphere) but not "key" ("seen" by the right hemisphere, which has no speech center). Conversely, upon request, the subject could grasp a key with the left hand, since the right brain controls the left side of the body—including its muscles, arm, and hand. Each half of a separated brain, Sperry reported, seemed to have separate and private sensations, with its own perceptions and impulses to carry out actions.

Additional research during the last 20 years has shown that in most people, specific sites in the left brain are in charge of language comprehension and speech production; the left brain seems to process information analytically and in sequences (as in formal logic and mathematics). Conversely, the right brain apparently specializes in spatial relations. For instance, the ability to identify an object by feeling its shape, contours, and texture is a right-brain function. The right brain also plays a key role in music recognition and in our ability to recognize faces. The physical differences between two faces may be so slight that they are almost impossible to describe verbally, yet most humans can unfailingly pick out the visage of a friend or celebrity, even from a crowd of look-alikes. If, however, the familiar visage is flashed to the right visual field only and so to the left brain, it remains completely unrecognized.

New work shows that the two hemispheres are not yet functionally wired in normal newborn rats. The memory of an odor can be established in the right hemisphere, for example, but not until about 12 days later can the

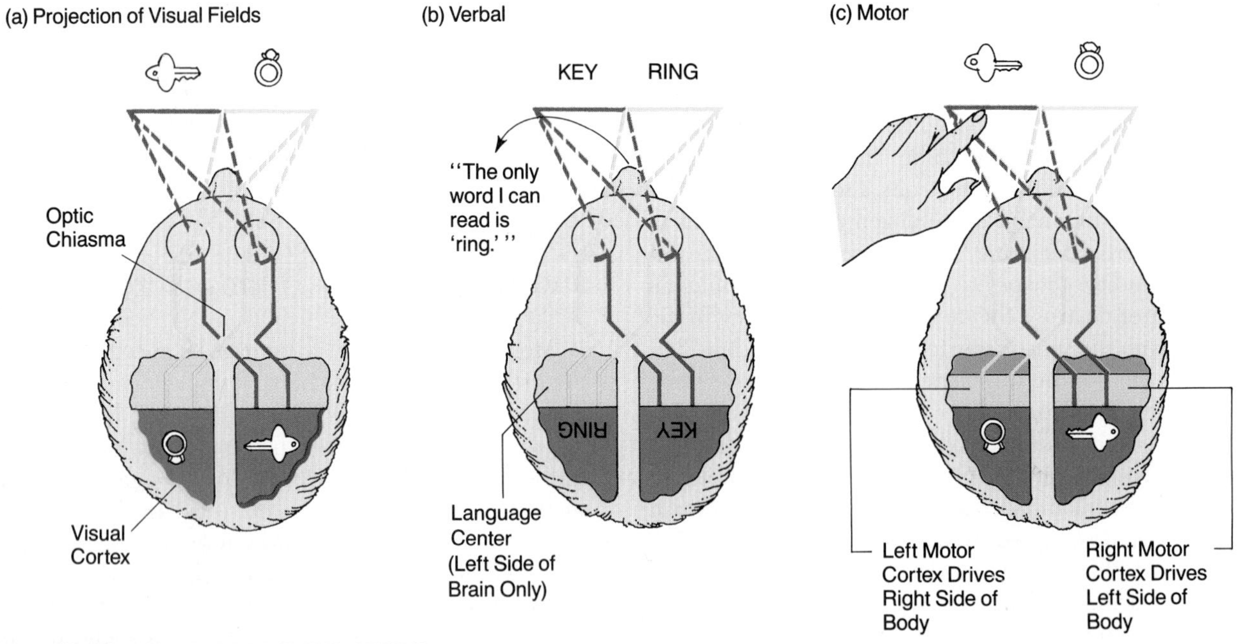

Figure 36-23 SPLIT BRAINS AND VISION.
(a) This drawing shows how an image in the left visual field of both eyes (here, a key) is conveyed to the right brain, and an image from the right (here, a ring) is conveyed to the left brain, seat of the language center. (b) A person with a split brain (whose corpus callosum has been severed) can see both the words "key" and "ring," but can only say "ring" because it is reported to the left hemisphere, with its speech center. (c) Although the person cannot utter the word "key" that is in the left visual field, he or she can point to a key with the left hand. The reason for this is that the right brain (which has "seen" and comprehended "key," but has no way of activating the speech center in the left brain to express it) controls the left hand.

baby rat recall the odor from both right and left hemispheres. During that time, new brain pathways apparently develop that can provide access to preexisting memories on the other side of the brain.

About 95 percent of right-handed people have speech centers located in the left hemisphere. About 70 percent of left-handed people do too, but 15 percent have speech centers in the right hemisphere, and 15 percent have bilateral speech centers. Just what are these centers? One is Broca's area, a portion of the left frontal lobe next to the part of the motor cortex that controls the muscles of the larynx (voice box), tongue, jaw, and lips—all involved in speech production (see Figure 36-21). A lesion in Broca's area can destroy a person's ability to speak, but not to write and read perfectly well, while damage to Wernicke's area can lead to speaking in jumbled, meaningless sentences. It appears that the form and meaning of speech originate inWernicke's area and go from there to Broca's area, where actual speech is generated.

Wernicke's area is also the site of language comprehension (see Figure 36-21). Spoken words are reported to the auditory cortex, but then the message goes to Wernicke's area before it can be understood. Damage to the link between these areas leaves the person capable of hearing but quite unable to comprehend spoken words. The words you are reading right now are sent from the visual cortex to a brain site called the angular gyrus, where they are translated into their equivalent as if they had been heard through the ears. Neurologists once thought that such information, in turn, goes to Wernicke's area, but newer work (described shortly) shows that the sensory input from seeing a printed word goes to a different site in the brain than hearing the same word.

The presence of the speech centers on the left side of most human brains helped establish the term *dominant hemisphere* for the left brain.

Electrical Stimulation of the Brain: Probing the Cortex

Brain researchers use a second major method for exploring nature's most complex organ: electrically stimulating specific sites in the brain. An experimental subject with a hair-fine electrode placed surgically in the neocortex and stimulated with tiny pulses of electricity can suddenly experience sights, sounds, and smells of a long-lost childhood event. If the neurosurgeon moves the electrode slightly, some other memory might arise, with all its unique sensations of odor, sound, color, mood, or perhaps uncontrollable rage or overwhelming pleasure.

Researchers can conduct such human studies only when a surgeon is already opening the skull because of injury or disease. More deliberate explorations of laboratory animals have therefore revealed many of the most profound insights in modern neurobiology. For example, when a researcher stimulates certain areas of a cat's limbic system, the animal immediately begins to stalk an imaginary prey and will pounce upon and energetically bite an object placed in the cage. By moving the site of electrical stimulation just a few millimeters, however, the cat will react to the same substitute "prey" with an arched back and a violent hiss of fear. Such results imply that the neural bases for all sorts of behaviors are present in the brain, in a sense prepackaged and ready to be called forth by a stimulus in normal daily life.

Electrical Recording to Explore the Brain: Sleep and Brain Waves

In a normal alert brain, the activities of billions of brain cells give rise to a "hum" of spontaneous electrical impulses that can be detected through the skull and scalp. Very recent "40-0" techniques allow direct observation of ion fluxes in different parts of single large neurons. With older technology that simultaneously records millions of such events, researchers can produce *electroencephalograms*, or *EEGs*. A person lying quietly with eyes closed produces *alpha waves*, regular, rhythmic electrical potentials of about 45 μV that occur about ten times a second (Figure 36-24a). They originate in the visual area of the occipital lobes. Mere opening of the eyes causes alpha waves to cease and *beta waves* to commence. Beta waves are irregular, rapid waves, dependent in part on visual input, and they are especially ac-

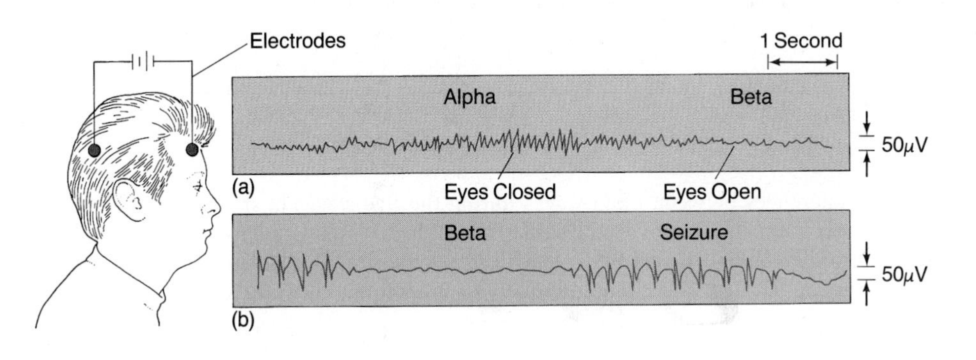

Figure 36-24 ELECTRICAL RECORDINGS OF THE BRAIN. Electroencephalograms (EEGs) measure brain waves. (a) A normal adult at rest with electrodes placed at the red dots shows alpha and beta waves. (b) An epileptic suffering a petit mal seizure has the waves shown here.

tive when one carries out mental activity, such as reading a textbook. Still other wave patterns appear in abnormal states, such as during an epileptic seizure or if a tumor or stroke (caused by a ruptured blood vessel in the brain) brings about local brain damage.

Sleep is the most profound change in behavior that most vertebrates regularly experience. If human subjects are confined to a windowless room and kept under constant environmental conditions, they sleep every 16 to 18 hours and remain oblivious to the world for about 8 or 9 hours. During a typical night's sleep, people experience five distinct sleep stages. Charted by an EEG, stages 1–4 are characterized by increasing amplitude and decreasing frequency of brain waves (Figure 36-25). This period of *slow-wave sleep* culminates in the deepest sleep stage, stage 4, with *delta waves*, about 1 hour after first falling asleep.

As stage 4 ends, the EEG registers fast, irregular beta waves, a trend that continues until the sleeper appears ready to waken. This is when the fifth and somewhat paradoxical state known as **rapid eye movement (REM)** sleep begins. During REM sleep, which tends to occur in periods of about 100 minutes each, the eyes dart rapidly back and forth beneath closed lids, as if the sleeping person were watching a frantic inner tennis match. Individuals awakened during REM sleep almost always report that they have been dreaming. Some evidence suggests that norepinephrine release in the reticular activating system (RAS) leads to nerve impulses going to the cortex; perhaps as a result, dreams occur; heart rate, respiration, and blood pressure all rise; and movements in the gastrointestinal tract cease. The likelihood that the sleeper will awaken spontaneously also increases. Oddly, however, throughout REM sleep, all of the body's skeletal muscles, except those of the eyes and ears, are virtually "paralyzed" by powerful inhibition of the spinal motor neurons. Work on rats and other mammals suggests that REM sleep is a time of processing the previous day's information and experiences and of establishing them in the so-called long-term memory (permanent memory). If an experimenter awakens an animal and prevents its REM sleep, tasks learned in the previous period of wakefulness will not be retained. Curiously, other evidence suggests that a person is likely to dream about a distinct occurrence on the first night after the event and then on the sixth night. While the meaning of this is still mysterious, good evidence suggests that "REM-on" and "REM-off" neurons in the hindbrain may be key players in regulating our essential periods of dreaming sleep.

Brain Studies at the Level of the Neuron

While the EEG is an excellent indicator of the brain's overall state, the data it provides are statistical averages much too general to reveal the intricate cell-to-cell communications by which the brain processes information and generates commands for actions. Several features of the mammalian nervous system simplify studies at the cellular level. First, many parts of the brain exhibit *par-*

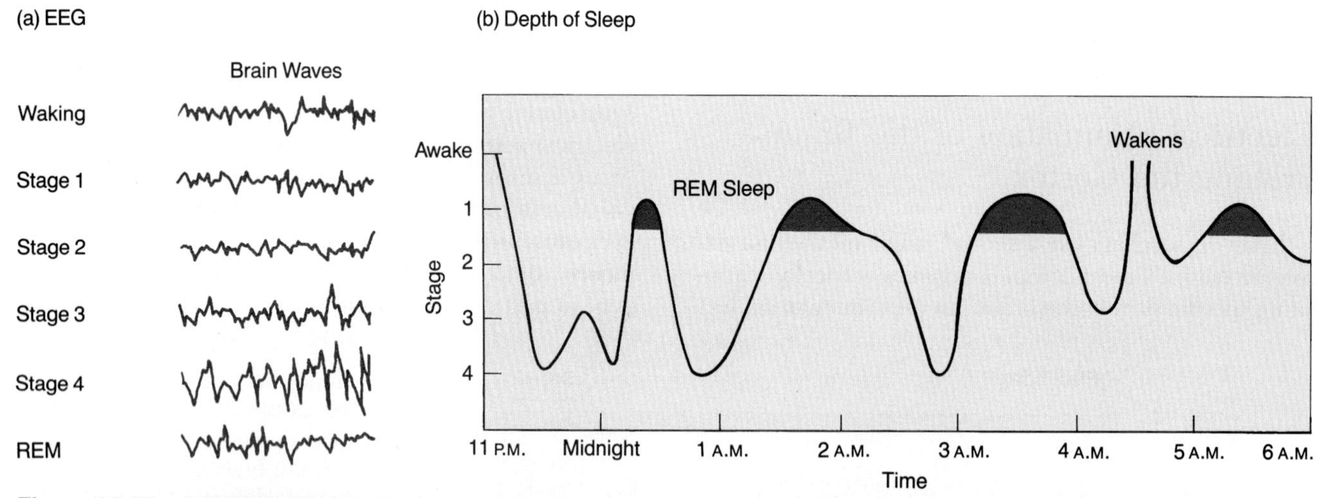

Figure 36-25 THE STAGES OF SLEEP.
(a) The EEG recordings reveal the remarkable differences between brain waves during the five stages of sleep (stages 1–4 and REM sleep). During stages 1–4, the amplitude rises and the frequency falls. The waves during REM sleep are reminiscent of wakefulness, but the eyes show very rapid movements behind closed lids. (b) During a typical night's sleep, the subject cycles through the various sleep stages, sometimes reaching REM sleep, sometimes not, and awakening once in the early morning.

allel processing of information: They are built of enormous numbers of similar circuits, each containing similar cell types. By examining just a few random subsets of the small circuits that connect the hundreds of millions of rods and cones in the eye's retina to the 1 million ganglion cells, biologists can draw some general conclusions about all those circuits.

The wiring of the neural retina also illustrates the *hierarchical organization* of the nervous system, built on the convergence pattern (see Chapter 34). Only a specific pattern of stimulation by light among the hundreds of rods or cones that converge on a given ganglion cell will activate that ganglion. For example, an *on-center* ganglion cell will generate an impulse only when illumination strikes the *center* of the set of rods or cones connected to that ganglion cell. The impulses from such ganglion cells are sent to a special area of the mammalian thalamus, and from there to layer IV of the visual cortex.

As we saw earlier, the visual cortex is organized into six layers, each running parallel to the brain's surface and each containing specifically shaped neurons different from those of the other layers. Hundreds of slabs run perpendicular to the brain's surface; and within each slab there are columns of cells also running perpendicular to the brain's surface and extending inward through the gray matter (see Figure 36-20). If an experimenter inserts an extremely slender electrode into a neuron in layer IV, then shines light in the animal's eye, the neuron will generate impulses, revealing that the cell receives input from the eye. Layer IV neurons in one column might prove to receive signals primarily from the right eye, while layer IV neurons in a nearby column receive input from the same spot in the left eye. The columns within the visual cortex thus have what is called *ocular dominance:* They are driven by either the right or the left eye.

To explore this hierarchy, researchers David Hubel and Torsten Wiesel of Harvard University recorded electrical responses from single cells in the cat's visual cortex as the researchers moved spots of light in the visual field. They found that while layer IV cells see dots of light, these cells in turn stimulate cells in layers V and VI, which see oriented lines, edges, and corners. Working together, the cells in the three layers convert the dot patterns that a cat's (or a person's) eyes perceive into a three-dimensional world of moving and stationary lines and shapes.

Further work on primate visual systems suggests that the visual cortex has three separate subdivisions. Two "parvo layers" receive information from large ganglion cells in the retina, and process data about objects' form and color, while a "magno layer" receives information from smaller ganglion cells and processes data on depth and movement.

Whereas vision is topographical—involved in locating the position of things—smell and most senses are recognitional—involved in identifying things. Smell, in particular, can be thought of as an associational process: We identify a smell because of the memories it evokes, just as Proust observed so gracefully in the nineteenth century. One pungent odor would immediately mean "orange rind"; another fragrance, "rose"; another, "bacon cooking." Recent work on the sense of smell by Gary Lynch of the University of California, Irvine, suggests that the brain's olfactory regions may contain circuits similar in operation to what computer scientists call interactive combinatorial networks. Within these networks, each nerve cell's long axon contacts a whole series of cells, one after another. This results in a grid with many feedback loops. The number of connections that any one neuron makes appear to be random. And the properties of synapses of such circuits may change if they are used repeatedly, growing and becoming more efficient with use.

Although biologists long believed that the adult brain contains a precisely wired set of permanent circuits, they now consider the brain to have great plasticity in both its neurons and its circuits.

Researchers have observed brain plasticity by transplanting pieces of cortex and cerebellum from embryos or young animals into lesions at corresponding sites in the adult brain. In studies of this kind, researchers have collected neural tissues from human fetuses, frozen it in liquid nitrogen, then implanted it into the brains of adult monkeys. There, the human cells survived and developed, and specific types of implanted neurons established precise, correct connections with the host brain cells. Researchers hope that the future application of similar techniques may help alleviate a number of severe human brain disorders, including Alzheimer's and Parkinson's diseases.

In still other promising transplant experiments, researchers substituted a piece of quail forebrain and midbrain for the corresponding parts of a chick brain. Upon hatching, the chicks vocalized like newly hatched quail, reflecting that the implanted tissue had established correct wiring in its new site (Figure 36-26).

How do the structures of neurons and the plasticity of synapses and dendrites relate to olfaction and memory? A rabbit might, for instance, smell a musky odor and learn, to its benefit, that the odor comes from a fox. The circuit, perhaps with its increased dendrites, synapses, and feedback loops, generates impulses only when the particular odor enters the nostrils; but when it does, the memory of "fox" is triggered. Finally, Gary Lynch suggests, the use of that circuit without the odor stimulus is really what we mean by a memory, or the idea "fox." Lynch's theories suggest an alternative to assuming that

Figure 36-26 BIRD BRAINS AND NEURAL PLASTICITY. The chicks on both sides received parts of a quail's brain, and after hatching, peeped like a baby quail instead of like a normal chick, the one in the middle.

the elegant order of the visual cortex column system, with its capacity to identify the positions of things, is also the basis for memory and learning.

MEMORY: MULTIPLE CIRCUITS AT WORK?

The hallmark of mammalian behavior is the ability to learn—good places to hunt or graze, dangerous predators to avoid, and so on; and learning is based on *memory*, the capacity to record and recall past events. Some memories involve the recall of the collective sensations and feelings surrounding some past event, such as high-school graduation, and the specific sights, sounds, and emotions it evokes. Other memories are less sensory and less direct—the appropriate way to behave during a wedding ceremony, for example, based on successful behavior at past weddings. This involves reasoning and is based on learning and recollection. How can such complex processes be explained in cellular, electrical, or molecular terms? A central mystery in the study of memory is whether storage and recall are special, unique processes or whether memory is simply a persistence—albeit a remarkable one—of information processing. In other words, are there circuits and synapses just for memory? Or does the basic circuitry for inputting and outputting data and orchestrating responses also somehow store records of past events?

The work of researchers like Wilder Penfield has revealed a great deal about memory. During an experiment in which he used tiny electrodes to stimulate specific sites on an epileptic patient's neocortex, the patient stated, "I just heard one of my children speaking . . . and I looked out of my kitchen window and saw Frankie in the yard." A phenomenon like this implies that much of our life—sights, smells, moods, and so on—is preserved in our cortex in incredible detail. We are unaware of that stored treasure, however, because we cannot recall it voluntarily. On the other hand, we can bring certain kinds of information voluntarily to consciousness—the value of pi, a line from Shakespeare, or even some facts from a course. So how are classes of memories differentiated and recorded?

Evidence from a variety of vertebrates suggests that there are at least three categories of memory: immediate, short-term, and long-term. Immediate memories—the parade of events as one rides to class or work on a bicycle, for example—are stored for just a few seconds. Virtually all these memories fade to obscurity in just seconds unless something unusual happens (a near collision, say) and calls greater attention to the incident. Other events are stored in short-term memory for minutes. If you look up a telephone number and make the call, you may remember the number for a while, but if you try to recall it a few days later, you probably won't be able to. Long-term memory persists for months or years and often involves repeated inputs or particularly vivid single events.

There is strong evidence that the hippocampus is involved in learning and memory. If a person's hippocampus is damaged badly or removed surgically, he or she will no longer be able to learn or to recall recent events or new facts and faces. The subject's recall of events that occurred prior to the accident or surgery, however, will remain quite normal. Stimulation of the hippocampus may cause the number of synapses and dendrites to rise and the properties of synapses to change, and both these alterations allow information to be stored for longer periods of time. Recall, too, that the hippocampus and limbic system receive direct input from the olfactory bulbs and that odors can trigger certain memories. Researchers have recently succeeded in localizing certain types of short-term memory to individual neurons in the anterior ventral part of the temporal cortex. These cells seem to be involved in the short-term retention of an object's shape, but not its size, orientation, color, or position in space. When an experimental subject is asked to recall a recently memorized object, these cells show sustained activity.

Various lines of evidence show that the cerebellum stores learned reflexes involving motor output. If a researcher directs a puff of air at a rabbit's eyeball, for example, the animal will blink. If a sound accompanies each puff of air as it is delivered, the rabbit becomes *conditioned;* its eyelid will blink simply in response to the sound. Regions deep within the cerebellum store the memory trace required to elicit this and similar conditioned reflexes.

In the 1940s, psychologist Karl Lashley trained animals to do a particular task and then removed one piece after another of the animals' cortices. Surprisingly, the obliteration of no single spot alone could destroy the learned behavior. The reason, he and others concluded, is that long-term memories appear to be *distributed* rather than localized—that is, they are stored in a redundant and diffuse way in the cortex, with probably no single site for a given memory. Biologists have found that the *consolidation* of long-term memory (its transfer to the neocortex) improves greatly if an event to be remembered has significant physiological consequences. When an animal is in a stressful, tense, or life-threatening situation, levels of epinephrine released from the adrenal gland and circulating in the bloodstream rise, and the events or things being experienced may be more likely to enter long-term memory. Events at the cellular level may gradually affect the number and properties of dendrites and synapses; this, in turn, may underlie memory storage.

Investigators have recently used positron emission tomography (see Chapter 2) to detect minute changes in blood flow to sites of neuron activity, thereby seeing, for the first time, what goes on in the neocortex as a person processes a single word. Intriguingly, hearing and reading the same word appear to activate different portions of the neocortex. This contradicts the older idea that visual inputs go first to the auditory region and then to other sites in the brain. In fact, when visual input occurs, Wernicke's area is not activated at all. Furthermore, visual and auditory information can directly influence motor programs (like those involved in speech) without going through the so-called semantic areas of the left frontal cortex. These results support the idea of parallel processing and distributed functions when we see or hear a word, interpret it, and then speak.

Biologists measuring biochemical changes in the brain neurons of invertebrates taught to perform learned behaviors have recorded specific activity by second messengers, various protein kinases, calcium ions, potassium channels, and newly synthesized receptors for neurotransmitters. Some evidence indicates that preexisting proteins are modified as part of a short-term memory, whereas new genes must be expressed for long-term memory to function. The memory process may depend on changes in neurons and their surfaces, along with changes in dendrites and synapses.

Still other studies suggest that for a memory to be stored in a mammal, new links must be forged between neurons in a special part of the brain, the hippocampus (see Figure 36-19), under the direction of a unique cell surface molecule, the NMDA receptor. Researchers have studied a remarkable neurology patient called R.B. This man could vividly describe television shows he had seen in the 1950s, but could not form new memories; he had to be reintroduced to his physician every few minutes! R.B. had suffered brain damage in an accident, but the damage was only to the hippocampus, an area of the cerebrum rich in NMDA receptors. These molecules, named for their ability to bind N-methyl-D-aspartate in the laboratory, are unusual because they respond both to chemical neurotransmitters and to electrical depolarization. When stimulated in either way, the NMDA receptor allows quantities of calcium ions to rush into the cell. The calcium rush changes the strength of the synapse by either increasing the activity of the sending neuron or increasing the sensitivity of the receiving neuron. This fascinating area of research is beginning to reveal new views of how we access memory—a capacity that lies at the very heart of learning in higher animals.

THE CHEMICAL MESSENGERS OF THE BRAIN

The brain's intercellular spaces are awash with an unexpected variety of chemicals, including neurotransmitters, hormones, and other substances.

In recent years, investigators have found that *neurotransmitters*—chemicals released from a presynaptic vesicle (see Chapter 34)—predominate in certain local regions of the brain. Norepinephrine, for instance, is found mainly in the reticular activating system, where it is involved in arousal, maintaining wakefulness, motivation, and other functions. Similarly, dopamine occurs exclusively in certain midbrain cells, where it takes part in neuronal control of complex muscle movements.

Not surprisingly, substances that interact with neurotransmitters also affect body functions. For example, certain antidepressant drugs act by inhibiting enzymes that degrade norepinephrine, so that levels of the neurotransmitter remain constant and a person remains alert and active. Caffeine in coffee, tea, and cola drinks inhibits the enzyme that degrades cyclic AMP; a high level of cAMP, in turn, excites various brain systems. And the addictive properties of cocaine may be based on the neurotransmitter dopamine, as the box on page 678 explains. Research is progressing rapidly on peptides that act on brain neurons, so-called **neuroactive peptides**.

Just a few decades ago, biologists had identified only seven neurotransmitters active in brain tissues and thought that they simply excited or inhibited postsynap-

DOPAMINE AND ADDICTION: THE COCAINE CONNECTION

North America is experiencing an epidemic of drug use. The most frightening part of the phenomenon, however, is surely the sharp rise in cocaine abuse since the mid-1980s. Forty percent of college age men and women have tried cocaine, and of those who use it regularly (12 times per year or more) fully one-quarter will eventually become addicted. Pushing that percentage even higher is the growing practice of freebasing—inhaling cocaine vapors from burning "crack" or other adulterated forms of the alkaloid directly into the lungs, where it can pass quickly into the bloodstream.

Recent research has revealed why cocaine is so addictive, and just how dangerous the drug can be. Because of its molecular structure, cocaine appears to directly stimulate a reward or pleasure system in the brain and lead to the "rush" people sometimes experience. Specifically, cocaine affects the neurons of the nucleus accumbens. In this small brain region, "sending" neurons give off the neurotransmitter dopamine, which in turn inhibits the spontaneous firing of receiving neurons, and leads to a neurologic reward—a sensation of pleasure. Cocaine blocks the normal reuptake of dopamine into the sending neuron, thereby increasing the level of dopamine in the nucleus accumbens and greatly expanding the sensations of pleasure produced in this part of the brain. Morphine, amphetamines, alcohol, nicotine, and certain other drugs appear to act in a similar way.

After repeated exposures, this biochemical pleasure connection stops working as efficiently as it did before. A cocaine user must take the drug in larger, more frequent doses to induce the same response, and, eventually, just to feel somewhat normal and avoid a plummeting mood, fatigue, and an intense craving for the alkaloid.

While cocaine addiction is ruinously expensive and notoriously difficult to overcome, even a single dose of the drug can be dangerous. Cocaine triggers the release of the "fight or flight" neurotransmitters, epinephrine and norepinephrine, then blocks their reuptake. This can cause the heart to beat so strongly and rapidly that if the natural beat is even slightly irregular, the heart can fibrillate and stop. What's more, a single dose of cocaine taken by a pregnant woman quickly accumulates in the amniotic fluid, and is converted there to a related compound, norcocaine. Norcocaine remains in the fluid for days, bathing the fetus and potentially causing neurological deficits in the child, which can be exhibited as poor attention span, irritability, or learning disabilities. Norcocaine can even cause brain-damaging strokes in the fetus before birth. Tragically, medical researchers in Boston recently reported that 18 percent of the new mothers in their study had used cocaine during pregnancy.

tic cell surfaces. By now, they have detected dozens of additional chemicals—most of which are peptides that act as neurotransmitters or as modulators of neuron activity. Perhaps most surprising is the fact that many of these "new" substances were identified long ago elsewhere in the body, where they were shown to act as hormones. Angiotensin II, for example, was originally isolated in the kidney and was believed to regulate blood pressure; it is now known to act on specific brain neurons to stimulate drinking behavior. Other brain neuroactive peptides include gastrin and cholecystokinin (CCK), which we encountered in the gut (see Chapter 32), as well as the traditional hormones prolactin, insulin, and thyroid-stimulating hormone (see Chapter 35).

Where do these powerful peptides come from? In a brain neuron, a neuroactive peptide is synthesized and transported down the neuron's axon to a synaptic terminal; it is then stored and secreted when nerve impulses travel down the axon. A neurotransmitter such as acetylcholine, on the other hand, is manufactured directly in the synaptic terminal. Some individual neurons appear to synthesize and release at synapses both a neuropeptide and a traditional neurotransmitter, such as dopamine or acetylcholine.

Compared to neurotransmitters, which are "hit-and-run" chemicals, affecting a target cell briefly before being quickly degraded by metabolic processes, neuroactive peptides resist destruction and thus act on target cells over prolonged periods of time. The **endorphins** and **enkephalins**—chemicals that mimic the effects of the opium-derived drug morphine and appear to serve as the body's natural painkillers—fall into this category.

In the mid-1970s, researchers found that certain brain cells have highly specific receptor sites for morphine, even though this substance is not normally present in the body. Reasoning that the brain must itself produce a chemical with an opiate-like molecular structure, J. T. Hughes and H. W. Kasterlitz found the first enkephalins in 1975. As predicted, these enkephalins, which are five amino acids long, are peptides that behave chemically like morphine. Within weeks after the enkephalins were discovered, three somewhat longer peptides, endorphins, were shown to have similar properties.

Endorphins, enkephalins, and other neuroactive peptides have continued to surprise biologists by the broad range of roles they appear to play in body functions. Biologists now know that the brain is dotted with sites where specific neurons synthesize specific peptides; moreover, these sites seem to be mutually exclusive—enkephalins are not found where endorphins exist, and other neuropeptides are similarly uniquely distributed. This highly specific arrangement suggests that such peptide substances may prove to regulate much of an animal's biology—from hormone secretion and blood pressure to body temperature and even movement. The activity of cholecystokinin (CCK) reflects the complexity of the situation; this biochemical is both a brain peptide and a gastrointestinal hormone, and it can also stimulate memory when a researcher injects it into an animal's body cavity. The vagus nerve reports the presence of CCK, and that enhances the memory of feeding; presumably, it also occurs when food enters the intestine.

Brain peptides may act by rendering target cells *less able to respond to other signals*. For example, when the peptide known as **substance P** reaches a target cell in the cerebral cortex, the action of a neurotransmitter that might otherwise excite the neuron may be blocked. Conversely, neuroactive peptides may prevent an inhibitory neurotransmitter from suppressing neuron activity. Substance P is of great interest medically because it is believed to be the chemical carrier of pain messages in the body's sensory system. When a person pricks a finger deeply with a thumbtack (Figure 36-27, step 1), pain receptors in the skin generate nerve impulses over a fast pathway up the spinal cord to the thalamus and then to the sensory and motor cortices (step 2). These impulses report the instantaneous, perhaps blinding pain. But in addition, slow-pathway fibers go to the reticular activating system and parts of both the hindbrain and the forebrain (step 3). Those pathways cause a prolonged perception of the pain, a "nagging pain" that allows the brain time for a better assessment of the real damage. It is the neuroactive peptide substance P that is released in spinal cord and brain cells, especially of the slow pathways. In the brain, substance P probably affects the

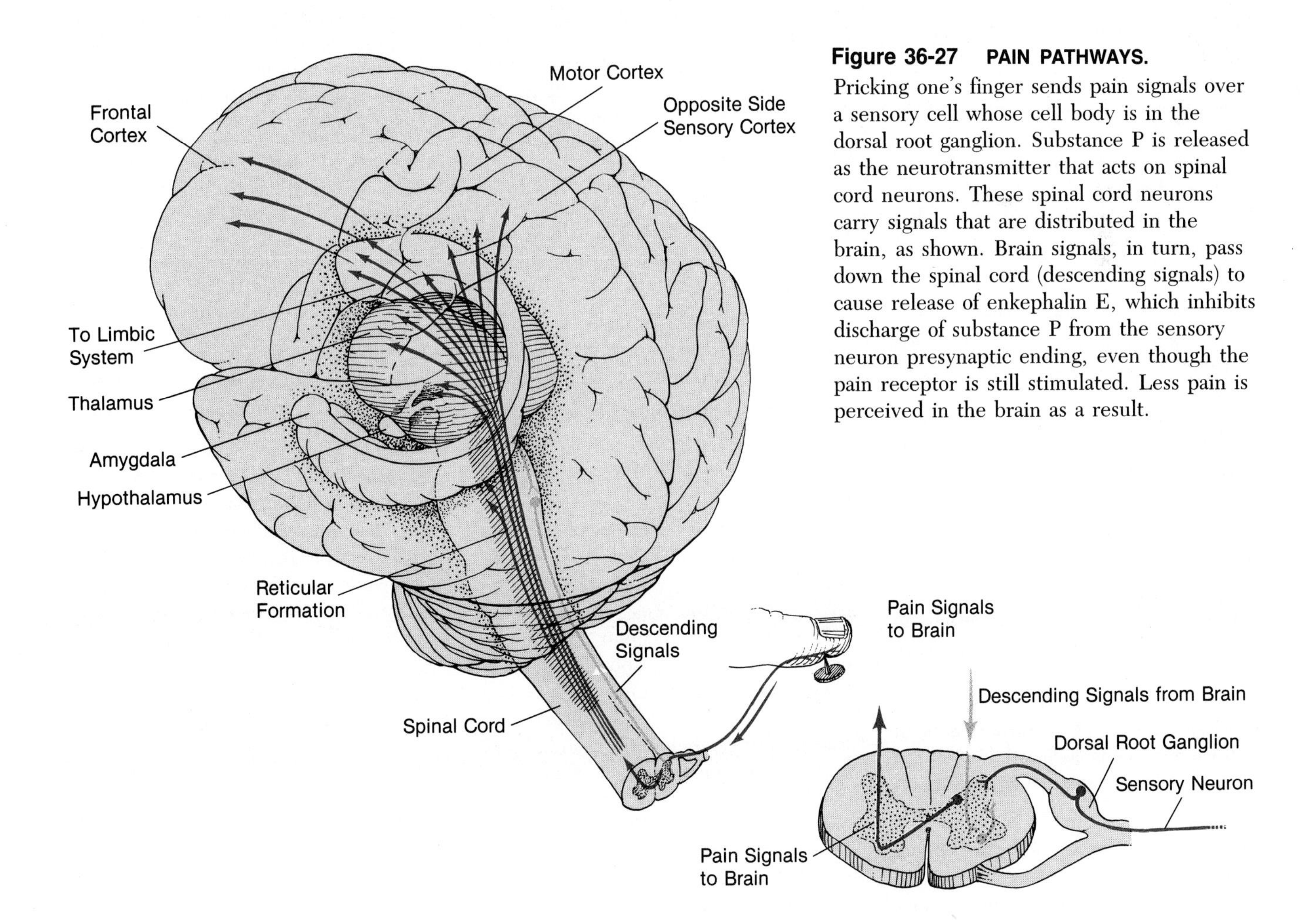

Figure 36-27 PAIN PATHWAYS. Pricking one's finger sends pain signals over a sensory cell whose cell body is in the dorsal root ganglion. Substance P is released as the neurotransmitter that acts on spinal cord neurons. These spinal cord neurons carry signals that are distributed in the brain, as shown. Brain signals, in turn, pass down the spinal cord (descending signals) to cause release of enkephalin E, which inhibits discharge of substance P from the sensory neuron presynaptic ending, even though the pain receptor is still stimulated. Less pain is perceived in the brain as a result.

emotions associated with pain. And it is there, too, that endorphins act. They modulate pain reception and emotion so that the organism can respond in an adaptive way to eliminate or cope with the causes of pain. For example, in the case of the thumbtack, descending signals from the brain cause spinal cord neurons to release enkephalin E, which inhibits the discharge of substance P by the sensory neuron. The family of endorphins probably acts analogously to modulate other emotions. Researchers propose, for instance, that endorphins may prevent extreme anxieties, irrational fears, and the like. Too little endorphin released at the right place in the brain may allow extreme fears of enclosed spaces (claustrophobia), heights, or snakes to develop.

It seems certain that one of the chief functions of neuropeptides is to modulate and refine the continuing flow of nerve activities that govern much of life. Thus, the distinction between the nervous and endocrine systems becomes increasingly blurred, so that it is best to consider them not as independent networks, but as parts of a highly organized and integrated continuum.

The nerve cell is remarkable for its electrical excitability, its secretory capacities (neurotransmitters and neuroactive peptides), and its intricate axonal and dendritic trees. From such cells, arrayed in networks and circuits ranging from the crayfish's tail-flip circuits to the human's cerebral cortex, come learning, language, and, indeed, the human condition.

SUMMARY

1. Sensory receptor cells transduce various types of energy into a form of electrical activity. Receptor cells may be true neurons, they may synapse with sensory neurons, and they may occur singly or be part of *sense organs*.

2. *Chemoreceptors*, the detectors partially responsible for *gustation* (taste) and *olfaction* (smell), appear to have receptor molecules on their surface. Binding of a specific stimulus molecule or ion leads to changes in a second messenger and produces graded changes in the cell membrane potential that ultimately alter nerve impulse rates to associated centers in the brain.

3. *Mechanoreceptors* respond to deformations of the receptor cell and its membrane by mechanical forces; the state of polarization rises or falls as a result, and in the same or an adjacent sensory neuron, corresponding changes take place in the rate of nerve impulse generation. Such receptor cells often have cilia or microvilli that deform the cell's plasma membrane when they are bent.

4. In invertebrates, sensory hair cells that detect body position and motion lie beneath *statoliths*. In vertebrates, *otoliths* shift and bend sensory hair cells in chambers of the inner ear so that the animal can sense body position. Sets of sensory hair cells associated with the *semicircular canals* detect angular acceleration and deceleration.

5. In many vertebrates, the initial transduction of sound takes place in the inner ear. Sound waves strike the *tympanic membrane*, or eardrum; resultant vibrations are then transmitted by a chain of tiny bones through the middle ear to the *oval window*, where they set up pressure waves in the inner-ear fluid.

6. Sound of a specific frequency sets up pressure waves in the cochlear chambers that displace maximally a specific site on the *basilar membrane*, thereby bending microvilli on sensory hair cells against the *tectorial membrane*. This generates activity in associated neurons, whose axons extend via the *auditory nerve* to the brain.

7. In some animals, mechanoreceptors derived from the lateral line have become specialized to detect weak electric currents.

8. Many types of animals and even some bacteria can "sense" the lines of force of the earth's magnetic field.

9. *Thermoreceptors* in the facial pits of certain snakes can detect and locate in space very small differences in infrared (heat) radiation.

10. Mollusks and vertebrates have camera eyes. Light enters through the transparent *cornea* and passes through the *iris's pupil* to the *lens*, which focuses light on the photoreceptor cells of the *neural retina*. These specialized cells, known as *rods* and *cones*, have visual pigment molecules *(rhodopsin)* that change shape when they absorb light. *Retinal*, a component of rhodopsin, changes shape and later dissociates from the protein *opsin*. These and other changes cause the cell to hyperpolarize and send nerve impulses to the brain via the *optic nerve*.

11. Color vision in mammals and some other vertebrates results from sensitivity of three types of cones to different wavelengths of light.

12. Much processing of visual information goes on in the retina prior to reaching the ganglion cells. Each ganglion cell has a *visual field* composed of the photoreceptors to which it is wired. Much more complex processing goes on in the brain, especially in the *visual cortices*.

13. Arthropod *compound eyes* have separate visual units, called *ommatidia*, which are sensitive to both visible and ultraviolet wavelengths.

14. The human brain contains at least 100 billion neurons that interact in extraordinarily complex ways.

15. The brain is divided into *hindbrain*, *midbrain*, and *forebrain*. The *cerebrospinal fluid* fills tissue spaces and cavities and transports vital materials.

16. Among the brain's "oldest" parts are the *medulla oblongata* and the *cerebellum*. The medulla controls basic functions, such as respiration and swallowing, while networks of giant *Purkinje cells* in the cerebellum regulate balance, stance, and some other motor functions.

17. The *cerebrum* and its *neocortex* (cortex) carry out more complex brain functions, including many complex reflexes and instinctive behaviors. The limbic system, including the hypothalamus, thalamus, amygdala, and hippocampus, controls hunger, thirst, anger, fear, and sexual desire.

18. The *reticular activating system*

(*RAS*), is the source of arousal and wakefulness and filters diverse sensory inputs.

19. The neocortex has six layers of neurons arranged into columns and, in the visual cortex, into orientation slabs. Vertically oriented columns of interconnecting neurons may be the basic units of operation in the cortex. Regions of the cortex are specialized for specific functions and are called the sensory, motor, visual, and auditory cortices.

20. Split-brain studies, in which the upper cortex of human brains were divided, reveal that each half of the brain is capable of independent functioning and that each is somewhat specialized for different cognitive tasks.

21. Broca's area in the cerebrum controls actual production of speech, while Wernicke's area controls the content.

22. EEG studies reveal that sleep is a complicated state consisting of at least five stages. REM, or dreaming, sleep may play a role in laying down long-term memories.

23. The perception of moving or still images and of depth arises from hierarchies of neurons in the mammalian visual cortex in which a large number of cells at each level drive a smaller number of cells in the next level.

24. Memories can be immediate, short-term, or long-term. The hippocampus and cerebellum are involved in immediate and short-term memory, whereas the cortex appears to be the site of long-term memory.

25. There is substantial plasticity of brain neurons, with dendrites, synapses, and even neurons changing with use and disuse.

26. Central nervous system neurons communicate with each other through neurotransmitters and *neuroactive peptides*.

KEY TERMS

auditory nerve
basilar membrane
bipolar cell
cerebellum
cerebrospinal fluid
cerebrum
chemoreceptor
cochlea
column
compound eye
cone
cornea
corpus callosum
echolocation
endolymph
endorphin
enkephalin
forebrain
fovea
ganglion cell
gustation
hindbrain
iris
lateral line organ
lens
limbic system
mechanoreceptor
medulla oblongata
midbrain
neocortex
neural retina
neuroactive peptide
olfaction
ommatidium
opsin
optic nerve
organ of Corti
otolith
oval window
pigmented retina
pons
proprioceptor
pupil
Purkinje cell
rapid eye movement (REM) sleep
reticular activating system (RAS)
reticular formation
retinal
rhabdomere
rhodopsin
rod
round window
semicircular canal
sense organ
statocyst
statolith
substance P
taste bud
tectorial membrane
thermoreceptor
transducin
trichromatic theory of color vision
tympanic membrane
visual cortex
visual field

QUESTIONS

1. What is the basic sequence of nerve cell activities during the conveyance of a sensory message from a receptor to the brain?

2. What is meant by transduction by a sensory receptor cell? Give an example.

3. What are the three major classes of sensory receptors?

4. Compare gustation in the human and the blowfly.

5. Briefly discuss the structure and function of a dog's olfactory epithelium.

6. Describe Pacinian corpuscles, and outline the role they play in mechanoreception.

7. Compare and contrast the mechanisms that vertebrates and invertebrates use to sense position of the body in space. What is the principle of sensory hair cell function in sensing position and in detecting angular or linear accelerations and decelerations?

8. How does the mammalian cochlea detect a specific frequency of sound?

9. Describe the sequence of activities that begin when a photon of light strikes molecules of visual pigment in the outer segment of a rod in the retina.

10. Compare and contrast the camera eyes of vertebrates and cephalopods with the compound eyes of arthropods.

11. What are the four hollow cavities of the brain called? What fluid fills them, and what is its function? What are meninges?

12. Draw and label a diagram of the human brain. What are some functions of each region?

13. What is the limbic system, and what does it do?

14. Where is the reticular formation, and what is its function? What is the reticular activating system?

15. Describe the structure of the human cerebral cortex.

16. What is a homunculus? Which regions of the human cortex have a homunculus, and what do its unusual proportions reflect?

17. What is the corpus callosum? What structures does it connect? Does severing it affect speech? Reading? Thinking? What is affected?

18. What is the function of Broca's area? Wernicke's area? Where are they located?

19. Does damage to the hippocampus affect long-term memory? Short-term memory? How does damage to the cerebellum affect memory?

20. How do neurotransmitters and neuroactive peptides differ in terms of production, longevity, and function? What is the significance of such peptides and of the view that the brain is a "neuroendocrine" organ?

ESSAY QUESTIONS

1. How are modified cilia or hair cells involved in the senses of chemoreception, mechanoreception, and photoreception in animals?

2. Explain how a curved series of dots in the retina is translated into a moving line—let's say the edge of a tennis ball rocketing through the air. Include the organization and functioning of the neural retina and visual cortex in your answer.

For additional readings related to topics in this chapter, see Appendix C.

37 Skeletons and Muscles

We made threads of the highly viscous new complex of actin and myosin, "actomyosin," and added boiled muscle juice. The threads contracted. To see them contract for the first time, and to have reproduced in vitro one of the oldest signs of life, motion, was perhaps the most thrilling moment of my life.

Albert Szent-Gyorgyi,
"Lost in the Twentieth Century,"
Review of Biochemistry (1963)

Tap your flexed knee with a rubber mallet, and your lower leg automatically kicks the air. Lift a load of books, and your biceps bulge. Run a marathon, and muscle fibers in your thighs work for hours to propel your body forward. Muscles and the bones they move accomplish most large-scale forms of work in a vertebrate's daily life. Along with glands and other organs, they are the body's *effectors*—organs that carry out movements and a variety of other actions that enable an animal to breathe, eat, walk, and go about its daily activities. Recall that sensory cells and sense organs send information to the central nervous system. Orders issuing from the CNS, in turn, control the body's effector organs to generate overt behavior or to regulate physiological processes.

The intricate activities of actin and myosin molecules under the control of calcium ions allow movement by and within individual eukaryotic cells (see Chapter 6).

Muscles and motion: This male red kangaroo flies across the arid ground of Australia's parched interior, muscles contracting and extending with each bound.

These movements include whole-cell locomotion, aspects of mitosis, and the shuttling of cytoplasmic organelles and fluids. Interestingly, these same molecules and ions also allow the much larger movements we see as the contraction of a biceps muscle, the beating of a heart, or the peristalsis in an intestine.

Two principles guide this universal contractile machinery: (1) Coordinated movements of precisely aligned molecules can generate force, and (2) this force results from the shortening of arrays of molecules, not from the contraction of individual molecules. As you will see, individual muscles virtually always work by pulling, rather than by pushing, and they usually pull against a skeleton, or against other muscles or tissues.

Our discussions in this chapter include:

- The various types of skeletons that have evolved in animals
- How nervous signals control muscle action
- The anatomy of muscle cells and muscles as organs
- The molecular events that underlie muscle contraction
- How muscle contractions are varied
- The three major classes of muscle and how they function in vertebrates and invertebrates

Figure 37-1 SCARAB BEETLE: ARMORED EXOSKELETON.
The hard, nonliving exoskeleton of the scarab beetle protects soft internal structures. Besides serving as a protective armor, the exoskeleton is impermeable and thereby prevents loss of water to the beetle's environment. The muscles that move parts of such a skeleton are located internally.

THE ANIMAL SKELETON: A LIVING SCAFFOLD

Animal skeletons have a variety of forms and functions, but all are of two basic types: fluid *hydroskeletons* and *hard skeletons* (see Chapter 24).

Hydroskeletons are volumes of fluid contained by the gut, pseudocoelom, coelom, vascular system, or water vascular system of various invertebrates. The earthworm is able to burrow through soil, for example, because waves of contractions in circular and longitudinal muscles compress fluid in the worm's gut and coelom, causing successive portions of the body to elongate, to press against the ground, and to propel the animal forward slowly.

Hard skeletons can be rigid, jointed, external encasements called *exoskeletons*, or they can be internal scaffolds called *endoskeletons*.

Exoskeletons

Exoskeletons serving as protective armor include the shells of clams and snails and the hard coverings of lobsters, spiders, and beetles (Figure 37-1). Besides providing support, these shells often block moisture loss or gain. Exoskeletons tend to be nonliving, acellular deposits of crystallized mineral salts, such as calcium carbonates, or a mixture of organic and inorganic substances, such as chitin. While a mussel shell or clamshell may increase in size as calcium salts are continually deposited at the edges, an arthropod's nonexpandable outer covering prevents much growth, and thus the lobster, insect, or other arthropod must periodically molt to expand in size.

All hard skeletons are composed of separate pieces that are *articulated*, or hinged, to permit movement. Movement of body parts is possible because joints composed of thinner, flexible exoskeletal materials are strategically located at points of relative movement in an antenna, claw, leg, wing, or tail and because muscles operate *within* the exoskeleton.

Two sets of muscles help move each skeletal part: The *flexors* bend a limb joint, and the *extensors* straighten the joint (Figure 37-2a). Figure 37-2b shows two typical arrangements, where one muscle elevates a fly's wing and another muscle depresses the wing. In a fly or bee, the contraction of the wing elevator muscles itself triggers the contraction of wing depressor muscles without any nerve signals. This *automatic stretch activation* explains why a fly, for example, can beat its wings hundreds of times per second.

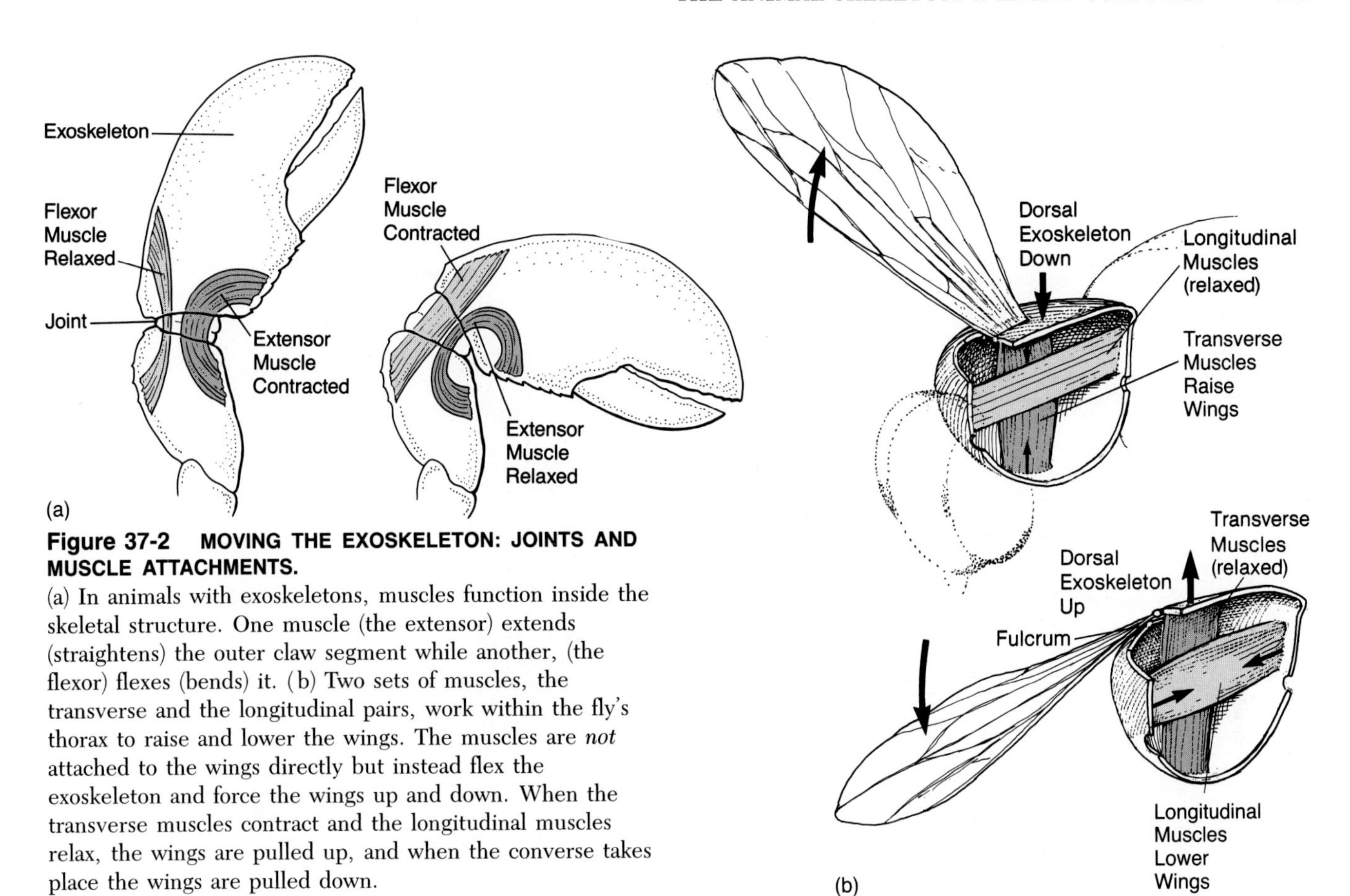

Figure 37-2 MOVING THE EXOSKELETON: JOINTS AND MUSCLE ATTACHMENTS.
(a) In animals with exoskeletons, muscles function inside the skeletal structure. One muscle (the extensor) extends (straightens) the outer claw segment while another, (the flexor) flexes (bends) it. (b) Two sets of muscles, the transverse and the longitudinal pairs, work within the fly's thorax to raise and lower the wings. The muscles are *not* attached to the wings directly but instead flex the exoskeleton and force the wings up and down. When the transverse muscles contract and the longitudinal muscles relax, the wings are pulled up, and when the converse takes place the wings are pulled down.

Endoskeletons

In animals such as rays and sharks, endoskeletons consist of cartilage, whereas in most vertebrates, they consist of bone plus cartilage. **Cartilage** is primarily the protein collagen and complex polysaccharides (see Figure 29-16), while **bone** is predominantly collagen in combination with a large amount of apatite, a calcium and phosphate salt. In contrast to acellular, mineralized exoskeletons, both bone and cartilage contain living, metabolizing cells.

Most vertebrates have two classes of bones in their endoskeletons: those that develop directly as bone within the skin (*dermal membrane* bones) and those that first develop as cartilage and are later transformed into true bone (*endochondral* bones). Outer skull bones are dermal membrane bones, whereas bones in the limbs and the pelvic girdle are endochondral bones. In addition, the skeleton itself is subdivided into the **axial skeleton**—which consists of the skull, the vertebral column, and, in animals that possess them, the ribs—and the **appendicular skeleton**—which is made up of the forelimb bones attached to the *pectoral girdle* and the hindlimb bones attached to the *pelvic girdle* (Figure 37-3).

Several structures enable an endoskeleton to move. The **joints,** such as the human knee joint (Figure 37-4), are the regions where individual bones meet. **Ligaments** are strong, flexible bands, composed primarily of collagen fibers, that help hold bones together at a joint. Although joint ligaments are fairly elastic, too great a stress—for example, a blow to the knee during a football tackle—may stretch or snap them or pull one end free of the bone. Within joints, where the bones rub together, cartilage acts as an internal pillow, softening the rubbing action. Finally, the whole joint cavity is surrounded by *synovial membranes,* which contain the shock-absorbing *synovial fluid* that acts as a lubricant.

Some junctions between bones are rigid. As an animal matures, individual bones of the vertebrate skull or of the pelvic girdle grow until their edges meet. Then *sutures*—stiff, fibrous, calcified regions—develop to weld the bones together. The fontanelles, or "soft spots," on the surface of a baby's head are areas in which the skull bones have yet to fuse.

The Inner Structure of Bone

Living bone is a dynamic tissue. Mammalian bones typically have an external layer of extremely hard *compact bone* around a softer center of *spongy bone* (Figure 37-5a). Compact bone is constructed of thousands of cy-

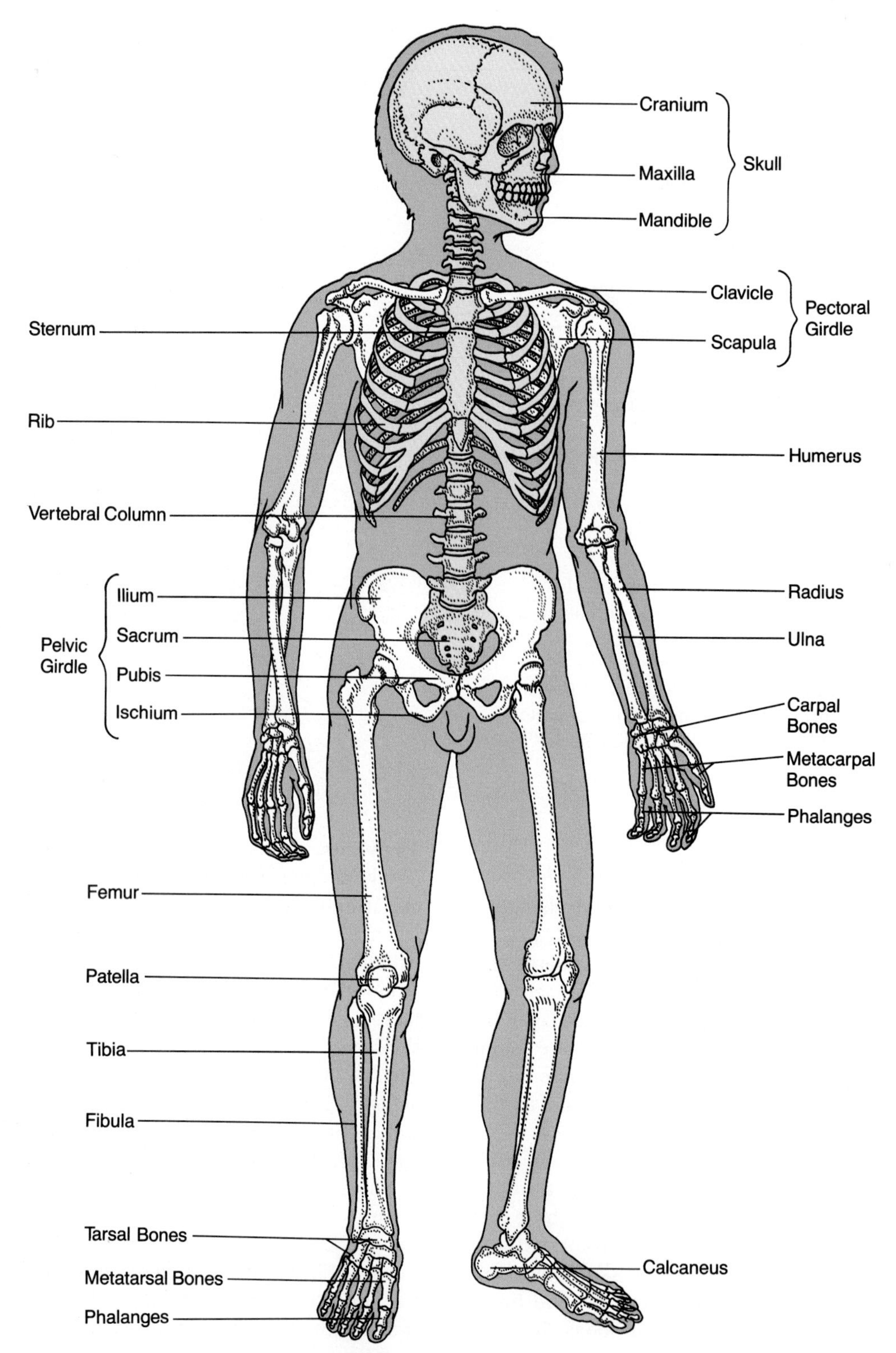

Figure 37-3 **THE HUMAN ENDOSKELETON.**
The axial skeleton, consisting of the skull, vertebral column, and ribs, is colored beige in the drawing. The pelvic girdle, the pectoral girdle, and bones peripheral to them compose the appendicular skeleton and are white. Both sets of bones are made of the same substances, primarily collagen and apatite.

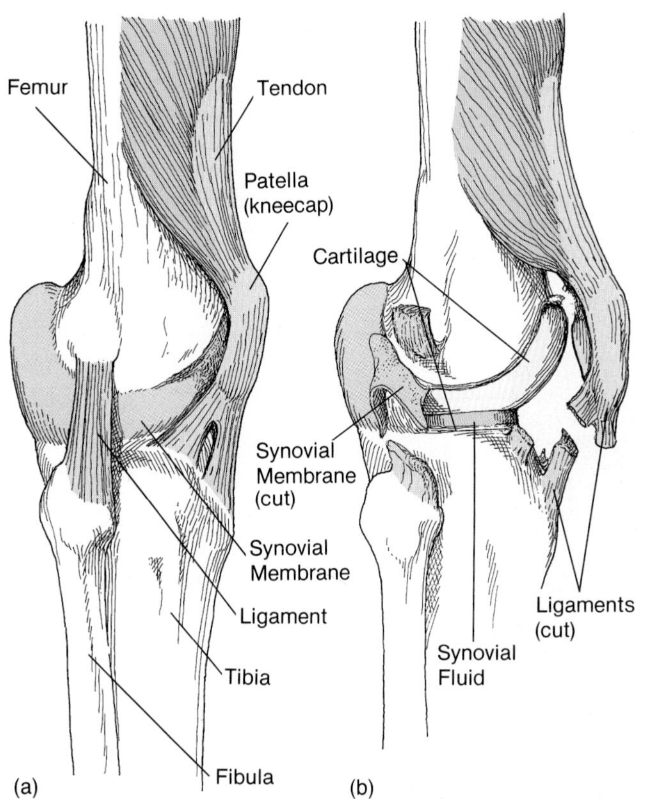

Figure 37-4 THE HUMAN KNEE JOINT: A LIVING HINGE OF CARTILAGE AND LIGAMENTS.
These drawings show a dissection of the human knee joint from the side. (a) The relationship of ligaments holding bones together and tendons attaching muscle to bone. (b) The knee with tendons, ligaments, and synovial membrane cut to reveal the cartilage cushions on the ends of the bones. Synovial fluid bathes the cartilage and also cushions the joint. Here, cartilage is shown as yellow; ligaments, manila; tendons, beige; synovial membrane, gray.

lindrical, densely packed **Haversian systems,** made of multiple layers of bone surrounding a single blood capillary (Figure 37-5b). Within the honeycomb-like layers are living bone cells, called *osteocytes;* each osteocyte resides in a separate chamber that is connected by tiny canals to the capillary, so that nutrients, wastes, and gases may pass to and from the cell. The outer surfaces of adjacent Haversian systems fuse totally, and together, many such layers give compact bone its hardness.

The less dense spongy bone is often richly endowed with blood vessels in its interior, the bone marrow region. Bony struts extend across the marrow cavity and serve as braces for the surrounding compact bone, and two kinds of cells fill the spaces in between: fat cells (so-called white marrow) and developing blood cells (so-called red marrow, where erythrocytes, lymphocytes, and macrophages arise).

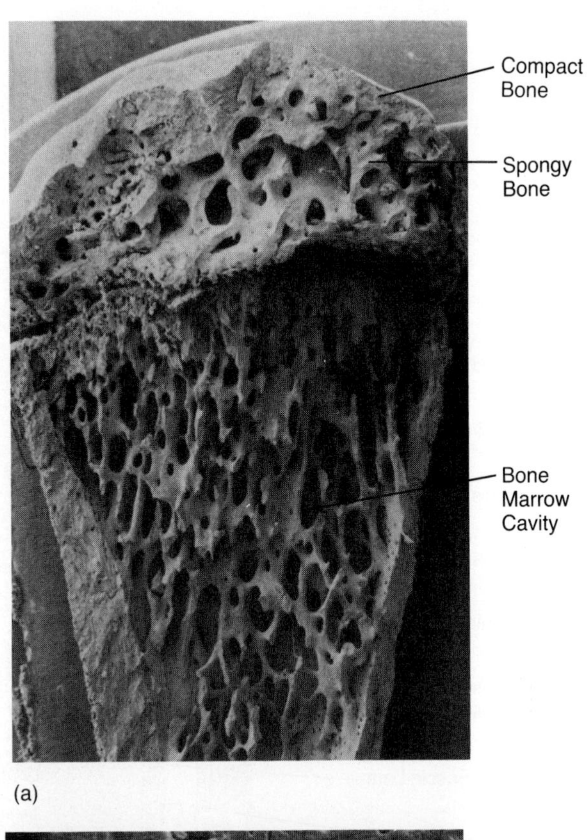

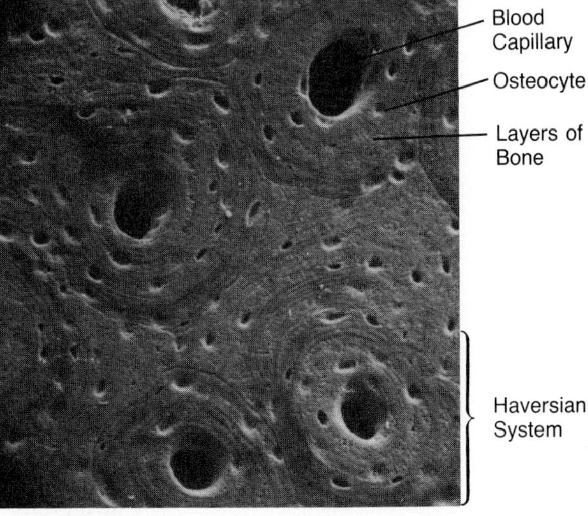

Figure 37-5 THE STRUCTURE OF BONE.
(a) The human femur (thighbone) in longitudinal section, magnified about 1.3 times. The meshwork at the top and the interior of the shaft is spongy bone. The dense bone of the exterior portion of the shaft is compact bone. The main bone marrow cavity can be seen at the center of the lower bone shaft in this photograph. (b) Electron micrograph showing a cross section of compact bone, magnified about 200 times. The Haversian systems are composed of concentric rings of bone around a blood capillary that occupied the large holes seen here. Individual living bone cells, osteocytes, resided in the small depressions when this bone was alive.

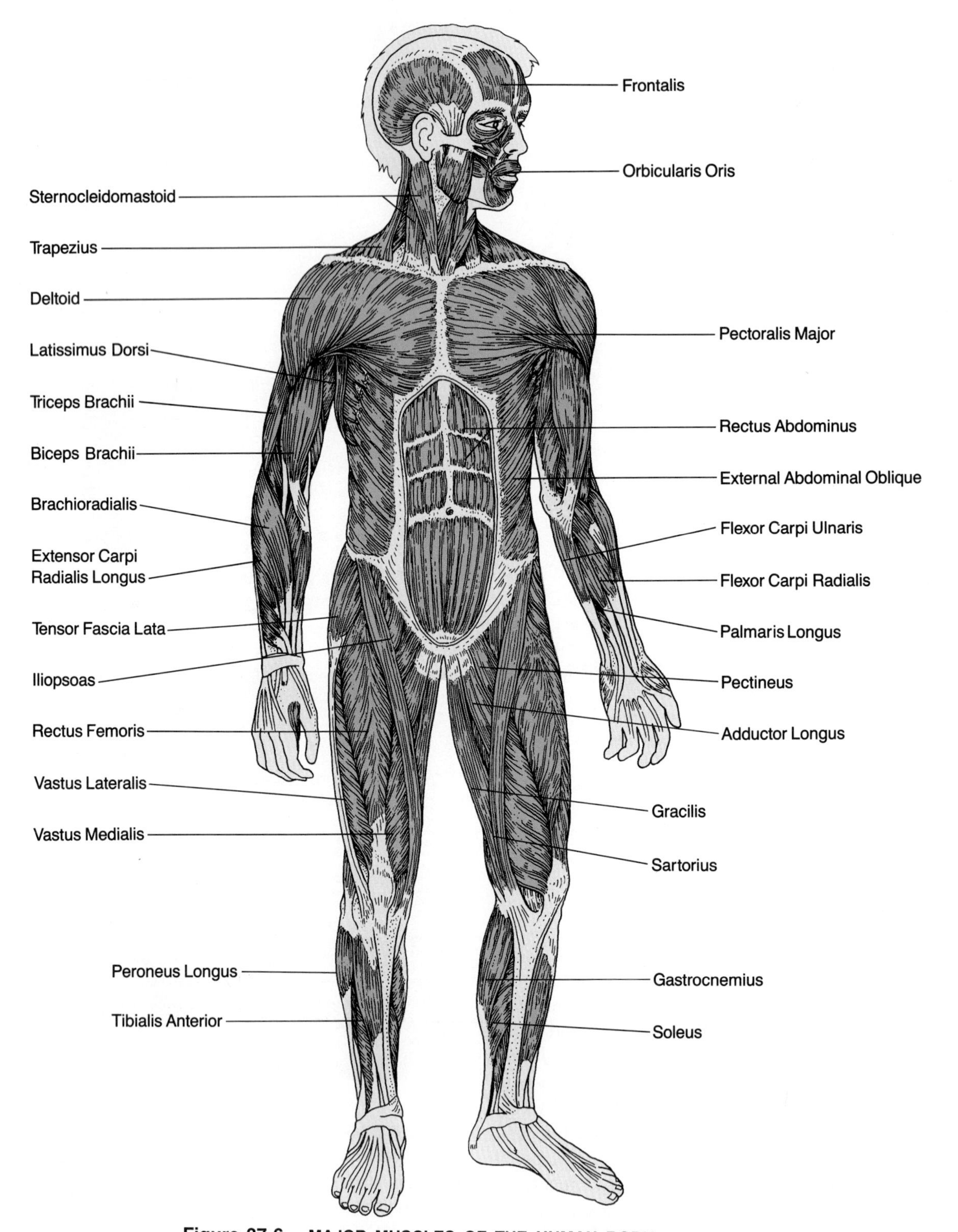

Figure 37-6 MAJOR MUSCLES OF THE HUMAN BODY. This diagram shows the major muscle groups. In some cases, tendons and ligaments are visible, as are the points of origin and insertion.

Some bones, such as the long bones of the limbs, must grow substantially as a vertebrate matures. Lengthening occurs in cartilaginous growth zones at each end of a bone, and the cartilage, in turn, is replaced by bone.

In living bone, varying levels of the hormones calcitonin and parathyroid hormone act on certain bone cells and cause them either to break down bone—thereby raising calcium and phosphate levels in the blood—or to deposit new bony material. Recently identified substances called bone morphogenetic proteins (BMPs) seem to help regulate cartilage and bone formation, and physicians have used them with some success to stimulate the repair of skull defects and broken bones. Researchers have also developed bioceramics, which they can insert into cracks or fissures in bone and which appear to promote infiltration by host bone cells. Others have invented numerous ingenious implants made of metal, plastic, or combinations of materials to replace damaged hip, ankle, elbow, or other joints.

A person's bones reach peak density and mass about age 35, then tend to become weaker and more porous with age. Extreme bone loss, or **osteoporosis,** can lead to brittle, easily broken bones, including the hip fractures so common in the elderly. Recent studies show that the decline of estrogen at menopause plays a role in bone loss, as can cigarette smoking, heavy drinking, and lack of exercise. To prevent osteoporosis, doctors are encouraging people to get plenty of calcium in the diet and to exercise regularly, and they are advising many older women to take estrogen replacement therapy and calcium supplements.

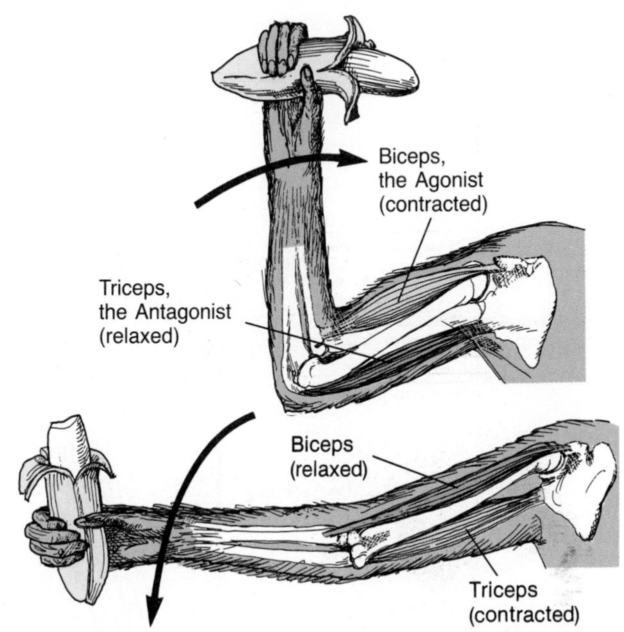

Figure 37-7 MUSCLE ANTAGONISM IN A MONKEY'S ARM.
When a monkey raises a banana to its mouth, the action of bending the arm at the elbow results from contraction of the biceps and relaxation of the triceps. In this motion, the biceps is the agonist, or prime mover, while the triceps is the antagonist. However, when the monkey straightens its arm, the triceps is contracted and thus represents the prime mover, while the biceps is relaxed and becomes the antagonist to this movement. In fact, several other muscles not shown here also participate in such movements.

The Skeleton-Muscle Connection

The speed and strength with which limbs, fingers, jaws, and other body parts can move depend on the connections and interactions of nerves, bones, and muscles. Figure 37-6 shows the major skeletal muscles of the human body. When a large animal moves a major limb—when a monkey picks up a banana, for example—the biceps muscle in its arm contracts, flexing the elbow, which acts as a hinge (Figure 37-7). The biceps is the *agonist*, or *prime mover*—the muscle primarily responsible for the banana-raising movement. Alternatively, if the triceps muscle—the *antagonist* of the biceps—contracts, it works against the agonist, pulling on the ends of the forearm bones, and causing the elbow to extend. In general, when a muscle contracts, its antagonist relaxes. Additional muscles may serve a joint and contract simultaneously with a prime mover, augmenting or modifying the direction in which a bone moves; these complementary muscles are known as *synergists*.

In more complex muscular operations—the measured finger manipulations of a pianist playing a sonata, for instance—the gradually increasing contraction of prime movers is coupled with balanced relaxation of antagonists, and the result is exquisite control. Thus, before a finger begins to move, both agonist and antagonist muscles are partially contracted; as one contracts more, the other contracts less.

In a sense, vertebrate skeletal muscles operate like systems of springs acting on levers. In the terminology of a mechanical engineer, one bone moves on another at a *fulcrum*, the support about which a lever turns (Figure 37-8). The length of bone between the point where a muscle is attached and the fulcrum is the *power arm*, and the length between the fulcrum and the site where work is done (such as a foot or hand) is the *load arm*. When the power arm is relatively long, as in a badger's limb, the load arm can be moved with substantial force, but not with great speed (Figure 37-8a). If the power arm is short, as in a cheetah's limb, the load arm can be moved quickly but with less force (Figure 37-8b).

Whether a bone can act as a lever depends, in large part, on where the muscles attach to it. Each muscle has two attachment sites: the *origin*, which acts as an anchor,

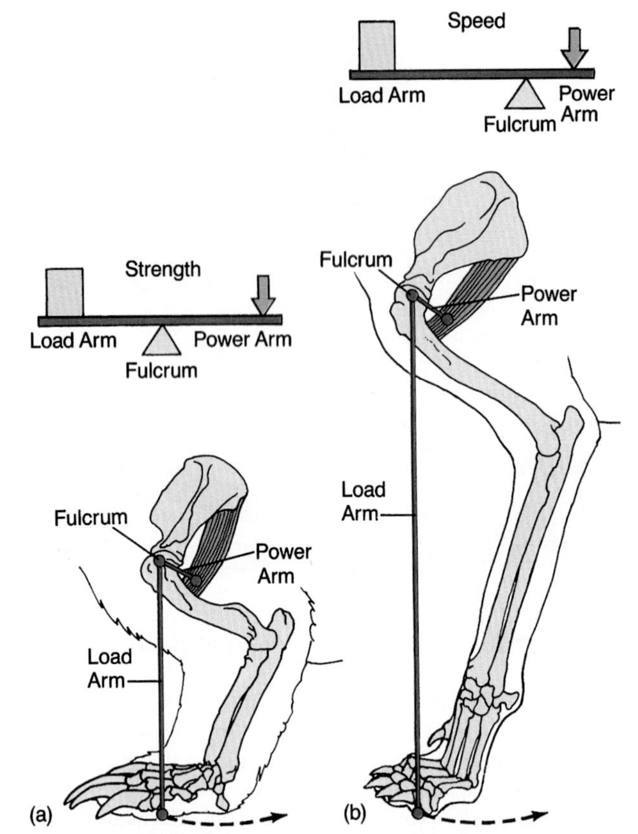

Figure 37-8 THE POWER ARM–LOAD ARM CONCEPT. The greater the length of the power arm (the distance from the fulcrum to the site of muscle attachment), relative to the length of the load arm (the distance from the fulcrum to where the work is done), the greater the power that can be applied to the bone. Conversely, the shorter the power arm relative to the load arm, the faster the bone (the foot, in this case) can be moved. (a) The badger's leg is slow but powerful. (b) The cheetah's longer leg sacrifices power to achieve great speed of movement. (The cheetah's leg is actually three to four times as long as the badger's.)

and the *insertion*, located near the region of movement. The insertion is the site where the force of the muscle's contraction is applied to the bone (Figure 37-9). Muscles attach to bones or other muscles by way of **tendons**—extremely tough bundles of collagen fibers.

A tendon can concentrate all the contractile force of a large or broad muscle at a small, specific site, usually one that provides greatest leverage for the motion in question. Tendons can also change the direction of the force generated by a muscle. For instance, muscles in a bird's lower chest raise its wings. This arrangement is possible because the tendons from the lower chest muscles travel over pulley joints of the shoulder bones and then attach to the upper surface of the wing bones. Therefore, when the chest muscles contract, the wing is pulled upward.

REGULATING MOVEMENT: FEEDBACK CONTROL OF MUSCLE ACTION

At every moment, the central nervous system keeps track of what hundreds of agonist, antagonist, and synergist muscles are doing and signals some motor neurons to generate impulses while others remain quiet. Instrumental to this management feat is a class of sense organs called *proprioceptors*.

Most complex animals have proprioceptors that report on the position of the body or the body's major

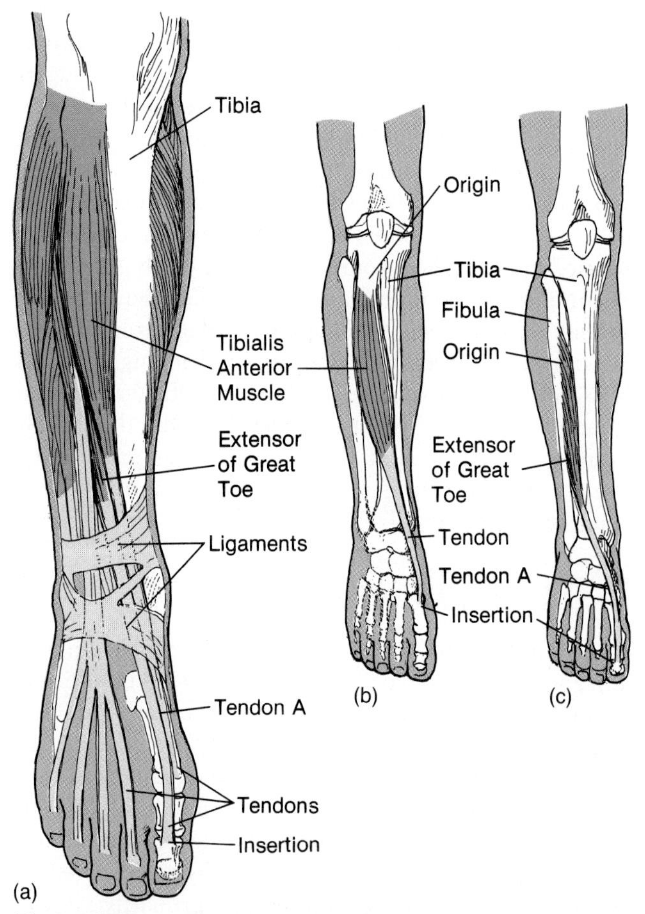

Figure 37-9 MUSCLE ATTACHMENTS AND TENDONS. (a) In a human leg, the muscles on each side of the tibia end in tendons running under straplike ligaments at the ankle. These then insert on the upper surface of the toes. (b) The tendon at the origin of the tibialis anterior muscle is quite broad; the tendon at the insertion on the upper surface of the base of the great toe is much narrower, concentrating the total force of the muscle's contraction at a small spot. (c) The extensor of the great toe has a long origin on the surface of the fibula and also a narrow insertion on top of the great toe's tip.

parts, such as the limbs or head. Vertebrates also have special proprioceptors called **muscle spindles** (or, sometimes, *stretch receptors*) in joints, tendons, and muscles. The muscle spindles are small but remarkable sets of special muscle cells found near the center of each skeletal muscle. Muscle spindle cells have highly ordered contractile systems at both ends, but not in their central region, where many endings of so-called *1a sensory neurons* reside (Figure 37-10). These 1a sensory neurons form the first part of a reflex arc, entering the spinal cord and forming synapses with *α motor neurons*. The axons of these cells pass to the very muscle in which the muscle spindle is located.

Suppose that a chimpanzee's triceps contracts, extending its elbow (Figure 37-10, step 1). The antagonist muscle, the biceps, is stretched as the forearm straightens (step 2), and so is the muscle spindle within the biceps. A 1a sensory neuron in the spindle generates nerve impulses at a slow rate when a muscle is at rest, but as the spindle is stretched, these neurons send a volley of nerve impulses to the spinal cord (step 3). Such neurons—in this case, coming from the biceps—usually synapse in the spinal cord directly with the α motor neurons serving the muscle cells around the muscle spindle—again, in this case, in the biceps (step 4). Thus, incoming sensory impulses resulting from a stretched biceps spindle trigger a reflex contraction of the biceps (step 5). That contraction resists the action of the triceps. As the biceps contracts, its muscle spindle is shortened somewhat, and its 1a sensory output to the spinal cord decreases; as a result, the motor output causing biceps contraction decreases. Feedback loops involving muscle spindles allow an animal to maintain normal body stance without falling over and underlie the knee-jerk response.

Certain centers in the brain can also cause the muscle spindle to contract and thus indirectly cause the muscle to contract as well. Innervating the contractile end regions of the muscle spindle cells are small motor neurons known as *γ motor neurons*. Motor centers in the brain or reflex circuits in the spinal cord activate the γ motor neurons, and they in turn generate impulses that cause the muscle spindle cells to contract, thereby stretching their central, noncontractile region (step 6). This stretch is signaled by the 1a sensory neuron serving the muscle spindle to the CNS; that information immediately triggers a reflex, and α motor neurons cause the muscle itself to contract. The synapse between a motor neuron fiber and a skeletal muscle cell is clearly a crucial structure and is called the neuromuscular junction. Recent studies show that a unique glycoprotein, s-laminin, marks this junction in such a way that growing nerve fibers recognize it, stop growing, and form the appropriate synapse.

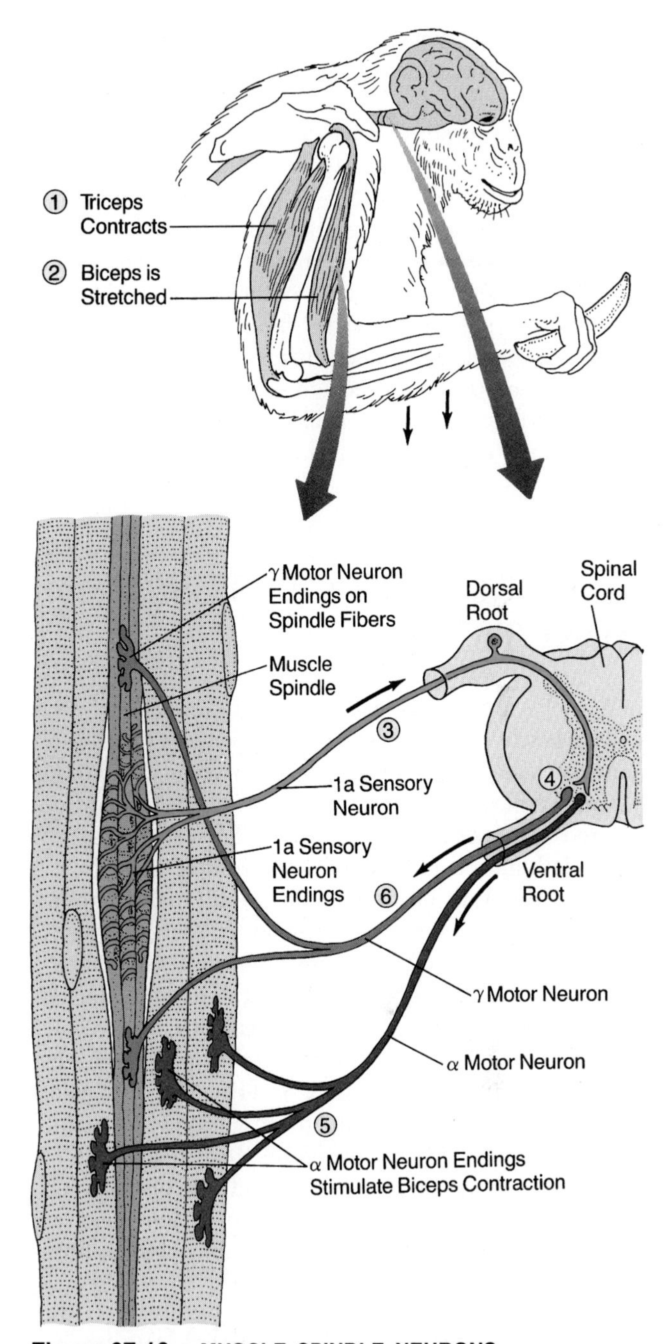

Figure 37-10 MUSCLE SPINDLE NEURONS.
When a chimpanzee's triceps contracts, its biceps stretches, and with it the muscle spindle inside the biceps. The 1a sensory neuron monitors the degree of stretch in the muscle spindle's central portion. It reports to the spinal cord and forms synapses on the α motor neuron, as shown. A reflex arc may thus involve sensory input over the 1a fiber and motor output over the α fiber. The result is contraction of the muscle and a shortening of the muscle spindle. Also notice that the γ motor neurons cause contraction of the ends of the muscle spindle fibers; that localized contraction stretches the central region of the spindle and activates the 1a sensory neurons, thereby triggering the reflex arc and contraction of the entire muscle.

The actions of γ motor neurons allow a beneficial anticipatory control of contraction. If a chimp's or person's eyes perceive that a large stone is about to be placed in one hand, the brain can activate γ motor neuron pathways so that he or she can tense appropriate finger and arm muscles ahead of time. The system also functions like a contraction "thermostat," which, when turned up or down, causes muscles to contract more or less and to maintain higher or lower states of tension. In addition to more direct signals from the brain (described later), these indirect control mechanisms allow the animal to keep its head or body erect, to carry an armful of textbooks, or to perform other tasks that require constant finely honed muscular control.

MUSCLE STRUCTURES AND FUNCTIONS

Muscles are organs made up of differentiated muscle cells. Muscle cells are also vital components of other organs, such as the gut, lungs, secretory glands, and blood vessels. Wherever they reside, muscle cells can be categorized as smooth, cardiac, or skeletal.

Smooth muscle consists of single contractile cells; it is found in the walls of the gut and the uterus, in the linings of certain ducts, and in blood vessels (Figure 37-11a). **Cardiac muscle** cells are also single and have a much more regular organization of the macromolecules responsible for contraction than do smooth muscle cells; cardiac muscle cells are linked together in special ways in the walls of the heart (Figure 37-11b). The third, most abundant, and best understood muscle type is striated, or **skeletal muscle** (Figure 37-11c). In it, each "cell" is really a fused set of dozens or hundreds of cells. The resultant muscle cells are usually very long and are known as **muscle fibers.** A large number of such fibers make up the organ we call a muscle.

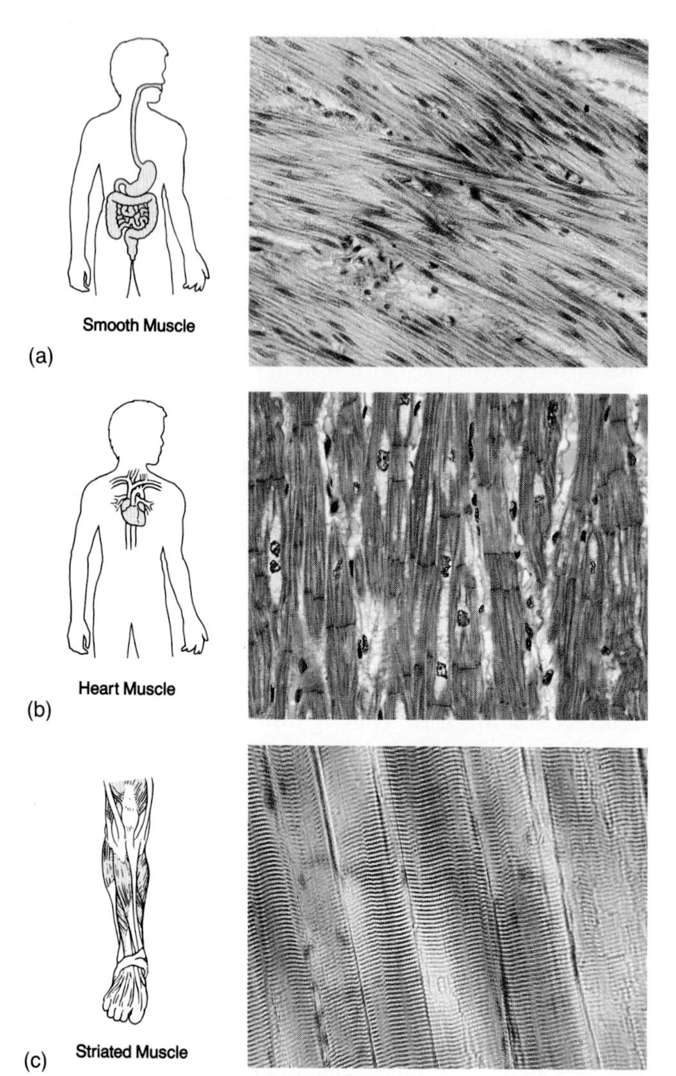

Figure 37-11 MUSCLE TYPES.
(a) Smooth muscle (magnified here about 60×) is composed of individual cells with contractile proteins that are not organized into regular units (sarcomeres). (b) Cardiac muscle (magnified here about 100×) is also composed of individual cells; these are interconnected at special junctions called intercalated disks and have striations owing to the ordering of the contractile proteins, actin and myosin. (c) Skeletal, or striated, muscle (magnified here about 400×) is composed of huge multinucleated cells (nuclei are visible between these striated regions); the large quantities of contractile proteins are assembled in sarcomeres aligned so precisely that the cytoplasm takes on this striped, or striated, appearance.

Special Features of Skeletal Muscle

Skeletal muscle fibers may be classified in several ways; one way is by color—*red* or *white.* Red muscle, sometimes called *slow-twitch muscle* (or tonic muscle), contains a large quantity of the reddish oxygen-binding compound myoglobin, as well as many blood capillaries that run through the muscle tissue. In addition, red muscle fibers store fat and glycogen, fuels for the manufacture of ATP by the large number of mitochondria that are present. And as we shall see, the action of myosin is quite slow in red muscle; thus, the tissue uses energy slowly. The numerous mitochondria can generate sufficient ATP to keep up with the cell's needs for prolonged periods, and as a result, red muscle can contract repeatedly and yet resist muscle "fatigue." The "dark meat" of a duck's breast is red muscle, allowing it to fly over long distances.

By contrast, white muscle (as in a domestic turkey's "white meat" breast muscles) is called *fast-twitch muscle.* It has little or no myoglobin, is traversed by relatively few capillaries, stores little fat and possibly glyco-

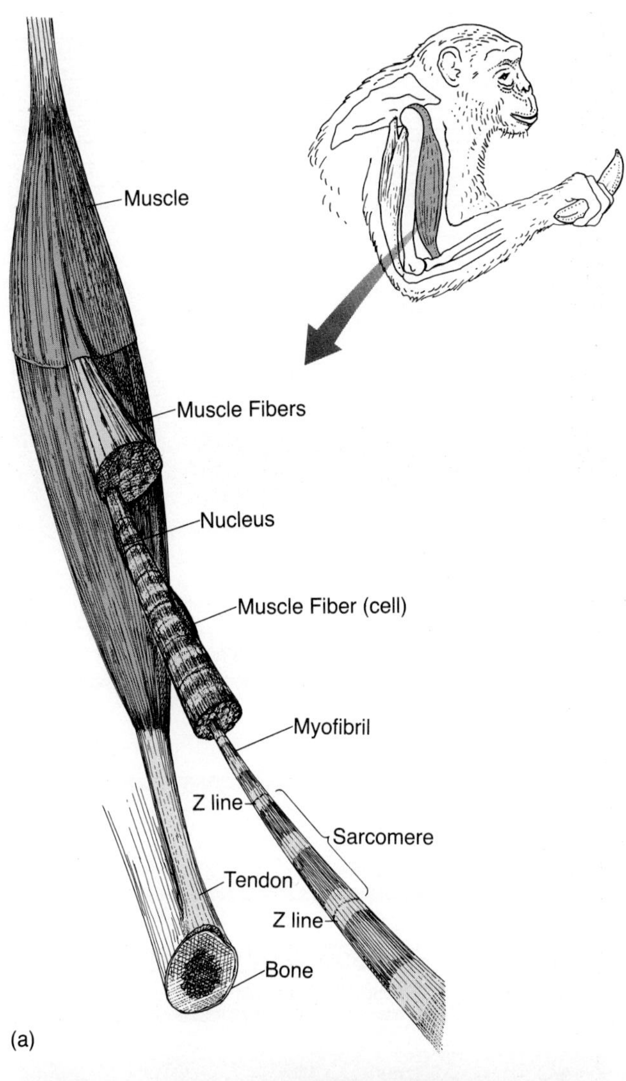

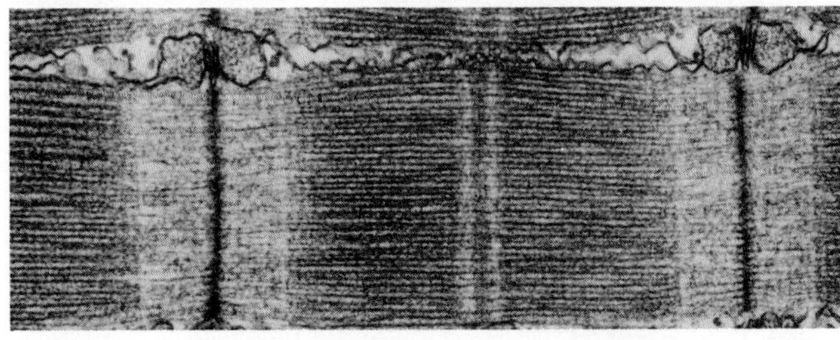

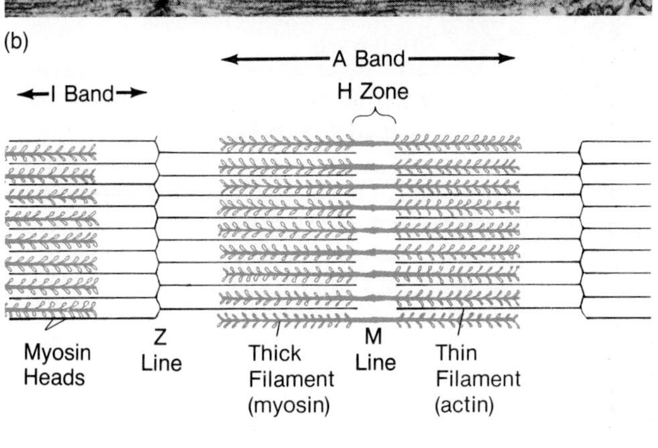

gen, and contains few mitochondria. Its myosin hydrolyzes ATP rapidly—too fast for the cells to produce sufficient ATP even by glycolysis. For these reasons, fast-twitch muscle is most active during brief, intensive flurries of contractions and soon becomes fatigued as lactic acid builds up from anaerobic metabolism.

The dichotomy of red versus white muscle is really an oversimplification. For instance, some white fibers have many mitochondria and fatigue slowly. Individual muscles in birds and mammals may have red fibers in one region and white fibers in another region. Within limits, sustained use can apparently alter a muscle's characteristics, even though both fiber types are present. A marathon runner, for example, will have more slow-twitch activity and a champion weight lifter more fast-twitch activity.

Both red and white muscles in vertebrates usually contain **creatine phosphate,** an organic molecule with a high-energy phosphate bond. This storage compound's phosphate group can be transferred enzymatically to ADP, resulting in the rapid formation of ATP. The presence of creatine phosphate thus allows muscle contractions to start and to persist (as ATP is replenished), even when aerobic metabolism is at a low level.

As a red or white muscle fiber continues to contract, all the high-energy phosphate stored in creatine may be transferred to ADP, and mitochondria may use up the oxygen stored in myoglobin. When this happens, the muscle cells generate some ATP through glycolysis and fermentation (see Chapter 7) but accumulate lactate and must eventually "pay off" an *oxygen debt* when the muscle stops working so hard and returns to a normal aerobic state.

Muscle Structure

Within your biceps, a sheath of connective tissue covers bundles of hundreds or thousands of cylindrical muscle fibers resembling wires in an underground telephone cable (Figure 37-12a). Each of these fibers is a giant sin-

◀ Figure 37-12 STRIATED MUSCLE: FROM ORGAN TO MOLECULES.
(a) A skeletal muscle is constructed of muscle fibers, each of which is a huge, multinucleate cell. Within every muscle fiber are many myofibrils, composed of sarcomeres. Sarcomeres, which are bounded at each end by Z lines, are built of actin, myosin, and several other proteins. (b) An electron micrograph (magnified about 22,500×) showing the orderly, repeating sarcomere units within a striated muscle cell. The membranous saclike structures at the end of each Z line are sites where calcium ions are released to trigger contraction. (c) Each sarcomere is composed of a series of thick filaments (myosin) and thin filaments (actin).

gle cell, fused from hundreds of embryonic precursor cells. Each can reach some 5–100 μm in diameter and lengths of centimeters to a meter and can contain many hundreds of functional nuclei.

Packed into the cytoplasm of the muscle fiber cell are long, cylindrical assemblies of molecules, called **myofibrils** (Figure 37-12b). Myofibrils are built from **sarcomeres;** these precisely repeating units of muscle proteins are the actual sites of contraction (Figure 37-12b and c). The sarcomeres of neighboring myofibrils tend to be aligned, so that under a microscope, the muscle fiber looks banded, or striated (hence the term *striated muscle*).

Biologists have labeled the various stripes and bands of sarcomeres for easy reference. The *Z lines* mark the ends of each sarcomere (see Figure 37-12c). Extending from each side of the Z lines are *I bands*, which appear lighter. Between the two I bands of a sarcomere, the dark *A band* is located; the A band, in turn, is subdivided by a light, centrally located *H zone*. Finally, the *M line* bisects the H zone, marking the center of the sarcomere. Let's see, now, how the structure of muscle cells allows them to contract.

THE MOLECULAR BASIS OF MUSCLE CONTRACTION

Biologists have observed that as a muscle fiber contracts, the sarcomere shortens and the I bands and central H zones become narrower. How and why do such changes in band width take place?

Each sarcomere contains two types of filament: *thin filaments*, which are composed primarily of the protein *actin*, and *thick filaments*, which are made of many *myosin* molecules that function both as a structural protein and as a mechanoenzyme (see Figure 37-12c and Chapter 6). Like slender molecular arms, actin thin filaments extend from each Z line into the adjacent sarcomeres, passing through the I bands and well into the A band regions. In contrast, myosin thick filaments are found only in the A band, where they straddle the H zone. The free ends of the myosin thick filaments and the actin thin filaments thus overlap—an organizational feature that is vital to the mechanics of muscle movement.

The individual myosin molecules that make up a thick filament function as mechanoenzymes partly because of their shape and partly because of their tendency to bind together. Individual myosin molecules are composed of two polypeptide chains, each shaped like a golf club; the "shafts," or "handles," of the two polypeptides are twisted around each other, and the "heads" are bent to the sides at hingelike sites (Figure 37-13a, b). In addi-

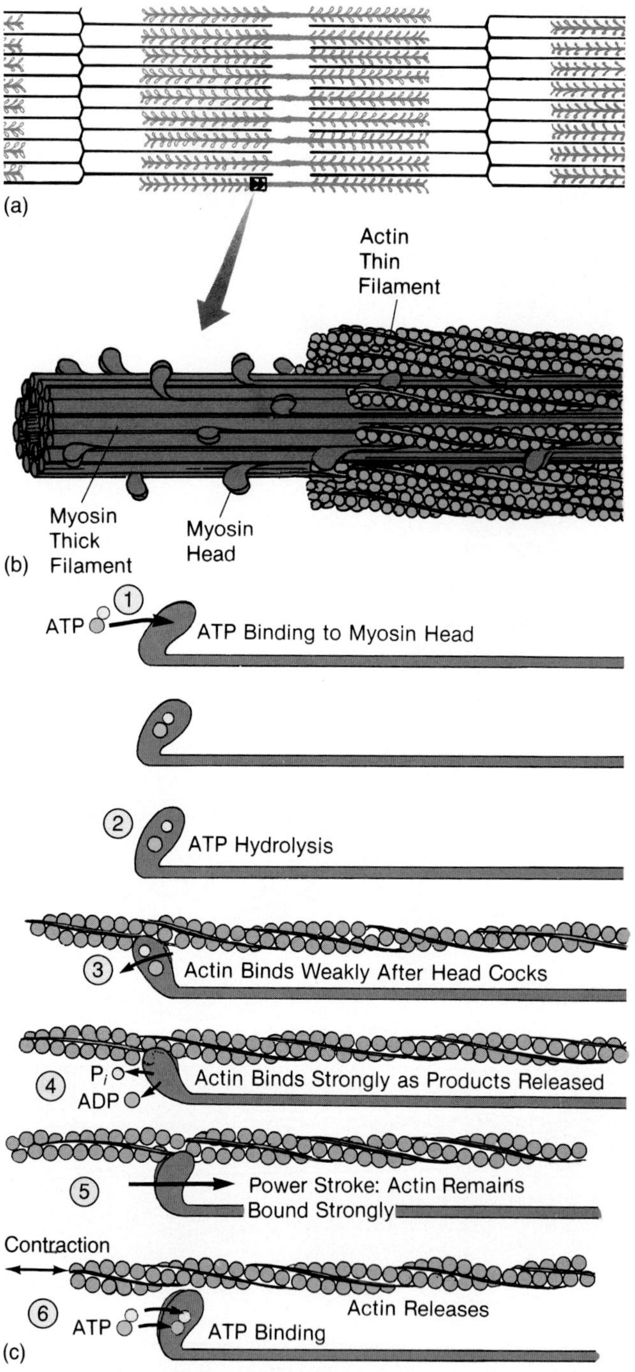

Figure 37-13 HOW ACTIN AND MYOSIN INTERACT TO CREATE A MYOSIN POWER STROKE.
Myosin molecules joined to form thick filaments (a), with myosin heads projecting, (b) allowing cross bridges to form with actin molecules. In a muscle cell, actin filaments surround such myosin filaments so that the myosin heads may contact the actin. (c) During a contraction, ATP binds to the myosin head and is hydrolyzed. The head then cocks and binds actin weakly. As ADP and inorganic phosphate (P_i) are released, binding of the actin head becomes strong, the power stroke is applied, and the actin filament moves. When another ATP binds, actin is released and the cycle repeats.

tion, sets of myosin molecules clump together (i.e., their "tails" bind to each other) in a staggered array, with many myosin heads jutting out in a spiral pattern at the surface of the thick filament. In a sarcomere, each myosin thick filament is surrounded by actin thin filaments in such a way that the protruding myosin heads and the actin filaments can come in contact with each other when the muscle contracts (Figure 37-13c).

In the mid-1950s, two British biologists developed the **sliding-filament theory** to explain how actin and myosin generate the force for muscle contraction. In this model, the myosin heads act as cross bridges between actin and myosin filaments, applying "power strokes," much like an oar pushing on water, that push the actin filaments inward toward the central H zone. As they slide, the actin filaments pull on the Z lines to which they are anchored, and the result is a shortened sarcomere. If the same process goes on in the thousands of sarcomeres along the whole muscle fiber cell, and in many muscle fiber cells simultaneously, the muscle itself becomes shorter—that is, it contracts.

The theory rests on the fact that myosin heads act as ATPase enzymes, not just structural proteins. Myosin's action begins as ATP binds to the active ATPase site—a site just 4 nm away from the actin binding site (Figure 37-13c, step 1). Then ATP is hydrolyzed into ADP plus inorganic phosphate, and both remain bound to the head (step 2). The energy released by the hydrolysis "charges" (activates) the myosin head into a "cocked" position, just as a person's thumb might cock the hammer on a pistol. In this cocked position, the head and actin bind weakly together (step 3). Then several events occur that help to bring about muscle contraction. The binding of actin to the head causes ADP and inorganic phosphate to be released. As those products exit, the head binds strongly to actin (step 4). Virtually simultaneously, the head rocks relative to the myosin backbone, thereby applying the power stroke to actin (step 5). The power stroke moves, or "slides," the actin filament a distance of 5–10 nm. The uncocked myosin head remains strongly bound to actin but in the uncharged state. Only a new ATP molecule can start the cycle once again by weakening the myosin-to-actin binding and allowing the head to be released (step 6).

ATP and ADP obviously play key roles in muscle contraction. First, ATP binding releases actin from the myosin head, allowing myosin to bind to a new position on the sliding actin filament. Second, the hydrolysis of ATP releases energy to cock the head; the head—a mechanoenzyme—uses that energy in turn by applying the power stroke to actin. This cycle occurs rapidly in thousands of heads along a sarcomere as long as ATP is available to weaken head-to-actin binding and to provide energy for the cocking process (and as long as calcium ions are present, as we shall see shortly).

Several recent studies support the basic sliding-filament model and the idea that a simple molecular shape change of the myosin heads is translated at higher levels into filament sliding. This, in turn, can bring about muscle cell contraction, the moving of an animal's limb, and facilitate an instinctive behavior that promotes survival—the dodging of a rabbit, for instance, to avoid a coyote's snapping jaws.

How Muscle Contraction Is Initiated

Muscle cells have a special property that they share with neurons: They are electrically excitable; that is, they can propagate action potentials (impulses). As Chapter 34 explained, a chemical neurotransmitter, such as acetylcholine, is released when action potentials reach the ending (synaptic terminal) of a motor nerve. When a threshold level of acetylcholine is released at the specialized junction of motor nerve and muscle cell membranes, called the neuromuscular junction, an action potential is initiated and propagates over the whole muscle cell surface. That rapid spread of the potential immediately triggers events *inside* the muscle cell that culminate in sarcomere contraction.

Skeletal muscle cells have a vital architectural feature: a system of infoldings of the plasma membrane that extends deep within the cell. These *transverse tubules*, or **T tubules,** surround the myofibrils at their Z lines (Figure 37-14). Between adjacent T tubules is a calcium reservoir, a network of **sarcoplasmic reticulum** (a version of smooth endoplasmic reticulum), whose hollow sacs contain calcium ions. When nerve impulses initiate an action potential in the muscle fiber, the T tubule membranes act like a maze of miniature electrical conduits, conducting the action potential to the terminal sacs of the sarcoplasmic reticulum and possibly causing the second messenger inositol triphosphate to be produced and to stimulate Ca^{2+} release. That release may occur through channels that span the junctions between T tubules and sarcoplasmic reticulum, and calcium ions quickly flood into the fluid bathing the actin and myosin filaments.

This chemical barrage of Ca^{2+} directly triggers a muscle contraction by interacting with components of the thin filaments. To understand the steps between the reception of motor nerve impulses by a skeletal muscle fiber and the resulting contraction, we need to consider the effects of two additional proteins associated with actin.

An actin thin filament is actually composed of two actin chains twisted into a helix (Figure 37-15a). In the grooves of this helix are molecules of **tropomyosin,** a long, thin protein. When a muscle is at rest, tropomyosin prevents myosin heads from binding actin, probably because it masks the sites at which binding occurs.

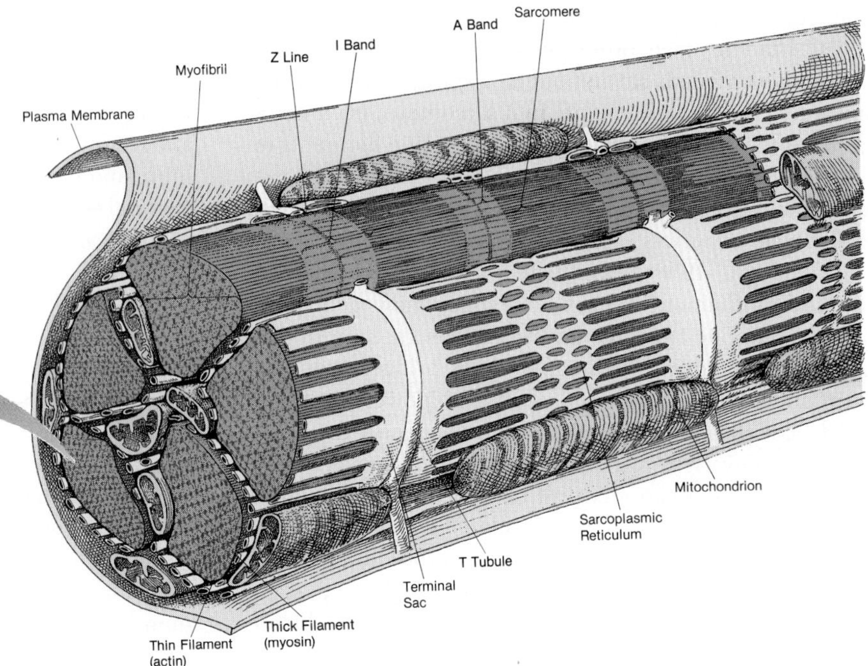

Figure 37-14 T TUBULES AND SARCOPLASMIC RETICULUM.
A T tubule immediately surrounds the Z line of every sarcomere. T tubules conduct action potentials from the cell surface inward and cause calcium ions to be released from the nearby terminal sacs of the sarcoplasmic reticulum. That release stimulates the myosin-actin interaction responsible for contraction and shortening of the muscle cell.

Another protein, **troponin,** is situated at regular intervals along the actin thin filament, where it binds to tropomyosin molecules and to actin. Troponin is in fact a complex of four proteins, one of which has a critical property: It can also bind calcium ions. Knowing this, we can trace the events that begin when motor nerve impulses reach a muscle.

As calcium ions bind to sites on the troponin component (Figure 37-15b, step 1), the complex changes shape. Because troponin is linked to tropomyosin, it, too, shifts position, unmasking the sites on actin that bind myosin heads (step 2). Then, like pegs tumbling into slots, the cocked myosin heads bind strongly to actin, and the molecular power strokes begin (step 3). The more Ca^{2+} released from the sarcoplasmic reticulum, the more Ca^{2+} troponin binds, the greater the sliding of filaments, and the stronger the muscle fiber contraction.

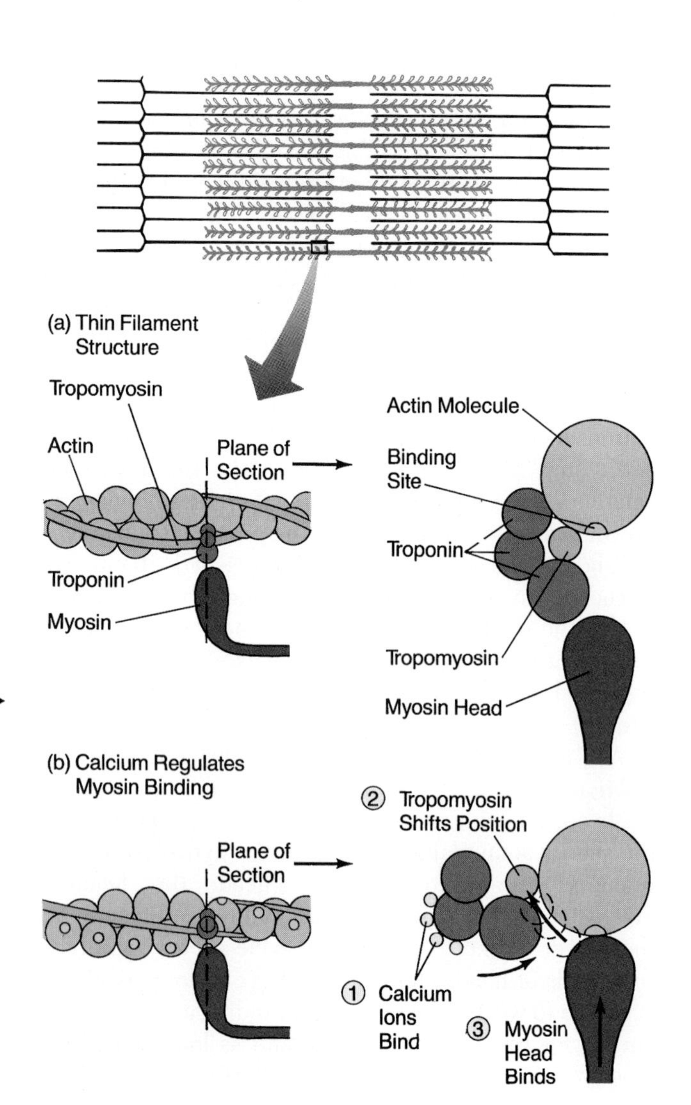

Figure 37-15 ACTIN THIN FILAMENTS, MYOSIN BINDING, AND MUSCLE CONTRACTION.
(a) Actin thin filaments are composed mainly of a helical chain of actin monomers. Tropomyosin and troponin associate with the helical actin structure to form actin thin filaments. (b) The myosin head does not bind to the actin molecule in the absence of calcium. Increased levels of calcium ions in the cytoplasm, however, result in the binding of calcium by one of the troponin component proteins. This binding results in a change in shape of troponin and a shift in the placement of tropomyosin on the actin molecule that uncovers the myosin-binding site. The myosin head then can bind and the muscle can contract. Not all of the subunits of troponin are shown here.

Turning Muscles Off

What, then, causes the muscle fiber to relax? When muscle action potentials stop being initiated, the release of calcium ions from the sarcoplasmic reticulum halts. At the same time, ATPase pump enzymes in the membrane of the sarcoplasmic reticulum pump calcium ions back into the terminal sacs. As calcium is withdrawn, troponin reverts to its resting configuration, tropomyosin immediately shifts to cover the myosin-binding sites on actin, and the contraction ends. Finally, actin filaments slide outward again, and the sarcomere returns to its resting length.

Muscles that lack a supply of ATP remain contracted. An extreme case develops after a vertebrate dies, when *rigor mortis* sets in. At death, calcium begins to leak from the sarcoplasmic reticulum, causing contractions; soon muscle cell ATP is fully hydrolyzed to ADP, and actin and myosin filaments remain locked together in their contracted position. Because aerobic and anaerobic metabolism cease when the animal dies, no new ATP is generated that could release myosin from actin or drive the pump enzyme that would lower calcium ion levels again. Only after some hours do the stiff and contracted muscles relax as other degeneration processes dominate over the final contraction of death.

MORE ON MUSCLE ACTION: GRADED RESPONSES AND MUSCLE TONE

In vertebrate animals, most *individual* striated muscle fiber cells contract, or "twitch," in an all-or-none fashion: Incoming motor nerve impulses exceeding a certain threshold trigger a muscle action potential and liberate calcium ions, but stronger, faster, or larger impulses produce only a slightly stronger contraction.

In contrast, whole vertebrate muscles exhibit *graded responses* to a stimulus. A rabbit's leg muscles, for example, might contract fully during a leap to escape a coyote, but twitch weakly when the animal shivers to generate heat for body warmth. The individual muscle fiber cells still contract in an all-or-none fashion, but the more fibers that are stimulated by motor neurons, the stronger the overall contraction, and the fewer the fibers stimulated, the weaker the contraction of the whole muscle.

Graded responses are also based on *summation*, a condition during which nerve impulses arrive at a muscle before its previous contraction has subsided and trigger a new contraction (Figure 37-16). The strength of summated contractions is always greater than the strength of individual twitches, since twitches are responses to less frequent stimulation.

Summation often culminates in **tetanus,** a state of sustained maximum contraction. Tetanus is a normal and crucial element of muscle function; in fact, most everyday muscle actions depend on the smooth, strong contractions of tetanus. For instance, when agonist and antagonist muscles in a typist's fingers counteract each other, providing precise control of movement, both muscles are in states of summation. The length of time that high levels of summation and tetanus can be maintained varies with muscle type—briefly in white muscle, much longer in red muscle—because of the differences in the rate at which each type utilizes ATP.

Another important characteristic of muscle function is the ability to achieve **tonus,** or "muscle tone," a condition in which a muscle (or sets of muscles) is kept partially contracted over a long period. Tonus is produced when first one set of fibers, then another, and finally yet another is briefly stimulated, so that in the muscle as a whole, some parts are always contracted, though most remain relaxed. The tonus of back, abdominal, neck, and limb muscles enables humans and other terrestrial animals to maintain normal posture in the presence of gravity. And tonus in the leg muscles helps squeeze blood out of the leg veins and aid its return to the heart.

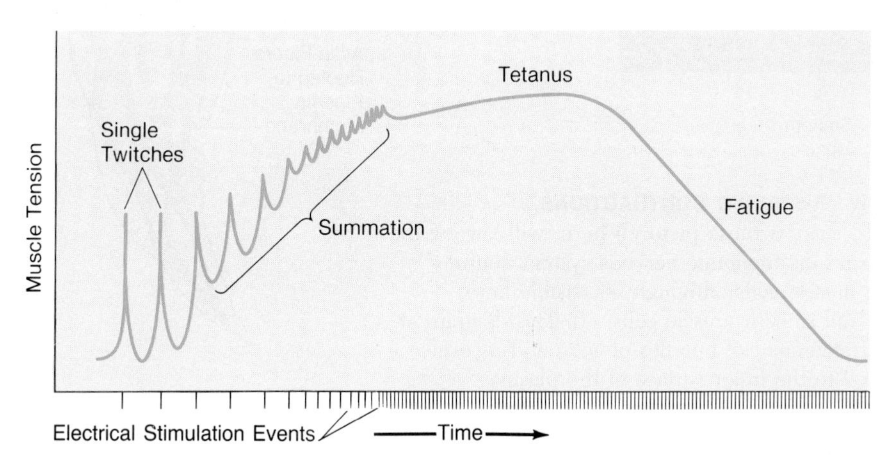

Figure 37-16 SUMMATION AND TETANUS DURING MUSCLE CONTRACTIONS.
Motor neurons electrically stimulate a muscle. When the rate of stimulation is slow, contraction of the muscle occurs by individual twitches. When the rate of electrical stimulation increases, the muscle no longer has time to relax between stimuli. After a period of summation of the individual twitches, a sustained, smooth tetanic contraction results. Ultimately, the muscle fatigues because it no longer has sufficient ATP to maintain contractile processes.

SMOOTH MUSCLE

Smooth muscle, a vertebrate's second most abundant muscle type, is found in the gut wall, the walls of blood vessels (Figure 37-17a), the iris of the eye, reproductive organs, and glandular ducts. Smooth muscle usually carries out sustained, slow contractions not under voluntary control.

Smooth muscle tissue is strikingly simple: Each slender, elongated cell lacks striations and has a single nucleus. The cells are arranged in sheets in the wall of the large intestine or as straps around blood vessels or ducts. Individual smooth muscle cells are linked to each other by abundant collagen fibers, by gap junctions that couple groups of cells electrically, and by peglike surface protrusions and socketlike depressions that couple adjacent cells. This linkage allows the contractile force generated in a single smooth muscle cell to spread to others, so that the whole muscle sheet gradually contracts. The peristalsis that propels food and fluid through the gut depends on this mechanism.

Although smooth muscle cells lack striations, they can nevertheless contract because the cytoplasm is crowded with actin thin filaments and also contains molecules in the myosin family organized into thick filaments. The myosin heads in smooth muscle probably function just as they do in the sarcomeres of skeletal muscle: When actin fibers rooted in the plasma membrane (and probably arrayed helically) are pulled inward, the cell shortens in a corkscrew fashion. Smooth muscle cells contract more slowly than do striated muscle cells and are able to sustain the contraction far longer, in part because metabolism can continuously supply the ATP needed to fuel each cell's contractile activity.

Since smooth muscle cells lack a sarcoplasmic reticulum, the large cell surface must itself act as the site of calcium ion entry from the tissue fluids bathing the cell's exterior. The second messengers cyclic AMP and diacylglycerol help stimulate the release of calcium. The cell surface is also the location of the ATPase that pumps Ca^{2+} back out of the cell to terminate contraction. Finally, in smooth muscles, calcium initiates a chain of events leading to the phosphorylation of myosin, and this is what starts the contractile cycles.

Although some smooth muscle cells show spontaneous, frequently rhythmic, contractile activity, in general, sympathetic nerves of the autonomic nervous system stimulate contractions, and parasympathetic nerves inhibit them (see Chapter 34). Various hormones can

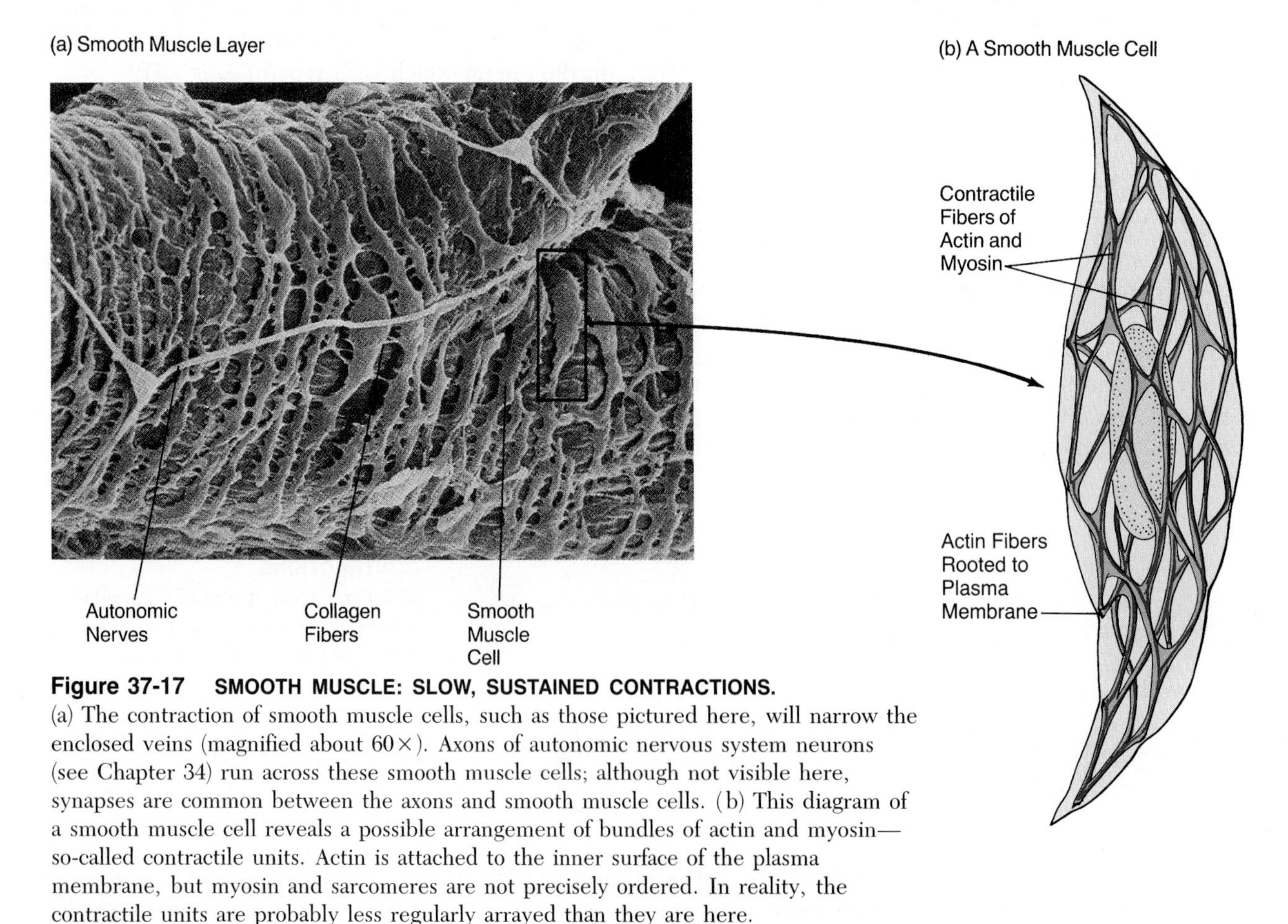

Figure 37-17 SMOOTH MUSCLE: SLOW, SUSTAINED CONTRACTIONS. (a) The contraction of smooth muscle cells, such as those pictured here, will narrow the enclosed veins (magnified about 60×). Axons of autonomic nervous system neurons (see Chapter 34) run across these smooth muscle cells; although not visible here, synapses are common between the axons and smooth muscle cells. (b) This diagram of a smooth muscle cell reveals a possible arrangement of bundles of actin and myosin—so-called contractile units. Actin is attached to the inner surface of the plasma membrane, but myosin and sarcomeres are not precisely ordered. In reality, the contractile units are probably less regularly arrayed than they are here.

RUNNING THE MILE: WHAT IS OUR SPEED LIMIT?

Experts predicted that no one would ever run a 4-minute mile—right up until 1954, when Roger Bannister did it. Since that feat, runners have shaved 13 seconds off Bannister's time and continue to push on to ever faster records in various Olympic events (Figure A). But how much faster can a human run the mile and other events?

Most exercise physiologists believe that, indeed, there *are* limits to human performance based on some fundamental physiological aspects of the human body. For example, measurements show that a world-class marathoner, such as Bill Rodgers, can use about 80 ml of oxygen per kilogram of body weight per minute compared with an amateur jogger's 45 ml. Some physiologists think that 80–90 ml/kg is the upper limit on aerobic performance, and thus, marathoners may already be nearing the limits to speed and endurance in their sport.

The strength of bone is another potential limit to human performance—but a distant one. Orthopedic surgeons removed a femur from a cadaver and subjected it to greater and greater compression. It splintered at 1,600 lb/in.^{2}, which is about twice the force exerted on the leg bones of a 160 lb runner and more than current weight lifters can heft. Even a very large athlete, such as a 7 ft basketball player, does not approach the limits of bone strength when pounding down the court or jumping for a basket.

Yet another physiological limitation on performance is the proportion of fast-twitch to slow-twitch muscle fibers. Marathoners like Rodgers have 80 to 90 percent slow-twitch (a type of red muscle), while sprinters, such as Carl Lewis and Evelyn Ashford, have about 70 percent fast-twitch (a type of white muscle). Record-breaking performances could well be expected from athletes with better ratios—say, 95 percent slow-twitch or 80 to 90 percent fast-twitch, if such a person is ever born and identified.

Based on all these considerations, some physiologists are now predicting that the fastest a person can ever run the mile will be about 3:34, or roughly 13 seconds faster than existing records. Only time will tell, however, whether they are right or whether some Roger Bannister of the future will smash that record, too.

Figure A
What are the limits? American runner Tim Bright, straining to the utmost, leaps a hurdle in a 110-meter section of the decathalon race at the 1988 Olympics.

also stimulate smooth muscle contractions directly, as when oxytocin stimulates uterine muscle during mammalian birth.

CARDIAC MUSCLE: STRIATED TISSUE WITH SMOOTH MUSCLE CHARACTERISTICS

The heart is built from the third type of muscle cell—cardiac muscle. Like smooth muscle cells, each cardiac muscle cell has a single nucleus, but unlike smooth muscle, cardiac muscle is striated. Cardiac muscle is constantly active throughout an organism's lifetime, and the cells are liberally supplied with mitochondria, myoglobin, and capillaries.

Cardiac muscle cells are often branched, with the branches of neighboring cells forming an interlocking system. Depending on the vertebrate species, there is a T-tubule system of varying complexity (Figure 37-18). There is also an extensive network of sarcoplasmic reticulum close to the plasma membrane that serves as the calcium ion reservoir for contraction. The ends of cardiac muscle cells are bound firmly to each other by **intercalated disks**—regions of folded, reinforced cell membrane that act both as welds holding cells together and as very leaky gap junctions, allowing ions or electric currents to flow easily (see Figure 37-18). Actin filaments of the terminal sarcomeres are also attached to the cell sur-

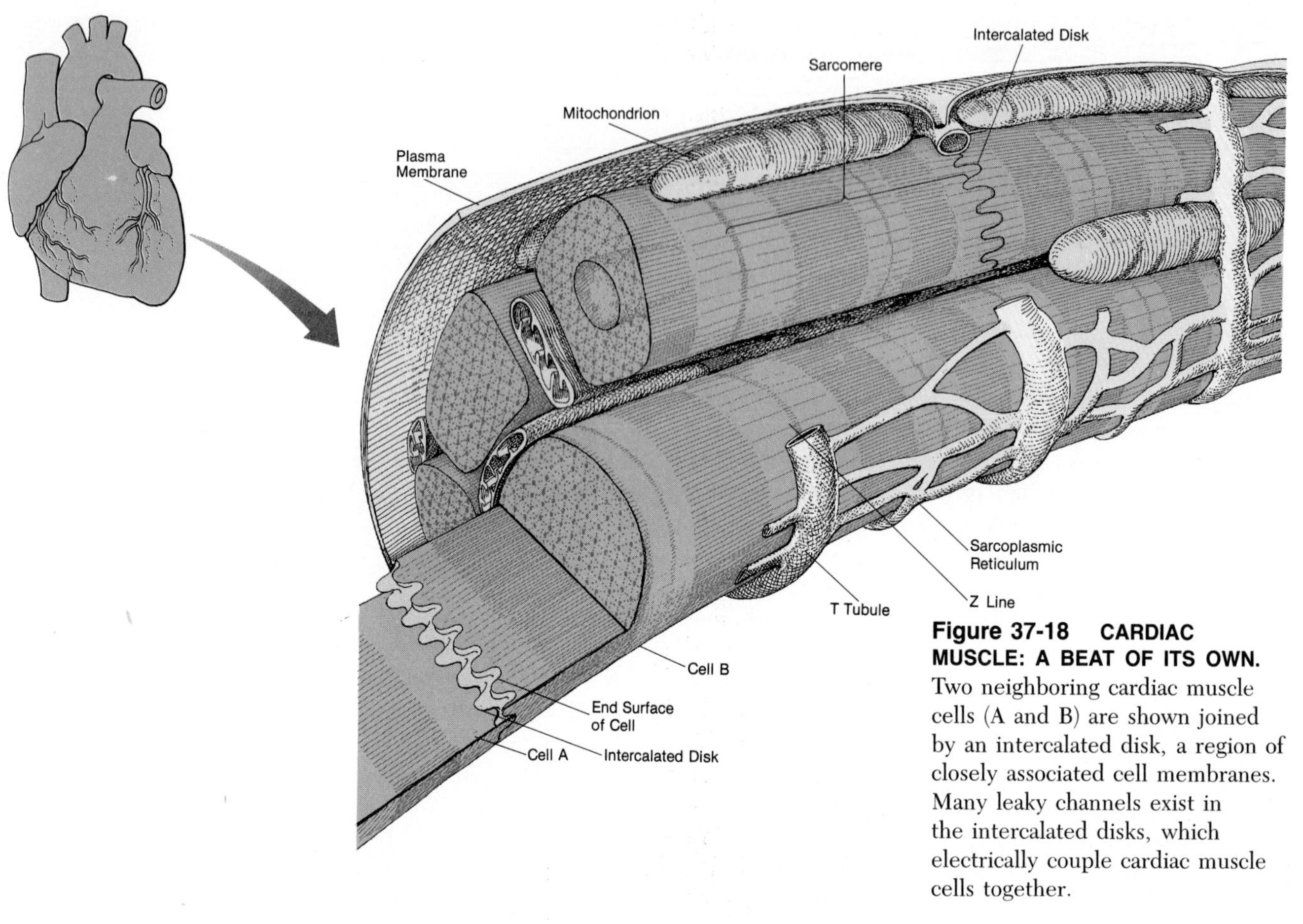

Figure 37-18 CARDIAC MUSCLE: A BEAT OF ITS OWN. Two neighboring cardiac muscle cells (A and B) are shown joined by an intercalated disk, a region of closely associated cell membranes. Many leaky channels exist in the intercalated disks, which electrically couple cardiac muscle cells together.

face at the intercalated disk, and thus they are somewhat analogous to Z lines, since the forces of sarcomere contraction are applied to them.

Placed in a culture dish, isolated single cardiac muscle cells contract spontaneously at periodic intervals: They truly "beat," just as the entire vertebrate heart, free of all nerves, will contract rhythmically. The organ's heartbeat is *myogenic* ("muscle-generated"). In contrast, the heartbeats of animals such as lobsters, crabs, and spiders are *neurogenic* ("nerve-driven"). An invertebrate's heart, if isolated, fails to contract spontaneously.

MUSCLES IN EVOLUTION: THE CONTRACTILE SYSTEMS OF INVERTEBRATES

While virtually all contractile systems in eukaryotes rely on actin, myosin, and control by calcium ions, there are distinct differences in the muscle cell architecture—and hence muscle cell function—of vertebrates and invertebrates.

The first muscle cells were likely quite simple—smooth muscle cells associated with protrusible organs, such as the proboscis and the penis, and later with the gut. These muscle cells are common in many modern

Figure 37-19 MOLLUSKS AND CATCH MUSCLES. This giant clam (*Tridachna gigantea*), as well as its smaller mollusk relatives, can remain tightly closed because of properties of its very thick myosin filaments and its muscle cells.

worms, for example, and in sponges, cnidarians, and mollusks.

The striated muscle cell—with its highly organized sarcomeres—apparently arose independently in many invertebrate lines. Barnacles and octopuses, for example, possess striated muscles in only certain parts of their bodies, but such cells are ubiquitous in insects, from the gut wall to the flight muscles.

Muscle cell biochemistry can also differ. Mollusk cells, for instance, generally lack troponin; myosin itself binds calcium ions and initiates the power stroke that causes actin to slide. And arthropod muscle cells lack the all-or-none mechanism and instead show a graded response in which the degree of contraction is proportional to the level of polarization of muscle cell membranes.

Invertebrate smooth muscles teem with myosin thick filaments, and this can produce some dramatic effects. For example, the tentacle muscles in a Portuguese man-of-war can stretch some 21 m (70 ft) and yet can contract to shorten the tentacle to 14 cm (5.5 in.)! The powerful catch muscle of a giant clam, another prime example, can hold the animal's shell firmly closed for up to 30 days at a time without fatigue (Figure 37-19). In this species, as in virtually all other animals, survival depends on functioning muscles.

SUMMARY

1. Exoskeletons—nonliving deposits of mineral salts and other materials—provide armor and permeability barriers around many invertebrate animals. Muscles can move the hinged external skeletons.

2. The vertebrate's endoskeleton contains *cartilage* and/or *bone*. Two major subdivisions of the endoskeleton are the *axial skeleton* and the *appendicular skeleton*. Individual bones meet at the joints, within which *ligaments* hold bones together.

3. Compact bone is especially hard because the surface regions are composed of densely packed, fused *Haversian systems*. Spongy bone is an open framework and may contain central regions of bone marrow.

4. The agonist, or prime mover, muscle acts in opposition to its antagonist muscle. Muscles grade into *tendons*, which attach to bones or other muscles.

5. The three classes of muscle tissue are *skeletal (striated) muscle*, *smooth muscle*, and *cardiac muscle*. Striated muscle may be red or white (or gradations thereof), reflecting each type's individual chemistry.

6. *Muscle spindles*, located near the center of each skeletal muscle, function to report the changing length of their parent muscle.

7. By way of γ motor neurons, the nervous system can indirectly cause muscle spindles to contract; this stretches the spindles' sensory region and triggers the same reflex arc that causes the parent muscle to contract. This is an alternative means for the brain and spinal cord to elicit muscle contraction.

8. Striated muscle is composed of giant, multinucleate cells called *muscle fibers*, which contain long assemblies of molecules known as *myofibrils*. Myofibrils are built from *sarcomeres*.

9. Sarcomeres shorten during contraction because actin thin filaments slide past myosin thick filaments. Myosin heads generate the force that drives this sliding mechanism.

10. Motor nerve impulses arriving at the neuromuscular junction trigger action potentials in the striated muscle cell's surface. The action potential spreads to the *T tubules* and *sarcoplasmic reticulum*. As a result, calcium ions are released into the cytoplasm to start the contraction event.

11. When calcium binds to the *troponin* protein complex, another protein, *tropomyosin*, is moved away from actin, thereby uncovering binding sites for myosin heads. As a result, the "power stroke" of contraction can begin. ATP must be present as well as calcium reserves for the cycle to continue.

12. Vertebrate muscle fibers respond to nerves in an all-or-none fashion, but display graded responses. Tetanus, the sustained maximum contraction of a muscle, is a vital element in normal muscle functioning.

13. Smooth muscle cells lack striations, have just one nucleus, and usually are interconnected by gap junctions. Contractions are involuntary, usually slow, and may be sustained for a relatively long period without muscle fatigue.

14. Cardiac muscle cells, whose contractions are involuntary, have only a single nucleus and typical sarcomeres and are tightly bound to each other by specialized junctions known as *intercalated disks*.

15. Vertebrate cardiac muscle has an intrinsic rhythmic contractile beat (myogenic beat).

KEY TERMS

appendicular skeleton
axial skeleton
bone
cardiac muscle
cartilage
creatine phosphate
Haversian system
intercalated disk
joint
ligament
muscle fiber
muscle spindle
myofibril
osteoporosis
sarcomere
sarcoplasmic reticulum

skeletal muscle
sliding-filament theory
smooth muscle
T tubule
tendon
tetanus
tonus
tropomyosin
troponin

QUESTIONS

1. Compare exoskeletons and endoskeletons: Which contain living cells? Which grow? Which are jointed? How does each type move?

2. What are the materials that make up bone? What materials are present in growth zones? Describe the structures of compact and spongy bone. What are osteocytes, and how do they exchange or obtain gases, wastes, and nutrients?

3. Name the major parts of the vertebrate endoskeleton. Explain what ligaments and tendons do and of what materials they are formed. Name the three major types of vertebrate muscles, their locations, and their general characteristics. Compare and contrast the three types.

4. What is a muscle spindle, and how does it work? What is proprioception?

5. Describe a skeletal muscle fiber. What are myofibrils and sarcomeres? Draw a sarcomere and label its components.

6. What is the function of T tubules in skeletal muscle fibers?

7. What is the role of sarcoplasmic reticulum in skeletal muscle fibers? What structure fulfills this function in smooth muscle cells? In cardiac muscle?

8. Explain how a muscle fiber contracts; include the roles of ATP, inositol triphosphate, calcium, troponin, actin, myosin heads, and tropomyosin.

9. Describe relaxation of skeletal muscles. Does this process use up calcium or ATP? Why does rigor mortis occur after death?

10. Does the way a single vertebrate skeletal muscle fiber responds to a stimulus differ from the way a whole muscle responds? Explain. What is tetanus? What is tonus?

For additional readings related to topics in this chapter, see Appendix C.

Part
SIX

POPULATION BIOLOGY

Most species live in populations—groups that confer survival value in numerous ways.

38 Evolution and the Genetics of Populations

Unlike physics, every generalization about biology is a slice in time; and it is evolution which is the real creator of originality and novelty in the universe.

Jacob Bronowski,
The Ascent of Man (1974)

In September of 1835, Charles Darwin stood for the first time on the shores of the Galápagos Islands, a cluster of small volcanic upwellings straddling the equator 600 miles west of Ecuador. Darwin became fascinated by the low, scrubby vegetation on these desolate islands and by the equally unique wildlife he observed: giant tortoises, prehistoric-looking marine iguanas, drab finches of many types, and numerous other animals. This collection of native organisms, so strikingly well suited to their remote islands, helped inspire Darwin years later to formulate a theory that would revolutionize the study of biology and forever alter society's traditional view of humankind's place in nature.

Darwin's theory of **evolution** by *natural selection* became a central unifying concept that allows biologists to probe and understand the structures, functions, and behaviors of modern organisms; to determine how new species may arise; to learn why some species may thrive

A dawn scene like this of giant tortoises in a pond on Isla Isabela, Galápagos, may have greeted Charles Darwin as he wandered the islands.

and diversify, while others die out; and to trace the historical links between groups of organisms over vast stretches of time. The major task in studying the formation and extinction of species is to separate chance events from causal factors—a difficult prospect, as we will see. In this chapter and the next two, we will focus on the precise mechanisms of evolutionary change. In particular, we will see that while natural selection operates at the level of phenotypes, the gene itself may be viewed as a unit of evolution and is amenable to detailed study with techniques borrowed from molecular genetics and other branches of modern biology.

Contemporary biologists are still asking questions much like those that Darwin pondered as he surveyed the strange life forms of the Galápagos Islands, fossils in Patagonia, and domestic animals in England. We will consider those questions in this chapter and see how the genetics of *populations*, or groups of interbreeding individuals within a species, as well as the study of changes in gene frequencies, can help biologists understand the mechanisms of evolution at work.

Our topics in this chapter include:

- The scientific and cultural underpinnings for the theory of evolution
- The raw material of natural selection: genetic variation
- The genetics of populations: the link between heredity and evolutionary change
- How preferential mating, mutation, migration, and other mechanisms lead to evolution

THE ORIGINS OF EVOLUTIONARY THOUGHT

As a youth, Charles Darwin was an avid amateur naturalist, and he held to the generally accepted scientific principles of his day. Most nineteenth-century people believed each of the millions of species to be immutable, created in its present form and remaining unchanged over the eons, although some natural philosophers, such as Georges-Louis Leclerc de Buffon, concluded that change over time was common (see Chapter 1). It was in the late eighteenth century that evidence for such change began to accumulate.

Much of this evidence came from rocky European hillsides and stream beds. There, geologists and amateur fossil hunters were turning up a huge number of ancient bones, shells, and fossilized plant parts that were clearly the remains of bygone forms. Like a parade of the earth's natural history frozen in layers of stone, plant and animal types unlike any currently alive seemed to appear, diversify, and become extinct. Inspired by the careful fossil reconstructions of eighteenth-century French paleontologist Georges Cuvier and by the work of British geologists James Hutton and William Smith, French zoologist Jean Baptiste Lamarck presented a new evolutionary theory in 1809, the year Darwin was born.

Lamarck and the Inheritance of Acquired Characteristics

Lamarck accepted the idea, put forth by Buffon and others, that life forms evolve. He proposed that the driving force of evolution is the **inheritance of acquired characteristics** (see Chapter 1). He believed that organisms change physically as they strive to meet the demands of their environment and that these changes are then passed to future generations by *pangenes* (elements the ancient Greeks believed were responsible for inheritance and were supposedly produced in every organ of the body). Lamarck also theorized that the inheritance of acquired characteristics is the mechanism by which lower life forms move up the ladder of life to become more complex forms through a series of heritable changes.

A classic example of Lamarckism is the elongation of the giraffe's long neck in response to stretching. Lamarck also believed that body parts can be lost because of disuse and cited the "loss" of eyes in moles—animals that spend most of their lives in dark, underground tunnels—as an example.

Today, we know Lamarck's theory to be incorrect because such things as stretching to feed in a tall tree or living in a dark cave have no effect on germ cells, gametes, or heredity. Still, Lamarck's theory does constitute a major landmark because it focused on evolution and, as Darwin later noted, declared that living things evolve according to natural laws, not divine intervention.

Darwin and Natural Selection

Charles Darwin was only 22 years old in 1831 when he boarded the HMS *Beagle*, a surveying ship that had been chartered to sail around the globe and map the coasts of South America (Figure 38-1). Five years later, when the ship returned to England, the theory that would forever change the way both scientists and the public perceive the living world was already taking shape in Darwin's mind and notebooks.

Even before Darwin's fact-finding journey as ship's naturalist, he had begun reading Charles Lyell's *Principles of Geology*. Lyell had observed that the earth's physical landscape underwent long, slow, continuous

Figure 38-1 CHARLES DARWIN AT AGE 31, IN A WATERCOLOR PAINTED BY GEORGE RICHRORD IN 1840.

(a) (b)

Figure 38-2 THE GALÁPAGOS FINCHES: EXEMPLARS OF EVOLUTION BY NATURAL SELECTION.
The dozen or so species of Galápagos finches show differences in body size and dramatic variation in beak shape. The specialized beaks of each species facilitate consumption of different foods, such as berries, hard seeds, or insects. Here, you can see a large ground finch (a) with a heavy, seed-cracking bill and a small ground finch (b) with a tiny, slender bill that can easily handle small seeds.

change. On the basis of the fossil record, Lyell speculated that animal and plant species arose, developed variations, and then became extinct as the ages passed. Darwin had also read an essay published nearly 50 years earlier by the clergyman and economist Thomas Malthus. In his *Essay on the Principles of Population*, Malthus set out the proposition that populations have the inevitable potential to grow faster than their food supplies. As a result, said Malthus, organisms are forced into a "struggle for existence."

Darwin's own observations, however, formed the basis for his theory. In Patagonia, Darwin found fossilized bones of extinct, cow-sized sloths and armadillos that resembled their smaller cousins still inhabiting Central and South America. And Darwin saw striking variation in the forms, habitats, and geographical distributions of plants and animals all during his travels.

Nowhere were the plants and animals more distinctive than in the Galápagos archipelago. Here Darwin discovered giant tortoises and a strange species of iguana—normally a terrestrial reptile in a desertlike environment—that on the Galápagos swam in the sea and ate seaweed in the surf. In what was probably his most significant observation, Darwin noted that similar animal types show distinctive variations in body form and functions from island to island. Once he had returned to England, two central concepts of his developing theory of evolution crystallized for Darwin.

First, he realized that the differences among related populations represent adaptations to differing environments. In biology, an **adaptation** is any genetically based feature that results in an individual or species being better suited to some aspect of its environment—jaws for eating particular types of food, insulation to resist freezing temperatures, behavioral instincts to avoid predators such as snakes, and so on. For example, there are obvious specializations in the body sizes and beak shapes of Galápagos finches; scientists have observed that the birds' adaptations correlate with each group's diet and feeding behaviors (Figure 38-2): Finches with strong beaks eat seeds; those with small, pointed beaks catch insects; and so on. What is the significance of adaptation? Darwin reasoned that individuals possessing advantageous adaptations are more likely to outreproduce individuals lacking the adaptations: *Survival to reproduce* is the key.

Second, Darwin recognized that in the Galápagos, organisms are geographically isolated on separate islands, and this provides an opportunity for *reproductive isolation*—the division of a population into groups that do not interbreed (and may eventually diverge into distinct species; see Chapter 39). Under such isolation, differences in form, function, or behavior accumulate among the members of the separated groups, and some of these differences may be adaptations.

Darwin was not the only scientist to formulate a theory of natural selection, however. Another young British naturalist, Alfred Russel Wallace, had studied plants and animals in Brazil and then in the Malay archipelago in Southeast Asia (Figure 38-3). Imagine Darwin's surprise and chagrin when he received a copy of Wallace's paper, which outlined the very theory that had been Darwin's primary focus for decades. Mutual friends arranged to have an abstract of Darwin's work read along with Wallace's paper at a meeting of the Linnaean Society in London in 1858. While Wallace emphasized competition for resources, Darwin focused on reproductive success—a concept that rapidly became the fundamental idea in evolutionary thought.

Two decades after the *Beagle*'s voyage, Darwin published *On the Origin of Species by Means of Natural*

THE FABLE AND FACTS OF DARWIN'S FINCHES

Galápagos finches—14 species of dusky birds with their variously shaped beaks and diets to match—were Darwin's final inspiration, crystallizing his great theory of evolution by natural selection. Or were they? Historians are no longer so sure that the finches were Darwin's turning point, because while Darwin describes variations among mockingbirds and other bird species in his magnum opus, nowhere does he make the briefest mention of the now-famous finches. Several researchers now conclude that Darwin's omission of the finches was based on rigorous scientific thinking and revolved around three crucial factors.

First, Darwin collected much of the data on the Galápagos finches secondhand from records made by shipmates. Given his penchant for accuracy, Darwin was probably leery of drawing conclusions from it.

Second, Darwin believed—mistakenly—that 11 of the 14 finch species ate similar food. He must therefore have missed the connection between each species' diet and its beak size and shape and thus the conclusion that new finch species arose as they became adapted to exploit different resources in the environment.

Finally, Darwin questioned whether all the Galápagos finches, with their very different characteristics, could have arisen from a single pioneering species. Indeed, not until well into the twentieth century did biologists have overwhelming evidence about the birds' evolution and adaptation from a common ancestor.

It seems time to lay an appealing but fanciful legend to rest: The Galápagos finches were probably not Darwin's ultimate inspiration, and it is unlikely that any single set of species served that role. Instead, Darwin reached his great insight by seeing order in the vast assemblage of biological and geological facts he collected so carefully and described so elegantly in *Origin of Species*. Darwin's precise, skeptical, and inquiring mind, however, will live on in the history of science.

Selection, or the Preservation of Favoured Races in the Struggle for Life. This seminal work integrated thousands of bits of evidence with the scholarly ideas that had influenced his theory. His eloquent summary reads:

> As more individuals are produced than can possibly survive, there must in every case be a struggle for existence, either one individual with another of the same species or with the individuals of a distant species, or with the physical conditions of life. . . . Can it therefore be thought improbable seeing that variations useful in some way to each being in the great and complex battle of life, should sometimes occur in the course of thousands of generations? If such do occur, can we doubt (remembering that many more individuals are born than can possibly survive) that individuals having any advantage, however slight, over others would have the best chance of surviving and of procreating their kind? On the other hand, we may feel sure that any variation in the least degree injurious would be rigidly destroyed. This preservation of favourable variations and the rejection of injurious variations, I call natural selection.*

Figure 38-3 ALFRED RUSSEL WALLACE, CODISCOVERER OF EVOLUTION BY NATURAL SELECTION. Working independently of Darwin, Wallace formulated his own theory of evolution by natural selection. Wallace stressed competition for limited resources as a main basis of natural selection, whereas Darwin emphasized competition within populations that tend to expand beyond their food supply. Wallace lived from 1823 to 1913. This photograph was taken when he was about 42.

The kernel of Darwin's argument for **natural selection,** then, is that organisms best adapted to their environment will have an edge in the battle of life, and this edge will tend to increase their chances to survive and reproduce.

Although many religious leaders and followers strenuously objected to Darwin's theories, his ideas were so powerful and reasonable that most scientists of the day accepted them enthusiastically. The concept of evolution by means of natural selection seemed a revelation that brought all living things into much closer harmony with

*Charles Darwin, *On the Origin of Species* . . . , 5th ed. (London: John Murray, 1869).

nature. It remained for later generations of biologists to explain the mechanisms of evolution and how inheritance and evolution intersect at the level of the gene.

VARIATIONS IN GENES: THE RAW MATERIAL OF NATURAL SELECTION

The key to Darwin's concept of a struggle for existence was the idea that some individuals in a species will arise with a longer neck, an enzyme that catalyzes a reaction faster, a leaf that gathers light better, or some other adaptation that increases an organism's chances for survival to reproductive age. If inherited variations in necks, enzymes, leaves, and other characteristics occur among members of species, then the genes responsible for them must also vary. Variations in both phenotypes and the genotypes that produce them are a good starting place for studying the mechanisms underlying evolution.

People are intriguingly diverse (Figure 38-4), and the individuals of most other species are equally distinctive. In light of such variation, evolutionary biologists wonder whether, for example, a tall person or pine tree is lofty because of "tall" genes or because the organism received favorable nutrients during crucial growth periods. Geneticists have discovered that variations in phenotype do not always reflect variations in genotype, and even when they do, it is usually difficult or impossible to determine the exact number of gene loci and alleles responsible for a particular variation. (Recall our discussion of pleiotropy in Chapter 11, for instance.)

Given such difficulties, how can one catalog the genetic variations that produce clear-cut differences among organisms, as well as understand the other subtle variations in phenotype that lack an easy genetic explanation? The techniques of molecular biology that followed the unraveling of the genetic code have allowed biologists to explore genotype and phenotype at more discriminating levels and to ascertain the degree of protein and genetic variation in natural, evolving groups of organisms.

Looking for Genetic Variation: Protein Electrophoresis

Implicit in Darwin's theory is the concept of variation of inherited characteristics. Researchers can measure such variation by analyzing either the sequences of amino acids in proteins or the sequences of nitrogenous bases in DNA. The preferred method for surveying a protein and its gene among many individuals in a population has been protein electrophoresis.

Electrophoresis—literally, the "carrying of electricity"—is a technique used to trace the movement of

(a)

(b)

(c)

(d)

Figure 38-4 THE RACES AND INDIVIDUALS OF *Homo sapiens:* ASTOUNDING VARIATION IN MORPHOLOGY, BUT ALSO IN CULTURE.

(a) In Tanzania, Masai men pausing for a picture. (b) In the People's Republic of China, a Beijing family going to the park. (c) In Scotland, an Edinburgh boy posing in his new piper's uniform. (d) In India, Agra women, three generations of an ancient civilization. All one species, all capable of interbreeding; yet, over the millenia, these and many other differences in morphology have arisen in human populations relatively isolated from each other.

electrically charged particles through a fluid medium. It depends on the fact that the amino acid subunits of proteins from plant and animal cells may carry a positive or negative charge or no charge at all. Since each kind of protein molecule has a characteristic combination of amino acids, each will tend to move at a characteristic rate in the electric field and in the electrophoretic medium.

Using protein electrophoresis, researchers can distinguish between the enzymes, other proteins, and other charged molecules they extract from tissues. Because a protein's primary structure is determined by the nucleotide sequence of its coding gene, protein variations revealed by electrophoresis serve as a kind of biological window on the genetic characteristics of populations.

To carry out electrophoresis, an experimenter places a protein sample at one site on a wet sheet or slab of coarse paper, starch, or various types of gels. Then he or she applies an electric current, running from a positive electrode (anode) to a negative electrode (cathode). Figure 38-5 shows how an electrophoretic gel is used to analyze protein differences in fruit flies.

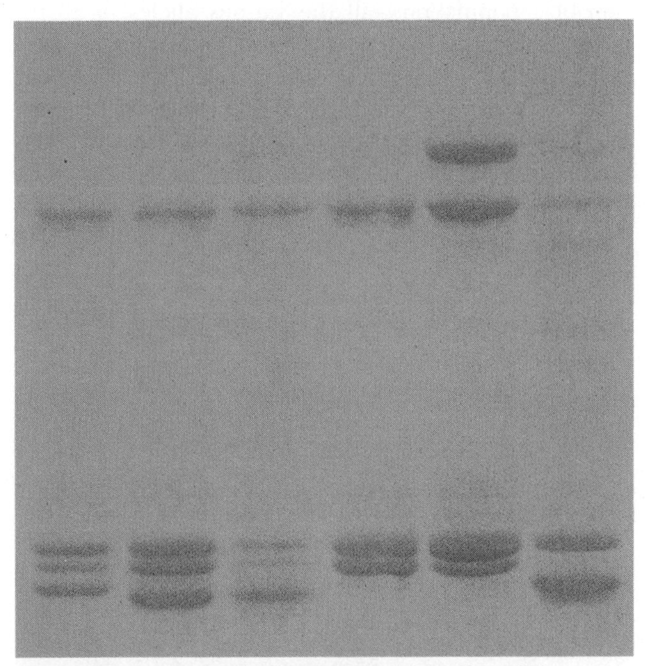

Figure 38-5 ELECTROPHORESIS: SORTING OUT PROTEINS IN AN ELECTRON FIELD.
Researchers can use electrophoresis to separate and sort proteins by size and properties. This photo shows the electrophoretic separation of yolk proteins in the blood of fruit flies from six different places in four different states (Oregon, California, Rhode Island, and Iowa). The bands reveal that the yolk proteins have travelled different distances, and thus are slightly different in their sizes and characteristics, based on genetic variability within the fly populations.

Using electrophoresis to analyze the protein product of a gene, biologists can get a relatively clear idea of how many individuals in a sample are homozygous and how many are heterozygous for each allele they assay. To translate these data into estimates of how much genetic variation actually exists in a population, a researcher usually surveys about 20 or more gene loci in each of perhaps 100 individuals by testing their respective proteins, then summarizes the result in a way that expresses the variability of the population.

One method for doing this is to figure the **average heterozygosity** (symbolized by H) of a population, that is, the average frequency of individuals that are heterozygous at each gene locus surveyed. Suppose you collect 100 fruit flies at random from a much larger population living in a pile of fallen apples at a local orchard. Now, suppose that your laboratory analysis of 20 different proteins in each fly—representing 20 different gene loci—turns up the following results. At 6 of the loci, 50 of the 100 flies are heterozygous. (You know this because two forms of each of these 6 proteins are present on the gel.) At 8 of the loci, 10 of the 100 flies are heterozygous; and at the remaining 6 loci, all the flies are homozygous for the alleles—that is, only one form is present for each of these 6 proteins.

To determine the average heterozygosity, you would note that 6 loci have a heterozygosity of $^{50}/_{100}$, or 0.50; 8 loci have a heterozygosity of $^{10}/_{100}$, or 0.10; and the remaining 6 loci have a heterozygosity of $^{0}/_{100}$, or 0. By summing and averaging over all 20 loci, you would get

$$H = \frac{(6 \times 0.50) + (8 \times 0.10) + (6 \times 0)}{20} = \frac{3 + 0.8 + 0}{20} = 0.19$$

or an average heterozygosity of 19 percent. Another way to view the genetic variation is through the population's **percent polymorphism.** If 6 + 8 of the 20 loci are heterozygous, as in this example, the percent polymorphism is 70 percent ($^{14}/_{20} = 0.7$).

Geneticists have used electrophoresis to estimate the amount of variation in a large number of loci in many species of plants and animals (Figure 38-6). They have concluded that such populations show high levels of genetic variation—the variation that makes natural selection possible. On the other hand, electrophoretic differences alone say nothing about actual differences in phenotype, and recent studies show that changes are likely to accumulate more slowly in the critical regions of proteins (such as the active site) and in the critical regions of DNA (such as the second base in a codon, which invariably leads to an amino acid change) than in other parts of the molecules.

Still, one interesting study implies that amino acid differences between proteins can represent true adaptations. Both langur monkeys and cows independently

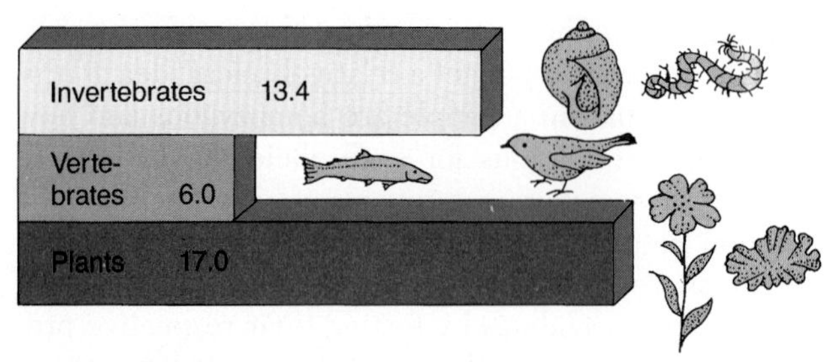

Figure 38-6 GENETIC VARIATION IN NATURAL POPULATIONS.
A substantial percentage of loci are heterozygous in all these major groups of sexually reproducing organisms. Recombination of those different alleles during sexual reproduction helps account for the variations on which evolution operates.

evolved special digestive chambers and processes that allow bacteria to ferment plant materials in their foreguts, and both have unique types of lysozyme (see Chapter 4). Forms of their enzyme resist digestion by pepsin, function best at low pH, and, significantly, have five amino acids that do not occur in other monkeys or domestic animals. As their special stomach chambers evolved, lysozyme evolution apparently proceeded as well, and these amino acid substitutions may allow the protein to function in an acid environment.

An extraordinary amount of variation exists within populations of our own species. Human beings have an average heterozygosity of 6.7 percent, and our DNA includes about 100,000 loci for structural genes (genes that code for proteins). An H value of 6.7 percent means that the average person is heterozygous at about 6,700 structural genes. Even considering linkage groups on chromosomes and the absence of complete independent assortment, people can still produce 2^{23} different gametes—an unimaginably large number—and aside from identical twins (derived from a single fertilized egg), no two gametes from different people are ever identical. Therefore, the chances that any two human beings will ever be genetically identical are just about zero.

The sheer number of genetically different gametes is only part of the biological insurance that organisms will vary. Crossing over during meiosis (see Chapter 9) and the random distribution of parental chromosomes to sperm and eggs provide a built-in guarantee that offspring will inherit new allele combinations. Further diversity is ensured by the fact that any sperm may chance to meet and fertilize any egg. And finally, this stockpile of genetic variability is continually replenished by random mutations. Clearly, large natural populations present countless, ever-changing genetic opportunities for natural selection to operate and for evolution to occur.

POPULATION GENETICS: THE LINKS BETWEEN GENETICS AND EVOLUTION

Knowing nothing of genes, Charles Darwin was never able to pin down the exact connection between heredity and differences in organisms' characteristics. Since the turn of the century, however, researchers in **population genetics** have used mathematical models to help them trace evolutionary trends within populations.

The Gene Pool and Gene Frequencies

Geneticists define a **population** as a group of organisms that interbreed or that have the potential to interbreed. All the marmots of a particular species living in a mountain meadow qualify as a population (Figure 38-7), as do the cottonwoods growing in a grove alongside a creek. In addition, biologists call the genetic composition of a population—all the various alleles of all the genes carried by individuals in a population—a **gene pool,** and this is the entity that changes over space and time and hence evolves. Related populations of trout living in unconnected river systems, for example, may gradually accumulate genetic differences that cause them to look, function, and behave differently.

There are two ways to depict the genetic makeup of a population or gene pool. One is to describe the types

Figure 38-7 POPULATIONS: GROUPS OF ORGANISMS THAT INTERBREED.
This colony of marmots (*Marmota flaviventris*) in Olympia National Park, Washington, shows typical watchful behavior for hawks or other predators. Three generations of this colony are seen here.

Table 38-1 PHENOTYPES, GENOTYPES, AND ALLELE FREQUENCIES FOR THE MN GENE LOCUS IN A U.S. CAUCASIAN POPULATION

Phenotypes (Number of Individuals with Each Blood Type)	M 1,787	MN 3,039	N 1,303	Total 6,129
Genotypes (Frequency)	$L^M L^M$ 1,787/6,129 = 0.292	$L^M L^N$ 3,039/6,129 = 0.496	$L^N L^N$ 1,303/6,129 = 0.213	Sum 1.0
Allele Frequency	L^M [(1,787 × 2) + 3,039]/(6,129 × 2) = 0.539		L^N [(1,303 × 2) + 3,039]/(6,129 × 2) = 0.461	Total 1.0

and relative number of different *genotypes* found in a given population. The other is to list the number and relative frequency of each *allele* in the gene pool. As we noted in Chapter 10, there may be many alleles of a given gene if there have been many mutations of the original gene.

Population geneticists sometimes define evolution as a change in gene frequencies. To measure evolution—to detect whether it is happening—they must therefore accurately quantify the frequencies of different alleles. Consider the MN gene locus in human blood groups, for example, which are codominant surface antigens analogous to the ABO markers (see Chapter 11). The three possible blood groups, M, N, and MN, are determined by two alleles, L^M and L^N, as shown in Table 38-1.

By testing 6,129 Caucasians living in the United States, geneticists found that 1,787 had the M phenotype and 1,303 had the N phenotype. The remaining 3,039 members of the sample were MN. To calculate the relative frequency of each genotype in the sample population, the geneticists divided the number of individuals with each phenotype by the number of people in the sample. The frequency of the $L^M L^M$ genotype, therefore, is $^{1,787}/_{6,129}$, or 0.292; likewise, the frequency of the $L^N L^N$ genotype is 0.213; and the frequency of the $L^M L^N$ genotype is 0.496. Of course, MM + MN + NN = 1 (although, because of rounding, the total will appear to be slightly greater).

To determine the allelic frequencies of the MN gene locus, the geneticists simply divided the number of alleles found by the total number of alleles in the sample. Now, $L^M + L^N = 1.0$. The total number of L^M alleles in the sample is arrived at by adding the number of alleles in the $L^M L^M$ individuals ($1,787 \times 2 = 3,574$) to the number of L^M alleles in the $L^M L^N$ individuals (3,039), giving a total of 6,613. Dividing this number by the total number of alleles in the sample (the diploid number is $2 \times 6,129 = 12,258$), we obtain 6,613/12,258, or 0.539. Because $L^M + L^N = 1$, it is then easy to calculate the frequency of the L^N allele: 0.461. Knowing these simple relationships allows us to calculate the proportion of alleles in a population. And as we shall see, numbers like these can lead to measures of real evolutionary change.

Genetic Equilibrium and the Hardy-Weinberg Law

To early geneticists, it seemed intuitively obvious that dominant phenotypes should be more abundant than recessive phenotypes in a population. This, however, did not correspond with the abundant evidence of persistent recessive phenotypes (and their underlying genotypes).

The English geneticist R. C. Punnett, whose squares we encountered in Chapter 10, was particularly intrigued by the fact that although blue eyes are recessive to brown eyes, there were a great many blue-eyed English people (Figure 38-8). He mentioned this to his

Figure 38-8 RECESSIVE TRAIT IN A POPULATION. The proportions of genotypes—say, for hair or eye color—in a population are the result of the frequency of alleles in that population, a frequency that is completely independent of the dominance or recessiveness of particular alleles. The great majority of blue-eyed blondes in England (such as these boys) and in northern Europe, for example, demonstrate that recessive traits can appear in a majority of a population.

cricket partner at lunch one day, and noted mathematician G. H. Hardy reportedly scribbled the explanation on a napkin. Hardy speculated that the proportions of genotypes within a population reflect the frequency of alleles, which is completely independent of the dominance or recessiveness of particular alleles. Furthermore, he suggested, if certain conditions are met, the frequency of alleles in a population will be stable—it will stay in *equilibrium*—from generation to generation. Because German physician W. Weinberg reached this same conclusion independently in the same year (1908), it is known today as the **Hardy-Weinberg law.**

Their "law" describes what happens to the frequencies of alleles and genotypes in a hypothetical ideal population that meets these requirements:

1. *Random mating:* No factors can influence the organisms' choice of mates.
2. *Large population size:* The laws of probability must be able to predict accurately.
3. *No mutations:* Alleles must not be changed to other alleles.
4. *Isolated population:* No exchange of genes with other populations, such as through migration.
5. *No natural selection:* No alleles can have reproductive advantage over others.

One could summarize the Hardy-Weinberg law this way: Allele frequencies and genotype frequencies remain constant over the generations in an infinitely large population with random matings but no mutation, migration, or selection. In any population that meets the five conditions, allele frequencies will not change over the generations, and the result is genetic equilibrium. But when one or more of these conditions is *not* met, changes may occur in gene frequencies. Since, by definition, changes in gene pools constitute evolution, then a deviation from the Hardy-Weinberg equilibrium tells us that a change has occurred in the gene pool and that evolution has taken place. Thus, the law is a means of detecting evolution.

Population geneticists use Punnett squares to demonstrate how a population might conform to the Hardy-Weinberg law. Suppose that a dominant allele *S* codes for straight hair and that this allele makes up 90 percent of the alleles at that gene locus in a population's gene pool. The other 10 percent of the alleles are the recessive *s*, which codes for curly hair. The Punnett square in Figure 38-9a represents random crosses between all the males and females in the population. Both the sperm population and the egg population have an *S*:*s* ratio of 90:10 (or 9:1). If the matings are truly random, any sperm may have a chance of fertilizing any egg, and we can calculate the frequencies of the resultant genotypes (*SS*, *Ss*, and *ss*) using the equations shown in Figure

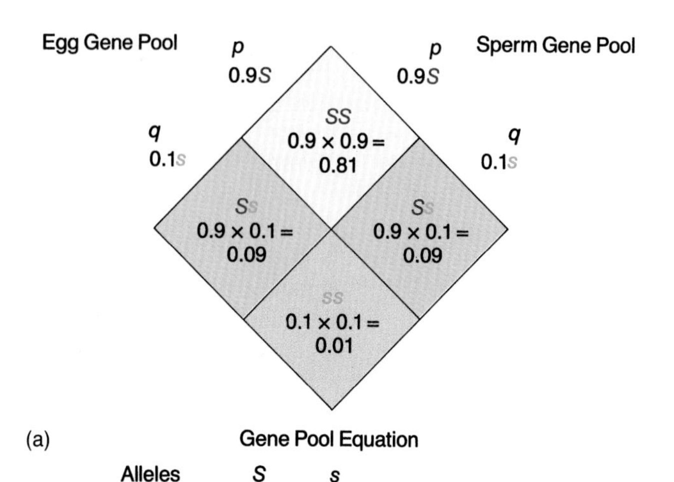

(a)

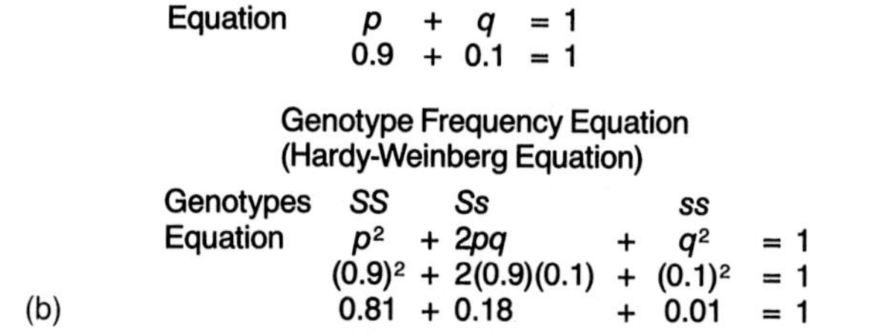

(b)

Figure 38-9 THE HARDY-WEINBERG LAW IN TERMS OF THE PUNNETT SQUARE.
The Punnett square can be used to indicate how many of each genotype to expect in a population. In this example, 90 percent of the alleles in eggs are dominant *S* and 10 percent are recessive *s*; the alleles in sperm show the same ratio. We can see from the Punnett square that 81 percent of the offspring through random mating will be *SS*, 18 percent will be *Ss*, and 1 percent will be *ss*.

38-9b. Note that 81 percent of the offspring will be homozygous for the dominant *SS* genotype, and only 1 percent will be homozygous for the recessive *ss* genotype. The remaining 18 percent of the organisms of the F_1 generation will be straight-haired heterozygotes—that is, they will show the dominant phenotype, since they carry only one recessive *s* allele.

Now, if the F_1 generation mates randomly, the genotypic frequencies remain unchanged; they could have arisen only from a set of gametes with the original ratio of 9:1 for *S*:*s*. The recessive alleles are not lost, but are "hidden" in F_1 heterozygotes. So the proportion of *S* to *s* stays the same from one generation to the next. Furthermore, we can see that even if the ratio were reversed, so that 10 percent of the alleles were *S* and 90 percent were *s*, the proportions of the genotypes *SS*, *Ss*, and *ss* in the F_1 generation would be 1, 18, and 81 percent, respectively.

Hardy and Weinberg showed that the stable gene frequencies in a population could be expressed by an algebraic equation:

$$p^2 + 2pq + q^2 = 1$$

where p and q stand for the frequencies of two alleles of a single gene. To see how this equation works, let us review our earlier calculations for the frequencies of alleles at the MN gene locus (see Table 38-1). Using p to represent the L^M allele frequency (0.539) and q to represent the L^N allele frequency (0.461), we obtain

$$(0.539)^2 + 2(0.539)(0.461) + (0.461)^2 = 1$$

$$0.29 + 0.50 + 0.21 = 1$$

Notice that the frequencies calculated according to the Hardy-Weinberg equation correspond rather closely to the actual frequencies of MN blood group genotype (0.292, 0.496, and 0.213) in the Caucasian populations in Table 38-1.

Based on this, we can state the consequences of the Hardy-Weinberg law in three straightforward principles:

1. If individuals in a sexually reproducing population mate at random, genotypic frequencies will be the product of the frequencies of their alleles.
2. If males and females in a population have the same frequencies of alleles, only one generation of random mating will be necessary to achieve equilibrium in genotypic frequencies.
3. In the absence of external factors that affect equilibrium, gene frequencies will remain constant from generation to generation.

MECHANISMS OF EVOLUTION: UPSETTING THE GENE POOL EQUILIBRIUM

In real populations of plants and animals, genes continually mutate, animals migrate, and natural selection imparts a steady pressure: The conditions for the Hardy-Weinberg equilibrium are never achieved. If the Hardy-Weinberg equilibrium is continually contradicted, implying that evolution is occurring, then the factors behind the contradictions are automatically the "causes" of evolution. Included among these factors are nonrandom mating, genetic drift, mutation, migration, and natural selection.

Nonrandom Mating

One of the most important requirements for gene pool equilibrium is *random mating*, independent of the individual's genetic makeup. Pollen from any one spruce tree, for instance, is equally likely to blow to any other spruce in a forest, irrespective of their genes. In **nonrandom mating,** on the other hand, the union of two organisms depends on their phenotypes (and indirectly on their genotypes). Lesser snow geese, for example, will preferentially mate with birds of their own feather color (white, silvery gray, or gray-brown; Figure 38-10). Nonrandom mating is commonplace in nature and usually involves inbreeding or assortative mating.

Inbreeding refers to mating between relatives at a greater frequency than would be expected by chance. The most extreme form of inbreeding is self-fertilization, a reproductive strategy found in some species of plants, including Mendel's peas, and in a few kinds of animals.

A variety of other organisms, including humans, show forms of inbreeding less extreme than self-fertilization. Although most human societies have incest taboos that reduce mating between siblings or between parents and siblings, among the Tuaregs, a Berber tribe that inhabits the Sahara, marriage is almost always between first cousins. Such nonrandom unions lead to deviations from the predicted Hardy-Weinberg gene frequencies. Animals that inbreed tend to live in small and generally isolated populations.

Self-fertilization and significant inbreeding can have either positive or negative consequences for a species. The major genetic results are an increase in the frequency of homozygotes and a decrease in the frequency

Figure 38-10 NONRANDOM MATING: A PREFERENCE FOR PHENOTYPE.
White and blue lesser snow geese were once believed to be different species. They are now known to be members of a single species (*Anser caerulescens hyperborea*). White is recessive, and both heterozygotes and homozygotes for blue coloration have silvery gray and grayish brown feathers. The geese mate preferentially with birds of their own color (assortative mating), so that there are more homozygotes and fewer heterozygotes than would be expected by the Hardy-Weinberg equation.

of heterozygotes within the population. With increased homozygosity comes a decrease in the diversity of genotypes, but that can be beneficial in a stable, unchanging environment—say, on the seafloor. Most environments vary physically or biologically, however, making inbreeding generally disadvantageous. Inbred wheat, for instance, which is highly homozygous, cannot survive a prolonged cold snap. Inbreeding can also bring out deleterious traits, since in homozygotes, recessive alleles are no longer masked. The small remaining populations of South African cheetahs are so inbred that they have 10 to 100 times less genetic variability at 200 protein loci than do other mammals, including artificially selected laboratory mice! Cheetahs have poor reproductive rates, high juvenile mortality, and great vulnerability to diseases, and up to 71 percent of their sperm cells are abnormal. We'll return to the cheetah's situation in a later section.

Assortative mating is a special case of nonrandom mating that involves union between unrelated individuals at a greater or lesser frequency than predicted by chance. Tall women, for instance, marry tall men more often than they marry short men; some white-crowned sparrows prefer to mate with individuals that sing in their local dialect; and lesser snow geese choose mates by color. Preferences like this are clear contradictions of the Hardy-Weinberg random mating principle and lead to changes in genotype proportions within populations.

Genetic Drift

Another of the Hardy-Weinberg conditions for genetic equilibrium is large population size—at least several thousand individuals. Why is size so important? It is because gene pool equilibrium depends on the laws of chance, and these laws apply to any situation where an event can produce more than one outcome, whether it is the flip of a coin or the union of gametes in sexual reproduction. When there are only a few cases, a phenomenon known as sampling error may occur; the smaller the sample, the greater the chance that actual frequencies will deviate from expectations.

Suppose that there are two alleles for a certain trait in a large population, with each allele present 50 percent of the time (a frequency of 0.50; Figure 38-11). If no external factors intervene, the gene frequencies in the next generation should also be 0.50, according to the Hardy-Weinberg law. But pure chance may produce deviations from this expected frequency. Such a chance change in gene frequency from generation to generation is called **genetic drift,** and it tends to increase the diversity be-[illegible] isolated populations.

[illegible] drift usually has a slight effect, and in very [illegible]lations it may be almost unnoticeable. In a

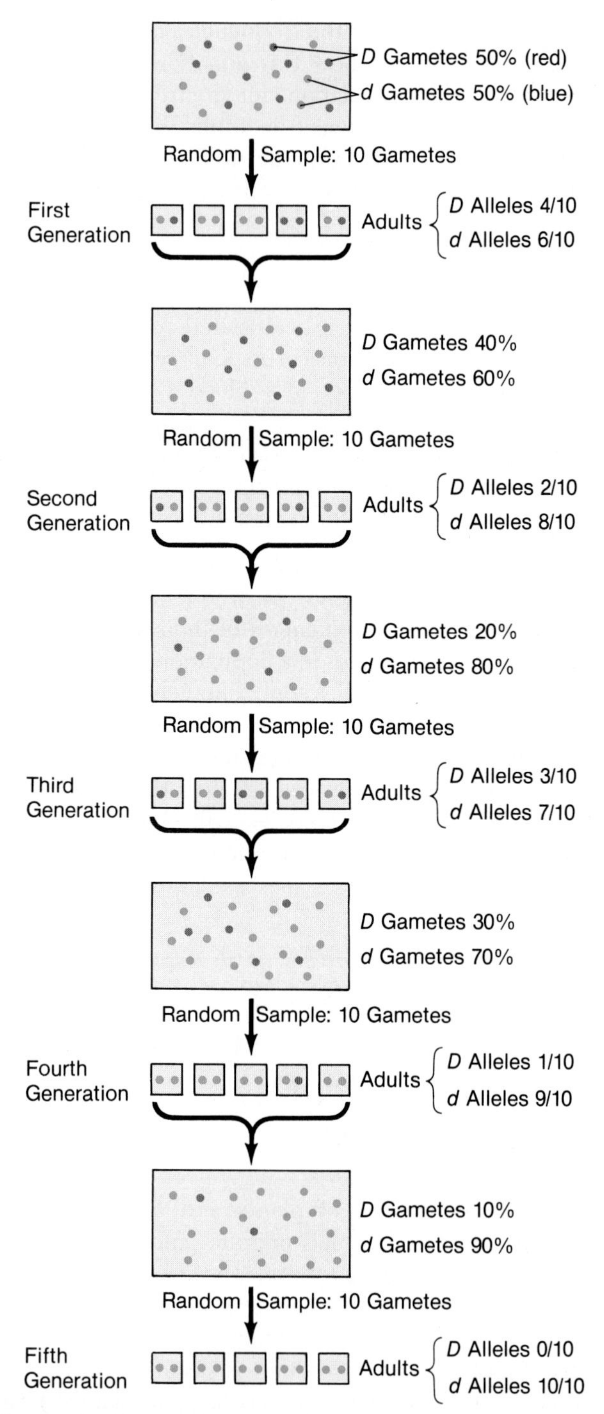

Figure 38-11 LOSS OF AN ALLELE IN A POPULATION OWING TO GENETIC DRIFT.
In this example, we start with an equal number of gametes with alleles *D* and *d*, and after five generations, all of the *D* alleles are eliminated as a result of genetic drift. For each generation, the gamete pool from the five parental adults is shown; then random selection of gametes from that pool yields the next generation. By this random process, the dominant allele happens to be eliminated, although the recessive allele could as easily have been lost.

small population, however, genetic drift can lead to substantial changes in gene frequencies and even the complete loss of an allele in a relatively short period of time. Let's say some natural disaster cuts a big population down to just ten individuals, and only one of them has a particular allele. If this individual fails to reproduce, the allele will be lost from the population, and genetic drift will have occurred. Figure 38-11 traces the way random genetic drift can cause an allele to be lost from a population.

Gene frequencies often change most dramatically when a few individuals leave a large population and take up residence away from other members of their own species. If these few "pioneers" succeed in starting a new population, its gene pool will reflect only the genotypes present by chance in the founders and may differ substantially from that of the original population, so that a rare allele is more frequent or a common one less frequent in the new group. This kind of change in gene frequencies is called the **founder effect** because the genetic makeup of the founder individuals determines the characteristics of the new gene pool.

Geologically isolated places, such as islands, provide an ideal setting for biologists to study the founder effect. For example, a small group of Yugoslavs moved to an isolated offshore island centuries ago, and one of the founders must have carried a rare recessive allele for the skin disease *mal de Meleda;* it is now quite common there, but remains very rare elsewhere in the world.

A well-documented example of the founder effect is evident among the Amish, a religious sect living in isolated communities in Pennsylvania, Ohio, and Indiana. One group of Amish has a high frequency of an otherwise rare genetic disease, the Ellis–van Creveld syndrome, a type of dwarfism in which the limbs are shortened, extra fingers or toes sometimes develop, and the victim usually dies within a few months of birth (Figure 38-12). In 1964, there were 43 cases of Ellis–van Creveld syndrome among the 8,000 Amish of Lancaster County, Pennsylvania, but none among the Amish of Indiana or Ohio. Those 43 cases were, in fact, more than the combined number of cases reported anywhere else in the world. The ancestry of all the people with this genetic syndrome can be traced back to a single couple, Mr. and Mrs. Samuel King, who emigrated to Pennsylvania in 1744.

In addition to the founder effect, random genetic drift was also at work among the Lancaster Amish. If members of the founding population had mated at random, the incidence of the disease would be expected to be about 1 in 400, yet it is 1 in 14. By chance, however, the Kings and their descendants had larger families than did the other founders, and the frequency of the Ellis–van Creveld allele thus drifted higher as time passed.

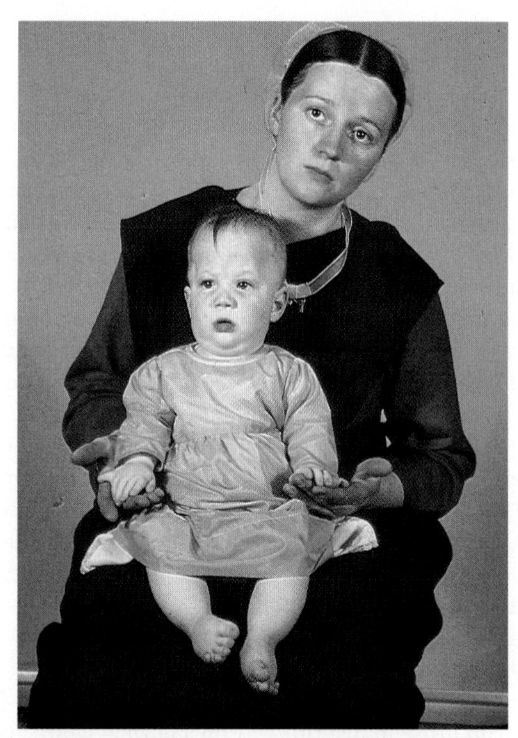

Figure 38-12 AN AMISH CHILD WITH ELLIS–VAN CREVELD SYNDROME.
The child has shortened limbs and six fingers on each hand. All the Amish with this syndrome are descendants of a single couple that helped found the Amish community in Lancaster County, Pennsylvania, in 1744. Because of inbreeding in the isolated community, the recessive trait is now common.

Another phenomenon, the **bottleneck effect,** is related to the founder effect and occurs when only a small portion of the original population survives to serve as the sole source of a new population. Among certain flies in New Zealand, for example, only a small subset of a very large summer population survives the harsh winter. Therefore, only a small proportion of the original population (and its gene pool) gives rise to the new generation each spring. The cheetahs of southern Africa *(Acinonyx jubatus jubatus)* apparently suffered a population bottleneck as well, perhaps through some environmental change thousands of years ago or from the pressure of disease or predation. Modern cheetahs arose from such a small surviving population that the animals show remarkable genetic uniformity. In fact, the immune system of one cheetah will not even reject a skin graft from a totally unrelated cheetah. While genetic variability usually declines after the bottleneck, some recent experiments on other species have shown unexpected increases in variability in the postbottleneck population—findings that have yet to be explained.

Mutation

Mutations—heritable changes in the chemical structure of genes—upset the Hardy-Weinberg equilibrium because they alter alleles. The concept of mutation has expanded in recent years to include chromosomal rearrangements of genes and DNA; gene duplication followed by divergence of extra copies; and the relocation of genes so that they come under the control of new regulatory elements. A simple point mutation, caused, let's say, by the ultraviolet rays in sunlight and resulting in a change in DNA base sequence, may have good or bad consequences. Such a mutation in an animal's somatic cell will not be passed on to the next generation of the population; but if the mutation occurs in an animal's germ line cells, so that sperm or eggs carry the altered allele, future populations may be affected. Plants lack a germ line (see Chapter 26); hence, if a point mutation strikes a somatic cell in a plant's growth zone—say, in the meristem in a branch that later forms a flower—the mutation may be passed on and show up as a single red petal on a pink chrysanthemum (Figure 38-13). Pink grapefruit arose in just this way on a normal white grapefruit tree.

Molecular research techniques have revealed how a duplicated copy of a gene may arise by chance (Figure 38-14a) and how this duplicated copy is then free, in a sense, to evolve. The original diploid set is still present to produce the phenotype and may still be influenced by natural selection, but the duplicated allele may undergo

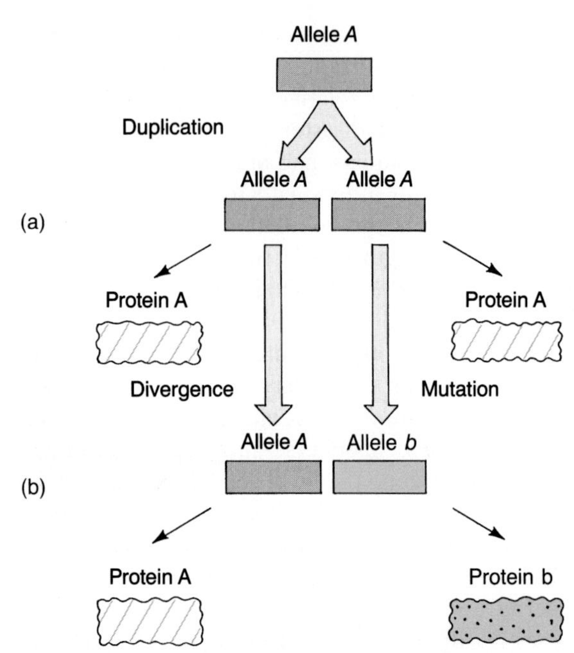

Figure 38-14 DUPLICATION AND DIVERGENCE: GENERATING NEW ALLELES.
The *A* allele is duplicated; each copy produces mRNA that codes for the A protein. Since two copies of the A allele are present, either one is "free" to evolve without the organism losing the A protein. Hence, mutation can generate the *b* allele, which codes for the b protein. This divergence process can generate families of related genes.

a mutation that alters its base sequence (Figure 38-14b) and yields a new protein with novel properties. This process is called *gene duplication and divergence.*

Recent research reveals that in most organisms, many different proteins share amino acid sequences. This implies that they share pieces of genes and that at one time, the genes may have duplicated or diverged. Recall from Chapter 36 that adult men can have one, two, or multiple copies of "green genes," all of which arose by duplication and divergence from the original gene coding for red-sensitive visual pigment.

Every time an allele mutates to a different allele, the gene frequencies are changed: The original allele decreases in frequency and the new mutated allele increases—that is, evolution occurs. By creating the new alleles that are a major source of genetic variation, mutation provides the raw material on which natural selection or other processes can act to change gene frequencies in a larger way.

As it happens, mutations are rare events; each human probably harbors an average of one new mutation. Most of these are recessive and are not expressed, but given enough individuals and enough time, every allele will eventually mutate, and ample genetic variability arises in populations through even low natural mutation rates. The time required can of course vary tremendously. African clawed toads have accumulated many point muta-

Figure 38-13 SOMATIC MUTATIONS.
...ce no germ line is present in plants, a somatic mutation, ... one red petal on this chrysanthemum, can be ...om one generation to the next if cells with the ...e to be included in tissue giving rise to

tions over the past 90 million years, but still resemble their ancestors closely, while in that same time period, placental mammals have diversified into the highly varied group we see today, from tree shrews to whales, antelopes, and people. Perhaps point mutation in mammalian regulatory genes made the difference.

Migration and Gene Flow

There are continual **migrations** into and out of populations: Pollen can blow for dozens or hundreds of kilometers; jellyfish can float on ocean currents to distant sites; and bull elk can invade a new territory and mate with local females. In each case, the migrant can introduce new genetic materials into the new population. When individuals (or gametes) migrate from one population to another and interbreed with the existing population, **gene flow** takes place, and new, often intermediate phenotypes can arise, as in the interracial family in Figure 38-15.

As a general rule, gene flow tends to reduce the genetic differences between populations—an effect just opposite to genetic drift, which tends to increase diversity between populations. For example, when African blacks were forced to immigrate to the United States beginning about 300 years ago, they probably had the *R* allele for the rhesus blood type at a frequency of 0.630, as do modern black Africans. The frequency among present-day black Americans, however, is 0.446, showing that gene flow has probably taken place between blacks and whites, who have a much lower frequency (0.028) for the *R* allele. Gene flow tends to be a constraining force in evolution, reducing genetic differences between populations. And as we will see in Chapter 39, the absence of gene flow between populations can be a major factor in the origin of new species, a major event in evolution.

Figure 38-15 GENE FLOW AS IT AFFECTS GENETIC DIVERSITY BETWEEN POPULATIONS.
Gene flow between races in the United States has resulted in an increased sharing of alleles. The result is often seen as a combination of phenotypes—as in skin and hair color and blood types. The individuals in this family illustrate some of the visible traits affected by gene flow.

LOOKING AHEAD

We have seen in this chapter that Hardy-Weinberg stability does not generally operate in natural populations, gene frequencies usually do change, and populations evolve unavoidably. We shall see in Chapter 39 that natural selection plays a role, along with nonrandom mating, genetic drift, mutation, and migration, in shaping the biological characteristics and evolutionary history of every species on earth.

SUMMARY

1. Jean Baptiste Lamarck presented a theory of evolution through the *inheritance of acquired characteristics.*

2. Charles Darwin and Alfred Russel Wallace independently proposed the theory of evolution by means of *natural selection.*

3. An *adaptation* is a genetically based characteristic that suits an organism to some aspect of the physical or biological environment and gives the organism a better chance of surviving to reproduce.

4. The more genetic variation in a population, the greater the opportunity for adaptive evolutionary change. Natural selection operates on phenotypic variations that result from underlying genetic variation.

5. The amount of variation in structural genes of natural populations is expressed as *average heterozygosity* or *percent polymorphism.*

6. *Population genetics* views genes as the units of evolution.

7. A *gene pool* is all the alleles carried by all the individuals in a *population,* or group of interbreeding or potentially interbreeding organisms. Changes in the gene pool over time, reflected in corresponding changes in the phenotype, constitute *evolution.*

8. The genetic makeup of a population can be characterized by frequencies of different alleles or different genotypes.

9. The *Hardy-Weinberg law* describes what happens to the frequencies of alleles and genotypes in an ideal population with random mating, large population size, no mutations, physical isolation, and no natural selection pres-

sure. The conditions for equilibrium are rarely, if ever, met.

10. *Nonrandom mating* occurs when the probability of mating depends on the phenotypes of the participating individuals. Nonrandom mating includes *inbreeding* and *assortative mating*.

11. Changes in gene frequency over generations that result from sampling error are called random *genetic drift*. Extreme cases of genetic drift occur as the result of the *founder effect* and the *bottleneck effect*.

12. Mutations, including point mutations in DNA, chromosomal rearrangements, and gene duplication and divergence, are a major source of genetic variation and serve as the raw material for evolution.

13. *Migration* of individuals from one population to another produces *gene flow* between the populations, which may tend to reduce the genetic diversity between the populations. Conversely, when isolation makes migration impossible, more diversity can accumulate. Gene flow can be a conservative or a creative force in evolution, depending on geographical, ecological, and demographic circumstances.

KEY TERMS

adaptation
assortative mating
average heterozygosity
bottleneck effect
electrophoresis
evolution
founder effect
gene flow
gene pool
genetic drift
Hardy-Weinberg law
inbreeding
inheritance of acquired characteristics
migration
natural selection
nonrandom mating
percent polymorphism
population
population genetics

QUESTIONS

1. What does adaptation mean? Give some examples.

2. Define and give examples of genetic drift, the founder effect, and the bottleneck effect.

3. How do biologists usually characterize a population's genetic makeup? How do they describe the amount of genetic variation in a population?

4. Suppose that the S allele in a population of organisms is present at an 80 percent frequency, and the *s* allele at a 20 percent frequency. After individuals in one generation mate randomly, what will be the frequencies of the genotypes SS, *Ss*, and *ss*? If the Hardy-Weinberg conditions apply, what will the genotypic frequencies be after five generations?

5. What conditions must operate to maintain the Hardy-Weinberg equilibrium? Do these conditions occur in nature? What can we conclude?

6. If for several generations the members of the population described in question 4 mate only with others of the same genotype rather than randomly, will the genotypic frequencies change? If so, which genotype(s) will become more frequent, and why? Will the frequencies of the S and *s* alleles change? If so, which allele will increase in frequency, and why?

7. Give some examples of nonrandom mating in nature and among humans.

8. Among the populations that Darwin studied on the Galápagos Islands, what phenomena might have affected gene frequencies when those islands were first colonized? When a species migrated to a neighboring island?

9. Calculate the average heterozygosity in a population of 1,000 sea urchins in which you analyze, using electrophoresis, 30 different proteins from each animal. At 5 of the 30 loci, 300 of the urchins are heterozygous: at 20 loci, 200 are heterozygous; at 4 loci, 400 are heterozygous; and at 1 locus, 100 are homozygous.

ESSAY QUESTION

1. What conditions would need to prevail to prevent evolution? What is the likelihood that such conditions could exist among natural populations on earth?

For additional readings related to topics in this chapter, see Appendix C.

39
Natural Selection

The essence of Darwin's theories is his contention that natural selection is the creative force of evolution—not just the executioner of the unfit.

Stephen Jay Gould,
Ever Since Darwin (1977)

The lungfish is a true "living fossil"—an air-breathing fish that has not changed substantially for 3 million centuries. In midsummer, a lungfish may lie just a few centimeters below the fractured surface of a dried-up pond, encased in a cocoon of damp earth and mucus. There it rests, breathing very slowly through a tiny air hole and sustained by a drastically slowed metabolism that will quicken only when rains return in autumn and the fish can safely free itself from its protective containment.

Although the seasonal shifts in the lungfish's environment are pronounced, the yearly pattern has remained essentially the same since the Carboniferous period, and as a result, biologists believe that *natural selection* has had a stabilizing influence on the Australian lungfish. The animal's anatomical and behavioral adaptations have served it so well that an ancestral form encased in the same pond bottom 300 million years ago would resemble its modern successor in most details.

Three hundred and fifty million years with remarkably little change: An African lungfish (*Protopterus dolloi*) awaits the drying of its lake prior to burrowing into the mud.

In marked contrast to such evolutionary stability are the nearly 500 species of *Drosophila* fruit flies on the Hawaiian Islands. Biologists believe that these insects evolved at an astonishing speed that reflects the availability of an extremely rich but relatively unoccupied set of environments—a chain of islands that arose from the seafloor in just the past 5 million years. Researchers have compared the spectrum of proteins in these Hawaiian flies and have found a remarkable degree of variability—apparently the result of selective pressures that were anything but stable.

In each of these examples, natural selection is acting on the organism's phenotype at all stages of the life cycle. In so doing, natural selection acts indirectly on the genotype, and it does so for all other organisms. Of the many factors that affect gene pools over time (see Chapter 38), natural selection best accounts for the broad range of adaptations that enable populations to survive in given environments. As Darwin recognized a century and a half ago, natural selection is a key mechanism in the evolution of life.

This chapter examines how natural selection can affect the phenotypes of plants, animals, and microbes over time to alter gene frequencies or maintain the genetic status quo, thus sculpting the way today's organisms look, act, and function. We consider:

- The concept of *fitness*—how an organism's characteristics can increase or decrease its chances of surviving to reproduce
- How phenotype affects genotype—that is, how natural selection acting on organisms affects gene frequencies in populations
- The various types of natural selection
- New views and lively debates on the mechanisms of evolution

THE DARWINIAN AND GENETIC MEANINGS OF FITNESS

Charles Darwin had four major points to his argument for evolution by natural selection: (1) Some organisms survive to reproduce, while others do not; (2) natural variations in traits affect the chances that a given individual will survive and reproduce; (3) offspring resemble their parents and thus tend to inherit those traits that gave their parents a reproductive advantage; and (4) therefore, traits that increase reproductive success tend to increase in frequency from generation to generation.

In light of what we know today about genes, chromosomes, and the mechanisms of heredity, we can restate Darwin's argument this way: (1) Phenotypic variations among organisms affect survival and reproductive success; (2) phenotypic variations are heritable (genetically determined), at least in part; and (3) genes contributing to variations that encourage reproductive success tend to increase in frequency over time.

Geneticists who study evolving populations use the term **fitness** to define the relative reproductive efficiency of various individuals or genotypes in a population. In essence, an individual's fitness (or an allele's or a genotype's fitness) depends on the likelihood that one individual, relative to other individuals in the population, will contribute its genetic information to the next generation. Fitness, then, includes an organism's ability to survive, to mate, and to reproduce successfully. The familiar phrase "survival of the fittest" relates to fitness, but the largest, strongest, and most aggressive organisms do not necessarily have the highest fitness. A much more subtle combination of structure, physiology, biochemistry, and behavior contributes to an individual's fitness and therefore the probability that it will leave offspring.

A good example of fitness is coat color in arctic hares. A pure-white hare sitting in the snow is less visible to a fox or an owl than a hare with a light tan coat. The allele for white coat color would therefore have a higher fitness than an allele for tan coat color and would allow white hares to survive to reproduce in higher frequency (Figure 39-1). In the same snowy environment, an allele for dark tan coat would have an even lower fitness and over time might be eliminated from the population. In a co-

Figure 39-1 FITNESS: A WHITE ARCTIC HARE IN THE SNOW.
Arctic hares (*Lepus timidus arcticus*) with pure-white fur are less visible against the snow than are hares with tan fur. For this reason, they are more likely to avoid predator attacks and survive to reproduce in higher frequency. Of course, this sort of adaptation is an indirect consequence of predators that use vision for finding prey; if foxes and birds hunted hares by sense of smell alone, the selection of fur color might not occur.

niferous forest, however, the allele for dark tan might confer a higher fitness than that for either white or lighter tan.

An allele coding for a particular enzyme form in the marine mussel *Mytilus edulis* (Figure 39-2) beautifully illustrates how sensitive an allele can be to natural selection. *Mytilus* species have multiple alleles and multiple forms of an aminopeptidase enzyme that cleaves amino acids from peptides. One particular allele, hap^{94}, cleaves amino acids so efficiently that the molecules build up and raise the internal osmotic pressure inside the mussel's cells to levels that match seawater.

Each year, ocean currents sweep little juvenile mussels from the open ocean into Long Island Sound between New York State and Connecticut, with its mixture of salt and fresh water. Measurements of gene frequency show that an intense selective pressure must be operating on the little animals, because the farther the young mussels live from the pure seawater, the lower the frequency of the allele and the more stringently hap^{94} is selected against. Survivors tend to have alleles that code for variant enzymes that release fewer amino acids; their body fluids thus have lower osmotic pressures, in balance with the surrounding brackish water. Clearly, the hap^{94} allele confers high fitness in the sea, but low fitness in more dilute waters.

Because varying alleles and proteins allow the mussels to survive under a wide range of environmental conditions, multiple alleles are preserved in the full *Mytilus edulis* population. The frequency of any given allele, however, will vary widely from site to site. In other organisms, alleles may be selected against and lost from a species permanently.

In cases where one allele of a pair is eliminated completely from a population, the remaining allele is said to be **fixed.** If the allele for tan coat color were lost, the white allele would be fixed in a population of arctic hares. Fixed alleles automatically have a high level of fitness because they are guaranteed representation in future generations. The "fixed" state of an allele is not permanent, however, because mutations are inevitable—mutations that could raise or lower the allele's fitness.

Figure 39-2 MUSSELS AND ALLELE FITNESS.
Blue mussels (here, *Mytilus edulis*) can live in various environments, such as on this exposed reef in the sea. Others live in brackish estuaries. Although the individuals look the same, organisms exposed to such different environments may have important differences in their physiology and biochemistry. Alleles such as hap^{94} (which codes for an enzyme that helps balance the mussel cells' osmotic pressure with that of surrounding water) vary accordingly in frequency, at least in mussels studied in Long Island Sound.

(a)

(b)

Figure 39-3 ALLEN'S RULE: EXTREMITIES, TEMPERATURE, AND EVOLUTION.
According to Allen's rule, animals living in cold climates tend to have shorter extremities (tails, limbs, and ears) than animals living in warm climates. This is illustrated by (a) the arctic fox (*Alopex lagopus*), with its short tail, ears, and legs, and (b) the desert fox (*Vulpes chama*), with its longer tail, ears, and legs. These characteristics have obvious adaptive value: Body heat is retained more efficiently by short extremities and is dissipated more efficiently by large extremities with their greater surface areas.

At the opposite end of the scale from fixed alleles are alleles for sterility or for lethal traits; alleles like these are unfit because the individuals carrying them would have no chance to reproduce. Between the two extremes are situations in which a number of alleles may exist for a given gene locus, as with the multiple eye color alleles in *Drosophila* and the numerous hemoglobin alleles in humans.

Biologists have recognized some intriguing general rules that show how a species' range can interact with the fitness of its allelic variations. **Bergmann's rule** states that individual warm-blooded animals living in the colder parts of their species' range tend to be larger than individuals living in warmer areas; **Allen's rule** observes that animals inhabiting cold regions tend to have shorter limbs, tails, and other extremities than do those in warmer regions (Figure 39-3). Both larger size and smaller extremities would aid heat retention. Clearly, environment can play a fundamental role in determining which of several alleles is optimum for a population's survival.

NATURAL SELECTION: HOW PHENOTYPE AFFECTS GENOTYPE

Since natural selection tends to eliminate less fit alleles, genotypes, and individuals, selective processes tend to increase a population's adaptation to its environment. Suppose, for instance, that fruit flies in a population can find the species' favorite food at the tops of trees and on the ground but nowhere in between. Can natural selection produce fruit flies that only fly up or down?

Researchers designed an experiment that would select for a genetically determined behavior known as geotaxis in fruit flies. *Geotaxis* is the movement of organisms in response to the force of gravity—either up (negative geotaxis) or down (positive geotaxis). The researchers devised a funnel-shaped maze with 105 "decision chambers" connected by one-way valves (Figure 39-4). Each chamber had two possible exits—one going up and one going down—and flies had to make an up or down decision 14 times. Flies that consistently took the upper passage ended up in the top three or four food vials at the end of the maze, while those that always tended toward the lower exits collected in the bottom few food vials. Flies with no particular tendency for either positive or negative geotaxis ended up in the middle vials.

Discarding the middle flies, the researchers allowed flies from the top and bottom vials to breed separately for a dozen more generations. The researchers eventually produced one population of flies that almost invariably took the up exits and another population of flies with a strong down preference. The researchers knew of no

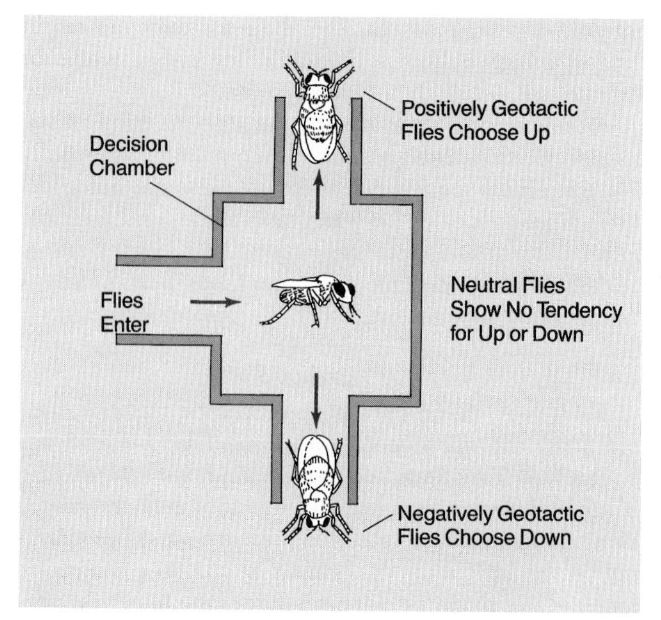

Figure 39-4 TESTING POSITIVE AND NEGATIVE GEOTAXIS IN *Drosophila*.
To test whether a fly's movement in response to gravity is genetically controlled, experimenters allowed flies to move through a maze consisting of many chambers like the one shown here. The offspring of flies that tended to move up through the maze of chambers also tended to walk up (and the offspring of flies that consistently chose lower chamber exits also tended to walk down). The workers concluded that geotaxis in flies is genetically controlled.

"new" genes for geotaxis in the populations due to mutations or gene flow. Instead, the selection pressure they provided acted on variation already present in the original population and caused a change in allele frequencies and so in phenotype (the flies' behavior).

What would happen if such selection pressure was removed? To answer this question, the experimenters allowed the "high" flies and the "low" flies to interbreed and reproduce. The original balance of alleles soon was restored: Some flies went up, some went down, and most took the route in between. However, when Jerry Hirsch at the University of Illinois continued the original selection for high and low fliers for over 600 generations, not just a dozen, then allowed interbreeding, the two populations remained fully stable as high or low fliers. The new, stable phenotype is associated with three major genetic loci and perhaps many more genes in them. The prolonged experimental selection caused (1) an evolutionary change involving stable, inherited differences in behavior (and in other aspects of phenotypes) and (2) changes in groups of genes that underlie those phenotypic characters. The capacity for genetic and hence evolutionary change was clearly present. Should the fruit flies' food supply diversify in height in nature, *Drosophila* populations might evolve in response.

TYPES OF SELECTIVE PROCESSES

Natural selection is not a generator of new genotypes and phenotypes, but a mechanism to ruthlessly eliminate the less fit and leave the more fit to compete in an unending quest for reproductive success. Let us consider the four major selective processes, which can preserve existing traits or produce evolutionary change.

Normalizing Selection: Preserving the Status Quo

Sometimes the effect of natural selection is simply to preserve a phenotype that is already present. This **normalizing selection** (also called *stabilizing selection*) maintains the status quo for the lungfish in its cocoon of crusted mud. Lungfish populations have survived the eons not because frequent genetic changes have helped them adapt to new environments, but because adaptations which evolved early in their history have remained adequate in their peculiar but extraordinarily stable environment. While there is no fossil proof, it seems likely that other lungfish phenotypes arising through the normal processes of meiotic recombination, mutation, and random mating were selected against.

Biologists documented a famous case of normalizing selection among a population of northern water snakes (genus *Nerodia,* formerly *Natrix*) living on the islands in Lake Erie. Members of this species normally have plain gray skin, but occasionally, some snakes are born with a banded phenotype (Figure 39-5). These patterned mutants are strongly selected against, however, because they make conspicuous targets for predatory birds. As a result, their chances of surviving to reproductive age and contributing their genes to the next generation are significantly less than are those of their duller compatriots.

Normalizing selection usually operates continuously over many generations, but an enterprising British scientist, E. B. Ford of Oxford University, documented a significant exception. An avid observer of moths and butterflies, Ford kept track of the phenotypes of a population of marsh fritillary butterflies for 15 years. When he began, the population was small and highly homogeneous in size, body shape, and wing color patterns. Suddenly, however, a 4-year population explosion took place, and during this time, butterflies showed great variation in both color patterns and basic body morphology. Then the population stabilized again, larger than before, with a "new" phenotype that was quite distinct from the original one. Although Ford could not establish the exact cause of the remarkable changes he had recorded, the implication of his observations seemed clear:

Figure 39-5 NORMALIZING SELECTION AND THE BANDED NORTHERN WATER SNAKE.
Some members of the species *Nerodia sipedon sipedon* have this striking banded pattern, but it leaves them more vulnerable to predators, and so is strongly selected against, while the phenotype status quo, the plain gray form, is selected for.

Normalizing selection pressure apparently relaxed for a time, then once again began to hold back variation.

Normalizing selection is especially common in environments that have remained stable through long periods of the earth's history, such as the black, cold, high-pressure regions of the ocean floor. In extremely stable environments, organisms resemble ancestors that inhabited such places for tens or hundreds of millions of years and had phenotypes already very well adapted for coping with the challenges of that environment. If a new predator, competitor, or parasite appears, however, adaptive changes may well arise in response. Normalizing selection also works in more variable locations—a mountain meadow, a desert sand dune, or other environments that may change over time spans that are short by geological standards but long when measured against the life of an individual.

Directional Selection: Changing Phenotypes

A second category of natural selection, **directional selection,** helps bring about change in an organism's phenotype. In 1915, for instance, oyster farmers hoisting their barrels in Malpeque Bay in the Gulf of St. Lawrence began to notice frightfully diseased individuals among the plump, healthy mollusks. These ailing shellfish were small and flabby and covered with yellowish pus-filled blisters, and by 1922, the Malpeque disease had all but wiped out the oyster beds. By 1925, fishermen found that they could again harvest a small catch, and by 1940, Malpeque Bay was producing more oysters

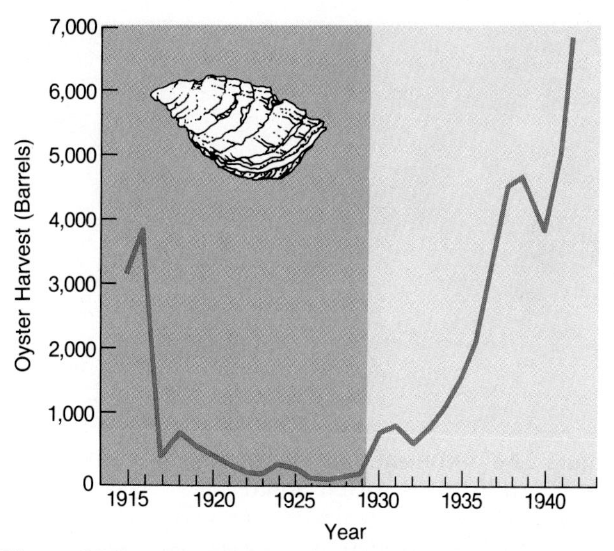

Figure 39-6 DIRECTIONAL SELECTION: OYSTER YIELDS FROM MALPEQUE BAY, 1915–1940.
Malpeque disease reduced the oyster population drastically, but disease-resistant oysters survived and, beginning in 1929, gave rise to the new, large population.

than ever in its history (Figure 39-6). The fishermen could also use the "new" Malpeque oysters to successfully repopulate other areas recently decimated by the disease. What had happened?

Every year, each adult oyster produced some 60 million offspring, a few of which carried an allele that conferred resistance to Malpeque disease. When environmental conditions favored the survival of individuals carrying the genetic variant, that variant increased in frequency, and the result was directional selection. ("Directional" here refers to a consistent trend over time, not to "direction" by some force.) A variant arose or existed completely by chance, and selective pressure operated to preserve it rather than other alleles.

Rapid and dramatic adaptive change often occurs in just a few generations in response to a sudden and substantial environmental change. One of the classic examples of this kind of directional selection is the case of *Biston betularia*, the peppered moth, and the phenomenon called *industrial melanism.*

In eighteenth-century England, butterfly collectors vied for the chance to snare a rare black form of this normally light-colored speckled moth—a variant that was dark owing to high levels of melanin pigment. Both forms of the moth spent a good deal of time resting and feeding on the trunks of trees that were often covered with pale, gray-green lichens. Birds and human collectors had difficulty spotting the light, peppered insects when they rested on the lichens (Figure 39-7a). The black moths, however, had no natural camouflage at all on the light-colored background and were easier to spot. For this reason, birds tended to pick them off (a strong selective pressure against the allele for the dark form), and live adults were hard for collectors to find.

As the Industrial Revolution progressed in mid-

(a)

(b)

Figure 39-7 THE PEPPERED MOTH: A CLASSIC EXAMPLE OF DIRECTIONAL SELECTION.
The peppered moth, *Biston betularia,* occurs naturally in two color variations: the light, speckled moth and the black variant. (a) The light, peppered type was much more common in England until the nineteenth century; when on the light-colored tree bark, it was less susceptible to predation than was its darker and more visible relative. (b) Then the Industrial Revolution darkened the vegetation of England with pollution; the darker moth came to have more of a selective advantage against predatory birds, and the frequency of the allele for dark pigment increased in the population. Today, as British pollution controls lead to a cleaner countryside, the frequency of the alleles for light, peppered moths is again increasing in the population.

nineteenth-century Europe, smoke and soot began to darken the vegetation around English factory towns. In heavily polluted regions, the dark form of *Biston betularia,* camouflaged on the sooty trunks, began to flourish, while the light form became increasingly rare (Figure 39-7b). Decades later, naturalist E. B. Ford suggested that either the light or dark form would have a selective advantage, depending on environmental conditions, and in the 1950s, entomologist H. B. D. Kettlewell of Oxford University designed several experiments to test Ford's ideas. First, Kettlewell released equal numbers of both light and dark moths in an area where 95 percent of local moths were light, and he was able to recapture twice as many light as dark moths. When he and his colleagues spied on the moths from a blind, they were able to directly observe birds picking off many more of the dark moths as the dark and light moths rested together on pale, lichen-covered tree trunks. Near heavily industrialized Birmingham, however, Kettlewell observed just the opposite. Clearly, predatory birds supplied the selective pressure, and their hunting habits led to higher frequencies of the light or dark moth phenotypes (and its underlying allele), depending on the color of the tree trunks. Interestingly, as British engineers have reduced factory emissions in industrialized regions of England, both lichens on tree trunks and light, peppered moths are making a comeback, and the century-long domination of the black moths may now be ending.

In recent years, biologists using the techniques of molecular biology have revealed novel, heretofore unsuspected genetic mechanisms that may underlie some types of directional selection. Many insect populations develop resistance to various insecticides, and this represents a kind of directional selection. For example, populations of *Culex* mosquitoes have a gene coding for an esterase, an enzyme that can detoxify small amounts of deadly organophosphate insecticides. In fields and swamps treated with the pesticides, most *Culex* mosquitoes die, but the few surviving individuals show an *amplification* of the gene, so that there are 250 or more copies present in each diploid body cell rather than just two genes, as in the cells of the original sensitive population. With this genetic arrangement, the mosquito can make huge quantities of the esterase enzyme, detoxify the organophosphates, and live.

In the case of the Malpeque Bay oysters, the peppered moths, and the *Culex* mosquitoes, directional selection took place in just a few decades in response to a relatively sudden environmental change. However, the selective process is usually believed to work very slowly, over immense periods of time. When this happens, the effect of selection can be detected only by identifying gradual—but clear—directional trends in the fossil record. The gradual decrease in the number of horses' toes is a classic example (Figure 39-8). Evolutionists suggest that much of nature's variety can be accounted for by this kind of gradual genetic change.

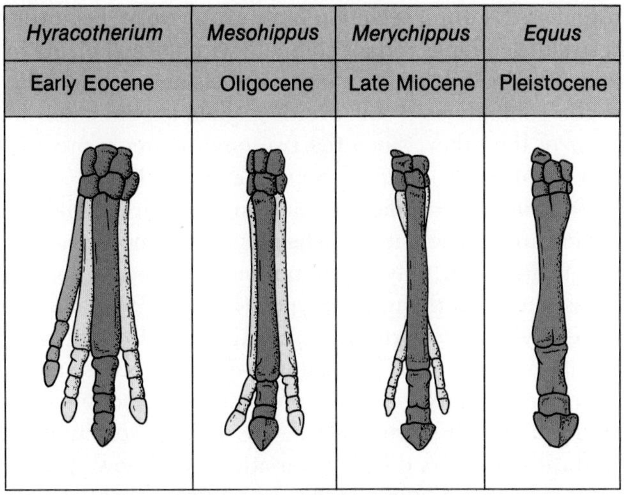

Hyracotherium	*Mesohippus*	*Merychippus*	*Equus*
Early Eocene	Oligocene	Late Miocene	Pleistocene

Figure 39-8 DIRECTIONAL SELECTION IN THE FOSSIL RECORD: THE EVOLUTION OF HORSES' FEET.
Horses evolved from four-toed and three-toed browsers to single-toed grazers over about 50 million years (see Figure 40-12). Fossil evidence reveals the trend in this directional selection as large horses capable of rapid running evolved from their slower-moving ancestors.

Diversifying Selection: Producing Variant Phenotypes

Sometimes, a population may be faced with new conditions so diverse that no single phenotype can exploit every extreme. If there is sufficient genetic and phenotypic variability to allow different selective pressures to operate, a third category of natural selection—**diversifying,** or **disruptive, selection**—may lead to two or more phenotypes, each adapted to a different specialized feature of the total environment.

Botanist A. D. Bradshaw found diversifying selection operating in a population of bent grass plants growing in the copper-mining regions of Wales. There, heaps of soil contaminated with copper and other metals dot the landscape. This tainted soil is lethal for most plant species in the area. While observing "normal" bent grass plants in clean soils nearby, however, Bradshaw and his coworkers discovered a variant strain of the plant thriving on adjacent piles of copper-laden dirt. This copper-resistant strain turned out to be highly specialized for its peculiar environment: It would barely grow at all on uncontaminated soil. Since resistant and nonresistant stands of bent grass are separated from each other by only a few hundred meters, and since bent grass is a cross-pollinating species, the two variants can easily exchange genes. As a result, hybrid seeds containing genes from both types of plant continually sprout in both environments. Natural selection is always at work, however; resistant seedlings cannot compete with normal plants in uncontaminated soil, whereas nonresistant plantlets die in the contaminated soil. Neither allele is eliminated in favor of the other, since the two environments provide two permanent "pockets" for the extreme types.

To summarize so far, normalizing and directional selection are "conservative": They tend to reduce variation and to maintain only a few optimally adaptive phenotypes and underlying genotypes (Figure 39-9a and b). Diversifying, or disruptive, selection has the opposite effect, maintaining or increasing variations because more than one phenotype can be optimal in the nonuniform environment (Figure 39-9c). Disruptive selection is "radical": It favors differing variations, and new species may arise (see Figure 39-9c). In nature, these various categories grade into each other.

Maintaining Genetic Variation

Two forms of natural selection can lead to the retention of an otherwise deleterious genetic variation: balanced selection and frequency-dependent selection.

Balanced Selection

Balanced selection, or **heterozygote advantage,** exists when a heterozygote *(Aa)* has a higher fitness than either homozygote *(AA* or *aa)*. One of the most famous examples of heterozygote advantage is the retention of the sickle-cell allele in certain human populations, since heterozygous individuals carrying the recessive allele are protected from malaria (see Chapter 11 and Figure 39-10).

Still another term for heterozygote advantage is **hybrid vigor,** or *heterosis.* Plant and animal breeders routinely exploit the genetic benefits of this phenomenon by breeding together two genetically distinct parental lines. If both parents are relatively homozygous at a number of sets of alleles—perhaps as a result of inbreeding to improve a sweet corn crop or a line of prize dogs—then the hybrid corn or the mongrel dog will be heterozygous for many of those sets of alleles. Because any deleterious alleles that have accumulated in the homozygous individuals are suddenly counteracted by other alleles, such offspring tend to be larger, healthier, and—in the case of the corn—more productive than the specialized inbred parents (Figure 39-11).

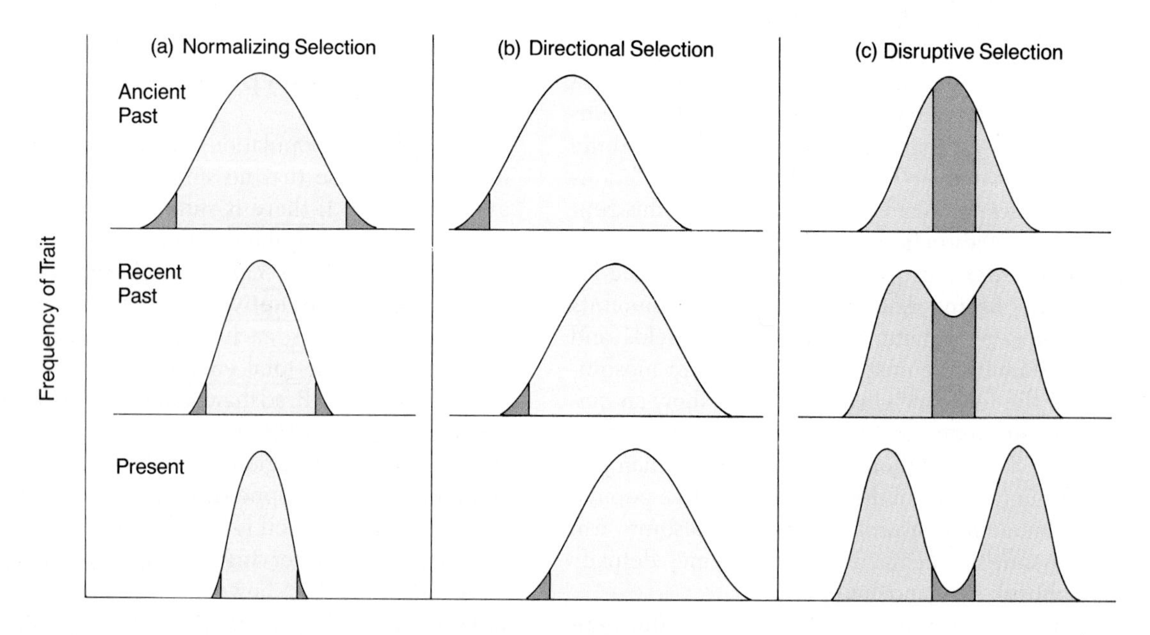

Figure 39-9 NORMALIZING, DIRECTIONAL, AND DIVERSIFYING SELECTION.
A polygenic character, such as height of a tree or giraffe, is subjected to different types of selection. (a) Normalizing, or stabilizing, selection tends to reduce variation, so that only the phenotypes best adapted to some set of circumstances are preserved in a population. (b) Directional selection leads to the establishment of certain phenotypes, especially as a trend over time. Here, the phenotypes at one end of the distribution are favored, and selection results in the distribution of phenotypes moving to that end. (c) Diversifying, or disruptive, selection favors individuals that are most diverse. If the selection is quite strong, two nonoverlapping populations may result, and new species might arise. The blue areas represent phenotypes that are being selected against and that are not reproducing as successfully as the others.

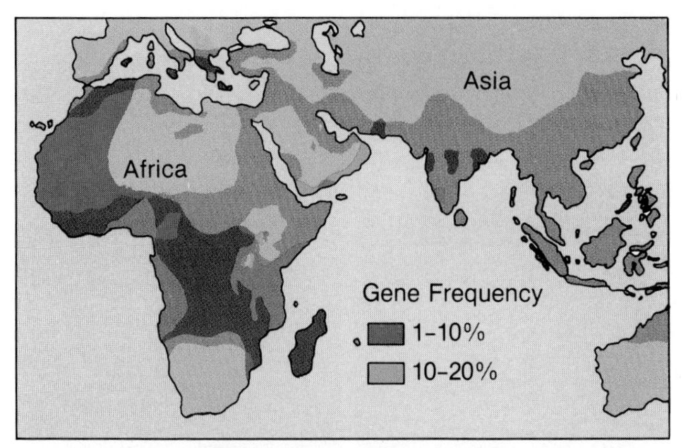

Figure 39-10 HETEROZYGOTE ADVANTAGE: MALARIA AND THE SICKLE-CELL ALLELE.
Heterozygotes carrying the recessive sickle-cell allele of the hemoglobin gene have protection against malaria. The distribution of malaria caused by the parasite *Plasmodium falciparum* in Africa, the Middle East, and India is shown in darker gray. The distribution of the sickle-cell allele, which confers protection against malaria, is shown in red and blue, depending on gene frequency. Note the overlap with malaria-infected areas. The recessive allele is selected for because individuals carrying it are more likely to survive malaria and live to reproduce.

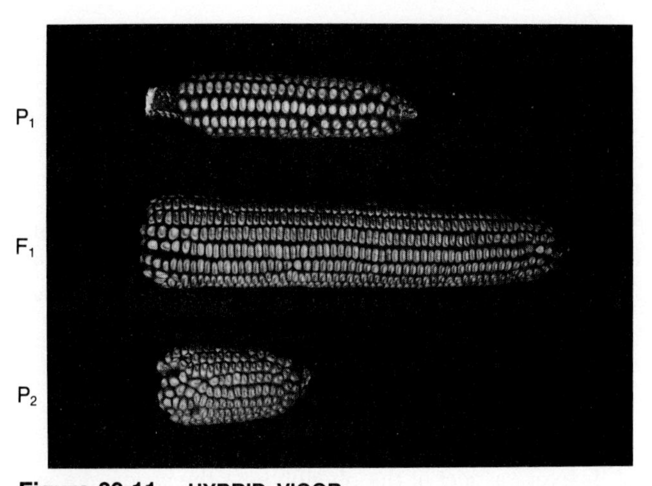

Figure 39-11 HYBRID VIGOR.
Two inbred lines of corn, P_1 and P_2, are crossed and yield an F_1 that is much larger as a plant and produces five times as much corn: The corn produced is also much larger, as this photo shows.

Heterozygote advantage, maintained by natural selection, may have indirect but very real biological "costs." The sickle-cell allele of the hemoglobin gene, for instance, provides resistance to malaria, but in homozygous form can lead to sickle-cell disease and even death. A population's **genetic load** is the sum total of alleles that, like the sickle-cell allele, yield some advantage when heterozygous but are lethal or deleterious when homozygous. Since at least a few organisms will be homozygous for the deleterious allele, genetic load is a price most kinds of organisms pay for the benefits the alleles confer to the heterozygote.

Frequency-Dependent Selection

A disadvantageous trait can be maintained not only by balancing selection but also by **frequency-dependent selection.** In such a case, a phenotype (and its genotype) will be selected for or against, depending on how rare or common it is in the population. One study revealed, for instance, that very common camouflaged snails in Uganda are subjected to much heavier predation by birds than are snails of the same species with rare conspicuous phenotypes.

Just as with birds feeding on the peppered moth in England, the Ugandan birds seem to concentrate on a particular phenotype—in this case, the commonest one. Even though the conspicuous snail is easier to spot, the birds concentrate on the more abundant phenotype, presumably because a given amount of foraging yields more of the common snails than the rare ones. A case like this demonstrates that the fitness of an individual with one particular genotype or allele is not independent of other individuals, which may attract predators, consume nutrients, absorb light (in plants), or have other positive or negative effects.

Biologist Lee Ehrman and her colleagues at the State University of New York demonstrated another example of frequency-dependent selection in fruit flies. The researchers collected two groups of flies of a single species, one group from Texas and another from California. When they mixed the groups in unequal proportions, males of the smaller group (the "rare mates") were more successful than the common males of the larger group at finding mates. Ehrman called this phenomenon **rare-mate advantage,** and it is temporary; once rare mates become more common owing to successful reproduction, they lose their special attraction. This phenomenon, which probably operates only in species with complex behaviors, provides insurance against the loss of alleles from a population's gene pool. Thus, the rarer the males or females of a species become (to a point), the better chance they have as individuals of being reproductively successful.

Thus far, our discussion of selective processes shows that natural selection in its many forms can produce diverse effects on living organisms. Selective pressures may favor the status quo in a phenotype, a genotype, or an allele's frequency; or pressures may produce trends in one direction or another or may produce such increased diversity that a new species forms.

NEW VIEWS ON EVOLUTIONARY MECHANISMS

Today's researchers are asking new questions and using the powerful techniques of molecular biology to answer them. Let's consider a few of the more intensely debated questions about evolution.

The Meaning of Adaptation

Is every aspect of the phenotype—the shape of a leaf on a tree or the nose on a human, the precise color and pattern on a butterfly wing or a flower blossom—adapted for some aspect of survival? One answer to this question is yes: No protein, no appendage, no process will persist or retain its special character in the absence of selective pressure. However, some evolutionary biologists argue against this view, saying that much of what is present in any one individual is not selected "for" or "against"; rather, some characteristics are neutral vis-à-vis natural selection and so cannot correctly be regarded as adaptations. Many regard the size and shape of the human nose or the utility of the small toes on a pig's leg (which do not touch the ground) as examples of neutral features that are not adaptations (Figure 39-12).

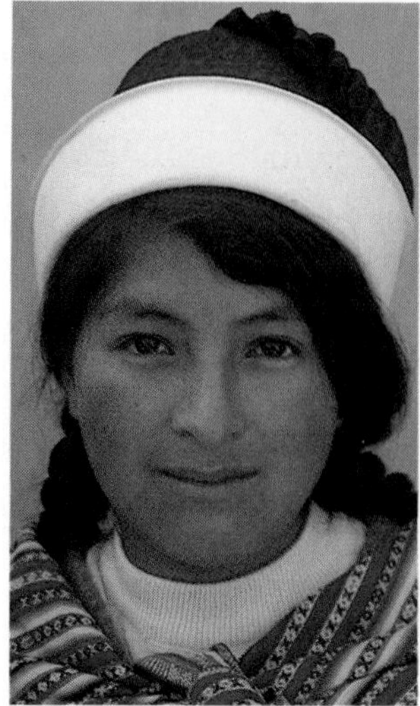

Figure 39-12 SOME ADAPTATIONS MAY NOT BE SUBJECT TO NATURAL SELECTION.
Genes and developmental processes may determine the size of a nose; yet as long as this, or any appendage, can carry out its basic function, the specific size and shape may not be subject to natural selection.

The Neutralist-Selectionist Debate and Molecular Evolution

A scientific argument called the *neutralist-selectionist debate* centers on whether all genetic variation is the result of natural selection. According to biologists in the "selectionist" camp, natural selection affects the many small variations in the amino acid sequences of protein molecules that show up during electrophoresis (see Chapter 38). They contend that such variation may contribute to heterozygote advantage and the resultant vigor of heterozygotes. On the other side are the "neutralists," who believe that changes in the amino acid sequences of protein molecules are mainly a matter of chance. Neutralists hypothesize that most of the random mutations in the genes coding for proteins are adaptively neutral; that is, they do not affect the function of the protein or the fitness of the organism carrying the mutations. If they are correct, natural selection would function neither to actively remove nor to accumulate such alleles in a population.

Biologists are designing and carrying out molecular-level studies to explore the merits of these alternative hypotheses. One intriguing set of observations involves cytochrome *c* (see Chapter 7) and other proteins in eukaryotic organisms. By comparing the amino acid sequences in cytochrome *c* molecules from horse and snake cells and from some other mammal-reptile pairs, researchers have detected about 15 differences between the sequences. This corresponds to roughly one amino acid change—presumably through mutation—per 17 million years, since mammals and reptiles diverged from a common ancestor in the Paleozoic era about 265 million years ago (Figure 39-13). Similar comparisons of cytochrome *c* sequences between two mammals, such as a horse and an ape, show about five amino acid changes—again, approximately one change per 17 million years, since the two orders of mammals diverged, perhaps 90 million years ago. These data imply that amino acid changes accumulated in cytochrome *c* at a slow and relatively constant rate. While the rate for other proteins—such as the small peptides clipped from fibrin during blood clotting (fibrinopeptides)—may be faster than that for cytochrome *c*, the rate for each protein is constant over time and can serve as an "evolutionary clock" for timing events during evolution (see Figure 39-13).

Studies on nucleotide substitutions in DNA rather than amino acid changes reveal some possible evolutionary mechanisms. In most animal species, a small proportion—only 5 to 10 percent—of the total DNA codes for protein and is clearly subject to natural selection. The rest of the DNA consists of introns and multiple copies of nucleotide sequences with no known functions.

Since so little of the DNA actually codes for proteins, it is no surprise that even closely related species show numerous differences in the nucleotide sequences in the remaining 90 to 95 percent of the DNA. Chimps and humans, for instance, have about 60 million DNA sequence differences, but they amount to just 1 percent of their bases, and the vast majority of these have no effect on phenotype. Biologist Roy Britten from the California Institute of Technology points out that the DNA from different people may vary by as many as 5 million nucleotides and that for every birth, hundreds of new nucleotides probably arise.

Britten and other molecular biologists have carefully studied the nucleotide changes in regions that do not code for proteins. One of their most surprising findings is that among different organisms, nucleotide substitution rates can vary fivefold: They found the fastest rates of change in sea urchins, insects, and rodents and the slowest in primates and some birds. They speculate that rates are fastest where large numbers of DNA replication events take place in the germ line, such as during the many generations of *Drosophila* each year or the sea urchin's annual huge gamete production. In mammals and birds, DNA repair mechanisms (which correct some mutational events) are probably more efficient and result in fewer changes accumulating in the nonprotein-coding portion of DNA. Studies such as Britten's show clearly that neutral change has gone on continually in all species' DNA and that the rate can vary among species. Biologists have yet to reconcile these findings of neutral mutation over evolutionary time and the seemingly constant rate at which amino acids are substituted in proteins with more traditional views of natural selection and the factors that enable it to operate, such as variations in climate, predators, competition, and local geology.

Returning to the neutralists and selectionists, neutralists who use the apparently constant rate of protein evolution as an "evolutionary clock" argue that random changes lead to alterations in gene frequencies (one per 17 million years or so for a molecule like cytochrome *c*). For them, selective processes play a minor role at best. But selectionists argue that evolutionary clocks based on amino acid sequences are not fine-tuned; that is, they do not reveal the huge number of mutations that occur in structural genes but that are selected against and thus do not persist. In fact, many selectionists interpret exceedingly slow rates of change—such as one per 17 million years—as evidence of continual, intense natural selec-

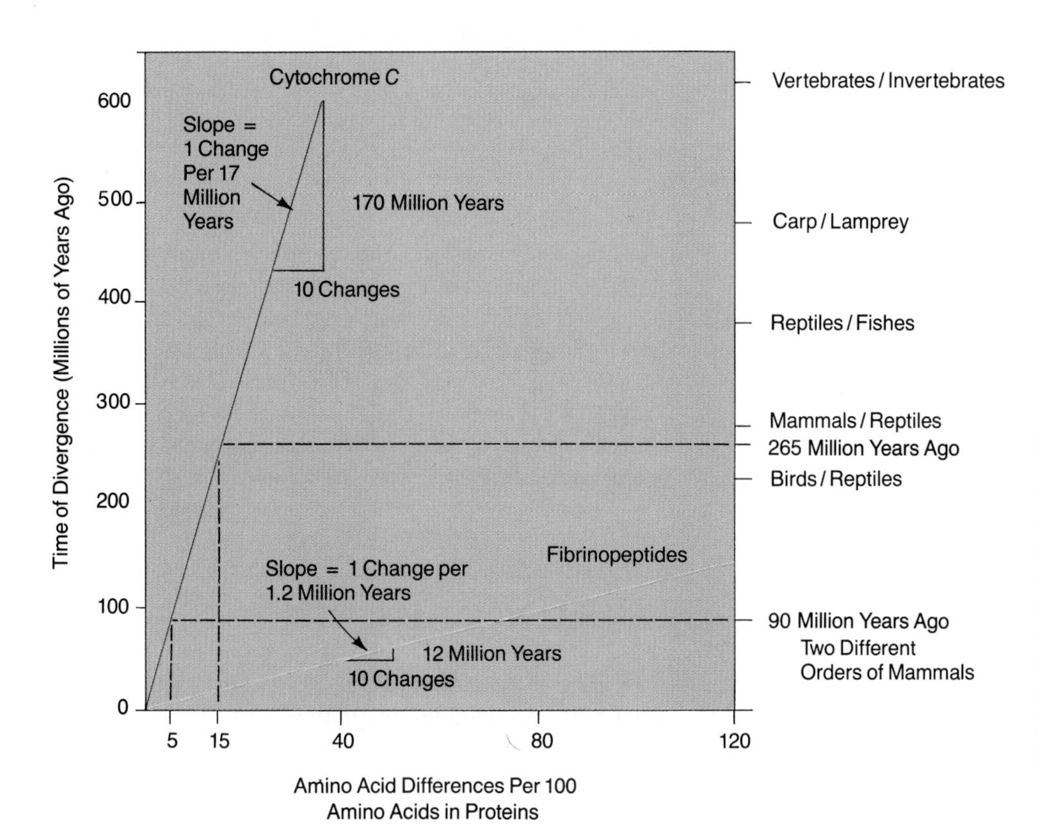

Figure 39-13
THE MOLECULAR EVOLUTIONARY CLOCK: MEASUREMENTS BY AMINO ACID AND NUCLEOTIDE SUBSTITUTIONS.
The number of amino acid changes per 100 amino acids in two proteins are plotted against geological time, with the approximate times the different pairs of animal types diverged from common ancestors indicated by arrows. Thus, the graph shows that there are about 30 amino acid differences in each 100 amino acids when cytochromes of vertebrates and of insects are compared; there are about 15 differences per 100 amino acids when cytochromes of mammals and of reptiles are compared.

tion: Most mutations yield abnormal proteins and so are lethal.

Groups of Genes as Units of Selection

Modern evolutionary biologists are weighing yet another fundamental question: In what sense are genes or groups of genes acted on by selection? Geneticists have assumed that the gene is the unit of evolution, yet the individual—with its full phenotype and underlying genotype—is widely considered the unit of natural selection.

In a case like sickle-cell anemia, natural selection seems to be operating at a single genetic locus. But in other instances, it seems to operate on groups of genes. For example, biologists have noted instances where combinations of alleles at two separate loci occur in the same genomes more often than would be expected by chance. Significantly, the two loci in each case code for enzymes that interact in biochemical pathways or for structural proteins that function together, such as actin and myosin, or tubulin and dynein. In such cases, selection may have favored a certain combination of alleles for the two genes because these two amino acid sequences allow the two interacting proteins to function optimally together.

A much different, more complex, and more puzzling case of multigene selection involves higher-order phenomena, such as animal behavior. In bird calls, for example, groups of genes *and* groups of organisms may serve as the units of selection. Many birds issue calls with specific meanings, and a feeding call, a flight call, or a warning call can be very similar from one species to the next (Figure 39-14). Such bird calls are inborn, or instinctive, in contrast to birds' mating "songs," which may be learned. Even a bird deafened as a hatchling can perfect calls of its species—having never heard them. Calls are obviously inherited in the same way as feather color or beak shape, and many genes must be required to code for the brain circuitry involved in a specific call.

Bird calls that warn other flock members of danger may be unfavorable to the individual that makes them, since they attract the attention of the predator to the caller himself. Many biologists argue that any set of alleles that causes its bearer to carry out such an "altruistic" act would likely be quickly eliminated if genes were the exclusive units of selection. The fact that such "unselfish" traits are common in many species of social animals requires, according to this view, the action of **group selection**—the selective favoring of gene combinations that enhance the survival chances of a group (a breeding population).

Despite this plausible set of arguments, other biologists interpret birds' alarm calls as favorable to the caller. Since the alarm caller enlists the aid of others to flee together, they suggest, the caller's chances of being snared by a predator are thereby reduced. In addition, the alarm caller may warn its offspring of danger, thereby improving their chances of surviving to propagate the caller's own genes. We return to these issues in Chapter 45. For now, the concept of natural selection acting on groups of genes or groups of organisms remains an intriguing but unproved hypothesis.

Figure 39-14 BIRD CALLS: INHERITED SIGNALS COMMON AMONG SPECIES.
These sound recordings show the great similarity of the alarm calls of three species of British birds in response to a hawk seen flying nearby. The high-pitched, lengthy calls are believed to be hard to trace to their source. Calls such as these are dependent on complex brain functions, yet they apparently are not learned in most species; they are present because of the way the brain circuits develop and so are attributable to the activities of many genes.

LOOKING AHEAD

Despite the unanswered questions, the facts at hand about natural selection, adaptation, and fitness help explain how organisms survive and evolve. In Chapter 40, we shall see how these processes can lead to the appearance of entirely new species.

SUMMARY

1. Natural selection can result in changes in gene frequencies because the individuals carrying variant alleles have differential reproductive success. Natural selection is a major contributor to evolution because it promotes the adaptations that permit differential survival and reproduction.

2. The *fitness* of an individual is a measure of the relative likelihood that the individual will contribute its genetic information to the next generation.

3. *Normalizing selection,* or stabilizing selection, works against atypical phenotypes that arise in populations and maintains the genotypic status quo.

4. *Directional selection* brings about changes in phenotype, such as darker pigmentation.

5. *Diversifying,* or *disruptive, selection* results in two or more phenotypes coexisting in the same population, often a result of strong pressure against the intermediate, or average, phenotypes.

6. Two mechanisms help maintain genetic variability in a population: *heterozygote advantage,* or *hybrid vigor,* and *frequency-dependent selection.* The latter occurs when the relative fitnesses of the genotypes in a population vary according to their frequency, often giving a reproductive advantage to rare phenotypes.

7. According to selectionists, all variations in protein molecules have evolved as the result of natural selection; according to neutralists, many variations may be the result of random mutations, with no effect on survival and reproductive success.

8. Recent evidence suggests that natural selection may operate on many levels of biological organization, from single allele to groups of genes, to individual organisms, and finally, to groups of individuals.

KEY TERMS

Allen's rule
balanced selection
Bergmann's rule
directional selection
diversifying (disruptive) selection
fitness
fixed allele
frequency-dependent selection
genetic load
group selection
heterozygote advantage
hybrid vigor
normalizing selection
rare-mate advantage

QUESTIONS

1. What is fitness?

2. List the various types of selection, and indicate whether each is likely to increase or decrease phenotypic diversity.

3. Contrast heterozygote advantage and genetic load. Why is one so advantageous and the other so disadvantageous?

4. Suppose you irradiate a large population of bacteria with ultraviolet (UV) light, using a dose that kills 90 percent of the bacteria. What determines which individual bacteria will survive and which will die? Repeating this procedure several times, you find that only 10 percent of the bacteria survive each irradiation. But eventually, with only a small remaining fraction of the original population, the same dose of UV light kills only 20 percent of the bacteria instead of 90 percent. What may have happened? If you grow the survivors into another large population and irradiate them, would you expect 10 percent survival, or higher? Explain.

5. Consider a population of mammals in which individuals homozygous for a lethal recessive allele die at birth. Assume that the phenotype of the heterozygotes is identical to that of homozygous normals and that the mutation frequency is very low. Would it take just one generation to eliminate this allele from the population? A few generations? Many generations? Why?

6. Consider a population of birds in which codominant alleles occur at one locus and in which only the heterozygotes can breed. The homozygotes of both kinds are shunned by all, including other homozygotes. What will the frequencies of the two alleles be in the population after one generation? After ten generations? What will the genotype frequencies be for the alleles A_1 and A_2?

7. Biologists usually think of selection as acting on the reproducing (or nonreproducing) individual and its phenotype. What raises the possibility that group selection occurs?

ESSAY QUESTION

1. How, as the agent of natural selection, could you obtain a population of brown arctic hares? Drought-resistant maple trees? Flowers pollinated by very small hummingbirds, but not large ones? Use other organisms or physical conditions in your selection regime.

For additional readings related to topics in this chapter, see Appendix C.

40 The Origin of Species

The ordinary naturalist is not sufficiently aware that, when dogmatizing on what species are, he is grappling with the whole question of the organic world and its connection with a time past and with man.

Charles Lyell, *Scientific Journals* (1856)

In Charles Darwin's day—the Victorian England of the nineteenth century—people liked to gather plant specimens, colored feathers, eggs, insects, minerals, and other natural objects. For many, this interest extended to fossil hunting, and devotees found hundreds of fossilized bones, teeth, shells, and exoskeletons in stratified rock beds around the British Isles.

Some observers concluded from the fossil evidence that thousands of species must have lived on the continents and in the ancient seas and become extinct. Just as the earth had been a continually changing geological panorama for millions of years, living things in that restless tableau seemed equally changeable rather than static. Such exciting fossil finds inspired Darwin and other biologists to wonder, How do species originate?

This chapter explores modern answers to the origin of species and lays out the principles that explain nature's

Frozen in time. Fifty million years ago, these herringlike fish swam in a shallow sea that covered what is now Wyoming.

astonishing diversity, past and present. To do this, the chapter builds on the background concepts covered in previous chapters, including phenotypic and genotypic variations, population genetics, and natural selection. Our current topics include:

- The modern biologist's definition of species
- The mechanisms that prevent gene exchange between populations
- How gene pools become isolated, making it possible for *speciation*—the origin of new species—to ultimately occur
- How gene changes contribute to speciation
- How modern evolutionary biologists explain macroevolution—the appearance of major taxa, such as families, orders, classes, phyla, and kingdoms

HOW BIOLOGISTS DEFINE A SPECIES

Carolus Linnaeus and other classical biologists distinguished a "species" as a group of organisms recognizable as a distinct and unique type by their morphological differences from all other life forms (see Chapter 19). According to this scheme, an individual's *morphological identity*, or physical correspondence with other individuals, suited it for membership in a given species.

Darwin added a powerful new insight into relationships in nature by suggesting that degrees of morphological similarity in general reflect degrees of *common ancestry*. Still, the physical traits taxonomists used to categorize organisms were often arbitrary. They might, for example, assign a butterfly to one species on the basis of wing pattern and another to a different species on the basis of body shape.

The introduction of population genetics in the early 1900s led to a new way of thinking about species. In particular, the concept of a shared gene pool (see Chapter 38) generated a biological definition of species that can be applied to a broad range of life forms. This definition describes a **species** as a group of actually or potentially interbreeding populations that are reproductively isolated from other such groups. The "actually" refers to organisms that are members of a population in which breeding is, in fact, taking place. And "potentially" means that individuals could exchange genes if given the opportunity, even though they might never actually do so in nature.

The key to today's biological definition of species is **reproductive isolation:** Members of a species can interbreed with each other but cannot breed with organisms belonging to another species. For instance, a crab and a spider, both arthropods, can never interbreed, nor can a rose and a cherry, both members of the rose family. Reproductive isolation thus provides a precise standard for determining whether related organisms belong to the same species. Note, however, that this standard applies only to sexually reproducing organisms; for the earth's asexually reproducing creatures, including nearly all prokaryotes, many plants, and some animals, species must still be defined by observable differences in physical traits, such as morphology or biochemistry.

Reproductive isolation can also serve as a biological rule of thumb for judging how far two populations have diverged. Researchers, for example, crossed similar-looking populations of a species of Latin American fruit fly *(Drosophila paulistorum)* and generated at least six subgroups of the tiny flies, all resembling each other but tending strongly to choose mates from within their own group. When members of different subgroups do mate, the result is an evolutionary compromise: The hybrid males are sterile, but the hybrid females are fertile. While gene flow among the subgroups is limited, it is still possible. So biologists still classify the six subgroups as *D. paulistorum*, although they may diverge into separate species some day.

PREVENTING GENE EXCHANGE

How does one subgroup of a species lose entirely the capacity to breed and exchange genes with other members of that species? Physical aspects of the environment as well as biological features of the organisms may lead to a restricted exchange of genes. Let's see how.

The Role of Isolating Mechanisms

Several mechanisms can prevent the effective exchange of genes between reproducing individuals. A **prezygotic isolating mechanism** blocks the formation of zygotes; for one reason or another, the male gamete never comes into contact with the female gamete or cannot successfully fertilize it. By contrast, a **postzygotic isolating mechanism** affects a zygote that has already formed. Such a zygote may develop into a nonviable embryo or into a hybrid adult that is sterile. Table 40-1 summarizes the many forms of these isolating mechanisms.

Table 40-1 REPRODUCTIVE ISOLATING MECHANISMS	
PREZYGOTIC	*Blockage of formation of viable zygotes*
Ecological Isolation	Life-styles using different parts of the environment prevent reproduction
Behavioral Isolation	Behavioral differences result in failure to mate successfully
Mechanical Isolation	Structural or molecular blockage of formation of zygote
Temporal Isolation	Reproduction at different times
POSTZYGOTIC	*Failure of development or of reproduction in individual or descendants*
Hybrid Inviability	Death of hybrids before reproductive capacity is attained
Hybrid Sterility	Inability of hybrids—even though vigorous—to reproduce
Hybrid Breakdown	Reduced viability or infertility in second or later generations

Prezygotic Isolating Mechanisms

Prezygotic isolating mechanisms that prevent successful reproduction can be ecological and involve environmental blocks to mating, or they can be behavioral and involve activities that influence the timing and physiology of reproduction. Two species of native bushes with small leathery leaves and vivid blue flowers grow in the California coast range. The geological upheavals that lifted and twisted these mountains into their current shapes left rocky alkaline soils and rich loams lying side by side in many places. Interestingly, the bushes called *Ceanothus jepsonii* grow only on the alkaline soils, while the species *C. vamulosa* can tolerate soils of many different types. The two species illustrate **ecological isolation;** each is adapted for a specific soil environment, and their genetic differences have grown so great over time that they can no longer successfully crossbreed, even though the plants may grow within a few feet of each other.

In Hawaii, two species of fruit flies—so closely related that an observer cannot tell the females of each group apart—have evolved ecological isolation *par excellence.* Both feed on a scarce resource, the oozing sap of *Myoporum* trees. Rather than one group of flies being selected against and eliminated over the years, one population of flies has taken the high ground, feeding and breeding on oozing tree trunks, while the other makes use of sap puddles on the forest floor. The evolutionary result is two separate, surviving species that have become genetically distinct, apparently because of their preferences for the different food locations.

Biologists have discovered a prime example of **behavioral isolation** in the complex light patterns that fireflies display (Figure 40-1). A female will mate only with a male whose flight path (straight, zigzagging, or looping) and whose flashing light pattern (short regular bursts of light, for example, or long sustained gleams) reveal the visual signaling pattern characteristic of her species. The male displays result from *sexual selection* leading to differential reproductive success (see the box on page 735). Evolution has produced an awesome collection of visual, chemical, and acoustic signals, all contributing to sexual selection and the greater reproductive success of certain individuals. The peacock's tail, the elk's antlers, insects' sex pheromones, frogs' mating croaks, and showy courtship displays such as those of the amazing riflebird (Figure 40-2) are all examples.

Mating signals like these restrict the exchange of genes to members of the same species or occasionally even to members of a single population. Populations of white-crowned sparrows around San Francisco, for example, sing their species' song in different "dialects," and females recognize and preferentially mate with males that sing with the correct "accent." Although the females could breed successfully with males from neigh-

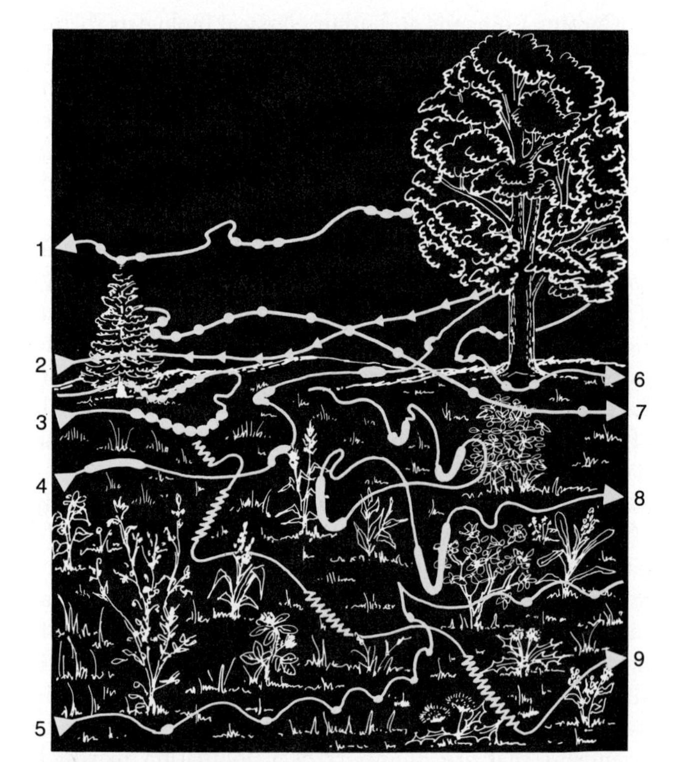

Figure 40-1 FIREFLY LIGHT PATTERNS: A BEHAVIORAL ISOLATING MECHANISM.
A female firefly will mate only with a male whose flight paths and flash patterns match those characteristic of her species. The patterns of the males of nine species of *Lampyridae* fireflies are illustrated. The dots, ovals, and elongated squiggles and curves represent the flashes; the interconnecting lines trace the flight path.

SORTING OUT SEXUAL SELECTION

In many animals, the males and females are so different they scarcely look like members of the same species. They can vary in size, color, and body parts (antlers, crests, hairy manes, showy tails, shoulder patches, and the like), and these contrasting features influence natural selection: Males compete for mates, and evolution favors those characteristics that enhance the tendency to attract the opposite sex.

In one type of sexual selection, called intrasexual sexual selection, males battle physically for the chance to mate with females. This phenomenon has produced elk with massive antlers and rams with powerful butting muscles. A bull elk's body must expend a great deal of extra energy to produce these heavy appendages each year and carry them around in the woods. But this investment can be repaid in full genetically when a bull with large antlers is able to defeat other males, copulate more often, and leave more offspring.

In the second type of sexual selection, called epigamic selection, the female is an active selective agent, choosing mates from a field of genetically variable males. An intriguing example is the "gardener" bowerbird of New Guinea. Both sexes are a drab brown, but the male builds an elaborate bower of sticks and twigs and adorns it with bits of fungi, colorful flower petals, bright fruits, iridescent butterfly wings, shimmering snail shells, even multihued plastic bottles and poker chips—all to entice a receptive female (Figure A).

Figure A
A male bowerbird (*Ptilonorhynchus violaceus*) at the front of his decorated bower in an eastern Australian forest.

Recent research suggests that epigamic selection—female choice—can also explain the huge array of extravagantly shaped male copulatory organs among worms, insects, mollusks, certain fish, snakes, rodents, and bats. As with many other physical traits, intricate, sometimes flamboyant, structures can arise from the drive for reproductive success.

Figure 40-2 SEXUAL SELECTION AND THE MALE RIFLEBIRD.
The dramatic courtship display of the male riflebird (*Ptiloris victoriae*) attracts females and is a form of behavioral isolation. Female riflebirds mate only with males that have the "correct" blue and green colors and that display wing or tail feathers expanded as shown here.

boring populations in Berkeley or Marin, they seldom do. By cutting off the flow of genes over hundreds or thousands of years, this kind of prezygotic behavioral isolating mechanism might lead to full reproductive isolation and the formation of new species.

Sometimes the differences that isolate populations or species prevent mating physically—so-called **mechanical isolation.** In flowering plants, different flower shapes can discourage cross-pollination by preventing an animal pollinator from accessing pollen-covered anthers or a sticky stigma. Other flowers—blooms that smell like rotten meat, for example—attract only specific pollinators whose instinctive behaviors ensure that appropriate gene exchange will take place. Animals that carry out external fertilization provide a good example of mechanical isolation at the molecular level. In the sea stars, sea urchins, sand dollars, and various other echinoderms, differences in the fit of the molecules that bind sperm and egg together may inhibit cross-fertilization. (Botanists have discovered equivalent differences in pollen binding in flowering plants.) In animals that carry out internal fertilization, differences in genital size or shape may prevent mating, or molecular incompatibilities may keep the sperm from successfully fertilizing the egg.

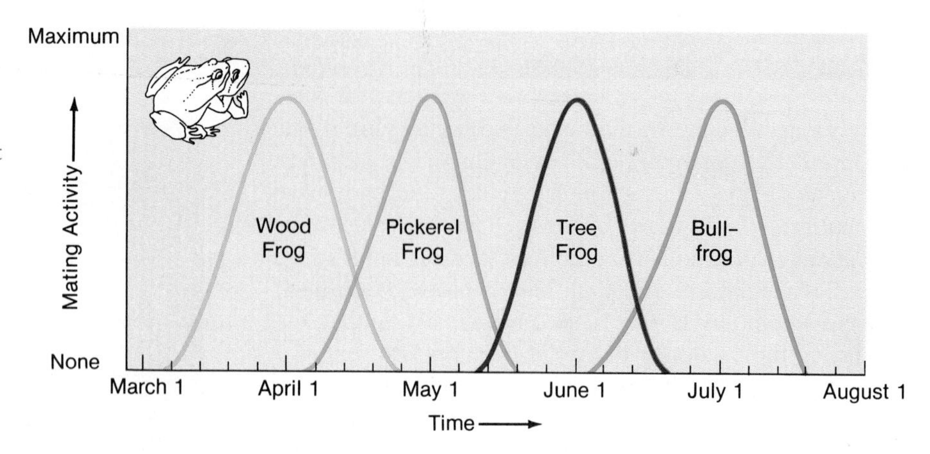

Figure 40-3 TEMPORAL ISOLATION IN BREEDING OF AMPHIBIANS. Four kinds of frogs are seen to have maximal reproductive behavior at different times; this helps to ensure that interbreeding is reduced or absent.

Finally, environmental cues that trigger mating and reproduction can be timed so differently that related species are prevented from interbreeding; this kind of prezygotic isolating mechanism is called **temporal isolation.** Figure 40-3 shows the peak breeding seasons for four kinds of frogs that may live in adjacent overlapping areas. Hormone surges that control egg maturation and sperm production come in four different months; thus, when wood frogs are primed for mating, tree frogs are unreceptive, and so on. Likewise, the *Pinus vadiata* and *Pinus muricata* trees growing on California's wind-swept Monterey peninsula have remained genetically distinct because one species sheds its pollen in early February, while the other does so in late April.

All of these prezygotic strategies work by preventing the members of diverging populations or distinct species from interbreeding. Occasionally, however, mating still occurs, and postzygotic isolating mechanisms usually come into play.

Postzygotic Isolating Mechanisms

When individuals from different species do interbreed, hybrids sometimes result, but they usually die before reproducing successfully (**hybrid inviability**) or are sterile (**hybrid sterility**). Both conditions function as postzygotic isolating mechanisms, preventing gene flow between species.

Sheep and goats can mate, for example, but the hybrid embryo that develops from the fertilized egg is inviable; it dies at an early stage of development because the parental genomes usually fail to coordinate. Alternatively, hybrid seeds may grow into a sterile plant that cannot produce mature flowers, or perhaps it makes flowers that produce no functional gametes. A mule is a hybrid that fails to produce viable gametes because its chromosomes (half from a donkey, half from a horse) do not pair and cross over correctly during meiosis (Figure 40-4).

While crosses between members of two genetically distant populations of a single species can result in *heterozygote advantage,* or *hybrid vigor* (see Chapter 39), crosses between closely related but distinct species can end up in *hybrid breakdown*—decreased reproductive success in later generations. When geneticists cross two fruit fly species (*Drosophila pseudoobscura* with *D. persimilis*), the first-generation females are healthy and deposit as many fertilized eggs as do purebred flies. Second-generation hybrid daughters, however, produce weak offspring that tend to be sterile.

The Role of Clines in Speciation

Just as reproductive isolating mechanisms can lead to new species, the chance distribution of populations

Figure 40-4 MULES AND HYBRID STERILITY. Mules, the hybrid offspring of a female horse and a male donkey, are invariably sterile. While mules are useful to humans for performing heavy hauling work, they cannot breed and become a true species.

throughout certain geographical ranges may also contribute to speciation.

Populations of a species are sometimes separated by a river or a deep canyon, but more often, they are spread out along a broad geographical range and display a gradual change in characteristics from one part of the range to another. Biologists call such gradual changes **clines.** Each local population along a cline seems to evolve adaptations for its particular environment. Gene flow from adjacent populations counteracts this specialization to local conditions; nevertheless, significant differences can still build up along a cline, as the following examples reveal.

The grass frog, *Rana pipiens*, displays a typical cline; eleven distinctive geographical races of the frogs inhabit localities between Vermont and Louisiana. Although members of adjacent frog races sometimes interbreed, more distant populations are reproductively isolated. Likewise, salamanders of the genus *Ensatina* in California's coastal mountains and Sierra Nevada range exist in seven subspecies (Figure 40-5). (A **subspecies** is a genetically distinct population of a species with at least some distinctive characteristics.) Each salamander subspecies displays its own slightly different body shape, skin color pattern, and adaptations to the specific ecological conditions in its part of the species' range. As in the grass frogs, subspecies of *Ensatina* sometimes interbreed successfully where their ranges overlap, but animals from opposite ends of the cline are reproductively isolated and cannot mate at all. They would appear, then, to be different species, but at least in theory they can still exchange genes, since interbreeding can occur anywhere along the cline. Like the interlocking gears in a gearbox, the "first" and "last" populations are linked by the intermediate groups between them.

Clines provide dramatic evidence of how species begin to form, and clines apparently arise as an original population spreads from its site of origin and adapts to new and different geographical and ecological settings. If one or more of the intervening groups in a cline were to leave or die out, geographical isolation would halt any gene flow between the adjacent groups, with full reproductive isolation an almost certain result.

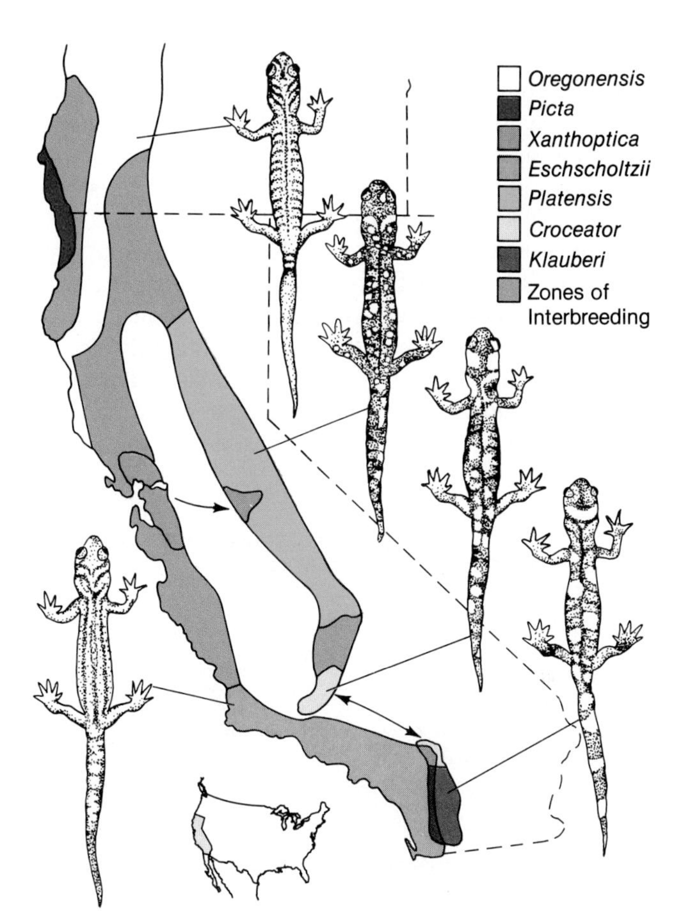

Figure 40-5 THE CLINE OF THE *Ensatina* SALAMANDERS.
Seven subspecies of this salamander form a cline; subspecies can interbreed with variants living in neighboring or overlapping ranges, but not with individuals from more distant parts of the cline. The cline extends through the coastal and Sierra Nevada mountains of California. The arrows indicate sites where subspecies have "jumped" across geographical regions in which no *Ensatina* subspecies live.

BECOMING A SPECIES: HOW GENE POOLS BECOME ISOLATED

How do gene pools become sufficiently isolated for new species to form? Ernst Mayr at the American Museum of Natural History has proposed an **allopatric** (geographical) **model of speciation** (Figure 40-6) to describe how species can originate. (*Allopatry* means "living in different places," whereas its opposite, *sympatry*, means "living in the same place.") Let us say that populations of an existing species of squirrels (Figure 40-6a) become separated, by chance, by a physical or geographic barrier, such as the Grand Canyon, carved by the Colorado River (Figure 40-6b and e). Once gene flow has been halted, more and more genetic differences accumulate between the two allopatric populations (Figure 40-6c). If another chance event removes the physical barrier, the two original populations could once again become sympatric. By now, however, the differences will have become so significant that at least some postzygotic isolation will occur: Hybrids of the two populations will not survive or will not reproduce (Figure 40-6d). Now, if hybrids produced by the two parent strains prove to be less fit, they will be selected against; prezygotic isolating mechanisms will arise, and the two former populations will function as distinct species (in this example, as the Albert and Kaibab squirrels).

Biologists have found a great deal of evidence to support Mayr's concept of allopatric speciation and its twin postulates: that physically separated populations diverge genetically and that they then become fully distinct by natural selection. Many questions remain, however, about the amount and kind of changes in the genetic material that are ultimately responsible for the differences among species.

THE GENETIC BASES OF SPECIATION

Researchers can use gel electrophoresis (see Chapter 39) to estimate how widely populations vary in their genetic makeup. They can also use the technique to compare variations in structural genes among different sorts of organisms, between newly formed species, or within members of one population. Such comparisons show that as one moves up the taxonomic ladder from subspecies to species to genus, pairs of organisms have decreased **genetic identity,** or a smaller proportion of

(a) One Initial Population

(b) A Geographical Barrier Creates Two Populations

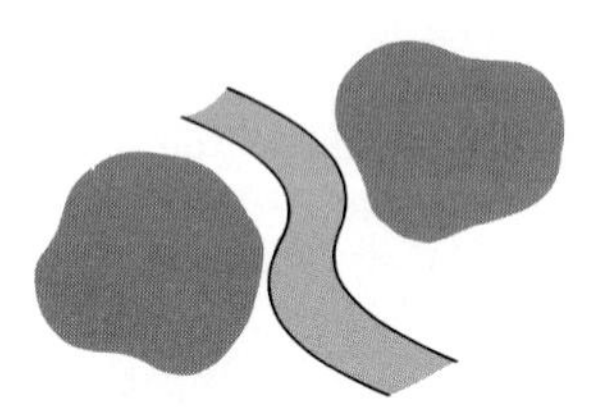

(c) The Two Populations Become Genetically Distinct

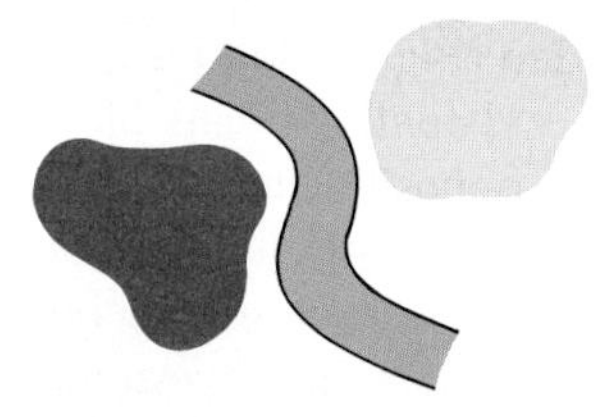

(d) Reproduction Isolation Continues Even When Geographical Barrier Disappears

(e)

Figure 40-6 ALLOPATRIC MODEL OF SPECIATION.
Allopatric speciation occurs when (a) a single population of a species becomes (b) separated by physical or geographical barriers, such as the formation of an island, a canyon, or a mountain range. (c) Without gene flow between the populations, genetic differences accumulate as natural selection operates on each group independently. Eventually, the two populations may become unable to interbreed. (d) If the physical barrier between the populations is removed, the two populations may coexist but not reproduce; they have become different species. (e) The Albert squirrel (*Sciurus alberti*; left of photo) and the Kaibab squirrel (*Sciurus kaibabensis*) are two different species believed to have been one population before the formation of the Grand Canyon. The two species now live on opposite sides of this deep chasm, which, as seen here, is indeed a forbidding barrier to cross. Allopatric speciation has occurred; the squirrels have developed differently and can no longer interbreed.

identical genes in the two groups. Such differences support Mayr's hypothesis that gene pools diverge gradually as populations slowly become reproductively isolated.

Within the primate order, humans, chimpanzees, gorillas, monkeys, and lemurs diverged from each other between about 8 and 30 million years ago, and the organisms look distinctly different from each other. The groups, however, remain amazingly alike in their structural genes. While there are 60 million nucleotide sequence differences between chimp and human DNA, for example, most or all of these are neutral mutations occurring in regions that do not code for proteins. Gel electrophoresis may overlook some actual genetic differences because it measures only the *products* of structural genes. More importantly, however, evidence increasingly suggests that small changes in special regulatory genes may underlie many of the major physical changes that lead to new species and higher taxonomic groups.

Regulatory Genes and Evolutionary Changes

During an organism's development, regulatory genes coordinate blocks of structural genes, ensuring that a bird's wing, for example, develops anterior, posterior, dorsal, and ventral surfaces in proper orientation to the body and that a crayfish or fruit fly develops the correct number of body segments and appendages. Furthermore, many biologists believe that changes in the activity of such regulatory genes may explain important evolutionary alterations in organisms and may perhaps explain speciation itself.

In both chimps and humans, fetuses develop at about the same rate and with amazingly parallel features. After birth, however, human development slows down, while chimps continue to undergo rapid and marked changes in their physical characteristics. A human skull, for example, changes relatively little as a person matures from infant to adult, but a chimpanzee's skull undergoes a striking transformation (Figure 40-7). It is as though such transformation has been truncated in humans, so that we retain the embryonic proportions in the adult stage. According to one hypothesis, the on and off timing of specific regulatory genes during the young animal's development may explain the disparities in adult chimp and human skull shapes.

Some biologists also think that chance mutations in regulatory genes might lead to whole new species—"hopeful monsters" that, although drastically altered, are still adapted to their surroundings and survive to reproduce.

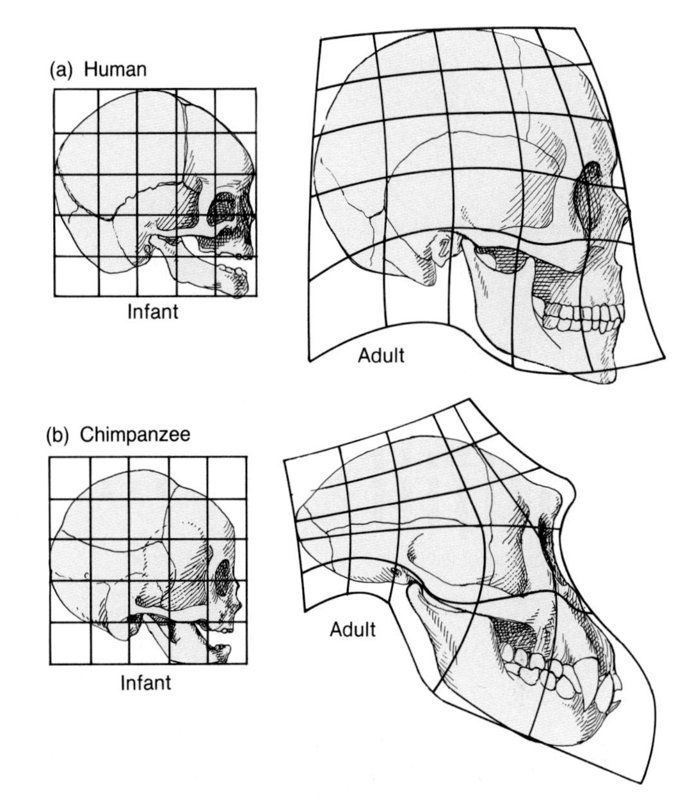

Figure 40-7 DEVELOPMENT OF HUMAN AND CHIMPANZEE SKULLS: REGULATORY GENES AND EVOLUTIONARY CHANGE.
Human and chimpanzee fetuses develop at about the same rate and have very similar skulls. (a) The human skull does not change a great deal in form from infancy to adulthood. (b) But the chimpanzee skull continues to change form as the animal matures, resulting in a markedly different morphology between infant and adult. Since humans and chimps share most of their structural genes and protein sequences, these differences are probably explained by the action of the regulatory genes during development.

Chromosomal Changes

Most biologists have traditionally believed that species arise as genetic changes (mutations) accumulate gradually over many generations and long stretches of time. Mounting evidence, however, shows that mechanisms other than mutation can split populations genetically at a much faster pace. In plants, for instance, isolating mechanisms may arise in a single generation through **polyploidization**—the sudden multiplication of an entire complement of chromosomes. The multiplication of chromosome sets in a single species (e.g., $2n$ going to $4n$ directly) produces *autopolyploids,* while the combination of chromosome sets from different parental species produces *allopolyploids* such as the new gilia species in

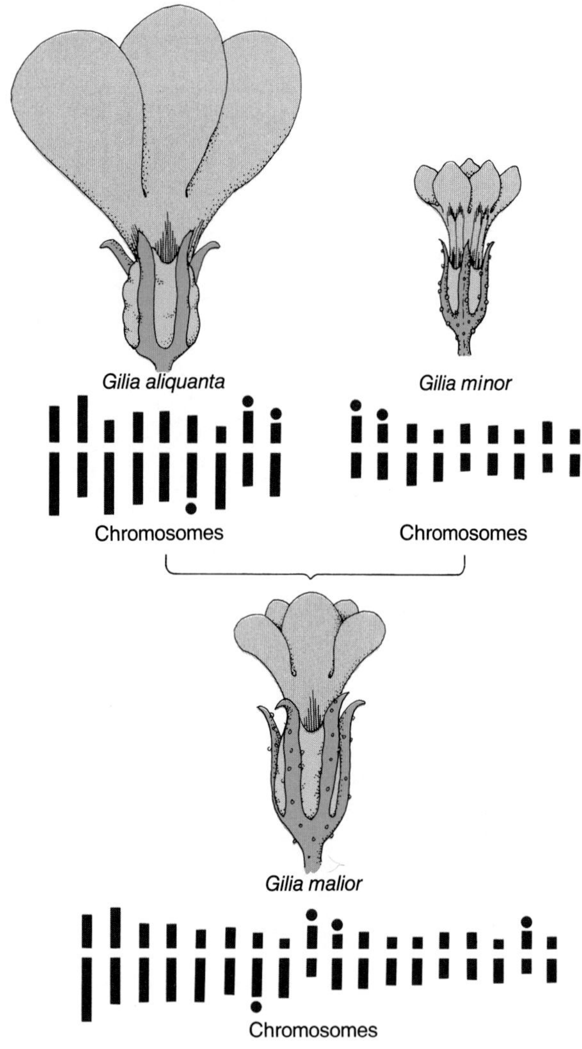

Figure 40-8 INSTANT SPECIATION: POLYPLOIDY IN *Gilia*.
Plants can form "instant species" through polyploidization, the duplication of an entire set of chromosomes. *Gilia malior*, for example, occasionally undergoes allopolyploidization, the doubling of chromosomes from two parent plants of different species. The 18 chromosomes in *G. malior* are no longer exact copies of the 9 from each parent species, reflecting changes in the chromosomes subsequent to the "speciation" event.

Figure 40-8. Polyploidization circumvents the problems of hybrid sterility: Since each chromosome set is doubled, matching pairs of chromosomes are present during meiosis. Once a polyploid population has been established, it automatically forms a new reproductively isolated species, with the increased chromosome number now functioning as the diploid state. Many of our economically important plants, including tobacco, cotton, wheat, sugarcane, and coffee, are polyploids.

Chromosome sets occasionally become polyploid in animals' somatic cells—for instance, human liver cells become polyploid during puberty—but polyploidization normally does not occur in germ line cells of the ovary or testis. Animal chromosomes can, however, become rearranged in ways that lead rapidly to new species. Among certain neighboring populations of flightless grasshoppers in Australia, for example, a random change in chromosome structure seems to have arisen, leaving those grasshoppers possessing the new structure more fit for a certain portion of the original species' range and effectively isolating two groups of grasshoppers into two species, even though no geographical barrier prevented gene flow. This is a case of **sympatric speciation,** speciation in populations inhabiting the same geographical range and not experiencing geographical isolation.

Evolutionary biologists also believe that the splitting of the giant panda from the bear lineage involved unusual chromosome rearrangements. While giant pandas resemble bears in many ways (Figure 40-9a), they eat primarily bamboo, rather than a variety of meat and vegetable matter as do bears. Also, unlike bears, pandas' forelimbs and shoulders are much larger than their hind limbs and hindquarters; they have five fingers and an opposable thumblike digit on their forepaws; and they "bleat" rather than roar. Significantly, giant pandas also have 42 chromosomes, each with the centromere located in the middle, while bears have 74 chromosomes, most with the centromere at one end (Figure 40-9b).

Researchers have been able to match every bear chromosome with an arm from a giant panda chromosome and have concluded that most giant panda chromosomes represent two bear chromosomes fused at their centromeres. Using DNA hybridization studies, electrophoretic analysis of over 50 proteins, and immunological comparisons of blood proteins, biologists have been able to deduce that chromosome fusion occurred some 20 million years ago as the giant panda line arose. Researchers have yet to prove, however, that this fusion was the key to the panda's divergence. Still, changes in chromosome structure and the process of polyploidization and sympatric speciation accentuate a fundamental biological fact: Species can originate in a variety of ways, and the alternative pathways leading to reproductive isolation can take thousands of years or just a single generation.

EXPLAINING MACROEVOLUTION: HIGHER-ORDER CHANGES

We have seen how the mechanisms of speciation can lead to new life forms, but how do new higher-order taxa, such as genera, families, or phyla, arise? Biologists

(a)

Centromeres at Ends

Bear Chromosomes

Centromere Fusion

Panda Chromosomes

Centromeres in Middle

(b)

Figure 40-9 THE GIANT PANDA: A CASE OF FUSED CHROMOSOMES.
(a) The giant panda *(Ailuropoda melanolenca)* resembles a bear, yet it has many distinctive features. (b) Most of the panda's chromosomes seem to be fused pairs of chromosomes that are found singly in brown bears and other bear species.

never speak of "phylogenization" or "genusation" to describe how these larger groups may form. One reason is that the higher-order taxa that encompass "all echinoderms," "all animals," and "all members of the genus *Felis*" are abstract concepts involving groups that do not possess a common gene pool or trade genes during reproduction, as do the tangible members of a species. Another reason is that since Darwin's day, most evolutionary biologists have believed that the principles that generate new species also underlie **macroevolution**—the major phenotypic changes that occur over evolutionary time and that result in the origin of new taxonomic levels above species. To distinguish between the two, biologists sometimes use the term **microevolution** to refer to the small-scale changes in gene frequencies that generate species.

Macroevolutionary events include major changes in physiology and alterations in basic body design. The acquisition of a closed circulatory system in animals, for example, allowed larger body size; the evolution of stomata and guard cells allowed plants to control gas exchange in leaves; vascular plants arose from nonvascular ones; the reptilian foreleg gave rise to the bird wing; simple flowers led to compound ones; and so on. One way biologists attempt to trace the course of such changes is by reconstructing the evolutionary past.

The Fossil Record

At best, the fossil record provides an incomplete image of the evolution of the earth's flora and fauna. Most fossils consist of the hard body parts left behind when an organism died, such as animal bones and shells and the tough cell wall materials of plants, while some fossils show burrows, tracks, or other impressions left in once-soft sediments. Some lines of descent, like those of the modern horse or the transition from reptile to mammal, are well documented by an abundant fossil record; but in many lineages, the evolutionary picture is cloudy because field workers have yet to find intermediate forms.

There are several reasons for the spottiness of the fossil record. First, probably two-thirds of all the organisms that have ever lived lacked readily fossilizable parts. Like the invertebrate inhabitants of ancient seas, they had soft bodies, no rigid skeleton, and no teeth or other hard structures, and they died out leaving no trace of what must have been a lavish array of life forms. Second, even an organism with hard parts—a long-extinct reptile, say, or a Paleozoic fern—would not have been preserved in stone unless its remains were quickly entombed by mud, silt, resin, or some other stable protective covering in which it could lie undisturbed for millenia. Finally, the fossils that did form were often destroyed, altered, or moved when the sedimentary rocks around them eroded, when overlying rocks crushed them with immense pressures, or when chemical reactions dissolved them after they were uplifted into the surface soils.

While paleontologists have found areas like Beartooth Butte, in southern Montana, where conditions were ideal for fossil formation (Figure 40-10), there is still the problem of interpreting the fossil record—in particular, determining the length of time represented in a given sample of fossils. The organisms fossilized at a single site could have lived hundreds, thousands, or even millions of years apart and thus represent a "time window" with huge and uncertain dimensions.

(a)

(b)

Figure 40-10 OPTIMAL SITES FOR FOSSIL FORMATION WERE RARE.
(a) Beartooth Butte, in southern Montana, is composed of sedimentary rocks that have yielded many early fish fossils, especially from the Devonian period (see Chapter 24). Conditions for fossilization were ideal when these rocks were laid down in an ancient inland sea. Now, all of the surrounding sedimentary rocks are eroded away, so that the butte stands alone above underlying Paleozoic era rocks that contain still older fossils. Erosion now exposes the well-preserved fossils to the keen-eyed paleontologist. (b) This fossil fish of the species *Priscara serrata* was discovered in the neighboring state of Wyoming in the Green River Formation.

Estimations of Fossil Age

The best technique for determining the relative ages of fossils within reasonably precise limits is *radioisotope dating,* in which isotopes serve as "radioactive clocks" for measuring the passage of time. To radioactively date a given fossil, a scientist must know: (1) the *half-life* of the isotope being measured (the length of time it takes for one-half of a given amount of the isotope to decay); (2) how much of the isotope was originally present in the fossil or the rock containing the fossil; and (3) how much of the isotope is left. Under ideal conditions, radioactive dating allows paleontologists to establish the age of a fossil with about 98 percent accuracy.

One of the best-known radioisotopes is carbon 14 (^{14}C). The ratio of ^{14}C to the more common ^{12}C makes up a low but constant fraction of the carbon atoms in living matter. Since scientists know the half-life of ^{14}C (5,730 years), they can accurately date a fossil by measuring how much ^{14}C it contains relative to ^{12}C. Unfortunately, the half-life of ^{14}C is short in geological terms, and little is left in a fossil more than 40,000 years old. Thus, to date more ancient fossils, the researcher must trace isotopes with much longer half-lives, such as uranium 235, with a half-life of 713 million years, or uranium 238, with a half-life of 4.5 billion years.

Reconstruction of Evolutionary Lines

Once a researcher has established the ages of groups of fossils, the next step is to build a **phylogeny**—a description of the *lineages,* or lines of descent, among the plants and animals as they lived from one era to the next. Paleontologists and biologists have constructed a remarkably detailed lineage for the horse (genus *Equus;* Figure 40-11). The early "horses," of the genus *Hyracotherium,* were about the size of a small German shepherd and had several toes (review Figure 39-8). For more than 50 million years, a series of evolutionary stages led to large animals with a single horny toenail (hoof). Altogether, scientists have reconstructed a complex evolutionary tree with more than 20 genera of the horse family represented (only one of which, *Equus,* remains today).

Unfortunately, a spotty fossil record makes it difficult to build such complete trees for most lineages. Where many descendants still survive, however, biologists can infer likely phylogenies by comparing morphological features and analyzing the genes and proteins of living forms.

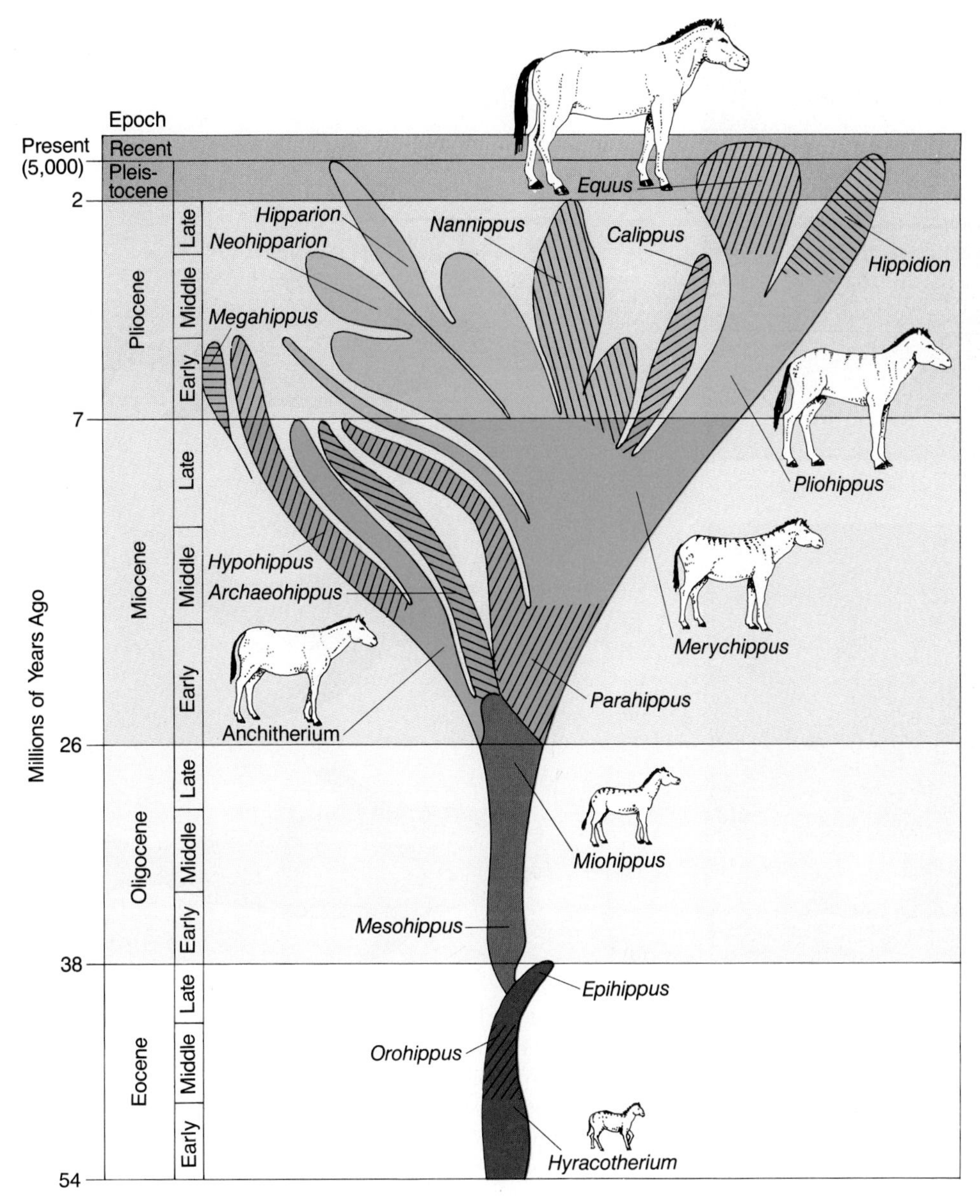

Figure 40-11 THE EVOLUTION OF HORSES. The fossil record for the phylogeny of horses is unusually complete. This figure shows the phylogenetic tree and an artist's renderings of what some of the ancient horses may have looked like. Although the overall trend is toward larger size, genera such as *Nannippus* became progressively smaller again. Note the great diversification of this family tree of horses, most branches of which became extinct.

Parallel, Convergent, and Divergent Evolution

To build a phylogeny, evolutionary biologists use similarities in body structures, biochemistry, reproductive strategies, and other features to trace lines of common descent. The typical patterns of evolution are called parallel, convergent, and divergent evolution, and they can be a bit confusing.

Porcupines are a good example of **parallel evolution.** Two separate species of the spiny mammals evolved independently in Africa and South America (Figure 40-12). More than 70 million years ago, when those two landmasses were joined, the porcupine's ancestral form scurried about, resembling a large, furry rat. When the two continents drifted apart, each rodent group evolved independently, but in parallel ways in generally similar environmental surroundings. Even though the two groups have been evolving separately for 70 million years, both New World and Old World porcupines are remarkably similar in size and shape and are covered by sharp, hollow spines.

In the pattern of descent called **convergent evolution,** two (or more) distantly related lineages grow *more* alike as they evolve similar adaptations. The porpoises, sea lions, seals, whales, and other marine mammals each evolved with a sleek, streamlined body; backswept dorsal appendages; and strong, flexible tails for generating

(a)

(b)

Figure 40-12 PARALLEL EVOLUTION IN PORCUPINES. (a) The American porcupine, *Coendou prehensilis,* and (b) the Old World porcupine, *Hystrix africaeaustralis,* have a common ancestor that lived 70 million years ago, before South America and Africa drifted apart. The porcupines have evolved independently on separate continents to modern forms that are amazingly similar. This is an example of parallel evolution.

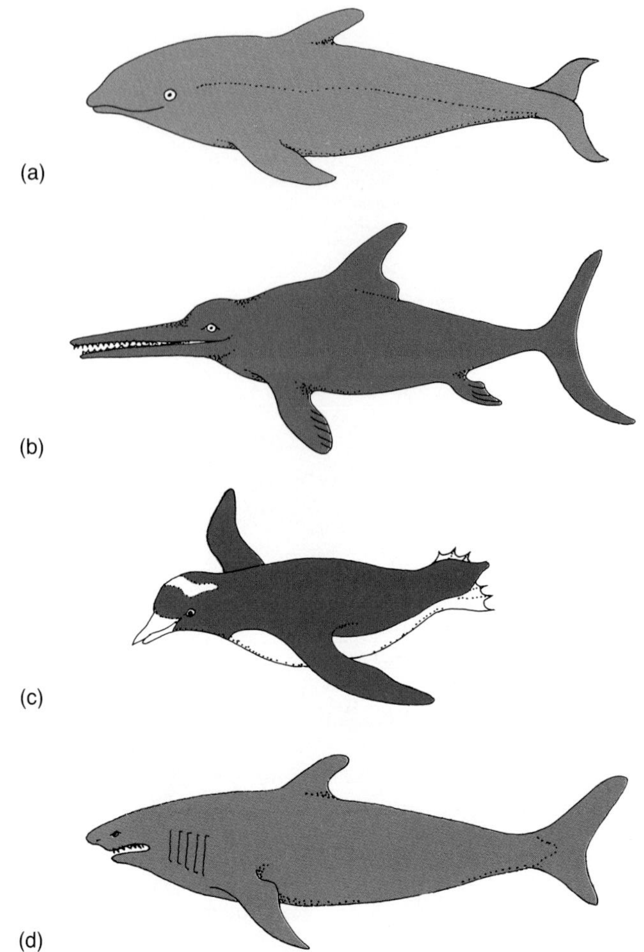

Figure 40-13 CONVERGENT EVOLUTION. Convergent evolution is particularly striking in the body types evolved for rapid swimming in an aquatic environment. (a) The porpoise, a mammal, (b) *Ichthyosaurus*, an extinct reptile, (c) the penguin, a bird, and (d) the shark, a fish, evolved independently, yet each has a shape that creates minimal drag, allows fast swimming, and employs the anterior pectoral appendage as a control surface during swimming. (See also Chapter 24.)

thrust in the water (Figure 40-13a). They share those traits with the extinct reptilian *Ichthyosaurus* (Figure 40-13b); with the penguins, auks, cormorants, and certain other aquatic birds (Figure 40-13c); and with powerful predatory fishes like the shark (Figure 40-13d).

In cases of parallel and convergent evolution, the evolutionary biologist must decide whether a particular similarity in form or function is an example of **homology**—derivation from a common ancestor, such as a human's arm, bat's wing, and whale's flipper from the forelimbs of early four-legged reptiles (Figure 40-14)—or an example of **analogy**—independent evolution of features with similar functions, such as the squid's gliding fin and the wings of birds and insects. Homologous structures in related organisms are built upon the same basic genetic program, beginning in the embryo. Analogous structures, in contrast, serve the same function, but have no common genetic basis; birds, flies, and bats can all glide through the air, but have no common winged ancestor and have traveled separate evolutionary paths.

One of the most common evolutionary patterns biologists reconstruct from the fossil record is a tree with branches splitting off from an ancestral trunk. This sort of **divergent evolution,** or **radiation,** has produced many (perhaps most) modern species and genera, including Hawaiian honeycreepers, with their distinctive bills and habits of feeding on nectar, fruits, insects, or seeds (depending on bill type; Figure 40-15), and the Hawaiian silversword plants (see the box on page 746). Radiation

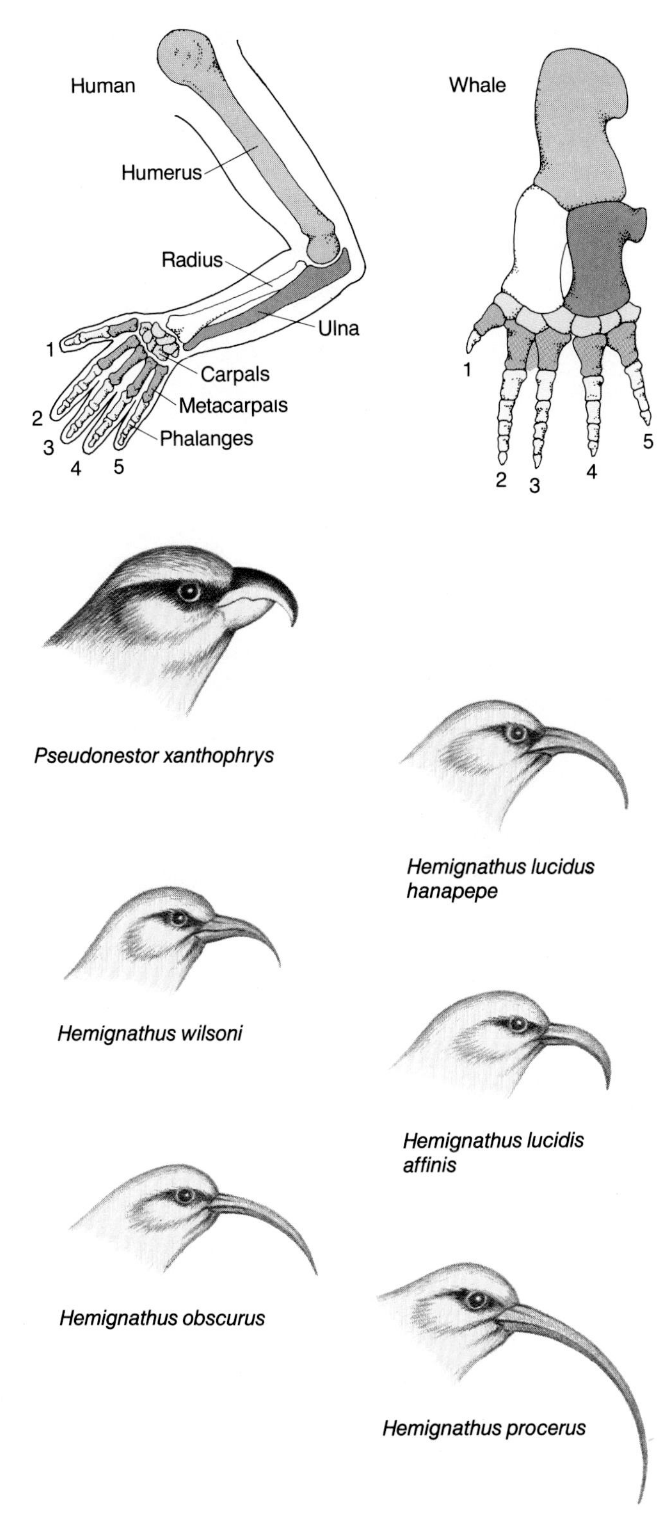

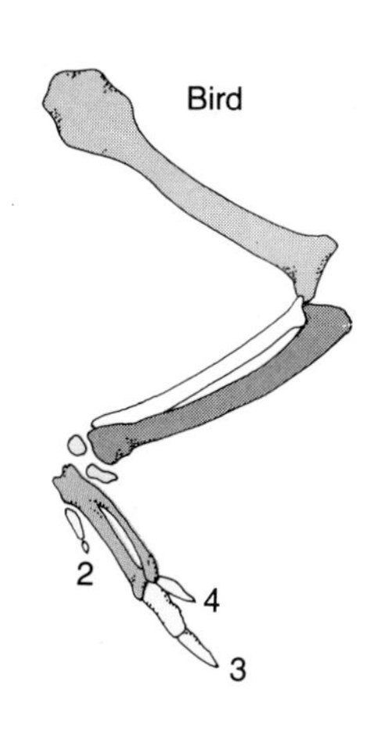

Bird
2
4
3

Figure 40-14 HOMOLOGOUS STRUCTURES.
The forelimbs of a human, whale, and bird are homologous. These limbs derived from that of a common ancestor. The homologous bones—the humerus, radius, ulna, carpals, metacarpals, and phalanges—can be compared in each animal, but the limbs serve vastly different functions in each.

Figure 40-15 ADAPTIVE RADIATION OF HAWAIIAN HONEYCREEPERS.
Divergent evolution of many branches off a single line is called radiation (or adaptive radiation, because the evolution is an adaptation to a variety of living conditions). More than 20 species of Hawaiian honeycreepers (6 are shown here) have evolved from only a few founders in the Hawaiian Islands. Each species has its own particular bill shape and feeding specialization.

almost always reflects an expanding repertoire of adaptations to new living conditions.

Extinction

Ten thousand years ago, vast herds of horses thundered across the inland prairies of North America. Suddenly, the horses vanished, perhaps hunted to extinction. Wild horses did survive in Europe, North Africa, and elsewhere, and Spanish explorers reintroduced the animals to both North and South America in the 1500s. The demise of populations in specific habitats has been documented. Fossil evidence suggests another example: About 340 million years ago, echinoderms called blastoids disappeared from shallow marine waters near present-day North America and Europe. Then, after millions of years of apparently diminished abundance and diversity worldwide, blastoids reappeared in shallow seas. If *all* the horses or blastoids had disappeared and no relatives survived elsewhere, **extinction**—the permanent loss of species (or higher taxa)—would have taken place.

Extinction has been the eventual fate of nearly all species—of the 4 billion species that have lived on earth, only about 2 million are present today. One well-studied extinction involved the saber-toothed tiger (Figure 40-16).

By analyzing all sorts of marine and terrestrial fossils, paleontologists have determined that during the earth's protracted history, both a continuous background level of individual extinctions and a number of *mass extinction* events have taken place. They recognize five mass extinctions, two of which resulted in enormous species loss; one at the end of the Permian period removed 96 percent of all then-living marine invertebrate species, and another at the end of the Cretaceous period claimed 60 to 75 percent of all marine species.

The Cretaceous event, lying at the boundary between the Mesozoic and Cenozoic eras, has been a key to un-

SILVERSWORDS: HAWAII'S EXEMPLARS OF ADAPTIVE RADIATION

Adaptive radiation is the height of evolutionary drama: A single ancestral lineage experiences a burst of change and gives rise to many new species. This concept helps explain the marvelous diversity among living things, but until recently, biologists had to rely mostly on fossils to trace instances of adaptive radiation. Now, however, researchers have found living proof of adaptive radiation in the 28 species of silverswords that grow in Hawaii's lush rain forests and on the islands' volcanic mountainsides.

The related species in the "silversword alliance" show a remarkable diversity of shapes adapted to virtually every type of Hawaiian habitat. The compact, rosettelike silversword *Argyroxiphium macrocephalum* (much like the plant in Figure A) grows in high, dry, barren volcanic soil at altitudes of nearly 10,000 ft on the massive Haleakala Crater on the island of Maui. The carpet of silvery hairs on each swordlike leaf probably inhibits water loss and screens out the intense sunlight at high altitudes. The silversword *Dubrutia menziesii* flourishes nearby but grows as densely branching shrubs, not as ground-hugging rosettes. A third silversword species, this one a denizen of moist lowlands, has large, broad leaves and a stem more than 1 ft in diameter, and it stands more than 20 ft tall!

Researchers studying the evolutionary relationships between silverswords have found that *all* of the species in the alliance can form hybrids, no matter how unlikely the mating of a 20 ft "tree" and a 2 ft shrub. What is more, all the hybrids are at least somewhat fertile, and slight differences in the degree of fertility help reveal evolutionary distance between the species.

In other studies, geneticists have found that two-thirds of the species have 28 chromosomes in each somatic cell, but the remaining third have only 26 chromosomes. Researchers Gerald D. Carr, Robert Robichaux, and Donald W. Kyhos identified one 28-chromosome species as the probable ancestor of the nine 26-chromosome species, all of which, interestingly, grow on the geologically younger islands of the Hawaiian chain.

New information on the evolution of fruit flies on the Hawaiian Islands has additional relevance for the study of silverswords. The new evidence suggests that as each new volcanic island in the chain emerged above the Pacific waves, flies from older islands must have been blown to it and formed colonies. Plants, too, could have colonized the first-formed islands, and then their seeds, carried by wind, birds, and ocean currents, might have landed on each new island as it emerged. This sort of serial colonization might have contributed to the silverswords' impressive radiation and suggests that the diversification may have taken much longer than the currently estimated 6 million years.

Figure A
A silversword adapted for the high, dry, sunbaked, and windswept volcanic heights on Hawaii.

Studies like these of adaptive radiation at work should give evolutionary biologists an unparalleled view of one primary pattern of change in the living world.

derstanding mass extinctions. One recent theory, for example, claims that the dinosaurs died out suddenly at this boundary, perhaps as a result of a giant meteorite hitting the earth. Newer studies on the earth's climate patterns, however, suggest that relatively abrupt climate transitions can arise and contribute to biotic crises and extinctions. In fact, there is considerable evidence that dinosaur extinctions began in earnest some 8 million years earlier and that 18 dinosaur genera had already disappeared before the Cretaceous boundary. More-

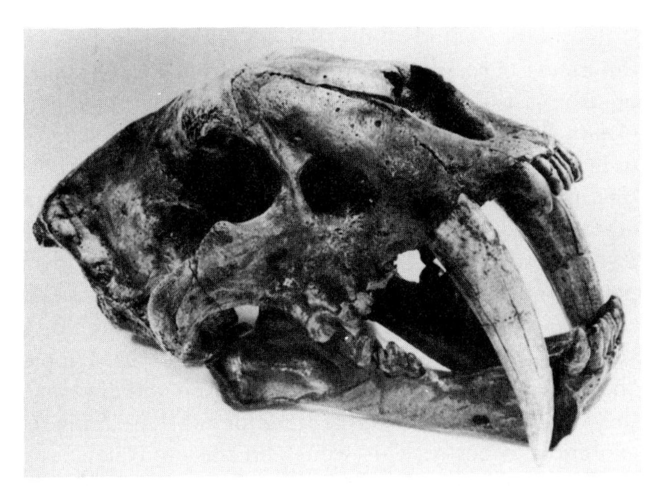

Figure 40-16 FOSSIL SKULL OF AN EXTINCT SABER-TOOTHED TIGER.
The saber-toothed tiger, *Smilodon californicus*, lived in what is now California until about 10,000 years ago. It weighed about 225 kg and probably preyed on animals such as mastodons and ground sloths. We can only speculate on the reason this fierce predator became extinct.

over, fossil remains show that 7 to 11 of the 12 surviving genera lived for at least another 40,000 years. The Cretaceous boundary is also marked by other climate-related events, including the appearance of ferns and then their replacement by many new angiosperm species. While many questions remain about mass extinctions, biologists are certain that they do clear the environment, in a sense, and allow many new species to evolve to fill the empty habitats.

Some observations raise the fascinating idea that mass extinctions may occur in cycles many millions of years in length (see Chapter 24). Other evidence and interpretation contradicts the idea. What is truly important is the realization that characteristics arising via natural selection and proving highly adaptive during normal times may become quite irrelevant in the face of global climate changes and subsequent mass extinctions. Species and their special adaptations may disappear not because of low fitness but because of a new environmental crisis never before experienced. For example, the widespread geographical distribution and richness of species within a biological group, which promote survival during normal times, offer no advantage during times of mass extinction. Extinction events tens of millions of years apart would involve such huge intervals that no individual could experience such events and be selected to survive them. Still, mass extinction events have undoubtedly had major effects on life's major evolutionary outlines. And ominously, we may be able to record the effects of the greatest mass extinction in history—the one our own species is causing by the continued cutting and burning down of large portions of the species-rich tropics. The box on page 747 describes the epidemic of extinctions our world now faces, and we will return to this serious concern once more in Chapter 41.

The Punctuated Equilibrium Theory

How fast do *major* evolutionary changes take place? Generations of scientists have believed that changes ac-

AN EPIDEMIC OF EXTINCTIONS: THE DISHEARTENING FACTS

- Biologists are not sure how many distinct species now inhabit the earth, but estimates range from 5 to 30 million.
- For the past 600 million years, an average of about one species has become extinct each year.
- Today's rate of extinction is at least 1,000 species per year.
- By the year 2000, the rate could rise to 36,500 per year, or 100 species per day.
- During our lifetimes, we could witness the extinction of one-third or more of all earth's species—one million or more.
- The planet's rain forests cover only 6 percent of the land surface, but contain 50 percent of the species. Human activities have already disturbed 55 percent of the original tropical rain forests to one degree or another, and an additional 100,000 square kilometers are cut and/or burned yearly to expose new agricultural land.
- While habitat preservation is the best single remedy for preventing extinctions, only 1 percent of the world's land area has been set aside so far as biological reserves to protect other species.
- During the next 10 years, the United States alone could lose up to 680 plant species, which is three times more than in the past 200 years combined.
- Poaching can also lead to extinctions. Between 1979 and 1987, ivory poachers cut populations of African elephants, the largest land animals, from 1.3 million animals to 750,000—a rate that will cause the species to disappear within 50 years. The rhinoceros, second largest land animal, is going the same way for similar reasons.
- To help prevent the epidemic of extinctions, learn about the human activities that destroy natural habitats and individual species, and avoid them when possible. Chapter 41 gives more details.

cumulate in species gradually over immense stretches of geological time. This theory has earned the name **gradualism.** In recent years, however, some biologists have suggested that evolution can proceed by "jumps"—radical changes over a short period of time, separated by long periods of stability. This new idea is called **punctuated equilibrium.**

Paleontologists have proposed these two contrasting phases—rapid change (the "punctuations") followed by equilibrium—to explain apparent gaps in the fossil record as well as instances in which certain life forms, such as the lungfish and the horseshoe crab, have remained essentially unchanged over hundreds of millions of years. The central tenet of punctuated equilibrium is that a lineage of organisms arises by some dramatic changes—say, the rapid acquisition of body segmentation in annelids—after which there is a lengthy period with far fewer radical changes taking place. The fossil history of one type of echinoderm, for example, the sea urchin, appears to reflect two "punctuations," one in the Ordovician period and another in the early Jurassic, with "equilibrium" times in between. To complement real examples from nature, researchers have created mathematical analyses showing that within small to moderate-sized populations undergoing typical Darwinian selection, random phenotypic variations can periodically lead to large transitions (punctuations) and thus to new types. This implies that special genetic, developmental, or ecological mechanisms are not needed to yield punctuations.

Many biologists have pointed out problems with the concept of punctuated equilibrium. First, the fossil record is frustratingly incomplete; are some or most "equilibrium periods" just plain gaps? Even with radioactive dating techniques, it is very hard to discriminate fossil ages precisely, and in some cases, errors of 5 million years in either direction might be as accurate as one can get. Such a time period may represent hundreds of thousands or even millions of generations in the life history of a group of organisms.

Interestingly, when an abundant fossil record is available, researchers tend not to see new groups arising as the result of macroevolutionary "bursts." There are, for example, an abundance of specimens representing trilobite evolution in the Ordovician period. A detailed study of about 15,000 fossils in eight genera over a period of 3 million years shows gradualistic patterns of change and no hints of punctuation points. An early trilobite fossil and a late one have very different morphologies and are classified as different species, yet the gradual change in intervening fossils suggests that the lineage is probably one species evolving slowly over time, not a second one arising from the first. Even to assign something as basic as a taxonomic name, a researcher must be fairly certain of rates and patterns of evolution.

In another noted study, P. G. Williamson examined thousands of fossil clams and snails from the Cenozoic era (the last 65 million years) and found that many species in the 13 lineages he studied remained unchanged in morphology over millions of years. "New" species did turn up where geological formations reflected the periodic drying up of lakes and other environmental stresses. Even these species changes, however, took place over periods of 5,000 to 50,000 years. Modern living snails and clams produce new generations in 6 months or a year; in the time Williamson allotted for speciation, about 20,000 generations could have lived and died, and this would have been ample time for substantial microevolution to occur. This would be the equivalent of a 1,000-year experiment with fruit flies or a 6,000-year mouse-breeding test in the laboratory! Clearly, the perspective we take influences the way we interpret facts. On the one hand, the equilibrium of some lineages of organisms may indeed be "punctuated" by environmental changes or other factors; on the other hand, gradualism characterizes much of the history of many species.

THE ROLE OF MICROEVOLUTION IN MACROEVOLUTION

Biologists are still grappling with key questions about macroevolution: Does some radical genetic process create the radically different new organisms that can lead to the evolution of whole new genera, families, orders, and higher taxa? Do punctuation points in the history of life, with their gross phenotypic changes, come at the time of a catastrophic environmental event that wipes out competing lines? Or do the known mechanisms leading to genetic variation and microevolutionary changes—nonrandom mating, genetic drift, migration, and the many types of gene and chromosomal change grouped as mutations—suffice to explain the founding of the major lineages we see represented in the earth's fossil history and among today's living organisms?

New fossil findings and reinterpretations of old ones continue to stimulate new theories about macroevolution. For example, studies show that marine invertebrates such as echinoderms and terrestrial mammals were far more diverse earlier in their evolutionary histories than they are now. Fossils from the Burgess shale deposits of British Columbia, for instance, show that during the early Paleozoic, there was far more diversity in body plans among animals with coeloms (true gut cavities) than among all of today's such animals combined. Does such early diversity suggest normal adaptive radiation of successful new forms? Or could there have been "ecological vacuums"—times of lessened competition during certain geological eras—that allowed a plethora of suboptimal species to survive for a while and then die out in later eras with stiffer competition and greater natural selection pressures?

Such questions remain as investigators, using traditional methods as well as the newer techniques of molecular biology, continue to study fossils, to explore novel genetic and developmental mechanisms, and to refine their understanding of microevolution. One day, no doubt, biologists will have a much clearer idea of how minute, steady changes, and perhaps "hopeful monsters" as well, contributed to the forming of whole new lineages.

LOOKING AHEAD

In the next five chapters, we move from evolutionary processes to the ecology of organisms and how they exist and interact in their specific environments. In the process, we will see what causes adaptations to persist, and how their presence explains so many basic biological characteristics of living species.

SUMMARY

1. *Species* are defined as groups of actually or potentially interbreeding populations that are reproductively isolated from other such groups.

2. *Reproductive isolation* can prevent gene flow and split interbreeding populations into genetically distinct species. *Prezygotic isolating mechanisms* block the formation of zygotes, while *postzygotic isolating mechanisms* result in nonviable or sterile hybrids.

3. Prezygotic isolating mechanisms include *ecological isolation*, *behavioral isolation*, *mechanical isolation*, and *temporal isolation*.

4. Postzygotic reproductive isolation may be the result of *hybrid inviability*, in which the organism dies as an embryo or before reproducing, and *hybrid breakdown*, in which hybrid offspring, often in later generations, are weak or sterile.

5. *Clines* are gradual changes in a species' phenotypic characteristics from one end of the range to the other. Local *subspecies* with distinctive characteristics may arise along a cline.

6. According to the model of *allopatric speciation*, a physical or geographical barrier separates two subgroups, and genetic differences accumulate until subgroups become reproductively isolated species.

7. Small changes in regulatory genes may be responsible for many of the major changes that mark the evolution of species and of higher taxonomic groups.

8. Rapid speciation in plants is typically due to *polyploidization*—the rapid duplication of chromosome sets. This can result in *sympatric speciation*—reproductive isolation arising even though there is no geographical barrier to prevent gene flow.

9. *Macroevolution* describes the major changes in phenotype that have occurred over evolutionary time and the origin of taxonomic groups above the level of species.

10. Researchers construct *phylogenies*, lines of organisms' descent over evolutionary time.

11. In *parallel evolution*, two or more lineages evolve along similar lines. In *convergent evolution*, similar adaptations evolve within very distantly related lineages. In *divergent evolution*, or *radiation*, lineages split off from a parental group and then diversify into related species. *Homology* refers to phenotypic characters based on the same basic genetic program in related lineages. *Analogy* refers to phenotypic characters that resemble each other in different lineages, but are not based on common descent or genetics.

12. *Extinction* is the complete loss of a species or group of species. At least five mass extinction events have occurred during the earth's history.

13. According to the *punctuated equilibrium* theory, evolution proceeds by jumps—radical changes over short periods of time—separated by long periods of stability. Gradualism refers to small-scale evolutionary changes in species, which ultimately might lead to reproductive isolation and speciation.

KEY TERMS

allopatric model of speciation
analogy
behavioral isolation
cline
convergent evolution
divergent evolution (radiation)
ecological isolation
extinction
genetic identity
gradualism
homology
hybrid inviability
hybrid sterility
macroevolution
mechanical isolation
microevolution
parallel evolution
phylogeny
polyploidization
postzygotic isolating mechanism
prezygotic isolating mechanism
punctuated equilibrium
reproductive isolation
species
subspecies
sympatric speciation
temporal isolation

QUESTIONS

1. Design a test to see whether two newly discovered populations of tropical rain forest orchids are distinct species, according to the modern biological definition.

2. Write a brief essay that gives examples of the various types of prezygotic and postzygotic isolating mechanisms.

3. Field biologists in Africa have discovered a cline of zebras with different stripe patterns. How could you test whether the obvious differences in color pattern correlate with differences that might result in speciation?

4. Distinguish between allopatric and sympatric speciation. Cite as examples populations suddenly separated by a Grand Canyon–sized chasm and populations on a remote Pacific island that is large enough to have a variety of local environments.

5. Define analogous structures and homologous structures, and give examples of both.

6. What is convergent evolution? What is parallel evolution? Give examples of both. How do the terms *analogy* and *homology* apply to convergent evolution and parallel evolution?

7. What is macroevolution? Give some examples.

8. How are fossils dated?

9. What is a radioactive half-life? The half-life of carbon 14 is nearly 6,000 years. In a particular sample, what fraction of the original radioactivity will remain after 12,000 years? After 24,000 years?

10. What kinds of observations does the punctuated equilibrium theory attempt to explain? Present some arguments for and against the theory. What kinds of data are needed to test the theory?

ESSAY QUESTION

1. Explain how gradualism and microevolutionary processes could lead to a group in a new taxonomic category higher than species—say, a different family.

For additional readings related to topics in this chapter, see Appendix C.

41

Ecosystems and the Biosphere

. . . Rachel Carson [in Silent Spring*] was alerting the world to what has been called the fundamental principle of ecology, namely: We can never do merely one thing, because the world is a system of fantastic complexity. Nothing stands alone. No intervention in nature can be focused exclusively on but one element of the system.*

Garrett Hardin, *Bulletin of the Atomic Scientists* (January 1970)

For many, the word *ecology* has come to connote "the environment," with particular emphasis on dwindling tropical forests; disturbed weather and climate; spoiled rivers and lakes; buildups of toxic waste, air pollution, and acid rain; and other evidences of global human disruption. The study of ecology does indeed include such very real concerns, and we will touch upon them repeatedly in this and the next two chapters. But it also has a more precise and scholarly meaning: **Ecology** is the study of how organisms interact with their physical and biological environments and how those interactions influence the distribution and abundance of living things.

Such interactions necessarily begin with a constant source of energy, and virtually all living organisms depend on the flow and harvesting of life-supporting solar energy through photosynthesis or food gathering. Eco-

Tahiti: A multitude of ecosystems, from coral reefs to mountain rain forests, each with its own abundance of species.

logical interactions then extend to chains of dependencies between various species and, ultimately, to complex communities of living things, all dependent for their very survival on particular aspects of the environment. Just as the earth's early landscapes helped shape the evolution of the first living cells (see Chapter 19), our planet's current climatic zones, latitudes, and soil and water types continue to influence how and where organisms can live. Life forms, in turn, alter their surroundings, and the mutual influences create regional and sometimes global cycles and a stable balance of nature.

To fully understand the interactions between organisms and their environment, we must take a systematic approach to ecology, moving from *ecosystem*, the broadest view of biological and physical interactions (covered in this chapter), to *communities*, the interacting populations of different species within an ecosystem (see Chapter 42), and finally to *populations*, individuals of a single species within the community (see Chapter 43).

In this chapter, we discuss:

- Ecosystems—communities of organisms interacting with their physical environments
- The biosphere—the earth's collective life-supporting environment and all its organisms
- The biomes—the major ecosystems, including deserts, rain forests, and tundra
- Each of the general habitats within the biosphere: air, land, and water
- How energy flows through ecosystems, including the food chains that operate within each community of organisms
- How water, carbon, nitrogen, and phosphorus cycle through living things and their environments
- How an aquatic ecosystem, such as a freshwater lake, operates

ECOSYSTEMS

The most complex level of biological organization is the ecosystem. An **ecosystem** is a complete life-supporting environment, including the entire community of interacting organisms, their physical and chemical environments, energy fluxes, and the types, amounts, and cycles of nutrients in the various habitats within the system. Ecosystems are both dynamic and incredibly complicated. An ecosystem can be as small as a single island, lake, or meadow within a forest, or it can be as large as the biosphere—the earth's life-supporting soils, seas, and atmosphere and all the organisms associated with them.

The boundaries between ecosystems are seldom sharp, and they tend to grade into each other (Figure 41-1). Nor are the ecosystems *truly* self-contained: Energy must enter; gases, nutrients, and water cycle in and out; and organisms come and go in space and time. This blurring of boundaries reflects the fact that every small ecosystem exists within larger ecosystems, which in turn exist within the biosphere itself.

Figure 41-1 GRADING OF ECOSYSTEMS INTO EACH OTHER.
Here, near the tree line high in the mountains, one can find a variety of ecosystems, including these grassy, flower-strewn meadows, open conifer forests, and higher rocky surfaces with little besides lichen surviving.

Nevertheless, ecologists tend to study smaller, circumscribed ecosystems so they can better analyze and understand the interplay of solar energy, moisture, tides, mineral nutrients, and other physical factors within the communities, populations, and individual organisms coexisting in space and time.

THE BIOSPHERE

The earth's surface region—its lands, its waters, and the air above them—forms the **biosphere.** This life-supporting environment for all living creatures is a fragile sheath of habitable regions about 14 miles thick surrounding the huge rocky globe (Figure 41-2) and providing the myriad **habitats,** or actual places where organisms live.

The earth and its biosphere are like an unbelievably large, complex machine with literally billions of "ingredients" and "parts" (carbon, nitrogen, minerals, liquid water, decomposers, photosynthetic organisms, etc.). As with any machine, remove one crucial factor—*energy*—and it all stops. Let us see how sunlight provides a constant flow of usable energy and establishes the climates that determine both the nature and the distribution of all living things.

Figure 41-2 THE BIOSPHERE: EARTH'S THIN FILM OF LIFE.
Earth's air, land, and water form a collective life-supporting environment for all living things.

The Sun as a Source of Energy

Most of the energy that flows through the earth's ecosystems emanates from the sun as light and heat and falls upon our atmosphere at a steady rate called the **solar constant,** equal to about 10^{10} calories of solar energy per square meter (or about 1 million calories per square centimeter) every year (Figure 41-3). Of that energy, approximately 9 percent is made up of short, invisible wavelengths in the ultraviolet (UV) part of the spectrum. These wavelengths are damaging to life, but fortunately, a protective screen of ozone (O_3) in our atmosphere filters out most of the shortest ultraviolet rays (see Chapter 19). Most terrestrial and shallow-water organisms in the biosphere would perish without this gaseous barrier. That is why in 1987 and 1989, ecologists became alarmed when scientists discovered two "holes" in the ozone layer over the South Pole. Figure 41-4 shows the situation.

Of the remaining 91 percent of the sun's energy reaching the earth's atmosphere, half is light—wavelengths in the visible spectrum—while the other half is heat—longer wavelengths in the infrared region. This may seem like an unlimited shower of electromagnetic radiation, but nearly all of the heat and light is either absorbed or reflected as it passes through the atmosphere. As much as 35 percent of visible light is simply reflected back into space by clouds, ice, snow, oceans, and other particles in the air and at the planet's surface (see Figure 41-3). And atmospheric carbon dioxide, water vapor, and the water droplets that form clouds absorb most of the infrared, which makes up nearly half of the solar constant. The net result is a drastic loss of energy, so that only about 5 percent of the energy first entering the atmosphere is available to do work at the planet's land and water surfaces (see Figure 41-3).

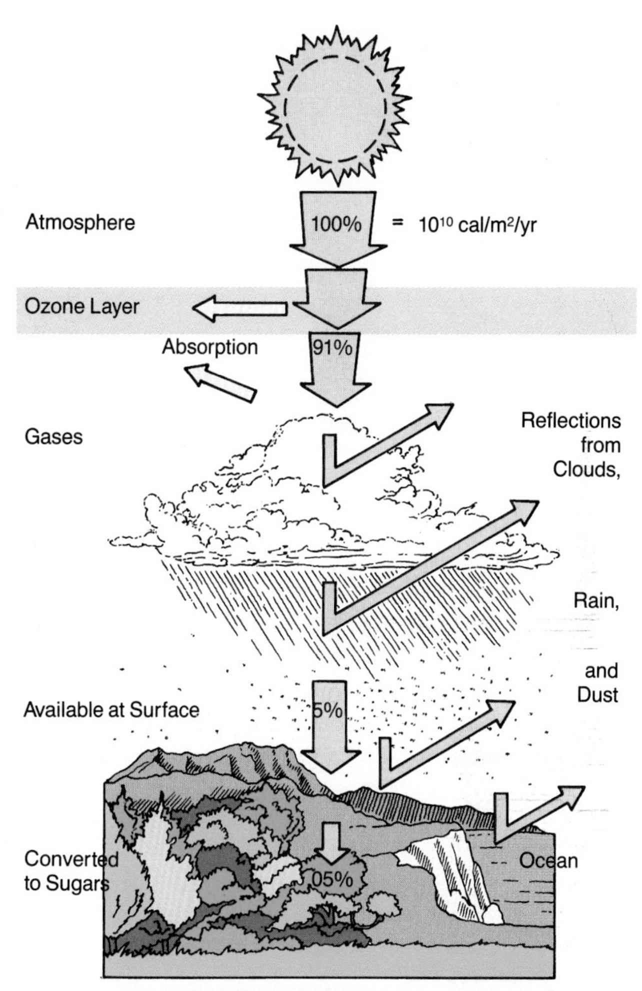

Figure 41-3 PROVIDING ENERGY TO SUPPORT LIFE: AN INEFFICIENT PROCESS.
Less than 5 percent of the solar energy first entering the atmosphere reaches the earth's surface. First, ultraviolet light is largely absorbed by ozone; then other wavelengths are partially absorbed by CO_2, O_2, and water vapor. Finally, reflection from clouds, rain, oceans, sand, and snow takes place. Gross production by all photosynthetic organisms is only about 0.05 percent of the solar constant; respiration then reduces that amount further, yielding the net productivity of plants shown here. Energy is expressed as calories per square meter of the earth's surface per year.

Of the solar energy—heat and light—that finally strikes a terrestrial environment, plants convert only 1.2

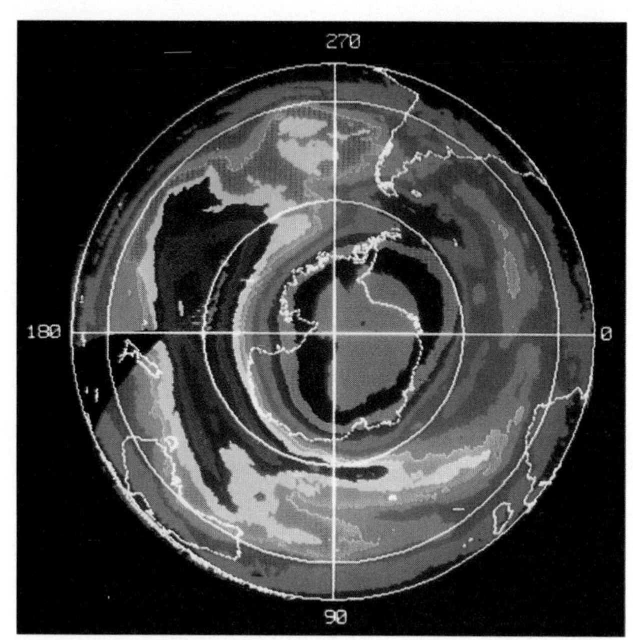

Figure 41-4 THE OZONE HOLE: A TEAR IN THE EARTH'S PROTECTIVE CANOPY.
Chemicals called chloroflurocarbons appear to play a major role in creating these holes in the ozone layer.

percent to sugars. The plants then use 15 percent of this converted energy to fuel their own aerobic respiration. In the end, the net conversion of solar energy to plant materials, the so-called *net productivity*, represents only about 0.05 percent of the solar constant. Other studies have shown that the use of solar energy during photosynthesis by aquatic plants in the earth's oceans is even less efficient. The finite amount of usable solar energy that plants convert and store in the chemical energy of organic compounds sets the upper limits of energy available to support all other organisms in the biosphere.

The Sun, Seasons, and Climate

The biosphere—the earth's thin zone of habitation—lies at the surface of a rotating, tilted sphere circling the sun. This fact has dramatic consequences for weather patterns, local climates, and the ability of different sorts of organisms to live at various places on the globe. The earth's rotation on its axis creates day and night, of course, and the tilt of that axis as the earth orbits the sun causes certain hemispheres to incline toward the sun and to heat up differentially (Figure 41-5). This uneven heating creates the seasons and global climate patterns and largely determines the mix of photosynthetic producers and, in turn, microbial, fungal, and animal consumers that can survive in a given region. In this and other ways, climate is a major evolutionary force, imposing constraints and establishing the ground rules for many plant and animal adaptations. Moreover, as the continents have drifted during the millennia relative to the fixed equator and the Tropics of Cancer and Capricorn (23.5° north and south latitude, respectively) local climate and weather have inevitably changed. Similarly, as huge mountain chains have pushed upward, only then to erode away, great changes in the continental topography have affected local weather conditions and thus organisms and their survival. Let us see, then, how temperature, wind, rainfall, and the earth's surface features create the climate patterns that so influence living things.

Temperature, Wind, and Rainfall

Because the earth is a sphere, different parts of the globe receive differing amounts of sunlight. In the middle zone, between 23.5° north and south of the equator, the earth's surface receives direct, perpendicular rays during certain times of the year. The highest average temperatures, therefore, are at and near the equator. In contrast, higher latitudes generally receive the sun's rays at an angle, and average temperatures are lower.

The northern latitudes receive the most direct sunlight near the time of the summer solstice (hence the long summer days and extended twilight), and simultaneously, the Southern Hemisphere experiences shortened days and winter. At the time of the winter solstice, the southern latitudes receive the most direct sunlight, and they experience summer as the north is in winter.

The uneven heating of the earth's surface not only causes temperatures to vary, but also sets air masses in motion as *wind*. Intense solar radiation at the equator heats the earth's surface. This, in turn, heats the air above, and it rises in massive air currents that ascend from the tropics and lift high into the atmosphere. The heated air spreads out north and south, and because it cools in the process, the air descends to earth again in circulating "cells" of alternately heated and cooled air (Figure 41-6).

As these giant atmospheric "pinwheels" spin, the earth is continually rotating at a speed (along the equator) of about 1,000 mph. This rotation applies a force on the atmosphere, oceans, and other fluids—a force named the **Coriolis effect.** This phenomenon skews the churning masses of air in a clockwise direction in the Northern Hemisphere and in a counterclockwise direction in the Southern Hemisphere. This, in turn, deflects

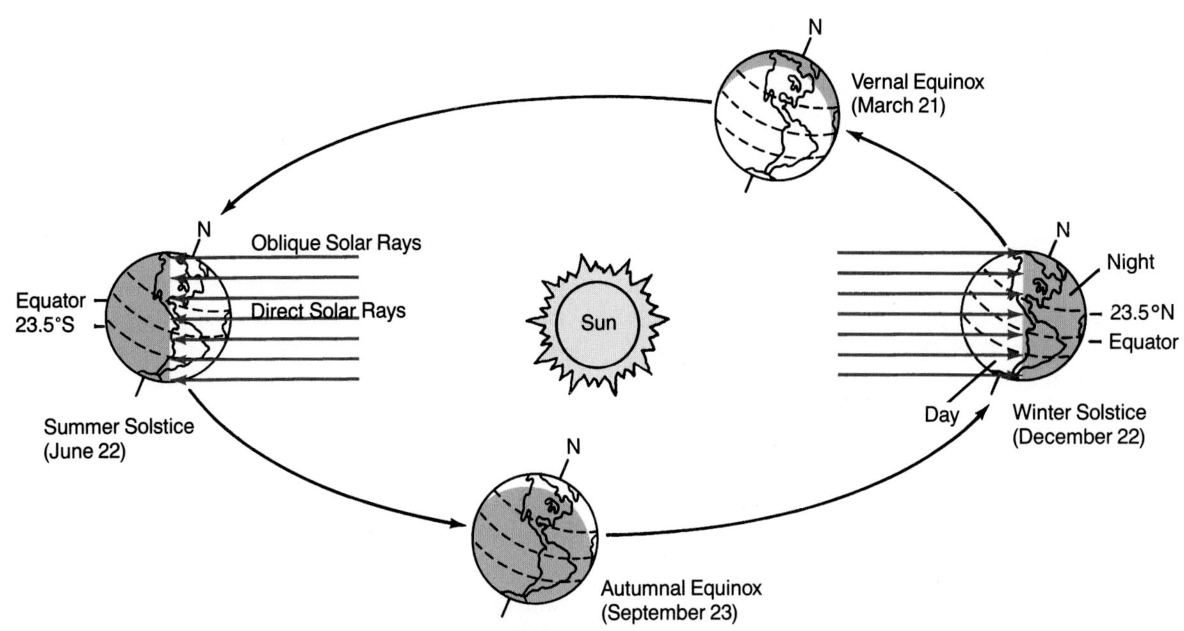

Figure 41-5 GLOBAL TILT AND THE SEASONS.
The earth is tilted on its axis as it revolves around the sun. The tilt leads to uneven heating of the earth's surface and, along with the earth's spherical shape, produces seasons and climate patterns around the globe. At the summer solstice (left), the sun's rays hit the earth perpendicularly in the Northern Hemisphere, heating it more and producing summer conditions there. That occurs in the Southern Hemisphere (right) at the winter solstice. The sun's rays fall perpendicularly on the equator at the vernal and autumnal equinoxes. Rotation of the earth about the axis each 24 hours produces day and night.

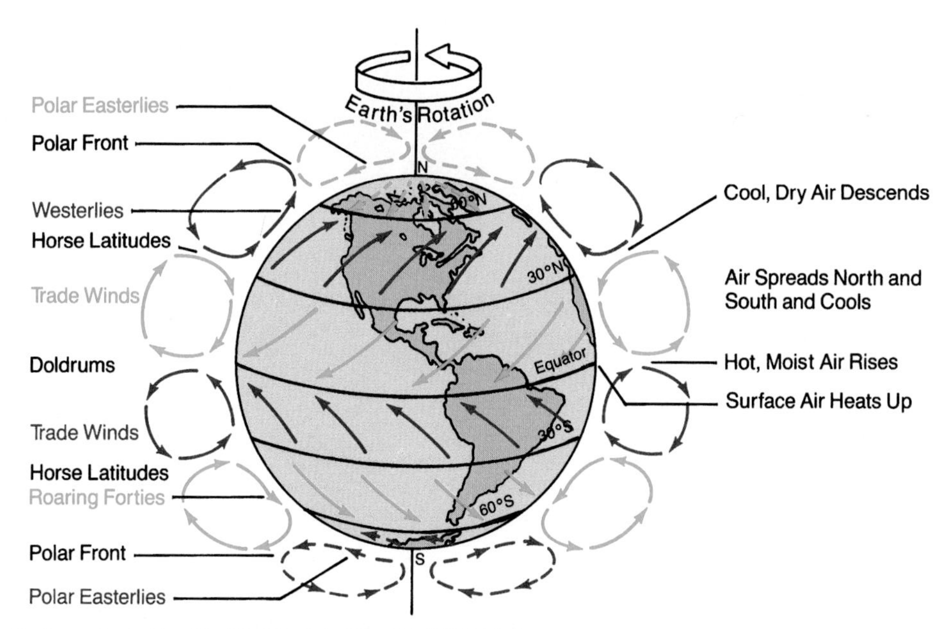

Figure 41-6 GLOBAL PATTERNS OF AIR CIRCULATION.
Uneven heating and the earth's rotation cause circulating patterns of air to whirl along the surfaces of land and ocean in the directions shown by the rows of curved arrows drawn on the globe itself. The circular patterns of arrows drawn outside the globe show that air masses are larger near the equator and smaller at higher latitudes. (These circles of air are shown in greatly exaggerated size in the drawing.) The circles near the equator show that air currents rise at the equator, spread out north and south, cool, then descend near the horse latitudes. The irregular polar wind patterns are indicated by broken lines. The common terms for regions of the earth's surface affected by these air patterns are listed to the left of the globe.

the wind currents near the Earth's surface away from routes that would have been due north and south. Thus, winds flow predominantly in prevailing directions (e.g., from west to east across much of North America), and move at normal speeds (slowly near the equator—the so-called "doldrums"—and faster at higher latitudes—the westerlies and "roaring forties"). These directional winds help generate the major ocean currents. Of course, the topography of landmasses, especially when high mountain ranges are present, may influence these wind patterns. Huge lakes, deserts, and other features of continents also may affect wind patterns, so we must include topography as a potentially important influence on climate.

Rainfall is yet another factor helping to determine a region's climate, and it, too, is determined by the uneven heating of the earth's surface. Warm air can hold much more moisture than cool air; because air carrying moisture rises and descends at different places on the globe, different regions receive differing amounts of precipitation (rain or snow). The tropics, for example, have abundant rainfall because air warmed near the earth's surface carries much water vapor; on rising, this saturated warm air cools until its water vapor condenses into droplets to form clouds, and ultimately, rain falls. When this same, now drier air moves through its "cell" and descends around latitudes 30° north and south, it is both compressed and dry. As a result, this low-humidity air creates a worldwide band of aridity and even deserts at these latitudes (Figure 41-7).

Yet another factor influences climate, and with it, communities of living organisms: altitude. Altitude is said to "mimic" latitude. Even in the Andes, which straddle the equator, temperature falls as one climbs, until finally, near the tops of the ancient volcanic peaks, one reaches polar climatic conditions. Equally important, moist air forced aloft when it blows against mountain ranges tends to drop its water content as rain or

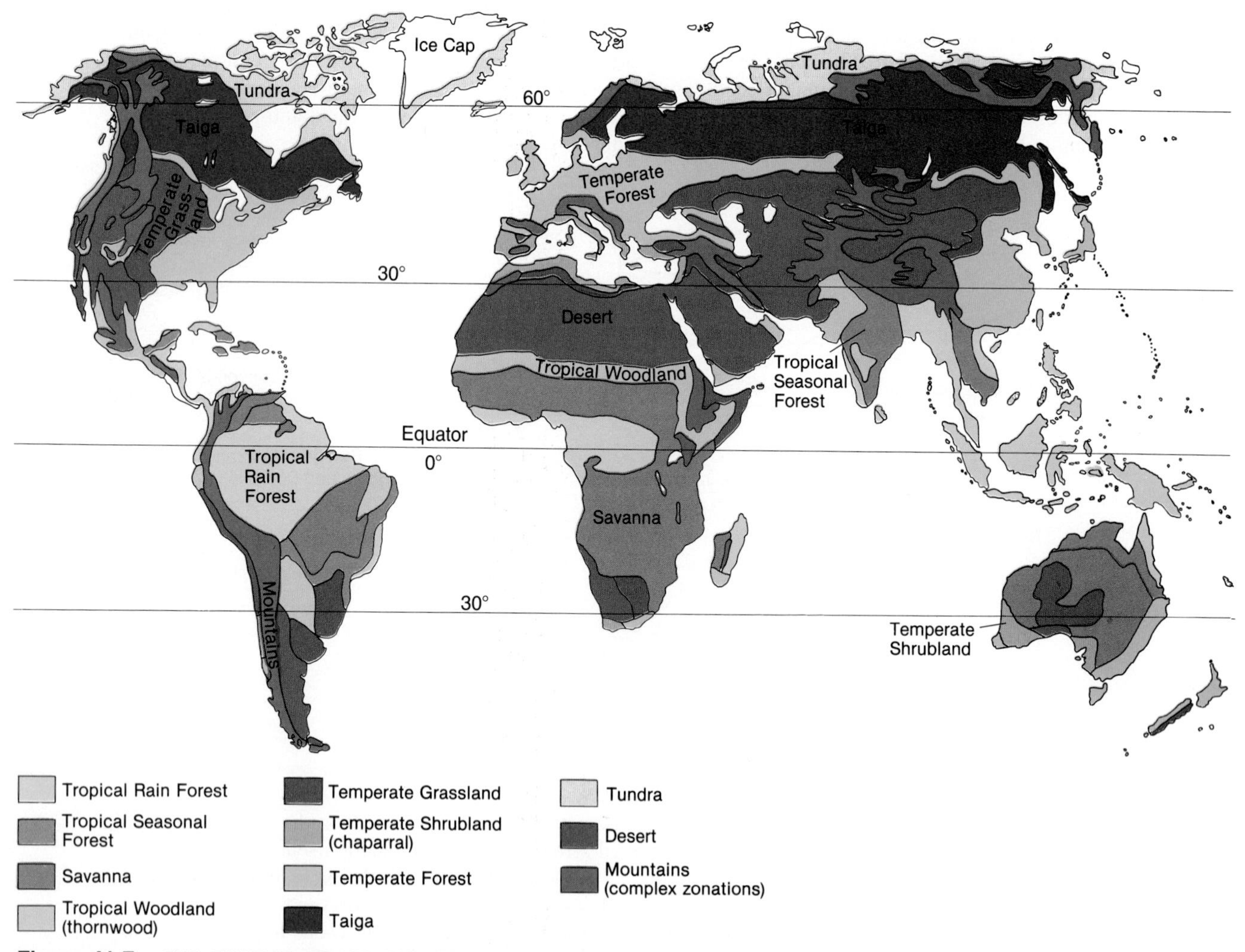

Figure 41-7 THE EARTH'S MAJOR BIOMES.
The major types of biome are shown in the map and described in the text. The biomes occur where they do because of global wind, water, and weather patterns in combination with topography and other factors.

snow on the mountains. If winds tend to hit a mountain range from one direction, that side of the range may be lush forest and the opposite, downwind side, may be a desert because it is in a "rain shadow." Winds traveling from east to west across the Andes, for example, dump moisture on the chain's eastern side, creating an arid zone in the rain shadow to the west. The parallels between altitude and latitude are reflected in the local fauna and flora; as one climbs from sea level to 7,000 m in the Andes, one moves from lush tropical rain forests to drier, treeless zones, and finally to the same kinds of low, scrubby plants one might see on the arctic tundra.

The similarity of organisms living in different parts of the world but under similar climatic conditions underscores the evolutionary principle that essentially similar adaptations must be present to allow survival in the face of similar sets of environmental conditions—whether that be on a mountain top, in a tundra, or in one of the world's desert zones.

To summarize, the earth's rotation, tilt, and orbit around the sun cause the dynamic air movements and ocean currents that in turn establish global climate and weather patterns. The locations and topography of the continents and oceans relative to these global patterns affect the climate of a given area and thus, indirectly, its living organisms. Temperature, precipitation, and wind patterns indirectly determine the **biomes**—the earth's major ecosystems, with their predominant vegetation types. Tropical thorn scrubs and savanna grasslands, for example, occur near latitudes 30° north and south, where rainfall is low and temperatures high; tropical rain forests grow at the equator, where both rainfall and temperatures are high; and so on. Let us tour earth's biomes now, one by one.

BIOMES

Our world's major visual features are the forests, deserts, grasslands, and bodies of water that ecologists subdivide into many biomes (Figure 41-7). By and large, every biome has characteristic types of organisms, as we will see shortly. Within a given biome, however, there may be local peculiarities in geography, climate, and biological history that give rise to isolated areas with unique groupings of small and large plants, animals, fungi, and microbes. At their edges, adjacent biomes usually grade into each other, as with the imperceptible merging of tall, lush vegetation at the foot of a mountain into zones of lower, drier, more sparse vegetation at higher altitudes.

Tropical rain forests occur at low latitudes wherever rain falls abundantly throughout the year (Figure 41-8). Dense tree foliage grows high above the forest floor in a "canopy" near the light, and epiphytes (such as orchids), lianas (vines of various types), and many kinds of animals inhabit this lush and lofty habitat. A few tall trees emerge above the canopy, and shorter trees and bushes create an understory that deeply shades the forest floor. Rain forests support an exceptional diversity of invertebrate wildlife, including butterflies and other insects, and most resident species of mammals, such as monkeys, and reptiles, such as snakes, live in the trees rather than on the damp ground. The box on page 758 describes the current and accelerating human destruction of tropical forests and its implications for global climate and our planet's diversity of species.

Tropical seasonal forests occur in slightly higher latitudes than rain forests or where topographical conditions cause locally drier climate (Figure 41-9). Trees in these

Figure 41-8 A TROPICAL RAIN FOREST IN TRINIDAD.
Tall trees, many vines, and moisture everywhere are typical of rain forests.

Figure 41-9 A TROPICAL SEASONAL FOREST IN AFRICA.
Photographed when leaves have fallen from deciduous trees, the tropical seasonal forest is much more open than rain forests.

ECOLOGICAL TROUBLE IN THE TROPICS

The human population explosion, the extinction crisis, and the threat of global warming are interrelated ecological calamities, with deep roots in the human use and misuse of tropical rain forests. While these dense, wet forests account for only 6 percent of the world's land area, they harbor many species of animals, 80,000 plant species, several million kinds of insects—in all, more than half of earth's species.

The trouble comes because the human population of the tropics has burgeoned, and by the year 2000, an additional 1.2 billion people will be born in these regions. Crowding and poverty are driving people to cut rain forests for the sale of hardwood timber, and to burn it to clear new land for agriculture. By 1989, about 55 percent of the world's tropical forests had been severely disrupted, and each year, people clear another area the size of Switzerland and the Netherlands combined. At this relentless pace, *all* tropical rain forests will have been altered or destroyed within 60 years.

The consequences of this habitat destruction are enormous. As we saw in Chapter 40, 1 million species or more could become extinct in the next 20 years—a catastrophic extinction rate without parallel in biological and geological history.

Perhaps the most tragic fact about tropical deforestation is that once tropical farmers have cleared the land, they can cultivate it or graze animals on it for just a few years. Many tropical soils can support lush forests only because most of the nutrients in the ecosystem are locked up in the trees and other vegetation. Slash-and-burn agriculture (Figure A), however, depletes the soil severely, and it takes about 80 years for the soil to regain its fertility and support a new forest. In the meantime, farmers must cut down other forests and start the cycle again.

Research shows that forest burning causes the rapid, direct extinction of some species, and dooms others indirectly. By fragmenting large stretches of tropical forest into islandlike patches separated by agricultural land, populations are separated into small units more susceptible to genetic drift, much as they are on oceanic islands.

Figure A
Slash-and-burn agriculture.

Tropical deforestation is certain to have consequences for global climate, as well as for global diversity. More than half of the rainfall in tropical forests is due to evaporation; thus, the precipitation as well as the temperature regime is disrupted severely when the forest is cut. What is more, deforested regions fix less carbon dioxide into plant tissues than they once did, and they release less oxygen into the air. Finally, tree burning adds to the high levels of atmospheric CO_2 that developed countries generate through the burning of fossil fuels. This promotes the greenhouse effect, which, as we have seen, creates global warming and itself has an impact on climate, wind, and rain patterns.

Ecologists have suggested various measures to combat the serious trouble in the tropics. These include the rapid cataloging of tropical sepecies before many are lost forever, the identification of areas in critical danger of destruction and in immediate need of preservation efforts; more reforestation in clear-cut areas; and the establishment of "extractive reserves," where people can harvest rubber, nuts, fruits, and other forest products but leave the trees standing. Ultimately, many think, the growing human population in these regions must be able to feed and support itself in stable, sustainable ways so people will not be forced by their own survival pressures to continue cutting virgin forests until they are gone.

forests lose many of their leaves during the dry season and so are deciduous. The uniformly warm tropics have wet and dry seasons rather than cold and hot seasons, and in the tropical seasonal forests in central India, for example, monsoon seasons can bring several inches of rain per day, while conditions can approach drought during the dry seasons.

Savannas are tropical grasslands with scattered trees that occur either where the climate is too dry for dense forests to grow or where soil conditions and/or periodic fires maintain a grassland system (Figure 41-10). Usually, the driest savannas have only grasses; where more rain falls, scattered trees grow; and with more rain still, savannas grade into tropical woodlands. The famous Af-

Figure 41-10 SAVANNA GRASSLAND OF THE SERENGETI PLAINS, TANZANIA.
Although the number of trees can be substantially greater than shown here, grasses remain the dominant vegetation in savanna grasslands.

Figure 41-12 A TEMPERATE GRASSLAND IN CENTRAL NORTH AMERICA.
This bison cow and calf are grazing on a classic rolling grassland biome in Custer Park, South Dakota.

rican savannas are grasslands with occasional flat-topped trees, where huge herds of grazing mammals and their predators roam the open landscape.

Tropical woodlands, or tropical thornwoods, occur in regions with somewhat more rain than savannas but less rain than tropical seasonal forests. Thornwoods—bushes and trees with spines, such as *Acacia* species in the Americas—dominate the vegetation.

Deserts occur in subtropical latitudes, where dry air descends (around 30° north and south; see Figure 41-7), or they occur where mountains block moist winds and cause rain shadows, as in the region west of the Andes. Although some deserts are extremely barren, others have seasonal rainfall and contain many species of succulents such as cacti and other dry-adapted vegetation (Figure 41-11). When low temperatures combine with low rainfall, simple deserts result, as in the Great Basin sagebrush-scrub desert of North America. Virtually all support vertebrates such as snakes and kangaroo rats, and may support bobcats, foxes, coyotes, badgers, and other large animals. Among the invertebrates are scorpions and many species of drought-tolerant insects.

Figure 41-11 DESERTS OF THE WORLD.
Deserts are always in dry regions, yet their conditions vary substantially, depending on local topography, soil, and temperature. A typical Arizona desert, shown here, has enough water to support a rich cover of cacti and drought-resistant flowering plants.

Temperate grasslands, such as the North American tall-grass *prairies* and short-grass *plains*, the Eurasian *steppes*, and the South African *veld*, are similar to tropical savannas, differing from them only in their lack of trees and cooler, high-latitude climate, which causes differences in soil (Figure 41-12). The climate is usually drier than in forested regions, but not as dry as in deserts. Like savannas, temperate grasslands are subject to frequent fires (often caused by lightning) that repress the growth of trees. Although they appear simple in biological structure, with a single layer of vegetation, the rich soils of temperate grasslands can sometimes support more plant species (including agricultural ones) than can temperate rain forests.

Temperate shrublands, such as the California chaparral and the Australian heath, grow in areas with "Mediterranean" climates—that is, hot, dry summers and cool, moist winters. Like grasslands, they are subject to frequent fires, and many indigenous plant species, such as manzanita bushes, are remarkably well adapted to periodic conflagrations. Rabbits, beetles, vultures, and deer mice are typical animal inhabitants.

Temperate forests, like grasslands and shrublands, occur to the north and south of subtropical latitudes in the two hemispheres (Figure 41-13). These forests are varied and include deciduous forests, evergreen forests, and even rain forests, depending on prevailing moisture

Figure 41-13 TRANQUILITY IN A TEMPERATE FOREST.
The woodlands along the Delaware River in Worthington State Park, New Jersey, are lovely examples of the temperate forest biome.

Figure 41-15 TUNDRA IN ALASKA, NOT FAR FROM NORTH AMERICA'S HIGHEST PEAK, DENALI (MT. McKINLEY).
Low bushes and soft, springy mosses cover the wet soil.

and temperature conditions. But temperate forests have far fewer species than do tropical rain forests. The coastal redwood forests of northern California contain the tallest trees in the world (up to 90 m high), and the deciduous forests covering much of the eastern United States and Canada are species-rich, with varied bird and mammal life.

Taiga is subarctic or subalpine needle-leaved forest growing in the coldest zones possible for trees (Figure 41-14). Taiga forms the great conifer forests of the northern United States, Canada, and northern Eurasia. The forests and woodlands of the taiga are simple and often dominated by just one or two tree species. Taiga animals include elk, moose, wolves, jays, grouse, and often an abundance of mosquitoes.

Tundra occurs in the Arctic as well as on tall mountains in the Andes, the Himalayas, the Rockies, and ranges throughout the world where high altitudes mimic arctic latitudes (Figure 41-15). Small shrubs sometimes grow on this treeless plain, but the predominant species are grasses, lichens, sedges, and mosses. Only the upper few inches of soil thaw each summer; the deeper soil remains frozen as *permafrost*. Some animal residents, such as the arctic hare and the snowy owl, inhabit the tundra year round despite the snows and bitter cold, while others, including many songbirds, are summer migrants.

Figure 41-14 TAIGA: NORTHERN CONIFEROUS FOREST.
This stand of conifers grows in Athabasca Valley at Jasper National Park in Alberta, Canada.

In addition to these terrestrial biomes, there are several aquatic biomes. *Freshwater biomes* include **lentic** (lake and pond) and **lotic** (stream) **communities** occurring worldwide, regardless of climate (Figure 41-16). Within lentic communities, shore, planktonic, and bottom communities tend to blend with one another. In lotic communities, there is less plankton than in lentic communities, but larger invertebrates live on the bottom or attached to rocks in fast-moving streams.

Marine biomes include sandy beaches, intertidal zones, marine mud flats, coral reefs in tropical waters, open-ocean communities, and bottom communities. While beaches have constantly shifting sand, they are nevertheless inhabited by many kinds of crustaceans, worms, and microbes. Tide pools are much more stable substrates than beaches, and a great diversity and abundance of sea stars, snails, algae, crabs, and other aquatic organisms inhabit these rocky intertidal areas. Marine mud flats are home to a range of invertebrate fauna, including crustaceans, worms, and clams. Coral reefs form fringes around tropical islands and are similar to tropical terrestrial communities in supporting a bewildering variety of plants and an even greater variety of animals (see Chapter 42 opening photo, page 775). **Pelagic** (open-ocean) **communities** are supported primarily by the pho-

Figure 41-16 A FRESHWATER COMMUNITY IN CALIFORNIA'S SIERRA NEVADA RANGE.
The sparkling Merced River flows through the spectacular glacial valley of Yosemite National Park.

tosynthesis of phytoplankton, yet innumerable diatoms, invertebrate larvae, fishes, and other buoyant creatures are also found there. **Benthic** (bottom) **communities** commonly contain diverse invertebrate types, even on the deep-ocean bottom.

THE HABITATS OF LIFE: AIR, LAND, AND WATER

The three very different states of matter—solid, liquid, and gas—that make up habitats impose constraints on organisms and affect their ways of living.

Air: The Atmosphere

Air is a vast, moving reservoir into which water evaporates and from which organisms' vital needs are met. Air temperatures can fluctuate rapidly and widely, and land-dwelling organisms have evolved numerous mechanisms for coping with environmental extremes (see Chapter 33). Air's molecular composition (79 percent nitrogen, 21 percent oxygen, and 0.03 percent carbon dioxide) also affects all plants and animals. Since the number of oxygen molecules per liter of air decreases as altitude increases, there are limits on where certain animals can live. People, for instance, have never established permanent settlements above 6,000 m.

The gaseous constituents of air provide important chemical raw materials for life. Carbon dioxide is, of course, the carbon source for photosynthesis. Nitrogen, reduced by nitrogen-fixing organisms, is a necessary component of proteins and nucleic acids. The amount of water vapor in the air (the relative humidity) largely determines the rates of evaporation from the skin of a frog, through the stomata of a plant, and so on. Hence, organisms that live in habitats with low relative humidities must have special adaptations to prevent extreme water loss. The American pygmy cedar, for example, can absorb enough moisture from airborne water vapor alone to sustain life in the parched deserts of the American southwest.

Land

Many plants and most animals acquire water and nongaseous mineral nutrients directly or indirectly from the *soil.* Without living things, the material we know as soil would not exist, for it is a product of the interplay of organisms and the earth's crust. Plant roots, burrowing mammals, earthworms, and fungal hyphae exploit the water, air, and space between soil particles and, in the process, add immense quantities of organic materials to it. The soil's chemical composition, texture, and porosity in turn affect its habitability.

Soil pH, for example, affects whether Ca^{2+}, PO^{4+}, and other essential ions are bound tightly to particles or are free to be absorbed by organisms or leached away in water runoff. Plants tend to be adapted to a specific soil pH; for example, the graceful rhododendrons and azaleas of northwestern and southeastern conifer forests thrive in acidic soils where many other plant species fail to grow.

Soils are composed of four main constituents in varying proportions: **Silt** and **sand** are pulverized rock and are intermediate in particle size between gravel and clay. **Clay** is a dense material made up of very fine particles produced by weathering and is composed of aluminum, silica, and other minerals. **Humus** consists of decomposing organic materials; it is a source of nutrients that are released slowly by decomposers. In a rich soil such as **loam,** which is optimal for plant growth, a mixture of silt, sand, clay, and humus particles creates spaces that contain life-sustaining water and air. Water remains between soil particles because of capillarity and **imbibition,** a process in which water enters the soil and binds to clay and humus particles. Clay-rich soils imbibe the most water and can become waterlogged, whereas soils with a high proportion of sand drain quickly and tend to hold less water. Humus particles act like tiny sponges; thus, in addition to providing nutrients, they keep soil from drying quickly, and from compacting.

While soil is a restrictive environment, hard to burrow through and often containing little food or oxygen, animals adapted to life underground can hide from predators and are relatively well insulated from both heat and cold. This insulation enables a small mammal such as the arctic hare to survive a bitter arctic winter nestled below the tundra.

Water

Aquatic life forms derive food, water, gases, support, protection from temperature extremes, and other basic needs from the surrounding water. Some waters contain few nutrients, and most provide less than 5 percent of the oxygen present in air. Water is denser than air and buoys up the bodies of aquatic plants and animals so effectively that they need less massive support structures than their terrestrial counterparts. Water moderates temperatures because of its large capacity for absorbing and holding heat, and this prevents the extremes found on land. Very deep waters may be cold—near freezing—yet their constancy has allowed organisms to evolve enzymes that function very rapidly at such low temperatures.

While animals can live in even the deepest oceans, plants can survive only in a top layer of water known as the **photic zone,** where enough light for photosynthesis can penetrate (Figure 41-17a). (Bacteria, present at levels of 10^7 to 10^9 per liter of seawater, tend to cluster near the phytoplankton in order to absorb nutrients from them.) Since water both reflects and absorbs light, even the clearest, cleanest ocean water has a photic zone only about 100 m deep. Most animals living deeper are either carnivores, which eat smaller animals that consume phytoplankton in the photic zones, or scavengers that depend on a steady supply of dead and dying plant and animal bodies drifting down from above.

True miniature ecosystems, however, that do not rely on energy arriving from the photic zone exist in the deep ocean abyss. Communities of giant tube worms, huge clams, crabs, a few fish, and other animals cluster around deep-sea vents, perforations in the seafloor where superheated, mineral-laden water pours out (Figure 41-17b). Biologists now believe that chemosynthetic bacteria that oxidize hydrogen sulfide or other compounds spewing from the vents are the ultimate food source supporting these vent communities and allowing them to harvest energy and build organic materials. These sulfur bacteria live symbiotically in the tissues of the tube worms and the gills of the clams, while crabs consume the producers directly.

In lakes and oceans, constant mixing and wave action help dissolve and distribute atmospheric oxygen. Circulation patterns can be horizontal (as in ocean currents) or vertical. Vertical ocean currents are called *upwellings;* vertical movements in lakes take place in fall and spring

(a)

(b)

Figure 41-17 MARINE COMMUNITIES: PHOTIC ZONE AND DEEP-SEA VENT.
(a) In the photic zone, light penetrates the seawater. Here, fish graze around large photosynthetic kelp plants moored to the shallow seafloor. (b) In this deep-sea vent community, scientists discovered many animal species, including yellow vent mussels, white crabs, and large tube worms with red plumes extended, all interacting in a previously unknown ecosystem based on heat from underwater vents and chemoautotrophic bacteria.

and are called *overturns*. This vertical movement is especially important because critical nutrients, such as phosphorus, are most soluble in the deeper, colder water; upwelling lifts this nutrient-rich cold water to the photic zone, where organisms can exploit it.

ENERGY FLOW IN ECOSYSTEMS

The habitability of every region, zone, and individual habitat depends on the availability and fluxes of energy—a fact that emerged when Charles Elton, a young biologist at Oxford University, began studying the ecology of Bear Island, off the northern coast of Norway, in the 1920s. Elton wanted to study how the local tundra animals divide up available food resources, and he noted how arctic foxes, for example, caught and ate tundra birds—sandpipers, buntings, and ptarmigan. These birds (which, like Elton himself, spent only summers in the area) in turn consumed the tender leaves and berries of tundra plants or ate insects that had fed on plant parts. Elton described the links between plant, insect, bird, and fox as a **food chain.** He defined the first link of the chain as the trapping of solar energy by a plant via photosynthesis and its conversion to stored chemical energy. The stored energy then passes from the plant to a herbivore and then to a carnivore. Each link in this chain represents a specific **trophic,** or feeding, **level.**

He went on to define levels of energy transfer in biological communities. Plants and other photosynthetic organisms are the **producers.** The solar energy they store as carbohydrates and other compounds is transferred to a series of **consumers** at successive trophic levels. An insect consuming a berry is a *primary consumer;* an indigo bunting eating an insect, a *secondary consumer;* and so on. At the end of every food chain, as well as at each of its levels, are the **decomposers**—mainly bacteria and fungi that break down the remnants of organic material—and the **detrivores**—bacteria, protists, and animals that feed on dead and decomposing organisms. As we saw in earlier chapters, these last two groups are nature's primary recyclers.

In his Bear Island studies, Elton also noticed that most food chains are part of more complex systems known as **food webs.** Arctic foxes did not limit their food intake to one species of tundra bird; they also ate sea gulls, auks, and other marine birds. These birds ate fish, which grazed on barnacles or other invertebrates, which in turn may have eaten photosynthetic algae. In winter, the birds migrated to warmer climates, and the foxes ate bits of dead seals left by polar bears, as well as polar bear dung. Figure 41-18 illustrates food webs, and the box on page 764 describes the consequences of human tampering with an Antarctic food web.

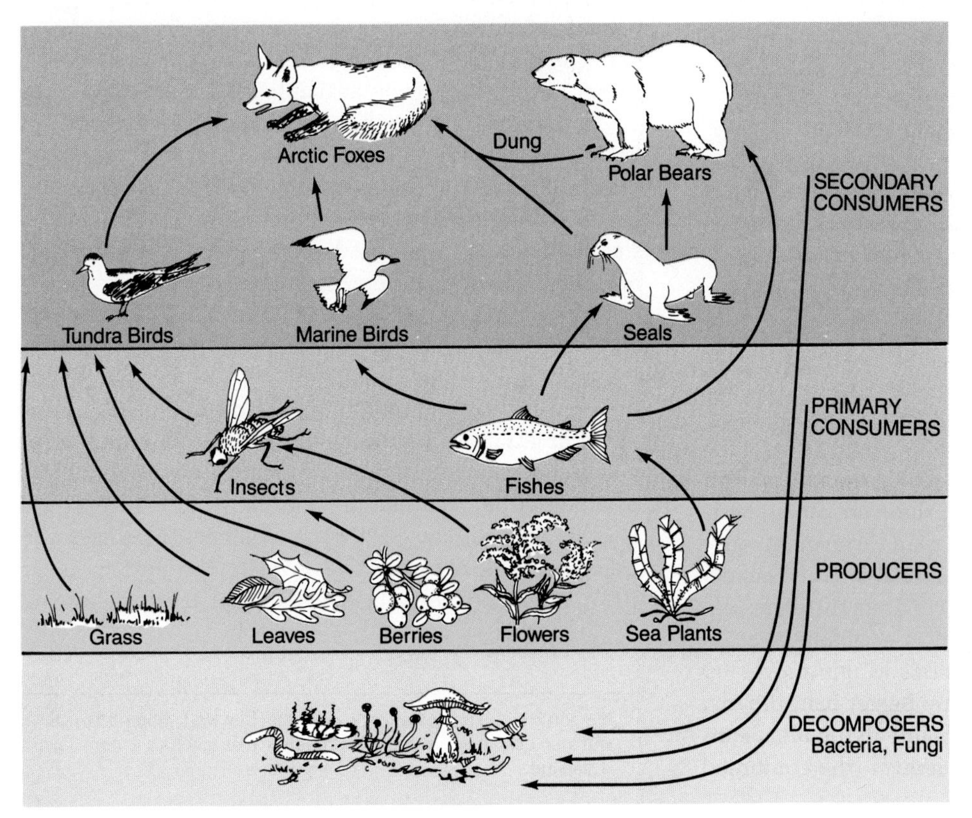

Figure 41-18 FOOD WEB OF ARCTIC BEAR ISLAND. A food web is a complex system of interlocking food chains. Each organism is positioned according to its trophic level; some, including tundra birds, are both primary consumers (of leaves, berries, and seeds) and secondary consumers (of insects that also feed on berries). Producers, primary consumers, and secondary consumers will ultimately be consumed by the decomposers. Arrows point from each food source to its consumers. This is a highly simplified version of the food web for this tundra and marine biome.

ANTARCTIC FOOD WEBS: AN UNCONTROLLED EXPERIMENT

Despite frigid temperatures and interminable winters, the Antarctic has always supported a rich and complex food web. Living in the icy waters that surround the earth's southernmost continent, phytoplankton—the main primary producers—support 750 to 1,350 million tons of krill each year (Figure Aa).

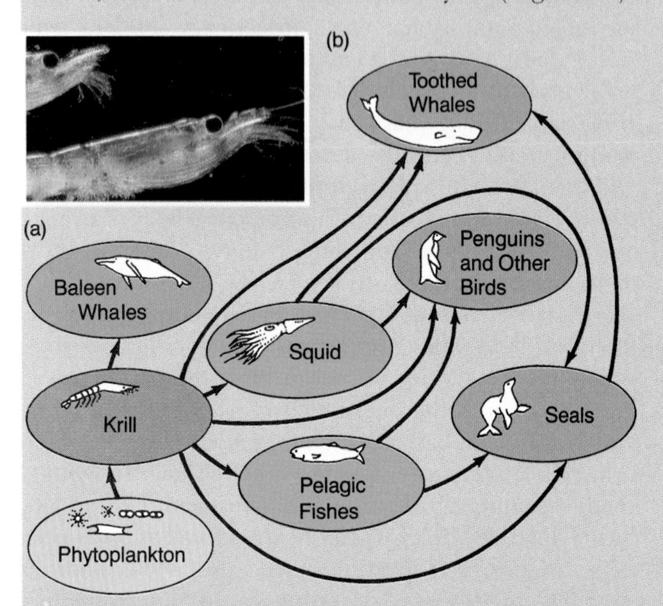

Figure A ANTARCTIC KRILL AND THE ANTARCTIC FOOD WEB.
(a) These primary consumers feed on algae and are themselves the huge "crop" that is grazed by fishes, marine mammals, and birds. (b) As a result of the extensive human harvesting of large whales, various seals and birds have become the main consumers of krill and of the squid and fishes that feed on krill. If populations of the large krill-eating baleen whales rebound, there will undoubtedly be disturbances in this pelagic, or ocean-surface, food web, but we cannot be sure what they will be.

Historically, this titanic krill population supported impressive numbers of filter-feeding baleen whales, squid, and fishes. And the squid and fishes, in turn, sustained the toothed whales, seals, penguins, and other birds (Figure Ab). During the past 150 years, however, whalers have reduced the baleen whale population to less than one-quarter of what it was, and in so doing have created a large-scale biological experiment.

Studies show that while the whales are consuming far less krill, the krill are still under intense natural harvesting pressure from other expanded animal populations. The 30 million crabeater seals of this region, for example, now represent about 66 percent of all the world's seals. Together with other seal species, crabeater seals annually harvest 130 million tons of krill (far more than all remaining whales combined), 10 million tons of squid, and 8 million tons of fish.

Populations of three penguin species have also burgeoned. These flightless birds make up some 60 million individuals and also feed on krill and squid by the millions of tons.

Finally, smaller whales have grown more numerous as the larger baleens have been killed off: As blue, humpback, right, and other large whales have disappeared, populations of small minke whales (which are krill feeders), for example, have expanded greatly. Clearly, removing major consumers from the "top" of one portion of the food web has allowed other consumers to exploit newly available food resources.

Commercial whaling has been more strictly regulated in recent years, and scientists now have an opportunity to monitor whether the food web is returning to its original balance or whether a new order is stabilizing—an order that largely excludes the great baleens from their former position.

Ecological Pyramids

Elton's observations revealed a basic principle of ecology: In virtually every community, there are more plants than herbivores, more herbivores than carnivores, more small carnivores than large ones, and so forth. What accounts for this **pyramid of numbers?** It might seem reasonable to assume that body size is a factor, since in many food webs, body size increases as one moves up the pyramid (little fish are eaten by bigger fish, they by still bigger fish, and so on). If we count all the organisms in an area and convert them to **biomass** (the combined dry or wet weight of all the organisms in a habitat measured in grams) or to energy (the calories or joules* represented by each group of organisms), we usually get a **pyramid of biomass** like the one shown in Figure 41-19. As the second law of thermodynamics predicts, the transfer of energy from one level to the next is never 100 percent efficient; in every one of life's millions of

*A joule is a unit of energy equal to the work done when a current of 1 ampere passes through a resistance of 1 ohm for 1 second.

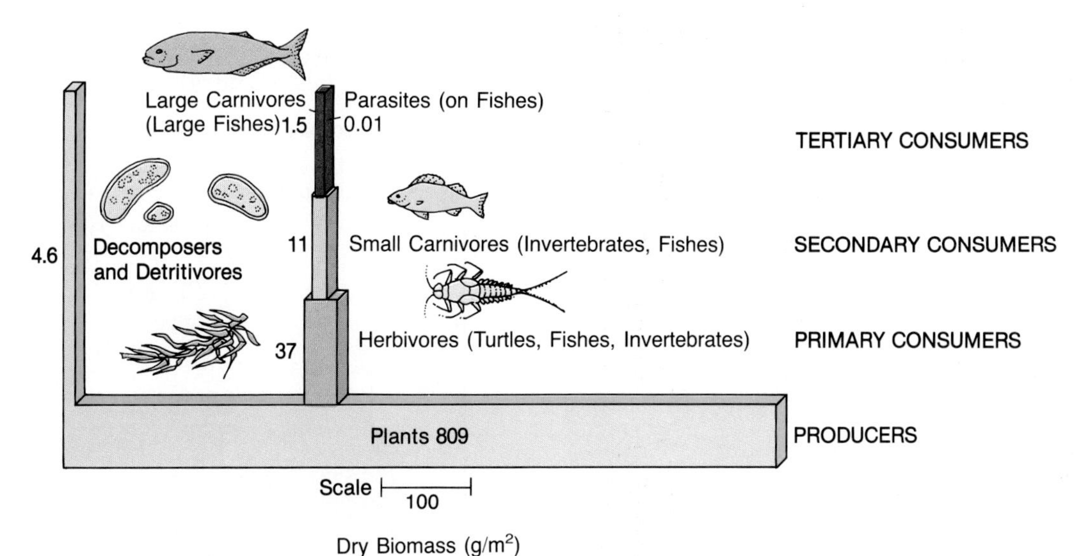

Figure 41-19 PYRAMID OF BIOMASS.
The transfer of energy from one level to the next is never 100 percent efficient, so that as we move from the producers (plants) to the primary consumers (herbivores) to the secondary consumers (small carnivores) to the tertiary consumers (large carnivores), the biomass of each successive group is less. Note the relatively constant biomass of detritus feeders (detritivores) and decomposers at all levels of the pyramid.

chemical reactions, there is always loss of heat and some gain in entropy. Hence, with less energy available for each successive trophic level in a food chain, less biomass in each succeeding level is not a surprising result.

Not all ecosystems appear to follow the pyramid-of-numbers pattern. In some aquatic systems, for example, ecologists have discovered *inverted pyramids*. If you take a sample of water from the English Channel, weigh the phytoplankton in it, and then weigh the zooplankton (the tiny animals that eat the phytoplankton), you will find that there are more primary consumers, by weight, than producers. At this point, what is being measured is the **standing crop**—the number of organisms existing at one specific moment. As it turns out, phytoplankton grow and reproduce quickly, but are consumed by the zooplankton as fast as they arise. Zooplankton, on the other hand, survive relatively longer and so accumulate a larger standing crop. Nevertheless, the total mass of phytoplankton consumed by a given mass of zooplankton in their lifetimes would far exceed the consumers' mass. The second law of thermodynamics wins again!

Generating Biomass: Energy Relationships in the Ecosystem

Within an ecosystem, there are two kinds of production. **Primary production** is the synthesis and storage of organic molecules during the growth and reproduction of photosynthetic organisms. Significant amounts of this energy pass directly to decomposers and detritivores. The rest is passed on to herbivores, which are primary consumers of photosynthesizers. Herbivores use some of that energy for life processes, while the rest is stored in their molecules. The processing and storage of energy by herbivores—which, like photosynthesizers, are living factories—is called **secondary production.** Some of the energy stored in the molecules of herbivores passes, again, to decomposers and detritivores, while the rest is passed on to secondary consumers.

Of course, not all the organic molecules that a photosynthesizer produces (so-called **gross production**) are available to primary consumers. Some are consumed in the primary producers' own respiration and metabolism. Therefore, **net production** equals gross production minus respiration and metabolism.

Plant ecologists have devised sophisticated ways to measure photosynthetic rates and the rate of net production in a range of ecosystems. Notice in Table 41-1 that production varies, depending on climatic conditions and rainfall. Thus, equatorial coral reefs are the most productive ecosystems on earth, followed closely by estuaries and typical rain forests. In contrast, the open ocean is more like a desert. Oceans cover 63 percent of the globe, but they account for only one-quarter of the world's net primary production. Why? The main reason is that most oceans lack nutrients—especially nitrogen and phosphorus—that are essential components of living systems. In the central Atlantic, for example, where horizontal currents are slow, there is little surface-to-bottom mixing, phytoplankton drift in a thin veil across great reaches of the ocean, and the region is the biosphere's

Table 41-1 NET PRIMARY PRODUCTION OF ECOSYSTEMS

Habitats	*Mean Net Primary Production (g/m²/year)*
Terrestrial	
Tropical rain forest	2,200
Temperate evergreen forest	1,300
Temperate deciduous forest	1,200
Taiga	800
Savanna	900
Temperate grassland	600
Tundra and alpine communities	140
Desert and semidesert scrub	90
Extreme desert	3
Cultivated land	650
Aquatic	
Lake and stream	250
Open ocean	125
Upwelling zones	500
Continental shelf	360
Algal beds and coral reefs	2,500
Estuaries	1,500

From R. H. Whittaker, *Communities and Ecosystems*, 2d ed., Macmillan, 1975.

least productive. But in places like the North Sea (fed by many nutrient-carrying rivers) and offshore from Antarctica and Peru, cold, upwelling water carries a high concentration of nutrients near the ocean surface. These nutrients, in turn, support high levels of phytoplankton productivity and some of the world's richest fishing grounds.

One shocking outgrowth of such ecological studies is the realization that humans are now appropriating about 40 percent of the primary productivity of our world's landmass. Figure 41-20 shows the components of net primary productivity (NPP) and how we direct them to our use. This large percentage is probably the greatest energy consumption by a single terrestrial species in life's history on earth. As our human populations double and redouble (see Chapter 43), our use of the earth's net primary productivity will no doubt increase substantially, leaving less and less for other organisms. In light of this, the predicted mass extinctions within tropical rain forests should be no surprise.

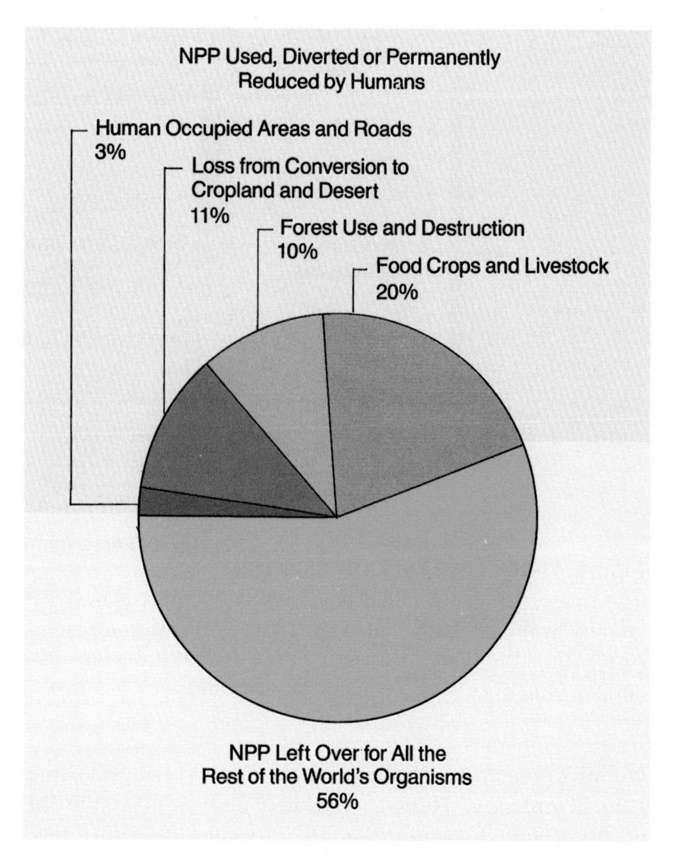

Figure 41-20 THE HUMAN USE OF THE EARTH'S RESOURCES.
As this graph shows, our single species uses 44 percent of the world's net primary productivity, leaving 56 percent for all the millions of other species combined. The net primary productivity of the earth's landmass is 132.1 units, where 1 unit = 10 g of organic matter per year.

The Efficiency of Energy Transfer

How much energy is transferred from the cytoplasm of organisms at one trophic level to the cytoplasm of organisms at the next? Ecologists have calculated that from 70 to 95 percent of the calories of net production at one level are lost by the time organisms at the next level have produced their cytoplasm (measured as caloric content) from that food. Or we can say that the efficiency of energy transfer between adjacent trophic levels, called **ecological efficiency,** amounts to between 5 and 30 percent with an average of about 10 percent. The efficiency of a particular level-to-level energy transfer depends on how successfully consumers at different trophic levels can find, capture, and eat the available plants or prey; how thoroughly they can digest food and absorb its molecules; and how much energy they expend in building new cytoplasm, maintaining cellular structure and function, and so on. As energy flows through an ecosystem, the ultimate effect of relatively low ecological efficiencies is to limit the number of trophic levels in a food chain. This limit is usually reached by the fourth or fifth level. Harvesting the energy from lower trophic levels (say, below phytoplankton) is too inefficient to be biologically worthwhile, while surviving as a supercarnivore—say, a predator of polar bears—would require an enormous total energy amount.

Ecologists try to understand how biological diversity arises and is maintained. The longer the food chain—the

ACID RAIN AND DUST: AIRBORNE RECIPE FOR DISASTER?

In recent years, ecologists have issued a series of stern warnings: Industry and government must tackle the problem of acid precipitation from the sky before the damage to plants, fishes, algae, and entire ecosystems is too great to reverse. At the center of the controversy is sulfur dioxide (SO_2), a gaseous pollutant spewed into the atmosphere by coal-burning power plants, factories, and other industrial sources. Ecologists worry that if not better controlled, SO_2 falling either as dry "acid dust" or as wet "acid rain" will damage living things in disastrous proportions.

The problem begins when airborne SO_2 is picked up by dust particles or is dissolved in water vapor and forms sulfuric acid. When this acid dust or rain contacts plant tissues, metals, cement, stone, or even some plastics, it quickly begins to corrode away the surface, as it has on this gravestone monument in a cemetery in Brooklyn, New York (Figure A).

In 1982, a West German government task force reported that fully 40 percent of that country's fir trees were either sick or dying as a result of acid rain. And a year earlier, in 1981, investigators at Cornell University discovered that more than half of the 400 lakes situated above 2,000 ft in the Adirondack Mountains of New York State were (and continue to be) so acidified that fish can no longer inhabit them without becoming stunted and deformed, as did this trout from Ontario, Canada (Figure B). Almost a decade later, researchers are estimating that some 50,000 lakes in the United States and Canada are in imminent danger of ceasing to support fish, algae, insect larvae, or aquatic life of any kind.

Figure B
Trout from the same lake in Northern Ontario, 1979 and 1982.

In addition to damaging lakes and forests, studies show that acid rain leaches minerals from soil; contaminates reservoirs and leaches lead and other heavy metals from pipes carrying human drinking water; increases respiratory illnesses causing, by some estimates, as many as 120,000 human deaths per year; and reduces agricultural yields by up to 50 percent per year.

Observers have noted that despite these environmental insults, the U.S. federal government and industries with the power to curb the problem are resisting change. Such groups have labeled efficient smokestack scrubbers and other new technologies capable of significantly reducing industrial emissions as "unnecessary" or "too expensive." Perhaps one reason that governments and industries have resisted solutions is that acid sulfates may be carried thousands of miles from their sources by prevailing winds to pristine wilderness areas as remote as the Arctic Circle. One region's pollution can therefore be another distant region's acid precipitation problem.

For many concerned scientists and laypeople alike, the question is no longer whether acid deposition represents a major ecological catastrophe. They question now whether the catastrophe can or will be stopped.

Figure A
Monument melting in the acid rain.

more kinds of organisms that depend on other organisms along the chain—the greater the diversity. What factors, then, contribute to longer food chains? Ecologists have assumed that total available energy is one such factor, but that view is changing. New results by ecologists at Rockefeller University, who studied 113 community food webs, show that the maximal length of food chains does not depend on the amount of primary productivity. Similarly, constant environments do not support longer chains than do variable environments. Instead, the environment's dimensionality is the key: Three-dimensional environments, such as thick forest canopies or ocean water columns, support longer chains than flat, two-dimensional environments, such as grassland, meadows, or lake bottoms.

Ecologists have learned to trace some individual substances (in addition to energy) through food chains. An unfortunate example involves the phenomenon of **biological magnification,** the process by which toxic materials present in trace amounts in the environment accumulate in organisms. For example, mercury, a poisonous metal, is an industrial by-product often released into rivers and lakes. Some bacteria can add mercury ions to organic groups, forming methyl mercury, which can easily be incorporated by photosynthetic cells. As these cells are eaten by small invertebrate animals, and they, in turn, are consumed by larger animals, the quantity of mercury tends to build up to higher and higher levels in certain animal tissues, such as fat cells; ultimately, what were extremely low concentrations in a lake are high toxic levels in the bodies of large carnivores.

CYCLES OF MATERIALS

While the earth receives a continuous supply of solar energy, the supply of minerals is limited to those already present on our planet. Living systems have adapted to this limited supply of minerals and have evolved complex processes for taking in and giving off the chemical constituents of life, thus participating in global cycles for water, carbon, nitrogen, and phosphorus.

The Water Cycle

Three-fourths of our planet's surface is covered by oceans, lakes, rivers, and ponds of water, the universal solvent of biological reactions, and this is endlessly recycled in a global **hydrologic cycle,** or water cycle (Figure 41-21). This begins when precipitation falls as rain or snow, with some evaporating as it falls, but most entering the earth's ecosystems in a variety of ways. Terrestrial plants and animals take up and retain water for varying periods. Other water runs off into lakes or streams or percolates down through layers of soil and rock until it reaches the *water table*—the lowermost layer of water in the earth's crust, underlain by solid rock. This groundwater can eventually seep into streams, flow in underground conduits to the sea, or be taken up by plant roots. Most of the water that falls as rain or snow ultimately flows back to the seas. The final leg of the global cycle is evaporation: Moisture evaporates from bodies of

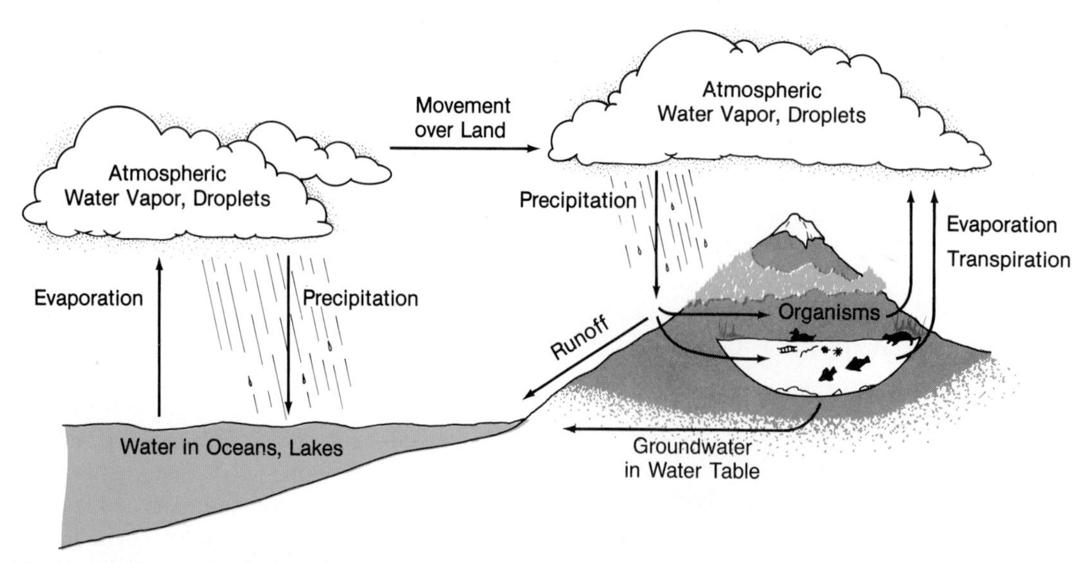

Figure 41-21 THE WATER CYCLE.
Water continuously cycles from oceans and other bodies of water to the atmosphere and then back again to the earth. Water then moves through plants, animals, and other organisms; from there it returns to the atmosphere. Water also enters the water table or returns to the sea and lakes as runoff.

water, moist soils, or animals' respiratory surfaces, or it is transpired through plant stomata; it then rises into the atmosphere and eventually returns to the earth as precipitation.

The Carbon Cycle

The cycling of carbon dioxide to organic compounds and back to CO_2 affects not only life itself, but also the earth's weather and geology. During photosynthesis, cyanobacteria, algae, and plants reduce carbon from gaseous CO_2 to generate sugars, which support most metabolic processes directly or indirectly (see Chapter 8). As living cells respire and liberate CO_2, some of these fixed carbon molecules are returned to the biosphere almost immediately. Other carbon passes from producers to consumers and then up the food chain; still more is eliminated as organic wastes into the environment. Collectively, this use and reuse of carbon is called the **carbon cycle** (review Figure 8-15), and carbon enters and leaves it from other sources as well. As leaves fall to the ground, and as fungi, animals, or plants die, carbon-rich residues build up in the soils or water systems. Carbon can be stored temporarily when such organic matter accumulates at the bottom of ocean basins and lakes or in stagnant swamps, and in some cases, it is compressed and chemically modified over time into oil and coal deposits.

In recent centuries, people have retrieved huge quantities of carbon stored in this way and have returned it to the biosphere by burning the fossil fuels. The result has been a general rise in the CO_2 content of the world's atmosphere; a typical record of this rise is seen in Figure 41-22. The spikes in the curve represent normal seasonal differences in photosynthesis, but the upward trend of the curve comes entirely from human activity. No one is sure what these rising CO_2 levels will mean to our planet's future climate, but predictions are alarming. Why? Because the layer of CO_2 allows light to pass through to strike the earth. Some of the light bounces off the earth as infrared radiation, and CO_2 then acts as a sort of heat blanket in the atmosphere, holding in some of the heat that would otherwise be radiated from the earth back toward space. Many scientists believe that the steady increase in atmospheric CO_2 is likely to raise surface temperatures by creating a *greenhouse effect* (much as the glass windows in a greenhouse or in a tightly closed automobile allow sunlight to enter but trap some of the infrared rays and heat inside). That could cause serious shifts in global climate patterns, including rainfall. In the 1980s, meteorologists recorded five of the nine warmest, driest years since 1850. While such records are still within normal expected temperature ranges, the forest fires and agricultural losses reminded many scientists of what could lie in store. Large land areas, including the

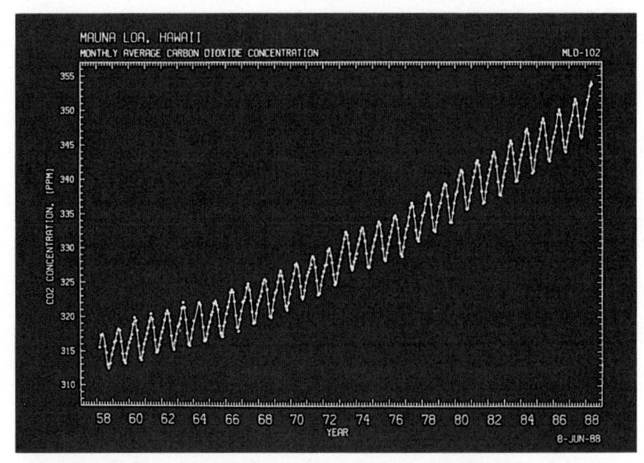

Figure 41-22 ELEVATING CARBON DIOXIDE LEVELS: A GROWING THREAT.
This 30-year record reveals a substantial increase in atmospheric CO_2, as measured in the air above the South Pole. The burning of fossil fuels worldwide is freeing CO_2 faster than plants can fix it or the sea can dissolve it. The zigzags in this curve reflect the annual cycling of CO_2 and seasonal differences in carbon fixation via photosynthesis. Ominously, however, the curve has been trending upward for three decades.

U.S. corn and wheat belts, might become hot, desertlike, and far less productive. Sea levels would rise from the melting of polar ice caps, inundating coastal cities. And the Great Lakes could dry into giant mud pans. A few scientists think that increased CO_2 will help produce a photosynthetic boon in global vegetation and that a similar phenomenon led to the lush polar forests that grew down to 85° south latitude during the Cretaceous period and early Cenozoic era. Even so, today's recorded rise in atmospheric CO_2 is huge and dangerously rapid—about 15 percent in 100 years.

Nature does have mechanisms for removing some excess carbon from the global carbon cycle: Atmospheric CO_2 tends to dissolve into cold seawater at high latitudes (a "sink"). Since CO_2 is less soluble at warmer temperatures, however, warm tropical seas act as a countervailing "source" of CO_2. In seawater, CO_2 forms the relatively insoluble compound calcium carbonate ($CaCO_3$), and this is deposited in the vast numbers of shells and exoskeletons of marine invertebrates or zooplankton. Carbon locked in rocks (such as limestone and marble) formed of such sediments can be returned to the cycle only when the rocks are uplifted by geological changes in the earth's crust, eroded by the action of wind and rain, and rendered soluble once again.

Ominously, our depletion of the ozone layer by the use of chlorofluorocarbons may be exacerbating the greenhouse effect. Experiments suggest that a thinner ozone layer and the resulting passage of more ultraviolet light can inhibit CO_2 fixation by phytoplankton, thus leaving a higher level of CO_2 in the atmosphere. What's more, the ozone hole could be allowing an even greater decrease in phytoplankton activity in the higher-latitude cold seas of the Antarctic. This would further disturb the natural "sink" effect we just discussed, and still more CO_2 would remain in the atmosphere.

The global carbon cycle is so complex that the best atmospheric scientists with the largest computers still find it nearly impossible to make realistic predictions. Right now, our future is truly unknown but the box below describes some of the steps we can take immediately, as individuals, to improve our prospects.

The Nitrogen Cycle

Like the carbon cycle, the **nitrogen cycle** involves chemical interactions between organisms and their environment (review Figure 27-8). First, atmospheric nitrogen is "fixed"—incorporated into nitrogen compounds—by certain bacteria and various cyanobacteria, by the action of lightning, and through industrial processes. Only when it is fixed can nitrogen be used by plants and, through them, by animals, fungi, and microbes. Ultimately, microorganisms dismantle the fixed compounds and return N_2 to the atmosphere.

Microbes play several crucial roles in the nitrogen cycle. During **nitrogen fixation,** specialized bacteria and cyanobacteria in soil and aquatic ecosystems reduce N_2 to ammonia (NH_3). Those organisms then use ammonia to synthesize nitrites (NO_2^-) and nitrates (NO_3) and then organic nitrogen compounds. Leguminous plants such as peas, with their symbiotic bacteria, not only fix nitrogen for direct use, but also produce enough that nitrates escape into the surrounding soil—the reason farmers often plant legumes in rotation with other crops, since with their rhizobial symbionts, they may fix several hundred pounds of fertilizing nitrogen per year in an acre of farmland. Furthermore, studies in Sweden of a legume-rhizobium crop called lucerne show that the nitrogen it fixes is not as easily leached from the soil as is nitrogen from around a heavily fertilized nonlegume like barley.

When a plant or animal dies and begins to decay, or when an animal excretes its wastes, decomposer organisms break down nitrogen compounds, forming ammonia (which forms NH_4^+, or ammonium ions). This is advantageous biologically, since the charged ion tends to remain bound in soil, in contrast to uncharged NO_2, which readily leaches away in water. Other soil bacteria then complete the cycle by carrying out **nitrification,** the oxidation of NH_4^+ to nitrite and then nitrate, which can be used by plants or microbes. Finally, still other prokaryotes carry out **denitrification,** the reduction of nitrate and nitrite to gaseous N_2, NO (nitric oxide), or N_2O (nitrous oxide, or laughing gas), all of which pass into the atmosphere, thereby completing the cycle.

HELP FIGHT GREENHOUSE WARMING

- Leave your car at home. Internal combustion engines are a big source of CO_2 emissions, and the United States—with its three cars for every four people and its power plants and industries—contributes one-quarter of the greenhouse gases even though it has only 5 percent of the world's population. Experts on global warming say we should all drive fewer miles, and, ideally, use vehicles that get 40 miles per gallon (instead of the U.S. average of 20 mpg).
- Plant a tree. The reforestation of cleared areas could mean the uptake of significant amounts of atmospheric CO_2 as the new, fast-growing trees photosynthesize. To remove about one-third of the global CO_2 released, humankind will have to plant new forests four and one-half times the area of California—a very tall order. To offset a single American family's contribution will require planting about 6 acres per year. Nevertheless, every tree helps.
- Use fewer propellants and coolants. The chlorofluorocarbon (CFC) compounds traditionally used in spray cans and in refrigerators and air conditioners float to the upper atmosphere. There, they trap 10,000 times more heat than CO_2, as well as destroying ozone and allowing more solar energy to reach earth's surface. The United States and a few other countries have banned the use of CFCs in spray cans, and are phasing them out as coolants. But manufacturers still use them to make those ubiquitous plastic foam fast food containers and other items. Some scientists see the total replacement of CFCs with benign substitutes as the first and easiest step in fighting the greenhouse effect.
- Support alternative energy sources. The United States still generates most of its electricity by burning coal and oil, and every year these fossil fuels contribute millions of tons of CO_2, SO_2, nitric oxides, and other pollutants to the atmosphere. Political and financial support for solar power, wind power, geothermal energy, and safe nuclear energy were never more essential.

The Phosphorus Cycle

Unlike nitrogen gas and carbon dioxide, molecules of phosphorus do not float about in the atmosphere; instead, all the phosphorus that living systems require for their ATP, nucleic acids, enzymes, structural proteins, and many membranes must enter biological pathways as soluble phosphate ions (PO_4^{3-}), dissolved originally by the action of water on rocks in the earth's crust. Plants, microorganisms, and phytoplankton incorporate these dissolved ions from soil moisture or from lake or ocean water, and thus the element enters the food chain. Figure 41-23 shows the **phosphorus cycle.**

Phosphate is a relatively scarce commodity—a state of affairs that limits the primary productivity of both land and aquatic environments. The most productive ocean regions, as we saw, are those where cold waters rich in phosphates upwell and support larger, phytoplankton populations in the photic zone, which, in turn, support more marine life (Figure 41-23, step 1).

How then, does phosphorus cycle back from the oceans to the continents? Over millions of years, phosphate-containing sediments in the sea form rocks; then those rocks must be thrust upward as mountains, only to erode away and once more yield PO_4^{3-} (steps 2–4). There are three shortcuts in the cycle, however. First, plants and microorganisms can take up PO_4^{3-} derived from animal wastes and dead organisms directly (step 5). Second, humans harvest some marine organisms directly for food and fertilizer and thus eventually return a small

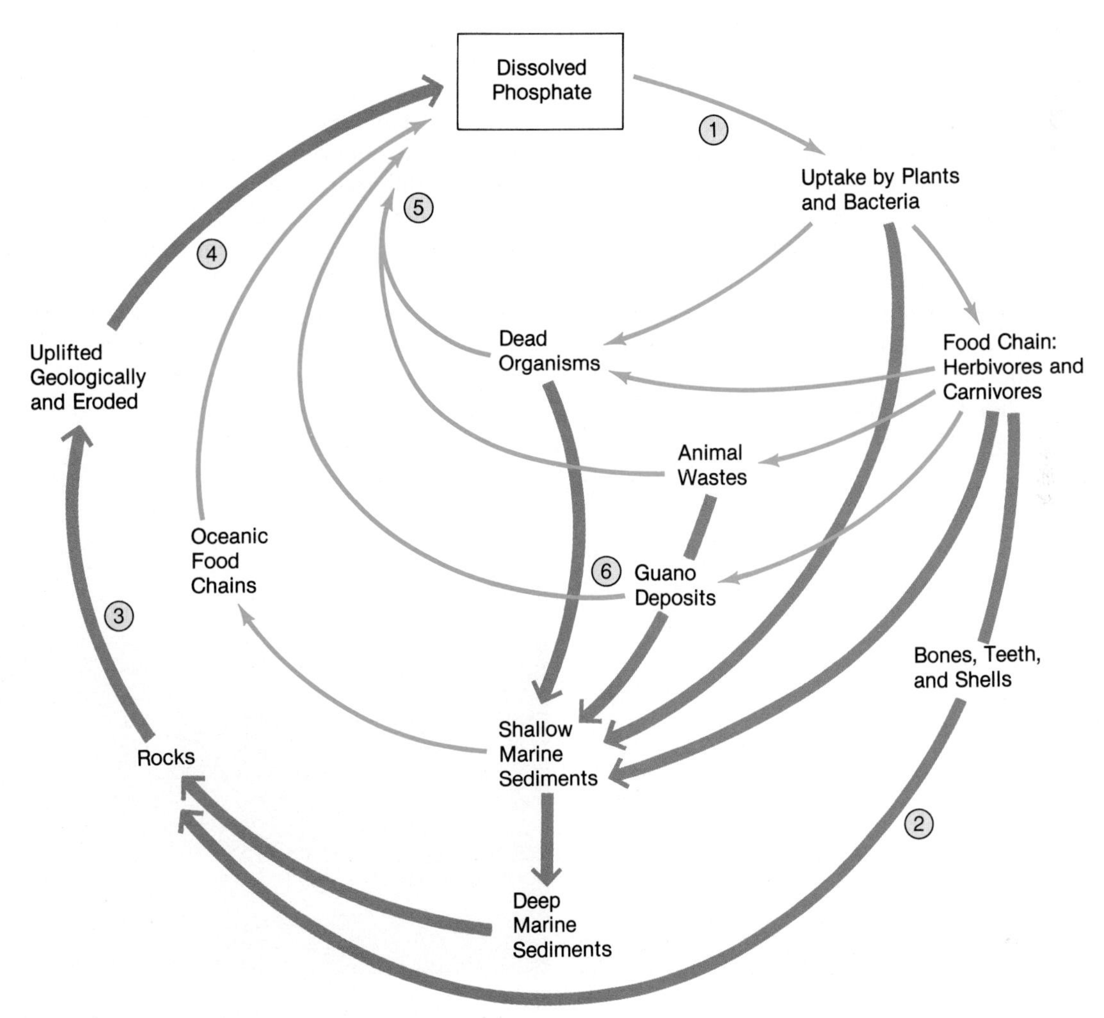

Figure 41-23 THE PHOSPHORUS CYCLE.
Phosphorus cycles in our biosphere as the phosphate ion (PO_4^{3-}). It cycles through water, soil, living systems, bones, some exoskeletons such as shells, and waste materials, including guano from birds and reptiles. The green arrows indicate the direct reutilization cycle; the blue arrows show PO_4^{3-} entering shallow marine sediments, from which some may be recaptured by the marine food chain (leading to fishes, birds); but most is lost to deep marine sediments (the "indirect cycle"). These sediments ultimately become rock that must emerge on the continents again in order to be eroded and once more yield phosphate. Some bones, teeth, and shells are degraded and reenter the direct reutilization cycle.

fraction of needed phosphate ions from the sea to the soil. Overall, however, this minor bit of recycling cannot begin to meet the needs of modern agriculture, which consumes huge quantities of phosphates to increase the productivity of food plants. The third, and greater, source of phosphates used in agriculture is *guano*—the droppings of marine birds that feed on phosphate-rich ocean plants and animals (step 6)—which people gather from bird colonies and process into fertilizer.

Despite these steps, the world's rivers discharge perhaps 14 million tons of phosphate each year, whereas seabirds, mining, and fisheries return only 70,000 tons per year back to the land.

CLOSE-UP OF A LAKE ECOSYSTEM

Research has revealed how a relatively self-contained ecosystem such as a lake operates. The physical and chemical properties of water, especially light penetration, govern much of what goes on in a lake. Rock and soil chemistry, both of the bottom and of the drainage flowing into the lake, also affects life there. Photosynthesis by phytoplankton or water plants occurs only down to a depth where approximately 5 percent of the sunlight hitting the surface can penetrate. Hence, free oxygen is produced only above that depth. The edges of the lake, where light reaches the bottom, make up the **littoral zone** (Figure 41-24); plants in the littoral zone are rooted in the lake bottom and grow to the surface, and whirligig beetles, water striders, and many protists and microbes inhabit the water or its surface.

The center of the lake, down to the depth where O_2 production by photosynthesis just equals respiratory utilization of O_2 by the collective life forms, is called the **limnetic zone.** The limnetic zone is inhabited by dinoflagellates; algae, such as *Volvox* and *Euglena;* nitrogen-fixing cyanobacteria, such as *Anabaena;* buoyant crustaceans, called copepods; and rotifers, which graze on the algal organisms. Below the depth of effective light penetration is the **profundal zone,** where decomposition exceeds production of organic materials. The energy input to these deeper lake waters comes in the form of organic materials or dead organisms that sink from the upper water levels. Various species, including bacteria, fungi, invertebrates, and fishes, decompose and/or consume it.

In winter, the coldest water in a lake is at the surface, where it may freeze into an ice layer. The densest water, at 4° C, is found near the lake bottom; there, trace heating from the underlying earth may occur. The layering of water at different temperatures is called *stratification.* As ice melts in the spring, the surface waters sink as they warm toward 4° C and become denser. Winds also churn the waters, the overall result being movement of deeper waters to the surface, in a process called *spring overturn.* An *autumn overturn* also occurs, and both redistribute nitrates, phosphates, oxygen, and other raw materials.

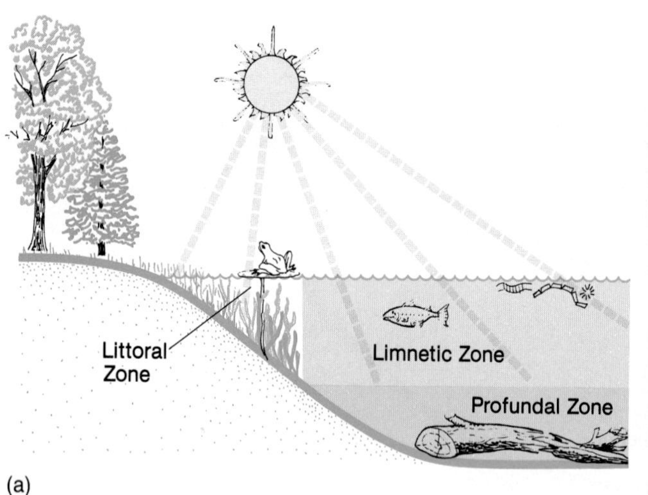

(a)

(b)

Figure 41-24 LAKE ZONES.
Lake zones are defined by the amount of light available and the amount of oxygen and biomass produced. (a) The littoral zone lies along the shoreline, where light reaches the bottom and rooted plants grow. The limnetic zone lies in the middle of the lake, beginning at the edge of the littoral zone, and extends to the depth beyond which light is insufficient to support a rate of photosynthesis greater than the rate of respiration. In the profundal zone, there is no effective light penetration.
(b) Reflection Lake, in Washington, has a typical shallow littoral zone, in the foreground, and the deep-water limnetic zone farther from the shore.

The clear, gorgeous blue waters of some deep mountain lakes reflect very low biological productivity because essential mineral nutrients are lacking. Low-productivity lakes are termed *oligotrophic*. In contrast, *eutrophic*, or "well-nourished," lakes are good producers of organic materials and organisms. These lakes tend to be shallow; to have high turnover rates of phosphorus, nitrogen, and other nutrients; and to be inhabited by a wide variety of microbes, algae, plants, and animals.

In the history of many lakes, a natural process called **eutrophication** takes place, during which a lake becomes richly productive, and more and more organic debris and silt accumulate. The lake becomes shallower, and the spectrum of organisms that live in it changes as it fills in and becomes a bog and then, perhaps, a meadow. Human activities often cause this natural succession to accelerate: If human sewage, drainage from fertilized farmlands, or water from homes or industries enters a lake, the normally minute amounts of phosphates that limit growth are supplemented massively. Huge populations of algae (especially cyanobacteria) and aquatic plants can spring up, and they choke the open-water areas. In time, the extra vegetation dies and rots, and the decomposition process consumes oxygen and makes it impossible for fishes and many invertebrates to survive.

LOOKING AHEAD

We have seen that ecosystems—from the biosphere, to the major biomes, to individual lakes—are incredibly complex tapestries of interacting organisms, energy transfer, and cycling minerals. To understand those complex interactions more fully, we must peel back another layer of the ecological onion and consider the workings of *communities* of organisms—our subject in Chapter 42.

SUMMARY

1. *Ecology* is defined as the study of the interactions that determine the distribution and abundance of organisms.

2. The largest and most complex level of biological organization is the *ecosystem*. The *biosphere* is the largest ecosystem, but the *biomes*, the major vegetation regions on our planet, as well as individual lakes, meadows, and forests, are also ecosystems.

3. The earth's land surface, the oceans and fresh water, and the atmosphere above them form the *biosphere*—the life-supporting environment for all living creatures.

4. The amount of solar radiation intercepted by our planet is known as the *solar constant*. Only about 5 percent of the energy first entering the atmosphere ever reaches the planet's land and water surfaces.

5. Day and night, seasons, different patterns of climate, and uneven heating of the planet's surface all result from the combination of the earth's tilt, its rotation each 24 hours, and its circling of the sun. A region's climate, in turn, determines the sorts of plant and animal life that may populate a given area.

6. The terrestrial biomes include: *tropical rain forests, tropical seasonal forests, savannas, deserts, temperate grasslands, temperate shrublands, temperate forests, taiga*, and *tundra*. There are also several aquatic biomes: freshwater biomes include *lentic* (lake and pond) *communities* and *lotic* (stream) *communities;* marine biomes include sandy beaches, intertidal zones, coral reefs, *pelagic* (open-ocean) *communities*, and *benthic* (bottom) *communities*.

7. On land, many plants and most animals must acquire water and nongaseous mineral nutrients directly or indirectly from the soil.

8. The food, physical support, gases, and water needed by aquatic organisms are provided by the surrounding medium. Photosynthesis takes place in the strongly illuminated upper water layers, or *photic zone*.

9. Energy flows through ecosystems by way of *food chains* that are part of larger interlocking *food webs*. Food chains are organized into *trophic levels;* photosynthetic organisms are the *producers* at the first level, and *consumers* are at successive trophic levels. Energy may also be transferred from each level to *decomposers* and *detritivores*.

10. Ecosystems often show *pyramids of numbers* and *pyramids of biomass* (total mass of living organisms). In typical food chains, there are decreasing numbers of larger and larger organisms as one goes up the chain. The number of organisms existing at any one time in a trophic level or population is referred to as the *standing crop*.

11. An ecosystem's *primary production* consists of the energy stored in molecules of photosynthetic organisms; *secondary production* is the energy stored in the molecules of consumers.

12. *Biological magnification* refers to the accumulation of trace materials in animal tissues, often to toxic levels, as one moves up a food chain.

13. The water cycle, or *hydrologic cycle*, is a global phenomenon; water that evaporates in one place is likely to fall to earth as rain in another part of the globe.

14. Carbon taken up from the atmosphere as CO_2 by photosynthesizing cells is returned to the air by respiration, by combustion, or as organic wastes in a massive *carbon cycle*.

15. In the *nitrogen cycle*, atmospheric nitrogen is fixed by bacteria and cyanobacteria. The fixed nitrogen is utilized by plants and animals and then is returned to the atmosphere through a complex series of chemical steps carried out by microorganisms.

16. As part of the *phosphorus cycle*, all the phosphorus required by living systems enters biological pathways as phos-

phate ions dissolved by water eroding rocks on the earth's crust. Phosphates enter the food chain when they are taken up from soil or water by producer organisms. There is continual loss, however, to deep marine sediments and, over eons, the earth's rocks.

17. A freshwater lake ecosystem can be divided into the *littoral zone*, the *limnetic zone*, and the *profundal zone*. *Eutrophication* is the natural process by which lakes become increasingly productive.

KEY TERMS

benthic community
biological magnification
biomass
biome
biosphere
carbon cycle
clay
consumer
Coriolis effect
decomposer
denitrification
desert
detritivore
ecological efficiency
ecology
ecosystem
eutrophication
food chain
food web
gross production
habitat
humus
hydrologic cycle
imbibition
lentic community
limnetic zone
littoral zone
loam
lotic community
net production
nitrification
nitrogen cycle
nitrogen fixation
pelagic community
phosphorus cycle
photic zone
primary production
producer
profundal zone
pyramid of biomass
pyramid of numbers
sand
savanna
secondary production
silt
solar constant
standing crop
taiga
temperate forest
temperate grassland
temperate shrubland
trophic level
tropical rain forest
tropical seasonal forest
tropical woodland
tundra

QUESTIONS

1. Explain briefly why each of the following is or is not an ecosystem: the earth; a biome; the biosphere; a lake; the moon.

2. What percentage of the solar constant actually arrives at the earth's surface? What percentage of *that* energy is then fixed during photosynthesis? What happens to the rest of the solar energy?

3. The sun's energy strikes the earth unevenly. Explain aspects of our planet and its motions that combine to produce what we call climate and seasons.

4. Ecologists recognize approximately sixteen major biomes. Which of the terrestrial and aquatic biomes described in the text can be found in the United States? Where?

5. Outline the challenges presented by each of the earth's three major habitats, and name the adaptations that help organisms overcome them.

6. Explain how energy flows through an ecosystem, using food webs and ecological pyramids in your answer.

7. In the Bear Island ecosystem, which organisms were producers? Which were primary consumers? Secondary consumers?

8. Of four natural cycles—water, carbon, nitrogen, and phosphorus—which depends on a number of microorganism species? Which depends (in the contemporary world) on the activity of seabirds? Which has been affected by human activity? Which may lead to a greenhouse effect?

9. Explain the difference between oligotrophic and eutrophic lakes. Into which category would a deep, high-mountain lake most likely fall? Why?

ESSAY QUESTION

1. List ten foods that you have eaten in the past week, and for each, describe the probable food chain (or food web) through which the sun's energy passed to you, the "top consumer" in the chain.

For additional readings related to topics in this chapter, see Appendix C.

42
The Ecology of Communities

It is interesting to contemplate a tangled bank, clothed with many plants of many kinds, with birds singing on the bushes, with various insects flitting about, and worms crawling through the damp earth, and to reflect that these elaborately constructed forms, so different from each other, and dependent upon each other in so complex a manner, have all been produced by laws acting around us.

Charles Darwin, *On the Origin of Species by Means of Natural Selection* (1859)

In the spring of 1985, a widespread fire broke out on Isabela Island, the largest of the Galápagos Islands. Firefighters battled the blaze for more than a month, but it eventually claimed more than 100,000 acres, or one-quarter of the island's vegetation, destroyed several stands of the unique Galápagos sunflower tree, and removed much of the habitat of a subspecies of giant tortoise and of a rare seabird, the dark-rumped petrel.

While the fire had marked effects on those individual species, research may well show that the greatest impact was on the Isabela Island **community:** the interacting populations of different species in an area. After a region is denuded by fire, there is a period of regrowth and reestablishment of plant and animal species. But which will come back, and in what order? Will the area eventually look just like it did before? Or will it be changed forever by the disaster? The answers depend on the

A community beneath the Red Sea surface: Soft and hard corals, kelp, sea stars, limpets, dozens of fish species, all interdependent in myriad ways.

complex interactions of all the community members, and in this chapter, we see how those interactions set limits on where and how organisms can live and how successfully they survive.

Members of species within communities sometimes interact in mutually beneficial ways, but most of the time, those organisms compete for limited resources or use members of other species for their own survival. Selective pressures, such as interspecies competition and predation, have led to the evolution of fascinating **adaptations**—genetically based characteristics that increase an organism's chances of survival. Adaptations for escape, defense, or intimate associations between species' members (so-called symbioses) are just as vital to a species' survival as their adaptations to temperature, habitat, water, or other physical aspects of the community's environment.

Some of the most dramatic adaptations help protect organisms against predators or help protect predators against their prey species' defenses. Prey and predator act as two sides in an endless evolutionary tug of war, with new modes of attack evolving for each defensive innovation. This reciprocal phenomenon, or *coevolution* (discussed in Chapter 23), has produced an arsenal of poisons, venoms, toxins, thorns, bold warning colors, and other devices. No species can evolve such predatory or defensive adaptations deliberately; they simply arise as natural selection alters gene frequencies and resultant phenotypes within population members.

Defining and studying communities and community interactions are vitally important processes as we confront our own species' enormous impact on the earth's ecosystems. How well we come to understand the causes and consequences of each interaction in the living world—from the countermeasures of prey and predator to natural fires on remote islands to the worldwide greenhouse effect—may well determine our collective futures and those of our descendants.

This chapter discusses five major topics:

- *Ecological succession,* or gradual changes in the mix of species that populate all land and many aquatic communities
- The structure of communities, including the kinds and numbers of species and their niches, or ecological roles
- Defensive adaptations among species in a community, including chemical arsenals and bright colors that warn of noxious flavors or behaviors
- Adaptations for escape, including locomotion, camouflage, and life cycle stages that diminish contact with potential predators
- Symbioses—intimate associations between different species that can be harmful, beneficial, or neutral to one or both

(a)

(b)

Figure 42-1 PRIMARY SUCCESSION: A BEGINNING WITH BARE ROCK. (a) A lava flow from the Pahoehoe vent on the Kilauea volcano, the island of Hawaii: About 10 years later, plants are taking root. (b) Fugitive species such as these daisylike composites tend to be poor competitors under crowded conditions, but good pioneers in harsh, newly opened environments.

ECOLOGICAL SUCCESSION: A BASIC FEATURE OF LIFE

As you travel inland from the North Carolina coast, you can see the clumps of grasses that dot the sand dunes along the coast giving way to wild oats and scrubby brush and then, farther inland, where the habitat is protected from ocean breezes and salt spray, to pine groves. The birds and other animal residents also change from one geographical area to the next. While the differences between communities of plants and animals can be obvious, it is less obvious that in any single area, the community may change over the years. A lake in a high mountain meadow, for example, may become a marsh, the marsh may become a thicket of willows, and eventually, the thicket may become a forest of pines.

A basic feature of life in the earth's ecosystems is that the types of species present in a given area, the *species composition*, are nearly always changing—sometimes imperceptibly, sometimes quite rapidly. Biologists call the gradual changes in a community's species composition **ecological succession,** and this alteration can result from natural changes in the physical environment or simply from changes that the living organisms bring on themselves. Lichens, for example, can erode and destroy their own rocky substrates, and soil-rooting plants can displace the lichens. And certain plants can so deplete the soil of nutrients that other plant species better adapted to poor soils can outcompete them.

Ecologists describe a succession of communities as either primary or secondary. **Primary succession** begins in areas with no true soil; the first plants must establish themselves on the base material, such as the lava rock spewed periodically from Kilauea Crater on the big island of Hawaii (Figure 42-1a). **Secondary succession,** on the other hand, begins on soils that have developed over millenia, but where a fire, a shifting river, lumbering, or some other physical disturbance has denuded the preexisting vegetation.

Primary Succession

The first colonizers in a newly available soil-free area, such as the lava flow from Kilauea Crater, are often the lichens and mosses that can subsist on bare rock. Both trap dead organic materials, creating a layer of humus, and some ecologists think that lichens carry out biochemical processes that slowly break down and erode rock surfaces.

As pockets of soil develop, plants that can survive with few soil-bound nutrients and little moisture take root. These hardy pioneers are often grasses and, in tropical areas, ferns. The seeds of pioneer species are blown long distances by the wind or are carried in the coats of animals far from the site where the parent plant grew. Because such species seem to specialize in precarious existences, ecologists have dubbed them **fugitive species.** Fugitive species are "weeds" in their respective habitats; they reproduce quickly but are quickly replaced by other organisms (Figure 42-1b).

As the pioneer plant species die and decay, their tissues add organic debris and nutrients to the decomposing and weathering base. As time passes, the accumulating soil deepens, traps more water, and allows less hardy species—other kinds of grasses, shrubs, and small trees, for example—to spread into the area. Soon, as resources are increasingly available, the habitat may host a population explosion of individuals from many species.

After time passes, the community consists of a mixture of middle and late species and is at its most diverse state. Succession will eventually create a rich *Acacia* and *Metrosideros* forest on the once-barren lava flows of Hawaii. Late-successional species depend on support structures (such as tree trunks) to carry photosynthetic leaves above competing plants toward the sunlight and on root structures deep into the soil, where moisture is more abundant. As the community undergoes successive changes, larger and longer-lived plant species tend to become dominant, and the biomass of the area increases. If left undisturbed, the community may eventually be-

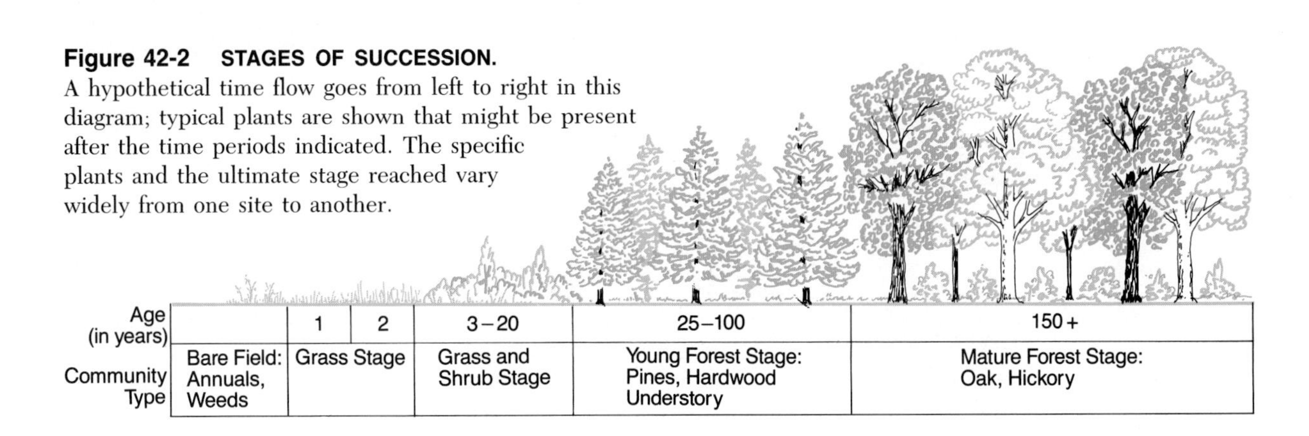

Figure 42-2 STAGES OF SUCCESSION. A hypothetical time flow goes from left to right in this diagram; typical plants are shown that might be present after the time periods indicated. The specific plants and the ultimate stage reached vary widely from one site to another.

come dominated by species that simply reproduce themselves and are not replaced by new arrivals. This kind of ecological constancy is known as a **climax community.** In theory, the climax community will endure as long as the environment remains stable (Figure 42-2).

People have sometimes introduced foreign species into an area that alter the area's primary succession. For example, Portuguese sailors introduced a nitrogen-fixing tree, *Myrica faya*, into Hawaii. This species can invade volcanic lava sites where no native plants can live; it has spread widely, and because it adds substantial nitrogen to the site, it seems to increase overall biological productivity.

A form of primary succession called *aquatic succession* occurs in lakes and ponds as silt and organic debris from decaying organisms accumulate on the lake bottom until the lake fills in and becomes a semisolid bog (Figure 42-3a, b and e). Shrubs may then invade, and, finally, as the soil dries even more, a grove of trees can grow (Figure 42-3c and d).

Secondary Succession

Primary succession is as rare as new lava flows, but secondary succession is common because developed soils have covered most of the earth's land surface for thousands of centuries. These soils can be an ideal "cradle" for new plant colonizers that replace others destroyed by ecological disturbances such as flooding rivers, forest fires, avalanches, farming, or logging.

When rivers change their channels, for example, soils can be eroded in one area and deposited in another. This process has dominated the vast upper Amazon regions where numerous rivers drain the Andes Mountains. Over the years, the rivers have meandered back and forth, with bends in the river eroding forest soils at the rate of 12 m per year. In time, this wipes out forests, leaves "islands" of surviving trees scattered as mosaics, or so thoroughly denudes an area that secondary succession must start, in which case it proceeds in the typical stages shown in Figure 42-2 from the riverbank inland. This process has broken the vast Amazon jungles into a mosaic of small forests of different successional stages on soil deposits of varying composition and has contributed to the ecosystem's great biological diversity.

Sometimes, farmers or loggers abandon a stretch of farmland or a hillside denuded of trees, and the area undergoes secondary succession. Often termed *old field succession*, this gradual recovery process can leave trees standing in once-plowed furrows (Figure 42-4).

In dry shrublands and grasslands, where floods or fires can be common, secondary succession begins when certain plants recover quickly by resprouting from roots.

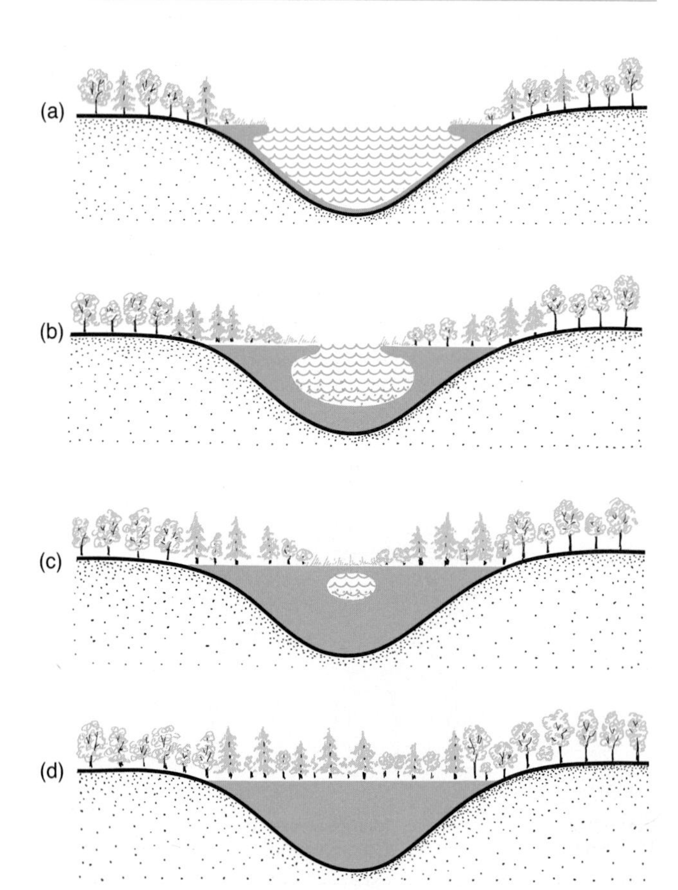

Figure 42-3 AQUATIC SUCCESSION: A DISAPPEARING LAKE.
(a) Organic material encroaches on the edges of a lake and is washed into the lake bottom. (b) A floating mat of organic material eventually closes off the surface, (c) forming a peat bog, which (d) then undergoes succession into forest stages. (e) Black Pond, in Norwell, Massachusetts, is at the stage of succession illustrated in (b). A mat of sphagnum moss can be found around the pond. Rooted in it are water willows, sedges, cranberry, grasses, orchids, and the carnivorous pitcher plant and sundew.

Figure 42-4 OLD FIELD SUCCESSION: A FOREST RETURNS.
In this photo of old field succession in North Carolina, plow furrows are still visible, even though the field is already in the pine forest stage of succession. When this "field" was still under yearly cultivation, natural succession was held in check.

Others, such as the jack pine, produce cones that open in response to heat and drop their seeds on the scorched earth. In the summer of 1988, after several exceptionally dry years, devastating fires swept Yellowstone National Park (Figure 42-5). While many people have bemoaned the fact that one-fifth of the scenic park burned, ecologists are seeing it more as a renewal than a disaster, and as an invaluable lesson in secondary succession. About 77 percent of Yellowstone's forests are composed of stock-straight lodgepole pines growing in dense, heavily shaded stands. Ecologists have begun dozens of studies of precisely which plant species will regenerate, in what order they will return, and which animals they will support.

Fugitive species returning to an area grow quickly, expend little energy developing "unnecessary" woody support tissue, and instead produce structures such as seeds, or storage tubers. In fact, many of our food crops, including wheat, oats, potatoes, sugarcane, and beets are early-stage fugitive species whose wild stocks required bare and sunny ground for seed germination. One of the prices we must pay for continually reaping this harvest is keeping the community in an early-successional stage by tilling, clearing, and weeding frequently (Figure 42-6). Late-successional plant species, in contrast, tend to develop large trunks, large branches, and roots and, as we have seen, tend to remain as stable, long-term community members.

Succession can result in "lateral" movement of vegetation types. Since the last ice age ended 10,000 years ago, the edges of various forest types have "migrated" north at rates of 1–8 km per generation. Seeds, of course, are the means by which tree species disperse and invade.

One might be tempted to conclude that succession proceeds in an orderly, preordained sequence, but the exact sequence of species during either secondary or primary succession can be highly variable, and early researchers attempted to understand this variability and to define the rules of succession.

Figure 42-5 CONFLAGRATION AND RENEWAL.
Young lodgepole pine trees grow at the base of scorched adults in Grand Teton National Park, just south of Yellowstone in Wyoming.

Figure 42-6 CULTIVATION: THE PREVENTION OF NATURAL SUCCESSION.
The prevention of natural succession requires an enormous investment of energy and labor, as any farmer can testify.

Looking for the "Rules" of Succession

In 1916, a young botanist from Nebraska, Frederick Clements, forged a general theory of succession that incorporated three simple assumptions: (1) In every community, climate alone determines the types of plants that occur in various stages of succession; (2) all successions eventually lead to a climax in a series of predictable stages; and (3) succession is governed by a sort of grand plan, each new group of species paving the way for the next, so that just one type of climax community is possible in any given area.

A core idea of Clements's theory was that climax communities seem to be more stable (i.e., they seem to persist longer) than other successional stages, *and thus they must be better adapted*—in fact, best adapted—to the climate of a particular region. Despite these conclusions, fieldwork often showed that sites with the same climate had different patterns of species replacement and that climax communities very often failed to appear.

In 1970, British ecologist Donald Walker tested Clements's rules by studying 66 lakes and ponds scattered throughout Great Britain for which the stages of aquatic succession—the progression from bog to shrubs and trees—were well documented. By comparing the sequences in each plant community, he found wide variation in the sequence of communities from one lake to another, rather than a predictable order to succession. Furthermore, the supposed climax species, willow and alder trees, were often replaced by long-lasting peat bogs.

New research shows that there are three mechanisms of succession: facilitation, inhibition, and tolerance. In *facilitation*, which may be fairly rare, early species make colonization or growth easier for later species. In *inhibition*, early species prevent the establishment of later ones until some disturbance, such as a destructive storm, extensive grazing, or deliberate bulldozing, removes the early-successional species, allowing the later plants to emerge and flourish. The mechanism of *tolerance* lies between these two extremes: Later species tolerate early ones, and the early species have no effect on the speed with which later species appear. At any one site, soils, local topography, weather, season, herbivores, the chance presence of certain seeds but not others, and a long list of additional factors can affect the three mechanisms.

Stable associations such as climax communities arise when the environment remains stable and gives the component species an adaptive edge, but they may disappear when the environment changes again. In the end, then, organisms' evolutionary adaptations determine when and how one stage in a succession will give way to the next. Certain types of algae, for example, are adapted to become established quickly on storm-cleared boulders, whereas other algal species are adapted to flourish only when herbivores reduce the cover.

Succession is one more reflection of the competition for light, nutrients, space, food, and other resources in the struggle of organisms to survive and reproduce.

Animal Succession

Some animals—such as herbivorous insects that breed, pupate, or feed on a particular species of flowering plant—are so closely associated with certain plants that when the plants are replaced in a succession, the animals disappear, too. Ecologist Victor Shelford studied a series of different-aged ponds near the southern shore of Lake Michigan in 1911 and found that the animal communities indeed undergo succession. He noted different species of fish in different ponds, as well as different species of insects in the dunes around each pond and varying types of mammals and reptiles inland from each pond.

Shelford's contemporary, L. L. Woodruff, observed animal succession in the laboratory by creating a murky artificial soup of pond water and hay in a glass flask. Over the next few weeks, species after species of protozoa—amoebae, paramecia, tiny flagellates, and other forms—replaced each other in "blooms" at the upper surface of the flask. Woodruff's experiment showed that while successions are hard to predict in nature, they always proceed in the same sequence in a simple laboratory culture.

THE STRUCTURE OF COMMUNITIES

Ecological successions are series of communities in time, but what is a community really? In their definition, ecologists generally consider (1) the physical appearance of the community—the *physiognomy*, or physical form that the component species take; (2) the *relative abundance* of rare and common species and individuals within those species; (3) the total number of species, or *species richness;* and (4) the *niches*, or "occupations," of each species in the community.

Physiognomy: What Does a Community Look Like?

Most ecologists depict a community's physical structure through its permanent affixed residents—its plants. Table 42-1 reviews the major growth forms of plants.

Table 42-1 GROWTH FORMS OF TERRESTRIAL PLANTS

Form	*Description*	*Example*
Thallophytes	Small plants without differentiated leaves	Mosses, lichens
Herbs	Plants without woody stems	Grasses, ferns
Shrubs	Small woody-stemmed plants less than 3 m tall	Yucca, cacti, chaparral shrubs, broad-leaved shrubs
Trees	Large woody plants more than 3 m tall	Conifers, oak and other hardwoods, bamboo, tree ferns, palms
Epiphytes	Whole plants that grow on other plants	Bromeliads
Lianas	Woody vines rooted in soil	Passion flower

Recall from Chapter 41 that the earth's major terrestrial communities of plants and animals are the biomes and that they are controlled by climate and geography. Lacking neat boundaries, these major communities merge into one another, often along climate gradients of moisture or temperature called **ecoclines.**

As the environment along a gradient becomes colder, drier, or otherwise harsher, community physiognomy changes from complex to simple, with a diversity of plants giving way to fewer, smaller forms and multiple layers of foliage becoming single. You can see just such an ecocline when you hike up a mountain from a forested valley to the alpine tundra above the tree line.

The particular *combination* of plant species growing in an area largely determines the community's physical form, and botanist R. H. Whittaker developed a quantitative technique called *gradient analysis* to record the actual distribution of plant species, along with data on moisture levels, temperature, and elevation gradients in "boundary regions."

Whittaker carried out his investigations in two mountainous regions in the western United States: the Siskiyou Mountains in Oregon and the Santa Catalina Mountains in Arizona. In both areas, a temperate forest biome gives way to an arid desert landscape along a moisture gradient, but Whittaker found that both plant and animal species occur at random with respect to most other members of the communities, and the communities intergrade—they lack boundaries with other communities. This conclusion rules out the notion that a community functions as a "superorganism" with a fixed group of species carrying out specific tasks.

Relative Abundance of Species

Every community contains a mixture of species in which certain individuals are common, others are fairly abundant, and a few are rare. The proportion of individuals of various species in a forest of firs in the Great Smoky Mountains of Tennessee ranges from 70 percent down to 0.001 percent of the total community. By contrast, most of the species in a tropical rain forest in Brazil have about equal population sizes; typical shares of the community range from 0.003 to 1 percent of the total. This feature of *equitability* is typical of tropical communities, whereas communities in other biomes, where one or a few species are often extremely common, are said to show a high degree of *dominance* (Figure 42-7). Full explanations for such patterns await future research.

Species Richness: Taking Count of Species

On a global scale, the number of species from different taxonomic groups varies tremendously between one community and the next. Biologists have identified more than 8,000 species of seed plants in the tropical rain forest biome of Costa Rica, for example, compared to only about 1,600 species in the British Isles, with six times the area of Costa Rica. Similarly, biologists have named about 120 species of mosquitoes in the whole of the

Figure 42-7 DOMINANCE IN A TEMPERATE FOREST. In contrast to the usual equitability in tropical forests, this lodgepole pine forest in the northern Rockies is composed predominantly of one species; the trees grow at incredible density and extend tall and straight as they seek the light above.

United States and Canada, but more than 150 mosquito species within a 10-mile radius in Colombia. What underlies such startling differences in the number of species in a community?

Looking at our planet as a whole, in most major taxonomic groups of land plants and animals, **species richness,** or *species diversity,* can be correlated with latitude. The farther one travels north or south of the equator, the fewer species they will find in various biomes. Tropical rain forests near the equator are the richest biological areas on earth, while the arctic tundra is one of the poorest (Figure 42-8a, b). Ecologists have reported up to 200 species of trees in a 5-acre patch of rain forest (eight times the number of types that might be found in a vast temperate forest), while entomologists armed with collecting nets have scooped up more than 500 species of tropical insects at a single site. Because of this species richness, the rapid destruction of tropical rain forests for lumber and agriculture is endangering millions of plant and animal species, some of which have not even been described yet (see Chapter 41).

(a)

(b)

Figure 42-8 SPECIES RICHNESS VARIES WITH LATITUDE.

A tropical rain forest, such as this one from LaSelva Biological Station in Costa Rica (a), shows a far broader diversity of species than a cold, dry tundra area such as this musk ox in Northern Canada (b).

Ecologists have devised many theories for the global trends in species richness. Ecologist George Stevens at the University of Alaska, for example, recently proposed that the species at high latitudes tend to live in broad geographical ranges and that fewer species persist there because members are subjected to wide fluctuations in climate and environment. At lower latitudes, however, species have much smaller ranges and their members are adapted precisely to specific microclimates that vary little from season to season. Stevens thinks that tropical species richness may result from a core group of stable species that is continuously replenished by an influx of less adapted colonists from other microclimates. Other researchers studying North American tree species conclude that solar energy, warmth, and water are keys to the degree of species richness at a given latitude, and, of course, the luxuriant tropics have plenty of all three.

The Species Equilibrium Model

In the 1960s, theoreticians Robert MacArthur and Edward O. Wilson began looking at the structures of communities on islands to try to understand why a given number of species occupies a particular area. Ecologists had long observed that large islands tend to have more species of a particular taxonomic group—perhaps birds or ferns—than do small islands. To explain such species richness, MacArthur and Wilson proposed the **species equilibrium model.**

Their model assumes that the number of species on an island is a balance between (1) the *immigration rate* of "new" species (from other inhabited areas) onto the island and (2) the *extinction rate* of established species (Figure 42-9). The model further assumes that the balance is dynamic, with new species continuing to enter the system, and others becoming extinct. Imagine an island that was formed recently, for example, by a volcanic eruption. As more time passes, species colonize the island and accumulate. With more species present, more can become extinct, so the extinction rate increases (Figure 42-9a). Further, as the number of species on an island increases, there is less likelihood that a newly arriving organism will be a member of a new species. Therefore, the rate of immigration decreases (see Figure 42-9a). Eventually, the rates of immigration and extinction will be equal, and that point will determine the equilibrium number of species. As the analysis in Figure 42-9b shows, the equilibrium number of species on a small island will be lower than on a large island. This is because (1) the extinction rate will increase more steeply on a small island, owing to each species' smaller population size, and (2) the immigration rate will be much lower than on a large island. The model also predicts that the farther away the island is from sources of

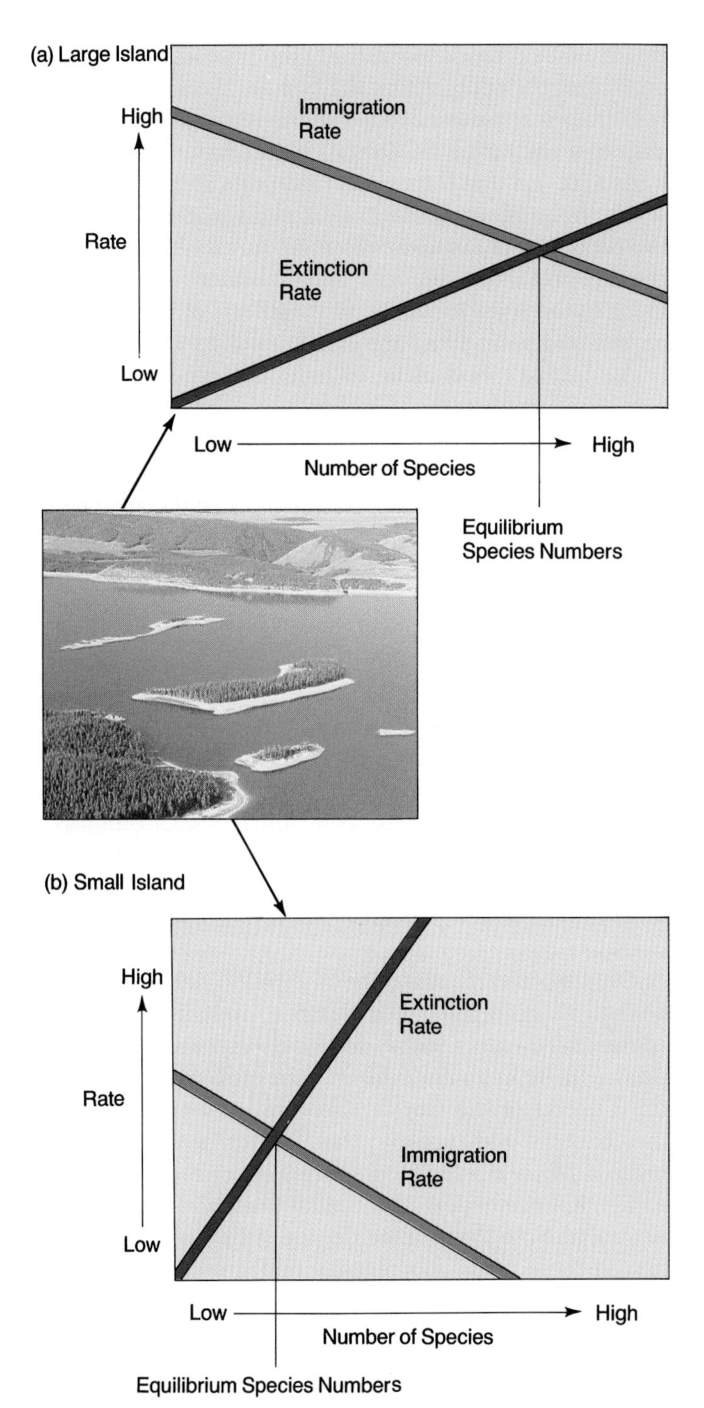

Figure 42-9 SPECIES EQUILIBRIUM, SPECIES RICHNESS, AND ISLAND SIZE.
The number of species on an island is determined by the balance between the immigration of new species and the extinction of species already present. At the junction of the two lines, the density of species is at equilibrium. Large islands have lower population extinction rates than small islands (compare parts a and b). For this reason, if all other factors are equal, a large island will have a higher equilibrium species number S than a small island. The islands in this photo lie in Lake Grandy, in Colorado's Big Thompson area.

new species (such as the mainland), the fewer species will be likely to immigrate and the lower will be the value of S, the number of species.

Researchers tested the species equilibrium model on a series of tiny islands of red mangrove trees off the coast of Florida, tallying all the invertebrate species on the islands—mostly insects—then sealing the islands off and fumigating them to kill all the animal life but not the trees. The experimenters then monitored the process of repopulation and found that in less than a year, the *same number* of species had repopulated each of the islands, including approximately equal numbers of species in each ecological category (herbivore, detritivore, etc.). Interestingly, the new communities often contained *different* species from those in the original communities, but the same number of species; species identity, in other words, was not predictable, but species richness was. They also noted that, as the model predicts, new immigrant species continued to balance the loss of recolonizers over time and that smaller islands reached equilibrium with fewer species than larger islands. A recent study of marsh grass in a Florida bay confirmed that islands nearest the mainland are indeed colonized by more species than islands farther away; and more recent work on island orb spiders suggests that a given island will have small fugitive populations that invade, become extinct, and reinvade, while larger populations are more stable and persistent.

Body size also appears to affect colonization: Tiny shrews lose too much body heat while swimming to reach islands more than 700 m from shore, whereas the larger voles can swim to islands 1 km out in Canada's St. Lawrence River. When the distances are very great, however, rafting on floating logs or debris is the means of transport, and then small size is favored. In the Galápagos, the Philippine Islands, and Australia, only rodent-sized placental, flightless mammal species arrived via rafts.

Ecologists now believe that most environments are composed of habitat "islands," such as an island of woodland in an ocean of grassland. The most likely colonists are plants and animals with adaptations that ease *dispersal* by wind or water or animal carriers. Some new arrivals will flourish, others may go through repeated invasions and extinctions like the orb spiders, and some established species will become extinct. The equilibrium number of species for a particular community will depend partly on the "island's" area and partly on its distance from other patches of woodland that serve as sources of colonists. The concept of habitat islands is assuming great practical importance in tropical countries where people are cutting rain forests and leaving only scattered islands of trees (see Chapter 41). The smaller the "island," the fewer species can survive in the long term. For this reason, ecologists recommend that cutting be minimized and that, where possible, large nature preserves be established as species "banks" and a hedge against future extinction.

Species Interactions and Physical Disturbance

A forest composed of Douglas fir (98 percent), spruce (1 percent), and pine (1 percent) shows a high degree of dominance. If each species composed one-third of the forest, however, the community would be characterized by a high degree of equitability (roughly equal proportions of species) and it would appear to have much more species diversity, even though it was still only a three-species community. Tropical rain forests appear so diverse not just because they have more species, but also because their relative abundance of species is more equitable than that in temperate forests.

The processes that determine species richness and relative abundance in communities on a local scale are **competition** between members of different species for limited resources, **physical disturbance,** and **predation.** Competition sometimes reduces species diversity. When a critical resource, such as food, water, or nesting sites, is limited, those species that can best secure the scarce resource keep it from the other species, members of which then die, emigrate, or become rare. Disturbances such as new lava flows and huge brush fires wipe out plants and thus also remove habitats and food from animals; as a result, they reduce species diversity. Minor physical disturbances, however, can increase species diversity by preventing dominant competitors from monopolizing resources. Predation can act similarly to physical disturbance. If predators or disease organisms reduce the populations of competitively superior species, other species that are competitively inferior may be able to persist in the community, thereby increasing species diversity. In later sections, we will consider the fascinating defenses that have evolved and that allow species to cope with predation and competition.

The Niche

Every creature has its habitat—in a sense, its "address" in the community. But it also has a **niche**—a role in the community and a way of life more equivalent to a "profession" than to an address. An organism's niche is determined both by physical factors—such as the amount of light, carbon dioxide, and oxygen it needs and the ranges of temperature and pH it can tolerate—and by biological factors—such as the kinds of food it needs, the diseases it tends to contract, the predators that feed on it, and the competitors that vie for the same limited resources it requires. Studies of the red squirrel, for example, a chattering denizen of North American conifer forests, reveal that the animal transports and eats spruce seeds, fungi, flowers, and other plant materials and in the process spreads them about the forest. All aspects of the squirrel's environment and behavior—from these foods to the trees it climbs, to the effects of its activities on soil and plant life—are elements of its niche.

An animal's **food niche** includes the types of prey it eats at different stages of its life cycle, the size range of its food items, and the places where it finds the food. The red tree mouse, for example, has a specialized food niche—the needles of fir trees—and inhabits only conifer forests, while the fly-catching swallow intercepts a wide variety of insects hovering in the air of varied habitats.

An animal's **habitat niche** includes physical factors—temperature, light, salinity, pH, and so on—as well as surrounding vegetation. Physical factors, such as light, temperature, soil types, soil moisture and nutrient content, and sizes of soil particles, largely determine a plant's habitat niche, but the biological factors of competition and herbivory also contribute. Organisms are adapted by form, physiology, and behavior to occupy their food and habitat niches. Therefore, these adaptations restrict where populations of a species may or may not survive.

One can also discuss a given type of niche independent of its occupants. For instance, rodents may fill a "desert seed-eating niche" in one desert, ants may fill a similar niche in another desert, and birds may fill it in a third desert of the world. Likewise, hummingbirds in the New World occupy the "flower nectar-feeding niche," whereas unrelated species of sunbirds do so in Africa, honeycreepers in Ecuador, and honey eaters in Australia. Note the striking degree of convergent evolution in those birds' morphology in Figure 42-10.

The niche concept is important to our understanding of communities because various species of organisms are adapted by form, physiology, and behavior to occupy their food and habitat niches, and these may overlap—that is, species may compete with each other for the same kinds of resources in a given area. This competition may limit where and how each species can live and thus influence the structure of the entire community.

Biologist Joseph Connell carried out a classic study of competition between two barnacle species that inhabit the shallow waters of a rocky coast. Connell studied one barnacle species, in the genus *Chthamalus*, that tends to live in the shallower waters of the marine intertidal zone (Figure 42-11, purple) while the other, in the genus *Balanus*, lives in deeper waters of that same zone (Fig-

Figure 42-10 THE NECTAR-FEEDING NICHE.
The availability of nectar-laden flowers in four parts of the world resulted in the convergent evolution of four unrelated birds: (a) Costa's hummingbird *(Calypte costae)* in the western United States. (b) The eastern spinebill *(Acanthorychus tenuirostris)*, a honey eater, in eastern Australia. (c) The Ecuadorian honeycreeper *(Cyanerpes cyaneus)*, called the liwi. (d) The sunbird *(Nectarinia mediocris)* in Africa.

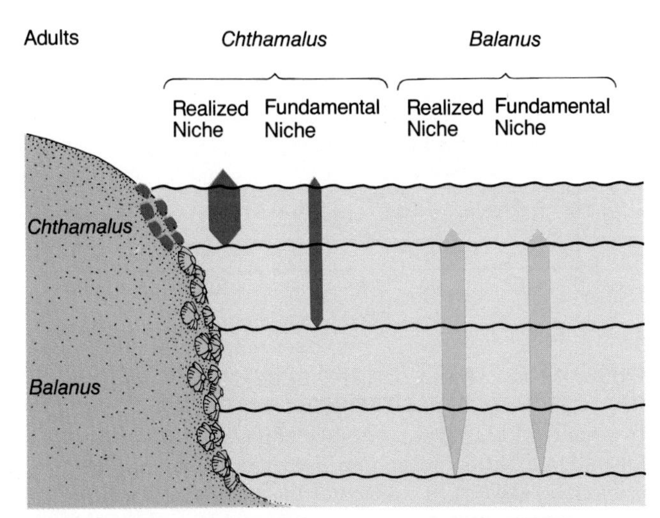

Figure 42-11 A CLASSIC CASE OF COMPETITION AFFECTING NICHE STRUCTURE.
Larvae of both *Chthamalus* and *Balanus* barnacle species can settle at any level on the intertidal rocks depicted on the left. *Balanus* fails to survive in the upper range of the intertidal zone because it cannot resist desiccation and heat. *Chthamalus* ends up restricted to the upper range of its fundamental niche because it grows more slowly than *Balanus*, which grows over it or displaces it from rocks in the lower reaches of the intertidal zone. The result is a greatly reduced realized niche for *Chthamalus* and near correspondence of the fundamental and realized niches for *Balanus*.

ure 42-11, tan). The two species appear to have clearly separate habitat niches: Although the tiny larvae of each species often settle in the other's area, adult *Chthamalus* and *Balanus* are always found at their characteristic depths.

Connell discovered that *Balanus* that settled in the upper part of the intertidal zone (areas usually inhabited by *Chthamalus*) dried up and died during low tides. Connell also observed that any *Chthamalus* gaining a foothold in the deeper areas were crushed by the larger, faster-growing *Balanus*. If Connell removed all *Balanus* individuals, however, *Chthamalus* flourished even in the deeper zones. Connell concluded that *Balanus* adults are indeed occupying their full potential niche, whereas *Chthamalus* adults—outcompeted by *Balanus* in areas where they could potentially survive—are not.

Experiments like these on barnacles and algae have led ecologists to formulate two definitions of niche that reflect interspecies competition. A **fundamental niche** is the full environmental range that a species can occupy if there is no direct competition from another species. *Balanus*, for example, occupies its full fundamental niche. *Chthamalus*, however, does not; it occupies a **realized niche,** a niche that is narrowed from the fundamental state by competition. The experiments also illustrate what is known today as the **competitive exclusion principle,** the postulate that no two species can occupy exactly the same niche at the same time in a particular locale if resources are limiting, since one species will always achieve a competitive edge and eventually exclude the other, at least from part of its fundamental niche.

In communities where species compete, every species uses a different aspect of the resource—a phenomenon called *resource partitioning.* For example, food is a limited resource to hawks, and competing hawk species differ from each other in body size and the size of prey species. Since different-sized hawks take different-sized prey, their food niches do not overlap greatly and competition is reduced. Likewise, African seed-cracking finches of the species *Pyrenestes ostrinus* display two strikingly different bill sizes. The large-billed birds feed on large, hard seeds; the small-billed members feed on small, soft seeds. Researchers are currently trying to determine whether the finches' bill size represents the evolution of resource partitioning within a single species. In contrast to these cases, species often seem to occupy very similar niches in communities with abundant resources and limited competition, such as the dozens of insect species that feed by sucking plant juices.

Some species evolve with narrow niches—they can use only one or a few types of food or growth sites—and they are called **specialists.** The blue-gray gnatcatcher is a habitat specialist (Figure 42-12) that catches insects of particular sizes flying at specific heights. In contrast, some species have broad niches—they are adapted to life over a wide range of conditions—and they are called **generalists.** The common crow is a generalist that occurs in almost every temperate biome and from wilderness to farm to urban area. Communities made up of competing generalist (broad-niched) species contain few types of organisms. In contrast, many specialist (narrow-niched) species may coexist in a community; in effect, they "share the wealth" by dividing the community into more discrete pieces, and the result may be greater species richness.

Generalists can easily exploit a fluctuating or sparsely available food supply. In the United States, eight broad-niched species of hummingbirds compete for the nectar in seasonal, scattered, temperate-zone flowers. Each species feeds on many types of flowers, and when the nectar supply shrinks in winter, most of the birds migrate south in search of more abundant food.

By contrast, more than 300 hummingbird species inhabit the tropical regions of Central and South America, where flowers grow in abundance year-round (Figure 42-13) and microclimates—and therefore habitats—are innumerable. Furthermore, the year-round availability of many types of food has allowed the evolution of highly specialized feeding niches; many hummingbird species can coexist in the community, each visiting only a few species of flowers.

(a)

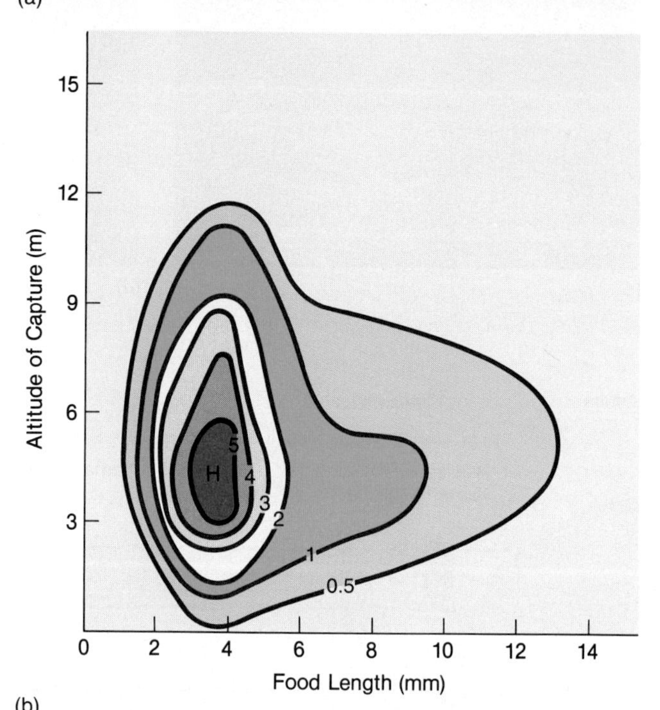

(b)

Figure 42-12 NICHE OF THE BLUE-GRAY GNATCATCHER: A HABITAT SPECIALIST.
In a study of blue-gray gnatcatchers (*Polioptila caerulea*) inhabiting a narrow niche in a particular community of California oakwood, it was found that they live primarily within a 12 m altitude range and ingest prey only in the size range of 1–13 mm in length. They most frequently capture insects that are 3–5 m above the ground and that measure 3.5–4.5 mm long. H represents this core part of the niche.

In general, the most diverse communities are those in which (1) resources are not limited, so that competition between species is absent or mild, or (2) resources are predictable in time and exploited by specialists. If food resources can always be counted on, specialists, as well as generalists, will be able to survive in the long term.

Figure 42-13 HUMMINGBIRD DIVERSITY AMID FLOWER DIVERSITY IN TROPICAL FORESTS.
These lovely birds show the range in size, bill structure, and coloration in birds occupying different food niches. This is an example of specialists in feeding niches; each species concentrates on a particular small set of flowers over the course of a year.
1. Violet-tailed sylph *(Aglaiocercus coelestis)*, Colombia, Ecuador. 2. Crimson topaz *(Topaza pella)*, northeastern South America. 3. Wire-crested thorntail *(Popelairia popelairii)*, Colombia, Ecuador, Peru. 4. Tufted coquette *(Lophornis ornata)*, northeastern South America. 5. Booted rackettail *(Ocreatus underwoodii)*, northwestern South America.

ADAPTATIONS FOR DEFENSE

Most plants and animals living and interacting in communities have evolved physical defenses against neighboring species, especially predators. These defenses can be either mechanical adaptations that serve as armor and weaponry, or chemical warfare devices to discourage attackers.

(a) (b)

Figure 42-14 PLANT DEFENSES.
(a) The sharp, curved spines of the mammillaria cactus *(Mammillaria microcarpa)* of the southwestern United States, known as the "fishhook cactus," can catch flesh as tenaciously as fishhooks. (b) Thorns and barbs are widely used plant defenses; one pointed example is the prickly stem and leaves of a thistle (*Cirsium* species).

Mechanical Defenses of Animals and Plants

Animals have an arsenal of *mechanical defenses*, including the elephant's large sharp tusks; the lobster's powerful pincers; the imposing size and tough skin of whales and hippos; and the bony, calcified, or chitinous armor of a turtle, clam, or pill bug. These weapons and armors work in concert with defensive behavior (described in Chapter 45).

Plants have equally effective mechanical defenses against herbivores, including abrasive materials, spines, thorns, hairs, and tough leathery seed coats. Heavily grazed grasses on the African savannas, for example, contain deposits of silica that wear away the teeth of grazing herbivores. And the cacti of North and South America have sharp, highly modified leaves that protrude from thick pads as wicked needles, hooks, or spines (Figure 42-14a). Thistles produce similar deterrents on leaves and stems (Figure 42-14b).

Chemical Defenses of Animals

Like hundreds of other marine animals, puffer fish harbor powerful nerve poisons, or *neurotoxins*, in their tissues. The puffer's chemical weapon is tetrodotoxin, a substance that interrupts the transmission of a predator's nerve impulses and can lead to paralysis and death. Tetrodotoxin is just one very effective example of the many protective chemicals that animals manufacture and harbor in their body cells or that they consume in foods and store for later defense against attackers.

In the bombardier beetle, oxygen released from hydrogen peroxide (H_2O_2) mixes with a solution of quinones in an abdominal chamber. The resulting exothermic reaction propels a boiling hot spray of foul-smelling or irritating chemicals, which the insect can aim like a water cannon (Figure 42-15). This barrage usually discourages a would-be predator, but the chemical defenses of other animals—including the poisons and venoms secreted by some snakes, toads, bees, and wasps—may inflict pain, cause illness, or even kill the predator.

Chemical defenses do not provide 100 percent protection: Some individuals in each generation are inevitably attacked, damaged, or consumed by predators that discover their prey's noxious qualities too late. Occasionally, though, the predator will quickly drop the animal and back off; in fact, defensive poisons may have been selected for during evolution because they taste bad and repel most predators quickly, *not* because they occasionally kill the predator outright.

Some animals borrow the chemical defenses of other species. Hedgehogs, for example, not only have a coat of tough, spinelike hairs, but sometimes also rub themselves with the bodies of poisonous toads—tipping their "arrows," if you will. Some species incorporate the protective chemicals of others into their own tissues. And larvae of the graceful monarch butterfly feed on milkweed plants, which contain *cardiac glycosides,* compounds that interfere with the heartbeats of vertebrates and many insects. After pupation, the tissues of the adult butterfly are saturated with the chemicals, and birds that eat a monarch vomit violently. As a result, experienced predators generally avoid eating monarchs or look-alike species.

In certain predators, clever countermeasures have evolved to circumvent a prey's chemical barricade. Some flycatcher birds have learned to remove bee stingers before dining, and the grasshopper mouse of the desert Southwest will wedge the backside of a tenebrionid beetle ("stink beetle"), with its nozzlelike chemical squirter, into the sand and eat the insect's head first.

Figure 42-15 ONE ANIMAL'S CHEMICAL DEFENSE. The bombardier beetle aiming its "cannon" forward and spraying its hot defensive secretion. The 100° C heat comes from an exothermic reaction that takes place as phenols are oxidized to quinones. The pressure for the spray comes from oxygen released suddenly from the splitting of hydrogen peroxide. All this occurs in a specialized abdominal chamber that contains the required enzymes and raw materials.

Chemical Defenses of Plants

Plants are particularly vulnerable to the world's many herbivores, since they are rooted in place. Perhaps the most widespread defense is the plant's fundamental growth and regeneration strategy: the ability to produce new limbs, leaves, roots, or reproductive parts throughout the life of the adult plant. But plants can also produce tissues of low nutrient value, as well as a marvelous assortment of defensive chemicals.

Some plants are simply not worth eating—a passive chemical defense called **nutrient exclusion**—because their tissues contain only a tiny amount of iron, sodium, or other nutrients many animals need in abundance. In addition, most herbivores (aside from termites) are adapted to consuming nonwoody foods, and so a related plant "defense" is the production of stems, trunks, bark, and mature leaves that are low in protein and high in indigestible cellulose.

The edible tissues of some plants are more nutritious but protected by nutrient-blocking compounds. Oak leaves, for example, have numerous vacuoles near the leaf surface that store *tannins,* phenolic compounds that bind leaf proteins into indigestible compounds. A caterpillar feeding on oak leaves will suffer a protein (amino acid) deficiency, decreasing the likelihood that it will survive to reproductive age. Not surprisingly, some herbivorous insects have evolved countermeasures; the larvae of leaf-mining beetles, for example, burrow into the nutritious inner layers and avoid rupturing the tannin vacuoles altogether (Figure 42-16).

Plants produce an impressive arsenal of defensive chemicals known as *secondary compounds* (as opposed to such primary compounds as proteins, nucleic acids, fats, and carbohydrates). Secondary compounds are commonly stored so as not to harm the plant cell that manufactures them; they are released or activated when an herbivore damages that cell within a leaf, a flower, a fruit, or a woody stem. There are six major types of secondary compounds. The *phenolics* include the tannins in

Figure 42-16 A DEFENSIVE STRATEGY AGAINST A PLANT'S OWN CHEMICAL DEFENSE.
Leaf-mining Amazon beetle larvae (*Galerucine* species) avoid tannin-filled vacuoles located near the surface of an oak leaf by burrowing inside the leaf and eating the inner layers.

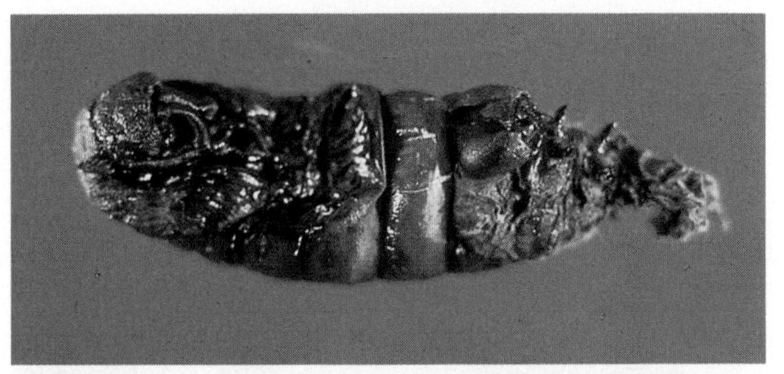

Figure 42-17 DEFENSE BY A TROPICAL LEGUME.
This abnormal larva ate leguminous plant tissue containing canavanine. Only an oxygen–for–CH_2 group substitution distinguishes the amino acid canavanine from the normal arginine that it replaces in proteins of the larva.

oak trees and the spicy substance in nutmeg trees that causes nerve disorders in herbivores. The *terpenes* in conifers include aromatic materials from which we derive turpentine and which interfere with the growth of insect larvae, as well as the fungicide α-pinene, produced in response to fungal infections. Balsam firs produce a terpene derivative, a *steroid* called juvabione, that interferes with metamorphosis in a large family of insects. Some legumes produce toxic *alkaloids*, including nicotine, strychnine, caffeine, colchicine (sometimes used as a treatment for gout), opium, and peyote. These alkaloids act on nerve, muscle, heart, or other vital cell types to do their damage. Walnut trees produce a *quinone* called juglone that inhibits the growth of most other plants. And among a miscellaneous collection of *nitrogen and sulfur compounds* are the cyanogenic glycosides, which cycads produce and which cause cancer or nerve degeneration in humans.

Botanists have discovered a most interesting example of chemical defense in certain species of tropical legumes, whose tissues contain *canavanine*, an amino acid not found in proteins and which mimics the common amino acid arginine. When most insects consume canavanine, the amino acid is incorporated into their proteins, resulting in severe developmental abnormalities (Figure 42-17) and a lower rate of survival in the insect population. Not surprisingly, however, one insect, the bruchid beetle, evolved with a counterstrategy that turns canavanine to great nutritional advantage. The beetle's cells contain an enzyme that distinguishes canavanine from arginine, so that only arginine is utilized to construct proteins. A different enzyme breaks down the canavanine molecules to yield large quantities of nitrogen for the beetle's metabolism. Likewise, in this coevolutionary game of move and countermove, the specialized cucumber beetles have evolved with the ability to metabolize cucurbitacin, a toxin that is poisonous to most insect generalists. And other insects can use plant toxins in their own defense. As we saw earlier, monarch butterflies are impervious to the ill effects of milkweed cardiac glycosides. And sawflies can store the sticky terpenes from pine needles or eucalyptus leaves and regurgitate droplets as a defense against predatory birds (Figure 42-18a).

Insect herbivores may be able to evolve an adaptation for bypassing or detoxifying one defensive chemical, but rarely do they develop several independent adaptations for avoiding a battery of chemical defenses. Thus, the plants in a given community tend to produce a range of toxin compounds, and the local herbivores tend to specialize on one or a few plant species whose chemical defenses are similar. This, in turn, naturally limits the herbivore population.

Plants may also inadvertently communicate with each other to trigger synthesis of defensive compounds.

According to some recent studies, trees step up production of defensive chemicals in response to attack, and some of the chemicals may actually "communicate" this attack and defense to other trees. Jack C. Schultz and Ian Balwin at Dartmouth College found that when larvae begin to eat leaves on a red oak tree, the remaining undamaged leaves produce 200 to 300 percent more protective tannin. Schultz and Balwin also found that untouched seedlings sharing an enclosed air space with trees that the team had deliberately injured, produced higher levels of defensive compounds within two to three days. While these results are still controversial, the researchers interpreted their findings to mean that the damaged trees gave off some airborne cue and inadvertently triggered their untouched neighbors to increase production of defensive chemicals.

(a)

(b)

Figure 42-18 ANIMAL DEFENSES: CHEMICAL AND VISUAL.

(a) The sawfly larvae in this circular mass have regurgitated yellow droplets around the periphery of the wriggling mass. The droplets contain a repellent substance that comes from a potentially harmful substance in their own food, eucalyptus leaves. (b) A skunk's bold stripes warn predators of danger—a potential olfactory shock. Here, a Canada goose chases a skunk away before it can come close enough to steal eggs or spray its powerful defensive chemicals.

One logical candidate for the message carrier between trees is ethylene gas, a substance produced by damaged plant tissues (as well as a plant hormone; see Chapter 28).

Recent advances in isolating, identifying, and synthesizing defensive chemicals are starting to pay off in developing life-saving human drugs and effective insecticides. For instance, leaf-cutting ants harbor a mutualistic partner in their underground nests—a species of fungus that can break down cellulose into simple nutrients. The ants then harvest a proportion of fungus from their "fungus gardens," thereby getting the nutrients they need. Researchers have found that leaf-cutting ants naturally avoid the leaves of certain tree species that contain a potent fungicide that kills the ants' ecological partner, the fungus. Pharmacologists are now testing that fungicide as a possible new antibiotic treatment for fungal infections in humans.

Fully one-half of the prescription drugs sold today contain such plant-derived substances as active agents, and several plant alkaloids have been found effective against hypertension and heart ailments. A major active ingredient in birth control pills, for example, is extracted from a species of yam. Researchers have also gathered the alkaloid compounds known as pyrethroids from chrysanthemums and related flowers and used them as principal ingredients in an array of commercially produced insecticides.

So far, we have seen mechanical and chemical defense strategies of animals and plants and the counterstrategies of predators. Let us turn now to very different kinds of defense based on appearance.

Warning Coloration and Mimicry

Many predators and herbivores use their eyes to seek and find food species, and prey species that produce chemical defenses are often marked with bright colors or striking patterns that warn off potential predators. These adaptations are called **warning coloration,** or **aposematic coloration.** For instance, vivid shades of red, orange, and yellow set off with black markings serve as signals to potential predators that bees, wasps, and monarch butterflies are undesirable as food. Similarly, a skunk's bold black and white stripes serve to warn dogs, foxes, and other predators away from a rude olfactory experience (Figure 42-18b).

Evidently, the bright, contrasting colors and patterns typical of aposematic coloration are so memorable to hunting animals that rely on visual cues that only a few negative experiences are required before the predator learns to eliminate such prey from its diet. The recognition and avoidance can even be instinctive. Some laboratory-raised birds of prey, for instance, recognized and avoided the skin patterns of the poisonous coral snake with no prior learning.

Interestingly, more than one species living in the same community may have identical or similar warning colors—a phenomenon called **Müllerian mimicry.** Ecologists have speculated that all the species resembling each other benefit because the learning process is simplified for predators. Birds that prey on similarly marked, noxious butterfly species such as those shown in Figure 42-19 learn one warning coloration pattern and subsequently tend to avoid all the noxious species. This, in turn, may lead to the selection for groups, or *complexes*, of Müllerian mimics.

A related adaptation is **Batesian mimicry.** Among groups of boldly patterned aposematic species that taste bad, there are sometimes palatable species that lack noxious chemicals but have evolved with the noxious species' warning coloration as a means of self-defense. The classic example of Batesian mimicry is the viceroy butterfly, which closely resembles the monarch (Figure 42-20). Studies show that blue jays readily consume viceroys before their first meal of monarchs. When a jay eats its first monarch, however, the bird becomes violently ill from the insect's cardiac glycosides and quickly learns to avoid the monarchs. Once a blue jay has had that violent gastronomic lesson, it will also avoid the perfectly palatable viceroys; thus, ecologists can prove that Batesian mimicry successfully protects "copycat species" from predation.

Some of nature's most colorful species are Batesian mimics. Several species of nonpoisonous milk snakes and the scarlet king snake, for example, mimic the appearance of the venomous coral snake (Figure 42-21). And female fireflies of the genus *Photuris* employ Batesian mimicry for aggressive ends. These fireflies flash their

Figure 42-20 BATESIAN MIMICS: THE MONARCH AND THE VICEROY.
The smaller viceroy butterfly *(Limenitis archippus)* mimics the larger monarch butterfly *(Danaus plexippus)*, which is characterized by a strong odor and unpleasant taste. A predator that consumes a monarch and subsequently becomes ill stays away from the viceroy as well.

Figure 42-19 MÜLLERIAN MIMICS: BAD-TASTING BUTTERFLY SPECIES.
Members of these four *Heliconius* butterfly species (clockwise from top left: *H. hewitsoni*, *H. sara theudela*, *H. doris f. viridis*, and *H. pachinus*) are quite distasteful. Because they resemble each other, a predator that consumes any one of them will learn to stay away from members of all four species.

(a)

(b)

Figure 42-21 REPTILIAN BATESIAN MIMICS.
(a) The nonpoisonous scarlet king snake *(Lampropeltis doliata)* is a Batesian mimic of (b) the poisonous coral snake *(Micrurus fulvius)*.

luminescent abdomens in the same pattern as females of the genus *Photinus*. Such flashing patterns attract males, and the *Photuris* females seize and consume any *Photinus* males that respond to this false advertising.

ADAPTATIONS FOR ESCAPE

In the grand cat-and-mouse game of survival, organisms use another very effective means of self-defense—escape. Some adaptations, such as fast running, flying, or swimming, enable potential prey to flee, while others separate predator and prey in space and time, and still other adaptations allow effective hiding.

Fast-moving animals are usually adapted for fleeing or chasing. African ungulates living in grasslands, such as the fleet-footed wildebeest and gazelle, have leg structures and muscles that permit high-speed running. Predators, of course, have responded by evolving an equal capacity for speed. No terrestrial animal is swifter than the cheetah over short distances. Its vertebral column, pelvic girdle, and limbs have become modified to permit exceptionally long, leaping strides and bursts of speed up to 116 km (72 miles) per hour (Figure 42-22). In fact, the cheetah's head and jaw are so small and streamlined that the animal sometimes has trouble killing the prey it runs down.

African ungulates may also protect themselves in their exposed habitat by *herding*. This is a safety-in-numbers strategy. If animals in a population stay together in a group, the chances are small that a solitary predator will single out any one particular individual. Schools of fish and flocks of birds employ a similar strategy. An important interpretation of herds, schools, and flocks came from W. D. Hamilton, who coined the term *selfish herd*. What looks like a group of cooperating animals, he theorized, is really a set of individuals that derive individual benefit from the evolved habit of group living.

Figure 42-22 ADAPTED FOR HIGH SPEED.
The body of the cheetah *(Acinonyx jubatus)*—from its musculature to its skeletal system—is designed for speed, allowing it to sprint up to 72 mph and thereby to catch fast-moving gazelles and antelope. Here a cheetah captures a baby gazelle.

Escape in Space and Time

A species can escape from its predators by avoiding them in space or time. Palatable insects, such as the periodic cicada, often spend most of their lives underground, remaining relatively inactive for many years. Then large numbers emerge simultaneously, breed, and produce a new generation of offspring that quickly tunnel underground to repeat the cycle. Although the birds and other animals that feed on "17-year locust" and similar cicadas enjoy a temporarily unlimited feast, the total number of insects is so overwhelming that predators have little effect on population density and tend not to evolve adaptations that enable them to specialize on the insects.

Plant life cycles can protect those organisms from predators in time and space. Bamboo, for example, flower once every 120 years, producing huge quantities of seeds each time. Since all potential seed eaters must feed more often than once every century, they are adapted to utilize other plants as food sources. This escape mechanism has a down side, however; bamboo reproduces sexually only once per century, and the intervening vegetative reproduction has traditionally been thought to limit genetic variation. Researchers have recently discovered, however, that both the rate of mutation and accumulation of total mutations is 25 times greater in long-lived trees such as mangroves than in short-lived annuals such as barley. Bamboo remains to be investigated for somatic mutations that chance to accumulate in meristems which ultimately give rise to flowers after one hundred years or so. The bamboo flower must be wind pollinated, since no animal pollinators adapt to a plant that flowers so rarely, and this, too, has disadvantages.

Wind pollination, however, can be its own adaptation for escaping predation in space: The wind-borne seeds of dandelions, grasses, and maple trees may travel for miles on a summer breeze, far from predators in the local area of parent plants.

Camouflage

Camouflage (also called *crypsis*) is one of nature's most fascinating escape mechanisms. Camouflage involves colors or patterns that allow organisms to blend

Figure 42-23 CAMOUFLAGE IN ACTION.
(a) The flower mantid *(Pseudocreobota ocellata)* resembles orchids so closely that the mantid's insect prey are caught unawares—and are consumed. (b) This tree frog *(Megophrys monticola)* closely resembles the leaves of the tree on which it sits. (c) The large thorn at the top is actually a treehopper that mimics thorns.

with their background or to appear to be inedible or nonthreatening objects. Biologists believe that these adaptations arose in prey because of selective pressure from predators that hunt by sight. Camouflage also arose in some predators, leaving them well concealed from their potential prey.

A common strategy for camouflaging bulk and body outlines is **cryptic coloration.** The lavender-hued flower mantid, for example, can await the arrival of its insect prey undetected (Figure 42-23a), while the coloring of tree frogs and treehoppers can render them almost indistinguishable from nearby foliage and thus protect them from predators (Figure 42-23b, c). In the deep ocean, numerous fish species are shades of pink and red, colors that are bright in sunlight but are perceived as gray or black in the watery depths, where there is little light. Even the vertical black and white stripes of a zebra—so conspicuous to us—are a form of camouflage because lions, which are color-blind, seem to have trouble discerning zebras against the vertical blacks, whites, and grays they see in the grassy savanna habitat.

Figure 42-24 CAMOUFLAGE BY SHAPE.
The African stoneplant *(Lithops villetti)* closely resembles nearby stones.

The organisms in Figure 42-23 are camouflaged by shape as well as by color. And the nubby *Lithops* plant in Figure 42-24 can easily be mistaken for smooth, gray-green pebbles.

While camouflaged animals often seem to bear an amazing resemblance to the objects and features they copy, merely looking like a leaf or a piece of wood is not enough. Organisms that possess such adaptations must also *behave* like the things they imitate. The insects commonly called walkingsticks, for example, spend much of their time on branches, frozen in the angular posture of a twig (Figure 42-25). And some species that normally may be quite inconspicuous, such as the *Automeris io* moth (Figure 42-26a), flash warning displays (in this case, show the false eyespots on its wings) that are believed to frighten predators (Figure 42-26b).

Like other protective adaptations, camouflage is not the result of conscious effort on the part of protected species; resemblances, colors, and many of the behaviors

Figure 42-25 CAMOUFLAGE BY BEHAVIOR.
This European walkingstick insect not only resembles a stick, but also behaves like one, spending almost all its time in branches, frozen into the position of a twig. Can you see the head just to the right of center?

are determined by genes and genetically regulated patterns of development. Natural selection can easily account for the spread of such genes after they arose by chance mutations: If a prey species resembles a twig even slightly, it will be partially camouflaged, and the genes for "twiglike" appearance will be selected for, as the predators find and eat "untwiglike" prey in greater numbers.

While many of the interactions between community members are competitive and predatory, some are beneficial, as the next section explains.

SYMBIOSIS

Sometimes, two or more species live together in **symbiosis,** intimate association that can involve unilateral or mutual benefit to the **symbionts** involved. *Parasitism* is one extreme, with one species benefiting greatly and the other being harmed. *Commensalism* also involves benefit to one species, but without identified harm to the other. Finally, *mutualism* describes a relationship in which both species benefit.

Parasitism

Ecologists have dubbed some organisms "live-in" predators: These *parasites* make their living by taking up residence on or within the bodies of their prey. Nearly all plants and animals are prey to parasitic animals or microbes at some time during their life cycle. Such parasitic organisms, including protozoa, roundworms, and tapeworms, are often highly specific in their choice of prey: The host provides nutrients, a protective environment, or other resources the parasite must have to reproduce effectively. Parasites often have lost organs during evolution; tapeworms, for example, have lost the

(a)

(b)

Figure 42-26 A FRIGHTENING SURPRISE: WARNING "EYES."
The *Automeris io* moth startles its predators with its prominent eyespots when disturbed. (a) The normal resting position. (b) When touched, the front wings move forward instantly, revealing the "eyes."

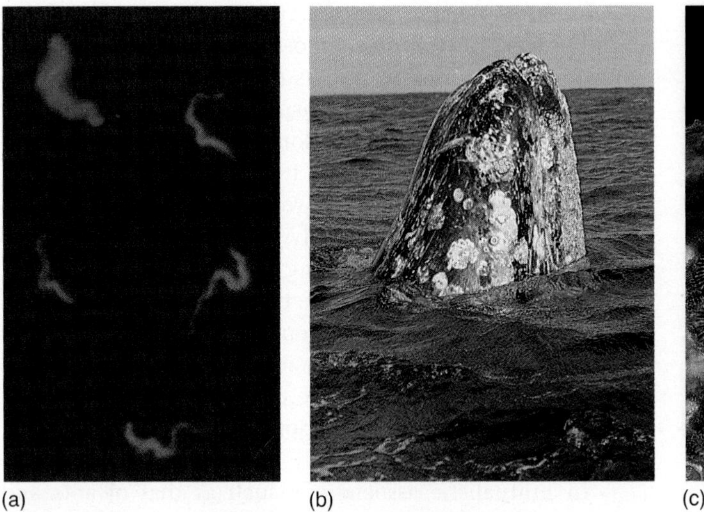
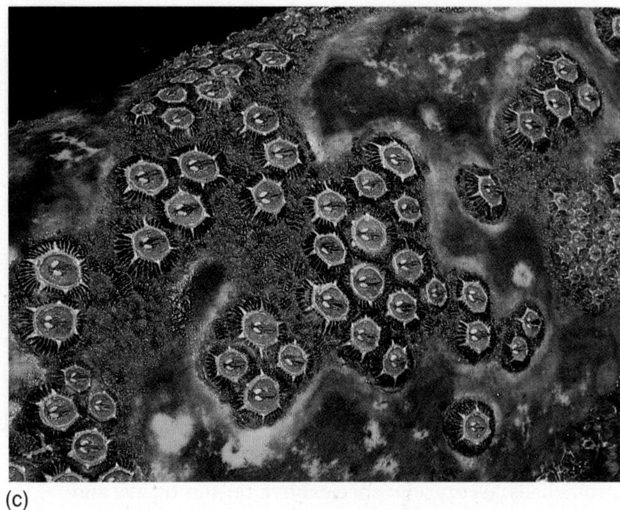

(a) (b) (c)

Figure 42-27 PARASITISM AND COMMENSALISM.
(a) Parasitic trypanosomes are moving targets for the immune system. The green-stained trypanosomes in this human blood sample are descended from cells that entered the bloodstream when this host was bitten by a tsetse fly. Just as the immune system mounts an attack against the "green" lineage, variant trypanosomes arise with different surface proteins (VSGs). The new "red" lineage, as stained by another antibody, persists; but it, too, will ultimately be killed off. Only if the "red" lineage gives rise to still another surface protein will the trypanosomes live on in this particular host. (b) A "spy-hopping" gray whale *(Eschrichtius robustus)* rises above the ocean's surface, exposing its eye. (c) "Hitchhiking" acorn barnacles *(Cyamus scammoni)* and lice on the skin of the gray whale take part in commensalism; the barnacles are carried through the water, filtering food along the way, with no apparent harm or benefit to the whale.

entire gut and acquired the ability to absorb nutrients from the host directly through its surface. The host, in turn, is often adapted to coexisting with the parasite or to fighting it off with a versatile immune system.

Among vertebrates, antibodies or T lymphocytes often attack parasites (see Chapter 30). Some parasitic species have, in turn, evolved their own countermeasures, including chemicals that suppress or evade the host's immune system. Trypanosomes, the protozoa that cause African sleeping sickness (see Chapter 21), change their surface antigens periodically so that the host's immune system cannot respond fast enough to kill all the invader cells (Figure 42-27a). Similarly, some invasive organisms are adapted to "borrow" a surface coat of host proteins immediately on entering their prey and thus elude immune recognition.

The most successful parasites do not kill their hosts, since that would destroy their living "hotel," perhaps before the parasite has reproduced successfully. This implies that natural selection tends to eliminate the most deadly parasites and pathogens and preserve those that do less harm.

Host-parasite relations that may be very stable for long periods of time can be altered radically if the ecological situation changes. In the mid-fourteenth century, the high density of population in European cities, unsanitary conditions, and increased commerce and transportation among cities created ideal conditions for bacteria of the genus *Pasturella,* which causes the black plague. The bacteria killed one-quarter of the population of Europe within 2 years and tens of thousands more before the fourteenth century ended.

Commensalism and Mutualism

The breaching, barnacle-encrusted gray whale in Figure 42-27b is a participant in the interspecific association known as **commensalism.** In this arrangement, one organism lives with or on another, neither harming nor benefiting its host. Thus, hitchhiking barnacles and lice obtain substrate and transportation while they filter microscopic food particles from passing ocean currents (Figure 42-27c).

While parasites derive nourishment at their host's expense, commensals do not. Bromeliads (plants related to cultivated pineapples), for example, live on the trunks and branches of tropical trees (Figure 42-28) and obtain water and nutrients from the air or bark surface without penetrating host tissues. Still, the collective weight may eventually harm the tree, causing branches to break during downpours.

The potentially negative effect of bromeliads may, in fact, have contributed to a relationship between certain ant and tree species that protects the trees against bromeliads. Ants of the genus *Azteca* commonly live in *Ce-*

Figure 42-28 COMMENSALISM IN PLANTS.
Bromeliads (*Neoregelia* species) live on the trunks and branches of their host trees without doing any apparent harm.

cropia trees, taking advantage of the hollow stems for their nests and feeding on glycogen-rich nodules, or "Müllerian bodies," found at the base of leaf petioles (Figure 42-29a). Plants, of course, normally store carbohydrate as starch; the production of glycogen, the typical storage carbohydrate of animals, is an evolutionary adaptation that lures the ants. The *Azteca*, in return for room and board, literally scavenge the *Cecropia* trees for new small bromeliads; the ants dissect even small mats of bromeliads tied experimentally to *Cecropia* limbs and drop them to the rain forest floor (Figure 42-29b).

The *Azteca-Cecropia* relationship clearly benefits both partners: It is an example of **mutualism.** The symbiotic relationships between legumes and nitrogen-fixing bacteria (see Chapter 41), between fungi and algae in lichens (see Chapter 22), and between cellulose-digesting microorganisms and the ruminant mammals (sheep and cattle) or termites whose guts they inhabit (see Chapter 32) are all examples of mutualism.

Some instances of mutualism are *obligatory* for one or both partners. House ants, for example, live in the hollow thorns of certain *Acacia* trees. The animals eat protein-rich droplets manufactured by the leaves and in return attack herbivores that try to nibble the trees. The ants also clear the surrounding area of competing vegetation. While the ants can live elsewhere, the trees grow slowly or die if the ants are removed. In other associated species, including other species of ants and acacias, the mutualism is *facultative*: The two species derive mutual benefits from their association, but can survive quite well independently.

In mutualistic associations such as that of ants and acacias, the host species trades food for reduced losses to some other, more damaging predator. Other forms of mutualism enhance the host's ability to reproduce or disperse progeny; prime examples are plant-pollinator and plant-disperser relationships, as we saw in Chapter 23.

LOOKING AHEAD

In every community of organisms, one finds plant, animals, fungi, and microbes coexisting, sometimes with highly evolved strategies for attack, defense, or mutual benefit. Some of these methods involve animals carrying out specific acts. In the next chapters, we consider animal behavior in greater detail, including some of the adaptations we have been discussing, plus other complex activities that together allow animals to survive in their physical environment.

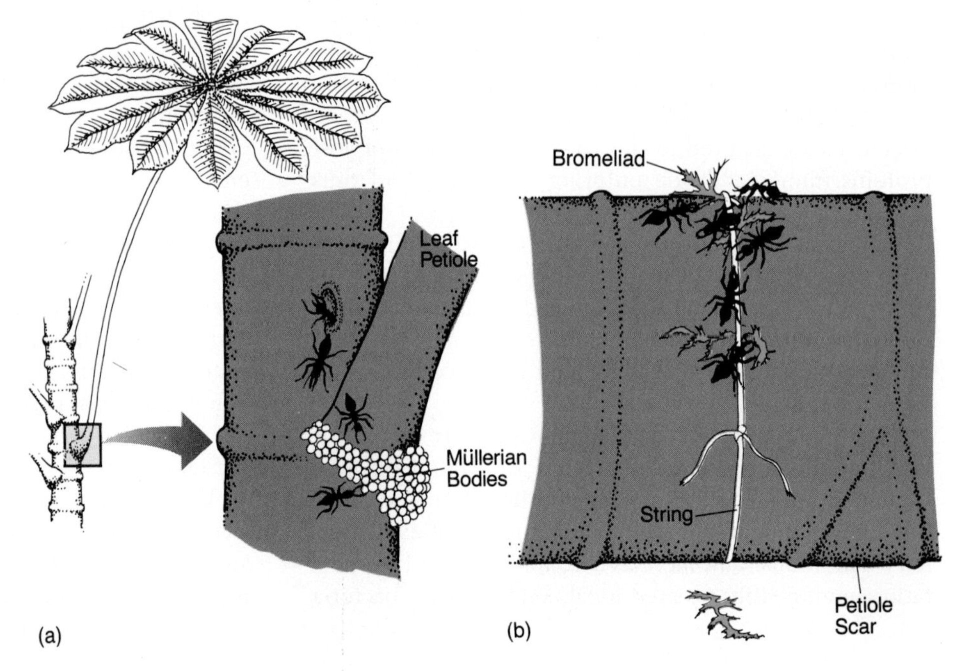

Figure 42-29 MUTUALISM: *Azteca* ANTS AND *Cecropia* TREES.
(a) Müllerian bodies provide glycogen to the *Azteca* ants that colonize the hollow stems of the *Cecropia* trees. (b) The ants, in turn, keep the trees cleared of bromeliads and other epiphytes, and prune away any mosses a researcher ties to the limb or that happen to sprout on a limb in natural conditions.

SUMMARY

1. A *community* is an association of the interacting members of populations of different species in a particular area.

2. *Ecological succession* is the gradual change in the species composition of communities over time. *Primary succession* begins on rock, lava, or sand, while *secondary succession* begins on developed soil. Members of *fugitive species* (weeds, grasses) disperse well, grow fast, and are commonly the early stages of secondary successions.

3. A *climax community* has reached a stable state in which new arrivals no longer replace existing species.

4. The species involved in each stage of a succession are often not predictable.

5. The number of species in a succession tends to be low in the earliest stage, highest in intermediate stages, and lower again in later stages.

6. Gradations that affect organisms along climatic gradients are known as *ecoclines*.

7. Different communities typically contain different numbers of species; a dramatic example is the higher *species richness* found in tropical rain forests.

8. The *species equilibrium model* states that larger islands are richer in species than smaller islands and that both immigration and extinction help account for species richness at a given site. In some ecosystems, sets of small ephemeral populations are present in addition to more persistent populations that are less likely to become extinct.

9. Species richness and relative abundance are affected by *competition*, *predation* (including herbivory), and *physical disturbances*.

10. A species' *niche* is its total way of life. The *fundamental niche* is the full range that a species can occupy, whereas the *realized niche* is the niche that results when competition narrows the fundamental niche.

11. If resources are limited and competition is severe, only one species can exist in a given niche at a given time. This is the *principle of competitive exclusion*.

12. The adaptive strategies of interacting species are often reciprocal; this phenomenon has been termed *coevolution*.

13. Animals and plants can employ mechanical defenses, such as horns, claws, spines, or thorns. Toxins, venoms, and bad-smelling sprays are examples of animal chemical defenses.

14. Plants have evolved a variety of chemical defenses against predation, including noxious and poisonous substances stored in plant tissue.

15. Coevolution has resulted in predators that are adapted to circumvent prey defenses, as by the recognition of potentially noxious prey. This has led to the evolution, among prey, of warning characteristics, such as *aposematic coloration (warning coloration)*, including *Müllerian mimicry* and *Batesian mimicry*.

16. Escape strategies include physical adaptations for rapid flying, schooling, herding, or flocking, which provides safety in numbers; escape mechanisms that avoid predators in space or time; and *camouflage*.

17. *Symbiosis* is an intimate relationship between members of different species. There is a range of symbiotic relationships, including *parasitism* (harm to one, benefit to the other); *commensalism* (benefit to one, no effect on the other); and *mutualism* (benefit to both partners).

KEY TERMS

adaptation
aposematic coloration
Batesian mimicry
camouflage
climax community
commensalism
community
competition
competitive exclusion principle
cryptic coloration
ecocline
ecological succession
food niche
fugitive species
fundamental niche
generalist
habitat niche
Müllerian mimicry
mutualism
niche
nutrient exclusion
physical disturbance
predation
primary succession
realized niche
secondary succession
specialist
species equilibrium model
species richness
symbiont
symbiosis
warning coloration

QUESTIONS

1. Define ecological community. What is the relationship between a community and an ecosystem?

2. Which of the following are examples of primary ecological succession? Secondary? Aquatic? Explain why.

a. A lake fills with vegetation, becoming first a bog, then a meadow, and finally a forest.

b. Grasses colonize sand dunes, followed by shrubs and finally trees.

c. Lichens begin to grow on bare lava, followed by mosses, grasses, and shrubs.

d. Mosses and lichens colonize rocks scraped bare by a receding glacier; grasses, perennials, and shrubs follow over the years.

e. An abandoned corn field sprouts weeds, then shrubs, and then trees.

3. Name the four categories that ecologists often use to define and describe

communities, and explain briefly what is included in each category.

4. How is the physiognomy of most communities described? Why?

5. In general, how does species richness differ between tropical and temperate communities? Briefly discuss possible explanations for this phenomenon.

6. How do competition, predation, and physical disturbances affect species richness and relative abundance? In most communities, are species at the point of competitive equilibrium? Why?

7. Why is it difficult to describe the complete niche of an organism?

8. Could a Batesian mimic be, at the same time, a member of a Müllerian mimic complex? Why or why not? Could a Batesian model be part of such a Müllerian complex? Why or why not?

9. How have animal species responded to secondary compounds in their food species?

10. Outline the sorts of features that an optimally camouflaged prey and equally well-camouflaged predator might display.

11. Is every structural feature of an organism's body necessarily an adaptation? Explain.

12. Define the three major types of symbiosis, pointing out advantages or disadvantages to the participants.

ESSAY QUESTION

1. Thinking of a city lawn as an ecological community, describe succession, climax, pioneer species, fugitive species, physiognomy, species richness, and competition in relation to that community.

For additional readings related to topics in this chapter, see Appendix C.

43
The Ecology of Populations

Through the animal and vegetable kingdoms nature has scattered the seeds of life abroad with the most profuse and liberal hand. She has been comparatively sparing in the room and the nourishment necessary to rear them.

Thomas Malthus, *Essays on the Principles of Population* (1798)

For several hundred years, North American fur trappers have sold the pelts of snowshoe hares and lynx to brokers of Canada's Hudson's Bay Company. The company has kept precise records of its purchases since 1800, and these tallies reveal that lynx populations reach a peak—followed by a severe decline, or "crash"—every 9 to 10 years. Populations of the lynx's prey, the snowshoe hare, also show a cycle, with peaks occurring just before those of their predators (Figure 43-1).

The swings are no great mystery: As the lynx population grows, these predators consume hares faster than the small prey can multiply. Then lynx begin to starve, and the predator population crashes. The hare population peaks and crashes for a different reason: The vegetation they consume has its own peak and crash cycles, based on rainfall, temperature, predatory insects, and the like. Therefore, more plants mean more hares and more lynx—but only until the predators overeat and crash once again!

As far as the eye can see: A wildebeest population grazing on a Kenyan plain.

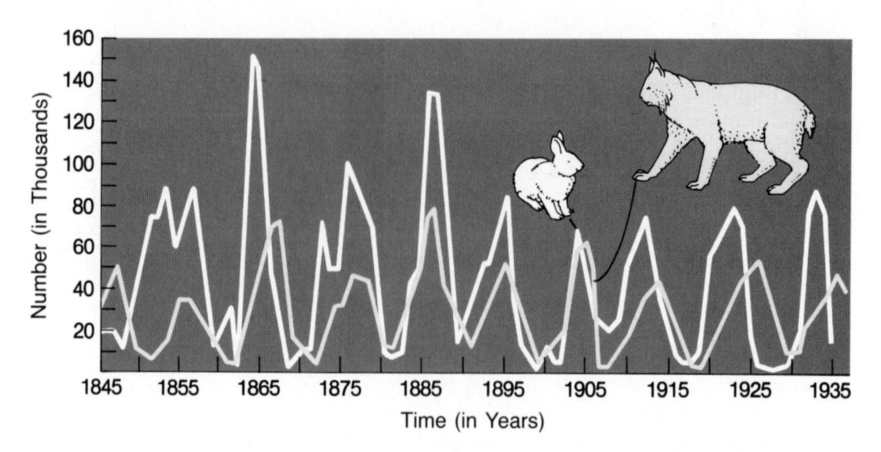

Figure 43-1 PEAKS AND "CRASHES" IN LYNX AND SNOWSHOE HARE POPULATIONS IN CANADA FROM 1845 TO 1935.
As the prey population cycles in numbers, the predators do also, but with a slight lag. These data come from records of pelts bought by the Hudson's Bay Company.

We turn in this chapter to *populations*, groups of individuals of the same species, and to the reasons why a particular population grows over time, remains steady, or declines in number. Ecologists use various models to explain the fluctuations, and they focus on reproductive strategies, competition, predation, and other factors that influence a population's success. Such principles allow them, for example, to predict whether an endangered population will become extinct and to determine how large to make nature preserves so that the populations within them can survive. Our own human population has been enormously successful—dangerously so, in fact—and we will consider the reasons for our population explosion and its serious implications for the future.

Our specific topics will include:

- Models for predicting and understanding population growth
- Limits on population size, including competition for resources and attack by predators
- How populations are distributed in space and why
- The problem of human overpopulation, and how the science of ecology can help predict our species' future and its possible impact on the biosphere

POPULATION GROWTH

While geneticists see populations as groups of interbreeding organisms, ecologists see them as groups that produce and consume resources and that provide or use up nest sites or other habitats. In some ways, a population is a dynamic entity, like a living thing: Over time, it grows larger; its growth may decrease and increase again; its composition shifts as old individuals die and new members arise; it occupies a certain range of habitats and performs specific ecological roles, just as individuals do; and a population eventually becomes extinct.

Populations also have statistical characteristics, such as per capita birthrate, known as **natality,** and a per capita death rate, or **mortality,** as well as a certain *population density*—the number of individuals per unit of area. Together, a birthrate, death rate, and density help determine a population's growth rate and changing size over time and help ecologists answer their central question: What determines the abundance and distribution of organisms?

Exponential Growth

Thomas Malthus, whose theories so influenced Charles Darwin, calculated that populations can grow in *exponential*, or geometric, leaps. Applying the principle to elephants, Darwin predicted that in 750 years, the descendants of a single original breeding pair would number almost 19 million pachyderms!

This sort of unrestrained exponential rate of growth in a population can be expressed in a mathematical equation:

$$\frac{dN}{dt} = r \times N$$

where dN stands for the change in a population's size, and dt equals the period of time over which the change occurs. Thus, dN/dt represents the *rate*—change divided by time—it takes for the change to take place, or the rate at which the population is growing. The growth rate, r, is the per capita growth rate (the birthrate minus the death rate), and N stands for the population size at the beginning of the time period under study. For example, if a population doubles from 2,000 to 4,000 individuals in 4 years, then

$$\frac{dN}{dt} = rN$$

$$\frac{4{,}000 - 2{,}000}{4} = r(2{,}000)$$

$$r = 0.25$$

The population, therefore, is expanding by one-quarter of its original size per year. Figure 43-2 shows the **exponential growth curve,** or steep rising line, for a rapidly growing population of bacteria.

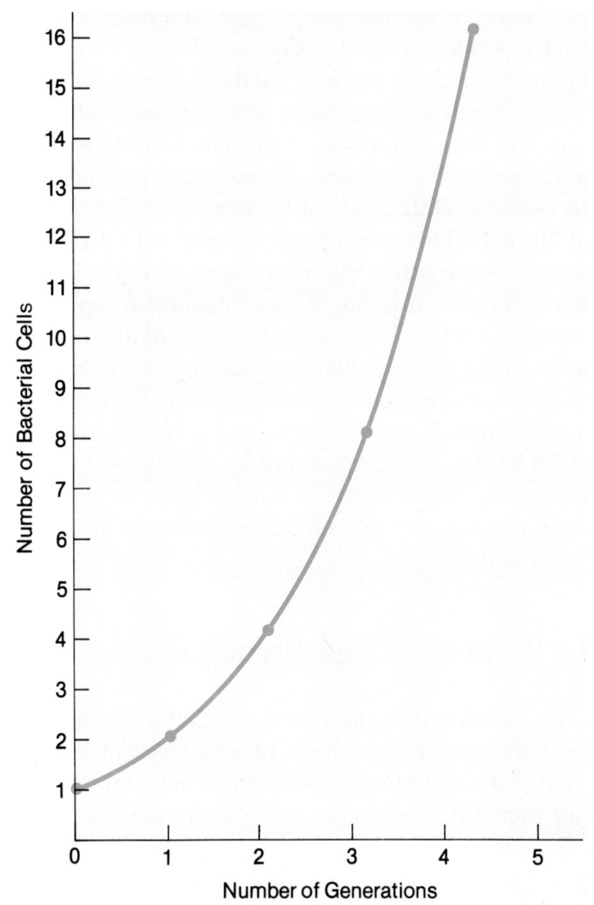

Figure 43-2 EXPONENTIAL GROWTH OF A BACTERIAL POPULATION.
The overall rate of population growth rises as the number of bacterial cells, each capable of dividing, continues to grow.

Sometimes, a population in nature grows exponentially when it colonizes a promising new area of habitat. Russian population biologist G. F. Gause showed this in 1930 when he cultured the protozoan ciliate *Paramecium caudatum* and added a plentiful, steady supply of bacteria to the culture medium. The paramecium population grew quickly from a few initial cells and followed a perfect exponential pattern. Eventually, despite a daily ration of bacteria, the number of protozoa leveled off.

Gause's experiment demonstrated that a population cannot grow exponentially for very long. In reality, rapidly growing populations always level off before they become absurdly large or before their size begins to fluctuate up and down, because limits on the food supply or other critical resources or a buildup of wastes inevitably slows population growth, causing more deaths or causing the reproductive rate to decline.

Sometimes, the population "boom" is followed by a "bust" (a sharp decline), not a leveling off. In 1944, U.S. Coast Guardsmen introduced 29 reindeer onto a small island in the Bering Sea off the coast of Alaska. With abundant food and no predators, the population rose steadily to 6,000 reindeer in 1963. At that point, however, the island's supply of winter forage—mostly reindeer moss—gave out, there was a calamitous population bust, and within 3 years, only 42 reindeer remained.

Carrying Capacity and Logistic Growth

As the reindeer example shows, the environment imposes an upper limit to population size—a limit called the **carrying capacity**, *K*, of the population's environment. The carrying capacity is simply the maximum number of individuals of a population that environmental resources will support. The growth limit imposed by the carrying capacity is expressed by simply joining the term $\frac{K-N}{K}$ to the previous growth equation:

$$\frac{dN}{dt} = rN\left(\frac{K-N}{K}\right)$$

Notice that the new term tends to decrease the growth rate; for example, if the current population size *N* (say, 990) is near to the environment's carrying capacity *K* (say, 1,000), then $\frac{K-N}{K}$ will be close to zero $\left(\frac{1{,}000 - 990}{1{,}000}\right)$, and the rate of growth will be small. When we apply the modified growth equation to a population, a **logistic growth curve,** with its characteristic S shape, results (Figure 43-3).

A logistic growth curve reflects various phases of a population's growth history. At the beginning, when resources are plentiful, growth is rapid and the size of the population increases. When the population reaches equilibrium with available environmental resources (i.e., when $K - N$ nears zero, and the carrying capacity is approached), the population may level off at a more or less constant size. The population can remain *stable* like this unless a crucial resource is reduced or depleted or the climate changes; these changes can mean that the environment, in effect, has a new *K* value—that is, a lower maximum carrying capacity, and the population in this area may decline or even become extinct.

Fluctuations in Population Size

Exponential and logistic growth curves are abstractions that only approximate real events. In fact, population sizes often shift, and over time, the number of individuals in any population always changes. For example, even a population that generally shows logistic growth

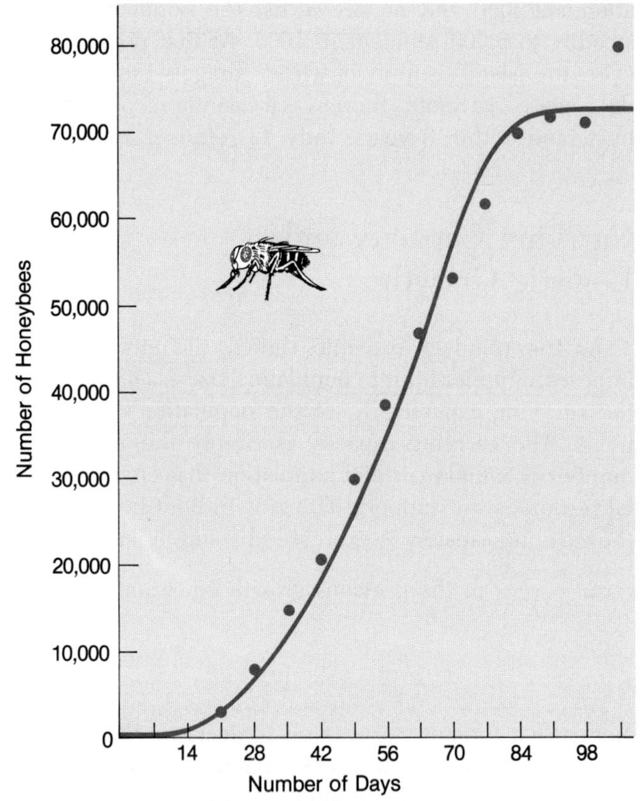

Figure 43-3 LOGISTIC GROWTH PATTERN OF A NATURAL POPULATION.
This population of an Italian strain of honeybees introduced near Baltimore shows a logistic pattern over a 3-month period. The ultimate population size, near *K*, the carrying capacity, is determined largely by available space within the hives.

may not level off smoothly when the environment's carrying capacity is reached. That's because it takes time for the birthrate to fall and the death rate to rise. This response time is known as a **reproductive time lag.** When growth eventually does decline, the population may begin a sort of ecological roller-coaster ride, alternately exceeding *K* and falling below it because of lagging reproductive responses. This is one factor that caused hares to cycle in our opening example.

An environment's carrying capacity and the population it supports may vary with the time of year. Seasonal fluctuations in the supply of food, water, hiding places, nesting sites, or other crucial environmental resources can be found in many habitats, such as temperate grasslands that are buried in snow in the winter and mountain streams fed by melting snow that run dry by August. As you might expect, many plant and animal populations are adapted to the seasonality of their habitats. They reproduce rapidly during the resource-abundant or climatically favorable season, grow to high densities, and then suffer high, but not complete, mortality during less favorable times. (Some species migrate, of course, to avoid such consequences.) Tiny amphibians called Couch's spadefoot toads (Figure 43-4), for example, emerge from their burrows in the Sonoran Desert of western North America only after a chance July rainstorm and then quickly feed and lay (or fertilize) eggs. Soon, thousands of tadpoles emerge and metamorphose into toadlets. Only a few of these will survive the heat and drought long enough to burrow into the caked desert soil, awaiting the next rains and the time to emerge from their hiding places to feed and reproduce. An ecologist trying to document the size of the toad population might get very different numbers from season to season or as the environment's carrying capacity rises and falls. In such cases, the researcher can discover how the population size is changing by simply studying the population's **age structure**—the relative numbers of young, mature, and old individuals in the population at a single point in time.

The Effects of Age Structure

The age structure can reveal whether a population is experiencing a boom, a bust, or a plateau. In Mexico's rapidly growing human population in 1985, for example, more than half the people were under age 20 (Figure 43-5a), while in the United States, with its slower population growth, the age structure was more equally distributed from young to old (Figure 43-5b). For many species with stable populations, the age structure is even more balanced, with deaths of older organisms about equally offset by births.

Another way to represent a population's age structure is with a **survivorship curve,** which shows the number of survivors in different age groups (Figure 43-6). Human

Figure 43-4 COUCH'S SPADEFOOT TOAD
A spadefoot toad *(Scophiepus couchi)* emerges from its earthen burrow in Comanche, Colorado, after a rain.

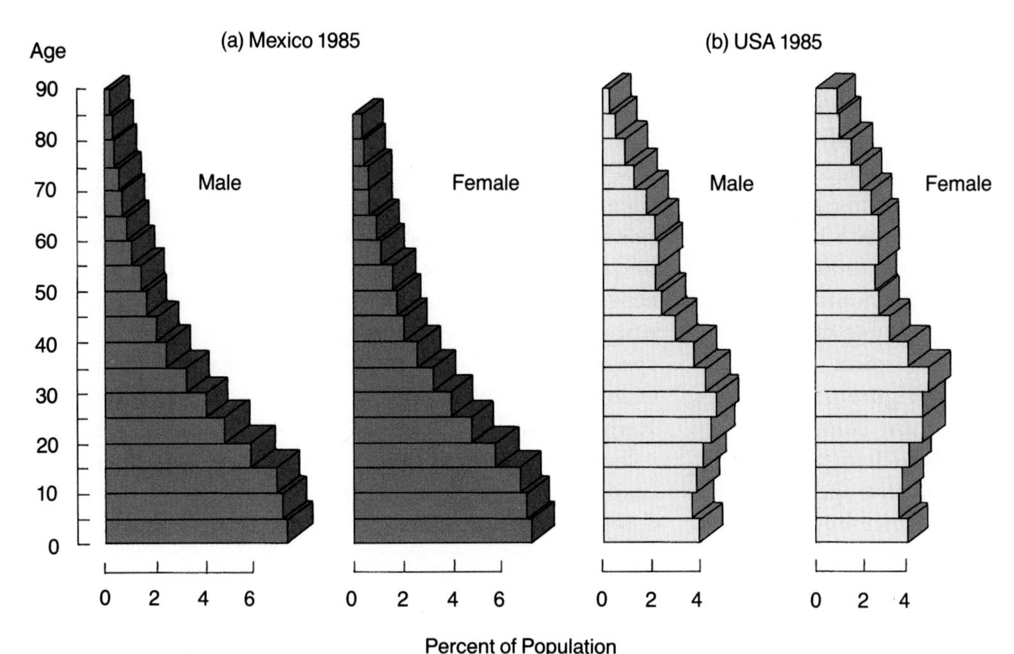

Figure 43-5 THE AGE STRUCTURES OF POPULATIONS.
The age composition is shown for both women and men in (a) a developing country, Mexico, and in (b) the United States. The large number and proportion of young people now approaching child-bearing age suggest that the Mexican population will continue to expand in the foreseeable future.

females have low rates of mortality in infancy, youth, and adulthood and are likely to die only after the seventh or eighth decade. Bony fishes and plants, in contrast, are likely to die before reproducing, but if they do survive the dangers of early life, will tend to live to old age (for their species). A bird has an intermediate survivorship curve (more or less diagonal); its chance of dying remains about the same at every age.

Survivorship curves are related not only to death rates, but also to important reproductive issues: How many offspring does an individual produce, and how much effort does the parent expend on each young?

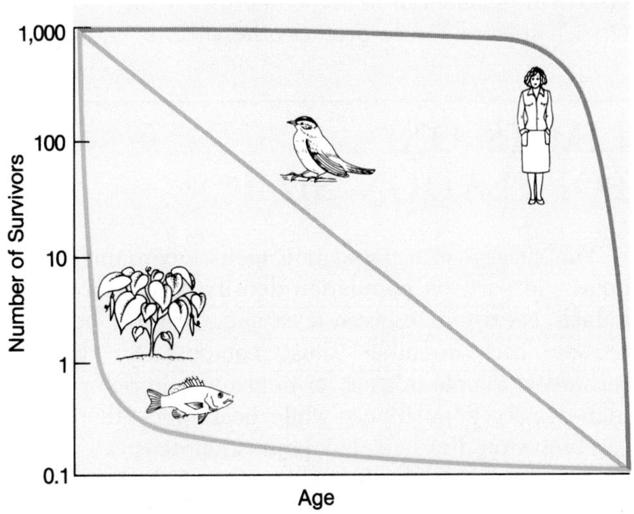

Figure 43-6 IDEALIZED SURVIVORSHIP CURVES.
The first type of survivorship curve (orange) is a rough approximation for healthy human females. The second curve (blue) illustrates survival in birds; it shows no obvious change in rate of mortality with age. The third type (green), for fishes and certain plants, shows massive death early in life and then long-term survival for those few organisms that make it through the early crisis years. In nature, the actual curves for human females and birds would show a greater degree of early mortality than do these idealized curves.

Reproductive Strategies: *r*- and *K*-Selection

In reproductive terms, each species has the same goal—to produce as many offspring as possible that survive to reproduce—and in each case, a complex adaptation called a **reproductive strategy** has evolved. For some species, the strategy is to produce a small number of offspring, each with a high chance of survival; humans, elephants, and whales fall into this category. For other species, the tactic is to generate a large number of young each time the adults reproduce; oysters, for example, release millions of eggs into the sea each season, but each offspring has only a small probability of surviving to reproduce. Yet another option, often associated with short-lived species such as bacteria, microbes, and fruit flies, is to reproduce early in the organism's life span so that many generations arise in a short period of time.

Ecologists categorize organisms into ***r*-selected** and ***K*-selected species:** *r*-selected species display rapid rates of reproduction despite risks to survival, while *K*-selected species have a slow rate of reproduction that allows the population to approach and remain at a density near their environment's carrying capacity (*K*). Females of most bony fish species (teleosts), for example, are *r*-

selected species; they lay hundreds, thousands, or millions of eggs, each containing only a modest amount of yolk, and each (when fertilized) producing a tiny hatchling that will probably be snatched up as food by another organism before it reaches adult size and reproductive maturity (Figure 43-7a). Sharks, on the other hand, are *K*-selected species; they usually produce a few very large, yolky eggs, which some species retain in the uterus for gestation periods of 6 to 20 months. These sharks give birth to sizable young quite able to fend for themselves and thus subject to much lower mortality rates than the young of bony fishes (Figure 43-7b).

The young of *r*-selected species are competitively inferior and are likely to survive only if they are dispersed to unexploited areas. The *r*-selected species tend to have adaptations for rapid dispersal, enabling them to rapidly colonize unexploited habitats and to reproduce before the habitat is claimed by a competitor. Fugitive species (see Chapter 42) are *r*-selected populations, and so are most organisms found in the early stages of ecological succession. By contrast, the competitive ability of *K*-selected offspring is much more important than are sheer numbers of young, since when a population reaches *K*, its members must compete intensely for limited resources. Each *K*-selected offspring represents a considerable investment of parental resources, and because of its larger size and greater ability to compete for space, food, and other necessities, it has the maximum chance of surviving to contribute to the next generation. Early dispersal to new habitats is not as crucial for *K*-selected species and they are more common in late successional stages, when community members use and compete more intensely for space, light, and other resources.

Logistic growth, age structure, and reproductive strategies are much more than just intellectual constructs for research ecologists. They are natural characteristics of populations that are propelling many of humankind's favorite organisms to extinction. Elephants, pandas, snow leopards, condors, and lady slipper orchids are all *K*-selected species living in areas of rapid habitat destruction. These species have so few offspring that their populations cannot recover from a crash as easily as an *r*-selected species, such as the snowshoe hare. Only by understanding the factors that underlie population growth or decline can we hope to preserve the remnants of such species and to protect others.

(a)

(b)

Figure 43-7 REPRODUCTIVE STRATEGIES: *r*- AND *K*-SELECTION.
(a) The grunion *(Leuresthes tenuis)* lays hundreds of eggs at a time, few of which survive—the *r*-selection strategy. (b) Sand sharks give birth to one live young (shown here with its yolk sac still attached)—a classic *K*-selection strategy.

LIMITS ON POPULATION SIZE

The success of a population in its environment depends, in part, on **population density:** how many individuals coexist in a given area and, in turn, how intensely each member must compete for limited resources. People in a city or nematodes in rich soil are high-density populations, while bears and other large carnivores tend to have low population densities and to be sprinkled sparsely throughout their habitats.

Whether the populations are dense or sparse, their distribution in space is usually uneven, and ecologists classify the spatial distributions of populations into three types of *dispersion patterns:* (1) clumped, such as a school of minnows, a stand of pines, or a herd of elephants (Figure 43-8a); (2) uniform, or evenly spaced, like hawks, bears, or other such territorial animals and some types of plants (Figure 43-8b); and (3) random, such as the spacing of dandelions on a lawn, which is entirely independent of the presence or absence of other dandelions (Figure 43-8c).

(a)

(b)

(c)

Figure 43-8 DISPERSION PATTERNS. Individuals within a habitat display three basic types of spatial dispersion: (a) clumped, as with elephants, minnows, and pines; (b) uniform, as with hawks and polar bears; and (c) random, as with dandelions.

Clumped species are most common, and the "clumps" (flocks or herds of animals, or groups of plants) tend to aggregate in places where the specific resources they need are abundant. Uniformly distributed populations occur when resources are spread thinly and evenly or when each individual aggressively defends its territory to ensure access to the resource. Populations disperse randomly if resources are uniformly distributed but individuals have little influence on each other's spacing or survival.

A single species can be densely distributed in one location or at one time and sparsely distributed in another. Breeding pairs of the great tit *(Parus major)*, for example, are distributed sparsely in the Netherlands but densely in England. Clumped species, such as lemmings or locusts, sometimes disperse across the landscape in a mass migration. For example, from about 1948 to the mid-1960s, much of sub-Saharan Africa was devastated by swarms of grasshoppers and locusts. Then, largely as a result of droughts, the infestations ceased. But now, in the late 1980s, rain has not only supported better harvests but has also triggered population explosions of the migratory insects (Figure 43-9). Densities and distributions clearly vary. So what controls these factors for any particular species?

Factors Controlling Density

Ecologists generally find that dense populations have lower birthrates, higher death rates, and slower growth rates than less dense populations, and they have identified four so-called **density-dependent factors** affecting population size: predation, parasitism, disease, and competition.

As a population of prey species grows denser, it is a more inviting target for predators; as predators increase in number, they may in turn exert greater control over the size of prey populations. Similarly, parasites and pathogens spread more easily among their hosts when the host population is dense (e.g., it's easier to catch the flu in a crowded city than on a farm), and crowding can lower an organism's defenses (a tree shaded out by competitors may become weak and more susceptible to fungal infestation). Overall, density-dependent mechanisms may limit a population to a level below its environmental carrying capacity.

Populations in nature often fluctuate much more erratically than one might expect, because **density-independent factors,** such as floods, fire, landslides, unseasonal weather, or other natural catastrophes, can operate independent of predation, competition, parasitism, and disease and can kill many or all the individuals in a population, regardless of its density. Rare killing frosts, for example, sometimes cause population crashes in areas where the winter temperatures are usually mild. In fact, plant and animal species living in places periodically visited by frosts, floods, or other disasters may be in a perpetual state of recovery from severe declines. Their

Figure 43-9 DRAMATIC FLUCTUATIONS IN POPULATION DENSITY.
Locusts and grasshoppers periodically show immense increases in population size. Following a 20-year respite, huge populations of grasshoppers and locusts are once again devastating large areas in Africa. Here, an Ethiopian farmer futilely swats at desert locusts devouring his millet crop. These migratory insects can eat their own weight in food each day, and a swarm can travel up to 200 miles every 24 hours.

evolution is likely to reflect such perturbations if the species continue to survive in the long term. In still other species, combinations of physical factors may trigger an unusual spurt in reproduction and corresponding population growth, as seen in the locusts of Tanzania, whose numbers rise and fall in relation to the area's rainfall.

Density-dependent and density-independent factors probably interact to some extent in all populations, and only through careful observation and experimentation can ecologists sort out the factors and their relative effects at a particular point in a population's life history. Researchers have focused considerable attention, for example, on how competition and predator-prey relationships affect populations.

Competition

A few years ago, ecologists working in the rain forests of Puerto Rico built tiny jungle nest sites of bamboo big enough to accommodate an unusual species of tropical frog, *Eleutherodactylus coqui* (Figure 43-10), whose members tend to be active only at night. Research suggested that the frogs' food sources were plentiful and did not limit population size, and thus the ecologists wondered if space could be the critical factor.

They divided the study area into 100 m^2 plots, built shelters in some plots, and left others unchanged as controls. They tallied frog populations in each plot at the beginning and end of the experiment, and the results were unequivocal: The original dense frog population had become even denser in the plots with bamboo houses, while no significant change had occurred in the control plots. Clearly, for this population, the number of available nest sites and hiding places determined K (carrying capacity), and the frogs competed with other members of the species for this limited resource.

In a case like this of **intraspecific** (within the species) **competition** the population may experience increased mortality, since frogs without hiding places are more vulnerable to predators. In other species, adaptations such as the production of fewer offspring may evolve. Among several bird species, including the Heermann's gulls nesting on an island in Mexico's Sea of Cortez (Figure 43-11), the higher the population density, the smaller the clutches, or groups of eggs, the females lay at a given time. The female can thus give extra care to a few young, and the offspring are more likely to survive to pass on the parents' genes. Similarly, limitations on light, water, or soil nutrients may cause individual plants in a population to produce less biomass or fewer flowers or seeds.

Sometimes, several species in a particular area compete for the same limited resource, and ecologists call this **interspecific competition.** As more individuals divide the resource pie, each competing population generally grows more slowly, and a very successful species can occasionally outcompete its neighbors so well that they become scarce or locally extinct. Eighteenth- and nineteenth-century sailors freed goats on one of the Galápagos Islands, for example, and these hungry mammals

Figure 43-10 ***Eleutherodactylus coqui:* COMPETITORS FOR SPACE.**
Given extra hiding places, the populations of this tropical frog can grow denser in a tropical forest. Among these frogs, only the males hear the "co" of the "coqui" song, while the females hear only the "qui" (see Chapter 45).

Figure 43-11 EFFECT OF POPULATION DENSITY ON REPRODUCTION IN BIRDS.
Population density at nesting sites of some bird species can grow quite high, as with these Heermann's gulls *(Larus heermanni)* nesting on an island in the Sea of Cortez, Mexico. In many birds and other species, fecundity, or production of eggs per breeding pair, drops off dramatically as population density rises.

consumed so many plants normally eaten by giant Galápagos tortoises that those indigenous reptiles became extinct on the island by about 1960.

Much of our modern theory of population ecology depends on the idea that intraspecific and interspecific competition largely determines the stability—or natural equilibrium—of wild populations. Ecologists believe that competition regulates population size and density in ecological communities in ways that help maintain a stable community structure. Density-independent factors, however, can cause populations to fluctuate dramatically, as when a hurricane rips through the plant and animal communities of a tropical island or when a grass fire destroys the food supply for a population of field mice. For many organisms, especially insects, the combined effects of predators, disease, and disastrous weather keep populations well below the levels where competition would operate.

Some ecologists have proposed that the higher a species' position in the food web (or trophic ladder) of its community, the more likely competition will be an important regulator of its population density. One reason is that predators at the top of the web—such as owls, hawks, and coyotes—tend to consume and therefore compete for the same classes of food. The role of competition in population ecology is an extremely active and controversial area of research today, and many of the studies try to distinguish between the intensity of competition and its ultimate importance to long-term ecological or evolutionary success. Certain plants in a community may be competing intensely for limited minimal nutrients, for example, and yet such intense competition may be relatively unimportant in the long run. Conversely, even weak competition may assume great importance in controlling populations evolving through time if other factors are not limiting.

Predation

Like competition, the interactions of predators and their prey can help determine a population's size and success in a given environment. Broadly speaking, any organism that feeds on another living creature can be termed a predator. Thus, predators include not only large carnivorous hunters, such as lions and eagles, but also grazing herbivores—from tiny planktonic swimmers to bison or giraffes—that consume plant parts but do not kill the plants themselves, and parasites, such as wasps that deposit their eggs in the tissues of living caterpillars (Figure 43-12).

In the early 1920s, two mathematicians, Alfred Lotka and Vittora Volterra, independently developed a mathematical model to describe the relationship between predator and prey populations. The *Lotka-Volterra equation*, as it is called, predicted that such populations

Figure 43-12 PREDATION IN PROGRESS.
Braconid wasps (such as *Rogas terminalis*) deposit their eggs on caterpillars, such as this larval sphinx moth *(Manduca quanquemaculatus)*. As the eggs mature, cocoons are formed, and wasp larvae eventually emerge and begin to eat the caterpillar alive.

would oscillate in a recurring cycle. As a prey population grows, its predators have more food available, and their density, after a reproductive time lag, increases as well. Eventually, the numerous predators eat prey faster than the prey can reproduce, and the prey population declines. After another time lag, predators begin to starve, and their populations become less dense. The classic, but much debated, real-world example of a regular predator-prey oscillation is the cycle involving lynx and hare populations (review Figure 43-1).

Experimenter G. F. Gause tested the Lotka-Volterra equation by setting up test-tube cultures with *Paramecium* as prey and *Didinium* (another protozoan) as predator. At first, the paramecia flourished; soon, however, the *Didinium* population expanded rapidly and wiped out its burgeoning food supply. Alas, the predators became extinct as well; no cycling of populations took place.

Gause tried the experiment again, this time periodically adding protozoan cells of each species as "immigrants." The populations soon began to cycle as the model predicted, and later studies confirmed that in the laboratory, oscillations occur as predicted, but only as long as the researcher provides immigrants, access to hiding places for prey, or other variable features.

In nature, it is often difficult to study predation because each prey species can have several predators (e.g., owls, snakes, and coyotes all eat desert kangaroo rats);

BIOLOGICAL PEST CONTROL: DOING WHAT COMES NATURALLY

With world hunger mounting as fast as public controversy over chemical pesticides, the stakes are high in the search for safer ways to control agricultural pests. As a result, dozens of research teams are actively identifying predators, parasites, and diseases that can be used to fight plant pests with biological rather than chemical warfare.

Ironically, the success of chemical pesticides helped agricultural scientists recognize the importance of biological pest control. Just after World War II, the chemical industry presented American farmers with what seemed like a revolution in pest control, and citrus growers, among others, raced to protect their crops with such "miracle" pesticides as DDD and DDT. In addition to dispatching the mites and scale insects that attack fruit trees, however, many chlorinated hydrocarbons like DDT also killed the natural enemies of the mites and scale insects. Released from such natural controls, populations of cottony-cushion scale, citrus red mites, and other pests exploded in outbreaks that sometimes defoliated trees severely or even killed them altogether (Figure A). Today, citrus growers still use some chemical sprays, but they also use insect pheromones to attract and survey populations of male red scale mites and thus to determine precisely when and how much to spray.

Biological control may be particularly promising where pests are inadvertently imported to a new area while their natural enemies are left behind. A case in point is the winter moth, a highly destructive insect introduced into northeastern Canada from Europe in the 1930s. Within 20 years, winter moth larvae, which feed on leaves and other tree parts, had begun to inflict millions of dollars in damage in Canadian hardwood forests. Then, in 1954, a team of ecologists tracked down two winter moth predators in Europe and imported them to the affected area. These insects—the tachinid fly and an ichneumonid wasp—attack the caterpillars, just like the wasps we discussed in Chapter 42. Within 6 years, the predators effectively held winter moth populations to acceptable densities. The ecology of the predatory parasites themselves was an important aspect of the program: The fly is most efficient at high moth densities, and the wasp is best when densities fall.

Biologists continue to study organisms with potential usefulness in biological control programs, and as they do, they will also be unraveling some of the fundamental ecological relationships that link organisms in the natural world.

Figure A
A lemon infested by red scale mites. Fruit growers now use pheromones to survey for red scale outbreaks.

and other factors impinge on prey populations (interspecific competition, disease, and emigration can all affect the numbers of kangaroo rats). Conversely, predators tend to eat several kinds of prey. (Coyotes trapped in Iowa, for example, had traces of grasshoppers, muskrats, deer, chickens, lizards, corn, snails, bark, crayfish, and even other coyotes in their stomachs.) And predators can shift from one species to another when times are lean: Arctic foxes prefer to eat lemmings, but when these small mammals suffer population crashes, the foxes switch to ground-nesting birds such as brant geese.

To complicate matters still more, what appears to be population regulation due to predation may actually be the result of some subtle, hard-to-determine fluctuation in the environment. Ecologists studying the 600 to 1,000 moose populating Isle Royale, a small island in Lake Superior, suggested that wolves and other predators were regulating the size of the moose population. Newer research, however, shows that sodium, an essential micronutrient in the moose diet, may be the crucial limiting factor for moose on the island. Researchers now believe that wolves cull from the herd mainly those animals that are old, sickly, or too young to defend themselves from attack (Figure 43-13).

Predation slows or stops the growth of a prey population only when many reproducing members are eliminated, and as Chapter 42 described, both predators and their plant and animal prey have evolved a fascinating variety of adaptations that increase each species' chances of survival.

Figure 43-13 FATE OF THE OLD, SICK, OR WEAK.
This moose calf *(Alces alces)* bears the attack wounds of wolves. It escaped once, but no doubt the predators will try again.

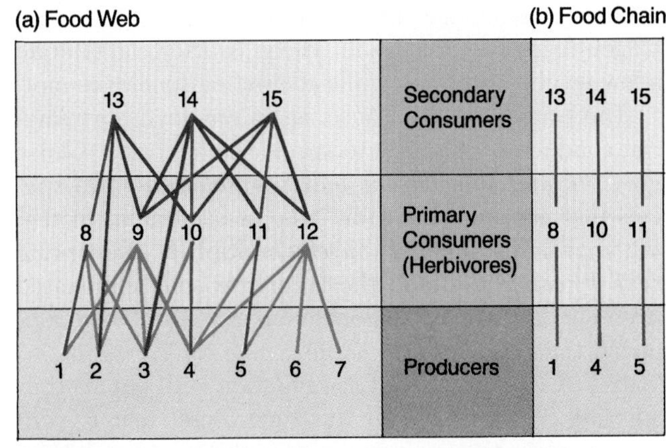

Figure 43-14 FOOD WEBS, FOOD CHAINS, AND COMMUNITY STABILITY.
If the earthworm (here, number 8) in the middle of this food chain (b) were to decline or become extinct owing to disease, let's say, then the mouse (13) specialized to eat those worms might die out. Plant number 2 might increase in size, since it is part of a simple food chain. But local extinction of the worm (8) in the food web (a) would not affect a sparrow (13), hawk (14), or snake (15), because they have alternative food sources within the complex web.

Population Fluctuation and Community Structure

Simple communities, such as a monoculture field of corn, are highly vulnerable to the attacks of pests, and this fact made ecologists wonder whether community structure, in general, could influence a population's growth or decline. In communities with many species and complex food webs, most herbivores feed on many plant species, and each carnivore exploits several herbivore species. As a result of these diversified food sources, if one species suffers a severe population decline or becomes extinct locally, the consumers above it in the community's food web still have an alternative food supply available. By contrast, in a simple community with linear food chains involving a few species, any drastic change in a population at one trophic level may have severe repercussions on the populations above and below it in the trophic hierarchy (Figure 43-14). Such a system seems inherently unstable.

Many ecologists, however, dispute the idea that stability results from community diversity. In nature, high diversity is not always found in combination with complex food webs. In fact, many plants and animals found in tropical regions have highly specialized adaptations for interacting that result in simple food webs (see Chapter 42). For the present, the idea that diversity begets stability seems, at best, oversimplified, and the reverse hypothesis may be closer to the truth: Stable environments may beget diversity because they allow rare species to persist.

HOW POPULATIONS ARE DISTRIBUTED

Some species can only live in very specific places; redwoods and desert yucca, for example, have climate-related adaptations that limit their populations to the foggy Pacific coast and hot, dry terrain, respectively. Other species, however, such as the North American coyote, can thrive from Mexico to Canada in a variety of environments. Even within its potential range, however, no species ever occupies all the area available to it; recall that a variety of factors, including interspecific competition, can reduce the fundamental niche to the realized niche (see Chapter 42). To understand why a population lives where it does and why it grows, declines, or stabilizes in that location, ecologists study the many factors that affect population distribution, including competition among species for food, breeding sites, or other resources that delimit the realized niche.

Interspecific competition is most intense when two or more species require many of the same resources (i.e., are adapted to similar or identical habitat niches). Among plants, one of the most effective forms of interspecific competition is **allelopathy,** in which chemical substances produced by one species inhibit the germination or growth of seedlings of another species. At least one allelopathic grass species *(Aristida oligantha)* exudes a phenolic acid that acts as an antibiotic and kills nitrogen-fixing bacteria in soil. The grass itself can tolerate a low level of nitrogen, while competing species cannot.

The brilliant G. F. Gause, again employing paramecia, conducted a test of interspecific competition. Gause grew *Paramecium aurelia* and *P. caudatum*—two species that consume the same type of bacteria as food—together in test tubes. *P. aurelia* multiply at a faster rate than *P. caudatum*, and the latter was soon eliminated from the "niche," as the population of *P. aurelia* monopolized the common food supply.

Next Gause combined *P. caudatum* with yet another species, *P. bursaria*, but this time found that *P. caudatum* fed on bacteria suspended in the culture solution, while *P. bursaria* consumed only those bacteria that were at the bottom of the test tube. By occupying different food niches, the two species were able to share a limited resource and coexist successfully.

This kind of **resource partitioning** is a common adaptive solution utilized by species that share similar or identical habitat niches. Figure 43-15 shows Robert MacArthur's famous study of how three species of warblers in a conifer forest partition resources by feeding in different parts of the tree and on insects and seeds of different sizes and types.

Often, closely related species show **character displacement:** They evolve physical differences in body structures, and these adaptations allow them to exploit a limited resource in different ways. For instance, two related species of earth-burrowing lizards in Africa eat a particular type of termite. Where the lizard species occupy the same small area, one species has evolved a larger head and body than the other and feeds on large termites. In places where the lizards' habitats are separate, each species has a small head and body and eats small termites exclusively.

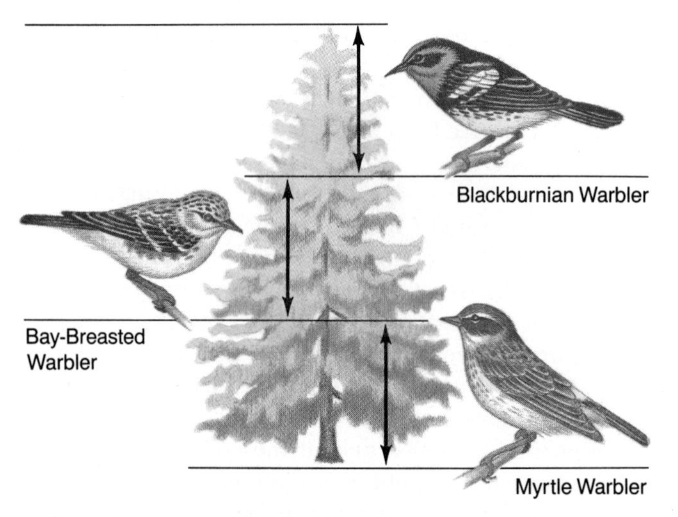

Figure 43-15 RESOURCE PARTITIONING IN FEEDING NICHES OF WARBLERS.
These three warbler species live in the same habitat but do not compete for food because they feed at different heights in the conifers of northeastern forests. Thus, their food niche has a height component, and there is little direct competition among the three species, at least for food.

Figure 43-16 BILL TYPES IN GALÁPAGOS FINCHES.
The labels indicate the types of food each finch eats predominantly. Notice how the bill size and shape correlate with the food type.

Perhaps the most famous case of character displacement is among the finches of the Galápagos Islands. British biologist David Lack described in detail how the birds' varying bill shapes correlate with their diet and ways of foraging (Figure 43-16). From the original seed-eating ground finch immigrant have evolved five species that feed on different sorts of seeds on the ground; two that feed on seeds in cactuses; six that feed in trees; one a vegetarian; and six insect eaters (one of which behaves like a woodpecker and even excavates insects by using a stick held in its bill). The other birds are warblerlike and generally insectivorous. Ecologists believe that isolation on the different Galápagos Islands and competition for food primarily led to this remarkable evolutionary diversification.

Predation or disease can limit a population's distribution as well as its size. Predatory sea urchins, for example, are very effective at restricting populations of algae to areas where the urchins are absent (Figure 43-17). And the Dutch elm disease we discussed in Chapter 22 has vastly reduced or eliminated elm populations, thereby leaving vacant land for other trees. Thus, both population and community structures are affected, just as with predators, chance storms, or competition among species.

HUMAN POPULATIONS: A CASE STUDY IN EXPONENTIAL GROWTH

The human population and its awesome explosion in the last century provides a real-world case study of population size, exponential growth, and the ecological limitations and consequences of such rapid expansion.

About 10,000 years ago, people began planting seeds,

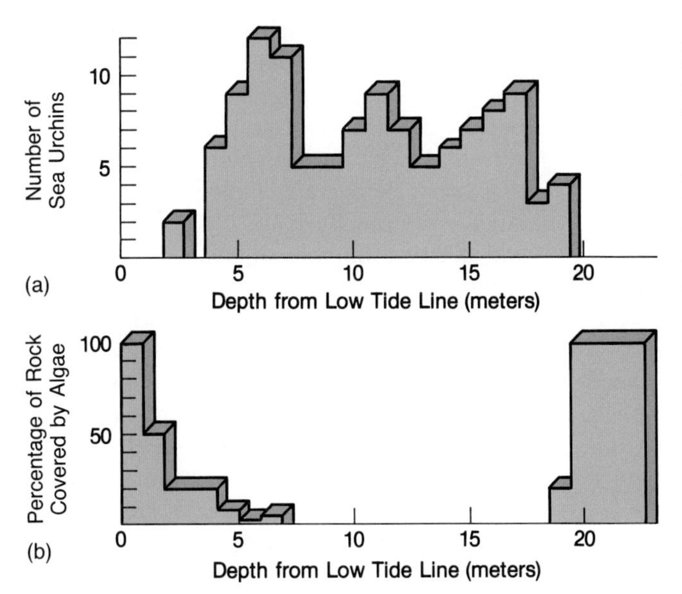

Figure 43-17 COMPETITION, PREDATION, OR DISEASE: LIMITATIONS ON SPECIES' LOCALE.
This graph shows the distribution of (a) sea urchins and (b) their algal food along a range of water depths. Notice that the urchins do not occur higher or lower than the depths indicated in part (a), and thus these areas are refuges for the vulnerable algae.

harvesting crops, and domesticating animals rather than hunting and gathering food exclusively. This so-called agricultural revolution had a stimulating effect on the size of the human population. Although increased agricultural efficiency allowed farmers to feed more people, it also required more hours per week than hunting and gathering, and it thus encouraged people to have larger families to provide extra hands for tending fields and herds. Human populations began to grow, increasing 25-fold between about 10,000 and 2,000 years ago. At the onset of the agricultural revolution, there were about 133 million people in the world (Figure 43-18); by 1650, there were about four times as many—roughly 500 million people. Human population doubled during the next 200 years, and then, with the coming of the industrial revolution in the eighteenth century, doubled again in only 80 years, reaching 2 billion people in 1930. The next doubling in world population size took only 45 years. In 1987, the world population reached 5 billion, and by the end of the 1990s, it will reach 6 billion. The staggering upsweep of the curve in Figure 43-18 should look familiar; it is a nearly perfect exponential curve of the sort that—as we saw earlier—cannot be sustained indefinitely without reaching and exceeding the carrying capacity of the environment and without a leveling off or a population crash.

Why did the industrial revolution so contribute to population growth? Among other things, while the enormous human population fanned out across the globe and occupied most of the arable (farmable) land, new forms of transportation allowed food to be widely distributed from low-density rural areas to high-density urban areas. Technological advances, such as urban sanitation systems, as well as medical discoveries, such as the role of bacteria in diseases, drugs to fight them, and immunizations to prevent infection, have helped diminish mortal-

Figure 43-18 WORLD POPULATION GROWTH SINCE THE AGRICULTURAL REVOLUTION.
Malthus was right. Human world population has grown geometrically. This graph shows the stupendous increase in human population beginning about 10,000 years ago, interrupted only briefly by the many deaths due to the black plague in the fourteenth century. As the inset photo suggests, human density will continue to increase along with human population.

ity and allowed more people to survive to reproductive age and beyond. At the end of the nineteenth century, for example, the life expectancy for males and females in the United States was below 50 years; today, it is 71 and 78, respectively. The ecological consequences of this are straightforward: Birthrates increased during the agricultural revolution, making large families the cultural standards; death rates decreased since the start of the industrial revolution. More births plus fewer deaths and longer life expectancies equals a rapidly expanding population.

In the latter half of the twentieth century, population pressures have started to ease in some industrialized nations; the birthrate has fallen as families have left farms and women have entered the workplace in unprecedented numbers. Between 1500 and 1700, English women commonly bore between 12 and 20 babies (30 was not unusual)! That has dropped steadily, and today in the United States, the rate is 1.8 births per woman, and in West Germany, Scandinavia, Canada, Switzerland, and certain other industrialized countries, the birthrate is even lower. Were it not for net immigration, the population size would be declining in most industrialized countries.

In stark contrast, many less-developed nations in Africa, South America, and Asia have birthrates of 4 children per woman or higher. Since most of the world's people live in those countries, and since the age structure is skewed toward young people in their reproductive years, demographers predict that the human population will grow by some 2 billion people between 1975 and 2000—a number equivalent to our species' entire population in 1930.

Demographers combine age structure diagrams for human males and females into **age pyramids** (Figure 43-19), which vividly depict our dilemma. The narrow bullet-shaped graph of population numbers in the developed world shows relatively small total populations, even distribution throughout the various age groups, and a modest projected increase, largely in the older age groups. In contrast, the broad pyramid-shaped graph for the developing countries shows huge total populations (current and projected), a largely young, reproductive-age population, and larger increases to come, mainly in the younger age brackets in the next few years. If these patterns persist, there will be nearly 10 billion people in 2030, when many readers of this book are young grand-

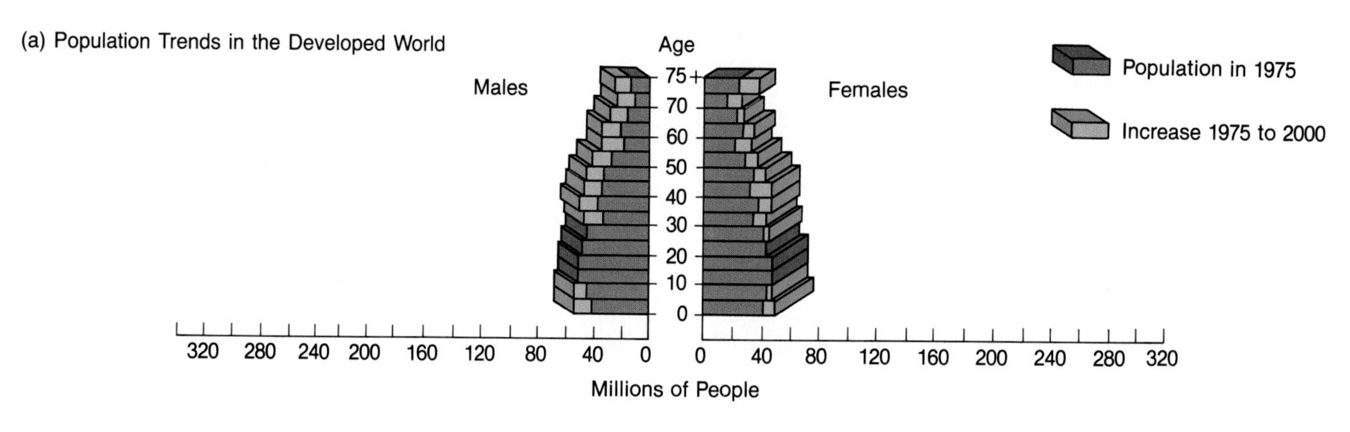

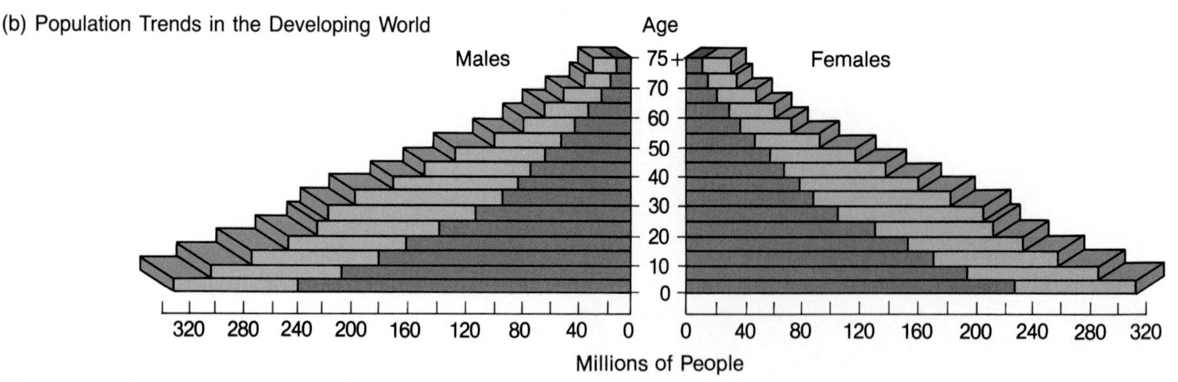

Figure 43-19 HUMAN AGE PYRAMIDS.
These graphs show the numbers of people in various age groups in (a) the developed world and (b) the less developed world. Males are shown on the left, females on the right. Purple areas show the populations in 1975, and blue areas the change between 1975 and 2000. In 1975, for example, there were 240 million boys between the ages of 0 and 5 years and 233 million girls in the same age range in the less developed world; by 2000 there will be 330 million boys between 0 and 5 years and 315 million girls. Note the greater total number of young people of both sexes in the less developed nations than in the developed world. Reproductive activity by those people will continue to drive upward the populations in less developed countries.

parents, and 30 billion people before the end of the twenty-first century.

Ecologists believe that the earth's ecosystems cannot support 30 billion people. The U.S. National Academy of Sciences has estimated that the earth's carrying capacity for human beings is not much more than the 10 billion we will reach in less than 40 years. This estimate assumes an intensively managed world that still preserves some degree of individual freedom. Even if our species completely relinquishes freedom of choice and we push our planet well beyond the carrying capacity, 30 billion people would be an upper limit, and human life could be uniformly fraught with hunger, crowding (Figure 43-20), disease—in short, with misery.

Humans have clearly become an overwhelming ecological factor during their short tenure on earth, and the details of our population explosion help explain specific physical impacts such as the greenhouse effect and the destruction of the tropics. Nearly three-quarters of earth's people inhabit developing nations—most of them in the tropics—and this figure will grow to about five-sixths within 30 years (Table 43-1). The pressure of population and poverty is immense—hence the cutting and burning of rain forests for agricultural land and wood and the added levels of CO_2 in the atmosphere.

Only in recent years have citizens, scientists, and government officials become fully aware of the profound ecological, economic, and political consequences that overpopulation in developing countries could have on our species' long-term survival. Clearly, humanity must begin to focus its huge capacity for invention and innovation upon this frightening problem.

Figure 43-20 AN URBAN FUTURE?
This crowded shopping street in Yokohama is duplicated in a thousand cities around the world. Faces everywhere, noise, smog, electric power to keep the city going—such a different life from that of our ancestors just a few generations ago. Is it the kind of future we will all face?

Table 43-1 POPULATION ESTIMATES (IN MILLIONS OF PEOPLE) AND PERCENTAGES OF WORLD POPULATION

	1950	*1984*	*2020*
Developed world	832 (33%)	1,166 (24.5%)	1,350 (16.7%)
China	557 (22%)	1,034 (21.7%)	1,545 (19.1%)
Less-developed world (excluding China)	1,136 (45%)	2,561 (53.8%)	5,191 (64.2%)
Total	2,525	4,761	8,086

Source: Population Reference Bureau, Inc., Washington, D.C.

LOOKING AHEAD

We turn in the next two chapters to the biological roots of animal behavior. We will discuss not only the kinds of survival drives that lead to reproduction and sometimes overpopulation, but the kinds of behavior patterns that enable social animals (including humans) to coexist in complex societies and to solve mutual problems.

SUMMARY

1. Populations—groups of individuals belonging to the same species—share a set of characteristics: *natality*, *mortality*, and *population density*.

2. In theory, a population can grow exponentially, or geometrically, if there are no limits on resources and no competition or predation, no disease, and no other potentially limiting factors. In nature, populations do not show exponential growth curves—at least not indefinitely.

3. Different populations are characterized by different intrinsic rates of increase (r), the maximum possible per capita growth rate. The maximum number of individuals that environmental resources can support is the *carrying capacity* (K) of the environment.

4. A logistic growth curve reflects various phases in a population's growth history.

5. If a population outstrips its available resources severely, the carrying capacity of the environment may be lowered permanently. In other instances, carrying capacity may fluctuate up and down owing to such factors as changes in season.

6. The rate at which a population grows can also be affected by its *age structure*—the relative numbers of young, mature, and old individuals in the group.

7. Populations show particular *reproductive strategies:* r-selected species grow to maturity quickly and produce many offspring, but the young suffer high mortality; K-selected species mature more slowly and produce few young, but each new individual is relatively well prepared to compete for food,

space, and other resources, and thus mortality is low.

8. *Density-dependent factors* that influence population size include *intraspecific competition, interspecific competition,* predation, parasitism, and disease. *Density-independent factors* include natural catastrophes, seasonal changes, bad weather, and accidents, all of which cause populations to fluctuate erratically.

9. Some populations exhibit regular cycles of decline in density and then recovery; interactions between predator and prey populations, combined with reproductive time lag, may cause such cycles.

10. Diverse ecological communities with complex food webs may be inherently more stable than simpler communities. Alternatively, stable environments beget diversity because they allow rare species to survive.

11. Distribution of a population depends on the food, habitat, and physiological requirements of its component individuals, and on interspecific competition and chance events, such as severe storms or disease.

12. Interspecific competition for resources can result in *resource partitioning.*

13. *Character displacement* reflects the evolution of adaptations for slightly different aspects of the niche among otherwise competing species.

14. Human population provides a case study in exponential growth, ecological limitations, and the consequences of rapid population expansion.

15. The agricultural revolution provided more food and hence more human survival, and encouraged larger family size. The industrial revolution saw improved transportation, sanitation, and medicine, with concommitant reduction in human mortality, increased life expectancy, and an even greater expansion of human population.

16. Age pyramids for human populations in developed and developing countries reveal stable population sizes in the former and rapid growth patterns in the latter.

17. We will reach earth's carrying capacity for humans (about 10 billion) in less than 40 years. The exploding human population is deeply entwined with global problems such as hunger, loss of species, and the greenhouse effect.

KEY TERMS

age pyramid
age structure
allelopathy
carrying capacity
character displacement
density-dependent factor
density-independent factor
exponential growth curve
interspecific competition
intraspecific competition
K-selected species
logistic growth curve
natality/mortality
population density
reproductive strategy
reproductive time lag
resource partitioning
r-selected species
survivorship curve

QUESTIONS

1. Theoretically, populations of organisms can increase in size exponentially. What does this mean? Why is this not seen in nature?

2. The equation $dN/dt = rN$ is the curve of exponential increase in an unlimited environment. How does the modification $dN/dt = rN(K - N/K)$ make the equation more closely reflect real-life conditions?

3. Give an example to show how the activities of certain members of a community may permanently alter the carrying capacity of a habitat.

4. Give examples to show what is meant by r selection and K selection.

5. Name the three dispersion patterns used to describe the spatial distribution of species, and give a species example for each pattern.

6. Density-dependent factors and density-independent factors limit population size. Define these terms. Give examples of limiting factors in each category.

7. Explain why it is difficult to study the effects of predator activity on the size and structure of prey populations.

8. Explain the apparent relationship between species diversity and community stability.

9. Explain how some plant species employ allelopathy to compete with other plant species.

10. Use examples to show what is meant by the adaptive strategies of resource partitioning and/or character displacement.

11. Where will most human population growth take place in the next few decades? What shape might the age pyramids for those countries be in 2000, 2020, 2040?

12. Are humans r- or K-selected species? What is likely to happen as we approach earth's carrying capacity for our species?

ESSAY QUESTION

1. Discuss the following generalization: Pioneer species that colonize disturbed habitats tend to be r-selected species, while the inhabitants of more mature and stable ecosystems tend to be K-selected species.

For additional readings related to topics in this chapter, see Appendix C.

44 Animal Behavior: Adaptations to the Environment

Beneath the varying behavior which animals learn lie unvarying motor patterns which they inherit. These behavior traits are as much a characteristic of a species as bodily structure and form.

Konrad Lorenz, *The Evolution and Modification of Behavior* (1965)

In the late nineteenth century, French naturalist Jean-Henri Fabre recorded the habits of bee-hunting insects called digger wasps. These animals always sting their prey in the same spot, then stuff their victim in a burrow where the wasps' developing offspring can eat the now-paralyzed bee alive. Curiously, the larvae seem to "know" just how to eat a paralyzed victim without causing its premature death. Fabre wondered whether the wasps acted out of pure inherited instinct or whether they learned their hunting and eating techniques.

While Fabre never discovered the answer, his work and that of other early naturalists helped set the stage for modern studies of *behavior:* the things that organisms, especially animals, do as they grow, reproduce, seek food, and otherwise interact with their environment. Behavior is a direct product of an animal's nervous and endocrine systems, employing, as it were, the body's arsenal of limbs, sensory systems, effectors, and other

Animal migration: Behavior for survival. In a "march to motherhood," these pregnant caribou *(Rangifer tarandus)* may travel 300 miles in the Alaskan Brooks Range.

features to act and react. Some animal behaviors are indeed *innate* (inborn); these instinctive actions are heritable and subject to natural selection. The ability to perform the behavior originates in the genes, arises during development, and can serve as an adaptation to the environment that leaves the organism better able to survive to reproductive age.

Digger wasp larvae do not "know" how to eat bees in any conscious sense; their instinctive feeding behaviors are *reflexes*, automatic actions brought about by direct links between the nervous system and the body's effectors. Nor do they learn the fine points of eating bees alive by watching other larvae or adults. Much animal behavior is, however, the result of *learning*, modifications based on experience, as well as on predispositions to certain types of learning. Learning is especially common in vertebrates. A young monkey chances to observe a small mammal being attacked by a snake and learns thereby to avoid these slithering predators. Again, the behavior is adaptive, but it is passed along indirectly, not directly through the genes. Most of the complex behaviors we see in birds and mammals involve a combination of reflexes, instincts, and learning intertwined in complicated patterns.

Such complexity makes the study of behavior both fascinating and difficult, whether in the laboratory or in nature. The study of animal's feeding, mating, and defense strategies in the wild is called **ethology.** All those who study behavior must contend with a real problem—our natural human tendency to *anthropomorphize,* or attribute human qualities to other species. There is no evidence to date, however, that nonhuman animals can "be anxious," "want something," "worry," or "love" in any way that resembles our own feelings or that motivates the behavior we see (or, for that matter, that one person's "anxiety" or "love" equals another's). Anthropomorphism has been a great stumbling block for the young science of animal behavior, and in this chapter, we describe the careful methodologies researchers use to avoid these pitfalls and to understand the evolutionary origins of behaviors and their adaptive roles in the lives of animals.

Our topics here include:

- The automatic behaviors called reflexes
- The broader set of instinctive behaviors that can be preset or modified and that can be triggered in various ways
- Learning, the flexibility that allows changed behavior based on experience and that enables a huge range of activities from a gull's simple feeding response to a person's understanding of biology
- How instinctive and learned behaviors underlie two important survival patterns: navigation and migration

REFLEX BEHAVIOR

At its most fundamental level, behavior is indistinguishable from the physiological processes that maintain life itself. A single cell such as *E. coli* moves backward or forward (simple behavior) as it detects various compounds by means of thousands of proteinaceous receptors in or on the cell surface. Studies show that such cellular organisms have receptor "systems" sensitive to dozens of chemicals and thus can respond by moving up or down concentration gradients for all of them. The activities of such receptors serve as a model for the sensory, nervous, and motor systems of multicellular animals and represent the simplest type of innate behavior—the reflex.

Reflexes are automatic, involuntary responses to external stimuli. Typically, a reflex behavior takes place when sensory receptor cells communicate a stimulus to the nervous system, which in turn sends a set of instructions to effector organs, such as muscles. The so-called startle reflex common to many animals illustrates this point: If you prod a garden snail with a stick, giant nerve fibers convey this touch stimulus simultaneously to nearly all muscle groups, producing an almost instant retreat of the snail into its shell (Figure 44-1). The neuronal circuits for some reflexes are very simple, with the sensory cell synapsing directly on the motor cell. Most reflexes, however, including the snail's fast contraction and the human knee-jerk response, involve more complex neuronal circuits, and these—whether simple or complex—can be the fundamental elements underlying other simple behaviors, including tropisms, taxes, and kineses.

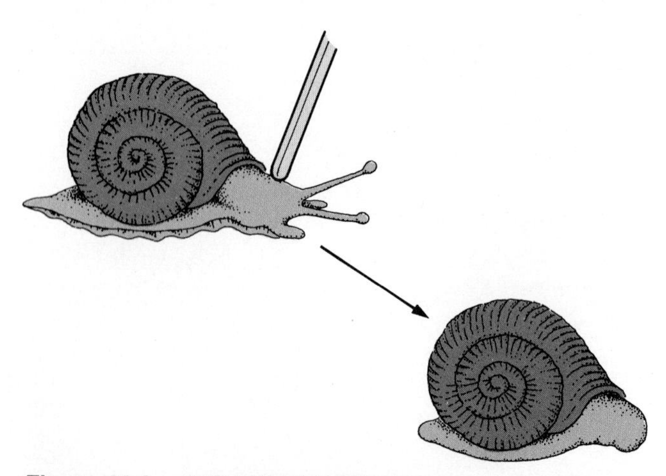

Figure 44-1 THE STARTLE REFLEX.
Touching a garden snail with a stick stimulates its nervous system to initiate muscle contraction automatically. The result is the animal's retreat into its shell.

Tropisms, Taxes, and Kineses

In the late nineteenth century, animal physiologist Jacques Loeb was intrigued by the orientation of plants toward stimuli such as light, gravity, and moisture, and he wondered if animals might also exhibit such orientations, or *tropisms*. Loeb found that the free-floating larvae of marine barnacles are phototropic—that is, attracted to light; that phototropic caterpillars presented with food in an unilluminated spot will starve rather than turn their heads into the shadow; and that other animals are attracted to or repelled by gravity, electric currents, and chemicals. In each case, the tropism was an instinctive reflex and did not involve learning. Modern physiologists call animal tropisms **taxes** (singular, *taxis*), to distinguish behavior produced by electrochemical activity of the animal nervous system from the analogous tropisms in plants produced by growth-regulating hormones (see Chapter 28).

The maggot of the common housefly, for example, exhibits *phototaxis*—movement in response to light. The maggot has primitive photoreceptors on its head, and when the larva is ready to pupate, it begins to crawl while moving its head from side to side, which allows the greatest light intensity to hit first one side of the head and then the other side. Each time, the maggot responds by turning its head (and body) toward the darker side, and it eventually faces completely away from the source of illumination and crawls into a dark place. This reflex taxis—this program of anatomical responses to simple stimuli—leaves the animal less vulnerable to its predators.

Kineses (singular, *kinesis*) are also simple reflexive behaviors, but ones in which the intensity of the animal's response is proportional to the intensity of the stimulus. The single-celled paramecium shows this type of response in the presence of carbon dioxide; at a certain distance from a bubble of carbon dioxide, the microscopic protozoan slows its forward movement in response to the acidity surrounding the bubble. If the paramecium moves too close to the bubble, the high acidity causes its cilia to beat in a way that rotates the animal through an angle of about 30 degrees; having changed course, the single-celled creature continues swimming away from the bubble. This process is repeated until the organism is in the zone of optimum pH once more (Figure 44-2). In nature, such adjustments to pH levels cause the paramecium to remain near its chief food source, decomposer bacteria, which produce some acid, but to avoid dangerously high acid levels.

In general, taxes and kineses are vital adaptations that enable simple animals to better cope with features of their environment and that have increased their chances of surviving to reproduce. In multicellular animals, seemingly automatic but much more complex behaviors called instincts also serve as important adaptations for survival.

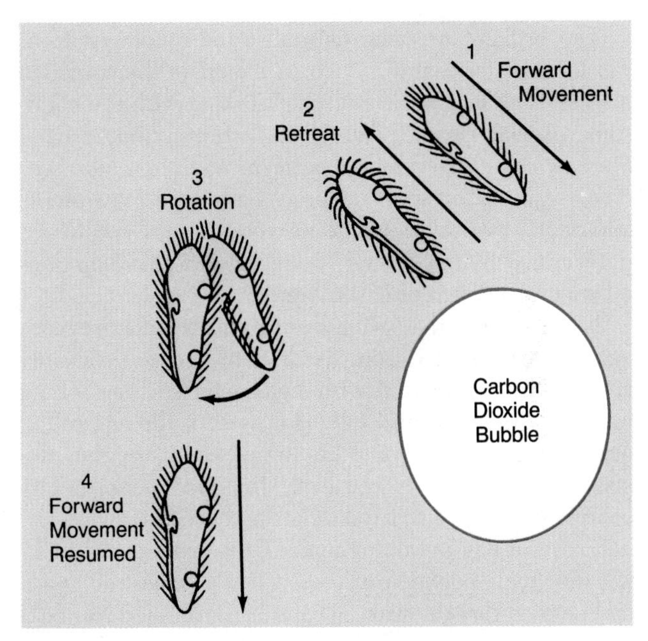

Figure 44-2 KINESIS IN A PROTOZOAN.
When this paramecium nears a noxious object, such as a CO_2 bubble, the cilia on its surface reverse their beat, causing the animal to back away (1–3). The organism then turns about 30 degrees to the left and advances (4). On encountering the object again, it repeats the reflexive backing and turning.

INSTINCTS: INHERITED BEHAVIORAL PROGRAMS

Biologists usually define an **instinct** as a stereotyped, inherited pattern of behavior that takes place in response to a specific environmental stimulus. When a fawn senses a disturbance in the environment caused by a predator, for example, it instinctively freezes; this instinct makes the fawn less likely to be noticed and attacked. The stinging and eating behavior of Fabre's digger wasps is also the result of instinct and does not depend on experience or learning as far as behaviorists now know. (Recent work, however, has revealed components of learning in supposedly instinctive bee behavior.) Beginning in the 1930s, behavioral psychologist B. F. Skinner triggered immense controversy by suggesting that nearly *all* of an animal's activities can be explained as learned—not instinctive—responses to environmental stimuli. More recent students of ethology have resoundingly rejected this view by providing a wealth of experimental evidence that genetically programmed activities can be important components of animal behavior.

The brilliant Austrian naturalist and ethologist Konrad Lorenz, for example, helped elucidate the nature of instinct with his classic studies. Working with the handsome greylag goose, (Figure 44-3), Lorenz noted what are now called **fixed motor patterns** or *fixed action patterns*—unvarying series of precise physical movements apparently "wired" into the nervous system and hence determined by the animal's genes and the developmental program that builds the brain.

Lorenz found that whenever he displaced a greylag's egg a short distance from its nest, the goose invariably attempted to retrieve it with a characteristic egg-rolling motion of the neck and bill. However, if the egg rolled out of reach entirely or if Lorenz picked it up after the goose had begun its retrieval, the bird nevertheless completed the entire fixed motor pattern, retrieving the nonexistent egg by rolling it along the ground and tucking the "egg" between its legs. Clearly, once the goose had noticed the wayward egg and had extended its neck, the fixed pattern, a complex type of instinct, was initiated and carried to conclusion.

Other researchers had discovered even earlier that such stereotyped behavior is a common feature among courting birds: Each species has its own characteristic pattern of genetically determined behavior that is followed faithfully from generation to generation. Figure 44-4, for example, shows the courtship display of the blue-footed booby. Another example of stereotyped behavior is the production of sounds by the 500 species of *Drosophila* fruit flies that evolved and diversified on the Hawaiian Islands. The sounds from the fly's wings and abdominal pulsations, as well as their locustlike and cricketlike clicks, can vary among species and differ radically from the sounds of mainland fruit flies. These differences demonstrate clearly that stereotyped behaviors can evolve: The genetic and physiological programs that produce these instincts contribute to diversification into new species.

Figure 44-4 FIXED MOTOR PATTERNS: A MATING PAIR OF BLUE-FOOTED BOOBIES.
These sometimes comical diving birds, appropriately called blue-footed boobies (after *bobos*, Spanish for "clowns"; *Sula nebouxii*), inhabit the Galápagos Islands and other areas of the Pacific. Here, they are executing the moves of a stereotyped courtship display before mating.

Figure 44-3 FIXED MOTOR PATTERNS AND IMPRINTING: THE GREYLAG GOOSE.
Konrad Lorenz studied this kind of greylag goose, *Anser anser*, for many years, gaining great insight into the concept of fixed motor patterns—seemingly innate behaviors initiated by the nervous system in response to specific stimuli.

Open and Closed Programs

Fixed motor patterns can be considerably more complicated than reflexes. Many of the actions of short-lived organisms, such as insects, are **closed programs**—that is, they are entirely "prewired" at birth. Larger, long-lived animals with more complex brains, however, often exhibit **open programs**—innate motor patterns in which certain elements can be modified by learning. This reflects the general rule of thumb that the lower an animal is on the phylogenetic scale, the greater the fraction of its behavior that depends directly on genes.

Kittens, for instance, inherit a basic "pounce and disembowel" behavior that can be modified to conform to the behavior of their parents, litter mates, or other cats and to fit the specific characteristics of available prey. Killing a tiny mouse thus requires different pounce and disembowel techniques than killing a strong ground squirrel or a large jay (Figure 44-5). Human facial expressions are also based on open programs: Most are strikingly similar in form and meaning throughout our species, and in fact, humans who are born blind still smile, frown, and grimace, without ever having seen another human face. People, however, often develop their own individual nuances and their own patterns of when and how they make particular expressions. Facial expressions are almost certainly inherited as fixed motor patterns, but are modified to the mood changes and life situations people encounter.

Open programs have adaptive advantages over closed programs, allowing an individual to cope better with the unpredictable opportunities and dangers that inevitably arise during its lifetime. Animals with prewired, closed programs simply cannot cope with the unforeseen or the rare, while animals with open programs can benefit from experiences, rare or common. Their expanded repertoire of motor patterns, in turn, can increase their chance of success in hunting or avoiding danger and ultimately in surviving to reproduce—the hallmark of evolutionary success.

Figure 44-5 OPEN PROGRAM: A CARNIVORE'S HUNTING BEHAVIOR.
An open program is an innate motor pattern that can be modified by learning and experience. The tendency of this white cat to pounce on a large bird is a good example of a fixed pattern that must be modified to meet the specific demands of a given hunting experience.

Triggering Behavior: Sign Stimuli and Innate Releasing Mechanisms

Most instinctive behaviors are triggered by some environmental event—a visual, auditory, olfactory, or tactile stimulus that prompts an animal to action. An event that sparks a behavioral response is called a **sign stimulus,** and such a stimulus sets in motion an *innate releasing mechanism* (an historical term sometimes avoided now) that results in a particular behavior.

Using herring gulls as subjects, ethologist Niko Tinbergen studied the sign stimuli involved in the feeding behavior of gull chicks. Such a sign stimulus, when produced by another member of an animal's species, has been known historically as a **releaser.** In nature, chicks peep in a stereotyped way that initiates the parents' search for food and tells the parent gulls where to deliver the food. When a parent returns to the nest with food, it holds its bill pointing downward and waves it back and forth in front of the chicks; chicks then peck at the tip of the parent's bill, which is marked with a species-specific red spot (Figure 44-6), and the parent releases the food.

To determine what the exact releaser for the chicks' pecking behavior might be, Tinbergen devised various cardboard and wood models of parent gulls. He found that chicks peck whenever the model has both a particular vertical orientation and a red spot like that on an

Figure 44-6 RELEASER: RED SPOT ON A BIRD'S BILL.
When gulls and their relatives, such as this black albatross (*Diomedea irrorata*), are hatched, they instinctively peck when hungry at moving red spots and vertical bars, features that form a caricature of their parents' bills. Adults, on the other hand, are driven to feed whatever pecks at the tips of their bills. This strategy of sign and countersign not only results in the chicks' getting fed, but also begins a process, similar to imprinting, wherein the parents and their chicks come to recognize each other as individuals.

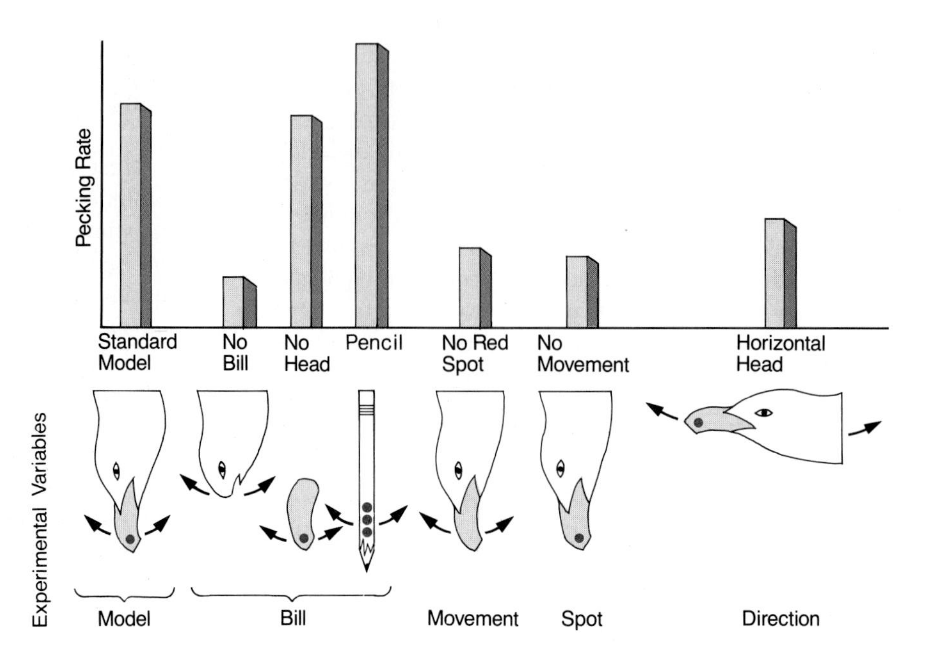

Figure 44-7 AN EXPERIMENT IN ANIMAL BEHAVIOR.
The bar graphs at the top show the rate at which herring gull chicks pecked at a model of an adult gull's head (left) and at other substituted variables (a head with no bill; a bill with no head; spots on a moving pencil; a head and bill with no spot; a complete head lacking movement; and a horizontal head). Arrows show the directions of movement. What conclusions would you draw about the necessary releasers for gull chick feeding behavior? (From P. Klopher and J. Hailman, *Control of Developmental Behavior* © 1972. Adapted by permission of Benjamin/Cummings Publishing Company.)

adult gull's bill. In a similar experiment, American ethologist Jack Hailman showed that, in fact, the chicks respond specifically to the red dot as it moves from side to side. As Figure 44-7 shows, gulls peck nearly as vigorously on a headless dotted bill, and even more so on a pencil with three dots, as long as both swing like a parent's head.

The colors, forms, odors, courtship displays, and other characteristics of many animals have been shown to release innate behaviors in other individuals. In fact, the reason that such colors, odors, and so on, are present at all may be their releaser function. And just as with the gulls' moving red dot, the *essential* characteristics of a behavioral trigger are as potent as the natural sign stimulus. Tinbergen described another example of a sign stimulus in the tiny minnowlike fish known as the three-spined stickleback.

When mating season approaches, male sticklebacks begin to develop a long red band on the belly (Figure 44-8), as well as to display changes in reproductive behavior. These changes in pigment cells and in the nervous system are triggered by environmental factors: The stickleback's nervous system monitors fluctuations in day length or water temperature, and resulting hormonal or nervous signals in turn initiate changes in coloration or behavior. As the male fish changes color, it also becomes highly aggressive, stakes out small territories, and constructs tunnel-like nests.

Male sticklebacks recognize other territorial males on the basis of a single essential sign stimulus: the red belly. Almost any long, narrow red object, in fact—no matter how unfishlike—provokes a reaction. Territorial males in laboratory aquaria have even been known to "attack" a passing red truck seen through the laboratory window! Very lifelike stickleback models lacking the red belly, however, are scarcely noticed. The fact that objects irrelevant to survival (like a red-spotted pencil or distant red truck) can act as sign stimuli emphasizes that these

Figure 44-8 RELEASER: THE MALE STICKLEBACK'S RED BELLY.
As the mating season nears for the three-spined stickleback (*Gasterosteus aculeatus*), environmental stimuli such as elevating water temperature and lengthening days cause physiological and behavioral changes. In the male stickleback, this includes the development of a bright red band on the belly as well as nest-building activity.

animals are behaving in ways programmed in their genes during evolution rather than as a result of learning.

Physiological Readiness for Behavior

What happens if a releaser, such as the red belly on a male stickleback, is presented to an "unprepared" organism? Researchers have discovered that the behavior in question will probably not occur. Indeed, for a sign stimulus to exert its effect, an animal must be *motivated*—that is, it must be in a physiological state that renders it receptive to the stimulus. A lion's predatory behavior, for example, normally requires a physiological state of hunger. Likewise, the seasonal aggression of a male stickleback to another male's red marking occurs only if hormones have already brought about the necessary state of readiness in the responding animal's brain.

Biologists probed the important relationship between internal physiological states and behavior in the ring dove, among other species. In this relative of the common pigeon, once a pair mate and build a nest, the female lays eggs and 16 days later begins to produce crop "milk," a cheesy substance, in response to the hormone prolactin (see Chapter 35). After the eggs hatch, the sight of the chicks triggers the female to feed the hatchlings crop milk. Finally, after a rigidly programmed waiting period of 3 days, the adult begins to feed the chicks solid food.

Experimenters were able to elucidate the actual triggers involved at each step. The presence of eggs appears to set an internal clock. Sixteen days after researchers placed eggs in a breeding pair's empty nest, the female produced crop milk, and if they quickly removed each egg the female laid, she never produced crop milk. In addition, the sight of baby chicks was the trigger for the parents' milk-feeding behavior. Experimenters placed partially incubated eggs under a newly nesting dove pair that had not seen eggs or chicks. These eggs hatched in fewer than 16 days, and on seeing the chicks, the parents attempted to feed the young—but with no prolactin-triggered crop milk to offer. Finally, the researchers found that 3 days must pass between crop milk feeding and solid food feeding. If they placed birds that lack crop milk with starving chicks, the adults failed to respond to the sight and sound of the hungry youngsters. The adults did not begin feeding the chicks with solid food until 3 days had passed after they first saw the chicks. Crop milk feeding in ring doves reveals that what appears to be devoted parental care is actually a series of automatic, preprogrammed steps, during which a sign stimulus triggers a physiological state, which leads to another sign stimulus triggering another physiological state, and so on.

Habituation

House cats, which descended from wild cats that fed on small birds and rodents, are innately sensitive to the stimulus "small, moving object." The opposite of such sensitization is a phenomenon known as **habituation,** in which repeated exposure to a sign stimulus lessens an animal's responsiveness to it. A sort of "behavioral boredom," habituation is common in higher animals and provides a safety valve that prevents them from repeating the same behavior over and over.

Habituation apparently lowers the sensitivity of the nervous system to sign stimuli. This occurs in the brain and its complex circuits. The sense organs usually continue to receive and report information, but the information is gradually ignored owing to habituation. The result is that the animal is numbed to the usual, while remaining alert to the special and novel. Young mallard ducks, for example, will instinctively flee and hide when a hawk's silhouette appears overhead. They eventually habituate to it, however, if it passes overhead repeatedly, and simply remain in place and eye the potential threat attentively.

While such responses are instinctive, habituation also has elements of what we define as learning. Eric Kandel at Columbia University, for example, studied the neurophysiological basis of habituation in *Aplysia*, the sluglike marine invertebrate also called the "sea hare" or "sea slug," and in the process helped provide perspective on the neuronal basis of learning. If some object or movement disturbs an *Aplysia*'s siphon, an instinctive reflex causes the animal to withdraw its muscular gill immediately and quite far into its mantle cavity (Figure 44-9a). If

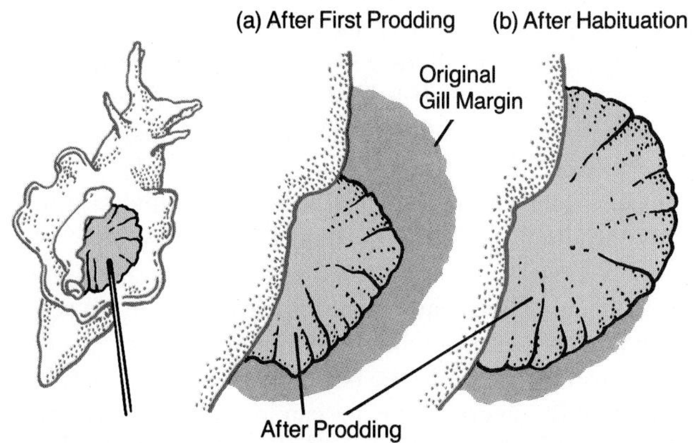

Figure 44-9 HABITUATION: A FORM OF SIMPLE LEARNING.
(a) When prodded, the marine invertebrate *Aplysia* initially contracts its gill almost 60 percent of the way into its mantle. (b) As learning takes place, the animal begins to ignore the stimulus, and gill contraction becomes less and less complete. The animal learns to suppress the gill contraction response.

an investigator repeatedly prods the siphon, however, after only a few such "training sessions," the slug responds less and less to the stimulus and may not respond at all for long periods of time (Figure 44-9b). Some biologists view this sort of habituation as a prototype of simple learning because a sea slug's nervous system must change with experience and remember being poked. Studies show that the portion of *Aplysia*'s nervous system that regulates the gill withdrawal reflex includes 24 sensory neurons, each synapsing on motor neurons in the gill; axons of the motor neurons, in turn, synapse on nearby muscle cells. Kandel's experiments showed that as habituation takes place, the sensory neurons release less and less neurotransmitter and so have a progressively weaker effect on the motor neurons that stimulate muscle contraction.

A long-term change apparently occurs in the ion permeability of the sensory cell's membrane; this change prevents calcium ions from entering and triggering the cell's secretion of neurotransmitter and is thus the chemical basis for an *Aplysia*'s nerve cell remembering being repeatedly poked with no ill effect. Many instinctive and learned behaviors can also be built on the chemical foundation of brain circuit plasticity and modulation by neuroactive peptides (see Chapter 36).

LEARNING

When a rat masters a maze, it is exhibiting the results of **learning:** behavior that can have reflexive and instinctive components but that is modified on the basis of experience. Most animals have nervous systems sufficiently complex to allow at least some degree of learning, but it is often difficult to distinguish true learning from complex series of reflexive and instinctive acts.

An experimenter might, for example, confuse simple physical maturation with learning. As an animal develops to puberty, altered hormonal levels trigger certain new behaviors, such as sexual play or male aggression. What's more, some animals can learn only during **sensitive periods**—times when permanent behavioral repertoires can be built into the nervous system. For instance, young birds of certain species can learn the precise songs of local adult males only during a brief "window" of brain development—between about 10 and 30 days after hatching. Finally, there are cases in which an animal really does learn something, yet fails to display it; the learning may emerge only in another context at some other time. Nevertheless, through careful experimentation, behaviorists are making real progress in their study of various types of learning.

Programmed Learning

Learning allows an animal to cope with highly complex phenomena or unpredictable features of its environment. In many animals, an early sensitive period of **programmed learning** initiated by instinct allows them to apply survival skills in unique settings.

Through a period of programmed learning, honeybees, for example, come to recognize the types of flowers their species visits at a particular time. They can communicate this learned information to other bees, but it can be "forgotten" in a period of hours and replaced by new learned information about other flowers as the day or the blooming season continues.

German zoologist Karl von Frisch and several generations of his students have shown that a forager bee can learn and relate to those back in the hive not only the color, odor, and shape of a flower, but also its location, as well as characteristics of nearby landmarks and the time of day associated with the bee's visit. In learning these flower parameters, the foraging bee follows an inflexible, genetically coded routine. She learns the color of a flower only during the last 3 seconds before she lands (Figure 44-10); she learns the odor while sitting on the flower itself; and she memorizes landmarks during the departing circling flight. Newly hatched, naive honeybees have no predilection for specific flower species. Nevertheless, the rate at which they learn a new flower species depends first on the geometric complexity of the

Figure 44-10 PROGRAMMED LEARNING: HONEYBEES AND FLOWERS.
As a honeybee *(Apis mellifera)* forager lands on a flower, she learns its color. This pollen collector (note the full pollen "basket" on her right rear leg) memorizes the odor of the sunflower only after landing and learns nearby landmarks only as she circles the flower before returning home. She also learns the shape of the flower, calculates its distance and direction from the hive, and registers the time of day when sunflowers are open.

THE FLEXIBILITY OF ANIMAL BEHAVIOR

A philosopher once said that a person never casts a stone twice in the same stream; and, indeed, all environments change, sometimes moment to moment. In light of this, how can an animal survive in a world with limited food and other resources, but a seemingly unlimited capacity to change and threaten? Behavioral ecologists study just this issue, and recent research reveals that even the simplest animals employ learning to a degree never before suspected.

Classical theory suggests that a great deal of animal behavior is instinctive, particularly in invertebrates and "lower" vertebrates, but how do such animals adjust to and survive new situations? In 1974, Ernst Mayr of Harvard University theorized that animals generate behavior not just as many separate actions, but as coordinated streams of actions that he called *behavior programs* and which include the open behavioral programs discussed on pages 818–819.

Mayr's theory involved a certain amount of flexibility, since a *completely* closed program would be too tight and inflexible to be useful in the real world, and a *completely* open program would be too loose to provide the advantages of advance preparation. Mayr argued that social or *communicative* behaviors like mating or parental care, are more likely to be relatively closed, to help ensure clear and unambiguous communication. He believed that nonsocial or *resource-directed* behaviors like food gathering or predator escape, on the other hand, are more likely to remain relatively open, because environments are so continually variable. Mayr also argued that long-lived animals, with their extensive opportunities for experience in the environment, should have behavioral programs that remain more open than those of short-lived animals.

In experiments at the University of Toronto, doctoral student Terrence Laverty lent powerful support to Mayr's theory. Like honeybees, bumblebees fly out from their nests to forage for flower nectar (Figure A). Unlike honeybees, however, they do not communicate information about foraging by executing a specific dance when they return to the nest. During its lifetime, an individual bumblebee forages from many species of flowers that vary greatly in shape and structure, or morphology. The bumblebee's "flower hopping" behavior is clearly a resource-directed behavior by a short-lived animal. So how much of it is instinctive and how much is learned? And to what extent is its foraging program open or closed?

Laverty reasoned that if the behavior is closed, practice will not help a bee learn to find new flower types. To test this, Laverty allowed bees raised in the laboratory to feed only by flying down long tubes to specific flowers, and he carefully chronicled their behavior.

Figure A Bumblebee *(Bombus impatiens)* foraging for nectar on a wild aster.

On their first foraging flight, the bumblebees took much longer to find the nectar on complex flowers like monkshood or lousewort than on simple flowers like daisies. It was as if the bees "knew" only to land and search "flowerlike" objects. They would usually find the nectar on a new flower eventually, but would sometimes leave without feeding. The more experience a bee had with a complex flower, however, the less time it took to find the nectar, until eventually the animal could fly directly to the food source with no exploration.

Most significantly, experience with flowers of one species improved performance on flowers of other species, as long as the flowers had the same general morphology. Experience with monkshood, for example, aided the bee in later feeding on lousewort.

Laverty's results provide strong evidence that bumblebees learn their foraging behavior not simply by memorizing details of specific flowers, but by learning something deeper and more abstract about how these details are organized.

The bumblebee experiment is just one of a growing number of similar studies. As Mayr predicted, most of these examples involve resource-directed behaviors and include activities of mammals, birds, fishes, reptiles, amphibians, and a variety of invertebrates. Together, the studies show animals to be amazingly flexible, to have a well-developed ability to learn, and to be more sensitive to their environments than anyone suspected only a few years ago.

flower and second on whether that individual bee has experienced that geometry (not that species) before (see the box on page 823). Thus, a kind of abstraction—geometric shape—is involved in learning even in the relatively simple brain of a honeybee.

For bees, then, learning is more than a simple, rigidly controlled behavioral operation. Although much is preprogrammed, the final ingredients—the distance, direction, color, and odor of a newly blooming set of flowers—can never be. They are each learned at different rates, and once assimilated, these elements form a coherent set of stored information in the honeybee's brain. The interplay of detailed information with that set appears to be a key feature of learning, evidenced by the fact that a honeybee can transfer what it has learned about one flower species to another differently shaped species it has never seen, as long as the relationship between the new details is the same as that in the first. All these studies show a remarkable level of abstraction in a relatively simple brain.

Recent work on cabbage butterflies *(Pieris rapae)* extends this new view of insect learning to yet another species. These insects must learn to extract nectar from a new flower over several attempts; this learning experience interferes with previous learning so that the butterfly must relearn how to gather nectar from flowers of different species.

Imprinting: A Form of Programmed Learning

Another form of learning is **imprinting.** Konrad Lorenz coined this term to describe a phenomenon long noted by naturalists: On hatching out of an egg, goslings, swans, and similar birds follow the first object they see that even approximates their species' natural sign stimulus—a walking parent. In his early experiments, Lorenz offered himself as a model parent to newly hatched birds; soon, flocks of geese, swans, and jackdaws (crowlike birds) were following him relentlessly about the grounds of his Max Planck Institute and would even ignore their natural parents in their pursuit of the imprinted object—Lorenz himself. As with programmed learning, such imprinting can occur only during a discrete sensitive period after hatching; and an appropriate stimulus must be present then if imprinting is to occur.

Lorenz discovered that this imprinting could also have lasting effects. For instance, on reaching maturity, goslings imprinted on Lorenz courted *him* rather than members of their own species, and in the wild, young birds reared by a different species will be more receptive to potential mates with their foster parents' plumage color than their own.

Quite a different kind of imprinting is apparent in young salmon. As the little fish mature into the smolt stage, they imprint on odors in their home stream during a sensitive period that may be only a few hours long. Perhaps 4 or 5 years later, the sexually mature adult fish migrate from the sea up long river systems, steering into correct tributaries as they wend their way upstream. To do so, they use the odor on which they imprinted years before—and which they have not smelled in the intervening time—as the guide to their spawning site (Figure 44-11).

Figure 44-11 IMPRINTING AND THE LONG RETURN HOME.
These red sockeye salmon *(Oncorhynchus nerka)* have been triggered by hormones to turn red, change their behavior, and return to the stream where they began life. It is at a specific time that young fingerling salmon apparently become imprinted on odors present in the stream. They retain the memory for years while growing to large size at sea. This photo shows adult salmon near the final stages of their return, as they almost fill a small stream in Alaska from bank to bank.

From an evolutionary perspective, imprinting provides a ready mechanism for recognizing one's own species or one's own parents or offspring. Such a strategy protects parents from wasting their energies on fostering offspring that may not bear their own alleles, helps wandering young identify their parents, and helps ensure mating between individuals of the same species.

Latent Learning

Latent learning is more complex than either programmed learning or imprinting. It involves a delay between when a behavior is learned and when it is used. The best examples of this process involve the elaborate songs of perching birds, such as finches, warblers, and sparrows. While some birds, such as the mourning dove or the roadrunner, have primitive bird songs that are entirely instinctive, the complex songs of most perching species are not wholly innate: The birds must learn important parts. The different dialects sung by white-crowned sparrows in the San Francisco Bay area (which we first encountered in Chapter 39) are a case in point. Local populations of white-crowned sparrows have evolved their own dialects—each a song built on the basic white-crowned sparrow set of notes, but each also possessing unique sound frequencies and timing (Figure 44-12).

Studies reveal that young nestlings seem to listen to and learn their local dialect when they are between 10 and 30 days old; they cannot learn song dialects they hear either before or after that critical time. During their 3-week sensitive period, both males and females "memorize" the group's song, even though females never normally sing as adults. In the listening and memorizing process, the young birds neither sing nor practice what they hear and apparently never actually try out the fixed motor pattern they will use later as adults. The following springtime, as male sex hormone levels rise in the blood of a young male white-crowned sparrow, he begins to sing. Over the course of the next few weeks, he refines his performance, apparently "comparing" his developing song with the pattern "recorded" the previous summer until he produces a perfect copy. He then sings the song thousands of times during the late spring and summer. Once assimilated through such latent learning, the dialect becomes a fixed motor pattern.

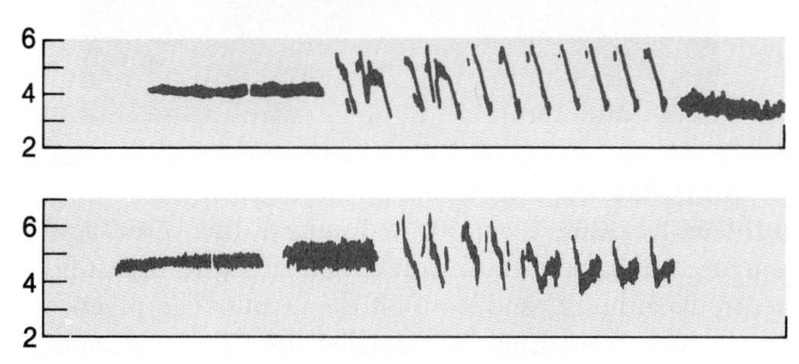

Figure 44-12 DIALECTS IN BIRD SONG.
Populations of white-crowned sparrows have developed their own variations on the basic song of the species. Here, a male white-crowned sparrow (this one from the Puget Sound area) delivers a melodious rift. Here are sound spectrograph recordings from sparrow populations that live fifty miles apart to the north and south of San Francisco Bay.

Newer work shows that some birds can still learn to sing after the sensitive period if a bird interacts with a living "tutor bird" (not a tape recorder, as used in the earlier studies). Moreover, the more aggressive the "tutor," the better the "pupil" learns. Clearly, social interactions play a role in latent learning.

Additional studies are probing the neuronal circuits involved in complex bird songs—circuits set up early in the animal's life and activated as sex hormone levels rise in spring and prepare the bird for migration, courtship, nest building, mating, or singing. Dendritic branching, numbers of synapses, and even numbers of neurons in a region of a canary's brain responsible for singing all increase each spring in response to sex hormones. And this circuitry is established even in birds that don't sing: A female white-crowned sparrow that receives a series of testosterone injections will sing the song only males warble in nature, and in the dialect that she heard as a nestling and memorized months or years earlier!

Complex Learning Patterns

In general, the more complex an animal's nervous system, the greater the proportion of its behaviors that involve learning and abstraction. Nevertheless, even humans and other primates exhibit a surprisingly large degree of instinct in their learning, as in the maternal behavior that follows childbirth and in the child's early development (learning to crawl, stand, walk, and talk). Let's see how instinct and complex learning are intertwined.

Many species exhibit **trial-and-error learning,** also called **feedback learning.** Thus, if a behavior elicits a favorable outcome, it is **reinforced,** and the animal will tend to repeat it. For example, a cat that discovers by accident that brushing against a screen door causes the door to open soon learns to carry out this behavior quite deliberately. Alternatively, if the action produces a neutral or negative outcome, it tends to be extinguished. As we saw in Chapter 42, after one taste of a noxious monarch butterfly, most birds never try to eat another one.

Feedback learning is also the basis for **operant conditioning,** or animal training based on rewards (reinforcers) and punishments—a method B. F. Skinner developed extensively. Russian physiologist Ivan Pavlov identified another kind of complex learning, known today as **classical** or **Pavlovian conditioning** or sometimes as **associative learning.** By ringing a bell immediately before each time he fed a dog, Pavlov found that he was eventually able to get the dog to salivate by simply ringing the bell, since the animal had come to associate the bell with being fed.

Insight learning, or **reasoning,** is the ability to use abstractions based on past learning in novel ways, combinations, or situations. This type of learned behavior is apparently available only to animals such as humans and certain other primates that possess highly complex nervous systems and perhaps also to porpoises, whales, and other cetaceans (although evidence is still insufficient). A chimpanzee placed in a room with bananas hanging from the ceiling and boxes on the floor demonstrates the classic example of insight learning. The chimp in this situation quickly figures out that it can reach the fruit by stacking the boxes one atop the other (Figure 44-13a). The long-term experimental efforts to teach American Sign Language and other forms of symbolic communication to chimps and orangutans are also based on insight learning (Figure 44-13b). An animal capable of using insight is able to generalize from previous experience, rather than relying on trial and error to cope successfully with a new situation. This explains why prosimians, monkeys, apes, and humans are more adaptable than other animals and more able to apply environmental information to their own needs.

(a)

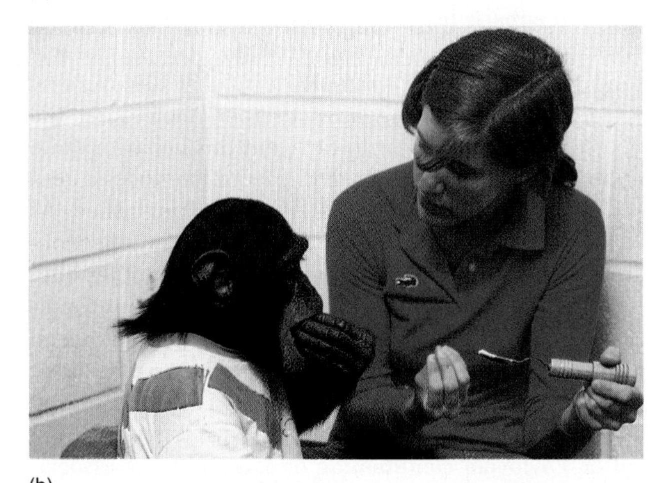

(b)

Figure 44-13 INSIGHT LEARNING AND PROBLEM SOLVING IN PRIMATES.
Chimpanzees display some evidences of insight learning. (a) When faced with the problem of bananas out of reach, a chimp can resolve the situation by building a tower of boxes. (b) Here a researcher teaches the chimpanzee "Nim Chimpsky" the American Sign Language symbol for "eat."

Learning and Cultural Transmission

Sometimes, one member of a population or species transmits learned behavior to the entire group. Human culture is based on this principle, but it occasionally shows up in other species as well. In the early 1920s, for example, a blue tit, a small English songbird, learned to puncture the foil caps on milk bottles sitting on doorsteps and to drink the cream (Figure 44-14). Before long, others of its species—and of a dozen other species throughout the British Isles and Europe—were pecking on milk bottle tops and drinking cream.

A second case of cultural spread involves a group of Japanese macaques in which a young, low-ranking female macaque discovered by accident that washing sweet potatoes removes sand from them. This novel

Figure 44-14 MILK BOTTLES AND THE TRANSMISSION OF CULTURAL INFORMATION.
In the early 1920s, a blue tit in England discovered how to puncture foil caps on milk bottles (delivered to doorsteps in the predawn hours) and skim off the cream. The practice spread throughout England, invaded the Continent, and was ultimately picked up by a dozen other species of birds, including this great tit *(Parus major)*. The rapid transmission of this piece of cultural information probably results from the fact that young birds closely watch older birds feeding and display a readiness to imitate them. Indeed, chicks that observe a hen trained to peck only green grains out of a mixture of colors will focus on the same color themselves without the slightest encouragement.

Figure 44-15 MOB ATTACKS ON PREDATORS. Nesting adult birds regularly mob or attack potential predators to drive them away. Here a common barn owl *(Tyto alba)* is harassed. As they attack, the smaller birds' mobbing calls alert other birds of the danger and serve to teach young birds to recognize enemies.

behavior, washing sweet potatoes, has become a permanent part of the troop's mealtime behavior that is now passed from generation to generation as a kind of cultural inheritance. We'll return to this subject in Chapter 45.

Many bird species show instinctive, not learned, responses to predators, but in starlings and certain others, the response can involve "mobbing calls" based on cultural inheritance—strident calls that, once learned, serve as releasers for members of a flock to join together and harass the intruder (Figure 44-15). In an experiment, researchers showed birds in one cage a stuffed owl (a natural predator) and birds in a second cage a milk bottle. Birds that saw the owl acted enraged, sounded the mobbing call, and attempted to attack the owl through the cage bars. This commotion set off the birds in the second cage, and they directed their activity to the only "predator" they could see—the milk bottle. In fact, the second group would continue to mob the milk bottle, teaching successive generations of offspring the same aversion by their example.

Cultural inheritance is turning out to be so widespread and important a phenomenon that ecologists may have to revise their concepts of generalists and specialists within feeding niches (see Chapter 43). Recent studies, for example, show that overall, a population of the classic generalist, the Cocos finch, will eat 17 kinds of fruit, nectar, and seeds from 29 species of flowers and innumerable crustacean, insect, mollusk, and lizard species. An individual Cocos finch, however, might consume only a few types of insects and a few types of seeds—a very specific diet that it learns as a youngster from observing and imitating the feeding techniques and choices of parents and other nearby adults. Ecologists and behaviorists were surprised by this extreme specialization of individuals within a generalist population, but cultural transmission does help explain how a young bird can avoid poisonous plants and animals among the thousands of prey species to choose from in a tropical area.

There are no genes in Cocos finches for choosing one insect species over another, in macaques for washing sweet potatoes, or in people for avoiding spinach, learning to ride a bicycle, or deciding to pursue a college degree; yet behaviors are passed on generation after generation. Examples such as these emphasize the prominent role of behavior in ecology, survival, and evolution.

Considering learning in its many forms, we can conclude that learning fits the biology of the living thing: Organisms once thought to be primarily instinct driven may possess substantial learning capabilities, and those able to learn in complex ways may be preprogrammed in some senses. Behaviorists James L. Gould and Peter Marler state that an organism is "innately equipped to recognize when it should learn, what cues it should attend to, how to store the new information, and how to refer to it in the future." A great distinguishing feature of the human species, for instance, the capacity for speech, seems to be preprogrammed to a great degree, with babbling, consonant learning, and subsequent stages unfolding quite automatically, even in deaf children. About the role of behavior in survival and evolution, Gould and Marler conclude, "Animals are smart in the ways natural selection has favored and stupid where life-style does not require a customized learning program."*

COMPLEX BEHAVIOR: NAVIGATION AND MIGRATION

Courtship, nest building, rearing of young, group hunting, and many other behaviors are built on an underlying base of reflex and instinct overlain with various types of learning. Such complex behaviors can be complex adaptations, enabling an animal to better cope with unfamiliar situations and to exploit entirely new relationships that may themselves contribute to greater success in surviving to reproduce.

The adaptive significance of complex behavior is clearly visible in navigation and migration, processes by

*J. L. Gould and P. Marler, "Learning by Instinct," *Scientific American*, January 1987, pp. 74–85.

which certain animals orient themselves in time and space and travel long distances with remarkable accuracy. The mixtures of innate and learned behaviors at work in these processes permit repeated journeys of hundreds or thousands of miles to sites where survival is likely in the different seasons.

Navigation

True **navigation**—the ability to find one's way over novel routes to precise spots—would not be necessary if a bird, let's say, had an innate program for migrating from one place to another. If a bird summers in Maine and winters in the Florida Everglades, it would simply have to fly the compass direction south-southwest for 10 days at 125 miles per day to travel from Maine to the Everglades. But what if the bird is blown or purposely carried off its normal course—say, to St. Louis, 950 miles west? To reach its normal wintering grounds, it must somehow compensate for its displacement by sensing that (1) it is in the wrong spot; (2) its new location is west of the right spot; and (3) its destination can be reached by flying to the east and south. These are the components of true navigation, which requires both an internal *compass* and a *map* and can involve olfaction, vision, and perception of the earth's magnetism.

Experiments show that honeybees memorize landmarks close to their hives, but when they forage farther away, they navigate through a remarkable series of "readings" involving the position of the sun. As a bee searches for food in a variety of directions, it keeps track of the direction back to the hive. To do so, it uses the sun for orientation (rather like north on a compass). But the bee must compensate for the sun's changing position in the sky; apparently, both the rate and the direction of the sun's movement are recorded. When the sun disappears from view (as on a cloudy day), the bee automatically switches to a backup system: the patterns of polarized light in the sky. These patterns bear a regular relationship to the sun's position. Finally, when both sun and sky are hidden, bees fall back on yet a third compass system, which enables them to orient to the earth's magnetic field.

Natural navigation involves *homing,* return of an individual to its nest site, feeding ground, and the like. Homing pigeons, for example, can be displaced in any direction from their roost; yet most will promptly return. And black bears and grizzly bears, after being captured, sedated, and moved in a vehicle over long distances, may nevertheless return home after release in a matter of days or weeks. The animals must have some compass sense to tell direction, and the earth's magnetic field and the sun are good candidates as the natural compasses, as researchers have clearly established with homing pigeons. Experiments reveal that a pigeon needs either a clear view of the sun or its own undisturbed internal magnetic source in order to home correctly. If the day is overcast, however, and its magnetic field is disrupted with tiny experimental magnets, the birds fly off in random directions. Despite this, having a compass sense that says which way is north is not enough; the animal must also have some map sense so it can avoid random wandering and follow a reasonably direct route home to safety, food, and (in some species) colony mates.

Figure 44-16 GREATER SHEARWATER: LONG-DISTANCE MIGRANT.
The seasonal migrations of birds, such as by this greater shearwater *(Puffinus gravis)*, are incredible examples of the relationship between programmed learning and innate behavior.

Migration

Nowhere is the ability of animals to navigate seen more dramatically than in the periodic journeys, or **migrations,** made by many species of insects and vertebrates. Animals as different as caribou (see chapter-opening photo, page 815) and monarch butterflies cross huge stretches of terrain, shifting residence from one region to another in response to environmental cues, such as seasonal changes in temperature, food supplies, and rainfall.

Tiny arctic terns, for example, migrate from Greenland and Alaska to Antarctica and back again, flying as much as 9,000 miles each way in a quest for perpetual summer. Another spectacular migrant is the greater shearwater, which nests on the tiny island of Tristan da Cunha in the South Atlantic and then flies in March to Newfoundland, Iceland, and Greenland (Figure 44-16). Finding such precise spots on the earth's surface is quite a feat. In one study of a related species, researchers transported a female Manx shearwater nesting on the Atlantic coast of England to Boston, tagged her, and set her free; less than 2 weeks later, she was found once

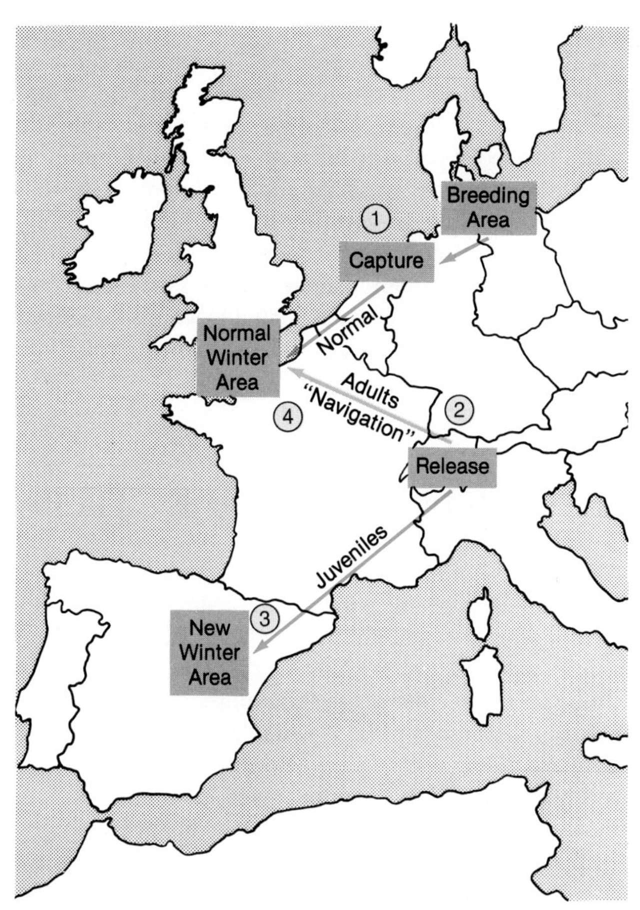

Figure 44-17 MIGRATORY BIRDS: LEARNING AND NAVIGATION.
Young starlings *(Sturnus vulgaris)* captured and displaced to Switzerland flew in the same fixed compass direction (southwest) on their first flight; but that carried them to Spain, not the French coast. Adult starlings captured and displaced to Switzerland returned to their normal wintering areas. Somehow, past learning of the wintering area or the flight paths was used to permit this true navigation; as adults, they are no longer behavioral slaves to fixed compass directions.

again in her original nest in the British countryside, despite the fact that no Manx shearwater ever normally flies east and west across the Atlantic. True navigation must have occurred in this spectacular feat of homing.

How do birds navigate and migrate with such precision? Experimenters have captured, banded, and released individuals of many species and have found that birds may use solar or stellar cues, landmarks on the earth's surface, their magnetic sense (see Chapter 36), and other cues.

In one famous experiment, researchers captured starlings flying south from northern Germany to their wintering grounds in southern England and France. They trapped the birds in Holland (Figure 44-17, step 1) and then banded and released them in Switzerland, about 460 miles southeast of their capture point (step 2). First-year migrants in the group were later recaptured; they had flown southwest parallel to their original course, although doing so had taken them into Spain—far south of their original destination (step 3). Clearly, they had been flying a fixed, "innate" compass direction and had not been navigating. Interestingly, however, when adults in the group (with experience of the route from previous years) were released in Switzerland, they did not blindly fly southwest toward Spain; instead, they corrected their courses and navigated successfully to the proper wintering grounds, flying over a new route as they went (step 4). They could navigate. Apparently, their navigational abilities had matured so that they could return to the wintering grounds learned on their first migration.

Experimenters have worked out the general strategy in annual bird migrations. During a bird's first migration, innate factors lead it to fly in a given direction and for a given distance, and it learns as it spends time at

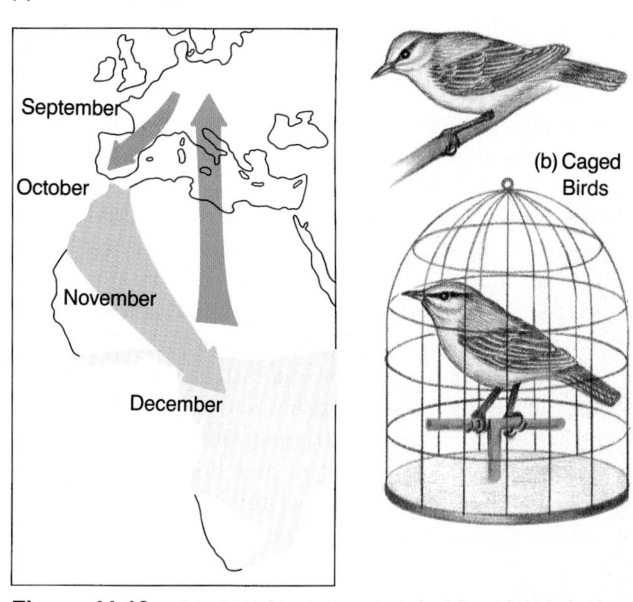

Figure 44-18 DIRECTION AND TIMING OF MIGRATIONS: INNATE MECHANISMS.
In autumn, willow warblers normally migrate from Germany to Spain and then to southern Africa. Even birds caged their entire lives will display migratory behavior each autumn, orienting their bodies first to the southwest (during August and September) and then to the southeast (during October through December). Thus, uncaged birds do not need true navigation to reach their destinations; they can use innate directions and times of flight to reach them. The shorter, more direct route that willow warblers follow back to Germany in the spring (blue arrow) is also innate. Interestingly, even experienced birds that have flown that shorter route in the spring still fly the longer route via Spain each autumn.

both summer and winter sites. If a storm (or an experimenter) later places the bird off its course, it is able to navigate to the correct destination. Evidence suggests that a bird acquires the information it needs for navigating only during a sensitive period in early life.

Animals must also have an internal clock to compensate for the changing position of the sun as the earth rotates, and they must be able to use the sun to establish compass direction. Chapter 35 described the pineal gland and suprachiasmatic nucleus of the brain, the probable sites of the biological clock in birds and mammals. But another clock operates in migrating animals. For instance, willow warblers raised in cages under a constant light/dark cycle for 7 years or more show intense migratory activity such as the flapping of wings and moving predominantly toward one side of the cage every spring and autumn, even though the birds have never experienced natural seasons as primarily reflected in day lengths or temperatures (Figure 44-18). These birds have a *circannual clock* that turns migratory behavior on and off twice each year.

The warblers not only face the parts of their cages nearest the appropriate migratory destination, but they do so for as long as it would take them to fly the normal route in the wild! First, at the time of autumn migration in Germany, they face southwest for about a month, long enough to fly from Germany to Spain. Then, although still locked securely in their cages, they face southeast for 3 months, long enough to fly from Spain to South Africa. Here, time equals distance: 30 days of flight, at so many miles per day, would take them to their first destination; another 90 days of flight, at a certain number of miles per day, would complete the migration from Spain to southern Africa. Clearly, their innate internal program specifies both direction and distance! How this fantastic capacity is built into genes and a developing nervous system remains an intriguing puzzle.

Perhaps the most important new finding about navigation and migration is that a number of animals can sense the lines of the earth's magnetic field. Birds, porpoises, honeybees, and even some humans possess tiny crystals of magnetite, a type of iron oxide, in a region near their brains (see Chapter 36). This magnetite may be involved in detection of movement relative to the magnetic lines of force. Moreover, different sites on the earth's surface apparently have special magnetic properties; thus, it is not too far-fetched to hypothesize that home for many higher vertebrates may be partly determined by cues involving magnetism and gravity. How such information about place—a map sense—is combined with learned aspects of spatial locations, as shown by foraging honeybees and other animals, is yet to be discovered.

BEHAVIOR IN PERSPECTIVE

We have seen in this chapter that animals can display innate and learned behaviors that are simple or complex and that correlate roughly with the organization of an animal's nervous system. This could conceivably reflect cellular and molecular plasticity in nervous systems. The old ideas that instinct and learning are separate and opposite are giving way to new evidence of intimate interplays between both types of behavior; even in animals as simple as insects, long considered to be primarily instinct driven and genetically programmed. Conversely, animals capable of great learning may be more limited than behaviorists once thought, perhaps being able to learn certain kinds of things only in certain ways. Overall, animal behavior in nature (and the underlying nervous system) has been powerfully influenced by natural selection, as individuals with certain behaviors have proved better able to meet the demands of the environment. Behaviors, then, may be adaptations of paramount importance, helping to explain evolution just like physiology and anatomy do.

We turn in our next chapter to behaviors that involve groups of individuals interacting in social situations—behaviors that can, in their own way, contribute to evolutionary success.

SUMMARY

1. *Ethology* is the study of animals' behavioral interactions with their environment.

2. Animal behavior can be divided into several broad categories: *Innate* (inborn) behaviors, which include *reflexes* and *instincts,* both built by genetic and developmental processes, yet occasionally modifiable by experience; and *learned* behaviors that are built on prior experiences.

3. The simplest behaviors seen in animals involve reflexes—automatic, involuntary responses to external stimuli.

4. Instincts are stereotyped, inherited patterns of behavior that take place in response to specific environmental stimuli.

5. The behavior of animals such as insects, which have relatively simple nervous systems, includes *closed programs,* which are innate. But some insects show varying degrees of experience-based behavior and exhibit *open programs,* behavioral patterns that can be modified by learning.

6. An event that sparks a behavioral response is called a *sign stimulus.* Sign stimuli can set in motion *innate releasing*

mechanisms that result in a particular behavior.

7. For a stimulus to exert its effect, an animal must be motivated—that is, physiologically receptive to the stimulus's message. Repeated exposure to a stimulus can result in an animal's decreased responsiveness to it, a phenomenon known as *habituation*.

8. *Programmed learning*, seen in bees, young birds, and other animals, follows a rigid pattern, much as innate behaviors do.

9. Species subject to imprinting pass through a *sensitive period*—a brief period after birth when the appropriate stimulus may be recorded in the animal's nervous system and attachments to it formed.

10. *Latent learning* is slightly more complex and involves learning at one particular time but behaving appropriately at later times.

11. Complex learning mechanisms include *trial-and-error learning*, *classical* or *Pavlovian conditioning (associative learning)*, and *insight learning*, the ability to use abstractions in novel ways, combinations, or situations.

12. *Navigation* involves learning, as well as the recognition of visual landmarks and use of a solar, stellar, or magnetic compass to travel novel routes to targets. Many animals show extremely complex innate migratory behavior in annual seasonal travels over great distances. These also require true navigation and a map sense.

KEY TERMS

associative learning
classical (Pavlovian) conditioning
closed program
ethology
feedback learning
fixed motor pattern
habituation
imprinting
innate releasing mechanism
insight learning
instinct
kinesis
latent learning
learning
migration
navigation
open program
operant conditioning
programmed learning
reasoning
reflex
reinforcement
releaser
sensitive period
sign stimulus
taxis
trial-and-error learning

QUESTIONS

1. What is a reflex? A taxis? A kinesis? Give a specific example of each.

2. What is meant by instinct? Can an instinct ever be modified? What is the difference between an open and a closed program?

3. Define the following terms, and give at least one example of each: sign stimulus; releaser; innate releasing mechanism; motivation; habituation.

4. How is habituation like learning?

5. Give at least one example of culturally transmitted behavior among nonhuman animals. What human behaviors are culturally transmitted?

6. What is imprinting? What is meant by the sensitive period? Give some examples.

7. What is latent learning? Is it possible to learn to perform a task without practicing it? Explain.

8. Explain each of the following types of complex learning: trial-and-error, or feedback, learning; associative learning; insight learning, or reasoning.

9. Many animals migrate hundreds or thousands of miles twice each year. How do they know where to go? Is migration completely instinctive, or are navigational skills required? What does it mean to say "time of flight equals distance" in migratory birds?

ESSAY QUESTION

1. In the springtime, male robins often attack their own reflection in a window. Is this bizarre action related to a "natural" behavior? Design an experiment to determine what feature of the reflection provokes the attack.

For additional readings related to topics in this chapter, see Appendix C.

45 Social Behavior

The cohesiveness and coordination of animal societies are often their most striking feature. . . . Natural selection operates upon individuals and the responses of a social animal to the other members of the group will evolve to its own best advantage.

Aubrey Manning, *An Introduction to Animal Behavior* (1979)

A pack of wolves. A hive of bees. A school of fish. A pride of lions. A bevy of quail. A gaggle of geese. Our language is full of such colorful labels for sets of animals, but the umbrella term for animal groups of the same species that communicate, interact, and cooperate with each other is a **society.**

Not all social groupings involve social interaction; caterpillars from a hatch, for example, all feeding on a single leaf, act entirely as individuals. But where there is true collective activity, the social behavior is adaptive; cooperation within a group tends to promote the survival of individuals, and with survival, the likelihood that they will reproduce successfully. A small fish is less likely to be eaten if it swims in a large school; a prairie dog in a colony can warn or be warned of approaching predators; a lion in a pride can hunt more effectively; and a person in a society can trade one specialized type of labor for food, clothing, shelter, and other necessities of survival. The individual animal's actions in social situations are

High-density social interactions: Plovers (*Pluviales* species) at rest between feedings.

built from the combinations of innate and learned behaviors we considered in Chapter 44. Hence, social behavior and its underlying innate and learned components are subject to natural selection, just as surely as morphological and physiological characteristics are.

The study of the biological basis of social behavior is a subfield of biology called **sociobiology.** Harvard biologist Edward O. Wilson coined this term while studying the extraordinary cooperative societies of certain insects and seeking to explain how their complex behavior may have evolved. Sociobiology addresses whether and how an individual's social behavior affects the propagation of its genes and those of its closest relatives—and, ultimately, its evolutionary fitness.

We survey the main kinds and characteristics of social behavior in this chapter, including discussions of:

- The key ecological factors that shape a species' social structure
- Communication, the cornerstone of information exchange, on which social interactions and systems of differing complexity are built
- Mating behavior and the specific social adaptations that promote successful reproduction
- *Altruism*—self-sacrificing behavior that benefits other members of a society and indirectly benefits the alleles of the "selfless" individual
- Insect societies—complex cooperative units based mainly on instinctive behavior
- Societies of vertebrates, characterized by the dominance hierarchy, or "pecking order"
- Mammal societies, such as herds of ungulates, packs of wolves, and troops of monkeys, and also touching on the evolutionary origins of our species' own cooperative behavior

Figure 45-1 MOTHER GORILLA WITH HER YOUNG. Gorillas *(Gorilla gorilla beringei)*, like all *K*-selected animals, invest much energy in the raising, care, and protection of their offspring. The young mature slowly and are few in number.

BEHAVIOR AND ECOLOGY

Behavior, like most other aspects of an animal's biology, can be shaped by natural selection. For a social system to evolve in a species, the social group must give its individual members competitive advantage in safety from predators, defending resources, finding food, and caring for young. The close association between social animals, however, can lead to more competition for resources, and a higher risk of disease.

Social behavior is often built around parent-offspring interactions. Thus, in a species with many small offspring and little parental care of young (*r*-selected species; see Chapter 43), social behavior is uncommon.

In contrast, among organisms that produce a small number of slow-maturing offspring (*K*-selected species), specific behaviors have evolved for elaborate parental care. Adult female gorillas, for example, interact with and care for their offspring for long periods (Figure 45-1). Not surprisingly, social behavior is often characteristic of such *K*-selected species.

Sociality can be an advantage when group membership offers the strength of numbers. Fish in schools, birds in flocks, and grazing animals in herds are usually less vulnerable to predators than the same individuals would be if isolated. Thus, the extra eyes, ears, and noses of fellow pronghorn antelope make the individual less likely to be caught unawares by a mountain lion or wolves. What works for prey works for predators, too, however: Wolves and hyenas hunt in groups, and this increases their chances of catching prey, even though they must share the food collectively caught. And coyotes hunt in large packs when their prey are too big for an individual coyote to kill, but as individuals when their prey is small and provides little meat per kill.

Clearly, social behaviors of varying complexity have evolved, and these behaviors primarily affect the individual and its survival to reproductive age or its parent-offspring relationships. Both cooperative survival techniques and parental care contribute to **inclusive fitness**—an individual's likelihood of passing its genes directly to its own offspring, as well as promoting those genes indirectly through the success of related individuals.

Clearly, too, a cornerstone of sociality is communication, whether by chemical signals, visual cues, or even spoken language.

COMMUNICATION

Communication—the transmission of information from one organism to another—is itself a kind of behavior, one that alters the behavior of one or more other individuals. Hence, communication must be mediated by the recipients' sensory organs as well as by the communicators' signaling devices. At its most fundamental level, communication consists of a sign stimulus that can release a behavior in another individual (see Chapter 44), but intricate learned behaviors are often part of the vast repertoire of socially useful signals.

Consider just a few of the astonishingly large number of ways by which animals communicate with members of their own or other species. The alarm calls of different bird species, for example, can have very similar sounds. Such calls are difficult for a predator such as a hawk to locate directionally, and since all such calls are so similar to each other, birds of a number of species can benefit from a single bird's alarm communication. Some birds even respond to alarms from other animals; hummingbirds in the mountains of California, for instance, respond immediately when a chipmunk or ground squirrel chatters an alarm. And the vocal sounds of both vertebrates and invertebrates convey more than simply alarm: The songs emitted by whales, cicadas, and birds often play a role in mating behavior; the lion's roar is an aggressive signal (Figure 45-2a); the snake's hiss is a warning; and the human's language can communicate feelings from love to hate, as well as complex information and abstract and aesthetic ideas and images (Figure 45-2b).

Sometimes the information content can be conveyed visually; the elaborate movements of male and female ducks that precede reproductive activity are visual display, and so are the baring of fangs by a carnivore or the laying back of its ears in a threatening way. Another communication channel is chemical; many animal species produce pheromones or other odor signals in specialized glands or excrete them in waste products, such as feces, urine, or perspiration. Thus, black bears, coyotes, and rabbits repeatedly mark the limits of their territories with urine containing pheromones, and many female insects are renowned for producing sexual attractants that can bring males running or flying from great distances.

Communication stimuli often have very specific meanings. A mother black bear, for example, sends her cubs up a tree with a particular open-mouthed tooth click. A large fish-eating hawk called an osprey (*Pandion haliaetus*) can communicate to colony mates where it found a particular fish, but the other birds will respond and take up the hunt only if the prey is a fish species that travels in schools (thus, the responding osprey is more likely to find another fish there). And small male tropical tree frogs of the species *Eleutherodactylus coqui* (see Figure 43-10) repeatedly emit the cry "coqui!" "coqui!" Careful analysis of frogs' sense of hearing reveals that the female's auditory system responds primarily to the "-qui" frequencies, while the males have almost no sensitivity to "-qui," but hear the "co-" quite well. Researchers discovered that the "co-" is a territorial warning call between males; the "-qui" sound attracts females to males; and each sex's simple brain interprets and responds only to appropriate sound "filtered" by the audi-

Figure 45-2 ANIMAL COMMUNICATION: DIFFERENT MODES FOR DIFFERENT MESSAGES.
Vocal messages can range from (a) the lion's aggression to (b) the complex and emotional song of opera singer Luciano Pavarotti playing Idomeneo in Mozart's opera of the same name.

(a)

(b)

tory system. Research on *Eleutherodactylus* indicates that specific properties of sensory systems and brain circuits are critical components in the specificity of communication. These results cause behaviorists to wonder what other specific communication links between land vertebrates are actually "hard-wired" in the brain, and research is still ongoing. In the meantime, researchers have discovered an analogous mechanism in red-winged blackbirds: Nonsinging females can discriminate between aspects of the males' songs that other males completely ignore.

Insect Communication: The "Waggle Dance" of the Honeybee

Like many kinds of vertebrates, honeybees employ chemical signals as trail markers, attractants, and warnings, but the insects also communicate extraordinarily complex information using tactile stimuli in a special dance language. A bee returning from foraging uses its "waggle dance" to describe food sources located between about 80 and 600 m from the hive to others in the colony.

Positioning herself on the vertical surface of a honeycomb inside the dark hive, the bee executes the dance to indicate the distance, direction, and richness of the food source. First, the length of a straight-line run, known as the "waggle run," specifies the distance from the hive to the food source; during the run, the bee waves her abdomen from side to side. After executing a semicircular turn, she makes another waggle run, turns in the opposite direction, and dances out her message several times (Figure 45-3).

The waggle run conveys more than just distance: The angle of the waggle run also specifies the direction of the food source from the hive. The insect translates her visual observation of the direction of the sun into a gravity-guided orientation of her dance on the comb—that is, the straight vertical direction up the wall of the comb represents the direction of the sun from the hive at that time. If the food source is toward the sun, she dances straight up; if it is away from the sun, she dances straight down; if it lies at an angle 30 degrees to the left of the sun, the straight-line run of the waggle dance points 30 degrees to the left of the vertical; and so on.

Recruits that attend the bee's dance in the dark hive feel her movements and learn not only the location of the food but its quality and its precise odor (which clings to the waxy hairs on the dancer's body or is present in regurgitated food). In addition, the frequency of dancing and the amplitude and rate of abdominal waggling communicate whether the forager found a rich or a weak food source. Variations in the waggle dances of several bee species communicate the location of water, of pollen sources, and even of possible new nest sites at the time of swarming.

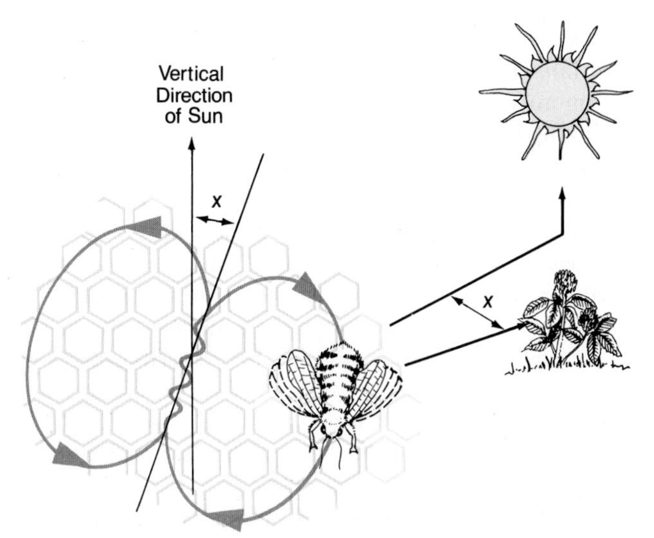

Figure 45-3 THE WAGGLE DANCE OF THE HONEYBEE. Voiceless organisms often develop elaborate methods of communication: The honeybee's waggle dance is a prime example. As the insect performs the straight-line run, her abdomen waggles; the number of waggles and the orientation of the straight-line run contain information about the distance and direction to a food source from the hive. After completing one waggle run, the bee makes a semicircular turn and does more waggle runs, to ensure that the feeding information is successfully conveyed to and received by the other forager bees in her hive. On a vertically situated honeycomb, straight up is toward the sun; the angle x is the angle between the sun and the food source.

Once a bee has visited a feeding site—perhaps originally as a result of a waggle-dance instruction—she learns a locale map of the area. If captured and released at some other site in the vicinity, the bee can fly directly to the food site without first returning to the hive, using landmark cues of the locale map to reach its goal—a rich source of sugar. This example, as well as those we discussed in Chapter 44, refutes the old notion that insect brains function solely on innate programs; learning is surely part of their repertoire and contributes to the precision of the communication and its survival value.

MATING BEHAVIOR

Mating behavior is a major component of inclusive fitness for individuals and species. While the male and female may spend very little actual time together for transfer of sperm and fertilization of eggs, some behavioral steps are likely to precede the mating itself. These can be very rudimentary: Among sea stars or sea urchins that inhabit the same reef, for example, the only social communication involves release of a substance by spawning females that triggers the males to shed sperm. Among salmon or frogs, specific behaviors bring the individuals close together, so that despite their external fertilization, sperm have a greater likelihood of meeting an egg. Species with internal fertilization—whether earthworms, dragonflies, salamanders, or warm-blooded vertebrates—usually display more complex mating rituals and means of transferring sperm or sperm packets to the female. These can include displays of territoriality and aggression among competing males, as well as ritualized series of body positions, movements, vocalizations, and other actions.

Mating behavior tends to be simpler in *monogamous* species than in *polygamous* ones. Among monogamous animals, single males and females may remain together during a full breeding season or longer, while polygamy involves multiple matings with different individuals. Monogamous species tend to have roughly equal numbers of males and females, and there may be little direct competition among males. The two sexes tend to look alike, and males rarely evolve special markings, such as antlers or showy tail feathers.

Things are quite different among polygamous species. Mating can involve one male and many females (**polygyny**) or one female and many males (**polyandry**). With so many mates on the scene, competition becomes commonplace, and members of the two sexes tend to look quite distinctive—that is, to display **sexual dimorphism.** Dimorphic males may have large canine teeth, huge antlers, luxuriant manes, ferocious pincers (Figure 45-4), or other recognition devices, while females are often smaller, drabber, and thus better camouflaged. There tends to be less genetic variability in polygamous groups than in monogamous ones, since one male may father all the young in the herd. But the female derives certain advantages from joining a harem and sharing a male, including better protection, social benefits of rearing young (built in "babysitters" and "nurse maids" among the other females), and a greater chance of gaining food through shared resources.

Figure 45-4 SEXUAL DIMORPHISM IN THE FIDDLER CRAB.
A male fiddler crab *(Uca pugnax)* threatens the photographer with its greatly enlarged pincer claw.

Territoriality and Aggression

One understandable offshoot from the competition for mates is **territoriality**, the defense of a particular feeding or breeding site, generally by one male against other males of the same species. Animals ranging from insects to lizards to frogs to monkeys establish territories of various types. Perhaps the most familiar examples of territorial animals are songbird species in which males claim areas for mating or nest building and advertise their boundaries by repeatedly singing the species-specific song (see Chapter 44).

The size of an animal's territory depends on the species' mobility, on which competitors tend to enter the territory, on the function of the territory (whether for nesting or foraging), and on other factors. A male songbird's territory is usually large enough to support one female and her nestlings. A single male redwing blackbird, however, can have a territory sufficient for a harem of ten breeding females. And a male vicuña—a relative of the camel that lives in the high Andes—can guard a patch of grassland large and rich enough to support up to 18 grazing females. Conversely, an animal may have only a symbolic territory—a *display court:* tropical birds called white-bearded manakins, for example, defend nothing more than a sapling surrounded by a small, bare area cleared of leaves and debris and situated very close to the display courts of other males (Figure 45-5). A female ultimately enters and mates with one of the males in his display court, but then flies away to nest elsewhere. For unknown reasons, it is the display court rather than the individual male that seems to attract female manakins!

Advertising a territory with scent, sound, or some other signal can have multiple effects. It may warn competitors away, and it may attract females of the species. Dividing the environment into territories can thus affect reproductive processes and influence their success, as well as improve the individual's chances of surviving. In some species, failing to establish a territory can prevent a male from maturing sexually, and in others, males never succeed in reproducing successfully until a territory becomes available.

Some species have not evolved the strategy of territoriality, and instead, individuals of such species may interact aggressively anyplace they face competition in

Figure 45-5 MATING TERRITORIES OF WHITE-BEARDED MANAKINS.
The cleared patch beneath this tree is the display court, or mating territory, of a male manakin. A female watches as a male jumps repeatedly from its perch to the ground and back, all the while emitting loud explosive sounds. Eventually, each female will select a male and enter his territory to mate.

obtaining a mate. Still others display aggression in day-to-day adult life as part of the ordering process of social groups. **Aggression** includes overt fighting, displaying, and posturing—sometimes between animals of different species but more often between members of the same species.

Singing, growling, depositing odors, and other behaviors used to mark off territories may constitute one form of aggression. Another frequently involves levels of fighting ranging from **threat displays**—including the baring of fangs or hair raising on the back of the neck (Figure 45-6)—to full-blown battles between bull moose, antlers clashing. Fortunately for the animals involved, most such fighting is carefully stylized, probably by instinctive programming, so that they do not inflict lethal damage on each other. And threat displays can play important roles in the establishment and maintenance of social hierarchies, as used by the dominant male chimpanzee in a large troop. Threat displays that do not culminate in actual violent physical contact confer a great benefit: One animal can warn off another without either being subjected to the dangers of real combat. Aggressive behaviors—all of which ultimately arise from competition—can be interpreted as adaptations that render the aggressive individual more likely to reproduce successfully. Once again, this increases inclusive fitness.

ALTRUISTIC BEHAVIOR

Very often, social groups are made up of genetically related individuals. Sister lionesses, for example, often stay with their mother and perhaps aunts and cousins in a pride. A pair or a few related male lions (brothers, half-brothers, or cousins) usually live with the female relatives; periodically, a new set of younger, stronger males will come in and displace the male relatives. Zebra colts remain in the herd with their mothers and fathers until reaching sexual maturity. Then, perhaps to decrease the likelihood of being inseminated by the father

Figure 45-6 THREAT DISPLAYS: SUBSTITUTES FOR BATTLE.
Aggressive gestures deter intruders. A cat's hiss, a dog's growl, and the bared canine teeth of this olive baboon *(Papio anubis)* are common threat displays. These behaviors are performed as warnings and often are not followed by actual combat.

(which would lead to inbreeding), the young female is separated and taken from the herd by an aggressive young stallion that is establishing his own herd.

In a surprisingly large proportion of social groups, individuals display so-called altruistic acts toward each other, and these may influence reproductive success. **Altruism** is defined as self-sacrifice by one member of an animal species that brings benefit to others of the species. A prime example is the sparrow that emits a shrill cry that serves as a signal to the rest of its flock that a hawk is swooping down; that cry is likely to focus the hawk's attention on the sparrow itself. What could account for potentially self-destructive behavior? At least some altruistic acts are reputed to stem from so-called selfish genes. Parents that work themselves ragged to feed insatiable offspring or go without food as long as a predator is near are probably carrying out genetically programmed behavior—behavior that increases the chances of parental genes within the offspring being passed on to yet another generation. These innate, instinctive responses to predators may seem "purposeful" to the human observer, but in fact they are behavioral programs triggered by sights, sounds, odors, and other cues. By increasing the statistical chances of one's own genes (or similar genes) being passed on to subsequent generations, altruistic behaviors contribute to inclusive fitness.

A leading behaviorist, R. L. Trivers, suggests that altruism can also have reciprocal benefits: The recipients may inadvertently "repay" the altruist at a later date. Thus, the alarm call that a bird emits one day may be a temporary source of danger for the caller, but the innumerable alarm warnings that other flock members give in the face of other dangers repay the original act of altruism with multiple dividends. So-called **reciprocal altruism** thereby raises the inclusive fitness of each participating member of the social group.

Kin Selection

Some cases of altruism are harder to account for, including those that lead to **kin selection.** The scorecard of evolutionary success registers only how many copies of a gene penetrate into succeeding generations, regardless of which individual transports them there. In the 1960s, geneticist W. D. Hamilton proposed that an individual's fitness can be measured by the total representation of his or her genes in future generations, whether they are contributed by children or by other kin.

Behaviorists have noted the clearest-cut cases of kin selection in certain social species of hymenopteran insects—ants, wasps, and bees. These insect societies contain a single fertile queen and her sterile daughters. The queen often becomes huge and serves as an immobile, living egg factory that female workers tend. Males leave the colony shortly after they emerge from pupation. Unlike males, which develop parthenogenetically from unfertilized eggs and are thus haploid, females develop from fertilized eggs and are diploid. Each time a queen produces an egg, meiosis distributes 50 percent of her genes into that egg (Figure 45-7a), the remainder lost in polar bodies. When an egg is fertilized and becomes a diploid female, that individual possesses 50 percent of

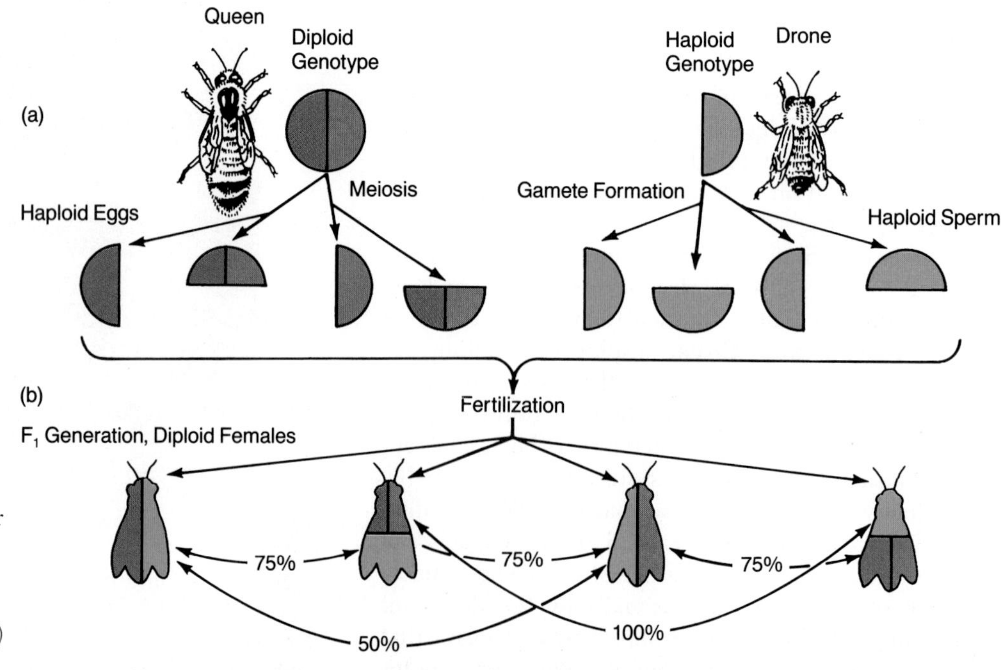

Figure 45-7 KIN SELECTION AND SHARED GENES IN ANTS, WASPS, AND BEES. (a) Genetic relatedness: mother to progeny. Meiosis yields four types of ova in a female, but just one type of sperm from the haploid males. (b) Females arising from those eggs share only half their chromosomes with the mother, since the other half comes from the father's sperm. Assuming only one male mates with a queen, the sisters share identical chromosomes from the father plus varying numbers from the mother. Each sister shares 50 percent to 100 percent (average, 75 percent) of her alleles with her other sisters.

the queen's genes (Figure 45-7b, F_1 generation). But Hamilton determined that the daughters themselves share 75 percent of their genes with one another (because the father was haploid, each daughter received 100 percent of his genes, as well as 50 percent of her mother's genes; see Figure 45-7b). The result of this peculiar arrangement is that the genes in a female worker are more likely to be passed on if she tends to the needs of the egg-factory queen and helps her mother raise more of her sisters than if she starts a new colony of her own offspring.

The sterile worker strategy has evolved independently in 11 types of hymenopterans and in termites. Clearly, the unusual degree of relatedness (75 percent) among sterile sister workers is a successful strategy for propagating alleles, and a worker's behavior helps genes similar to her own to be passed on to new generations, thus substantially increasing her inclusive fitness.

Despite these plausible arguments, kin selection and sterile sisters were probably not the main factors in the evolution of complex insect societies. For one thing, termites have complex societies, yet both males and females are diploid. For another, the queens of most insect societies mate with several males and store sperm from them all. A honeybee queen, for example, copulates between seven and ten times on her single mating flight. This means that sisters produced by such a queen may have any of several fathers, and thus their genetic relatedness averages out to be considerably less than the 75 percent calculated by Hamilton. Perhaps the queen's many daughters receive sufficient evolutionary benefit by working to increase their mother's inclusive fitness.

Let's look more deeply, now, at other aspects of insect societies.

Figure 45-8 COORDINATED BEHAVIOR IN HUGE INSECT COLONIES.
This is a swarm of honeybees *(Apis mellifera)* with thousands of workers, drones, and guards, all of which have specific functions in the colony. Here the buzzing swarm blankets the fork of a tree while scouts search for a new site for a hive.

INSECT SOCIETIES

Insect societies can be quite complex, involving frequent contact, division of labor, and interdependence among members. Nevertheless, the animals' behavior is highly mechanical, showing that great behavioral complexity can be built from simple elements.

The Social Organization of the Honeybee

As with colonies of other social insects, a honeybee colony consists of 1 queen, 10,000 to 50,000 sterile female workers, and, in spring and summer, 1,000 to 5,000 male drones (Figure 45-8). The tasks within the colony are rigidly divided; depending on their age, workers clean the hive, tend larvae and the queen, build honeycomb cells and serve as guards, or forage for nectar and pollen. In contrast with this life of constant drudgery, drones (the males) have a single function—to mate. Periodically, a virgin queen leaves a hive, mates with drones—thereby receiving and storing a lifetime supply of sperm—and then forms a new hive.

Unlike other bee and wasp species, in which entire populations may die in winter (except for a mated queen), honeybees are protected from the elements by the positioning of their nest sites in hollow trees or in walls. They begin the spring with an exploding population of workers. When the hive becomes critically overcrowded, it splits in two: The old queen and half the workers leave for a new nest site.

The social coordination necessary for this buzzing corporation to function is achieved through communication. A queen is recognized as such by her pheromones (sometimes called "queen substance"); those molecules consist of a mixture of at least five substances related to the aromatic substance benzene. Besides identifying the queen, the odors suppress worker bee ovary development and egg laying, influence worker bee orientation during swarming, and inhibit queen rearing. The substance also makes females develop into workers. Any female can develop into either a queen or a worker, depending primarily on the presence or absence of queen substance acting on the genome. As long as a queen is alive and producing queen substance, workers do not build cells for rearing potential rival queens.

When the queen dies or the colony becomes too large for all members of the group to receive a sufficient quantity of queen substance, the colony automatically begins preparations for swarming. Workers build half a dozen queen cells and feed the larvae within them a protein-rich diet that causes these otherwise ordinary diploid individuals to develop as queens.

As time for swarming approaches, the old queen begins to produce a pulsating sound signal known as "quacking," which in turn elicits a "tooting" sound from any developing queens nearing maturity. This exchange apparently tells the colony that a second queen is available; in addition, the old queen's quacking warns new queens to remain in the safety of their cells. Then a silence descends over the hive until a worker returns and does the waggle dance that communicates the distance and direction to a site for the new hive. The old queen and a swarm leave the hive, whereupon a new queen emerges and stings the remaining queen candidates to death. A day or two later, she flies out to mate, returns to the hive, and begins her career of laying eggs.

Behaviorists are beginning to believe that genetics may underlie the division of labor in bee societies. Researchers have discovered, for instance, that workers descended from one father that mated with the queen carry out the activity of removing corpses from a hive, whereas those with a different father but the same queen mother guard the hive. Similarly, genes inherited from the father also seem to determine grooming of other workers, whereas all workers carry out mutual feeding to some extent, and so behaviorists consider this to be unrelated to the father's genes. Finally, workers show "honeybee nepotism" in their care of queen larvae. A given worker will tend and feed only larvae that share their father's genes and will ignore larvae containing genes from a different father. This is a clear instance of kin selection at work.

Queens mate with up to 17 males—a polyandry that produces genetic diversity among the queen's progeny and, in turn, differences in at least some worker behaviors. These findings clearly contradict the older notion that colonies are "superorganisms" composed of identical individuals capable of all roles and simply responding to environmental cues to develop or behave in their characteristic ways. The division of labor and, in turn, the smooth functioning of the entire complex society—with its protection against intruders, temperature regulation, nest site selection, rearing of young, collective food gathering, waste disposal, and undertaking of corpses—are at least partly determined by genes.

SOCIAL SYSTEMS OF VERTEBRATES

Vertebrate social groups can be as complex as insect societies and then some, but they have an additional feature: a **social hierarchy** (sometimes called **dominance hierarchy**), in which members possess ranks with varying privileges and responsibilities. These hierarchies are usually established and reinforced by aggressive behaviors or threats. An example is the **pecking order** in chickens (Figure 45-9), in which there is a clear-cut line of dominance from the bird at the top, which can peck all others in the coop and is pecked by none, to the hapless chicken at the bottom, which can peck none and is pecked by all.

Often, the dominant individual in a social hierarchy is a male whose privileges include preferential access to food and mates. In return, this individual may lead and coordinate group activities, such as where and how far the group travels, the length of rest and activity periods, and attacks on prey. Most of the time, the dominant animal must also act as the primary defender of the group or its territory, and that can be a real cost. For animals at the bottom of the dominance heap, the story is quite different. Such individuals may reproduce infrequently or not at all, starve if food is in short supply, or be excluded from the group altogether. Still, there is often genetic benefit for the low-ranking animal. While the dominant male lion in a pride may mate some three and a half times as often as the most subordinate male, the latter is better off remaining with the pride rather than going it alone or attempting to join another pride. This is because he shares many genes with relatives that rank above him in the hierarchy. Therefore, even though he may not sire many cubs, he contributes to the

Figure 45-9 PECKING ORDER: TOUGHEST ON TOP. Animals display social hierarchies, such as the pecking order of chickens. Those at the top of the hierarchy can peck (and hence dominate) all the others, while those at the bottom can peck no one.

Figure 45-10 BELDING GROUND SQUIRREL "ON GUARD."
Female belding ground squirrels (*Spermophilus beldingi*) are likely to emit alarm calls when they spot hawks or coyotes. This behavior endangers the caller, but the warning benefits both their own young and their genetic relatives. This is an example of kin selection.

welfare of the pride and increases his own inclusive fitness through cooperative hunting and other social activities.

Habitats can have strong, indirect influences on how social systems have evolved in vertebrates. In a tropical rain forest, for example, with its continuously warm climate, many female primates experience estrus (see Chapter 18) in any month of the year, and this helps explain why primate societies have year-round dominance hierarchies: The males must compete again and again as individual females enter estrus independently. This is in sharp contrast to the pattern displayed by many animals in temperate regions. Female elk, for example, mate only at times that ensure that young will arrive in a season when their survival is most likely. In the elk society, males often do not stay with the females for most of the year, and they compete aggressively with other males mainly at the time of mating.

Social structure can also reflect an animal's niche. In colonies of ground squirrels living on the plains in the western United States, one can often observe a number of individuals standing high on their haunches and peering about, literally "on watch" for hawks or ground-based predators. This is a good example of kin selection; among Belding ground squirrels, for example, the females, not the males, are on guard and emit alarm calls. This endangers them, but their offspring and genetic relatives receive benefit from the warning (Figure 45-10). Put simply, an organism's place in the community is reflected in both its individual and its social behavior. This is true not just for insects and for vertebrates in general, but for mammals, possessing as they do the most complex societies in the animal kingdom.

SOCIETIES OF MAMMALS

Mammals display a vast array of social organizations, but perhaps the most instructive are the herds of ungulates (hooved animals), packs of wolves, and troops of primates.

Life in the Herd

Most large ungulates live in herds that can number either a few individuals or hundreds or thousands and tend to remain (1) in monogamous pairs in exclusive territories; (2) as single males in a matrix of territories, across which females roam at will to graze; or (3) in wandering groups with a strict male dominance hierarchy. In different settings, each of these social structures can be ecologically adaptive, as the African antelope illustrate.

(a)

(b)

Figure 45-11 VARIATION IN ANTELOPE SOCIETIES.
(a) Small duikers (such as *Cephalophus zebra*) live in pairs in forest territories. (b) African buffalo (*Synceros caffer*) travel in large herds with a dominant male and a typical dominance hierarchy.

Small antelope known as duikers, for example, live in forests, where the stability of the habitat favors permanent territories (Figure 45-11a). They don't migrate, they pair-bond, and the pairs in a herd divide up territorial resources. Wildebeest, typical of the second group, roam the open, grassy savanna in herds as a defense against predation. Even so, males persist in defending exclusive territories during the mating season, contesting for the lushest patches of grass and facing off along their mutual borders with display rituals (see Chapter 43 opening photo). The resident male has undisputed mating rights with any female in estrus coming onto his turf.

The buffalo, the largest of the African ruminants, eats tough, poor-quality grass (Figure 45-11b) and must be constantly on the move to find sufficient forage—a fact of life that makes holding territories impossible. Within each herd, males are organized in a strict dominance hierarchy, with the top, or "alpha," male having the exclusive right to mate with any receptive female in the group. The "fittest" (here, strongest, most aggressive) is therefore most likely to contribute genes to the next generation. Clearly, these three herd strategies fit the animals' different environments well.

Figure 45-12 THE WOLF PACK: DOMINANCE HIERARCHIES IN ACTION.
Note the lowered ears, cowering posture, and gentle pawing of the wolf *(Canis lupus)* on the left as it interacts with a male higher in the pecking order.

The Social Organization of Wolves

One of the most intensely studied of all social mammals is *Canis lupus*, the wolf, which lives in North America (as well as Eurasia) and feeds on deer, elk, caribou, and moose. For a predator to hunt such large mammals successfully, it either must weigh about as much as its prey—the strategy of tigers and polar bears—or must hunt in groups, like the wolf pack.

Wolf-pack size varies greatly, from a pair to several dozen individuals. Pack size is influenced in part by *economic limits:* on the one hand, by the smallest number of animals required to find and kill prey efficiently, and, on the other hand, by the largest number that could be fed by a kill. There are also *behavioral* limits: There appears to be a limit to the number of social bonds that individual animals can form and to the degree of competition that each pack member can tolerate.

Males and females are organized in separate dominance orders, with the original mating pair of the pack accorded the alpha position in each ranking; an established pack typically consists entirely of successive generations of their offspring. Within the group, the alpha animal displays its dominance by raising tail and ears, standing broadside to an inferior pack mate, and urinating on territorial markers and even food. The subordinate animal, on the other hand, may show a range of submissive behaviors (Figure 45-12), from shifting its ear position subtly to whining conspicuously and rolling on its back. Wolves also carry out a striking group greeting, crowding together, nuzzling the alpha male, and often howling together in an eerie chorus.

Wolf behavior is intimately entwined with group hunting of large game. Scientists have found that packs grow, divide, and dissolve because the pack leadership changes as a result of deaths, pair bonding, or fluctuations in the population size and availability of prey. In the isolated environment of Isle Royale, for example (see Chapter 42), packs grew in good times and shrank and even dissolved as the supply of game declined. In addition, surviving wolves were forced to change their usual group hunting style, splitting up and concentrating on beaver, hare, and other small prey as solitary hunters. Like the coyotes mentioned earlier, wolves probably inherit one lifestyle, but it is flexible.

Primate Societies

Gorillas, chimpanzees, orangutans, and other primates are our closest living relatives in the animal kingdom, and their complex societies hint at the origins of our own. Nearly all primates are tree-dwelling herbivores and insectivores (Figure 45-13). Among semiarboreal and ground-living species—in particular, gibbons, baboons, macaques, and the highly social chimpanzees and gorillas—several types of stable, complex social systems have been described.

Asian gibbons are among the least social of the higher primates. Living nearly full-time in dense tropical forest, mated pairs and their offspring form the primary social unit. The food resources for these animals, fruits and parts of green plants, are abundant and relatively predictable, and a gibbon family typically defends a terri-

Figure 45-13 PRIMATES: SOCIAL AND OFTEN ARBOREAL.
Nonhuman primates, such as these common langurs *(Presbytis entellus)*, are usually tree dwellers that feed on arboreal plant and insect life. Their societies are complex, with a strict social hierarchy and division of labor.

tory of about 250 acres. Males and females are nearly identical in size and markings, and both parents take part in territorial defense and care of the young. Like duiker antelope pairs, their social structure is correlated with their stable environment and food supplies.

In striking contrast, the hamadryas baboons *(Papio hamadryas)* of the desert grasslands and savannas of northeastern Africa are dog-faced monkeys with extreme sexual dimorphism. Hamadryas males may weigh more than 50 kg, twice the weight of the average female, and they have a characteristic bright-red face wreathed with a lionlike mane. The basic social unit of the hamadryas baboon is the **harem,** which consists of one male and up to ten adult females and their offspring. The male "overlord" is an autocrat that, because of his superior size and strength, is able to exert virtually total control over the harem's activities. Harems combine to form **bands,** the basic unit for food gathering and defense. Finally, bands unite in a third level of organization, the **troop,** to spend the night sleeping on cliffs or in trees. Troops may contain more than 700 individuals.

Behaviorists speculate that this elaborate social structure is an adaptation to the savanna's uncertain food supplies. In lean years, fewer females are impregnated per harem, and the population remains stable relative to available food, whereas in good years, more offspring can be produced.

Nonhuman primates have displayed some fascinating instances of cultural evolution (first described in Chapter 44). Observers of a captive group of Japanese macaque monkeys, for example, had for months watched individual animals pick up sand-covered sweet potatoes from a beach on their island and eat them, grit and all. One day, one observer saw a 1½-year-old female, dubbed Imo, holding a potato in one hand, cleaning off the sand in the water of a brook with the other hand, and then eating the potato. She washed her other potatoes, and soon two infant members of the troop were doing the same. Over several years, washing sweet potatoes spread among closely related families and groups of young playmates, and finally mothers passed the habit along regularly to their children as they trained them (Figure 45-14). The monkeys later switched to seawater, perhaps because of its salty taste. Now, the apparent salt lovers dip their potatoes into the sea between each bite!

Imo invented another novel cultural trait when she was 4 years old: She learned to gather handfuls of wheat and sand and drop them into the sea, where the heavier sand settles out rapidly and the buoyant wheat can be gathered together in handfuls for easy eating. This troop of macaques now washes its sweet potatoes and separates its wheat from the sand, while other troops do not.

Figure 45-14 CULTURAL TRANSMISSION: MACAQUES AND SWEET POTATOES.
Here a Japanese macaque (*Macaca* species) washes its potatoes. The behavior, invented by a young female, Imo, is now passed on from one generation to the next. This and another novel feeding behavior, also invented by Imo, are not transmitted genetically, but must be learned by each young macaque.

There is no gene for washing sweet potatoes or wheat; instead, such learned behaviors are culturally transmitted from one generation to the next—precisely as humans pass on information about the arrowhead, the spear, the wheel, or the computer.

Animal Behavior and *Homo sapiens*

Which of the lessons of animal sociality apply to our own species? As Chapter 46 explains in more detail, perhaps 5 million years or so ago, our ancestors were ground-dwelling hunters who probably still spent some time in the trees; they fed on both animals and plants and lived in small groups in the bush and on the plains of Africa. Since that time, humans began using weapons and tools, and civilizations sprang up all over the world. However, while cultural evolution has become the dominant feature of human behavior, there is no reason to think that it has obliterated the original biological features of human sociality or that our biological evolution has ceased.

A good place to see the residuum of biological evolution in humans is in the development of human infants, in which early behavior is genetically and developmentally programmed. Infants only a few days old cling tenaciously; "walk" if supported; paddle and hold their breath if submerged; search for a nipple and suck rhythmically; and, of course, perform that well-coordinated, communicative motor behavior—crying (Figure 45-15). This list of accomplishments goes on well into childhood, structuring both cognitive and motor development. As a result, human children the world over pass through a strikingly stereotyped sequence of developmental stages.

Figure 45-15 INSTINCTIVE—AND EFFECTIVE—HUMAN COMMUNICATION.
Fixed motor patterns are found in people as well as in other vertebrates. Many aspects of infant behavior are thought to be programmed. The baby's loud cry of distress—whether from hunger or a wet diaper—is a good example.

Considered in the light of our increasing understanding of neuronal plasticity and of how peptides and hormones operate in subtle ways in the brain (see Chapter 36), it seems plausible that many of the behaviors that humans blithely attribute to free will may not be so free after all. We are, in the end, an extraordinary example of primate evolution, capable of remarkable feats of learning, thought, creative expression, and foible. Nevertheless, much of what we do may ultimately be strongly influenced by the circuits that arise in our brains, by chemicals that act in profound and powerful ways, and by the genes that are responsible for our development. Recall the remarkable behavioral similarities between human twins reared apart. But cultural evolution—from the invention of agriculture and the wheel to the invention of the computer—is uniquely important in the process that has led to contemporary *Homo sapiens*, as our next and final chapter explains.

SUMMARY

1. A *society* is a group of animals of the same species in which the members communicate and interact in cooperative ways. Social behavior can be adaptive—that is, it may evolve in situations in which being social tends to promote survival.

2. All social behavior involves *communication* that alters the behavior of others and is mediated through senses and signals.

3. The waggle dance of honeybees is an extraordinary example of complex insect communication that relies on tactile stimuli. Recruits that attend a forager bee's dance feel her movements and learn not only the distance and direction of a food source from the hive, but also its odor and quality.

4. Mating behavior is a key component of behavior in social animals. *Polygamy* (either *polyandry* or *polygyny*) has profound effects on social structure for a species. It also affects whether males and females are *sexually dimorphic*.

5. *Territoriality* is the defense of a particular feeding or breeding site, generally against other individuals of the same species.

6. *Aggression* refers to behaviors such as fighting and *threat displays* between animals of the same or different species.

7. *Altruism* is self-sacrifice or restraint that is exhibited by some members of a social group and that benefits other members who are not direct genetic relatives. Altruism may increase *inclusive fitness* and may result from *kin selection*—behavior that promotes the survival of genes similar to one's own, whether they are carried by children or by other kin.

8. The social organization of insects is extremely complex, involving frequent physical contact, division of labor, and interdependence among members. The behavioral repertoire of insect societies includes both innate, closed-program behavior and learned, open-program behaviors.

9. A typical feature of vertebrate social systems is the *social hierarchy*, or *dominance hierarchy*, in which members possess ranks with varying privileges and responsibilities.

10. Mammals display a vast array of ecologically adaptive social organizations, including ungulate herds of varying types and highly social wolf packs—an organization that enables the predators to capture prey much larger than themselves. Under other circumstances of food availability, wolves and coyotes may hunt in small groups or alone.

11. Primate societies range from small family groups to huge assemblies of individuals organized in multilevel social structures. Cultural transmission of information and behaviors is especially prominent in monkeys, apes, and humans.

12. It is likely that some aspects of human behavior are programmed by the nervous and endocrine systems and so have some genetic basis.

KEY TERMS

aggression
altruism
band
communication
dominance hierarchy
harem
inclusive fitness
kin selection
pecking order
polyandry
polygyny
reciprocal altruism
sexual dimorphism
social hierarchy
society
sociobiology
territoriality
threat display
troop

QUESTIONS

1. To improve your inclusive fitness, what sorts of behavior should you show, and toward whom?

2. What are some advantages and disadvantages of sociality?

3. How does a worker bee communicate in the darkness of the hive? What information can she pass to other bees? What role does learning play once a worker has left the hive en route to a food source?

4. What are the various ways in which mammals can communicate?

5. What is sexual dimorphism, and under what conditions is it seen?

6. What roles do territoriality and aggression play for various species? Does either increase inclusive fitness?

7. What is reciprocal altruism, and how can it benefit an individual?

8. Describe common forms of kin selection in social insects.

9. How are changing ecological circumstances coped with behaviorally by coyotes or wolves?

10. What roles do communication, aggression, and cultural evolution play in primate social groups?

For additional readings related to topics in this chapter, see Appendix C.

46 Human Origins

We are the products of editing, rather than of authorship.

George Wald

Paleontologist Mary Leakey watched with anticipation as her co-workers carefully chipped and brushed the last layers of solidified volcanic ash away from three sets of footprints—footprints that were to reveal a visible trail, set in stone, of our early human ancestors (Figure 46-1). From this exciting find in the Laetoli region of Tanzania in 1976, and from other fossils gathered in the same region, Leakey and her colleagues were able to reconstruct an image of the animals that had tracked

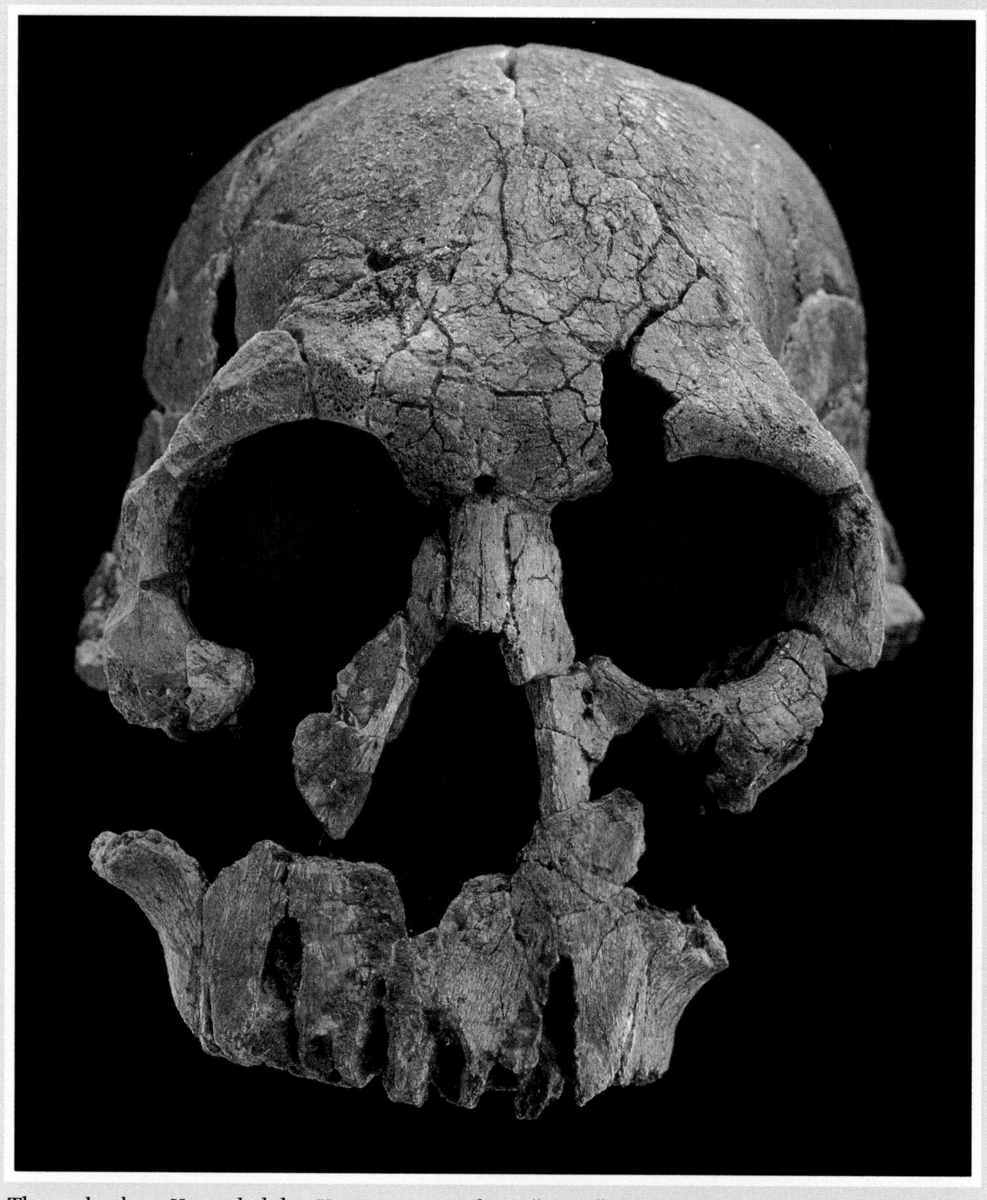

The toolmaker: *Homo habilis.* Known to us only as "1470," this ancestor of modern humans was probably a brown-eyed, hairy, social individual living in the savanna and forests of East Africa some 2 million years ago.

through the warm ash that covered an ancient riverbank. About 3.6 million years ago, three figures walked at that site. The largest may have been about 1.2 m (4 ft) tall; the other two were shorter and smaller—perhaps a female and offspring. Their skin or fur was probably dark brown; their hair was long and shaggy; and heavy brow ridges may have jutted forward above their eyes. Where was the group going? How did they live? What did they eat? Were organisms like this really distant ancestors of modern human beings? These are just some of the kinds of questions that paleontologists try to answer as they search Africa, Asia, and Europe for fossil remains of our early ancestors and for links between them and the apes, monkeys, and other mammals.

In his *Systema Naturae* Carolus Linnaeus (see Chapter 19) placed humans within the mammalian order Primates, next to the apes and monkeys, and gave us a genus and species, *Homo sapiens*. As we survey the primates in this chapter and explore our own species' fascinating evolutionary history, we shall see that all primates display adaptations for life in the trees. A few species returned to life on the ground, but only one lineage, our own, evolved erect bipedalism, the upright balanced stance on two limbs that allows a mode of locomotion unique among all vertebrates. Erect bipedalism, the remarkable dextrous hand, and an interplay of biological and cultural evolution are believed to have led to the remarkably rapid expansion of the brain's size and complexity in the line leading to modern *Homo sapiens*. Fossils in the human lineage are relatively rare and notoriously difficult to interpret. This leads to controversy about the evolutionary tree of *Homo sapiens* and about the factors that contributed to our evolution. Nevertheless, paleontologists are making great advances in finding promising fossil beds, in determining the ages of fossils, and in interpreting the fossils in ecological, social, and functional terms. As they do, they are putting into perspective our recent past as hunter-gatherers and agriculturists, our present population explosion, our impact on the biosphere, and our technological resources for meeting the current challenges we face.

Figure 46-1 SEQUENCE OF FOOTPRINTS OF THREE INDIVIDUALS OF *Australopithecus afarensis*.
Mary Leakey and her associates discovered these footprints, along with those of many other types of animals, in 1976 at Laetoli in Tanzania. The footprints were filled with black sand to make them stand out. They were made about 3.6 million years ago, when early humans walked across a fresh ash fall from a nearby volcano; a brief rainfall shortly afterward solidified the ash layer, preserving the footprints. The footprints were made by humanlike feet applying weight to the ground as humans do today; and their regular progression indicates a bipedal striding gait basically similar to our own.

Our topics in this chapter include:

- The characteristics and evolution of the monkeys, apes, and other nonhuman primates
- The transition from ape to hominid, or early human
- The evolution of the hominids and the emergence of modern *Homo sapiens*
- The origin and diversification of recent humans
- Our species and the future of the biosphere

THE PRIMATES

Modern **primates** (Table 46-1) exhibit most of the body forms also found in the fossil record for our mammalian order. **Prosimians** (meaning "before apes") include lemurs, lorises, and tarsiers (Figure 46-2a). Lemurs, and lorises, with their foxlike snout, are usually tree dwellers and are closest in structure to the most ancient placental mammals, including insectivores such as shrews. Tarsiers are vertical climbers and leapers, active nocturnally, and usually about the size of a small rat. The head is bent on the vertebral column, so that the face and eyes point forward when the tarsier sits upright in the trees. Humans have this same head orientation. The tarsier's snout is shortened, and the visual fields of its enormous eyes overlap for binocular depth perception. Finally, the hand has taken over the tactile functions carried out by the snout in lemurs and more primitive primates.

The **anthropoids** include the monkey, ape, and human lineages (see Table 46-1). All possess larger brains than do prosimians, and the cerebral hemispheres are particularly large. Spider and howler monkeys and

Table 46-1 CLASSIFICATION OF SOME LIVING MEMBERS OF THE ORDER PRIMATES

Suborder	Family	Common Name(s)
Prosimii		Lower primates
	Lemuridae	Lemur
	Lorisidae	Loris
		Galago (bush baby)
	Tarsiidae	Tarsier
Anthropoidea		Higher primates
Platyrrhini		New World higher primates
	Cebidae	New World monkeys
		Howler monkey
		Spider monkey
		Capuchin monkey
	Callithricidae	Marmoset
Catarrhini		Old World higher primates
	Cercopithecidae	Old World monkeys
	Cercopithecinae	Baboon
		Vervet
		Macaque
	Colobinae	Colobus monkeys
		Langur
Hominoidea		
	Hylobatidae	Lesser apes
		Gibbon
	Pongidae	Great apes
		Chimpanzee
		Gorilla
		Orangutan
	Hominidae	Humans

other **New World monkeys** are *arboreal* or tree dwelling, running about the branches, leaping wildly from tree to tree, and occasionally swinging by the arms. Many possess a prehensile tail capable of gripping branches (Figure 46-2b), and some lack a thumb that can be fully opposed against the other digits for gripping. The **Old World monkeys** live in Africa and Asia, tend to be larger than their New World relatives, have tails that are not prehensile, usually have a fully opposable thumb, and possess sexual cycles closely resembling those of apes and humans. The **apes**, all tailless, include gibbons, a so-called lesser ape; these fantastic acrobats swing through the trees suspended by one forelimb at a time—a mode of locomotion called **brachiation.** The

(a) (b) (c)

Figure 46-2 TYPICAL LIVING PRIMATES.
(a) This small tarsier *(Tarsius syrichta)* tends to sit upright in trees; its head is bent at a right angle to the vertebral column as it sits. The large eyes, with the flattened snout, allow good binocular vision. And the elongated fingers and toes allow grasping of branches. (b) Red howler monkeys *(Alouatta seniculus)* grip branches with fingers, toes, and tail. (c) A chimpanzee *(Pan troglydytes)* has long, powerful arms, and exceptional intelligence for a nonhuman primate.

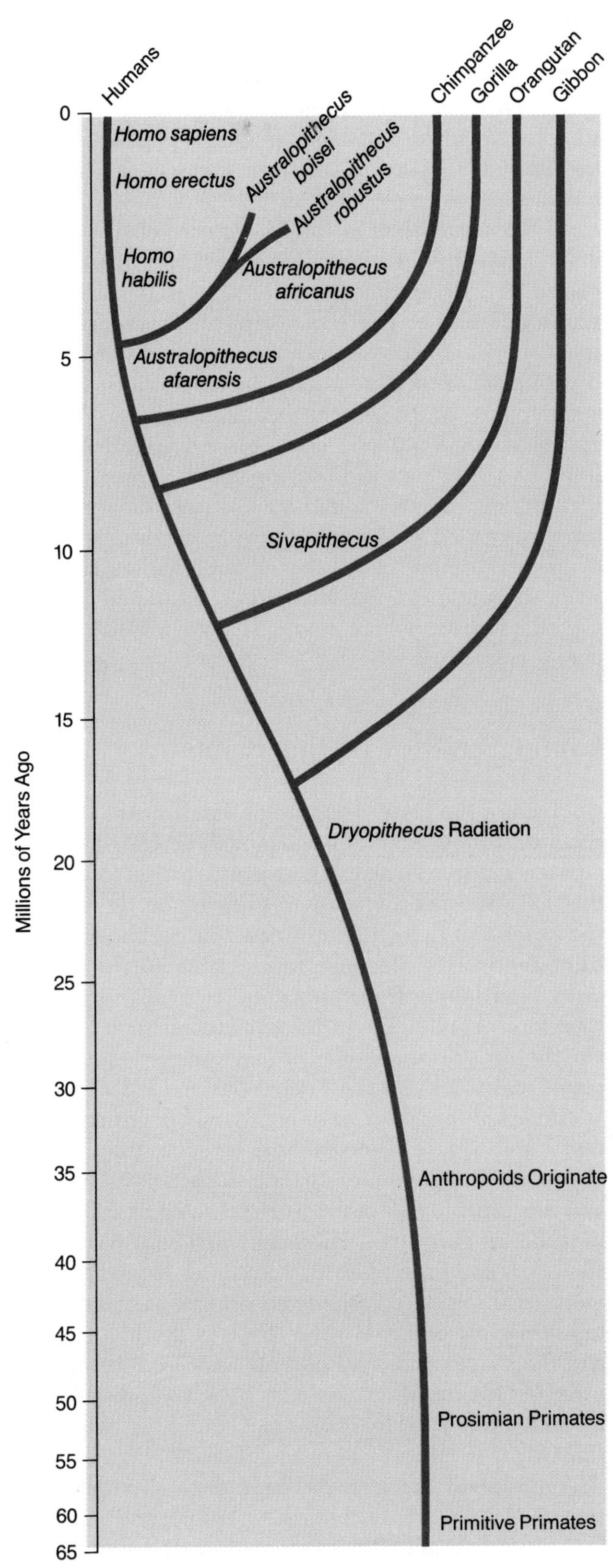

Figure 46-3 THE PRIMATE PHYLOGENETIC TREE. The major types of primates and times at which different lines diverged are shown here. This tree depicts just one interpretation of events of the past 4 million years.

great apes include chimpanzees, gorillas, and orangutans, the first two of which, like humans, have adaptations for life on the ground (Figure 46-2c). But reflecting their former life in the trees, they have enlarged, elongate forelimbs that prop up the front of the body when the animals are on the ground.

Our prehistoric ancestors, living millions of years ago, were never identical to today's chimpanzees, baboons, lemurs, or any other modern primate. However, we can use these living primates as a mirror for our past, particularly by studying primate anatomy, genetics, and behavior combined with analyses of the fossils preserved in geological strata.

Early Evolution of the Primates

The primates are one of the oldest orders of placental mammals, extending back in time more than 65 million years to the Cretaceous period (Figure 46-3). The earliest primates were small, shrewlike animals contemporary with at least six species of dinosaurs, and they were similar in many ways to today's living tree shrews, which are insectivores rather than primates (Figure 46-4).

Sixty-five million years ago, during the Paleocene epoch, there was an evolutionary radiation of primitive primates that still looked like small rodents and insectivores but had some telltale skeletal features of the skull and teeth, as well as limbs that indicated a life in the trees. At the beginning of the Eocene epoch, about 54 million years ago, the first primates with bodily characteristics of modern prosimians appeared. Some of these Eocene fossils closely resemble living lemurs and tarsiers, and huge numbers of Eocene prosimians lived in

Figure 46-4 THE MODERN TREE SHREW. This Southeast Asian resident *(Tupaia tana)* is generally representative of the types of insectivores from which the earliest primates evolved.

most of the earth's tropical areas. Prosimian fossils, in fact, are among the most common animal remains found in some European and North American fossil beds of the Eocene epoch. By the end of the Eocene, however, most of these prosimian groups had become extinct, probably as a result of climatic cooling, continental drift, and competition from rodents and other well-adapted animals. Some Eocene prosimians did survive, however, and gave rise to the various modern prosimian groups, as well as the earliest members of the Anthropoidea.

The earliest anthropoids appeared in Africa during the Oligocene epoch, about 30 million years ago (see Figure 46-3), and spread across the Old World into South America. Their ancient geographical separation led to the establishment of the New World monkey lineages, on the one hand, and the Old World monkeys, apes, and humans, on the other. One of the earliest apelike creatures was *Aegyptopithecus* (meaning "Egyptian ape"); it was about the size of a large house cat, lived in the forests of northern Africa, resembled a small New World monkey (such as a squirrel monkey), and may have given rise to later apes.

Much as the Eocene epoch was the heyday of the prosimians, the Miocene epoch saw the flourishing of the apes and Old World monkeys, from 26 to 5 million years ago. During the first half of the Miocene, apes or apelike animals occupied much of the tropical and subtropical Old World. Paleontologists have identified them

Figure 46-5 ***Dryopithecus:* A MIOCENE APE.** *Dryopithecus* was the first genus of the group of apes that occupied much of the tropical and subtropical Old World from about 25 million to 15 million years ago. They filled niches similar to those of modern Old World monkeys. (From Eli C. Minkoff, *Evolutionary Biology* © 1984. Adapted by permission of Benjamin/Cummings Publishing Company.)

primarily on the basis of tooth crowns that resemble those of later apes. These fossils are generally classified as dryopithecines, after ***Dryopithecus*** (meaning "forest ape"). *Dryopithecus* lacked the long, large forelimbs so apparent in modern apes (Figure 46-5), but was on the lineage that led to modern **hominoids** (apes and humans).

In the second half of the Miocene, monkeys largely indistinguishable from modern Old World monkeys replaced the dryopithecines across most of Africa and southern Eurasia. These monkeys could eat and digest the abundant grasses and leaves, whereas the dryopithecines were restricted to softer fruits, berries, and tender shoots. By the beginning of the Pliocene epoch, about 5 million years ago, the apes descended from dryopithecines had become greatly limited in their distribution, as are modern apes, and monkeys had taken over most of the apes' previous niches. A paucity of ape fossils more recent than about 5 million years old suggests that apes were increasingly restricted to tropical rain forests, where bones seldom fossilize in the damp, acidic soil.

Characteristics of the Primates

Much of the body structure of the primates that distinguishes them from other mammals is related to their arboreal nature. Virtually all primates, except for the few that have secondarily adapted to living on the ground, spend most of their time in trees—eating buds, leaves, fruit, and insects; sleeping; reproducing; interacting socially; and hiding from predators. It is not surprising, therefore, that the first major primate radiation followed closely on the radiation of angiosperms—flowering plants and trees—in the Cretaceous and Paleocene.

Almost all primates have highly mobile arms, shoulders, and legs for locomoting through trees. Their thumbs and big toes are separate from their other digits and are capable of grasping objects such as cylindrical branches (Figure 46-6). The small, light body has narrow shoulders and hips, allowing the center of gravity to be more easily balanced when the primate is standing on top of narrow branches. The shape of the tooth arcade and the shapes and sizes of teeth vary from primate to primate (Figure 46-6), and the brain's cerebral cortex increased dramatically in size and complexity from prosimians to monkeys, apes, and humans (Figure 46-6).

Anthropoids living in the trees depend on acute vision as they leap from branch to branch and reach for fruit or insects. With their flattened snout and their eyes facing forward, monkeys and apes have binocular, stereoscopic vision that permits accurate depth perception—especially useful when grabbing small mobile prey such as insects.

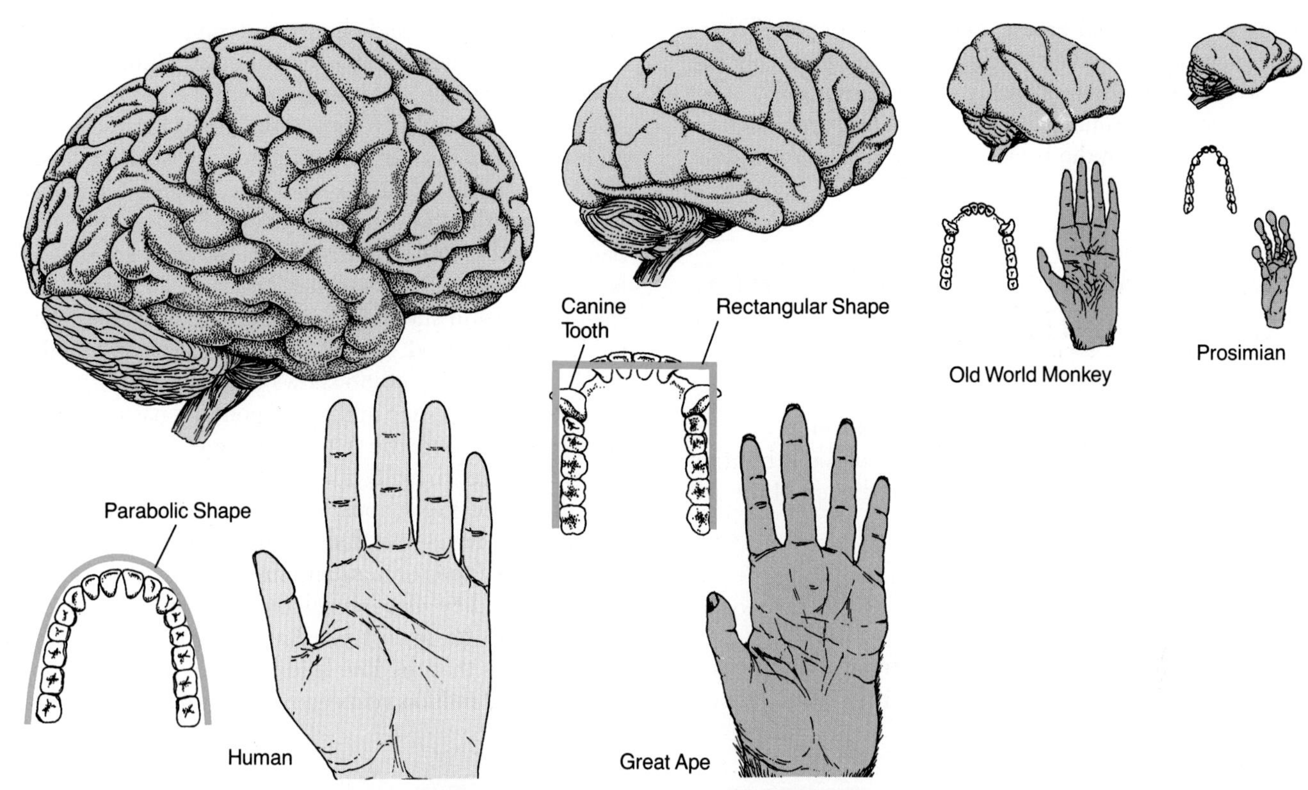

Figure 46-6 HANDS, TEETH, AND BRAINS OF HUMANS AND OTHER PRIMATES.
The opposable thumb (and large toe) arose early, allowing the gripping of branches from above. The long fingers of monkeys and apes can be wrapped around branches as the body hangs or swings beneath. Note how the shape of the upper jaw varies among these primates, as do the shape and size of different types of teeth. The large canines of monkeys and apes have been lost in the human lineage. The great increase in surface area of the brain's cerebral cortex is reflected by the folding and fissures in the surface of the ape and human brains. (The hands and brains are not drawn to the same scale; the human brain is much larger in real size than the monkey and prosimian brains.)

Early apes apparently carried out a kind of slow vertical climbing, with the full body weight suspended from the upward-reaching forelimbs and with the feet of the hind limbs gripping the trunk or branch. Contemporary tarsiers do the same thing (Figure 46-7a). Some apes also developed brachiation, swinging beneath the branches with the full weight suspended from one arm at a time (Figure 46-7b). The forelimbs of apes became longer and stronger to support the weight and also became longer than the hind limbs. The shoulders widened and became highly mobile, too. When modern apes walk quadrupedally on the ground, their long forelimbs prop up the front of the body; the animals carry out so-called **knuckle walking,** with the greatly elongate fingers, evolved for gripping branches during climbing and brachiation, curled inward so that the weight is applied to the knuckles of the fist (see Figure 46-9b).

Anthropoids usually have single births; the infants are largely helpless, and they commence a long period of growth and maturation that requires constant parental care. This means that the primates invest much time and energy in each offspring, and offspring have considerable time to learn the complex behaviors necessary to get

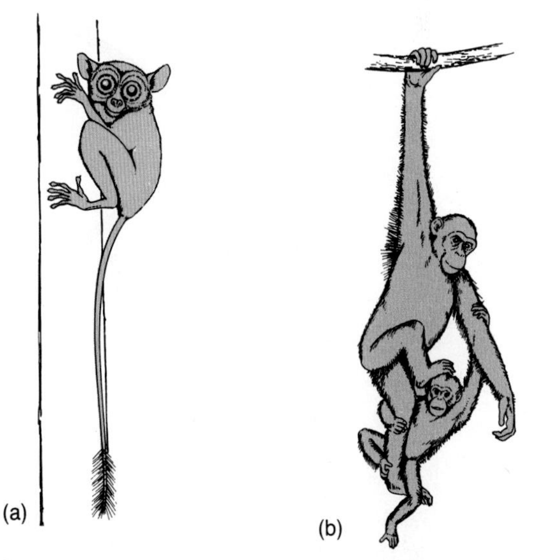

Figure 46-7 ACCENTUATION OF THE FORELIMBS.
Primates living in the trees frequently hold their forelimbs in raised positions and use them in tension-bearing ways. Imagine the stresses and forces you would need to exert with your legs and arms if (a) you were in the position of this tarsier or (b) you had to carry a little friend the way this chimpanzee is doing.

along within a social group. This is an example of *K* selection, a reproductive strategy based on optimizing the survival of individual offspring (see Chapter 43).

The general primate anatomy, physiology, and behavioral patterns form the background for human evolution.

THE TRANSITION FROM APE TO HOMINID

In Charles Darwin's *The Descent of Man* (1871), he predicted that the earliest human fossils would eventually be found in Africa, and a century of fossil hunting worldwide has supported Darwin's prediction of the African origin.

Molecular Relatedness

Humans share a number of anatomical traits with chimpanzees, including details of the brain, teeth, shoulder and arm, and thorax and some aspects of the legs and feet. Remarkable biochemical similarities between chimpanzee and human proteins also bespeak our close relatedness. In studies of the genes for one type of globin protein in various primates, researchers using DNA-DNA hybridization and sequencing techniques have found that humans are genetically closer to chimpanzees and pygmy chimps than those apes are to gorillas (Figure 46-8a), but are far closer to all three kinds than apes are to sheep or than dogs are to wolves. Data on DNA and blood groups demonstrate corresponding similarities. Figure 46-8b shows the banding patterns of a human chromosome and corresponding patterns in the great apes. In fact, the *X* chromosomes and four of the autosomes are virtually identical among *Homo sapiens* and the three great apes. Comparisons of some 1,000 genetic loci have revealed a variety of inversions and duplications (note orangutan chromosome 10 in Figure 46-8b), but they also reveal that humans and chimps clearly have the fewest banding differences. Evidence now suggests that the line leading to the orangutan split off at least 12 million years ago; the gorilla line separated

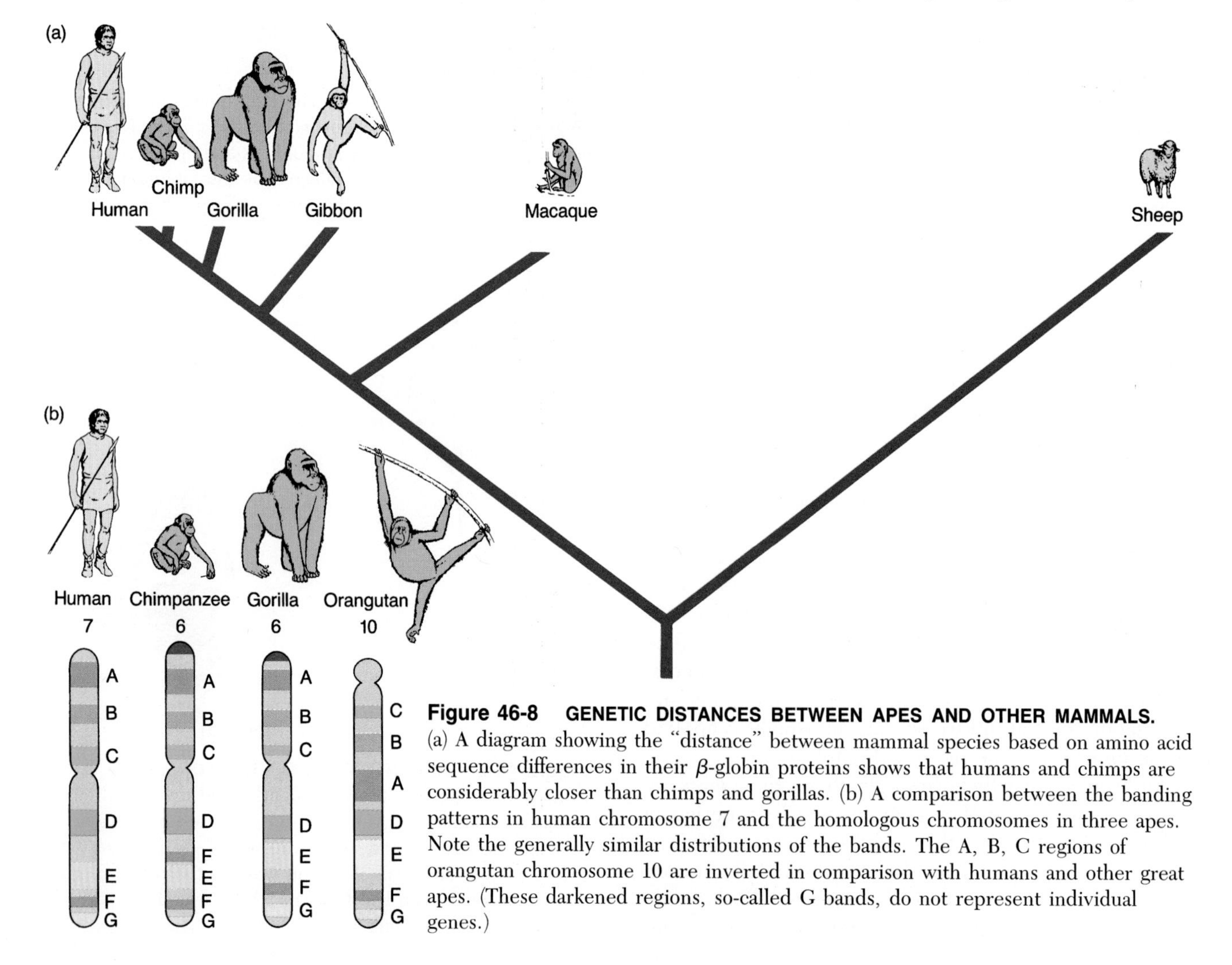

Figure 46-8 GENETIC DISTANCES BETWEEN APES AND OTHER MAMMALS. (a) A diagram showing the "distance" between mammal species based on amino acid sequence differences in their β-globin proteins shows that humans and chimps are considerably closer than chimps and gorillas. (b) A comparison between the banding patterns in human chromosome 7 and the homologous chromosomes in three apes. Note the generally similar distributions of the bands. The A, B, C regions of orangutan chromosome 10 are inverted in comparison with humans and other great apes. (These darkened regions, so-called G bands, do not represent individual genes.)

about 9 million years ago; and, about 6 million years ago, the lines leading to the chimpanzee (family Pongidae) and to humans (family Hominidae) diverged (review Figure 46-3).

The best evidence suggests that the common ancestor of chimpanzees and humans probably resembled chimpanzees in overall appearance, especially in the skull. But it was undoubtedly smaller than a modern chimpanzee, standing only .9 or 1.2 m (3 or 4 ft) high and weighing perhaps 23 kg (50 lb), whereas an adult chimpanzee weighs about 45 kg (100 lb). Some paleontologists conclude that these animals probably lived in open woodland savannas, ate fruit, nuts, insects, and small game, and walked on all fours or upright. Other scientists point out, however, that humans lack sun-reflecting fur and have subcutaneous fat for insulation, two characteristics never seen in savanna-dwelling animals. We also sweat a great deal and lose much water and sodium to keep cool and also cannot greatly concentrate our urine. In these and other respects, we need much more water than do savanna-adapted mammals and other primates, and thus, some believe that our ancestors lived in moist African forests, not open savannas.

Bipedalism

What one feature most distinguished the earliest hominid from other primates? The answer, say many anthropologists, is an erect, bipedal stance. That is because this stance and the striding bipedal gait appear to have evolved well before the large brain, the dexterous hand, and many other human anatomical features. Mary Leakey's footprint find at Laetoli (review Figure 46-1) tends to confirm this. The hominid bipedal stance depends on a special anatomy of the legs and feet, of the pelvic girdle, and of the vertebral column. Only when such features had evolved could the early hominid (or, later, the modern human) have moved with the body's center of gravity balanced between the two feet when the legs extend straight down, allowing the human to stand or stride easily and efficiently (Figure 46-9a). A chimpanzee or monkey, in contrast, must exert great muscular effort to stand and is still out of balance (Figure 46-9b–d).

This energy-efficient habitual bipedalism among early hominids would have allowed them to cover long distances; carry food, infants, and perhaps other objects in their arms; and, later in hominid evolution, use the hands for wielding weapons or simple tools. Many scientists believe that the hand and its myriad uses contributed greatly to the remarkably rapid expansion of the cerebral cortex in the human lineage.

Reproductive Behavior and Human Evolution

Many anthropologists have noted that several sexual characteristics and mating patterns appear to be uniquely human: (1) Females have permanently enlarged mammary glands (breasts), and males have a prominent penis; (2) copulation is usually performed face to face; (3) male and female body and facial hair and fat distributions are quite distinctive; and (4) females can

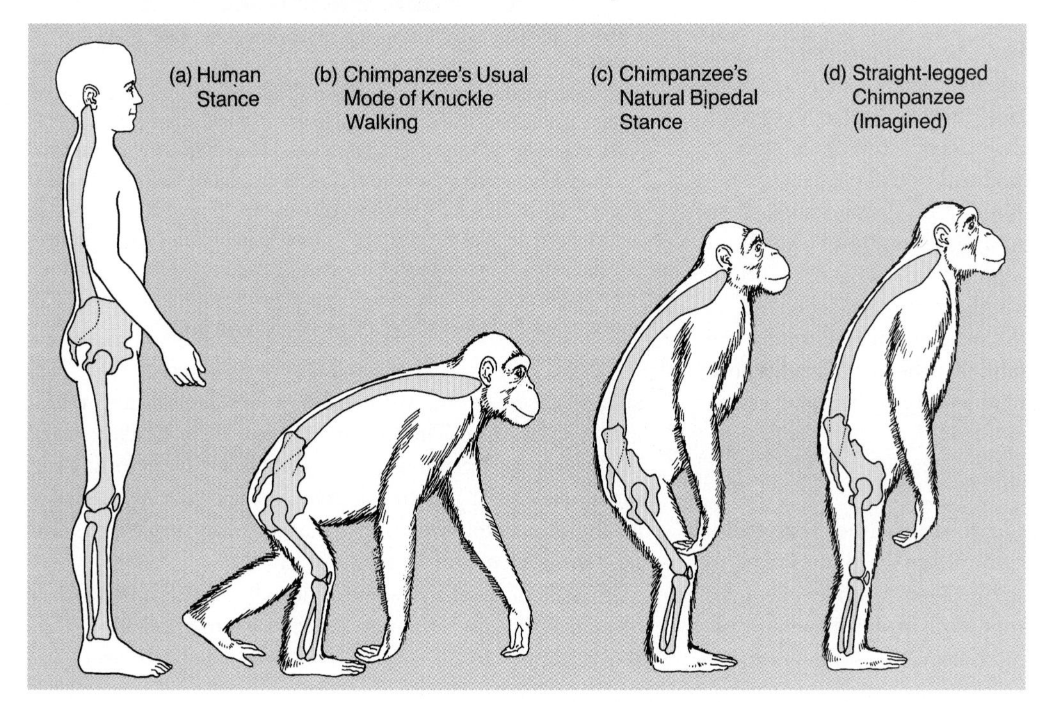

Figure 46-9 ATTAINMENT OF ERECT BIPEDALISM. (a) The erect human's legs are considerably longer than a chimp's, and the backbone bends backward in the pelvic region, bringing the shoulders and head directly above the feet. The chimpanzee usually knuckle walks (b), but can assume a bipedal stance with flexed legs and bent back (c). If a chimp's legs were straightened (d), its body would tilt forward in an unbalanced state never seen in nature.

copulate independently of ovulation, thus separating copulation from reproduction (see Chapter 18). Frequent copulation may well have contributed to the male-female pair bonding that is a key feature of human family structure and to the availability of both parents to rear and teach offspring.

THE EVOLUTION OF THE HOMINIDS

Hominid evolution is a fascinating story, pieced together, literally, by fossil shards and other evidence. By studying fossils of the human lineage that diverged from the chimpanzee line between 5 and 10 million years ago, researchers have discovered that bipedalism preceded brain enlargement and that even after the balanced, bipedal skeleton had evolved, our ancestors retained characteristics appropriate for retreating and climbing into the trees.

Paleontologists divide the past 4 million years of human evolution into two overlapping periods, each represented by different early humans. Members of the genus *Australopithecus* lived during the earlier period, which extended from before 4 million years ago to about 1.3 million years ago. The later period of human evolution began at least 2 million years ago and includes individuals in our own genus, *Homo*.

Australopithecus

Early members of the genus ***Australopithecus*** (meaning "southern ape") survived on earth for nearly 3 million years. While there is considerable controversy over the species in this genus, we accept the view that at least four distinct species lived and became extinct: *Australopithecus afarensis* (which lived from about 4 to about 3 million years ago), *A. africanus* (about 3 to about 2.5 million years ago), *A. boisei* (about 2.5 to about 1.3 million years ago), and *A. robustus* (about 2 to about 1.3 million years ago). These early humans are known from only eastern and southern Africa, where they appear to have inhabited wooded grassland areas.

Australopithecus afarensis

The fossil remains of the earliest hominids, *Australopithecus afarensis*, were first found in 1978 at Hadar, in the Afar region of Ethiopia, by Donald Johanson, Yves Coppens, and their co-workers. One of these specimens, "Lucy" (Figure 46-10), named after the song "Lucy in the Sky with Diamonds," is the most complete early fossil yet discovered, aside from a much younger *Homo erectus* skeleton. The footprints that Mary Leakey found at Laetoli in Tanzania are also believed to belong to *A. afarensis*. Most paleontologists now regard *A. afarensis* as the stem stock leading to later *Australopithecus* and *Homo* species (see Figure 46-12).

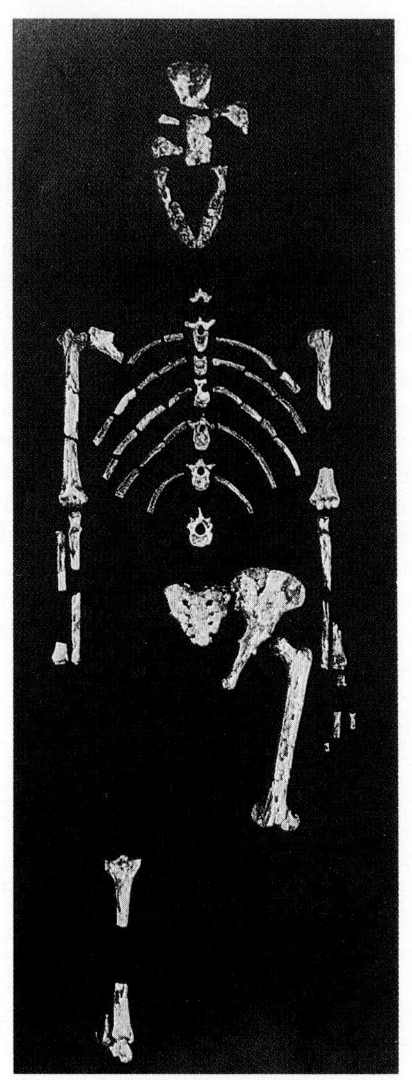

Figure 46-10 ***Australopithecus afarensis*: THE FIRST HUMAN.**
This remarkably complete 3.5-million-year-old skeleton is of a young female of the species *Australopithecus afarensis*. This skeleton is popularly known as Lucy. It is from fossils such as Lucy that paleontologists are able to reconstruct the overall body size and proportions of *A. afarensis*. The pelvis of this female has the typical human shape; the thigh bone is relatively shorter than our own. Finger and foot remains from other individuals indicate similarities, but not identical configurations, with those of modern humans.

Fossils of *A. afarensis* individuals exhibit a mixture of human and ape characteristics (Figure 46-11a). The skulls appear to have been very apelike, with large projecting faces and small brains—about 450 cm^3 (about

1 pt) in volume, which is only slightly larger than a chimpanzee's brain (about 400 cm^3), but about one-third the size of a modern human's brain (average about 1,350 cm^3). The fossilized teeth and jaws are also quite apelike, with the molars and premolars similar to those of chimpanzees. The front teeth (incisors and canines) are large and slightly projecting, with spaces between the upper canines and incisors to accommodate the projecting canines from the lower jaw. In contrast to these apelike characteristics, the *foramen magnum*, the hole at the base of the skull through which the spinal cord exits, points downward. It thus indicates that the head sat on top of the vertebral column rather than in front of it, as in all quadrupedal apes. This means that these creatures could stand upright.

The arms of the *A. afarensis* fossilized skeleton have a mixture of apelike and humanlike features. The shoulder and elbow joints could have held the arms overhead and supported the weight of the animal hanging in a tree. The hands are human in overall proportions; the fingers are shorter than a modern ape's, yet they are somewhat curved like an ape's and could have been used for gripping branches. Indeed, the finger joints are very similar to a chimpanzee's, suggesting that *A. afarensis* lacked the later humans' extensive manipulative abilities.

The fossil remains of *A. afarensis* also indicate that this group possessed legs adapted for a balanced bipedal gait. The pelvis is short and widened, indicating the same hip-stabilizing mechanism that we have today; the knees angle in toward the midline, and the feet lack the divergent, grasping big toe that other primates use to grasp branches. Nevertheless, the four outside toes are longer and more curved than our own and suggest that Lucy and her relatives still had adaptations for grasping branches. The fossil *A. afarensis* thighbone and lower leg bones are also substantially shorter than those of later hominids, reflecting a closeness to the apes, with their short legs and long arms. Only after the leg lengthened during later hominid evolution could humans attain the fully modern stride.

A. afarensis individuals stood only 1 to 1.4 m (3.5 to 4.5 ft) high and must have weighed only 18–23 kg (40–50 lb; Figure 46-11a). They probably spent much of their time in the trees, perhaps when feeding and sleeping and when fleeing danger. The brain size and surface folding pattern are little larger than a chimpanzee's, and so the early hominid's intelligence was probably little greater than a chimp's. *A. afarensis* may have made and used simple tools at least as often as chimpanzees do today. Field workers have not yet found definite stone or bone tools associated with *A. afarensis* fossils, nor evidence of group hunting or butchering of prey.

Australopithecus africanus

The next hominids in the fossil record belong to the species *Australopithecus africanus*. In 1924, Raymond Dart identified the first australopithecine, the Tuang child, in South Africa, and this fossil shares many anatomical features with those of *A. afarensis*. *A. africanus* individuals were relatively small bipedal humans with a protruding face, with a brain only slightly larger (450–500 cm^3) than that of *A. afarensis* (Figure 46-11b), and

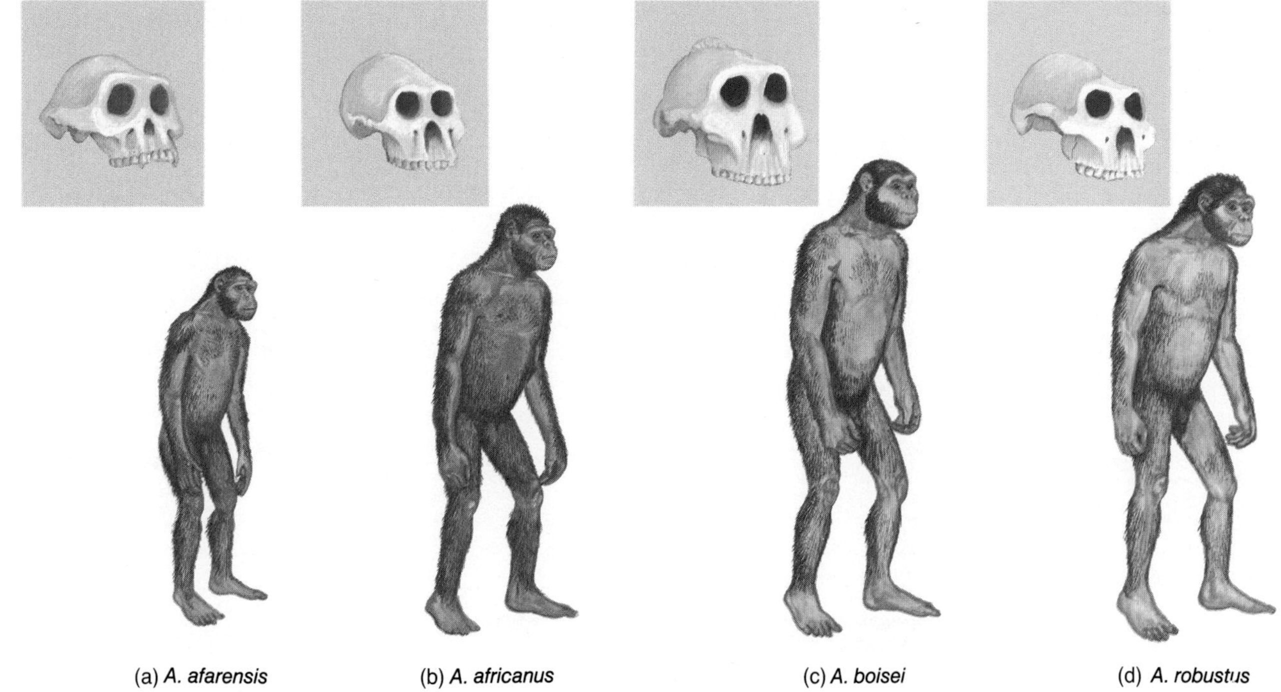

Figure 46-11 THE AUSTRALOPITHECINES: EARLY HUMAN ANCESTORS.

with a body that differed little from the earlier hominid. New analyses of hand and foot structure suggest that these hominids, too, spent time in the trees—perhaps while sleeping or feeding, perhaps for safety.

The teeth, however, had started to change. The front teeth lost the slight apelike projection seen in *A. afarensis* and became entirely human in size and shape, and the row of teeth was more curved than square. The cheek teeth, however—the flat grinding and crushing molars and premolars—became exaggerated in *A. africanus*, no doubt for processing coarse, abrasive food, such as tubers, and they were set in heavily built jaws.

Australopithecus boisei and *Australopithecus robustus*

Members of *A. africanus* are known to have existed from about 3 to 2.5 million years ago. Between 2.5 and 2 million years ago, two later species of *Australopithecus* appeared in East Africa: *A. boisei* (Figure 46-11c), named for a donor of funds to the Leakey expedition, and *A. robustus* (Figure 46-11d). In 1986, researchers found a 2.5-million-year-old skull at Olduvai Gorge; dubbed the "black skull" because of mineral discoloration, the specimen seems to be a precursor of *A. boisei*. Fossils of both species show large flat teeth and jaws. However, the limb skeletons were similar to earlier species of *Australopithecus*, although their brains were slightly bigger, as were their bodies. There is evidence that a form of *A. robustus* called *Paranthropus robustus*, dating back to 1.8 million years ago, had stone tools and the fingers and thumbs to use them.

Anthropologists disagree rather strenuously over the australopithecine's possible relationships to each other and to the genus *Homo*. Figure 46-12 shows one possible phylogenetic tree. Whatever the relationships turn out to be, the australopithecines were highly successful in evolutionary terms. They occupied much of sub-Saharan Africa for at least 3 million years, far longer than the genus *Homo* has been around.

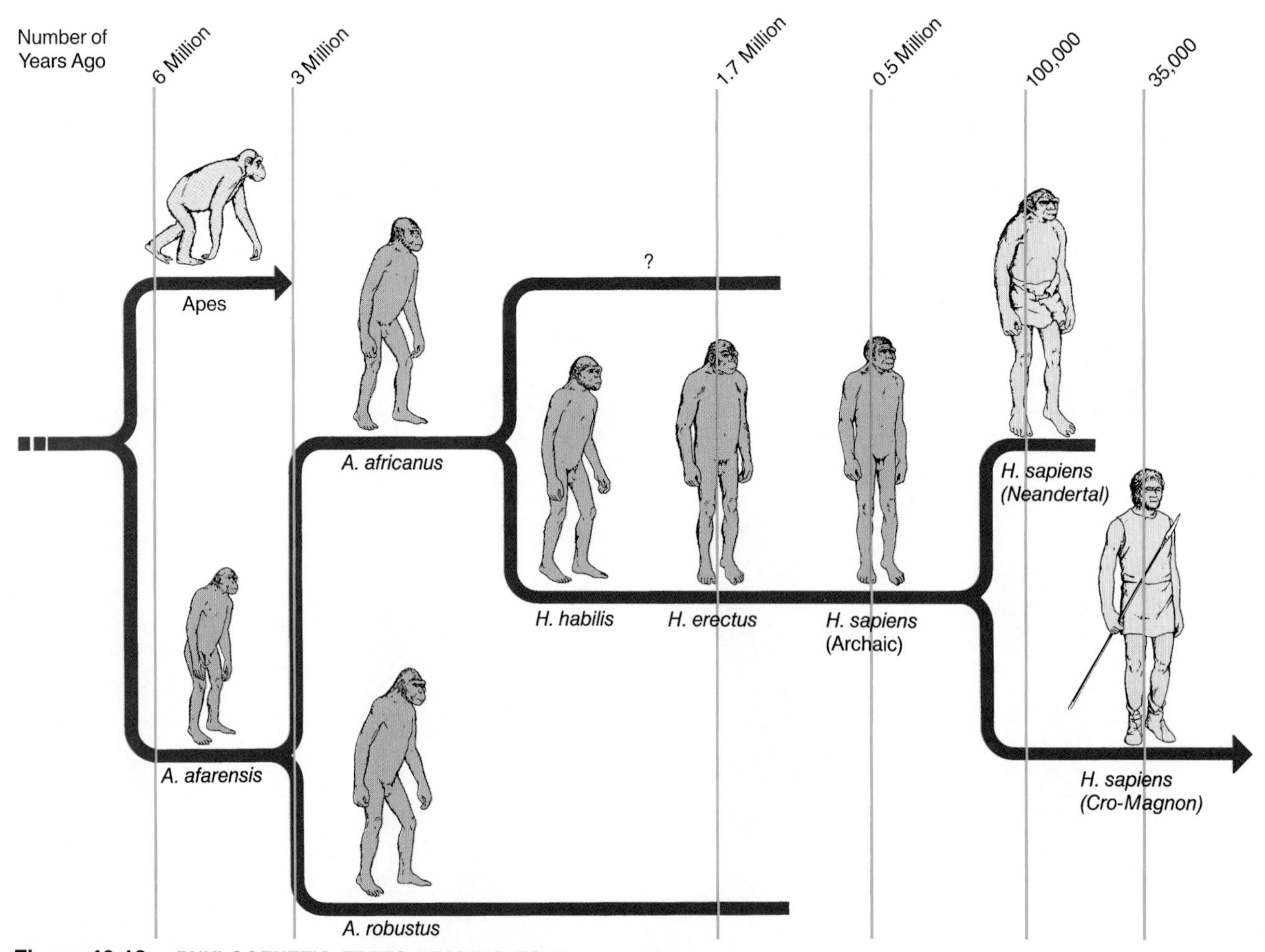

Figure 46-12 PHYLOGENETIC TREES LEADING TO *Homo sapiens*.
Many paleontologists favor the scheme shown here, but this is just one of several suggested trees for our species' ancestry.

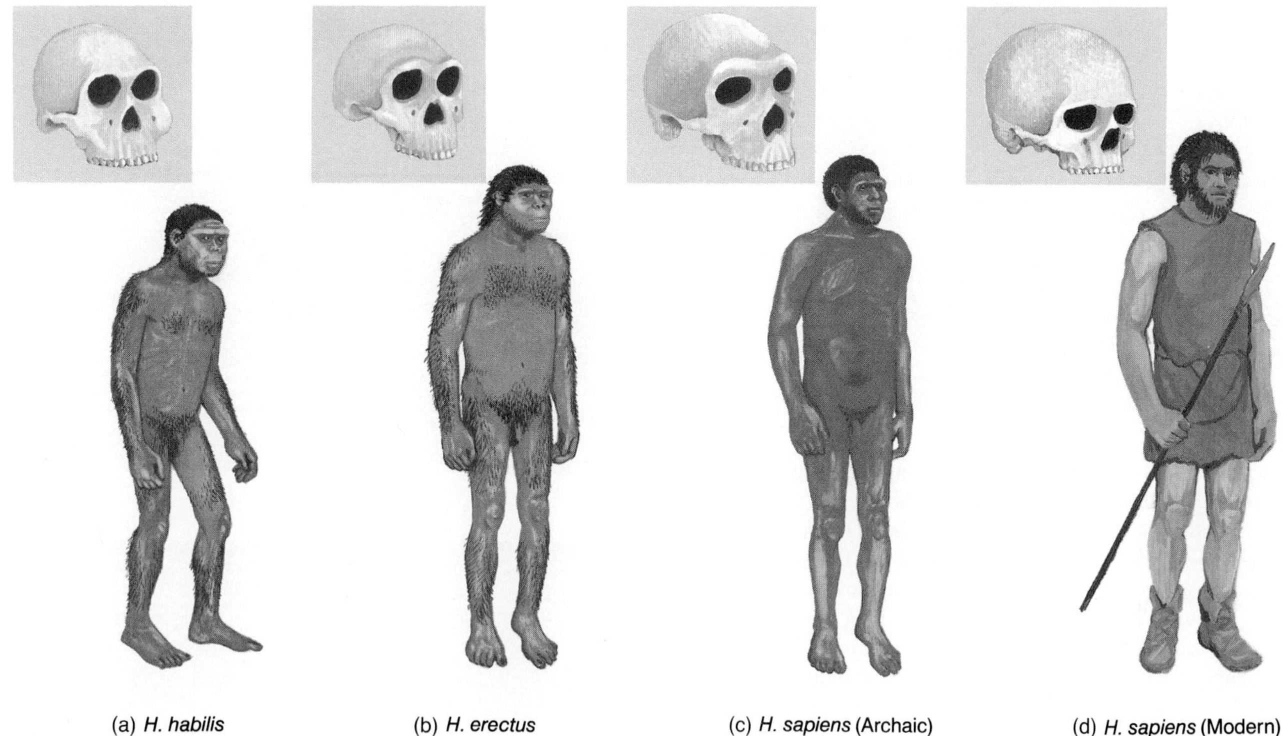

(a) *H. habilis* (b) *H. erectus* (c) *H. sapiens* (Archaic) (d) *H. sapiens* (Modern)

Figure 46-13 THE GENUS *Homo:* HUMANS EMERGE.

The Genus *Homo*

By about 2 million years ago, at the beginning of the Pleistocene epoch, a new type of hominid had appeared in the prehistoric record of Africa: the first members of our own biological group, the genus *Homo*.

Homo habilis

The earliest humans have been placed in the species ***Homo habilis*** (meaning "handy man"; Figure 46-13a).

Homo habilis superficially resembled its forebear, *Australopithecus*, but was different in several important aspects. The *H. habilis* brain was about 750 cm^3 on the average and thus somewhat more than one-half the size of the modern human brain instead of the australopithecine's one-third. Some individuals may have also had larger bodies, although a fossil analyzed in 1987 had a typical enlarged *H. habilis* skull on a body no larger than an *A. afarensis* individual. An increase in brain size relative to body mass indicates the presence of additional neurons over and above those needed to operate basic body functions. The marked increase in the *H. habilis* brain relative to that of *Australopithecus* thus implies heightened intelligence.

Anthropologists dispute the evolutionary origins of *H. habilis*. The species appears in the fossil record of around 2 million years ago, and some derive *H. habilis* directly from *A. afarensis* at about 3 million years ago. Others regard the origin of *H. habilis* as more recent and think that *A. africanus* was the immediate progenitor. Only a more complete fossil record will resolve the disputes over the origin and early history of *H. habilis*.

The trunks and limbs of *H. habilis* were fully adapted for upright, bipedal locomotion, but the hand represents a curious mosaic: Fossilized thumb and fingertip bones are similar to recent humans', suggesting improved manipulative skills; yet the bones of its digits were heavy and somewhat curved, permitting the powerful grasping seen in chimpanzees and gorillas. Researchers have interpreted these characteristics to mean that *H. habilis* lived on the ground during the day, but retreated to the trees at night to sleep.

Homo habilis apparently performed two characteristic human activities: toolmaking and butchering of large animals. The earliest stone tools were found among *H. habilis* fossils dated at 2.4 million years old, suggesting that the "handy men" were already "handy" that long ago. The tools are rather crude fractured rocks that do not look particularly like deliberately fashioned tools (Figure 46-14a). Yet, their shapes are not randomly fractured, but form patterns of breakage, leaving little doubt that people manufactured them. Anthropologists called their creation and use the **Oldowan industry,** after Olduvai Gorge.

Homo habilis definitely butchered both large and small game, as cut marks on the bones indicate. A major addition of meat to their diet would have set them apart

(a)

(b)

(c)

Figure 46-14 PRIMITIVE TOOL "INDUSTRIES" OF HUMAN ANCESTORS.
(a) An Oldowan tool excavated in Olduvai Gorge, Tanzania. A *Homo habilis* individual probably made this coarse cutting and chopping tool from a lava pebble by knocking off flakes on a small block of stone. (b) An Acheulian hand ax from France—a general-purpose chopping and cutting tool probably fashioned by *Homo erectus*. The oldest known hand axes are from deposits in East Africa about 1.5 million years old, and they continued to appear in stone tool industries in Africa, Europe, and much of Asia as recently as 100,000 years ago. (c) A selection of Mousterian tools, mostly from archeological sites in western Europe, where *Homo sapiens* archaic (Neandertals) lived and died. The tools shown are, from left to right, a backed knife, a piercer, a scraper, and a hand ax.

from other primates and probably from most australopithecines. The use of tools and animal carcasses may well have started the shift in human adaptive patterns that was to dominate subsequent human evolution.

Homo erectus

It seems unlikely that *Homo habilis* could have lived outside the savanna and savanna-woodland belt that extended from Ethiopia to South Africa, a zone that offered steady, mild temperatures and abundant plant and animal resources.

Some time around 1.5 million years ago, however, prehumans passed an adaptive threshold. Within a short period of time, these early peoples extended their geographical range to include all the tropical and subtropical areas of the Old World, excluding the inaccessible islands. Evidence of their presence, in the form of either fossils or primitive tools, appears in rock strata dated to that time in mountainous areas of Africa, across the North African coast, and across southern Asia into China and Indonesia. In general, the Chinese and Indonesian fossils date from 1.2 to 1 million years ago and thus are younger than the African stock, which is at least 1.6 million years old.

This major geographical expansion, an important indicator of ecological and evolutionary success, was accompanied by anatomical and behavioral changes in the human lineage. A new form of *Homo*, called ***Homo erectus*** (Figure 46-13b), appears in the fossil record around 1.6 million years ago, and a new stone industry, called the **Acheulian** (named after the town of St. Acheul, in France), appears in the archeological record (Figure 46-14b).

Early *Homo erectus*—at least the males—may have been substantially taller than *H. habilis*. The most complete early human fossil yet found is that of a 12-year-old boy (as judged by the teeth and the bone growth areas) who stood about 1.6 m (5.5 ft) tall and would have been perhaps 1.8 m (6 ft) tall if he had lived to maturity (Figure 46-15). The fossilized skull is quite massive, considerably more so than many female *H. erectus* skulls,

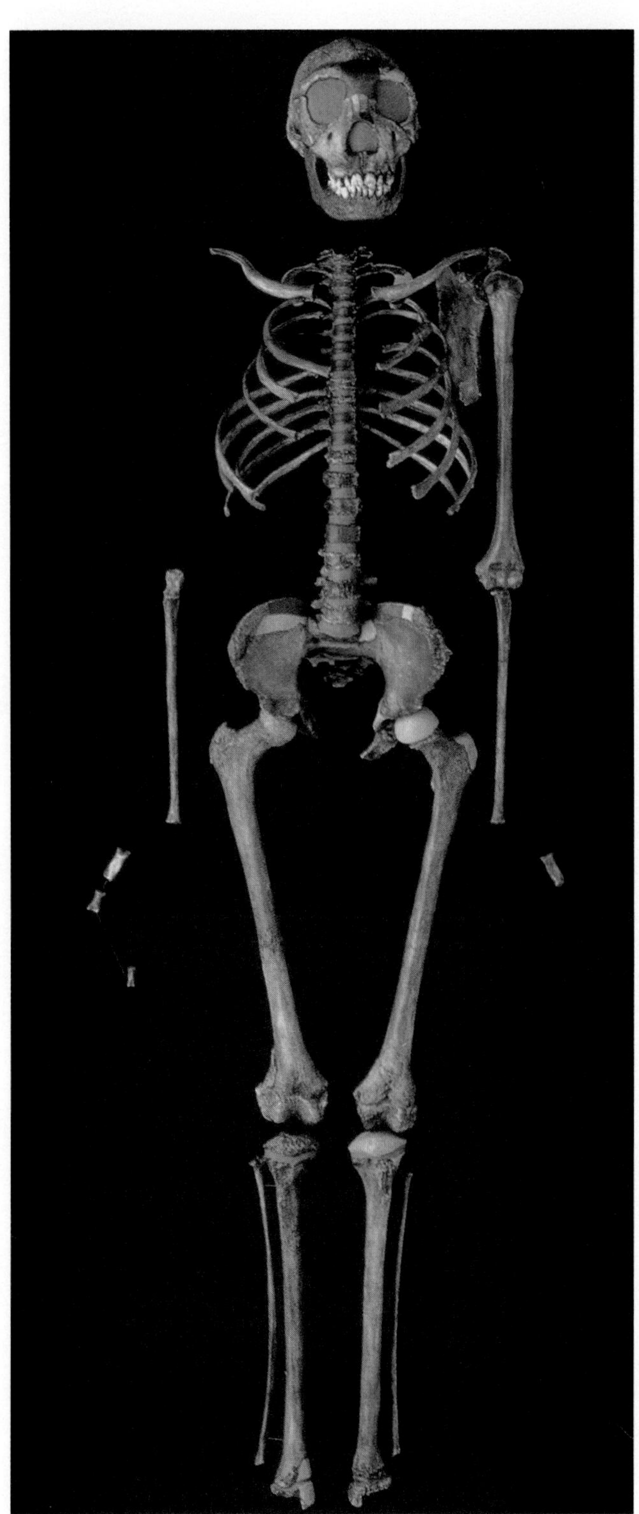

Figure 46-15 ***Homo erectus:*** **THE MOST COMPLETE FOSSIL HUMAN SKELETON EVER FOUND.**
This 12-year-old boy lived 1.6 million years ago. Note the narrow hips, yet very long heads on the two femur bones. The knees could be held in, near each other, when this boy stood. In contrast, an ape's legs are directed straight down or slightly lateral when it stands.

which implies considerable sexual dimorphism in the species. The arms and legs were quite similar to those of modern humans, although they were considerably more robust, indicating greater strength and endurance. An intriguing aspect of the boy's fossilized skeleton is the extreme narrowness of the pelvic girdle and especially the hole that, in females, serves as the birth canal. This may mean that *H. erectus* had extremely small newborns, and perhaps this necessitated a longer period for rearing and passing on learning and culture. The brain of *H. erectus* was much larger than that of *Australopithecus*, and the Acheulian industry shows a greater variety of tools and finer workmanship than the Oldowan industry, as evidenced by the hand ax.

There is evidence that *H. erectus* hunted, indicating that the species learned to fully exploit nontropical regions, where plant foods are seasonably scarce in the winter. These early humans also learned to control and use fire, perhaps for warmth and cooking.

During the tenure of *H. erectus* (from 1.6 to 0.3 million years ago), the species underwent evolutionary change. The face and teeth decreased in size and massiveness, while the brain increased in size spectacularly, from about 800 cm^3 around 1.5 million years ago to about 1,200 cm^3 500,000 years ago. Anthropologists cannot yet explain this great expansion, but cultural evolution and even the beginnings of language may have played roles. The gradual alteration in form of the head and face led imperceptibly into the anatomical configuration that we identify as that of archaic forms of *Homo sapiens*, our own species. There is no distinct boundary between late *Homo erectus* and early *Homo sapiens*, yet they were very different organisms.

Homo sapiens

The last 300,000 years of human evolution have been dominated by two groups of humans: an early group known as archaic ***Homo sapiens*** (Figure 46-13c), which includes the well-known **Neandertals** (also spelled "Neanderthal"), among others; and a late group of *Homo sapiens* who were physically indistinguishable from modern humans (see Figure 46-13d). Archaic *H. sapiens* occupied most of the Old World from about 300,000 years ago to about 75,000 years ago. In the following 40,000 years, a transition took place to yield modern *H. sapiens*.

The Neandertals lived in Europe and the Near East from about 130,000 to about 35,000 years ago, but Neandertal-like peoples preceded them in the same regions and occupied other areas of the Old World during earlier and the same times. Like *H. erectus*, all these Neandertal and Neandertal-like people had large faces with projecting brow ridges. Yet, their brains were similar in size (perhaps even larger) and probably similar in organization to those of modern humans. The face, although large and with a projecting brow ridge, was less massive than that of *H. erectus*. And the limb skeletons, while heavily built, were somewhat less robust than those of their hominid ancestors. The large brain and powerful limbs suggest an increasing ability to deal with the environment, as do the more specialized stone tools the Neandertals left behind, the **Mousterian industry** (after Le Moustier, France; see Figure 46-14c). The Neandertals routinely built shelters, hunted large game (anthropologists even found a wooden spear between the ribs of a fossil elephant in northern Germany), cared for the aged and infirm for long periods of time, and buried their dead (Figure 46-16). Thus, for the first time in the human fossil record, we find among the Neandertals a significant number of elderly individuals, several of whom had suffered broken bones that had healed with deformities years, if not decades, before the individuals died. Even so, the oldest Neandertals still lived only to about age 50, and few lived more than about 35 years. The Neandertals not only buried their dead, but also made a few simple body ornaments—both indications of growing consciousness, socialization, and even spirituality.

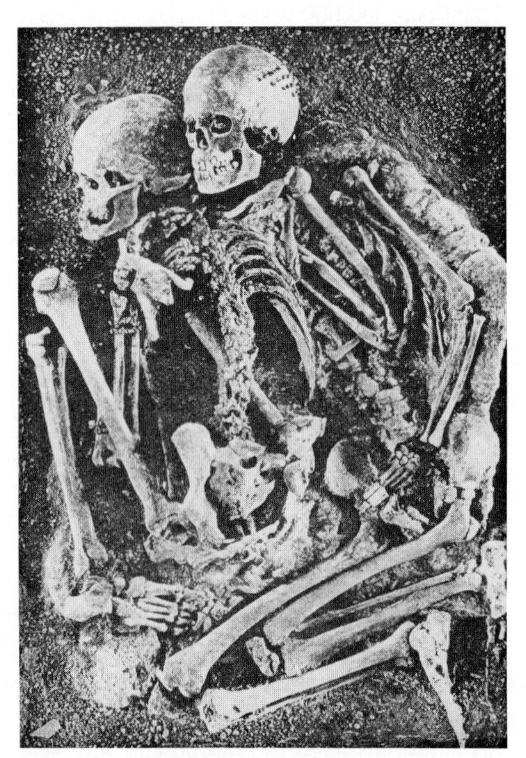

Figure 46-16 NEANDERTAL BURIAL.
The skeletons of an old woman and an adolescent male as they were discovered in the Grotte des Enfants in France, near Monaco on the Mediterranean coast. These two individuals were buried around 60,000 years ago, at the beginning of the Upper Paleolithic. This is one of many burials from this time in human evolution, some with elaborate grave offerings.

While Neandertals are included in the species *Homo sapiens*, paleoanthropologists do not know whether they could have successfully exchanged genes with the first modern humans. Where did the transition take place from archaic Neandertals to modern *Homo sapiens*? Both the fossil record and genetic evidence suggest that Africa was the source of all anatomically modern humans and that these people spread worldwide and did not interbreed with preexisting older populations. Analyses of mitochondrial DNA from modern populations across the world imply that all contemporary human mitochondria are derived from women who lived in Africa from about 290,000 to 140,000 years ago (mitochondria are passed on solely through eggs, never human sperm). Some observers think that founder populations left Africa 180,000 to 90,000 years ago. Their descendants in Asia show no evidence on the basis of mitochondrial DNA of having interbred with preexisting Asian *H. erectus* individuals. The fossils of early modern *Homo sapiens* (dubbed proto-Cro-Magnons) some 90,000 years ago in what is now Israel indicates an unexpectedly early date for our species. This population was present in the Middle East as long as 30,000 years before Neandertals, and this has led anthropologists to conclude that Neandertals were not a stem stock for modern humans.

THE ORIGIN AND DIVERSIFICATION OF RECENT HUMANS

The earliest humans of modern appearance, the Cro-Magnons, represented a new phase in human evolution, with far greater ability than their predecessors to adapt to varied and often demanding environments. They were different from Neandertals in several anatomical features. The brain case was higher and rounder, the brow ridge had all but disappeared (see Figure 46-13), and the whole body, but particularly the limbs, was athletic but markedly less massive. Physical changes continued to their current state as people gradually became less nomadic, adopted agriculture, and became urbanized during the past 10,000 to 15,000 years.

With the arrival of modern-appearing humans, several important behavioral changes took place. There was an elaboration of technology, reflected primarily in tools: They created many more kinds of tools from varied materials, and through more sophisticated techniques than in earlier cultures. Art appeared in the form of sculp-

(a) (b)

Figure 46-17 ART: AN ANCIENT FORM OF ABSTRACTION.
(a) Form, perspective, and shadowing all appear in this sophisticated work created by a Stone Age artist some 12,000 to 15,000 years ago in the Lascaux Caves in France. (b) The famous Venus of Willendorf, carved from limestone, was carved as a fertility symbol during the Pleistocene.

ture, painting, and engraving, suggesting a major development in symbolic communication (Figure 46-17). The use of art, a form of abstraction, was probably accompanied by an increase in verbal communication. The encampments and associated structures became larger and more complex, and human burials became more common and often included elaborate offerings.

These cultural manifestations were accompanied by massive migrations and a general population explosion (as we saw in Chapter 43). This period saw the first major increase in territorial occupation since *Homo erectus* had spread outward from Africa 1.5 million years ago. For the first time, people inhabited the extremely harsh arctic regions of Eurasia, following the herds of mammoth, woolly rhinoceros, reindeer, and other large game. They expanded across the Bering land bridge into North America, and they crossed open water to occupy New Guinea and Australia. By 12,000 years ago, humans had occupied virtually all the inhabitable regions of the earth (Figure 46-18).

Perhaps 5 million people lived at that time, only about 400 generations ago, and population sizes remained roughly constant or increased very slowly. People existed then in small nomadic bands of hunter-gatherers who followed game and collected diverse, seasonally available plants. The modern Lapps of northern Scandinavia probably still live much as those hunter-gatherers did, following giant herds of reindeer on the animals' yearly migrations; using reindeer flesh, hide, and bones for food, clothing, shelter, tools, and means of transportation; and gathering tundra plants for food, medication, and other purposes.

One of the great events of human cultural evolution marked the transition from hunter-gatherer to agriculturalist. About 10,000 years ago, the so-called agricultural revolution took place; that is, people began planting seeds, harvesting crops, and domesticating animals. New evidence suggests that agriculture arose independently at many sites and that the earliest farmers planted wheat, barley, and alfalfa in the Fertile Crescent, which extended from the Nile to the Tigris and Euphrates rivers in what is now eastern Iraq. Simultaneously, agriculture arose in the New World—with the growing of corn, potatoes, and tomatoes—and in the tropical and subtropical Far East—with the cultivation of rice, millet, soybeans, and other crops. Eurasian peoples began to subject camels, dogs, horses, sheep, cattle, fowl, and other animals to the breeding regimens that ultimately led to the animal varieties we use today for food, transportation, work, and companionship.

The agricultural revolution, with its widespread environmental modifications, had far-reaching consequences, including stable communities at specific sites; career specialization; writing; written numbers; and population expansion (as we saw in Chapter 43).

In a species as numerous and widely dispersed geographically as *Homo sapiens*, it is only natural to expect genetic differences to have arisen among populations, and the product of adaptation to different environmental pressures was the formation of subspecies, known popu-

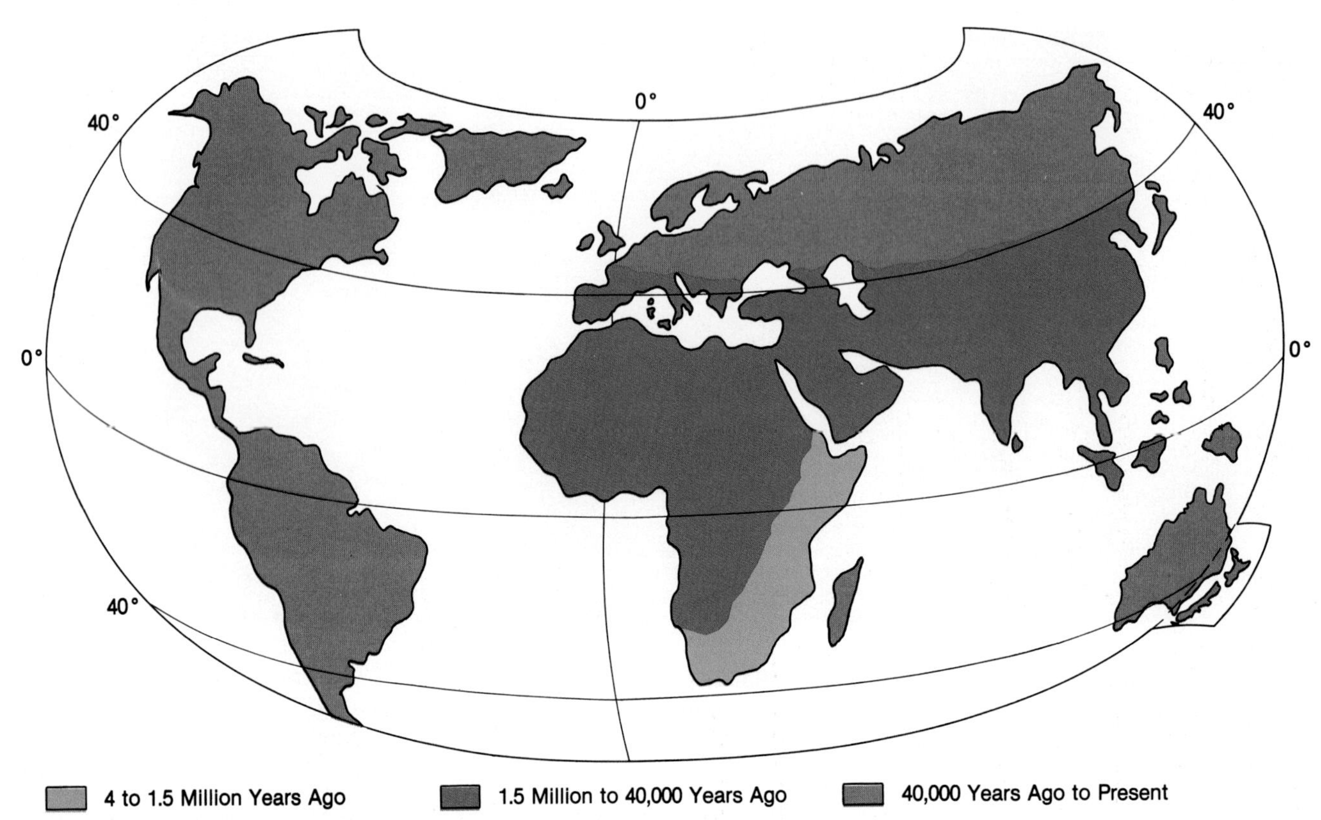

Figure 46-18 HUMAN MIGRATION OVER 4 MILLION YEARS IN THE OLD WORLD.
The successive periods of human geographical expansion are based on current fossil evidence. From about 4 million to 1.5 million years ago, humans were restricted to eastern and southern Africa (blue). From 1.5 million to 35,000 or 40,000 years ago, humans spread rapidly across the tropical and subtropical Old World and then slowly moved into more temperate zones (purple). About 35,000 to 40,000 years ago, humans began migrating into arctic regions, the Americas, New Guinea and Australia, and outlying islands (orange).

larly as *races*. Nevertheless, Linnaeus recognized long ago the basic unity of living humanity when he placed all people in our single species. We now know that this biological unity is due to our common evolutionary origins that lead back through geological time to our nonhuman primate ancestors.

HUMANKIND AND THE FUTURE OF THE BIOSPHERE

The emergence of modern *Homo sapiens* from primate ancestors was an odyssey of biological and cultural evolution. Cultural development, in particular, has changed the human past and will, no doubt, affect its future. Our bodies, brains, and biochemistry remain subject to the same processes of variation and natural selection as those that influenced our ancestors; yet, our capacity to intentionally alter some aspects of the environment in which we live—delivering water, heat, and electricity to a home, for example—permits a way of life unknown to other organisms and relaxes certain constraints of natural selection.

As we have seen throughout this book, our unique relationship to the environment presents us with both problems and opportunities. Our human population has grown exponentially, and with it, our impact on the earth's land, air, and water. Just to feed our still expanding populations over the next 30 years, we will need to produce as much food as has been produced in all of human history. This will require a second agricultural revolution, a green revolution based on genetic engineering techniques (Figure 46-19). We will need to develop:

Crops that can withstand extremes of temperature and moisture and that can thrive in an expanded range of habitats
Crops that are resistant to insects, fungi, and nematodes, allowing yields per acre to increase and the need for applying pesticides to decrease
Crops that fix their own nitrogen, so they can thrive on less fertilizer
Crops with heightened nutritive value, so we can de-

Figure 46-19 GENETIC ENGINEERING OF CROP PLANTS: A GLOWING FUTURE.
Genetic engineers transferred a gene for the enzyme luciferase from fireflies to tobacco cells, using methods described in Chapter 14, and then grew plants from the hybrid cells. The adult tobacco plants glow in the dark when watered with a solution containing the enzyme's substrate, luciferin, proving that an animal protein can be both manufactured and made to function in plant cells.

crease our reliance on fish, beef, poultry, and other animal foods

Crops that are living factories capable of harboring the genes for, and producing, insulin, interferon, or other animal proteins of critical importance in medicine

Once available, such high-yield genetically engineered plants will still require proper soils, sufficient water; beneficial weather patterns, and other benign environmental conditions, as well as the availability of energy. As we saw in Chapter 41, rising CO_2 levels and the greenhouse effect could well raise global temperatures and change rainfall patterns in ways that disrupt the very same temperate-zone farm belts we will depend on so heavily to feed our people.

Simultaneously, other human activities are causing the world's present agricultural lands to deteriorate from erosion, loss of nutrients and soil compaction, rise in salt content, pollution, and desertification—the transformation of farmland into desert. To add to these problems, in both developed and less-developed countries, people and their roads, parking lots, tract houses, cities, and factories are covering arable land at an increasing rate.

In addition to addressing these problems, we will also need more energy-intensive technology to maintain high agricultural productivity. We will need to produce four times more fertilizer, pesticides, and herbicides than we now manufacture, and yet energy costs are rising and supplies of fossil fuels (coal and oil) are finite.

The overall picture looks grim—yet the stakes are nothing short of human survival and the maintenance of a life-sustaining biosphere for ourselves and the world's other species. We argued in Chapter 1 that biology is one of the liberal arts, worth studying for its own sake.

(a)

(b)

Figure 46-20 ELECTRICITY FROM SUN AND WIND.
(a) "Solar One," one of the first commercial solar power plants, is an array of solar collectors that transduce the "free" energy of the sun into the electrical power for civilization. (b) Huge blades await the winds of the California desert to power generators and send electricity to western cities and agriculture. Solar power plants and wind generators are built where there is sun and wind, respectively, most days of the year.

Beyond that very real value, however, the problems facing humankind demand biological solutions. Without question, if we are to live in a tolerable world, the new biology will need to assert itself and insert itself farther and farther into the fabric of commerce and society. And we will require rapid new developments in energy production, especially alternative forms, such as solar energy, wind energy, fusion, fission, and hydroelectric generation (Figure 46-20).

It is indeed good fortune that the major new steps in human cultural evolution—the development of the computer and the ability to manipulate and control genes—have come at just the time when our species' reproductive capacity is straining the planet's ecosystems so severely. But while scientists and technologists can produce such great advances, informed world leaders and citizens must *exploit* the advances to help the earth, with its more than 5 billion human inhabitants, to continue on as the marvelous spaceship it has been for the past 4.5 billion years. Science and technology can provide possibilities, but sociopolitical processes within and among nations will dictate what actually occurs. Preservation and use of land for food production; protection of water quality and water availability; and limitation of air pollutants are concerns that transcend traditional national boundaries.

The great problems confronting humankind will be overcome, however, only if educated people work together and make the personal commitment to invest lives and careers in the areas of greatest global need (Figure 46-21). The biological and cultural evolution of *Homo sapiens* has reached new heights with the powerful tools of the computer and genetic engineering. Now it is time to apply these and other tools in an effort to forge a new and different way of living that will protect the huge human populations of the twenty-first century from abject hunger, poverty, and disease. As E. Barton Worthington wrote in 1983: "The eighteenth century was marked by enlightenment, and the nineteenth by industry. The twentieth may go down in history as the ecological century. . . . What the twenty-first century may be labelled—progress or chaos—will depend on how well mankind learns to ensure a proper balance between himself and his environment."*

Figure 46-21 TRANSFERRING NEW CROPS TO THE FIELD: THE HARDEST PART OF THE NEW AGRICULTURE. Here, scientists and a farmer in Bangladesh confer about a new rice strain. Molecular agriculture will only realize its potential if this last step is successful.

*E. Barton Worthington, *The Ecological Century: A Personal Appraisal.* New York: Oxford University Press, 1983.

SUMMARY

1. The *primates* include the *prosimians* and the *anthropoids: New World monkeys, Old World monkeys, apes*, and the human lineage.

2. The earliest primates appeared more than 65 million years ago. The earliest anthropoids appeared around 30 million years ago.

3. Primates are characterized by adaptations to an arboreal way of life. These adaptations include arms and legs adapted to locomotion through the trees, hands and feet adapted for grasping branches, and stereoscopic vision.

4. Humans and the African apes, the chimpanzee and gorilla, share a common ancestor that lived after the time of the evolutionary split with the ancestors of the Asian orangutan.

5. The earliest members of the family *Hominidae*—humans and their direct ancestors—were of the genus *Australopithecus*, including *A. afarensis, A. africanus, A. boisei*, and *A. robustus*.

6. *Homo habilis* appeared about 2 million years ago and was characterized by a significant increase in brain size, habitual tool use, and butchering of animal carcasses.

7. The first spread of humans outside the African source land began about 1.5 million years ago with *Homo erectus*. During the course of its existence, *H. erectus* experienced a gradual decrease in face and tooth size and a spectacular increase in brain size.

8. The past 300,000 years of hominid evolution have been occupied by two groups of humans: archaic *Homo sapiens* (including *Neandertals*) and modern humans. They used more sophisticated stone tools (the *Mousterian industry*) than did earlier cultures, built shelters, cared for the aged and infirm, and buried their dead.

9. The humans who evolved from archaic *Homo sapiens* appeared by about 75,000 years ago in Africa, and they rapidly spread across the inhabitable earth.

10. The wide geographical spread of

recent humans led to the agricultural revolution, as well as the evolution of human racial differences.

11. We face severe challenges in the future, including overpopulation, pollution, alteration of the world's weather patterns, and the biological sciences will play a crucial role in addressing those problems.

KEY TERMS

Acheulian industry
anthropoid
ape
Australopithecus
brachiation
Dryopithecus
Homo erectus
Homo habilis
Homo sapiens
knuckle walking
Mousterian industry
Neandertal
New World monkey
Old World monkey
Oldowan industry
primate
prosimian

QUESTIONS

1. No field of modern biology is more beset by controversy than primate paleontology. Suggest some reasons why.

2. How could life in the trees have contributed to the anatomy of monkeys, apes, and humans?

3. What evidence of bipedalism can be found in the fossil lineage leading to modern *Homo sapiens?*

4. Trace the dramatic expansion of the brain in the human ancestral lineage.

5. What is the significance of the different tool industries of the human lineages?

6. What are the possible relationships of Neandertals to modern *Homo sapiens?*

7. What does the earliest art imply about the early humans?

8. Does evidence of molecular relatedness between primates support deductions based on the fossil record? Explain.

For additional readings related to topics in this chapter, see Appendix C.

Appendix A
The Classification of Organisms

Organisms are classified by taxonomic rules originated in part by Linnaeus and modified each year at international meetings of taxonomists. The following list follows, in general, R. H. Whittaker's five-kingdom scheme (*Science* 163 [1969]: 150–160), but with the addition of the kingdom Archaebacteria. The text discussion explains our reasons for placing groups in specific kingdoms in cases where there is lack of agreement on such assignments (e.g., placement of the slime molds).

KINGDOM MONERA

Prokaryotic, mainly single-celled bacteria and cyanobacteria; reproduce by asexual fission, but also have genetic recombination processes.

Subkingdom Schizomycete: Bacteria; unicellular or chains; most are heterotrophs; some with a solid molecular flagellum that rotates; include mycoplasmae, the smallest living cells.

Subkingdom Cyanobacteria: Photosynthetic bacteria; chlorophyll on membranes; some also fix nitrogen; single cells and colonies; ancient oxygen liberators.

Subkingdom Chloroxybacteria (Prochlorophyta): Photosynthetic marine bacteria; may be source of chloroplasts.

KINGDOM ARCHAEBACTERIA

Prokaryotes with unique nucleic acid and biochemical properties.

Phylum Methanobacteria: Methane-producing archaebacteria; obligatory anaerobes.

Phylum Sulfolobales: Sulfur-reducing archaebacteria; some also tolerate very high temperatures (called thermoproteales).

KINGDOM PROTISTA*

Eukaryotic, single-celled organisms (some colonies); heterotrophs and autotrophs; reproduce sexually; protozoa and some algae included; flagella, cilia of 9+2 microtubule type.

Plantlike Autotrophs

Phylum Euglenophyta: Euglenoid, usually photosynthetic organisms with chloroplasts; most unicellular, but colonies, too.

Phylum Pyrrophyta: Dinoflagellates; photosynthetic phytoplankton, usually single cells with flagella.

Phylum Chrysophyta: Golden-brown and yellow-green algae; photosynthetic single cells and colonies; include diatoms that contribute to sedimentary rocks.

Funguslike Heterotrophs

Phylum Gymnomycota: Slime molds; phagocytic; sexual and asexual reproduction; stages with single cells, cell aggregates, and spore-forming structures; sometimes classified as fungi.

CLASS MYXOMYCETE: Acellular slime molds; plasmodium multinucleate; also single ameoboid cells.

CLASS ACRASIOMYCETE: Cellular slime molds; plasmodium composed of separate cells. Also single amoeboid cells.

Animal-Like Heterotrophs

Group Mastigophora: Flagellated protozoa; usually reproduce asexually; some parasitic (e.g., *Trypanosoma*, cause of African sleeping sickness).

Group Sarcodina: Amoebae; single cells; creeping movement using pseudopods; many marine; foraminiferans and radiolarians contribute to sedimentary rocks.

Group Sporozoa: Single-celled, nonmotile cellular organisms that parasitize animals (e.g., *Plasmodium*, cause of malaria); sporelike stage in life cycle.

Phylum Ciliphora: Unicellular organisms with many cilia; carry out conjugation.

Phylum Caryoblastea: Amoeboid cellular organism lacking mitochondria; may represent early stage in eukaryotic evolution.

KINGDOM FUNGI

Filamentous heterotrophic organisms with many nuclei in mycelium; nuclei sometimes separated by septae; the

*Some taxonomists place the Chlorophyta, Rhodophyta, and Phaeophyta in this kingdom.

slime molds, Gymnomycota, are sometimes classified as fungi.

Division Mycota (or Eumycota):

CLASS CHYTRIDIOMYCETE: Chytrids; single-celled or small mycelia; gametes and spores with flagella; parasites or saprobes; thought to be most ancient mycota.

CLASS OOMYCETE: Water molds; uni- or multicellular organisms that may be terrestrial or aquatic; asexual spores with flagella; produce large oocytes; parasites or saprobes.

CLASS ZYGOMYCETE: Bread molds; consume decaying organic materials; chitin in cell walls; form mycorrhizae with plants.

CLASS ASCOMYCETE: Cup or sac fungi, molds, yeasts, and relatives; form asci during sexual reproduction; most common fungi in lichens.

CLASS BASIDIOMYCETE: Common mushrooms, shelf fungi, rusts, and relatives; conidial spores; carry out sexual reproduction; form mycorrhizae with plants.

CLASS DEUTEROMYCETE: Fungi imperfecti; sexual reproductive structures not known; many are probably ascomycetes.

KINGDOM PLANTAE/ALGAE AND LOWER SEEDLESS PLANTS

Photosynthetic, eukaryotic, and usually multicellular organisms; some vascular.

***Division Chlorophyta:** Green algae; both multi- and unicellular photosynthetic and usually aquatic organisms; some with flagella.

***Division Rhodophyta:** Red algae; mostly multicellular and marine organisms that lack flagella; photosynthetic; unique kind of storage starch; coralline algae have calcareous deposits.

***Division Phaeophyta:** Brown algae; multicellular photosynthesizers that are mostly marine; some very large; may have vascular tissues; food stored as laminarin.

Division Bryophyta: Liverworts, hornworts, and mosses; largely nonvascular; alternation of sexual (gametophyte) and asexual (sporophyte) generations; sporophytes derive food from gametophytes.

CLASS HEPATOPHYTA (OR HEPATICOPSIDA): Liverworts; terrestrial with photosynthetic haploid gametophyte as dominant stage.

CLASS ANTHOCEROTOPHYTA (OR ANTHOCERATOPSIDA): Hornworts; terrestrial with photosynthetic haploid gametophyte that is leafy.

CLASS MUSCI (OR MUSCOPSIDA): Mosses; contain some vascular tissue; photosynthetic haploid gametophyte as dominant stage of terrestrial and semiaquatic plants.

Division Psilophyta: Whisk ferns; vascular phloem and xylem tissue; motile sperm must swim in water.

Division Lycophyta: Club mosses, quill worts; vascular plants with leaves, stems, roots; motile sperm move in water.

Division Sphenophyta: Horsetails; vascular plants with jointed stems and nonphotosynthetic leaves; tiny gametophytes with sperm that travel in water.

Division Pteridophyta: Ferns; vascular sporophyte dominant stage with feathery fronds; small gametophyte with sperm that travel in water.

KINGDOM PLANTAE/HIGHER SEED PLANTS

Photosynthetic, vascular, nonmotile, with cellulose cell walls; alternation of generations between haploid and diploid stages of life cycle.

Gymnosperms: Seed plants without flowers; naked seeds (no fruits).

Division Cycadophyta: Cycads; palmlike; individuals with either seed- or pollen-producing cones (dioecious); flagellated sperm move through pollen tube.

Division Ginkgophyta: Ginkgoes; large pollen- and seed-producing trees (dioecious); sperm swim in pollen tube.

Division Gnetophyta: Gnetinas; Mormon tea; vines, shrubs; nonmotile sperm.

Division Coniferophyta: Conifers; leaves reduced to "needles"; usually monoecious, with both pollen and seed cones on single trees; gametophyte stage greatly reduced to cells in cone; no fruits.

Angiosperms: Flowering plants with seeds enclosed in fruit; nonmotile sperm move through pollen tube; double fertilization; unique endosperm nourishes embryo.

Division Anthophyta: Flowering plants; monocotyledon or dicotyledon seed leaves; annuals, biannuals, perennials; herbaceous or woody plants; gametophyte stage only a few cells.

KINGDOM ANIMALIA

Animals; eukaryotic multicellular heterotrophs; sexual reproduction with meiosis; some reproduce asexually; many motile, but some stationary.

*Some taxonomists consider green, red, and brown algae to be protists, not true plants.

Phylum Porifera (sometimes subkingdom Parazoa): Sponges; colonial or solitary, nonmotile, aquatic; filter feeders; phagocytosis.

Phylum Placozoa (sometimes in subkingdom Parazoa): *Trichoplax;* multicellular but lacks tissues; ciliated; changes shape like amoeba; may be early stage of multicellularity.

Subkingdom Eumetazoa: Animals with radial or bilateral symmetry and true tissues.

Phylum Cnidaria: Hydroids, jellyfish (medusae), corals, anemones; marine, are radially symmetrical, with two tissue layers; cnidoblast stinging cells; extracellular digestion.

CLASS HYDROZOA: Hydras, Portuguese men-of-war; alternation of generations in life cycle; may lack polyp stage or may have only short medusa stage.

CLASS SCYPHOZOA: Jellyfish; mostly medusa stage in life cycle.

CLASS ANTHOZOA: Sea anemones, corals; no medusa stage in life cycle.

Phylum Ctenophora: Comb jellies; marine, radially symmetrical with eight comb rows.

Phylum Platyhelminthes: Flatworms; terrestrial, marine, and freshwater; flat dorsoventrally; some three tissue layers; one opening for digestive system; acoelomate; asexual and sexual reproduction.

CLASS TURBELLARIA: Planarians; bilaterally symmetrical; organ systems; asexual fission of body, plus hermaphroditic sexual reproduction.

CLASS TREMATODA: Flukes; parasites, with two or more animal hosts; have digestive tract.

CLASS CESTODA: Tapeworms; parasites that absorb nutrients and lack digestive tract; complex life cycles.

Phylum Nemertina (Rhynchocoela): Ribbon worms; one-way digestive tract with two openings; marine; lack coelom; protrusible proboscis; circulatory system with blood.

Phylum Nematoda: Roundworms; unsegmented, pseudocoelomate worms with complete gut and longitudinal muscles; ubiquitous soil organisms that parasitize many plants and animals.

Phylum Rotifera: Rotifers; anterior end with ciliated crown; posterior end a foot. Microscopic; freshwater, some marine, some inhabitants of mosses.

Phylum Annelida: Segmented worms; protostomes with coelom, complete gut, closed circulatory system, and septae between body segments.

CLASS OLIGOCHAETA: Earthworms; terrestrial and freshwater aquatic worms; lack distinct head or external sense organs.

CLASS POLYCHAETA: Marine worms; free-swimming; parapodia used in swimming and respiration; well-developed head with sense organs and mandibles.

CLASS HIRUDINEA: Leeches; flattened with reduced segmentation; external parasites or scavengers with suckers.

Phylum Pogonophora: Pogonophorans; deep-water marine animals; long body within chitinous tube; lack digestive tract.

Phylum Tardigrada: Water bears; microscopic segmented animals with four pairs of stubby legs; freshwater and terrestrial in lichens and mosses; a few marine.

Phylum Arthropoda: Segmented animals with exoskeleton of chitin and paired, jointed appendages; protostomes; with coelom, complete gut, open circulatory system, and dorsal brain; phylum with largest number of species.

SUBPHYLUM TRILOBITA: Trilobites; extinct, highly diverse group of segmented marine arthropods with well-developed compound eyes.

SUBPHYLUM CHELICERATA: Spiders, ticks, horseshoe crabs, sea spiders; six pairs of appendages, including chelicerae (fanglike biting appendages); two main body regions.

SUBPHYLUM CRUSTACEA: Crabs, shrimps, barnacles, water fleas; aquatic; with two-branched (biramous) appendages; often three pairs of legs; legs on both thorax and abdomen; mandibles.

SUBPHYLUM UNIRAMIA: Centipedes, millipedes, and insects; terrestrial and aquatic with single-branched appendages; up to 200 body segments.

CLASS INSECTA: Largest number of species of any group of organisms; includes bees, ants, termites, dragonflies; 900,000 species described so far; three body segments; three pairs of walking legs; many have wings; varied mouthparts; compound eyes.

Phylum Onychophora: Peripatus; wormlike; segmented; appendages with claws; tracheae for respiration; characters of both annelids and uniramia.

Phylum Mollusca: Mollusks; protostomes, usually with open circulatory system; muscular foot and visceral mass common; often with external shell; both coelom and segmentation greatly reduced.

CLASS MONOPLACOPHORA: Neopilina; coelomate, possibly segmented; characters of both annelids and mollusks.

CLASS POLYPLACOPHORA (AMPHINEURA): Chitons; eight skeletal plates embedded in dorsal body; long foot; many gills; simple body plan; marine.

CLASS GASTROPODA: Snails, abalone, sea slugs, land slugs; large muscular foot; radula for feeding; often with coiled shell; marine, freshwater, and terrestrial.

CLASS BIVALVIA: Oysters, clams, scallops; two hinged shells; filter feeders (no radula); reduced head.

CLASS CEPHALOPODA: Octopuses, squid; marine predators; enlarged head that includes tentacles; large, complex brains and eyes; complex behavior; closed circulatory system and skeleton that may be outside, inside, or absent.

Phylum Echinodermata: Echinoderms; coelomate deuterostomes; fivefold radial symmetry; water vascular system and tube feet; calcareous plates and sometimes spines; marine.

CLASS ECHINOIDEA: Sea urchins and sand dollars; rigid skeleton; spines; tube feet.

CLASS HOLOTHUROIDEA: Sea cucumbers; long cylindrical body; lack tube feet and spines.

CLASS CRINOIDEA: Sea lilies; cup-shaped body with feathery arms used in filter feeding; mouth and anus on upper disk.

CLASS ASTEROIDEA: Sea stars; five or more arms; tube feet; mouth and arms on ventral surface.

CLASS CONCENTRICYCLOIDEA: Sea daisies; deep-marine-water inhabitants in rotting wood; tiny, disk-shaped; spines at edge like daisy petals; unique two rings of tube feet; no gut; embryos develop within gonads; discovered 1986.

CLASS OPHIUROIDEA: Brittle stars; long, spiny arms that are very flexible horizontally.

Phylum Hemichordata: Acorn worms and pterobranchs; coelomate deuterostomes; gill slits and dorsal nerve cord; soft bodied; larvae like those of echinoderms; marine.

Phylum Chordata: Deuterostome vertebrates and relatives; coelomate; gill slits; hollow dorsal nerve cord; notochord endoskeleton; postanal tail; segmental muscles for locomotion.

SUBPHYLUM UROCHORDATA: Sea squirts, tunicates; ciliary filter feeders; gill slits; open circulatory system; tadpole larvae with tail, notochord, and dorsal nerve cord; adults usually sessile; marine.

SUBPHYLUM CEPHALOCHORDATA: Lancelet (amphioxus); fishlike body; ciliary filter feeder; segmental body muscles; marine.

SUBPHYLUM VERTEBRATA: Adults with notochord (ancient fishes, amphibians) or vertebral column; prominent head with major sense organs; closed circulatory system; many with two pairs of appendages.

CLASS AGNATHA: Ostracoderms, lampreys, hagfish; jawless; ancient forms with bone; contemporary forms with cartilage endoskeletons; parasites or scavengers; mostly marine.

CLASS ACANTHODII: First jawed vertebrates; paired fins; marine and later freshwater; extinct.

CLASS CHONDRICHTHYES: Sharks, skates; jawed, cartilaginous fishes; urea osmoregulators; oils for buoyancy; cartilage endoskeleton; pectoral fins for maneuvering or locomotion; mostly marine; internal fertilization and lengthy development.

CLASS OSTEICHTHYES: Jawed bony fishes; bony endoskeleton.

SUBCLASS ACTINOPTERYGII: Ray-finned teleost fishes such as trout; swim bladder for buoyancy; many eggs, embryos.

SUBCLASS SARCOPTERYGII: Lungfish (Dipnoi); coelacanths; rhipidistians, the source of amphibians; most with lungs; few eggs, embryos.

CLASS AMPHIBIA: Frogs (Anura), salamanders (Urodela), and wormlike caecilians (Apoda); jawed tetrapods; bony endoskeleton; lungs plus moist skin for respiration; external fertilization with aquatic embryos and larvae; metamorphosis to adult terrestrial form; terrestrial and aquatic.

CLASS REPTILIA: Turtles, snakes, lizards, crocodiles; many extinct aquatic, aerial, and terrestrial types, including dinosaurs; jawed tetrapods with dry skin; bony endoskeleton; lungs for respiration; cleidoic amniotic egg; uric acid osmoregulators; terrestrial, but some marine or freshwater.

CLASS AVES: Birds; jawed tetrapods; feathers; light bony endoskeleton; wings for flight; cleidoic eggs; endothermic homeotherms; uric acid osmoregulators; high-pressure circulatory system.

CLASS MAMMALIA: Mammals, including monotremes (Prototheria), marsupials (Metatheria), and placentals (Eutheria); latter group includes 26 orders, including primates; jawed tetrapods; bony endoskeleton; hair; endothermic homeotherms; mammary glands to feed newborn; eggs laid by monotremes; others nourish embryo in uterus, usually with placenta; large brains and complex behavior.

Appendix B
Some Useful Chemical Measurements

All precise science rests on quantitation and numbers, for mathematics is a language common to all science. The chemist analyzing hereditary material, the biologist measuring water and salt loss in a sweating desert rodent, or the physician interpreting the action of a drug on a patient with kidney disease, all must be concerned with the amounts of substances and the concentrations of solutions. This appendix summarizes the key quantitative measures for such scientists.

DALTON

The dalton is a special unit of measurement for the weights of atoms and atomic particles. These structures are so tiny that expressing their weights in minute fractions of grams is impractical. The atomic weight of the most common isotope of carbon is arbitrarily set at exactly 12 atomic mass units, or 12 daltons. One *dalton* is, therefore, one-twelfth the mass of an atom of ^{12}C.

ATOMIC WEIGHT

Each element occurs not in one form and at one weight, but in the form of isotopes of different weights. Therefore, the standard weight established for atoms of an element is an average figure, called the *atomic weight*. For example, a typical sample of chlorine is 75.4 percent $^{35}_{17}Cl$ and 24.6 percent $^{37}_{17}Cl$. Based on the weights of these isotopes and their percentages in a typical sample, the atomic weight of Cl is calculated to be 35.45.

MOLE

A *mole* is Avogadro's number (6.023×10^{23}) of particles of a substance, or 1 gram molecular weight of the substance. A gram molecular weight is the weight in grams equal to the atomic weight of each atom in a molecule multiplied by the number of those atoms. Thus the molecular weight of methane, CH_4, is

$$\underbrace{4(1.01)}_{\text{atomic weight of H}} + \underbrace{1(12.01)}_{\text{atomic weight of C}} = 16.05 \text{ g}$$

Each mole of methane, 16.05 g, contains 6.023×10^{23} molecules.

MOLARITY

Biologists often are interested in the amount of a substance per unit of a solution, such as how much salt is dissolved in the blood. A 1 molar, or $1M$, solution is 1 mole of solute in 1 liter of solution. A $1M$ solution of NaCl is, therefore, 1 gram molecular weight of NaCl in 1 liter of solution:

$$1M = \underbrace{23.0}_{\text{atomic weight of Na}} + \underbrace{35.5}_{\text{atomic weight of Cl}} = 58.5 \text{ g NaCl/liter solution}$$

A $2M$ solution of NaCl is $2(58.5 \text{ g}) = 117.0$ g NaCl/liter solution.

A $0.5M$ solution of NaCl is $0.5(58.5 \text{ g}) = 29.25$ g NaCl/liter solution.

Appendix C
Suggested Readings

CHAPTER 1

DARWIN, C. *On the Origin of Species: A Facsimile of the First Edition.* Cambridge, Mass.: Harvard University Press, 1975.

GARDNER, E. J. *History of Biology.* 3d ed. Minneapolis: Burgess, 1972.

HARRE, R. *Great Scientific Experiments.* New York: Oxford University Press, 1983.

KARP, G. *Cell Biology.* 2d ed. New York: McGraw-Hill, 1984.

MAYER, W. V. "Wallace and Darwin." *The American Biology Teacher,* Nov./Dec. 1987, pp. 406–410.

WALKER, W. "Study Sees Big Leap in Science Jobs by 2000." *The Scientist,* June 1988, p. 21.

CHAPTER 2

BRODY, J. E. "Natural Chemicals Now Called Major Force of Destruction." *New York Times,* April 26, 1988, p. 7, section B.

HENDERSON, L. S. *The Fitness of the Environment.* Boston: Beacon Press, 1958.

HILL, J. W., and D. M. FEIGL. *Chemistry and Life: An Introduction to General, Organic, and Biological Chemistry.* New York: Macmillan, 1987.

MARX, J. L. "Oxygen Free Radicals Linked to Many Diseases." *Science* 235 (1987): 529–531.

RADDA, G. K. "The Use of NMR Spectroscopy for the Understanding of Disease." *Science* 233 (1986): 640–645.

CHAPTER 3

CALVIN, M., and W. A. PRYOR. *Organic Chemistry of Life: Readings from "Scientific American."* San Francisco: Freeman, 1973.

DICKERSON, R. E., and I. GEIS. *The Structure and Action of Proteins.* New York: Harper & Row, 1970.

HILL, J. W., and D. M. FEIGL. *Chemistry and Life: An Introduction to General, Organic, and Biological Chemistry.* 3d ed. New York: Macmillan, 1987.

KARPLUS, M., and J. A. MCCAMMON. "The Dynamics of Proteins." *Scientific American,* April 1986, pp. 42–51.

REGAN, L., and W. F. DEGRADO. "Characterization of a Helical Protein Designed from First Principles." *Science* 241 (1988): 976–978.

STRYER, L. *Biochemistry.* 3d ed. San Francisco: Freeman, 1988.

CHAPTER 4

ALBERTS, B., et al. *Molecular Biology of the Cell.* 2d ed. New York: Garland, 1989.

BECKER, W. M. *Energy and the Living Cell.* Philadelphia: Lippincott, 1977.

BLUM, H. F. *Time's Arrow and Evolution.* New York: Harper & Row, 1962.

CHRISTENSEN, H. H., and R. A. CELLARUS. *Introduction to Bioenergetics: Thermodynamics for the Biologist.* San Francisco: Freeman, 1972.

KRAUT, J. "How Do Enzymes Work?" *Science* 242 (1988): 533–540.

SRIVASTAVA, D. K., and S. A. BERNHARD. "Metabolite Transfer via Enzyme-Enzyme Complexes." *Science* 234 (1986): 1081–1086.

STRYER, L. *Biochemistry.* 3d ed. San Francisco: Freeman, 1988.

CHAPTER 5

ALBERTS, B. et al. *Molecular Biology of the Cell.* 2d ed. New York: Garland, 1989.

BOATMAN, E. S., et al. "Today's Microscopy." *Bioscience* 37 (1987): 384.

CUTTER, E. G. *Plant Anatomy: Experiment and Interpretation.* Reading, Mass.: Addison-Wesley, 1969.

FAWCETT, D. *The Cell.* Philadelphia: Saunders, 1981.

LEWIN, R. "New Horizons for Light Microscopy." *Science* 230 (1985): 1258.

SINGER, S. J., and G. NICOLSON. "The Fluid-Mosaic Model of the Structure of Cell Membranes." *Science* 175 (1972): 720–731.

VERNER, K., and G. SCHATZ. "Protein Translocation Across Membranes." *Science* 241 (1988): 1307–1313.

CHAPTER 6

ALBERTS, B., et al. *Molecular Biology of the Cell.* 2d ed. New York: Garland, 1989.

ALLEN, R. D. "The Microtubule Is an Intracellular Engine." *Scientific American,* February 1987, pp. 42–49.

BRETSCHER, M. S. "How Animal Cells Move." *Scientific American,* December 1987, pp. 72–80.

KARP, G. *Cell Biology.* 2d ed. New York: McGraw-Hill, 1984.

LANE, M. D., et. al. "The Mitochondrion Updated." *Science* 234 (1986): 526–527.

ROTHMAN, J. E. "The Compartmental Organization of the Golgi Apparatus." *Scientific American,* September 1985, pp. 74–89.

CHAPTER 7

ALBERTS, B., et al. *Molecular Biology of the Cell.* 2d ed. New York: Garland, 1989.

HINKLE, P. C., and R. E. MCCARTY. "How Cells Make ATP." *Scientific American,* March 1978, pp. 104–123.

KREBS, H. A. "The History of the Tricarboxylic Acid Cycle." *Perspectives in Biology and Medicine* 14 (1970): 154–170.

LANE, M. D., P. L. PEDERSEN, and A. S. MILDVAN. "The Mitochondrion Updated." *Science* 234 (1986): 526–527.

MARX, J. L. "Ubiquitin Move, to the Cell Surface." *Science* 231 (1986): 796–797.

RACKER, E. "From Pasteur to Mitchell, a Hundred Years of Bioenergetics." *Federation Proceedings* 39 (1980): 210–215.

SRIVASTAVA, D. K., and S. A. BERNHARD. "Metabolite Transfer via Enzyme-Enzyme Complexes." *Science* 234 (1986): 1081–1086.

CHAPTER 8

BARBER, J. "Signals from the Reaction Center." *Nature* 332 (1988): 111–112.

BIDWELL, R. G. S. *Plant Physiology.* 2d ed. New York: Macmillan, 1979.

BOGORAD, L. "Chloroplasts." *Journal of Cell Biology* 91 (1981): 256s–270s.

BONNER, J., and J. E. VARNER. *Plant Biochemistry.* 3d ed. New York: Academic Press, 1976.

LEWIN, R. "Membrane Protein Holds Photosynthetic Secrets." *Science* 242 (1988): 672–673.

MILLER, K. R. "The Photosynthetic Membrane." *Scientific American,* October 1979, pp. 102–113.

YOUVAN, D. C., and B. L. MARRS. "Molecular Mechanisms of Photosynthesis." *Scientific American,* June 1987, pp. 42–48.

CHAPTER 9

ALBERTS, B., et al. *Molecular Biology of the Cell.* 2d ed. New York: Garland, 1989.

BOY DE LA TOUR, E., and U. K. LAEMMLI. "The Metaphase Scaffold Is Helically Folded." *Cell* 55 (1988): 937–944.

GORBSKY, G. J., PAUL J. SAMMAK, and G. G. BORISY. "Chromosomes Move Poleward in Anaphase Along Stationary Microtubules That Coordinately Disassemble from Their Kinetochore Ends." *Journal of Cell Biology* 104 (1987): 9–18.

JOHN, B., and K. R. LEWIS. *The Meiotic Mechanism.* New York: Oxford Biology Readers, 1976.

JOHN, P. C. L. *The Cell Cycle.* New York: Cambridge University Press, 1981.

KOSHLAND, D. E., et al. "Polewards Chromosome Movement Driven by Microtubule Depolymerization in Vitro." *Nature* 331 (1988): 449–504.

PRESCOTT, D. M. *Reproduction of Eukaryotic Cells.* New York: Academic Press, 1976.

ROBERTS, L. "Chromosomes: The Ends in View." *Science* 240 (1988): 982–983.

WEINERT, T. A., and L. H. HARTWELL. "The RAD9 Gene Controls the Cell Cycle Response to DNA Damage in *Saccharomyces cerevisiae.*" *Science* 241 (1988): 317–322.

WHITEHOUSE, H. L. *Towards an Understanding of the Mechanism of Heredity.* 3d ed. New York: St. Martin's Press, 1973.

ZIMMERMAN, A. M., and A. FORER, eds. *Mitosis/Cytokinesis.* New York: Academic Press, 1981.

CHAPTER 10

AYALA, F. J., and J. A. KIGER, JR. *Modern Genetics.* 2d ed. Redwood City, Calif.: Benjamin/Cummings, 1984.

GAFFNEY, B., and S. P. CUNNINGHAM. "Estimation of Genetic Trend in Racing Performance of Thoroughbred Horses." *Nature* 332 (1988): 722–724.

HARRISON, D. *Problems in Genetics with Notes and Examples.* Reading, Mass.: Addison-Wesley, 1970.

MENDEL, G. *Experiments in Plant-Hybridisation.* Cambridge, Mass.: Harvard University Press, 1965.

SRB, A. S., K. D. OWEN, and R. S. EDGAR. *General Genetics.* 2d ed. San Francisco: Freeman, 1965.

CHAPTER 11

AYALA, F. J., and J. A. KIGER, JR. *Modern Genetics.* 2d ed. Redwood City, Calif.: Benjamin/Cummings, 1984.

BOWER, B. "Schizophrenia: Genetic Clues and Caveats." *Science News,* Nov. 1988, p. 308.

ROBINSON, R. *Genetics for Cat Breeders.* New York: Pergamon Press, 1977.

STRICKBERGER, M. W. *Genetics.* 2d ed. New York: Macmillan, 1976.

STURTEVANT, A. H. *A History of Genetics.* New York: Harper & Row, 1965.

CHAPTER 12

AYALA, F. J., and J. A. KIGER, JR. *Modern Genetics.* 2d ed. Redwood City, Calif.: Benjamin/Cummings, 1984.

JUDSON, H. F. *The Eighth Day of Creation: The Makers of the Revolution in Biology.* New York: Simon & Schuster, 1979.

SCHMECK, H. M. "DNA Pioneer to Tackle Biggest Gene Project Ever." *New York Times,* Oct. 4, 1988, B1, B8.

STRYER, L. *Biochemistry.* Salt Lake City: Freeman, 1988.

WATSON, J. D. *The Double Helix.* New York: Atheneum, 1968.

WATSON, J. D., and F. H. C. CRICK. "Molecular Structure of Nucleic Acids: A Structure for Deoxyribose Nucleic Acid." *Nature* 171 (1953): 737–738.

CHAPTER 13

CASKEY, C. T. "Peptide Chain Termination." *Trends in Biochemical Science* 5 (1980): 234–237.

CLARK, B. "The Elongation Step of Protein Biosynthesis." *Trends in Biochemical Science* 5 (1980): 207–210.

CRICK, F. H. C. "The Genetic Code." *Scientific American,* October 1962, pp. 66–74.

______. "The Genetic Code III." *Scientific American,* October 1966, pp. 55–62.

HUNT, T. "The Initiation of Protein Synthesis." *Trends in Biochemical Science* 5 (1980): 178–188.

NIRENBERG, M. W. "The Genetic Code II." *Scientific American,* March 1963, pp. 80–94.

SCHULMAN, L. H., and J. ABELSON. "Recent Excitement in Understanding Transfer RNA Identity." *Science* 240 (1988): 1591–1592.

STRYER, L. *Biochemistry.* 3d ed. New York: Freeman, 1981.

YANOFSKY, C. "Gene Structure and Protein Structure." *Scientific American,* May 1967, pp. 80–90.

CHAPTER 14

ALBERTS, B., et al. *Molecular Biology of the Cell.* New York: Garland, 1989.

AYALA, F. J., and J. A. KIGER, JR. *Modern Genetics.* 2d ed. Redwood City, Calif.: Benjamin/Cummings, 1984.

CHAMBON, P. "Split Genes." *Scientific American,* May 1981, pp. 60–71.

COCKING, E. C., and M. R. DAVEY. "Gene Transfer in Cereals." *Science* 236 (1987): 1252–1262.

FEDOROFF, N. V. "Transposable Elements in Maize." *Scientific American,* June 1984, pp. 84–98.

GLOVER, D. M. *Genetic Engineering.* New York: Methuen, 1980.

JACOB, F., and J. MONOD. "Genetic Regulatory Mechanisms in the Synthesis of Proteins." *Journal of Molecular Biology* 33 (1961): 318.

LEWIS, R. "Fish: New Focus for Biotechnology." *Bioscience* 38 (1988): 225–227.

NOVICK, R. "Plasmids." *Scientific American,* December 1980, pp. 102–127.

RUBY, S. W., and J. ABELSON. "An Early Hierarchical Role of U1 Small Nuclear Ribonuclearprotein in Spliceosome Assembly." *Science* 242 (1988): 1028–1035.

CHAPTER 15

DIAMOND, J., and J. I. ROTTER. "Observing the Founder Effect in Human Evolution." *Nature* 329 (1987): 105–106.

EPSTEIN, C., and M. GOLBUS. "Prenatal Diagnosis of Genetic Diseases." *American Scientist* 65 (1977): 703–711.

HARTL, D. *Our Uncertain Heritage: Genetics and Human Diversity.* 2d ed. New York: Harper & Row, 1984.

HOLDEN, C. "The Genetics of Personality." *Science* 237 (1987): 598–601.

NATHANS, J. "The Genes for Color Vision." *Scientific American,* February 1989, pp. 42–49.

ROTTER, J. I., and DIAMOND, J. "What Maintains the Frequencies of Human Genetic Diseases?" *Nature* 329 (1987): 289–290.

STERN, C. *Principles of Human Genetics.* 3d ed. San Francisco: Freeman, 1973.

SUTTON, H. E. *An Introduction to Human Genetics.* 3d ed. Philadelphia: Saunders, 1980.

WALTERS, L. "The Ethics of Human Gene Therapy." *Nature* 320 (1986): 225–227.

WHITE, R., and J-M. LALOUEL. "Chromosome Mapping with DNA Markers." *Scientific American,* February 1988, pp. 40–48.

CHAPTER 16

BALINSKY, B. I. *An Introduction to Embryology.* 6th ed. Philadelphia: Saunders, 1982.

BARNES, D. M. "Orchestrating the Sperm-Egg Summit." *Science* 239 (1988): 1091–1092.

BROWDER, L. W. *Developmental Biology.* 2d ed. Philadelphia: Saunders, 1984.

COHN, J. P. "The Molecular Biology of Aging." *Bioscience* 37 (1987): 99–102.

COOKE, J. "The Early Embryo and the Formation of Body Pattern." *American Scientist* 76 (1988): 35–42.

EPEL, D. "The Program of Fertilization." *Scientific American,* Nov. 1977, pp. 128–140.

WASSARMAN, P. M. "The Biology and Chemistry of Fertilization." *Science* 235 (1987): 553–560.

CHAPTER 17

BISHOP, J. M. "The Molecular Genetics of Cancer." *Science* 235 (1987): 305–311. See also *Nature* 322 (1988): 203.

DAVIDSON, E. *Gene Activity in Early Development.* 2d ed. New York: Academic Press, 1976.

GEHRING, W. J. "Homeo Boxes in Development." *Science* 236 (1987): 1245–1252. See also *Science* 238 (1987): 1675–1681.

GURDON, J. *The Control of Gene Expression in Animal Development.* Cambridge, Mass.: Harvard University Press, 1974.

HOPPER, A., and N. HART. *Foundations of Animal Development.* Oxford: Oxford University Press, 1985.

MANIATIS, T., S. GOODBOURN, and J. A. FISCHER. "Regulation of Inducible and Tissue-Specific Gene Expression." *Science* 236 (1987): 1237–1244. See also *Nature* 322 (1986): 697–701 and 329 (1987): 79–80.

RAFF, R. A., and T. C. KAUFMAN. *Embryos, Genes, and Evolution.* New York: Macmillan, 1983.

TRINKAUS, J. P. *Cells into Organs.* Englewood Cliffs, N.J.: Prentice-Hall, 1984.

WESSELLS, N. K. *Tissue Interactions in Development.* Menlo Park, Calif.: Benjamin/Cummings, 1977.

WOLFFE, A. P., and D. D. BROWN. "Developmental Regulation of Two 5S Ribosomal RNA Genes." *Science* 241 (1988): 1626–1632.

WOLPERT, L. "Pattern Formation in Biological Development." *Scientific American*, April 1978, p. 154.

CHAPTER 18

AUSTIN, C. R., and R. V. SHORT. *Embryonic and Fetal Development.* New York: Cambridge University Press, 1972.

BEACONSFIELD, P., G. BIRDWOOD, and R. BEACONSFIELD. "The Placenta." *Scientific American*, August 1980, pp. 94–102.

DIAMOND, J. "Double Trouble." *Discover*, August 1988, pp. 65–71.

GOLANTY, E. *Human Reproduction.* New York: Holt, Rinehart, & Winston, 1975.

KATCHADOURIAN, H. *Human Sexuality: Sense and Nonsense.* New York: Norton, 1979.

KOLATA, G. "Maleness Pinpointed on Y Chromosome." *Science* 234 (1986): 1076–1077.

MASTERS, W. H., and V. E. JOHNSON. *Human Sexual Response.* Boston: Little, Brown, 1966.

MOORE, K. L. *Before We Are Born: Basic Embryology and Birth Defects.* 2d ed. Philadelphia: W.B. Saunders, 1983.

RAYMOND, C. "A Miracle Goes Sour." *Discover*, January 1989, p. 72.

WASSARMAN, P. M. "Fertilization in Mammals." *Scientific American*, December 1988, pp. 78–84.

WEISS, R. "Shape-Inducing Chemical Identified." *Science News*, June 1987, p. 406.

CHAPTER 19

BENNER, S.A., and A. D. ELLINGTON. "Return of the Last Ribo-Organism." *Nature* 332 (1988): 688–689.

DICKERSON, R. E. "Chemical Evolution and the Origin of Life." *Scientific American*, September 1978, pp. 70–86.

DOBZHANSKY, T., F. J. AYALA, G. L. STEBBINS, and J. W. VALENTINE. *Evolution.* San Francisco: Freeman, 1977.

GOULD, S. J. "The Ediacaran Experiment." *Natural History*, February 1984, pp. 14–23.

LEWIN, R. "RNA Catalysis Gives Fresh Perspective on the Origin of Life." *Nature* 319 (1986): 545–546.

MAYR, E. "Biological Classification: Toward a Synthesis of Opposing Methodologies." *Science* 214 (1981): 510–516.

NISBET, E. G. "RNA and Hot-Spring Waters." *Nature* 322 (1986): 206.

NORTH, G. "Back to the RNA World . . . and Beyond." *Nature* 328 (1987): 18–19.

SCHIDLOWSKI, M. "A 3,800-Million Year Isotopic Record of Life from Carbon in Sedimentary Rocks." *Nature* 333 (1988): 313–318.

WESTHEIMER, F. H. "Polyribonucleic Acids as Enzymes." *Nature* 319 (1986): 534–535.

WHITTAKER, R. H. "New Concepts of Kingdoms of Organisms." *Science* 163 (1969): 150–160.

CHAPTER 20

HIRSCH, M. S., and J. C. KAPLAN, "Antiviral Therapy." *Scientific American*, April 1987, pp. 76–85.

KLUYVER, A. J., and C. B. VAN NIEL. *The Microbe's Contribution to Biology.* Cambridge, Mass.: Harvard University Press, 1956.

LURIA, S. E., J. DARNELL, D. BALTIMORE, and A. CAMPBELL. *General Virology.* New York: Wiley, 1978.

MAGASANIK, B. "Research on Bacteria in the Mainstream of Biology." *Science* 240 (1988): 1435–1440.

MARGULIS, L., and D. SAGAN. *Microcosmos: Four Billion Years of Microbial Evolution.* New York: Summit Books, 1986.

SHAPIRO, J. A. "Bacteria as Multicellular Organisms." *Scientific American*, June 1988, pp. 82–89.

STANIER, R., and M. DOUDOROFF. *The Microbial World.* 4th ed. Englewood Cliffs, N.J.: Prentice-Hall, 1984.

VARMUS, H. "Retroviruses." *Science* 240 (1988): 1427–1434.

WOESE, C. "Archaebacteria." *Scientific American*, June 1981, pp. 98–122. See also R. A. Garrett, "The Uniqueness of Archaebacteria," *Nature* 318 (1985): 233–235.

YANG, D., et al. "Mitochondrial Origins." *Proceedings of the National Academy of Science, U.S.A.* 82 (1985): 4443–4447.

CHAPTER 21

CORLISS, J. "The Kingdom Protista and Its 45 Phyla." *Biosystems* 17 (1984): 87–126.

DOBELL, C. *Anton van Leeuwenhoek and His "Little Animals."* New York: Dover, 1962.

FARMER, J. N. *The Protozoa.* St. Louis: Mosby, 1980.

HANSON, E. D. *Understanding Evolution.* New York: Oxford University Press, 1981.

HARDY, A. *The Open Sea.* Part 1, *The World of Plankton.* Boston: Houghton Mifflin, 1965.

WALSH, J. "Human Trials Begin for Malaria Vaccine." *Science* 235 (1987): 1319–1320.

CHAPTER 22

ALEXOPOULOS, C. J., and C. W. MIMS. *Introductory Mycology.* 3d ed. New York: Wiley, 1979.

BEARD, J. "Spores Get in Your Eyes." *New Scientist*, January 1987, p. 26.

CHRISTENSEN, C. M. *The Molds and Man: An Introduction to the Fungi.* 3d ed. Minneapolis: University of Minnesota Press, 1965.

"Cracking the Mold." *Scientific American*, August 1987, p. 26.

KLEM, P. S., et al. "A Chemoattractant Receptor Controls Development in *Dictyostelium discoideum*." *Science* 241(1988): 1467–1472.

RAVEN, P. H., R. F. EVERT, and S. EICHHORN. *Biology of Plants.* 4th ed. New York: Worth, 1986.

RAY, P. M., T. STEEVES, and S. A. FULTZ. *Botany.* Philadelphia: Saunders, 1983.

CHAPTER 23

BARRETT, S. C. H. "Mimicry in Plants." *Scientific American*, September 1987, pp. 76–83.

DUDDINGTON, C. L. *Evolution in Plant Design.* London: Faber & Faber, 1969.

DYER, A. F., ed. *The Experimental Biology of Ferns.* New York: Academic Press, 1979.

EAMES, A. J. *Morphology of the Angiosperms.* New York: McGraw-Hill, 1961.

FOSTER, A. S., and E. M. GIFFORD. *Comparative Morphology of the Vascular Plants.* San Francisco: Freeman, 1974.

GENSEL, P. G., and H. N. ANDREWS. "The Evolution of Early Land Plants." *American Scientist* 75 (1987): 478–490.

GRAHAM, L. E. "The Origin of the Life Cycle of Land Plants." *American Scientist* 73 (1985): 178–186.

KUOLL, A. H., and N. J. BUTTERFIELD. "New Window on Proterozoic Life." *Nature* 337 (1989): 602–603.

MILNE, L., and M. MILNE. *Living Plants of the World.* New York: Random House, 1975.

MIROV, N. T., and J. HASBROUCK. *The Story of Pines.* Bloomington: Indiana University Press, 1976.

NORSTOG, K. "Cycads and the Origin of Insect Pollination." *American Scientist* 75 (1987): 270–278.

RAY, P. M., T. STEEVES, and S. A. FULTZ. *Botany.* Philadelphia: Saunders, 1983.

WATSON, E. V. *Mosses.* Oxford Biology Reader, no. 29. Burlington, N.C.: Carolina Biological Supply, 1972.

CHAPTER 24

ALEXANDER, R. M. *The Chordates.* Cambridge, England: Cambridge University Press, 1975.

BARNES, R. D. *Invertebrate Zoology.* 5th ed. Philadelphia: Saunders, 1987.

BARRINGTON, E. J. *Invertebrate Structure and Function.* Boston: Houghton Mifflin, 1967.

BUCHSBAUM, R. *Animals Without Backbones.* 2d ed. Chicago: University of Chicago Press, 1976.

DALY, H. V., J. T. DOYEN, and P. R. EHRLICH. *Introduction to Insect Biology and Diversity.* New York: McGraw-Hill, 1978.

FIELD, K. G., et al. "Molecular Phylogeny of the Animal Kingdom." *Science* 239 (1988): 748–752.

GANS, C., ed. *Biology of the Reptiles.* 9 vols. New York: Academic Press, 1969–1979.

The Life of Vertebrates. 3d ed. Oxford: Oxford University Press, 1981.

MILLER, J. A. "Ecology of a New Disease." *BioScience* 37 (1987): 11–15.

OFFICER, C. B., A. HALLAM, C. L. DRAKE, and J. D. DEVINE. "Late Cretaceous and Paroxysmal Cretaceous/Tertiary Extinctions." *Nature* 306 (1987): 143–149.

RAUP, D. M., and J. J. SEPKOSKI. "Testing for Periodicity of Extinction." *Science* 241 (1988): 94–99.

ROMER, A. S. *Vertebrate Paleontology.* Chicago: University of Chicago Press, 1966.

ROPER, C. F. E., and K. J. BOSS. "The Giant Squid." *Scientific American,* April 1982, pp. 96–105.

WICKELGREN, I. "At the Drop of a Tick." *Science News* 135 (March 25, 1989): 184–187.

WOLBACH, W. S. "Global Fire at the Cretaceous-Tertiary Boundary." *Nature* 334 (1988): 665–669.

YOUNG, J. Z. *The Life of Mammals: Their Anatomy and Physiology.* Oxford: Oxford University Press (Clarendon Press), 1975.

CHAPTER 25

BALANDRIN, M. F., et al. *Science* 228 (1985): 1154–1159.

CRONSHAW, J. *Support and Protection in Plants: Topics in the Study of Life.* New York: Harper & Row, 1971.

EAMES, A. J. *Morphology of the Angiosperms.* New York: McGraw-Hill, 1961.

EPSTEIN, E. "Roots." *Scientific American,* May 1973, pp. 48–55.

FELDMAN, L. J. "The Habits of Roots." *BioScience* 38 (1988): 612–618.

GUNNING, B. E. S., and M. W. STEER. *Ultrastructure and Biology of Plant Cells.* London: Arnold, 1975.

HARCOMBE, P. A. "Tree Life Tables." *BioScience* 37 (1987): 557–568.

ROBACKER, D. C., et al. "Floral Aromas." *BioScience* 38 (1988): 380–398.

WIENS, D. "Secrets of a Cryptic Flower." *Natural History,* April 1985, pp. 71–77.

WILSON, C., W. LOOMIS, and T. STEEVES. *Botany.* New York: Holt, Rinehart & Winston, 1971.

WOODING, F. P. B. *Phloem.* Oxford Biology Reader, no. 15. Burlington, N.C.: Carolina Biological Supply, 1977.

CHAPTER 26

BEWLEY, J. D., and M. BLACK. *Physiology and Biochemistry of Seeds.* New York: Springer-Verlag, 1978.

BRISTOW, A. *The Sex Life of Plants.* New York: Holt, Rinehart & Winston, 1978.

CLOWES, F. A. L. *Morphogenesis of the Shoot Apex.* Oxford Biology Reader, no. 23. Burlington, N.C.: Carolina Biological Supply, 1972.

ESAU, K. *Anatomy of Seed Plants.* 2d ed. New York: Wiley, 1977.

MOSES, P. B., and N.-H. CHUA. "Light Switches for Plant Genes." *Scientific American,* April 1988, pp. 88–93.

TREFIL, J. S. "Concentric Clues from Growth Rings Unlock the Past." *Smithsonian,* July 1985, pp. 47–52.

WALBOT, V. "On the Life Strategies of Plants and Animals." *Trends in Genetics,* June 1985, pp. 165–169.

CHAPTER 27

BAKER, D. A. *Transport Phenomena in Plants.* New York: Chapman & Hall, 1978.

BEEVERS, L. *Nitrogen Metabolism in Plants.* London: Arnold, 1976.

HESLOP-HARRISON, Y. "Carnivorous Plants." *Scientific American,* February 1987, pp. 104–115.

HEWITT, E. J., and T. A. SMITH. *Plant Mineral Nutrition.* New York: Wiley, 1975.

RAY, P. M., T. STEEVES, and S. A. FULTZ. *Botany.* Philadelphia: Saunders, 1983.

SCHULZE, E.-D., et al. "Plant Water Balance." *BioScience* 37 (1987): 30–37.

URIBE, E. G., and U. LUTTAGE. "Solute Transport and the Life Functions of Plants." *American Scientist* (November/December 1984): 567–573.

ZIMMERMAN, M. H., and C. L. BROWN. *Trees: Structure and Function.* New York: Springer-Verlag, 1971.

CHAPTER 28

ALBERSHEIM, P., and A. G. DARVILL. "Oligosaccharins." *Scientific American,* September 1985, pp. 58–64.

BLEECKER, A. B., et al. "Insensitivity to Ethylene Conferred by a Dominant Mutation in *Arabidopsis thaliana.*" *Science* 241 (1988): 1086–1088.

EVANS, L. T. *Daylength and Flowering of Plants.* Redwood City, Calif.: Benjamin/Cummings, 1975.

EVANS, M. L., R. MOORE, and K.-H. HASENSTEIN. "How Roots Respond to Gravity." *Scientific American,* December 1986, pp. 112–119.

GASSER, CHARLES S., and ROBERT T. FRALEY. "Genetically Engineering Plants for Crop Improvement." *Science* 244 (1989): 1293–1299.

GOODAVAGE, MARCIA. "Food Frankensteins?" *San Francisco Chronicle,* March 8, 1989, Food Section, p. 1.

KENDRICK, R. E., and B. FRANKLAND. *Phytochrome and Plant Growth.* Studies in Biology, no. 68. Baltimore: University Park Press, 1976.

MOORE, T. C. *Biochemistry and Physiology of Plant Hormones.* New York: Springer-Verlag, 1979.

MOSES, P. B., and N.-H. CHUA. "Light Switches for Plant Genes." *Scientific American,* April 1988, pp. 88–93.

NICKELL, L. G. *Plant Growth Regulators—Agricultural Uses.* New York: Springer-Verlag, 1982.

RAAB, M. M., and R. E. KONING. "How is Floral Expansion Regulated?" *BioScience* 38 (1988): 670–674.

RALOFF, JANET. "Cornucopious Nutrition." *Science News,* 134, August 13, 1988, p. 104–105.

CHAPTER 29

BOURNE, G. H., ed. *Hearts and Heart-Like Organs.* Vol. 1. New York: Academic Press, 1980.

CLARK, S. C., and R. KAMEN. "The Human Hematopoietic Colony-Stimulating Factors." *Science* 236 (1987): 1229–1236.

FOX, S. I. *Human Physiology.* 2d ed. Dubuque, Iowa: Wm. C. Brown, 1987.

KLOOG, Y., et al. "Sarafotoxin, a Novel Vasoconstrictor Peptide." *Science* 242 (1988): 268–270.

ROBINSON, T. F., S. M. FACTOR, and E. H. SONNENBLICK. "The Heart as a Suction Pump." *Scientific American,* June 1986, pp. 84–91.

SACHS, L. "The Molecular Control of Blood Cell Development." *Science* 238 (1987): 1374–1379.

CHAPTER 30

BARNES, D. M., et al. "Blood Forming Stem Cells Purified." *Science* 241 (1988): 24–25.

DING-E. YOUNG, J., and Z. A. COHN. "How Killer Cells Kill." *Scientific American,* January 1988, pp. 38–44.

GALLO, R. C., and L. MONTAGNIER. "AIDS in 1988." *Scientific American,* October 1988, pp. 41–48.

HALL, S. S. "A Molecular Code Links Emotions, Mind, and Health." *Smithsonian,* June 1989, pp. 62–71.

HOOD, L. E., I. L. WEISSMANN, and W. B. WOOD. *Immunology.* 3d ed. Menlo Park, Calif.: Benjamin/Cummings, 1984.

MACQUEEN, G., et al. "Pavlovian Conditioning of Rat Mucosal Mast Cells to Secrete Rat Mast Cell Protease II." *Science* 243 (1989): 83–85.

MARX, J. L. "What T Cells See and How They See It." *Science* 242 (1988): 863–865.

OLD, L. J. "Tumor Necrosis Factor." *Scientific American,* May 1988, pp. 59–75.

ROITT, I. *Essential Immunology.* 4th ed. Oxford, England: Blackwell Scientific, 1980.

TONEGAWA, S. "The Molecules of the Immune System." *Scientific American,* October 1985, pp. 122–131.

CHAPTER 31

BROOKS, G. A. and T. D. FARNEY. *Exercise Physiology.* New York: Wiley, 1984.

ECKERT, R., and D. RANDALL. *Animal Physiology.* 2d ed. New York: Freeman, 1983.

FEDER, M. E., and W. W. BURGGREN. "Skin Breathing in Vertebrates." *Scientific American,* November 1985, pp. 126–134.

FISHMAN, A. P., section ed. "Respiratory Physiology." *Annual Review of Physiology* 45 (1983): 391–451.

NAGAI, K., et al. "Distal Residues in the Oxygen-Binding Site of Hemoglobin Studied by Protein Engineering." *Nature* 329 (1987): 858–860.

RANDALL, D. J. "Gas Exchange in the Fish." In *Fish Physiology,* vol. 4, edited by W. S. Hoar and D. J. Randall. New York: Academic Press, 1970.

SCHMIDT-NIELSEN, K. *How Animals Work.* Cambridge, England: Cambridge University Press, 1972.

STEEN, J. B. "The Physiology of the Swim Bladder." *Acta Physiologica Scandinavica* 58 (1963): 124–137.

CHAPTER 32

BARNES, R. D. *Invertebrate Zoology.* 4th ed. Philadelphia: Saunders, 1980.

DAVENPORT, H. W. *Physiology of the Digestive Tract.* 4th ed. Chicago: Year Book Medical Publishers, 1978.

ECKERT, R., and D. RANDALL. *Animal Physiology.* 3d ed. New York: Freeman, 1988.

MILNER, A. "Flamingos, Stilts, and Whales." *Nature* 289 (1981): 347.

MOOG, F. "The Lining of the Small Intestine." *Scientific American,* May 1981, pp. 154–176.

PURSEL, V. G., et al. "Genetic Engineering of Livestock." *Science* 244 (1989): 1281–1288.

WEISS, R. "Bulimia's Binges Linked to Hormone." *Science News,* September 17, 1988, p. 182.

WHITNEY, E. N., and E. HAMILTON. *Understanding Nutrition.* St. Paul, Minn.: West, 1987.

WURTMAN, R. J., and J. J. WURTMAN. "Carbohydrates and Depression." *Scientific American,* January 1989, pp. 68–75.

CHAPTER 33

BARNES, B. M. "Freeze Avoidance in a Mammal; Body Temperature Below 0° C in an Arctic Hibernator." *Science* 244 (1989): 1593–1595.

BEAUCHAMP, G. K. "The Human Preference for Excess Salt." *American Scientist* 75 (1987): 27–33.

ECKERT, R., and D. RANDALL. *Animal Physiology.* 3d ed. New York: Freeman, 1988.

FLAM, F. "Antifreezes in Fish Work Quite Similarly." *Science News* 135 (June 30, 1989): 102.

FRENCH, A. R. "The Patterns of Mammalian Hibernation." *American Scientist* 76 (1988): 569–575.

HAINSWORTH, F. R. *Animal Physiology: Adaptations in Function.* 2d ed. Reading, Mass.: Addison-Wesley, 1984.

HOCHACHKA, P., and G. SOMERO. *Biochemical Adaptations.* Princeton, N.J.: Princeton University Press, 1984.

LEE, R. E., Jr. "Insect Cold-Hardness: To Freeze or Not to Freeze." *BioScience* 39, May 1989, pp. 308–313.

MARSHALL, E. "Testing Urine for Drugs." *Science* 241 (1988): 150–151.

STRICKER, E. M., and J. G. VERBALIS. "Hormones and Behavior: The Biology of Thirst and Sodium Appetite." *American Scientist* 76 (1988): 261–268.

VOGEL, S. "Cold Storage." *Discover,* February 1988, pp. 52–53.

WEISBURD, S. "Death-Defying Dehydration." *Science News,* February 13, 1988, pp. 107–110.

CHAPTER 34

ALBERTS, B., et al. *The Molecular Biology of the Cell.* 2d ed. New York: Garland, 1989.

"The Brain." *Scientific American,* September 1979. Whole issue.

DAVIS, L., G. A. BANKER, and O. STEWARD. "Selective Dendritic Transport of RNA in Hippocampal Neurons in Culture." *Nature* 330 (1987): 477–479.

ECKERT, R., and D. RANDALL. *Animal Physiology.* 3d ed. New York: Freeman, 1988.

HICKEY, W. F., and H. KIMURA. "Perivascular Microglial Cells of the CNS Are Bone Marrow Derived and Present in Antigen in Vivo." *Science* 239 (1988): 290–292.

KUFFLER, S., and J. NICHOLS. *From Neuron to Brain.* Sunderland, Mass.: Sinauer, 1976.

PURVES, D., et al. "Nerve Terminal Remodeling Visualized in Living Mice by Repeated Examination of the Same Neuron." *Science* 238 (1987): 1122–1126.

WEISS, R. "Neurons Regenerate into Spinal Cord." *Science News,* November 1988, p. 324.

CHAPTER 35

ALBERTS, B., et al. *The Molecular Biology of the Cell.* 2d ed. New York: Garland, 1989.

AUTELITANO, D. J., et al. "Hormonal Regulation of POMC Gene Expression." *Annual Review of Physiology* 51 (1989): 715–726.

BRODY, J. "Quest for Lean Meat Prompts New Approach." *New York Times,* May 13, 1988, page B1.

FROHMAN, L. A. "CNS Peptides and Glucoregulation." *Annual Review of Physiology* 45 (1983): 95–107.

MARSHALL, ELIOT. "The Drug of Champions." *Science* 242 (1988): 183–184.

NISHIZUKA, Y. "Turnover of Inositol Phospholipids and Signal Transduction." *Science* 225 (1985): 1365–1369.

ORCI, L., J.-D. YASSALLI, and A. PERRELET. "The Insulin Factory." *Scientific American,* September 1988, pp. 85–94.

POPE, H. G., JR., and D. L. KATZ. "Affective and Psychotic Symptoms Associated with Anabolic Steroid Use." *American Journal of Psychiatry* 145 (1988): 487–490.

SAMUELS, H. H., et al. "Regulation of Gene Expression by Thyroid Hormone." *Annual Review of Physiology* 51 (1989): 623–640.

SAUNDERS, D. S. "The Biological Clock of Insects." *Scientific American,* February 1976, pp. 114–120.

WHITE, J. D., et al. "Biochemistry of Peptide-Secreting Neurons." *Physiological Reviews* 65 (1985): 553–597.

CHAPTER 36

ALTMAN, J. "A Quiet Revolution in Thinking." *Nature* 328 (1987): 572–573.

AOKI, C., and P. SIEKEVITZ. "Plasticity in Brain Development." *Scientific American,* December 1988, pp. 56–64.

ATTWELL, D. "Phototransduction Changes Focus." *Nature* 317 (1985): 14–15.

BLOOM, F. E., A. LAZERSON, and L. HOFSTADTER. *Brain, Mind, and Behavior.* New York: Freeman, 1985.

BOTSTEIN, D. "The Molecular Biology of Color Vision." *Science* 232 (1986): 142–143.

BRODY, J. E. "Cocaine: Litany of Fetal Risks Grows." *New York Times,* September 6, 1988, p. B11.

CLARK, S. A., et al. "Receptive Fields in the Body Surface Map in Adult Cortex Defined by Temporally Correlated Inputs." *Nature* 332 (1988): 444–445.

ENGEN, T. "Remembering Odors and Their Names." *American Scientist* 75 (1987): 497–503.

FINE, A. "Transplantation in the Central Nervous System." *Scientific American,* August 1986, pp. 52–58.

GAZZANIGA, M. "Organization of the Human Brain." *Science* 245 (1989): 947–952.

KING, A. J., et al. "Developmental Plasticity in the Visual and Auditory Representations in the Mammalian Superior Colliculus." *Nature* 332 (1988): 73–76.

KOOB, G. F., and F. E. BLOOM. "Cellular and Molecular Mechanisms of Drug Dependence." *Science* 242 (1988): 715–723.

KORETY, J. F., and G. H. HANDELMAN. "How the Human Eye Focuses." *Scientific American,* July 1988, pp. 92–99.

KUCHARSKI, D., and W. G. HALL. "New Routes to Early Memories." *Science* 238 (1987): 786–787.

LIVINGSTONE, M. S. "Art, Illusion, and the Visual System." *Scientific American,* January 1988, pp. 92–99.

LIVINGSTONE, M. S., and D. HUBEL. "Segregation of Form, Color, Movement, and Depth: Anatomy, Physiology, and Perception." *Science* 240 (1988): 740–749.

MARTIN, G. R. "How Do Birds Accommodate?" *Nature* 328 (1988): 383.

MAYASHITA, Y., and H. S. CHANG. "Neuronal Correlate of Pictorial Short Term Memory in the Primate Temporal Cortex." *Nature* 331 (1988): 68–70.

MISHKIN, M., and T. APPENZELLER. "The Anatomy of Memory." *Scientific American*, June 1987, pp. 80–89.

NATHANS, J., et al. "Molecular Genetics of Human Blue Cone Monochromacy." *Science* 245 (1989): 831–838.

NILSSON, D. E. "A New Type of Imaging Optics in Compound Eyes." *Nature* 332 (1988): 76–78.

PETERSON, S. E., et al. "Positron Emission Tomographic Studies of the Cortical Anatomy of Single Word Processing." *Nature* 331 (1988): 585–588.

RAKIC, P. "Specification of Cerebral Cortical Areas." *Science* 241 (1988): 170–176.

SEMM, P., and C. DEMAINE. "Neurophysiological Properties of Magnetic Cells in the Pigeon's Visual System." *Journal of Comparative Physiology* 159 (1989): 619–625.

SHEPHERD, G. "Welcome Whiff of Biochemistry." *Nature* 316 (1985): 214–215. Also see *Nature* 324 (1986): 17–18; and *Nature* 325 (1987): 389.

SQUIRE, L. R. "Mechanisms of Memory." *Science* 232 (1986): 1612–1619.

STRYER, L. "The Molecules of Visual Excitement." *Scientific American*, July 1987, pp. 42–50.

CHAPTER 37

ADELSTEIN, R. S., and E. EISENBERG. "Regulation and Kinetics of the Actin-Myosin-ATP Interaction." *Annual Review of Biochemistry* 49 (1980): 921–956.

ALBERTS, B., et al. *The Molecular Biology of the Cell.* 2d ed. New York: Garland, 1989.

BRODY, J. E. "Dozens of Factors Critical in Bone Loss Among Elderly." *New York Times*, February 17, 1987, p. B17.

HANKER, J. S., and B. L. GIAMMARA. "Biomaterial and Biomedical Devices." *Science* 242 (1988): 885–892.

HUXLEY, A. F. *Reflections on Muscle.* Princeton, N.J.: Princeton University Press, 1980.

LAI, F. A., et al. "Purification and Reconstitution of the Calcium Release Channel from Skeletal Muscle." *Nature* 331 (1988): 315–319.

MURPHY, R. A. "Contraction in Smooth Muscle Cells." *Annual Review of Physiology* 51 (1989): 275–349.

WOZNEY, J. M., et al. "Novel Regulators of Bone Formation: Molecular Cones and Activities." *Science* 242 (1988): 1528–1534.

CHAPTER 38

KOEHN, R. K., and T. J. HILBISH. "The Adaptive Importance of Genetic Variation." *American Scientist* 75 (1987): 134–141.

LACK, D. *Darwin's Finches.* New York: Cambridge University Press, 1983.

MCCONNELL, T. J., et al. "The Origin of MHC Class II Gene Polymorphism Within the Genus *Mus.*" *Nature* 332 (1988): 651–653.

SLATKIN, M. "Gene Flow and the Geographic Structure of Natural Populations." *Science* 236 (1987): 787–792.

STRICKBERGER, M. W. *Genetics.* 3d ed. New York: Macmillan, 1985.

WILSON, A. C. "The Molecular Basis of Evolution." *Scientific American*, October 1985, pp. 164–173.

CHAPTER 39

BRITTEN, R. J. "Rates of DNA Sequence Evolution Differ Between Taxonomic Groups." *Science* 231 (1986): 1393–1398.

DIAMOND, J. "Competition Among Different Taxa." *Nature* 326 (1988): 241–242.

JOHNSON, C. *Introduction to Natural Selection.* Baltimore: University Park Press, 1976.

KOEHN, R. K., and T. J. HILBISH. "The Adaptive Importance of Genetic Variation." *American Scientist* 75 (1987): 134–141.

MAYR, E. "Evolution." *Scientific American*, September 1978, pp. 46–55.

SLATKIN, M. "Gene Flow and the Geographic Structure of Populations." *Science* 236 (1987): 787–792.

STEBBINS, G. H., and F. J. AYALA. "The Evolution of Darwinism." *Scientific American*, July 1985, pp. 72–82.

WALLACE, A. R. *The Malay Archipelago.* New York: Oxford University Press, 1987. Reprint of original work from 1850.

CHAPTER 40

AUSICH, W. I., et al. "Middle Mississippian Blastoid Extinction Event." *Science* 240 (1988): 796–798.

BOOTH, WILLIAM. "Africa Is Becoming an Elephant Graveyard." *Science* 243 (1989): 732.

BUSH, L. "Modes of Animal Speciation." *Annual Review of Ecology and Systematics* 6 (1975): 339–364.

COHN, JEFFREY P. "Halting the Rhino's Demise." *BioScience* 38 (1988): 740–744.

CROWLEY, T. J., and G. R. NORTH. "Abrupt Climate Changes and the Extinction Events in Earth History." *Science* 240 (1988): 996–1002.

GOULD, S. J. *Ontogeny and Phylogeny.* Cambridge, Mass.: Harvard University Press, 1977.

HALLAN, A. "End-Cretaceous Mass Extinction Event: Argument for Terrestrial Causation." *Science* 238 (1987): 1237–1242.

MAYR, E. *Populations, Species, and Evolution.* Cambridge, Mass.: Harvard University Press, 1970. Also see Mayr's article on panda evolution in *Nature* 323 (1986): 769–771.

MYERS, N. "Extinction Rates Past and Present." *BioScience* 39 (1989): 39–41.

POOL, ROBERT. "Hard Chores Ahead in Biodiversity." *Science* 241 (1988): 1759–1762.

RAUP, D. M. "Biological Extinction in Earth History." *Science* 231 (1986): 1528–1533.

SMITH, J. M. "Darwinism Stays Unpunctuated." *Nature* 330 (1987): 516. Related article on pp. 561–563.

STANLEY, S. M. *Macroevolution.* San Francisco: Freeman, 1979. Also see Stanley's article "Mass Extinctions in the Ocean," *Scientific American*, June 1984, pp. 64–72.

WHITE, M. J. D. *Modes of Speciation.* San Francisco: Freeman, 1977.

CHAPTER 41

BRIAND, F., and J. E. COHEN. "Environmental Correlates of Food Chain Length." *Science* 238 (1987): 956–960.

COLINVAUX, P. A. "The Past and Future Amazon." *Scientific American*, May 1989, pp. 102–108.

DIAMOND, J. "Human Use of World Resources." *Nature* 328 (1987): 479–480.

FEARNSIDE, P. M. "Extractive Reserves in Brazilian Amazonia." *BioScience* 39 (1989): 387–393.

LONG, S. P., and D. O. HALL. "Nitrogen Cycles in Perspective." *Nature* 329 (1987): 584–585.

MARSHALL, E. "EPA's Plan for Cooling the Global Greenhouse." *Science* 243, March 24, 1989, pp. 1544–1545.

ODUM, E. P. *Basics of Ecology.* Philadelphia: Saunders, 1983.

RALOFF, J. "CO_2: How Will We Spell Relief?" *Science News*, December 24 and 31, 1988, pp. 411–414.

REVKIN, A. C. "Endless Summer: Living with the Greenhouse Effect." *Discover*, October 1988, pp. 50–61.

SCHNEIDER, S. H. "The Changing Climate." *Scientific American*, September 1989, pp. 70–79.

SHABECOFF, P. "Major Greenhouse Impact Is Unavoidable." *New York Times*, July 19, 1988, p. B5.

VITOUSEK, P., et al. "Net Primary Production: Original Calculations." *BioScience* 36 (1986): 368–373.

VITOUSEK, P. M., et al. "Net Primary Production: Original Calculations." *Science* 235 (1987): 730.

WALTER, H. *Vegetation of the Earth in Relation to Climate and the Eco-physiological Conditions.* Trans, J. Wieser. London: English Universities Press, 1973.

WHITTAKER, R. H. *Communities and Ecosystems.* 2d ed. New York: Macmillan, 1975.

WILSON, E. O. "Threats to Biodiversity." *Scientific American*, September 1989, pp. 108–116.

CHAPTER 42

BROWER, L. "Ecological Chemistry." *Scientific American*, February 1969, pp. 22–30.

CASE, T. J., and M. L. CODY. "Testing Theories of Island Biogeography." *American Scientist* 75 (1987): 402–410.

CURRIER, D. J., and V. PAQUIN. "Large Scale Biogeographical Patterns of Species Richness of Trees." *Nature* 329 (1987): 326–328.

DRURY, W. H., and I. C. T. NISBET. "Succession." *Journal of the Arnold Arboretum* 54 (1973): 331–368.

EHRLICH, P., and P. H. RAVEN. "Butterflies and Plants." *Scientific American*, June 1967, pp. 104–112.

KAREIVA, P. "Habitat Fragmentation and the Stability of Predator Prey Interactions." *Nature* 326 (1987): 388–390.

KLEKOWSKI, E. J., and P. J. GODFREY. "Aging and Mutation in Plants." *Nature* 340 (1989): 389–391.

KLOPFER, P. J., and P. J. HAILMAN. *Behavioral Aspects of Ecology*. Englewood Cliffs, N.J.: Prentice-Hall, 1973.

LEWIN, R. "Latitude Gradients in Geographic Ranges." *American Naturalist*, February 1989, pp. 240–256.

LEWONTIN, R. C. "Adaptation." *Scientific American*, September 1978, pp. 213–230.

MACARTHUR, R., and E. O. WILSON. *The Theory of Island Biogeography*. Princeton, N.J.: Princeton University Press, 1967.

MONASTERSKY, R. "Lessons from the Flames." *Science News* 134 (1988): 314–317 and 330–332.

RICKLEFS, R. E. *The Economy of Nature*. 2d ed. Newton, Mass.: Chiron Press, 1983.

ROSENTHAL, G. A. "The Chemical Defenses of Higher Plants." *Scientific American*, January 1986, pp. 94–99.

SCHOENER, T. W., and D. A. SPILLER. "High Populations Persistence in a System with High Turnover." *Nature* 330 (1987): 474–477.

STEVENS, G. "The Latitudinal Gradient in Geographical Range: How So Many Species Coexist in the Tropics." *American Naturalist* 133 (February 1989): 240–256. Also see *Science* 244 (1989): 527–528.

WHITTAKER, R. H. *Communities and Ecosystems*. 2d ed. New York: Macmillan, 1975.

CHAPTER 43

CONNELL, J. H. "Diversity in Tropical Rainforests and Coral Reefs." *Science* 199 (1978): 1302.

HUSTON, M., et al. "New Computer Models Unify Ecological Theory." *BioScience* 38 (1988): 628–691.

LACK, D. *Darwin's Finches*. New York: Cambridge University Press, 1983.

———. *Ecological Isolation in Birds*. London: Methuen, 1971.

MCLAREN, I., ed. *Natural Regulation of Animal Populations*. New York: Atherton Press, 1971.

OSMOND, C. B., et al. "Stress Physiology and the Distribution of Plants." *BioScience* 37 (1987): 38–48.

PEET, K., and N. L. CHRISTENSEN. "Competition and Tree Death." *BioScience* 37 (1987): 586–594.

WELDEN, C. W., and W. L. SLANSON. "The Intensity of Competition Versus Its Importance." *Quarterly Review of Biology* 61(1986): 23–44.

WILSON, E. O., and W. H. BOSSERT. *A Primer of Population Biology*. Sunderland, Mass.: Sinauer, 1971.

CHAPTER 44

ALCOCK, J. *Animal Behavior, an Evolutionary Approach*. 2d ed. Sunderland, Mass.: Sinauer, 1979.

DIAMOND, J. M. "Learned Specializations of Birds." *Nature* 330 (1987): 16–17. Also see T. K. Werner and T. W. Sherry, *Proceedings of the National Academy of Sciences* 84 (1987): 5506–5510.

GOULD, J. L., and P. Marler. "Learning by Instinct." *Scientific American*, January 1987, pp. 74–85.

GWINNER, E. "Internal Rhythms in Bird Migration." *Scientific American*, April 1986, pp. 84–92.

HOY, R. R., et al. "Hawaiian Courtship Songs: Evolutionary Innovation in Communication Signals of *Drosophila*." *Science* 240 (1988): 217–218.

LORENZ, K. *King Solomon's Ring*. New York: Crowell, 1952.

MARLER, P., and W. J. HAMILTON. *Mechanisms of Animal Behavior*. New York: Wiley, 1966.

TINBERGEN, N. *The Animal and Its World: Explorations of an Ethologist*. Cambridge, Mass.: Harvard University Press, 1973.

CHAPTER 45

ALCOCK, J. *Animal Behavior: An Evolutionary Approach*. 3d ed. Sunderland, Mass.: Sinauer, 1984.

ALEXANDER, R. D. "The Evolution of Social Behavior." *Annual Review of Ecology and Systematics* 5 (1974): 325–383.

BREED, M. "Genetics and Labour in Bees." *Nature* 333 (1988): 229. Also see G. D. Robinson and R. E. Page, *Nature* 333 (1988): 356–358; and P. C. Frumhoff and J. Baker, *Nature* 333 (1988): 358–360.

HAMILTON, W. D. "The Genetical Evolution of Social Behavior." *Journal of Theoretical Biology* 7 (1964): 1–52.

PAGE, R. E., JR., et al. "Genetic Specialists, Kin Recognition, and Nepotism in Honey Bee Colonies." *Nature* 338 (1989): 576–579.

SLESSOR, K. N., et al. "Semiochemical Basis of the Retinue Response to Queen Honeybees." *Nature* 332 (1988): 354–355.

WILSON, E. O. *Sociobiology: The New Synthesis*. Cambridge, Mass.: Harvard University Press, 1975.

CHAPTER 46

CARTMILL, M., D. PILBEAM, and G. ISAAC. "One Hundred Years of Paleoanthropology." *American Scientist* 74 (1986): 410–420.

DAY, M. H. *Guide to Fossil Man*. 4th ed. Chicago: University of Chicago Press, 1986.

EDMONDS, J., and J. M. REILLY. *Global Energy: Assessing the Future*. New York: Oxford University Press, 1985.

GOODMAN, M., R. E. TASHIAN, and J. H. TASHIAN. *Molecular Anthropology*. New York: Plenum Press, 1976.

LOVEJOY, C. O. "Evolution of Human Walking," *Scientific American*, November 1988, pp. 118–125.

MIYAMOTO, M. M., et al. "Phylogenetic Relations of Humans and African Apes from DNA Sequences in Globin." *Science* 238 (1987): 369–372. See also *Science* 238 (1987): 273–275.

MYERS, N. *The Primary Source*. New York: Norton, 1984.

RICHARD, A. E. *Primates in Nature*. San Francisco: Freeman, 1985.

SIMONS, E. L. "Human Origins." *Science* 245 (1989): 1343–1350.

STRINGER, C. B., and P. ANDREWS. "Genetic and Fossil Evidence for the Origin of Modern Humans." *Science* 239 (1988): 1263–1268.

VALLADAS, H., et al. "Thermoluminescence Dating of Mousterian Proto-CroMagnon Remains from Israel and the Origins of Modern Man." *Nature* 331 (1988): 614–616. See also *Nature* 331 (1988): 565–566.

WEISS, M. L., and A. E. MANN. *Human Biology and Behavior*. 4th ed. Boston: Little, Brown, 1985.

WYMER, J. *The Paleolithic Age*. New York: St. Martin's Press, 1982.

Glossary

abscisic acid: A plant hormone that suppresses growth.

abscission: The normal separation of a leaf, fruit, or flower from a plant.

abscission layer: A special zone of cells in the stem of a leaf or fruit that forms in response to hormonal signals and allows the leaf or fruit to fall at the appropriate time.

absorption: The taking up of liquids by root hairs or other biological structures.

absorption spectrum: The region of the spectrum of electromagnetic energy (usually visible light) that is absorbed by a particular molecule or atom.

Acheulian: A stone tool industry associated with the fossil record of *Homo erectus* and characterized by bifacial hand axes.

acid: Any substance that gives up or donates H^+ ions in solution, thereby increasing the H^+ content of the solution.

acrosome reaction: A two-part event that marks the first stage of fertilization of an egg by a sperm, and which permits the sperm to penetrate the egg's outer protective layers.

actin: A structural protein that with myosin carries out contraction; also called microfilaments.

actinomycete (ak-ti-no-**my**-seet)**:** Any of a diverse group of filamentlike bacteria.

action potential (nerve impulse): A temporary depolarization of the potential across the membrane of a nerve or skeletal muscle cell.

action spectrum: The range of light wavelengths that triggers a chemical process, as of photosynthesis.

activation energy: The amount of energy needed to break the bonds between a molecule's constituent atoms.

active immunity: Immunity to disease based on prior exposure to an antigenic pathogen.

active site: The small area (or areas) on an enzyme molecule, often a groove or pocket, into which the enzyme's substrate fits in a "lock-and-key" arrangement.

active transport: The carrier-mediated, energy-dependent movement of materials into or out of cells, especially against a concentration gradient. Active transport differs from diffusion in that it requires expenditure of energy, usually from ATP.

adaptation: (1) The accumulation of inherited characteristics that make an organism specifically suited to its environment and way of life. (2) A particular genetically based feature, characteristic, or behavior that results in an individual or species being better suited to some aspect of its environment.

adaptive radiation: The development from an ancestral group of a variety of forms adapted for different habitats and ways of life.

adenine: One of four nucleotide bases that are fundamental structural components of DNA molecules; classed as a purine and complementary in size to thymine in the DNA double helix.

adenosine triphosphate (ATP): The primary energy-storage molecule in cells. Large amounts of energy are stored in bonds of the three phosphate groups that make up an ATP molecule's tail.

adhesion: The tendency of unlike molecules to cling together.

adrenal gland: Either of two endocrine glands near the vertebrate kidney that secretes a variety of steroid hormones.

adrenocorticotropic hormone (ACTH): A hormone secreted by the pituitary gland that regulates the adrenals.

adventitious root: Any new root that arises from an organ other than an existing root.

aerial hypha (plural: **hyphae**)**:** Fungal structures that grow vertically and discharge spores into the air in various ways.

aerobic cell: A cell that requires oxygen for its metabolism.

afterbirth: The placenta and attached membranes which are expelled from the uterus after the birth of a newborn.

age structure: The relative numbers of young, mature, and old individuals in a population at a single point in time.

aggression: Overt fighting, displaying, posturing, etc., sometimes between animals of different species, but more often between members of the same species.

air sac: In birds, one of numerous balloonlike sacs that are filled with air and lighten the body.

albinism: A genetically caused deficiency of pigment in the hair, eyes, and skin.

aldosterone: A steroid hormone released by cells in the cortex of the adrenal gland that regulates salt reabsorption in the kidney.

aleurone: A layer of living cells that surrounds the food stores of a monocot seedling.

alga (plural: **algae**)**:** Any of a broad group of simple, mostly aquatic plants; seaweeds are prime examples.

alimentary canal: A two-ended gut that extends from mouth to anus.

allantois: A membrane of embryos of land vertebrates. In reptiles and birds, the role of the allantois is to store nitrogenous wastes generated by the embryo's metabolism.

allele: An alternate form of a particular gene.

allelopathy: Among plants, a form of interspecific competition in which chemical substances produced by one species inhibit the germination or growth of seedlings of another species.

Allen's rule: The observation that animals inhabiting cold regions tend to have shorter limbs and other extremities than do those in warmer regions, as an adaptation to minimize convective heat loss.

allergy: A chronic immune-system response induced by a usually benign environmental stimulant.

allopatric speciation: The evolution of two or more new species from a single ancestral species as the result of chance geographical separation.

allosteric enzyme: An enzyme that undergoes reversible changes in shape and in catalytic activity when "control" substances bind.

α (alpha) helix: A secondary folding pattern of polypeptide chains, in which the protein's amino-acid residues are wrapped in a helical "spiral staircase" shape.

alternation of generations: The alternating life cycle of plants, which includes both gametophyte (haploid) and sporophyte (diploid) phases.

altruism: Self-sacrifice by one member of an animal species for the benefit of others of the species.

alveolus (plural: **alveoli**)**:** Any of the tiny, blind-ended cavities in lungs in which gas exchange takes place.

Ames test: A test that usually employs *Salmonella* bacteria to screen chemicals for possible mutagenic effects.

amino (uh-**mee**-no) **acid:** An organic molecule consisting of a central carbon atom, an amino group ($-NH_2$), a carboxyl group ($-COOH$), and a distinctive side chain (R). Amino acids are the monomers from which proteins are built.

amino-acid activation: The process prior to protein synthesis in a cell in which amino acids are joined to tRNA by high energy bonds.

aminoacyl attachment site: The position on a tRNA molecule where a specific amino acid attaches.

aminoacyl-tRNA synthetases: Enzymes which catalyze a reaction that attaches a specific amino acid to an appropriate tRNA molecule.

amniocentesis (am-nee-oh-sen-**tee**-sis)**:** The removal of amniotic fluid from a pregnant woman's womb, and analysis of fetal cells in the fluid for chromosomal defects.

amnion: A protective membrane which is filled with fluid and forms a cushionlike sac around a developing embryo.

anabolism: The chemical reactions involved in the synthesis of essential large and small biological molecules in cells.

analogy: The independent evolution in unrelated life forms of similar structures that carry out similar functions.

anaphase: The third stage of mitosis, in which the two sets of chromatids separate and move to opposite ends of the dividing cell.

anaphylactic shock: A life-threatening allergic reaction during which certain cells of the immune system discharge their contents simultaneously, causing rapid fluid loss from blood vessels throughout the body and reduced blood flow to the brain and heart.

anchorage dependence: An animal cell's need to adhere to a solid surface.

androgen: The collective term for certain male sex hormones.

aneuploidy: A genetic condition marked by the absence of one or more chromosomes or the presence of extra chromosomes.

angiosperm: A flowering plant.

anisogametes: Gametes of unequal size that are produced by a single species, such as human egg and sperm.

annual plant: A plant whose life cycle—germination from seed, growth, reproduction, and death—is completed within a single growing season.

antheridium: In liverworts and related plants, a sperm-producing chamber.

anthropoid: Pertaining to the suborder of primates that includes the monkey, ape, and human lineages.

antibody: Globular blood proteins that are produced by B lymphocytes and that bind specifically to foreign antigenic materials in the body.

anticodon: A set of three unpaired bases on a tRNA molecule that binds to a complementary codon on mRNA.

antidiuretic hormone (ADH): A posterior pituitary gland hormone that regulates the amount of water allowed to pass from the kidney as urine.

antigen: Any substance that elicits an immune response.

aorta: The largest artery in the vertebrate body; carries oxygenated blood away from the heart.

aortic arch: Any of the paired blood vessels running through the gill arches of vertebrate embryos and adult fish.

apical dominance: The tendency of the main shoot of a plant to predominate over all others.

apical meristem: The undifferentiated, actively dividing cells at the growing tip of a plant shoot; such tissue is the source of a plant's leaves, stems, branches, and flowers.

apoplastic pathway: In a plant root, the "compartment" made up of all extracellular spaces, along with the spaces within cell walls that water can traverse without crossing any plasma membranes.

aposematic (warning) coloration: Bright colors or striking patterns used as a warning by organisms that also possess chemical or other defenses.

appendicular skeleton: The portion of the vertebrate skeleton made up of the pectoral girdle and forelimb bones, and the pelvic girdle and hindlimb bones.

archaebacteria: A phylum or subkingdom of monerans that includes methane producers, sulfur-dependent species, and cells that tolerate very salty or hot environments, and is thought to be very ancient.

arteriole: A small blood vessel branching off an artery.

artery: A vessel that carries blood to tissues.

ascus: In some fungi, a small sac in which sexually produced spores develop.

A site: One of two groovelike sites on a ribosome where aminoacyl tRNA, the carrier of amino acids during protein synthesis, can bind to mRNA. When both sites are occupied (see also *P site*), the two attached amino acids can be joined together.

associative learning: Also known as classical or Pavlovian conditioning, learned behavior based on the association of a particular activity with a particular reward, punishment, or other event.

assortative mating: The union of unrelated individuals at a greater or lesser frequency than that predicted by chance.

atmospheric pressure: The weight of the atmosphere surrounding the Earth, equal to about 14.7 pounds per square inch at sea level.

atom: The smallest unit of matter that still displays the characteristic properties of an element.

atomic number: The particular number of protons in the nucleus of an atom of an element.

atomic orbital: A cloudlike three-dimensional zone around the nucleus of an atom, in which electrons are confined.

atomic weight: The sum of an atom's neutrons and protons.

ATP synthetase: Large, complex enzyme molecules that catalyze the synthesis of ATP and through which protons flow down an electrochemical gradient.

atrionatriuretic factor (ANF): Any of several small peptide hormones produced by heart atrial cells that increase water and sodium loss in the urine.

atrioventricular (A-V) node: A bundle of cells that receives electrical impulses from the heart's "pacemaker" (the sinoatrial node) and that in turn triggers waves of contraction in the heart ventricles.

auditory nerve: The eighth cranial nerve, which carries impulses to the brain and makes hearing possible.

auto-: From Greek, meaning "self."

autoimmune response: An abnormal process in which the immune system attacks the body's own cells or substances.

autonomic nervous system: The portion of the peripheral nervous system that controls physiological functions such as heart rate, respiration, and digestion.

autosome: A nonsex chromosome.

autotroph: Any organism that uses light energy or chemical energy to manufacture the sugars, fats, and proteins required for cellular metabolism.

auxin: Any of a group of hormones that stimulate growth of various plant parts.

auxotroph: A mutant bacterium that, unlike its normal relatives, is unable to manufacture all the materials necessary for its growth on a simple nutrient solution.

average heterozygosity: A measure of the average frequency of individuals that are heterozygous at each gene locus in a particular genome.

axial skeleton: That portion of the vertebrate skeleton made up of the skull, the vertebral column, and, in animals having them, the ribs.

axillary bud: A bud that develops at the junction (the axil) of a leaf primordium and the shoot apical meristem.

axon: The portion of a neuron that transmits action potentials (nerve impulses).

bacteroid: A nitrogen-fixing bacterium that is able to survive only within a plant root nodule.

balanced selection: The process that operates to counteract the loss of variant alleles in a population.

band: Among primate societies, the basic unit for food gathering and defense.

Barr body: Any of one or more inactive X chromosomes present in the cells of a normal female mammal.

basal body: One of several types of microtubule-organizing centers in animal and other kinds of flagellated cells; similar to a centriole.

basal lamina: A feltlike layer attached to epithelial tissues which consists of highly ordered sugar–protein complexes, along with a specialized form of collagen.

base: (1) A substance that upon dissociation in water, forms hydroxyl ions (OH^-); such ions commonly combine with hydrogen ions (H^+), thereby raising the pH of the solution. The opposite of an acid, a base has a pH of more than 7. (2) A ring structure composed of carbon and nitrogen that serves as one of the chemical building blocks of nucleotides. The nucleotide bases are adenine, guanine, cytosine, uridine, and thymine.

basidiocarp: The dense mass of hyphae that forms the main body of a mushroom.

basidium (plural: **basidia**): One of many club-shaped structures that lines the surfaces of gills on the underside of a mushroom.

basilar membrane: A thin sheet of tissue that lies over the inner cochlear canal in the vertebrate ear, and upon which rest numerous sensory hair cells.

Batesian mimicry: A defensive adaptation in which a species lacking chemical defenses evolves the same warning coloration or patterns as a noxious species.

B cell (B lymphocyte): One of two major types of white blood cells of the immune system; B cells synthesize antibody molecules.

behavioral isolation: A prezygotic reproductive isolating mechanism that keeps different species from mating successfully due to differences in mating behavior.

benthic community: An ocean-bottom marine biome.

Bergmann's rule: Individuals living in colder parts of their species' range tend to be larger than individuals living in warmer areas.

***β* (beta)-pleated sheet:** A configuration for the folding of polypeptide chains in which two or more polypeptides lying side by side become crosslinked by hydrogen bonds and form an accordianlike sheet of connected molecules.

biennial plant: A plant whose life cycle from seed germination to the production of a new generation of seeds requires two growing seasons.

Big Bang: A theory of how the universe began with the explosion of a giant ball of gases about 18 billion years ago.

bilateral symmetry: The animal body plan in which an organism's right and left sides are mirror images.

bilayer (lipid): The thin, two-layered arrangement of lipid molecules that make up membranes within and at the surface of cells.

bile: A greenish fluid produced in the liver which emulsifies fats and aids in fat digestion and absorption.

binary fission: The division of prokaryotic cells into two virtually identical daughter cells.

binomial system of nomenclature: The assignment of names to organisms using two Latin words, the first denoting the genus and the pair denoting the species, e.g., *Homo sapiens*.

biological magnification: The process by which toxic materials present in trace amounts in the environment accumulate in organisms.

biology: The study of life.

biomass: The combined dry or wet weight of all the organisms in a habitat.

biome: Any of the Earth's major ecosystems.

biosphere: The land and waters and the air above them which make up the life-supporting region of the Earth's surface.

biosynthesis: The process of building metabolic proteins, fats, carbohydrates, and other biological molecules from precursor components.

bipolar cell: Any of the specialized neurons in the eye which synapse with rods and cones, and also with ganglion cells which transmit information along the optic nerve to the brain.

bladder: In animals, a balloonlike storage organ for urine.

blade: The broad, usually flattened portion of a leaf.

blastomere: Early embryonic cells of animals.

blastula: The usually spherelike arrangement of cells around a central cavity or mass of yolk that results when a zygote undergoes cleavage during embryonic development.

blood: A dynamic, life-sustaining solution in animals with closed circulatory systems containing ions, nutrients, waste products, hormones, other substances, and cells. The cells—blood cells and platelets—are suspended in plasma.

blood–brain barrier: In the central nervous system, the system of tightly joined, highly impermeable capillary walls that acts to prevent most blood-borne substances from passing easily into the cerebrospinal fluid that bathes CNS neurons. Also inhibits movement in the opposite direction. Lipid-soluble substances can pass the barrier.

blood pressure: The hydrostatic pressure exerted by the blood in an animal's circulatory system as a result of the rhythmic contractions of the heart and peristaltic waves of contraction in some blood vessels.

Bohr effect: A decrease in the affinity of hemoglobin for O_2 as pH falls, and an increase in affinity as pH rises.

bolus: A moistened lump created by chewing food.

bond energy: The amount of energy needed to break a chemical bond.

bone: A hard, living tissue consisting primarily of collagen and apatite, a calcium and phosphate salt.

bottleneck effect: A form of genetic drift that occurs when only a small portion of an original population provides the gene pool for a new population.

brachiation: A mode of locomotion that involves swinging through the trees suspended by one forelimb at a time.

bronchus (plural: **bronchi**): One of the branched passageways through which air enters the lungs.

brown fat: A special type of fat in mammals that is darkened by increased numbers of mitochondria; as these organelles oxidize fatty acids, they give off extra heat.

bud primordium: A group of meristematic cells that has begun to develop into a bud and will, in turn, grow into a branch or form flowers.

buffer: A substance that binds hydrogen ions when concentrations of H^+ are high and releases hydrogen ions when concentrations of H^+ are low.

bulb: A compact, subterranean conical stem having modified leaves in which carbohydrate is stored; participates in vegetative reproduction.

bulk flow: The movement of a mass of fluid.

C_3 plants: Plants showing decreased carbohydrate production in hot, dry weather and in which three-carbon sugars are the first stable intermediates in the Calvin-Benson cycle of photosynthesis.

C_4 plants: Plants that can photosynthesize in hot, dry climates at a faster rate than C_3 plants due to special leaf anatomy and a unique biochemical pathway which begins with a stable, four-carbon sugar intermediate.

calcitonin: A peptide hormone that lowers blood calcium levels; produced in the thyroid gland.

callus tissue: A disorganized lump of plant tissue.

calorie: The amount of heat required to raise the temperature of 1 gram of water by 1°C.

Calvin-Benson cycle: The dark reactions of photosynthesis.

calyx: The ring of sepals, the outer, usually green parts of a flower.

camouflage: Colors or patterns that allow organisms to blend with their background or to appear to be either inedible or nonthreatening.

capillarity: The tendency of molecules to move upward in a narrow space against the tug of gravity.

capillary: Any of the tiny, thin-walled blood vessels interwoven throughout body tissues.

capillary action: The tendency of water in a thin tube to move upward.

carbohydrate: One of a group of carbon compounds including sugars, starches, and cellulose and consisting of a carbon backbone with various functional groups attached. Carbohydrates are the most abundant organic compounds found in living organisms.

carbon cycle: The natural cycle established by the activities of photosynthesis and respiration in various life forms. Basic steps of the cycle are the fixing of atmospheric CO_2 into carbohydrates via photosynthesis, the use of carbohydrates as fuel by organisms, and the release of waste CO_2 back to the atmosphere.

carcinogen: An agent that has been shown to cause cancer in a laboratory animal or person.

carcinoma: A tumor arising in one of the epithelial sheets that cover the outer and inner surfaces of the body.

cardiac muscle: The specialized striated muscle tissue of the heart.

cardiac output: A measure of the amount of blood pumped by the heart per unit of time.

carnivore: An animal that eats only meat.

carnivorous plant: A plant such as the Venus's-flytrap or the sundew that traps and digests insects as a source of nitrogen.

carotenoid pigment: A pigment related to vitamin A that appears red, orange, or yellow to the human eye.

carrier-facilitated diffusion: A process in which specialized proteins act as carriers that transport substances across a cell's plasma membrane.

carrying capacity (*K*): The maximum size population that can be supported by the resources in the population's environment.

cartilage: A fibrous connective tissue consisting primarily of collagen and complex polysaccharides.

Casparian strip: Any of the waterproof, suberin-coated walls of endodermal cells in a plant.

catabolism: The energy-yielding processes in cells, in which molecules are broken down to obtain structural elements, to release energy, or to digest waste products.

catabolite repression: A mechanism whereby the formation of enzymes that catabolize certain sugars (such as lactose) is repressed by the presence of a simple sugar (such as glucose) that is easier to break down.

catalyst: Any molecule that increases the rate of a chemical reaction without being used up during that reaction. Biological catalysts are primarily protein molecules known as enzymes.

cecum (plural: **ceca**): A blind-ended pouch that extends from the intestine and holds food for an extended period of time to enhance digestion and absorption of nutrients.

cell: The basic structural unit of living organisms; see also *prokaryotic cell* and *eukaryotic cell*.

cell cycle: The regular sequence during which a cell grows, prepares for division, and divides to form two daughter cells.

cell death: A developmental mechanism for sculpting digits, shaping body parts, and generating other tissues and organs that involves the death of cells in specific developing structures.

cell-mediated immunity: The direct attack on foreign cells or substances by T lymphocytes.

cell plate: The layer of membranous sacs that arises to separate the two new daughter cells when a plant cell divides.

cell theory: A set of statements encapsulating the essential characteristics of cells, including that cells are the basic units of life, make up all organisms, and arise from preexisting cells.

cellular respiration: The oxygen-dependent metabolic process by which cells derive energy (ATP) from glucose and other fuel molecules.

cellular slime mold: A type of slime mold in which separate amoeboid cells aggregate to form a multicellular slug.

cellulose: The fibrous structural material of plant cells and wood; a high-molecular-weight polysaccharide composed of long chains of glucose units.

cell wall: The rigid outer structure that surrounds the plasma membranes of plant cells, bacteria, and some fungi.

central nervous system (CNS): The brain and spinal cord of vertebrates.

centriole: A cellular organelle composed of nine microtubular triplets. Centrioles serve as assembly sites for spindle microtubules used in cell division (similar to *basal bodies*).

centromere: The constricted area at which two chromatids are attached to one another and to which spindle microtubules are attached; site of satellite DNA.

cephalization: The formation of a front end or "head" where organs specialized for sensing and feeding are located.

cerebellum: The region of the vertebrate hindbrain that regulates balance, stance, and some locomotor movements.

cerebrospinal fluid: The fluid filling the hollow ventricular chambers of the vertebrate central nervous system, as well as the intercellular spaces in the walls of the brain and spinal cord and the spaces between the meninges.

cerebrum: The anterior dorsal part of the vertebrate forebrain; the site where most sensory processing and complex behaviors are coordinated.

cervix: The base of the uterus, which also serves as the upper end of the vagina.

chalone: A possible inhibitor of cell division in animal cells that is hypothesized to be produced by differentiated tissue cells.

character displacement: In closely related species, the evolution of slight physical differences in structures used to exploit a limited resource.

Chargaff's rules: The constant and variable features of DNA codified by geneticist Erwin Chargaff into two rules: first, A and T occur in equal amount as do G and C; second, the ratio of A and T to G and C is constant within a species but varies among different species.

chemical energy: Potential energy stored in atoms and molecules and in their bonds.

chemical evolution: The evolution of simple cells from nonliving substances on the early Earth.

chemical-gated channel: An ion-transport channel in the membrane of a neuron that opens when a specific chemical is present.

chemical reaction: The interaction of atoms, ions, or molecules to form new substances.

chemiosmotic coupling hypothesis: A model for cellular respiration in which the movement of electrons through the electron transport chain in mitochondria is accompanied by a proton-pumping mechanism, which in turn sets up an "energy gradient." Energy released from this gradient is conserved in the form of ATP.

chemoautotroph: A bacterium capable of oxidizing inorganic compounds to gain energy for its life processes.

chemoreceptor: A sensory receptor cell that detects specific molecules and ions; active in taste and smell perception.

chemotaxis: Movement toward or away from the source of a diffusing chemical.

chitin: A substance built from nitrogen-containing polysaccharides that is the main ingredient of the hard exoskeletons of lobsters, spiders, houseflies, and their relatives.

chlorophyll: The principal green pigment active in photosynthesis.

chloroplast: The chlorophyll-containing plastid organelle in plant cells that is the site of photosynthesis.

chorion: One of the four extraembryonic membranes surrounding the embryos of land vertebrates. The chorion and the allantois fuse to become the chorioallantois.

chorionic villi: The thousands of finger-like projections that expand outward from the fluid-filled chorion and provide a huge surface area for the exchange of materials.

chrom-, -chrome: From Greek, meaning "color" or "pigmented."

chromaffin cell: Any of the endocrine cells of the adrenal medulla that secretes epinephrine and norepinephrine.

chromatid: One of the two copies of a chromosome in a cell undergoing division.

chromatin: The chromosomal substance within the nucleus of a nondividing cell which consists of DNA, histones, and nonhistone proteins.

chromosomal mutation: A change in the physical structure of a chromosome.

chromosome: A long strand of coiled DNA (deoxyribonucleic acid) that is the site of genes, the genetic information for most organisms. Eukaryotic chromosomes also include many proteins.

chyle (kile): A whitish, watery solution of partially digested food material produced by the neutralization (by bile and pancreatic juices) of chyme.

chyme (kyme): The semifluid mass of food material produced by the action of digestive juices in the stomach; the material that passes from the stomach to the small intestine.

ciliate: A single-celled organism characterized by the many cilia on its surface.

cilium (plural: **cilia**): A short, hairlike organelle that has a microtubular skeleton and a capacity to beat, thereby propelling a cell through fluid or a fluid past a cell.

circulatory system: A transport system in animals, consisting of specialized structures such as vessels and a heart and blood, which delivers nutrients and other essential materials to cells and carries away metabolic wastes.

clade: A group of organisms that shares sets of specific characteristics that reflect common descent; hence, organisms belonging to a single taxonomic unit.

cladistics: Classification of organisms based on the historical sequences by which they have diverged from common ancestors.

class: A taxonomic grouping of organisms belonging to related orders.

clay: A dense soil material made up of very fine particles of aluminum, silica, and other minerals and produced by the weathering of rocks.

cleavage: A specialized form of mitosis (cell division) in embryos in which no cell growth occurs between divisions.

climax community: A stable ecological community dominated by species that tend not to be replaced by new species.

cline: A gradual change in one or more characteristics among members of a species over a broad geographical range.

clonal selection theory: The theory stating

that many precommitted types of T and B lymphocytes are present in an organism, and that a newly invading antigen binds to one, thereby "selecting" it for activation.

clone: An offspring that arises mitotically or asexually and is genetically identical to its parent.

closed circulatory system: A circulatory system in which blood moves through a continuous set of interconnected vessels.

closed (behavioral) program: An innate motor pattern that cannot be altered easily by learning.

co-: From Greek, meaning "a shared condition."

coacervate: A polymer-rich droplet having certain cell-like properties.

coated pit: Indentations in the cell's plasma membrane that are coated or lined with a protein called clathrin and that can pinch off coated vesicles; these vesicles are thought to be involved in the importation of substances across the cell surface.

cochlea: A set of spiraled tubes in the vertebrate inner ear; in mammals, the primary structure involved in hearing.

codominance: When both alleles in a heterozygotic individual are expressed in the organism's phenotype.

codominant gene: A gene with several alleles, two of which are equally dominant. Both dominant alleles are fully expressed when they appear together, and both phenotypic traits are present.

codon: A set of three consecutive nucleotides that code for a single amino acid.

coelom: A fluid-filled body cavity that is lined by mesoderm and in which a variety of body organs are suspended.

coenocytic (see-no-**sit**-ik): Having multiple nuclei in one mass of cytoplasm.

coevolution: A process in which different life forms undergo interrelated, often complementary, evolutionary change.

cofactor: A special enzyme substrate that binds temporarily to a site on an enzyme and participates in chemical reactions and the formation of products.

cohesion: The tendency of like molecules to cling to one another.

cohesion-adhesion-tension theory: The idea that water is pulled up through the xylem due to transpiration from the plant's leaves, the adhesion of water to plant vessel walls, and the cohesion of water molecules to each other rather than being pushed upward due to root pressure.

coleoptile: A specialized sheath that protects the growing shoot tissue of a monocot embryo.

collecting duct: The duct associated with a nephron in the vertebrate kidney in which water is removed from the filtrate and returned to the blood.

collenchyma: A plant cell type having especially strong primary walls that lend extra support to stem tissue.

colloidal osmotic pressure: Osmotic pressure created by the presence of proteins (colloids) in blood or body fluids.

colostrum: An antibody- and protein-rich fluid synthesized and stored prior to birth in the mammary glands of a pregnant mammal; when fed to the newborn, it gives immediate resistance to a variety of potentially dangerous microbes.

commensalism: A form of symbiosis in which members of two species live in intimate association, with one partner benefiting while the other is unharmed.

communication: The transmission of information from one organism to another.

community: All of the interacting populations of different species in a particular area.

companion cell: In a flowering plant (angiosperm), a specialized cell associated with sieve tube elements, the cells that make up the transport tissue phloem.

compatible osmolyte strategy: The presence of nonharmful organic molecules in body fluids of organisms to raise total osmotic pressure without inhibiting proteins.

compensatory hypertrophy: A growth response sometimes seen in damaged organs, in which cells in remaining healthy tissue divide extremely rapidly so that the affected organ increases in mass and cell number.

competition: Fighting for limited resources among members of the same or different species.

competitive exclusion principle: The postulate that no two species can occupy the same niche at the same time in a particular locale.

competitive inhibition: A system of enzyme control in which the active site is occupied by a compound other than the normal substrate, thereby preventing the binding of the substrate.

complementary gene: A gene whose protein product must act with the product of another gene to produce a given phenotype. Human albinism is produced by complementary genes.

complement response: In the immune system, a process whereby proteins bind to antibody-invader cell complexes and perforate the invader's membrane, killing the invader.

compound: An aggregate of atoms of more than one element.

compound eye: An eye composed of many separate optic units (ommatidia).

concentration gradient: A variation in the concentration of a substance, as from higher to lower.

condensation reaction: A chemical process which takes place when two monomers join; typically, as a covalent bond forms between them, one monomer loses an –OH group and the other loses a hydrogen atom.

cone: One of two types of photoreceptors in the eyes of many vertebrates; responsive to higher light intensities, cones provide high visual acuity and color vision.

conidium: A spore generated during asexual reproduction in certain fungi.

conifer: A cone-bearing plant.

conjugation: An exchange of genes in which DNA is transferred from one (usually) bacterial cell to another by way of a cytoplasmic bridge.

connective tissue: The mesodermally derived, collagen-fiber-containing tissue that surrounds most animal organs and literally connects them together.

constant region: That portion of an antibody protein that has the same amino-acid sequence as other, different antibodies.

consumer: An animal that derives energy from eating all or parts of other animals and plants.

continental drift: The movement of the Earth's massive tectonic plates over the underlying mantle, with formation or subduction of land.

control experiment: As part of the scientific method, the step in which all of the experimental factors are kept constant except for the single parameter in question.

convergent evolution: An evolutionary pattern in which distantly related organisms become more alike as they evolve similar adaptations.

coralline alga: A red alga that can deposit hard calcium-carbonate crystals in its cell walls, creating a coral reef.

core (of the Earth)**:** The extremely dense mixture of materials, primarily iron and nickel, that makes up the innermost region of the Earth.

Coriolis effect: Earth's rotation applies a force on the atmosphere and oceans that skews wind and ocean currents in different directions depending on the hemisphere.

cork: A tough surface tissue in woody plants produced by cell divisions in the cork cambium; cork replaces the epidermis, becomes impregnated with suberin, and serves as a waterproof barrier.

cork cambium: A layer of cells just beneath the epidermis of a woody plant which produces cork, the outer, nonliving component of bark.

corm: A solid, subterranean stem structure that serves as a site for carbohydrate storage and vegetative reproduction.

cornea: In the eye, the transparent portion of the sphere through which light enters.

corolla: Collectively, the petals of a flower.

corpus callosum: A giant bundle of nerve axons that links the two hemispheres of the vertebrate brain.

corpus luteum: Literally, "yellow body"; a cell mass that forms from an ovarian follicle after the egg has been released, and which secretes progesterone.

cortex: (1) In a plant, the region of the stem underlying the epidermis and composed mainly of parenchyma cells. (2) The outer region of some animal organs, as the kidney, adrenal gland, or brain; see *neocortex*.

cortical reaction: A series of events occurring immediately after an egg is fertilized by a sperm and during which the egg's internal electrical charge is reversed long enough to prevent the entry of additional sperm.

corticosteroid: Any of the steroid hormones produced in the adrenals; divided into glucocorticoids, mineralocorticoids, and sex steroids.

corticotropin: Also known as adrenocorticotropic hormone (ACTH), this substance regulates corticosteroid production by the adrenal cortex.

co-transport: A form of active transport in which sodium ions and a sugar or amino-acid molecule bind to a carrier that transports them together and discharges them inside the cell.

cotyledon: An embryonic seed leaf that stores nutrients to sustain the growth of a newly germinated plant.

counteracting osmolyte strategy: The maintenance of osmotic pressure in body fluids through the presence of pairs of inhibitory and stimulatory substances such as urea and trimethylamine oxide.

countercurrent exchange system: The flow of two fluids, which are separated by a permeable interface, in opposite directions; the result is much more efficient exchange of commodities that can pass through the interface (heat, O_2, etc.).

coupled reactions: Exergonic reactions are energetically linked to endergonic reactions in such a way that energy freed during a exergonic reaction drives an endergonic one.

coupling factor: Large, complex molecules that are sites where the flow of protons through usually impermeable membranes drives the production of high-energy compounds (as ATP).

covalent (coe-vay-lent) **bond:** A chemical bond between atoms characterized by the sharing of electrons.

creatine phosphate: An organic molecule with a high-energy phosphate bond, contained in the red and white muscle tissues of vertebrates and elsewhere.

crop: In some animals, a thick-walled chamber that receives and temporarily stores food materials.

crossing over: The exchange of corresponding pieces of genetic material between a maternal chromatid and a paternal one.

crust (of the Earth): The rocky outermost layer of the Earth.

cryptic coloration: Having the same color or pattern as the background; a form of defensive camouflage.

cultural evolution: The transfer of information from generation to generation of animals by nongenetic means.

cuticle: The waxy outer layer of cutin on leaves and stems that inhibits the loss of moisture from within.

cutin: A transparent, waxy substance secreted by the epidermal cells of leaves and other structures that helps seal moisture into the tissues.

cyanobacteria: Formally called blue-green algae, single rod-shaped or spherical prokaryotic cells that occur in clusters or in long filamentous chains and carry out photosynthesis by means of chlorophyll a, carotenoids, and red and blue pigments.

cycad: A primitive seed plant; a member of the division Cycadophyta.

cyclic AMP (cAMP): Cyclic adenosine monophosphate; a compound related to ATP which acts as an intermediate or second messenger in the transmission of signals within target cells.

cyclic GMP (cGMP): Cyclic guanosine monophosphate; a second-messenger compound.

cyclic photophosphorylation: A recurring ATP-generating process in photosynthesis that is driven solely by light energy.

-cyte, cyto-: From Greek, meaning "hollow vessel"; refers to cells.

cytochrome: Any of a class of iron-containing proteins that acts as carriers in the electron transport chain of cellular respiration.

cytokinesis: The separation of the cytoplasm of a dividing cell into two daughter cells.

cytokinin: Any of a class of plant hormones that regulates cell division.

cytoplasm (**site**-oh-plazm): The semifluid substance that makes up the nonnuclear part of a eukaryotic cell, or the nonnucleoid part of a prokaryotic cell.

cytoplasmic streaming: The rapid movement of cell cytoplasm to circulate nutrients, proteins, pigments, or other cellular materials or organelles.

cytosine: A single-ring pyrimidine base that forms a DNA nucleotide; complementary to guanine in the DNA double helix.

cytoskeleton: The three-dimensional weblike structure that fills the cytoplasm of a cell, and within which organelles are suspended.

dark reactions: The second-stage reactions of photosynthesis, which do not require light energy to proceed and in which CO_2 is reduced to carbohydrate.

deciduous: In plants, the property of shedding leaves at the end of the growing season.

decomposer: An organism that derives energy for its life processes by breaking down remnants of organic materials.

dedifferentiation: A developmental process during which cells lose their functional phenotype and divide rapidly to generate a population of cells that can, for instance, reform lost parts.

deductive reasoning: A process of predicting new facts or processes for which new experiments can be designed and new information collected.

dendrite: Any of the short, multibranched extensions of the neuron surface that receives input to the cell.

denitrification: The reduction of nitrate and nitrite to gaseous N_2, NO, or N_2O.

density-dependent factor: Any factor, such as disease or competition, whose effect on a population's numbers is related to population density and size.

density-independent factor: Any factor, such as a natural calamity, that regulates population size independent of population density and size.

depolarization: A change in a cell's electrical state toward a nonpolarized condition.

desert: A very dry, often rather barren biome found in subtropical latitudes or where mountains block moisture-bearing winds.

desmosome: An intercellular junction thought to "glue together" the plasma membranes of adjacent cells.

determination: The final commitment of cells to a single developmental pathway, from among several alternatives.

detritivore: An organism that obtains food energy by consuming disintegrated organic matter.

deuterostome: Any member of the lineage of animals in which the blastopore of the developing embryo becomes the anus, while a second opening becomes the mouth.

di-: From Greek, meaning "two."

diacylglycerol (DG): A second-messenger compound; see *cyclic AMP*.

diaphragm: A muscular sheet that separates the thoracic cavity from the abdominal cavity in mammals; active in the inhalation-exhalation mechanism of breathing.

diastole (die-**ast**-uh-lee): The phase of relaxation of the heart muscle.

diatom: A phytoplankton organism having sculptured, glasslike cell walls containing silica.

diatomaceous earth: The hard silica deposits that remain after billions of diatoms die and their glassy "shells" accumulate.

dicotyledon: An angiosperm in which embryos have two seed leaves.

differential gene activity: The turning on and functioning of some genes during the synthesis and processing of RNA, while other genes are turned off and inactive.

differentiation: During development, the phenotypic maturation of cells in both structure and function.

diffusion: The process by which a dissolved substance moves passively through a fluid.

digestion: The chemical breakdown of foods into compounds that can be used for cellular metabolism.

dihybrid cross: In genetics, a cross between parents that are heterozygous for two characteristics.

dikaryon: A specialized cell type in fungal hyphae which contains two nuclei and plays a role in spore formation.

dinoflagellate: A single-celled phytoplankton, often photosynthetic, having a set of flagella that causes the organism to spin as it swims.

dioecious (dy-**ee**-shus): Plant species having individuals that are either male or female; hence individuals produce only one of the two types of gametophytes.

diploid: Literally, "paired"; having two sets of homologous chromosomes in each cell.

directional selection: Natural selection in which the result is a change from one phenotype to another.

disaccharide: A sugar made up of two monosaccharides, such as glucose and fructose, joined together by a glycosidic bond.

disruptive coloration: Having body parts of contrasting colors which break up an organism's silhouette.

disruptive selection: Natural selection in which a population may be separated into discrete groups because the "average" phenotype for a particular trait is strongly selected against.

divergent evolution (radiation): The evolutionary pattern in which one or more phylogenetic lines branch and rebranch as organisms acquire an expanding repertoire of adaptations to new environmental circumstances.

diversifying selection: A form of natural selection that helps to generate distinctive phenotypes adapted to specialized features of a portion of the total environment.

diving reflex: In marine mammals, a complex set of responses to swimming downward in the sea during which the heart rate slows and blood flow is diverted from other body regions mainly to the heart muscles, the brain, and skeletal muscles.

division: In botany, a taxonomic grouping of organisms belonging to similar classes; the equivalent of a phylum.

DNA ligase: Any of a class of enzymes which can rejoin complementary cohesive ends of DNA fragments.

DNA polymerase: An enzyme that catalyzes the synthesis of a DNA strand complementary to an original strand of the parent molecule.

DNA replication: The faithful formation of daughter molecules after the helically entwined chains of a DNA molecule separate and act as templates for the newly forming molecules.

DNA/RNA hybridization: A technique for creating a molecular probe in order to locate a specific gene sequence.

dominance hierarchy: A pecking order within the ranks of a social animal species, often reinforced by aggressive behaviors or threats, that confers varying privileges and responsibilities upon the members.

dominant (allele): An allele whose action masks that of another (recessive) allele of the same gene and thereby determines the phenotype of a heterozygous offspring.

dormancy: A resting state, as of seeds, in which an organism or cell respires at a very low rate and carries out only a small amount of metabolic work.

double fertilization: In angiosperms, the process in which one sperm from a pollen grain fuses with the small egg cell of the megagametophyte, while the second sperm penetrates the adjoining large endosperm cell containing the two polar nuclei.

double helix: The "twisted ladder" structure of linked double strands of nucleotides that comprises a DNA molecule.

Down syndrome: A genetic disorder, based on nondisjunction of chromosome 21, that is characterized by mental retardation, heart defects, a short, stocky body, and distinctive eyelid folds, among other traits.

DPG (2,3-diphosphoglycerate): A substance in mammalian red blood cells that decreases the binding affinity of O_2 to hemoglobin.

dynein: A mechanoenzyme, associated with microtubules, that cleaves ATP. Dynein arms may generate the forces that cause cilia and flagella to beat.

early wood: Xylem cells formed in the early part of a tree's growing season.

ecdysone: Any of a family of steroid hormones that regulates molting in insects and other arthropods; alpha ecdysone is the precursor of the substance that initiates insect metamorphosis.

echolocation: A sensory system in which high-frequency sound pulses are used to produce echoes that convey information on the location of objects in an animal's environment.

ecocline: Variation in the composition of plant or animal communities along a climate gradient—usually of moisture or temperature.

ecological efficiency: The efficiency of energy transfer between adjacent trophic (feeding) levels in a food chain.

ecological isolation: The presence of genetic differences between populations that prevent interbreeding and that have arisen as adaptations to particular features of the environment.

ecological succession: Gradual changes in the composition of the species that make up a community.

ecology: The study of the interplay of organisms and their environments.

ecosystem: Any community of interacting organisms, including their physical and chemical environment, energy fluxes, and the types, amounts, and cycles of nutrients in the various habitats within the system.

ecto-: From Greek, meaning "outer."

ectoderm: The outer layer of cells that arises during cleavage, and later develops, in vertebrates, into the epidermis, the nervous system, and the sense organs.

ectotherm: An animal that derives most of its body heat from the external environment.

effector (motor) neuron: Any of the class of neurons that transmits messages to muscles and glands.

electron: A subatomic particle that bears a negative electrical charge and moves continuously about the nucleus.

electronegativity: The ability of an atom within a molecule to attract electrons from other atoms, which then spend more time near the nucleus of the electronegative atom.

electron transport chain: The final phase of cellular respiration, in which the compounds NADH and $FADH_2$ are oxidized and their electrons passed along a chain of oxidation-reduction steps.

electrophoresis: A technique used to trace the movement of electrically charged molecules through a fluid medium.

element: A substance that cannot be decomposed by chemical processes into simpler substances.

elongation: (1) In all cells, the stage of protein synthesis in which amino acids are added sequentially to the lengthening polypeptide chain. (2) In plant cells, the increase in the length of cells to cause stem or root growth.

enantiomer: Mirror-image arrangements of atoms within molecules of identical chemical make-up. Enantiomers of a compound share the same chemical properties, but they will rotate light passing through them in opposite directions.

end-, endo-: From Greek, meaning "inside."

endergonic reaction: Any chemical reaction in which the products have more total energy and more free energy than did the reactants. Endergonic reactions thus require the input of energy from another source before they can take place.

endocrine system: The animal organs, tissues, and cells which secrete hormones.

endocytosis: A process through which materials are taken into a cell.

endoderm: The innermost layer of cells arising during cleavage and gastrulation, and ultimately forming the lining of the gut and other internal epithelial tissues.

endodermis: In a plant, the innermost layer of the cortex tissue.

endolymph: The fluid that fills the semicircular canals of the vertebrate ear.

endometrium: The lining of the uterus.

endoplasmic reticulum (ER): An array of membranous sacs, tubules, and vesicles within the cell cytoplasm. ER may be rough or smooth; the rough form is studded with ribosomes and is the site of synthesis of proteins that will be secreted or built into membranes.

endorphin: A neuroactive peptide in the brain that acts as a natural painkiller; see also *enkephalin*.

endospore: A heavily encapsulated resting cell formed within many types of bacterial cells during times of environmental stress.

endosymbiont theory: A theory explaining how eukaryotic cells may have arisen from earlier prokaryotes as progenitor cells became "colonized" by other simple prokaryotes, and mitochondria, chloroplasts, and other cell organelles evolved.

endotherm: An animal that produces its own body heat.

endotoxin: Toxin released when bacterial cells die and burst.

enkephalin: A five-amino-acid neuroactive peptide synthesized in the brain; along with endorphins, enkephalins are natural opiates.

enterogastric reflex: A reflex that inhibits the release of *chyme* into the small intestine, thereby allowing the small intestine more time to digest the material it already contains.

entropy: The amount of disorder in a system.

enzyme: A member of the class of proteins (and certain RNAs) that catalyze chemical reactions.

enzyme saturation: A condition in which all of the active sites on one or more available enzyme molecules are occupied by substrate molecules most of the time.

enzyme-substrate (ES) complex: A complex consisting of an enzyme and its reactant (substrate) which is held together by weak bonds. The formation of an ES is the crucial first step in enzyme catalysis.

epicotyl: The structure in a plant embryo that is the future stem of the seedling.

epidermis: In plants, a surface layer of cells—usually one cell thick—that reduces moisture loss (from leaves and stems), or is the site of water uptake (in roots). In animals, the outermost tissue.

epinephrine: A substance produced by the adrenal medulla which can function both as a neurotransmitter and as a hormone; also known as adrenalin.

epiphyte: Any of the so-called air plants that usually grow on other plants rather than being rooted in soil.

episome: A genetic element that can replicate in a bacterial host independently of the chromosome, or that can integrate into the chromosome and replicate with it.

epistasis (eh-**pee**-stay-sis)**:** A type of gene interaction in which the effects of one gene override or mask the effects of other, entirely different genes.

epithelial tissue: A single or multilayered sheet of cells bounded on one side by a ruglike basement membrane.

epithelium (plural: **epithelia**)**:** A population of animal cells arranged in a sheet. Epithelial tissue lines or covers a variety of internal and external body surfaces.

equilibrium (chemical): The point in a chemical reaction at which no further net conversion of reactants takes place. When equilibrium is reached, the combined free energy of reactants equals the combined free energy of reaction products.

ergot: The dark purple or black spore case of the fungus *Claviceps purpurea*, which often grows on grains such as rye and is the source of the hallucinogenic drug LSD.

erythrocyte: A red blood cell.

erythropoetin: A peptide hormone secreted by the vertebrate kidney that stimulates the production of red blood cells in bone marrow.

esophagus: The tubular passageway through which food travels from the pharynx to the stomach.

essential amino acid: Any of the eight amino acids that adult humans cannot synthesize and must obtain from food.

ester bonds: Chemical bonds which form between any alcohol and any carboxylic acid.

estrogen: One of the gynogens, or female sex hormones.

estrus: The period during which a female mammal is receptive to sexual activity.

ethology: The study of an animal's feeding, mating, and defense strategies in the wild.

ethylene: The plant hormone responsible for ripening in fruit.

eubacteria: Literally, "true bacteria"; by far the most abundant group of prokaryotes.

eukaryotic cell: A cell that possesses a membrane-enclosed nucleus, chromosomes built of DNA and protein, a cytoskeleton, and a variety of membrane-bound organelles.

eutrophication: The gradual accumulation of organic debris and silt in a lake as the result of high productivity of the organisms that inhabit it.

evolution: Change in a gene pool (and corresponding phenotype) over time.

ex-, exo-: From Greek, meaning "outside" or "external to."

excitatory synapse: A synapse in which the synaptic transmission between neurons depolarizes the receiving neuron.

excitement phase: The first phase of human sexual response in which sex organs become engorged with blood, and blood pressure, heart rate, and breathing rate increase.

excretory system: Organs such as the kidney, bladder, and associated structures that regulate osmotic balance in body fluids and eliminate metabolic wastes from the body.

exergonic reaction: An energy-liberating chemical reaction in which the products contain less total energy and less usable energy than existed in the original reactants.

exocytosis: A process, often involving a vacuole or vesicle, through which materials are expelled from a cell.

exon: Any of the protein-coding segments of a gene.

exoskeleton: The tough, rigid, outer covering (cuticle) that encloses and protects the body of an arthropod.

exotoxin: Toxin secreted by intact bacterial cells of certain species.

exponential growth curve: The steeply climbing curve describing unchecked population growth, in which population size repeatedly doubles.

external intercostal: A diagonal set of muscles between each pair of ribs in humans that contract to draw air into the lungs.

extinction: The permanent loss of a species.

extracellular digestion: Digestion carried out by enzymes secreted outside of cells, as into an organism's gut cavity.

extracellular matrix: Material made up primarily of space-filling sugar polymers and collagen fibers that contributes to the bulk of bones, cartilage, the eye cavities, and other body parts.

extrinsic protein: A protein molecule that is attached to the outer surface of the cell's plasma membrane.

F_1 (first filial) generation: In a genetic cross, the first generation of offspring from a set of parents.

F_2 (second filial) generation: The offspring from a genetic cross between members of the same F_1 generation.

facultative anaerobe: A bacterium that can grow with or without the presence of free oxygen.

FAD (flavin adenine dinucleotide): Along with NAD^+, a coenzyme that carries electrons and hydrogen in a variety of metabolic oxidations and reductions, such as those of the Krebs cycle.

Fallopian tube (oviduct): One of the two hollow canals that extends from the ovaries to the uterus (in mammals), and that conduct ovulated eggs toward the uterus.

family: A taxonomic grouping of similar genera.

fat-soluble vitamin: A vitamin, such as A, D, E, or K, that is transported in the blood as a complex linked to lipids or proteins.

fatty acid: A molecule consisting of a long chain of carbon atoms attached to an acidic carboxyl group (–COOH). Fatty acids are basic units of fats and oils.

feedback inhibition: A type of metabolic pathway control that regulates the rate at which cells synthesize amino acids (or other monomers) and use them in building proteins (or other polymers).

fermentation: A set of anaerobic reactions in which pyruvate generated by glycolysis is modified to ethanol, lactate, or some other organic end product.

fertilization: The process that unites the nuclei of male and female gametes and initiates development.

fetus: A human embryo during its main growth period, the second two trimesters of pregnancy.

F factor: A DNA fragment in male bacterial cells that codes for a cytoplasmic bridge through which genes are transferred to other bacterial cells.

F factor plasmid: A self-replicating circle of bacterial DNA that carries the F factor genes, which are transferred to a recipient cell during conjugation.

fiber: A long sclerenchyma cell found in bundles or cylinders, often toward the periphery of a stem or root.

fibrous root system: A root system consisting of many equal-sized roots.

filamentous alga: Any of a group of green algae with a specialized, threadlike body form.

filter feeder: An invertebrate organism that feeds by straining microscopic organisms from water pumped through its hollow body. Some insects, birds, and mammals also have structures to filter food particles.

first law of thermodynamics: The physical law stating that energy can change from one form to another form, but can never be created or destroyed. Also termed the law of conservation of energy.

fitness: The relative reproductive efficiency of various individuals or genotypes in a population.

5′ to 3′ direction: The direction in which one strand of the DNA molecule is oriented and in which DNA lengthens during replication.

fixed allele: An allele that is the sole allele for a given gene in a population, because other alleles have been lost.

fixed motor pattern: An unvarying series of precise physical movements, thought to have a genetic basis.

flagellum (plural: **flagella**): A fine, whiplike, microtubule-containing organelle that undulates to move eukaryotic cells.

florigen: A plant hormone that stimulates the formation and opening of flowers.

flower: The reproductive structure of a flowering plant. Flowers attract animals that inadvertently help disperse pollen to fertilize other plants or disperse the seeds encased in the fruit or ovary.

flower bud: A bud produced by an apical meristem that will develop into a flower.

fluid-mosaic model: A model explaining the properties of cell membranes. The membrane structure includes a lipid bilayer with several types of proteins embedded and protruding. At normal biological temperatures, the plasma membrane acts like a thin layer of fluid across which proteins move freely, like icebergs in a lipid sea.

follicle-stimulating hormone (FSH): A gonadotropin hormone secreted by the pituitary gland that supports sperm production in males and stimulates egg maturation in females.

food chain: The chain of events in which plants in a community convert solar energy to a stored chemical form, and that energy (and material) is sequentially transferred to herbivores, carnivores, and decomposers.

food niche: A major division of the factors defining an organism's niche or ecological role and function. The food niche includes the types and size ranges of food or prey as well as its distribution.

food web: A complex system of numerous food chains.

foraminiferan: A marine protozoan that secretes a calcium-containing shell.

forebrain: The anterior region of the vertebrate brain, encompassing the cerebrum, pituitary gland, thalamus, hypothalamus, and olfactory bulb.

founder effect: An increased or decreased frequency of certain alleles in a new population because the founding members chanced to differ in those alleles from the original population.

fovea: In primates and some other vertebrates, a region in the neural retina where maximal resolution of images is achieved.

frameshift mutation: A genetic mutation that arises from a shift in the normal reading frame of a nucleotide sequence.

free energy: The energy available to do work as a result of a chemical reaction.

frequency-dependent selection: A process that operates when the relative fitnesses of the genotypes in a population vary according to their frequency.

frond: The leaf of a fern or large alga.

fruit: A mature ovary or group of ovaries that surrounds and protects a plant seed, and aids in its dispersal.

fruiting body: In fungi, the structure that carries sexually produced spores.

fugitive species: A plant species that typically is one of the first to colonize harsh terrain, and is capable of maturing and reproducing quickly.

functional group: A cluster of atoms that imparts a similar chemical behavior to all molecules to which it is attached.

fundamental niche: The full environmental range that a species can occupy if the proper physical and biological conditions are met and if there is not direct competition from another species.

fusiform initial cell: In a vascular plant, the source of xylem and phloem cells.

G_1 phase: The period of normal metabolism (gap 1) that is the first phase of the cell cycle.

G_2 phase: The brief period of cell metabolism and growth (gap 2) that follows the S (DNA synthesis) phase in the cell cycle and is a prelude to cell division.

gallbladder: A small organ near the liver in which bile is stored.

gametangium (plural: **gametangia**): In certain fungi and other organisms, a structure that contains gametes.

gametophyte: A plant in the haploid, gamete-producing stage of its life cycle.

ganglion (plural: **ganglia**): An aggregation of nerve cells.

ganglion cell: Any of the neurons associated with rods and cones whose axons extend via the optic nerve to the vertebrate brain.

gap junction: A perforated channel which serves as the primary electrical, ionic, and molecular communication junction between animal cells.

gastrin: A digestive hormone (also a neuropeptide) secreted in the stomach that causes the secretion of other digestive juices.

gastrocoel: The central gut (previously called the coelenteron) of cnidarians.

gastrula: The three-layered early embryo that carries out gastrulation.

gastrulation: The infolding process in the gastrula-stage embryo that creates a complex, three-dimensional organism from the simpler blastula.

gene: The basic unit of heredity. Residing on chromosomes, genes consist of linked sequences of nucleotides that provide the blueprints for the construction of all cellular proteins and RNAs.

gene amplification: A process in which extra copies of a gene are manufactured, such as those needed to generate large numbers of ribosomes during oogenesis.

gene flow: A change in gene frequencies as the result of interbreeding between members of two populations.

gene map: A map that shows the relative positions of and distances between genes on a chromosome.

gene pool: All the alleles of all the genes carried by individuals in a population.

generalist: A species that is adapted to a wide range of living conditions.

gene regulation: The process by which a special class of genes called regulatory genes controls the shape and pattern of body structures coded by structural genes.

genetic code: The molecular "grammar" in genes that relates nucleic-acid bases in nucleotides to amino acids. The functional units of the code, codons, consist of sets of three nucleotides.

genetic counseling: Advice provided to couples who are at risk for producing a genetically defective child or who have had such a child.

genetic drift: A change in gene frequency from generation to generation as the result of chance events.

genetic engineering: The science of manipulating genes between chromosomes and between organisms using recombinant DNA technology. These techniques have revolutionized biological research and have many applications to medicine, agriculture, and other fields.

genetic identity: A statistical measure of the degree of similarity among structural genes in related organisms. The number of shared structural genes decreases proportionally in larger taxonomic groups.

genetic load: The sum total of those alleles in a genome that yield some advantage when they are heterozygous, but are lethal or deleterious when homozygous.

genome (genotype): The full set of genes on an organism's chromosomes.

genotype: The complete genetic make-up of a cell or organism, part of which is expressed as the organism's phenotype or physical appearance.

genus (plural: **genera**): A taxonomic grouping of very similar organisms considered to be closely related species.

germination: The reactivation and subsequent growth of a pollen grain or of a plant seed.

germ plasm theory: A theory stating that only hereditary information in the "germ plasm" of the gametes transmits traits to the progeny of a multicellular organism, so other body cells (somatic cells) do not make a contribution.

gibberellin: Any of a class of growth-promoting plant hormones that acts to increase stem length.

gill: Any of the specialized gas-exchange organs of many aquatic animals.

gill slit: An opening of the pharynx through which water passing over an aquatic animal's gills leaves the body. Gill slits are an ancient feature of chordates and still occur in adult fish and other vertebrate embryos.

ginkgo: The maidenhair tree, last remaining member of the division Ginkgophyta.

gizzard: In birds and some invertebrates, a large muscular chamber in which food particles are pulverized to aid digestion before passing to the stomach or intestines.

glial cell (glia): A cell in the animal nervous system that supports neurons metabolically and serves as electrical insulating material around axons.

glomerular filtrate: The solution of wastes and other substances that passes from the blood into the cavity of a nephron.

glucagon: A hormone that, along with insulin, regulates glucose levels in blood.

glucocorticoid: Any of a group of steroid hormones produced in the adrenal gland that stimulates the production of glucose from proteins and carbohydrates.

glucose: A six-carbon monosaccharide which is a universal cellular fuel.

glycerol: A water soluble organic substance that is one of the basic components of fats and oils.

glycocalyx: Literally, "sugar coat"; an agglomeration of complexes of sugar polymers, proteins, and sometimes lipids that are found on the surface of animal cells.

glycogen: A polysaccharide sugar used by animals to store glucose.

glycolysis: The first phase of energy metabolism in cells. By way of the multistep glycolysis pathway, a single six-carbon glucose molecule is broken down to yield two molecules of the three-carbon compound pyruvate, two molecules of NADH, and two molecules of ATP.

glycosidic bond: A covalent bond linking two monosaccharides via an oxygen atom.

gnetina (net-**eye**-na): Any member of a small, diverse group of gymnosperms, including certain tropical vines and arid region shrubs.

golden-brown alga: Any of a group of phytoplankton having a golden color created by the presence of the carotenoid pigment fucoxanthin.

Golgi complex: A membranous organelle in the eukaryotic cell cytoplasm that is specialized for the modification, transport, storage, or secretion of proteins.

gonial cell: Any of the specialized cells in the gonads that gives rise to sperm or to eggs.

gradualism: The theory that evolutionary changes accumulate in species gradually over immense stretches of geological time.

gram molecular weight: The sum of the atomic weights of the atoms in a molecule, equal to the weight of one mole of the molecule.

gram-negative bacteria: Bacteria, such as *E. coli*, whose cell walls are surrounded by a lipid bilayer and hence do not take up iodine dye (crystal violet) during staining.

gram-positive bacteria: Bacteria having a peptidoglycan cell wall, which takes up crystal violet dye and hence appears to stain purple under the light microscope.

grand postsynaptic potential (GPSP): The summation of all the excitatory and inhibitory PSPs active on a neuron over any short time span.

granulocyte: A type of white blood cell that holds granules containing enzymes or other substances involved in inflammatory or allergic reactions.

granum (plural: **grana**): A stack of thylakoids arranged like a pile of coins within a chloroplast.

gravitropism: Growth of plant roots and shoots toward and away from, respectively, the direction of gravitational force.

gross production: The total amount of organic matter produced via photosynthesis, including energy consumed by the plants' own metabolism and energy available to other members of the community.

ground parenchyma: In a monocot plant, an epidermal tissue in which bundles of vascular tissue are scattered.

group selection: The theory that natural selection favors traits or gene combinations which contribute to the survival of an entire group of the same species.

growth hormone (GH): A hormone secreted by the anterior pituitary lobe that acts to stimulate the growth of long bones during maturation.

guanine: A double-ring purine base that is a fundamental structural element of some DNA nucleotides.

guard cell: One of a pair of cells which borders each stoma (pore) on a leaf and regulates the opening and closing of the pore.

gustation: The sense of taste.

guttation: The formation of water droplets on pores at the edge of a leaf.

gymnosperm: Any of the broad group of nonflowering seed plants, such as pines and spruces, in which both ovules and seeds are borne on the surface of the sporophyte.

gynogen: Any of the female sex hormones.

habitat: The actual place where an organism lives.

habitat niche: The physical parameters of an organism's niche—soil, vegetation, light, and other factors—that restrict where an organism can live.

habituation: A phenomenon in which repeated exposure to an environmental stimulus lessens an animal's responsiveness to it.

halobacteria: Members of the archaebacteria that live in conditions of high salt or very high temperatures, as in volcanoes.

haploid: The condition of a cell that contains only one set of parental chromosomes, or of an entire organism whose cells have only one set.

Hardy-Weinberg law: A series of mathematical statements that describe the conditions necessary for the frequencies of alleles (genes) in a population to remain stable over time.

harem: A basic social unit in some mammals that consists of one male, several females, and their offspring.

haustorium (plural: **haustoria**): In fungi, a feeding structure that penetrates the living cells of other organisms and absorbs nutrients.

Haversian system: The basic component of compact bone, in which cylindrical layers of bone surround a single blood capillary.

heart: A region or organ made up of specialized muscle tissue that acts as a pump to propel blood or hemolymph.

heartwood: Dark wood at the center of a tree trunk, composed of dead xylem which no longer conducts water or nutrients.

heat of vaporization: The amount of heat needed to turn a given amount of liquid water into gas (water vapor).

heavy chain: In an antibody molecule, one of two polypeptide chains connected to two light chains forming a Y-shaped molecule.

hemocoel: A blood-filled cavity.

hemolymph: The extracellular fluid in the body cavity that bathes the tissue cells in animals with open circulatory systems. It is grossly similar in composition to intracellular fluid, containing mostly water with dissolved nutrients, ions, gases, and some macromolecules.

hemophilia: A genetic disease linked to the X chromosome in which the body lacks a protein necessary for normal blood clotting.

hemoskeleton: A noncompressible, blood-filled space (sinus) against which the muscles of a mollusk push to generate movement.

herbaceous: In a plant, the property of having a relatively thin, soft, nonwoody stem.

herbivore: An animal that eats only plant material.

hermaphrodite: An individual born possessing both testes and ovaries.

hetero-: From Greek, meaning "different."

heterokaryon: A single cytoplasm with dissimilar nuclei.

heterokaryosis: A process that increases the genetic variability of fungi by forming heterokaryons, in which genetically different nuclei reside in the same cell.

heterospore: Two different types of spores in lycophytes and certain other plants that differentiate into either male or female gametophytes.

heterosporous: The property in certain plants of having two types of spores, which yield male and female gametophytes respectively.

heterotroph: An organism that obtains energy for cellular processes by taking in food consisting of whole autotrophs or other heterotrophs, their parts, or their waste products.

heterozygote advantage: A condition said to exist when a heterozygote has a higher reproductive fitness than homozygotes.

heterozygous (hetero-**zye**-gus): The quality of having two different alleles for a particular genetic trait.

Hfr cell: A bacterial cell with a high frequency of recombination, formed by integrating an F factor plasmid and a bacterial chromosome during conjugation.

hibernation: A temporary physiological state in which an animal's set point (normal body temperature) falls and metabolism slows dramatically.

hindbrain: The posterior region of the vertebrate brain that encompasses the cerebellum, pons, and lower portion of the brain stem (medulla oblongata).

histone: A protein that bears a net positive electrical charge. The DNA molecules in eukaryotic chromosomes coil around clusters of histones.

holdfast: An algal cell or tissue specialized to anchor the plant to a substrate.

homeostasis: The maintenance of a relatively constant internal environment despite fluctuations in the external world.

homeotherm: An animal that has a relatively constant body temperature.

homo-: From Greek, meaning "same or similar."

Homo erectus: An extinct hominid which appeared in Africa around 1.6 million years ago and spread into Europe and Asia.

Homo habilis: The earliest species of genus *Homo*, which originated in Africa around 2 million years ago and is distinguished from earlier hominids by its larger brain.

homologous pair: Identical pairs of autosomes, or non-sex chromosomes, derived one each from the egg and sperm.

homology: The appearance in related life forms of similar structures or functions, based on the inheritance of the same basic genetic program.

Homo sapiens: The single species of living humans, divided into an archaic group that lived in most of the Old World from about 300,000 to 75,000 years ago and a subsequent modern group.

homospore: Identical spores in certain club mosses and other plants that give rise to gametophytes, producing antheridia and archegonia.

homosporous: The property in certain plant species of producing a single type of spore, which gives rise to gametophytes having both male and female structures.

homozygous (homo-**zye**-gus): The quality of having two identical alleles for a particular genetic trait.

hormone: A substance that is manufactured in minute quantities in one part of an organism and that produces effects in other parts of the same organism.

human chorionic gonadotropin (HCG): A hormone that stimulates the corpus luteum to produce progesterone and estrogen during pregnancy.

humoral immunity: The immunity conferred by antibodies carried in the blood.

humus: One of four main constituents of soil, humus consists of decomposing organic materials that release nutrients and prevent soil from compacting.

Huntington's disease: A currently incurable genetic disease caused by a lethal, dominant mutation usually expressed in midlife that leads to brain cell deterioration, muscle spasms, and death.

hybrid inviability: A postzygotic reproductive isolating mechanism in which a hybrid offspring dies before reproducing successfully.

hybrid sterility: A postzygotic reproductive isolating mechanism in which the hybrid offspring survives but is unable to reproduce.

hybrid vigor: A condition in which heterozygotes for one or several genetic traits exhibit a more desirable phenotype than do homozygotes.

hydr-, hydro-: From Greek, meaning "fluid" or "water."

hydrogen bond: A weak bond formed as a result of the attraction between the oxygen atom of one molecule and a hydrogen atom of another.

hydrologic cycle: The global cycling of water from the atmosphere to the Earth's surface and back, including precipitation, evaporation, and transpiration.

hydrolysis (high-**drol**-i-sis): A "splitting with water" process in which one larger molecule is split into two monomers by the addition of parts of water molecules.

hydrophilic: The property of being "water-loving." Hydrophilic compounds tend to form hydrogen bonds with water molecules, and thus readily dissolve in water.

hydrophobic: The property of being "water-hating." Hydrophobic compounds have nonpolar covalent bonds which prevent them from forming bonds with hydrogen and from being electrically attracted to water molecules. Thus such compounds tend to be insoluble in water.

hydroskeleton: A volume of water trapped within an animal's tissues that is noncompressible and serves as a firm mass against which opposing sets of muscles can act.

hyper-: From Greek, meaning "over" or "more."

hyperpolarization: An increase in the polarization of a cell (as from −50 to −80 mV).

hypertonic: A condition of a solution reflecting the presence of a solute concentration that is higher than that of some other solution.

hypervariable region: Any of several areas in the variable portions of antibody protein chains that account for the different binding site (antigen) specificities of antibodies.

hypha (plural: **hyphae**): The cellular filaments that are the basic structural units of a fungus.

hypo-: From Greek, meaning "under" or "lower."

hypocotyl: The initial length of stem that emerges from a germinating seed.

hypothesis: A precisely constructed proposition to explain a scientific question. The systematic testing of hypotheses is the fundamental process that sets science apart from other disciplines and enables scientists to produce accurate, enduring explanations of natural phenomena.

hypotonic solution: A solution in which the salt concentration is lower than that of another solution.

imaginal disk: Little sacs of cells in fruit flies that develop into a specific part of the adult fly.

imbibition: A process in which water enters soil and binds to clay and humus particles.

immune deficiency: The inability of an immune system to produce enough antibodies to ward off infection, due to the lack of some component of the system.

immune system: White blood cells, antibodies, and organs such as the lymph nodes, spleen, and thymus which work together to protect the body from foreign substances.

immunoglobulin: A protein antibody molecule.

implantation: The process during which a newly formed mammalian embryo becomes embedded in the inner wall of the uterus.

imprinting: A type of learning during a brief, early sensitive period of an animal's life in which a behavioral response is learned to a stimulus with specific characteristics.

inbreeding: Mating between relatives at a greater or lesser (negative inbreeding) frequency than would be predicted by chance.

inclusive fitness: A concept in evolutionary theory in which an individual's total fitness includes both its likelihood of passing on its own genes and the likelihood that it will contribute to the successful passing on of the genes of its relatives.

incomplete dominance: In genetics, a situation in which both alleles of a heterozygous pair exert an effect, jointly producing a phenotype intermediate between the two.

incomplete penetrance: In genetics, a situation in which a dominant allele is present but is not expressed at all in certain individuals.

induced fit: An adjustment in an enzyme's shape which improves the fit between the enzyme molecule's active site and its substrate.

inducible enzyme: A protein synthesized in a cell in response to the presence of a particular inducer substance.

inductive reasoning: How scientists proceed from specific observations and facts to a general hypothesis proposing to explain the observations and answer questions arising from them.

inflammatory response: The initial defensive response to a wound; characterized by the release of histamine, increased blood flow to the wound site, the arrival of macrophages, and other related events.

ingestion: The taking of food pieces into the body to be digested.

inheritance of acquired characteristics: Lamarck's early evolutionary theory that organisms change physically as they strive to meet the demands of their environment and that these changes are passed to future generations so that the organisms grow more complex.

inhibitory synapse: A synapse in which synaptic transmission acts to prevent function of the receiving cell, often by inhibiting depolarization.

initiation: The beginning of protein synthesis from a transcribed gene. Initiation takes place when a ribosome, an mRNA, and two tRNAs bearing specific amino acids bind together.

inositol triphosphate (IP_3): A second-messenger compound (see *cyclic AMP*) that helps trigger cell responses to a range of neurotransmitters, hormones, and growth factors.

inside cell: One of two plant cell types during the globular embryonic stage which is an accelerated period of growth.

insight learning (reasoning): The ability to respond correctly to a novel situation by applying information garnered from past experience.

instinct: A stereotyped, inherited pattern of behavior that occurs in response to a specific environmental stimulus.

insulin: A hormone produced in the vertebrate pancreas that acts to lower the concentration of glucose in the blood; also found in the brain.

integral protein: Individual protein molecules that are embedded within the lipid bilayer of a cell's plasma membrane.

integument: A tissue that serves as a covering layer, such as the outer covering of a seed.

intercalated disk: A region of cell membrane in a cardiac-muscle cell that serves to hold adjoining cells together; it also serves as a site through which ion or electric currents can flow.

interferon: A naturally occurring protein liberated by a mammalian cell after it has been infected by certain viruses. Interferons bind to nearby infected cells, stimulating the production of antiviral proteins.

intermediate filament: A threadlike structure made of protein which is thought to lend tensile strength to animal cells.

internal intercostal: A set of human muscles between each pair of ribs which contracts and forcibly expels residual air from the lungs during active exercise.

interneuron: Any of a class of neurons that receive and process input from receptor, sensory, or other interneurons and send commands to other interneurons or to effector neurons.

internode: The stem tissue between leaves of a plant.

interphase: The longest portion of the cell cycle, divided into G_1, S, and G_2 phases, during which the nucleus does not divide.

intersex: An individual with both male and female sex characteristics, produced by errors in the hormonal control of development.

interspecific competition: Competition between members of several species for a limited resource.

intestine: The long, tubelike section of the digestive tract where most food digestion and absorption take place.

intraspecific competition: Competition for one or more resources among members of the same species.

intrinsic protein: A protein molecule in which part or all of its peptide chain is embedded in the lipid bilayer of a cell's plasma membrane.

intron: In a eukaryotic gene, a segment of DNA which does not code for protein.

invertebrate: An animal that lacks a backbone.

ion (eye-on): An atom that bears a net electrical charge.

ionic bond: A chemical bond formed when one atom gives up a valence electron and another atom adds the free electron to its outermost orbital. An ionic bond holds its atoms together in an energetically stable unit.

iris: A pigmented ring of tissue in the vertebrate eye that opens or closes in response to changing light levels.

islets of Langerhans: Clusters of endocrine cells in the pancreas in which the hormones insulin and glucagon are produced.

isogamete: In certain plants, a gamete of one of two mating types, identical in size and appearance to gametes of the other mating type.

isomer: One of two chemical compounds which have identical formulas, but different spatial arrangements of atoms. Isomers often have very different chemical properties as well.

isotonic solution: A solution which has the same salt concentration as that of a comparison solution.

isotope: An atom of an element which contains the normal number of protons for atoms of that element, but a different number of neutrons. Thus isotopes of an element have different atomic weights.

joint: A region where two or more individual bones meet, often associated with ligaments and cartilage.

juvenile hormone (JH): A hormone whose presence at high levels in insects prevents metamorphosis.

karyotype: A display of the number, sizes, and shapes of an organism's chromosomes.

kidney: A blood-filtering and waste-excreting organ; found in all vertebrates and some invertebrates.

kilocalorie: The amount of heat energy required to raise the temperature of one kilogram of water by 1°C.

kinesis (cih-nee-sis): A simple reflexive behavior in which the intensity of the animal's response is proportional to the intensity of the stimulus.

kinetic energy: The energy of motion.

kinetochore: The portion of a chromosome to which centromere fibers attach during mitosis and meiosis.

kingdom: The most inclusive taxonomic grouping, such as the classification of all plants into the kingdom Plantae.

kin selection: Behavior that promotes the reproductive success of closely related individuals.

Klinefelter's syndrome: A type of sex-chromosome nondisjunction producing usually sterile individuals with male external sex characteristics.

knuckle walking: A form of quadruped locomotion in modern apes, during which the animals curl their fingers and apply weight to their knuckles.

Krebs cycle: The fundamental metabolic pathway in cellular respiration. The cycle consists of a series of chemical reactions in which pyruvate (the end product of glycolysis) is oxidized to carbon dioxide, and ATP is generated. Also known as the citric acid cycle.

***K*-selected species:** A species whose members reach reproductive age relatively slowly, and produce small numbers of young that are well-prepared to compete for food and other resources.

lactation: The production and secretion of milk by mammary glands.

large intestine: The sizable, nonconvoluted portion of the mammalian intestine that extends from the small intestine to the anus.

larynx: The boxlike entrance to the mammalian trachea and lung system.

latent learning: A process in which a learned behavior is repeated after some time delay, as in the distinct dialects learned and sung by certain songbirds.

lateral line: A series of small sensory organs on the sides of fish and amphibians that contain specialized sensory cells that respond to movements, sound vibrations, electric fields, and other stimuli in the water.

lateral meristem: Either of two tissues, the cork cambium and the vascular cambium, that generates the stem-thickening secondary growth in a nonherbaceous plant.

lateral root: Branches of existing plant roots which arise in the pericycle and grow through the cortex and out from the surface of the root.

late wood: Secondary xylem cells that arise in the latter part of a tree's growing season.

law of independent assortment: Mendel's second law, which states that different characters are inherited independently of one another.

law of segregation: Mendel's first law, which states that individuals carry two discrete hereditary units (alleles) for a given character, receiving one from each parent. When sperm or ova are produced, each allele pair is separated and the alleles are distributed on chromosomes to different gametes.

leaf: A plant organ specialized for collecting light for photosynthesis.

leaf primordium: A flattened mound on the side of a plant meristem that will eventually develop into a leaf.

learning: The modification of behavior on the basis of experience.

lens: A structure composed of highly ordered cells in the eyes of vertebrates, octopuses, and some other animals that focuses light images.

lentic community: The fresh-water biome represented by a lake or pond.

lenticel: One of numerous porelike sites in the cork layer of bark at which gas exchange can take place.

lethal allele: A dominant or recessive gene that, when expressed, can cause death by interfering with some basic biological function.

leukemia: A type of cancer in which undeveloped white blood cells proliferate uncontrollably in bone marrow.

leukocyte: A white blood cell.

lichen: A composite organism consisting of one fungus species and one or more species of algae.

life: A particular set of processes that results from the organization of matter and energy and is passed on from an organism to its progeny, not generated anew on the modern Earth.

ligament: A strong, flexible band of collagen fibers that connects bones.

light chain: One of a pair of short polypeptide chains in an antibody molecule.

light-dependent reaction: The first stage of photosynthesis, which uses light energy to oxidize water and release oxygen.

light-independent reaction: The second stage of photosynthesis, in which carbon dioxide is reduced to carbohydrate and which can proceed whether light is present or not.

light reactions: The first of the two distinctive sets of reactions in photosynthesis in which light energy is required to oxidize water and O_2 is released.

limbic system: The seat of the emotions and short-term memory in the vertebrate forebrain.

limnetic zone: The center of a lake, extending to the depth where O_2 production by photosynthesis equals the uptake of oxygen in the respiration of organisms in the lake.

linkage: A measure of the degree to which genes on chromosomes are inherited together.

lipid: A member of the class of organic compounds that includes oils, fats, waxes, and other fatlike substances. The main categories of lipids are fats, oils, phospholipids (integral components of cell membranes), and steroids.

liposome: A spherical lipid bilayer formed in the laboratory from phospholipids in aqueous solution; a possible model for the precursor of true cells.

lipotropin (LPH): An anterior-pituitary hormone that causes the hydrolysis of fat into free fatty acids and glycerol.

littoral zone: The relatively shallow edges of a lake, where light reaches the bottom.

liver: A large, lobed organ in vertebrates that regulates the nutrient content of the blood, absorbs and degrades hormones, serves as a reservoir for glycogen, and carries out a variety of other functions vital to health.

loam: A type of soil composed of silt, sand, clay, and humus particles.

local circuit current: In a neuron, any of the minute, internal electrical currents that flow from the initial point of depolarization and which are involved in propagation of impulses.

locus: The specific site at which a gene or allele appears on a chromosome.

logistic growth curve: An S-shaped curve that represents the phases of a population's growth history (rapid initial growth followed by an equilibrium phase which may in turn be followed by a plateau or a decline).

loop of Henle: The long, looped portion of a mammalian kidney tubule that helps form a concentrated (hypertonic) extracellular brine bath.

lotic community: The fresh-water biome represented by a stream.

lung: One of a pair of hollow, usually branched, internal respiratory organs; connected through passageways to the outside air and also interfacing with the circulatory system.

luteinizing hormone (LH): A gonadotropin hormone produced in the pituitary which stimulates the production of testosterone in male infants and embryos, and triggers ovulation in mature females.

lymphatic system: A system of vessels that drains excess extracellular fluid from the spaces around cells and houses important parts of the immune system.

lymph node: A mass of tissue that contains lymphocytes through which lymph is filtered.

lymphocyte: A cell of the immune system which responds to foreign substances; some lymphocytes secrete antibodies.

lymphokine: A lymphocyte-activating protein secreted by macrophages and helper T cells.

lymphoma: A type of cancer which arises in the blood-forming cells of the lymph nodes.

lysogeny: The process in which viral DNA is replicated along with a host cell chromosome each time the host cell passes through a growth and division cycle.

lysosome: A spherical, membrane-bound sac which contains digestive enzymes that can digest most known biological macromolecules.

lytic pathway: The serial events in which viral genes within a host cell begin to replicate independently, mature virus particles assemble, and the host cell bursts, releasing the particles, which may then infect other host cells.

macro-: From Greek, meaning "large."

macroevolution: Major phenotypic changes occurring over evolutionary time.

macromolecule: An extremely large molecule, with a molecular weight of about 10,000 daltons or more, that is an aggregation of smaller molecules and contributes to the diversity of organic structure.

macronutrient: A nutrient required by a heterotroph in large amounts, such as proteins, fats, carbohydrates, and some minerals.

macrophage: Any of a class of white blood cells that engulfs and digests foreign particles.

major histocompatibility (MHC) complex: A group of genes coding for protein markers that enables body tissues to be recognized by the immune system as self or nonself.

Malpighian tubule: The blind-ended sac that is the functional unit of an insect's excretory system.

mammary glands: Glands evolved in mammals that are specialized in mature females for the production and secretion of milk.

mantle: The hot, semisolid region of the Earth that surrounds the dense core and underlies the lighter, solid crust.

marsupial: A mammal, such as the kangaroo, that possesses an external pouch in which young are nurtured.

mass flow theory: A model for the large-scale movement of phloem fluid in plants.

matrix: A semifluid mix of ribosomes, DNA, and enzymes which surrounds the cristae inside mitochondria.

mechanical isolation: A prezygotic reproductive isolating mechanism involving structural or molecular differences between organisms of different species.

mechanoenzyme: An enzyme that exerts mechanical forces, such as myosin in muscle.

mechanoreceptor: A sensory receptor cell sensitive to pressure, stretching, and bending forces.

mechanoreceptor reflex: During exercise, the changing of the breathing rate by special receptors in limb tendons and joints which ensure that blood gases remain within normal levels.

medulla: The central portion of organs, as of the kidney, which contains the loops of Henle, or of the adrenal gland, where epinephrine is made.

medulla oblongata (medulla): The area of the vertebrate hindbrain where the brain and the spinal cord merge.

medusa: The radial, free-floating form of a jellyfish.

mega-: From Greek, meaning "large."

megagametogenesis: The process in angiosperms during which the female gamete is produced and readied for fertilization.

megagametophyte: A female gametophyte.

megasporangium: The structure that contains megaspores.

megaspore: A large spore that differentiates into a female gametophyte.

megaspore mother cell: A diploid cell in the ovule of conifers that produces four haploid megaspores, one of which develops into a female gametophyte.

megasporogenesis: The meiotic division of a megaspore mother cell that generates the first stage of a female gamete, the megaspore.

meiosis: The specialized form of nuclear division in which, following chromosome duplication in a reproductive cell, the diploid parent nucleus divides twice and four haploid daughter cells are formed.

melanophore-stimulating hormone (MSH): An anterior-pituitary hormone that affects skin pigmentation and fur color.

melatonin: A hormone secreted by the pineal gland; in some animals, affects the internal biological rhythms associated with light and dark ("day" and "night").

memory cell: One of two cell types formed by dividing B cells, providing a secondary immune response against future exposure to a specific antigen.

menarche: The first menstruation at puberty in human females.

menopause: The ending of monthly menstrual cycles in middle-aged women, when the ovaries stop making normal amounts of progesterone and estrogen.

menstrual cycle: The cyclic preparation of the mammalian uterus to receive an embryo.

menstruation: The shedding of the uterine lining that begins the menstrual cycle each month if the egg is not fertilized.

meristem: In plants, an organizing center of undifferentiated, actively dividing cells forming zones where new organs can be generated throughout the life of the plant.

meristematic region: Major areas of growth and development in plant roots and stems, behind which cells elongate and allow primary growth.

meso-: From Greek, meaning "middle."

mesoderm: The midlayer (between ectoderm and endoderm) that arises during gastrulation in an embryo. Mesodermal cells give rise to the skeleton, muscles, and circulatory and immune systems, among other structures.

mesophyll: The major photosynthetic tissue in a leaf.

messenger RNA (mRNA): The type of RNA (ribonucleic acid) that encodes information from DNA and is translated into corresponding protein structures (amino-acid sequences).

meta-: From Greek, meaning "change" or "posterior."

metabolic pathway: A linked series of chemical reactions by which various enzymes catalyze the specific steps needed to construct or break down biological compounds.

metabolism: The combination of simultaneous, interrelated chemical reactions taking place in a cell at any given time.

metamorphosis: Among insects, amphibians, and other animals, the developmental transformation from the larval to the adult body plan.

metaphase: The second stage of mitosis, in which the fully condensed chromosomes become associated with the spindle.

metaphase plate: A plane on which chromosomes align perpendicular to the spindle fibers during metaphase in mitosis.

metastasize (muh-**tas**-tuh-size)**:** To break away from a tumor and spread to a distant site in the body.

methanobacteria: A type of anaerobic archaebacteria that live in decaying vegetation, cattle intestines, and certain other places and produce methane.

micro-: From Greek, meaning "small."

microbody: A small, membranous vesicle containing enzymes that break down the waste products of eukaryotic cells.

microevolution: Small-scale changes in gene frequencies hypothesized to lead to the evolution of distinct species.

microfilament: Any of the many threadlike actin fibers that make up much of a cell's cytoskeleton.

microgametogenesis: The second stage in pollen formation, in which microspores differentiate into functional pollen grains.

microgametophyte: A male gametophyte.

micronutrient: A nutrient required by a heterotroph in small amounts, such as vitamins and trace inorganic minerals.

micropyle: The opening at one end of the megasporangium in the ovule of a conifer.

microsporangium: The structure that holds microspores.

microspore: A small spore that differentiates into a male gametophyte.

microspore mother cell: A diploid cell in the microsporangia of pine cones that produces haploid microspores, which develop into male gametophytes.

microsporogenesis: The initial step in the development of pollen grains, in which a diploid microspore mother cell divides meiotically.

microtubule: A long cylindrical tube that serves as bonelike scaffolding to help stabilize the shape of a cell; also, sites along which transport may occur.

microvillus (plural: **microvilli**)**:** A tiny hairlike projection of epithelial cells that has an actin skeleton.

midbrain: A centrally located region in the vertebrate brain; in fish and amphibians, the primary site for processing visual input.

midrib: The largest, centrally located vein in a dicot leaf.

migration: Periodic journeys by animals, often over long distances and to precise destinations, in response to environmental and probably internal cues.

mineral: An inorganic element or compound.

mineralocorticoid: Any of a group of steroid hormones produced in the adrenal glands that regulates the retention and excretion of minerals, such as sodium and potassium.

mitochondrion (my-toe-**kon**-dree-on; plural: **mitochondria**)**:** A cellular organelle specialized to harvest energy from food molecules and store that energy in ATP.

mitosis (my-**toe**-sis)**:** The process of nuclear division in which genetic information is distributed equally to two identical daughter cells.

molecular formula: A shorthand system used in chemistry to show how many atoms of each type are present in a particular molecule, and whether any of the atoms occur in certain common groups. The molecular formula of water is H_2O.

molecular orbital: A stable arrangement of paired electrons that forms a strong bond between atoms of a molecule such as hydrogen.

molecule: Two or more atoms that have been bound together by chemical bonds.

mon-, mono-: From Greek, meaning "single."

monoclonal antibody: A highly specific antibody produced by a single clone of B cells, from which huge quantities of a single antibody can be obtained for research and medical diagnosis and treatment.

monocotyledon: An angiosperm in which the embryo has only one seed leaf.

monocyte: A type of white blood cell which matures into a scavenger cell called a macrophage.

monoecious (mon-**ee**-shus)**:** In plants, having both male and female sexual parts on each individual of a species.

monomer: A simple molecule that can be linked with other monomers to form a more complex polymer.

monophyletic: In taxonomy, the property ascribed to taxons which share a single ancestral source.

monosaccharide: A sugar monomer that is the basic carbohydrate subunit of more complex sugars.

monotreme: One of a primitive group of modern mammals, including the duckbilled platypus, which lays leathery eggs rather than giving birth directly to live young.

morpho-, -morph: From Greek, meaning "structure" or "form."

morphogenesis: The developmental process that generates changes in the shapes of cells and cell populations.

mortality: The per capita death rate of a population.

mosaicism: A genetic condition in which some cells of an organism express the phenotype of one chromosome, while other cells express the phenotype of the complementary homologous chromosome.

Mousterian industry: The more specialized stone tools, such as knives, scrapers, and piercers, associated with the Neandertals.

M phase: The period of mitosis in the cell cycle, in which the chromosomes become equally apportioned in the two new nuclei.

Müllerian duct: The tube in a female mammalian embryo which develops into the Fallopian tubes, uterus, and upper vagina.

Müllerian inhibiting substance (MIS): A hormone that inhibits female reproductive structures from forming in a developing male embryo.

Müllerian mimicry: A phenomenon in which different aposematic species have identical or similar warning colors or patterns; see *aposematic coloration.*

multiple allelic series: A group of alleles of a single gene capable of determining many forms of the same trait.

muscle fiber: A fused set of dozens or hundreds of muscle cells; the basic unit of muscle tissue.

muscle spindle: Any of the specialized stretch receptors (proprioreceptors) in the joints, tendons, and muscles of a vertebrate.

mutagen: Any agent capable of causing a mutation.

mutation: A change in the chemical structure of a gene; see also *chromosomal mutation.*

mutation rate: The frequency with which mutations arise naturally in a given population.

mutualism: An arrangement in which two species live in intimate association, to the benefit of both partners; a form of symbiosis.

mycelium: The dense network of filaments (hyphae) that form a fungus.

mycoplasma: A minute prokaryote that is the smallest known living cell.

mycorrhiza: An association of a plant root and fungal filaments which aid the plant in obtaining nutrients.

myelin sheath: The insulating covering of nerve cell axons, produced by glial cells.

myofibril: A cylindrical assembly of contractile muscle proteins arranged as sarcomeres in which actin and myosin overlap, found within the cytoplasm of muscle-fiber cells.

myoglobin: A respiratory pigment abundant in muscle and some other cells.

myosin: A mechanoenzyme protein that, in the form of thick filaments, interacts with actin to bring about the contraction of muscle cells.

myotomes: Blocks of lateral muscles in chordates, which contract as a unit and are associated with the evolution of swimming in chordates.

myx-, myxo-: From Greek, meaning "mucus" or "slime."

myxamoeba: A single, motile haploid cell of the true slime molds.

myxobacteria: Small, unflagellated, rod-shaped bacterium that moves by gliding along slime tracks.

NAD^+ (nicotinamide adenine dinucleotide): One of two important coenzymes (the other is FAD) that serves as an electron and hydrogen carrier in the metabolic oxidations and reductions of the Krebs cycle and other cell processes.

naive B cell: In the developing immune system, a resting cell found in the lymph nodes and blood that constitutes a stage of clonal selection theory.

natality: A statistic for the per capita birth rate in a population.

natural killer (NK) cell: Any of a class of large, granular, white blood cells that can directly attack certain viruses and virus-infected tumor cells, as well as some bacteria.

natural law: A theory, such as evolution, which has been tested and corroborated in diverse ways, until it has become generally accepted as scientific fact.

natural selection: The differential survival to reproduce of some genotypes; heritable adaptations account for the difference in success.

navigation: A complex animal behavior involving movement over novel routes to precise spots using numerous sensory processes.

Neandertal: A robust group of archaic *Homo sapiens* that inhabited Europe and the Near East from about 130,000 to 35,000 years ago.

negative feedback: A regulatory system in which an increase in the concentration of a substance inhibits the continued synthesis of that substance, and vice versa.

neocortex (cortex): The outer, sheetlike wall of the mammalian cerebrum, which consists of highly ordered arrangements of neurons that carry out many higher brain functions.

neotony: The retention of larval or juvenile traits in sexually mature forms of organisms, or larval reproduction in amphibians.

nephridium (plural: **nephridia**): The excretory organ of an earthworm.

nephron: A tubule that is the functional unit of the vertebrate kidney.

nerve: A bundle of many neuronal axons.

nerve cord: The spinal cord, located just dorsal to the notochord, which is present in all chordates and which coordinates sequential muscle action.

nerve impulse: See *action potential.*

nerve net: A primitive, noncentralized nervous system, as in radially symmetrical cnidarians, that includes several types of nerve cells.

net production: The amount of organic molecules available to primary consumers in an ecosystem, equal to gross production minus respiration and metabolism by primary producers.

neural retina: The epithelial layer in the eye that contains photoreceptors and on which an image is focused.

neuroactive peptide: A protein that acts as a neurotransmitter or that modulates neuron activity.

neuron: A nerve cell.

neurotransmitter: A chemical messenger between nerve cells.

neurulation: The embryonic process in chordates of neural tube formation; the brain and spinal cord develop from the neural tube.

neutron: A subatomic particle that carries no electrical charge.

New World monkey: An arboreal monkey, such as a spider monkey, that occurs in the Americas and usually has a prehensile tail.

niche: The total way of life of a species; includes every facet of its ecological, physiological, and behavioral roles in nature.

nitrification: The oxidation of ammonia (NH_3) to nitrogen oxides.

nitrogen cycle: The cycling of nitrogen between organisms and the Earth; crucial steps are the fixation of atmospheric nitrogen by bacteria and later the release of N_2 back to the atmosphere through the action of microbes.

nitrogen fixation: The conversion in plants of atmospheric N_2 to a usable form, NH_4^+ (ammonium ion).

node: The position at which a leaf arises on a stem.

node of Ranvier: Any of the tiny, nonmyelinated sections of a nerve cell axon where Na^+ channels are abundant.

noncompetitive inhibition: A general system for activating or inhibiting enzyme function in which a substance binds to an enzyme in a way that alters the shape of the active site. It is "noncompetitive" because the control substance does not compete with the substrate to occupy the enzyme's active site.

noncyclic photophosphorylation: A system of electron flow driven by light energy during photosynthesis in which the electrons pass in a one-way sequence through a series of pigments, proteins, and energy carriers.

nondisjunction: During meiosis, the failure of the members of a homologous chromosome pair to separate.

nonlinkage: Independent assortment of gene pairs during meiosis into gametes with equal proportions of four kinds of genotypes.

nonrandom mating: A situation in which the probability of two organisms mating depends on their phenotypes.

nonvascular plant: A plant lacking vascular tissue for conducting water and nutrients.

norepinephrine: A common neurotransmitter of the sympathetic nervous system; also called noradrenalin.

normalizing selection: Natural selection whose effect is to preserve the status quo in the genetic makeup of a population.

notochord: The stiff but flexible rod that runs the length of a chordate, just ventral to the nerve cord.

nuclear envelope: A flattened, double-layered sac that separates the nucleus from the cytoplasm in a eukaryotic cell.

nuclear transplantation: A technique for studying the determination of specific developmental pathways in cells that involves the replacement of a frog zygote nucleus with the nuclei of other cells.

nucleic acid: A polymer chain made up of nucleotide subunits that are arranged in a specific linear sequence. The two types of nucleic acids are deoxyribonucleic acid (DNA) and ribonucleic acid (RNA).

nucleoid: A dense, unbounded area within a prokaryotic cell that encompasses the cell's single chromosome and serves much like a nucleus.

nucleolus (plural: **nucleoli**)**:** An organelle associated with a specific chromosome and in which are found the genes coding for the major ribosomal RNAs. It is the site of ribosome manufacture.

nucleoside: The molecule that is the central component of DNA nucleotides, consisting of a nitrogenous base and a five-carbon sugar.

nucleosome: A beadlike complex consisting of a portion of a DNA molecule coiled around a cluster of eight histone protein molecules.

nucleotide: The building block of nucleic acid, made up of a nitrogen-containing base, a five-carbon sugar, and a phosphate group.

nucleus: Present only in eukaryotic cells, it encloses the chromosomes, is bounded by a membranous envelope, and serves as the source of genetic information for most cellular proteins.

nutrient exclusion: A passive chemical defense mechanism in plants during which the amounts of certain nutrients required by animals are limited in the tissues of the plants they consume.

obligate aerobe: An organism, generally a bacterium, which must have oxygen for its metabolic processes.

obligate anaerobe: A bacterium which can grow only in the absence of free oxygen, usually carrying out fermentation to generate ATP.

Okazaki fragment: In replicating DNA, a small, independent fragment of DNA that is synthesized on one of the two separated strands (the lagging strand). Ultimately each Okazaki fragment is joined to its predecessor as the new DNA strand lengthens.

Oldowan industry: The earliest stone tool industry, associated with *Homo habilis* and consisting of coarse cutting and chopping tools.

Old World monkey: A monkey found in Africa and Asia, such as a baboon, that is relatively large, has a nonprehensile tail, and possesses an opposable thumb and a sexual cycle similar to that in apes and humans.

olfaction: The chemical perception of odor molecules.

oligodendrocyte: Any of the myelin-producing glial cells in the brain and spinal cord.

ommatidium (plural: **ommatidia**)**:** Any of the individual optic units that make up a compound eye.

omnivore: An animal that consumes both plant and animal matter.

oncogene: A cancer-causing gene.

one gene–one enzyme hypothesis: The idea that one gene might code for a particular enzyme in each step of a biosynthetic pathway. This hypothesis provided the foundation for learning how genes determine a phenotype.

one gene–one polypeptide hypothesis: The basis for current knowledge of the link between a genotype and its phenotype. This idea states that each gene codes for the formation of one polypeptide chain in an enzyme or structural protein and thus influences an organism's physical traits.

1,25-Dihydroxycholecalciferol: A hormone formed by kidney enzymes, which regulates calcium uptake in the small intestine.

oocyte: A large, immobile egg cell.

oogamy: The development of true oocytes.

oogenesis: The process of egg production.

open circulatory system: A circulatory system in which the circulating fluid is not entirely enclosed within a continuous set of interconnected vessels.

open program: An innate motor pattern in which certain elements can be modified by learning.

operant conditioning: Behavior modification, developed by B.F. Skinner, that involves learning by feedback from rewards and punishments for certain behaviors.

operator: A regulatory segment of DNA whose role is to receive signals in the form of a repressor substance. Such signals in turn regulate expression of a nearby target gene.

operon: A unit which regulates prokaryotic gene activity, consisting of an operator plus the protein-coding genes it controls.

opsin: A lipoprotein that is one component of the visual pigment rhodopsin.

optic nerve: The nerve of sight, which connects the retina and the brain.

orbital: A three-dimensional path taken by electrons around the nucleus of an atom.

order: A taxonomic grouping or organisms belonging to similar families.

organ: A body part composed of several tissues that operate in concert to perform specific functions within the organism.

organelle: A structure within a cell that carries out specific functions; for example, a mitochondrion, ribosome, or microtubule.

organic compound: Any chemical compound that contains one or more carbon atoms.

organ of Corti: A structure composed of thousands of sensory hair cells situated on the basilar membrane in the vertebrate inner ear; displacement of the cells by sound pressure waves generates the impulses carried by the auditory nerve.

organogenesis: The developmental stage during which discrete organs and tissues form.

organ system: A collection of organs that are coordinated to carry out a specific biological function such as respiration or excretion.

orgasmic phase: The peak of sexual excitement, marked by ejaculation in males and orgasm in females.

orientation slab: Any of the functional units of the visual cortex sensitive to lines oriented at different angles.

osmolyte: Any substance that causes osmotic pressure to rise.

osmoregulation: The process of maintaining a stable internal fluid environment by regulating osmolyte concentrations in body fluids.

osmoregulatory system: An organ system that governs the levels of water and salt in body fluids by regulating osmolyte concentrations.

osmosis: The diffusion of water through a semipermeable membrane in response to distribution of osmolytes.

osmotic potential: The tendency of water to move to areas of lower solute concentration across a semipermeable membrane.

osmotic pressure: The pressure exerted by a solution separated by a semipermeable membrane from pure water; practically measured as the pressure that must be applied to such a solution to prevent it from gaining additional water through the membrane.

osteoporosis: Extreme bone loss, usually with age, that leads to brittle, easily broken bones.

otolith: Literally "ear stone"; any of numerous crystalline structures in the inner ear of vertebrates that impinge on sensory hairs to register changes in the position of an animal's head.

outside cells: One of the first two differentiated plant cell types, produced during the globular embryonic stage and developing into the embryonic epidermis.

oval window: A thin, taut, inner-ear membrane.

ovarian cycle: The cycle during which eggs mature and ovulation occurs.

ovary: The female sex organ in which eggs (ova) are generated.

oviduct: The tube through which a mature animal egg travels after leaving the ovary.

ovulation: The release of an egg from an ovary.

ovule: In a seed plant, the structure within which an egg cell forms.

ovum (plural: **ova**)**:** An egg, or a female gamete produced in an ovary.

oxidation: The removal of electrons from an atom or compound.

oxidation-reduction reaction: A chemical reaction in which one molecule loses electrons (oxidation) while another molecule simultaneously gains electrons (reduction).

oxidative phosphorylation: The process by which energy released during oxidation reactions is stored in high-energy phosphate bonds.

oxygen debt: The condition in which reduced metabolic products (such as lactic acid) comprising the "debt" accumulate due to the inability of oxidative metabolism to function rapidly enough. The debt is paid off when the metabolism that produces reduced products slows.

oxygen dissociation curve: A graphic measure

of the binding of O_2 to hemoglobin under different oxygen partial pressures.

oxytocin: A peptide hormone released by the female posterior pituitary lobe that causes uterine contractions during sexual intercourse and childbirth; also a neuroactive peptide.

ozone layer: A high-altitude layer of O_3 which shields the earth from much of the sun's ultraviolet radiation.

P_1 (parental) generation: In a genetic cross, the two parents which produce the first generation of offspring.

palisade parenchyma: A tightly packed layer of rod-shaped, chloroplast-filled cells just below the upper epidermis of a leaf.

pancreas: A long, slender organ near the stomach that secretes digestive enzymes.

Pangaea: A single giant land mass joining all of today's continents, formed about 250 million years ago by the convergence of Laurasia and Gondwana.

pangenesis: A theory first proposed by Hippocrates, that each body part produces a characteristic "semen," or "seed," that passes to reproductive organs and combines with seed from a mate after copulation to form respective body parts in offspring.

parallel evolution: The evolution in separate but related groups of similar characteristics apparently as a result of exposure to similar environmental or selective conditions.

parasite: An organism that feeds on its living host, often rendering harm but usually not causing the host's death.

parasympathetic nervous system: The portion of the autonomic nervous system that, in general, acts to inhibit cell or organ function.

parathyroid gland: Small endocrine gland located on the surface of the thyroid; the source of parathyroid hormone (PTH).

parathyroid hormone (PTH): A hormone that raises calcium levels in the blood.

parenchyma cell: Any of the thin-walled plant cells with large vacuoles that make up leaf mesophyll tissue.

parthenogenesis: A form of reproduction in which no fertilization of an egg takes place. Instead, the mature egg is spontaneously activated and subsequently undergoes normal development.

partial linkage: When gametes form during meiosis bearing two parental and two recombinant genotypes in unequal proportions. Partial linkage has led to many developments in gene mapping.

partial pressure: The pressure exerted by a gas in air; corresponds to the percentage volume of the gas in a given volume of air.

passive immunity: Immunity to disease provided indirectly, as in the transfer of antibodies from mother to fetus across the placenta.

pathogen: An agent that causes a specific disease.

pattern formation: During development, the gradual emergence of the body plan and allocation of organs to their relative positions.

Pavlovian conditioning: Also known as associative learning; a type of conditioning developed by physiologist Ivan Pavlov, who provoked a dog to salivate upon hearing a bell, which the animal had come to associate with the arrival of food.

pecking order: An example of vertebrate social hierarchy involving a clear-cut line of dominance, with each position established and reinforced by aggressive behavior.

pedigree: A formal representation of a set of traits for all members of a family lineage.

pelagic: The quality of drifting about freely in water.

pelagic community: A near-surface, open ocean community supported primarily by the photosynthesis of phytoplankton, although diatoms, invertebrate larvae, fishes, and other buoyant creatures are also found in these upper layers.

pelvis: (1) The central cavity of the kidney; (2) the hip bone of land vertebrates.

penicillin: The first antibiotic created from a species of *Penicillium* mold.

penis: The male copulatory organ.

pepsin: A digestive enzyme.

peptide bond: A chemical bond which results from a condensation reaction between the carboxyl group of one amino acid and the amino group of another.

peptidoglycan polymer: A huge carbohydrate and protein chain that makes up bacterial cell walls, within which short peptides stabilize the main chains of sugar molecules.

peptidyl transferase: An enzyme that catalyzes the formation of peptide bonds between amino acids during the synthesis of a polypeptide chain.

percent polymorphism: The fraction of individuals in a population heterozygous at given gene loci; a measure of the genetic variation in a population.

perennial plant: A plant whose life cycle typically lasts for a number of years.

perianth: The showy, exterior parts of a flower.

pericycle: A circular zone of cells that surrounds the xylem and phloem tissue of a plant root.

peripheral lymphoid tissue: Organs including the lymph nodes, spleen, gut-associated appendix, tonsils, adenoids, and intestinal Peyer's patches, where B lymphocytes from the bone marrow are localized.

peripheral nervous system (PNS): The entire portion of the vertebrate nervous system other than the brain and spinal cord.

peristalsis: Wavelike muscular contractions, such as a mechanism that propels food within the digestive tract.

petiole: The stemlike portion of a leaf which connects the leaf blade to the stem of the plant.

pH: An expression of the concentration of hydrogen ions in a solution. The pH scale runs from 0 to 14, with acidic solutions having a pH of less than 7 (neutral), and basic (alkaline) solutions having a pH of more than 7.

phagocytosis: Literally, "cell eating"; the engulfing of particulate matter by cells.

pharynx: The region of the vertebrate gut immediately posterior to the mouth cavity.

phenotype: The observable expression of the genetic makeup (genotype) of a cell or organism.

phenylketonuria (PKU): A genetic condition in newborns caused by abnormal metabolism leading to excess amounts of this amino acid in the blood and severe retardation.

pheromone: A chemical signal produced by an organism that stimulates a physiological or behavioral response in another member of the same species.

phloem: A plant vascular tissue specialized for carbohydrate transport. Unlike the other plant vascular material, xylem, phloem cells must be alive in order to function.

phospholipid: A lipid molecule in which the glycerol is linked to at least one other molecule containing a phosphate group. Phospholipids are fundamental components of cell membranes.

phosphorus cycle: The cycling of phosphate between sources in soil, rocks, or water and living organisms.

phosphorylation: The transfer of a phosphate group.

photic zone: The top layer of a body of water, where enough light is present for photosynthesis.

photo-: From Greek, meaning "light."

photoautotroph: A bacterium that contains a modified form of chlorophyll and derives its food energy from photosynthesis.

photon: A particle of visible light.

photoperiod: The number of hours of light in a given day.

photophosphorylation: The production of ATP through the transport of electrons excited by light energy down an electron transport chain.

photoreceptor: A sensory receptor cell sensitive to visible or ultraviolet light.

photorespiration: An inefficient form of the dark reactions of photosynthesis in which O_2 accumulates, CO_2 is depleted, and no carbohydrates are generated.

photosynthesis: A metabolic process in which light energy is converted to chemical energy stored in chemical compounds. Photosynthesis takes place in green plants, algae, and certain protists and bacteria.

photosystems I and II: The two basic molecular systems for converting light to chemical energy during photosynthesis. Photosystem I tends to absorb light with a wavelength near 700 nm. Photosystem II most strongly absorbs light with a wavelength near 680 nm.

phototropism: Movement toward light.

phycobilin: A pigment in red algae that gives the algae their color and enables them to capture green and blue wavelengths of light in deep water.

phylogeny: A description of the line of descent of a particular organism.

phylum: A broad taxonomic grouping of organisms belonging to similar classes; in botany a phylum is termed a division.

physical disturbance: Natural event (fire, flood, etc.) that can determine local species richness in a community and can affect ecological succession and either increase or decrease species diversity.

phytochrome: A light-absorbing plant pigment.

phytoplankton: Microscopic, usually photosynthetic organisms that float near the surface of fresh and salt waters.

pigment: A natural coloring substance in plant or animal tissues which absorbs light.

pigmented retina: A layer surrounding the eye's neural retina that absorbs or reflects light.

piloerection: The muscular erection of fur or feathers as a mechanism for conserving body heat.

pineal gland: A pea-sized endocrine gland situated in the roof of the brain; the source of melatonin.

pinocytosis (pie-no-sy-**toe**-sis): A process of "cell drinking" in which a fluid is taken up in a cell vesicle or is discharged from a cell vesicle to the outside.

pith: The innermost zone of cells in a plant stem; a storage tissue, especially in a young plant.

pituitary gland: A small mass of endocrine tissue attached to the lower surface of the brain; made up of the anterior and the posterior pituitary lobes, which together produce at least ten hormones having a variety of effects.

placenta: The organ in sharks and mammals that connects a developing embryo to surrounding maternal tissue, and through which the fetus may obtain nutrients, give off wastes, and exchange O_2 and CO_2.

placental: Referring to mammals of the infraclass Eutheria, in which adult females produce a placenta that nourishes the embryo.

plakula: A three-layered marine worm that is placed in its own phylum and may be the most primitive free-living multicellular animal.

planula (plural: **planulae**): The ciliated larva produced by the medusa (sexual) stage of various invertebrates, such as by a cnidarian—for instance a hydra.

plasma: The fluid portion of the blood, consisting of water that contains a variety of dissolved substances including gases, ions, proteins, and antibodies.

plasma cell: A mature B cell that is actively secreting an antibody in response to a particular antigen.

plasma membrane: The thin bilayer of lipid and protein molecules that surrounds the cytoplasm of cells.

plasmid: A small circle of bacterial DNA that is separate from the organism's single chromosome and can replicate independently.

plasmodesma (plural: **plasmodesmata**): A bridge of membrane-enclosed cytoplasm connecting adjacent plant cells.

plasmodium: A coenocytic mass of cytoplasm, either branched or solid, that forms the multinucleate body of a true slime mold.

plastid: An organelle in plant cells that is the site of photosynthesis and of storage of sugars in the form of starch. There are two main categories of plastid: leucoplasts and pigment-containing chromoplasts.

plateau phase: The second phase of human sexual response, marked in women by retracting of the clitoris and swelling of the vagina, and in men by increased blood pressure, heart rate, and breathing rate.

platelet: A living, motile mass of membrane-bound cytoplasm; platelets circulate in the blood and play an important role in blood clotting.

plate tectonics: A theory describing the building and drifting of crustal plates that bear Earth's continental surface masses. Plate tectonics has been corroborated from numerous lines of evidence and has revolutionized a number of fields including geology and biogeography.

pleiotropy (**ply**-o-trow-pee): A situation in which an individual gene affects several traits.

poikilotherm: Any animal having a variable body temperature that tends to track the temperature of the surrounding environment.

point mutation: A mutation that alters a single site in a gene and creates a new allele.

polar bond: A chemical bond in which atoms share electrons only partially. The electrical charge from the cloud of moving electrons is asymmetrical, and tends to be found near the nucleus of the negative atom in the pair.

polarization: In a cell, the creation through ion flow of an internal environment having a net negative electrical charge, contrasted with a net positive charge outside the cell.

polar nuclei: In a megagametophyte of a flowering plant, the two nuclei that migrate to the center of the multinucleate cell prior to the processes that yield a haploid egg cell ready for fertilization.

pollen grain: A mature male gametophyte.

pollen tube: After pollination, a pipelike structure extended by the developing male gametophyte toward the egg cell. When the tube reaches the egg, some of its cytoplasm and the sperm are transferred into the egg.

pollination: The reception of a pollen grain by a female ovule.

pollinators: Animals, such as bees, birds, or bats, that pollinate plants by inadvertently dispersing pollen between flowers.

poly-: From Greek, meaning "many."

polyandry: Mating between one female and many males.

polygenic: Controlled by several genes.

polygyny: Mating between one male and many females.

polymer: A large chainlike molecule made up of many simpler units (monomers) which are linked in a specific sequence by covalent bonds.

polyp: The sessile stage of hydras and some other cnidarians, characterized by a hollow, elongated body.

polypeptide: A long chain of amino acids linked by peptide bonds.

polyploidization: The sudden multiplication of an entire complement of chromosomes.

polysaccharide: A long-chain carbohydrate made up of large numbers of monosaccharides linked by glycosidic bonds.

polysome: A complex consisting of a single molecule of messenger RNA, plus several ribosomes.

polytene chromosome: A giant chromosome, consisting of as many as 1,000 DNA double helix molecules, that arises during a prolonged S phase in certain insect cells.

pons: A region of the vertebrate hindbrain that forms part of the brain stem; along with the medulla, helps control respiration, circulation, swallowing, and vomiting.

population: A group of individuals of a species that can or do interbreed.

population density: The number of individuals in a population per unit of area.

population genetics: The study of evolutionary trends within populations using mathematical models of genetic phenomena such as frequency distribution and mutation.

postsynaptic potential (PSP): A transient depolarization of a postsynaptic nerve cell upon receipt of a synaptic transmission.

postzygotic isolating mechanism: A type of reproductive isolation that affects a zygote, which will become either an inviable embryo or adult or a sterile adult.

potential energy: Stored energy with the capacity to do work later.

predation: The habitual feeding on some organism by another living organism.

prezygotic isolating mechanism: A type of reproductive isolation that blocks the formation of zygotes between two species via ecological, behavioral, or structural phenomena.

primary growth: In a plant seedling, the initial elongation of cells at the tips of the root and shoot.

primary production: The synthesis and storage of organic molecules during the growth and reproduction of photosynthetic organisms.

primary succession: The initial establishment of plant life in an area, generally on bare rock.

primate: The most highly developed order of living mammals, including humans, apes, monkeys, and prosimians.

primitive streak: A thickened region in reptile, bird, and mammal embryos that serves as the site of gastrulating cell movements.

probe: A radioactively labeled stretch of DNA or RNA that hybridizes to any complementary piece of DNA, becoming a useful tool in numerous types of molecular and genetic research.

procambium: In a young plant, a layer of cells that separates the xylem from the phloem in the vascular bundles of stems.

producer: In ecology, a plant or other photosynthetic organism that converts solar energy to a stored chemical form usable as food by other creatures.

product: The substance formed as the result of a chemical reaction.

profundal zone: In a lake, the deep region where little light penetrates, and hence decomposition of organic material exceeds production.

progenote: A primitive cell-like structure that has been proposed as a hypothetical ancestor to the prokaryotes.

progesterone: A female sex hormone.

programmed learning: An early learning experience initiated by instinct, such as imprinting by newborn animals or recognition of specific flowers by bees.

progress zone: In vertebrates, a special set of mesodermal cells located just beneath the tip of an elongating limb bud, from which developing limbs arise.

prokaryotic cell: A cell that lacks a membrane-bound nucleus and a cytoskeleton. All prokaryotes, including bacteria and cyanobacteria, are single-celled organisms.

prolactin: A pituitary hormone that stimulates the production of milk in female mammals; it has other functions in such females and in other vertebrates.

promoter: A specific DNA sequence that serves as a binding site for RNA polymerase near each gene. Transcription proceeds from these sites.

promoter site: In gene regulation, the site adjacent to an operator where RNA polymerase binds to initiate mRNA synthesis.

prophase: The first phase of mitosis, in which the chromatin begins to condense and individual chromosomes become visible.

proprioreceptor: A type of stretch receptor found in joints, tendons, and muscles.

prop roots: Roots that sprout downward from the stem to anchor a tree in waterlogged soil, such as in mangroves, which live in swampy habitats.

prosimian: The most primitive living primates, including lemurs, lorises, and tarsiers.

prostaglandin: Any of a class of fatty-acid hormones present in semen and also produced by virtually all body cells. Prostaglandins stimulate smooth-muscle contractions.

prot-, proto-: From Greek, meaning "first."

protein: Any member of a class of diverse polymer macromolecules constructed of linked amino acids.

proteinoid: A proteinlike polypeptide made up of linked amino acids that can form spontaneously in clays and micas.

proteinoid microsphere: A sphere of proteinlike polymers that exhibits cell-like activity and is formed by exposing a heated mixture of dry amino acids to water.

protoderm: The embryonic epidermis covering a plant embryo; also present in the growth centers of roots and shoots.

proton: A subatomic particle that bears a positive electrical charge.

proto-oncogene: A normal gene that can be changed in state to that of an oncogene, as by a mutation or chromosome translocation.

protostome: Any member of a lineage of animals in which the blastopore of the developing embryo becomes the mouth.

protozoan: Literally, a "first animal"; any of the single-celled organisms that is primarily animallike in its method of obtaining food and in other characteristics.

pseudo-: From Greek, meaning "false."

pseudocoelom: The "false," fluid-filled body cavity that is a characteristic of nematodes.

pseudohermaphrodite: A person having the gonads of one sex and the external genitals of the other.

pseudoplasmodium: A sluglike mass of independent cells that makes up the "body" of a cellular slime mold.

pseudopod: A retractable extension of the cell surface used for locomotion.

P site: The grooved site on a ribosome in which an amino-acid–bearing initiator tRNA can bind. Once the P site and its adjacent A site are occupied, protein synthesis on the ribosome can begin.

pulmonary (lung) circulation: Blood flow from the right heart, to the lungs, and to the left heart, during which the blood picks up oxygen and removes carbon dioxide.

punctuated equilibrium: A model of evolutionary change which proposes that evolution proceeds in radical bursts over short periods of time, separated by long periods of stability.

Punnett square: A box diagram for displaying the possible outcomes of a genetic cross; all of the possible combinations of the parental alleles are shown.

pupil: The central opening of the iris in the eye, through which light passes to the lens.

purine: A nitrogen-containing base with a double-ring structure, such as adenine or guanine. Purines are structural elements of nucleic acids.

Purkinje cell: A giant, many-branched cell in the cerebellum that relays nervous signals across synapses from the brain down the spinal cord and to motor neurons in the body.

pyramid of biomass: The relationship between the total masses of various groups of organisms in a food chain, in which there is usually less mass—and hence less stored energy—present at each successive trophic level.

pyramid of numbers: The numerical relationship between groups of organisms in a food chain, in which plants typically outnumber herbivores, herbivores outnumber carnivores, and so on.

pyrimidine (pie-rim-ih-deen)**:** A single-ring nitrogen-containing base that is a structural component of nucleic acids. The bases cytosine, thymine, and uracil are pyrimidines.

pyruvate: A three-carbon compound formed by the breakdown of glucose during glycolysis.

radial symmetry: A body plan that looks circular when viewed from above or below, and in which certain structures radiate outward in all directions from the center.

radial transport: The lateral transfer of nutrients across the tissues of a tree, conducted by square-shaped cells called vascular rays.

radicle: An embryonic root.

radiolarian: A type of marine protozoan which secretes a delicately patterned, silicone-containing shell.

rapid eye movement (REM) sleep: A sleep state characterized by a high rate of brain activity, in which the eyes dart rapidly back and forth beneath closed lids.

rare-mate advantage: As observed in fruit flies and certain other organisms, the temporary ability of a smaller alien group of individuals to mate more successfully than a larger group of the same species. This apparently ensures against the loss of alleles from a population's gene pool.

ray initial cell: Any of the cells in a woody plant that gives rise to vascular rays, spokelike conduits through which nutrients travel from more central parts of the stem toward the periphery.

reactant: One of the substances that interacts with another in a chemical reaction to form some product.

reaction center chlorophyll: A specialized form of the pigment chlorophyll to which light energy must be transferred during photosynthesis.

reading frame: The specific unit of three nucleotides that is deciphered when a gene is expressed; each reading frame codes for one amino acid in a newly synthesized protein.

realized niche: The actual portion of a species' potential niche that can be occupied.

receptor (sensory) neuron: Any of the class of neurons sensitive to a particular type of stimulus, such as light, pressure, heat, or specific chemicals.

recessive (allele): An allele whose expression is blocked by another, dominant allele of the same gene in a heterozygous individual.

reciprocal altruism: A theory proposed to explain the evolution of altruistic behavior, stating that single altruistic acts which place individuals at risk are repaid in kind, thus raising the genetic fitness of each participating member of a social group.

recombinant DNA molecule: A new genetic entity created by cutting and splicing genes from different organisms by means of specific genetic techniques.

recombinant DNA technology: The basis for genetic engineering, using recombinant DNA molecules to manipulate and transfer genes from one organism to another in the laboratory and produce new genomes and new organisms.

recombinant genotype (or **phenotype**): A genotype (or phenotype) that appears in an offspring but was not present in either parent.

recombination frequency: A measure of how often crossing over occurs between specific gene loci.

reduction: The addition of electrons to an atom or compound.

reflex: An automatic, involuntary response to a stimulus.

reflex arc: A neuronal circuit in which input from sensory neurons is relayed to motor neurons, generally via interneurons.

refractory period: The time period following passage of an action potential during which the nerve cell membrane is unable to react to additional stimulation.

regeneration: The regrowth of lost tissue in an adult organism.

region of elongation: A zone in plant roots that allows vertical growth, located just behind the meristem and containing mitotically active precursors of root epidermal cells and a central core of cells forming vascular tissues and cortex.

region of maturation and differentiation: A major zone in plant roots that allows differentiation and development into individual cell types, such as xylem and epidermal cells.

reinforcement: The strengthening of a learned behavior because it produces a favorable outcome.

relaxin: A hormone from the ovaries and placenta which acts to loosen the junction of the pubic bones of a woman's pelvis to ease childbirth.

release-inhibiting hormone (RIH): Any of the small peptide hormones secreted in the hypothalamus that acts in turn to inhibit the secretion of anterior pituitary lobe hormones.

releaser: A sign stimulus produced by a member of an animal's own species.

releasing hormone (RH): Any of the hypothalamic hormones that stimulates secretion of hormones of the anterior pituitary lobe.

replication fork: A site on replicating DNA at which the two component strands of the helix separate, each becoming a template for the construction of a daughter strand.

repressible enzyme: An enzyme whose synthesis is repressed when the end product of its pathway is present.

repressor: A substance which blocks the expression of a gene or genes.

reproductive isolation: The inability of members of a species to breed with organisms belonging to other species.

reproductive strategy: A strategy, such as *r* selection and *K* selection, by which a species produces as many offspring as possible that survive to reproductive age; it usually involves a compromise between the number of young produced and the amount of parental care required.

reproductive time lag: The period between when a population reaches its environmental carrying capacity and the subsequent fall in birth rate and rise in death rate.

resolution: The final phase of human sexual response, marked in women by a sensation of warmth and a return to normal physiological processes, and in men by the detumescence of the penis and a resumption of normal heart rate and breathing.

resource partitioning: The division of a limited resource among competing species in a community, in which each species uses a different aspect of the resource.

respiration: The process by which organisms exchange gases with the environment; see also *cellular respiration.*

respiratory pigment: A pigment molecule that circulates in the blood or resides within red blood cells and is specialized to bind O_2 reversibly.

resting potential: A state of electrical charge in which the inside of a cell is electrically negative relative to the outside of the cell.

restriction endonuclease: Any of a class of enzymes that recognizes specific nucleotide sequences along DNA molecules and cuts both complementary DNA strands at those specific sites.

restriction fragment polymorphism (RFLP): DNA fragments of variable length formed by restriction endonucleases cleaving DNA. These fragments are used in DNA fingerprinting and other genetic techniques.

rete mirabile: A network of blood vessels in which fluids flow in a countercurrent arrangement so that the efficiency of commodity (O_2, heat, etc.) exchange is facilitated.

reticular activating system (RAS): A network of neurons and nerve tracts in the brain that governs awareness of sensory stimuli and wakefulness.

reticular formation: The diffuse network of neurons in the mammalian medulla and cerebrum that receives information from sensory systems.

retina: See *neural retina.*

retinal: The light-receptor molecule that in part makes up the visual pigment rhodopsin and converts a photon of light into molecular motion that leads to vision.

reverse genetics: A procedure permitted by DNA sequencing that works from chromosomes toward phenotypic traits by identifying specific genes, their associated proteins, and the proteins' roles in forming the phenotype.

R factor: A bacterial plasmid gene that is resistant to antibiotics and is able to be transmitted to other bacteria by conjugation.

R factor plasmid: A DNA plasmid molecule that carries the R factor genes and has been found to convey multiple drug resistance.

rhabdomere: In the compound eyes of arthropods, a set of cellular microvilli on which visual pigment molecules are arrayed.

rhizoid: A specialized filament that serves to anchor fungi and other nonvascular plants to a substrate.

rhizome: An underground plant stem that grows laterally from the main shoot.

rhodopsin: The visual pigment present in the rods and cones of vertebrate eyes that is activated by light energy, beginning a series of events that result in vision.

ribosomal RNA (rRNA): Molecules that, along with a variety of proteins, make up ribosomes.

ribosome: Any cytoplasmic organelle that serves as a site for the synthesis of amino acids into proteins.

ribozyme: A piece of RNA that splices itself from a larger RNA molecule and acts as an enzymatic catalyst for the building of new RNA molecules.

ribulose biphosphate (RuBP): A short-lived precursor to phosphoglyceric acid that acts as an electron acceptor for atmospheric carbon dioxide and helps catalyze the light-independent reactions of photosynthesis.

ribulose biphosphate carboxylase: A key enzyme that catalyzes the first reaction in the metabolic pathway leading to the reduction of CO_2 in the dark reactions of photosynthesis; probably the most abundant protein found in nature.

rickettsiae: Among the tiniest prokaryotes, they live as parasites in two alternating hosts: arthropods, and birds or mammals. They cause typhus and Rocky Mountain spotted fever in mammals.

rod: One of two types of photoreceptors in the vertebrate eye; unlike cones, rods are responsive to low light intensities.

root: The part of a plant which anchors it to a substrate or in soil and absorbs water, minerals, and oxygen.

root cap cell: Any of a group of cells at the tip of a root meristem that acts as a protective shield and is positively gravitropic.

root hair: A hairlike extension of an epidermal cell on a plant root.

root meristem: In a plant embryo, the precursor tissue that will grow and mature into functioning root tissue.

root nodule: A small hard lump on the root surface that encapsulates nitrogen-fixing microorganisms.

root-pressure theory: A theory that water pressure builds up in roots and pushes upward toward the leaves, as a result of mineral uptake and transfer to the root xylem.

root sucker: A horizontal root from which new stems and roots can emerge.

round window: A thin membrane at the base of the lower chamber of the cochlea that flexes to relieve pressure created by sound waves.

***r*-selected species:** A species in which individuals reach reproductive age quickly and produce many offspring, relatively few of which survive to reproduce.

rumen: The largest of the three storage chambers in the four-chambered stomach of a ruminant, such as cattle and sheep.

runner: A stem capable of vegetative reproduction that grows horizontally out from the base of a plant and runs along the ground.

salivary glands: The paired submandibular, sublingual, and parotid glands that produce saliva in the mouth and moisten food.

saltatory propagation: The spread of an action potential along a myelinated axon by "jumping" from one node of Ranvier to another.

sand: One of the main constituents of soil, sand is pulverized rock that is intermediate in grain size between gravel and clay.

saprobe: An organism, such as a fungus, that derives food molecules by decomposing dead organic matter.

sapwood: The younger xylem that lies near the periphery of a tree trunk or woody stem.

sarcoma: Cancerous tumors which arise in connective tissue and other non-epithelial tissues.

sarcomere: A precisely repeating unit of actin, myosin, and other proteins; the functional contractile unit of skeletal and cardiac muscle myofibrils.

sarcoplasmic reticulum: A type of smooth membranous network in muscle, in which hollow terminal sacs serve as reservoirs for calcium ions that trigger muscle contractions.

savanna: A tropical grassland biome.

Schwann cell: Any of the myelin-producing glial cells in peripheral nerves.

scientific method: The "organized common sense" approach to the study of natural phenomena. The first step in the scientific method, observation, is followed by the development of a hypothesis, systematic experimentation to test the hypothesis, and eventually the formulation of an explanation based on experimental results.

sclereid: A type of unusually hard plant sclerenchyma cell having a strong secondary wall that is sometimes impregnated with the non-carbohydrate polymer lignin.

sclerenchyma: A type of plant cell having strong secondary walls of cellulose that enhances the ability of plant organs to withstand physical stresses.

scutellum: The single, cylindrical cotelydon in a monocot.

secondary growth: The thickening of the stem in a maturing nonherbaceous plant.

secondary production: The processing and storage of energy by herbivores; see *primary production*.

secondary sex characteristic: A sex-related feature arising as a result of hormonal changes around puberty, such as growth of body hair and breast and penis enlargement.

secondary succession: The initial establishment of plant life on developed soil, as in an area cleared of preexisting vegetation by a physical disturbance.

second law of thermodynamics: The physical law stating that the energy in any system always decreases as energy conversions take place. The "lost" energy is dissipated as heat—a by-product of every energy conversion.

second messenger: A chemical go-between that transfers regulatory messages from non-steroid hormones to target cells.

seed: The complex unit that includes a plant embryo and stored nutrients.

segmentation: The organization of an animal body into a series of segments, either containing the same structures or having specialized structures in each segment. Segmentation is associated with the evolution of increased body size and specialization.

semen: The sperm-containing fluid produced in the male reproductive system.

semicircular canal: A fluid-filled, tubelike channel within the inner ear of a vertebrate that measures angular accelerations and decelerations of the head and body.

semiconservative replication: DNA replication in which each newly formed double-stranded DNA molecule contains one strand conserved intact from the parent molecule.

sense organ: A cluster of sensory receptor cells, as in eyes, ears, and so on.

sensitive period: A discrete time during which an animal must learn something to be incorporated into its later behavioral repertoire; similar learning cannot occur after the sensitive period.

septum (plural: **septa**): In fungi, cross walls in hyphae that segregate independent cells, each with at least one nucleus.

sessile: In animals, the quality of being permanently attached to a fixed surface.

sessile blade: A leaf blade that is joined directly to the stem; seen in plants where the petiole is reduced or absent.

set point: A point on a physiological scale about which regulation occurs, as body temperature is maintained near its set point by a bird.

sex chromosome: An X, Y (or equivalent) chromosome that carries genes involved in sex determination.

sex-linked trait: A hereditary trait carried on only one sex chromosome, such as the X-linked traits of hemophilia and color blindness.

sexual dimorphism: A situation in which members of the two sexes in a species have highly distinctive appearances.

sieve plate: One of the two end walls of a sieve tube element that are perforated by pores.

sieve tube: A pipelike arrangement of sieve tube elements in the phloem vascular tissue of an angiosperm.

sieve tube element: In angiosperms, a cell that is stacked vertically and forms tube-like columns of phloem within a stem or other structure.

sign stimulus: An event that triggers a specific behavioral response in an animal.

silt: Fine-grained, pulverized rock that is intermediate in grain size between sand and clay and which is a major constituent of soil.

sinoatrial (S-A) node: A lump of modified heart muscle cells that is spontaneously electrically excitable; the "pacemaker" that governs the basic rate of heart contractions in vertebrates.

siphonous alga: A type of green alga having cells that contain many nuclei within one large mass of cytoplasm.

skeletal (striated) muscle: Muscle tissue composed of greatly elongated muscle fibers (each of which is built from many fused cells).

sliding-filament theory: The model to explain contraction stating that myosin molecules exert force on adjacent actin filaments and cause thin filaments to slide past thick filaments.

small intestine: The lengthy, convoluted portion of mammalian gut between the stomach and the large intestine; site of most digestion of food and absorption of nutrients and water.

smooth muscle: Muscle tissue consisting of single contractile cells and having actin and myosin but no sarcomeres.

social hierarchy: A feature of animal social groups in which members possess ranks with varying privileges and responsibilities. Sometimes called a dominance hierarchy.

social insect: A member of any of the insect groups, such as ants, bees, and termites, that lives in colonies of related individuals and exhibits complex social behaviors.

society: A group of animals of the same species in which the members communicate and interact in cooperative ways.

sociobiology: The study of the biological basis of social behavior, or the study of behavior from an evolutionary perspective.

solar constant: The amount of solar energy (measured in calories) that falls on a square meter of the Earth's atmosphere in one year.

solubility: A measure of the amount of a solute that will dissolve in a specified volume of solvent at specified temperatures and a pressure of one atmosphere.

solute: A substance that will dissolve in a solvent; sodium chloride is a solute in water.

solute potential: A measure of the concentration of solutes either inside or outside a cell; used in regard to plant cells.

solvent: A substance capable of forming a homogeneous mixture with molecules of another substance.

somatic cell genetics: A method of human gene mapping which bypasses human reproduction as a means of producing new genotypes by relying on somatic cells instead of gametes.

somatic nervous system: The division of the peripheral nervous system that controls muscle movements thought of as voluntary.

somatomedin: Any of several small peptide hormones secreted in the liver; one somatomedin stimulates rapid bone growth during vertebrate adolescence.

somatostatin: A hormone produced in the pancreas that inhibits the secretion of insulin and glucagon and the absorption of glucose.

sorus (plural: **sori**): A cluster of sporangia on the underside of a fern frond.

specialist: A species that has evolved a narrow niche in which it uses only one or a few types of food or growth sites.

species: A group of actually or potentially interbreeding individuals that are reproductively isolated from other such individuals.

species equilibrium model: A model which assumes that species richness on an island is a balance between the immigration rate of "new" species to the island and the extinction rate of established species.

species richness (species diversity): A measure of the total numbers of different species within a community.

specific heat: The amount of heat needed to raise the temperature of 1 gram of water by 1°C.

sperm: Haploid male sex cells, or gametes; these are often formed in the testes and released by ejaculation to fertilize eggs.

spermatogenesis: The process of sperm production.

S phase: The stage in the cell cycle during which DNA is replicated and histones are synthesized as a prelude to cell division.

spindle: A set of microtubules that stretches across a cell from opposite poles during mitosis and that is involved in the separation and movement of the chromosomes.

spirochete (**spy**-row-keet): Any of the bacteria that has a spiral or corkscrew shape.

spleen: An organ whose functions include the storage of red blood cells; as part of the immune system, the spleen also contains lymphocytes and macrophages.

spongy parenchyma: The layer of loosely arranged cells below the palisade parenchyma in leaf tissue, in which much gas exchange is carried out.

spontaneous generation: The belief that life arises anew spontaneously under the right conditions; refuted by experiments by Redi and Pasteur.

sporangium: A spore case.

sporophyll: A leaf bearing spore cases.

sporophyte: A plant in the spore-producing, diploid stage of its life cycle.

spring wood: Also known as early wood, it is denoted by newly produced, large diameter xylem cells which form when abundant moisture is available.

stamen: The pollen-bearing organ of flowers that is surrounded by petals and includes the anther and filament.

standing crop: The number of (photosynthetic) organisms existing in a defined place at a particular moment.

starch: A mixture of polysaccharides (amylose and amylopectin) that serves as the primary nutrient reserve of plants.

Starling's law: A rule stating that the more the cardiac muscle is stretched, the more vigorously it responds and the larger the volume of blood that can be pumped per contraction.

statocyst: In invertebrates, a fluid-filled sac containing sensory hairs that register changes in the organism's body position.

statolith: Any of the movable masses of calcium carbonate crystals or other dense materials in an invertebrate that rests on the sensory hairs of statocysts (organs of balance).

stele: The central cylinder of vascular tissue in a root.

stem: The central, often elongated part of a plant that is composed mainly of vascular tissue and serves both as a structural support for leaves and as a conduit for the transport of water and nutrients.

stem cell: A cell that serves as a continuing source of a differentiated cell type, such as a bone marrow stem cell that generates red blood cells.

stereoisomer: A compound in which the constituent atoms may be arranged in one of two ways, each a mirror image of the other. All stereoisomers of a compound share the same chemical properties.

steroid: Any member of a class of lipid compounds that is composed of four interconnected rings of carbon atoms linked with various functional groups. Some steroids act as vitamins, others as hormones.

stipe: The stemlike structure that provides vertical support to an alga.

stolon: A lateral branch of an aerial plant stem that is capable of putting out roots and new stems where it touches the ground.

stoma (plural: **stomata**): A pore in the epidermis of a leaf, through which gases, including water vapor, diffuse.

stomach: An expandable, elastic-walled sac of the gut that receives food from the esophagus.

strobilus: A spore-producing organ common to club mosses and quillworts.

stroma: The matrix that surrounds the grana within a chloroplast. Among other constituents, the stroma contains enzymes used in photosynthesis.

stromatolite: A closely layered rock mound that consists mainly of the remains of ancient photosynthetic cyanobacteria.

strong bond: A stable chemical bond that can only be broken by the input of a large amount of energy. All covalent and ionic bonds are strong bonds.

structural formula: A diagrammatic representation of the approximate spatial arrangement of atoms in a molecule that also shows the number of bonds between atoms.

structural isomer: A compound that has the same molecular formula as another compound, but has different structural properties.

suberin: A substance impermeable to water that lines the walls of plant root endodermis cells (the innermost layer of the cortex), where two such cells adjoin.

subspecies: Any subpopulation of a species set apart from the original population by at least some distinctive, genetically derived characteristics.

substance P: A peptide that is able to block the action of a neurotransmitter upon a neuron and is believed to be the chemical carrier of pain messages in the body's sensory system.

substrate: The compound or pair of reacting compounds on which a given enzyme can act.

sulfobale: Sulfur-dependent archaebacteria which live by oxidizing or reducing sulfur, often living at high temperatures.

summation: The cumulative effect of individual postsynaptic potentials in a neuron.

summer wood: Also known as late wood, it is denoted by small diameter xylem cells that form during dry summers.

suprachiasmatic nucleus (SCN): Either of the paired regions of the hypothalamus thought to govern the innate rhythm of biological activities (the "biological clock") of mammals.

surface tension: The tendency of a liquid to minimize its surface area.

surface-to-volume ratio: The relationship between surface area and volume of a cell or body, whereby area is squared and volume cubed with each doubling of length.

survivorship curve: A curve which shows the number of survivors in different age groups of an animal population as a way to represent the age structure.

suspensor: A column of cells that attaches a plant embryo to the ovule wall.

symbiont: An organism that lives in a state of symbiosis, or intimate association, with an organism of a different species.

symbiosis: An arrangement in which organisms of more than one species live together in intimate association.

sympathetic nervous system: The division of the autonomic nervous system that, in general, stimulates cell or organ function.

sympatric speciation: Speciation in populations that overlap in geographical range and are not isolated from one another.

symplastic pathway: In a plant root, a "compartment" consisting of the collective cytoplasms of all the cells, and the channels (plasmodesmata) between them; one of two pathways for the movement of water and ions.

synapse: The site where a neuron communicates electrically or chemically with a target cell.

synapsis: The pairing of homologous chromosomes during the prophase I stage of meiosis.

synaptic terminal: In a neuron, the branching structure at the end of an axon across which impulses are transmitted to other neurons or cells.

synaptic transmission: The sending of nerve impulses or action potentials across a synapse between synaptic terminal and target cell.

synaptinemal complex: A bridgelike structure of proteins and RNA that aligns corresponding regions of homologous chromosomes during the prophase I stage of meiosis.

systemic (body) circulation: The portion of the circulatory system in which oxygenated blood is pumped by the heart to body tissues and returns through veins to the heart.

systole (sis-toe-lee): The phase when the heart muscle is contracted.

taiga: The sub-Arctic or subalpine biome consisting of needle-leaved forests; the coldest zone in which trees may live.

taproot: In some types of plants, a root that includes one main axis extending underground.

taste bud: A cluster of taste (gustatory) receptor cells capable of detecting relatively high concentrations of dissolved substances.

taxis: A reflex movement that orients an animal's body with respect to an external stimulus, such as light.

taxon: One category in a system of taxonomy.

taxonomy: The classification of organisms.

Tay-Sachs disease: A lethal human genetic disorder in which afflicted infants are born homozygous for a recessive allele that prevents the production of an enzyme, hexoaminidase A, required for normal lipid metabolism.

T cell (T lymphocyte): One of the two major types of white blood cells in the immune system; T cells congregate at sites of infection and directly attack foreign substances, organisms, or tissues.

tectorial membrane: In the vertebrate inner ear, the membrane which overlies the sensory hair cells of the organ of Corti.

telophase: The final stage of mitosis, in which the two nuclei form and cytokinesis takes place.

temperate forest: Any of the forest biomes that occurs north or south of subtropical latitudes; characterized by varied populations of evergreen and deciduous trees.

temperate grassland: A grassland biome characterized by the absence of trees, a relatively cool, dry climate, and deep, rich soil.

temperate shrubland: Any of the shrub-dominated biomes typical of areas having hot, dry summers and moist winters.

temporal isolation: A prezygotic reproductive isolating mechanism that prevents interbreeding by the different timing of environmental cues that trigger mating in a species.

tendon: An extremely tough bundle of collagen fibers.

tensile strength: A measure of the resistance of molecules of a substance to being pulled apart.

terminal bud: A bud that is produced by the apical meristem and is situated at the end of a branch.

termination: The completion of protein synthesis that began with transcription of a gene. Termination occurs when a specific codon is reached on an mMRA molecule.

territoriality: The defense of a particular feeding or breeding site by an animal, usually against other members of the same species.

test cross: The crossing of an organism of known phenotype but unknown genotype to one that is homozygous recessive for some trait to determine the unknown genotype.

testis (plural: **testes**): The male sex organ in which sperm are generated.

testosterone: One of the androgens, or male sex hormones.

tetanus: A state of sustained maximum muscle contraction.

thalassemia: A group of genetically caused anemias characterized by low or absent synthesis of either alpha or beta hemoglobin.

theory: A general statement that is usually based on a number of substantiated hypotheses and is designed to explain a range of observations. One feature of a theory is that it can be used to account for future observations as well as for existing findings.

therapsid: One of a diverse group of ancient reptiles that included the direct ancestors of mammals.

thermiogenesis: The generation of body heat.

thermoproteale: A type of archaebacteria that lives at very high temperatures in volcanoes and similar habitats.

thermoreceptor: A sensory receptor cell sensitive to infrared heat waves or temperature changes in the skin or brain.

thermoregulation: Maintaining homeostasis by regulating body temperature using behavior and elements of the nervous, endocrine, circulatory, and other systems.

thermoregulatory center: The brain site (the hypothalamus) that carries out thermoregulation by governing physiological responses to changes in body temperature.

thermoregulatory system: A system that can involve parts of numerous organ systems to maintain an organism's body temperature and/or govern the organism's response to environmental temperature shifts.

threat display: A generally instinctive behavioral response, such as the baring of fangs, that serves as a warning to perceived competitors or predators.

threshold: A level of electrical voltage (polarization) that must be reached before an action potential is initiated in a neuron.

thylakoid: Any of the flattened sacs within grana, which in turn are the most prominent internal structures in chloroplasts.

thymine: A single-ring pyrimidine base that is a component of a nucleotide in DNA.

thymus: An organ located behind the breastbone in mammals; a component of the immune system in which T lymphocytes differentiate.

thyroid gland: The thyroxine-producing endocrine organ located near the esophagus in air-breathing vertebrates.

thyrotropic hormone (TH): A hormone produced in the anterior pituitary lobe that regulates the secretion of thyroid hormones.

thyroxine (T_4; also T_3): A thyroid hormone characterized by the presence of atoms of iodine; regulates metabolic rate.

tidal volume: The amount of air taken in during normal inhalation in birds and mammals, roughly equivalent to ten percent of total tracheal and lung volume.

tight junction: A seal that encompasses the lateral surfaces of cells in epithelia; they act as barriers to fluid leakage.

tissue: A grouping of cells that function collectively.

tissue interaction: Also known as induction, a developmental strategy for controlling cell differentiation that was discovered by transplanting embryonic mesodermal cells that induced new structures to form.

tonus: A condition in which a muscle is kept partially contracted over a long period.

torpor: A state in which the body's metabolic rate and activity are lowered temporarily.

trachea: In animals, the windpipe; in insects and most terrestrial arthropods, it is the air-filled, hollow, branching tube that carries air to and from the animal's tissues.

tracheid (tray-kee-id): An elongated hollow cell having thick, rigid, pitted walls; a basic unit of vascular tissue in plants.

trans-: From Latin, meaning "across" or "through."

transcription: The process in which the genetic information encoded in DNA is deciphered to yield a variety of types of RNA.

transdetermined: Cells which shift from one state of determination to another after extensive division, as in the disk cells of fruit fly larvae.

transducin: A protein that interacts with the light-sensitive pigment rhodopsin in a series of light-triggered events leading to visual perception.

transduction: A mode of indirect gene transfer in the laboratory, in which DNA is carried from one bacterium to another by a virus.

transfer RNA (tRNA): The form of RNA that is responsible for transporting individual amino acids to sites of protein elongation on a ribosome–mRNA complex.

transformation: A laboratory process of indirect gene transfer, in which DNA that has been released from one bacterium into the surrounding medium is taken up by another bacterial cell.

translation: The process in which proteins are synthesized based on information transcribed from DNA.

translocation: The transport of solutes in the phloem of a plant.

transpiration: Evaporative water loss through leaves.

transpiration-pull theory: Also called the cohesion-adhesion-tension theory, which states that water is pulled up through a plant's xylem from the top of the plant rather than pushed up from the roots.

trial-and-error learning: Also called feedback learning; a form of learning in which a behavior that elicits a favorable outcome will be reinforced and repeated.

trichromatic theory: The accepted theory of color vision, which states that there are three classes of cones, each maximally sensitive to either blue, green, or yellow or red wavelengths of visible light. The theory was supported by recombinant DNA experiments that discovered separate genes for the various cone opsins.

triglyceride: A lipid compound in which three fatty acids are joined to a single molecule of glycerol.

triiodothyronine (T_3): A hormone derived from tyrosine that is secreted by the thyroid gland in response to TH secretions and which seems to act on target cells.

trimester: A period of three months, one of three such periods in a human pregnancy.

trochophore: The ciliated larva of an aquatic polychaete worm.

trophic level: A feeding level in a food chain.

trophoblast: The outer layer of cells of a mammalian blastocyst-stage embryo.

tropical rain forest: A biome typified by a warm, wet climate, abundant vegetation, and exceptionally diverse animal life.

tropical seasonal forest: A forest biome characterized by wet–dry seasonality and in which trees are deciduous during the dry season.

tropism: Any movement in response to an environmental signal, such as light or gravity.

tropomyosin: A protein thought to mask the actin binding sites of myosin in a noncontracting muscle.

troponin: A protein complex that affects tropomyosin function and so helps regulate actin–myosin interaction in muscle cells.

true slime mold: Any of the species of Myxomycota, a class of funguslike protists that take the form of giant multinucleate cells.

T tubule: In a skeletal-muscle cell, any of numerous deep infoldings of the plasma membrane that serve as conduits for action potentials that can trigger muscle contractions.

tuber: A modified plant stem that can reproduce vegetatively, sprouting new stems and roots from budlike "eyes."

tubulin: The globular protein that is the structural unit of microtubules.

tumor: A solid mass of cancer cells.

tundra: In Arctic and high-mountain regions, a treeless plain in which the predominant plant types are grasses, lichens, sedges, and occasional small shrubs. Deep tundra soil remains frozen year-round as permafrost.

Turner's syndrome: A genetic disorder caused by nondisjunction in either parent and resulting in an XO genotype. Individuals have external female organs but degenerate ovaries that lack germ cells, so puberty does not occur.

tympanic membrane: A tightly stretched membrane separating the outer and middle ear which vibrates when struck by sound waves; the eardrum.

tyrosine phosphokinase: The protein products of oncogenes that usually act as enzymes. These catalyze the addition of phosphate groups to tyrosine residues in cellular proteins, and thus can trigger many cellular events and help cancerous cells spread.

umbilical cord: The thick, ropelike structure that connects the abdomen of a developing fetus to the placenta, and in which a fetus's umbilical arteries and vein spiral about each other.

uracil: A pyrimidine base which takes the place of thymine in RNA molecules.

urea: A nitrogenous waste that can be formed during the breakdown of proteins and other nitrogen-containing substances.

uric acid: A nitrogenous waste product; excreted typically by birds, land reptiles, and insects.

uterus: The muscular reproductive organ in female mammals in which an embryo may implant and the developing embryo be maintained during pregnancy.

vacuole: A saclike membrane-bound organelle found within both animal and plant cells. Vacuoles are typically filled with fluids and soluble molecules.

vagina: The muscular tube that leads from the uterus to the outside of the female mammal's body.

variable expressivity: Differences in the way a gene is expressed among individuals with identical genotypes, as in the gene for polydactyly.

variable region: Either of the two ends of the arms of a Y-shaped antibody protein. The characteristic amino-acid sequences in the variable regions of an antibody determine the antibody's specificity to a particular antigen.

vascular bundle: The distinctive grouping of plant vascular tissues, xylem and phloem, that appears as a ring just inside the cortex layer of the stem.

vascular cambium: In a mature plant stem, the layer of cells that separates xylem and phloem tissues in each vascular bundle in a stem.

vascular plant: A plant having tissues (xylem, phloem) specialized to conduct water and other materials and to provide vertical support.

vascular ray: A spokelike line of parenchyma cells that serves as a channel for the transport of nutrients from the center of a woody plant stem outward toward the periphery.

vascular tissue: The stem zone lying just interior to a plant's cortex, made up of xylem and phloem which support the plant and conduct water and nutrients.

vasoactive intestinal peptide (VIP): A digestive hormone secreted by the duodenum when fats are present in its lumen; controls stomach and pancreatic secretions during digestion.

vasoconstriction: Contraction of the capillary sphincters and the walls of veins in response to action potentials or certain hormones.

vasodilation: The relaxation of capillary sphincters and walls of veins in response to action potentials, certain hormones, or, in muscles, local small molecules.

vasomotor center: An area of the hypothalamus that operates to keep blood pressure, blood gases, pH, and other physiological indicators within normal ranges, generally by regulating blood flow through the heart and vessels.

vegetative reproduction: A process in which new plants genetically identical to each other and to the parent emerge from the parent's body.

vein: (1) A blood vessel that carries blood toward the heart. (2) In a leaf, any of the threadlike channels that provides structural support to leaf tissue and serves as transport pathways for the movement of water and nutrients.

ventilation: The process of filling and emptying internal respiratory organs, such as lungs; critical to the evolution of terrestrial life.

venule: A tiny vein.

vertebrate: An animal that has an internal bony skeleton and usually a backbone.

vertical transport: In plants, the transport of water and nutrients between roots and leaves via xylem and phloem tissues.

vessel element: One of two types of cells in the xylem tissue of an angiosperm, with structural features that permit the relatively free flow of water.

virion: A virus particle outside a living cell; composed of an inner core of nucleic acid surrounded by a protein coat.

viroid: A minute particle of RNA that lacks a protein coat and is capable of causing disease in both plants and animals.

virus: A particle of genetic material that can invade a living cell and utilize its metabolic machinery to reproduce.

visual cortex: The part of the brain that is the center of visual processing.

visual field: The full set of rods or cones linked to each ganglion cell in the eye.

vital capacity: The measurement of a volume of air equivalent to a forced exhalation; usually equals eighty percent of total lung capacity.

vitamin: An organic molecule that functions as a coenzyme or cofactor of enzymes.

voltage-gated channel: A channel through which ions may pass into a neuron that opens in response to changes in membrane potential.

volvocine alga: Any of the colonial green algae.

vulva: The external genitals of a female human.

warning coloration: See *aposematic coloration.*

water potential: A measure of the tendency of a plant cell to gain or lose water.

water-soluble vitamin: A vitamin, such as vitamin C and the B vitamins, that is transported as a free compound in the blood.

weak bond: A chemical bond that is easily broken by the input of a small amount of energy.

wild-type allele: In *Drosophila,* the most common unmutated allele for any genetic characteristic.

Wolffian duct: The embryonic structure that develops into the epididymis and the vas deferens if the embryo is male (XY).

wood: The hard, cellulose-containing secondary xylem of a nonherbaceous plant.

x-ray diffraction: A technique that allows determination of the spatial arrangement of atoms in molecules by bouncing x-rays off of parallel planes of the atoms.

xylem: In plants, a pipelike transport tissue made up of dead, cellulose- and lignin-reinforced cells stacked end to end.

yolk: Material made up of proteins, carbohydrates, nucleic acids, and lipids that is stored in an egg to serve as a nutrient supply for a developing embryo.

yolk sac: One of the membranes that supports embryos of land vertebrates. In reptile and bird eggs the yolk sac completely encases the large sphere of yolk that serves as the embryo's food supply.

zone of polarizing activity (ZPA): At the posterior junction between the limb bud and the body wall, the region where digits in limbs originate.

zonulae adherens: Sites of firm physical contact between cells that are literally "zones of adhesion."

zooflagellate: An animallike protozoan that can move about by means of its whiplike flagellum.

zooplankton: Nonphotosynthetic marine protozoans that live at or near the ocean surface.

zoospore: The motile reproductive cell that matures into a haploid adult during asexual reproduction in certain algae.

zygospore: The diploid zygote produced by the sexual fusion of two isogametes in certain algae.

zygote: A fertilized egg.

zymogen: Any of numerous inactive precursors of digestive enzymes.

Credits and Acknowledgments

Chapter 1 Page 1: W. H. Hodge/Peter Arnold, Inc. Figure 1-1: Walter E. Harvey/National Audubon Society—Photo Researchers. Figure 1-2: (a) Manfred Kage/Peter Arnold; (b) S. J. Krasemann/Peter Arnold, Inc. Figure 1-3: Richard H. Smith/Photo Researchers. Figure 1-4: (a) E. R. Degginger/Earth Scenes; (b) Leonard Lee Rie III/Photo Researchers. Figure 1-5: (a) Jerome Wexler/Photo Researchers; (b) Tom McHugh/Photo Researchers. Figure 1-6: John Colwell/Grant Heilman. Figure 1-7: Michael C. T. Smith/National Audubon Society—Photo Researchers. Figure 1-8: Fred Bavendam/Peter Arnold, Inc. Figure 1-10: The Bettmann Archive. Figure 1-11: Warner Lambert/Park, Davis & Company. Figure 1-12: Jeff Foott/Bruce Coleman. Figure 1-13: Jonathan D. Eisenback/Phototake. Figure 1-14: David Leah/Science Photo Library—Photo Researchers. Figure 1-17: (a) Tom McHugh/Photo Researchers; (b) Animals Animals. Figure 1-18: Pat and Tom Leeson/Photo Researchers. Figure 1-20: Wide World Photos. Figure 1-21: (a) Steve Kaufman/Peter Arnold, Inc.; (b) Hank Morgan/Photo Researchers. Figure 1-22: Verna R. Johnston/Photo Researchers. Figure 1-23: J. R. Simon/Photo Researchers.

Part One Page 17: R. B. Taylor/Science Photo Library—Photo Researchers.

Chapter 2 Page 18: Bruce Matheson/Aberdeen, Washington. Figure 2-1: (left) Richard Hutchings/Photo Researchers; (right) Scott Blackman/Tom Stack & Associates. Page 21, Figure A: Science Photo Library; Figure B: Phillipe Plailly/Photo Researchers. Figure 2-6: Yoav/Phototake. Page 25, Figure A: Bruno J. Zehnder/Peter Arnold, Inc. Figure 2-10: Spencer Swanger/Tom Stack & Associates. Figure 2-11: G. I. Bernard/Animals Animals. Figure 2-14: Mark McKenna.

Chapter 3 Page 35: Tripos Associates/Peter Arnold, Inc. Figure 3-8: (a) Biophoto Associates/Photo Researchers; (c) Don Fawcett/Photo Researchers; (d) Science Source/Photo Researchers. Figure 3-9: (e) Stephen J. Krasemann/Peter Arnold, Inc. Figure 3-17: D. Cauagnaro/Peter Arnold, Inc. Figure 3-18: (b) Tony Brain/Science Photo Library—Photo Researchers. Figure 3-19: Mark McKenna. Figure 3-21: K. R. Porter/Photo Researchers.

Chapter 4 Page 58: Carol Hughes/Bruce Coleman. Figure 4-1: (a) Diane Rawson/Photo Researchers; (b) Gary Milburn/Tom Stack & Associates; (c) Farrell Grehan/Photo Researchers. Figure 4-11: (a) Brian W. Matthews/University of Oregon. Page 70, Figure A: A. Kerstitch/Tom Stack & Associates.

Chapter 5 Page 77: Biological Photo Service. Figure 5-1: Brian Ford. Figure 5-2: (a and b) S. E. Frederick & John N. A. Lott/Biological Photo Service; (c) John N. A. Lott/Biological Photo Service. Page 79, Figure A: IBM Almaden; Figure B: Edwin S. Boatman, University of Washington, Seattle. Figure 5-8: (a) W. Rosenberg/Biological Photo Service. Figure 5-13: From Richard G. Kessel and Randy H. Kardon, *Tissues and Organs: A Text-Atlas of Scanning Electron Microscopy.* W. H. Freeman and Company. © 1979. Figure 5-14: (a–c) Courtesy of Michael P. Sheetz, Department of Physiology, University of Connecticut Health Center, Farmington, Connecticut. Figure 5-15: (b, top) Norman K. Wessells; (b, bottom) Biophoto Associates/Photo Researchers. Figure 5-16: (a) Biological Photo Service. Figure 5-17: Courtesy of Audrey M. Glavert and Geoffrey M. W. Cook, Strangeways Research Lab, Cambridge, England. Figure 5-19: Courtesy of Norton B. Gilula, Rockefeller University. Figure 5-20: Micrograph by W. P. Wergin, Courtesy of E. H. Newcomb, University of Wisconsin/Biological Photo Service. Figure 5-21: From *Tissues and Organs: A Text-Atlas of Scanning Microscopy* by Richard G. Kessel and Randy Kardon. W. H. Freeman and Company. © 1979.

Chapter 6 Page 98: Normal K. Wessells. Figure 6-1: (a) G. F. Leedale/Biophoto Associates—Photo Researchers; (b) Science Photo Library International/Taurus. Figure 6-2: (b) Don Fawcett and B. Gilula—Photo Researchers. Figure 6-3: (b) Norman K. Wessells. Figure 6-4: (b) D. W. Fawcett/Photo Researchers. Figure 6-6: D. W. Fawcett/Photo Researchers. Figure 6-7: (b) H. T. Bonner & E. H. Newcomb/Biological Photo Service. Figure 6-9: J. Burgess, Department of Cell Biology, John Innes Institute, Norwich, England. Figure 6-10: M. M. Perry and A. B. Gilbert, *J. Cell Sci.* 39 (1979): 357–372. Figure 6-11: (b) D. W. Fawcett/Photo Researchers. Figure 6-13: (b) Courtesy of Daniel S. Friend, University of California School of Medicine, San Francisco. Figure 6-14: (b) G. F. Leedale/Photo Researchers. Figure 6-15: (c–e) D. W. Fawcett & John Heuser/Photo Researchers. Figure 6-16; L. E. Roth/Biological Photo Service. Figure 6-18: (a) Norman K. Wessells; (b) Courtesy of Lloyd A. Culp, Department of Molecular Biology and Microbiology, School of Medicine, Case Western Reserve University, Cleveland. Figure 6-19: Courtesy of Sidney Tamm. Figure 6-21: (b) Diane Woodrum, North Hills, West Virginia, and Richard Linck, University of Minnesota, Minneapolis.

Chapter 7 Page 117: Stephen J. Krasemann/Peter Arnold, Inc. Figure 7-1: (c) Herve Chaumeton/Bruce Coleman.

Chapter 8 Page 135: Art Wolfe/The Image Bank. Figure 8-4: (b) Courtesy of Lewis K. Shumway, Washington State University. Figure 8-7: (b) Mark McKenna. Figure 8-13: (a and b) K. Bendo.

Part Two Page 153: Norman Myers/Bruce Coleman.

Chapter 9 Page 159: Andrew Bajer, Department of Biology, University of Oregon, Eugene. Figure 9-2: (a) Andrew Bajer, Department of Biology, University of Oregon, Eugene; (b) W. E. Engler, Courtesy of G. F. Bahr, Washington, D. C. Figure 9-3: (g) E. Boy de la Tour and U. K. Laemmli, "The Metaphase Scaffold Is Helically Folded: Sister Chromatids Have Predominantly Opposite Helical Handedness," *Cell* 55 (1988): 937–944. Figure 9-4: Science Source/Photo Researchers. Figure 9-6: K. R. Porter/Science Source—Photo Researchers. Figure 9-7: (a–f) Carolina Biological Supply Company. Figure 9-8: (a) T. E. Schroeder, University of Washington/Biological Photo Service; (b) B. A. Palevitz/Biological Photo Service. Figure 9-11: (a) Alan Detrick/Photo Researchers; (b) Science Source/Photo Researchers.

Chapter 10 Page 170: Dale Darwin/Photo Researchers. Figure 10-2: The Bettmann Archive. Figure 10-3: (b) Stephen J. Krasemann/Peter Arnold, Inc. Page 176, Figure A: Jerry Cooke/Animals Animals. Figure 10-8: (b) Courtesy of W. Atlee Burpee Company.

Chapter 11 Page 185: Anne Cumbers/Spectrum Colour Library. Figure 11-3: (a) Courtesy of C. Firling, Department of Biology, University of Minnesota, Duluth; (b) Courtesy of George Lefevre, Jr., from *The Genetics and Biology of Drosophila,* Vol. 1a, M. Ashburne and E. Novitski, eds. (Academic Press, 1976), Ch. 2. Figure 11-8: Victor A. McKusick, Johns Hopkins University; from Mange/Mange, *Human Genetics: Human Aspects* (W. B. Saunders and Co., 1980). Figure 11-11: Courtesy of Bruce F. Cameron and Robert Zucker, Miami Comprehensive Sickle Cell Center, Donald R. Harkness, Director.

Chapter 12 Page 199: R. Lagridge & D. McCoy/Rainbow. Figure 12-1: (c) Courtesy of Ronald L. Phillips, University of Minnesota, St. Paul. Figure 12-4: (a) Maclyn McCarty, M. D., The Rockefeller University. Figure 12-5: (a) Science Source/Photo Researchers; (b) Lee D. Simon/Photo Researchers. Figure 12-8: (a) Courtesy of the Biophysics Department, King's College, London; (b) The Bettmann Archive. Figure 12-9: (c) Dan McCoy/Rainbow. Page 209, Figure A: Michael Shavel/*New York Times.* Figure 12-12: (b) From John Cairns, Imperial Cancer Research Cancer Fund, London, England; permission from *Cold Spring Harbor Symposium on Quantitative Biology* 28 (1963): 43. Figure 12-14: (a) Courtesy of David R. Wolstenholm from *Chromosoma* 43 (1973): 1.

Chapter 13 Page 215: O. L. Miller, B. A. Hamkalo, and C. A. Thomas, Jr., "Visualization of Bacterial Genes in Action," *Science* 169 (1970): 392–395. Figure 13-4: (b) O. L. Miller and Barbara R. Beatty, "Portrait of a Gene," *Journal of Cell Physiology* 74 (Suppl. 1): 225–232. Figure 13-6: (b) Courtesy of Alexander Rich, Department of Biology, Massachusetts Institute of Technology. Figure 13-7: O. L. Miller, B. A. Hamkalo, and C. A. Thomas, Jr., "Visualization of Bacterial Genes in Action," *Science* 169 (1970): 392–395. Page 226, Figure A: Nik Kleinberg/Picture Group.

Chapter 14 Page 229: Courtesy of R. L. Brinster and R. E. Hammer, School of Veterinary Medicine, University of Pennsylvania, Philadelphia. Figure 14-3: (a) Courtesy of Charles C. Brinton, Jr., and Judith Carnahan, Department of Biological Sciences, University of Pittsburgh. Figure 14-6: Stanley N. Cohen/Photo Researchers. Figure 14-12: (a) Courtesy of P. Leder, Genetics Department, Harvard Medical School. Figure 14-15: (a) Courtesy of R. L. Brinster and Myrna Trumbauer, University of Pennsylvania, School of Veterinary Medicine, Philadelphia. Page 248, Figure A: Paul Hosefras/*New York Times.*

Chapter 15 Page 251: Wendell Metzen/Bruce Coleman. Figure 15-1: (b) Statens Museum for Kunst, Copenhagen. Figure 15-2: (a) Terry McKay/The Picture Cube; (b) Courtesy of Theodore T. Puck, E. Roosevelt Institute for Cancer Research, University of Colorado Medical Center. Figure 15-6: Karl Fredga, Department of Genetics, University of Uppsala, Uppsala, Sweden. Figure 15-9: (a and b) Courtesy of Murray L. Barr, Health Science Center, University of Western Ontario, London, Ontario, Canada. Figure 15-12: E. M. Pembrey, R. M. Winter, and Kay E. Davis, "Fragile X Mental Retardation: Current Controversies," *TINS* 2 (1986): 59, Fig. 2. Figure 15-14: (b) Gernsheim Collection, Harry Ransom Humanities Research Center, University of Texas at Austin. Figure 15-16: (a and b) The New York Historical Society; (c) Library of Congress; (d) Massachusetts Historical Society. Figure 15-17: (b) Armed Forces Institute of Pathology.

Chapter 16 Page 269: Norman K. Wessells. Figure 16-7: (a and c) Courtesy of Mia Tegner, Scripps Institution of Oceanography; (b) Courtesy of Gerald Schatten, Department of Biological Sciences, Florida State University, Tallahassee. Figure 16-13: (f) From *Tissues and Organs: A Text-Atlas of Scanning Electron Microscopy* by Richard G. Kessel and Randy H. Kardon, W. H. Freeman and Company. © 1979. Figure 16-15: Courtesy of R. J. Goss, from *Principles of Regeneration* (Academic Press).

Chapter 17 Page 287: Walter J. Gehring, from *Nature* (January 1985), MacMillan Journals Ltd., London. Figure 17-1: (a–c) Courtesy of Carnegie Institution of Washington. Figure 17-5: (e) Norman K. Wessels. Figure 17-6: (a) Courtesy of Kathryn Tosney, Biological Sciences, The University of Michigan, Ann Arbor. Figure 17-10: Courtesy of Lewis Wolpert from "Pattern Formation in Biological Development," *Scientific American* (Oct. 1978). Figure 17-11: (a) Courtesy of Cheryl Tickle, Department of Biology as Applied to Medicine, Middlesex Hospital Medical School, London. Figure 17-12: (a and b) John Postlethwait, University of Oregon. Figure 17-13: Courtesy of E. Firling, Department of Biology, University of Minnesota, Duluth. Figure 17-17: (a) Science Source/Photo Researchers; (b) Mark McKenna.

Chapter 18 Page 307: Science Source/Photo Researchers. Figure 18-1: (a) Charlie Ott/Photo Researchers; (b) John Chellman/Animals Animals; (c) Richard H. Chesher/Photo Researchers. Figure 18-4: Robert Apcarian, Yerkes Regional Primate Center, Emory University, Atlanta. Figure 18-8: (a–e) Petit Format/Nestle/Science Source—Photo Researchers; (f) Lennart Nilsson from *A Child Is Born* (Dell Publishing Company); (g and h) Lennart Nilsson, *Behold Man* (Little, Brown, and Company). Figure 18-13: Bob Marinace.

Part Three Page 325: C. W. Perkins/Animals Animals.

Chapter 19 Page 326: Lee McElfresh/Auscape International. Figure 19-1: Helmut K. Wimmer, American Museum of Natural History. Figure 19-2: David Weintraub/Photo Researchers. Figure 19-3: (a and b) Courtesy of Sidney W. Fox, Institute for Molecular and Cellular Evolution, University of Miami, Coral Gables. Figure 19-6: A. H. Knoll and E. S. Barghoorn, Harvard University. Figure 19-11: Robert Noonan, from "New Genuses Found at Neblina," *National History,* vol. 95, no. 4, April 1986, pp. 93–95.

Chapter 20 Page 340: Grant Heilman. Figure 20-2: A. S. M. LeBeau/Biological Photo Service. Figure 20-3: Courtesy of J. D. Ellar, D. G. Lundgren, and R. A. Slepecky from *Journal of Bacteriology* 94 (1967): 1189, Fig. 20. Figure 20-4: J. P. Duguid, E. S. Anderson, and I. Campbell, *J. Path. Bact.*, 1966, 92, 107. Figure 20-5: (b) Courtesy of Julius Adler, Department of Biochemistry, University of Wisconsin, Madison, from M. L. DePamphilis and Julius Adler, *Journal of Bacteriology* 105 (1971). Figure 20-6: (b) Courtesy of Yasuo Mizugushi, Department of Microbiology, University of Occupational and Environmental Health, from Takade et al., *J. Gen. Microbiol.* 129 (1983): 2315. Figure 20-7: T. J. Beveridge/Biological Photo Service. Figure 20-8: From *Living Images* by Gene Shih and Richard G. Kessel, Science Books International, 1982. Reprinted by permission of the present publisher, Jones and Bartlett Publishers. Figure 20-9: (a) David M. Phillips/Visuals Unlimited; (b) From *Living Images* by Gene Shih and Richard G. Kessel, Science Books International, 1982. Reprinted by permission of the present publisher, Jones and Bartlett Publishers. Figure 20-10: Centers for Disease Control, Atlanta. Figure 20-11: *Scientific American,* June 1988; micrograph made by Hans Reichenbach, Society for Biotechnological Research, Braunschweig, Germany. Figure 20-12: Eli Lilly and Company. Figure 20-13: (a) John D. Cunningham/Visuals Unlimited; (b) Ray Simons/Photo Researchers. Figure 20-14: (a) Tony Brain/Science Source—Photo Researchers; (b) Norma J. Lang, Department of Biology, University of California, Davis. Figure 20-15: The Bettmann Archive. Figure 20-16: (a) Mahl/Photo Researchers; (b) Marc Adrian, Jacques Dubochet, Jean Lapault, and Alasdair W. McDowall, *Nature* 308 (March 1–7, 1984); (c) Courtesy of H. G. Pereira, Ministério de Saúde Fundacao Oswaldo Cruz, from R. C. Valentine and H. G. Pereira, *Journal of Molecular Biology* 13 (1965): 13–20.

Chapter 21 Page 359: M. I. Walker/Photo Researchers. Figure 21-2: (b) P. L. Walne and J. H. Arnott, *Planta* 77 (1967): 325–354. Figure 21-3: Courtesy of F. J. R. Taylor and G. Gaines, Institute of

Oceanography, University of British Columbia, Vancouver. Figure 21-4: (a and b) Biophoto Associates/Photo Researchers. Figure 21-6: James P. Jackson/Photo Researchers. Figure 21-7: (b) From W. F. Loomis, *Dictyostelium discoidem* (Academic Press, 1975). Figure 21-8: Eric V. Gravé. Figure 21-9: Biological Photo Service. Figure 21-10: Eric V. Gravé. Figure 21-11: Bruce Coleman Limited, London. Figure 21-13: Courtesy of Gary Grimes, Hofstra University.

Chapter 22 Page 373: Hans Reinhard/Bruce Coleman. Figure 22-1: G. F. Leedale/Biophoto Associates—Photo Researchers. Figure 22-2: (b) C. E. Bracker, Department of Botany and Plant Pathology, Purdue University, West Lafayette, Indiana. Figure 22-3: William J. Weber/ Visuals Unlimited, Figure 22-4: (a) P. A. Hinchliffe/Bruce Coleman; (b) From *Living Images* by Gene Shih and Richard G. Kessel, Science Books International, 1982. Reprinted by permission of the present publisher, Jones and Bartlett Publishers. Figure 22-5: Noble Proctor/ Photo Researchers. Figure 22-6: D. Gotelli, Department of Biological Science, California State College Stanislaus, Turlock. Figure 22-7: (b) Biophoto Associates/Photo Researchers. Page 380, Figure A and Figure B: John D. Cunningham/Visuals Unlimited. Figure 22-9: (a) Peter Katsaros/Photo Researchers. Figure 22-10: Robert P. Carr/ Bruce Coleman. Figure 22-11: (a) John D. Cunningham/Visuals Unlimited; (b) Charlie Ott/Photo Researchers. Figure 22-12: Biophoto Associates/Photo Researchers. Figure 22-13: (a) Russ Kinne/Photo Researchers; (b) Robert Ashworth/Photo Researchers.

Chapter 23 Page 388: Alan Pitcarin/Grant Heilman. Figure 23-2: Kim Taylor/Bruce Coleman. Figure 23-3: (a) D. P. Wilson/Eric and David Hosking. Figure 23-4: Bruce Coleman. Figure 23-5: (a) Jeff Foott/Bruce Coleman. Figure 23-8: (b) Steve Solum/Bruce Coleman. Figure 23-9: (a) G. R. Roberts/Documentary Photos; (b) John D. Cunningham/Visuals Unlimited. Figure 23-10: Robert Waaland/Biological Photo Service. Figure 23-11: (b) Biophoto Associates/Photo Researchers. Figure 23-12: Biophoto Associates/Photo Researchers. Figure 23-13: (a) E. R. Degginger/Bruce Coleman; (b) C. R. Roberts/Documentary Photos. Figure 23-14: (a) Michael Giannechini/Photo Researchers; (b) Norman K. Wessells. Table 23-2: (left) Manuel Rodriguez; (right) John D. Cunningham/Visuals Unlimited. Figure 23-17: (a) Merlin Tuttle/Photo Researchers; (b) Spencer C. H. Barrett, Department of Botany, University of Toronto. Figure 23-18: (a) Gregory K. Scott/ Photo Researchers; (b) S. N. Postlethwait/Rainchild Gardens. Figure 23-19: (a) F. Collet/Photo Researchers; (b) John S. Flannary/Bruce Coleman; (c) Robert H. Potts, Jr./Photo Researchers.

Chapter 24 Page 409: W. E. Harvey/Photo Researchers. Figure 24-2: Carl Rossler/Tom Stack & Associates. Figure 24-3: (a) Kim Taylor/Bruce Coleman; (b) Charles A. Jacoby, Harbor Branch Oceanographic Institute, Fort Pierce, Florida. Figure 24-5: Gary Milburn/ Tom Stack & Associates. Figure 24-6: Kathie Atkinson/Oxford Scientific Films. Figure 24-7: Chris Newbert/Bruce Coleman. Figure 24-10: (b and c) Biophoto Associates/Photo Researchers. Figure 24-11: (b) Kjell A. Sandved/Bruce Coleman. Figure 24-14: (a) E. R. Degginger/Bruce Coleman; (b) Carl Roessler/Tom Stack & Associates. Figure 24-17: Michael Fogden/Bruce Coleman. Figure 24-18: J. MacGregor/Peter Arnold. Figure 24-19: James H. Robinson/Photo Researchers. Figure 24-20: (b) A. Kerstitch/Tom Stack & Associates. Figure 24-21: (a) J. H. Robinson/Photo Researchers; (b) Peter Ward/ Bruce Coleman; (c) Kjell B. Sandved/Photo Researchers; (d) Jeff Lepore/Photo Researchers. Figure 24-22: (a) Mike Newmann/Photo Researchers; (b) Gary Milburn/Tom Stack & Associates. Figure 24-24: (a) Jeff Foott/Bruce Coleman; (b) E. R. Degginger; (c) Edward Robinson/Tom Stack & Associates. Figure 24-29: P. Morris/Ardea, London. Figure 24-31: (a) Breck P. Kent; (b) Fred Bavendam/Peter Arnold, Inc. Figure 24-32: (a) Shostal Associates; (b) Fred Bavendam/Peter Arnold, Inc.; (c) A. Kerstitch/Tom Stack & Associates. Figure 24-33: Steve Martin/Tom Stack & Associates. Figure 24-34: (a) Gregory G. Dimijian/Photo Researchers; (b) John R. MacGregor/ Peter Arnold, Inc. Figure 24-36: (a) Breck P. Kent; (b) Hans D. Dossenbach/Ardea, London; (c) H. E. Vible/Photo Researchers. Figure 24-37: (a) Steven C. Kaufman/Peter Arnold, Inc.; (b) G. Ziesler/Peter Arnold, Inc.; (c) Brian Rogers/Biophotos. Figure 24-38: Jean-Paul Ferrero/Auscape International. Figure 24-39: (a) Joyce Photographics/ Photo Researchers; (b) Tom McHugh/Photo Researchers; (c) Carl Purcell/Photo Researchers. Figure 24-40: (a) Mickey Gibson/Tom Stack & Associates; (b) Stephen J. Krasemann/Photo Researchers; (c) David Hosking.

Part Four Page 447: Chris O'Riley/The Stock Broker.

Chapter 25 Page 448: William M. Harlow/Photo Researchers. Page 450, Figure A: Edwards/FPG. Figure 25-2: (a) S. D. Halperin/Earth Scenes; (b) Barry Lopez/Photo Researchers; (c) Biological Photo Service. Figure 25-3: (a) Long Ashton Research Station, Bristol, England. Figure 25-4: Norman K. Wessells. Figure 25-5: G. I. Bernard/Earth Scenes. Figure 25-6: (a) Barbara J. Miller/Biological Photo Service; (b) John Gerlach/Tom Stack & Associates. Figure 25-7: (a and b) Ray F. Evert, University of Wisconsin, Madison. Figure 25-8: (b) Courtesy of J. H. Troughton, Department of Scientific and Industrial Research/ Physics and Engineering Laboratory, Lower Hutt, New Zealand. Figure 25-10: Francois Gohier/Photo Researchers. Figure 25-11: (b) From *Living Images* by Gene Shih and Richard G. Kessel, Science Books International, 1982. Reprinted by permission of the present publisher, Jones and Bartlett Publishers. Figure 25-13: Harry Howard, Department of Biology, University of Oregon, Eugene. Figure 25-14: (a and b) William Ferguson; (c) Patti Murray/Earth Scenes.

Chapter 26 Page 463: Robert L. Dunne/Bruce Coleman. Figure 26-1: (a) Gilbert Grant/Photo Researchers; (b) Runk/Shoenberger/ Grant Heilman; (c) Jerome Wexler/National Audubon Society—Photo Researchers. Figure 26-2: Plant Genetics, Inc. Page 466, Figure A: BOWN; Figure B: Delbert Wiens. Figure 26-3: (a) Biological Photo Service; (b) L. West/Photo Researchers; (c) Alford W. Cooper/Photo Researchers. Figure 26-5: (d) Norman K. Wessells. Figure 26-7: (c) David J. Mulcahy, Department of Botany, University of Massachusetts at Amherst. Figure 26-9: (a) Lois and George Cox/Bruce Coleman; (b) Forest W. Buchanan/Visuals Unlimited; (c) William J. Weber/ Visuals Unlimited. Figure 26-11: (b) J. Robert Waaland/Biological Photo Service. Figure 26-12: (a) J. Robert Waaland/Biological Photo Service; (b) Paul Green, Biology Department, Stanford University. Figure 26-13: (a) Mary M. Thatchek/Photo Researchers; (b) Richard Shiell/Earth Scenes. Figure 26-14: (d) Ardea, London. Figure 26-15: (a) John D. Cunningham/Visuals Unlimited; (b) Robert E. Lyons/ Visuals Unlimited; (c) Tom Myers/Photo Researchers.

Chapter 27 Page 483: John Shaw/Bruce Coleman. Figure 27-3: William J. Weber/Visuals Unlimited. Figure 27-6: Norman K. Wessells. Figure 27-8: Hugh Spencer/Photo Researchers. Figure 27-9: Harry E. Calvert/Battelle-Kettering Research Laboratory. Figure 27-10: (a and b) Jack Dermid.

Chapter 28 Page 498: Breck Kent/Earth Scenes. Figure 28-4: (a and b) J. P. Nitsch, *American Journal of Botany* 37 (1950): 3. Figure 28-5: (b) Michael Evans. Figure 28-6: Courtesy of Sylvan H. Wittwer, Agricultural Experiment Station, Michigan State University, East Lansing. Figure 28-8: (a) Lowell Georgia/Science Source—Photo Researchers; (b) Devere Patton. Page 505, Figure A: DNAP. Figure 28-9: (b) Bill Ross/Westlight.

Part Five Page 513: George Holton/Photo Researchers.

Chapter 29 Page 514: CNRI/Science Photo Library/Photo Researchers. Figure 29-8: Thomas Eisner, Cornell University. Figure 29-9: Thomas Eisner, Cornell University. Figure 29-10: (a) D. W. Fawcett/Photo Researchers. Figure 29-14: From *Tissues and Organs:*

A Text-Atlas of Scanning Electron Microscopy by Richard G. Kessel and Randy H. Kardon, W. A. Freeman and Company. © 1979. Figure 29-15: (b) Lennart Nilsson, © Boehringer Ingelheim International GmbH/Bonnier Fakta, Stockholm, Sweden.

Chapter 30 Page 533: Lennart Nilsson, © Boehringer Ingelheim International GmbH/Bonnier Fakta, Stockholm, Sweden. Figure 30-1: Don Fawcett/Photo Researchers. Figure 30-5: (c) Courtesy of R. R. Dourmashkin, from R. R. Parkhouse et al., *Immunology* 18 (1970): 575. Figure 30-10: From O. Carper, I. Virtaner, and E. Saksela, *Journal of Immunology* 128 (1982): 2691. Figure 30-11: Science Source/Photo Researchers. Figure 30-14: (e) Lennart Nilsson, © Boehringer Ingelheim International GmbH/Bonnier Fakta, Stockholm, Sweden.

Chapter 31 Page 552: Bruce M. Wellman/Animals Animals. Page 554, Figure A: John Roskelley/Photo Researchers. Figure 31-4: Tom McHugh/Photo Researchers. Figure 31-5: E. R. Degginger. Figure 31-6: (b) Warren Burggren, Department of Zoology, University of Massachusetts, Amherst. Figure 31-19: William E. Townsend, Jr./Photo Researchers.

Chapter 32 Page 570: Mark J. Jones/Bruce Coleman. Figure 32-2: Lennart Nilsson from *Close to Nature* (New York: Pantheon Books). Figure 32-4: (a and b) Francois Gohier. Figure 32-10: From *Tissues and Organs: A Text-Atlas of Scanning Electron Microscopy* by Richard G. Kessel and Randy H. Kardon, W. H. Freeman and Company. © 1979. Figure 32-11: (a) From *Tissues and Organs: A Text-Atlas of Scanning Electron Microscopy* by Richard G. Kessel and Randy H. Kardon, W. H. Freeman and Company. © 1979. Figure 32-12: (b) R. D. Specian, Department of Anatomy, Louisiana State University, Medical Center, Shreveport. Page 586, Figure A: (left) David Madison/Bruce Coleman; (right) Dianne Duchlin/Bruce Coleman.

Chapter 33 Page 592: Spencer Swanger/Tom Stack & Associates. Page 594, Figure A: John Crowe, Department of Anatomy, University of California, Davis. Figure 33-2: Heather Angel. Figure 33-9: Mark McKenna. Figure 33-10: Christopher Crowley/Tom Stack & Associates. Figure 33-11: Robert Harding Picture Library. Page 610, Figure A: Brian M. Barnes/Institute of Arctic Biology.

Chapter 34 Page 612: Biophoto Associates/Photo Researchers. Figure 34-1: (b) Norman K. Wessells. Figure 34-3: (b) A. C. Hodgkin and R. D. Keynes from *Journal of Physiology* 131 (1956). Figure 34-8: (d) Micrograph produced by John E. Heuser of Washington University School of Medicine, St. Louis. Figure 34-10: (b) From E. R. Lewis et al., "Studying Neural Organization in Aplysia with the Scanning Electron Microscope," *Science* 165 (September 12, 1969): 1140–1143.

Chapter 35 Page 629: Heather Angel. Figure 35-3: (a) Walter E. Stumpf, Department of Anatomy, University of North Carolina, Chapel Hill. Page 633, Figure A: Wide World Photos. Figure 35-6: E. R. Degginger/Animals Animals. Figure 35-7: 1965 CIBA Pharmaceutical Company, Division of CIBA-GEIGY Corporation, reproduced with permission from CIBA Collection of Medical Illustrations by Frank H. Netter, M. D., all rights reserved. Figure 35-10: Custom Medical Stock Photo. Figure 35-12: (b) Courtesy of W. W. Douglas, F. R. S., Yale University, from Douglas, Nagasawa, and Shulz, "Sub Cellular Organization and Function in Endocrine Tissues," *Memoirs of the Society of Endocrinology* 19 (Cambridge University Press).

Chapter 36 Page 649: Barbara Kirk/The Stock Market. Figure 36-1: (b) SIU/Photo Researchers. Figure 36-2: J. H. Carmichael/Photo Researchers. Figure 36-8: (d) From *Tissues and Organs: A Text-Atlas of Scanning Electron Microscopy* by Richard G. Kessel and Randy H. Kardon, W. H. Freeman and Company. © 1979. Figure 36-9: Merlin D. Tuttle/Photo Researchers. Figure 36-10: Brian Parker/Tom Stack & Associates. Figure 36-12: (b) Sjur Olsnes, Institute of Cancer Research, Norwegian Radium Hospital, Oslo, Norway. Figure 36-14: (b) John Rogers. Figure 36-16: (a) Science Photo Library/Photo Researchers. Figure 36-18: D. W. Tank. Figure 36-21: (c) Fritz Goro. Figure 36-26: Nicole Le Davarin.

Chapter 37 Page 683: Alan Jones/Westlight. Figure 37-1: Kjell B. Sandved/Photo Researchers. Figure 37-5: (a) From *Living Images* by Gene Shih and Richard G. Kessel, Science Books International, 1982. Reprinted by permission of the present publisher, Jones and Bartlett Publishers; (b) From *Tissues and Organs: A Text-Atlas of Scanning Electron Microscopy* by Richard G. Kessel and Randy H. Kardon, W. H. Freeman and Company. © 1979. Figure 37-11: (a) Science Source/Photo Researchers; (b) G. W. Willis/Biological Photo Service; (c) Tom Stack & Associates. Figure 37-12: (b) C. Franzini Armstrong, *Journal of Cell Biology* 47 (1970): 488. Figure 37-17: (a) Fawcett/Vehara/Photo Researchers. Page 699, Figure A: Focus on Sports. Figure 37-19: Jeff Rotman.

Part Six Page 703: Adam Woolfitt/Woodfin Camp & Associates.

Chapter 38 Page 704: Frans Lanting. Figure 38-1: The Bettmann Archive. Figure 38-2: (a and b) Miguel Castro/Photo Researchers. Figure 38-3: The Granger Collection. Figure 38-4: (a) Luis Villota/The Stock Market; (b) Joan Lebold Cohen/Photo Researchers; (c) Susan McCartney/Photo Researchers; (d) Luis Villota/The Stock Market. Figure 38-5: Yi-lin Yan, Department of Biology, University of Oregon, Eugene. Figure 38-7: Charles Krebs, Issaguah, Washington. Figure 38-8: Manuel Grossberg/Photo Researchers. Figure 38-10: C. Allan Morgan. Figure 38-12: Victor McKusick, Johns Hopkins University. Figure 38-13: Tau-San Chou. Plant Geneticist, George J. Ball, Inc., West Chicago. Figure 38-15: Joel Gordon.

Chapter 39 Page 719: Zig Leszczynski/Animals Animals. Figure 39-1: Richard P. Smith/Tom Stack & Associates. Figure 39-2: Zig Leszczynski/Animals Animals. Figure 39-3: (a) Breck P. Kent; (b) Anthony Bannister/Animals Animals. Figure 39-5: Robert A. Lubeck/Animals Animals. Figure 39-7: (a and b) J. L. Mason/Ardea, London. Figure 39-11: Courtesy of J. M. Poehlman, from J. M. Poehlman, *Breeding Field Crops*, Third Edition (Van Nostrand-Reinhold, 1987), Figure 12-4b. Figure 39-12: (Peruvian) Owen Franken/Stock Boston; (African) Ivan Massar/Photo Nats; (Italian) Luis Villota/The Stock Market.

Chapter 40 Page 732: Jeff Foott/Tom Stack & Associates. Page 735, Figure A: C. B. Frith/Bruce Coleman. Figure 40-2: Hans and Juoy Beste/Ardea, London. Figure 40-4: Kenneth W. Fink/Bruce Coleman. Figure 40-6: (e, Albert squirrel) Larry Brock/Tom Stack & Associates; (Kaibab squirrel) Ian Beames/Ardea, London; (Grand Canyon) Tom Bean/DRK Photo. Figure 40-9: (a) Art Wolfe/Wildlife Photobank. Figure 40-10: (a) P. G. Wessells; (b) John Cahcalosi/Tom Stack & Associates. Figure 40-12: (a) E. R. Degginger; (b) Anthony Bannister/Animals Animals. Page 746, Figure A: Gerald D. Carr, Department of Botany, University of Hawaii, Honolulu. Figure 40-16: Reprinted with permission of Macmillan Publishing Company from *The Science of Evolution*, by William D. Stansfield, © 1977.

Chapter 41 Page 751: Nicholas deVore III/Bruce Coleman. Figure 41-1: C. A. Morgan. Figure 41-2: NASA. Figure 41-4: NASA. Figure 41-8: Christopher Crowley/Tom Stack & Associates. Figure 41-9: Pierre C. Fischer. Page 758, Figure A: Jacques Jangoux/Peter Arnold, Inc. Figure 41-10: E. E. Kingsley/Photo Researchers. Figure 41-11: E. R. Degginger. Figure 41-12: Charles Palek/Animals Animals. Figure 41-13: E. R. Degginger/Earth Scenes. Figure 41-14: Breck P. Kent/Earth Scenes. Figure 41-15: Dale Johnson/Tom Stack & Associates. Figure 41-16: William Ferguson. Figure 41-17: (a) Tom Bean/ Aperture; (b) R. Hessler, Research Institute of Scripps Clinic. Page 764, Figure A: (a) I. Everson, British Antarctic Survey. Page 767, Figure A: Ray Pfortner/Peter Arnold, Inc.; Figure B: D. W. Schindler et al., "Effects of Years of Experimental Acidification on a Small Lake," *Science*, vol. 228. Figure 41-22: Charles Keeling, Scripps Institution of Oceanography, University of California. Figure 41-24: (b) Jim Corwin/ Aperture.

Chapter 42 Page 775: Stephen Frink/The WaterHouse. Figure 42-1: (a) Sydney Thompson/Earth Scenes; (b) Tom Bean/DRK Photo. Figure 42-3: (e) B. Lund/The Nature Conservancy, Massachusetts/ Rhode Island Field Office, Boston. Figure 42-4: Jack Dermid. Figure 42-5: Diana L. Stratton/Tom Stack & Associates. Figure 42-6: Grant Heilman. Figure 42-7: Frans Lanting. Figure 42-8: (a) G. G. Dimijian/ Photo Researchers; (b) Stephen Krasemann/Photo Researchers. Figure 42-9: John Mutrux. Figure 42-10: (a) Bob and Clara Calhoun/Bruce Coleman; (b) Taounson & Clampett/Ardea; (c) Kjell B. Sandved; (d) Dale and Marian Zimmerman/Bruce Coleman. Figure 42-12: (a) Bill Dyer/Photo Researchers. Figure 42-14: (a) Lynn M. Stone/ Bruce Coleman; (b) Bill Everitt/Tom Stack & Associates. Figure 42-15: Thomas Eisner, Cornell University, and D. Aneshansley. Figure 42-16: Kjell B. Sandved. Figure 42-17: Gerald Rosenthal, University of Kentucky, Lexington. Figure 42-18: (a) Thomas Eisner, Cornell University; (b) Thomas Kitchen/Tom Stack & Associates. Figure 42-19: James L. Nation, Jr., Courtesy of Thomas C. Emmel, University of Florida, Gainesville. Figure 42-20: Breck P. Kent. Figure 42-21: (a) E. R. Degginger; (b) Breck P. Kent. Figure 42-22: E. R. Degginger/ Animals Animals. Figure 42-23: (a) K. G. Preston-Malham/Animals Animals; (b) E. R. Degginger; (c) Breck P. Kent. Figure 42-24: Heather Angel. Figure 42-25: Heather Angel. Figure 42-26: (a and b) J. L. Lapore/Photo Researchers. Figure 42-27: (a) Klaus Esser, Department of Cell Biology, Smith Kline and French Laboratories, Swedeland, Pennsylvania; (b) Richard Kolar/Animals Animals; (c) Francois Gohier/Photo Researchers. Figure 42-28: Zig Leszczynski/ Earth Scenes/Animals Animals.

Chapter 43 Page 799: G. Schaller/Bruce Coleman. Figure 43-4: Phil A. Dotson/Photo Researchers. Figure 43-7: (a) Jeff Frost/Bruce Coleman; (b) Jeff Rotman/Peter Arnold, Inc. Figure 43-9: Gianni Tortem/Photo Researchers. Figure 43-10: Ray Pfortner/Peter Arnold. Figure 43-11: C. Allen Morgan. Figure 43-12: Larry D. Slocum/ Animals Animals. Page 808, Figure A: U. U. Department of Agriculture. Figure 43-13: Wayne Lankiner/DRK Photo. Figure 43-18: Alan Oddie/PhotoEdit. Figure 43-20: Bruno J. Zehnder/Peter Arnold, Inc.

Chapter 44 Page 815: Entheos. Figure 44-3: J. C. Carton/Bruce Coleman. Figure 44-4: (a) Michael & Barbara Reed/Animals Animals; (b) Carl Purcell/Photo Researchers. Figure 44-5: Allen Russel/Profiles West. Figure 44-6: E. R. Degginger. Figure 44-8: Kim Taylor/Bruce Coleman. Figure 44-10: Don and Pat Valenti/DRK Photo. Page 823, Figure A: Dwight Kuhn/Bruce Coleman. Figure 44-11: Marty Cordano/DRK Photo. Figure 44-12: J. Van Wormer/Bruce Coleman. Figure 44-13: (b) H. S. Terrace/Animals Animals. Figure 44-14: G. I. Bernard/Animals Animals. Figure 44-15: A. Morris/Academy of Natural Sciences, Philadelphia/Virco. Figure 44-16: M. Hyett/Academy of Natural Sciences, Philadelphia/Virco.

Chapter 45 Page 832: Charles Krebs. Figure 45-1: Peter Veit/DRK Photo. Figure 45-2: (a) Arthur Bertrand/Photo Researchers; (b) Bettmann Newsphotos. Figure 45-4: Tony Floria/Photo Researchers. Figure 45-6: Fawcett/Tom Stack & Associates. Figure 45-8: Hans Pfletschinger/Peter Arnold, Inc. Figure 45-10: Paul W. Sherman, Cornell University. Figure 45-11: (a) Wardene Weisser/Bruce Coleman; (b) Stephen J. Krasemann/DRK Photo. Figure 45-12: Gary Milburn/ Tom Stack & Associates. Figure 45-13: G. Ziesler/Peter Arnold. Figure 45-14: Tom Cajacob/Minnesota Zoo. Figure 45-15: Tom Stack/Tom Stack & Associates.

Chapter 46 Page 846: *Discover* (Sept. 1986), p. 920. Figure 46-1: John Reader, Surrey, England; Courtesy of Dr. Mary Leakey. Figure 46-2: (a) Kenneth Fink/Bruce Coleman; (b) Sullivan and Rogers/Bruce Coleman; (c) Zig Leszczynski/Animals Animals. Figure 46-4: Rod Williams/Bruce Coleman. Figure 46-10: Photo Courtesy of the Cleveland Museum of Natural History. Figure 46-14: (a) Tim D. White; (b) Collection Musée de l'Homme; (c) American Museum of Natural History. Figure 46-15: David L. Brill, Courtesy of the National Geographic Society. Figure 46-16: M. Boule and H. V. Vallois from *Fossil Men* (New York: Dryden Press, 1957), p. 286, Courtesy of University of California Library, Berkeley. Figure 46-17: (a) Shelly Grossman/ Woodfin Camp & Associates; (b) Jim Cartier/Photo Researchers. Figure 46-19: Courtesy of Donald R. Helinski, University of California, La Jolla, photo by Keith V. Wood. Figure 46-20: (a) Jeff Hunter/The Image Bank; (b) David Madison/Bruce Coleman. Figure 46-21: John Paul Kay/Peter Arnold.

Index

Pages on which definitions or main discussions of topics appear are indicated by **boldface.**
Pages containing illustrations or tables are indicated by *italics.*

Abalone, 428
A band, *693*, 694
Abdomen, 423
Abies grandis, *451*
ABO blood group, 84, 191–192, *191*
Abomasum, 574–575, *574*
Abortion, 322, *323*
 spontaneous, 257, 550
Abscisic acid, 489, *500*, **505–508**
Abscission, **505–506**
Abscission layer, **507–508**, *507*
Absorption, by root, **458**
Absorption spectrum, chlorophyll, *139*, **140**
Abstraction, 861
Abzyme, 542
Acacia, 759, 777, 796
Acanthocephala, *410*
Acanthodii, 433–434
Acanthorychus tenuirostris, *785*
Accessory pigment, 140
Accommodation (eye), 660
Accutane, 297, 317
Acer, *507*
Acetabularia, *155*, 156
Acetic acid, *27*, 38, 350
Acetylcholine, 526, 621, 627, 678, 695
Acetyl-CoA, *126*, 127–128, *131–132*, 132
Acheulian industry, **858–859**, *858*
Achondroplastic dwarfism, 252, *253*
Acid, **32–33**
Acid dust, 767
Acidic solution, 32
Acid rain, 439, 767
Acinonyx jubatus, 715, *792*
Acne medication, 297
Acoelomate, 415
cis-Aconitic acid, *126*
Acorn, *4*
Acorn worm, 428, 430
Acquired immunodeficiency syndrome, *See* AIDS
Acrasiomycota, **364–365**, *365*
Acromegaly, 642
Acrosome, 270, *271*
Acrosome reaction, **274**, *274*
ACTH, **318**, *319*, *638*, *641*, **642**, 647
Actin, **110**, 111–112, *112*, 114, 274–275, 292, 683, *693–694*, 694–700, *696*, *698*
 in cytokinesis, 163–164, *163*
Actinomyces, *346*
Actinomycete, *346*, **348**, *348*
Actinopterygii, 434
Action potential
 all-or-none character, 618
 muscle, 695–696, *696*
 nerve, **613**
  generation, 615–618, *616–617*
  propagation, 618–619, *618–619*
  transmission between cells, 620–622, *620–621*
 plant, 495, *495*
Action spectrum, **141**
Activated molecule, 64
Activation energy, **64**, *64*, 66, *66*
Active immunity, **546–547**, *547*
Active site, **67**, *67*
Active transport, **89**, *90*
Acyclovir, 357
Adams family, pattern baldness, *264*
Adaptation, 4, *5*, 11, **706**, 728, *728*, **776**
 amphibian, 436
 angiosperm (flowering plant), 403–404
 defense, 787–792
 escape, 792–794, *792–794*
 leaves, 453–454
 mammals, 442
 reptile, 437
 roots, 460–461
 stems, 456–457, *457*
Adaptive radiation, *441*, **744–746**, *745–746*
Addison's disease, 638
Adenine, 54, *55*, 118, *119*, **206**, *206*
Adenine deaminase, 266
Adenoids, *535*
Adenosine monophosphate, *See* AMP
Adenosine triphosphate, *See* ATP
Adenovirus, 351, *352*
Adenyl cyclase, 633, *634*
ADH, *See* Antidiuretic hormone
Adhesion, water molecules, **28–29**, 30–31
Adipose tissue, 516
Adolescent, growth, 283–284
ADP, 118–119, *118–119*
 hydrolysis, 118, *119*
Adrenal cortex, 638–639, *638*, 644
Adrenal gland, 603, *603*, 638–639, *638*
Adrenal medulla, *638*, 639
Adrenocorticotropic hormone, *See* ACTH
Adrostenol, 382
Adventitious root, **460**, *460*
Aegyptopithecus, 850
Aerial hyphae, **376**
Aerobe, **333**
 obligate, **345**
 origin, 333
Aerobic performance, limits, 699
Aerobic respiration, 121, *122*
 compared to photosynthesis, 135–136, *136*
Aflatoxin, 384
African sleeping sickness, 366, *366*, 795
Afterbirth, **319**
Agaiocercus coelestis, *787*
Agar, 230, 392
Age of Reptiles, 437–439
Age pyramid, **812–813**, *812*
Age structure, population, **802–803**, *803*
Agglutination, 539
Aggregate, formation on primitive earth, 329–330
Aggressive signal, 834
Aging, 285
 immune system, 285
 plant, 504, *504*
 skin, 285
Agnatha, **432–433**, *432–433*
Agonist muscle, 689, *689*
Agricultural researcher, *13*
Agricultural revolution, 811–812, 861
Agrobacterium tumefaciens, 248
AIDS, 354, *355*, 533–534, *533*, 545–546, *546*
Ailuropoda melanolenca, *741*
Air, pathway in human, 559–560, *559*
Air capillary, 561, *562*
Air quality indicator, 385
Air sac, 561–562, *562*
Alarm call, 730, 834, 841, *841*
Alarm caller, 730, *730*, 838
Albatross, *441*
 black, *819*
Albinism, 193, *193*, 195, **262–263**, *263*
Albumin, 273, 581
Alcohol, 38–39
Alcohol dehydrogenase gene, 248
Alcoholic beverage, 124, *125*, 382
Aldehyde, *38*, 39
Alder tree, 494, 780
Aldosterone, *602*, **603**, 635, **638–639**
Aleurone, **475**
Alfalfa, 345, *454*, 457, *465*, 494, 861
Algae, *389*, **389–394**, 760, A2
 golden-brown, *360–361*, 361
 in lichens, 384–385, *385*
Alginic acid, 393
Alimentary canal, **571**
Alkaline solution, 32
Alkaloid, 790
Alkaptonuria, 263
Allantois, **282**, *282*, 317, *318*
Allard, H.A., 508
Allele, **174**, 711
 fixed, **721–722**
 lethal, **194–195**, *194*
 multiple allelic series, **191–193**, *191–192*, 722
 wild-type, **190**
Allele frequency, 710–712, *711*
Allelopathy, **809**
Allen's rule, *721*, **722**
Allergy, **548–549**, *548*
Alligator, 439–440, 572

Allium, *476*
Allopatric speciation, **737–738**, *738*
Allopolyploid, 739–740, *740*
Allosteric effector, 73–74, *73–74*
Allosteric enzyme, **73–74**, *73–74*, 132, *133*
Allosteric site, 73–74, *73–74*
Alouatta seniculus, *848*
Alpaca, *568*
Alpha animal, 842
Alpha carbon, 47, *47*
α cells, 637, *637*
α factor, 646
α-helix, **50**, *50*
Alpha wave, 673, *673*
Alternation of generations, **390**, *390*, 412–414, *412–413*
Altitude
adaptation to high altitude, 567–568, *568*
climate and, 756–757
effect on brain, 554
partial pressure of oxygen and, 553–554, *553*
Altitude sickness, 567
Altruistic behavior, 730, **837–839**
Alveoli, *559*, **560**
Alvarez, Luis, 440
Alvarez, Walter, 440
Alzheimer's disease, 25, 252, 675
Amanita muscaria, *383*
Amblyospiza albifrons, *441*
Amblyrhynchus oristosis, *606*
Ames test, **197**
Amine hormone, 631, *631*, 633–635, *634–635*
Amino acid, **46**
absorption in intestine, 579–580, *580*
activation, 221–222, *222*
essential, **587**
metabolism, 131–132, *131–132*, 596–604, *596*
structure, 47–48, *47*
uptake by cells, 89, *90*
Aminoacyl-tRNA, *223*, *225*
Aminoacyl-tRNA synthetase, **222**, *222*, 226
Amino group, *38*, 39, 47, *47*
Aminopeptidase, *581*, 582, *583*, 721
Amish, 715, *715*
Ammonia, *27*, 596, *596–597*, 770
Amniocentesis, **266**, *267*
Amnion, **282**, *282*, *316*, 317, *318*, *320*
Amniotic fluid, 282
Amoeba, *359–361*
Amoeba proteus, 366–367, *367*
Amoebocyte, 411, *411*
AMP, 118–119, *119*
Amphibian, *432*, **436–437**, *437*
adaptations, 436
body temperature, 606–607
embryonic development, 277, *277*
evolution, *10*
gastrulation, *280*
heart, 519, *520*
kidney, 598
metamorphosis, 645–646, *646*
nitrogenous waste, *597*
osmotic balance, 595
ventilation, 560, *560*
Amphioxus, *See* Lancelet
Amphitetras antediluviana, *363*
Amplification, 529
Ampulla
ear, 655, *655*
echinoderm, 428, *429*
Amygdala, *668*, 669
Amylase, 475, 581, *581–582*
α-Amylase, 508
Amylopectin, 41, *42*
Amyloplast, 502, *502*
Amylose, 41, *42*
Anabaena, *341*, *348–349*, 494, *495*, 722
Anabolic steroid, 633
Anabolism, **72**, 132
Anaerobe, 124
facultative, 124, **345**
obligate, **345**
Analogy, **744**, *745*
Anal pore, 571
Anaphase
meiosis I, *164*, 165
meiosis II, *165*, 166
mitotic, *154*, **161–163**
Anaphylactic shock, **548–549**
Anaphylaxis, **548–549**
Anchorage dependence, **93**, 111–112, *112*
Androgen, **310**, 639
Anemia, 263
pernicious, *588*
sickle-cell, *See* Sickle-cell anemia
Anesthetic, 88
Aneuploidy, **257–258**, *257–258*
ANF, *See* Atrionatriuretic factor
Anger, 669
Angiography, 21
Angiosperm (flowering plant), *389*, 399, **403–407**, 449, *449*
adaptations, 403–404
evolution, *394*
life cycle, *404*
Angiotensin I, 602–603, *602*
Angiotensin II, *602*, 603, 637–638, *638*, 678
Angular gyrus, 673
Anhidrotic ectodermal dysplasia, 259, *259*
Animal
adaptations to land, 436
evolution, *410*
Animal breeding, 10, *10*, 176, *176*, 726, 861
genetic engineering, 247–248
Animalia, **338**
Animal-like heterotroph, 361
Animal pole, 277–278, *277*
Anion, 31
Annelida, *410*, *415*, **420–422**, *421–422*, 426, 594
Annual plant, 406–407, **479–480**, *480*
Annual rhythm, 644
Anopheles, 367, *368*
Anopodium, *110*
Anorexia nervosa, 585
Anser anser, *818*
Anser caerulescens hyperborea, *713*
Ant, 426, 795–796, *796*, 838, *838*
house, 796
leaf-cutting, 790
Antagonist muscle, 689, *689*
Antarctic food web, 764, *764*
Anteater, spiny, 441, 443
Antelope, 841–842, *841*
pronghorn, 833
Antenna complex, 141–142, *141*, *143*, 149, *149*
Antennae, insect, 652, *652*, 660
Antennapedia gene complex, 298–299, *298*
Anterior vena cava, 520
Anther, 172, *173*, 404, *404*, 468, *468*
Antheridium, *378*, **395**, *396*, 398, *398*
Anthoceros, *395*
Anthocerotae, 389, **395**, *395*
Anthophyta, *389*, **399**, **403–407**
Anthopleura, *413*
Anthozoa, **413–414**
Anthrax, 346
Anthropoid, **847–848**, *848*, 850
Anthropomorphism, 816
Antibiotic, 89, 348, 351, 382, 658
Antibiotic resistance, 230, *230*
Antibody, 192, 299, 528, 533, **537–543**, *538–541*, 795
binding site, 538
constant region, **538–539**, *538–539*
diversity, 541–542, *541*
engineered, 357
heavy chain, **538–539**, *538–539*
hypervariable region, **538–539**, *538–539*
light chain, **538–539**, *538–539*
monoclonal, **542–543**, *543*
specificity, 539–540, *540*
structure, 538–539, *538–539*
variable region, **538–539**, *538–539*
Anticodon, **220**, *220*, *223*
Antidepressant drug, 677
Antidiuretic hormone (ADH), *602*, **603**, **642**, 668
"Antifreeze," 248, 605
Antigen, 192, **537**
Antigenic determinant, 537
Antigen receptor, 540
Antihistamine, 549
Antimycotic agent, 380
Antiport, 89
Antivenin, 548
Antiviral drug, 357
Antler, *3*
Antriopeptin, *See* Atrionatriuretic factor
Antrozous pallidus, *658*
Anura, 436
Anus, *576*, 580
Anxiety, 680
Aorta, *519*, **520–521**
Aortic body, 566, *567*
Apatite, 433, 685
Ape, *10*, 308, 443, 847–848, *848*, 850
Aphid, 490, *490*, *666*
Apical dominance, **500**, *501*
Apical meristem, **472**, *472*, 475–477, *476*
Apis mellifera, *822*
Aplysia, 821–822, *821*
Apoda, 436
Apoderus giraffa, *424*
Apoplastic pathway, **458–459**, *458*
Aposematic coloration, *See* Warning coloration

Appendage, jointed, 422, *424*
Appendicular skeleton, **685**, *686*
Appendix, *535*, *576*, 580
Appetite, 585
Apple tree, *406*, *451*, 473, 503
 red Delicious, 465
Aquatic plant, *451*
Aquatic succession, 778, *778*, 780
Aqueous humor, 660, *661*
Arabidopsis, 504, 507
Arachnid, 423, *423*
Archaebacteria, 338, 340, **349–350**, A1
Archaeoglobus, 350
Archaeopteryx, 440
Archean eon, *333*
Archegonium, 395, *396*, 398, *398*, *401*, 402–403
Archenteron, *280*, 420
Arctium minus, *473*
Arginine vasopressin, 642
Argyroxiphium macrocephalum, 746
Arisaema astrorubens, *467*
Aristida oligantha, 809
Armadillo, nine-banded, 319
Arousal, 669, 677
Art, 860–861, *860*
Artemesia, *460*
Artery, 514, *516*, **517**, 521–523, *521*
Arthritis, 25, 545, *545*
Arthropoda, *410*, *415*, **422–426**, *422–425*
 evolution, 426
 eye, 664–665, *666*
 nervous system, *623*
Artificial selection, 10, *10*
Artiodactyl, *336*, 444
Ascending colon, *576*, 580
Asclepias syriaca, *473*
Ascogonium, *381*
Ascomycota, 377, *377*, **380–382**, *381–382*, *385*
Ascorbic acid, *See* Vitamin C
Ascospore, 200, *201*, 381–382, *381*
Ascus, 200, *201*, 381–382
Asexual reproduction, 166–167, *168*, *See also* Vegetative reproduction
 bacteria, 343–344
Aspen tree, 465
Aspergillus, *377*, 384
Aspirin, 646
Associational cortex, 671
Associative learning, **825**
Assortative mating, *713*, **714**
Aster chilensis, *468*
Asthma, 548
Atherosclerosis, 522, *522*
Athlete's foot, 374
Atmosphere, 753, *753*, 761
 composition, 761
 primitive earth, 7–8, 327–328
Atmospheric pressure, **553**, *553*
Atom, 18, **19**, 33
 structure, 20–23, *20–23*
Atomic mass, 20
Atomic number, **20**
Atomic orbital, 20–23
Atomic weight, **20**, *20*, A5
ATP, 8, *35*, **118**, *118*
 formation, 119
 hydrolysis, 118–119, *119*
 on primitive earth, 329
 production in cellular respiration, *122*, 124–130, *125–131*
 production in fermentation, *122*
 production in glycolysis, *122–123*, 124
 production in photosynthesis, 136–150
 structure, 118–119, *119*
 turnover, *133*
 use in active transport, 89, *90*
 use in amino acid activation, 222, *222*
 use in biosynthesis, 119, *120*
 use in Calvin-Benson cycle, *145*
 use in cell movement, 113–114
 use in glycolysis, 123
 use in muscle contraction, 692–697, *696*
 use in nerve action, 618
ATPase, 89, *90*
ATP synthetase, *129*, **130**, *143*
Atrazine, 149
Atrionatriuretic factor (ANF), 247, *602*, **603**, 638
Atrioventricular node (A-V node), **525–526**, *526*
Atrium
 heart, 518
 sponge, *411*
Auditory cortex, 669
Auditory nerve, **658**
Auk, 744, 763
Australopithecus, **854–856**, *856*
Australopithecus afarensis, *847*, 854–855, *854–856*
Australopithecus africanus, 855–856, *855–856*
Australopithecus boisei, *855–856*, 856
Australopithecus robustus, *855–856*, 856
Autocrine cells, *630*, 631
Autoimmune disorder, 544–545, 614
Autoimmune response, 285, **545**, *545*
Autonomic nervous system, *624–626*, **627**, 698, *698*
Autopolyploid, 739
Autoradiography, 81
Autosome, **157**, *158*
 abnormal number, 257–258, *257–258*
Autotroph, **81**, 345
 origin, 332–333, *333*
 plantlike, 360
Autumn overturn, 772–773
Auxin, **498–507**, *500*, *507*
 effect, 500–501, *501*
 mechanism of action, 501–502, *502*
Auxotroph, **231**, *231*
Average heterozygosity, **709**
Avery, Oswald T., 204
Aves, *432*, **440–441**, *441*
A-V node, *See* Atrioventricular node
Avogadro's number, 26
Axial filament, 343
Axial skeleton, **685**, *686*
Axil, 477
Axillary bud, 477
Axon, *613*, 614, *616*
Axon hillock, *613*, 614
Azolla caroliniana, *495*
Azotobacter, 493
AZT, 357
Azteca, 795–796, *796*

Baboon
 hamadryas, 843
 olive, *837*
Baby tears, 453
Bacillus, *341*, 346
Bacillus anthracis, 346
Bacteria, 340
 detection of phenotype, 230–231
 in food production, 350
 gene exchange, 230–234
 gene regulation, 234–240
 Gram staining, *93*, 342, *342*
 intestinal, 580–581, 587
 luminous, 70, *70*
 metabolism, 345
 movement, 342–343, *342–343*
 photosynthetic, 149
 reproduction, 343–345, *344*
 size, *84*, 341, *341*
 structure, 341–342, *342*
Bacterial disease, 350–351, *350–351*
Bacteriochlorophyll, 345
Bacteriophage, 204–206, *205*, 233, *233*, 351, *352*
Bacteroid, **493–494**
Badger, *690*
BAL, *38*
Balanced selection, **726–727**, *727*
Balanus, 784–785, *785*
Baldness, 264, *264*
Baleen, 573, *573*
Bamboo, 404, 464, 466, *781*, 792
Banana, *405*
Band (society), **843**
Bandicoot, 443
Barb, *787*
Barbiturate, 103
Bark, **478–479**, *478*
Barley, 404, 475, 861
Barnacle, 423, 701, 784–785, *785*, 795, 817
 acorn, *795*
Barr body, **259**, *259*
Basal body, **113–114**, *113*
Basal lamina, 93
Basal metabolic rate, 609
Base, **32–33**
Basement membrane, 515, *515*, 600
Base pairing, 208, *208*, 210, *210*
Basidia, 384
Basidiocarp, **382**, *383*
Basidiomycota, **382–384**, 377, *377*, *383*, *385*
Basidiospore, 382, *383*, 384
Basilar membrane, **656**, *657*, 658
Bat, 443, *444*, 610, 658, *659*
 fruit-eating, *405*
 horseshoe, *444*
Batesian mimicry, **791**, *791*
Bayliss, W.M., 584
B cells, *292*, **534**, 537, 540, *540*
Beach, 760
Beadle, George W., 200

Bean, 450, 471, 473, *474*, 493, 587
string, 469
Bear, 443, 740, *741*, 804
black, 828, 834
grizzly, 828
polar, 763, *763*, *805*
Beartooth Butte, Montana, *742*
Beaumont, William, 66
Beaver, 443
Bee, 426, 558, 607, 822–824, *822*, 838, *838*
Bee sting, 788
Beet, 461, 779
Beetle, 607, 684, 759
bark, 380, *380*
bombardier, 788, *788*
bruchid, 789
cucumber, 789
giraffe, *424*
leaf-mining, 788, *789*
scarab, *684*
stink, 788
whirligig, 772
Behavior, 673
ecology and, 833, *833*
heart disease and, 522
innate, 816
physiological readiness for, 821
reflex, **816–817**, *816–817*
selection for, 730
stereotyped, 426
Behavioral isolation, **734**, *734*
Behavior program, 823
Benson, Andrew, 145
Bentgrass, 725–726
Benthic community, **761**
Bergmann's rule, **722**
Beriberi, *588*
Bermudagrass, 449
Bernstein, Julius, 615
β cells, 636–637, *637*
β-pleated sheet, **50**, *51*
Beta wave, 673–674, *673*
Biceps muscle, 689, *689*, 691, *691*
Bicuspid, *See* Premolar
Biennial plant, **479–480**, *480*
Big Bang, **327**
Bilateral symmetry, 414–420, 422
Bile, **579**, 581
Bile pigment, **581**
Bile salt, *583*
Bilirubin, 581
Biliverdin, 581
Binary fission, **343–344**, *344*
Bindin, 275
Binomial nomenclature, **335**
Bioceramic, 689
Biochemist, *13*
Biological clock, 644–645, 669, 830
Biological control, 808
Biological magnification, **768**
Biology, **1–2**
Bioluminescence, *118*
Biomass, **764**
generation, 765–766
Biome, *756*, 757–761
Biosphere, 752–757, *753–755*
future, 862–864
Biotin, *588*
Bipedalism, 437, 441, 847, *847*, 853, *853*
Bipolar cells, **663–664**, *664*
Birch tree, 506
Bird, **440–441**, *441*
body temperature, 607–610
embryonic development, *4*, 277, *277*, *282*, 283
evolution, *438*, *441*
gastrulation, *280*
heart, *520*
lungs, 558
migration, 828–830, *828–829*
nitrogenous waste, *597*
osmotic balance, 595–596
respiratory system, 561–562, *562*
Bird calls, 730
Bird of prey, 790
Bird song, 734, 824–825, *825*
Birth control, 322, *323*
Birth control pill, 790
Birth defect, drug-induced, 297
Birth process, 318–319, *319*
Birthrate, *See* Natality
Bison, *336*, 807
Biston betularia, 724
Bithorax gene complex, 298
Bitter taste, 651, *651*
Bivalvia, *427*, **428**
Blackberry, 406, *406*, 457
Blackbird, *730*
Black bread mold, *377*, 379
Black plague, 795
Bladder, *See* Swim bladder; Urinary bladder
Blade, leaf, **450**, *451*
Blastocoele, 276–277, *277*, *280*, 420
Blastocyst, 315, *315–316*
Blastodisk, 277, *277*
Blastoid, 745
Blastomere, **276**, *277*, 278
Blastopore, 279, *280*, 420
Blastula, **276**, *277*, *280*
Blind spot, 660
Bloating, 594
Blood, 518, **528–529**
pathways, 520–521, *521*
pH, 564–566, *564*
temperature, 564
transport of gases, 562–565
Blood cells, 518, 528–529, *528*
Blood clotting, 529, *529*
Blood flow
regulation, 525–527
vessel size and, 525
Blood fluke, 416
Blood group, 84, 191–193, *191–192*
ABO, 191–192, *191*
MN, 711, *711*
Rh, 192, *192*
Blood pressure, **517**, 518, 524, 603, 642
Blood transfusion, 191–192, *191*
Blood vessel, evolution, 417–418
Bloom, dinoflagellate, 363
Blowfly, 584
Blubber, 442
Blueberry, 463–464
Bluegill, 592
BMP, *See* Bone morphogenetic protein
Body cavity, evolution, 418–419, *419*
Body fluid
freshwater organism, 593–594, *593*
land animals, 595–596, *595*
marine organisms, *593*, 594–595
pH, 603
regulation, 593–596
Body plan
arthropod, 423–424
mollusk, 427–428, *427*
Body position, 654–655, *655*
Body size, 442, 783
Body surface, gas exchange, 555–556
Body symmetry, 410
bilateral, 414–420, 422
evolution, *415*
radial, 412–414, 428–430
Bog, 773, 778, *778*
Bohr effect, *564*, 565
Bolete, 382
Boletus puritus, *373*
Bolinopsis chuni, *414*
Bolus, **576**, *577*
Bond, 24–26, *See also* specific types of bonds
strength, 26–27
Bond energy, **26**, *27*
Bone, 96, 516, **685**
evolution, 433
strength, 699
structure, 685–689, *687*
Bone marrow, 528–529, 535, *535*, 540, 687, *687*
Bone morphogenetic protein (BMP), 689
Bonner, James, 508
Bony fish, 434–435
Bony-skinned fish, 432–433
Booby, blue-footed, 818, *818*
Book lung, 423
Booster shot, 546
Bordetella pertussis, *351*
Boron, 492, *492*
Bottleneck effect, **715**
Botulism, 351
Boveri, Theodor, 180
Bowman's capsule, 599–600, *599*, *601*
Boxwood shrub, 501
Boyer, Herbert, 243
Brachiation, **848**, *848*, 851
Brachiopoda, *410*
Brachyury mutant, 194
Bracket fungi, *377*, 382
Bract, 402
Bradshaw, A.D., 725
Bradycardia, 568
Brain, *623–625*, 625, *630*, 649–650, 665–666, *666–667*
chemical messengers, 677–680, *679*
effect of altitude, 554
electrical recording, 673–674
electrical stimulation, 673
plasticity, 675
primate, 850, *851*
size, 443
split-brain studies, 671–673, *672*

studies at neuron level, 674–676
temperature, 608
transplantation, 675
visual processing, 663–664, *664*
Brain damage, 554, 674
Brain hormone, 635, *636*
Brain peptide, 646
Brain stem, 667
Brain wave, 673–674
Branchial arch, 433, *433*
Brazil nut, 473
Bread dough, 124, *125*, 382
Bread mold, 374
Breath holding, 566
Breathing, 565–567, *567*
Bridges, Calvin, 182
Briggs, Robert, 289
Britten, Roy, 729
Brittle star, 429, 605
Broca's area, 673
Broccoli, 456
Bromeliad, *781*, 795–796, *796*
Bronchi, *559*, **560**
Bronchiole, *559*, 560
Brown, Donald, 273, 301
Brown, Louise, 320
Brown algae, **393–394**, *393*
Brown fat, **608–609**
Brussels sprouts, 453
Bryophyta, 389, *389*, **395–396**
Bryopsis, 391
Bryum capillae, *396*
Bud
axillary, 477
terminal, **477**
Budding
bacteria, 343–344
hydra, 167, *168*
yeast, 382
Bud primordium, *476*, **477**
Buffalo, *336*, *841*, 842
Buffer, **32–33**
blood, 565, *566*
Bufo periglenes, *437*
Bulb (plant), 457, *464*, **465**
Bulblet, 465
Bulimia nervosa, 585
Bulk flow, 490, **516**
Bullfrog, *736*
Bumblebee, 607, 823
Bundle of His, 525, *526*
Bundle-sheath cells, 147, *148*
Bunting, 763
reed, *730*
Buoyancy, 562, *563*
Burdock, *473*
Burial ritual, 860–861, *860*
Burkitt's lymphoma, 302
Burnet, Sir Macfarlane, 539
Butchering, 857–858, *858*
Butter, 43, 350
Butterfly, *558*, 635, 757, 791
cabbage, 824
marsh fritillary, 723
monarch, *424*, 788–789, 791, *791*, 828
viceroy, *791*
Butterfly fish, *435*
Buttermilk, 350
Butyric acid, 400

C_3 plant, **147–149**, *147–148*
C_4 plant, **147–149**, *147–148*
Cabbage, 453, 477, 503, *503*
Cacodylic acid, 380
Cactus, *457*, *759*, *781*, 787
Christmas, 467, *467*
mammillaria, *787*
Caecum, *576*
Caffeine, 677, 789
Caiman, 440
Calciferol, *See* Vitamin D
Calcitonin, **639–640**, *639*, 689
Calcium, 20, 433, 588, *589*, 689
in cell cycle, 161
in cleavage, 276
in cortical reaction, 275–276
in muscle contraction, 695, *696*, 698–700
in plants, 492, *492*
regulation of cell cycle, 159
as third messenger, 634–635, *635*
Calcium carbonate, 769
Calithricidae, *848*
Callorhinus ursinus, *444*
Callus tissue (plant), 465, *465*, **504**
Calorie, **586**
Calvin, Melvin, 145
Calvin-Benson cycle, **145–146**, *145*
Calypte costae, *785*
Calyptra, *396*
Calyx, **467**
Cambium
cork, 478–479, *478*
vascular, *454*, 459, 478–479, *478*
Cambrian period, 426
Camel, 444, 595
Camera eye, 428, **660**
Camouflage, **792–794**, *793–794*
cAMP, *See* Cyclic AMP
Canary, 825
Canavanine, 789
Cancer, **93**, 159, *159*, 302
causes, 302
free radicals and, 25
lung, 302–303
oncogenes and, 303–304
proto-oncogenes and, 302–303
viruses and, 302–304, *354*, 355
Canine (tooth), 442, 572, *573*, 575
Canus lupus, 842
Capillarity, water, **28–29**
Capillary, 514, *516*, **517**
fluid transport across, 105–106, *106*
Capillary action, **487**
Capillary bed, 523–524, *523*
Capsid, viral, 351
Capsule, bacterial, 203, 342, *342*
Carapace, 423, 440
Carbohydrate, 35, **39–42**
breakdown, 117
digestion, 581–583, *582*
energy content, 586
export, 103–105
metabolism, 39, *39*
production in photosynthesis, 136–150
structure, 39–42
Carbohydrate craving, 585
Carbon, *19*, 20
atomic structure, *22–23*, 23
chemical bonding, 24
organic compounds, 36–39
Carbonaceous chondrite, 329
Carbon chain, *See also* Organic compound
changing conformations, 37, *37*
Carbon cycle, **149–150**, *150*, 333, **769–770**, *769–770*
Carbon dioxide, 564, 769–770, *769*
fixation in Calvin-Benson cycle, *145*, 146
production in Krebs cycle, 124–130, *126–133*
transport in blood, 565, *566*
use in photosynthesis, 136–150
Carbonic acid, 564–565, *566*
Carbonic anhydrase, 69, 565, *566*
Carboniferous period, 397–399, 436
Carbon monoxide, 564
Carbon tetrachloride, 103
Carboxyl group, 38–39, *38*, 47, *47*
Carboxylic acid, 38–39
Carboxypeptidase, *581*, 582, *583*
Carcharhinus maculipinnis, *434*
Carcinogen, **196–197**, 303
Cardiac glycoside, 788–789, 791
Cardiac muscle, **692**, *692*, 699–700, *700*
Cardiac output, 525–526
Career, applied biology, *13*
Caribou, *815*, 828
Carnivore, 443, **572–573**, *573*
Carnivorous plant, *483*, **495–496**
β-Carotene, *140*
Carotenoid, 108, **141**, 349, 362, 392
Carotid body, 566, *567*
Carp, 248, *248*
Carpel, 404, 467, *467*
Carpobrotus edulis, *451*
Carr, Gerald R., 746
Carrageenan, 392
Carrier
color blindness, 261, *261*, 663
Huntington's disease, 264–265
identification, 254–255
sickle-cell anemia, 263
Tay-Sachs disease, 254–255
Carrier-facilitated diffusion, **88–89**, *89*
Carrier protein, 88–89, *89*
Carrot, 458, 460–461
Carrying capacity, **801**
Cartilage, 96, 433, 516, **685**, *687*
Cartilaginous fish, 434
Caryoblastea, *360–361*, **371**
Casparian strip, *458*, **459**
Caste, insect society, 426
Cat, 443, 572, 587, 818–819, *819*, 825
agouti, 193–194, *193*
blotched-tabby, 193–194, *193*
coat color, 193–194, *193*
mackerel-tabby, 193–194, *193*
Manx, 185, *185*, 195
Catabolism, **72**, 117
Catabolite gene activator protein, 236–237, *238*

Catabolite repression, **236–237**, *238*
Catalase, 107
Catalyst, **65–66**, *66*
Catarrhini, *848*
Caterpillar, 426, 635, 789
Catfish, 654
bearded, 338
Cation, 31
Cat's cry syndrome, *See* Cri du chat syndrome
Cattail, 464
Cattle, 44, *336*, 444, 574
Brahman, *10*
Scottish Highlanders, *10*
Cave painting, *861*
CCK, *See* Cholecystokinin
Ceanothus jepsonii, 734
Ceanothus vamulosa, 734
Cebidae, *848*
Ceca, **572**, 574
Cech, Thomas, 331
Cecropia, 796
Cecum, 580
Cedar
pygmy, 761
white, 402
Celery, 455
Cell, **78**
characteristics, 81–84
discovery, 77
first, 7–8, *8*, 332–333, *332–333*
methods of studying, 78–81
types, 81
Cell adhesion, 93–94, 293
Cell body, neuron, 613–614, *613*
Cell-cell communication, 94–95, *94–95*
Cell cycle, **158–159**, *158*
cancer cells, 159, *159*
control, 159, *159*
stages, 158–159, *158*
Cell death, **295**, *295*
Cell division, 154, 292
Cell fractionation, 80, *80*
Cell-free system, 218, 225
Cell hybrid, 255
Cell integrity, 90–91
Cell junction, 94–95
Cell-mediated immunity, **543–544**
Cell membrane, 45–46
Cell migration, 279, *280*
single-cell, 292–293, *293*
Cell movement
creeping, 111–112, *112*
extracellular substances in, 293, *294*
gliding, 111–112, *112*
internal, 114
populations of cells, 294, *294*
swimming, 112–114
Cell plate, *163*, **164**
Cell shape, 110, *110*, 292
Cell size, limits, 83–84, *84–85*
Cell surface, 84–92
Cell theory, **81**
Cellular protein, 101
Cellular respiration, **124–128**, *See also* Aerobic respiration
in carbon cycle, 150, *150*
energy score, 130, *131*
origin, 333
Cellulase, 574
Cellulose, **41**, *42*, 92, *92*
degradation, 41, 366, *366*, 574–575
Cell volume, 91, *91*
Cell wall, 41, 83, *83*, **92–93**, *111*
bacterial, 92–93, *93*, 341–342, *342*
biosynthesis, 103
in cell division, *163*, 164
diatom, 364, *364*
fungi, 93, 374
plant, 92, *92*
primary, 92, *92*
secondary, 92, *92*
Cenozoic era, *333*, 334, 406, 441
Centimorgan, 188
Centipede, 423
Central nervous system (CNS), **624–627**, *624–626*
Central theme of molecular biology, 215
Centrifugation, 80, *80*
Centriole, *82*, *111*, **114**, *114*, *160*, 163, *164*, *271*
Centromere, **156**, *156*, *160*, 161, 165–166, *165*
Centrum, 436
Cephalization, **414**
Cephalochordata, **432**, *432*
Cephalophus zebra, *841*
Cephalopoda, *427*, **428**
eye, 664–665, *666*
Cephalothorax, 423
Cercopithecidae, *848*
Cerebellum, 665, *666–667*, **667**, 676–677
Cerebral cortex, 669
Cerebral ganglia, 624
Cerebrospinal fluid, 615, **665–667**
Cerebrum, 665, *666–667*, **668–670**
Cervical cap, 322
Cervix, **311**, *311*, 318, *319*
Cesium chloride density gradient, 210, *211*
Cestoda, **416**
Cetacea, 443
CGD, 256
cGMP, *See* Cyclic GMP
Chaetodon semilarvatus, *435*
Chailakhian, M.H., 510, *510*
Chalone, **159**
Chameleon, *439*
Chaparral, 759, *781*
Character displacement, **810**, *810*
Chargaff, Erwin, 206–207
Chargaff's rules, **206–207**
Chase, Martha, 204–206, *205*
Cheese, 124, *125*, 350, 382
Cheetah, *690*, 714–715, *792*
Chelicerae, 423
Chelicerata, **423**, 426
Chelonia, 440
Chemical bond, *See* Bond
Chemical communication, 834
Chemical defense
animal, 787–788, *788*
plant, 788–790, *789*
Chemical energy, **59**, *59–60*
Chemical equation, 27–28
Chemical evolution, **326**
Chemical formula, 27–28
Chemical-gated channel, **617**
Chemical reaction, **27–28**, 58
coupled, **62–63**
endergonic, *61*, **62**
energetics, 59–61
at equilibrium, 62–63, *63*
exergonic, *61*, **62**, *63*
rate, 59, 63–64, *See also* Enzyme
catalysts and, 65–66, *66*
reactant concentration and, 65, *65*
spontaneous, 62
temperature effect, 64–65, *65*, 604
Chemical synapse, *620*, 621
Chemiosmotic coupling hypothesis, **128–130**, *129*
Chemoautotroph, **345**
Chemoreceptor, 566, *567*, **650–652**, *651*
Chemotaxis, **112**
Cherry tree, 406, *406*, 473
Chert, 368
Chestnut blight, 381
Chewing, 576
Chiasmata, 166, *167*
Chicken, 840, *840*
Chicken pox, 546
Chief cells, 577, *578*
Child, growth, 283–284
Childbirth, *See* Birth process
Chimera, 319–321, *321*
Chimpanzee, 729, 739, 826, 837, 842, 849, 852–853, *852*
pygmy, 852
Chipmunk, 609, 834
Chiroptera, 443
Chitin, 41, **92**, 422
Chiton, 428
Chlamydiae, *346*, **348**
Chlamydomonas, 391, *391*
Chloride, 492, *492*, 588, *589*, 593–596
Chlorophyll, 136, 138, **139**
absorption spectrum, *139*, **140**
reaction center, **141**, *141*
structure, *140*
Chlorophyll *a*, *139–141*, 140, 349, 362–363, 391–392, 394
Chlorophyll *b*, 140, *140*, 362, 391, 394
Chlorophyll *c*, 363
Chlorophyll *d*, 392
Chlorophyta, 389, *389*, **391–392**, *391–392*
Chloroplast, *78*, 83, *83*, **108**, *135*, 138, *451*
evolution, 355, *356*
structure, 108–109, *108*, *138*
Chloroquine, 367
Chloroxybacteria, **349**
Choanocyte, 411, *411*
Choanoflagellate, 414
Chocolate, 124, *125*
Cholecystokinin (CCK), 584–585, *584*, 646, 678–679
Cholera, *351*
Cholesterol, 46, *46*
blood, 522, *522*

dietary, 522
membrane, *86*, 87
Chondrichthyes, *432*, **434**, *434*
Chondrite, carbonaceous, 329
Chondromyces crocatus, *347*
Chordata, *410*, 428, **430**
Chordate, *415*
characteristics, 430–431, *430*
nonvertebrate, 431–432
Chorioallantois, *282*, 283
Chorion, **282**, *282*, 317, *318*, *320*
Chorionic villi, **317**, *318*
Chorionic villi sampling, **266**
Choroid, 660, *661*
Chromaffin cells, **639**
Chromatid, *156*, **159**, 161, 165
Chromatin, **157**, *157*, 162, *164*
Chromodoris banksi, *409*
Chromoplast, 108
Chromosomal abnormalities, human, 257–260
Chromosomal mutation, **196**, *196*
Chromosomal scaffold, 157, *157*
Chromosome, **99**, 100, *111*, 155, **156**
circular, 212, *212*
eukaryotic, *156*
evolution and, 739–740
fruit fly, 181, *181*
genetics and, 180–183
homologous, **157**, *164*, 165–166
lampbrush, *272*, 273
movement, 162, *162*
polytene, **189**, *189*, 299, *300*
primate, *852*
sex, *See* Sex chromosome
structure, 156–158, *156–158*
Chromosome fusion, 740, *741*
Chromosome imprinting, 315
Chromosome puff, 299, *300*
Chromosome theory of heredity, 180
Chronic myelogenous leukemia, 260
Chrysanthemum, 470, 508, 510, *716*, 790
Chrysophyta, *360–361*, 361, **363–364**, *363–364*
Chthamalus, 784–785, *785*
Chylomicron, 580
Chyme, **578**
Chymotrypsin, 67, *581*, 582, *583*
Chytridiomycota, *377*
Cicada, *423*, 834
periodic, 792
Cilia, *82*, 83, *98*, 110, *111–112*, **112–114**
ciliate, 369
evolution, 355, *356*
fallopian tube, *311*
respiratory tract, 113, 560
Ciliary body, 660, *661*
Ciliary muscle, 660, *661*
Ciliate, **369–371**, *369–370*
Ciliophora, *360–361*, 365, **369–371**, *369–370*
Circadian rhythm, 644–645
Circannual clock, 830
Circular chromosome, 212, *212*
Circulatory system, **514**
brain regulation, 527
closed, 420–421, *421*, *516*, **517**, *518*
control, 667
direction of fluid flow, 517, *517*
mollusk, 427
open, **516–517**, *516*, *518*
vertebrate, 518–525
Cirrhosis, 581
Cirsium, *787*
Cisternae, between RER membrane, 101, *102*
Citric acid, *126*, 128, 384
Citric acid cycle, *See* Krebs cycle
Citrus tree, 587, 808, *808*
C-jun gene, 304
Clade, **336**
Cladistics, **336**
Cladogram, 336, *336*
Clam, 427–428, *427*, 595, 684, 748, 762, 787
giant, *700*, 701
Class (taxonomic), **335**, *335*
Classification
fungi, 376–384
plants, 389–390, *389*
protists, 360–361, *360–361*
seed plant, 398–399
Clathrin, *105*, 106
Claustrophobia, 680
Clavaria, *383*
Claviceps purpurea, 382
Clay, **761**
catalysis of polymerization, 329
Cleavage, **276**
cytoplasmic distribution during, 278–279, *278*
patterns, 276–278, *277*, 420, *420*
radial, 278, 420, *420*
spiral, 278, 420, *420*
Cleavage furrow, *163*, 164
Cleidoic egg, 437
Clements, Frederick, 780
Climate, 754, *755*
Climax community, **778**, 780
Cline, **737**
in speciation, 736–737, *737*
Clitoris, *311*, 312
Clonal deletion, 545
Clonal expansion, 540
Clonal selection theory, **539–540**, *540*, 543
Clone, **167**, **246**, 540, *540*
Closed circulatory system, *516*, **517**, *518*
Closed program, **818–819**, 823
Clostridium, 493
Clostridium botulinum, *346*, 351, *351*
Clostridium perfringens, *351*
Clostridium tetani, *351*
Cloud, 753, *753*
Clover, 345
Clown fish, 412
Club moss, 396–397
Cnidarian, 410, *410*, **412–414**, *412–413*, *415*, *418–419*, 516, 571, 594
nervous system, *623*
Cnidocyte, 412, *413*
CNS, *See* Central nervous system
Coacervate, **330**, *330*
Coated pit, *105*, **106**
Coated vesicle, 106
Cobalt, 492, *492*, *589*
Coccus, 346, *347*
Cochlea, 654, *655–657*, **656**
Cocklebur, 407, 508, *509*
Cockroach, 426
Cocksonia, 394
Coconut palm, 406–407, *406*, *463*, 470
Cocoon, *424*, 635
Cocos nucrifera, *406*
Codominance, **192**
Codon, **216**
nonsense, 218
start, *223*, *227*
stop, 224, *224*, 226
triplet nature, 218
Coelacanth, 435, *435*
Coelenteron, *See* Gastrocoel
Coelom, 410, *410*, *415*, 420–421, *421*
Coendou prehensilis, *744*
Coenocytic fungi, **375**
Coenzyme, 67, 120, 587
Coenzyme A, 127
Coenzyme Q, *129*, 130
Coevolution, 14, 405, *405*, 776, 789
Cofactor, **67**, *67*, 587
Coffee, 740
Cohen, Stanley, 243
Cohesion, water molecules, **28–29**, 30–31
Cohesion-adhesion theory, *See* Transpiration-pull theory
Colchicine, 789
Cold receptor, 609, 659
Cold sore, 354, 357
Cold-tolerant plant, 88
Cold virus, 354
Coleoptile, *472*, *474*, **475**
Coleus, *451*
Collagen, 52, *52*, 93, 285, *292*, 293, 433, 516
Collagen gene, 241
Collateral axon, *613*, 614
Collecting duct, **601**, *601*
Collenchyma, *449*, **454–455**
Colloidal osmotic pressure, **524**, *524*
Colobinae, *848*
Colon, 580
Colonial theory, origin of multicellularity, 414
Color blindness, 255, 261, *261*, 663
Color vision, 661–663, *662*
Colostrum, **321**, 547
Column, neocortex, 669
Columnar epithelium, 95
Comb jelly, 412, 414, *414*
Comet shower, 328, 440
Commensalism, **794–796**, *795*
Commitment, **288–290**
Common ancestor, 9, *9*, 733
Common cold, *355*, 357
Common intermediate, 119, *120*
Communication, **834–835**, *834–835*
Communicative behavior, 823
Community, 752
physiognomy, 780–781

Community (*Cont.*)
relative abundance of species, 781
species interactions, 784
species richness, 781–784
structure, 809, *809*
Compact bone, 685–689, *687*
Companion cell, 456, *456*, 491, *491*
Compass sense, 828–830
Compensatory hypertrophy, **284**
Competition, **784–785**, 805–807, *806*, *811*
interspecific, **806–807**, 809–810
intraspecific, **806–807**, *806*
Competitive exclusion principle, **785–786**
Competitive inhibition, **74**, *74*
Complementary genes, **193**, *193*
Complement response, **539**
Complete linkage, **186**, *187*
Complete metamorphosis, 425–426
Compound, **23**
Compound eye, 426, **665**, *666*
Computerized tomography (CT), 21, *21*
Concentration, 31
Concentration gradient, **89**
Concentricycloidea, 430
Condensation reaction, **39**, *39*, 48, *48*, 58
Conditioned reflex, 677
Conditioning, 677
Condom, 322, *323*
Cone (plant)
ovulate, *401*, 402
pine, 401–402, *401*
pollen, *401*, 402
Cones (eye), **661–664**, *661*, *664*, 675
Conidia, **381**, *381*, 384
Conidiophore, *374*, 381, *381*
Conifer, **399–403**, 484, *781*
Coniferophyta, *389*, **399–403**
Conjugation, 344
bacteria, **232**, *232*
ciliate, 370–371, *370*
Connective tissue, 95–96, *96*, *515*, **516**, 521, *521*
Connell, Joseph, 784
Connexon, 94
Conodont, 433
Consciousness, 669
Conservative replication, 210
Consolidation, long-term memory, 677
Constipation, 580
Constitutive mutant, 236
Consumer, 361, 365–371, **763**
primary, 763, *763*, *765*
secondary, 763, *763*, *765*
tertiary, *765*
Contact dependence, 159
Continental drift, **334**, *334*, 586, 754
Continental shelf, *766*
Contraception, 322, *323*
Contractile vacuole, 105, 362–363, *362*, *369*, 597
Control experiment, **11**
Conus arteriosus, 518
Convection, 607
Convergent evolution, 428, **743–744**, *744*, 784, *785*
Convolution, brain, 669
Copepod, 772
Copper, 492, *492*, *589*
Coprinus comatus, *383*
Coquette, tufted, *787*
Coral, 412, 414, *775*
Coral fungus, 382, *383*
Coral reef, *751*, 760
Core temperature, 605
Corey, Robert, 50
Coriolis effect, **754–756**
Cork, *478*, **479**
Cork cambium, 478–479, *478*
Corm, *464*, **465**
Cormel, 465
Cormorant, 605, 744
Corn, *78*, 147, *240*, *335*, 450, *453–454*, 470–471, *474*, 475, 477, *504*, 509, 587, 861
genetics, 226
kernel color, 226
origin, 402
Cornea, **660**, *661*
Corn oil, 43
Corolla, 467, *467*
Coronary artery, 520
Corpora allata, 635
Corpora cavernosa, 309
Corpus callosum, **671–672**
Corpus luteum, **310**, *311*, 312, 317
Corpus spongiosum, 309
Correns, Carl, 180
Cortex
brain, *666*, 673
kidney, **599**, *599*
root, 457–458, *457–458*
stem, **454–455**, *454*
Cortical lobe, *670*, 671
Cortical reaction, **275–276**, *276*
Corticosteroid, **638–639**, *638*
Corticotropic-releasing hormone (CRH), *643*
Cortisol, 638
Cortisone, 638
Corynebacterium diphtheriae, *351*
Coryza virus, *355*
Co-transport, **89**, *90*
Cotton, 156, 506, 740
Cotyledon, **403**, 471–472, *472*, *474*, 475
Cotylosaur, 436–437, *438*
Counteracting osmolyte strategy, **595**
Countercurrent exchange system, **556**, *557*, 595, 601, *608*
Coupled reactions, **62–63**
Coupling factor, *See* ATP synthetase
Courtship behavior, 734, *735*, 818, *818*
Covalent bond, **24–25**
Cowpox, 547
Coxiella burneti, *351*
Coyote, *11*, 759, 807–809, 833–834
Crab, 422, 760, 762
fiddler, *836*
horseshoe, 423, 665, *666*
Crabgrass, 147, *147*
Cramp, leg, 124
Cranberry, *778*
Cranial nerve, 432–433, 625, 627
Crayfish, 422, 605, *606*, 665
Creatine phosphate, **693**
Creeping movement, 111–112, *112*
Cretaceous period, *334*, 439, 441, 745, 849
Cretinism, 640
CRH, *See* Corticotropic-releasing hormone
Crick, Francis, 200, 207–209, *207*, 216, 218
Cri du chat syndrome, 260
Crimson topaz, *787*
Cristae, mitochondrial, 107, *107*
Crocodile, 440
salt-water, *439*
Crocodilia, *438*, 439–440
Crocodylus prorosus, *439*
Crocus, 465, 470, 508
saffron, *464*
Crocus saticus, *464*
Cro-magnon, *856*, 860
Crop, earthworm, **571**, *572*
Crop milk, 821
Cross-fertilization, 735
Crossing over, **166**, *167*, 186–188, *187*, 710
Cross-pollination, *175*, 735
Crow, 786
Crowe, John, 594
Crowe, Lois, 594
Crustacean, **422–423**, *424*, 426, 635
Crustose lichen, 384, *385*
Crying, 844, *844*
Crypsis, *See* Camouflage
Cryptic coloration, **793**, *793*
CT, *See* Computerized tomography
Ctenophora, *410*, **412**, 414, *414–415*, 701
Cuboidal epithelium, 95
Cucumber, 509
Cucurbitacin, 789
Cud, *574*, 575
Culex, 725
Cultivated land, *766*
Cultural evolution, **15**, 843–844, 860–862, 864
Cultural inheritance, 826–827, *826*
Cunningham, E.P., 176
Cup fungi, *377*, 381
Cupula, 654, *654–655*
Cuspid, *575*
Cuticle
arthropod, 422, 425
flatworm, 416
plant, 147, 401, *451*, *454*
Cutin, 396, **452**
Cuttlebone, *427*, 428
Cuvier, Georges, 705
Cyamus scammoni, *795*
Cyanerpes cyaneus, *785*
Cyanide, 91, 618
Cyanobacteria, *326*, 332, 340, **348–349**, *348–349*, 493–494, *495*
Cyanocobalamine, *See* Vitamin B_{12}
Cyanogenic glycoside, 789
Cyanophyta, 348
Cycad, 398, 789
Cycadophyta, *389*, **398–400**, *399*
Cyclic AMP (cAMP), 112, 237, *238*, 365, 633–634, *634*, 677, 698
as second messenger, 637
Cyclic GMP (cGMP), **634**

Cyclic nucleotide, as second messenger, 633–634, *634*
Cyclic photophosphorylation, **143–144**, *144*
Cyclostomata, 433
Cyprus tree, bald, 400
Cysteine, 48, *48*
Cystic fibrosis, 265
Cytochrome, **128**, *129*
Cytochrome *b-f*, *143–144*
Cytochrome *c*, *129*, 130, 336, 728–729, *729*
Cytochrome *c* oxidase, *129*
Cytochrome *c* reductase, *129*
Cytogenetic (gene) map, 189–190, *189*
Cytokinesis, 159, *161*, **163–164**, *163*, *165*, 166
Cytokinin, 159, *500*, **503–505**, *504*, 507
Cytoplasm, 82, *82*, **99**
Cytoplasmic inheritance, 371
Cytoplasmic membrane, *104*
Cytoplasmic streaming, **114**
Cytosine, 54, *55*, **206**, *206*
Cytoskeleton, 81–82, *82*, 99, **109–111**, *109–111*, 292
Cytostome, 369, *369*

d[2] factor, 380
Daffodil, 449, *464*
Dalton, A5
Dalton, John, 19
Damsel fly, *308*
Danaus plexippus, 791, *791*
Dandelion, *406*, 407, 792, 804, *805*
Danielli, J.F., 84
Dart, Raymond, 855
Darwin, Charles, 8–11, 499, *499*, 704–708, *704*, *706*, 720
Darwin, Francis, 499, *499*
Davson, H., 84
Day-neutral plant, 508
Deafness, 658
Deamination, 587, 596, *596*
Death, 285
Death rate, *See* Mortality
de Buffon, Georges-Louis Leclerc, 8–9, 705
Deciduous tree, **400**
Decomposer, *150*, 361, 364–365, *364–365*, 371, 373, **763**, *763*, 765
Dedifferentiation, **284**, 288
Deductive reasoning, **12**
Deep-sea vent, 345, 762, *762*
Deer, *336*, 444, 574, 759, 817
Defense adaptation, 787–792, *787*
DeGrado, William, 54
Dehydration, 594–595
3-Dehydrocholesterol, 586
Dehydrogenase, 128
Deletion, 196, *196*, *217*, 218, 259–260
δ cells, 637, *637*
Delta wave, 674
Denaturation
  enzyme, *71*
  protein, 52–53, *53*
Dendragapus canadensis, *441*
Dendrite, **613**, *613*
Dendrobates histrionicus, *595*
Dengue fever, *355*
Denitrification, **770**
Deoxyribonuclease, *581*
Deoxyribonucleic acid, *See* DNA
Deoxyribose, 54, *55*, 206, *206*
Depolarized cell, **616**
Dermal membrane bone, 685
Dermal tissue, plant, 449, *449*
Dermatitis, *588*
Descending colon, *576*, 580
Desert, *756*, **759**, *759*, *766*, 781
Desmid, *16*
Desmosome, **94**, *94*, 111
Determination, **288–290**
  changes, 288–289
  dependence on cytoplasm, 289–290
Dethier, Vincent, 584
Detoxification, 103, 581
Detritivore, **763**, 765
Deuteromycota, 377, **384**, *385*
Deuterostome, 410, *410*, **420**, *420*, 428–430
Development, 3, *4*, 14, *269*, 299–302, 739
Devonian period, *334*, 394, 397–398, 426, 435, *742*
de Vries, Hugo, 180
Dextrin, *582*
D form, 37
Diabetes insipidus, 642
Diabetes mellitus, 637
Diacylglycerol, **634–635**, *635*, 698
Dialysis, kidney, 603
Diaphragm (birth control), 322, *323*
Diaphragm (muscle), **561**, *561*
Diarrhea, 580
Diatom, *3*, *360–361*, 361, 363–364
Diatomaceous earth, **363**
Dicotyledon, **403**, *403*, 449, *449*
  embryonic development, *472*
  germination, 474–475, *474*
  vascular bundles, *454*
  vein pattern, 453, *453*
Dictostelium, *365*
Didinium, *369*, 807
Diencephalon, 665, *666*
Diet, heart disease and, 522
Dietary guidelines, 589, *590*
Differentiated cells, 291–292, *292*
Differentiation, **279–282**, 290–291, *291*
Diffusion, **87**, 554
  carrier-facilitated, **88–89**, *89*
  limitations, 516
  lipid, 88
  simple, 87–88, *87*, *89*
Digestion, **570**
  animal strategies, 571–575
  chemistry, 581–584, *581–583*
  coordination with ingestion, 584–585, *584*
  extracellular, **374**, **412**, 571–575, *572–575*
  intracellular, 106–107, 571
Digestive enzyme, 67, 106–107
Digestive juice, 66
Digestive system, 571
  earthworm, 571–572, *572*
  human, 575–581, *576*
  insect, 572, *572*
  one-way, 417, *417*
Digit reversal, 296, *297*
Digitalis purpurea, *480*
Dihybrid cross, **177–179**, *178*
Dihydrotestosterone, 314
Dihydroxyacetone, 40
Dihydroxyacetone phosphate, *38*, *122*, 123
1,25-Dihydroxycholecalciferol, **637**
Dikaryotic hyphae, **382**, *383*
Dill weed, 508
Dinoflagellate, *360–362*, **361–363**, 772
Dinosaur, 437, 441, 746
  extinction, 439–440
Dioecious plant, **400**
Diomedea, *441*, *819*
Dipeptidase, *581*, 582, *583*
Dipeptide, 48
1,3-Diphosphoglycerate (DPGA), 123, *123*, *145*, 146
Diphtheria, *351*, 547
Diplococcus pneumoniae, *240*
Diploid, **157**
Directional selection, **723–725**, *724–726*
Disaccharide, **40–41**, *40*
Disease, 805, 810, *811*
  bacterial, 350–351, *350–351*
  fungal, 374
  transmission, *351*
  viral, 353–355, *355*, 357
Dispersal spore, 376
Dispersive replication, 210
Distal convoluted tubule, 599–600, *599*, *601*
Disulfide bond, 48, *48*, *53*, 538–539, *538–539*
Disulfide group, 39
Diuretic, 603
Divergent evolution, **744–745**, *745*
Diversifying selection, **725–726**, *726*
Diving bird, 605
Diving mammal, 605
Diving reflex, **568**
Division (taxonomic), **335**, *335*
Dizygotic twins, 319, *320*
DNA, 8, *8*
  chloroplast, 109, 138
  in chromosomes, 156–158, *156–158*
  composition, 206–207, *206*
  content of haploid genome, 240, *240*
  dissociation-reassociation experiment, 240, *241*
  highly repeated, 240, *241*
  hybridization, 240, *241*
  mitochondrial, 108, 860
  moderately repeated, 240, *241*
  nonprotein-coding, 719
  origin, 331
  processing, 299
  proof that it is genetic material, 203–206, *204–205*
  replication, **209–213**
    continuous, 212, *213*
    direction, 212, *213*
    discontinuous, 212, *213*
    Escherichia coli, 211–212, *212–213*
    eukaryote, 212–213, *213*
    fidelity, 211
    lagging strand, 212, *213*

DNA, replication (*Cont.*)
leading strand, 212, *213*
semiconservative nature, **210–211**, *211*
satellite, *See* DNA, highly repeated
sticky ends, 243, *244*
structure, 54–56, *199*, 206–212, *206*, *209–210*
template, 210
transcription, *See* Transcription
viral, 353
x-ray diffraction pattern, 207–208, *207*
DNA fingerprint, 255–256, *255*
DNA ligase, **243**, *244*
DNA polymerase, **211**, 212, 357
DNA-RNA hybridization, **246**
DNA-RNA synthesizer, 242
Dog, 443, 572, 587, 595
husky, *607*
variation, *4*
Doldrums, *755*, 756
Dolphin, 443, *608*
Dominance, incomplete, **179**, *179*
Dominance hierarchy, *See* Social hierarchy
Dominant hemisphere, 673
Dominant species, 781, *781*
Dominant trait, **172–174**, 190, 252, *253*
human, 264–265, *264–265*
Dopamine, 621, 677
Dormancy, **473**
plant, 506
seed, 473–474
Dorsal root ganglia, 625, *625*
Double bond, 24–25, *25*, *36*, 44
Double fertilization, *404*, **470**, *471*
Double helix, *55*, 56, **208**, *208*
Douche, *323*
Dove
mourning, 824
ring, 821
Down syndrome, **252**, *253*, 258, *258*, 260, *260*
Downy mildew, *377*, 379
DPGA, *See* 1,3-Diphosphoglycerate
Dragonfly, 425, 607
Dreaming, 674
Drinking behavior, 678
Drinking center, 603
Drone, 839–840, *839*
Drosera rotundifolia, *483*
Drosophila, *See* Fruit fly
Drosophila paulistorum, 733
Drosophila pseudoobscura, 736
Drosophila similis, 736
Drought tolerance, 594
Drug resistance, 351, 367
Dryopithecine, 850
Dryopithecus, 850, *850*
Dubrutia menziesii, 746
Duck, 295, *295*, 565, 692, 834
mallard, 821
Duckweed, 403
Duiker, *841*, 842
Duodenum, *576*, 579, *630*
Duplication, 196, *196*, 259–260
Duplication and divergence, gene, 716, *716*
Dutch elm disease, 380, *380*, 810
Dynein, **113–114**, *113*
Eagle, 807
Ear, 656–658, *656–657*
Eardrum, *See* Tympanic membrane
Early wood, *478*, **479**
Earth
air circulation, *755*
global tilt, 754, *755*
primitive, 7–8, 327–330, *330*
Earth's crust, **327**
continental drift, 334, *334*
elements in, 19–20, *19*
Earth star, *376*
Earthworm, 420, *421*, *516*, 517, 555, 571–572, *572*, *598*, 684
Eccentric cells, 665, *666*
α-Ecdysone, **635**, *636*
Echidna, 441
Echinodermata, *410*, **428–430**, *429*
Echolocation, **658–659**, *659*
Ecocline, **781**
Ecological efficiency, **766**
Ecological isolation, **734**, *734*
Ecological pyramid, 764–765, *765*
Ecology, **751**
behavior and, 833, *833*
Ecosystem, **752**, *752*
energy flow, 763–768
natural cycles, 768–772
Ectoderm, **279**, *280*, 290–291, 298, 415, *415*
fate, *281*, 288, *289*
Ectoprocta, *410*
Ectotherm, **605**
Edidin, M., 86
EEG, *See* Electroencephalogram
Eel, 434
conger, *240*
Effector, 132, 613, 683
Effector neuron, *See* Motor neuron
EGF, *See* Epidermal growth factor
Egg
bird, 277, 441
cleidoic, 437
monotreme, 442
reptile, 437
Egg cells, 271
manufacture of molecules by, 273
plant, 469–470, *469*
production, 310–311, *311*
sperm penetration, 274–275, *274–275*
structure, 271–272
transport, 310–311, *311*
Egg shell, 273, 282
Egg white, 273
Egg yolk, *105*, **273**, *282*, 283
cleavage and, 276–278, *277*
Ehyman, Lee, 727
Ejaculation, 310
Eland, 608
Elastic recoil, 561
Elastin, 516
Elderberry, 465
Electrical energy, 59
Electrical synapse, 620, *620*
Electric fish, 654
Electrocardiogram, 526
Electroencephalogram (EEG), 673, *673*
Electron, **20**, *20*
energy level, 20–23
valence, 24
Electron carrier, 120, *121*, *129*, 130
Electronegativity, **26**
Electron microscope, 80
scanning, *78*, 79–80
scanning tunneling, 79, *79*
transmission, 79–80
Electron transport chain
photosynthetic, 142–144, *142–143*
respiratory, **124–125**, *126*, 128, *129*
Electrophoresis, 202–203, *203*, *255*, **708–710**, *709*
Electroreceptor, 654
Element, **19**
earth's crust, 19–20, *19*
human body, 19–20, *19*
inert, 23
Elephant, 443, 803, *805*
Eleutherodactylus coqui, 804, *806*, 834–835
Elk, *3*, *11*, 308, 717, 734, 760, 841
Ellis-van Creveld syndrome, 715, *715*
Elm tree, 380, *380*, 406, 473, 810
Elton, Charles, 763
Embolotherapy, intracranial, 21
Embryonic membranes, 282–283, *282*
Embryonic coverings, 282–283, *282*
Embryonic development, 44, 283
bird, *4*
plant, 470–472, *472*
Embryo sac, *469*, 470
Emotion, 680
Emulsifying agent, *44*
Enantiomer, **37**, *38*
Encephalitis, *355*
Endergonic reaction, *61*, **62**
Endochondral bone, 685
Endocrine function, invertebrates, 635–636, *636*
Endocrine gland, *630*, 631
Endocrine system, **629**, 636–644
Endocytosis, **106**
Endoderm, **279**, *280*, 415, *415*, 420
fate, *281*, *289*
Endodermis, root, **459**, *459*
Endolymph, **655**, *655*, *657*, 658
Endometrium, **310**, *311*, *313*, *315–316*, 317
Endopeptidase, 582, *583*
Endoplasmic reticulum, 82, *82*, **101–103**, *111*
rough, 101–103, *102*, *104*, 105
smooth, 103, *103–104*
Endorphin, 647, **678–680**
Endoskeleton, 684–690
Endosperm, 404, *404*, 470–471, *471–472*, 475, 508
Endospore, **344–345**, *344*
Endosymbiont theory, **355**, *356*, 371
Endothelin, 527
Endothelium, blood vessel, 521, *521*
Endothelium-derived relaxing factor, 527
Endotherm, **605**
Endotoxin, **351**
Energetics, chemical reactions, 59–61
Energy, *See also* specific types of energy
conservation, 60

Energy currency, 118–119
Energy flow, ecosystem, 763–768
Energy reserve, 144
carbohydrates, 41, *42*
lipids, 43–46
Enhancer, 300
Enkephalin, **678–680**
Enoplometopus vanuato, 424
Ensatina, 737, *737*
Enteropneust, 430
Enthalpy, *61*, 62
Entropy, **60–62**, *60–61*
Envelope, viral, 352
Environment, effect on phenotype, 256–257, *256–257*
Enzyme, 41, **46**, 58–59, 66
allosteric, **73–74**, *73–74*, 132, *133*
control, 72–74, 132–133
covalent modification, 132–133
denaturation, *71*
function, 67–69, *68–69*
changes in enzyme shape, 68, *68*
changes in substrate concentration, 68, *68*
completing catalysis, 68–69
distortion of substrate, 68, *68*
orientation of reactants, 68, *68*
inducible, 235, *237*
pH optimum, 67
reaction rate and, 70–71, *71*
repressible, **238–240**
saturation with substrate, **71**, *71*
specificity, 66
structure, 67, *67*
temperature effect, 70–71, *71*, 604
Enzyme-substrate complex, **67–68**, *68*
Eocene epoch, *334*, 849–850
Ephrussi, Boris, 200
Epidermal cells, *292*
Epidermal growth factor (EGF), 159, *303*, 304
Epidermal growth factor (EGF) receptor, *303*
Epidermis
cnidarian, 412, *412*
leaf, **452**
plant, 394, *449*, *451*, 484
root, 457–458, *457–458*
stem, 454, *454*
Epididymis, 309
Epiglottis, 560, 576, *577*
Epinephrine, 527, 621, 631, *631*, 633–635, *634*, *638*, **639**, 644
Epiphyte, **461**, 757, *781*
Episome, **232**
Epistasis, **193–194**, *193*
Epithelial cells, *90*, 95–96, *96*, *110*, *292*
columnar, 95
cuboidal, 95
gap junctions, *95*
squamous, 95
Epithelial tissue, **515**, *515*
Epitheliomuscular cells, 412, *413*
Epithelium, 93
cnidarian, *413*
sponge, 411, *411*
Epstein-Barr virus, *355*
Equilibrium (balance), 654–655, *655*
Equilibrium (chemical reaction), **62**
free energy and, 62–63, *63*
Equisetum, 397, *397*
Equus, 742, *743*
Erection, 310
Eregata minor, *629*
Ergot, **382**
Ergot poisoning, 382
Erysiphe graminis, *374*
Erythroblastosis, 192, *192*
Erythrocytes, *See* Red blood cells
Erythromycin, 348
Erythropoietin, 283, **637**
Escape adaptation, 792–794, *792–794*
Escherichia coli, 204–206, *205*, *240*, *341–343*, *346*, *351*, 580
DNA replication, 211–212, *212–213*
gene map, 234, *235*
Eschrichtius robustus, *795*
Esophagus, **571**, *572*, 576, *576*
Essential amino acid, **587**
Essential fatty acid, 587
Ester bond, *43*, **44–45**
Estradiol-17β, 631–632, *631*
Estrogen, 45–46, **312**, *313*, 314, 317, 322, 633, 639, 689
Estrus, **308**, 841
Estuary, 765, *766*
Ethanol, *27*, 37–38, *38*, *122*, 124, *125*
Ether, 88
Ethics, 12–13
Ethology, **816**
Ethylene, 24–25, *25*, *27*, *500*, **506–508**, *507*
etr gene, 507
Eubacteria, **346–347**, *346–347*
Eucalyptus, 789, *790*
Eucalyptus marginata, 403
Euglena, 362, *362*
Euglenoid, *360–361*, 361–362
Euglenophyta, *360–362*, **361–362**
Eukaryote, **81**
chromosomes, *156*
components, 81–83
DNA replication, 212–213, *213*
evolution, 355–356, *356*
gene regulation, 240–243
genetic center, 99–100, *99*
origin, 333, *333*
ribosomes, 221
transcription, 221
translation, 221
Eumetazoa, A3
Eustachian tube, *656–657*, 658
Eutheria, 443
Eutrophication, **773**
Eutrophic lake, 773
Evaporation, 768, *768*
respiratory, 595
Evolution, 14, **704**
amphibians, *10*
animals, *410*
ape to hominid, 852–854
arthropods, 426
birds, *438*, *441*
blood vessels, 417–418
body plan, *415*
bone, 433
chemical, **326**
chloroplast, 355, *356*
chromosomal changes and, 739–740
cilia, 355, *356*
constraints, **754**
convergent, 428, **743–744**, *744*, 784, *785*
cultural, *See* Cultural evolution
divergent, **744–745**, *745*
eukaryotes, 355–356, *356*
extremities, *721*, 722
flagella, 355, *356*
flatworms, 416
flowers, *10*, 405, *406*
forelimbs, 744–745, *745*
fruit flies, 720, 746
fungi, 385–386, *385*
heart, 518–520, *521*
hominid, 854–860
horse, 725, *725*, 742, *743*
humans, 586, 846–864
intestine, 574
land plants, *10*, 394–398, *394*
limbs, 9, *9*, 442
mammals, *10*, *438*, 442–443
mechanism, 713–717
mitochondria, 108, 355, *356*
molecular, 728–730, *729*
moneran, 355–356, *356*
muscle, 700–701, *700*
origins of evolutionary thought, 705–708
parallel, **743**, *744*
photosynthesis, 332–333, *333*
plants, 448–449
plastid, 109
primates, *10*, 849–850
protists, 371
regulatory genes and, 739, *739*
reptiles, *10*, *438*
seed plants, 399
taxonomy and, 335–337
teeth, 442
theory, 8–11
vertebrates, *10*
viruses, 355–356, *356*
Evolutionary clock, 729, *729*
Evolutionary line, reconstruction, 742, *743*
Evolutionary tree, *6*
Excitable cells, 617
Excitatory synapse, **621–622**
Excretory pore, *598*
Excretory system
invertebrate, 597–598, *598*
plant, 484, *485*
vertebrate, 598–604
Excurrent pore, 411
Exercise
aerobic, 124
breathing rate and, 566–567
heart rate and, 527
osteoporosis and, 689
Exergonic reaction, *61*, **62**, *63*
Exhalation, 561
Exocrine gland, 630, *630*
Exocytosis, **106**, *354*
Exon, **241**, *242*, 301, 331, *331*, 541
Exopeptidase, 582, *583*
Exoskeleton, **422**, *423*, **684**, *684–685*

Exotoxin, **351**
Experience, 823
Experiment
control, **11**
reproducibility, 12
Exponential growth, **800–801**, 810–813, *811–812*
Exponential growth curve, **800**, *801*
Exportable protein, 101–105
Extensor, 684, *685*
External auditory canal, *656*
External ear, 656, *656*
External fertilization, 274
External intercostal muscle, **561**, *561*
Extinction, mass, 439–440, 745–747
Extinction rate, 782, *783*
Extracellular digestion, **374**, **412**, 571–575, *572–575*
Extracellular fluid, 87, 517
Extracellular matrix, **295**
Extraembryonic membrane, 282–283, *282*
Extremity
evolution, *721*, 722
heat loss, 608–609, *608*
Eye
arthropod, 664–665, *666*
camera, 428, **660**
cephalopod, 664–665, *666*
color, 711–712, *711*
compound, 426, **665**, *666*
embryonic development, 290–291, *291*, 293, *294*
focusing, 660–661
vertebrate, 660–664, *661–665*
Eye (tuber), 465
Eyespot
protist, 660
turbellaria, 416, *416*

F_1 generation, *See* First filial generation
F_2 generation, *See* Second filial generation
Fabre, Jean-Henri, 815
Facial expression, 819
Facial pit, *659*, 660
Facilitation mechanism, of succession, 780
Facultative anaerobe, 124, **345**
FAD, **120**, *121*, 124–130, *126–131*
Fallopian tube, **273**, *292*, **310–311**, *311*, *315–316*
False morel, 381–382
Family (taxonomic), **335**, *335*
Family tree, *See* Pedigree
Farsightedness, 660–661
Fast-twitch muscle, 692–693
Fat, 43–45
body, 605, 607
metabolism, 131–132, *131–132*
stored, 442
Fat-soluble vitamin, **587**, *588*
Fatty acid, **44**
essential, 587
metabolism, 131–132, *131–132*
saturated, *43*, 44
unsaturated, *43*, 44
Feather, 440–441, 605, 607, *607*, 652
color, *140*, 713, *713*
Feces, 580–581
Fecundity, *806*
Feedback inhibition, *72–73*, **73**, 132
Feedback learning, *See* Trial-and-error learning
Feedforward response, 489
Feeding, control, 584–585, *584*
Feeding behavior, 819–821, *819–820*
Feeding call, 730
Feeding current, 411
Feeding niche, *810*
Feeding style
meat-eating vertebrate, 572–573, *573*
plant-eating vertebrate, 573–575, *574–575*
Feinberg, Gerald, 20
Femaleness, 312–315, *314*
Female reproductive system, 310–312, *311*
Femur, *687*
Fermentation, 121, *122*, **124**, *125*
Fern, 156, 396, 398, 455, 777, *781*
Ferredoxin, 142–144, *144*
Fertile shoot, 398
Fertilization, *272*, **274**, *315*
barriers to other sperm, 275–276, *276*
completion of meiosis, 275
double, *404*, **470**, *471*
external, 274
internal, 274, 437
plant, 403, 470, *471*
sperm penetration of egg, 274–275, *274–275*
in vitro, 320
Fertilization membrane, 275–276, *276*
Fescue, meadow, *460*
Festuca etatior, *460*
Fetal hemoglobin, *563*, 564
Fetus, *307*, **315**
monitoring, *13*
Feulgen, Robert, 203
F factor, 232, *232*
F factor plasmid, **232**, *232*
Fiber
dietary, 41
plant, **455**
Fibrin, 529, *529*
Fibrinogen, 529, *529*
Fibrinopeptide, 728–729, *729*
Fibroblasts, *292*, *515*, 516
Fibronectin, 293, *294*, 301
Fibronectin gene, 241
Fibrous root system, **460**, *460*
Filament
flagellar, 343, *343*
stamen, 468, *468*
Filamentous algae, 391–392, *392*
Filarial worm, 419
Filter feeder, **411**, 573, *573*
Finch, 824
Cocos, 827
Galápagos, 706–707, *706*, 810, *810*
seed-cracking, 786
weaver, *441*
woodpecker, *706*
Fins, 434–435, *434–435*
Fire, 759, 778–779, *779*, 805
Firefly, 70, 791
light pattern, 734, *734*
Fire plant, 362
First filial generation (F_1), **174**, *175*
First law of thermodynamics, **59–60**, *60*
Fir tree, *451*, 767
balsam, 789
Douglas, *135*, 399, 784
Fish
bony, 434–435
bony-skinned, 432–433
cartilaginous, 434
cold-water, 554
fleshy-finned, 434
heart, 518, *520*
jawed, 433–436, *433*
jawless, 432–433, *433*
lobed-fin, 435
luminescent, 118, *118*
marine versus freshwater, *593*
nitrogenous waste, *597*
spiny-finned, 434
Fish oil, 586
Fission, 230
Fissure, brain, 669
Fitness, genetic meaning, **720–722**, *720–721*
Fixed action pattern, *See* Fixed motor pattern
Fixed allele, **721–722**
Fixed motor pattern, **818–819**, *818*
Flagella, *111*, **112–114**
bacterial, 113–114, 342–343, *342–343*
dinoflagellate, 362, *362*
euglenoid, 362, *362*
evolution, 355, *356*
sperm, 270
Flagellate, 660
Flame cell, 416, 597, *598*
Flashlight fish, 70, *70*
Flatworm, 415–416, *415–416*, 516, 571
evolution, 416
excretory system, *598*
free-living, 415–416, *415–416*
gas exchange, 555
nervous system, *623*
parasitic, 416
tiger, *415*
Flavin adenine dinucleotide, *See* FAD
Flax, 455
Flea, 348
Fleming, Alexander, 12, 382
Fleshy-finned fish, 434
Flexor, 684, *685*
Flight call, 730
Flight-or-fight response, *638*, **639**
Flock, 805
Flood, 805
Floridean starch, 392
Florigen, **510**, *510*
Flower, **403–405**, 467, *467*
anatomy, 467, *467*
color, 108, *140*, 179, *179*, 190–191, 405, *405*, 466–467, 470
evolution, 405, *406*
odor, 405, *405*, 466, 470
Flowering, 11–12, *12*, 470
control, 508–510, *509–510*
Flowering plant, *See* Angiosperm
Fluid balance, 524–525

Fluid-mosaic model, **85–86**, *86*
Fluke, 416
Fluorine, *589*
Flutter, *112*
Fly, 607
 tachnid, 808
Flycatcher, 788
Flying fish, 435
Folic acid, *588*
Foliose lichen, 384
Follicle cells, 272, *273*, 310
Follicle-stimulating hormone (FSH), **310**, 312, *313*, 317, 640, *641*
Fontanel, 685
Food
 production processes using bacteria, 350
 spoilage, 350, 374
Food chain, **763**, 809, *809*
Food niche, **784**, 810
Food poisoning, *351*
Food vacuole, 105, *369*, 571
Food web, **763**, 809, *809*
 Antarctic, 764, *764*
Foolish seedling disease, 502
Foot, mollusk, 427–428, *427*
Foramen magnum, 855
Foraminiferan, *360–361*, **368–369**
Force pump system, 560, *560*
Ford, E.B., 723, 725
Forebrain, **665**, *666*, 668–669
Forelimb, evolution, 744–745, *745*
Forest, 777, *777*
Fossil, 705, 732, *732*
 dating, 742
 first cells, 332, *332*
 formation, 741–742, *742*
Fossil fuel, burning, 150, *150*, 767, 769, *769*
Fossil record, 9, *10*, 12, 706, 741–742
 gaps, 748
Founder effect, 252, **715**
Fovea, **660–661**, *661*
Fox
 arctic, *721*, 763, *763*, 808
 desert, *721*
 flying, 443
Fox, Sidney, 329, *330*
Foxglove, 480, *480*
Fragaria ananassa, *487*
Fragile X syndrome, 260, *260*
Frameshift mutation, **218**
Francisella tularensis, *351*
Franklin, Rosalind, 207
Fraternal twins, *See* Dizygotic twins
Free energy, 59
 changes in chemical reactions, 61–63, *61*
 equilibrium and, 62–63, *63*
Free radical, 24–25
Freeze-fracture preparation, 80
Frequency-dependent selection, **727**
Freshwater biome, 760, *761*
Frigate bird, *629*
Frisch, Karl von, 822
Frog, 277, 436, 555, 606, 645–646, *646*, 734, 736, *736*, 836
 black, 338
 gas exchange, 556, *556*
 grass, 737
 Lake Titicaca, *556*
 lungs, 558
 pickerel, *736*
 poison arrow, *595*
 tree, 736, *736*, 793, *793*, 834–835
 wood, 736, *736*
Frond, algal, **393**, *393*
Frontal lobe, *666*, *670*, 671
Fructose, 40, *40*
Fructose-1,6-diphosphate, *122*, 123, *145*, 146
Fructose-6-phosphate, *122*, 123
Fruit, **404**, 406–407, *406*, 473, *473*
 color, *140*
 development, 501, *501*
 ripening, 506
Fruit fly, *170*, 644, 733
 body color, 186–187
 chromosomes, 181, *181*
 DNA, *240*
 embryonic development, *287*, 288–289, *290*
 evolution, 720, 746
 eye, *666*
 eye color, 181–183, *182–183*, 186–189, *189*, 191, *191*, 200
 gene mapping, 188–190, *188–189*
 geotaxis, 722, *722*
 mutation rate, 196
 polytene chromosomes, 189, *189*, 299, *300*
 sounds, 818
 wing mutants, 188–189, *189*
Fruiting body, 347, *347*, **381**
Fruticose lichen, 384, *385*
Frye, L.D., 86
FSH, *See* Follicle-stimulating hormone
Fucoxanthin, 363, 393
Fucus, 393–394
Fugitive species, **777**, 779, 804
Fulcrum, 689, *690*
Fumaric acid, *127*
Functional group, **38–39**, *38*
Fundamental niche, **785**
Fungal disease, 374
Fungi, **338**, 373, A1–A2
 characteristics, 374–376
 classification, 376–384
 evolution, 385–386, *385*
 growth, 375
 in lichens, 384–385, *385*
 life cycle, 376
Fungi Imperfecti, *See* Deuteromycota
Fungus garden, 790
Funguslike heterotroph, 360–361
Fur, *444*, 605, 607
 color, 720–721, *720*
Fused gene, 246
Fusiform initial cells, *478*, **479**

G_1 phase, **158**, *158*
G_2 phase, *158*, **159**
Gaffney, B., 176
Galactic cloud, 329
Galactosemia, *267*
β-Galactosidase, 235
Galago, *848*
Gall bladder, *576*, **579**
Gametangia, *378*, 379
Gamete, formation, 164–166, *164–165*
Gametocyte, *Plasmodium*, *368*, 369
Gametophyte, **390**, *390*, *392*, 394, 399, 464
Gamma globulin, 547
Ganglia, **416**, *623*, **624**
Ganglion cells, **663–664**, *664–665*
 on-center, 675
Gap junction, **94–95**, *94–95*
Garner, W.W., 508
Garrod, Archibald, 200, 263
Gas
 diffusion, 554
 partial pressure, **553–554**
 physics, 553–555
 solubility, 554
 transport in blood, 562–565
Gas exchange, 436
 leaf, 452, *452*
 plants, 484, *484*
 respiratory organs, 555–559, *555–558*
Gas gangrene, *351*
Gastric juice, 578, 584
Gastric pit, 577–578, *578*
Gastrin, **584**, *584*, 678
Gastrocoel, **412**, *412*, 420
Gastrodermis, *413*, *416*
Gastropoda, *427*, **428**
Gastrovascular cavity, *416*, 571
Gastrula, **279**, *280*
Gastrulation, **279**, *280*, 317
Gause, G.F., 801, 807, 810
Gazania, 467, *467*
Gazelle, 792
Geastrum triplex, *376*
Gemma cup, 395
Gene, **99**, **174**
 action, 190–191, *190*
 arrangement on chromosomes, 186–190
 coding for proteins, 200–203
 colinearity with protein, 216–217, *216*
 complementary, **193**, *193*
 epistatic, **193–194**, *193*
 eukaryotic, 241
 fused, 246
 jumping, 226
 multiple alleles, 191–193, *191–192*
 proof that they are DNA, 203–206, *204–205*
 structural, 235
Gene activity, differential, **299**
Gene amplification, **273**, 299, 303, 725
Gene duplication and divergence, 716, *716*
Gene flow, **717**, *717*
Gene frequency, **710–711**
Gene map, **188–190**
 bacteria, 232, 234, *235*
 cytogenetic, 189–190, *189*
 human, *254–255*, 255–256
 linearity, 189
Gene pool, **710–711**
 isolated, 737–738, *738*
Generalist, **786**, 827
Generative cell, 468–469, *468*
Generative nucleus, 470, *471*

Gene rearrangement, **299**
in development, 299
immunoglobulin genes, 541–542, *541*
Gene regulation, **234**
bacteria, 234–240
eukaryotes, 240–243
Gene therapy, 248, 263, 266
Genetic code, **217**, *227*
colinearity, 216–217, *216*
cracking, 224–226
degeneracy, 218, 226, *227*
nonoverlapping nature, 217, *217*
reading frame, 217–218, *217*
second, 226
Genetic counseling, **266**
Genetic disease, 209
biochemical analysis, 254, *254*
treatment, 266
Genetic drift, **714–715**, *714–715*
Genetic engineering, 13, **229**, *229*, 243–247
ethics, 249
plants, 149, 505, 863, *863*
prospects, 247–249
safety, 249
Genetic equilibrium, **711–713**
Genetic identity, **738–739**
Genetic language, *14*, 15
Genetic load, **727**
Genetic marker, 255
Genetics, *See also* Inheritance
reverse, **256**
Genetic variation, 196–197, 708–710, *708–710*
Genital herpes, 354, 357
Genotype, **174**, *175*, *180*, 711
parental, 186, *187*
recombinant, **186**, *187–188*
Genotype frequency, 712
Genotypic ratio, *175*
Gentamicin, 658
Gentiana lutea, *477*
Genus, **335**, *335*
Geochelone elephantopus abingdoni, *570*
Geological era, 334
Geological period, 334
Geotaxis, 722, *722*
German measles, *See* Rubella
Germ cells, 171, *171*, 288
Germination
pollen, **403**, 470, *471*
seed, 474–475, *474*, 508
Germ layer, 279, *280–281*
Germ plasm theory, **171**, *171*
Gestation, 442
Geyser, *7*, *28*
GH-release-inhibiting hormone (GRIH), *643*
GH-releasing hormone (GRH), *643*
Giant tube worm, 762
Gibberella fujikuori, 502
Gibberellic acid, *500*, 503, 508
Gibberellin, **501–503**, *503*, 507–508
Gibbon, 842–843, 848, *852*
Gibbs equation, 61–62
Gill filament, *557*
Gills, **556**, *556*
arthropod, 423
fish, 434, 556, *557*, 594–597
gas exchange, 556, *556–557*
mollusk, 427, *427*
Gill slit, *316*, 430, *433*
Gill withdrawal reflex, 821–822, *821*
Ginkgo tree, 398–400, *400*
Ginkgolide B, 400
Ginkgophyta, *389*, **398**, **400**, *400*
Giraffe, 9, *336*, 595, 614, 705, 807
Girdling (tree), 479
Gizzard
bird, 573
earthworm, **571**, *572*
Gladiolus, 465
Glans penis, 310
Glial cells, **614**
Gliding movement, 111–112, *112*
Globin, 301
Globin gene, 263
Globular protein, 51, 53, 67
Globular-stage embryo, 471, *472*
Glomerular filtrate, **600**, *601*
Glomerulus, 599, *599*
Glowing wave, 363
Glucagon, **636–637**, *637*, 644
Glucocorticoid, **638–639**, *638*, 644
Gluconeogenesis, 644
Glucose, **40**
blood, 636–638, *637*, 644
degradation, 121–124, **122–123**, *See also* Glycolysis
structure, 40, *40*
Glucose-1-phosphate, 119, *120*
Glucose-6-phosphate, *122*, 123
Glyceraldehyde, 40
Glyceraldehyde-3-phosphate (PGAL), *38*, *122*, 123, 132, *145*, 146
Glycerol, **43–44**, *43*, 132, 605
Glycerol phosphoric acid, 45
Glycocalyx, 82, *82*, **93**, *93*, *111*
Glycogen, **41**, *42*
Glycogen phosphorylase, 634, *634*
Glycogen synthetase, 634, *634*
Glycolipid, 84, *86*
Glycolysis, **121–124**, *122–123*
control, 132–133, *133*
Glycoprotein, *86*
Glycosaminoglycan, 41, 295
Glycosidic bond, **40**, *40*
Glyoxysome, 107
Gnatcatcher, blue-gray, *786*
Gnetina, 398, **400**
Gnetophyta, *389*, **398**, **400**
GnRH, *See* Gonadotropin-releasing hormone
Goat, 736
Goiter, **640**, *640*
Golden-brown algae, *360–361*, 361, 363–364
Goldfish, 248
Golgi, Camillo, 103
Golgi complex, 82, *82*, **103–105**, *104*, *111*
Gonad, indifferent, 313, *314*
Gonadotropin, 310, 669
Gonadotropin-releasing hormone (GnRH), *643*, 647
Gondwana, *334*
Gonial cells, **270**
Gonium, 391
Gonorrhea, *351*
Goose, 824
graylag, 818, *818*
Gorilla, 739, 833, *833*, 842, 849, 852, *852*
Gorilla gorilla beringei, *833*
Gould, James, 827
Gould, S.J., 338
Gout, 597
G-protein, 633, *634*
GPSP, *See* Grand postsynaptic potential
Gradient analysis, 781
Gradualism, **748**
Grafting, plant, 94, 465
Gram molecular weight, **26**
Gram-negative bacteria, **342**, *342*
Gram-positive bacteria, *93*, **342**, *342*
Gram staining, 342, *342*
Grana, *108*, 109, 138, *138*
Grand postsynaptic potential (GPSP), **622–623**, *622*
Granulocytes, **528**, 534, 536, *556*
Grape, 407
Grapefruit
Marsh, 465
pink, 716
Grape hyacinth, 465
Graptopetalum paraguageuse, *476*
Grass, *147*, 405, 460, 465, 470–471, 480, 777, *777–778*, *781*, 787, 792, 809–810
Grasshopper, *424*, 426, *516*, 558, *598*, *623*, *805*
flightless, 740
Gravitropism, 474, **501–502**, *502*
Gray matter, 625, *625*
Green algae, 391–392, *392*
Greenhouse effect, 149, 769–770, 863
Green sulfur bacteria, 345
GRH, *See* GH-releasing hormone
Griffith, Frederick, 203–204, *204*
GRIH, *See* GH-release-inhibiting hormone
Gross production, **765**
Ground meristem, *478*
Ground parenchyma, **455**
Ground substance, 293
Ground tissue, 449, *449*
Groundwater, 768, *768*
Group selection, **730**, *730*
Grouse, 760
spruce, *441*
Growth, 3, *4*
embryo, 283
fungi, 375
juvenile and adolescent, 283–284
localized relative, 295
plant
primary, **475–478**
secondary, 478–479, *478*
Growth factor, 304, *304*
Growth hormone (GH), 243–248, *245–246*, **640–642**, *641*, 644
Growth ring, *448*, *478*, 479
GTP, *127*, 128, 633, *634*
Guanine, 54, *55*, **206**, *206*

Guano, 437, 597, *771*, 772
Guard cells, *451*, **452**, 488–489, *488–489*
Guard hair, 452, *452*
Guinea pig, 609
Gull, 763
  Heermann's, 806, *806*
  herring, 819–820, *820*
Gullet, 571
Gurdon, John, 289
Gusella, James, 264
Gustation, **651**, *651*
Gut, 571
Guttation, **487**, *487*
Gymnodinium, 363
Gymnomycota, *360–361*, **364**
Gymnosperm, *389*, **399–403**
  evolution, *394*
Gynogen, **312**
Gyri, *See* Convolution

Habitat, **752**
Habitat island, 782–784
Habitat niche, **784**, 810
Habituation, **821–822**, *821*
Haemmerling, Joachim, *155*, 156
Hagfish, 433
Hailman, Jack, 820
Hair cells, 654–655, *654–655*
Half-life, 742
Halobacteria, **350**
Hamilton, W.D., 838
Hamner, Karl, 508
Hand
  development, *288*
  primate, 847, 850, *851*
Haploid, **157**
Hard skeleton, 684
Hardy, G.H., 712
Hardy-Weinberg law, **711–713**, *712*
Hare, 443
  arctic, 720–721, *720*, 760
  snowshoe, 799, *800*
Harem, **843**
Harris, Albert, 293
Harvey, William, 518
Hatching, 283
  bird, *4*
  reptile, 437
Haustoria, **374**, *375*
Haversian system, **687**, *687*
Hawk, 804, *805*, 807, 838
HCG, *See* Human chorionic gonadotropin
Head
  arthropod, 423
  mollusk, 427, *427*
  vegetable, 453
Hearing, 656–658, *656–657*
Hearing loss, 658
Heart, 514, **517–520**, *519–520*, *630*
  endocrine functions, 638
  evolution, 518–520, *521*
Heart attack, 522, 529
Heartbeat, 525–526, *526*, 700
Heart disease, 46, 522
Heart sounds, 526
Heartwood, *478*, **479**
Heat energy, 59–60, *60*
Heath, 759
Heat of vaporization, **28**, 30
Heat receptor, 609, 659
Hedgehog, 443, 788
Height, body, 256, *257*
Heliconius, *791*
Helium, *22–23*, 23
Helix-turn-helix structure, 300
Heme, *52*, 128, 565, *565*
Hemicellulose, 92, *92*
Hemichordata, *410*, 428, **430**
Hemin, 301
Hemocoel, **422**
Hemocyanin, 427, 563
Hemoglobin, *292*, 528, 563
  affinity for oxygen, 564
  control, 564, *564*
  fetal, *563*, 564
  oxygen transport, 563–564, *563–564*
  saturated, 563
  size, *84*
  structure, *49*, 50, 52, *52*
Hemoglobin S, 195, *195*, 202–203, *203*
Hemolymph, 423, **517**, 558, 597, *598*
Hemophilia, **261**, *262*, 529
Hemoskeleton, 427
Hemp, 455
Heparin, 548
Hepatica, 406, *406*
Hepatic portal vein, 520, *579*
Hepatic vein, 520
Hepatitis, *355*
Hepatocyte, 581
Hepatophyta, 389, **395**
HER-2/*neu* gene, 303
Herb, 480, *781*
Herbaceous plant, **454**
Herbal medicine, 400
Herbicide resistance, 248
Herbivore, **573–575**, *574–575*, 765, 807
Herd, 792, 804–805, 833, 841–842, *841*
Heredity, *See* Inheritance
Hermaphrodite, **314**, 416, 422
Herpes simplex virus, 354
Hershey, Alfred, 204–206, *205*
Heterocentrotus mammillatus, *429*
Heterocyst, 349, *349*
Heterokaryon, *254*, 255, **375**
Heterokaryosis, **375**
Heterosis, *See* Hybrid vigor
Heterospore, **397**
Heterotherm, 607
Heterotroph, **81**, 345, 374, 410
  animal-like, 361
  funguslike, 360–361
  origin, 332–333, *333*
Heterozygote, **174**, *175*, 190–191, *190*
Heterozygote advantage, **726–727**, *727*, 736
Hexasomy, *258*
Hexokinase, 123
Hexosaminidase A, 254, *254*
Hfr cell, **232**
Hibernation, 44, 585, 605, **608–610**, *609*
Hickory tree, *777*
  butternut, *448*
Hierarchical organization, nervous system, 675
High-energy group, 118
High-energy phosphate group, 118
Hindbrain, **665**, *666*, 667
Hippocampus, *668*, 669, 676–677
Hippocrates, 171
Hippopotamus, *336*, 787
Hirsch, Jerry, 722
Hirudinean, 420
Histamine, 536, *536*, 548
Histocompatibility antigen, 549
Histocompatibility genes, 549–550
Histone, **157**, *157*, 240, *300*, 301
Hodgkin, Alan L., 615
Holdfast, **393**, *393*
Homarus, *424*
Homeobox, 299–300
Homeostasis, **514–515**, **592**
Homeotherm, 441–443, **604–605**, *604*
Homing behavior, 659, 828
Hominid evolution, 852–860
Hominidae, *848*, 853
Hominoidea, *848*
Homo erectus, *856–859*, 858–859
Homo habilis, *846*, *856–858*, 857–858
Homo sapiens, 852, *856–857*, **859–860**
Homogentisic acid, 200
Homologous chromosome, **157**, *164*, 165–166
Homology, **744**, *745*
Homospore, **397**, 398
Homozygote, **174**, *175*
Homunculus, 670–671
Honey, 40
Honeybee, *335*, *802*, 822–824, *822*, 828, 830, 835, *835*, 839–840, *839*
Honeycreeper, 784, *785*
  Hawaiian, 744, *745*
Honey eater, 784, *785*
Honeysuckle, *476*
Hooke, Robert, 77–78
Hookworm, 419
Hopeful monster, 739
Hormone, 72, 310, 629
  characteristics, 630–631
  female reproductive, 312
  male reproductive, 310
  mechanism of action, 631–635
  plant, **498**
  protein, 46
Hormone receptor, 300, 631
Hornwort, 389, **395**, *395*
Horse, 335, 443, 575, 745
  evolution, 725, *725*, 742, *743*
  foot, 725, *725*
  palomino, 179
  thoroughbred, 176, *176*
Horsefly, 572
Horse latitudes, *755*
Horsetail, 396–398, *397*, 455
Hughes, J.T., 678
Human
  chromosomes, 156
  classification, *335*, 443
  DNA, *240*

Human (*Cont.*)
elemental composition, 19–20, *19*
evolution, *10*, 586, 846–864
migration, 861, *862*
modern, origin and diversification, 860–862
races, *708*, 862
society, 844, *844*
use of earth's resources, 766, *766*
Human chorionic gonadotropin (HCG), *313*, **317**
Human genetics
chromosomal abnormalities, 257–260
gene flow, 717, *717*
genetic variation, 710
methods of studying, 252–257
survey of traits, 260–265
Human Genome Project, 209
Human immunodeficiency virus, 546, *546*
Human population
age pyramid, **812–813**, *812*
age structure, 802, *803*
birthrate, 812
growth, 810–813, *811–812*
size, 861
twenty-first century, 813, *813*
Humidity, 761
Hummingbird, 405, 610, 784, *785*, 786, *787*, 834
violet-eared, 2, 562
Humoral immunity, **537–543**, *538–541*
Humors, 547
Humus, **761**
Hunger, 584–585, *584*, 669
Hunter, John, 629
Hunter-gatherer, 450, 811, 861
Hunter syndrome, *267*
Hunting, 859–860
Huntington's disease, **264–265**, *265*
Hurler syndrome, *267*
Hutton, James, 705
Huxley, Andrew F., 615
Hyalophora cecropia, 635
Hyaluronic acid, 293, *294*
Hyaluronidase, *294*
Hybrid, 733
Hybrid breakdown, *734*, 736
Hybrid inviability, *734*, **736**
Hybridization, in situ, 256
Hybridoma, 542–543, *543*
Hybrid sterility, *734*, **736**, *736*
Hybrid vigor, **726**, *727*, 736
Hydra, 167, *168*, 412, *413*, 571, *623*
Hydration, 594
Hydration sphere, 31, *31*
Hydrochloric acid, 578, *578*
Hydrogen, *19*, 20, 22, *27*, 120, *120*
atomic structure, *22–23*, 23
chemical bonding, 24, *24*
Hydrogen bond, 27, 53, 207–208
water molecules, **20–21**, *20*
Hydrogen ion, 32
Hydrogen peroxide, *146*
Hydrogen sulfide, *27*, 762
Hydroid, 396
Hydrologic cycle, *See* Water cycle
Hydrolysis, **39**, *39*, 58
Hydronium ion, 32
Hydrophilic compound, **31**
Hydrophobic compound, **31**, *32*
Hydrophobic interactions, 53
Hydroskeleton, **412**, 421, 684
20-Hydroxyecdysone, 635
Hydroxyl group, 38–39, *38*
Hydroxyl ion, 32
Hydrozoa, **413**
Hyena, 833
Hylobatidae, *848*
Hymenoptera, 426
Hypertension, 527
Hyperthyroidism, 640
Hypertonic solution, **91**, *91*
Hypertrophy, compensatory, **284**
Hyperventilation, 566
Hypervitaminosis, 586
Hyphae, **374**, *374–375*
aerial, **376**
dikaryotic, **382**, *383*
septate, 375, *375*
Hypocotyl, **474**, *474*
Hypoglycemia, 644
Hypothalamus, 310, 584–585, *602*, 603, 609–610, *630*, *641*, 645, 665, *666–668*, 668–669
hypothalamus-pituitary connection, 642–644, *643*
Hypothesis, **11**
testing, 11–12, *12*
Hypothyroidism, 640
Hypotonic solution, **91**, *91*, 486
Hypoxia, 554
Hyracotherium, 742, *743*
Hystrix africaeaustralis, *744*
H zone, *693*, 694

I band, *693*, 694
Ice, *18*, 29–30, *30*, 772
Ice plant, *451*, 453
Ichthyosaur, *438*, 439
Ichthyosaurus, 744, *744*
Icthyostegid, 436
Identical twins, *See* Monozygotic twins
Ig, *See* Immunoglobulin
Iguana, 440
marine, 606, *606*
Ileum, *576*, 579
Imaginal disk, **288–289**, *290*
Imbibition, 474
Immediate memory, 676
Immigration rate, 782, *783*
Immune deficiency, **545–546**
Immune response, 537
diversity, 537
memory, 537
primary, 537
secondary, 537
Immune system, **533**
aging, 285
components, 534–536, *535*
Immune tolerance, 537, 544–545
Immunity
active, **546–547**, *547*
cell-mediated, **543–544**
humoral, **537–538**, *538–541*
passive, **547–548**, *547*
Immunoglobulin (Ig), *292*, **538–539**, *539*
Immunoglobulin (Ig) genes, 541–542, *541*
Immunoglobulin A (IgA), 538–539
Immunoglobulin D (IgD), 538–540, *540*
Immunoglobulin E (IgE), 538–539, 548, *548*
Immunoglobulin G (IgG), 538–539, 548
Immunoglobulin M (IgM), 538–540, *540*
Implantation, *315*, **317**
Imprinting, **824**, *824*
chromosome, 315
Inbreeding, 176, **713–714**, *715*
Incest taboo, 713
Incisor, 442, 573, *573–575*, 575
Inclusive fitness, **833**
Incomplete dominance, **179**, *179*
Incomplete metamorphosis, 426
Incomplete penetrance, **264**, *265*
Incurrent pore, 411
Incus, 656, *656*
Independent assortment, 166, *167*, *180*, 186
law of, 177–179, *178*
Indeterminate plant, *See* Day-neutral plant
Indifferent stage, 313, *314*
Induced fit, **68**, *68*
Inducer, 235–236, *237*
Inducible enzyme, 235, *237*
Induction, 290
Inductive reasoning, **11**
Industrial melanism, 724–725, *724*
Industrial revolution, 811–812
Infant behavior, 844, *844*
Infection thread, 493
Inferior colliculi, 665
Inferior vena cava, *519*, 520
Infertility
female, 320
male, 113, 320
Inflammatory response, **536–537**, *536*
Inflorescence, *406*
Influenza, 354, *355*
Infrared detector, *659*, 660
Ingestion, **570**, 584–585, *584*
animal strategies, 571–575
Ingram, Vernon, 202
Inhalation, 561
Inheritance, 3
chromosomes theory, 180–183
cytoplasmic, 371
early theories, 171
heart disease, 522
human, *See* Human genetics
Mendel's experiments, 172–180
Inheritance of acquired characteristics, **705**
Inhibin, 310, 312
Inhibitory synapse, **622**
Initiator tRNA, 222, *223*
Innate behavior, 816
Inner cell mass, *277*, 315, *315*, 317, *320*
Inner ear, 654–655, *655*
Inorganic compound, 36
Inositol triphosphate, **634–635**, *635*

Insect, 423, *424*, **425–426**, 701
 body temperature, 606–607
 circulatory system, 516, *516*
 digestive tract, 572, *572*
 excretory system, 597–598, *598*
 eye, 665
 gas exchange, 558, *558*
 pollination by, *1*, 405
 social, **426**, 838–840, *838–839*
Insecticide, 622, 790
Insecticide resistance, 725
Insectivora, 443
Insertion, *217*, 218
Insertion sequence, 234
Inside cells, **471**
Insight learning, **825–826**, *826*
Instinct, 667–668, **817–822**
Insulation, 607, *607*
Insulin, **49**, 54, 159, *292*, 631, **635–637**, *637*, 644, 678
 genetically engineered, 243
 microbial production, 247
Integral protein, **85**, *86*
Integration, 612
Integument (plant), 399, **402**, 404
Intercalated disk, 700, *700*
Interferon, 247, 357
Intermediate filament, *109*, **110–111**
Internal cell movement, 114
Internal fertilization, 274, 437
Internal intercostal muscle, **561**
Interneuron, **614–615**, *615*
Internode, 503, *503*
Interphase, *158*, **159**, *160*, *164–165*, 166
Interrupted mating experiment, 232, *233*
Intersexes, 314
Interspecific competition, **806–807**, 809–810
Interstitial cells, 310
Intervertebral disk, 436
Intestinal cell, *90*
Intestinal gland, *579*
Intestine, **572**, *572*, 574
Intracellular digestion, 571
Intracellular fluid, 87, 517
Intraspecific competition, **806–807**, *806*
Intrauterine device (IUD), 322, *323*
Intron, **241–242**, *242*, 301, 331, *331*, 356, 541
Invagination, in gastrulation, 279, *280*
Inversion, 196, *196*
Invertebrate, **409**, *418–419*
 body temperature, 606–607
 contractile system, 700–701, *701*
 endocrine function, 635–636, *636*
 excretory system, 597–598, *598*
 marine, 594
Inverted ecological pyramid, 765
Iodine, 589, *589*, 640
Ion, 31
 hydration, 31, *31*
Ion channel, 616–618, *617*
Ionic bond, **24–26**, *26*, 53
Ionic compound, dissociation, 26
Ionophore, 89, *89*
Ir genes, 550
Iris (eye), **660**, *661*
Iris (plant), 465
Iron, in plants, 492, *492*, *589*
Irritability, 623
Island, 782–784, *783*
Islets of Langerhans, **636–637**, *637*, 644
Isocitric acid, *127*
Isogamete, 391, *392*
Isolating mechanism, 733–736
 postzygotic, **733**, *734*, 736
 prezygotic, **733–736**, *734–736*
Isoleucine, 586
Isomer, **37**
Isoosmotic animal, 594
Isoptera, 426
Isotonic solution, **91**, *91*
Isotope, **20**
Isotretinoin, 297
Istiophorus platypterus, *435*
IUD, *See* Intrauterine device
Ivy, 461

Jack-in-the-pulpit, 467, *467*
Jacob, Francois, 235
Jawed fish, 433–436, *433*
Jawless fish, 432–433, *433*
Jay, 760
 blue, 791
Jeffreys, Alec, 255–256
Jejunum, *576*, 579
Jelly coat, egg, 274, *274*
Jellyfish, 412, 414, 571, 654, 717
Jenner, Edward, 547
Jerne, Niels, 539
JH, *See* Juvenile hormone
Johanson, Donald, 854
Johnson, Virginia, 310
Joint, **684–685**, *685*
Juglone, 789
Jumping gene, 226
Juniper, 402
Jute, 455
Juvabione, 789
Juvenile hormone (JH), **635**, *636*

Kalanchoë daigremontiana, *464*, 465
Kandel, Eric, 821
Kangaroo, 442–443
 red, *443*
Karyotype, **157**, *158*, **252**, *253*
Kasterlitz, H.W., 678
Kelp, 393–394, *393*, *775*
Keratin, 50, *50*, *292*
α-Ketoglutaric acid, *127*, 128
Ketone, *38*, 39
Kettlewell, H.B.D., 725
Kidney, **598–604**, *599–602*, *630*
 artificial, 603
 concentration, 600–602
 endocrine functions, 637–638
 filtration, 600
 mesonephric, 598
 reabsorption, 600
 regulation, 602–604, *602*
 secretion, 600
 transplantation, 549–550
Kilocalorie, **26**, 586
Kinesin, 114
Kinesis, **817**, *817*
Kinetic energy, **59**, *59*
Kinetochore, **161–163**, *161*
King, Thomas, 289
Kingdom, **335**, *335*, *337*, 338
Kin selection, **838–841**, *838*
Kirk, David, 71
Klinefelter's syndrome, **258–259**, *258*
Knee, 685, *687*
Knee-jerk reflex, 624
Knuckle walking, **851**
Koala, 443, *443*
Koch, Robert, 350, *350*
Kohler, Georges, 542
Krause's end bulb, 652, *653*
Krebs, Sir Hans, 128
Krebs cycle, *122*, **124**, *126–127*, 128
 in fat metabolism, 131–132, *131–132*
 in protein metabolism, 131–132, *131–132*
Krill, 573, *573*, 764, *764*
K-selected species, **803–804**, *804*, 833, *833*
Kudzu, 464, *464*
Kyhos, Donald W., 746

Labia majora, *311*, 312
Labia minora, *311*, 312
Lack, David, 810
Lactase, 581, *581*
Lactation, **321–322**, *321*
Lacteal, 580, *583*
Lactic acid, 39, *122*, 124, *125*, 350, 568, 693
Lactose, 40
Lactose intolerance, 581
Lactose operon, 235–236, *237*
Lagena, 654, *655*
Lagomorpha, 443
Lake, 762, *766*, 767, 771–772, *772*, 777–778, *778*
Lamarck, Jean Baptiste, 8–9, 705
Laminaria, 393
Laminarin, 393
Laminin, 293
s-Laminin, 691
Lampbrush chromosome, *272*, 273
Lamprey, *51*, 433, *433*
Lampropeltis doliata, *791*
Lampyridae, *734*
Lancelet, 279, *280*, 432, *432*, 518
Land animal, body temperature, 606–608
Land plant, *389*
 evolution, *10*, 394–398, *394*
Landsteiner, Karl, 537–538
Language, *14*, 15, 671, 673, 834
Language center, 672–673, *672*
Langur, *843*
Laproides phthirophagus, *435*
Larch tree, 400, *478*, 487
Lard, 43
Large intestine, *576*, **580–581**
Larus heermanni, *806*
Larva, 283, 411, 426, *636*
Larynx, *559*, **560**
Lashlev, Karl, 677
Latent learning, **824–825**, *825*

Lateral geniculate body, *665*
Lateral line organ, 436, **653**, *654*
Lateral meristem, **478**
Lateral root, *457*, **459–460**
Late wood, *478*, **479**
Latimaria chalumnae, *435*
Latitude, 782
Latrodectus mactans, *423*
Laurasia, *334*
Lava flow, 777
Laverty, Terrence, 823
Lavoisier, Antoine, 19
Law of independent assortment, *See* Independent assortment
Law of segregation, *See* Segregation
Leaf, *78*, **449**, **452**
 adaptations, 453–454
 color, 108, *139–140*
 conifer, 401, *401*
 embryonic, *474*, 475
 energy conversions, *60*
 falling, 507–508, *507*
 gas exchange, 452, *452*
 opposite pattern, 477, *477*
 shape, *451*
 spiral pattern, 477, *477*
 structure, *148*, 450–453, *451–453*
Leaf primordium, *476*, **477**
Leakey, Mary, 846–847, 854
Learning, 671, 676, 816, **822–827**
 associative, **825**
 complex pattern, 825–826
 cultural transmission, 826–827, *826*
 insight, **825–826**, *826*
 latent, **824–825**, *825*
 programmed, **822–824**, *822*
 trial-and-error, **825**
Leather, 52
Lecithin, *44*
Lederberg, Joshua, 231
Leech, 420
Leeuwenhoek, Anton van, 78, *78*
Leg, *690*
Leghemoglobin, 493–494
Legume, 345, 493–494, *494*, 770, 796
Lemming, 805, 808
Lemur, 443, 739, 847, *848*, 849
Lemuridae, *848*
Lens, **660**, *661*
Lens field, 291
Lentic community, **760**, *761*
Lenticel, **479**, 484
Leprosy, 348
Leptoid, 396
Leptospira, *346*
Lepus timidus arcticus, *720*
Lesch-Nyhan syndrome, *267*
Lethal allele, **194–195**, *194*, 722
Leucine, 587
Leucoplast, 83, *83*, 108
Leukemia, **302**
 chronic myelogenous, 260
Leukocytes, *See* White blood cells
Levene, P.A., 206
L form, 37
LH, *See* Luteinizing hormone
Liana, 757, *781*
Lice, 348
Lichen, 384–385, *385*, 777, *781*, 796
Life, **2**
 characteristics, 2–5
  adaptation, 4, *5*
  development, 3, *4*
  growth, 3, *4*
  metabolism, 3
  organization, 2–3, *3*
  reproduction, 3, *4*
  responsiveness, 4–5, *5*
  variation, 3–4, *4*
 history of, 5, *See also* Evolution
 origin, *See* Origin of life
Life cycle, 157–158
Life expectancy, human, 812
Life span
 differentiated cells, 292, *292*
 plant, 479–480
Life-style
 heart disease and, 522
 plant, 479–480, *485*
Ligament, **685**, *687*
Light, nature of, 139
Light-dependent reactions, photosynthesis, **137**, *137*, 142–144, *143–144*
Light energy, 59, *60*
Light-independent reactions, photosynthesis, **137**, *137*, 144–146, *145–146*
Light microscope, 78–80
Lightning, 493, *494*
Lignin, 92, *92*, **455**
Lilac, 465
 western mountain, 494
Lily, *240*
 Arum, 466, *466*
Limb
 evolution, 9, *9*, 442
 primate, 850–851, *851*
 vertebrate, development, 295–298, *296–297*
Limbic system, *668*, **669**
Limenitis archippus, *791*
Limestone, 368–369
Limnetic zone, **772**, *772*
Limpet, *775*
Linkage, **186**, *187*
 complete, **186**, *187*
 partial, **186**, *187*
Linnaeus, Carolus, 335, 733
Linoleic acid, 587
Lion, 807, 821, 834, 837, 840–841
Lipase, *581*, 582, *583*
Lipid bilayer, 45, *45*, 84–87, *86*
Lipid diffusion, 88
Lipids, 35, **43–46**
 absorption in intestine, 580, *580*, 582
 digestion, 581–583, *583*
 energy content, 586
 hydrophobicity, 31, *32*
 oxidation, 44
 structure, 39
Lipofucsin, *107*, 285
Lipopolysaccharide, 341–342
Lipoprotein, 522, 647
Liposome, **330**
Lipotropin, *641*, **642**
Lips, 575
Lithium, *22–23*, 23
Lithium fluoride, 25–26, *26*
Lithops villetti, *793*
Littoral zone, **772**, *772*
Liver, *84*, *95*, *576*, **581**
Liverleaf, *406*
Liverwort, 389, **395**
Living things, composition, 46, *46*
Liwi, *785*
Lizard, *438*, 440, 604–606, 642, 810
Llama, 567, *568*
Loach, 248
Load arm, 689, *690*
Loam, **761**
Lobed-fin fish, 435
Lobster, 283, 422–423, *424*, 594, 684, 787
 American, *424*
Locus, **186**
Locust, 805, *805*
Locust tree, 473
Loeb, Jacques, 817
Loewenstein, Werner, 655
Logging, 778
Logistic growth, **801**
Logistic growth curve, **801**, *802*
Long-day plant, 508
Long-term memory, 674, 676–677
Lonicera, *476*
Loop of Henle, **598–602**, *599*, *601*
Lophornis ornata, *787*
Lorenz, Konrad, 818, 824
Loris, 847, *848*
Lotic community, **760**, *761*
Lotka-Volterra equation, 807
Louse, 795, *795*
Lovastatin, 522
LSD, *See* Lysergic acid diethylamide
Lucerne, 770
Luciferase, 70, *863*
Luciferin, 70
"Lucy" (fossil), 854, *854*
Lumbricus, *421*
Lumen, 571
 of chloroplast, 138, *138*
Lung, 436, **558–559**, *559*
 book, 423
 cancer, 302–303
Lungfish, 435, 556, 558, 560, 719, *719*, 723
Lupus erythematosus, 545
Luteinizing hormone (LH), **310**, 312, *313*, 317, 640, *641*
Lycoperdon giganteum, *376*
Lycophyta, *389*, **396–397**
Lycopodium, 397
Lyell, Charles, 705–706
Lyme disease, 425
Lymph, 530, 535
Lymphatic system, **530**, *530*
Lymphatic vessel, 530, 535, *535*, *579–580*, 580
Lymph node, 530, *530*, **535–536**, *535*
Lymphocytes, **528**, 534, *534*

Lymphokine, **543**
Lymphoma, **302**
Burkitt's, 302
Lynch, Gary, 675
Lysergic acid diethylamide (LSD), 382
Lysimachia quadrifolia, *477*
Lysine, 587
Lysogeny, **352**, *353*
Lysosome, 82, *82*, 104, *104*, **106–107**, *107*, *111*
Lysozyme, 536, 710
mechanism of action, 69–70, *69*
structure, 50
Lytic pathway, **352**, *353*

Macaque, *852*
Japanese, 826–827, 843–844, *843*
MacArthur, Robert, 782, 810
Mackerel, 556, 565
MacLeod, Colin, 204
Macroevolution, **740–749**
Macromolecule, 33, **36**
Macronucleus, *369*
Macronutrient, 492, *492*, **585**, 586–587
Macrophages, 528, **534**, *534*, 536–537, *536*, 543, 546, 550
Macropus rufus, *443*
Maggot, 5, *6*, 426, 817
Magnesium, 492, *492*, *589*
Magnetic field, detection, 659
Magnetic sense, 828–830
Magnetite, 659, 830
Magnetobacteria, 659
Magnolia, *453*
Maidenhair tree, 400, *400*
Major histocompatibility (MHC) complex, 543, **549–550**
Malaria, 195, 263, 367, *368*, 369, 547, 726
Mal de Meleda, 715
Maleness, 312–315, *314*
Male reproductive system, 309–310, *309*
Malic acid, *127*, 147–149, *148*
Malleus, 656, *656*
Malpeque disease, 723–725, *724*
Malpighian tubule, **598**, *598*
Maltase, 581, *581*
Malthus, Thomas, 10–11, 706, 800
Maltose, 40–41
Mammal, *432*, *438*, **441–442**
adaptations, 442
body temperature, 607–610
embryonic development, 276–277, *277*
evolution, *10*, *438*, 442–443
heart, *520*
lungs, 558
marine, 568
modern, 443–444
nitrogenous waste, *597*
social system, 841–844
Mammary gland, **321**, 630, *630*
Mammillaria microcarpa, *787*
Mandible, arthropod, 423, 572
Manduca quanquemaculatus, *807*
Manganese, 492, *492*, 588, *589*
Mangold, Hilde, 290
Mangrove tree, red, *460*
Mantid, flower, 793, *793*
Mantle, 427, *427*
Mantle cavity, 427, *427*
MAP 1C, 114
Maple sugar urine disease, *267*
Maple syrup, 456
Maple tree, *3*, 449, 453, 473, *507*, 792
sugar, 456
Map sense, 828–830
Marathon runner, 699
Marine biome, 760
Marine worm, 605, *606*
Marler, Peter, 827
Marmot, 609, *710*
Marmota flaviventris, *710*
Marrs, Barry, 149
Marsh, 777
Marsh grass, 783
Marsupial, *438*, **442–443**, *443*, 574
Marsupium, 442, *443*
Masked messenger, 301
Mass extinction, 439–440, 745–747
Mass flow theory, **491–492**, *491*
Mast cells, 539, 548–549, *548*
Masters, William, 310
Mastigophora, *360–361*, **365–369**, *366–367*
Maternal age, Down syndrome and, *258*
Maternal gene, 273–74
Mating behavior, 274, 308, *308*, 820–821, *820*, 834–837, *836–837*
Mating signal, 734
Matrix, mitochondrial, *107*, **108**, 125, *126*
Matthaei, J. Heinrich, 225
Mauthner cells, 619
Mayr, Ernst, 823
McCarty, Maclym, 204
McClintock, Barbara, 226, *226*, 234
Meadow, 773
Measles, *355*, 546
Meat-eating vertebrate, 572–573, *573*
Mechanical defense, 787
Mechanical isolation, *734*, **735–736**
Mechanoenzyme, **110**
Mechanoreceptor, **650**, 652–659
Mechanoreceptor reflex, **567**, *567*
Median eminence, 642
Median fissure, 671
Medical adhesive, 54
Medical diagnostics, 542
Medoglea, *415*
Medulla, 565
kidney, **599**, *599*
Medulla oblongata, 665, *666–667*, **667**
Medusa, **412–414**, *412*
Megagametogenesis, **469**, *469*
Megagametophyte, **403**, *404*, *469*, 470
Megakaryocytes, 529
Megasporangium, 399, *401*, **402**, 404
Megaspore, **397**, *401*, **402**, *404*
Megaspore mother cell, *401*, **402**, 469, *469*
Megasporogenesis, **469**, *469*
Megophrys monticola, *793*
Meiosis, **164–166**, *164–165*, 180, *180*
recombination during, 166, *167*
Meiosis I, 164–166, *164*
Meiosis II, 164–166, *165*
Meiotic arrest, 271, *272*
Meissner's corpuscle, 652, *653*
Melanin, 263, *263*, 586, 642, 724–725, *724*
Melanophore, 642
Melanophore-stimulating hormone (MSH), *641*, **642**, 647
Melatonin, 585, **645**, 659
Melosira sulcata, *363*
Melting, 60, *60*
Membrane
fluidity, 85, 87, 604
semipermeable, 88
Membrane protein, 101–104, *104*
Memory, 671–673, 675–677, 679
immediate, 676
long-term, 674, 676–677
short-term, 669, 676–677
Memory cells, **540**, *540*, 546–547, *547*
Menarche, **312**
Mendel, Gregor, 170, 172–180, *172*
Meninges, 665
Meningitis, *351*
Menopause, **312**
Menstrual cycle, **312**, *313*
Menstruation, **311**, 312
Mental retardation, 262
Mercury, 768
Meristem
apical, **472**, *472*, 475–477, *476*
ground, *478*
lateral, **478**
root, **472**, 475
Meristematic region, root, **475**, *476*
Merozoite, 367, *368*, 369
Meselson, Matthew, 210–211, *211*
Mesencephalon, 665, *666*
Mesentery, *415*, 576, *577*
Mesochyme, sponge, 411
Mesoderm, **279**, *280*, 290–291, 415, *415*, 420
fate, *281*, *289*
Mesoglea, 412, *412–413*
Mesonephric kidney, 598
Mesophyll, 96, **452–453**
Mesophyll cells, 147, *148*, 149
Mesosome, 342
Mesozoa, *410*
Mesozoic era, 333, 334, *334*, 437–441, 586
Messenger RNA (mRNA), 99, 101, *101*, *215*, **220**, 241, *See also* Translation
capping, 242, *242*
lampbrush chromosomes, 273
masked, 301
poly-A tail, 242, *242*
processing, 241–243, *242*, 301
regulation of production, 299–301, *300*
synthesis, *See* Transcription
stability, 301
storage, 301
Metabolic pathway, 59, **71–72**, *72*
control of enzymes, 72–74
Metabolic rate, arthropod, 423
Metabolism, 3, **72**
bacterial, 345
control, 132–133, *133*
origin, 331–332

Metafemale, *258*
Metallothionen, 248
Metamorphosis, **283**
amphibian, 645–646, *646*
complete, 425–426
incomplete, 426
insect, 635, *636*
Metanephros, 598
Metaphase
meiosis I, *164*, 165
meiosis II, *165*, 166
mitotic, *154*, *160*, **161**
Metaphase plate, *160*, **161**, *164–165*, 166
Metastasis, **302**
Metatheria, 443
Metazoa, 411
Metencephalon, 665, *666*
Meteorite, 328–329
Meteor shower, 440
Methane, 24, *24*, 27, *27*, 37, 350, 575
Methanobacteria, **350**
Methanol, *27*
Methionine, 587
Methyl ether, 37–38, *38*
Methyl group, *38*, 39
Metrosideros, 777
MHC complex, *See* Major histocompatibility complex
Microbody, **107**
Microevolution, **741**, 748–749
Microfibril, cell wall, 92
Microfilament, 82, *82*, *109*, **110**
in cell movement, 111, *112*
cytoskeletal, 114
in morphogenesis, 294, *294*
Microfossil, 332, *332*
Microgametogenesis, **468**, *468*
Microgametophyte, **400**, *404*
Microglial cells, 614
Micronucleus, *369*, 370, *370*
Micronutrient, 492, *492*, **585**, 587
Micropyle, *401*, **402**, 404, 470
Microscope, 78–80
electron, *See* Electron microscope
light, 78–80
Microspora, *360–361*, **371**
Microsporangium, *401*, **402**, 468, *468*
Microspore, **397**, *401*, *404*, 468, *468*
Microspore mother cell, *401*, **402**, 468, *468*
Microsporogenesis, **468**, *468*
Microtubule, 82, *82*, *84*, *109–110*, **110**
in cell movement, 111, *112*
in chromosome movement, 162–163, *162*
in cilia and flagella, 112–114
cytoskeletal, 114
doublet, 113, *113*
in morphogenesis, 294, *294*
Microtubule-organizing center, 113–114, 163
Microvilli, *82*, 83, *90*, *96*, **572**, 578, *579*, 651–652, *651*
Micrurus fulvius, *791*
Micturition, 601
Midbrain, **665**, *666–667*, 667–668
Middle lamella, 92, *92*
Midge, gall, 425
Midrib, **453**, *453*
Miescher, Johann Friedrich, 203
Migration, 44, 585, 608, 644, 659, 668, 717, 805, *815*, **828–830**, *828–829*
Milk
microbial production, 247
production, 321–322
Milk chocolate, *44*
Milk sugar, *See* Lactose
Milkweed, 473, *473*, 788–789
Miller, Stanley, 328
Millet, 861
Millipede, 423
Milstein, Cesar, 542
Mimicry, **790–792**, *791*
Mimosa pudica, 495–496, *495*
Mineral, **402**, 587 588, *589*
requirement in plants, 492, *492*
uptake by plants, 492–493
Mineralocorticoid, **638–639**, *638*
Minimal medium, 200–201, *202*
Miocene epoch, 850
MIS, *See* Müllerian inhibiting substance
Mitchell hypothesis, 128–130
Mite, 423, 808
red scale, *808*
Mitochondria, 82, *82*, **107–108**, *107*, *111*, *135*, *See also* Electron transport chain, respiratory
DNA, 860
evolution, 108, 355, *356*
respiration in, 125, *126*
size, *84*
trypanosome, 366
Mitochondrial membrane
inner, 125, *126*, 130
outer, 125, *126*
Mitosis, **159–163**, *160–161*
M line, *693*, 694
MN blood group, 711, *711*
Mobbing call, 827, *827*
Molar, 442, 572–573, *573–575*, 575
Molarity, A5
Mole, A5
Molecular chaperone, 352–353
Molecular evolution, 728–730, *729*
Molecular formula, **27**, *27*
Molecular orbital, **24**, *24*
Molecular relatedness, apes and hominids, 852–853
Molecule, **23**, 33
Mollusk, *410*, *415*, *426–427*, **427–428**
body plan, 427–428, *427*
circulatory system, 516
contractile system, 700–701, *700*
Molting, 422, *423*, 635, *636*, 684
Molybdenum, 492, *492*
Monera, **338**, 341–351, A1, *See also* Bacteria
evolution, 355–356, *356*
Monkey, 308, 443, 739, 816, 847
New World, **848**, *848*, 850
Old World, **848**, *848*, 850
red howler, *848*
Monoclonal antibody, **542–543**, *543*
Monocotyledon, **403**, *403*, 449, *449*
embryonic development, *472*
germination, 474–475, *474*
vascular bundles, *454*
vein pattern, 453, *453*
Monocytes, **528**
Monod, Jacques, 235
Monoecious plant, **402**
Monogamous species, 836
Monomer, **39**, *39*
formation on primitive earth, 328
Mononucleosis, *355*
Monophyletic organism, **336**
Monoplacophora, **427–428**
Monosaccharide, **40**
Monosomy, 257–258, *257–258*
Monotreme, *438*, **441–443**, *442*
Monozygotic twins, 319, *320*
Moose, 760, 808, *809*, 837
Moral choices, 12–13
Morel, 374, *377*, 381–382
Morgan, Thomas Hunt, 181, 188–189
Mormyrid, 654
Morphine, 678
Morphogen, 296–297
Morphogenesis, **279–282**, *281*, 292–295
traction-induced, 293
Morphological identity, 733
Mortality, **800**
Morula, 276, *277*, 315, *315*
Mosaicism, **259**, *259*, 261
Mosquito, 572, *572*
Moss, 389, **395–396**, *396*, 777, *781*
sphagnum, *778*
Moth, 607, 635
Automeris, 793, *794*
hawk, 405
peppered, 724–725, *724*
sphinx, 607, *807*
winter, 808
Mother of thousands, *464*
Motivation, 677, 821
Motor cortex, 669–671, *671–672*, 673
Motor neuron, **614–615**, *615*, 691
α, 691, *691*
γ, 691–692, *691*
Mountain climber, 554, *554*
Mouse, *240*, 443, *558*, 759
chimeric, 319–321, *321*
giant, 243–247, *245–246*
grasshopper, 788
red tree, 784
tail length, 194–195, *194*
Mousterian industry, *858*, **860**
Mouthparts, insect, 572, *572*
Movement
bacterial, 342–343, *342–343*
cell, *See* Cell movement
regulation, 690–692, *691*
sensing, 654–655, *655*
M phase, *158*, **159**
MRIH, *See* MSH-release-inhibiting hormone
mRNA, *See* Messenger RNA
MSH, *See* Melanophore-stimulating hormone
MSH-release-inhibiting hormone (MRIH), *643*
Mucigel, 475
Mucosa, digestive tract, 576, *577*

Mucous membrane, 536
Mucous secretion, 539
Mucus, 536, 576
Mud flat, 760
Mudpuppy, *556*
Mule, 736, *736*
Muller, Herman I., 196
Müllerian body, 796, *796*
Müllerian duct, **313–314**, *314*
Müllerian inhibiting substance (MIS), **314**
Müllerian mimicry, **791**, *791*
Multicellular organization, 95–96
  origin, 414
Multichannel cochlear implant, 658
Multiple allelic series, **191–193**, *191–192*, 722
Multiple birth, 319–3221
Multiple drug resistance, 234
Multiple-gene family, 240
Multiple sclerosis, 614
Mumps, *355*, 546
Musci, 389, **395–396**, *396*
Muscle
  contraction, 694–697, *694–696*
    heat production, 608
    initiation, 695–696, *696*
  evolution, 700–701, *700*
  feedback control, 690–692, *691*
  function, 692–694, *693*
  graded response, 697
  insertion, 690, *690*
  myoglobin, 564–565, *565*
  origin, 689, *690*
  relaxation, 697
  skeleton-muscle connection, 689–690, *689–690*
  structure, 692–694, *693*
  types, 692, *692*
Muscle block, 431
Muscle cell, 96, 110, *292*, *515*
  fermentation, 124
Muscle fiber, *693*
Muscle spindle, **691**, *691*
Muscle tone, 697
Muscular dystrophy, Duchenne, 265
Muscularis externa, 576, *577*
Mushroom, *335*, *373*, 374, 376, *377*, *383*
Mussel, 428, 721, *721*
Mustard, white, 508
Mutagen, **196**
Mutant, 181
  detection, 201, *202*
Mutation, 168, 181, **196-197**, *196*, 200, *202*, 210, 216-217, *216*, 716-717, *716*, 721, 739-740
  chromosomal, **196**, *196*
  frameshift, **218**
  neutral, 728–729
  point, **196**, *196*, 217, *217*, *227*
  somatic, 260, 466, 716, *716*
Mutation rate, **196**
Mutualism, 794, **795–796**, *796*
  facultative, 796
  obligatory, 796
Myasthenia gravis, 545
Mycelium, **375**, *375*
myc gene, 303
Mycobacterium tuberculosis, *346*, *351*
Mycoplasma, 341, *341*, *346*, **348**
Mycorrhizae, **379–380**, *379*, 394, 494–495
Mycota, **377**
Myelencephalon, 665
Myelin, 614
Myelin sheath, **619**
Myeloma, 542, *543*
Myofibril, *693*, **694**
Myogenic heartbeat, 525–526, *526*, 700
Myoglobin, **564–565**, *565*, 692
Myoneme, 369
Myosin, **110–112**, 114, 292, *292*, 683, *693–694*, 694–697, *696*, 698, *698*, 700–701
  in cytokinesis, 163–164, *163*
Myotome, **431**
Myrica faya, 778
Myrmecia pilosula, 426
Mytilus edulis, 721, *721*
Myxamoebae, **364**, *364*
Myxobacteria, 343, *346–347*, **347**
Myxomycota, **364**, *364*
Myxospore, 347, *347*

NAD^+, **120**, *121*
  regeneration in fermentation, 124, *125*
  use in glycolysis, *122*, 123
  use in Krebs cycle, 124–130, *126–131*
NADH, 120, *121*
  production in glycolysis, *122*, 124
  production in photosynthesis, 137
NADH-coenzyme Q reductase, *129*
$NADP^+$, use in photosynthesis, 142, *143*
NADPH, use in Calvin-Benson cycle, *145*
NADP reductase, *143*
Naive B cells, **540**, *540*
Naked-seed plant, 399–403
Naphthaquinone, *See* Vitamin K
Napp, Cyrill, 171
Narcissus pseudonarcissus, *464*
Nares, 435
Nasal cavity, 559, *559*, 595, 608, 652, *653*
Natality, **800**
Natix, 723
Natural catastrophe, 805
Naturalist-selectionist debate, 728–730, *728*
Natural killer (NK) cells, **534**, 537, 544, *544*
Natural law, **12**
Natural selection, **8–11**, *11*, **704–708**, *706*, 719–720
  directional, **723–725**, *724–726*
  diversifying, **725–726**, *726*
  frequency-dependent, **727**
  groups of genes, 730, *730*
  normalizing, **723**, *723*, *726*
  phenotype effects genotype, 722, *722*
Nature-nurture question, 256–257, *256–257*
Nautilus, 428
Navicula perpusilla, *363*
Navigation, **828**
Neandertal, *856*, *858*, **859–860**, *860*
Nearsightedness, 660–661
Neat's-foot oil, 44
Nectar, *483*, 823
Nectarinia mediocris, *785*
Necturus masulosus, *556*
Needle, *451*
  conifer, 401
  pine, 453
Negative control, 238
Neisseria gonorrhoeae, *351*
Neisseria meningitidis, *351*
Nematocyst, 412, *413*
Nematoda, *410*, *415*, **418–420**, *418–419*
Nemertina, *410*, **416–418**, *417–419*
Neocortex, 443, **669–670**
Neomycin, 658
Neopallium, *See* Neocortex
Neotony, **646**
Nephridia, 421–422, *421*, **597**, *598*
Nephridiopore, 597, *598*
Nephron, **598–604**, *599–602*
Nephrostome, 597, *598*
Nereis, *420*
Nerve, 614
  myelinated, 619, *619*
Nerve cell, *103*, *See also* Neuron
  cnidarian, 412–413, *413*
Nerve cord, *430*, **431**, *623*, 624
Nerve gas, 622
Nerve impulse, **613**
Nerve net, 412–414, *413*, **623–624**, *623*
Nerve tract, 667
Nervous system
  complex circuits, 624
  simple circuits, 623–624, *623*
Nervous tissue, 96, *515*
Net primary production, 766, *766*
Net production, **765**
Net productivity, 754
Neural arch, 436
Neural crest, *281*
Neural crest cells, 293–294, *293–294*
Neural fold, *281*
Neural plate, 281
Neural tube, 281, *281*, 294
Neuritis, *588*
Neuroactive peptide, **677–680**
Neurogenic heartbeat, 700
Neuromuscular junction, 691, 695
Neuron, *103*, *612*, **613–615**
  structure, 613–614, *613*
  types, 614–615, *615*
Neurospora crassa, *377*
  Beadle and Tatum experiments, 200–201, *201–202*
  life cycle, 200, *200*
Neurotoxin, 787
Neurotransmitter, 614, *620–621*, **621–622**, 677–678
Neurulation, **281**, *281*
Neutral mutation, 728–729
Neutral solution, 32
Neutron, **20**, *20*
Newt, 436
Niacin, *588*
Niche, **784–786**, *785–786*
Nicolson, Garth, 85
Nicotinamide adenine dinucleotide, *See* NAD^+
Nicotine, 789
Night blindness, *588*
Night vision, *662*
Nirenberg, Marshall, 225

Nitrate, 770
Nitrate reductase, 493
Nitrification, **493**, *494*, **770**
Nitrobacter, 493
Nitrogen, 20, 492–494, *492*
Nitrogenase, 349, 493
Nitrogen-containing base, 54, *55*
Nitrogen cycle, **770**
Nitrogen fixation, **493–494**, *494*, **770**
  bacterial, 345, *345*, 348, 493, 796
  bryophyte, 395
  cyanobacteria, 349
Nitrogenous waste
  excretion in animals, 596–604
  excretion in plants, 484, *485*
Nitrosomonas, 493
NK cells, *See* Natural killer cells
NMDA receptor, 677
NMR imaging, *See* Nuclear magnetic resonance imaging
Nocturnal creature, 661
Node of Ranvier, **619**, *619*
Noncompetitive inhibition, **74**, *74*
Noncyclic photophosphorylation, **143**, *143*
Nondisjunction, **182–183**, *183*, **252**, 258, *258*
Nonlinkage, **186**, *187*
Nonoverlapping code, 217, *217*
Nonrandom mating, **713–714**, *713*
Nonsense codon, 218
Nonshivering thermogenesis, 608
Nonspecific defense mechanism, 536
Nonvascular plant, **389**, *389*, 395–396
Norepinephrine, 526, 608, 621, 627, *638*, **639**, 677
Normalizing selection, **723**, *723*, *726*
Nostoc, 384
Nostril, 559, *559*
Notochord, 428, **430**, *430*
Notophthalmus viridescens, *437*
Nuclear envelope, *99–100*, **100**, 103, *104*, *160–161*, 162, *164*, 165–166
Nuclear lamina, 157, 162
Nuclear magnetic resonance imaging (NMR), 21, *21*
Nuclear pore, *99–100*, 100, *104*
Nuclease, *581*
Nucleic acid, 36, **54–56**, *See also* DNA; RNA
  discovery, 203
  metabolism, 597
  structure, 54–56
Nucleocapsid, 352, *352*
Nucleoid, *99*, **100**, 342, *342*
Nucleolus, *82*, *99*, **100**, *160–161*, 162, 221
Nucleoside, 54, *55*
Nucleosome, **157**, *157*, *300*
Nucleotide, **54**, *55*, 118, **206**, *206*
Nucleus, *82*, 83, 99, **99–100**, *99–100*, *111*
  atomic, 20
  brain, 667
  role, 155–156
  transplantation, **289–290**
Nudibranch, *409*, *426*
Nurse cells, 272, *273*
Nutmeg, 789
Nutrient
  absorption in digestive tract, 578–580, *580*
  transport in animals, *See* Circulatory system
  transport in plants, 484, 489–492, *490–491*
Nutrient exclusion, **788**
Nutrition, human, 585–589
Nutritional mutant, 200–201, *202*, 230–231
Nutshell, 455
Nymph, 426
Nystagmus, 263

Oak tree, 405, 452–453, *452*, *777*, *781*, 788, *789*
  red, *4*
Oats, 779
Obesity, 585
Obligate aerobe, **345**
Obligate anaerobe, **345**
Occipital lobe, *666*, *670*, 671
Ocean, 762, 765–766, *766*
Ocreatus underwoodii, *787*
Octopus, *426*, 427–428, 517, 664, 701
Ocular dominance, *670*, 675
Odor, 652
Odorant-binding protein, 652
Oils, 43–45
Okazaki fragment, **212**, *213*
Oken, Lorenz, 81
Oldowan industry, **857**, *858*
Oleic acid, *43*
Olfaction, **651–652**, *652–653*, 675
Olfactory bulb, 652, *653*, 665, *666*, 668
Olfactory center, 675
Olfactory epithelium, 652, *653*
Olfactory hair, *653*
Olfactory receptor cells, 652, *653*
Oligocene epoch, 850
Oligochaete, 420–422
Oligodendrocytes, **619**
Oligotrophic lake, 773
Olive, ripe, 350
Olive oil, 43
Omasum, *574*, 575
Ommatidia, **665**, *666*
Omnivore, **575**
On-center ganglion cell, 675
Oncogene, **302**, 303–304, *304*, 355
Oncorhynchus nerka, *824*
One gene–one enzyme hypothesis, **200–201**, *201–202*
One gene–one polypeptide hypothesis, **202–203**, *203*
One gene–several polypeptides hypothesis, 301
Onion, 453, 465, *476*, 509
Onychophora, *410*, **422**
Oocyte, 271, *272–273*
  completion of meiosis, 275
  maturation, 273
Oogamy, 391
Oogenesis, **270–274**, *272–273*, 315
Oogonia, 271, *272*, *378*
Oomycota, 377, *377–378*, **379**, *385*
Oospore, *378*
Oparin, Aleksandr I., 330
Open behavior program, 823
Open circulatory system, **516–517**, *516*, *518*
Open program, **818–819**
Operant conditioning, **825**
Operator, **236**, *237*
Operculum, 556
Operon, **235–236**
Ophiostoma ulmi, 380, *380*
Ophrys speculum, 405, *405*
Opium, 789
Opossum, 442–443
  brush tail, *443*
Opsin, 660, **662–663**, *663*
Optic chiasma, 664, *665*, *672*
Optic cup, *291*
Optic disk, 660, *661*, 663
Optic nerve, **660**, *661*, 663, *664–665*
Optic tectum, 665, 667–668
Optic vesicle, 290–291, *291*
Opuntia, *457*
Oral cavity, 575–576, *575*
Oral contraceptive, 322, *323*
Orangutan, *444*, 826, 842, 849, 852, *852*
Orchid, *1*, 405, *405–406*, 461, 509, 757, *778*
Order (taxonomic), **335**, *335*
Ordovician, 422
Organ, 81, **95–96**
  bilateral symmetry and, 414–420
Organelle, 33, **81**, 87, 98
Organic compound, 35, **36**
  branched chain, *36*
  extraterrestrial, 329
  fused rings, *36*
  primitive earth, 7, *7*, 328–330, *330*
  ring, *36*
  sources and sinks in plants, 490
  unbranched, *36*
Organizer, 290
Organizer-induction experiment, 290, *291*
Organ of Corti, *657*
Organogenesis, **279–292**, *281*, 317
Organ system, 81, **95–96**
Organ transplant, 549–550, 675
Orgasmic platform, 312
Orientation, 659, 828
Orientation slab, *670*
Origin of life
  early beliefs, 5–7, *6–7*
  modern view, 7–8, *7–8*
Origin of replication, 212, *212*
Ornithischian, *438*
Orthomyxovirus, *355*
Osculum, 411, *411*
Osmolyte, **593**
Osmolyte strategy, **593**
Osmoregulation, **593**
Osmosis, **90–91**, *91*, 486, *486*, 524, 593
Osmotic potential, **486**
Osmotic pressure, **91**, 486, *486*, 524
Osprey, 834
Ossicle, 429
Osteichthyes, *432*, **434–435**, *435*
Osteoblasts, 640
Osteoclasts, 640
Osteocytes, 433, 687, *687*

Osteoporosis, **689**
Ostracoderm, 432–433
Otolith, **654**, *655*
Outgassing, 327
Outside cells, **471**
Ovalbumin, *292*
Oval gland, 562, *563*
Oval window, **656**, *656–657*
Ovarian cycle, **312**
Ovary, **271**, *272*, 310, *311*, *630*
development, 313, *314*
plant, 404, *404*, 469, *469*
Overhydration, 668
Overlapping code, 217, *217*
Overproduction, 11
Overturn (lake), 762–763
Oviduct, *See* Fallopian tube
Ovulate, 400
Ovulation, 271, *272*, 275, **310**, *311*, 312, *315*, 322
Ovule, **402**, 404, *404*, 469, *469*
Owl
barn, *827*
snowy, 760
Oxaloacetic acid, *126*, 128, 147, *148*
Oxalosuccinic acid, *127*
Oxidation, **120**, *120*
Oxidation-reduction reaction, **119–121**, *120–121*
Oxidative phosphorylation, **128**, *129*, 130
Oxygen, *19*, 20
atomic structure, 23, *23*
inhibition of photosynthesis, 146–147
primitive earth, 333, *333*
production in photosynthesis, 136–150
roots and, 460
use in cellular respiration, 128, *129*
Oxygen debt, **568**, 693
Oxygen dissociation curve, hemoglobin, **563–564**, *563–564*, 567–568, *568*
Oxygen-poor environment, 567–568, *568*
Oxyhemoglobin, 563
Oxytocin, **311**, 319, *319*, 322, *584*, 585, 631, *631*, **642**, 646, 668
Oyster, 428, 723–725, *724*, 803
Ozone hole, 753, *754*, 769–770
Ozone layer, **333**, 439, 753, *753*, 769–770

P_1 generation, *See* Parental generation
P680, 141–142, *143*
P700, 141–142, *143–144*, 144
Pacinian corpuscle, 652, *653*
Pack (animal), 833
Pain, 669
Pain pathway, 679–680, *679*
Pain receptor, 652, *653*, 679–680, *679*
Palade, George, 101, 105
Palate, *559*, 560
Paleocene epoch, 849
Paleozoic era, *333–334*, 334, 423, 437
Palisade parenchyma, **452–453**
Palmitic acid, *43*
Palm tree, 449, *781*
Pancreas, 282, *292*, *576*, 579, *581*, 582, *630*, 636–637, *637*
Pancreatic juice, 579
Panda, giant, 740, *741*
Pandion haliaetuc, 834
Pangaea, **334**, *334*
Pangenes, 705
Pangenesis, **171**
Panting, 607, 668
Pantothenic acid, *588*
Papaver, *480*
Papillae, taste, 651, *651*
Papio anubis, *837*
Papio hamadryas, 843
Para-aminobenzoic acid, *588*
Paracrine cells, *630*, 631
Paracucumana tricolor, *429*
Parallel evolution, **743**, *744*
Parallel processing, 675
Paramecium, 112, *112*, 369–370, *369*, 571, 807, 817, *817*
Paramecium aurelia, 810
Paramecium bursaria, 810
Paramecium caudatum, 801, 810
Paramylum, 362
Paramyxovirus, *355*
Parapodia, 421
Parasitism, **794–795**, *795*, 805
Parasympathetic nervous system, *624*, *626*, **627**
Parathyroid gland, *630*, **639–640**, *639*
Parathyroid hormone (PTH), *639*, **640**, 689
Parazoa, 411
Parenchyma, 415, *449*, *451*, **452**, 454, *454*, 458
ground, **455**
palisade, **452–453**
spongy, **452–453**
Parental generation (P_1), **174**, *175*
Parental genotype, 186, *187*
Parental-type gamete, *180*, 181
Parietal cells, 577, *578*
Parietal lobe, *666*, *670*, 671
Parkinson's disease, 675
Parotid gland, *576*
Parsley, 480
Parthenogenesis, **276**
Partial linkage, **186**, *187*
Partial pressure, **553–554**
Parus major, 805, *826*
Passion flower, *781*
Passive immunity, **547–548**, *547*
Passive transport, 87–89, *89*
Pastan, Ira, 106
Pasteur, Louis, 5–7, *7*, 350
Pasteur flask, 6, *7*
Pasturella, 795
Pasturella pestis, *351*
Patella, *687*
Pathogen, **350**
Pattern baldness, 264, *264*
Pattern formation, **295**
regulatory genes in, 298–299, *298*
vertebrate limb, 295–298, *296–297*
Pauling, Linus, 50, 202, 207
Pavlov, Ivan, 825
Pavlovian conditioning, *See* Associative learning
Pea, 345, 471, 493, 509
Mendel's experiments, 172–180
Peach leaf curl, 381
Peach tree, *406*, 407, 503
Peacock, 734, 834
Peanut, 384, 473
Pea pod, 469, 473
Pear tree, 473
Pecking order, 840, *840*
Pectin, 92, 363
Pectoral girdle, 434, 685, *686*
Pedigree, **252**
autosomal dominant trait, 252, *253*
dominant trait, *265*
dominant trait with incomplete penetrance, *265*
X-linked trait, *261–262*
Pedipalp, 423
Pelagic animal, **412**
Pelagic community, **760–761**
Pellagra, *588*
Pellicle
ciliate, *369*
euglenoid, 362, *362*
Pelomyxa carolinensis, *367*
Pelomyxa palustris, 371
Pelvic girdle, *311*, 434, 685, *686*
Penfield, Wilder, 676
Penguin, 44, 607, 744, *744*, 764, *764*
emperor, *5*, *607*
Penicillin, 351, **382**
Penicillium, *377*
Penis, 274, **309**
Pentasomy, *258*
PEP, *See* Phosphoenolpyruvate
Pepsin, 67, *581*, 582, *583*
Pepsinogen, 577–578, *578*
Peptic ulcer, 578
Peptide bond, **48**, *48*, 101
Peptide hormone, 633–635, *634–635*
Peptidoglycan polymer, 92–93, **341–342**, 349
Peptidyl transferase, *223–224*
Percent polymorphism, **709**
Perennial plant, 407, **479–480**, *480*
Perforin, 539, 544
Perianth, **467**, *467*
Pericarp, 406
Pericycle, 457–458, *457–459*, **459**
Perilymph, *655*, 656, *657*
Peripatus, 422, *422*, 426
Peripheral cells, 468, *468*
Peripheral lymphoid tissue, **534–535**
Peripheral nervous system, **624–627**, *624–626*
Peripheral protein, **85**, *86*
Perissodactyla, 443–444
Peristalsis, 517, **576**, 698
Peritoneum, 576, *577*
Periwinkle, 428
Permafrost, 760
Permian period, *334*, 399, 437, 745
Pernicious anemia, *588*
Peroxisome, 107, *146*, 147
Personality, 257, 671
Pesticide, 808
Pesticide resistance, 248, 367

PET, *See* Positron-emission tomography
Petal, 404, *404*, 467, *467*
Petiole, **450**, *451*
Peyer's patch, *535*
Peyote, 789
PFK, *See* Phosphofructokinase
2PGA, *See* 2-Phosphoglycerate
3PGA, *See* 3-Phosphoglycerate
PGAL, *See* Glyceraldehyde-3-phosphate
pH, **32**, *32*
 blood, 564–566, *564*
 body fluids, 603
 soil, 761
Phaeophyta, 389, *389*, **393–394**, *393*
Phagocytosis, **105**, *105*
Phallus, 313, *314*
Phanerozoic eon, *333*
Pharynx, *559*, **560**, 571, *572*, 576, *576*
Phascolarctos cinereus, *443*
Phenolic, 788–789, 810
Phenotype, **174**, *175*
 environmental influences, 256–257, *256–257*
 recombinant, **186**
Phenotypic ratio, *175*
Phenylalanine, 262, *263*, 587
Phenylalanine hydroxylase, 262, *263*
Phenylketonuria (PKU), **262**, *263*
Phenylpyruvate, 262, *263*
Pheophytin, *143*
Pheromone, **380**, 652, *652*, 734, 834, 839–840
Phloem, 396, *449*, 452, *454*, **455–456**, *456*, *459*, *478*, **479**, 484, *485*, 489–492, *490–491*
Phoronida, *410*
Phosphate, 433
 inorganic, 118, *118–119*
Phosphate group, *38*, 39, 54, *55*, 118, *119*, 206, *206*
Phosphatidylinositol, 85
Phosphodiesterase, 634, 662, *663*
Phosphoenolpyruvate (PEP), *123*, 124, 149
Phosphoenolpyruvate (PEP) carboxylase, 147, *148*
Phosphofructokinase (PFK), 132, *133*
2,3-Phosphoglycerate, *145*
2-Phosphoglycerate (2PGA), *123*
3-Phosphoglycerate (3PGA), 123–124, *123*, 145–146
Phosphoglycolate, *146*, 147
Phospholipid, *44–45*, **45**
 hydrophilic head, *44*
 hydrophobic tail, *44*
 in membranes, 84–87, *86*
 in water, 45, *45*
Phosphorus, 20, 492, *492*, 588, *589*, 689
Phosphorus cycle, **771–772**, *771*
Phosphorylation, **118**, 304, 634, *634*
Photic zone, **762**, *762*
Photinus, 792
Photoautotroph, **345**
Photon, **139**, *139*
Photooxidation, 146
Photoperiodism, **508–509**, *509*
Photophobia, *588*
Photophosphorylation, **142–144**
 cyclic, **143–144**, *144*
 noncyclic, **143**, *143*
Photoreceptor, **650**, 660–665, 817
Photorespiration, **146**, *146*
Photosynthesis, *135*, **136**, 452–453, 754, 765, *See also* Leaf
 bacterial, 149, 348–349
 basic reactions, 136–138
 in carbon cycle, 150, *150*
 compared to cellular respiration, 135–136, *136*
 euglenoid, 362
 evolution, 332–333, *333*
 inhibition by oxygen, 146–147
 light-dependent reactions, **137**, *137*, 142–144, *143–144*
 light-independent reactions, **137**, *137*, 144–146, *145–146*
 overview, 136–138
Photosynthetic pigment
 complexes, 141–142, *141*
 excited, 140
 light absorption, 139–141
 types, 140–141
Photosystem, **141**
Photosystem I, 141–144, *143–144*
Photosystem II, 141–144, *143–144*
Phototaxis, 817
Phototropism, **499–500**, *499*, 817
Photuris, 791–792
Phycobilin, **392**
Phycocyanin, 349
Phycoerythrin, 349, 392
Phyllium, *424*
Phylogenetic tree, 336, *336*
 primate, *849*
 protists, *361*
Phylogeny, **742**
Phylum, **335**, *335*
Physical disturbance, **784**
Physical law, 14, *14*
Phytochrome far-red, 509, *509*
Phytochrome pigment system, **509–510**, *509*
Phytochrome red, 509, *509*
Phytophthora infestans, *377*, 379
Phytoplankton, 764–765, *764*
Pickled food, 124, *125*, 350
Pig, *336*
Pigeon, homing, 828
Pigment, **139**
 aging, 285
 photosynthetic, *See* Photosynthetic pigment
Pill bug, 787
Piloerection, **607**
Pincer, 787, *836*
Pineal gland, *630*, **645**, 659
Pineapple, 465
α-Pinene, 789
Pine tree, *3*, 399, 401–402, *401*, 500, *777*, 784
 bristlecone, 400
 jack, 779, *779*
 lodgepole, *781*
 sugar, 402
Pin mold, 379
Pinna, *656*
Pinnipedia, 443
Pinocytosis, **105–106**, *106*, 523, *523*
 reverse, 106, *106*
Pinus banksiana, *779*
Pinus longaeva, 400
Pinus muricata, 736
Pinus vadiata, 736
Pinworm, 419
Pioneer species, 777
Pisaster ochraceus, *429*
Pistil, 469, *469*
Pit (seed), 469
Pitcher plant, 495, *778*
Pith, *454*, **455**, 459, *478*
Pittendrigh, C.S., 644
Pituitary dwarfism, 642
Pituitary gigantism, 642
Pituitary gland, 585, *630*, **640–641**, 644, 665, *666*, 668, *668*
 anterior, 310, 640–642, *641*
 hypothalamus-pituitary connection, 642–644, *643*
 posterior, *602*, 603, **642**
PKU, *See* Phenylketonuria
Placenta, **317–318**, *318*, 442, *630*
Placental mammal, *438*, **442–444**, *444*
Placoderm, 434
Placozoa, *410*, **414**
Plague, *351*
Plain, 759
Plakula, **414**
Planaria, 415, 516, 571
Planarian, *598*
Plankton, 760
Plant, **338**
 classification, 389–390, *389*
 dormancy, 44
 evolution, 448–449
 genetic engineering, 149
 life cycle, 390, *390*
Plant breeding, 247–248, 402, 450, 726
Plant cell, 83, *83–84*
Plant-eating vertebrate, 573–575, *574–575*
Plant growth regulator, **498**
Plantlet, *464*, 465
Plantlike autotroph, 360
Planulae, **413**
Plaque, atherosclerotic, 522, *522*
Plasma, 518, **528**
Plasma cells, **540**, *540*
Plasma membrane, 81–82, *82*, **84**, *111*
 bacterial, 342, *342*
 movement of materials across, 87–89, *87–90*
 structure, 84–87, *86*
Plasmid, **232**, *232*, *240*, 243, 248
Plasmodesmata, 83, *83*, **94–95**, *95*, 458
Plasmodium, **364**, *364*
Plasmodium falciparum, 369
Plasmodium vivax, *368*
Plasmolysis, 486, *486*
Plasmopara viticola, *377*, 379
Plastid, 83, *83*, **107–109**, *108*, *111*
Plastiquinone, 142, *143–144*, 144

Plastocyanin, *143–144*
Platelets, **529**, *529*
Plate tectonics, **334**, *334*
Platyhelminthes, *410*, **415–416**, *415–416*, *418–419*
Platypus, duck-billed, 441, *442*, 443
Platyrrhini, *848*
Pleiotropy, **194–195**
Pleodorina, 391
Plesiosaur, *438*
Pleuropneumonia, 348
Pliocene epoch, 850
Plover, *832*
Plum tree, *480*
Pluviales, *832*
Pneumococcus, rough versus smooth, 203–204, *204*
Pneumonia, *351*, *355*
Poikilotherm, **604–605**, *604*
Point mutation, **196**, *196*, 217, *217*, *227*, 716
Polar body
 first, *272*
 second, *272*, 275
Polar bond, 24, **26**, *26*
Polar easterlies, *755*
Polar front, *755*
Polarized cell, **616**
Polar lobe, *278*, 279
Polar nucleus, *469*, **470**
Polar transport, 502
Pole, dividing cell, 161
Pole plasm, 288
Poliomyelitis, *355*, 357, 547
Polioptila caerulea, *786*
Pollen, 399, 466, 717
 production, 468–469, *468*
Pollen grain, *401*, **402**, *404*
Pollen sac, 468, *468*
Pollen tube, *401*, **403–404**, 469–470, *471*
Pollination, *1*, 400, **402–406**, *405–406*, 470, *471*
Pollinator, 405–406, *405–406*, 453, 466, 470
Polyadenine, 225
Polyandry, **836**, 840
Polychaete, 420, *421*
Polycytosine, 225
Polydactyly, 264, *265*
Polygamous species, 836
Polygyny, **836**
Polymer, **39**
 formation on primitive earth, 328–329
Polynucleotide ligase, 212, *213*
Polyp, **412–414**, *412*
Polypeptide, **48–49**
Polyplacophora, **428**
Polyploidization, 739–740, *740*
Polyploidy, 480–481
Polypodium vulgare, *398*
Polysaccharide, **40–42**, *42*
Polysome, **101**, *101–102*, *104*, 221, *221*
Polytene chromosome, **189**, *189*, 299, *300*
Polyuridylic acid, 225
Pond scum, *348*
Pond skater, 29, *29*
Pongidae, *848*, 853
Pongo pygmaeus abelii, *444*
Pons, 665, *666–667*, **667**
Popcorn, 470
Popelairia popelairii, *787*
Poplar tree, 465
Poppy, *480*
Population, 705, **710**, *710*, 752, 800
 age structure, **802–803**, *803*
 dispersion pattern, 804–805, *805*
 distribution, 809–810, *810*
 fluctuations, 809, *809*
Population density, 800, **804**
 factors controlling, 805–808, *806–807*
Population genetics, **710–713**
Population growth, 800–804
Population size, 11, 714
 fluctuations, 801–802
 limits, 804–809
Porcupine, 443, 743, *744*
Pore cell, 411, *411*
Pore mushroom, 382
Porifera, *410–411*, **411**, *418–419*, 516
Porpoise, 443, 658, 743–744, *744*, 826, 830
Portal vessel, 520
Porter, Keith, 101
Portuguese man-of-war, 413, 701
Position, body, 690–691
Positive control, 238
Positron-emission tomography (PET), 21
Posterior vena cava, 520
Postsynaptic neuron, 620, *620*
Postsynaptic potential (PSP), **622**
Posture, 697
Postzygotic isolating mechanism, **733**, *734*, 736
Potassium, 492, *492*, 588, *589*, 593–596
Potato, *42*, 449–450, 456–457, 465, 779, 861
Potato blight, *377*, 379
Potential energy, **59**, *59*
Powder puff, *375*
Powdery mildew, *377*
Power arm, 689, *690*
Prairie, 759
Pre-B cells, 540, *540*
Precambian period, 334
Precipitation, 539
Predation, **784**, 805, 807–808, *807*, *809*, 810, *811*
Predator, 807–808, *807*, *809*
Predatory behavior, 821
Pregnancy, 315–318
 first trimester, 315–317, *315–316*
 Rh disease, 192, *192*
 second trimester, *316*, 317
 third trimester, 317
Pregnancy test, 317, 542
Premature withdrawal, *323*
Premolar, 442, *573*, 575, *575*
Prenatal development, 315–318
Prenatal diagnosis, 12–13, *13*, 266, *267*
Presbytis entellus, *843*
Pressure receptor, 566, *567*, 652, *653*
Presynaptic neuron, 620, *620*
Pretonema, *396*
Prezygotic isolating mechanism, **733**, 734–736, *734–736*
PRIH, *See* Prolactin-release-inhibiting hormone
Primary host, 366
Primate, 443, **847–852**
 characteristics, 850–852, *851*
 evolution, *10*, 849–850
 society, 842–844, *843*
Prime mover (muscle), 689, *689*
Primitive streak, **279**, *280*, 317
Primordial soup, 328
Probability, 174–176
Probe, molecular, **244**, *244*
Proboscidea, 443
Proboscis, 572, *572*
Procambium, **455**, 471, *472*, *478*
Producer, 360–364, 371, **763**, *763*, *765*
 aquatic, 390–394
Product, **28**, 59
Production
 primary, **765**
 secondary, **765**
Profundal zone, **772**, *772*
Progenote, **332**
Progesterone, 275, **312**, *313*, 317, 319, 322, *632*, 633
Proglottid, 416, *417*
Programmed learning, **822–824**, *822*
Progress zone, **296**, *296*, 298
Prokaryote, **81**, **341**
 genetic center, 99–100, *99*
 ribosomes, 221
 transcription, 221, **222**, *225*
 translation, 221, **222**, *225*
Prolactin, **321–322**, **640–641**, *641*, 645, 678
Prolactin-release-inhibiting hormone (PRIH), *643*
Promoter, 220, **236**, *237*, 300–301, *300*
Pro-opiomelanocortin, 647
Prophase
 meiosis I, *164*, 165
 meiosis II, *165*, 166
 mitotic, **160–161**, *160*
Proplastid, 109
Proprioceptor, **652–654**, 690–691, *691*
Prop root, **460–461**, *460*
Prosimian, **847**, *848*, 849
Prostaglandin, **309**, 312, 318–319, *319*, **646**
Prostate gland, 309
Prosthetic group, 48, 67, *67*
Protea, ground-flowering, 466, *466*
Protein, 36, **46**
 artificial, 54, *54*
 biosynthesis, 100–101, *101*
 carrier, 88–89, *89*
 cellular, 101
 colinearity with gene, 216–217, *216*
 content of living things, 46, *46*
 degradation, 302
 denaturation, 52–53, *53*, 604
 differentiated cells, 291–292, *292*
 digestion, 581–583, *583*
 electrophoresis, 708–710, *709*
 energy content, 586
 exportable, 101–105

Protein (*Cont.*)
factors causing specific three-dimensional structure, 52–54
globular, 51, 53, 67
hormone, 46
integral, **85**, *86*
membrane, 85–86, *96*, 101–104, *104*
metabolism, 131–132, *131–132*, 596–604, *596*
modification, 101–103, *102*, 301
peripheral, **85**, *86*
primary structure, 48–50, *49*
quaternary structure, *49*, 52, *52*
regulator, 190
ribosomal, 221, *221*
secondary structure, *49–51*, 50–51
self-assembly, 52–53, *53*
structural, 46, 52, *52*, 190
synthesis, *See* Translation
tertiary structure, *49*, 51–52, *51–52*
transport, 46, 101–103, *102*
Protein hormone, 631, *631*
Protein kinase A, 634, *634*
Protein kinase C, 634–635, *635*
Proteinoid, **329**, 330, 332
Proteinoid microsphere, **329**, *330*
Protein sequencer, 242
Proterozoic eon, *333*
Prothallus, 398, *398*
Prothoracic gland, 635, *636*
Prothrombin, 529, *529*, 581
Protist, **338**, 359, A1
animal-like, 365–371
classification, 360–361, *360–361*
evolution, 371
funguslike, 364–365, *364–365*
plantlike, 361–364
Proto-Cro-Magnon, 860
Protoderm, **471**, *472*
Protogonyaulax, *362*, 363
Proton, **20**, *20*
Proton pump, 130, *143*
Proto-oncogene, **302–304**, *303*, 355
Protoplast
fungal, *378*, 379
plant, 504, *504*
Protopterus dolloi, *719*
Protostome, 410, *410*, **420–428**, *420*
Prototheria, 443
Protozoan, **365**, 571
Provirus, 352, *353*
Proximal convolutes tubule, 599–600, *599*, *601*
Pruning (plant), 501, *501*
Prunus, *406*
Prunus cerasifera, *480*
Pseudocoelom, *415*, **419**, *419*
Pseudocreobota ocellata, *793*
Pseudohermaphrodite, **314**
Pseudomonas syringae, 380
Pseudoplasmodium, **365**, *365*
Pseudopod, *359*, **365–367**, *367*
Psilophyta, *389*, **396–397**, *397*
Psittacosis, 348
PSP, *See* Postsynaptic potential
Ptarmigan, 763
Pteridophyta, *389*, **396**, 398, *398*
Pterobranch, 430
Pterosaur, *438*, 439
PTH, *See* Parathyroid hormone
Ptiloris vistoriae, *735*
Puberty, 639
female, 312
male, 310
Pueraria lobata, *464*
Puerperal fever, *351*
Puffball, 376, *376–377*, 382
Puffer fish, 787
Puffinus gravis, *828*
Pulmonary artery, *519*, 520, *559*
Pulmonary circulation, **519**, *519*, *521*
Pulmonary vein, *519*, *559*
Punctuated equilibrium theory, **747–749**
Punnett, R.C., 711
Punnett square, **174–176**, *175*, 712, *712*
Pupa, 283, *636*
Pupil, **660**, *661*
Purine, **206**, *206*
Purkinje cells, **667**, *667*
Purple bacteria, 345
Pus, 537
Pyloric sphincter, 578
Pyramid of biomass, **764–765**, *765*
Pyrenestes ostrinus, 786
Pyrenoid, 391
Pyrethroid, 790
Pyridoxine, *See* Vitamin B_6
Pyrimidine, **206**, *206*
Pyrogen, 609
Pyrrophyta, *360–362*, **361–363**
Pyruvate, **121**, *122–123*, 124, *125–126*, 127, *131–132*, 132

Q fever, *351*
Quadruplets, 319–321
Queen bee, 426, 838–840, *838–839*
Queen substance, 839–840
Queen Victoria (England), 261, *262*
Quercus agrifolia, *452*
Quillwort, 396–397
Quinone, 789
Quinone-binding protein, 149

Rabbit, 443, 675, 697, 759, 834
Rabies, *355*
Rackettail, booted, *787*
Radial cleavage, 278, 420, *420*
Radial symmetry, 412–414, 428–430
Radial transport, **479**
Radiation, of heat, 607
Radicle, **474**, *474*
Radioactive isotope, 20, 80
Radioisotope dating, 742
Radiolarian, *360–361*, **368–369**
Radish, *457*, 461
Radula, *427*, 428
Rainbow wrasse, *435*
Rainfall, 754–757, 768–769, *768*
Rain forest, *751*
Rain shadow, 757
Rana pipiens, 737
Random coil, 51, *51*
Random mating, 713
Rangifer tarandus, *815*
Rapid eye movement (REM) sleep, **674**, *674*
Rare-mate advantage, **727**
Rat, 443
kangaroo, 592, 595, 759
Rattler, 660
Raup, David, 440
Ray (animal), 595, 685
Ray initial cells, *478*, **479**
Reactant, **28**, 59
concentration, 65, *65*
Reaction center chlorophyll, **141**, *141*, 149, *149*
Reading frame, 217–218, *217*
Realized niche, **785**
Reasoning, *See also* Insight learning
deductive, **12**
inductive, **11**
Receptacle, flower, 467, *467*
Receptor-mediated endocytosis, 106
Receptor neuron, *See* Sensory neuron
Recessive allele, 190
Recessive trait, **172–174**
autosomal human, 262–263, *263*
Reciprocal altruism, **838**
Recombinant DNA technology, **229**, *229*, 243–247
Recombinant genotype, **186**, *187–188*
Recombinant phenotype, **186**
Recombinant-type gamete, *180*, 181
Recombination, 166, *167*
crossing over and, 186–187
Recombination analysis, 188
Recombination frequency, **188**
Rectal gland, 595
Rectum, *311*, *576*, 581
Red algae, **392**, *393*
Red blood cells, *91*, *292*, 518, 528, *528*, 563, 637, *See also* Hemoglobin
monitoring shelf life, 70
replacement, 283
sickled, 195, *196*
size, *84*
Red bread mold, *377*
Red gland, 562, *563*
Redi, Francesco, 5, *6*
Red marrow, 687
Red muscle, 565, 692
Red tide, 362–363
Reduction, **120**, *120*
Redwood tree, 399, 760, 809
coast, 400
dawn, 400
Reflex, 667–668, 816, **816**
Reflex arc, **624**
Reflex behavior, **816–817**, *816–817*
Regeneration, **284**, *284*
Acetabularia, *155*, 156
echinoderm, 429
nemertines, 418
Region of differentiation, root, **475**, *476*
Region of elongation, root, **475**, *476*
Regulator protein, 190
Regulatory gene
evolution and, 739, *739*
pattern formation and, 298–299, *298*
Reindeer, 44, 801, 861

Relative humidity, 761
Relaxin, **318**
Release factor, 224, *224*
Release-inhibiting hormone (RIH), **643–644**, *643*
Releaser, **819–820**, *819–820*
Releasing hormone (RH), **643–644**, *643*
REM sleep, *See* Rapid eye movement sleep
Renal pelvis, 599, *599*
Renin, 602–603, *602*, 637–638, *638*
Replica plating, 231, **321**
Replication fork, **212**, *212–213*
Repressible enzyme, **238–240**, *239*
Repressor, **236**, *237*, 300
Reproduction, 3, *4*
 annelid, 422
 asexual, 166–167, *168*
 bacterial, 343–345, *344*
 origin, 330–331
 sexual, 167–168, 308
 vertebrate, 308–309
 virus, 352–353
Reproductive behavior, human evolution and, 853–854
Reproductive isolation, 336, 706, **733**
Reproductive strategy, **803–804**, *804*
Reproductive success, 11, 720
Reproductive system
 female, 310–312, *311*
 male, 309–310, *309*
Reproductive time lag, **802**
Reptile, *432*, **436–440**, *438–439*
 adaptations, 437
 body temperature, 606–607, *606*
 evolution, *10*, *438*
 gas exchange, 555
 heart, 519, *520*
 kidney, 598
 lungs, 558
 modern, 439–440
 nitrogenous waste, *597*
 osmotic balance, 595–596
Resin, 484, *485*, 488
Resource-directed behavior, 823
Resource partitioning, 786, **810**, *810*
Respiration, **552**
 aerobic, *See* Aerobic respiration
 control, 667
Respiratory center, 565–567, *567*
Respiratory organ, 555–559, *555–558*
Respiratory pigment, **562–565**
Respiratory system
 birds, 561–562, *562*
 human, 559–561, *559–561*
Responsiveness, 4–5, *5*, 282
Resting potential, **615–616**
Resting spore, *See* Survival spore
Restriction endonuclease, **243**, *244*, 255, *255*, 264
Restriction fragment length polymorphism (RFLP), **256**, 264–265
Rete mirabile, **605**, *606*, 607–608
Reticular activating system, *668*, **669**
Reticular formation, *668*, **669**
Reticulum (stomach), 574, *574*
Retina, **661–662**, *662*, *665*, 675
 neural, **660**, *661*
 pigmented, **660**, *661*
 visual processing, 663–664, *664*
Retinal, **662**, *663*
11-*cis*-Retinal, 662, *663*
all-*trans*-Retinal, 662, *663*
Retinoic acid, 297
Retrovirus, 303, 353–354
Reverse genetics, **256**
Reverse pinocytosis, 106, *106*
Reverse transcriptase, 331, *331*, 353, *354*, 357
R factor, **234**, *234*, 351
R factor plasmid, **234**, *234*
RFLP, *See* Restriction fragment length polymorphism
R group, 47, *47*
RH, *See* Releasing hormone
Rhabdomere, **665**
Rhabdovirus, *355*
Rh blood group, 192, *192*
Rheumatic fever, *351*
Rheumatoid arthritis, 545
Rhinoceros, 443
 Indian, *308*
Rhinolophus ferrum-equinum, *444*
Rhinovirus, *355*, 357
Rhipidistan, 435
Rhizobium, 345, *345*, 493–494, *494*
Rhizoid, **374**, *378*, 379, 395, *398*
Rhizome, 396, 398, **464**
Rhizopus, *377*, 379
Rhodophyta, 389, *389*, **392**, *393*
Rhodopsin, **662**, *663*
Rhodospeudomonas capsulata, 149
Rhubarb, 456
Rhythm method, 322, *323*
Ribbonworm, 416–418, *417*
Rib cage, expandable, 560–561, *561*
Riboflavin, *See* Vitamin B_2
Ribonuclease, *581*
 structure, 51–53, *51*, *53*
Ribose, 54, *55*, 118, *119*
Ribosomal protein, 221, *221*
Ribosomal RNA (rRNA), **220–221**, *221*, 240, 508
Ribosome, 82, *82*, *84*, **100–101**, *101–102*, *111*, *215*, *See also* Translation
 A site, **222**, *223*
 egg cells, 273
 mitochondrial, 108
 prokaryotic versus eukaryotic, 221
 in protein synthesis, 100–101, *101*
 P site, **222**, *223*
Ribozyme, 242, **331–332**
Ribulose bisphosphate (RuBP), **145–147**, *145–146*
Ribulose bisphosphate carboxylase, 146–147, *146*
Ribulose-5-phosphate, *145*
Rice, 450, 502, 504, 508, 861
Rickets, 586, *588*
Rickettsia, *341*, **348**
Rickettsia rickettsii, *351*
Rickettsia typhi, *351*
Rickettsiae, *346*
Riflebird, 734, *735*
Rigor mortis, 697
Ringworm, 374, *377*
River, 778
RNA, 218–219
 catalytic, 330–331, *330–331*
 messenger, *See* Messenger RNA
 origin, 330–331, *330–331*
 processing, 220
 ribosomal, *See* Ribosomal RNA
 5S, 300–301, *300*
 structure, 54–56
 synthesis, *See* Transcription
 transfer, *See* Transfer RNA
 types, 220–221
 viral, 353
RNA polymerase, *219*, 220, 236–237, *238*, 301
RNA transcript, primary, 220, 241, *242*
Roadrunner, 824
Roaring forties, *755*, 756
Robichaux, Robert, 746
Rocky Mountain spotted fever, 348, *351*
Rodentia, 443
Rod eye, 661–662
Rods, **661–664**, *662*, *664*, 675
Rogas terminalis, *807*
Röntgen, Wilhelm, 21
Root, 397, **457**
 adaptations, 460–461
 adventitious, **460**, *460*
 embryonic, 474, *474*
 function, 457
 lateral, *457*, **459–460**
 nutrient uptake, 492–494
 oxygen and, 460
 primary growth, 475, *476*
 prop, **460–461**, *460*
 structure, 457–460, *457–459*
 types, 460, *460*
Root cap cells, **475**, *476*
Root cells, *104*
Root hair, *457*, **458**, 475, *476*, *485*
Root meristem, **472**, 475
Root nodule, **493**, *494*
Root pressure, **487**
Root sucker, 465
Root tip, *95*, *99*, 475, *476*
Rose, 457, 465
 twisted stalk, *473*
Rosette (plant), 503
Rotifer, 772
Rough endoplasmic reticulum, 101–103, *102*, *104*, 105
Round window, **656**, *656–657*
Roundworm, 794
Rous sarcoma virus, *354*, 357
Roux, E., 537
Royal jelly, 426
rRNA, *See* Ribosomal RNA
r-selected species, **803–804**, *804*, 833, *833*
Rubella, 317, *355*, 546
RuBP, *See* Ribulose bisphosphate
Rubus rubrisetus, *406*
Ruffini's corpuscle, 652, *653*
Ruffle, 111, *112*
Rumen, **574–575**, *574*
Ruminant, *336*, 574, *574*, 796
Runner (plant), *168*, 464

Running, 792
human speed limit, 699
Rusts, *377*, 382

Saccharomyces cerevisiae, 382
Sacculus, 654, *655–656*
Sage, 405
Sago palm, 399, *399*
Sailfish, *435*
Sake, 384
Salamander, 277, 284, *284*, 436, 555, 558, 606, *623*, 645–646, 737, *737*
spotted, *437*
Saliva, 539, 576, 584
Salivary gland, **576**, *576*, *581*, 630–631
Salmon, 248, 274, 308, 605, 652, 824, *824*, 836
Salmonella, 197, *351*
Salmonella anatom, *342*
Salmonella typhi, *351*
Salt
dietary, 603
iodized, 640
Saltatory propagation, **619**
Salt balance, 593
Salt gland, 597
Salt-secreting cells, 595
Salty taste, 651, *651*
Sampling error, 714
Sand, **761**
Sandpiper, 763
Sand sage, *460*
Sanger, Frederick, 49
S-A node, *See* Sinoatrial node
Sap, 40, 456
Saprobe, **374**, 379
Saprolegnia, *378*, 379
Sapwood, *478*, **479**
Sarcodina, *360–361*, **365–369**, *367*
Sarcoma, **302**
Sarcomere, 692, *693*, **694**, 701
Sarcoplasmic reticulum, **695**, *696*, 699, *700*
Sarcopterygian, 434–435
Sargassum, 393, *393*
Sarischian, *438*
Saturated fatty acid, *43*, 44
Sauerkraut, 350, 587
Savanna, *756*, **758–759**, *759*, *766*
Sawfly, 789, *790*
Scale insect, 808
Scallop, 427–428, 664
Scanning electron microscope (SEM), *78*, 79–80
Scanning tunneling microscope (STM), 79, *79*
Scarlet fever, *351*
Scarlet gilia, 405
Schistosoma, 416
Schistosomiasis, 416
Schizophyta, **346–348**, *346–348*
Schleiden, Matthias, 81
Schlumbergira, *467*
Schneiderman, Howard, 249
School (fish), 833
Schwann, Theodor, 81
Schwann cells, **619**
Scientific method, **11–12**
Sciurus, *738*
Sclera, 660, *661*
Sclerenchyma, *449*, **454–455**
Scleroid, **455**
Scolex, 416, *417*
Scorpion, 423, 759
Scouring rush, 397, *397*
Scrotum, 309
Scurvy, *588*
Scutellum, **472**, *472*
Scyphozoa, **413–414**
Sea anemone, 412, *413*, 414
Sea apple, *429*
Sea bass, 434, 562
Sea cucumber, 428–429, *429*, 605
Sea daisy, 430
Sea horse, 434
Seal, 443, 568, 605, 743–744, *763–764*, 764
Alaskan fur, *444*
Sea lettuce, 392, *392*
Sea lily, 428–429
Sea lion, 443, 743–744
Sea otter, *393*
Sea slug, *426*, 821–822, *821*
Seasonal affective disorder syndrome, 585
Seasons, 754, *755*
Sea squirt, 278, *278*, 431
Sea star, 420, 428–429, *429*, 594, 760, *775*, 836
Sea urchin, *275*, *308*, 428–429, *429*, 748, 810, *811*, 836
Seaweed, 393–394, *393*
Secondary compound, 788–789, *789*
Secondary host, 366
Secondary sexual characteristics
female, 312
male, **310**
Second filial generation (F_2), **174**, *175*
Second genetic code, 226
Second law of thermodynamics, **60–61**, *60*
Second messenger, **633–635**, *634–635*
Secretin, 584, *584*
Secretory granule, 104, *104*
Sedge, *778*
Sediment, 769
Seed, **399**, *401*, *404*, 471
conifer, 403
dispersal, 406–407, *406*, 472–473, *473*
dormancy, 44, 473–474
germination, 474–475, *474*, 508
hybrid, 736
maturation, 473
round versus wrinkled, 190, *190*
starch reserves, 41, *42*
synthetic, 465, *465*
Seed coat, 403, 472–473, *472*, 787
Seed fern, 399
Seedless plant, *389*, 396–398
Seedling, *401*
Seed plant, *389*, 448
classification, 398–399
evolution, 399
Segmentation, **421**, 422
Segmentation gene, 298
Segmented worm, **420–422**, *421–422*
Segregation, law of, *175*, **176–177**
Selaginella, 397
Selector gene, 298
Self-assembly, protein, 52–53, *53*
Self-fertilization, 713
Selfish gene, 838
Selfish herd, 792
Self-pollination, 172, *173*, 405
SEM, *See* Scanning electron microscope
Semen, **309**
Semicircular canal, **654–655**, *655–656*
Semiconservative replication, **210–211**, *211*
Seminal vesicle, 309
Seminiferous tubule, 270–271, *270–271*, 309
Semipermeable membrane, 88
Semliki Forest virus, 352, *355*
Senescence, plant, 504, *504*, 506
Sense organ, **650**
Senses, 649–650, *649*
Sensitive period, **822**, 824
Sensitive plant, 495–496, *495*
Sensory cortex, 669–671, *671*
Sensory ganglia, 625
Sensory neuron, **614–615**, *615*
1a, 691, *691*
Sensory receptor, 650
baseline level, 650
Sepal, 404, *404*, 467, *467*
Sepkoski, John, 440
Septum, fungal, **374**, *375*
Sequoia, giant, 400
Sequoiadendron giganteum, 400
Sequoia sempervirens, 400
Serosa, digestive tract, 576, *577*
Serotonin, 585, 621
Sertoli cells, 270, *271*, 310
Sessile animal, 412
Setae, 421
Set point, 585, **609**
Severe combined immunodeficiency disease, 266
Sex chromosome, **157**, *158*, 181, *181*
abnormal numbers, 258–259, *258–259*
Sex differences, origin, 312–315, *314*
Sex hormone, *629*
Sex-influenced trait, **264**, *264*
Sex-linked trait, **182–183**, *182*
human, 261–262, *261–262*
Sex pilus, 232, *232*
Sex steroid, *638*, 639
Sex switch, 314
Sexual behavior, 633, 669
Sexual development, 312–315, *314*
chromosome imprinting and, 315
hormonal control, 313–314
Sexual dimorphism, **836**, *836*
Sexual process, bacterial, 344–345
Sexual reproduction, 167–168, 308
fungi, 200, *201*
meiosis, 164–166, *164–165*
plants, 466–470, *467–469*
Sexual response
excitement phase, **310**, 312
female, 312
male, 310

orgasmic phase, **310**
plateau phase, **310**, 312
refractory period, 312
resolution phase, **310**, 312
Sexual selection, 734, *734*
Shade leaf, 453
Shaggy mane fungus, *383*
Shark, 565, 592, 595, 654, 665, 685, 744, 804
black-tip, *434*
mako, 605–606
Shearwater, greater, 828–829, *828*
Sheep, 444, *568*, 574, 736
Merino, 607
Shelford, Victor, 780
Shell, 41, *See also* Egg shell
arthropod, 684
clam, 684
mollusk, *426–427*, 427–428
sarcodine, 367–368, *367*
snail, 684
turtle, 440
Shigella dysenteriae, 234, *351*
Shigellosis, *351*
Shingles, 357
Shivering, 608
Shock, anaphylactic, **548–549**
Shoot, primary growth, 475–478, *476*
Short-day plant, 508
Short-term memory, 669, 676–677
Shrew, 443, 658, 783
tree, 849, *849*
Shrimp, 422–423, 605, *606*, 635
nautilid, 665
Shrub, *777*, 778, *781*
Sickle-cell anemia, 50, 195–196, *195*, 202–203, *203*, 263, 726–727, *727*, 730
Sieve plate, 491, *491*
Sieve tube, **456**, *456*
Sieve tube element, **456**, *456*, 459, 491, *491*
Signal peptide, 102–103, *102*
Sign stimulus, **819–821**, *819–820*
Siliceous rock, 368
Silicon, 492, *492*
Silk, 50, **51**
Silk gland, 423
Silk moth, 283, 635, 652, *652*
Silks (corn), 470
Silt, **761**
Silversword plant, 744, 746, *746*
Simple diffusion, 87–88, *87*, *89*
Singer, Jonathan, 85
Single bond, 24, *24*
Single-cell movement, 292–293, *293*
Sinoatrial node (S-A node), **525–526**, *526*
Sinus, *516*, 517
Sinus venosus, 518
Siphonous algae, 391
Sisal, 455
Skate, 595
Skeletal muscle, **692**, *692*
characteristics, 692–693
structure, 693–694, *693*
Skeleton, 684–690
articulated, 684
hard, advantages, 422
skeleton-muscle connection, 689–690, *689–690*
Skin
aging, 285
amphibian, 436
in body defense, 536
color, 586
gas exchange, 555–556
reptile, 437
Skinner, B.F., 817, 825
Skull, development, 739, *739*
Skunk, 443, 609, 790
Slab, neocortex, 669
Sleep, 673–674, *674*
Sliding-filament theory, *694*, **695**
Slime mold, 111–112, *360–361*
cellular, **364–365**, *365*
true, **364**, *364*
Slime track, 343, 347
Slip (plant), 465
Sloth, 574
Slow-twitch muscle, 692–693
Slow-wave sleep, 674
Slug, 427
Small intestine, 573, *576*, **578–581**, *579–581*
Smallpox, *355*, 547
Smell, 651–652, *652–653*
Smilodon californicus, *747*
Smith, William, 705
Smoking, 522, 689
Smooth endoplasmic reticulum, 103, *103–104*
Smooth muscle, **692**, *692*, 698–699, *698*
blood vessel, 521, *521*
Smuts, *377*, 382
Snail, *278*, 427–428, *427*, 684, 748, 760, 816, *816*
Snake, *438*, 440, 659, *659*, 834
coral, 790, *791*
milk, 791
scarlet king, 791, *791*
Snake venom, 548, 788
Snapdragon, 179, *179*, 190–191
Snow goose, lesser, 713–714, *713*
Social behavior, 832–844
Social hierarchy, 837, **840–841**
Social insect, **426**
Social system
biology and, 12–13
mammals, 841–844
primate, 842–844, *843*
vertebrate, 840–841
Sociobiology, **833**
Sodium, 20, *22*, 492, *492*, 588, *589*, 593–596
Sodium chloride, 26
Sodium-potassium pump, 89, *90*, 91, 608
neuron, 616, *616*, 618
Soil, 493, 761–762
erosion, 778
formation, 777
pH, 761
Solar constant, **753**
Solar energy, 3, *3*, 753–754, *753*
in photosynthesis, 138–142
"Solar One", *863*
Solar power, *863*
Solar system, formation, 327–328, *327*
Solenoid, 157, *157*
Solubility, **554**
Solute, **31**
Solution, 31–32, *31*
Solvent, **31**
Somatic cell genetics, *254–255*, **255–256**
Somatic cells, 171, *171*
Somatic mutation, 260, 466, 716, *716*
Somatic nervous system, *624–625*, **625–627**
Somatomedin, 159, 581, **641–642**
Somatostatin, **636–637**, *637*
Songbird, 822, 824–825, *825*
Sorbitol, 605
Sorus, **398**, *398*
Sound
locating, 658
loudness, 658
Sour cream, 350
Sourdough bread, 124, *125*
Sour taste, 651, *651*
Sow bug, 423
Soybean, 149, *494*, 509, 861
Soy sauce, 124, *125*, 384
Spark chamber, 328
Sparrow, 824, 838
white-crowned, 734, 825, *825*
Spatial summation, 623
Spawning, 274, *308*, 824, *824*, 836
Specialist, **786**, *786–787*, 827
Speciation
allopatric, 737–738, *738*
genetic bases, 738–740, *739–740*
sympatric, **740**
Species, **335**, *335*
definition, 336, **733**
Species composition, 777
Species diversity, 782
Species equilibrium model, **782–784**, *783*
Species richness, 781–784, *783*
Specific heat, **28**, 30
Speech, 673, 827
Speed, human, 699
Spemann, Hans, 290
Sperm, 112–113, *270*
penetration of egg, 274–275, *274–275*
production, 309–310, *309*
transport, 309–310, *309*
Spermatid, 270, *270–271*
Spermatocyte, 270, *270*
Spermatogenesis, **270–271**, *270–271*, 315
Spermatogonia, 270, *270–271*
Spermatophore, 274
Spermophilus beldingi, *841*
Sperry, Roger, 672
S phase, **158**, *158*, 164
Sphenophyta, *389*, **396–398**, *397*
Sphincter muscle, 517, 525–526, *526*
Spicula, 411, *411*
Spider, 423, 665, 684
black widow, *423*
orb, 783
Spinach, 508

Spinal cord, 431, *623–624*, 625, *666*
Spinal nerve, *625*, 627
Spindle, 110, *160–161*, **161–163**, 162, 166
Spine, plant, 787, *787*
Spinebill, eastern, *785*
Spiny-finned fish, 434
Spiral cleavage, 278, 420, *420*
Spirillum, *341*, 346–347, *347*
Spiro branchus, *421*
Spirochete, *341*, *346–347*, **347**
Spiroplasma, 355, *356*
Spleen, *535*, **536**, *576*
Spliceosome, 241, *242*
Split-brain study, 671–673, *672*
Sponge, 410–411, *411*, 516, 555, 605, *606*
Spongy bone, 685–689, *687*
Spongy parenchyma, **452–453**
Spontaneous generation, **5**
Spontaneous reaction, 62
Sporangiophore, 364, *378*, 379
Sporangium, *378*, 379, 396
Spore
 bacterial, 343–345, *344*
 dispersal, 376
 fungal, 374, 376, *376*
 slime mold, 364–365, *364–365*
 survival, 376
Sporophore, 365, *365*
Sporophyte, **390**, *390*, *392*, 394, 399, *404*, 464
Sporozoa, *360–361*, 365, *368–369*, **369–371**
Sporozoite, 367, 369
Spring overturn, 772–773
Spruce tree, 399, 784
 blue, 452
Sputum, 536
Squamata, 440
Squamous epithelium, 95
Squid, 274, 427–428, *427*, 517, 605, *606*
 boreal, *5*
 giant, 614–615, *616*
Squirrel, 443, 610
 Albert, 737–738, *738*
 ground, 609, *609*, 834, 841, *841*
 Kaibab, 737–738, *738*
 red, 784
src gene, 304, *354*
Stahl, Franklin W., 210–211, *211*
Stamen, 404, **467–468**, *467–468*
Standing crop, **765**
Stanley, Wendell, 351
Stapes, 656, *656*
Staphylococcus, *341–342*, 346, *346*, *351*
Starch, **41**, *42*, 144, 581
Starch granule, *42*, *78*
Starfish, *See* Sea star
Starling, 829, *829*
Starling, E.H., 584
Start codon, *223*, *227*
Startle reflex, 816, *816*
Statocyst, **654**, *655*
Statolith, **654**
Stele, **457–459**, *457–458*
Stem, **454**
 adaptations, 456–457, *457*
 structure, 454–455, *454*
 vascular tissue, *454–456*, **455–456**
Stem cells, **283**, 528–529
Stentor, 371
Steppe, 759
Stereotyped behavior, 426
Sternum, 561
Steroid, **45–46**, *46*
Steroid hormone, 631, *631*
 anabolic, 633
 mechanism of action, 632–633, *632*
Stevens, George, 782
Stickleback, three-spined, 820–821, *820*
Stigma, 172, *173*
 algal, 391
 euglenoid, 362, *362*
 flower, 404, *404*, 469, *469*
Stingray, black-spotted, *434*
Stipe, **393**, *393*
STM, *See* Scanning tunneling microscope
Stolon, *378*, 379, **464**
Stomach, **573**, 576–578, *576–577*, *581*, *630*
Stomata, 147, 394, 401, *451*, **452**, 484, 487–489, *488–489*
Stoneplant, African, 793, *793*
Stone tools, 857–860, *858*
Stopak, David, 293
Stop codon, 224, *224*, 226
Stratification
 lake, 772
 seed, 506
Strawberry, 167, *168*, 464, 466, 473, *487*, 501, *501*
Strep throat, 347
Streptococcus, *346*, 347, *351*
Streptococcus lacti, 350
Streptococcus mutans, 346
Streptomyces, *346*
Streptomycin, 230, 348, 658
Streptopus roseus, *473*
Stretch receptor, 567, *567*, 691, *691*
Strobilus, **397**, 400
Stroke, 522, 529
Stroma, *108*, **109**
 chloroplast, 138, *138*, 144, *146*
Stromatolite, *326*, **332–333**, 349
Strong bond, **27**
Structural formula, **27**, *27*
Structural gene, 235
Structural isomer, **37–38**
Structural protein, 46, 52, *52*, 190
Strychnine, 789
Sturnus vulgaris, *829*
Sturtevant, Alfred H., 188–189
Style, 469, *469*
Stylet, 490, *490*
Subatomic particle, 20, *20*, 33
Suberin, **459**, 479
Sublingual gland, *576*
Submandibular gland, *576*
Submucosa, digestive tract, 576, *577*
Subphylum, *335*
Subspecies, **737**, *737*
Substance P, **679–680**, *679*
Substrate, **66**
 binding to enzyme, 68
 concentration, 68, *68*
Subtilisin, 54
Succession, 773, **777**, *777*
 animal, 780
 old field, 778, *779*
 primary, *776*, **777–778**
 rules, 780
 secondary, **777–779**, *779*
Succinic acid, *127*, 128
Succinic dehydrogenase, 128
Succinyl-CoA, *127*
Succulent, 457, *476*
Sucker, plant, 461
Sucrase, 70, *581*
Sucrose, 40, *40*, 119, *120*, 490
Suction pump system, 560–561, *561*
Sudden infant death syndrome, 567
Sugar, 39
 absorption in intestine, 579–580, *580*
 uptake by cells, 89, *90*
Sugarcane, 147, 456, 465, 740, 779
Sulci, *See* Fissure
Sulfhydryl group, *38*, 39
Sulfobale, **350**
Sulfonolipid, 343
Sulfur, 20, 492, *492*, *589*
Sulfur dioxide, 767
Summation, **622–623**, 697, *697*
 spatial, 623
 temporal, 623
Summer solstice, 754, *755*
Sunbird, 784, *785*
Sundew, *483*, 495, *778*
Sunflower, 449
Sun leaf, 453
Sunlight, *See* Light-dependent reactions; Solar energy
Superfish, 248, *248*
Superior colliculi, 665
Superior vena cava, *519*, 520
Suprachiasmatic nucleus, **645**
Supraoptic nucleus, 668
Surface tension, water, **29–31**, *29*
Surface-to-volume ratio, **84**, *84*
Surfactant, lung, 560
Surrogate mother, 320
Survival spore, 376
Survivorship curve, 802–803, *803*
Suspensor, **471**, *472*
Sutherland, Earl W., 633
Sutton, Walter, 180
Suture, 685
Swallow, fly-catching, 784
Swallowing, 576, *577*, 667
Swamp gas, 350
Swan, 824
Swarming, *839*, 840
Sweat gland, 607, 630
Sweating, 603, 607, 609, 668
Sweet gale, 494
Sweet potato, 458, 461
Sweet taste, 651, *651*
Swim bladder, 434–435, 562, *563*
Swimming, 112–114, 431
Sylph, violet-tailed, *787*
Symbiont, **794**
Symbiosis, 345, 355, **794–796**, *795–796*
Symbolic communication, 861
Sympathetic nervous system, *624*, *626*, **627**
Sympatric speciation, **740**

Symplastic pathway, **458–459**, *458*, 493
Symport, 89
Synapse, *615*, **620–622**, *620*
 nonrectifying, 621
 rectifying, 621
Synapsis, **165**
Synaptic cleft, *620*, 621
Synaptic terminal, *613*, **614**
Synaptic transmission, **620**
Synaptic vesicle, *620*, 621
Synaptinemal complex, *164*, **165–166**
Synceros caffer, *841*
Syncytial theory, origin of multicellularity, 414
Synergistic muscles, 689
Synomone, 412
Synovial fluid, 685, *687*
Synovial membrane, 685, *687*
Syphilis, 347, *347*, *351*
Systemic circulation, **519**, *519*, *521*

T_3, *See* Triiodothyronine
T_4, *See* Thyroxine
Table sugar, *See* Sucrose
Tadpole, 431, *431*, 436, 645–646, *646*
Taenia serrata, *417*
Taeniura lymma, *434*
Taiga, *756*, 760, **760**, *766*
Tail
 postanal, *430*
 prehensile, 848
Tandem scanning reflected-light microscope (TSRLM), 79
Tannin, 484, *485*, 788, *789*
Tapetum, 468, *468*
Tapeworm, 416, *417*, 571, 794
Tapir, 443
Taproot system, **460**, *460*
Tar, 488
Taraxacum officinale, *406*
Tardigrade, 594, *594*
Target cells, 631
Tarsier, 443, 847, *848*, 849
Tarsiidae, *848*
Tarsius syrichta, *848*
Taste, 651, *651*
Taste bud, 651, *651*
Taste pore, *651*
Taste receptor cells, 651, *651*
Tatum, Edward, 200, 231
Taxis, **817**
Taxon, **335**
Taxonomic group, 335, *335*
Taxonomy, **335**
 evolution and, 335–337
 molecular, 336
Taylor, F.J.R., 355
Tay-Sachs disease, **254–255**, *254*, *267*
T cells, **534–537**, 543–544, *544*, 550, 795
 helper, *533*, 543, *544*, 546
 killer, 543, *544*
 suppressor, 543–545, *544*
TDF, *See* Testis determining factor
T-DNA, 248
Tears, 536, 539
Tectorial membrane, *657*, **658**
Teeth, 572–575, *573–576*
 evolution, 442
 primate, 850, *851*
Telencephalon, 665
Teleost, 434–435, *435*, *593*, 803–804
Telmatobius culcus, *556*
Telomere, 157
Telophase
 meiosis I, *164*, 165–166
 meiosis II, *165*, 166
 mitotic, *161*, **162**
TEM, *See* Transmission electron microscope
Temperate forest, *756*, **759–760**, *760*, *766*, 781–782
Temperate grassland, *756*, **759**, *766*
Temperate shrubland, *756*, **759**
Temperature
 body, 564
  Allen's rule, *721*, 722
  amphibian, 606–607
  aquatic organism, 605, *606*
  birds, 607–610
  insects, 606–607
  mammals, 607–610
  regulation, 604–610
  reptiles, 606–607
 earth's, 754–757
 effect on chemical reactions, 64–65, *65*
 effect on enzymes, 70–71, *71*
Temporal isolation, *734*, **736**, *736*
Temporal lobe, *666*, *670*, 671
Temporal summation, 623
Tendon, 52, 516, **690**, *690*
Tensile strength, 111
 water, **28–31**
Tentacle, 428
Teosinte, 402, *402*
Teratogen, 297
Terminal bud, **477**
Termination codon, *See* Stop codon
Termite, 366, *366*, 426, 572, 574, 796, 839
Tern, arctic, 828
Terpene, 789
Territoriality, **836–837**, *837*
Territorial warning, 834
Test cross, **177**, *177*
Testes, **270–271**, *270–271*, 313, *314*, *630*
Testis determining factor (TDF), 314
Testosterone, 45–46, *46*, 264, **310**, 314, 631, *631*
Test-tube baby, 320
Testudo elephantopus, *439*
Tetanus, *351*, 547, **697**, *697*
Tetrad, *164*, 165
Tetraploid, *257*, 258
Tetrasomy, *258*
Tetrodotoxin, 787
TH, *See* Thyrotropic hormone
Thalamus, 665, *666–668*, 669
Thalassemia, 202–203, **263**
Thalidomide, 297, 317
Thallophyte, *781*
Thallus, 374, 392, *392*
Thecodont, 437, *438*
Theory, **12**
Therapsid, 437, *438*, 441
Thermal spring, 345, 349
Thermode, 609
Thermodynamics
 first law, **59–60**, *60*
 second law, **60–61**, *60*
Thermogenesis, **608–609**
Thermogenin, 130
Thermoproteales, **350**
Thermoreceptor, **650**, 659–660, *659*
Thermoregulatory center, **609–610**
Thiamine, *See* Vitamin B_1
Thicket, 777
Thick filament, *693–694*, 694–698, *696*
Thin filament, *693–694*, 694–698, *696*
Thirst, 603, 669
Thistle, 787, *787*
Thoracic cage, 561, *561*
Thoracic duct, 530, *535*, *579*
Thorax, 423
Thorn, 787, *787*
Thorntail, wire-crested, *787*
Thoroughbred, 176, *176*
Threat display, **837**, *837*
TH-releasing hormone, *643*
Threonine, *38*, 587
Thrombin, 529, *529*
Thromboplastin, *529*
Thumb, opposable, 848, *851*
Thylakoid, **138**, *138*, 349
Thylakoid membrane, 138, *138*, 142, *143*
Thymine, 54, *55*, **206**, *206*
Thymus, 534, *535*, **536**, *630*, 642
Thyroglobulin, *292*
Thyroid gland, *292*, *630*, **639–640**, *639*
Thyroid-stimulating hormone, 678
Thyrotropic hormone-releasing hormone (TRH), 645, *646*
Thyrotropic hormone (TH), *639*, **640**, *641*
Thyroxine (T_4), 608–609, 633, **639–640**, *639*, 645, *646*
Tibecen, *423*
Tick, 348, 423
 deer, 425
Tide pool, 760
Tiger, saber-toothed, 745, *747*
Tight junction, **94**, *94*
Timothy, *95*
Tinbergen, Niko, 819
Tissue, 81, **95–96**
Tissue compatibility, 549–550
Tissue culture, plant, 504–505, *504*
Tissue factor, 529
Tissue interaction, in differentiation, **290–291**, *291*
Tissue plasminogen activator, 247, 529
Tit
 blue, 826, *826*
 great, 805, *826*
Titmouse, great, *730*
Tmesipteris, 396
TMV, *See* Tobacco mosaic virus
Toad, 436
 Couch's spadefoot, 802, *802*
 golden, *437*
Tobacco, 248, 740, *863*
Tobacco mosaic virus (TMV), 351, *352*
Tobramycin, 658
Tocopherol, *See* Vitamin E

Togavirus, *355*
Tolerance mechanism, of succession, 780
Toll gene, 298
Tomato, 465, 473, 487, 861
Tongue, 575–576, *576–577*, 651, *651*
Tonicity, 91, 105
Tonsillitis, *351*
Tonsils, 530, *535*
Tonus, **697**
Tool making, 860–861
Tooth enamel, 575–576
Topaza pella, *787*
Topoisomerase, 157, *157*
Torgor, 486, *486*
Tornaria larva, 429
Torpor, 605, **610**
Tortoise, 440
 Galápagos, *439*, *570*
 giant, *704*, 706
Totipotent cells, 478, 504
Touch receptor, 652, *653*
Toxic waste, 54
Toxin
 dinoflagellate, 363
 plant, 484
Trachea, 113, *559*, **560**
Tracheae, 423, 517, **558**, *558*
Tracheid, 396, *455*, **456**, 459
Tracheole, 558, *558*
Tracheophyta, 390
Trachoma, 348
Traction-induced morphogenesis, 293
Trade winds, *755*
Transcription, *215*, **219**
 eukaryotic, 221
 overview, 219–220, *219*
 prokaryotic, **221–222**, *225*
 regulation, 299–301, *300*
Transcription complex, 300–301
Transcription factor, 300–301, *300*
Transdetermination, **289**, *290*
Transduction, **232–234**, *233*, 344
 phage-mediated, *233*, 234
Transduction machinery, 631
Transfer RNA (tRNA), **220–222**, *220*, *222*
 amino acid attachment site, 220, *220*
 anticodon, **220**, *220*
 initiator, 222, *223*
Transformation, 204, *204*, **232–234**, *233*, 344
Transition state, 64, *64*
Translation, *215*, **219**
 amino acid activation, 221–222, *222*
 elongation stage, 221–224, *223*
 eukaryotic, 221
 initiation stage, **221–222**, *223*
 origin, 330–331
 prokaryotic, **221–222**, *225*
 regulation, 301–302
 termination stage, **221**, 224, *224*
Translocation, 196, *196*, 259–260, *260*
Translocation (nutrient transport), **489–492**
Transmission electron microscope (TEM), 79–80
Transpiration, **487–488**, *488*
Transpiration-pull theory, **487–488**, *488*
Transplantation, nuclear, **289–290**
Transport protein, 46
Transport system, plant, 484
Transport vesicle, *104*
Transposable element, 226
Transposon, 234
Transverse colon, *576*, 580
Transverse tubule, **695**, *696*, 699, *700*
Trap tree, 380
Traveler's diarrhea, *351*
Trebouxia, 384
Tree, *781*
Tree fern, *781*
Treehopper, 793, *793*
Trehalose, 594, 605
Trematoda, **416**
Trentepohlia, 384
Treponema pallidum, *346–347*, 347, *351*
TRH, *See* Thyrotropic hormone-releasing hormone
Trial-and-error learning, **825**
Tricarboxylic acid cycle, *See* Krebs cycle
Triceps muscle, 689, *689*, 691, *691*
Trichinella spiralis, 419–420
Trichinosis, 419–420
Trichocyst, 369
Trichonympha, 366, *366*
Trichoplax, 414
Trichosurus vulpecula, *443*
Trichromatic theory, **662–663**
Tridachnia gigantea, *700*
Triglyceride, *43*, **44**
Triiodothyronine (T_3), **639–640**, *639*
Trilobita, **423**
Trilobite, 748
Trimester, **315**
Trimethylamine oxide, 595
Tripeptidase, *581*
Tripeptide, 48
Triple bond, 24–25
Triplets, 319–321
Triploid, *257*, 258
Trisomy, *257–258*, 258
Trisomy 13, 258
Trisomy 18, 258
Trisomy 21, *See* Down syndrome
Trivers, R.L., 838
tRNA, *See* Transfer RNA
Trochophore, **422**, *422*, 427
Troop, **843**
Trophic level, **763**
Trophoblast, **315**, *315*, 317
Tropical rain forest, *756–757*, **757**, 765, *766*
Tropical seasonal forest, *756–757*, **757–758**
Tropical thornwood, *See* Tropical woodland
Tropical woodland, *756*, **759**
Tropism, **499**, **817**
Tropomyosin, **695–696**, *696*
Troponin, **696**, *696*
Trout, 434, 554, 556, 562, 604, 665, 767
 rainbow, 248, *248*
trpA gene, 216, *216*
trp operon, 238–240, *239*
True-breeding strain, 172
Truffle, *377*, 382
Trypanosoma gambiense, 366, *366*
Trypanosome, 795, *795*
Trypsin, 67, *292*, *581*, 582, *583*
Trypsin inhibitor, 248
Tryptophan, 587
Tryptophan synthetase, 216, *216*
Tsetse fly, 366, *366*
TSRLM, *See* Tandem scanning reflected-light microscope
T tubule, *See* Transverse tubule
Tuang child, 855
Tubal ligation, 322, *323*
Tube cell, 468–469, *468*
Tube feet, 428–429, *429*
Tube nucleus, 470, *471*
Tuber (plant), **465**
Tuberculosis, 255, 348, *351*
Tubulin, **110**, *113*, 114, 292, *292*
Tularemia, *351*
Tulip, 465
Tumor, **302**
 benign, 302
 malignant, 302
Tumor necrosis factor, 544
Tuna, 434, 562, 605–606, *606*
Tundra, *756*, **760**, *760*, 763
Tunic, 431, *431*
Tupaia tana, *849*
Turbellaria, **415–416**, *415–416*
Turgor, 91, 105
Turgor pressure, 486, *486*, 491, 495–496, *495*
Turkey, 156, 565, 692
Turner's syndrome, **258**, *258*
Turnip, 458, 460–461
Turtle, *438*, 440, 787
Tusk, 787
Twins, 319–321
 dizygotic, 319, *320*
 monozygotic, 319, *320*
Twin study, 256–257, *256–257*
Tympanic membrane, **656**, *656–657*
Typhoid fever, *351*
Typhus, *351*
Tyrosinase, 263, *263*, *292*
Tyrosine phosphokinase, **304**
Tyto alba, *827*

Ubiquinone, *See* Coenzyme Q
Uca pugnax, *836*
Ulcer, peptic, 578
Ulothrix, 391
Umbilical artery, *318*
Umbilical cord, *307*, *316*, **317**, *318*
Umbilical vein, *318*
Ungulate, 443–444, 792
Uniport, 89
Uniramian, **423**, 426
Unsaturated fatty acid, *43*, 44
Unwinding protein, 211
Upwelling, 762–763, *766*
Uracil, 54, *55*, **219**
Urea, 27, *27*, 39, 581, 595, *596–597*, **597**
Urease, 71
Ureter, 599, *599*
Urethra, *311*, 599, *599*

Urey, Harold, 328
Uric acid, *596–597*, **597**
Urinary bladder, *311*, **595**, 597, 599, *599*
Urination, *See* Micturition
Urine, 594–595, 601
Urine testing, 600
Urkaryote, 350
Urochordata, **431**, *431*
Urodela, 436
Urokinase, 247
Uterine artery, *318*
Uterine vein, *318*
Uterus, **310**, *311*, 632
 contraction, 318–319, *319*
Utriculus, 654, *655–656*

Vaccination, 351, 546–547
Vaccine, 247, 546–547
 malaria, 367, 547
 synthetic, 357, 547
Vacuole, 83, *83*, **105–106**, *105*, *111*
 contractile, 105, 362–363, *362*, *369*, 597
 food, 105, *369*, 571
Vagina, **311**
Vaginal foam, *323*
Vaginal jelly, *323*
Vagus nerve, 584–585, 679
Valence electron, 24
Valine, 73, *73*, 587
Valve, blood vessel, 517, *517*
van der Waals forces, 27, 53
van Niel, C.B., 136–137
van Niel's equations, 137, *137*
Variable expressivity, **264**, *265*
Variation, 3–4, *3*, 11
 genetic, 166–168, 196–197, 308
 maintenance, 726–727, *727*
Vascular bundle, *454*, **455**
Vascular cambium, *454*, 459, 478–479, *478*
Vascular plant, 389–390, *389*, **396**
 seedless, 396–398
Vascular tissue, plant, 449, *449*, *454–456*, **455**
Vas deferens, 309
Vasectomy, 309, 322, *323*
Vasoactive intestinal peptide (VIP), **583**, *584*, 646
Vasoconstriction, **527**, *527*, 607, *608*
Vasodilation, **527**, *527*, 607, *608*
Vasomotor center, **527**
Vasopressin, *See* Antidiuretic hormone
Vector, 366
 molecular, 243, *244*
Vegetable oil, 587
Vegetal pole, 277–278, *277*
Vegetative reproduction, **463–466**
 advantages and disadvantages, 465–466
 in agriculture, 465, *465*
 in nature, 464–465, *464*
Vein, 514, *516*, **517**
 leaf, *451*, 453, *453*
 valves, *517*
Veld, 759
Venereal disease, 348
Ventilation, **560–561**
Ventricle
 heart, 518–519, *519–520*
 brain, 665, *666*
Venus flytrap, 495
Vertebra, 436
Vertebral column, 430, 434
Vertebrate
 circulatory system, 518–525
 classes, *432*
 endocrine system, 636–644
 evolution, *10*
 excretory system, 598–604
 meat-eating, 572–573, *573*
 nervous system, 624–627, *624–626*
 osmotic balance, 595
 plant-eating, 573–575, *574–575*
 reproduction, 308–309
 social systems, 840–841
Vertical transport, **479**
Vessel cells, 393, 404
Vessel element, *455*, **456**, 459
Vestigial structure, 11
Vibrio cholerae, *351*
Villi, intestinal, 578, *579*
Vimentin, 111
Vine, 457, 461
Vinegar, 38, 350
Violet
 African, 452
 woodland, 453
VIP, *See* Vasoactive intestinal peptide
Viper
 bamboo, *659*
 pit, *659*, 660
Viperfish, Sloan's, *118*
Viral disease, 353–355, *355*, 357
Virchow, Rudolf, 81, 95
Viroid, **351**, 356
Virulence, 203
Virus, *240*, 341, **351**
 cancer and, 302–304, *354*, 355
 dormant, 354
 evolution, 355–356, *356*
 nonvirulent, 352
 reproduction, 352–353
 size, *84*
 structure, 351–352, *352*
 temperate, 352
 virulent, 352, *353*
Virus receptor, 354
Visceral mass, 427, *427*
Visible light, 139, *139*
Visual acuity, 660–661, *662*, 665
Visual communication, 834
Visual cortex, **664**, *665*, 669, *670*, 671, *672*, 675
Visual field, **664**, 672, *672*
Vital capacity, **561**
Vitamin, 67, **587**, *588*
 fat-soluble, **587**, *588*
 water-soluble, **587**, *588*
Vitamin A, *588*, 662
Vitamin B_1, *588*
Vitamin B_2, *588*
Vitamin B_6, *588*
Vitamin B_{12}, 587, *588*
Vitamin C, 587, *588*
Vitamin D, 586, *588*, 637
Vitamin E, *588*
Vitamin K, *588*
Vitella, *108*
Vitelline layer, 274, *274–276*, 276
Vitreous humor, 660, *661*
Viviparity, 437
Vocal cord, 560
Vole, 783
Voltage-gated channel, 616–619, *617–618*, 621, *621*
Volume reservoir, 523
Volvocine algae, 391, *391*
Volvox, *77*, 391, *391*, 772
Vomiting, 667
VSG protein, 366, *795*
Vulture, 759
Vulva, **312**
Vulvourethral gland, 310

Waggle dance, 835, *835*
Walker, Donald, 780
Walkingstick, 793, *794*
Wallace, Alfred Russel, 8, 706, *707*
Walnut tree, 789
Walrus, 443, 607
Warbler, 668, 824
 bay-breasted, *810*
 blackburnian, *810*
 myrtle, *810*
 willow, *829*, 830
Warm-blooded, *See* Homeotherm
Warning call, 834
Warning coloration, **790–792**, *791*
Warthog, *336*
Wasp, 405, *405*, 426, 838, *838*
 braconid, *807*
 digger, 815–817
 ichneumonid, 808
Water
 absorption in intestine, 580
 content of living things, 28, *28*
 dissociation, 32–33
 as habitat, 762–763
 human body, *19*
 hydrogen bonding, **20–21**, *20*
 molecular structure, 29–32, *29–32*
 movement across plasma membrane, 90–91, *91*
 physical properties, 28–29, *29*
 polarity, 26, *26*
 production in respiration, 128, *129*
 properties, *14*
 as solvent, 31–32, *31*
 transport in plants, 484–489, *485–489*
 use in photosynthesis, 136–150
 walking on, 29, *29*
Water bear, 594, *594*
Water cycle, **768–769**, *768*
Water fern, 494, *495*
Water hog, *336*
Water hyacinth, 464
Water lily, *388*
Water mold, 377, *377*
Water potential, **486**, *486*

Water snake, 723
Water-soluble vitamin, **587**, *588*
Water strider, 772
Water table, 768, *768*
Water vascular system, 428, *429*
Watson, James, 200, 207–209, *207*, *209*
Wavelength, 139, *139*
Wax, 45, 422
Weak bond, **27**
Weasel, 443
Weaverbird, 668
Webbed feet, 295, *295*
Weeds, *777*
Weight, body, 256, *257*, 585
Weinberg, W., 712
Weismann, August, 171
Weissman, Irving, 535
Welwitschia mirabilis, 400
Went, F.W., 499–500, *499*
Wernicke's area, 673, 677
Westerlies, *755*
Whale, 443, 568, 658, 743–744, 787, 803, 826, 834
baleen, 573, 764, *764*
blue, 283, 442, 573, *573*
gray, *795*
minke, 764, *764*
toothed, 572, 764, *764*
Wheat, 714, 740, 779, 861
winter, 88, 509
Whelk, 428
Whisk fern, **396–397**, *397*, 455
White blood cells, 528, *528*, 535–536, *535*
White marrow, 687
White matter, 625, *625*
White muscle, 565, 692–693
Whittaker, R.H., 338, 781
Whooping cough, *351*, 547
Wildebeest, 792, *799*, *841*, 842
Wild-type allele, **190**
Wilkins, Maurice, 207
Williamson, P.G., 748
Willingham, Mark, 106
Willow tree, 780
water, *778*
Wilson, Edward O., 782, 833
Wilting, 486
Wind, 754–757
Wind pollination, 405, 470, 792
Wind power, *863*
Wings
bird, 441
insect, 423, 425, 684, *685*
Winter solstice, 754, *755*
Wolf, 760, 808, 833
social organization, 842, *842*
Wolffia microscopica, 403
Wolffian duct, **313–314**, *314*
Wolpert, Lewis, 296
Wombat, 443
Wood, 41, 92, *92*, **478–479**, *478*
Woodruff, L.L., 780
Woody plant, *455*
Worker bee, 839–840, *839*
Wrinkling, 25

X chromosome, 157, *158*, 181–182, *182*, 312–313, 852
inactivation, 259, *259*
Xenopus, 240, *272*, 273, 289
5S RNA, 300–301, *300*
X-linked trait, human, 261–262, *261–262*
X-ray, 21
X-ray diffraction, **207–208**, *207*
Xylem, 396, *449*, 453, *454–455*, 455, **456**, *459*, *478*, **479**, 484, *485*, 487–488, *488*

Yam, 790
Yanofsky, Charles, 216
Y chromosome, 157, *158*, 261, 312–313
Yeast, 164, 247, *377*, 382, 646
Yellow fever, *355*
Yersin, D., 537
Yogurt, 124, *125*, 350
Yolk sac, 277, **282**, *282*, 317, *318*
Young, J.Z., 615
Youvan, Douglas, 149
Yucca, *781*, 809

Zamia pumila, 400
Zebra, *153*, 335, 443, 837
Zebra fish, 793
Zinc, 492, *492*, *589*
Zinc finger, 300
Z line, *693*, 694
Zona pellucida, *273*
Zone of polarizing activity (ZPA), **296–298**, *297*
Zonulae adherens, **94**, *94*
Zooflagellate, *360–361*, **365–369**, *366–367*
Zooplankton, 765
Zoospore, *378*, **391–392**, *392*
ZPA, *See* Zone of polarizing activity
Zygomycota, 377, *377–379*, **379–380**, *385*
Zygosporangium, *378*, 379
Zygospore, **391**
Zygote, **275**, 470, *471*
Zymogen, **582–584**